Handbook of Cell Signaling

Volume 3

Handbook of Cell Signaling

Volume 3

Editors-in-Chief

Ralph A. Bradshaw
Department of Physiology and Biophysics
University of California Irvine
Irvine, California

Edward A. Dennis
Department of Chemistry and Biochemistry
University of California San Diego
La Jolla, California

ACADEMIC PRESS

An imprint of Elsevier Science

Amsterdam Boston Heidelberg London New York Oxford
Paris San Diego San Francisco Singapore Sydney Tokyo

This book is printed on acid-free paper. ♾

Academic Press
An imprint of Elsevier Science
525 B Street, Suite 1900, San Diego, California 92101-4495, USA
http://www.academicpress.com

Academic Press
84 Theobald's Road, London WC1X 8RR, UK
http://www.academicpress.com

Library of Congress Catalog Card Number: 2003103352

International Standard Book Number: 0-12-124546-2 (Set)
International Standard Book Number: 0-12-124547-0 (Volume 1)
International Standard Book Number: 0-12-124548-9 (Volume 2)
International Standard Book Number: 0-12-124549-7 (Volume 3)

PRINTED IN THE UNITED STATES OF AMERICA
03 04 05 06 07 7 6 5 4 3 2 1

Contents

VOLUME 3

Section D: Post-Translational Control

Nahum Sonenberg, Editor

CHAPTER 279

Orphan Receptor COUP-TFII and Vascular Development

Fabrice G. Petit, Sophia Y. Tsai, and Ming-Jer Tsai
Department of Molecular and Cellular Biology, Baylor College of Medicine, Houston, Texas

Introduction

Nuclear receptors belong to a large family of transcription factors, which can modulate gene expression [1,2]. Since the discovery of the steroid/thyroid hormone receptor superfamily, an increasing number of transcription factors with similar structural motif have been added. A majority of these newly added members have unknown ligands, thus they are classified as orphan nuclear receptors. Like the other members of the steroid/thyroid hormone receptor superfamily, the orphan receptors are involved in development, differentiation, and homeostasis processes. The absence of ligand makes it more difficult to dissect the functional role of orphan receptors. However, the analysis of orphan receptor null mice will provide new insights into the physiological function of these receptors during development, organogenesis, and homeostasis. In this review, we attempt to describe the functional role of orphan receptors during vascular development.

Vascular Development

During embryogenesis, the vascular system plays an important role. It is in charge of the transport of oxygen and nutrients, which are indispensable to the growth and development of the different tissues in vertebrate embryos. For that reason, the development of the vascular system is one of the first events to occur. The vascular network formation requires two distinct phenomena, vasculogenesis and angiogenesis [3,4].

Vasculogenesis

Vasculogenesis is the formation of the early vascular plexus from mesodermal cells (hemangioblasts), which proliferate and differentiate to form the precursors of the endothelial cells of the vessel wall. During early development, vasculogenesis gives rise to a primary vascular plexus including the major vessels of the embryo, such as the aorta and major veins [3]. Factors such as fibroblast growth factors (FGFs), vascular endothelial growth factor (VEGF), and VEGFR-2 are important for this process.

Angiogenesis

The next step in formation of the vascular system is *angiogenesis,* which corresponds to the formation of new blood vessels from preexisting vessels. Two different processes characterize angiogenesis: angiogenic remodeling and angiogenic sprouting. The remodeling process consists of the modification of the initial vessel plexus in order to form interconnecting branching patterns present in the mature vascular network. The sprouting process corresponds to the formation of new capillaries originating from small venules or from other capillaries. It occurs both in the yolk sac and in the embryo (neural tube and retina). It also involves many factors responsible for the proteolytic degradation of the extracellular matrix, for the chemotactic migration and proliferation of endothelial cells, for the formation of a lumen, and for functional maturation of the endothelium. The involvement of factors such as VEGF, VEGFR-2, Tie1, Tie2, angiopoietin-1 (Ang1), Ang2, EphrinB2, platelet-derived growth factor (PDGF), and PDGFR is primordial for the formation of a mature vascular network [3,4].

Transcription Factors Involved in Vasculogenesis and Angiogenesis

Recently, Roman and Weinstein [4] provided a concise overview of the molecules involved in the development and maintenance of the vertebrate embryonic vascular network. It appears that COUP-TFII is the only orphan nuclear receptor with a well-described role in angiogenesis during embryonic development in mice [5]. Some recent studies suggest that the orphan receptor, PPARγ, may also play a role in the vascular development.

PPARγ: Inhibitor of Angiogenesis

The peroxisome proliferator-activated receptor (PPAR, NR1C) group contains three subtypes: α, β (also known as δ or NUCI), and γ. Since the cloning of PPAR members, many pharmacological activators have been identified [6,7]. PPARs are known to regulate genes involved in lipid and glucose metabolism. Despite the ubiquitous expression of PPARδ, its physiological function is not well known. In PPARα knock-out mice, the liver accumulates droplets of lipid. PPARα ligands can inhibit inflammatory responses in vascular smooth muscle cells *in vivo*, suggesting that PPARα agonists may have a positive effect on vasculature in atherosclerosis [6,7]. The specific ligand for PPARγ, 15d-PGJ_2, can inhibit endothelial tube formation and proliferation, as well as the induction of VEGF receptors *in vitro*. Moreover, 15d-PGJ_2 was shown to be an inhibitor of angiogenesis in the cornea *in vivo* [8]. It is not clear whether PPARγ is involved because we do not know whether the ligand used acts through PPARγ or another nuclear receptor [6,9]. The study of PPARγ knock-out mice has revealed that the establishment and maintenance of the fetal and maternal vascular networks are abnormal in the placenta. However, all other vascular processes appear normal in the null embryos. Whether other PPAR members can compensate for PPARγ has not yet been defined. Therefore, it appears that the vascularization of the placenta is dictated by PPARγ-dependent trophoblast functions [10].

COUP-TFII: Positive Effector in Angiogenesis

COUP-TF (NR2F) is one of the best characterized orphan receptor among the nuclear receptor superfamily. Two members have been identified in mammals, COUP-TFI (also called v-ErbA related protein 3, EAR-3) and COUP-TFII (also called apolipoprotein regulating protein 1, ARP-1). COUP-TF members share a high degree of homology within and between species, so it is assumed they must play important physiological functions [11].

In the developing mouse embryo, COUP-TFI and COUP-TFII exhibit overlapping, but distinct, expression patterns. They are also differentially expressed in the nervous system and during organogenesis [12]. The expression pattern of COUP-TFII in the nervous system is more restricted than that of COUP-TFI. During organogenesis, COUP-TFII is expressed in the mesenchymal compartment. Expression of COUP-TFII is lower than that of COUP-TFI in the nasal septum, tongue, follicles of vibrissae, and cochlea. Expression of COUP-TFII is higher than that of COUP-TFI in the developing salivary gland, atrium of the heart, lung, stomach, pancreas primordium, mesonephros, kidney, and prostate [12]. Because COUP-TFI and COUP-TFII have different expression patterns, they may have different physiological functions.

To study the physiological function of COUP-TFI and -II *in vivo*, we have generated null mice for these genes. COUP-TFI seems to be important for the neuronal development and differentiation. Accordingly, COUP-TFI null mutant mice die perinatally from starvation due to defects in formation of the glossopharyngeal nerve [12]. In contrast, the knock-out mice for COUP-TFII die early during embryonic development at 9.5–10.5 days from heart and vasculature defects [5]. COUP-TFII mutants probably die as a result of malformation of the vasculature, leading to extensive hemorrhage and edema in the brain and heart. Analyses of the mutant embryos revealed defects in angiogenesis, in the development of the atrium chamber and sinus venosus, and malformations of the cardinal veins.

Detailed observation of COUP-TFII null mice revealed that the major vessels, which arise through vasculogenesis, were formed and expressed the specific markers VEGF-R1, VEGF-R2, and PECAM, suggesting that COUP-TFII does not play an essential role in vasculogenesis. However, the remodeling of the vascular plexus was impaired. Therefore, the sprouting and branching from preexisting vessels was defective in COUP-TFII mutants, suggesting an angiogenesis defect. The defect in vasculature development observed in COUP-TFII mutants was similar to the one observed in mice deficient for Ang1 or its receptor, Tie2 [3]. Moreover, the Ang1 expression pattern is very similar to that of COUP-TFII. In fact, Ang1 mRNA expression is down-regulated 10-fold in the mesenchymal compartments of the brain, eyes, somites, and heart in COUP-TFII mutants. Because COUP-TFII may function through the Ang1-Tie2 signaling pathway, it is likely that it acts on the vasculature formation via mesenchymal–endothelial interactions. Further experiments, using subtraction library screening or P19 cells overexpressing COUP-TFII, also showed that Ang1 is a downstream target gene of COUP-TFII (Petit, F., Tsai, M.-J., and Tsai, S.Y., unpublished data). In the COUP-TFII mutants, there is no obvious alteration in the expression of Tie2, which is regulated by its own ligand, Ang1. This observation suggests that another signaling pathway must be affected by COUP-TFII. Taken together these data suggest that, in mesenchymal cells, COUP-TFII is the upstream regulator of Ang1 expression, which in turn activates its receptor, Tie2, to exert the differentiation of endothelial cells (Fig. 1).

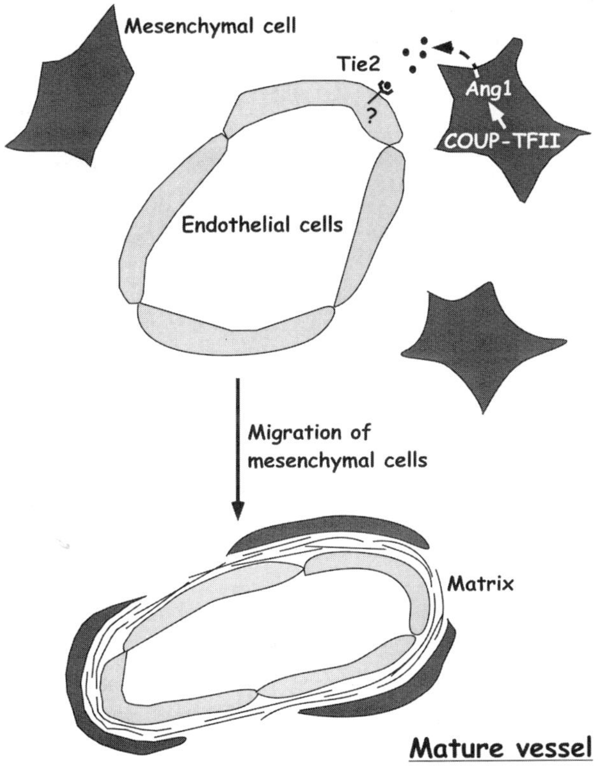

Figure 1 Role of COUP-TFII in the development of vessels. COUP-TFII induces the expression of Ang1, which is released by the mesenchymal cells, Ang1 activates its own receptor, Tie2, located on the endothelial cells, which in turn allow the migration of mesenchymal cells and subsequently the differentiation of mesenchymal and endothelial cells.

Conclusion

So far, among the orphan nuclear receptor group, COUP-TFII is the only member shown to play a role in vascular development. The involvement of COUP-TFII in angiogenesis suggests that it might play a role in tumor growth and progression. Indeed, COUP-TFII is highly expressed in several tumor cell lines. The discovery of artificial or natural ligands for COUP-TFII will provide important tools for the therapeutic treatment of disease involving angiogenesis or vascular modeling. Moreover, these putative ligands will allow us to further study the function of COUP-TFII in other signaling pathways.

References

1. Mangelsdorf, D. J., Thummel, C., Beato, M., Herrlich, P., Schutz, G., Umesono, K., Blumberg, B., Kastner, P., Mark, M., and Chambon, P. (1995). The nuclear receptor superfamily: The second decade. *Cell* **83**, 835–839.
2. Tsai, M.-J. and O'Malley, B. W. (1994). Molecular mechanisms of action of steroid/thyroid receptor superfamily members. *Annu. Rev. Biochem.* **63**, 451–486.
3. Patan, S. (2000). Vasculogenesis and angiogenesis as mechanisms of vascular network formation, growth and remodeling. *J. Neurooncol.* **50**, 1–15.
4. Roman, B. L. and Weinstein, B. M. (2000). Building the vertebrate vasculature: research is going swimmingly. *Bioessays* **22**, 882–893.
5. Pereira, F. A., Qiu, Y., Zhou, G., Tsai, M. J., and Tsai, S. Y. (1999). The orphan nuclear receptor COUP-TFII is required for angiogenesis and heart development. *Genes Dev.* **13**, 1037–1049.
6. Bishop-Bailey, D. (2000). Peroxisome proliferator-activated receptors in the cardiovascular system. *Br. J. Pharmacol.* **129**, 823–834.
7. Giguere, V. (1999). Orphan nuclear receptors: From gene to function. *Endocr. Rev.* **20**, 689–725.
8. Xin, X., Yang, S., Kowalski, J., and Gerritsen, M. E. (1999). Peroxisome proliferator-activated receptor gamma ligands are potent inhibitors of angiogenesis *in vitro* and *in vivo*. *J. Biol. Chem.* **274**, 9116–9121.
9. Lazar, M. A. (2001). Progress in cardiovascular biology: PPAR for the course. *Nat. Med.* **7**, 23–24.
10. Barak, Y., Nelson, M. C., Ong, E. S., Jones, Y. Z., Ruiz-Lozano, P., Chien, K. R., Koder, A., and Evans, R. M. (1999). PPAR gamma is required for placental, cardiac, and adipose tissue development. *Mol. Cell* **4**, 585–595.
11. Tsai, S. Y. and Tsai, M.-J. (1997). Chick ovalbumin upstream promoter-transcription factors (COUP-TFs): coming of age. *Endocr. Rev.* **18**, 229–240.
12. Pereira, F. A., Tsai, M. J., and Tsai, S. Y. (2000). COUP-TF orphan nuclear receptors in development and differentiation. *Cell. Mol. Life Sci.* **57**, 1388–1398.

VOLUME 1

VOLUME 2

PART II
TRANSMISSION: EFFECTORS AND CYTOSOLIC EVENTS (CONTINUED FROM VOLUME 1)

Section C: Calcium Mobilization
Michael J. Berridge, Editor

Contributors

Numbers in parentheses indicate the pages on which the authors' contributions begin.

John M. Abrams (3:213)
Department of Cell Biology, University of Texas Southwestern Medical Center, Dallas, Texas 75390

John P. Adelman (1:227)
Oregon Health Sciences University, Vollum Institute, Portland, Oregon 97201

Joseph L. Alcorn (3:509)
Department of Pediatrics, University of Texas Health Science Center, Houston, Texas 77030

Dario R. Alessi (1:513; 2:193)
Department of Biochemistry, MRC Unit, University of Dundee, Dundee DD1 5EH, Scotland, United Kingdom

Emil Alexov (1:11)
Department of Biochemistry and Molecular Biophysics, Columbia University, Howard Hughes Medical Institute, New York, New York 10032

Simon Alford (2:663)
Department of Biological Sciences, University of Illinois at Chicago, Chicago, Illinois 60637

Kari Alitalo (2:855)
Molecular/Cancer Biology Laboratory and Ludvig Institute for Cancer Research, Haartman Institute and Helsinki University Central Hospital, Biomedicum Helsinki, University of Helsinki, FIN-00014 Helsinki, Finland

James P. Allison (1:355)
Department of Molecular and Cell Biology, Howard Hughes Medical Institute, University of California, Berkeley, Berkeley, California 94720

Steven C. Almo (1:74)
Department of Biochemistry, Albert Einstein College of Medicine, Bronx, New York 10461-1975

Christelle Alory (2:689)
Departments of Cell and Molecular Biology and Institute for Childhood and Neglected Disease, The Scripps Research Institute, La Jolla, California 92037

Aymen Al-Shamkhani (1:319)
University of Southampton General Hospital, Tenovus Research Lab, Cancer Sciences Division, School of Medicine, Southampton SO16 6YD, United Kingdom

Sally A. Amundson (3:179)
Gene Response Section, Center for Cancer Research, National Cancer Institute, National Institutes of Health, Bethesda, Maryland 20892

Carl W. Anderson (3:237)
Biology Department, Brookhaven National Laboratory, Upton, New York 11973

Jannik N. Andersen (1:659)
Cold Spring Harbor Laboratory, Cold Spring Harbor, New York 11724

Peter Angel (3:99)
Department of Signal Transduction and Growth Control, Deutsches Krebsforschungszentrum, 69120 Heidelberg, Germany

Ettore Appella (3:237)
Laboratory of Cell Biology, National Cancer Institute, National Institutes of Health, Bethesda, Maryland 20892

William J. Arendshorst (3:573)
Department of Cell and Molecular Physiology, University of North Carolina at Chapel Hill, Chapel Hill, North Carolina 27599

Steve Arkinstall (1:703)
Departments of Microbiology and Medicine, Serono Reproductive Biology Institute, Rockland, Massachusetts 02370

Anjon Audhya (2:177)
Division of Cellular and Molecular Medicine, The Howard Hughes Medical Institute, University of California San Diego, School of Medicine, La Jolla, California 92093

Joseph Avruch (1:523)
Harvard Medical School, Massachusetts General Hospital, Boston, Massachusetts 02114

Gary D. Bader (2:311)
Samuel Lunenfeld Research Institute, Mount Sinai Hospital, Toronto, Ontario M5G 1X5, Canada

Cinzia Bagala (3:393)
Center for Molecular Medicine, Maine Medical Center Research Institute, Scarborough, Maine 04074

William E. Balch (2:689)
Departments of Cell and Molecular Biology and Institute for Childhood and Neglected Disease, The Scripps Research Institute, La Jolla, California 92037

Jesus Balsinde (2:261)
Institute of Molecular Biology and Genetics, University of Valladolid School of Medicine, E-47005 Valladolid, Spain

Utpal Banerjee (2:827)
Department of Molecular Cell and Developmental Biology, Biological Chemistry, Human Genetics, Molecular Biology Institute, University of California at Los Angeles, Los Angeles, California 90095

David Barford (1:601)
Chester Beatty Laboratories, Section of Structural Biology, Institute of Cancer Research, London, SW3 6JB, United Kingdom

Dafna Bar-Sagi (2:711)
Department of Molecular Genetics and Microbiology, State University of New York at Stony Brook, Stony Brook, New York 11794

Perry F. Bartlett (3:625)
Institute for Brain Research, The University of Queensland, Brisbane QLD 4072, Queensland, Australia

Philippe I. H. Bastiaens (2:305)
Cell Biology and Cell Biophysics Program, European Molecular Biology Laboratory, 69120 Heidelberg, Germany

Chiara Battelli (3:393)
Center for Molecular Medicine, Maine Medical Center Research Institute, Scarborough, Maine 04074

Linnea M. Baudhuin (2:253)
Department of Cancer Biology, The Lerner Research Institute, Cleveland Clinic Foundation, Cleveland, Ohio, and Department of Chemistry, Cleveland State University, Cleveland, Ohio 44195

Andrew J. Beavil (1:45)
The Randall Centre, King's College London, New Hunt's House, Guy's Campus, London SE1 1UL, United Kingdom

Rebecca L. Beavil (1:45)
The Randall Centre, King's College London, New Hunt's House, Guy's Campus, London SE1 1UL, United Kingdom

Joseph A. Beavo (2:431; 2:525)
Department of Pharmacology, University of Washington, Seattle, Washington 98195

Elsa Bello-Reuss (3:573)
Department of Internal Medicine, Division of Nephrology, and Department of Physiology and Biophysics, The University of Texas Medical Branch, Galveston, Texas 77555

Stephen Bellum (3:393)
Center for Molecular Medicine, Maine Medical Center Research Institute, Scarborough, Maine 04074

Juan Carlos Izpisúa Belmonte (2:799)
The Salk Institute for Biological Studies, Gene Expression Laboratory, La Jolla, California 92037

Craig B. Bennett (3:203)
Department of Surgery, Duke University Medical Center, Durham, North Carolina 27710

Jeffrey L. Benovic (3:441)
Cell Biology and Signaling Program, Kimmel Cancer Center, Thomas Jefferson University, Philadelphia, Pennsylvania 19107

Michael J. Berridge (2:51)
The Babraham Institute, Babraham, Cambridge CB2 4AT, United Kingdom

Penny J. Beuning (3:185)
Biology Department, Massachusetts Institute of Technology, Cambridge, Massachusetts 02139

Rashna Bhandari (1:343)
Departments of Molecular and Cell Biology and Chemistry, HHMI, University of California, Berkeley, Berkeley, California 94720-3202

Ananya Bhattacharya (2:779)
Waksman Institute, Department of Molecular Biology and Biochemistry, and Cancer Institute of New Jersey, Rutgers University, Piscataway, New Jersey 08901

Martin Biel (2:515)
Department Pharmzaie-Zentrum für Pharmaforschung, Ludwig-Maximilians-Universität München, 81377 München, Germany

Vincent A. Bielinski (2:405)
Department of Pharmacology, University of Texas Southwestern Medical Center, Dallas, Texas 75392

Hana Bilak (2:333)
National Center of Scientific Research, Institute of Molecular and Cellular Biology, 67084 Strasbourg, France

Lutz Birnbaumer (2:557)
Laboratory of Signal Transduction, National Institute of Environmental Health Sciences, Research Triangle Park, North Carolina 27709

Geoff Birrell (3:225)
Queensland Cancer Fund Research Laboratory, Queensland Institute of Medical Research, Brisbane 4029, Australia

Gail A. Bishop (1:315)
Departments of Microbiology and Internal Medicine, University of Iowa and the VA Medical Center, Iowa City, Iowa 52242

Trillium Blackmer (2:663)
Department of Biological Sciences, University of Illinois at Chicago, Chicago, Illinois 60637

Perry J. Blackshear (3:319)
Office of Clinical Research and Laboratory of Signal Transduction, National Institute of Environmental Health Sciences, Research Triangle Park, North Carolina 27709

and Departments of Medicine and Biochemistry, Duke University, Durham, North Carolina 27710

Christine Blattner (3:257)
Institute of Genetics and Toxicology, Forschungszentrum Karlsruhe, 76021 Karlsruhe, Germany

Mordecai P. Blaustein (2:63)
Department of Physiology, University of Maryland, School of Medicine, Baltimore, Maryland 21201

Gary M. Bokoch (2:705)
Departments of Immunology and Cell Biology, The Scripps Research Institute, La Jolla, California 92037

Lynda F. Bonewald (3:497)
University of Missouri at Kansas City, School of Dentistry, Kansas City, Missouri 64108

Marco Bonomi (1:161)
Faculty of Medicine, University of Brussels, Institut de Recherche Interdisciplinaire, Brussels B-1070, Belgium

Michelle A. Booden (2:675)
Lineberger Comprehensive Cancer Center, University of North Carolina at Chapel Hill, Chapel Hill, North Carolina 27599

Charles Boone (2:311)
Banting and Best Department of Medical Research and Department of Molecular and Medical Genetics, University of Toronto, Toronto, Ontario M5G 1X5, Canada

Martin D. Bootman (2:51)
The Babraham Institute, Babraham, Cambridge CB2 4AT, United Kingdom

Johannes L. Bos (2:521)
Department of Physiological Chemistry and Center for Biomedical Genetics, University Medical Center Utrecht, Utrecht, The Netherlands

Jane M. Bradbury (3:219)
Wellcome Trust and Cancer Research Campaign, Institute of Cancer and Developmental Biology, Cambridge, United Kingdom

Ralph A. Bradshaw (1:1; 1:361; 3:453)
Department of Physiology and Biophysics, Department of Anatomy and Neurobiology, University of California, Irvine, Irvine, California 92697-4560

Anne R. Bresnick (1:74)
Center for Synchrotron Biosciences, Albert Einstein College of Medicine, Bronx, New York 10461-1975

Lena Brevnova (1:259)
Department of Microbiology and Immunology, Department of Structural Biology, Stanford University School of Medicine, Stanford, California 94305-5124

Ross I. Brinkworth (2:495)
Department of Biochemistry and Molecular Biology, Institute for Molecular Bioscience, University of Queensland, St. Lucia, Brisbane QLD 4072, Queensland, Australia

Michael S. Brown (3:353)
Department of Molecular Genetics, University of Texas Southwestern Medical Center at Dallas, Dallas, Texas 75390

Steven A. Brown (3:139)
Department of Molecular Biology, Sciences II, University of Geneva, CH-1211 Geneva, Switzerland

Anne Brunet (3:83)
Division of Neuroscience, Children's Hospital and Department of Neurobiology, Harvard Medical School, Boston, Massachusetts 02115

Robert Bucki (2:209)
Department of Physiology, Institute of Medicine and Engineering, University of Pennsylvania, Philadelphia, Pennsylvania 19104

Robert D. Burgoyne (2:79)
The Physiological Laboratory, The University of Liverpool, Liverpool L69 3BX, United Kingdom

Janice E. Buss (2:675)
Department of Biochemistry, Biophysics, and Molecular Biology, Iowa State University, Ames, Iowa 50011

Ronald A. Butow (3:365)
Department of Molecular Biology, University of Texas Southwestern Medical Center, Dallas, Texas 75390

Javier Capdevila (2:799)
The Salk Institute for Biological Studies, Gene Expression Laboratory, La Jolla, California 92037

Ernesto Carafoli (2:52)
Department of Biological Chemistry, University of Maryland School of Medicine, Baltimore, Maryland 21201

Cathrine R. Carlson (2:501)
Department of Medical Biochemistry, Institute of Basic Medical Sciences, University of Oslo, N-0316 Oslo, Norway

Graham Carpenter (2:5)
Department of Biochemistry, Vanderbilt University School of Medicine, Nashville, Tennessee 37232

Juan J. Carrillo (2:619)
Molecular Pharmacology Group, Division of Biochemistry and Molecular Biology, University of Glasgow, Glasgow G12 8QQ, Scotland, United Kingdom

Patrick J. Casey (2:601)
Department of Pharmacology and Cancer Biology, Duke University Medical Center, Durham, North Carolina 27710

William A. Catterall (2:23)
Department of Pharmacology, University of Washington, Seattle, Washington 98195

Richard A. Cerione (2:715; 2:721)
Department of Molecular Medicine and Department of Chemistry and Chemical Biology, Cornell University, Ithaca, New York 14853

Gianni Cesareni (2:311)
Department of Biology, University of Rome Tor Vergata, 00153 Rome, Italy

Andrew C. Chan (1:475)
Department of Immunology, Genentech, Inc., South San Francisco, California 94080

Geoffrey Chang (1:115)
Department of Molecular Biology, The Scripps Research Institute, La Jolla, California 92037

Moses V. Chao (2:839)
Molecular Neurobiology Program, Skirball Institute of Biomolecular Medicine, New York University School of Medicine, New York, New York 10016

Harry Charbonneau (1:697)
Department of Biochemistry, Purdue University, West Lafayette, Indiana 47907-1153

Philip Chen (3:225)
Queensland Cancer Fund Research Laboratory, Queensland Institute of Medical Research, Brisbane 4029, Australia

Alan Cheng (1:729)
McGill Cancer Center, Department of Biochemistry, McGill University, Montreal, Quebec H3G 1Y6, Canada

Chris Chiu (1:475)
Department of Immunology, Genentech, Inc., South San Francisco, California 94080

Dar-chone Chow (1:259)
Department of Microbiology and Immunology, Department of Structural Biology, Stanford University School of Medicine, Stanford, California 94305-5124

Ted D. Chrisman (2:427)
Cecil H. and Ida Green Center for Reproductive Biology Sciences, Howard Hughes Medical Institute and Department of Pharmacology, University of Texas Southwestern Medical Center, Dallas, Texas 75390

Anne Elisabeth Christensen (2:549)
Department of Anatomy and Cell Biology, University of Bergen, N-5009 Bergen, Norway

Jee Y. Chung (1:311)
Department of Biochemistry, Cornell University, Weill Medical College, New York, New York 10021

Grant C. Churchill (2:15)
Department of Pharmacology, University of Oxford, Oxford OX1 2JD, United Kingdom

Aaron Ciechanover (3:129)
Department of Biochemistry, Bruce Rappaport Faculty of Medicine, Technion-Israel Institute of Technology, Haifa, Israel

Gino Cingolani (3:419)
Department of Cell Biology, The Scripps Research Institute, La Jolla, California 92037

Sylvie Claeysen (1:161)
Faculty of Medicine, University of Brussels, Institut de Recherche Interdisciplinaire, Brussels B-1070, Belgium

Jean Closset (3:591)
Department of Biochemistry, Faculty of Medicine, Institute of Pathology, University of Liège, B 4000 Liège, Belgium

Shamshad Cockcroft (2:225)
Department of Physiology, University College London, London WC1E 6JJ, United Kingdom

Patricia T. W. Cohen (1:593)
Medical Research Council Protein Phosphorylation Unit, School of Life Sciences, University of Dundee, Dundee DD1 5EH, Scotland

Philip Cohen (1:547)
MRC Protein Phosphorylation Unit, Department of Biochemistry, University of Dundee, Dundee DD1 5EH, Scotland

Roger J. Colbran (2:397)
Department of Molecular Physiology and Biophysics and the Center for Molecular Neuroscience, Vanderbilt University School of Medicine, Nashville, Tennessee 37232

Clay E. S. Comstock (2:545)
Department of Biochemistry and Molecular Biology, University of North Dakota School of Medicine and Health Sciences, Grand Forks, North Dakota 58203

Marco Conti (2:437)
Department of Gynecology and Obstetrics, Stanford University Medical Center, Stanford, California 94305

Jackie D. Corbin (2:447; 2:465)
Department of Molecular Physiology and Biophysics, Vanderbilt University School of Medicine, Nashville, Tennessee 37232

Daniela Corda (2:613)
Department of Cell Biology and Oncology, Instituto di Ricerche Farmacologiche "Mario Negri," Santa Maria Imbaro, Chieti, Italy

Sabine Costagliola (1:161)
Faculty of Medicine, University of Brussels, Institut de Recherche Interdisciplinaire, Brussels B-1070, Belgium

Rick H. Cote (2:453; 2:465)
Department of Biochemistry and Molecular Biology, University of New Hampshire, Durham, New Hampshire 03824

Shaun R. Coughlin (1:167)
Departments of Medicine and Cellular and Molecular Pharmacology, Cardiovascular Research Institute, University of California San Francisco, San Francisco, California 94920-0130

L. Ashley Cowart (2:257)
Department of Biochemistry and Molecular Biology, Medical University of South Carolina, Charleston, South Carolina 29425

Adrienne D. Cox (2:681; 2:737)
Department of Pharmacology, and Lineberger Comprehensive Cancer Center, University of North Carolina at Chapel Hill, Chapel Hill, North Carolina 27599

Mark S. Cragg (1:319)
University of Southampton General Hospital, Tenovus Research Laboratory, Cancer Sciences Division, School of Medicine, Southampton SO16 6YD, United Kingdom

José L. Crespo (3:299)
Division of Biochemistry, Biozentrum, University of Basel, 4056 Basel, Switzerland

Claudia Crosio (3:91)
Institute of Genetic and Molecular and Cellular Biology, 67084 Strasbourg, France

Christopher Daly (2:849)
Regeneron Pharmaceuticals, Inc., Tarrytown, New York 10591

Sami Damak (2:657)
Department of Physiology and Biophysics, The Mount Sinai School of Medicine, New York, New York 10029

Mary Dasso (2:695)
Laboratory of Gene Regulation and Development, National Institute of Child Health and Human Development, National Institutes of Health, Bethesda, Maryland 20892

Michael David (3:145)
Molecular Biology Section, Division of Biology, University of California San Diego, La Jolla, California 92093

Anthony J. Davis (2:405)
Department of Pharmacology, University of Texas Southwestern Medical Center, Dallas, Texas 75390

Roger J. Davis (2:365)
Howard Hughes Medical Institute and Program in Molecular Medicine, University of Massachusetts Medical School, Worcester, Massachusetts 01605

Richard N. Day (3:135)
Departments of Medicine and Cell Biology, NSF Center for Biological Timing, University of Virginia Health Sciences Center, Charlottesville, Virginia 22908

Eva Degerman (2:441)
Section for Molecular Signaling, Department of Cell and Molecular Biology, Lund University, SE-221 00 Lund, Sweden

Warren L. DeLano (1:51)
Sunesis Pharmaceuticals, Inc., San Francisco, California 94080

Mark L. Dell'Acqua (2:377)
Department of Pharmacology, University of Colorado Health Sciences Center, Denver, Colorado 80262

Emmanuèle Délot (2:833)
Department of Human Genetics, Department of Pediatrics, Department of Orthopedic Surgery, The David Geffen School of Medicine at the University of California Los Angeles, Los Angeles, California 90095

Bruce Demple (3:191)
Department of Cancer Cell Biology, Harvard School of Public Health, Boston, Massachusetts 02115

Edward A. Dennis (1:1; 2:261)
Department of Chemistry and Biochemistry, School of Medicine, University of California San Diego, La Jolla, California 92093

John M. Denu (1:653)
Department of Biochemistry and Molecular Biology, Oregon Health and Science University, Portland, Oregon 97201-3098

Anna A. DePaoli-Roach (1:613)
Department of Biochemistry and Molecular Biology, Indiana University School of Medicine, Indianapolis, Indiana 46202-5122

Channing J. Der (2:681)
Department of Pharmacology, and Lineberg Comprehensive Cancer Center, University of North Carolina at Chapel Hill, Chapel Hill, North Carolina 27599

Johan de Rooij (2:521)
Department of Physiological Chemistry and Center for Biomedical Genetics, University Medical Center Utrecht, Utrecht, The Netherlands

Frederic de Sauvage (2:793)
Department of Molecular Biology, Genentech, Inc., South San Francisco, California 94080

Peter N. Devreotes (2:645)
Department of Cell Biology, Johns Hopkins School of Medicine, Baltimore, Maryland 21201

Valérie Dewaste (2:11)
Interdisciplinary Research Institute (IRIBHN), Université Libre de Bruxelles, Brussels, Belgium

Robert B. Dickson (3:565)
Department of Oncology, Lombardi Cancer Center, Georgetown University, Washington, DC

Becky A. Diebold (2:705)
Department of Immunology, The Scripps Research Institute, La Jolla, California 92037

Pier Paolo Di Fiori (3:411)
Department of Experimental Oncology, Istituto Europeo di Oncologia, 20141 Milan, Italy, and Medical School, University of Milan, Milan Italy, and FIRC Institute for Molecular Oncology, 20134 Milan, Italy

Maria Di Girolamo (2:613)
Department of Cell Biology and Oncology, Instituto di Ricerche Farmacologiche "Mario Negri," Santa Maria Imbaro, Chieti, Italy

Julie Diplexcito (1:607)
School of Life Sciences, University of Dundee, MRC Protein Phosphorylation Unit, Dundee, Scotland DD1 5EH, United Kingdom

Jack E. Dixon (2:143)
The Life Science Institute and Department of Biological Chemistry, University of Michigan Medical School, Ann Arbor, Michigan 48109

Jack E. Dixon (1:591)
Department of Biological Chemistry, University of Michigan, Ann Arbor, Michigan 48109

Robert W. Doms (1:191)
Department of Microbiology, University of Pennsylvania, Philadelphia, Pennsylvania 19104

Daniel J. Donoghue (2:861)
Department of Chemistry and Biochemistry, Center for Molecular Genetics, University of California San Diego, La Jolla, California 92093

Russell F. Doolittle (1:119)
Center for Molecular Genetics, University of California San Diego, La Jolla, California 92093-0634

Stein Ove Døskeland (2:549)
Department of Anatomy and Cell Biology, University of Bergen, N-5009 Bergen, Norway

Wolfgang R. G. Dostmann (2:487)
Department of Pharmacology, University of Vermont, College of Medicine, Burlington, Vermont 05405

Matthias K. Dreyer (1:289)
Aventis Pharma Deutschland GmbH, Structural Biology, Frankfurt 65926, Germany

Guo Guang Du (2:45)
Banting and Best Department of Medical Research, University of Toronto, Charles H. Best Institute, Toronto, Ontario M5G 1X5, Canada

Keyong Du (3:115)
Peptide Biology Laboratories, The Salk Institute for Biological Studies, La Jolla, California 92037

Michael R. Duchen (2:73)
Department of Physiology and UCL Mitochondrial Biology Group, University College London, London WC1E 6JJ, United Kingdom

William G. Dunphy (1:693)
California Institute of Technology, Howard Hughes Medical Institute, Division of Biology, Pasadena, California 91125

Joanne Durgan (2:389)
Cancer Research UK, London Research Institute, Lincoln's Inn Fields Laboratories, WC2A 3PX London, United Kingdom

Michael L. Dustin (1:79)
NYU School of Medicine, Department of Pathology, Skirball Institute of Biomolecular Medicine, New York, New York 10016

Peter A. Edwards (2:287)
Departments of Biological Chemistry and Medicine, Molecular Biology Institute, University of California, Los Angeles, California 90095

Jackson G. Egen (1:355)
Department of Molecular and Cell Biology, Howard Hughes Medical Institute, University of California, Berkeley, Berkeley, California 94720

Lee E. Eiden (3:375)
Section on Molecular Neuroscience, Laboratory of Cellular and Molecular Regulation, National Institute of Mental Health Intramural Research Program, Bethesda, Maryland 20892

Elaine A. Elion (2:357)
Department of Biological Chemistry and Molecular Pharmacology, Harvard Medical School, Boston, Massachusetts 02115

Scott Emr (2:177)
Division of Cellular and Molecular Medicine, The Howard Hughes Medical Institute, University of California San Diego, School of Medicine, La Jolla, California 92093

Othmar G. Engelhardt (2:771)
Abteilung Virologie, Institut für Medizinische Mikrobiologie und Hygiene, Universität Freiburg, 79085 Freiburg, Germany

Christophe Erneux (2:11)
Interdisciplinary Research Institute (IRIBHN), Université Libre de Bruxelles, Brussels, Belgium

Peter J. Espenshade (3:353)
Department of Molecular Genetics, University of Texas Southwestern Medical Center at Dallas, Dallas, Texas 75390

Edward D. Esplin (2:405)
Department of Pharmacology, University of Texas Southwestern Medical Center, Dallas, Texas 75390

B. Mark Evers (3:477)
Department of Surgery, The University of Texas Medical Branch, Galveston, Texas 77555

Joanne L. Eyles (1:431)
The Walter and Eliza Hall Institute of Medical Research, The Cooperative Research Centre for Cellular Growth Factors, Victoria 3050, Australia

Sheelagh Fame (1:547)
MRC Protein Phosphorylation Unit, Department of Biochemistry, University of Dundee, Dundee DD1 5EH, Scotland

Marilyn Farquhar (3:351)
Department of Cellular and Molecular Medicine, University of California at San Diego, La Jolla, California 92093

Robert Feil (2:511)
Institut für Pharmakologie und Toxikologie, TU München, D-80802 München, Germany

Gui-Jie Feng (2:619)
Molecular Pharmacology Group, Division of Biochemistry and Molecular Biology, University of Glasgow, Glasgow G12 8QQ, Scotland, United Kingdom

Stanley Fields (2:311)
Howard Hughes Medical Institute, Departments of Genome Sciences and Medicine, University of Washington, Seattle, Washington 98195

James J. Fiordalisi (2:737)
Departments of Radiation Oncology and Pharmacology, and the Lineberger Comprehensive Cancer Center, University of North Carolina at Chapel Hill, Chapel Hill, North Carolina 27599

Richard A. Firtel (2:217)
Section of Cell Developmental Biology, Division of Biological Sciences, and Center for Molecular Genetics, University of California San Diego, La Jolla, California 92093

Garret A. Fitzgerald (2:265)
Center for Experimental Therapeutics, University of Pennsylvania, Philadelphia, Pennsylvania 19104

Andrew Flint (1:671)
CEPTYR, Inc., Bothell, Washington 98021

Marco Foiani (3:197)
Dipartimento di Genetica e Biologia dei Microrganismi, Universita' di Milano, 20133 Milano, Italy and Instituto FIRC di Oncologia Molecolare, 20139 Milano, Italy

Barry Marc Forman (3:47)
The Beckman Research Institute, City of Hope National Medical Center, Division of Molecular Medicine, The Gonda Diabetes and Genetic Research Center, Duarte California 91010

Albert J. Fornace, Jr. (3:179)
Gene Response Section, Center for Cancer Research, National Cancer Institute, Bethesda, Maryland 20892

Sharron H. Francis (2:447; 2:465)
Department of Molecular Physiology and Biophysics, Vanderbilt University School of Medicine, Nashville, Tennessee 37232

Günter Fritz (2:87)
Department of Pediatrics, Division of Clinical Chemistry and Biochemistry, University of Zürich, CH-8032 Zürich, Switzerland

David A. Fruman (2:135)
Department of Molecular Biology and Biochemistry, University of California Irvine, Irvine, California 92697

Antony Galione (2:15)
Department of Pharmacology, University of Oxford, Oxford OX1 2JD, United Kingdom

Chris S. Gandhi (1:209)
Department of Molecular and Cell Biology, University of California, Berkeley, Berkeley, California 94720

David L. Garbers (2:427)
Cecil H. and Ida Green Center for Reproductive Biology Sciences, Howard Hughes Medical Institute and Department of Pharmacology, University of Texas Southwestern Medical Center, Dallas, Texas 75390

K. Christopher Garcia (1:259)
Department of Microbiology and Immunology, Department of Structural Biology, Stanford University School of Medicine, Stanford, California 94305-5124

Benjamin Geiger (2:317)
Department of Molecular Cell Biology, Weizmann Institute of Science, Rehovot 76100, Isreal

Larry Gerace (3:419)
Department of Cell Biology, The Scripps Research Institute, La Jolla, California 92037

Andrea Gerstner (2:515)
Department Pharmzaie-Zentrum für Pharmaforschung, Ludwig-Maximilians-Universität München, 81377 München, Germany

Amato J. Giaccia (3:277)
Department of Radiation Oncology, Stanford University School of Medicine, Stanford, California 94305

Michele Giannattasio (3:197)
Dipartimento di Genetica e Biologia dei Microrganismi, Universita' di Milano, 20133 Milano, Italy

Vincent Giguère (3:35)
Molecular Oncology Group, McGill University Health Centre, Montréal, Québec H3G 1Y6, Canada

Christopher K. Glass (3:25)
Department of Cellular and Molecular Medicine, School of Medicine, University of California San Diego, La Jolla, California 92093

Martin J. Glennie (1:319)
University of Southampton General Hospital, Tenovus Research Laboratory, Cancer Sciences Division, School of Medicine, Southampton SO16 6YD, United Kingdom

Jennifer L. Glick (2:431)
Department of Pharmacology, University of Washington, Seattle, Washington 98195

Joseph L. Goldstein (3:353)
Department of Molecular Genetics, University of Texas Southwestern Medical Center at Dallas, Dallas, Texas 75390

Venkatesh Gopal (2:465)
Department of Molecular Physiology and Biophysics, Vanderbilt University School of Medicine, Nashville, Tennessee 37232

Myriam Gorospe (3:283)
Laboratory of Cellular and Molecular Biology, Gerontology Research Center, National Institute on Aging – IRP, National Institutes of Health, Baltimore, Maryland 21224

Cedric Govaerts (1:161)
Faculty of Medicine, University of Brussels, Institut de Recherche Interdisciplinaire, Brussels B-1070, Belgium

Paul R. Graves (2:301)
Department of Pharmacology and Cancer Biology, Center for Chemical Biology, Duke University, Durham, North Carolina 27710

Patrick W. Gray (1:149)
Macrogenics, Inc., Seattle, Washington 98103

Irene Graziani (3:393)
Center for Molecular Medicine, Maine Medical Center Research Institute, Scarborough, Maine 04074

Douglas R. Green (3:431)
Division of Cellular Immunology, La Jolla Institute for Allergy and Immunology, San Diego, California 92093

Michael E. Greenberg (3:83)
Division of Neuroscience, Children's Hospital and Department of Neurobiology, Harvard Medical School, Boston, Massachusetts 02115

Iva Greenwald (2:809)
Howard Hughes Medical Institute, Department of Biochemistry and Molecular Biophysics, Columbia University, New York, New York 10032

Haihua Gu (1:707)
Beth Israel-Deaconess Medical Center, Cancer Biology Program, Division of Hematology-Oncology, Harvard Medical School, Boston, Massachusetts 02215

Nuri Gueven (3:225)
Queensland Cancer Fund Research Laboratory, Queensland Institute of Medical Research, Brisbane 4029, Australia

J. Silvio Gutkind (2:589)
Oral and Pharyngeal Cancer Branch, National Institute of Dental and Craniofacial Research, National Institutes of Health, Bethesda, Maryland 20892

Jesper Z. Haeggström (2:275)
Department of Medical Biochemistry and Biophysics, Division of Chemistry II, The Scheele Laboratory, Karolinska Institutes, S-171 77 Stockholm, Sweden

Alan Hall (2:701)
Department of Biochemistry and Molecular Biology, University College London, London WC1E 6BT, United Kingdom

Michael N. Hall (3:299)
Division of Biochemistry, Biozentrum, University of Basel, 4056 Basel, Switzerland

Otto Haller (2:771)
Abteilung Virologie, Institut für Medizinische Mikrobiologie und Hygiene, Universität Freiburg, 79085 Freiburg, Germany

Heidi E. Hamm (1:335)
Vanderbilt University Medical Center, Department of Pharmacology, Nashville, Tennessee 37232-6600

Yusef A. Hannun (2:257)
Department of Biochemistry and Molecular Biology, Medical University of South Carolina, Charleston, South Carolina 29425

Carl A. Hansen (2:623)
Department of Biological and Allied Health Sciences, Bloomsburg University, Bloomsburg, Pennsylvania 17815

T. Kendall Harden (2:631)
Department of Pharmacology, Lineberger Comprehensive Cancer Center, and UNC Neuroscience Center, The University of North Carolina at Chapel Hill, Chapel Hill, North Carolina 27599

D. Grahame Hardie (1:535)
University of Dundee, Wellcome Trust Biocentre, Devision of Molecular Physiology, Dundee DD1 I, Scotland, United Kingdom

Kiminori Hasegawa (1:475)
Department of Immunology, Genentech, Inc., South San Francisco, California 94080

Phillip T. Hawkins (2:203)
Inositide Laboratory, Signaling Program, Babraham Institute, Cambridge CB2 4AT, United Kingdom

Timothy A. J. Haystead (2:301)
Serenex, Inc., Research Triangle Park, North Carolina 27709

Xiao-lin He (1:259)
Department of Microbiology and Immunology, Department of Structural Biology, Stanford University School of Medicine, Stanford, California 94305-5124

Claus W. Heizmann (2:87)
Department of Pediatrics, Division of Clinical Chemistry and Biochemistry, University of Zürich, CH-8032 Zürich, CH-8032 Switzerland

Carl-Henrik Heldin (1:391)
Biomedical Center, Ludwig Institute for Cancer Research, Uppsala S-751 24, Sweden

Michelle L. Hermiston (1:689)
Departments of Medicine, Microbiology and Immunology, Howard Hughes Medical Institute and Department of Pediatrics, University of California-San Francisco, San Francisco, California 94143-0795

Peter Herrlich (3:61; 3:257)
Forschungszentrum Karlsruhe, Institute of Genetics and Toxicology, 76021 Karlsruhe, Germany

Elizabeth A. Hewat (1:95)
Institut de Biologie Structurale J-P Ebel, Grenoble 38027, France

Bertil Hille (1:215)
Department of Physiology and Biophysics, University of Washington School of Medicine, Seattle, Washington 98195-7290

Douglas J. Hilton (1:431)
The Walter and Eliza Hall Institute of Medical Research, The Cooperative Research Centre for Cellular Growth Factors, Victoria 3050, Australia

K. A. Hinchliffe (2:123)
Department of Pharmacology, University of Cambridge, Cambridge CB2 1GA, United Kingdom

Steffan N. Ho (3:125)
Departments of Pathology and Cellular and Molecular Medicine, University of California San Diego, La Jolla, California 92093

Su-Chin Ho (1:161)
Faculty of Medicine, University of Brussels, Institut de Recherche Interdisciplinaire, Brussels B-1070, Belgium

Mark Hochstrasser (2:347)
Department of Molecular Biophysics and Biochemistry, Yale University, New Haven, Connecticut 06520

Franz Hofmann (2:511)
Institut für Pharmakologie und Toxikologie, TU München, D-80802 München, Germany

Christopher W. Hogue (2:311)
Department of Biochemistry, University of Toronto, Toronto, Ontario M5G 1X5, Canada

Wim G. J. Hol (2:525)
Department of Pharmacology, University of Washington, Seattle, Washington, and Department of Biochemistry and Biological Structure, Howard Hughes Medical Institute, University of Washington, Seattle, Washington 98195

Jocelyn Holash (2:849)
Regeneron Pharmaceuticals, Inc., Tarrytown, New York 10591

Robert A. Holmgren (3:167)
Department of Biochemistry Molecular Biology and Cell Biology, Robert H. Lurie Comprehensive Cancer Center, Northwestern University, Evanston, Illinois 60208

Barry Honig (1:11)
Department of Biochemistry and Molecular Biophysics, Howard Hughes Medical Institute, Columbia University, New York, New York 10032

Bruce S. Hostager (1:315)
Department of Microbiology, University of Iowa, Iowa City, Iowa 52242

Stevan R. Hubbard (1:299)
New York University, School of Medicine, Skirball Institute of Biomolecular Medicine, New York, New York 10016

Michael Huber (1:327)
Department of Molecular Immunology, Biology III, University of Freiburg and Max-Planck Institute for Immunobiology, Freiburg 79108, Germany

Tony Hunter (1:369; 1:373; 1:685)
Molecular and Cell Biology Laboratory, The Salk Institute for Biological Sciences, La Jolla, California 92037-1099

Anna Huttenlocher (2:105)
Departments of Pediatrics and Pharmacology, University of Wisconsin, Madison, Wisconsin 53716

Sarah G. Hymowitz (1:305)
Department of Protein Engineering, Genentech, Inc., South San Francisco, California 94080

James N. Ihle (1:427)
Department of Biochemistry, Howard Hughes Medical Institute, St. Jude Children's Research Hospital, Memphis, Tennessee 38105-2729

Jean-Luc Imler (2:333)
National Center of Scientific Research, Institute of Molecular and Cellular Biology, 67084 Strasbourg, France

R. F. Irvine (2:123)
Department of Pharmacology, University of Cambridge, Cambridge CB2 1GA, United Kingdom

Ehud Y. Isacoff (1:209)
Department of Molecular and Cell Biology, University of California, Berkeley, Berkeley, California 94720

Xavier Iturrioz (2:389)
Cancer Research UK, London Research Institute, Lincoln's Inn Fields Laboratories, WC2A 3PX London, United Kingdom

Lars F. Iversen (1:659)
Protein Chemistry, Novo Nordisk, Bagsvaerd DK 2880, Denmark

Ravi Iyengar (2:605)
Department of Pharmacology and Biological Chemistry, Mount Sinai School of Medicine, New York, New York 10029

Stephen P. Jackson (1:557; 3:219)
Wellcome Trust and Cancer Research Campaign, Institute of Cancer and Developmental Biology, Cambridge, United Kingdom, and Department of Zoology, University of Cambridge, Cambridge CB2 1GA, United Kingdom

Lily Yeh Jan (1:187; 2:667)
Howard Hughes Medical Institute, Departments of Physiology and Biochemistry, University of California San Francisco, San Francisco, California 94143

Fabiola Janiak-Spens (1:563)
Department of Chemistry and Biochemistry, University of Oklahoma, Norman, Oklahoma 73019

Paul A. Janmey (2:209)
Department of Physiology, Institute of Medicine and Engineering, University of Pennsylvania, Philadelphia, Pennsylvania 19104

Peter Gildsig Jansen (1:659)
Scientific Computing, Novo Nordisk, Måløv DK 2760, Denmark

Sophie Jarriault (2:809)
Howard Hughes Medical Institute, Department of Biochemistry and Molecular Biophysics, Columbia University, New York, New York 10032

Jonathan A. Javitch (1:155)
Columbia University Center for Molecular Recognition, and Departments of Psychiatry and Pharmacology, New York, New York 10032

Elwood V. Jensen (3:7)
Department of Cell Biology, University of Cincinnati College of Medicine, Cincinnati, Ohio 45267

Kristen Jepsen (3:29)
Howard Hughes Medical Institute, Department and School of Medicine, University of California San Diego, La Jolla, California 92093

E. Yvonne Jones (1:57)
Division of Structural Biology, Wellcome Trust Centre for Human Genetics, Cancer Research United Kingdom Receptor Stucture Research Group, Oxford, OX3 7BN, United Kingdom

Katherine A. Jones (3:161)
Regulatory Biology Laboratory Salk Institute for Biological Studies, La Jolla, California 92037

J. Dedrick Jordan (2:605)
Department of Pharmacology and Biological Chemistry, Mount Sinai School of Medicine, New York, New York 100329

Jomon Joseph (2:695)
Laboratory of Gene Regulation and Development, National Institute of Child Health and Human Development, National Institutes of Health, Bethesda, Maryland 20892

Louis B. Justement (3:555)
Division of Developmental and Clinical immunology, Department of Microbiology, University of Alabama at Birmingham, Birmingham, Alabama 35294

Yariv Kafri (2:317)
Department of Physics of Complex Systems, Weizmann Institute of Science, Rehovot 76100, Israel

Richard A. Kahn (2:727)
Department of Biochemistry, Emory University School of Medicine, Atlanta, Georgia 30322

Shin W. Kang (1:475)
Department of Immunology, Genentech, Inc., South San Francisco, California 94080

Arthur Karlin (1:223)
Departments of Biochemistry and Molecular Biophysics, College of Physicians and Surgeons, Center for Molecular Recognition, Columbia University, New York, New York 10032

Heidi R. Kast-Woelbern (2:287)
Department of Biological Chemistry, University of California, Los Angeles, California 90095

Randal J. Kaufman (3:311)
Department of Biological Chemistry, Howard Hughes Medical Institute, University of Michigan Medical Center, Ann Arbor, Michigan 48109

Andrius Kazlauskas (1:397)
Schepens Eye Research Institute, Harvard Medical School, Boston, Massachusetts 02114

James H. Keen (3:441)
Cell Biology and Signaling Program, Kimmel Cancer Center, Thomas Jefferson University, Philadelphia, Pennsylvania 19107

Rolf Kemler (2:789)
Department of Molecular Embryology, Max-Planck Institute of Immunobiology, 79085 Freiburg, Germany

Bruce E. Kemp (2:495)
St. Vincent's Institute of Medical Research, Fitzroy, Victoria, Australia

Bruce E. Kemp (1:539)
St. Vincent's Institute of Medical Research, Fitzroy, Victoria 3065, Australia

Mary B. Kennedy (1:543; 2:329)
Division of Biology, California Institute of Technology, Pasadena, California 91125

Matthew A. Kennedy (2:287)
Department of Biological Chemistry, University of California, Los Angeles, California 90095

Ushio Kikkawa (2:119)
Biosignal Research Center, Kobe University, Kobe, Japan

Albert H. Kim (2:839)
Molecular Neurobiology Program, Skirball Institute of Biomolecular Medicine, New York University School of Medicine, New York, New York 10016

Soo-A Kim (2:143)
The Life Science Institute and Department of Biological Chemistry, University of Michigan Medical School, Ann Arbor, Michigan 48109

Sung-Hou Kim (1:197)
Department of Chemistry and Lawrence Berkeley National Laboratory, University of California, Berkeley, Berkeley, California 94720

Youngjoo Kim (1:653)
Department of Biochemistry and Molecular Biology, Oregon Health and Science University, Portland, Oregon 97201-3098

Kirst King-Jones (3:69)
Howard Hughes Medical Institute, University of Utah, Salt Lake City, Utah 84112

Chris Kintner (2:813; 3:149)
Molecular Neurobiology Laboratory, Salk Institute for Biological Studies, La Jolla, California 92037

Saul Kivimäe (1:575)
Laboratory of Genetics, The Rockefeller University, New York, New York 10021

Claude B. Klee (1:631)
Laboratory of Biochemistry, National Cancer Institute, National Institutes of Health, Bethesda, Maryland 20892-4255

Rüdiger Klein (1:421)
Department of Molecular Neurobiology, Max-Planck Institute of Neurobiology, Martinsried D-82152, Germany

Thomas Kleppisch (2:511)
Institut für Pharmakologie und Toxikologie, TU München, D-80802 München, Germany

Steven A. Kliewer (3:53)
Nuclear Receptor Discovery Research, GlaxoSmithKline Research Triangle Park, North Carolina 27709

Richard A. Klinghoffer (2:845)
CEPTYR, Inc., Bothell, Washington 98021

Juergen A. Knoblich (2:571)
Research Institute of Molecular Pathology (I.M.P.), A-1030 Vienna, Austria

Bostjan Kobe (1:539; 2:495)
Department of Biochemistry and Molecular Biology, Institute for Molecular Bioscience, University of Queensland, St. Lucia, Brisbane QLD 4072, Queensland, Australia

George Kochs (2:771)
Abteilung Virologie, Institut für Medizinische Mikrobiologie und Hygiene, Universität Freiburg, 79085 Freiburg, Germany

Monica Kong-Beltran (2:861)
Genentech, Inc., San Francisco, California 94080

Rolf König (3:545)
Department of Microbiology and Immunology, Sealy Center for Molecular Science, The University of Texas Medical Branch, Galveston, Texas 77555

Albert C. Koong (3:277)
Division of Radiation Biology, Department of Radiation Oncology, Stanford University School of Medicine, Stanford, California 94305

Murray Korc (3:471)
Division of Endocrinology, Diabetes and Metabolism, University of California Irvine, Irvine, California 92697

Daniel Kornitzer (3:29)
Department of Molecular Microbiology, Bruce Rappaport Faculty of Medicine, Technion-Israel Institute of Technology, Haifa, Israel

Anthony A. Kossiakoff (1:241)
Cummings Life Sciences Center, University of Chicago, Department of Biochemistry and Molecular Biology, Chicago, Illinois 60637

Jun Kotera (2:465)
Department of Molecular Physiology and Biophysics, Vanderbilt University School of Medicine, Nashville, Tennessee 37232

M. V. Kovalenko (1:397)
Schepens Eye Research Institute, Harvard Medical School, Boston, Massachusetts 02114

Tohru Kozasa (2:639)
Department of Pharmacology, University of Illinois at Chicago, Chicago, Illinois 60637

Sergei Kozlov (3:25)
Queensland Cancer Fund Research Laboratory, Queensland Institute of Medical Research, Brisbane 4029, Australia

Keith G. Kozminski (2:733)
Departments of Biology and Cell Biology, University of Virginia, Charlottesville, Virginia 22904

Sonja Krugmann (2:203)
Inositide Laboratory, Signalling Programme, Babraham Institute, Cambridge CB2 4AT, United Kingdom

John Kuriyan (1:343; 1:387)
Physical Biosciences Division, Lawrence Berkeley National Laboratory, Berkeley, California 94720-3202

Riki Kurokawa (3:25)
Department of Cellular and Molecular Medicine, School of Medicine, University of California San Diego, La Jolla, California 92093

Peter D. Kwong (1:99)
National Institute of Allergy and Infectious Diseases, National Institute of Health Vaccine Research Center, Bethesda, Maryland 20892

Wi S. Lai (3:319)
Office of Clinical Research and Laboratory of Signal Transduction, National Institute of Environmental Health Sciences, Research Triangle Park, North Carolina 27709

Elise Lamar (3:149)
Molecular Neurobiology Laboratory, Salk Institute for Biological Studies, La Jolla, California 92037

Millard H. Lambert (3:21)
Nuclear Receptor Discovery Research, GlaxoSmithKline, Research Triangle Park, North Carolina 27709

David G. Lambright (2:575)
Program in Molecular Medicine and Department of Biochemistry and Molecular Pharmacology, University of Massachusetts Medical School, Worcester, Massachusetts 01605

Doron Lancet (1:145)
Department of Molecular Genetics, Weizmann Institute of Science, Rehovot 76100, Israel

Reiko Landry (3:161)
Regulatory Biology Laboratory Salk Institute for Biological Studies, La Jolla, California 92037

Wallace Y. Langdon (1:483)
Department of Pathology, University of Western Australia, Crawley, Western Australia 6009, Australia

Lorene K. Langeberg (2:383)
Howard Hughes Medical Institute, Vollum Institute, Oregon Health and Sciences University, Portland, Oregon 97239

Paul Lasko (3:327)
Department of Biology, McGill University, Montréal, Québec H3G 1Y6, Canada

Vaughn Latham (3:293)
Department of Anatomy and Structural Biology, Albert Einstein College of Medicine, Bronx, New York, and Dana-Farber Cancer Institute, Boston, Massachusetts 02115

Martin F. Lavin (3:225)
Queensland Cancer Fund Research Laboratory, Queensland Institute of Medical Research, Brisbane 4029, Australia, and Department of Surgery, University of Queensland, Brisbane QLD 4072, Australia

Kevin A. Lease (1:579)
Department of Biology, University of Virginia, Charlottesville, Virginia 22904

Hakon Leffler (1:87)
Section MIG, Institute of Laboratory Medicine, University of Lund, SE 223 62 Lund, Sweden

Mark A. Lemmon (2:161)
Department of Biochemistry and Biophysics, University of Pennsylvania Medical Center, Philadelphia, Pennsylvania 19104

Ann E. Leonard (2:883)
Gene Expression Laboratory, The Salk Institute for Biological Studies, La Jolla, California 92037

Alexander Levitzki (1:451)
Unit of Cell Signaling, Department of Biological Chemistry, The Alexander Silberman Institute of Life Sciences, The Hebrew University of Jerusalem, Jerusalem 91904, Israel

Hong-Jun Liao (2:5)
Department of Biochemistry, Vanderbilt University School of Medicine, Nashville, Tennessee 37232

Lucy Liaw (3:455)
Center for Molecular Medicine, Maine Medical Center Research Institute, Scarborough, Maine 04074

Giordano Liberi (3:197)
Dipartimento di Genetica e Biologia dei Microrganismi, Universita' di Milano, 20133 Milano, Italy and Instituto FIRC di Oncologia Molecolare, 20139 Milano, Italy

Heiko Lickert (2:789)
Department of Molecular Embryology, Max-Planck Institute of Immunobiology, 79085 Freiburg, Germany

Robert C. Liddington (1:123)
Program on Cell Adhesion, The Burnham Institute, La Jolla, California 92037

Thomas M. Lincoln (2:479)
Department of Pathology, University of Alabama at Birmingham, Birmingham, Alabama 35294

Jürgen U. Linder (2:535)
Department of Pharmaceutical Biochemistry, Pharmaceutical Institute, University of Tübingen, 72074 Tübingen, Germany

Maurine E. Linder (1:331; 2:585)
Department of Cell Biology and Physiology, Washington University School of Medicine, St. Louis, Missouri 63110

Hui Liu (2:171)
Center for Cancer Research, Massachusetts Institute of Technology, Cambridge, Massachusetts 02139

Zhengchang Liu (3:365)
Department of Molecular Biology, University of Texas Southwestern Medical Center, Dallas, Texas 75390

Marja K. Lohela (2:855)
Molecular/Cancer Biology Laboratory and Ludvig Institute for Cancer Research, Haartman Institute and Helsinki University Central Hospital, Biomedicum Helsinki, University of Helsinki, FIN-00014 Helsinki, Finland

Sarah H. Louie (1:177)
Department of Pharmacology and Center for Developmental Biology, Howard Hughes Medical Institute, University of Washington School of Medicine, Seattle, Washington 98195–7750

Deirdre K. Luttrell (2:609)
Department of High Throughput Biology, GlaxoSmithKline, Research Triangle Park, North Carolina 27709

Louis M. Luttrell (2:609)
The Geriatrics Research, Education and Clinical Center, Durham Veterans Affairs Medical Center, Durham, North Carolina, and Department of Medicine, Duke University Medical Center, Durham, North Carolina 27710

Karen M. Lyons (2:833)
Department of Molecular, Cell and Developmental Biology, Department of Biological Chemistry, Department of Orthopaedic Surgery, The David Geffen School of Medicine at the University of California Los Angeles, Los Angeles, California 90095

S. Lance Macaulay (1:293)
Health Sciences and Nutrition, CSIRO, Parkville, Victoria 3052, Australia

Michael Maceyka (2:247)
Department of Biochemistry, Medical College of Virginia Campus, Virginia Commonwealth University, Richmond, Virginia 23284

Thomas Maciag (3:393; 3:455)
Center for Molecular Medicine, Maine Medical Center Research Institute, Scarborough, Maine 04074

Fernando Macian (3:119)
Center for Blood Research and Department of Pathology, Harvard Medical School, Boston, Massachusetts 02115

Carol MacKintosh (1:607)
MRC Protein Phosphorylation Unit, School of Life Sciences, University of Dundee, Dundee, Scotland DD1 5EH, United Kingdom

David H. MacLennan (2:45)
Banting and Best Department of Medical Research, University of Toronto, Charles H. Best Institute, Toronto, Ontario M5G 1X5, Canada

Nadir A. Mahmood (2:405)
Department of Pharmacology, University of Texas Southwestern Medical Center, Dallas, Texas 75390

Craig C. Malbon (1:177)
Department of Molecular Pharmacology, Diabetes and Metabolic Research Center, University Medical Center, SUNY/Stony Brook, Stony Brook, New York 11794-8651

Sohail Malik (3:11)
Laboratory of Biochemistry and Molecular Biology, Rockefeller University, New York, New York 10021

Orna Man (1:145)
Department of Molecular Genetics, Weizmann Institute of Science, Rehovot 76100, Israel

Carol L. Manahan (2:645)
Department of Cell Biology, Johns Hopkins School of Medicine, Baltimore Maryland 21201

Anna Mandinova (3:393)
Center for Molecular Medicine, Maine Medical Center Research Institute, Scarborough, Maine 04074

Vincent C. Manganiello (2:441)
Department of Cell and Molecular Biology, Lund University, SE-22100 Lund, Sweden, and Pulmonary-Critical Care Medicine Branch, National Heart, Lung, and Blood Institute, National Institutes of Health, Bethesda, Maryland 20892

James L. Manley (3:331)
Department of Biological Sciences, Columbia University, New York, New York 10032

Matthias Mann (2:295)
Protein Interaction Laboratory, Center for Experimental Bioinformatics, University of Southern Denmark, DK-5230 Odense M, Denmark

Gerald Manning (1:373)
SUGEN, Inc., South San Francisco, California 94080

Ed Manser (1:499)
Institute of Molecular and Cell Biology, Glaxo–IMCB Group, Singapore 117609, Singapore

Marta Margeta-Mitrovic (1:187)
Department of Pathology, University of California, San Francisco, San Francisco, California 94143-0450

Robert F. Margolskee (2:657)
Department of Physiology and Biophysics, Howard Hughes Medical Institute, The Mount Sinai School of Medicine, New York, New York 10029

Julia Marinissen (2:589)
Oral and Pharyngeal Cancer Branch, National Institute of Dental and Craniofacial Research, National Institutes of Health, Bethesda, Maryland 20892

Roy A. Mariuzza (1:39)
Center for Advanced Research, W.M. Keck Laboratory for Structural Biology, University of Maryland, Biotechnology Institute, Rockville, Maryland 20850

Mina D. Marmor (1:405)
Department of Biological Regulation, Weizmann Institute of Science, Rehovot 76100, Israel

G. Steven Martin (1:441)
Department of Molecular and Cell Biology, University of California at Berkeley, Berkeley, California 94720-3204

Karen H. Martin (1:463)
Department of Microbiology, University of Virginia Health System, Charlottesville, Virginia 22908-0734

Sergio E. Martinez (2:525)
Department of Pharmacology, University of Washington, Seattle, Washington 98195

Michael B. Mathews (3:335)
Department of Biochemistry and Molecular Biology, New Jersey Medical School and the Graduate School of Biomedical Sciences, University of Medicine and Dentistry of New Jersey, Newark, New Jersey 07107

Bruce J. Mayer (1:471)
Department of Genetics and Developmental Biology, University of Connecticut Health Center, Farmington, Connecticut 06030-3301

Mark L. Mayer (1:219)
Laboratory of Cellular and Molecular Neurophysiology, National Institute of Child Health and Human Development, National Institutes of Health, Bethesda, Maryland 20892-4495

Maria R. Mazzoni (1:335)
Department of Pharmacology, Vanderbilt University Medical Center, Nashville, Tennessee 37232-6600

Frank McCormick (2:671)
Cancer Research Institute, University of California Comprehensive Cancer Center, San Francisco, California 94080

Clare H. McGowan (3:401)
Department of Molecular Biology and Department of Cell Biology, The Scripps Research Institute, La Jolla, California 92037

Melissa M. McKay (2:727)
Department of Biochemistry, Emory University School of Medicine, Atlanta, Georgia 30322

Wallace L. McKeehan (1:265)
Institute of Biosciences, Texas A&M University System Health Science Center, Center for Cancer Biology and Nutrition, Houston, Texas 77030-3303

Alison J. McLean (2:619)
Molecular Pharmacology Group, Division of Biochemistry and Molecular Biology, University of Glasgow, Glasgow G12 8QQ, Scotland, United Kingdom

Anthony R. Means (2:83)
Department of Pharmacology and Cancer Biology, Duke University Medical Center, Durham, North Carolina 27710

Ruedi Meili (2:217)
Section of Cell Developmental Biology, Division of Biological Sciences, and Center for Molecular Genetics, University of California San Diego, La Jolla, California 92093

Jingwei Meng (2:601)
Department of Pharmacology and Cancer Biology, Duke University Medical Center, Durham, North Carolina 27710

Mark Merchant (2:793)
Department of Molecular Biology, Genentech, Inc., South San Francisco, California 94080

Frank Mercurio (3:107)
Signal Research Division, Celgene, San Diego, California 92037

Graeme Milligan (2:619)
Molecular Pharmacology Group, Division of Biochemistry and Molecular Biology, University of Glasgow, Glasgow G12 8QQ, Scotland, United Kingdom

Guo-Li Ming (2:871)
The Salk Institute, La Jolla, California 92037

Daniel L. Minor (1:203)
Department of Biochemistry and Biophysics, Cardiovascular Research Institute, University of California, San Francisco, San Francisco, California 94143-0450

Nadeem Moghal (2:805)
HHMI and Division of Biology, California Institute of Technology, Pasadena, California 91125

Neils Peter H. Møller (1:659)
Signal Transduction, Novo Nordisk, Bagsvaerd, DK 2880, Denmark

Marco Mongillo (2:459)
Venetian Institute of Molecular Medicine, Department of Biomedical Studies, University of Padua, 35129 Padua, Italy

Marc Montminy (3:115)
Peptide Biology Laboratories, The Salk Institute for Biological Studies, La Jolla, California 92037

Randall T. Moon (1:177)
Department of Pharmacology and Center for Developmental Biology, Howard Hughes Medical Institute, University of Washington, Seattle, Washington 98195-7750

Richard I. Morimoto (3:269)
Department of Biochemistry, Molecular Biology and Cell Biology, Rice Institute for Biomedical Research, Northwestern University, Evanston, Illinois 60208

Stephen E. Moss (2:101)
Division of Cell Biology, Institute of Ophthalmology, University College London, London WC1E 6JJ, United Kingdom

Helen R. Mott (2:745)
Department of Biochemistry, University of Cambridge, Cambridge CB2 1GA, United Kingdom

Carla Mouta (3:455)
Center for Molecular Medicine, Maine Medical Center Research Institute, Scarborough, Maine 04074

Marco Muda (1:703)
Serono Reproductive Biology Institute, Rockland, Massachusetts 02370

Marc C. Mumby (2:405)
Department of Pharmacology, University of Texas Southwestern Medical Center, Dallas, Texas 75390

Gretchen A. Murphy (2:681)
Department of Pharmacology, and Lineberg Comprehensive Cancer Center, University of North Carolina at Chapel Hill, Chapel Hill, North Carolina 27599

Marco Muzi-Falconi (3:197)
Dipartimento di Genetica e Biologia dei Microrganismi, Universita' di Milano, 20133 Milano, Italy

Raghavendra Nagaraj (2:827)
Department of Molecular Cell and Developmental Biology, Biological Chemistry, Human Genetics, Molecular Biology Institute, University of California at Los Angeles, Los Angeles, California 90095

Stefan R. Nahorski (2:19)
Department of Cell Physiology and Pharmacology, University of Leicester, Leicester LE2 3QE, United Kingdom

Angus C. Nairn (1:567)
Department of Psychiatry, Yale University School of Medicine, New Haven, Connecticut 06508

Piers Nash (1:379)
Samuel Lunenfeld Research Institute, Mount Sinai Hospital, Toronto, ONT M5G 1X5, Canada

Benjamin G. Neel (1:707)
Cancer Biology Program, Division of Hematology-Oncology, Beth Israel Deaconess Medical Center, Harvard Medical School, Boston, Massachusetts 02215

Alexandra C. Newton (1:551; 2:187)
Department of Pharmacology, University of California at San Diego, La Jolla, California 92093

Yasutomi Nishizuka (2:119)
Biosignal Research Center, Kobe University, Kobe 657-8501, Japan

Joseph P. Noel (1:685)
Structural Biology Lab, The Salk Institute for Biological Sciences, La Jolla, California 92037

Ellen A. A. Nollen (3:269)
Department of Biochemistry, Molecular Biology and Cell Biology, Rice Institute for Biomedical Research, Northwestern University, Evanston, Illinois 60208

Irene M. A. Nooren (1:21)
European Bioinformatics Institute, Wellcome Trust Genome Campus, Hinxton, Cambridge, CB10 1SD, United Kingdom

Rodney O'Connor (2:51)
The Babraham Institute, Babraham, Cambridge CB2 4AT, United Kingdom

Stefan Offermanns (2:581)
Pharmakologisches Institut, Universität Heidelberg, 69120 Heidelberg, Germany

Tsviya Olender (1:145)
Department of Molecular Genetics, Weizmann Institute of Science, Rehovot 76100, Israel

Shao-En Ong (2:295)
Protein Interaction Laboratory, Center for Experimental Bioinformatics, University of Southern Denmark, DK-5230 Odense M, Denmark

Darerca Owen (2:745)
Department of Biochemistry, University of Cambridge, Cambridge CB2 1GA, United Kingdom

Lisa J. Pagliari (3:431)
Division of Cellular Immunology, La Jolla Institute for Allergy and Immunology, San Diego, California 92037

Lily Pao (1:707)
Cancer Biology Program, Division of Hematology-Oncology, Beth Israel Deaconess Medical Center, Harvard Medical School, Boston, Massachusetts 02215

John Papaconstantinou (3:515)
University of Texas Medical Branch, Department of Human Biological Chemistry and Genetics, Galveston, Texas 77555

Leonardo Pardo (1:161)
Unitat de Bioestadistica, Facultat de Medicina, Laboratori de Medicina Computacional, Universitat Autonoma de Barcelona, Bellaterra 08193, Spain

Hay-Oak Park (2:733)
Department of Molecular Genetics, The Ohio State University, Columbus, Ohio 43210

Young Chul Park (1:311)
Department of Biochemistry, Cornell University, Weill Medical College, New York, New York 10021

Peter J. Parker (2:389)
Cancer Research UK, London Research Institute, Lincoln's Inn Fields Laboratories, WC2A 3PX London, United Kingdom

J. Thomas Parsons (1:463)
Department of Microbiology, University of Virginia Health Systems Charlottesville, Virginia 22908-0734

J. M. Passner (2:531)
Department of Physiology and Biophysics, Mount Sinai School of Medicine, New York, New York 10029

Tony Pawson (1:379)
Samuel Lunenfeld Research Institute, Mount Sinai Hospital, Toronto, ONT M5G 1X5, Canada

Achille Pelliccioli (3:197)
Dipartimento di Genetica e Biologia dei Microrganismi, Universita' di Milano, 20133 Milano, Italy and Instituto FIRC di Oncologia Molecolare, 20139 Milano, Italy

J. Regino Perez-Polo (3:485)
The University of Texas Medical Branch, Galveston Texas 77555

Norbert Perrimon (2:783)
Department of Genetics, Harvard Medical School, Howard Hughes Medical Institute, Boston, Massachusetts 02115

Fabrice G. Petite (3:57)
Department of Molecular and Cellular Biology, Baylor College of Medicine, Houston, Texas 77030

Emmanuel Petroulakis (3:343)
Department of Biochemistry and McGill Cancer Research Center, McGill University, Montréal, Québec H3G 1Y6, Canada

Samuel L. Pfaff (2:883)
Gene Expression Laboratory, The Salk Institute for Biological Studies, La Jolla, California 92037

Jacob Piehler (1:27)
Institute of Biochemistry, Goethe-University Frankfurt, Frankfurt am Main 60439, Germany

Linda J. Pike (1:323)
Department of Biochemistry and Molecular Biophysics, Washington University School of Medicine, St. Louis, Missouri 63110

Michael J. Pinkoski (3:431)
Division of Cellular Immunology, La Jolla Institute for Allergy and Immunology, San Diego, California 92037

Fiona J. Pixley (3:615)
Department of Developmental and Molecular Biology, Albert Einstein College of Medicine, Yeshiva University, Bronx, New York 10461

Paolo Plevani (3:197)
Dipartimento di Genetica e Biologia dei Microrganismi, Universita' di Milano, 20133 Milano, Italy

Mu-ming Poo (2:871)
Division of Neurobiology, Department of Molecular and Cell Biology, University of California Berkeley, Berkeley, California 94720

Tullioi Pozzan (2:459)
Venetian Institute of Molecular Medicine, Department of Biomedical Studies, University of Padua, 35129 Padua, Italy

Stephen M. Prescott (2:243)
Huntsman Cancer Institute, University of Utah, Salt Lake City, Utah 84112

Igor Prudovsky (3:393)
Center for Molecular Medicine, Maine Medical Center Research Institute, Scarborough, Maine 04074

James W. Putney, Jr. (2:31)
Calcium Regulation Section, Laboratory of Signal Transduction, National Institute of Environmental Health Sciences, National Institute of Health, Research Triangle Park, North Carolina 27709

Thomas Radimerski (3:305)
Friedrich Miescher Institute for Biomedical Research, 4056 Basel, Switzerland

Elzbieta Radzio-Andzelm (2:471)
Department of Chemistry and Biochemistry and the Howard Hughes Medical Institute, University of California San Diego, La Jolla, California 92093

Prahlad T. Ram (2:605)
Department of Pharmacology and Biological Chemistry, Mount Sinai School of Medicine, New York, New York 10029

Lucia Rameh (2:129)
Boston Biomedical Research Institute, Watertown, Massachusetts 02115

Danica Ramljak (3:565)
Department of Oncology, Lombardi Cancer Center, Georgetown University, Washington, DC 20057

Barbara Ranscht (2:889)
The Burnham Institute, Neurobiology Program, La Jolla, California 92037

Anjana Rao (3:119)
Center for Blood Research and Department of Pathology, Harvard Medical School, Boston, Massachusetts 02115

Carol J. Raport (1:149)
ICOS corporation, Bothell, Washington

Jacqueline D. Reeves (1:191)
Department of Microbiology, University of Pennsylvania, Philadelphia, Pennsylvania 19104

Holger Rehman (2:521)
Department of Physiological Chemistry and Center for Biomedical Genetics, University Medical Center Utrecht, Utrecht, The Netherlands, and Max-Planck Institute for Molecular Physiology, 44227 Dortmund, Germany

Trevor W. Reichman (3:335)
Department of Biochemistry and Molecular Biology, New Jersey Medical School and the Graduate School of Biomedical Sciences, University of Medicine and Dentistry of New Jersey, Newark, New Jersey 07107

Eric Reiter (3:591)
Station de Physiologie de la Reproduction et des Comportements, INRA/CNRS, University of Tours, 37380 Nouzilly, France

Michael A. Resnick (3:203)
Laboratory of Molecular Genetics, National Institute of Environmental Health Sciences, National Institutes of Health, Research Triangle Park, North Carolina 27709

Michael Reth (1:327)
Department of Molecular Immunology, Biology III, University of Freiburg and Max-Planck Institute for Immunobiology, Freiburg 79108, Germany

Sue Goo Rhee (2:113)
Laboratory of Cell Signaling, National Heart, Lung, and Blood Institute, National Institutes of Health, Bethesda, Maryland 20892

Joel D. Richter (3:323)
Program in Molecular Medicine, University of Massachusetts Medical School, Worcester, Massachusetts 01605

Rodney L. Rietze (3:625)
Institute for Brain Research, The University of Queensland, Brisbane QLD 4072, Queensland, Australia

James M. Rini (1:87)
Departments of Molecular and Medical Genetics and Biochemistry, University of Toronto, Toronto, ONT M5S 1A8, Canada

Jürgen A. Ripperger (3:139)
Department of Molecular Biology, Sciences II, University of Geneva, CH-1211 Geneva, Switzerland

Josep Rizo (2:95)
Departments of Biochemistry and Pharmacology, The University of Texas Southwestern Medical Center, Dallas, Texas 75390

Janet D. Robishaw (2:623)
Weis Center for Research, Geisinger Clinic, Danville, Pennsylvania 17822

H. Llewelyn Roderick (2:51)
The Babraham Institute, Babraham, Cambridge CB2 4AT, United Kingdom

Robert G. Roeder (3:11)
Laboratory of Biochemistry and Molecular Biology, Rockefeller University, New York, New York 10021

Larry R. Rohrschneider (2:147)
Fred Hutchinson Cancer Research Center, Division of Basic Sciences, Seattle, Washington 98109

David Ron (3:263; 3:359)
Skirball Institute of Biomolecular Medicine, New York University School of Medicine, New York, New York 10016

Michael G. Rosenfeld (3:29)
Howard Hughes Medical Institute, Department and School of Medicine, University of California San Diego, La Jolla, California 92093

Hans Rosenfeldt (2:589)
Oral and Pharyngeal Cancer Branch, National Institute of Dental and Craniofacial Research, National Institutes of Health, Bethesda, Maryland 20892

Kent L. Rossman (2:751)
Department of Pharmacology, Department of Biochemistry and Biophysics, and Lineberger Comprehensive Cancer Center, The University of North Carolina at Chapel Hill, Chapel Hill, North Carolina 27599

Christopher B. Roth (1:115)
Department of Molecular Biology, The Scripps Research Institute, La Jolla, California 92037

Markus G. Rudolph (1:63)
Department of Molecular Biology, Skaggs Institute for Chemical Biology, The Scripps Research Institute, La Jolla, California 92037

Anja Ruppelt (2:501)
Department of Medical Biochemistry, Institute of Basic Medical Sciences, University of Oslo, N-0316 Oslo, Norway

Lino Saez (1:575)
Laboratory of Genetics, The Rockefeller University, New York, New York 10021

Thomas P. Sakmar (1:139)
Laboratory of Molecular Biology and Biochemistry, Howard Hughes Medical Institute, Rockefeller University, New York, New York 10021

Guy S. Salvesen (2:351)
Program in Apoptosis and Cell Death Research, Burnham Institute, San Diego, California 92037

Paolo Sassone-Corsi (3:91)
Institute of Genetic and Molecular and Cellular Biology, 67084 Strasbourg, France

Charles L. Saxe (2:323)
Department of Cell Biology, Emory University School of Medicine, Atlanta, Georgia 30322

Beat W. Schäfer (2:87)
Department of Pediatrics, Division of Clinical Chemistry and Biochemistry, University of Zürich, CH-8032 Zürich, Switzerland

Ueli Schibler (3:139)
Department of Molecular Biology, Sciences II, University of Geneva, CH-1211 Geneva, Switzerland

Christian W. Schindler (3:77)
Department of Microbiology and Medicine, College of Physicians and Surgeons, Columbia University, New York, New York 10032

Tobias Schmelzle (3:299)
Division of Biochemistry, Biozentrum, University of Basel, 4056 Basel, Switzerland

Sandra L. Schmid (2:763)
Department of Cell Biology, The Scripps Research Institute, La Jolla, California 92037

Anja Schmidt (2:701)
MRC Laboratory for Molecular Cell Biology, CRC Oncogene and Signal Transduction Group, University College London, London WC1E 6BT, United Kingdom

Eric F. Schmidt (2:877)
Interdepartmental Science Program, Yale University School of Medicine, New Haven, Connecticut 06520

Gideon Schreiber (1:27)
Department of Biological Chemistry, Weizmann Institute of Science, Rehovot 76100, Israel

Joachim E. Schultz (2:535)
Department of Pharmaceutical Biochemistry, Pharmaceutical Institute, University of Tübingen, 72074 Tübingen, Germany

Beat Schwaller (2:67)
Division of Histology, Department of Medicine, University of Fribourg, Perolles, Fribourg, Switzerland

Klaus Schwamborn (3:107)
Signal Research Division, Celgene, San Diego, California 92037

Thue Schwartz (1:173)
Laboratory of Molecular Endocrinology, Ringshospitalet Copenhagen DK-2100, Denmark

William F. Schwindinger (2:623)
Weis Center for Research, Geisinger Clinic, Danville, Pennsylvania 17822

Giorgio Scita (3:411)
Department of Experimental Oncology, Istituto Europeo di Oncologia, 20141 Milan, Italy, and IFOM, The FIRC Institute of Molecular Oncology, 20134 Milan, Italy

John D. Scott (2:291; 2:383)
Howard Hughes Medical Institute, Vollum Institute, Oregon Health and Sciences University, Portland, Oregon 97239

Shaun Scott (3:225)
Queensland Cancer Fund Research Laboratory, Queensland Institute of Medical Research, Brisbane 4029, Australia

Thomas Seebeck (2:539)
Institute of Cell Biology, University of Bern, CH-3012 Bern, Switzerland

Charles N. Serhan (2:281)
Center for Experimental Therapeutics and Reperfusion Injury, Department of Anesthesiology, Perioperative and Pain Medicine, Brigham and Women's Hospital and Harvard Medical School, Boston, Massachusetts 02115

John B. Shabb (2:501; 2:545)
Department of Biochemistry and Molecular Biology, University of North Dakota School of Medicine and Health Sciences, Grand Forks, North Dakota 58203

Andrey S. Shaw (2:339)
Department of Pathology and Immunology, Washington University School of Medicine, St. Louis, Missouri 63110

Stephen B. Shears (2:233)
Laboratory of Signal Transduction, National Institute of Environmental Health Sciences, Research Triangle Park, North Carolina 27709

Shirish Shenolikar (1:627)
Department of Pharmacology and Cancer Biology, Duke University Medical Center, Durham, North Carolina 27710

Lei Shi (1:155)
Center for Molecular Recognition, Columbia University, New York, New York 10032

Chanseok Shin (3:331)
Department of Biological Sciences, Columbia University, New York, New York 10032

Kazuhiro Shiozaki (1:637)
Section of Microbiology, University of California, Davis, California 95616

Kevan M. Shokat (1:583)
Department of Cellular and Molecular Pharmacology, Department of Chemistry, University of California, Berkeley, California 94143-0450

Trevor J. Shuttleworth (2:35)
Department of Pharmacology and Physiology, University of Rochester Medical Center, Rochester, New York 14642

David P. Siderovski (2:631)
UNC Neuroscience Center, University of North Carolina at Chapel Hill, Chapel Hill, North Carolina 27599

Steven A. Siegelbaum (1:233)
Department of Pharmacology, Center for Neurobiology and Behavior, Howard Hughes Medical Institute, Columbia University, New York, New York 10032

Adam M. Silverstein (2:405)
Department of Pharmacology, University of Texas Southwestern Medical Center, Dallas, Texas 75390

Robert H. Singer (3:293)
Department of Anatomy and Structural Biology, Albert Einstein College of Medicine, Bronx, New York 10461

Michael K. Skinner (3:531)
Center for Reproductive Biology, School of Molecular Biosciences, Washington State University, Pullman, Washington 99164

Jill K. Slack-Davis (1:463)
Department of Microbiology, University of Virginia Health System, Charlottesville, Virginia 22908-0734

Stephen J. Smerdon (1:505; 2:171)
Division of Protein Structure, National Institute for Medical Research, London NW7 1AA, United Kingdom

Graeme C. M. Smith (1:557)
KuDOS Pharmaceuticals, Ltd, Cambridge CB4 0WG, United Kingdom

Guillaume Smits (1:161)
Faculty of Medicine, University of Brussels, Institut de Recherche Interdisciplinaire, Brussels B-1070, Belgium

Sarah M. Smolik (3:167)
Department of Medicine, Oregon Health Sciences University, Portland, Oregon 97239

Jessica E. Smotrys (2:585)
Department of Cell Biology and Physiology, Washington University School of Medicine, St. Louis, Missouri 63110

Emer M. Smyth (2:265)
Center for Experimental Therapeutics, University of Pennsylvania, Philadelphia, Pennsylvania 19104

Jason T. Snyder (2:751)
Department of Pharmacology, Department of Biochemistry and Biophysics, and Lineberger Comprehensive Cancer Center, The University of North Carolina at Chapel Hill, Chapel Hill, North Carolina 27599

Naoko Sogame (3:213)
Department of Cell Biology, University of Texas Southwestern Medical Center, Dallas, Texas 75390

Raffaella Soldi (3:393)
Center for Molecular Medicine, Maine Medical Center Research Institute, Scarborough, Maine 04074

John Sondek (2:751)
Department of Pharmacology, Department of Biochemistry and Biophysics, and Lineberger Comprehensive Cancer Center, The University of North Carolina at Chapel Hill, Chapel Hill, North Carolina 27599

Nahum Sonenberg (3:343)
Department of Biochemistry and McGill Cancer Research Center, McGill University, Montréal, Québec H3G 1Y6, Canada

Erica Dutil Sonneberg (1:685)
Structural Biology Lab, The Salk Institute for Biological Sciences, La Jolla, California 92037

Lindsay G. Sparrow (1:293)
Health Sciences and Nutrition, CSIRO, Parkville, Victoria 3052, Australia

Sarah Spiegel (2:247)
Department of Biochemistry, Medical College of Virginia Campus, Virginia Commonwealth University, Richmond, Virginia 23284

Stephen R. Sprang (1:127; 1:275)
Howard Hughes Medical Institute and Department of Biochemistry, The University of Texas Southwestern Medical Center, Dallas, Texas 75235-9050

Deepak Srivastava (3:463)
Department of Pediatrics and Molecular Biology, University of Texas Southwestern Medical Center, Dallas, Texas 75390

Robyn L. Stanfield (1:33)
Department of Molecular Biology, The Scripps Research Institute and The Skaggs Institute for Chemical Biology, La Jolla, California 92037

E. Richard Stanley (3:615)
Department of Developmental and Molecular Biology, Albert Einstein College of Medicine, Yeshiva University, Bronx, New York 10461

Deborah J. Stauber (1:251)
Department of Molecular Biology, The Scripps Research Institute and The Skaggs Institute for Chemical Biology, La Jolla, California 92037

Christopher Stefan (2:177)
Division of Cellular and Molecular Medicine, The Howard Hughes Medical Institute, University of California San Diego, School of Medicine, La Jolla, California 92093

Lena Stenson-Holst (2:441)
Section for Molecular Signaling, Department of Cell and Molecular Biology, Lund University, SE-22100 Lund, Sweden

Len Stephens (2:203)
Inositide Laboratory, Signalling Programme, Babraham Institute, Cambridge CB2 4AT, United Kingdom

Paul W. Sternberg (2:805)
HHMI and Division of Biology, California Institute of Technology, Pasadena, California 91125

Paul C. Sternweis (2:237)
Department of Pharmacology, University of Texas Southwestern Medical Center, Dallas, Texas 75390

Ruth Steward (2:779)
Waksman Institute, Department of Molecular Biology and Biochemistry, and Cancer Institute of New Jersey, Rutgers University, Piscataway, New Jersey 08901

John T. Stickney (2:675)
Department of Cell Biology, Neurobiology, and Anatomy, University of Cincinnati, Cincinnati, Ohio 45221

Andrew W. Stoker (2:867)
Neural Development Unit, Institute of Child Health, University College London, London WC1N 61EH, United Kingdom

Stephen M. Strittmatter (2:877)
Department of Neurology and Section of Neurobiology, Yale University School of Medicine, New Haven, Connecticut 06520

Beth E. Stronach (2:783)
Department of Biological Sciences, University of Pittsburgh, Pittsburgh, Pennsylvania 15261

Roland K. Strong (1:83)
Basic Sciences Division Fred Hutchinson Cancer Research Center, Seattle, Washington 98109-1024

Robert M. Stroud (1:239)
Department of Biochemistry and Biophysics, University of California, San Francisco, San Francisco, California 94143-0448

Thomas C. Südhof (2:95)
Center for Basic Neuroscience, Department of Molecular Genetics, and Howard Hughes Medical Institute, University of Texas Southwestern Medical Center, Dallas, Texas 75390

Roger K. Sunahara (2:419)
Department of Pharmacology, University of Michigan Medical School, Ann Arbor, Michigan 48109

Brian J. Sutton (1:45)
The Randall Centre, King's College London, New Hunt's House, Guy's Campus, London SE1 1UL, United Kingdom

Sipeki Szabolcs (2:389)
Cancer Research UK, London Research Institute, Lincoln's Inn Fields Laboratories, WC2A 3PX London, United Kingdom

Xiao-Bo Tang (2:525)
Department of Pharmacology, University of Washington, Seattle, Washington 98195

Kjetil Taskén (2:501)
Department of Medical Biochemistry, Institute of Basic Medical Sciences, University of Oslo, N-0316 Oslo, Norway

Hisashi Tatebe (1:637)
Section of Microbiology, University of California, Davis, California 95616

Servane Tauszig-Delamasure (2:333)
National Center of Scientific Research, Institute of Molecular and Cellular Biology, 67084 Strasbourg, France

Colin W. Taylor (2:41)
Department of Pharmacology, University of Cambridge, Cambridge CB2 1PD, United Kingdom

Garry L. Taylor (1:105)
Centre for Biomolecular Sciences, University of St. Andrews, St. Andrews, Fife KY16 9UA, Scotland

Laura J. Taylor (2:711)
Department of Molecular Genetics and Microbiology, State University of New York at Stony Brook, Stony Brook, New York 11794

Susan S. Taylor (2:471)
Department of Chemistry and Biochemistry and the Howard Hughes Medical Institute, University of California San Diego, La Jolla, California 92093

George Thomas (3:305)
Friedrich Miescher Institute for Biomedical Research, 4056 Basel, Switzerland

Robert P. Thomas (3:477)
Department of Surgery, The University of Texas Medical Branch, Galveston, Texas 77555

E. Brad Thompson (3:451)
Department of Human Biological Chemistry and Genetics, The University of Texas Medical Branch, Galveston, Texas 77555

Michael J. Thompson (1:15)
Protein Pathways, Inc., Woodland Hills, California 91367

Janet M. Thornton (1:21)
Wellcome Trust Genome Campus, European Bioinformatics Institute, Hinxton, Cambridge CB10 1SD, United Kingdom

Carl S. Thummel (3:69)
Howard Hughes Medical Institute, University of Utah, Salt Lake City, Utah 84112

Hideaki Togashi (2:877)
Department of Neurology and Section of Neurobiology, Yale University School of Medicine, New Haven, Connecticut 06520

Amy Hin Yan Tong (2:311)
Banting and Best Department of Medical Research and Department of Molecular and Medical Genetics, University of Toronto, Toronto, Ontario M5G 1X5, Canada

Nicholas K. Tonks (1:641)
Cold Spring Harbor Laboratory, Cold Spring Harbor, New York 11724-2208

Peter Tontonoz (3:39)
Howard Hughes Medical Institute, Department of Pathology and Laboratory Medicine, UCLA School of Medicine, Los Angeles, California 90095

M. K. Topham (2:243)
Huntsman Cancer Institute, University of Utah, Salt Lake City, Utah 84112

Knut Martin Torgersen (2:143)
The Life Science Institute and Department of Biological Chemistry, University of Michigan Medical School, Ann Arbor, Michigan 48109

Hien Tran (3:83)
Division of Neuroscience, Children's Hospital and Department of Neurobiology, Harvard Medical School, Boston, Massachusetts 02115

Michel L. Tremblay (1:729)
McGill Cancer Center, McGill University, Department of Biochemistry, Montreal, Quebec H3G 1Y6, Canada

Ming-Jer Tsai (3:57)
Department of Molecular and Cellular Biology, Baylor College of Medicine, Houston, Texas 77030

Sophia Y. Tsai (3:57)
Department of Molecular and Cellular Biology, Baylor College of Medicine, Houston, Texas 77030

Susan Tsunoda (1:349)
Department of Biology, Boston University, Boston, Massachusetts 02215

Stewart Turley (2:525)
Department of Biochemistry and Biological Structure, Howard Hughes Medical Institute, University of Washington, Seattle, Washington 98195

Darren Tyson (1:361)
Department of Physiology and Biophysics, University of California, Irvine, Irvine, California 92697-4560

Robert L. Van Etten (1:733)
Department of Chemistry, Purdue University, West Lafayette, Indiana 47907-1393

Gilbert Vassart (1:161)
Faculty of Medicine, University of Brussels, Institut de Recherche Interdisciplinaire and Department of Medical Genetics, Brussels B-1070, Belgium

Peter J. Verveer (2:305)
Cell Biology and Cell Biophysics Program, European Molecular Biology Laboratory, 69120 Heidelberg, Germany

Virginie Vlaeminck (1:161)
Faculty of Medicine, University of Brussels, Institut de Recherche Interdisciplinaire, Brussels B-1070, Belgium

Abraham M. de Vos (1:281; 1:285; 1:305)
Department of Protein Engineering, Genentech, Inc., South San Francisco, California 94080-4990

Ty C. Voss (3:135)
Departments of Medicine and Cell Biology, NSF Center for Biological Timing, University of Virginia Health Sciences Center, Charlottesville, Virginia 22908

Robert Walczak (3:39)
Howard Hughes Medical Institute, Department of Pathology and Laboratory Medicine, UCLA School of Medicine, Los Angeles, California 90095

Graham C. Walker (3:185)
Biology Department, Massachusetts Institute of Technology, Cambridge, Massachusetts 02139

John C. Walker (1:579)
Division of Biological Sciences, University of Missouri, Columbia, Missouri 65211

Gernot Walter (1:621)
Department of Pathology, University of California, San Diego, La Jolla, California 92093-0612

Mark R. Walter (1:271)
Department of Microbiology and Center for Macromolecular Crystallography, University of Alabama at Birmingham, Birmingham, Alabama 35294-0005

Fen Wang (1:265)
Center for Cancer Biology and Nutrition, Institute of Biosciences and Technology, Texas A&M University System Health Science Center, Houston, Texas 77030–3303

Jean Y. J. Wang (3:249)
Division of Biological Sciences and The Cancer Center, University of California at San Diego, La Jolla, California 92093

Weiru Wang (1:197)
Department of Chemistry and Lawrence Berkeley National Laboratory, University of California, Berkeley, Berkeley, California 94720

Richard J. Ward (2:619)
Molecular Pharmacology Group, Division of Biochemistry and Molecular Biology, University of Glasgow, Glasgow G12 8QQ, Scotland, United Kingdom

Philip Wedegaertner (2:651)
Department of Microbiology and Immunology, Kimmel Cancer Institute, Thomas Jefferson University, Philadelphia, Pennsylvania 19104

Christian Wehrle (2:789)
Department of Molecular Embryology, Max-Planck Institute of Immunobiology, 79085 Freiburg, Germany

Arthur Weiss (1:689)
Departments of Medicine, Microbiology and Immunology, Howard Hughes Medical Institute, University of California-San Francisco, San Francisco, California 94143-0795

Jamie L. Weiss (2:79)
The Physiological Laboratory, The University of Liverpool, Liverpool L69 3BX, United Kingdom

Alan Wells (2:105)
Department of Pathology, University of Pittsburgh, Pittsburgh, Pennsylvania 15261

Claudia Werner (2:511)
Institut für Pharmakologie und Toxikologie, TU München, D-80802 München, Germany

Ann H. West (1:563)
Department of Chemistry and Biochemistry, University of Oklahoma, Norman, Oklahoma 73019

Marie C. Weston (2:441)
Pulmonary-Critical Care Medicine Branch, National Heart, Lung, and Blood Institute, National Institutes of Health, Bethesda, Maryland 20892

John K. Westwick (3:107)
Signal Research Division, Celgene, San Diego, California 92037

Anders Wetterholm (2:275)
Department of Medical Biochemistry and Biophysics, Division of Chemistry II, The Scheele Laboratory, Karolinska Institutes, S-17177 Stockholm, Sweden

Morris F. White (1:409)
Joslin Diabetes Center, Howard Hughes Medical Institute, Harvard Medical School, Boston, Massachusetts 02215

Malcolm Whitman (3:171)
Department of Cell Biology, Harvard Medical School, Boston, Massachusetts 02115

Matt R. Whorton (2:419)
Department of Pharmacology, University of Michigan Medical School, Ann Arbor, Michigan 48109

Christian Wiesmann (1:281; 1:285)
Protein Engineering Department, Genetech, Inc., South San Francisco, California 94080

Christian Wiesmann (1:281; 1:285)
Sunesis Pharmaceuticals, South San Francisco, California 94080

Roger L. Williams (2:153)
Medical Research Council, Laboratory of Molecular Biology, Cambridge CB2 1GA, United Kingdom

William D. Willis (3:607)
Department of Anatomy and Neurosciences, Marine Biomedical Institute, The University of Texas Medical Branch, Galveston, Texas 77555

Timothy M. Willson (3:53)
Nuclear Receptor Discovery Research, GlaxoSmithKline Research Triangle Park, North Carolina 27709

Ian A. Wilson (1:33; 1:63; 1:251)
Department of Molecular Biology, The Scripps Research Institute and the Skaggs Institute for Chemical Biology, La Jolla, California 92037

Ofer Wiser (2:667)
Howard Hughes Medical Institute, Departments of Physiology and Biochemistry, University of California San Francisco, San Francisco, California 94080

Matthew J. Wishart (1:741)
Biological Chemistry Department, University of Michigan, Ann Arbor, Michigan 48109-0606

Alfred Wittinghofer (2:757)
Max-Planck Institute for Molecular Physiology, 44227 Dortmund, Germany

James R. Woodgett (1:493)
Ontario Cancer Institute, Toronto, Ontario M5G 2M9, Canada

David K. Worthylake (2:751)
Department of Pharmacology, Department of Biochemistry and Biophysics, and Lineberger Comprehensive Cancer Center, The University of North Carolina at Chapel Hill, Chapel Hill, North Carolina 27599

Jeffrey L. Wrana (1:487)
Samuel Lunenfeld Research Institute, Mount Sinai Hospital, Program in Molecular Biology and Cancer and Department of Medical Genetic and Microbiology, University of Toronto, Toronto, Ontario M5G 1X5, Canada

Hao Wu (1:311)
Department of Biochemistry, Cornell University, Weill Medical College, New York, New York 10021

Yijin Xiao (2:253)
Department of Cancer Biology, The Lerner Research Institute, Cleveland Clinic Foundation, Cleveland, Ohio 44195

H. Eric Xu (3:21)
Nuclear Receptor Discovery Research, GlaxoSmithKline, Research Triangle Park, North Carolina 27709

Yan Xu (2:253)
Department of Cancer Biology, The Lerner Research Institute, Cleveland Clinic Foundation, Cleveland, Ohio 44195

Zheng Xu (1:689)
Departments of Medicine, Microbiology and Immunology, Howard Hughes Medical Institute, University of California-San Francisco, San Francisco, California 94143-0795

Michael B. Yaffe (2:171)
Center for Cancer Research, Massachusetts Institute of Technology, Cambridge, Massachusetts 02139

Michael B. Yaffe (1:505)
Center for Cancer Research, Massachusetts Institute of Technology, Cambridge, Massachusetts 02139

Kenneth M. Yamada (2:317)
Craniofacial Developmental Biology and Regeneration Branch, National Institute of Dental and Craniofacial Research, National Intitutes of Health, Bethesda, Maryland 20892

Seun-Ah Yang (1:631)
Division of Hematology-Oncology, University of Pennsylvania School of Medicine, Philadelphia, Pennsylvania 19104

Wannian Yang (2:715)
Department of Molecular Medicine, Cornell University, Ithaca, New York 14853

Yosef Yarden (1:405)
Department of Biological Regulation, Weizmann Institute of Science, Rehovot 76100, Israel

Hong Ye (1:311)
Department of Biochemistry, Cornell University, Weill Medical College, New York, New York 10021

Weilan Ye (2:793)
Department of Molecular Biology, Genentech, Inc., San Francisco, California 94080

Todd O. Yeates (1:15)
UCLA-DOE Center for Genomics and Proteomics and UCLA Molecular Biology Institute, University of California, Los Angeles, 611 Charles Young Dr. East, Los Angeles, California 90095-1569

Helen L. Yin (2:209)
Department of Physiology, University of Texas Southwestern Medical Center, Dallas, Texas 75390

John D. York (2:229)
Departments of Pharmacology and Cancer Biology, and Biochemistry, Howard Hughes Medical Institute, Duke University Medical Center, Durham, North Carolina 27710

Edgar C. Young (1:233)
Department of Pharmacology, Center for Neurobiology and Behavior, Howard Hughes Medical Institute, Columbia University, New York, New York 10032

Kenneth W. Young (2:19)
Department of Cell Physiology and Pharmacology, University of Leicester, Leicester LE2 3QE, United Kingdom

Matthew A. Young (1:387)
Departments of Molecular and Cell Biology and Chemistry, Howard Hughes Medical Institute, University of California, Berkeley, Berkeley, California 94720-3202

Michael W. Young (1:575)
Laboratory of Genetics, The Rockefeller University, New York, New York 10021

Minmin Yu (1:251)
Department of Molecular Biology, The Scripps Research Institute and The Skaggs Institute for Chemical Biology, La Jolla, California 92037

Nathan R. Zaccai (1:57)
Division of Structural Biology, Wellcome Trust Centre for Human Genetics, Cancer Research United Kingdom Receptor Stucture Research Group, Oxford OX3 7BN United Kingdom

Manuela Zaccolo (2:459)
Venetian Institute of Molecular Medicine, and Dulbecco Telethon Institute, University of Padua, 35129 Padua, Italy

Eli Zamir (2:317)
Department of Molecular Cell Biology, Weizmann Institute of Science, Rehovot 76100, Israel

Mark von Zastrow (1:181)
Psychiatry, Box 0984 LPPI F A014, University of California, San Francisco, 513 Parnassus, Box 0450, San Francisco, California 94143-0450

Chao Zhang (1:583)
Department of Cellular and Molecular Pharmacology, University of California, San Francisco, California 94143-0450

Xuewu Zhang (1:74)
Department of Cell Biology, Albert Einstein College of Medicine, Bronx, New York 10461

Zhong-Yin Zhang (1:677)
Department of Molecular Pharmacology, Albert Einstein College of Medicine, Bronx, New York 10461

Wenhong Zhou (3:545)
Department of Microbiology and Immunology, Sealy Center for Molecular Science, The University of Texas Medical Branch, Galveston, Texas 77555

Roya Zoraghi (2:539)
Institute of Cell Biology, University of Bern, CH-3012 Bern, Switzerland

PART III

Nuclear and Cytoplasmic Events: Transcriptional and Post-Transcriptional Regulation

Michael Karin, Editor

Introduction

Michael Karin

All living organisms, whether unicellular or multicellular, animals or plants, need to be able to constantly adapt to changing environmental conditions and be capable of successfully handling stressful and life-threatening challenges. This requirement for constant adjustment and adaptation is the major evolutionary driving force underlying the development of signal transduction systems that control and modulate the use of genetic information stored in the cell's DNA in response to extracellular conditions and stimuli. Thus the *raison d'etre* for the various receptors, channels, adaptors, GTP binding proteins, protein kinases, and protein phosphatases discussed in Parts I and II is to allow the regulation of the cellular gene expression program in response to the cell's environment, neighborhood, and history. Part III of the Handbook is therefore focused on the different mechanisms through which signal transduction systems regulate the utilization of genetic information.

The process of gene expression in both prokaryotes and eukaryotes can be divided into several mechanistically and temporally distinct steps: transcription and generation of the primary RNA transcript, processing and transport of the RNA transcript, translation of the mRNA into protein, turnover (decay) of the mRNA, and, finally, turnover of its protein product. In a eukaryotic cell, the first two steps (transcription and RNA processing) take place in the nucleus, whereas the last three steps (translation, mRNA turnover, protein turnover) take place in the cytoplasm (although nuclear proteins are turned over in the nucleus). Regardless of locale, all of these processes are subject to regulation by signal transduction pathways.

The first two sections of Part III deal with transcriptional control mechanisms that are targeted by signal transduction pathways. The most common form of regulation is based on phosphorylation of either sequence-specific transcription factors or proteins that directly interact with such transcription factors. The phosphorylation event can either affect the subcellular distribution of the transcription factors (e.g., NFAT, NFκB), their ability to bind DNA, or their ability to activate or repress transcription (e.g., CREB, c-Jun). As mentioned earlier, regulation can be achieved through phosphorylation of the transcription factor itself (e.g., CREB) or through phosphorylation of an interacting protein, for instance, an inhibitor (e.g., IκB) that regulates the activity or subcellular distribution of the transcription factor. Rather than being extensive and cover every single transcription factor known to be regulated via phosphorylation, Section B of Part III includes authoritative reviews that highlight the general aspects and principles governing this form of regulation.

In addition to posttranslational modifications, such as protein phosphorylation, sequence-specific transcription factors are also subject to allosteric regulation. In this case the binding of a small ligand has a dramatic effect on either the DNA binding or transcriptional activation properties of the transcription factor. This type of regulation is most relevant to members of the nuclear receptor family that play very important roles in signal transduction and cell regulation. Members of this family of ligand-regulated transcription factors, reviewed in Section A, form the core of signal transduction pathways that regulate gene expression in response to steroid and thyroid hormones, fatty acids, bile acids, and cholesterol metabolites as well as certain xenobiotic compounds. Unlike the multicomponent pathways that control transcription in response to activation of cell surface receptors, nuclear receptors are multifunctional proteins that incorporate signal detection, amplification, and execution in one molecule.

Although sequence-specific transcription factors represent the most common target for signal transduction pathways, as far as transcriptional regulation is concerned, new evidence is emerging that some of the coactivators, corepressors, or mediators with which these factors interact may also be subject to regulation. Coactivators, corepressors, and mediators are often large multicomponent protein complexes that are recruited to promoters or enhancers through interactions with sequence-specific transcription factors. We have much more to learn about the regulation of these protein complexes, which may act either through chromatin modifications or direct interactions with the RNA polymerase holoenzyme. In addition to modulation of chromatin structure via recruitment of chromatin modifiers to sequence-specific transcription factors, evidence also exists to show that signal-responsive protein kinases may directly phosphorylate histones and regulate chromatin structure via a more direct route. All of these modes of transcriptional regulation are discussed in Sections A and B. Evidence is also growing to indicate that the activity of various transcription factors and chromatin components can be regulated via additional posttranslational modifications, such as acetylation and methylation. However, it remains to be demonstrated that signal transduction pathways can directly regulate the activity of protein acetylases or methylases in a way that does not involve protein phosphorylation. Finally, another important form of regulation discussed in Section B is regulated proteolysis, which is responsible for both activation and inactivation of transcription factors.

While transcriptional regulation has received most of the attention, it is quite clear that every single step in the gene expression process, including RNA processing, RNA transport, RNA translation, and RNA and protein turnover can be subject to regulation. Therefore a section of Part III is dedicated to posttranscriptional regulation of gene expression. Because much of the available information pertains to translational control and regulation of mRNA turnover, these topics receive most of the coverage. However, regulated RNA processing (splicing) and transport are also discussed. Although the currently available information regarding posttranscriptional regulation is far from being extensive, it is our hope that this largely neglected area of gene regulation will receive more attention in the near future and that, one day, it will be as well understood as transcriptional regulation.

Given the important role of transcriptional and posttranscriptional control of gene expression in adaptation to adverse environmental conditions, an entire section of this volume is dedicated to the signaling mechanisms underlying various stress responses. Among the topics covered are the heat-shock and UV responses and the different responses that are elicited by DNA damage, which have provided valuable basic information relevant to many other aspects of cell regulation and signal transduction. In addition to metabolic control, these stress responses are evolutionarily ancient and are conserved in many eukaryotic orders.

Like all of the other parts of this Handbook, the prevailing approach was thematic in spirit, intended to provide a state-of-the-art understanding of fundamental principles rather than detailed coverage of the entire field. One remaining challenge for the future is to be able to identify all of the transcription factors and their target genes that fall under the influence of a given cell surface receptor or a signal responsive protein kinase. Such information will be essential both for the ultimate understanding of biological specificity and for being able to predict the biological and pathophysiological outcomes of activation or inhibition of a particular signaling molecule.

SECTION A

Nuclear Receptors

Michael G. Rosenfeld, Editor

CHAPTER 270

History of Nuclear Receptors

Elwood V. Jensen
Department of Cell Biology,
University of Cincinnati College of Medicine,
Cincinnati, Ohio

Introduction

Nuclear receptors comprise a group of intracellular proteins found either in the cytosol, weakly bound in the nucleus, or both, that, in combination with small molecule ligands, bind tightly in the nucleus, recruit appropriate cofactors, and regulate transcription in target genes. Included in this "superfamily" are receptors for the steroid hormones (estrogens, progestins, androgens, glucocorticoids, mineralocorticoids, ecdysteroids, vitamin D), for thyroid hormone and retinoids, and for an ever-increasing number of "orphan" receptors, which show structural similarity to the other receptors but for which ligands are only now being gradually identified.

Progress in the understanding nuclear receptors and their action is summarized in Table I. Because studies with estrogens began earlier than those for the other agents, the first six discoveries, identifying receptor proteins and establishing the basic action pattern, as well as the preparation of specific antibodies, were first made with estrogen receptors and later extended to other types of cellular regulators. For the most part, the remaining findings were first obtained with progestins or glucocorticoids and then shown to apply to estrogens. Thus, this short history of the receptor concept focuses on the earlier studies with estrogens that provide the general model, with a brief summary of more recent discoveries that have expanded our understanding. For the most part, only the initial reports, which stimulated other investigations, are cited for each phenomenon, in some cases along with relevant review articles. Detailed references to the many studies that have contributed to our current knowledge of hormone-receptor interaction can be found in comprehensive review articles or monographs [1–13].

Discovery of Receptors and Shift in Research Direction

The synthesis in 1957 of estradiol labeled with carrier-free tritium provided a means for determining what actually happens to the steroid itself as it exerts its hormone action. Knowledge about what the target cell does to the steroid, taken together with the influence of the hormone on cellular processes, has provided a better understanding than would have been possible from either approach alone [14].

The demonstration that rat uterus and other estrogen-responsive tissues contain characteristic hormone-binding components [15], with which estradiol associates to induce growth without itself undergoing chemical alteration [16], directed thinking away from earlier concepts that had linked estrogen action to participation in the enzymatic processes of steroid metabolism. It became apparent that interaction with the steroid must in some way alter the properties of macromolecules involved in RNA and protein biosynthesis, processes that were known to be enhanced by hormone action [17]. But in the 1960s the idea that steroid hormones must interact with enzymes was so firmly entrenched that many biochemists questioned whether the observed binding had any direct role in hormone action. The finding that the progressive inhibition of estradiol uptake in the rat uterus by increasing amounts of the antiestrogen, nafoxidine, closely parallels its inhibition of uterine growth [18] provided evidence that hormone binding actually is involved in uterotrophic action and that binding components could be regarded as true receptors.

Receptor Forms and Physiological Action

When tritiated estradiol of high specific activity became commercially available in late 1962, other investigators

Table I Steps in Development of the Receptor Concept

A. Conceptual stepping stones	
	1. Hormone-binding components in target cells
	2. Hormonal action without metabolism
	3. Binding substance is true receptor
	4. Two receptor forms–cytosol and nuclear bound
	5. Nuclear receptor derived from cytosol receptor
	6. Composition of the two receptor forms
	7. Hormone-induced receptor "transformation"
	8. Transcriptional enhancement by transformed receptor
B. Detailed understanding of receptor structure and function	
	9. Antibodies to receptor proteins
	10. Cloning and sequencing
	11. Domain content and function
	12. Interaction of transformed receptor with target genes
	a. Hormone-response elements
	b. Recruitment of coactivators and corepressors

confirmed its specific uptake in target tissues of various species. Cell fractionation and autoradiographic studies from several laboratories showed that most of the hormone becomes tightly bound in the nucleus, with a smaller amount present in the cytosol fraction, whereas, before hormone treatment, the binding substance appears entirely in the cytosol fraction. Similar results were obtained with excised uteri exposed to hormone at physiological temperature *in vitro*.

The application of sucrose gradient centrifugation to characterize the cytosol receptor complex as an 8 S entity (originally called 9.5 S) was an especially valuable contribution [19], because it provided a means to show that the complex bound in the nucleus is different from that in the cytosol. After exposure of rat uteri to estradiol *in vivo* or at physiological temperature *in vitro*, extraction of their nuclei with 400 m*M* KCl was found to solubilize the bound hormone as a 5 S estradiol–receptor complex [20], readily distinguished [21] from the 4 S entity produced from the cytosol complex by exposure to similar salt concentrations. After several studies suggested that the 5 S nuclear receptor is not present as such in target cells but is derived from the 8 S cytosol receptor [22,23], direct evidence was provided by the fact that simply warming cytosol to physiological temperature in the presence, but not the absence, of estradiol converts the receptor to the 5 S form [24], a process that is accelerated by the presence of DNA [25].

Based on observations first made with progestin and glucocorticoid receptors, the 8 S cytosol complex was shown to be a conglomerate of the 4 S receptor protein associated with a variety of molecular chaperones, including heat-shock proteins and immunophilin [10], which is dissociated by interaction with the hormone, as well as by exposure to salt. The 5 S "transformed" receptor was recognized as a dimer of the 4 S protein thus produced [26,27]. When it was found that the 5 S, but not the 4 S, complex stimulates RNA synthesis in isolated nuclei from hormone-responsive cells [28], the biochemical role of the hormone became clear, namely, to produce a functional transcription factor from an inactive precursor [29].

After analogous receptor systems were identified for other classes of steroid hormones (summarized in [7]), similar hormone-induced conversion of the native receptor to a nuclear binding form was observed. Subsequent molecular biological studies demonstrated that all hormone-transformed receptors react in dimeric form with their target genes to regulate transcription [30,31], but only with the estrogen receptor is the dimer stable enough to permit its identification in the absence of target cell DNA.

Subsequent Discoveries Relevant to Receptor Structure and Function

As summarized in Table I, more recent investigations, especially those involving techniques of molecular biology, have greatly increased our understanding of the processes by which nuclear receptors serve as regulators of cellular processes. After several investigators had tried unsuccessfully to obtain specific antibodies to estrogen receptors by conventional immunochemical techniques, application of sucrose density sedimentation to identify the nonprecipitating immune complexes formed with tritiated estradiol–receptor complexes provided specific polyclonal [32] and monoclonal [33] antibodies. Subsequently, antibodies were obtained for all classes of steroid hormone receptors, permitting the cloning and sequence determination, first of the glucocorticoid receptor [34], followed by the estrogen receptor [35,36] and other members of the nuclear receptor family.

Early studies with the glucocorticoid receptor indicated that it contained at least three distinct functional domains [37]. After cloning and sequencing, it became evident that receptors for steroid and thyroid hormones, vitamin D, and retinoids represent a superfamily of regulatory agents [4], consisting of DNA-binding (C) and ligand-binding (E) domains, a small, highly conserved hinge region (D) joining these two units, and, at the amino acid terminus, a large region (A/B) that shows little homology among the different receptors. Receptors for estrogens and thyroid hormones contain a small additional region (F) at the carboxy terminus, the function of which is not entirely clear.

For the regulation of transcription, it was found that the transformed hormone–receptor complex does not interact with the site of transcription but with a *hormone response element*, an enhancer located at the promoter area in the 5′-flanking area of the target gene [38,39]. Palindromic sequences appear to make up the binding site for the hormone–receptor complex. This interaction leads to the recruitment of coactivators or corepressors that modulate the resulting transcription [40–43].

References

1. Gorski, J. and Gannon, F. (1976). Current models of steroid hormone action: A critique. *Ann. Rev. Physiol.* **38**, 425–450.
2. Yamamoto, K. R. and Alberts, B. M. (1976). Steroid receptors: Elements for modulation of eukaryotic transcription. *Ann. Rev. Biochem.* **45**, 721–746.
3. Jensen, E. V., Greene, G. L., Closs, L. E., DeSombre, E. R., and Nadji, M. (1982). Receptors reconsidered: A 20-year perspective. *Recent. Progr. Hormone Res.* **38**, 1–34.
4. Evans, R. M. (1988). The steroid and thyroid hormone superfamily. *Science* **240**, 889–895.
5. Beato, M. (1989). Gene regulation by steroid hormones. *Cell* **56**, 335–344.
6. Carson-Jurica, M. A., Schrader, W. T., and O'Malley, B. W. (1990). Steroid receptor superfamily. *Endocr. Rev.* **11**, 201–220.
7. Jensen, E. V. (1991). Steroid hormone receptors. *Curr. Topics Pathol.* **83**, 365–431.
8. Tsai, M. J. and O'Malley, B. W. (1994). Molecular mechanisms of action of steroid/thyroid superfamily members. *Ann. Rev. Biochem.* **63**, 451–486.
9. Parker, M., Ed. (1994). *Nuclear Hormone Receptors.* Academic Press, New York.
10. Pratt, W. B. and Toft, D. O. (1997). Steroid receptor interactions with heat shock protein and immunophilin chaperones. *Endocr. Rev.* **18**, 306–360.
11. Freedman, L., Ed. (1998). *Molecular Biology of Steroid and Nuclear Hormone Receptors.* Birkhäuser, Boston.
12. Enmark, E. and Gustafsson, J.-Å. (1996). Orphan nuclear receptors—the first eight years. *Mol. Endocrinol.* **10**, 1293–1307.
13. Giguére, V. (1999). Orphan nuclear receptors: From gene to function. *Endocr. Rev.* **20**, 689–725.
14. Gorski, J. (1992). Remembrance: The introduction of molecular biology and receptors into the study of hormone action. *Endocrinology* **131**, 1583–1584.
15. Jensen, E. V. and Jacobson, H. I. (1960). Fate of steroid estrogens in target tissues, in Pincus, G., and Vollmer, E. P., Eds., *Biological Activities of Steroids in Relation to Cancer*, pp. 161–178. Academic Press, New York.
16. Jensen, E. V. and Jacobson, H. I. (1962). Basic guides to the mechanism of estrogen action. *Recent Progr. Hormone Res.* **18**, 387–414.
17. Mueller, G. C., Herranen, A. M., and Jervell, K. F. (1958). Studies on the mechanism of action of estrogens. *Recent Progr. Hormone Res.* **14**, 95–129.
18. Jensen, E. V. (1965). Mechanism of estrogen action in relation to carcinogenesis. *Can. Cancer Conf.* **6**, 143–165.
19. Toft, D. and Gorski, J. (1966). A receptor molecule for estrogens. Isolation from the rat uterus and preliminary characterization. *Proc. Natl. Acad. Sci. USA* **55**, 1574–1581.
20. Jensen, E. V., DeSombre, E. R., Hurst, D. J., Kawashima, T., and Jungblut, P. W. (1967). Estrogen-receptor interactions in target tissues. *Arch. Anat. Microsc. Morphol. Exp.* **56**(Suppl), 547–569.
21. Jensen, E. V., Mohla, S., Gorell, T., Tanaka, S., and DeSombre, E. R. (1972). Estrophile to nucleophile in two easy steps. *J. Steroid Biochem.* **3**, 445–458.
22. Gorski, J., Toft., D., Shyamala, G., Smith, D., and Notides, A. (1968). Hormone receptors: Studies on the interaction of estrogen with the uterus. *Recent Progr. Hormone Res.* **24**, 45–80.
23. Jensen, E. V., Suzuki, T., Kawashima, T., Stumpf, W. E. *et al.* (1968). A two-step mechanism for the interaction of estradiol with rat uterus. *Proc. Natl. Acad. Sci. USA* **59**, 632–638.
24. Jensen, E. V., Numata, M., Brecher, P. I., and DeSombre, E. R. (1971). Hormone-receptor interaction as a guide to biochemical mechanism. *Biochem Soc. Symp.* **32**, 133–159.
25. Yamamoto, K. R. and Alberts, B. M. (1972). *In vitro* conversion of estradiol–receptor protein to its nuclear form: Dependence on hormone and DNA. *Proc. Natl. Acad. Sci. USA* **69**, 2105–2109.
26. Little, M., Szendro, P. I., and Jungblut, P. W. (1973). Hormone-mediated dimerization of microsomal estradiol receptor. *Hoppe-Seiler's Z. Physiol. Chem.* **354**, 1599–1610.
27. Notides, A. C., Hamilton, D. E., and Auer, H. E. (1975). A kinetic analysis of the estrogen receptor transformation. *J. Biol. Chem.* **250**, 3945–3950.
28. Mohla, S., DeSombre, E. R., and Jensen, E. V. (1972). Tissue-specific stimulation of RNA synthesis by transformed estradiol-receptor complex. *Biochem. Biophys. Res. Commun.* **46**, 661–667.
29. Jensen, E. V. and DeSombre, E. R. (1973). Estrogen-receptor interaction: Estrogenic hormones effect transformation of specific receptor proteins to a biochemically functional form. *Science* **182**, 126–134.
30. Kumar, V. and Chambon, P. (1988). The estrogen receptor binds tightly to its responsive element as a ligand-induced homodimer. *Cell* **55**, 145–156.
31. Tsai, S. Y., Carlstedt-Duke, J., Weigel, N. L., Dahlman, K. *et al.* (1988). Molecular interactions of steroid hormone receptor with its enhancer element: Evidence for receptor dimer formation. *Cell* **55**, 361–369.
32. Greene, G. L., Closs, L. E., Fleming, H., DeSombre, E. R., and Jensen, E. V., (1977). Antibodies to estrogen receptor: immunochemical similarity of estrophilin from various mammalian species. *Proc. Natl. Acad. Sci. USA* **74**, 3681–3685.
33. Greene, G. L., Nolan, C.. Engler, P., and Jensen, E. V. (1980). Monoclonal antibodies to human estrogen receptor. *Proc. Natl. Acad. Sci. USA* **77**, 5115–5119.
34. Hollenberg, S. M., Weinberger, C., Ong, E. S., Cerelli, G. *et al.* (1985). Primary structure and expression of a functional human glucocorticoid receptor cDNA. *Nature* **318**, 635–641.
35. Green, S. Walter, P., Kumar, V., Krust, A. *et al.* (1986). Human estrogen receptor cDNA: Sequence, expression, and homology to v-erb-A. *Nature* **320**, 134–139.
36. Greene, G. L., Gilna, P., Waterfield, M., Baker, A. *et al.* (1986). Sequence and expression of human estrogen receptor complementary DNA. *Science* **231**, 1150–1154.
37. Carlstedt-Duke, J., Okret, S., Wrange, Ö., and Gustafsson, J.-Å. (1982). Immunochemical analysis of the glucocorticoid receptor: identification of a third domain separate from the steroid-binding and DNA-binding domains. *Proc. Natl. Acad. Sci. USA* **79**, 4260–4264.
38. Payvar, F., Firestone, G. L., Ross, S. R., Chandler, V. L. *et al.* (1982). Multiple specific binding sites for purified glucocorticoid receptors on mammary tumor virus DNA. *J. Cell. Biochem.* **19**, 241–247.
39. Geisse, S., Scheidereit, C., Westphal, H. M., Hynes, N. E., Groner, B., and Beato, M. (1982). Glucocorticoid receptors recognize DNA sequences in and around murine mammary tumour virus DNA. *EMBO J.* **1**(12), 1613–1619.
40. Onate, S. A., Tsai, S. Y., Tsai, M. J., and O'Malley, B. W. (1995). Sequence and characterization of a coactivator for the steroid receptor superfamily. *Science* **270**, 1354–1357
41. Kamei, Y., Xu, I., Heinzel, T., Torchia, J. *et al.* (1996). A CBP integrator complex mediates transcriptional activation and AP-1 inhibition by nuclear receptors. *Cell* **85**, 403–414.
42. Rosenfeld, M. G. and Glass, C. K. (2001). Coregulator codes of transcriptional reputation by nuclear receptors *J. Biol. Chem.* **276**, 36865–36868.
43. McKenna, N. J. and O'Malley, B. W. (2002). Combinational control of gene expression by nuclear receptors and coregulators. *Cell* **108**, 465–474.

CHAPTER 271

Regulation of Basal Transcription by RNA Polymerase II

Sohail Malik and Robert G. Roeder
Laboratory of Biochemistry and Molecular Biology, Rockefeller University, New York

Introduction

Transcription of protein-encoding target genes by RNA polymerase II (Pol II) represents the end point of many signal transduction pathways. A wide variety of cell type- and gene-specific transcriptional regulators (activators and repressors), including nuclear receptors, are the conduits through which these signals are transmitted to Pol II. These regulators typically function by binding to cognate promoter or enhancer elements located distal to the transcription start sites of target genes. From there they relay their regulatory effects to Pol II, which is directly controlled by the general transcription factors (GTFs: TFIIA, TFIIB, TFIID, TFIIE, TFIIF, and TFIIH) (Table I) [1,2]. A primary function of the GTFs is to position and orient Pol II on core promoter elements that determine the transcription start site and to provide the active site of Pol II with initial access to the transcribed strand of the template. Despite the potential of Pol II and GTFs to support low levels of accurately initiated transcription *in vitro* (*basal transcription*), as well as the structural complexity (>40 polypeptides) of this machinery, responsiveness to regulatory factors is dependent on additional cofactors (coactivators and corepressors, see later discussion) [3]. The most general cofactors include TATA box binding protein (TBP)-associated factors (TAF_{II}s) in TFIID [4,5], positive cofactors (PC1, PC2, PC3, PC4, and PC52) derived from the USA cofactor fraction [3,6], and a multiprotein Mediator complex [7–9]. Together, Pol II, GTFs, and these cofactors constitute an intricate piece of machinery that makes it possible for the signals borne by activators and repressors to be translated into the appropriate genetic response on specific target genes.

This chapter discusses structural and functional aspects of the basal transcription apparatus and also addresses the role of regulatory cofactors that in turn, ultimately appear to function by directly or indirectly impacting Pol II and GTFs.

The Preinitiation Complex

Structural Organization

Multiple approaches, including X-ray crystallography and protein–DNA cross-linking, have led to a detailed description of the Pol II- and GTF-containing preinitiation complex (PIC) (Fig. 1), the key protein–DNA intermediate whose efficient assembly (and subsequent function) dictates, to a large extent, the level of expression of target genes. The crystal structure of Pol II from yeast cells reveals a complex machine that can effectively carry out RNA synthesis from double-stranded DNA templates [10–12]. Like their bacterial counterparts, with which they share extensive sequence homology [13], the two largest subunits (RPB1 and RPB2) of Pol II constitute the core of the structure and are surrounded by various smaller subunits. The catalytic site contains (at least) one magnesium ion that is buried deep within the cleft of the jaw-like assembly, from where it can drive templated incorporation of nucleotides into the growing RNA chain [12].

Pol II has not yet been visualized in the context of the PIC. However, subsets of the PIC components have been analyzed by X-ray crystallography. Together with other biophysical and biochemical approaches that have probed the fully assembled PIC, it has been possible to construct a working model of this multiprotein complex. On the paradigmatic TATA-containing core promoter, the TBP subunit of TFIID

Table I Subunit Composition of RNA Polymerase II and Associated Factors

Pol II	TFIIA	TFIIB	TFIID	TFIIE	TFIIF	TFIIH	Mediator
RPB1	α	35 kDa	TBP	α	RAP30	ERCC3	TRAP240
RPB2	β		TAF250	β	RAP74	ERCC2	TRAP230
RPB3	γ		TAF150			P62	TRAP220
RPB4			TAF135			p52	TRAP170/RGR1
RPB5			TAF100			p44	TRAP150β/hSUR2
RPB6			TAF80			cdk7	TRAP100
RPB7			TAF55			cyclinH	TRAP97
RPB8			TAF43			p34	TRAP95
RPB9			TAF31			MAT1	TRAP93
RPB10			TAF30				TRAP80
RPB11			TAF28				p78/CRSP70
							SRB10/CDK8
							p37/CRSP34
							p36
							MED7
							MED6
							SRB11/Cyc C
							TRFP
							TRAP25
							SRB7
							SOH1
							NUT2
							p12
							p10

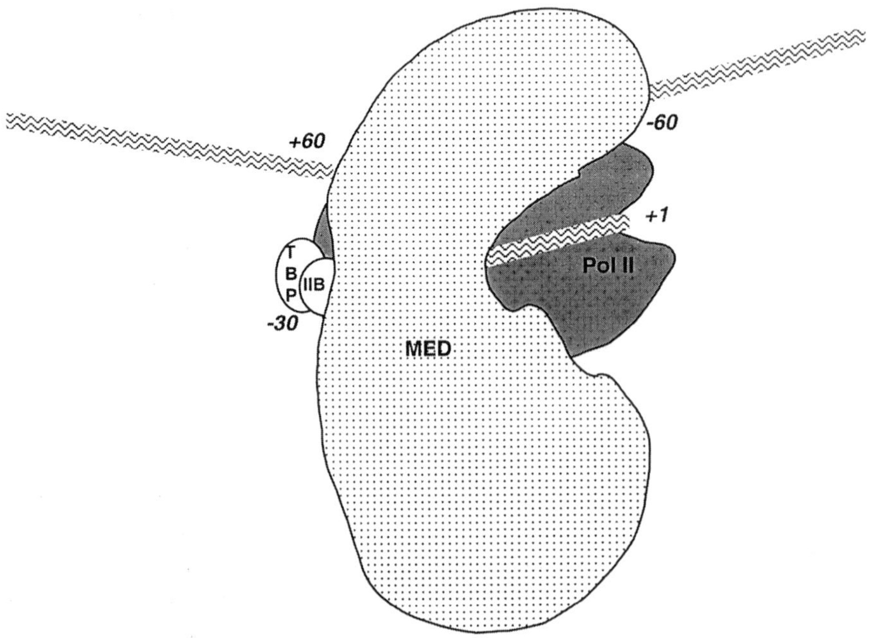

Figure 1 A schematic drawing showing a partial preinitiation complex. See text for details. For clarity, the TFIID-associated $TAF_{II}s$, TFIIF, TFIIE, and TFIIH are not shown. The figure is meant to emphasize the interaction of Pol II with Mediator and to highlight how the PIC likely consists of a central protein core wrapped within the template DNA (wavy lines). Although the positions of the factors are conjectural, care has been taken to ensure that they conform with previously published structural data. Activators and other cofactors, which are predicted to impinge on this machinery, have also been left out.

provides the main recognition contacts with promoter DNA. X-ray crystal structures of the TBP–promoter [14,15] and the TBP–TFIIB–promoter [16] complexes have revealed that the saddle-shaped TBP molecule interacts with the minor groove of DNA through its concave underside and induces both an unusual distortion of the central TATA element and a sharp bend in the template DNA. This arrangement leaves available large areas on the relatively convex upper surface of TBP for potential interactions with TAF_{II}s within TFIID and other factors.

The interaction of TBP with promoter DNA is stabilized by TFIIA, which binds upstream of TBP and makes contacts both with TBP and with phosphates in the DNA backbone upstream of the TATA box [17,18]. The TFIIA requirement in transcription, which appears to be variable [2], is also reflected in its ability to alleviate repression by negatively acting cofactors such as NC2 [6] (see later discussion also) and may be related to how TFIIA is positioned within the complex.

The TBP–promoter complex is also stabilized by interactions with the C-terminal core repeat domains of TFIIB, which essentially slots into the TBP–promoter complex and makes contacts both with the underside of TBP and with promoter DNA upstream and downstream of the TATA element [16]. Fluorescence anisotropy measurements suggest the presence of a conserved element (TFIIB recognition element, BRE) through which TFIIB makes specific contacts with the DNA [19]. These conclusions are directly confirmed by the crystal structure of a promoter complex containing archaeal TBP and TFB, its TFIIB ortholog [20]. Together these results establish the underlying basis for polarity in transcription start site selection. The N-terminal zinc-ribbon region of TFIIB has not been structurally analyzed in the context of the PIC, but the well-documented role of TFIIB in transcription start selection [21] suggests that it must be positioned toward the start site in order to make necessary contacts with Pol II and the RAP30 subunit of TFIIF [21–23].

Although high-resolution three-dimensional structures are available for segments of TFIIE [24] and TFIIF [25,26], currently it has only been possible to deduce the positions and function of these and other GTFs (including TFIIH) in the PIC using different approaches. Photo cross-linking analyses, in which the trajectory of the promoter DNA was mapped onto the PIC components [27,28], suggest that TFIIF plays a critical role in the fully assembled complex both in recruiting Pol II and in compacting the complex by tightly wrapping the promoter DNA around Pol II. This wrapping involves DNA contacts both upstream and downstream of the start site and may be further enhanced by TFIIE, which appears to be primarily positioned close to the transcription site [27,28]. Consistent with these results, a low-resolution three-dimensional structural analysis [29] has localized TFIIE close to the DNA binding cleft of Pol II. An important structural role for TFIIE is to recruit TFIIH to the PIC [30–32]. However, it has been suggested that, at the structural level, TFIIE also might contribute to the establishment of the PIC by a TFIIH-independent mechanism involving direct stabilization of TBP binding to the TATA element [33].

PICs on Nonstandard Core Promoters

Most of our understanding of PIC structure is based on analyses of TATA-containing promoters, but these may represent only a minority of cellular gene promoters [34]. Additional distinct elements that dictate core promoter function have been identified. These include the initiator (Inr), which spans the transcription site [35], and the downstream promoter element (DPE) [36], located approximately 30 nucleotides downstream of the start site. As expected, PIC assembly on promoters with these elements requires additional factors. Thus, transcription from Inr-containing promoters has been shown to require distinct TAF_{II}s in TFIID [37,38]; uncharacterized activities designated TIC1, TIC2, and TIC3 [39]; and possibly other transcription factors as well [40]. DPE promoter activity also appears to require TAF_{II} interactions [36] and, intriguingly, the cofactor NC2 [41], which was previously thought to possess only a negative modulatory function [6] (see later discussion also). An altogether novel strategy may be employed by core promoters that utilize the various TBP-related factors (TRFs) [42] in place of TBP-containing TFIID, and whose target sites differ considerably from the canonical TATA box [43,44].

In all of these cases of divergent core promoter usage, the precise composition of the final PIC that is competent for transcription initiation remains to be determined. Inasmuch as the function of the basal machinery is to enable selective transcription initiation by Pol II, these elements would be expected, *a priori*, to nucleate a PIC whose overall architecture would be rather similar to that described earlier for PICs formed on canonical TATA boxes. However, the primary recognition would be unlikely to be mediated by a TBP–promoter interaction. This expectation is strengthened by the observation that the TATA binding function of TBP is dispensable for transcription of an Inr-containing promoter [38] and by the presence of TBP in initiation factors required by Pol I and Pol III for transcription of promoters that contain no discernible TATA-like sequences [45].

Global Mechanisms of PIC Function

Promoter Melting

Kinetic and structural analyses have revealed a series of well-orchestrated events that precede processive RNA synthesis by Pol II [46,47]. In the broadest terms, the PIC described earlier, which is experimentally isolated in the absence of nucleotides, can be thought of as corresponding to the “closed complex” originally described for *Escherichia coli* RNA polymerase [48]. The isomerization of the eukaryotic closed complex to the open complex (in which the double-stranded promoter DNA around the start site is partially unwound) is dependent on energy derived from the hydrolysis of ATP [46,47,49,50]. Although the details are still not clear, initial promoter melting appears to entail multiple transitions involving TFIIE and the helicase activity of the ERCC3 subunit of TFIIH [51,52]. In the first step, the promoter region between positions –9 and +2 relative to the start site is unwound. In the next transition, the downstream edge of the unwound region is extended to +4 [46]. Templated phosphodiester bond formation ensues, but, perhaps because of the intrinsic thermodynamic instability of the complex, appears to result only in abortive synthesis of short RNA products [46].

Two distinct models have been proposed for how the TFIIH helicase catalyzes promoter melting [53]. As a conventional helicase, it could simply peel the two DNA strands apart until intrinsic Pol II functions, which are also involved in this process, take over. Alternatively, coupled with ATP hydrolysis, TFIIH could twist the DNA and crank it into an unwound state by strongly gripping DNA surrounding the melted region [54]. Although the role of TFIIE in this process is, in part, to deliver TFIIH, there also are indications that under some circumstances it directly contributes to promoter melting independently of TFIIH and ATP hydrolysis [50]. Consistent with the primary roles of TFIIE and TFIIH in promoter melting, the requirement for these factors can be partially obviated either with supercoiled templates [55] or with templates containing nonpaired regions at the start site [51,56].

Promoter Clearance

As the melted region is extended (in an ATP hydrolysis-independent fashion) to +9, longer RNA chains begin to form [46,50]. Concomitantly, the upstream DNA abruptly reanneals, whereas the downstream DNA is melted in a continuous motion into a more or less stable *transcription bubble* that is further propagated as Pol II moves into an elongation mode. This process, referred to as *promoter clearance*, prepares the promoter for a second incoming Pol II molecule for additional rounds of transcription [57].

Promoter clearance involves extensive rearrangements of transcription factor interactions, including relinquishing of contacts that hold Pol II in a stable structure. Thus, in addition to the changes in the DNA template structure and the Pol II movements discussed earlier, the PIC undergoes loss of a subset of GTFs. Early studies (of basal transcription only) suggested that TFIID remains bound to the promoter [58,59], whereas other GTFs are sequentially released as Pol II advances along the template; TFIIB and TFIIE are released by register +10 and TFIIH after +30. On the other hand, TFIIF appears to travel with Pol II. A more recent study analyzing activator-dependent transcription in extracts from yeast cells points to a somewhat different mechanism [60]. In this case, a "scaffold" consisting of TFIIA, TFIID, TFIIH, TFIIE, and Mediator (see later discussion) remains behind at the promoter after the Pol II departs. In both situations, the residual PIC is thought to serve as an efficient intermediate for a second round of transcription (reinitiation), requiring, in the latter case, only TFIIB and TFIIF [60].

Phosphorylation of the Pol II CTD

Cyclical changes in the phosphorylation status of the CTD repeats of the largest subunit (RPB1) of Pol II have long been correlated with promoter escape by Pol II [61,62]. It has been proposed that hypophosphorylated Pol II (form IIA) preferentially associates with the PIC [63]. Following promoter melting, and either concomitant with or immediately subsequent to promoter escape, the CTD is phosphorylated at specific serine residues by the cdk7 kinase subunit of TFIIH [62]. Thus, in the cell, the hyperphosphorylated form of Pol II (IIO) is predominantly engaged in transcription elongation [61]. The observation that the CTD is dispensable for transcription of many promoters in purified cell-free systems [64,65] has suggested that its phosphorylation is unlikely to be a causative factor in the initiation-to-elongation transition. However, in less-purified systems and *in vivo,* other constraining factors could impose a CTD phosphorylation requirement for this transition [66]. A role for CTD phosphorylation in transcription elongation has also been suggested, consistent with the identification of an elongation factor (pTEFb) that actively phosphorylates the CTD [67]. However, new evidence strongly suggests that CTD phosphorylation also might be a device to couple transcription and RNA processing [68,69]. Importantly, CTD phosphorylation does not constitute a unidirectional pathway because Pol IIO may be converted back to the IIA form by a variety of phosphatases, such as the TFIIF-regulated FCP [70], thereby allowing it to reinitiate.

Gene-Specific Regulation of PIC Function by Transcriptional Activators

Each of the steps in PIC assembly and function is a potential target for transcriptional regulators, especially activators. Early work emphasized direct interactions of activators with GTFs. Beginning with the original observations that TFIID could be directly targeted by activators [71], interactions of various regulators with each of the GTFs have now been reported [72–76]. However, GTFs do not suffice to support activator function *in vitro* [77] and mutations in some GTF residues that are critical for activator interactions *in vitro* do not affect transcriptional activation in functional tests [78].

Identification of the Mediator as a Key Coactivator

As mentioned, initial biochemical analyses in metazoan systems strongly suggested that TAF_{II} subunits of TFIID [4] and various USA-derived PCs [3,6] are essential coactivators for activated transcription from DNA templates. In yeast, concurrent genetic studies aimed at delineating the role of the Pol II CTD and various regulatory pathways, as well as biochemical studies aimed at eliciting activator function *in vitro*, led to the identification of SRB/Mediator (or Mediator) as the key coactivator [8,9,79]. Although the yeast

Mediator was isolated as a holoenzyme complex with Pol II, this association is reversible [80]. Variable levels of subsets of GTFs were also found to be present in the Pol II holoenzyme [80,81]. A genome-wide analysis has further suggested that, like Pol II, the Mediator is required for essentially all genes [82]. These observations have suggested that Mediator function in transcriptional regulation might be intimately associated with basal transcriptional control.

Subsequently, in metazoan systems, parallel efforts to recapitulate thyroid hormone receptor activity *in vitro* and to purify complexes containing orthologs of yeast Mediator subunits converged with the description of the TRAP [83] and SMCC [65] coactivator complexes, which proved to be identical [84]. Similar complexes were independently identified based on different criteria [7]. These include DRIP [85], ARC [86], and hMediator [87], which were isolated on the basis of their ability to interact with distinct activators. Furthermore, USA-derived PC2 also was identified as a Mediator-like complex [88]. The variously isolated complexes may all represent the same cellular entity or its derivatives [7], which by analogy to the yeast complex is also referred to here as the Mediator.

The metazoan Mediator complex consists of about 25 polypeptides (Table I) that include RGR1, SRB10/CDK8, MED7, MED6, SRB11/cyclin C, SRB7, SOH1, and NUT2, which are phylogenetically conserved from yeast to human [7]. Other polypeptides that are specific to the metazoan complex include the various TRAP polypeptides, as well as additional less-well characterized polypeptides. As for the yeast Mediator, which the metazoan complex resembles in its overall three-dimensional structure [89], the metazoan Mediator subunits are predicted to be organized in a modular fashion. There appears to be a relatively stable core (likely represented by the subunits found in PC2) surrounded by somewhat more reversibly associating polypeptides such as the kinase-cyclin pair that consists of SRB10/CDK8 and SRB11/cyclin C [7,88].

Potential Mechanisms of Mediator Function

Although no stable, stoichiometric Mediator–Pol II complex akin to the yeast holoenzyme has been reported in metazoans, there is evidence for direct interactions of metazoan Mediator with Pol II [7]. Furthermore, activators appear to contact distinct subunits within the Mediator. Thus, nuclear receptors primarily target TRAP220 [90], whereas activation domains of the viral activator VP16 and the tumor suppressor p53 interact with TRAP80 [84]. Based in part on these data, as well as similar data from yeast [80,81], a simple model for the mechanism of action of the Mediator postulates that it functions as an adaptor that links activators with the basal transcription machinery. Nonetheless, it remains to be established whether the initial PIC-forming entity is the Pol II holoenzyme or whether, as some recent reports suggest, PIC formation involves temporally separable recruitment of Mediator and Pol II [91,92].

It is possible that Mediator effects are exerted both at the level of recruitment (that is, by enhancing PIC formation) and by alterations in the function of the PIC (Fig. 1). In this regard, note that both the yeast and metazoan Mediators directly enhance basal (activator-independent) transcription [80,93,94], especially in the context of a complement of other nuclear factors. Direct interactions with the basal transcription machinery that could stabilize the PIC may be involved. Thus, in yeast, SOH1 displays genetic interactions with Pol II and TFIIB [95]. (Note, however, that although SOH1 appears to be an integral subunit of the human Mediator [7,65], it has not yet been directly implicated as such in yeast.) Similarly, the GAL11 subunit of the yeast Mediator (which has no apparent ortholog in the metazoan complex) interacts both physically and genetically with TFIIE [96]. Related, *in vitro* recruitment experiments have revealed that the Mediator-containing yeast holoenzyme is dependent on both TBP and TFIIB for PIC formation in some assays [97]. Functionally, the Mediator can also modulate the TFIIH CTD kinase activity and, thus, potentially also intervene at a postinitiation stage [80]. On the other hand, it has been proposed that the SRB10/CDK8 kinase subunit down-regulates TFIIH activity by phosphorylating its cyclin H subunit [98]. Effects at the level of promoter opening have not been reported. However, a role for the Mediator in efficient multiple-round transcription is likely because, in a yeast transcription system, the scaffold that remains after Pol II clears the promoter contains Mediator [60].

It is intriguing that multiple subunits in the Mediator have been identified as targets of activators [7,84,85,87,90,99]. Given that many activator signals emanate from enhanceosome-type assemblies that are composed of various activators [100], and that distinct activators may simultaneously interact with the Mediator [84], this provides a basis for synergistic activation mechanisms. Furthermore, these observations are consistent with coactivators (especially the Mediator) being devices for processing diverse signals into an integrated output for the Pol II machinery.

Mediator in Relation to Other Coactivators

An unresolved issue related to the mechanisms by which the various coactivators affect the basal machinery is whether they act synergistically or whether they are functionally redundant and thus reflect alternative activation pathways. In the case of USA-derived positive cofactors, we have previously found that Mediator-like PC2 functionally synergizes with PC4 and PC3/topoisomerase I [88]. Here, PC4 and PC3/topoisomerase I are believed to fulfill a structural role in stabilizing the PIC. In the case of the TAF_{II} subunits of TFIID, the relative contributions of these components and the Mediator in activated transcription have been harder to dissect. Earlier biochemical studies in yeast systems demonstrated Mediator function, including transcriptional activation, in reactions that contained TBP rather than the intact TFIID [80,81]. Moreover, genetic analysis revealed that at least a subset of TFIID-specific TAF_{II}s was dispensable for general activator function in yeast cells [101,102]. In the small number of cases in which a strong TAF_{II} dependence was demonstrable, the TAF_{II} requirement appeared to be dictated by the core-promoter sequences [103]. However, more recent genome-wide analyses in yeast have identified a subset of genes whose expression is critically dependent on histone-like TAF_{II}s [104] (including yTAF_{II} 17, a component shared by both the TFIID and SAGA complexes [105]). These genes display a requirement of their upstream activating sequences for this dependence [104,106].

In metazoan systems, differential requirements for TAF_{II}s have been observed depending on the experimental system employed [107,108]. Nonetheless, under some conditions in unfractionated extracts, TAF_{II}s within TFIID and Mediator appear to be jointly required both for activator-dependent and activator-independent transcription [94]. An interesting possibility is also suggested from a recent study in which the relative contributions of TAF_{II}s and Mediator in activation of several *Drosophila* promoters were examined [109]. It was observed that TBP sufficed to support activation from a TATA-box containing promoter (albeit still in a Mediator-dependent manner). However, activation through another promoter, which lacks a TATA box but contains a DPE, required intact TFIID (and simultaneously also displayed a Mediator dependence). These results are consistent with the idea that TAF_{II}s are required primarily for core promoter recognition, whereas Mediator is required for both activator-dependent and activator-independent effects. Therefore, it may be that, ultimately, both the TAF_{II} and Mediator requirements are intertwined, and reflect, at least in part, how each of these complexes interacts with the basal transcription machinery.

Although factors and mechanisms involved in transcription from DNA templates have been stressed in the foregoing discussion, it is important to keep in mind that in the cell transcription takes place within the constraints of a chromatin template. This necessitates other coactivator functions such as those associated with histone acetyl transferases like CBP/p300 and the p160 family of nuclear receptor coactivators [105]. A sequential mode of action of the factors, entailing an initial activator-mediated chromatin remodeling step followed by the recruitment of Pol II, GTFs, and Mediator, has been proposed [3,7]. Thus the mechanisms discussed earlier are likely to be relevant regardless of the nature of the template (chromatin or DNA). Indeed, in a sense, the PIC, with promoter DNA wrapped around a central protein core, can be viewed as being in competition with the structurally similar nucleosomes (Fig. 1) [110].

Negative Control of GTF Function

In addition to constraints imposed by chromatin structure, the basal machinery is subject also to other kinds of negative control. This introduces antirepression as an additional mode of gene activation [3]. It is noteworthy that some of the USA-derived PCs contain intrinsic repressive capabilities that are not only reversed during transcriptional activation but also are exploited to effect net transcription levels that exceed the basal (unregulated) levels [3]. Similarly, TFIID-associated TAF_{II}s, especially TAF_{II} 250, can also be repressive in the absence of an activator [111]. Part of the mechanism by which TAF_{II} 250 represses basal transcription involves molecular mimicry. Thus, the TAF_{II} 250 N-terminal domain resembles the unwound DNA structure of the TBP–TATA complex and makes repressive contacts with the underside of the TBP saddle that normally interacts with the TATA sequences [112]. In the presence of an activator, this interaction can be eliminated, thus allowing PIC formation to proceed. NC2 also is a negative cofactor that interacts directly with the basal transcription machinery [6]. A recent crystal structure of a ternary complex containing TBP, NC2, and promoter DNA nicely illustrates how a potential steric clash with TFIIB would preclude productive PIC assembly [113]. While the precise mechanisms are unknown, antagonism of this repression could be an important regulatory step in the cell.

Other negatively acting cofactors that have functional links to TBP and that have been identified both in yeast and metazoans include MOT1 [114,115] and the various NOT proteins [116,117]. In yeast, MOT1 may function to "remodel" TBP complexes in an ATPase-dependent fashion. Although regarded as a negative cofactor, MOT1 also has been localized to active genes [118]. This situation is somewhat reminiscent of the apparent functional duality associated with NC2 [119], and it is not immediately apparent how the various observations can be reconciled. The NOT proteins form heterogeneous complexes that include the CCR4 protein [120]. Their mechanism of action also is unclear, although one recent report has identified an ubiquitin ligase activity associated with the metazoan ortholog of one of the NOT proteins [121].

GTF Regulation by Covalent Modification

GTFs also undergo covalent modifications that might have regulatory consequences, but which at present are largely uncharacterized. The effect of the SRB10/CDK8 kinase on TFIIH was mentioned earlier. Other modifications include TAF_{II} 250-mediated phosphorylation of the RAP74 subunit of TFIIF [122] and acetylation of the large subunit of TFIIE by p300 [123]. In some cases, such as for the extensive phosphorylation of TFIID subunits (TBP and TAF_{II}s) during mitosis [124], these modifications may not reflect gene-specific controls in the context of the PIC but, instead, global control mechanisms that operate, for example, during progression through the cell cycle.

Conclusion

As summarized here, strong evidence linking activator function to the basal transcription machinery has been obtained. At the same time, structural, mechanistic and kinetic details of how this machinery carries out its basic functions have begun to emerge. Therefore, a main task for the future is to understand more completely how the various subprocesses that constitute PIC function are modulated by activators and coactivators. The description in recent years of the Mediator as a key transcriptional regulator has opened up the possibility of understanding these mechanisms in greater depth.

References

1. Roeder, R. G. (1996). The role of general initiation factors in transcription by RNA polymerase II. *Trends Biochem. Sci.* **21**, 327–335.
2. Orphanides, G., Lagrange, T., and Reinberg, D. (1996). The general transcription factors of RNA polymerase II. *Genes Dev.* **10**, 2657–2683.
3. Roeder, R. G. (1998). Role of general and gene-specific cofactors in the regulation of eukaryotic transcription. *Cold Spring Harb. Symp. Quant. Biol.* **63**, 201–218.
4. Verrijzer, C. P. and Tjian, R. (1996). TAFs mediate transcriptional activation and promoter selectivity. *Trends Biochem. Sci.* **21**, 338–342.
5. Burley, S. K. and Roeder, R. G. (1996). Biochemistry and structural biology of transcription factor IID (TFIID). *Annu. Rev. Biochem.* **65**, 769–799.
6. Kaiser, K. and Meisterernst, M. (1996). The human general co-factors. *Trends Biochem. Sci.* **21**, 342–345.
7. Malik, S. and Roeder, R. G. (2000). Transcriptional regulation through Mediator-like coactivators in yeast and metazoan cells. *Trends Biochem. Sci.* **25**, 277–283.
8. Lee, T. I. and Young, R. A. (2000). Transcription of eukaryotic protein-coding genes. *Annu. Rev. Genet.* **34**, 77–137.
9. Myers, L. C. and Kornberg, R. D. (2000). Mediator of transcriptional regulation. *Annu. Rev. Biochem.* **69**, 729–749.
10. Cramer, P., Bushnell, D. A., Fu, J., Gnatt, A. L., Maier-Davis, B., Thompson, N. E., Burgess, R. R., Edwards, A. M., David, P. R., and Kornberg, R. D. (2000). Architecture of RNA polymerase II and implications for the transcription mechanism. *Science* **288**, 640–649.
11. Gnatt, A. L., Cramer, P., Fu, J., Bushnell, D. A., and Kornberg, R. D. (2001). Structural basis of transcription: An RNA polymerase II elongation complex at 3.3 A resolution. *Science* **292**, 1876–1882.
12. Cramer, P., Bushnell, D. A., and Kornberg, R. D. (2001). Structural basis of transcription: RNA polymerase II at 2.8 angstrom resolution. *Science* **292**, 1863–1876.
13. Zhang, G., Campbell, E. A., Minakhin, L., Richter, C., Severinov, K., and Darst, S. A. (1999). Crystal structure of *Thermus aquaticus* core RNA polymerase at 3.3 A resolution. *Cell* **98**, 811–824.
14. Kim, Y., Geiger, J. H., Hahn, S., and Sigler, P. B. (1993). Crystal structure of a yeast TBP/TATA-box complex. *Nature* **365**, 512–520.
15. Kim, J. L., Nikolov, D. B., and Burley, S. K. (1993). Co-crystal structure of TBP recognizing the minor groove of a TATA element. *Nature* **365**, 520–527.
16. Nikolov, D. B., Chen, H., Halay, E. D., Usheva, A. A., Hisatake, K., Lee, D. K., Roeder, R. G., and Burley, S. K. (1995). Crystal structure of a TFIIB-TBP-TATA-element ternary complex. *Nature* **377**, 119–128.
17. Geiger, J. H., Hahn, S., Lee, S., and Sigler, P. B. (1996). Crystal structure of the yeast TFIIA/TBP/DNA complex. *Science* **272**, 830–836.
18. Tan, S., Hunziker, Y., Sargent, D. F., and Richmond, T. J. (1996). Crystal structure of a yeast TFIIA/TBP/DNA complex. *Nature* **381**, 127–151.
19. Lagrange, T., Kapanidis, A. N., Tang, H., Reinberg, D., and Ebright, R. H. (1998). New core promoter element in RNA polymerase II-dependent transcription: sequence-specific DNA binding by transcription factor IIB, *Genes Dev.* **12**, 34–44.
20. Littlefield, O., Korkhin, Y., and Sigler, P. B. (1999). The structural basis for the oriented assembly of a TBP/TFB/promoter complex. *Proc. Natl. Acad. Sci. USA* **96**, 13668–13673.
21. Hampsey, M. (1998). Molecular genetics of the RNA polymerase II general transcriptional machinery. *Microbiol. Mol. Biol. Rev.* **62**, 465–503.
22. Ha, I., Roberts, S., Maldonado, E., Sun, X., Kim, L. U., Green, M., and Reinberg, D. (1993). Multiple functional domains of human transcription factor IIB: distinct interactions with two general transcription factors and RNA polymerase II. *Genes Dev.* **7**, 1021–1032.
23. Malik, S., Lee, D. K., and Roeder, R. G. (1993). Potential RNA polymerase II-induced interactions of transcription factor TFIIB. *Mol. Cell Biol.* **13**, 6253–6259.
24. Okuda, M., Watanabe, Y., Okamura, H., Hanaoka, F., Ohkuma, Y., and Nishimura, Y. (2000). Structure of the central core domain of TFIIEbeta with a novel double-stranded DNA-binding surface. *EMBO J.* **19**, 1346–1356.
25. Groft, C. M., Uljon, S. N., Wang, R., and Werner, M. H. (1998). Structural homology between the Rap30 DNA-binding domain and linker histone H5: implications for preinitiation complex assembly. *Proc. Natl. Acad. Sci. USA* **95**, 9117–9122.
26. Kamada, K., De Angelis, J., Roeder, R. G., and Burley, S. K. (2001). Crystal structure of the C-terminal domain of the RAP74 subunit of human transcription factor IIF. *Proc. Natl. Acad. Sci. USA* **98**, 3115–3120.
27. Robert, F., Douziech, M., Forget, D., Egly, J. M., Greenblatt, J., Burton, Z. F., and Coulombe, B. (1998). Wrapping of promoter DNA around the RNA polymerase II initiation complex induced by TFIIF. *Mol. Cell.* **2**, 341–351.
28. Coulombe, B. and Burton, Z. F. (1999). DNA bending and wrapping around RNA polymerase: a "revolutionary" model describing transcriptional mechanisms. *Microbiol. Mol. Biol. Rev.* **63**, 457–478.
29. Leuther, K. K., Bushnell, D. A., and Kornberg, R. D. (1996). Two-dimensional crystallography of TFIIB- and IIE-RNA polymerase II complexes: Implications for start site selection and initiation complex formation. *Cell* **85**, 773–779.
30. Flores, O., Lu, H., and Reinberg, D. (1992). Factors involved in specific transcription by mammalian RNA polymerase II. Identification and characterization of factor IIH. *J. Biol. Chem.* **267**, 2786–2793.
31. Maxon, M. E., Goodrich, J. A., and Tjian, R. (1994). Transcription factor IIE binds preferentially to RNA polymerase IIa and recruits TFIIH: A model for promoter clearance. *Genes Dev.* **8**, 515–524.
32. Ohkuma, Y., Hashimoto, S., Wang, C. K., Horikoshi, M., and Roeder, R. G. (1995). Analysis of the role of TFIIE in basal transcription and TFIIH-mediated carboxy-terminal domain phosphorylation through structure-function studies of TFIIE-alpha. *Mol. Cell Biol.* **15**, 4856–4866.
33. Yokomori, K., Verrijzer, C. P., and Tjian, R. (1998). An interplay between TATA box-binding protein and transcription factors IIE and IIA modulates DNA binding and transcription. *Proc. Natl. Acad. Sci. USA* **95**, 6722–6727.
34. Kutach, A. K. and Kadonaga, J. T. (2000). The downstream promoter element DPE appears to be as widely used as the TATA box in *Drosophila* core promoters. *Mol. Cell Biol.* **20**, 4754–4764.
35. Smale, S. T. and Baltimore, D. (1989). The "initiator" as a transcription control element, *Cell* **57**, 103–113.
36. Burke, T. W. and Kadonaga, J. T. (1997). The downstream core promoter element, DPE, is conserved from *Drosophila* to humans and is recognized by TAFII60 of *Drosophila*. *Genes Dev.* **11**, 3020–3031.
37. Verrijzer, C. P., Chen, J. L., Yokomori, K., and Tjian, R. (1995). Binding of TAFs to core elements directs promoter selectivity by RNA polymerase II. *Cell* **81**, 1115–1125.
38. Martinez, E., Zhou, Q., L'Etoile, N. D., Oelgeschlager, T., Berk, A. J., and Roeder, R. G. (1995). Core promoter-specific function of a mutant transcription factor TFIID defective in TATA-box binding. *Proc. Natl. Acad. Sci. USA* **92**, 11864–11868.
39. Martinez, E., Ge, H., Tao, Y., Yuan, C. X., Palhan, V., and Roeder, R. G. (1998). Novel cofactors and TFIIA mediate functional core promoter selectivity by the human TAFII150-containing TFIID complex. *Mol. Cell Biol.* **18**, 6571–6583.
40. Smale, S. T. (2001). Core promoters: Active contributors to combinatorial gene regulation. *Genes Dev.* **15**, 2503–2508.
41. Willy, P. J., Kobayashi, R., and Kadonaga, J. T. (2000). A basal transcription factor that activates or represses transcription. *Science* **290**, 982–985.

42. Dantonel, J. C., Wurtz, J. M., Poch, O., Moras, D., and Tora, L. (1999). The TBP-like factor: An alternative transcription factor in metazoa? *Trends Biochem. Sci.* **24**, 335–339.
43. Takada, S., Lis, J. T., Zhou, S., and Tjian, R. (2000). A TRF1:BRF complex directs *Drosophila* RNA polymerase III transcription. *Cell* **101**, 459–469.
44. Holmes, M. C. and Tjian, R. (2000). Promoter-selective properties of the TBP-related factor TRF1. *Science* **288**, 867–870.
45. Hernandez, N. (1993). TBP, a universal eukaryotic transcription factor? *Genes Dev.* **7**, 1291–1308.
46. Holstege, F. C., Fiedler, U., and Timmers, H. T. (1997). Three transitions in the RNA polymerase II transcription complex during initiation. *EMBO J.* **16**, 7468–7480.
47. Yan, M. and Gralla, J. D. (1997). Multiple ATP-dependent steps in RNA polymerase II promoter melting and initiation. *EMBO J.* **16**, 7457–7467.
48. McClure, W. R. (1985). Mechanism and control of transcription initiation in prokaryotes. *Annu. Rev. Biochem.* **54**, 171–204.
49. Bunick, D., Zandomeni, R., Ackerman, S., and Weinmann, R. (1982). Mechanism of RNA polymerase II—specific initiation of transcription *in vitro*: ATP requirement and uncapped runoff transcripts. *Cell* **29**, 877–886.
50. Holstege, F. C., van der Vliet, P. C., and Timmers, H. T. (1996). Opening of an RNA polymerase II promoter occurs in two distinct steps and requires the basal transcription factors IIE and IIH. *EMBO J.* **15**, 1666–1677.
51. Holstege, F. C., Tantin, D., Carey, M., van der Vliet, P. C., and Timmers, H. T. (1995). The requirement for the basal transcription factor IIE is determined by the helical stability of promoter DNA. *EMBO J.* **14**, 810–819.
52. Tirode, F., Busso, D., Coin, F., and Egly, J. M. (1999). Reconstitution of the transcription factor TFIIH: Assignment of functions for the three enzymatic subunits, XPB, XPD, and cdk7. *Mol. Cell.* **3**, 87–95.
53. Fiedler, U. and Timmers, H. T. M. (2000). Peeling by binding or twisting by cranking: Models for promoter opening and transcription initiation by RNA polymerase II. *Bioessays* **22**, 316–326.
54. Kim, T. K., Ebright, R. H., and Reinberg, D. (2000). Mechanism of ATP-dependent promoter melting by transcription factor IIH. *Science* **288**, 1418–1422.
55. Parvin, J. D. and Sharp, P. A. (1993). DNA topology and a minimal set of basal factors for transcription by RNA polymerase II. *Cell* **73**, 533–540.
56. Pan, G. and Greenblatt, J. (1994). Initiation of transcription by RNA polymerase II is limited by melting of the promoter DNA in the region immediately upstream of the initiation site. *J. Biol. Chem.* **269**, 30101–30104.
57. Jiang, Y., Yan, M., and Gralla, J. D. (1996). A three-step pathway of transcription initiation leading to promoter clearance at an activation RNA polymerase II promoter. *Mol. Cell Biol.* **16**, 1614–1621.
58. Van Dyke, M. W., Roeder, R. G., and Sawadogo, M. (1988). Physical analysis of transcription preinitiation complex assembly on a class II gene promoter. *Science* **241**, 1335–1338.
59. Zawel, L., Kumar, K. P., and Reinberg, D. (1995). Recycling of the general transcription factors during RNA polymerase II transcription. *Genes Dev.* **9**, 1479–1490.
60. Yudkovsky, N., Ranish, J. A., and Hahn, S. (2000). A transcription reinitiation intermediate that is stabilized by activator. *Nature* **408**, 225–229.
61. Dahmus, M. E. (1996). Reversible phosphorylation of the C-terminal domain of RNA polymerase II. *J. Biol. Chem.* **271**, 19009–19012.
62. Oelgeschlager, T. (2002). Regulation of RNA polymerase II activity by CTD phosphorylation and cell cycle control. *J. Cell Physiol.* **190**, 160–169.
63. Lu, H., Flores, O., Weinmann, R., and Reinberg, D. (1991). The nonphosphorylated form of RNA polymerase II preferentially associates with the preinitiation complex. *Proc. Natl. Acad. Sci. USA* **88**, 10004–10008.
64. Akoulitchev, S., Makela, T. P., Weinberg, R. A., and Reinberg, D. (1995). Requirement for TFIIH kinase activity in transcription by RNA polymerase II. *Nature* **377**, 557–560.
65. Gu, W., Malik, S., Ito, M., Yuan, C. X., Fondell, J. D., Zhang, X., Martinez, E., Qin, J., and Roeder, R. G. (1999). A novel human SRB/MED-containing cofactor complex, SMCC, involved in transcription regulation. *Mol. Cell.* **3**, 97–108.
66. Li, Y. and Kornberg, R. D. (1994). Interplay of positive and negative effectors in function of the C-terminal repeat domain of RNA polymerase II. *Proc. Natl. Acad. Sci. USA* **91**, 2362–2366.
67. Jones, K. A. (1997). Taking a new TAK on tat transactivation. *Genes Dev.* **11**, 2593–2599.
68. Bentley, D. (1999). Coupling RNA polymerase II transcription with pre-mRNA processing. *Curr. Opin. Cell Biol.* **11**, 347–351.
69. Hirose, Y. and Manley, J. L. (2000). RNA polymerase II and the integration of nuclear events. *Genes Dev.* **14**, 1415–1429.
70. Kobor, M. S., Archambault, J., Lester, W., Holstege, F. C., Gileadi, O., Jansma, D. B., Jennings, E. G., Kouyoumdjian, F., Davidson, A. R., Young, R. A., and Greenblatt, J. (1999). An unusual eukaryotic protein phosphatase required for transcription by RNA polymerase II and CTD dephosphorylation in *S. cerevisiae. Mol. Cell.* **4**, 55–62.
71. Horikoshi, M., Hai, T., Lin, Y. S., Green, M. R., and Roeder, R. G. (1988). Transcription factor ATF interacts with the TATA factor to facilitate establishment of a preinitiation complex. *Cell* **54**, 1033–1042.
72. Kobayashi, N., Boyer, T. G., and Berk, A. J. (1995). A class of activation domains interacts directly with TFIIA and stimulates TFIIA-TFIID-promoter complex assembly. *Mol. Cell Biol.* **15**, 6465–6473.
73. Lin, Y. S. and Green, M. R. (1991). Mechanism of action of an acidic transcriptional activator *in vitro*. *Cell* **64**, 971–981.
74. Sauer, F., Fondell, J. D., Ohkuma, Y., Roeder, R. G., and Jackle, H. (1995). Control of transcription by Kruppel through interactions with TFIIB and TFIIE beta. *Nature* **375**, 162–164.
75. Joliot, V., Demma, M., and Prywes, R. (1995). Interaction with RAP74 subunit of TFIIF is required for transcriptional activation by serum response factor. *Nature* **373**, 632–635.
76. Xiao, H., Pearson, A., Coulombe, B., Truant, R., Zhang, S., Regier, J. L., Triezenberg, S. J., Reinberg, D., Flores, O., Ingles, C. J. *et al.* (1994). Binding of basal transcription factor TFIIH to the acidic activation domains of VP16 and p53. *Mol. Cell Biol.* **14**, 7013–7024.
77. Meisterernst, M., Roy, A. L., Lieu, H. M., and Roeder, R. G. (1991). Activation of class II gene transcription by regulatory factors is potentiated by a novel activity. *Cell* **66**, 981–993.
78. Tansey, W. P. and Herr, W. (1995). The ability to associate with activation domains *in vitro* is not required for the TATA box-binding protein to support activated transcription *in vivo*. *Proc. Natl. Acad. Sci. USA* **92**, 10550–10554.
79. Carlson, M. (1997). Genetics of transcriptional regulation in yeast: Connections to the RNA polymerase II CTD. *Annu. Rev. Cell Dev. Biol.* **13**, 1–23.
80. Kim, Y. J., Bjorklund, S., Li, Y., Sayre, M. H., and Kornberg, R. D. (1994). A multiprotein mediator of transcriptional activation and its interaction with the C-terminal repeat domain of RNA polymerase II. *Cell* **77**, 599–608.
81. Thompson, C. M., Koleske, A. J., Chao, D. M., and Young, R. A. (1993). A multisubunit complex associated with the RNA polymerase II CTD and TATA-binding protein in yeast. *Cell* **73**, 1361–1375.
82. Holstege, F. C., Jennings, E. G., Wyrick, J. J., Lee, T. I., Hengartner, C. J., Green, M. R., Golub, T. R., Lander, E. S., and Young, R. A. (1998). Dissecting the regulatory circuitry of a eukaryotic genome. *Cell* **95**, 717–728.
83. Fondell, J. D., Ge, H., and Roeder, R. G. (1996). Ligand induction of a transcriptionally active thyroid hormone receptor coactivator complex. *Proc. Natl. Acad. Sci. USA* **93**, 8329–8333.
84. Ito, M., Yuan, C. X., Malik, S., Gu, W., Fondell, J. D., Yamamura, S., Fu, Z. Y., Zhang, X., Qin, J., and Roeder, R. G. (1999). Identity between TRAP and SMCC complexes indicates novel pathways for the function of nuclear receptors and diverse mammalian activators. *Mol. Cell.* **3**, 361–370.
85. Rachez, C., Lemon, B. D., Suldan, Z., Bromleigh, V., Gamble, M., Naar, A. M., Erdjument-Bromage, H., Tempst, P., and Freedman, L. P. (1999). Ligand-dependent transcription activation by nuclear receptors requires the DRIP complex. *Nature* **398**, 824–828.

86. Naar, A. M., Beaurang, P. A., Zhou, S., Abraham, S., Solomon, W., and Tjian, R. (1999). Composite co-activator ARC mediates chromatin-directed transcriptional activation. *Nature* **398**, 828–832.
87. Boyer, T. G., Martin, M. E., Lees, E., Ricciardi, R. P., and Berk, A. J. (1999). Mammalian Srb/Mediator complex is targeted by adenovirus E1A protein. *Nature* **399**, 276–279.
88. Malik, S., Gu, W., Wu, W., Qin, J., and Roeder, R. G. (2000). The USA-derived transcriptional coactivator PC2 is a submodule of TRAP/SMCC and acts synergistically with other PCs. *Mol. Cell.* **5**, 753–760.
89. Dotson, M. R., Yuan, C. X., Roeder, R. G., Myers, L. C., Gustafsson, C. M., Jiang, Y. W., Li, Y., Kornberg, R. D., and Asturias, F. J. (2000). Structural organization of yeast and mammalian mediator complexes. *Proc. Natl. Acad. Sci. USA* **97**, 14307–14310.
90. Yuan, C. X., Ito, M., Fondell, J. D., Fu, Z. Y., and Roeder, R. G. (1998). The TRAP220 Component of a thyroid hormone receptor-associated protein (TRAP) coactivator complex interacts directly with nuclear receptors in a ligand-dependent fashion. *Proc. Natl. Acad. Sci. USA* **95**, 7939–7944.
91. Bhoite, L. T., Yu, Y., and Stillman, D. J. (2001). The Swi5 activator recruits the Mediator complex to the HO promoter without RNA polymerase II. *Genes Dev.* **15**, 2457–2469.
92. Cosma, M. P., Panizza, S., and Nasmyth, K. (2001). Cdk1 triggers association of RNA polymerase to cell cycle promoters only after recruitment of the mediator by SBF. *Mol. Cell.* **7**, 1213–1220.
93. Mittler, G., Kremmer, E., Timmers, H. T., and Meisterernst, M. (2001). Novel critical role of a human Mediator complex for basal RNA polymerase II transcription. *EMBO Rep.* **2**, 808–813.
94. Baek, H. J., Malik, S., Qin, J., and Roeder, R. G. (2002). Requirement of TRAP/Mediator for Both Activator-Independent and Activator-Dependent Transcription in Conjunction with TFIID-Associated TAF(II)s. *Mol. Cell Biol.* **22**, 2842–2852.
95. Fan, H. Y., Cheng, K. K., and Klein, H. L. (1996). Mutations in the RNA polymerase II transcription machinery suppress the hyper-recombination mutant hpr1 delta of *Saccharomyces cerevisiae. Genetics* **142**, 749–759.
96. Sakurai, H., Kim, Y. J., Ohishi, T., Kornberg, R. D., and Fukasawa, T. (1996). The yeast GAL11 protein binds to the transcription factor IIE through GAL11 regions essential for its *in vivo* function. *Proc. Natl. Acad. Sci. USA* **93**, 9488–9492.
97. Ranish, J. A., Yudkovsky, N., and Hahn, S. (1999). Intermediates in formation and activity of the RNA polymerase II preinitiation complex: Holoenzyme recruitment and a postrecruitment role for the TATA box and TFIIB. *Genes Dev.* **13**, 49–63.
98. Akoulitchev, S., Chuikov, S., and Reinberg, D. (2000). TFIIH is negatively regulated by cdk8-containing mediator complexes. *Nature* **407**, 102–106.
99. Stevens, J. L., Cantin, G. T., Wang, G., Shevchenko, A., and Berk, A. J. (2002). Transcription control by E1A and MAP kinase pathway via Sur2 Mediator subunit. *Science* **296**, 755–758.
100. Carey, M. (1998). The enhanceosome and transcriptional synergy. *Cell* **92**, 5–8.
101. Moqtaderi, Z., Bai, Y., Poon, D., Weil, P. A., and Struhl, K. (1996). TBP-associated factors are not generally required for transcriptional activation in yeast. *Nature* **383**, 188–191.
102. Walker, S. S., Reese, J. C., Apone, L. M., and Green, M. R. (1996). Transcription activation in cells lacking TAFIIS. *Nature* **383**, 185–188.
103. Shen, W. C. and Green, M. R. (1997). Yeast TAF(II)145 functions as a core promoter selectivity factor, not a general coactivator. *Cell* **90**, 615–624.
104. Apone, L. M., Virbasius, C. A., Holstege, F. C., Wang, J., Young, R. A., and Green, M. R. (1998). Broad, but not universal, transcriptional requirement for yTAFII17, a histone H3-like TAFII present in TFIID and SAGA. *Mol. Cell.* **2**, 653–661.
105. Brown, C. E., Lechner, T., Howe, L., and Workman, J. L. (2000). The many HATs of transcription coactivators. *Trends Biochem. Sci.* **25**, 15–19.
106. Michel, B., Komarnitsky, P., and Buratowski, S. (1998). Histone-like TAFs are essential for transcription *in vivo. Mol. Cell.* **2**, 663–673.
107. Ryu, S., Zhou, S., Ladurner, A. G., and Tjian, R. (1999). The transcriptional cofactor complex CRSP is required for activity of the enhancer-binding protein Sp1. *Nature* **397**, 446–450.
108. Oelgeschlager, T., Tao, Y., Kang, Y. K., and Roeder, R. G. (1998). Transcription activation via enhanced preinitiation complex assembly in a human cell-free system lacking TAFIIs. *Mol. Cell.* **1**, 925–931.
109. Park, J. M., Gim, B. S., Kim, J. M., Yoon, J. H., Kim, H. S., Kang, J. G., and Kim, Y. J. (2001). *Drosophila* Mediator complex is broadly utilized by diverse gene-specific transcription factors at different types of core promoters. *Mol. Cell Biol.* **21**, 2312–2323.
110. Workman, J. L. and Roeder, R. G. (1987). Binding of transcription factor TFIID to the major late promoter during *in vitro* nucleosome assembly potentiates subsequent initiation by RNA polymerase II. *Cell* **51**, 613–622.
111. Guermah, M., Malik, S., and Roeder, R. G. (1998). Involvement of TFIID and USA components in transcriptional activation of the human immunodeficiency virus promoter by NF-kappaB and Sp1. *Mol. Cell Biol.* **18**, 3234–3244.
112. Liu, D., Ishima, R., Tong, K. I., Bagby, S., Kokubo, T., Muhandiram, D. R., Kay, L. E., Nakatani, Y., and Ikura, M. (1998). Solution structure of a TBP-TAF(II)230 complex: Protein mimicry of the minor groove surface of the TATA box unwound by TBP. *Cell* **94**, 573–583.
113. Kamada, K., Shu, F., Chen, H., Malik, S., Stelzer, G., Roeder, R. G., Meisterernst, M., and Burley, S. K. (2001). Crystal structure of negative cofactor 2 recognizing the TBP-DNA transcription complex. *Cell* **106**, 71–81.
114. Auble, D. T., Hansen, K. E., Mueller, C. G., Lane, W. S., Thorner, J., and Hahn, S. (1994). Mot1, a global repressor of RNA polymerase II transcription, inhibits TBP binding to DNA by an ATP-dependent mechanism. *Genes Dev.* **8**, 1920–1934.
115. van der Knaap, J. A., Borst, J. W., van der Vliet, P. C., Gentz, R., and Timmers, H. T. (1997). Cloning of the cDNA for the TATA-binding protein-associated factorII170 subunit of transcription factor B-TFIID reveals homology to global transcription regulators in yeast and *Drosophila. Proc. Natl. Acad. Sci. USA* **94**, 11827–11832.
116. Collart, M. A. and Struhl, K. (1994). NOT1(CDC39), NOT2(CDC36), NOT3, and NOT4 encode a global-negative regulator of transcription that differentially affects TATA-element utilization. *Genes Dev.* **8**, 525–537.
117. Albert, T. K., Lemaire, M., van Berkum, N. L., Gentz, R., Collart, M. A., and Timmers, H. T. (2000). Isolation and characterization of human orthologs of yeast CCR4-NOT complex subunits. *Nucleic Acids Res.* **28**, 809–817.
118. Dasgupta, A., Darst, R. P., Martin, K. J., Afshari, C. A., and Auble, D. T. (2002). Mot1 activates and represses transcription by direct, ATPase-dependent mechanisms. *Proc. Natl. Acad. Sci. USA* **99**, 2666–2671.
119. Geisberg, J. V., Holstege, F. C., Young, R. A., and Struhl, K. (2001). Yeast NC2 associates with the RNA polymerase II preinitiation complex and selectively affects transcription *in vivo. Mol. Cell Biol.* **21**, 2736–2742.
120. Bai, Y., Salvadore, C., Chiang, Y. C., Collart, M. A., Liu, H. Y., and Denis, C. L. (1999). The CCR4 and CAF1 proteins of the CCR4-NOT complex are physically and functionally separated from NOT2, NOT4, and NOT5. *Mol. Cell Biol.* **19**, 6642–6651.
121. Albert, T. K., Hanzawa, H., Legtenberg, Y. I., de Ruwe, M. J., van den Heuvel, F. A., Collart, M. A., Boelens, R., and Timmers, H. T. (2002). Identification of a ubiquitin-protein ligase subunit within the CCR4-NOT transcription repressor complex. *EMBO J.* **21**, 355–364.
122. Dikstein, R., Ruppert, S., and Tjian, R. (1996). TAFII250 is a bipartite protein kinase that phosphorylates the base transcription factor RAP74. *Cell* **84**, 781–790.
123. Imhof, A., Yang, X. J., Ogryzko, V. V., Nakatani, Y., Wolffe, A. P., and Ge., H. (1997). Acetylation of general transcription factors by histone acetyltransferases. *Curr. Biol.* **7**, 689–692.
124. Segil, N., Guermah, M., Hoffmann, A., Roeder, R. G., and Heintz, N. (1996). Mitotic regulation of TFIID: Inhibition of activator-dependent transcription and changes in subcellular localization. *Genes Dev.* **10**, 2389–2400.

CHAPTER 272

Structural Mechanisms of Ligand-Mediated Signaling by Nuclear Receptors

H. Eric Xu and Millard H. Lambert
Nuclear Receptor Discovery Research, GlaxoSmithKline
Research Triangle Park, North Carolina

Introduction

Nuclear receptors (NRs) contain three major domains: an N-terminal domain, a central DNA-binding domain (DBD), and a C-terminal ligand-binding domain (LBD). The N-terminal domain contains, in some NRs, an "activation function 1" (AF-1) but little is known about its structure. The DBD, consisting of two highly conserved zinc-finger motifs, recognizes specific DNA sequences in the target promoters, and is responsible for the specificity of gene regulation. The LBD, as its name suggests, is responsible for ligand recognition and subsequent signaling cascades. In most NRs, the LBD also contains dimerization functions and a C-terminal "activation function 2" (AF-2) that promotes ligand-dependent transcription. The LBD is thus the key to the ligand-dependent regulation of NR signaling and, as such, has been the focus of intense structural studies. In this brief review, we focus on structural insights into the molecular basis of ligand binding and signaling by nuclear receptors.

Overall Structure of the LBD

All of the NR LBD structures determined to date share a similar fold, where about a dozen α-helices are packed into a three-layer sandwich with a cavity for binding ligands (Fig. 1). The C-terminal AF-2 region also adopts an α-helical conformation, but can pack onto alternative sites on the core structure depending on the presence and nature of the ligand. Aside from the AF-2 helix, the core helical structure is relatively stable and well preserved among different receptors, particularly in the top half of the LBD. The conservation of the overall structure suggests that the LBD fold is selected through evolution for the binding of small molecule ligands despite a wide divergence in their amino acid sequences.

Ligand-Binding Pockets

Helices 3, 5, 6, 7 and 10 enclose a ligand-binding pocket in the bottom half of the LBD. This pocket is covered by a β-sheet on one side, and it can be covered by the C-terminal AF-2 helix on the other side. Although different NR LBDs have similar overall structures, their ligand-binding pockets are remarkably diverse, with pocket volumes ranging from 390 to 1440 $Å^3$ (Table I). The receptors with smaller pocket volumes, such as the retinoid and steroid receptors, seem to recognize specific ligands, whereas the receptors with larger pockets, such as the pregnane X receptor (PXR) and the peroxisome proliferator-activated receptors (PPARs), are able to bind a broader range of ligands. The ligand promiscuity in the PXR and the PPARs may be essential for their physiological functions as sensors of diverse xenobiotic substances and diverse fatty acids and eicosanoids, respectively.

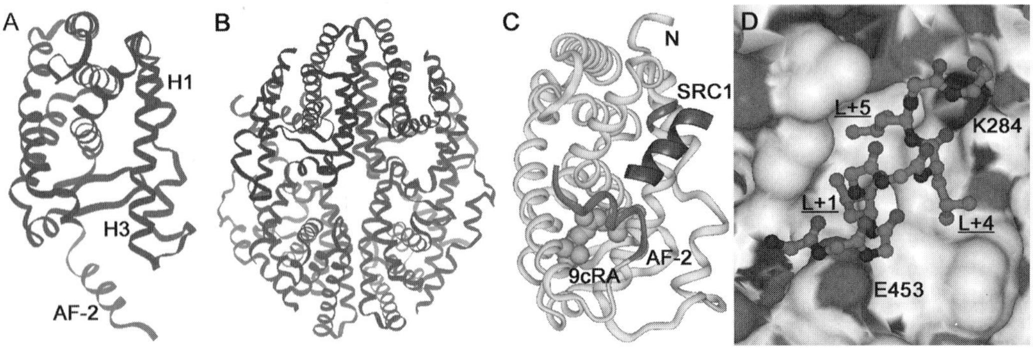

Figure 1 Structures of the RXR LBD in different conformational states. A, an overview of the apo-RXR LBD where the key helices are noted. B, the self-repressed tetrameric form consisting of two symmetric dimers packed together (dark and gray ribbons). C, RXR bound to 9 cis-RA (gray worm) with the SRC-1 LxxLL motif (dark) and D, a close-up view of the RXR charge-clamp groove where the leucine residues of the LXXLL motif are underlined.

Table I Volumes (Å^3) of NR Ligand Binding Pockets

Receptor	Ligand	Pocket vol	Ligand vol	PDB ID
ERβ	Genistein	390	236	1QKM
AR	DHT	422*	319*	1I37
RARγ	Retinoic acid	429	335*	2LBD
ERα	Estradiol	450	245	1ERE
RXRα	9-cis RA	470	350	1FM6
TRα	T3	600	530	–
PR	Progesterone	603	349*	1A28
VDR	Vitamin D	697	390	1DB1
RORβ	Stearic acid	766	–	1K4W
PXR	SR12813	1150	552*	1ILH
USP	Phospholipid	1256	878*	1G2N
PPARδ	EPA	1300	300	3GWX
PPARα	GW409544	1400	524*	1K7L
PPARγ	GI262570	1440	580	1FM6

Volumes indicated with an asterisk were calculated using the methodology of Gampe *et al.* [1]. Other values were compiled from the literature.

The shape of the ligand-binding pocket also varies greatly, allowing different NRs to accommodate ligands with different shapes. An interesting example of ligand selectivity between different subfamilies of receptors is provided by the retinoid X receptor (RXR) and the retinoic acid receptor (RAR), where 9-*cis* retinoic acid can bind to both receptors, but the all-*trans* retinoic acid binds only RAR. RXR has an L-shaped pocket, allowing it to accommodate the bent shape of the 9-*cis* isomer but not that of the extended all-*trans* isomer, [1]. By contrast, the RAR pocket is relatively straighter, and can accommodate both isomers [2]. The role of the pocket shape in ligand selectivity can also be found in receptors from the same subfamily. The three PPARs (α, γ, and δ) show different ligand preferences for fibrates, a class of lipid-lowering drugs, or glitazones, a class of diabetes drugs. A single residue variation that alters the shape of the PPAR headgroup pockets has been shown to be responsible for the differential binding of these ligands [3]. These examples show that the pocket shape is a key factor in determining ligand specificity.

Ligand-Mediated Activation: Mouse Trap versus Charged Clamp

Activation of nuclear receptors depends on the precise positioning of the AF-2 helix, which in turn depends on the nature of the bound ligand. In the apo-RXR structure, the AF-2 helix was extended away from the main body of the domain, leaving an opening into the ligand-binding pocket [4,5]. In contrast, the AF-2 helix is packed over this opening in the agonist-bound structures. This observation led to a "mouse-trap" model for activation, in which the ligand enters the pocket through this opening, causing the AF-2 helix to close over the ligand like a mouse trap closes over a mouse. The mouse-trap model was confirmed, for RXR, by the subsequent PPARγ/RXR heterodimer structure, in which the RXR monomer was bound to 9-*cis* retinoic acid [1]. However, the activation process has proven to be more complex than originally proposed because RXR activation requires the dissociation from its self-repressed tetrameric form (Fig. 1B) [4]. The apo structures of PPARγ, PPARδ, and PXR have shown that the AF-2 helix can be packed against the receptor in the active conformation even in the absence of ligand [6–8]. With the AF-2 opening blocked, the ligand instead enters the PPAR pocket through a large channel between helix 3 and the β-sheet. Ultraspiracle (the insect homolog of RXR) also has an alternate channel for ligand entry [9,10].

In contrast to multiple pathways of ligand entry, a conserved mechanism is employed by nuclear receptors for ligand-dependent activation. In the apo state, the lower half of the LBD is structurally unstable due to a large unfilled cavity inside the protein. Ligand binding, which partly fills this cavity, stabilizes the protein and allows the AF-2 helix to dock into the active conformation. In PPARs, the AF-2 helix is further reinforced in the active conformation by a direct hydrogen bond with the acidic headgroup of the bound ligand. In this conformation, the AF-2 helix positions a conserved glutamate so as to form a charge clamp with a lysine from helix 3. Coactivators such as SRC-1 contain an LxxLL motif that forms two turns of α-helix, orienting the leucine side chains into a hydrophobic cleft between the two charge-clamp residues (Fig. 1). The charge-clamp residues make a network of hydrogen bonds with the ends of the LxxLL helical motif, further stabilizing the docking of the coactivator helix. This mode of coactivator interactions appears to be common for ligand-mediated activation because residues that contact coactivator are highly conserved among NRs.

Ligand-Mediated Repression

Some nuclear receptors can actively repress the transcription of target genes by recruiting corepressors such as *nuclear corepressor* (N-CoR) and *silencing mediator for retinoid and thyroid hormone receptors* (SMRT) [11,12]. The recent structure of a ternary complex of the PPARα LBD bound to an antagonist and a segment from SMRT reveals that the corepressor interacts with the receptor through an α-helical LxxxIxxxL/I consensus sequence that packs into the same general site as the LxxLL motif of the coactivator [13]. However, the LxxxIxxxI/L helix is longer than the LxxLL helix, with three turns of α-helix instead of two (Fig. 2). Whereas the coactivator interacts with a glutamate residue in the AF-2 helix, the corepressor helix protrudes into volume normally occupied by the AF-2 helix, effectively displacing the AF-2 helix from its active position. The net effect is that coactivators recognize and bind to NRs when the AF-2 helix is in its "active" conformation, whereas corepressors recognize and bind to NRs when the AF-2 helix is in other, "inactive" conformations.

The antagonist in the PPARα/SMRT structure was specifically designed, based on an earlier agonist-bound PPAR structure, to protrude into volume normally occupied by the AF-2 helix. The structure confirms that the designed antagonist displaces the AF-2 helix out of the active position. However, different from the antagonist-bound estrogen receptor structures, the AF-2 helix does not occupy the coactivator-binding groove but is loosely packed against helix 3 (Fig. 2). The displacement of the AF-2 helix out of its active position by the antagonist opens up a larger groove composed by helices 3, 4 and 5 and further reinforces the binding of the three-turn corepressor helix. These structural observations highlight the conformational flexibility of the AF-2 helix, which functions as a molecular switch to turn on or turn off NR transcriptional activities in response to the binding of an agonist or an antagonist.

Dimerization

As DNA-binding transcriptional factors, most NRs function as dimers. X-ray structures of the estrogen receptor (ER) homodimer and two different RXR heterodimers reveal a common mode of dimerization, in which the helix 10 from one monomer packs against the helix 10 from the other monomer in a parallel manner [4,5]. The dimerization involves only the N-terminal (upper) half of helix 10, with the lower half of the helix curving away from the partner monomer. Although the homodimers have perfect twofold symmetry, the PPARγ/RXR heterodimer displays significant asymmetry. This can be seen in the contacts made by the AF-2 helix. In the homodimers, both AF-2 helices fail to contact the partner, whereas in the heterodimer, the PPARγ AF-2 helix contacts RXR, but the RXR AF-2 helix fails to contact PPARγ [1]. The RXR/RAR complex has a similar asymmetry, although the RAR AF-2 helix cannot make analogous interactions with RXR because the C-terminal end of its AF-2 helix is not conserved, and because the AF-2 helix lies in the coactivator pocket. Interestingly, the RXR/PPARγ heterodimer is "permissive" in that it can be activated by RXR ligand alone, whereas the RXR/RAR heterodimer is nonpermissive, and cannot be activated by the RXR ligand alone. The asymmetric interactions between RXR and its partner's AF-2 helix may provide a structural basis for the permissive activation by the RXR ligand.

The steroid receptors of progesterone (PR) and androgen (AR) reveal a distinct mode of dimerization involving the AF-2 helix (in its active conformation) and the C-terminal (lower) half of helix 10 [14,15]. The dimer interface in these steroid receptors is significantly smaller than that seen in the RXR heterodimer and ER homodimer, and its physiological relevance remains to be determined.

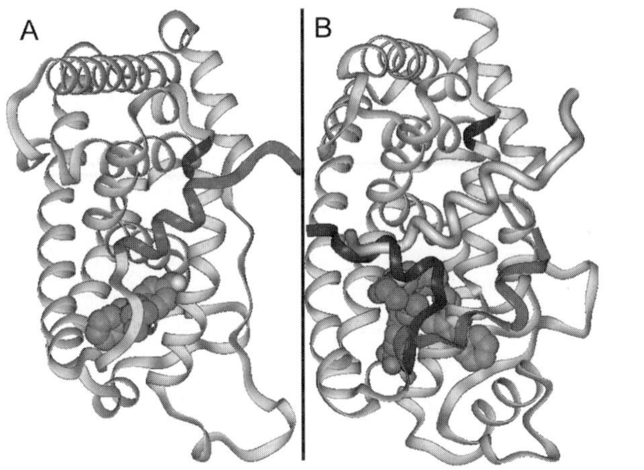

Figure 2 Antagonist-bound LBD structures. A, ERα bound to tamoxifen (spheres), with the AF-2 helix shown in a darker shade of gray; and B, the ternary complex of PPARα with GW6471 (spheres), with the AF-2 helix and SMRT peptide in darker and lighter shades of gray, respectively. The active position of the AF-2 is shown with a very dark gray to illustrate how it would bump the SMRT corepressor helix.

Summary

X-ray structures have now been solved for more than a dozen NR LBDs, bound to agonists and antagonists, coactivators and corepressors, and as monomers, homodimers, heterodimers, and tetramers. These structures have illustrated the details of ligand binding, the conformational changes induced by agonists and antagonists, the basis of NR dimerization, and the mechanism of coactivator and corepressor binding. All of the NRs studied to date have broadly similar structures and mechanisms of activation, but functionally significant differences have arisen in different NRs over the course of evolution. We can expect more surprises as structural work continues on the remaining NRs, and as crystallographers tackle higher order complexes involving the LBD with the AF-1 domain, the DBD, and other proteins and nucleic acids involved in gene transcription.

References

1. Gampe, R. T., Jr., Montana, V. G., Lambert, M. H., Miller, A. B., Bledsoe, R. K., Milburn, M. V., Kliewer, S. A., Willson, T. M., and Xu, H. E. (2000). Asymmetry in the PPARγ/RXRalpha crystal structure reveals the molecular basis of heterodimerization among nuclear receptors. *Mol. Cell* **5**, 545–555.
2. Renaud, J. P., Rochel, N., Ruff, M., Vivat, V., Chambon, P., Gronemeyer, H., and Moras, D. (1995). Crystal structure of the RAR-gamma ligand-binding domain bound to all-*trans* retinoic acid. *Nature* **378**, 681–689.
3. Xu, H. E., Lambert, M. H., Montana, V. G., Plunket, K. D., Moore, L. B., Collins, J. L., Oplinger, J. A., Kliewer, S. A., Gampe, R. T., Jr., McKee, D. D., Moore, J. T., and Willson, T. M. (2001). Structural determinants of ligand binding selectivity between the peroxisome proliferator-activated receptors. *Proc. Natl. Acad. Sci. USA* **98**, 13919–13924.
4. Gampe, R. T., Jr., Montana, V. G., Lambert, M. H., Wisely, G. B., Milburn, M. V., and Xu, H. E. (2000). Structural basis for autorepression of retinoid X receptor by tetramer formation and the AF-2 helix. *Genes Dev.* **14**, 2229–2241.
5. Bourguet, W., Ruff, M., Chambon, P., Gronemeyer, H., and Moras, D. (1995). Crystal structure of the ligand-binding domain of the human nuclear receptor RXR-alpha. *Nature* **375**, 377–382.
6. Xu, H. E., Lambert, M. H., Montana, V. G., Parks, D. J., Blanchard, S. G., Brown, P. J., Sternbach, D. D., Lehmann, J. M., Wisely, G. B., Willson, T. M., Kliewer, S. A., and Milburn, M. V. (1999). Molecular recognition of fatty acids by peroxisome proliferator-activated receptors. *Mol. Cell* **3**, 397–403.
7. Nolte, R. T., Wisely, G. B., Westin, S., Cobb, J. E., Lambert, M. H., Kurokawa, R., Rosenfeld, M. G., Willson, T. M., Glass, C. K., and Milburn, M. V. (1998). Ligand binding and co-activator assembly of the peroxisome proliferator-activated receptor-gamma. *Nature* **395**, 137–143.
8. Watkins, R. E., Wisely, G. B., Moore, L. B., Collins, J. L., Lambert, M. H., Williams, S. P., Willson, T. M., Kliewer, S. A., and Redinbo, M. R. (2001). The human nuclear xenobiotic receptor PXR: Structural determinants of directed promiscuity. *Science* **292**, 2329–2333.
9. Clayton, G. M., Peak-Chew, S. Y., Evans, R. M., and Schwabe, J. W. (2001). The structure of the ultraspiracle ligand-binding domain reveals a nuclear receptor locked in an inactive conformation. *Proc. Natl. Acad. Sci. USA* **98**, 1549–1554.
10. Billas, I. M., Moulinier, L., Rochel, N., and Moras, D. (2001). Crystal structure of the ligand-binding domain of the ultraspiracle protein USP, the ortholog of retinoid X receptors in insects. *J. Biol. Chem.* **276**, 7465–7474.
11. Horlein, A. J., Naar, A. M., Heinzel, T., Torchia, J., Gloss, B., Kurokawa, R., Ryan, A., Kamei, Y., Soderstrom, M., Glass, C. K. *et al.* (1995). Ligand-independent repression by the thyroid hormone receptor mediated by a nuclear receptor co-repressor. *Nature* **377**, 397–404.
12. Chen, J. D. and Evans, R. M. (1995). A transcriptional co-repressor that interacts with nuclear hormone receptors. *Nature* **377**, 454–457.
13. Xu, H. E. *et al.* (2002). Structural basis of antagonist-mediated recruitment of nuclear corepressors by PPARalpha. *Nature* **415**, 813–817.
14. Williams, S. P. and Sigler, P. B. (1998). Atomic structure of progesterone complexed with its receptor. *Nature* **393**, 392–396.
15. Matias, P. M., Donner, P., Coelho, R., Thomaz, M., Peixoto, C., Macedo, S., Otto, N., Joschko, S., Scholz, P., Wegg, A., Basler, S., Schafer, M., Egner, U., and Carrondo, M. A. (2000). Structural evidence for ligand specificity in the binding domain of the human androgen receptor. Implications for pathogenic gene mutations. *J. Biol. Chem.* **275**, 26164–26171.

Nuclear Receptor Coactivators

Riki Kurokawa and Christopher K. Glass
Department of Cellular and Molecular Medicine,
School of Medicine, University of California
at San Diego, La Jolla, California

Introduction

Nuclear receptors (NRs) comprise a family of DNA-binding transcription factors that regulate programs of gene expression related to cellular growth, differentiation, and homeostasis in a ligand-dependent manner [1]. In addition to receptors for ligands such as steroid hormones, retinoids, and metabolites of fatty acids, the NR family also includes several so-called "orphan receptors," representing members for which regulatory ligands are not known [2]. NRs activate transcription by binding as monomers, dimers, or heterodimers to specific hormone response elements (HREs) within target genes [3]. The central highly conserved DNA-binding domain (DBD) mediates sequence-specific recognition of HREs. Two distinct domains contribute to transcriptional activation functions, referred to as AF-1 and AF-2 [1]. The AF-1 domain is located in the N-terminal region and is poorly conserved among NR family members. In some NRs, particularly steroid hormone receptors, AF-1 plays a quantitatively important role in transcriptional activation. AF-2 resides in the C-terminal ligand-binding domain (LBD), which in addition to conferring the specific ligand-binding properties of each receptor, also contains a dimerization interface and ligand-regulated transcriptional activation and repression functions. Several lines of evidence have emerged in recent years indicating that ligand-dependent transcriptional activation involves the recruitment of a series of coactivator proteins to the NR LBD [1,4]. Several of these proteins are components of complexes that remodel chromatin, modify histone tails, or act to recruit core transcription factors (Fig. 1). In this chapter, we discuss general molecular mechanisms responsible for ligand-dependent interactions among NRs and coactivator proteins and provide examples of different classes of coactivators that have been linked to NR function.

Mechanism of Coactivator Recruitment

The ligand-dependent transactivation function of NRs has been demonstrated to depend on a short conserved sequence within the C terminus of the LBD [5,6]. Mutations within this region have been identified that have little or no effect on ligand binding, but abolish ligand-dependent transcription. Crystallographic analysis of several nuclear receptor LBDs has revealed a conserved structure in which a three-layered antiparallel α-helical sandwich encloses a central, hydrophobic ligand-binding pocket [7,8]. In the unliganded RXR structure, the AF-2 helix extends away from the ligand-binding domain [9]. In contrast, in the agonist-bound RAR, TR, and ER LBD structures, the AF2 helix is tightly packed against the body of the LBD and makes direct contact with ligand [10–13]. In concert, these studies are consistent with the idea that ligand-dependent changes in the conformation of the AF-2 helix result in the formation of a surface that facilitates coactivator interactions. Intriguingly, the structures of the estrogen receptor LBD bound to the antagonists raloxifene or dihydroxytamoxifen (OHT) demonstrate a distortion in the position of the AF-2 helix that prevents the binding of coactivators [12,13].

Extensive analysis of the amino acid sequences of several nuclear receptor coactivators revealed that a short helical sequence LXXLL (L: leucine and X: any amino acid) within the nuclear receptor interaction domain was necessary for ligand-dependent recruitment to the NR LBD [14–16]. LXXLL motifs have been identified within nearly all of the putative coactivators that interact with NRs in a ligand-dependent manner, including members of the p160/SRC family (Fig. 2A), CBP/p300, TRAP220/DRIP205/PBP, and PGC-1. Structural studies of the PPARγ LBD complexed to a fragment of the SRC-1 nuclear receptor interaction domain containing two LXXLL motifs revealed that these motifs

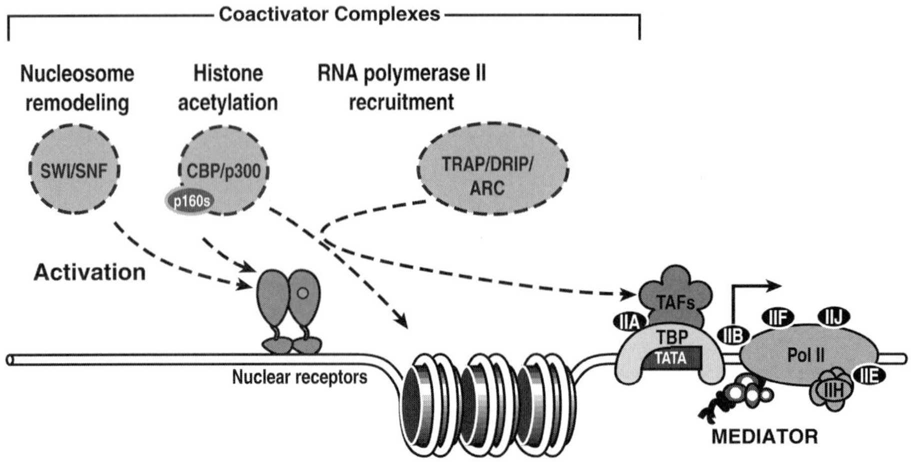

Figure 1 Ligand-dependent transcriptional activation of nuclear receptor target genes. Nuclear receptors bind to hormone response elements in promoter or enhancer elements and recruit one or more coactivator complexes in response to activating ligands. Many coactivator proteins have been identified that are components of multiprotein complexes, such as Swi/Snf complexes involved in nucleosome remodeling, p160/CBP complexes containing histone acetyltransferase activities, and TRAP/DRIP/ARC complexes involved in recruitment of basal transcription factors.

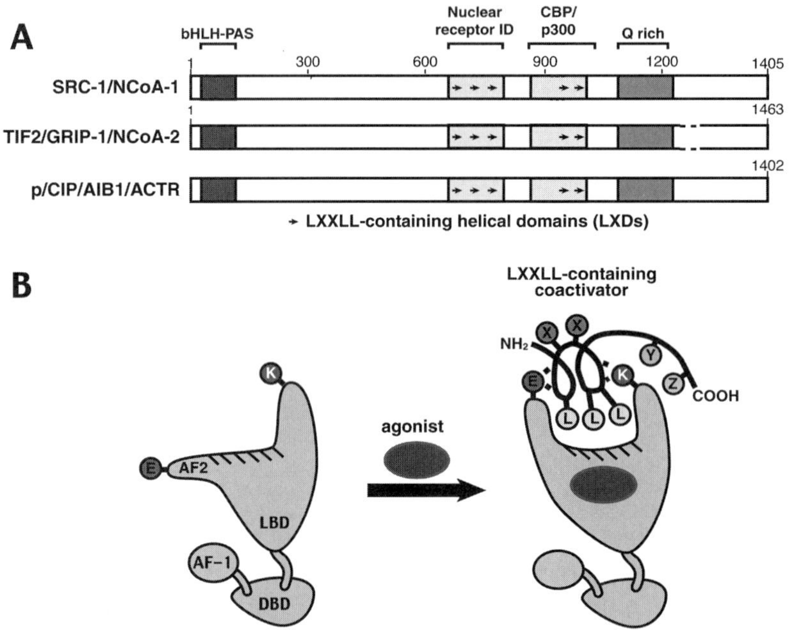

Figure 2 Mechanism of coactivator recruitment. (A) Alignment of members of the p160/SRC family of nuclear receptor coactivator illustrating the positions of LXXLL motifs mediating NR interaction and the CBP/p300 interaction domain. (B) Binding of agonist to the NR LBD induces a conformational change in the C-terminal AF-2 helix. This results in the formation of a charge-clamp that grips the ends of the LXXLL helix and allows the leucine residues to pack into an intervening hydrophobic cavity.

form short α-helices [17]. The two LXXLL motifs within this fragment of SRC-1 interacted with each subunit of a LBD dimer, suggesting a mechanism for cooperative interactions among coactivators and nuclear receptor dimers or heterodimers. The LXXLL helix was gripped at each end by a charge-clamp consisting of a conserved lysine in helix 3 of the LBD and a conserved glutamate in the AF-2 helix (Fig. 2B). This positioned the LXXLL helix so that the leucine residues could pack into a hydrophobic pocket between the end of helix 3 and the AF-2 helix, stabilizing the interaction. Structures of the thyroid hormone receptor and estrogen receptor LBDs complexed to LXXLL peptides from other p160 family members exhibit the same structural basis for binding [13,18,19]. Although the LXXLL helix is necessary for ligand-dependent interactions, additional residues N and C terminal to the helix contribute to binding specificity [14,20].

General Classes of Coactivator Complexes

Since the initial biochemical characterization of p160 and p140 proteins as ligand-dependent estrogen receptor-associated proteins [21,22], dozens of additional proteins have been identified that interact with nuclear receptors in a ligand-dependent manner [1]. Functional assignment of these proteins as NR coactivators has generally been based on co-immunoprecipitation experiments demonstrating that they interact with liganded nuclear receptors in cells and that forced expression results in increased ligand-dependent transcriptional responses. The development of chromatinized *in vitro* transcription assays has allowed biochemical approaches to functional analysis of coactivator function [23]. More recently, demonstration of ligand-dependent recruitment of putative coactivators to NR target genes by chromatin immunoprecipitation assays has been considered as an additional line of evidence for physiological relevance [24]. Relatively few coactivators have been demonstrated to be required for NR action by loss of function (e.g., gene knock-out) experiments. Gene knock-out experiments are in some cases difficult to interpret because of the potential for coactivators to serve redundant functions. In many cases, proteins initially identified to function as nuclear receptor coactivators have proven to function as coactivators for other classes of transcription factors.

Coactivator proteins are generally considered to function by modifying chromatin architecture or by serving as adapters that recruit core transcription factors to the promoter [1,4]. Many coactivator proteins have been found to be components of stable, multiprotein complexes. Modifications of histone tails are becoming increasingly recognized to constitute a "code" that specifies overall chromatin architecture and the potential of genes to be silenced or transcribed [25]. Several proteins shown to function as nuclear receptor coactivators harbor histone acetyltransferase activity, including CBP and p300. Conversely, the nuclear receptor corepressors N-CoR and SMRT are components of protein complexes that contain histone deacetylase activities [4]. Histone methylation has also emerged as a modification that is associated with either transcriptional activation or gene silencing, depending on the specific histone and amino acid residue that is modified [26]. The nuclear receptor coactivator CARM-1 methylates histone H3 at arginine 3, and this modification is correlated with transcriptional activation [27]. Histone phosphorylation has been linked to transcriptional activation, but NR coactivators with histone kinase activities have not yet been described. A distinct type of chromatin modification is provided by ATP-dependent chromatin remodeling complexes, exemplified by the BRG/BAF complexes, which are highly related to SWI/SNF complexes originally identified genetically in yeast and *Drosophila.* SWI/SNF complexes mediate the repositioning of nucleosomes, presumably facilitating access to DNA-binding transcription factors. Although these proteins may not interact directly with NRs, there is evidence that they are recruited to promoters in a ligand-dependent manner and cooperate with other classes of coactivators [28].

The p160/SRC and TRAP220/DRIP205/PBP proteins are examples of nuclear receptor coactivators that appear to function as adapter proteins. SRC-1 and related proteins interact with liganded NRs via LXXLL motifs, as described earlier, and with CBP and p300 via a C-terminal domain (Fig. 2A). The ability of SRC-1 to function as a coactivator is based in part on its ability to recruit CBP/p300 [15,29,30]. TRAP220/DRIP205/PBP is a component of the TRAP/DRIP/ARC complex identified biochemically as a coactivator complex required for *in vitro* transcriptional activities of the thyroid hormone receptor, vitamin D receptor, and SREBP [31–33]. The TRAP220/DRIP205/PBP subunit contains two LXXLL motifs, which allow the entire complex consisting of more than a dozen proteins to be recruited to nuclear receptors in a ligand-dependent manner [34]. The TRAP/DRIP/ARC complex is proposed to function by recruiting RNA polymerase II holoenzyme to the promoter in a receptor- and ligand-dependent manner [35].

Coactivators as Targets of Signal Transduction Pathways

Components of coactivator complexes are emerging as targets for various signal transduction pathways, providing an additional level at which these pathways may be integrated. An example is provided by PGC-1, initially reported as a fat cell-specific coactivator for PPARγ [36]. Recently, PGC-1 has been found to enhance the expression of key gluconeogenic enzymes, including phosphoenolpyruvate carboxykinase (PEPCK) and glucose-6-phosphatase in the livers of fasting mice, leading to increased glucose synthesis [37]. These genes are under combinatorial control of several sequence-specific transcription factors, including the glucocorticoid receptor and the orphan nuclear receptor HNF-4, which collectively utilize PGC-1 as a coactivator. The expression of PGC-1 is stimulated by glucagon and suppressed by insulin, suggesting that it may be a key target of these hormones in the maintenance of blood glucose homeostasis. These observations are of particular interest because they suggest that the expression of a specific coactivator is required for a specific program of gene expression that is regulated by transcription factors that have multiple physiologic roles.

Conclusion

The identification and functional characterization of nuclear receptor coactivators represent major current efforts being undertaken in the nuclear receptor field. Many fundamental questions remain to be answered, including why so many of these proteins exist, and how they exert their transcriptional effects. The study of these proteins could have practical benefits in that it may be possible to selectively regulate patterns of gene expression by developing synthetic NR ligands that differentially recruit coactivators to the LBD.

It is likely that differential recruitment of coactivators and corepressors underlies the tissue-specific actions of selective estrogen receptor modulators, and that ligands with similar properties can be developed for other members of the NR family. Such ligands would have potential therapeutic applications in a variety of human diseases, including metabolic syndromes, atherosclerosis, diabetes, and chronic inflammatory diseases.

References

1. McKenna, N. J. and O'Malley, B. W. (2002). Combinatorial control of gene expression by nuclear receptors and coregulators. *Cell* **108**(4), 465–474.
2. Chawla, A., Repa, J., Evans, R. *et al.* (2001). Nuclear receptors and lipid physiology: Opening the X-files. *Science* **294**, 1866–1870.
3. Glass, C. K. (1994). Differential recognition of target genes by nuclear receptor monomers, dimers and heterodimers. *Endocrin. Rev.* **15**, 1503–1519.
4. Glass, C. K. and Rosenfeld, M. G. (2000). The coregulator exchange in transcriptional functions of nuclear receptors. *Genes Dev.* **14**, 121–141.
5. Durand, B., Saunders, M., Gaudon, C. *et al.* (1994). Activation function 2 (AF-2) of retinoic acid receptor and 9-cis retinoic acid receptor: presence of a conserved autonomous constitutive activating domain and influence of the nature of the response element on AF-2 activity. *EMBO J.* **13**, 5370–5382.
6. Danielian, P. S., White, R., Lees, J. A. *et al.* (1992). Identification of a conserved region required for hormone-dependent transcriptional activation by steroid hormone receptors. *EMBO J.* **11**, 1025–1033.
7. Moras, D. and Gronemeyer, H. (1998). The nuclear receptor ligand-binding domain: structure and function. *Curr. Opin. Cell Biol.* **10**, 384–391.
8. Bourguet, W., Germain, P., and Gronemeyer, H. (2000). Nuclear receptor ligand-binding domains: Three-dimensional structures, molecular interactions and pharmacological implications. *Trends Pharm. Sci.* **21**, 381–388.
9. Bourguet, W., Ruff, M., Chambon, P. *et al.* (1995). Crystal structure of the ligand-binding domain of the human nuclear receptor RXR-α. *Nature* **375**, 377–382.
10. Wagner, R. L., Apriletti, J. W., McGrath, M. E. *et al.* (1995). A structural role for hormone in the thyroid hormone receptor. *Nature* **378**, 690–697.
11. Renaud, J.-P., Rochel, N., Ruff, M. *et al.* (1995). Crystal structure of the RAR-γ ligand-binding domain bound to all-*trans* retinoic acid. *Nature* **378**, 681–689.
12. Brzozowski, A. M., Pike, A. C. W., Dauter, Z. *et al.* (1997). Molecular basis of agonism and antagonism in the oestrogen receptor. *Nature* **389**, 753–758.
13. Shiau, A. K., Barstad, D., Loria, P. M. *et al.* (1998). The structural basis of estrogen receptor/coactivator recognition and the antagonism of this interaction by tamoxifen. *Cell* **95**(7), 927–937.
14. Heery, D. M., Kalkhoven, E., Hoare, S. *et al.* A signature motif in transcriptional co-activators mediates binding to nuclear receptors. *Nature* **387**, 733–736.
15. Torchia, J., Rose, D. W., Inostroza, J. *et al.* The transcriptional co-activator p/CIP binds CBP and mediates nuclear-receptor function. *Nature* **387**, 677–684.
16. Ding, X. F., Anderson, C. M., Ma, H. *et al.* (1998). Nuclear receptor-binding sites of coactivators glucocorticoid receptor interacting protein 1 (GRIP1) and steroid receptor coactivator 1 (SRC-1): Multiple motifs with different binding specificities. *Mol. Endo.* **12**(2), 302–313.
17. Nolte, R. T., Wisely, G. B., Westin, S. *et al.* (1998). Ligand binding and co-activator assembly of the peroxisome proliferator-activated receptor-γ. *Nature* **395**, 137–143.
18. Feng, W., Ribeiro, R. C. J., Wagner, R. L. *et al.* Hormone-dependent coactivator binding to a hydrophobic cleft on nuclear receptors. *Science* **280**, 1747–1749.
19. Darimont, B. D., Wagner, R. L., Apriletti, J. W. *et al.* (1998). Structure and specificity of nuclear receptor-coactivator interactions. *Genes Dev.* **12**, 3343–3356.
20. McInerney, E. M., Rose, D. W., Flynn, S. E. *et al.* (1998). Determinants of coactivator LXXLL motif specificity in nuclear receptor transcriptional activation. *Genes Dev.* **12**, 3357–3368.
21. Cavailles, V., Dauvois, S., L'Horset, F. *et al.* (1995). Nuclear factor RIP140 modulates transcriptional activation by the estrogen receptor. *EMBO J.* **14**, 3741–3751.
22. Halachmi, S., Marden, E., Martin, G. *et al.* (1994). Estrogen receptor-associated proteins: Possible mediators of hormone-induced transcription. *Science* **264**, 1455–1458.
23. Kraus, W. L. and Kadonaga, J. T. (1998). p300 and estrogen receptor cooperatively activate transcription via differential enhancement of initiation and reinitiation. *Genes Dev.* **12**(3), 331–342.
24. Shang, Y., Hu, X., DiRenzo, J. *et al.* (2000). Cofactor dynamics and sufficiency in estrogen receptor-regulated transcription [in process citation]. *Cell* **103**(6), 843–852.
25. Jenuwein, T. and Allis, C. D. (2001). Translating the histone code. *Science* **293**, 1074–1080.
26. Zhang, Y. and Reinberg, D. (2001). Transcription regulation by histone methylation: Interplay between different covalent modifications of the core histone tails. *Genes Dev.* **15**(18), 2343–2360.
27. Strahl, B. D., Briggs, S. D., Brame, C. J. *et al.* (2001). Methylation of histone H4 at arginine 3 occurs *in vivo* and is mediated by the nuclear receptor coactivator PRMT1. *Curr. Biol.* **11**(12), 996–1000.
28. DiRenzo, J., Shang, Y., Phelan, M. *et al.* (2000). BRG-1 is recruited to estrogen-responsive promoters and cooperates with factors involved in histone acetylation. *Mol. Cell Biol.* **20**, 7541–7549.
29. Yao, T.-P., Ku, G., Zhou, N. *et al.* (1996). The nuclear hormone receptor coactivator SRC-1 is a specific target of p300. *Proc. Natl. Acad. Sci. USA* **93**, 10626–10631.
30. Li, J., O'Malley, B., and Wong, J. (2000). p300 requires its histone acetyltransferase activity and SRC-1 interaction domain to facilitate thyroid hormone receptor activation in chromatin. *Mol. Cell Biol.* **20**, 2031–2042.
31. Rachez, C., Suldan, Z., Ward, J. *et al.* (1998). A novel protein complex that interacts with the vitamin D_3 receptor in a ligand-dependent manner and enhances VDR transactivation in a cell-free system. *Genes Dev.* **12**, 1787–1800.
32. Fondell, J. D., Ge, H., and Roeder, R. G. (1996). Ligand induction of a transcriptionally active thyroid hormone receptor coactivator complex. *Proc. Natl. Acad. Sci. USA* **93**, 8329–8333.
33. Naar, A M., Beaurang, P. A., Zhou, S. *et al.* (1999). Composite co-activator ARC mediates chromatin-directed transcriptional activation. *Nature* **398**, 828–832.
34. Rachez, C., Gamble, M., Chang, C. *et al.* (2000). The DRIP complex and SRC-1/160 coactivators share similar nuclear receptor binding determinants but constitute functionally distinct complexes. *Mol. Cell Biol.* **20**, 2718–2726.
35. Chiba, N., Suldan, Z., Freedman, L. *et al.* (2000). Binding of liganded vitamin D receptor to the vitamin D receptor interacting protein coactivator complex induces interaction with RNA polymerase II holoenzyme. *J. Biol. Chem.* **275**, 10719–10722.
36. Puigserver, P., Wu, Z., Park, C. W. *et al.* (1998). A cold-inducible coactivator of nuclear receptors linked to adaptive thermogenesis. *Cell* **92**, 829–839.
37. Yoon, J. C., Puigserver, P., Chen, G. *et al.* (2001). Control of hepatic gluconeogenesis through the transcriptional coactivator PGC-1. *Nature* **413**, 131–138.

CHAPTER 274

Corepressors in Mediating Repression by Nuclear Receptors

Kristen Jepsen and Michael G. Rosenfeld
Howard Hughes Medical Institute, Department and School of Medicine, University of California at San Diego, La Jolla, California

Introduction

Activation of gene transcription has long been recognized as an essential component of regulation of gene expression by nuclear receptors, but repression of transcription plays an equally important role in developmental and homeostatic gene regulation. Transcriptional repression by nuclear receptors requires the actions of the closely related receptor-associated corepressors, nuclear receptor corepressor (N-CoR) and silencing mediator of retinoic acid and thyroid hormone receptors (SMRT), which were identified based on their ability to associate with and mediate transcriptional repression by nuclear receptors. These corepressors recruit many classes of transcription factors as components of multiprotein complexes, some containing specific histone deacetylases (Fig. 1A). This association with histone deacetylase activity provides one component of the mechanism that allows DNA-binding proteins interacting with N-CoR or SMRT to repress transcription of specific target genes. However, both N-CoR and SMRT are components of additional complexes that can result in permanent repression. The biological importance of these corepressors has emerged from genetic studies both of humans and mice. In turn, various signaling pathways regulate levels of expression, subcellular localization, and function of these corepressors.

N-CoR and SMRT in Repression by Nuclear Receptors

Thyroid hormone and retinoic acid receptors (T_3R and RAR) actively repress transcription in the absence of their cognate ligands via transferable repression domains [1] that can associate with N-CoR [2] and SMRT [3–5]. These corepressors contain a conserved bipartite nuclear receptor interaction domain [6–8] and three independent repressor domains that are capable of transferring active repression to a heterologous DNA-binding domain (DBD) [2–5] (Fig. 1B). The interaction domains contain related putative helical motifs LXXXIXXXI/L [9–11], one helical turn longer than the LXXLL recognition motif present in nuclear receptor coactivators [12,13] that binds in a hydrophobic pocket in the hormone binding domain, similar to coactivator binding. Specific sequences in the nuclear receptor interaction motif are suggested to account for the preference of RAR for SMRT and T_3R for N-CoR [14].

The mechanism by which N-CoR and SMRT function proved, in part, to reflect their association with mRpd3 and mSin3A and B, mammalian homologs of the yeast proteins Rpd3p/histone deacetylase 1 and Sin3p [15,16]. Previous observations that acetylation of specific lysine residues in the N termini of histones correlates with increased transcription, and that heterochromatic regions are generally hypoacetylated [17,18], led to studies of the role of HDAC proteins in transcriptional repression. There are now 10 HDACs, classified on the basis of their homology to two closely related yeast HDACs, RPD3 and Hda-1 [19,20], as class I or class II HDACs, respectively [21]. In some assays, specific HDAC proteins appear to require association with N-CoR or SMRT for full enzymatic activity [22,23]. Additionally, there is evidence of a combinatorial code based on distinct histone modifications, indicating a complex relationship between various recruited enzymatically active complexes [24].

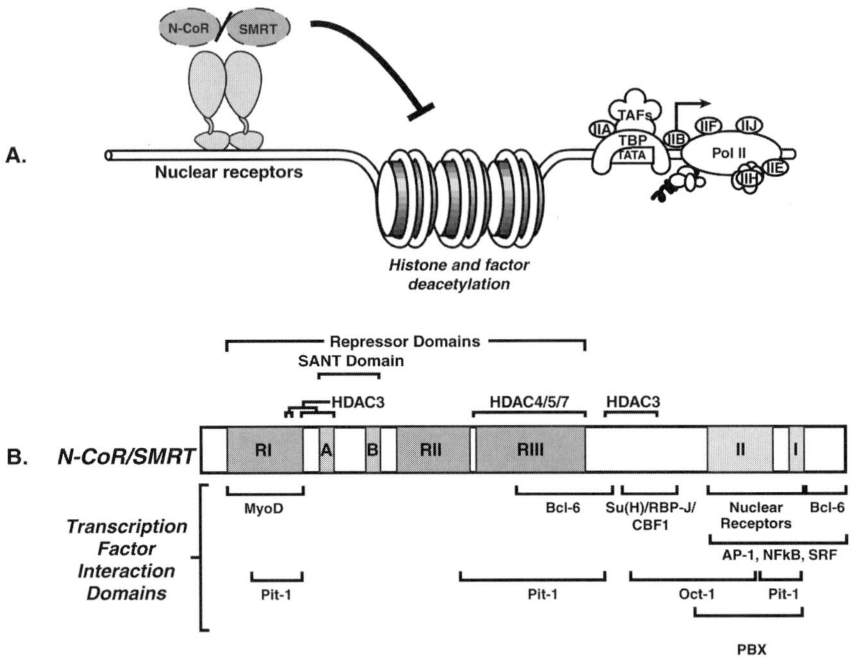

Figure 1 (A) Transcriptional repression by nuclear receptors is regulated by recruitment of the corepressors N-CoR and/or SMRT. (B) Schematic of domains of N-CoR/SMRT. Repression domains (RI, RII, RIII) and SANT domains (A and B) are indicated, as are interaction domains for HDACs, nuclear receptors (I and II), and other transcription factors.

Purification of Corepressor Complexes

Biochemical purification of complexes using anti-N-CoR or anti-HDAC3 antibodies revealed several corepressor-associated complexes, one which contains HDAC3, N-CoR, or SMRT, and transducin (beta)-like protein 1 (TBL1) [23,25–27]. TBL-1 has six WD-40 repeats [28], a motif also present in the Tup1 and Groucho corepressors, and is homologous to the *Drosophila* protein ebi, which is involved in epidermal growth factor receptor signaling pathways [29]. Under different conditions, an N-CoR–SMRT–HDAC3 complex can also contain Krab associated protein 1 (KAP-1), a TSA-sensitive corepressor that interacts with members of the heterochromatin protein 1 (HP1) family, and several members of the Swi/Snf ATP-dependent chromatin-remodeling complex family, which is reminiscent of the ATP-dependent chromatin-remodeling proteins found in the NURD complex (Fig. 1B) [27]. A third N-CoR/SMRT complex, which appears to be among the weakest, shares common components (HDAC1, HDAC2, and mSin3) with the Sin-associated protein (SAP) complex [30,31]. Several groups have also shown that the third repressor domain of N-CoR and SMRT can directly interact *in vitro* with class II HDACs, including HDAC4, HDAC5, and HDAC7 [32,33], suggesting that the full range of complexes has yet to be purified.

Other Nuclear Receptor and Transcription Factor Partners of N-CoR/SMRT

Although cloned based on their interactions with unliganded RAR and T_3R, N-CoR and SMRT are capable of conferring transcriptional repression to an ever increasing number of transcription factors (Fig. 1B). N-CoR or SMRT serve as corepressors for several other members of the nuclear receptor superfamily, including v-ErbA, RevErb, COUP-transcription factors, PPARα and DAX1 [34]. Although steroid hormone receptors do not appear to interact with N-CoR or SMRT in the absence of ligand [2,3], both the estrogen receptor (ER) and progesterone receptor (PR) can interact with these corepressors in the presence of their respective antagonists to repress transcription [35–38]. In chromatin immunoprecipitation assays (ChIP) performed in the breast tumor derived cell line MCF-7, N-CoR and SMRT were present on the estrogen-responsive cathepsin D and pS2 promoters in the presence of the antagonist tamoxifen but not estrogen [39]. These data suggest a role for N-CoR and SMRT in mediating the antagonist-associated effects of steroid hormone receptors [2,3] and provide a mechanism for the clinical application of antagonistic ligands.

In addition to their interactions with members of the nuclear receptor family, N-CoR and SMRT have been implicated as corepressors for a variety of unrelated transcription factors that regulate diverse cellular processes, including various homeodomain factors, MAD, SRF, BCL-6, MyoD, STAT5, HERP, and Su(H) [34].

Multiple Mechanisms of N-CoR/SMRT Regulation

The actions of N-CoR and SMRT are highly regulated. The N terminus of N-CoR interacts with mSiah2, the mammalian homolog of *Drosophila Seven in absentia* [40]. Consistent with evidence that mSiah2 regulates proteasomal degradation of proteins [41], cotransfection of N-CoR and mSiah2 resulted in a dramatic decrease of N-CoR protein levels, an effect abolished by inclusion of a proteasome inhibitor [40].

Stimulation of the MAPK and PKC pathways resulted in decreased association of N-CoR and ER in the presence of the antagonist tamoxifen [38]. In one example, EGF-dependent phosphorylation of ER converted tamoxifen from an antagonist to an agonist [38]. Activation of the mitogen-activated protein kinase (MAPK) pathway by L-thyroxine (T_4) results in serine phosphorylation of TRβ1 and dissociation of SMRT in a hormone-independent manner [42], while phosphorylation of SMRT by MAPK-extracellular signal-regulated kinase 1 (MEK-1) and MEK-1 kinase (MEKK-1) is suggested to inhibit interactions between SMRT and nuclear receptors or PLZF [43]. In contrast, phosphorylation of SMRT by the protein kinase casein kinase II (CK2) is reported to stabilize the SMRT/nuclear receptor interaction [44]. Thus, different cell signaling pathways can effect different transcriptional outcomes.

Signaling pathways have also been shown to cause changes in subcellular distribution of the corepressors, as CamKIV, MEK-1, and MEKK-1 signaling are suggested to result in a redistribution of SMRT from the nucleus to the perinucleus or cytoplasm [43,45]. Indeed, specific signaling events influence the subcellular distribution of HDAC proteins [46–48].

Roles in Development and Disease

During normal development N-CoR is required for normal progression of specific developmental stages in erythrocyte and thymocyte development and in neural maturation [49] based on evidence from gene deletion studies. Expression of a dominant-negative N-CoR protein in hepatocytes of transgenic mice causes an increased proliferation of hepatocytes and derepression of T_3-regulated hepatic target genes [50], while expression of a dominant-negative N-CoR protein in lactotropes abolishes long-term repression of the growth hormone gene [51]. In *Xenopus*, expression of a dominant-negative N-CoR protein can result in embryos exhibiting phenotypes similar to those treated by RA, namely, reduction of anterior structures such as forebrain and cement gland, suggesting that RAR-mediated repression of target genes is critical for head formation [52]. Resistance to thyroid hormone (RTH), characterized by an impaired physiological response to thyroid hormone, is associated with mutations in the T_3R-β gene that fail to release N-CoR or SMRT upon hormone treatment [53–55]. Gene repression appears to be a critical component of normal thyroid hormone physiology, and deletion of all known thyroid hormone receptors results in a phenotype less severe than that of thyroid-hormone-deficient mice [56], implying the role for repression by the unliganded T_3R. Roles for N-CoR and SMRT in several types of leukemia are also well characterized [57].

Other Mediators of Nuclear Receptor Repression

The p140 factor, RIP 140, has proven to be an example of an inhibitor of ligand-occupied receptor based on displacement of other activators [58–62], and perhaps based on its interactions with other corepressors [63]. In addition, other corepressors, acting both in the presence and absence of ligand, may act through nuclear receptors to exert critical biological roles [64,65].

Conclusion

Nuclear receptors can serve as repressors of transcription, characterized by association with proteins with enzymatic functions. In addition to a ligand-dependent switch, various signal transduction pathways can modulate interactions of corepressors with nuclear receptors or mediate their activity or distribution between nuclear or cytoplasmic compartments.

References

1. Baniahmad, A., Kohne, A. C., and Renkawitz, R. (1992). A transferable silencing domain is present in the thyroid hormone receptor, in the v-erbA oncogene product and in the retinoic acid receptor, *EMBO J.* **11**, 1015–1023.
2. Horlein, A. J., Naar, A. M., Heinzel, T., Torchia, J., Gloss, B., Kurokawa, R., Ryan, A., Kamei, Y., Soderstrom, M., Glass, C. K. *et al.* (1995). Ligand-independent repression by the thyroid hormone receptor mediated by a nuclear receptor co-repressor. *Nature* **377**, 397–404.
3. Chen, J. D. and Evans, R. M. (1995). A transcriptional co-repressor that interacts with nuclear hormone receptors. *Nature* **377**, 454–457.
4. Ordentlich, P., Downes, M., Xie, W., Genin, A., Spinner, N. B., and Evans, R. M. (1999). Unique forms of human and mouse nuclear receptor corepressor SMRT. *Proc. Natl. Acad. Sci. USA* **96**, 2639–2644.
5. Park, E. J., Schroen, D. J., Yang, M., Li, H., Li, L., and Chen, J. D. (1999). SMRTe, a silencing mediator for retinoid and thyroid hormone receptors-extended isoform that is more related to the nuclear receptor corepressor. *Proc. Natl. Acad. Sci. USA* **96**, 3519–3524.
6. Seol, W., Mahon, M. J., Lee, Y. K., and Moore, D. D. (1996). Two receptor interacting domains in the nuclear hormone receptor corepressor RIP13/N-CoR. *Mol. Endocrinol.* **10**, 1646–155.
7. Zamir, I., Harding, H. P., Atkins, G. B., Horlein, A., Glass, C. K., Rosenfeld, M. G., and Lazar, M. A. (1996). A nuclear hormone receptor corepressor mediates transcriptional silencing by receptors with distinct repression domains. *Mol. Cell Biol.* **16**, 5458–5465.
8. Li, H., Leo, C., Schroen, D. J., and Chen, J. D. (1997). Characterization of receptor interaction and transcriptional repression by the corepressor SMRT. *Mol. Endocrinol.* **11**, 2025–2037.
9. Perissi, V., Staszewski, L. M., McInerney, E. M., Kurokawa, R., Krones, A., Rose, D. W., Lambert, M. H., Milburn, M. V., Glass, C. K., and Rosenfeld, M. G. (1999). Molecular determinants of nuclear receptor-corepressor interaction. *Genes Dev.* **13**, 3198–3208.
10. Hu, X. and Lazar, M. A. (1999). The CoRNR motif controls the recruitment of corepressors by nuclear hormone receptors. *Nature* **402**, 93–96.
11. Nagy, L., Kao, H. Y., Love, J. D., Li, C., Banayo, E., Gooch, J. T., Krishna, V., Chatterjee, K., Evans, R. M., and Schwabe, J. W. (1999). Mechanism of corepressor binding and release from nuclear hormone receptors. *Genes Dev.* **13**, 3209–3216.

12. Heery, D. M., Kalkhoven, E., Hoare, S., and Parker, M. G. (1997). A signature motif in transcriptional co-activators mediates binding to nuclear receptors. *Nature* **387**, 733–736.
13. McInerney, E. M., Rose, D. W., Flynn, S. E., Westin, S., Mullen, T. M., Krones, A., Inostroza, J., Torchia, J., Nolte, R. T., Assa-Munt, N., Milburn, M. V., Glass, C. K., and Rosenfeld, M. G. (1998). Determinants of coactivator LXXLL motif specificity in nuclear receptor transcriptional activation. *Genes Dev.* **12**, 3357–3368.
14. Cohen, R. N., Putney, A., Wondisford, F. E., and Hollenberg, A. N. (2000). The nuclear corepressors recognize distinct nuclear receptor complexes. *Mol. Endocrinol.* **14**, 900–914.
15. Heinzel, T., Lavinsky, R. M., Mullen, T. M., Soderstrom, M., Laherty, C. D., Torchia, J., Yang, W. M., Brard, G., Ngo, S. D., Davie, J. R., Seto, E., Eisenman, R. N., Rose, D. W., Glass, C. K., and Rosenfeld, M. G. (1997). A complex containing N-CoR, mSin3 and histone deacetylase mediates transcriptional repression. *Nature* **387**, 43–48.
16. Nagy, L., Kao, H. Y., Chakravarti, D., Lin, R. J., Hassig, C. A., Ayer, D. E., Schreiber, S. L., and Evans, R. M. (1997). Nuclear receptor repression mediated by a complex containing SMRT, mSin3A, and histone deacetylase, *Cell* **89**, 373–380.
17. Turner, B. M. (1993). Decoding the nucleosome. *Cell* **75**, 5–8.
18. Grunstein, M. (1990). Histone function in transcription. *Annu. Rev. Cell Biol.* **6**, 643–678.
19. Vidal, M. and Gaber, R. F. (1991). RPD3 encodes a second factor required to achieve maximum positive and negative transcriptional states in *Saccharomyces cerevisiae. Mol. Cell Biol.* **11**, 6317–6327.
20. Rundlett, S. E., Carmen, A. A., Kobayashi, R., Bavykin, S., Turner, B. M., and Grunstein, M. (1996). HDA1 and RPD3 are members of distinct yeast histone deacetylase complexes that regulate silencing and transcription. *Proc. Natl. Acad. Sci. USA* **93**, 14503–14508.
21. Gray, S. G. and Ekstrom, T. J. (2001). The human histone deacetylase family. *Exp. Cell Res.* **262**, 75–83.
22. Guenther, M. G., Barak, O., and Lazar, M. A. (2001). The smrt and n-cor corepressors are activating cofactors for histone deacetylase 3. *Mol. Cell Biol.* **21**, 6091–6101.
23. Wen, Y. D., Perissi, V., Staszewski, L. M., Yang, W. M., Krones, A., Glass, C. K., Rosenfeld, M. G., and Seto, E. (2000). The histone deacetylase-3 complex contains nuclear receptor corepressors. *Proc. Natl. Acad. Sci. USA* **97**, 7202–7207.
24. Jenuwein, T. and Allis, C. D. (2001). Translating the histone code. *Science* **293**, 1074–1080.
25. Li, J., Wang, J., Nawaz, Z., Liu, J. M., Qin, J., and Wong, J. (2000). Both corepressor proteins SMRT and N-CoR exist in large protein complexes containing HDAC3. *EMBO J.* **19**, 4342–4350.
26. Guenther, M. G., Lane, W. S., Fischle, W., Verdin, E., Lazar, M. A., and Shiekhattar, R. (2000). A core SMRT corepressor complex containing HDAC3 and TBL1, a WD40-repeat protein linked to deafness. *Genes Dev.* **14**, 1048–1057.
27. Underhill, C., Qutob, M. S., Yee, S. P., and Torchia, J. (2000). A novel nuclear receptor corepressor complex, N-CoR, contains components of the mammalian SWI/SNF complex and the corepressor KAP-1. *J. Biol. Chem.* **275**, 40463–40470.
28. Bassi, M. T., Ramesar, R. S., Caciotti, B., Winship, I. M., De Grandi, A., Riboni, M., Townes, P. L., Beighton, P., Ballabio, A., and Borsani, G. (1999). X-linked late-onset sensorineural deafness caused by a deletion involving OA1 and a novel gene containing WD-40 repeats. *Am. J. Hum. Genet.* **64**, 1604–1616.
29. Dong, X., Tsuda, L., Zavitz, K. H., Lin, M., Li., S., Carthew, R. W., and Zipursky, S. L. (1999). ebi regulates epidermal growth factor receptor signaling pathways in *Drosophila. Genes Dev.* **13**, 954–965.
30. Zhang, Y., Sun, Z. W., Iratni, R., Erdjument-Bromage, H., Tempst, P., Hampsey, M., and Reinberg, D. (1998). SAP30, a novel protein conserved between human and yeast, is a component of a histone deacetylase complex. *Mol. Cell* **1**, 1021–1031.
31. Zhang, Y., Iratni, R., Erdjument-Bromage, H., Tempst, P., and Reinberg, D. (1997). Histone deacetylases and SAP18, a novel polypeptide, are components of a human Sin3 complex. *Cell* **89**, 357–364.
32. Huang, E. Y., Zhang, J., Miska, E. A., Guenther, M. G., Kouzarides, T., and Lazar, M. A. (2000). Nuclear receptor corepressors partner with class II histone deacetylases in a Sin3-independent repression pathway. *Genes Dev.* **14**, 45–54.
33. Kao, H. Y., Downes, M., Ordentlich, P., and Evans, R. M. (2000). Isolation of a novel histone deacetylase reveals that class I and class II deacetylases promote SMRT-mediated repression. *Genes Dev.* **14**, 55–66.
34. Glass, C. K. and Rosenfeld, M. G. (2000). The coregulator exchange in transcriptional functions of nuclear receptors. *Genes Dev.* **14**, 121–141.
35. Xu, J., Nawaz, Z., Tsai, S. Y., Tsai, M. J., and O'Malley, B. W. (1996). The extreme C terminus of progesterone receptor contains a transcriptional repressor domain that functions through a putative corepressor. *Proc. Natl. Acad. Sci. USA* **93**, 12195–12199.
36. Jackson, T. A., Richer, J. K., Bain, D. L., Takimoto, G. S., Tung, L., and Horwitz, K. B. (1997). The partial agonist activity of antagonist-occupied steroid receptors is controlled by a novel hinge domain-binding coactivator L7/SPA and the corepressors N-CoR or SMRT. *Mol. Endocrinol.* **11**, 693–705.
37. Smith, C. L., Nawaz, Z., and O'Malley, B. W. (1997). Coactivator and corepressor regulation of the agonist/antagonist activity of the mixed antiestrogen, 4-hydroxytamoxifen. *Mol. Endocrinol.* **11**, 657–666.
38. Lavinsky, R. M., Jepsen, K., Heinzel, T., Torchia, J., Mullen, T. M., Schiff, R., Del-Rio, A. L., Ricote, M., Ngo, S., Gemsch, J., Hilsenbeck, S. G., Osborne, C. K., Glass, C. K., Rosenfeld, M. G., and Rose, D. W. (1998). Diverse signaling pathways modulate nuclear receptor recruitment of N-CoR and SMRT complexes. *Proc. Natl. Acad. Sci. USA* **95**, 2920–2925.
39. Shang, Y., Hu, X., DiRenzo, J., Lazar, M. A., and Brown, M. (2000). Cofactor dynamics and sufficiency in estrogen receptor-regulated transcription. *Cell* **103**, 843–852.
40. Zhang, J., Guenther, M. G., Carthew, R. W., and Lazar, M. A. (1998). Proteasomal regulation of nuclear receptor corepressor-mediated repression. *Genes Dev.* **12**, 1775–1780.
41. Della, N. G., Hu, Y., Holloway, A. J., Wang, D., and Bowtell, D. D. (1995). A combined genetic and biochemical approach to mammalian signal transduction. *Aust. NZ J. Med.* **25**, 845–851.
42. Davis, P. J., Shih, A., Lin, H. Y., Martino, L. J., and Davis, F. B. (2000). Thyroxine promotes association of mitogen-activated protein kinase and nuclear thyroid hormone receptor (TR) and causes serine phosphorylation of TR. *J. Biol. Chem.* **275**, 38032–38039.
43. Hong, S. H. and Privalsky, M. L. (2000). The SMRT corepressor is regulated by a MEK-1 kinase pathway: inhibition of corepressor function is associated with SMRT phosphorylation and nuclear export. *Mol. Cell Biol.* **20**, 6612–6625.
44. Zhou, Y., Gross, W., Hong, S. H., and Privalsky, M. L. (2001). The SMRT corepressor is a target of phosphorylation by protein kinase CK2 (casein kinase II). *Mol. Cell Biochem.* **220**, 1–13.
45. Jang, M. K., Goo, Y. H., Sohn, Y. C., Kim, Y. S., Lee, S. K., Kang, H., Cheong, J., and Lee, J. W. (2001). Ca2+/calmodulin-dependent protein kinase IV stimulates nuclear factor-kappa B transactivation via phosphorylation of the p65 subunit. *J. Biol. Chem.* **276**, 20005–20010.
46. McKinsey, T. A., Zhang, C. L., Lu, J., and Olson, E. N. (2000). Signal-dependent nuclear export of a histone deacetylase regulates muscle differentiation. *Nature* **408**, 106–111.
47. McKinsey, T. A., Zhang, C. L., and Olson, E. N. (2000). Activation of the myocyte enhancer factor-2 transcription factor by calcium/calmodulin-dependent protein kinase-stimulated binding of 14–3–3 to histone deacetylase 5. *Proc. Natl. Acad. Sci. USA* **97**, 14400–14405.
48. Grozinger, C. M. and Schreiber, S. L. (2000). Regulation of histone deacetylase 4 and 5 and transcriptional activity by 14–3–3-dependent cellular localization. *Proc. Natl. Acad. Sci. USA* **97**, 7835–7840.
49. Jepsen, K., Hermanson, O., Onami, T. M., Gleiberman, A. S., Lunyak, V., McEvilly, R. J., Kurokawa, R., Kumar, V., Liu, F., Seto, E., Hedrick, S. M., Mandel, G., Glass, C. K., Rose, D. W., and Rosenfeld, M. G. (2000). Combinatorial roles of the nuclear receptor corepressor in transcription and development. *Cell* **102**, 753–763.

50. Feng, X., Jiang, Y., Meltzer, P., and Yen, P. M. (2001). Transgenic targeting of a dominant negative corepressor to liver blocks basal repression by thyroid hormone receptor and increases cell proliferation. *J. Biol. Chem.* **276**, 15066–15072.
51. Scully, K. M., Jacobson, E. M., Jepsen, K., Lunyak, V., Viadiu, H., Carriere, C., Rose, D. W., Hooshmand, F., Aggarwal, A. K., and Rosenfeld, M. G. (2000). Allosteric effects of Pit-1 DNA sites on long-term repression in cell type specification. *Science* **290**, 1127–1131.
52. Koide, T., Downes, M., Chandraratna, R. A., Blumberg, B., and Umesono, K. (2001). Active repression of RAR signaling is required for head formation. *Genes Dev.* **15**, 2111–2121.
53. Kopp, P., Kitajima, K., and Jameson, J. L. (1996). Syndrome of resistance to thyroid hormone: insights into thyroid hormone action. *Proc. Soc. Exp. Biol. Med.* **211**, 49–61.
54. Yoh, S. M., Chatterjee, V. K., and Privalsky, M. L. (1997). Thyroid hormone resistance syndrome manifests as an aberrant interaction between mutant T3 receptors and transcriptional corepressors. *Mol. Endocrinol.* **11**, 470–480.
55. Safer, J. D., Cohen, R. N., Hollenberg, A. N., and Wondisford, F. E. (1998). Defective release of corepressor by hinge mutants of the thyroid hormone receptor found in patients with resistance to thyroid hormone. *J. Biol. Chem.* **273**, 30175–30182.
56. Gothe, S., Wang, Z., Ng, L., Kindblom, J. M., Barros, A. C., Ohlsson, C., Vennstrom, B., and Forrest, D. (1999). Mice devoid of all known thyroid hormone receptors are viable but exhibit disorders of the pituitary-thyroid axis, growth, and bone maturation. *Genes Dev.* **13**, 1329–1341.
57. Behre, G., Zhang, P., Zhang, D. E., and Tenen, D. G. (1999). Analysis of the modulation of transcriptional activity in myelopoiesis and leukemogenesis. *Methods* **17**, 231–237.
58. Cavailles, V., Dauvois, S., L'Horset, F., Lopez, G., Hoare, S., Kushner, P. J., and Parker, M. G. (1995). Nuclear factor RIP140 modulates transcriptional activation by the estrogen receptor. *EMBO J.* **14**, 3741–3751.
59. Lee, C. H., Chinpaisal, C., and Wei, L. N. (1998). Cloning and characterization of mouse RIP140, a corepressor for nuclear orphan receptor TR2. *Mol. Cell Biol.* **18**, 6745–6755.
60. Lee, C. H. and Wei, L. N. (1999). Characterization of receptor-interacting protein 140 in retinoid receptor activities. *J. Biol. Chem.* **274**, 31320–31326.
61. Sugawara, T., Abe, S., Sakuragi, N., Fujimoto, Y., Nomura, E., Fujieda, K., Saito, M., and Fujimoto, S. (2001). RIP 140 modulates transcription of the steroidogenic acute regulatory protein gene through interactions with both SF-1 and DAX-1. *Endocrinology* **142**, 3570–3577.
62. Treuter, E., Albrektsen, T., Johansson, L., Leers, J., and Gustafsson, J. A. (1998). A regulatory role for RIP140 in nuclear receptor activation. *Mol. Endocrinol.* **12**, 864–881.
63. Vo, N., Fjeld, C., and Goodman, R. H. (2001). Acetylation of nuclear hormone receptor-interacting protein RIP140 regulates binding of the transcriptional corepressor CtBP. *Mol. Cell Biol.* **21**, 6181–6188.
64. Lutz, M., Baniahmad, A., and Renkawitz, R. (2000). Modulation of thyroid hormone receptor silencing function by co-repressors and a synergizing transcription factor. *Biochem. Soc. Trans.* **28**, 386–389.
65. Dressel, U., Thormeyer, D., Altincicek, B., Paululat, A., Eggert, M., Schneider, S., Tenbaum, S. P., Renkawitz, R., and Baniahmad, A. (1999). Alien, a highly conserved protein with characteristics of a corepressor for members of the nuclear hormone receptor superfamily. *Mol. Cell Biol.* **19**, 3383–3394.

CHAPTER 275

Steroid Hormone Receptor Signaling

Vincent Giguère
Molecular Oncology Group, McGill University Health Centre, Montréal, Québec, Canada

Introduction

Steroid hormones are essential regulators of key physiological processes such as reproduction, glucose metabolism, and the response to stress and salt balance. The biological effects of steroid hormones are transduced by intracellular receptors that directly mediate the action of their cognate hormone [1]. Steroid hormone receptors (SHRs) were the first recognized members of the steroid/thyroid/retinoid nuclear receptor superfamily, a class of transcription factors whose activity is regulated by small lipid-soluble molecules. The SHR subgroup includes the receptors for estradiol: estrogen receptor α and β [ERα (NR3A1) and ERβ (NR3A2)]; cortisol: glucocorticoid receptor [GR (NR3C1)]; aldosterone: mineralocorticoid receptor [MR (NR3C2)]; progesterone: PR (NR3C3); and dihydrotestosterone: androgen receptor [AR (NR3C4)]. In addition, the SHR subgroup contains three orphan nuclear receptors closely related to the ERs: the estrogen-related receptors α, β, and γ [ERRα (NR3B1), β (NR3B2), and γ (NR3B3)] for which a natural ligand remains to be identified. SHRs share a common modular structure composed of independent functional domains [2]. The DNA-binding domain (DBD) is centrally located, well conserved among SHRs, and comprised of two zinc-finger motifs involved in both protein–DNA and protein–protein contacts. The ligand-binding domain (LBD), located at the carboxy terminal of the receptor, is moderately conserved and folds into a canonical α-helical sandwich generally consisting of 12 α-helices (H1 to H12) [3]. The LBD can also contain a ligand-dependent nuclear translocation signal, determinants to bind chaperone proteins, dimerization interfaces, and a potent ligand-dependent activation domain referred to as AF-2. A ligand-independent activation domain (AF-1) is encoded within the nonconserved amino-terminal region of the receptors.

Activation by the Hormone

The classic model of SHR action dictates that SHRs interact with chaperones in the cytoplasm and be dissociated in a ligand-dependent fashion, leading to the reorganization of the receptor and exposure of nuclear localization signal(s) and translocation to the nucleus. Although this model is widely accepted for SHRs, there are important exceptions [4]. The ERs are clearly localized to the nucleus despite being part of a complex with chaperone proteins. Perhaps more striking is the finding that the two forms of PR, PR-A and PR-B, that differ only in the length of their amino-terminal domains have distinct cellular localization. PR-A is found predominantly in the nucleus, whereas PR-B is mainly located in the cytoplasm in the absence of hormone [5]. Given that the nuclear localization and chaperone binding functions are identical for both forms of PR, these data strongly suggest that the interactions of SHRs with complexes containing coregulatory proteins might influence the intracellular distribution of SHRs.

Intracellular redistribution of ligand-bound SHRs is accompanied by the recognition of specific sites on chromatin, referred to as hormone response elements (HREs) [6]. The HREs are short *cis*-acting sequences located within the promoters or enhancers of target genes. SHRs bind DNA as homodimers to HREs composed of inverted repeats of AGGTCA (for the ERs and ERRs) or AGAACA (for the GR,

MR, PR, and AR) motifs spaced by three nucleotides. SHR isoforms such as ERα and ERβ or PR-A and PR-B can also form functional heterodimeric complexes on DNA [7–9]. In the case of the ERs, the functional properties of each isoform are retained in the heterodimeric complex [10], whereas PR-A is dominant over PR-B [11].

The transcriptional activity of the SHRs is mediated by the independent AF-1 and AF-2 [12]. AF-1 is ligand independent and constitutive since it can activate transcription in the absence of the ligand when fused to a heterologous DBD. The structural determinants and mode of action of the distinct AF-1 domains found in each SHR have yet to be characterized. In contrast, the ligand-dependent AF-2 is well defined: a short amphipathic α-helix (H12) located at the carboxy-terminal end of the LBD is repositioned on hormone binding into a hydrophobic cleft formed mainly by residues from helices 3, 4, and 12. The resulting structural change provides a functional interface for coactivator recruitment by the receptor [13]. AF-1 and AF-2 recruit both common and specific cofactors. These regulatory proteins possess various enzymatic activities such as acetylase, deacetylase, methylase, kinase, and ubiquitinase functions [14]. One RNA molecule known as SRA has also been characterized as a SHR coactivator [15]. The cofactors participate in the remodeling of chromatin, the formation of a stable transcription initiation complex, the association or dissociation of other cofactors within the SHR-cofactor complex as well as recycling and degradation of the receptor [16].

Hormone-Independent Activation

SHRs are phosphoproteins and targets of kinase cascades involved in the response to growth factors and cytokines [17]. Many aspects of SHR function can be modulated in this way, including dimerization, DNA binding, and both ligand-independent and -dependent activation [18]. SHRs are the targets of protein kinase A, mitogen-activated protein kinase (MAPK), cyclin-dependent kinases, casein kinase, and glycogen-synthase kinase. The molecular mechanisms underlying modulation of SHR activity on phosphorylation have yet to be elucidated in most instances. However, in the case of ERβ, phosphorylation of two serine residues within AF-1 promotes the recruitment of steroid receptor coactivator-1 (SRC-1) both *in vivo* and *in vitro* [19]. The physiological relevance of hormone-independent activation pathways was clearly demonstrated using the uterus of the ERα mouse knock-out as a model [20]. In wild-type animals, the uterus displays growth responses to epidermal growth factor (EGF) and insulin-like growth factor 1 (IGF-1). In the ERα knockout animals, the uterus is unresponsive to the mitogenic actions of EGF and IGF-1, demonstrating an essential requirement for ERα in these biological responses [21, 22]. In cellular and *in vitro* models, EGF, IGF-1, and other agents can induce the phosphorylation of serine 118 in the ERα amino-terminal region [23]. Mutation of this residue abolishes the response of ERα to EGF and considerably reduces the ability of the receptor to respond to its cognate ligand.

Cross-Talk with Other Transcription Factors

Steroid hormone regulation of a number of target genes does not require the presence of a HRE within the transcriptional unit of those genes. Instead, SHRs can tether to a transcription factor and thus modify its activity. The formation of a SHR–transcription factor complex can lead to changes in cellular localization, DNA binding activity, and enzymatic function within the transcription initiation complex, or conversely can provide a hormone-dependent transcriptional activation function to a non-HRE site [24, 25]. The biological actions of SHRs have been linked to binding sites for more than a dozen transcription factors, as well as to non-HRE sites for unidentified factors. The *in vivo* relevance of the HRE-independent pathways was highlighted by the observation that, in contrast to GR-deficient mice, genetically engineered mice expressing a non-DNA binding form of the GR are viable [26].

Nongenomic Action of Steroid Hormones

Steroid hormones have been shown to elicit biological responses too rapid to involve gene transcription and subsequent synthesis of new proteins. Rapid effects of steroid hormones have been reported in blood vessels, bone, breast cancer cells, nervous system, sperm, and maturating oocytes [27–31]. These effects involve changes in the activities of enzymes such as phospholipase C, PI3 kinase, and adenylate cyclase, leading to increases in intracellular calcium levels, second messengers, and activation of kinase cascades [32,33]. It is not clear whether the membrane SHRs are identical to the nuclear isoforms, but responses to pharmacological agents and transfection experiments with SHR expression vectors indicate that the membrane SHRs must share common determinants with their nuclear counterparts [34, 35] and thus originate from the same genes. The ER has been shown to physically interact with c-SRC, the insulin-like growth factor, the p85 subunit of PI3 kinase and calveolin-1, whereas the PR possesses a proline-rich domain within its amino-terminal region that mediates direct hormone-dependent interaction with the SH3 domain of a variety of cytoplasmic signaling proteins, including c-SRC [35].

Estrogen Related Receptors

The ERRs were the first orphan nuclear receptors identified through a search for genes related to ERα [36]. Initial studies of the ERRs showed that they did not bind estradiol or other physiologically relevant steroid hormones and indicated that their biological roles could be quite distinct from those of the classic ERs [37–40]. However, the recent observation that the ERs and ERRs share target genes and coactivators [41, 42], coupled with the striking discovery that diethylstilbestrol, a potent synthetic estrogen, and 4-hydroxytamoxifen, a mixed estrogen agonist/antagonist, are ERR ligands [43–45], suggests that the ERRs are bona fide SHRs [46].

Selective Steroid Hormone Receptor Modulators

Pharmacological and toxicological studies of synthetic SHR ligands have led to the realization that certain drugs can have distinct effects on the same SHR depending on the target tissue. This concept has mainly emerged from the study of tamoxifen, a molecule that acts as an estrogen antagonist in the breast but as an estrogen agonist in bone and uterus [47]. The term *selective estrogen receptor modulator* (SERM) is now being used to describe estrogenic drugs that display cell type- and context-dependent actions, and this concept has now been extended to all SHRs. The mechanisms by which selective SHR modulators exert tissue-specific effects are not yet understood and are likely to be distinct for each class of compounds, targeted receptors, and site of action. However, our better understanding of the modes of action of SHR at the molecular level indicate that these mechanisms may include selective activation of receptor isoforms (e.g., ERα versus ERβ), distinct behavior of the drug-receptor complex in the presence of different corepressor/coactivator ratios, preferential recruitment of specific coregulatory proteins (including other transcription factors involved in cross-talk with SHRs such as AP-1 and NF-κB), and possibly selective activation of nongenomic pathways [48–50]. The future development of selective SHR modulators with improved therapeutic indexes and no undesirable side effects is likely to constitute the most significant outcome of the vast effort dedicated to understanding how SHRs work, and basic molecular and genetic studies of SHR action will continue to be the main driving force behind this process.

Acknowledgments

The literature on steroid hormone receptor signaling is extensive and only selected studies and reviews were cited in this article due to strict space limitation. I apologize to all investigators whose work was not included. Many thanks to members of my group for revising the manuscript. My laboratory is funded by the Canadian Institute of Health, the Canadian Breast Cancer Research Initiative and the National Cancer Institute of Canada.

References

1. Evans, R. M. (1988). The steroid and thyroid hormone receptor superfamily. *Science* **240**, 889–895.
2. Giguère, V., Hollenberg, S. H., Rosenfeld, M. G., and Evans, R. M. (1986). Functional domains of the human glucocorticoid receptor. *Cell* **46**, 645–652.
3. Wurtz, J. M., Bourguet, W., Renaud, J. P., Vivat, V., Chambon, P., Moras, D., and Gronemeyer, H. (1996). A canonical structure for the ligand-binding domain of nuclear receptors. *Nature Struct. Biol.* **3**, 87–94.
4. Hager, G. L., Lim, C. S., Elbi, C., and Baumann, C. T. (2000). Trafficking of nuclear receptors in living cells. *J. Steroid Biochem. Molec. Biol.* **74**, 249–254.
5. Lim, C. S., Baumann, C. T., Htun, H., Xian, W., Irie, M., Smith, C. L., and Hager, G. L. (1999). Differential localization and activity of the A- and B-forms of the human progesterone receptor using green fluorescent protein chimeras. *Mol. Endocrinol.* **13**, 366–375.
6. Glass, C. K. (1994). Differential recognition of target genes by nuclear receptors monomers, dimers, and heterodimers. *Endocr. Rev.* **15**, 391–407.
7. Cowley, S. M., Hoare, S., Mosselman, S., and Parker, M. G. (1997). Estrogen receptors α and β form heterodimers on DNA. *J. Biol. Chem.* **272**, 19858–19862.
8. Pace, P., Taylor, J., Suntharalingam, S., Coombes, R. C., and Ali, S. (1997). Human estrogen receptor β binds DNA in a manner similar to and dimerizes with estrogen receptor α. *J. Biol. Chem.* **272**, 25832–25838.
9. Pettersson, K., Grandien, K., Kuiper, G. G. J. M., and Gustafsson, J.-Å. (1997). Mouse estrogen receptor β forms estrogen response element-binding heterodimers with estrogen receptor α. *Mol. Endocrinol.* **11**, 1486–1496.
10. Tremblay, G. B., Tremblay, A., Labrie, F., and Giguère, V. (1999). Dominant activity of activation function 1 (AF-1) and differential stoichiometric requirements for AF-1 and -2 in the estrogen receptor α-β heterodimer complex. *Mol. Cell. Biol.* **19**, 1919–1927.
11. Mohamed, M. K., Tung, L., Takimoto, G. S., and Horwitz, K. B. (1994). The leucine zippers of c-fos and c-jun for progesterone receptor dimerization: A-dominance in the A/B heterodimer. *J. Steroid Biochem. Molec. Biol.* **51**, 241–250.
12. Aranda, A. and Pascual, A. (2001). Nuclear hormone receptors and gene expression. *Physiological Review* **81**, 1269–1304.
13. Feng, W., Ribeiro, R. C. J., Wagner, R. L., Nguyen, H., Apriletti, J. W., Fletterick, R. J., Baxter, J. D., Kushner, P. J., and West, B. L. (1998). Hormone-dependent coactivator binding to a hydrophobic cleft on nuclear receptors. *Science* **280**, 1747–1749.
14. McKenna, N. J., Lanz, R. B., and O'Malley, B. W. (1999). Nuclear receptor coregulators: Cellular and molecular biology. *Endocr. Rev.* **20**, 321–344.
15. Lanz, R. B., McKenna, N. J., Onate, S. A., Albrecht, U., Wong, J., Tsai, S. Y., Tsai, M. J., and O'Malley, B. W. (1999). A steroid receptor coactivator, SRA, functions as an RNA and is present in an SRC-1 complex. *Cell* **97**, 17–27.
16. Hermanson, O., Glass, C. K., and Rosenfeld, M. G. (2002). Nuclear receptor coregulators: Multiple modes of modification. *Trends Endocrinol. Metab.* **13**, 55–60.
17. Weigel, N. L. and Zhang, Y. (1998). Ligand-independent activation of steroid hormone receptors. *J. Mol. Med.* **76**, 469–479.
18. Shao, D. and Lazar, M. A. (1999). Modulating nuclear receptor function: May the phos be with you. *J. Clin. Invest.* **103**, 1617–1618.
19. Tremblay, A., Tremblay, G. B., Labrie, F., and Giguère, V. (1999). Ligand-independent recruitment of SRC-1 by estrogen receptor β through phosphorylation of activation function AF-1. *Mol. Cell* **3**, 513–519.
20. Couse, J. F. and Korach, K. S. (1999). Estrogen receptor null mice: What have we learned and where will they lead us? *Endocr. Rev.* **20**, 358–417.
21. Curtis, S. W., Washburn, T., Sewall, C., DiAugustine, R., Lindzey, J., Couse, J. F., and Korach, K. S. (1996). Physiological coupling of growth factor and steroid receptor signaling pathways: Estrogen receptor knockout mice lack estrogen-like response to epidermal growth factor. *Proc. Natl. Acad. Sci. USA* **93**, 12626–12630.
22. Klotz, D. M., Hewitt, S. C., Ciana, P., Raviscioni, M., Lindzey, J. K., Foley, J., Maggi, A., DiAugustine, R. P., and Korach, K. S. (2002). Requirement of estrogen receptor α in insulin-like growth factor-1 (IGF-1)-induced uterine responses and *in vivo* evidence for IGF-1/estrogen receptor cross-talk. *J. Biol. Chem.* **277**, 8531–8537.
23. Kato, S., Endoh, H., Masuhiro, Y., Kitamoto, T., Uchiyama, S., Sasaki, H., Masushige, S., Gotoh, Y., Nishida, E., Kawashima, H., Metzger, D., and Chambon, P. (1995). Activation of the estrogen receptor through phosphorylation by mitogen-activated protein kinase. *Science* **270**, 1491–1494.
24. Diamond, M. I., Miner, J. N., Yoshinaga, S. K., and Yamamoto, K. R. (1990). Transcription factor interactions: Selectors of positive or negative regulation from a single DNA element. *Science* **249**, 1266–1272.
25. Nissen, R. M. and Yamamoto, K. R. (2000). The glucocorticoid receptor inhibits NFκB by interfering with serine-2 phosphorylation of the RNA polymerase II carboxy-terminal domain. *Genes Dev.* **14**, 2314–2329.

26. Reichardt, H. M., Kaestner, K. H., Tuckermann, J., Kretz, O., Wessely, O., Bock, R., Gass, P., Schmid, W., Herrlich, P., Angel, P., and Schütz, G. (1998). DNA binding of the glucocorticoid receptor is not essential for survival. *Cell* **93**, 531–541.
27. Kousteni, S., Bellido, T., Plotkin, L. I., O'Brien, C. A., Bodenner, D. L., Han, L., Han, K., DiGregorio, G. B., Katzenellenbogen, J. A., Katzenellenbogen, B. S., Roberson, P. K., Weinstein, R. S., Jilka, R. L., and Manolagas, S. C. (2001). Nongenotropic, sex-nonspecific signaling through the estrogen or androgen receptors: Dissociation from transcriptional activity. *Cell* **104**, 719–730.
28. Revelli, A., Massobrio, M., and Tesarik, J. (1998). Nongenomic actions of steroid hormones in reproductive tissues. *Endocr. Rev.* **19**, 3–17.
29. Watson, C. S. and Gametchu, B. (1999). Membrane-initiated steroid actions and the proteins that mediate them. *Proc. Soc. Exp. Biol. Med.* **220**, 9–19.
30. Ferrell, J. E., Jr. (1999). *Xenopus* oocyte maturation: New lessons from a good egg. *BioEssays* **21**, 833–842.
31. McEwen, B. S. (1994). Steroid hormone actions on the brain: When is the genome involved? *Horm. Behav.* **28**, 396–405.
32. Moggs, J. G. and Orphanides, G. (2001). Estrogen receptors: Orchestrators of pleiotropic cellular responses. *EMBO Rep.* **2**, 775–781.
33. Kelly, M. J. and Levin, E. R. (2001). Rapid actions of plasma membrane estrogen receptors. *Trends Endocrinol. Metab.* **12**, 152–156.
34. Migliaccio, A., Piccolo, D., Castoria, G., Di Domenico, M., Bilancio, A., Lombardi, M., Gong, W., Beato, M., and Auricchio, F. (1998). Activation of the Src/p21^{ras}/Erk pathway by progesterone receptor via cross-talk with estrogen receptor. *EMBO J.* **17**, 2008–2018.
35. Boonyaratanakornkit, V., Scott, M. P., Ribon, V., Sherman, L., Anderson, S. M., Maller, J. L., Miller, W. T., and Edwards, D. P. (2001). Progesterone receptor contains a proline-rich motif that directly interacts with SH3 domains and activates c-Src family tyrosine kinases. *Mol. Cell* **8**, 269–280.
36. Giguère, V., Yang, N., Segui, P., and Evans, R. M. (1988). Identification of a new class of steroid hormone receptors. *Nature* **331**, 91–94.
37. Wiley, S. R., Kraus, R. J., Zuo, F., Murray, E. E., Loritz, K., and Mertz, J. E. (1993). SV40 early-to-late switch involves titration of cellular transcriptional repressors. *Genes Dev.* **7**, 2206–2219.
38. Sladek, R., Bader, J.-A., and Giguère, V. (1997). The orphan nuclear receptor estrogen-related receptor α is a transcriptional regulator of the human medium-chain acyl coenzyme A dehydrogenase gene. *Mol. Cell. Biol.* **17**, 5400–5409.
39. Vega, R. B. and Kelly, D. P. (1997). A role for estrogen-related receptor α in the control of mitochondrial fatty acid β-oxidation during brown adipocyte differentiation. *J. Biol. Chem.* **272**, 31693–31699.
40. Luo, J., Sladek, R., Bader, J.-A., Rossant, J., and Giguère, V. (1997). Placental abnormalities in mouse embryos lacking orphan nuclear receptor ERRβ. *Nature* **388**, 778–782.
41. Vanacker, J. M., Pettersson, K., Gustafsson, J., and Laudet, V. (1999). Transcriptional targets shared by estrogen receptor- related receptors (ERRs) and estrogen receptor (ER) α, but not by ERβ. *EMBO J.* **18**, 4270–4279.
42. Lu, D., Kiriyama, Y., Lee, K. Y., and Giguère, V. (2001). Transcriptional regulation of the estrogen-inducible pS2 beast cancer marker gene by the ERR family of orphan nuclear receptors. *Cancer Res.* **61**, 6755–6761.
43. Tremblay, G. B., Bergeron, D., and Giguère, V. (2001). 4-Hydroxytamoxifen is an isoform-specific inhibitor of orphan estrogen-receptor-related (ERR) nuclear receptors β and γ. *Endocrinology* **142**, 4572–4575.
44. Tremblay, G. B., Kunath, T., Bergeron, D., Lapointe, L., Champigny, C., Bader, J.-A., Rossant, J., and Giguère, V. (2001). Diethylstilbestrol regulates trophoblast stem cell differentiation as a ligand of orphan nuclear receptor ERRβ. *Genes Dev.* **15**, 833–838.
45. Coward, P., Lee, D., Hull, M. V., and Lehmann, J. M. (2001). 4-Hydroxytamoxifen binds to and deactivates the estrogen-related receptor γ. *Proc. Natl. Acad. Sci. USA* **98**, 8880–8884.
46. Giguère, V. (2002). To ERR in the estrogen pathway. *Trends In Endocr & Metab.* **13**, 220–225.
47. Park, W. C. and Jordan, V. C. (2002). Selective estrogen receptor modulators (SERMS) and their roles in breast cancer prevention. *Trends Mol. Med.* **8**, 82–88.
48. Tremblay, G. B., Tremblay, A., Copeland, N. G., Gilbert, D. J., Jenkins, N. A., Labrie, F., and Giguère, V. (1997). Cloning, chromosomal localization and functional analysis of the murine estrogen receptor β. *Mol. Endocrinol.* **11**, 353–365.
49. Norris, J. D., Paige, L. A., Christensen, D. J., Chang, C. Y., Huacani, M. R., Fan, D., Hamilton, P. T., Fowlkes, D. M., and McDonnell, D. P. (1999). Peptide antagonists of the human estrogen receptor. *Science* **285**, 744–746.
50. Paech, K., Webb, P., Kuiper, G. G. J. M., Nilsson, S., Gustafsson, J.-Å., Kushner, P. J., and Scanlan, T. S. (1997). Differential ligand activation of estrogen receptors ERα and ERβ at AP1 sites. *Science* **277**, 1508–1510.

CHAPTER 276

PPARγ Signaling in Adipose Tissue Development

Robert Walczak and Peter Tontonoz
Howard Hughes Medical Institute,
Department of Pathology and Laboratory Medicine,
UCLA School of Medicine, Los Angeles, California

Introduction

Adipose tissue plays a central role in maintaining lipid homeostasis and energy balance in vertebrates by storing triglycerides or releasing free fatty acids in response to changing energy demands. Adipocytes also perform an important endocrine function by secreting numerous signaling molecules, which are involved in the regulation of food intake, energy balance, and whole-body lipid homeostasis. Excessive accumulation of adipose tissue leads to obesity and often to insulin resistance, whereas its absence is associated with lipodystrophic disorders. In contrast to rodents, which have no white adipose tissue until after birth [1], human preadipocytes begin to differentiate during late embryonic development, although the majority of the differentiation occurs after birth [2]. All vertebrates have the ability to induce adipose tissue differentiation throughout their lives in response to the body's energy storage demands.

During the past 10 years, work from a number of groups has outlined a pivotal role for PPARγ in adipocyte development. Evidence from both cell culture and animal models indicates that this protein is absolutely required for adipogenesis. A member of the nuclear hormone receptor superfamily, PPARγ is a transcription factor whose activity is regulated by the binding of specific small molecule ligands. Like many other nuclear receptors, PPARγ binds to DNA in a heterodimeric complex with the retinoid X receptor (RXR). The discovery that the natural ligands for PPARγ are likely to be native and modified polyunsaturated fatty acids provided a mechanism whereby changes in cellular or systemic lipid homeostasis might be translated into specific changes in adipocyte gene expression and differentiation. Obesity, the development of excess adipose tissue, is invariably associated with insulin resistance. PPARγ has also been identified as a key regulator of insulin sensitivity, an activity that is likely to be directly related to its function in adipose tissue biology. Synthetic ligands for this receptor are widely used for the treatment of type II diabetes.

PPARγ: A Dominant Regulator of Adipose Tissue Development

Early insights into the function of PPARγ in adipocyte differentiation came from studies of the aP2 gene. Through a combination of polymerase chain reaction (PCR)-mediated cloning and biochemical purification, PPARγ was identified as the factor responsible for directing adipocyte-specific expression of the aP2 enhancer [3–5]. It was shown to be highly expressed in both white and brown adipose tissue and to be induced early during the time course of adipocyte differentiation [5,6]. Subsequent studies demonstrated that expression and activation of PPARγ in fibroblastic cells is sufficient to trigger the adipocyte gene expression cascade and lead to the development of the adipose phenotype [7]. In addition to the fatty acid binding protein aP2, PPARγ target genes in adipose tissue include phosphoenol pyruvate carboxykinase [8], acyl-CoA synthetase [9,10], fatty acid transport protein 1 (FATP-1) [11,12], CD36 [13], lipoprotein lipase (LPL) [14] and liver X receptor α (LXRα) [15]. Although activation of PPARγ leads ultimately to the induction of all of the genes necessary to define an adipocyte, not all of these genes are direct targets for PPARγ. Execution of the differentiation program now appears likely to involve the sequential and coordinated action of PPARγ and members of the CAAT/enhancer binding protein (C/EBP) and sterol regulatory element binding protein (SREBP) transcription factor families (see later discussion).

Although PPARγ was once considered an orphan receptor, considerable evidence has emerged to suggest that the endogenous ligands of the PPARs are likely to be native or modified polyunsaturated fatty acids [16–18]. Thus, it appears that the substrates for energy storage in adipose tissue, fatty acids, are also signaling molecules that direct the differentiation of new adipocytes. The naturally occurring prostanoid 15-deoxyprostaglandin J2 binds to PPARγ with high affinity; however, its role as a signaling molecule *in vivo* is not yet clear [19,20]. Of particular pharmacological importance was the discovery that PPARγ is the biological receptor for a class of insulin-sensitizing agents known as the thiazolidinediones [19,21–23]. Two of these drugs, rosiglitazone (Avandia) and pioglitazone (Actos), are in widespread clinical use for the treatment of type II diabetes. The availability of potent and highly specific ligands has greatly facilitated analysis of PPARγ function. For example, both naturally occurring and synthetic PPARγ ligands are potent stimulators of adipogenesis in cultured systems [19,23]. The thiazolidinediones have also been shown to increase adipose tissue mass *in vivo* [24].

Analysis of PPARγ Function in Animal Models

The human and mouse PPARγ genes extend over more than 100 kb of genomic DNA (Fig. 1). Each is composed of nine exons that give rise to three distinct mRNA molecules, PPARγ1, PPARγ2, and PPARγ3, by alternate promoter usage and differential splicing [25–28]. Fat and large intestine are the only tissues, where considerable amounts of PPARγ2 and PPARγ3 transcripts have been reported [26,27]. In rodents, PPARγ appears late during embryonic development and is restricted to the brown adipose tissue (day 18.5), and only transiently appears in the central nervous system [29–31]. Rodents do not develop the white adipose tissue until after birth.

Homozygous inactivation of the PPARγ gene in mice results in embryonic lethality [32]. PPARγ-deficient fetuses can only survive until midgestation and die by day 10 of development. At day 9.5, PPARγ–/– fetuses are alive and are similar to their wild-type counterparts; however, at day 10, embryos show serious vascular abnormalities in the placenta. Barak and colleagues [32] utilized a tetraploid-rescue approach to generate a single PPARγ null pup that survived to term. This PPARγ-deficient animal was completely devoid of both white and brown adipose tissues. Studies using embryonic stem cells have also pointed to an absolute requirement for PPARγ in adipose tissue development. Injection of PPARγ-deficient cells into wild-type blastocysts produces chimeric mice in which the adipose tissue is composed exclusively of PPARγ+/+ cells [33]. Similarly, embryonic fibroblasts

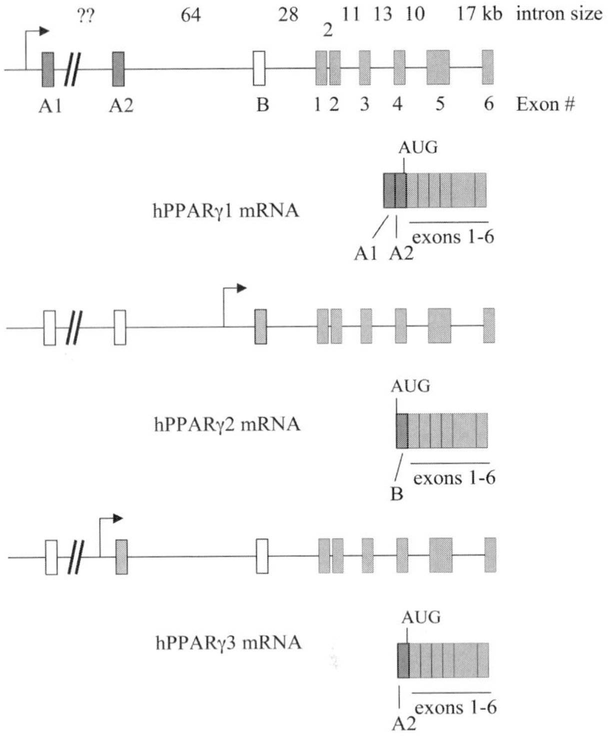

Figure 1 Structure of the murine and human PPARγ genes. Three PPARγ isoforms are produced by the differential use of three promoters and alternative splicing of the three 5'-exons (A1, A2, B). Exons 1–6 are common to all three transcripts. In the human PPARγ1 and PPARγ3 genes, the AUG translation initiation codon is located six nucleotides upstream of the one used in other species. Therefore, the human PPARγ1 protein has two additional amino acids at its N terminus compared with the rodent PPARγ1 protein. Note that transcription from the promoters 1 and 3 results in the same protein of 477 amino acids. The PPARγ2 protein of 505 amino acids is produced by transcription from the promoter 2. Distances between exons are based on the alignment of PPARγ mRNA sequences with the Human Genome build #28.

derived from PPARγ-deficient fetuses fail to differentiate into adipocytes *in vitro* [32,34]. Very recently, studies using PPARγ-deficient fibroblasts have shown that it is the PPARγ2 isoform that is specifically required for the differentiation program [35].

A limited amount of data on genetic polymorphisms in human populations corroborate studies in animal models. Mutations that result in permanent activation of PPARγ protein are associated with obesity [36]. In contrast, mutations that render PPARγ less active are generally associated with a lower body mass index (BMI) and ameliorated insulin sensitivity [37–41].

Transcriptional Networks in Adipose Tissue Development

Adipocytic differentiation has been primarily studied *in vitro*. The 3T3-L1 and 3T3-F422A cell lines [42], are the most widely used adipocyte culture models. These cells are morphologically similar to the fibroblastic preadipose cells

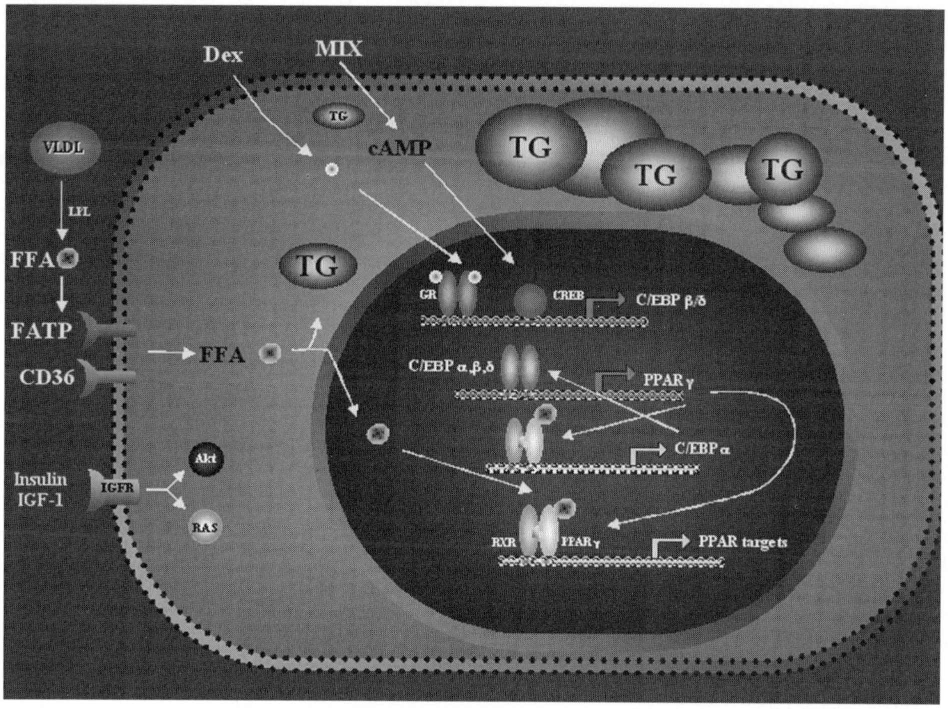

Figure 2 Transcriptional networks in adipocyte differentiation. Insulin and IGF act on the IGF receptor to activate signaling pathways involving ras and Akt. C/EBPβ and C/EBPδ are induced following the hormonal stimulation of 3T3-L1 cells with MDI. C/EBPβ expression is regulated by glucocorticoids through the glucocorticoid receptor (GR), whereas C/EBPδ expression is induced by cAMP via CREB. Both C/EBPβ and C/EBPδ activate the expression of the PPARγ gene. PPARγ in turn induces the expression of C/EBPα. PPARγ and C/EBPα cross-regulate each other to maintain their high level of expression as C/EBPβ and C/EBPδ levels decline. Finally, PPARγ and C/EBPα cooperate to induce expression of genes specific for the mature adipocyte phenotype.

found in the stroma of adipose tissue and, once differentiated, they exhibit virtually all of the characteristics associated with adipocytes present within the adipose tissue. When injected into mice, 3T3-L1 cells differentiate into adipocytes and form fat pads that are indistinguishable from normal adipose tissue [43]. Using these model systems a number of the molecular pathways that control adipogenesis have been elucidated (Fig. 2).

Confluent 3T3-L1 cells can be converted in adipocytes by a treatment with a mixture of insulin, dexamethasone, methyl-isobutyl-xanthine (MIX), and fetal bovine serum, commonly referred to in literature as MDI [44]. Insulin activates the insulin-like growth factor 1 (IGF-1) receptor in 3T3-L1 cells; therefore, IGF-1 can be substituted for insulin in the adipogenic cocktail [45]. Dexamethasone activates the glucocorticoid receptor pathway, which results in a rise in the intracellular cAMP levels. MIX, a cAMP-phosphodiesterase inhibitor, prevents cAMP hydrolysis and is used to stimulate the cAMP-dependent protein kinase pathway.

The developmental program of the 3T3-L1 preadipocytes can be divided into four distinct stages: (1) preconfluent proliferation, (2) confluence/growth arrest, (3) hormonal induction/clonal expansion, and (4) permanent growth arrest/terminal differentiation (Table I). Studies during the past 20 years have defined specific patterns of gene expression associated with each of these stages. Approximately 48 hours after induction by MDI, 3T3-L1 cells complete the postconfluent mitosis and enter into an unusual growth arrest phase, called GD [46,47]. Following the growth arrest, cells are committed to become adipocytes and begin to express late markers of differentiation by day 3. These late markers consist of lipogenic and lipolytic enzymes, as well as other proteins responsible for modulating the mature adipocyte phenotype. Finally, the cells round up, accumulate triglyceride droplets, and become terminally differentiated adipocytes by days 5–7.

At confluence, within 1 hour after the addition of MDI, the transient expression of c-fos, c-jun, junB, c-myc is observed [48]. These factors have not been implicated directly in any differentiation-specific events but they are believed to have mitogenic properties. C/EBPβ and C/EBPδ are the first transcription factors induced following the hormonal stimulation [49–52]. C/EBPβ expression is regulated by glucocorticoids, whereas the promoter of C/EBPδ is responsive primarily to MIX. The activity of C/EBPβ and C/EBPδ is believed to directly induce the expression of PPARγ2 via interaction with a C/EBP site in the PPARγ2 promoter [49,53,54]. PPARγ2 can be first detected on day 2 after addition of MDI and its expression acquires the maximum by days 3–4. In turn, PPARγ2 induces the expression of C/EBPα, which acquires its highest expression level on day 5 [55]. Finally, PPARγ and C/EBPα cross-regulate each other to maintain their high level of expression when C/EBPβ and C/EBPδ levels decline [53].

Table I Temporal Pattern of Gene Expression during *In Vitro* Adipocyte Differentiation[a]

		JunB							aP2
		c-jun	Wnt-10b						SCD
		c-fos	CUP	ADD1	C/EBPβ				Adipsin
	DAY	c-myc	Pref-1	SREBP1c	C–/EBPδ	C/EBPα	PPARγ2	LPL	PEPCK
Proliferation	–5	–	+++	–	–	–	–	–	–
Confluence	–4	–	+++	–	–	–	–	++	–
MDI simulation clonal expansion	0	+++	+++	–	–	–	–	+++	–
	0.5	–	+	–	++	–	–	+++	–
	1	–	–	++	+++	–	+	+++	–
Permanent growth arrest terminal differentiation	2	–	–	+++	+++	+	++	+++	+
	2.5	–	–	+++	++	++	++	+++	++
	3	–	–	+++	–	++	+++	+++	+++
	4	–	–	+++	–	++	+++	+++	+++
	5	–	–	+++	–	+++	+++	+++	+++
	7	–	–	+++	–	+++	+++	+++	+++

[a]Stages of differentiation are depicted in the first column: (1) preconfluent proliferation, (2) confluence/growth arrest, (3) clonal expansion, and (4) permanent growth arrest/terminal differentiation. The expression pattern of genes associated with each step of differentiation is shown.

It is now clear that PPARγ and C/EBPα cooperate to activate gene expression and the full program of adipogenesis. Although, retroviral expression of either PPARγ [7] or C/EBPα [53,56] alone is sufficient to induce *in vitro* adipogenesis in 3T3-L1 cells, this procedure is much more efficient if both factors are coexpressed [7,57,58].

Another protein implicated in the adipocyte differentiation pathway is a member of the basic helix–loop–helix protein family of transcription factors, called ADD-1/SREBP-1c (adipocyte differentiation and determination factor1 or sterol regulatory element binding protein 1c) [59]. Ectopic expression of ADD1 alone renders 3T3-L1 cells differentiation competent [60], whereas the expression of the ADD1 dominant-negative form strongly inhibits 3T3-L1 differentiation [60]. Coexpression of both ADD1 and PPARγ2 results in a synergistic effect on adipocyte differentiation [60]. The mechanism by which ADD1/SREBP1c promotes differentiation is still unclear, however, it has been postulated that genes activated by this transcription factor lead to synthesis of endogenous ligands for PPARγ [61]. Paradoxically, however, transgenic overexpression of ADD1/SREBP-1c in adipose tissue of mice was found to cause severe lipodystrophy [62]. The mechanism underlying this effect is not yet clear.

Negative Regulation of Adipocyte Differentiation

Exposure of preadipocytes to differentiating agents, such as the MDI cocktail, releases the cells from a number of inhibitory pathways that are antagonistic to terminal differentiation. Examples of such inhibitory factors include the CUP/AP-2α transcriptional repressor, Pref-1, and members of the Wnt family. CUP/AP-2α is a C/EBP family member abundantly expressed in preadipocytes, where it has been proposed to mediate repression of the C/EBPα gene through direct binding to specific sequences in the C/EBPα promoter. Expression of the CUP/AP-2α factor in preadipocytes is blocked upon hormonal stimulation [63,64]. Another postulated inhibitor of adipogenesis inhibition, Pref-1, is also highly expressed in 3T3-L1 preadipocytes. Pref-1 is a small, membrane associated, EGF-repeat-containing protein [65,66]. Pref-1 is elevated in the mouse model of congenital generalized lipodystrophy, a condition characterized by poorly developed white and brown adipose tissue [62]. Cleavage of the membrane associated Pref-1 in its extracellular domain results in the generation of a soluble signaling mediator that may interact with an as yet unidentified receptor. Pref-1 expression in preadipocytes is blocked by treatment with glucocorticoids [67]. Finally, the Wnt family of signaling factors has recently been implicated in the control of adipose differentiation. Forced expression of Wnt-1 in 3T3-442A cells inhibited the formation of adipose tissue when these cells were grafted into nude mice [68]. Furthermore, 3T3-L1 cells, which are engineered to express a dominant-negative form of the TCF4, a transcriptional mediator of the Wnt pathway, undergo adipogenesis without any hormonal induction. Activation of the Wnt pathway in these studies repressed expression of both the PPARγ and C/EBPα genes, and this has been postulated to be the mechanism by which Wnts inhibit adipogenesis.

PPARγ, TNF-α Signaling Antagonism and Insulin Resistance

Insulin resistance is a common feature of obesity, lipodystrophy, and non-insulin-dependent diabetes mellitus (NIDDM). A possible connection between PPARγ and insulin sensitivity was first suggested by the critical role of PPARγ in adipocyte development. However, the discovery that PPARγ is the biologic target for the thiazolidinedione class of antidiabetic drugs provided the definitive link [19,21–23]. In rodents, PPARγ gene expression is down-regulated by fasting and by insulin-deficient diabetes, whereas exposure to a high-fat diet increases PPAR expression in adipose tissue and may result in obesity and insulin resistance (reviewed in [24]).

The cytokine tumor necrosis factor α (TNF-α) has long been recognized for its capacity to impair normal adipocyte physiology by inducing lipolysis, blocking insulin signaling, and antagonizing adipogenesis. Elevated TNF-α levels are observed in several rodent models of obesity and NIDDM [69]. Recent studies utilizing gene disruption in mice confirmed that absence of TNF-α or its receptor results in significantly improved insulin sensitivity [70,71]. It has been suggested that TNF-α blocks insulin signaling in adipocytes, at least in part, by down-regulating expression of the C/EBPα and PPARγ genes. Recent studies have shed light on potential mechanisms for this repression. One mechanism by which TNF-α could inhibit adipogenesis is through activation of the MEK/MAPK signaling pathway [72,73]. It has been previously shown that MAP kinase mediated phosphorylation of the PPARγ protein at Ser82 (Ser112 for PPARγ2) results in repression of its transcriptional activity [74]. Other groups have demonstrated that administration of TNF-α early during 3T3-L1 adipocyte differentiation could block clonal expansion due to the erroneous regulation of two retinoblastoma protein family members, p130 and p107. These proteins regulate cell cycle events through interactions with the E2F transcription factors. TNF-α disrupts the normal pattern of expression of both p130 and p107, leading to a complete block in mitotic clonal expansion and the activation of apoptosis [75].

TZDs, potent and specific PPARγ ligands, appear to antagonize TNF-α-mediated inhibition of insulin signaling at several different levels. First, TZD treatment was shown to decrease circulating TNF-α levels in rodents [76,77]. Second, TNF-α affects insulin action through inhibition of the insulin receptor (IR) tyrosine kinase activity [78]. TZDs can antagonize this effect of TNF-α on IR itself as well as on IR downstream targets, such as IRS proteins, and restore insulin sensitivity in fat [79]. This effect has not been observed in tissues that do not express PPARγ. Third, TZDs rapidly increase expression of genes that are crucial for insulin signaling in adipocytes, such as c-Cbl associated protein (CAP), IRS-2, and the p85 subunit of the PI3 kinase [80–83]. The question of whether these genes are direct PPARγ targets has not yet been addressed. The concept that TNF-α is an important modulator of insulin sensitivity and adipocyte development is based primarily on data in animal models. The role of this factor in human adipose tissue biology is not yet clear.

PPARγ and Cell Cycle Regulation

Given the strong correlation between growth arrest and terminal differentiation, it is not surprising that PPARγ has been linked to cell cycle regulation. For many specialized cell types, differentiation and division are mutually exclusive choices. In adipocytes, induction of the differentiation program by PPARγ is accompanied by cell cycle withdrawal. Studies using tumor cell lines have shown that binding of the E2F/DP-1 complex to its target promoters, an event usually associated with cell cycle reentry, is significantly decreased on stimulation with the PPARγ-specific ligand. PPARγ is believed to participate in the silencing of the E2F/DP-1 transcriptional activity by down-regulating the PP2A protein phosphatase [84]. Another study revealed that PPARγ, in concert with C/EBPα, may mediate cell growth arrest during adipogenesis by up-regulating the cyclin-dependent kinase inhibitors p18 and p21 [85]. It was also demonstrated that treatment with PPARγ ligand could effectively bypass the block in adipocyte differentiation imposed by the retinoblastoma protein deficiency [86]. These authors proposed that the adipocyte differentiation observed in RB null embryonic fibroblasts is related to a severely reduced level of C/EBPβ dependent transactivation and that pRB may participate in a pathway leading to the production of an endogenous PPARγ ligand.

References

1. Ailhaud, G., Grimaldi, P., and Negrel, R. (1992). Cellular and molecular aspects of adipose tissue development. *Annu. Rev. Nutr.* **12**, 207–233.
2. Burdi, A. R., Poissonnet, C. M., Garn, S. M., Lavelle, M., Sabet, M. D., and Bridges, P. (1985). Adipose tissue growth patterns during human gestation: A histometric comparison of buccal and gluteal fat depots. *Int. J. Obes.* **9**, 247–256.
3. Graves, R. A., Tontonoz, P., Platt, K. A., Ross, S. R., and Spiegelman, B. M. (1992). Identification of a fat cell enhancer: Analysis of requirements for adipose tissue-specific gene expression. *J. Cell Biochem.* **49**, 219–224.
4. Tontonoz, P., Graves, R. A., Budavari, A. I., Erdjument-Bromage, H., Lui, M., Hu, E., Tempst, P., and Spiegelman, B. M. (1994). Adipocyte-specific transcription factor ARF6 is a heterodimeric complex of two nuclear hormone receptors, PPAR γ and RXR alpha. *Nucleic Acids Res.* **22**, 5628–5634.
5. Tontonoz, P., Hu, E., Graves, R. A., Budavari, A. I., and Spiegelman, B. M. (1994). mPPAR γ 2: Tissue-specific regulator of an adipocyte enhancer. *Genes Dev.* **8**, 1224–1234.
6. Chawla, A., Schwarz, E. J., Dimaculangen, D. D., and Lazar, M. A. (1994). Peroxisome proliferator-activated receptor γ: Adipose-predominant expression and induction early in adipocyte differentiation. *Endocrinology* **135**, 798–800.
7. Tontonoz, P., Hu, E., and Spiegelman, B. M. (1994). Stimulation of adipogenesis in fibroblasts by PPAR γ 2, a lipid-activated transcription factor. *Cell* **79**, 1147–1156.
8. Tontonoz, P., Hu, E., Devine, J., Beale, E. G., and Spiegelman, B. M. (1995). PPAR γ 2 regulates adipose expression of the phosphoenolpyruvate carboxykinase gene. *Mol. Cell Biol.* **15**, 351–357.
9. Schoonjans, K., Staels, B., Grimaldi, P., and Auwerx, J. (1993). Acyl-CoA synthetase mRNA expression is controlled by fibric-acid derivatives, feeding and liver proliferation. *Eur. J. Biochem.* **216**, 615–622.
10. Schoonjans, K., Watanabe, M., Suzuki, H., Mahfoudi, A., Krey, G., Wahli, W., Grimaldi, P., Staels, B., Yamamoto, T., and Auwerx, J. (1995). Induction of the acyl-coenzyme A synthetase gene by fibrates and fatty acids is mediated by a peroxisome proliferator response element in the C promoter. *J. Biol. Chem.* **270**, 19269–19276.
11. Martin, G., Schoonjans, K., Lefebvre, A. M., Staels, B., and Auwerx, J. (1997). Coordinate regulation of the expression of the fatty acid transport protein and acyl-CoA synthetase genes by PPARalpha and PPARγ activators. *J. Biol. Chem.* **272**, 28210–28217.

12. Frohnert, B. I., Hui, T. Y., and Bernlohr, D. A. (1999). Identification of a functional peroxisome proliferator-responsive element in the murine fatty acid transport protein gene. *J. Biol. Chem.* **274**, 3970–3977.
13. Tontonoz, P., Nagy, L., Alvarez, J. G., Thomazy, V. A., and Evans, R. M. (1998). PPARγ promotes monocyte/macrophage differentiation and uptake of oxidized LDL. *Cell* **93**, 241–252.
14. Schoonjans, K., Peinado-Onsurbe, J., Lefebvre, A. M., Heyman, R. A., Briggs, M., Deeb, S., Staels, B., and Auwerx, J. (1996). PPARalpha and PPARγ activators direct a distinct tissue-specific transcriptional response via a PPRE in the lipoprotein lipase gene. *EMBO J.* **15**, 5336–5348.
15. Laffitte, B. A., Joseph, S. B., Walczak, R., Pei, L., Wilpitz, D. C., Collins, J. L., and Tontonoz, P. (2001). Autoregulation of the human liver X receptor alpha promoter. *Mol. Cell Biol.* **21**, 7558–7568.
16. Forman, B. M., Chen, J., and Evans, R. M. (1997). Hypolipidemic drugs, polyunsaturated fatty acids, and eicosanoids are ligands for peroxisome proliferator-activated receptors alpha and delta. *Proc. Natl. Acad. Sci. USA* **94**, 4312–4317.
17. Kliewer, S. A., Sundseth, S. S., Jones, S. A., Brown, P. J., Wisely, G. B., Koble, C. S., Devchand, P., Wahli, W., Willson, T. M., Lenhard, J. M., and Lehmann, J. M. (1997). Fatty acids and eicosanoids regulate gene expression through direct interactions with peroxisome proliferator-activated receptors alpha and γ. *Proc. Natl. Acad. Sci. USA* **94**, 4318–4323.
18. Krey, G., Braissant, O., L'Horset, F., Kalkhoven, E., Perroud, M., Parker, M. G., and Wahli, W. (1997). Fatty acids, eicosanoids, and hypolipidemic agents identified as ligands of peroxisome proliferator-activated receptors by coactivator-dependent receptor ligand assay. *Mol. Endocrinol.* **11**, 779–791.
19. Forman, B. M., Tontonoz, P., Chen, J., Brun, R. P., Spiegelman, B. M., and Evans, R. M. (1995). 15-Deoxy-delta 12, 14-prostaglandin J2 is a ligand for the adipocyte determination factor PPAR γ. *Cell* **83**, 803–812.
20. Kliewer, S. A., Lenhard, J. M., Wilson, T. M., Patel, I., Morris, D. C., and Lehmann, J. M. (1995). A prostaglandin J2 metabolite binds peroxisome proliferator-activated receptor gamma and promotes adipocyte differentiation. *Cell* **83**, 813–819.
21. Kletzien, R. F., Foellmi, L. A., Harris, P. K., Wyse, B. M., and Clarke, S. D. (1992). Adipocyte fatty acid-binding protein: Regulation of gene expression *in vivo* and *in vitro* by an insulin-sensitizing agent. *Mol. Pharmacol.* **42**, 558–562.
22. Ibrahimi, A., Teboul, L., Gaillard, D., Amri, E. Z., Ailhaud, G., Young, P., Cawthorne, M. A., and Grimaldi, P. A. (1994). Evidence for a common mechanism of action for fatty acids and thiazolidinedione antidiabetic agents on gene expression in preadipose cells. *Mol. Pharmacol* **46**, 1070–1076.
23. Lehmann, J. M., Moore, L. B., Smith-Oliver, T. A., Wilkinson, W. O., Willson, T. M., and Kliewer, S. A. (1995). An antidiabetic thiazolidinedione is a high affinity ligand for peroxisome proliferator-activated receptor γ (PPAR γ). *J. Biol. Chem.* **270**, 12953–12956.
24. Walczak, R. and Tontonoz, P. (2002). PPARadigms and PPARadoxes. Expanding roles for PPARγ in the control of lipid metabolism. *J. Lipid Res.* **43**, 177–186.
25. Beamer, B. A., Negri, C., Yen, C. J., Gavrilova, O., Rumberger, J. M., Durcan, M. J., Yarnall, D. P., Hawkins, A. L., Griffin, C. A., Burns, D. K., Roth, J., Reitman, M., and Shuldiner, A. R. (1997). Chromosomal localization and partial genomic structure of the human peroxisome proliferator activated receptor-γ (hPPAR γ) gene. *Biochem. Biophys. Res. Commun.* **233**, 756–759.
26. Fajas, L., Auboeuf, D., Raspe, E., Schoonjans, K., Lefebvre, A. M., Saladin, R., Najib, J., Laville, M., Fruchart, J. C., Deeb, S., Vidal-Puig, A., Flier, J., Briggs, M. R., Staels, B., Vidal, H., and Auwerx, J. (1997). The organization, promoter analysis, and expression of the human PPARγ gene. *J. Biol. Chem.* **272**, 18779–18789.
27. Fajas, L., Fruchart, J. C., and Auwerx, J. (1998). PPARγ3 mRNA: A distinct PPARγ mRNA subtype transcribed from an independent promoter. *FEBS Lett.* **438**, 55–60.
28. Zhu, Y., Qi, C., Korenberg, J. R., Chen, X. N., Noya, D., Rao, M. S., and Reddy, J. K. (1995). Structural organization of mouse peroxisome proliferator-activated receptor γ (mPPAR γ) gene: Alternative promoter use and different splicing yield two mPPAR γ isoforms. *Proc. Natl. Acad. Sci. USA* **92**, 7921–7925.
29. Kliewer, S. A., Forman, B. M., Blumberg, B., Ong, E. S., Borgmeyer, U., Mangelsdorf, D. J., Umesono, K., and Evans, R. M. (1994). Differential expression and activation of a family of murine peroxisome proliferator-activated receptors. *Proc. Natl. Acad. Sci. USA* **91**, 7355–7359.
30. Braissant, O., Foufelle, F., Scotto, C., Dauca, M., and Wahli, W. (1996). Differential expression of peroxisome proliferator-activated receptors (PPARs): Tissue distribution of PPAR-alpha, -beta, and -γ in the adult rat. *Endocrinology* **137**, 354–366.
31. Braissant, O. and Wahli, W. (1998). Differential expression of peroxisome proliferator-activated receptor-alpha, -beta, and -γ during rat embryonic development. *Endocrinology* **139**, 2748–2754.
32. Barak, Y., Nelson, M. C., Ong, E. S., Jones, Y. Z., Ruiz-Lozano, P., Chien, K. R., Koder, A., and Evans, R. M. (1999). PPARγ is required for placental, cardiac, and adipose tissue development. *Mol. Cell* **4**, 585–595.
33. Rosen, E. D., Sarraf, P., Troy, A. E., Bradwin, G., Moore, K., Milstone, D. S., Spiegelman, B. M., and Mortensen, R. M. (1999). PPARγ is required for the differentiation of adipose tissue in vivo and in vitro. *Mol. Cell* **4**, 611–617.
34. Kubota, N., Terauchi, Y., Miki, H., Tamemoto, H., Yamauchi, T., Komeda, K., Satoh, S., Nakano, R., Ishii, C., Sugiyama, T., Eto, K., Tsubamoto, Y., Okuno, A., Murakami, K., Sekihara, H., Hasegawa, G., Naito, M., Toyoshima, Y., Tanaka, S., Shiota, K., Kitamura, T., Fujita, T., Ezaki, O., Aizawa, S., Kadowaki, T., *et al.* (1999). PPAR γ mediates high-fat diet-induced adipocyte hypertrophy and insulin resistance. *Mol. Cell* **4**, 597–609.
35. Rosen, E. D., Hsu, C. H., Wang, X., Sakai, S., Freeman, M. W., Gonzalez, F. J., and Spiegelman, B. M. (2002). C/EBPalpha induces adipogenesis through PPARγ: A unified pathway. *Genes Dev.* **16**, 22–26.
36. Ristow, M., Muller-Wieland, D., Pfeiffer, A., Krone, W., and Kahn, C. R. (1998). Obesity associated with a mutation in a genetic regulator of adipocyte differentiation. *N. Engl. J. Med.* **339**, 953–959.
37. Beamer, B. A., Yen, C. J., Andersen, R. E., Muller, D., Elahi, D., Cheskin, L. J., Andres, R., Roth, J., and Shuldiner, A. R. (1998). Association of the Pro12Ala variant in the peroxisome proliferator-activated receptor-γ2 gene with obesity in two Caucasian populations. *Diabetes* **47**, 1806–1808.
38. Deeb, S. S., Fajas, L., Nemoto, M., Pihlajamaki, J., Mykkanen, L., Kuusisto, J., Laakso, M., Fujimoto, W., and Auwerx, J. (1998). A Pro12Ala substitution in PPARγ2 associated with decreased receptor activity, lower body mass index and improved insulin sensitivity. *Nat. Genet.* **20**, 284–287.
39. Hara, K., Okada, T., Tobe, K., Yasuda, K., Mori, Y., Kadowaki, H., Hagura, R., Akanuma, Y., Kimura, S., Ito, C., and Kadowaki, T. (2000). The Pro12Ala polymorphism in PPAR γ2 may confer resistance to type 2 diabetes. *Biochem. Biophys. Res. Commun.* **271**, 212–216.
40. Vigouroux, C., Fajas, L., Khallouf, E., Meier, M., Gyapay, G., Lascols, O., Auwerx, J., Weissenbach, J., Capeau, J., and Magre, J. (1998). Human peroxisome proliferator-activated receptor-γ2: Genetic mapping, identification of a variant in the coding sequence, and exclusion as the gene responsible for lipoatrophic diabetes. *Diabetes* **47**, 490–492.
41. Yen, C. J., Beamer, B. A., Negri, C., Silver, K., Brown, K. A., Yarnall, D. P., Burns, D. K., Roth, J., and Shuldiner, A. R. (1997). Molecular scanning of the human peroxisome proliferator activated receptor γ (hPPAR γ) gene in diabetic Caucasians: Identification of a Pro12Ala PPAR γ 2 missense mutation. *Biochem. Biophys. Res. Commun.* **241**, 270–274.
42. Green, H. and Kehinde, O. (1975). An established preadipose cell line and its differentiation in culture. II. Factors affecting the adipose conversion. *Cell* **5**, 19–27.
43. Green, H. and Kehinde, O. (1979). Formation of normally differentiated subcutaneous fat pads by an established preadipose cell line. *J. Cell Physiol.* **101**, 169–171.

44. Student, A. K., Hsu, R. Y., and Lane, M. D. (1980). Induction of fatty acid synthetase synthesis in differentiating 3T3-L1 preadipocytes. *J. Biol. Chem.* **255**, 4745–4750.
45. Smith, P. J., Wise, L. S., Berkowitz, R., Wan, C., and Rubin, C. S. (1988). Insulin-like growth factor-I is an essential regulator of the differentiation of 3T3-L1 adipocytes. *J. Biol. Chem.* **263**, 9402–9408.
46. Bernlohr, D. A., Bolanowski, M. A., Kelly, T. J., Jr., and Lane, M. D. (1985). Evidence for an increase in transcription of specific mRNAs during differentiation of 3T3-L1 preadipocytes. *J. Biol. Chem.* **260**, 5563–5567.
47. Scott, R. E., Florine, D. L., Wille, J. J., Jr., and Yun, K. (1982). Coupling of growth arrest and differentiation at a distinct state in the G1 phase of the cell cycle: GD. *Proc. Natl. Acad. Sci. USA* **79**, 845–849.
48. Cornelius, P., MacDougald, O. A., and Lane, M. D. (1994). Regulation of adipocyte development. *Annu. Rev. Nutr.* **14**, 99–129.
49. Wu, Z., Xie, Y., Bucher, N. L., and Farmer, S. R. (1995). Conditional ectopic expression of C/EBP beta in NIH-3T3 cells induces PPAR γ and stimulates adipogenesis. *Genes Dev.* **9**, 2350–2363.
50. Yeh, W. C., Cao, Z., Classon, M., and McKnight, S. L. (1995). Cascade regulation of terminal adipocyte differentiation by three members of the C/EBP family of leucine zipper proteins. *Genes Dev* **9**, 168–181.
51. Wu, Z., Bucher, N. L., and Farmer, S. R. (1996). Induction of peroxisome proliferator-activated receptor γ during the conversion of 3T3 fibroblasts into adipocytes is mediated by C/EBPbeta, C/EBPdelta, and glucocorticoids. *Mol. Cell Biol.* **16**, 4128–4136.
52. Tanaka, T., Yoshida, N., Kishimoto, T., and Akira, S. (1997). Defective adipocyte differentiation in mice lacking the C/EBPbeta and/or C/EBPdelta gene. *EMBO J.* **16**, 7432–7443.
53. Wu, Z., Rosen, E. D., Brun, R., Hauser, S., Adelmant, G., Troy, A. E., McKeon, C., Darlington, G. J., and Spiegelman, B. M. (1999). Cross-regulation of C/EBP alpha and PPAR γ controls the transcriptional pathway of adipogenesis and insulin sensitivity. *Mol. Cell* **3**, 151–158.
54. Clarke, S. L., Robinson, C. E., and Gimble, J. M. (1997). CAAT/enhancer binding proteins directly modulate transcription from the peroxisome proliferator-activated receptor γ 2 promoter. *Biochem. Biophys. Res. Commun.* **240**, 99–103.
55. Wu, Z., Xie, Y., Morrison, R. F., Bucher, N. L., and Farmer, S. R. (1998). PPARγ induces the insulin-dependent glucose transporter GLUT4 in the absence of C/EBPalpha during the conversion of 3T3 fibroblasts into adipocytes. *J. Clin. Invest.* **101**, 22–32.
56. Freytag, S. O., Paielli, D. L., and Gilbert, J. D. (1994). Ectopic expression of the CCAAT/enhancer-binding protein alpha promotes the adipogenic program in a variety of mouse fibroblastic cells. *Genes Dev.* **8**, 1654–1663.
57. Brun, R. P., Tontonoz, P., Forman, B. M., Ellis, R., Chen, J., Evans, R. M., and Spiegelman, B. M. (1996). Differential activation of adipogenesis by multiple PPAR isoforms. *Genes Dev.* **10**, 974–984.
58. Hu, E., Tontonoz, P., and Spiegelman, B. M. (1995). Transdifferentiation of myoblasts by the adipogenic transcription factors PPAR γ and C/EBP alpha. *Proc. Natl. Acad. Sci. USA* **92**, 9856–9860.
59. Tontonoz, P., Kim, J. B., Graves, R. A., and Spiegelman, B. M. (1993). ADD1: A novel helix–loop–helix transcription factor associated with adipocyte determination and differentiation. *Mol. Cell Biol.* **13**, 4753–4759.
60. Kim, J. B. and Spiegelman, B. M. (1996). ADD1/SREBP1 promotes adipocyte differentiation and gene expression linked to fatty acid metabolism. *Genes Dev.* **10**, 1096–1107.
61. Kim, J. B., Wright, H. M., Wright, M., and Spiegelman, B. M. (1998). ADD1/SREBP1 activates PPARγ through the production of endogenous ligand. *Proc. Natl. Acad. Sci. USA* **95**, 4333–4337.
62. Shimomura, I., Hammer, R. E., Richardson, J. A., Ikemoto, S., Bashmakov, Y., Goldstein, J. L., and Brown, M. S. (1998). Insulin resistance and diabetes mellitus in transgenic mice expressing nuclear SREBP-1c in adipose tissue: Model for congenital generalized lipodystrophy. *Genes Dev.* **12**, 3182–3194.
63. Jiang, M. S., Tang, Q. Q., McLenithan, J., Geiman, D., Shillinglaw, W., Henzel, W. J., and Lane, M. D. (1998). Derepression of the C/EBPalpha gene during adipogenesis: Identification of AP-2alpha as a repressor. *Proc. Natl. Acad. Sci. USA* **95**, 3467–3471.
64. Tang, Q. Q., Jiang, M. S., and Lane, M. D. (1997). Repression of transcription mediated by dual elements in the CCAAT/enhancer binding protein alpha gene. *Proc. Natl. Acad. Sci. USA* **94**, 13571–13575.
65. Smas, C. M. and Sul, H. S. (1993). Pref-1, a protein containing EGF-like repeats, inhibits adipocyte differentiation. *Cell* **73**, 725–734.
66. Smas, C. M., Chen, L., and Sul, H. S. (1997). Cleavage of membrane-associated pref-1 generates a soluble inhibitor of adipocyte differentiation. *Mol. Cell Biol.* **17**, 977–988.
67. Smas, C. M., Chen, L., Zhao, L., Latasa, M. J., and Sul, H. S. (1999). Transcriptional repression of pref-1 by glucocorticoids promotes 3T3-L1 adipocyte differentiation. *J. Biol. Chem.* **274**, 12632–126341.
68. Ross, S. E., Hemati, N., Longo, K. A., Bennett, C. N., Lucas, P. C., Erickson, R. L., and MacDougald, O. A. (2000). Inhibition of adipogenesis by Wnt signaling. *Science* **289**, 950–953.
69. Hotamisligil, G. S., Shargill, N. S., and Spiegelman, B. M. (1993). Adipose expression of tumor necrosis factor-alpha: Direct role in obesity-linked insulin resistance. *Science* **259**, 87–91.
70. Uysal, K. T., Wiesbrock, S. M., and Hotamisligil, G. S. (1998). Functional analysis of tumor necrosis factor (TNF) receptors in TNF-alpha-mediated insulin resistance in genetic obesity. *Endocrinology* **139**, 4832–4838.
71. Uysal, K. T., Wiesbrock, S. M., Marino, M. W., and Hotamisligil, G. S. (1997). Protection from obesity-induced insulin resistance in mice lacking TNF-alpha function. *Nature* **389**, 610–614.
72. Xing, H., Northrop, J. P., Grove, J. R., Kilpatrick, K. E., Su, J. L., and Ringold, G. M. (1997). TNF alpha-mediated inhibition and reversal of adipocyte differentiation is accompanied by suppressed expression of PPARγ without effects on Pref-1 expression. *Endocrinology* **138**, 2776–2783.
73. Font de Mora, J., Porras, A., Ahn, N., and Santos, E. (1997). Mitogen-activated protein kinase activation is not necessary for, but antagonizes, 3T3-L1 adipocytic differentiation. *Mol. Cell Biol.* **17**, 6068–6075.
74. Hu, E., Kim, J. B., Sarraf, P., and Spiegelman, B. M. (1996). Inhibition of adipogenesis through MAP kinase-mediated phosphorylation of PPARγ. *Science* **274**, 2100–2103.
75. Lyle, R. E., Richon, V. M., and McGehee, R. E., Jr. (1998). TNFalpha disrupts mitotic clonal expansion and regulation of retinoblastoma proteins p130 and p107 during 3T3-L1 adipocyte differentiation. *Biochem. Biophys. Res. Commun.* **247**, 373–378.
76. Hamann, A., Benecke, H., Le Marchand-Brustel, Y., Susulic, V. S., Lowell, B. B., and Flier, J. S. (1995). Characterization of insulin resistance and NIDDM in transgenic mice with reduced brown fat. *Diabetes* **44**, 1266–1273.
77. Hofmann, C., Lorenz, K., Braithwaite, S. S., Colca, J. R., Palazuk, B. J., Hotamisligil, G. S., and Spiegelman, B. M. (1994). Altered gene expression for tumor necrosis factor-alpha and its receptors during drug and dietary modulation of insulin resistance. *Endocrinology* **134**, 264–270.
78. Paz, K., Hemi, R., LeRoith, D., Karasik, A., Elhanany, E., Kanety, H., and Zick, Y. (1997). A molecular basis for insulin resistance. Elevated serine/threonine phosphorylation of IRS-1 and IRS-2 inhibits their binding to the juxtamembrane region of the insulin receptor and impairs their ability to undergo insulin-induced tyrosine phosphorylation. *J. Biol. Chem.* **272**, 29911–29918.
79. Peraldi, P., Xu, M., and Spiegelman, B. M. (1997). Thiazolidinediones block tumor necrosis factor-alpha-induced inhibition of insulin signaling. *J. Clin Invest* **100**, 1863–1869.
80. Ribon, V., Johnson, J. H., Camp, H. S., and Saltiel, A. R. (1998). Thiazolidinediones and insulin resistance: Peroxisome proliferator-activated receptor γ activation stimulates expression of the CAP gene. *Proc. Natl. Acad. Sci. USA* **95**, 14751–14756.
81. Smith, U., Gogg, S., Johansson, A., Olausson, T., Rotter, V., and Svalstedt, B. (2001). Thiazolidinediones (PPARγ agonists) but not PPARalpha agonists increase IRS-2 gene expression in 3T3-L1 and human adipocytes. *FASEB J.* **15**, 215–220.
82. Rieusset, J., Auwerx, J., and Vidal, H. (1999). Regulation of gene expression by activation of the peroxisome proliferator-activated receptor γ with rosiglitazone (BRL 49653) in human adipocytes. *Biochem. Biophys. Res. Commun.* **265**, 265–271.

83. Rieusset, J., Chambrier, C., Bouzakri, K., Dusserre, E., Auwerx, J., Riou, J. P., Laville, M., and Vidal, H. (2001). The expression of the p85alpha subunit of phosphatidylinositol 3-kinase is induced by activation of the peroxisome proliferator-activated receptor γ in human adipocytes. *Diabetologia* **44**, 544–554.
84. Altiok, S., Xu, M., and Spiegelman, B. M. (1997). PPARγ induces cell cycle withdrawal: Inhibition of E2F/DP DNA-binding activity via down-regulation of PP2A. *Genes Dev.* **11**, 1987–1998.
85. Morrison, R. F. and Farmer, S. R. (1999). Role of PPARγ in regulating a cascade expression of cyclin-dependent kinase inhibitors, p18(INK4c) and p21(Waf1/Cip1), during adipogenesis. *J. Biol. Chem.* **274**, 17088–17097.
86. Hansen, J. B., Petersen, R. K., Larsen, B. M., Bartkova, J., Alsner, J., and Kristiansen, K. (1999). Activation of peroxisome proliferator-activated receptor γ bypasses the function of the retinoblastoma protein in adipocyte differentiation. *J. Biol. Chem.* **274**, 2386–2393.

CHAPTER 277

Orphan Nuclear Receptors

Barry Marc Forman
The Beckman Research Institute,
City of Hope National Medical Center,
Division of Molecular Medicine,
The Gonda Diabetes & Genetic Research Center,
Duarte, California

Classical Receptors versus Orphan Receptors

Endocrine Signaling Paradigm

Nuclear hormone receptors comprise a superfamily of ligand-modulated transcription factors that regulate gene transcription in response to their cognate ligands. The transcriptional effects of these receptors are mediated via two critical functional domains: the DNA-binding domain (DBD) and the ligand-binding domain (LBD). The DBD recognizes specific target genes, whereas the LBD modulates the gene transcription by interacting with transcriptional regulatory complexes in a ligand-dependent fashion.

Nuclear acting hormones such as steroid and thyroid hormones were identified nearly a century ago. These signaling molecules are secreted into the bloodstream at nanomolar concentrations and ultimately regulate transcription in distant target tissues. Tissues that respond to these hormones express receptors whose ligand-binding affinity closely matches the circulating concentration of the hormone. The combination of secreted hormones and high-affinity targets is a characteristic that defines the classical endocrine signaling system. This signaling paradigm provides an efficient means to simultaneously regulate gene expression at multiple sites throughout the body.

To date, 49 human nuclear receptors have been identified [1]. Although virtually all of these proteins possess the characteristic LBD, approximately 34 had not originally been associated with a cognate ligand. These proteins are known as *orphan receptors* and their existence implies that additional signaling molecules remain to be identified. Because more than 30 years had passed since a classical steroid hormone had been identified, it is reasonable to imagine that ligands for orphan receptors may have escaped detection because they are not generated via classical endocrine paradigms.

Nature of Orphan Receptor Ligands

Specialized endocrine signaling molecules are common in higher organisms, but in lower organisms, intermediary metabolites commonly possess both metabolic and signaling functions. For example, in bacteria and yeast, numerous gene networks are regulated by sugars, amino acids, fatty acids, and adenosine triphosphate (ATP) derivatives [2–4]. These regulatory networks sense specific nutrients and modulate their levels by regulating, in a coordinated manner, the expression of genes required for their synthesis, degradation, or transport. In addition to metabolic signals, single-cell organisms have also developed networks to detect and minimize exposure to exogenous toxins such as antibiotics (e.g., tetracycline) [5] and heavy metal derivatives (e.g., mercury, arsenic) [6,7]. Because higher organisms must also regulate their exposure to nutrients and foreign toxins, it is possible that they possess receptors that respond to similar signals. Indeed, studies during the past 5–6 years have established a definitive role for orphan receptors in these processes.

An important component of all signaling networks is the ability to respond to signaling molecules at physiologically relevant concentrations. In contrast to endocrine hormones that are secreted at nanomolar concentrations, intermediary metabolites are by necessity present at much higher (micromolar) concentrations. Thus, an important distinction between endocrine and metabolite/toxin-controlled receptors is that the former are high-affinity sensors, whereas the latter bind their ligands with an appropriately lower affinity.

ɲhan Receptors and Metabolite-Derived Signals

RXR, A Common Heterodimeric Partner

Many receptors function as an obligate heterodimer with the nuclear receptor RXR [8]. Although the identity of the endogenous RXR ligand remains controversial, it is clear that this receptor can be activated by synthetic rexinoids [9], 9-*cis* retinoic acid [10], and very high concentrations of phytanic [11,12] and docosahexaenoic acids [13]. Receptors that heterodimerize with RXR fall into two classes: those that are "permissive" and respond to RXR ligands and those that are "silent" and fail to respond [14–16]. The permissive receptors include the PPARs, LXR, bile acid receptor (BAR), and constitutive androstane receptor (CAR). Although these receptors respond to their specific ligands, the potential also exists for cross-activation via RXR agonists.

PPARs: Regulators of Lipid and Glucose Homeostasis

The PPARs (peroxisome proliferator-activated receptors) include three different receptor genes (PPARα, β/δ, and γ) whose ligands are lipid-derived compounds. PPARα binds to endogenous polyunsaturated fatty acids and regulates hepatic genes involved in fatty acid oxidation. PPARβ/δ also binds polyunsaturated fatty acids and increases HDL cholesterol and reverse cholesterol transport from peripheral tissues. PPARγ is expressed at high levels in adipose tissue where it regulates lipid storage and glucose homeostasis. These receptors (discussed elsewhere in this series) provide excellent examples of the nexus between metabolism and transcription in higher animals.

LXR, A Sterol and Fatty Acid Sensor

Two related LXR genes have been identified: LXRα and LXRβ. The β-subtype is expressed ubiquitously and its physiological function is unclear [17]. In contrast, LXRα is expressed in a variety of tissues that contribute to lipid homeostasis including hepatocytes, intestinal epithelia, macrophages, and adipose tissue (Table I) [18]. LXRα exhibits partial constitutive activity that reflects a combination of ligand-independent activity [19] and activation by an endogenous mevalonic acid-derived metabolite [20]. LXRα can be further activated by a variety of oxysterols. The endogenous LXR ligand is unclear and different sterol ligands may be acting in different tissues, for example, 24(*S*),25-epoxycholesterol in the liver [18] and 27-hydroxycholesterol in macrophages [21]. Whereas oxysterols activate LXRα, the transcriptional activity of this receptor can be repressed by polyunsaturated fatty acids [22] and geranylgeranyl pyrophosphate [20]. LXRα is therefore a dynamic "integrator" of multiple lipid metabolites.

Studies using knock-out mice and synthetic agonists have demonstrated that LXR is not only a lipid sensor but also an effector of cholesterol and fatty acid homeostasis. LXRα activation results in increased hepatic cholesterol degradation and a decrease in intestinal cholesterol absorption [23,24]. The increase in cholesterol turnover is due to enhanced transcription of cyp7a1, the rate-limiting step in the conversion of cholesterol to bile acids. Cyp7a1 is regulated in rodents but not humans because the LXR response element is defective in the human promoter. In contrast, the decrease in intestinal cholesterol absorption is due to activation of the ABCA1 (ATP binding cassette A1) transporter in both humans and rodents. LXRα also activates expression of ABCG1 [25] and of ABCG5/G8 [26], which have been implicated in the efflux of sitosterol and other sterols.

In macrophages, LXRα coordinately activates expression of ABCA1 [27] and apolipoprotein E (apoE) [28], a secreted cholesterol-acceptor protein. The net effect is enhanced removal of cholesterol from atherosclerotic lesions (macrophage foam cells) and its transfer to the liver via HDL cholesterol. Interestingly, this process appears to be under feed-forward control in humans as a macrophage-specific LXRα promoter is activated by PPARγ and autoregulated by LXRα [29–31].

The ability to reduce cholesterol absorption and to stimulate cholesterol efflux makes LXRα an attractive target for the treatment of atherosclerosis. Unfortunately, LXRα activation is also associated with an increase in circulating triglycerides [32]. This is likely a result of enhanced expression of SREBP-1c (sterol response element binding protein), a transcription factor that activates a battery of genes required for saturated fatty acid synthesis [33]. One of these genes, fatty acid synthase (FAS), is also a direct target of LXRα [34]. Because elevated triglycerides may be counterproductive in atherosclerosis, an ideal LXRα-based agent may be one that selectively activates certain genes (e.g., ABCA1, apoE) but not others (SREBP-1c, FAS). The identification of selective estrogen receptor modulators [35] raises the possibility that similar reagents may eventually be identified for other nuclear receptors.

FXR, A Bile Acid Receptor

As early as 1957 it had been appreciated that a biliary component could inhibit bile acid production [36]. A key feature of this negative feedback loop is the ability of bile acids to inhibit Cyp7a transcription [36]. The nature of the bile acid sensor remained elusive until 1999 when it was identified as the nuclear receptor FXR or BAR [37–39]. It is now clear that BAR plays a broader role in regulating bile acid homeostasis. In addition to inhibiting hepatic expression of CYP7A1, the rate-limiting enzyme in bile acid biosynthesis, BAR also inhibits expression of CYP8B [40], which acts downstream in the bile acid biosynthetic pathway. BAR also stimulates expression of the ABC cassette protein BSEP (bile-salt export protein, ABCB11) [41], which promotes biliary secretion of bile acids. Thus, the net effect of BAR activation is to limit bile acid accumulation in the liver. Indeed, BAR-deficient mice are highly sensitive to the toxic effects of excessive bile acids [42].

Table I Expression Patterns, Key Target Genes, Ligands and Physiological Functions of Selected Orphan Nuclear Receptors

H_2N— DNA BINDING DOMAIN ———— LIGAND BINDING DOMAIN AF2 —COOH

Receptor	Gene symbol[a]	Key tissues of expression	Critical target genes[b]	Ligands[c]	RXR ligand response?	Function
LXRα	NR1H3	Hepatocytes, Intestinal Epithelium, Macrophage, Fat	cypa7a1 (rodent), ABCA1, apoE, ABCG1/G4/G5/G8, SREBP-1c, FAS, LXRa	Oxysterols, polyunsaturated fatty acids (–)	Yes	Cholesterol, triglyceride & bile acid homoestasis
BAR FXR	NR1H2	Hepatocytes, Intestinal Epithelium	BSEP, SHP, IBABP, PLTP, apoCII ↓CYP7A1, ↓CYP8B	*Bile Acids* Chenodeoxycholic,cholic & deoxycholic acids	Yes	Cholesterol, triglyceride & bile acid homoestasis
SXR PXR	NR1I2	Hepatocytes, Intestinal Epithelium	CYP3A4, CYP2C8, CYP2C9, CYP2B10, MDR1, MRP2, OATP2	*Xenobiotics* *Human*: rifampicin,ritonavir, taxol, clotrimazole, hyperforin *Mouse*: PCN[d]	No	Clearance of xenobiotics & endogenous toxins
CAR	NR1I3 NR1I4	Hepatocytes, Intestinal Epithelium	cyp2b10, cyp3a11, MRP2	*Xenobiotics* *Human*: ? *Mouse*: androstanol(–) TCPOBOP[e]	Yes	Clearance of xenobiotics & endogenous toxins

A schematic representation of an orphan nuclear receptor is shown above the table. Target gene specificity is determined by the DNA binding domain; ligand binding and transcriptional activation are mediated by the ligand binding domain.

[a]Gene symbols are according to the unified nomenclature system for the nuclear receptor superfamily (*www.ens-lyon.fr/LBMC/laudet/nomenc.html*).

[b]↓ indicates genes that are down-regulated by agonist ligands; other genes up-regulated by agonist ligands.

[c](–) indicates an antagonist, inverse agonist or inhibitory ligand; other ligands are agonists or activating ligands.

[d]PCN, pregnenolone-16α-carbonitrile.

[e]TCPOBOP, 1,4-bis[2-(3,5-dichloropyridyloxy)]benzene.

Bile acids are produced in the liver, secreted into the bile, and then efficiently absorbed via an ileal bile acid transporter. Thus, the ileal epithelium is exposed to extremely high levels of bile acids. The cytosolic ileal bile acid binding protein (IBABP) is thought to act as a shuttle protein that aids in transport and protects the cell from high levels of free bile acids. Interestingly, BAR is an extremely effective inducer of IBABP [43], which is consistent with the role of this receptor in protecting the body from excess bile acids.

The ability to inhibit CYP7A1 transcription implies that an antagonist of BAR could be a useful cholesterol-lowering agent. However, like LXRα, BAR induces a reciprocal relationship between cholesterol and triglyceride levels [42,44]. This is potentially related to BAR-mediated regulation of apolipoprotein CII (apoCII) [45] or phospholipid transfer protein (PLTP) [46]. Regardless of the mechanism, LXRα and BAR share similar caveats with regard to their potential use as targets for cholesterol lowering.

Orphan Receptors and Xenobiotic Signals

SXR, A Master Regulator of Drug Clearance

The ligand-binding properties of the steroid and xenobiotic receptor SXR (also known as PXR, PAR, PRR, NR1I2) are different from most nuclear receptors. Whereas most receptors possess a high degree of ligand-binding specificity, SXR is activated by a very large number of structurally unrelated pharmaceutical agents including rifampicin, clotrimazole, SR12813 [47–51], HIV protease inhibitors [52], taxol [53], and hyperforin [54], the active ingredient in the herbal remedy called St. John's wort. Many of these compounds bind directly to the receptor LBD with affinities in the micromolar range. In an effort to understand the basis for SXR's extreme promiscuity in ligand binding, Watkins and colleagues [55] determined the three-dimensional structure of human SXR. They found that SXR has a larger ligand binding cavity than most nuclear receptors as well as a flexible loop, which accommodates both small and large ligands. These studies highlight the broad ligand-binding specificity of SXR and confirm its designation as a xenobiotic receptor.

Many SXR activators were previously described as inducers of the hepatic cytochrome *P*-450 monooxygenase CYP3A (CYP3A4 in human, cyp3a11 in mouse, CYP3A23 in rat, CYP3A6 in rabbit). A variety of biochemical and genetic studies have demonstrated that SXR directly regulates CYP3A induction via SXR response elements in the CYP3A promoter. CYP3A4 is the most abundant cytochrome *P*-450 in human liver and is responsible for the metabolism of approximately 50% of all drugs [56]. Subsequent studies have expanded the number of SXR target genes to include other *P*-450 enzymes including CYP2C8, CYP2C9, and CYP2B [52,53,57]. These enzymes metabolize a variety of drugs including nonsteroidal anti-inflammatory drugs, hypoglycemic agents, anti-epileptic agents, and taxol, an antineoplastic agent [56]. Thus, SXR is not only a xenobiotic sensor, but a direct effector of xenobiotic degradation.

In addition to being expressed in the liver, SXR is also highly expressed in the intestine [47–50], which raises the question as to the function of SXR in this organ. We have found that SXR regulates expression of MDR-1/P-glycoprotein, a broad-specificity efflux pump located on the lumenal surface of intestinal enterocytes [58]. P-glycoprotein can severely limit the oral absorption of its substrates [58,59]. Because most drugs are administered orally, the SXR/P-glycoprotein linkage may limit the bioavailability of many pharmaceutical agents.

Other transporters including OATP2 (Na+-independent organic anion transporter 2) [60] and the ABC protein MRP2 (multidrug resistance protein 2, ABCC2) [52] also act as SXR target genes. OATP2 is present on the sinusoidal membrane of hepatocytes and promotes uptake of organic anions (e.g., bile acids and xenobiotics) into the liver. MRP2 resides on the canalicular membrane of hepatocytes and mediates the biliary excretion of its substrates. Taken together, these studies indicate that SXR is a master regulator of a diverse array of xenobiotic clearance pathways.

The ability to activate drug clearance has important implications. In particular, drugs that activate SXR may decrease their own bioavailability as well as the availability of coadministered drugs [61–65]. In principle, this type of drug–drug interaction could be overcome by screening for analogs that retain therapeutic activity but fail to activate SXR. An example of this is illustrated by the use of taxotere, a taxol analog that retains antineoplastic activity but fails to activate SXR [53]. The identification of such "SXR-transparent" drugs would be predicted to decrease the likelihood of drug–drug interactions. Finally, it is important to note that the LBDs of human and rodent SXR possess distinct ligand-response profiles (see Table I) [51]. This divergence is consistent with the observation that rodent models are poor predictors of drug metabolism in humans.

CAR, A Close Relative of SXR

The nuclear receptor CAR is a close relative of SXR. Both are highly expressed in the liver and intestine and these receptors share sequence identity in their DNA (66%) and ligand-binding domains (41%) [66,67]. Mouse CAR was originally identified as a constitutively active receptor whose activity could be repressed by androstanol [68]. Subsequent studies demonstrated that CAR was a receptor for TCPOBOP (1,4-bis[2-(3,5-dichloropyridyloxy)]benzene) [69–72], a highly potent inducer of cyp2b10 expression in mouse hepatocytes. TCPOBOP had previously been demonstrated to be a member of the phenobarbital class of hepatic *P*-450 inducing agents suggesting that CAR played a role in mediating this xenobiotic response. This link became definitive with the demonstration that CAR knock-out mice fail to induce cyp2b10 in response to either TCPOBOP or phenobarbital [67].

Although CAR is a strong inducer of cyp2b10, it also serves as a weaker regulator of SXR target genes [73]. Moreover, these receptors recognize a similar array of ligands, albeit with distinct dose-response profiles [74]. These findings suggest that SXR and CAR may be partially redundant in their functions. This notion is confirmed by the observation that mice lacking either SXR or CAR are highly sensitive to the paralytic effects of zoxazolamine [67,75]. This presumably reflects a shared defect in the clearance of this drug. It is important to note that effective ligands have been identified for mouse CAR but not for its human counterpart. In the future, it will be important to test drugs for their ability to modulate human CAR and to elucidate the contribution of this receptor to drug metabolism in humans.

Future Directions

Orphan nuclear receptors provide convincing evidence that higher organisms regulate gene transcription in response to metabolic and xenobiotic cues. As metabolic disease (atherosclerosis, hypercholesterolemia, diabetes) represents the leading cause of death in industrialized societies, these receptors provide important targets for drug discovery. At the same time, discovery efforts will be enhanced by dual-screening strategies that select for therapeutic activity while minimizing interactions with xenobiotic receptors. Finally, orphan receptors represent a valuable source for future exploration because 26 orphans have yet to be associated with specific ligands.

References

1. Maglich, J. M., Sluder, A., Guan, X., Shi, Y., McKee, D. D., Carrick, K., Kamdar, K., Willson, T. M., and Moore, J. T. (2001). *Genome Biol.* **2**, research0029.1–0029.7.
2. Yang, S. Y. and Schulz, H. (1983). *J. Mol. Chem.* 258, 9780–9785.
3. Bykowski, T., Ploeg, J. R., Iwanicka-Nowicka, R., and Hryniewicz, M. M. (2002). *Mol. Microbiol.* **43**, 1347–1358.
4. Dworkin, J. and Losick, R. (2001). *Genes Dev.* **15**, 1051–1054.
5. Hillen, W., Klock, G., Kaffenberger, I., Wray, L. V., and Reznikoff, W. S. (1982). *J. Biol. Chem.* **257**, 6605–6613.
6. Prithivirajsingh, S., Mishra, S. K., and Mahadevan, A. (2001). *Mol. Biol. Rep.* **28**, 63–72.
7. Walsh, C. T., Distefano, M. D., Moore, M. J., Shewchuk, L. M., and Verdine, G. L. (1988). *FASEB J.* **2**, 124–130.
8. Mangelsdorf, D. J. and Evans, R. M. (1995). *Cell* **83**, 841–850.
9. Boehm, M. F., Zhang, L., Zhi, L., McClurg, M. R., Berger, E., Wagoner, M., Mais, D. E., Suto, C. M., Davies, J. A., Heyman, R. A. *et al.* (1995). *J. Med. Chem.* **38**, 3146–3155.
10. Heyman, R. A., Mangelsdorf, D. J., Dyck, J. A., Stein, R. B., Eichele, G., Evans, R. M., and Thaller, C. (1992). *Cell* **68**, 397–406.
11. Kitareewan, S., Burka, L. T., Tomer, K. B., Parker, C. E., Deterding, L. J., Stevens, R. D., Forman, B. M., Mais, D. E., Heyman, R. A., McMorris, T., and Weinberger, C. (1996). *Mol. Biol. Cell* **7**, 1153–1166.
12. Lemotte, P. K., Keidel, S., and Apfel, C. M. (1996). *Eur. J. Biochem.* **236**, 328–333.
13. de Urquiza, A. M., Liu, S., Sjoberg, M., Zetterstrom, R. H., Griffiths, W., Sjovall, J., and Perlmann, T. (2000). *Science* **290**, 2140–2144.
14. Forman, B. M., Umesono, K., Chen, J., and Evans, R. M. (1995). *Cell* **81**, 541–550.
15. Kurokawa, R., DiRenzo, J., Boehm, M., Sugarman, J., Gloss, B., Rosenfeld, M. G., Heyman, R. A., and Glass, C. K. (1994). *Nature* **371**, 528–531.
16. Germain, P., Iyer, J., Zechel, C., and Gronemeyer, H. (2002). *Nature* **415**, 187–192.
17. Alberti, S., Schuster, G., Parini, P., Feltkamp, D., Diczfalusy, U., Rudling, M., Angelin, B., Bjorkhem, I., Pettersson, S., and Gustafsson, J. A. (2001). *J. Clin. Invest.* **107**, 565–573.
18. Lu, T. T., Repa, J. J., and Mangelsdorf, D. J. (2001). *J. Biol. Chem.* **276**, 37735–37738.
19. Wiebel, F. F., Steffensen, K. R., Treuter, E., Feltkamp, D., and Gustafsson, J. A. (1999). *Mol. Endocrinol.* **13**, 1105–1118.
20. Forman, B. M., Ruan, B., Chen, J., Schroepfer, G. J., Jr., and Evans, R. M. (1997). *Proc. Natl. Acad. Sci. USA* **94**, 10588–10593.
21. Fu, X., Menke, J. G., Chen, Y., Zhou, G., MacNaul, K. L., Wright, S. D., Sparrow, C. P., and Lund, E. G. (2001). *J. Biol. Chem.* **276**, 38378–38387.
22. Ou, J., Tu, H., Shan, B., Luk, A., DeBose-Boyd, R. A., Bashmakov, Y., Goldstein, J. L., and Brown, M. S. (2001). *Proc. Natl. Acad. Sci. USA* **98**, 6027–6032.
23. Peet, D. J., Turley, S. D., Ma, W., Janowski, B. A., Lobaccaro, J. M., Hammer, R. E., and Mangelsdorf, D. J. (1998). *Cell* **93**, 693–704.
24. Repa, J. J., Turley, S. D., Lobaccaro, J. A., Medina, J., Li, L., Lustig, K., Shan, B., Heyman, R. A., Dietschy, J. M., and Mangelsdorf, D. J. (2000). *Science* **289**, 1524–1529.
25. Kennedy, M. A., Venkateswaran, A., Tarr, P. T., Xenarios, I., Kudoh, J., Shimizu, N., and Edwards, P. A. (2001). *J. Biol. Chem.* **276**, 39438–39447.
26. Repa, J. J., Berge, K. E., Pomajzl, C., Richardson, J. A., Hobbs, H. H., and Mangelsdorf, D. J. (2002). *J. Biol. Chem.* **277**(21), 18793–18800.
27. Venkateswaran, A., Laffitte, B. A., Joseph, S. B., Mak, P. A., Wilpitz, D. C., Edwards, P. A., Tontonoz, P., Kast, H. R., and Anisfeld, A. M. (2000). *Proc. Natl. Acad. Sci. USA* **97**, 12097–12102.
28. Laffitte, B. A., Repa, J. J., Joseph, S. B., Wilpitz, D. C., Kast, H. R., Mangelsdorf, D. J., and Tontonoz, P. (2001). *Proc. Natl. Acad. Sci. USA* **98**, 507–512.
29. Chawla, A., Boisvert, W. A., Lee, C. H., Laffitte, B. A., Barak, Y., Joseph, S. B., Liao, D., Nagy, L., Edwards, P. A., Curtiss, L. K., Evans, R. M., and Tontonoz, P. (2001). *Mol. Cell* **7**, 161–171.
30. Whitney, K. D., Watson, M. A., Goodwin, B., Galardi, C. M., Maglich, J. M., Wilson, J. G., Willson, T. M., Collins, J. L., and Kliewer, S. A. (2001). *J. Biol. Chem.* **276**, 43509–43515.
31. Laffitte, B. A., Joseph, S. B., Walczak, R., Pei, L., Wilpitz, D. C., Collins, J. L., and Tontonoz, P. (2001). *Mol. Cell Biol.* **21**, 7558–7568.
32. Schultz, J. R., Tu, H., Luk, A., Repa, J. J., Medina, J. C., Li, L., Schwendner, S., Wang, S., Thoolen, M., Mangelsdorf, D. J., Lustig, K. D., and Shan, B. (2000). *Genes Dev.* **14**, 2831–2838.
33. Repa, J. J., Liang, G., Ou, J., Bashmakov, Y., Lobaccaro, J. M., Shimomura, I., Shan, B., Brown, M. S., Goldstein, J. L., and Mangelsdorf, D. J. (2000). *Genes Dev.* **14**, 2819–2830.
34. Joseph, S. B., Laffitte, B. A., Patel, P. H., Watson, M. A., Matsukuma, K. E., Walczak, R., Collins, J. L., Osborne, T. F., and Tontonoz, P. (2002). *J. Biol. Chem.* **277**, 11019–11025.
35. Shang, Y. and Brown, M. (2002). *Science* **295**, 2465–2468.
36. Chiang, J. Y. L. (1998). *Front Biosci* **3**, D176–193.
37. Makishima, M., Okamoto, A. Y., Repa, J. J., Tu, H., Learned, R. M., Luk, A., Hull, M. V., Lustig, K. D., Mangelsdorf, D. J., and Shan, B. (1999). *Science* **284**, 1362–1365.
38. Parks, D. J., Blanchard, S. G., Bledsoe, R. K., Chandra, G., Consler, T. G., Kliewer, S. A., Stimmel, J. B., Willson, T. M., Zavacki, A. M., Moore, D. D., and Lehmann, J. M. (1999). *Science* **284**, 1365–1368.
39. Wang, H., Chen, J., Hollister, K., Sowers, L. C., and Forman, B. M. (1999). *Mol. Cell* **3**, 543–553.
40. del Castillo-Olivares, A. and Gil, G. (2000). *J. Biol. Chem.* **275**, 17793–17799.
41. Ananthanarayanan, M., Balasubramanian, N., Makishima, M., Mangelsdorf, D. J., and Suchy, F. J. (2001). *J. Biol. Chem.* **276**, 28857–28865.
42. Sinal, C. J., Tohkin, M., Miyata, M., Ward, J. M., Lambert, G., and Gonzalez, F. J. (2000). *Cell* **102**, 731–744.
43. Kanda, T., Foucand, L., Nakamura, Y., Niot, I., Besnard, P., Fujita, M., Sakai, Y., Hatakeyama, K., Ono, T., and Fujii, H. (1998). *Biochem. J.* **330**, 261–265.

44. Maloney, P. R., Parks, D. J., Haffner, C. D., Fivush, A. M., Chandra, G., Plunket, K. D., Creech, K. L., Moore, L. B., Wilson, J. G., Lewis, M. C., Jones, S. A., and Willson, T. M. (2000). *J. Med. Chem.* **43**, 2971–2974.
45. Kast, H. R., Nguyen, C. M., Sinal, C. J., Jones, S. A., Laffitte, B. A., Reue, K., Gonzalez, F. J., Willson, T. M., and Edwards, P. A. (2001). *Mol. Endocrinol.* **15**, 1720–1728.
46. Urizar, N. L., Dowhan, D. H., and Moore, D. D. (2000). *J. Biol. Chem.* **275**, 39313–39317.
47. Bertilsson, G., Heidrich, J., Svensson, K., Asman, M., Jendeberg, L., Sydow-Backman, M., Ohlsson, R., Postlind, H., Blomquist, P., and Berkenstam, A. (1998). *Proc. Natl. Acad. Sci. USA* **95**, 12208–12213.
48. Blumberg, B., Sabbagh, W., Jr., Juguilon, H., Bolado, J., Jr., van Meter, C. M., Ong, E. S., and Evans, R. M. (1998). *Genes Dev.* **12**, 3195–3205.
49. Kliewer, S. A., Moore, J. T., Wade, L., Staudinger, J. L., Watson, M. A., Jones, S. A., McKee, D. D., Oliver, B. B., Willson, T. M., Zetterstrom, R. H., Perlmann, T., and Lehmann, J. M. (1998). *Cell* **92**, 73–82.
50. Lehmann, J. M., McKee, D. D., Watson, M. A., Willson, T. M., Moore, J. T., and Kliewer, S. A. (1998). *J. Clin. Invest.* **102**, 1016–1023.
51. Jones, S. A., Moore, L. B., Shenk, J. L., Wisely, G. B., Hamilton, G. A., McKee, D. D., Tomkinson, N. C., LeCluyse, E. L., Lambert, M. H., Willson, T. M., Kliewer, S. A., and Moore, J. T. (2000). *Mol. Endocrinol.* **14**, 27–39.
52. Dussault, I., Lin, M., Hollister, K., Wang, E. H., Synold, T. W., and Forman, B. M. (2001). *J. Biol. Chem.* **276**(36), 33309–33312.
53. Synold, T. W., Dussault, I., and Forman, B. M. (2001). *Nature Med.* **7**, 584–590.
54. Moore, L. B., Goodwin, B., Jones, S. A., Wisely, G. B., Serabjit-Singh, C. J., Willson, T. M., Collins, J. L., and Kliewer, S. A. (2000). *Proc. Natl. Acad. Sci. USA* **97**, 7500–7502.
55. Watkins, R. E., Wisely, G. B., Moore, L. B., Collins, J. L., Lambert, M. H., Williams, S. P., Willson, T. M., Kliewer, S. A., and Redinbo, M. R. (2001). *Science* **292**, 2329–2333.
56. Anzenbacher, P. and Anzenbacherova, E. (2001). *Cell Mol. Life Sci.* **58**, 737–747.
57. Gerbal-Chaloin, S., Pascussi, J. M., Pichard-Garcia, L., Daujat, M., Waechter, F., Fabre, J. M., Carrere, N., and Maurel, P. (2001). *Drug Metab. Dispos.* **29**, 242–251.
58. Ambudkar, S. V., Dey, S., Hrycyna, C. A., Ramachandra, M., Pastan, I., and Gottesman, M. M. (1999). *Annu. Rev. Pharmacol. Toxicol.* **39**, 361–398.
59. Sparreboom, A., van Asperen, J., Mayer, U., Schinkel, A. H., Smit, J. W., Meijer, D. K., Borst, P., Nooijen, W. J., Beijnen, J. H., and van Tellingen, O. (1997). *Proc. Natl. Acad. Sci. USA* **94**, 2031–2035.
60. Staudinger, J. L., Goodwin, B., Jones, S. A., Hawkins-Brown, D., MacKenzie, K. I., LaTour, A., Liu, Y., Klaassen, C. D., Brown, K. K., Reinhard, J., Willson, T. M., Koller, B. H., and Kliewer, S. A. (2001). *Proc. Natl. Acad. Sci. USA* **98**, 3369–3374.
61. Ernst, E. (2002). *Arch. Surg.* **137**, 316–319.
62. Turton-Weeks, S. M., Barone, G. W., Gurley, B. J., Ketel, B. L., Lightfoot, M. L., and Abul-Ezz, S. R. (2001). *Prog. Transplant* **11**, 116–120.
63. Hennessy, M., Kelleher, D., Spiers, J. P., Barry, M., Kavanagh, P., Back, D., Mulcahy, F., and Feely, J. (2002). *Br. J. Clin. Pharmacol.* **53**, 75–82.
64. Barone, G. W., Gurley, B. J., Ketel, B. L., and Abul-Ezz, S. R. (2001). *Transplantation* **71**, 239–241.
65. Markowitz, J. S., DeVane, C. L., Boulton, D. W., Carson, S. W., Nahas, Z., and Risch, S. C. (2000). *Life Sci.* **66**, L133–139.
66. Choi, H. S., Chung, M., Tzameli, I., Simha, D., Lee, Y. K., Seol, W., and Moore, D. D. (1997). *J. Biol. Chem.* **272**, 23565–23571.
67. Wei, P., Zhang, J., Egan-Hafley, M., Liang, S., and Moore, D. D. (2000). *Nature* **407**, 920–923.
68. Forman, B. M., Tzameli, I., Choi, H. S., Chen, J., Simha, D., Seol, W., Evans, R. M., and Moore, D. D. (1998). *Nature* **395**, 612–615.
69. Honkakoski, P., Zelko, I., Sueyoshi, T., and Negishi, M. (1998). *Mol. Cell Biol.* **18**, 5652–5658.
70. Honkakoski, P., Moore, R., Washburn, K. A., and Negishi, M. (1998). *Mol. Pharmacol.* **53**, 597–601.
71. Kawamoto, T., Sueyoshi, T., Zelko, I., Moore, R., Washburn, K., and Negishi, M. (1999). *Mol. Cell Biol.* **19**, 6318–6322.
72. Tzameli, I., Pissios, P., Schuetz, E. G., and Moore, D. D. (2000). *Mol. Cell Biol.* **20**, 2951–2958.
73. Xie, W., Barwick, J. L., Simon, C. M., Pierce, A. M., Safe, S., Blumberg, B., Guzelian, P. S., and Evans, R. M. (2000). *Genes Dev.* **14**, 3014–3023.
74. Moore, L. B., Parks, D. J., Jones, S. A., Bledsoe, R. K., Consler, T. G., Stimmel, J. B., Goodwin, B., Liddle, C., Blanchard, S. G., Willson, T. M., Collins, J. L., and Kliewer, S. A. (2000). *J. Biol. Chem.* **275**, 15122–15127.
75. Xie, W., Barwick, J. L., Downes, M., Blumberg, B., Simon, C. M., Nelson, M. C., Neuschwander-Tetri, B. A., Brunt, E. M., Guzelian, P. S., and Evans, R. M. (2000). *Nature* **406**, 435–439.

CHAPTER 278

Identification of Ligands for Orphan Nuclear Receptors

Steven A. Kliewer and Timothy M. Willson
Nuclear Receptor Discovery Research, GlaxoSmithKline
Research Triangle Park, North Carolina

Introduction

The nuclear receptors comprise a superfamily of transcription factors that are activated by small, lipophilic chemicals including the classic steroid, retinoid, and thyroid hormones. There are 48 members of this family encoded by the human genome [1], far more receptors than established lipophilic hormones. The work of a number of laboratories has shown that many of the remaining "orphan" members of the family function as metabolic sensors, linking physiological levels of intracellular metabolites to the transcriptional regulation of key biochemical pathways. In this brief review, we highlight recent progress that has been made in the identification of ligands for orphan nuclear receptors. With the exception of the hepatocyte nuclear receptor 4 (HNF4, NR2A) and estrogen related receptor (ERR; NR3B) subtypes, all of the receptors discussed in this review bind to their cognate DNA response elements as heterodimers with the 9-*cis* retinoic acid receptors (RXR, NR2B).

PPARs: Fatty Acid Sensors

A wealth of evidence now exists to support roles for the peroxisome proliferator-activated receptors (PPARs; NR1C) in lipid and carbohydrate metabolism [2]. Three PPAR subtypes are encoded by distinct genes: PPARα, PPARγ, and PPARδ. Each PPAR subtype has a unique biology. PPARα is abundantly expressed in the liver, kidney, heart, and muscle and plays a central role in fatty acid oxidation. PPARγ is expressed at high levels in adipose and has a prominent role in fat cell formation and fatty acid storage. PPARδ is expressed in most tissues and was recently shown to regulate systemic cholesterol and lipid homeostasis.

Synthetic ligands have provided tremendous insight into the functions of the PPARs [2]. PPARα is the molecular target for the fibrates (Fig. 1), a class of drugs used to lower triglyceride levels in dyslipidemic patients. PPARγ is the target for the thiazolidinedione drugs (Fig. 1), which enhance insulin sensitivity and lower glucose levels in patients with type II diabetes. Although there are no marketed drugs targeted against PPARδ, a potent, synthetic PPARδ agonist was recently shown to increase serum levels of high-density lipoprotein and to decrease fasting triglyceride and insulin levels in insulin resistant nonhuman primates [3]. Thus, PPARδ agonists may have utility in the treatment of cardiovascular disease associated with metabolic syndrome X.

All three PPAR subtypes are activated in cell-based reporter assays by a variety of fatty acids and fatty acid metabolites [2]. Many of these fatty acids bind to the receptors at low micromolar concentrations, which are consistent with their levels in serum. These data suggest that rather than having a single, high-affinity ligand, the PPARs may function as more generalized sensors of free fatty acid levels. Recently, a higher affinity (K_d~40 nM) PPARγ ligand, hexadecyl azelaoyl phosphatidylcholine, was extracted from oxidized low-density lipoprotein [4]. Thus, higher affinity natural ligands may also exist for the PPARs.

LXRs: Cholesterol Sensors

Cholesterol is an essential component of cell membranes and serves as the precursor to the steroid hormones and

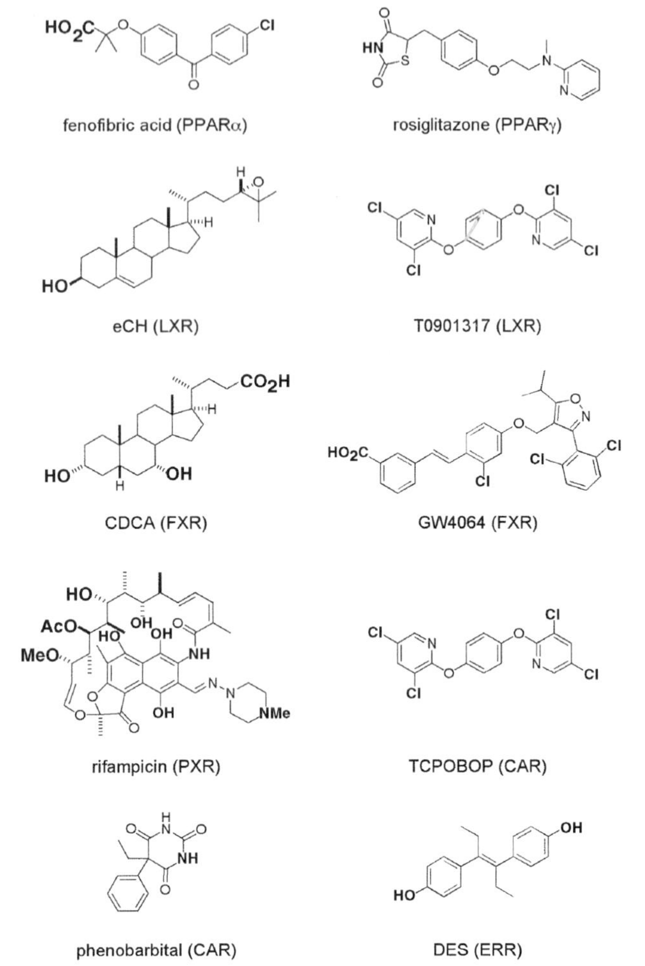

Figure 1 Orphan nuclear receptor ligands. The chemical structures of representative ligands for orphan nuclear receptors are shown.

bile acids. However, excess cholesterol is associated with diseases such as atherosclerosis and the formation of gallstones. Thus, a delicate balance must be maintained between the ingestion and synthesis of cholesterol and its elimination from the body. It is now established that the two liver X receptor (LXR; NR1H) subtypes, LXRα and LXRβ, play important roles in the elimination of excess cholesterol from the body [5]. The LXRs regulate a variety of genes involved in cholesterol metabolism including CYP7A1, which catalyzes the rate-limiting step in the conversion of cholesterol to bile acids in the liver, and the ATP-binding cassette (ABC) proteins A1 and G1, which mediate the efflux of cholesterol from enterocytes and macrophages for elimination from the body. Thus, the LXRs are components of a physiological regulatory system that senses excess cholesterol and modifies gene expression accordingly.

The natural ligands for the LXRs are oxidized cholesterol derivatives including 24(*S*),25-epoxycholesterol (eCH) (Fig. 1), 22(*R*)-hydroxycholesterol, and 24(*S*)-hydroxycholesterol [5,6]. Each of these compounds has a distinct tissue distribution profile. eCH is produced in the liver from a shunt of the mevalonate pathway in which squalene epoxide is metabolized to the diepoxide prior to undergoing cyclization. eCH is present in extracts from rat livers at 1–5 μM concentrations and rises upon cholesterol feeding. 22(*R*)-hydroxycholesterol is present at micromolar concentrations in extracts of the adrenals. Interestingly, 24(*S*)-hydroxycholesterol is generated exclusively in the brain. Its abundance in this tissue has led to it being termed *cerebrosterol.* Recently, the first potent, selective synthetic LXR agonist (T0901317) (Fig. 1) was described [7]. T0901317 binds and activates both LXR subtypes at nanomolar concentrations.

FXR: Bile Acid Sensor

Bile acids are essential for the elimination of cholesterol from the body and the solubilization and transport of lipids in the intestine. However, bile acids are also detergents that are extremely toxic at high concentrations. Thus, their levels must be tightly regulated. In 1998, several groups reported that the farnesoid X receptor (FXR, NR1H4) is a bile acid receptor [5,8]. Several different naturally occurring bile acids bind and activate FXR at physiological concentrations including chenodeoxycholic acid (Fig. 1), cholic acid, deoxycholic acid, and lithocholic acid and/or their glycine or taurine conjugates. The first potent, selective nonsteroidal FXR agonist, GW4064, was recently described (Fig. 1) [9]. FXR is highly expressed in tissues that are critical in bile acid biology including the liver, intestine, and kidney, and it regulates the expression of genes involved in bile acid homeostasis including the intestinal bile acid binding protein, the bile salt export pump, and CYP7A1 [10].

PXR and CAR: Xenobiotic Sensors

The body must protect itself against myriad xenobiotics ingested in the diet or otherwise absorbed. Two nuclear receptors, the pregnane X receptor (PXR, NR1I2) and the constitutive androstane receptor (CAR, NR1I3), have evolved to detect xenobiotics as part of the body's detoxification machinery [11]. Both receptors are highly expressed in the liver and intestine and regulate the expression of a number of genes involved in the hydroxylation, conjugation, and transport of xenobiotics. PXR is activated by a remarkable diverse collection of natural and synthetic chemicals including macrocyclic antibiotics (e.g., rifampicin; Fig. 1), antimycotics (e.g., clotrimazole), steroids (e.g., dexamethasone), and the herbal antidepressant St. John's wort. Because it is activated by many widely used prescription medicines, PXR activation represents the basis for a common class of drug–drug interaction, in which one drug stimulates the metabolism of another. Although CAR can also function as a xenobiotic sensor, it differs from PXR in two important respects. First, CAR is sequestered in the cytoplasm in its inactive state. Second, once it has entered the nucleus, CAR has a high basal level of transcriptional activity even in the absence of an exogenous ligand. CAR can be activated by compounds such as phenobarbital (Fig. 1), which promote its translocation from the cytoplasm into the nucleus without binding directly to CAR. However, once in the nucleus, CAR can also be modulated by compounds that bind directly to the receptor such as the androstanols, which suppress CAR's high basal activity, or TCPOBOP (Fig. 1), which further increases CAR's transcriptional activity [12].

Ligands for Other Orphan Nuclear Receptors

Recently, ligands have been identified for the three ERR subtypes (α, β, γ) [13,14] and two mammalian HNF4 subtypes (α, γ) [15]. All of these receptors have high basal activity in cell-based reporter assays in the absence of exogenous ligands. The three ERR subtypes are expressed in a number of different tissues and bind to DNA efficiently as monomers. The basal transcriptional activity of all three ERR subtypes can be directly suppressed by the estrogen receptor agonist diethylstilbestrol (Fig. 1), and the estrogen receptor antagonist 4-hydroxytamoxifen can efficiently deactivate ERRγ [13,14]. These data suggest that the ERRs may function as receptors for estrogens or related steroids. The HNF4s are known to bind to DNA as homodimers and regulate genes involved in lipid and carbohydrate homeostasis [16]. X-ray crystallography and mass spectroscopy revealed that both HNF4 subtypes bind to saturated and monounsaturated fatty acids with chain lengths of 14–16 carbon atoms [15]. The transcriptional activity of HNF4α has also been reported to be higher in the presence of fatty acids in cell-based reporter assays [17]. However, unlike other nuclear receptors, the HNF4 subtypes do not appear to readily exchange their ligands [15]. Thus, HNF4 may define a new class of constitutively active nuclear receptors in which the lipophilic ligand serves as an integral component of the receptor complex rather than a reversible switch.

Conclusion

During the past decade, ligands have been identified for more than a dozen of the orphan nuclear receptors. Many of these are now known to be receptors for key intracellular metabolites including fatty acids and cholesterol derivatives. Others serve as sentinels in the body's defense mechanism against xenobiotics. Studies of the ERRs suggest that additional steroid hormones may yet exist to be discovered. Finally, recent studies with HNF4 raise the possibility that some orphan nuclear receptors have high constitutive levels of transcriptional activity because they bind irreversibly to their lipophilic ligands. The availability of natural and synthetic ligands will aid tremendously in unraveling the biological functions of these orphan nuclear receptors.

References

1. Maglich, J. M., Sluder, A., Guan, X., Shi, Y., McKee, D. D., Carrick, K., Kamdar, K., Willson, T. M., and Moore, J. T. (2001). *Genome Biol.* 2, RESEARCH0029.
2. Willson, T. M., Brown, P. J., Sternbach, D. D., and Henke, B. R. (2000). *J. Med. Chem.* **43**, 527–550.
3. Oliver, W. R., Jr., Shenk, J. L., Snaith, M. R., Russell, C. S., Plunket, K. D., Bodkin, N. L., Lewis, M. C., Winegar, D. A., Sznaidman, M. L., Lambert, M. H., Xu, H. E., Sternbach, D. D., Kliewer, S. A., Hansen, B. C., and Willson, T. M. (2001). *Proc. Natl. Acad. Sci. USA* **98**, 5306–5311.
4. Davies, S. S., Pontsler, A. V., Marathe, G. K., Harrison, K. A., Murphy, R. C., Hinshaw, J. C., Prestwich, G. D., Hilaire, A. S., Prescott, S. M., Zimmerman, G. A., and McIntyre, T. M. (2001). *J. Biol. Chem.* **276**, 16015–16023.
5. Lu, T. T., Repa, J. J., and Mangelsdorf, D. J. (2001). *J. Biol. Chem.* **17**, 17.
6. Schroepfer, G. J., Jr. (2000). *Physiol. Rev.* **80**, 361–554.
7. Schultz, J. R., Tu, H., Luk, A., Repa, J. J., Medina, J. C., Li, L., Schwendner, S., Wang, S., Thoolen, M., Mangelsdorf, D. J., Lustig, K. D., and Shan, B. (2000). *Genes Dev.* **14**, 2831–2838.
8. Russell, D. W. (1999). *Cell.* **97**, 539–542.
9. Maloney, P. R., Parks, D. J., Haffner, C. D., Fivush, A. M., Chandra, G., Plunket, K. D., Creech, K. L., Moore, L. B., Wilson, J. G., Lewis, M. C., Jones, S. A., and Willson, T. M. (2000). *J. Med. Chem.* **43**, 2971–2974.
10. Sinal, C. J., Tohkin, M., Miyata, M., Ward, J. M., Lambert, G., and Gonzalez, F. J. (2000). *Cell* **102**, 731–744.
11. Waxman, D. J. (1999). *Arch. Biochem. Biophys.* **369**, 11–23.
12. Tzameli, I. and Moore, D. D. (2001). *Trends Endocrinol. Metab.* **12**, 7–10.
13. Coward, P., Lee, D., Hull, M. V., and Lehmann, J. M. (2001). *Proc. Natl. Acad. Sci. USA* **98**, 8880–8884.
14. Tremblay, G. B., Kunath, T., Bergeron, D., Lapointe, L., Champigny, C., Bader, J. A., Rossant, J., and Giguere, V. (2001). *Genes Dev.* **15**, 833–838.
15. Miller, A. B., Wisely, G. B., Davis, R. G., Thornquest, A. D., Johnson, R., Spitzer, T., Sefler, A., Shearer, B., Moore, J. T., Miller, A. B., Willson, T. M., and Williams, S. (2001). Submitted.
16. Hayhurst, G. P., Lee, Y. H., Lambert, G., Ward, J. M., and Gonzalez, F. J. (2001). *Mol. Cell Biol.* **21**, 1393–1403.
17. Hertz, R., Magenheim, J., Berman, I., and Bar-Tana, J. (1998). *Nature* **392**, 512–516.

Chapter 280

Cross-Talk between Nuclear Receptors and Other Transcription Factors

Peter Herrlich
Forschungszentrum Karlsruhe, Institute of Toxicology and Genetics, Karlsruhe, Germany

Introduction

All cellular processes depend on extracellular cues ranging from nutrient and oxygen supply, through cell density and contact with adverse agents, to highly specific stimuli originating within the multicellular organism. The extracellular cues are sensed by receptors and converted into responses through the intracellular signaling network. To maintain homeostasis of the cell and of the multicellular organism, the reactions to extracellular cues must be well controlled. Any change of condition requires finding a new balance of life processes achieved by constant readjustment of the cellular signaling network. Each stimulatory pathway is, therefore, counterbalanced by inhibitory components. For instance, receptor tyrosine kinases are associated with negatively regulating protein tyrosine phosphatases, which keep receptor autophosphorylation at a low level in the absence of ligand and limit the duration of activity on activation of the receptors. Similarly, dual-specificity kinases within pathways are balanced by dual-specificity phosphatases whose abundance or activities are up-regulated together with the pathway, thus limiting the extent and duration of stimulation. Upon mitogenic stimulation, cells enter and proceed through the cell cycle, itself an ordered circuit of consecutive positive and negative elements.

One can easily imagine what would happen if the control circuits did not operate: A wound-healing reaction would not stop once the damage was repaired; it would instead form excessive keloid material (scar tissue). Excessive expression and activation of metalloproteases in response to inflammatory cytokines would cause various types of tissue damage, as seen in rheumatoid arthritis. The expansion of T and B lymphocytes upon antigen-specific stimulation would lead to leukemia-like cell numbers unless negative control mechanisms and ultimately induced apoptosis limited the expansion. A bacterial invasion would lead to uncontrolled and long-lasting release of tumor necrosis factor alpha (TNF-α), for example, by macrophages, which results in septic shock and death of the organism.

This chapter deals with a particular organismic control mechanism established by hormones. Glucocorticoid hormones or retinoic acid as well as vitamin D counteract proliferative, inflammatory, and immune responses. This is why glucocorticoids are widely used in treating unwanted inflammatory reactions, asthma, autoimmune diseases, and leukemia. Vitamin D is applied to psoriatic skin to block proliferation. Retinoids are being used for certain types of cancer. The presumable mechanistic principles were discovered in 1990 and in the years thereafter. The nuclear receptors, which are activated by the hormone ligands, interfere with several of the signal transduction and transcription steps that are relevant for the panel of reactions mentioned. The mechanisms of interference with components of signal transduction are briefly reviewed here before describing in more detail the most interesting mechanism: *cross-talk* between transcription factors.

Proliferative and Proinflammatory Pathways

If a process is to interrupt the immune response or unwanted proliferation, or even inhibit cancer cells, which signaling pathways and which gene expressions need to be addressed? Obviously, we need to address pathways and genes that turn on the cell cycle, which cause the release of cytokines and which control cellular migration, adhesion, and invasiveness. It is these properties that lymphocytes, macrophages, dendritic cells, and other inflammatory cells acquire in order to "squeeze" through endothelial barriers, to invade tissues, and to home into lymphoid organs. Cancer cells share many of these phenotypic features. Knock-out technology has revealed some of the pathways and transcription factors relevant for these processes in immune cells (see, e.g., [1]). Members of the Rel family of transcription factors, of the AP-1 family as well as NF-AT, Oct-1, and C/EBP, fulfill central functions in immune cells upon activation. These factors or the signaling pathways leading to their activation need to be targeted in order to inhibit the responses. AP-1 and NF-κB are driven also by oncogenic pathways. Inhibiting their action or activation would be antiproliferative [2–5].

For instance, components of the bacterial cell wall such as lipopolysaccharides (LPS) activate toll-like receptors that are linked to the protein kinase complex (IKK) phosphorylating I-kBα [6,7]. I-κBα keeps NF-κB in an inactive cytoplasmic state [5,8,9]. Upon phosphorylation, I-κBα is subjected to degradation and NF-κB is released and transported into the nucleus, where it acts on the promoters of numerous cytokine genes. In the case of macrophages the toll-like receptors trigger the synthesis and release of TNF-α. Important for the mechanism discussed here, nonlethal (to wild-type mice) doses of LPS become lethal in adrenalectomized mice (unable to produce glucocorticoid hormone [10]). Inflammatory cytokines can pass through the blood–brain barrier and exert central nervous system effects [11,12], including the stimulation of the hypothalamic–pituitary–adrenal axis, which leads to release of glucocorticoids from the adrenal gland if present [13].

Another central pathway involved in growth factor transcription as well as in transcription of cyclin D1, the entry step into the cell cycle, and of adhesion and migration components, originates from the small G protein Ras. The Ras– MEK–Erk pathway regulates a wide variety of features essential for the immune response as well as for a proliferative and invasiveness program. For instance, the pathway leads to expression of AP-1 subunits as well as to their activation. AP-1 and NF-κB control the synthesis of several cytokines in T cells and macrophages. Interestingly, the Ras–MEK–Erk pathway controls also alternative splicing [14], yielding different CD44 transcripts and subsequently CD44 proteins that, in turn, are required for specific activation of certain receptor tyrosine kinases [15]. Thus, interference with inflammation, proliferation, and immune responses should address predominantly the pathways leading to AP-1 and NF-κB activity.

Nuclear Receptors

The mediators of hormone-dependent interference with proliferative and inflammatory responses belong to the family of nuclear receptors that share several structural features [16,17]. Nuclear receptors are bona fide transcription factors whose action is turned on by the specific hormone ligand. The C-terminal half carries the ligand-binding domain and a transactivation domain (AF-2) whose accessibility depends on the presence of the proper ligand. AF-2 assembles coactivator complexes necessary for initiation of transcription. The monomeric nuclear receptor does not bind sufficiently to DNA; rather homodimerization (e.g., in the case of steroid hormone receptors) or heterodimerization with a common subunit, RXR (of the receptors for vitamin D, for thyroid hormone, and for retinoids) is required. A conserved structure in the middle of the nuclear receptor molecule is folded by coordinating zinc atoms into two finger-like structures, the N terminal of which confers, together with its dimer partner, binding to the specific promoter element. The C-terminal zinc finger mediates protein interactions; for example, through the extended so-called D-loop, it mediates homodimerization of steroid hormone receptors. The N-terminal half carries another transactivation domain (AF-1) that is ligand independent but requires modifications by other pathways. Recent evidence suggests that modifications by Erk or Rsk-1 enhance the transcriptional function of some nuclear receptors (e.g., ERα carries an Erk site at amino acid 118 [18] and a pp90rsk1 site at amino acid 167 [19,20]). Steroid hormone receptors are associated with chaperones that, in addition to other functions, are needed to keep the ligand-binding domain in a steroid hormone accepting conformation. Often the steroid hormone receptors are cytoplasmic in the absence of ligand [21].

Induced Expression of Inhibitory Molecules

The most straightforward explanation for an inhibitory role of nuclear receptors would be the hormone-induced synthesis of an inhibitory molecule. To what extent the transcription of such effector molecules plays a physiological role is still a matter of debate. Removal of the adrenal gland leads to thymic hypertrophy suggestive of reduced cell death because the thymus is the organ of high cellular apoptosis. Glucocorticoids induce apoptosis in isolated thymocytes for which proper transcriptional function of the GR is required [22,23]. Also several other nuclear receptors appear to exert functions in the thymus [24]. It is therefore plausible that nuclear receptors are involved in the control of synthesis of proapoptotic transcripts. Nevertheless, it is still unclear to what extent this is physiologically relevant [24].

Induced inhibitors of transcription and of signal transduction have been identified. In several types of cells, the abundance of the inhibitor of NF-κB, I-κBα, is enhanced upon treatment of cells with glucocorticoid hormone [10,23,25]. As we will see, the induction requires the presence of a dimerization-competent glucocorticoid receptor. One could imagine that enhanced expression of I-κBα terminates the action of NF-κB and thus the release of cytokines. (For a recent discussion of NF-κB:I-κB shuttling between cytoplasm and nucleus see [5].) Another glucocorticoid receptor target is MKP-1, the phosphatase acting on Erk. The hormone-enhanced abundance (by both increased transcription and reduced turnover [26]) of MKP-1 counteracts the activation state of Erk. As one would expect for induced new synthesis, the inhibition occurs with a delay of more than 16 hr. Thus, the induced expression of inhibitory proteins represents one type of mechanism explaining how a hormone can terminate inflammatory processes. The induced synthesis will not lead to an instantaneous block.

Immediate Hormone Responses

Most of the anti-inflammatory and antiproliferative action of nuclear receptors does not depend on prior protein synthesis. Such immediate action implies that the liganded receptor exerts a molecular function by direct interaction with another component, for example, DNA or protein. The function could be exerted in the cytoplasm or in the nucleus. The steroid hormone receptors are associated with chaperones that keep the proteins in ligand-susceptible conformation, and they share prolonged periods of cytoplasmic localization. Thus, a cytoplasmic action would be feasible although difficult to reconcile with hormone-dependent activation of cytoplasmic components: The bulk of ligand-activated receptor is located in the nucleus, whereas signaling components including JNK remain by and large cytoplasmic [27,28]. Controversy still surrounds questions as to what extent and how nuclear receptors affect signal transduction pathways directly. Both inhibitory and stimulatory actions have been reported, the kinetics of which are often too slow to be taken as direct actions (e.g., 50% inhibition of JNK activity [29–32], Erk [33,34], p38 [35]). One of the signaling molecules addressed is JNK, the enzyme activating the AP-1 subunit Jun. Interference with JNK is possibly a nuclear event [28]. The more convincing data, which are based on the use of gene disruptions and coprecipitations, classify interactions with components of the signaling network as costimulatory. The associations, for instance, of the androgen or estrogen receptors with Src, PI3Kp85, Fyn, and other signaling molecules appear to enhance nuclear receptor function in transcription [36,37].

The inhibitory action of nuclear receptors is by and large nuclear. Two types of mechanisms for immediate inhibition appear to emerge: blocking of transcription by binding to a promoter element, and modulation of other transcription factors without obligatory binding of the nuclear receptor to DNA. The repressor function at the promoter element recognized by the nuclear receptor has been best documented for the thyroid receptor. In the nonliganded state, the receptor recruits corepressor molecules such as N-CoR and histone deacetylases [38]. Hormone binding converts the receptor into the transactivating molecule. Less well understood is the action of the glucocorticoid receptor on so-called nGREs, negative glucocorticoid response elements [39]. nGREs seem to play roles in the autoregulatory circuit of the hypothalamus–pituitary–adrenal axis, for example, in the transcription of the genes for promelanocorticotropin and prolactin. The binding to nGREs appears to require dimerization or multimerization of the glucocorticoid receptor [39]. nGREs possibly induce a conformational change such that the GR cannot recruit the complex of components required for transcriptional initiation and instead binds inhibitory cofactors.

The modulation of transcription factors without direct binding to DNA is the more predominant mechanism relevant for the control of proliferation and of the immune reactions. It is also the most intriguing mechanism and presumably the evolutionarily more ancient one [40]. All nuclear receptors (perhaps with the exception of the mineralocorticoid receptor) appear to share this ability [41–48]. The evidence for the glucocorticoid receptor for which most data have been collected is described next.

Direct Modulation of Transcription Factors

Mutual Modulation, A Method of Maintaining Homeostasis

Rapid repression of phorbol-ester-induced transcription of collagenase (MMP-13) absence of phorbol synthesis led to the discovery of transcription factor modulation by nuclear receptors [41]. The collagenase promoter fragments sufficient for induced transcription and hormone-dependent repression comprised neither GRE or nGRE sequence elements. Similarly, the repressible promoter for interleukin-2 carried no GRE or nGRE [10]. These data led to the recognition that nuclear receptors interfere with other transcription factors. The glucocorticoid receptor modulates negatively the transcription by CREB [4,49], by the Jun:Fos heterodimer (AP-1) [41–43], and by, for example, Oct-1 [50], Spi-1/PU-1 [51], GATA-1 [52], and NF-κB [53]. This list is probably not complete. As introduced earlier, interference with AP-1 and NF-κB function is certainly most relevant in the inhibition of inflammation and of proliferation [3,54]. Interestingly, the modulation is mutual, liganded glucocorticoid receptor down-modulating AP-1- and NF-κB-dependent transcriptions, activated AP-1 or p65 (NF-κB) inhibiting GRE-dependent transcription [41]. Interestingly, less hormone was required for repression of AP-1 than for induction of a GRE-containing promoter [41]. Accordingly, the program of gene expression can be modeled by a fine-tuned transition from totally AP-1 (NF-κB, or other factor) dependent transcription and complete inhibition of GRE promoters, through increasing repression of AP-1 and decreasing repression of GRE promoters to a totally hormone-dependent program [55]. This adjustment would be created by the relative concentrations of hormone activating the nuclear receptor versus growth factors activating AP-1 or NF-κB [55].

Tethering

The mutual modulation of transcription factors AP-1 or NF-κB by GR and the reverse is based on protein–protein interactions. By means of coimmunoprecipitations, low-affinity interactions between nuclear receptors and AP-1 were detected [41–43,56]. Although initially thought that mutual binding would interfere with DNA binding, this assumption turned out to be wrong. According to genomic footprints [4,57] and chromatin immunoprecipitations [58], the inhibited AP-1 (at the MMP-13 promoter) or NF-κB (at the promoters for interleukins 2 and 8, as well as for ICAM-1) remain bound to their promoter elements [4,57,58]. Chromatin immunoprecipitation proved that the glucocorticoid receptor was associated with the complex. The molecules "tethered" together apparently exert molecular properties different from those as individual factors; that is, they "forget" to be positive transcription factors [59,60].

This hypothesis is not far-fetched. Theoretical and experimental arguments support that the binding to a specific DNA element determines the conformation of the partners: of DNA as well as of the protein factor [60]. It is thus not too surprising that protein–protein tethering also determines protein function. Synergy has been found for the tethering of nuclear receptors with AP-1 or NF-κB in certain tissues [4,47]. Interestingly Jun : Jun homodimer function can be enhanced by glucocorticoids [4,56]. It is not clear, however, when Jun homodimers are formed under physiological conditions. In the organism, however, the negative interference of GR with AP-1 and NF-κB appears to predominate.

Dissociation of GRE Promoter Activity from Tethering Action

I turn now to a discussion of the dissociation of GRE promoter activity from the tethering action, mainly by the dimerization-defective GR, and to cross-talk as the major mechanism of action for the anti-inflammatory and immune-suppressive functions of the glucocorticoid receptor.

What was already suggested by the different hormone concentration requirements became obvious by mutations created in the glucocorticoid receptor GR [61]. One mutation in particular was most informative: a point mutation in the D-loop responsible for subunit dimerization at the GRE element. This mutation was introduced into the mouse by knock-in strategy. The resulting $GR^{dim/dim}$ mice carried a receptor that was unable to bind to GREs and to activate GRE promoters, but could perfectly repress AP-1 and NF-κB promoters [22,23]. The $GR^{dim/dim}$ mice were viable, suggesting that the induction of hormone-dependent genes was not essential for life under normal animal house conditions. The repressive action was, however, essential, a conclusion that could be derived from the nonviability of total GR knock-outs [62]. The dimerization-defective GR suffices to inhibit most proinflammatory responses [23] with the only exception being thymocyte apoptosis [63], which is counteracted by Bcl-2 in transgenic mice [64] and may involve the activation of caspase-9 [65]. Thus cross-talk between GR and AP-1 and NFκB is possibly the decisive mechanism in the control of the immune system and most relevant for the medical use of glucocorticoid hormones as anti-inflammatories. For instance, inhibitions by hormone of phorbol ester induced inflammation, of the acute phase response or of LPS-induced TNF-α release, were perfectly normal in the mutant mouse. The synthesis and release of macrophage and T-cell cytokines as well as transcription of the cyclooxygenase-2 gene were blocked to the same degree as in wild-type mice. The role of GR in thymocyte development, however, has not been resolved and remains controversial [23,24].

Transcriptional Initiation Steps That Could Be GR Targets

Ever since the discovery of transcription factors and their binding to elements at a fair distance to the TATA box or to the site of RNA polymerase and initiation complex assembly, researchers have been puzzled about how they might activate transcription. According to an early hypothesis, the DNA between the enhancing factor and the initiation complex loops out to bring all partners into proximity, where they provide an unknown start stimulus. Consequently, the modulation of AP-1 by GR was seen as interrupting the contact between AP-1 and the basic transcription complex. When additional factors were found—coactivators—competition for shared coactivators was considered as a mechanism for how a nuclear receptor could affect another transcription factor [47]. One of the coactivators, CBP/p300, is of low abundance, making competition for CBP a plausible proposal [66]. CBP indeed collaborates with both AP-1 and nuclear receptors. However, other CBP-binding transcription factors such as STATs or NF-κB should also compete for CBP, but do not repress AP-1. Even more convincingly, however, the activation domains of the receptors that recruit coactivators could be destroyed without any effect on cross-talk [67].

The discovery of corepressors and their association with histone deacetylase activity primed the hypothesis that recruitment of these negative counterplayers of coactivators may be responsible for hormone-dependent repression. However, inhibitors of histone deacetylase could not release repression [4,58,68].

What then is arrested and prevents transcription after the preinitiation complex has been formed? In chromatin immunoprecipitations, the GR-arrested preinitiation complex lacked one of the phosphorylations of the RNA polymerase II C-terminal tail [58]. This phosphorylation is thought to be important for the start of transcription and for elongation. Because the data are based on the specificity of an antibody, further confirmation is needed.

A current area of intensive research concerning gene expression addresses the state of the chromatin and its regulation. Transcribed genes are less densely packed, a state suggestive of better accessibility for protein factors. Regulated changes of chromatin accessibility occur prior to assembly of the preinitiation complex, during or after assembly. This order of remodeling may differ among genes [69,70]. It is therefore possible that the nuclear receptor prevents the remodeling step at genes that require such a process after formation of the preinitiation complex. Studies with chromatin immunoprecipitations have yet not explored this possibility.

Whatever the mechanism, yet unknown additional factors are likely to be involved. A cofactor isolated by two-hybrid screens promises to reveal such a factor (tentatively named protein #198; unpublished data). The protein carries three LIM domains, zinc-finger-like structures that mediate protein–protein interactions. One of the LIM domains binds specifically to either Fos (one of the AP-1 subunits) or p65 (the NF-κB subunit), while nuclear receptors interact with another LIM domain, offering the possibility that the protein forms a trimeric complex. Protein #198 carrying only one LIM domain acts as dominant-negative factor abolishing cross-talk between AP-1 and either GR, thyroid hormone receptor, or retinoic acid receptor. Loss of protein #198 by morpholino antisense oligonucleotide technique also prevented the cross-talk. What exactly this new cofactor does in conferring cross-talk modulation of transcription is currently being explored.

Conclusion

The cross-talk function of nuclear receptors is essential for adult life. It is exerted by interference with the activity of another transcription factor at a step after formation of the preinitiation complex. The cross-talk is mutual and offers a way to adjust pro- and antiproliferative transcription in an elegant and fine-tuned way. A new class of cofactors appears to be involved in the cross-talk. Because long-term systemic therapy with glucocorticoids can result in severe side effects such as osteoporosis and joint necroses, current efforts are directed toward distinguishing whether the side effects are caused by GRE-dependent genetic programs or whether they are intimately connected to the cross-talk function of GR, which is unavoidably used for these therapies. It may be possible to select ligands that activate only the cross-talk function.

References

1. Köntgen, F., Grumont, R. J., Strasser, A., Metcalf, D., Li, R., Tarlinton, D., and Gerondakis, S. (1995). Mice lacking the c-rel proto-oncogene exhibit defects in lymphocyte proliferation, humoral immunity, and interleukin-2 expression. *Genes Dev.* **9**, 1965–1977.
2. Angel, P. and Karin, M. (1991). The role of Jun, Fos and the AP-1 complex in cell proliferation and transformation. *Biochim. Biophys. Acta* **1072**, 129–157.
3. Barnes, P. J. and Karin, M. (1997). Nuclear factor-κB–A pivotal transcription factor in chronic inflammatory diseases. *New Engl. J. Med.* **336**, 1066–1071.
4. Herrlich, P. (2002). Cross-talk between glucocorticoid receptor and AP-1. *Oncogene* **20**, 2465–2475.
5. Gosh, S. and Karin, M. (2002). Missing pieces in the NF-κB puzzle. *Cell* **109**, S81–S96.
6. Hoshino, K., Takeuchi, O., Kawai, T., Sanjo, H., Ogawa, T., Takeda, Y., Takeda, K., and Akira, S. (1999). Cutting edge: Toll-like receptor 4 (TLR4)-deficient mice are hyporesponsive to lipopolysaccharide: Evidence for TLR4 as the LPS gene product. *J. Immunol.* **162**, 3749–3752.
7. Zhang, G. and Gosh, S. (2001). Toll-like receptor-mediated NF-κB activation: A phylogenetically conserved paradigm in innate immunity. *J. Clin. Invest.* **107**, 13–19.
8. Baldwin, A. S. (2001). Series introduction: The transcription factor NF-κB and human disease. *J. Clin. Invest.* **107**, 3–6.
9. Silverman, N. and Maniatis, T. (2001). NF-κB signaling pathways in mammalian and insect innate immunity. *Genes Dev.* **15**, 2321–2342.
10. Auphan, N., DiDonato, J. A., Rossette, C., Helmberg, A. and Karin, M. (1995). Immunosuppression by glucocorticoids: Inhibition of NF-κB activity through induction of I-κB synthesis. *Science* **270**, 286–290.
11. Besedovsky, H., del Rey, A., Sorkin, E., and Dinarello, C. A. (1986). Immunoregulatory feedback between interleukin-1 and glucocorticoid hormones. *Science* **233**, 652–654.
12. Karin, M. and Chang, L. (2001). AP-1-glucocorticoid receptor crosstalk taken to a higher level. *J. Endocrinol.* **169**, 447–451.
13. Chrousos, G. P. (1995). The hypothalamic–pituitary–adrenal axis and immune-mediated inflammation. *New Engl. J. Med.* **332**, 1351–1362.
14. König, H., Ponta, H., and Herrlich, P. (1988). Coupling of signal transduction to alternative pre-mRNA splicing by a composite splice enhancer. *EMBO J.* **17**, 2904–2013.
15. Orian-Rousseau, V., Chen, L. Sleeman, J. P., Herrlich, P., and Ponta, H. (2002). CD44 is required for two consecutive steps in HGF/c-Met signaling. *Genes Dev.* **16**(23), 3074–3086.
16. Mangelsdorf, D. J. and Evans, R. M. (1995). The RXR heterodimers and orphan receptors. *Cell* **83**, 841–850.
17. Beato, M., Herrlich, P., and Schütz, G. (1995). Steroid hormone receptors: Many actors in search of a plot. *Cell* **83**, 851–857.
18. Kato, S., Endoh, H., Masuhiro, Y., Kitamoto, T., Uchiyama, S., Sasaki, H., Masushige, S., Gotoh, Y., Nishida, E., Kawashima, H., Metzger, D., and Chambon, P. (1995). Activation of the estrogen receptor through phosphorylation by mitogen-activated protein kinases. *Science* **270**, 1491–1494.
19. Joel, P. B., Smith, J., Sturgill, T. W., Fisher, T. L., Blenis, J., and Lannigan, D. A. (1998). pp90^{rsk1} regulates estrogen receptor-mediated transcription through phosphorylation of Ser-167. *Mol. Cell Biol.* **18**, 1978–1984.
20. Tremblay, A., Tremblay, G. B., Labrie, F., and Giguère, V. (1999). Ligand-independent recruitment of SRC-1 to Estrogen Receptor β through phosphorylation of activation function AF-1. *Mol. Cell* **3**, 513–519.
21. Pratt, W. B. (1997). The role of the hsp90-based chaperone system in signal transduction by nuclear receptors and receptors signaling via MAP kinase. *Annu. Rev. Pharmacol. Toxicol.* **37**, 297–326.
22. Reichardt, H. M., Kaestner, K. H., Tuckermann, J., Kretz, O., Wessely, O., Bock, R., Gass, P., Schmid, W., Herrlich, P., Angel, P., and Schütz, G. (1998). DNA binding of the glucocorticoid receptor is not essential for survival. *Cell* **93**, 531–541.
23. Reichardt, H. M., Tuckermann, J. P., Göttlicher, M., Vujic, M., Weih, F., Angel, P., Herrlich, P., and Schütz, G. (2001). Repression of inflammatory responses in the absence of DNA binding by the glucocorticoid receptor. *EMBO J.* **20**, 7168–7173.
24. Winoto, A. and Littman, D. R. (2002). Nuclear hormone receptors in T lymphocytes. *Cell* **109**, S57–66.
25. Scheinman, R. I., Cogswell, P. C., Lofquist, A. K., and Baldwin, A. S. (1995). Role of transcriptional activation of I kappa B alpha in mediation of immunosuppression by glucocorticoids. *Science* **270**, 283–286.
26. Kassel, O., Sancono, A., Krätzschmar, J., Kreft, B., Stassen, M., and Cato, A. C. B. (2001). Glucocorticoids inhibit MAP kinase via increased expression and decreased degradation of MKP-1. *EMBO J.* **20**, 7108–7116.
27. Cavigelli, M., Dolfi, F., Claret, F. X., and Karin, M. (1995). Induction of c-fos expression through JNK-mediated TCF/Elk-1 phosphorylation. *EMBO J.* **14**, 5957–5964.

28. González, M. V., Jimenez, B., Berciano, M. T., González-Sancho, J. M., Caelles, C., Lafarga, M., and Munoz, A. (2000). Glucocorticoids antagonize AP-1 by inhibiting the activation/phosphorylation of JNK without affecting its subcellular distribution. *J. Cell Biol.* **150**, 1199–1208.
29. Caelles, C., González-Sancho, J. M., and Munoz, A. (1997). Nuclear hormone receptor antagonism with AP-1 by inhibition of the JNK pathway. *Genes Dev.* **11**, 3351–3364.
30. Swantek, J. L., Cobb, M. H., and Geppert, T. D. (1997). Jun N-terminal kinase/stress-activated protein kinase (JNK/SAPK) is required for lipopolysaccharide stimulation of tumor necrosis factor α (TNFα) translation: Glucocorticoids inhibit TNFα translation by blocking JNK/SAPK. *Mol. Cell Biol.* **17**, 6274–6282.
31. Rogatsky, I., Logan, S. K., and Garabedian, M. J. (1998). Antagonism of glucocorticoid receptor transcriptional activation by the c-Jun N-terminal kinase. *Proc. Natl. Acad. USA* **95**, 2050–2055.
32. Hirasawa, N., Sato, Y., Fujita, Y., Mue, S., and Ohuchi, K. (1998). Inhibition by dexamethasone of antigen-induced c-Jun N-terminal kinase activation in rat basophilic leukemia cells. *J. Immunol.* **161**, 4939–4943.
33. Rider, L. G., Hirasawa, N., Santini, F., and Beaven, M. A. (1996). Activation of the mitogen-activated protein kinase cascade is suppressed by low concentrations of dexamethasone in mast cells. *J. Immunol.* **157**, 2374–2380.
34. Hulley, P. A., Gordon, F., and Hough, F. S. (1998). Inhibition of mitogen-activated protein kinase activity and proliferation of an early osteoblast cell line (MBA 15.4) by dexamethasone: Role of protein phosphatases. *Endocrinology* **139**, 2423–2431.
35. Lasa, M., Brook, M., Saklatvala, J., and Clark, A. R. (2001). Dexamethasone destabilizes cyclooxygenase 2 mRNA by inhibiting mitogen-activated protein kinase p38. *Mol. Cell Biol.* **21**, 771–780.
36. Kousteni, S., Bellido, T., Plotkin, L. I., O'Brien, C. A., Bodenner, D. L., Han, L., Han, K., DiGregorio, G. B., Katzenellenbogen, J. A., Katzenellenbogen, B. S., Roberson, P. K., Weinstein, R. S., Jilka, R. L., and Manolagas, S. C. (2001). Nongenotropic, sex-nonspecific signaling through the estrogen or androgen receptors: Dissociation from transcriptional activity. *Cell* **104**, 719–730.
37. Song, R. X.-D., McPherson, R. A., Adam, L., Bao, Y., Shupnik, M., Kumar, R., and Santen, R. J. (2002). Linkage of rapid estrogen action to MAPK activation by ERα-Shc association and Shc pathway activation. *Mol. Endocrinol.* **16**, 116–127.
38. Glass, C. K. and Rosenfeld, M. G. (2000). The coregulator exchange in transcriptional functions of nuclear receptors. *Genes Dev.* **14**, 121–141.
39. Drouin, J., Sun, Y. L., Chamberland, M., Gauthier, Y., De, L. A., Nemer, M., and Schmidt, T. J. (1993). Novel glucocorticoid receptor complex with DNA element of the hormone-repressed POMC gene. *EMBO J.* **12**, 145–156.
40. Karin, M. (1998). New twists in gene regulation by glucocorticoid receptor: Is DNA binding dispensable? *Cell* **93**, 487–490.
41. Jonat, C., Rahmsdorf, H. J., Park, K. K., Cato, A. C. B., Gebel, S., Ponta, H., and Herrlich, P. (1990). Antitumor promotion and anti-inflammation: Downmodulation of AP-1 (Fos/Jun) activity by glucocorticoid hormone. *Cell* **62**, 1189–1204.
42. Schüle, R., Rangarajan, P., Kliewer, S., Ransone, L. J., Bolado, J., Yang, N., Verma, I. M., and Evans, R. M. (1990). Functional antagonism between oncoprotein c-Jun and the glucocorticoid receptor. *Cell* **62**, 1217–1226.
43. Yang-Yen, H. F., Chambard, J. C., Sun, Y. L., Smeal, T., Schmidt, T. J., Drouin, J., and Karin, M. (1990). Transcriptional interference between c-Jun and the glucocorticoid receptor: Mutual inhibition of DNA-binding due to direct protein–protein interaction. *Cell* **62**, 1205–1215.
44. Lafyatis, R., Kim, S. J., Angel, P., Roberts, A. B., Spors, M. B., Karin, M., and Wilder, R. L. (1990). Interleukin-1 stimulates and all-*trans*-retinoic acid inhibits collagenase gene expression through its 5'activator protein-1-binding site. *Mol. Endocrinol.* **4**, 973–980.
45. Nicholson, R. C., Mader, S., Nagpal, S., Leid, M., Rochette-Egly, C., and Chambon, P. (1990). Negative regulation of the rat stromelysin gene promoter by retinoic acid is mediated by an AP1 binding site. *EMBO J.* **9**, 4443–4454.
46. Doucas, K. V., Spyrou, G., and Yaniv, M. (1991). Unregulated expression of c-Jun or c-Fos proteins but not JunD inhibits oestrogen receptor activity in human breast cancer derived cells. *EMBO J.* **10**, 2237–2245.
47. Shemshedini, L., Knauthe, R., Sassone Corsi, P., Pornon, A., and Gronemeyer, H. (1991). Cell-specific inhibitory and stimulatory effects of Fos and Jun on transcription activation by nuclear receptors. *EMBO J.* **10**, 3839–3849.
48. Zhang, X. K., Wills, K. N., Husmann, M., Hermann, T., and Pfahl, M. (1991). Novel pathway for thyroid hormone receptor action through interaction with jun and fos oncogene activities. *Mol. Cell Biol.* **11**, 6016–6025.
49. Akerblom, I. E., Slater, E. P., Beato, M., Baxter, J. D., and Mellon, P. L. (1988). Negative regulation by glucocorticoids through interference with a cAMP responsive enhancer. *Science* **241**, 350–353.
50. Kutoh, E., Stromstedt, P. E., and Poellinger, L. (1992). Functional interference between the ubiquitous and constitutive octamer transcription factor 1 (OTF-1) and the glucocorticoid receptor by direct protein-protein interaction involving the homeo subdomain of OTF-1. *Mol. Cell Biol.* **12**, 4960–4969.
51. Gauthier, J. M., Bourachot, B., Doucas, V., Yaniv, M., and Moreau-Gachelin, F. (1993). Functional interference between the Spi-1/PU.1 oncoprotein and steroid hormone or vitamin receptors. *EMBO J.* **12**, 5089–5096.
52. Chang, T. J., Scher, B. M., Waxman, S., and Scher, W. (1993). Inhibition of mouse GATA-1 function by the glucocorticoid receptor: Possible mechanism of steroid inhibition of erythroleukemia cell differentiation. *Mol. Endocrinol.* **7**, 528–542.
53. Ray, A. and Prefontaine, K. E. (1994). Physical association and functional antagonism between the p65 subunit of transcription factor NF-κB and the glucocorticoid receptor. *Proc. Natl. Acad. Sci. USA* **91**, 752–754.
54. Ip, Y. T. and Davis, R. J. (1998). Signal transduction by the c-Jun N-terminal kinase (JNK)–from inflammation to development. *Curr. Opin. Cell Biol.* **10**, 205–219.
55. Herrlicfh, P., Jonat, C., Ponta, H., and Rahmsdorf, H. J. (1990). A major decision: Proliferation or differentiation. Interactions between pathway-specific transcription factors exemplify a molecular mechanism for the decision. In Bartram, C. R., Munk, K., and Schwab, M., Eds, *Oncogenes in Cancer Diagnosis,* pp. 1–9. Karger, Basel.
56. Diamond, M. A., Miner, J. N., Yoshinaga, S. K., and Yamamoto, K. R. (1990). Transcription factor interactions: Selectors of positive or negative regulation from a single DNA element. *Science* **249**, 1266–1272.
57. König, H., Ponta, H., Rahmsdorf, H. J., and Herrlich, P. (1992). Interference between pathway-specific transcription factors: Glucocorticoids antagonize phorbol ester-induced AP-1 activity without altering AP-1 site occupation *in vivo*. *EMBO J.* **11**, 2241–2246.
58. Nissen, R. M. and Yamamoto, K. R. (2000). The glucocorticoid receptor inhibits NFκB by interfering with serine-2 phosphorylation of the RNA polymerase II carboxy-terminal domain. *Genes Dev.* **14**, 2314–2329.
59. Saatcioglu, F., Claret, F. X., and Karin, M. (1994). Negative transcriptional regulation by nuclear receptors. *Sem. Cancer Biol.* **266**, 1719–1723.
60. Lefstin, J. A. and Yamamoto, K. R. (1998). Allosteric effects of DNA on transcriptional regulators. *Nature* **392**, 885–888.
61. Heck, S., Kullmann, M., Gast, A., Ponta, H., Rahmsdorf, H. J., Herrlich, P., and Cato, A. C. (1994). A distinct modulating domain in glucocorticoid receptor monomers in the repression of activity of the transcription factor AP-1. *EMBO J.* **13**, 4087–4095.
62. Cole, T. J., Blendy, J. A., Monaghan, A. P., Krieglstein, K., Schmid, W., Aguzzi, A., Fantuzzi, G., Hummler, E., Unsicker, K., and Schütz, G. (1995). Targeted disruption of the glucocorticoid receptor gene blocks adrenergic chromaffin cell development and severely retards lung maturation. *Genes Dev.* **9**, 1608–1621.
63. Godfrey, D. I., Purton, J. F., Boyd, R. L., and Cole, T. J. (2001). Glucocorticoids and the thymus: The view from the middle of the road. *Trends Immunol.* **22**, 243.

64. Chao, D. T., Linette, G. P., Boise, L. H., White, L. S., Thompson, C. B., and Korsmeyer, S. J. (1995). Bcl-XL and Bcl-2 repress a common pathway of cell death. *J. Exp. Med.* **182**, 821–828.
65. Kuida, K., Haydar, T. F., Kuan, C. Y., Gu, Y., Taya, C., Karasuyama, H., Su, M. S., Rakic, P., and Flavell, R. A. (1998). Reduced apoptosis and cytochrome c-mediated caspase activation in mice lacking caspase 9. *Cell* **94**, 325–337.
66. Kamei, Y., Xu, L., Heinzel, T., Torchia, J., Kurokawa, R., Gloss, B., Lin, S. C., Heyman, R. A., Rose, D. W., Glass, C. K., and Rosenfeld, M. G. (1996). A CBP integrator complex mediates transcriptional activation and AP-1 inhibition by nuclear receptors. *Cell* **85**, 403–414.
67. Schneider, S., Heilbock, C., Göttlicher, M., Herrlich, P., and Kassel, O. unpublished.
68. Herrlich, P and Göttlicher, M. (2002). The anti-inflammatory action of glucocorticoid hormones. *Ernst Schering Res Found Workshop* **40**, 297–304.
69. Soutoglou, E. and Talianidis, I. (2002). Coordination of PIC assembly and chromatin remodeling during differentiation-induced gene activation. *Science* **295**, 1901–1904.
70. Fry, C. J. and Peterson, C. L. (2002). Unlocking the gates to gene expression. *Science* **295**, 1847–1848.

CHAPTER 281

Drosophila Nuclear Receptors

Kirst King-Jones and Carl S. Thummel
Howard Hughes Medical Institute, University of Utah, Salt Lake City, Utah

Introduction

The completion of the *Drosophila* genome sequence in early 2000 revealed 21 nuclear receptor (NR) genes [1], about half of the 49 NR genes identified to date in the human genome [2]. In striking contrast, ~270 NR genes have been described in *Caenorhabditis elegans* [3]. It seems likely that the ~255 nematode-specific NRs reflect an adaptation unique to the nematode group. The 21 *Drosophila* NRs represent all major subfamilies described previously [4], and 16 receptors appear to have orthologs in either vertebrates or nematodes, or both (Table I).

Only five receptors have not been identified outside the arthropod branch. Knirps, Knirps-related, and Eagle form a unique class of highly related proteins (often referred to as the Knirps group) that lack a ligand-binding domain (LBD). The other two receptors, DHR39 and DHR83, are closely related to FTZ-F1 and dFAX-1, respectively, and probably represent—in evolutionary terms—relatively recent splits within the arthropod lineage. In general, it appears that most *Drosophila* NRs are more similar to their human counterparts than to the corresponding *C. elegans* homolog (Table I). The fruit fly, *Drosophila melanogaster*, provides an ideal system to analyze NR function in the context of a developing organism, with a wide range of genetic tools that allow the dissection of regulatory pathways and in-depth characterization of gene function.

Nuclear Receptors and Embryonic Pattern Formation

Drosophila embryogenesis is characterized by a succession of events that transform the early embryo into a segmented first instar larva. Maternally deposited transcripts create opposing protein gradients along the anterior-posterior axis of the early embryo that trigger spatially restricted activation of the so-called gap genes, establishing more defined subdivisions. The gap genes, in turn, activate the next set of genes, the pair-rule genes, that divide the embryo into regions representing the precursors of the body segments.

The NR genes *tailless* (*tll*) and *knirps* (*kni*) are gap genes. Embryos mutant for *tll* exhibit deletions in the terminal domains of the embryo [5]. Once Tll expression is activated through the Ras signaling pathway, it sets up a posterior boundary, restricting the expression of other gap genes like *kni*, *krüppel,* and *giant* [6]. The *tll* gene also plays a role in establishing the optic lobe primordium, which, at a later stage, develops into the optic lobe neuropiles and the larval photoreceptor [7]. The *kni* gene is required for the proper development of A1–A7 abdominal segments [8]. Closely related to *kni*, *knirps-related* (*knrl*) has partially overlapping functions with *kni* in development of the nervous system [9]. Kni and Knrl also play a critical role in cell migration and branch morphogenesis during embryonic tracheal development [10].

The NR gene *FTZ-F1* encodes two isoforms with distinct functions. During embryogenesis, maternally supplied αFTZ-F1 interacts with the Fushi tarazu (FTZ) homeodomain protein. These two proteins constitute mutually dependent cofactors required for correct segmentation [11,12]. FTZ contacts the AF-2 domain of αFTZ-F1 via an LXXLL motif, a signature typically found in NR cofactors and previously unknown to exist in homeodomain proteins [13,14]. The other isoform, βFTZ-F1, which plays a crucial role in molting and metamorphosis at later stages, is first expressed in the late embryo and appears to be required for the completion of embryogenesis [15].

Ecdysone Regulatory Hierarchies

Following embryogenesis, *Drosophila* progresses through three larval stages and metamorphosis, a dramatic process in

Table I ***Drosophila* Nuclear Receptors**

D. melanogaster	Cytology	IF	*C. elegans*	*H. sapiens*	Ecd	Lethal phase	Function
dERR	66B12-13	1	N/A	ERRβ (88/39)	?	?	?
dFAX-1	51F7	1	FAX-1 (93/22)	PNR (83/51)	?	?	?
dHNF4	29E3-4	2	NHR-64 (69/37)	HNF4α (92/65)	?	?	gut formation?
DHR3	46F4-5	2	CHR3 (92/32)	ROR (79/39)	+	embryo	pupariation, activates βFTZ-F1, PNS development
DHR4	2C1-2	1	NHR-91 (68/27)	GCNF (63/20)	+	?	?
DHR38	38E2-3	1 (2?)	NHR-6 (73/22)	NURR1 (92/58)	-	pharate adult[1]	adult cuticle formation, dimerizes with Usp
DHR39	39B1-3	1	NHR-25 (63/25)	LRH-1 (65/32)	+	viable	?
DHR78	78D6-7	1	NHR-41 (73/20)[2]	TR2 (76/25)	(+)	3rd instar	required in mid 3rd instar, trachea molting
DHR83	83E4	1	FAX-1 (67/20)	PNR (62/25)	?	?	?
DHR96	96B14-15	1	DAF-12 (68/29)	VDR (59/25)	(+)	?	?
Dissatisfaction	26A6-7	1	NHR-67 (71/19)	TLX (78/43)	?	sterile	sexual behaviour, serotonin neuron development
E75	75A9-B6	3	NHR-85 (82/12)	REVERBα (82/37)	+	1st instar	oogenesis, embryonic midgut constrictions
E78	78C4-7	2	SEX-1 (76/26)	REVERBα (73/34)	+	viable	follicle cell migration/patterning
Eagle	79A3-4	1	N/A	N/A	?	embryo, 1st instar	neuroblast determination
EcR	42A10	3	N/A	FXR (76/31), LXR (69/38)	+	embryo	20E receptor, molting, pupariation
FTZ-F1	75D4-6	2	NHR-25 (83/24)	LRH-1 (88/38)	+	embryo	FTZ cofactor, mid-prepupal competence factor
Knirps	77E1-2	1	N/A	N/A	?	embryo	embryonic pattern formation, gap gene
Knirps-related	77D1-4	1	N/A	N/A	?	?	embryonic tracheal development
Seven-up	87B8-9	2	UNC-55 (69/24)	COUP-TF1 (95/92)	?	embryo	photoneuron determination
Tailless	100B1	1	NHR-67 (80/16)	TLX (83/36)	?	embryo	embryo termini (gap gene), optic lobe cell fate
Ultraspiracle	2C1-2	1	N/A	RXR (88/44)	+	2nd instar	EcR-partner in 20E receptor, molting, pupariation

Cytology: Chromosomal location was based on published polytene chromosome *in situ* hybridizations and on estimated positions by the genome project. IF: isoforms, only mRNA isoforms encoding different proteins are considered. Number is based on published cDNA and EST data. *C. elegans* and *H. sapiens*: homologs of corresponding *D. melanogaster* NR and percent identity in the DBD/LBD. N/A = no homolog was found, closest match was not significantly more similar than matches to other members of the NR family. Ecd: regulated by ecdysone, tested by organ culture experiments, (+) indicates a weak induction. lethal phase: The zygotic lethal phase of known null mutations. function: summary of key functions

[1]T. Kozlova, personal communication.

[2]The NHR-41 sequence was kindly provided by A. Sluder and C. Gissendanner.

which the crawling larva is transformed into a reproductively active adult fly. These postembryonic developmental transitions are triggered by pulses of the steroid hormone ecdysone. An ecdysone pulse at the end of larval development signals the onset of metamorphosis and initiation of prepupal development, followed by another ecdysone pulse ~12 hr later that triggers the prepupal–pupal transition. The ecdysone signal is transduced through a heterodimer of two NRs: the EcR ecdysone receptor and an ortholog of vertebrate RXR, USP, which appears to function as an obligatory partner [16–18]. USP can also dimerize with DHR38, a NR involved in adult cuticle formation [19,20]. The EcR/USP and DHR38/USP dimers can recognize similar DNA sequences and may directly compete for binding to common regulatory targets [21]. Upon binding ecdysone, the EcR/USP receptor directly induces a set of primary-response genes, some of which encode transcription factors that, in turn, induce a larger set of secondary-response target genes. These gene cascades not only transduce and amplify the ecdysone signal, but also diversify it to achieve temporal and spatial specificity [22].

A surprising number of the key players induced in the ecdysone-regulatory hierarchies at the onset of metamorphosis are members of the NR family [23]. At least 10 NR genes are regulated by ecdysone (Ecd, Table I); most are induced by the hormone, whereas βFTZ-F1 is repressed by ecdysone [24]. The *E75* primary-response gene encodes three NR isoforms, designated E75A, E75B, and E75C [25]. *E75* plays a critical role in oogenesis [26] and is required for proper midgut constrictions during embryonic development [27]. *DHR3* is induced by ecdysone at puparium formation and is required for induction of βFTZ-F1 in the mid-prepupa [28]. E75B, a NR without a functional DNA-binding domain (DBD), can bind directly to DHR3, blocking its ability to induce βFTZ-F1 [29]. The timing of βFTZ-F1 expression thus appears to be controlled by induction of the DHR3 activator and decay of the E75B repressor. βFTZ-F1 provides competence for prepupal responses to ecdysone and is required for larval salivary gland cell death through stage-specific induction of the *E93* death gene [24,30]. *E78* and *DHR39*, two ecdysone-inducible NR genes, are expressed during the early stages of metamorphosis but do not carry out essential functions in the fly [31,32]. Among *Drosophila* NR genes, *E78* is most similar to *E75*. Intriguingly, one of the *E78* isoforms, E78B, is expressed in synchrony with E75B and is also missing its DBD, suggesting possible functional redundancy. DHR39 is closely related to FTZ-F1 and is almost as similar to DHR4, indicating that this NR group might share overlapping functions. *DHR4* mutations have not yet been described, although *DHR4* expression in early prepupae and its induction by ecdysone in organ culture suggest a role at the onset of metamorphosis (K. King-Jones, unpublished results).

Null mutations in the ecdysone-inducible *DHR78* NR gene lead to breaks in the tracheal respiratory system of the larva and lethality during the third instar [33]. Another *Drosophila* NR gene known to be regulated by ecdysone, *DHR96* [34], is an ortholog of *C. elegans daf-12*, which functions at the intersection of pathways that regulate dauer larva formation and adult longevity [35,36]. No specific mutations have yet been reported in *DHR96*, or in two other *Drosophila* NR genes, *dERR* and *dHNF4*. Recent studies, however, have shown that vertebrate HNF4α transcriptional activity can be modulated by long-chain fatty acids, and that diethylstilbestrol inactivates ERRβ [37,38]. It will be interesting to determine whether dERR and dHNF4 are capable of binding similar ligands and if *in vivo* ligands for these NRs can be identified.

The Neuronal Connection

Several NRs are involved in neuronal development, and their expression is often limited to specific neuronal cells. In embryos, *eagle* (*eg*) is almost exclusively expressed in four neuronal lineages that arise from a specific neuroblast (NB 7-3), which ultimately develops into a set of serotonin neurons. Mutant alleles of *eg* exhibit the correct number of *eg*-expressing NB 7-3 progeny, but fail to establish the correct projections, suggesting that the specification of cell identity in this lineage is altered [39,40]. *seven-up* (*svp*) encodes a COUP-TFI-like NR and represents the most conserved member of the NR family between humans and flies. SVP is expressed specifically in four of the eight photoreceptor neurons in the developing eye (R1, R3, R4, R6). *svp* mutations cause a transformation of these cells toward an R7 cell fate [41]. Conversely, ectopic expression of SVP in R7 photoneurons drives them toward an R1–R6 cell fate [42]. Another NR gene known to be expressed in the nervous system is *dissatisfaction* (*dsf*), a *tll*-like gene. Curiously, the LBD of the human Tll ortholog, TLX, is more similar to Dsf than to Tll itself, suggesting that both *dsf* and *tll* are orthologs of the *TLX* gene. *Dsf* is required for appropriate courtship behaviors and sex-specific neural development, and is expressed in a few cells in the larval and adult brain. Sexually dimorphic transcripts of *dsf* have not yet been identified, suggesting that its distinct roles in males and females may be mediated by a sex-specific dimerization partner and/or ligand(s) [43,44].

The genome project revealed two previously unknown NR genes that we refer to here as *DHR83* and *dFAX-1*. Both are closely related, but *DHR83* appears to have no orthologs in either humans or worms. dFAX-1 was named after its *C. elegans* counterpart, FAX-1, which is required for axonal pathfinding [45]. Whether dFAX-1 plays a similar role in *Drosophila* is unknown, although the absence of any ESTs, as well as the fact that this gene was not identified before the completion of the genome project, argues that its expression is either highly cell-type specific or of low abundance. *DHR83* appears to encode a rare transcript as well, but it remains to be seen whether this gene has functions similar to *dfax-1*. Finally, *DHR3* mutants display defects in embryonic peripheral nervous system development, but it is not clear whether this contributes to the early lethality seen in these mutants [46].

References

1. Adams, M. D. *et al.* (2000). The genome sequence of *Drosophila melanogaster. Science* **287**, 2185–2195.
2. Robinson-Rechavi, M., Carpentier, A., Duffraisse, M., and Laudet, V. (2001). How many nuclear hormone receptors are there in the human genome? *Trends Genet.* **17**, 554–556.
3. Sluder, A. E. and Maina, C. V. (2001). Nuclear receptors in nematodes: themes and variations. *Trends Genet.* **17**, 206–213.
4. Nuclear Receptors Nomenclature Committee (1999). A unified nomenclature system for the nuclear receptor superfamily. *Cell* **97**, 161–163.
5. Pignoni, F. *et al.* (1990). The *Drosophila* gene *tailless* is expressed at the embryonic termini and is a member of the steroid receptor superfamily. *Cell* **62**, 151–163.
6. Steingrimsson, E., Pignoni, F., Liaw, G. J., and Lengyel, J. A. (1991). Dual role of the *Drosophila* pattern gene *tailless* in embryonic termini. *Science* **254**, 418–421.
7. Daniel, A., Dumstrei, K., Lengyel, J. A., and Hartenstein, V. (1999). The control of cell fate in the embryonic visual system by *atonal, tailless* and EGFR signaling. *Development* **126**, 2945–2954.
8. Nauber, U. *et al.* (1988). Abdominal segmentation of the *Drosophila* embryo requires a hormone receptor-like protein encoded by the gap gene *knirps. Nature* **336**, 489–492.
9. Gonzalez-Gaitan, M., Rothe, M., Wimmer, E. A., Taubert, H., and Jackle, H. (1994). Redundant functions of the genes *knirps* and *knirps-related* for the establishment of anterior *Drosophila* head structures. *Proc. Natl. Acad. Sci. USA* **91**, 8567–8571.
10. Chen, C. K. *et al.* (1998). The transcription factors KNIRPS and KNIRPS RELATED control cell migration and branch morphogenesis during *Drosophila* tracheal development. *Development* **125**, 4959–4968.
11. Guichet, A. *et al.* (1997). The nuclear receptor homologue Ftz-F1 and the homeodomain protein Ftz are mutually dependent cofactors. *Nature* **385**, 548–552.
12. Yu, Y. *et al.* (1997). The nuclear hormone receptor Ftz-F1 is a cofactor for the *Drosophila* homeodomain protein Ftz. *Nature* **385**, 552–555.
13. Schwartz, C. J. *et al.* (2001). FTZ-Factor1 and Fushi tarazu interact via conserved nuclear receptor and coactivator motifs. *EMBO J.* **20**, 510–519.
14. Yussa, M., Lohr, U., Su, K., and Pick, L. (2001). The nuclear receptor Ftz-F1 and homeodomain protein Ftz interact through evolutionarily conserved protein domains. *Mech. Dev.* **107**, 39–53.
15. Yamada, M. *et al.* (2000). Temporally restricted expression of transcription factor βFTZ-F1: Significance for embryogenesis, molting and metamorphosis in *Drosophila melanogaster. Development* **127**, 5083–5092.
16. Yao, T. P. *et al.* (1993). Functional ecdysone receptor is the product of *EcR* and *Ultraspiracle* genes. *Nature* **366**, 476–479.
17. Yao, T. P., Segraves, W. A., Oro, A. E., McKeown, M., and Evans, R. M. (1992). *Drosophila ultraspiracle* modulates ecdysone receptor function via heterodimer formation. *Cell* **71**, 63–72.
18. Riddiford, L. M., Cherbas, P., and Truman, J. W. (2000). Ecdysone receptors and their biological actions. *Vitam. Horm.* **60**, 1–73.
19. Kozlova, T. *et al.* (1998). Drosophila hormone receptor 38 functions in metamorphosis: A role in adult cuticle formation. *Genetics* **149**, 1465–1475.
20. Sutherland, J. D., Kozlova, T., Tzertzinis, G., and Kafatos, F. C. (1995). *Drosophila* hormone receptor 38: A second partner for *Drosophila* USP suggests an unexpected role for nuclear receptors of the nerve growth factor-induced protein B type. *Proc. Natl. Acad. Sci. USA* **92**, 7966–7970.
21. Crispi, S., Giordano, E., D'Avino, P. P., and Furia, M. (1998). Cross-talking among *Drosophila* nuclear receptors at the promiscuous response element of the *ng-1* and *ng-2* intermolt genes. *J. Mol. Biol.* **275**, 561–574.
22. Thummel, C. S. (1996). Flies on steroids—*Drosophila* metamorphosis and the mechanisms of steroid hormone action. *Trends Genet.* **12**, 306–310.
23. Thummel, C. S. (1995). From embryogenesis to metamorphosis: The regulation and function of *Drosophila* nuclear receptor superfamily members. *Cell* **83**, 871–877.
24. Woodard, C. T., Baehrecke, E. H., and Thummel, C. S. (1994). A molecular mechanism for the stage specificity of the *Drosophila* prepupal genetic response to ecdysone. *Cell* **79**, 607–615.
25. Segraves, W. A. and Hogness, D. S. (1990). The *E75* ecdysone-inducible gene responsible for the 75B early puff in *Drosophila* encodes two new members of the steroid receptor superfamily. *Genes Dev.* **4**, 204–219.
26. Buszczak, M. *et al.* (1999). Ecdysone response genes govern egg chamber development during mid-oogenesis in *Drosophila. Development* **126**, 4581–4589.
27. Bilder, D. and Scott, M. P. (1995). Genomic regions required for morphogenesis of the *Drosophila* embryonic midgut. *Genetics* **141**, 1087–1100.
28. Lam, G. T., Jiang, C., and Thummel, C. S. (1997). Coordination of larval and prepupal gene expression by the DHR3 orphan receptor during *Drosophila* metamorphosis. *Development* **124**, 1757–1769.
29. White, K. P., Hurban, P., Watanabe, T., and Hogness, D. S. (1997). Coordination of *Drosophila* metamorphosis by two ecdysone-induced nuclear receptors. *Science* **276**, 114–117.
30. Broadus, J., McCabe, J. R., Endrizzi, B., Thummel, C. S., and Woodard, C. T. (1999). The *Drosophila* βFTZ-F1 orphan nuclear receptor provides competence for stage-specific responses to the steroid hormone ecdysone. *Mol Cell* **3**, 143–149.
31. Russell, S. R., Heimbeck, G., Goddard, C. M., Carpenter, A. T., and Ashburner, M. (1996). The *Drosophila Eip78C* gene is not vital but has a role in regulating chromosome puffs. *Genetics* **144**, 159–170.
32. Horner, M. A. and Thummel, C. S. (1997). Mutations in the *DHR39* orphan receptor gene have no effect on viability. *DIS* **80**, 35–37.
33. Fisk, G. J. and Thummel, C. S. (1998). The DHR78 nuclear receptor is required for ecdysteroid signaling during the onset of *Drosophila* metamorphosis. *Cell* **93**, 543–555.
34. Fisk, G. J. and Thummel, C. S. (1995). Isolation, regulation, and DNA-binding properties of three *Drosophila* nuclear hormone receptor superfamily members. *Proc. Natl. Acad. Sci. USA* **92**, 10604–10608.
35. Antebi, A., Yeh, W. H., Tait, D., Hedgecock, E. M., and Riddle, D. L. (2000). *daf-12* encodes a nuclear receptor that regulates the dauer diapause and developmental age in *C. elegans. Genes Dev.* **14**, 1512–1527.
36. Antebi, A., Culotti, J. G., and Hedgecock, E. M. (1998). *daf-12* Regulates developmental age and the dauer alternative in *Caenorhabditis elegans. Development* **125**, 1191–1205.
37. Tremblay, G. B. *et al.* (2001). Diethylstilbestrol regulates trophoblast stem cell differentiation as a ligand of orphan nuclear receptor ERRβ. *Genes Dev.* **15**, 833–838.
38. Hertz, R., Magenheim, J., Berman, I., and Bar-Tana, J. (1998). Fatty acyl-CoA thioesters are ligands of hepatic nuclear factor-4α. *Nature* **392**, 512–516.
39. Higashijima, S., Shishido, E., Matsuzaki, M., and Saigo, K. (1996). *eagle,* A member of the steroid receptor gene superfamily, is expressed in a subset of neuroblasts and regulates the fate of their putative progeny in the *Drosophila* CNS. *Development* **122**, 527–536.
40. Lundell, M. J. and Hirsh, J. (1998). *eagle* is required for the specification of serotonin neurons and other neuroblast 7–3 progeny in the *Drosophila* CNS. *Development* **125**, 463–472.
41. Mlodzik, M., Hiromi, Y., Weber, U., Goodman, C. S., and Rubin, G. M. (1990). The *Drosophila seven-up* gene, a member of the steroid receptor gene superfamily, controls photoreceptor cell fates. *Cell* **60**, 211–224.
42. Hiromi, Y., Mlodzik, M., West, S. R., Rubin, G. M., and Goodman, C. S. (1993). Ectopic expression of *seven-up* causes cell fate changes during ommatidial assembly. *Development* **118**, 1123–1135.

43. Finley, K. D., Taylor, B. J., Milstein, M., and McKeown, M. (1997). *dissatisfaction*, A gene involved in sex-specific behavior and neural development of *Drosophila melanogaster. Proc. Natl. Acad. Sci. USA* **94**, 913–918.
44. Finley, K. D. *et al.* (1998). *dissatisfaction* Encodes a Tailless-like nuclear receptor expressed in a subset of CNS neurons controlling *Drosophila* sexual behavior. *Neuron* **21**, 1363–1374.
45. Much, J. W., Slade, D. J., Klampert, K., Garriga, G., and Wightman, B. (2000). The *fax-1* nuclear hormone receptor regulates axon pathfinding and neurotransmitter expression. *Development* **127**, 703–712.
46. Carney, G. E., Wade, A. A., Sapra, R., Goldstein, E. S., and Bender, M. (1997). *DHR3*, an ecdysone-inducible early-late gene encoding a *Drosophila* nuclear receptor, is required for embryogenesis. *Proc. Natl. Acad. Sci. USA* **94**, 12024–12029.

SECTION B

Transcription Factors

Marc Montminy, Editor

CHAPTER 282

JAK-STAT Signaling

Christian. W. Schindler
Department of Microbiology and Medicine,
College of Physicians and Surgeons,
Columbia University, New York, New York

Introduction

Characterization of the ability of interferon α (IFN-α) to rapidly induce genes led to the discovery of the JAK-STAT pathway. Subsequent studies determined that JAKs (Janus kinases) and STATs (signal transducers and activators of transcription) represent a families of proteins that play a critical role in mediating the biological response to the ~50 hematopoietins, all members of the mammalian cytokine family of ligands (see Table I). Although components of this pathway can be found in the most primitive metazoans, the pathway has grown in complexity as multicellular organisms evolved. In mammals, this pathway regulates pivotal aspects of cellular homeostasis and inflammatory response. This review will provide a brief overview of the significant progress that has been made in understanding this pathway since it was first discovered a decade ago. For a more detailed discussion, readers are directed to more comprehensive reviews [1–4].

The JAK-STAT Paradigm

Hematopoietins represent a family of secreted ligands, which exert their potent biological effects on target cells that express the corresponding receptors. These receptors can be divided into five subgroups based on their structures and the STATs that they activate (see Table I). Of note, a number of receptor tyrosine kinases and G-protein-coupled receptors have also been shown to activate STATs, but these signals do not appear to be as critical to their biological response [2]. Upon binding ligand, the dimeric receptors undergo a conformational change, bringing together and thereby activating associated JAKs (see Fig. 1; [5]). Activated JAKs in turn phosphorylate receptor tyrosine motifs, leading to the recruitment of specific STATs (Table I). At the receptor, these STATs become activated by a single tyrosine phosphorylation event.

After release from the receptor, they dimerize through reciprocal phosphotyrosine–SH2 domain interactions ("classical dimerization"; [6,7]). Intriguingly, recent evidence indicates that STATs may actually form stable dimers prior to activation, but through a different mechanism (J. Braunstein and C. Schindler, unpublished observation). "Classically dimerized" STATs are competent for both DNA binding and robust nuclear translocation. Although structural studies have provided detailed insight into STAT DNA binding activity [6,7], little is known about the structural motifs that regulate nuclear import [2]. Once in the nucleus, STAT dimers bind to members of the gamma-activated site (GAS) family of enhancers, which culminates in the activation of target genes [2,8]. One well-established exception is the Stat1-Stat2 heterodimer. This unique heterodimer, which forms in response to stimulation with type I IFNs, associates with a DNA binding protein, IFN regulatory factor 9 (IRF-9), to form the transcription factor IFN stimulating gene factor 3 (ISGF-3). This factor binds to the IFN stimulated response element (ISRE) subset of the IRF family of enhancers [2].

STAT signals decay within a period of hours and the STATs are re-exported back to the cytoplasm, effectively resetting the cell for the next round of signaling. This appears to entail nuclear dephosphorylation and a Crm1/Ran-GTPase-dependent nuclear export process [2,9]. Some activated STATs may also undergo targeted degradation [10]. Intriguingly, recent studies suggest that the predominantly cytoplasmic distribution of STATs found in resting cells reflects a steady state, which represents a balance between basal (i.e., in the absence of tyrosine phosphorylation) nuclear import and export (S. Bhattacharya and C. Schindler, unpublished observation). This raises the possibility that STATs may regulate the expression of some genes in unstimulated cells [11].

Table I Hematopoietins Signal through JAKs and STATs

Ligands	JAKs	STATs
IFN family		
IFN-α's†**/β/ω/Limitin**	**Jak1**, Tyk2	**Stat1**, **Stat2**, (Stats3–6)
IFN-γ	**Jak1**, **Jak2**	**Stat1**, (Stat5)
IL-10*	**Jak1**, Tyk2	**Stat3**
IL-19	?	?
IL-20	?	Stat3
IL-22	?	Stat3, (Stat5)
IL-24	?	Stat3
gp130 family		
IL-6	**Jak1**, (Jak2)	**Stat3**, Stat1
IL-11	Jak1	**Stat3**, Stat1
OSM	Jak1, (Jak2)	**Stat3**, Stat1
LIF	**Jak1**, (Jak2)	**Stat3**, Stat1
CNTF	Jak1, (Jak2)	**Stat3**, Stat1
NNT-1/BSF-3	Jak1, (Jak2)	**Stat3**, Stat1
G-CSF	Jak1, (Jak2)	**Stat3**
CT-1	Jak1, (Jak2)	**Stat3**
Leptin	Jak2	**Stat3**
IL-12	**Tyk2**, Jak2	**Stat4**
IL-23	?	**Stat4**
γC family		
IL-2	**Jak1**, **Jak3**	**Stat5**, (Stat3)
IL-7	**Jak1**, **Jak3**	**Stat5**, (Stat3)
TSLP¶	?	Stat5
IL-9	Jak1, Jak3	**Stat5**, Stat3
IL-15	**Jak1**, Jak3	**Stat5**, (Stat3)
IL-21	Jak3, (Jak1),	**Stat3**, Stat5, (Stat1)
IL-4	**Jak1**, **Jak3**	**Stat6**
IL-13¶	Jak1, Jak2	Stat6, (Stat3)
IL-3 family		
IL-3	**Jak2**	**Stat5**
IL-5	**Jak2**	**Stat5**
GM-CSF	**Jak2**	**Stat5**
Single Chain family		
Epo	**Jak2**	**Stat5**
GH	Jak2	**Stat5**, (Stat3)
Prl	Jak2	**Stat5**
Tpo	**Jak2**	**Stat5**

The JAKs and STATs activated by members of the hematopoietin family are indicated. Confidence in these assignments vary: very strong (based on knockout studies and shown in bold), reasonably strong (plain lettering) and least strong (shown in brackets).

†There are 12 IFNα's.

*IL-10 homologue AK155 has not yet been functionally characterized.

¶Bind to a related, but γc independent receptor.

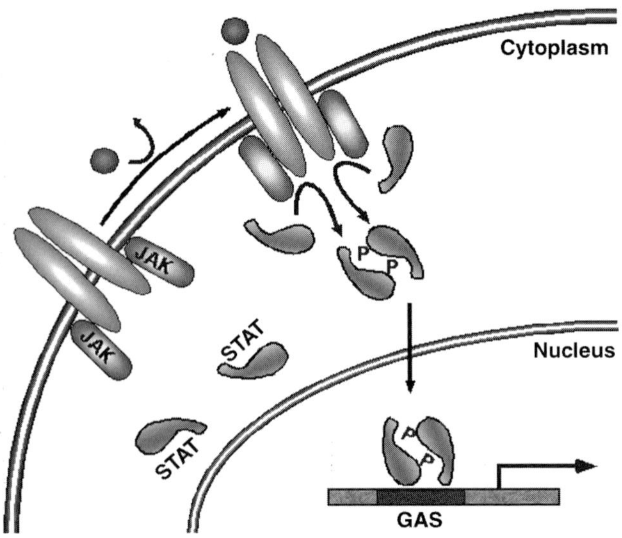

Figure 1 The JAK-STAT signaling paradigm. See text for details.

The JAK Family

Four members of the mammalian JAK family, Jak1, Jak2, Jak3, and Tyk2, are associated with a proline-rich, membrane proximal receptor domain [1]. JAKs range in molecular weight from 120 to 130 kDa. Jak1, Jak2 and Tyk2 are expressed ubiquitously, but the expression of Jak3 is more restricted [3]. Seven conserved JAK homology (JH) domains have been identified. The JH1 and JH2 domains constitute the carboxy-terminal kinase and pseudokinase domains, which distinguish JAKs from most other protein tyrosine kinases [1]. The amino-terminal JAK homology domains, JH3–JH7, constitute a FERM (four-point-one, Ezrin, Radixin, Moesin) domain, which mediates receptor association [2].

Although biochemical studies have implicated differing sets of JAKs for each hematopoietin receptor subgroup, it appears they can be interchanged without affecting signaling specificity [1,2]. Gene targeting studies have demonstrated that Jak1 plays an important role in the biological response to members of the interleukin (IL)-2, IL-6, and IFN/IL-10 receptor families [12]. Because the Jak1–/– mice exhibit a perinatal lethal phenotype, more complete analysis awaits adoptive transfer and tissue-specific knockout studies. The Jak2 knock-out mice exhibited a midgestational lethal phenotype (i.e., day 12.5), which has been attributed to a block in definitive erythropoiesis [13,14]. *Ex vivo* studies with Jak2–/– fetal liver cells indicate defects in their responses to Tpo, IL-3, members of the IL-2 family, IFN-γ, but not IL-6 or IFN-α. Consistent with biochemical studies demonstrating a robust association of Jak3 with γC [3], both the Jak3–/– and γC[–/–] mice demonstrate severe SCID-like defects in lymphopoiesis [3,15–17]. The phenotype of the Tyk2 knock-out mice is the most surprising of all. In contrast to earlier biochemical and genetic studies, these mice exhibit only relatively modest defects in their response to type I IFNs and IL-10. Defects in their response to IL-12 and LPS are considerably more remarkable [18,19].

The STAT Family

Mammals express seven STAT proteins, which range in size from 750 to 900 amino acids. Both the chromosomal distribution of these STATs and the identification of homologs in more primitive eukaryotes suggest that this family arose from a single primordial gene [2,20]. Duplications of this locus appear to reflect an increasing need for cell-to-cell communication. Homologs, most closely related to Stat3 and Stat5, have been identified in model eukaryotes, including *Dictyostelium*, *Caenorhabditis elegans*, and *Drosophila* [2]. Whereas in *Drosophila* the single STAT transduces signals through a "classical" JAK-STAT pathway, the STAT homologs in *Dictyostelium* and *C. elegans* appear to signal through different pathway(s).

STATs can be divided into five structurally and functionally conserved domains (see Fig. 2; [6,7]). The amino-terminal domain (NH_2; ~125 amino acids) is well conserved and promotes cooperativity in DNA binding [2,21]. This domain has also been implicated in nuclear translocation [2]. The coiled-coil domain (amino acids ~135 to ~315) consists of a four-helix bundle that protrudes laterally (~80 Å) from the core. This domain has been shown to associate with a number of potentially important regulatory modifiers including IRF-9 and StIP [2,4]. It has also been implicated in nuclear export [2]. The DNA-binding domain (DBD; amino acids ~320 to ~480) recognizes the palindromic GAS element, upon dimerization, and may also participate in nuclear export [6–9]. However, the Stat2 DBD, despite its sequence conservation, is unable to bind DNA directly. The linker domain (amino acids ~480 to ~575) structurally translates the dimerization signal to the DNA binding motif. Recent studies suggest it regulates a basal (i.e., in resting cells) nuclear export process [22]. The SH2 domain (amino acids ~575 to ~680) is the most highly conserved motif. It mediates both specific recruitment to the appropriate receptor, as well as STAT dimerization [2]. The tyrosine activation motif represents a conserved tyrosine near residue 700 that is activated by phosphorylation. This phosphotyrosyl residue is specifically recognized by a SH2 domain of the partner STAT [2]. Each STAT, except Stat2, has been shown to homodimerize *in vivo*. The carboxy termini of all STATs vary considerably in both length and sequence. They encode the transcriptional activation domain (TAD), which is conserved between mouse and man for every STAT, except Stat2 [2,23].

Gene targeting studies have provided important insight into the functional specificity of the STATs. Consistent with its identification during the purification of ISGF-3 and an IFN-γ induced factor, Stat1 knock-out mice exhibit defects in their biological response to both type I and type II IFNs, but not other ligands [2,24,25]. Likewise, Stat2 knock-out mice are only defective in their response to type I IFNs [23]. Analysis of these mice has highlighted the role type I IFNs play in regulating both innate and acquired immunity ([23]; E. Cha and C. Schindler, unpublished observation). The Stat3 knock-out mice exhibit an embryonic lethal phenotype, reflecting the pleiotropism of Stat3 activating ligands (see Table I). Tissue-specific Stat3 knock-outs have identified important defects in responses to IL-6, IL-10, KGF, CNTF, and LIF [2]. The diverse role Stat3 plays is reminiscent of the more pervasive role STATs play in primitive metazoans. The Stat4 and Stat6 knock-out mice exhibit specific defects in their response to the cytokines that regulate the polarization of naïve T-helper cells into effector Th1 and Th2 subsets, where IL-12 (as well as closely related IL-23) promotes Th1 cells and IL-4 (as well as closely related IL-13) promotes Th2 cells [26–30]. Finally, the Stat5 knock-out mice have yielded the most unexpected findings [1,31–33]. Stat5a and Stat5b, which are 96% identical, are not completely redundant, as earlier biochemical studies had suggested. Stat5a null mice exhibit a defect in PRL-dependent mammary gland development. In contrast, Stat5b single knock-out mice exhibit defects in their response to GH. The Stat5a/b double knock-out mice have the most severe defects, yet these mice still develop a full complement of hematopoietic lineages. Defects in the response to GH and PRL are striking, while those to IL-2, IL-3, IL-5, IL-7, GM-CSF, G-CSF, and Tpo are more subtle [1,2]. The double knock-out mice have also been found to be defective in TEL/Jak2-dependent leukemogenesis [1].

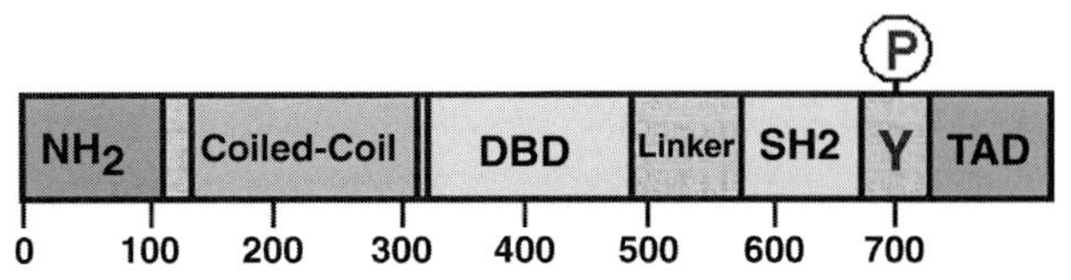

Figure 2 STAT structure. See text for details.

A Promising Future

STATs, which were first identified as IFN-dependent transcription factors, are now known to transduce vital signals for all ~50 hematopoietins (see Table I). Although the basic outline for these JAK-STAT signaling cascades has been established, much work remains to be done in understanding how this pathway is regulated and how it mediates its potent biological effects. Future developments in STAT regulation are likely to include further insight into STAT modifications and signal decay. Likewise, studies on the wide variety of biological responses mediated by STATs are likely to include more precise and comprehensive identification of STAT target genes, as well as a better understanding as to how STATs interact with other transcription factors to achieve desired gene regulation.

The most compelling evidence that STATs are modified at residues in addition to the activation tyrosine comes from work on serine phosphorylation, which appears to enhance transcriptional activity of several STATs [34]. Gene targeting studies, which are under way, are likely to provide valuable insight into this potentially important regulatory loop. Data on other modifications, including acetylation, arginine-methylation, ubiquitination, and potentially SUMO-lation, are more limited, but promising avenues for future study [2,4,10,35–37].

Studies on STAT signal decay have made considerable progress with the recent characterization of the SOCS family of counter-regulatory proteins. Several SOCS family members are themselves STAT target genes, thereby establishing a classic negative feedback loop [2,38]. Another important focus has been on phosphatases that dephosphorylate receptors, JAKs and STATs, but many remain to be identified [1,2]. Other studies have suggested that some signal decay is achieved by targeting activated STATs for ubiquitin-dependent degradation [10]. In addition, the role that structural changes play in both the activation and decay of STAT signaling has not been explored, but is likely to be important.

Lastly, there is increasing evidence that STATs interact with other transcription factors to regulate the expression of target genes. Although, a detailed characterization of STAT target gene promoters has not been completed, ongoing studies will provide important insight into the nature of these interactions and how they may regulate acetylation, nucleosome phasing, and stimulation of the basal transcriptional machinery [39,40]. Thus, it seems likely that the coming decade will continue to bring important developments in our understanding of how STATs mediate their potent biological responses.

References

1. Ihle, J. N. (2001). The Stat family in cytokine signaling. *Curr. Opin. Cell Biol.* **13**, 211–217.
2. Kisseleva, T., Bhattacharya, S., Schröder-Braunstein, J., and Schindler, C. W. (2002). Signaling through the JAK/STAT pathway, recent advances and future challenges. *Gene* **285**, 1–24.
3. Leonard, W. and O'Shea, J. J. (1998). JAKS and STATS: Biological implications. *Ann. Rev. Immunol.* **16**, 293–322.
4. Horvath, C. M. (2000). STAT proteins and transcriptional responses to extracellular signals. *Trends Biochem. Sci.* **25**, 496–502.
5. Remy, I., Wilson, I. A., and Michnick, S. W. (1999). Erythropoietin receptor activation by a ligand-induced conformation change. *Science* **283**, 990–993.
6. Becker, S., Groner, B., and Müller, C. W. (1998). Three-dimensional structure of the Stat3β homodimer bound to DNA. *Nature* **394**, 145–151.
7. Chen, X., Vinkemeier, U., Zhao, Y., Jeruzalmi, D., Darnell, J. E., Jr., and Kuriyan, J. (1998). Crystal structure of a tyrosine phosphorylated STAT-1 dimer bound to DNA. *Cell* **93**, 827–839.
8. Decker, T., Kovarik, P., and Meinke, A. (1997). GAS elements: A few nucleotides with a major impact on cytokine-induced gene expression. *J. Interferon Cytokine Res.* **17**, 121–134.
9. McBride, K. M., McDonald, C., and Reich, N. C. (2000). Nuclear export signal located within the DNA-binding domain of the STAT1transcription factor. *EMBO J.* **19**, 6196–6206.
10. Wang, D., Moriggl, R., Stravopodis, D., Carpino, N., Marine, J. C. *et al.* (2000). A small amphipathic alpha-helical region is required for transcriptional activities and proteasome-dependent turnover of the tyrosine-phosphorylated Stat5. *EMBO J.* **19**, 392–399.
11. Kumar, A., Commane, M., Flickinger, T. W., Horvarth, C. M., and Stark, G. R. (1997). Defective TNF-α induced apoptosis in Stat1 null cells due to low constitutive levels of caspases. *Science* **278**, 1630–1632.
12. Rodig, S. J., Meraz, M. A., White, J. M., Lampe, P. A., Riley, J. K. *et al.* (1998). Disruption of the Jak1 gene demonstrates obligatory and nonredundant roles of the Jaks in cytokine-induced biologic responses. *Cell* **93**, 373–383.
13. Neubauer, H., Cumano, A., Mueller, M., Wu, H., Huffstadt, U., and Pfeffer, K. (1998). Jak2 deficiency defines an essential developmental checkpoint in definitive hematopoiesis. *Cell* **93**, 397–409.
14. Parganas, E., Wang, D., Stravopidis, D., Topham, D., Marine, J.-C. *et al.* (1998). Jak2 is essential for signaling through a variety of cytokine receptors. *Cell* **93**, 385–395.
15. Nosaka, T., van Deursen, J. M., Tripp, R. A., Thierfelder, W. E., Witthuhn, B. A. *et al.* (1995). Defective lymphoid development in mice lacking Jak3. *Science* **270**, 800–802.
16. Park, S. Y., Saijo, K., Takahashi, T., Osawa, M., Arase, H. *et al.* (1995). Developmental defects of lymphoid cells in Jak3 kinase-deficient mice. *Immunity* **3**, 771–782.
17. Thomis, D. C., Gurniak, C. B., Tivol, E., Sharpe, A. H., and Berg, L. J. (1995). Defects in B lymphocyte maturation and T lymphocyte activation in mice lacking Jak3. *Science* **270**, 794–797.
18. Karaghiosoff, M., Neubauer, H., Lassnig, C., Kovarik, P., Schindler, H. *et al.* (2000). Partial impairment of cytokine responses in Tyk2-deficient mice. *Immunity* **13**, 549–560.
19. Shimoda, K., Kato, K., Aoki, K., Matsuda, T., Miyamoto, A. *et al.* (2000). Tyk2 plays a restricted role in IFN alpha signaling, although it is required for IL-12-mediated T cell function. *Immunity* **13**, 561–571.
20. Miyoshi, K., Cui, Y., Riedlinger, G., Lehoczky, J., Zon, L. *et al.* (2001). Structure of the mouse stat 3/5 locus: Evolution from *Drosophila* to zebrafish to mouse. *Genomics* **71**, 150–155.
21. Vinkemeier, U., Moarefi, I., Darnell, JE., Jr., and Kuriyan, J. (1998). Structure of the amino-terminal protein interaction domain of STAT-4. *Science* **279**, 1048–1052.
22. Bhattacharya, S. and Schindler, C. (2003). Regulation of Stat3 nuclear export. *J. Clin. Invest.* **111**, 553–559.
23. Park, C., Li, S., Cha, E., and Schindler, C. (2000). Immune response in Stat2 knockout mice. *Immunity* **13**, 795–804.
24. Durbin, J. E., Hackenmiller, R., Simon, M. C., and Levy, D. E. (1996). Targeted disruption of the mouse Stat1 gene results in compromised innate immunity to viral disease. *Cell* **84**, 443–450.
25. Meraz, M. A., White, J. M., Sheehan, K. C., Bach, E. A., Rodig, S. J. *et al.* (1996). Targeted disruption of the Stat1 gene in mice reveals unexpected physiological specificity in the JAK-STAT pathway. *Cell* **84**, 431–442.
26. Kaplan, M. H., Sun, Y.-L., Hoey, T., and Grusby, M. J. (1996). Impaired IL-12 responses and enhanced development of Th2 cells in Stat4-deficient mice. *Nature* **382**, 174–177.
27. Thierfelder, W. E., vanDeursen, J. M., Yamamoto, K., Tripp, R. A., Sarawar, S. R. *et al.* (1996). Requirement for Stat4 in interleukin-12 mediated response of natural killer and T cells. *Nature* **382**, 171–174.
28. Kaplan, M. H., Schindler, U., Smiley, S. T., and Grusby, M. J. (1996). Stat6 is required for mediating responses to IL-4 and for the development of Th2 cells. *Immunity* **4**, 313–319.
29. Shimoda, K., vanDeursen, J., Sangster, M. Y., Sarawar, S. R., Carson, R. T. *et al.* (1996). Lack of IL-4 induce Th2 response in IgE class switching mice with disrupted Stat6 gene. *Nature* **380**, 630–633.
30. Takeda, K., Tanaka, T., Shi, W., Matsumoto, M., Minami, M. *et al.* (1996). Essential role of Stat6 in IL-4 signaling. *Nature* **380**, 627–630.
31. Liu, X., Robinson, G. W., Wagner, K. U., Garrett, L., Wynshaw-Boris, A., and Hennighausen, L. (1997). Stat5a is mandatory for adult mammary gland development and lactogenesis. *Genes Devel.* **11**, 179–186.
32. Teglund, S., McKay, C., Schuetz, E., VanDeursen, J. M., Stravopodis, D. *et al.* (1998). Stat5a and Stat5b proteins have essential roles and nonessential, or redundant, roles in cytokine responses. *Cell* **93**, 841–850.
33. Udy, G. B., Towers, R. P., Snell, R. G., Wilkins, R. J., Park, S. H. *et al.* (1997). Requirement of Stat5b for sexual dimorphism of body growth rates and liver gene expression. *Proc. Natl. Acad. Sci. USA.* **94**, 7239–7244.
34. Decker, T. and Kovarik, P. (2000). Serine phosphorylation of STATs. *Oncogene* **19**, 2628–2637.
35. Sachdev, S., Bruhn, L., Sieber, H., Pichler, A., Melchior, F., and Grosschedl, R. (2001). PIASy, a nuclear matrix-associated SUMO E3 ligase, represses LEF1 activity by sequestration into nuclear bodies. *Genes Dev.* **15**, 3088–3103.

36. Shuai, K. (2000). Modulation of STAT signaling by STAT-interacting proteins. *Oncogene* **19**, 2638–2644.
37. Hochstrasser, M. (2001). Sp-ring for sumo. new functions bloom for a ubiquitin-like protein. *Cell* **107**, 5–8.
38. Kile, B. T., Nicola, N. A., and Alexander, W. S. (2001). Negative regulators of cytokine signaling. *Int. J. Hematol.* **73**, 292–298.
39. Kim, H., Kelly, J., and Leonard, W. J. (2001). The basis for IL-2-induced IL-2 receptor alpha chain gene regulation: Importance of two widely separated IL-2 response elements. *Immunity* **15**, 159–172.
40. Zhang, X., Wrzeszczynska, M. H., Horvath, C. M., and Darnell, J. E., Jr. (1999). Interacting regions in Stat3 and c-Jun that participate in cooperative transcriptional activation. *Mol. Cell Biol.* **19**, 7138–7146.

CHAPTER 283

FOXO Transcription Factors: Key Targets of the PI3K-Akt Pathway That Regulate Cell Proliferation, Survival, and Organismal Aging

Anne Brunet, Hien Tran, and Michael E. Greenberg

Division of Neuroscience,
Children's Hospital and Department of Neurobiology,
Harvard Medical School, Boston, Massachusetts

Introduction

The development and integrity of multicellular organisms depend on the ability of each cell to integrate a wide range of external and internal cues and to trigger appropriate cellular responses. In recent years, Forkhead transcription factors of the FOXO subfamily have emerged as key players in controlling cell cycle progression, cell survival, detoxification, and DNA damage repair by integrating extracellular signals and triggering changes in gene expression. At the organismal level, FOXO transcription factors may play a critical role in the control of life span, a function that may be conserved throughout evolution.

Identification of the FOXO Subfamily of Transcription Factors

Forkhead transcription factors all share a conserved DNA-binding domain (DBD) of 100 amino acids that folds into a winged helix structure termed the *forkhead box*. The Forkhead family is evolutionarily conserved from yeast to mammals, and in humans comprises around 30 members that have been divided into 17 subgroups (FOX for "Forkhead Box" A to Q) [1]. FOX transcriptional regulators play a wide range of roles during development, from organogenesis (FOXC) [2] to language and speech acquisition (FOXP) [3].

Among the large Forkhead family, the FOXO subgroup first attracted interest because all three members of this family, namely, FOXO1/FKHR, FOXO3a/FKHRL1, and FOXO4/AFX, were identified at the site of chromosomal translocations in human tumors. The chromosomal translocations involving FOXO family members result in the generation of a chimeric protein in which the transactivation domain of FOXOs is fused to the DBD of another transcription factor, creating a dysregulated and highly active transcriptional fusion protein (Table I) [4–7]. These initial findings suggested that, when mutated, FOXO transcription factors might play a role in tumor development.

Table I FOXO Family Members

New name	Old name	Chromosomal translocation	Fusion protein	Type of cancer	Expression
FOXO1	FKHR	2–13 1–13	PAX3-FOXO1 PAX7-FOXO1	Alveolar rhabdomyosarcoma	Ubiquitous but high in ovary
FOXO2 FOXO3a	AF6q21 FKHRL1	6–21	MLL-FOXO3a	Acute myeloblastic leukemia	Ubiquitous but high in brain and kidney
FOXO4	AFX	X-11	MLL-FOXO4	Acute myeloblastic leukemia	Placenta and muscle

Regulation of FOXO Transcription Factors by the PI3K-Akt Pathway

In recent years, a series of genetic and biochemical studies revealed that FOXO family members are regulated by the PI3K-Akt pathway in response to growth factor stimulation. Binding of growth factors or insulin to their receptors triggers the activation of the phosphoinositide kinase (PI3K), which in turn is responsible for the activation of several serine/threonine kinases, including Akt or SGK (serum glucocorticoid inducible kinase) [8]. The PI3K-Akt pathway is conserved in the nematode *Caenorhabditis elegans*, where this pathway is activated by an insulin-like signal [9,10]. Inactivation of the PI3K-Akt pathway during development of the nematode results in arrest in dauer, a long-lived stress-resistant larval stage, whereas in the adult, inactivation of this pathway increases life span by two- to threefold. Strikingly, two independently conducted genetic screens found that all suppressor alleles of the PI3K-Akt pathway mutants map to the gene encoding a FOXO transcription factor termed DAF-16 [11,12]. These results indicate that DAF-16 is a key target of the PI3K-Akt pathway.

Several studies conducted in mammalian cells provide a molecular mechanism that appears to explain the genetic link between the PI3K-Akt pathway and FOXOs. These studies show that the protein kinase Akt phosphorylates all FOXO family members at three key regulatory sites (Thr32, Ser253, and Ser315 for FOXO3a) that are conserved from *C. elegans* to mammals and are part of a perfect consensus sequence for Akt phosphorylation [RXRXX(S/T)] (Fig. 1) [13–17]. Akt is not the only protein kinase that phosphorylates FOXOs at these regulatory sites; SGK also phosphorylates FOXO3a. However, the efficacy with which SGK and Akt phosphorylate the three phosphorylation sites of FOXO transcription factors. Whereas Thr32 is phosphorylated by both kinases, Akt preferentially phosphorylates Ser253, and SGK favors the phosphorylation of Ser315 [18].

The three FOXO regulatory sites are phosphorylated in response to a wide range of stimuli that activate Akt and SGK, including insulin-like growth factor 1 (IGF-1) [13], insulin [19], interleukin 3 [20], erythropoietin [21], epidermal growth factor [22], and transforming growth factor β [23]. The phosphorylation of FOXO transcription factors by Akt and SGK in response to growth factors triggers a change in the subcellular localization of FOXOs [13,14] (Fig. 2). In the absence of growth factors, when Akt and SGK are inactive, FOXOs are localized within the nucleus. When cells are exposed to growth factors, the PI3K-Akt/SGK cascade is activated and leads to the relocalization of FOXOs from the nucleus to the cytoplasm, away from FOXO target genes (Fig. 2). Mutation analyses of the three regulatory sites have shown that the phosphorylation of each site contributes to the relocalization of FOXOs from the nucleus to the cytoplasm [18]. This apparent redundancy of phosphorylation may represent a way of modulating the extent of the relocalization of FOXOs to the cytoplasm in different cell types or in response to different combinations of signals.

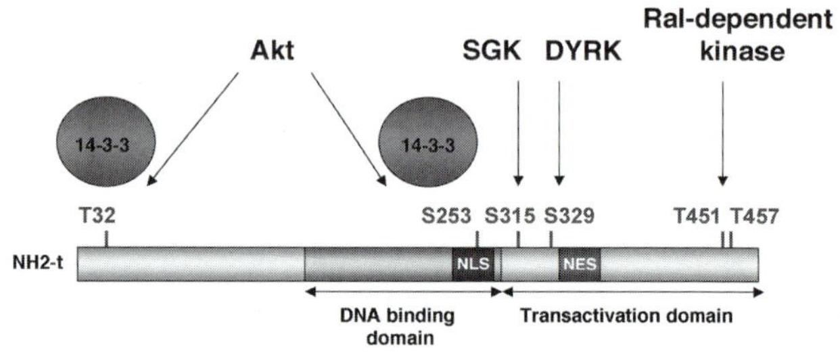

Figure 1 Domain structure and phosphorylation sites within FOXO family members. Numbering for FOXO3a: T32, S253, S315; numbering for FOXO1, S329; numbering for FOXO4: T451, T457.

Other Regulatory Phosphorylation Sites in FOXOs

In addition to the sites on FOXOs that are phosphorylated in response to growth factors, several other sites of phosphorylation have recently been identified (Fig. 1). FOXO4 (AFX) is phosphorylated at two residues, Thr 447 and Thr 451, in response to the activation of the small G protein termed Ral, although the kinase directly responsible for the phosphorylation of these two threonines is not yet known [24]. Phosphorylation of Thr 447 and Thr 451 does not affect the subcellular localization of FOXO4, but instead appears to abolish FOXO4 transcriptional activity. Another study has reported that the MAP kinase family member DYRK catalyzes the phosphorylation of Ser 329 in FOXO1, a residue that is conserved in all FOXO family members [25]. However, the phosphorylation of Ser 329 is not modulated by growth factors and its role in the regulation of FOXO function is still unclear.

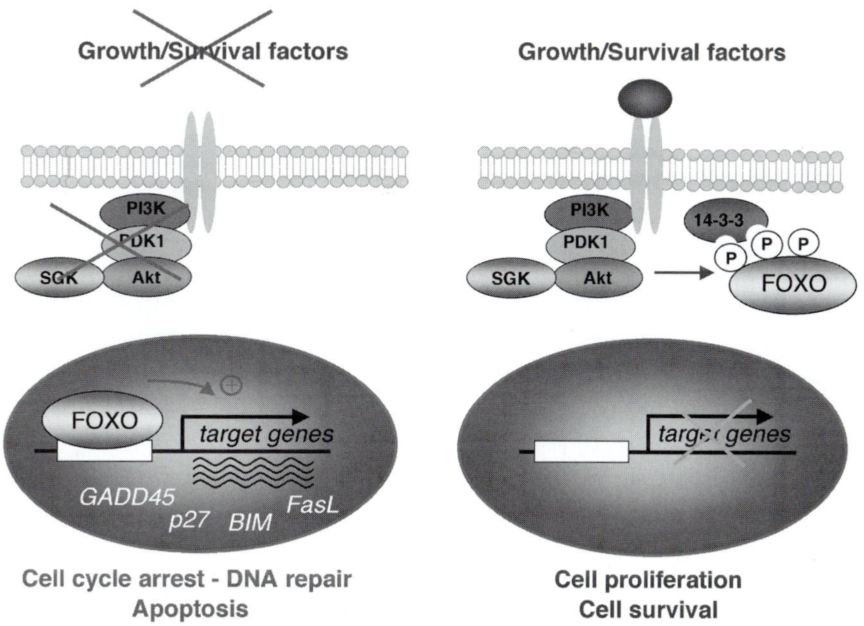

Figure 2 Regulation of FOXO family members by growth and survival factors. In the absence of growth/survival factors, FOXO family members induce the expression of death genes, cell cycle arrest genes or stress response genes, including Fas ligand, the proapoptotic Bcl-2 family member BIM, the cell cycle inhibitor p27KIP1, and the growth arrest and DNA damage response gene GADD45. In the presence of growth/survival factors, the PI3K-Akt/SGK pathway is activated. Akt and SGK promote survival and cell cycle progression by phosphorylating and inhibiting FOXOs.

Mechanism of the Exclusion of FOXOs from the Nucleus in Response to Growth Factor Stimulation

The translocation of FOXOs from the nucleus to the cytoplasm appears to be the primary mechanism of regulation of these transcription factors in response to growth factors. Experiments using leptomycin B, a specific inhibitor of nuclear export, show that the translocation of FOXO transcription factors from the nucleus to the cytoplasm is mediated by a nuclear export signal (NES)-dependent mechanism [14,26,27]. Mutation analyses have revealed that one (FOXO1) or two (FOXO3a) leucine-rich domains in the conserved C-terminal region of FOXOs function as NES [14,26,27] (Fig. 1). Furthermore, the phosphorylated forms of FOXOs have been shown to specifically interact with 14-3-3 proteins [13], which appear to serve as chaperone molecules. The phosphorylation of the first FOXO regulatory site (Thr 32 in FOXO3a) creates a perfect 14-3-3 binding site, and the phosphorylation of the second FOXO regulatory site (Ser 253 in FOXO3a)—although it does not create a canonical 14-3-3 binding site—also participates in 14-3-3 binding [13]. Several mechanisms have been proposed to explain how 14-3-3 binding to FOXOs may promote the relocalization of FOXOs from the nucleus to the cytoplasm. One study suggests that 14-3-3 binding decreases the ability of FOXOs to bind DNA, releasing FOXOs from a nuclear DNA anchor and allowing the relocalization of FOXOs to the cytoplasm [28]. A second study suggests that 14-3-3 binding to FOXOs occurs within the nucleus and actively promotes the nuclear export of FOXOs, perhaps by inducing a conformational change in FOXOs that would expose the NES and allow interaction with Exportin/Crm1 [27].

Finally 14-3-3 binding to FOXOs has been shown to prevent the nuclear reimport of FOXOs by masking FOXOs' nuclear localization signal (NLS) [26,29], consistent with a known function of 14-3-3 in regulating the subcellular localization of several of its other binding partners. These various mechanisms for regulating the translocation of FOXOs from the nucleus to the cytoplasm may serve as a fail-safe mechanism to ensure a complete sequestration of FOXOs in the cytoplasm, away from FOXO target genes in the nucleus. The sequestration of a transcription factor in the cytoplasm upon phosphorylation is a mechanism by which the activity of a variety of transcriptional regulators, including NFAT [30,31] and the yeast transcription factor Pho4 [32], are controlled, and this mechanism may have been selected by evolution because it represents an efficient way to inhibit the function of a transcription factor.

Transcriptional Activator Properties of FOXOs

Like other FOX family members, FOXOs bind to DNA as a monomer via the Forkhead box, a 100-amino-acid region located in the central part of the molecule (Fig. 1). The NMR structure of the DBD of FOXO4 has been solved and shows that this domain adopts a winged helix structure similar to those of the other FOX family members that have been solved, FOXA (HNF3 γ) and FOXD (Genesis) [33]. The consensus recognition site for FOXOs on DNA has been determined by three independent groups [34–36] and the core motif, GTAAA(C/T)A, is very similar to that of other Forkhead transcription factors [37]. Both more extensive DNA sequence motifs in promoters of FOXO target genes and regions of FOXO proteins outside the Forkhead box are likely involved in conferring binding specificity to particular FOXO family members under different conditions of stimulation.

When present in the nucleus and bound to DNA, FOXOs act as potent activators of transcription. The phosphorylation of FOXOs by Akt or SGK appears to regulate the transcriptional activity of FOXOs primarily by excluding this transcription factor from the nucleus [13,15,17,18,38]. When the phosphorylated form of FOXOs is artificially maintained within the nucleus either by leptomycin B treatment or by mutation of the FOXOs NES, FOXOs are transcriptionally active even in the presence of growth factors, indicating that the main function of phosphorylation at the growth factor-regulated sites is to relocalize FOXOs to the cytoplasm rather than to directly affect the DNA binding or transcriptional activity of FOXOs [26]. Deletion analyses have revealed that the transactivation domain of FOXOs spans the C-terminal region of the molecule (Fig. 1). In the chimeric protein PAX3-FOXO1 present in some human rhabdomyosarcomas, it is the transactivation domain of FOXO1 that is fused to PAX3 and that confers potent transactivation properties to PAX3 [39]. FOXOs' ability to bind to the transcriptional coactivator CBP (CREB binding protein) may provide a connection with the basal transcriptional machinery resulting in FOXO/CBP-dependent changes in target gene expression [40].

FOXOs and the Regulation of Apoptosis

The expression of constitutively active mutants of FOXOs triggers cell death in several different cell types, ranging from primary neurons to lymphocytes [13,17,41–43]. FOXO-induced apoptosis is dependent on the ability of FOXOs to bind to DNA and to induce transcription, indicating that a subset of target genes that FOXOs transactivate consists of death genes (Table II). Thus, one way in which Akt and SGK promote cell survival is by sequestering FOXOs away from death genes. Several death genes contain FOXO binding sites in their promoters, including the genes encoding for death cytokines (Fas ligand, TNF α, CD30 ligand) or death cytokine receptors (Fas, TNFR). In particular, FOXO3a has been shown to bind to and activate the promoter for the Fas ligand gene [13,44]. Consistent with the hypothesis that FOXOs regulate Fas ligand transcription, the inactivation of Akt has recently been shown to induce Fas ligand expression [45]. Because FOXO3a-induced apoptosis is reduced in cells in which Fas ligand signaling is blocked [13], Fas ligand may relay in a paracrine manner the effects of FOXOs

Table II FOXO Target Genes

Genes	Function	Gene induction	Binding to promoter	Promoter activation	FOXO binding sites
Cell Death					
Fas ligand	Death cytokine	ICC	EMSA; ChIP	+	+
Bim	BH3-only Bcl-2 family member	NB; WB			
bNIP3	BH3-only Bcl-2 family member	Microarray			
Legumain	Cysteine protease	Microarray			+
Bcl-6	Transcriptional repressor	Microarray; NB	EMSA	+	+
Cell Cycle					
p27KIP1	Cdk inhibitor	WB; NB		+	+
WIP1	Phosphatase	Microarray			+
EXT1	Tumor suppressor	Microarray			+
Cyclin G_2	G_2 delay	Microarray			+
Polo-like kinase (PLK)	Mitosis control	WB	ChIP	+	+
Cyclin B	Cdc2 activator	WB	ChIP	+	+
Antioxidant					
Selenoprotein P	Detoxification of ROS	Microarray; NB			+
SOD3		NB			+
DNA Repair					
GADD45	Growth arrest and DNA damage response	Microarray; NB; WB		+	+
PA26		Microarray; NB			+
Metabolism					
Glucose-6-phosphatase	Glucose metabolism	NB	EMSA	+	+
Phosphoenolpyruvate carboxykinase	Glucose metabolism	NB		+	
IGF-BP1	Regulation of IGF-I levels	NB	EMSA	+	+

Key: ICC, immunocytochemistry; NB, Northern blot; WB, Western blot; EMSA, electrophoretic mobility shift assay; ChIP, chromatin immunoprecipitation.

by triggering cellular pathways that culminate in apoptosis. FOXOs also appear to control the expression of several members of the Bcl-2 family, which are known to play a critical role in the balance between cell survival and cell death. FOXO3a induces the expression of Bim, a potent prodeath Bcl-2 family member, and apoptosis induced by the inactivation of the PI3K-Akt/SGK pathway is reduced in lymphocytes from Bim-deficient mice, indicating that Bim may be an important target gene of FOXO3a that mediates cell death [43,46]. In addition, FOXO4 indirectly downregulates the expression of the prosurvival Bcl-2 family member Bcl-xL via the induction of a transcriptional repressor Bcl-6 [47]. Thus, one way in which FOXOs appear to trigger apoptosis is by modulating the ratio of prodeath and prosurvival members of the Bcl-2 family.

FOXOs Are Key Regulators of Several Phases of the Cell Cycle

In cycling cells such as fibroblasts, the main effect of the expression of FOXO family members is not apoptosis but cell cycle arrest at the G_1/S boundary [48]. One target gene that mediates FOXO-induced cell cycle arrest is the Cdk inhibitor p27KIP1, which inhibits Cdk2 activity, thereby blocking cell cycle progression at the G_1/S transition [20,48,49] (Table II). FOXOs induce p27 promoter activity, and FOXOs' ability to induce G_1 arrest is diminished in p27-deficient fibroblasts, suggesting that p27 is a critical FOXO target that mediates G_1 arrest [48]. The FOXO-mediated arrest at the G_1/S transition may be a critical first step in the differentiation process because transgenic mice expressing a dominant-negative form of FOXO1 have impaired T-cell differentiation [36].

Endogenous FOXO3a is localized to the nucleus in cells passing through the G_2 phase of the cell cycle, suggesting a role for FOXO3a at the G_2/M checkpoint [50]. Cells in which FOXO3a is activated in the S phase display a delay in their progression through the G_2 phase of the cell cycle. Microarray analyses led to the identification of several FOXO3a target genes that may mediate FOXOs' effect at the G_2/M boundary (Table II) [50]. As with the G_1/S checkpoint, the G_2/M checkpoint is critical in the cellular response to stress, in part to allow time for detoxification and the repair of damaged DNA. Indeed, when active, FOXO3a induces DNA repair [50]. One of the genes that may mediate both FOXO3a-induced arrest at the G_2/M checkpoint and DNA repair is the growth arrest and DNA damage response gene 45 (GADD45) since the GADD45 promoter is activated by FOXO3a, and FOXO3a-induced DNA repair is diminished in GADD45-deficient fibroblasts [50].

FOXO3a also promotes the exit from the M phase and allows the transition to the following G_1 phase [44]. FOXO3a binds to and induces the promoter of two genes that play a critical role in the exit from M phase, cyclin B and polo-like kinase (PLK) (Table II) [44]. These findings are consistent with the observation that the activation of PI3K and Akt prevents the completion of the M phase [44]. As the two yeast Forkhead transcription factors Fkh1 and Fkh2 have been found to play a role in the completion of M phase [51], this function of FOXOs may have been conserved throughout evolution.

How can FOXOs induce both an arrest in G_1 and G_2 and promote the exit from the M phase? One possible explanation is that during normal cell cycles, FOXOs would be recruited to promoters of genes that promote the exit from the M phase. In contrast, under conditions of stress or absence of growth factors, FOXOs may be recruited to promoters of genes controlling cell cycle arrest at the G_1/S and G_2/M checkpoints, allowing repair of damaged DNA. Thus, FOXOs may integrate different extracellular cues, possibly by undergoing various posttranslational modifications, and elicit the appropriate cell cycle response by triggering a specific array of target genes.

FOXOs in Cancer Development: Potential Tumor Suppressors

In mammals, FOXOs' ability to induce a G_1 arrest, a G_2 delay, DNA repair, and apoptosis makes it an attractive candidate as a tumor suppressor. Loss of FOXO function may lead to a decreased ability to induce cell cycle arrest, leading to tumor development. A decreased ability to repair damaged DNA due to the absence of FOXOs may result in genomic instability. Finally, in the absence of FOXOs, abnormal cells that would normally die may instead survive, resulting in tumor formation. FOXOs' ability to induce cell cycle arrest, DNA repair, and apoptosis are reminiscent of the functions of the tumor suppressor protein p53. In that respect, it is interesting to note that genes such as GADD45, WIP1, and PA26 (Table II) that are induced in response to FOXOs have also been found to be regulated by p53 [52–54]. These observations raise the possibility that FOXOs and p53 may under some circumstances function in a cooperative manner.

One further link between the FOXO family and cancer is that all FOXO members had been initially characterized because of their presence at chromosomal breakpoints in cells from human tumors (Table I). However the expression of these human tumor fusion proteins in transgenic mice is not sufficient to promote cancer [55]. This finding suggests that the human tumors may have arisen not only as a result of the chimeric molecules but perhaps also as a consequence of the loss of one FOXO allele. If FOXOs function as tumor suppressors, then the haploinsufficiency of one FOXO family member could be a contributing factor to the development of the tumor.

Role of FOXOs in the Response to Stress and Organismal Aging

All mutants of the nematode that lead to the activation of the FOXO transcription factor DAF-16 are not only long lived but also display resistance to oxidative stress, heat shock, and UV [9,10,56,57]. This observation suggests that one way in which DAF-16 activity may lead to an increase in organismal life span is by augmenting the resistance of cells to various stresses. In the nematode, expression of the manganese-dependent superoxide dismutase SOD3 is induced when DAF-16 is active [58]. The promoter of SOD3 contains a Forkhead binding site that is in a region conserved across species, suggesting that SOD may be a conserved target of FOXOs that detoxify cells of reactive oxygen species (ROS), thereby conferring resistance to oxidative stress [34]. Another gene that is induced when DAF-16 is active and contains FOXO binding sites in its promoter is the tyrosine kinase receptor OLD-1, which is distantly related to the mammalian PDGF receptor. Expression of OLD-1 increases nematode longevity in part by increasing the resistance to oxidative stress [59].

In mammals, FOXOs induce cell cycle arrest at two critical checkpoints that allow repair of damaged DNA [48,50]. An organism's ability to respond to stress and in particular to induce detoxification and repair damaged DNA, has been shown to correlate with an increased longevity in many organisms [60]. The GADD45 gene, which in part mediates FOXO-induced arrest and DNA repair [50], is up-regulated in old mice [61]. These correlative evidence suggest that this pathway may be up-regulated to protect against oxidative stress produced by metabolism as an organism ages. Thus, by up-regulating a program of gene expression that protects cells against both internal and external cellular stresses, FOXO transcriptional regulators might promote longevity in mammals as well as in *C. elegans*.

Mosaic analyses in *C. elegans* indicate that the longevity phenotype of the nematode mutants that leads to DAF-16 activation is noncell autonomous [62]. In addition, neurons are the cell types that appear to be critical regulators of the aging phenotype in the nematode [63]. These findings, combined with the indications that DAF-16 may protect cells against stress, raise the possibility that a subset of neurons in the central nervous system may represent a "longevity control center" that might be more sensitive than other organs to oxidative stress [63]. It is tempting to speculate that factor(s) released from these neurons would convey the life span signal to the whole organism. It will be interesting to determine if a central neuronal regulator of life span also operates in mammals.

FOXOs and the Regulation of Metabolism in Relation to Organismal Aging

Reduction in nutrient intake consistently correlates with an increase in life span from *C. elegans* to mammals [60]. In the nematode, nutrient deprivation induces the relocalization of the DAF-16 to the nucleus [56], suggesting that one way starvation may result in increased life span is by relocalizing FOXO transcription factors to the nucleus where they may in turn induce target genes that lead to the cellular resistance to stress.

In addition, FOXOs appear to themselves regulate target genes that are involved in the control of metabolism (Table II). FOXO1 induces the expression of glucose-6-phosphatase, a gene that catalyzes the hydrolysis of glucose-6-phosphate into glucose [64,65]. FOXO1 also regulates the gene encoding phosphoenolpyruvate carboxykinase (PEPCK), which mediates the conversion of pyruvate to glucose, although the mechanism by which FOXO1 up-regulates PEPCK expression is probably not via direct binding of FOXO1 to the promoter of the PEPCK gene [66,67]. The regulation of glucose metabolism by FOXOs may be one indirect mechanism by which FOXOs increase organismal longevity. Indeed, switching the cell metabolism toward glucose production and away from the oxidative phosphorylation at the mitochondria may decrease the production of ROS by this organelle. Thus, FOXOs may extend organismal life span both by decreasing the production of ROS and by increasing detoxification and repair of DNA damage caused by ROS.

Conclusion

A series of recent studies have shown that FOXO transcription factors are regulated in a conserved manner throughout evolution by the PI3K-Akt pathway. In mammalian cells, FOXOs' functions include cell cycle regulation at various key checkpoints, apoptosis, repair of damaged DNA, and the regulation of glucose metabolism. How can the same transcription factor induce such a range of biological responses, responses that appear in some cases even to be antagonistic? Depending on their posttranslational modifications or interaction with protein partners, FOXOs may be recruited to particular subsets of promoters. It is possible that FOXOs may act as rheostats depending on the stress level: In low stress conditions, FOXOs may promote cell cycle arrest and DNA repair, whereas in high levels of stress, FOXOs may induce apoptosis. This graded response to stress stimuli would protect cells from damage, but also facilitate the elimination of terminally damaged cells. The result would be a suppression of tumor formation or an increase in the life span of the organism.

Understanding the molecular mechanisms that generate specificity among the FOXO family members will help to define the role of this family in various cellular responses. In addition, the generation of mice models in which FOXO family members are either inactivated or constitutively activated will help to uncover the respective contribution of the FOXO family members to tumor development and the control of life span in mammals.

Acknowledgments

We thank members of the Greenberg lab, in particular S. R. Datta, S. E. Ross, and A. J. Shaywitz, for their helpful comments on the manuscript. This work was supported by a Senior Scholars Award from the Ellison Foundation, NIH grant PO1-HD24926, and Mental Retardation Research Center grant NIHP30-HD18655 (MEG). AB was supported by a Goldenson Berenberg Fellowship. MEG acknowledges the generous contribution of the FM Kirby Foundation to the Division of Neuroscience.

References

1. Kaestner, K. H., Knochel, W., and Martinez, D. E. (2000). Unified nomenclature for the winged helix/forkhead transcription factors. *Genes Dev.* **15**, 142–146.
2. Kume, T., Deng, K., and Hogan, B. L. (2000). Murine forkhead/winged helix genes Foxc1 (Mf1) and Foxc2 (Mfh1) are required for the early organogenesis of the kidney and urinary tract. *Development* **127**, 1387–1395.
3. Lai, C. S., Fisher, S. E., Hurst, J. A., Vargha-Khadem, F., and Monaco, A. P. (2001). A forkhead-domain gene is mutated in a severe speech and language disorder. *Nature* **413**, 519–523.
4. Galili, N. *et al.* (1993). Fusion of a fork head domain gene to PAX3 in the solid tumour alveolar rhabdomyosarcoma. *Nat. Genet.* **5**, 230–235.
5. Borkhardt, A. *et al.* (1997). Cloning and characterization of AFX, the gene that fuses to MLL in acute leukemias with a t(X;11) (q13;q23). *Oncogene* **14**, 195–202.
6. Hillion, J., Le Coniat, M., Jonveaux, P., Berger, R., and Bernard, O. A. (1997). AF6q21, a novel partner of the MLL gene in t(6;11) (q21;q23), defines a forkhead transcriptional factor subfamily. *Blood* **90**, 3714–3719.
7. Davis, R. J., D'Cruz, C. M., Lovell, M. A., Biegel, J. A., and Barr, F. G. (1994). Fusion of PAX7 to FKHR by the variant t(1;13) (p36;q14) translocation in alveolar rhabdomyosarcoma. *Cancer Res.* **54**, 2869–2872.
8. Datta, S. R., Brunet, A., and Greenberg, M. E. (1999). Cellular survival: A play in three Akts. *Genes Dev.* **13**, 2905–2927.
9. Morris, J. Z., Tissenbaum, H. A., and Ruvkun, G. (1996). A phosphatidylinositol-3-OH kinase family member regulating longevity and diapause in *Caenorhabditis elegans*. *Nature* **382**, 536–539.
10. Kimura, K. D., Tissenbaum, H. A., Liu, Y., and Ruvkun, G. (1997). daf-2, an insulin receptor-like gene that regulates longevity and diapause in *Caenorhabditis elegans*. *Science* **277**, 942–946.
11. Lin, K., Dorman, J. B., Rodan, A., and Kenyon, C. (1997). daf-16: An HNF-3/forkhead family member that can function to double the life-span of *Caenorhabditis elegans*. *Science* **278**, 1319–1322.
12. Ogg, S. *et al.* (1997). The Fork head transcription factor DAF-16 transduces insulin-like metabolic and longevity signals in *C. elegans*. *Nature* **389**, 994–999.
13. Brunet, A. *et al.* (1999). Akt promotes cell survival by phosphorylating and inhibiting a Forkhead transcription factor. *Cell* **96**, 857–868.
14. Biggs, W. H. I., Meisenhelder, J., Hunter, T., Cavenee, W. K., and Arden, K. C. (1999). Protein kinase B/Akt-mediated phosphorylation promotes nuclear exclusion of the winged helix transcription factor FKHR1. *Proc. Natl. Acad. Sci. USA* **96**, 7421–7426.
15. Kops, G. J. *et al.* (1999). Direct control of the Forkhead transcription factor AFX by protein kinase B. *Nature* **398**, 630–634.
16. Rena, G., Guo, S., Cichy, S., Unterman, T. G., and Cohen, P. (1999). Phosphorylation of the transcription factor forkhead family member FKHR by protein kinase B. *J. Biol. Chem.* **274**, 17179–17183.
17. Tang, E. D., Nunez, G., Barr, F. G., and Guan, K.-L. (1999). Negative regulation of the Forkhead transcription factor FKHR by Akt. *J. Biol. Chem.* **274**, 16741–16746.
18. Brunet, A. *et al.* (2001). The protein kinase SGK mediates survival signals by phosphorylating the Forkhead transcription factor FKHRL1/FOXO3a. *Mol. Cell Biol.* **21**, 952–965.
19. Nakae, J., Park, B. C., and Accili, D. (1999). Insulin Stimulates Phosphorylation of the Forkhead Transcription Factor FKHR on Serine 253 through a Wortmannin-sensitive Pathway. *J. Biol. Chem.* **274**, 15982–15985.
20. Dijkers, P. F. *et al.* (2000). Forkhead transcription factor FKHR-L1 modulates cytokine-dependent transcriptional regulation of p27(KIP1). *Mol. Cell Biol.* **20**, 9138–9148.
21. Kashii, Y. *et al.* (2000). A member of Forkhead family transcription factor, FKHRL1, is one of the downstream molecules of phosphatidylinositol 3-kinase-Akt activation pathway in erythropoietin signal transduction. *Blood* **96**, 941–949.
22. Jackson, J. G., Kreisberg, J. I., Koterba, A. P., Yee, D., and Brattain, M. G. (2000). Phosphorylation and nuclear exclusion of the forkhead transcription factor FKHR after epidermal growth factor treatment in human breast cancer cells. *Oncogene* **19**, 4574–4581.
23. Shin, I., Bakin, A. V., Rodeck, U., Brunet, A., and Arteaga, C. L. (2001). Transforming growth factor beta enhances epithelial cell survival via Akt-dependent regulation of FKHRL1. *Mol. Cell Biol.* **12**, 3328–3339.
24. De Ruiter, N. D., Burgering, B. M., and Bos, J. L. (2001). Regulation of the Forkhead transcription factor AFX by Ral-dependent phosphorylation of threonines 447 and 451. *Mol. Cell Biol.* **21**, 8225–8235.
25. Woods, Y. L. *et al.* (2001). The kinase DYRK1A phosphorylates the transcription factor FKHR at Ser329 in vitro, a novel in vivo phosphorylation site. *Biochem. J.* **355**, 597–607.
26. Brownawell, A. M., Kops, G. J., Macara, I. G., and Burgering, B. M. (2001). Inhibition of nuclear import by protein kinase B (Akt) regulates the subcellular distribution and activity of the forkhead transcription factor AFX. *Mol. Cell Biol.* **21**, 3534–3546.
27. Brunet, A. *et al.* (2002). 14-3-3 Transits to the nucleus and actively participates in dynamic nucleo-cytoplasmic transport. *J. Cell. Biol.* **156**, 817–828.
28. Cahill, C. M. *et al.* (2000). PI-3 kinase signaling inhibits DAF-16 DNA binding and function via 14-3-3 dependent and 14-3-3 independent pathways. *J. Biol. Chem.* **276**, 13402–13410.
29. Rena, G., Prescott, A. R., Guo, S., Cohen, P., and Unterman, T. G. (2001). Roles of the forkhead in rhabdomyosarcoma (FKHR) phosphorylation sites in regulating 14–3–3 binding, transactivation and nuclear targeting. *Biochem. J.* **354**, 605–612.
30. Beals, C. R., Sheridan, C. M., Turck, C. W., Gardner, P., and Crabtree, G. R. (1997). Nuclear export of NF-ATc enhanced by glycogen synthase kinase-3. *Science* **275**, 1930–1934.
31. Chow, C. W., Rincon, M., Cavanagh, J., Dickens, M., and Davis, R. J. (1997). Nuclear accumulation of NFAT4 opposed by the JNK signal transduction pathway. *Science* **278**, 1638–1641.
32. Komeili, A. and O'Shea, E. K. (1999). Roles of phosphorylation sites in regulating activity of the transcription factor Pho4. *Science* **284**, 977–980.
33. Weigelt, J., Climent, I., Dahlman-Wright, K., and Wikstrom, M. (2001). Solution structure of the DNA binding domain of the human forkhead transcription factor AFX (FOXO4). *Biochemistry* **40**, 5861–5869.
34. Furuyama, T., Nakazawa, T., Nakano, I., and Mori, N. (2000). Identification of the differential distribution patterns of mRNAs and consensus binding sequences for mouse DAF-16 homologues. *Biochem. J.* **349**, 629–634.
35. Biggs, W. H., III, and Cavenee, W. K. (2001). Identification and characterization of members of the FKHR (FOXO) subclass of winged-helix transcription factors in the mouse. *Mamm. Genome* **12**, 416–425.
36. Leenders, H., Whiffield, S., Benoist, C., and Mathis, D. (2000). Role of the forkhead transcription family member, FKHR, in thymocyte differentiation. *Eur. J. Immunol.* **30**, 2980–2990.
37. Pierrou, S., Hellqvist, M., Samuelsson, L., Enerback, S., and Carlsson, P. (1994). Cloning and characterization of seven human forkhead proteins: binding site specificity and DNA bending. *EMBO J.* **13**, 5002–5012.
38. Guo, S. *et al.* (1999). Phosphorylation of serine 256 by protein kinase B disrupts transactivation by FKHR and mediates effects of insulin on IGF binding protein-1 promoter activity through a conserved insulin response sequence. *J. Biol. Chem.* **274**, 17184–17192.
39. Fredericks, W. J. *et al.* (1995). The PAX3-FKHR fusion protein created by the t(2;13) translocation in alveolar rhabdomyosarcomas is a more potent transcriptional activator than PAX3. *Mol. Cell Biol.* **15**, 1522–1535.
40. Nasrin, N. *et al.* (2000). DAF-16 recruits the CREB-binding protein coactivator complex to the insulin-like growth factor binding protein 1 promoter in HepG2 cells. *Proc. Natl. Acad. Sci. USA* **97**, 10412–10417.

41. Takaishi, H. *et al.* (1999). Regulation of nuclear translocation of forkhead transcription factor AFX by protein kinase B. *Proc. Natl. Acad. Sci. USA* **96**, 11836–11841.
42. Dijkers, P. F. *et al.* (2000). Forkhead transcription factor FKHR-L1 modulates cytokine-dependent transcriptional regulation of p27(KIP1). *Mol. Cell Biol.* **20**, 9138–9148.
43. Dijkers, P. F. *et al.* (2002). FKHR-L1 can act as a critical effector of cell death induced by cytokine withdrawal: protein kinase B-enhanced cell survival through maintenance of mitochondrial integrity. *J. Cell Biol.* **156**, 531–542.
44. Alvarez, B., Martinez, A. C., Burgering, B. M., and Carrera, A. C. (2001). Forkhead transcription factors contribute to execution of the mitotic programme in mammals. *Nature* **413**, 744–747.
45. Suhara, T., Kim, H. S., Kirshenbaum, L. A., and Walsh, K. (2002). Suppression of Akt signaling induces Fas ligand expression: involvement of caspase and Jun kinase activation in Akt-mediated Fas ligand regulation. *Mol. Cell Biol.* **22**, 680–691.
46. Dijkers, P. F., Medemadagger, R. H., Lammers, J. J., Koenderman, L., and Coffer, P. J. (2000). Expression of the pro-apoptotic bcl-2 family member bim is regulated by the forkhead transcription factor FKHR-L1. *Curr. Biol.* **10**, 1201–1204.
47. Tang, T. T. *et al.* (2002). The forkhead transcription factor AFX activates apoptosis by induction of the BCL-6 transcriptional repressor. *J. Biol. Chem.* **2**, 2.
48. Medema, R. H., Kops, G. J., Bos, J. L., and Burgering, B. M. (2000). AFX-like Forkhead transcription factors mediate cell-cycle regulation by Ras and PKB through p27kip1. *Nature* **404**, 782–787.
49. Nakamura, N. *et al.* (2000). Forkhead transcription factors are critical effectors of cell death and cell cycle arrest downstream of PTEN. *Mol. Cell Biol.* **20**, 8969–8982.
50. Tran, H. *et al.* (2002). DNA repair pathway stimulated by the Forkhead transcription factor FOXO3a (FKHRL1) through the GADD45 protein. *Science* In Press.
51. Zhu, G. *et al.* (2000). Two yeast forkhead genes regulate the cell cycle and pseudohyphal growth. *Nature* **406**, 90–94.
52. Fiscella, M. *et al.* (1997). Wip1, a novel human protein phosphatase that is induced in response to ionizing radiation in a p53-dependent manner. *Proc. Natl. Acad. Sci. USA* **94**, 6048–6053.
53. Velasco-Miguel, S. *et al.* (1999). PA26, a novel target of the p53 tumor suppressor and member of the GADD family of DNA damage and growth arrest inducible genes. *Oncogene* **18**, 127–137.
54. Kastan, M. B. *et al.* (1992). A mammalian cell cycle checkpoint pathway utilizing p53 and GADD45 is defective in ataxia-telangiectasia. *Cell* **71**, 587–597.
55. Anderson, M. J., Shelton, G. D., Cavenee, W. K., and Arden, K. C. (2001). Embryonic expression of the tumor-associated PAX3-FKHR fusion protein interferes with the developmental functions of Pax3. *Proc. Natl. Acad. Sci. USA* **98**, 1589–1594.
56. Henderson, S. T. and Johnson, T. E. (2001). daf-16 integrates developmental and environmental inputs to mediate aging in the nematode *Caenorhabditis elegans*. *Curr. Biol.* **11**, 1975–1980.
57. Larsen, P. L. (1993). Aging and resistance to oxidative damage in *Caenorhabditis elegans*. *Proc. Natl. Acad. Sci. USA* **90**, 8905–8909.
58. Honda, Y. and Honda, S. (1999). The daf-2 gene network for longevity regulates oxidative stress resistance and Mn-superoxide dismutase gene expression in *Caenorhabditis elegans*. *FASEB J.* **13**, 1385–1393.
59. Murakami, S. and Johnson, T. E. (2001). The OLD-1 positive regulator of longevity and stress resistance is under DAF-16 regulation in *Caenorhabditis elegans*. *Curr. Biol.* **11**, 1517–1523.
60. Kirkwood, T. B. L. and Austad, S. N. (2000). Why do we age? *Nature* **408**, 233–238.
61. Lee, C.-K., Klopp, R. G., Weindruch, R., and Prolla, T. A. (1999). Gene expression profile of aging and its retardation by caloric restriction. *Science* **285**, 1390–1393.
62. Apfeld, J. and Kenyon, C. (1998). Cell nonautonomy of C. elegans daf-2 function in the regulation of diapause and life span. *Cell* **95**, 199–210.
63. Wolkow, C. A., Kimura, K. D., Lee, M. S., and Ruvkun, G. (2000). Regulation of C. elegans life-span by insulinlike signaling in the nervous system. *Science* **290**, 147–150.
64. Schmoll, D. *et al.* (2000). Regulation of glucose-6-phosphatase gene expression by protein kinase Balpha and the forkhead transcription factor FKHR. Evidence for insulin response unit-dependent and -independent effects of insulin on promoter activity. *J. Biol. Chem.* **275**, 36324–36333.
65. Nakae, J., Kitamura, T., Silver, D. L., and Accili, D. (2001). The forkhead transcription factor Foxo1 (Fkhr) confers insulin sensitivity onto glucose-6-phosphatase expression. *J. Clin. Invest.* **108,** 1359–1367.
66. Yeagley, D., Guo, S., Unterman, T., and Quinn, P. G. (2001). Gene- and activation-specific mechanisms for insulin inhibition of basal and glucocorticoid-induced insulin-like growth factor binding protein-1 and phosphoenolpyruvate carboxykinase transcription. Roles of forkhead and insulin response sequences. *J. Biol. Chem.* **276**, 33705–33710.
67. Hall, R. K. *et al.* (2000). Regulation of phosphoenolpyruvate carboxykinase and insulin-like growth factor-binding protein-1 gene expression by insulin. The role of winged helix/forkhead proteins. *J. Biol. Chem.* **275**, 30169–30175.

92

Multiple Signaling Routes to Histone Phosphorylation

Claudia Crosio and Paolo Sassone-Corsi
Institut de Génétique et de Biologie Moléculaire et Cellulaire Strasbourg, France

Introduction

The eukaryotic genome is organized in a nucleoprotein structure that enables essential biological processes such as regulation of transcription, DNA repair, apoptosis, and cell division [1–3]. A number of remodeling events lead to dynamic changes in the architecture of chromatin. These appear to be prerequisite steps to evolve from a condensed to a decondensed state, and vice versa, each state being coupled to specific cellular functions. Various mechanisms have been implicated in inducing modifications in chromatin structure, some involving the active participation of ATP-dependent remodeling factors such as Swi/Snf and NURF [4–7]; others implicate enzymatic activities that elicit a number of histone modifications, including acetylation, methylation, and phosphorylation. In the past years evidence has accumulated indicating that these covalent modifications occur on histones H2A, H2B, H3, and H4, all assembled with DNA to form nucleosomes, and on H1, the linker histone. In this chapter, we focus on histone phosphorylation and the influence that phosphorylation may have on other modifications.

Although the nucleosome is a highly compacted structure [8], the N-terminal regions, or tails, of the histones are exposed [9]. Indeed, the N-terminal tails of core histones are not essential for maintaining the integrity of nucleosomes since removal of these tails by trypsin treatment does not diminish nucleosome stability [10,11]. Instead, histone tails are thought to confer secondary and more flexible contacts with DNA that allow for dynamic changes in the accessibility of the underlying genome. In addition, the disposition of the tails renders these domains accessible to modifications that could reversibly modulate chromatin structure [12]. The concept that chromatin modifications had to be reversible in nature was already implicit in some important early observations. For example, specific lysine residues were found to be acetylated and/or methylated on the histone H3 tail [13], and it was proposed that acetylation was a hallmark of active chromatin [14,15]. The cAMP-dependent kinase PKA was found to phosphorylate at least *in vitro* various chromatin substrates [16], and phosphrylation of H3 was thought to promote interactions with other chromatinic proteins [17]. In recent years fundamental advances have been made in the identification and characterization of histone modifications. These have pushed forward a unifying concept: the presence of a *histone code*, established by distinct combinations of histone modifications that can then be translated into different nuclear responses and dictate a particular biological outcome [18]. This model is strongly supported by experimental evidence indicating that posttranslational modifications are involved in gene-specific regulation [19]. Whether histone modifications modulate chromatin structure directly, by altering the contacts between histones and DNA, or indirectly, by presenting a special surface for interaction with other regulatory proteins, is still a subject of investigation.

Histone phosphorylation has been associated with extensive alteration of chromatin structure, and phosphorylation of the five individual histones has been shown. Although we focus here primarily on the role of histone phosphorylation, it is important to note that the central concepts are also applicable to other covalent modifications known to influence histone structure and function.

Histone Phosphorylation and Gene Activation

Phosphorylation of histones occurs by serine-threonine kinases at specific sites on the N-terminal tails. With the

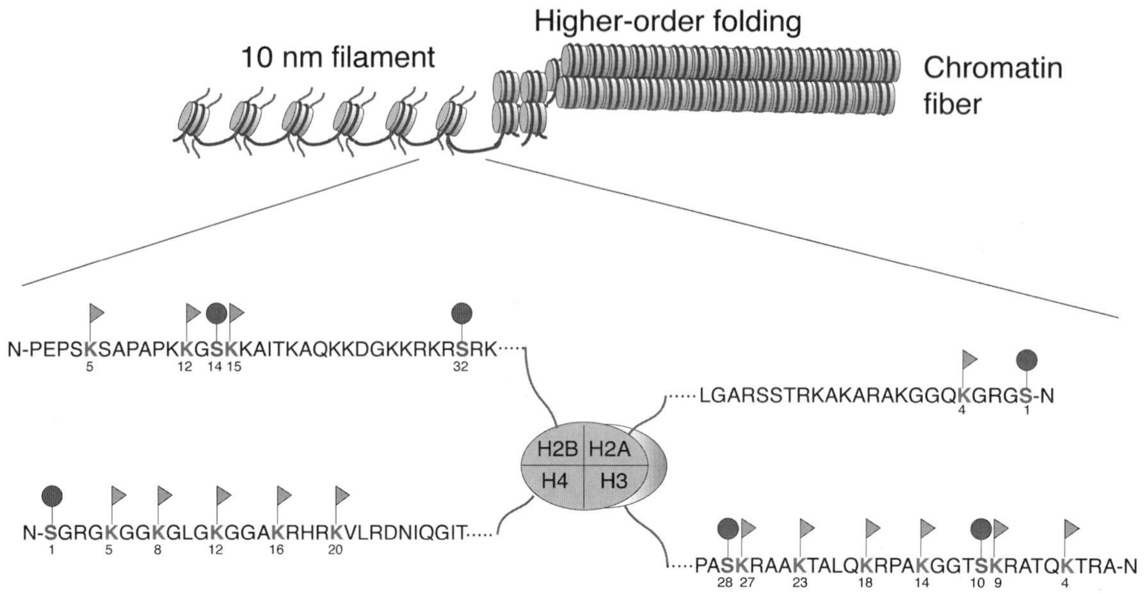

Figure 1 Chromatin organization and amino-terminal tails of core histones. Simplified representation of general chromatin organization. Histone tails (represented in fuchsia) are likely to be exposed or extended outward from chromatin fibers. The amino acid sequence of the N-terminal tails of the four human core histones is represented: red dots indicate known phosphorylation sites; whereas green flags are acetylation sites.

exception of H2A, where only one site has been demonstrated, for each histone tail there are two phosphoacceptor sites: histone H2A is phosphorylated on Ser1 (and possibly on Ser18), H2B on Ser14 and Ser32, H3 on Ser10 and Ser28, and H4 on Ser1 and Ser18. Strikingly, the distance between phosphoacceptor sites is conserved among all histones: It is always 18 residues (Fig. 1). The significance of this constant spacing is unclear, but further supports the above-mentioned hypothesis of the histone code. Of all of these modifications, the best characterized is phosphorylation of H3 on Ser10 (also because of the very successful generation of phosphospecific antibodies directed to the site [20]).

The Remarkable Case of Histone H3

Since the first description of histone H3 phosphorylation upon mitogenic stimulation of quiescent fibroblasts [21], significant evidence has contributed to establishing a direct link between this posttranslational modification and gene activation. In fact, in a variety of cellular systems, it has clearly been shown that a given stimulation causes a rapid and transient induction in Ser10 phosphorylation, with kinetics that are reminiscent of the transcriptional induction of immediate-early genes (IEGs). Importantly, only a subset of histone H3 molecules becomes phosphorylated upon mitogenic stimulation [22], generating a speckled pattern of phospho-H3 immunostaining in the nucleus. The possibility that these speckles correspond to physical sites of active transcription in fascinating. This hypothesis seems supported by chromatin-immunoprecipitation (ChIP) assays demonstrating that the induced expression of a number of IEGs is associated with Ser10 transient phosphorylation [23]. Table I summarizes the described cases of Ser10 H3 transient phosphorylation induced by various treatments and through different signaling pathways.

Although most of our knowledge of histone modifications is based on studies in cell culture systems, an important challenge is to reach a deeper understanding of the role that chromatin modifications play *in vivo* in physiological systems. We have observed that chromatin remodeling occurs in the vertebrate nervous system in response to a photic stimulation in neurons that constitute the mammalian circadian clock (SCN, suprachiasmatic nucleus) and after GABAergic stimulation in the supraoptic nucleus (SON) [24].

In most of the cellular systems analyzed, phosphorylation of Ser10 H3 involves the ERK/MAPK and p38 cascades, as shown by drug sensitivity studies (Fig. 2). The MAPK-activated RSK-2 (ribosomal S6-kinase) kinase has been identified as a candidate for EGF-induced H3 phosphorylation [25]. Interestingly, RSK-2 is also involved in EGF-induced CREB phosphorylation and c-*fos* expression in fibroblasts [26], establishing a direct link between signaling to chromatin and induction of IEGs. Also MSK-1 (mitogen- and stress-activated kinase-1) appears to function as H3 kinase at the same Ser10 site [23,27]. Both MAP and SAP kinase pathways can activate MSK-1; therefore, it is likely that different signaling pathways converge on a single kinase to phosphorylate serine 10 of histone H3. Salvador and collegues [28], on the other hand, have described Ser10 H3 phosphorylation upon stimulation with follicle-stimulating hormone (FSH) in rat ovarian granulosa cells. FSH-stimulated H3 phosphorylation seems correlated with cAMP-dependent protein kinase A (PKA) activation. More recently the first Ser10 histone H3 kinase complex from *Saccharomyces cerevisiae* was

Table I Association of Phosphorylated Histone H3 with Promoters of IEGs in Response to Various Stimuli[a]

Stimulus	Cell type	Activated gene(s)	Signaling cascaede	Ref.
Oncogenic transformation	Mouse fibroblasts	c-*myc* c-*fos*	Ras-MAPK	22
UV	10T1/2 fibroblast	c-*fos* c-*jun*	p38	23
EGF	10T1/2 fibroblast	c-*fos* c-*jun*	ERK/MAPK	24,25
Heat shock	*Drosophila* larvae	Heat shock genes		26
Arsenite	10T1/2 fibroblast	MKP-1		27
FSH	Granulosa cells	Serum glucocorticoid kinase Inhibin α c-*fos*	PKA	28
	Yeast	INO1 ACT1	Snf1/AMPK	29
Inflammatory stimuli (lipopolysaccharide)	Primary human dendridic cells	IL-6 IL-8 IL-12p40 Macrophage chemoattractant protein 1	p38	30
Retinoic acid	P19 cells	Retinoic acid recptor β	Not tested	31
TPA	10T1/2 fibroblast	c-*fos*	ERK/MAPK	32

[a]ChIP assays were performed on different cell types stimulated in various ways, using an antibody that recognizes histone H3 when phosphorylated on serine 10.

purified and identified as the catalytic subunit of Snf1. The Snf1/AMPK kinase phosphorylated Ser10 H3 to enhance the transcription of the INO1 and ACT1 genes [29].

Interestingly, all of the above-mentioned kinases belong to the AGC branch of cyclic nucleotide-regulated protein kinase, underscoring their fundamental role in transducing the signal to the nucleosomes. However, how one single class of kinases activated by separate pathways can elicit different responses remains to be clarified.

Coupling Acetylation and Phosphorylation

How phosphorylation at Ser10 of H3 affects gene expression is unclear. One possibility could be that—similar to what has been proposed for histone acetylation—the addition of negatively charged phosphate groups to the H3 tail may disrupt electrostatic interactions with the negatively charged DNA backbone, thereby increasing the accessibility of nuclear factors to underlying regulatory promoter elements. Another important observation concerns studies that have have shown that H3 phosphorylation and acetylation may be coupled in response to EGF stimulation [30–32]. Indeed, nucleosomes containing phosphorylated and acetylated H3 are preferentially associated with the EGF-activated genes, suggesting that both modifications may operate in concert to elicit induced transcription. Kinetics studies suggest that H3 phosphorylation precedes acetylation in the formation of multiply modified H3 [31].

A mechanistic link between the enzymatic reactions of phosphorylation and acetylation is suggested by the fact that various coactivators involved in signaling-dependent activation of early response genes act as histone acetyltransferases (HATs) and display strong preferences for H3 phosphorylated at serine 10 as substrate [29,31]. Therefore, it appears that a significant interplay exists between the different modifications occuring on the same histone tail.

As mentioned earlier, mitogen-induced H3 phosphorylation is involved in the regulation of only a selected set of genes, hinting at the existence of a highly regulated mechanism that targets the H3 kinase (and probably acetyltransferase) to the appropriate gene loci. These notions indicated that coupling of phosphorylation and acetylation at the level of the histone substrates could possibly be paralleled by a physical and functional interplay of the respective effectors, kinases, and acetylases. At least one case has been described that supports this view and concerns the coactivator CBP (CREB-binding protein), which has HAT activity [33]. In quiescent cells, CBP and RSK-2 proteins are found associated in a complex with low kinase and HAT activities. In serum-starved cells the basal phosphorylation levels of two RSK-2 substrates, CREB and histone H3, are also low. Upon EGF mitogenic stimulation, and consequent activation by phosphorylation of RSK-2 at Ser227, CBP and RSK-2 dissociate. Thus, in contrast to the well-characterized CREB–CBP interaction, where association requires CREB phosphorylation at Ser133, formation of the RSK-2–CBP complex is efficient only when RSK-2 is not phosphorylated. The dissociation results in increased Ser133 CREB and Ser10 H3 phosphorylation. Indeed, following the dissociation, RSK-2 becomes available to phosphorylate CREB in response to EGF. At the same time, CBP

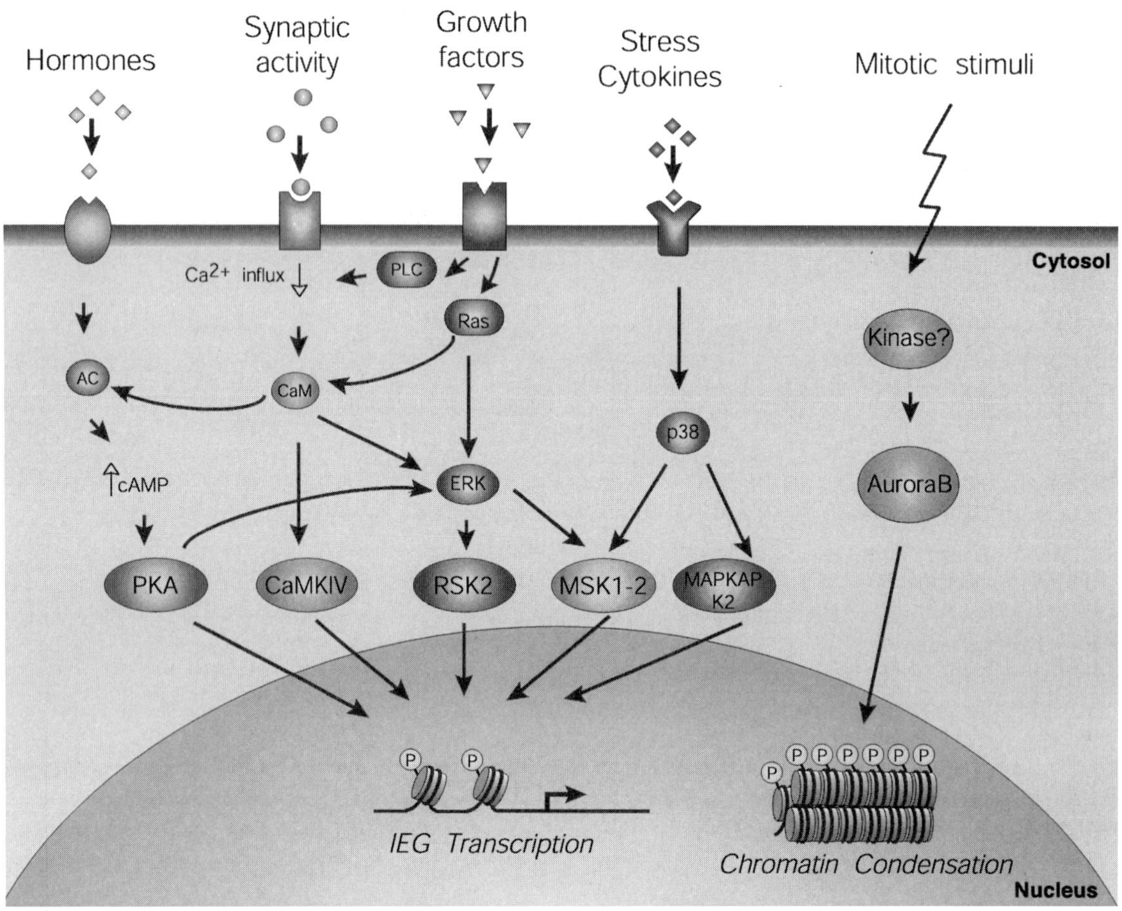

Figure 2 Duality of histone H3 phosphorylation in mammalian cells. A rapid and transient phosphorylation of hitone H3 molecules, located in the nucleosomes of IEG promoters, is achieved after activation of different signaling pathways, upon stimulation of cells with hormones, cytokines, growth factors, or UV light. During mitosis, yet ill-defined cell cycle signals induce a robust histone H3 phosphorylation, which is likely to trigger the chromosome condensation process.

becomes available to interact with the newly phosphorylated CREB. The dynamics of this tripartite regulation fits well the kinetics of early gene transcriptional activation [33].

Histone Phosphorylation and DNA Repair

In addition to transcription regulation, histone phosphorylation is also linked to other cellular processes. For example, several reports have shown that following exposure to various agents that cause double-strand DNA breaks, the histone H2A.X, one of the H2A variants, is rapidly phosphorylated at its C-terminal tail, at serine 139 in mammalian cells [34,35], and at serine 129 in yeast [36]. The number of H2A.X molecules becoming phosphorylated is directly proportional to the amount of double breaks induced and it spans the megabase of DNA around the lesion [34]. H2A.X phosphorylation is not limited to repair of DNA breaks in somatic cells, but it is involved in a number of programmed genomic rearrangements, including recombination in lymphocytes [37,38] and germ cells [39].

Recently mice lacking H2AX have been generated [40]. Mutant mice show pleiotropic phenotypes, including radiation sensitivity, growth retardation, immune deficiency, and male infertility. Those defects seem to be associated with chromosomal instability, repair defects, and impaired recruitment of many DNA repair factors to irradiation-induced foci, implying that H2AX is critical for facilitating the assembly of specific DNA repair complexes on damaged DNA.

Members of the phosphatidylinositol-3-OH kinase-related kinase (PIKK) family have been largely implicated in the response of mammalian cells to double-strand breaks [41], thus making them ideal candidates for the kinase responsible for H2A.X phosphorylation. Although PIKK family members are able to phosphorylate H2A.X *in vitro* and inhibtion of PIKK kinase activity prior to double-strand break abolished H2A.X phosphorylation, repair-deficient cells still display normal phosphorylation of H2A.X, indicating functional redundancy with other as yet ill-defined kinases [42].

Histone Phosphorylation and Apopotis

Chromatin condensation is a morphological hallmark of apoptosis and although commonly used as an indicator of apoptosis, its biochemical basis remains unclear. Different lines of evidence indicate that histone phosphorylation occurs during programmed cell death, but at present there is little evidence of a mechanistic involvement of histone modification in triggering apoptosis in mammalian cells.

Using different mammalian cell lines and a wide range of apoptosis-inducing agents, H2A.X and H2B have been shown to be phosphorylated during apoptosis at, respectively, serine 139 and serine 32 [35,43]. These phosphorylation events initiate around the time of nucleosomal DNA fragmentation at early stages of apoptosis and seem to be dependent on activation of caspases, suggesting that caspase-induced pathways result in phosphorylation of multiple histone species.

More controversial are the data concerning the phosphorylation of histones H3 and H1 during apoptosis. Phosphorylation of histone H3 at serine 10 and histone H1 was observed upon treatment of rat thymocytes with phosphatase inhibitors [44]. Analogous results were obtained using gliotoxin, but not dexamethasone or thapsigargin [45]. In contrast to these findings, others have reported that apoptotic cells show no H1 and H3 phosphorylation [43,46]. The reasons for this discrepancy are not clear.

An analysis by mass spectrometry of posttranslational modification of histones after induction of apoptosis by okadaic acid (OA), a serine-threonine phosphatase inhibitor, has provided some important information [47]. Indeed, very soon (1 hr) after OA treatment, H2A becomes phosphorylated, whereas H3 and H4 phosphorylation became evident only 18 hr after treatment. Interestingly, H4 phosphorylation was not observed when apoptosis was induced by UV light, indicating that this event it is unlikely to occur as a secondary effect of apoptosis. Therefore, the identities of the residues phosphorylated in association with apoptosis—and the roles these modifications may play—still need to be clarified.

Histone Phosphorylation and Mitosis

Phosphorylation of histones H1 and H3 has been linked to mitotic chromatin condensation [48]. H1 hyperphosphorylation is temporally associated with entry into mitosis and depends on Cdc2 kinase activity [49]. Recent studies, however, have revealed that chromatin condensation can occur without H1 hyperphosphorylation [50] or even without H1 itself [51]. Therefore, the biological significance of mitotic H1 hyperphosphorylation remains undefined.

Histone H3 is phosphorylated during mitosis on at least two serine residues located on the N-terminal tail, serines 10 and 28 (for a review, see [52]). Thus, at least for Ser10, two distinct natures characterize its phosphorylation—one linked to gene transcription and local chromatin decondensation (see earlier section), and one coupled to condensation during mitosis. Whereas mitogenically induced Ser10 phosphorylation concerns only a subset of H3 molecules—generating the nuclear speckles mentioned previously—mitotic phosphorylation appears to involve all H3 proteins.

Phosphorylation at serine 10 begins in early G_2 in the pericentrometric heterochromatin of each chromosome and by metaphase spreads throughout all chromosomes, whereas phosphorylation on serine 28 starts to be evident only in early mitosis. The different timing of serine 10 and 28 phosphorylation suggests that diverse signaling routes are utilized for the two sites. Mitotic H3 phosphorylation at serine 10 is required for proper execution of mitosis in *Tetrahymena*, whereas *S. cerevisiae* strains bearing serine>alanine mutations at positions 10 and/or 28 in the H3 tail have no apparent mitotic defects [53]. This may indicate redundancy with modifications on other histones in some specific systems versus others. It has been proposed that H2B phosphorylation can compensate for loss of serine 10 of histone H3 in yeast [53,54]. Moreover, there is evidence for histone H2B as essential for chromosome condensation in *Xenopus* egg extracts [55]. Indeed, phosphorylation on H3 serine 10 appears to be involved in the initiation, but the maintenance, of mammalian chromosome condensation [56]. Additional evidence indicates that the N-terminal tail of histone H3 is able to stably bind its mitotic kinase, inhibiting chromatin condensation in an *in vitro* system [57]. In this study an H3 kinase activity could be sequestered *in vitro* by nucleosome complexes in peptide competition experiments.

Solid evidence has accumulated showing that in yeast, *Drosophila*, nematodes, *Xenopus* and mammals [53,58–61] members of the Aurora kinase family are responsible for mitotic phosphorylation at Ser 10 of histone H3. These findings have not been confirmed in *Aspergillus*, where NIMA was found to act as H3 kinase [62].

The role of Aurora kinases in phosphorylating H3 has also been explored by different approaches. RNAi experiments in *Drosophilla* and *Coenirhabcitis elegans* indicate that targeting Aurora-B causes a reduction of histone H3 phosphorylation during mitosis [36,49], whereas at least in *C. elegans* Aurora-A does not appear to be implicated [49]. In *Xenopus*, both Aurora-A [59] and Aurora-B [58] were identified as potential histone H3 mitotic kinases. In mammals both Aurora-A and Aurora-B can directly phosphorylate H3 Ser 10 *in vitro* and *in vivo*. However, the spatiotemporal expression of Aurora-A and Aurora-B during the mitotic cell cycle of mammalian cells shows that the two kinases have a mutually exclusive localization: While Aurora-A is present in the centromeres, Aurora-B colocalizes with the phosphorylated form of histone H3. These results strongly implicate Aurora-B as the mitotic kinase for H3 (Fig. 2).

Conclusions

This review focused on a major histone modification, phosphorylation, and on its implication in different cellular process (Fig. 3). Increasing evidence shows that several types of modifications are linked and, in particular, one modification may influence the presence of another nearby modification. This has been demonstrated for acetylation at Ser10 and phosphorylation at Lys14 on the histone H3 tail

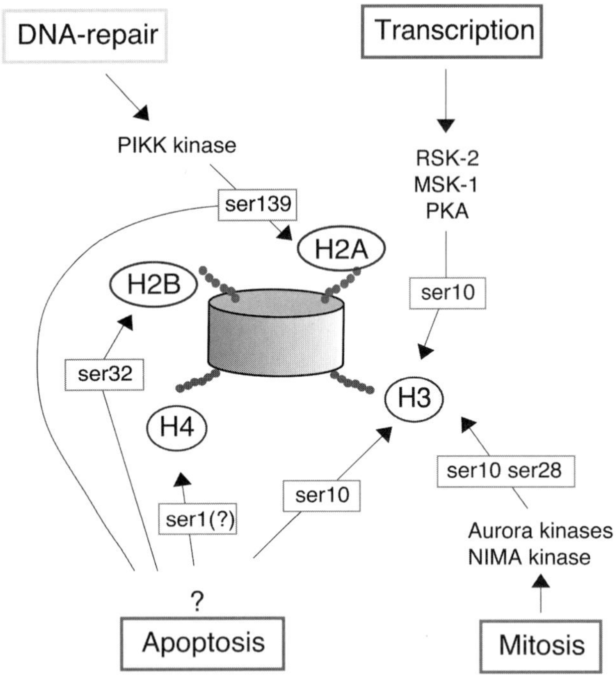

Figure 3 Multiple cellular processes are associated with histone phosphorylation. DNA damage induces phosphorylation in H2A-Ser132; transcription, upon mitogenic stimulation, on H3-Ser10; mitosis on H3-Ser10 and Ser28; apoptosis, depending on the stimulus used, on H4-Serl, H3-Ser10, H2B-Ser32, and H2A-Ser32. The putative kinases responsible for these phosphorylation events are indicated.

(see earlier section) and for phosphorylation at Ser10 and methylation at Lys9 on the same tail [63]. While the first pair modifications have been coupled to activation of gene expression, increasing evidence indicates that methylation at Lys9 results in decreased Ser10 phosphorylation and is thereby responsible for silencing. In addition to an intrahistonic *histone code* operating on the individual tail of a specific histone, interhistone codes may exist. An area of exciting research will be to study the influence that phosphorylation on a specific histone may have on the phosphorylation (or other modifications) on a tail of a different histone. The large number of possible signaling pathways involved in the many physiological scenarios discussed here (Fig. 2) provides a picture in which an impressive number of combinatorial possibilities exist. How can the same histone modification be linked with such a wide range of "chromatin states"? We favor a scenario in which histone tails function to integrate upstream signals that, in turn, impart regulatable dynamic changes to the chromatin structure of localized regions of the genome. Mitogenically induced phosphorylation of histone H3 on serine 10 is thought to cause a reduction of the overall positive charge, inducing a local decondensation. This decondensation event may be crucial for the binding of regulatory factors that could induce either chromatin condensation or transcription. Gaining further understanding of the molecules involved in signaling to chromatin and how phosphorylation can be targeted in a determined locus will be an exciting area for future investigations.

Note added in proofs

The field of histone modifications is moving very fast and some additional references need to be pointed out.

1. Two recent reports indicate that phosphorylation of histone H3 at Ser10 may occur in response to cytokine stimulation by IKK-α. This event was coupled to activation of NF-κB-responsive genes:
Y. Yamamoto, U. N. Verma, S. Prajapati, Y. T. Kwak, and R. B. Gaynor, *Nature* **423**, 655 (2003); V. Anest, J. L. Hanson, P. C. Cogswell, K. A. Steinbrecher, B. D. Strahl, and A. S. Baldwin, *Nature* **423**, 659 (2003).
2. An histone modification event that is uniquely associated with apoptotic chromatin, in species ranging from frogs to humans, is H2B phosphorylation at position Ser14. The kinase involved in this event was recently identified as Mst1 (mammalian sterile twenty), whose function is directly regulated by caspase-3.
W. L. Cheung, K. Ajiro, K. Samejima, M. Kloc, P. Cheung, C. A. Mizzen, A. Beeser, L. D. Etkin, J. Chernoff, W. C. Earnshaw, and C. D. Allis, *Cell* **113**, 507 (2003).
3. A case of regulation of inter-histone modifications has been reported, suggestive that several others may exist, all that will contribute to the 'histone code'.
Sun, Z. W. and Allis, C. D. *Nature* **418**, 104 (2002).

References

1. Peterson, C. L. (2001). Chromatin: Mysteries sovled? *Biochem. Cell Biol.* **79**, 219–225.
2. Narlikar, G. J., Fan, H. Y., and Kingston, R. E. (2002). Cooperation between complexes that regulate chromatin structure and transcription. *Cell* **108**, 475–487.
3. Berger, S. L. (2002). Histone modifications in transcriptional regulation. *Curr. Opin. Genet. Dev.* **12**, 142–148.
4. Peterson, C. L. and Workman, J. L. (2000). Promoter targeting and chromatin remodeling by the SWI/SNF complex. *Curr. Opin. Genet. Dev.* **10**, 187–192.
5. Becker, P. B. and Horz, W. (2002). ATP-dependent nucleosome-remodelling factors. *Nat. Rev. Mol. Cell Biol.* **3**, 422–429.
6. Emerson, B. M. (2002). Specificity of gene regulation. *Cell* **109**, 267–270.
7. Tsukiyama, T. (2002). The *in vivo* functions of ATP-dependent chromatin-remodeling factors. *Nat. Rev. Mol. Cell Biol.* **3**, 422–429.
8. Richmond, T. J., Rechsteiner, T., and Luger, K. (1993). Studies of nucleosome structure. *Cold Spring Harb. Symp. Quant. Biol.* **58**, 265–272.
9. Wolffe, A. P. (1998). *Chromatin: Structure and Function.* Academic Press, San Diego.
10. Whitlock, J. P., Jr. and Simpson, R. T. (1977). Localization of the sites along nucleosome DNA which interact with NH2-terminal histone regions. *J. Biol. Chem.* **252**, 6516–6520.
11. Ausio, J., Dong, F., and van Holde, K. E. (1989). Use of selectively trypsinized nucleosome core particles to analyze the role of the histone "tails" in the stabilization of the nucleosome. *J. Mol. Biol.* **206**, 451–463.
12. Luger, K., Mader, A. W., Richmond, R. K., Sargent, D. F., and Richmond, T. J. (1997). Crystal structure of the nucleosome core particle at 2.8 A resolution. *Nature* **389**, 251–260.
13. Dixon, G. H., Candido, E. P. M., Honda, B. M., Louie, A. J., MacLeod, A. R., and Sung, M. T. (1975). The structure and function of chromatin. *Ciba Foundation Symp.* **28**, 229–256.

14. Allegra, P., Sterner, R., Clayton, D. F., and Allfrey, V. G. (1987). Affinity chromatographic purification of nucleosomes containing transcriptionally active DNA sequences. *J. Mol. Biol.* **196**, 379–388.
15. Tazi, J. and Bird, A. (1990). Alternative chromatin structure at CpG islands. *Cell* **60**, 909–920.
16. Taylor, S. S. (1982). The *in vitro* phosphorylation of chromatin by the catalytic subunit of cAMP-dependent protein kinase. *J. Biol. Chem.* **257**, 6056–6063.
17. Mazen, A., Hacques, M. F., and Marion, C. (1987). H3 phosphorylation-dependent structural changes in chromatin. Implications for the role of very lysine-rich histones. *J. Mol. Biol.* **194**, 741–745.
18. Jenuwein, T. and Allis, C. D. (2001). Translating the histone code. *Science* **293**, 1074–1080.
19. Rice, J. C. and Allis, C. D. (2001). Code of silence. *Nature* **414**, 258–261.
20. Hendzel, M. J., Wei, Y., Mancini, M., Van Hooser, A., Ranalli, T., Brinkley, B. R., Bazett-Jones, D. P., and Allis, C. D. (1997). Mitosis-specific phosphorylation of histone H3 initiates primarily within pericentromeric heterochromatin during G_2 and spreads in an ordered fashion coincident with mitotic chromosome condensation. *Chromosoma* **106**, 348–360.
21. Mahadevan, L. C., Willis, A. C., and Barratt, M. J. (1991). Rapid histone H3 phosphorylation in response to growth factors, phorbol esters, okadaic acid, and protein synthesis inhibitors. *Cell* **65**, 775–783.
22. Chadee, D. N., Hendzel, M. J., Tylipski, C. P., Allis, C. D., Bazett-Jones, D. P., Wright, J. A., and Davie, J. R. (1999). Increased Ser-10 phosphorylation of histone H3 in mitogen-stimulated and oncogene-transformed mouse fibroblasts. *J. Biol. Chem.* **274**, 24914–24920.
23. Thomson, S., Clayton, A. L., Hazzalin, C. A., Rose, S., Barratt, M. J., and Mahadevan, L. C. (1999). The nucleosomal response associated with immediate-early gene induction is mediated via alternative MAP kinase cascades: MSK1 as potential histone H3/HMG-14 kinase. *EMBO J.* **18**, 4779–4793.
24. Crosio, C., Cermakian, N., Allis, C. D., and Sassone-Corsi, P. (2000). Light induces chromatin modification in cells of the mammalian circadian clock. *Nat. Neurosci.* **3**, 1241–1247.
25. Sassone-Corsi, P., Mizzen, C. A., Cheung, P., Crosio, C., Monaco, L., Jacquot, S., Hanauer, A., and Allis, C. D. (1999). Requirement of Rsk-2 for epidermal growth factor-activated phosphorylation of histone H3. *Science* **285**, 886–891.
26. De Cesare, D., Jacquot, S., Hanauer, A., and Sassone-Corsi, P. (1998). Rsk-2 activity is necessary for epidermal growth factor-induced phosphorylation of CREB protein and transcription of c-*fos* gene. *Proc. Natl. Acad. Sci. USA* **95**, 12202–12207.
27. Arthur, J. S. and Cohen, P. (2000). MSK1 is required for CREB phosphorylation in response to mitogens in mouse embryonic stem cells. *FEBS Lett.* **482**, 44–48.
28. Salvador, L. M., Park, Y., Cotton, J., Maizels, E. T., Jones, J. C., Schillace, R. V., Carr, D. W., Cheung, P., Allis, C. D., Jameson, J. L. *et al.* (2001). Follicle-stimulating hormone stimulates protein kinase A-mediated histone H3 phosphorylation and acetylation leading to select gene activation in ovarian granulosa cells. *J. Biol. Chem.* **276**, 40146–40155.
29. Lo, W. S., Duggan, L., Tolga, N. C., Emre, Belotserkovskya, R., Lane, W. S., Shiekhattar, R., and Berger, S. L. (2001). Snf1—a histone kinase that works in concert with the histone acetyltransferase Gcn5 to regulate transcription. *Science* **293**, 1142–1146.
30. Barratt, M. J., Hazzalin, C. A., Zhelev, N., and Mahadevan, L. C. (1994). A mitogen- and anisomycin-stimulated kinase phosphorylates HMG-14 in its basic amino-terminal domain *in vivo* and on isolated mononucleosomes. *EMBO J.* **13**, 4524–4535.
31. Cheung, P., Tanner, K. G., Cheung, W. L., Sassone-Corsi, P., Denu, J. M., and Allis, C. D. (2000). Synergistic coupling of histone H3 phosphorylation and acetylation in response to epidermal growth factor stimulation. *Mol. Cell* **5**, 905–915.
32. Clayton, A. L., Rose, S., Barratt, M. J., and Mahadevan, L. C. (2000). Phosphoacetylation of histone H3 on c-*fos*- and c-*jun*-associated nucleosomes upon gene activation. *EMBO J.* **19**, 3714–3726.
33. Merienne, K., Pannetier, S., Harel-Bellan, H., and Sassone-Corsi, P. (2001). Mitogen-regulated RSK2–CBP interaction controls reciprocal kinase and acetylase activities. *Mol. Cell Biol.* **21**, 7089–7096.
34. Rogakou, E. P., Pilch, D. R., Orr, A. H., Ivanova, V. S., and Bonner, W. M. (1998). DNA double-stranded breaks induce histone H2AX phosphorylation on serine 139. *J. Biol. Chem.* **273**, 5858–5868.
35. Rogakou, E. P., Nieves-Neira, W., Boon, C., Pommier, Y., and Bonner W. M. (2000). Initiation of DNA fragmentation during apoptosis induces phosphorylation of H2AX histone at serine 139. *J. Biol. Chem.* **275**, 9390–9395.
36. Downs, J. A., Lowndes, N. F., and Jackson, S. P. (2002). A role for *Saccharomyces cerevisiae* histone H2A in DNA repair. *Nature* **408**, 1001–1004.
37. Chen, H. T., Bhandoola, A., Difilippantonio, M. J., Zhu, J., Brown, M. J., Tai, X., Rogakou, E. P., Brotz, T. M., Bonner, W. M., Ried, T. *et al.* (2000). Response to RAG-mediated VDJ cleavage by NBS1 and gamma-H2AX. *Science* **290**, 1962–1965.
38. Petersen, S., Casellas, R., Reina-San-Martin, B., Chen, H. T., Difilippantonio, M. J., Wilson, P. C., Hanitsch, L., Celeste, A., Muramatsu, M., Pilch, D. R. *et al.* (2001). AID is required to initiate Nbs1/gamma-H2AX focus formation and mutations at sites of class switching. *Nature* **414**, 660–665.
39. Mahadevaiah, S. K., Turner, J. M., Baudat, F., Rogakou, E. P., de Boer, P., Blanco-Rodriguez, J., Jasin, M., Keeney, S., Bonner, W. M., and Burgoyne, P. S. (2001). Recombinational DNA double-strand breaks in mice precede synapsis. *Nat. Genet.* **27**, 271–276.
40. Celeste, A., Petersen, S., Romanienko, P. J., Fernandez-Capetillo, O., Chen, H. T., Sedelnikova, O. A., Reina-San-Martin, B., Coppola, V., Meffre, E., Difilippantonio, M. J. *et al.* (2002). Genomic instability in mice lacking histone H2AX. *Science.* Published online 4 April 2002; 10.1126/science.1069398.
41. Zhou, B. B. and Elledge, S. J. (2000). The DNA damage response: Putting checkpoints in perspective. *Nature* **408**, 433–439.
42. Modesti, M. and Kannar, R. (2001). DNA repair. Spot(light)s on chromatin. *Curr. Biol.* **11**, R229–R232.
43. Ajiro, K. (2000). Histone H2B phosphorylation in mammalian apoptotic cells. An association with DNA fragmentation. *J. Biol. Chem.* **275**, 439–443.
44. Lee, E., Nakatsuma, A., Hiraoka, R., Ishikawa, E., Enomoto, R., and Yamauchi, A. (1999). Involvement of histone phosphorylation in thymocyte apoptosis by protein phosphatase inhibitors. *IUBMB Life* **48**, 79–83.
45. Waring, P., Khan, T., and Sjaarda, A. (1997). Apoptosis induced by gliotoxin is preceded by phosphorylation of histone H3 and enhanced sensitivity of chromatin to nuclease digestion. *J. Biol. Chem.* **272**, 17929–17936.
46. Hendzel, M. J., Nishioka, W. K., Raymond, Y., Allis, C. D., Bazett-Jones, D. P., and Th'ng, J. P. (1998). Chromatin condensation is not associated with apoptosis. *J. Biol. Chem.* **273**, 24470–24478.
47. Galasinski, S. C., Louie, D. F., Gloor, K. K., Resing, K. A., and Ahn, N. G. (2002). Global regulation of post-translational modifications on core histones. *J. Biol. Chem.* **277**, 2579–2588.
48. Hirano, T. (2000). Chromosome cohesion, condensation, and separation. *Annu. Rev. Biochem.* **69**, 115–144.
49. Langan, T. A., Gautier, J., Lohka, M., Hollingsworth, R., Moreno, S., Nurse, P., Maller, J., and Sclafani, R. A. (1989). Mammalian growth-associated H1 histone kinase: A homolog of cdc2+/CDC28 protein kinases controlling mitotic entry in yeast and frog cells. *Mol. Cell Biol.* **9**, 3860–3868.
50. Guo, X. W, Th'ng, J. P., Swank, R. A., Anderson, H. J., Tudan, C., Bradbury E. M., and Roberge, M. (1995). Chromosome condensation induced by fostriecin does not require p34cdc2 kinase activity and histone H1 hyperphosphorylation, but is associated with enhanced histone H2A and H3 phosphorylation. *EMBO J.* **14**, 976–985.
51. Ohsumi, K., Katagiri, C., and Kishimoto, T. (1993). Chromosome condensation in *Xenopus* mitotic extracts without histone H1. *Science* **262**, 2033–2035.

52. Hans, F. and Dimitrov, S. (2001). Histone H3 phosphorylation and cell division. *Oncogene* **20**, 3021–3027.
53. Hsu, J. Y., Sun, Z. W., Li, X., Reuben, M., Tatchell, K., Bishop, D. K., Grushcow, J. M., Lin, R., Mitchell Smith, M., and Allis, C. D. (2000). Mitotic phosphorylation of histone H3 is governed by Ipl1/aurora kinase and Glc7/PP1 phosphatase in budding yeast and nematodes. *Cell* **102**, 279–291.
54. Cheung, P., Allis, C. D., and Sassone-Corsi, P. (2000). Signaling to chromatin through histone modifications. *Cell* **103**, 263–271.
55. de la Barre, A. E., Angelov, D., Molla, A., and Dimitrov, S. (2001). The N-terminus of histone H2B, but not that of histone H3 or its phosphorylation, is essential for chromosome condensation. *EMBO J.* **20**, 6383–6393.
56. Van Hooser, A., Goodrich, D. W., Allis, C. D., Brinkley, B. R., and Mancini, M. A. (1998). Histone H3 phosphorylation is required for the initiation, but not maintenance, of mammalian chromosome condensation. *J. Cell. Sci.* **111**, 3497–3506.
57. de la Barre, A. E., Gerson, V., Gout, S., Creaven, M., Allis, C. D., Dimitrov, S. (2000). Core histone N-termini play an essential role in mitotic chromosome condensation. *EMBO J.* **19**, 379–391.
58. Giet, R. and Glover, D. M. (2001). *Drosophila* Aurora B kinase is required for histone H3 phosphorylation and condensin recruitment during chromosome condensation and to organize the central spindle during cytokinesis. *J. Cell Biol.* **152**, 669–682.
59. Scrittori, L., Hans, F., Angelov, D., Charra, M., Prigent, C., and Dimitrov, S. (2001). pEg2 Aurora-a kinase, histone H3 phosphorylation and chromosome assembly in xenopus egg extract. *J. Biol. Chem.* **11**, 11.
60. Murnion, M. E., Adams, R. R., Callister, D. M., Allis, C. D., Earnshaw, W. C., and Swedlow, J. R. (2001). Chromatin-associated protein phosphatase 1 regulates Aurora-B and histone H3 phosphorylation. *J. Biol. Chem.* 26656–26665.
61. Crosio, C., Fimia, G. M., Loury, R., Kimura, M., Okano, Y., Zhou, H., Sen, S., Allis, C. D., and Sassone-Corsi, P. (2002). Mitotic phosphorylation of histone H3: Spatio-temporal regulation by mammalian Aurora kinases. *Mol. Cell Biol.* **22**, 874–885.
62. De Souza, C. P., Osmani, A. H., Wu, L. P., Spotts, J. L., and Osmani, S. A. (2000). Mitotic histone H3 phosphorylation by the NIMA kinase in *Aspergillus* nidulans. *Cell* **102**, 293–302.
63. Rea, S., Eisenhaber, F., O'Carroll, D., Strahl, B. D., Sun, Z. W., Schmid, M., Opravil, S., Mechtler, K., Ponting, C. P., Allis, C. D. *et al.* (2000). Regulation of chromatin structure by site-specific histone H3 methyltransferases. *Nature* **406**, 593–599.

CHAPTER 285

Multigene Family of Transcription Factor AP-1

Peter Angel

Division of Signal Transduction and Growth Control, Deutsches Krebsforschungszentrum, Heidelberg, Germany

Introduction

Much of our present knowledge about transcription factors comes from the discovery and study of the activating protein 1 (AP-1) family. AP-1 (and the transcription factor NFκB) has served to detect one of the decisive DNA-binding motifs required for regulation of both basal and inducible transcription of several genes containing AP-1 sites (5'-TGAG/CTCA-3′), also known as TPA-responsive elements (TRE), by a variety of extracellular signals. These include growth factors; cytokines; tumor promoters, such as the phorbol ester TPA (12-*O*-tetradecanoyl-phorbol-13-acetate) and carcinogens; UV irradiation; and other DNA-damaging agents. The core of the AP-1 family of transcription factors is formed by heterodimeric associations of members of the Jun (c-Jun, JunB, and JunD), Fos (with c-Fos, FosB, Fra-1, and Fra-2), and ATF (ATFa, ATF-2, and ATF-3) proteins. In this chapter we summarize our current knowledge of (1) the characteristic structural features of AP-1 subunits, (2) mechanisms involved in modulation of AP-1 activity, and (3) function of AP-1 proteins in cellular processes using mouse genetics.

General Structure of AP-1 Subunits

According to their function in controlling gene expression, AP-1 subunits are composed of a region responsible for binding to a specific DNA recognition sequence (DNA-binding domain, DBD) and a second region that is required for transcriptional activation (transactivation domain) once the protein is bound to DNA. The DBD of AP-1 proteins, also known as the *bZip* region, can be divided into two evolutionarily conserved, independently acting domains: the *basic domain* ("b"), which is rich in basic amino acids and responsible for contacting the DNA, and the *leucine zipper* ("Zip") region, which is characterized by heptad repeats of leucines forming a coiled-coil structure that is responsible for dimerization (Fig. 1). In addition to the leucines, other hydrophobic and charged amino acid residues within the leucine zipper region are responsible for specificity and stability of homo- or heterodimer formation between the various Jun, Fos, or ATF proteins. Jun-Jun and Jun-Fos dimers preferentially bind to the 7-base-pair (bp) motif 5′-TGAG/CTCA-3′ whereas Jun-ATF dimers or ATF homodimers prefer to bind to a related, 8-bp consensus sequence 5′-TTACCTCA-3′ [1,2]. Therefore, the characteristics of the AP-1 DNA-binding sites in promoters as well as the abundance of the individual AP-1 subunits are decisive for the selection of target genes.

In addition to the "classical" AP-1 members (Jun, Fos, ATFs), other bZip proteins have been discovered, some of which can heterodimerize with the core AP-1 subunits, for example, Maf, Maf-related proteins, Nrl, Smads, and Jun-dimerizing partners (JDPs). Binding of AP-1 to DNA may also support binding of other transcription factors to adjacent or overlapping binding sites (composite elements) to allow the formation of "quaternary" complexes. The interaction of NF-AT and Ets proteins with the interleukin 2 (IL-2) and collagenase promoters, respectively, may serve as paradigms for this type of protein–protein interaction [3,4]. AP-1 has been reported to associate with NFκB [5], the glucocorticoid receptor [6,7], and hypoxia-inducible factor 1α [8].

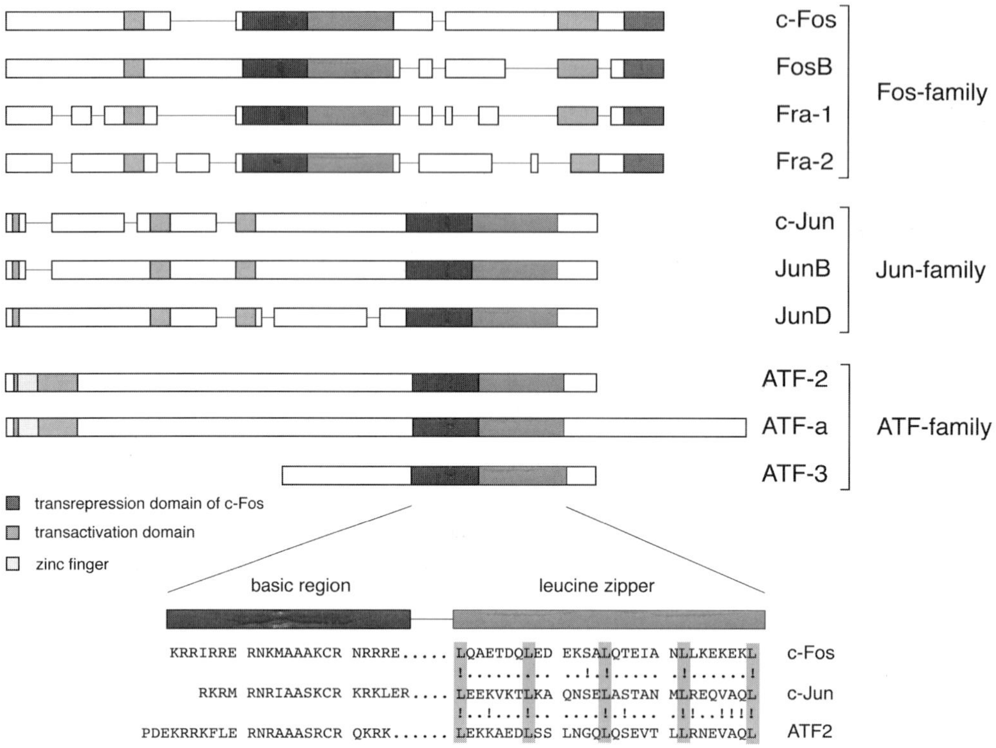

Figure 1 Structural organization of the Fos, Jun, and ATF proteins. Conserved regions are highlighted. On the bottom, the amino acid sequences of the bZip region of c-Fos, c-Jun, and ATF-2, the most extensively analyzed members of the Fos, Jun, and ATF protein families, are shown.

In contrast to the well-defined DBDs of AP-1 proteins, the structural properties of the domains mediating transcriptional activation of target genes (transactivation domain, TAD) are still ill defined. The TAD (and its function) can be transferred to heterologous DBDs that do not require heterodimerization, for example, to that of the yeast transcription factor GAL4. Such chimeric proteins are permitted to identify critical amino acids in TADs and to recognize that the TADs of individual Jun, Fos, and ATF proteins differ in their transactivation potential. Under specific circumstances, some subunits, such as Fra-1 or JunB, may even act as repressors of AP-1 activity by competitive binding to AP-1 sites, or by forming inactive heterodimers with other AP-1 members [9–11].

Transcriptional and Posttranslational Control of AP-1 Activity

Given the aspects of subunit heterogeneity, their dimerization and DNA binding rules, the control of transcription of AP-1 subunit genes, the time course of subunit synthesis, and the regulation of their function become important issues for an evaluation of the effect on target gene expression and on dependent phenotypes. The *jun* and *fos* genes represent the prototype of the class of "immediate-early" genes that are characterized by a rapid and transient activation of transcription in response to changes of environmental conditions, such as growth factors, cytokines, tumor promoters, carcinogens, and expression of certain oncogenes [12]. Because this type of regulation of promoter activity is also observed in the absence of ongoing protein synthesis, it is generally accepted that preexisting factors, whose activity gets altered by changes in posttranslational modification (in particular phosphorylation), are responsible for the regulation of promoter activity.

Most of our current knowledge on transcriptional activation of immediate early genes is derived from studies on deletion and point mutations the *c-fos* and *c-jun* promoters, combined with *in vitro* and *in vivo* footprinting analyses. The serum-response element (SRE), which is bound by a ternary complex containing the transcription factors p67-SRF and p62-TCF, is required for the majority of extracellular stimuli including growth factors and phorbol esters. Changes in the phosphorylation pattern of SRF and, predominantly, TCF regulate *c-fos* promoter activity by these stimuli. Other elements include the c-AMP response element (CRE) and the Sis-inducible enhancer (SIE), which is recognized by the STAT group of transcription factors. The very transient induction of promoter activity by extracellular stimuli can be explained by increased phosphatase activity counteracting the activity of upstream protein kinases, which address promoter-associated transcription factors. On the other hand, negative autoregulation by newly synthesized c-Fos may also play an important role [13,14].

Analysis of deletion mutants of the *c-jun* promoter identified two AP-1-like binding sites (Jun1, Jun2), which are recognized by Jun/ATF heterodimers or ATF homodimers and which are involved in transcriptional regulation in response to the majority of extracellular stimuli affecting *c-jun* transcription. In response to G-protein-coupled receptor activation, EGF, and other growth factors, the AP-1 sites and an additional element in the *c-jun* promoter recognized by MEF2 proteins cooperate in transcriptional control of the *c-jun* gene. Similar to the factors binding to the *c-fos* promoter, the activity of factors binding to the *c-jun* promoter is regulated by their phosphorylation status (for review, see [12,15]).

The most critical members of the class of protein kinases regulating the activity of AP-1 in response to extracellular stimuli are mitogen-activated protein kinases (MAPKs). Depending on the type of stimuli, these proline-directed kinases can be dissected into three subgroups: the extracellular signal-regulated kinases (Erk-1, Erk-2), which are robustly activated by growth factors and phorbol esters, but only weakly activated by cytokines and cellular stress-inducing stimuli (UV irradiation, chemical carcinogens). In contrast Jun-N-terminal kinases (JNK-1, -2, -3), also known as stress-activated kinases (SAPKs), and a structurally related class, p38 MAP kinases (p38α, -β, -γ), are strongly activated by cytokines and environmental stress, but are poorly activated by growth factors and phorbol ester. These kinases themselves are under strict control of upstream kinases and phosphatases, which are part of individual signaling pathways initiated by specific classes of extra- and intracellular stimuli (growth factors, DNA-damaging agents, oncoproteins). This network, which exhibits a high degree of evolutionary conservation between yeast, *Drosophila*, and mammals is, however, far too complex to be discussed in greater detail in this review (for in-depth information on this subject, see [16,17]).

Erk1 and Erk2 mediate mitogen-stimulated phosphorylation of sites in the TAD of TCF proteins, resulting in enhanced transactivation activity. The JNK/SAPKs were originally identified by their ability to specifically phosphorylate c-Jun at two positive regulatory sites (Ser-63 and Ser-73) residing within the TAD [18]. Hyperphosphorylation of both sites is observed in response to stress stimuli and oncoproteins. The JNKs can also phosphorylate and stimulate the transcriptional activity of ATF-2 [19,20]. The same positive sites on ATF-2 also serve as phosphoacceptor sites for p38, whereas Ser-63 and -73 of c-Jun are not affected by p38 [21,22]. Most likely, hyperphosphorylation of Jun, ATF, and TCF proteins results in a conformational change of the TAD, allowing more efficient interaction with cofactors, which facilitate and stabilize the connection with the RNA polymerase II/initiation complex to enhance transcription of target genes. Interestingly, in addition to enhanced transactivation, hyperphosphorylation of the TAD of c-Jun also regulates the stability of c-Jun by reducing ubiquitin-dependent degradation of c-Jun [23,24]. Similarly, phosphorylation-dependent changes in the half-life of c-Fos have been observed [25].

The DBD of c-Jun is phosphorylated at multiple sites by GSK-3 and/or casein kinase II (CK-II), which results in reduced DNA binding. In response to extracellular stimuli, such as UV, phosphorylation is reduced, thus enhancing DNA binding [26,27]. The mechanism (reduced activity of the kinase or enhanced activity of a phosphatase) has not yet been definitively clarified.

In addition to phosphorylation, other posttranslational modifications regulate AP-1 activity [28]. Oxidation of a cysteine in the basic region inhibits DNA binding and there is pronounced regulation of nuclear localization. Moreover, positive and negative interference of other cellular proteins with AP-1 activity (in addition to the protein kinases and coactivators described earlier) has been identified in, for example, Maf, MyoD, YY1, STAT, SMADs and Menin [2]. The mutual interference between AP-1 and nuclear receptors, particularly the glucocorticoid receptor (GR), which may be involved in the anti-inflammatory and immunosuppressive activities of glucocorticoids, represents the most extensively analyzed example for this type of cross-talk [6,7].

Function of Mammalian AP-1 Subunits: Lessons from Loss-of-Function Approaches in Mice

As described earlier, AP-1 activity is enhanced when cells are stimulated by agents that promote cell proliferation. Moreover, oncogenic versions of c-Jun and c-Fos have been isolated from retroviruses, and various membrane-associated or cytoplasmic oncogenes (Ras, Src, Raf) permanently upregulate AP-1 abundance as part of their transforming capacity, suggesting that AP-1 members play an important role in cell proliferation and transformation [28–30]. Initial evidence for this assumption has been obtained by overexpression studies in mice and various cell lines, by blocking AP-1 activity either through expression of a transdominant-negative c-Jun mutants, by expression of antisense sequences, or by microinjection of Jun- and Fos-specific antibodies. Under these conditions cell cycle progression was disturbed and the efficiency of oncoprotein-mediated cell transformation was reduced. However, already in these systems, different lines of evidence, including variations in the expression pattern and phosphorylation status of AP-1 members during the cell cycle [31,32], suggested that the members of the Jun and Fos families play specific roles during these processes or may even antagonize each other. The generation of mice harboring genetic disruption and/or transgenic overexpression, as well as the availability of genetically defined mutant cells isolated from these animals, represents a major breakthrough in our understanding of the regulatory functions of AP-1 subunits. The distinct phenotypes of the individual knock-out mice (Table I), induced by defects in cells or tissues in which the subunit was particularly important or where its absence became first limiting, support the notion that AP-1 dimers exhibit specific and independent functions *in vivo*. As a general rule derived from all studies, the AP-1 subunits must be present in a complementary and coordinated manner to ensure proper development or physiology of the organism.

Loss of c-Jun, JunB, or Fra-1 results in embryonic lethality. c-Jun null embryos die at midgestation (E12.5 to 13.5 [33,34]) due to dysregulation in liver and heart development [35]. Mice carrying a conditional (floxed) allele of *c-jun*, survive

Table I

AP-1 member	Knockout/Knockin mice	References
c-Jun	Complete loss: embryonic lethal E13, hepatic failure, heart defect	33–35
	Liver-specific loss: viable, impaired postnatal hepatocyte proliferation and liver regeneration	36
junB in *c-jun* locus	Rescue of *c-jun*$^{-/-}$ phenotype by JunB (cardiac and liver defects); rescue of c-Jun/c-Fos-, but not c-Jun/ATF-dependent gene expression	39
JunB	Embryonic lethal E8.5-E10.0, placentation defect, Adult: myeloproliferative disease	40, 41
JunD	Viable, reduced postnatal growth, age-dependent defects in male reproductive function	50
c-Fos	Viable, defective bone remodelling, osteopetrosis, light-induced apoptosis of photoreceptors	43–46
fra-1 in *c-fos* locus	Dose-dependent rescue of the *c-fos*$^{-/-}$ phenotype (bone remodeling, photoreceptor apoptosis)	48
FosB	Viable, nurturing defect	50
Fra-1	Embryonic lethal E10, placenta defect	42
ATF-2	Hypomorth allele: decreased postnatal viability and growth, defects in endochondral ossification, ataxic gait, hyperactivity, decreased hearing, decreased number of Purkinje cells	53
	Complete loss: Neonatal lethality, Meconium aspiration syndrome, decreased cytotrophoblasts in placenta	54

the specific postnatal gene inactivation in hepatocytes but show an impaired liver regeneration in response to partial hepatectomy [36].

Knock-in mice expressing a mutated c-Jun protein in which the N-terminal phosphorylation sites at serines 63 and 73 were changed into alanines (JunAA mice) developed normally and were viable and fertile as adults [37]. However, these JunAA mice are resistant to epileptic seizures and neuronal apoptosis induced by the excitatory amino acid kainate, and JunAA fibroblasts show proliferation defects as well as apoptotic defects on stress induction [37]. By contrast, T-cell proliferation and differentiation appear to be independent of c-Jun N-terminal phosphorylation, whereas efficient T-cell-receptor-induced thymocyte apoptosis is affected [38].

Interestingly, the JunB protein, when expressed at sufficient levels, can substitute at least for some of c-Jun's functions. Knock-in mice having the *c-jun* allele replaced by *junB* undergo normal embryogenesis and develop a normal liver and heart. Analysis of indicative marker genes showed that expression of genes regulated by Jun/Fos, but not those regulated by Jun/ATF dimers, are restored, thereby rescuing c-Jun-dependent defects *in vivo* as well as in primary fibroblasts and fetal hepatoblasts *in vitro* [39].

Targeted inactivation of JunB resulted also in embryonic lethality between days 8.5 and 10.0 due to multiple defects in extraembryonic tissues [40]. Affected cell types/organs comprise the trophoblast giant cells, yolk sac mesenthelium and placental labyrinth. The observed phenotypes in JunB null embryos appear to result from severe impairment of general vasculogenic and angiogenic processes resulting in a failure to establish proper vascular interactions with the maternal circulation. The lethal phenotype can be rescued by an ubiquitously expressed *junB* transgene [40]. In such rescued mice lacking JunB in the myeloid lineage, a transplantable myeloproliferative disease resembling human chronic myeloid leukemia was observed [41].

Fra-1 is so far the only Fos member essential for embryonic development. Fra-1 null embryos died between E10.0 and 10.5 [42]. Similar to JunB null embryos, the vascularization of the placental labyrinth was impaired, suggesting that JunB and Fra-1, possibly as heterodimers, address common target genes responsible for the generation of a functional placental labyrinth. Fra-1-deficient embryos can be rescued from embryonic death by JunB overexpression [42].

Fos-null mice are viable and fertile but suffer from severe osteopetrosis caused by lack of mature osteoclasts [43–45]. Moreover, light-induced apoptosis of photoreceptors is lost in c-Fos-deficient mice [46]. The c-Fos-dependent functions in bone cells can be substituted by Fra-1, when the *fos* locus is deleted and replaced by the *fosl1* gene encoding Fra-1. *Fosl1* is a transcriptional target of c-Fos during osteoclast differentiation [47] and can fully complement for the lack of c-Fos in bone development in a gene-dosage-dependent manner. However, Fra-1 is not able to induce expression of c-Fos target genes in fibroblasts derived from the knock-in mice [48], suggesting the need for additional tissue-specific factors.

The embryonic development of FosB-deficient mice is normal [49]. Adult *fosB*$^{-/-}$ females, however, nurture insufficiently [50]. JunD-null mice develop normally but postnatal growth of homozygous *junD*$^{-/-}$ animals is reduced. JunD null males develop age-dependent defects in reproduction, hormone imbalance, and impaired spermatogenesis [51]. ATF-2 mutant mice that express small amounts of a mutant ATF-2 protein (ATF-2$^{m/m}$; [52]) are chondrodysplastic and neurologically abnormal [53], which might be explained by reduced expression of cyclin D1 in chondrocytes [52]. ATF-2-null mutant mice die shortly after birth and suffer from a disease resembling a severe type of human meconium aspiration syndrome [54].

Knock-out mice represent an excellent system for the isolation of genetically defined mutant cells. Primary and immortalized fibroblasts can be isolated from almost all mice lacking individual AP-1 members. Analysis of these cells revealed that c-Jun acts as positive regulator of the cell cycle by suppressing p53 and, indirectly, the p53 target gene p21 [55]. Moreover, loss of c-Jun results in reduced cyclin D1 activity [56]. JunB, on the other hand, serves as a negative regulator of cell cycle progression by induction of the cyclin-CDK inhibitor p16 and down-regulation of c-Jun and cyclin D1 expression [32,57]. Data from fibroblasts lacking both c-Fos and FosB established a critical role for these AP-1 subunits in cyclin D expression [58]. Moreover, fibroblasts lacking either *c-jun* or *c-fos* cannot be transformed by oncogenes, such as Ras and Src [59]. JunD-deficient fibroblasts exhibit specific alterations in cell proliferation, depending on p53 and p19-ARF expression [60].

In addition to these cell-autonomous effects, critical and antagonistic functions of c-Jun and JunB in cell proliferation and differentiation *in trans* were observed. In an *in vitro* skin equivalent model system (which mimics many characteristics of cutaneous wound healing) composed of primary keratinocytes and mouse embryonic fibroblasts of wild-type, *c-jun*$^{-/-}$ or *junB*$^{-/-}$ genotype, c-Jun- and JunB-dependent expression of critical cytokines (KGF, GM-CSF) could be demonstrated [61,62].

As described before, AP-1 activity is greatly enhanced upon treatment of cells with genotoxic agents, implying that AP-1 target genes are involved in the cellular "stress response," such as DNA repair, induction of survival functions, or initiation of the apoptotic program. Indeed, analysis of fibroblasts from mice lacking specific AP-1 subunits provided experimental proof for this assumption. Fibroblasts lacking c-Fos are hypersensitive to UV irradiation when compared to wild-type cells, which is caused by a higher rate of apoptosis. In contrast to cells lacking c-Fos, the ability of *c-jun*-deficient fibroblasts to undergo apoptosis is greatly reduced. In contrast, overexpression of c-Jun induced apoptosis in fibroblasts. Reduced apoptosis in response to genotoxic agents was also observed in mice lacking members of the JNK/SAPK family of protein kinases, suggesting that c-Jun and ATF proteins are the major substrates of JNK/SAPKs to mediate the cellular stress response [30,63]. There is evidence that JunD participates in an anti-apoptotic pathway [60], whereas JunB appears to be part of a pro-apoptotic pathway, at least in myeloid cells, through negative regulation of anti-apoptotic genes [41]. However, it is important to note that AP-1 members depending on the cell type and extracellular stimuli may be involved in both apoptotic and anti-apoptotic responses. For example, primary liver cell cultures and erythroblasts derived from *c-jun*$^{-/-}$ embryos exhibit increased apoptotic rates. On the other hand, the lack of c-Fos results in the loss of light-induced apoptosis of photoreceptors in retinal degeneration [30]. Obviously, in addition to the abundance of individual AP-1 subunits and other cellular proteins modulating AP-1 activity on the posttranslational level some of the specificity of the function of a given AP-1 subunit is presumably based on the choice of the heterodimeric partner, dictating sequence specificity and, in turn, the subsets of AP-1 target genes to be addressed [64]. A shift in the equilibrium of gene expression of such distinct classes of c-Jun target genes, in conjunction with alterations in c-Jun-independent pathways, will contribute to the decision of the cell to either proliferate, to activate survival factors or to induce the genetic program of cell death in response to extracellular signals.

Despite the fact that AP-1 has been identified more than a decade ago, it still maintains a lot of its mystery. Further research on tissue-specific inactivation of AP-1 members and the identification of subunit-specific target genes may yield an even more complex picture of function and regulation of AP-1 than exists at present.

Acknowledgements

I would like to thank Axel Szabowski for his help with preparing the figure and table. The laboratory is supported by grants from the Deutsche Forschungsgemeinschaft, by the Cooperation Program in Cancer Research of the DKFZ and Israel's Ministry of Science, and by the BioMed-2 and Training and Mobility of Researchers (TMR) Programs of the European Community.

References

1. Angel, P. and Karin, M. (1991). The role of Jun, Fos and the AP-1 complex in cell-proliferation and transformation. *Biochim. Biophys. Acta* **1072**, 129–157.
2. Chinenov, Y. and Kerppola, T. K. (2001). Close encounters of many kinds: Fos-Jun interactions that mediate transcription regulatory specificity. *Oncogene* **20**, 2438–2452.
3. Macian, F., Lopez-Rodriguez, C., and Rao, A. (2001). Partners in transcription: NFAT and AP-1. *Oncogene* **20**, 2476–2489.
4. Gutman, A. and Wasylyk, B. (1990). The collagenase gene promoter contains a TPA and oncogene-responsive unit encompassing the PEA3 and AP-1 binding sites. *EMBO J.* **9**, 2241–2246.
5. Stein, B., Baldwin, A. S., Jr., Ballard, D. W., Greene, W. C., Angel, P., and Herrlich, P. (1993). Cross-coupling of the NF-kappa B p65 and Fos/Jun transcription factors produces potentiated biological function. *EMBO J.* **12**, 3879–3891.
6. Karin, M. (1998). New twists in gene regulation by glucocorticoid receptor: is DNA binding dispensable? *Cell* **93**, 487–490.
7. Herrlich, P. (2001). Cross-talk between glucocorticoid receptor and AP-1. *Oncogene* **20**, 2465–2475.
8. Laderoute, K. R., Calaoagan, J. M., Gustafson-Brown, C., Knapp, A. M., Li, G. C., Mendonca, H. L., Ryan, H. E., Wang, Z., and Johnson, R. S. (2002). The response of c-jun/AP-1 to chronic hypoxia is hypoxia-inducible factor 1 alpha dependent. *Mol. Cell Biol.* **22**, 2515–2523.
9. Chiu, R., Angel, P., and Karin, M. (1989). Jun-B differs in its biological properties from, and is a negative regulator of, cJun. *Cell* **59**, 979–986.
10. Deng, T. and Karin, M. (1993). JunB differs from c-Jun in its DNA-binding and dimerization domains, and represses c-Jun by formation of inactive heterodimers. *Genes Dev.* **7**, 479–490.
11. Wagner, E. F. (2001). AP-1. *Oncogene* **20**, 2336–2497.
12. Mechta-Grigoriou, F., Gerald, D., and Yaniv, M. (2001). The mammalian Jun proteins: redundancy and specificity. *Oncogene* **20**, 2378–2389.
13. Nordheim, A., Janknecht, R., and Hipskind, R. A. (1994). Transcriptional regulation of the human c-fos proto-oncogene, in Angel, P., and Herrlich, P., Eds., *The Fos and Jun Families of transcription Factors,* pp. 97–113. CRC Press, Boca Raton, FL, 1994.
14. Treisman, R. (1995). Journey to the surface of the cell: Fos regulation and the SRE. *EMBO J.* **14**, 4905–4913.

15. Mechta, F. and Yaniv, M. (1994). Structure and regulation of the c-jun promoter, in Angel P. E., and Herrlich, P. A., Eds., *The Fos and Jun Families of transcription Factors*, pp. 115–129. CRC Press, Boca Raton, FL, 1994.
16. Minden, A. and Karin, M. (1997). Regulation and function of the JNK subgroup of MAP kinases. *Biochim. Biophys. Acta* **1333**, 85–104.
17. Wilkinson, M. G. and Millar, J. B. (1998). SAPKs and transcription factors do the nucleocytoplasmic tango. *Genes Dev.* **12**, 1391–1397.
18. Hibi, M., Lin, A., Smeal, T., Minden, A., and Karin, M. (1993). Identification of an oncoprotein- and UV-responsive protein kinase that binds and potentiates the c-Jun activation domain. *Genes Dev.* **7**, 2135–2148.
19. Gupta, S., Campbell, D., Derijard, B., and Davis, R. J. (1995). Transcription factor ATF2 regulation by the JNK signal transduction pathway. *Science* **267**, 389–393.
20. van Dam, H., Wilhelm, D., Herr, I., Steffen, A., Herrlich, P., and Angel, P. (1995). ATF-2 is preferentially activated by stress-activated protein kinases to mediate c-jun induction in response to genotoxic agents. *EMBO J.* **14**, 1798–1811.
21. Gupta, S., Barrett, T., Whitmarsh, A. J., Cavanagh, J., Sluss, H. K., Derijard, B., and Davis, R. J. (1996). Selective interaction of JNK protein kinase isoforms with transcription factors. *EMBO J.* **15**, 2760–2770.
22. Enslen, H., Brancho, D. M., and Davis, R. J. (2000). Molecular determinants that mediate selective activation of p38 MAP kinase isoforms. *EMBO J.* **19**, 1301–1311.
23. Treier, M., Staszewski, L. M., and Bohmann, D. (1994). Ubiquitin-dependent c-Jun degradation in vivo is mediated by the delta domain. *Cell* **78**, 787–798.
24. Musti, A. M., Treier, M., and Bohmann, D. (1997). Reduced ubiquitin-dependent degradation of c-Jun after phosphorylation by MAP kinases. *Science* **275**, 400–402.
25. Papavassiliou, A. G., Treier, M., Chavrier, C., and Bohmann, D. (1992). Targeted degradation of c-Fos, but not v-Fos, by a phosphorylation-dependent signal on c-Jun. *Science* **258**, 1941–1944.
26. Boyle, W. J., Smeal, T., Defize, L. H., Angel, P., Woodgett, J. R., Karin, M., and Hunter, T. (1991). Activation of protein kinase C decreases phosphorylation of c-Jun at sites that negatively regulate its DNA-binding activity. *Cell* **64**, 573–584.
27. Lin, A., Frost, J., Deng, T., Smeal, T., al-Alawi, N., Kikkawa, U., Hunter, T., Brenner, D., and Karin, M. (1992). Casein kinase II is a negative regulator of c-Jun DNA binding and AP-1 activity. *Cell* **70**, 777–789.
28. Vogt P. K. (2001). Jun, the oncoprotein. *Oncogene* **20**, 2365–2377.
29. Karin, M., Liu, Z., and Zandi, E. (1997). AP-1 function and regulation. *Curr. Opin. Cell Biol.* **9**, 240–246.
30. Jochum, W. E. P. and Wagner, E. F. (2001). AP-1 in mouse development and tumorigenesis. *Oncogene* **20**, 2401–2412.
31. Lallemand, D., Spyrou, G., Yaniv, M., and Pfarr, C. M. (1997). Variations in Jun and Fos protein expression and AP-1 activity in cycling, resting and stimulated fibroblasts. *Oncogene* **14**, 819–830.
32. Bakiri, L., Lallemand, D., Bossy-Wetzel, E., and Yaniv, M. (2000). Cell cycle-dependent variations in c-Jun and JunB phosphorylation: A role in the control of Cyclin D1 expression. *EMBO J.* **19**, 2056–2068.
33. Hilberg, F., Aguzzi, A., Howells, N., and Wagner, E. F. (1993). c-jun is essential for normal mouse development and hepatogenesis. *Nature* **365**, 179–181.
34. Johnson, R. S., van Lingen, B., Papaioannou, V. E., and Spiegelman, B. M. (1993). A null mutation at the c-jun locus causes embryonic lethality and retarded cell growth in culture. *Genes Dev.* **7**, 1309–1317.
35. Eferl, R., Sibilia, M., Hilberg, F., Fuchsbichler, A., Kufferath, I., Guertl, B., Zenz, R., Wagner, E. F., and Zatloukal, K. (1999). Functions of c-Jun in liver and heart development. *J. Cell Biol.* **145**, 1049–1061.
36. Behrens, A., Sibilia, M., David, J. P., Mohle-Steinlein, U., Tronche, F., Schutz, G., and Wagner, E. F. (2002). Impaired postnatal hepatocyte proliferation and liver regeneration in mice lacking c-jun in the liver. *EMBO J.* **21**, 1782–1790.
37. Behrens, A., Sibilia, M., and Wagner, E. F. (1999). Amino-terminal phosphorylation of c-Jun regulates stress-induced apoptosis and cellular proliferation. *Nat. Genet.* **21**, 326–329.
38. Behrens, A., Sabapathy, K., Graef, I., Cleary, M., Crabtree, G. R., and Wagner, E. F. (2001). Jun N-terminal kinase 2 modulates thymocyte apoptosis and T cell activation through c-Jun and nuclear factor of activated T cell (NF-AT). *Proc. Natl. Acad. Sci. USA* **98**, 1769–1774.
39. Passegue, E., Jochum, W., Behrens, A., Ricci, R., and Wagner, E. F. (2002). JunB can substitute for Jun in mouse development and cell proliferation. *Nat. Genet.* **30**, 158–166.
40. Schorpp-Kistner, M., Wang, Z. Q., Angel, P., and Wagner, E. F. (1999). JunB is essential for mammalian placentation. *EMBO J.* **18**, 934–948.
41. Passegue, E., Jochum, W., Schorpp-Kistner, M., Mohle-Steinlein, U., and Wagner, E. F. (2001). Chronic myeloid leukemia with increased granulocyte progenitors in mice lacking junB expression in the myeloid lineage. *Cell* **104**, 21–32.
42. Schreiber, M., Wang, Z. Q., Jochum, W., Fetka, I., Elliott, C., and Wagner, E. F. (2000). Placental vascularization requires the AP-1 component fra1. *Development* **127**, 4937–4948.
43. Wang, Z. Q., Ovitt, C., Grigoriadis, A. E., Mohle-Steinlein, U., Ruther, U., and Wagner, E. F. (1992). Bone and hematopoietic defects in mice lacking c-fos. *Nature* **360**, 741–745.
44. Johnson, R. S., Spiegelman, B. M., and Papaioannou, V. (1992). Pleiotropic effects of a null mutation in the c-fos proto-oncogene. *Cell* **71**, 577–586.
45. Grigoriadis, A. E., Wang, Z. Q., Cecchini, M. G., Hofstetter, W., Felix, R., Fleisch, H. A., and Wagner, E. F. (1994). c-Fos: A key regulator of osteoclast-macrophage lineage determination and bone remodeling. *Science* **266**, 443–448.
46. Hafezi, F., Steinbach, J. P., Marti, A., Munz, K., Wang, Z. Q., Wagner, E. F., Aguzzi, A., and Reme, C. E. (1997). The absence of c-fos prevents light-induced apoptotic cell death of photoreceptors in retinal degeneration *in vivo. Nat. Med.* **3**, 346–349.
47. Matsuo, K., Owens, J. M., Tonko, M., Elliott, C., Chambers, T. J., and Wagner, E. F. (2000). Fosl1 is a transcriptional target of c-Fos during osteoclast differentiation. *Nat. Genet.* **24**, 184–187.
48. Fleischmann, A., Hafezi, F., Elliott, C., Reme, C. E., Ruther, U., and Wagner, E. F. (2000). Fra-1 replaces c-Fos-dependent functions in mice. *Genes Dev.* **14**, 2695–2700.
49. Gruda, M. C., van Amsterdam, J., Ruizzo, C. A., Durham, S. K., Lira, S., and Bravo, R. (1996). Expression of FosB during mouse development: Normal development of FosB knockout mice. *Oncogene* **12**, 2177–2185.
50. Brown, J. R., Ye, H., Bronson, R. T., Dikkes, P., and Greenberg, M. E. (1996). A defect in nurturing in mice lacking the immediate early gene fosB. *Cell* **86**, 297–309.
51. Thepot, D., Weitzman, J. B., Barra, J., Segretain, D., Stinnakre, M. G., Babinet, C., and Yaniv, M. (2000). Targeted disruption of the murine junD gene results in multiple defects in male reproductive function. *Development* **127**, 143–153.
52. Beier, F., Lee, R. J., Taylor, A. C., Pestell, R. G., and LuValle, P. (1999). Identification of the cyclin D1 gene as a target of activating transcription factor 2 in chondrocytes. *Proc. Natl. Acad. Sci. USA* **96**, 1433–1438.
53. Reimold, A. M., Grusby, M. J., Kosaras, B., Fries, J. W., Mori, R., Maniwa, S., Clauss, I. M., Collins, T., Sidman, R. L., Glimcher, M. J., and Glimcher, L. H. (1996). Chondrodysplasia and neurological abnormalities in ATF-2-deficient mice. *Nature* **379**, 262–265.
54. Maekawa, T., Bernier, F., Sato, M., Nomura, S., Singh, M., Inoue, Y., Tokunaga, T., Imai, H., Yokoyama, M., Reimold, A., Glimcher, L. H., and Ishii, S. (1999). Mouse ATF-2 null mutants display features of a severe type of meconium aspiration syndrome. *J. Biol. Chem.* **274**, 17813–17819.
55. Schreiber, M., Kolbus, A., Piu, F., Szabowski, A., Mohle-Steinlein, U., Tian, J., Karin, M., Angel, P., and Wagner, E. F. (1999). Control of cell cycle progression by c-Jun is p53 dependent. *Genes Dev.* **13**, 607–619.
56. Wisdom, R., Johnson, R. S., and Moore, C. (1999). c-Jun regulates cell cycle progression and apoptosis by distinct mechanisms. *EMBO J.* **18**, 188–197.

57. Passegue, E. and Wagner, E. F. (2000). JunB suppresses cell proliferation by transcriptional activation of p16 (INK4a) expression. *EMBO J.* **19**, 2969–2979.
58. Brown, J. R., Nigh, E., Lee, R. J., Ye, H., Thompson, M. A., Saudou, F., Pestell, R. G., and Greenberg, M. E. (1998). Fos family members induce cell cycle entry by activating cyclin D1. *Mol. Cell Biol.* **18**, 5609–5619.
59. Johnson, R., Spiegelman, B., Hanahan, D., and Wisdom, R. (1996). Cellular transformation and malignancy induced by ras require c-jun. *Mol. Cell Biol.* **16**, 4504–4511.
60. Weitzman, J. B., Fiette, L., Matsuo, K., and Yaniv, M. (2000). JunD protects cells from p53-dependent senescence and apoptosis. *Mol. Cell* **6**, 1109–1119.
61. Szabowski, A., Maas-Szabowski, N., Andrecht, S., Kolbus, A., Schorpp-Kistner, M., Fusenig, N. E., and Angel, P. (2000). c-Jun and JunB antagonistically control cytokine-regulated mesenchymal-epidermal interaction in skin. *Cell* **103**, 745–755.
62. Angel, P., Szabowski, A., and Schorpp-Kistner, M. (2001). Function and regulation of AP-1 subunits in skin physiology and pathology. *Oncogene* **20**, 2413–2423.
63. Shaulian, E. and Karin, M. (2001). AP-1 in cell proliferation and survival. *Oncogene* **20**, 2390–2400.
64. van Dam, H., Huguier, S., Kooistra, K., Baguet, J., Vial, E., van der Eb, A. J., Herrlich, P., Angel, P., and Castellazzi, M. (1998). Autocrine growth and anchorage independence: two complementing Jun-controlled genetic programs of cellular transformation. *Genes Dev.* **12**, 1227–1239.

CHAPTER 286

108

NFκB: A Key Integrator of Cell Signaling

John K. Westwick, Klaus Schwamborn, and Frank Mercurio
Signal Research Division, Celgene, San Diego, California

Eukaryotic cells possess a number of distinct signal transduction pathways that couple environmental stimuli to specific changes in gene expression. One such pathway regulates the transcription factor NFκB, which is known to orchestrate the expression of a diverse array of genes essential in host defense, inflammatory, cell survival, and immune responses. NFκB is activated by a surprisingly broad array of cellular stimuli [1]. Likewise, NFκB has been shown to induce the expression of a large functionally diverse array of genes [2]. The seemingly promiscuous nature of NFκB biology raises the question as to how NFκB proteins elicit a specific transcriptional program in response to a given environmental challenge. The discovery of several key regulatory proteins within this pathway has provided insight into the mechanism by which the NFκB achieves specific coupling of these distinct cellular processes. As would be predicted of such a central signaling pathway, aberrant regulation of NFκB activation has been associated with the pathogenesis of several diseases, including autoimmunity, arthritis, asthma, and cancer [3]. This chapter summarizes the mechanisms of NFκB regulation and discusses emerging opportunities for target-based therapeutic intervention in various disease settings.

NFκB was originally identified as a transcription factor required for B-cell-specific gene expression [4]. However, it was quickly recognized that NFκB activity could be induced in most cell types in response to myriad stimuli, including proinflammatory cytokines, bacterial lipopolysaccharides, viral infection, DNA damage, oxidative stress, and chemotherapeutic agents (Fig. 1) [1]. The classic experiment by Baeuerle and Baltimore, which demonstrated that NFκB exists in latent form in the cytosol of unstimulated cells and undergoes rapid translocation to the nucleus upon stimulation, set the stage for what is now the hallmark of NFκB regulation [5]. This paradigm provides a mechanism for NFκB to undergo rapid induction in response to cellular stress, resulting in the up-regulation of NFκB target genes. A wide variety of genes are regulated by NFκB, including those encoding cytokines, chemokines, adhesion molecules, acute phase proteins, inducible effector enzymes, antimicrobial peptides, adaptive immune response, regulators of apoptosis, and cell proliferation (Fig. 1). Remarkably, specific subsets of NFκB responsive genes can be activated in a cell- and stimulus-specific fashion. Biochemical and genetic characterization of molecular components that impinge on the NFκB signaling pathway has greatly facilitated our understanding of this process.

NFκB exists as a multigene family of proteins that can form stable homo- and heterodimeric complexes that vary in their DNA-binding specificity and transcriptional activation potential. Five proteins belonging to the NFκB family have been identified in mammalian cells: RelA (p65), c-Rel, RelB, NFκB1 (p50 and its precursor p105), and NFκB2 (p52 and its precursor p100) (Fig. 2) [1]. NFκB/Rel proteins share a highly conserved 300-amino-acid N-terminal Rel homology domain (RHD) responsible for DNA binding, dimerization, and association with the IκB inhibitory proteins. The prototype NFκB complex is comprised of p50 and p65, but a variety of NFκB/Rel-containing dimers are also known to exist. The p50/p65 complex displays strong transcriptional activation, whereas the p50/p50 and p52/p52 homodimers function to repress transcription of NFκB target genes [1]. Thus, the existence of a multigene family provides one tier of regulation by which the cell can fine-tune NFκB-mediated gene expression. Moreover, knock-out mice lacking distinct NFκB subunits display distinct phenotypes, further implicating

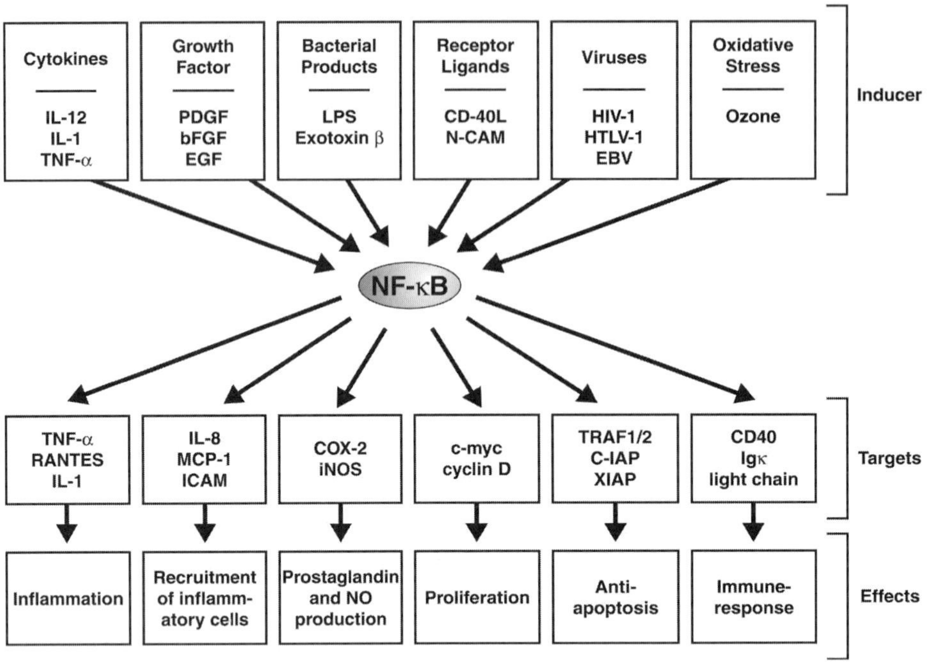

Figure 1 NFκB plays a central role in gene regulation: Stimulus-dependent activation of NFκB results in the modulation of specific subsets of NFκB target genes that are involved in distinct cellular processes. IL, interleukin; TNF, tumor necrosis factor; bFGF, basic fibroblast growth factor; PDGF, platelet-derived growth factor; EGF, epidermal growth factor; LPS, lipopolysaccharide; CD-40L, CD 40 ligand; N-CAM, neural cell adhesion molecule; HIV, human immunodeficiency virus; HTLV, human T-cell leukemia/lymphoma virus; EBV, Epstein-Barr virus; RANTES, regulated on activation, normal T cells expressed and secreted; MCP, monocyte chemoattractant protein; COX-2, cyclooxygenase-2; iNOS, inducible nitric oxide synthetase; TRAF1/2, TNF receptor associated factor; c-IAP, cellular inhibitors of apoptosis protein; XIAP, X-chromosome-linked inhibitor of apoptosis protein.

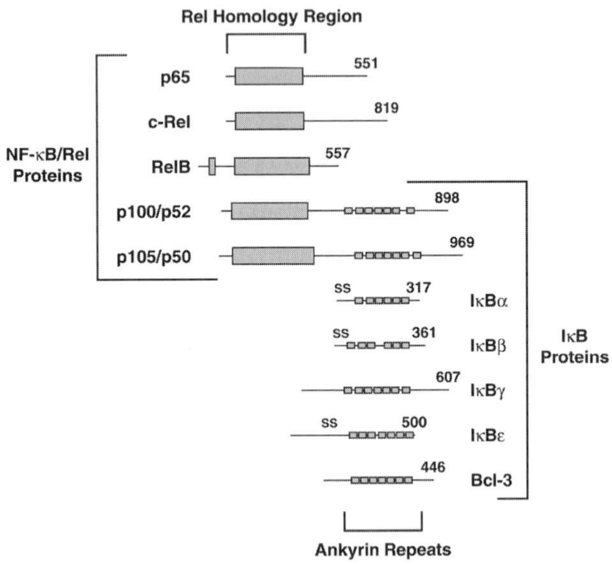

Figure 2 Schematic representation of members of the NFκB/Rel and IκB family of proteins. The position of the Rel homology region and the ankyrin repeat domain are indicated in the figure. p100/p52 and p105/p50 Precursor proteins comprise a unique subgroup of proteins that contain both a Rel homology and an ankyrin repeat domain. p100 and p105 Proteins have been shown to function in an IκB-like capacity. The number of amino acids of each protein is indicated to the right. For RelB, the shaded box N terminal to the Rel homology region represents a putative leucine zipper region.

specific NFκB dimers as activators of distinct sets of NFκB target genes [6–10].

NFκB resides in the cytoplasm in an inactive form by virtue of its association with a class of inhibitory proteins termed *IκBs*. Seven IκBs have been identified: IκBα, IκBβ, IκBε, IκBγ, Bcl3, NFκB1 precursor (p105), and the NFκB2 precursor (p100) (Fig. 2) [11]. The IκB family members, having in common ankyrin repeat domains, regulate the subcellular localization, and hence the DNA-binding and transcriptional activity, of NFκB proteins. The basis for the cytoplasmic localization of the inactive NFκB:IκB complex is thought to be due to masking of the nuclear localization signals (NLS) on the NFκB subunits by the IκB proteins. Thus, IκB degradation would lead to unmasking of the NLS, allowing NFκB to undergo translocation to the nucleus (Fig. 3). The IκBs display a preference for specific NFκB/Rel complexes, which may provide a means to regulate the activation of distinct Rel/NFκB complexes. Interestingly, NFκB induces the expression of IκBα; the newly synthesized IκB molecules enter the nucleus and remove NFκB from DNA [12–14]. The NFκB:IκB is then expelled from the nucleus as a result of potent nuclear export signals on IκB and p65 [15]. The IκBs provide yet another tier of regulatory complexity to modulate NFκB-mediated gene expression.

Activation of NFκB is achieved primarily through the signal-induced proteolytic degradation of IκB that is mediated by

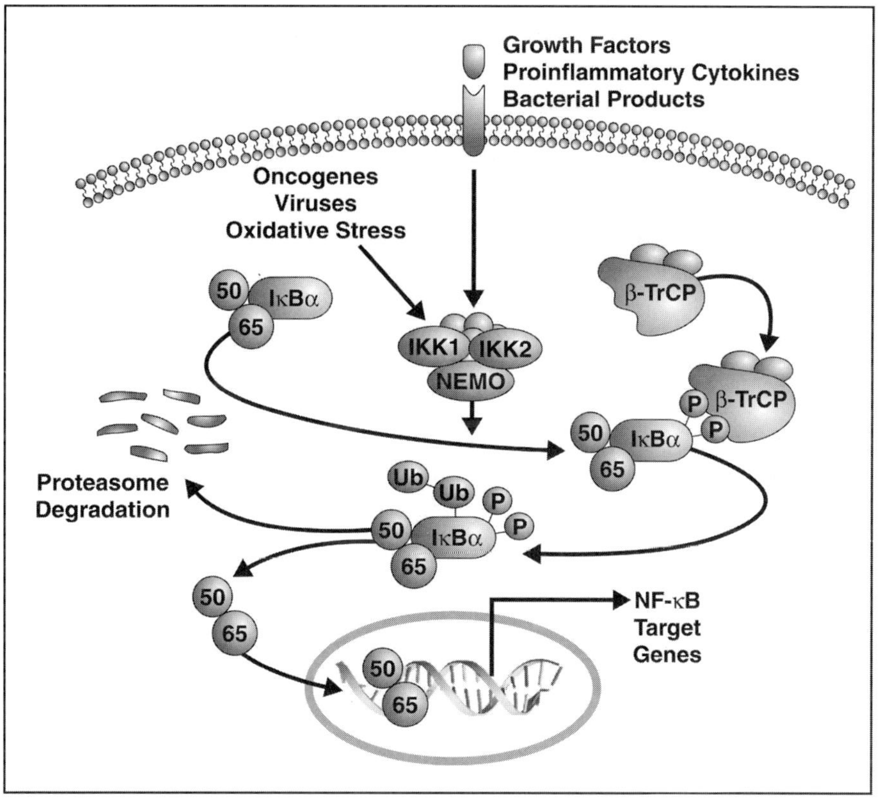

Figure 3 Schematic representation of components of the classical NFκB signal transduction pathway. In response to extracellular signals, the IKK complex becomes activated by a poorly defined series of membrane proximal events. The activated IKK complex phosphorylates (P) IκBα at serines 32 and 36, leading to site-specific ubiquitination (Ub) and degradation by the proteasome. The released NFκB complex (p50/p65) is then free to undergo nuclear translocation and activate transcription of NFκB target genes.

the ubiquitin/proteasome system [2]. The critical event that initiates the degradation of IκB is the stimulus-dependent phosphorylation of IκB at specific N-terminal residues (S32/S36 for IκBα, S19/S23 for IκBβ) [16,17]. Mutation of IκBα at either of these residues was found to block stimulus-dependent IκB phosphorylation, thereby preventing IκB degradation and subsequent activation of NFκB. Phosphorylation of IκB serves as a molecular tag leading to its rapid ubiquitination and subsequent degradation by the proteasome. β-TrCP, an F-box/WD-containing component of the Skp1-Cullin-F-box (SCF) class of E3 ubiquitin ligases, is required for recognition of phosphorylated IκB and recruitment of the degradation machinery [18,19].

Phosphorylation of IκBα on serines 32 and 36 is mediated by IκB kinases (IKKs), whose activity is induced by activators of the NFκB pathway [20–24]. IKK activity exists as large cytoplasmic multisubunit complex (700–900 kDa) containing two kinase subunits, IKK1 (IKKα) and IKK2 (IKKβ), and a regulatory subunit, NEMO (IKKγ, IKKAP1, FIP3) [25–28]. Hence the "core" IKK complex is comprised of IKK1, IKK2, and NEMO. IKK1 and IKK2 are highly homologous kinases, both containing an N-terminal kinase domain and a C-terminal region with two protein interaction motifs, a leucine zipper (LZ), and a helix–loop–helix (HLH) motif. The LZ domain is responsible for dimerization of IKK1 and IKK2 and is essential for activity of the IKK complex. The IKK1/2 complex associates with NEMO through a short interaction motif located at the very C terminus of either catalytic subunit. Short peptides derived from the interaction motif can be used to disrupt the IKK complex and prevent its activation [29]. NEMO is thought to serve an important regulatory function by connecting the IKK complex to upstream activators through its C terminus, which contains a Zn finger motif (Fig. 4). Further support for the regulatory function of NEMO is found by the observation that NEMO undergoes stimulus-dependent interaction with components of the TNF receptor complex [30]. More recently, a report described two potentially novel components of the IKK complex, namely, Cdc37 and Hsp90 [31]. Apparently, formation of the core IKK complex with Cdc37/Hsp90 is required for TNF-induced activation and recruitment of the core IKK complex from the cytoplasm to the membrane.

Sequence analysis revealed that both IKK1 and IKK2 contain a canonical MAP kinase kinase (MAPKK) activation loop motif. This region contains specific sites whose phosphorylation induces a conformational change that results in kinase activation. Phosphorylation within the activation loop typically occurs through the action of an upstream kinase or through transphosphorylation enabled by regulated proximity between two kinase subunits. IKK2 activation loop mutations,

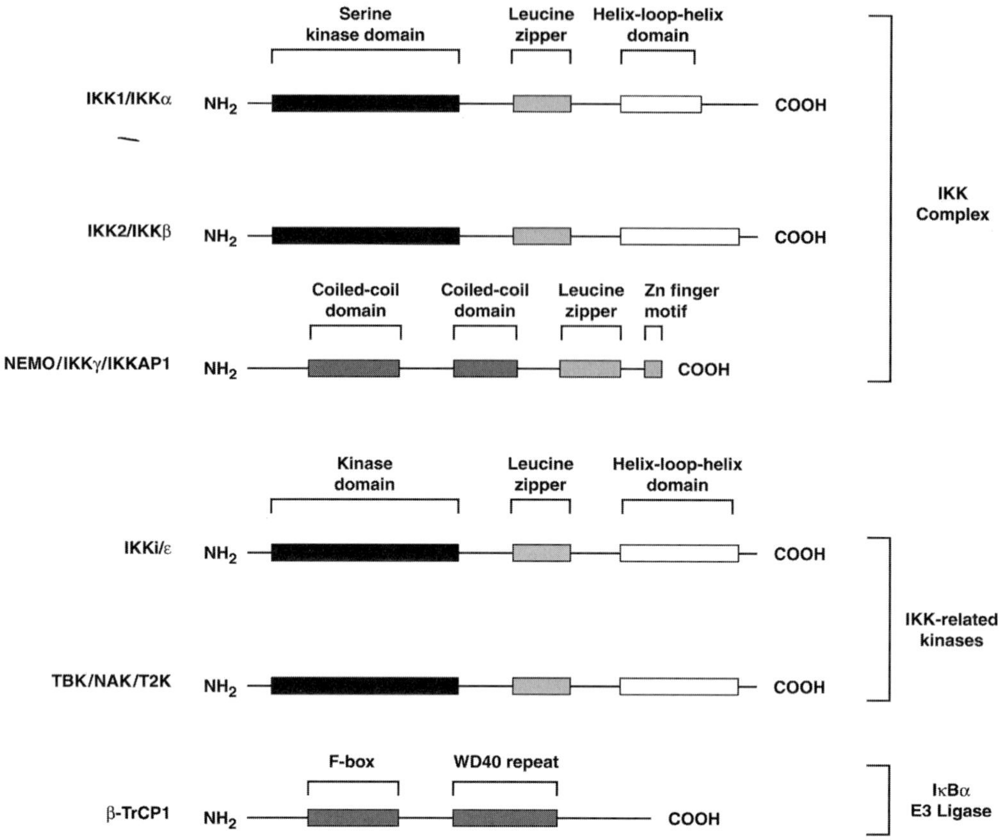

Figure 4 Schematic diagram showing the known subunits of the IKK complex, IKK-related kinases and β-TrCP, the IκBα E3 ubiquitin ligase. The name(s) of the protein is indicated to the left and the putative structural and functional motifs are indicated on top.

in which serines 177 and 181 were replaced with alanine, render the kinase refractory to stimulus-dependent activation. In contrast, replacement of serine 177 and 181 with glutamic acid, to mimic phosphoserine, yielded a constitutively active kinase, and was capable of inducing NFκB-mediated gene expression in the absence of cell stimulation [21]. The corresponding mutations in IKK1 did not interfere with NFκB activation in response to IL-1 or TNF, providing the first data suggesting that IKK2 plays a more prominent role in NFκB activation in response to proinflammatory cytokines [21,23,24]. Subsequent studies with IKK knock-out mice further validate that IKK2, and not IKK1, is required for NFκB activation in response to most proinflammatory stimuli [32–35]. The mechanism by which diverse stimuli converge to activate the IKK complex remains unresolved. For example, a number of kinases have been reported to function in the capacity of an IKK kinase, including NIK, IKKι/ε, NAK/T2K/TBK, MEKK1, MEKK3, and Cot/TPL2 [36]. However, with the exception of MEKK3, no effect was observed on either IKK activation or induction of NFκB DNA-binding activity in mice devoid of these kinases [37].

Biochemical and genetic analyses demonstrate that IKK1 and IKK2 have distinct cellular functions [2]. Disruption of the IKK2 gene results in embryonic lethality due to extensive liver apoptosis [32,34,38]. This phenotype is remarkably similar to that seen previously for RelA$^{-/-}$ mice [39]. Interestingly, IKK2$^{-/-}$- and RelA-mice could be rescued by ablation of the TNF receptor I gene, which is consistent with the role of NFκB in preventing TNF-induced hepatocyte apoptosis [34,40]. In addition, IKK2$^{-/-}$ mouse embryonic fibroblasts were demonstrated to be refractory to activation of NFκB in response to inducers of NFκB, including TNF, IL-1, LPS, and dsRNA [34,38]. In contrast, the IKK1$^{-/-}$ mice were born alive but died within 30 min, and IKK1$^{-/-}$-derived mouse embryonic fibroblasts display normal activation of IKK activity and induction of NFκB DNA binding in response to proinflammatory stimuli. These mice exhibit a plethora of developmental defects, the most striking of which is defective epidermal differentiation [32,33]. Interestingly, IKK1 was found to play a prominent role in regulating keratinocyte differentiation, which is independent from its kinase activity or modulation of NFκB [32,41].

More recently, a series of eloquent studies revealed a unique function of IKK1 in the lymphoid system [42]. Specifically, lethally irradiated mice that were reconstituted with IKK$^{-/-}$ hematopoietic stem cells displayed a diminution in B-cell maturation, germinal center formation, and antibody production, as well as defective splenic microarchitecture. Interestingly, it was discovered that IKK1$^{-/-}$ B cells exhibit a specific deficiency in NFκB2/p100 processing [42]. Moreover, independent studies demonstrated that NIK induces ubiquitin-dependent processing of NFκB2/p100 [43]. Subsequently, it

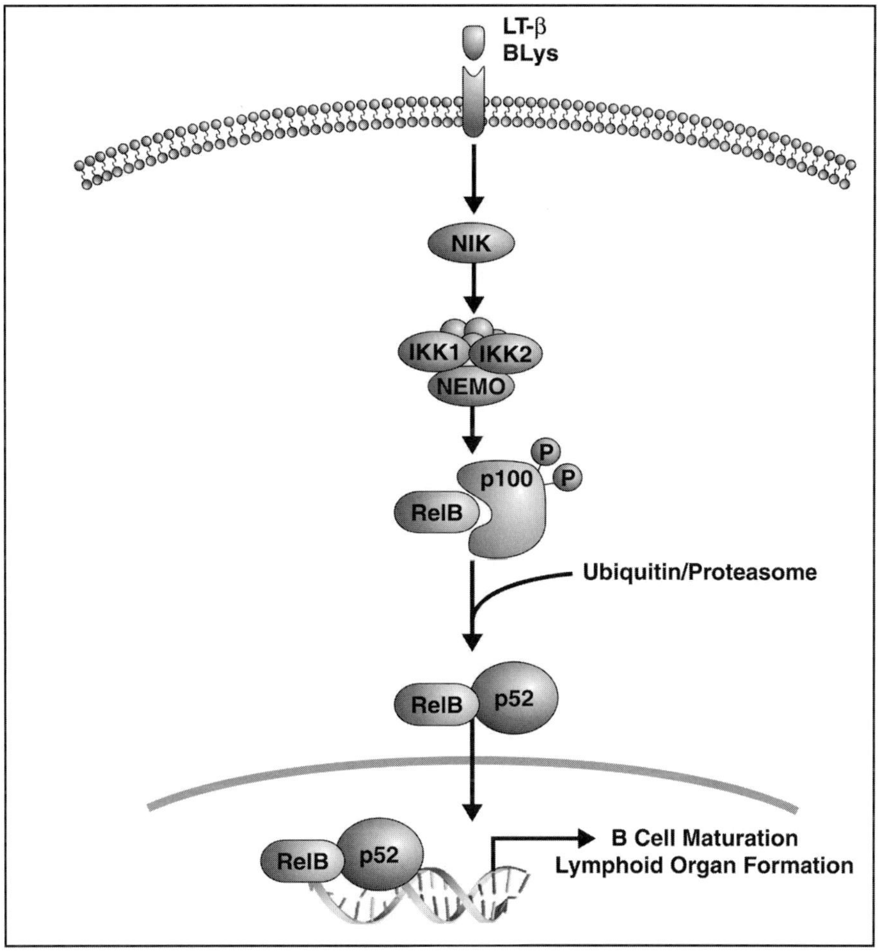

Figure 5 Activation by IKK1 is involved in an alternative NFκB signaling pathway. Select members of the TNF family, namely, lymphotoxin-beta (LT-beta) and B-lymphocyte stimulator (BLyS), activate IKK1 through NIK (NFκB-inducing kinase). Activation of this pathway leads to IKK1-dependent phosphorylation (P) and processing of NFκB2/p100. The active p52:RelB heterodimer undergoes nuclear translocation and regulates specific NFκB target genes involved in B-cell maturation and lymphoid organ formation.

was demonstrated that an inactive variant of IKK1 blocked NIK-induced NFκB2/p100 processing. Together, these studies suggest that NIK-mediated activation of IKK1 may lead to the phosphorylation and ubiquitin-dependent processing of NFκB2/p100 (Fig. 5). NFκB2/p100 is unique in that it contains both a Rel homology domain and an ankyrin domain and, consequently, possesses an intrinsic IκB-like capacity. Thus, enhanced proteolytic removal of the inhibitory C-terminal ankyrin domain will generate increased levels of transcriptionally active p52- containing NFκB complexes, perhaps leading to up-regulation of a specific subset of NFκB target genes. It is quite intriguing that, although IKK1 and IKK2 display high similarity and form stable heterodimers, they modulate distinct signaling pathways with unique biological consequences.

NFκB proteins may play a role in any disease possessing one of the cytokines, including TNF, IFNγ, IL-2, IL-6, IL-8, and IL-12, as a component of their pathophysiology (Fig. 1). The likely diseases for an NFκB-targeted therapeutic strategy may be those with a chronic, unresolved inflammatory component characterized by constitutively elevated systemic or local cytokine levels. Rheumatoid and osteoarthritis, inflammatory bowel disease, atherosclerosis, diabetes, multiple sclerosis and cachexia, among other conditions, fall into this category. Note that the chronic inhibition of NFκB likely to be necessary to ameliorate these conditions may have unknown repercussions. However, agents such as glucocorticoids inhibit NFκB and are used chronically, so we may predict that long-term NFκB-based therapy can be tolerated [44].

Despite the emphasis on NFκB's role as a regulator of cytokine gene expression and immune function, the initial application of therapeutic modulators of NFκB is likely to be in the field of oncology. If this is indeed the case, it will recapitulate the initial discovery of v-Rel proteins as transforming oncogenes nearly 20 years ago [45]. While the genetics of viral Rel proteins were being worked out, simultaneous work was proceeding on the transcription factors (including NFκB) regulating immunoglobulin and MHC genes [4,46]. It was several years before these divergent fields of research coalesced with the discovery that NFκB proteins are the cellular homologs of v-Rel [47,48]. The functional roles of NFκB proteins, of course, extend far beyond regulation of immunoglobulins and are quite widely expressed.

More recent information has shown a wide variety of genetic aberrations of these genes in human neoplasms, from overexpression to rearrangement and amplification. The list of such observations is beyond the scope of this review, and details can be found elsewhere [49,50]. Transforming viral proteins of SV40, EBV, adenovirus E1A, and HTLV-1 activate NFκB [51–54]. Regardless of the mechanism, it is clear that the NFκB transcription factor is constitutively active in a wide variety of human cancers, including but not limited to colon, gastric, pancreatic, ovarian, hepatocellular, breast, head and neck carcinomas, melanoma, lymphoblastic leukemias, and Hodgkin's disease [55–65].

As described earlier, NFκB activity is controlled to a large extent by the activity of the IKKs, and this represents the most tractable point for therapeutic intervention. However, other mechanisms for activation and avenues for intervention exist. For example, the DNA-binding and transcriptional activity of NFκB proteins can be controlled by phosphorylation of multiple sites in both the Rel homology and transactivation domains [66–69]. Well-characterized oncogenes such as Ras, Rac, and Bcr-Abl as well as integrin-activated signals induce NFκB signaling both through the traditional IKK/IκB pathway and through direct phosphorylation of the p65 subunit of the transcription factor [66–70]. Recent results with agents selectively inhibiting the IKKs (e.g., PS-1145) or acting in part on other aspects of NFκB signaling (e.g., proteasome inhibitors, NSAIDs, and thalidomide) are beginning to show great promise in the treatment of various cancers [50,71,72].

From a functional standpoint, NFκB proteins have been suggested to play a role in cellular transformation via two general routes: direct stimulation of growth (cell cycle dysregulation) and/or inhibition of differentiation, and inhibition of programmed cell death (apoptosis). Both are likely to be involved to varying degrees based on the type of neoplasm, but the protection from apoptosis has garnered the most attention of late and may be the most important aspect from a therapeutic standpoint. With respect to direct cell cycle control, NFκB has been shown to up-regulate cyclin D1 transcription, a crucial step in the G_1 to S phase cell cycle transition [73,74]. NFκB activity peaks at multiple points in the mammalian cell cycle, including G_0/G_1, late G_1, and S phase [75]. In support of this concept, analysis of tumor cells treated with selective inhibitors of IKKs reveals a G_2/M phase blockade as well (unpublished observation), suggesting the existence of additional cell cycle targets.

NFκB activates central players in the programmed cell death pathways, including c-IAP-1 and -2, Bcl-Xl, and A1/Bfl-1 [76,77]. NFκB activation of protective proteins holds the key to the promise of NFκB inhibitors as chemotherapeutic agents. Of particular importance is the observation that diverse classes of chemotherapeutic agents are potent activators of NFκB, a response that proceeds at least in part through the activation of IKKs. Agents known to activate NFκB include etoposide, CPT-11, and TRAIL [78–80]. *In vitro* and *in vivo* experiments using the IκB super-repressor, a mutant version of IκB resistant to stimulus-induced degradation, have shown that inhibition of NFκB activity acts synergistically with chemotherapeutic agents to decrease tumor cell growth and tumor burden. Small molecule inhibitors of IKK2 show the same effect *in vitro* (unpublished observations). It is likely that agents previously used in chemotherapeutic regimens, such as glucocorticoids and proteasome inhibitors in multiple myeloma, are working at least in part through the inhibition of NFκB. It is interesting to note that specific inhibitors of IKK are only partially efficacious when used as monotherapy compared to a general proteasome inhibitor, suggesting other proteasome targets contribute to efficacy [71]. On the other hand, the side effect profile of more specific inhibitors is likely to be superior.

The development of potent and selective inhibitors of NFκB, targeting multiple aspects of NFκB signal transduction, is imminent. An important task for the near future is to determine the human diseases most likely to be resolved with these agents. Chronic inflammatory conditions are intriguing possibilities, but we must first define the consequences of long-term NFκB inhibition. The optimal tumor types to approach have not been completely defined, but are likely to include multiple myeloma [71] and one of the many neoplasms with chronically elevated NFκB (as described earlier). Tumors in which traditional chemotherapy elicits a robust (and presumably activating or protective) induction of NFκB will be likely targets for combination therapy. The molecular correlates of sensitivity to NFκB inhibitors, alone or in combination with other agents, need to be identified. For example, tumors lacking the PTEN tumor suppressor may be more sensitive to NFκB inhibitors [81]. It is clear that continued work on the basic aspects of NFκB biology is necessary. In concert with development of potent and specific modulators, these efforts have the potential to yield novel therapeutics for a wide variety of human diseases.

References

1. Grossmann, M., O'Reilly, L. A., Gugasyan, R. *et al.* (2000). The anti-apoptotic activities of Rel and RelA required during B-cell maturation involve the regulation of Bcl-2 expression. *EMBO J.* **19**, 6351–6360.
2. Karin, M. and Ben-Neriah, Y. (2000). Phosphorylation meets ubiquitination: the control of NFκB activity. *Annu. Rev. Immunol.* **18**, 621–663.
3. Yamamoto, Y. and Gaynor, R. B. (2001). Therapeutic potential of inhibition of the NFkappaB pathway in the treatment of inflammation and cancer. *J. Clin. Invest.* **107**, 135–142.
4. Sen, R. and Baltimore, D. (1986). Inducibility of kappa immunoglobulin enhancer-binding protein Nf-kappa B by a posttranslational mechanism. *Cell* **47**, 921–928.
5. Baeuerle, P. A. and Baltimore, D. (1988). Activation of DNA-binding activity in an apparently cytoplasmic precursor of the NFkappa B transcription factor. *Cell* **53**, 211–217.
6. Sha, W. C., Liou, H. C., Tuomanen, E. I. *et al.* (1995). Targeted disruption of the p50 subunit of NFkappa B leads to multifocal defects in immune responses. *Cell* **80**, 321–330.

7. Kontgen, F., Grumont, R. J., Strasser, A. *et al.* (1995). Mice lacking the c-rel proto-oncogene exhibit defects in lymphocyte proliferation, humoral immunity, and interleukin-2 expression. *Genes Dev.* 1995. **9**, 1965–1977.
8. Beg, A. A. and Baltimore, D. (1996). An essential role for NFkappaB in preventing TNF-alpha-induced cell death. *Science* **274**, 782–784.
9. Caamano, J. H., Rizzo, C. A., Durham, S. K. *et al.* (1998). Nuclear factor (NF)-kappa B2 (p100/p52) is required for normal splenic microarchitecture and B cell-mediated immune responses. *J. Exp. Med.* **187**, 185–196.
10. Franzoso, G., Carlson, L., Poljak, L. *et al.* (1998). Mice deficient in nuclear factor (NF)-kappa B/p52 present with defects in humoral responses, germinal center reactions, and splenic microarchitecture. *J. Exp. Med.* **187**, 147–159.
11. Tam, W. F. and Sen, R. (2001). IkappaB family members function by different mechanisms. *J. Biol. Chem.* **276**, 7701–7704.
12. Arenzana-Seisdedos, F., Thompson, J., Rodriguez, M. S. *et al.* (1995). Inducible nuclear expression of newly synthesized I kappa B alpha negatively regulates DNA-binding and transcriptional activities of NFkappa B. *Mol. Cell Biol.* **15**, 2689–2696.
13. Chiao, P. J., Miyamoto, S., and Verma, I. M. (1994). Autoregulation of I kappa B alpha activity. *Proc. Natl. Acad. Sci. USA* **91**, 28–32.
14. Sun, S. C., Ganchi, P. A., Ballard, D. W. *et al.* (1993). NFkappa B controls expression of inhibitor I kappa B alpha: Evidence for an inducible autoregulatory pathway. Science **259**, 1912–1915.
15. Harhaj, E. W. and Sun, S. C. Regulation of RelA subcellular localization by a putative nuclear export signal and p50. *Mol. Cell Biol.* **19**, 7088–7095.
16. Brown, K., Park, S., Kanno, T. *et al.* (1993). Mutual regulation of the transcriptional activator NFkappa B and its inhibitor, I kappa B-alpha. *Proc. Natl. Acad. Sci. USA* **90**, 2532–2536.
17. DiDonato, J., Mercurio, F., Rosette, C. *et al.* (1996). Mapping of the inducible IkappaB phosphorylation sites that signal its ubiquitination and degradation. *Mol. Cell Biol.* **16**, 1295–1304.
18. Yaron, A., Gonen, H., Alkalay, I. *et al.* (1997). Inhibition of NFkappa-B cellular function via specific targeting of the I-kappa-B-ubiquitin ligase. *EMBO J.* **16**, 6486–6494.
19. Yaron, A., Hatzubai, A., Davis, M. *et al.* (1998). Identification of the receptor component of the IkappaBalpha-ubiquitin ligase. *Nature* **396**, 590–594.
20. DiDonato, J. A., Hayakawa, M., Rothwarf, D. M. *et al.* (1997). A cytokine-responsive IkappaB kinase that activates the transcription factor NFkappaB. *Nature* **388**, 548–554.
21. Mercurio, F., Zhu, H., Murray, B. W. *et al.* (1997). IKK-1 and IKK-2: Cytokine-activated IkappaB kinases essential for NFkappaB activation. *Science* **278**, 860–866.
22. Regnier, C. H., Song, H. Y., Gao, X. *et al.* (1997). Identification and characterization of an IkappaB kinase. *Cell* **90**, 373–383.
23. Woronicz, J. D., Gao, X., Cao, Z. *et al.* (1997). IkappaB kinase-beta: NFkappaB activation and complex formation with IkappaB kinase-alpha and NIK. *Science* **278**, 866–869.
24. Zandi, E., Rothwarf, D. M., Delhase, M. *et al.* (1997). The IkappaB kinase complex (IKK) contains two kinase subunits, IKKalpha and IKKbeta, necessary for IkappaB phosphorylation and NFkappaB activation. *Cell* **91**, 243–252.
25. Yamaoka, S., Courtois, G., Bessia, C. *et al.* (1998). Complementation cloning of NEMO, a component of the IkappaB kinase complex essential for NFkappaB activation. *Cell* **93**, 1231–1240.
26. Rothwarf, D. M., Zandi, E., Natoli, G. *et al.* (1998). IKK-gamma is an essential regulatory subunit of the IkappaB kinase complex. *Nature* **395**, 297–300.
27. Mercurio, F., Murray, B. W., Shevchenko, A. *et al.* (1999). IkappaB kinase (IKK)-associated protein 1, a common component of the heterogeneous IKK complex. *Mol. Cell Biol.* **19**, 1526–1538.
28. Li, Y., Kang, J., Friedman, J. *et al.* (1999). Identification of a cell protein (FIP-3) as a modulator of NFkappaB activity and as a target of an adenovirus inhibitor of tumor necrosis factor alpha-induced apoptosis. *Proc. Natl. Acad. Sci. USA* **96**, 1042–1047.
29. May, M. J., D'Acquisto, F., Madge, L. A. *et al.* (2000). Selective inhibition of NFkappaB activation by a peptide that blocks the interaction of NEMO with the IkappaB kinase complex. *Science* **289**, 1550–1554.
30. Zhang, S. Q., Kovalenko, A., Cantarella, G. *et al.* (2000). Recruitment of the IKK signalosome to the p55 TNF receptor: RIP and A20 bind to NEMO (IKKgamma) upon receptor stimulation. *Immunity* **12**, 301–311.
31. Chen, G., Cao, P., and Goeddel, D. V. (2002). TNF-induced recruitment and activation of the IKK complex require Cdc37 and Hsp90. *Mol. Cell* **9**, 401–410.
32. Hu, Y., Baud, V., Delhase, M. *et al.* (1999). Abnormal morphogenesis but intact IKK activation in mice lacking the IKKalpha subunit of IkappaB kinase. *Science* **284**, 316–320.
33. Li, Q., Lu, Q., Hwang, J. Y. *et al.* (1999). IKK1-deficient mice exhibit abnormal development of skin and skeleton. *Genes Dev.* **13**, 1322–1328.
34. Li, Q., Van Antwerp, D., Mercurio, F. *et al.* (1999). Severe liver degeneration in mice lacking the IkappaB kinase 2 gene. *Science* **284**, 321–325.
35. Takeda, K., Takeuchi, O., Tsujimura, T. *et al.* (1999). Limb and skin abnormalities in mice lacking IKKalpha. *Science* **284**, 313–316.
36. Ghosh, S. and Karin, M. (2002). Missing pieces in the NFkappaB puzzle. *Cell* **109**(Suppl), S81–S96.
37. Yang, J., Lin, Y., Guo, Z. *et al.* (2001). The essential role of MEKK3 in TNF-induced NFkappaB activation. *Nat. Immunol.* **2**, 620–624.
38. Li, Z. W., Chu, W., Hu, Y. *et al.* (1999). The IKKbeta subunit of IkappaB kinase (IKK) is essential for nuclear factor kappaB activation and prevention of apoptosis. *J. Exp. Med.* **189**, 1839–1845.
39. Beg, A. A., Sha, W. C., Bronson, R. T. *et al.* (1995). Embryonic lethality and liver degeneration in mice lacking the RelA component of NFkappa B. *Nature* **376**, 167–170.
40. Alcamo, E., Mizgerd, J. P., Horwitz, B. H. *et al.* (2001). Targeted mutation of TNF receptor I rescues the RelA-deficient mouse and reveals a critical role for NFkappa B in leukocyte recruitment. *J. Immunol.* **167**, 1592–1600.
41. Hu, Y., Baud, V., Oga, T. *et al.* IKKalpha controls formation of the epidermis independently of NFkappaB. *Nature* **410**, 710–714.
42. Senftleben, U., Cao, Y., Xiao, G. *et al.* (2001). Activation by IKKalpha of a second, evolutionary conserved, NFkappa B signaling pathway. *Science* **293**, 1495–1499.
43. Xiao, G., Harhaj, E. W., and Sun, S. C. NFkappaB-inducing kinase regulates the processing of NFkappaB2 p100. *Mol. Cell* **7**, 401–409.
44. Auphan, N., DiDonato, J. A., Rosette, C. *et al.* (1995). Immunosuppression by glucocorticoids: Inhibition of NFkappa B activity through induction of I kappa B synthesis. *Science* **270**, 286–290.
45. Wilhelmsen, K. C., Eggleton, K., and Temin, H. M. (1984). Nucleic acid sequences of the oncogene v-rel in reticuloendotheliosis virus strain T and its cellular homolog, the proto-oncogene c-rel. *J. Virol.* **52**, 172–182.
46. Baldwin, A. S., Jr., and Sharp, P. A. (1998). Two transcription factors, NFkappa B and H2TF1, interact with a single regulatory sequence in the class I major histocompatibility complex promoter. *Proc. Natl. Acad. Sci. USA* **85**, 723–727.
47. Kieran, M., Blank, V., Logeat, F. *et al.* The DNA binding subunit of NFkappa B is identical to factor KBF1 and homologous to the rel oncogene product. *Cell* **62**, 1007–1018.
48. Ghosh, S., Gifford, A. M., Riviere, L. R. *et al.* (1990). Cloning of the p50 DNA binding subunit of NFkappa B: Homology to rel and dorsal. *Cell* **62**, 1019–1029.
49. Rayet, B. and Gelinas, C. (1999). Aberrant rel/nfkb genes and activity in human cancer. *Oncogene* **18**, 6938–6947.
50. Yamamoto, Y. and Gaynor, R. B. (2001). Role of the NFkappaB pathway in the pathogenesis of human disease states. *Curr. Mol. Med.* **1**, 287–296.
51. Hiscott, J., Kwon, H., and Genin, P. (2001). Hostile takeovers: Viral appropriation of the NFkappaB pathway. *J. Clin. Invest.* **107**, 143–151.
52. Chu, Z. L., DiDonato, J. A., Hawiger, J. *et al.* (1998). The tax oncoprotein of human T-cell leukemia virus type 1 associates with and

persistently activates IkappaB kinases containing IKKalpha and IKKbeta. *J. Biol. Chem.* **273**, 15891–15894.

53. Yamaoka, S., Inoue, H., Sakurai, M. *et al.* (1996). Constitutive activation of NFkappa B is essential for transformation of rat fibroblasts by the human T-cell leukemia virus type I Tax protein. *EMBO J.* **15**, 873–887.
54. Mosialos, G. (1997). The role of Rel/NFkappa B proteins in viral oncogenesis and the regulation of viral transcription. *Semin. Cancer Biol.* **8**, 121–129.
55. Izban, K. F., Ergin, M., Huang, Q. *et al.* (2001). Characterization of NFkappaB expression in Hodgkin's disease: Inhibition of constitutively expressed NFkappaB results in spontaneous caspase-independent apoptosis in Hodgkin and Reed-Sternberg cells. *Mod. Pathol.* **14**, 297–310.
56. Romieu-Mourez, R., Landesman-Bollag, E., Seldin, D. C. *et al.* (2001). Roles of IKK kinases and protein kinase CK2 in activation of nuclear factor-kappaB in breast cancer. *Cancer Res.* **61**, 3810–3818.
57. Arlt, A., Vorndamm, J., Breitenbroich, M. *et al.* (2001). Inhibition of NFkappaB sensitizes human pancreatic carcinoma cells to apoptosis induced by etoposide (VP16) or doxorubicin. *Oncogene* **20**, 859–868.
58. Arlt, A., Vorndamm, J., Muerkoster, S. *et al.* (2002). Autocrine production of interleukin 1beta confers constitutive nuclear factor kappaB activity and chemoresistance in pancreatic carcinoma cell lines. *Cancer Res.* **62**, 910–916.
59. Sasaki, N., Morisaki, T., Hashizume, K. *et al.* (2001). Nuclear factor-kappaB p65 (RelA) transcription factor is constitutively activated in human gastric carcinoma tissue. *Clin. Cancer Res.* **7**, 4136–4142.
60. Kordes, U., Krappmann, D., Heissmeyer, V. *et al.* (2000). Transcription factor NFkappaB is constitutively activated in acute lymphoblastic leukemia cells. *Leukemia* **14**, 399–402.
61. Tai, D. I., Tsai, S. L., Chang, Y. H. *et al.* (2000). Constitutive activation of nuclear factor kappaB in hepatocellular carcinoma. *Cancer* **89**, 2274–2281.
62. Hinz, M., Loser, P., Mathas, S. *et al.* (2001). Constitutive NFkappaB maintains high expression of a characteristic gene network, including CD40, CD86, and a set of antiapoptotic genes in Hodgkin/Reed-Sternberg cells. *Blood* **97**, 2798–2807.
63. Dhawan, P. and Richmond, A. (2002). A novel NFkappa B-inducing kinase-MAPK signaling pathway up-regulates NFkappa B activity in melanoma cells. *J. Biol. Chem.* **277**, 7920–7928.
64. Lind, D. S., Hochwald, S. N., Malaty, J. *et al.* (2001). Nuclear factor-kappa B is upregulated in colorectal cancer. *Surgery* **130**, 363–369.
65. Tamatani, T., Azuma, M., Aota, K. *et al.* (2001). Enhanced IkappaB kinase activity is responsible for the augmented activity of NFkappaB in human head and neck carcinoma cells. *Cancer Lett.* **171**, 165–172.
66. Finco, T. S., Westwick, J. K., Norris, J. L. *et al.* (1997). Oncogenic Ha-Ras-induced signaling activates NFkappaB transcriptional activity, which is required for cellular transformation. *J. Biol. Chem.* **272**, 24113–24116.
67. Zhong, H., SuYang, H., Erdjument-Bromage, H. *et al.* (1997). The transcriptional activity of NFkappaB is regulated by the IkappaB-associated PKAc subunit through a cyclic AMP-independent mechanism. *Cell* **89**, 413–424.
68. Zhong, H., Voll, R. E., and Ghosh, S. (1998). Phosphorylation of NFkappa B p65 by PKA stimulates transcriptional activity by promoting a novel bivalent interaction with the coactivator CBP/p300. *Mol. Cell* **1**, 661–671.
69. Zhong, H., May, M. J., Jimi, E. *et al.* (2002). The Phosphorylation Status of Nuclear NFkappaB Determines Its Association with CBP/p300 or HDAC-1. *Mol. Cell* **9**, 625–636.
70. Reuther, J. Y., Reuther, G. W., Cortez, D. *et al.* (1998). A requirement for NFkappaB activation in Bcr-Abl-mediated transformation. *Genes Dev.* **12**, 968–981.
71. Hideshima, T., Chauhan, D., Richardson, P. *et al.* (2002). NFkappa B as a Therapeutic Target in Multiple Myeloma. *J. Biol. Chem.* **277**, 16639–16647.
72. Baldwin, A. S., Jr. (2001). Series introduction: The transcription factor NFkappaB and human disease. *J. Clin. Invest.* **107**, 3–6.
73. Guttridge, D. C., Albanese, C., Reuther, J. Y. *et al.* (1999). NFkappaB controls cell growth and differentiation through transcriptional regulation of cyclin D1. *Mol. Cell Biol.* **19**, 5785–5799.
74. Joyce, D., Albanese, C., Steer, J. *et al.* (2001). NFkappaB and cell-cycle regulation: The cyclin connection. *Cytokine Growth Factor Rev.* **12**, 73–90.
75. Ansari, S. A., Safak, M., Del Valle, L. *et al.* (2001). Cell cycle regulation of NFkappa b-binding activity in cells from human glioblastomas. *Exp. Cell Res.* **265**, 221–233.
76. Wang, C. Y., Mayo, M. W., Korneluk, R. G. *et al.* (1998). NFkappaB antiapoptosis: Induction of TRAF1 and TRAF2 and c-IAP1 and c-IAP2 to suppress caspase-8 activation. *Science* **281**, 1680–1683.
77. Grumont, R. J., Rourke, I. J., O'Reilly, L. A. *et al.* (1998). B lymphocytes differentially use the Rel and nuclear factor kappaB1 (NFkappaB1) transcription factors to regulate cell cycle progression and apoptosis in quiescent and mitogen-activated cells. *J. Exp. Med.* **187**, 663–674.
78. Cusack, J. C., Jr., Liu, R., Houston, M. *et al.* (2001). Enhanced chemosensitivity to CPT-11 with proteasome inhibitor PS-341: Implications for systemic nuclear factor-kappaB inhibition. *Cancer Res.* **61**, 3535–3540.
79. Wang, C. Y., Cusack, J. C., Jr., Liu, R. *et al.* (1999). Control of inducible chemoresistance: Enhanced anti-tumor therapy through increased apoptosis by inhibition of NFkappaB. *Nat. Med.* **5**, 412–417.
80. Wang, C. Y., Guttridge, D. C., Mayo, M. W. *et al.* (1999). NFkappaB induces expression of the Bcl-2 homologue A1/Bfl-1 to preferentially suppress chemotherapy-induced apoptosis. *Mol. Cell Biol.* **19**, 5923–5929.
81. Gustin, J. A., Maehama, T., Dixon, J. E. *et al.* (2001). The PTEN tumor suppressor protein inhibits tumor necrosis factor-induced nuclear factor kappa B activity. *J. Biol. Chem.* **276**, 27740–27744.

Transcriptional Regulation via the cAMP Responsive Activator CREB

Marc Montminy and Keyong Du
Peptide Biology Laboratories, The Salk Institute for Biological Studies, La Jolla, California

The Transcriptional Response to cAMP

Phosphorylation constitutes a predominant mechanism by which transcription factors are regulated in response to extracellular signals. The first phosphorylation-dependent activator to be described, the cAMP responsive factor CREB, belongs to the basic region/leucine zipper family of activators, which includes, among others, jun, fos, and myc [1,2]. CREB binds as a homodimer to a conserved cAMP responsive element (CRE), which often appears as a palindromic sequence (TGACGTCA), but also occurs as a lower affinity half site (TGACG) [3]. Treatment with cAMP agonist stimulates cellular gene expression by promoting phosphorylation of CREB at Ser133 [4]. A characteristic burst-attenuation pattern of transcription ensues, with maximal rates of transcription and Ser133 phosphorylation evident after 30 min [5]. This burst in transcription is rate limited by nuclear entry of a PKA catalytic subunit (PKAc), a passive process that is triggered by liberation of PKAc from the inactive holoenzyme in response to cAMP. The amplitude of target gene expression achieved in response to cAMP agonist is proportional to signal strength [5]. Following induction, target gene expression declines during the next 2–4 hr, reflecting the PP-1 and PP-2a mediated dephosphorylation of CREB [6,7].

Mechanism of Transcriptional Activation via CREB

Although phosphorylation can regulate activators by modulating their nuclear targeting or DNA-binding activities, CREB belongs to a group whose transactivation potential is specifically affected. The CREB transactivation domain is bipartite, consisting of kinase inducible (KID) and constitutive (Q2) activators that act synergistically in response to cAMP stimulus [8,9]. Following phosphorylation at Ser133, KID promotes transcription via an association with the KIX domain of the histone acetylase (HAT) paralogs CBP and P300 [10–12]. By contrast, Q2 stimulates target gene expression via an interaction with the $hTAF_{II}130$ component of TFIID [13–15].

The mechanism by which the KID and Q2 domains promote transcription has been studied extensively by *in vitro* transcription analysis on naked DNA and chromatin templates. CREB behaves as a constitutive activator when assayed with crude nuclear extracts on a naked DNA template; and Ser133 phosphorylation has only marginal effects on transcription [15,16]. Indeed, the glutamine-rich Q2 domain is highly active in this context; and reconstitution studies using purified basal transcription factors indicate that Q2 stimulates target gene expression by recruiting TFIID to the promoter [13,17].

By contrast, basal CREB activity is strongly repressed on a chromatin template [18]. And recruitment of P300/CBP in response to Ser133 phosphorylation is required for target gene activation. CBP/P300 HAT activity is essential for transcriptional induction in this setting [19]; disruption of CBP/P300 HAT activity either by mutagenesis or by addition of CBP/P300 specific inhibitor (LysCoA) impairs transcription [18]. Coincident with this observation, promoter bound histones H4 and H3 are inducibly acetylated *in vivo* on CREB target promoters in response to cAMP agonist.

Recruitment of CBP HAT activity to the promoter does not appear sufficient for transcriptional induction, however. CREB polypeptides lacking the Q2 domain are not competent for target gene activation despite wild-type CBP binding

..vity in the Ser133 phosphorylated state [18]. These experiments suggest that the activity of the Q2 domain, and more specifically its ability to associate with TFIID, is impaired in the chromatin context under basal conditions. The recruitment of CBP HAT activity to the promoter via association with Ser133 phosphorylated KID may promote target gene expression by disrupting the chromatin imposed barrier to the Q2–TFIID interaction.

A key aspect to this regulatory model is the presence of repressive complexes that inhibit CREB activity under basal conditions. In this regard, histone deacetylases (HDACs) have been shown to repress transcription from a variety of promoters, often by associating with certain activators and blocking recruitment of the transcriptional apparatus. Recent studies suggest that CREB associates with type I HDAC complexes, and these complexes appear important for maintaining low levels of histone acetylation over the promoter [20]. Indeed, treatment with HDAC inhibitors such as butyrate or trichostatin A (TSA) has been found to potentiate target gene expression in part by prolonging Ser133 phosphorylation of CREB during the attenuation phase [21]. The mechanism by which TSA and butyrate potentiate CREB phosphorylation is unclear, but current data suggest that HDACs inhibit CREB activity via an association with Ser/Thr phosphatases.

Signal Discrimination via CREB

Although first characterized as a cAMP-dependent activator, CREB also promotes cellular gene expression in response to a wide range of signals including growth factors, cytokines, UV irradiation, and neuronal depolarization [22]. At least three major intracellular pathways have been implicated in this process: cAMP, MAPK, and calcium. Despite inducing Ser133 phosphorylation with comparable kinetics and stoichiometry, non-cAMP pathways appear far less potent in stimulating target gene expression [3].

Activation of the MAPK pathway in response to growth factors promotes Ser133 phosphorylation of CREB in large part via the mitogen and stress-activated kinase MSK1 [23]. Highly related to pp90RSK, MSK1 and its paralog MSK2 appear to phosphorylate CREB with high efficiency in response to most if not all non-cAMP stimuli [23]. Importantly, CREB phosphorylation in response to non-cAMP stimuli is severely impaired in MSK1/2 null cells [24].

Despite the high stoichiometry of CREB phosphorylation via MSK1 and MSK2, CREB target gene activation via these kinases is much lower than that via PKA. Using a CREB–CBP complex specific antiserum, Wagner and colleagues [25] found that this association was strongly induced in response to cAMP but not other stimuli, suggesting that signal discrimination via CREB reflects the relative ability of CBP to be recruited to the promoter. Using fluorescence resonance energy transfer (FRET), Mayr and colleagues [26] found that relevant interaction domains in CREB and CBP, referred to as KID and KIX, respectively, are sufficient for signal discrimination in response to cAMP and non-cAMP stimuli. The process of signal discrimination appears to reside in the nucleus; cytoplasmic KID and KIX peptides showed no preference for various stimuli.

Secondary Phosphorylation of CREB: Ser142

The nature of the discriminatory signal remains elusive; but current work has drawn considerable attention to a second phosphorylation site in CREB: Ser142. In earlier studies, Sun and colleagues [27] and Sun and Maurer [28] had observed that the calcium/calmodulin-dependent kinase (CaMKII) strongly induces Ser133 phosphorylation of CREB. But target gene induction via this kinase was blocked due to concomitant phosphorylation of CREB at Ser142. Indeed, mutagenesis of Ser142 to alanine rescued target gene induction by CaMKII, whereas a Ser142Asp substitution, which mimicked Ser142 phosphorylation, reduced CREB activation in response to cAMP.

The solution structure of the KID/KIX complex has provided insight into the mechanism by which Ser142 phosphorylation blocks CREB activity [29,30]. The structure reveals that KID undergoes a random coil-to-helix transition upon binding to KIX. Ser133 appears to nucleate this interaction by forming direct contacts with residues in KIX. Hydrophobic residues lining one face of the inducible helix in KID further stabilize the complex via interactions with a hydrophobic groove in KIX. Remarkably, Ser142 projects into this groove and appears to destabilize the KID–KIX interaction by electrostatic repulsion [30].

Whether Ser142 is phosphorylated *in vivo* and whether this site explains signal discrimination via CREB are somewhat controversial questions. Substitution of Ser142 with alanine does not appear to rescue target gene in response to certain non-cAMP stimuli [26]; and Ser142 phosphorylation is not detectable using two-dimensional tryptic mapping studies on ^{32}p-labeled cells [31]. Using phosphospecific antisera, however, Kornhauser and colleagues [31] found that Ser142 and Ser143 were indeed phosphorylated in neurons following depolarization in response to KCl or glutamate. Despite their inhibitory effect on the CREB–CBP interaction, Ser142/143 phosphorylation appeared essential, in conjunction with Ser133 phosphorylation, for target gene activation in response to these stimuli. Confirming this notion, Gau and colleagues [32] found that light-induced shifts in locomotion and in CREB target gene activation in the suprachiasmatic nucleus were compromised in Ser142/Ala knock-in mice. In contrast with these findings, Impey and colleagues [33] found that CBP was actually critical for CREB target gene expression in neurons following exposure to glutamate, and that inducible phosphorylation of CBP at Ser301 was important in this process. One clear limitation in the former studies is the absence of data concerning the stoichiometry of Ser142/143 phosphorylation, particularly since phosphorylation at this site is not detectable using standard 2-D tryptic mapping experiments.

Methylation of the KIX Domain

In addition to secondary phosphorylation of KID at Ser142, formation of the KID/KIX complex is also modulated by modification of the KIX domain in CBP. In the process of evaluating the mechanism by which the methyltransferase CARM1 potentiates nuclear hormone receptor activity, Xu and coworkers [34] found unexpectedly that CARM1 blocked target gene activation via CREB in response to cAMP stimulus. Biochemical studies reveal that CARM1 disrupts CREB activity by catalyzing the methylation of CBP at Arg600 within the KIX domain. Indeed, Arg600 was first identified in a genetic screen for mutants in KIX that were unable to bind to phospho (Ser133) CREB *in vitro* [35]. NMR studies reveal that Arg600 does not participate directly in surface interactions with KID; rather, modification of this residue appears to disrupt KIX activity by destabilizing its structure. Thus Arg600 may serve as a methylation-sensitive switch that regulates the ability of CBP to associate with phospho (Ser133) CREB. The development of methylation-specific antisera should reveal under what conditions this switch is activated.

Cooperative Binding with MLL

In addition to covalent modification, CREB–CBP complex formation also appears to be regulated via allosteric mechanisms. In an elegant three-hybrid screen, Ernst and colleagues [36] found that the *Drosophila* trithorax homolog mixed lineage leukemia (MLL) protein stabilized CREB–CBP complex formation following its association with the KIX domain. NMR studies indicate the MLL stabilizes the CREB–CBP complex upon binding to a surface in KIX that is distinct from the CREB binding surface [37]. Whether MLL potentiates CREB activity *in vivo*, however, remains unclear.

The presence of several distinct mechanisms by which CREB–CBP complex formation is modulated suggests that signaling pathways may impose specific constraints on target gene activation by employing a combination of modifications and allosteric regulators. A complete characterization of these modifications will be critical in understanding how CREB has such different effects on cellular activity in response to distinct stimuli.

References

1. Hoeffler, J. P. *et al.* (1988). Cyclic-AMP-responsive DNA-binding protein: Structure based on a cloned placental cDNA. *Science* **242**, 1430–1432.
2. Gonzalez, G. A. *et al.* (1989). A cluster of phosphorylation sites on the cyclic AMP-regulated nuclear factor CREB predicted by its sequence. *Nature* **337**, 749–752.
3. Mayr, B. and Montminy, M. (2001). Transcriptional regulation by the phosphorylation dependent factor CREB. *Nat. Rev. Mol. Cell Biol.* **2**, 599–609.
4. Gonzalez, G. A. and Montminy, M. R. (1989). Cyclic AMP stimulates somatostatin gene transcription by phosphorylation of CREB at serine 133. *Cell* **59**, 675–680.
5. Hagiwara, M. *et al.* (1993). Coupling of hormonal stimulation and transcription via cyclic AMP-responsive factor CREB is rate limited by nuclear entry of protein kinase A. *Mol. Cell Biol.* **13**, 4852–4859.
6. Hagiwara, M. *et al.* (1992). Transcriptional attenuation following cAMP induction requires PP-1-mediated dephosphorylation of CREB. *Cell* **70**, 105–113.
7. Wadzinski, B. *et al.* (1993). Nuclear protein phosphatase 2A dephosphorylates protein kinase A-phosphorylated CREB and regulates CREB transcriptional stimulation. *Mol. Cell Biol.* **13**, 2822–2834.
8. Brindle, P., Linke, S., and Montminy, M. (1993). Analysis of a PK-A dependent activator in CREB reveals a new role for the CREM family of repressors. *Nature* **364**, 821–824.
9. Quinn, P. G. (1993). Distinct activation domains within cAMP response element-binding protein (CREB) mediate basal and cAMP-stimulated transcription. *J. Biol. Chem.* 1993. **268**, 16999–17009.
10. Chrivia, J. C. *et al.* (1993). Phosphorylated CREB binds specifically to the nuclear protein CBP. *Nature* **365**(6449), 855–859.
11. Kwok, R. *et al.* (1994). Nuclear protein CBP is a coactivator for the transcription factor CREB. *Nature* **370**, 223–226.
12. Arias, J. *et al.* (1994). Activation of cAMP and mitogen responsive genes relies on a common nuclear factor. *Nature* **370**, 226–228.
13. Ferreri, K., Gill, G., and Montminy, M. (1994). The cAMP regulated transcription factor CREB interacts with a component of the TFIID complex. *Proc. Natl. Acad. Sci. USA* **91**, 1210–1213.
14. Nakajima, T. *et al.* (1997). Analysis of a cAMP-responsive activator reveals a two-component mechanism for transcriptional induction via signal-dependent factors. *Genes Dev.* **11**, 738–747.
15. Felinski, E. A. *et al.* (2001). Recruitment of an RNA polymerase II complex is mediated by the constitutive activation domain in CREB, independently of CREB phosphorylation. *Mol. Cell Biol.* **21**(4), 1001–1010.
16. Gonzalez, G. A. *et al.* (1991). Characterization of motifs which are critical for activity of the cAMP-responsive transcription factor CREB. *Mol. Cell Biol.* **11(3)**, 1306–1312.
17. Felinski, E. A. and Quinn, P. G. (1999). The CREB constitutive activation domain interacts with TATA-binding protein-associated factor 110 (TAF110) through specific hydrophobic residues in one of the three subdomains required for both activation and TAF110 binding. *J. Biol. Chem.* **274**(17), 11672–11678.
18. Asahara, H. *et al.* (2001). Chromatin dependent cooperativity between constitutive and inducible activation domains in CREB. *Mol. Cell Biol.* **21**, 7892–7900.
19. Korzus, E. *et al.* (1998). Transcription factor-specific requirements for coactivators and their acetyltransferase functions. *Science* **279**(5351), 703–707.
20. Canettieri, G. *et al.* (2003). *Nat. Struct. Biol.* 2003 (in press).
21. Michael, L. F. *et al.* (2000). The phosphorylation status of a cyclic AMP-responsive activator is modulated via a chromatin-dependent mechanism. *Mol. Cell Biol.* **20**, 1596–1603.
22. Lonze, B. E. and Ginty, D. D. (2002). Function and regulation of CREB family transcription factors in the nervous system. *Neuron* **35**(4), 605–623.
23. Deak, M. *et al.* (1998). Mitogen- and stress-activated protein kinase-1 (MSK1) is directly activated by MAPK and SAPK2/p38, and may mediate activation of CREB. *EMBO J.* **17**, 4426–4441.
24. Wiggin, G. R. *et al.* (2002). MSK1 and MSK2 are required for the mitogen- and stress-induced phosphorylation of CREB and ATF1 in fibroblasts. *Mol. Cell Biol.* **22**(8), 2871–2881.
25. Wagner, B. *et al.* (2000). Stimulus-specific interaction between activator-coactivator cognates revealed with a novel complex-specific antiserum. *J. Biol. Chem.* **275**, 8263–8266.
26. Mayr, B., Canetierri, L., and Montminy, M. (2001). Distinct effects of cAMP and mitogenic signals on CBP recruitment impart specificity to target gene activation via CREB. *Proc. Natl. Acad. Sci. USA* **98**, 10936–11041.
27. Sun, P. *et al.* (1994). Differential activation of CREB by Ca^{2+}/calmodulin-dependent protein kinase type II and type IV involves phosphorylation of a site that negatively regulates activity. *Genes Dev.* **8**, 2527–2539.

28. Sun, P. and Maurer, R. (1995). An inactivating point mutation demonstrates that interaction of cAMP response element binding protein (CREB) with the CREB binding protein is not sufficient for transcriptional activation. *J. Biol. Chem.* **270**, 7041–7044.
29. Radhakrishnan, I. *et al.* (1997). Solution structure of the KIX domain of CBP bound to the transactivation domain of CREB: A model for activator:coactivator interactions. *Cell* **91**(6), 741–752.
30. Parker, D. *et al.* (1998). Analysis of an activator:coactivator complex reveals an essential role for secondary structure in transcriptional activation. *Mol. Cell* **2**, 353–359.
31. Kornhauser, J. M. *et al.* (2002). CREB transcriptional activity in neurons is regulated by multiple, calcium-specific phosphorylation events. *Neuron* **34**(2), 221–233.
32. Gau, D. *et al.* (2002). Phosphorylation of CREB Ser142 regulates light-induced phase shifts of the circadian clock. *Neuron* **34**(2), 245–253.
33. Impey, S. *et al.* (2002). Phosphorylation of CBP mediates transcriptional activation by neural activity and CaM kinase IV. *Neuron* **34**(2), 235–244.
34. Xu, W. *et al.* (2001). A transcriptional switch mediated by cofactor methylation. *Science* **294**(5551), 2507–2511.
35. Parker, D. *et al.* (1996). Phosphorylation of CREB at Ser133 induces complex formation with CBP via a direct mechanism. *Mol. Cel. Biol.* **16**, 694–703.
36. Ernst, P. *et al.* (2001). MLL and CREB bind cooperatively to the nuclear coactivator CREB-binding protein. *Mol. Cell Biol.* **21**(7), 2249–2258.
37. Goto, N. K. *et al.* (2002). Cooperativity in transcription factor binding to the coactivator CREB-binding protein (CBP). The mixed lineage leukemia protein (MLL) activation domain binds to an allosteric site on the KIX domain. *J. Biol. Chem.* **277**(45), 43168–43174.

The NFAT Family: Structure, Regulation, and Biological Functions

Fernando Macian and Anjana Rao
Center for Blood Research and Department of Pathology, Harvard Medical School, Boston, Massachusetts

Introduction

Originally identified as regulators of the immune response in T cells, the NFAT family of transcription factors controls various biological processes in diverse tissues and stages of development. NFAT proteins share a high degree of structural homology with the Rel/NFκB family of transcription factors. Four calcium-regulated family members, NFAT1–4, mediate gene transcription in response to calcium/calcineurin signals, while NFAT5 transduces the hypertonic stress response and is also regulated by signals through an integrin receptor and the T-cell receptor. Once activated, NFAT proteins cooperate with other transcriptional factors and induce specific patterns of gene expression that control distinct biological programs. This chapter reviews recent data on NFAT structure and regulation and summarizes some biological functions controlled by NFAT transcription factors.

NFAT is a small but important family of transcription factors originally described in T cells [1–6]. It is now well established that NFAT proteins direct specific biological programs in a variety of cells and tissues. The NFAT family consists of five members. The primordial family member NFAT5 (TonEBP) is found in *Drosophila,* and the genes encoding the four calcium-regulated NFAT proteins, NFAT1–4 (also known as NFATc1–c4), appear to have emerged simultaneously early in the course of vertebrate evolution. The five NFAT proteins are classified into one family based on the sequence and structural similarity of their DNA-binding domains: the degree of sequence identity is ~60% to 70% when NFAT1–4 are compared among themselves, and ~40% to 50% when NFAT1–4 are compared with NFAT5.

The primordial NFAT family member NFAT5/TonEBP is expressed ubiquitously in mammalian cells and regulates the response to hypertonic stress [4,7]. NFAT5 is also likely to be involved in regulating diverse other biological programs. Genetic analysis in *Drosophila* suggests that NFAT5 plays a role in the Ras signal transduction pathway, and this function may be conserved in evolution [8]. NFAT5 protein levels are increased by antigen receptor stimulation of T cells, suggesting a role for this protein in lymphocyte responses downstream of the TCR [9,10]. NFAT5 is also activated by $\alpha6\beta4$ integrin stimulation of carcinoma cells, suggesting a role in tumor metastasis [11].

Except for NFAT2, which is the only family member expressed at a specific stage of development of cardiac valves [12,13], one or more calcium-regulated NFAT proteins are expressed redundantly in many cell types of the embryo and the adult [5,6]. Three structural features are common to all four calcium-regulated NFAT proteins: an N-terminal transactivation domain, a highly phosphorylated regulatory domain that binds and is dephosphorylated by the calcium/calmodulin-regulated phosphatase calcineurin, and a DNA-binding domain that is monomeric in solution, forms dimers on certain κB-like DNA elements, and interacts with Fos and Jun proteins on "composite" NFAT:AP-1 DNA elements [1–5].

Structure and DNA-Binding

The NFAT family is evolutionarily related to the Rel/NFκB family [10,14,15]. The level of sequence identity between these two families is marginal (~17%) but the structural similarity is remarkable. The DNA-binding domains of all NFAT and NFκB/Rel family members have two domains, an N-terminal specificity domain (~180 amino acids) involved in making base-specific DNA contacts, and a C-terminal domain (~100 amino acids) involved in dimer formation [14,16,17]. Together these domains constitute the Rel homology region common to all members of the extended NFAT/NFκB/Rel family. Comparison of the structures of NFAT5/TonEBP and NFκB p50 dimers on DNA illustrates the striking structural similarity between the NFAT and NFκB families [15–17]. The C-terminal domains of NFAT5 and NFκB p50 utilize a similar interface to form the dimer contacts, but NFAT5 has an additional surface for dimer formation, which involves the N-terminal domain [15–17]. NFAT5 and Rel proteins are stable dimers in the absence of DNA, whereas the calcium-regulated NFAT proteins are monomeric [10,18]. Nevertheless, NFAT1–4 can also form dimers on certain κB-like sites, a finding now confirmed by crystallography of two NFAT1 dimer–DNA complexes [S. Harrison and L. Chen, personal communication]. An interesting point is that all NFAT proteins, including NFAT5, prefer monomeric binding sites on DNA, whereas Rel proteins bind loosely palindromic DNA elements [19].

Regulation

NFAT1–4

NFAT1–4 proteins are activated by ligand binding to a variety of cell-surface receptors [1–3,5,6]. The common feature of the receptors is their ability to activate phosphatidylinositol-specific phospholipase C (PLC), thereby inducing calcium influx across the plasma membrane. In the immune system, the ability of immunoreceptors (T- and B-cell antigen receptors, Fcε receptors on mast cells, and Fcγ receptors on NK cells and monocytes) to activate NFAT is well documented [1]. Stimulation of immune cells through these receptors activates a cascade of several tyrosine kinases, leading to tyrosine phosphorylation and activation of PLC-γ. In other cell types, NFAT activation has been shown to result from stimulation of G-protein-coupled receptors leading to PLCβ activation, or stimulation of receptor tyrosine kinases leading to PLCγ activation. This leads to PIP2 hydrolysis and generation of IP3, which by binding to the IP3 receptor and depleting intracellular endoplasmic reticulum (ER) calcium stores, initiates the process of store-operated calcium entry through the plasma membrane [20,21].

NFAT-dependent gene transcription is exquisitely sensitive to changes in intracellular calcium concentration ($[Ca^{2+}]_I$). Even in the continuous presence of stimulus, $[Ca^{2+}]_I$ levels may oscillate depending on specific parameters of receptor occupancy and desensitization [20]. $[Ca^{2+}]_I$ levels are thus modulated at two levels—amplitude and oscillation frequency—and NFAT activation is sensitive to both types of modulation [22–24]. This is well illustrated by T cells from two severe combined immunodeficiency patients with a primary defect in store-operated calcium entry [25]. When activated, these cells show a small $[Ca^{2+}]_I$ spike resulting from store depletion, but they lack the ability to sustain increased $[Ca^{2+}]_I$ levels for several hours as observed in wild-type T cells. The brief increase in $[Ca^{2+}]_I$ suffices for transient dephosphorylation and nuclear localization of NFAT, manifested by transcription of a small number of NFAT target genes [25,26]. However, optimal activation of NFAT-dependent genes requires sustained calcium/calcineurin signals [25,26], which are most effectively elicited by slow $[Ca^{2+}]_I$ oscillations or low sustained $[Ca^{2+}]_I$ increases [22–24]. The differing sensitivities of NFAT target genes to intracellular free calcium levels are likely to reflect the multiple configurations of NFAT sites in gene regulatory regions; for instance, the requirement for sustained activation could be either because NFAT itself needs to remain within a transcription complex for many hours, or because formation of a cooperative NFAT:AP-1 complex is important for gene transcription (*de novo* synthesis of Fos proteins is required for optimal AP-1 activation).

Interaction with calcineurin is central to the calcium responsiveness of NFAT1–4. The major calcineurin docking site on NFAT is located at the N terminus of the regulatory domain and has the consensus sequence PxIxIT [27]. Substitution of the PxIxIT sequence of NFAT1 with a higher affinity version obtained by peptide selection increases the basal $[Ca^{2+}]_I$ sensitivity of NFAT [28]. The surface of NFAT–calcineurin interaction is unlikely to be limited to the PxIxIT motif, however; a second interacting region is present at the C terminus of the regulatory domain [29,30]. By mass spectrometric analysis, NFAT1 isolated from resting cells was shown to contain at least 21 phosphoserine residues, of which 14 were located in characteristic conserved sequence motifs in the regulatory domain [31]. Following cell stimulation, 13 of these residues were dephosphorylated by calcineurin. Dephosphorylation of 5 phosphoserines in one of these conserved motifs (SRR-1) sufficed for exposure of a nuclear localization signal in the regulatory domain, while dephosphorylation of all 13 residues was required for masking of a nuclear export sequence, complete nuclear localization, and full transcriptional activity [31]. In NFAT2, dephosphorylation of the conserved SPxx motifs has been correlated with increased DNA binding [32]. It is interesting that the gene for an endogenous calcineurin inhibitor, MCIP1, is itself an NFAT target, implying the existence of a negative feedback loop that down-modulates NFAT activity [33].

When calcium entry is prevented or calcineurin activity is inhibited in stimulated cells, NFAT is rephosphorylated by NFAT kinases and rapidly leaves the nucleus. A variety of constitutive and inducible kinases, including casein kinase 1, casein kinase 2, GSK3, and the MAP kinases p38 and JNK, have been suggested as NFAT kinases ([32,34–40] and reviewed in [3]). Different NFAT proteins can be differentially regulated, even in a single cell type under identical cell culture or stimulation conditions, a finding most likely explained by differential susceptibility of individual NFAT proteins to signal-activated kinases [41,42].

Mass spectrometric analysis indicated that in stimulated cells, NFAT1 becomes phosphorylated on a serine residue in its transactivation domain [31,43]. It remains to be determined if this modification is necessary for transcriptional function in the context of the full-length protein. Pim1, Cot, and protein kinase C zeta have been suggested as candidate kinases, because when overexpressed, they enhance reporter activity dependent on the NFAT transactivation domain [44–46]. However, as previously suggested in a similar study using calmodulin-dependent kinase II, such enhancement could arise from indirect effects on coactivator function rather than through direct NFAT phosphorylation [43].

NFAT5

The mechanism of NFAT5 regulation is not yet fully understood. In most cell types examined, NFAT5 is present in both the nucleus and the cytoplasm; however, in at least one subline of Jurkat T cells, NFAT5 is cytoplasmic and is translocated to the nucleus following hypertonic stimulation [7,19]. NFAT5 is basally phosphorylated in resting cells and hyperphosphorylated following stimulation [10], but it is not known how (or whether) phosphorylation contributes to its function. The kinases responsible for inducible NFAT5 phosphorylation have not been identified; in particular, NFAT5 does not appear to be regulated by p38, a kinase known to be responsive to osmotic stress (C. Lopez-Rodriguez and A. Rao, unpublished). Dimerization of NFAT5 is required for transcriptional activity; however, hypertonic stress does not regulate dimerization, since NFAT5 is already a dimer in resting cells [10]. Direct binding studies have not provided evidence for interaction of NFAT5 with any of the known IκBs, despite the fact that this protein conserves many IκB-interacting residues common to Rel proteins (C. Lopez-Rodriguez and A. Rao, unpublished). Potentially, the DNA binding or nuclear localization of NFAT5 could be regulated intramolecularly, by masking interactions with other protein domains, in a manner sensitive to phosphorylation. Alternatively, NFAT5 function might be regulated by cooperation with nuclear partner proteins whose expression or function was modulated in a signal-dependent manner.

Transcriptional Functions

All NFAT proteins contain intrinsic transactivation domains, and thus are bona fide transcription factors that can independently induce gene transcription. The large C-terminal region of NFAT5 contains a transactivation domain that functions in reporter assays when fused to the GAL4 DNA-binding domain [10]. By the same criterion, NFAT1 and NFAT4 possess two transactivation domains, at their N and C termini, respectively [47,48]; both are strongly acidic and contain a DELDF[S/K] sequence that may serve as a docking site for the SAGA histone acetyltransferase complex [49]. Similar motifs are found in the corresponding regions of NFAT3 and the constitutive isoforms of NFAT2. The transactivation domain of NFAT2 has surprisingly been mapped to the beginning of its regulatory domain, overlapping a sequence utilized as the major docking site for calcineurin [50]. Constitutively active NFAT1 and NFAT2, which mimic the corresponding dephosphorylated proteins, have been shown to induce or potentiate target gene expression in resting cells [31,51,52].

NFAT-Fos-Jun cooperation constitutes a major mechanism of NFAT-dependent gene transcription. This subject has been recently reviewed [4] and we will only briefly elaborate on it here. Whereas the residues required for Fos-Jun contact are almost completely conserved in the DNA-binding domains of all four calcium-regulated NFAT proteins [53], they are absent from NFAT5, suggesting that the ability to cooperate with Fos and Jun was a relatively late evolutionary development [19]. NFAT has also been shown to cooperate functionally with many other families of transcription factors including GATA [54,55], MEF2 [56], Maf [57], IRF4 [58], and PPARγ [59]. Unlike the NFAT:AP-1 interaction, which cannot be detected in solution, these interactions appear to be mediated by direct protein–protein contact and are not cooperative on DNA. In at least one case (the NFAT–GATA interaction), the surface of interaction of these other transcription factors with NFAT is known not to overlap the surface of NFAT:AP-1 interaction; a mutant NFAT1 protein engineered to lack completely the ability to cooperate with AP-1 [60] was as or more effective than wild-type NFAT1 in its ability to synergize functionally with GATA3 in a transient reporter assay in T cells [61].

The shortest isoform of NFAT2 (NFATc/A) is induced in a CsA-sensitive manner by NFAT itself, in a process suggested to constitute a positive autoregulatory loop [62]. This protein is generated through utilization of a distinct inducible promoter that is preferentially coupled to the most proximal polyadenylation site [63]. As a result, NFATc/A lacks the entire C-terminal domain and contains an alternate N-terminal domain that is not highly acidic and lacks the SAGA interaction sequence DELDF[S/K]. It is not clear whether this protein would be transcriptionally active in the absence of partner proteins such as AP-1.

Biological Programs Regulated by NFAT

NFAT1–4

The calcium-regulated NFAT proteins have been implicated in a variety of gene expression programs in diverse cell types and tissues. These include cardiac hypertrophy, slow- and fast-twitch fiber differentiation, cardiac valve development, vascular patterning during embryogenesis, chondrocyte development, and adipose differentiation among others. These developmental and differentiation processes have been recently reviewed and are not discussed further here [5,6].

Cooperation of NFAT proteins with members of the AP-1 family of transcription factors is known to play an important role in the establishment of a productive immune response. Activation of both transcription factors is achieved upon full stimulation of B and T cells through their antigen receptors and costimulatory molecules. Composite sites for NFAT and AP-1 have been described in promoters and enhancers of many cytokine genes [1]. Paradoxically, NFAT without AP-1 induces an unresponsive or tolerant state in lymphocytes [51]. This response is induced in T cells by TCR stimulation that is caused by stimulation through the TCR without engagement of costimulatory receptors. This results in increased $[Ca^{2+}]_I$ levels with minimal MAP kinase or IκB kinase activation, thereby leading to NFAT activation without appreciable AP-1 or NFκB activation. Under these conditions NFAT initiates a distinct gene expression profile associated with a profound block in TCR signaling [51]. Anergic T cells are unable to respond to a subsequent complete stimulation by producing IL-2 or initiating a proliferative response. Thus, NFAT proteins play a central role in the control of two opposite aspects of T-cell function by activating two different genetic programs depending on the presence or absence of AP-1 cooperation.

NFAT5

NFAT5 has been shown or suggested to play a role in several cellular responses including the osmoprotective response, the TCR response, Ras signaling, and tumor invasion. It was independently cloned in a yeast one-hybrid system as TonEBP (tonicity element binding protein) [7]. NFAT5 activates a large number of target genes implicated in osmoprotective responses, including those encoding aldose reductase, the betaine transporter, and the inositol transporter [64]. In *Drosophila,* NFAT5 has been implicated in modulating Ras signaling: NFAT5 overexpression affects Ras-1-dependent photoreceptor cell development, suppressing a rough-eye phenotype produced by activated Ras [8]. In carcinoma cells, NFAT5 activity is induced in response to α6β4 clustering and specific inhibition of NFAT5 results in reduced carcinoma invasion [11], suggesting a novel role for NFAT5 as an inducer of integrin-mediated tumor metastasis. In hyperosmotically stressed lymphocytes, NFAT5 up-regulates transcription of TNFα and LTβ and may play a role in organogenesis, apoptosis, and various pathological responses [10]. In human T cells, activation through the TCR increases NFAT5 levels and activates transcription from NFAT5-dependent reporters. The signal transduction pathways that induce NFAT5 in response to TCR engagement seem to be calcineurin dependent and, therefore, different from the mechanisms controlling NFAT5 activation in response to hyperosmotic stimuli [9,10]. These results suggest an additional function for NFAT5 as a mediator of responses downstream of the TCR.

Perspectives

Although a great deal has been discovered about the NFAT family in the 10 years since the founding family members were cloned, much still remains to be understood. It is clear that the individual family members can be independently regulated even in the same cell types, but the basis of this specificity is unclear. The genes regulated by NFAT1–4 in immune cells and by NFAT5 under hypertonic conditions in kidney cells are by large known, but the target genes for these proteins and their cell specific biological functions in other cell types remain to be identified.

References

1. Rao, A., Luo, C., and Hogan, P. G. (1997). Transcription factors of the NFAT family: Regulation and function. *Annu. Rev. Immunol.* **15**, 707–747.
2. Crabtree, G. R. (1999). Generic signals and specific outcomes: Signaling through Ca^{2+}, calcineurin, and NF-AT. *Cell* **96**, 611–614.
3. Kiani, A., Rao, A., and Aramburu, J. (2000). Manipulating immune responses with immunosuppressive agents that target NFAT. *Immunity* **12**, 359–372.
4. Macian, F., Lopez-Rodriguez, C., and Rao, A. (2001). Partners in transcription: NFAT and AP-1. *Oncogene* **20**, 2476–2489.
5. Crabtree, G. R., and Olson, E. N. (2002). NFAT signaling: Choreographing the social lives of cells. *Cell* **109**, S67–S79.
6. Horsley, V., and Pavlath, G. K. (2002). NFAT: Ubiquitous regulator of cell differentiation and adaptation. *J. Cell. Biol.* **156**, 771–774.
7. Miyakawa, H., Woo, S. K., Dahl, S. C., Handler, J. S., and Kwon, H. M. (1999). Tonicity-responsive enhancer binding protein, a rel-like protein that stimulates transcription in response to hypertonicity. *Proc. Natl. Acad. Sci. USA.* **96**, 2538–2542.
8. Huang, A. M., and Rubin, G. M. (2000). A misexpression screen identifies genes that can modulate RAS1 pathway signaling in *Drosophila melanogaster. Genetics* **156**, 1219–1230.
9. Trama, J., Lu, Q., Hawley, R. G., and Ho, S. N. (2000). The NFAT-related protein NFATL1 (TonEBP/NFAT5) is induced upon T cell activation in a calcineurin-dependent manner. *J. Immunol.* **165**, 4884–4894.
10. Lopez-Rodriguez, C., Aramburu, J., Jin, L., Rakeman, A. S., Michino, M., and Rao, A. (2001). Bridging the NFAT and NF-kappaB families: NFAT5 dimerization regulates cytokine gene transcription in response to osmotic stress. *Immunity* **15**, 47–58.
11. Jauliac, S., Lopez-Rodriguez, C., Shaw, L. M., Brown, L. F., Rao, A., and Toker, A. (2002). The role of NFAT transcription factors in the regulation of integrin-mediated carcinoma invasion. *Nat. Cell. Biol.* **4** (in press).
12. de la Pompa, J. L., Timmerman, L. A., Takimoto, H., Yoshida, H., Elia, A. J., Samper, E., Potter, J., Wakeham, A., Marengere, L., Langille, B. L., Crabtree, G. R., and Mak, T. W. (1998). Role of the NF-ATc transcription factor in morphogenesis of cardiac valves and septum. *Nature* **392**, 182–186.
13. Ranger, A. M., Grusby, M. J., Hodge, M. R., Gravallese, E. M., de la Brousse, F. C., Hoey, T., Mickanin, C., Baldwin, H. S., and Glimcher, L. H. (1998). The transcription factor NF-ATc is essential for cardiac valve formation. *Nature* **392**, 186–190.
14. Chen, L., Rao, A., and Harrison, S. C. (1999). Signal integration by transcription-factor assemblies: Interactions of NF-AT1 and AP-1 on the IL-2 promoter. *Cold Spring Harb. Symp. Quant. Biol.* **64**, 527–531.
15. Stroud, J. C., Lopez-Rodriguez, C., Rao, A., and Chen, L. (2002). Structure of a TonEBP-DNA complex reveals DNA encircled by a transcription factor. *Nat. Struct. Biol.* **9**, 90–94.
16. Ghosh, G., van Duyne, G., Ghosh, S., and Sigler, P. B. (1995). Structure of NF-kappa B p50 homodimer bound to a kappa B site. *Nature* **373**, 303–310.
17. Muller, C. W., Rey, F. A., Sodeoka, M., Verdine, G. L., and Harrison, S. C. (1995). Structure of the NF-kappa B p50 homodimer bound to DNA. *Nature* **373**, 311–317.
18. Hoey, T., Sun, Y. L., Williamson, K., and Xu, X. (1995). Isolation of two new members of the NF-AT gene family and functional characterization of the NF-AT proteins. *Immunity* **2**, 461–472.

19. Lopez-Rodriguez, C., Aramburu, J., Rakeman, A. S., and Rao, A. (1999). NFAT5, a constitutively nuclear NFAT protein that does not cooperate with Fos and Jun. *Proc. Natl. Acad. Sci. USA* **96**, 7214–7219.
20. Berridge, M. J., Lipp, P., and Bootman, M. D. (2000). The versatility and universality of calcium signaling. *Nat. Rev. Mol. Cell. Biol.* **1**, 11–21.
21. Lewis, R. S. (2001). Calcium signaling mechanisms in T lymphocytes. *Annu. Rev. Immunol.* **19**, 497–521.
22. Dolmetsch, R. E., Lewis, R. S., Goodnow, C. C., and Healy, J. I. (1997). Differential Activation of Transcription Factors Induced By Ca^{2+} Response Amplitude and Duration. *Nature* **386**, 855–858.
23. Dolmetsch, R. E., Xu, K., and Lewis, R. S. (1998). Calcium oscillations increase the efficiency and specificity of gene expression. *Nature* **392**, 933–936.
24. Li, W., Llopis, J., Whitney, M., Zlokarnik, G., and Tsien, R. Y. (1998). Cell-permeant caged InsP3 ester shows that Ca^{2+} spike frequency can optimize gene expression. *Nature* **392**, 936–941.
25. Feske, S., Giltnane, J., Dolmetsch, R., Staudt, L. M., and Rao, A. (2001). Gene regulation mediated by calcium signals in T lymphocytes. *Nat. Immunol.* **2**, 316–324.
26. Feske, S., Draeger, R., Peter, H. H., Eichmann, K., and Rao, A. (2000). The duration of nuclear residence of NFAT determines the pattern of cytokine expression in human SCID T cells. *J. Immunol.* **165**, 297–305.
27. Aramburu, J., Garcia-Cozar, F., Raghavan, A., Okamura, H., Rao, A., and Hogan, P. G. (1998). Selective inhibition of NFAT activation by a peptide spanning the calcineurin targeting site of NFAT. *Mol. Cell* **1**, 627–637.
28. Aramburu, J., Yaffe, M. B., Lopez-Rodriguez, C., Cantley, L. C., Hogan, P. G., and Rao, A. (1999). Affinity-driven peptide selection of an NFAT inhibitor more selective than cyclosporin A. *Science* **285**, 2129–2133.
29. Liu, J., Masuda, E. S., Tsuruta, L., Arai, N., and Arai, K. (1999). Two independent calcineurin-binding regions in the N-terminal domain of murine NF-ATx1 recruit calcineurin to murine NF-ATx1. *J. Immunol.* **162**, 4755–4761.
30. Park, S., Uesugi, M., and Verdine, G. L. (2000). A second calcineurin binding site on the NFAT regulatory domain. *Proc. Natl. Acad. Sci. USA* **97**, 7130–7135.
31. Okamura, H., Aramburu, J., Garcia-Rodriguez, C., Viola, J. P. B., Raghavan, A., Tahiliani, M., Zhan, X. L., Qin, J., Hogan, P. G., and Rao, A. (2000). Concerted dephosphorylation of the transcription factor NFAT1 induces a conformational switch that regulates transcriptional activity. *Mol. Cell* **6**, 539–550.
32. Neal, J. W. and Clipstone, N. A. (2001). Glycogen synthase kinase-3 inhibits the DNA binding activity of NFATc. *J. Biol. Chem.* **276**, 3666–3673.
33. Yang, J., Rothermel, B., Vega, R. B., Frey, N., McKinsey, T. A., Olson, E. N., Bassel-Duby, R., and Williams, R. S. (2000). Independent signals control expression of the calcineurin inhibitory proteins MCIP1 and MCIP2 in striated muscles. *Circ. Res.* **87**, E61–E68.
34. Beals, C. R., Sheridan, C. M., Turck, C. W., Gardner, P., and Crabtree, G. R. (1997). Nuclear export of NF-ATc enhanced by glycogen synthase kinase-3. *Science* **275**, 1930–1934.
35. Chow, C. W., Rincon, M., Cavanagh, J., Dickens, M., and Davis, R. J. (1997). Nuclear accumulation of NFAT4 opposed by the JNK signal transduction pathway. *Science* **278**, 1638–1641.
36. Zhu, J., Shibasaki, F., Price, R., Guillemot, J. C., Yano, T., Dotsch, V., Wagner, G., Ferrara, P., and McKeon, F. (1998). Intramolecular masking of nuclear import signal on NF-AT4 by casein kinase I and MEKK1. *Cell* **93**, 851–861.
37. Graef, I. A., Mermelstein, P. G., Stankunas, K., Neilson, J. R., Deisseroth, K., Tsien, R. W., and Crabtree, G. R. (1999). L-type calcium channels and GSK-3 regulate the activity of NF-ATc4 in hippocampal neurons. *Nature* **401**, 703–708.
38. Gomez del Arco, P., Martinez-Martinez, S., Maldonado, J. L., Ortega-Perez, I., and Redondo, J. M. (2000). A role for the p38 MAP kinase pathway in the nuclear shuttling of NFATp. *J. Biol. Chem.* **275**, 13872–13878.
39. Porter, C. M., Havens, M. A., and Clipstone, N. A. (2000). Identification of amino acid residues and protein kinases involved in the regulation of NFATc subcellular localization. *J. Biol. Chem.* **275**, 3543–3551.
40. Yang, T. T., Xiong, Q., Enslen, H., Davis, R. J., and Chow, C. W. (2002). Phosphorylation of NFATc4 by p38 mitogen-activated protein kinases. *Mol. Cell. Biol.* **22**, 3892–3904.
41. Loh, C., Carew, J. A., Kim, J., Hogan, P. G., and Rao, A. (1996). T-cell receptor stimulation elicits an early phase of activation and a later phase of deactivation of the transcription factor NFAT1. *Mol. Cell. Biol.* **16**, 3945–3954.
42. Abbott, K. L., Friday, B. B., Thaloor, D., Murphy, T. J., and Pavlath, G. K. (1998). Activation and cellular localization of the cyclosporine A-sensitive transcription factor NF-AT in skeletal muscle cells. *Mol. Biol. Cell* **9**, 2905–2916.
43. Garcia-Rodriguez, C., and Rao, A. (2000). Requirement for integration of phorbol 12-myristate 13-acetate and calcium pathways is preserved in the transactivation domain of NFAT1. *Eur. J. Immunol.* **30**, 2432–2436.
44. de Gregorio, R., Iniguez, M. A., Fresno, M., and Alemany, S. (2001). Cot kinase induces cyclooxygenase-2 expression in T cells through activation of the nuclear factor of activated T cells. *J. Biol. Chem.* **276**, 27003–27009.
45. Rainio, E. M., Sandholm, J., and Koskinen, P. J. (2002). Cutting edge: Transcriptional activity of NFATc1 is enhanced by the Pim-1 kinase. *J. Immunol.* **168**, 1524–1527.
46. San-Antonio, B., Iniguez, M. A., and Fresno, M. (2002). Protein kinase C zeta phosphorylates nuclear factor of activated T cells and regulates its transactivating activity. *J. Bio. Chem.* **277** (in press).
47. Luo, C., Burgeon, E., and Rao, A. (1996). Mechanisms of transactivation by nuclear factor of activated T cells-1. *J. Exp. Med.* **184**, 141–147.
48. Imamura, R., Masuda, E. S., Naito, Y., Imai, S., Fujino, T., Takano, T., Arai, K., and Arai, N. (1998). Carboxyl-terminal 15-amino-acid sequence of NFATx1 is possibly created by tissue-specific splicing and is essential for transactivation activity in T cells. *J. Immunol.* **161**, 3455–3463.
49. Massari, M. E., Grant, P. A., Pray-Grant, M. G., Berger, S. L., Workman, J. L., and Murre, C. (1999). A conserved motif present in a class of helix–loop–helix proteins activates transcription by direct recruitment of the SAGA complex. *Mol. Cell* **4**, 63–73.
50. Chuvpilo, S., Avots, A., Berberich-Siebelt, F., Glockner, J., Fischer, C., Kerstan, A., Escher, C., Inashkina, I., Hlubek, F., Jankevics, E., Brabletz, T., and Serfling, E. (1999). Multiple NF-ATc isoforms with individual transcriptional properties are synthesized in T lymphocytes. *J. Immunol.* **162**, 7294–7301.
51. Macian, F., Garcia-Cozar, F., Im, S.-H., Horton, H. F., Byrne, M. C., and Rao, A. (2002). Transcriptional mechanisms underlying lymphocyte tolerance. *Cell* **109**, 719–731.
52. Porter, C. M. and Clipstone, N. A. (2002). Sustained NFAT signaling promotes a Th1-like pattern of gene expression in primary murine CD^{4+} T cells. *J. Immunol.* **168**, 4936–4945.
53. Chen, L., Glover, J. N., Hogan, P. G., Rao, A., and Harrison, S. C. (1998). Structure of the DNA-binding domains from NFAT, Fos and Jun bound specifically to DNA. *Nature* **392**, 42–48.
54. Molkentin, J. D., Lu, J. R., Antos, C. L., Markham, B., Richardson, J., Robbins, J., Grant, S. R., and Olson, E. N. (1998). A calcineurin-dependent transcriptional pathway for cardiac hypertrophy. *Cell* **93**, 215–228.
55. Musaro, A., McCullagh, K. J., Naya, F. J., Olson, E. N., and Rosenthal, N. (1999). IGF-1 induces skeletal myocyte hypertrophy through calcineurin in association with GATA-2 and NF-ATc1. *Nature* **400**, 581–585.
56. Youn, H. D., Chatila, T. A., and Liu, J. O. (2000). Integration of calcineurin and MEF2 signals by the coactivator p300 during T-cell apoptosis. *EMBO J.* **19**, 4323–4331.
57. Ho, I. C., Hodge, M. R., Rooney, J. W., and Glimcher, L. H. (1996). The proto-oncogene c-maf is responsible for tissue-specific expression of interleukin-4. *Cell* **85**, 973–983.
58. Rengarajan, J., Mowen, K. A., McBride, K. D., Smith, E. D., Singh, H., and Glimcher, L. H. (2002). Interferon regulatory factor 4 (IRF4) interacts with NFATc2 to modulate interleukin 4 gene expression. *J. Exp. Med.* **195**, 1003–1012.

59. Yang, X. Y., Wang, L. H., Chen, T., Hodge, D. R., Resau, J. H., DaSilva, L., and Farrar, W. L. (2000). Activation of human T lymphocytes is inhibited by peroxisome proliferator-activated receptor gamma (PPARgamma) agonists. PPARgamma co-association with transcription factor NFAT. *J. Biol. Chem.* **275**, 4541–4544.
60. Macian, F., Garcia-Rodriguez, C., and Rao, A. (2000). Gene expression elicited by NFAT in the presence or absence of cooperative recruitment of Fos and Jun. *EMBO J.* **19**, 4783–4795.
61. Avni, O., Lee, D., Macian, F., Szabo, S. J., Glimcher, L. H., and Rao, A. (2002). T(H) cell differentiation is accompanied by dynamic changes in histone acetylation of cytokine genes. *Nat. Immunol.* **10**, 10.
62. Zhou, B., Cron, R. Q., Wu, B., Genin, A., Wang, Z., Liu, S., Robson, P., and Baldwin, H. S. (2002). Regulation of the murine Nfatc1 gene by NFATc2. *J. Biol. Chem.* **277**, 10704–10711.
63. Chuvpilo, S., Zimmer, M., Kerstan, A., Glockner, J., Avots, A., Escher, C., Fischer, C., Inashkina, I., Jankevics, E., Berberich-Siebelt, F., Schmitt, E., and Serfling, E. (1999). Alternative polyadenylation events contribute to the induction of NF-ATc in effector T cells. *Immunity* **10**, 261–269.
64. Burg, M. B., Kwon, E. D., and Kultz, D. (1997). Regulation of gene expression by hypertonicity. *Annu. Rev. Physiol.* **59**, 437–455.

CHAPTER 289

Transcriptional Control through Regulated Nuclear Transport

Steffan N. Ho
Departments of Pathology and Cellular and Molecular Medicine, University of California at San Diego, La Jolla, California

Introduction

A defining characteristic of eukaryotic cells is the compartmentalization of the genome into the nucleus, a membrane-bound subcellular organelle that is essentially impermeable to the passive diffusion of macromolecules. This spatial segregation of genomic DNA within the cell provides a means to control the expression of genome information that takes place during cell growth and differentiation through the regulated localization of transcriptional control proteins to the nucleus. The regulation of nuclear transport has thus been widely utilized in a variety of biological contexts to regulate transcription in a signal-dependent manner.

Regulated Nuclear Transport: Overview

The flow of information (i.e., proteins, RNA) into and out of the nucleus represents a dynamic balance between nuclear import and nuclear export. The signal-dependent localization of a protein to the cytoplasm or the nucleus can be altered by regulating either import or export, or by regulating both of these processes in a coordinated manner (Fig. 1). Transport occurs through interactions with soluble transport receptors of the karyopherin/importin family that interact or "dock" with components of the nuclear pore complex [1,2]. The interaction between a transport receptor and a target protein is mediated through specific nuclear localization (i.e., import) sequences (NLSs) or nuclear export sequences (NESs) present within the transported protein [3]. Directionality of transport is established in part by the differential interaction of transport receptors with the small GTPase, Ran, which is present in a GDP-bound form in the cytoplasm and a GTP-bound state in the nucleus [1].

Nuclear transport can be regulated by three distinct mechanisms (Fig. 2). First, the nuclear targeting signal itself, represented schematically as a black box in Fig. 2, can be directly modified to alter the binding affinity between the signal sequence and the relevant transport receptor. This is achieved by phosphorylation of residues typically located within or adjacent to the NLS or NES that are directly involved in the interaction with a transport receptor. One well-characterized example of direct phosphorylation influencing the interaction between an NLS or NES and a transport receptor is provided by the yeast Pho4 transcription factor, which translocates to the nucleus upon dephosphorylation in response to phosphate-free growth conditions. Phosphorylation within the Pho4 NLS inhibits interaction with the import receptor Pse1, whereas phosphorylation at other sites is necessary for interaction with the export receptor Msn5 in the nucleus [4–6].

The second and perhaps more common means of regulating nuclear transport is through regulation of protein–protein interactions, either intramolecular or intermolecular, which alter the accessibility of the transport signal sequence to the transport receptor. This strategy is often referred to as the *masking* or *unmasking* of the signal sequences. The interaction between the NFκB transcription factor and IκBα provides a structurally defined example of intermolecular masking in which association of IκBα blocks accessibility of the NFκB nuclear localization signals [7,8]. The induced degradation of IκBα that occurs upon receptor-dependent signaling results in the unmasking of these NLSs and subsequent nuclear translocation of NFκB.

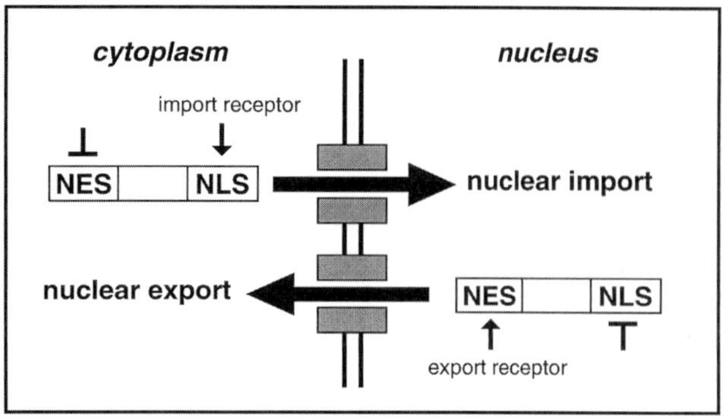

Figure 1 Dynamics of nuclear transport. Nuclear import results from association between a NLS and an import receptor or from the inhibition or disruption of the interaction between a NES and an export receptor. Conversely, nuclear export results from association between an NES and an export receptor or from the inhibition or disruption of the interaction between an NLS and an import receptor. Nuclear transport occurs through the nuclear pore complex. The coordinated regulation of both nuclear import and export provides a rapid, sensitive, and highly responsive means to dynamically translate intracellular signaling events into transcriptional responses in the nucleus.

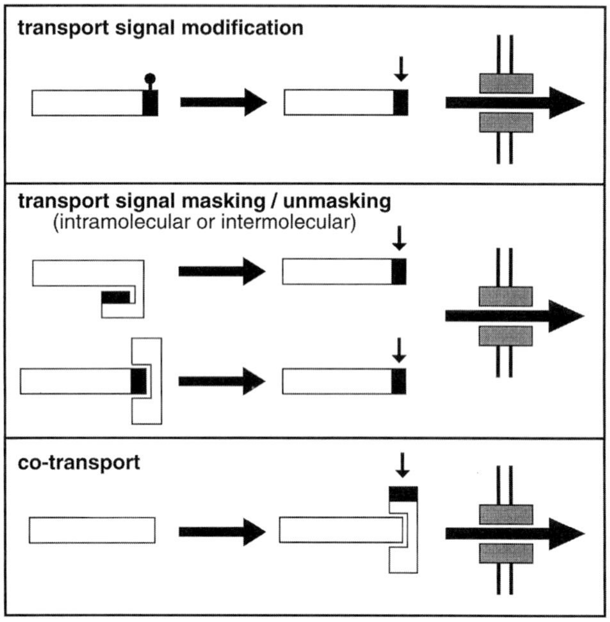

Figure 2 Mechanisms regulating nuclear transport. Nuclear import and export can be regulated by either direct modification of transport signals, masking or unmasking of transport signals through intramolecular or intermolecular protein interactions, or by interaction and cotransport with a carrier protein that contains a transport signal. The black box represents a transport signal. An arrow directed to the transport signal reflects accessibility and association with a transport receptor. The black circle represents protein phosphorylation. The direction of transport (import versus export) is not shown because the indicated mechanisms may affect either import (via NLS) or export (via NES).

The third strategy involves cotransport, whereby the import or export of a given protein is dictated by the interaction and cotransport with another protein that is itself differentially localized or expressed in a signal-dependent manner. Regulation of the NFκB transcription factor also provides an example of cotransport. The termination of signaling events that activate NFκB via IκB degradation results in reexpression of IκB. In addition to masking NLSs of NFκB and preventing nuclear import, IkBα also contains a functional NES that allows cotransport of NFκB out of the nucleus, thus providing a mechanism to actively terminate NFκB-dependent transcription upon cessation of upstream signaling [9–11].

Coordinate Regulation of Nuclear Import and Export: Calcium-Dependent Nuclear Localization of NFATc Transcription Factors

The coordinate regulation of both nuclear import and nuclear export in a signal-dependent manner is clearly illustrated in studies of the regulated nuclear transport of the NFATc family of transcription factors. NFATc transcription factors link calcium-dependent signaling to transcriptional responses that mediate a wide spectrum of biological processes, ranging from cardiac and vascular development to T-cell development and activation [12–14]. NFATc proteins localize to the cytosol in unstimulated cells and undergo rapid nuclear translocation in response to receptor-dependent increases in intracellular free calcium, which activate the calcium/calmodulin-dependent phosphatase calcineurin. In the absence of calcium-dependent signaling (i.e., at basal concentrations of cytoplasmic-free calcium), cytoplasmic NFATc proteins are constitutively and heavily phosphorylated, predominantly within a conserved regulatory domain consisting of a serine-rich region (SRR) and three SP repeat regions. The calcium-dependent activation of calcineurin results in NFATc dephosphorylation and nuclear localization.

Mutations of residues that are the target of calcineurin-dependent phosphatase activity result in constitutive, calcium-independent nuclear localization mediated by functionally defined NLSs [15–17]. Similarly, export of NFATc proteins from the nucleus requires phosphorylation-dependent interaction between the Crm1 export receptor and variably defined NESs, mutation of which results in impaired nuclear export [17–20]. The state of NFATc protein phosphorylation thus dictates the accessibility of both NLSs and NESs through phosphorylation-dependent alterations in protein conformation, hence, representing a mechanism of signal-dependent intramolecular masking and unmasking of nuclear transport signals [15,17].

Several kinases have been identified that are capable of phosphorylating NFATc proteins and functionally regulating the balance of nuclear import/export by antagonizing calcineurin-dependent phosphatase activity, thereby effecting NFATc-dependent transcription. These include glycogen synthase kinase 3, casein kinase I, JNK, ERK, p38, and casein kinase II [16,21–24]. Thus, in addition to calcium-dependent activation of calcineurin, signal-dependent alterations in NFATc kinase activities may also influence NFATc-dependent transcriptional responses by altering the dynamic balance between NFATc nuclear import and export. Calcineurin and NFATc protein kinases also form stable complexes with

NFATc proteins [20,24,25], providing an additional means of regulating the state of NFATc protein phosphorylation and the accessibility of transport sequences. The coordinated, calcium-dependent regulation of both the nuclear import and export of NFATc proteins thus provides a rapid, sensitive, and highly responsive mechanism through which the frequency and duration of receptor-dependent calcium signals can be differentially translated into appropriate transcriptional responses [26–28].

Regulated Nuclear Transport of Non-DNA-Binding Transcriptional Regulatory Proteins

Although regulated nuclear transport of site-specific DNA-binding transcription factors represents an important means of regulating transcription of those target genes that contain appropriate DNA-binding sites, more recent studies have demonstrated that non-DNA-binding transcriptional regulatory proteins that influence chromatin structure are also subject to signal-dependent nuclear transport, thus revealing a novel perspective on signal-dependent transcriptional control mechanisms. The epigenetic modification of histones, proteins that function to package DNA into chromatin, represents an important means of regulating transcription. Histone acetylation is a particularly important modification regulated by a dynamic balance between histone acetyltransferase (HAT) and histone deacetylase (HDAC) activity, with histone acetylases typically associated with active transcription and histone deacetylases associated with repressed transcription [29]. The expression of muscle-specific genes that are induced during myocyte differentiation is controlled by the MyoD and MEF2 family of transcription factors [30]. The class II histone deacetylases HDAC4 and HDAC5 bind to MEF2 and repress MEF2-dependent transcription [31–33]. Myogenic signals induce not only the dissociation of HDACs from MEF2 but also the export of the HDACs from the nucleus [34,35]. Export of HDAC5 from the nucleus is dependent on phosphorylation at Ser-259 and Ser-498 and mediated by activation of the calcium/calmodulin-dependent protein kinase (CaMK) pathway, specifically CaMKI and CaMKIV [34,35]. Alanine substitutions at these positions result in the constitutive nuclear localization of HDAC5 and inhibition of myocyte differentiation [35,36].

The regulated nuclear export of HDAC4 and HDAC5 involves the phosphorylation-dependent association with the intracellular chaperone protein 14-3-3 [36–38]. Phosphorylation-site mutants that constitutively localize to the nucleus also fail to interact with 14-3-3, suggesting that binding of 14-3-3 is necessary for nuclear export [36]. The recent identification of an NES present at the extreme carboxy terminus of HDAC5 (also present in HDAC4 and HDAC7) capable of mediating phosphorylation-dependent nuclear export suggests a model of intramolecular masking in which amino-terminal sequences mask the carboxy-terminal NES [39]. Thus, induction of calcium-dependent signaling activates CaMK, which phosphorylates HDAC5 at Ser-259 and Ser-498, allowing the association of 14-3-3 and unmasking of the carboxy-terminal NES. 14-3-3 has also been shown to decrease the affinity of HDAC4 for the import receptor importin α [37], suggesting that phosphorylation-dependent association of 14-3-3 may regulate HDAC nuclear transport by simultaneously masking an NLS and unmasking an NES.

Nuclear export of HDAC5 may not be essential for activation of MEF2-dependent transcription given that a HDAC5 nuclear export mutant that is competent to bind 14-3-3 continues to exhibit CaMK-dependent activation of MEF2 [36]. Thus, the phosphorylation-dependent binding of 14-3-3 to HDAC5 resulting in the dissociation of MEF2 and HDAC5 appears to be sufficient to allow activation of MEF2-dependent transcription independent of the nuclear export of HDAC. What then is the function of HDAC nuclear export? As suggested by McKinsey et al. [36], localization of HDACs out of the nucleus may allow for a more sustained activation of MEF2-dependent transcription. Alternatively, perhaps the signal-dependent nuclear export of HDACs, by altering the balance between acetylase and deacetylase activity in the nucleus, results in a more global reduction in transcriptional thresholds that facilitates the optimal execution of subsequent steps in a contingent program of transcriptional responses regulating cellular differentiation.

Conclusion

The regulated nuclear transport of DNA-binding transcription factors as well as non-DNA-binding transcriptional regulatory proteins represents an important means of regulating transcription in response to receptor-mediated signaling events. Both nuclear import and export can be regulated by either direct modification of transport signals, masking or unmasking of transport signals through intramolecular or intermolecular protein interactions, or by interaction and cotransport with a carrier protein that contains a transport signal. The coordinated regulation of both nuclear import and export by various combinations of these mechanisms provides a rapid, sensitive, and highly responsive means to dynamically translate intracellular signaling events into transcriptional responses in the nucleus.

References

1. Gorlich, D. and Kutay, U. (1999). Transport between the cell nucleus and the cytoplasm. *Annu. Rev. Cell Dev. Biol.* **15**, 607–660.
2. Rout, M. P. and Aitchison, J. D. (2001). The nuclear pore complex as a transport machine. *J. Biol. Chem.* **276**, 16593–16596.
3. Jans, D. A., Xiao, C. Y., and Lam, M. H. (2000). Nuclear targeting signal recognition: A key control point in nuclear transport? *Bioessays* **22**, 532–544.
4. Kaffman, A., Rank, N. M., O'Neill, E. M., Huang, L. S., and O'Shea, E. K. (1998). The receptor Msn5 exports the phosphorylated transcription factor Pho4 out of the nucleus. *Nature* **396**, 482–486.
5. Kaffman, A., Rank, N. M., and O'Shea, E. K. (1998). Phosphorylation regulates association of the transcription factor Pho4 with its import receptor Pse1/Kap121. *Genes Dev.* **12**, 2673–2683.

6. Komeili, A. and O'Shea, E. K. (1999). Roles of phosphorylation sites in regulating activity of the transcription factor Pho4. *Science* **284**, 977–980.
7. Huxford, T., Huang, D. B., Malek, S., and Ghosh, G. (1998). The crystal structure of the IkappaBalpha/NF-kappaB complex reveals mechanisms of NF-kappaB inactivation. *Cell* **95**, 759–770.
8. Jacobs, M. D. and Harrison, S. C. (1998). Structure of an IkappaBalpha/NF-kappaB complex. *Cell* **95**, 749–758.
9. Arenzana-Seisdedos, F., Turpin, P., Rodriguez, M., Thomas, D., Hay, R. T., Virelizier, J. L., and Dargemont, C. (1997). Nuclear localization of I kappa B alpha promotes active transport of NF-kappa B from the nucleus to the cytoplasm. *J. Cell Sci.* **110**, 369–378.
10. Johnson, C., Van Antwerp, D., and Hope, T. J. (1999). An N-terminal nuclear export signal is required for the nucleocytoplasmic shuttling of IkappaBalpha. *EMBO J.* **18**, 6682–6693.
11. Rodriguez, M. S., Thompson, J., Hay, R. T., and Dargemont, C. (1999). Nuclear retention of IkappaBalpha protects it from signal-induced degradation and inhibits nuclear factor kappaB transcriptional activation. *J. Biol. Chem.* **274**, 9108–9115.
12. Rao, A., Luo, C., and Hogan, P. G. (1997). Transcription factors of the NFAT family: Regulation and function. *Annu. Rev. Immunol.* **15**, 707–747.
13. Crabtree, G. R. (1999). Generic signals and specific outcomes: signaling through Ca^{2+}, calcineurin, and NF-AT. *Cell* **96**, 611–614.
14. Graef, I. A., Chen, F., and Crabtree, G. R. (2001). NFAT signaling in vertebrate development. *Curr. Opin. Genet. Dev.* **11**, 505–512.
15. Beals, C. R., Clipstone, N. A., Ho, S. N., and Crabtree, G. R. (1997). Nuclear localization of NF-ATc by a calcineurin-dependent, cyclosporin-sensitive intramolecular interaction. *Genes Dev.* **11**, 824–834.
16. Zhu, J., Shibasaki, F., Price, R., Guillemot, J. C., Yano, T., Dotsch, V., Wagner, G., Ferrara, P., and McKeon, F. (1998). Intramolecular masking of nuclear import signal on NF-AT4 by casein kinase I and MEKK1. *Cell* **93**, 851–861.
17. Okamura, H., Aramburu, J., Garcia-Rodriguez, C., Viola, J. P., Raghavan, A., Tahiliani, M., Zhang, X., Qin, J., Hogan, P. G., and Rao, A. (2000). Concerted dephosphorylation of the transcription factor NFAT1 induces a conformational switch that regulates transcriptional activity. *Mol. Cell* **6**, 539–550.
18. Kehlenbach, R. H., Dickmanns, A., and Gerace, L. (1998). Nucleocytoplasmic shuttling factors including Ran and CRM1 mediate nuclear export of NFAT *in vitro*. *J. Cell Biol.* **141**, 863–874.
19. Klemm, J. D., Beals, C. R., and Crabtree, G. R. (1997). Rapid targeting of nuclear proteins to the cytoplasm. *Curr. Biol.* **7**, 638–644.
20. Zhu, J. and McKeon, F. (1999). NF-AT activation requires suppression of Crm1-dependent export by calcineurin. *Nature* **398**, 256–260.
21. Beals, C. R., Sheridan, C. M., Turck, C. W., Gardner, P., and Crabtree, G. R. (1997). Nuclear export of NF-ATc enhanced by glycogen synthase kinase-3. *Science* **275**, 1930–1934.
22. Chow, C. W., Rincon, M., Cavanagh, J., Dickens, M., and Davis, R. J. (1997). Nuclear accumulation of NFAT4 opposed by the JNK signal transduction pathway. *Science* **278**, 1638–1641.
23. Gomez del Arco, P., Martinez-Martinez, S., Maldonado, J. L., Ortega-Perez, I., and Redondo, J. M. (2000). A role for the p38 MAP kinase pathway in the nuclear shuttling of NFATp. *J. Biol. Chem.* **275**, 13872–13878.
24. Porter, C. M., Havens, M. A., and Clipstone, N. A. (2000). Identification of amino acid residues and protein kinases involved in the regulation of NFATc subcellular localization. *J. Biol. Chem.* **275**, 3543–3551.
25. Shibasaki, F., Price, E. R., Milan, D., and McKeon, F. (1996). Role of kinases and the phosphatase calcineurin in the nuclear shuttling of transcription factor NF-AT4. *Nature* **382**, 370–373.
26. Dolmetsch, R. E., Lewis, R. S., Goodnow, C. C., and Healy, J. I. (1997). Differential activation of transcription factors induced by Ca^{2+} response amplitude and duration. *Nature* **386**, 855–858.
27. Timmerman, L. A., Clipstone, N. A., Ho, S. N., Northrop, J. P., and Crabtree, G. R. (1996). Rapid shuttling of NF-AT in discrimination of Ca^{2+} signals and immunosuppression. *Nature* **383**, 837–840.
28. Neilson, J., Stankunas, K., and Crabtree, G. R. (2001). Monitoring the duration of antigen-receptor occupancy by calcineurin/glycogen-synthase-kinase-3 control of NF-AT nuclear shuttling. *Curr. Opin. Immunol.* **13**, 346–350.
29. Struhl, K. (1998). Histone acetylation and transcriptional regulatory mechanisms. *Genes Dev.* **12**, 599–606.
30. McKinsey, T. A., Zhang, C. L., and Olson, E. N. (2001). Control of muscle development by dueling HATs and HDACs. *Curr. Opin. Genet. Dev.* **11**, 497–504.
31. Lemercier, C., Verdel, A., Galloo, B., Curtet, S., Brocard, M. P., and Khochbin, S. (2000). mHDA1/HDAC5 histone deacetylase interacts with and represses MEF2A transcriptional activity. *J. Biol. Chem.* **275**, 15594–15599.
32. Lu, J., McKinsey, T. A., Zhang, C. L., and Olson, E. N. (2000). Regulation of skeletal myogenesis by association of the MEF2 transcription factor with class II histone deacetylases. *Mol. Cell* **6**, 233–244.
33. Youn, H. D., Grozinger, C. M., and Liu, J. O. (2000). Calcium regulates transcriptional repression of myocyte enhancer factor 2 by histone deacetylase 4. *J. Biol. Chem.* **275**, 22563–22567.
34. Lu, J., McKinsey, T. A., Nicol, R. L., and Olson, E. N. (2000). Signal-dependent activation of the MEF2 transcription factor by dissociation from histone deacetylases. *Proc. Natl. Acad. Sci. USA* **97**, 4070–4075.
35. McKinsey, T. A., Zhang, C. L., Lu, J., and Olson, E. N. (2000). Signal-dependent nuclear export of a histone deacetylase regulates muscle differentiation. *Nature* **408**, 106–111.
36. McKinsey, T. A., Zhang, C. L., and Olson, E. N. (2000). Activation of the myocyte enhancer factor-2 transcription factor by calcium/calmodulin-dependent protein kinase-stimulated binding of 14-3-3 to histone deacetylase 5. *Proc. Natl. Acad. Sci. USA* **97**, 14400–14405.
37. Grozinger, C. M. and Schreiber, S. L. (2000). Regulation of histone deacetylase 4 and 5 and transcriptional activity by 14-3-3-dependent cellular localization. *Proc. Natl. Acad. Sci. USA* **97**, 7835–7840.
38. Wang, A. H., Kruhlak, M. J., Wu, J., Bertos, N. R., Vezmar, M., Posner, B. I., Bazett-Jones, D. P., and Yang, X. J. (2000). Regulation of histone deacetylase 4 by binding of 14-3-3 proteins. *Mol. Cell Biol.* **20**, 6904–6912.
39. McKinsey, T. A., Zhang, C. L., and Olson, E. N. (2001). Identification of a signal-responsive nuclear export sequence in class II histone deacetylases. *Mol. Cell Biol.* **21**, 6312–6321.

CHAPTER 290

Proteasome/Ubiquitination

Daniel Kornitzer[1] and Aaron Ciechanover[2]
Departments of Molecular Microbiology[1] and Biochemistry[2], Bruce Rappaport Faculty of Medicine, Technion-Israel Institute of Technology, Haifa, Israel

Protein Degradation and the Ubiquitin/Proteasome System

Although most cellular proteins are long lived, a large class exists of normally short-lived proteins. Within this class one finds mainly regulatory factors, of which transcription factors constitute the largest group. One explanation for the short half-life of many regulatory factors is that it enables the rapid modulation of the steady-state concentration of the protein: The levels of constitutively short-lived proteins will respond much faster to changes in their rate of synthesis than that of long-lived ones. Additionally, many factors, rather than being constitutively short lived, can be conditionally stabilized or degraded in response to various stimuli. The question of how extracellular stimuli ultimately can determine the stability of specific transcription factors is central to the understanding of many signaling pathways.

The ubiquitin/proteasome pathway is the principal cellular system for selective protein degradation of normally short-lived proteins and of damaged or unfolded proteins (see [1] for a more extensive review). The ubiquitin system covalently ligates the conserved, 76-amino-acid protein ubiquitin to target proteins by creating an isopeptide bond linking the terminal carboxyl group of the ubiquitin polypeptide to an ε amino group of an internal lysine of the target polypeptide (or, occasionally, to the α amino group of the target polypeptide chain). An internal lysine residue of the first ubiquitin adduct can then serve as acceptor for an additional ubiquitin moiety, eventually yielding a cross-linked polyubiquitin chain that can contain tens of molecules. The polyubiquitin chain serves as a tag for recognition and destruction of the target protein by the 26S proteasome, a large, multicatalytic cytoplasmic protease. The ubiquitination reaction requires a number of enzymes and recognition factors that act sequentially: the ubiquitin-activating enzyme, or E1; an ubiquitin-conjugating enzyme, or E2; and an ubiquitin/protein ligase, or E3. Ubiquitin is initially conjugated in an ATP-requiring reaction to E1 via a thiolester bond, then transferred via a transesterification reaction to a cysteine residue in the active site of the E2, which finally transfers the activated ubiquitin moiety to an amino group of the target protein (Fig. 1). The E3—or ubiquitin/protein ligase—is responsible for substrate recognition; it serves to bring together the E2 and the substrate in a single complex, thereby allowing ubiquitination to occur. Ubiquitin ligases form a heterogeneous group of proteins that appears to consist of two main subgroups: RING finger domain-containing ligases (the larger group) and HECT domain-containing ligases. With the HECT domain ligases, the catalytic cascade of ubiquitination includes an additional transthiolation step in which the activated ubiquitin is transferred from the E2 to a cysteine residue on the E3 prior to its conjugation to the target.

The main site of regulation of the various ubiquitination reactions is at the level of the ubiquitin ligase–substrate interaction, which can be modulated either by modification of the substrate, or by modulation of ubiquitin ligase activity. In the next sections, we review pathways in which the ubiquitin system plays a role in the regulation of transcription factor activity in response to extracellular signals.

Regulation of Ubiquitination by Substrate Modification

Stimulation of Ubiquitination by Substrate Phosphorylation

Modification of proteins by the addition of phosphate groups can modulate their function in a multitude of ways, including by enabling interactions with other proteins. In the event that such an interaction occurs with a proteolytic system,

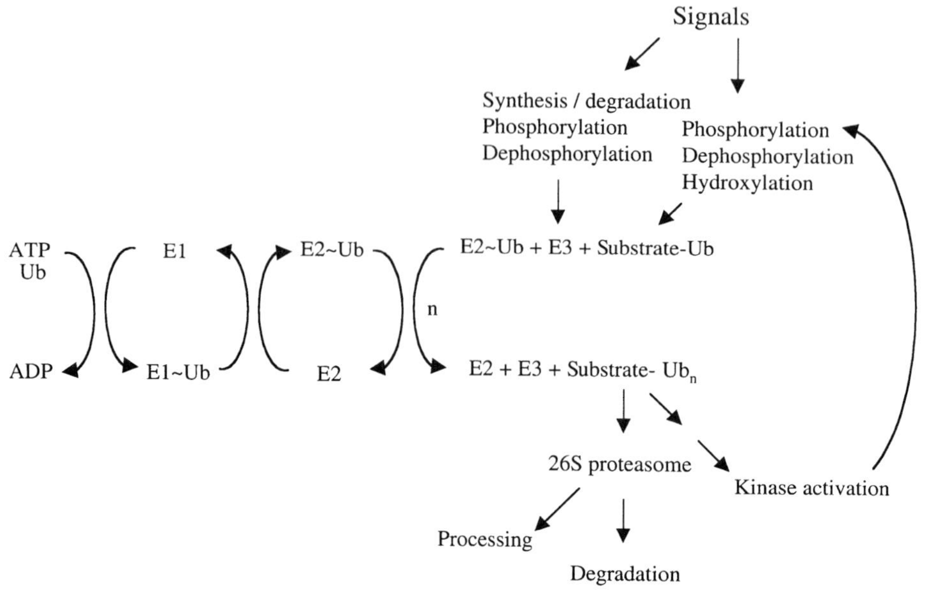

Figure 1 Schematic representation of the ubiquitination reaction cycle. Known points of regulation of the ubiquitin system by signaling pathways are indicated. See text for details.

phosphorylation will result in degradation of the protein. A number of ubiquitin ligases were described that require phosphorylation of the substrate prior to ubiquitination. The known ubiquitin ligases requiring prior phosphorylation of the substrate are primarily of the SCF class. SCF ubiquitin ligases form a complex of at least four proteins: the three components that were initially isolated, which gave the complex its name (SCF stands for *S*kp1, *C*dc53 or *C*ullin, and *F*-box protein [2,3]), and a subsequently identified RING finger domain-containing component, Rbx1 or Roc1 (reviewed in [4]). The F-box protein is the variable component of the complex. Given that SCF complexes with different F-box proteins exhibit different substrate affinities, the F-box proteins are thought to carry the substrate recognition function of the complex (see [5] for a review). The SCF complex containing a specific F-box protein is marked with a superscript, for example, SCF^{CDC4}.

Several transcription factors are known to date to be regulated by SCF-mediated ubiquitination, including Gcn4 [6] and Met4 [7] in yeast, and NFκB [8], β-catenin [9], ATF4 [10], and E2-F1 [11] in mammalian cells. Some of these factors were shown to be regulated at the level of protein stability by signaling pathways that modulate the phosphorylation state of the substrate, and therefore recognition by the SCF complex.

IκBα

One of the most thoroughly investigated signaling pathways is that of NFκB activation (reviewed in [8]). IκBα is an inhibitor of NFκB, which normally sequesters it in the cytoplasm. Upon stimulation of the pathway, IκBα becomes phosphorylated at two specific serine residues, whereupon it becomes a substrate for ubiquitination by the $SCF^{\beta\text{-TrCP}}$ complex. Interestingly, ubiquitination plays additional roles in the NFκB pathway: at the level of processing of the NFκB precursor, p105 (see later section), and at the level of activation of the IκBα kinase (see later section).

Gcn4

Another example of phosphorylation being required for degradation is that of the yeast transcription factor Gcn4. Gcn4 degradation depends on SCF^{CDC4} and on phosphorylation at a single specific site by a cyclin-dependent kinase, Pho85 [6] in conjunction with the Pho85 cyclin Pc15 [12], or, alternatively, on phosphorylation on an undefined site by another cyclin-dependent kinase, Srb10 [13]. Regulation of Gcn4 turnover by the availability of nutrients or, more generally, by the protein biosynthetic capacity of the cell, is mediated by regulation of Pho85 activity [6,12].

β-Catenin

As a last example, turnover of the mammalian transcriptional coactivator β-catenin depends on phosphorylation by glycogen synthase kinase 3β (GSK3β), which leads to ubiquitination of β-catenin by $SCF^{\beta\text{-TrCP}}$ (reviewed in [9]). Activation of the Wnt pathway leads to inhibition of the kinase and, consequently, to stabilization of β-catenin. Interestingly, another pathway of β-catenin destruction was recently identified that also depends on an SCF complex, but in this case the interaction with the substrate is phosphorylation independent and regulation occurs at the level of the ubiquitination complex (see later section).

Inhibition of Ubiquitination by Substrate Phosphorylation: p53

The tumor suppressor protein p53 is regulated at several levels, including degradation by the ubiquitin system. p53 Ubiquitination requires mdm2, a RING finger-containing ubiquitin ligase [14]. Genotoxic stress leads to phosphorylation of p53 at several serine residues, including serine 20.

Phosphorylation of serine 20 was found to stabilize the protein, probably by preventing its interaction with mdm2 [15]. Thus, in the case of p53, phosphorylation exerts a stabilizing effect on the protein.

Stimulation of Ubiquitination by Substrate Hydroxylation: HIF-1α

Although phosphorylation is the most common protein modification, other types of modification could, in principle, regulate the interaction between a protein and the ubiquitin system. The global transcriptional regulator of the hypoxic response, HIF-1α, provides an example of regulation by an alternative type of modification. HIF-1α is rapidly degraded under normoxic conditions, but stabilized under hypoxic conditions. Degradation of HIF-1α depends on ubiquitination by an ubiquitin ligase complex, the VBC complex (*V*on Hippel-Lindau protein–Elongin *B*–and Elongin *C*), which resembles the SCF complex in general architecture, and which contains the Von Hippel-Lindau tumor suppressor protein as a substrate recognition component [16–18]. However, rather than phosphorylation, hydroxylation of a specific proline residue in HIF-1α was recently found to be responsible for the interaction of the protein with the VBC complex [19,20]. This hydroxylation occurs only under normoxic conditions, thereby explaining the dependence of HIF-1α degradation on the oxygen concentration. A specific proline hydroxylase, the activity of which is directly regulated by the oxygen concentration, is responsible for the modification of HIF-1α [21,22]. This pathway thus represents a case where a very short signaling cascade links the stimulus (oxygen concentration) to the response (HIF-1α degradation).

Regulation of Ubiquitin Ligase Activity

Signaling pathways can modulate degradation of proteins by modification of the substrate, as shown earlier. In principle, the same effect could be achieved by modification of the ubiquitination system in response to extracellular signals. The ubiquitin/protein ligase, being the substrate recognition component of the ubiquitin system, would *a priori* be the preferred site of regulation. Regulation of the ubiquitin ligase is best established in the case of the E3 anaphase promoting complex/cyclosome (APC/C), which is responsible for the degradation of the regulators that govern, among other processes, the various mitotic transitions. Regulation of APC/C activity by signals emanating from the cell cycle machinery is essential for orderly cell cycle progression. However, it is not known whether the APC/C also responds to extracellular signals (reviewed in [23,24]).

F-Box Protein Stability

Activity of SCF-type ligases can be modulated at the level of regulation of the F-box component, the substrate recognition component of the complex. Many F-box proteins were found to be intrinsically unstable [25] and, therefore, could be tightly regulated at the level of their synthesis. The transcription factor E2F-1 accumulates in the late G_1 phase and is rapidly degraded in the S phase, following ubiquitination by the SCF^{SKP2} complex. Regulation of E2F-1 degradation was shown to be due to the cell-cycle-dependent synthesis of the F-box protein SKP2 [11]. Another example is that of the yeast transcription factor Met4, which is regulated by the SCF^{MET30} ubiquitin ligase [7]. Met30 is transcriptionally regulated by Met4 [7], but it is also regulated at the level of protein stability by the availability of methionine, so that in the absence of methionine, Met30 is destabilized and disappears from the cell, leading to increased Met4 activity [26]. Here again, regulation appears to be exerted by the concentration of an intracellular metabolite, rather than by extracellular signals.

Regulation of SCF Complex Activity by NEDD8

The recent convergence of several lines of investigation raises the possibility that SCF activity may be subject to direct regulation by cellular signaling pathways. First, the Cullin subunit of the SCF was found to be subject to modification by the addition of an ubiquitin-like protein, Rub1 (in yeast)/NEDD8 (in mammalians) to a specific lysine residue. This modification (henceforth "neddylation") appears to activate the SCF [27]; it is essential for SCF activity in fission yeast [28], although not in budding yeast [29]. The enzymatic machinery of neddylation, which consists of dedicated, Rub1/NEDD8-specific E1 and E2 enzymes, has been characterized, but it is not known whether it is subject to regulation [30]. A protein complex that contains a deneddylation activity, the COP9 signalosome, was also identified [31]. The COP9 signalosome is a conserved protein complex first identified in plants, where it plays a role in photomorphogenesis [32]. The signalosome complex is structurally similar to a part (the "lid") of the regulatory subcomplex (19S) of the 26S proteasome [33]. Although the functional significance of this homology is not yet clear, the signalosome and the "lid" subcomplex of the proteasome include homologous subunits, Jab1 and Rpn11, that appear to carry the catalytic activity of the deneddylation and deubiquitination activities of the respective complexes [34,35]. In mammalian cells, the NEDD8 pathway was recently shown to participate in the p105 processing pathway by affecting the activity of the $SCF^{\beta\text{-}TrCP}$ ubiquitination complex [36]. In *Arabidopsis*, COP9 activity modulates the activity of at least one specific ubiquitination complex, SCF^{TIR1} [31], previously implicated in the cellular response to the plant hormone auxin. Thus, the deneddylation function of the signalosome might connect extracellular signals (in the case of *Arabidopsis*, light) to cellular responses via modulation of the activity of the ubiquitin system.

Protein Processing by the Ubiquitin System

The p50 subunit of the transcription factor NFκB is generated by proteolytic processing of a 105-kDa precursor, p105.

Processing is mediated by the proteasome, in what constituted the first instance of partial degradation of an ubiquitinated protein [38]. The proteasome degrades the C-terminal half of the precursor, but arrests degradation and releases the mature, N-terminal half because of the presence of a "stop-transfer" signal of ill-defined nature in the middle of the precursor [39]. The only salient feature of the signal is that it is rich in glycine residues; how such a sequence would inhibit complete degradation of a protein is unclear, but it is notable that a similar glycine-rich region is found in the viral protein EBNA-1, where in contrast to p105, it confers complete protection of the protein against degradation by the proteasome [34]. p105 Processing can occur in two modes at least, constitutive and stimulated. The constitutive mode probably depends on signals near the glycine-rich region and on an undefined ubiquitin ligase [39]. The second, stimulated mode of processing, depends on phosphorylation of residues in the C terminus of p105 by the IκB kinase. The phosphorylated precursor is then recognized and ubiquitinated by $SCF^{\beta\text{-}TrCP}$, and processed or degraded by the proteasome [40,41]. This second mode ensures that upon activation of the NFκB pathway, not only are NFκB molecules formerly sequestered in the cytosol recruited to the nucleus upon degradation of IκB, but new NFκB molecules are rapidly generated.

The yeast transcription factors Spt23 and Mga2 present another example of processing by the ubiquitin/proteasome system [42]. These two proteins are required for expression of *OLE1*, encoding a fatty acid desaturase essential for the synthesis of the monounsaturated fatty acids, palmitoleic and oleic acid. Regulated synthesis of these compounds is essential for the maintenance of proper membrane fluidity. Spt23 and Mga2 are normally membrane bound due to a C-terminal transmembrane domain. The N-terminal domain of the transcription factor is released from its membrane anchor via ubiquitination and selective degradation of the C-terminal domain by the proteasome. This processing is inhibited by addition of unsaturated fatty acids to the medium. The simplest hypothesis is that in this case, the signal for processing of the precursor is not extracellular, but rather consists of variations in membrane fluidity or thickness that depend on the ratio of unsaturated to saturated fatty acids.

Modulation of Kinase Activity by Ubiquitination

The ubiquitin system is best known for its role in protein degradation; however, examples are emerging of nondestructive roles for ubiquitin. In signal transduction, a recent example for such a nondestructive role is the function of ubiquitination in the activation of the IκB kinase (IKK) [43]. The NFκB pathway is activated by proinflammatory factors such as interleukin 1 that bind to their cognate receptors and ultimately lead to phosphorylation and degradation of IκB, thereby releasing NFκB to the nucleus [8].

One of the first proteins in the NFκB pathway is TRAF6, a signal transducer that links receptor activation to activation of IKK. TRAF6 activation depends on a complex of two ubiquitin conjugating enzymes, Ubc13/Uev1A, that catalyze the polyubiquitination of TRAF6 following receptor activation and (probably) TRAF6 oligomerization [43]. IKK activation depends on phosphorylation by a complex that includes the kinase TAK1. TAK1, in turn, is activated by binding to activated TRAF6. The fact that polyubiquitinated TRAF6 escapes destruction is probably due to the nature of the ubiquitin chain: Whereas "classical" ubiquitin chains are cross-linked via lysine 48 of ubiquitin, the Ubc13/Uev1A ubiquitin conjugating enzymes catalyze the formation of a lysine 63-cross-linked chain, which is probably not recognized by the proteasome. The mechanistic details of how this polyubiquitination of TRAF6 activates TAK1, however, remain obscure.

Conclusion

In the interplay between cellular signal transduction and the ubiquitin system, the latter was often thought of as lying near the bottom of the signaling cascade, in the role of executioner of phosphorylated proteins, merely carrying out the verdict of the kinases. However, not only have we seen instances where the ubiquitination complex is itself the recipient of the signal, but in one case so far, that of the NFκB pathway, the ubiquitin system can even interfere with signaling by directly modulating the activity of kinases. It is likely that the NFκB pathway, which exemplifies the versatility of the ubiquitin system, will serve as a useful paradigm for the role of the ubiquitin system in other signal transduction pathways.

Acknowledgments

Work in the authors' laboratories is supported by grants from the Israel Science Foundation and the Israeli Ministry of Science (to DK), the Israel Science Foundation, the German–Israeli Foundation for Scientific R&D, the U.S.–Israel Binational Science Foundation, the German–Israeli Project Cooperation, a European Community TMR grant, CapCure Israel, and the VPR Fund–Hedson Fund for Medical Research (to AC), and a Mars Pittsburgh Award to DK and AC.

References

1. Hershko, A. and Ciechanover, A. (1998). The ubiquitin system. *Annu. Rev. Biochem.* **67**, 425–479.
2. Skowyra, D., Craig, K. L., Tyers, M., Elledge, S. J., and Harper, J. W. (1997). F-box proteins are receptors that recruit phosphorylated substrates to the SCF ubiquitin-ligase complex. *Cell* **91**, 209–219.
3. Feldman, R. M., Correll, C. C., Kaplan, K. B., and Deshaies, R. J. (1997). A complex of Cdc4p, Skp1p, and Cdc53p/cullin catalyzes ubiquitination of the phosphorylated CDK inhibitor Sic1p. *Cell* **91**, 221–230.
4. Deshaies, R. J. (1999). SCF and Cullin/ring H2-based ubiquitin ligases. *Annu. Rev. Cell Dev. Biol.* **15**, 435–467.
5. Patton, E. E., Willems, A. R., and Tyers, M. (1998). Combinatorial control in ubiquitin-dependent proteolysis: Don't Skp the F-box hypothesis. *Trends Genet.* **14**, 236–243.
6. Meimoun, A., Holtzman, T., Weissman, Z., McBride, H. J., Stillman, D. J., Fink, G. R., and Kornitzer, D. (2000). Degradation of the transcription factor Gcn4 requires the kinase Pho85 and the SCF^{CDC4} ubiquitin–ligase complex. *Mol. Biol. Cell* **11**, 915–927.

7. Rouillon, A., Barbey, R., Patton, E. E., Tyers, M., and Thomas, D. (2000). Feedback-regulated degradation of the transcriptional activator Met4 is triggered by the SCF^{Met30} complex. *EMBO J.* **19**, 282–294.
8. Karin, M. and Ben-Neriah, Y. (2000). Phosphorylation meets ubiquitination: The control of NFκB activity. *Annu. Rev. Immunol.* **18**, 621–663.
9. Polakis, P. (2001). More than one way to skin a catenin. *Cell* **105**, 563–566.
10. Lassot, I., Segeral, E., Berlioz-Torrent, C., Durand, H., Groussin, L., Hai, T., Benarous, R., and Margottin-Goguet, F. (2001). ATF4 degradation relies on a phosphorylation-dependent interaction with the $SCF^{\beta\text{-TrCP}}$ ubiquitin ligase. *Mol. Cell Biol.* **21**, 2192–2202.
11. Marti, A., Wirbelauer, C., Scheffner, M., and Krek, W. (1999). Interaction between ubiquitin-protein ligase SCF^{SKP2} and E2F-1 underlies the regulation of E2F-1 degradation. *Nat. Cell Biol.* **1**, 14–19.
12. Shemer, R., Meimoun, A., Holtzman, T., and Kornitzer, D. (2002). Regulation of the transcription factor Gcn4 by Pho85 cyclin PCl5. *Mol. Cell. Biol.* **22**, 5395–5404.
13. Chi, Y., Huddleston, M. J., Zhang, X., Young, R. A., Annan, R. S., Carr, S. A., and Deshaies, R. J. (2001). Negative regulation of Gcn4 and Msn2 transcription factors by Srb10 cyclin-dependent kinase. *Genes Dev.* **15**, 1078–1092.
14. Haupt, Y., Maya, R., Kazaz, A., and Oren, M. (1997). Mdm2 promotes the rapid degradation of p53. *Nature* **387**, 296–299.
15. Unger, T., Juven-Gershon, T., Moallem, E., Berger, M., Vogt Sionov, R., Lozano, G., Oren, M., and Haupt, Y. (1999). Critical role for Ser20 of human p53 in the negative regulation of p53 by Mdm2. *EMBO J* **18**, 1805–1814.
16. Kamura, T., Sato, S., Iwai, K., Czyzyk-Krzeska, M., Conaway, R. C., and Conaway, J. W. (2000). Activation of HIF-1α ubiquitination by a reconstituted von Hippel-Lindau (VHL) tumor suppressor complex. *Proc. Natl. Acad. Sci. USA* **97**, 10430–10435.
17. Maxwell, P. H., Wiesener, M. S., Chang, G. W., Clifford, S. C., Vaux, E. C., Cockman, M. E., Wykoff, C. C., Pugh, C. W., Maher, E. R., and Ratcliffe, P. J. (1999). The tumour suppressor protein VHL targets hypoxia-inducible factors for oxygen-dependent proteolysis. *Nature* **399**, 271–275.
18. Ohh, M., Park, C. W., Ivan, M., Hoffman, M. A., Kim, T. Y., Huang, L. E., Pavletich, N., Chau, V., and Kaelin, W. G. (2000). Ubiquitination of hypoxia-inducible factor requires direct binding to the β-domain of the von Hippel-Lindau protein. *Nat. Cell Biol.* **2**, 423–427.
19. Ivan, M., Kondo, K., Yang, H., Kim, W., Valiando, J., Ohh, M., Salic, A., Asara, J. M., Lane, W. S., and Kaelin, W. G., Jr. (2001). HIF-1α targeted for VHL-mediated destruction by proline hydroxylation: Implications for O_2 sensing. *Science* **292**, 464–468.
20. Jaakkola, P., Mole, D. R., Tian, Y. M., Wilson, M. I., Gielbert, J., Gaskell, S. J., Kriegsheim, A., Hebestreit, H. F., Mukherji, M., Schofield, C. J., Maxwell, P. H., Pugh, C. W., and Ratcliffe, P. J. (2001). Targeting of HIF-α to the von Hippel-Lindau ubiquitylation complex by O_2-regulated prolyl hydroxylation. *Science* **292**, 468–472.
21. Epstein, A. C., Gleadle, J. M., McNeill, L. A., Hewitson, K. S., O'Rourke, J., Mole, D. R., Mukherji, M., Metzen, E., Wilson, M. I., Dhanda, A., Tian, Y. M., Masson, N., Hamilton, D. L., Jaakkola, P., Barstead, R., Hodgkin, J., Maxwell, P. H., Pugh, C. W., Schofield, C. J., and Ratcliffe, P. J. (2001). C. elegans EGL-9 and mammalian homologs define a family of dioxygenases that regulate HIF by prolyl hydroxylation. *Cell* **107**, 43–54.
22. Bruick, R. K., and McKnight, S. L. (2001). A conserved family of prolyl-4-hydroxylases that modify HIF. *Science* **294**, 1337–1340.
23. Fang, G., Yu, H., and Kirschner, M. W. (1999). Control of mitotic transitions by the anaphase-promoting complex. *Philos. Trans. R. Soc. Lond. B Biol. Sci.* **354**, 1583–1590.
24. Hershko, A. (1999). Mechanisms and regulation of the degradation of cyclin B. *Philos. Trans. R. Soc. Lond. B Biol. Sci.* **354**, 1571–1575.
25. Zhou, P. and Howley, P. M. (1998). Ubiquitination and degradation of the substrate recognition subunits of SCF ubiquitin-protein ligases. *Mol. Cell* **2**, 571–580.
26. Smothers, D. B., Kozubowski, L., Dixon, C., Goebl, M. G., and Mathias, N. (2000). The abundance of met30p limits SCF^{Met30p} complex activity and is regulated by methionine availability. *Mol. Cell Biol.* **20**, 7845–7852.
27. Read, M. A., Brownell, J. E., Gladysheva, T. B., Hottelet, M., Parent, L. A., Coggins, M. B., Pierce, J. W., Podust, V. N., Luo, R. S., Chau, V., and Palombella, V. J. (2000). Nedd8 modification of cul-1 activates $SCF^{\beta\text{-TrCP}}$-dependent ubiquitination of IκBα. *Mol. Cell Biol.* **20**, 2326–2333.
28. Osaka, F., Saeki, M., Katayama, S., Aida, N., Toh, E. A., Kominami, K., Toda, T., Suzuki, T., Chiba, T., Tanaka, K., and Kato, S. (2000). Covalent modifier NEDD8 is essential for SCF ubiquitin-ligase in fission yeast. *EMBO J.* **19**, 3475–3484.
29. Lammer, D., Mathias, N., Laplaza, J. M., Jiang, W., Liu, Y., Callis, J., Goebl, M., and Estelle, M. (1998). Modification of yeast Cdc53p by the ubiquitin-related protein rub1p affects function of the SCF^{Cdc4} complex. *Genes Dev.* **12**, 914–926.
30. Liakopoulos, D., Doenges, G., Matuschewski, K., and Jentsch, S. (1998). A novel protein modification pathway related to the ubiquitin system. *EMBO J.* **17**, 2208–2214.
31. Lyapina, S., Cope, G., Shevchenko, A., Serino, G., Tsuge, T., Zhou, C., Wolf, D. A., Wei, N., and Deshaies, R. J. (2001). Promotion of NEDD-CUL1 conjugate cleavage by COP9 signalosome. *Science* **292**, 1382–1385.
32. Wei, N., Chamovitz, D. A., and Deng, X. W. (1994). Arabidopsis COP9 is a component of a novel signaling complex mediating light control of development. *Cell* **78**, 117–124.
33. Glickman, M. H., Rubin, D. M., Coux, O., Wefes, I., Pfeifer, G., Cjeka, Z., Baumeister, W., Fried, V. A., and Finley, D. (1998). A subcomplex of the proteasome regulatory particle required for ubiquitin-conjugate degradation and related to the COP9-signalosome and eIF3. *Cell* **94**, 615–623.
34. Cope, G. A., Suh, G. S., Aravind, L., Schwarz, S. E., Zipursky, S. L., Koonin, E. V., and Deshaies, R. J. (2002). Role of predicted metalloprotease motif of Jab1/Csn5 in cleavage of Nedd8 from Cull. *Science* **298**, 608–611.
35. Verma, R., Aravind, L., Oania, R., McDonald, W. H., Yates, J. R., 3rd, Koonin, E. V., and Deshaies, R. J. (2002). Role of Rpn11 metalloprotease in deubiquitination and degradation by the 26S proteasome. *Science* **298**, 611–615.
36. Amir, R. E., Iwai, K., and Ciechanover, A. (2002). The NEDD8 pathway is essential for $SCF^{\beta TrCP}$-mediated ubiquitination and processing of the NF-κB precursor p105. *J. Biol. Chem.* **277**, 23253–23259.
37. Schwechheimer, C., Serino, G., Callis, J., Crosby, W. L., Lyapina, S., Deshaies, R. J., Gray, W. M., Estelle, M., and Deng, X. W. (2001). Interactions of the COP9 signalosome with the E3 ubiquitin ligase SCF^{TIRI} in mediating auxin response. *Science* **292**, 1379–1382.
38. Palombella, V. J., Rando, O. J., Goldberg, A. L., and Maniatis, T. (1994). The ubiquitin-proteasome pathway is required for processing the Nf-κB1 precursor protein and the activation of Nf-κB. *Cell* **78**, 773–785.
39. Orian, A., Schwartz, A. L., Israel, A., Whiteside, S., Kahana, C., and Ciechanover, A. (1999). Structural motifs involved in ubiquitin-mediated processing of the NFκB precursor p105: Roles of the glycine-rich region and a downstream ubiquitination domain. *Mol. Cell Biol.* **19**, 3664–3673.
40. Orian, A., Gonen, H., Bercovich, B., Fajerman, I., Eytan, E., Israel, A., Mercurio, F., Iwai, K., Schwartz, A. L., and Ciechanover, A. (2000). $SCF^{\beta\text{-TrCP}}$ ubiquitin ligase-mediated processing of NFκB p105 requires phosphorylation of its C-terminus by IκB kinase. *EMBO J.* **19**, 2580–2591.
41. Heissmeyer, V., Krappmann, D., Hatada, E. N., and Scheidereit, C. (2001). Shared pathways of IκB kinase-induced $SCF^{\beta\text{-TrCP}}$-mediated ubiquitination and degradation for the NFκB precursor p105 and IκBα. *Mol. Cell Biol.* **21**, 1024–1035.
42. Hoppe, T., Matuschewski, K., Rape, M., Schlenker, S., Ulrich, H. D., and Jentsch, S. (2000). Activation of a membrane-bound transcription factor by regulated ubiquitin/proteasome-dependent processing. *Cell* **102**, 577–586.
43. Wang, C., Deng, L., Hong, M., Akkaraju, G. R., Inoue, J., and Chen, Z. J. (2001). TAK1 is a ubiquitin-dependent kinase of MKK and IKK. *Nature* **412**, 346–351.

CHAPTER 291

Fluorescence Resonance Energy Transfer Microscopy and Nuclear Signaling

Ty C. Voss and Richard N. Day
Departments of Medicine and Cell Biology, NSF Center for Biological Timing, University of Virginia Health Sciences Center, Charlottesville, Virginia

Introduction

Signal transduction pathways are typically represented as linear, connect-the-dot diagrams that terminate within the cell nucleus. The dots represent different protein complexes, each functioning sequentially to convey information that the cell will interpret into changing patterns of gene expression. Understanding how the cell assembles these signaling protein complexes is critically important to unraveling disease processes, and to the design of therapeutic strategies. This knowledge is being gained through the combination of biochemical, genetic, and molecular approaches. Importantly, these *in vitro* approaches are now being complemented by noninvasive techniques that allow direct visualization of dynamic protein interactions in their natural environment within the living cell.

Monitoring the social interactions of proteins inside the cell became possible with the cloning of fluorescent proteins (FP) from marine organisms [1,2]. Extensive mutagenesis of the jellyfish green fluorescent protein (GFP) and the cloning of a second FP from coral have yielded genetically encoded tags that emit light from the blue to red range of the visible spectrum [3,4]. These different color protein tags can be used in combinations that are distinguishable by multispectral imaging [4–7]. Because the FPs have no intrinsic intracellular targeting, they usually adopt the subcellular localization of the linked protein. Therefore, the intracellular distribution of several different labeled proteins relative to one another can be monitored in real time in the same living cell, and overlap in their distributions can be determined. The colocalization of the labeled proteins can be an indication that the proteins are assembled in common signaling complexes.

Unfortunately, the diffraction of light limits the resolution of the microscope to approximately 0.2 μm (2000 Å), and objects that are closer together than this will appear as a single object. For perspective, the basal transcription factor IID is an ensemble of three different interacting proteins forming a complex approximately 200 Å across. This complex was visualized *in vitro* by electron microscopy [8], but falls well below the detection limit of the light microscope (Fig. 1). Therefore, considerable distances may actually separate proteins that appear to be colocalized by fluorescence microscopy, and detecting the physical interactions between proteins within a signaling complex would appear to be beyond the reach of the light microscope.

FRET Microscopy Improves Spatial Resolution

Fortunately, it is possible to achieve this degree of spatial resolution by conventional light microscopy by using the technique of fluorescence resonance energy transfer (FRET) microscopy (Fig. 1). FRET has been used as a spectroscopic technique for many years [9,10], but the emergence of the genetically encoded FP tags has dramatically increased the utility of FRET microscopy of living cells [11]. FRET is the radiationless transfer of energy from a donor fluorophore to a nearby acceptor fluorophore resulting in sensitized

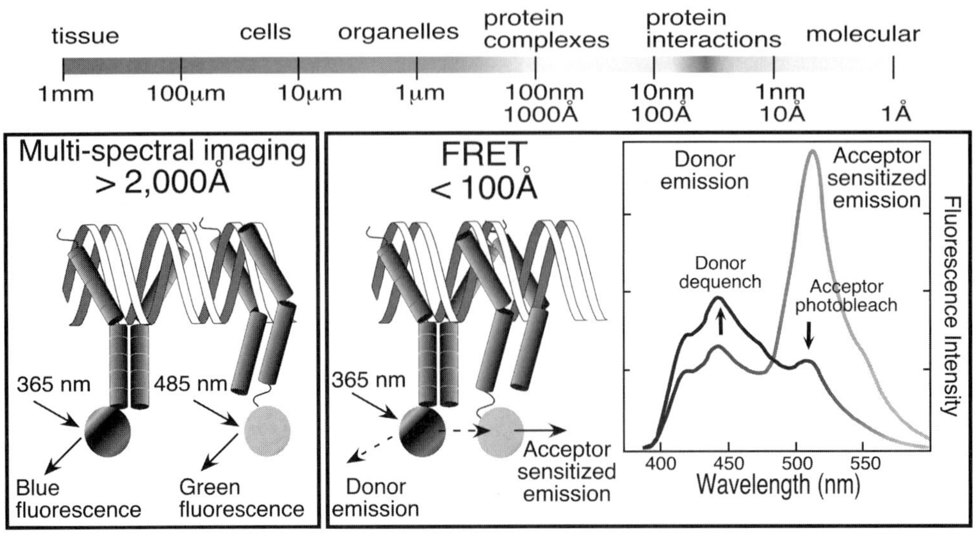

Figure 1 The optical resolution of the microscope is limited to approximately 2000 Å, but this resolution can be improved 50-fold by using FRET microscopy. The detection of sensitized fluorescence emission from acceptor fluorophores upon illumination of the donors can indicate that the tagged proteins are less than 100 Å apart. When energy transfer occurs, the donor fluorescence signal is quenched. Thus, FRET can also be measured by detecting the dequenching of the donor following acceptor photobleaching.

fluorescence emission from the acceptor (Fig. 1). Because the efficiency of the energy transfer drops precipitously as the inverse of the sixth power of the distance separating the donor and acceptor fluorophores, FRET is limited to distances of less than 100 Å. Therefore, the detection of sensitized fluorescence emission from acceptor fluorophores upon illumination of the donors can provide strong evidence for the physical interaction of the tagged proteins.

Limitations of FRET Microscopy

In practice, the detection of FRET signals from living cells expressing independent donor- and acceptor-tagged proteins is difficult because of limited control over the relative amounts of the expressed proteins in any given cell. This produces a highly variable spectral background signal, contributed by both the donor and acceptor fluorophores, from which weak FRET signals must be extracted [12]. Importantly when energy transfer occurs, the donor fluorescence signal is quenched because of the direct transfer of excitation energy to neighboring acceptors (Fig. 1). By measuring donor dequenching following acceptor photobleaching, FRET signals can be mapped to precise subcellular locations [13]. Here, we demonstrate how acceptor photobleaching FRET microscopy can detect the localized interactions of the transcription factor C/EBPα in specific subnuclear domains in the living pituitary cell.

The C/EBP family members are determinants of cell differentiation, and they regulate the expression of genes involved in energy metabolism. The basic region-leucine zipper of C/EBPα forms dimers through contacts in the leucine zipper and binds to specific DNA elements via the basic region. Our studies show that GFP-C/EBPα localizes to regions pericentromeric chromatin in pituitary GHFT1-5 and preadipocyte 3T3-L1 mouse cell lines. This pattern of subnuclear localization is identical to the location of endogenous C/EBPα in differentiated mouse adipocyte cells [14]. In Fig. 2, we show FRET signals originating from these subnuclear sites occupied by C/EBPα tagged with the yellowish (YFP) and blue (BFP) color variants. This fluorescent probe combination was selected for three reasons. First, despite its low quantum yield and sensitivity to photobleaching, the BFP variant used here provides a sufficient signal from the nucleus, where there is a low autofluorescence background. The much higher autofluorescence signal in the cytoplasm makes the BFP variant a poor choice for studies outside the nucleus. Second, the overlap of the BFP emission and YFP excitation spectra is suitable for energy transfer, but this pair has a reduced spectral background when compared to the CFP/YFP combination commonly used for these types of studies [13]. Third, the YFP variant is sensitive to photobleaching [13,15], making it a good choice for acceptor photobleaching FRET.

Detecting C/EBPα Intersections

The results shown in Fig. 2 illustrate the application of acceptor photobleaching FRET to detect the physical association of C/EBPα proteins in specific subnuclear domains. A GHFT1-5 cell coexpressing YFP- and BFP-CEBPα was selected and a reference image was acquired using the YFP filter set (for details, see [13]). Images were then acquired using the FRET filter set (donor excitation, acceptor emission) and the donor filter set (donor excitation, donor emission) at the same focal plane and under identical conditions (Figs. 2B and C, prebleach). The acceptor (YFP-CEBPα) was then

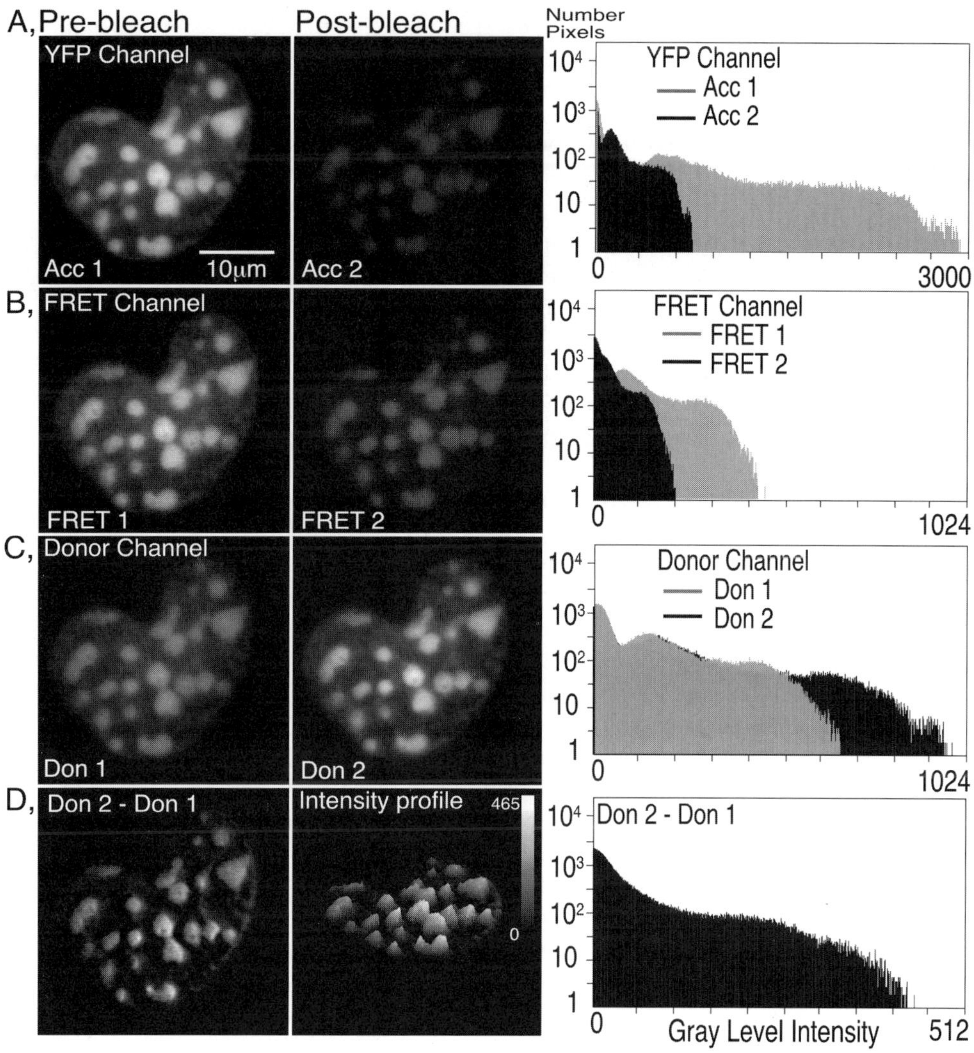

Figure 2 Acceptor photobleaching FRET microscopy of a pituitary GHFT1-5 cell expressing BFP- and YFP-C/EBPα. *Prebleach panels:* (A) The reference image showing subnuclear foci occupied by YFP-C/EBPα (Acc 1); bar = 10 μm. (B) Acceptor emission with donor excitation (FRET 1) and (C) the donor emission (Don 1) obtained under identical conditions. *Postbleach panels:* (A) YFP signal after 5-min bleach period obtained using the same conditions as for the first image. (B) The effect of selective acceptor photobleaching on the FRET signal (FRET 2) and (C) the donor fluorescence intensity (Don 2). (D, left panel) The dequenching of the donor was quantified by subtracting the Don 1 digital image from the Don 2 image, and (D, right panel) The subnuclear location where FRET occurred was mapped in the intensity profile; the calibration bar indicates the gray-level intensity.

photobleached by exposure to 500-nm excitation light for 5 min. A second YFP reference image was then acquired under identical conditions (Fig. 2A, postbleach), showing that the acceptor signal was reduced by approximately 70% (see histogram, Fig. 2A). This selective acceptor photobleaching abolished the FRET signal and revealed the contribution of the donor spectral background signal to the FRET signal (see histogram, Fig. 2B). In contrast, the donor fluorescence intensity was increased following acceptor photobleaching because of donor dequenching (Don 2, Fig. 2C, compare histograms). The donor dequenching was quantified by digital subtraction of the Don 1 image from the Don 2 image (Fig. 2D, left), and the subnuclear location of the dequenched donor signal was mapped by plotting the intensity profile (Fig. 2D, right). Together, the demonstration of a FRET signal above the expected spectral cross-talk background and the observed increased donor signal after acceptor photobleaching provide strong evidence for the physical association of these proteins at discrete subnuclear sites in these living pituitary cells.

Concluding Remarks

The detection of FRET in the living cell is limited by the accuracy of quantifying the intensity and position of the fluorescence signals in space. Accuracy is most severely compromised when measuring low signal intensities near the background noise of the system. Therefore, microscope systems used for FRET studies must be optimized to increase sensitivity and reduce noise. The FRET efficiency will be improved by selecting donor and acceptor fluorophores with substantial spectral overlap, but the cost of the improved efficiency is an increased spectral cross-talk background from which FRET signals must be extracted. The FRET signal can also be improved by optimizing the ratio of the expressed donor- and acceptor-tagged proteins to favor productive interactions. In this regard, however, artifacts can arise from fusion protein overexpression, and control experiments confirming the normal function of the labeled proteins are essential. Further, there are many potential reasons FRET may not be detected from fusion protein partners that interact, including competition by endogenous proteins for the tagged proteins or incorrect orientation of the donor or acceptor fluorophore relative to one another. Therefore, the failure to detect FRET from a pair of labeled proteins carries no intrinsic information regarding the association of the proteins.

Finally, the mobility of higher order protein complexes within the three-dimensional space of the living cell interior represents a major problem for intensity-based measurements of sensitized acceptor emission and donor dequenching. The example shown here is a dimerized transcription factor that localizes to discrete and relatively immobile structures in the cell nucleus. Detecting FRET from signaling protein complexes that are highly mobile requires a temporal resolution that is difficult to achieve using intensity-based measurements. The technique of fluorescence lifetime imaging (FLIM) microscopy, however, can achieve both the spatial and temporal resolution necessary to detect highly dynamic protein interactions. FLIM measures the nanosecond duration of the excited state of a fluorophore, and when combined with FRET, this approach can provide direct evidence for dynamic protein interactions in four dimensions [16,17].

References

1. Prasher, D. C., Eckenrode, V. K., Ward, W. W., Prendergast, F. G., and Cormier, M. J. (1992). Primary structure of the *Aequorea victoria* green-fluorescent protein. *Gene* **111**, 229–233.
2. Chalfie, M., Tu, Y., Euskirchen, G., Ward, W. W., and Prasher, D. C. (1994). Green fluorescent protein as a marker for gene expression. *Science* **263**, 802–805.
3. Tsien, R. Y. (1998). The green fluorescent protein. *Annu. Rev. Biochem.* **67**, 509–544.
4. Patterson, G., Day, R. N., and Piston, D. (2001). Fluorescent protein spectra. *J. Cell Sci.* **114**, 837–838.
5. Day, R. N., Nordeen, S. K., and Wan, Y. (1999). Visualizing protein–protein interactions in the nucleus of the living cell. *Mol. Endocrinol.* **13**, 517–526.
6. Ellenberg, J., Lippincott-Schwartz, J., and Presley, J. F. (1998). Two-color green fluorescent protein time-lapse imaging. *Biotechniques* **25**, 838–842.
7. Finley, K. R., Davidson, A. E., and Ekker, S. C. (2001). Three-color imaging using fluorescent proteins in living zebrafish embryos. *Biotechniques* **31**, 66–72.
8. Andel, F., III, Ladurner, A. G., Inouye, C., Tjian, R., and Nogales, E. (1999). Three-dimensional structure of the human TFIID-IIA-IIB complex. *Science* **286**, 2153–2156.
9. Stryer, L. (1978). Fluorescence energy transfer as a spectroscopic ruler. *Annu. Rev. Biochem.* **47**, 819–846.
10. Wu, P. and Brand, L. (1994). Review—Resonance energy transfer: Methods and applications. *Anal. Biochem.* **218**, 1–13.
11. Periasamy, A. and Day, R. N. (1999). Visualizing protein interactions in living cells using digitized GFP imaging and FRET microscopy. *Methods Cell Biol.* **58**, 293–314.
12. Mahajan, N. P., Linder, K., Berry, G., Gordon, G. W., Heim, R., and Herman, B. (1998). Bcl-2 and Bax interactions in mitochondria probed with green fluorescent protein and fluorescence resonance energy transfer. *Nat. Biotechnol.* **16**, 547–552.
13. Day, R. N., Periasamy, A., and Schaufele, F. (2001). Fluorescence resonance energy transfer microscopy of localized protein interactions in the living cell nucleus. *Methods* **25**, 4–18
14. Schaufele, F., Enwright, J. F., III, Wang, X., Teoh, C., Srihari, R., Erickson, R., MacDougald, O. A., and Day, R. N. (2001). CCAAT/enhancer binding protein alpha assembles essential cooperating factors in common subnuclear domains. *Mol. Endocrinol.* **15**, 1665–1674.
15. Miyawaki, A. and Tsien, R. Y. (2000). Monitoring protein conformations and interactions by fluorescence resonance energy transfer between mutants of green fluorescent protein. *Methods Enzymol.* **327**, 472–500.
16. Wouters, F. S., Verveer, P. J., and Bastiaens, P. I. (2001). Imaging biochemistry inside cells. *Trends Cell Biol.* **11**, 203–211.
17. Elangovan, M., Day, R. N., Periasamy, A. (2002). Nanosecond fluorescence resonance energy transfer-fluorescence lifetime imaging microscopy to localize the protein interactions in a single living cell. *J. Microsc.* **205**, 3–14.

CHAPTER 292

The Mammalian Circadian Timing System

Ueli Schibler, Steven A. Brown, and Jürgen A. Ripperger
Department of Molecular Biology Sciences II, University of Geneva, Geneva, Switzerland

Introduction

Many aspects of mammalian physiology follow daily oscillations controlled by a circadian pacemaker. Clock-regulated processes include sleep–wake cycles, heartbeat, blood pressure, body temperature, most functions of the digestive tract, renal physiology, liver metabolism, acuity of sensorial systems, and many endocrine functions [1]. Remarkably, this clock produces such daily oscillations indefinitely in the absence of external time cues such as changes in light intensity or temperature. Nevertheless, it can measure a 24-hour period only approximately, and thus it has to be readjusted every day by a separate input pathway in order to remain in resonance with geophysical time [2].

The central mammalian pacemaker resides in the suprachiasmatic nuclei (SCN) of the hypothalamus and drives all known overt circadian rhythms in physiology and behavior [3]. The photoperiod is the major *Zeitgeber* (timing cue) for the SCN clock, and SCN neurons receive photic information directly via the retino-hypothalamic tract [4]. The oscillations generated in SCN neurons are translated into overt circadian rhythms via mechanisms that are as yet poorly understood. Surprisingly, circadian oscillators with a molecular makeup similar to that of the SCN oscillator are also operative in most peripheral body cells, but these peripheral clocks dampen after a few days if they do not receive periodic phase-resetting inputs from the SCN pacemaker [5,6].

In this chapter we briefly summarize recently acquired information about the molecular circuitry generating daily oscillations, the input pathways by which light synchronizes these oscillations, and the output pathways by which the SCN clock governs circadian fluctuations in physiology and behavior. Most of the observations described here have been made in the mouse; however, the major conclusions are likely to apply to most mammalian organisms.

The Molecular Oscillator

During the past few years, seven mammalian clock genes have been isolated and studied by genetic and biochemical approaches: *Clock*, *Bmal1/Mop3*, *Cry1*, *Cry2*, *Per1*, *Per2*, and *CK1ε*. With the exception of *Clock*, which was identified and cloned via a forward genetics approach in the mouse, all mammalian pacemaker genes were identified by their sequence homology with known *Drosophila* clock genes (see reviews in [7,8]).

Circadian oscillations are generated by feedback loops in gene expression that involve both transcriptional and post-transcriptional mechanisms (Fig. 1). The two PAS domain helix–loop–helix transcription factors CLOCK and BMAL1 (also termed MOP3 in some publications) heterodimerize and bind to E-box sequences to activate transcription of the *Per* and *Cry* genes [9,10]. PER and CRY proteins are translocated into the nucleus, where they repress transcription

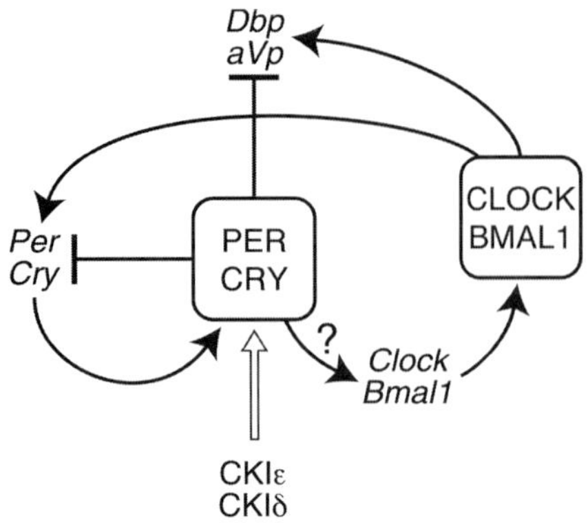

Figure 1 Simplified model of the mammalian circadian oscillator. At the heart of the oscillator is a multi-protein complex composed of period (PER) and cryptochrome (CRY) proteins. Transcription of the genes encoding these proteins is enhanced by CLOCK:BMAL1 heterodimers. When the CRY/PER proteins reach a critical threshold concentration, these proteins enter into the nucleus and repress transcription of their own genes by a yet unknown mechanism. Concomitantly, they provoke an anticyclic accumulation of CLOCK and BMAL1 proteins, probably via an indirect mechanism (?) (see note added in proof and reference 12). Upon decay of the CRY and PER proteins, transcription of the *Per* and *Cry* genes is reestablished, and a new cycle can occur. Casein kinase Iε and δ (CKIε, CKIδ) affect the nuclear translocation and the stability of the CRY/PER multi-protein complex by specific phosphorylations. The transcription of some clock-controlled genes, such as albumin D-element binding protein (*Dbp*) or arginine vasopressin (*aVp*), is regulated by the same mechanism as *Per* and *Cry* genes.

of their own genes as multisubunit complexes. Once the concentration of PER:CRY complexes falls below the critical threshold required for autorepression, *Per* and *Cry* transcription can resume (see review in [2]). This transcriptional feedback loop is assisted by several post-transcriptional control mechanisms [11]. For example, PER1, PER2, BMAL1, and CLOCK undergo robust daily cycles in protein phosphorylation, and these posttranslational modifications may affect their stability and function. CK1ε and CK1δ are among the protein kinases that phosphorylate PER and BMAL1 proteins, and the existence of protein complexes containing PER1, CRY2, CLOCK/BMAL1, CK1ε, and CK1δ was demonstrated by coimmunoprecipitation [11].

In liver, all of the clock genes specifying transcriptional regulatory components are transcribed in a circadian fashion, although the amplitudes of the various mRNA accumulation cycles vary greatly between genes (from less than 2-fold for *Cry2* to more than 20-fold for *Bmal1* and *Per2* [11]; J. A. Ripperger and U. Schibler, unpublished results). The circadian accumulation of *Cry* and *Per* mRNA is roughly in antiphase with that of *Bmal1* and *Clock* mRNA, suggesting that the rhythmic expression of positively and negatively acting clock components is governed by different mechanisms. Indeed, while CLOCK and BMAL1 stimulate the transcription of *Cry* and *Per* genes [9], they down-regulate transcription of their own genes via an indirect mechanism [12].

Photic Entrainment of the Central Pacemaker

SCN neurons receive photic information from the retina through the retino-hypothalamic tract. Glutamate is the major neurotransmitter used in this synaptic communication [13,14]. Surprisingly, rare light-sensitive ganglion cells in the inner retina layer, rather than the classical rods and cones in the outer retina layer, could be the photoreceptor cells required for the photic entrainment of the circadian pacemaker. These ganglion cells express the photopigment melanopsin and establish direct synaptic connections with SCN neurons as well as with other brain regions [15,16]; however, a disruption of the melanopsin gene will be needed to adequately assess its candidacy for the locus that encodes the circadian photopigment (see note added in proof and references 41 and 42). Light resets the phase of circadian rhythms in a highly gated manner. Thus, in nocturnal laboratory rodents kept in constant darkness, only light pulses delivered during the subjective night can elicit phase shifts. Furthermore, exposure to light during the first half of the subjective night provokes phase delays, whereas exposure during the second half of the subjective night causes phase advances [17].

Light pulses capable of phase shifting also elicit a transient activation of immediate early genes. These include *Per1*, *Per2*, and about a dozen other genes encoding transcriptional regulatory proteins (e.g., *cFos*, and *JunB*) [18]. cAMP serves as a second messenger both in the photic activation of immediate early gene transcription and in phase shifting ([19] and references therein). Increased cAMP levels activate the protein kinase PKA, which in turn phosphorylates the transcription factor CREB at serine residues S133 and S142 ([19] and references therein). Phosphorylated CREB then stimulates the transcription of immediate early genes, such as *Per1* and *c-Fos*. Light also provokes extensive histone H3 phosphorylation in the nuclei of SCN neurons, a phenomenon that is usually associated with changes in chromatin structure [20]. Whether these changes are causally related to immediate early genes has not yet been determined.

Outputs of the SCN Pacemaker

Most probably, the SCN controls overt circadian rhythms in behavior and physiology through neuronal and humoral outputs. In fact, SCN neurons project into various other brain centers (see reviews in [21,22]). These projections may affect circadian outputs directly through electrical signaling or via the rhythmic secretions of hormones of the hypothalamic-pituitary gland axis and other endocrine glands (e.g., the pineal gland). Some of these oscillating hormones, especially those of the hypothalamus-pituitary gland axis like glucocorticoids and thyroid hormones, may serve as phase-resetting cues for oscillators in non-neuronal cell types (see later sections). Cyclically secreted

SCN-derived neuropeptides such as arginine vasopressin and vasoactive intestinal peptide (VIP) may also participate in the regulation of rhythmic behavior and physiology [23,24].

Transforming growth factor-α (TGF-α), which signals through epidermal growth factor receptors, has recently been identified as a key regulator in the control of circadian locomotor activity. TGF-α is rhythmically expressed in SCN neurons, and constant infusion of this cytokine into the third ventricle abolishes circadian sleep–wake cycles in hamsters [25].

Outputs via Subsidiary Clocks

A second facet of the SCN's regulation of circadian physiology involves peripheral clocks. Circadian oscillators of a molecular makeup similar to that of SCN neurons exist in most peripheral tissues. Nevertheless, circadian gene expression in these tissues is not maintained in SCN-lesioned animals [26]. Hence, in intact animals, ongoing oscillations in peripheral organs depend on periodic inputs from the SCN. In immortalized tissue culture cells, circadian transcriptional oscillations can also be achieved by "jump starting" the cells with a serum shock or with substances that activate known signaling pathways. These involve tyrosine receptor kinases, G-protein-coupled receptors, nuclear hormone receptors (e.g., the glucocorticoid and retinoic acid receptors), protein kinases (e.g., PKA, PKC, and MAPK), and Ca^{2+} channels [27–31]. It is noteworthy that the immediate-early genes activated by these pathways before circadian gene expression is initiated include the same genes that are light induced in SCN neurons. This observation suggests that phase shifting by light in the SCN and signaling to the periphery by the SCN could share similar elements.

A similar conservation may exist in the direct regulation of output genes by the molecular clock. E-box sequences, similar to those used by CLOCK and BMAL1 proteins to activate *Per* gene transcription (see earlier section), are probably also important for the circadian expression of output genes such as DBP or arginine vasopressin [32,33]. Hence, these genes may be directly regulated by peripheral clock components in their cognate tissues.

Synchronization and Functions of Peripheral Oscillators: Time and Food

A functional SCN is required for the maintenance of circadian oscillations in peripheral cell types [6,26]. The output pathways by which the SCN synchronizes peripheral clocks are still poorly understood, but they are likely to be complex, given the panoply of signaling pathways that can affect circadian gene expression in peripheral cell types (see earlier section). Important clues on the phase entrainment of peripheral clocks have recently come from restricted feeding experiments [34,35]. Although feeding time does not affect the phase of the SCN pacemaker, it is the dominant *Zeitgeber* for many peripheral oscillators. Hence, the SCN may control the phase of peripheral timekeepers simply by determining the activity phase—and thus feeding time. Food entrainment may involve hormones related to metabolism, such as glucocorticoids [36] or metabolite-induced changes in the NAD(P)H/NAD(P)+ redox potential [37].

The dominant role of feeding time for peripheral phase resetting suggests that the anticipation of food uptake and processing may be a major purpose of peripheral clocks, at least in tissues such as the liver. This conjecture is strongly supported by recently published transcriptome profiling studies [38]. Indeed, a large fraction of cycling liver transcripts encodes proteins that are related to pathways implicated in sugar and fat metabolism, redox potential, and the detoxification of potentially harmful substances that may be present in the food.

Conclusions and Perspectives

Figure 2 summarizes our current view of the mammalian circadian timing system. Although impressive progress has been made in the analysis of all its aspects, much remains to be done. For example, the definitive identification of melanopsin as the key circadian photopigment will require genetic studies (see note added in proof and references 41 and 42). In addition, the presynaptic and postsynaptic light-signaling components downstream of melanopsin will have to be characterized. The possible role of mammalian cryptochrome as an additional circadian photopigment still remains a matter of debate, and compelling genetic evidence for or against such a function is still lacking. Considerable efforts will also have to be invested in the biochemical dissection of the mechanisms by which CRY and PER proteins repress CLOCK:BMAL1 mediated gene expression. Furthermore, studies aimed at the signaling pathways by which the SCN governs overt rhythms in physiology and behavior are still in their infancy. Finally, and perhaps most surprisingly, irrefutable experimental evidence supporting a strong benefit of circadian clocks is still scarce. Two studies, in which the survival of arrhythmic SCN-lesioned chipmunks and antelope ground squirrels in natural habitats have been recoded, showed only moderate advantages for the intact control animals [39,40]. Further studies conducted under conditions in which the survival of a species depends on predator avoidance, competition for mating partners and food, and resistance to extreme environmental variations will be required to evaluate the selective advantage that the circadian timing system may provide to its owners.

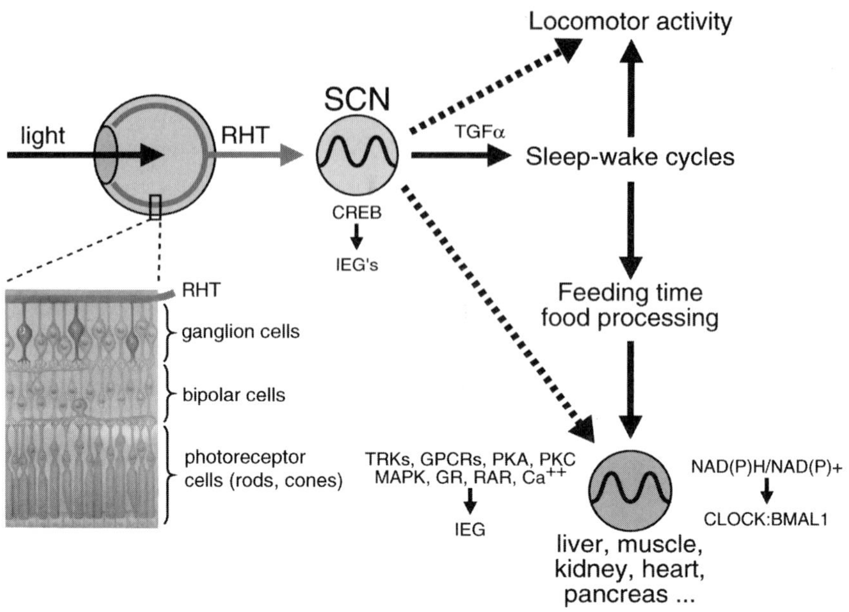

Figure 2 Phase-entrainment of central and peripheral circadian oscillators. Recent evidence suggests that photic cues are perceived by specific, melanopsin-expressing ganglion cells within the eye's retina. This light information is transmitted to the SCN via the retino-hypothalamic tract (RHT) and provokes the transcriptional stimulation of immediate early genes (IEG's, see text) via activation of the transcription factor CREB. Immediate early gene expression is thought to change the phase of the SCN pacemaker. In addition, specific diffusible molecules are secreted by SCN neurons to regulate physiological processes. One of these, TGFα, appears to regulate sleep-wake cycles. In turn, these cycles may affect locomotor activity and food uptake. Peripheral oscillators may be linked to the SCN directly via specific signaling cascades activating tyrosine receptor kinases (TRKs), G-protein coupled receptors (GPCRs), nuclear hormone receptors like the glucocorticoid (GR) and retinoic acid receptors (RAR), protein kinases like PKA, PKC, and MAPK, and Ca^{2+} channels. Alternatively, the SCN may control the phase of peripheral oscillators indirectly via food uptake and/or processing. The metabolism of nutrients may affect the redox-potential (the ratio of NADH to NAD+) within the cells, which in turn may regulate the DNA-binding activity of CLOCK:BMAL1 heterodimers.

Notes Added in Proof

Two important issues that had been unanswered (or unpublished) at the time of writing have been solved during the preparation of this book. First, the orphan nuclear receptor REV-ERBα has been identified as the transcriptional regulator connecting the positive and negative limbs of the circadian oscillator (see Reference 12). Second, genetic loss-of-function studies clearly revealed that melanopsin plays an important role in the photic entrainment of the mouse circadian clock. Thus, mice deficient in the conventional rod and cone photoreceptor cells and without a functional melanopsin allele are unable to light-entrain the phase of their circadian oscillator (see references 41 and 42).

References

1. Brown, S. A. and Schibler, U. (1999). The ins and outs of circadian timekeeping. *Curr. Opin. Genet. Dev.* **9**, 588–594.
2. Dunlap, J. C. (1999). Molecular bases for circadian clocks. *Cell* **96**, 271–290.
3. Ralph, M. R., Foster, R. G., Davis, F. C., and Menaker, M. (1990). Transplanted suprachiasmatic nucleus determines circadian period. *Science* **247**, 975–978.
4. Rusak, B. and Zucker, I. (1979). Neural regulation of circadian rhythms. *Physiol. Rev.* **59**, 449–526.
5. Balsalobre, A., Damiola, F., and Schibler, U. (1998). A serum shock induces circadian gene expression in mammalian tissue culture cells. *Cell* **93**, 929–937.
6. Yamazaki, S., Numano, R., Abe, M., Hida, A., Takahashi, R., Ueda, M., Block, G. D., Sakaki, Y., Menaker, M., and Tei, H. (2000). Resetting central and peripheral circadian oscillators in transgenic rats. *Science* **288**, 682–685.
7. Chang, D. C. and Reppert, S. M. (2001). The circadian clocks of mice and men. *Neuron* **29**, 555–558.
8. Ripperger, J. A. and Schibler, U. (2001). Circadian regulation of gene expression in animals. *Curr. Opin. Cell Biol.* **13**, 357–362.
9. Gekakis, N., Staknis, D., Nguyen, H. B., Davis, F. C., Wilsbacher, L. D., King, D. P., Takahashi, J. S., and Weitz, C. J. (1998). Role of the CLOCK protein in the mammalian circadian mechanism. *Science* **280**, 1564–1569.
10. Kume, K., Zylka, M. J., Sriram, S., Shearman, L. P., Weaver, D. R., Jin, X., Maywood, E. S., Hastings, M. H., and Reppert, S. M. (1999). mCRY1 and mCRY2 are essential components of the negative limb of the circadian clock feedback loop. *Cell* **98**, 193–205.
11. Lee, C., Etchegaray, J. P., Cagampang, F. R., Loudon, A. S., and Reppert, S. M. (2001). Posttranslational mechanisms regulate the mammalian circadian clock. *Cell* **107**, 855–867.
12. Preitner, N., Damiola, F., Luis Lopez, M., Zakany, J., Duboule, D., Albrecht, U., and Schibler, U. (2002). The orphan nuclear receptor REV-ERBα controls circadian transcription within the positive limb of the mammalian circadian oscillator, *Cell* **110**, 251–260.

13. Ding, J. M., Faiman, L. E., Hurst, W. J., Kuriashkina, L. R., and Gillette, M. U. (1997). Resetting the biological clock: Mediation of nocturnal CREB phosphorylation via light, glutamate, and nitric oxide. *J. Neurosci.* **17**, 667–675.
14. Ding, J. M., Buchanan, G. F., Tischkau, S. A., Chen, D., Kuriashkina, L., Faiman, L. E., Alster, J. M., McPherson, P. S., Campbell, K. P., and Gillette, M. U. (1998). A neuronal ryanodine receptor mediates light-induced phase delays of the circadian clock. *Nature* **394**, 381–384.
15. Berson, D. M., Dunn, F. A., and Takao, M. (2002). Phototransduction by retinal ganglion cells that set the circadian clock. *Science* **295**, 1070–1073.
16. Hattar, S., Liao, H. W., Takao, M., Berson, D. M., and Yau, K. W. (2002). Melanopsin-containing retinal ganglion cells: Architecture, projections, and intrinsic photosensitivity. *Science* **295**, 1065–1070.
17. Pittendrigh, C. S. and Daan, S. (1976). A functional analysis of circadian pacemakers in nocturnal animals. IV. Entrainment: Pacemaker as clock. *J. Comp. Physiol. A* **106**, 291–331.
18. Morris, M. E., Viswanathan, N., Kuhlman, S., Davis, F. C., and Weitz, C. J. (1998). A screen for genes induced in the suprachiasmatic nucleus by light. *Science* **279**, 1544–1547.
19. Gau, D., Lemberger, T., von Gall, C., Kretz, O., Le Minh, N., Gass, P., Schmid, W., Schibler, U. and Korf, H. W. (2002). Phosphorylation of CREB Ser-142 regulates light-induced phase shifts of the circadian clock. *Neuron* **34**, 245–253.
20. Crosio, C., Cermakian, N., Allis, C. D., and Sassone-Corsi, P. (2000). Light induces chromatin modification in cells of the mammalian circadian clock. *Nat. Neurosci.* **3**, 1241–1247.
21. Ibata, Y., Tanaka, M., Tamada, Y., Hayashi, S., Kawakami, F., Takamatsu, T., Hisa, Y., and Okamura, H. (1997). The suprachiasmatic nucleus: a circadian oscillator. *Neuroscientist* **3**, 215–225.
22. Ibata, Y., Okamura, H., Tanaka, M., Tamada, Y., Hayashi, S., Iijima, N., Matsuda, T., Munekawa, K., Takamatsu, T., Hisa, Y., Shigeyoshi, Y., and Amaya, F. (1999). Functional morphology of the suprachiasmatic nucleus. *Front Neuroendocrinol.* **20**, 241–268.
23. Reppert, S. M., Schwartz, W. J., and Uhl, G. R. (1987). Arginine vasopressin: A novel peptide rhythm in cerebrospinal fluid. *Trends Neurosci.* **10**, 76–80.
24. Fahrenkrug, J. (1993). Transmitter role of vasoactive intestinal peptide. *Pharmacol. Toxicol.* **72**, 354–363.
25. Kramer, A., Yang, F. C., Snodgrass, P., Li, X., Scammell, T. E., Davis, F. C., and Weitz, C. J. (2001). Regulation of daily locomotor activity and sleep by hypothalamic EGF receptor signaling. *Science* **294**, 2511–2515.
26. Sakamoto, K., Nagase, T., Fukui, H., Horikawa, K., Okada, T., Tanaka, H., Sato, K., Miyake, Y., Ohara, O., Kako, K., and Ishida, N. (1998). Multitissue circadian expression of rat period homolog (rPer2) mRNA is governed by the mammalian circadian clock, the suprachiasmatic nucleus in the brain. *J. Biol. Chem.* **273**, 27039–27042.
27. Akashi, M. and Nishida, E. (2000). Involvement of the MAP kinase cascade in resetting of the mammalian circadian clock. *Genes Dev.* **14**, 645–649.
28. Balsalobre, A., Marcacci, L., and Schibler, U. (2000). Multiple signaling pathways elicit circadian gene expression in cultured Rat-1 fibroblasts. *Curr. Biol.* **10**, 1291–1294.
29. Motzkus, D., Maronde, E., Grunenberg, U., Lee, C. C., Forssmann, W., and Albrecht, U. (2000). The human PER1 gene is transcriptionally regulated by multiple signaling pathways. *FEBS Lett.* **486**, 315–319.
30. Yagita, K. and Okamura, H. (2000). Forskolin induces circadian gene expression of rPer1, rPer2 and dbp in mammalian rat-1 fibroblasts. *FEBS Lett.* **465**, 79–82.
31. McNamara, P., Seo, S. P., Rudic, R. D., Sehgal, A., Chakravarti, D., and FitzGerald, G. A. (2001). Regulation of CLOCK and MOP4 by nuclear hormone receptors in the vasculature: A humoral mechanism to reset a peripheral clock. *Cell* **105**, 877–889.
32. Jin, X., Shearman, L. P., Weaver, D. R., Zylka, M. J., de Vries, G. J., and Reppert, S. M. (1999). A molecular mechanism regulating rhythmic output from the suprachiasmatic circadian clock. *Cell* **96**, 57–68.
33. Ripperger, J. A., Shearman, L. P., Reppert, S. M., and Schibler, U. (2000). CLOCK, an essential pacemaker component, controls expression of the circadian transcription factor DBP. *Genes Dev.* **14**, 679–689.
34. Damiola, F., Le Minh, N., Preitner, N., Kornmann, B., Fleury-Olela, F., and Schibler, U. (2000). Restricted feeding uncouples circadian oscillators in peripheral tissues from the central pacemaker in the suprachiasmatic nucleus. *Genes Dev.* **14**, 2950–2961.
35. Stokkan, K. A., Yamazaki, S., Tei, H., Sakaki, Y., and Menaker, M. (2001). Entrainment of the circadian clock in the liver by feeding. *Science* **291**, 490–493.
36. Le Minh, N., Damiola, F., Tronche, F., Schutz, G., and Schibler, U. (2001). Glucocorticoid hormones inhibit food-induced phase-shifting of peripheral circadian oscillators. *EMBO J.* **20**, 7128–7136.
37. Rutter, J., Reick, M., Wu, L. C., and McKnight, S. L. (2001). Regulation of clock and NPAS2 DNA binding by the redox state of NAD cofactors. *Science* **293**, 510–514.
38. Kornmann, B., Preitner, N., Rifat, D., Fleury-Olela, F., and Schibler, U. (2001). Analysis of circadian liver gene expression by ADDER, a highly sensitive method for the display of differentially expressed mRNAs. *Nucleic Acids Res.* **29**, E51–51.
39. Dark, J., Pickard, G. E., and Zucker, I. (1985). Persistence of circannual rhythms in ground squirrels with lesions of the suprachiasmatic nuclei. *Brain Res.* **332**, 201–207.
40. DeCoursey, P. J., Krulas, J. R., Mele, G., and Holley, D. C. (1997). Circadian performance of suprachiasmatic nuclei (SCN)-lesioned antelope ground squirrels in a desert enclosure. *Physiol. Behav.* **62**, 1099–1108.
41. Hattar, S., Lucas, R. J., Mrosovsky, N., Thompson, S., Douglas, R. H., Hankins, M. W., Lem, J., Biel, M., Hofmann, F., Foster, R. G., and Yau, K. W. (2003). Melanopsin and rod-cone photoreceptive systems account for all major accessory visual functions in mice. *Nature* **424**, 75–81.
42. Panda, S., Provencio, I., Tu, D. C., Pires, S. S., Rollag, M. D., Castrucci, A. M., Pletcher, M. T., Sato, T. K., Wiltshire, T., Andahazy, M., *et al.* (2003). Melanopsin is required for non-image-forming photic responses in blind mice. *Science* **301**, 525–527.

CHAPTER 293

Protein Arginine Methylation

Michael David
Molecular Biology Section, Division of Biology,
University of California at San Diego,
La Jolla, California

Introduction

Posttranslational modification of proteins is a hallmark of signal transduction, allowing cells to alter existing proteins to react rapidly to changes and events in their extracellular environment. Protein phosphorylation on serine, threonine, or tyrosine residues as a mechanism of intracellular signal transduction has been extensively studied. In contrast, the role of protein methylation at lysine, arginine, or histidine residues is much less understood. The subject of this chapter, methylation of arginine residues, was discovered more than 30 years ago, but has only recently received appreciation as a novel mechanism of signal transduction, transcriptional regulation, and protein sorting. As more arginine methylated substrates are discovered, and as novel functions of arginine methylation are reported at a very rapid pace, it appears that this protein modification may be of equal importance to protein phosphorylation in the regulation of cellular functions. This chapter summarizes the current knowledge on protein arginine methyltransferases and the function of arginine methylation in signal transduction.

Arginine Methylation and Arginine-Methyltransferases

The transfer of a methyl group from S-adenosylmethionine (SAM) onto arginine can occur on either or both of the guanidino nitrogen atoms. As such, arginine methylation, which is restricted to eukaryotic cells, can be found in three distinct forms—N^G-monomethylarginine, N^GN^G (asymmetric) dimethylarginine, and $N^GN'^G$ (symmetric) dimethylarginine (Fig. 1)—frequently in the context of RGG tripeptides [1,2].

Two classes of protein arginine methyltransferases have been characterized and classified based on the symmetry of their reaction products. Both, type I and type II enzymes generate N^G-monomethylarginine. However, whereas type I protein arginine methyltransferases account for the formation of asymmetric N^G,N^G-dimethylarginine, type II enzymes catalyze the formation of symmetric N^G,N'^G-dimethylarginine (Fig. 1). The best conserved region within this enzyme family is the SAM-interacting methyltransferase motif, which is closely related to that found in other methyltransferases that use SAM as a methyl donor [3].

Five related enzymes that catalyze asymmetric arginine methylation have thus far been identified. The majority of type I arginine methyltransferase activity in eukaryotic cells appears to be accounted for by PRMT1 and its functional yeast homolog Hmt1/Rmt1 [4]. These enzymes contain only a few residues outside the methyltransferase core domain, and their enzymatic activity appears to correlate with dimer formation. In contrast, PRMT2 contains an N-terminal SH3 domain, whereas PRMT3 harbors an N-terminal zinc-finger motif that modulates its substrate specificity [5]. The crystal structures of the conserved core regions of Hmt1 and human PRMT3 revealed significant structural similarity between these enzymes [6]. PRMT4 (Carm1) contains both N- and C-terminal extensions to the methyltransferase core region [7]. The recently identified PRMT6 resembles PRMT1 most closely in that it is comprised solely of the catalytic core without any additional domains [8]. PRMT5 (JBP), which was identified as a Janus kinase binding protein, is the only arginine methyltransferase identified to date that has type II enzyme activity [9,10]. As such, it is capable of catalyzing the symmetric arginine methylation of myelin basic protein, which was one of the first arginine-methylated proteins identified. In addition, two other symmetrically dimethylated RNA-binding proteins, spliceosomal snRNP proteins SmD1 and SmD3, can function as substrates for PRMT5 *in vitro* [11].

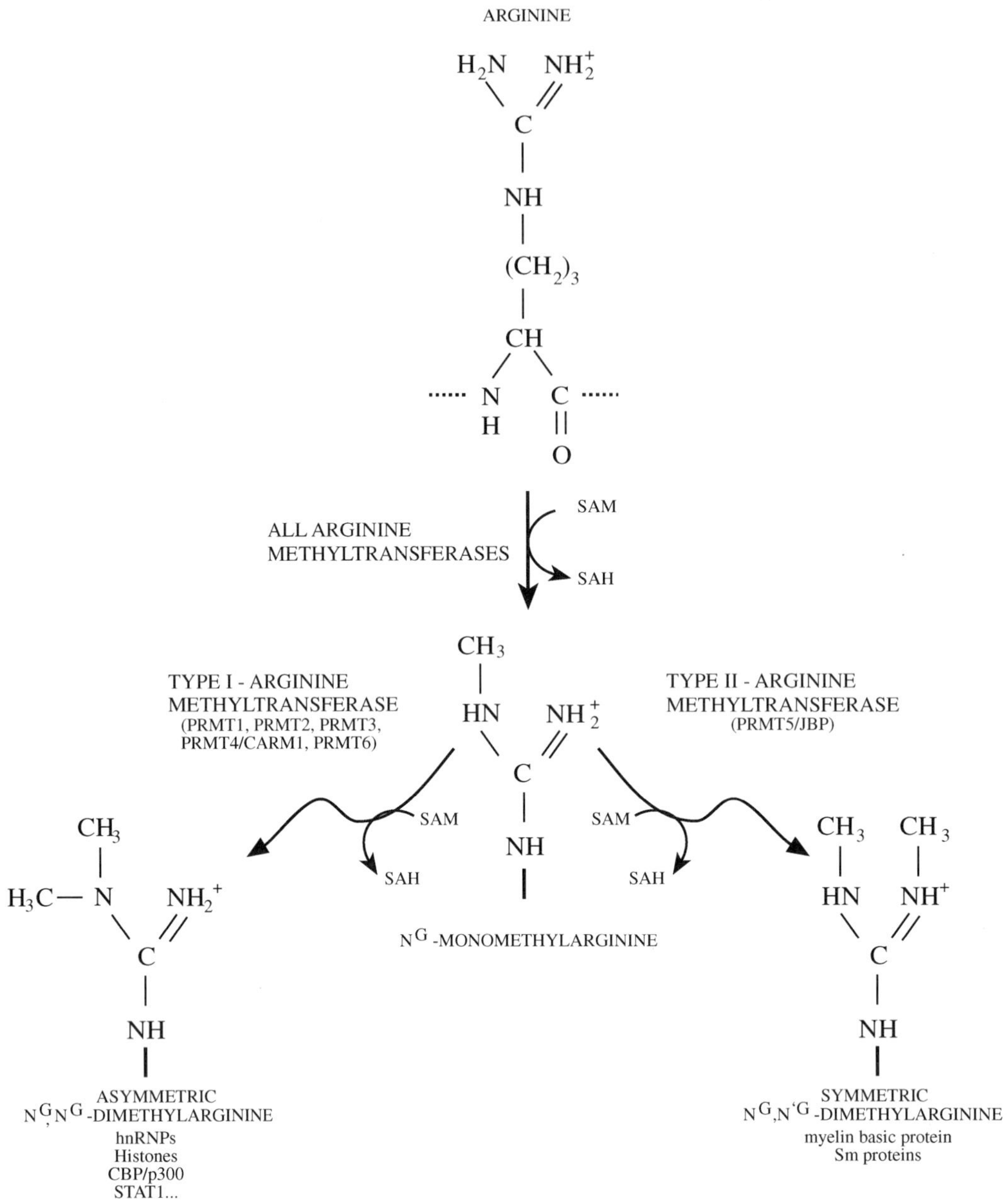

Figure 1 Protein arginine methyltransferases use SAM as methyl donor to catalyze mono- and dimethylation of the guanidino nitrogen atoms, generating SAH in the process.

Function of Arginine Methylation

Addition of methyl groups to the guanidino nitrogen atoms of arginine residues increases steric hindrance and reduces the ability to form hydrogen bonds without altering the overall charge of the amino acid. This suggests a role for methylation in the modulation of intra- or intermolecular interactions. Although the heterogeneous nuclear ribonucleoproteins (hnRNPs) involved in nucleocytoplasmic RNA transport comprise the majority of dimethylarginine containing proteins in the nucleus, their methylation appears to have no effect on their affinity for their respective RNA targets [12]. In contrast, several recent studies have illustrated an important function of arginine methylation in the regulation of protein–protein interactions. Methylation of Sam68, a proline-rich src-kinase substrate known to interact with WW- or SH3-domain containing signaling proteins, decreases its affinity for SH3 domains, but does not alter binding to WW-motif containing proteins [13]. Similarly, arginine methylation of the STAT1 (signal transducer and activator of transcription) transcription factor decreases its affinity for its inhibitor PIAS1 (protein inhibitor of activated STAT 1). In the absence of STAT1 arginine methylation, PIAS1 associates with STAT1 and prevents its binding to DNA, thereby modulating interferon induced gene transcription [14]. A positive modulatory effect of arginine methylation on protein–protein interaction is observed with the type II arginine methyltransferase substrates SmD1 and SmD3, which require methylation for efficient binding to the spinal muscular atrophy gene product, SMN [11,15].

Role of Arginine Methylation in Signal Transduction

Very little is still known about the regulation of enzymatic activity and the biological functions of arginine methyltransferases. PRMT1, the first mammalian enzyme, was discovered through its physical interaction with the immediate-early proteins TIS21 and the related BTG1, which have the ability to modulate the activity and substrate specificity of PRMT1 [16]. Deletion of PRMT1 in mice through gene targeting resulted in early embryonic lethality, demonstrating an important role for the enzyme in mammalian development. Interestingly, PRMT1 deficiency does not cause lethality on a cellular level [17]. PRMT1 has been implicated in intracellular signaling by its ability to bind the cytoplasmic domain of the type I interferon receptor [18] and to methylate an arginine residue conserved between the N-terminal regions of all mammalian STAT proteins. Indeed, methylation of the STAT1 transcription factor was shown to be essential for interferon α/β-mediated transcriptional induction [14].

Arginine methylation of histones H2A and H3 [7], as well as the transcriptional coactivators CBP/p300 [19], supports a model in which protein methylation contributes to chromatin remodeling and initiation of transcription [20]. PRMT4 or *c*oactivator-*a*ssociated a*r*ginine *m*ethyltransferase (Carm1) cooperates with p160 family coactivators to stimulate transcriptional activation by nuclear receptors [7]. Furthermore, PRMT4 interacts with myocyte enhancer factor 2C (MEF2C) to promote myocyte differentiation [21].

PRMT5 or Jak-binding protein 1 (JBP1) participates in the assembly of a large arginine methyltransferase complex that targets several spliceosomal Sm proteins of the snRNPs [22]. Direct interaction of survival of motor neurons (SMN), the spinal muscular atrophy gene product, with the arginine-rich domain of the Sm proteins to form functional snRNP core particles increased upon symmetric dimethylation of the Sm proteins. Interestingly, anti-Sm autoantibodies detected in lupus erythematosus patients specifically recognize symmetric arginine-dimethylated Sm proteins [11]. Autoimmune responses against myelin basic protein (MBP), the only other protein known to be symmetrically dimethylated on arginine residues, account for the development of multiple sclerosis. These observations raise the possibility that posttranslational modification by arginine methylation plays an important role in the development of autoimmune disorders.

With the number of identified arginine methyltransferases and arginine methylated proteins rapidly increasing, the question remains whether this modification is similarly reversible as phosphorylation. Thus far, no arginine demethylase activity has been reported. Although it is possible that this modification persists over the lifetime of the protein, other enzymatic reactions such as protein arginine diimination could in theory catalyze the removal of methylated nitrogens.

Much additional work is still required to elucidate the regulation of arginine methylation and its role in cellular function. Arginine methylation has just recently begun to be appreciated as a posttranslational modification that contributes significantly to signal transduction, protein sorting, transcriptional regulation, chromatin remodeling, and perhaps disease progression.

References

1. Gary, J. D. and Clarke, S. (1998). RNA and protein interactions modulated by protein arginine methylation. *Prog. Nucleic Acid Res. Mol. Biol.* **61**, 65–131.
2. Najbauer, J., Johnson, B. A., Young, A. L., and Aswad, D. W. (1993). Peptides with sequences similar to glycine, arginine-rich motifs in proteins interacting with RNA are efficiently recognized by methyltransferase(s) modifying arginine in numerous proteins. *J. Biol. Chem.* **268**, 10501–10509.
3. McBride, A. E. and Silver, P. A. (2001). State of the arg: Protein methylation at arginine comes of age. *Cell* **106**, 5–8.
4. Tang, J., Frankel, A., Cook, R. J., Kim, S., Paik, W. K., Williams, K. R., Clarke, S., and Herschman, H. R. (2000). PRMT1 is the predominant type I protein arginine methyltransferase in mammalian cells. *J. Biol. Chem.* **275**, 7723–7730.
5. Frankel, A. and Clarke, S. (2000). PRMT3 is a distinct member of the protein arginine N-methyltransferase family. Conferral of substrate specificity by a zinc-finger domain. *J. Biol. Chem.* **275**, 32974–32982.
6. Zhang, X., Zhou, L., and Cheng, X. (2000). Crystal structure of the conserved core of protein arginine methyltransferase PRMT3. *EMBO J.* **19**, 3509–3519.
7. Chen, D., Ma, H., Hong, H., Koh, S. S., Huang, S. M., Schurter, B. T., Aswad, D. W., and Stallcup, M. R. (1999). Regulation of transcription by a protein methyltransferase. *Science* **284**, 2174–2177.
8. Frankel, A., Yadav, N., Lee, J., Branscombe, T. L., Clarke, S., and Bedford, M. T. (2002). The novel human protein arginine N-methyltransferase PRMT6 is a nuclear enzyme displaying unique substrate specificity. *J. Biol. Chem.* **277**, 3537–3543.
9. Pollack, B. P., Kotenko, S. V., He, W., Izotova, L. S., Barnoski, B. L., and Pestka, S. (1999). The human homologue of the yeast proteins Skb1 and Hsl7p interacts with Jak kinases and contains protein methyltransferase activity. *J. Biol. Chem.* **274**, 31531–31542.
10. Branscombe, T. L., Frankel, A., Lee, J. H., Cook, J. R., Yang, Z., Pestka, S., and Clarke, S. (2001). PRMT5 (Janus kinase-binding protein 1) catalyzes the formation of symmetric dimethylarginine residues in proteins. *J. Biol. Chem.* **276**, 32971–32976.
11. Brahms, H., Raymackers, J., Union, A., de Keyser, F., Meheus, L., and Luhrmann, R. (2000). The C-terminal RG dipeptide repeats of the spliceosomal Sm proteins D1 and D3 contain symmetrical dimethylarginines, which form a major B-cell epitope for anti-Sm autoantibodies. *J. Biol. Chem.* **275**, 17122–17129.
12. Valentini, S. R., Weiss, V. H., and Silver, P. A. (1999). Arginine methylation and binding of Hrp1p to the efficiency element for mRNA 3′-end formation. *RNA* **5**, 272–280.
13. Bedford, M. T., Frankel, A., Yaffe, M. B., Clarke, S., Leder, P., and Richard, S. (2000). Arginine methylation inhibits the binding of proline-rich ligands to Src homology 3, but not WW, domains. *J. Biol. Chem.* **275**, 16030–16036.
14. Mowen, K. A., Tang, J., Zhu, W., Schurter, B. T. , Shuai, K., Herschman, H. R., and David, M. (2001). Arginine methylation of STAT1 modulates IFNalpha/beta-induced transcription. *Cell* **104**, 731–741.
15. Brahms, H., Meheus, L., de Brabandere, V., Fischer, U., and Luhrmann, R. (2001). Symmetrical dimethylation of arginine residues in spliceosomal Sm protein B/B' and the Sm-like protein LSm4, and their interaction with the SMN protein. *RNA* **7**, 1531–1542.
16. Lin, W. J., Gary, J. D., Yang, M. C., Clarke, S., and Herschman, H. R. (1996). The mammalian immediate-early TIS21 protein and the leukemia-associated BTG1 protein interact with a protein-arginine N-methyltransferase. *J. Biol. Chem.* **271**, 15034–15044.
17. Pawlak, M. R., Scherer, C. A., Chen, J., Roshon, M. J., and Ruley, H. E. (2000). Arginine N-methyltransferase 1 is required for early postimplantation mouse development, but cells deficient in the enzyme are viable. *Mol. Cell Biol.* **20**, 4859–4869.

18. Abramovich, C., Yakobson, B., Chebath, J., and Revel, M. (1997). A protein-arginine methyltransferase binds to the intracytoplasmic domain of the IFNAR1 chain in the type I interferon receptor. *EMBO J.* **16**, 260–266.
19. Xu, W., Chen, H., Du, K., Asahara, H., Tini, M., Emerson, B. M., Montminy, M., and Evans, R. M. (2001). A transcriptional switch mediated by cofactor methylation. *Science* **8**, 8.
20. Stallcup, M. R. (2001). Role of protein methylation in chromatin remodeling and transcriptional regulation. *Oncogene* **20**, 3014–3020.
21. Chen, S. L., Loffler, K. A., Chen, D., Stallcup, M. R., and Muscat, G. E. (2002). The coactivator-associated arginine methyltransferase is necessary for muscle differentiation. Car,1 coactivates myocyte enhancer factor-2. *J. Biol. Chem.* **277**, 4324–4333.
22. Friesen, W. J., Paushkin, S., Wyce, A., Massenet, S., Pesiridis, G. S., Van Duyne, G., Rappsilber, J., Mann, M., and Dreyfuss, G. (2001). The methylosome, a 20S complex containing JBP1 and pICln, produces dimethylarginine-modified Sm proteins. *Mol. Cell Biol.* **21**, 8289–8300.

CHAPTER 294

Transcriptional Activity of Notch and CSL Proteins

Elise Lamar and Chris Kintner
Molecular Neurobiology Laboratory,
Salk Institute for Biological Studies,
La Jolla, California

Introduction

The Notch receptor controls diverse cell fate decisions developmentally and functions in a variety of processes in adults (for reviews, see [1–8]). Because the Notch intracellular domain (ICD) was shown to translocate to the nucleus following receptor activation and cleavage [9,10], its transcriptional activity has been actively investigated. In this capacity, ICD interacts with its DNA binding partner and alters its function from that of a repressor to an activator (for reviews, see [11,12]). This protein is known variously as Su(H) (for *Su*ppressor of *H*airless), CBF1, RBP-J, and LAG-1, depending on the species, and all orthologs are collectively referred to as CSL proteins [for *C*BF1, *Su(H)*, and *L*ag-1]. [For simplicity, we use the designation Su(H) in this review.] Here we present a summary of the biochemical functions of Su(H), ICD, and proteins participating in the Notch transcriptional complex, followed by evidence supporting models of Notch transcriptional activity *in vivo*. Although Su(H)-independent pathways of Notch signaling have been described [13–17], the transcriptional role of Notch has been examined largely in relationship to Su(H); therefore we address Notch activity only in that context.

Components of the Notch Transcriptional Complex

Architecture of the Notch ICD

The domain structure of the Notch ICD is shown in Fig. 1, which uses the 110-kDa *Xenopus* Notch1 ICD as a model. The number of Notch subtypes varies from organism to organism, that is, *Drosophila* expresses one receptor subtype, *Caenorhabditis elegans* two, and vertebrates four (for a review, see [18]). Data derived from several species and receptor subtypes suggests that all Notch intracellular domains exhibit the conserved structural motifs shown in Fig. 1 and that these domains mediate common functions.

The notable exception to this rule is the activity of the Notch3 ICD, which is in some contexts a poor transcriptional activator and antagonistic to Notch1 ICD [19]. The N-terminal RAM domain of ICD interacts directly with Su(H) and is required for high-affinity binding to ICD *in vitro* [20]. Su(H) also binds weakly *in vitro* to the ankyrin repeats immediately downstream of the RAM domain [21]. All Notch ICDs contain six ankyrin repeats; however, structural studies of *Drosophila* ICD indicate a motif resembling a seventh [22]; human Notch1 mutant in this region binds Su(H) but is nonfunctional, suggesting that the motif is required for transactivation [23]. Immediately downstream of the ankyrin repeats is a conserved region known both as the ICD TAD (for transactivation domain), containing one of the two nuclear localization signals [9] and sequences required for transactivation; this region is discussed in detail later. The ICD C terminus contains a glutamine-rich opa repeat domain and a C-terminal PEST domain associated with protein turnover [24] (see later discussion).

Notch Signaling Antagonizes Su(H)-Mediated Repression

Notch ICD does not bind DNA but activates transcription through interaction with Su(H), which binds to the motif YGTGRGAA *in vitro* [225]. Sites identical or highly

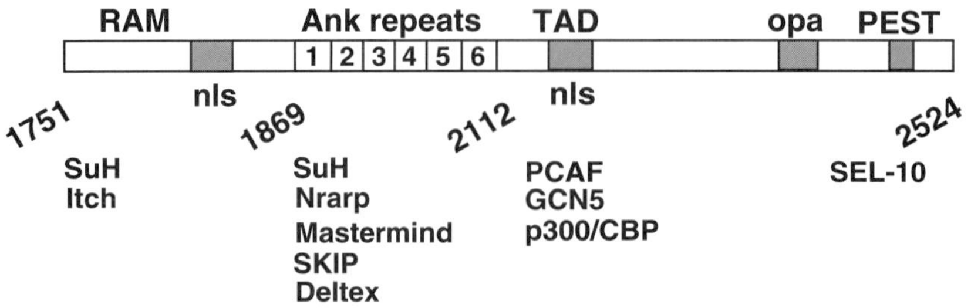

Figure 1 Domain structure of the 773 amino acid intracellular domain (ICD) of *Xenopus laevis* Notch1. Structural motifs shown are conserved in most Notch proteins, and numbering corresponds to amino acid residues of the unprocessed Notch receptor. Interacting proteins are listed below (for their functions see Table I). In some cases, namely Itch and SEL-10, potential interaction domains are fairly broad. Itch interacts with constructs of ICD containing the RAM domain and ankyrin repeats, while SEL-10 interacts with the C-terminus of ICD. SuH protein interacts with both the RAM domain and the reiterated ankyrin repeats, and both regions are required for Notch function. Six ankyrin repeats are shown, although there is some evidence for the existence of a seventh (see text). Sequences required for transactivation, designated the TAD, lie C-terminal to the ankyrin repeats and as indicated are known to bind several histone acetylases. Downstream from the TAD is a glutamine-rich domain designated opa and a PEST domain, which may be involved in protein turnover. ICD has two nuclear localization signals (nls): one C-terminal to the RAM domain and the other in the TAD.

homologous to this site (designated here as high-affinity sites) have been mapped in Notch-responsive promoters *in vivo* [26–31], as have divergent sites [32]. In the absence of Notch signaling, Su(H) has been shown to be a transcriptional repressor, by blocking interactions between TFIIA and TAF110 of TFIID on a viral promoter [33], but more generally by recruiting corepressors containing histone deacetylase (HDAC) activity. In vertebrates, Su(H) interacts with the corepressors SMRT [34], CIR [35], and a third repressor KyoT2 [36], and antagonizes Notch activation of Su(H)-dependent reporters in transient transfection assays. In *Drosophila*, Hairless, a nuclear protein that binds Su(H), genetically antagonizes Notch signaling in *Drosophila* [37–39] and mediates Su(H) repressor functions by acting as an adaptor protein between Su(H) and the corepressors CtBP [12,40] and Groucho [12,41]. Together, all of these corepressors can recruit both Class I and Class II HDACs (for a review, see [42]).

The implication of these findings is that in the absence of Notch signaling Su(H) acts as a local repressor; biochemical evidence for this hypothesis comes from chromatin immunoprecipitation assays showing that Su(H) binds to target promoters in the absence of Notch signaling [43]. *In vivo* support for a Su(H) repressor function comes from further analysis of both artificial enhancers and enhancers derived from Notch target genes, as described later [29,44].

Notch Signaling Derepresses Su(H) and Activates Transcription

Significantly, the physical interaction of corepressors and ICD with Su(H) is mutually exclusive, suggesting that Notch signaling leads to changes in gene expression in part by derepression. However, a large body of evidence also indicates that ICD activates transcription by recruiting additional proteins that contain histone acetyltransferase (HAT) activity. One line of evidence suggests that these HAT-containing proteins are recruited by directly binding to sites on ICD; specifically, to a TAD located in a 200-amino-acid stretch downstream of the ankyrin repeats (see Fig. 1) [45–47]. Deleting or mutating this region in ICD attenuates Notch-mediated transcription in frog embryos [438], *in vitro* [49], in cell culture models [46], and in *in vivo* models of ICD-mediated oncogenesis [50]. This region has been shown to bind the HAT-containing proteins, PCAF and GCN5, in both mouse Notch1 and *Drosophila* ICD, while a similar region in mouse Notch1 has been shown to bind the HAT-containing coactivator, p300/CBP.

Another line of evidence suggests that HAT-containing proteins are recruited indirectly by ICD through interactions with large, glutamine-rich nuclear proteins referred to here as the Mastermind (mam) proteins [49,51–53]. Loss-of-function analysis of the mam proteins in *Drosophila* [54], *C. elegans* (where it is known as Lag-3) [51], and vertebrates [49,55] suggests that these proteins are required for Notch signaling and the activation of Notch target genes. The mam proteins form a tight ternary complex with Su(H) and the ankyrin repeats of ICD [51–53] and this complex appear to potentiate ICD transcriptional activity in transient transfection assays [52,53], suggesting that mam is a Notch coactivator. Indeed, when assayed *in vitro* on chromatin templates, ICD requires mam to initiate transcription [49]. Activity in this assay requires a region of mam between amino acids 74 and 301 that binds tightly to CBP/p300; disrupting this binding markedly decreases the ability of ICD to initiate transcription *in vitro* and to promote the expression of Notch target genes *in vivo* [49]. CBP/p300 binds to mam under conditions that are more stringent than those that allow binding to ICD, suggesting that mam is the key element in recruiting HAT-containing proteins to the Notch transcriptional complex [49].

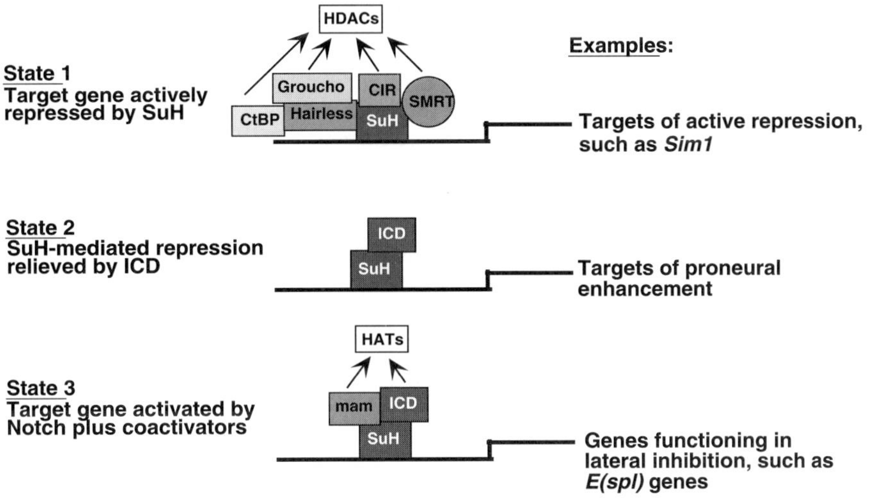

Figure 2 The SuH/ICD transcriptional switch can exist in several states. In State 1, SuH functions to actively repress transcription by interacting with several factors, such as the SMRT/NCor, CIR, CtBP, and Groucho co-repressors. In Drosophila, both CtBP and Groucho interact with SuH through the adaptor protein Hairless. For the sake of simplicity not all of the numerous proteins interacting with SuH in a repressive state are depicted. All of these co-repressors function to recruit HDACs, suggesting that chromatin remodeling occurs in the course of changes in Notch signaling. Active repression has been demonstrated in Drosophila not only on artificial enhancers but in vivo as being required for regulation of expression of *sim1*. State 3 represents the opposite activity; through interaction with ICD, co-repressors and accompanying HDACs are dismissed from SuH, and several HATs (See Figure 1 and Table 1) are recruited to the promoter through ICD and its dedicated co-activator mastermind (mam). Transcriptional activation by Notch is thought to be required for processes such as lateral inhibition and is best represented by upregulation of *E(spl)* genes in Drosophila. Evidence for an intermediate state (State 2), in which repression is off but activation through Notch is not required, comes from studies of eye development in SuH-files, which lack both SuH-mediated repression and Notch signaling. These studies indicate that relief from SuH-mediated repression is sufficient to activate, or "enhance", expression of putative targets, such as proneural genes, which function in differentiation.

The current model suggests that the Su(H)/ICD repressor/activator switch is mediated by HDAC/HAT recruitment, respectively, and strongly suggests that changes in the acetylation state of nucleosomes occurs in the course of Notch signaling [56]. This model posits several states in which Su(H) can exist, thus regulating gene expression in an ICD-regulated manner. In the absence of ICD, Su(H) acts as a local transcriptional repressor by recruiting HDAC-containing corepressors (state 1; Fig. 2). By associating with Su(H), ICD relieves this repression and may weakly recruit coactivators (state 2). The combination of Su(H) and ICD recruits additional coactivators such as mam, resulting in strong transcriptional activation (state 3).

This model will certainly undergo revision as various components of the Notch transcriptional complex are revealed. For example, we already know that ICD is more than a tether for mam because its TAD acts cooperatively with mam to promote transcription both *in vivo* and *in vitro*. How this TAD acts is unknown. In addition, mam contains a second TAD within the carboxy terminus that is required for Notch transcription *in vivo* but not *in vitro* [49], suggesting interactions with yet another group of transcriptional cofactors.

Additional Proteins Interacting with Su(H)/ICD

A number of proteins have been identified that interact with ICD and/or Su(H) (see Table I). In some cases, it remains unclear whether these proteins function in transcription per se, or whether they are regulatory subunits that modulate the activity of the transcriptional complex.

NRARP

Nrarp (for *N*otch *r*egulated *a*nkyrin *r*epeat *p*rotein) is expressed in many cases where Notch signaling regulates cell fate in the vertebrate embryo, and its expression is dependent on Notch signaling [57–59]. Nrarp binds to ICD and Su(H) but only as part of a ternary complex, interacting with the ICD ankyrin repeats. On the one hand, Nrarp promotes Notch transcription in transient transfection assays presumably by promoting the formation of a complex between ICD and Su(H). On the other hand, when misexpressed in frog embryos, Nrarp blocks Notch signaling and promotes the loss of ICD protein, suggesting that it inhibits Notch signaling by destabilizing ICD. Further genetic analysis is required to determine whether Nrarp is primarily a positive or negative factor in the Notch transcriptional complex, and whether it is required for Notch signaling in some or all cases *in vivo*.

SKIP

The v-*ski*-*i*nteracting *p*rotein SKIP interacts physically with both Su(H)/corepressor complexes, through interaction with both Su(H) and SMRT, and with ICD, through the fourth

Table I Proteins Interacting with ICD/Su(H) or Su(H)[a]

Interactor	Function	References
Interactors with ICD/Su(H)		
Mastermind	Coactivator	51–53
Nrarp	Unknown; mediates ICD degradation	57–59
SKIP	Unknown; potential coactivator	60
Deltex	Unknown; may antagonize activators	63–66
GCN5, PCAF	HATs; coactivators	46
CBP/p300	HAT; coactivator	47
Itch	E3 ligase; ICD degradation	83
SEL-10	E3 ligase; ICD degradation	86
Interactors with Su(H)		
SKIP	Unknown	60
SMRT/NCoR	Corepressor	34
CIR	Corepressor	35
KyoT2	Corepressor	36
CtBP	Corepressor	12, 40
Hairless	Adaptor; interacts with corepressors	37–39
TFIIA, TFIID	Coactivators; antagonized by Su(H)	33

[a]These proteins interact with either ICD/Su(H) or Su(H) alone in a manner suggesting they modulate or regulate transcriptional activity.

ankyrin repeat [60]. When expressed as Gal4 fusion proteins, human and *Drosophila* SKIP repress transcription. However, evidence that SKIP's primary role is to facilitate ICD transcription is more compelling. First, SKIP potentiates Su(H)-dependent ICD activity in transient transfection assays. Furthermore, antisense SKIP can rescue a block in muscle cell differentiation promoted by ICD [60], suggesting that SKIP is required for ICD function in this context. Finally, SKIP interacts with and transcriptionally coactivates several other proteins, including viral transactivators [60], nuclear receptors [61], and Smads [62].

DELTEX

Found in flies and vertebrates, Deltex proteins interact through their N terminus with the ICD ankyrin repeats [63]. Deltex proteins also contain a proline-rich domain and a C-terminal ring-finger motif, which promotes oligomerization [64]. What role Deltex plays in Notch signaling is currently unclear; furthermore, assigning a role for Deltex in transcription is speculative, because it is predominantly cytoplasmic [14,65] and also plays a role in Su(H)-independent Notch pathways [16,66]. On the one hand, genetic screens in *Drosophila* indicate that Deltex positively regulates Notch [67]. However, in functional assays Deltex proteins [66,68,69] can block ICD/Su(H) transcriptional activity, an activity requiring the ICD TAD and PEST sequences [69]. Similarly, Deltex inhibits transactivation of Gal4 reporters by both ICD and full-length mam fused to Gal4 [69]. This study, taken together with evidence that both ICD [47] and mam [49] interact with p300/CBP and that Deltex inhibits p300/CBP activity [66], suggests that in some contexts Deltex may negatively regulate Su(H)-dependent Notch activity by interfering with p300/CBP recruitment.

Modification of ICD

Notch ICD, like several transcription factors, is modified posttranslationally, primarily by phosphorylation but also by ubiquitination. Although ICD modification is likely to have transcriptional consequences, those consequences are not well understood at this time.

PHOSPHORYLATION

Following receptor activation and cleavage ICD becomes hyperphosphorylated [70–73]. Biochemical analysis of phosphorylation of Notch1 indicates that it occurs in two kinetically separable steps, in which ICD is phosphorylated in the cytoplasm and then hyperphosphorylated in the nucleus [74]. Although correlated with receptor activation, the mechanism and role of ICD phosphorylation remain unclear. Highly phosphorylated forms of *Drosophila* ICD immunoprecipitate with Su(H), suggesting that phosphorylation does not preclude inclusion in a transcriptional complex [70,72]; however, dephosphorylated ICD also immunoprecipitates with Su(H), indicating that ICD's phosphorylation state is unlikely to regulate complex formation.

It is not known whether phosphorylation of ICD is required for Su(H)-dependent transcription per se. Differential phosphorylation in response to growth factors alters Notch2 ICD function, although the transcriptional consequences are unknown [75]. In labeling experiments, much of the phosphate incorporated by ICD maps to the region required for transactivation [72], suggesting a transcriptional role, although the C-terminal half of mouse Notch1 ICD, containing both the TAD and PEST sequences, incorporates phosphate at multiple sites [72].

The activity of the coactivator mam suggests that phosphorylation regulates ICD turnover. Mam promotes hyperphosphorylation of *Xenopus* ICD in a Su(H)-dependent manner, an event correlating with loss of ICD protein from cell extracts [49]. Truncated ICD containing the TAD but lacking the C-terminal PEST domain is neither hyperphosphorylated nor destabilized in the presence of mam; likewise, full-length ICD is not hyperphosphorylated and destabilized by mam truncations containing only the N terminus and N-terminal TAD. Given that truncated forms of ICD initiate transcription in the presence of mam [49], these observations support the following model: that ICD/Su(H)/mam interaction favors recruitment of a kinase to catalyze the nuclear phosphorylation of sequences in the ICD C terminus, and that such hyperphosphorylation may function in a downstream transcriptional event such as elongation, in ICD turnover, or in both. Some ubiquitin ligases such as SEL-10 target phosphorylated ICD substrates for degradation (see later discussion); although loss of ICD protein from cell extracts accompanies mam-mediated hyperphosphorylation, it is not known whether a kinase activity potentially recruited by mam is required for ICD recognition by SEL-10.

An opposite role for phosphorylation in regulating ICD turnover comes from reports that the serine kinase GSKβ (Shaggy) phosphorylates the Notch1 ICD *in vitro* and thereby enhances the stability of ICD protein [76]. *In vivo* cells lacking GSKβ activity exhibit less extensively phosphorylated forms of ICD, which is attributable to either direct or indirect interaction with GSKβ [76]. Given that ICD is so highly phosphorylated, these data and the activity of mam indicate that phosphorylation, which is likely mediated by several kinases, plays multiple roles in regulating ICD stability and potentially transcriptional activity.

UBIQUITINATION

Ubiquitination not only marks proteins for degradation but is required for activation of some transcriptional factors (for a review, see [77]). Numerous studies indicate a role for ubiquitination in the turnover of proteins in the Notch pathway [74,78–84]; furthermore, disabling the proteasome stabilizes misexpressed ICD in both fly embryos and cultured cells [81,85]. As yet, however, there is no evidence for an activating role of ubiquitination in Notch transcription, although the observation of paradoxical degradation/activation activity by proteins such as Nrarp and mam suggests that ICD activation could be linked to turnover.

The F-box E3 ligase, SEL-10, was first identified as a suppressor of hypomorphic Notch alleles in *C. elegans* genetic screens [86]. Vertebrate Sel-10 homologs promote the ubiquitination of ICD, preferentially recognizing hyperphosphorylated forms of ICD [74,81,82]. Interaction with SEL-10 requires sequences downstream of the ICD ankyrin repeats, the domain shown to be highly phosphorylated as noted earlier [72]. Ubiquitination by SEL-10 apparently leads to turnover given that dominant-negative forms of SEL-10 stabilize ICD protein [81]. Dominant-negative SEL-10 potentiates ICD activity on Su(H)-dependent reporters [81,74], while full-length SEL-10 reduces transcriptional activity [82]. These observations coupled with genetic evidence suggest that SEL-10-mediated ubiquitination does not activate Notch transcription but primarily acts to promote ICD turnover.

A HECT-type ubiquitin ligase known in flies as Suppressor of Deltex (Su(dx)) and in mouse as Itch also ubiquitinates Notch ICD [83]. Although genetic evidence indicates that Su(dx) negatively regulates Notch signaling [87], the phenotype of $Itch^{-/-}$ mice, which exhibit inflammation and immune defects [88,89], does not immediately suggest a role in Notch signaling. Therefore the relevance of Itch to Notch signaling *in vivo* in vertebrates remains unclear.

Notch Transcriptional Activity *In Vivo*

Biochemical analysis of the Notch transcriptional complex has revealed the mechanism by which ICD converts Su(H) from a transcriptional repressor to an activator. We next consider how this transcriptional complex alters gene expression during Notch signaling based on *in vivo* analysis of Notch-responsive enhancers. This analysis has been primarily pioneered in *Drosophila* where a large number of Notch-responsive enhancers have been characterized molecularly. One issue addressed by these studies is whether ICD regulates transcription by simply promoting derepression of Su(H), or whether it also promotes transcriptional activation by recruiting coactivators. These two possibilities can be distinguished genetically in *Drosophila* because they predict that Su(H) loss-of-function mutants will exhibit increased Notch signaling phenotypes in the former case, but decreased Notch signaling phenotypes in the latter. We review here cases in which Notch acts by derepression, although even in these cases, Notch signaling also acts by switching Su(H) to a transcriptional activator.

A second issue addressed here is how Notch activates expression of targets in a context-dependent fashion. For example, Notch signaling directly regulates related genes that encode bHLH repressors in the *Enhancer of Split* locus even though these genes differ dramatically in both their spatial and temporal expression patterns during development. This finding implies that specificity in Notch signaling is derived from other factors, which act in combination with ICD. We discuss here support for such a combinatorial mechanism, which has been observed in *Drosophila* on both native and artificial Notch-responsive enhancers. Finally, we conclude with a brief discussion of vertebrate Notch response elements and enhancers, some of which reflect the molecular organization and/or activities seen in flies. Table II provides a summary of Notch response elements including bona fide Notch enhancers.

Table II Notch Responsive Enhancers/Promoters[a]

Target gene	Function	References
Dm Pax2	Enhancer	30
Dm m4	Enhancer	29
Dm E(spl)mγ	Enhancer	28
Dm E(spl)m8	Enhancer	95, 99
Dm sim	Enhancer	44, 96
Dm yan	Enhancer	100
Dm vestigial	Enhancer	31
Xenopus hairy2	Enhancer	26
Human, mouse *pre-TCRα*	Enhancer	101
Mouse *Lunatic fringe*	Enhancer	97, 98
Mouse *HES1*	Response element	102
Mouse *HES5*	Response element	103
Mouse *HES7*	Response element	104
Mouse *HERP1*	Response element	105
Mouse *HRT2*	Response element	106
Mouse, human *Hey1*	Response element	107
Human *NFkappaB2*	Response element	108
Human *ERBB-2*	Response element	109
Human *IL-6*	Response element	110, 111
Mouse *p21*	Response element	43
Rat, human *Cyclin D1*	Binding Site	112
Human β-globin LCR	Binding Site	32

[a]Function is defined as follows: enhancer, tissue specific enhancer responsive to Su(H)-dependent Notch signaling *in vivo*; response element, genomic fragment from a likely Notch target gene containing Su(H) binding sites and responsive to ICD activation, Su(H)-mediated repression, or both in cultured cells; binding site, element DNA binding Su(H) *in vitro* and derived from genomic DNA flanking a potential Notch and/or Su(H) target gene.

Repression by Su(H)

The role of Notch in relieving Su(H)-mediated repression is supported by the analysis of the *Drosophila singleminded* (*sim*) gene, which is expressed in a single row of cells in the embryonic ventral mesectoderm. High levels of *sim* expression within this single row of cells require Su(H)-dependent Notch signaling [44]. However, expression of *sim* in Su(H) null mutants is not only reduced, as one would predict, but also ectopic, suggesting that in the absence of Notch signaling Su(H) represses *sim* expression in cells adjacent to the stripe. This possibility was tested by characterizing sequences in the *sim* enhancer required for proper transgene expression. The *sim* enhancer contains 10 Su(H) sites, 6 of which are high-affinity sites, and several in proximity to binding sites for two other positive regulators of *sim* expression, Dorsal and Twist, and to binding sites for a negative regulator, Snail [44]. When all 10 Su(H) sites are mutated, the enhancer no longer drives high levels of expression at the midline but is also expressed ectopically in adjacent cells in a manner analogous to the *sim* expression pattern in Su(H) null flies. These data suggest strongly that Su(H) not only mediates activation of target gene expression in areas of Notch signaling, but also repression of genes outside this area where Notch signaling is silent.

Transcriptional repression mediated by Su(H) is also revealed by analysis of a synthetic enhancer containing different combinations of binding sites for the bHLH protein Grainyhead and for Su(H) [90]. In general, reporters driven by Grainyhead sites alone are expressed widely and at moderate levels. When Grainyhead sites are combined with Su(H) sites, expression of the enhancer is inhibited in tissues where Notch signaling is absent, suggesting that Su(H) without ICD represses the activity of Grainyhead. However, transgene expression driven by the same enhancer is robust in regions where Notch signaling and Grainyhead expression coincide, suggesting that ICD not only depresses Su(H) but cooperatively interacts with Grainyhead to activate transcription. These findings support those observed with the *sim* enhancer, revealing a role of Su(H) mediated repression in preventing ectopic expression of Notch target genes.

Importantly, in both of these examples, mutating Su(H) binding sites, or genetically removing Su(H), is insufficient to drive high levels of Notch target gene expression in the normal pattern, indicating that transcriptional activation by ICD is also needed. The only case in which Notch signaling has been proposed to act solely by promoting derepression is in a process known as *proneural enhancement*, in which Notch signaling is required to promote proneural gene expression in the developing eye [91]. Notch signaling in this case can be phenocopied by removing Su(H) genetically, indicating that Notch is required only for derepression [91]. However, this interpretation may be premature until more is known about whether or not this regulation is direct instead of a complex outcome of Su(H) regulation of other genes.

ICD Activates Target Genes in Combination with Other Activators

In the majority of Notch-responsive enhancers examined thus far, Notch signaling in combination with Su(H) is required for high levels of tissue-specific expression. Significantly, data from these *in vivo* studies also indicate that many target genes are fairly refractory to signaling by Notch alone. For example, transgenic flies expressing reporters driven by multimerized Su(H) sites exhibit little transgene expression in *Drosophila* larval tissues, even in areas positive for Notch signaling, suggesting that target activation *in vivo* requires input from Su(H)/ICD in combination with other factors [11,12,92]. Analysis of this combinatorial input in a number of different enhancers reveals synergistic transcriptional interactions among ICD/Su(H) and a variety of different transcriptional activators.

Drosophila m4

An extensively characterized family of Notch target genes in *Drosophila* is clustered within the *E(spl)* complex; these genes are closely related and likely arose by recent

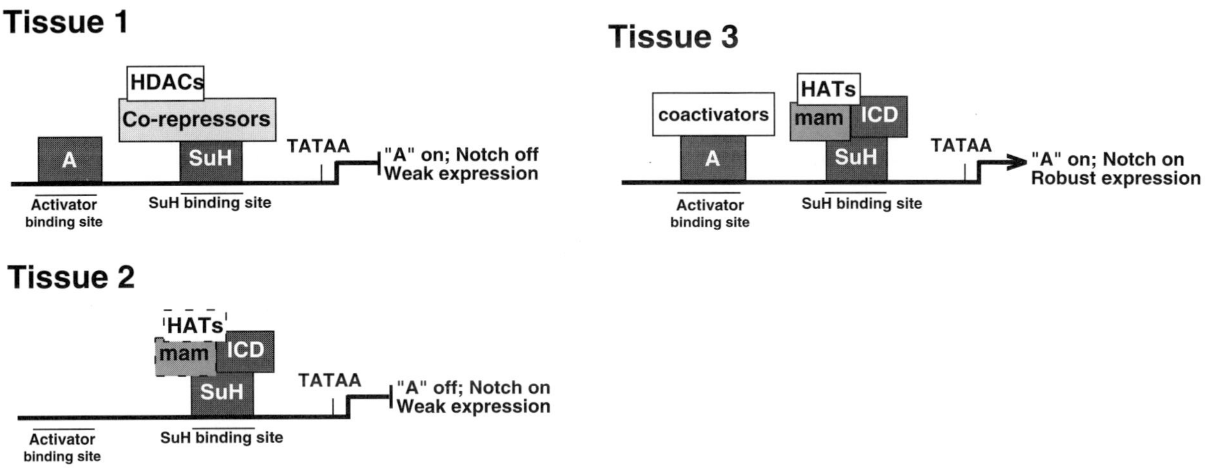

Figure 3 Hypothetical Notch responsive enhancer requiring cooperative interactions. Interactions on this element are shown in three different tissues: Tissue 1, in which Notch signaling is off, and Tissues 2 and 3, in which Notch signaling is on. For simplicity one heterologous activator (A) is depicted, although interaction with SuH/ICD by more than one cooperating activator is not precluded (see text). Likewise, SuH could bind to several sites in the enhancer. In Tissue 1 the target gene is either not expressed or expressed very weakly because potential activity mediated by A bound to the enhancer is repressed by corepressors interacting with SuH. Despite the interaction of ICD with SuH in Tissue 2, the absence of heterologous activators prevents cooperative interaction required for robust activation of transcription in vivo. Whether the co-activator mastermind (mam) or HATs interact with ICD in this condition is unknown; therefore, those potential interactions are represented by dashed lines. Robust activation of transcription can only occur in Tissue 3, where Notch signaling is on, heterologous activator proteins are bound to the enhancer, and coactivators functioning cooperatively are likely recruited by both. Such a model based on cooperativity does not rule out hypothetical activation states represented in figure 2; for example, proneural enhancement could occur in Tissue 1 if Notch signaling relieved SuH-mediated repression, allowing A and accompanying co-activators to activate transcription.

gene duplication [29,93,94]. The response of these genes to Notch signaling depends on Su(H) binding sites, many of which are found in a so-called paired arrangement consisting of two inverted Su(H) binding sites, separated by 30 bp and containing a conserved hexamer motif (GAAAGT). Mutational analysis of the enhancers for several *E(spl)* genes, including *m4* and *m*γ, shows that both the Su(H) binding sites and the hexamer are required for tissue-specific gene expression. However, the *m4* and *m*γ enhancers, as well as those of other *E(spl)* complex genes, also contain E box motifs recognized by proneural bHLH proteins in proximity to both paired and isolated Su(H) binding sites [28,29,95]. Mutating these sites also results in a marked reduction in enhancer activity [29]. Enhancer activity is similarly reduced in flies mutant in either candidate proneural proteins or Su(H), indicating further that both proteins cooperate to promote expression of these Notch target genes [28,29,94]. A model for a Notch enhancer exhibiting such cooperative activity is shown in Fig. 3.

Drosophila Pax2

Combinatorial activation by Notch has also been demonstrated by analyzing the *Drosophila* eye-specific enhancer of the gene encoding the transcription factor Pax2 (DPax2). This ~350-bp enhancer accounts for the *DPax2* expression in cone cell precursors in the larval eye disc and contains binding sites for three potential effectors; namely, three binding sites for the runt-domain transcription factor Lozenge, six Ets factor binding sites, and eight consensus Su(H) binding sites. Many of these sites are clustered within 100 base pairs of one another [30]. Disabling all three pathways independently indicates that Lozenge binding, EGF signaling (required to activate the Ets factor Pnt), and Notch signaling through the Su(H) sites are all required for eye-specific expression of *DPax2*. The marked loss in enhancer activity seen when one pathway is disrupted suggests that the activity of Lozenge, Pnt, and ICD/Su(H) is not only combinatorial but synergistic.

Drosophila sim

Cooperativity between Su(H)/ICD and the transcription factor Dorsal has also been demonstrated on the same *sim* enhancer element described earlier [96]. Just as ICD is required for high levels of *sim* expression [44], disabling mutations in Dorsal block *sim* expression in the mesectodermal stripe [96]. Ectopic expression of high levels of activated Notch rescues *sim* expression in Dorsal mutant embryos but very weakly; however, high expression of ectopic ICD in fly embryos expressing low levels of Dorsal robustly induces *sim* expression, indicating cooperation between ICD and Dorsal. This finding is remarkable given that the *sim* enhancer, like the eye-specific *DPax2* enhancer, contains many Su(H) binding sites, most of them high affinity and likely functional, and shows that without Dorsal activity even high levels of ectopic ICD are insufficient to activate transcription through these sites. Clearly, low levels of ICD and Dorsal pack more transcriptional punch than high levels of ICD alone, either due to architecture of the enhancer, a requirement for ICD/Dorsal cooperativity, or both.

Vertebrate Notch Response Elements

Compared to *Drosophila*, relatively little is known about enhancers that drive the expression of vertebrate Notch target genes. Nonetheless, likely targets of Notch signaling have been identified by isolating flanking sequences of vertebrate homologs of genes regulated by *Drosophila* Notch, such as those that encode the bHLH repressors in the Enhancer of Split Complex (*E(spl)-C*). One such mouse gene, called *HES1*, is likely regulated by Notch in part, and the sequence upstream of the start of transcription contains high-affinity, paired Su(H) binding sites similar to that found in the *Drosophila E(spl)-C* genes. In transient transfection assays, this element responds to activated Notch. However, whether these binding sites constitute part of a Notch-responsive *HES1* enhancer element *in vivo* has not been determined.

One characterized Notch-responsive enhancer in vertebrates is a *Xenopus* gene called *hairy2*, which encodes a homolog of the bHLH encoded in the ESPL-C [26]. *Hairy2* is expressed in the frog neural plate and transiently in a narrow anterior stripe of the presomitic mesoderm (PSM) immediately prior to formation of a somite. Reminiscent of the *m4* enhancer and *HES1* proximal promoter, the *hairy2* proximal promoter contains paired Su(H) binding sites flanking a hexameric motif. Frogs transgenic for a reporter driven from this element express the transgene in a pattern recapitulating *hairy2* expression in the neural plate and PSM. Mutation of either Su(H) site or the hexamer alone abolishes the PSM stripe pattern of transgene expression. At first glance, the fact that Su(H) consensus sites in the 90-bp element function as a tissue-specific *hairy2* enhancer seems to argue against the cooperativity model, because consensus binding sites for activators other than ICD/Su(H) are not apparent based on sequence analysis of the region. Nonetheless, unidentified factors binding to the hexameric element or elsewhere could cooperate with or be required for Su(H)/ICD activation of this enhancer.

A direct Notch target in vertebrates that is not a *HES/E(spl)* gene is the gene encoding the glycosyltransferase Lunatic fringe (Lfng), which modifies and modulates the activity of the Notch receptor. In the mouse and chick presomitic mesoderm, the expression of Lfng oscillates periodically in synchrony with somite formation [113], and a 110-bp upstream element from the *Lfng* 5′ flanking sequences containing both E-box motifs and a single Su(H) binding site drives oscillating expression of a reporter gene in transgenic mice [97,98]. Mutation of either the E boxes [97] or the Su(H) binding site [98] disrupts the cyclic expression pattern of the reporter in the presomitic mesoderm, strongly suggesting that input from both E-box recognizing proteins and Notch signaling is required to up-regulate periodic *Lfng* expression. As is the case with *sim*, *DPax2*, and a number of *E(spl)* genes noted earlier, these observations strongly suggest that in vertebrates as well as *Drosophila* Notch signaling functions in a combinatorial fashion with other transcription factors to up-regulate target genes.

Conclusion

It is clear from the study of Notch signaling in development that Notch transcription is often extremely dynamic as well as versatile in the activation of targets. In this chapter, we have reviewed recent literature analyzing the biochemical activities that underlie Notch transcription, with the hope that themes would emerge to account for such behavior. One such theme is that a multiplicity of cofactors is associated with the Notch transcriptional complex. Some of these cofactors such as mam are required for ICD-mediated transcription in many if not most cases. Others such as SKIP, Nrarp, HATs, and modifying activities such as kinases may be required for transcriptional activity *in vivo* but also may play important regulatory roles in both turning on and turning off the transcriptional complex. A second theme is that Notch seems to function as an on/off switch in that it converts Su(H) from a repressor into an activator. This switch presumably accounts for the large dynamic range over which Notch regulates expression of its targets during development. A third theme is that the up-regulation of Notch target genes is in many cases the result of combinatorial activity between ICD/Su(H) and other transcription factors. This feature of Notch transcription presumably accounts for the diversity of Notch target genes in different developmental contexts. Future studies will need to combine an evaluation of biochemical activities associated with Notch transcription *in vitro* with the analysis of Notch target genes *in vivo* to account for both dynamic and versatile activity.

Acknowledgments

The authors thank Drs. Anne Bang, Christy Fryer, Naoko Koyano-Nakagawa, and Don van Myel for critical input on this manuscript.

References

1. Gridley, T. (1997). Notch signaling in vertebrate development and disease. *Mol. Cell Neurosci.* **9**, 103–108.
2. Kimble, J. and Simpson, P. (1997). The LIN12/Notch signaling pathway and its regulation. *Ann. Rev. Cell Dev. Biol.* **13**, 333–361.
3. Artavanis-Tsakonas, S., Rand, M., and Lake, R. (1999). Notch signaling: Cell fate control and signal integration in development. *Science* **284**, 770–776.
4. Mumm J. and Kopan R. (2000). Notch signaling: From the outside in. *Dev. Biol.* **228**, 151–165.
5. Deftos, M. L. and Bevan, M. J. (2000). Notch signaling in T cell development. *Curr. Opin. Immunol.* **12**, 166–172.
6. Kopan, R. (2002). Notch: A membrane-bound transcription factor. *J. Cell Sci.* **115**, 1095–1097.
7. Allman, D., Punt, J. A., Izon, D. J., Aster, J. C., and Pear, W. S. (2002). An invitation to T and more: Notch signaling in lymphopoiesis. *Cell,* **109**, S1–11.
8. Justice, N. J. and Jan, Y. N. (2002). Variations on the Notch pathway in neural development. *Curr. Opin. Neurobiol.* 2002;**12**, 64–70.
9. Struhl, G. and Adachi, A. (1998). Nuclear access and action of Notch *in vivo*. Cell **93**, 649–660.
10. Schroeter, E., Kisslinger, J., and Kopan, R. (1998). Notch-1 signalling requires ligand-induced proteolytic release of intracellular domain. *Nature* **393**, 382–386.
11. Bray, S. and Furriols, M. (2001). Notch pathway: Making sense of Suppressor of Hairless. *Curr. Biol.* **11**, R217–R221.

12. Barolo, S. and Posakony, J. W. (2002). Three habits of highly effective signaling pathways: Principles of transcriptional control by developmental cell signaling. *Genes Dev.* **16**, 1167–1181.
13. Shawber, C., Nofziger, D., Hsieh, J., Lindsell, C., Bogler, O., Hayward, D., and Weinmaster, G. (1996). Notch signalling inhibits muscle cell differentiation through a CBF1-independent pathway. *Development* **122**, 3765–3773.
14. Ordentlich, P., Lin, A., Shen, C., Blaumueller, C., Matsuno, K., Artavanis-Tsakonas, S. and Kadesch, T. (1998). Notch inhibition of E47 supports the existence of a novel signaling pathway. *Mol. Cell Biol.* **18**, 2230–2239.
15. Bush, G., diSibio, G., Miyamoto, A., Denault, J. B., Leduc, R., and Weinmaster, G. (2001). Ligand-induced signaling in the absence of furin processing of Notch1. *Dev. Biol.* **229**, 494–502.
16. Ramain, P., Khechumian, K., Seugnet, L., Arbogast, N., Ackermann, C., and Heitzler, P. (2001). Novel Notch alleles reveal a Deltex-dependent pathway repressing neural fate. *Curr. Biol.* **11**, 1729–1738.
17. Ross, D. A. and Kadesch, T. (2001). The Notch intracellular domain can function as a coactivator for LEF-1. *Mol. Cell Biol.* **21**, 7537–7544.
18. Weinmaster, G. (1997). The ins and outs of Notch signaling. *Mol. Cell Neurosci.* **9**, 91–102.
19. Beatus, P., Lundkvist, J., Oberg, C., and Lendahl, U. (1999). The Notch 3 intracellular domain represses Notch 1-mediated activation through *Hairy/Enhancer of split (HES)* promoters. *Development* **126**, 3925–3935.
20. Tamura, K., Taniguchi, Y., Minoguchi, S., Sakai, T., Tun, T., Furukawa, T., and Honjo, T. (1995). Physical interaction between a novel domain of the receptor Notch and the transcription factor RBP-Jκ/Su(H). *Curr. Biol.* **5**, 1416–1423.
21. Kato, H., Taniguchi, Y., Kurooka, H., Minoguchi, S., Sakai, T., Nomura-Okazaki, S., Tamura, K., and Honjo, T. (1997). Involvement of RBP-J in biological functions of mouse Notch1 and its derivatives. *Development* **124**, 4133–4141.
22. Zweifel, M. E. and Barrick, D. (2001). Studies of the ankyrin repeats of the Drosophila melanogaster Notch receptor. 1. Solution conformational and hydrodynamic properties. *Biochemistry.* **40**, 14344–14356.
23. Jeffries, S., Robbins, D. J., and Capobianco, A. J. (2002). Characterization of a high-molecular weight Notch complex in the nucleus of Notch(ic)-transformed RKE cells and in a human T-cell leukemia cell line. *Mol. Cell Biol.* **22**, 3927–3941.
24. Rechsteiner, M. and Rogers, S. W. (1996). PEST sequences and regulation by proteolysis. *Trends Biochem. Sci.* **21**, 267–271.
25. Tun, T., Hamaguchi, Y., Matsunami, N., Furukawa, T., Honjo, T., and Kawaichi, M. (1994). Recognition sequence of a highly conserved DNA binding protein RBP-Jκ. *Nucl. Acids Res.* **22**, 965–971.
26. Davis, R., Turner, D., Evans, L., and Kirschner, M. (2001). Molecular targets of vertebrate segmentation: Two mechanisms control segmental expression of Xenopus *hairy2* during somite formation. *Dev. Cell* **1**, 553–565.
27. Jarriault, S., Brou, C., Logeat, F., Schroeter, E. H., Kopan, R., and Israel, A. (1995). Signalling downstream of activated mammalian Notch. *Nature* **377**, 355–358.
28. Nellesen, D. T., Lai, E. C., and Posakony, J. W. (1999). Discrete enhancer elements mediate selective responsiveness of *Enhancer of split* Complex genes to common transcriptional activators. *Dev. Biol.* **213**, 33–53.
29. Bailey, A. and Posakony, J. W. (1995). Suppressor of Hairless directly activates transcription of *Enhancer of split* Complex genes in response to Notch receptor activity. *Genes Dev.* **9**, 2609–2622.
30. Flores, G. V., Duan, H., Yan, H., Nagaraj, R., Fu, W., Zou, Y., Noll, M., and Banerejee, U. (2000). Combinatorial signaling in the specification of unique cell fates. *Cell,* **103**, 75–85.
31. Kim, J., Sebring, A., Esch, J., Kraus, M. E., Vorwerk, K., Magee, J., and Carroll, S. B. (1996). Integration of positional signals and regulation of wing formation and identity by *Drosophila vestigial* gene. *Nature* **382**, 133–138.
32. Lam, L. T. and Bresnick, E. H. (1998). Identity of the β-globin loculs control region binding protein HS2NF5 as the mammalian homolog of the Notch-regulated transcription factor Suppressor of Hairless. *J. Biol. Chem.* **273**, 24223–24231.
33. Olave, I., Reinberg, D., and Vales, L. D. (1998). The mammalian transcriptional repressor RBP (CBF1) targets TFIID and TFIIA to prevent activated transcription. *Genes Dev.* **12**, 1621–1637.
34. Kao, H.-Y., Ordentlich, P., Koyano-Nakagawa, N., Tang, Z., Downes, M., Kintner, C., Evans, R., and Kadesch, T. (1998). A histone deacetylase corepressor complex regulates the Notch signal transduction pathway. *Genes Dev.* **12**, 2269–2277.
35. Hsieh, J.-D., Zhou, S., Chen, L., Young, D. B., and Hayward, S. D. (1999). CIR, a corepressor linking the DNA binding factor CBF1 to the histone deacetylase complex. *Proc. Natl. Acad. Sci. USA* **96**, 23–28.
36. Taniguchi, Y., Furukawa, T., Tun, T., Han, H., and Honjo, T. (1998). LIM protein KyoT2 negatively regulates transcription by association with the RBP-J DNA-binding protein. *Mol. Cell Biol.* **18**, 644–654.
37. Bang, A. G. and Posakony, J. W. (1992). The *Drosophila* gene *Hairless* encodes a novel basic protein that controls alternative cell fates in adult sensory organ development. *Genes Dev.* **6**, 1752–1769.
38. Bang, A. G., Bailey, A. M., and Posakony, J. W. (1995). Hairless promotes stable commitment to the sensory organ precursor cell fate by negatively regulating the activity of the Notch signlaing pathway. *Dev. Biol.* **172**, 479–494.
39. Maier, D., Stumm, G., Kuhn, K., and Preiss, A. (1992). *Hairless*, a *Drosophila* gene involved in neural development, encodes a novel, serine rich protein. *Mech. Dev.* **38**, 143–156.
40. Morel, V., Lecourtois, M., Massiani, O., Maier, D., Preiss, A., and Schweisguth, F. (2001). Transcriptional repression by Suppressor of hairless involves the binding of a Hairless-dCtBP complex in *Drosophila. Curr. Biol.* **11**, 789–792.
41. Barolo, S., Stone, T., Bang, A. G., and Posakony, J. W. (2002). Hairless acts as an adaptor to recruit the corepressors Groucho and dCtBP to Suppressor of Hairless. *Genes Dev.* **16**, 1964–1976.
42. Courey, A. J. and Jia, S. (2001). Transcriptional repression: The long and the short of it. *Genes Dev.* **15**, 2786–2796.
43. Rangarajan, A., Talora, C., Okuyama, R., Nicolas, M., Mammucari, C., Oh, H., Aster, J., Krishna, S., Metzger, D., Chambon, P., Miele, L., Aguet, M., Radtke, F., and Dotto, G. (2001). Notch signaling is a direct determinant of keratinocyte growth arrest and entry into differentiation. *EMBO J.* **20**, 3427–3436.
44. Morel, V. and Schweisguth, F. (2000). Repression by Suppressor of hairless and activation by Notch are required to define a single row of single-minded expressing cells in the *Drosophila* embryo. *Genes Dev.* **14**, 377–388.
45. Kurooka, H., Kuroda, K., and Honjo, T. (1998). Roles of the ankyrin repeats and C-terminal region of the mouse Notch1 intracellular region. *Nucl. Acids Res.* **26**, 5448–5455.
46. Kurooka, H. and Honjo, T. (2000). Functional interaction between the mouse Notch 1 intracellular region and histone acetyltransferases PCAF and GCN5. *J. Biol. Chem.* **275**, 17211–17220.
47. Oswald, F., Tauber, B., Dobner, T., Bourteele, S., Kostezka, U., Adler, G., Liptay, S., and Schmid, R. M. (2001). p300 Acts as a transcriptional coactivator for mammalian Notch-1. *Mol. Cell Biol.* **21**, 7761–7774.
48. Wettstein, D., Turner, D., and Kintner, C. (1997). The *Xenopus* homolog of *Drosophila Suppressor of Hairless* mediates Notch signaling during primary neurogenesis. *Development* **124**, 3693–3702.
49. Fryer, C. J., Lamar, E., Turbachova, I., Kintner, C., and Jones, K. A. (2002). Mastermind mediates chromatin-specific transcription and turnover of the Notch enhancer complex. *Genes Dev.* **16**, 1397–1411.
50. Aster, J. C., Xu, L., Karnell, F. G., Patriub, V., Pui, J. C., and Pear, W. S. (2000). Essential roles for ankyrin repeat and transactivation domains in induction of T-cell leukemia by notch1. *Mol. Cell Biol.* **20**, 7505–7515.
51. Petcherski, A. G. and Kimble, J. (2000). Mastermind is a putative activator for Notch. *Curr. Biol.* **10**, R471–473.

52. Wu, L., Aster, J., Blacklow, S., Lake, R., Artavanis-Tsakonas, S., and Griffin, J. (2000). MAML1, a human homologue of *Drosophila* Mastermind is a transcriptional coactivator for NOTCH receptors. *Nature Genet.* **26**, 484–489.
53. Kitagawa, M., Oyama, T., Kawashima, T., Yedvobnick, B., Kumar, A., Matsuno, K., and Harigaya, K. (2001). A human protein with sequence similarity to *Drosophila* Mastermind coordinates the nuclear form of Notch and a CSL protein to build a transcriptional activator complex on target promoters. *Mol. Cell Biol.* **21**, 4337–4346.
54. Helms, W., Lee, H., Ammerman, M., Parks, A. L., Muskavitch, M. A., and Yedvobnick, B. (1999). Engineered truncations in the *Drosophila* mastermind protein disrupt Notch pathway function. *Dev. Biol.* **215**, 358–374.
55. Giraldez, A. J., Perez, L., and Cohen, S. M. (2002). A naturally occurring alternative product of the *mastermind* locus that represses notch signalling. *Mech. Dev.* **115**, 101–105.
56. Bresnick, E. H., Chu, J., Christensen, H., Lin, B., and Norton, J. (2000). Linking Notch signaling, chromatin remodeling, and T-cell leukemogenesis. *J. Cell Biochem. Supp.* **35**, 46–53.
57. Lamar, E., Deblandre, G., Wettstein, D., Gawantka, V., Pollet, N., Niehrs, C., and Kintner, C. (2001). Nrarp is a novel intracellular component of the Notch signaling pathway. *Genes Dev.* **15**, 1885–1899.
58. Krebs, L., Deftos, M., Bevan, M., and Gridley, T. (2001). The *Nrarp* gene encodes an ankyrin-repeat protein that is transcriptionally regulated by the Notch signaling pathway. *Dev. Biol.* **238**, 110–119.
59. Lahaye, K., Kricha, S., and Bellefroid, E. (2002). XNAP, a conserved ankyrin repeat-containing protein with a role in the Notch pathway during *Xenopus* primary neurogenesis. *Mech. Dev.* **110**, 113–124.
60. Zhou, S., Fujimuro, M., Hsieh, J., Chen, L., and Hayward, S. D. (2000). A role for SKIP in EBNA-2 activation of CBF1-repressed promoters. *J. Virol.* **74**, 1939–1947.
61. Zhang, C., Baudino, T. A., Dowd, D. R., Tokumaru, H., Wang, W., and MacDonald, P. N. (2001). Ternary complexes and cooperative interplay between NCoA-62/Ski-interacting protein and steroid receptor coactivators in vitamin D receptor-mediated transcription. *J. Biol. Chem.* **276**, 40614–40620.
62. Leong, G. M., Subramaniam, N., Figueroa, J., Flanagan, J. L., Hayman, M. J., Eisman, J. A., and Kouzmenko, A. P. (2001). Ski-interacting protein interacts with Smad proteins to augment transforming growth factor-β-dependent transcription. *J. Biol. Chem.* **276**, 18243–18248.
63. Matsuno, K., Diederich, R., Go, M., Blaumueller, C., and Artavanis-Tsakonas, S. (1995). Deltex acts as a positive regulator of Notch signaling through interactions with the Notch ankyrin repeats. *Development* **121**, 2633–2644.
64. Matsuno, K., Ito, M., Hori, K., Miyashita, F., Suzuki, S., Kishi, N., Artavanis-Tsakonas, S., and Okano, H. (2002). Involvement of a proline-rich motif and RING-H2 finger of Deltex in the regulation of Notch signaling. *Development* **129**, 1049–1059.
65. Diederich, R. J., Matsuno, K., Hing, H., and Artavanis-Tsakonas, S. (1994). Cytosolic interaction between deltex and Notch ankyrin repeats implicates deltex in the Notch signaling pathway. *Development* **120**, 473–481.
66. Yamamoto, N., Yamamoto, S., Inagaki, F., Kawaichi, M., Fukamizu, A., Kishi, N., Matsuno, K., Nakamura, K., Weinmaster, G., Okano, H., and Nakafuku, M. (2001). Role of Deltex-1 as a transcriptional regulator downstream of the Notch receptor. *J. Biol. Chem.* **276**, 45031–45040.
67. Xu, T., Rebay, I., Fleming, R., Scottgale, T., and Artavanis-Tsakonas, S. (1990). The *Notch* locus and the genetic circuitry involved in early *Drosophila* neurogenesis. *Genes Dev.* **4**, 464–475.
68. Sestan, N., Artavanis-Tsakonas, S., and Rakic, P. (1999). Contact-dependent inhibition of cortical neurite growth mediated by Notch signaling. *Science* **286**, 741–746.
69. Izon, D. J., Aster, J. C., He, Y., Weng, A., Karnell, F. G., Patriub, V., Xu, L., Bakkour, S., Rodriguez, C., Allman, D., and Pear, W. S. (2002). Deltex1 redirects lymphoid progenitors to the B cell lineage by antagonizing Notch1. *Immunity* **16**, 231–243.
70. Kidd, S., Lieber, T., and Young, M. W. (1998). Ligand-induced cleavage and regulation of nuclear entry of Notch in *Drosophila melanogaster* embryos. *Genes Dev.* **12**, 3728–3740.
71. Shimizu, K., Chiba, S., Hosoya, N., Kumano, K., Saito, T., Kurokawa, M., Kanda, Y., Hamada, Y., and Hirai, H. (2000). Binding of Delta1, Jagged1, and Jagged2 to Notch2 rapidly induces cleavage, nuclear translocation, and hyperphosphorylation of Notch2. *Mol. Cell Biol.* **20**, 6913–6922.
72. Foltz, D. R. and Nye, J. S. (2001). Hyperphosphorylation and association with RBP of the intracellular domain of Notch1. *BBRC* **286**, 484–492.
73. Redmond, L., Oh, S.-R., Hicks, C., Weinmaster, G., and Ghosh, A. (2000). Nuclear Notch1 signaling and the regulation of dendritic development. *Nature Neurosci.* **3**, 30–40.
74. Gupta-Rossi, N., LeBail, O., Gonen, H., Brou, C., Logeat, F., Six, E., Ciechanover, A., and Israel, A. (2001). Functional interaction between SEL-10, an F-box protein, and the nuclear form of activated Notch1 receptor. *J. Biol. Chem.* **276**, 34371–34378.
75. Ingles-Esteve, J., Espinosa, L., Milner, L. A., Caelles, C., and Bigas, A. (2001). Phosphorylation of Ser^{2078} modulates the Notch2 function in 32D cell differentiation. *J. Biol. Chem.* **276**, 44873–44880.
76. Foltz, D. R., Santiago, M. C., Berechid, B. E., and Nye, J. S. (2002). Glycogen synthase kinase-3β modulates Notch signaling and stability. *Curr. Biol.* **12**, 1006–1011.
77. Conaway, R. C., Brower, C. S., and Conaway, J. W. (2002). Emerging roles of ubiquitin in transcription regulation. *Science* **296**, 1254–1258.
78. Yeh, E., Dermer, M., Commisso, C., Zhou, L., McGlade, C. J., and Boulianne, G. L. (2001). Neuralized functions as an E3 ubiquitin ligase during *Drosophila* development. *Curr. Biol.* **11**, 1675–1679.
79. Deblandre, G., Lai, E., and Kintner, C. (2001). *Xenopus* Neuralized is a ubiquitin ligase that interacts with X-Delta1 and regulates Notch signaling. *Dev. Cell* **1**, 795–806.
80. Lai, E. C., Deblandre, G., and Kintner, C. (2001). *Drosophila* Neuralized is a ubiquitin ligase that interacts with X-Delta1 and regulates Notch signaling. *Dev. Cell* **1**, 795–806.
81. Wu, G., Lyapina, S., Das, I., Li, J., Gurney, M., Pauley, A., Chui, I., Deshaies, R. J., and Kitajewski, J. (2001). SEL-10 is an inhibitor of Notch signaling that targets Notch for ubiquitin-mediated protein degradation. *Mol. Cell Biol.* 7403–7415.
82. Oberg, C. J. L., Li, J., Pauley, A., Wolf, E., Gurney, M., and Lendahl, U. (2001). The Notch intracellular domain is ubiquitinated and negatively regulated by the mammalian Sel-10 homolog. *J. Biol. Chem.* **276**, 35847–35853.
83. Qiu, L., Joazeiro, C., Fang, N., Wang, H.-Y., Elly, C., Altman, Y., Fang, D., Hunter, T., and Liu, Y.-C. (2000). Recognition and ubiquitination of Notch by Itch, a Hect-type E3 ubiquitin ligase. *J. Biol. Chem.* **46**, 35734–35737.
84. Lai, E. (2002). Protein degradation: Four E3s for the Notch pathway. *Curr. Biol.* **12**, R74-R78.
85. Schweisguth, F. (1999). Dominant-negative mutation in the beta2 and beta6 proteasome subunit genes affect alternative cell fate decisions in the *Drosophila* sense organ lineage. *Proc. Natl. Acad. Sci. USA* **96**, 11382–11386.
86. Hubbard, E. J., Wu, G., Kitajewski, J., and Greenwald, I. (1997). *sel-10*, a negative regulator of *lin-12* activity in *Caenorhabditis elegans*, encodes a member of the CDC4 family of proteins. *Genes Dev.* **11**, 3182–3193.
87. Fostier, M., Evans, D. A., Artavanis-Tsakonas, S., and Baron, M. (1998). Genetic characterization of the *Drosophila melanogaster* Suppressor of deltex gene: A regulator of notch signaling. *Genetics* **150**, 1477–1485.
88. Hustad, C. M., Perry, W. L., Siracusa, L. D., Rasberry, C., Cobb, L., Cattanach, B. M., Kovatch, R., Copeland, N. G., and Jenkins, N. A. (1995). Molecular genetic characterization of six recessive viable alleles of the mouse agouti locus. *Genetics* **140**, 255–265.
89. Fang, D., Elly, C., Gao, B., Fang, N., Altman, Y., Joazeiro, C., Hunter, T., Copeland, N., Jenkins, N., and Liu, Y.-C. (2002). Dysregulation of T lymphocyte function in itchy mice: A role for Itch in TH2 differentiation. *Nature Immunol.*

90. Furriols, M. and Bray, S. (2001). A model Notch response element detects Suppressor of Hairless-dependent molecular switch. *Curr. Biol.* **11**, 60–64.
91. Li, Y. and Baker, N. E. (2001). Proneural enhancement by Notch overcomes Suppressor-of-Hairless repressor function in the developing *Drosophila* eye. *Curr. Biol.* **11**, 330–338.
92. Guss, K. A., Nelson, C. E., Hudson, A., Kraus, M. E., and Carroll, S. B. (2001). Control of a genetic regulatory network by a selector gene. *Science* **292**, 1164–1167.
93. Knust, E., Schrons, H., Grawe, F., and Campos-Ortega, J. A. (1992). Seven genes of the *Enhancer of split* complex of *Drosophila melanogaster* encode helix-loop-helix proteins. *Genetics* **132**, 505–518.
94. Cooper, M. T. D., Tyler, D. M., Furriols, M., Chaldiadaki, A., Delidakis, C., and Bray, S. (2000). Spatially restricted factors cooperate with Notch in the regulation of *Enhancer of split* genes. *Dev. Biol.* **221**, 390–403.
95. Singson, A., Leviten, M., Bang, A., Hua, X. H., and Posakony, J. W. (1994). Direct downstream targets of proneural activators in the imaginal disc include genes involved in lateral inhibitory signaling. *Genes Dev.* **8**, 2058–2071.
96. Cowden, J. and Levine, M. (2002). The Snail repressor positions Notch signaling in the *Drosophila* embryo. *Development* **129**, 1785–1793.
97. Cole, S. E., Levorse, J. M., Tilghman, S. M., and Vogt, T. F. (2002). Clock regulatory elements control cyclic expression of *Lunatic fringe* during somitogenesis. *Dev. Cell* **3**, 75–84.
98. Morales, A. V., Yasuda, Y., and Ish-Horowicz, D. (2002). Periodic *Lunatic fringe* expression is controlled during segmentation by a cyclic transcriptional enhancer responsive to Notch signaling. *Dev. Cell* **3**, 63–74.
99. Klambt, C., Knust, E., Tietze, K., and Campos-Ortega, J. (1989). Closely related transcripts encoded by the neurogenic gene complex *Enhancer of split* of *Drosophila melanogaster. EMBO J.* **8**, 203–210.
100. Rohrbaugh, M., Ramos, E., Nguyen, D., Price, M., Wen, Y., and Lai, Z.-C. (2002). Notch activation of *yan* expression is antagonized by RTK/Pointed signaling in the *Drosophila* eye. *Curr. Biol.* **12**, 576–581.
101. Reizis, B. and Leder, P. (2002). Direct induction of T lymphocyte-specific gene expression by the mammalian Notch signaling pathway. *Genes Dev.* **16**, 292–300.
102. Takebayashi, K., Sasai, Y., Sakai, Y., Watanabe, T., Nakanishi, S., and Kageyama, R. (1994). Structure, chromosomal locus, and promoter analysis of the gene encoding the mouse helix-loop-helix factor HES-1. *J. Biol. Chem.* **269**, 5150–5156.
103. Takebayashi, K., Akazawa, C., Nakanishi, S., and Kageyama, R. (1995). Structure and promoter analysis of the gene encoding the mouse helix-loop-helix factor HES-5. *J. Biol. Chem.* **270**, 1342–1349.
104. Bessho, Y., Miyoshi, G., Sakata, R., and Kageyama, R. (2001). *Hes7*: A bHLH type repressor gene regulated by Notch and expressed in the presomitic mesoderm. *Genes Cells* **6**, 175–185.
105. Iso, T., Chung, G., Hamamori, Y., and Kedes, L. (2002). *HERP1* is a cell type-specific primary target of Notch. *J. Biol. Chem.* **277**, 6598–6607.
106. Nakagawa, O., McFadden, D. G., Nakagawa, M., Yanagisawa, H., Hu, T., Srivastava, D., and Olson, E. N. (2000). Members of the HRT family of basic-loop-helix proteins act as transcriptional repressors downstream of Notch signaling. *Proc. Natl. Acad. Sci. USA* **97**, 13655–13660.
107. Maier, M. and Gessler, M. (2000). Comparative analysis of the human and mouse *Hey1* promoter: *Hey* genes are new Notch target genes. *Biochem. Biophys. Res. Commun.* **275**, 652–660.
108. Oswald, F., Liptay, S., Adler, G., and Schmid, R. M. (1998). NF-kappaB2 is a putative target gene of activated Notch-1 via RBP-Jkappa. *Mol. Cell Biol.* **18**, 2077–2088.
109. Chen, Y., Fischer, W. H., and Gill, G. N. (1997). Regulation of the ERBB-2 promoter by RBPJkappa and NOTCH. *J. Biol. Chem.* **272**, 14110–14114.
110. Plaisance, S., Vanden Berghe, W., Boone, E., Fiers, W., and Haegeman, G. (1997). Recombination signal sequence binding protein Jkappa is constitutively bound to the NF-kappaB site of the interleukin-6 promoter and acts as a negative regulatory factor. *Mol. Cell Biol.* **17**, 3733–3743.
111. Kannabiran, C., Zeng, X., and Vales, L. D. (1997). The mammalian transcriptional repressor RBP (CBF1) regulates interleukin-6 gene expression. *Mol. Cell Biol.* **17**, 1–9.
112. Ronchini, C. and Capobianco, A. J. (2001). Induction of cyclin D1 transcription and CDK2 activity by Notch^{icL}: Implication for cell cycle diruption in transformation by Notch^{ic}. *Mol. Cell Biol.* **21**, 5925–5934.
113. Palmeirim, I., Henrique, D., Ish-Horowicz, D., and Pourquie, O. (1997). Avian hairy gene expression identifies a molecular clock linked to vertebrate segmentation and somitogenesis. *Cell* **91**, 639–648.

CHAPTER 295

The β-Catenin: LEF/TCF Signaling Complex: Bigger and Busier than Before

Reiko Landry and Katherine A. Jones
Regulatory Biology Laboratory, Salk Institute for Biological Studies, La Jolla, California

Introduction

The Wnt/Wingless pathways regulate genes that control cell fate, proliferation, and apoptosis, as well as tissue and organ identity in many organisms. The best characterized Wnt pathway targets the armadillo-repeat protein, β-catenin (β-cat), which functions as a dedicated coactivator of the LEF-1/TCF HMG enhancer-binding proteins. In this pathway, Wnt signaling through the Frizzled:LRP (lipoprotein-related protein)-5/8 receptors inactivates the APC:Axin:GSK3 complex that would otherwise phosphorylate β-cat, marking it for degradation through the ubiquitin-proteasome machinery. The newly stabilized hypophosphorylated β-cat enters the nucleus and interacts with LEF/TCF proteins to induce downstream target genes in a manner dependent on a strong transcription activation domain located at the C terminus of β-cat.

Interestingly, recent genetic and molecular studies in *Drosophila* reveal that nuclear Wnt transcription complexes also incorporate at least two other coactivators: Legless (human Bcl9) and a PHD-domain protein termed Pygopus. Both of these factors appear to be required universally for the induction of Wnt target genes, and function in a Wnt/Wg pathway-specific manner. The different components of the Wnt enhancer complex are likely to control many different steps in transcription, from the initial binding of the complex to chromatin, leading to activation and the subsequent disassembly of the complex or its reassembly with corepressors to return to the repressed state. Importantly, the failure to properly control β-cat underlies the oncogenic activation of the Wnt transcription pathway in colon cancers, melanomas, and other human cancers arising from mutations in β-cat or the Axin:APC complex. In this chapter we examine recent advances in understanding β-cat regulation and its function as a dedicated transcriptional coactivator in Wnt signaling.

Regulated Proteolytic Turnover of β-Cat

The level of free cytosolic β-cat is tightly controlled by phosphorylation-directed ubiquitination and proteolysis carried out by several different destruction pathways, only one of which is sensitive to Wnt signaling. The Wnt-responsive destruction pathway is outlined schematically in Fig. 1 (for recent reviews, see [1–5]). As part of a multiprotein scaffolding complex that contains the adenomatous polyposis coli (APC) tumor suppressor and Axin, β-cat undergoes a series of sequential N-terminal phosphorylation events that are essential for its recognition by the F-box β-TrCP ubiquitin ligase complex (see review in [6]). First, phosphorylation by casein kinase 1 (CK1α or CK1ε) at Ser-45 in the amino terminus [7,8] primes β-cat for sequential phosphorylation at residues 41, 37, and 33 within the amino terminal "destruction box" motif (D-**S**33-G-I-H-**S**37-G-A-T-**T**41-T-A-P-**S**45-L-S) by glycogen synthase kinase 3β (GSK3β). The β-TrCP ubiquitin ligase specifically recognizes S33- and S37-phosphorylated β-cat for ubiquitination [9]. In support of this two-kinase mechanism, oncogenic mutations in β-cat have been described at both the CK1 and GSK3β-phosphorylation sites [5,6]. However, important questions remain as to whether CK1α or CK1ε is the relevant isoform,

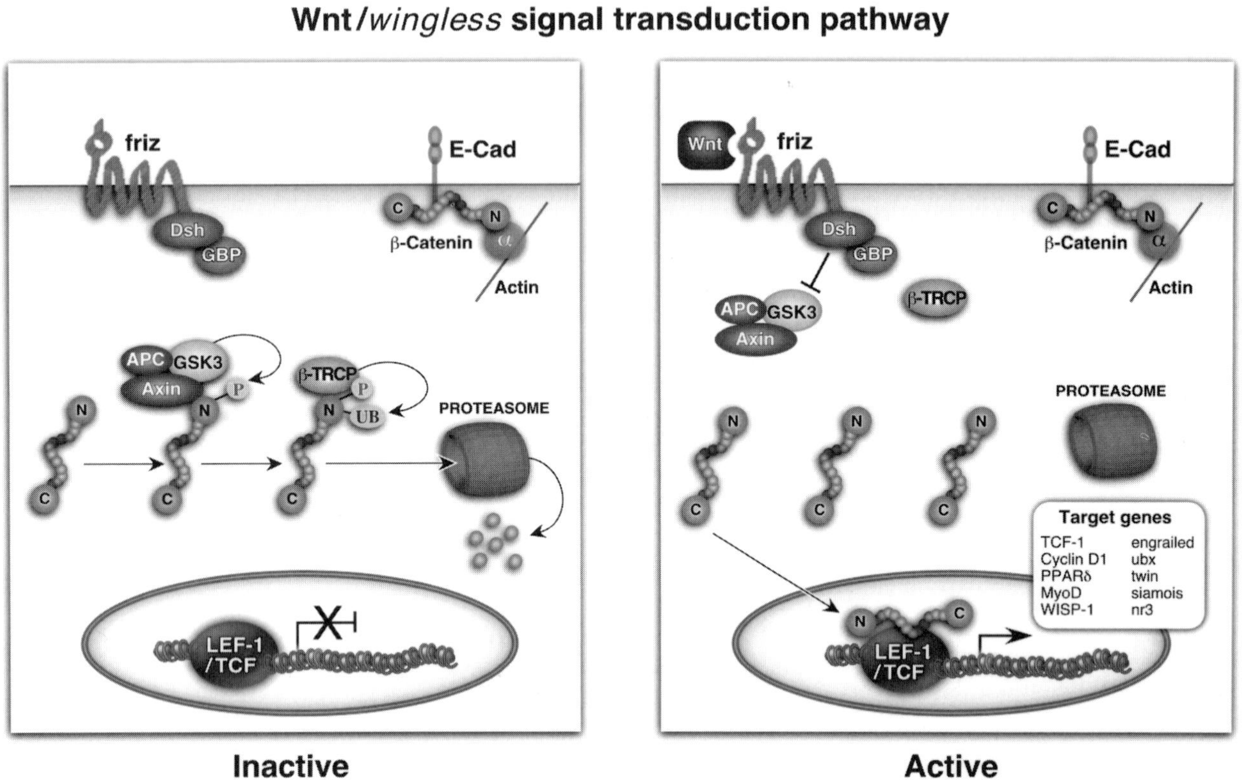

Figure 1 Overview of the canonical Wnt/Wg signaling pathway. A fraction of β-cat is tethered to the plasma membrane by E-cadherin and α-actin. Cytoplasmic free β-cat is rapidly phosphorylated by CK1 and GSK3β in the multiple-subunit Axin:APC complex, ubiquitinated by β-TrCP, and degraded by the ubiquitin-dependent proteosome degradation pathway (Inactive). The Wnt class of secretory glycoproteins binds to the Frizzled transmembrane receptors, which signals through Disheveled to inhibit the GSK3β phosphorylation. The stabilized and underphosphorylated β-cat enters the nucleus, binds to LEF1/TCF proteins, and activates the transcription from target genes (Active).

and how this protein kinase may become recruited to the Axin: APC complex. For example, CK1 has been reported to interact directly with Axin [7], or to be recruited to Axin through the ankyrin repeat protein Diversin [10], and alternative scenarios not involving Axin have also been suggested [6].

The central event in Wnt signaling is the inactivation of the APC:Axin scaffolding complex upon binding of Wnt ligands to the Frizzled/LRP coreceptor complex. The Disheveled (Dsh) protein plays a key intermediary role at this step, moving to the membrane to bind Axin and inhibit the activity of the complex. At present it is not clear whether the complex actually disassembles, or even if GSK3β is the direct target for inactivation within the complex. Binding of Dsh to GBP/Frat-1 might displace GSK3β from the complex or, alternatively, the mechanism may involve the Dsh-associated PP2C or Axin:APC-associated PP2A enzymes. Interestingly, phosphorylation at S45 is not affected by Wnt signaling, and S45-phosphorylated β-cat can enter the nucleus [7,8], indicating that β-cat can associate with the Axin:APC complex and undergo phosphorylation by CK1 during Wnt signaling, but fail to undergo the step that couples CK1 with GSK3β.

Additional mechanisms also contribute to the rapid proteolytic degradation of free β-cat in cells. Unexpectedly, a recent study found that presenilin (PS1) can also function to couple S45-phosphorylated forms of β-cat with GSK-3β, thereby modifying β-cat as a substrate for degradation [11]. In this case, phosphorylation at S45 is thought to be carried out by protein kinase A (PKA). Loss of presenilin enhances β-cat stability *in vivo* and disrupts phosphorylation by GSK3β, and activation of PKA by forskolin strongly enhances phosphorylation at S45 *in vivo* and *in vitro* [11]. In addition, the p53-inducible F-box protein Siah-1 can regulate nuclear β-cat levels in DNA-damaged cells (reviewed in [12]). However, neither of these destruction pathways is subject to regulation by Wnt signaling. Rather, these pathways seem to serve as alternative mechanisms to control β-cat molecules that escape the Axin:APC axis.

The subcellular distribution of the various regulatory components also plays a role in the dynamics of Wnt signaling. APC, like β-cat, is present in the nucleus and cytoplasm, and can actively shuttle between the two compartments (for recent reviews, see [2,13,14]). Phosphorylation by CK1ε may enhance binding of β-cat to LEF/TCF [7], whereas phosphorylation by GSK3β inhibits its transcriptional activity *in vivo* and *in vitro* [15,16]. Although most models indicate that degradation occurs in the cytoplasm, the β-TrCP ubiquitin ligase complex is predominantly nuclear. Nuclear β-TrCP exists in a stable complex with hnRNP-U, but dissociates in the presence of high-affinity substrates [17]. Thus β-TrCP must shuttle between compartments to ubiquitinate β-cat in the cytoplasm. In addition, β-cat may resemble other short-lived activators and undergo phosphorylation and ubiquitination in the nucleus as a consequence of transcriptional activation (for a review, see ref [18]). One possibility is that the differently modified forms of β-cat may arise separately in the nuclear and cytoplasmic compartments, yet undergo a similar proteolytic fate in the cytoplasm.

Regulation of the Wnt-Assembled Enhancer Complex in the Nucleus

Upon Wnt signaling, β-cat enters the nucleus and associates with LEF-1/TCF proteins (lymphocyte enhancer factor/T-cell factor; LEF-1, TCF1, TCF3, TCF4) (reviewed in [19,20]). These factors are monomeric high mobility group (HMG) sequence-specific DNA-binding proteins, frequently classified as architectural owing to their ability to bend DNA and recognize highly distorted DNA sequences. The LEF-1/TCF proteins are transcriptionally inactive on their own, but activate transcription strongly in association with β-cat. In T cells, the LEF/TCF proteins synergize with other T-cell factors to activate specific T-cell enhancers, but function independently of β-cat in this context [19,20].

In the absence of Wnt signaling, the LEF-1/TCF proteins repress Wnt gene expression through assembly with corepressors such as Groucho or the C-terminal binding protein (CtBP) [21–23], both of which recruit histone deacetylases to maintain chromatin in a repressed state (Fig. 2). Osa-containing Brahma chromatin remodeling complexes are also required for repression of Wnt target genes *in vivo* [24]. Interestingly, different LEF-1/TCF gene products and spliced isoforms appear to play dedicated roles in activation, inhibition, or repression of Wnt target genes [22,25,26]. For example, TCF3 functions as a potent repressor of Wnt signaling, even though it retains the ability to bind β-cat, potentially reflecting a higher intrinsic affinity for the CtBP corepressor [26]. The LEF/TCF proteins are also differentially regulated by Wnt signaling in transformed cells. Activation of the Wnt pathway up-regulates the expression of LEF-1 without affecting transcription from an adjacent promoter that directs expression of a truncated LEF-1 protein termed ΔNLEF. The ΔNLEF protein is unable to bind β-cat, and functions as a feedback inhibitor of Wnt signaling *in vivo* [27,28]. Therefore, the relative levels of different LEF/TCF gene products or isoforms can significantly affect the level of expression of Wnt target genes.

The activity of the Wnt enhancer components in the nucleus is also subject to regulation by posttranslational modifications and small regulatory peptides. For example, LEF-1 was recently shown to undergo modification by small ubiquitin-like modifier (SUMO) peptides, which causes it to become sequestered in an inactive form within PML/ND10 nuclear bodies [29]. Phosphorylation of LEF-1/TCF proteins by mitogen-activated protein kinases has been shown to block binding to DNA and may also affect proteolysis (reviewed in [2–4]). A small peptide inhibitor, ICAT, interacts directly with β-cat ARM repeats and is a strong and selective inhibitor of LEF-1: β-cat activity *in vivo* [30] and *in vitro* [16,30,31].

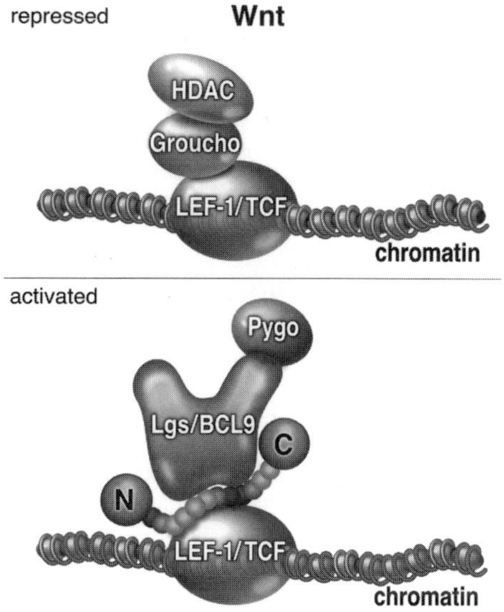

Figure 2 Repression of Wnt target genes is mediated by LEF/TCF isoforms that preferentially bind or function with Groucho or CtBP corepressors, and involves the recruitment of HDACs (top panel) as well as chromatin remodeling complexes (not shown). Upon signaling through the Wnt pathway, β-catenin assembles with LEF/TCF proteins and supplies a strong constitutive activation domain. The complex that forms on chromatin may also include Legless (human BCL9) and Pygopus [41–44].

ICAT can block the ability of β-cat to interact with LEF-1 [30,31] and can also interfere with transcriptional activation subsequent to binding of the complex to chromatin, potentially through interfering with the recruitment of transcriptional coactivators to the LEF-1:β-cat complex [16]. The ability of LEF-1/TCF to bind DNA can also be disrupted by the inhibitor of myogenic basic helix–loop–helix proteins, I-mfa [32], however this regulator, unlike ICAT, is not specific to the Wnt pathway.

Upon binding to LEF/TCF proteins, β-cat supplies a strong activation domain located within the C terminus of the protein. Although the specific target for the activation domain has not yet been identified, putative coactivators for β-cat include CBP/p300 [33–36], Brg-1-containing SWI/SNF chromatin remodeling complexes [37], pontin 52 [38], and the TATA-binding protein, TBP [39]. The central core of the β-cat molecule contains 12 armadillo (ARM) repeats that mediate mutually exclusive binding interactions with LEF/TCF, E-cadherin, APC, and other proteins required for Wnt signaling [40]. *In vitro*, recombinant β-cat interacts specifically with LEF-1 and the two proteins bind in a highly cooperative manner to nucleosomal templates [16]. In addition, β-cat strongly enhances transcription by LEF-1 in a chromatin-dependent manner on a minimal enhancer composed of reiterated LEF/TCF binding sites *in vitro*. An extended region of the C terminus of β-cat, including ARM repeats 11–12, appears to constitute the activation domain *in vitro*, and transcription *in vitro* is potently and selectively inhibited by a fragment of the CT-ARM domain, as well as by the physiological inhibitor ICAT. Although CBP/p300 is a positive coactivator of LEF-1:β-cat activity *in vivo* and *in vitro* and has been reported to interact directly with β-cat [34], nuclear p300 does not interact directly with β-cat in pull-down experiments, and, moreover, the purified LEF-1:β-cat complex is insufficient to direct p300-mediated histone acetylation on nucleosomal templates *in vitro* (R. Landry and K. A. Jones, unpublished data). Moreover, although binding of LEF-1:β-cat to chromatin requires ATP-dependent chromatin remodeling, purified SWI/SNF is unable to carry out this step alone *in vitro* [16]. Consequently the identity and ordered recruitment of specific coactivators by the β-cat transactivation domain remain to be clearly established.

Enter Pygopus and Legless (hBcl9)

Among the most important and exciting advances in the past year were the identification of Legless (Lgs) and Pygopus (Pygo) as two new essential nuclear components of the Wnt/Wg signaling pathway [41–44]. Basler and colleagues [41] identified Lgs as a genetic suppressor of an eye phenotype that arises from high levels of Wg signaling in *Drosophila*. Molecular cloning of Lgs revealed it to be a homolog of the Bcl-9 oncogene, which was originally identified by the translocation t(1;14) (q21;q32) from a patient with precursor-B-cell acute lymphoblastic leukemia, and has been implicated to play a role in the development of non-Hodgkin's lymphomas. Pygo was identified by several groups genetically as a suppressor of an activated ARM phenotype in flies [42–44], and was independently uncovered by the Basler group [41] in a two-hybrid screen for proteins that interact with Lgs. Mutant fly embryos that lack either Lgs or Pygo generate phenotypes that resemble loss of wingless. The functions of Pygo and Lgs appear to be largely specific to the Wg signaling pathway in flies, and both proteins are universally required in Wg-responsive cells and tissues throughout development [41–44]. Thus a Wnt/Wg signal cannot be transmitted in the absence of either protein *in vivo*.

Further genetic analyses revealed that both Lgs and Pygo function downstream of ARM in flies. Although nuclear levels of ARM/β-cat are somewhat reduced in cells lacking Pygo, the primary defect appears to be a loss in transcriptional activation mediated through the β-cat:LEF/TCF protein complex. Interestingly, Pygo is a strong activator of β-cat:LEF/TCF activity in transient expression experiments, whereas Lgs is comparatively inactive [41,42]. Therefore, Pygo may be more limiting than Lgs in cells or, alternatively, the normal function of Lgs may be effectively bypassed when Pygo is expressed at very high levels in the cell.

Pygopus is a large nuclear protein that contains an N-terminal nuclear localization sequence and C-terminal plant homeotic domain (PHD) motif, also known as a leukemia-associated protein (LAP) domain [41–44]. The PHD domain is essential for Pygo function, although its role in the transactivation mechanism remains unclear. PHD domains coordinate zinc through a C4HC3 motif that is distantly related to the RING and FYVE finger domains. PHD domains are frequently found in chromatin-binding proteins; however, no specific conserved function has been ascribed to this motif. Interestingly, a PHD-containing protein, Pf1, appears to link Groucho/TLE corepressor proteins with the Sin3A histone deacetylase complex [45]. The PHD domain of Pf1 is critical for its function, but does not mediate binding to Sin3A, and its role is unclear. Similarly, the PHD domain of Cti6 is required for switching of coactivator and corepressor complexes, but does not mediate protein–protein interactions between the two complexes [46]. The PHD domain of the unrelated MEKK1 protein functions enzymatically as an E3 ubiquitin ligase required for the proteolytic turnover of ERK1 [47], whereas the PHD domain of p300 is required for the structural integrity of the histone acetyltransferase domain [48].

In the model proposed by Basler *et al.* [41], the PHD domain of Pygo mediates a direct interaction with Lgs. Moreover, the N-terminal half of Lgs was reported to interact both with Pygo and with ARM/β-cat, through interactions involving the HD1 and HD2 domains of Lgs, respectively (Fig. 3). A mutant form of Lgs protein lacking HD2 functions as a strong dominant-negative inhibitor of Wg signaling *in vivo*. These observations indicate that Pygo may be tethered to the β-cat:LEF/TCF complex through binding to Lgs [41]. In this scenario, Pygo is postulated to stimulate transcription through an activation domain (NHD) located at the amino terminus of the protein [41]. An alternative possibility, proposed by Bienz *et al.* [42], suggests that the PHD domain in Pygo facilitates the binding or transcription activation function of the β-cat:LEF/TCF complex on chromatin. The Bienz group [42] suggests that Pygo may enter the complex indirectly or through binding to TCF proteins, and they further indicate that the conserved amino terminal sequences of Pygo most likely constitute a nuclear localization sequence. Consequently, the exact details concerning the function and mode of recruitment of Lgs and Pygo to the β-cat:LEF-1 complex remain, for the moment, open to debate. Taken together, these recent findings constitute a major advance in the field, particularly because they radically alter previous notions about the composition of the multiple-subunit enhancer complex that assembles on chromatin in response to Wnt signaling. The discovery of Lgs and Pygo also provides exciting new targets for anticancer drugs that may be able to selectively block Wnt signaling at target genes. Looking back at these remarkable accomplishments, we can anticipate that further mechanistic studies of Lgs, Pygo, and β-cat will continue to lead the field in interesting and unanticipated new directions.

Perspectives

We have presented a summary of the current understanding of the key regulated events in the Wnt/Wg pathway. One of the key challenges for the future will be to characterize biochemically the different multiple-subunit complexes that engage β-catenin, from the Axin:APC complex

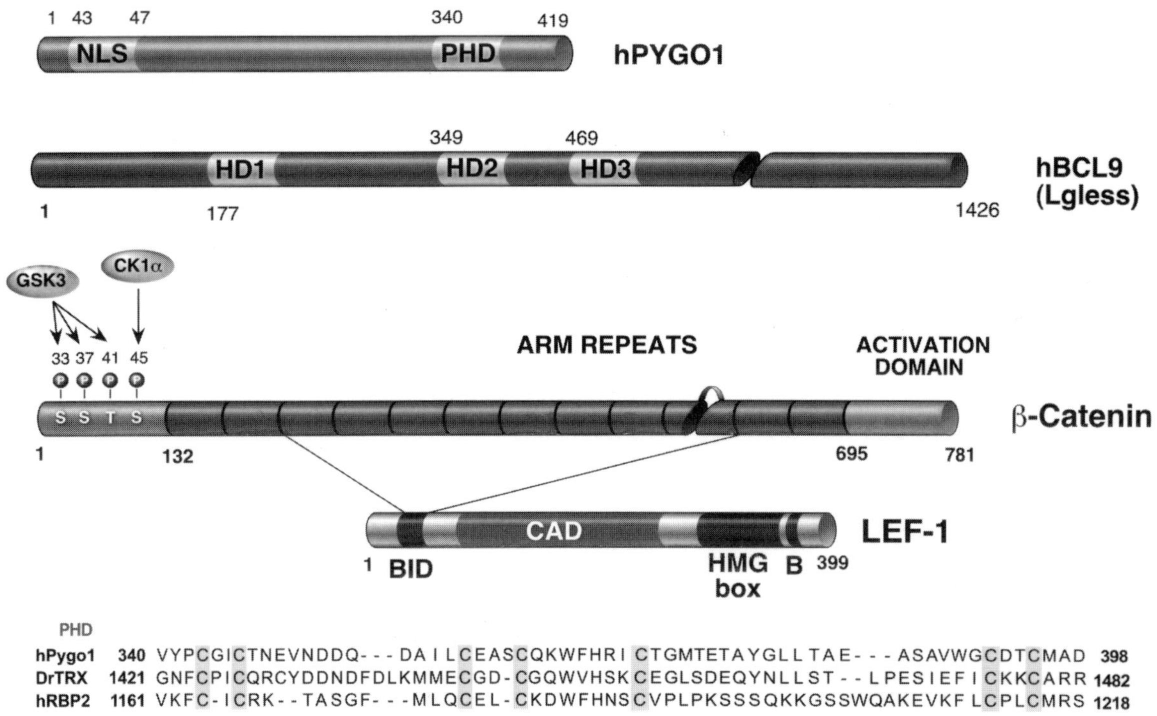

Figure 3 Domain structures of the regulatory components of the Wnt enhancer complex. The LEF-1/TCF proteins mediate sequence-specific binding to DNA through their HMG domain, and recruit β-catenin through the N-terminal β-cat interaction domain (BID). LEF-1 recognizes Armadillo (ARM) repeats 3–10, and structural studies indicate that the interaction does not significantly alter the conformation of β-cat [40]. Emerging studies indicate a critical role for the nuclear Pygo and Lgs in Wnt-directed transcription; however, it is not clear how these proteins are recruited to the β-cat:LEF/TCF complex. A comparative alignment of the critical PHD finger domain of human Pygopus 1 with *Drosophila* Trithorax and human RBP2 is shown schematically at the bottom. Sequences were aligned by the Clustal V method of multiple sequence alignment using DNASTAR.

to the nuclear shuttling complexes and the functional transcription complexes that assemble on Wnt target genes. Biochemical studies of the functions of these various regulatory proteins would also contribute to a deeper understanding of the dynamics of Wnt-initiated transcription. Analyses of the ordered recruitment of regulatory components by chromatin immunoprecipitation and other approaches will also reveal important facets of the dynamics of activation and repression. Continued efforts to build on the outstanding structural information already available for β-catenin and some of the components of the Wnt pathway will define the critical protein contacts in detail and are essential for efforts to develop specific chemical inhibitors of Wnt signaling. It will be interesting to learn whether these regulatory subunits are modified posttranslationally, and how the turnover of β-cat is regulated in the nucleus. Clearly many more surprises await in the future for those interested in understanding the mechanism of the Wnt/Wg regulatory network.

Acknowledgments

We apologize to our colleagues whose work could not be covered or cited directly due to the space constraints of this review, and we thank Kim Emerson for administrative help and Jamie Simon for assistance with the figures. Studies on Wnt transcription in our laboratory are supported by CA054418.

References

1. Polakis, P. (2000). Wnt signaling and cancer. *Genes Dev.* **14**, 1837–1851.
2. Bienz, M. and Clevers, H. (2000). Linking colorectal cancer to Wnt signaling. *Cell* **103**, 311–320.
3. Huelsken, J. and Birchmeier, W. (2001). New aspects of Wnt signaling pathways in higher vertebrates. *Curr. Opin. Genet. Dev.* **11**, 547–553.
4. Cadigan, K. M. (2002). Wnt signaling—20 years and counting. *Trends Genet.* **18**, 340–342.
5. Moon, R. T., Bowerman, B., Boutros, M., and Perrimon, N. (2002). The promise and perils of Wnt signaling through β-catenin. *Science* **296**, 1644–1646.
6. Ding, Y. and Dale, T. (2002). Wnt signal transduction: Kinase cogs in a nano-machine? *Trends Biochem. Sci.* **27**, 327–329.
7. Amit, S., Hatzubai, A., Birman, Y., Andersen, J. S., Ben-Shushan, E., Mann, M., Ben-Neriah, Y., and Alkalay, I. (2002). Axin-mediated CKI phosphorylation of β-catenin at Ser 45: A molecular switch for the Wnt pathway. *Genes Dev.* **16**, 1066–1076.
8. Liu, C., Li, Y., Semenov, M., Han, C., Baeg, G. H., Tan, Y., Zhang, Z., Lin, X., and He, X. (2002). Control of β-catenin phosphorylation/degradation by a dual-kinase mechanism. *Cell* **108**, 837–847.
9. Winston, J. T., Strack, P., Beer-Romeron, P., Chu, C. Y., Elledge, S. J., and Harper, J. W. (1999). The SCFβ-TRCP-ubiquitin ligase complex associates specifically with phosphorylation destruction motifs in IκBα and β-catenin and stimulates IκBα ubiquitination *in vitro*. *Genes Dev.* **13**, 270–283.
10. Schwarz-Romond, T., Asbrand, C., Bakkers, J., Kuhl, M., Schaeffer, H. J., Huelsken, J., Behrens, J., Hammerschmidt, M., and Birchmeier, W. (2002). The ankyrin repeat protein Diversin recruits Casein kinase I epsilon to the β-catenin degradation complex and acts in both canonical Wnt and Wnt/JNK signaling. *Genes Dev.* **16**, 2073–2084.

11. Kang, D. E., Soriano, S., Xia, X., Eberhart, C. G., De Strooper, B., Zheng, H., and Koo, E. H. (2002). Presenilin couples the paired phosphorylation of β-catenin independent of Axin: Implications for β-catenin activation in tumorigenesis. *Cell* **110**, 751–762.
12. Polakis, P. (2001). More than one way to skin a catenin. *Cell* **105**, 563–566.
13. Henderson, B. R. and Fagotto, F. (2002). The ins and outs of APC and β-catenin nuclear transport. *EMBO Rep.* **3**, 834–839.
14. Bienz, M. (2002). The subcellular destinations of APC proteins. *Nat. Rev. Mol. Cell. Biol.* **3**, 328–338.
15. Staal, F. J., van Noort, M., Strous, G. J., and Clevers, H. C. (2002). Wnt signals are transmitted through N-terminally dephosphorylated β-catenin. *EMBO Rep.* **3**, 63–68.
16. Tutter, A. V., Fryer, C. J., and Jones, K. A. (2001). Chromatin-specific regulation of LEF-1-β-catenin transcription activation and inhibition *in vitro*. *Genes Dev.* **15**, 3342–3354.
17. Davis, M., Hatzubai, A., Andersen, J. S., Ben-Shushan, E., Fisher, G. Z., Yaron, A., Bauskin, A., Mercurio, F., Mann, M., and Ben-Neriah, Y. (2002). Pseudosubstrate regulation of the SCF(beta-TrCP) ubiquitin ligase by hnRNP-U. *Genes Dev.* **16**, 439–451.
18. Conaway, R. C., Brower, C. S., and Conaway, J. W. (2002). Emerging roles of ubiquitin in transcription regulation. *Science* **296**, 1254–1258.
19. Hurlstone, A. and Clevers, H. (2002). T-cell factors: Turn-ons and turn-offs. *EMBO J.* **21**, 2303–2311.
20. van de Wetering, M., de Lau, W., and Clevers, H. (2002). WNT signaling and lymphocyte development. *Cell* **109**, S13–S19.
21. Cavallo, R. A., Cox, R. T., Moline, M. M., Roose, J., Polevoy, G. A., Clevers, H., Peifer, M., and Bejsovec, A. (1998). *Drosophila* Tcf and Groucho interact to repress Wingless signalling activity. *Nature* **395**, 604–608.
22. Roose, J., Molenaar, M., Peterson, J., Hurenkamp, J., Brantjes, H., Moerer, P., van de Wetering, M., Destree, O., and Clevers, H. (1998). The *Xenopus* Wnt effector XTcf-3 interacts with Groucho-related transcriptional repressors. *Nature* **395**, 608–612.
23. Brannon, M., Brown, J. D., Bates, R., Kimelman, D., and Moon, R. T. (1999). XCtBP is a XTcf-3 corepressor with roles throughout *Xenopus* development. *Development* **126**, 3159–3170.
24. Collins, R. and Treisman, J. (2000). Osa-containing Brahma chromatin remodeling complexes are required for the repression of Wingless target genes. *Genes Dev.* **14**, 3140–3152.
25. Kim, C. H., Oda, T., Itoh, M., Jiang, D., Artinger, K. B., Chandrasekharappa, S. C., Driever, W., and Chitnis, A. B. (2000). Repressor activity of Headless/Tcf3 is essential for vertebrate head formation. *Nature* **407**, 913–916.
26. Merrill, B. J., Gat, U., DasGupta, R., and Fuchs, E. (2001). Tcf3 and Lef1 regulate lineage differentiation of multipotent stem cells in skin. *Genes Dev.* **15**, 1688–1705.
27. Hovanes, K., Li, T., Munguia, J., Truong, T., Milovanovic, T., Lawrence Marsh, J., Holcombe, R., and Waterman, M. L. (2001). Beta-catenin-sensitive isoforms of lymphoid enhancer factor-1 are selectively expressed in colon cancer. *Nat. Genet.* **28**, 53–57.
28. Roose, J., Huls, G., van Beest, M., Moerer, P., van der Horn, K., Goldschmeding, R., Logtenberg, T., and Clevers, H. (1999). Synergy between tumor suppressor APC and the β-catenin-Tcf4 target Tcf1. *Science* **285**, 1923–1926.
29. Sachdev, S., Bruhn, L., Sieber, H., Pichler, A., Melchior, F., and Grosschedl, R. (2001). PIASy, a nuclear matrix-associated SUMO E3 ligase, represses LEF1 activity by sequestration into nuclear bodies. *Genes Dev.* **15**, 3088–3103.
30. Tago, K., Nakamura, T., Nishita, M., Hyodo, J., Nagai, S., Murata, Y., Adachi, S., Ohwada, S., Morishita, Y., Shibuya, H., and Akiyama, T. (2000). Inhibition of Wnt signaling by ICAT, a novel β-catenininteracting protein. *Genes Dev.* **14**, 1741–1749.
31. Graham, T. A., Clements, W. K., Kimelman, D., and Xu, W. (2002). The crystal structure of the β-catenin/ICAT complex reveals the inhibitory mechanism of ICAT. *Mol. Cell* **10**, 563–571.
32. Snider, L., Thirlwell, H., Miller, J., Moon, R., Groudine, M., and Tapscott, S. (2001). Inhibition of Tcf3 binding by I-mfa domain proteins. *Mol. Cell. Biol.* **21**, 1866–1873.
33. Hecht, A., Vleminckx, K., Stemmler, M., van Roy, F., and Kemler, R. (2000). The p300/CBP acetyltransferases function as transcriptional coactivators of β-catenin in vertebrates. *EMBO J.* **19**, 1839–1850.
34. Miyagishi, M., Fujii, R., Hatta, M., Yoshida, E., Araya, N., Nagafuchi, A., Ishihara, S., Nakajima, T., and Fukamizu, A. (2000). Regulation of Lef-mediated transcription and p53-dependent pathway by associating β-catenin with CBP/p300. *J. Biol. Chem.* **275**, 35170–35175.
35. Sun, Y., Kolligs, F., Hottiger, M., Mosavin, R., Fearon, E., and Nabel, G. (2000). Regulation of β-catenin transformation by the p300 transcriptional coactivator. *Proc. Natl. Acad. Sci. (USA)* **97**, 12613–12618.
36. Takemaru, K. and Moon, R. (2000). The transcriptional coactivator CBP interacts with β-catenin to activate gene expression. *J. Cell Biol.* **149**, 249–254.
37. Barker, N., Hurlstone, A., Musisi, H., Miles, A., Bienz, M., and Clevers, H. (2001). The chromatin remodeling factor Brg-1 interacts with β-catenin to promote target gene activation. *EMBO J.* **20**, 4935–4943.
38. Bauer, A., Huber, O., and Kemler, R. (1998). Pontin52, an interaction partner of beta-catenin, binds to the TATA box binding protein. *Proc. Natl. Acad. Sci. USA* **95**, 14787–14792.
39. Hecht, A., Litterst, C. M., Huber, O., and Kemler, R. (1999). Functional characterization of multiple transactivating elements in beta-catenin, some of which interact with the TATA-binding protein *in vitro*. *J. Biol. Chem.* **274**, 18017–18025.
40. Daniels, D. L., Eklof Spink, K., and Weis, W. I. (2001). β-catenin: Molecular plasticity and drug design. *Trends Biochem. Sci.* **26**, 672–678.
41. Kramps, T., Peter, O., Brunner, E., Nellen, D., Froesch, B., Chatterjee, S., Murone, M., Zullig, S., and Basler, K. (2002). Wnt/wingless signaling requires BCL9/legless-mediated recruitment of Pygopus to the nuclear β-catenin-TCF complex. *Cell* **109**, 47–60.
42. Thompson, B., Townsley, F., Rosin-Arbesfeld, R., Musisi, H., and Bienz, M. (2002). A new nuclear component of the Wnt signalling pathway. *Nat Cell Biol.* **4**, 367–373.
43. Parker, D. S., Jemison, J., and Cadigan, K. M. (2002). Pygopus, a nuclear PHD-finger protein required for Wingless signaling in *Drosophila*. *Development* **129**, 2565–2576.
44. Belenkaya, T. Y., Han, C., Standley, H. J., Lin, X., Houston, D. W., Heasman, J., and Lin, X. (2002). Pygopus encodes a nuclear protein essential for wingless/Wnt signaling. *Development* **129**, 4089–4101.
45. Yochum, G. S. and Ayer, D. E. (2001). Pf1, a novel PHD zinc finger protein that links the TLE corepressor to the mSin3A-histone deacetylase complex. *Mol. Cell. Biol.* **21**, 4110–4118.
46. Papamichos-Chronakis, M., Petrakis, T., Ktistaki, E., Topalidou, I., and Tzamarias, D. (2002). Cti6, a PHD domain protein, bridges the Cyc8-Tup1 corepressor and the SAGA coactivator to overcome repression at GAL1. *Mol. Cell.* **9**, 1297–1305.
47. Lu, Z., Xu, S., Joazeiro, C., Cobb, M. H., and Hunter, T. (2002). The PHD domain of MEKK1 acts as an E3 ubiquitin ligase and mediates ubiquitination and degradation of ERK1/2. *Mol Cell.* **9**, 945–956.
48. Kalkhoven, E., Teunissen, H., Houweling, A., Verrijzer, C. P., and Zantema, A. (2002). The PHD type zinc finger is an integral part of the CBP acetyltransferase domain. *Mol. Cell Biol.* **22**, 1961–1970.

CHAPTER 296

Cubitus Interruptus

Sarah M. Smolik[1] and Robert A. Holmgren[2]
[1]*Department of Medicine, Oregon Health Sciences University, Portland, Oregon*
[2]*Department of Biochemistry Molecular Biology and Cell Biology, Robert H. Lurie Comprehensive Cancer Center, Northwestern University, Evanston, Illinois*

Introduction

A striking example of multiple signaling pathways eliciting their effects through the regulation of a single transcription factor is seen in the coregulation of embryonic segmental patterning and limb development in Drosophila by the *hedgehog* (*hh*) and protein kinase A (PKA) signal transduction pathways. The Hh signaling cascade activates a series of genes through the transcription factor Cubitus interruptus (Ci). The PKA signaling pathway suppresses the expression of *hh*-responsive genes in cells that do not receive a Hh signal by phosphorylating the 155-kDa form of Ci and targeting it for proteolysis to a 75-kDa repressor form (Ci[rep]). Paradoxically, PKA can further activate Ci-mediated transcription in cells that do receive a Hh signal. Thus both the activation and repression of *hh*-responsive transcription is modulated by the PKA signaling pathway.

Protein Structure and Expression Patterns of Ci

A summary of the Ci functional domains is illustrated in Fig. 1. Both the *ci* gene and its vertebrate homologs *Gli1–3*, encode proteins with five highly homologous, unique, and tandemly repeated zinc fingers that bind DNA through a consensus DNA-binding site [1]. The crystal structure of the Gli1 zinc fingers bound to the DNA-binding site shows that zinc finger 1 does not contact the DNA, whereas the remaining four fingers wrap around the DNA, with fingers 4 and 5 making most of the base contact with the DNA [2]. This conformation might explain the higher homology shared by zinc fingers 3 through 5 among the members in the Gli gene family.

The N terminus of Ci has a repressor function; expression of Ci that is truncated C-terminal to the zinc-finger region in the anterior compartment of the fly embryonic segments causes the down-regulation of *ci* target genes [3]. Ci deleted for the N terminus can function as a transcriptional activator of *ci* target genes when introduced into Drosophila embryos [3]. Mutations in Ci that delete amino acids 611–760 result in proteins that cannot be proteolyzed in response to PKA phosphorylation [4,5]. Likewise, the region between amino acids 675 and 860 is required to maintain Ci in the cytoplasm [6].

Expression of the *ci* gene is zygotic and has a complex pattern during development that includes all embryonic tissues [7]. For this chapter, it is important to note that coincident with the onset of ectodermal segmentation, both the message and the protein are restricted to the anterior compartment of each segment by the *engrailed* (*en*) gene, resulting in 15 broad metamerically repeating stripes. Just prior to germ band shortening, the protein and message expression patterns no longer coincide. Although transcripts are still uniformly distributed throughout the anterior compartment of each segment, the stripe of Ci 155 within each segment is graded such that higher levels of protein are detected in the rows of cells adjacent to the Hh-producing cells of the posterior compartment. Lower levels are seen in the middle of the Ci expressing stripe. This pattern of expression also exists in imaginal disks where *ci* is uniformly transcribed in the entire anterior compartment and higher levels of CI 155 are detected along the anterior/posterior (A/P) boundary.

Regulation of Ci by Hedgehog

Genetic epistasis experiments and cell culture experiments from a variety of laboratories have provided substantial evidence that *hh* signaling establishes A/P polarity in the

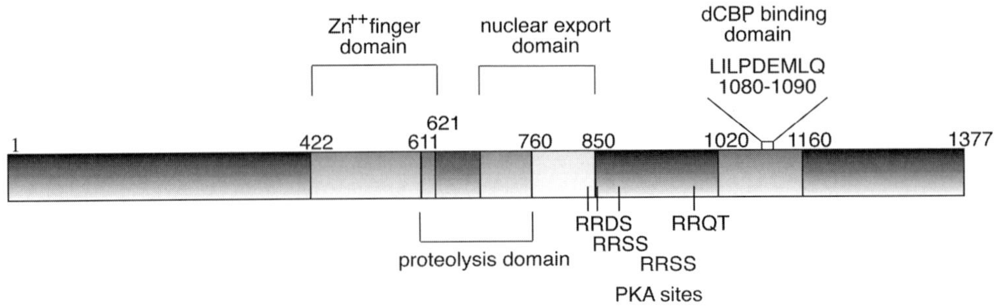

Figure 1 Cubitus interruptus.

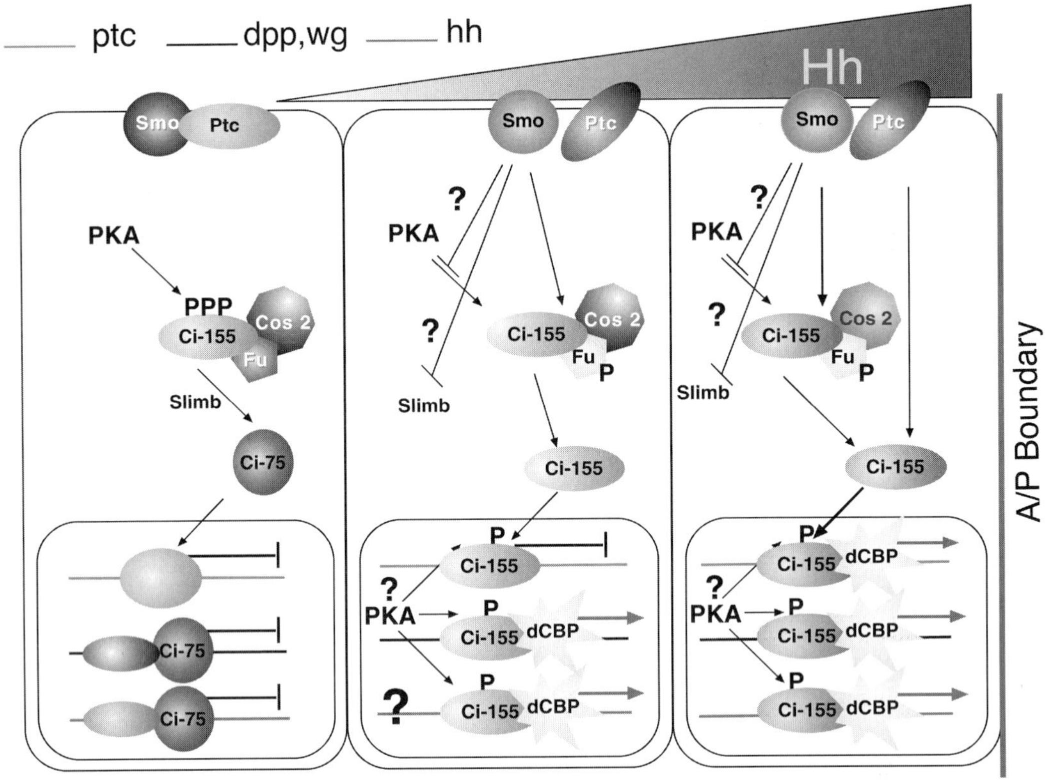

Figure 2 Hedgehog and PKA signaling regulate Ci activity. (A color representation of this figure is available on the CD version of the *Handbook of Cell Signaling*.)

embryonic segments and the developing wing primordia by regulating the transcriptional activity of Ci [3,8–10]. A model summarizing this work is illustrated in Fig. 2. The evidence to date shows that the transmembrane protein Ptc is the receptor for Hh and, in the absence of Hh binding, inhibits the *hh*-signaling pathway [11,12]. The binding of Hh to Ptc relieves the inhibition of Ptc on Smo, a protein that spans the bilayer seven times and the positive transducer of *hh* signaling [13,14]. Hh signaling antagonizes the proteolysis of Ci 155 and converts it to Ci[act], an activator of transcription, in the anterior cells abutting the A/P boundary [4–6]. In the anterior cells outside the range of Hh activity, proteolysis of the 155-kDa Ci generates the 75-kDa repressor, Ci[rep], by cleaving the sequences C-terminal to zinc-finger domain [4–6]. Antibodies that detect both Ci 155 and Ci[rep] stain the entire anterior compartment of the embryonic segments and imaginal disks [4]. At the cellular level, the antibody staining is distributed evenly in both the nucleus and cytoplasm. In contrast, Ci 155 resides primarily in the cytoplasm. The levels are very low in the cells of the anterior compartment that do not receive a *hh* signal, and very high along the A/P boundary that contacts the *hh*-producing cells [7].

The mechanism involved in the *hh* regulation of full-length Ci levels is currently under intense investigation. It has been shown that a domain C-terminal to the zinc-finger domain and N-terminal to the PKA phosphorylation sites is responsible for the retention of the Ci 155 in the cytoplasm [15]. In the absence of Hh signaling, a complex containing the Cos2, Fu, and Su(fu) proteins sequesters Ci in the cytoplasm. In response to Hh signaling, Ci appears to be released from this complex and is free to shuttle in and out of the nucleus [6,15–18]. The integrity of the complex appears critical to the retention of

Ci in the cytoplasm. Either Fu or Su(fu) can be removed without releasing Ci, but if both are missing, Ci is no longer sequestered [19]. The Fu protein plays two opposing roles in this process. The regulatory domain is in part responsible for retaining Ci in the cytoplasm, whereas the kinase function is required for the release of Ci in the presence of Su(fu) [19]. The Ci protein contains both a nuclear localization site (NLS) and a nuclear export signal, both of which account for its shuttling [15,20]. Processing of Ci 155 into the Ci 75 repressor removes the nuclear export signal but not the NLS, resulting in the stable accumulation of Ci 75 in the nucleus.

Regulation of Ci by PKA

PKA negatively regulates *ci* activity in cells that do not receive an *hh* signal by targeting Ci 155 for proteolysis to Ci[rep] [21–23]. Loss of PKA function or inhibition of PKA activity in these cells increases the levels of Ci 155 in both embryos and disks and results in the misregulation of some *hh*-responsive genes [24–28]. Ci has four consensus PKA phosphorylation sites in its C terminus that can be phosphorylated by PKA (Fig. 1). Substitution of the serine with alanine in any one of the first three PKA sites inhibits Ci proteolysis and increases the levels of Ci 155, suggesting that Ci is the direct target of PKA regulation. Recent evidence suggests that phosphorylation of Ci by PKA is a prerequisite for subsequent phosphorylation of Ci by glycogen synthase kinase 3(GSK3) and casein kinase 1 (CK1) and that all three phosphorylation events target Ci for proteolysis to Ci[rep] [29,30]. The *Drosophila slimb* gene is thought to mediate Ci proteolysis and encodes an F-box/WD40-repeat protein that is related to the yeast protein cdc4p, which is involved in targeting cell cycle regulators to the ubiquitination-mediated degradation pathway [31].

Although PKA promotes the generation of Ci[rep] and is thus a negative regulator of *hh*-responsive gene activation, it is an activator of Ci[act] in cells that receive an Hh signal [18,32]. It is not known how the *hh* signal transduction cascade inhibits the phosphorylation or proteolysis of Ci 155 in Hh-stimulated cells, but it is clear that the suppression of proteolysis is not sufficient to activate Ci 155 for all of its target genes [5,20]. For example, Ci that is mutant for the proteolytic cleavage site remains full length but cannot transactivate the Ci target gene *ptc* in the wing disk without an Hh signal. Ci that is mutant for the three PKA sites is a poor activator of *ptc* expression in the wing disk. However, in embryos, Ci that is mutant for the three PKA sites can activate the Ci target gene *wg* and is independent of Hh signaling [22]. Thus, PKA is required to activate Ci[act] for *ptc* expression, whereas Ci 155 is sufficient for *wg* expression in embryos.

Ci Transcriptional Regulation

The Ci target genes identified to date include the following: *wg*, a morphogen that signals to the cells posterior to the A/P boundary and stimulates *hh* expression; *decapentaplegic* (*dpp*), a long-range signaling morphogen of the TGF-β family; and *ptc*, a component of a negative feedback loop and *hh* itself. The context of the Ci target gene promoters defines overlapping but nonidentical sets of targets for Ci[rep] and Ci[act] even though they bind to and regulate the same consensus sequence [33]. Thus, although Ci[rep] and Ci 155 regulate both *wg* and *dpp*, Ci[rep] regulates *hh* and Ci[act] regulates *ptc* [5]. Furthermore, the different Ci targets are sensitive to the level of Hh signaling. Ci[act] is only found in the cells that receive the highest levels of Hh signaling. In the wing imaginal disk, where the targets of Hh signal transduction have been studied in most detail, this occurs in anterior cells directly adjacent to the compartment boundary. These boundary cells have high-level *ptc* expression, late anterior compartment expression of *en*, and activation of *collier*. On the other hand, expression of *dpp*, which is regulated by Ci 155 and Ci[rep], is observed in a much broader stripe. Why do certain targets require Ci[act] while others respond to Ci 155? This is most likely a consequence of the other components assembled on the particular enhancers.

The interaction of Ci with other transcription factors is also responsible for regional and tissue-specific responses to Hh signal transduction. For example the regulation of *dpp* imaginal disk expression is remarkably complex. The continuous *dpp* stripe is actually composed of a series of subelements, each of which is regulated by a distinct enhancer [34]. The *HO* enhancer directs *dpp* expression in the wing pouch because it pairs binding of Ci with that of Arm/dTCF and Vg/Sd [35]. The combined action of these factors is responsible for the final expression pattern. For efficient transactivation, Ci binding must be coupled with the binding of other factors [36]. This makes very good sense, because Hh signaling is utilized in many different contexts and must elicit distinct responses in different tissues. Ci's requirement for additional bound factors may be a consequence of its shuttling out of the nucleus. Even in cells receiving high-level Hh signaling, the nuclear concentration of Ci is low. Therefore, high-level occupancy of an enhancer by Ci may only occur when a prebound factor is present to facilitate Ci binding. Alternatively, it is known that activation of transcription requires the recruitment of multiple components. It is possible that binding of Ci brings certain set of factors to the promoter while the regional and tissue-specific transcription factors recruit complementary components. One coactivator that Ci 155 requires to activate *wg* is Drosophila CREB-binding protein, dCBP [37]. The binding domain for dCBP is in the activation domain and it is thought that Ci[rep] is a repressor, at least in part, because it cannot recruit dCBP to the promoter. However, it is not known whether Ci[act] or all Ci-activated promoter complexes require dCBP for their activity.

A number of questions regarding Ci regulation remain. How does Hh signaling antagonize the proteolysis of Ci? Does it inhibit the different kinases or does it activate a phosphatase? What are the biochemical events that control the release of Ci? Do they involve modification of Ci or the other components of the complex? Finally, what are the specific interactions of Ci[rep], Ci 155, and Ci[act] at enhancer complexes, and how do these interactions lead to distinct, Hh-signaling responses? Further studies will certainly bring us closer to an understanding of the intricacies involved in the regulation of development by signal transduction cascades.

References

1. Kinzler, K. W. and Vogelstein, B. (1990). The GLI gene encodes a nuclear protein which binds specific sequences in the human genome. *Mol. Cell Biol.* **10**, 634–642.
2. Pavletich, N. P. and Pabo, C. O. (1993). Crystal structure of a five-finger GLI-DNA complex: new perspectives on zinc fingers. *Science* **261**, 1701–1707.
3. Hepker, J. *et al.* (1997). *Drosophila* cubitus interruptus forms a negative feedback loop with patched and regulates expression of Hedgehog target genes. *Development* **124**, 549–558.
4. Aza-Blanc, P. *et al.* (1997). Proteolysis that is inhibited by Hedgehog targets Cubitus interruptus protein to the nucleus and converts it to a repressor. *Cell* **89**, 1043–1053.
5. Methot, N. and Basler, K. (1999). Hedgehog controls limb development by regulating the activities of distinct transcriptional activator and repressor forms of Cubitus interruptus. *Cell* **96**, 819–831.
6. Chen, C. H. *et al.* (1999). Nuclear trafficking of Cubitus interruptus in the transcriptional regulation of Hedgehog target gene expression. *Cell* **98**, 305–316.
7. Motzny, C. K. and Holmgren, R. (1995). The *Drosophila* Cubitus interruptus protein and its role in the wingless and hedgehog signal transduction pathways. *Mech. Dev.* **52**, 137–150.
8. Alexandre, C., Jacinto, A., and Ingham, P. W. (1996). Transcriptional activation of hedgehog target genes in *Drosophila* is mediated directly by the Cubitus interruptus protein, a member of the GLI family of zinc finger DNA-binding proteins. *Genes & Development* **10**, 2003–2013.
9. Dominguez, M. *et al.* (1996). Sending and receiving the hedgehog signal: Control by the *Drosophila* Gli protein cubitus interruptus. *Science* **272**, 1621–1625.
10. Methot, N. and Basler, K. (2001). An absolute requirement for Cubitus interruptus in Hedgehog signaling. *Development* **128**, 733–742.
11. Chen, Y. and Struhl, G. (1996). Dual roles for patched in sequestering and transducing Hedgehog. *Cell* **87**, 553–563.
12. Chen, Y. and Struhl, G. (1998). *In vivo* evidence that Patched and Smoothened constitute distinct binding and transducing components of a Hedgehog receptor complex. *Development* **125**, 4943–4948.
13. Alcedo, J. *et al.* (1996). The *Drosophila* smoothened gene encodes a seven-pass membrane protein, a putative receptor for the hedgehog signal. *Cell* **86**, 221–232.
14. van den Heuvel, M. and Ingham, P. W. (1996). Smoothened encodes a receptor-like serpentine protein required for hedgehog signalling. *Nature* **382**, 547–551.
15. Wang, G. *et al.* (2000). Interactions with Costal2 and suppressor of fused regulate nuclear translocation and activity of cubitus interruptus. *Genes Dev.* **14**, 2893–2905.
16. Methot, N. and Basler, K. (2000). Suppressor of fused opposes hedgehog signal transduction by impeding nuclear accumulation of the activator form of cubitus interruptus. *Development* **127**, 4001–4010.
17. Stegman, M. A. *et al.* (2000). Identification of a tetrameric hedgehog signaling complex. *J. Biol. Chem.* **275**, 21809–21812.
18. Wang, Q. T. and Holmgren, R. A. (2000). Nuclear import of cubitus interruptus is regulated by hedgehog via a mechanism distinct from Ci stabilization and Ci activation. *Development* **127**, 3131–3139.
19. Lefers, M. A., Wang, Q. T., and Holmgren, R. A. (2001). Genetic dissection of the *Drosophila* cubitus interruptus signaling complex. *Dev. Biol.* **236**, 411–420.
20. Wang, Q. T. and Holmgren, R. A. (1999). The subcellular localization and activity of *Drosophila* cubitus interruptus are regulated at multiple levels. *Development* **126**, 5097–5106.
21. Chen, Y. *et al.* (1998). Protein kinase A directly regulates the activity and proteolysis of cubitus interruptus. *Proc. Natl. Acad. Sci. USA* 2349–2354.
22. Chen, Y. *et al.* (1999). Mutants of cubitus interruptus that are independent of PKA regulation are independent of hedgehog signaling. *Development* **126**, 3607–3616.
23. Price, M. A. and Kalderon, D. (1999). Proteolysis of cubitus interruptus in *Drosophila* requires phosphorylation by protein kinase A. *Development* **126**, 4331–4339.
24. Lepage, T. *et al.* (1995). Signal transduction by cAMP-dependent protein kinase A in *Drosophila* limb patterning. *Nature* **373**, 711–715.
25. Li, W. *et al.* (1995). Function of protein kinase A in hedgehog signal transduction and *Drosophila* imaginal disk development. *Cell* **80**, 553–562.
26. Pan, D. and Rubin, G. M. (1995). cAMP-dependent protein kinase and hedgehog act antagonistically in regulating decapentaplegic transcription in *Drosophila* imaginal disks. *Cell* **80**, 543–552.
27. Jiang, J. and Struhl, G. (1995). Protein kinase A and hedgehog signaling in *Drosophila* limb development. *Cell* **80**, 563–572.
28. Johnson, R. L., Grenier, J. K., and Scott, M. P. (1995). Patched overexpression alters wing disk size and pattern: Transcriptional and post-transcriptional effects on hedgehog targets. *Development* **121**, 4161–4170.
29. Jia, J. *et al.* (2002). Shaggy/GSK3 antagonizes Hedgehog signalling by regulating cubitus interruptus. *Nature* **416**, 548–552.
30. Price, M. A. and Kalderon, D. (2002). Proteolysis of the Hedgehog signaling effector cubitus interruptus requires phosphorylation by glycogen synthase kinase 3 and casein kinase 1. *Cell* **108**, 823–835.
31. Jiang, J. and Struhl, G. (1998). Regulation of the Hedgehog and Wingless signalling pathways by the F-box/WD40-repeat protein Slimb. *Nature* **391**, 493–496.
32. Ohlmeyer, J. T. and Kalderon, D. (1997). Dual pathways for induction of wingless expression by protein kinase A and Hedgehog in *Drosophila* embryos. *Genes Dev.* **11**, 2250–2258.
33. Muller, B. and Basler, K. (2000). The repressor and activator forms of Cubitus interruptus control Hedgehog target genes through common generic gli-binding sites. *Development* **127**, 2999–3007.
34. Blackman, R. K. *et al.* (1991). An extensive 3' *cis*-regulatory region directs the imaginal disk expression of decapentaplegic, a member of the TGF-beta family in *Drosophila. Development* **111**, 657–666.
35. Hepker, J., Blackman, R. K., and Holmgren, R. (1999). Cubitus interruptus is necessary but not sufficient for direct activation of a wing-specific decapentaplegic enhancer. *Development* **126**, 3669–3677.
36. Guss, K. A. *et al.* (2001). Control of a genetic regulatory network by a selector gene. *Science* **292**, 1164–1167.
37. Chen, Y., Goodman, R. H., and Smolik, S. M. (2000). Cubitus interruptus requires *Drosophila* CREB-binding protein to activate wingless expression in the *Drosophila* embryo. *Mol. Cell. Biol.* **20**, 1616–1625.

CHAPTER 297

The Smads

Malcolm Whitman
Department of Cell Biology,
Harvard Medical School,
Boston, Massachusetts

Introduction

The transforming growth factor ß (TGF-ß) superfamily of ligands regulates a broad range of biological functions in both developing embryos and adult organisms, and the mechanisms by which these ligands regulate cell-type-specific effects on transcription has been a central problem in TGF-ß biology. The Smads were identified in the mid-1990s as a family of intracellular transducers of TGF-ß signals, and the study of how Smads mediate transcriptional regulation by TGF-ßs has begun to reveal how specificity is achieved for the broad range of biological activities of these ligands (for reviews, see [1–4]).

The founding member of the Smad family, MAD, was identified in *Drosophila* as a modifier of the effects of *dpp*, a *Drosophila* bone morphogenetic protein (BMP) homolog [5]. A related set of genes, Sma1, 2, and 3, were identified as required for signaling by a TGF-ß homolog in *Caenorhabditis elegans* [6], and both homology-based searches and functional screens in vertebrates subsequently identified vertebrate MAD/Sma homologs (reviewed in [7]), defining a broadly conserved family, renamed Smads [8], as transducers of TGF-ß signals.

Families: R-Smads, Co-Smads, and I-Smads

In both vertebrates and invertebrates, the Smads fall into three functional categories (reviewed in [9]): (1) the receptor-regulated R-Smads, which act downstream of specific subsets of TGF-ß superfamily ligands and are directly phosphorylated by Type I receptors; (2) the co-Smads, which are not pathway specific, but which interact with R-Smads in response to ligand stimulation and are required for R-Smad signaling; and (3) the inhibitory I-Smads, which are often induced as early responses to TGF-ß stimuli and which appear to act primarily as feedback inhibitors of signaling. A total of eight Smads have been identified in vertebrates (reviewed in [9]): Smads 1, 2, 3, 5, and 8 are R-Smads, Smad4 (and the closely related Smad10/Smad4ß, to date identified only in frogs) are co-Smads, and Smads 6 and 7 are I-Smads. All the Smads share two distinct regions of homology, an N-terminal MH1 domain and a C-terminal MH2 domain, separated by a linker that varies in size among the different Smad family members. The MH2 of Smad4 crystallizes as a trimer, and biochemical data indicate that Smads can homo-oligomerize via the MH2 domain *in vivo* in the presence or absence of ligand stimulation (reviewed in [10]). The extent to which Smads homo-oligomerize *in vivo* may differ for each Smad and has not been fully worked out [10].

Smad Oligomerization and Regulation by Receptors

Following the binding of TGF-ß ligand to specific type I and type II transmembrane receptors, signaling is initiated when the type I receptor phosphorylates an R-Smad at a conserved C-terminal SSXS motif (Fig. 1) (reviewed in [7]; see also the chapter by J. Wrana in this volume). Smad2 and Smad3 are activated downstream of activin, TGF-ß, and nodal family ligands, whereas Smad1, Smad5, and Smad8 are activated primarily downstream of BMP ligands. The interaction of Smad2 and Smad3 with their upstream type I receptors is mediated by a coupling protein, SARA [11]; whether a comparable Smad-receptor coupling protein exists for the BMP-Smad1/5/8 signaling pathway is not known.

Following phosphorylation by the type I receptor, the R-Smads heterodimerize with the co-Smad Smad4 [12,13] (Fig. 1). Whether the resulting complex is a simple heterodimer or a higher order oligomer as a result of homomeric

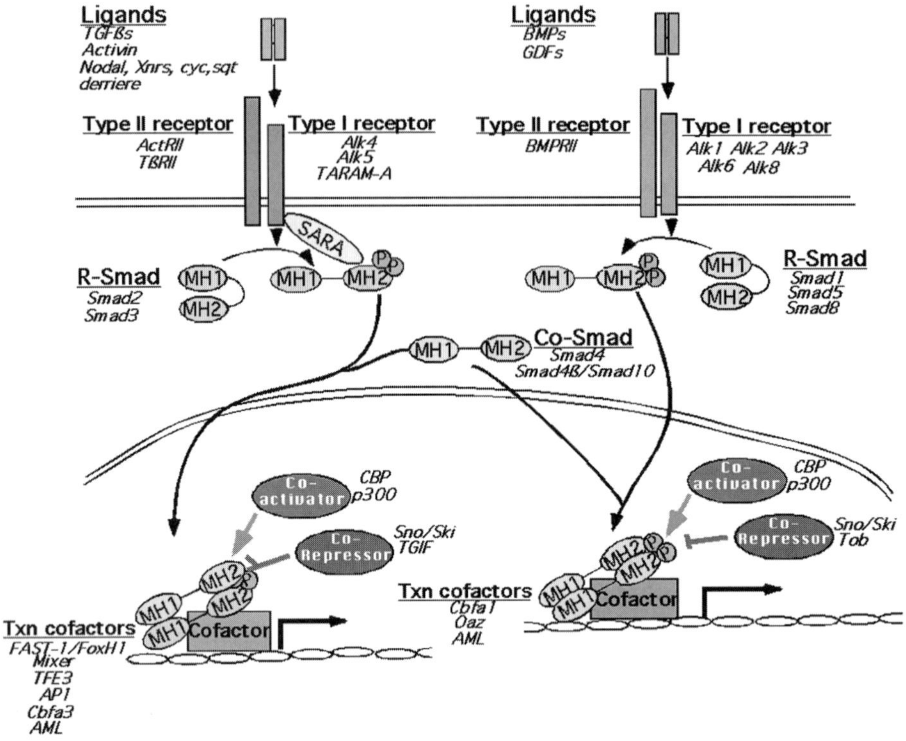

Figure 1 Basics of Smad signaling. The general pathway for transcriptional regulation by Smads is shown. Specific examples of vertebrate components of the Smad pathway are indicated in italics. These examples are not intended to be exhaustive and do not include invertebrate components of Smad pathways. For reviews of invertebrate Smad signaling, see [59,60].

interactions among the Smad MH2 domains (see earlier discussion) is not clear (reviewed in Shi, 2001[10]). The activated R-Smad/Smad4 complex then translocates to the nucleus. This nuclear translocation may be driven by two distinct properties of the R-Smads: (1) a canonical nuclear localization signal (NLS) in the MH1 domain that is unmasked following receptor phosphorylation at the C terminus [14] and (2) nuclear localization of the MH2 domain independent of a canonical NLS, which is inhibited in the unstimulated state by association (in the case of Smad2 and Smad3) with membrane bound SARA [15]. Smad4 is not required for nuclear localization, and in fact contains a nuclear export signal (NES) that may serve to rapidly remove Smad4 from the nucleus following dissociation from the R-Smads [16].

Transcriptional Regulation by Smads

Following translocation to the nucleus, the R-Smad/Smad4 complex can interact with DNA by means of two distinct mechanisms: (1) direct, site-specific binding of DNA by the MH1 domain [17] and (2) interaction with additional site-specific transcription factors [18,19] (Fig. 1). R-Smads and co-Smads preferentially recognize the motif GTCT, although binding to other target sites, particularly GC-rich sequences, has also been reported [10]. Although tandem repeats of the GTCT motif are sufficient to mediate TGF-ß-regulated transcriptional reporter activation, and these motifs are present in naturally occurring TGF-ß responsive promoters, recognition of this 4-bp motif by Smads does not seem sufficient to account for highly specific patterns of transcription in response to TGF-ßs *in vivo*, or for the differing biological effects of activation of different R-Smad pathways. This specificity in transcriptional responsiveness may be achieved primarily by interaction of the R-Smad/Smad4 complex with additional site-specific DNA-binding proteins. Interactions of R-Smad complexes with a wide variety of transcription factors have now been reported [2,3]. Many of these transcription factors are cell type specific, and many interact preferentially with either Smad2/Smad3 [18,20] or with Smad1 [21], providing an explanation for both ligand-specific and cell-type-specific transcriptional responses to TGF-ß superfamily ligands. Binding of the R-Smad/Smad4 complex and additional transcription factors to DNA is often cooperative, further enhancing the specificity and affinity of target site recognition [22–24]. Smad binding to transcription factors is often mediated through the MH2 domain; in one of the best characterized examples of Smad–transcription factor binding, a short α-helical region in the MH2 domain of Smad2 was shown to be a necessary and sufficient determinant of specificity in the binding of Smad2, but not Smad1, to the transcription factor FAST-1 (FoxH1) [25]. In other cases, however, regions outside the MH2 domain have also been shown to participate in the interaction of R-Smads with additional transcription factors.

Once bound to DNA, complexes of Smads and additional transcription factors regulate transcriptional activation by a number of mechanisms. R-Smads have been shown to interact with the histone acetyltransferase (HAT) p300 [22,26–28]. Additional transcriptional coactivators, which may act at least in part as bridges linking Smads and HATs, have also been shown to interact with both R-Smads and co-Smads, although their specific mechanism of action is not known [29,30]. In addition to binding coactivators, several examples have been reported in which Smads may activate transcription by displacing or sequestering transcriptional inhibitors (reviewed in [19]).

Although mechanisms of transcriptional activation have been the major focus of study of Smad function, several transcriptional corepressors have also been shown to interact with Smads and to be recruited to specific promoters by DNA-bound Smad complexes (reviewed in [31]). While interaction with corepressors may be a general mechanism for terminating or opposing Smad transcriptional activation, it is also possible that complex-specific interaction of Smads with corepressors results in the ligand-regulated inhibition of transcription at specific promoters. Specificity determinants that distinguish coactivator and corepressor interactions in different Smad–transcription factor complexes remain to be worked out, but may be an important component of understanding the full range of Smad effects on transcription.

Down-Regulation and Cross-Regulation of Smads

Various mechanisms have been identified by which Smad signaling can be down-regulated, including induction of I-Smads, specific protein degradative pathways, and kinase-mediated inhibition by other signaling pathways. Expression of I-Smads is often induced by TGF-ß superfamily signals, suggesting that they function as negative feedback components (reviewed in [32]). I-smads can act by functionally sequestering receptors or Smad4, or by targeting receptors for degradation, thus inhibiting TGF-ß signals at several distinct steps.

Specific degradation of R-Smads is regulated by a family of ubiquitin E3 ligases called Smurfs, which interact with R-Smads via the linker region that lies between the MH1 and MH2 domains [33]. Ubiquitin-mediated degradation of R-Smads may be either ligand independent or stimulated by TGF-ß ligands [33–37], whether there are additional points of regulation of ubiquitin/Smurf-mediated degradation of Smads remains to be elucidated.

In addition to the receptor-catalyzed activating phosphorylations at the C termini of the R-Smads, Smads are regulated by phosphorylation at additional sites. Best characterized are phosphorylations by MAP kinase in the linker domains of Smad1 and Smad2, which can inhibit Smad translocation to the nucleus and therefore inhibit signaling [38,39]. These MAP kinase catalyzed phosphorylations provide a mechanism by which receptor tyrosine kinase signaling can antagonize TGF-ß superfamily signals. Ca/CaM-dependent kinase phosphorylation of the Smad2 linker domain has also been reported to inhibit signaling without inhibiting nuclear translocation; the mechanism of this inhibition is not clear [40]. In addition to regulation by kinases, Smads can also stably interact with components of other signaling pathways, including LEF-1 [41], which acts downstream of wnt signals, and the estrogen receptor [42]. While it is not yet clear how these interactions are used *in vivo*, they do provide potential mechanisms for the integration of multiple signals in the determination of cell fate and behavior.

Function *In Vivo*: Gain of Function, Loss of Function

Both loss-of-function and gain-of-function experiments *in vivo* have confirmed a crucial role for Smads in TGF-ß signaling during a variety of biological processes [43–46]. Ectopic expression of Smad1 and Smad2 in frog embryos recapitulates the developmental effects of BMPs and nodals, respectively [47–49]. Conversely, the loss of Smad5 function in zebrafish phenocopies the loss of BMP ligands in early dorsal-ventral patterning [50], and Smad2 and Smad4 knock-outs in mice have severe developmental defects at gastrulation consistent with proposed roles for TGF-ß superfamily signals in early patterning [51–54]. Interestingly, however, cells derived from embryos lacking Smad4 do not show a complete loss of TGF-ß signaling, indicating that some biological effects of TGF-ßs are mediated independent of Smads in these cells [55].

Because of the complex roles TGF-ß signals play in different cell types, the simple association of phenotypes resulting from the loss of Smad function and known effects of TGF-ß ligands becomes more problematic as development progresses. Functional redundancy among different Smads also complicates the interpretation of Smad knock-out phenotypes [56]. Compound Smad knock-outs, and the substitution of one Smad for another at endogenous loci, will ultimately be necessary to sort out the distinctive role of each Smad *in vivo*. Antibodies directed against the C-terminally phosphorylated, activated forms of the R-Smads provide an additional tool for investigating exactly where and when Smads function *in vivo* [57].

While Smads have emerged as key mediators of TGF-ß signals, a growing body of data indicate that TGF-ßs utilize Smad-independent as well as Smad-dependent signaling pathways in the regulation of multiple biological processes (reviewed in [3,58]), and the relative contribution of each pathway to specific processes *in vivo* remains to be determined. The identification and characterization of the Smads has provided a coherent framework for understanding how specificity is generated in the wide range of biological actions of TGF-ßs. Future work will flesh out the full range of transcriptional targets for Smad interaction, the mechanisms and significance of Smad interaction with other signaling pathways, and the relationship between Smad-dependent and Smad-independent signals in mediating TGF-ß action.

Acknowledgments

Due to space constraints, reviews have been extensively cited at the expense of primary references. The author apologizes to workers whose work is not directly cited due to these constraints. The author is supported by grants from the NICHD.

References

1. Attisano, L. and Wrana, J. L. (2000). Smads as transcriptional co-modulators. *Curr. Opin. Cell Biol.* **12**, 235–243.
2. Derynck, R., Zhang, Y., and Feng, X. H. (1998). Smads: Transcriptional activators of TGF-β responses. *Cell* **95**, 737–740.
3. Massague, J. (2000). How cells read TGF-beta signals. *Nat. Rev. Mol. Cell Biol.* **1**, 169–178.
4. Moustakas, A., Souchelnytskyi, S., and Heldin, C. H. (2001). Smad regulation in TGF-beta signal transduction. *J. Cell Sci.* **114**, 4359–4369.
5. Sekelsky, J. J., Newfeld, S. J., Raftery, L. A., Chartoff, E. H., and Gelbart, W. M. (1995). Genetic characterization and cloning of Mothers against dpp, a gene required for decapentaplegic function in *Drosophila melanogaster. Genetics* **139**, 1347–1358.
6. Savage, C., Das, P., Finelli, A. L., Townsend, S. R., Sun, C. Y., Baird, S. E., and Padgett, R. W. (1996). *Caenorhabditis elegans* genes Sma2, Sma-3, and Sma-4 define a conserved family of transforming growth factor beta pathway components. *Proc. Nat. Acad. Sci. USA* **93**, 790–794.
7. Massague, J. (1998). TGF-beta signal transduction. *Annu. Rev. Biochem.* **67**, 753–791.
8. Derynck, R., Gelbart, W. M., Harland, R. M., Heldin, C.-H., Kern, S. E., Massagué, J., Melton, D. A., Moldzik, M., Padgett, R. W., Roberts, A. B. *et al.* (1996). Nomenclature: Vertebrate mediators of TGF-beta family signals. *Cell* **87**, 173.
9. Attisano, L. and Tuen Lee-Hoeflich, S. (2001). The Smads. *Genome Biol.* **2**.
10. Shi, Y. (2001). Structural insights on Smad function in TGFbeta signaling. *Bioessays* **23**, 223–232.
11. Tsukazaki, T., Chiang, T. A., Davison, A. F., Attisano, L., and Wrana, J. L. (1998). SARA, a FYVE domain protein that recruits Smad2 to the TGFbeta receptor. *Cell* **95**, 779–791.
12. Lagna, G., Hata, A., Hemmati-Brivanlou, A., and Massagué, J. (1996). Partnership between DPC4 and SMAD proteins in TGF-beta signalling pathways. *Nature* **383**, 832–836.
13. Zhang, Y., Feng, X., We, R., and Derynck, R. (1996). Receptor-associated Mad homologues synergize as effectors of the TGF-beta response. *Nature* **383**, 168–172.
14. Xiao, Z., Liu, X., Henis, Y. I., and Lodish, H. F. (2000). A distinct nuclear localization signal in the N terminus of Smad 3 determines its ligand-induced nuclear translocation. *Proc. Natl. Acad. Sci. USA* **97**, 7853–7858.
15. Xu, L., Chen, Y. G., and Massague, J. (2000). The nuclear import function of Smad2 is masked by SARA and unmasked by TGFbeta-dependent phosphorylation. *Nat. Cell Biol* **2**, 559–562.
16. Pierreux, C. E., Nicolas, F. J., and Hill, C. S. (2000). Transforming growth factor beta-independent shuttling of Smad4 between the cytoplasm and nucleus. *Mol. Cell. Biol* **20**, 9041–9054.
17. Kim, J., Johnson, K., Chen, H., Carroll, S., and Laughon, A. (1997). *Drosophila* Mad binds to DNA and directly mediates activation of vestigial by Decapentaplegic. *Nature* **388**, 304–308.
18. Chen, X., Rubock, M. J., and Whitman, M. (1996). A transcriptional partner for Mad proteins in TGF-β signalling. *Nature* **383**, 691–696.
19. Massagué, J. and Wotton, D. (2000). Transcriptional control by the TGF-β/Smad signaling system. *EMBO J.* **19**, 1745–1754.
20. Germain, S., Howell, M., Esslemont, G. M., and Hill, C. S. (2000). Homeodomain and winged-helix transcription factors recruit activated Smads to distinct promoter elements via a common Smad interaction motif. *Genes Dev.* **14**, 435–451.
21. Hata, A., Seoane, J., Lagna, G., Montalvo, E., Hemmati-Brivanlou, A., and Massagué, J. (2000). OAZ uses distinct DNA- and protein-binding zinc fingers in separate BMP-Smad and Olf signaling pathways. *Cell* **100**, 229–240.
22. Feng, X. H., Zhang, Y., Wu, R. Y., and Derynck, R. (1998). The tumor suppressor Smad4/DPC4 and transcriptional adaptor CBP/p300 are coactivators for smad3 in TGF-beta-induced transcriptional activation. *Genes Dev.* **12**, 2153–2163.
23. Hua, X., Liu, X., Ansari, D. O., and Lodish, H. F. (1998). Synergistic cooperation of TFE3 and smad proteins in TGF-beta-induced transcription of the plasminogen activator inhibitor-1 gene. *Genes Dev.* **12**, 3084–3095.
24. Yeo, C.-Y., Chen, X., and Whitman, M. (1999). The role of FAST-1 and Smads in transcriptional regulation of by activin during early *Xenopus* embryogenesis. *J. Biol. Chem.* **274**, 26584–26590.
25. Chen, Y.-G., Hata, A., Lo, R. S., Wotton, D., Shi, Y., Pavletich, N., and Massagué, J. (1998). Determinants of specificity of TGF-β signal transduction. *Genes Dev.* **12**, 2144–2152.
26. de Caestecker, M. P., Yahata, T., Wang, D., Parks, W. T., Huang, S., Hill, C. S., Shioda, T., Roberts, A. B., and Lechleider, R. J. (2000). The Smad4 activation domain (SAD) is a proline-rich, p300-dependent transcriptional activation domain. *J. Biol. Chem.* **275**, 2115–2122.
27. Janknecht, R., Wells, N. J., and Hunter, T. (1998). TGF-beta-stimulated cooperation of smad proteins with the coactivators CBP/p300. *Genes Dev.* **12**, 2114–2119.
28. Pouponnot, C., Jayaraman, L., and Massagué, J. (1998). Physical and functional interaction of SMADs and p300/CBP. *J. Biol. Chem.* **273**, 22865–22868.
29. Bai, R. Y., Koester, C., Ouyang, T., Hahn, S. A., Hammerschmidt, M., Peschel, C., and Duyster, J. (2002). SMIF, a Smad4-interacting protein that functions as a co-activator in TGFbeta signalling. *Nat. Cell Biol.* **4**, 181–190.
30. Shimizu, K., Bourillot, P. Y., Nielsen, S. J., Zorn, A. M., and Gurdon, J. B. (2001). Swift is a novel BRCT domain coactivator of Smad2 in transforming growth factor beta signaling. *Mol. Cell Biol.* **21**, 3901–3912.
31. Wotton, D. and Massague, J. (2001). Smad transcriptional corepressors in TGF beta family signaling. *Curr. Top. Microbiol. Immunol.* **254**, 145–164.
32. Miyazono, K., ten Dijke, P., and Heldin, C. H. (2000). TGF-beta signaling by Smad proteins. *Adv. Immunol.* **75**, 115–157.
33. Zhu, H., Kavsak, P., Abdollah, S., Wrana, J. L., and Thomsen, G. H. (1999). A SMAD ubiquitin ligase targets the BMP pathway and affects embryonic pattern formation. *Nature* **400**, 687–693.
34. Kavsak, P., Rasmussen, R. K., Causing, C. G., Bonni, S., Zhu, H., Thomsen, G. H., and Wrana, J. L. (2000). Smad7 binds to Smurf2 to form an E3 ubiquitin ligase that targets the TGF beta receptor for degradation. *Mol. Cell* **6**, 1365–1375.
35. Lo, R. S. and Massagué, J. (1999). Ubiquitin-dependent degradation of TGF-beta-activated Smad2. *Nat. Cell Biol.* **1**, 472–478.
36. Podos, S. D., Hanson, K. K., Wang, Y. C., and Ferguson, E. L. (2001). The DSmurf ubiquitin-protein ligase restricts BMP signaling spatially and temporally during *Drosophila* embryogenesis. *Dev. Cell* **1**, 567–578.
37. Zhang, Y., Chang, C., Gehling, D. J., Hemmati-Brivanlou, A., and Derynck, R. (2001). Regulation of Smad degradation and activity by Smurf2, an E3 ubiquitin ligase. *Proc. Natl. Acad. Sci. USA* **98**, 974–979.
38. Kretzschmar, M., Doody, J., and Massagué, J. (1997). Opposing BMP and EGF signalling pathways converge on the TGF-ß family mediator Smad1. *Nature* **389**, 618–622.
39. Kretzschmar, M., Doody, J., Timokhina, I., and Massagué, J. (1999). A mechanism of repression of TGF-ß/Smad signaling by oncogenic Ras. *Genes Dev.* **13**, 804–816.
40. Abdel-Wahab, N., Wicks, S. J., Mason, R. M., and Chantry, A. (2002). Decorin suppresses transforming growth factor-beta-induced expression of plasminogen activator inhibitor-1 in human mesangial cells through a mechanism that involves Ca2+-dependent phosphorylation of Smad2 at serine-240. *Biochem. J.* **362**, 643–649.

41. Labbe, E., Letamendia, A., and Attisano, L. (2000). Association of Smads with lymphoid enhancer binding factor 1/T cell-specific factor mediates cooperative signaling by the transforming growth factor-beta and wnt pathways. *Proc. Natl. Acad. Sci. USA* **97**, 8358–8363.
42. Matsuda, T., Yamamoto, T., Muraguchi, A., and Saatcioglu, F. (2001). Cross-talk between transforming growth factor-beta and estrogen receptor signaling through Smad3. *J. Biol. Chem.* **276**, 42908–42914.
43. Lu, C. C., Brennan, J., and Robertson, E. J. (2001). From fertilization to gastrulation: Axis formation in the mouse embryo. *Curr. Opin. Genet. Dev.* **11**, 384–392.
44. Weinstein, M., Yang, X., and Deng, C. (2000). Functions of mammalian Smad genes as revealed by targeted gene disruption in mice. *Cytokine Growth Factor Rev.* **11**, 49–58.
47. Baker, J. C. and Harland, R. M. (1996). A novel mesoderm inducer, Madr2, functions in the activin signal transduction pathway. *Genes Dev.* **10**, 1880–1889.
48. Graff, J. M., Bansal, A., and Melton, D. A. (1996). *Xenopus* Mad proteins transduce distinct subsets of signals for the TGFβ superfamily. *Cell* **85**, 479–487.
49. Thomsen, G. H. (1996). *Xenopus* mothers against decapentaplegic is an embryonic ventralizing agent that acts downstream of the Bmp-2/4 Receptor. *Development* **122**, 2359–2366.
50. Hild, M., Dick, A., Rauch, G. J., Meier, A., Bouwmeester, T., Haffter, P., and Hammerschmidt, M. (1999). The *smad5* mutation *somitabun* blocks Bmp2b signaling during early dorsoventral patterning of the zebrafish embryo. *Development* **126**, 2149–2159.
51. Nomura, M. and Li, E. (1998). Roles for Smad2 in mesoderm formation, left right patterning, and craniofacial development in mice. *Nature* **393**, 786–789.
52. Sirard, C., Pompa, J., de la Pompa, J. L., Elia, A., Itie, A., Mirtsos, C., Cheung, A., Hahn, S., Wakeham, A., Schwartz, L., Kern, S. *et al.* (1998). The tumor suppressor gene Smad4/Dpc4 is required for gastrulation and later for anterior development of the mouse embryo. *Genes Dev.* **12**, 107–119.
53. Waldrip, W., Bikoff, E., Hoodless, P., Wrana, J., and Robertson, E. (1998). Smad2 signaling in extraembryonic tissues determines anterior-posterior polarity of the early mouse embryo. *Cell* **92**, 797–808.
54. Weinstein, M., Yang, X., Li, C., Xu, X., Gotay, J., and Deng, C. X. (1998). Failure of egg cylinder elongation and mesoderm induction in mouse embryos lacking the tumor suppressor smad2. *Proc. Natl. Acad. Sci. USA* **95**, 9378–9383.
55. Sirard, C., Kim, S., Mirtsos, C., Tadich, P., Hoodless, P. A., Itie, A., Maxson, R., Wrana, J. L., and Mak, T. W. (2000). Targeted disruption in murine cells reveals variable requirement for Smad4 in transforming growth factor beta-related signaling. *J. Biol. Chem.* **275**, 2063–2070.
55. Whitman, M. (1998). Smads and early developmental signaling by the TGFβ superfamily. *Genes Dev.* **12**, 2443–2453.
56. Brennan, J., Lu, C. C., Norris, D. P., Rodriguez, T. A., Beddington, R. S., and Robertson, E. J. (2001). Nodal signalling in the epiblast patterns the early mouse embryo. *Nature* **411**, 965–969.
56. Whitman, M. (2001). Nodal signaling in early vertebrate embryos: themes and variations. *Dev. Cell* **1**, 605–617.
57. Lee, M., Heasman, J., and Whitman, M. (2001). Timing of endogenous activin-like signals and regional specification of the *Xenopus* embryo. *Development* **128**, 2939–2952.
58. von Bubnoff, A. and Cho, K. W. (2001). Intracellular BMP signaling regulation in vertebrates: Pathway or network? *Dev. Biol.* **239**, 1–14.
59. Raftery, L. A. and Sutherland, D. J. (1999). TGF-beta family signal transduction in *Drosophila* development: From Mad to Smads. *Dev. Biol.* **210**, 251–268.
60. Zimmerman, C. M. and Padgett, R. W. (2000). Transforming growth factor beta signaling mediators and modulators. *Gene* **249**, 17–30.

SECTION C

Damage/Stress Responses

Albert J. Fornace, Jr., Editor

CHAPTER 298

Complexity of Stress Signaling and Responses

Sally A. Amundson and Albert J. Fornace, Jr.
Gene Response Section, National Cancer Institute, Bethesda, Maryland

Introduction: A Variety of Stresses

Life has evolved in an environment fraught with potential dangers. Organisms have survived at least in part by evolving complex responses to meet the challenges posed by their environments. Many different kinds of stress can be encountered by a cell or organism. One important category of stressors is genotoxic agents that cause damage to DNA. These include ultraviolet and ionizing radiations as well as many chemical mutagens and carcinogens. Organisms must also defend against a myriad of physical stresses acting mainly through mechanisms other than DNA damage. Such stresses may include shear stress, wounding, infection, nutrient deprivation, osmotic stress, hypoxia, and heat shock. These last two stresses are discussed in detail in chapters by Koong and Giaccia, and Rick Morimoto, respectively, in this volume.

The cellular strategies used to combat these assaults on equilibrium are equally diverse, but relatively well conserved through evolution (Table I). Much has been learned from the study of lower organisms including bacteria, yeast, and *Drosophila* as discussed in later chapters by Walker, Demple, Bennett and Resnick, Muzi-Falconi, and Abrams. In all organisms, sensor systems are needed to detect stress and its resultant damage. Signals are then exchanged between different cellular compartments, and even between cells, resulting in changes in expression and function of specific transcripts and proteins (Fig. 1). This in turn impacts various cellular processes such as cell cycle progression, DNA repair, and activation of the apoptotic program, ultimately resulting in either recovery and repair or in death. This chapter attempts to provide an overview of these interwoven signaling pathways and processes, many of which will be presented in greater detail in the chapters that follow.

Origin of Signals

The Nucleus

Damaged DNA has long been recognized as originating a stress signal. For instance, tracts of single-stranded DNA at damaged sites in *Escherichia coli* are coated by the RecA protein, which initiates the bacterial SOS response to be discussed in detail in the chapter by Walker. In mammalian cells, double-strand breaks are bound by DNA-PK, a complex of the DNA-PKcs (catalytic subunit) and the Ku70-Ku80 heterodimer. DNA-PK binding to broken DNA ends may initiate signaling to the cellular apoptosis machinery [1]. The ATM kinase has also been shown to bind to broken DNA ends [2,3] and in combination with various cofactors may initiate a DNA damage signal (reviewed in [4]). Activation of ATM results in phosphorylation of a number of downstream targets involved in signal transduction, including p53, Chk1, Chk2, Brca1, and Mdm2 [5–8]. Signaling through ATM is discussed further in the chapter by Lavin *et al.*, whereas the cascades of events triggered specifically by double-strand breaks are covered in chapters by Bennett and Resnick and by Jackson and Bradbury.

Single-strand breaks and damaged bases in DNA also result from many types of genotoxic stress. When a damaged site cannot be bypassed by the polymerase, stalled replication forks can result. This in turn initiates a stress signal to prevent initiation of further DNA synthesis and to arrest cell cycle progression (reviewed in [9,10]), possibly signaling through ATM and Chk2 [11].

Table I Examples of Conserved Signaling Pathways Activated by Different Stress Stimuli

Stress	Signaling pathways	Drosophilla homologs	S. cerevisiae
Ionizing Radiation	ATM, CHK2	Mei-41	Mec1, Tel1, Rad53
	DNA-PKcs	Mei-41	Tor1
UV Radiation	ATM, CHK2	Mei-41	Mec1, Tel1
	ATR, CHK1	Mei-41, grp	Esr1
	p38	p38b-P1	Hog1
UPR	PEK/PERK	Dmpek (E2K3)	Gcn2
	IRE1, ATF6, JNK	Bsk	Ire1
	HSF1	Hsf	Hsf1
Hypoxia	HIF1A	Sima	Swi1
	VHL	d-VHL	
Heat Shock	HSF1	Hsf	Hsf1
	Raf-1/ERK	Draf-1	Bck1

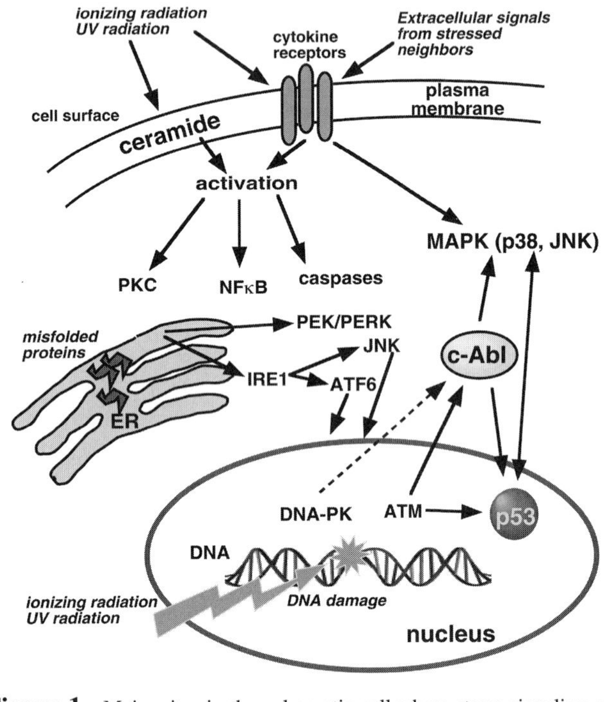

Figure 1 Major sites in the eukaryotic cell where stress signaling can originate, and some of the key signaling pathways activated in response. Many of these pathways will be discussed in detail in the following chapters in this section.

The Cytoplasm

Signaling can also be initiated by non-DNA-damage cellular stress. For instance, conditions including nutrient deprivation and exposure to certain toxins can cause accumulation of unfolded or misfolded proteins in the endoplasmic reticulum, triggering a complex unfolded protein response (UPR) (reviewed in [12]). This response has two major components, the first of which shuts down new protein synthesis, and appears to be mediated by the PEK/PERK protein kinase [13,14]. The second component of the UPR up-regulates expression of genes whose products are involved in protein folding and in the degradation of misfolded proteins. This pathway is initiated by activation of the IRE1 kinase, and the signal is transduced through both ATF6 and the JNK pathway [15]. These signal transduction pathways are discussed in the chapter by David Ron. Stresses that disrupt the normal function of other organelles, including the cytoskeleton and mitochondria, can also initiate signals leading to apoptosis [16].

The Plasma Membrane

Exposure to toxins or infection can give rise to a stress signal originating in the plasma membrane. Engagement of cell surface receptors, such as the tumor necrosis factor (TNF) receptor superfamily or the interleukin 1 (IL-1) receptor may lead to activation of NFκB or initiation of the caspase cascade (reviewed in [17]). Activation of TNF receptors initiates hydrolysis of sphingomyelin to produce ceramide, which acts as a second messenger. Irradiation of enucleated cells can also initiate ceramide signaling in the plasma membrane, clearly indicating signaling of radiation damage in the absence of DNA-damage-specific signals [18].

Extracellular Signals

Cells can also respond to stress or injury at a distance through communication with other cells. Cytokines, growth factors, and hormones are released in response to internal cues or external stressors, and their internalization or binding to receptors of other cells can set in motion signal cascades and alter transcriptional programs within the target cell. Such extracellular communication appears to be a basis of the so-called "bystander effect," where DNA damage in one cell can result in altered protein expression in undamaged cells in the same culture [19,20].

Signal Transduction

Once a stress situation has been recognized and signaling is initiated, numerous pathways exist for the amplification and direction of the message. Redox signaling, to be discussed in the chapter by Bruce Demple, is a common mechanism for signal amplification in cells. Reactive nitrogen (RNS) or oxygen species (ROS) can be generated by the p450 system or the electron transport machinery in the mitochondria in response to stress. Alteration of the cellular redox environment may directly affect protein function, such as ROS-mediated oxidation of Tyr phosphatase 1B, which inhibits its function and leads to activation of target proteins including ErbB1 [21]. Nitric oxide similarly has been shown to modulate the function of p21ras [22] through nitrosylation.

Protein Modification and Signaling

Protein modification is a major mechanism for stress signal propagation within the cell, and perhaps the most common modification is alteration of phosphorylation state. Many kinase cascades are activated by various stresses; one of the best studied kinase pathways associated with DNA damaging stress is that involving the ATM kinase. Activation of ATM by DNA damage leads to phosphorylation of both p53 and the c-Abl protein tyrosine kinase. In the case of c-Abl activation, this leads to activation of p73 and promotes apoptosis [23]. Stress signaling through the c-Abl kinase is discussed in a later chapter by Jean Wang, while the chapter by Lavin *et al.* focuses on ATM signaling.

The complexities of p53 activation by DNA damage and other stresses have been studied extensively, as reviewed in detail in the chapter by Anderson and Appella. Briefly, at least 18 sites of regulatory posttranslational modification of p53 have been described, including sites of phosphorylations, acetylations, and sumoylation. Modifications of these sites vary in response to different stresses and the activation of different upstream regulatory pathways. For instance, distinct patterns of p53 modification have been reported in ionizing radiation-treated, UV-irradiated, or senescent human primary fibroblasts [24,25]. Agents inducing double-strand breaks, such as ionizing radiation, activate the ATM kinase pathway, resulting in phosphorylation of p53 at Ser15, Ser20, Ser9, and Ser46 [26], as well as activating a number of additional kinases phosphorylating further sites in p53. In contrast, agents such as UVC that induce bulky lesions in DNA activate the ATR [27], p38 MAPK [28], and JNK [29] kinase pathways, resulting in phosphorylations at Ser15 and Ser37, Ser33 and Ser46, and Thr81, respectively. Such modifications of p53 subsequently affect its interactions with other proteins and its sequence-specific DNA-binding and transactivation activities, resulting in differential gene transcription profiles (Fig. 2). Thus p53 activation can promote such diverse outcomes as apoptosis [30,31], reversible cell cycle arrest [32], or a permanent

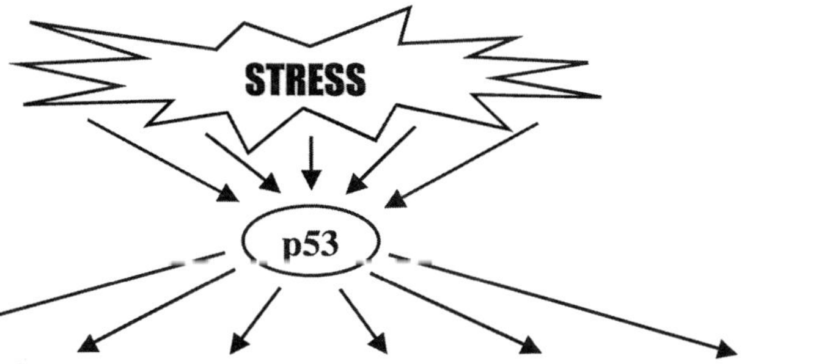

Cell Cycle	Apoptosis	DNA Repair	Senescence	Anti-angiogenesis	Other
CIP1/WAF1	*BAX*	*XPC*	*CIP1/WAF1*	*VEGF*	*MDM2*
ClnG	*BCL-X*	*DDB2*	type I PACAP	*IL-8*	*FRA1*
ClnD1	*PAG608*	*GADD45A*		*HIF1α*	*ATF3*
WIP1	*FAS/APO1*	*PCNA*		*MMP2*	*14-3-3σ*
EGF-R	*KILLER/DR5*	*p53R2*		*BAI-1*	*Rb*
Rb	*TRUNDD*			*TSP1 & 2*	*c-MYC*
BTG2	*TRID*				*ninjurin*
PCNA	*SIAH*				*MAP4*
GADD45A	*IGF-BP3*				*amyloid*
14-3-3σ	*PIG1 to PIG14*				*PIR121*
TGF-α	*14-3-3σ*				*WIG1*
SIAH	*MAP4*				
IGF-β	*TSP1 & 2*				
PIG1 to PIG14	*bFGF*				
inhibin-β	*GML*				
GML					

Figure 2 Diverse stress signals can be transduced by a variety of pathways through p53. This, in turn, results in the execution of different cellular programs, mediated in part by the activation or repression of transcription of a variety of genes, including the examples listed here.

cell cycle arrest indistinguishable from senescence [33,34]. The many different possible modifications and binding interactions of p53 could serve as a means to tailor the specificity of p53 responses to suit the originating stress or the specific cellular environment.

MAP Kinase Pathways

Protein activation by phosphorylation is a common mechanism for intracellular signal transduction, and the mitogen-activated protein kinase (MAPK) cascades provide a major example of this mechanism. The MAPK signaling cascades are three distinct but analogous pathways comprising the ERK-1/ERK-2 mitogen-activated pathway and the stress-activated protein kinases (SAPK), p38 and c-Jun N-terminal kinase (JNK). The MAPK/ERK pathway is activated primarily by mitogenic stimuli such as extracellular growth factors, as well as during physiological processes such as T-cell activation in immune response [35]. Conversely, the SAPK/JNK and p38 pathways are activated predominantly by stresses, such as UV or ionizing radiation, heat shock, or inflammatory cytokines. There is also a regulatory phosphatase system for the down-regulation and inactivation of each of these signaling cascades. The degree to which the different signaling cascades are activated and the duration of the activation appear to shape the cellular response and can result in altered gene transcription, cell cycle arrest, or apoptosis. The interplay of the various MAPK pathways is discussed in greater detail in the chapter by Blattner and Herrlich.

Functional Genomics and Proteomics Approaches

In addition to, and often consequent to, the modification of proteins, the cellular response to stress also alters the abundance of specific mRNA transcripts. This can occur both through alteration of message stability (discussed in the chapter by Miriam Gorosope) and via transcription factor modification leading to alteration of transcription programs. By changing the transcripts and subsequently the proteins manufactured in response to specific stress situations, the cell makes components available to promote or protect against cell death, to repair damaged DNA, or to halt cell cycle progression and maintain damage checkpoints.

The advent of functional genomics has vitalized interest in the transcriptional basis of stress response. Techniques such as serial analysis of gene expression (SAGE) and microarray analysis allow assessment of expression of thousands of genes in a single experiment [36,37]. For instance, SAGE has been used to identify around 30 transcripts apparently induced by p53 activation [38], while microarray analysis has been widely applied to the study of responses to stresses including ionizing [39,40] and UV [41] irradiation, hypoxia [42], alkylating agents [43], and toxic metals [44–46]. The data being collected in such experiments are extremely complex. Although some common factors emerge, such as increased expression of genes such as *CDKN1A* indicating activation of the p53 pathway, there is a profound heterogeneity in stress-responsive genes. Different "profiles" are beginning to emerge, however, based both on the initiating stressor (Fig. 3) and the responding cell type (Fig. 4). Mathematical approaches are being developed for analysis of this sort of data, including hierarchical clustering, principal component analysis, and multidimensional scaling analysis. Meanwhile, techniques for similar high-throughput analysis of protein changes occurring after stress are also being developed. Through the combination of these sophisticated approaches, we may soon be able to unravel much more of the complex interplay of signals involved in cellular stress response.

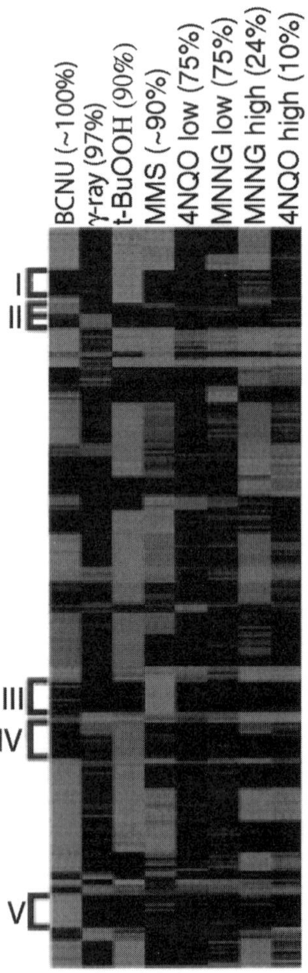

Figure 3 Global transcription profiles in *S. cerevisiae* exposed to different damaging agents. Log-phase cells were exposed to the indicated agents for 1 hr. For each treatment, the percent survival (in parentheses) was determined by colony formation. Expression of 2324 genes scored as responsive on Affymetrix arrays has been grouped into 30 clusters by K-means clustering. The six indicated clusters contain members appearing to respond only to a particular agent (Clusters I through IV). Treatments were as follows: 200 μM BCNU, 300 Gy γ-ray, 5 mM *t*-BuOOH, 0.1% MMS, 4NQO at 2 (low) or 8 (high) μg/ml, and MNNG at 8 (low) or 27 (high) μg/ml. (Reproduced with permission from [43].)

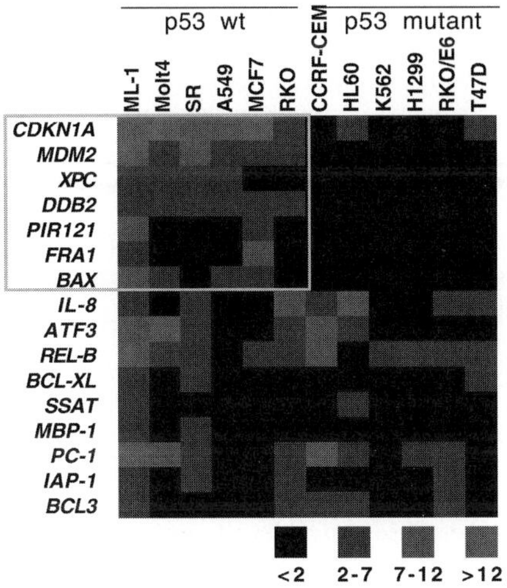

Figure 4 Variability of induction of a small number of genes in cell lines differing in functional p53 status. Cells were exposed to 20 Gy γ-rays, RNA was harvested 4 hr later, and relative gene induction was determined by quantitative single-probe hybridization [47]. The relative fold induction is indicated as a shade of gray according to the scale at the bottom of the figure. A black square represents no significant induction of a particular gene in that cell line. The box highlights genes dependent on wild-type p53 for robust induction by ionizing radiation.

References

1. Wang, S., Guo, M., Ouyang, H., Li, X., Cordon-Cardo, C., Kurimasa, A., Chen, D. J., Fuks, Z., Ling, C. C., and Li, G. C. (2000). The catalytic subunit of DNA-dependent protein kinase selectively regulates p53-dependent apoptosis but not cell-cycle arrest. *Proc. Natl. Acad. Sci. USA* **97**, 1584–1588.
2. Smith, G. C., Cary, R. B., Lakin, N. D., Hann, B. C., Teo, S. H., Chen, D. J., and Jackson, S. P. (1999). Purification and DNA binding properties of the ataxia-telangiectasia gene product ATM. *Proc. Natl. Acad. Sci. USA* **96**, 11134–11139.
3. Suzuki, K., Kodama, S., and Watanabe, M. (1999). Recruitment of ATM protein to double strand DNA irradiated with ionizing radiation. *J. Biol. Chem.* **274**, 25571–25575.
4. Durocher, D. and Jackson, S. P. (2001). DNA-PK, ATM and ATR as sensors of DNA damage: Variations on a theme? *Curr. Opin. Cell Biol.* **13**, 225–231.
5. Banin, S., Moyal, L., Shieh, S., Taya, Y., Anderson, C. W., Chessa, L., Smorodinsky, N. I., Prives, C., Reiss, Y., Shiloh, Y., and Ziv, Y. (1998). Enhanced phosphorylation of p53 by ATM in response to DNA damage. *Science* **281**, 1674–1677.
6. Canman, C. E., Lim, D. S., Cimprich, K. A., Taya, Y., Tamai, K., Sakaguchi, K., Appella, E., Kastan, M. B., and Siliciano, J. D. (1998). Activation of the ATM kinase by ionizing radiation and phosphorylation of p53. *Science* **281**, 1677–1679.
7. Khosravi, R., Maya, R., Gottlieb, T., Oren, M., Shiloh, Y., and Shkedy, D. (1999). Rapid ATM-dependent phosphorylation of MDM2 precedes p53 accumulation in response to DNA damage. *Proc. Natl. Acad. Sci. USA* **96**, 14973–14977.
8. Cortez, D., Wang, Y., Qin, J., and Elledge, S. J. (1999). Requirement of ATM-dependent phosphorylation of brca1 in the DNA damage response to double-strand breaks. *Science* **286**, 1162–1166.
9. Rowley, R., Phillips, E. N., and Schroeder, A. L. (1999). The effects of ionizing radiation on DNA synthesis in eukaryotic cells. *Int. J. Radiat. Biol.* **75**, 267–283.
10. Lowndes, N. F. and Murguia, J. R. (2000). Sensing and responding to DNA damage. *Curr. Opin. Genet. Dev.* **10**, 17–25.
11. Falck, J., Mailand, N., Syljuasen, R. G., Bartek, J., and Lukas, J. (2001). The ATM-Chk2-Cdc25A checkpoint pathway guards against radioresistant DNA synthesis. *Nature* **410**, 842–847.
12. Welihinda, A. A., Tirasophon, W., and Kaufman, R. J. (1999). The cellular response to protein misfolding in the endoplasmic reticulum. *Gene Expr.* **7**, 293–300.
13. Shi, Y., Vattem, K. M., Sood, R., An, J., Liang, J., Stramm, L., and Wek, R. C. (1998). Identification and characterization of pancreatic eukaryotic initiation factor 2 alpha-subunit kinase, PEK, involved in translational control. *Mol. Cell. Biol.* **18**, 7499–7509.
14. Harding, H. P., Zhang, Y., and Ron, D. (1999). Protein translation and folding are coupled by an endoplasmic-reticulum-resident kinase. *Nature* **397**, 271–274.
15. Urano, F., Bertolotti, A., and Ron, D. (2000). IRE1 and efferent signaling from the endoplasmic reticulum. *J. Cell Sci.* **113**, Part 21, 3697–3702.
16. Bratton, S. B. and Cohen, G. M. (2001). Apoptotic death sensor: An organelle's alter ego? *Trends Pharmacol. Sci.* **22**, 306–315.
17. Bowie, A. and O'Neill, L. A. (2000). Oxidative stress and nuclear factor-kappaB activation: A reassessment of the evidence in the light of recent discoveries. *Biochem. Pharmacol.* **59**, 13–23.
18. Haimovitz-Friedman, A., Kan, C. C., Ehleiter, D., Persaud, R. S., McLoughlin, M., Fuks, Z., and Kolesnick, R. N. (1994). Ionizing radiation acts on cellular membranes to generate ceramide and initiate apoptosis. *J. Exp. Med.* **180**, 525–535.
19. Iyer, R. and Lehnert, B. E. (2000). Factors underlying the cell growth-related bystander responses to alpha particles. *Cancer Res.* **60**, 1290–1298.
20. Mothersill, C. and Seymour, C. (2001). Radiation-induced bystander effects: Past history and future directions. *Radiat. Res.* **155**, 759–767.
21. Lee, S. R., Kwon, K. S., Kim, S. R., and Rhee, S. G. (1998). Reversible inactivation of protein-tyrosine phosphatase 1B in A431 cells stimulated with epidermal growth factor. *J. Biol. Chem.* **273**, 15366–15372.
22. Lander, H. M., Ogiste, J. S., Teng, K. K., and Novogrodsky, A. (1995). p21ras as a common signaling target of reactive free radicals and cellular redox stress. *J. Biol. Chem.* **270**, 21195–21198.
23. Wang, J. Y. (2000). Regulation of cell death by the Abl tyrosine kinase. *Oncogene* **19**, 5643–5650.
24. Webley, K., Bond, J. A., Jones, C. J., Blaydes, J. P., Craig, A., Hupp, T., and Wynford-Thomas, D. (2000). Posttranslational modifications of p53 in replicative senescence overlapping but distinct from those induced by DNA damage. *Mol. Cell. Biol.* **20**, 2803–2808.
25. Bulavin, D. V., Demidov, O. N., Saito, S., Kauraniemi, P., Phillips, C., Amundson, S. A., Ambrosino, C., Sauter, G., Nebreda, A. R., Anderson, C. W., Kallioniemi, A., Fornace, A. J. J., and Appella, E. (2002). Amplification of PPM1D in human tumors abrogates p53 tumor-suppressor activity. *Nat. Genet.* **31**, 210–215.
26. Saito, S., Goodarzi, A. A., Higashimoto, Y., Noda, Y., Lees-Miller, S. P., Appella, E., and Anderson, C. W. (2002). ATM mediates phosphorylation at multiple p53 sites, including Ser(46), in response to ionizing radiation. *J. Biol. Chem.* **277**, 12491–12494.
27. Tibbetts, R. S., Brumbaugh, K. M., Williams, J. M., Sarkaria, J. N., Cliby, W. A., Shieh, S. Y., Taya, Y., Prives, C., and Abraham, R. T. (1999). A role for ATR in the DNA damage-induced phosphorylation of p53. *Genes Dev.* **13**, 152–157.
28. Bulavin, D., Saito, S. I., Hollander, M. C., Sakaguchi, K., Anderson, C. W., Appella, E., and Fornace, A. J., Jr. (1999). Phosphorylation of human p53 by p38 kinase coordinates N-terminal phosphorylation and apoptosis in response to UV radiation. *EMBO J.* **18**, 6845–6854.
29. Buschmann, T., Potapova, O., Bar-Shira, A., Ivanov, V. N., Fuchs, S. Y., Henderson, S., Fried, V. A., Minamoto, T., Alarcon-Vargas, D., Pincus, M. R., Gaarde, W. A., Holbrook, N. J., Shiloh, Y., and Ronai, Z. (2001). Jun NH2-terminal kinase phosphorylation of p53 on Thr-81 is important for p53 stabilization and transcriptional activities in response to stress. *Mol. Cell. Biol.* **21**, 2743–2754.
30. Yonish-Rouach, E., Resnitzky, D., Lotem, J., Sachs, L., Kimchi, A., and Oren, M. (1991). Wild-type p53 induces apoptosis of myeloid leukaemic cells that is inhibited by interleukin-6. *Nature* **352**, 345–347.
31. Lowe, S. W., Ruley, H. E., Jacks, T., and Housman, D. E. (1993). p53-Dependent apoptosis modulates the cytotoxicity of anticancer agents. *Cell* **74**, 957–967.

32. Agarwal, M. L., Agarwal, A., Taylor, W. R., and Stark, G. R. (1995). p53 Controls both the G2/M and the G1 cell cycle checkpoints and mediates reversible growth arrest in human fibroblasts. *Proc. Natl. Acad. Sci. USA* **92**, 8493–8497.
33. Bond, J. A., Wyllie, F. S., and Wynford-Thomas, D. (1994). Escape from senescence in human diploid fibroblasts induced directly by mutant p53. *Oncogene* **9**, 1885–1889.
34. Gire, V. and Wynford-Thomas, D. (1998). Reinitiation of DNA synthesis and cell division in senescent human fibroblasts by microinjection of anti-p53 antibodies. *Mol. Cell. Biol.* **18**, 1611–1621.
35. Weg-Remers, S., Ponta, H., Herrlich, P., and Konig, H. (2001). Regulation of alternative pre-mRNA splicing by the ERK MAP-kinase pathway. *EMBO J.* **20**, 4194–4203.
36. Schena, M., Shalon, D., Heller, R., Chai, A., Brown, P. O., and Davis, R. W. (1996). Parallel human genome analysis: Microarray-based expression monitoring of 1000 genes. *Proc. Natl. Acad. Sci. USA* **93**, 10614–10619.
37. Schena, M., Shalon, D., Davis, R. W., and Brown, P. O. (1995). Quantitative monitoring of gene expression patterns with a complementary DNA microarray. *Science* **270**, 467–470.
38. Polyak, K., Xia, Y., Zweler, J. L., Kinzler, K. W., and Vogelstein, B. (1997). A model for p53-induced apoptosis. *Nature* **389**, 300–306.
39. Amundson, S. A., Bittner, M., Chen, Y. D., Trent, J., Meltzer, P., and Fornace, A. J., Jr. (1999). cDNA microarray hybridization reveals complexity and heterogeneity of cellular genotoxic stress responses. *Oncogene* **18**, 3666–3672.
40. Amundson, S. A., Shahab, S., Bittner, M., Meltzer, P., Trent, J., and Fornace, A. J., Jr. (2000). Identification of potential mRNA markers in peripheral blood lymphocytes for human exposure to ionizing radiation. *Radiat. Res.* **154**, 342–346.
41. Murakami, T., Fujimoto, M., Ohtsuki, M., and Nakagawa, H. (2001). Expression profiling of cancer-related genes in human keratinocytes following non-lethal ultraviolet B irradiation. *J. Dermatol. Sci.* **27**, 121–129.
42. Scandurro, A. B., Weldon, C. W., Figueroa, Y. G., Alam, J., and Beckman, B. S. (2001). Gene microarray analysis reveals a novel hypoxia signal transduction pathway in human hepatocellular carcinoma cells. *Int. J. Oncol.* **19**, 129–135.
43. Jelinsky, S. A., Estep, P., Church, G. M., and Samson, L. D. (2000). Regulatory networks revealed by transcriptional profiling of damaged *Saccharomyces cerevisiae* cells: Rpn4 links base excision repair with proteasomes. *Mol. Cell. Biol.* **20**, 8157–8167.
44. Momose, Y. and Iwahashi, H. (2001). Bioassay of cadmium using a DNA microarray: Genome-wide expression patterns of *Saccharomyces cerevisiae* response to cadmium. *Environ. Toxicol. Chem.* **20**, 2353–2360.
45. Bouton, C. M., Hossain, M. A., Frelin, L. P., Laterra, J., and Pevsner, J. (2001). Microarray analysis of differential gene expression in lead-exposed astrocytes. *Toxicol. Appl. Pharmacol.* **176**, 34–53.
46. Chen, H., Liu, J., Merrick, B. A., and Waalkes, M. P. (2001). Genetic events associated with arsenic-induced malignant transformation: applications of cDNA microarray technology. *Mol. Carcinog.* **30**, 79–87.
47. Koch-Paiz, C. A., Momenan, R., Amundson, S. A., Lamoreaux, E., and Fornace, A. J., Jr. (2000). Estimation of relative mRNA content by filter hybridization to a polyuridylic probe. *Biotechniques* **29**, 708–714.

CHAPTER 299

Signal Transduction in the *Escherichia coli* SOS Response

Penny J. Beuning and Graham C. Walker
Biology Department, Massachusetts Institute of Technology, Cambridge, Massachusetts

SOS Response

Inducing Signals and Transcriptional Control

An organism's DNA is often exposed to potential mutagens such as chemicals and UV light. *Escherichia coli* has a regulated system, the SOS response, that is induced in response to damage to its genome [1–4]. When the cell's DNA suffers damage or the normal replication process is stalled, a region of single-stranded DNA is formed. The RecA protein coats this single-stranded DNA, forming a nucleoprotein filament and in turn activating RecA for its coprotease function. The RecA/ssDNA filament serves as the signal for the SOS response, which governs the cellular response to the damage. The RecA/ssDNA filament stimulates the self-cleavage of the LexA repressor, thus inactivating it as a repressor (Fig. 1). This leads to expression of the SOS-regulated genes, many of which are involved in DNA repair. Many of the genes induced in the SOS response are involved in DNA damage repair and tolerance and cell cycle control, such as *polB*, *sulA*, and *uvrA* [1,5]. Among the genes induced are those that code for two members of the newly termed *Y-family* of low-fidelity DNA polymerases: *dinB*, also known as DNA Pol IV, and *umuDC*, also called DNA Pol V [6].

A recent genomic analysis demonstrated more than 40 LexA-regulated loci in *E. coli* (Table I) [7]. By including an Ind$^-$ allele of *lexA* in this analysis, several genes were identified that are induced following UV irradiation in a LexA-independent manner, including genes coding for proteins involved in replication, the heat-shock response, and purine, pyrimidine, and RNA metabolism [7]. Decreased levels of some messages were also found following UV irradiation, including some genes involved in carbon and lipid metabolism, and in septum formation prior to cell division. These results suggest yet another level of regulation of this system.

Posttranscriptional Control in the SOS Response

After induction of the SOS response, UmuD and UmuC are translated and UmuD persists as the homodimer of the full-length protein for about 20 min, at which point it undergoes a RecA/ssDNA nucleoprotein filament-facilitated self-cleavage reaction similar to LexA cleavage in which the Cys24–Gly25 peptide bond is severed to form UmuD'. The UmuD'$_2$C complex (DNA Pol V) is a low-fidelity lesion-bypass polymerase; the RecA/ssDNA filament and single-stranded DNA-binding protein (SSB) are also required *in vitro* for the polymerase activity of UmuD'$_2$C [8,9]. Uncleaved UmuD also seems to have an important role in protecting the cell against excessive mutations as a result of DNA damage by serving as a DNA damage checkpoint effector [10,11]. Also, UmuD'$_2$C is able to inhibit homologous recombination, whereas UmuD$_2$C is not, an interaction that may serve to prevent recombination in favor of translesion synthesis [12].

The SOS-induced *dinI* gene is able to inhibit the RecA/ssDNA-mediated cleavage of UmuD [13]. Recent evidence suggests that the role of DinI is to modulate the rate of UmuD cleavage [13]. Earlier *in vivo* work showed that overexpression of DinI inhibited LexA cleavage [14]. DinI binds to the RecA/ssDNA filament more strongly than UmuD does but less strongly than LexA *in vitro*, a finding that supports the idea that competition for binding sites on RecA/ssDNA filaments allows DinI to attenuate the rate of UmuD cleavage while LexA cleavage is apparently less affected [13].

Ending the SOS Response and Return to the Uninduced State

As the cell recovers from DNA damage, the amount of ssDNA decreases and so the inducing signal abates, thereby resulting in the accumulation of intact LexA repressor and

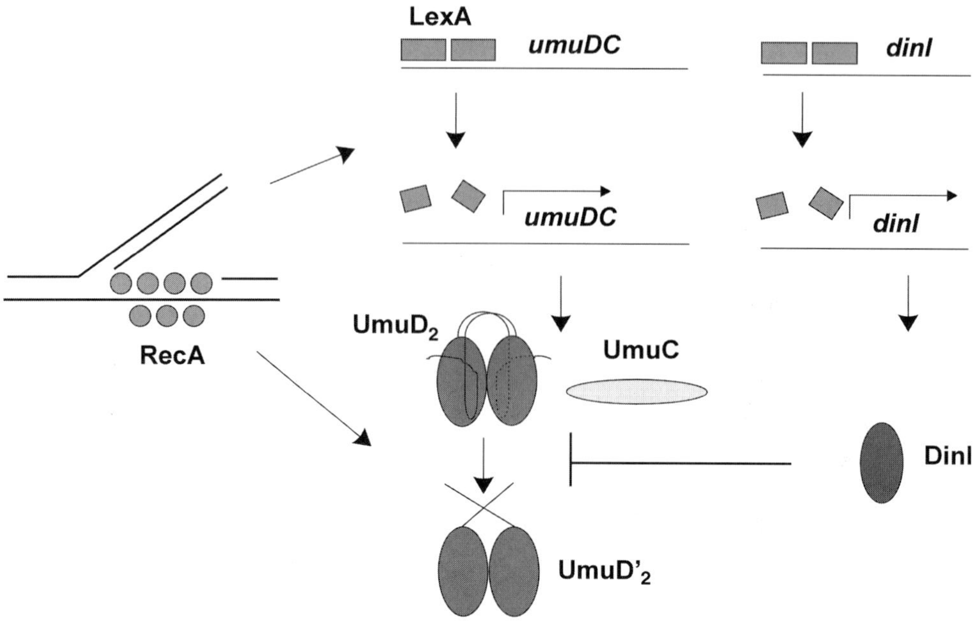

Figure 1 Induction of the SOS response showing two of the operons under LexA control. After DNA damage, RecA forms a filament on the single-stranded DNA that presumably results from the cell's failed attempts at replication (left), initiating a cascade of events including cleavage of the LexA repressor.

thus halting the general SOS response [1]. Because Pol V is a low-fidelity DNA polymerase, its activity is tightly regulated to prevent general use as a DNA polymerase. An additional level of control seems to be provided by the degradation of UmuD$_2$ and UmuC by the Lon protease and the selective digestion of UmuD' in the UmuD·D' heterodimer by the ClpXP protease [15,16]. While both UmuD and UmuD' form stable homodimers, the UmuD·D' heterodimer is more stable in solution than either homodimer [17]. The sequestration of UmuD' in the heterodimer may provide a way to target UmuD' for proteolysis [18].

LexA Cleavage and Other Self-cleavage Reactions Regulating the SOS Response

Researchers have known for some time that similarities exist between the self-cleavage reactions of LexA repressor, phage λ cI repressor, and UmuD [1]. All three self-cleave in a RecA/ssDNA nucleoprotein filament-dependent fashion, yet all three have autocatalytic capabilities in the absence of the RecA/ssDNA nucleoprotein filament at high pH [1]. These proteins utilize a Ser/Lys dyad active site for cleavage, and cleave either between Ala-Gly or Cys-Gly sequences. Additionally, cleavage of each of these three proteins results in distinct physiological events [1]. Thus, when the LexA repressor is cleaved, it loses its repressor activity and the genes it regulates become induced. Similarly, in the case of λ cI repressor protein, cleavage results in induction of phage genes. When UmuD undergoes cleavage, it converts itself, together with UmuC, from a DNA damage checkpoint effector to the lesion bypass polymerase, DNA Pol V.

A striking similarity between the structures of *E. coli* signal peptidase and UmuD was identified by Paetzel and Strynadka [19]. They found that the root mean square deviation (RMSD) for 69 common Cα carbon atoms between UmuD and signal peptidase was only 1.6 Å, despite the fact that there is only about 17% sequence identity between the aligned residues [19]. Signal peptidases contain a Ser/Lys dyad active site and their structures consist mainly of β-strands [19]. Several other recent structural studies involving this class of proteins have given significant insights into the control and mechanism of their cleavage reactions.

Crystal Structure of the LexA Repressor

Recently, the crystal structures of four different mutant forms of LexA were solved: two that are noncleavable due to mutations in the cleavage site, one that is noncleavable due to an active site mutation, and one that contains mutations to create a "hypercleavable" LexA coupled with an active site mutation to eliminate cleavage and thus facilitate crystallization [20]. The solved structures demonstrate that the C-terminal catalytic domain, like signal peptidase, consists mainly of β-strand structural elements [20]. Additionally, LexA also exhibits a high degree of structural similarity to UmuD', giving an RMSD of 0.96 Å for 81 C-terminal catalytic domain Cα atoms [20]. The several different solved structures of LexA exhibit two different but structured conformations in the cleavage site region: one in which the cleavage site lies nestled in the active site and the other in which the cleavage site is about 20 Å away from the active site [20].

It is proposed that, consistent with an earlier hypothesis based on a mutational analysis, the interconversion between the two structured forms is the basis for regulating the

Table I SOS-Induced Genes in *E. coli*[1]

Gene	Function/Possible function
lexA	Transcriptional repressor; regulator of SOS regulon
recA	Recombination, regulation of SOS response, role in SOS mutagenesis
umuC	DNA pol V, SOS mutagenesis
umuD	Forms complex with UmuC, role in DNA damage checkpoint
dinB/dinP	DNA pol IV
dinA/polB	DNA pol II
uvrA	Nucleotide excision repair, involved in damage recognition
uvrB	Nucleotide excision repair, involved in damage excision
uvrD	Nucleotide excision repair, helicase
ruvA	Daughter strand gap repair; Holliday junction helicase
recN	Recombinational double-strand break repair
ssb	Single-stranded binding protein
ydjQ/dinM/sosD	Putative excinuclease; Homolgous to *uvrC*
dinI	Modulates the rate of RecA/ssDNA-dependent UmuD cleavage
fis	Transcription factor
suhB	Transcriptional regulator
ttk	Putative transcriptional regulator
ydeO	Putative ARAC-type regulatory protein
ftsK	ATP-dependent DNA translocase, involved in chromosome segregation during cell division [36,37]
grxA	Glutaredoxin coenzyme for ribonucleotide reductase
ycgH	Putative ATP-binding transport system component
glvB	PTS system, arbutin-like IIB component
ybeW	Putative DnaK homolog
ibpB	Heat-shock protein
sulA	Suppressor of Lon, inhibits cell division
dinD/pcsA	DNA-damage inducible protein, cold-sensitive mutant
oraA	Regulator
dinF	Putative membrane protein [38], immediately downstream of *lexA*
sbmC	Resistance to Microcin B17
molR	Molybdate regulator
dinS	Transposase
dinG	Putative helicase
hokE/ydbY	Host killing protein, ensures stable maintenance of plasmids harboring it

[1]Largely from reference 7, with a cutoff of two-fold (LexA-dependent) induction to be included in this table. See also references 1, 2, and 5. Also noted as induced are the following proteins of unknown function: *yebG, yigF, arpB, yafO, yafN, yebF, yafP, ydiY, ydjM, yebE, ybiN, yceP, yjiW, yifL, ydeT, ysdAB, dinQ, ybfE.*

cleavage reaction [20,21]. In the two forms that are observed in the crystal structures, the catalytic core is mostly the same but there is a shift in the position of the cleavage site region. However, both conformations are stabilized by distinct sets of hydrophobic and hydrophilic interactions [20]. It is suggested that the RecA/ssDNA filament facilitates cleavage by binding preferentially to the cleavable conformation of LexA, thereby altering the equilibrium between the two forms. Extrapolation from cryoelectron microscopy experiments indicates that LexA may nestle in a groove in the RecA/ssDNA filament, thereby stabilizing the less stable cleavage-competent conformation [20,22]. Recent experiments with UmuD suggest that it may be stabilized in a cleavable conformation in a similar manner by repositioning the active site via interactions with the RecA/ssDNA filament [23,24]. The LexA side-chain ε-amino group of Lys156 is responsible for deprotonating the catalytic residue Ser119, but the ε-amino group itself is protonated in the noncleavable form of LexA [20]. The energetic cost of deprotonating and burying the ε-amino group of Lys156 in order to lower the pK_a sufficiently for catalysis likely provides an additional barrier, and thus level of control, to cleavage [20].

Structural Insights into the Cleavage of $UmuD_2$

NMR and crystal structures of $UmuD'_2$ show that the protein consists of a globular C-terminal domain that contains the dimerization domain, the catalytic residues Ser60 and Lys97, and flexible N-terminal arms [23,25]. A recent NMR study determined the structure of the $UmuD'_2$ homodimer, but the $UmuD_2$ homodimer was not amenable to NMR structure determination [23]. However, biochemical data indicate that the UmuD·D' heterodimer is a good model for the structure of the $UmuD_2$ homodimer, and NMR analyses yielded some insights into its structure [26,27]. Whereas secondary structural elements in the $UmuD'_2$ and UmuD·D' dimers are largely similar to each other and to those in the X-ray structure, the RMSD between $UmuD'_2$ homodimers in the NMR and X-ray structures is quite different at 4.59 Å [23,28]. Much of this difference can be accounted for in the more mobile exterior regions of the protein in solution [23]. In this solution structure of $UmuD'_2$, the active site Ser60 and Lys97 residues are not near enough to affect catalysis, whereas in the crystal structure, the ε-amino group of Lys97 points directly at the side-chain of Ser60 [23,25]. This suggests that the conformation of the C-terminal domain of UmuD' in the crystal may approximate its conformation when it is in contact with the RecA/ssDNA filament, which is required for cleavage [23]. The conformation of the N-terminal arm of UmuD as determined by cross-linking is similar to that of the linker region of LexA in the cleavable conformation [20,27]. Cleavage of $UmuD_2$ occurs intermolecularly, while the crystal structure of LexA indicates that cleavage in this case is intramolecular [20,23]. This may indicate an idiosyncratic feature of this family, but it is possible that subtle changes in the conformation of the linker region on LexA could produce intermolecular cleavage in this case as well [20].

Structures of Y-Family Polymerases

There are now crystal structures of the Y-family polymerases *Sulfolobus solfataricus* DinB and *Saccharomyces cerevisiae* Polη [29–32]. There are four reported structures of two different *S. solfataricus* DinBs (Dbh and Dpo4), including Dpo4 with DNA substrates bound [29,30,32]. In general, the Y-family polymerases possess the canonical shape of a right hand that is characteristic of other polymerases, but with shorter finger and thumb domains [33,34]. All of the structures determined thus far are observed in the "closed" conformation, whether they have substrate bound or not [33,34]. Thus, it appears that the PolY family polymerases do not use the same type of induced fit mechanism as the replicative polymerases, which alternate from open to closed conformations upon DNA binding.

In the structure of the complex, Dpo4 makes limited and nonspecific contacts with the newly forming base pair [29]. Moreover, the bound dsDNA is B-form, which does not allow the protein to access the minor groove to check the fidelity of the new base pair [29]. The minor groove of the DNA substrate bound to Dpo4 is solvent accessible, whereas in classical DNA polymerases it is bound by the protein. The DNA substrate bound to classical DNA polymerases is typically A-form, allowing the protein to bind the minor groove to check the fidelity of replication [35]. Some of these structures also show the presence of appended domains that are likely to play a role in providing additional DNA binding energy [29–31]. One of these, the "little finger" domain in the Dpo4 structure, is observed binding the double-stranded DNA in the major groove [29]. Finally, Dpo4 accommodates a bulged template base in its active site, which also provides enough space to bind a cyclobutane thymine dimer and may in part explain its substrate specificity [29].

Conclusions

The ultimate result of DNA damage, whether accurate repair such as nucleotide excision repair, homologous recombination, or translesion synthesis, is governed by several levels of control including transcriptional activation and posttranslational modification. Translesion synthesis itself is accomplished by a cascade of signaling events achieved through noncovalent interactions as well as covalent modification (self-cleavage) of several of the proteins involved. Translesion synthesis is executed by a specialized class of polymerases that appear to be uniquely suited to their role as low-fidelity DNA polymerases. It seems certain that there are many more insights to be gained into when and with which tools cells respond to an assault on their genetic material.

Acknowledgments

We acknowledge the financial support of Public Health Service grant CA21615 from the National Cancer Institute to GCW and Postdoctoral Fellowship DRG1648 from the Damon Runyon Cancer Research Foundation to PJB, and Dr. Veronica Godoy for helpful suggestions.

References

1. Friedberg, E. C., Walker, G. C., and Siede, W. (1995). *DNA Repair and Mutagenesis.* ASM Press, Washington, DC.
2. Koch, W. H. and Woodgate, R. (1998). The SOS response, in Nickoloff, J. A., and Hoekstra, M. F., Eds., *DNA Damage and Repair, Vol 1: DNA Repair in Prokaryotes and Lower Eukaryotes,* pp. 107–134. Humana Press, Totowa, NJ.
3. Sutton, M. D., Smith, B. T., Godoy, V. G., and Walker, G. C. (2001). The SOS response: Recent insights into *umuDC*-dependent mutagenesis and DNA damage tolerance. *Ann. Rev. Genet.* **34**, 479–497.
4. Walker, G. C., Smith, B. T., and Sutton, M. D. (2000). The SOS response to DNA damage, in Storz, G., and Hengge-Aronis, R., Eds., *Bacterial Stress Responses,* pp. 131–144. ASM Press, Washington, DC.
5. Fernandez de Henestrosa, A. R., Ogi, T., Aoyagi, S., Chafin, D., Hayes, J. J., Ohmori, H., and Woodgate, R. (2000). Identification of additional genes belonging to the LexA regulon in Escherichia coli. *Mol. Microbiol.* **35**, 1560–1572.
6. Ohmori, H., Friedberg, E. C., Fuchs, R. P. P., Goodman, M. F., Hanaoka, F., Hinkle, D., Kunkel, T. A., Lawrence, C. W., Livneh, Z., Nohmi, T., Prakash, L., Prakash, S., Todo, T., Walker, G. C., Wang, Z., and Woodgate, R. (2001). The Y-family of DNA polymerases. *Mol. Cell,* **8**, 7–8.
7. Courcelle, J., Khodursky, A., Peter, B., Brown, P. O., and Hanawalt, P. C. (2001). Comparative gene expression profiles following UV exposure in wild-type and SOS-deficient *Escherichia coli. Genetics* **158**, 41–64.
8. Reuven, N. B., Arad, G., Maor-Shoshani, A., and Livneh, Z. (1999). The mutagenesis protein UmuC is a DNA polymerase activated by UmuD', RecA, and SSB and is specialized for translesion replication. *J. Biol. Chem.* **274**, 31763–31766.
9. Tang, M., Shen, X., Frank, E. G., O'Donnell, M., Woodgate, R., and Goodman, M. F. (1999). UmuD'(2)C is an error-prone DNA polymerase, *Escherichia coli* pol V. *Proc. Natl. Acad. Sci. USA* **96**, 8919–8924.
10. Opperman, T., Murli, S., Smith, B. T., and Walker, G. C. (1999). A model for a umuDC-dependent prokaryotic DNA damage checkpoint. *Proc. Natl. Acad. Sci. USA* **96**, 9218–9223.
11. Sutton, M. D., Opperman, T., and Walker, G. C. (1999). The *Escherichia coli* SOS mutagenesis proteins UmuD and UmuD' interact physically with the replicative DNA polymerase. *Proc. Natl. Acad. Sci. USA* **96**, 12373–12378.
12. Sommer, S., Bailone, A., and Devoret, R. (1993). The appearance of the UmuD'C protein complex in *Escherichia coli* switches repair from homologous recombination to SOS mutagenesis. *Mol. Microbiol.* **10**, 963–971.
13. Yasuda, T., Morimatsu, K., Kato, R., Usukura, J., Takahashi, M., and Ohmori, H. (2001). Physical interactions between DinI and RecA nucleoprotein filament for the regulation of SOS mutagenesis. *EMBO J.* **20**, 1192–1202.
14. Yasuda, T., Morimatsu, K., Horii, T., Nagata, T., and Ohmori, H. (1998). Inhibition of *Escherichia coli* RecA coprotease activities by DinI. *EMBO J.* **17**, 3207–3216.
15. Frank, E. G., Ennis, D. G., Gonzalez, M., Levine, A. S., and Woodgate, R. (1996). Regulation of SOS mutagenesis by proteolysis. *Proc. Natl. Acad. Sci. USA* **93**, 10291–10296.
16. Gonzalez, M., Frank, E. G., Levine, A. S., and Woodgate, R. (1998). Lon-mediated proteolysis of the *Escherichia coli* UmuD mutagenesis protein: in vitro degradation and identification of residues required for proteolysis. *Genes Dev.* **12**, 3889–3899.
17. Battista, J. R., Ohta, T., Nohmi, T., Sun, W., and Walker, G. C. (1990). Dominant negative umuD mutations decreasing RecA-mediated cleavage suggest roles for intact UmuD in modulation of SOS mutagenesis. *Proc. Natl. Acad. Sci. USA* **87**, 7190–7194.
18. Gonzalez, M., Frank, E. G., McDonald, J. P., Levine, A. S., and Woodgate, R. (1998). Structural insights into the regulation of SOS mutagenesis. *Acta Biochim. Pol.* **45**, 163–172.

19. Paetzel, M. and Strynadka, N. C. (1999). Common protein architecture and binding sites in proteases utilizing a Ser/Lys dyad mechanism. *Protein Sci.* **8**, 2533–2536.
20. Luo, Y., Pfuetzner, R. A., Mosimann, S., Paetzel, M., Frey, E. A., Cherney, M., Kim, B., Little, J. W., and Strynadka, N. C. J. (2001). Crystal structure of LexA: A conformational switch for regulation of self-cleavage. *Cell* **106**, 585–594.
21. Roland, K. L., Smith, M. H., Rupley, J. A., and Little, J. W. (1992). *In vitro* analysis of mutant LexA proteins with an increased rate of specific cleavage. *J. Mol. Biol.* **228**, 395–408.
22. Yu, X. and Egelman, E. H. (1993). The LexA repressor binds within the deep helical groove of the activated RecA filament. *J. Mol. Biol.* **231**, 29–40.
23. Ferentz, A. E., Walker, G. C., and Wagner, G. (2001). Converting a DNA damage checkpoint effector ($UmuD_2C$) into a lesion bypass polymerase ($UmuD'_2C$). *EMBO J.* **20**, 4287–4298.
24. Sutton, M., Kim, M., and Walker, G. C. (2001). Genetic and biochemical characterization of a novel *umuD* mutation: Insights into a mechanism for UmuD self-cleavage. *J. Bact.* **183**, 347–357.
25. Peat, T. S., Frank, E. G., McDonald, J. P., Levine, A. S., Woodgate, R., and Hendrickson, W. A. (1996). Structure of the UmuD' protein and its regulation in response to DNA damage. *Nature* **380**, 727–730.
26. Lee, M. H., Ohta, T., and Walker, G. C. (1994). A monocysteine approach for probing the structure and interactions of the UmuD protein. *J. Bacteriol.* **176**, 4825–4837.
27. Sutton, M. D., Guzzo, A., Narumi, I., Costanzo, M., Altenbach, C., Ferentz, A. E., Hubbell, W. L., and Walker, G. C. (2002). A model for the structure of the *Escherichia coli* SOS-regulated $UmuD_2$ protein. *DNA Repair* **1**, 77–93.
28. Peat, T. S., Frank, E. G., McDonald, J. P., Levine, A. S., Woodgate, R., and Hendrickson, W. A. (1996). The UmuD' protein filament and its potential role in damage induced mutagenesis. *Structure* **4**, 1401–1412.
29. Ling, H., Boudsocq, F., Woodgate, R., and Yang, W. (2001). Crystal structure of a Y-family DNA polymerase in action: A mechanism for error-prone and lesion-bypass replication. *Cell* **107**, 91–102.
30. Silvian, L., Toth, E., Pham, P., Goodman, M. F., and Ellenberger, T. (2001). Crystal structure of a DinB family error-prone DNA polymerase from *Sulfolobus solfataricus. Nature Struct. Bio.* **8**, 984–989.
31. Trincao, J., Johnson, R. E., Escalante, C. R., Prakash, S., Prakash, L., and Aggarwal, A. K. (2001). Structure of the catalytic core of *S. cerevisiae* DNA polymerase eta: Implications for translesion DNA synthesis. *Mol. Cell* **8**, 417–426.
32. Zhou, B. L., Pata, J. D., and Steitz, T. A. (2001). Crystal structure of a DinB lesion bypass DNA polymerase catalytic fragment reveals a classic polymerase catalytic domain. *Mol. Cell* **8**, 427–437.
33. Ellenberger, T. and Silvian, L. F. (2001). The anatomy of infidelity. *Nature Struct. Bio.* **8**, 827–828.
34. Friedberg, E. C., Fischhaber, P. L., and Kisker, C. (2001). Error-prone DNA polymerases: Novel structures and the benefits of infidelity. *Cell* **107**, 9–12.
35. Doublie, S., Sawaya, M. R., and Ellenberger, T. (1999). An open and closed case for all polymerases. *Structure Fold Des* **7**, R31–R35.
36. Aussel, L., Barre, F.-X., Aroyo, M., Stasiak, A., Stasiak, A. Z., and Sherratt, D. (2002). FtsK is a DNA motor protein that activates chromosome dimer resolution by switching the catalytic state of the XerC and XerD recombinases. *Cell* **108**, 195–205.
37. Donachie, W. D. (2002). FtsK: Maxwell's Demon. *Mol. Cell* **9**, 206–207.
38. Brown, M. H., Paulsen, I. T., and Skurray, R. A. (1999). The multidrug efflux protein NorM is a prototype of a new family of transporters. *Mol. Microbiol.* **31**, 394.

CHAPTER 300

Oxidative Stress and Free Radical Signal Transduction

Bruce Demple
Department of Cancer Cell Biology
Harvard School of Public Health, Boston, Massachusetts

Introduction: Redox Biology

During the past two decades, considerable attention has been focused on issues related to oxidative stress and the biology of free radicals. *Oxidative stress* refers to excess production of various reactive species (H_2O_2, superoxide, singlet oxygen, etc.) or to insufficient antioxidant defenses in the cell [glutathione (GSH), superoxide dismutase, catalase, etc.] [1]. Environmental exposure produces oxidative stress through airway injury, irradiation, and exposure to compounds that disrupt cellular electron transfer pathways. Actively generated free radicals can play physiological roles, including in cellular signaling pathways [2–5]. High-level production of superoxide and another free radical, nitric oxide, contributes to immunological attack on pathogens and tumor cells [6,7]. The main cellular components are all subject to damage during oxidative stress, so it is not surprising to find that cells have evolved pathways to respond dynamically to oxidative stress by altering gene expression [2,8].

Oxidative Stress Responses in Bacteria: Well-Defined Models of Redox Signal Transduction

Study of oxidative stress responses has focused especially on *Escherichia coli* and *Salmonella enterica* (*typhimurium*) as model systems amenable to combined genetic and biochemical analysis [5,9,10]. Studies in the Gram-positive bacterium *Bacillus subtilis* [11], or in medically or agriculturally important organisms such as *Pseudomonas aeruginosa* [12] and *Xanthomonas oryzae* [13], are now beginning to yield similar fundamental insights in those systems.

Response to Superoxide Stress and Nitric Oxide: SoxR Protein

The SoxR regulatory system of *E. coli* (Fig. 1) was discovered by genetic analysis of strains with elevated defenses against superoxide-generating agents such as paraquat and quinones [14,15]. The system works in two stages of transcriptional control; only the first stage involves direct redox signal transduction [16,17]. As described later, SoxR protein is activated in cells exposed to oxidative stress or nitric oxide; activated SoxR then stimulates expression of just one gene, *soxS*. The resulting increase in the level of SoxS protein, itself a transcription activator in the AraC/XylS family [18], leads to the up-regulation of 42 genes and the repression of 58 genes [19]. Many of the induced genes have obvious functions in oxidative defense (e.g., Mn-containing superoxide dismutase, or endonuclease IV for oxidative DNA damage), but many others (e.g., the *acrAB*-encoded efflux pumps) serve broader functions [20].

The redox-sensing SoxR protein is a homodimer of 17-kDa subunits, each of which contains a single [2Fe-2S] iron-sulfur cluster anchored to the protein's only four cysteine residues near the C terminus [21]. In resting SoxR, these clusters are in the reduced (+2) state, which can be detected *in vivo* using electron paramagnetic resonance spectroscopy [22,23]. Upon cellular exposure to agents such as paraquat, which sets up a heavy flux of superoxide in the cell and consumes NADPH by cyclic reduction and autoxidation [24], the [2Fe-2S] centers are oxidized to the +3 state. This one-electron oxidation converts SoxR to a powerful activator of *soxS* transcription, which elevates the soxS mRNA level up to 100-fold [4].

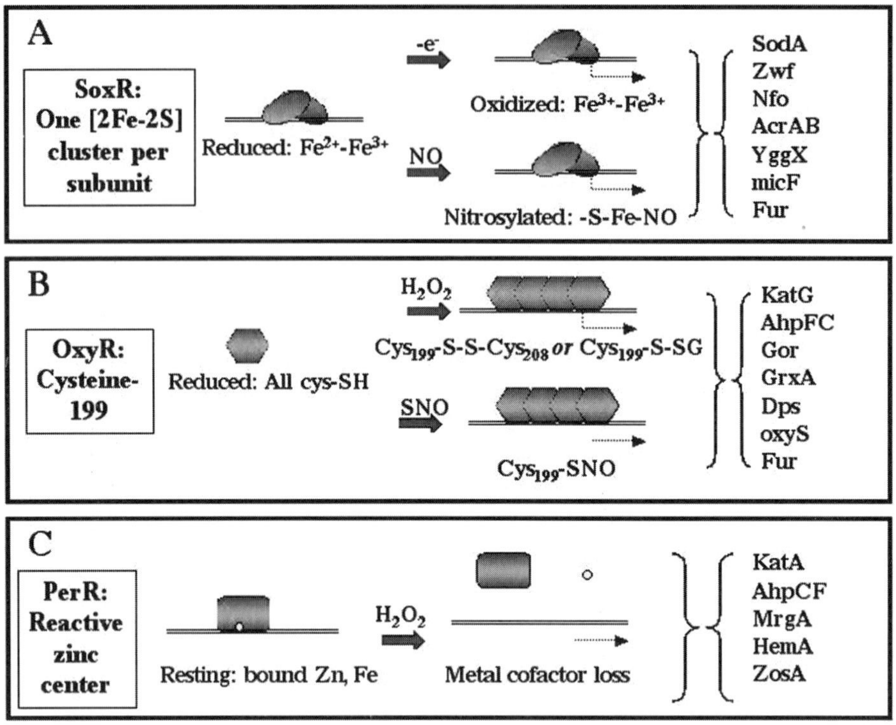

Figure 1 Representative systems of redox regulation. Each box depicts the main aspects of one of the regulatory systems described in the text. In all three cases, transcriptional activation is indicated by a dotted, right-angle arrow at the target DNA (thin parallel lines). (A) The SoxR system of *E. coli*. Each 17-kDa subunit of the homodimer contains a [2Fe-2S] cluster, which is in the reduced state in the absence of free-radical stress. Nonactivated SoxR binds its DNA site in the *soxS* promoter with high affinity. Oxidative stress exerted by superoxide-generating agents (such as paraquat) one-electron oxidizes the centers, which converts SoxR to a potent transcriptional activator without changing its DNA-binding affinity. Alternatively, nitric oxide activates SoxR by reacting with the iron-sulfur centers to form dinitrosyl-iron complexes anchored via cysteine residues in the protein. Activation of SoxR triggers expression of the *soxS* gene, whose protein product (not shown) then activates the many genes of the regulon. Among the many functions stimulated by activated SoxR are manganese-containing superoxide dismutase (SodA), glucose-6-phosphate dehydrogenase (Zwf), the DNA repair enzyme endonuclease IV (Nfo), the AcrAB efflux pump, and the YggX protein involved in maintenance of iron-sulfur clusters (P. Pomposiello, A. Koutsolioutsou, and B. Demple, submitted). In addition, secondary regulators are under SoxR control: the micF antisense RNA that down-regulates OmpF expression; and Fur protein, which controls many genes involved in iron uptake. (B) The OxyR system of *E. coli*. Resting OxyR is in the reduced state and does not bind most of its DNA targets. Oxidation by H_2O_2 causes structural changes that convert the protein to a DNA-binding tetramer. Cysteine-199 is the critical residue for this response and forms a disulfide, although there is controversy (see text) as to whether this involves cysteine-208 or a molecule of glutathione. A form of OxyR with a sulfenic acid based on cysteine-199 (not shown) has also been reported. In addition, nitrosothiols produce a different activated form of OxyR containing S-nitrosylated cysteine-199. Activation of OxyR stimulates the expression of many genes not activated by SoxR, although there is a limited overlap (e.g., Fur protein). The functions include catalase (KatG), alkyl hydroperoxide reductase (AhpFC), GSH reductase (Gor), glutaredoxin-1 (GrxA), and the protective DNA-binding protein Dps. As seen with SoxR, secondary regulators control still more genes; in addition to the Fur repressor, the oxyS RNA acts posttranscriptionally on the expression of several other genes. (C) The PerR system of *B. subtilis*. The Fur homolog PerR is a repressor of genes that respond to H_2O_2. The active (repressing) form of PerR contains zinc and iron (small circle in the cartoon); oxidation causes these metals to be lost from the protein and DNA-binding affinity is lowered dramatically. Functions regulated by PerR include catalase (KatA), AhpFC, the protective DNA-binding protein MrgA, HemA and other heme biosynthetic enzymes, and the ZosA protein involved in zinc uptake.

These events can be reproduced *in vitro* using purified SoxR protein [25,26]. SoxR isolated under air has iron-sulfur clusters in the oxidized state, and the protein has full transcriptional activity specific for *soxS*. SoxR stimulates transcription by increasing the rate of open complex formation by RNA polymerase [27]. Chemical reduction using sodium dithionite yields the +2 state of the [2Fe-2S] clusters, which switches off the transcriptional activity. The activity is restored by reoxidation [25,26].

SoxR binds tightly to its *soxS* target site independently of the iron-sulfur centers [28] or their oxidation state [26]. The SoxR binding site is centered between the –35 and –10 elements of the *soxS* promoter [27]. This promoter has an unusually long spacer of 19 bp (versus the usual 17±1 bp) between the RNA polymerase recognition elements, which are otherwise good matches to consensus sequences. The *soxS* promoter has very weak basal activity, but mutant promoters with spacers of 18, 17, or 16 bp are highly expressed *in vivo* independent of SoxR [29]. The model for SoxR transcription activation is analogous to that for the Hg^{2+}-sensing MerR protein [30]: DNA structural remodeling that compensates for the excessively long spacer. Recent crystallographic analysis of the SoxR-related BmrR protein [31] is consistent with this view.

SoxR is also activated by nitric oxide, and the chemistry of this process is fundamentally different from the one-electron oxidation just described. Nitric oxide directly modifies the [2Fe-2S] clusters to form dinitrosyl-iron-dicysteine complexes anchored to the protein [32]. These nitrosylated complexes are quite stable *in vitro*, which allowed NO-modified SoxR to be repurified for analysis. SoxR activation by NO may play a role in bacterial resistance to the cytotoxic attack of NO-generating macrophages [33,34].

Both the oxidation and the nitrosylation of SoxR are very rapidly reversed in intact bacteria when the oxidative stress or NO exposure is stopped [32]. The former observation, together with maintenance of SoxR in the reduced form in the absence of stress, implies that the [2Fe-2S] centers are actively reduced, probably enzymatically. Although a report of a SoxR reductase activity [35] has not been followed up, a recent paper presents genetic evidence of a SoxR-reducing system [54].

The reversal of the dinitrosyl complexes in NO-treated SoxR to regenerate unmodified [2Fe-2S] clusters may be of even more general significance. Nitrosylation of various types of iron-sulfur clusters is a general feature of NO toxicity in all cell types [36]. Delineating the mechanism by which such protein damage is corrected (possibly by removal of the nitrosylated clusters and the resynthesis of new ones) should provide general insights into how cells handle this basic problem.

Response to H_2O_2 and Nitrosothiols: OxyR Protein

The *oxyR* regulatory system (Fig. 1) was identified by genetic analysis of *S. enterica* and *E. coli* stains in which inducible resistance to H_2O_2 [37] was activated constitutively [38]. OxyR protein both senses oxidative stress and activates the >60 genes of the *oxyR* regulon [39]. OxyR protein is a member of the LysR family of transcriptional regulators [40], and close homologs of the *oxyR* gene are widely represented in bacterial genomes [5].

Like SoxR, OxyR is a positive regulator that exists in a resting state in nonstressed cells [4,5]. Upon cellular exposure to H_2O_2 or nitrosothiols [41,42], the protein is activated to stimulate transcription of the regulon genes. Unlike SoxR, the activity of OxyR is regulated largely by increased DNA binding [43,44], as is the case for many prokaryotic signal transduction systems. The positioning of the protein upstream of the RNA polymerase binding site stimulates transcription through protein–protein contacts with the C-terminal domain of the polymerase α subunit [45]. The *oxyR* regulon includes genes such as *katG* (HP-I catalase-hydroperoxidase), *gor* (GSH reductase), and the *ahpFC* operon (alkyl hydroperoxide reductase) [39].

The molecular basis for OxyR activation has been explored to yield a detailed model of both the redox activation and the stimulation of transcription [46]. However, the mechanism of redox sensing is still subject to some debate [41]. At the DNA-binding level, redox activation converts OxyR from a predominantly dimeric form to a tetramer, which binds consecutive repeats of a recognition element and bends the DNA [43].

The proposed mechanism of redox activation of OxyR by H_2O_2 is the formation of an intramolecular disulfide in the protein [46], and genetic and biochemical evidence supports this view. For example, of the six cysteine residues in the OxyR polypeptide, one (cysteine-199) is absolutely required for full activity in response to oxidative stress, and the other (cysteine-208) has a significant effect. Peptides have been isolated from oxidized OxyR that contain a disulfide bond linking cysteines 199 and 208 (although the other four cysteine residues of the protein had been replaced by alanines for this experiment). These observations led to a model in which oxidative activation of OxyR occurs through the formation of an intramolecular disulfide bond linking cysteine-199 and cysteine-208. The structural basis for this specificity is proposed to be both the proximity of the two cysteine residues, and a basic residue near cysteine-199 that increases its reactivity with oxidants [44].

A recent report challenges this view [41]. That work indicates that there are various activated forms of OxyR, none of which contains the cysteine-199/cysteine-208 disulfide. Instead, Kim *et al.* [41] identified a form with cysteine-199 as a sulfenic acid (-SOH) in air-oxidized OxyR, or with a mixed disulfide linking cysteine-199 and GSH [41]. Such mixed disulfides were detected biochemically as a prominent product in OxyR activated *in vivo* [41]. However, note that GSH is not necessary to activate OxyR *in vivo* [46].

GSH and glutaredoxin-1 (a small, thioredoxin-like protein with a redox-active cysteine pair) evidently play physiological roles in restoring reduced OxyR following oxidative stress [3,47]. The overall view is that these proteins participate in thiol/disulfide exchange reactions that yield reduced OxyR, ultimately at the expense of NADPH through the activity of GSH reductase. This control would establish a negative feedback loop in the system, since the *gor* and *grx1* (glutaredoxin-1) genes are activated by OxyR [3,47]. However, *gor* and *grx1* are not required to maintain OxyR in the reduced state in unstressed cells [46].

Nitrosothiols have also been shown to activate OxyR, which was proposed to occur by generating a cysteine nitrosothiol (-SNO) in the protein [42]. New data support this view [41], but also show that different activated forms (-SNO, -SOH, and –S-SG) have variable affinity and transcription-activating capacity for different promoters. The discrepancy with earlier studies [46] was rationalized by pointing out that they had employed mutant forms of OxyR with four of the six cysteines substituted by other amino acids [41]. These observations point to a general phenomenon in which different stress signals give rise to different patterns of gene activation even while operating on the same sensor/transducer protein. For the SoxR/SoxS system, current data indicate that different activated forms of SoxR would merely influence the *level* of SoxS protein. Differences in the pattern of gene expression would then reflect the degree of SoxR activation, rather than changes in its specificity.

Parallels in Redox and Free-Radical Sensing

Systems that seem to share the thiol/disulfide mechanism of redox sensing used by OxyR have now been identified in several contexts. A particularly interesting example is the *E. coli* Hsp33 chaperone, which is activated by H_2O_2 through the formation of intramolecular disulfide bonds [48]. This activation might provide for increased cellular capacity to handle oxidatively damaged proteins. The four cysteines involved are ligands for a zinc atom in the resting protein; oxidation eliminates the metal and activates the chaperone function [49]. A similar mechanism may govern the redox activation of the *B. subtilis* Fur homolog PerR, which governs an oxidative stress regulon (Fig. 1) [50]. Reversible disulfide bond formation (whether intramolecular or mixed) makes physiological sense as a regulatory mechanism because it links redox sensing to overall metabolism through NADPH and the oxidation state of intermediaries such as thioredoxin and glutaredoxin [5,47].

Disulfide bond formation in eukaryotic regulatory proteins has also been proposed. A comprehensive review of this and related questions for many mammalian systems has appeared recently [2], and the topic will not be further covered here. In plants, mRNA translation in *Chlamydomonas reinhardtii* chloroplasts is governed by disulfide bond formation in a regulatory protein [51,52]. Other examples will surely emerge, though we await a eukaryotic example of *direct activation* occurring through disulfide bond formation, and the role of mixed disulfides needs further exploration.

Examples analogous to the SoxR mechanism remain to be described, although the facile nature of one-electron oxidation and reduction of an iron-sulfur cluster should lend itself to redox signaling in various contexts. Another prokaryotic example of redox sensing using an iron-sulfur cluster is the *E. coli* Fnr protein, which acts as a repressor of genes involved in aerobic metabolism. Fnr harbors a [4Fe-4S] cluster that is sensitive to oxygen; exposure to O_2 converts the single cluster to a pair of [2Fe-2S] clusters and releases the protein from the DNA [53].

Themes in Redox Sensing

The ability of cellular proteins to detect changes in redox status and the presence of potentially toxic free radicals is likely a fundamental aspect of biology. Known examples of redox-regulated responses already include transcriptional, translational, and enzymatic control. Other key features might include feedback loops that return the system to a "ground state" (e.g., the role of reduced glutathione and glutaredoxin-1 in OxyR regulation) and the use of molecular "canaries in the coal mine"; that is, reactions that damage and inactivate most proteins actually switch on sensors such as OxyR and SoxR.

Acknowledgments

Work in the author's laboratory has been supported by the National Cancer Institute (grants R01-CA37831 and R01-CA82737) and the Kresge Center for Environmental Health (P30-ES00002).

References

1. Sies, H. (1991). Oxidative stress: Introduction, in Sies, H., Ed., *Oxidative Stress: Oxidants and Antioxidants,* pp. xv–xxii. Academic Press, London.
2. Droge, W. (2002). Free radicals in the physiological control of cell function. *Physiol. Rev.* **82**, 47–95.
3. Carmel-Harel, O. and Storz, G. (2000). Roles of the glutathione- and thioredoxin-dependent reduction systems in the *Escherichia coli* and *Saccharomyces cerevisiae* responses to oxidative stress. *Annu. Rev. Microbiol.* **54**, 439–461.
4. Hidalgo, E. and Demple, B. (1996). Adaptive responses to oxidative stress: The *soxRS* and *oxyR* regulons, in Lin, E. C. C., and Lynch, A. S., Eds., *Regulation of Gene Expression in Escherichia coli,* pp. 435–452. R. G. Landes, Austin, TX.
5. Zheng, M. and Storz, G. (2000). Redox sensing by prokaryotic transcription factors. *Biochem. Pharmacol.* **59**, 1–6.
6. Babior, B. M. (1992). The respiratory burst oxidase. *Enzymol. Rel. Areas Mol. Biol.* **65**, 49–65.
7. MacMicking, J., Xie, Q. W., and Nathan, C. (1997). Nitric oxide and macrophage function. *Annu. Rev. Immunol.* **15**, 323–350.
8. Finkel, T. (2000). Redox-dependent signal transduction. *FEBS Lett.* **476**, 52–54.
9. Pomposiello, P. J. and Demple, B. (2001). Redox-operated genetic switches: The SoxR and OxyR transcription factors. *Trends Biotechnol.* **19**, 109–114.
10. Pomposiello, P. J. and Demple, B. (2002). Global adjustment of microbial physiology during free radical stress, in Poole, R. K., Ed., *Advances in Microbial Physiology,* Vol. 46, pp. 319–341. Academic Press, London.
11. Hecker, M. and Volker, U. (2001). General stress response of *Bacillus subtilis* and other bacteria. *Adv. Microb. Physiol.* **44**, 35–91.
12. Hassett, D. J., Ma, J. F., Elkins, J. G., McDermott, T. R., Ochsner, U. A., West, S. E., Huang, C. T., Fredericks, J., Burnett, S., Stewart, P. S., McFeters, G., Passador, L., and Iglewski, B. H. (1999). Quorum sensing in *Pseudomonas aeruginosa* controls expression of catalase and superoxide dismutase genes and mediates biofilm susceptibility to hydrogen peroxide. *Mol. Microbiol.* **34**, 1082–1093.
13. Sukchawalit, R., Loprasert, S., Atichartpongkul, S., and Mongkolsuk, S. (2001). Complex regulation of the organic hydroperoxide resistance gene (*ohr*) from *Xanthomonas* involves OhrR, a novel organic peroxide-inducible negative regulator, and posttranscriptional modifications. *J. Bacteriol.* **183**, 4405–4412.
14. Greenberg, J. T., Monach, P., Chou, J. H., Josephy, P. D., and Demple, B. (1990). Positive control of a global antioxidant defense regulon activated by superoxide-generating agents in *Escherichia coli. Proc. Natl. Acad. Sci. USA* **87**, 6181–6185.
15. Tsaneva, I. R. and Weiss, B. (1990). *soxR,* A locus governing a superoxide response regulon in *Escherichia coli* K-12. *J. Bacteriol.* **172**, 4197–4205.

16. Wu, J. and Weiss, B. (1992). Two-stage induction of the *soxRS* (superoxide response) regulon of *Escherichia coli. J. Bacteriol.* **174**, 3915–3920.
17. Nunoshiba, T., Hidalgo, E., Amabile Cuevas, C. F., and Demple, B. (1992). Two-stage control of an oxidative stress regulon: The *Escherichia coli* SoxR protein triggers redox-inducible expression of the *soxS* regulatory gene. *J. Bacteriol.* **174**, 6054–6060.
18. Li, Z. and Demple, B. (1994). SoxS, an activator of superoxide stress genes in *Escherichia coli.* Purification and interaction with DNA. *J. Biol. Chem.* **269**, 18371–18377.
19. Pomposiello, P. J., Bennik, M. H., and Demple, B. (2001). Genome-wide transcriptional profiling of the *Escherichia coli* responses to superoxide stress and sodium salicylate. *J. Bacteriol.* **183**, 3890–3902.
20. White, D. G., Goldman, J. D., Demple, B., and Levy, S. B. (1997). Role of the *acrAB* locus in organic solvent tolerance mediated by expression of MarA, SoxS, or RobA in *Escherichia coli. J. Bacteriol.* **179**, 6122–6126.
21. Bradley, T. M., Hidalgo, E., Leautaud, V., Ding, H., and Demple, B. (1997). Cysteine-to-alanine replacements in the *Escherichia coli* SoxR protein and the role of the [2Fe-2S] centers in transcriptional activation. *Nucleic Acids Res.* **25**, 1469–1475.
22. Gaudu, P., Moon, N., and Weiss, B. (1997). Regulation of the *soxRS* oxidative stress regulon. Reversible oxidation of the Fe-S centers of SoxR *in vivo. J. Biol. Chem.* **272**, 5082–5086.
23. Hidalgo, E., Ding, H., and Demple, B. (1997). Redox signal transduction: Mutations shifting [2Fe-2S] centers of the SoxR sensor-regulator to the oxidized form. *Cell* **88**, 121–129.
24. Kappus, H. and Sies, H. (1981). Toxic drug effects associated with oxygen metabolism: Redox cycling and lipid peroxidation. *Experientia* **37**, 1233–1241.
25. Ding, H., Hidalgo, E., and Demple, B. (1996). The redox state of the [2Fe-2S] clusters in SoxR protein regulates its activity as a transcription factor. *J. Biol. Chem.* **271**, 33173–33175.
26. Gaudu, P. and Weiss, B. (1996). SoxR, a [2Fe-2S] Transcription factor is active only in its oxidized form. *Proc. Natl. Acad. Sci. USA* **93**, 10094–10098.
27. Hidalgo, E., Bollinger, J. M., Jr., Bradley, T. M., Walsh, C. T., and Demple, B. (1995). Binuclear [2Fe-2S] clusters in the *Escherichia coli* SoxR protein and role of the metal centers in transcription. *J. Biol. Chem.* **270**, 20908–20914.
28. Hidalgo, E. and Demple, B. (1994). An iron-sulfur center essential for transcriptional activation by the redox-sensing SoxR protein. *EMBO J.* **13**, 138–146.
29. Hidalgo, E. and Demple, B. (1997). Spacing of promoter elements regulates the basal expression of the *soxS* gene and converts SoxR from a transcriptional activator to a repressor. *EMBO J.* **16**, 1056–1065.
30. Summers, A. O. (1992). Untwist and shout: A heavy metal-responsive transcriptional regulator. *J. Bacteriol.* **174**, 3097–3101.
31. Heldwein, E. E. and Brennan, R. G. (2001). Crystal structure of the transcription activator BmrR bound to DNA and a drug. *Nature* **409**, 378–382.
32. Ding, H. and Demple, B. (2000). Direct nitric oxide signal transduction via nitrosylation of iron-sulfur centers in the SoxR transcription activator. *Proc. Natl. Acad. Sci. USA* **97**, 5146–5150.
33. Nunoshiba, T., deRojas-Walker, T., Wishnok, J. S., Tannenbaum, S. R., and Demple, B. (1993). Activation by nitric oxide of an oxidative-stress response that defends *Escherichia coli* against activated macrophages. *Proc. Natl. Acad. Sci. USA* **90**, 9993–9997.
34. Nunoshiba, T., deRojas-Walker, T., Tannenbaum, S. R., and Demple, B. (1995). Roles of nitric oxide in inducible resistance of *Escherichia coli* to activated murine macrophages. *Infect. Immun.* **63**, 794–798.
35. Kobayashi, K. and Tagawa, S. (1999). Isolation of reductase for SoxR that governs an oxidative response regulon from *Escherichia coli. FEBS Lett.* **451**, 227–230.
36. Drapier, J.-C. (1997). Interplay between NO and [Fe-S] clusters: Relevance to biological systems. *Methods* **11**, 319–329.
37. Demple, B. and Halbrook, J. (1983). Inducible repair of oxidative DNA damage in *Escherichia coli. Nature* **304**, 466–468.
38. Christman, M. F., Morgan, R. W., Jacobson, F. S., and Ames, B. N. (1985). Positive control of a regulon for defenses against oxidative stress and some heat-shock proteins in *Salmonella typhimurium. Cell* **41**, 753–762.
39. Zheng, M., Wang, X., Templeton, L. J., Smulski, D. R., LaRossa, R. A., and Storz, G. (2001). DNA microarray-mediated transcriptional profiling of the *Escherichia coli* response to hydrogen peroxide. *J. Bacteriol.* **183**, 4562–4570.
40. Christman, M. F., Storz, G., and Ames, B. N. (1989). OxyR, a positive regulator of hydrogen peroxide-inducible genes in *Escherichia coli* and *Salmonella typhimurium,* is homologous to a family of bacterial regulatory proteins. *Proc. Natl. Acad. Sci. USA* **86**, 3484–3488.
41. Kim, S. O., Merchant, K., Nudelman, R., Beyer, W. F., Keng, T., DeAngelo, J., Hausladen, A., and Stamler, J. S. (2002). OxyR, a molecular code for redox-related signaling. *Cell* **109**, 383–396.
42. Hausladen, A., Privalle, C. T., Keng, T., DeAngelo, J., and Stamler, J. S. (1996). Nitrosative stress: Activation of the transcription factor OxyR. *Cell* **86**, 719–729.
43. Toledano, M. B., Kullik, I., Trinh, F., Baird, P. T., Schneider, T. D., and Storz, G. (1994). Redox-dependent shift of OxyR-DNA contacts along an extended DNA-binding site: a mechanism for differential promoter selection. *Cell* **78**, 897–909.
44. Choi, H., Kim, S., Mukhopadhyay, P., Cho, S., Woo, J., Storz, G., and Ryu, S. (2001). Structural basis of the redox switch in the OxyR transcription factor. *Cell* **105**, 103–113.
45. Tao, K., Zou, C., Fujita, N., and Ishihama, A. (1995). Mapping of the OxyR protein contact site in the C-terminal region of RNA polymerase alpha subunit. *J. Bacteriol.* **177**, 6740–6744.
46. Zheng, M., Åslund, F., and Storz, G. (1998). Activation of the OxyR transcription factor by reversible disulfide bond formation. *Science* **279**, 1718–1721.
47. Åslund, F. and Beckwith, J. (1999). The thioredoxin superfamily: Redundancy, specificity, and gray-area genomics. *J. Bacteriol.* **181**, 1375–1379.
48. Jakob, U., Muse, W., Eser, M., and Bardwell, J. C. (1999). Chaperone activity with a redox switch. *Cell* **96**, 341–352.
49. Graumann, J., Lilie, H., Tang, X., Tucker, K. A., Hoffmann, J. H., Vijayalakshmi, J., Saper, M., Bardwell, J. C., and Jakob, U. (2001). Activation of the redox-regulated molecular chaperone Hsp33—a two-step mechanism. *Structure (Camb.)* **9**, 377–387.
50. Bsat, N., Herbig, A., Casillas Martinez, L., Setlow, P., and Helmann, J. D. (1998). *Bacillus subtilis* contains multiple Fur homologues: Identification of the iron uptake (Fur) and peroxide regulon (PerR) repressors. *Mol. Microbiol.* **29**, 189–198.
51. Kim, J. and Mayfield, S. P. (1997). Protein disulfide isomerase as a regulator of chloroplast translational activation. *Science* **278**, 1954–1957.
52. Fong, C. L., Lentz, A., and Mayfield, S. P. (2000). Disulfide bond formation between RNA binding domains is used to regulate mRNA binding activity of the chloroplast poly(A)-binding protein. *J. Biol. Chem.* **275**, 8275–8278.
53. Khoroshilova, N., Popescu, C., Münck, E., Beinert, H., and Kiley, P. J. (1997). Iron-sulfur cluster disassembly in the FNR protein of *Escherichia coli* by O_2: [4Fe-4S] to [2Fe-2S] conversion with loss of biological activity. *Proc. Natl. Acad. Sci. USA* **94**, 6087–6092.
54. Koo, M. S., Lee, J. H., Rah, S. Y., Yeo, W. S., Lee, J. W., Lee, K. L., Koh, Y. S., Kang, S. O., and Roe, J. H. (2003). A reducing system of the superoxide sensor SoxR in Escherichia coli. *Embo. J.* **22**, 2614–2622.

CHAPTER 301

Budding Yeast DNA Damage Checkpoint: A Signal Transduction-Mediated Surveillance System

Marco Muzi-Falconi,[1] Michele Giannattasio,[1] Giordano Liberi,[1,2] Achille Pellicciolo,[1,2] Paolo Plevani,[1] and Marco Foiani[1,2]

[1]*Dipartimento di Genetica e Biologia dei Microrganismi, Universita' di Milano, Milano, Italy*
[2]*Istituto FIRC di Oncologia Molecolare, Milano, Italy*

Introduction

The control of cell cycle progression is of fundamental importance for all cells. Alterations in the mechanisms that oversee the coordination between the various DNA metabolic processes and the alternation of the different cell cycle phases frequently lead to uncontrolled proliferation or cell death [1]. *Checkpoints* are genetically controlled systems that are deeply involved in these coordination efforts and are activated upon sensing the presence of altered structures in the cell. The molecular effect of activating these pathways is not yet fully understood. Upon detection of DNA damage and triggering of the checkpoint, a transcriptional activation response can be observed, which involves several genes acting in various DNA repair pathways, and cell cycle progression is delayed (Fig. 1A). This is supposed to provide the cell with enough time to repair the lesions before replicating or segregating the chromosomes, although no formal proof of this hypothesis has been produced yet [2]. Detection of damaged DNA during the replication phase of the cell cycle leads to an expansion of the S phase by controlling origin firing and by intervening in the stability of the replication forks [3–7]. During the S phase, tight control over the integration of the different repair pathways with the progression of the replication fork is critical for preventing potentially lethal events, such as replication fork collapse and unscheduled recombination processes. Recent work has shown that the checkpoint response indeed acts by protecting the integrity of stalled replication forks and possibly by favoring alternative replication strategies, thus allowing the DNA polymerases to overcome the lesions [5,7–10]. We can predict that mutations in these surveillance pathways would have profound effects and, in fact, are responsible for several predispositions to genome instability and tumor development.

Checkpoints are thought to be similar to signal transduction cascades with proteins, devolved to sensing the abnormalities that elicit a signal that, through a kinase cascade, is relayed to effectors that directly or indirectly control cell cycle progression and DNA metabolic processes (Fig. 1B) [11]. The basics of cell cycle checkpoints are well conserved by evolution, while most of the information obtained on the molecular details of these mechanisms derives from the study of genetically amenable simple organisms such as budding and fission yeasts. In this chapter, we provide a short summary of the state of the art with regard to the checkpoint-mediated signal transduction pathways, with special emphasis on budding yeast.

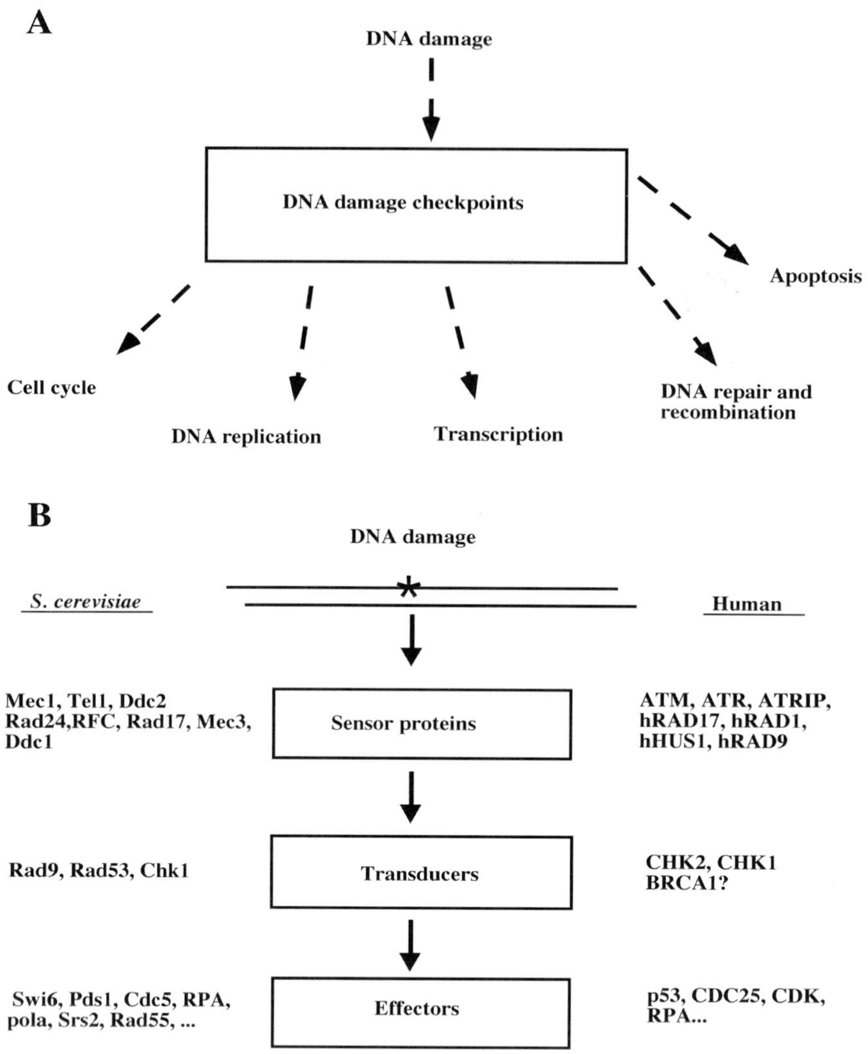

Figure 1

Sensing

Cells can recognize chromosome alterations and stalled replication forks, and this ability triggers a particular signal that results in a cascade of phosphorylation events. Very little is known about how this happens. How does the cell realize that its DNA has been damaged and how do different types of damage trigger the same response? Much work has been devoted to solve these questions, in the last few years, and there is evidence to suggest that single-stranded DNA might represent the checkpoint activating signal [7,12–14]. Genetic analysis has allowed the identification of some factors that are likely upstream in the signaling pathway. For the DNA damage checkpoint, these are represented by the Rad24, Rad17, Mec3, Ddc1, Mec1, and Ddc2/Lcd1 proteins of *Saccharomyces cerevisiae* [15–17]. All of the corresponding genes are present in all other eukaryotic cells, even though the names are different and confusing (Fig. 2).

Rad24 is a protein that bears some degree of homology to the five replication factor C (RFC) subunits and can be purified to RFC2-5 [18,19]. Rad17, Mec3, and Ddc1 are associated in a heterotrimeric complex, which *in silico* studies was predicted to have a toroidal structure very similar to that of proliferating cell nuclear antigen (PCNA) [20,21]. RFC1-5 is required to load PCNA at the primer-template interface during DNA replication; it is thus possible that a Rad24-modified RFC complex could be involved in loading the Rad17 complex onto damaged DNA. Moreover, conflicting evidence has suggested that at least one of the three subunits composing the heterotrimeric toroid might have exonucleolytic activity and thus be involved in DNA damage processing [22–24].

Mec1 is a protein kinase, of the PI3 family that is absolutely essential for checkpoint activity and is associated with the product of the *DDC2/LCD1* gene, which has been reported to mediate Mec1 binding to DNA [25–27]. The molecular details of its function and regulation have not been completely elucidated yet. All available data show that Mec1 targets are phosphorylated in response to genotoxic treatment or replication fork arrest, but this could happen in one of two ways. The kinase could be normally inactive and its catalytic activity could be triggered after the detection of some chromosomal structure (e.g., ssDNA). Because Mec1 is fairly similar to a

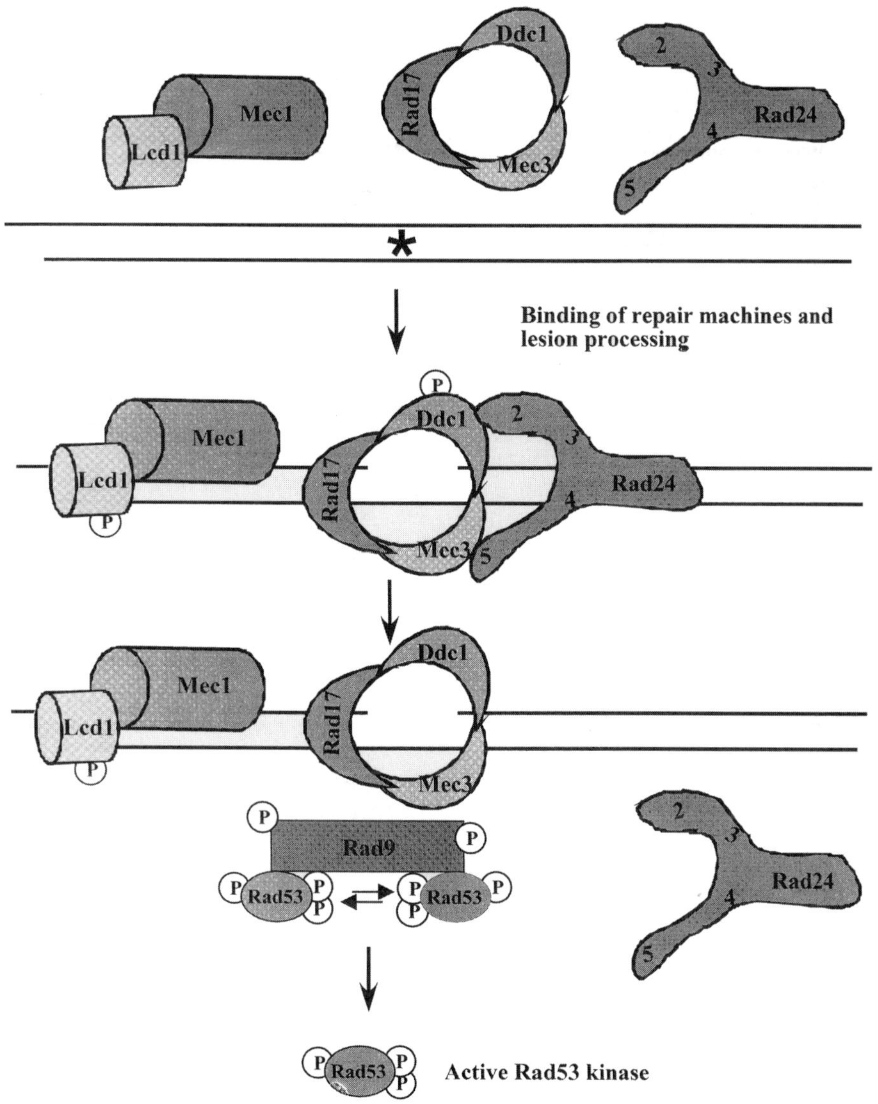

Figure 2

well-studied mammalian protein kinase (DNA PK), which binds to and is activated by DNA, one could imagine that binding of Mec1 to damaged DNA could lead to the activation of its catalytic function. On the other hand, it is also possible that the kinase could be constitutively active and only after DNA damage recognition does it encounter its substrates. In this second option, it is not the kinase catalytic activity that is regulated but rather the availability of the substrate.

Several groups, working on different organisms, have tried to obtain a clear demonstration explaining how all this could happen, although the results are somewhat contrasting [27–30]. Mec1 has several known targets and it has been demonstrated that the genetic requirements for their Mec1-dependent phosphorylation may be different. In fact, Rad53, a downstream kinase, is activated by Mec1-induced phosphorylation in a pathway that requires *RAD24* and the heterotrimeric complex [11]. On the other hand, Ddc2/Lcd1, which is presumably constitutively associated with Mec1, is phosphorylated in response to DNA damage, also in a strain lacking the *RAD24*, *RAD17*, *MEC3*, *DDC1* genes [15,26]. Thus, either we imagine a different regulation of Mec1 catalytic activity toward different targets, or we have to believe that the kinase is always active (or it could be activated upon DNA binding), and the substrates become available only following DNA damage.

Several groups have suggested that recruitment of the substrates could be, directly or indirectly, mediated by the Rad17 complex (Fig. 3) [31–34]. It is still not known how Mec1 and its substrates are brought in proximity to DNA after genotoxic treatment. Intriguingly, recently it has been shown that Mec1-Lcd1 and the Rad17 complex can be independently recruited onto damaged DNA [31–33]. It is hard to imagine that Mec1 or Lcd1 could themselves detect any kind of DNA alteration at any site of the genome. One possibility would be that DNA damage might alter the topological state of a chromosomal domain, thus unmasking the checkpoint signal and restricting the search for the lesion to limited portions of the genome. Another option, which is not mutually

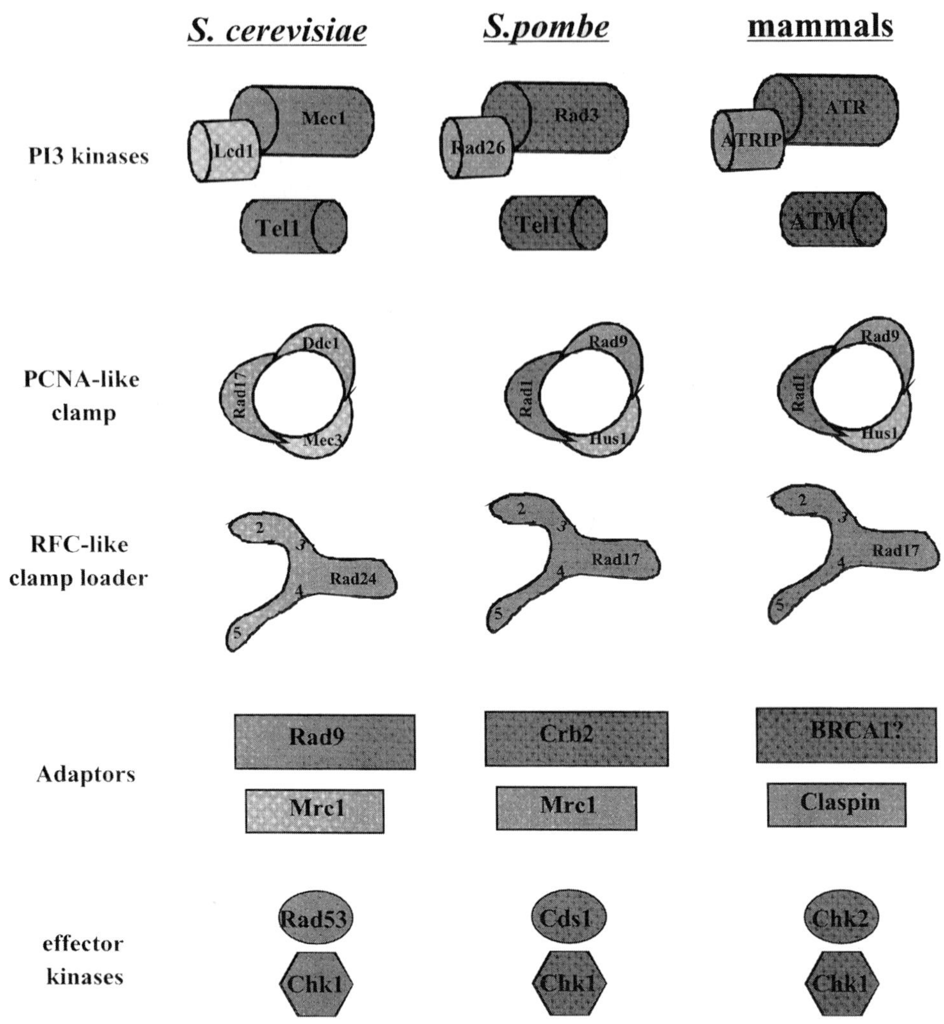

Figure 3

exclusive with the former one, predicts that it is the DNA repair proteins themselves that recognize DNA damages and by doing so trigger the checkpoint response, either by directly recruiting checkpoint proteins or by generating some intermediate DNA structure that can be recognized by the checkpoint. In this latter view the checkpoint response would be tightly connected to DNA repair functions. In this view it is intriguing that the Rad50/Xrs2/Mre11 complex, involved in double-strand break processing, has been reported to be required for DNA damage checkpoints triggered by double-strand breaks [35–37].

Downstream Events

Another important substrate of Mec1 kinase is the product of the *RAD9* gene, whose most likely human homolog is BRCA1, an oncosuppressor involved in breast cancer. The role of these modifications and how they are interconnected is not clear yet. Mec1-dependent hyperphosphorylation of Rad9 seems to be required for checkpoint activity: Hyperphosphorylated Rad9 captures Rad53 kinase molecules, promotes their transphosphorylation by Mec1 and, by increasing their local concentration, allows them to undergo autophosphorylation events that result in activation of Rad53 kinase activity. Rad9, thus functioning as a solid-state catalyst for Rad53 activation, works as an adaptor between Mec1 and Rad53 leading to signal amplification [38–42].

What happens after Rad53 has been activated is still unclear; in fact, very little is known about Rad53 targets. In G2, one possible substrate is Cdc5, a protein required for mitotic exit and for adaptation following double-strand break formation. Cdc5 is phosphorylated following genotoxic treatment in a Rad53-dependent manner; moreover genetic data suggest that Cdc5 could be a critical Rad53 target [43–45]. Swi6, a factor involved in cell-cycle-dependent transcription of several genes, among which are cyclins, has been shown to be a direct substrate of Rad53 *in vitro*, and this phosphorylation inactivates this protein [46]. Furthermore Rad53 activity causes modifications of the phosphorylative state of several enzymes involved in DNA replication, recombination, and repair [47–50].

Another kinase, Chk1, is activated by Mec1, even though the molecular details of this event are not completely clear.

It has also been reported that Chk1 phosphorylation is dependent on Rad9 and that the two proteins physically interact; this suggests the existence of a mechanism similar to the one described for Rad53 activation. Chk1 is important for the function of the G_2/M checkpoint and regulates the anaphase inhibitor Pds1. Upon phosphorylation by Chk1, Pds1 is stabilized and prevents loss of cohesion between sister chromatids, a fundamental step for the metaphase/anaphase transition, causing cells to arrest in mitosis. In G_2 cells two branches departing from Mec1 seem to amplify and transduce a signal in order to make sure that the cell cycle is arrested on detection of DNA damage [43,51].

In response to genotoxic treatments, cells also activate the so-called "DNA damage regulon" (DDR), which includes several genes from different repair pathways. Transcriptional induction of the DDR is genetically controlled through activation of the checkpoint pathway. Hence, in this way, checkpoint activation increases the intracellular concentrations of repair enzymes. Among the DDR-controlled genes is *RNR2*, which encodes a subunit of ribonucleotide reductase and whose transcription is cell cycle regulated, showing a peak at the beginning of the S phase. This is an enzyme that carries out the rate-limiting step in deoxyribonucleotides (dNTPs) synthesis and, thus, controls the pool of dNTPs within the cell. In recent years it has been shown that a deletion of *MEC1* or *RAD53* is lethal in budding yeast, but this lethality can be rescued by overproducing *RNR2* or by mutating Sml1, an inhibitor of ribonucleotide reductase.

These results indicate that the level of the intracellular pool of dNTP is limiting in yeast cells and that in order to cope with problems in DNA replication during the S phase or repair synthesis, cells have to increase dNTPs levels by upregulating *RNR2*. Intriguingly, the checkpoint feeds into the regulation of this critical activity in atleast two ways. In fact, as mentioned earlier, activation of the DNA damage response signal transduction cascade triggers transcription of the DDR genes. This is achieved through the activation, mediated by Rad53, of yet another kinase, namely, Dun1 [52]. Dun1 also phosphorylates the above-mentioned Sml1 protein and by doing this it causes Sml1 degradation by the ubiquitin pathway [53–55]. Thus, by activating Dun1, the checkpoint not only brings about more Rnr2 protein, through an increase in transcription, but also activates ribonucleotide reductase by extinguishing the Sml1 inhibitor.

The DNA damage checkpoint response is thus a very complicated signal transduction pathway, with several branches that likely interconnect and coordinate all DNA metabolic processes in the cell. In the last 5 years much has been learned about the molecular mechanisms underlying its function, but many important details are still unknown. The study of simple eukaryotes, like budding and fission yeast, thanks to their amenability to genetics studies, will be critical for the complete understanding of such complex phenomenon.

References

1. Weinert, T. and Lydall, D. (1993). Cell cycle checkpoints, genetic instability and cancer. *Semin. Cancer Biol.* **4**, 129–140.
2. Zhou, B. B. and Elledge, S. J. (2000). The DNA damage response: Putting checkpoints in perspective. *Nature* **408**, 433–439.
3. Santocanale, C. and Diffley, J. F. (1998). A Mec1- and Rad53-dependent checkpoint controls late-firing origins of DNA replication. *Nature* **395**, 615–618.
4. Shirahige, K., Hori, Y, Shiraishi, K., Yamashita, M., Takahashi, K., Obuse, C., Tsurimoto, T., and Yoshikawa, H. (1998). Regulation of DNA-replication origins during cell-cycle progression. *Nature* **395**, 618–621.
5. Lopes, M., Cotta-Ramusino, C., Pellicioli, A., Liberi, G., Plevani, P., Muzi-Falconi, M., Newlon, C. S., and Foiani, M. (2001). The DNA replication checkpoint response stabilizes stalled replication forks. *Nature* **412**, 557–561.
6. Tercero, J. A. and Diffley, J. F. (2001). Regulation of DNA replication fork progression through damaged DNA by the Mec1/Rad53 checkpoint. *Nature* **412**, 553–557.
7. Sogo, J. M., Lopes, M., and Foiani, M. (2002). Fork reversal and ssDNA accumulation at stalled replication forks de to checkpoint defects. *Science* (in press).
8. Foiani, M., Ferrari, M., Liberi, G., Lopes, M., Lucca, C., Marini, F., Pellicioli, A., Falconi, M. M., and Plevani, P. (1998). S-phase DNA damage checkpoint in budding yeast. *Biol. Chem.* **379**, 1019–1023.
9. Rhind, N. and Russell, P. (2000). Checkpoints: It takes more than time to heal some wounds. *Curr. Biol.* **10**, R908–911.
10. Higgins, N. P., Kato, K., and Strauss, B. (1976). A model for replication repair in mammalian cells. *J. Mol. Biol.* **101**, 417–425.
11. Lowndes, N. F. and Murguia, J. R. (2000). Sensing and responding to DNA damage. *Curr. Opin. Genet. Dev.* **10**, 17–25.
12. Garvik, B., Carson, M., and Hartwell, L. (1995). Single-stranded DNA arising at telomeres in cdc13 mutants may constitute a specific signal for the RAD9 checkpoint. *Mol. Cell. Biol.* **15**, 6128–6138.
13. Lee, S. E., Moore, J. K., Holmes, A., Umezu, K., Kolodner, R. D., and Haber, J. E. (1998). Saccharomyces Ku70, mre11/rad50 and RPA proteins regulate adaptation to G2/M arrest after DNA damage. *Cell* **94**, 399–409.
14. Pellicioli, A., Lee, S. E., Lucca, C., Foiani, M., and Haber, J. E. (2001). Regulation of Saccharomyces Rad53 checkpoint kinase during adaptation from DNA damage-induced G2/M arrest. *Mol. Cell.* **7**, 293–300.
15. Edwards, R. J., Bentley, N. J., and Carr, A. M. (1999). A Rad3-Rad26 complex responds to DNA damage independently of other checkpoint proteins. *Nat. Cell. Biol.* **1**, 393–398.
16. Longhese, M. P., Foiani, M., Muzi-Falconi, M., Lucchini, G., and Plevani, P. (1998). DNA damage checkpoint in budding yeast. *EMBO J.* **17**, 5525–5528.
17. Melo, J. and Toczyski, D. (2002). A unified view of the DNA-damage checkpoint. *Curr. Opin. Cell. Biol.* **14**, 237–245.
18. Green, C. M., Erdjument-Bromage, H., Tempst, P., and Lowndes, N. F. (2000). A novel Rad24 checkpoint protein complex closely related to replication factor C. *Curr. Biol.* **10**, 39–42.
19. Shimomura, T., Ando, S., Matsumoto, K., and Sugimoto, K., (1998). Functional and physical interaction between Rad24 and Rfc5 in the yeast checkpoint pathways. *Mol. Cell. Biol.* **18**, 5485–5491.
20. Kondo, T., Matsumoto, K., and Sugimoto, K., (1999). Role of a complex containing Rad17, Mec3, and Ddc1 in the yeast DNA damage checkpoint pathway. *Mol. Cell. Biol.* **19**, 1136–1143.
21. Thelen, M. P., Venclovas, C., and Fidelis, K., (1999). A sliding clamp model for the Rad1 family of cell cycle checkpoint proteins. *Cell,* **96**, 769–770.
22. Parker, A. E., Van de Weyer, I., Laus, M. C., Oostveen, I., Yon, J., Verhasselt, P., and Luyten, W. H. (1998). A human homologue of the *Schizosaccharomyces pombe* rad1+ checkpoint gene encodes an exonuclease. *J. Biol. Chem.* **273**, 18332–18339.
23. Freire, R., Murguia, J. R., Tarsounas, M., Lowndes, N. F., Moens, P. B., and Jackson, S. P. (1998). Human and mouse homologs of *Schizosaccharomyces pombe* rad1(+) and saccharomyces cerevisiae Rad17: Linkage to checkpoint control and mammalian meiosis. *Genes Dev.* **12**, 2560–2573.
24. Bessho, T. and Sancar, A. (2000). Human DNA damage checkpoint protein hRad9 is a 3' to 5' exonuclease. *J. Biol. Chem.* **275**, 7451–7454.

25. Rouse, J. and Jackson, S. P. (2000). *Lcd1:* An essential gene involved in checkpoint control and regulation of the *MEC1* signalling pathway in *Saccharomyces cerevisiae. EMBO J.* **19**, 5801–5812.
26. Paciotti, V., Clerici, M., Lucchini, G., and Longhese, M. P. (2000). The checkpoint protein Ddc2, functionally related to *S. pombe* Rad26, interacts with Mec1 and is regulated by Mec1-dependent phosphorylation in budding yeast. *Genes Dev.* **14**, 2046–2059.
27. Rouse, J. and Jackson, S. P. (2002). Lcd1p recruits Mec1p to DNA lesions in vitro and in vivo. *Mol. Cell.* **9**, 857–869.
28. Costanzo, V., Robertson, K., Ying, C. Y., Kim, E., Avvedimento, E., Gottesman, M., Grieco, D., and Gautier, J. (2000). Reconstitution of an ATM-dependent checkpoint that inhibits chromosomal DNA replication following DNA damage. *Mol. Cell.* **6**, 649–659.
29. Smith, G. C., Cary, R. B., Lakin, N. D., Hann, B. C., Teo, S. H., Chen, D. J., and Jackson, S. P. (1999). Purification and DNA binding properties of the ataxia-telangiectasia gene product ATM. *Proc. Natl. Acad. Sci. USA* **96**, 11134–11139.
30. Mallory, J. C. and Petes, T. D. (2000). Protein kinase activity of Tel1p and Mec1p, two *Saccharomyces cerevisiae* proteins related to the human ATM protein kinase. *Proc. Natl. Acad. Sci. USA* **97**, 13749–13754.
31. Melo, J. A., Cohen, J., and Toczyski, D. P. (2001). Two checkpoint complexes are independently recruited to sites of DNA damage *in vivo. Genes Dev.* **15**, 2809–2821.
32. Kondo, T., Wakayama, T., Naiki, T., Matsumoto, K., and Sugimoto, K. (2001). Recruitment of Mec1 and Ddc1 checkpoint proteins to double-strand breaks through distinct mechanisms. *Science* **294**, 867–870.
33. Zou, L., Cortez, D., and Elledge, S. J. (2002). Regulation of atr substrate selection by Rad17-dependent loading of Rad9 complexes onto chromatin. *Genes Dev.* **16**, 198–208.
34. Giannattasio, M., Sommariva, E., Vercillo, R., Lippi-Boncambi, F., Liberi, G., Foiani, M., Plevani, P., and Muzi-Falconi, M. (2002). A dominant negative *mec3* mutant uncovers new functions for the Rad17-complex and Tel1. Submitted.
35. D'Amours, D. and Jackson, S. P. (2001). The yeast Xrs2 complex functions in S phase checkpoint regulation. *Genes Dev.* **15**, 2238–2249.
36. Grenon, M., Gilbert, G., and Lowndes, N. F. (2001). Checkpoint activation in response to double-strand breaks requires the Mre11/Rad50/Xrs2 complex. *Nat. Cell. Biol.* **3**, 844–847.
37. Usui, T., Ogawa, H., and Petrini, J. H. (2001). A DNA damage response pathway controlled by Tel1 and the Mre11 complex. *Mol. Cell.* **7**, 1255–1266.
38. Soulier, J. and Lowndes, N. F. (1999). The BRCT domain of the S. cerevisiae checkpoint protein Rad9 mediates a Rad9-Rad9 interaction after DNA damage. *Curr. Biol.* **9**, 551–554.
39. Gilbert, C. S., Green, C. M., and Lowndes, N. F. (2001). Budding yeast Rad9 is an ATP-dependent Rad53 activating machine. *Mol. Cell,* **8**, 129–136.
40. Vialard, J. E., Gilbert, C. S., Green, C. M., and Lowndes, N. F. (1998). The budding yeast Rad9 checkpoint protein is subjected to Mec1/Tel1-dependent hyperphosphorylation and interacts with Rad53 after DNA damage. *EMBO J.* **17**, 5679–5688.
41. Emili, A. (1998). Mec1-dependent phosphorylation of Rad9p in response to DNA damage. *Mol. Cell,* **2**, 183–189.
42. Sun, Z. X., Hsiao, J., Fay, D. S., and Stern, D. F. (1998). Rad53 FHA domain associated with phosphorylated Rad9 in the DNA damage checkpoint. *Science* **281**, 272–274.
43. Sanchez, Y., Bachant, J., Wang, H., Hu, F., Liu, D., Tetzlaff, M., and Elledge, S. J. (1999). Control of the DNA damage checkpoint by Chk1 and Rad53 protein kinases through distinct mechanisms. *Science* **286**, 1166–1171.
44. Cheng, L., Hunke, L., and Hardy, C. F. (1998). Cell cycle regulation of the *Saccharomyces cerevisiae* polo-like kinase Cdc5p. *Mol. Cell. Biol.* **18**, 7360–7370.
45. Toczyski, D. P., Galgoczy, D. J., and Hartwell, L. H. (1997). CDC5 and CKII control adaptation to the yeast DNA damage checkpoint. *Cell* **90**, 1097–1106.
46. Sidorova, J. M. and Breeden, L. L. (1997). Rad53-dependent phosphorylation of Swi6 and down-regulation of CLN1 and CLN2 transcription occur in response to DNA damage in Saccharomyces cerevisiae. *Genes Dev.* **11**, 3032–3045.
47. Pellicioli, A., Lucca, C., Liberi, G., Marini, F., Lopes, M., Plevani, P., Romano, A., Di Fiore, P. P., and Foiani, M. (1999). Activation of Rad53 kinase in response to DNA damage and its effect in modulating phosphorylation of the lagging strand DNA polymerase. *EMBO J.* **18**, 6561–6572.
48. Liberi, G., Chiolo, I., Pellicioli, A., Lopes, M., Plevani, P., Muzi-Falconi, M., and Foiani, M. (2000). Srs2 DNA helicase is involved in checkpoint response and its regulation requires a functional Mec1-dependent pathway and Cdk1 activity. *EMBO J.* **19**, 5027–5038.
49. Brush, G. S., Morrow, D. M., Hieter, P., and Kelly, T. J. (1996). The ATM homologue MEC1 is required for phosphorylation of replication protein A in yeast. *Proc. Natl. Acad. Sci. USA* **93**, 15075–15080.
50. Bashkirov, V. I., King, J. S., Bashkirova, E. V., Schmuckli-Maurer, J., and Heyer, W. D. (2000). DNA repair protein Rad55 is a terminal substrate of the DNA damage checkpoints. *Mol. Cell. Biol.* **20**, 4393–4404.
51. Martinho, R. G., Lindsay, H. D., Flaggs, G., DeMaggio, A. J., Hoekstra, M. F., Carr, A. M., and Bentley, N. J. (1998). Analysis of Rad3 and Chk1 protein kinases defines different checkpoint responses. *EMBO J.* **17**, 7239–7249.
52. Huang, M., Zhou, Z., and Elledge, S. J. (1998). The DNA replication and damage checkpoint pathways induce transcription by inhibition of the Crt1 repressor. *Cell,* **94**, 595–605.
53. Zhao, X. and Rothstein, R. (2002). The Dun1 checkpoint kinase phosphorylates and regulates the ribonucleotide reductase inhibitor Sml1. *Proc. Natl. Acad. Sci. USA* **99**, 3746–3751.
54. Zhao, X., Chabes, A., Domkin, V., Thelander, L., and Rothstein, R. (2001). The ribonucleotide reductase inhibitor Sml1 is a new target of the Mec1/Rad53 kinase cascade during growth and in response to DNA damage. *EMBO J.* **20**, 3544–3553.
55. Zhao, X., Muller, E. G., and Rothstein, R. (1998). A suppressor of two essential checkpoint genes identifies a novel protein that negatively affects dNTP pools. *Mol. Cell,* **2**, 329–340.

CHAPTER 302

Finding Genes That Affect Signaling and Toleration of DNA Damage, Especially DNA Double-Strand Breaks

Craig B. Bennett[1,2] and Michael A. Resnick[2]
[1]*Department of Surgery, Duke University Medical Center, Durham, North Carolina*
[2]*Laboratory of Molecular Genetics, National Institute of Environmental Health Sciences, National Institutes of Health, Research Triangle Park, North Carolina*

Introduction

The total response of an organism to DNA perturbations is determined by many factors, including the nature of the lesions, ability to detect and tolerate the lesions, and status of the cell cycle at the time the damage occurs. The responses require transduction of damage information into cellular changes as well as processes for directly modifying the damaging effects of the lesions. The information can be transduced into a series of effectors that could direct a host of processes including apoptosis, replication inhibition, cell cycle arrest, lesion bypass, and repair. Some examples of elaborate DNA damage transducing systems are the ATM-p53 network of mammalian cells [1] or the RecA-LexA system from bacteria ([2] and elsewhere in this volume). Lesions are sensed through the ATM or RECA pathway, leading to posttranslation modifications of the p53 or the LEXA *transducer* proteins. Modifications of the transducer result in an orchestrated change in expression of more than 20 effector genes involved in many aspects of cell growth, repair, and even cell death in the case of p53.

The signaling elements could be the DNA lesions or their processed intermediates. DNA double-strand breaks (DSBs) are among the most biologically relevant lesions [3]. As indicated in Fig. 1, they are induced directly by a variety of agents or as a result of processing of single-strand lesions. DSBs can arise spontaneously during replication and they are an intrinsic part of meiosis leading to recombination [4]. Genetic and biochemical studies in yeast were the first to reveal that they can be repaired via recombination, a process with features common to nearly all organisms. End joining is another type of DSB repair common to most eukaryotes, although the nature of breaks may determine its efficiency [5].

Studies of DSBs have provided insights into signaling networks that deal with DNA lesions. Many kinds of lesions can produce signals that are transduced into biological effects and the responses are often lesion dependent (Fig. 1). Signaling targets could be genes that are relevant to repair and DNA metabolism. Repair may or may not require signal transduction depending on whether the full set of proteins is always available and active (constitutive), as appears to be the case for excision repair [6] and photoreactivation [7] of UV-induced lesions. DNA damage responses may also be *d*amage-*in*ducible (*DIN*) requiring the *de novo* production of proteins involved in repair or replication [7,8]. Alternatively, posttranslational modification such as phosphorylation may be required for activation of existing proteins, as is the case for many damage-induced cell cycle arrest phenomena.

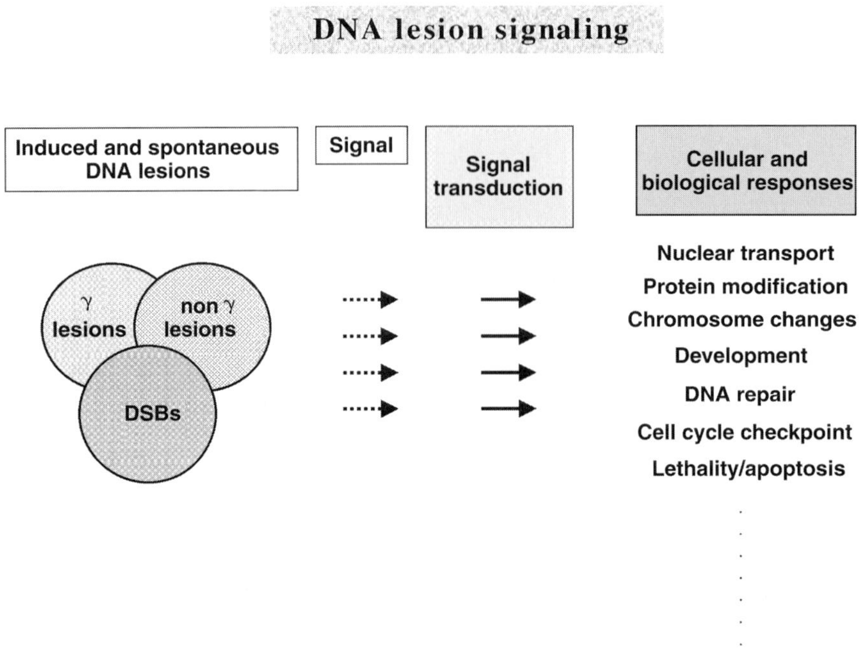

Figure 1 DNA lesion signaling in the yeast *S. cerevisiae*. DNA damage from agents such as ionizing radiation (gamma rays) are capable of inflicting DNA, protein, and lipid damage; however, DSBs appear to be the most biologically relevant lesions and they are responsible for most gamma-ray-induced lethality. Other DSB lesions such as those induced by MMS or HU (also see text) may result from unresolved replication intermediates that occur during the S phase of the cell cycle. Following the induction of DNA damage, yeast cells utilize a sensitive signaling system that can detect even a single DSB, transduce information to the cell cycle apparatus and in turn effect efficient repair of the lesion. This process requires gene products that act as (1) "sensors" of the type and magnitude of damage and (2) signal "transducers" that activate a wide range of cellular and biological responses, which are in turn "effectors" of cell cycle arrest and lesion repair or lesion toleration. Failure in any of these signaling processes can result in genome instability such as aneuploidy, deletions, and gross chromosomal rearrangements. In higher eukaryotes this instability can initiate the multistep process leading to cancer. Persistence of unrepaired DSB damage in nonessential (dispensable) DNA can also lead to cell lethality even in the presence of an intact DNA damage signaling system.

Studies using microarray expression approaches should reveal more of the network involved in *DIN* gene expression.

The budding yeast *Saccharomyces cerevisiae* has been an excellent model system for addressing the relationship between DNA lesions, signaling, and biological consequences. One of the great strengths of this organism is the opportunity to address genetically the components that connect lesion production to cellular changes. Moreover, there is the unique opportunity with budding yeast to follow morphologically cell cycle events in individual cells (Fig. 2). As cells transition from the G_1 to S phase, they develop buds and prior to mitosis (G_2/M) the resulting cells have a doublet appearance [9].

DNA damage could signal a variety of modifications including chromosome condensation, altered nuclear morphology, changes in cell cycle, and apoptosis. Because cell morphology reflects stages in the cell cycle in *S. cerevisiae*, yeast has been ideal for characterizing damage-induced cell cycle changes and for the isolation of mutants that affect the underlying processes. Cell cycle arrest is a hallmark signaling response to DNA damage. Cells monitor the status of chromosome integrity and possess elaborate, genetically controlled feedback networks, or checkpoints, to prevent cell cycle progression if there are chromosomal alterations [10] (Fig. 2). Checkpoints provide additional time for repair prior to cellular division. As discussed later, cell cycle arrest checkpoints have two facets: STOP (arrest) and GO (resumption of cell cycle progression). In yeast, genes regulating the STOP (solid lines, Fig. 2) and GO (dashed arrows, Fig. 2) functions have been identified at both the G_1/S or the G_2/M transitions, whereas only genes regulating STOP functions have been identified during replication (S-phase) arrest.

The strengths of yeast in addressing damage and cell signaling responses have been enhanced through global opportunities to examine genetic relationships. These include total analysis of gene expression [11] combined with functional evaluation of all genes that contribute to resistance of DNA damage [12]. Complete genome mRNA expression reveals the circuitry of responses to perturbations and may suggest participants in signaling pathways. Knowledge of which genes contribute to sensitivity to an agent may lead to identification of genes that play a critical role in damage signaling responses. In this chapter, we focus on approaches to the isolation of genes that play a role in signaling responses to DNA damage, particularly DSBs.

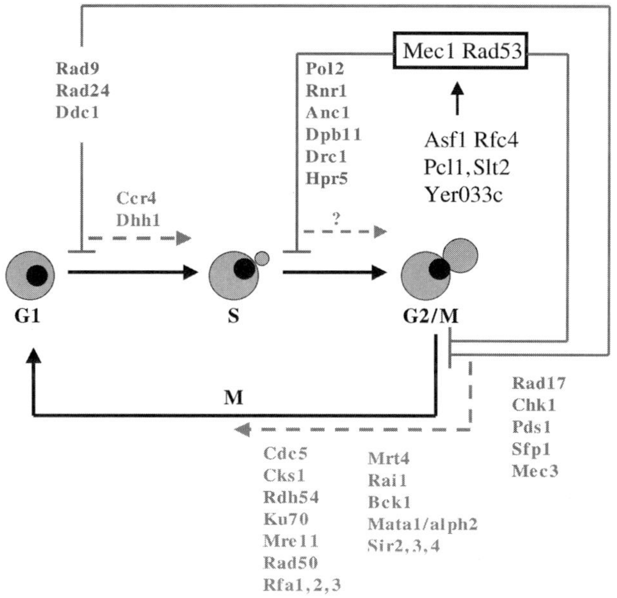

Figure 2 Genetic controls responsible for cell cycle arrest following DNA damage in the yeast *S. cerevisiae*. The G_1, S and G_2 stages of the cell cycle can be identified morphologically. Single unbudded G_1 cells progress through the cell cycle (solid arrows) and give rise to cells with small buds corresponding to the initiation of S phase. Cells progress into G_2 where the size of the daughter cell increases to approximately half the size of the mother cell. In the presence of unrepaired DSB damage, logarithmically growing cells can signal cell cycle arrest (⊥) at any of the three stages in the cell cycle: G_1, intra-S, and G_2/M transition. These damage-induced checkpoints are controlled by many genes that check for genetic integrity. Checkpoint genes such as *RAD9* cause cell cycle arrest, allowing greater time for repair. Mutations in the checkpoint genes can result in a lack of cell cycle arrest and reduced viability following DNA damage. There are also adaptation genes that enable cells to resume cell cycle progression, even in the presence of unrepaired damage (--->). Mutations in these genes cause prolonged cell cycle arrest and loss of viability following DNA damage. *CCR4* and *DHH1* are newly identified adaptation genes required for the resumption of cell cycling following DSB-induced G_1 arrest (T. J. Westmoreland *et al. J. Surg. Kes.*, in press). Genes such as *RAD9, RAD24,* and *DDC1* are required for both the G_1 and G_2/M checkpoints. *MEC1* and *RAD53* provide intra-S and G_2/M checkpoints. The roles of some genes (such as *ASF1, PCL1, SLT2,* and *YER033C*) that are high-copy suppressors of the temperature-sensitive growth phenotype of *mrc1rad53* double mutants are less clear (see [53] for details).

Nature of DSB and Repair and Genetic Consequences

DSBs can appear spontaneously or result from a variety of physical and chemical agents. They can vary markedly, both in structure and in biological responses. They are observed during DNA replication or meiosis and are intrinsic to mating-type switching in yeast. The ends of DSBs come in many forms. Random, radiation-induced DSBs have *dirty* ends often associated with base or sugar damage [13]. The *in vivo* expression of endonucleases such as the mating-type switching enzyme HO or restriction enzymes produces clean, site-specific breaks with blunt or recessed cohesive ends [5]. Many anticancer agents such as bleomycin are described as radiomimetic although the actual structure of the resulting DSBs differs from those produced by radiation [14]. Also, agents that interfere with DNA replication such as hydroxyurea (HU), the single-strand alkylating agent methylmethane sulfonate (MMS), or topoisomerase inhibitors such as camptothecin produce DSBs and/or unresolved replication intermediates that are specific to the S phase of the cell cycle.

DSBs can be repaired by recombination with homologous chromosomes or sister chromatids or by direct end joining [5,15,16]. In the absence of recombination, DSBs may lead to loss of essential chromosomal material via large deletions, rearrangements, and aneuploidy. However, in yeast, lethality can also be indirect as shown for a persistent unrepaired DSB in dispensable DNA [17–19]. The biological responses to specific types of DSBs vary in yeast, suggesting that the nature of the break influences recognition, signaling, and repair. In our recent study, many gamma-ray-sensitive mutants were sensitive to one or more of the following DSB-inducing agents: bleomycin, camptothecin, MMS, or HU, indicating that a variety of repair/checkpoint pathways are required to tolerate a wide spectrum of DSB lesions [12]. Furthermore, DSBs induced by ionizing radiation are almost exclusively repaired by recombination, which explains why radiation resistance in Rad^+ haploid cells is G_2 dependent [20]. An undamaged homolog is available for recombinational repair via the *RAD52* group of repair genes. Surprisingly, *in vivo* expression of *Eco*RI results in little or no lethality in recombination-deficient *rad52* strains in spite of the large number of DSBs. Unlike radiation-induced DSBs, these breaks are repaired almost exclusively by end joining [21].

Checkpoint Activation and Adaptation as Signaling Responses to DSBs

Because bud status identifies cell cycle stage, *S. cerevisiae* is ideal for examining the signaling impact of DNA damage, particularly DSBs, on cell cycle progression using time-lapse photomicroscopy. Following induction of radiation or HO-induced DSBs, cells arrest (STOP) predominantly in the G_2/M phase of the cell cycle [9,17]. This ensures both greater opportunities to repair lesions prior to division and accurate chromosome segregation. Inactivation of many of the genes within these checkpoint pathways, such as *RAD9*, renders cells sensitive to ionizing radiation [9,22]. Damaging agents also induce arrest at G/S preventing entry into S phase [23–25], and there is also an S-phase checkpoint that monitors successful completion of replication [26]. Cells arrested at G_2/M can, after a period of time, resume cell cycling even if the damage remains unrepaired. This GO process has been termed *adaptation* [27]. The duration of the checkpoint arrest at G_2/M may in part depend on the strength or persistence of the original inducing signal. Yeast have a robust recognition system for detecting DSBs. A single, site-specific HO-induced DSB is sufficient to signal prolonged G_2/M arrest, suggesting an exquisitely sensitive system for sensing a break and amplifying and transducing the signal to effectors that arrest the cell cycle machinery [9,17].

The DSB signaling response can occur even if it is induced in a completely dispensable plasmid [17] or yeast artificial chromosome (YAC) vector [18,19], under conditions where the break may or may not be repairable. In the absence of repair the break leads to loss of the vector. If there is rapid repair, there is no checkpoint response suggesting that induction of a DSB is not sufficient *per se* to signal a checkpoint response [18]. Surprisingly, in the absence of repair, the DSB can signal not only cell cycle arrest but also cell death even though there is no loss of genetically relevant material. It is interesting that some nonrepairable DSBs do not cause arrest or lethality [19].

DNA Damage Signaling Networks

Extensive DNA damage signaling networks have been revealed through genetic screens with yeast. DNA damage-induced checkpoint arrest and adaptation is a complex multicomponent signaling process involving more than 40 genes that can affect the G_1/S, S, and G_2/M phases of the cell cycle (Table I, Fig. 2). As described later, many genetic and biochemical strategies have been used to screen for checkpoint signaling mutants. They extend from signaling responses elicited by random radiation-induced DSBs or site-specific HO-induced breaks to DNA replication stress induced by agents such as MMS or HU. Knowledge of genes involved in damage-induced checkpoints has enabled the development of secondary screens that do not rely directly on DNA damage responses.

Identifying Checkpoint Defects by Screening Radiation-Sensitive Mutants

The first checkpoint mutant, *rad9*, was identified in a screen of radiation-sensitive *S. cerevisiae* mutants sensitive to ionizing radiation [9]. This *d*amage *s*ensitive *mu*tant screening approach (*DSMU*; see Table I), assumes that checkpoint defects will sensitize cells to damage. Therefore, simple visual screening of DNA damage-sensitive mutants for cell cycle progression has been successful in identifying many checkpoint and adaptation mutants (Table I). Recently, a broad

Table I Screening for Genes Involved with DNA Damage Checkpoint and Adaptation Responses in the Yeast *Saccharomyces Cerevisiae*

Gene	Essential	Phosphorylated	Checkpoint function	Screening method	First report of checkpoint mutant
RAD9	No	Yes	G2/M checkpoint arrest following DNA damage or replicative arrest	*DSMU*	[9]
			G1/S checkpoint after damage		[23]
RAD17	No	No	G2/M arrest following DNA damage or replicative arrest	*DSMU & DRAC*	[28]
MRT4	No	?	G2/M checkpoint arrest following DNA damage as diploid and adaptation to the G2 checkpoint as haploids	*DSMU*	[12]
RAI1	No	?			
BCK1	No	Yes?	Adaptation to G2/M checkpoint as haploid		
MRE11	No	Yes	Complex is required for phosphorylation of Rad53 and Chk1 to activate the G2/M checkpoint following damage	*DSMU*	[29]
XRS2	No	Yes			
RAD50	No	No			
KU70	No	No	Adaptation to G2/M checkpoint	*DSMU & DECP*	[30]
MRE11	No	Yes			
RAD50	No	No			
RPA(RFA1,2,3)	Yes	Yes			
TID1(RDH54)	No	?	Adaptation to G2/M checkpoint	*DSMU*	[31]
MRC1	No	Yes	S phase checkpoint	*DSMU*	[32]
DUN1	No	Yes	G2/M checkpoint following damage; mutant defective in DNA damage-induced transcription of *RNR2* and *RNR3*	*DITR*	[33]
HRR25	No	Yes?	Defective in DNA damage-induced transcription of *RNR2* and *RNR3*	*DITR*	[34]
SWI4	No	Yes?			
SWI6	No	Yes			
SFP1	No	?	G2/M checkpoint arrest following DNA damage	*DITR*	[35]
CRT1	No	Yes	Effector of DNA damage and replication arrest checkpoint pathway; required for transcriptional activation after damage	*DITR*	[36]

Table I *continued*

Gene	Essential	Phosphorylated	Checkpoint function	Screening method	First report of checkpoint mutant
RAD24	No	Yes	G2/M damage checkpoint S + G2/M	*DRAC*	[37]
MEC1,	Yes	No	incomplete replication and damage		
RAD53	Yes	Yes	checkpoint		
MEC3	No	No	G2/M damage checkpoint		
PDS1	No	Yes	G2/M checkpoint arrest following DNA damage	*DECM*	[38]
CDC5	Yes	Yes	Adaptation to G2/M checkpoint	*DECP*	[27]
CKB2	No	?			
POL2	Yes	No	Defective in S phase checkpoint and *DUN1* dependent transcription	*SRA*	[39]
CKS1	Yes	?	Suppressor of *RAD53*-dependent replication (S-phase) checkpoint	*SRA*	[26]
RFC5	Yes	?	Suppressed by overexpression of *RAD53*	*SRA*	[40]
DDC1	No	Yes	G1/S and G2/M checkpoints following DNA damage	*SRA*	[41]
RNR1	Yes	?	Suppression of the *mec1* and *rad53*	*SRA*	[42]
ANC1	No	?	replication defect		
DPB11	Yes	?	Suppression of *pol2* and *dph2* replication defect	*SRA*	[43]
DRC1	Yes	?	Suppression of *dpb11* replication defect	*SRA*	[44]
SML1	No	Yes	Suppression of *mec1* and *rad53* replication defect	SRA	[45]
CHK1	No	Yes	G2/M checkpoint arrest following DNA damage	*HOMS*	[46]
CHL12(CTF18)	No	?	Defective in S phase checkpoint when combined with *rad24*	*HOMS*	[47]
RAD55	No	Yes	*RAD53*-dependent activation by phosphorylation	*CHDP*	[48]
HPR5(SRS2)	No	Yes	S-phase checkpoint	*CHDP*	[49]
SIR2,	No	?	Adaptation to G2/M checkpoint	*ENCHR*	[50]
SIR3,	No	Yes?			
SIR4,	No	Yes?			
MAT	No	?			
HUG1	No	?	*MEC1*-dependent arrest following DNA damage or replication	*SAGE*	[51]
RFC4	Yes	?	S + G2/M incomplete replication and damage checkpoint	*SYNL*	[52]
YER033C	No	?	S + G2/M incomplete replication	*SSYNL*	[53]
PCL1	No	Yes?	and damage checkpoints		
SLT2	No	Yes?			
ASF1	No	?			
Total = 50					

Phosphorylation status: Yes? = phosphorylated but not yet examined following DNA damage; ? = not examined; No = not detected

Approaches for identifying checkpoint and adaptation defective mutants:

DSMU-*d*amage *s*ensitive *mu*tants that have a posttreatment checkpoint defect
DITR-mutants with altered *d*amage *i*nduced *tr*anscription
DRAC-mutants that are *d*efective in *r*eplication *a*ssociated *c*heckpoint
DECM-checkpoint mutants among mutants that are *de*fective in *c*hromosome *m*echanics
DECP-direct screen for mutants *de*fective in *c*ell *p*rogression (adaptation) following induction of DNA damage
CHOM-mutants with *c*heckpoint *hom*ology, *i.e.,* homology to known checkpoint genes in the same and/or other species
ENCHR-genes that *en*hance *ch*eckpoint *r*esponse to DNA damage when expressed in high copy
SRA-*S*uppression of *r*eplicative *a*rrest and/or lethality by deletion or overexpression.
SAGE-*S*erial *a*nalysis of *g*ene *e*xpression in the presence of HU
SYN-*Syn*thetic *l*ethality with the *rfa1-Y29H* ts allele
SSYNL-*S*uppression of *syn*thetic *l*ethality of a *mc11rad53* double mutant
CHDP-Direct screen for *ch*eckpoint *d*ependent *p*hosporylation of known recombination gene products

functional genomics approach was used in *diploid* yeast that identified 107 new genes that enhanced sensitivity to gamma radiation when deleted. Using *DSMU* screening, the mutants *bck1*, *mrt4*, and *rai1* were found to be defective in postirradiation G_2 arrest, similar to *rad9* [12]. Surprisingly, these mutations resulted in an adaptation defect in haploids where they experienced prolonged G_2 arrest after irradiation. This is unlike *RAD9* mutants, which have a defective checkpoint response in both haploid and diploid cells. These effects suggest modifiers of checkpoint signaling that may result from differences in transcriptional controls associated with mating type (*MAT*) and/or ploidy.

Checkpoint Mutants Revealed through Screening DNA Replication Mutants

Signaling defective mutants have been identified through secondary screens with replication mutants. Some temperature-sensitive (*ts*) mutants of replication-essential genes arrest in late S phase at high temperature. The arrest is due to activation of a checkpoint that detects replication-induced lesions (possibly DSBs). Therefore, mutations in damage-sensing checkpoint genes, when combined with *ts* mutants, should also be *d*efective *i*n *r*eplication *a*ssociated *c*heckpoints (*DIRAC*; Table I). Four *ts* cell division cycle (*cdc*) mutants including *cdc13* were not viable in combination with *rad9* or *rad17* mutations if held at the restrictive temperature for a short time [28]. Both RAD mutants were required for checkpoint arrest following UV or gamma-ray damage, suggesting they failed to sense DSBs, caused by replication arrest of the *CDC* mutants. The *cdc13* mutant was used in another *DIRAC* screen to identify the *m*itosis *e*ntry *c*heckpoint mutants *mec1, mec2 (rad53), mec3,* and *rad24* [37]. Further analysis with various *CDC* mutants and other damaging agents including hydroxyurea established that the *RAD17, RAD24,* and *MEC3* were specific to the G_2/M checkpoint, whereas *RAD9* was active at both G_1/S and G_2/M checkpoints. The essential *MEC1* and *MEC2* genes were required for both the S phase and the ionizing radiation-induced G_2/M arrests, suggesting two overlapping damage signaling checkpoint pathways (Fig. 2).

Using direct *s*uppression of *r*eplication *a*rrest (*SRA*; Table I), *S*-phase *a*rrest *d*efective (*sad*) mutants including *sad1/rad53* and *sad2/pol2* were initially isolated based on HU sensitivity [26]. In a second-site suppressor screen, the HU sensitivity of *sad1/rad53* was suppressed by mutations in *CKS1*, a subunit of the Cdc28p kinase required for normal G_1-to-S and G_2-to-M cell cycle transitions. The Cdc28p kinase appears to be a downstream target of Cdc5p kinase implicated in checkpoint adaptation; therefore, it appears to be involved in both S-phase and G_2/M checkpoints [26].

Because *RAD53* is central to damage-induced signaling responses and it is essential, high-copy suppressors of lethality were sought. Interestingly, overexpression of the ribonucleotide reductase gene *RNR1* completely suppressed the lethal effects of *rad53*Δ, and *mec1*Δ mutants but did not suppress the HU or UV-induced lethality or checkpoint defects [42]. Furthermore, deletion of the *SML1* gene was found to suppress the replication defects of *mec1* or *rad53* deletions without affecting their checkpoint functions [45]. These results indicated that two separate checkpoint functions exist in S phase, either dealing with damage or replication stress. Surprisingly, overexpression of the transcriptional regulator *ANC1* was found to be a weak suppressor of the *mec1*Δ replication defect, possibly through the transcriptional regulation of other genes [42]. Using *SRA* screens, both Dpb11p [43] and Drc1p [44] have also been identified as participants in the S-phase replication checkpoint. These proteins appear to interact with the C-terminal domain of DNA Pol2p. This domain participates in the sensing of DNA damage to prevent entry into mitosis when DNA is not replicated [43].

Recently synthetic lethality genetic screens (*SYNL*; Table I) have further expanded the role of replication factors in checkpoint controls. For example, the *RFC* complex (*RFC1-5*), which is a processivity factor for DNA polymerases 2 and 3, plays a role in checkpoint signaling. The *rfc4-2* mutant was lethal in combination with a mutant *(rfa-t11)* of *RFA1* [52] that encodes a subunit of the DNA-binding protein RPA required for G_2/M adaptation [30]. Both the *rfc4-2* and *rfa1-t11* mutants were defective in the G_2/M checkpoint, as determined by decreased arrest of *cdc13* or *cdc15* mutants at the nonpermissive temperature. The *rfc4-2* cells were found to have a defective S-phase checkpoint in response to MMS damage [52].

Screening for Checkpoint Defects

Screening for checkpoint defects associated with damage-induced changes in transcription also identifies kinase mutants. *D*amage-*i*nduced *tr*anscription (*DITR*; Table I) is an important signaling response to radiation. Genes associated with checkpoint function frequently influence gene expression by altering transcription. Among the *d*amage-*in*ducible (*DIN*) genes are those required for synthesis of ribonucleotide reductase (i.e., *RNR1*, *RNR2,* and *RNR3*). By fusing *RNR3* to a β-galactosidase color reporter gene, 5 complementation groups of DNA *d*amage-*un*inducible mutants were isolated [8,33]. One of these transcriptionally defective signaling mutants (*dun1)* is partially defective in the G_2/M checkpoint response [54]. It is a nuclear protein kinase that autophosphorylates following DNA damage in a *RAD53*-dependent manner and subsequently is required for inactivating the Crt1p transcriptional repressor (by hyperphosphorylation), which is required to tolerate replicative stress and maintain genome stability [36]. Thus, *DUN1* (through inactivation of *CRT1)* derepresses *RNR2* and *RNR3* following damage. Other candidate G_1/S and S-phase checkpoint genes include *HRR25*, *SWI6,* and *SWI4.* Mutations in these genes result in sensitivity to HU and MMS treatment as well as defective damage-induced (*DIN*) transcriptional signaling responses [34].

Central components of the damage-induced signaling pathway are the Mec1p and Mec2p/Rad53p kinases, which are essential for replication and are involved in both S and

G_2/M responses. Phosphorylation is clearly important in checkpoint responses because more than half of the associated gene products are phosphorylated, many in a damage-specific manner (Table I). The G_2/M checkpoint arrest involves two parallel *MEC1*-dependent pathways that contribute equally to arrest [55]. These pathways utilize upstream damage "sensor" functions attributed to the combined actions of the *RAD9, RAD17, RAD24, MEC3,* and *DDC1*gene products [41,56]. Both Rad9p and Ddc1p appear to be phosphorylated following DNA damage. Furthermore, Rad53p is also phosphorylated following damage in a Mec1p-dependent process. Rad53p in turn phosphorylates downstream effectors [46] such as the *CDC5* gene product required for checkpoint adaptation [27] and the *RAD55* gene product required for recombinational repair [48]. In a second parallel pathway, the checkpoint protein Chk1p functions in a Mec1-dependent manner to maintain abundant levels of Pds1, an anaphase inhibitor when DNA damage is present [46,55]. The Cdc5p, Rad55p, Chk1p, and Pds1p effectors are all activated by a *MEC1*-dependent phosphorylation cascade in response to damage.

Screen for Altered Checkpoint and Adaptation Responses to a single DSB

A single, persistent HO-induced DSB in a nonessential plasmid or YAC can induce a prolonged G_2 arrest followed by lethality. However, some strains do not exhibit this prolonged arrest and quickly adapt to the damage and resume cell cycle progression. Using such a strain, a high-copy library was screened for genes that would confer *en*hanced G_2 *ch*eckpoint *r*esponses (*ENCHR*; Table I) and lethality following induction of a HO break in dispensable DNA. Multiple copies of the silent mating type locus *HMRa* resulted in the *ENCHR* phenotype and this was attributed to the locus being able to titrate proteins involved in chromatin silencing [50]. This led to the finding that deletion of the multifunctional chromatin modification genes *SIR2, SIR3,* or *SIR4* has the same effect. Both the mating type status and the *SIR* genes determined the checkpoint and lethal responses to the single DSB. These results suggest that the transcriptional status of the cell (in this case subject to mating-type controls) can greatly influence checkpoint responses or lethality arising from a persistent DSB. Furthermore, since a *rad9Δ sir4Δ* mutant did not show prolonged G_2 arrest, but did exhibit DSB-induced lethality, the two responses to a DSB appear to involve separate pathways [50].

A single unrepairable DSB system has also been used to address the ability of cells to reenter the cell cycle, that is, to adapt to a break, and to screen for mutants that are *de*fective in *c*ell *p*rogression (*DECP* screen; Table I). A site-specific break in *MAT* was created in a dispensable copy of chromosome III in cells that could not repair the DSB because they were deficient in recombination (i.e., *rad52*). Cells mutated in *cdc5* and *cdk2* were permanently arrested as budded cells in a *RAD9*-dependent manner [27]. Furthermore, in diploid cells lacking recombinational repair, adaptation is a prerequisite of genomic rearrangement or break-induced replication [57]. Therefore, adaptation promotes the survival of diploid cells carrying one or a few persistent DSBs. Cells lacking Yku70p also permanently arrested as budded cells following induction of a single DSB. This was suppressed by mutations in *RAD50, MRE11*, or the ssDNA-binding protein complex *RPA* [30]. Using a derivation of the *DSMU* screening approach, the *RAD54* homolog *RDH54/TID1* was identified as an adaptation gene [31]. Following induction of a single HO-induced DSB, *tid1Δ* cells showed prolonged arrest in G_2/M that was *RAD9* dependent. This prolonged arrest was similar to that seen for *yku70Δ* cells; however, extensive resection of the DSB, characteristic of the *yku70Δ* mutant, was not seen in the *tid1Δ* strain. Recently, the Rad50/Mre11/Xrs2 complex has been shown to be required for both the S-phase and G_2/M checkpoints [29]. Specifically this complex is required for damage-specific phosphorylation of Rad9p, Rad53p, and Chk1p. Because Rad9p is also required for G_1/S arrest following damage, deletions of *RAD50, MRE11*, or *XRS2* may affect this checkpoint as well.

Other Screens for DNA Damage Checkpoint Pathway Genes

Several other screens have also identified genes associated with checkpoint control. For example, *CHK1* was identified in *S. cerevisiae* based on a *hom*ology *s*earch (*HOMS*; Table I) against the mammalian and *Schizosaccharomyces pombe* Chk1p [46]. Similarly, the yeast Chl12p is related to the yeast Rad24p; it provides a redundant replication checkpoint function [47].

The *PDS1* (*p*recocious *d*isassociation of *s*ister chromatids) checkpoint mutant was first identified by its *de*fect in *c*hromosome *m*echanics (*DECM*; Table I) [58]. Subsequently, it was shown to have a damage-associated checkpoint function [38]. Another approach, serial analysis of gene expression (*SAGE*; Table I), revealed a nonannotated transcribed gene *HUG1* that was highly expressed following HU treatment. Subsequently it was shown to affect checkpoint responses [51]. Using a variation of the *DSMU* screen and examining all *RAD52* group genes for damage-inducible *ch*eckpoint-*d*ependent *p*hosphorylation (*CHDP*; Table I), Rad55p was found to be a *terminal effector* of the *MEC1*-dependent checkpoint pathway since it was phosphorylated following DNA damage in a *RAD53*-dependent fashion [48]. In a similar manner, the *CHDP* approach identified Hpr5p (Srs2p) as being phosphorylated in response to damage in a *MEC1, RAD53*, and *DUN1*-dependent manner [49].

High-copy *s*uppressors of *syn*thetic *l*ethality (*SSYNL*; Table I) were identified that rescued growth of a *mrc1 rad53 mec1* triple mutant at a nonpermissive temperature. These mutants also suppressed the HU sensitivity of *rad53* mutants. One of them, *ASF1*, which is involved in chromatin remodeling, was later shown to have additional interesting checkpoint-related functions [46]. The mutant is gamma-ray sensitive, spontaneously arrests in G_2, and exhibits high levels of spontaneous transcription of a damage-responsive reporter, suggesting a role in replication or S-phase checkpoint response ([12]; unpublished).

Implications of DNA Damage Checkpoint Signaling

Genomic instability has been implicated as an early event in cancer progression. Defects in genes that regulate DNA damage checkpoints have been suggested to be predisposing factors for the onset of cancer because of their influence on genome stability. Mutations in many of the yeast genes described earlier, particularly those associated with the S-phase checkpoint, result in destabilization of the genome. The conservation of gene function among most eukaryotes predicts that many of these yeast genes that are critical for damage signaling and checkpoint control will have orthologs in human cells with similar roles in checkpoint signaling pathways. For example, Mec1 is the yeast ortholog of the human checkpoint gene ATR [11]. Mutations within ATR are responsible for sensitivity to DNA damaging agents and a predisposition for cancer. Also, mutations in the C-terminal domain of the breast cancer gene BRCA1 abrogate mammalian checkpoint response at G_2/M following DNA damage [59]. A large superfamily with repeats of the conserved BRCT domain first identified in the C terminus of BRCA1 have also been found in many damage-responsive checkpoint proteins including yeast *RAD9*, *DPB11*, and the large subunit of RFC [60]. Thus, characterization of new checkpoint genes in yeast will undoubtedly lead to the identification of new cancer gene targets and the elucidation of their underlying molecular role(s) in neoplastic progression as well as other diseases.

References

1. Khanna, K. K. (2000). Cancer risk and the ATM gene: A continuing debate. *J. Natl. Cancer Inst.* **92**, 795–802.
2. Sutton, M. D., Smith, B. T., Godoy, V. G., and Walker, G. C. (2000). The SOS response: Recent insights into umuDC-dependent mutagenesis and DNA damage tolerance. *Annu. Rev. Genet.* **34**, 479–497.
3. Khanna, K. K. and Jackson, S. P. (2001). DNA double-strand breaks: Signaling, repair and the cancer connection. *Nat. Genet.* **27**, 247–254.
4. Wu, T. C. and Lichten, M. (1994). Meiosis-induced double-strand break sites determined by yeast chromatin structure. *Science* **263**, 515–518.
5. Lewis, L. K. and Resnick, M. A. (2000). Tying up loose ends: Nonhomologous end-joining in *Saccharomyces cerevisiae. Mutat. Res.* **451**, 71–89.
6. Friedberg, E. C. (1988). Deoxyribonucleic acid repair in the yeast *Saccharomyces cerevisiae. Microbiol. Rev.* **52**, 70–102.
7. Sancar, G. B. (2000). Enzymatic photoreactivation: 50 years and counting. *Mutat. Res.* **451**, 25–37.
8. Ruby, S. W. and Szostak, J. W. (1985). Specific *Saccharomyces cerevisiae* genes are expressed in response to DNA-damaging agents. *Mol. Cell. Biol.* **5**, 75–84.
9. Weinert, T. A. and Hartwell, L. H. (1988). The RAD9 gene controls the cell cycle response to DNA damage in *Saccharomyces cerevisiae. Science* **241**, 317–322.
10. Foiani, M., Pellicioli, A., Lopes, M., Lucca, C., Ferrari, M., Liberi, G., Muzi Falconi, M., and Plevani, P. (2000). DNA damage checkpoints and DNA replication controls in *Saccharomyces cerevisiae. Mutat. Res.* **451**, 187–196.
11. Gasch, A. P., Huang, M., Metzner, S., Botstein, D., Elledge, S. J., and Brown, P. O. (2001). Genomic expression responses to DNA-damaging agents and the regulatory role of the yeast ATR homolog Mec1p. *Mol. Biol. Cell* **12**, 2987–3003.
12. Bennett, C. B., Lewis, L. K., Karthikeyan, G., Lobachev, K. S., Jin, Y. H., Sterling, J. F., Snipe, J. R., and Resnick, M. A. (2001). Genes required for ionizing radiation resistance in yeast. *Nat. Genet.* **29**, 426–434.
13. Ward, J. F. (1985). Biochemistry of DNA lesions. *Radiat. Res. Suppl.* **8**, S103–S111.
14. Povirk, L. F. and Goldberg, I. H. (1987). A role of oxidative DNA sugar damage in mutagenesis by neocarzinostatin and bleomycin. *Biochimie* **69**, 815–823.
15. Resnick, M. A. (1975). The repair of double-strand breaks in chromosomal DNA of yeast. *Basic Life Sci.,* 549–556.
16. Szostak, J. W., Orr-Weaver, T. L., Rothstein, R. J., and Stahl, F. W. (1983). The double-strand-break repair model for recombination. *Cell* **33**, 25–35.
17. Bennett, C. B., Lewis, A. L., Baldwin, K. K., and Resnick, M. A. (1993). Lethality induced by a single site-specific double-strand break in a dispensable yeast plasmid. *Proc. Natl. Acad. Sci. USA* **90**, 5613–5617.
18. Bennett, C. B., Westmoreland, T. J., Snipe, J. R., and Resnick, M. A. (1996). A double-strand break within a yeast artificial chromosome (YAC) containing human DNA can result in YAC loss, deletion or cell lethality. *Mol. Cell. Biol.* **16**, 4414–4425.
19. Bennett, C. B., Snipe, J. R., and Resnick, M. A. (1997). A persistent double-strand break destabilizes human DNA in yeast and can lead to G2 arrest and lethality. *Cancer Res.* **57**, 1970–1980.
20. Brunborg, G., Resnick, M. A., and Williamson, D. H. (1980). Cell-cycle-specific repair of DNA double strand breaks in *Saccharomyces cerevisiae. Radiat. Res.* **82**, 547–558.
21. Lewis, L. K., Kirchner, J. M., and Resnick, M. A. (1998). Requirement for end-joining and checkpoint functions, but not RAD52-mediated recombination, after EcoRI endonuclease cleavage of *Saccharomyces cerevisiae* DNA. *Mol. Cell. Biol.* **18**, 1891–1902.
22. Weinert, T. and Hartwell, L. (1989). Control of G2 delay by the rad9 gene of *Saccharomyces cerevisiae. J. Cell. Sci. Suppl.* **12**, 145–148.
23. Siede, W., Friedberg, A. S., and Friedberg, E. C. (1993). RAD9-dependent G1 arrest defines a second checkpoint for damaged DNA in the cell cycle of *Saccharomyces cerevisiae. Proc. Natl. Acad. Sci. USA* **90**, 7985–7989.
24. Alic, N., Higgins, V. J., and Dawes, I. W. (2001). Identification of a *Saccharomyces cerevisiae* gene that is required for G1 arrest in response to the lipid oxidation product linoleic acid hydroperoxide. *Mol. Biol. Cell* **12**, 1801–1810.
25. Gerald, J. N., Benjamin, J. M., and Kron, S. J. (2002). Robust G1 checkpoint arrest in budding yeast: Dependence on DNA damage signaling and repair. *J. Cell Sci.* **115**, 1749–1757.
26. Allen, J. B., Zhou, Z., Siede, W., Friedberg, E. C., and Elledge, S. J. (1994). The SAD1/RAD53 protein kinase controls multiple checkpoints and DNA damage-induced transcription in yeast. *Genes Dev.* **8**, 2401–2415.
27. Toczyski, D. P., Galgoczy, D. J., and Hartwell, L. H. (1997). CDC5 and CKII control adaptation to the yeast DNA damage checkpoint. *Cell* **90**, 1097–1106.
28. Weinert, T. A. and Hartwell, L. H. (1993). Cell cycle arrest of cdc mutants and specificity of the RAD9 checkpoint. *Genetics* **134**, 63–80.
29. Grenon, M., Gilbert, C., and Lowndes, N. F. (2001). Checkpoint activation in response to double-strand breaks requires the Mre11/Rad50/Xrs2 complex. *Nat. Cell Biol.* **3**, 844–847.
30. Lee, S. E., Moore, J. K., Holmes, A., Umezu, K., Kolodner, R. D., and Haber, J. E. (1998). *Saccharomyces* Ku70, mre11/rad50 and RPA proteins regulate adaptation to G2/M arrest after DNA damage. *Cell* **94**, 399–409.
31. Lee, S. E., Pellicioli, A., Malkova, A., Foiani, M., and Haber, J. E. (2001). The *Saccharomyces* recombination protein Tid1p is required for adaptation from G2/M arrest induced by a double-strand break. *Curr. Biol.* **11**, 1053–1057.
32. Alcasabas, A. A., Osborn, A. J., Bachant, J., Hu, F., Werler, P. J., Bousset, K., Furuya, K., Diffley, J. F., Carr, A. M., and Elledge, S. J. (2001).

Mrc1 transduces signals of DNA replication stress to activate Rad53. *Nat. Cell Biol.* **3**, 958–965.

33. Zhou, Z. and Elledge, S. J. (1993). DUN1 encodes a protein kinase that controls the DNA damage response in yeast. *Cell* **75**, 1119–1127.
34. Ho, Y., Mason, S., Kobayashi, R., Hoekstra, M., and Andrews, B. (1997). Role of the casein kinase I isoform, Hrr25, and the cell cycle-regulatory transcription factor, SBF, in the transcriptional response to DNA damage in *Saccharomyces cerevisiae*. *Proc. Natl. Acad. Sci. USA* **94**, 581–586.
35. Xu, Z. and Norris, D. (1998). The SFP1 gene product of *Saccharomyces cerevisiae* regulates G2/M transitions during the mitotic cell cycle and DNA-damage response. *Genetics* **150**, 1419–1428.
36. Huang, M., Zhou, Z., and Elledge, S. J. (1998). The DNA replication and damage checkpoint pathways induce transcription by inhibition of the Crt1 repressor. *Cell* **94**, 595–605.
37. Weinert, T. A., Kiser, G. L., and Hartwell, L. H. (1994). Mitotic checkpoint genes in budding yeast and the dependence of mitosis on DNA replication and repair. *Genes Dev.* **8**, 652–665.
38. Yamamoto, A., Guacci, V., and Koshland, D. (1996). Pds1p, an inhibitor of anaphase in budding yeast, plays a critical role in the APC and checkpoint pathway(s). *J. Cell Biol.* **133**, 99–110.
39. Navas, T. A., Zhou, Z., and Elledge, S. J. (1995). DNA polymerase epsilon links the DNA replication machinery to the S phase checkpoint. *Cell* **80**, 29–39.
40. Sugimoto, K., Ando, S., Shimomura, T., and Matsumoto, K. (1997). Rfc5, a replication factor C component, is required for regulation of Rad53 protein kinase in the yeast checkpoint pathway. *Mol. Cell Biol.* **17**, 5905–5914.
41. Longhese, M. P., Paciotti, V., Fraschini, R., Zaccarini, R., Plevani, P., and Lucchini, G. (1997). The novel DNA damage checkpoint protein ddc1p is phosphorylated periodically during the cell cycle and in response to DNA damage in budding yeast. *Embo J.* **16**, 5216–5226.
42. Desany, B. A., Alcasabas, A. A., Bachant, J. B., and Elledge, S. J. (1998). Recovery from DNA replicational stress is the essential function of the S-phase checkpoint pathway. *Genes Dev.* **12**, 2956–2970.
43. Araki, H., Leem, S. H., Phongdara, A., and Sugino, A. (1995). Dpb11, which interacts with DNA polymerase II(epsilon) in *Saccharomyces cerevisiae*, has a dual role in S-phase progression and at a cell cycle checkpoint. *Proc. Natl. Acad. Sci. USA* **92**, 11791–11795.
44. Wang, H., and Elledge, S. J. (1999). DRC1, DNA replication and checkpoint protein 1, functions with DPB11 to control DNA replication and the S-phase checkpoint in *Saccharomyces cerevisiae*. *Proc. Natl. Acad. Sci. USA* **96**, 3824–3829.
45. Zhao, X., Muller, E. G., and Rothstein, R. (1998). A suppressor of two essential checkpoint genes identifies a novel protein that negatively affects dNTP pools. *Mol. Cell* **2**, 329–340.
46. Sanchez, Y., Bachant, J., Wang, H., Hu, F., Liu, D., Tetzlaff, M., and Elledge, S. J. (1999). Control of the DNA damage checkpoint by chk1 and rad53 protein kinases through distinct mechanisms. *Science* **286**, 1166–1171.
47. Naiki, T., Kondo, T., Nakada, D., Matsumoto, K., and Sugimoto, K. (2001). Chl12 (Ctf18) forms a novel replication factor C-related complex and functions redundantly with Rad24 in the DNA replication checkpoint pathway. *Mol. Cell Biol.* **21**, 5838–5845.
48. Bashkirov, V. I., King, J. S., Bashkirova, E. V., Schmuckli-Maurer, J., and Heyer, W. D. (2000). DNA repair protein Rad55 is a terminal substrate of the DNA damage checkpoints. *Mol. Cell Biol.* **20**, 4393–4404.
49. Liberi, G., Chiolo, I., Pellicioli, A., Lopes, M., Plevani, P., Muzi-Falconi, M., and Foiani, M. (2000). Srs2 DNA helicase is involved in checkpoint response and its regulation requires a functional Mec1-dependent pathway and Cdk1 activity. *Embo J.* **19**, 5027–5038.
50. Bennett, C. B., Snipe, J. R., Westmoreland, J. W., and Resnick, M. A. (2001b). SIR functions are required for the toleration of an unrepaired double-strand break in a dispensable yeast chromosome. *Mol. Cell. Biol.* **21**, 5359–5373.
51. Basrai, M. A., Velculescu, V. E., Kinzler, K. W., and Hieter, P. (1999). NORF5/HUG1 is a component of the MEC1-mediated checkpoint response to DNA damage and replication arrest in *Saccharomyces cerevisiae*. *Mol. Cell. Biol.* **19**, 7041–7049.
52. Kim, H. S. and Brill, S. J. (2001). Rfc4 interacts with Rpa1 and is required for both DNA replication and DNA damage checkpoints in *Saccharomyces cerevisiae*. *Mol. Cell. Biol.* **21**, 3725–3737.
53. Hu, F., Alcasabas, A. A., and Elledge, S. J. (2001). Asf1 links Rad53 to control of chromatin assembly. *Genes Dev.* **15**, 1061–1066.
54. Pati, D., Keller, C., Groudine, M., and Plon, S. E. (1997). Reconstitution of a MEC1-independent checkpoint in yeast by expression of a novel human fork head cDNA. *Mol. Cell. Biol.* **17**, 3037–3046.
55. Gardner, R., Putnam, C. W., and Weinert, T. (1999). RAD53, DUN1 and PDS1 define two parallel G2/M checkpoint pathways in budding yeast. *Embo J.* **18**, 3173–3185.
56. Lydall, D. and Weinert, T. (1995). Yeast checkpoint genes in DNA damage processing: Implications for repair and arrest. *Science* **270**, 1488–1491.
57. Galgoczy, D. J. and Toczyski, D. P. (2001). Checkpoint adaptation precedes spontaneous and damage-induced genomic instability in yeast. *Mol. Cell. Biol.* **21**, 1710–1718.
58. Yamamoto, A., Guacci, V., and Koshland, D. (1996b). Pds1p is required for faithful execution of anaphase in the yeast, *Saccharomyces cerevisiae*. *J. Cell Biol.* **133**, 85–97.
59. Xu, B., Kim, S., and Kastan, M. B. (2001). Involvement of Brca1 in S-phase and G(2)-phase checkpoints after ionizing irradiation. *Mol. Cell. Biol.* **21**, 3445–3450.
60. Bork, P., Hofmann, K., Bucher, P., Neuwald, A. F., Altschul, S. F., and Koonin, E. V. (1997). A superfamily of conserved domains in DNA damage-responsive cell cycle checkpoint proteins. *FASEB J.* **11**, 68–76.

CHAPTER 303

Radiation Responses in *Drosophila*

Naoko Sogame and John M. Abrams
Department of Cell Biology,
University of Texas Southwestern Medical Center,
Dallas, Texas

Introduction

Four decades after the discovery of X rays, H. J. Muller established *Drosophila* as a preeminent model for an emerging field of *radiation biology*, receiving a Nobel Prize for showing a cause-and-effect relationship between radiation and heritable mutations [1]. We now understand that radiation exposure promotes genotoxic damage, causing wholesale chromosomal alterations that jeopardize viability and reproductive capacity. Consequently, all organisms have evolved sophisticated mechanisms that respond to and counteract these effects. In eukaryotes, radiation responses can be divided into three phases: damage sensors, signal transmitters, and effectors that promote (1) cell cycle arrest, (2) DNA repair, and (3) apoptosis. Except where specified we use the term *radiation* in the broad sense, not necessarily distinguishing UV rays versus X rays.

Sensors and Transmitters

Initial Detection of DNA Damage

Sensors must have the ability to discriminate damaged from normal DNA, and they must show a defect in the DNA damage checkpoint when mutated. A current model for the sensors in *Schizosaccharomyces pombe*, describes a multi-component complex composed of Hus1, Rad1, and Rad9 that is recruited to damaged DNA by R17 associated replication factor C (RFC) [2–6]. Though, we do not yet know which gene products act as direct sensors of DNA damage in flies, the best candidates are *Drosophila* orthologs of *S. pombe*'s gene products recruited to damaged DNA via the RFC.

Signal Transmission by Kinases

Once DNA damage is recognized, signal transmitters including the ATM and ATR kinases are activated. In mammals, ATM responds exclusively to ionizing radiation (IR), whereas ATR responds to IR, UV radiation, and stalled replication forks [7]. The *Drosophila* ATM/ATR homolog, *mei-41*, was isolated as a meiotic mutant four decades ago [8] and has been well characterized as a pleiotropic locus affecting DNA repair, chromosome instability, and defective replication control [9]. A second ATM/ATR like gene is found in the fly genome (CG6535) and, from sequence alignments, this is a likely ortholog of ATM, whereas Mei-41 more closely resembles ATR [10]. In mammals, ATM and ATR activate the downstream serine/threonine kinases Chk1 and Chk2, depending on the signals they receive [11] and, similarly, in flies, genetic studies suggest *mei-41* (ATR) functions upstream of *grapes*, the fly homolog of *chk1* [12]. Though *mei-41* and *chk2* mutant flies exhibit similar damage checkpoint defects [13], tests for genetic interactions between these have not been reported. Like their mammalian counterparts, *Drosophila* Chk1 (Grapes) and Chk2 kinases probably function to modulate the activity of damage response effectors.

Effectors

Cell Cycle Arrest

Cell division cycles in early *Drosophila* embryos are divided into two stages, defined by the midblastula transition (MBT), which occurs 120 min after fertilization. Cell cycle divisions prior to this (referred to as *pre-MBT*) are controlled by maternally derived transcripts, whereas the *post-MBT* phase is controlled by zygotically produced transcripts.

The first 13 nuclear cycles do not contain G_1 and G_2 stages, and proceed as a syncytium (without cytokinesis). A G_2 phase is added to the cell cycle during the 14th embryonic division, which is accompanied by cytokinesis. Syncytial embryos (1–13 cycles, pre-MBT) and cellular embryos (14–16 cycles, post-MBT) show very different responses to X-ray irradiation (Table I). In addition, syncytial embryos can eliminate damaged nuclei by incorporation into the yolk, and these are quickly replaced through rapid nuclear divisions of syncytial embryos. This elimination process cannot be accomplished in cellular embryos (owing to an emergence of cell membranes); therefore, alternative methods, such as programmed cell death, are involved. Su and Jaklevic [14] speculate that these different radiation responses reflect an emphasis on velocity of replication in syncytial embryo, and replication fidelity in cellular post-MBT embryos.

When post-MBT embryos are exposed to X-ray irradiation, delays are observed at both G_2/M and metaphase–anaphase transitions. These responses are evidently important for reducing the incidence of broken chromosomes [14]. In the wing discs, flies mutated for *mei-41*, *grapes*, *chk2*, and *14-3-3ε* exhibit G_2/M checkpoint defects of varying degrees, indicating these genes act in a common pathway whose elements are outlined in Fig. 1 [9,15,16]. The onset of a G_2/M transition checkpoint during normal development involves a Cdc25 homolog, String, and Cdks. String may activate Cdks by removing inhibitory phosphates, and activated Cdks drive the G_2/M transition together with mitotic cyclins. The inhibitory phosphorylation of Cdks may be necessary during damage-induced G_2/M arrest, as previously demonstrated from mammalian studies [19,20]. Induction of p21 is a major regulator of damage-induced G_1 cell cycle arrest in mammals [21–23], but interestingly, Dacapo, the fly ortholog of p21, is not responsive to radiation [17,18], though it can induce G_1 cell cycle arrest when ectopically expressed in eye discs [18,24]. Therefore, in flies, current evidence suggests that one or more p21-independent pathways specify radiation-induced cell cycle arrest.

DNA Repair

Some of the first mutants in DNA repair genes were identified in the 1970s in flies as mutagen-sensitive (mus) or meiotic recombination mutants (mei) [25–27]. Currently, approximately 30 mus genes and 2 mei genes are known or inferred to comprise the DNA repair pathway [28]. (For a comprehensive review, see [10,29].)

Apoptosis

The propensity toward radiation-induced apoptosis varies dramatically during development and appears to correlate with the proliferative potential. Tissues in early embryos and larval imaginal discs are particularly sensitive to radiation-induced apoptosis, whereas later stage embryos and adults are not. This pattern is also reflected in an organism's viability [30]. Figure 2 outlines an emerging model for radiation-induced apoptosis in *Drosophila*. In early embryos, *reaper* (*rpr*) is transcriptionally activated in response to radiation via an enhancer region that contains a binding site for Dp53 [31].

In functional studies, both Dp53 and the *rpr* region are required for radiation-induced apoptosis [24,31]. Another gene mapping near the *reaper* region, *sickle* (*skl*), was recently identified as a radiation-responsive gene capable of inducing apoptosis [32–34]. Rpr and Skl antagonize the function of inhibitor of apoptosis proteins (IAPs) by releasing activated caspases from IAP-mediated inhibition [35]. Another regulator of caspases, Dark, is required for radiation-induced apoptosis like its human counterpart, Apaf-1[36]. This gene is induced by X-ray and UV exposure [37–39], but whether it is a target of Dp53 has not been determined. In mammals, Chk2 phosphorylates and activates p53 in response to irradiation [40]. Because mutations in the *Drosophila chk2* gene also exhibit defects in both cell cycle arrest and apoptosis, it is reasonable to hypothesize that at least the apoptotic defects reflect p53-dependent functions. Requirements for the *Drosophila* homologs of ATM and ATR in apoptosis have not been tested.

Conclusions: What Can We Learn from the *Drosophila* Model?

It has been nearly a full century since T. H. Morgan used the fruit fly as a model to launch the science of genetics.

Table I Damage Induced Checkpoints Associated with Onset of MBT

	Cleavage cycles	Phases	Radiation induced arrest: G2	Radiation induced arrest: Metaphase/Anaphase	Other responses
Precellularized embryos	1–13	S ⟳ M	No	No	Metaphase chromosome align improperly Abortion of mitosis
Cellularized embryos	14–16	S → G2 → M ⟳	Yes	Yes	—

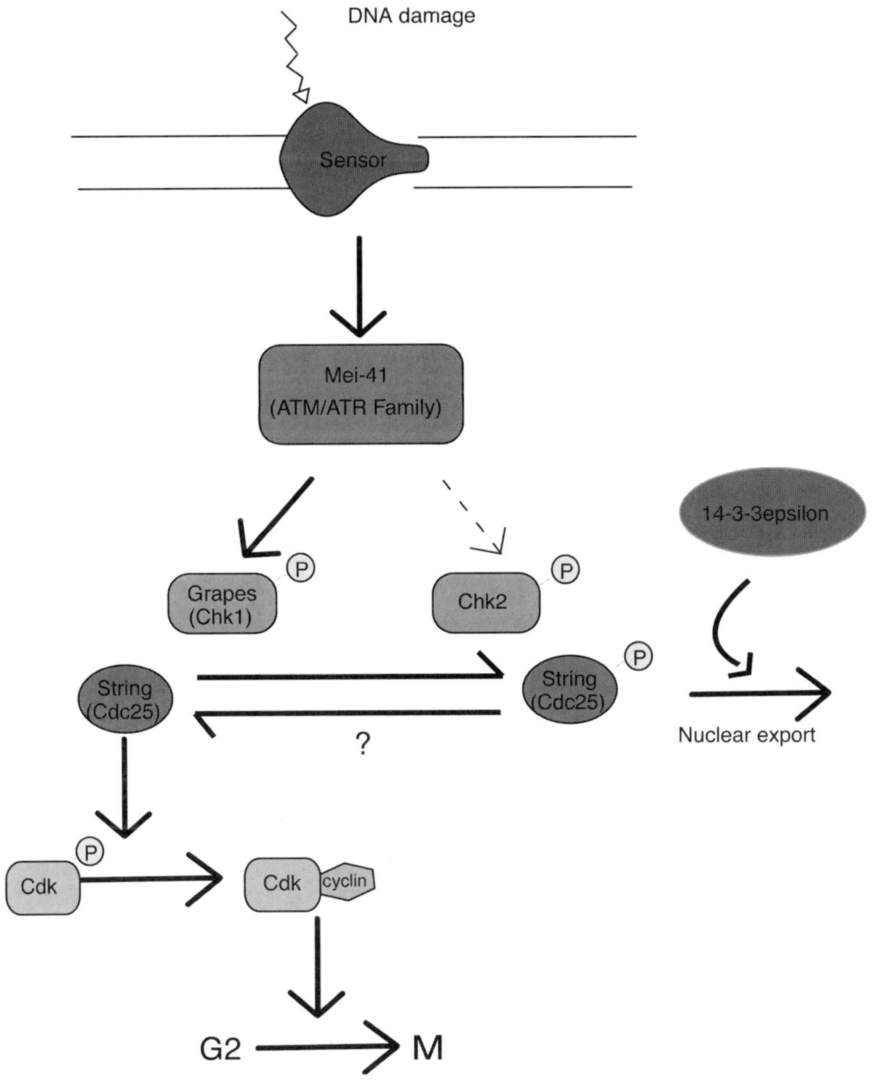

Figure 1 Organization of damaged-induced cell cycle arrest in *Drosophila*. The dashed arrow indicates the inferred orthologous relationship from mammalian studies. Phosphorylation of target proteins is also inferred. Upon radiation-induced damage, MEI-41 activates Grapes or Chk2, leading to the nuclear export of the phosphatase, String. The phosphorylated form of Cdks prevails due to nuclear export of String and, thus, a G2/M transition does not take place.

As a metazoan, *Drosophila* offers fundamental insights regarding the ways in which developmental programs exert constraints on adaptive responses to radiation damage. Although preferential attention is often focused on pathways that are shared by common descent, equally important insights can be gained by an awareness of what has not been conserved. For instance, p53 earned status as a "guardian of the genome," specifying a cell cycle arrest and a DNA repair program that can switch to an apoptotic option if the repair program fails. Current evidence from studies in *Drosophila* contradicts this notion since a robust cell cycle arrest program is triggered in irradiated flies, but *Drosophila* p53 evidently plays no part in this response [24,31]. Similarly, the requirement for induction of p21 as the predominant p53 effector of G_1/S arrest is widely documented in mammals, yet the *Drosophila* homolog of p21, Dacapo, is dispensable for radiation-induced G_1/S arrest (though it is competent to do so [18,24]) and is unlikely to be a radiation-responsive Dp53 target gene [17]. Together, these observations suggest that members of the p53 family can specify the apoptotic fate without first engaging a cell cycle arrest/DNA repair program. An evolutionary corollary related to these observations is that, among the damage responses governed by p53, cell cycle arrest could be a more recently acquired pathway. Continued elucidation of shared and distinct features of radiation-induced damage responses in this model should uncover important insights that help us understand, and perhaps treat, human disease.

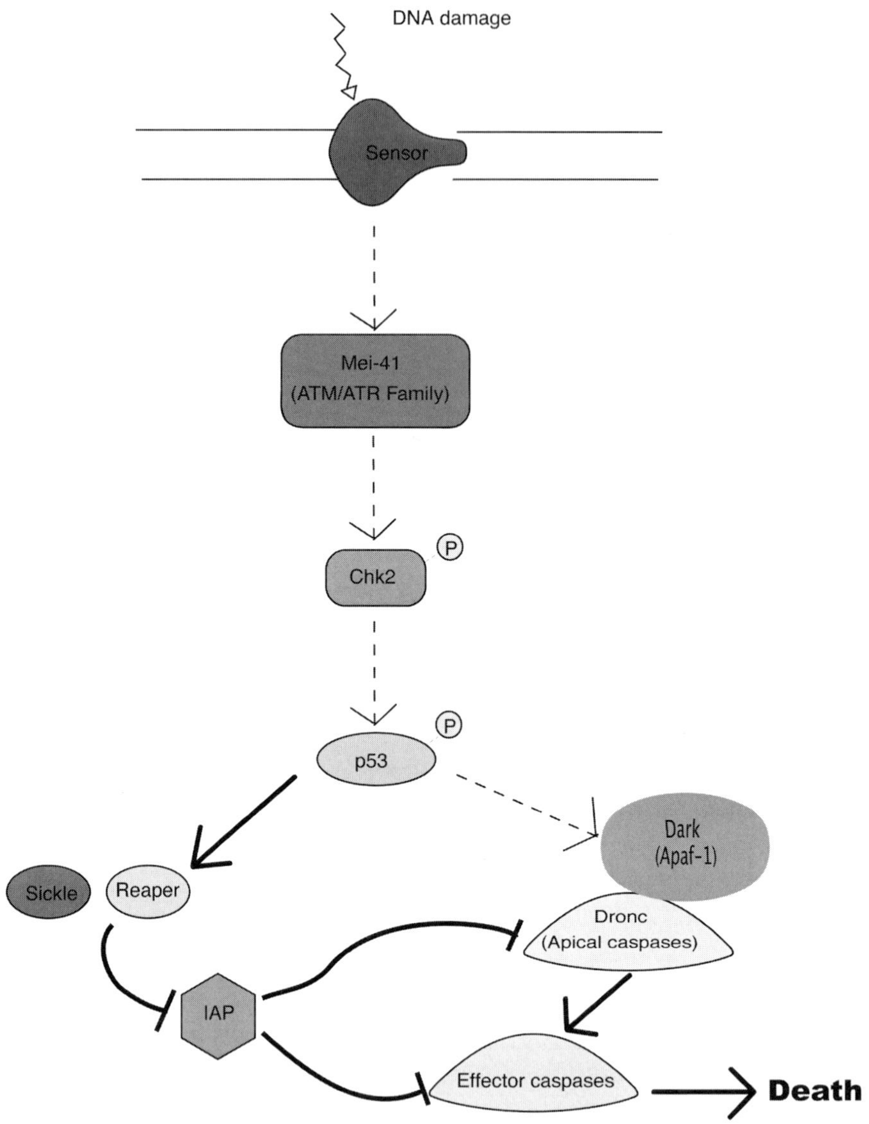

Figure 2 Organization of damage-induced apoptosis in *Drosophila*. The dashed arrows and phosphorylation of target proteins are inferred from orthologous relationships. Upon radiation-induced damage, Dp53 activates IAP antagonists Reaper and possibly Sickle, releasing IAP inhibition from caspases and leading to cell death.

References

1. Muller, H. J. (1927). Artificial transmutation of the gene. *Science.* **46**, 84–87.
2. Dean, F. B., Lian, L., and O'Donnell, M. (1998). cDNA cloning and gene mapping of human homologs for *Schizosaccharomyces pombe rad17, rad1,* and *hus1* and cloning of homologs from mouse, *Caenorhabditis elegans,* and *Drosophila melanogaster. Genomics.* **54**, 424–436.
3. Griffiths, D. J., Barbet, N. C., McCready, S., Lehmann, A. R., and Carr, A. M. (1995). Fission yeast rad17: A homologue of budding yeast RAD24 that shares regions of sequence similarity with DNA polymerase accessory proteins. *EMBO J.* **14**, 5812–5823.
4. Kaur, R., Kostrub, C. F., and Enoch, T. (2001). Structure–function analysis of fission yeast hus1-rad1-rad9 checkpoint complex. *Mol. Biol. Cell.* **12**, 3744–3758.
5. Thelen, M. P., Venclovas, C., and Fidelis, K. (1999). A sliding clamp model for the Rad1 family of cell cycle checkpoint proteins. *Cell* **96**, 769–770.
6. Venclovas, C. and Thelen, M. P. (2000). Structure-based predictions of Rad1, Rad9, Hus1 and Rad17 participation in sliding clamp and clamp-loading complexes. *Nucleic Acids Res.* **28**, 2481–2493.
7. Shiloh, Y. (2001). ATM and ATR: Networking cellular responses to DNA damage. *Curr. Opin. Genet. Dev.* **11**, 71–77.
8. Baker, B. S. and Carpenter, A. T. (1972). Genetic analysis of sex chromosomal meiotic mutants in *Drosophila melanogaster. Genetics.* **71**, 255–286.
9. Hari, K. L., Santerre, A., Sekelsky, J. J., McKim, K. S., Boyd, J. B., and Hawley, R. S. (1995). The mei-41 gene of *D. melanogaster* is a structural and functional homolog of the human ataxia telangiectasia gene. *Cell* **82**, 815–821.
10. Sekelsky, J. J., Brodsky, M. H., and Burtis, K. C. (2000). DNA repair in *Drosophila:* Insights from the *Drosophila* genome sequence. *J. Cell Biol.* **150**, F31–36.
11. Zhou, B. B. and Elledge, S. J. (2000). The DNA damage response: Putting checkpoints in perspective. *Nature* **408**, 433–439.

12. Sibon, O. C., Laurencon, A., Hawley, R., and Theurkauf, W. E. (1999). The *Drosophila* ATM homologue Mei-41 has an essential checkpoint function at the midblastula transition. *Curr. Biol.* **9**, 302–312.
13. Xu, J., Xin, S., and Du, W. (2001). *Drosophila* Chk2 is required for DNA damage-mediated cell cycle arrest and apoptosis. *FEBS Lett.* **508**, 394–398.
14. Su, T. T. and Jaklevic, B. (2001). DNA damage leads to a Cyclin A-dependent delay in metaphase-anaphase transition in the *Drosophila gastrula. Curr. Biol.* **11**, 8–17.
15. Brodsky, M. H., Sekelsky, J. J., Tsang, G., Hawley, R. S., and Rubin, G. M. (2000). mus304 Encodes a novel DNA damage checkpoint protein required during *Drosophila* development. *Genes Dev.* **14**, 666–678.
16. Su, T. T., Parry, D. H., Donahoe, B., Chien, C. T., O'Farrell, P. H., and Purdy, A. (2001). Cell cycle roles for two 14–3–3 proteins during *Drosophila* development. *J. Cell Sci.* **114**, 3445–3454.
17. Christich, A. and Abrams, J. M. (2002). Unpublished data.
18. de Nooij, J. C., Letendre, M. A., and Hariharan, I. K. (1996). A cyclin-dependent kinase inhibitor, Dacapo, is necessary for timely exit from the cell cycle during *Drosophila* embryogenesis. *Cell* **87**, 1237–1247.
19. Su, T. T., Walker, J., and Stumpff, J. (2000). Activating the DNA damage checkpoint in a developmental context. *Curr. Biol.* **10**, 119–126.
20. Jin, P., Gu, Y., and Morgan, D. O. (1996). Role of inhibitory CDC2 phosphorylation in radiation-induced G2 arrest in human cells. *J. Cell Biol.* **134**, 963–970.
21. Deng, C. X., Zhang, P. M., Harper, J. W., Elledge, S. J., and Leder, P. (1995). Mice lacking p21(C/P1/WAF1) undergo normal development, but are defective in G1 checkpoint control. *Cell* **82**, 675–684.
22. Harper, J. W., Adami, G. R., Wei, N., Keyomarsi, K., and Elledge, S. J. (1993). The p21 Cdk-interacting protein Cip1 is a potent inhibitor of G1 cyclin dependent kinases. *Cell* **75**, 805–816.
23. el-Deiry, W. S., Tokino, T., Velculescu, V. E., Levy, D. B., Parsons, R., Trent, J. M., Lin, D., Mercer, W. E., Kinzler, K. W., and Vogelstein, B. (1993). WAF1, a potential mediator of p53 tumor suppression. *Cell* **75**, 817–825.
24. Ollmann, M., Young, L. M., Di Como, C. J., Karim, F., Belvin, M., Robertson, S., Whittaker, K., Demsky, M., Fisher, W. W., Buchman, A., Duyk, G., Friedman, L., Prives, C., and Kopczynski, C. (2000). *Drosophila* p53 is a structural and functional homolog of the tumor suppressor p53. *Cell* **101**, 91–101.
25. Boyd, J. B., Golino, M. D., Nguyen, T. D., and Green, M. M. (1976). Isolation and characterization of X-linked mutants of *Drosophila melanogaster* which are sensitive to mutagens. *Genetics* **84**, 485–506.
26. Smith, P. D. (1976). Mutagen sensitivity of *Drosophila melanogaster*. III. X-linked loci governing sensitivity to methyl methanesulfonate. *Mol. Gen. Genet.* **149**, 73–85.
27. Boyd, J. B., Golino, M. D., Shaw, K. E., Osgood, C. J., and Green, M. M. (1981). Third-chromosome mutagen-sensitive mutants of *Drosophila melanogaster. Genetics* **97**, 607–623.
28. de Buendia, P. G. (1998). Search for DNA repair pathways in *Drosophila melanogaster. Mutat. Res.* **407**, 67–84.
29. Sekelsky, J. J., Burtis, K. C., and Hawley, R. S. (1998). Damage control: The pleiotropy of DNA repair genes in *Drosophila melanogaster. Genetics* **148**, 1587–1598.
30. Ashburner, M. (1989). *Drosophila: A Laboratory Manual.* Cold Spring Harbor Laboratory Press, Cold Spring Harbor, New York.
31. Brodsky, M. H., Nordstrom, W., Tsang, G., Kwan, E., Rubin, G. M., and Abrams, J. M. (2000). *Drosophila* p53 binds a damage response element at the reaper locus. *Cell* **101**, 103–113.
32. Christich, A., Kauppila, S., Chen, P., Sogame, N., Ho, S. I., and Abrams, J. M. (2002). The damage-responsive *Drosophila* gene sickle encodes a novel IAP binding protein similar to but distinct from reaper, grim, and hid. *Curr. Biol.* **12**, 137–140.
33. Wing, J. P., Karres, J. S., Ogdahl, J. L., Zhou, L., Schwartz, L. M., and Nambu, J. R. (2002). *Drosophila* sickle is a novel grim-reaper cell death activator. *Curr. Biol.* **12**, 131–135.
34. Srinivasula, S. M., Datta, P., Kobayashi, M., Wu, J. W., Fujioka, M., Hegde, R., Zhang, Z., Mukattash, R., Fernandes-Alnemri, T., Shi, Y., Jaynes, J. B., and Alnemri, E. S. (2002). sickle, a novel *Drosophila* death gene in the reaper/hid/grim region, encodes an IAP-inhibitory protein. *Curr. Biol.* **12**, 125–130.
35. Goyal, L. (2001). Cell death inhibition: Keeping caspases in check. *Cell* **104**, 805–808.
36. Rodriguez, A., Oliver, H., Wang, X., and Abrams, J. M. (2002). Unrestrained caspase dependent cell death caused by loss of Diap1 function requires the *Drosophila* Apaf-1 homolog, Dark. *EMBO J.* **21**(9), 2189–2197.
37. Rodriguez, A., Oliver, H., Zou, H., Chen, P., Wang, X. D., and Abrams, J. M. (1999). Dark is a *Drosophila* homologue of Apaf-1/CED-4 and functions in an evolutionarily conserved death pathway. *Nat. Cell Biol.* **1**, 272–279.
38. Zhou, L., Song, Z. W., Tittel, J., and Steller, H. (1999). HAC-1, a *Drosophila* homolog of APAF-1 and CED-4 functions in developmental and radiation-induced apoptosis. *Mol. Cell* **4**, 745–755.
39. Kanuka, H., Sawamoto, K., Inohara, N., Matsuno, K., Okano, H., and Miura, M. (1999). Control of the cell death pathway by Dapaf-1, a *Drosophila* Apaf-1/CED-4-related caspase activator. *Mol. Cell* **4**, 757–769.
40. Hirao, A., Kong, Y. Y., Matsuoka, S., Wakeham, A., Ruland, J., Yoshida, H., Liu, D., Elledge, S. J., and Mak, T. W. (2000). DNA damage-induced activation of p53 by the checkpoint kinase Chk2. *Science* **287**, 1824–1827.

CHAPTER 304

Double-Strand Break Recognition and Its Repair by Nonhomologous End Joining

Jane M. Bradbury[1] and Stephen P. Jackson[1,2]

[1]*Wellcome Trust and Cancer Research Center, Cambridge, United Kingdom*
Institute of Cancer and Developmental Biology, Cambridge, United Kingdom
[2]*Department of Zoology, University of Cambridge, Cambridge, United Kingdom*

Introduction

Many different agents can damage DNA, often with disastrous consequences. The most dangerous form of DNA damage is the double-strand break (DSB)—a single unrepaired DSB can induce a cell to undergo apoptosis [1]. DSBs can also directly inactivate key genes, lead to chromosomal translocations, or generate unstable chromosomal abnormalities [2], which may ultimately lead to cancer development. DSBs are induced by ionizing radiation and radiomimetic chemicals but can also be generated during site-specific recombination events (for example, during V(D)J recombination in the immune system and meiotic recombination) and during DNA replication.

Because the effects of DNA DSBs are potentially so serious, eukaryotes have elaborate systems to repair the damage rapidly or to stop DNA replication and cell division until the damage has been repaired (Fig. 1; [3]). Here, we consider nonhomologous end joining (NHEJ), a major DSB repair system, drawing on data obtained from the mammalian system and the highly conserved *Saccharomyces cerevisiae* system.

Repair of DSBs: Homologous Recombination and NHEJ

Eukaryotic cells have two major DSB repair systems: homologous recombination (HR) and NHEJ [4,5]. HR involves 5′ resection of the damaged DNA; invasion of the 3′ end of the broken DNA into a homologous DNA duplex molecule—a sister chromatid or a homologous chromosome; and replication of the damaged DNA using the homologous DNA as a template. Because no genetic material is normally lost during HR, this is the DNA repair pathway of choice throughout most of the life cycle of *S. cerevisiae* and other lower eukaryotes.

NHEJ, by contrast, does not require homologous DNA. Instead, the broken DNA ends are brought together and religated, often after limited nuclease digestion to remove damaged bases. Thus, repair of DSBs via NHEJ generally results in loss of some genetic material. Nevertheless, NHEJ is a major DSB repair pathway in mammals, possibly because it is an efficient way to remove potentially lethal DSBs.

Several of the principal players in NHEJ (Fig. 2) were originally identified by studying mutant mammalian cell lines that are hypersensitive to ionizing radiation and defective in DSB rejoining and V(D)J recombination (reviewed in [4,6]). Cells of complementation groups IR5 and IR7 lack Ku80 and DNA-PKcs, respectively. These proteins, together with Ku70, form a DNA-activated nuclear serine/threonine kinase called DNA-dependent protein kinase (DNA-PK; [7]). DNA-PKcs itself is a large protein that is a member of the phosphoinositol 3-kinase-like family of kinases (PIKK; see Smith and Jackson elsewhere in this series). There are yeast homologs of the two small Ku subunits but no obvious DNA-PKcs homolog. Cells of complementation group IR4

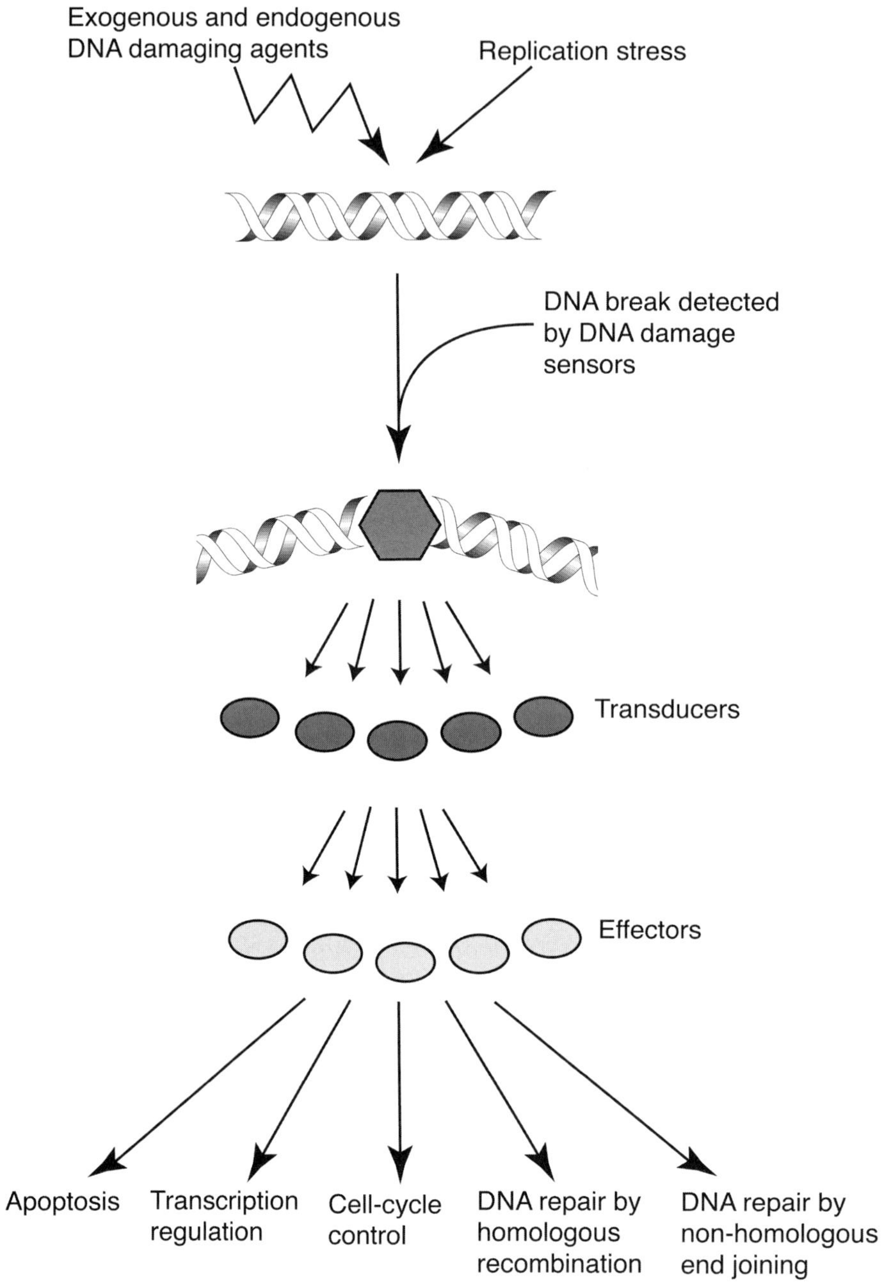

Figure 1 General outline of the cellular response to DNA double-strand breaks. For simplicity, responses are shown as an arbitrary number of separate arrows. In reality, the number of pathways and their interactions with each other are much more complicated, allowing the cell to deal with each and every DNA double-strand break that occurs rapidly and efficiently.

lack the small nuclear phosphoprotein XRCC4, a protein that forms a tight, specific complex with DNA ligase IV [8,9]. This is the ligase responsible for rejoining DNA at DSBs and thus the effector of the NHEJ pathway. Again, close homologs of XRCC4 and DNA ligase IV exist in yeast (Lif1p and Dnl4p, respectively; [10–12]).

Other evidence linking DNA-PK and ligase IV to NHEJ comes from studies on spontaneous mouse mutants and knock-out mice. The severe-combined immunodeficient (Scid) mouse strain has no mature circulating T or B lymphocytes and is hypersensitive to ionizing radiation and radiomimetic drugs. Scid mice lack active DNA-PKcs [13,14], thus indicating the importance of DNA-PKcs in both DNA repair and V(D)J recombination. More recently, knock-out mice have been derived in which DNA-PK components are missing (reviewed in [7]). Most of these are immunodeficient and hypersensitive to ionizing radiation, but, interestingly, mice lacking DNA ligase IV or XRCC4 [15–17] are embryonic lethal (see later discussion).

In addition to these players, the human Mre11/Rad50/Nbs1 complex (Mre11p/Rad50p/Xrs2p in *S. cerevisiae*) has been implicated in NHEJ. This complex may be involved in

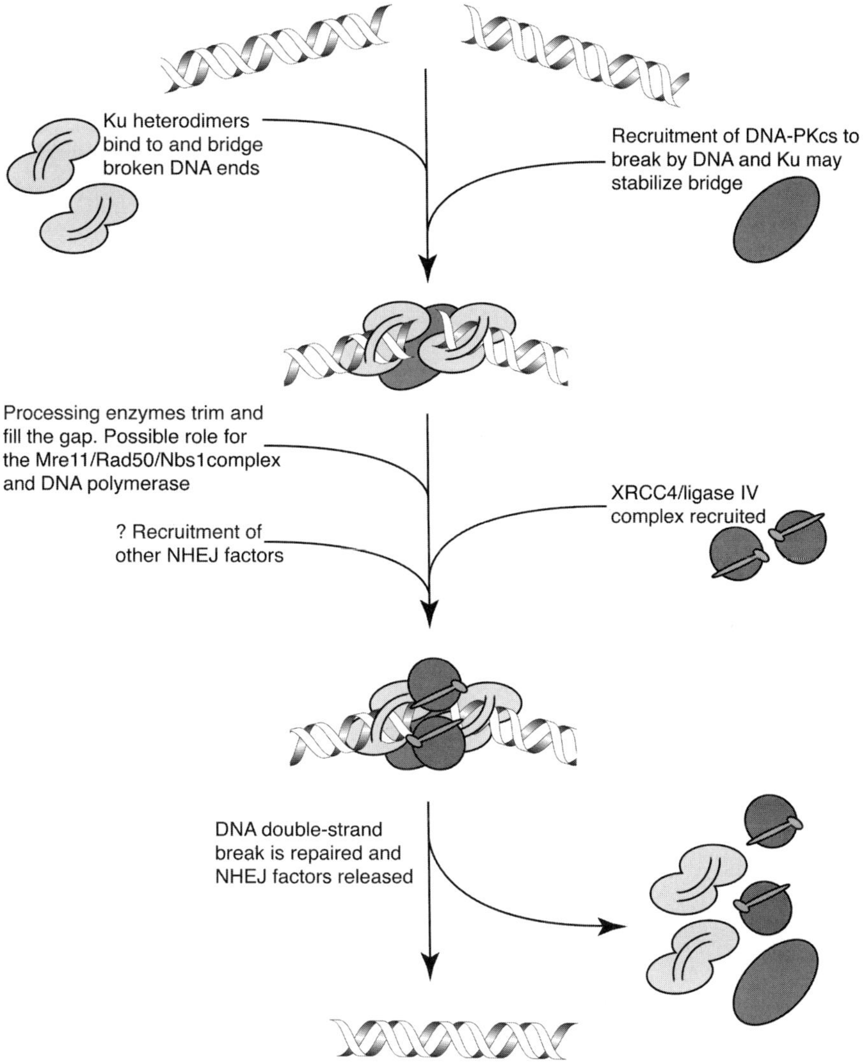

Figure 2 Schematic showing the major known players in NHEJ in mammalian cells. Some of the potential additional NHEJ components are described in the text where attention is also drawn to aspects of this model for which we have recently gained new insights.

preparing the DNA for ligation via the nuclease activity of Mre11p [18]. However, it also has a role in checkpoint signaling in mammalian cells (reviewed in [5]) and in yeast [19–21]. Finally, two new NHEJ factors have been identified recently (see later discussion).

Recognition of DNA DSBs

Substantial *in vitro* evidence indicates that the ends of a DNA DSB are initially recognized by two Ku heterodimers [6,7,22], which may interact to bring together the DNA ends [23,24]. Subsequent recruitment of DNA-PKcs may help to stabilize this bridging complex and to recruit or activate other factors needed for NHEJ (Fig. 2). The recent publication of the crystal structure of Ku bound to DNA provides support for this model [25]. In this structure, the Ku heterodimer forms an open ring that can be threaded onto a DNA end. One side of this ring forms a cradle that protects one surface of the DNA double helix. The other side of the ring is considerably more open, a finding that suggests how other NHEJ factors can access the DSB, including the processing enzymes needed to prepare the DNA for ligation. The structure also helps to explain many of the *in vitro* properties of Ku. For example, the observation that Ku makes contacts almost exclusively with the sugar-phosphate backbone of the DNA explains why Ku binds with high affinity to DNA ends irrespective of sequence or structure [22].

The structure also explains why Ku can only be loaded onto free DNA ends (reviewed in [7]) and becomes trapped if linear DNA is subsequently circularized [26]. However, the data beg the question: How is Ku released from DNA once the DSB has been repaired? One possibility is that Ku is released before or concomitant with ligation. Alternatively, Ku's removal from ligated DNA might require its structural rearrangement or proteolytic digestion.

Signal Transduction

DNA-PKcs can bind to DNA by itself [27–29] and is activated by binding to single-stranded DNA [30,31], possibly through a conformational change [32,33]. Nevertheless, mammalian DNA-PKcs can be regarded as a signaling kinase that links DNA damage sensing by Ku to DNA repair—the kinase activity of DNA-PKcs is certainly necessary for NHEJ [34]. *In vivo*, DNA-PKcs probably never binds alone to DNA. Instead, the weak intrinsic DNA-end binding activity of DNA-PKcs is greatly enhanced and stabilized by the Ku70–Ku80 heterodimer [27,29,33,35].

The DNA-PKcs–Ku complex preferentially phosphorylates proteins attached to the same DNA molecule [33] although targeting of non-DNA-bound substrates can also occur. Numerous *in vitro* substrates for DNA-PKcs have been defined, including several transcription factors, chromatin components, and replication protein A (reviewed in [7]). Although it remains unclear which, if any, of these are true physiological substrates, it seems reasonable to assume that other NHEJ components will be phosphorylated by DNA-PKcs to increase their intrinsic activity and to encourage the formation of multiprotein complexes. Indeed, XRCC4 is a phosphoprotein *in vivo* and is an effective substrate for DNA-PK *in vitro* [8,36]. The Ku subunits and DNA-PKcs itself are also phosphorylated *in vitro* by DNA-PKcs, modifications that may help in the disassembly of the DNA-PK complex at the time of or after DNA ligation [37].

DNA Repair

Once the DNA ends on either side of a DSB have been brought together and tidied up, the break is repaired by DNA ligase IV in mammalian cells and by Dnl4p in yeast. These enzymes form tight complexes with XRCC4 and Lif1p, respectively, interactions that probably regulate their activity [38]. Certainly, XRCC4–DNA ligase IV promotes DNA end joining *in vitro* [39], a reaction that is stimulated by the addition of Ku [23,40] and by DNA-PK [41].

The recent determination of the crystal structure of XRCC4, both alone [42] and in a complex with a peptide from DNA ligase IV [43], yields some mechanistic insights into NHEJ. XRCC4 alone forms an elongated dumb-bell-like tetramer, but in the XRCC4–DNA ligase IV complex, the helical tail of XRCC4 is partly unwound to produce a flat interaction site. The sites of contact within this complex are highly conserved from yeast to man, suggesting that the mechanism of interaction between XRCC4 and DNA ligase IV has been maintained throughout evolution.

Other Sensors and Transducers of DNA Damage

Although Ku is an important DNA sensor in the NHEJ system, it is not the only sensor for DNA DSBs. Indeed, signaling to the cell cycle apparatus seems to occur mainly through related PIKKs: ATM and ATR in mammalian cells [44,45] and Tel1p and Mec1p in *S. cerevisiae* [3]. For these signaling kinases, we know less about how the damage is initially sensed because no Ku homolog has been found for these enzymes, although a number of proteins have been proposed as possible damage sensors [3]. Recent work in yeast in our laboratory and elsewhere has yielded some information about how Mec1p, the yeast ATR homolog, might be recruited to DNA damage by the protein Lcd1p (also called Ddc2p and Pie1p; [46–49]). A human homolog of Lcd1p has also been identified ([50]; P. Reaper and S. P. Jackson, 2001, unpublished data) and its properties are likely to match those of yeast Lcd1p.

Interestingly, as mentioned earlier, transgenic animals lacking functional XRCC4 or DNA ligase IV are embryonic lethal, whereas mice lacking Ku are viable. One explanation for this apparent inconsistency is that in Ku-defective mice, the broken DNA ends can be detected by other DNA damage sensors, including those that signal to the checkpoint apparatus and to other DNA repair mechanisms. In mice that have defective XRCC4 or DNA ligase IV, however, Ku can still bind to the DSBs and this may result in a nonproductive complex that prevents other cellular responses to DNA damage being activated. Thus the cell, and indeed the whole organism, is unable to avoid lethal genomic instability. Further evidence for this proposition comes from recent work by Adachi *et al.* [51] in DT40 cells, which shows that LIG4$^{-/-}$ cells are more X-ray sensitive than Ku70$^{-/-}$ cells, but that Ku70$^{-/-}$/LIG4$^{-/-}$ double-mutant cells have a similar sensitivity to Ku70$^{-/-}$ cells.

New Factors in NHEJ

We now know a considerable amount about NHEJ but there may still be unknown factors in both the yeast and the mammalian system. For example, Ooi *et al.* [52] recently developed a microarray-based screen to look for additional NHEJ factors in *S. cerevisiae*. Their screen reidentified several known NHEJ components, but also identified a new factor required for NHEJ, Nej1p, which has been identified independently by three other groups using different approaches [53–55]. Nej1p (also called Lif2p) interacts with Lif1p but it does not stimulate NHEJ simply by regulating Lif1p levels because it is required for efficient NHEJ even when Lif1p is overexpressed in the nucleus. More work is needed to discover Nej1p's exact mode of action and the identity of any mammalian homologs of this yeast protein. Interestingly, Nej1p is expressed only in haploid yeast cells, an observation that provides an elegant explanation for why NHEJ is very inefficient in yeast diploid cells where HR is the preferred means of DSB repair.

Another new NHEJ component was recently identified by looking for the gene defect in a family of human SCID individuals who were also radiosensitive. Unlike Scid mice, human beings who lack T and B cells do not have defective DNA-PKcs. Instead, Artemis, a novel DNA double-strand break repair/V(D)J recombination protein, is mutated in this subset of SCID patients [56]. Artemis belongs to the metallo-β-lactamase superfamily so it may be involved in DNA processing. However, the precise function of Artemis in NHEJ remains to be elucidated.

Future Prospects

Although we have learned much about NHEJ, many questions remain. For example, is the conformation of DNA-PKcs altered on binding DNA and Ku? What are the *in vivo* physiological substrates for DNA-PKcs? How is the DNA-PK complex removed from the DNA repair site? In addition, mammalian and yeast cells possess multiple DNA damage sensing systems that must be tightly integrated, but we are a long way from understanding how this is brought about. Recent work also indicates that intimate links exist between NHEJ and the control of chromatin structure [57], telomere maintenance [6,58], retroviral integration [59,60], and innate immunity [61]. By understanding all of these aspects of the NHEJ machinery, we should gain insights into how genomic stability is maintained in the face of constant endogenous and exogenous insults, knowledge that should lead to better therapeutic and preventive measures for cancer and other human diseases.

Acknowledgments

Research in the S. P. J. laboratory is funded by the Cancer Research Campaign, the Association for International Cancer Research and the A-T Medical Research Trust.

References

1. Rich, T., Allen, R. L., and Wyllie, A. H. (2000). Defying death after DNA damage. *Nature* **407**, 777–783.
2. van Gent, D. C., Hoeijmakers, J. H. J., and Kanaar, R. (2001). Chromosomal stability and the DNA double-stranded break connection. *Nat. Rev. Genet.* **2**, 196–206.
3. Zhou, B. B. S. and Elledge, S. J. (2000). The DNA damage response: Putting checkpoints in perspective. *Nature* **408**, 433–439.
4. Critchlow, S. E. and Jackson, S. P. (1998). DNA-end joining: From yeast to man. *Trends Biochem. Sci.* **23**, 394–398.
5. Khanna, K. K. and Jackson, S. P. (2001). DNA double-strand breaks: Signaling, repair and the cancer connection. *Nat. Genet.* **27**, 247–254.
6. Featherstone, C. and Jackson, S. P. (1999). Ku, a DNA repair protein with multiple cellular functions? *Mutat. Res. DNA Repair* **434**, 3–15.
7. Smith, G. C. M. and Jackson, S. P. (1999). The DNA-dependent protein kinase. *Genes Dev.* **13**, 916–934.
8. Critchlow, S. E., Bowater, R. P., and Jackson, S. P. (1997). Mammalian DNA double-strand break repair protein XRCC4 interacts with DNA ligase IV. *Curr. Biol.* **7**, 588–598.
9. Grawunder, U., Wilm, M., Wu, X. T., Kulesza, P., Wilson, T. E., Mann, M., and Lieber, M. R. (1997). Activity of DNA ligase IV stimulated by complex formation with XRCC4 protein in mammalian cells. *Nature* **388**, 492–495.
10. Herrmann, G., Lindahl, T., and Schär, P. (1998). *Saccharomyces cerevisiae* LIF1: A function involved in DNA double-strand break repair related to mammalian XRCC4. *EMBO J.* **17**, 4188–4198.
11. Teo, S.-H. and Jackson, S. P. (1997). Identification of *Saccharomyces cerevisiae* DNA ligase IV: Involvement in DNA double-strand break repair. *EMBO J.* **15**, 4788–4795.
12. Schär, P., Herrmann, G., Daly, G., and Lindahl, T. (1997). A newly identified DNA ligase of *Saccharomyces cerevisiae* involved in *RAD52*-independent repair of DNA double-strand breaks. *Genes Dev.* **11**, 1912–1924.
13. Blunt, T., Finnie, N. J., Taccioli, G. E., Smith, G. C. M., Demengeot, J., Gottlieb, T. M., Mizuta, R., Varghese, A. J., Alt, F. W., Jeggo, P. A., and Jackson, S. P. (1995). Defective DNA-dependent protein kinase activity is linked to V(D)J recombination and DNA repair defects associated with the murine scid mutation. *Cell* **80**, 813–823.
14. Danska, J. S., Holland, D. P., Mariathasan, S., Williams, K. M., and Guidos, C. J. (1996). Biochemical and genetic defects in the DNA-dependent protein kinase in murine scid lymphocytes. *Mol. Cell. Biol.* **16**, 5507–5517.
15. Barnes, D. E., Stamp, G., Rosewell, I., Denzel, A., and Lindahl, T. (1998). Targeted disruption of the gene encoding DNA ligase IV leads to lethality in embryonic mice. *Curr. Biol.* **8**, 1395–1398.
16. Frank, K. M., Sekiguchi, J. M., Seidl, K. J., Swat, W., Rathbun, G. A., Cheng, H. L., Davidson, L., Kangaloo, L., and Alt, F. W. (1998). Late embryonic lethality and impaired V(D)J recombination in mice lacking DNA ligase IV. *Nature* **396**, 173–177.
17. Gao, Y. J., Sun, Y., Frank, K. M., Dikkes, P., Fujiwara, Y., Seidl, K. J., Sekiguchi, J. M., Rathbun, G. A., Swat, W., Wang, J. Y., Bronson, R. T., Malynn, B. A., Bryans, M., Zhu, C. M., Chaudhuri, J., Davidson, L., Ferrini, R., Stamato, T., Orkin, S. H., Greenberg, M. E., and Alt, F. W. (1998). A critical role for DNA end-joining proteins in both lymphogenesis and neurogenesis. *Cell* **95**, 891–902.
18. Boulton, S. J. and Jackson, S. P. (1998). Components of the Ku-dependent non-homologous end-joining pathway are involved in telomeric length maintenance and telomeric silencing. *EMBO J.* **17**, 1819–1828.
19. D'Amours, D. and Jackson, S. P. (2001). The yeast Xrs2 complex functions in S phase checkpoint regulation. *Genes Dev.* **15**, 2238–2249.
20. Usui, T., Ogawa, H., and Petrini, J. H. J. (2001). A DNA damage response pathway controlled by Tel1 and the Mre11 complex. *Mol. Cell* **7**, 1255–1266.
21. Grenon, M., Gilbert, C., and Lowndes, N. F. (2001). Checkpoint activation in response to double-strand breaks requires the Mre11/Rad50/Xrs2 complex. *Nat. Cell Biol.* **3**, 844–847.
22. Dynan, W. S. and Yoo, S. (1998). Interaction of Ku protein and DNA-dependent protein kinase catalytic subunit with nucleic acids. *Nucleic Acids Res.* **26**, 1551–1559.
23. Ramsden, D. A. and Gellert, M. (1998). Ku protein stimulates DNA end joining by mammalian DNA ligases: A direct role for Ku in repair of DNA double-strand breaks. *EMBO J.* **17**, 609–614.
24. Bliss, T. M. and Lane, D. P. (1997). Ku selectively transfers between DNA molecules with homologous ends. *J. Biol. Chem.* **272**, 5765–5773.
25. Walker, J. R., Corpina, R. A., and Goldberg, J. (2001). Structure of the Ku heterodimer bound to DNA and its implications for double-strand break repair. *Nature* **412**, 607–614.
26. Paillard, S. and Strauss, F. (1991). Analysis of the mechanism of interaction of simian Ku protein with DNA. *Nucleic Acids Res.* **19**, 5619–5624.
27. Hammarsten, O. and Chu, G. (1998). DNA-dependent protein kinase: DNA binding and activation in the absence of Ku. *Proc. Natl. Acad. Sci. USA* **95**, 525–530.
28. Yaneva, M., Kowalewski, T., and Lieber, M. R. (1997). Interaction of DNA-dependent protein kinase with DNA and with Ku: Biochemical and atomic-force microscopy studies. *EMBO J.* **16**, 5098–5112.
29. West, R. B., Yaneva, M., and Lieber, M. R. (1998). Productive and nonproductive complexes of Ku and DNA-dependent protein kinase at DNA termini. *Mol. Cell. Biol.* **18**, 5908–5920.
30. Hammarsten, O., DeFazio, L. G., and Chu, G. (2000). Activation of DNA-dependent protein kinase by single-stranded DNA ends. *J. Biol. Chem.* **275**, 1541–1550.
31. Mårtensson, S. and Hammarsten, O. (2002). DNA-dependent protein kinase catalytic subunit: Structural requirements for kinase activation by DNA ends. *J. Biol. Chem.* **277**, 3020–3029.

32. Lees-Miller, S. P., Chen, Y. R., and Anderson, C. W. (1990). Human cells contain a DNA-activated protein kinase that phosphorylates simian virus 40 T antigen, mouse p53, and the human Ku autoantigen. *Mol. Cell. Biol.* **10**, 6472–6481.
33. Gottlieb, T. M. and Jackson, S. P. (1993). The DNA-dependent protein kinase: Requirement for DNA ends and association with Ku antigen. *Cell* **72**, 131–142.
34. Kurimasa, A., Kumano, S., Boubnov, N. V., Story, M. D., Tung, C. S., Peterson, S. R., and Chen, D. J. (1999). Requirement for the kinase activity of human DNA-dependent protein kinase catalytic subunit in DNA strand break rejoining. *Mol. Cell. Biol.* **19**, 3877–3884.
35. Dvir, A., Stein, L. Y., Calore, B. L., and Dynan, W. S. (1993). Purification and characterization of a template-associated protein kinase that phosphorylates RNA polymerase II. *J. Biol. Chem.* **268**, 10440–10447.
36. Leber, R., Wise, T. W., Mizuta, R., and Meek, K. (1998). The *XRCC4* gene product is a target for and interacts with the DNA-dependent protein kinase. *J. Biol. Chem.* **273**, 1794–1801.
37. Chan, D. W. and Lees-Miller, S. P. (1996). The DNA-dependent protein kinase is inactivated by autophosphorylation of the catalytic subunit. *J. Biol. Chem.* **271**, 8936–8941.
38. Teo, S. H. and Jackson, S. P. (2000). Lif1p targets the DNA ligase Lig4p to sites of DNA double-strand breaks. *Curr. Biol.* **10**, 165–168.
39. Grawunder, U., Zimmer, D., Kulesza, P., and Lieber, M. R. (1998). Requirement for an interaction of XRCC4 with DNA ligase IV for wild type V(D)J recombination and DNA double strand break repair *in vivo*. *J. Biol. Chem.* **273**, 24708–24714.
40. McElhinny, S. A. N., Snowden, C. M., McCarville, J., and Ramsden, D. A. (2000). Ku recruits the XRCC4-ligase IV complex to DNA ends. *Mol. Cell. Biol.* **20**, 2996–3003.
41. Chen, L., Trujillo, K., Sung, P., and Tomkinson, A. E. (2000). Interactions of the DNA ligase IV-XRCC4 complex with DNA ends and the DNA-dependent protein kinase. *J. Biol. Chem.* **275**, 26196–26205.
42. Junop, M. S., Modesti, M., Guarné, A., Ghirlando, R., Gellert, M., and Yang, W. (2000). Crystal structure of the Xrcc4 DNA repair protein and implications for end joining. *EMBO J.* **19**, 5962–5970.
43. Sibanda, B. L., Critchlow, S. E., Begun, J., Pei, X. Y., Jackson, S. P., Blundell, T. L., and Pellegrini, L. (2001). Crystal structure of an Xrcc4–DNA ligase IV complex. *Nat. Struct. Biol.* **8**, 1015–1019.
44. Abraham, R. T. (2001). Cell cycle checkpoint signaling through the ATM and ATR kinases. *Genes Dev.* **15**, 2177–2196.
45. Durocher, D. and Jackson, S. P. (2001). DNA-PK, ATM and ATR as sensors of DNA damage: Variations on a theme? *Curr. Opin. Cell Biol.* **13**, 225–231.
46. Rouse, J. and Jackson, S. P. (2000). *LCD1*: An essential gene involved in checkpoint control and regulation of the *MEC1* signalling pathway in *Saccharomyces cerevisiae*. *EMBO J.* **19**, 5801–5812.
47. Paciotti, V., Clerici, M., Lucchini, G., and Longhese, M. P. (2000). The checkpoint protein Ddc2, functionally related to *S. pombe* Rad26, interacts with Mec1 and is regulated by Mec1-dependent phosphorylation in budding yeast. *Genes Dev.* **14**, 2046–2059.
48. Wakayama, T., Kondo, T., Ando, S., Matsumoto, K., and Sugimoto, K. (2001). Pie1, a protein interacting with Mec1, controls cell growth and checkpoint responses in *Saccharomyces cerevisiae*. *Mol. Cell. Biol.* **21**, 755–764.
49. Clerici, M., Paciotti, V., Baldo, V., Romano, M., Lucchini, G., and Longhese, M. P. (2001). Hyperactivation of the yeast DNA damage checkpoint by *TEL1* and *DDC2* overexpression. *EMBO J.* **20**, 6485–6498.
50. Cortez, D., Guntuku, S., Qin, J., and Elledge, S. J. (2001). ATR and ATRIP: Partners in checkpoint signaling. *Science* **294**, 1713–1716.
51. Adachi, N., Ishino, T., Ishii, Y., Takeda, S., and Koyama, H. (2001). DNA ligase IV-deficient cells are more resistant to ionizing radiation in the absence of Ku70: Implications for DNA double-strand break repair. *Proc. Natl. Acad. Sci. USA* **98**, 12109–12113.
52. Ooi, S. L., Shoemaker, D. D., and Boeke, J. D. (2001). A DNA microarray-based genetic screen for nonhomologous end-joining mutants in *Saccharomyces cerevisiae*. *Science* **294**, 2552–2556.
53. Kegel, A., Sjöstrand, J. O. O., and Åström S. U. (2001). Nej1p, a cell type-specific regulator of nonhomologous end joining in yeast. *Curr. Biol.* **11**, 1611–1617.
54. Frank-Vaillant, M. and Marcand, S. (2001). NHEJ regulation by mating type is exercised through a novel protein, Lif2p, essential to the ligase IV pathway. *Genes Dev.* **15**, 3005–3012.
55. Valencia, M., Bentele, M., Vaze, M. B., Herrmann, G., Kraus, E., Lee, S. E., Schär, P., and Haber, J. E. (2001). *NEJ1* controls non-homologous end joining in *Saccharomyces cerevisiae*. *Nature* **414**, 666–669.
56. Moshous, D., Callebaut, I., de Chasseval, R., Corneo, B., Cavazzana-Calvo, M., Le Deist, F., Tezcan, I., Sanal, O., Bertrand, Y., Philippe, N., Fischer, A., and de Villartay, J. P. (2001). Artemis, a novel DNA double-strand break repair/V(D)J recombination protein, is mutated in human severe combined immune deficiency. *Cell* **105**, 177–186.
57. Downs, J. A., Lowndes, N. F., and Jackson, S. P. (2000). A role for *Saccharomyces cerevisiae* histone H2A in DNA repair. *Nature* **408**, 1001–1004.
58. Gilley, D., Tanaka, H., Hande, M. P., Kurimasa, A., Li, G. C., and Chen, D. J. (2001). DNA-PKcs is critical for telomere capping. *Proc. Natl. Acad. Sci. USA* **98**, 15084–15088.
59. Downs, J. A. and Jackson, S. P. (1999). Involvement of DNA end-binding protein Ku in Ty element retrotransposition. *Mol. Cell. Biol.* **19**, 6260–6268.
60. Daniel, R., Katz, R. A., and Skalka, A. M. (1999). A role for DNA-PK in retroviral DNA integration. *Science* **284**, 644–647.
61. Chu, W. M., Gong, X., Li, Z. W., Takabayashi, K., Ouyang, H. H., Chen, Y., Lois, A., Chen, D. J., Li, G. C., Karin, M., and Raz, E. (2000). DNA-PKcs is required for activation of innate immunity by immunostimulatory DNA. *Cell* **103**, 909–918.

CHAPTER 305

Role of ATM in Radiation Signal Transduction

Martin F Lavin,[1,2] Shaun Scott,[1] Philip Chen,[1] Sergei Kozlov,[1] Nuri Gueven,[1] and Geoff Birrell[1]

[1]*Queensland Cancer Fund Research Laboratory, Queensland Institute of Medical Research, Brisbane, Australia,*

[2]*Department of Surgery, University of Queensland, Brisbane, Australia*

Introduction

To maintain the integrity of the genome, cells have evolved several mechanisms that recognize DNA damage and signal this to the DNA repair machinery, to cell cycle checkpoints, and to transcriptional control. Although there have been exhaustive reports on the nature of the lesions arising in DNA in response to a variety of damaging agents and on the mechanisms of repair, the ability of the cell to recognize these lesions and signal to the appropriate cellular machinery has only recently begun to be unraveled. The description of a number of human genetic disorders characterized by chromosomal instability and cancer predisposition has accelerated our understanding of the process of DNA damage recognition. One such syndrome, ataxia-telangiectasia (A-T), has been a focal point because of the universal sensitivity to ionizing radiation and because of the central role the gene product involved plays in radiation signal transduction. This syndrome is characterized by immunodeficiency, neurodegeneration, radiosensitivity, meiotic defects, and cancer predisposition [1,2]. Radiosensitivity is observed in patients exposed to therapeutic radiation [3,4] and in cells in culture [5,6]. The basis of the radiosensitivity remains unknown but can be explained by a failure of A-T cells to respond appropriately to double-strand breaks in DNA and the presence of residual breaks at longer times postirradiation [7]. It has been suggested that residual chromosomal breaks in A-T cells lead to oxidative stress for which there is evidence in both human cells and in *Atm*-gene disrupted mice [8–10]. Failure of mutant ATM to respond to double-strand breaks in DNA appears to account for the multiple cell cycle checkpoint defects in A-T cells [11,12]. Of these, radioresistant DNA synthesis (RDS) was the first to be described [13,14]. RDS may be explained by a defect in radiation-induced degradation of Cdc25A phosphatase that normally prevents dephosphorylation of Cdc2 kinase and a transient block in DNA replication [15]. Alternatively, it could be due to failure to inhibit cyclin-dependent kinase activity [16] or by defective signaling through BRCA1 [17].

The defect in regulating the passage of cells from the G_1 to S phase can be explained primarily by a defective p53 response in irradiated A-T cells [18,19] but it is also evident that ATM controls the G_1/S checkpoint through several intermediate phosphorylation reactions [20]. The exact basis for the G_2/M checkpoint defect in A-T cells remains unresolved but there is evidence that defective inhibition of Cdc2 cyclin B kinase may at least be partially responsible [16].

ATM is a member of a family of proteins that share a phosphatidylinositol 3-kinase (PI3K) domain [21]. This group includes the catalytic subunit of DNA-dependent protein kinase (DNA-PKcs), A-T and rad3-related protein (ATR), and proteins in other organisms responsible for DNA damage recognition or cell cycle control [22]. ATM kinase is rapidly activated by ionizing radiation to phosphorylate a series of substrates involved in radiation signaling [23]. However, evidence also exists to show that ATM can be regulated at both the transcriptional and translational levels [24–26]. A more widespread role for ATM in events other than DNA damage recognition exists including receptor signaling, cellular proliferation, K^+ channel activity, and insulin signaling pathways [27–30]. It is possible that ATM plays a direct role in these processes, or that in its absence cellular homeostasis is altered by oxidative stress or some other form of perturbation, leading to the myriad of defects described in A-T cells. Here we initially focus on the function of ATM; its role in DNA damage recognition and its relationship to other DNA damage recognition systems; intermediates phosphorylated; and pathways activated to help coordinate the cellular response to radiation. A possible role in more general signaling is discussed later.

Sensing Radiation Damage in DNA

Exposure of cells to ionizing radiation causes direct damage to DNA as well as ionized and excited species through interaction with water, also capable of damaging DNA [31]. A multitude of damage results, including base modifications and single- and double-strand breaks (DSBs) in the backbone of DNA [32]. Of these the DSB is the most lethal lesion. DNA DSBs also arise normally during the process of V(D)J recombination in B- and T-cell ontogeny and during meiosis and mitosis [33,34].

In both yeast and mammalian cells these breaks are repaired by three mechanisms: single-strand annealing, homologous recombination (HR), and nonhomologous end joining (NHEJ) [35]. The relative importance of these mechanisms was thought to vary between yeast and mammals but recent data suggest this may not be the case [36]. NHEJ is primarily responsible for repair of DSBs in G_1 and early S phase, whereas HR predominates in G_2/M [37]. The appearance of breaks in DNA is rapidly detected by several enzyme systems as outlined in Fig. 1. Poly(ADP-ribose) polymerase (PARP) responds rapidly to single-strand breaks in DNA by transferring ADP-ribosyl from NAD^+ to proteins involved in chromatin structure (histone proteins) and in DNA metabolism (topoisomerases, DNA replication factors, and PARP itself) [38]. Addition of chains of negatively charged poly(ADP-ribose) to these proteins alters their capacity to bind DNA and is responsible for their inactivation.

These protein alterations would be expected to increase the access of proteins involved in DNA repair. Concurrently with the recognition of single-strand breaks by PARP, DSBs are detected by a group of sensor proteins. In *Schizosaccharomyces pombe*, Rad3, a homolog of ATM and ATR, recognizes damage in DNA in conjunction with Rad1, Rad9, Rad17, Rad26, and Hus1 [39]. The corresponding proteins in human cells form a complex, and reduced extractability from nuclei after DNA damage points to tighter binding to damaged DNA. Recent results suggest that after DNA damage

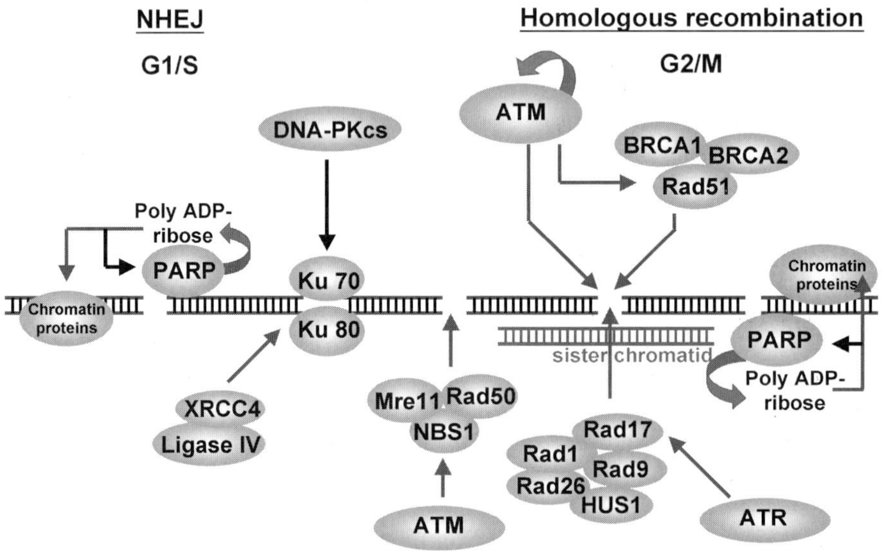

Figure 1 Detection of strand breaks in DNA. PARP is rapidly activated by single- and double-strand breaks in DNA to poly ADP ribosylate a number of proteins including itself. This modification alters their ability to bind to DNA and in turn assists in the processing of the break. DSBs are repaired by two major mechanisms: NHEJ and HR. Recognition of the DSBs involves the coordinated action of several proteins and complexes including DNA-PKcs and the Ku heterodimer that recognize free ends in DNA followed by recruitment of the DNA ligase IV/XRCC4 complex to seal the break. This occurs in concert with the Mre11/Rad50/Nbs1 complex, at least for some of these breaks. ATM also senses breaks in DNA through the HR pathway and phosphorylates and activates a number of downstream effector proteins including Nbs1 and BRCA1. Evidence for indirect involvement of ATM through Rad51 and Mre11 also exists.

Rad17 recruits the Rad1/Rad9/Hus1 complex, which is required for ATR phosphorylation of Rad17 and triggering of the checkpoint signaling cascade [40]. NHEJ is responsible for initiating the process of DSB repair primarily in G_1 phase cells. Characterization of radiation-sensitive mammalian mutants and the generation of gene-disrupted mutant mice have identified DNA-PKcs, a heterodimer of Ku70 and Ku80, ligase IV, and XRCC4 as central players in this process [41]. Mutants in any of these genes are hypersensitive to ionizing radiation, defective in DSB repair and V(D)J recombination, and have abnormalities in telomere maintenance [42–44]. The Ku heterodimer binds to the ends of a DSB and recruits and activates DNA-PKcs leading to the phosphorylation of a number of substrates implicated in DNA repair. There is also evidence that Ku recruits the XRCC4-ligase IV complex to DNA ends to complete the process of end joining [45].

Another complex, Mre11/Rad50/Xrs2, was also shown to participate in NHEJ of breaks in *Saccharomyces cerevisiae* [46]. Homologs of Mre11 and Rad50 have been identified in mammalian cells and Nbs1 (nibrin), the gene defective in Nijmegen breakage syndrome (NBS), appears to be the functional counterpart of Xrs2 [47]. In response to radiation, the Mre11/Rad50/Nbs1 complex is localized rapidly to sites of DNA DSBs and associates in discrete foci [48,49]. The Mre11 complex is also involved in homologous recombination, meiotic recombination, and telomere maintenance [50]. Cross-talk between the Mre11 complex and ATM, in recognizing DSB in DNA, is supported by overlap in phenotype between the three syndromes that arise when these genes are mutated: A-T (ATM), NBS (Nbs1), and A-T like syndrome (Mre11) [51]. Further support for this is derived from the observation that ATM phosphorylates Nbs1 to mediate cell cycle control and minimize sensitivity to radiation [52]. Thus it is evident that several mechanisms have evolved to recognize and repair DSBs in DNA. The relative importance of NHEJ and HR in this process is governed by cell cycle position but it is not clear how the various recognition systems coordinate their activities in the recognition and repair processes.

The Mre11 complex together with the Rad52 epistasis group of genes also participates in HR in organisms ranging from *S. cerevisiae* to mammals [53,54]. In response to DNA damage, Rad52 binds to free DNA ends, which may facilitate processing by an exonuclease activity of the Mre11 complex, which localizes rapidly to the break [55]. The resulting 3' overhangs represent a recognition structure for RPA binding and Rad51 multimerization that, in the presence of the Rad51 paralogues, BRCA1 and BRCA2, invades the homologous duplex DNA to achieve exchange between damaged and intact DNAs. DNA polymerase fills in the resulting gap followed by ligation and strand resolution. Colocalization of Rad51 and the BRCA proteins to nuclear foci at the sites of DNA damage suggests that those structures participate in the HR process [56]. This association, induced by ionizing radiation, is accompanied by increased phosphorylation of BRCA1 and, as discussed later, at least some of this is carried out by ATM kinase [52,57]. ATM may also participate at the level of Rad51 because radiation exposure causes tyrosine phosphorylation of Rad51, which is both ATM and c-Abl dependent and this facilitates complex formation between Rad51 and Rad52 [58]. In chicken DT40 cells, characterized by high homologous recombination frequencies, disruption of the ATM gene altered the kinetics of Rad51 and Rad54 focus formation [59]. $ATM^{-/-}$ $Ku^{-/-}$ (NHEJ-deficient) DT40 cells were more sensitive to radiation and displayed high levels of radiation-induced chromosome aberrations whereas the $ATM^{-/-}$ $Rad54^{-/-}$ (HR-deficient) cells showed only slightly increased aberrations suggesting that ATM is on the HR pathway.

ATM Signaling: Recognition of Breaks in DNA

ATM is primarily activated as a preexisting protein by ionizing radiation and radiomimetic agents [60,61]. The exact mechanism of activation is as yet unclear but it is clear that this molecule undergoes autophosphorylation, which is expected to be responsible for or contribute to activation [52]. Agents that break DNA activate ATM so that the initiating signal may be a relaxation of the superhelical density of chromatin, which will be rapidly transmitted to ATM in complexes associated with chromatin. The initiating event may be a conformational change or dissociation from other members of a protein complex such as BRCA1-associated genome surveillance complex (BASC) [62], which would allow access to ATP and protein substrates. In this context recent evidence points to a greater retention of a portion of ATM in nuclear aggregates, postirradiation [63].

This retained ATM colocalizes with the phosphorylated form of histone H2AX (γ-H2AX) and in foci with Nbs1, suggesting that it is being targeted to sites of DSBs. Phosphorylation of H2AX occurs with rapid kinetics after irradiation at sites of DSBs and there is evidence that this is dependent on both ATM and ATR [64,65]. A host of proteins are either phosphorylated directly by ATM or their phosphorylation is ATM-dependent during radiation signal transduction (Table I). To identify potential substrates Kim *et al.* [66] employed a rapid peptide screening assay and showed that the sequence S/TQ is a minimal essential requirement and that the presence of hydrophobic or negatively charged amino acids N terminal of S/T were positive determinants for activity. Candidate substrates identified included ATM itself, Nbs1, BRCA1, p53, Mre11, Rad17, and Werner's syndrome protein (WRN).

Oriented peptide library analyses were also used to define preferred amino acid motifs for ATM [67]. As will be discussed later, it was not surprising that key substrates predicted for ATM would be involved in DNA damage response and repair as well as cell cycle checkpoint activation. Because ATM phosphorylates proteins involved in both DNA damage recognition and cell cycle control, an important question is whether these substrates are simultaneously phosphorylated or are distinguished by their distribution in different subcompartments in the nucleus. Andegeko *et al.* [63] have

Table I Substrates for ATM Kinase and ATM-dependent Modification of Proteins

Protein	Site (amino acid)	Direct	Indirect	Function
p53	Ser15	+		G1/S checkpoint
	Ser20		+	G1/S and G2/M checkpoints
	Ser376 (dephosphorylation)		+	G1/S checkpoint
Mdm2	Ser395	+		G1/S checkpoint
Chk2	Thr68	+		G1/S, S and G2/M checkpoints
53BP1	Several (unidentified)	+		cell cycle (developmentally regulated)
BRCA1	ser 1387	+		G2/M checkpoint/radioresistance
	ser 1423	+		
	ser 1457	+		
	ser 1524	+		
CtIP	ser 664	+		DNA damage response/Induction of GADD45
Nbs1	ser 278	+		S phase checkpoint/
(nibrin)	ser 343	+		Mre11/Rad50/Nbsi foci/radioresistance
Rad51	ser 645		+	Assembly of rad51 complex
c-Jun	ser 63		+	Oxidative stress signaling
	ser 73		+	
Pin2/TRF1	ser 219	+		Telomere maintenance/G2/M checkpoint/radioresistance
BLM	Thr 99	+		chromosome integrity
	Thr 122	+		
I KappaB	?		+	gene expression via NF-kappa B
c-Abl	ser 465	+		cellular response to radiation
Histone H1	desphosphorylation			chromatin remodelling
SMC1	ser 957	+		sister chromatid cohesion
	ser 966	+		

addressed this issue to some extent by revealing that a slow migrating (phosphorylated) form of Nbs1 was found with the retained form of ATM, but in contrast neither the basal nor phosphorylated forms of the cell cycle checkpoint kinase, Chk2, were detected in the retained fraction. However, it is evident that only a small proportion of activated ATM is confined to the retained nuclear fraction [63].

While double-strand breaks arise in DNA in response to radiation and other agents, breaks occur normally in some cells in specialized processes such as V(D)J recombination in lymphoid cells and during meiosis in germ cells [68,69]. Evidence for an involvement of ATM in recognizing these breaks arises from hypogonadism in A-T patients [70] and severe meiotic disruption in prophase 1 in $Atm^{-/-}$ mice [71,72]. A high incidence of chromosomal rearrangements involving both the T-cell receptor and Ig heavy chain genes is characteristic of A-T [73]. This does not appear to be due to defective V(D)J recombination in A-T cells [74,75]. However, both A-T patients and *Atm* gene-disrupted mice develop lymphoid tumours characterized by translocations involving antigen receptor genes [76,77]. Blocking of V(D)J recombination by crossing $Atm^{-/-}$ mice with recombinant activating gene (RAG) deficient mice delays the onset of tumors in these mice [78,79]. Atm is localized to breaks associated with V(D)J, being recruited at approximately the same time as RAG [80].This report also provides evidence that Ser18 phosphorylated p53, a product of ATM kinase activity, is also localized to the breaks with Atm.

These data together with results demonstrating direct interaction between ATM and p53 [81,82], suggest that ATM is being activated by V(D)J recombination breaks and that it is phosphorylating a known substrate in the vicinity of the break. Although ATM may not be essential for V(D)J recombination, it appears to play a surveillance role for intermediates, suppressing aberrant rearrangements [80]. These data also are supportive of phosphorylation of substrates in the immediate vicinity of the DNA damage. Double-strand breaks in DNA also up-regulate the nonreceptor tyrosine kinase, c-Abl, in an ATM-dependent manner [83,84]. ATM interacts with c-Abl through its SH3 group, and in A-T cells radiation induction of c-Abl kinase is defective. c-Abl also interacts with and phosphorylates DNA-PKcs, causing it to dissociate from the Ku/DNA complex [85], and inhibition of c-Abl kinase prevented down-regulation of DNA-PK [86].

Thus, there is evidence that ATM regulates the activity of DNA-PK postirradiation providing support for a convergence of these pathways of DNA damage repair.

Checkpoint Activation

G_1/S-Phase Checkpoint

Perhaps the best described signaling pathway initiated by ATM postirradiation is that through p53 to arrest the passage of cells from G_1 to S phase [18,19,87]. In this pathway ATM phosphorylates p53 on Ser15 and is required for Ser20 phosphorylation and Ser376 dephosphorylation [88]. p53 Becomes transcriptionally activated to induce the cyclin kinase inhibitor p21/WAF1, which associates with cyclin E-Cdk2 to block its activity, to prevent phosphorylation of downstream substrates, and to delay the passage of cells from G_1 to S phase [89–91; Fig. 2]. ATM phosphorylates Chk2 kinase on Thr68, which depends on the integrity of the forkhead associated (FHA) domain in Chk2 [92]. This phosphorylation is required for the subsequent autophosphorylation and activation of Chk2 [93], which enables it to exert its control on the G_1/S, S, and G_2/M checkpoints. This activation of Chk2 by ATM also requires Nbs1 since activation is defective in NBS cells [94]. Whereas wild-type Nbs1 could complement this defect in NBS cells, a construct mutated in the ATM phosphorylation site on Nbs1 (ser 343) failed to do so as well as a mutant form that abrogated the formation of the Mre11/Rad50/Nbs1 complex. Phosphorylation of p53 on Ser20 by Chk2, in response to radiation, stabilizes the protein and increases its availability to induce downstream effector genes such as p21/WAF1 [95,96]. Dephosphorylation of p53 at Ser376 occurs in response to radiation treatment and this is also ATM dependent [97]. The significance of this change to p53 stabilization/activation is not known.

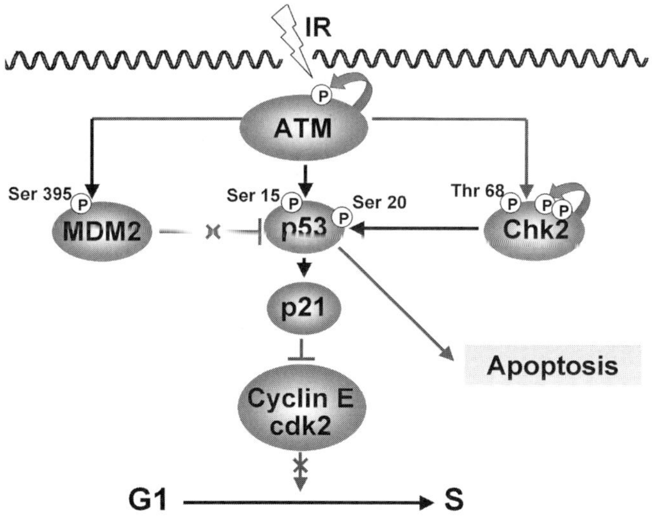

Figure 2 ATM activates the G_1/S checkpoint in response to ionizing radiation damage to DNA. ATM is rapidly activated by an unknown mechanism, in response to DNA breaks, to phosphorylate itself and three downstream effector proteins (p53, MDM2, and Chk2) to achieve arrest of cells at the G_1/S checkpoint. It directly phosphorylates p53 on ser15, which may alter its transcriptional capacity. Stabilization of p53 is achieved by phosphorylating Chk2 on thr 68, which leads to its subsequent activation to, in turn, phosphorylate p53 on ser20 leading to its stabilization. Negative regulation of MDM2 by ATM through phosphorylation on ser395 reduces its ability to bind to and mediate the ubiquitination and degradation of p53. Thus stabilization of p53 is ensured through two separate steps. When stabilized p53 is capable of inducing p21 and other downstream genes to inhibit cyclin E-Cdk2 kinase activity and bring about G_1 arrest.

The involvement of ATM signaling at multiple levels in a single pathway is further illustrated by the capacity of activated ATM to phosphorylate Mdm2, a negative regulator of p53, on Ser395 [98]. Expression of Mdm2 is controlled by p53, but Mdm2 in turn has a feedback effect, binding to p53, exporting it from the nucleus and promoting its degradation in the proteasome pathway [99]. ATM-dependent phosphorylation of Mdm2 is observed prior to p53 accumulation. Decreased reactivity of Mdm2 from irradiated cells to an anti-Mdm2 antibody, directed against an epitope containing Ser395, suggests that this is a site of phosphorylation *in vivo*, and phosphorylation at this site may reduce the capacity of Mdm2 to translocate p53 to the cytoplasm and ensure its stabilization [100]. Thus, it is evident that ATM acts directly on three different substrates and also has indirect effects on some of these, providing complex control of the G_1/S checkpoint.

S-Phase Checkpoint

Exposure of mammalian cells to radiation leads to a biphasic pattern of inhibition of DNA synthesis reflecting the greatest sensitivity in replicon initiation [101]. A much reduced inhibition of DNA synthesis in A-T cells in response to radiation, which was termed *radioresistant DNA synthesis* (RDS), represented the first report of a cell cycle anomaly [13,14]. This phenomenon is still not fully understood but may be explained by defects in more than one signaling pathway controlled by ATM. Beamish *et al.* [16] have shown that exposure of lymphoblastoid cells to radiation in S phase causes a rapid inhibition of cyclin A-Cdk2 accompanied by markedly increased binding of p21/WAF1 (Fig. 3). In contrast, radiation did not inhibit cyclin kinase activity in A-T cells in S phase nor was there any significant change in cdk-associated p21/WAF1 compared to unirradiated cells [16]. These observations are compatible with a reduced effect of radiation on DNA replication in A-T cells. A functional link has also been established between ATM, the checkpoint signaling kinase, Chk2, and Cdc25A phosphatase and its downstream target for activation, Cdk2, in S phase [15]. Radiation-induced degradation of Cdc25A requires both ATM and Chk2-mediated phosphorylation of Cdc25A on Ser123 and this prevents dephosphorylation of Cdk2, leading to a transient block in DNA replication (Fig. 3). Exposure of A-T cells to radiation failed to cause an increase in Cdk2 Tyr 15 dephosphorylation or inhibition of cyclin E-Cdk2 kinase activity, consistent with the radioresistant DNA synthesis phenotype. Furthermore, Chk2 alleles defective in catalytic activity or ability to interact with Cdc25A had a dominant interfering effect and abrogated the S-phase checkpoint [15].

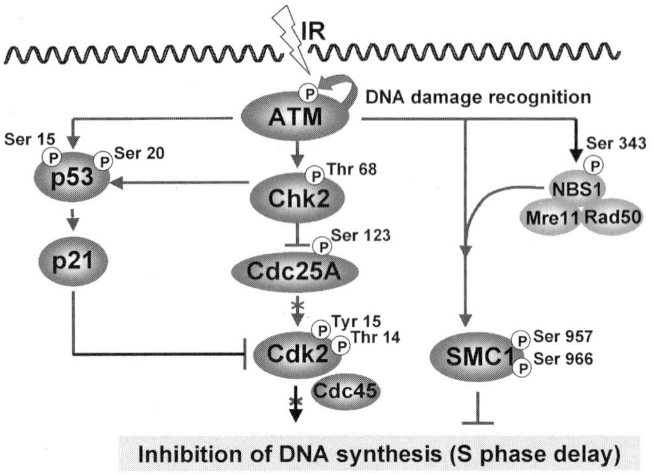

Figure 3 ATM activates the S-phase checkpoint. Radioresistant DNA synthesis was first described in cells from A-T patients lacking ATM. This phenomenon is also observed in NBS where the ATM substrate Nbs1 is defective and in the A-T like syndrome (ATLD) characterized by hypomorphic mutants in the Mre11 gene, a component of the Mre11/Rad50/Nbs1 complex. Radioresistant DNA synthesis is also observed when Chk2 is disrupted on its downstream pathway through Cdc25A/Cdk2. These observations point to an important role for ATM in inhibiting DNA synthesis through several separate but parallel pathways. In the first case ATM activates Chk2 as described for the G_1/S checkpoint, which in this case destabilizes Cdc25A by phosphorylation on Ser123, preventing its ability to remove the inhibitory Tyr 15/Thre 14 phosphorylations for activation of Cdk2, leading to inhibition of DNA synthesis. In a parallel but less well-described pathway, ATM activates the Mre11 complex and SMC1, a downstream target to inhibit DNA synthesis. Finally, the p53 activation pathway might also contribute through inhibition of Cdk2 kinase.

Similarities in phenotype between A-T and NBS including RDS provided another approach to understanding the basis of this phenomenon. After ionizing radiation exposure, Nbs1 is phosphorylated on Ser343 both *in vitro* and *in vivo* in an ATM-dependent manner [102–104].

In normal cells transiently transfected with Nbs1, radiation caused inhibition of DNA synthesis but when the Ser343 site was mutated prior to transfection, a significant reduction was seen in the extent of inhibition of DNA synthesis [104], indicating abrogation of the S-phase checkpoint by dominant interference. Furthermore, RDS was rescued by retroviral constructs expressing wild-type Nbs1 but not Ser343 mutants. Thus ATM phosphorylation of Nbs1 is an important part of the mechanism to ensure inhibition of DNA replication in response to radiation damage. Recent data suggest that the S-phase checkpoint is regulated by parallel pathways through Nbs1-Mre11 on the one hand and Chk2 on the other [105]. Concomitant interference with both of these pathways gave rise to RDS after exposure of cells to ionizing radiation. Thus it appears likely that ATM, by phosphorylating both Nbs1 and Chk2, triggers two pathways that inhibit distinct steps in DNA replication.

Evidence for parallel pathways is further supported by the demonstration that the structural maintenance of chromosomes protein, SMC1, is a downstream effector in the ATM/Nbs1 branch of the S-phase checkpoint [106,107]. It has been suggested that Nbs1 ceases to be a sensor of DSBs after phosphorylation by ATM and becomes an adaptor to facilitate the phosphorylation of SMC1 by ATM. Once phosphorylated SMC1 may participate in the S-phase checkpoint in a postreplicative role. A delay in radiation-induced phosphorylation of replication protein A (RPA), a component of the single-strand DNA binding protein complex, has also been implicated [108]. Furthermore UV-induced hyperphosphorylation of RPA (p34 subunit) is ATM dependent, and ATM kinase phosphorylates sites on p34 that are phosphorylated *in vivo* in response to UV irradiation [109]. The transitory delay in DNA synthesis postirradiation also appears to be mediated through a calmodulin-dependent regulatory cascade [110] and this pathway is defective in A-T.

G_2/M Checkpoint

Arrest of cells in G_2 phase in response to radiation damage leads to suppression of the mitotic index to protect cell viability [111]. Consequently, abrogation of G_2 arrest sensitizes cells to radiation [112]. This checkpoint is defective in A-T cells as evidenced by a lesser delay of cells, irradiated in G_2 phase, in progressing into mitosis [113]. Defective G_2-phase delay in A-T cells was confirmed by labeling cells in S phase with BrdU, blocking their passage into mitosis with nocodazole prior to irradiation (in G_2 phase), and scoring for ability to enter the next G_1 phase [11]. This was also achieved by distinguishing G_2 phase cells from mitotic cells using histone H3 phosphospecific antibody [114]. As observed with cells irradiated in S phase, when A-T cells were exposed to radiation in G_2 phase, there was no inhibition of Cdc2-cyclin B kinase activity and, unlike that obtained with irradiated control cells, no increase in p21/WAF1 associated with Cdc2 was observed [16]. Again this checkpoint defect is compatible with a lack of ATM signaling in A-T cells to the cyclin-kinase complex. As with S and G_1 phase checkpoints, it appears that ATM influences G_2 arrest by more than one pathway. It is well established that BRCA1 is involved in the cellular response to radiation and this involves phosphorylation on several sites mediated by both ATM and ATR [52,57,115]. ATM interacts with BRCA1 and this molecule is a substrate for ATM kinase both *in vitro* and *in vivo* [52]. Deletion of exon 11 of BRCA1 leads to a defective G_2 checkpoint and extensive chromosomal abnormalities [116].

A G_2/M checkpoint abnormality, similar to that seen in A-T cells, has been reported for the BRCA1-null cell line HCC1937 and a normal checkpoint was restored to these cells with BRCA1 transfection [17]. Transfection of HCC1937 cells with a mutant form of a major ATM phosphorylation site in BRCA1 (S1423A) failed to complement the defective G_2/M checkpoint. Thus, it appears likely that radiation-induced phosphorylation of BRCA1 on ser1423 by ATM is important in the regulation of the G_2/M checkpoint (Fig. 4). Because cells from NBS patients exhibited a normal radiation-induced G_2/M checkpoint, it appears that ATM phosphorylation of Nbs1 does not play a significant role for this checkpoint. BRCA1-mediated induction of GADD45 also leads to

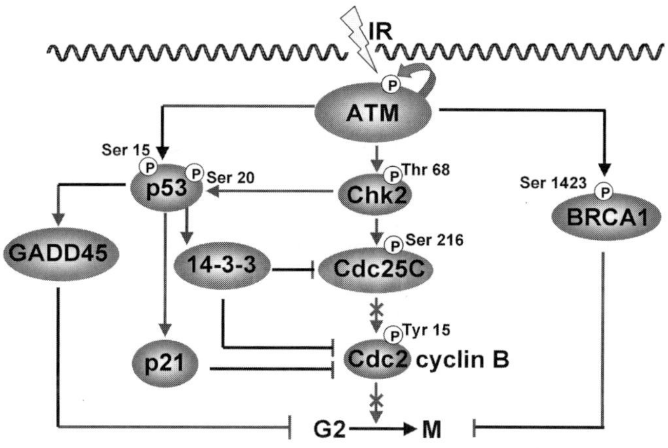

Figure 4 ATM involvement in G_2/M checkpoint control. The activation of the G_2/M checkpoint is no less complex than the S-phase checkpoint. Again ATM is a central player, via chk2 phosphorylation of Cdc25C on ser216 in S phase, and in response to DNA damage this site is maintained in a phosphorylated state. Inactivation appears to be mediated by binding to 14-3-3 protein, a p53 downstream effector. Again in this case activation of the p53 pathway is prominent to induce 14-3-3 protein and p21/WAF1, which can bind to and inhibit Cdc2-cyclin B kinase blocking cells at the G_2/M checkpoint. Induction of GADD45 in this pathway may also interfere with Cdc2-cyclin B kinase. To add to the complexity mutations in BRCA1, including an ATM phosphorylation site (ser 1423), leads to a defective G_2/M checkpoint indicating that ATM also controls the G_2/M checkpoint through BRCA1.

G_2/M phase delay but only in response to microtubule poisons, not DNA damaging agents, revealing that it is ATM independent [117].

Under normal proliferative conditions, the dual specificity Cdc25 phosphatase removes the phosphate group from tyrosine 15 on Cdc2 to facilitate mitotic entry [118]. As part of the G_2/M checkpoint activation, Cdc25C is phosphorylated on Ser216 and inactivated by binding to 14-3-3 protein and subsequently sequestered to the cytoplasm [119,120]. Both Chk1 and Chk2 phosphorylate Cdc25C on ser216 *in vitro* [121–124]. Since Cdc25C is already phosphorylated on ser216 in S phase it appears that DNA damage prolongs this state of phosphorylation [125]. As in the case of the G_1/S checkpoint it appears likely that ATM activates Chk2 to phosphorylate Cdc25C as part of the G_2/M checkpoint activation.

Role of ATM in More General Signaling

The presence of ATM in the nucleus is consistent with its role in sensing damage in DNA and signaling to cell cycle checkpoints. However, ATM is also extranuclear in proliferating cells with evidence for localization to endosomes [126] and peroxisomes [81]. In postmitotic human and mouse Purkinje cells and other neurons, ATM is predominantly a cytoplasmic protein [127,128]. A likely explanation for ATM activation by double-strand breaks in DNA is the binding either directly or indirectly to the lesion, which is supported by data revealing binding of ATM to DNA ends [129,130] and the recruitment of this protein to DNA double-strand breaks after ionizing radiation damage [131]. However, it is difficult to envisage the commonality that might occur when the same molecule is activated in the cytoplasm or in a subcellular organelle outside the nucleus. Support for a more general signaling role for ATM comes from a number of observations [132]. One of the earliest observations supporting a non-DNA damage role for ATM was the failure to record a cytoplasm to nucleus signaling after internalization of PHA in A-T lymphocytes [133]. Kondo *et al.* [134] demonstrated that Ca^{2+} mobilization, in response to both PHA and CD3 cross-linking, was defective in lymphocytes from A-T patients, and Khanna *et al.* [28] have provided evidence for defective signal transduction and Ca^{2+} mobilization after B-cell receptor ligation in A-T lymphoblastoid cells. Rhodes *et al.* [29] have shown that the ability of A-T fibroblasts to depolarize in response to increasing concentration of extracellular K^+ is significantly reduced compared to controls, and that the outward rectifier K^+ currents are largely absent in these cells.

Electrophysiological anomalies were also reported by Chiesa *et al.* [135] who showed that there was a significant decrease in the duration of calcium and sodium firing in Purkinje cells from $Atm^{-/-}$ mice. Other evidence for defective signaling in A-T includes greater demand for growth factors in A-T fibroblasts [136,137], poor growth capacity of fibroblasts from $Atm^{-/-}$ mice and from A-T patients [87], only partial protection against cell death in peripheral blood mononuclear cells from A-T patients by serum and added cytokines [138], and defective signaling through the EGF-receptor in A-T cells [25,26]. A more direct role for non-DNA damage activation of ATM kinase has been described in response to insulin [30]. Insulin activated ATM kinase to the same extent, but it did so somewhat slower than ionizing radiation, which led to the phosphorylation of 4E-BP1 (PHAS-1) on Ser111, while bound to eIF-4E. This appears to be the priming event for other phosphorylations on 4E-BPI, which lead to dissociation of the complex and availability of eIF-4E for the initiation of translation of mRNA. Furthermore, there is evidence for a reduction in the extent of dissociation of the complex in A-T cells in response to insulin. Thus while agents other than DNA damage can activate ATM kinase, it remains unclear whether the nuclear or extranuclear form is being activated. In addition unless ATM is present in soluble form in the cytoplasm, the additional problem of bringing substrates such as 4E-BPI-eIF-4E together with ATM in subcellular organelles (endosomes and peroxisomes) exists.

Putting aside the example of insulin-induced signaling through ATM, it might well be argued that all or most of the signaling defects present in A-T cells are indirect as a consequence of the loss of ATM. A possible unifying explanation for the myriad of defects at the level of the membrane in A-T cells might be oxidative stress for which there is accumulating evidence. It has been suggested that perturbation of the cellular balance of reactive oxygen species leads to oxidative stress in A-T, which may be responsible for several features of the phenotype including neurodegeneration [139]. Evidence for a state of oxidative stress in A-T is based on abnormal response of these cells to agents that cause stress [140]; reduced antioxidant capacity [141]; increased oxidative damage in the cerebellum of $Atm^{-/-}$ mice [8,81]; and the capacity of antioxidants to reduce constitutively high levels of p53, p21/WAF1, and other proteins in A-T cells [9]. Increased oxidative stress in the absence of ATM could be due to a failure to sense and respond to reactive oxygen species (ROS) levels, deregulation of enzymes involved in preventing oxidative stress, or the presence of unrepaired breaks in DNA that activate components of the DNA damage response to deplete NAD [139]. Thus at least some of the series of non-DNA damage signaling defects in A-T cells might be due to indirect consequences of loss of ATM.

Perspective

The cellular response to radiation exposure is complex, involving nuclear-, cytoplasmic-, and membrane-mediated events. As outlined earlier, ATM plays a central role in the recognition and response to the major radiation-induced lesion in DNA, the double-strand break. In the nucleus a series of rapid protein phosphorylations occurs that is controlled by ATM, but it remains unclear how ATM is activated as an existing protein to achieve this. It is likely that alterations in the superhelicity of chromatin loops may rapidly signal the presence of breaks to ATM, dislodging it from an inhibition or inhibitory complex associated with chromatin to become autophosphorylated and, in turn, phosphorylate key substrates such as p53, BRCA1, and Chk2. Evidence exists for ATM autophosphorylation, postirradiation, but it has not been established how many such sites there are in ATM and how this is related to its capacity to phosphorylate specific substrates. Compartmentalization of different ATM substrates in different complexes such as BASC may determine accessibility and rapidity of phosphorylation. This may help to explain how it is that ATM can activate several pathways for DNA damage recognition and cell cycle control almost simultaneously.

To add to this complexity it is evident that ATM has a fine-tuning role in these pathways by phosphorylating more than one pathway intermediate directly or mediating phosphorylation through other protein kinases. A good example of this is the G_1/S checkpoint where the downstream targets include p53, mdm2, and Chk2. It is not difficult to envisage a central role for ATM in radiation signaling in the nucleus where the DNA damage is located and where the enzyme is present in abundance in proliferating cells. However, being able to establish a similar mechanism of activation in the absence of chromatin for extranuclear (nonmitochondrial) ATM is somewhat more difficult. It has been reported that the majority of ATM is extranuclear in postmitotic cells, and even in proliferating cells up to 30% of ATM is outside the nucleus. Insulin has been shown to activate ATM kinase, but it is not clear whether this involves nuclear or extranuclear ATM and what the mechanism is. It is possible that constitutive activation of ATM in the nucleus is prevented by association with other proteins in a complex and it becomes activated as referred to earlier by a reorganization of chromatin. Similarly, ATM may be kept inactive or residually active by interaction with a different protein outside the nucleus that dissociates in response to a different stimulus. We should point out that no one has demonstrated active ATM kinase in extranuclear ATM to date.

What happens when ATM protein is mutated or absent by the description of the A-T syndrome was established as long ago as 1926. In the absence of ATM, patients undergo a progressive cerebellar degeneration, they are prone to infection, and up to 30% die from mainly lymphoid tumors. Failure of appropriate DNA damage recognition leads to genomic instability, which is related to the propensity to develop tumors. What is of particular interest is the penetration of subtle but significant aspects of the phenotype in carriers of the defective gene. For some time it has been demonstrated that A-T heterozygotes are intermediate in their sensitivity to ionizing radiation; they have an increased risk for the development of cancer, especially breast cancer, and their relative risk for dying from other conditions is increased. The challenge ahead requires a greater understanding of how ATM is activated, what pathways it controls, and how it minimizes genomic instability and protects against cancer.

References

1. Sedgwick, R. P. and Boder, E. (1991). Hereditary neuropathies and spinocerebellar atrophies, in Vianney De Jong, J. M. B., Ed., pp. 347–423. Alan R Liss, New York.
2. Lavin, M. F. and Shiloh, Y. (1997). The genetic defect in ataxia-telangiectasia, *Ann. Rev. Immunol.* **15**, 177–202.
3. Gotoff, S. P., Amirmokri, E., and Liebner, E. J. (1967). Ataxia telangiectasia. Neoplasia, untoward response to x-irradiation, and tuberous sclerosis. *Am. J. Dis. Child.* **114**, 617–625.
4. Morgan, J. L., Holcomb, T. M., and Morrissey, R. W. (1968). Radiation reaction in ataxia telangiectasia. *Am. J. Dis. Child* **116**, 557–558.
5. Taylor, A. M., Harnden, D. G., Arlett, C. F., Harcourt, S. A., Lehmann, A. R., Stevens, S., and Bridges, B. A. (1975). Ataxia telangiectasia: A human mutation with abnormal radiation sensitivity. *Nature* **258**, 427–429.
6. Chen, P. C., Lavin, M. F., Kidson, C., and Moss, D. (1978). Identification of ataxia telangiectasia heterozygotes, a cancer prone population. *Nature* **274**, 484–486.
7. Foray, N., Priestley, A., Alsbeih, G., Badie, C., Capulas, E. P., Arlett, C. F., and Malaise, E. P. (1997). Hypersensitivity of ataxia telangiectasia fibroblasts to ionizing radiation is associated with a repair deficiency of DNA double-strand breaks. *Int. J. Radiat. Biol.* **72**, 271–283.
8. Barlow, C., Hirotsune, S., Paylor, R., Liyanage, M., Eckhaus, M., Collins, F., Shiloh, Y., Crawley, J. N., Ried, T., Tagle, D., and Wynshaw-Boris, A. (1996). Atm-deficient mice: A paradigm of ataxia telangiectasia. *Cell* **86**, 159–171.
9. Gatei, M., Shkedy, D., Khanna, K. K., Uziel, T., Shiloh, Y., Pandita, T. K., Lavin, M. F., and Rotman, G. (2001). Ataxia-telangiectasia: Chronic activation of damage-responsive functions is reduced by alpha-lipoic acid. *Oncogene* **20**, 289–294.
10. Rotman, G. and Shiloh, Y. (2001). ATM: A mediator of multiple responses to genotoxic stress. *Oncogene* **18**, 6135–6144.
11. Beamish, H., and Lavin, M. F. (1994). Radiosensitivity in ataxia-telangiectasia: Anomalies in radiation-induced cell cycle delay. *Int. J. Radiat. Biol.* **65**,175–184.
12. Morgan, S. E. and Kastan, M. B. (1997). p53 and ATM: Cell cycle, cell death, and cancer. *Adv. Cancer Res.* **71**, 1–25.

13. Houldsworth, J. and Lavin, M. E. (1980). Effect of ionizing radiation on DNA synthesis in ataxia telangiectasia cells. *Nucleic Acids Res.* **8**, 3709–3720.
14. Painter, R. B. and Young, B. R. (1980). Radiosensitivity in ataxia-telangiectasia: A new explanation. *Proc. Natl. Acad. Sci. USA* **77**, 7315–7317.
15. Falck, J., Mailand, N., Syljuasen, R. G., Bartek, J., and Lukas, J. (2001). The ATM-Chk2-Cdc25A checkpoint pathway guards against radioresistant DNA synthesis. *Nature* **410**, 842–847.
16. Beamish, H., Williams, R., Chen, P., and Lavin, M. F. (1996). Defect in multiple cell cycle checkpoints in ataxia-telangiectasia postirradiation. *J. Biol. Chem.* **271**, 20486–20493.
17. Xu, B., Kim, S. T., and Kastan, M. B. (2001). Involvement of Brca1 in S-phase and G2-phase checkpoints after ionizing radiation. *Mol. Cell Biol.* **21**, 3445–3450.
18. Kastan, M. B., Zhan, O., el-Deiry, W. S., Carrier, F., Jacks, T., Walsh, W. V., Plunkett, B. S., Vogelstein, B., and Fornace, A. J. Jr. (1992). A mammalian cell cycle checkpoint pathway utilizing p53 and GADD45 is defective in ataxia-telangiectasia. *Cell* **71**, 587–597.
19. Khanna, K. K. and Lavin, M. F. (1993). Ionizing radiation and UV induction of p53 protein by different pathways in ataxia-telangiectasia cells. *Oncogene* **8**, 3307–3312.
20. Shiloh, Y. (2001). ATM (ataxia telangiectasia mutated): Expanding roles in the DNA damage response and cellular homeostasis. *Biochem Soc. Trans.* **29**, 661–666.
21. Zakian, V. A. (1995). ATM-related genes: What do they tell us about functions of the human gene? *Cell* **82**, 685–687.
22. Abrahams, R. T. (2001). Cell cycle checkpoint signaling through the ATM and ATR kinases. *Genes Dev.* **15**, 2177–2196.
23. Shiloh, Y. and Kastan, M. B. (2001). ATM: Genome stability, neuronal development, and cancer cross paths. *Adv. Cancer. Res.* **83**, 209–254.
24. Fukao, T., Kaneko, H., Birrell, G., Gatei, M., Tashata, H., Kasahara, K., Cross, S., Kedar, P., Watters, D., Khanna, K. K., Misko, I., Kondo, N., and Lavin, M. F. (1999). ATM is upregulated during the mitogenic response in peripheral blood mononuclear cells. *Blood* **94**,1998–2006.
25. Gueven, N., Keating, K. E., Chen, P., Fukao, T., Khanna, K. K., Watters, D., Rodemann, P. H., Lavin, M. F. (2001). Epidermal growth factor sensitizes cells to ionizing radiation by down-regulating protein mutated in ataxia-telangiectasia. *J. Biol. Chem.* **276**, 8884–8891.
26. Keating, K. E., Gueven, N., Watters, D., Rodemann, H. P., and Lavin, M. F. (2001). Transcriptional downregulation of ATM by EGF is defective in ataxia-telangiectasia cells expressing mutant protein. *Oncogene* **20**, 4281–4290.
27. Shiloh, Y., Tabor, F., and Becker, Y. (1982). The response of ataxia-telangiectasia homozygous and heterozygous skin fibroblasts to neocarzinostatin. *Carcinogenesis.* **3**, 815–820.
28. Khanna, K. K., Yan, J., Watters, D., Hobson, K., Beamish, H., Spring, K., Shiloh, Y., Gatti, R. A., and Lavin, M. F. (1997). Defective signaling through the B cell antigen receptor in Epstein-Barr virus-transformed ataxia-telangiectasia cells. *J. Biol. Chem.* **272**, 9489–9495.
29. Rhodes, N., D'Souza, T., Foster, C. D., Ziv, Y., Kirsch, D. G., Shiloh, Y., Kastan, M. B., Reinhart, P. H., and Gilmer, T. M. (1998). Defective potassium currents in ataxia telangiectasia fibroblasts. *Genes Dev.* **12**, 3686–3692.
30. Yang, D. O. and Kastan, M. B. (2000). Participation of ATM in insulin signaling through phosphorylation of EIF-4E binding protein. *Nat. Cell Biol.* **2**, 893–898.
31. Ward, J. F. (1975). Molecular mechanisms of radiation-induced damage to nucleic acids. *Radiat. Res.* **86**, 185–195.
32. Collins, A. (1987). Cellular responses to ionizing radiation: Effects of interrupting DNA repair with chemical agents. *Int. J. Radiat. Biol. Relat. Stud. Phys. Chem. Med.* **51**, 971–983.
33. Jeggo, P. A., Taccioli, G. E., and Jackson, S. P. (1995). Menage a trois: Double-strand break repair, V(D)J recombination and DNA-PK. *Bioessays* **17**, 949–957.
34. Haber, J. E. (2000). Partners and pathways: Repairing a double-strand break. *Trends Genet.* **16**, 259–264
35. Jeggo, P., Singleton, B., Beamish, H., and Priestley, A. (1999). Double-strand break rejoining by the Ku-dependent mechanism of non-homologous end-joining. *CR Acad. Sci. III* **322**, 109–112.
36. Johnson, R. D. and Jasin, M. (2000). Sister chromatid gene conversion is a prominent double-strand break repair pathway in mammalian cells. *EMBO J.* **19**, 3398–3407.
37. Sonoda, E., Morrison, C., Yamashita, Y. M., Takata, M., and Takeda, S. (2001). Reverse genetic studies of homologous DNA recombination using the chicken B-lymphocyte line, DT40. *Philos. Trans. R. Soc. Lond. B Biol. Sci.* **356**, 111–117.
38. Althaus, F. R. and Richter, C. (1987). ADP-ribosylation of proteins. Enzymology and biological significance. *Mol. Biol. Biochem. Biophys.* **37**, 1–237.
39. O'Connell, M. J., Walworth, N. C., and Carr, A. M. (2000). The G2-phase DNA damage checkpoint. *Trends Cell Biol.* **10**, 296–303.
40. Zou, L., Cortez, D., and Elledge, S. J. (2002). Regulation of ATR substrate selection by Rad17-dependent loading of Rad9 complexes onto chromatin. *Genes Dev.* **16**, 198–208.
41. Zdzienicka, M. Z. (1995). Mammalian mutants defective in the response to ionizing radiation-induced DNA damage. *Mutat Res.* **336**, 203–213.
42. Errani, A., Smider, V., Rathmell, W. K., He, D. M., Hendrickson, E. A., Zdzienicka, M. Z., and Chu, G. (1996). Ku86 defines the genetic defect and restores X-ray resistance and V(D)J recombination to complementation group 5 hamster cell mutants. *Mol. Cell Biol.* **16**, 1519–1526.
43. Jackson, S. P. (1997). DNA-dependent protein kinase. *Int. J. Biochem. Cell Biol.* **29**, 935–938.
44. Grawunder, U., Zimmer, D., Fugmann, S., Schwarz, K., and Lieber, M. R. (1998). DNA ligase IV is essential for V(D)J recombination and DNA double-strand break repair in human precursor lymphocytes. *Mol. Cell.* **2**, 477–484.
45. McElhinny, N., Snowden, C. M., McCarville, J., and Ramsden, D. A. (2000). Ku recruits the XRCC4-ligase IV complex to DNA ends. *Mol. Cell Biol.* **20**, 2996–3003.
46. Moore, J. K. and Haber, J. E. (1996). Capture of retrotransposon DNA at the sites of chromosomal double-strand breaks. *Nature* **383**, 644–646.
47. Petrini, J. H. (2000). The Mre11 complex and ATM: Collaborating to navigate S phase. *Curr. Opin. Cell Biol.* **12**, 293–296.
48. Maser, R. E., Monsen, K. J., Nelms, B. E., and Petrini, J. H. (1997). hMre11 and hRad50 nuclear foci are induced during the normal cellular response to DNA double-strand breaks. *Mol. Cell Biol.* **17**, 6087–6096.
49. Nelms, B. E., Maser, R. S., MacKay, J. F., Lagally, M. G., and Petrini, J. H. (1998). *In situ* visualization of DNA double-strand break repair in human fibroblasts. *Science* **280**, 590–592.
50. Haber, J. E. (1998). The many interfaces of Mre11. *Cell* **95**, 583–586.
51. Stewart, G. S., Maser, R. S., Stankovic, T., Bressan, D. A., Kaplan, M. I., Jaspers, N. G., Raams, A., Byrd, P. J., Petrini, J. H., and Taylor, A. M. (1999). The DNA double-strand break repair gene hMRE11 is mutated in individuals with an ataxia-telangiectasia-like disorder. *Cell* **99**, 577–587.
52. Gatei, M., Scott, S. P., Filippovitch, I., Soronika, N., Lavin, M. F., Weber, B., and Khanna, K. K. (2000). Role for ATM in DNA damage-induced phosphorylation of BRCA1. *Cancer Res.* **60**, 3299–3304.
53. Sung, P., Trujillo, K. M., and Van Komen, S. (2000). Recombination factors of *Saccharomyces cerevisiae*. *Mutat. Res.* **451**, 257–275.
54. Thompson, L. H. and Schild, D. (2001). Homologous recombinational repair of DNA ensures mammalian chromosome stability. *Mutat. Res.* **477**, 131–153.
55. van Gent, D. C., Hoeijmakers, J. H., and Kanaar, R. (2001). Chromosomal stability and the DNA double-stranded break connection. *Nat. Rev. Genet.* **2**, 196–206.
56. Scully, R., Chen, J., Ochs, R. L., Keegan, K., Hoekstra, M., Feunteun, J., and Livingston, D. M. (1997). Dynamic changes of BRCA1 subnuclear location and phosphorylation state are initiated by DNA damage. *Cell* **90**, 425–435.

57. Cortez, D., Wang, Y., Qin, J., and Elledge, S. J. (1999). Requirement of ATM-dependent phosphorylation of brca1 in the DNA damage response to double-strand breaks. *Science* **286**, 1162–1166.
58. Chen, G., Yuan, S. S., Liu, W., Xu, Y., Trujillo, K., Song, B., Cong, F., Goff, S. P., Wu, Y., Arlinghaus, R., Baltimore, D., Gasser, P. J., Park, M. S., Sung, P., and Lee, E. Y. (1999). Radiation-induced assembly of Rad51 and Rad52 recombination complex requires ATM and c-Abl. *J. Biol. Chem.* **274**, 12748–12752.
59. Morrison, C., Sonoda, E., Takao, N., Shinohara, A., Yamamoto, K., and Takeda, S. (2000). The controlling role of ATM in homologous recombinational repair of DNA damage. *EMBO J.* **19**, 463–471.
60. Banin, S., Moyal, L., Shieh, S., Taya, Y., Anderson, C. W., Chessa, L., Smorodinsky, N. I., Prives, C., Reiss, Y., Shiloh, Y., and Ziv, Y. (1998). Enhanced phosphorylation of p53 by ATM in response to DNA damage. *Science* **281**, 1674–1677.
61. Canman, C. E., Lim, D. S., Cimprich, K. A., Taya, Y., Tamai, K., Sakaguchi, K., Appella, E., Kastan, M. B., and Siliciano, J. D. (1998). Activation of the ATM kinase by ionizing radiation and phosphorylation of p53. *Science* **281**, 1677–1679.
62. Wang, Y., Cortez, D., Yazdi, P., Neff, N., Elledge, S. J and Qin, J. (2000). BASC, a super complex of BRCA1-associated proteins involved in the recognition and repair of aberrant DNA structures. *Genes Dev.* **14**, 927–939.
63. Andegeko, Y., Moyal, L., Mittelman, L., Tsarfaty, I., Shiloh, Y., Rotman, G. (2001). Nuclear retention of ATM at sites of DNA double-strand breaks. *J. Biol. Chem.* **276**, 38224–38230.
64. Paull, T. T., Rogakou, E. P., Yamazaki, V., Kirchgessner, C. U., Gellert, M., and Bonner, W. M. (2000). A critical role for histone H2AX in recruitment of repair factors to nuclear foci after DNA damage. *Curr. Biol.* **10**, 886–895.
65. Burma, S., Chen, B. P., Murphy, M., Kurimasa, A., and Chen, D. J. (2001). ATM phosphorylates Histone H2AX in response to DNA double-strand breaks. *J. Biol. Chem.* **276**, 42462–42467.
66. Kim, S. T., Lim, D. S., Canman, C. E., and Kastan, M. B. (1999). Substrate specificities and identification of putative substrates of ATM kinase family members. *J. Biol. Chem.* **274**, 37538–37543.
67. O'Neill, T., Dwyer, A. J., Ziv, Y., Chan, D. W., Lees-Miller, S. P., Abraham, R. H., Lai, J. H., Hill, D., Shiloh, Y., Cantley, L. C., and Rathbun, G. A. (2000). Utilization of oriented peptide libraries to identify substrate motifs selected by ATM. *J. Biol. Chem.* **275**, 22719–22727.
68. Gellert, M. (1997). Recent advances in understanding V(D)J recombination. *Adv. Immunol.* **64**, 39–64.
69. Stack, S. M. and Anderson, L. K. (2001). A model for chromosome structure during the mitotic and meiotic cell cycles. *Chromosome Res.* **9**, 175–198.
70. Boder, E. (1985). Ataxia-telangiectasia: An overview. *Kroc. Found. Ser.* **19**, 1–63.
71. Barlow, C., Liyanage, M., Moens, P. B., Tarsounas, M., Nagashima, K., Brown, K., Rottinghaus, S., Jackson, S. P., Tagle, D., Ried, T., and Wynshaw-Boris, A. (1998). Atm deficiency results in severe meiotic disruption as early as leptonema of prophase I. *Development* **125**, 4007–4017.
72. Plug, A. W., Peters, A. H., Xu, Y., Keegan, K. S., Hoekstra, M. F., Baltimore, D., de Boer, P., and Ashley, T. (1997). ATM and RPA in meiotic chromosome synapsis and recombination. *Nat. Genet.* **17**, 457–461.
73. Aurias, A., Dutrillaux, B., Buriot, D., and Lejeune, J. (1980). High frequencies of inversions and translocations of chromosomes 7 and 14 in ataxia telangiectasia. *Mutat. Res.* **69**, 369–374.
74. Hsieh, C. L., Arlett, C. F., and Lieber, M. R. (1993). V(D)J recombination in ataxia telangiectasia, Bloom's syndrome, and a DNA ligase I-associated immunodeficiency disorder. *J. Biol. Chem.* **268**, 20105–20109.
75. Kirsch, I. R. (1994). V(D)J recombination and ataxia-telangiectasia: A review. *Int. J. Radiat. Biol.* **66**, S97–108.
76. Barlow, C., Hirotsune, S., Paylor, R., Liyanage, M., Eckhaus, M., Collins, F., Shiloh, Y., Crawley, J. N., Ried, T., Tagle, D., and Wynshaw-Boris, A. (1996). Atm-deficient mice: A paradigm of ataxia telangiectasia. *Cell* **86**, 159–171.
77. Taylor, A. M., Metcalfe, J. A., Thick, J., and Mak, Y. F. (1996). Leukemia and lymphoma in ataxia telangiectasia. *Blood* **87**, 423–438.
78. Liao, M. J. and Van Dyke, T. (1999). Critical role for Atm in suppressing V(D)J recombination-driven thymic lymphoma. *Genes Dev.* **13**, 1246–1250.
79. Petiniot, L. K., Weaver, Z., Barlow, C., Shen, R., Eckhaus, M., Steinberg, S. M., Ried, T., Wynshaw-Boris, A., and Hodes, R. J. (2000). Recombinase-activating gene (RAG). 2-mediated V(D)J recombination is not essential for tumorigenesis in Atm-deficient mice. *Proc. Natl. Acad. Sci. USA* **97**, 6664–6669.
80. Perkins, E. J., Nair, A., Cowley, D. O., Van Dyke, T., Chang, Y., and Ramsden, D. A. (2002). Sensing of intermediates in V(D)J recombination by ATM. *Genes Dev.* **16**, 159–164.
81. Watters, D., Kedar, P., Spring, K., Bjorkman, J., Chen, P., Gatei, M., Birrell, G., Garrone, B., Srinivasa, P., Crane, D. I., and Lavin, M. F. (1999). Localization of a portion of extranuclear ATM to peroxisomes. *J. Biol. Chem.* **274**, 34277–34282.
82. Khanna, K. K., Keating, K. E., Kozlov, S., Scott, S., Gatei, M., Hobson, K., Taya, Y., Gabrielli, B., Chan, D., Lees-Miller, S. P., and Lavin, M. F. (1998). ATM associates with and phosphorylates p53: Mapping the region of interaction. *Nat. Genet.* **20**, 398–400.
83. Baskaran, R., Wood, L. D., Whitaker, L. L., Canman, C. E., Morgan, S. E., Xu, Y., Barlow, C., Baltimore, D., Wynshaw-Boris, A., Kastan, M. B., and Wang, J. Y. (1997). Ataxia telangiectasia mutant protein activates c-Abl tyrosine kinase in response to ionizing radiation. *Nature* **387**, 516–519.
84. Shafman, T., Khanna, K. K., Kedar, P., Spring, K., Kozlov, S., Yen, T., Hobson, K., Gatei, M., Zhang, N., Watters, D., Egerton, M., Shiloh, Y., Kharbanda, S., Kufe, D., and Lavin, M. F. (1997). Interaction between ATM protein and c-Abl in response to DNA damage. *Nature* **387**, 520–523.
85. Kumar, S., Pandley, P., Bharti, A., Jim, S., Weichselbaum, R., Weaver, D., Kufe, D., and Kharbanda, S. (1998). Regulation of DNA-dependent protein kinase by the Lyn tyrosine kinase. *J. Biol. Chem.* **273**, 25654–25658.
86. Shangary, S., Brown, K. D., Adamson, A. W., Edmonson, S., Ng, B., Pandita, T. K., Yalowich, J., Taccioli, G. E., and Baskaran, R. (2000). Regulation of DNA-dependent protein kinase activity by ionizing radiation-activated abl kinase is an ATM-dependent process. *J. Biol. Chem.* **275**, 30163–30168.
87. Xu, Y. and Baltimore, D. (1996). Dual roles of ATM in the cellular response to radiation and in cell growth control. *Genes. Dev.* **10**, 2401–2410.
88. Giaccia, A. J. and Kastan, M. B. (1998). The complexity of p53 modulation: Emerging patterns from divergent signals. *Genes Dev.* **12**, 2973–2983.
89. El-Deiry, W. S., Tokino, T., Velculescu, V. E., Levy, D. B., Parsons, R., Trent, J. M., Lin, D., Mercer, W. E., Kinzler, K. W., and Vogelstein, B. (1993). WAF1, a potential mediator of p53 tumor suppression. *Cell* **75**, 817–825.
90. Gu Y., Turck, C.W., and Morgan, D. O. (1993). Inhibition of Cdk2 activity *in vivo* by an associated 20K regulatory subunit. *Nature* **366**, 707–710.
91. Harper, J. W., Adami, G. R., Wei, N., Keyomarsi, K., and Elledge, S. J. (1993). The p21 Cdk-interacting protein Cip1 is a potent inhibitor of G1 cyclin-dependent kinases. *Cell* **75**, 805–816.
92. Bartek, J. and Lukas, J. (2001). Pathways governing G1/S transition and their response to DNA damage. *FEBS Lett.* **490**, 117–122.
93. Lee, C. H. and Chung J. H. (2001). The hCds1 (Chk2)-FHA domain is essential for a chain of phosphorylation events on hCds1 that is induced by ionizing radiation. *J. Biol. Chem.* **276**, 30537–30541.
94. Buscemi, G., Savio, C., Zannini, L., Micciche, F., Masnada, D., Nakanishi, M., Tauchi, H., Komatsu, K., Mitzutani, S., Khanna, K., Chen, P., Concannon, P., Chessa, L., and Delia, D. (2001). Chk2 activation dependence on Nbs1 after DNA damage. *Mol. Cell Biol.* **21**, 5214–5222.

95. Chehab, N. H., Malikzay, A., Stavridi, E. S., and Halazonetis, T. D. (1999). Phosphorylation of Ser-20 mediates stabilization of human p53 in response to DNA damage. *Proc. Natl. Acad. Sci. USA* **96**, 13777–13782.
96. Shieh, S. Y., Ahn, J., Tamai, K., Taya, Y., and Prives, C. (2000). The human homologs of checkpoint kinases Chk1 and Cds1 (Chk2). phosphorylate p53 at multiple DNA damage-inducible sites. *Genes Dev.* **14**, 289–300.
97. Waterman, M. J., Stavridi, E. S., Waterman, J. L., and Halazonetis, T. D. (1998). ATM-dependent activation of p53 involves dephosphorylation and association with 14–3–3 proteins. *Nat. Genet.* **19**, 175–178.
98. Khosravi, R., Maya, R., Gottlieb, T., Oren, M., Shiloh, Y., and Shkedy, D. (1999). Rapid ATM-dependent phosphorylation of MDM2 precedes p53 accumulation in response to DNA damage. *Proc. Natl. Acad. Sci. USA* **96**, 14973–14977.
99. Oren, M. (1999). Regulation of the p53 tumor suppressor protein. *J. Biol. Chem.* **274**, 36031–36034.
100. Maya, R., Balass, M., Kim, S. T., Shkedy, D., Leal, J. F., Shifman, O., Moas, M., Buschmann, T., Ronai, Z., Shiloh, Y., Kastan, M. B., Katzir, E., and Oren, M. (2001). ATM-dependent phosphorylation of 5: Role in p53 activation by DNA damage. *Genes Dev.* **15**, 1067–1077.
101. Painter, R. B. (1996). Inhibition of mammalian cell DNA synthesis by ionizing radiation. *Int. J. Radiat. Biol. Relat. Stud. Phys. Chem. Med.* **49**, 771–781.
102. Gatei, M., Young, D., Cerosaletti, K. M., Desai-Mehta, A., Spring, K., Kozlov, S., Lavin, M. F., Gatti, R. A., Concannon, P., and Khanna, K. (2000). ATM-dependent phosphorylation of nibrin in response to radiation exposure. *Nat. Genet.* **25**, 115–119.
103. Zhao, S., Weng, Y. C., Yuan, S. S., Lin, Y. T., Hsu, H. C., Lin, S. C., Gerbino, E., Song, M. H., Zdzienicka, M. Z., Gatti, R. A., Shay, J. W., Ziv, Y., Shiloh, Y., and Lee, E. Y. (2000). Functional link between ataxia-telangiectasia and Nijmegen breakage syndrome gene products. *Nature* **405**, 473–477.
104. Lim, D. S., Kim, S. T., Xu, B., Maser, R. S., Lin, J., Petrini, J. H., and Kastan, M. B. (2000). ATM phosphorylates p95/nbs1 in an S-phase checkpoint pathway. *Nature* **404**, 613–617.
105. Falck, J., Petrini, J. H., Williams, B. R., Lukas, J., and Bartek, J. (2002). The DNA damage-dependent intra-S phase checkpoint is regulated by parallel pathways. *Nat. Genet.* **30**, 290–294.
106. Kim, S., Xu, B., and Kastan, M. B. (2002). Involvement of the cohesion protein, Smc1, in Atm-dependent and independent responses to DNA damage. *Genes Dev.* **16**, 560–570.
107. Yazdi, P. T., Wang, Y., Zhao, S., Patel, N., Lee, E. Y.-H. P., and Qin, J. (2002). SMC1 is a downstream effector in ATM/NBS1 branch of the human S-phase checkpoint. *Genes Dev.* **16**, 571–582.
108. Liu, V. F. and Weaver, D. T. (1993). The ionizing radiation-induced replication protein A phosphorylation response differs between ataxia telangiectasia and normal human cells. *Mol. Cell Biol.* **13**, 7222–7331.
109. Oakley, G. G., Loberg, L. I., Yao, J., Risinger, M. A., Yunker, R. L., Zernik-Kobak, M., Khanna, K. K., Lavin, M. F., Carty, M. P., and Dixon, K. (2001). UV-induced hyperphosphorylation of replication protein A depends on DNA replication and expression of ATM protein. *Mol. Biol. Cell.* **12**, 1199–1213.
110. Mirzanyans, R., Famulski, K. S., Enns, L., Fraser, M., and Paterson, M. C. (1995). Characterization of the signal transduction pathway mediating gamma ray-induced inhibition of DNA synthesis in human cells: Indirect evidence for involvement of calmodulin but not protein kinase C nor p53. *Oncogene* **11**, 1597–1605.
111. Zampetti-Besseler, F. and Scott, D. (1981). Cell death, chromosome damage and mitotic delay in normal human, ataxia telangiectasia and retinoblastoma fibroblasts after x-irradiation. *Int. J. Radiat. Biol. Relat. Stud. Phys. Chem. Med.* **39**, 547–558.
112. Bache, M., Pigorsch, S., Dunst, J., Wurl, P., Meye, A., Bartel, F., Schmidt, H., Rath, F. W., and Taubert, H. (2001). Loss of G2/M arrest correlates with radiosensitization in two human sarcoma cell lines with mutant p53. *Int. J. Cancer* **96**, 110–117.
113. Scott, D., and Zampetti-Bosseler, F. (1982). Cell cycle dependence of mitotic delay in X-irradiated normal and ataxia-telangiectasia fibroblasts. *Int. J. Radiat. Biol. Relat. Stud. Phys. Chem. Med.* **42**, 679–683.
114. Xu, B., Kim, S. T., Lim, D. S., and Kastan, M. B. (2002). Two molecularly distinct g(2)/m checkpoints are induced by ionizing irradiation. *Mol. Cell Biol.* **22**, 1049–1059.
115. Gatei, M., Zhou, B. B., Hobson, K., Scott, S., Young, D., and Khanna, K. K. (2001b). Ataxia telangiectasia mutated (ATM) kinase and ATM and Rad3 related kinase mediate phosphorylation of Brca1 at distinct and overlapping sites. In vivo assessment using phospho-specific antibodies. *J. Biol. Chem.* **276**, 17276–17280.
116. Xu, X., Wagner, K. U., Larson, D., Weaver, Z., Li, C., Ried, T., Hennighausen, I., Wynshaw-Boris, A., and Deng, C. X. (1999). Conditional mutation of Brca1 in mammary epithelial cells results in blunted ductal morphogenesis and tumour formation. *Nat. Genet.* **22**, 37–43.
117. Mullan, P. B., Quinn, J. E., Gilmore, P. M., McWilliams, S., Andrews, H., Gervin, C., McCabe, N., McKenna, S., White, P., Song, Y. H., Maheswaran, S., Liu, E., Haber, D. A., Johnston, P. G., and Harkin, D. P. (2001). BRCA1 and GADD45 mediated G2/M cell cycle arrest in response to antimicrotubule agents. *Oncogene* **20**, 6123–6131.
118. Nurse, P. (1997). Regulation of the eukaryotic cell cycle. *Eur. J. Cancer* **33**, 1002–1004.
119. Peng, C. Y., Graves, P. R., Thoma, R. S., Wu, Z., Shaw, A. S., and Piwnica-Worms, H. (1997). Mitotic and G2 checkpoint control: regulation of 14-3-3 protein binding by phosphorylation of Cdc25C on Ser-216. *Science* **277**, 1501–1505.
120. Dalal, S. N., Schweitzer, C. M., Gan, J., and DeCaprio, J. A. (1999). Cytoplasmic localization of human cdc25C during interphase requires an intact 14–3–3 binding site. *Mol. Cell. Biol.* **19**, 4465–4479.
121. Sanchez, Y., Wong, C., Thoma, R. S., Richman, R., Wu, Z., Piwnica-Worms, H., and Elledge, S. J. (1997). Conservation of the Chk1 checkpoint pathway in mammals: Linkage of DNA damage to Cdk regulation through Cdc25. *Science* **277**, 1497–1501.
122. Matsuoka, S., Huang, M., and Elledge, S. J. (1998). Linkage of ATM to cell cycle regulation by the Chk2 protein kinase. *Science* **282**, 1893–1897.
123. Ogg, S., Gabrielli, B., and Piwnica-Worms, H. (1994). Purification of a Ser kinase that associates with and phosphorylates human Cdc25C on Ser 216. *J. Biol. Chem.* **269**, 30461–30469.
124. Brown, A. L., Lee, C. H., Schwarz, J. K., Mitiku, N., Piwnica-Worms, H., and Chung, J. H. (1999). A human Cds-1-related kinase that functions downstream of ATM protein in the cellular response to DNA damage. *Proc. Natl. Acad. Sci. USA* **96**, 3745–3750.
125. Zhou, B. B. and Elledge, S. J. (2000). The DNA damage response: Putting checkpoints in perspective. *Nature* **408**, 435–439.
126. Lim, D. S., Kirsch, D. G., Canman, C. E., Ahn, J. H., Ziv, Y., Newman, L. S., Darnell, R. B., Shiloh, Y., and Kastan, M. B. (1998). ATM binds to beta-adaptin in cytoplasmic vesicles. *Proc. Natl. Acad. Sci. USA* **95**, 10146–10151.
127. Oka, A. and Takashima, S. (1998). Expression of the ataxia-telangiectasia gene (ATM) product in human cellular neurons during development. *Neurosci. Lett.* **252**, 195–198.
128. Barlow, C., Ribaut-Barassin, C., Zwingman, T. A., Pope, A. J., Brown, K. D., Owens, J. W., Larson, D., Harrington, E. A., Haeberle, A. M., Mariani, J., Eckhaus, M., Herrup, K., Bailly, Y., and Wynshaw-Boris, A. (2000). ATM is a cytoplasmic protein in mouse brain required to prevent lysosomal accumulation. *Proc. Natl. Acad. Sci. USA* **97**, 871–876.
129. Smith, G. C., Cary, R. B., Lakin, N. D., Hann, B. C., Teo, S. H., Chen, D. J., and Jackson, S. P. (1999). Purification and DNA binding properties of the ataxia-telangiectasia gene product ATM. *Proc. Natl. Acad. Sci. USA* **96**, 11134–11139.
130. Chan, D. W., Khanna, K. K., Lavin, M. F., and Lees-Miller, S. P. (1999). Purified ATM has manganese dependent, wortmannin sensitive Ser/thr kinase activity that is independent of DNA. *J. Biol. Chem.* **275**, 7803–7810.

131. Suzuki, K., Kodama, S., and Watanabe, M. (1999). Recruitment of ATM protein to double-strand DNA irradiated with ionizing radiation. *J. Biol. Chem.* **274**, 25571–25575.
132. Lavin, M. F. (2000). An unlikely player joins the ATM signaling network. *Nat. Cell Biol.* **2**, E215–217.
133. O'Connor, R. D. and Scott-Linthicum, D. S. (1980). Mitogen receptor redistribution defects and concomitant absence of blastogenesis in ataxia-telangiectasia T lymphocytes. *Clin. Immunol. Immunopathol.* **15**, 66–75.
134. Kondo, N., Inoue, R., Nishimura, S., Kasahara, K., Kameyama, T., Miwa, Y., Lorenzo, P. R., and Orii, T. (1993). Defective calcium-dependent signal transduction in T lymphocytes of ataxia-telangiectasia. *Scand. J. Immunol.* **38**, 45–48.
135. Chiesa, N., Barlow, C., Wynshaw-Boris, A., Strata, P., and Tempia, F. (2000). Atm-deficient mice Purkinje cells show age-dependent defects in calcium spike bursts and calcium currents. *Neuroscience* **96**, 575–583.
136. Elmore, E. and Swift, M. (1976). Growth of cultured cells from patients with ataxia-telangiectasia. *J. Cell Physiol.* **89**, 429–431.
137. Shiloh, Y., Tabor, E., and Becker, Y. (1982). Colony-forming ability of ataxia-telangiectasia skin fibroblasts is an indicator of their early senescence and increased demand for growth factors. *Exp. Cell Res.* **140**, 191–199.
138. Lynn, W. S. and Wong, P. K. (1997). Possible control of cell death pathways in ataxia telangiectasia. A case report. *Neuroimmunomodulation* **4**, 277–284.
139. Barzilai, A., Rotman, G., and Shiloh, Y. (2002). ATM deficiency and oxidative stress, a new dimension of defective response to DNA damage. *DNA Repair* **1**, 3–26.
140. Roisin, M. Y. and Anderson, C. K. (1990). Response of fibroblast cultures from ataxia-telangiectasia patients to oxidative stress. *Cancer Lett.* **54**, 43–50.
141. Reichenbach, J., Schubert, R., Schwan, C., Muller, K., Bohles, H. J., and Zielen, S. (1999). Anti-oxidative capacity in patients with ataxia telangiectasia. *Clin. Exp. Immunol.* **117**, 535–539.

CHAPTER 306

Signaling to the p53 Tumor Suppressor through Pathways Activated by Genotoxic and Nongenotoxic Stresses

Carl W. Anderson[1] and Ettore Appella[2]
[1]Biology Department, *Brookhaven National Laboratory, Upton, New York*
[2]*Laboratory of Cell Biology, National Cancer Institute, National Institutes of Health, Bethesda, Maryland*

Introduction

The product of the human p53 tumor suppressor gene (*TP53*) is a 393-amino-acid polypeptide that functions as a homotetrameric transcription factor. p53 regulates the expression of genes that control cell cycle progression, the induction of apoptosis, DNA repair, and other functions involved in cellular responses to stress. Loss of p53 function, either directly through mutation or indirectly through several mechanisms, plays a central role in the development of cancer [1,2]. While the p53 protein normally is short lived and is present at low levels in unstressed mammalian cells, in response to both genotoxic and nongenotoxic stresses it accumulates in the nucleus where it binds to specific DNA sequences [2,3]. Genomic approaches have shown that p53 induces or inhibits the expression of more than 150 genes including *p21*, *GADD45*, *MDM2*, *IGFBP3,* and *BAX* [4], many of which mediate cell cycle arrest or apoptosis. p53 modulates DNA repair processes [5,6], and the arrest of cell cycle progression may provide time for the repair of DNA damage. In some circumstances, cell cycle arrest is permanent and indistinguishable from senescence [7]. Alternatively, stress signaling may initiate p53-dependent apoptosis [2]. The biochemical links between p53, G_1 arrest, senescence, and apoptosis are cell- and stress-type dependent. These observations suggest that specific posttranslational modifications to the p53 protein, at least in part, determine cellular fate. In turn, these modifications reflect the specific pathways that become activated in response to any particular stress condition. In this chapter, we highlight recent studies on the pathways that modulate p53 stability and activity in response to genotoxic and nongenotoxic stresses through covalent posttranslational modifications to p53 including the phosphorylation of serine and threonines and the acetylation of lysines.

p53 Protein Structure

The p53 polypeptide can be divided into three functionally distinct regions: an amino-terminal region (1–101, numbering for human p53) that interacts with regulatory proteins and the transcriptional machinery; a central, sequence-specific DNA-binding domain (102–292); and a carboxyl-terminal tetramerization and regulatory domain (293–393) (Fig. 1). In the N-terminal domain, residues 1–42 are required for transactivation activity and interact with the transcription factors TFIID, TFIIH, several TAFs, the histone acetyltransferases CBP/p300, and possibly PCAF, as well as the MDM2 ubiquitin ligase. Residues 17–28 form an amphipathic helix that

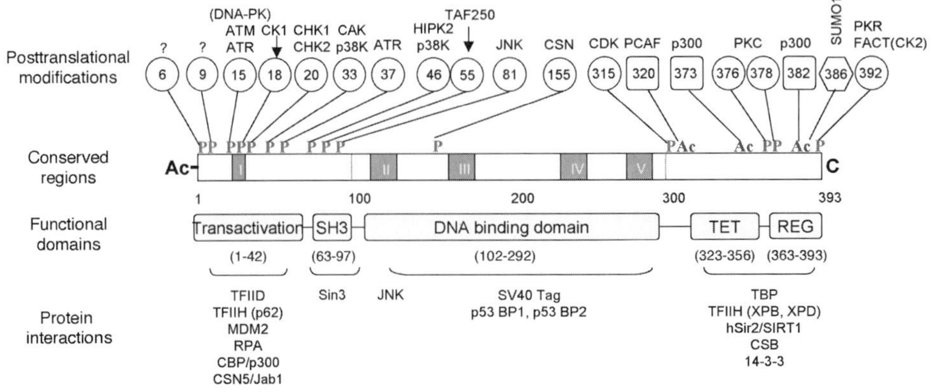

Figure 1 Protein domains, posttranslational modification sites, and proteins that interact with human p53. The 393-amino-acid human p53 polypeptide is represented schematically (box) with the five most highly conserved regions marked (I–V); postulated function regions and domains also are indicated. Residues ~1–42 comprise the transactivation domain; residues ~63–97 constitute a Src homology 3-like (SH3) domain that overlaps a poorly conserved proline- and alanine-rich segment (33–80); residues ~102–292 contain the central, sequence-specific, DNA-binding core region; residues ~300–323 contain the primary nuclear localization signal (NLS); residues 324–356 comprise the tetramerization domain (TET), which contains a nuclear export signal [86]; and residues 363–393 (REG) negatively regulate DNA binding by the central core to consensus recognition sites in oligonucleotides and interact in a sequence-independent manner with single- and double-stranded nucleic acids. Interaction regions for selected proteins are indicated below the polypeptide, and posttranslational modification sites (P, phosphorylation; Ac, acetylation) are indicated above the peptide together with enzymes that can accomplish the modifications *in vitro*. Lys386 may be modified by conjugation with SUMO1, an ubiquitin-like peptide. References are found in the text and recent reviews.

interacts directly with a hydrophobic cleft in the N-terminal domain of MDM2 [8], while residues 11–26 are reported to function as a secondary nuclear export signal [9]. Residues 63–97 comprise a proline-rich SH3 domain required for interaction with the Sin3 corepressor [10] and other proteins required for the induction of apoptosis. While residues 1–31 and 80–101 are highly conserved, especially among mammals, residues 32–79 are poorly conserved in sequence even among mammals.

The majority of tumor-derived p53 mutations affect the central domain and block or alter sequence-specific DNA binding either by destabilizing the domain or through changes to residues that directly contact DNA. The three-dimensional structure of the DNA-binding domain (DBD) bound to DNA was determined by X-ray crystallography [11]. The consensus DNA-binding site is composed of two 10-bp segments (RRRCWWGYYY) separated by 0–13 bp. The carboxy-terminal region contains the nuclear localization signal (312–324), a tetramerization domain (323–356), and a basic segment that binds certain DNA structures, including short single strands, four-way junctions, and insertions/deletions in a sequence-independent manner [12]. Only tetrameric p53 appears to be active as a transcription factor.

Posttranslational Modifications to p53

p53 activity is thought to be regulated largely through numerous posttranslational modifications that occur mainly in the N- and C-terminal regions (Fig. 1). Using monoclonal or affinity-purified polyclonal antibodies produced by several laboratories and companies that recognize specific, modified sites in human or mouse p53, increased phosphorylation at most of the 15 known phosphorylation sites has been demonstrated in response to the treatment of cells with DNA damage-inducing agents in Western immunoblot experiments (reviewed in [13]). Seven serines and two threonines in the N-terminal domain of human p53, specifically Ser6, 9, 15, 20, 33, 37, and 46 and Thr18 and 81, are phosphorylated in response to exposing cells to ionizing radiation or UV light. Recently, Thr55 was found to be phosphorylated in unstressed cells [14] and dephosphorylated after DNA damage (X. Liu, personal communication, 2001). Thus, all N-terminal serines and threonines in the first 89 residues of human p53 may be phosphory-lated or dephosphorylated in response to one or more stress conditions. In the C-terminal regulatory domain, Ser315 and Ser392 are phosphorylated; Lys320, 373, and 382 are acetylated; and Lys386 is sumoylated in response to DNA damage. Ser376 and 378 were reported to be constitutively phosphorylated in unstressed cells, whereas Ser376 was dephosphorylated in response to ionizing radiation [15]. Thr155 and Thr150 or Ser149 in the central site-specific DBD recently were reported to be phosphorylated by the COP9 signalosome (CSN)-associated kinase [16]; so far, these are the only sites in the central domain that have been reported to be posttranslationally modified.

Regulation of p53 Activity

The biochemical mechanisms that regulate p53 activity are complex and incompletely understood, but it is widely believed that activation of p53 as a transcription factor involves two stages, as discussed next.

p53 Stabilization

First, in response to stress-activated signaling pathways, p53 is stabilized and accumulates in the nucleus. In unstressed cells, p53 protein is maintained at low steady-state levels and has a short half-life due to rapid, ubiquitin-dependent degradation through the 26S proteosome. Ubiquitin is a 76-amino-acid polypeptide that is transferred to lysine residues in proteins by ubiquitin ligases; multiple ubiquitination targets the protein to the 26S proteosome complex where it is degraded. At least three cellular systems that target p53 for ubiquitination have been described. In the G_0 phase of the cell cycle, the Jun N-terminal kinase (JNK) binds p53 and targets it for ubiquitination [17]. Activation of JNK in response to DNA damage allows JNK to phosphorylate p53 on Thr81, enhancing its transactivation potential and releasing JNK from p53. The COP9 signalosome (SSN) recently was shown to bind p53, phosphorylate it on Thr155 and nearby residues, and promote p53 degradation by targeting it for ubiquitination [16]. Inhibition of the CSN-associated kinase activity or mutation of Thr155 to valine resulted in increased p53 stability and its accumulation. The CSN kinase is thought to be constitutively active; thus, CSN is believed to contribute to the normal turnover of p53 in cells, but it is not known if modulation of the CSN kinase activity in response to stress contributes to p53 stability.

In dividing cells, the primary system that ubiquitinates p53 is the MDM2 E3 ubiquitin ligase. This activity of MDM2 is vital as shown by the rescue of Mdm2 knock-out mice from embryonic lethality by deletion of p53. Interestingly, loss of MdmX expression also results in embryonic lethality that is rescued by deletion of p53 [18]. MdmX is a structurally related protein that interacts with Mdm2 and blocks p53 degradation. A cleft in the N-terminal domain of MDM2 (amino acids 25–109) binds to an amphipathic helix (amino acids 17–29) in the transactivation domain at the N terminus of p53 [8], and binding is required for subsequent ubiquitination at multiple C-terminal p53 lysines. p53 is phosphorylated at several N-terminal sites that reside in or near the MDM2 binding site by kinases activated through several stress response pathways; this led to the hypothesis that phosphorylation might stabilize p53 by preventing its interaction with MDM2 [19]. Initially it was proposed that phosphorylation of Ser15 and 37 in response to DNA damage induced a conformational change in p53 that prevented its interaction with MDM2, thus inhibiting p53 ubiquitination [19]. Subsequently, phosphorylation of Thr18 and Ser20 were reported to negatively regulate the interaction of p53 with MDM2 [20–23]. Both Thr18 and Ser20 lie within the p53 N-terminal amphipathic helix that directly interacts with the N terminus of MDM2. Thr18 makes several hydrogen bonds with neighboring residues that stabilize the helix and that would be disrupted by phosphorylation. Consistent with the structural data, phosphorylation of Thr18, but not phosphorylation of Ser15, Ser20, or Ser37, was found to interfere directly with the interaction of an N-terminal p53 peptide with the N-terminal domain of MDM2 [22,24]. Nevertheless, changing Ser15 to alanine was shown to significantly decrease the ability of p53 to activate transcription and induce apoptosis in both human [19,25] and mouse [26] systems, and changing Ser20 to alanine in human p53 abrogated stabilization in response to DNA damage [20]. Taken together, these results suggest that phosphorylation of p53 Ser15 and Ser20 may indirectly affect complex formation with Mdm2.

One mechanism that would be consistent with the present results is increased competition for binding to the N terminus of phosphorylated p53 by other factors. For example, CBP/p300 interacts with the N terminus of p53, and binding is dramatically enhanced by phosphorylation of Ser15 [27]. Consistent with this finding, changing Ser6, Ser9, Ser15, or Thr18 (but not more distal serines) to alanine reduced acetylation of Lys382 [28]. Surprisingly, however, changing murine Ser23, the equivalent of human Ser20, to alanine had no measurable effect on transactivation, apoptosis, and Mdm2 binding in mouse ES cells, fibroblasts, or thymocytes [29]. Thus, in spite of the high degree of sequence conservation between human and mouse p53 in this region, the two species appear to regulate p53 stabilization and MDM2 binding somewhat differently. Several other proteins have been shown to interact with the N terminus of p53 (see Fig. 1), but the effect of p53 phosphorylation on their binding has not been examined.

Recent results suggest that other mechanisms also play an important role in regulating MDM2 activity. First, $p14^{ARF}$ can inhibit the activity of MDM2 by sequestering it in the nucleolus or through other mechanisms [30]. Second, MDM2 itself is a target for DNA damage-induced posttranslational modifications that may positively or negatively regulate its activity [31]. Third, several studies have shown that the C-terminal regulatory region of p53 is important for its stabilization. The circumstances under which each mechanism operates and their relative importance in regulating p53 stability have not been fully characterized, but the existence of several mechanisms may explain reported differences in results relating to different phosphorylation sites.

A fourth system that targets p53 for ubiquitin-mediated degradation is the E6 protein of human papillomaviruses. Papillomaviruses are small, DNA tumors viruses, and certain serotypes are strongly associated with cervical cancer. Like other DNA tumor viruses, papillomaviruses inactivate p53, both to allow cells to enter a state in which the DNA viral genome can replicate and to avoid virus-induced apoptosis. The papillomavirus E6 protein, in conjunction with a ~100-kDa cellular protein E6AP, functions as an ubiquitin ligase in a manner similar to MDM2 [32]. As a consequence, papillomavirus transformed human cells (e.g., HeLa) frequently have wild-type p53 genes but are functionally deficient for p53 activity.

p53 Activation

p53 accumulation alone is not sufficient to fully activate p53-dependent transcription [33]. Early studies by Hupp and

Lane [34] indicated that p53 is synthesized in a latent form that is incompetent for sequence-specific DNA binding; they proposed that subsequent to its synthesis p53 was activated to a DNA-binding competent state through posttranslational modifications that were postulated to induce a conformational change in the DBD of p53. Indeed, several modifications to the C terminus, including phosphorylation [34], acetylation [35,36], binding of antibodies, and truncation of the C-terminal 30 amino acids, were shown to result in increased p53 binding to oligonucleotides containing the p53 consensus recognition site. These results suggested that the C-terminal domain negatively regulated sequence-specific binding by the central domain. The C-terminal regulatory domain of p53 contains numerous basic residues and, subsequently, it was found to bind single-stranded DNA and RNA, as well as several DNA structures.

An alternative model to explain latency was proposed by Anderson *et al.* [37], who showed that the sequence-specific DNA binding of p53 was inhibited by long DNAs but not by short oligonucleotides (Fig. 2, top). These results suggested that strong non-sequence-specific binding by the C-terminal domain of tetrameric p53 prevented sequence-specific binding. Furthermore, relief of inhibition of sequence-specific DNA binding through removal of the C terminus only occurred in the presence of long DNA molecules. The interference model recently received indirect support from NMR-based structural studies on a dimeric derivative of p53 and its C-terminal truncated form [38]. No significant differences in the structure of the central DBD were observed between the two forms, suggesting that increased DNA binding did not result from an allosteric conformational change in p53 structure. However, the concept of a latent DNA binding form of p53 recently was challenged by the finding that unmodified p53 binds well to long DNAs containing consensus recognition sites [39]. Indeed, p53 was reported to be constitutively bound to chromatin at some recognitions sites *in vivo*, including sites in the *CDKN1A* (*p21, WAF1, CIP1*), and *MDM2* promoters [40]. Furthermore, genotoxic stress caused only a small increase in the amount of p53 bound to chromatin at these sites.

If DNA/chromatin binding is not the rate-limiting step for activation of p53, then what might this step be? Although direct attempts to demonstrate a role for p53 acetylation in

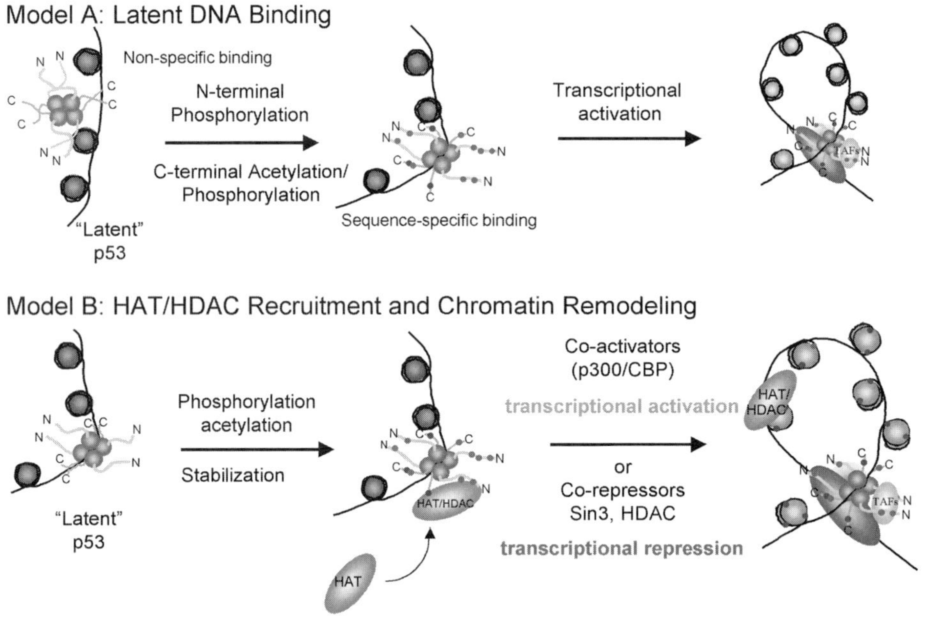

Figure 2 Models for the activation of p53 as a transcription factor. *Top:* Model A: Regulation of DNA binding by the C-terminal domain. Unmodified p53 exists as a latent tetramer incapable of sequence-specific DNA binding. Upon modification of C-terminal residues by phosphorylation or acetylation, latent p53 is converted into an active, sequence-specific, DNA-binding protein that then recruits elements of the transcription apparatus (e.g., TAFs). The conversion of latent p53 into active p53 might occur through an allosteric transition in the DBD [34] or as a consequence of the inhibition of nonspecific DNA binding to the C-terminal regulatory domain [37]. Recent NMR studies [38] argue against activation through an allosteric transition. *Bottom:* Model B: Recent studies suggest that p53 may be constitutively bound at some p53 consensus sites; thus, latency of DNA binding may be an artifact of the *in vitro* conditions employed [40]. Other studies have shown that both histone acetyltransferase (HAT) coactivators, such as CBP/p300, and histone deacetylase (HDAC) corepressors are recruited to p53 through stress-induced posttranslational modifications to the p53 N terminus. Although CBP/p300 and PCAF acetylate lysines at the C terminus of p53, mutation of these sites failed to demonstrate a strong effect on p53-mediated transcription [38]. Instead, recruitment of HATs and HDACs to promoter sites may alter chromatin structure in the vicinity of p53 promoters, suggesting that activation of p53-mediated transcription may result from the action of these coactivators and repressors on the modification state and structure of chromatin in the vicinity of p53 promoters [39].

transcriptional activation *in vivo* have been unsuccessful [41], the importance of histone acetyltransferases (HATs) for p53 transcriptional activity was demonstrated by overexpression of the histone deacetylases HDAC-1, -2, or -3, or hSir2, which deacetylate p53 and inhibit the transcription of p53-target genes [42–46]. These and other recent results suggest an alternative model for p53-mediated transcriptional activation in which p53 serves to target HATs to the promoters of p53 activated target genes (Fig. 2, bottom). Likewise, the targeting of HDACs by p53 to p53 repressed genes has been suggested as a mechanism of p53-mediated gene repression [47]. The mechanisms by which HATs (and HDACs) associate with p53 are not completely understood. As noted earlier, we previously proposed that DNA damage-induced N-terminal phosphorylations promote the association of p300/CBP with p53 [36]. However, other targeting mechanisms may exist. Recently it was suggested that acetylated residues in the p53 C terminus may serve to recruit coactivators [48].

A role for p53 in targeting the coactivator p300 to chromatin is consistent with recent *in vitro* transcription studies by Espinosa and Emerson [39]. They found that a plasmid containing the *CDKN1A* (*p21*) promoter, which contains two p53 binding sites located 2.3- and 1.5-kb proximal to the transcription start site, was not transcribed efficiently when assembled into chromatin. Transcription was strongly enhanced, however, by addition of both p53 and p300 but not by either factor alone. Importantly, unacetylated p53 bound efficiently to the plasmid DNA and with even higher affinity to the nucleosomal template, but not to a 25-bp oligonucleotide containing the 5′ *p21* promoter consensus site. Transcriptional activation required the C terminus of p53 but not acetylation of C-terminal sites. Furthermore, p53 was shown to induce acetylation of nucleosomal histones preferentially in the promoter region of the chromosomal template, consistent with the model depicted in Fig. 2 (bottom). The authors [39] suggested that binding of p53 to consensus sites in long DNA might be facilitated by the ability of long DNA, but not short oligonucleotides, to form non-B-DNA structures [49] or to bend DNA [50], as suggested previously. Surprisingly, however, several p53 activated promoters including *KARP1* [51], p21, and 14-3-3σ [51a] were found to be in an "open" conformation as judged by the presence of DNAse I hypersensitive sites in these promoters prior to p53 activation by treatment of cells with DNA damage-inducing agents. DNAse I chromatin hypersensitivity is a hallmark of promoter regions of actively transcribed genes, whereas the promoter regions of inactive genes usually are insensitive to DNAse I.

Activation of p53 by Genotoxic Stresses

Mammalian cells appear to have at least two largely independent signaling pathways for activating p53 in response to genotoxic stress; one is activated by the presence of DNA double-strand breaks, the other in response to bulky lesions such as pyrimidine dimers and base adducts.

Ionizing Radiation

Treatment of cells with ionizing radiation or several radiomimetics (e.g., neocarzinostatin, bleomycin) activates several kinases that phosphorylate p53 at multiple sites (Fig. 3). Although the molecular mechanism(s) by which DNA strand breaks are recognized are still obscure, key among the kinases activated in response to DNA breaks is ATM, a protein kinase member of the phosphatidylinositol-3-kinase (PI3K) kinase family encoded by the gene responsible for the human genetic disorder ataxia-telangiectasia (A-T). ATM directly phosphorylates p53 at Ser15 and activates several other protein kinases that phosphorylate the N-terminal transactivation domain including Chk1 and Chk2, which

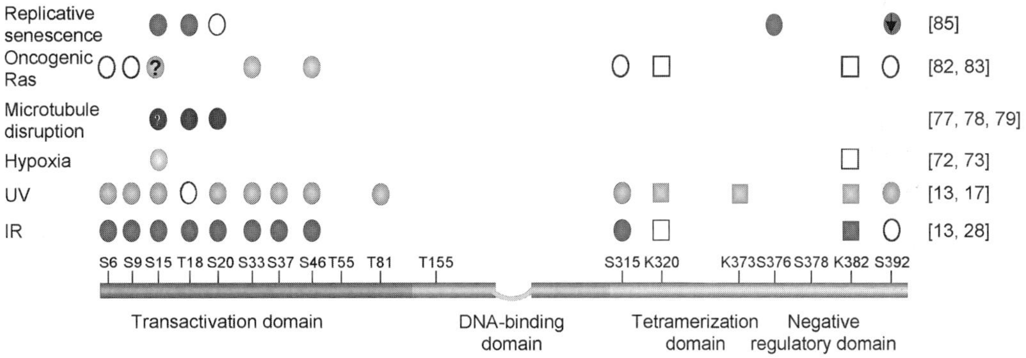

Figure 3 Posttranslational modifications to p53 in response to genotoxic and nongenotoxic stress. The bar at the bottom represents the human 393-amino-acid p53 polypeptide; functional regions are indicated. Reported posttranslational modification sites are indicated above the bar; S, serine; T, threonine; K, lysine. Filled circles (phosphorylation) or squares (acetylation) indicate modification in response to the indicated stress (left); open symbols indicate no change in modification in response to stress. No symbol indicates the site has not been examined. Selected references are given at the right. A "?" indicates conflicting literature reports; a down arrow indicates that treatment induced a decrease in site modification. Thr55 [14] and Ser376 and 378 [15] were reported to be constitutively phosphorylated; Thr55 and Ser376 are dephosphorylated in response to DNA damage. Ser6, 9, 33, 315, and 392 may be constitutively phosphorylated at low levels in some cell lines (Saito *et al.* unpublished).

phosphorylate p53 at Ser20 and perhaps other residues (see Fig. 1). ATM appears to activate other, unidentified protein kinases that directly phosphorylated Ser9 and Ser46 [28]. Protein kinases that directly phosphorylate Ser6, Ser33, Ser37, and Ser315 appear to be activated in response to DNA double-strand breaks by ATM-independent pathways, although inactivation of possible site-specific phosphatases for these residues in response to DNA strand breaks cannot be ruled out. Induction of phosphorylation through substrate modification also is possible. Casein kinase 1 (CK1) phosphorylates sites two residues distal to a previously phosphorylated serine or threonine. *In vitro* p53 previously phosphorylated at Ser6 and Ser15 is phosphorylated by CK1 at Ser9 or Thr18, respectively [22,52]. Thus, CK1, or a similar activity, may provide a mechanism for phosphorylation site signal amplification, at least in some cells. Several kinases, including CAK (Ser33), CDK (Ser315), PKA, and PKC (Ser376, Ser378), which are capable of phosphorylating p53 *in vitro*, have been identified (see Fig. 1), but if and how these might be activated in response to DNA strand breaks is unknown (reviewed in [13]).

DNA-PK, another PI3K family member that phosphorylates p53 at both Ser15 and Ser37 *in vitro*, is directly activated by DNA strand breaks through targeting to DNA ends by its Ku subunit, but its role *in vivo* remains uncertain. A recent study reported that p53 is found in a complex that contains nucleotide analog-modified DNA and activated DNA-PK [53]. Treatment of cells with the analog induced DNA-PK protein accumulation phosphorylation of p53 on Ser15, resulting in its stabilization and activation; and triggered apoptosis. Consistent with this report, Woo *et al.* [54] recently reported that DNA-PK is required for the IR-mediated apoptosis of mouse embryonic fibroblasts. However, DNA-PK is not required for p53-mediated cell cycle arrest in G_1 or the induction of p53 activated transcription in response to ionizing radiation [55].

UV Light

A second DNA damage response pathway is activated in response to bulky lesions, such as the pyrimidine dimers caused by UV-C (see Fig. 3). As for IR, the molecular mechanism(s) responsible for sensing UV-damaged DNA are not known, although the pathway is thought to involve sensing a block to transcriptional elongation [56] and activation of a third PI3K family member, ATR (A-T and Rad3-related) [57]. Recently, it was reported that ATR exhibits preferential binding to UV-damaged DNA; thus, under some circumstance, it may act directly as a sensor of UV damage [58]. *In vitro* ATR phosphorylates p53 on Ser15 and Ser37. UV-C also induces p53 phosphorylation on Ser6, 9, 20, 33, 46, 315, 392, and on Thr18 and 81 (see Fig. 3); thus exposure of cells to UV light must activate or induce a number of other kinases that phosphorylate p53. Among these are p38 MAPK, which targets Ser33 and Ser46 [59]; HIPK2, which also targets Ser46 [60,61]; JNK, which phosphorylates Thr81 [17]; and a complex containing casein kinase 2 (CK2) and the transcription elongation factor FACT that phosphorylates Ser392 [62]. In response to UV light, each of these sites is phosphorylated in cells lacking ATM ([28]; Saito *et al.*, unpublished).

The dependence of phosphorylation at other sites on ATR has not been carefully investigated since ATR-deficient cells are not viable. Phosphorylation at Ser46 correlates with the induction of apoptosis and the transcriptional induction of *AIP1* (p53 regulated *a*poptosis *i*nducing *p*rotein *1*) by UV-C [63]. Substitution of Ser46 with alanine inhibited p53-mediated apoptosis and the induction of *AIP1*. Evidence has been presented that the majority of p53 protein can be phosphorylated at Ser315 by cdc2/Cdk2 in response to UV-induced DNA damage and that the cyclin-dependent kinases play a role in stimulating p53 function [64]. In contrast to UV-C, UV-A triggers activation of p53 through activation of the ATM kinase activity, apparently through the production of reactive oxygen species [65].

Carboxy-Terminal DNA Damage-Induced p53 Modifications

As noted earlier, two C-terminal PKC sites, Ser376 and 378, were reported to be constitutively phosphorylated in the absence of DNA damage [15]. Thr55 also may be constitutively phosphorylated and then dephosphorylated in response to DNA damage. Treatment of MCF7 cells with IR led to the ATM-dependent dephosphorylation of Ser376 and to the association of p53 with a 14-3-3 protein, which increased its *in vitro* affinity for sequence-specific DNA. A confounding observation is the fact that p53 from unstressed cells is recognized by the monoclonal antibody PAb421, which recognizes an epitope including Lys372 to Lys382, but recognition is inhibited by phosphorylation of Ser376 or Ser378 (and acetylation of Lys382). Recently H7, an inhibitor of PKC, was reported to prevent p53 ubiquitination and to stimulate p53 accumulation in cells without inducing its activation, suggesting that the constitutive phosphorylation of p53 at Ser376 or Ser378 contributes to its degradation [66]. The function of UV-mediated phosphorylation of Ser392 is not clear, because previously Ser392 was shown not to be required for the suppression of cell growth or the activation of transcription [67]. *In vitro*, phosphorylation of Ser392 stimulated formation of p53 tetramers, while phosphorylation of Ser315 reversed this effect [68]; however, these effects are seen at low p53 concentrations and may have been masked in cells overexpressing exogenous p53.

The acetylation of several C-terminal p53 lysines is stimulated by several forms of stress including the genotoxic damage caused by both UV and IR [36,69]. These residues include Lys320, which is acetylated by PCAF *in vitro*, and Lys373 and Lys382, which can be acetylated by p300/CBP. Prior to the discovery of DNA damage-mediated p53 acetylation, CBP/p300 and PCAF were shown to be coactivators of p53-mediated transcription. As noted earlier, C-terminal acetylation is promoted by the phosphorylation of N-terminal sites, especially Ser15, but acetylation at Lys382 also was enhanced by phosphorylation at nearby residues, including Ser6, Ser9, and Thr18 [28]. However, in A549 cells acetylation was not induced equivalently by IR and UV. Acetylation, especially of Lys320, was much more strongly induced by UV than IR. Potential roles for p53 acetylation include contributions to p53 stability through interference with C-terminal ubiquitination and the modulation of p53's interactions with other proteins and DNA. Possible roles for p53 in targeting HATs and HDACs to chromatin were discussed earlier and illustrated in Fig. 2.

Other Genotoxic Agents

Because of their convenience, ionizing radiation and UV light are commonly used in the laboratory to produce two different forms to genotoxic damage: DNA double-strand breaks and pyrimidine dimers. Many other environmental, physiological, and therapeutic agents cause genotoxic damage that activate p53 through one or more signaling pathways. These include anticancer drugs such as adriamycin; topoisomerase inhibitors such as camptothecin, etoposide, and quercetin; DNA synthesis and transcription inhibitors including aphidicolin, actinomycin D and 5,6-dichloro-1-β-D-ribofuranosylbenzimidazole (DRB); DNA cross-linking agents, such as cisplatin and mitomycin C; and environmental chemicals including arsenite, cadmium, and chromate. Each of these agents have been shown to induces p53 accumulation and its phosphorylation on Ser15 except DRB, which interferes with phosphorylation of the CTD domain of RNA polymerase II rather than with elongation [70]. For most of these agents, however, modifications to p53 at other sites have not been investigated.

Activation of p53 by Nongenotoxic Stresses

p53 activation occurs in response to several physiological processes that are not associated with frank DNA damage, including hypoxia, nucleotide deprivation, microtubule inhibitors, oncogene activation, and senescence potentiated by telomere erosion. Some of these processes may be mimicked by pharmacological agents that may or may not induce identical responses.

Hypoxia

Tumor hypoxia occurs in most solid tumors from abnormal vasculature development; hypoxia is also an important pathophysiological feature of ischemic disorders. Hypoxia and several hypoxia mimetics have been shown to induce p53 accumulation as a result of the down-regulation of MDM2 [71] with concomitant phosphorylation of Ser15, but not acetylation of Lys382 [72] (see Fig. 3). However, in contrast to IR, hypoxia treatment failed to induce the transcription of downstream effector mRNAs including GADD45, Bax, and p21 [72]. Hypoxia does not induce detectable DNA damage, and, in contrast to DNA damage-inducing agents, primarily caused an association of p53 with mSin3A rather than p300. Consistent with this finding, p53-mediated transrepression was induced. In hypoxia-treated human papillomavirus HPV-16 transformed cells, p53 was resistant to E6-mediated degradation, and its association with E6AP was reduced [71]. Interestingly, a recent study has shown that inhibition of ATR kinase activity reduced the hypoxia-induced phosphorylation of p53 protein on Ser15 as well as p53 protein accumulation [73]. These data suggest that hypoxia could select for the loss of ATR-dependent checkpoint controls, thus promoting cell transformation.

Ribonucleotide Depletion

Studies by Linke *et al.* [74] showed that p53 is activated in normal human fibroblasts by the *N*-phosphoacetyl-L-aspartate-induced (PALA) depletion of ribonucleotides in the absence of detectable DNA damage. In contrast to the G_1 arrest induced by DNA damage, that induced by PALA was readily reversible. PALA treatment induced a pattern of gene expression that was distinct from that induced by IR. Some of these gene expressions, such as *MDGI*, a mammary-derived growth inhibitor gene, were induced independent of p53, whereas for others, such as *TSG6*, a tumor necrosis factor stimulated gene, induction was p53 dependent [75]. The modification status of p53 after PALA treatment has not been addressed.

Microtubule Disruption

Activation of p53 also occurs in response to factors such as colcemid, nocodazole, and taxol that deregulate cell adhesion or microtubule architecture and dynamics. Taxol (Paclitaxel), which inhibits microtubule depolymerization, is one of the newer chemotherapy drugs commonly used to treat ovarian, breast, and head and neck cancers. After nocodazole treatment, which depolymerizes microtubules, quiescent human fibroblasts accumulated transcriptionally active p53 and arrested in G_1 with a 4 N DNA content [76]. Activation of p53 after colcemid treatment was accompanied by a moderate increase in phosphorylation at Ser15 and correlated with activation of Erk1/2 MAP kinases and the development of focal adhesions rather than disruption of the microtubule system [77]. Curiously, murine fibroblasts did not undergo the same response. Taxol and vincristine, but not nocodazole, were found to induced multisite phosphorylation of p53 in several tumor-derived human cell lines, including HCT-116 and RKO cells, and the pattern of p53 phosphorylation was distinct from that observed after DNA damage [78]. Nevertheless, both nocodazole and taxol increased phosphorylation at Ser15 (Fig. 3).

Interestingly, microtubule inhibitor-induced p53 stabilization and Ser15 phosphorylation did not occur in ATM-deficient fibroblasts nor in normal human dermal fibroblasts. Studies with ectopically expressed p53 phosphorylation site mutants indicated that several p53 amino-terminal residues, including Ser15 and Thr18, were required for the taxol-mediated phosphorylation of p53 [78]. In contrast, Damia *et al.* [79] reported that taxol induced p53 phosphorylation at Ser20 but not at Ser15 in HCT-116 cells. Phosphorylation at Ser20 was accompanied by increased Chk2 activity and was not inhibited in A-T cells lines nor by wortmannin treatment. Thus, the signaling pathways that impinge on p53 after hypoxia, ribonucleotide depletion, or microtubule disruption, although still not well defined, appear distinct from those induced by genotoxic stresses.

Oncogene Activation

Oncogenes, such as Ras, c-Myc, or E1a, when activated or overexpressed, stabilize and activate p53, and, depending on the cell type, induce senescence (Ras) or apoptosis (c-Myc or E1a) through hyperproliferative signaling pathways that activate ARF, the product of the alternative reading frame of the cell cycle regulatory gene INK4a/CDKN2a [30]. ARF, in turn, inhibits p53 degradation by MDM2. Based on the observation that p53 Ser15 was not phosphorylated in response to adenovirus E1A expression, it was concluded that oncogenic activation of p53 occurs in the absence of DNA damage [80]. However, in normal fibroblasts brief c-Myc overexpression induced DNA damage prior to S phase that correlated with the induction of reactive oxygen species [81], raising the question of whether oncogenes activate p53 through DNA damage and whether the ability of oncogenes to promote either apoptosis or senescence correlates with different p53 posttranslational modifications. Ferbeyre *et al.* [82] reported that expression of oncogenic Ras induced phosphorylation of Ser15 in IMR90 cells. In contrast, Bulavin *et al.* [83] found that p53 was phosphorylated at Ser33 and Ser46 but not at other N- or C-terminal sites, nor was it acetylated at Lys382 (see Fig. 3). Interestingly, a similar induction of permanent cell cycle arrest resembling cellular senescence was produced in murine fibroblasts engineered to express the MAP kinase Mek1 [84]. The induction of senescence by Ras required wild-type p53 and ARF, but p53 was not required to maintain the senescent state. These data indicate that other signals may influence the outcome of p53 activation, likely by changing its association with various coactivators; however, whether this leads to the expression of different p53 target genes remains to be determined.

Replicative Senescence

Replicative senescence in human fibroblasts correlates with activation of p53-dependent transcription and was shown to be associated with increased phosphorylation at Ser15, Thr18, and probably Ser376, and decreased phosphorylation at Ser392 [85] (see Fig. 3). It was inferred from the finding that no change occurred in staining with the DO-1 monoclonal antibody, the epitope for which includes Ser20, that phosphorylation on Ser20 was not induced. These results, in conjunction with findings showing that changes in p53 phosphorylation are abrogated in cells that have been immortalized by overexpression of telomerase, indicate that the preceding modifications may be the product of telomere erosion. Shorten or disrupted telomere structures may signal to p53 via pathways partially shared with DNA damage responses.

Conclusions

Multiple, distinct signal transduction pathways clearly activate and modulate p53-dependent transcription in response to both genotoxic and nongenotoxic stresses. Although key protein kinases that are likely to phosphorylate p53 in response to DNA damage have been identified, the identities of kinases that phosphorylate several important sites are still unknown. Furthermore, several sites may be phosphorylated by more than one protein kinase. This complexity is augmented further by the facts that signaling pathway activation may be cell-type and cell-cycle dependent and that many signaling initiation events activate more than one pathway.

A fundamental question that remains unanswered is what mechanism(s) contribute to the ability of different cells to interpret p53 activation in different ways. The activation of p53 by hypoxia or oncogenes clearly induces different effects than the response to genotoxic stresses. Therefore, the pattern of posttranslational modifications may determine the selection of the subsets of target genes regulated in response to p53 activation, but a precise understanding of the mechanisms is not yet in hand. It is clear that the p53 protein forms complexes with many other cellular components and with particular nuclear structures. This characteristic may influence the degree of its activation and contribute to the heterogeneity of p53-dependent responses observed within a specific tissue. The analysis of the modification patterns in different mouse tissues of knock-in mutants should give insights as to the role played by individual phosphorylations and acetylations sites in eliciting a molecular signaling outcome. While there is still much to learn, substantial progress in understanding the causes and effects of p53 responses is being made.

Acknowledgments

We thank Sharlyn J. Mazur for constructive suggestions. We apologize to those whose publications could not be cited due to space limitations. C.W.A. was supported in part by a CRADA funded by the Laboratory Technology Research Program in the Office of Science of the U.S. Department of Energy.

References

1. Hollstein, M., Sidransky, D., Vogelstein, B., and Harris, C. C. (1991). p53 mutations in human cancers. *Science* **253**, 49–53.
2. Bálint, É. E. and Vousden, K. H. (2001). Activation and activities of the p53 tumour suppressor protein. *Br. J. Cancer* **85**, 1813–1823.
3. Wahl, G. M. and Carr, A. M. (2001). The evolution of diverse biological responses to DNA damage: Insights from yeast and p53. *Nat. Cell Biol.* **3**, E277–286.
4. Zhao, R., Gish, K., Murphy, M., Yin, Y., Notterman, D., Hoffman, W. H., Tom, E., Mack, D. H., and Levine, A. J. (2000). Analysis of p53-regulated gene expression patterns using oligonucleotide arrays. *Genes Dev.* **14**, 981–993.

5. Hanawalt, P. C. (2001). Controlling the efficiency of excision repair. *Mutat. Res.* **485**, 3–13.
6. Smith, M. L. and Seo, Y. R. (2002). p53 regulation of DNA excision repair pathways. *Mutagenesis* **17**, 149–156.
7. Campisi, J. (2001). Cellular senescence as a tumor-suppressor mechanism. *Trends Cell. Biol.* **11**, S27–31.
8. Kussie, P. H., Gorina, S., Marechal, V., Elenbaas, B., Moreau, J., Levine, A. J., and Pavletich, N. P. (1996). Structure of the MDM2 oncoprotein bound to the p53 tumor suppressor transactivation domain. *Science* **274**, 948–953.
9. Zhang, Y. and Xiong, Y. (2001). A p53 amino-terminal nuclear export signal inhibited by DNA damage-induced phosphorylation. *Science* **292**, 1910–1915.
10. Zilfou, J. T., Hoffman, W. H., Sank, M., George, D. L., and Murphy, M. (2001). The corepressor mSin3a interacts with the proline-rich domain of p53 and protects p53 from proteasome-mediated degradation. *Mol. Cell. Biol.* **21**, 3974–3985.
11. Cho, Y., Gorina, S., Jeffrey, P. D., and Pavletich, N. P. (1994). Crystal structure of a p53 tumor suppressor-DNA complex: Understanding tumorigenic mutations. *Science* **265**, 346–355.
12. Wolkowicz, R. and Rotter, V. (1997). The DNA binding regulatory domain of p53: See the C. *Pathol. Biol. (Paris)* **45**, 785–796.
13. Appella, E. and Anderson, C. W. (2001). Post-translational modifications and activation of p53 by genotoxic stresses. *Eur. J. Biochem.* **268**, 2764–2772.
14. Gatti, A., Li, H.-H., Traugh, J. A., and Liu, X. (2000). Phosphorylation of human p53 on Thr-55. *Biochemistry* **39**, 9837–9842.
15. Waterman, M. J. F., Stavridi, E. S., Waterman, J. L. F., and Halazonetis, T. D. (1998). ATM-dependent activation of p53 involves dephosphorylation and association with 14–3–3 proteins. *Nat. Genet.* **19**, 175–178.
16. Bech-Otschir, D., Kraft, R., Huang, X., Henklein, P., Kapelari, B., Pollmann, C., and Dubiel, W. (2001). COP9 signalosome-specific phosphorylation targets p53 to degradation by the ubiquitin system. *EMBO J.* **20**, 1630–1639.
17. Buschmann, T., Potapova, O., Bar-Shira, A., Ivanov, V. N., Fuchs, S. Y., Henderson, S., Fried, V. A., Minamoto, T., Alarcon-Vargas, D., Pincus, M. R., Gaarde, W. A., Holbrook, N. J., Shiloh, Y., and Ronai, Z. (2001). Jun NH_2-terminal kinase phosphorylation of p53 on Thr-81 is important for p53 stabilization and transcriptional activities in response to stress. *Mol. Cell. Biol.* **21**, 2743–2754.
18. Parant, J., Chavez-Reyes, A., Little, N. A., Yan, W., Reinke, V., Jochemsen, A. G., and Lozano, G. (2001). Rescue of embryonic lethality in *Mdm4*-null mice by loss of *Trp53* suggests a nonoverlapping pathway with MDM2 to regulate p53. *Nat. Genet.* **29**, 92–95.
19. Shieh, S.-Y., Ikeda, M., Taya, Y., and Prives, C. (1997). DNA damage-induced phosphorylation of p53 alleviates inhibition by MDM2. *Cell* **91**, 325–334.
20. Chehab, N. H., Malikzay, A., Stavridi, E. S., and Halazonetis, T. D. (1999). Phosphorylation of Ser-20 mediates stabilization of human p53 in response to DNA damage. *Proc. Natl. Acad. Sci. USA* **96**, 13777–13782.
21. Craig, A. L., Burch, L., Vojtesek, B., Mikutowska, J., Thompson, A., and Hupp, T. R. (1999). Novel phosphorylation sites of human tumour suppressor protein p53 at Ser^{20} and Thr^{18} that disrupt the binding of mdm2 (mouse double minute 2) protein are modified in human cancers. *Biochem. J.* **342**, 133–141.
22. Sakaguchi, K., Saito, S., Higashimoto, Y., Roy, S., Anderson, C. W., and Appella, E. (2000). Damage-mediated phosphorylation of human p53 threonine 18 through a cascade mediated by a casein 1-like kinase. Effect on Mdm2 binding. *J. Biol. Chem.* **275**, 9278–9283.
23. Unger, T., Juven-Gershon, T., Moallem, E., Berger, M., Vogt Sionov, R., Lozano, G., Oren, M., and Haupt, Y. (1999). Critical role for Ser20 of human p53 in the negative regulation of p53 by Mdm2. *EMBO J.* **18**, 1805–1814.
24. Lai, Z., Auger, K. R., Manubay, C. M., and Copeland, R. A. (2000). Thermodynamics of p53 binding to hdm2(1–126): Effects of phosphorylation and p53 peptide length. *Arch. Biochem. Biophys.* **381**, 278–284.
25. Fiscella, M., Ullrich, S. J., Zambrano, N., Shields, M. T., Lin, D., Lees-Miller, S. P., Anderson, C. W., Mercer, W. E., and Appella, E. (1993). The carboxy-terminal serine 392 phosphorylation site of human p53 is not required for wild-type activities. *Oncogene* **8**, 1519–1528.
26. Chao, C., Saito, S., Anderson, C. W., Appella, E., and Xu, Y. (2000). Phosphorylation of murine p53 at Ser-18 regulates the p53 responses to DNA damage. *Proc. Natl. Acad. Sci. USA* **97**, 11936–11941.
27. Lambert, P. F., Kashanchi, F., Radonovich, M. F., Shiekhattar, R., and Brady, J. N. (1998). Phosphorylation of p53 serine 15 increases interaction with CBP. *J. Biol. Chem.* **273**, 33048–33053.
28. Saito, S., Goodarzi, A. A., Higashimoto, Y., Noda, Y., Lees-Miller, S. P., Appella, E., and Anderson, C. W. (2002). ATM mediates phosphorylation at multiple p53 sites, including Ser46, in response to ionizing radiation. *J. Biol. Chem.* **277**, 12491–12494.
29. Wu, Z., Earle, J., Saito, S., Anderson, C. W., Appella, E., and Xu, Y. (2002). Mutation of mouse p53 Ser23 and the response to DNA damage. *Mol. Cell. Biol.* **22**, 2441–2449.
30. Sherr, C. J. (2001). The *INK4a/ARF* network in tumour suppression. *Nat. Rev. Mol. Cell Biol.* **2**, 731–737.
31. Michael, D. and Oren, M. (2002). The p53 and Mdm2 families in cancer. *Curr. Opin. Genet. Dev.* **12**, 53–59.
32. Talis, A. L., Huibregtse, J. M., and Howley, P. M. (1998). The role of E6AP in the regulation of p53 protein levels in human papillomavirus (HPV) positive and HPV-negative cells. *J. Biol. Chem.* **273**, 6439–6445.
33. Agarwal, M. L., Taylor, W. R., Chernov, M. V., Chernova, O. B., and Stark, G. R. (1998). The role of E6AP in the regulation of p53 protein levels in human papillomavirus (HPV)-positive and HPV-negative cells. *J. Biol. Chem.* **273**, 1–4.
34. Hupp, T. R. and Lane, D. P. (1994). Allosteric activation of latent p53 tetramers. *Curr. Biol.* **4**, 865–875.
35. Gu, W. and Roeder, R. G. (1997). Activation of p53 sequence-specific DNA binding by acetylation of the p53 C-terminal domain. *Cell* **90**, 595–606.
36. Sakaguchi, K., Herrera, J. E., Saito, S., Miki, T., Bustin, M., Vassilev, A., Anderson, C. W., and Appella, E. (1998). DNA damage activates p53 through a phosphorylation-acetylation cascade. *Genes Dev.* **12**, 2831–2841.
37. Anderson, M. E., Woelker, B., Reed, M., Wang, P., and Tegtmeyer, P. (1997). Reciprocal interference between the sequence-specific core and nonspecific C-terminal DNA binding domains of p53: Implications for regulation. *Mol. Cell. Biol.* **17**, 6255–6264.
38. Ayed, A., Mulder, F. A. A., Yi, G.-S., Lu, Y., Kay, L. E., and Arrowsmith, C. H. (2001). Latent and active p53 are identical in conformation. *Nat. Struct. Biol.* **8**, 756–760.
39. Espinosa, J. M. and Emerson, B. M. (2001). Transcriptional regulation by p53 through intrinsic DNA/chromatin binding and site-directed cofactor recruitment. *Mol. Cell* **8**, 57–69.
40. Kaeser, M. D. and Iggo, R. D. (2002). Chromatin immunoprecipitation analysis fails to support the latency model for regulation of p53 DNA binding activity *in vivo*. *Proc. Natl. Acad. Sci. USA* **99**, 95–100.
41. Nakamura, S., Roth, J. A., and Mukhopadhyay, T. (2000). Multiple lysine mutations in the C-terminal domain of p53 interfere with MDM2-dependent protein degradation and ubiquitination. *Mol. Cell. Biol.* **20**, 9391–9398.
42. Juan, L.-J., Shia, W.-J., Chen, M.-H., Yang, W.-M., Seto, E., Lin, Y.-S., and Wu, C.-W. (2000). Histone deacetylases specifically down-regulate p53-dependent gene activation. *J. Biol. Chem.* **275**, 20436–20443.
43. Luo, J., Su, F., Chen, D., Shiloh, A., and Gu, W. (2000). Deacetylation of p53 modulates its effect on cell growth and apoptosis. *Nature* **408**, 377–381.
44. Luo, J., Nikolaev, A. Y., Imai, S., Chen, D., Su, F., Shiloh, A., Guarente, L., and Gu, W. (2001). Negative control of p53 by Sir2a promotes cell survival under stress. *Cell* **107**, 137–148.
45. Vaziri, H., Dessain, S. K., Eaton, E. N., Imai, S. I., Frye, R. A., Pandita, T. K., Guarente, L., and Weinberg, R. A. (2001). *hSIR2*^{SIRT1} functions as an NAD-dependent p53 deacetylase. *Cell* **107**, 149–159.

46. Langley, E., Pearson, M., Faretta, M., Bauer, U.-M., Frye, R. A., Minucci, S., Pelicci, P. G., and Kouzarides, T. (2002). Human SIR2 deacetylates p53 and antagonizes PML/p53-induced cellular senescence. *EMBO J.* **21**, 2383–2396.
47. Murphy, M., Ahn, J., Walker, K. K., Hoffman, W. H., Evans, R. M., Levine, A. J., and George, D. L. (1999). Transcriptional repression by wild-type p53 utilizes histone deacetylases, mediated by interaction with mSin3a. *Genes Dev.* **13**, 2490–2501.
48. Barlev, N. A., Liu, L., Chehab, N. H., Mansfield, K., Harris, K. G., Halazonetis, T. D., and Berger, S. L. (2001). Acetylation of p53 activates transcription through recruitment of coactivators/histone acetyltransferases. *Mol. Cell* **8**, 1243–1254.
49. Kim, E., Rohaly, G., Heinrichs, S., Gimnopoulos, D., Meißner, H., and Deppert, W. (1999). Influence of promoter DNA topology on sequence-specific DNA binding and transactivation by tumor suppressor p53. *Oncogene* **18**, 7310–7318.
50. Nagaich, A. K., Zhurkin, V. B., Durell, S. R., Jernigan, R. L., Appella, E., and Harrington, R. E. (1999). p53-induced DNA bending and twisting: p53 tetramer binds on the outer side of a DNA loop and increases DNA twisting. *Proc. Natl. Acad. Sci. USA* **96**, 1875–1880.
51. Braastad, C. D., Leguia, M., and Hendrickson, E. A. (2002). Ku86 autoantigen related protein-1 transcription initiates from a CpG island and is induced by p53 through a nearby p53 response element. *Nucleic Acids Res.* **30**, 1713–1724.
51a. Hendrickson, E. A. (2003). Constitutive DNase I hypersensitivity of p53-regulated promoters. *J. Biol. Chem.* **278**, 8261–8268.
52. Higashimoto, Y., Saito, S., Tong, X.-H., Hong, A., Sakaguchi, K., Appella, E., and Anderson, C. W. (2000). Human p53 is phosphorylated on serines 6 and 9 in response to DNA damage-inducing agents. *J. Biol. Chem.* **275**, 23199–23203.
53. Achanta, G., Pelicano, H., Feng, L., Plunkett, W., and Huang, P. (2001). Interaction of p53 and DNA-PK in response to nucleoside analogues: Potential role as a sensor complex for DNA damage. *Cancer Res.* **61**, 8723–8729.
54. Woo, R. A., Jack, M. T., Xu, Y., Burma, S., Chen, D. J., and Lee, P. W. K. (2002). DNA damage-induced apoptosis requires the DNA-dependent protein kinase, and is mediated by the latent population of p53. *EMBO J.* **21**, 3000–3008.
55. Jimenez, G. S., Bryntesson, F., Torres-Arzayus, M. I., Priestley, A., Beeche, M., Saito, S., Sakaguchi, K., Appella, E., Jeggo, P. A., Taccioli, G. E., Wahl, G. M., and Hubank, M. (1999). DNA-dependent protein kinase is not required for the p53-dependent response to DNA damage. *Nature* **400**, 81–83.
56. Ljungman, M., Zhang, F., Chen, F., Rainbow, A. J., and McKay, B. C. (1999). Inhibition of RNA polymerase II as a trigger for the p53 response. *Oncogene* **18**, 583–592.
57. Abraham, R. T. (2001). Cell cycle checkpoint signaling through the ATM and ATR kinases. *Genes Dev.* **15**, 2177–2196.
58. Ünsal-Kaçmaz, K., Makhov, A. M., Griffith, J. D., and Sancar, A. (2002). Preferential binding of ATR protein to UV-damaged DNA. *Proc. Natl. Acad. Sci. USA* **99**, 6673–6678.
59. Bulavin, D. V., Saito, S., Hollander, M. C., Sakaguchi, K., Anderson, C. W., Appella, E., and Fornace, A. J. Jr. (1999). Phosphorylation of human p53 by p38 kinase coordinates N-terminal phosphorylation and apoptosis in response to UV radiation. *EMBO J.* **18**, 6845–6854.
60. D'Orazi, G., Cecchinelli, B., Bruno, T., Manni, I., Higashimoto, Y., Saito, S., Gostissa, M., Coen, S., Marchetti, A., Del Sal, G., Piaggio, G., Fanciulli, M., Appella, E., and Soddu, S. (2002). Homeodomain-interacting protein kinase-2 phosphorylates p53 at Ser 46 and mediates apoptosis. *Nat. Cell Biol.* **4**, 11–19.
61. Hofmann, T. G., Möller, A., Sirma, H., Zentgraf, H., Taya, Y., Dröge, W., Will, H., and Schmitz, M. L. (2002). Regulation of p53 activity by its interaction with homeodomain-interacting protein kinase-2. *Nat. Cell Biol.* **4**, 1–10.
62. Keller, D. M., Zeng, X., Wang, Y., Zhang, Q. H., Kapoor, M., Shu, H., Goodman, R., Lozano, G., Zhao, Y., and Lu, H. (2001). A DNA damage-induced p53 serine 392 kinase complex contains CK2, hSpt16, and SSRP1. *Mol. Cell* **7**, 283–292.
63. Oda, K., Arakawa, H., Tanaka, T., Matsuda, K., Tanikawa, C., Mori, T., Nishimori, H., Tamai, K., Tokino, T., Nakamura, Y., and Taya, Y. (2000). *p53AIP1* a potential mediator of p53-dependent apoptosis, and its regulation by Ser-46-phosphorylated p53. *Cell* **102**, 849–862.
64. Blaydes, J. P., Luciani, M. G., Pospisilova, S., Ball, H. M.-L., Vojtesek, B., and Hupp, T. R. (2001). Stoichiometric phosphorylation of human p53 at Ser^{315} stimulates p53-dependent transcription. *J. Biol. Chem.* **14**, 4699–4708.
65. Zhang, Y., Ma, W.-Y., Kaji, A., Bode, A. M., and Dong, Z. (2002). Requirement of ATM in UVA-induced signaling and apoptosis. *J. Biol. Chem.* **277**, 3124–3131.
66. Chernov, M. V., Bean, L. J. H., Lerner, N., and Stark, G. R. (2001). Regulation of ubiquitination and degradation of p53 in unstressed cells through C-terminal phosphorylation. *J. Biol. Chem.* **276**, 31819–31824.
67. Fiscella, M., Zambrano, N., Ullrich, S., Unger, T., Lin, D., Cho, B., Mercer, W., Anderson, C., and Appella, E. (1994). Mutation of the serine 15 phosphorylation site of human p53 reduces the ability of p53 to inhibit cell cycle progression. *Oncogene* **9**, 3249–3257.
68. Sakaguchi, K., Sakamoto, H., Xie, D., Erickson, J. W., Lewis, M. S., Anderson, C. W., and Appella, E. (1997). Effect of phosphorylation on tetramerization of the tumor suppressor protein p53. *J. Protein Chem.* **16**, 553–556.
69. Ito, A., Lai, C.-H., Zhao, X., Saito, S., Hamilton, M. H., Appella, E., and Yao, T.-P. (2001). p300/CBP-mediated p53 acetylation is commonly induced by p53-activating agents and inhibited by MDM2. *EMBO J.* **20**, 1331–1340.
70. Ljungman, M., O'Hagan, H. M., and Paulsen, M. T. (2001). Induction of ser15 and lys382 modifications of p53 by blockage of transcription elongation. *Oncogene* **20**, 5964–5971.
71. Alarcón, R., Koumenis, C., Geyer, R. K., Maki, C. G., and Giaccia, A. J. (1999). Hypoxia induces p53 accumulation through MDM2 down-regulation and inhibition of E6-mediated degradation. *Cancer Res.* **59**, 6046–6051.
72. Koumenis, C., Alarcon, R., Hammond, E., Sutphin, P., Hoffman, W., Murphy, M., Derr, J., Taya, Y., Lowe, S. W., Kastan, M., and Giaccia, A. (2001). Regulation of p53 by hypoxia: Dissociation of transcriptional repression and apoptosis from p53-dependent transactivation. *Mol. Cell. Biol.* **21**, 1297–1310.
73. Hammond, E. M., Denko, N. C., Dorie, M. J., Abraham, R. T., and Giaccia, A. J. (2002). Hypoxia links ATR and p53 through replication arrest. *Mol. Cell. Biol.* **22**, 1834–1843.
74. Linke, S. P., Clarkin, K. C., Di Leonardo, A., Tsou, A., and Wahl, G. M. (1996). A reversible, p53-dependent G_0/G_1 cell cycle arrest induced by ribonucleotide depletion in the absence of detectable DNA damage. *Genes Dev.* **10**, 934–947.
75. Seidita, G., Polizzi, D., Costanzo, G., Costa, S., and Di Leonardo, A. (2000). Differential gene expression in p53-mediated G_1 arrest of human fibroblasts after γ-irradiation or *N*-phosphoacetyl-L-aspartate treatment. *Carcinogenesis* **21**, 2203–2210.
76. Khan, S. H. and Wahl, G. M. (1998). p53 and pRb prevent rereplication in response to microtubule inhibitors by mediating a reversible G_1 arrest. *Cancer Res.* **58**, 396–401.
77. Sablina, A. A., Chumakov, P. M., Levine, A. J., and Kopnin, B. P. (2001). p53 activation in response to microtubule disruption is mediated by integrin-Erk signaling. *Oncogene* **20**, 899–909.
78. Stewart, Z. A., Tang, L. J., and Pietenpol, J. A. (2001). Increased p53 phosphorylation after microtubule disruption is mediated in a microtubule inhibitor- and cell-specific manner. *Oncogene* **20**, 113–124.
79. Damia, G., Filiberti, L., Vikhanskaya, F., Carrassa, L., Taya, Y., D'incalci, M., and Broggini, M. (2001). Cisplatinum and taxol induce different patterns of p53 phosphorylation. *Neoplasia* **3**, 10–16.
80. de Stanchina, E., McCurrach, M. E., Zindy, F., Shieh, S. Y., Ferbeyre, G., Samuelson, A. V., Prives, C., Roussel, M. F., Sherr, C. J., and Lowe, S. W. (1998). E1A signaling to p53 involves the p19(ARF) tumor suppressor. *Genes Dev.* **12**, 2434–2442.
81. Vafa, O., Wade, M., Kern, S., Beeche, M., Pandita, T. K., Hampton, G. M., and Wahl, G. M. (2002). c-myc Can induce DNA damage,

increase reactive oxygen species, and mitigate p53 function. A mechanism for oncogene-induced genetic instability. *Mol. Cell* **9**, 1031–1044.

82. Ferbeyre, G., de Stanchina, E., Querido, E., Baptiste, N., Prives, C., and Lowe, S. W. (2000). PML is induced by oncogenic *ras* and promotes premature senescence. *Genes Dev.* **14**, 2015–2027.
83. Bulavin, D. V., Demidov, O. N., Saito, S., Kauraniemi, P., Phillips, C., Amundson, S. A., Ambrosino, C., Sauter, G., Nebreda, A. R., Anderson, C. W., Kallioniemi, A., Fornace, A. J. Jr., and Appella, E. (2002). Amplification of *PPM1D* in human tumors abrogates p53 tumor-suppressor activity. *Nat. Genet.* **31**, 210–215.
84. Ferbeyre, G., de Stanchina, E., Lin, A. W., Querido, E., McCurrach, M. E., Hannon, G. J., and Lowe, S. W. (2002). Oncogenic *ras* and p53 cooperate to induce cellular senescence. *Mol. Cell. Biol.* **22**, 3497–3508.
85. Webley, K., Bond, J. A., Jones, C. J., Blaydes, J. P., Craig, A., Hupp, T., and Wynford-Thomas, D. (2000). Posttranslational modifications of p53 in replicative senescence overlapping but distinct from those induced by DNA damage. *Mol. Cell. Biol.* **20**, 2803–2808.
86. Stommel, J. M., Marchenko, N. D., Jimenez, G. S., Moll, U. M., Hope, T. J., and Wahl, G. M. (1999). A leucine-rich nuclear export signal in the p53 tetramerization domain: Regulation of subcellular localization and p53 activity by NES masking. *EMBO J.* **18**, 1660–1672.

Chapter 307

Abl in Cell Signaling

Jean Y. J. Wang
Division of Biological Sciences and the Cancer Center, University of California at San Diego, La Jolla, California

Introduction

Abl encodes a nonreceptor tyrosine kinase that is conserved in metazoan. The c-*abl* gene was discovered by virtue of its identity to the oncogene of Abelson murine leukemia virus, which expresses a Gag-Abl oncoprotein. The oncogenic potential of Abl is also activated in the BCR-ABL fusion protein of human chronic myelogenous leukemia (CML). The oncogenic function of Gag-Abl and BCR-ABL requires the Abl tyrosine kinase activity. This is best demonstrated by the clinical success in treating CML with an Abl kinase inhibitor, GLEEVEC [1]. The BCR-ABL tyrosine kinase can activate many signaling pathways in transformed cells. However, because overproduction of c-Abl does not transform cells, the signaling capabilities of BCR-ABL may not be identical to those of c-Abl. This chapter focuses on the signaling functions of mammalian c-Abl.

In the mouse and human genome, an *abl-r*elated *g*ene (*arg*) has been identified. In mice, mutation of the c-*abl* gene causes neonatal lethality and several low-penetrant phenotypes including lymphopenia and osteoporosis [2,3]. In addition, splenocytes from *abl*-knock-out mice do not respond to bacterial lipopolysaccharide [4]; and retinas from *abl*-knock-out mice are resistant to hyperoxia-induced retinopathy [5]. The *arg*-knock-out mice, by contrast, are healthy and fertile [6]. The double knock-out of *abl* and *arg* causes early embryonic lethality [6]. Thus, *abl* and *arg* have redundant functions during early embryonic development, whereas *abl* has unique functions later in development. The present understanding of *abl* and *arg* cannot provide a coherent explanation for the various knock-out phenotypes. Nevertheless, these phenotypes suggest that *abl* is required for the proper development and function of different tissues and cell types.

Functional Domains of Abl

N-Terminal Region: Kinase Function

The mammalian c-Abl protein contains modular functional domains, consistent with its role as a signal transducer (Fig. 1). The N terminus of c-Abl is variable, encoded by two alternative 5′-exons. This variable domain is followed by the Src homology (SH) domains 3, 2, and the tyrosine kinase domain. The Abl SH3 domain exerts a negative effect on the catalytic activity. This inhibitory effect is mediated by an intramolecular interaction between the SH3 and the kinase domains [7], similar to that found in the inhibited conformation of the Src tyrosine kinase [8]. The inhibited conformation of Src involves a second intramolecular interaction between its SH2 domain and a C-terminal phosphotyrosine [8]. An SH2-dependent intramolecular interaction with the kinase domain also exists in Abl. This SH2/kinase interaction does not involve an internal phosphotyrosine ligand for the SH2 domain. However, this interaction depends on myristate binding to the kinase C-lobe [7]. The Abl SH3 and SH2 domains can also engage in intermolecular interactions with other cellular proteins, which are activators, inhibitors, or substrates of the Abl kinase (see later discussion).

C-Terminal Region: Localization Cues

The C-terminal region of c-Abl specifies its subcellular localization. The mammalian c-Abl contains three nuclear localization signals (NLSs) [9]. Mutations of all three NLSs are required to prevent c-Abl from entering the nucleus [9]. The mammalian c-Abl also contains a nuclear export signal (NES) [10]. Mutation of the NES causes the nuclear accumulation of c-Abl [10]. In proliferating fibroblasts, c-Abl shuttles between the cytoplasmic and the nuclear compartments [10].

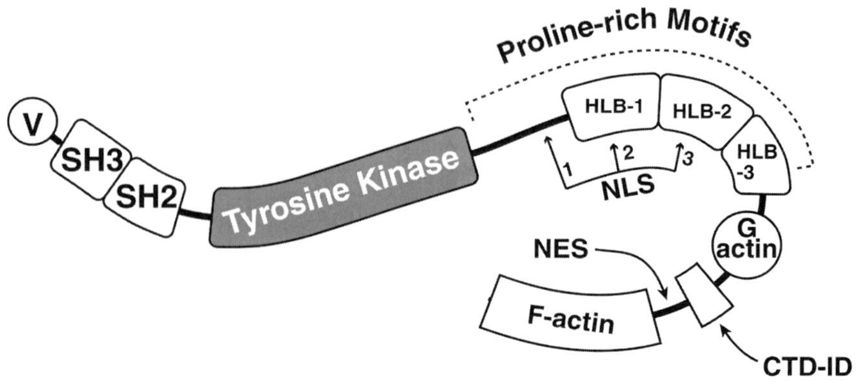

Figure 1 Functional domains of mammalian Abl. The variable N terminus (V) is encoded by two different exons. The murine type I (human Ia) exon encodes 26 amino acids. The murine type IV (human Ib) exon encodes 45 amino acids. SH3, Src homology 3; SH2, Src homology 2; NLS, nuclear localization signal; NES, nuclear export signal; HLB, high-mobility group (HMG)-like box. The three HLBs cooperatively bind to A/T-rich DNA with a distorted structure. Proline-rich motifs that bind to SH3 domains of several adaptor proteins are distributed in the C-terminal region. G-actin, binding site for actin monomers; CTD-ID, binding site for the CTD of RNA polymerase II. The CTD-ID is conserved in mammalian Abl and Arg. An F-actin binding consensus sequence is located at the very C terminus of Abl and this sequence is conserved in mammalian and *Drosophila* Abl.

In the cytoplasm, c-Abl associates with the actin cytoskeleton because it contains binding sites for F-actin and G-actin [11,12]. The C terminus of Abl contains an F-actin binding consensus sequence, which is conserved in Arg and the *Drosophila* Abl [11]. Purified mammalian c-Abl protein is an active kinase, which can be inhibited by F-actin *in vitro* [13]. This inhibition is abolished by mutation of the F-actin binding site [13]. The cytoplasmic c-Abl plays a role in actin-dependent cellular processes, including cell adhesion, cell migration, and axon guidance [14], possibly because Abl can regulate the actin dynamics [15].

In the nucleus, c-Abl associates with the chromatin [16]. The mouse and human c-Abl protein contains three HMG-like domains that bind to A/T-rich DNA with distorted structures [17,18]. The nuclear c-Abl plays a role in the regulation of transcription. This is suggested by the interaction between c-Abl and several transcription factors, including E2F-1 [19], p53 [20], p73 [21,22], c-Jun [23], RXFI [24], and the catalytic subunit of RNA polymerase II [25].

Functional Interactions between the N- and C-Terminal Regions of Abl

The C-terminal region plays an important role in the biological function of Abl kinase. A mutant mouse strain that expresses a C-terminal truncated Abl protein exhibits phenotypes similar to the Abl-null mice [26]. A functional interaction between the N- and the C-terminal regions of Abl is also supported by other observations. For example, F-actin binding to the C terminus can exert a negative effect on Abl kinase activity, suggesting an interaction between the C terminus with the kinase domain [13]. The C-terminal region of Abl also contains a binding site for the C-terminal repeated domain (CTD) of mammalian RNA polymerase II [25,27]. This CTD-binding site is required for Abl to phosphorylate RNA polymerase II [27,28], again, suggesting juxtaposition of the C-terminal region to the kinase domain.

The C-terminal region of Abl is not well conserved through evolution. Consequently, the C-terminal region of the murine Abl cannot substitute for that of the *Drosophila* Abl; however, the kinase region could be exchanged between mouse and fly [29]. Taken together, these observations support the notion that interaction between the two regions of Abl is important to the biological function of the Abl kinase.

Proteins that Interact with Abl

The multiple functional domains of Abl direct its interaction with a large number of proteins, which are listed in Table I.

SH3 Binding Proteins

The SH3 domain binds proteins with the PXXP motif [30]. Most of the Abl SH3-binding proteins are substrates of the Abl kinase (Table I). Some of these substrates were thought to be "inhibitors" of Abl kinase. Those conclusions could have been biased by the experiments, which were based on kinase assays where substrates competed with each other for Abl. Another cause for the biased conclusion might be the use of truncated proteins that could have exerted a dominant negative effect.

As mentioned earlier, the SH3 domain engages in an intramolecular interaction to impose an inactive conformation on Abl kinase [7]. Intermolecular interaction through the SH3 domain should disrupt this intramolecular interaction to cause an increase in catalytic activity. Thus, SH3 binding by a substrate may promote its phosphorylation by the Abl kinase. It is conceivable that a protein may bind to the SH3 domain without disrupting the intramolecular interaction.

Table I Proteins that Bind to the Mammalian c-Abl

Binding domain in Abl	Protein	Function	Functional interaction with Abl	Ref.
SH3	3BP-1	GAP of Rac	?	[62, 63]
	3BP-2	adaptor	?	[62]
	Abi-1, Abi-2	adaptor	Abl substrate	[64, 65]
	APP-1	unknown	Abl inhibitor	[66]
	ATM	DNA damage signaling	Abl activator	[50]
	Cables	adaptor	Abl substrate	[67]
	Cbl	adaptor	Abl substrate	[68]
	DNA-PK	DNA repair	Abl substrate	[69]
	NMDA receptor NR2D subunit	neuronal signaling	Abl inhibitor	[70]
	Mena	F-actin assembly	Abl substrate	[71]
	PAG	peroxiredoxin	Abl inhibitor	[39]
	Scramblase-1	lipid bilayer remodeling	Abl substrate	[72]
	WAV E-1	F-actin assembly	?	[42]
SH2	Ptyr-Cbl	adaptor	Abl substrate	[68]
	Ptyr-Eph receptor	axon guidance	Abl substrate	[37]
	Ptyr-p130cas	adaptor	Abl substrate	[36]
	Ptyr-CTD of RNA polymerase II	transcription	Abl substrate	[35]
Kinase	PAG	peroxiredoxin	Abl inhibitor	[39]
	RB	transcription co-repressor	Abl inhibitor	[38]
Proline-rich motifs	Abi-1, Abi-2	adaptor	Abl substrate	[64, 65]
	Crk, CrkL	adaptor	Abl substrate	[40, 73]
	Grb2	adaptor	?	[40]
	Nck	adaptor	Abl activator	[68]
	PSTPIP1	adaptor	Abl substrate	[43]

If this SH3-binding protein further stabilizes the inactive conformation, then it would qualify as an inhibitor of Abl kinase. Inhibitors of Abl should not become phosphorylated, despite their binding to the SH3 domain. Given these considerations, proteins that interact with the Abl SH3 domain can be substrates, activators, or inhibitors of Abl kinase.

The SH3 domain is deleted in the Gag-Abl oncoprotein, and it does not contribute to the oncogenic function of BCR-ABL. Mutation of the SH3 domain can convert c-Abl into an oncogenic kinase, albeit of weaker potency than Gag-Abl and BCR-ABL [31,32]. Thus, the SH3 domain is not required for the signaling functions of oncogenic Abl tyrosine kinase. Instead, the Abl SH3 domain is likely to constrain the kinase function toward specific substrates.

SH2 Binding Proteins

The SH2 domain of Abl interacts with phosphorylated tyrosine in the consensus sequence (P) YXXP [33]. The kinase domain of Abl also prefers the YXXP motif in its substrates [34]. The coordinated recognition of substrates by the kinase and the SH2 domain is important for catalysis for two reasons. First, the SH2 domain can mediate a continued association of Abl with a substrate. This is particularly important in the processive phosphorylation of the multiple tyrosines in the p130cas adaptor protein or the CTD of RNA polymerase II [35,36]. Second, the SH2 domain may increase the catalytic efficiency by extracting the product from the kinase domain [35]. The Abl SH2 domain plays an essential role in the oncogenic functions of Gag-Abl and BCR-ABL, in keeping with its positive role in catalysis. In addition, the Abl SH2 domain could serve as an adaptor to bring Abl kinase to specific signaling complexes. For example, the Abl SH2 domain binds to tyrosine phosphorylated EphB2 receptors [37]. Thus, Abl SH2-binding proteins can be substrates or recruiters of Abl kinase in signal transduction.

Kinase Domain Binding Proteins

The substrates of Abl are expected to bind the kinase domain. In addition, two protein inhibitors of the Abl kinase have been shown to bind Abl through its kinase domain. These are the retinoblastoma tumor suppressor (RB) and PAG [38,39]. RB binds to the ATP-binding lobe of the Abl kinase domain, and this interaction inhibits Abl kinase activity [38]. *In vitro*, RB can inhibit Abl, Gag-Abl, and BCR-ABL kinase [38]. *In vivo*, only the nuclear Abl is inhibited by RB, most likely because RB is an exclusively nuclear protein [38]. PAG is inducibly expressed under oxidative stress, and it catalyzes the destruction of H_2O_2. PAG binds Abl through the SH3 domain and the kinase domain to inhibit Abl kinase [39]. Thus, Abl substrates as well as Abl inhibitors can bind to the kinase domain.

Proline-Rich Motif Binding Proteins

The C-terminal region of Abl contains a series of proline-rich motifs that bind to proteins with SH3 domains [40]. A number of adaptor proteins have been shown to bind Abl through the proline-rich motifs (Table I). Most of these proteins are also phosphorylated by Abl. Thus, the proline-rich motifs appear to function as substrate binding sites. In addition to the recruitment of substrates, the C-terminal region of Abl may have an adaptor function of its own and can contribute to the assembly of signaling complexes. A kinase-defective mutant has been shown to rescue some of the phenotypes of *abl*-deficient *Drosophila* [29]. This observation supports the idea that Abl may have a scaffolding function besides being a tyrosine kinase.

Abl in Signal Transduction

Cytoplasmic Signaling Function of Abl

In the cytoplasm, Abl can transduce extracellular signals to regulate F-actin assembly (Fig. 2). This signaling function is consistent with Abl's ability to interact with G-actin and F-actin [14].

Platelet-derived growth factor (PDGF) activates Abl kinase to stimulate membrane ruffling in fibroblasts [41]. The Src tyrosine kinase and phospholipase C-α1 are involved in the activation of Abl kinase by the PDGF receptor [41,41a]. In response to PDGF, a complex of Abl/WAVE-1/PKA (cyclic AMP-dependent protein kinase) is recruited to the sites of membrane ruffling [42]. PDGF also stimulates the tyrosine phosphorylation of PSTPIP1, a scaffolding protein, and this phosphorylation event requires the Abl kinase [43]. PSTPIP1 can bring Abl to a tyrosine phosphatase (PTP-PEST), which is thought to dephosphorylate and inactivate Abl [43]. PSTPIP1 also binds WASP [44]. Both WAVE-1 and WASP can interact with the Arp2/3 complex to promote cortical F-actin polymerization [45]. Taken together, these observations suggest that Abl can transduce PDGF signal to regulate the assembly of actin filaments (Fig. 2A).

Cell adhesion to fibronectin activates Abl kinase to stimulate the formation of F-actin microspikes [15]. The Abl tyrosine kinase is inactive in cells deprived of extracellular matrix (ECM) signals [13,46]. Upon cell adhesion to fibronectin, Abl kinase is activated to stimulate the formation of F-actin microspikes [15]. Interestingly, Abl-dependent F-actin microspikes can form under conditions when the Rho family of GTPases is inhibited by the bacterial C3 toxin [15]. These observations suggest that c-Abl tyrosine kinase can stimulate F-actin assembly in response to cell adhesion signals through a pathway that is independent of the Rho/Rac/Cdc42 family of small G proteins (Fig. 2A).

Hepatocyte growth factor (HGF) can activate Abl kinase in epithelial cells [47]. Inhibition of Abl kinase by GLEEVEC causes an increase in the chemotaxis response to HGF, suggesting a role for Abl in the reduction of cell migration [47]. The negative effect of Abl kinase on cell motility is also observed in fibroblasts [48]. In transient transfection experiments, Abl is found to reduce motility by phosphorylating and inactivating Crk, which is a rate-limiting factor in the motile response [48].

Taken together, current evidence suggests that the cytoplasmic Abl kinase can transduce a variety of extracellular signals to regulate F-actin assembly [14]. The phenotypic effect of Abl kinase on F-actin appears to be dependent on the signals as well as the cell context. It is interesting to note that F-actin itself can exert a negative effect on the Abl kinase activity [13]. The reciprocal regulation between Abl and F-actin may provide a self-limiting mechanism to control the dynamics of actin filament formation (Fig. 2A).

Nuclear Signaling Function of Abl

In the nucleus, the Abl kinase can transduce cell cycle and DNA damage signals to regulate transcription. The nuclear Abl kinase is inhibited in quiescent and early G_1 cells, through the action of RB [38]. Upon RB phosphorylation, nuclear Abl kinase is released from RB and can be further activated by DNA damage. Ionizing radiation (IR) activates nuclear Abl kinase and this requires the function of ATM, which phosphorylates a SQ motif that is conserved in the kinase domain of mammalian Abl and Arg [49,50]. Cisplatin, a DNA cross-linking agent, also activates the nuclear Abl kinase and this requires the function of mismatch repair (MMR) proteins [22,51]. The DNA lesions induced by IR and cisplatin are different. Hence, nuclear Abl may be activated by different types of DNA lesions through alternative signaling pathways.

As discussed earlier, Abl phosphorylates the CTD of RNA polymerase II [25,28]. Indeed, DNA damage can increase the tyrosine phosphorylation of RNA polymerase II through c-Abl [49,52]. The CTD is composed of heptad repeats with the consensus sequence YSPTSPS [53]. Phosphorylation of the CTD repeats converts the transcription initiation complex to an elongation complex [53]. The Abl-mediated tyrosine phosphorylation of the CTD can be correlated with increased transcription elongation [27]. Moreover, the Abl protein binds A/T-rich DNA with distorted structure, which resembles a transcription bubble [17]. Thus, the nuclear Abl kinase is likely to stimulate transcription elongation during cell cycle progression and following DNA damage (Fig. 2B). In addition to RNA polymerase II, nuclear Abl kinase also phosphorylates p73, a p53-related transcription factor [54]. Activated Abl also contributes to the stabilization of p73 [22],

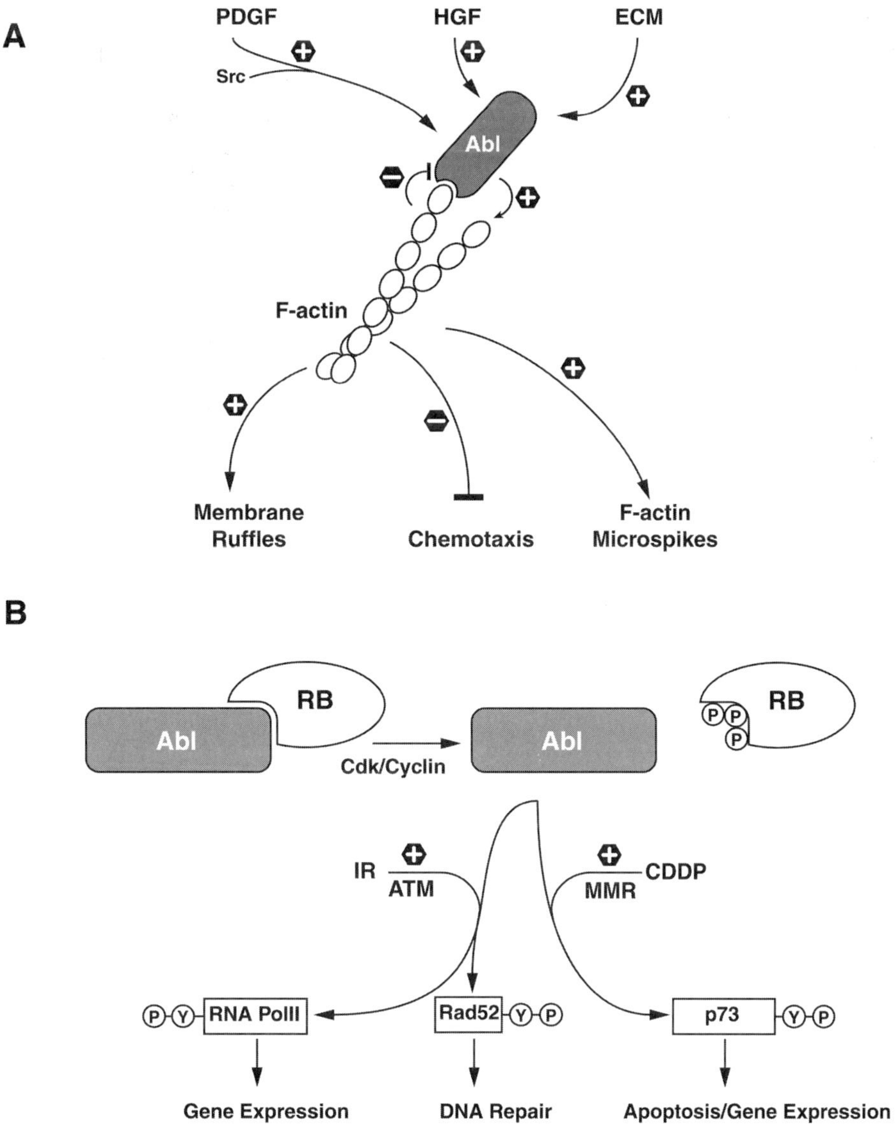

Figure 2 Signaling functions of Abl kinase. (A) Cytoplasmic Abl transduces extracellular signals to F-actin. Growth factors (PDGF, HGF) and extracellular matrix (ECM) can activate Abl kinase to regulate the assembly of actin filaments. Phenotypic outcomes resulting from each signaling pathway appear to be different. Abl stimulates membrane ruffling in response to PDGF. Abl inhibits chemotaxis in response to HGF. Abl stimulates F-actin microspikes in response to ECM. The Abl kinase activity can be inhibited by F-actin. The reciprocal regulation between Abl and F-actin may provide a self-limiting mechanism in regulating the dynamic structure of F-actin. (B) Nuclear Abl transduces cell cycle and DNA damage signals. The nuclear Abl interacts with the retinoblastoma tumor suppressor (RB), which inhibits the Abl kinase activity. Phosphorylation of RB during cell cycle progression releases Abl and allows it to be further activated by DNA damage signals. Ionizing radiation (IR) activates Abl kinase through ATM, leading to an increased tyrosine phosphorylation of RNA polymerase II and Rad52 and contributing to gene expression and possibly DNA repair. Cisplatin (CDDP) activates Abl kinase through the mismatch repair proteins (MMR), leading to the stabilization and activation of p73, a transcription factor related to the p53 tumor suppressor. The interaction between activated Abl kinase and p73 causes apoptosis.

as well as p53 [55], through mechanisms that have not been completely elucidated.

The "biological consequence" of the Abl function in the nucleus is likely to be determined by its interactions with other factors. For example, the nuclear Abl can be recruited to the E2F-1 transcription complex through RB [19]. In this capacity, Abl may stimulate transcription of E2F-regulated genes upon the inactivation of RB. Nuclear Abl has also been detected in transcription complexes that contain c-Jun, p53, p73, or RXF1. Thus, the nuclear signaling function of Abl is likely to be varied and determined by its recruitment to different protein complexes (Fig. 2B).

One of the biological consequences resulting from the activation of nuclear Abl kinase is apoptosis [56]. Inducers of

DNA damage, for example, cisplatin and doxorubicin, can activate nuclear Abl kinase, which in turn activates the pro-apoptotic function of p73 [22,57]. The Abl kinase is an essential upstream activator of p73, because ectopic expression of p73 does induce apoptosis in Abl-null cells [21,22,57]. The reverse is also true, in that the pro-apoptotic function of nuclear Abl kinase is greatly compromised in p73-null cells [74]. Thus, the nuclear Abl kinase can transduce DNA damage signal to p73 to activate apoptosis (Fig. 2B).

The BCR-ABL tyrosine kinase has the ability to inhibit apoptosis. The discrepancy between the anti-apoptotic function of BCR-ABL and the pro-apoptotic function of c-Abl has been resolved [58]. The BCR-ABL tyrosine kinase is an exclusively cytoplasmic protein. Its signaling functions in the cytoplasm culminate in the inhibition of apoptosis. The BCR-ABL protein can enter the nucleus when its tyrosine kinase is inhibited by GLEEVEC [58]. The nuclear BCR-ABL can be trapped by a second drug, leptomycin B, which inhibits the nuclear export of BCR-ABL [58]. Upon removal of GLEEVEC, the nuclear BCR-ABL regains activity to cause apoptosis [58]. These results demonstrate that the pro-apoptotic function of Abl kinase is dependent on it being in the nucleus of a cell. Moreover, these results show that BCR-ABL retains the pro-apoptotic function of Abl. Hence, conversion of Abl into an oncoprotein requires not only the activation of its tyrosine kinase but also the inhibition of its nuclear import to abolish the pro-apoptotic function.

The nuclear Abl can transduce death-inducing signals other than those generated by DNA damage [56]. For example, Abl tyrosine kinase contributes to apoptosis induced by tumor necrosis factor [59]. Abl tyrosine kinase may also contribute to apoptosis caused by oxidative stress [60]. The signaling pathways linking these death inducers to the activation of nuclear Abl kinase are under investigation.

While the nuclear Abl kinase can promote apoptosis, this may not be the only biological consequence of Abl activation. For example, the maximal activation of nuclear Abl tyrosine kinase is achieved by IR at a dose (2 Gy) that does not induce apoptosis. Thus, activation of the nuclear Abl kinase is necessary but not sufficient to cause cell death. The Abl kinase has been shown to phosphorylate Rad52, suggesting a possible role for Abl in DNA repair [61]. Nuclear Abl kinase can also phosphorylate MyoD to inhibit myogeniz differentiation in response to DNA damage [75]. Hence, the signaling functions of the nuclear Abl kinase are likely to be modulated by the nature of the signal and the cell context.

Future Prospects

A framework for the signaling functions of Abl has emerged from (1) the delineation of its modular domains and (2) the identification of cellular phenotypes associated with Abl kinase activation. The cytoplasmic Abl transduces extracellular signals to regulate F-actin. The nuclear Abl transduces cell cycle and DNA damage signals to regulate transcription. With this framework, it is now feasible to further elucidate the mechanistic steps in Abl-dependent signaling pathways.

The accumulated knowledge on Abl has not provided a unifying mechanism for the biological effects associated with Abl kinase activation. Instead, the current evidence suggests a highly context-dependent biological function for Abl kinase. The biological effect of Abl is strongly modulated by its association with other signaling molecules. For example, cytoplasmic Abl kinase stimulates membrane ruffling in response to PDGF, but it stimulates F-actin microspikes in response to fibronectin [14]. Although both phenotypic outcomes may involve the ability of Abl to regulate F-actin assembly, the pathways linking Abl to ruffles or microspikes might not be identical. Likewise, nuclear Abl kinase can be recruited to several different transcription complexes, which are likely to influence the biological output from activated Abl kinase.

Because the signaling functions of Abl may be modulated by the cell context, it is important to study the Abl function in a defined biological system. In addition, a biological effect of Abl found in one system might not be easily extrapolated or generalized to another experimental system. Despite these possible complexities, it is clear that Abl can regulate F-actin and transcription. The precise mechanisms by which these functions of Abl are utilized under specific biological conditions will await further investigation.

Acknowledgment

Work in the author's lab has been supported by grants from the National Institutes of Health (CA43054, CA58320 and HL57900). The author thanks Dr. Jiangyu Zhu for the artwork.

References

1. Druker, B. J. *et al.* (2001). Efficacy and safety of a specific inhibitor of the BCR-ABL tyrosine kinase in chronic myeloid leukemia. *N. Engl. J. Med.* **344**(14), 1031–1037.
2. Tybulewicz, V. L. J. *et al.* (1991). Neonatal lethality and lymphopenia in mice with a homozygous disruption of the *c-abl* proto-oncogene. *Cell* **65**, 1153–1163.
3. Li, B. J. *et al.* (2000). Mice deficient in Abl are osteoporotic and have defects in osteoblast maturation. *Nat. Genet.* **24**(3), 304–308.
4. Hardin, J. D. *et al.* (1996). Abnormal peripheral lymphocyte function in c-abl mutant mice. *Cell Immunol.* **172**(1), 100–107.
5. Nunes, I. *et al.* (2001). c-abl Is required for the development of hyperoxia-induced retinopathy. *J. Exp. Med.* **193**(12), 1383–1391.
6. Koleske, A. J. *et al.* (1998). Essential roles for the Abl and Arg tyrosine kinases in neurulation. *Neuron* **21**(6), 1259–1272.
7. Nagar, B., *et al.* (2003). Structural basis for the autoinhibition of c-Abl tyrosine kinase. *Cell* **112**(6), 859–871.
8. Xu, W., Harrison, S., and Eck, M. (1997). Three-dimensional structure of the tyrosine kinase c-Src. *Nature* **385**, 595–602.
9. Wen, S.-T., Jackson, P. K., and Van Etten, R. A. (1996). The cytostatic function of c-Abl is controlled by multiple nuclear localization signals and requires the *p53* and *Rb* tumor suppressor gene products. *EMBO J.* **15**(7), 1583–1595.
10. Taagepera, S. *et al.* (1998). Nuclear-cytoplasmic shuttling of c-Abl tyrosine kinase. *Pro. Natl. Acad. Sci. USA,* **98**, 7457–7462.
11. McWhirter, J. R. and Wang, J. Y. J. (1993). An actin-binding function contributes to transformation by the Bcr-Abl oncoprotein of Philadelphia chromosome-positive human leukemias. *EMBO J.* **12**(4), 1533–1546.

12. Van Etten, R. A. *et al.* (1994). The COOH terminus of the c-Abl tyrosine kinase contains distinct F- and G-actin binding domains with bundling activity. *J. Cell Biol.* **124**(3), 325–340.
13. Woodring, P. J., Hunter, T. and Wang, J. Y. J. (2001). Inhibition of c-Abl tyrosine kinase activity by filamentous actin. *J. Biol. Chem.* **276**(29), 27104–27110.
14. Woodring, P. J., Hunter, T., and Wang, J. Y. J. (2003). Regulation of F-actin-dependent processes by the Abl family of tyrosine kinases. *J. Cell Sci.* **116**(13), 2613–2626.
15. Woodring, P. J. *et al.* (2002). Modulation of the F-actin cytoskeleton by c-Abl tyrosine kinase in cell spreading and neurite extension. *J. Cell Biol.* **156**(5), 879–892.
16. Kipreos, E. T. and Wang, J. Y. J. (1992). Cell cycle-regulated binding of c-Abl tyrosine kinase to DNA. *Science* **256**, 382–385.
17. Miao, Y. and Wang, J. Y. J. (1996). Binding of A/T-rich DNA by three high mobility group like domains in c-Abl tyrosine kinase. *J. Biol. Chem.* **271**, 22823–22830.
18. David-Cordonnier, M.-H., *et al.* (1998). The DNA binding domain of the human c-Abl tyrosine kinase preferentially binds to DNA sequences containing an AAC motif and to distorted DNA structures. *Biochemistry* **37**, 6065–6076.
19. Welch, P. and Wang, J. Y. J. (1995). Disruption of retinoblastoma protein function by coexpression of its C pocket fragment. *Genes Dev.* **9**, 31–46.
20. Goga, A. *et al.* (1995). p53 dependent growth suppression by the c-Abl nuclear tyrosine kinase. *Oncogene* **11**(4), 791–799.
21. Agami, R. *et al.* (1999). Interaction of c-Abl and p73 alpha and their collaboration to induce apoptosis. *Nature* **399**(6738), 809–813.
22. Gong, J. G. *et al.* (1999). The tyrosine kinase c-Abl regulates p73 in apoptotic response to cisplatin-induced DNA damage. *Nature* **399**(6738), 806–809.
23. Barila, D. *et al.* (2000). A nuclear tyrosine phosphorylation circuit: c-Jun as an activator and substrate of c-Abl and JNK. *EMBO J.* **19**(2), 273–281.
24. Agami, R. and Shaul, Y. (1998). The kinase activity of c-Abl but not v-Abl is potentiated by direct interaction with RFXI, a protein that binds the enhancers of several viruses and cell-cycle regulated genes. *Oncogene* **16**(14), 1779–1788.
25. Baskaran, R., Chiang, G. G., and Wang, J. Y. J. (1996). Identification of a binding site in c-Abl tyrosine kinase for the C-terminal repeated domain of RNA polymerase II. *Mol. Cell. Biol.* **16**, 3361–3369.
26. Schwartzberg, P. L. *et al.* (1991). Mice homozygous for the abl^{m1} mutation show poor viability and depletion of selected B and T cell populations. *Cell* **65**, 1165–1175.
27. Baskaran, R., Escobar, S. R., and Wang, J. Y. J. (1999). Nuclear c-Abl is a COOH-terminal repeated domain (CTD)-tyrosine kinase-specific for the mammalian RNA polymerase II: Possible role in transcription elongation. *Cell Growth Differentiation* **10**(6), 387–396.
28. Baskaran, R., Dahmus, M., and Wang, J. Y. J. (1993). Tyrosine phosphorylation of mammalian RNA polymerase II carboxyl-terminal domain. *Proc. Natl. Acad. Sci. USA* **90**, 11167–11171.
29. Henkemeyer, M. *et al.* (1990). A novel tyrosine kinase-independent function of Drosophila *abl* correlates with proper subcellular localization. *Cell* **63**, 949–960.
30. Ren, R. *et al.* (1993). Identification of a ten-amino acid proline-rich SH3 binding site. *Science* **259**, 1157–1161.
31. Franz, W., Berger, P. and Wang, J. Y. J. (1989). Deletion of an N-terminal regulatory domain of the c-*abl* tyrosine kinase activates its oncogenic potential. *EMBO J.* **8**(1), 137–147.
32. Jackson, P. and Baltimore, D. (1989). N-terminal mutations activate the leukemogenic potential of the myristoylated form of *c-Abl. EMBO J.* **8**(2), 449–456.
33. Songyang, Z. *et al.* (1993). SH2 domains recognize specific phosphopeptide sequences. *Cell* **72**, 767–778.
34. Songyang, Z. *et al.* (1994). Use of an oriented peptide library to determine the optimal substrates of protein kinases. *Curr. Biol.* **4**(11), 973–982.
35. Duyster, J., Baskaran, R., and Wang, J. Y. J. (1995). Src homology 2 domain as a specificity determinant in the c-Abl mediated tyrosine phosphorylation of the RNA polymerase II carboxyl-terminal repeated domain. *Proc. Natl. Acad. Sci. USA* **92**, 1555–1559.
36. Mayer, B. J., Hirai, H., and Sakai, R. (1995). Evidence that SH2 domains promote processive phosphorylation by protein-tyrosine kinases. *Curr. Biol.* **5**(3), 296–305.
37. Yu, H. H. *et al.* (2001). Multiple signaling interactions of Abl and Arg kinases with the EphB2 receptor. *Oncogene* **20**(30), 3995–4006.
38. Welch, P. J. and Wang, J. Y. J. (1993). A C-terminal protein-binding domain in the retinoblastoma protein regulates nuclear c-Abl tyrosine kinase in the cell cycle. *Cell* **75**, 779–790.
39. Wen, S.-T. and Etten, R. A. V. (1997). The PAG gene product, a stress-induced protein with antioxidant properties, is an Abl SH3-binding protein and a physiological inhibitor of c-bl tyrosine kinase activity. *Genes Dev.* **11**(19), 2456–2467.
40. Ren, R., Ye, Z.-S., and Baltimore, D. (1994). Abl protein-tyrosine kinase selects the Crk adapter as a substrate using SH3-binding sites. *Genes Dev.* **8**, 783–795.
41. Plattner, R. *et al.* (1999). c-Abl is activated by growth factors and Src family kinases and has a role in the cellular response to PDGF. *Genes Dev.* **13**(18), 2400–2411.
41a. Plattner, R. *et al.* (2003). A new link between the c-Abl tyrosine kinase and phosphoinositide signaling through PLC-γ1. *Nat. Cell Biol.* **5**(4), 309–319.
42. Westphal, R. S. *et al.* (2000). Scar/WAVE-1, a Wiskott-Aldrich syndrome protein, assembles an actin-associated multi-kinase scaffold. *EMBO J.* **19**(17), 4589–4600.
43. Cong, F. *et al.* (2000). Cytoskeletal protein PSTPIP1 directs the PEST-type protein tyrosine phosphatase to the c-Abl kinase to mediate Abl dephosphorylation. *Mol. Cell* **6**(6), 1413–1423.
44. Cote, J. F. *et al.* (2002). PSTPIP is a substrate of PTP-PEST and serves as a scaffold guiding PTP-PEST toward a specific dephosphorylation of WASP. *J. Biol. Chem.* **277**(4), 2973–2986.
45. Welch, M. D. (1999). The world according to Arp: Regulation of actin nucleation by the Arp2/3 complex. *Trends Cell Biol.* **9**(11), 423–427.
46. Lewis, J. *et al.* (1996). Integrin regulation of c-Abl tyrosine kinase activity and cytoplasmic-nuclear transport. *Proc. Natl. Acad. Sci. USA* **93**, 15174–15179.
47. Frasca, F. *et al.* (2001). Tyrosine kinase inhibitor STI571 enhances thyroid cancer cell motile response to Hepatocyte Growth Factor. *Oncogene* **20**(29), 3845–3856.
48. Kain, K. H. and Klemke, R. L. (2001). Inhibition of cell migration by Abl family tyrosine kinases through uncoupling of Crk-CAS complexes. *J. Biol. Chem.* **276**(19), 16185–16192.
49. Baskaran, R. *et al.* (1997). Ataxia telangiectasia mutant protein activates c-Abl tyrosine kinase in response to ionizing radiation. *Nature* **387**(6632), 516–519.
50. Shafman, T. *et al.* (1997). Interaction between ATM protein and c-Abl in response to DNA damage. *Nature* **387**(6632), 520–523.
51. Nehme, A. *et al.* (1997). Differential Induction of c-Jun NH2-terminal kinase and c-Abl kinase in DNA mismatch repair-proficient and deficient cells exposed to cisplatin. *Cancer Res.* **57**(15), 3253–3257.
52. Liu, Z. *et al.* (1996). Three distinct signalling responses by murine fibroblasts to genotoxic stress. *Nature* **384**, 273–276.
53. Dahmus, M. E. (1995). Phosphorylation of the C-terminal domain of RNA polymerase II. *Biochim. Biophys. Acta* **1261**, 171–182.
54. Levrero, M. *et al.* (1999). Structure, function and regulation of p63 and p73. *Cell Death Differentiation* **6**, 1146–1153.
55. Sionov, R. V. *et al.* (2001). c-Abl regulates p53 levels under normal and stress conditions by preventing its nuclear export and ubiquitination. *Mol. Cell Biol.* **21**(17), 5869–5878.
56. Wang, J. Y. J. (2000). Regulation of cell death by the Abl tyrosine kinase. *Oncogene* **19**(49), 5643–5650.
57. Costanzo, A. *et al.* (2002). DNA damage-dependent acetylation of p73 dictates the selective activation of apoptotic target genes. *Mol. Cell* **9**(1), 175–186.
58. Vigneri, P. and Wang, J. Y. J. (2001). Induction of apoptosis in chronic myelogenous leukemia cells through nuclear entrapment of BCR-ABL tyrosine kinase. *Nat. Med.* **7**(2), 228–234.

59. Dan, S. *et al.* (1999). Activation of c-Abl tyrosine kinase requires caspase activation and is not involved in JNK/SAPK activation during apoptosis of human monocytic leukemia U937 cells. *Oncogene* **18**, 1277–1283.
60. Sun, X. G. *et al.* (2000). Interaction between protein kinase C delta and the c-Abl tyrosine kinase in the cellular response to oxidative stress. *J. Biol. Chem.* **275**(11), 7470–7473.
61. Chen, G. *et al.* (1999). Radiation-induced assembly of Rad51 and Rad52 recombination complex requires ATM and c-Abl. *J. Biol. Chem.* **274**(18), 12748–12752.
62. Cicchetti, P. *et al.* (1992). Identification of a protein that binds to the SH3 region of Abl and is similar to Bcr and GAP-rho. *Science* **257**, 803–806.
63. Cicchetti, P. *et al.* (1995). 3BP-1, an SH3 domain binding protein, has GAP activity for Rac and inhibits growth factor-induced membrane ruffling in fibroblasts. *EMBO J.* **14**(13), 3127–3135.
64. Dai, Z. and Pendergast, A. M. (1995). Abi-2, a novel SH3-containing protein interacts with the c-Abl tyrosine kinase and modulates c-Abl transforming activity. *Genes Dev.* **9**, 2569–2582.
65. Shi, Y., Alin, K., and Goff, S. (1995). Abl-interactor-1, a novel SH3-protein binding to the carboxy-terminal portion of the v-Abl protein, suppresses v-*abl* transforming activity. *Genes Dev.* **9**, 2583–2597.
66. Zhu, J. and Shore, S. K. (1996). c-ABL tyrosine kinase activity is regulated by association with a novel SH3-domain-binding protein. *Mol. Cell Biol.* **16**(12), 7054–7062.
67. Zukerberg, L. R. *et al.* (2000). Cables links Cdk5 and c-Abl and facilitates Cdk5 tyrosine phosphorylation, kinase upregulation, and neurite outgrowth. *Neuron* **26**(3), 633–646.
68. Miyoshi-Akiyama, T. *et al.* (2001). Regulation of Cbl phosphorylation by the Abl tyrosine kinase and the Nck SH2/SH3 adaptor. *Oncogene* **20**(30), 4058–4069.
69. Kharbanda, S. *et al.* (1997). Functional interaction between DNA-PK and c-Abl in response to DNA damage. *Nature* **386**, 732–735.
70. Glover, R. T. *et al.* (2000). Interaction of the N-methyl-D-aspartic acid receptor NR2D subunit with the c-Abl tyrosine kinase. *J. Biol. Chem.* **275**(17), 12725–12729.
71. Gertler, F. B. *et al.* (1996). Mena, a relative of VASP and Drosophila Enabled, is implicated in the control of microfilament dynamics. *Cell* **87**(2), 227–239.
72. Sun, J. *et al.* (2001). c-Abl tyrosine kinase binds and phosphorylates phospholipid scramblase 1. *J. Biol. Chem.* **276**(31), 28984–28990.
73. Feller, S., Knudsen, B., and Hanafusa, H. (1994). c-Abl kinase regulates the protein binding activity of c-Crk. *EMBO J.* **13**(10), 2341–2351.
74. Vella, V. *et al.* (2003). Exclusion of c-Abl from the nucleus restrains the p73 tumor suppression function. *J. Biol. Chem.*, In Press (e-pub, April 24, 2003).
75. Puri, P. L. *et al.* (2002). A myogenic differentiation checkpoint activated by genotoxiz stress. *Nat, Genet.* **32**(4), 585–593.

CHAPTER 308

Radiation-Induced Cytoplasmic Signaling

Christine Blattner and Peter Herrlich
Institute of Genetics and Toxicology, Forschungszentrum Karlsruhe, Karlsruhe, Germany

Introduction

Radiation initiates a plethora of cellular alterations including inflammation, skin reddening, and cancer and also activates several intrinsic cellular programs such as cell cycle arrest, DNA repair, and apoptosis. Many of these changes result from radiation-induced chromosomal damage, while others are a result of lesions in other cellular macromolecules that link into signal transduction pathways. A frequent consequence is the activation of mitogen activated protein (MAP) kinases resulting in phosphorylation of their substrates, for example, transcription factors. In this chapter we summarize our current knowledge of the buildup of the signaling network and of radiation-induced cytoplasmic signaling in mammalian cells.

Cytoplasmic Signaling Network

All living cells require extracellular stimuli for survival, for apoptosis, for differentiation—in essence for any cellular achievement, for any change in gene expression. Physiologically relevant extracellular cues are perceived primarily by cell surface receptors, each of which feeds into signal transduction chains to the nucleus. One class of these signaling chains involves a series of successively acting protein kinases, the most distal of which in the direction of signal flow and the closest to the nucleus are the MAP-kinases. (Fig. 1 shows the principle of a protein kinase cascade.) Some of the upstream protein kinases as well as the MAP-kinases can shuttle between cytoplasm and the nucleus (reviewed in [1]) and phosphorylate transcription and splice regulatory factors directly [2–4] as well as factors that control RNA and protein turnover [5–8]. Some target molecules are addressed by MAP-kinases through still another effector protein kinase (e.g. Msk, Rsk) [9–11].

The linear signal transduction cascades as diagrammed in Fig. 1 are simplified abstractions of reality. In fact, signaling pathways form an elaborate communication network with an enormous degree of cross-talk between pathways (reviewed in [12]). In addition, signaling components are likely ordered in preformed arrays rather than being freely diffusible. To this end, the protein kinases carry, in addition to their catalytic domain, a specific interaction surface. Such interaction domains have been characterized, for instance, for the binding of JNK to its transcription factor substrate c-Jun [13] and for the interaction between MEK and ERK [14]. In addition, specialized scaffolding proteins may contribute to establish preformed arrays of protein kinases [15–17].

A large number of cell surface receptors activate a given MAP-kinase pathway, seemingly to the same extent. How then is specificity established? One possibility is that the combination of several signaling pathways determines the exact phenotypic outcome provided each receptor induces a specific pathway in addition to the common one. Or the duration of signaling, which could be organized by receptor-specific negative regulators such as protein tyrosine phosphatases [18], may be the determining parameter. An interesting aspect concerns the specific adaptation of the signaling components to the receptor. Based on studies of the receptor tyrosine kinase Met, an additional class of factors has been proposed to create specific interactions: A closely associated coreceptor forms a link to the actin cytoskeleton, which is absolutely

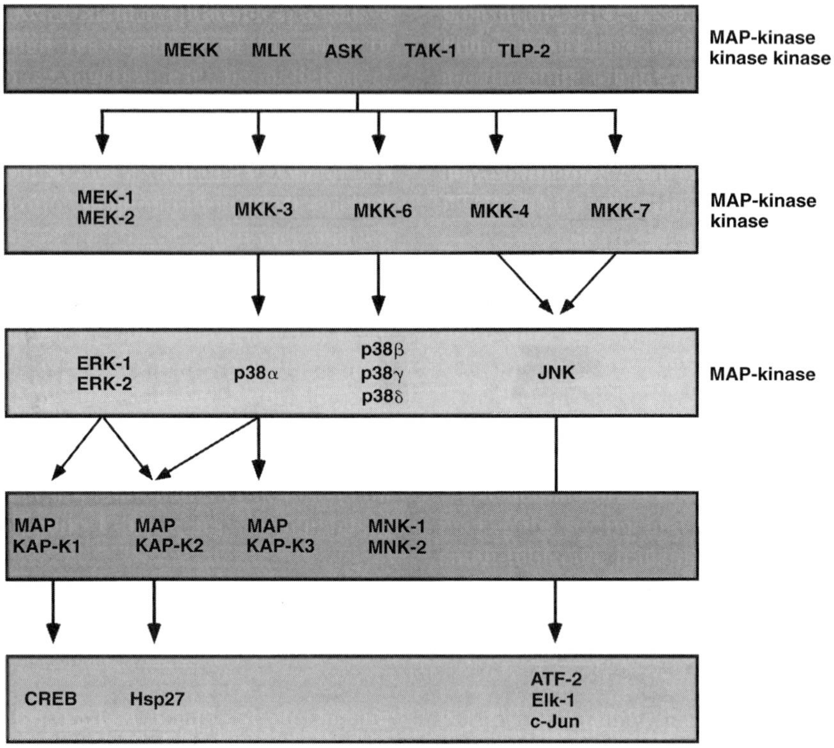

Figure 1 MAP-kinase signaling cascades. MAP-kinases are activated by dual phosphorylation caused by members of the MAP-kinase kinase group, which are in turn activated by phosphorylation mediated by members of the group of MAP-kinase kinase kinases. On activation, MAP-kinases phosphorylate transcription factors either directly or through other effector protein kinases.

required for any signal transfer from Met to the level of the MAP-kinase kinase MEK [19].

As touched on earlier, stimulation of a signaling pathway must be terminated after a "reasonable" period of time. On all levels, means of counterregulation are built in. Protein kinases are associated with protein phosphatases. The stimulation of a pathway often leads to synthesis or activation of a negative terminating factor, for example, of a phosphatase such as MKP-1 [20–22] or of the inhibitor of NFκB, IκB [23]. Inactivation of a phosphatase derepresses signal transfer and enhances transcription [24], suggesting that there is constant firing through signaling components and that specific negative regulators counteract the firing in order to keep the noise level low. Also interference with the actin cytoskeleton, which occurs physiologically during mitosis, disrupts signal transfer from the plasma membrane to the nucleus [19,25–27].

Redox Sensitivity and Metal Toxicity: Toxic Agents Activate Signaling Pathways

Obviously individual cells as well as cells in the multicellular organism are not only exposed to physiological stimuli but also to various adverse agents, ranging from heat and radiation through hypoxia and osmotic shock to many types of toxic chemicals. Low to moderate doses induce elaborate cellular responses—known as UV response, heat-shock response, osmotic shock response, and so on—all of which involve changes in gene expression.

To trigger an active response, the toxic agent is required to find access to the signaling process, which addresses transcription, splicing, or else. Statistically it is by far more likely that a toxic agent destroys the function of a molecule rather than conferring on it a gain of function. The structure and function of most cellular macromolecules depends on charge (and therefore on pH and reductive microenvironment). Moreover, the macromolecules carry reactive chemical groups relevant for function. In particular, redox-sensitive structures can be addressed by adverse agents leading to, for instance, inactivation of an enzyme. A direct gain of function can, in the rare case, be achieved by an induced conformational change or by the generation of a cross-link of two proteins leading to an increase in their activity [28]. More frequently, an indirect gain of function can result from inactivation by the toxic agent of a negative control element and thus derepression of, for example, some step in the signal transduction process.

Indeed, adverse agents of a very different kind have been reported to stimulate signaling in the cell. Here we first briefly review the evidence showing that individual components of the intracellular communication network are activated in response to an insult. This does not indicate that the agent acts directly on the component in question, but rather that a pathway has been addressed into which the component is integrated.

We then ask where the signaling comes from and what the primary target of interaction with the adverse agent is.

Activation of Signaling Components

Adverse agents alter cellular parameters, such as proliferation and gene expression, and induce apoptosis, differentiation, and senescence. Alterations in the program of gene expression are reflected by, for instance, induction of *c-fos*, *c-jun*, uPA, collagenase, metallothionein, and many other genes. These changes in gene expression result from agent-induced signaling activity, which in turn enhances the function of several transcription factors. Indeed numerous components of the signaling network have been found to be activated on cellular contact with adverse agents. Both UV and IR cause activation of IκB kinase and Akt/PKB, or Ras and Raf kinase [29–33]. The predominant link between signal flow and transcriptional activity is provided by the large group of MAP-kinases.

MAP-kinases are parts of five distinct signaling cascades including the mitogenic ERK-1/2 cascade, the stress-activated JNK and p38 cascades, and the ERK-3 and ERK-5 cascades (Fig. 1). These cascades are typically arranged in a three-kinase architecture comprising a MAP-kinase, an activator of the MAP-kinase (MAP-kinase kinase), and an activator of the MAP-kinase activator (MAP-kinase kinase kinase). MAP-kinase cascades have been conserved through evolution and similar cascades are found in yeast, *Drosophila*, and plants [34].

Activation by adverse agents has been reported for the ERK 1/2 branch of the MAP-kinases and for p38 and JNK. Different members of the MAP-kinase family are activated to a different extent. MAP-kinase signaling is habitually transient in nature, because neutralizing principles are addressed by the same signal flow. Indeed, ERK-kinase specific phosphatase (MKP) is rapidly induced following stimuli that provoke MAP kinase signaling [20–22,35]. Activation of JNK (and p38) is counteracted by, for example, heat-shock protein 72 [36] and JIP-1 (JNK-interacting protein). The latter retains JNK in the cytoplasm, hence interfering with subsequent JNK activity [37].

MAP-kinases are normally activated by numerous mitogenic stimuli, and one of the as yet unresolved mysteries of MAP-kinase signaling is that of where the specificity of the response comes from after mitogenic stimuli and cellular stress. One explanation would be that degree or duration of activation accounts for the specificity of the response. p38/JNKs are usually efficiently and stably stimulated in response to cellular stress but only weakly and transiently in response to growth factors such as EGF, PDGF, FGF, and phorbol esters. In contrast, ERKs are heavily activated in response to mitogenic stimuli and marginally after UV or ionizing irradiation. A more convincing, but not well explored idea, as introduced earlier, postulates that the cross-talk between pathways and the actual integration of all of the stimuli hitting cells at a given point in time defines the phenotypic outcome.

Primary Radiation Targets: DNA Damage versus Cytoplasmic Signaling

The previous paragraphs summarized evidence for the activation of signaling chains by adverse agents. For that discussion, we ignored the mechanism of activation. This mechanism is now considered. What are the primary targets that "emit" a signal and lead to activation of, for instance, JNK or ERK?

Radiation injures DNA, RNA, proteins, and other biomolecules. Lesions result either from energy absorption [for example, DNA photoproducts; UVB (280–320 nm) and UVC (200–280 nm) are absorbed by nucleic acid and protein] or from oxidation by free radicals, which are formed by radiation of various wavelengths including UVA (320–400 nm), UVB, UVC, and IR. The finding that treatment of cells with radical scavengers prior to stimulation with adverse agents prevents subsequent signaling and gene transcription provides indirect evidence for the generation of reactive oxygen intermediates [38,39]. Reactive oxygen intermediates such as superoxide anion radicals (O^{2-}), singlet oxygen ($^1O_2$), and hydroxyl radicals (OH^-) and other adverse agents may damage several types of macromolecules, which could be relevant for the subsequent activation of signaling and gene expression. Ionizing radiation may cause damage to DNA and inactivate numerous proteins in the cytoplasm at the same time. With increasing dose, certain targets may be damaged earlier than others. For instance, arsenite has been found to activate AP-1 by inhibiting JNK-phosphatase [40], although it is clear that arsenite must interact with numerous proteins.

With the notion in mind that adverse agents target numerous macromolecules at the same time, dissecting the pathways and defining the primary signal-generating targets is not a trivial task. To find the primary target molecules relevant for signal transduction, "retrograde" analyses were performed. Interestingly, both nuclear DNA damage as well as oxidative lesions of cytoplasmic macromolecules have been found to activate preexisting signaling chains.

Activation of JNK appears to reveal the complexity of finding the source of radiation-induced signaling. There is good evidence that JNK activation is triggered by prior membranal receptor tyrosine kinase activation in that it works in cytoblasts [41] and is addressed by the TNFR activated by UV [42]. Others have, however, reported on JNK activation by damaged DNA [43]. In addition, JNK is activated during the ribotoxic stress response. This illustrate the point that adverse agents have usually targets.

DNA Damage-Dependent Signaling

In recent years it turned out that nuclear and cytoplasmic signaling cascades are not completely separated branches of the cellular signaling network. Signals go in and out of the nucleus and influence the signal flow in both compartments after exposure to adverse agents. Transcription of collagenase, metallothionein, and plasminogen activator is enhanced in response to UV-DNA lesions in transcribed regions of the genome [44,45]. If these are repaired proficiently, higher doses of UV are required to obtain the same response, suggesting strongly that the lesions generate signals to the transcription factors acting on these genes. The relevant factors are probably members of the AP-1 family. The DNA-lesion-dependent signals thus likely feed into MAP-kinase pathways. How the DNA lesion is recognized and how a signal is generated is processes linked to not precisely understood. However, as lesions in transcribed units provoke collagenase and metallothionein induction, one might speculate that signaling originates from stalled RNA-polymerase [44].

As mentioned earlier, it has been reported that JNK activation comprises a DNA-damage-sensitive component in that irradiated or sonicated DNA added to lysates of UV-treated cells stimulated JNK activity in a dose-dependent manner [43]. An alternative mechanism has been proposed by the finding that JNK can be activated by binding of the protein GADD45 to its NH_2-terminal domain [46]. The synthesis of GADD45 is induced in response to DNA damage and GADD45 may mediate sustained JNK activation. A third example is provided by the activation of NFκB in response to IR. Activated IKK phosphorylates IκB in the cytoplasm. This activity appears to depend on DNA double-strand breaks and ATM-kinase whose activity is regulated by double-strand breaks [47].

Activation of Receptor Tyrosine Kinases

One of the earliest measurable responses of mammalian cells to UV irradiation of all wavelengths is tyrosine phosphorylation of membrane-bound growth factor receptors [24,42,48,49]. The autophosphorylation of growth factor receptors is ligand independent and is an essential condition for transcription of the downstream genes *c-fos* and *c-jun,* which is completely ablated on inhibition of tyrosine kinase activity [24,49]. Activators of the receptors for EGF, IL-1, and bFGF account for most of the UV response in HeLa cells [49]. UV also activates the receptor for tumor necrosis factor (TNF) in a ligand-independent manner and to an extent comparable to that induced by the natural ligand TNF-α [50]. Fas (CD95) can serve as the third example of ligand-independent receptor activation by UV [51] or by the organometal TBT (Own unpublished data).

Growth factor receptors are usually phosphorylated in response to growth factor binding. The interaction with their cognate ligand animates the receptors to adopt a phosphorylated state achieved by the mutual action of two neighboring receptor molecules exhibiting intrinsic receptor tyrosine kinase activity. UV-mediated receptor phosphorylation occurs in the absence of ligand [49], yet is the result of protein tyrosine phosphatase (PTP) inhibition. PTPs are particularly sensitive to oxidative stress because each carries a conserved cysteine in its catalytic center, which is prone to oxidation [18,52]. The UV-induced inhibition of membrane-associated PTP activity [24] and of individual candidate PTPs (SHP-1, RPTPα, RPTPσ, and DEP-1; [48,53]) could account for the increase in receptor phosphorylation after UV irradiation provided that the intrinsic receptor tyrosine kinase is constitutively active. PTP inhibition is also achieved by treatments of cells in metal toxins, for example, vanadate [54], arsenite [40], or the organometal TBT (Unpublished data). Although plausible, PTP inactivation must not be the only mechanism responsible for RTK activation. Inactivation of other negatively controlling components and possibly a direct stabilization of receptor oligomers are conceivable options.

Phosphorylation of growth factor receptors upon UV irradiation seems to be sufficient to render them fully functional. Proteins such as Grb-2, phospholipase C-γ and Shc coprecipitate with the receptors after UV irradiation, similar to their response after growth factor stimulation [24,55,56]. Translocation of Grb-2/Sos complexes from the cytoplasm to activated growth factor receptors at the cell periphery leads to the activation of Ha-ras. UV and IR exposure increases the low basal activity of Ha-ras and enhance its association with Grb2 and Sos, followed by activation of Raf-1 and Raf-2 [33,38,57].

Ribotoxic Stress Response

The ribotoxic stress response is a cellular reaction to toxic agents that is initiated by modification or damage to the functional center of the 28S ribosomal RNA. This response is characterized by three features: (1) interference with translational elongation, (2) interference with enzymatic activities of the 28S ribosomal RNA (aminoacyl-tRNA, peptidyl transfer, or ribosomal translocation), and (3) activation of stress kinases [58]. UV irradiation provokes all of these features, presumably elicited by UV-mediated damage to the 28S ribosomal RNA. Lesions in the ribosomal RNA after UV irradiation resemble UV-induced photoproducts in the DNA, and they likewise involve adjacent pyrimidine nucleotides [59]. UV-induced inhibition of translational elongation results in activation of JNK-1 and p38 [60].

Other Signaling-Initiating Principles

Signaling initiated at the cell periphery can also be based on the activation of voltage-gated K^+ ion channels. Changes in K^+ channel activity are normally used to stabilize the membrane potential and maintain the salt and water balance in the cell. Accordingly, channels are regulated by growth factors, but they also respond to a variety of chemical and physical stresses. By using the patch-clamp technique, UVC irradiation as low as 45 J/m^2 was found to increase the K^+ current within seconds by an as yet unknown mechanism [61]. When K^+ channel activity was suppressed by specific inhibitors, UV-dependent activation of MKK-4 and JNK was ablated and apoptosis largely abrogated, at least in myeloblastic cells. The degree of JNK inhibition correlated with the potency of the K^+ channel inhibitors [61].

UV and ionizing irradiation also enhance phospholipid turnover in the plasma membrane and stimulate the generation of diacylglycerol (DAG) and ceramide. Both DAG and ceramide are potent activators of PKC, and several isoforms of PKC are activated in response to UV and ionizing irradiation [62,63]. Ionizing irradiation-mediated JNK activation was furthermore prevented by induced PKC degradation or by treating cells with the kinase inhibitor H7, implying that induction of JNK in response to IR depends at least partially on PKC [64].

Conclusions

Radiation affects the function of numerous macromolecules. Among these are components that result in elevated signaling and subsequent changes in gene expression. Extranuclear signals are predominantly generated by the inactivation of negative control elements, for example, PTPs. They are, however, supplemented by stimuli generated by nuclear DNA damage. How nuclear DNA damage is converted into signal transduction is as yet not well understood. The final phenotype of irradiated cells results from the mix of stimuli—radiation-induced intranuclear and extranuclear signals and physiological stimuli—that address cells at the same time. Radiation of cells induces apoptosis, triggered either through MAP-kinases, or through DNA damage and p53, or through direct Fas (CD95) activation. Modulated by the sum of all stimuli, radiation can act as a tumor promoter, is mitogenic, or induces differentiation apoptosis and senescence.

References

1. Karin, M. and Hunter, T. (1995). Transcriptional control by protein phosphorylation: Signal transmission from the cell surface to the nucleus. *Curr. Biol.* **5**, 747–757.
2. Davis, R. J. (2000). Signal transduction by the JNK group of MAP kinases. *Cell* **103**, 239–252.
3. Matter-König, N., Herrlich, P., and König, H. (2002). Signal dependent regulation of splicing via phosphorylation of Sam68. *Nature* **420**, 691–695.
4. Price, M. A., Cruzalegui, F. H., and Treisman, R. (1996). The p38 and ERK MAP kinase pathways cooperate to activate ternary complex factors and c-fos transcription in response to UV light. *EMBO J.* **15**, 6552–6563.
5. Bulavan, D. V., Saito, S., Hollander, M. C., Sakaguchi, K., Anderson, C. W., Appella, E., and Fornace, A. J. Jr. (1999). Phosphorylation of human p53 by p38 kinase coordinates N-terminal phosphorylation and apoptosis in response to UV radiation. *EMBO J.* **23**, 6845–6854.
6. Kim, M., Lee, W., Park, J., Kim, J. B., Jang, Y. K., Seong, R. H., Choe, S. Y., and Park, S. D. (2000). The stress-activated MAP kinase Styl/Spcl and a 3′-regulatory element mediate UV-induced expression of the uvi15(+) gene at the post-transcriptional level. *Nucleic Acids Res.* **28**, 3392–402.
7. Musti, A. M., Treier, M., and Bohman, D. (1997). Reduced ubiquitin-dependent degradation of c-Jun after phosphorylation by MAP kinases. *Science* **275**, 400–402.
8. Zhou, B. P., Liao, Y., Xia, W., Zou, Y., Spohn, B., and Hung, M. C. (2001). Her-2/neu induces p53 ubiquitination via Akt-mediated Mdm2 phosphorylation. *Nat. Cell Biol.* **3**, 973–982.
9. Arthur, J. S. and Cohen, P. (2000). MSK1 is required for CREB phosphorylation in response to mitogens in mouse embryonic stem cells. *FEBS Lett.* **482**, 44–48.
10. Bruning, J. C., Gillette, J. A, Zhao, Y., Bjorbaeck, C., Kotzka, J., Knebel, B., Avci, H., Hanstein, B., Lingohr, P., Moller, D. E., Krone, W., Kahn, C. R., and Muller-Wieland, D. (2000). Ribosomal subunit kinase-2 is required for growth factor-stimulated transcription of the c-Fos gene. *Proc. Natl. Acad. Sci. USA* **97**, 2462–2467.
11. Shimamura, A., Ballif, B. A., Richards, S. A., and Blenis, J. (2000). Rsk1 mediates a MEK-MAP kinase cell survival signal. *Curr. Biol.* **10**, 127–135.
12. Hunter, T. (2000). Signaling-2000 and beyond. *Cell* **100**, 113–127.
13. Hibi, M., Lin, A., Smeal, T., Minden, A., and Karin, M. (1993). Identification of an oncoprotein- and UV-responsive protein kinase that binds and potentiates the c-Jun activation domain. *Genes Dev.* **7**, 2135–2148.
14. Bardwell, A. J., Flatauer, L. J., Matsukuma, K., Thorner, J., and Bardwell, L. (2001). A conserved docking site in MEKs mediates high-affinity binding to MAP kinases and cooperates with a scaffold protein to enhance signal transmission. *J. Biol. Chem.* **276**, 10374–10386.
15. Schaeffer, H. J., Catling, A. D., Eblen, S. T., Collier, L. S., Krauss, A., and Weber, M. J. (1998). MP1: A MEK binding partner that enhances enzymatic activation of the MAP kinase cascade. *Science* **281**, 1668–1671.
16. Whitmarsh, A. J., Cavanagh, J., Tournier, C., Yasuda, J., and Davis, R. J. (1998). A mammalian scaffold complex that selectively mediates MAP kinase activation. *Science* **281**, 1671–1674.
17. Yasuda, J., Whitmarsh, A. J., Cavanagh, J., Sharma, M., and Davis, R. J. (1999). The JIP group of mitogen-activated protein kinase scaffold proteins. *Mol. Cell Biol.* **19**, 7245–7254.
18. Ostman, A. and Bohmer, F. D. (2001). Regulation of receptor tyrosine kinase signaling by protein tyrosine phosphatases. *Trends Cell Biol.* **11**, 258–266.
19. Orian-Rousseau, V., Chen, L., Sleeman, J., Herrlich, P., and Ponta, H. (2002). CD44 is required for two consecutive step in HGF/c-Met signaling. *Genes Dev.* **16**, 3074–3086.
20. Erhardt, A., Hassan, M., Heintges, T., and Haussinger, D. (2002). Hepatitis C virus core protein induces cell proliferation and activates ERK, JNK, and p38 MAP kinases together with the map kinase phosphatase MKP-1 in a HepG2 tet-off cell line. *Virology* **292**, 272–284.
21. Kusari, A. B., Byon, J., Bandyopadhyay, D., Kenner, K. A., and Kusari, J. (1997). Insulin-induced mitogen-activated protein (MAP) kinase phosphatase-1 (MKP-1) attenuates insulin-stimulated MAP kinase activity: A mechanism for the feedback inhibition of insulin signaling. *Mol. Endocrinol.* **11**, 1532–1543.
22. Sun, H., Charles, C. H., Lau, L. F., and Tonks, N. K. (1993). MKP-1 (3CH134), an immediate early gene product, is a dual specificity phosphatase that dephosphorylates MAP kinase *in vivo*. *Cell* **75**, 487–493.
23. Ito, C. Y., Kazantsev, A. G., and Baldwin, A. S. Jr. (1994). Three NF-kappa B sites in the I kappaB-alpha promoter are required for induction of gene expression by TNF alpha. *Nucleic Acids Res.* **22**, 3787–3792.
24. Knebel, A., Rahmsdorf, H. J., Ullrich, A., and Herrlich, P. (1996). Dephosphorylation of receptor tyrosine kinases as target of regulation by radiation, oxidants or alkylating agents. *EMBO J.* **15**, 5314–5325.
25. Lallemand, D., Ham, J., Garbay, S., Bakiri, L., Traincard, F., Jeannequin, O., Pfarr, C. M., and Yaniv, M. (1998). Stress-activated protein kinases are negatively regulated by cell density. *EMBO J.* **17**, 5615–5626.
26. Morrison, H., Sherman, L. S., Legg, J., Banine, F., Isacke, C., Haipek, C. A., Gutmann, D. H., Ponta, H., and Herrlich, P. (2001). The NF2 tumor suppressor gene product, merlin, mediates contact inhibition of growth through interactions with CD44. *Genes Dev.* **15**, 968–980.
27. Sotiropoulos, A., Gineitis, D., Copeland, J., and Treisman, R. (1999). Signal-regulated activation of serum response factor is mediated by changes in actin dynamics. *Cell* **98**, 159–169.
28. Delhase, M., Hayakawa, M., Chen, Y., and Karin, M. (1999). Positive and negative regulation of IkappaB kinase activity through IKKbeta subunit phosphorylation. *Science* **284**, 309–313.
29. Cataldi, A., Zauli, G., Di Pietro, R., Castorina, S., and Rana, R. (2001). Involvement of the pathway phosphatidylinositol-3-kinase/AKT-1 in the establishment of the survival response to ionizing radiation. *Cell Signal.* **13**, 369–375.
30. Huang, C., Li, J., Ding, M., Leonard, S. S., Wang, L., Castranova, V., Vallyathan, V., and Shi, X. (2001). UV Induces phosphorylation of

protein kinase B (Akt) at Ser-473 and Thr-308 in mouse epidermal Cl 41 cells through hydrogen peroxide. *J. Biol. Chem.* **276**, 40234–40240.
31. Kasid, U., Suy, S., Dent, P., Ray, S., Whiteside, T. L., and Sturgill, T. W. (1996). Activation of Raf by ionizing radiation. *Nature* **382**, 813–816.
32. Li, N. and Karin, M. (1998). Ionizing radiation and short wavelength UV activate NF-kappaB through two distinct mechanisms. *Proc. Natl. Acad. Sci. USA* **95**, 13012–13017.
33. Radler-Pohl, A., Sachsenmaier, C., Gebel, S., Auer, H. P., Bruder, J. T., Rapp, U., Angel, P., Rahmsdorf, H. J., and Herrlich, P. (1993). UV-induced activation of AP-1 involves obligatory extranuclear steps including Raf-1 kinase. *EMBO J.* **12**, 1005–1012.
34. Ulm, R., Revenkova, E., DiSansebastiano, G.-P., Bechtold, N., and Paszkowski, J. (2001). Mitogen-activated protein kinase phosphatase is required for genotoxic stress relief in *Arabidopsis. Genes Dev.* **15**, 699–709.
35. Liu, Y., Gorospe, M., Yang, C., and Holbrook, N. J. (1995). Role of mitogen-activated protein kinase phosphatase during the cellular response to genotoxic stress. Inhibition of c-Jun N-terminal kinase activity and AP-1-dependent gene activation. *J. Biol. Chem.* **270**, 8377–8380.
36. Gabai, V. L., Meriin, A. B., Mosser, D. D., Caron, A. W., Rits, S., Shifrin, V. I., and Sherman, M. Y. (1997). Hsp70 prevents activation of stress kinases. A novel pathway of cellular thermotolerance. *J. Biol. Chem.* **272**, 18033–18037.
37. Dickens, M., Rogers, J. S., Cavanagh, J., Raitano, A., Xia, Z., Halpern, J. R., Greenberg, M. E., Sawyers, C. L., and Davis, R. J. (1997). A cytoplasmic inhibitor of the JNK signal transduction pathway. *Science* **277**, 693–696.
38. Devary, Y., Gottlieb, R. A., Smeal, T., and Karin, M. (1992). The mammalian ultraviolet response is triggered by activation of Src tyrosine kinases. *Cell* **71**, 1081–1091.
39. Huang, R. P., Wu, J. X., Fan, Y., and Adamson, E. D. (1996). UV activates growth factor receptors via reactive oxygen intermediates. *J. Cell Biol.* **133**, 211–220.
40. Cavigelli, M., Li, W. W., Lin, A., Su, B., Yoshioka, K., and Karin, M. (1996). The tumor promoter arsenite stimulates AP-1 activity by inhibiting a JNK phosphatase. *EMBO J.* **15**, 6269–6279.
41. Devary, Y., Rosette, C., DiDonato, J. A., and Karin, M. (1993). NF-kappa B activation by ultraviolet light not dependent on a nuclear signal. *Science* **261**, 1442–1445.
42. Rosette, C. and Karin, M. (1996). Ultraviolet light and osmotic stress: Activation of the JNK cascade through multiple growth factor and cytokine receptors. *Science* **274**, 1194–1197.
43. Adler, V., Fuchs, S. Y., Kim, J., Kraft, A., King, M. P., Pelling, J., and Ronai, Z. (1995). Jun-NH2-terminal kinase activation mediated by UV-induced DNA lesions in melanoma and fibroblast cells. *Cell Growth Differ.* **6**, 1437–1446.
44. Blattner, C., Bender, K., Herrlich, P., and Rahmsdorf, H. J. (1998). Photoproducts in transcriptionally active DNA induce signal transduction to the delayed UV-responsive genes for collagenase and metallothionein. *Oncogene* **16**, 2827–2834.
45. Miskin, R. and Ben-Ishai, R. (1981). Induction of plasminogen activator by UV light in normal and xeroderma pigmentosum fibroblasts. *Proc. Natl. Acad. Sci. USA* **78**, 6236–6240.
46. Takekawa, M. and Saito, H. (1998). A family of stress-inducible GADD45-like proteins mediate activation of the stress-responsive MTK1/MEKK4 MAPKKK. *Cell* **95**, 521–530.
47. Li, N., Banin, S., Ouyang, H., Li, G. C., Courtois, G., Shiloh, Y., Karin, M., and Rotman, G. (2001). ATM is required for IkappaB kinase (IKKk) activation in response to DNA double strand breaks. *J. Biol. Chem.* **276**, 8898–8903.
48. Gross, S., Knebel, A., Tenev, T., Neininger, A., Gaestel, M., Herrlich, P., and Bohmer, F. D. (1999). Inactivation of protein-tyrosine phosphatases as mechanism of UV-induced signal transduction. *J. Biol. Chem.* **274**, 26378–26386.
49. Sachsenmaier, C., Radler-Pohl, A., Zinck, R., Nordheim, A., Herrlich, P., and Rahmsdorf, H. J. (1994). Involvement of growth factor receptors in the mammalian UVC response. *Cell* **78**, 963–972.
50. Sheikh, M. S., Antinore, M. J., Huang, Y., and Fornace, A. J. Jr. (1998). Ultraviolet-irradiation-induced apoptosis is mediated via ligand independent activation of tumor necrosis factor receptor 1. *Oncogene* **17**, 2555–2563.
51. Aragane, Y., Kulms, D., Metze, D., Wilkes, G., Poppelmann, B., Luger, T. A., and Schwarz, T. (1998). Ultraviolet light induces apoptosis via direct activation of CD95 (Fas/APO-1) independently of its ligand CD95L. *J Cell Biol.* **140**, 171–182.
52. Meng, T. C., Fukada, T., and Tonks, N. K. (2002). Reversible oxidation and inactivation of protein tyrosine phosphatases *in vivo. Mol. Cell* **9**, 387–399.
53. Gulati, P., Klohn, P. C., Krug, H., Gottlicher, M., Markova, B., Bohmer, F. D., and Herrlich, P. (2001). Redox regulation in mammalian signal transduction. *IUBMB Life* **52**, 25–28.
54. Denu, J. M., Lohse, D. L., Vijayalakshmi, J., Saper, M. A., and Dixon, J. E. (1996). Visualization of intermediate and transition-state structures in protein-tyrosine phosphatase activity. *Proc. Natl. Acad. Sci. USA* **93**, 2493–2498.
55. Coffer, P. J., Burgering, B. M., Peppelenbosch, M. P., Bos, J. L., and Kruijer, W. (1995). UV activation of receptor tyrosine kinase activity. *Oncogene* **11**, 561–569.
56. Pomerance, M., Multon, M. C., Parker, F., Venot, C., Blondeau, J. P., Tocque, B., and Schweighoffer, F. (1998). Grb2 interaction with MEK-kinase 1 is involved in regulation of Jun-kinase activities in response to epidermal growth factor. *J. Biol. Chem.* **273**, 24301–24304.
57. Suy, S., Anderson, W. B., Dent, P., Chang, E., and Kasid, U. (1997). Association of Grb2 with Sos and Ras with Raf-1 upon gamma irradiation of breast cancer cells. *Oncogene* **15**, 53–61.
58. Iordanov, M. S., Pribnow, D., Magun, J. L., Dinh, T. H., Pearson, J. A., Chen, S. L., and Magun, B. E. (1997). Ribotoxic stress response: Activation of the stress-activated protein kinase JNK1 by inhibitors of the peptidyl transferase reaction and by sequence-specific RNA damage to the alpha-sarcine/ricin loop in the 28S rRNA. *Mol. Cell Biol.* **17**, 3373–3381.
59. Iordanov, M. S., Pribnow, D., Magun, J. L., Dinh, T. H., Pearson, J. A., and Magun, B. E. (1998). Ultraviolet radiation triggers the ribotoxic stress response in mammalian cells. *J. Biol. Chem.* **273**, 15794–15803.
60. Iordanov, M. S. and Magun, B. E. (1999). Different mechanisms of c-Jun NH(2)-terminal kinase-1 (JNK1) activation by ultraviolet-B radiation and by oxidative stressors. *J. Biol. Chem.* **274**, 25801–25806.
61. Wang, L., Xu, D., Dai, W., and Lu, L. (1999). An ultraviolet-activated K+ channel mediates apoptosis of myeloblastic leukemia cells. *J. Biol. Chem.* **274**, 3678–3685.
62. Carsberg, C. J., Warenius, H. M., and Friedmann, P. S. (1994). Ultraviolet radiation-induced melanogenesis in human melanocytes. Effects of modulating protein kinase C. *J. Cell Sci.* **107**, 2591–2597.
63. Cridland, N. A., Martin, M. C., Stevens, K., Baller, C. A., Pearson, A. J., Driscoll, C. M., and Saunders, R. D. (2001). Role of stress responses in human cell survival following exposure to ultraviolet C radiation. *Int. J. Radiat. Biol.* **77**, 365–374.
64. Hara, T., Namba, H., Yang, T. T., Nagayama, Y., Fukata, S., Kuma, K., Ishikawa, N., Ito, K., and Yamashita, S. (1998). Ionizing radiation activates c-Jun NH2-terminal kinase (JNK/SAPK) via a PKC-dependent pathway in human thyroid cells. *Biochem. Biophys. Res. Commun.* **244**, 41–44.

Chapter 309

Endoplasmic Reticulum Stress Responses

David Ron
Skirball Institute of Biomolecular Medicine,
New York University School of Medicine,
New York, New York

Introduction

Eukaryotic cells must adapt to wide variation in the load of client proteins that their endoplasmic reticulum (ER) must process. They do so by modulating the synthesis of client proteins and the expression of genes whose products determine the folding capacity of the organelle. The signaling pathways activated to effect these adaptations are part of the unfolded protein response (UPR), which is similar in many ways to the cytoplasmic heat-shock response. In mammalian cells, signaling is initiated by three known ER resident transmembrane proteins (IRE1, ATF6, and PERK) that can sense the stress caused by imbalance between capacity and demand in the ER. IRE1 and PERK are transmembrane protein kinases whose activity is repressed by free chaperones in the ER lumen.

During ER stress, as free chaperones are recruited to serve client proteins, these transducers are activated, resulting in downstream signaling. IRE1 effector functions include endonucleolytic processing of a substrate mRNA (HAC1 in yeast and XBP-1 in metazoans) and activation of JUN N-terminal kinase, resulting in activated gene expression. PERK phosphorylates the translation initiation factor eIF2 and mediates both translational repression of client protein biosynthesis and activation of specific mRNA translation. The latter culminates in increased expression of the transcription factor ATF4 and activates a gene expression program that is also accessed by other forms of stress and is referred to as the *integrated stress response.* ATF6 activation is mediated by proteolytic processing and liberation of an active N-terminal protein fragment that serves as a transcription factor for activating gene expression in the UPR.

ER Stress Defined

Proteins destined for secretion and membrane insertion are synthesized on ribosomes associated with the endoplasmic reticulum. Such client proteins of the ER are translocated into the organelle's lumen where they undergo posttranslational modifications specific to that environment. Chaperone proteins specific to the ER assist in the folding and posttranslational processing of the client proteins, and quality-control mechanisms operating within the organelle clear properly folded client proteins for transport to the Golgi apparatus or earmark others for degradation by the ER associated degradation apparatus (ERAD).

The ER is thus a dynamic cellular compartment whose activity level varies among different cell types and between physiological states. In complex metazoans it is most developed in cells devoted to secretion, such as those of the exocrine and endocrine glands, immunoglobulin-secreting plasma cells, and cells with extensive membranous processes such as neurons and myelin-producing glial cells. In other words, the quantity and intrinsic properties of client proteins synthesized in response to developmental programs and physiological states set the level of *demand* made of the ER. The ability of the ER to *supply* this demand is impacted by the size of the organelle, by the complement of chaperones and other intrinsic protein and lipid components at its disposal, and by the physiological state of the cell. The latter is particularly important in pathological states associated with nutrient limitation, hypoxia, or toxins that degrade the ER's ability to process client proteins (reviewed in [1–3]).

Specific signaling pathways are activated by an imbalance between the demand on the ER and the ability of the organelle to meet the demand, so-called "ER stress." Activation of these pathways occurs under both pathological conditions and in response to normal physiological demands. However, much of what we have learned about them is derived from experiments in which cells were treated with toxins, for example, tunicamycin, an inhibitor of N-linked glycosylation. Because such toxins cause ER stress by interfering with protein folding, the signaling pathways that they activate came to be known as the *unfolded protein response* (UPR) [4,5]. This is a somewhat misleading term; because the pathways in question appear to be activated by the threat of malfolded proteins and do not necessarily require the accumulation of measurable amounts of malfolded proteins.

The UPR in Yeast

The UPR was first studied in yeast where treatment with tunicamycin or mutations that affect protein folding in the ER were noted to activate transcription of the *KAR2* gene encoding the yeast homolog of the mammalian ER chaperone BiP [6,7]. Genetic analysis of signaling from the yeast ER to the *KAR2* promoter unveiled a truly novel pathway. Signaling in response to ER stress is initiated by a type 1 ER resident transmembrane protein encoded by the *IRE1* gene [8,9]. The N-terminal lumenal domain of Ire1p responds to the ER stress signal and the signal is transmitted to the C-terminal, cytoplasmic effector domain in which two distinct activities reside: a protein kinase activity for which the only known substrate is Ire1p itself [8,9] and, remarkably, an endonuclease activity that removes an intron from the *HAC1* mRNA [10–12]. Ire1p activation thus entails transautophosphorylation that unmasks the endonucleolytic activity [13]. The cleaved ends of the *HAC1* mRNA are joined by tRNA ligase to generate an activated, spliced form of the *HAC1* mRNA [14]. The unspliced *HAC1* mRNA is poorly translated, whereas the protein encoded by the spliced form is well expressed. Thus, noncanonical splicing controls expression of Hac1p, a transcription factor that activates downstream target genes of the yeast UPR [15,11].

In yeast the IRE1->HAC1 pathway is a simple linear one that is required for activation of all known target genes of the UPR. These have been the subjects of detailed analysis by expression microarrays. Surprisingly, the genes induced by the UPR in yeast include not only the expected ER chaperones but also components of the apparatus for degrading ER proteins, the machinery required for membrane biosynthesis and non-ER components of the secretory apparatus [16,17]. The yeast UPR, therefore, appears to up-regulate the capacity of the entire endomembrane system to deal with client proteins. In this context the UPR should be viewed as a pathway concerned both with adapting to stress in the short term and with growth and anabolism [18]. It is also clear, however, that the ability to up-regulate target genes of the UPR is an important adaptation, because yeast defective in this function are hypersensitive to ER stress [16,17,19].

IRE1 proteins are conserved in higher eukaryotes [20,21] and their mechanism of action appears to be conserved too (see later discussion), however their preeminent role in regulating all aspects of the UPR has been superseded by new components that are lacking in yeast. One important difference between higher eukaryotes and yeast is that the former rapidly and reversibly attenuate protein biosynthesis in response to ER stress. This adaptation decreases the quantity of client proteins that are delivered to the stressed ER and reduces the demand on the organelle.

The UPR is Metazoans

Translation control by ER stress is mediated by PERK (also known as PEK or EIF2AK3), a type 1 transmembrane ER resident protein kinase related to IRE1 in its stress-sensing lumenal domain. The C-terminal, cytoplasmic effector domain of PERK is a protein kinase that phosphorylates the alpha subunit of translation initiation factor 2 [22]. Phosphorylation of eIF2α inhibits nucleotide exchange on the eIF2 complex and blocks the initiation step of translation. An important component of the regulation of PERK and IRE1 activity depends on an interaction between their related N-terminal lumenal domains and the ER chaperone BiP. In unstressed cells, dispensable BiP protein is available to bind the lumenal domains of PERK and IRE1. BiP binding maintains these proteins in a monomeric, inactive state. When the folding reserve of the ER is challenged, either by increased demand or reduced function, levels of dispensable BiP decrease and PERK-BiP and IRE1-BiP complexes dissociate. Freed from the inhibitory effects of BiP binding, PERK and IRE1 oligomerize, transautophosphorylate, and initiate downstream signaling [23,24].

Cells lacking PERK are unable to reduce translation when experiencing ER stress and consequently are exquisitely sensitive to agents and conditions that cause such stress [25]. Humans and mice with *PERK* mutations develop early onset diabetes mellitus caused by destruction of the insulin-producing pancreatic β cells [26,27]. Protein synthesis in β cells is strongly activated by plasma glucose; PERK modulates this increase, ensuring that the quantity of proteins translocated into the ER does not exceed the capacity of the organelle to fold them. In $PERK^{-/-}$ islets one finds cells with distended ER, and the presence of these correlates with high rates of programmed cell death in the mutant islets. It is clear, therefore, that PERK signaling is activated under normal physiological circumstances, notably in professional secretory cells.

The hypothesis whereby loss of translational control leads to stuffing of the ER with client proteins goes a long way toward explaining the phenotype of the *PERK* mutation. However, this simple hypothesis does not take into account the role of PERK in regulating gene expression in the UPR. eIF2α phosphorylation decreases the efficiency of translation of most mRNAs; however, some mRNAs are translated more efficiently in such circumstances. The mechanism for this has been worked out in great detail in the case of the yeast *GCN4* mRNA, which encodes a transcription factor that activates genes involved in amino acid biosynthesis [28]. Translation of the mammalian *ATF4* mRNA is similarly up-regulated by eIF2α phosphorylation [29]. Thus, ATF4 translation during ER stress and up-regulation of its downstream target genes require PERK. Because this signaling pathway is activated both by ER stress and other conditions associated with eIF2α phosphorylation (e.g., amino acid starvation, arsenite treatment), we propose referring to it as the *integrated stress response* (ISR). Preliminary results suggest that the ISR may contribute substantially to cell survival during ER stress and to developmental programs [30].

GADD34 is an interesting target gene of PERK (and the ISR). The encoded protein binds the catalytic subunit of protein phosphatase 1 and recruits it to dephosphorylate eIF2α. Thus, the integrated stress response induces a negative feedback loop that serves to terminate the response [31]. It seems likely that eIF2α dephosphorylation plays a role in the recovery of protein biosynthesis, which, in turn, is required for the translation of mRNAs that are up-regulated as part of the gene expression program activated during the UPR (and other stresses associated with an ISR). According to this model, the first line of defense against ER stress in complex metazoans is translational repression. It is followed, soon thereafter, by remodeling of the internal milieu; the latter process requires new protein synthesis that is favored by GADD34.

Higher eukaryotes possess a third signaling pathway from the ER to the nucleus (Fig. 1), which is mediated in mammals by two ATF6 genes, α and b, encoding very similar type 2 transmembrane ER resident proteins. The N-terminal cytoplasmic portion of these proteins is a bZIP transcription factor that binds and activates ER stress response elements (ERSEs) in the promoter of such classic UPR target genes as BiP and CHOP [32–34]. In unstressed cells ATF6 is maintained in an inactive state by being tethered to the ER membrane. In response to ER stress, the protein is proteolytically cleaved, generating a free N-terminal fragment that migrates to the nucleus and activates transcription through ERSEs [35,33]. This mode of activation resembles that of another ER-tethered bZIP protein, the sterol-regulated transcription factor SREBP. This is more than a superficial resemblance, because the proteases that process and activate SREBP were found to be essential also for the activation of ATF6 [36]. These proteases are found in the Golgi apparatus. It is likely, therefore, that ATF6 activity is controlled by ER retention, and migration of ATF6 to the Golgi apparatus during ER stress likely leads to processing and liberation of the active transcription factor.

The comprehensive analysis of the yeast UPR, carried out by expression microarrays, has not yet been extended to mammalian cells. Consequently, we do not have detailed information on how the different signaling components of the UPR interact to control gene expression in higher eukaryotes. However, based on even the limited information available, it is clear that the mammalian UPR has diverged considerably from that of the yeast. The ATF6 pathway appears to have assumed a major role in controlling chaperone gene expression. This is revealed by the severe defect in *BiP*

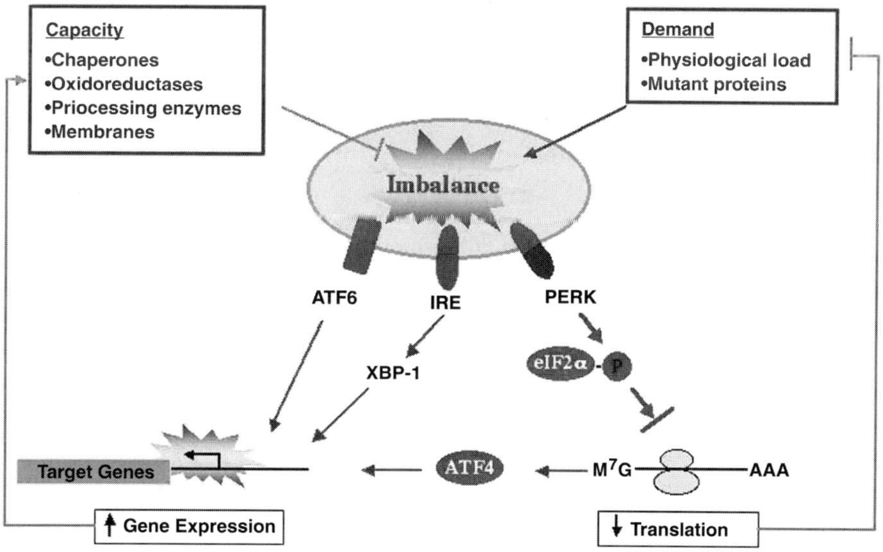

Figure 1 Depiction of the major aspects of the endoplasmic reticulum stress response and its mediators in mammalian cells. An imbalance between the load of client proteins imposed on the ER (demand) and capacity of the ER to fold client proteins is sensed by the three ER stress transducers: ATF6, IRE1, and PERK. These coordinate two functional responses: activation of the expression of genes whose products increase ER folding capacity and inhibition of protein synthesis that attenuated demand on the ER. IRE1 activity is transduced though XBP-1 to activate gene expression, whereas ATF6 directly mediates ER to nuclear signaling. PERK signaling is mediated by eIF2α phosphorylation modifying translation initiation rates. The latter results in a general decrease in demand on the ER and at the same time activates the translation of transcription factors such as ATF4 that contribute to activated gene expression.

expression manifested by cells that lack ATF6-processing protease activity [36]. PERK and the integrated stress response also contribute significantly to gene expression in the UPR [29,30]. Some target genes of the UPR, such as *GADD34* and *CHOP*, are highly dependent on the ISR, whereas other genes, such as *BiP*, are less dependent on PERK signaling and the ISR. IRE1 genes, of which mammals have two, appear to have been diverted from exclusive control of a broad set of UPR genes (as is the case in yeast) toward certain specific cellular functions.

We have found that mouse cells lacking *IRE1* are virtually unimpaired in activating most target genes of the UPR. Furthermore, such mutant cells are not manifestly hypersensitive to pharmacological manipulations that cause ER stress. One measure of the divergence of IRE1 function from yeast to higher eukaryotes is the finding that mammalian IRE1 activates stress-activated protein kinases during ER stress. It does so by recruiting TRAF2 to the ER membrane [37]. The physiological significance of this signaling pathway is not fully understood, but it may link ER stress and programmed cell death. The association between ER stress and programmed cell death is a poorly understood but potentially important subject, particularly in light of recent findings suggesting that ER stress may accompany common pathological conditions [38]. Several signaling pathways leading to CHOP [39], JNK [37], CASPASE-12 [40], and cABL [41] activation may be involved, but the manner in which these are integrated into the UPR remains to be studied in greater detail.

Further clues to the role of mammalian IRE1 have been provided by the identification of a direct downstream target mRNA (analogous to yeast *HAC1*). The *XBP-1* mRNA is processed by mammalian IRE1 in an ER-stress-dependent manner and IRE1 thus controls the expression of this important transcription factor [42–44]. The target genes of XBP-1 have not been identified, but analysis of *XBP-1* knock-out mice reveals its essential role in development of B cells to plasma cells [45]. XBP-1 mediated signaling presumably up-regulates the capacity of the endomembrane system to deal with the load of immunoglobulins that must be secreted by the developing B cell. The finding that XBP-1 is controlled by IRE1 suggests that in mammalian cells the IRE1 genes have specialized to link the load on the ER to certain long-term "developmental" adaptations of the ER. This hypothesis is further supported by the observation that both XBP-1 and IRE1 are enriched in cells that specialize in secretion.

Conclusion

In summary, the mammalian UPR consists of rapid and transient translational repression, which is completely PERK dependent. This short-term adaptation is supplanted by up-regulation of genes that enhance the capacity of ER to deal with increased load. All three known signaling components of the UPR (PERK, IRE1, and ATF6) participate in this latter adaptation. Additional signaling downstream of these and other ER stress responders effects important and poorly understood life and death decisions.

References

1. Lee, A. S. (1992). Mammalian stress response: Induction of the glucose-regulated protein family. *Curr. Opin. Cell Biol.* **4**, 267–273.
2. Aridor, M. and Balch, W. E. (1999). Integration of endoplasmic reticulum signaling in health and disease. *Nat. Med.* **5**, 745–751.
3. Kaufman, R. J. (1999). Stress signaling from the lumen of the endoplasmic reticulum: Coordination of gene transcriptional and translational controls. *Genes Dev.* **13**, 1211–1233.
4. Kozutsumi, Y., Segal, M., Normington, K., Gething, M. J., and Sambrook, J. (1988). The presence of malfolded proteins in the endoplasmic reticulum signals the induction of glucose-regulated proteins. *Nature* **332**, 462–464.
5. Gething, M. J. and Sambrook, J. (1992). Protein folding in the cell. *Nature* **355**, 33–45.
6. Normington, K., Kohno, K., Kozutsumi, Y., Gething, M. J., and Sambrook, J. (1989). S. cerevisiae encodes an essential protein homologous in sequence and function to mammalian BiP. *Cell* **57**, 1223–1236.
7. Kohno, K., Normington, K., Sambrook, J., Gething, M. J., and Mori, K. (1993). The promoter region of the yeast KAR2 (BiP) gene contains a regulatory domain that responds to the presence of unfolded proteins in the endoplasmic reticulum. *Mol. Cell Biol.* **13**, 877–890.
8. Cox, J. S., Shamu, C. E., and Walter, P. (1993). Transcriptional induction of genes encoding endoplasmic reticulum resident proteins requires a transmembrane protein kinase. *Cell* **73**, 1197–1206.
9. Mori, K., Ma, W., Gething, M. J., and Sambrook, J. (1993). A transmembrane protein with a cdc2+/CDC28-related kinase activity is required for signaling from the ER to the nucleus. *Cell* **74**, 743–756.
10. Cox, J. S. and Walter, P. (1996). A novel mechanism for regulating activity of a transcription factor that controls the unfolded protein response. *Cell* **87**, 391–404.
11. Kawahara, T., Yanagi, H., Yura, T., and Mori, K. (1997). Endoplasmic reticulum stress-induced mRNA splicing permits synthesis of transcription factor Hac1p/Ern4p that activates the unfolded protein response. *Mol. Biol. Cell* **8**, 1845–1862.
12. Sidrauski, C. and Walter, P. (1997). The transmembrane kinase Ire1p is a site-specific endonuclease that initiates mRNA splicing in the unfolded protein response. *Cell* **90**, 1031–1039.
13. Shamu, C. E. and Walter, P. (1996). Oligomerization and phosphorylation of the Ire1p kinase during intracellular signaling from the endoplasmic reticulum to the nucleus. *Embo J* **15**, 3028–3039.
14. Sidrauski, C., Cox, J. S., and Walter, P. (1996). tRNA ligase is required for regulated mRNA splicing in the unfolded protein response. *Cell* **87**, 405–413.
15. Chapman, R. E. and Walter, P. (1997). Translational attenuation mediated by an mRNA intron. *Curr. Biol.* **7**, 850–859.
16. Casagrande, R., Stern, P., Diehn, M., Shamu, C., Osario, M., Zuniga, M., Brown, P. O., and Ploegh, H. (2000). Degradation of proteins from the ER of S. cerevisiae requires an intact unfolded protein response pathway. *Mol. Cell* **5**, 729–735.
17. Travers, K. J., Patil, C. K., Wodicka, L., Lockhart, D. J., Weissman, J. S., and Walter, P. (2000). Functional and genomic analyses reveal an essential coordination between the unfolded protein response and ER-associated degradation. *Cell* **101**, 249–258.
18. Cox, J. S., Chapman, R. E., and Walter, P. (1997). The unfolded protein response coordinates the production of endoplasmic reticulum protein and endoplasmic reticulum membrane. *Mol. Biol. Cell* **8**, 1805–1814.
19. Ng, D. T., Spear, E. D., and Walter, P. (2000). The unfolded protein response regulates multiple aspects of secretory and membrane protein biogenesis and endoplasmic reticulum quality control. *J. Cell Biol.* **150**, 77–88.
20. Tirasophon, W., Welihinda, A. A., and Kaufman, R. J. (1998). A stress response pathway from the endoplasmic reticulum to the nucleus

requires a novel bifunctional protein kinase/endoribonuclease (Ire1p) in mammalian cells. *Genes Dev.* **12**, 1812–1824.
21. Wang, X. Z., Harding, H. P., Zhang, Y., Jolicoeur, E. M., Kuroda, M., and Ron, D. (1998). Cloning of mammalian Ire1 reveals diversity in the ER stress responses. *EMBO J* **17**, 5708–5717.
22. Harding, H., Zhang, Y., and Ron, D. (1999). Translation and protein folding are coupled by an endoplasmic reticulum resident kinase. *Nature* **397**, 271–274.
23. Bertolotti, A., Zhang, Y., Hendershot, L., Harding, H., and Ron, D. (2000). Dynamic interaction of BiP and the ER stress transducers in the unfolded protein response. *Nature Cell Biology* **2**, 326–332.
24. Liu, C. Y., Schroder, M., and Kaufman, R. J. (2000). Ligand-independent dimerization activates the stress-response kinases IRE1 and PERK in the lumen of the endoplasmic reticulum. *J. Biol. Chem.* **275**, 17680–17687.
25. Harding, H., Zhang, Y., Bertolotti, A., Zeng, H., and Ron, D. (2000). *Perk* is essential for translational regulation and cell survival during the unfolded protein response. *Mol. Cell* **5**, 897–904.
26. Delepine, M., Nicolino, M., Barrett, T., Golamaully, M., Lathrop, G. M., and Julier, C. (2000). EIF2AK3, encoding translation initiation factor 2-alpha kinase 3, is mutated in patients with Wolcott-Rallison syndrome. *Nat. Genet.* **25**, 406–409.
27. Harding, H., Zeng, H., Zhang, Y., Jungreis, R., Chung, P., Plesken, H., Sabatini, D., and Ron, D. (2001). Diabetes Mellitus and exocrine pancreatic dysfunction in Perk$^{-/-}$ mice reveals a role for translational control in survival of secretory cells. *Mol. Cell* **7**, 1153–1163.
28. Hinnebusch, A. G. (1997). Translational regulation of yeast GCN4. A window on factors that control initiator-tRNA binding to the ribosome. *J. Biol. Chem.* **272**, 21661–21664.
29. Harding, H., Novoa, I., Zhang, Y., Zeng, H., Wek, R. C., Schapira, M., and Ron, D. (2000). Regulated translation initiation controls stress-induced gene expression in mammalian cells. *Mol. Cell* **6**, 1099–1108.
30. Scheuner, D., Song, B., McEwen, E., Gillespie, P., Saunders, T., Bonner-Weir, S., and Kaufman, R. J. (2001). Translational control is required for the unfolded protein response and in-vivo glucose homeostasis. *Mol. Cell* **7**, 1165–1176.
31. Novoa, I., Zeng, H., Harding, H., and Ron, D. (2001). Feedback inhibition of the unfolded protein response by GADD34-mediated dephosphorylation of eIF2α. *J. Cell Biol.* **153**, 1011–1022.
32. Yoshida, H., Haze, K., Yanagi, H., Yura, T., and Mori, K. (1998). Identification of the cis-acting endoplasmic reticulum stress response element responsible for transcriptional induction of mammalian glucose-regulated proteins. Involvement of basic leucine zipper transcription factors. *J. Biol. Chem.* **273**, 33741–33749.
33. Yoshida, H., Okada, T., Haze, K., Yanagi, H., Yura, T., Negishi, M., and Mori, K. (2000). ATF6 activated by proteolysis binds in the presence of NF-Y (CBF) directly to the *cis*-acting element responsible for the mammalian unfolded protein response. *Mol. Cell Biol.* **20**, 6755–6767.
34. Yoshida, H., Okada, T., Haze, K., Yanagi, H., Yura, T., Negishi, M., and Mori, K. (2001). Endoplasmic reticulum stress-induced formation of transcription factor complex ERSF including NF-Y (CBF) and activating transcription factors 6alpha and 6beta that activates the mammalian unfolded protein response. *Mol. Cell Biol.* **21**, 1239–1248.
35. Haze, K., Yoshida, H., Yanagi, H., Yura, T., and Mori, K. (1999). Mammalian transcription factor ATF6 is synthesized as a transmembrane protein and activated by proteolysis in response to endoplasmic reticulum stress. *Mol. Biol. Cell* **10**, 3787–3799.
36. Ye, J., Rawson, R. B., Komuro, R., Chen, X., Dave, U. P., Prywes, R., Brown, M. S., and Goldstein, J. L. (2000). ER stress induces cleavage of membrane-bound ATF6 by the same proteases that process SREBPs. *Mol. Cell* **6**, 1355–1364.
37. Urano, F., Wang, X., Bertolotti, A., Zhang, Y., Chung, P., Harding, H., and Ron, D. (2000). Coupling of stress in the endoplasmic reticulum to activation of JNK protein kinases by transmembrane protein kinase IRE1. *Science* **287**, 664–666.
38. Lee, A. S. (2001). The glucose-regulated proteins: Stress induction and clinical applications. *Trends Biochem. Sci.* **26**, 504–510.
39. Zinszner, H., Kuroda, M., Wang, X., Batchvarova, N., Lightfoot, R. T., Remotti, H., Stevens, J. L., and Ron, D. (1998). CHOP is implicated in programmed cell death in response to impaired function of the endoplasmic reticulum. *Genes Dev.* **12**, 982–995.
40. Nakagawa, T., Zhu, H., Morishima, N., Li, E., Xu, J., Yankner, B. A., and Yuan, J. (2000). Caspase-12 mediates endoplasmic-reticulum-specific apoptosis and cytotoxicity by amyloid-beta. *Nature* **403**, 98–103.
41. Ito, Y., Pandey, P., Mishra, N., Kumar, S., Narula, N., Kharbanda, S., Saxena, S., and Kufe, D. (2001). Targeting of the c-Abl tyrosine kinase to mitochondria in endoplasmic reticulum stress-induced apoptosis. *Mol. Cell Biol.* **21**, 6233–6242.
42. Shen, X., Ellis, R. E., Lee, K., Liu, C.-Y., Yang, K., Solomon, A., Yoshida, H., Morimoto, R., Kurnit, D. M., Mori, K., and Kaufman, R. J. (2001). Complementary signaling pathways regulate the unfolded protein response and are required for *C. elegans* development. *Cell* **107**, 893–903.
43. Yoshida, H., Matsui, T., Yamamoto, A., Okada, T., and Mori, K. (2001). XBP1 mRNA is induced by ATF6 and spliced by IRE1 in response to ER stress to produce a highly active transcription factor. *Cell* **107**, 881–891.
44. Calfon, M., Zeng, H., Urano, F., Till, J. H., Hubbard, S. R., Harding, H. P., Clark, S. G., and Ron, D. (2002). IRE1 couples endoplasmic reticulum load to secretory capacity by processing the *XBP-1* mRNA. *Nature* **415**, 92–96.
45. Reimold, A. M., Iwakoshi, N. N., Manis, J., Vallabhajosyula, P., Szomolanyi-Tsuda, E., Gravallese, E. M., Friend, D., Grusby, M. J., Alt, F., and Glimcher, L. H. (2001). Plasma cell differentiation requires the transcription factor XBP-1. *Nature* **412**, 300–307.

CHAPTER 310

The Heat-Shock Response: Sensing the Stress of Misfolded Proteins

Richard I. Morimoto and Ellen A. A. Nollen
Department of Biochemistry, Molecular Biology and Cell Biology, Rice Institute for Biomedical Research, Northwestern University, Evanston, Illinois

Introduction

The heat-shock response is an inducible molecular response to physiological, environmental, and biochemical stress conditions (Fig. 1) that results in the elevated expression of heat-shock genes. These can be classified into four categories: (1) environmental stresses, such as heat shock, amino acid analogs, drugs, oxidative stress, toxic chemicals, heavy metals, and pharmacologically active small molecules; (2) nonstress conditions, such as the cell cycle, growth factors, serum stimulation, development, differentiation, and activation by certain oncogenes; (3) physiological stress and disease states, such as neuroendocrine hormones, tissue injury and repair, fever, inflammation, infection, ischemia and reperfusion, and cancer; and (4) diseases of protein aggregation, such as Huntington's disease, Alzheimer's disease, Parkinson's disease, and ALS. For each of these categories, the various conditions indicated are typically associated with the overexpression of one or more heat-shock proteins through activation of heat-shock factor (HSF) and the heat-shock response.

Common to these stresses are challenges to protein homeostasis that influence folding, translocation, assembly, and degradation events. Consequently, an increased flux of non-native intermediates, if left unprotected, will have an increased propensity to misfold and self-associate to form protein aggregates and other toxic protein species. The heat-shock response, through the elevated synthesis of molecular chaperones and proteases, responds rapidly and precisely to the intensity and duration of specific environmental and physiological stress signals to reestablish protein homeostasis and prevent protein damage [1–4]. Transient exposure to intermediate elevated temperatures or lower levels of chemical and environmental stress has cytoprotective effects against sustained, normally lethal, exposures to stress [5]. This reveals a valuable survival strategy that "a little stress is good."

Transcriptional Regulation of the Heat-Shock Response

The stress-induced regulation of the heat-shock response in vertebrates occurs by activation of a family of heat shock transcription factors (HSFs) from an inert state under normal steady-state conditions to a transcriptionally competent DNA-binding state [6–9]. Four HSF genes (HSF1–4) are expressed in vertebrates whereas yeast, *Caenorhabditis elegans*, and *Drosophila* encode a single HSF [9–15] with a high degree of conservation in the helix–loop–helix DNA-binding domain [8,16,17], adjacent 80-amino-acid hydrophobic repeat (HR-A/B) necessary for trimer formation [12,18,19], the centrally localized negative regulatory domain, and the carboxyl-terminal transcriptional transactivation domain [20–25].

Stress-induced activation of HSF1 involves a multistep process (Fig. 2) including translocation from the cytosol and nuclear relocalization, oligomerization to a DNA-binding competent state, binding to heat-shock promoter elements, hyperphosphorylation at serine residues, and elevated transcription of heat-shock genes [8,26–32]. These steps can be uncoupled; for example, nonsteroidal anti-inflammatory drugs induce HSF1 trimers that are bound *in vivo* to promoter sites and lack transcriptional activity, yet can be subsequently

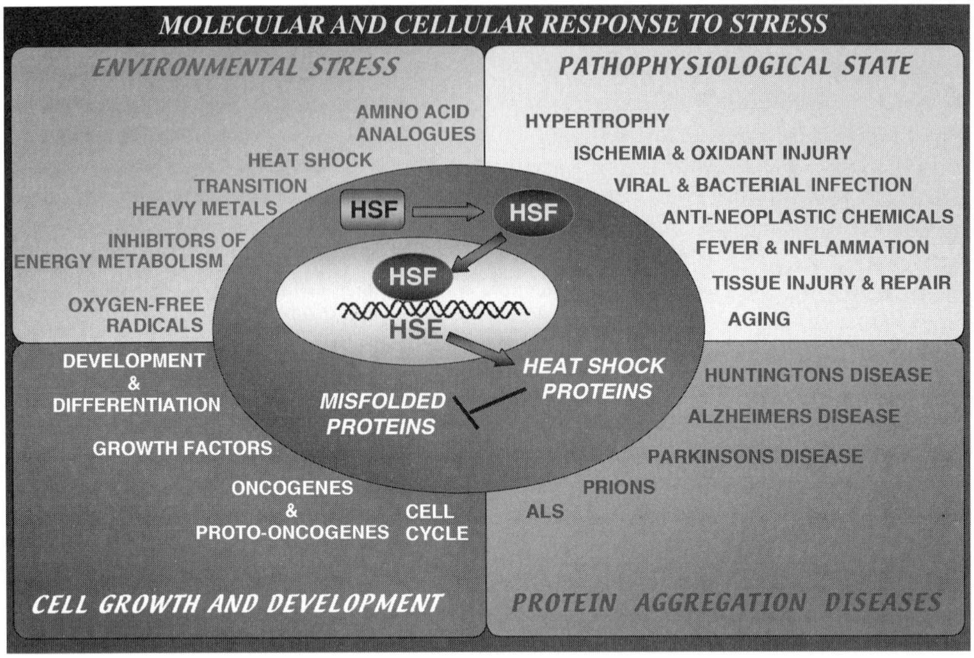

Figure 1 Conditions that activate the heat-shock response. Heat-shock gene expression is induced by environmental and physiological stress, nonstressful conditions including cell growth and development, pathophysiological states, and protein aggregation diseases, that is, conditions that are associated with the appearance of misfolded proteins. Activation of the heat-shock transcription factor leads to the expression of heat-shock proteins, which function to prevent and repair protein damage through interaction with the misfolded proteins.

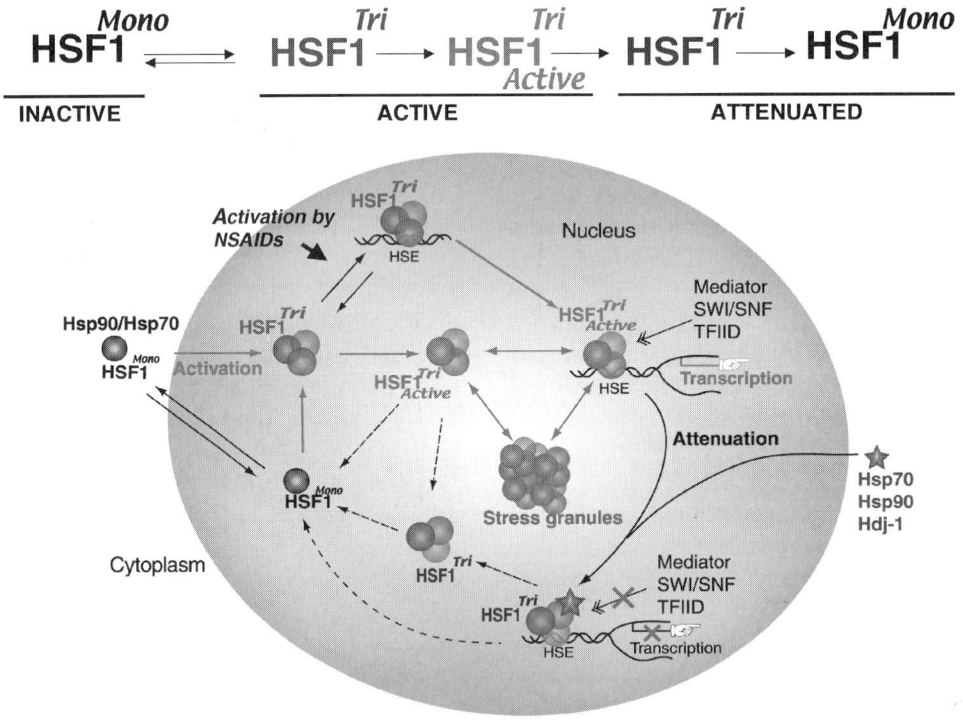

Figure 2 Regulation of the heat-shock response. Activation of HSF1 is linked to the appearance of non-native proteins and the requirement for molecular chaperones (Hsp90, Hsp70, and Hdj1) to prevent the appearance of misfolded proteins. HSF1 exists in the control state as an inert monomer (shown as intramolecularly negatively regulated for DNA binding and transcriptional activity) and undergoes stepwise activation to a DNA-binding competent state that is transcriptionally inert, acquisition of inducible phosphorylation resulting in complete activation and inducible transcription of heat-shock genes, and attenuation of HSF1 activity. HSF1 activity is negatively regulated by heat-shock factor binding protein 1 (HSBP1), which binds to the region of HSF1 corresponding to the heptad repeat, and by Hsp70 and Hdj1, which bind to the transcriptional transactivation domain thus repressing HSF1 activity.

activated to the fully active HSF1 by subsequent exposures to intermediate levels of stress [33–35]. During continued exposure to heat shock, HSF1 activity attenuates leading to the arrest of heat-shock gene transcription. HSF1 associates with molecular chaperones Hsp70, Hdj1, and Hsp90 during attenuation, and these events correlate with transcriptional arrest, dissociation of HSF1 from DNA, and refolding of the active trimer to the inert monomer [36–42]. Conditional overexpression of Hsp70 is sufficient to inhibit the induction of heat-shock gene transcription with little or no effect on the oligomeric state or level of phosphorylation of HSF1 [38]. These results are also supported by genetic observations that the heat-shock response is autoregulated in the yeast *Saccharomyces cerevisiae*, but interactions of the yeast Hsp70s and HSF [43–46] of HSF1 in a repressed state are delicately balanced and easily disrupted; for example, slight changes in the endogenous levels of HSF1 or expression of point mutants in the negative regulatory domain result in constitutive activity [31,47]. Collectively, these results reveal that multiple chaperones are involved in different aspects of HSF1 regulation and moreover that differences in specificity and kinetics may influence different aspects of the heat-shock transcriptional response.

Another member of the HSF family, HSF2, has been suggested to have a role in heat-shock gene expression during cell growth, differentiation, and upon deregulation of the protein degradative machinery. Activation of HSF2 was first observed in human erythroleukemia (K562) cells exposed to the differentiation agent hemin [48–50], and subsequently detected during murine spermatocyte differentiation [51] and embryogenesis [52,53]. The stress-sensing pathway for HSF2 activation is associated with changes in protein degradation, in a cell-type-independent manner, following exposure to inhibitors of the ubiquitin-dependent proteasome or in cells harboring conditional mutations in components of the proteasome machinery [54]. The function of the ubiquitin-dependent proteasome machinery is to degrade short-lived and misfolded proteins; therefore, the relationship between HSF2 activity and the protein degradative machinery reveals a requirement for heat-shock proteins, perhaps to prevent misfolding and aggregation of non-native proteins targeted for degradation.

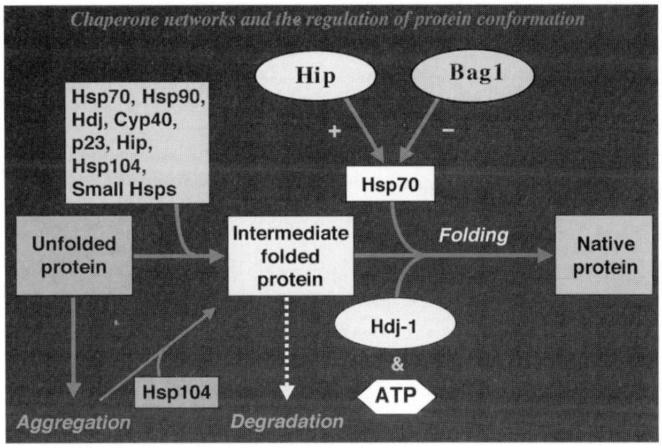

Figure 3 Chaperone networks and the regulation of protein conformation. The biochemical fate of an unfolded protein is schematically presented. In absence of chaperones, unfolded proteins are prone to aggregation. The presence of the molecular chaperones Hsp70, Hsp90, Hdj, Cyp40, p23, Hip, Hsp104, or small Hsps results in an intermediate folded state. The chaperones prevent aggregation or target the unfolded proteins for degradation. Proteins in an intermediate folded state can be refolded to the native state by Hsp70, ATP, and the cochaperone Hdj-1. Refolding can be positively (+) or negatively (–) influenced by Hsp70 cochaperones such as Hip and Bag1.

Molecular Chaperones: Folding, Misfolding, and the Assembly of Regulatory Complexes

Although some proteins can refold spontaneously, *in vitro*, larger, multidomain proteins often misfold and aggregate. The challenge, *in vivo*, within the densely packed environment of the cell is to ensure that nascent polypeptides fold, translocate, and assemble as multimeric complexes, and that non-native intermediates that accumulate during normal biosynthesis or are enhanced due to mutations or environmental stress are efficiently captured, refolded, or degraded. Molecular chaperones of the Hsp104, Hsp90, Hsp70, Hsp60, and small Hsp class accomplish these activities by capturing non-native intermediates and, together with cochaperones and ATP, facilitate the folded native state (Fig. 3) [3]. The Hsp70 chaperones recognize hydrophobic residue-rich stretches in polypeptides that are transiently exposed in folded intermediates and typically confined to the hydrophobic core in the native state [55]. This contrasts with the Hsp60/GroEL chaperonin, which creates a protected environment with the properties of a "protein-folding test tube" in which non-native proteins undergo rounds of binding and release to acquire the native state [56,57]. Common to these chaperone interactions is their ability to shift the equilibrium of protein folding toward on-pathway events and to minimize the appearance of off-pathway, aggregation-prone species.

The family of molecular chaperones is large, diverse in size and apparent structure, yet highly conserved [1,3,58]. Chaperones are abundant in growing cells and can attain concentrations of 5–20% of total cell protein. They are classified according to molecular size, that is, Hsp100, Hsp90, Hsp70, Hsp60, Hsp40, and small Hsps distributed in all subcellular compartments (Table I). Biochemical studies on chaperones have established common properties of Hsp104, Hsp90, Hsp70, the small Hsps, immunophilins (FKPB52 and CyP40), the steroid aporeceptor protein p23, and Hip (Hsp70 and Hsp90 interacting protein) to prevent the *in vitro* aggregation of model protein substrates and to maintain the substrate in an intermediate folded state competent for subsequent refolding to the native state [59–62]. A distinction among proteins that exhibit the properties of chaperones is that refolding to the native state requires the activity of a specific subset of chaperones such as Hsp90, Hsp70, or Hsp60/GroEL and nucleotide (Fig. 2). Cycles of nucleotide

TABLE I Brief Summary of the Nomenclature, Location, and Function of the Major Heat-Shock Protein and Molecular Chaperone Families

Family	Organism	Chaperones	Location	Functions
HSP 100	*E. coli*	ClpA,B,C	Cytosol	Role in stress tolerance; helps the resolubilization of heat-inactivated proteins from insoluble aggregates
	S. cerevisiae	HSP104	Cytosol	
HSP90	*E. coli*	HtpG	Cytosol	Role in signal transduction (e.g., interaction with steriod hormone receptors, tyrosine kinases, serine/threonine kinases); refolds and maintains proteins *in vitro;* autoregulation of the heat-shock response; role in cell cycle and proliferation
	S. cerevisiae	HSP83	Cytosol	
	Mammals	HSP90	Cytosol	
		GRP94	ER	
HSP70	*E. coli*	DnaK	Cytosol	Roles in lambda phage replication; autoregulation of the heat-shock response; interaction with nascent chain polypeptides; functions in interorganellar transport; roles in signal transduction; refolds and maintains denatured proteins *in vitro;* role in cell cycle and proliferation; anti-apoptotic activity; potential antigen-presenting molecule in tumor cells
	S. cerevisiae	Ssa 1-4	Cytosol	
		Ssb 1,2	Cytosol	
		Kar2	ER	
		Ssc1	Mitochondria	
	Mammals	HSC70	Cytosol/nucleus	
		HSP70	Cytosol/nucleus	
		BIP	ER	
		mHSP70	Mitochondria	
HSP60	*E. coli*	groEL	Cytosol	Refolds and prevents aggregation of denatured proteins *in vitro;* may facilitate protein degradation by acting as a cofactor in proteolytic systems; role in the assembly of bacteriophages and Rubisco (an abundant protein in the chloroplast)
	S. cerevisiae	HSP60	Mitochondria	
	Plants	Cpn60	Chloroplasts	
	Mammals	HSP60	Mitochondria	
HSP40	*E. coli*	dnaJ	Cytosol	Essential co-chaperone activity with HSP70 proteins to enhance rate of ATPase activity and substrate release
	S. cerevisiae	Ydh1	Cytosol/nucleus	
	Malmmals			
Small HSPs	*E. coli*	lbp A and B	Cytosol	Suppresses aggregation and heat inactivation of protein *in vitro;* confers thermotolerance through stabliization of microfilaments; anit-apoptotic activity
	S. cerevisiae	HSP27	Cytosol	
	Mammals	αA and αB-crystallin	Cytosol	
		HSP27	Cytosol	

binding and hydrolysis are regulated by a large family of cochaperones such as p23 or the immunophilins which enhance Hsp90, dnaJ/Hsp40, and Hip, or Bag proteins which associate with Hsp70, or Hsp10/groES which stimulates Hsp60/groEL [59,63,64]. The association of chaperones with different partner proteins thus influences the folded state of the substrate and consequently its activities (Fig. 2).

The ability of a cell to know whether to grow, divide, differentiate, or die depends on extracellular signals and the ability to properly recognize and respond to these signals. The cell may receive these extracellular signals in different forms such as soluble hormones, small peptides, or proteins attached to neighboring cells. Cellular receptors receiving the signals transmit the extracellular information to the nucleus through cascades of protein–protein interactions and biochemical reactions. Chaperones of the Hsp90 and Hsp70 family and their cochaperones have been found to interact with key cell signaling molecules, including intracellular receptors, tyrosine- and serine/threonine kinases, cell cycle regulators, and cell death regulators [42,65–67]. Decreasing the levels of functional Hsp90 in *Drosophila,* by genetic mutation or by treatment with Hsp90 inhibitor geldanamycin, causes developmental abnormalities [68]. Likewise, increasing the levels of Hsp70, by overexpression or upon heat shock, has growth inhibitory effects on mammalian tissue culture cells and in *Drosophila* salivary gland cells, whereas expression of a dominant-negative form of Hsp70 causes developmental defects in *Drosophila* [69–71]. However, what remains less well understood is whether this represents a general strategy of the cell to link specific signaling pathways with cell stress-sensing events.

Interestingly, cells that have lost their ability to properly regulate cell growth, such as tumor cells, often express high levels of multiple heat-shock proteins compared to their normal parental cells [72]. Depletion of Hsp90 by geldanamycin, or of Hsp70 by antisense methodology in transformed cells, but not in their nontransformed counterparts, causes either an arrest of cell growth or progression into cell death [73,74]. Apparently, tumor cells have become dependent on increased levels of Hsps, although the beneficial reasons for this have yet to be clearly established. One possibility is the ability of chaperones to suppress and buffer mutations that accumulate during the transformation process, thus allowing cell viability and even enhanced cell growth of otherwise mutant cells. This is exemplified by the relationship between p53 and Hsp90, where mutant forms of p53, but not wild-type p53, depend on Hsp90 for their normal levels and function [75].

As a consequence of stress and physiological change, the equilibrium between substrates, heat-shock proteins, and cochaperones within a cell may be perturbed, with potentially profound consequences for the cellular phenotype. Changes in the abundance and relative levels of chaperones and cochaperones could result in novel combinations of heat-shock proteins, which, in turn, could redirect information flow through the intracellular pathways and change the overall response to signals. Whereas some pathways may become favored because of an increase in a particular cochaperone that is specifically required for its regulation, other pathways might be shut down entirely by the lack of critical chaperone components. The overall effect of changes in chaperone or cochaperone levels on cellular and organismal phenotypes will depend on which chaperone or cochaperone is affected. For example, changes in the levels of Hsp90 by exposure of cells or organisms to geldanamycin can have strong pleiotropic effects on development [68,76]. Likewise, the Ras/Raf-1 signaling pathway in mammalian cells is arrested when the levels of Hsp70 increase in response to stress. The inhibition of MAP kinase is due to the sequestration of Bag1, a Raf1 positive effector, by Hsp70, thus disrupting cell growth [71,77].

Altogether, heat-shock proteins have evolved as integrated components of signal transduction networks, in which they can function in the maturation, activation, and inactivation of signaling molecules. Their involvement in a particular pathway within the network is determined by the availability and relative abundance of partner specific cochaperones, which will influence, in a cell-type-specific manner, the natural response to physiological intracellular and extracellular signals. Consequently, we suggest that altered levels of heat-shock proteins and cochaperones in response to stress or disease states alters how organisms integrate and respond to the flow of their normal physiological signals. Future studies in multicellular model systems will help to elucidate with greater detail the molecular basis for the pervasive role of molecular chaperones in organismal development and disease and how they respond to altered chaperone and cochaperone levels associated with fluctuating environmental conditions and disease.

Neurodegenerative Diseases: When Aggregation-Prone Proteins go Awry

Human neurodegenerative disorders appear to be associated with the accumulation of misfolded proteins and the appearance of protein aggregates, fibrils, or plaques. These include inherited disorders caused by CAG/polyglutamine-expansion as occurs in Huntington's disease (HD), Kennedy disease, the spinocerebellar ataxias SCA1, SCA2, MJD (SCA3), SCA6, and SCA7, dentorubral-pallidoluysian atrophy (DPRLA), Alzheimer's disease, and prion disease [78]. A hallmark of these diseases is the expression of abnormal proteins that form a predominant β-sheet conformation that self-associates to form insoluble aggregates in neuronal cells or in the extracellular space [79–81].

Prion diseases from yeast to humans are associated with changes in the folded state of the prion, which correlates with conversion from a noninfective to an infective state. Among the fascinating features of prion diseases is the epigenetic transmission of an altered protein conformation, thus endowing the infectious prion with unique biological properties. The association of chaperones with prion disease has been suggested based on studies in the yeast *S. cerevisiae*. A subunit of a translation termination factor, Sup35p, can form self-seeded fibers in [psi+] strains, in a prion-like manner which is prevented if Hsp104 is deleted or overproduced [82]. Hsp104 can directly interact with the prion-like domain of Sup35p and also with the human prion protein, which is suggestive for a conserved mechanism [83]. The conserved properties of heat-shock proteins and the striking similarity in the characteristics of *in vitro*-formed fibers of Sup35p, polyglutamine repeat, and prion proteins have further strongly suggested the involvement of heat-shock proteins in aggregate formation in other neurodegenerative disorders.

Indeed the heat-shock response and heat-shock proteins have been implicated in these polyglutamine-expansion misfolding diseases. Studies with mammalian tissue culture cells and the nematode *C. elegans* have established that the heat-shock response is activated in cells expressing polyglutamine-expansion-containing proteins [84,85]. Moreover, the colocalization of several heat-shock proteins, including the Hsp40 family members Hdj-1 and Hdj-2, Hsp70, and ubiquitin, with polyglutamine aggregates in mouse tissues and tissue culture cells has suggested a direct relationship between these heat-shock proteins and polyglutamine diseases [86]. Recently, evidence has accumulated that heat-shock proteins could play important roles in cytoprotection based on observations that overexpression of Hdj-1, Hdj-2, or Hsp70 reduces polyglutamine aggregate formation and prevents cellular degeneration [87–89]. These observations offer interesting possibilities to develop therapeutic strategies based on activation of the stress response or selectively increasing the levels of individual chaperones.

Acknowledgments

R.M. was supported by a research grant from the National Institutes of General Medical Sciences (NIH), the Gollub Foundation, the Huntington's Disease Society of America Coalition for the Cure, and the Daniel F. and Ada L. Rice Foundation. E. A. A. N. was supported by a postdoctoral fellowship from the Netherlands Organization for Scientific Research and an EMBO Long Term Fellowship. We thank K. Veraldi and S. Fox for their invaluable assistance in the preparation of this manuscript and G. Matsumoto for Fig. 3.

References

1. Morimoto, R. I., Jurivich, D. A., Kroger, P. E., Mathur, S. K., Murphy, S. P., Nakai, A., Sarge, K., Abravaya, K., and Sistonen, L. T. (1994). In Morimoto, R. I., Tissieres, A., and Georgopoulos, C., Eds., *The Biology of Heat-Shock Proteins and Molecular Chaperones,* pp. 417–455. Cold Spring Harbor Laboratory Press, Cold Spring Harbor, NY.
2. Bukau, B. (1999). *Molecular Chaperones and Folding Catalyst, Regulation, Cellular Function and Mechanism.* Harwood Academic Publishers, Amsterdam.
3. Hartl, F. U. (1996). Molecular chaperones in cellular protein folding. *Nature* **381**, 571–579.
4. Downes, C. P., Wolf, C. R., and Lane, D. P. (1999). Portland Press, London.

5. Parsell, D. A. and Lindquist, S. (1994). In Morimoto, R. I., Tissieres, A., and Georgopoulos, C., Eds., *The Biology of Heat-Shock Proteins and Molecular Chaperones,* pp. 457–494. Cold Spring Harbor Laboratory Press, Cold Spring Harbor, NY.
6. Lis, J. and Wu, C. (1993). Protein traffic on the heat shock promoter: Parking, stalling, and trucking along. *Cell* **74**, 1–4.
7. Morimoto, R. I. (1993). Cells in stress: Transcriptional activation of heat shock genes. *Science* **259**, 1409–1410.
8. Wu, C. (1995). Heat shock transcription factors: Structure and regulation. *Annu. Rev. Cell Dev. Biol.* **11**, 441–469.
9. Morimoto, R. I. and Santoro, M. G. (1998). Stress-inducible responses and heat shock proteins: New pharmacologic targets for cytoprotection. *Nat. Biotechnol.* **16**, 833–838.
10. Wiederrecht, G., Seto, D., and Parker, C. S. (1988). Isolation of the gene encoding the *S. cerevisiae* heat shock transcription factor. *Cell* **54**, 841–853.
11. Sorger, P. K. and Pelham, H. R. (1988). Yeast heat shock factor is an essential DNA-binding protein that exhibits temperature-dependent phosphorylation. *Cell* **54**, 855–864.
12. Clos, J., Westwood, J. T., Becker, P. B., Wilson, S., Lambert, K., and Wu, C. (1990). Molecular cloning and expression of a hexameric *Drosophila* heat shock factor subject to negative regulation. *Cell* **63**, 1085–1097.
13. Rabindran, S. K., Giorgi, G., Clos, J., and Wu, C. (1991). Molecular cloning and expression of a human heat shock factor, HSF1. *Proc. Natl. Acad. Sci. USA* **88**, 6906–6910.
14. Sarge, K. D., Zimarino, V., Holm, K., Wu, C., and Morimoto, R. I. (1991). Cloning and characterization of two mouse heat shock factors with distinct inducible and constitutive DNA-binding ability. *Genes Dev.* **5**, 1902–1911.
15. Nakai, A. and Morimoto, R. I. (1993). Characterization of a novel chicken heat shock transcription factor, heat shock factor 3, suggests a new regulatory pathway. *Mol. Cell Biol.* **13**, 1983–1997.
16. Harrison, C. J., Bohm, A. A., and Nelson, H. C. (1994). Crystal structure of the DNA binding domain of the heat shock transcription factor. *Science* **263**, 224–227.
17. Vuister, G. W., Kim, S. J., Orosz, A., Marquardt, J., Wu, C., and Bax, A. (1994). Solution structure of the DNA-binding domain of *Drosophila* heat shock transcription factor. *Nat. Struct. Biol.* **1**, 605–614.
18. Sorger, P. K. and Nelson, H. C. (1989). Trimerization of a yeast transcriptional activator via a coiled-coil motif. *Cell* **59**, 807–813.
19. Peteranderl, R. and Nelson, H. C. (1992). Trimerization of the heat shock transcription factor by a triple-stranded alpha-helical coiled-coil. *Biochemistry* **31**, 12272–12276.
20. Nakai, A., Tanabe, M., Kawazoe, Y., Inazawa, J., Morimoto, R. I., and Nagata, K. (1997). HSF4, a new member of the human heat shock factor family which lacks properties of a transcriptional activator. *Mol. Cell Biol.* **17**, 469–481.
21. Green, M., Schuetz, T. J., Sullivan, E. K., and Kingston, R. E. (1995). A heat shock-responsive domain of human HSF1 that regulates transcription activation domain function. *Mol. Cell Biol.* **15**, 3354–3362.
22. Shi, Y., Kroeger, P. E., and Morimoto, R. I. (1995). The carboxyl-terminal transactivation domain of heat shock factor 1 is negatively regulated and stress responsive. *Mol. Cell Biol.* **15**, 4309–4318.
23. Zuo, J., Rungger, D., and Voellmy, R. (1995). Multiple layers of regulation of human heat shock transcription factor 1. *Mol. Cell Biol.* **15**, 4319–4330.
24. Wisniewski, J., Orosz, A., Allada, R., and Wu, C. (1996). The C-terminal region of *Drosophila* heat shock factor (HSF) contains a constitutively functional transactivation domain. *Nucleic Acids Res.* **24**, 367–374.
25. Farkas, T., Kutskova, Y. A., and Zimarino, V. (1998). Intramolecular repression of mouse heat shock factor 1. *Mol. Cell Biol.* **18**, 906–918.
26. Morimoto, R. I. (1998). Regulation of the heat shock transcriptional response: Cross talk between a family of heat shock factors, molecular chaperones, and negative regulators. *Genes Dev.* **12**, 3788–3796.
27. Larson, J. S., Schuetz, T. J., and Kingston, R. E. (1988). Activation in vitro of sequence-specific DNA binding by a human regulatory factor. *Nature* **335**, 372–375.
28. Mosser, D. D., Kotzbauer, P. T., Sarge, K. D., and Morimoto, R. I. (1990). In vitro activation of heat shock transcription factor DNA-binding by calcium and biochemical conditions that affect protein conformation. *Proc. Natl. Acad. Sci. USA* **87**, 3748–3752.
29. Zimarino, V., Tsai, C., and Wu, C. (1990). Complex modes of heat shock factor activation. *Mol. Cell Biol.* **10**, 752–759.
30. Westwood, J. T., Clos, J., and Wu, C. (1991). Stress-induced oligomerization and chromosomal relocalization of heat-shock factor. *Nature* **353**, 822–827.
31. Knauf, U., Newton, E. M., Kyriakis, J., and Kingston, R. E. (1996). Repression of human heat shock factor 1 activity at control temperature by phosphorylation. *Genes Dev.* **10**, 2782–2793.
32. Kline, M. P. and Morimoto, R. I. (1997). Repression of the heat shock factor 1 transcriptional activation domain is modulated by constitutive phosphorylation. *Mol. Cell Biol.* **17**, 2107–2115.
33. Jurivich, D. A., Sistonen, L., Kroes, R. A., and Morimoto, R. I. (1992). Effect of sodium salicylate on the human heat shock response. *Science* **255**, 1243–1245.
34. Giardina, C. and Lis, J. T. (1995). Sodium salicylate and yeast heat shock gene transcription. *J. Biol. Chem.* **270**, 10369–10372.
35. Cotto, J. J., Kline, M., and Morimoto, R. I. (1996). Activation of heat shock factor 1 DNA binding precedes stress-induced serine phosphorylation. Evidence for a multistep pathway of regulation. *J. Biol. Chem.* **271**, 3355–3358.
36. Abravaya, K., Myers, M. P., Murphy, S. P., and Morimoto, R. I. (1992). The human heat shock protein hsp70 interacts with HSF, the transcription factor that regulates heat shock gene expression. *Genes Dev.* **6**, 1153–1164.
37. Baler, R., Welch, W. J., and Voellmy, R. (1992). Heat shock gene regulation by nascent polypeptides and denatured proteins: Hsp70 as a potential autoregulatory factor. *J. Cell Biol.* **117**, 1151–1159.
38. Rabindran, S. K., Wisniewski, J., Li, L., Li, G. C., and Wu, C. (1994). Interaction between heat shock factor and hsp70 is insufficient to suppress induction of DNA-binding activity *in vivo*. *Mol. Cell Biol.* **14**, 6552–6560.
39. Nunes, S. L. and Calderwood, S. K. (1995). Heat shock factor-1 and the heat shock cognate 70 protein associate in high molecular weight complexes in the cytoplasm of NIH-3T3 cells. *Biochem. Biophys. Res. Communications* **213**, 1–6.
40. Shi, Y., Mosser, D. D., and Morimoto, R. I. (1998). Molecular chaperones as HSF1-specific transcriptional repressors. *Genes Dev.* **12**, 654–666.
41. Marchler, G. and Wu, C. (2001). Modulation of *Drosophila* heat shock transcription factor activity by the molecular chaperone DROJ1. *EMBO J.* **20**, 499–509.
42. Morimoto, R. I. (2002). Dynamic remodeling of transcription complexes by molecular chaperones. *Cell* **110**, 1–4
43. Stone, D. E. and Craig, E. A. (1990). Self-regulation of 70-kilodalton heat shock proteins in *Saccharomyces cerevisiae*. *Mol. Cell Biol.* **10**, 1622–1632.
44. Boorstein, W. R. and Craig, E. A. (1990). Transcriptional regulation of SSA3, an HSP70 gene from *Saccharomyces cerevisiae*. *Mol. Cell Biol.* **10**, 3262–3267.
45. Nelson, R. J., Heschl, M. F., and Craig, E. A. (1992). Isolation and characterization of extragenic suppressors of mutations in the SSA hsp70 genes of *Saccharomyces cerevisiae*. *Genetics* **131**, 277–285.
46. Halladay, J. T. and Craig, E. A. (1995). A heat shock transcription factor with reduced activity suppresses a yeast HSP70 mutant. *Mol. Cell Biol.* **15**, 4890–4897.
47. Sarge, K. D., Murphy, S. P., and Morimoto, R. I. (1993). Activation of heat shock gene transcription by heat shock factor 1 involves oligomerization, acquisition of DNA-binding activity, and nuclear localization and can occur in the absence of stress. *Mol. Cell Biol.* **13**, 1392–1407.
48. Theodorakis, N. G., Zand, D. J., Kotzbauer, P. T., Williams, G. T., and Morimoto, R. I. (1989). Hemin-induced transcriptional activation of the HSP70 gene during erythroid maturation in K562 cells is due to a heat shock factor-mediated stress response. *Mol. Cell Biol.* **9**, 3166–3173.

49. Sistonen, L., Sarge, K. D., Phillips, B., Abravaya, K., and Morimoto, R. I. (1992). Activation of heat shock factor 2 during hemin-induced differentiation of human erythroleukemia cells. *Mol. Cell Biol.* **12**, 4104–4111.
50. Sistonen, L., Sarge, K. D., and Morimoto, R. I. (1994). Human heat shock factors 1 and 2 are differentially activated and can synergistically induce hsp70 gene transcription. *Mol. Cell Biol.* **14**, 2087–2099.
51. Sarge, K. D., Park-Sarge, O. K., Kirby, J. D., Mayo, K. E., and Morimoto, R. I. (1994). Expression of heat shock factor 2 in mouse testis: Potential role as a regulator of heat-shock protein gene expression during spermatogenesis. *Biol. Reprod.* **50**, 1334–1343.
52. Mezger, V., Rallu, M., Morimoto, R. I., Morange, M., and Renard, J. P. (1994). Heat shock factor 2-like activity in mouse blastocysts. *Dev. Biol.* **166**, 819–822.
53. Rallu, M., Loones, M., Lallemand, Y., Morimoto, R., Morange, M., and Mezger, V. (1997). Function and regulation of heat shock factor 2 during mouse embryogenesis. *Proc. Natl. Acad. Sci. USA* **94**, 2392–2397.
54. Mathew, A., Mathur, S. K., and Morimoto, R. I. (1998). Heat shock response and protein degradation: Regulation of HSF2 by the ubiquitin-proteasome pathway. *Mol. Cell Biol.* **18**, 5091–5098.
55. Rudiger, S., Germeroth, L., Schneider-Mergener, J., and Bukau, B. (1997). Substrate specificity of the DnaK chaperone determined by screening cellulose-bound peptide libraries. *EMBO J.* **16**, 1501–1507.
56. Horwich, A. L., Low, K. B., Fenton, W. A., Hirshfield, I. N., and Furtak, K. (1993). Folding *in vivo* of bacterial cytoplasmic proteins: Role of GroEL. *Cell* **74**, 909–917.
57. Mayhew, M., da Silva, A. C., Martin, J., Erdjument-Bromage, H., Tempst, P., and Hartl, F. U. (1996). Protein folding in the central cavity of the GroEL-GroES chaperonin complex. *Nature* **379**, 420–426.
58. Gething, M. J. (1997). *Molecular Chaperones and Protein-Folding Catalysts.* Oxford University Press, New York.
59. Freeman, B. C., Toft, D. O., and Morimoto, R. I. (1996). Molecular chaperone machines: Chaperone activities of the cyclophilin Cyp-40 and the steroid aporeceptor-associated protein p23. *Science* **274**, 1718–1720.
60. Jakob, U., Lilie, H., Meyer, I., and Buchner, J. (1995). Transient interaction of Hsp90 with early unfolding intermediates of citrate synthase. Implications for heat shock *in vivo. J. Biol. Chem.* **270**, 7288–7294.
61. Pratt, W. B. and Welsh, M. J. (1994). Chaperone functions of the heat shock proteins associated with steroid receptors. *Semin. Cell Biol.* **5**, 83–93.
62. Smith, D. F. (1997). In Gething, M. J., Ed., *Molecular Chaperones* and *Protein-Folding Catalysts,* pp. 518–521. Oxford University Press, New York.
63. Freeman, M. L., Borrelli, M. J., Syed, K., Senisterra, G., Stafford, D. M., and Lepock, J. R. (1995). Characterization of a signal generated by oxidation of protein thiols that activates the heat shock transcription factor. *J. Cell. Physiol.* **164**, 356–366.
64. Goloubinoff, P., Gatenby, A. A., and Lorimer, G. H. (1989). GroE heat-shock proteins promote assembly of foreign prokaryotic ribulose bisphosphate carboxylase oligomers in *Escherichia coli. Nature* **337**, 44–7.
65. Xu, Y. and Lindquist, S. (1993). Heat-shock protein hsp90 governs the activity of pp60v-src kinase. *Proc. Natl. Acad. Sci. USA* **90**, 7074–7078.
66. Dittmar, K. D., Banach, M., Galigniana, M. D., and Pratt, W. B. (1998). The role of DnaJ-like proteins in glucocorticoid receptor. hsp90 heterocomplex assembly by the reconstituted hsp90.p60.hsp70 foldosome complex. *J. Biol. Chem.* **273**, 7358–7366.
67. Sato, S., Fujita, N., and Tsuruo, T. (2000). Modulation of Akt kinase activity by binding to Hsp90. *Proc. Natl. Acad. Sci. USA* **97**, 10832–10837.
68. Rutherford, S. L. and Lindquist, S. (1998). Hsp90 as a capacitor for morphological evolution [see comments]. *Nature* **396**, 336–342.
69. Elefant, F. and Palter, K. B. (1999). Tissue-specific expression of dominant negative mutant *Drosophila* HSC70 causes developmental defects and lethality. *Mol. Biol. Cell* **10**, 2101–2117.
70. Feder, J. H., Rossi, J. M., Solomon, J., Solomon, N., and Lindquist, S. (1992). The consequences of expressing hsp70 in *Drosophila* cells at normal temperatures. *Genes Dev.* **6**, 1402–1413.
71. Song, J., Takeda, M., and Morimoto, R. I. (2001). Bag1-Hsp70 mediates a physiological stress signalling pathway that regulates Raf-1/ERK and cell growth. *Nat. Cell Biol.* **3**, 276–282.
72. Jaattela, M. (1999). Heat shock proteins as cellular lifeguards. *Ann. Med.* **31**, 261–271.
73. Nylandsted, J., Rohde, M., Brand, K., Bastholm, L., Elling, F., and Jaattela, M. (2000). Selective depletion of heat shock protein 70 (Hsp70) activates a tumor-specific death program that is independent of caspases and bypasses Bcl-2. *Proc. Natl. Acad. Sci. USA* **97**, 7871–7876.
74. Whitesell, L., Mimnaugh, E. G., De Costa, B., Myers, C. E., and Neckers, L. M. (1994). Inhibition of heat shock protein HSP90-pp60v-src heteroprotein complex formation by benzoquinone ansamycins: Essential role for stress proteins in oncogenic transformation. *Proc. Natl. Acad. Sci. USA* **91**, 8324–8328.
75. Blagosklonny, M. V., Toretsky, J., Bohen, S., and Neckers, L. (1996). Mutant conformation of p53 translated *in vitro* or *in vivo* requires functional HSP90. *Proc. Natl. Acad. Sci. USA* **93**, 8379–8383.
76. Queitsch, C., Sangster, T. A., and Lindquist, S. (2002). hsp90 as a capacitor of phenotypic variation. *Nature* **417**, 618–624.
77. Nollen, E. A. A. and Morimoto, R. I. (2002). Chaperoning signaling pathways: Molecular chaperones as stress-sensing "heat shock" proteins. *J. Cell Sci.* **115**, 2809–2816.
78. Kakizuka, A. (1998). Protein precipitation: A common etiology in neurodegenerative disorders? *Trends Genet.* **14**, 396–402.
79. Jackson, G. S., Hosszu, L. L., Power, A., Hill, A. F., Kenney, J., Saibil, H., Craven, C. J., Waltho, J. P., Clarke, A. R., and Collinge, J. (1999). Reversible conversion of monomeric human prion protein between native and fibrinogenic conformations. *Science* **283**, 1935–1937.
80. Jimenez, J. L., Guijarro, J. I., Orlova, E., Zurdo, J., Dobson, C. M., Sunde, M., and Saibil, H. R. (1999). Cryo-electron microscopy structure of an SH3 amyloid fibril and model of the molecular packing. *EMBO J.* **18**, 815–821.
81. Scherzinger, E., Lurz, R., Turmaine, M., Mangiarini, L., Hollenbach, B., Hasenbank, R., Bates, G. P., Davies, S. W., Lehrach, H., and Wanker, E. E. (1997) Huntington-encoded polyglutamine expansions form amyloid-like protein aggregates *in vitro* and *in vivo. Cell* **90**, 549–558.
82. Chernoff, Y. O., Lindquist, S. L., Ono, B., Inge-Vechtomov, S. G., and Liebman, S. W. (1995). Role of the chaperone protein Hsp104 in propagation of the yeast prion-like factor [psi+]. *Science* **268**, 880–884.
83. Schirmer, E. C. and Lindquist, S. (1997). Interactions of the chaperone Hsp104 with yeast Sup35 and mammalian PrP. *Proc. Natl. Acad. Sci. USA* **94**, 13932–13937.
84. Satyal, S. H., Schmidt, E., Kitagawa, K., Sondheimer, N., Lindquist, S., Kramer, J. M., and Morimoto, R. I. (2000). Polyglutamine aggregates alter protein folding homeostasis in *Caenorhabditis elegans. Proc. Natl. Acad. Sci. USA* **97**, 5750–5755.
85. Wyttenbach, A., Carmichael, J., Swartz, J., Furlong, R. A., Narain, Y., Rankin, J., and Rubinsztein, D. C. (2000). Effects of heat shock, heat shock protein 40 (HDJ-2), and proteasome inhibition on protein aggregation in cellular models of Huntington's disease. *Proc. Natl. Acad. Sci. USA* **97**, 2898–2903.
86. Cummings, C. J., Mancini, M. A., Antalffy, B., DeFranco, D. B., Orr, H. T., and Zoghbi, H. Y. (1998). Chaperone suppression of aggregation and altered subcellular proteasome localization imply protein misfolding in SCA1. *Nat. Genet.* **19**, 148–154.
87. Krobitsch, S. and Lindquist, S. (2000). Aggregation of Huntington in yeast varies with the length of the polyglutamine expansion and the expression of chaperone proteins. *Proc. Natl. Acad. Sci. USA* **97**, 1589–1594.
88. Warrick, J. M., Chan, H. Y., Gray-Board, G. L., Chai, Y., Paulson, H. L., and Bonini, N. M. (1999). Suppression of polyglutamine-mediated neurodegeneration in *Drosophila* by the molecular chaperone HSP70. *Nat. Genet.* **23**, 425–428.
89. Kazemi-Esfarjani, P. and Benzer, S. (2000). Genetic suppression of polyglutamine toxicity in *Drosophila. Science* **287**, 1837–1840.

CHAPTER 311

Hypoxia-Mediated Signaling Pathways

Albert C. Koong and Amato J. Giaccia
Division of Radiation Biology, Department of Radiation Oncology, Stanford University School of Medicine, Stanford, California

Introduction

Oxygen homeostasis is a tightly regulated process. Disruption of oxygen delivery pathways can lead to cardiovascular disease, cerebrovascular disorders, and chronic pulmonary disease. Within solid tumors, poorly oxygenated regions develop because the tumor vasculature is unable to meet the demands of tumor expansion, and because of aberrant blood vessel formation within the tumor. Numerous clinical studies have documented the presence of low oxygen levels within most solid tumors and correlated the presence of these hypoxic regions with poor clinical outcome [1–4].

Hypoxic cells are relatively resistant to conventional radiation therapy and chemotherapy. When extremely hypoxic cells are irradiated, three times the radiation dose is required to achieve the same level of killing as required under fully aerobic conditions. Furthermore, most hypoxic cells are not actively undergoing cell division, thus impeding the efficacy of conventional chemotherapeutic agents that are targeted to actively dividing cells [5]. Numerous studies have indicated that hypoxia selects for tumors with an increased malignant phenotype by promoting angiogenesis [6], increasing metastases [7], and decreasing apoptotic potential [8].

HIF-1 Signaling

At the molecular level, transcriptional control of hypoxia-regulated genes occurs most prominently through hypoxia-induced factor (HIF-1). Although other transcription factors have also been reported to be activated by hypoxia (AP-1 [9–11], NFκB [12,13], Egr-1 [14,15], and NF-IL6 [16,17]), HIF-1 controls the majority of hypoxia-regulated genes. HIF-1-regulated genes play critical roles in a variety of cellular processes including angiogenesis, energy metabolism, iron homeostasis, vascular regulation, cell proliferation/survival, and erythropoiesis. To date, more than 30 genes have been reported to be transcriptionally regulated by HIF-1 [18]. Analysis of HIF-1 staining from clinical specimens suggests a relationship between HIF-1 protein accumulation, malignant progression, and outcome after therapy [19–22].

HIF-1 is a heterodimer consisting of HIF-1α and the aryl hydrocarbon receptor nuclear translocator (ARNT or HIF-1β). HIF-1 belongs to the basic helix–loop–helix (bHLH) family of transcription factors in which the HLH motifs mediate dimerization and the basic regions bind to DNA. In contrast to the constitutively expressed HIF-1β subunit, HIF-1α is an oxygen-labile protein that becomes stabilized in response to hypoxia, iron chelators, and divalent cations. These inducers of HIF-1α are all related to the metabolism of heme, leading some investigators to postulate that the cellular oxygen sensor is a heme-associated protein [23,24]. To date, no definitive identification of such a protein has been made. Interestingly, under hypoxic conditions, HIF-1a mRNA levels do not change in most cell lines, but HIF-1a protein levels increase [25].

The stability of HIF-1α protein is one of the critical elements in its regulation. HIF-1α interaction with the von Hippel-Lindau tumor suppressor gene product targets it for ubiquitin-mediated proteosomal degradation under normoxic conditions. Under hypoxic conditions, HIF-1α degradation is inhibited, and it translocates to the nucleus where it dimerizes with HIF-1β to form an active HIF-1 complex. The HIF-1 heterodimer then binds to the promoter region of target genes that contain specific recognition sequence(s) known as the hypoxia response element (HRE).

Tumors with mutations in the Von Hippel-Lindau (VHL) gene, such as renal cell carcinomas, exhibit high levels of HIF-1α protein under normoxic conditions. When wild-type is reintroduced into these cells, the expression of -1α protein significantly decreases during normoxia. VHL binds to a region in HIF-1α known as the *oxygen dependent degradation domain*. Several groups have reported that HIF-1α binding to VHL is regulated by proline hydroxylation at residue 564. This enzymatic modification is dependent on oxygen and requires iron as a cofactor. Epstein *et al.* [26] demonstrated that HIF-1α posttranslational modification is mediated by a novel prolyl-4-hydroxylase that they have termed HIF-α prolyl-hydroxylase (HIF-PH). Under hypoxic conditions or in the presence of an iron chelator (desferrioxamine, DFO), this interaction is disrupted and HIF-1α is stabilized (Fig. 1).

Direct phosphorylation of HIF-1α also appears to play a role in the regulation of HIF-1 activity. Wang and Semenza [27] were the first to report that HIF-1 was a phosphoprotein. They demonstrated by electrophoretic mobility shift assays that the HIF-1/DNA complex could be disrupted by treating nuclear extracts of hypoxic cells with a phosphatase [27]. More recently, Suzuki and coworkers [28] reported that the phosphorylated form of HIF-1α is the predominant form found to associate with HIF-1β, whereas the dephosphorylated form of HIF-1α mediates binding to p53. Although the phosphorylation sites for HIF-1α are not known, these investigators correlated dephosphorylation of HIF-1α with increased apoptosis. These studies suggest that HIF-1α may have different cellular functions depending on its phosphorylation status.

The most widely studied downstream target gene of HIF-1 is vascular endothelial growth factor (VEGF), a potent angiogenic factor that is involved in the progression of a variety of cancers. Transcriptional regulation of VEGF is mediated by binding of HIF-1 to an HRE within the VEGF promoter [29]. In addition, under hypoxic conditions when cap-dependent translation may be limited, several groups have reported indirect evidence that VEGF gene induction may be mediated by an internal ribosome entry site mechanism to maintain a high level of gene expression during stress conditions [30,31].

Several lines of evidence suggest that the PI3-kinase/Akt pathway is involved in hypoxia-regulated VEGF expression and that the tumor suppressor gene PTEN can function as a negative regulator of the PI3-kinase/HIF pathway (Fig. 2). PTEN is mutated or inactivated in a large percentage of glioblastomas [32] and direct measurement of pO_2 with the Eppendorf microelectrode within these tumors demonstrates that they are hypoxic relative to normal tissue. Using PTEN-deficient glioblastoma cell lines, Zundel *et al.* [33] demonstrated that PTEN inhibits HIF-1 activation and VEGF induction by hypoxia to the same extent as wortmannin, a potent PI3-kinase inhibitor. Interestingly, PTEN overexpression inhibited the hypoxia-stimulated accumulation of HIF-1α, indicating that PTEN regulates growth factor signaling to HIF-1. Similar observations have been made in other cell types such as human breast cancer cells [34,35] and human prostate cancer cells [36,37]. These observations suggest that tumor cells deficient in PTEN tend to have elevated HIF-1 activity and thus increased production of pro-angiogenic factors.

Downstream of PTEN, Mazure and colleagues [38] demonstrated that in *ras*-transformed cells, transcriptional induction of VEGF is mediated by HIF-1 and occurs through the activation of the PI3-kinase/Akt pathway. Furthermore, the PI3-kinase/Akt pathway preferentially regulates HIF-1α without apparent effect on HIF-2α [34] or HIF-1β [37]. However, another report [39] indicated that the effect of the PI3-kinase/Akt pathway on hypoxia-induced activation of HIF-1α may be cell-type specific. For example, in some cell types (HepG2 and HEK293T), no correlation is seen between the activation of the PI3-kinase/Akt pathway and HIF-1α activity during hypoxia. Taken together, these seemingly inconsistent findings suggest that multiple and perhaps redundant regulatory mechanisms modulate HIF-1α activity

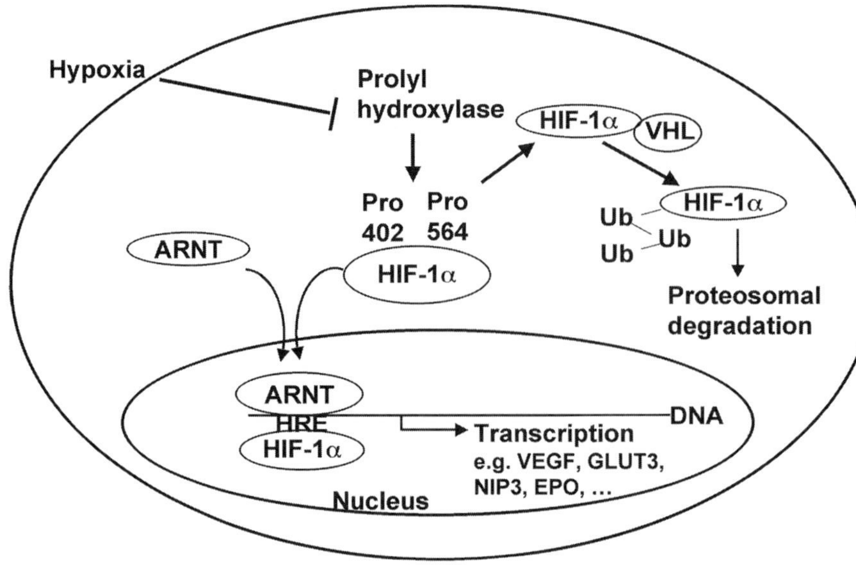

Figure 1 Diagram depicting our current understanding of HIF-1 regulation.

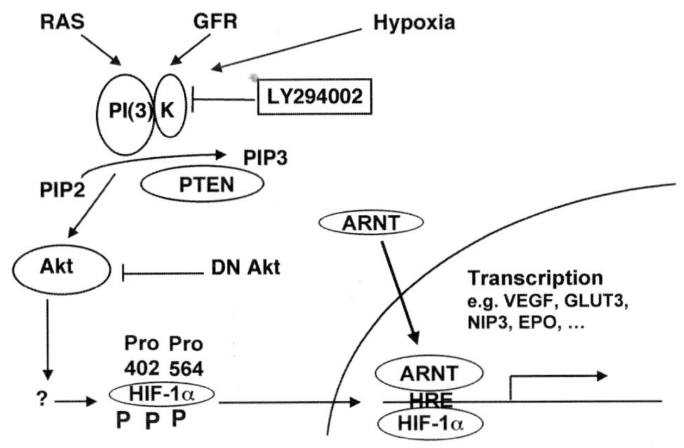

Figure 2 Involvement of a growth factor receptor/PI3-kinase/Akt pathway in HIF-1 regulation. In this figure, an activated growth factor receptor (GFR) binds the lipid kinase PI3K and results in the activation of Akt. Akt itself does not phosphorylate HIF-1α but through a presumed intermediate kinase results in HIF-1α phosphorylation (P). The phosphorylated form of HIF-1α then binds to ARNT and activates transcription.

during hypoxia. Therefore, it will be important to delineate specific pathways leading to the modulation of HIF-1α activities under hypoxia in any given cell type.

Involvement of the Raf-1/ERK pathway has also been implicated in the hypoxic induction of VEGF. With the use of dominant negatives of c-*src* and Raf-1 in 293 cells, Mukhopadhyay and coworkers [40] inhibited VEGF induction by hypoxia. However, this observation may also be dependent on the cell type, because other investigators do not report involvement of the Raf-1/ERK signaling cascade in hypoxia-mediated VEGF induction in *ras*-transformed NIH3T3 cells [38]. Similarly, Gleadle and colleagues [41] did not observe any modulation of HIF-1 DNA binding or downstream effector gene activation (GLUT-1, VEGF) using *src* dominant negatives or $src^{-/-}$ fibroblasts. More recently, however, Pal *et al.* [42] demonstrated increased c-*src* activity in breast cancer cells after exposure to hypoxia and inhibition of VEGF induction with exposure to antisense *src*. This area remains controversial and further work will help to delineate the role of *src* in the regulation of HIF-1/VEGF.

Unfolded Protein Response

Although HIF-1 is the major transcriptional regulator of the hypoxic stress response, during exposure to chronic hypoxia, other non-HIF-1 signaling pathways are also activated. For example, induction of the glucose regulated proteins (GRP78, GRP94, GRP170, GRP58) protects cells against hypoxia and other environmental stresses. The function of these endoplasmic reticulum (ER) resident proteins is to aid in protein folding, trafficking, and secretion. These chaperone proteins are induced during ER, stress in some cell types to help maintain proteins in their native conformation. The process, referred to as the *unfolded protein response* (UPR), involves the activation of a signaling cascade from the ER to the nucleus. Although this pathway has been well studied in yeast, many of the mammalian signaling pathways leading to the UPR have only recently been elucidated.

With the use of microarray technology, Travers and colleagues [43] have characterized global transcriptional changes during exposure to UPR-inducing agents in yeast. In this study, they found that the genes responsible for the yeast UPR were involved not only in protein folding in the ER but also in regulating secretion. They concluded that activation of the UPR results in a complete remodeling of the secretory pathway to minimize the accumulation of unfolded proteins in the ER [43]. Some of the mammalian orthologs to these yeast genes are known and characterization of the mammalian response to ER stress is an area of active investigation.

In yeast, the Ire1p and Hac1p genes are essential elements of the UPR. Ire1p, an ER transmembrane protein, has been implicated as a sensor for the accumulation of unfolded proteins. The exact role of Ire1p in the UPR is unknown, but in response to ER stress, Ire1p is autophosphorylated, leading to activation of its endonuclease activity. Activated Ire1p cleaves a 252-nucleotide inhibitory intron from HAC1 mRNA, resulting in an efficiently translated protein that up-regulates transcription of ER chaperone proteins [44,45].

The mammalian orthologs of Ire1 were identified as Ire1α and Ire1β [46,47]. Ire1α is ubiquitously expressed and Ire1β is expressed predominantly in the gut. Two different groups have independently shown that Ire1α endonuclease activity is required for proper processing of XBP1 mRNA, which results in transcriptional activation of the UPR [48,49]. Furthermore, during ER stress, ATF6 cleavage is mediated by site-2 protease (S2P), releasing a transcriptionally active 50-kDa protein. UPR target genes are then up-regulated by these transcription factors.

Analysis of mammalian GRP promoters revealed a novel consensus sequence responsible for the induction of these genes. This element, designated as the ER stress response element (ERSE), consists of $CCAATN_9CCACG$ [50]. Previously, Li *et al.* [51] showed that the CCAAT portion of the ERSE is constitutively occupied, and later studies by Yoshida and coworkers [52] identified the general transcription factor NF-Y (also known as CBF) to bind constitutively to this portion of the ERSE. Activated transcription factor 6 (ATF6) and X-box binding protein 1 (XBP1) were identified in a yeast 1 hybrid screen as two inducible transcription factors that can bind to the ERSE [50]. The transcriptional activity of activated ATF6 is enhanced by interactions with TFII-I [53], NF-Y [54], and YY1 [54].

The current model proposes that in response to ER stress, the cytoplasmic domain of ATF6 is cleaved, translocates to the nucleus, binds to the ERSE, and promotes the induction of UPR target genes. XBP1 also contains an ERSE in its promoter and is regulated by ATF6 activation, which ultimately leads to increased transcription of XBP1 mRNA (Fig. 3). During normal growth conditions, unspliced XBP1 mRNA is produced at low levels. However, during ER stress, XBP1 mRNA is spliced and the spliced form of XBP1 enhances transcription of ERSE-regulated

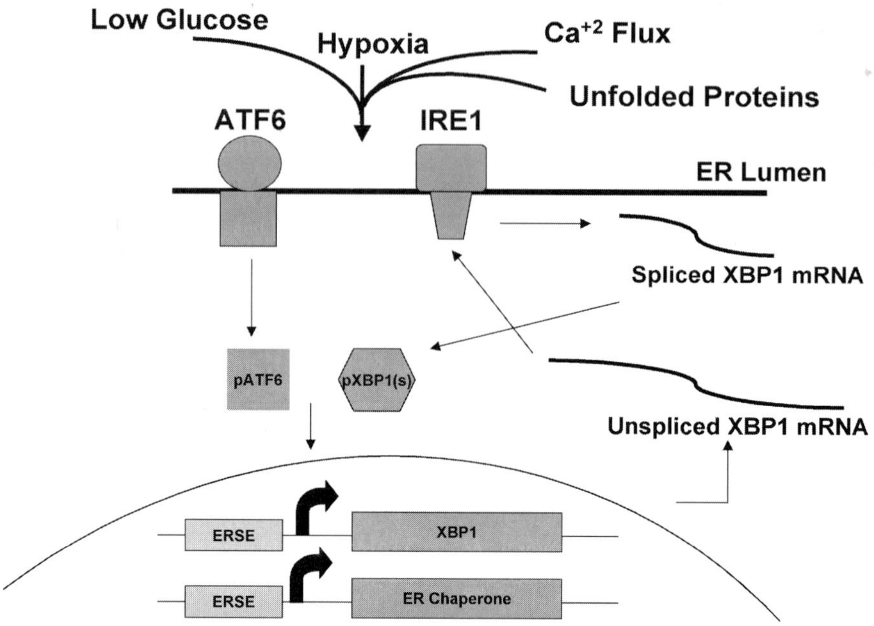

Figure 3 Model depicting the relationship between hypoxia and ER stress.

ER chaperone proteins. XBP-1 is postulated to be responsible for maintaining a prolonged UPR during exposure to chronic ER stress [48].

Song and coworkers [55] reported that GRP78 induction by hypoxia is mediated through a protein kinase C (PKC) epsilon/ERK/AP-1 signaling cascade. They showed that transfection with a dominant negative of PKC epsilon blocks the hypoxia-regulated signaling pathway leading to GRP78 production. GRP78 induction by hypoxia can also be blocked by pretreating cells with PKC inhibitors such as staurosporine. Furthermore, as evidenced by gel mobility shift assays, hypoxia activated the binding of a protein complex to a specific sequence in the promoter region of GRP78 containing the ERSE, and this binding was abolished in the presence of PKC inhibitors [56]. These results suggest that PKC mediates a portion of the signaling cascade, leading to induction of GRP78 during hypoxia. ER stressors such as glucose deprivation, protein accumulation, glycosylation inhibition, and calcium flux are all initiating agents of the UPR. Similarly, hypoxia is another stress that can illicit the UPR through similar and overlapping pathways. The chaperone proteins appear to serve a protective function during ER stress. Experiments have shown that antisense inhibition of GRP78 during hypoxia results in decreased cell survival under these conditions [57]. Another important function of the ER chaperone proteins was demonstrated by Ozawa and colleagues [58], who showed that antisense inhibition of oxygen-regulated protein (ORP150), also known as glucose-regulated protein 170 (GRP170), resulted in intracellular accumulation of VEGF within the ER and decreased VEGF secretion. These results suggest that the chaperone proteins may be an important component of angiogenesis.

Conclusions

Much progress has been made in characterizing the signaling pathways that are activated during hypoxia; however, many questions remain unanswered. Ultimately, by understanding the signaling pathways that are regulated by hypoxia, novel anticancer therapies may be developed to selectively target hypoxic tumor cells.

References

1. Brizel, D. M. *et al.* (19xx). Tumor hypoxia adversely affects the prognosis of carcinoma of the head and neck. *Int. J. Radiat. Oncol. Biol. Phys.* **38**, 285–289.
2. Nordsmark, M., Overgaard, M., and Overgaard, J. (1996). Pretreatment oxygenation predicts radiation response in advanced squamous cell carcinoma of the head and neck. *Radiother. Oncol.* **41**(1), 31–39.
3. Brizel, D.M. *et al.* (1996). Tumor oxygenation predicts for likelihood of distant metastasis in human soft tissue sarcoma. *Cancer Res.* **56**(5), 941–943.
4. Hockel, M. *et al.* (1996). Association between tumor hypoxia and malignant progression in advanced cancer of the uterine cervix. *Cancer Res.* **56**(19), 4509–4515.
5. Brown, J. M. and Giaccia, A. J. (1998). The unique physiology of solid tumors: Opportunities (and problems) for cancer therapy. *Cancer Res.* **58**(7), 1408–1416.
6. Harris, A. L. (2001). Hypoxia—A key regulatory factor in tumour growth. *Nat. Rev. Cancer* **2**(1), 38–47.
7. Cairns, R. A., Kalliomaki, T., and Hill, R. P. (2001). Acute (cyclic) hypoxia enhances spontaneous metastasis of KHT murine tumors. *Cancer Res.* **61**(24), 8903– 8908.
8. Graeber, T. G. *et al.* (1996). Hypoxia-mediated selection of cells with diminished apoptotic potential in solid tumours. *Nature* **379**(6560), 88–91.
9. Yao, K. S. *et al.* (1994). Activation of AP-1 and of a nuclear redox factor, Ref-1, in the response of HT29 colon cancer cells to hypoxia. *Mol. Cell Biol.* **14**(9), 5999–6003.

10. Ausserer, W. A. *et al.* (1994). Regulation of c-jun expression during hypoxic and low-glucose stress. *Mol. Cell Biol.* **14**(8), 5032–5042.
11. Laderoute, K. R. *et al.* (2002). The response of c-jun/AP-1 to chronic hypoxia is hypoxia-inducible factor 1 alpha dependent. *Mol. Cell Biol.* **22**(8), 2515–2523.
12. Koong, A. C., Chen, E. Y., and Giaccia, A. J. (1994). Hypoxia causes the activation of nuclear factor kappa B through the phosphorylation of I kappa B alpha on tyrosine residues. *Cancer Res.* **54**(6), 1425–1430.
13. Koong, A. C. *et al.* (1994). Hypoxic activation of nuclear factor-kappa B is mediated by a Ras and Raf signaling pathway and does not involve MAP kinase (ERK1 or ERK2). *Cancer Res.* **54**(20), 5273–9.
14. Bae, S. K. *et al.* (1999). Egr-1 mediates transcriptional activation of IGF-II gene in response to hypoxia. *Cancer Res.* **59**(23), 5989–94.
15. Nishi, H., Nishi, K. H., and Johnson, A. C. (2002). Early Growth Response-1 gene mediates up-regulation of epidermal growth factor receptor expression during hypoxia. *Cancer Res.* **62**(3), 827–34.
16. Matsui, H. *et al.* (1999). Induction of interleukin (IL)-6 by hypoxia is mediated by nuclear factor (NF)-kappa B and NF-IL6 in cardiac myocytes. *Cardiovasc. Res.* **42**(1), 104–12.
17. Hehlgans, T. *et al.* (2001). Hypoxic upregulation of TNF receptor type 2 expression involves NF-IL-6 and is independent of HIF-1 or HIF-2. *J. Interferon Cytokine Res.* **21**(9), 757–62.
18. Semenza, G. L. (2001). Hypoxia-inducible factor 1: Oxygen homeostasis and disease pathophysiology. *Trends Mol. Med.* **7**(8), 345–350.
19. Zhong, H. *et al.* (1999). Overexpression of hypoxia-inducible factor 1alpha in common human cancers and their metastases. *Cancer Res.* **59**(22), 5830–5.
20. Talks, K. L. *et al.* (2000). The expression and distribution of the hypoxia-inducible factors HIF-1alpha and HIF-2alpha in normal human tissues, cancers, and tumor-associated macrophages. *Am. J. Pathol.* **157**(2), 411–21.
21. Zagzag, D. *et al.* (2000). Expression of hypoxia-inducible factor 1alpha in brain tumors: Association with angiogenesis, invasion, and progression. *Cancer* **88**(11), 2606–18.
22. Aebersold, D. M. *et al.* (2001). Expression of hypoxia-inducible factor-1alpha: A novel predictive and prognostic parameter in the radiotherapy of oropharyngeal cancer. *Cancer Res.* **61**(7), 2911–6.
23. Goldberg, M. A., Dunning, S. P., and Bunn, H. F. (1988). Regulation of the erythropoietin gene: Evidence that the oxygen sensor is a heme protein. *Science* **242**(4884), 1412–5.
24. Goldberg, M. A. and Schneider, T. J. (1994). Similarities between the oxygen-sensing mechanisms regulating the expression of vascular endothelial growth factor and erythropoietin. *J. Biol. Chem.* **269**(6), 4355–9.
25. Huang, L. E. *et al.* (1996). Activation of hypoxia-inducible transcription factor depends primarily upon redox-sensitive stabilization of its alpha subunit. *J. Biol. Chem.* **271**(50), 32253–9.
26. Epstein, A. C. *et al.* (2001). *C. elegans* EGL-9 and mammalian homologs define a family of dioxygenases that regulate HIF by prolyl hydroxylation. *Cell* **107**(1), 43–54.
27. Wang, G. L. and Semenza, G. L. (1993). Characterization of hypoxia-inducible factor 1 and regulation of DNA binding activity by hypoxia. *J. Biol. Chem.* **268**(29), 21513–18.
28. Suzuki, H., Tomida, A., and Tsuruo, T. (2001). Dephosphorylated hypoxia-inducible factor 1alpha as a mediator of p53-dependent apoptosis during hypoxia. *Oncogene* **20**(41), 5779–88.
29. Forsythe, J. A. *et al.* (1996). Activation of vascular endothelial growth factor gene transcription by hypoxia-inducible factor 1. *Mol. Cell Biol.* **16**(9), 4604–13.
30. Stein, I. *et al.* (1998). Translation of vascular endothelial growth factor mRNA by internal ribosome entry: Implications for translation under hypoxia. *Mol. Cell Biol.* **18**(6), 3112–9.
31. Huez, I. *et al.* (1998). Two independent internal ribosome entry sites are involved in translation initiation of vascular endothelial growth factor mRNA. *Mol. Cell Biol.* **18**(11), 6178–90.
32. Cantley, L. C. and Neel, B. G. (1999). New insights into tumor suppression: PTEN suppresses tumor formation by restraining the phosphoinositide 3-kinase/AKT pathway. *Proc. Natl. Acad. Sci. USA* **96**(8), 4240–5.
33. Zundel, W. *et al.* (2000). Loss of PTEN facilitates HIF-1-mediated gene expression. *Genes Dev.* **14**(4), 391–6.
34. Blancher, C. *et al.* (2001). Effects of ras and von Hippel-Lindau (VHL) gene mutations on hypoxia-inducible factor (HIF)-1alpha, HIF-2alpha, and vascular endothelial growth factor expression and their regulation by the phosphatidylinositol 3′-kinase/Akt signaling pathway. *Cancer Res.* **61**(19), 7349–55.
35. Laughner, E. *et al.* (2001). HER2 (neu) signaling increases the rate of hypoxia-inducible factor 1alpha (HIF-1alpha) synthesis: Novel mechanism for HIF-1-mediated vascular endothelial growth factor expression. *Mol. Cell Biol.* **21**(12), 3995–4004.
36. Zhong, H. *et al.* (2000). Modulation of hypoxia-inducible factor 1alpha expression by the epidermal growth factor/phosphatidylinositol 3-kinase/PTEN/AKT/FRAP pathway in human prostate cancer cells: Implications for tumor angiogenesis and therapeutics. *Cancer Res.* **60**(6), 1541–5.
37. Jiang, B. H. *et al.* (2001). Phosphatidylinositol 3-kinase signaling controls levels of hypoxia-inducible factor 1. *Cell Growth Differ.* **12**(7), 363–9.
38. Mazure, N. M. *et al.* (1997). Induction of vascular endothelial growth factor by hypoxia is modulated by a phosphatidylinositol 3-kinase/Akt signaling pathway in Ha-ras-transformed cells through a hypoxia inducible factor-1 transcriptional element. *Blood* **90**(9), 3322–31.
39. Alvarez-Tejado, M. *et al.* (2002). Lack of evidence for the involvement of the phosphoinositide 3-kinase/Akt pathway in the activation of hypoxia-inducible factors by low oxygen tension. *J. Biol. Chem.* **277**(16), 13508–17.
40. Mukhopadhyay, D. *et al.* (1995). Hypoxic induction of human vascular endothelial growth factor expression through c-Src activation. *Nature* **375**(6532), 577–81.
41. Gleadle, J. M. and Ratcliffe, P. J. (1997). Induction of hypoxia-inducible factor-1, erythropoietin, vascular endothelial growth factor, and glucose transporter-1 by hypoxia: Evidence against a regulatory role for Src kinase. *Blood* **89**(2), 503–9.
42. Pal, S. *et al.* (2001). Central role of p53 on regulation of vascular permeability factor/vascular endothelial growth factor (VPF/VEGF) expression in mammary carcinoma. *Cancer Res.* **61**(18), 6952–7.
43. Travers, K. J. *et al.* (2000). Functional and genomic analyses reveal an essential coordination between the unfolded protein response and ER-associated degradation. *Cell* **101**(3), 249–58.
44. Mori, K. (2000). Tripartite management of unfolded proteins in the endoplasmic reticulum. *Cell* **101**(5), 451–4.
45. Ma, Y. and Hendershot, L. M. (2001). The unfolding tale of the unfolded protein response. *Cell* **107**(7), 827–30.
46. Wang, X. Z. *et al.* (1998). Cloning of mammalian Ire1 reveals diversity in the ER stress responses. *EMBO J.* **17**(19), 5708–17.
47. Tirasophon, W., Welihinda, A. A., and Kaufman, R. J. (1998). A stress response pathway from the endoplasmic reticulum to the nucleus requires a novel bifunctional protein kinase/endoribonuclease (Ire1p) in mammalian cells. *Genes Dev.* **12**(12), 1812–24.
48. Yoshida, H. *et al.* (2001). XBP1 mRNA is induced by ATF6 and spliced by IRE1 in response to ER stress to produce a highly active transcription factor. *Cell* **107**(7), 881–91.
49. Lee, K. *et al.* (2002). IRE1-mediated unconventional mRNA splicing and S2P-mediated ATF6 cleavage merge to regulate XBP1 in signaling the unfolded protein response. *Genes Dev.* **16**(4), 452–66.
50. Yoshida, H. *et al.* (1998). Identification of the cis-acting endoplasmic reticulum stress response element responsible for transcriptional induction of mammalian glucose-regulated proteins. Involvement of basic leucine zipper transcription factors. *J. Biol. Chem.* **273**(50), 33741–9.
51. Li, W. W. *et al.* (1994). Stress induction of the mammalian GRP78/BiP protein gene: In vivo genomic footprinting and identification of p70CORE from human nuclear extract as a DNA-binding component specific to the stress regulatory element. *Mol. Cell Biol.* **14**(8), 5533–46.

52. Yoshida, H. *et al.* (2000). ATF6 activated by proteolysis binds in the presence of NF-Y (CBF) directly to the cis-acting element responsible for the mammalian unfolded protein response. *Mol. Cell Biol.* **20**(18), 6755–67.
53. Parker, R. *et al.* (2001). Identification of TFII-I as the endoplasmic reticulum stress response element binding factor ERSF: Its autoregulation by stress and interaction with ATF6. *Mol. Cell Biol.* **21**(9), 3220–33.
54. Li, M. *et al.* (2000). ATF6 as a transcription activator of the endoplasmic reticulum stress element: Thapsigargin stress-induced changes and synergistic interactions with NF-Y and YY1. *Mol. Cell Biol.* **20**(14), 5096–106.
55. Song, M. S. *et al.* (2001). Induction of glucose-regulated protein 78 by chronic hypoxia in human gastric tumor cells through a protein kinase C-epsilon/ERK/AP-1 signaling cascade. *Cancer Res.* **61**(22), 8322–8330.
56. Koong, A. C., Auger, E. A., Chen, E. Y., and Giaccia, A. J. (1994). The regulation of GRP78 and messenger RNA levels by hypoxia is modulated by protein kinase C activators and inhibitors. *Radiat. Res.* **138**(1 Suppl.), S60–63.
57. Koong, A. C., Chen, E. Y., Lee, A. S., Brown, J. M., and Giacci, A. J. (1994). Increased cytotoxicity of chronic hypoxic cells by molecular inhibition of GRP78 induction. *Int. J. Radiat. Oncol. Biol. Phys.* **28**(3), 661–666.
58. Ozawa, K. *et al.* (2001). Regulation of tumor angiogenesis by oxygen-regulated protein 150, an inducible endoplasmic reticulum chaperone. *Cancer Res.* **61**(10), 4206–4213.

284

Regulation of mRNA Turnover by Cellular Stress

Myriam Gorospe
Laboratory of Cellular and Molecular Biology, Gerontology Research Center, National Institute on Aging–IRP, Baltimore, Maryland

Introduction

Changes in gene expression occurring in response to stressful stimuli are controlled at multiple levels, through a variety of regulatory mechanisms. Although we have extensive knowledge of the transcriptional mechanisms involved in altering gene expression, the posttranscriptional regulatory events are poorly understood. Alterations in mRNA stability constitute a major mechanism of posttranscriptional gene regulation, effecting rapid changes in gene expression (reviewed in [1–3]). In mammalian cells, the stability of mRNA fluctuates widely, ranging from 20 min to 24 hr [4]. In the absence of altered gene transcription, even small differences in half-life provide a very effective means of dramatically altering the abundance of a given mRNA and consequently the amount of protein expressed [2]. Different types of mRNA turnover have been identified, some promoting mRNA stability, others enhancing mRNA degradation. Among the genes whose expression is regulated through altered mRNA turnover are many encoding T-cell activation molecules, cell cycle regulators, cell survival proteins, and stress-response factors. Given their pivotal role in many physiological and pathological conditions, including conditions of stress, there is increasing interest in understanding the mechanisms that govern mRNA stability and mRNA degradation.

Messenger mRNA stabilization is indeed emerging as a major mechanism of *coordinate* regulation of stress-response gene expression. For example, the growth arrest and DNA-damage-inducible (*gadd*) genes are synchronously up-regulated in response to genotoxic stress, at least in part, through mRNA stabilization [5]; likewise, mRNAs encoding immediate-early genes (c-fos, c-myc, c-jun, IκB, etc.) are coordinately stabilized following irradiation with short-wavelength ultraviolet light [6]. Although the precise mechanisms determining mRNA turnover during stress are poorly understood, they are generally believed to involve RNA-binding proteins recognizing specific RNA sequences [7] (Fig. 1). During the past two decades some of the *cis* elements (RNA sequences) and *trans*-acting factors (participating proteins) involved in these processes have been described. However, the specific stress-triggered events that regulate the process of mRNA turnover are only now beginning to be characterized. In this chapter, I focus on our current understanding of how stress-triggered signaling pathways influence mRNA stability and regulate the expression of labile mRNAs.

mRNA Stability

Cis Elements Regulating mRNA Turnover

Certain mRNA stability-promoting structures are ubiquitous, like the 5′ *end cap*, a 7 methylguanosine triphosphate (m^7GpppG) structure that protects messages from general 5′→3′ exonucleases, and the 3′ *end polyadenylate* or *poly(A) tail*, which protects the 3′ terminus from degradation by 3′→5′ exonucleases. Other mRNA stability determinants are specific and they can be found throughout the 5′ untranslated region (UTR), as described for interleukin 2 (IL-2) mRNA [8], and the coding region, as shown for the c-fos

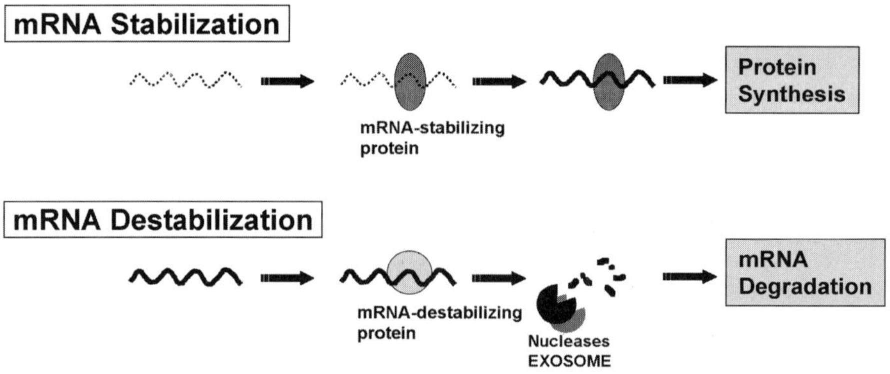

Figure 1 Schematic of mRNA Turnover. *mRNA Stabilization*: Labile mRNAs can be stabilized through mechanisms involving their association with proteins that promote enhanced transcript integrity. Examples include mRNAs encoding p21, VEGF, c-Fos, TNFα, and several interleukins. *mRNA Destabilization*: Labile mRNAs can be destabilized through mechanisms involving their association with proteins that promote transcript degradation, including RNases. Examples of mRNAs whose stability can decrease in a regulated manner include cyclin D1, cyclooxygenase-2, and APP.

and c-myc mRNAs [9,10]. However, the majority of specific regulatory sequences reside within the 3′UTR. The best understood examples of 3′UTR-mediated stability are the IRE (iron-responsive element), which confers stability to transferrin mRNA, and the ARE (AU-rich elements) present in the 3′UTR (and occasionally the 5′UTR) of many mRNAs, such as those encoding cytokines (interferons, interleukins, tumor necrosis factor [TNF] α), cell cycle regulatory proteins (p21, cyclin A, cyclin B1, cdc25), growth factors (GM-CSF, VEGF), and oncoproteins (c-fos, c-myc) [11–13]. AREs consist of one or several copies of the pentamer AUUUA, generally within a U-rich sequence.

Trans Factors Regulating mRNA Turnover

Many RNA-binding proteins have been described that selectively recognize and bind to AU-rich regions of labile mRNAs, including AU-A, AU-B, AU-C, AUBF, AUF1 (hnRNP D), AUH, GAPDH, Hel-N1 (HuB), hnRNP A0, hnRNP A1, hnRNP C, HuR (HuA), tristetraprolin, and TIAR [14–25]. Among these, the influence of HuR and AuF1 on the stability of ARE-containing mRNAs has been studied most extensively.

HuR is an ubiquitously expressed member of the *elav* family of RNA-binding proteins, which also comprises the neuronal-specific proteins HuD, HuC, and Hel-N1 [23,26–29]. Elav proteins contain three RNA recognition motifs (RRM) and bind with high affinity and specificity to AREs in a variety of mRNAs, among which are those encoding VEGF, p21, Id, cyclin A, cyclin B1, GLUT-1, and c-Fos [20,26,30–37]. Although HuR is believed to increase the stability and/or translation of target mRNAs, the precise mechanism(s) regulating HuR function remain largely unknown. HuR can shuttle between the nucleus and the cytoplasm through a motif named HNS [38]. Because HuR is predominantly localized in the nucleus of resting cells, the stabilizing influence of HuR is believed to require its translocation to the cytoplasm [38–40].

HuR's mRNA stabilizing effect is illustrated in two recent studies from our laboratory. Exposure of cells to short-wavelength ultraviolet light (UVC) resulted in stabilization of the p21 mRNA, increased HuR abundance in the cytoplasm, and increased binding of HuR to the p21 ARE [41,42]. In another study, HuR was shown to stabilize target mRNAs encoding cyclin A and cyclin B1. In synchronous cultures, HuR was found to be almost exclusively nuclear during early G_1 phase, but increased in the cytoplasm during the S and G_2 phases. This cell-cycle-dependent HuR localization correlated closely with HuR's increased binding to, and stabilization of, target mRNAs encoding cyclin A and cyclin B1 [43]. In both studies, cells expressing reduced HuR levels exhibited decreased HuR binding to target transcripts and diminished half-lives for the corresponding mRNAs.

AUF1 consists of four alternatively spliced isoforms: p37, p40, p42, and p45 [44]. All isoforms contain two RRMs and can be found in both the nucleus and the cytoplasm [18,45]. Overexpression of AUF1 accelerates the decay of ARE-containing mRNAs [46–49], although the specific pathways and/or cellular events that regulate AUF1 activity remain to be defined. We recently reported that prostaglandin A_2 (PGA_2), a model stress agent and experimental chemotherapeutic compound capable of eliciting a potent inhibition of cell growth, decreased cyclin D1 expression by rendering the cyclin D1 mRNA unstable. PGA_2 treatment also increased the association of AUF1 with an instability element within the cyclin D1 3′UTR, suggesting that AUF1 may participate in the degradation of cyclin D1 mRNA [50,51].

Generally speaking, the RNA sequence requirements for recognition by a given RNA-binding protein appear to be substantially more flexible than those governing, for example, transcription factor–DNA interactions. A given RNA-binding protein can bind to a collection of RNA substrates of related sequences with varying relative affinities, at least *in vitro*. Conversely, a given mRNA sequence can be the target of many RNA-binding proteins *in vitro*. These observations have led several groups to postulate that the half-life of a

CHAPTER 312

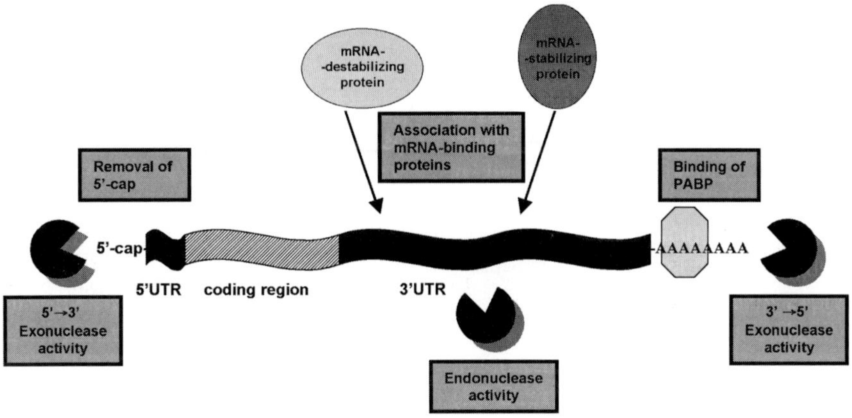

Figure 2 Potential levels of regulation of mRNA turnover. mRNA turnover can be influenced by a number of events, such as the enzymatic action of endonucleases and exonucleases, the association of PABP with the transcript's poly(A) tail, the cleavage of the 5′ cap, and the formation of complexes with RNA-binding proteins that either enhance or reduce transcript stability. Each of these regulatory mechanisms, indicated in shaded squares, is a potential target of stress-modulated events.

given labile mRNA is likely to reflect the *net* effect of its interaction with RNA-binding proteins that promote stabilization and with RNA-binding proteins that promote degradation. Although this hypothesis remains to be proven experimentally, we anticipate that the relative association of proteins that alter mRNA turnover will depend on whether cells are proliferating or not, and on the type of stress to which they are exposed.

Models of Regulation of mRNA Turnover

Although we have a considerable amount of information concerning the *cis* and *trans* determinants of mRNA turnover, we know relatively little about the different levels at which mRNA stability can be regulated. For example, the 5′ end cap that renders the mRNA resistant to general 5′→3′ exonucleases can be removed in a potentially regulatable fashion by a decapping nuclease (DCP1), an enzyme activity that has been purified from yeast and mammalian cells. The 3′ poly(A) tracts can be bound, and therefore protected, by a poly(A) RNA-binding protein (PABP), with the steps leading to poly(A) removal also being subject to regulation. Moreover, poly(A) shortening appears to be a rate-limiting step in the decay of many mammalian mRNAs. Nevertheless, regulation of the turnover of specific mRNAs is largely exerted through AREs. Among the plausible mechanisms whereby associating RNA-binding proteins, with target AREs (or with other stability-influencing mRNA sequences) can affect mRNA turnover are the following: (1) RNA-binding proteins may mask/unmask sites of endonucleolytic cleavage; (2) they may enhance or reduce the activity of potential decapping nucleases; (3) they may elevate or diminish the protective influence of PABPs; (4) they may direct mRNAs to other subcellular locations, such as polysomes, which can also influence mRNA translation and turnover, although the precise influence that these two processes exert on each other remains controversial; and (5) they may target given mRNAs to the *exosome*, a cellular complex of proteins with exonuclease activities (similar to the proteasome, but specialized in degradation of nucleic acid). Such possible regulatory events are shown in Fig. 2.

These possible mechanisms remain to be investigated in depth, but two general mRNA turnover pathways have been described. *Deadenylation-independent mRNA decay* is initiated through the endonucleolytic cleavage of a mRNA, followed by subsequent exonuclease digestion of each fragment through the combined action of 5′→3′ and 3′→5′ exonucleases. *Deadenylation-dependent decay*, believed to be the major pathway, is initiated by the removal of the poly(A) tract, followed by decapping and degradation by both 5′→3′ and 3′→5′ exonucleases. This is the pathway of degradation of most ARE-containing mRNAs. For a review on this topic, see [52].

Stress-Activated Signalling Molecules that Regulate mRNA Turnover

Only a limited number of studies reported to date have addressed specifically how stress stimuli and stress-triggered signaling influence mRNA turnover. However, the available evidence provides support for the notion that mRNA turnover is regulated through mechanisms akin to those controlling gene transcription, that is, by triggering signal transduction pathways that involve phosphorylation cascades. I next discuss our current knowledge of stress-triggered signaling events that regulate mRNA turnover, including those altering the levels of mRNA-binding proteins, their association with target mRNAs, and the degradation rates of labile mRNAs. Importantly, mRNA turnover changes occurring during other cellular processes, such as T-cell activation, neoplastic transformation, and cell cycle progression, may be regulated through the very same stress- and/or growth factor-activated pathways discussed here. A schematic representation of

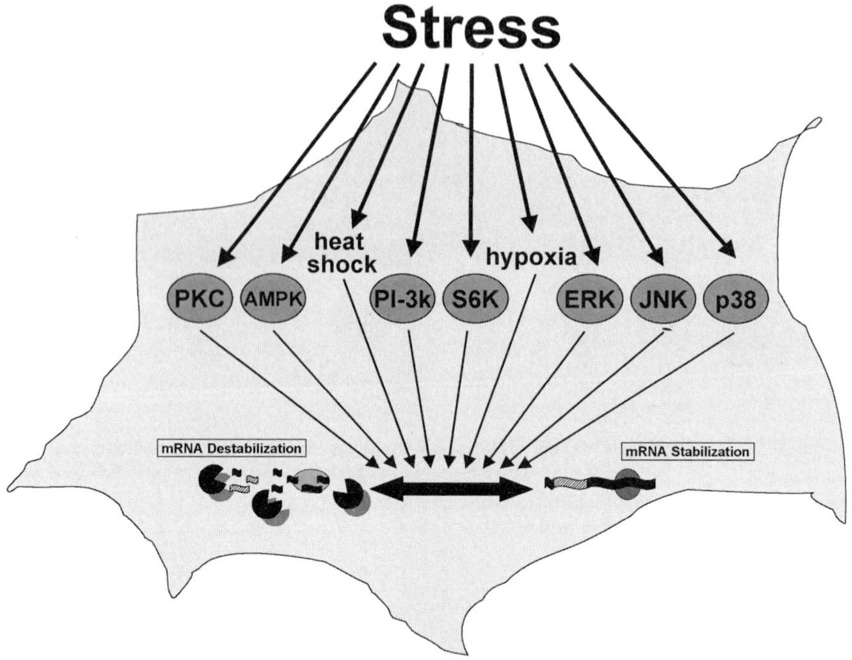

Figure 3 Stress-triggered signaling pathways that modulate mRNA turnover. Schematic representation of stress-activated signaling events that either enhance or reduce mRNA stability. Detailed knowledge of the molecules and mechanisms that influence mRNA turnover along each stress-triggered pathway awaits further investigation. However, we anticipate that the precise regulatory events will vary depending on factors such as the type of stress agent, the particular labile transcript targeted, the cell type, and the cell's metabolic status.

stress-regulated pathways known to influence mRNA stability, either positively or negatively, is shown in Fig. 3.

Protein Kinase C

Among the earliest reports describing regulated mRNA stabilization were several studies describing the role of protein kinase C (PKC) in the stabilization of ARE-containing mRNAs. In response to a variety of stimuli, including phorbol esters, calcium ionophores, diacylglycerol, antibodies recognizing CD3/CD28 surface receptors, TNFα, oxidants, NGF, and interleukins, PKC was implicated in the enhanced stability of many labile mRNAs, such as those encoding IL-1, IL-2, interferon γ, nitric oxide synthase, GAP-43, GM-CSF, HSP70, HSP90, ribonucleotide reductase R1, bcl-2, uPA, and p21 [53–66]. PKC activity was implicated in the stabilization of TNFα mRNA by preventing removal of its poly(A) tail [67]. PKC activity was also shown to influence binding of an adenosine-uridine binding factor (AUBF) to target labile mRNAs, leading to their enhanced stability [16]. More recently, PKC was proposed to increase binding of hnRNP C to the uPA mRNA ARE [64]. In addition, PKC-dependent neurite outgrowth was shown to require HuD-mediated stabilization of the GAP-43 mRNA [68].

Mitogen-Activated Protein Kinases (MAPKs)

Stress-Activated Protein Kinases (SAPKs)

In mast cells treated with ionomycin (a Ca^{2+} ionophore), the c-Jun N-terminal kinase (JNK) pathway was found to participate in the stabilization of the IL-3 mRNA through an ARE present in its 3′UTR [69]. The JNK pathway was also found to be involved in the stabilization and increased translation of IL-2 mRNA in Jurkat cells. In this paradigm, the JNK-regulated response was mediated through a JNK-response element (JRE) present in its 5′UTR, and involved the proteins Nucleolin and YB-1 [8,70]. In response to anisomycin, an inhibitor of protein translation, the JNK pathway was also found to be involved in stabilizing VEGF mRNA, through an ARE in its 3′UTR [71].

Another SAPK, p38, was also shown to participate in the anisomycin-induced stabilization of VEGF mRNA [71]. Exposure to stress stimuli including LPS, cytokines, dexamethasone, and serum deprivation similarly revealed a requirement for p38 and its downstream target MAPK-activated protein kinase 2, in the ARE-dependent turnover of mRNAs encoding IL-8, c-fos, GM-CSF, TNFα, and cyclooxygenase-2 [72–76].

Extracellular Signal-Regulated Kinases (ERKs)

ERK proteins have also been implicated in the stabilization of mRNAs encoding p21, VEGF, and nucleolin [53,71,77,78]. The ERK-controlled up-regulation of nucleolin has been proposed to mediate the destabilization of the β-amyloid precursor protein (APP) mRNA [79].

Heat-Shock-Regulated Events

The heat-shock pathway has been linked to changes in mRNA turnover at many levels. The mRNA encoding the major heat-shock protein, hsp70, has long been known to be stabilized by heat shock [80]. Laroia and colleagues [49] showed that heat shock also stabilizes mRNAs encoding cytokine and protooncogene mRNAs. This stabilization mechanism involves hsp70-mediated sequestration of the AUF1 protein, which would otherwise promote the degradation of ARE-containing cytokine and protooncogene mRNAs [49]. These studies also directly link the rapid degradation of cytokine mRNAs to the ubiquitin-proteasome system. Two additional roles for heat shock and hsp70 have been postulated, although they have not been fully explored: hsp70 can influence mRNA turnover by modulating mRNA decapping in yeast [81], and hsp70 has been found to preferentially bind to, and protect, ARE-containing mRNAs [82].

Heat shock was also shown to decrease HuR's ability to bind poly(A)-containing RNA in the cytoplasm and increase it in the nucleus [83], suggesting that heat shock causes HuR to retain the bulk of its target mRNAs in the nucleus. Interestingly, however, HuR-mediated nuclear export of hsp70 mRNA, and consequently hsp70 expression, increases with heat shock, revealing the specific influence of HuR on the transport and enhanced expression of a particular target mRNA [84]. The authors of this study propose that heat-shock-triggered specific binding of HuR to hsp70 mRNA involves the association of HuR with protein ligands pp32, SETα, and SETβ [84].

AMP-Activated Protein Kinase

Recent studies from our laboratory have identified signaling events that regulate HuR function. Because HuR activity appears to depend largely on its subcellular localization, with binding to target mRNAs occurring in the cytoplasm, we sought to identify signaling events regulating its nucleocytoplasmic transport. Exposure to many different stresses, including UVC, H_2O_2, and PGA_2, elevated HuR's presence in the cytoplasm. Surprisingly, however, neither the major stress-activated signaling cascades (ERK, JNK, p38 pathways) nor other more general signaling pathways (phosphatidylinositol 3-kinase [PI-3K], PKA, PKC) appeared to influence HuR's subcellular localization. Instead, we found evidence that the subcellular compartmentalization of HuR is potently influenced by the activity of a lesser known kinase that is primarily involved in sensing and responding to metabolic stresses: the AMP-activated protein kinase (AMPK). Specific AMPK activators, including treatment with A1CA riboside and infection with an adenovirus that expresses constitutively active AMPK, promoted the nuclear localization of HuR. By contrast, inhibition of AMPK activity, including that achieved through infection with adenoviruses expressing dominant-negative AMPK, caused cells to elevate cytoplasmic HuR levels [85]. Whereas the precise mechanisms of AMPK-regulated subcellular localization of HuR remains to be identified, *in vivo* phosphorylation experiments indicate that HuR itself is not phosphorylated. Instead, we propose that one or several of the HuR protein ligands [87] may be phosphorylated, directly or indirectly, by AMPK-triggered mechanisms, and that these phosphorylation events, in turn, affect the transport of HuR to the cytoplasm [86].

Hypoxia-Triggered Events

Among the response mechanisms implemented by the cell in response to low oxygen levels (hypoxia) is the stabilization of critical mRNAs (reviewed in [88]). VEGF, a key hypoxia-induced angiogenic factor, contains many instability sequences (generally AU-rich) in its 5′UTR, coding region, and 3′UTR. Several hypoxia-inducible proteins that bind to the VEGF mRNA and appear to enhance the stability of its mRNA have been described, among them HuR and hnRNP L [89,90]. However, sequences in the 5′UTR, coding region and 3′UTR have been found to be required for fully effective mRNA degradation under normoxic conditions, and for mRNA stabilization during hypoxia [91]. The mRNA encoding erythropoietin (EPO), another major hypoxia-regulated gene, also contains many turnover determinants. The EPO mRNA is targeted by several cytoplasmic proteins, including the poly(C)-binding proteins $PCBP_1$ and $PCBP_2$, although these associations are not hypoxia inducible [92]. Tyrosine-hydroxylase (TH) mRNA is also stabilized by hypoxia. A pyrimidine-rich sequence in the TH 3′UTR that was identified as important for basal and hypoxia-regulated mRNA stability constitutes a binding site of PCBPs. Significantly, hypoxia induces the stability of the mRNA encoding the hypoxia-inducible factor 1α (HIF-1α), a protein pivotal for the transcriptional up-regulation of hypoxia-inducible genes [93]. Less well understood are the mechanisms controlling the turnover of other mRNAs regulated by hypoxia, including Glut-1, PAI-1, eNOS, MnSOD, Cu, ZnSOD, and the iron-regulatory proteins IRP1 and IRP2.

Other Signaling Molecules and Pathways

Phosphatidylinositol-3 Kinase

The PI3K pathway was recently implicated in the stability of cyclin D1 mRNA [94]. However, the precise cyclin D1 mRNA elements and RNA-binding proteins remain to be identified.

S6 Kinase

The stability of cyclin D1 mRNA is also dependent on the activity of the S6 kinase pathway, because treatment with the S6 kinase inhibitor rapamycin stabilizes the mRNA and enhances cyclin D1 expression [95].

Conclusions

Given that steady-state mRNA levels are critically important for determining cellular protein levels, the regulation of mRNA turnover is crucial in determining the cell's ability to respond to situations of stress. The importance of mRNA turnover in determining the expression of individual genes is being investigated on a case-by-case basis. My group recently initiated efforts to investigate the "global" contribution of mRNA turnover to the stress-induced changes in expressed gene patterns by using cDNA arrays. Following treatment of cells with stress agents (heat shock, PGA_2, or UVC), we compared *changes in total mRNA levels* (after hybridization of arrays with standard radiolabeled cDNA probes obtained through reverse transcription of total cellular RNA) with *changes in gene transcription* using nascent, radiolabeled RNA from nuclear run-on reactions to hybridize the same cDNA arrays. We estimated that approximately *one-half* of all transcripts whose expression was altered by stress was subject to a significant change in mRNA turnover, either mRNA stabilization or destabilization [96].

The role that cellular stress plays in regulating mRNA half-life is thus gaining increasing attention. Some common mechanisms regulating mRNA stability are currently emerging in the form of related *cis* elements and even shared *trans*-acting factors. However, the complete elucidation of how mRNA turnover is controlled requires that complex events regulating the expression and function of RNA-binding and RNA-degrading proteins be thoroughly investigated. Considering the great relevance of many stress-regulated genes for cellular growth and homeostasis, it is not surprising that such regulatory events have begun to be the focus of intense investigation in recent years. It is widely anticipated that an in-depth understanding of the stress-triggered processes that control mRNA stability will contribute novel insight into basic mechanisms serving to regulate gene expression and will help identify new targets of clinical intervention.

References

1. Sachs, A. B. (1993). Messenger RNA degradation in eukaryotes. *Cell* **74**, 413–421.
2. Ross, J. (1995). mRNA stability in mammalian cells. *Microbiol. Rev.* **59**, 423–450.
3. Brennan, C. M. and Steitz, J. A. (2001). HuR and mRNA stability. *Cell. Mol. Life Sci.* **58**, 266–277.
4. Peltz, S. W., Brewer, G., Bernstein, P., Hart, P. A., and Ross, J. (1991). Regulation of mRNA turnover in eukaryotic cells. *Crit. Rev. Eukaryot. Gene Expr.* **1**, 99–126.
5. Jackman, J., Alamo, I. Jr., and Fornace A. J. Jr. (1994). Genotoxic stress confers preferential and coordinate messenger RNA stability on the five gadd genes. *Cancer Res.* **54**, 5656–5662.
6. Blattner, C., Kannouche, P., Litfin, M., Bender, K., Rahmsdorf, H. J., Angulo, J. F., and Herrlich, P. (2000). UV-induced stabilization of c-fos and other short-lived mRNAs. *Mol. Cell. Biol.* **20**, 3616–3625.
7. Kenan, D. J., Query, C. C., and Keene, J. D. (1991). RNA recognition: Towards identifying determinants of specificity. *Trends Biochem. Sci.* **16**, 214–220.
8. Chen, C.-Y., Del Gatto-Konczak, F., Wu, Z., and Karin, M. (1998). Stabilization of Interleukin-2 mRNA by the c-Jun NH_2-Terminal Kinase Pathway. *Science* **280**, 1945–1949.
9. Shyu, A. B., Greenberg, M. E., and Belasco, J. G. (1989). The c-fos transcript is targeted for decay by two distinct mRNA degradation pathways. *Genes Dev.* **3**, 60–72.
10. Wisdom, R. and Lee, W. (1991). The protein-coding region of c-myc mRNA contains a sequence that specifies rapid mRNA turnover and induction by protein synthesis inhibitors. *Genes Dev.* **5**, 232–243.
11. Chen, C. Y. and Shyu, A. B. (1995). AU-rich elements: Characterization and importance in mRNA degradation. *Trends Biochem. Sci.* **20**, 465–470.
12. Lagnado, C. A., Brown, C. Y., and Goodall, G. J. (1994). AUUUA is not sufficient to promote poly(A) shortening and degradation of an mRNA: The functional sequence within AU-rich elements may be UUAUUUA(U/A) (U/A). *Mol. Cell. Biol.* **14**, 7984–7995.
13. Xu, N., Chen, C.-Y. A., and Shyu, A.-B. (1997). Modulation of the fate of cytoplasmic mRNA by AU-rich elements: Key sequence features controlling mRNA deadenylation and decay. *Mol. Cell. Biol.* **17**, 4611–4621.
14. Bohjanen, P. R., Petryniak, B., June, C. H., Thompson, C. B., and Lindsten, T. (1991). An inducible cytoplasmic factor (AU-B) binds selectively to AUUUA multimers in the 3′ untranslated region of lymphokine mRNA. *Mol. Cell Biol.* **11**, 3288–3295.
15. Bohjanen, P. R., Petryniak, B., June, C. H., Thompson, C. B., and Lindsten, T. (1992). AU RNA-binding factors differ in their binding specificities and affinities. *J. Biol. Chem.* **267**, 6302–6309.
16. Malter, J. S. and Hong, Y. (1991). A redox switch and phosphorylation are involved in the posttranslational up-regulation of the adenosine-uridine binding factor by phorbol ester and ionophore. *J. Biol. Chem.* **266**, 3167–3171.
17. Zhang, W., Wagner, B. J., Ehrenman, K., Schaefer, A. W., DeMaria, C. T., Crater, D., DeHaven, K., Long, L., and Brewer, G. (1993). Purification, characterization, and cDNA cloning of an AU-rich element RNA-binding protein, AUF1. *Mol. Cell. Biol.* **13**, 7652–7665.
18. Nakagawa, J., Waldner, H., Meyer-Monard, S., Hofsteenge, J., Jeno, P., and Moroni, C. (1995). AUH, a gene encoding an AU-specific RNA binding protein with intrinsic enoyl-CoA hydratase activity. *Proc. Natl. Acad. Sci. USA* **92**, 2051–2055.
19. Nagy, E. and Rigby, W. F. (1995). Glyceraldehyde-3-phosphate dehydrogenase selectively binds AU-rich RNA in the NAD(+)-binding region (Rossmann fold). *J. Biol. Chem.* **270**, 2755–2763.
20. Levine, T. D., Gao, F., King, P. H., Andrews, L. G., and Keene, J. D. (1993). Hel-N1: An autoimmune RNA-binding protein with specificity for 3′ uridylate-rich untranslated regions of growth factor mRNAs. *Mol. Cell. Biol.* **13**, 3494–3504.
21. Myer, V. E. and Steitz, J. A. (1995). Isolation and characterization of a novel, low abundance hnRNP protein: A0. *RNA* **1**, 171–182.
22. Hamilton, B. J., Nagy, E., Malter, J. S., Arrick, B. A., and Rigby, W. F. (1993). Association of heterogeneous nuclear ribonucleoprotein A1 and C proteins with reiterated AUUUA sequences. *J. Biol. Chem.* **268**, 8881–8887.
23. Ma, W. J., Cheng, S., Campbell, C., Wright, A., and Furneaux, H. (1996). Cloning and characterization of HuR, a ubiquitously expressed Elav-like protein. *J. Biol. Chem.* **271**, 8144–8151.
24. Carballo, E., Lai, W. S., and Blackshear, P. J. (1998). Feedback inhibition of macrophage tumor necrosis factor-alpha production by tristetraprolin. *Science* **281**, 1001–1005.
25. Gueydan, C., Droogmans, L., Chalon, P., Huez, G., Caput, D., and Kruys, V. (1999). Identification of TIAR as a protein binding to the translational regulatory AU-rich element of tumor necrosis factor alpha mRNA. *J. Biol. Chem.* **274**, 2322–2326.
26. Chung, S., Jiang, L., Cheng, S., and Furneaux, H. (1996). Purification and properties of HuD, a neuronal RNA-binding protein. *J. Biol. Chem.* **271**, 11518–11524.
27. Good, P. J. (1995). A conserved family of Elav-like genes in vertebrates. *Proc. Natl. Acad. Sci. USA* **89**, 4557–4561.
28. Marusich, H. M., Furneaux, H., Henion, P., and Weston, J. A. (1994). Hu neuronal proteins are expressed in proliferating neurogenic cells. *J. Neurobiol.* **25**, 143–155.
29. Szabo, A., Dalmau, J., Manley, G., Rosenfeld, M. R., Wong, E., Henson, J., Posner, J. B., and Furneaux, H. (1991). HuD, a paraneoplastic encephalomyelitis antigen, contains RNA-binding domains and is homologous to Elav and Sex-lethal. *Cell* **67**, 325–333.
30. Antic, D., Lu, N., and Keene, J. D. (1999). ELAV tumor antigen hel-N1 increases translation of neurofilament M mRNA and induces formation of neurites in human teratocarcinoma cells. *Genes Dev.* **13**, 449–461.
31. Fan, X. C. and Steitz, J. A. (1998). Overexpression of HuR, a nuclear-cytoplasmic shuttling protein, increases the in vivo stability of ARE-containing mRNAs. *EMBO J.* **17**, 3448–3460.
32. Gao, F. B., Carson, C. C., Levine, C. T., and Keene, J. (1994). Selection of a subset of mRNAs from combinatorial 3′ untranslated region libraries using neuronal RNA-binding protein Hel-N1. *Proc. Natl. Acad. Sci. USA* **91**, 11207–11211.

33. Jain, R. G., Andrews, L. G., McGowan, K. M., Pekala, P. H., and Keene, J. D. (1997). Ectopic expression of Hel-N1, an RNA-binding protein, increases glucose transporter (GLUT1) expression in 3T3-L1 adipocytes. *Mol. Cell. Biol.* **17**, 954–962.
34. Joseph, B., Orlian, M., and Furneaux, H. (1998). p21^{waf1} mRNA contains a conserved element in its 3′-untranslated region that is bound by the Elav-like mRNA-stabilizing proteins. *J. Biol. Chem.* **273**, 20511–2051.
35. King, P. H., Levine, T. D., Fremeau, R. T. Jr., and Keene, J. D. (1994). Mammalian homologs of Drosophila ELAV localized to a neuronal subset can bind in vitro to the 3′UTR of mRNA encoding the Id transcriptional repressor. *J. Neurosci.* **14**, 1943–1952.
36. Ma, W. J., Chung, S., and Furneaux, H. (1997). The Elav-like proteins bind to AU-rich elements and to the poly(A) tail of mRNA. *Nucleic. Acids. Res.* **25**, 3564–3569.
37. Myer, V. E., Fan, X. C., and Steitz, J. A. (1997). Identification of HuR as a protein implicated in AUUUA-mediated mRNA decay. *EMBO J.* **16**, 2130–2139.
38. Fan, X. C. and Steitz, J. A. (1998). HNS, a nuclear-cytoplasmic shuttling sequence in HuR. *Proc. Natl. Acad. Sci. USA* **95**, 15293–15298.
39. Atasoy, U., Watson, J., Patel, D., and Keene, J. D. (1998). ELAV protein HuA (HuR) can redistribute between nucleus and cytoplasm and is upregulated during serum stimulation and T cell activation. *J. Cell Sci.* **111**, 3145–3156.
40. Peng, S. S., Chen, C. Y., Xu, N., and Shyu, A. B. (1998). RNA stabilization by the AU-rich element binding protein HuR, an ELAV protein. *EMBO J.* **17**, 3461–3470.
41. Gorospe, M., Wang, X., and Holbrook, N. J. (1998). p53-dependent elevation of p21^{Waf1}expression by UV light is mediated through mRNA stabilization and involves a vanadate-sensitive regulatory system. *Mol. Cell. Biol.* **18**, 1400–1407.
42. Wang, W., Furneaux, H., Cheng, H., Caldwell, M. C., Hutter, D., Liu, Y., Holbrook, N. J., and Gorospe, M. (2000). HuR regulates p21 mRNA stabilization by UV light. *Mol. Cell. Biol.* **20**, 760–769.
43. Wang, W., Lin, S., Caldwell, C. M., Furneaux, H., and Gorospe, M. (2000). HuR regulates cyclin A and cyclin B1 mRNA stability during the cell division cycle. *EMBO J.* **19**, 2340–2350.
44. Wagner, B. J., DeMaria, C. T., Sun, Y., Wilson, G. M., and Brewer, G. (1998). Structure and genomic organization of the human AUF1 gene: Alternative pre-mRNA splicing generates four protein isoforms. *Genomics* **48**, 195–202.
45. Wilson, G. M., Sun, Y., Sellers, J., Lu, H., Penkar, N., Dillard, G., and Brewer, G. (1999). Regulation of AUF1 expression via conserved alternatively spliced elements in the 3′ untranslated region. *Mol. Cell. Biol.* **19**, 4056–4061.
46. DeMaria, C. T. and Brewer, G. (1996). AUF1 binding affinity to A+U-rich elements correlates with rapid mRNA degradation. *J. Biol. Chem.* **271**, 12179–12184.
47. DeMaria, C. T., Sun, Y., Long, L., Wagner, B. J., and Brewer, G. (1997). Structural determinants in AUF1 required for high affinity binding to A+U-rich elements. *J. Biol. Chem.* **272**, 27635–27643.
48. Loflin, P., Chen, C. Y., and Shyu, A.-B. (1999). Unraveling a cytoplasmic role for hnRNP D in the in vivo mRNA destabilization directed by the AU-rich element. *Genes Dev.* **13**, 1884–1897.
49. Laroia, G., Cuesta, R., Brewer, G., and Schneider, R. J. (1999). Control of mRNA decay by heat shock-ubiquitin-proteasome pathway. *Science* **284**, 499–502.
50. Gorospe, M., Liu, Y., Xu, Q., Chrest, F. J., and Holbrook, N. J. (1996). Inhibition of G_1 cyclin-dependent kinase activity during prostaglandin A_2-mediated growth arrest. *Mol. Cell. Biol.* **16**, 762–770.
51. Lin, S., Wang, W., Wilson, G. M., Yang, X., Brewer, G., Holbrook, N. J., and Gorospe, M. (2000). Downregulation of cyclin D1 expression by prostaglandin A_2 is mediated by enhanced cyclin D1 mRNA turnover. *Mol. Cell. Biol.* **20**, 7903–7913.
52. Mitchell, P. and Tollervey, D. (2000). mRNA stability in eukaryotes. *Curr. Opin. Genet. Dev.* **10**, 193–198.
53. Esposito, F., Cuccovillo, F., Vanoni, M., Cimino, F., Anderson, C. W., Appella, E., and Russo, T. (1997). Redox-mediated regulation of p21(waf1/cip1) expression involves a post-transcriptional mechanism and activation of the mitogen-activated protein kinase pathway. *Eur. J. Biochem.* **245**, 730–737.
54. Fenton, M. J., Vermeulen, M. W., Clark, B. D., Webb, A. C., and Auron, P. E. (1988). Human pro-IL-1 beta gene expression in monocytic cells is regulated by two distinct pathways. *J. Immunol.* **140**, 2267–2273.
55. Yamato, K., el-Hajjaoui, Z., and Koeffler, H. P. (1989). Regulation of levels of IL-1 mRNA in human fibroblasts. *J. Cell. Physiol.* **139**, 610–616.
56. Kowalski, J. and Denhardt, D. T. (1989). Regulation of the mRNA for monocyte-derived neutrophil-activating peptide in differentiating HL60 promyelocytes. *Mol. Cell. Biol.* **9**, 1946–1957.
57. Park, J. W., Jang, M. A., Lee, Y. H., Passaniti, A., and Kwon, T. K. (2001). p53-independent elevation of p21 expression by PMA results from PKC-mediated mRNA stabilization. *Biochem. Biophys. Res. Commun.* **280**, 244–248.
58. Carpenter, L., Cordery, D., and Biden, T. J. (2001). Protein kinase Cdelta activation by interleukin-1beta stabilizes inducible nitric-oxide synthase mRNA in pancreatic beta-cells. *J. Biol. Chem.* **276**, 5368–5374.
59. Shaw, G. and Kamen, R. (1986). A conserved AU sequence from the 3′ untranslated region of GM-CSF mRNA mediates selective mRNA degradation. *Cell* **46**, 659–667.
60. Gorospe, M., Kumar, S., and Baglioni, C. (1993). Tumor necrosis factor increases stability of interleukin-1 mRNA by activating protein kinase C. *J. Biol. Chem.* **268**, 6214–6220.
61. Lindstein, T., June, C. H., Ledbetter, J. A., Stella, G., and Thompson, C. B. (1989). Regulation of lymphokine messenger RNA stability by a surface-mediated T cell activation pathway. *Science* **244**, 339–343.
62. Lafuse, W. P., Alvarez, G. R., and Zwilling, B. S. (2000). Regulation of Nramp1 mRNA stability by oxidants and protein kinase C in RAW264.7 macrophages expressing Nramp1(Gly169). *Biochem. J.* **351**, 687–696.
63. Schiavone, N., Rosini, P., Quattrone, A., Donnini, M., Lapucci, A., Citti, L., Bevilacqua, A., Nicolin, A., and Capaccioli, S. (2000). A conserved AU-rich element in the 3′ untranslated region of bcl-2 mRNA is endowed with a destabilizing function that is involved in bcl-2 down-regulation during apoptosis. *FASEB J.* **14**, 174–184.
64. Nanbu, R., Montero, L., D'Orazio, D., and Nagamine, Y. (1997). Enhanced stability of urokinase-type plasminogen activator mRNA in metastatic breast cancer MDA-MB-231 cells and LLC-PK1 cells down-regulated for protein kinase C–correlation with cytoplasmic heterogeneous nuclear ribonucleoprotein C. *Eur. J. Biochem.* **247**, 169–174.
65. Jacquier-Sarlin, M. R., Jornot, L., and Polla, B. S. (1995). Differential expression and regulation of hsp70 and hsp90 by phorbol esters and heat shock. *J. Biol. Chem.* **270**, 14094–14099.
66. Chen, F.-Y., Amara, F. M., and Wright, J. A. (1994). Regulation of mammalian ribonucleotide reductase R1 mRNA stability is mediated by a ribonucleotide reductase R1 mRNA 3′-untranslated region cis-trans interaction through a protein kinase C-controlled pathway. *Biochem. J.* **302**, 125–132.
67. Lieberman, A. P., Pitha, P. M., and Shin, M. L. (1992). Poly(A) removal is the kinase-regulated step in tumor necrosis factor mRNA decay. *J. Biol. Chem.* **267**, 2123–2126.
68. Mobarak, C. D., Anderson, K. D., Morin, M., Beckel-Mitchener, A., Rogers, S. L., Furneaux, H., King, P., and Perrone-Bizzozero, N. I. (2000). The RNA-binding protein HuD is required for GAP-43 mRNA stability, GAP-43 gene expression, and PKC-dependent neurite outgrowth in PC12 cells. *Mol. Biol. Cell.* **11**, 3191–3203.
69. Ming, X. F., Kaiser, M., and Moroni, C. (1998). c-jun N-terminal kinase is involved in AUUUA-mediated interleukin-3 mRNA turnover in mast cells. *EMBO J.* **17**, 6039–6048.
70. Chen, C.-Y., Gherzi, R., Andersen, J. S., Gaietta, G., Jurchott, K., Royer, H. D., Mann, M., and Karin, M. (2000). Nucleolin and YB-1 are required for JNK-mediated interleukin-2 mRNA stabilization during T-cell activation. *Genes Dev.* **14**, 1236–1248.
71. Pagès, G., Berra, E., Milanini, J., Levy, A. P., and Pouyssegur, J. (2000). Stress-activated protein kinases (JNK and p38/HOG) are essential for vascular endothelial growth factor mRNA stability. *J. Biol. Chem.* **275**, 26484–26491.

72. Winzen, R., Kracht, M., Ritter, B., Wilhelm, A., Chen, C. Y., Shyu, A. B., Muller, M., Gaestel, M., Resch, K., and Holtmann, H. (1999). The p38 MAP kinase pathway signals for cytokine-induced mRNA stabilization via MAP kinase-activated protein kinase 2 and an AU-rich region-targeted mechanism. *EMBO J.* **18**, 4969–4980.
73. Lasa, M., Mahtani, K. R., Finch, A., Brewer, G., Saklatvala, J., and Clark, A. R. (2000). Regulation of cyclooxygenase 2 mRNA stability by the mitogen-activated protein kinase p38 signaling cascade. *Mol. Cell Biol.* **20**, 4265–4274.
74. Kotlyarov, A., Neininger, A., Schubert, C., Eckert, R., Birchmeier, C., Volk, H.-D., and Gaestel, M. (1999). MAPKAP kinase 2 is essential for LPS-induced TNF-α biosynthesis. *Nat. Cell Biol.* **1**, 94–97.
75. Dean, J. L. E., Brook, M., Clark, A. R., and Saklatvala, J. (1999). p38 Mitogen-activated Protein Kinase Regulates Cyclooxygenase-2 mRNA stability and Transcription in Lipopolysaccharide-treated Human Monocytes. *J. Biol. Chem.* **274**, 264–269.
76. Jang, B. C., Sanchez, T., Schaefers, H. J., Trifan, O. C., Liu, C. H., Creminon, C., Huang, C. K., and Hla, T. (2000). Serum withdrawal-induced post-transcriptional stabilization of cyclooxygenase-2 mRNA in MDA-MB-231 mammary carcinoma cells requires the activity of the p38 stress-activated protein kinase. *J. Biol. Chem.* **275**, 39507–39515.
77. Park, J. S., Qiao, L., Gilfor, D., Yang, M. Y., Hylemon, P. B., Benz, C., Darlington, G., Firestone, G., Fisher, P. B., and Dent, P. (2000). A role for both Ets and C/EBP transcription factors and mRNA stabilization in the MAPK-dependent increase in p21 (Cip-1/WAF1/mda6) protein levels in primary hepatocytes. *Mol. Biol. Cell.* **11**, 2915–2932.
78. Westmark, C. J. and Malter, J. S. (2001). Up-regulation of nucleolin mRNA and protein in peripheral blood mononuclear cells by extracellular-regulated kinase. *J. Biol. Chem.* **276**, 1119–1126.
79. Westmark, C. J. and Malter, J. S. (2001). Extracellular-regulated kinase controls beta-amyloid precursor protein mRNA decay. *Brain Res. Mol. Brain Res.* **90**, 193–201.
80. Theodorakis, N. G. and Morimoto, R. I. (1987). Posttranscriptional regulation of hsp70 expression in human cells: Effects of heat shock, inhibition of protein synthesis, and adenovirus infection on translation and mRNA stability. *Mol. Cell Biol.* **7**, 4357–4368.
81. Zhang, S., Williams, C. J., Hagan, K., and Peltz, S. W. (1999). Mutations in VPS16 and MRT1 stabilize mRNAs by activating an inhibitor of the decapping enzyme. *Mol. Cell Biol.* **19**, 7568–7576.
82. Henics, T., Nagy, E., Oh, H. J., Csermely, P., von Gabain, A., and Subjeck, J. R. (1999). Mammalian Hsp70 and Hsp110 proteins bind to RNA motifs involved in mRNA stability. *J. Biol. Chem.* **274**, 17318–17324.
83. Gallouzi, I.-E., Brennan, C. M., Stenberg, M. G., Swanson, M. S., Eversole, A., Maizels, N., and Steitz, J. A. (2000). HuR binding to cytoplasmic mRNA is perturbed by heat shock. *Proc. Natl. Acad. Sci. USA* **97**, 3073–3078.
84. Gallouzi, I.-E., Brennan, C. M., and Steitz, J. A. (2001). Protein ligands mediate the CRM1-dependent export of HuR in response to heat shock. *RNA* **7**, 1348–1361.
85. Woods, A., Azzout-Marniche, D., Foretz, M., Stein, S. C., Lemarchand, P., Ferre, P., Foufelle, F., and Carling, D. (2000). Characterization of the role of AMP-activated protein kinase in the regulation of glucose-activated gene expression using constitutively active and dominant negative forms of the kinase. *Mol. Cell. Biol.* **20**, 6704–6711.
86. Wang, W., Fan, J., Yang, X., Fürer, S., López de Silanes, I., von Kobbe, C., Guo, J., Georas, S. N., Foufelle, F., Hardie, D. G., Carling, D., and Gorospe, M. (2002). AMP-activated kinase regulates cytoplasmic HuR. *Mol. Cell. Biol.* **22**, 3425–3436.
87. Brennan, C. M., Gallouzi, I. E., and Steitz, J. A. (2000). Protein ligands to HuR modulate its interaction with target mRNAs in vivo. *J. Cell. Biol.* **151**, 1–14.
88. Paulding, W. R. and Czyzyk-Krzeska, M. F. (2000). Hypoxia-induced regulation of mRNA stability. *Adv. Exp. Med. Biol.* **475**, 111–121.
89. Levy, A. P., Levy, N. S., Wegner, S., and Goldberg, M. A. (1996). post-transcriptional regulation of vascular endothelial growth factor by hypoxia. *J. Biol. Chem.* **271**, 2746–2753.
90. Shih, S. C. and Claffey, K. P. (1999). Regulation of human vascular endothelial growth factor mRNA stability in hypoxia by heterogeneous nuclear ribonucleoprotein L. *J. Biol. Chem.* **274**, 1359–1365.
91. Dibbens, J. A., Miller, D. L., Damert, A., Risau, W., Vadas, M. A., and Goodall, G. J. (1999). Hypoxic regulation of vascular endothelial growth factor mRNA stability requires the cooperation of multiple RNA elements. *Mol. Biol. Cell.* **10**, 907–919.
92. Czyzyk-Krzeska, M. F. and Bendixen, A. C. (1999). Identification of the poly(C) binding protein in the complex associated with the 3′ untranslated region of erythropoietin messenger RNA. *Blood* **93**, 2111–2120.
93. Zou, A. P., Yang. Z. Z., Li, P. L., and Cowley, A. W. Oxygen-dependent expression of hypoxia-inducible factor-1α in renal medullary cells of rats. *Physiol. Genomics* **6**, 159–168.
94. Dufourny, B., van Teefelen H. A. A. M., Hamelers, I. H. L., Sussenbach, J. S., and Steenbergh, P. H. (2000). Stabilization of cyclin D1 mRNA via the phosphatidylinositol 3-kinase pathway in MCF-7 human breast cancer cells. *J. Endocrinol.* **166**, 329–338.
95. Hashemolhosseini, S., Nagamine, Y., Morley, S. J., Desrivieres, S., Mercep, L., and Ferrari, S. (1998). Rapamycin inhibition of the G1 to S transition is mediated by effects on cyclin D1 mRNA and protein stability. *J. Biol. Chem.* **273**, 14424–14429.
96. Fan, J., Yang, X., Wang, W., Wood III, W. H., Becker, K. G., and Gorospe, M. (2002). Global analysis of stress-regulated mRNA turnover using cDNA arrays. *Proc. Natl. Acad. Sci. USA* **99**, 10611–10616.

SECTION D

Post-Translational Control

Nahum Sonenberg, Editor

CHAPTER 313

RNA Localization and Signal Transduction

Vaughan Latham[1,2] and Robert H. Singer[1]

[1]*Department of Anatomy and Structural Biology, Albert Einstein College of Medicine, Bronx, New York*
[2]*Dana-Farber Cancer Institute, Boston, Massachusetts*

Introduction

Sorting of mRNA to specific compartments of the cell determines cell asymmetry. This sorting occurs in oocytes and embryos as well as somatic cells such as fibroblasts and neurons (for reviews see [1–6]). Translation of localized mRNAs spatially directs protein synthesis. The cellular signals that direct specific RNA sequences to particular cellular compartments have recently been examined in fibroblasts, neurons, and *Drosophila* embryos. This chapter examines the regulation of mRNA localization through signal transduction pathways in organisms and their tissues.

Growth Factors Induce mRNA Localization

In chicken embryo fibroblasts (CEFs), β-actin mRNA has been shown to be localized toward the leading edge where it plays a role in cell motility and asymmetry [7,8]. A specific sequence in the 3′ untranslated region (UTR) of the β-actin mRNA, termed the *zipcode* [9], is necessary for the localization of the mRNA and specific *trans*-acting factors, termed *zipcode binding proteins* [10]. Growth factors can affect the site of synthesis of β-actin in the cytoplasm by inducing rapid localization of β-actin mRNA toward the leading edge in CEFs [11,12]. This localization can be induced after serum starvation by serum as well as growth factors such as LPA and PDGF and is inhibited by tyrosine kinase inhibitors herbimycin and genistein [11].

β-Actin mRNA localization can also be induced in growth cones of forebrain neurons by neurotrophin-3 (NT-3), forskolin, or db-cAMP [13]. Growth factors can act on growth cones to induce path finding in neuronal development [14]. In another study, NT-3 also induced localization of mRNA granules into dendrites and was inhibited by K252a, an inhibitor of tyrosine kinase receptors [15]. Brain-derived neurotrophic factor (BDNF) induced dendritic mRNA localization and translation of a reporter GFP with the 5′ and 3′UTR of CAMKII-α [16]. Growth factors such as NT-3 and BDNF may induce the movement of mRNAs into dendrites and induce the localized translation of these mRNAs at their target cellular compartment and affect changes in actin cytoskeletal reorganization within growth cones.

Tetanic stimulation of hippocampal neurons induces an increase in the concentration of CaMKII-α in the dendrites of postsynaptic neurons within 5 min after a tetanus [17]. This is the result of local synthesis on CaMKII mRNA within dendrites rather than the slower transport of CaMKII protein and was blocked by protein synthesis inhibitors [17–19]. In the case of *Arc* mRNA being targeted to active synapses, NMDA receptor activation is necessary for this localization [18,20]. The expression of *Arc* mRNA was induced by electroconvulsive seizure, and newly synthesized *Arc* mRNA was targeted to synapses in the dentate gyrus, but not when NMDA receptor antagonists were present [20]. Ca^{2+} signaling may be also be involved in localization and subsequent translation as binding of the growth factors to their receptors induced tyrosine phosphorylation [21] and subsequent increase in intracellular Ca^{2+} [22]. Increased [Ca^{2+}] may induce stability of some mRNAs and subsequent translation [23,24].

Drosophila oogenesis the localization of mRNA ... translated is a signal to surrounding cells. *Gurken* mRNA is localized in the posterior oocyte cortex and then translated into an EGF similar growth factor, which then induces the follicle cells to form the polarity of the egg chamber [25,26]. This localization of *Gurken* mRNA is necessary for signaling via the EGF receptor pathway, as *maelstrom* mutants that do not localize the mRNA develop follicle cells with anterior rather than posterior fates [27]. A similar localization of the mRNA of *wingless* determines the apical localization of the signaling protein [28,29].

Signaling from the Extracellular Matrix Induces mRNA Localization

Another cellular signaling system is that from the extracellular matrix, usually integrins [30]. A study using microbeads coated with focal adhesion complex (FAC) proteins such as β_1 integrin talin, vinculin, and the RGD peptide induced "halos" of poly $(A)^+$ RNA and ribosomes within a few minutes [31]. These are presumably the mRNAs of FAC proteins although this was not determined. These halos also formed when tension was exerted on adhered beads.

mRNAs Localized via the Cytoskeleton

Numerous investigations have examined which cytoskeletal components are involved in localizing mRNA. In fibroblasts it appears that the actin cytoskeleton is involved in localizing β-actin mRNA. Early work demonstrated that β-actin mRNA was not localized in the presence of cytochalasin D, but was localized with colchicine [32]. More recently the rapid (<5 min) localization induced by serum or growth factors was shown to be inhibited by cytochalasin D, but not nocodazole [33]. Conversely, in neurons, mRNA localization is predominantly microtubule based and is inhibited by colchicine but not cytochalasin D [4].

mRNA Granule Movement in Neurons

Movement of mRNA in neurons was first tracked with ^{3}H-uridine in pulse-chase experiments [34]. More recently a membrane-permeable nucleic acid stain, SYTO-14, has been used to track mRNA movement in living cells [35]. Messenger RNA granules move into neurites within 15 min after NT-3 stimulation [15]. These granules are enriched in the RNA binding protein Staufen [36,37] and are complexed with polyribosomes [38]. After depolarization, mRNAs shift to a less dense granule fraction but are not translationally competent [38]. The movement of mRNA granules is responsive to neurotrophins in neurons [39] and growth factors in fibroblasts [11,12].

Regulation of mRNA Localizing Proteins

Many proteins necessary for localization of mRNAs such as ZBP1 [10], ZBP2 [40] in fibroblasts and neurons, and Egalitarian (Egl) and BicaudalD (Bic D) in *Drosophila* [41] may be targets of signal transduction pathways regulating mRNA localization. These proteins may be directly phosphorylated through signaling pathways or through their interactions with motor molecules such as myosins, dyneins, or kinesins. One such protein, zipcode binding protein 1, is required for the localization of β-actin mRNA to growth cones [39]. Neurons transfected with EGFP-ZBP exhibited rapid bidirectional movements of granules that required ZBP1 binding to RNA.

GTPase Signals Regulating Actomyosin Interactions Are Involved in mRNA Localization

β-Actin mRNA localization near the leading edge in CEFs is dependent on the actin cytoskeleton and not the microtubule-based cytoskeleton [32]. Similarly the rapid induction of β-actin mRNA localization by serum and growth factors is also dependent on actin and not microtubules [33]. Consistent with the Rho GTPases regulation of the actin cytoskeleton [42,43], β-actin mRNA localization at the leading edge has been shown to be regulated by the Rho GTPase [33]. This signaling pathway goes through the Rho-associated kinases, because the specific inhibitorY-27632 inhibited localization and transfected Rho-kinase constructs induced localization. Because β-actin mRNA localization can lead to the development of cell polarity [7,44], this would allow for signaling regulation of cell polarity. Similarly in *Drosophila*, Rho-associated kinase (Drock) regulates planar cell polarity through the actin cytoskeleton [45]. Also in *Drosophila*, Rho regulatory proteins such as rhoGEF may play a role in the cytoskeletal changes during development [46] and may be involved in mRNA localization. The Rho signaling pathway may involve the formin protein p140mDia [47]. Because mDia can regulate the cytoskeleton through stabilization of microtubules [48] and the actin stress fibers [49,50] it could participate in mRNA localization on either actin or microtubules perhaps via anchoring of RNA by EF1α [51].

Cytoskeletal motors involved in mRNA localization have been demonstrated during development in *Xenopus* [52], and Drosophila and more recently a myosin V has been found to be involved in *ASH 1* mRNA localization in budding yeast [53–58]. Some studies using pharmacological inhibition of myosin showed inhibition of β-actin mRNA localization to the leading edge [33], and the inhibition of poly$(A)^+$ RNA movement into "halos" around integrin-coated beads with myosin ATPase inhibitor BDM [31] and the myosin light chain kinase inhibitor ML-7 [33]. Because Rho and Rho-kinase can regulate myosin phosphatase by inhibiting the phosphatase in the presence of ongoing myosin light chain kinase (MLCK) activity [59,60], the induction of mRNA localization by activation of this pathway acts through myosin. Recently mouse embryo fibroblasts (MEFs) from myosin IIB knock-out mice were shown to be deficient in localizing β-actin mRNA to the leading edge [33]. MEFs from homozygous knock-out mice showed no movement of β-actin mRNA to the leading edge, whereas wild-type and the heterozygous MEFs responded within minutes [33]. The two-headed myosin II filaments can translocate on polarized bundles of actin filaments or stress fibers toward the leading edge associated with an mRNA complex that can bind the myosin. These actin bundles have a polarity with barbed ends increasingly directed toward the lamellipodium and thus could constrain activated myosin II-B movement only toward the leading edge [61,62]. Rho kinase can lead to phosphorylation of the light chains internally near the nucleus, where myosin filament assembly, stress fiber formation, and motility occur [63]. Growth cones from IIB knock-out mice neurons have been analyzed and showed differences in growth cone shape, actin organization, and reduced filipodial traction force [64], as does a localized knockout of myosin IIB using lazer inactivation.

Conclusion

There is no question now that motors move localized RNAs to their ultimate destination in the periphery of the polarized cell. This then allows proteins to be sorted asymmetrically to the specific region of the cell where they can exert their functionality. But this movement of the RNA is in response to the polarized nature of the asymmetric cell and is not causal for the asymmetry. The synthesis of specific protein components at the periphery enhances and stabilizes the polarity of the cell. How this polarity all gets established is a result of the signaling molecules impinging on the cell directionally from the outside. The transduction of these signals into polarity, through actin polymerization, for instance, provides a substrate for the trafficking of the RNA. In this way, the extracellular environment determines the distribution of specific protein synthesis within the cell. Figure 1 summarizes this process for asymmetric cells.

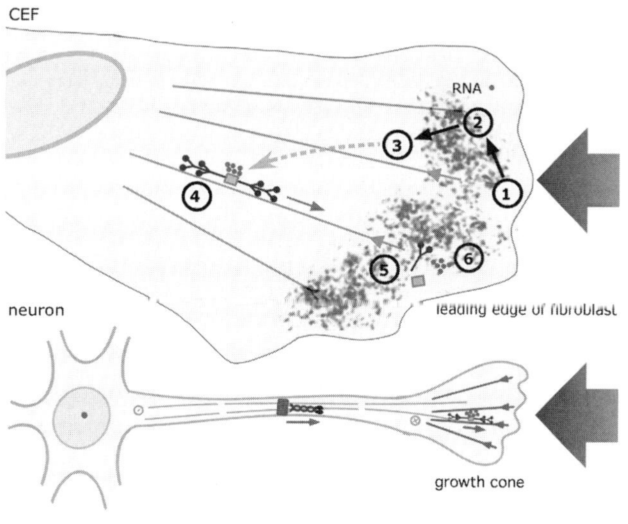

Figure 1 Model for how mRNA may move in response to a signal transduction mechanism on polarized actin bundles, associated with myosin II-B in fibroblasts (top) or with kinesin in neurons (bottom). Incoming signals (arrow) binding receptors (1) activate the RhoA cascade (2) through Rho-kinase (3), which activates myosin II-B light chains and myosin assembly with an mRNA complex. (4) The complex moves over polarized actin bundles to disassemble toward the leading edge as a result of heavy chain phosphorylation (5) and the RNA then anchors (6). The speed of myosin movement predicts that the mRNA would only be transported for 20 sec before it becomes anchored. Hence the steady-state distribution of the mRNA population would be at the lamella [33]. For neurons, an analogous cascade would be initiated by neurotrophins, possibly leading to the activation of the kinesin light chains. Kinesin would then move the ZBP1-RNA complex down the process and then onto microfilament bundles in the growth cone where myosin would take over [39].

References

1. Palacios, I. M. and Johnston, D. S. (2001). Getting the message across: The intracellular localization of mRNAs in higher eukaryotes. *Annu. Rev. Cell Dev. Biol.*17, 569–614.
2. Job, C. and Eberwine, J. (2001). Localization and translation of mRNA in dendrites and axons. *Nat. Rev. Neurosci.* **2**, 889–898.
3. Steward, O. and Schuman, E. M. (2001). Protein synthesis at synaptic sites on dendrites. *Annu. Rev. Neurosci.* **24**, 299–325.
4. Bassell, G. J. and Singer, R. H. (2001). Neuronal RNA localization and the cytoskeleton. *Results Probl. Cell Differ.* **34**, 41–56.
5. Hazelrigg, T. (1998). The destinies and destinations of RNAs. *Cell* **95**, 451–460.
6. Kloc, M., Zearfoss, N. R., and Etkin, L. D. (2002). Mechanisms of subcellular mRNA localization. *Cell* **108**, 533–544.
7. Kislauskis, E. H., Zhu, X., and Singer, R. H. (1997). β-Actin messenger RNA localization and protein synthesis augment cell motility. *J. Cell Biol.* **136**, 1263–1270.
8. Kislauskis, E. H., Zhu, X., and Singer, R. H. (1994). Sequences responsible for intracellular localization of beta-actin messenger RNA also affect cell phenotype. *J. Cell Biol.* **127**, 441–451.
9. Kislauskis, E. H. and Singer, R. H. (1992). Determinants of mRNA localization. *Curr. Opin. Cell Biol.* **4**, 975–978.
10. Ross, A. F., Oleynikov, Y., Kislauskis, E. H., Taneja, K. L., and Singer, R. H. (1997). Characterization of a beta-actin mRNA zipcode-binding protein. *Mol. Cell Biol.* **17**, 2158–2165.
11. Latham, V. M., Jr., Kislauskis, E. H., Singer, R. H., and Ross, A. F. (1994). Beta-actin mRNA localization is regulated by signal transduction mechanisms. *J. Cell Biol.* **126**, 1211–1219.
12. Hill, M. A., Schedlich, L., and Gunning, P. (1994). Serum-induced signal transduction determines the peripheral location of beta-actin mRNA within the cell. *J. Cell Biol.* **126**, 1221–1229.
13. Zhang, H. L., Singer, R. H., and Bassell, G. J. (1999). Neurotrophin regulation of beta-actin mRNA and protein localization within growth cones. *J. Cell Biol.* **147**, 59–70.
14. Song, H. J. and Poo, M. M. (1999). Signal transduction underlying growth cone guidance by diffusible factors. *Curr. Opin. Neurobiol.* **9**, 355–363.
15. Knowles, R. B. and Kosik, K. S. (1997). Neurotrophin-3 signals redistribute RNA in neurons. *Proc. Natl. Acad. Sci. USA* **94**, 14804–14808.
16. Aakalu, G., Smith, W. B., Nguyen, N., Jiang, C., and Schuman, E. M. (2001). Dynamic visualization of local protein synthesis in hippocampal neurons. *Neuron* **30**, 489–502.
17. Ouyang, Y., Rosenstein, A., Kreiman, G., Schuman, E. M., and Kennedy, M. B. (1999). Tetanic stimulation leads to increased accumulation of Ca(2+)/calmodulin-dependent protein kinase II via dendritic protein synthesis in hippocampal neurons. *J. Neurosci.* **19**, 7823–7833.
18. Steward, O. and Worley, P. F. (2001). A cellular mechanism for targeting newly synthesized mRNAs to synaptic sites on dendrites. *Proc. Natl. Acad. Sci. USA* **98**, 7062–7068.

19. Steward, O. and Worley, P. (2001). Localization of mRNAs at synaptic sites on dendrites. *Results Probl. Cell Differ.* **34**, 1–26.
20. Steward, O. and Worley, P. F. (2001). Selective targeting of newly synthesized Arc mRNA to active synapses requires NMDA receptor activation. *Neuron* **30**, 227–240.
21. Ip, N. Y., Li, Y., Yancopoulos, G. D., and Lindsay, R. M. (1993). Cultured hippocampal neurons show responses to BDNF, NT-3, and NT-4, but not NGF. *J. Neurosci.* **13**, 3394–3405.
22. Gao, F. B. (1998). Messenger RNAs in dendrites: Localization, stability, and implications for neuronal function. *Bioessays* **20**, 70–78.
23. Berninger, B., Garcia, D. E., Inagaki, N., Hahnel, C., and Lindholm, D. (1993). BDNF and NT-3 induce intracellular Ca2+ elevation in hippocampal neurones. *Neuroreport* **4**, 1303–1306.
24. Nikcevic, G., Perhonen, M., Boateng, S. Y., and Russell, B. (2000). Translation is regulated via the 3′ untranslated region of alpha-myosin heavy chain mRNA by calcium but not by its localization. *J. Muscle Res. Cell Motil.* **21**, 599–607.
25. Roth, S., Neuman-Silberberg, F. S., Barcelo, G., and Schupbach, T. (1995). cornichon and the EGF receptor signaling process are necessary for both anterior-posterior and dorsal-ventral pattern formation in *Drosophila. Cell* **81**, 967–978.
26. Gonzalez-Reyes, A., Elliott, H., and St. Johnston, D. (1995). Polarization of both major body axes in *Drosophila* by gurken-torpedo signalling. *Nature* **375**, 654–658.
27. Clegg, N. J., Frost, D. M., Larkin, M. K., Subrahmanyan, L., Bryant, Z., and Ruohola-Baker, H. (1997). Maelstrom is required for an early step in the establishment of *Drosophila* oocyte polarity: Posterior localization of grk mRNA. *Development* **124**, 4661–4671.
28. Simmonds, A. J., dosSantos, G., Livne-Bar, I., and Krause, H. M. (2001). Apical localization of wingless transcripts is required for wingless signaling. *Cell* **105**, 197–207.
29. Manseau, L. J. (2001). RNA localization meets wingless signaling. *Sci. STKE* **2001**, PE1.
30. Schwartz, M. A. (2001). Integrin signaling revisited. *Trends Cell Biol.* **11**, 466–470.
31. Chicurel, M. E., Singer, R. H., Meyer, C. J., and Ingber, D. E. (1998). Integrin binding and mechanical tension induce movement of mRNA and ribosomes to focal adhesions. *Nature* **392**, 730–733.
32. Sundell, C. L. and Singer, R. H. (1991). Requirement of microfilaments in sorting of actin messenger RNA. *Science* **253**, 1275–1277.
33. Latham, V. M., Yu, E. H., Tullio, A. N., Adelstein, R. S., and Singer, R. H. (2001). A Rho-dependent signaling pathway operating through myosin localizes beta-actin mRNA in fibroblasts. *Curr. Biol.* **11**, 1010–1016.
34. Davis, L., Banker, G. A., and Steward, O. (1987). Selective dendritic transport of RNA in hippocampal neurons in culture. *Nature* **330**, 477–479.
35. Knowles, R. B., Sabry, J. H., Martone, M. E., Deerinck, T. J., Ellisman, M. H., Bassell, G. J., and Kosik, K. S. (1996). Translocation of RNA granules in living neurons. *J. Neurosci.* **16**, 7812–7820.
36. Kiebler, M. A., Hemraj, I., Verkade, P., Kohrmann, M., Fortes, P., Marion, R. M., Ortin, J., and Dotti, C. G. (1999). The mammalian staufen protein localizes to the somatodendritic domain of cultured hippocampal neurons: Implications for its involvement in mRNA transport. *J. Neurosci.* **19**, 288–297.
37. Kohrmann, M., Luo, M., Kaether, C., DesGroseillers, L., Dotti, C. G., and Kiebler, M. A. (1999). Microtubule-dependent recruitment of Staufen-green fluorescent protein into large RNA-containing granules and subsequent dendritic transport in living hippocampal neurons. *Mol. Biol. Cell* **10**, 2945–2953.
38. Krichevsky, A. M. and Kosik, K. S. (2001). Neuronal RNA granules: A link between RNA localization and stimulation-dependent translation. *Neuron* **32**, 683–696.
39. Zhang, H. L., Eom, T., Oleynikov, Y., Shenoy, S. M., Liebelt, D. A., Dictenberg, J. B., Singer, R. H., and Bassell, G. J. (2001). Neurotrophin-induced transport of a beta-actin mRNP complex increases beta-actin levels and stimulates growth cone motility. *Neuron* **31**, 261–275.
40. Gu, W., Pan, F., Zhang, H., Bassell, G. J., and Singer, R. H. (2002). A predominantly nuclear protein affecting cytoplasmic localization of beta-actin mRNA in fibroblasts and neurons. *J. Cell Biol.* **156**, 41–51.
41. Bullock, S. L. and Ish-Horowicz, D. (2001). Conserved signals and machinery for RNA transport in *Drosophila* oogenesis and embryogenesis. *Nature* **414**, 611–616.
42. Hall, A. (1998). Rho GTPases and the actin cytoskeleton. *Science* **279**, 509–514.
43. Ridley, A. J. (2001). Rho family proteins: Coordinating cell responses. *Trends Cell Biol.* **11**, 471–477.
44. Shestakova, E. A., Singer, R. H., and Condeelis, J. (2001). The physiological significance of beta-actin mRNA localization in determining cell polarity and directional motility. *Proc. Natl. Acad. Sci. USA* **98**, 7045–7050.
45. Winter, C. G., Wang, B., Ballew, A., Royou, A., Karess, R., Axelrod, J. D., and Luo, L. (2001). *Drosophila* Rho-associated kinase (Drok) links Frizzled-mediated planar cell polarity signaling to the actin cytoskeleton. *Cell* **105**, 81–91.
46. Werner, L. A. and Manseau, L. J. (1997). A *Drosophila* gene with predicted rhoGEF, pleckstrin homology and SH3 domains is highly expressed in morphogenic tissues. *Gene* **187**, 107–114.
47. Watanabe, N., Madaule, P., Reid, T., Ishizaki, T., Watanabe, G., Kakizuka, A., Saito, Y., Nakao, K., Jockusch, B. M., and Narumiya, S. (1997). p140mDia, a mammalian homolog of *Drosophila* diaphanous, is a target protein for Rho small GTPase and is a ligand for profilin. *EMBO J.* **16**, 3044–3056.
48. Palazzo, A. F., Cook, T. A., Alberts, A. S., and Gundersen, G. G. (2001). mDia mediates Rho-regulated formation and orientation of stable microtubules. *Nat. Cell Biol.* **3**, 723–729.
49. Nakano, K., Takaishi, K., Kodama, A., Mammoto, A., Shiozaki, H., Monden, M., and Takai, Y. (1999). Distinct actions and cooperative roles of ROCK and mDia in Rho small G protein-induced reorganization of the actin cytoskeleton in Madin-Darby canine kidney cells. *Mol. Biol. Cell* **10**, 2481–2491.
50. Satoh, S. and Tominaga, T. (2001). mDia-interacting protein acts downstream of Rho-mDia and modifies Src activation and stress fiber formation. *J. Biol. Chem.* **276**, 39290–39294.
51. Liu, G., Grant, W. M., Persky, D., Latham, V. M., Jr., Singer, R. H., and Condeelis, J. (2002). Interactions of elongation factor 1alpha with F-actin and beta-actin mRNA: Implications for anchoring mRNA in cell protrusions. *Mol. Biol. Cell* **13**, 579–592.
52. Bashirullah, A., Cooperstock, R. L., and Lipshitz, H. D. (1998). RNA localization in development. *Annu. Rev. Biochem.* **67**, 335–394.
53. Kwon, S. and Schnapp, B. J. (2001). RNA localization: SHEdding light on the RNA-motor linkage. *Curr. Biol.* **11**, R166–R168.
54. Long, R. M., Singer, R. H., Meng, X., Gonzalez, I., Nasmyth, K., and Jansen, R. P. (1997). Mating type switching in yeast controlled by asymmetric localization of ASH1 mRNA. *Science* **277**, 383–387.
55. Long, R. M., Gu, W., Lorimer, E., Singer, R. H., and Chartrand, P. (2000). She2p is a novel RNA-binding protein that recruits the Myo4p-She3p complex to ASH1 mRNA. *EMBO J.* **19**, 6592–6601.
56. Bertrand, E., Chartrand, P., Schaefer, M., Shenoy, S. M., Singer, R. H., and Long, R. M. (1998). Localization of ASH1 mRNA particles in living yeast. *Mol. Cell* **2**, 437–445.
57. Takizawa, P. A. and Vale, R. D. (2000). The myosin motor, Myo4p, binds Ash1 mRNA via the adapter protein, She3p. *Proc. Natl. Acad. Sci. USA* **97**, 5273–5278.
58. Bohl, F., Kruse, C., Frank, A., Ferring, D., and Jansen, R. P. (2000). She2p, a novel RNA-binding protein tethers ASH1 mRNA to the Myo4p myosin motor via She3p. *EMBO J.* **19**, 5514–5524.
59. Kimura, K., Ito, M., Amano, M., Chihara, K., Fukata, Y., Nakafuku, M., Yamamori, B., Feng, J., Nakano, T., Okawa, K., Iwamatsu, A., and Kaibuchi, K. (1996). Regulation of myosin phosphatase by Rho and Rho-associated kinase (Rho-kinase). *Science* **273**, 245–248.
60. Somlyo, A. P. and Somlyo, A. V. (2000). Signal transduction by G-proteins, rho-kinase and protein phosphatase to smooth muscle and non-muscle myosin II. *J. Physiol.* **522** (Pt 2), 177–185.

61. Cramer, L. P., Siebert, M., and Mitchison, T. J. (1997). Identification of novel graded polarity actin filament bundles in locomoting heart fibroblasts: Implications for the generation of motile force. *J. Cell Biol.* **136**, 1287–1305.
62. Sellers, J. R. and Kachar, B. (1990). Polarity and velocity of sliding filaments: Control of direction by actin and of speed by myosin. *Science* **249**, 406–408.
63. Totsukawa, G., Yamakita, Y., Yamashiro, S., Hartshorne, D. J., Sasaki, Y., and Matsumura, F. (2000). Distinct roles of ROCK (Rho-kinase) and MLCK in spatial regulation of MLC phosphorylation for assembly of stress fibers and focal adhesions in 3T3 fibroblasts. *J. Cell Biol.* **150**, 797–806.
64. Bridgman, P. C., Dave, S., Asnes, C. F., Tullio, A. N., and Adelstein, R. S. (2001). Myosin IIB is required for growth cone motility. *J. Neurosci.* **21**, 6159–6169.
65. Diefenbach, T. J., Latham, V. M., Yimlamai, D., Liv, C. A., Herman, I. M., Vay, D. G. (2002). Myosin IC and myosin IIB serve opposing roles in lamellipodial dynamics of the neuronal growth cone. *J. Cell Biol.* **158**, 1207–1217.

CHAPTER 314

Translational Control by Amino Acids and Energy

Tobias Schmelzle, José L. Crespo, and Michael N. Hall

Division of Biochemistry, Biozentrum, University of Basel, CH-4056 Basel, Switzerland

Introduction

Translation is possibly the most energy-consuming process of the cell. Organisms have therefore developed broad control mechanisms that adjust the rate of translation to available nutrient and energy resources. The control of translation rapidly and directly sets a cell growth rate that is appropriate to prevailing environmental conditions. In general, translational control involves the transduction of a signal through a pathway that at one end senses environmental conditions and at the other end impinges on the translational machinery. Among many other signals, translation is controlled by amino acids—the building blocks that are required for the translation process itself—and by the energy status of the cell. Here we summarize recent findings on translational control by amino acids and energy, focusing on two signal transduction pathways—the GCN system and the TOR signaling pathway—that control at least four different steps of protein synthesis. The yeast GCN system, in response to amino acid availability, controls the translation initiation factor eIF2. The mammalian TOR (mTOR) signaling pathway, in response to amino acids and energy, controls translation by regulating the initiation factor eIF4E, ribosomal protein S6, and elongation factor eEF2. mTOR controls the latter two via p70 S6 kinase (S6K) (Fig. 1). Depending on the signal and the specific control mechanism, the global rate of protein synthesis or only the translation rate of specific mRNAs is affected. In general, the changes in global translation are rather modest, but the effects on specific mRNA populations are dramatic.

GCN System

The GCN signaling pathway has been characterized in yeast. The kinase GCN2 responds to amino acid or purine deprivation by phosphorylating the initiation factor subunit eIF2α at Ser51. This results in a down-regulation of initiation-competent 40S ribosomal subunits, which in turn leads to increased translation of a specific mRNA encoding GCN4 (see later discussion). GCN4 is a transcriptional activator of several genes encoding amino acid biosynthetic enzymes [1,2]. Thus, as a consequence of GCN-mediated down-regulation of translation, amino acid synthesis is transiently boosted to compensate for a limited supply of exogenous amino acids. Mammalian cells also make extensive use of eIF2α Ser51 phosphorylation to down-regulate translation initiation, employing different kinases in response to different stresses. These kinases include PKR (dsRNA viral infection), HRI (heme/iron metabolism), PERK (unfolded protein response in the ER), and a mammalian GCN2 homolog. It remains to be determined if mGCN2, like its yeast counterpart, is activated by amino acid deprivation.

How is GCN2 activated in response to amino acid deprivation? GCN2 contains a domain that is similar to histidyl-tRNA synthetase (HisRS) and which interacts with a tRNA acceptor stem. The activating ligand for GCN2 is therefore likely uncharged (deacylated) tRNA—especially since levels of uncharged tRNA significantly increase upon amino acid limitation. In mammalian cells, the phosphorylation of eIF2α typically observed upon amino acid depletion can also be induced by treatment with histidinol, a competitive inhibitor

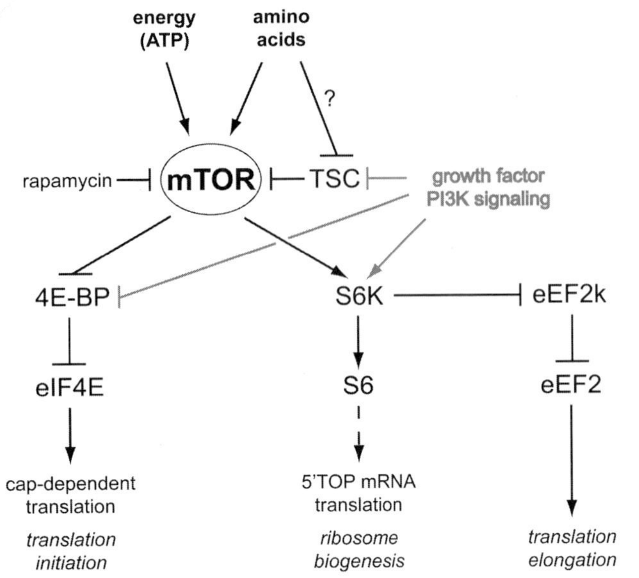

Figure 1 Model of mTOR controlling translation in response to amino acids and ATP. mTOR controls translation in conjunction with growth factor signaling via PI3K. The input of amino acids into mTOR may be direct or via inhibition of the TSC complex. The dashed arrow between S6 and 5′TOP mRNA translation reflects the uncertainty of this link (see text). TSC indicates the TSC1-TSC2 complex. Arrows indicate activation; bars indicate repression.

of HisRS, indicating that the mechanism of GCN2 activation may be conserved [2,3].

How does eIF2α Ser51phosphorylation lead to a down-regulation of translation initiation? As one of the early steps of initiation, GTP-loaded eIF2 (of which eIF2α is a subunit) delivers Met-tRNA$_i$ to the 40S ribosomal subunit. GTP loading and activation of the eIF2 complex is catalyzed by the GDP/GTP exchange factor eIF2B. Phosphorylation of Ser51 interferes with the activation (GTP-loading) of eIF2 by eIF2B. This results in decreased formation of the eIF2-GTP-Met-tRNA$_i$ ternary complex required for assembly of initiation-competent ribosomes and, thus, a global down-regulation of translation initiation [2]. Despite the global down-regulation of translation initiation, translation of GCN4 is up-regulated due to a mechanism that depends on the presence of short open reading frames (uORFs) in the leader region of the *GCN4* mRNA [1,2].

TOR Signaling Pathway

In mammalian cells, amino acid deprivation causes rapid dephosphorylation and activation of eIF-4E binding protein(s) (4E-BP; an inhibitor of eIF4E, also known as PHAS-I) and dephosphorylation and inhibition of the kinase S6K. This results in a down-regulation of translation initiation (see the following two sections). There are several 4E-BPs, but most experimental work has focused on 4E-BP1. Similarly, S6K comprises S6K1 and the more recently discovered S6K2, but most studies have involved only S6K1. The effect of amino acids on 4E-BP and S6K phosphorylation is mediated by the kinase mTOR (Fig. 1). TOR, initially identified in yeast, is the target of the immunosuppressive and anticancer drug rapamycin. The conserved TOR kinases, found in yeast, plants, worm, fly, and mammals, control a large and diverse set of cell growth-related readouts, including translation initiation [4,5]. TOR positively controls translation initiation and is thus essential for growth. This is in contrast to the nonessential GCN system, which negatively regulates translation as a homeostatic response to fluctuations in amino acid availability.

Inactivation of mTOR by treatment with rapamycin elicits the same changes in 4E-BP and S6K phosphorylation as amino acid deprivation. Furthermore, a truncated version of S6K that is resistant to rapamycin is also resistant to inhibition by amino acid withdrawal [6,7]. These findings indicate that amino acids signal to eIF4E and S6K via mTOR. mTOR phosphorylates 4E-BP and S6K *in vitro*, suggesting that mTOR may act on these two proteins directly. In yeast, TOR controls the phosphorylation of several downstream targets by negatively regulating the type-2A-related phosphatase SIT4. Recent evidence indicates that such a mechanism may also apply in mammalian cells to regulate the phosphorylation and thus the activity of mTOR effectors such as S6K [4].

The phosphorylation of 4E-BP and S6K is also controlled by the phosphatidylinositol 3-kinase (PI3K) signaling pathway, which is activated in response to growth factors such as insulin. Downstream of PI3K, this growth-controlling pathway includes 3′-phosphoinositide-dependent protein kinase 1 (PDK1) and protein kinase B (PKB/Akt). How are mTOR and the PI3K signaling pathway linked? First, mTOR and PI3K signaling are linked via their common targets 4E-BP and S6K, although they respond separately to amino acids (and energy) and growth factors (mitogens), respectively. The inhibition of 4E-BP and the activation of S6K require simultaneous inputs from mTOR and PI3K. The complex and hierarchical phosphorylation of several sites in 4E-BP [8,9] and S6K [10] reflects the dual control of these proteins by mTOR and PI3K (Fig. 1). Second, in addition to mTOR and PI3K controlling common targets in response to different signals, the PI3K pathway may also be linked to mTOR via the TSC complex (see later discussion) (Fig. 1).

Branched-chain amino acids (Leu, Ile, Val) have been shown to play an important role in the regulation of protein synthesis, in particular, in skeletal muscle. Of the branched-chain amino acids, leucine is the most potent in stimulating translation. Leucine as the sole exogenously added amino acid is sufficient to drive phosphorylation of 4E-BP or S6K in diverse cells, such as skeletal muscle, pancreatic β cells, or adipocytes. For leucine to be effective, mTOR must be active, since pretreatment with rapamycin abolishes the leucine-mediated stimulatory effects on the translation apparatus. However, evidence also suggests a so far uncharacterized, mTOR-independent mechanism by which leucine stimulates translation. It is also important to note that the regulatory role of leucine may be tissue specific (reviewed in [11]).

How is mTOR signaling regulated in response to amino acids? For S6K, it has been proposed that, similar to the

GCN system, the aminoacylation state of tRNA may be responsible for the regulation of S6K phosphorylation [12]. It is not known how mTOR would sense the aminoacylated tRNA and, despite the possible similar regulatory mechanism, a connection between GCN2 and TOR has not been demonstrated. Dennis *et al.* [13] argue that amino acid pools rather than the amount of aminoacylated tRNA are important to mTOR signaling. How mTOR would sense amino acid pools is also unknown.

The control of translation initiation by energy is less characterized than the control by amino acids. Moreover, one must define the term *energy*—is it "nutrients" in general (and if so, which?) or is it an "energy-rich" intracellular product? Intracellular levels of ATP or related nucleotides may indicate the energy status of the cell. Along these lines, mTOR has been suggested recently to respond to intracellular levels of ATP. The mechanism by which ATP acts on TOR appears to be different from that used by amino acids, but both ATP and amino acids are required for mTOR signaling [13] (Fig. 1). AMP-activated protein kinase (AMPK) is also involved in controlling cellular metabolism in response to the energy status (AMP concentration or ATP/AMP ratio) of the cell. In contrast to mTOR, AMPK is activated in response to low energy levels, leading to increased glucose uptake and fatty acid oxidation in muscle. AMPK activation by treatment of cells with the compound AICAR (phosphorylated in the cell to form an AMP analog) has recently been demonstrated to result in inhibition of protein synthesis in skeletal muscle and hepatocytes. Interestingly, phosphorylation of the mTOR targets 4E-BP and S6K decreases upon AMPK activation [14,15]. The nature of the relationship between AMPK activity and mTOR signaling, also in terms of mTOR responsiveness to ATP levels (see earlier discussion), remains to be determined.

In terms of energy, another nutrient to be considered is carbohydrates. Patel *et al.* [16] have recently reported that glucose exerts a permissive effect on the ability of insulin to promote the phosphorylation of 4E-BP and the formation of translation-competent initiation complexes. The effect may be mediated by a product of hexose metabolism and seems to be restricted to 4E-BP, since glucose did not allow full activation of S6K. Thus, 4E-BP in particular appears to be regulated in response to the availability of both amino acids (especially leucine) and an energy source, such as glucose [16].

In yeast, TOR signaling controls cap-dependent translation initiation in response to nutrients, such as nitrogen and carbon sources. The mechanism by which TOR controls translation initiation remains to be determined, even though the TOR dependence of cap-dependent translation initiation was first shown in yeast [17]. TOR also controls translation by ensuring the availability of amino acids, via stabilization of plasma membrane amino acid transporters [4]. Recent evidence argues that TOR responds to the amino acid glutamine, suggesting that glutamine is a particularly important indicator of nutrient status [18]. Indeed, glutamine is a preferred nitrogen source and controls TCA cycle-mediated carbon metabolism. TOR must also respond to other yet-to-be-identified nutrients because glutamine starvation affects only a subset of TOR readouts [18]. In mammalian cells, a recent transcriptional profiling has revealed that rapamycin treatment mimics, in part, glutamine or leucine starvation [19]. It is also interesting that, similarly to rapamycin treatment, a decrease in blood glutamine levels causes immunosuppression in humans and mice [20].

The recent identification of TOR interacting proteins as part of large TOR complexes in yeast and mammalian cells may contribute to the understanding of how TOR signaling responds to nutrient availability [21–23]. In particular, the mTOR interactor raptor binds 4E-BP and S6K, and the raptor-mTOR interaction is required for the mTOR-catalyzed phosphorylation of 4E-BP and S6K. Raptor may thus serve as a scaffold linking mTOR to its substrates [21]. Kim *et al.* [22] have proposed that raptor, in a nutrient-sensitive complex with mTOR, has both a positive and a negative role in the control of S6K. Finally, the yeast homolog of raptor, KOG1, interacts with yeast TOR, and depletion of KOG1 mimics rapamycin treatment, including a decrease in protein synthesis [23].

The TSC1–TSC2 complex, composed of the tumor suppressor proteins TSC1 and TSC2 (tuberous sclerosis 1 and 2, also known as hamartin and tuberin, respectively), also physically associates with TOR, as shown in *Drosophila* [24]. Epistasis analysis examining the effects of *TSC* and *TOR* mutations on cell viability and cell size in *Drosophila* places the TSC complex upstream of TOR [24]. Loss of TSC1–TSC2 results in a TOR-dependent increase of S6K activity in both *Drosophila* and mammalian cells and renders these cells resistant to amino acid starvation [24]. Thus, the TSC complex appears to antagonize amino acid signaling to TOR [24]. Recent findings have also shown that TSC2 is phosphorylated and inactivated by the PI3K effector PKB/Akt in response to growth factors [25–27]. Phosphorylation of TSC2 causes dissociation of the TSC complex [25,27] and stimulation of TOR-dependent phosphorylation of 4E-BP and S6K [25,26,28]. Furthermore, TSC1–TSC2 inhibits mTOR kinase activity [25]. These findings suggest that growth factor signaling to mTOR requires PKB-dependent inactivation of the TSC complex, and thereby provide a so far elusive link between growth factor signaling and mTOR. However, because PKB/Akt activity is not controlled by amino acids, it is difficult to reconcile a role for TSC in both amino acid and growth factor signaling to TOR. It would be of interest to determine how amino acid availability inactivates the TSC1-TSC2 complex. The nature of the apparently multiple roles of TSC in TOR signaling remains to be determined.

4E-BP: Regulation of eIF4E

Binding of the 40S ribosomal subunit to mRNA is mediated by the eIF4F complex, which is composed of the mRNA 5′ cap (m^7GpppN, m = methyl, N = nucleotide) binding protein eIF4E, the helicase eIF4A, and the scaffold protein eIF4G. After binding to the 5′ cap, eIF4F (via eIF4A and the associated factor eIF4B) unwinds inhibitory secondary structure present in the 5′ untranslated leader region (UTR). Growing cells exhibit high eIF4F activity, whereas starved or stressed cells display low eIF4F activity. In mammalian cells, formation of the eIF4F complex is regulated by the 4E-BP family of translational repressors. 4E-BP competes with eIF4G for binding to eIF4E, and binding of 4E-BP to eIF4E is regulated by the phosphorylation state of 4E-BP. Hyperphosphorylation of 4E-BP inhibits the 4E-BP–eIF4E interaction, whereas hypophosphorylation, observed upon amino acid starvation or mTOR inactivation (see earlier discussion), promotes this interaction. Low 4E-BP phosphorylation thus results in the down-regulation of initiation-competent eIF4F complex due to the displacement of eIF4G from the eIF4F complex. Dephosphorylation of 4E-BP occurs not only upon amino acid deprivation, but also in response to a wide variety of other environmental stresses.

Although modulation of eIF4E activity affects global (cap-dependent) protein synthesis, translation of selective mRNAs is particularly affected. For example, the mRNA for ornithine decarboxylase possesses an unusually long 5′UTR predicted to form extensive secondary structure, and therefore falls into a class of mRNAs that are particularly dependent on the helicase activity of the eIF4F complex for their efficient translation. Many mRNAs encoding oncoproteins or other gene products related to cell growth and proliferation display such large, highly structured 5′UTRs and are thus predicted to be particularly subject to control by 4E-BP [5].

S6K: Regulation of S6 and eEF2

The main substrate of the mTOR-controlled kinase S6K is the 40S ribosomal protein S6. Based on its location within the ribosome and its interactions with translation initiation factors, S6 has been proposed to affect protein synthesis at the level of initiation. Increased S6 phosphorylation upon S6K activation may mediate, in particular, the initiation of translation of mRNAs with a 5′ terminal oligopyrimidine tract (TOP) [10,29] (Fig. 1). However, it is controversial whether 5′TOP mRNA translation is indeed controlled via S6K and S6 [30]. Most 5′TOP mRNAs encode components of the translational machinery, such as ribosomal proteins (RP), elongation factors, and the poly(A)-binding protein. Thus, ribosome biogenesis and synthesis of other components of the translational machinery are controlled at the translational level. The biosynthesis of ribosomes consumes a large part of the cell's energy and therefore constitutes an ideal target of translational control by mTOR.

TOR also controls ribosome biogenesis in yeast, but by a mechanism independent of a S6K homolog or 5′TOP regulatory elements because both of these are absent in yeast. Inhibition of TOR leads to a rapid down-regulation of transcription of rRNA (RNA polymerase I [Pol I] and RNA polymerase III [Pol III]) and RP genes, and thus of ribosome biogenesis [31, 32]. Furthermore, treatment of yeast cells with rapamycin blocks tRNA synthesis (Pol III) and processing of 35S rRNA [31,32]. The conserved TOR signaling pathway may thus act in all eukaryotic organisms to control ribosome biogenesis in response to nutrients (amino acids and energy).

mTOR-controlled S6K also positively regulates translation elongation. The eukaryotic elongation factor 2 (eEF2) is phosphorylated and inactivated by the kinase eEF2k. S6K phosphorylates and inhibits eEF2k. Thus, S6K positively controls eEF2, and thus elongation, by inhibiting eEF2k [33] (Fig. 1).

Note Added in Proof

Cherkasova and Hinnebusch have recently demonstrated that the GCN and TOR systems in yeast are functionally connected. Inhibition of TOR by treatment with rapamycin causes dephosphorylation of Ser 577 in GCN2, GCN2-mediated phosphorylation of eIF2α, and induction of *GCN4* translation. Thus, TOR negatively regulates GCN2. TOR controls GCN2 through TAP42, a regulator of yeast type 2A-related phosphatases [34].

References

1. Hinnebusch, A. G. (1997). Translational regulation of yeast GCN4. A window on factors that control initiator tRNA-binding to the ribosome. *J. Biol. Chem.* **272**, 21661–21664.
2. Hinnebusch, A. G. (2000). Mechanism and regulation of initiator methionyl-tRNA binding to ribosomes, in Sonenberg, N., Hershey, J. W. B., and Mathews, M. B., Eds., *Translational Control of Gene Expression*, pp. 185–243. Cold Spring Harbor Laboratory Press, Cold Spring Harbor, NY.
3. Kimball, S. R. and Jefferson, L. S. (2000). Regulation of translation initiation in mammalian cells by amino acids, in Sonenberg, N., Hershey, J. W. B., and Mathews, M. B., Eds., *Translational Control of Gene Expression*, pp. 561–579. Cold Spring Harbor Laboratory Press, Cold Spring Harbor, NY.
4. Schmelzle, T. and Hall, M. N. (2000). TOR, a central controller of cell growth. *Cell* **103**, 253–262.
5. Gingras, A. C., Raught, B., and Sonenberg, N. (2001). Regulation of translation initiation by FRAP/mTOR. *Genes Dev.* **15**, 807–826.
6. Hara, K., Yonezawa, K., Weng, Q. P., Kozlowski, M. T., Belham, C., and Avruch, J. (1998). Amino acid sufficiency and mTOR regulate p70 S6 kinase and eIF-4E BP1 through a common effector mechanism. *J. Biol. Chem.* **273**, 14484–14494.
7. Wang, X., Campbell, L. E., Miller, C. M., and Proud, C. G. (1998). Amino acid availability regulates p70 S6 kinase and multiple translation factors. *Biochem. J.* **334**, 261–267.
8. Mothe-Satney, I., Yang, D., Fadden, P., Haystead, T. A., and Lawrence, J. C. (2000). Multiple mechanisms control phosphorylation of PHAS-I in five (S/T)P sites that govern translational repression. *Mol. Cell. Biol.* **20**, 3558–3567.
9. Gingras, A. C., Raught, B., Gygi, S. P., Niedzwiecka, A., Miron, M., Burley, S. K., Polakiewicz, R. D., Wyslouch-Cieszynska, A., Aebersold, R., and Sonenberg, N. (2001). Hierarchical phosphorylation of the translation inhibitor 4E-BP1. *Genes Dev.* **15**, 2852–2864.
10. Fumagalli, S. and Thomas, G. (2000). S6 phosphorylation and signal transduction, in Sonenberg, N., Hershey, J. W. B., and Mathews, M. B., Eds., *Translational Control of Gene Expression*, pp. 695–717. Cold Spring Harbor Laboratory Press, Cold Spring Harbor, NY.
11. Kimball, S. R. and Jefferson, L. S. (2002). Control of protein synthesis by amino acid availability. *Curr. Opin. Clin. Nutr. Metab. Care* **5**, 63–67.
12. Iiboshi, Y., Papst, P. J., Kawasome, H., Hosoi, H., Abraham, R. T., Houghton, P. J., and Terada, N. (1999). Amino acid-dependent control of $p70^{s6k}$. Involvement of tRNA aminoacylation in the regulation. *J. Biol. Chem.* **274**, 1092–1099.

13. Dennis, P. B., Jaeschke, A., Saitoh, M., Fowler, B., Kozma, S. C., and Thomas, G. (2001). Mammalian TOR: A homeostatic ATP sensor. *Science* **294**, 1102–1105.
14. Bolster, D. R., Crozier, S. J., Kimball, S. R., and Jefferson, L. S. (2002). AMP-activated protein kinase suppresses protein synthesis in rat skeletal muscle through down-regulated mammalian target of rapamycin (mTOR) signaling. *J. Biol. Chem.* **277**, 23977–23980.
15. Dubbelhuis, P. F. and Meijer, A. J. (2002). Hepatic amino acid-dependent signaling is under the control of AMP-dependent protein kinase. *FEBS Lett.* **521**, 39–42.
16. Patel, J., Wang, X., and Proud, C. G. (2001). Glucose exerts a permissive effect on the regulation of the initiation factor 4E binding protein 4E-BP1. *Biochem. J.* **358**, 497–503.
17. Barbet, N. C., Schneider, U., Helliwell, S. B., Stansfield, I., Tuite, M. F., and Hall, M. N. (1996). TOR controls translation initiation and early G1 progression in yeast. *Mol. Biol. Cell* **7**, 25–42.
18. Crespo, J. L., Powers, T., Fowler, B., and Hall, M. N. (2002). The TOR-controlled transcription activators GLN3, RTG1, and RTG3 are regulated in response to intracellular levels of glutamine. *Proc. Natl. Acad. Sci. USA* **99**, 6784–6789.
19. Peng, T., Golub, T. R., and Sabatini, D. M. (2002). The immunosuppressant rapamycin mimics a starvation-like signal distinct from amino acid and glucose deprivation. *Mol. Cell. Biol.* **22**, 5575–5584.
20. Calder, P. C. and Yaqoob, P. (1999). Glutamine and the immune system. *Amino Acids* **17**, 227–241.
21. Hara, K., Maruki, Y., Long, X., Yoshino, K., Oshiro, N., Hidayat, S., Tokunaga, C., Avruch, J., and Yonezawa, K. (2002). Raptor, a binding partner of target of rapamycin (TOR), mediates TOR action. *Cell* **110**, 177–189.
22. Kim, D.-H., Sarbassov, D. D., Ali, S. M., King, J. E., Latek, R. R., Erdjument-Bromage, H., Tempst, P., and Sabatini, D. M. (2002). mTOR interacts with raptor to form a nutrient-sensitive complex that signals to the cell growth machinery. *Cell* **110**, 163–175.
23. Loewith, R., Jacinto, E., Wullschleger, S., Lorberg, A., Crespo, J. L., Bonenfant, D., Oppliger, W., Jenoe, P., and Hall, M. N. (2002). Two TOR complexes, only one of which is rapamycin sensitive, have distinct roles in cell growth control. *Mol. Cell* **10**, 457–468.
24. Gao, X., Zhang, Y., Arrazola, P., Hino, O., Kobayashi, T., Yeung, R. S., Ru, B., and Pan, D. (2002). Tsc tumour suppressor proteins antagonize amino-acid-TOR signalling. *Nat. Cell Biol.* **4**, 699–704.
25. Inoki, K., Li, Y., Zhu, T., Wu, J., and Guan, K.-L. (2002). TSC2 is phosphorylated and inhibited by Akt and suppresses m TOR signalling. *Nat. Cell Biol.* **4**, 648–657.
26. Manning, B. D., Tee, A. R., Logsdon, M. N., Blenis, J., and Cantley, L. C. (2002). Identification of the tuberous sclerosis complex-2 tumor suppressor gene product tuberin as a target of the phosphoinositide 3-kinase/Akt pathway. *Mol. Cell* **10**, 151–162.
27. Potter, C. J., Pedraza, L. G., and Xu, T. (2002). Akt regulates growth by directly phosphorylating Tsc2. *Nat. Cell. Biol.* **4**, 658–665.
28. Kwiatkowski, D. J., Zhang, H., Bandura, J. L., Heiberger, K. M., Glogauer, M., el-Hashemite, N., and Onda, H. (2002). A mouse model of TSC1 reveals sex-dependent lethality from liver hemangiomas, and up-regulation of p70S6 kinase activity in Tsc1 null cells. *Hum. Mol. Genet.* **11**, 525–534.
29. Meyuhas, O. and Hornstein, E. (2000). Translational control of TOP mRNAs, in Sonenberg, N., Hershey, J. W. B., and Mathews, M. B., Eds., *Translational Control of Gene Expression*, pp. 671–693. Cold Spring Harbor Laboratory Press, Cold Spring Harbor, NY.
30. Tang, II., Hornstein, E., Stolovich, M., Levy, G., Livingstone, M., Templeton, D., Avruch, J., and Meyuhas, O. (2001). Amino acid-induced translation of TOP mRNAs is fully dependent on phosphatidylinositol 3-kinase-mediated signaling, is partially inhibited by rapamycin, and is independent of S6K1 and rpS6 phosphorylation. *Mol. Cell. Biol.* **21**, 8671–8683.
31. Zaragoza, D., Ghavidel, A., Heitman, J., and Schultz, M. C. (1998). Rapamycin induces the G0 program of transcriptional repression in yeast by interfering with the TOR signaling pathway. *Mol. Cell. Biol.* **18**, 4463–4470.
32. Powers, T. and Walter, P. (1999). Regulation of ribosome biogenesis by the rapamycin-sensitive TOR-signaling pathway in *Saccharomyces cerevisiae*. *Mol. Biol. Cell* **10**, 987–1000.
33. Wang, X., Li, W., Williams, M., Terada, N., Alessi, D. R., and Proud, C. G. (2001). Regulation of elongation factor 2 kinase by $p90^{RSK1}$ and p70 S6 kinase. *EMBO J.* **20**, 4370–4379.
34. Cherkasova, V. A. and Hinnebusch, A. G. (2003). Translational control by TOR and TAP42 through dephosphorylation of eIF2α kinase GCN2. *Genes Dev.*, **17**, 859–872.

Translational Control and Insulin Signaling

Thomas Radimerski and George Thomas
Friedrich Miescher Institute for Biomedical Research, Basel, Switzerland

In mammalian systems the primary function of insulin is to control glucose homeostasis by regulating glucose utilization in peripheral tissues and its own production and secretion in pancreatic β cells. Insulin controls these responses through intracellular signal transduction pathways, which directly modulate major anabolic responses, one of which is the activation and maintenance of elevated rates of protein synthesis. Because protein synthesis is the most energy-consuming anabolic process in the cell, insulin must also serve to sustain elevated levels of energy production required for the continuance of high rates of protein synthesis. To do so insulin acts to stimulate nutrient uptake, to meet the increased demand for energy. Here we describe the insulin-induced signal transduction pathways that coordinate increased protein synthesis rates with increased energy demand.

To maintain glucose homeostasis, insulin must control a number of major anabolic responses in specific target tissues such as liver, adipose tissue, and skeletal and cardiac muscle. Among these responses, the activation and maintenance of high rates of protein synthesis are the most energy-consuming anabolic processes. Therefore, it is not surprising that mechanisms have evolved to precisely modulate translation rates as a function of time, space, and demand [1]. In addition, regulation of gene expression at the level of translation allows the cell to respond more rapidly to sudden changes in the external milieu [2,3]. Insulin controls protein synthesis at the level of both initiation and elongation. Furthermore insulin enhances the cells' translational capacity by stimulating the translation of specific subsets of mRNAs, such as those encoding ribosomal proteins and translation elongation factors [4,5].

The use of mammalian cell culture systems and animal models has led to the identification of a number of signaling components involved in insulin-mediated translational control. In the quiescent state, basal translation rates are required to maintain appropriate levels of housekeeping proteins that turnover at a finite rate. In addition, the translation of certain classes of mRNAs are repressed under these conditions [6,7]. The effects of insulin on the protein synthetic machinery are mediated by signal transduction cascades (Fig. 1), which in turn modulate the specific activity of key translation factors [8]. Moreover, insulin stimulates global protein synthesis by increasing both the rates of translation initiation and elongation as well as by triggering ribosome biogenesis to increase translational capacity [5,9].

Activation of the insulin signal transduction pathway is triggered by the binding of the ligand to its tyrosine kinase receptor, leading to the inter-autophosphorylation of the two β-chains of the receptor at specific tyrosine residues. These phosphorylated residues act as docking sites for proteins containing phosphotyrosine binding domains, such as PTB and SH2 domains. Among these adapter proteins are the insulin receptor substrates (IRSs) [10]. IRSs are recruited to the insulin receptor and phosphorylated on tyrosine residues, which act as docking sites for additional signaling components. One of these is the GRB2/SOS complex, which through the small GTPase Ras stimulates the MAP kinase pathway, a pathway mainly involved in the control of transcription and proliferation [11], but which has a significant impact on translation (see later discussion). Through a distinct set of phosphorylation sites, IRSs recruit the p85 adapter of the p110 catalytic subunit of the phosphatidylinositol 3-kinase (PI3K). PI3K signaling is involved in the regulation of cell

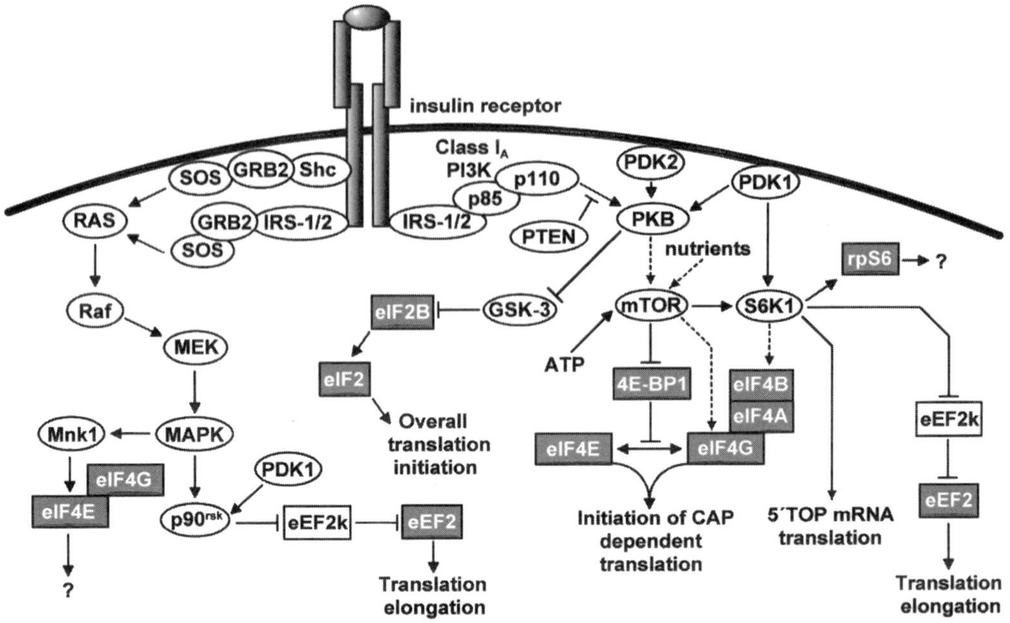

Figure 1 Schematic representation of insulin-induced molecular pathways signaling to the translational control machinery in mammalian cells. Arrows represent a positive input, while bars represent a negative input. Broken arrows indicate unclear and/or possibly indirect relationships. Translation factors targeted by insulin signaling components are depicted as shaded boxes. See text for detailed description of the individual signaling cassettes.

growth, survival, and metabolism and, in the latter case, particularly protein synthesis [12]. The recruitment of PI3K to IRS stimulates the production of the lipid second messenger phosphatidylinositol 3,4,5-trisphosphate [PtdIns(3,4,5)P3] from PtdIns(4,5)P2, leading to the recruitment and activation of protein kinase B (PKB/Akt) by the phosphatidylinositol-dependent kinases, PDK1 and PDK2 [13]. PtdIns(3,4,5)P3 production is counteracted by the lipid phosphatase PTEN (*p*hosphatase and *ten*sin homolog deleted from chromosome 10), a tumor suppressor gene that is either absent or mutated in a large number of cancers [14]. Insulin also leads to the activation of ribosomal protein S6 kinase 1 (S6K1) via *m*ammalian *t*arget *o*f *r*apamycin (mTOR), as well as through PI3K and PDK1 signaling inputs [15].

The primary target of translational control is the initiation step of protein synthesis. The first step of translation initiation is the binding of the ternary eIF2-Met-tRNA$_i$-GTP complex to the 40S ribosomal subunit, containing initiation factors eIF3 and eIF1A, to form the 43S preinitiation complex [16]. After formation of the initiation complex at the translation start site (see later discussion), the GTP bound to the γ subunit of eIF2, is hydrolyzed and eIF2, in a complex with GDP, is released. To begin a new cycle of initiation, eIF2 must interact with the guanine nucleotide-exchange factor, eIF2B, to efficiently exchange GDP for GTP. The guanine nucleotide exchange activity of eIF2B itself is subject to insulin regulation [4]. Insulin treatment promotes eIF2B activation by inducing the dephosphorylation of its ε subunit. This subunit is phosphorylated by GSK-3 [17], a key enzyme involved in regulating glycogen synthase [18]. Insulin induces GSK-3 phosphorylation and inactivation by PKB [19]. Separately, insulin can also promote dephosphorylation of eIF2 α [20]. Phosphorylation of this subunit of eIF2 prevents GDP exchange for GTP, resulting in a strong inhibition of initiation. Despite this knowledge, the insulin signaling pathway controlling eIF2 α dephosphorylation has yet to be elucidated.

Insulin also acts on eIF4F complex formation to exert translational control. The eIF4F complex is composed of three proteins, eIF4G, eIF4A, and eIF4E, which mediate the binding of the mRNA to the 43S preinitiation complex to form the 48S initiation complex [21]. eIF4G is a scaffold protein, which binds eIF4E and eIF4A and directly interacts with eIF3 of the 43S preinitiation complex [3]. eIF4A serves as an ATP-dependent helicase, whereas the function of eIF4E is to bind the 5′ terminal 7-methyl-guanosine, or cap, found on most nuclear encoded mRNAs. eIF4A interacts with a cofactor, eIF4B, which enhances eIF4A helicase activity [22], to drive unwinding of inhibitory secondary structures present within the 5′ untranslated regions (UTRs) of certain mRNAs, until the initiator AUG is encountered during ribosome scanning. eIF4E appears to be the limiting factor in eIF4F complex formation [23] and to be regulated by insulin by two distinct signal transduction pathways. The first involves phosphorylation by MAP kinase activated protein kinase 1 (Mnk1) [24], an event requiring binding of the kinase to the carboxyl terminus of eIF4G [25]. The role of eIF4E phosphorylation is controversial. eIF4E phosphorylation has been reported to increase its affinity toward the mRNA cap structure [26]; however, it has recently been argued that it is neither required for cap-dependent translation nor for eIF4F complex formation [27]. The second involves a family of small repressor proteins termed the eIF4E binding proteins (4E-BPs), which compete with eIF4G for binding to eIF4E, inhibiting translation [28]. Insulin-induced hierarchical phosphorylation of 4E-BP1 disrupts its binding to eIF4E, favoring eIF4F complex formation [29].

Insulin-induced 4E-BP1 phosphorylation has been proposed to be dependent on PKB activation and to be mediated via mTOR [30]. However, in PDK1$^{-/-}$ embryonic stem cells, which lack PKB activity, growth-factor-induced 4E-BP1 phosphorylation was found to be intact and still dependent on mTOR [31]. eIF4G [32] and eIF4B [3] are also phosphorylated in both mTOR-dependent and -independent manners, however, the functional consequences of these events is yet to be established.

mTOR is emerging as a critical player in insulin-controlled translation, having been placed at the interphase between nutrient and energy sensing to serve as a gatekeeper to balance protein synthesis versus autophagy [2,33], which is a process responsible for bulk lysosomal degradation of proteins and organelles during periods of nutrient limitation. Despite the recent finding that mTOR is phosphorylated by PKB [34], the role of PI3K/PKB in mTOR activation is controversial ([3] and see later discussion). For example, insulin does not cause a change in mTOR kinase activity measured against S6K1 as an *in vitro* substrate [35], suggesting that insulin might instead influence the composition and signaling capabilities of factors that are bound to mTOR rather than mTOR kinase activity itself [35,36].

Interestingly, recent reports have demonstrated constitutive activation of the mTOR/S6K1 pathway in cells lacking tuberous sclerosis complex (TSC) tumor suppressor function [37–39]. Importantly, the TSC tumor suppressor might provide the link between growth factor and nutrient signaling and the mTOR/S6K1 pathway [40,41]. In support of this notion, the PI3K/PKB pathway can promote TSC phosphorylation, which is proposed to lead to inactivation of the tumor suppressor, and overexpression of TSC inhibits nutrient and growth factor induced S6K1 activation and 4E-BP1 phosphorylation [40,42]. However, note that neither Ca^{2+}, amino acids, nor TPA treatment of cells affect PKB activation despite leading to acute S6K1 activation.

Insulin specifically affects the translation of some classes of mRNAs [3]. Among these, are transcripts containing a 5′ oligopyrimidine tract (TOP) that have been shown to encode for components of the translational apparatus, such as ribosomal proteins and elongation factors [7]. The 5′TOP acts as a *cis*-repressing element in quiescent cells [43], with insulin induction relieving this repression by an unknown mechanism that allows efficient translation of 5′TOP mRNAs. The mTOR/S6K1 pathway has been shown to be in part required for recruitment of the 5′TOP mRNAs into polysomes [44,45]. It has been speculated that the effects of S6K1 on 5′TOP translation are mediated through the phosphorylation of the 40S ribosomal subunit protein S6 [46]. However, recent data suggest that S6 phosphorylation may not be required for 5′TOP mRNA translation in response to insulin (S. Fumagalli and G. T., unpublished work), suggesting the involvement of other targets. Alternatively, S6 phosphorylation may protect ribosomes from autophagic degradation [47], which is concomitantly down-regulated following induction of cell proliferation by mitogens [48,49].

Another class of mRNAs, whose translation is strongly induced by insulin, are those that contain long and complex secondary structures within their 5′UTRs. Such transcripts encode for proteins involved in proliferation, including c-Myc, Bcl3, and Cyclin D1 and their translation may involve regulation through the mTOR/S6K/4E-BP pathway [3]. Whether the translation of these transcripts and of 5′TOP mRNAs is constitutively derepressed in cells lacking TSC function remains to be determined.

The elongation phase of protein synthesis is also subject to insulin control. Peptide-chain elongation is mediated by the two elongation factors, eEF1 and eEF2 [1]. In particular, insulin has been shown to regulate phosphorylation of eEF2 [4]. eEF2 forms a complex with GTP and is required for the translocation step during elongation. Phosphorylation of eEF2 by eEF2 kinase occurs within the GTP-binding domain and renders the factor inactive [50]. Insulin induces the rapid inactivation of eEF2 kinase and eEF2 dephosphorylation (by protein phosphatase-2A), leading to a doubling of translation elongation rates [51]. Recently, S6K1 and p90^{rsk} were shown to mediate eEF2 kinase phosphorylation and inactivation [31], providing a link between insulin signaling and the elongation phase of translation.

In parallel with increasing translation rates, insulin also acutely stimulates nutrient uptake [52] to provide the energy required for anabolic responses. Several of the components of the insulin-signaling pathway involved in regulating protein synthesis are also involved in the control of cellular energy homeostasis. For example, insulin-induced glucose uptake by GLUT-4 translocation at the plasma membrane is mediated by PKB, atypical PKCs, and phospholipase D [53–55]. PKB has also been implicated in mediating increased mitochondria-associated hexokinase activity and thus the coupling of glycolysis to oxidative phosphorylation [56]. Moreover, PI3K and PDK1 have been demonstrated to be involved in the activation of wortmannin-sensitive and insulin-stimulated kinase (WISK), which phosphorylates and activates heart 6-phosphofructo-2-kinase [57], leading to increased rates of glycolysis [58].

As with glucose uptake, insulin positively regulates amino acid uptake [59–61]. Interestingly, branched-chain amino acids and, in particular, leucine exert a potent stimulatory effect on protein synthesis, an effect most probably mediated by mTOR [62]. Indeed, regulation by mTOR in response to insulin, of both S6K1 and 4E-BP phosphorylation, requires the presence of amino acids [3,63]. In addition, it has been shown that mTOR activity is directly controlled by homeostatic levels of ATP [35]. Thus mTOR is suggested to regulate protein synthesis according to the level of translational and energy precursors [2,35]. Consistent with this model, mTOR and also PI3K have been proposed to negatively regulate autophagy [47,64,65], whereas PTEN has been suggested to have a positive effect [66].

In conclusion, in the future it will be important to obtain a deeper knowledge of the mechanisms that link the control of translation to metabolism in order to have a clearer understanding of insulin-mediated protein synthesis.

References

1. Hershey, J. W. B. (1991). Translational control in mammalian cells. *Annu. Rev. Biochem.* **60**, 717–755.
2. Dennis, P. B., Fumagalli, S., and Thomas, G. (1999). Target of rapamycin (TOR): Balancing the opposing forces of protein synthesis and degradation. *Curr. Opin. Genet. Dev.* **9**, 49–54.
3. Gingras, A. C., Raught, B., and Sonenberg, N. (2001). Regulation of translation initiation by FRAP/mTOR. *Genes Dev.* **15**(7), 807–826.
4. Proud, C. G. and Denton, R. M. (1997). Molecular mechanisms for the control of translation by insulin. *Biochem. J.* **328**(Pt 2), 329–341.
5. Thomas, G. (2000). An "encore" for ribosome biogenesis in cell proliferation. *Nat. Cell Biol.* **2**, E71–E72.
6. Meyuhas, O., Avni, D., and Shama, S. (1996). Translational control of ribosomal protein mRNAs in eukaryotes, in Hershey, J. W. B., Mathews, M. B., and Sonenberg, N., Eds., *Translational Control*, pp. 363–388. Cold Spring Harbor Laboratory Press, Cold Spring Harbor, NY.
7. Meyuhas, O. (2000). Synthesis of the translational apparatus is regulated at the translational level. *Eur. J. Biochem.* **267**(21), 6321–6330.
8. Peterson, R. T. and Schreiber, S. L. (1998). Translation control: connecting mitogens and the ribosome. *Curr. Biol.* **8**, R248–250.
9. Antonetti, D. A. *et al.* (1993). Regulation of rDNA transcription by insulin in primary cultures of rat hepatocytes. *J. Biol. Chem.* **268**(34), 25277–25284.
10. Whitehead, J. P. (2000). Signalling through the insulin receptor. *Curr. Opin Cell Biol.* **12**(2), 222–228.
11. Frodin, M. and Gammeltoft, S. (1999). Role and regulation of 90 kDa ribosomal S6 kinase (RSK) in signal transduction. *Mol. Cell Endocrinol.* **151**(1–2), 65–77.
12. Leevers, S. J., Vanhaesebroeck, B., and Waterfield, M. D. (1999). Signalling through phosphoinositide 3-kinases: The lipids take centre stage. *Curr. Opin Cell Biol.* **11**(2), 219–225.
13. Brazil, D. P. and Hemmings, B. A. (2001). Ten years of protein kinase B signalling: a hard Akt to follow. *Trends Biochem. Sci.* **26**(11), 657–664.
14. Maehama, T. and Dixon J. E. (1999). PTEN: A tumour suppressor that functions as a phospholipid phosphatase. *Trends Cell Biol.* **9**(4), 125–128.
15. Pullen, N. and Thomas, G. (1997). The modular phosphorylation and activation of $p70^{s6k}$. *FEBS Lett.* **410**, 78–82.
16. Pain, V. M. (1996). Initiation of protein synthesis in eukaryotic cells. *Eur. J. Biochem.* **236**(3), 747–771.
17. Welsh, G. I. and Proud, C. G. (1993). Glycogen synthase kinase-3 is rapidly inactivated in response to insulin and phosphorylates eukaryotic initiation factor eIF-2B. *Biochem. J.* **294**, 625–629.
18. Skurat, A. V. and Roach P. J. (1995). Phosphorylation of sites 3a and 3b (Ser640 and Ser644) in the control of rabbit muscle glycogen synthase. *J. Biol. Chem.* **270**(21), 12491–12497.
19. Cross, D. A. E. *et al.* (1995). Inhibition of glycogen synthase kinase-3 by insulin mediated by protein kinase B. *Nature* **378**, 785–789.
20. Towle, C. A. *et al.* (1984). Insulin promoted decrease in the phosphorylation of protein synthesis initiation factor eIF-2. *Biochem. Biophys. Res. Commun.* **121**, 134–140.
21. Pestova, T. V. *et al.* (2001). Molecular mechanisms of translation initiation in eukaryotes. *Proc. Natl. Acad. Sci. USA* **98**(13), 7029–7036.
22. Rozen, F. *et al.* (1990). Bidirectional RNA helicase activity of eucaryotic translation initiation factors 4A and 4F. *Mol. Cell Biol.* **10**(3), 1134–1144.
23. Duncan, R., Milburn, S. C. and Hershey, J. W. B. (1987). Regulated phosphorylation and low abundance of HeLa cell initiation factor eIF-4F suggest a role in translational control: Heat shock effects eIF-4F. *J. Biol. Chem.* **262**, 380–388.
24. Waskiewicz, A. J. *et al.* (1999). Phosphorylation of the cap-binding protein eukaryotic translation initiation factor 4E by protein kinase Mnk1 *in vivo*. *Mol. Cell Biol.* **19**(3), 1871–1880.
25. Pyronnet, S. *et al.* (1999). Human eukaryotic translation initiation factor 4G (eIF4G) recruits mnk1 to phosphorylate eIF4E. *EMBO J.* **18**(1), 270–279.
26. Minich, W. B. *et al.* (1994). Chromatographic resolution of in vivo phosphorylated and nonphosphorylated eukaryotic translation initiation factor eIF-4E: Increased cap affinity of the phosphorylated form. *Proc. Natl. Acad. Sci. USA* **91**, 7668–7672.
27. Knauf, U., Tschopp, C. and Gram, H. (2001). Negative regulation of protein translation by mitogen-activated protein kinase-interacting kinases 1 and 2. *Mol. Cell Biol.* **21**(16), 5500–5511.
28. Pause, A. *et al.* (1994). Insulin-dependent stimulation of protein synthesis by phosphorylation of a regulator of 5′-cap function. *Nature* **371**, 762–767.
29. Lin, T. A. *et al.* (1995). Control of PHAS-I by insulin in 3T3-L1 adipocytes. *J. Biol. Chem.* **270**, 18531–18538.
30. Gingras, A. C. *et al.* (2001). Hierarchical phosphorylation of the translation inhibitor 4E-BP1. *Genes Dev.* **15**(21), 2852–2864.
31. Wang, X. *et al.* (2001). Regulation of elongation factor 2 kinase by p90(RSK1) and p70 S6 kinase. *EMBO J.* **20**(16), 4370–4379.
32. Raught, B. *et al.* (2000). Serum-stimulated, rapamycin-sensitive phosphorylation sites in the eukaryotic translation initiation factor 4GI. *EMBO J.* **19**(3), 434–444.
33. Dennis, P. B. and Thomas, G. (2002). Quick guide: Target of rapamycin. *Curr. Biol.* **12**(8), R269.
34. Sekulic, A. *et al.* (2000). A direct linkage between the phosphoinositide 3-kinase-AKT signaling pathway and the mammalian target of rapamycin in mitogen-stimulated and transformed cells. *Cancer Res.* **60**(13), 3504–3513.
35. Dennis, P. B. *et al.* (2001). Mammalian TOR: A homeostatic ATP sensor. *Science* **294**(5544), 1102–1105.
36. Oldham, S. *et al.* (2000). Genetic and biochemical characterization of dTOR, the Drosophila homolog of the target of rapamycin. *Genes Dev.* **14**(21), 2689–2694.
37. Kwiatkowski, D. J. *et al.* (2002). A mouse model of TSC1 reveals sex-dependent lethality from liver hemangiomas, and up-regulation of p70S6 kinase activity in Tsc1 null cells. *Hum Mol. Genet.* **11**(5), 525–534.
38. Jaeschke, A. *et al.* (2002). Tuberous sclerosis complex tumor suppressor-mediated S6 kinase inhibition by phosphatidylinositide-3-OH kinase is mTOR independent. *J. Cell Biol.* **159**(2), 217–224.
39. Kenerson, H. L. *et al.* (2002). Activated mammalian target of rapamycin pathway in the pathogenesis of tuberous sclerosis complex renal tumors. *Cancer Res.* **62**(20), 5645–5650.
40. Inoki, K. *et al.* (2002). TSC2 is phosphorylated and inhibited by Akt and suppresses mTOR signalling. *Nat. Cell Biol.* **4**(9), 648–657.
41. Gao, X. *et al.* (2002). Tsc tumour suppressor proteins antagonize amino-acid-TOR signalling. *Nat. Cell Biol.* **4**(9), 699–704.
42. Manning, B. D. *et al.* (2002). Identification of the tuberous sclerosis complex-2 tumor suppressor gene product tuberin as a target of the phosphoinositide 3-kinase/akt pathway. *Mol. Cell* **10**(1), 151–162.
43. Avni, D. *et al.* (1994). Vertebrate mRNAs with a 5′-terminal pyrimidine tract are candidates for translation repression in quiescent cells. Characterization of the translational cis-regulatory element (TLRE). *Mol. Cell. Biol.* **14**, 3822–3833.
44. Jefferies, H. B. J. *et al.* (1994). Rapamycin selectively represses translation of the "polypyrimidine tract" mRNA family. *Proc. Natl. Acad. Sci. USA* **91**, 4441–4445.
45. Jefferies, H. B. J. *et al.* (1997). Rapamycin suppresses 5′TOP mRNA translation through inhibition of $p70^{s6k}$. *EMBO J.* **12**, 3693–3704.
46. Fumagalli, S. and Thomas, G. (2000). S6 phosphorylation and signal transduction, in Sonenberg, N., Hershey, J. W. B., Mathews, M. B., Eds., *Translational Control of Gene Expression*. Cold Spring Harbor Laboratory Press, Cold Spring Harbor, NY.
47. Blommaart, E. F. *et al.* (1995). Phosphorylation of ribosomal protein S6 is inhibitory for autophagy in isolated rat hepatocytes. *J. Biol. Chem.* **270**(5), 2320–2326.

48. Seglen, P. O. and Bohley, P. (1992). Autophagy and other vacuolar protein degradation mechanisms. *Experientia* **48**(2), 158–172.
49. Blommaart, E. F., Luiken, J. J., and Meijer, A. J. (1997). Autophagic proteolysis: control and specificity. *Histochem J.* **29**(5), 365–385.
50. Price, N. T. *et al.* (1991). Identification of the phosphorylation sites in elongation factor-2 from rabbit reticulocytes. *FEBS Lett.* **282**(2), 253–258.
51. Redpath, N. T., Foulstone, E. J., and Proud, C. G. (1996). Regulation of translation elongation factor-2 by insulin via a rapamycin-sensitive signalling pathway. *EMBO J.* **15**, 2291–2297.
52. Saltiel, A. R. and Kahn, C. R. (2001). Insulin signalling and the regulation of glucose and lipid metabolism. *Nature* **414**(6865), 799–806.
53. Wang, Q. *et al.* (1999). Protein kinase B/Akt participates in GLUT4 translocation by insulin in L6 myoblasts. *Mol. Cell Biol.* **19**(6), 4008–4018.
54. Kotani, K. *et al.* (1998). Requirement of atypical protein kinase clambda for insulin stimulation of glucose uptake but not for Akt activation in 3T3-L1 adipocytes. *Mol. Cell Biol.* **18**(12), 6971–6982.
55. Kristiansen, S. *et al.* (2001). GLUT-4 translocation in skeletal muscle studied with a cell-free assay: involvement of phospholipase D. *Am. J. Physiol. Endocrinol. Metab.* **281**(3), E608–618.
56. Gottlob, K. *et al.* (2001). Inhibition of early apoptotic events by Akt/PKB is dependent on the first committed step of glycolysis and mitochondrial hexokinase. *Genes Dev.* **15**(11), 1406–1418.
57. Deprez, J. *et al.* (2000). Partial purification and characterization of a wortmannin-sensitive and insulin-stimulated protein kinase that activates heart 6-phosphofructo-2-kinase. *Biochem J.* **347**(Pt 1), 305–312.
58. Bertrand, L. *et al.* (1999). Heart 6-phosphofructo-2-kinase activation by insulin results from Ser-466 and Ser-483 phosphorylation and requires 3-phosphoinositide-dependent kinase-1, but not protein kinase B. *J. Biol. Chem.* **274**(43), 30927–30933.
59. McDowell, H. E., Eyers, P. A., and Hundal, H. S. (1998). Regulation of System A amino acid transport in L6 rat skeletal muscle cells by insulin, chemical and hyperthermic stress. *FEBS Lett.* **441**(1), 15–19.
60. Munoz, M., Sweiry, J. H., and Mann, G. E. (1995). Insulin stimulates cationic amino acid transport activity in the isolated perfused rat pancreas. *Exp. Physiol.* **80**(5), 745–753.
61. Jefferson, L. S. and Kimball, S. R. (2001). Amino acid regulation of gene expression. *J. Nutr.* **131**(9 Suppl), 2460S-6S; discussion 2486S–2487S.
62. Kimball, S. R. *et al.* (1999). Leucine regulates translation of specific mRNAs in L6 myoblasts through mTOR-mediated changes in availability of eIF4E and phosphorylation of ribosomal protein S6. *J. Biol. Chem.* **274**(17), 11647–11652.
63. Schmelzle, T. and Hall, M. N. (2000). TOR, a central controller of cell growth. *Cell* **103**(2), 193–200.
64. Petiot, A. *et al.* (2000). Distinct classes of phosphatidylinositol 3′-kinases are involved in signaling pathways that control macroautophagy in HT-29 cells. *J. Biol. Chem.* **275**(2), 992–998.
65. Shigemitsu, K. *et al.* (1999). Regulation of translational effectors by amino acid and mammalian target of rapamycin signaling pathways. Possible involvement of autophagy in cultured hepatoma cells. *J. Biol. Chem.* **274**(2), 1058–1065.
66. Arico, S. *et al.* (2001). The tumor suppressor PTEN positively regulates macroautophagy by inhibiting the phosphatidylinositol 3-kinase/protein kinase B pathway. *J. Biol. Chem.* **276**(38), 35243–35246.

Unfolded Protein Response: An Intracellular Signaling Pathway Activated by the Accumulation of Unfolded Proteins in the Lumen of the Endoplasmic Reticulum

Randal J. Kaufman
Department of Biological Chemistry, Howard Hughes Medical Institute, University of Michigan Medical Center, Ann Arbor, Michigan

Introduction

The endoplasmic reticulum (ER) is a membranous network extending throughout the cytoplasm of the eukaryotic cell and is contiguous with the nuclear envelope. The ER is the site of cholesterol and steroid biosynthesis, lipid biosynthesis, assembly of core-asparagine linked oligosaccharides, and membrane and secreted protein biosynthesis. The ER has evolved as a protein-folding machine and a major intracellular signaling organelle. Numerous posttranslational modification reactions occur at the ER and many of these are required for proteins to attain their final-folded functional conformation. The quality of protein folding is strictly monitored by protein chaperones that prevent aberrant folding and aggregation. These chaperones permit only properly folded proteins to exit the ER, a process termed *quality control.* The ER lumen also contains catalysts of protein-folding reactions, such as *cis-trans* polylisomerases and protein disulfide-bond isomerases (PDIs). The oxidizing environment of the ER provides a constant demand for the cell to monitor and promote correct disulfide bond formation.

Perturbations that alter ER homeostasis disrupt protein folding and lead to the accumulation of unfolded proteins and protein aggregates, which are detrimental to cell survival. These perturbations include disturbances in calcium homeostasis or redox status, an increased demand for chaperone function due to elevated secretory protein synthesis, the expression of mutant, misfolded, or unassembled proteins, and nutrient deprivation [1]. When homeostasis in the ER is altered, signaling pathways are activated to elicit an adaptive response. These pathways are collectively termed the *unfolded protein response* (UPR). If adaptation is not sufficient, then the cell activates a cell death response. It is important to note that all newly synthesized proteins are produced in an unfolded state. Therefore, high-level expression of wild-type proteins can also activate the UPR [2]. One enigma in cell biology is how the cell chooses between survival and death upon activation of the UPR. This chapter summarizes the diverse mechanisms for UPR signaling.

Pastan and colleagues [3] first reported that viral transformation of eukaryotic cells induces a set of genes. Because the transcription of the same set of genes was induced upon glucose deprivation, the products of these genes were termed *glucose-regulated proteins* (GRPs) [4]. Many of the GRPs are now known to function as molecular chaperones to either promote productive protein folding or prevent protein aggregation in the ER [5]. Productive protein folding in the oxidizing environment of the ER requires extensive energy, and glucose deprivation reduces the amount of energy that is available for such reactions. A number of ER chaperones, including BiP, use the energy from ATP hydrolysis to promote folding and prevent aggregation of proteins within the ER. In addition, glucose is directly required for the synthesis of asparagine-linked oligosaccharide core structures. The cell monitors the glucosylation status of N-linked oligosaccharides on glycoproteins as a mechanism for quality control. The protein chaperones calnexin and calreticulin bind monoglucosylated asparagine-linked core oligosaccharides on unfolded glycoproteins and prevent their exit to the Golgi compartment [6]. Individual proteins require different levels of ATP for protein folding prior to ER export [7]. It is probable that different levels of glucose deprivation can differentially induce the UPR in different cell types, depending on the amount of secretory proteins they need to fold.

In yeast, the accumulation of unfolded proteins in the ER lumen induces the transcription of 381 genes, many encoding functions to increase the volume or capacity for ER protein folding, or to increase the degradation of misfolded proteins [8]. Transcriptional induction of the ER protein chaperone KAR2/BiP/GRP78 is a classical marker for UPR activation in yeast and mammalian cells [9]. In higher eukaryotic cells, protein synthesis at the level of initiation is inhibited in order to decrease the protein-folding load on the ER [10].

UPR in *Saccharomyces Cerevisiae*

The original description of the UPR signaling pathway in the budding yeast *S. cerevisiae* occurred less than 10 years ago [11,12]. The only ER-stress-sensing protein in yeast *S. cerevisiae* is the kinase/endoribonuclease, IRE1. IRE1 is similar to class I receptor-transmembrane protein kinases consisting of an ER-lumenal extracellular, transmembrane and a serine/threonine kinase intracellular domain. In addition, IRE1 has a carboxy-terminal cytoplasmic endoribonuclease (RNase) domain with strong homology to mammalian RNaseL that is essential to signal UPR transcriptional activation. Accumulation of unfolded proteins in the ER lumen promotes dimerization of IRE1 with subsequent transautophosphorylation to activate its RNase function. Activated IRE1 cleaves the 5′ and 3′ exon–intron splice site junctions in *HAC1* messenger RNA, encoding a basic leucine zipper (bZIP)-containing transcription factor [13,14]. The 5′ and 3′ exons are then joined by transfer RNA ligase [15]. The unconventional *HAC1*-mRNA processing reaction removes a 252-nucleotide intron and thereby replaces the carboxy-terminal 10 amino acids in HAC1 ($HAC1^u$) with a novel 19-amino-acid segment ($HAC1^i$). This splicing reaction regulates UPR transcriptional activation in two ways. First, the new carboxyl terminus on $HAC1^i$ converts $HAC1^u$ into a 10-fold more potent transcriptional activator [16]. Second, removal of the intron by IRE1 and tRNA ligase increases the translational efficiency of *HAC1* mRNA. It was proposed that translation elongation is attenuated by base pairing between the 5′ untranslated region of *HAC1* mRNA and sequences within the intron [17]. Therefore, the *HAC1*-mRNA splicing reaction removes the intron to stimulate HAC1 protein translation. The *HAC1* mRNA splicing reaction occurs independent of *HAC1* transcription, supporting the supposition that it occurs in the cytoplasm [17]. However, mammalian IRE1 was localized primarily to the inner nuclear envelope, suggesting that nuclear splicing may occur in mammals [19,20].

$HAC1^i$ binds to a DNA sequence motif termed the UPR element (UPRE) (consensus motif: CAGCGTG) in the promoter regions of about one-half of UPR-responsive genes in yeast. In addition to activating the transcription of ER-chaperone genes, $HAC1^i$ activates transcription of genes involved in ER-associated protein degradation (ERAD) [8,21,22] and phospholipid biosynthesis (for example, *INO1*) [23,24]. Because there are no readily recognizable UPR elements in the promoter regions of ERAD genes, the molecular basis for their transcriptional activation by $Hac1^i$ is not understood. However, there is a close relationship between the UPR and ERAD. Deletion of the ERAD genes *UBC7*, *DER1*, *HRD1*, and *HRD3* moderately activates the UPR, indicating that the unfolded protein load in the ER of these ERAD-defective cells is increased [8,13,22]. Inositol starvation also activates IRE1-mediated splicing of *HAC1* mRNA [24]. It was proposed that $Hac1^i$ binds the transcriptional repressor OPI1 to relieve repression of *INO1* in order to promote inositol biosynthesis [24].

Not only does the UPR coordinate membrane biosynthesis with ER expansion, it also regulates nutrient-deprivation-induced differentiation responses in yeast. The budding yeast *S. cerevisiae* undergo distinct morphological transitions when starved of nitrogen. Nitrogen starvation in the presence of fermentable carbon sources promotes invasive pseudohyphal or filamentous growth [25,26]. In contrast, nitrogen starvation in the absence of fermentable carbon sources promotes meiosis and sporulation [27]. Genetic and physiological data demonstrated that nitrogen activates IRE1-mediated *HAC1* mRNA splicing to repress both pseudohyphal growth and meiosis [28].

UPR Transcriptional Activation in Metazoan Species

The UPR in higher eukaryotic cells has conserved many of the essential and unique properties from *S. cerevisiae*, but has also superimposed additional sensors and control mechanisms to mediate more diverse responses. In metazoan cells, the complex network of physiological responses to ER stress is regulated by three ER transmembrane proteins: IRE1, PERK, and ATF6 [29]. If these adaptive responses are not sufficient to relieve the unfolded protein load (so-called "ER stress"), the cell enters one of the cell-death pathways: apoptosis or necrosis. The UPR in metazoan cells not only signals differentiation, but also regulates energy homeostasis and differentiation responses.

UPR Transcriptional Activation by ER Stress-Activated Splicing of *XBP1* mRNA

The mammalian genome contains two homologs of yeast *IRE1*: *IRE1α* and *IRE1β*. Whereas *IRE1α* is expressed in most cells and tissues, with high-level expression in the pancreas and placenta [19], *IRE1β* expression is primarily restricted to the intestinal epithelial cells [30]. Both IRE1α and IRE1β molecules respond to the accumulation of unfolded proteins in the ER to activate their kinase and subsequent RNase activities (Fig. 1B). The cleavage-site specificities of IRE1α and IRE1β are similar, suggesting they do not recognize different sets of substrates, but rather generate temporal- and tissue-specific splicing [31].

Although overexpression of either *IRE1α* or *IRE1β* activates transcription from a BiP promoter reporter construct [32], deletion of *IRE1α* and/or *IRE1β* in the mouse did not interfere with transcriptional activation of the endogenous *BiP* gene in cultured cells [20,33,34]. Therefore, at least one additional mechanism for UPR transcriptional activation exists. However, deletion of *IRE1α* caused lethality at embryonic day E10.5 [20,33]. In contrast, no developmental defect was detected on deletion of *IRE1β*, although the mice did show increased susceptibility to experimental-induced colitis [35]. These studies demonstrate that although *IRE1α* and *IRE1β* are not essential for the UPR, *IRE1α* is required for mammalian embryonic development.

The only known substrate for mammalian IRE1 is the mRNA encoding the bZIP transcription factor X-box binding protein 1 (XBP1) [20,36–38]. XBP1 was isolated as a protein that interacts with the mammalian ER stress-response element [ERSEI; CCAAT (N_9)CCACG] in a yeast one-hybrid screen [39]. *XBP1* mRNA has two conserved overlapping open reading frames (ORFs). On activation of the UPR, *XBP1* mRNA is cleaved by IRE1 to initiate a splicing reaction that removes a 26-nucleotide intron and generates a translation frame shift to produce a fusion protein encoded from the two ORFs. The new carboxyl terminus on the product from spliced *XBP1* mRNA (XBP1-s) converts XBP1 into a potent transcriptional activator, similar to that described for yeast HAC1. Although deletion of *IRE1α* and/or *IRE1β* did not diminish the transcriptional activation of several UPR-responsive genes, analysis of a minimal UPR reported did uncover a transcriptional defect in *IRE1α*-null mouse embryo fibroblasts (MEFs). This defect was complemented by expression of spliced *XBP1* mRNA [20]. Therefore, it is possible that a subset of UPR targets, those composed of minimal UPR motifs, requires IRE1.

UPR Transcriptional Activation by ER Stress-Induced Proteolysis of ATF6

Activating transcription factor 6 (ATF6) was identified as another ERSEI-binding protein in the same yeast one-hybrid screen that detected XBP1 [39]. There are two ATF6 genes, ATF6α (90 kDa) and ATF6β (110 kDa; also known as CREB-RP), that encode type II ER transmembrane proteins. On activation of the UPR, both forms of ATF6 are processed to generate 50- to 60-kDa cytosolic bZIP-containing fragments that migrate to the nucleus (Fig. 1A). In the presence of the CCAAT-binding factor (CBF; also called NF-Y [40]), both forms bind to the ERSEI in the promoter regions of UPR-responsive genes to activate transcription [41–44]. The DNA-binding specificities of ATF6 and XBP1 are similar; however, XBP1 can bind the ERSEI half-site (CCACG) in the absence of CBF/NFY [20,45,46]. Whereas CBF binds the CCAAT motif, ATF6α/β binds the CCACG motif.

Processing of ATF6 occurs within the transmembrane segment and at an adjacent site that is exposed to the ER lumen. The ATF6 processing enzymes S1P and S2P are the same processing enzymes that are activated upon cholesterol deprivation to cleave the ER-associated transmembrane sterol-response element binding protein (SREBP) [47]. However, only the UPR elicits ATF6 cleavage and sterol deprivation alone induces SREBP cleavage. The mechanism for ATF6 processing is similar to that of SREBP, involving the ER-stress-induced translocation of ATF6 from the ER to the Golgi compartment [48]. The sterol cleavage activating protein (SCAP) confers specificity for SREBP cleavage in response to sterol deprivation [49]. It is possible that another cleavage-activating protein confers S1P and S2P specificity for ATF6 upon activation of the UPR.

IRE1 and ATF6 Signaling Converge at XBP1

Signaling through ATF6 and IRE1 merge to induce *XBP1* transcription and mRNA splicing, respectively (Fig. 1). Transcription of *XBP1* is directly activated by the UPR, as well as by the cleaved form of ATF6. Increased *XBP1* transcription produces more substrate for IRE1-mediated splicing to generate more active XBP1, providing a positive feedback for UPR activation. Cells that lack either IRE1α or IRE1β or are defective for ATF6 cleavage can induce *XBP1* mRNA [20], suggesting that these two pathways provide complementary mechanisms to activate *XBP1* transcription. Alternatively, another pathway—possibly mediated by the ER-localized protein kinase PERK/PEK—may also contribute to *XBP1* mRNA induction.

UPR Transcriptional Activation through Translational Regulation by PERK

Stimuli that activate the transcriptional components of the UPR also transiently attenuate protein translation. This response is coupled with phosphorylation on the α-subunit of heterotrimeric eukaryotic translation initiation factor 2 (eIF2α). eIF2 is a GTPase required for bringing initiator tRNA to the 40S ribosomal subunit and for using GTP hydrolysis to promote 60S ribosomal subunit joining. To perform another round of initiation, GTP exchange for bound GDP is required. Phosphorylation of eIF2α at Ser51 inhibits

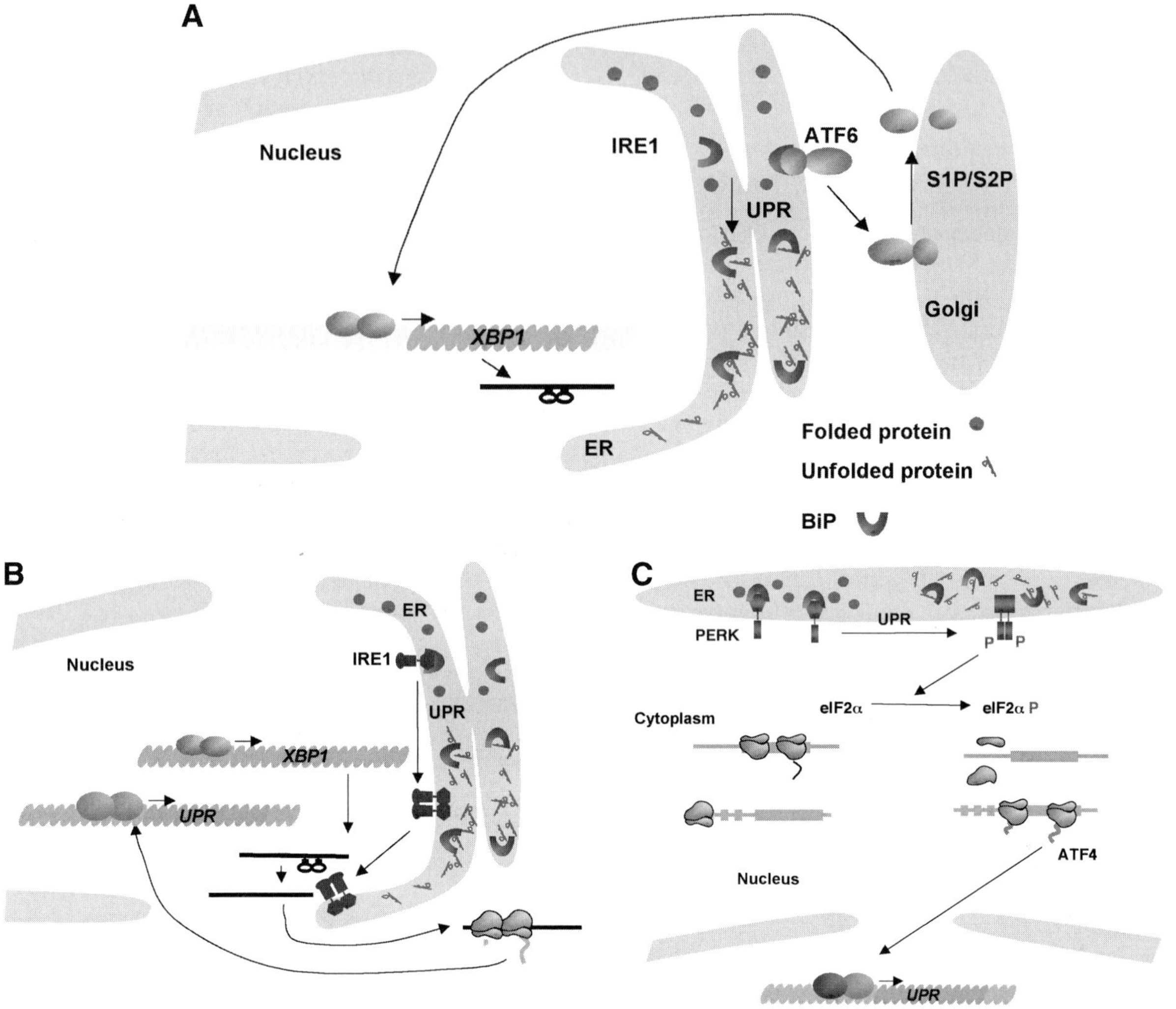

Figure 1 Signaling the unfolded protein response in eukaryotes. The signaling pathways for the three proximal sensors of the UPR are depicted. Whereas only IRE1 is present in *S. cerevisiae*, all three transducers are present in metazoan cells. *Caenorhabditis elegans* only requires IRE1 and PERK for survival to accumulation of unfolded proteins in the endoplasmic reticulum (ER) [37]. Although IRE1 and PERK may be dispensable for most transcriptional regulation of the UPR, ATF6 cleavage is required for UPR transcriptional induction [23,43]. Therefore, ATF6 is the most significant signaling pathway in mammalian cells. BiP negatively regulates these pathways. (A) Upon accumulation of unfolded proteins in the ER lumen, ATF6 transits to the Golgi compartment. In the Golgi, ATF6 is cleaved by S1P and S2P proteases to yield a cytosolic fragment that migrates to the nucleus to activate transcription of responsive genes, including *XBP1*. (B) In parallel, IRE1 protein kinase/RNase dimerizes and activates its RNase activity. (C) Simultaneously, PERK is activated to phosphorylate eIF2 on the α-subunit at Ser51 to reduce the frequency of AUG codon recognition.

this exchange reaction so translation initiation is prevented. The UPR activates a transmembrane ER-localized kinase, PERK/PEK, to phosphorylate Ser51 in eIF2α and to inhibit translation [50,51] (Fig. 1C).

The best characterized translational response to eIF2α phosphorylation is the one that occurs on amino acid starvation in yeast. On amino acid deprivation, uncharged histidyl tRNA activates the eIF2α protein kinase called GCN2. Phosphorylation of eIF2α (yeast SUI2) reduces the rate of polypeptide-chain initiation and induces the transcription of a limited set of genes through preferential translation of *GCN4* mRNA. *GCN4* mRNA encodes a bZIP transcription factor that is required for induction of amino acid biosynthetic genes [52]. The 5′ untranslated region of *GCN4* mRNA contains four upstream ORFs that inhibit the ability of the ribosome to scan through the 5′ end of the mRNA and reach the correct AUG initiation codon. Phosphorylation of eIF2α limits 60S ribosomal subunit joining to allow the 40S ribosomal subunit to scan through the ORFs and initiate polypeptide chain synthesis at the authentic initiation AUG codon for GCN4.

Phosphorylation of eIF2α is a major translational-control mechanism conserved in all nucleated cells. On accumulation of unfolded proteins in the ER, PERK-mediated phosphorylation of eIF2α is required to attenuate translation, to induce transcription of one-third of the UPR-responsive genes, and for survival [53,54]. Interestingly, transcriptional induction of the ubiquitous glucose transporter-1 (GLUT1) by low glucose is coincident with induction of the UPR marker BiP, and may contribute to survival in low glucose [55]. The mechanistic requirement for eIF2α phosphorylation in UPR gene induction is unknown. However, in the absence of eIF2α phosphorylation or PERK, the mRNA encoding the bZiP transcription factor ATF4 was not translated [53,56]. It was proposed that translation of *ATF4* mRNA, by virtue of ORFs within the 5′UTR, requires phosphorylation of eIF2α. Analogous to the amino-acid-starvation-induced translation of *GCN4* mRNA in yeast, *ATF4* mRNA may be one of many mRNAs that is preferentially translated when eIF2 levels are limiting. ATF4 is another member of the C/EBP family of transcription factors implicated in metabolic control regulated through cyclic AMP [57]. Nutrient-deprivation-induced regulation of translation and transcription of the C/EBP family members may facilitate the interconversion and use of different energy sources; that is, fat, carbohydrate and protein.

Free BiP Inhibits the Activation of IRE1, PERK, and ATF6

One interesting question concerning the UPR is how IRE1, PERK, and ATF6 are simultaneously activated by accumulation of unfolded proteins in the ER. Dimerization and transautophosphorylation are required to activate yeast IRE1 [58,59]. In addition, the lumenal domain of yeast IRE1 is required for UPR signaling and the lumenal domain of IRE1 can dimerize *in vitro* [60,61]. These results support the hypothesis that the lumenal domain promotes dimerization with subsequent kinase activation and autophosphorylation. Interestingly, the amino terminus of the lumenal domain of PERK is homologous to the amino terminus of the IRE1 lumenal domain. Although yeast do not have PERK, the lumenal domain of PERK, as well as the lumenal domain of human IRE1, can functionally replace the lumenal domain of yeast IRE1 to provide UPR-dependent activation in yeast [60]. Therefore, mammalian PERK and IRE1 have conserved the same sensing mechanism as yeast IRE1. Surprisingly, a bZIP dimerization domain can also functionally replace the yeast IRE1 lumenal domain and this chimeric molecule also responds to the UPR [60]. These findings led to the hypothesis that IRE1 is actively held in a monomeric state in the cell and on UPR activation, dimerization is permitted.

What might prevent IRE1 dimerization when the UPR is off? Overexpression of BiP, but not other ER protein chaperones, inhibits PERK and IRE1 activation and also suppresses both the translational attenuation and transcriptional activation that occurs on accumulation of unfolded proteins in the ER [62,63]. Therefore, it was proposed that the level of free BiP negatively regulates UPR activation. In the absence of unfolded proteins, IRE1 and PERK are in a complex with BiP [64]. As unfolded proteins accumulate, their exposed hydrophobic surfaces may bind BiP, thereby reducing the pool of free BiP that is available to bind IRE1 and PERK. As a consequence, PERK and IRE could spontaneously dimerize for transautophosphorylation.

The level of free BiP also regulates ATF6 cleavage. A BiP binding site within the lumenal domain of ATF6 was recently identified that when deleted produced constitutive transport of ATF6 to the *cis*-Golgi compartment [65]. In addition, overexpression of BiP prevents ATF6 transit to the Golgi and subsequent cleavage [65]. Therefore, when the UPR is activated and the level of free BiP decreases, BiP release from ATF6 would permit ATF6 transit to the *cis*-Golgi to gain access to S1P and S2P proteases at the *cis*-Golgi compartment. The BiP negative-regulation model permits the simultaneous activation of all three ER stress sensors on activation of the UPR.

Physiological Role for the UPR in Mammals

The physiological role for UPR signaling has been elucidated through genetic modification in the mouse. To date, mice deleted in *IRE1α*, *IRE1β*, and *PERK* as well as mice deleted of the IRE1 substrate XBP1 and mice with targeted knock-in mutation at the PERK phosphorylation site Ser51 in eIF2α have been analyzed. The observations from these mice support that IRE1/XBP1 provides a major role in cell differentiation, possibly restricted to cells designed to have a specialized secretory compartment. In contrast, PERK/eIF2α may provide a major control over *in vivo* glucose metabolism.

IRE1α/XBP1 Signaling Required for Mammalian Development

Although deletion of *IRE1α* did not interfere with UPR transcriptional activation of many UPR-responsive genes in MEFs, deletion of *IRE1α* caused embryonic lethality at day E10.5. Therefore, although IRE1 is not essential for the UPR, *IRE1α* is required in mammalian development. In contrast, developmental defects were not detected in *IRE1β*-null mice.

Deletion of *XBP1*, encoding the only known mammalian substrate of IRE1, produced embryonic lethality at days 12–14 due to multiple defects in hematopoiesis, cardiomyocyte development, and hepatocyte differentiation [66,67]. However, the *XBP1*-deleted mice studied lack both the unspliced and spliced forms of XBP1 so it is not possible to conclude these defects are due to a loss of IRE1 signaling. Analysis of chimeric mice produced between $XBP1^{-/-}$ mice and immunoincompetent $RAG1^{-/-}$ (i.e., mice deficient in T- and B-cell maturation) demonstrated that *XBP1* is required for plasma cell differentiation and high-level immunoglobulin production [68]. Interestingly, immunoglobulin heavy-chain and light-chain gene rearrangement and surface expression of immunoglobulin μ in B cells occurred normally in *XBP1*-null mice. However, high immunoglobulin-secreting plasma cells were not detected, suggesting a role in plasma cell differentiation and/or survival. On B-cell differentiation into a plasma cell, the ER compartment expands approximately fivefold to accommodate the large increase in immunoglobulin synthesis [68]. Interestingly, plasma-cell differentiation is stimulated *in vivo* by activation of the innate immune response—by treatment with lipopolysaccharide (LPS), for example, or by ligation of CD40 receptors. Treatment of lymphoid cells with LPS induced *XBP1* mRNA splicing [38], suggesting that activation of IRE1, at least indirectly, might signal an inflammatory response. These observations with the *XBP1*-null mice are consistent with the UPR providing a programmed response to expand ER capacity to accommodate protein secretion needs in highly specialized cells.

PERK/eIF2α Required for Glucose Homeostasis

Strong evidence supports the supposition that PERK signaling through eIF2α phosphorylation provides major control over glucose homeostasis *in vivo*. Two independent human families were identified with an autosomal-recessive infancy-onset insulin-requiring diabetes, termed Wolcott–Rallison syndrome, resulting from loss-of-function mutations in the *PERK* gene [69]. *PERK*-null mice have pancreatic dysfunction and develop diabetes several weeks after birth [70]. Mice with a homozygous Ser51 to Ala mutation in the *eIF2α* gene develop a more severe β-cell dysfunction that appears *in utero* [53]. The more severe phenotype resulting from Ser51Ala mutation in *eIF2α* compared to the *PERK* knock-out suggests that additional eIF2α kinases partially complement for β-cell function in the absence of PERK. In addition, homozygous Ser51Ala eIF2α mice also have a defect in liver gluconeogenesis, suggesting a broader role for eIF2α phosphorylation in glucose production and storage [53].

Why is phosphorylation of eIF2α required for β-cell differentiation and/or survival? Glucose stimulates both the secretion of insulin and the translation of proinsulin [71,72]. It was proposed that blood glucose levels not only affect transcription and translation of glucose-regulated genes but also the protein-folding status in the ER [53]. As glucose levels decline, the cellular energy supply decreases, protein folding becomes inefficient, and IRE1 and PERK are activated to induce UPR transcription and inhibit translation. The downstream effects of activating ER stress kinases in the β cell are largely unknown, but the transcriptional induction of glucose-regulated proteins may have a protective function by increasing the cellular capacity for the uptake and use of glucose. Prolonged activation of IRE1 may activate the c-Jun amino-terminal kinase (JNK) to induce an apoptotic state [33]. PERK activation in response to low blood glucose would inhibit protein synthesis, thereby limiting translation of proinsulin mRNA. In contrast, as blood glucose levels rise, PERK would be turned off so that translation would accelerate, allowing translocation of proinsulin into the ER. In this manner, eIF2α phosphorylation by PERK would mediate glucose regulation of insulin translation. Thus, the signaling mechanisms that β cells use for sensing glucose levels, triggering insulin secretion, and rapidly controlling insulin biosynthesis may have coevolved with the ER signaling pathway to support these specialized functions,

Why do β cells uniquely respond to physiological fluctuations in blood glucose (i.e., between 3 and 10 m*M*)? In contrast to other cell types, β cells express very little, if any, low Km-hexokinases. Therefore, the production of glucose-6-phosphate and glucose utilization for ATP production is controlled by glucokinase, which has a weaker affinity for glucose [73]. Consequently, in β cells the cellular [ATP/ADP] ratio is tightly coupled with blood glucose levels [74]. During intermittent feeding, periodic decreases in the ATP/ADP ratio may compromise protein folding to activate the UPR in β cells. Therefore, the ER lumen of the β cell would be more susceptible to fluctuations in glucose levels within the physiological range. The β cell may require the UPR for survival during intermittent decreases in the blood glucose level. Additionally, the very high level of PERK and IRE1α expression in the pancreas may lower the activation threshold by predisposing them to dimerization.

Future Directions

In conclusion, IRE1, PERK, and ATF6 are proximal sensors that regulate the quantity and/or quality of a class of bZIP-containing transcription factors that may form homo- and heterodimers. Combinatorial interactions of these factors may generate diversity in responses for different subsets of UPR-responsive genes. Each of these pathways contributes to a different subset of responses that are collectively termed the UPR. Future studies should identify under what physiological and pathological states these pathways are activated *in vivo* and how they influence disease outcome. It is likely that under some conditions their activation will be beneficial and under different conditions their activation will be detrimental. As we gain a greater understanding of the three subpathways, it should eventually be possible to target them for intervention in the pathogenesis of different disease states, such as conformational diseases of protein folding (Alzheimer's disease), viral infection, diabetes, cancer, and autoimmune diseases.

Acknowledgements

I gratefully thank present and past members of the Kaufman Laboratory.

References

1. Kozutsumi, Y., Segal, M., Normington, K., Gething, M. J., and Sambrook, J. (1988). *Nature* **332**, 462–464.
2. Dorner, A. J., Wasley, L. C., and Kaufman, R. J. (1989). *J. Biol. Chem.* **264**, 20602–20607.
3. Pouyssegur, J., Shiu, R., and Pastan, I. (1977). *Cell* **11**, 941–947.
4. Lee, A. S. (1992). *Curr. Opin. Cell Biol.* **4**, 267–273.
5. Lee, A. S. (2001). *Trends Biochem. Sci.* **26**, 504–510.
6. Ellgaard, L. and Helenius, A. (2001). *Curr. Opin. Cell Biol.* **13**, 431–437.
7. Dorner, A. J., Wasley, L. C., and Kaufman, R. J. (1990). *Proc. Natl. Acad. Sci. USA* **87**, 7429–7432.
8. Travers, K. J., Patil, C. K., Wodicka, L., Lockhart, D. J., Weissman, J. S., and Walter, P. (2000). *Cell* **101**, 249–528.
9. Kaufman, R. J. (1999). *Genes Dev.* **13**, 1211–1233.
10. Prostko, C. R., Brostrom, M. A., Malara, E. M., and Brostrom, C. O. (1993). *J. Biol. Chem.* **267**, 16751–16754.
11. Mori, K., Ma, W., Gething, M.-J., and Sambrook, J. (1993). *Cell* **74**, 743–756.

12. Cox, J. S., Shamu, C. E., and Walter, P. (1993). *Cell* **73**, 1197–1206.
13. Cox, J. S. and Walter, P. (1996). *Cell* **87**, 391–404.
14. Kawahara, T., Yanagi, H., Yura, T., and Mori, K. (1997). *Mol. Biol. Cell* **8**, 1845–1862.
15. Sidrauski, C., Cox, J. S., and Walter, P. (1996). *Cell* **87**, 405–413.
16. Mori, K., Ogawa, N., Kawahara, T., Yanagi, H., and Yura, T. (2000). *Proc. Natl. Acad. Sci. USA* **97**, 4660–4665.
17. Ruegsegger, U., Leber, J., and Walter, P. (2001). *Cell* **107**, 103–114.
18. Kuhn, K. M., DeRisi, J. L., Brown, P. O., and Sarnow, P. (2001). *Mol. Cell Biol.* **21**, 916–927.
19. Tirasophon, W., Welihinda, A. A., and Kaufman, R. J. (1998). *Genes Dev.* **12**, 1812–1824.
20. Lee, K., Tirasophon, W., Shen, X., Michalak, M., Prywes. R., Okada, T., Yoshida, H., Mori, K., and Kaufman, R. J. (2002). *Genes Dev.* **16**, 452–466.
21. Casagrande, R., Stern, P., Diehn, M., Shamu, C., Osario, M., Zuniga, M., Brown, P. O., and Ploegh, H. (2000). *Mol. Cell* **5**, 729–735.
22. Friedlander, R., Jarosch, E., Urban, J., Volkwein, C., and Sommer, T. (2000). *Nat. Cell Biol.* **2**, 379–384.
23. Nikawa, J. I. and Yamashita, S. (1992). *Mol. Microbiol.* **6**, 1441–1446.
24. Cox, J. S., Chapman, R. E., and Walter, P. (1997). *Mol. Biol. Cell* **8**, 1805–1814.
25. Madhani, H. D. and Fink, G. R. (1998). *Trends Cell Biol.* **8**, 348–353.
26. Pan, X., Harashima, T., and Heitman, J. (2000). *Curr. Opin. Microbiol.* **3**, 567–572.
27. Kupiec, M. B. B., Esposito, R. E., and Mitchell, A. P. (1997). *The Molecular and Cellular Biology of the Yeast Saccharomyces,* Vol. **3**, pp. 889–1036. Cold Spring Harbor Press, Cold Spring Harbor, New York.
28. Schröder, M., Chang, J. S., and Kaufman, R. J. (2000). *Genes Dev.* **14**, 2962–2975.
29. Mori, K. (2000). *Cell* **101**, 451–454.
30. Wang, X. Z., Harding, H. P., Zhang, Y., Jolicoeur, E. M., Kuroda, M., and Ron, D. (1998). *EMBO J.* **17**, 5708–5717.
31. Niwa, M., Sidrauski, C., Kaufman, R. J., and Walter, P. (1999). *Cell* **99**, 691–702.
32. Tirasophon, W., Lee, K., Callaghan, B., Welihinda, A., and Kaufman, R. J. (2000). *Genes Dev.* **14**, 2725–2736.
33. Urano, F., Wang, X., Bertolotti, A., Zhang, Y., Chung, P., Harding, H. P., and Ron, D. (2000). *Science* **287**, 664–666.
34. Urano, F., Bertolotti, A., and Ron, D. (2000). *J. Cell Sci.* **113** (Pt 21), 3697–3702.
35. Bertolotti, A., Wang, X., Novoa, I., Jungreis, R., Schlessinger, K., Cho, J. H., West, A. B., and Ron, D. (2001). *J. Clin. Invest.* **107**, 585–593.
36. Yoshida, H., Matsui, T., Yamamoto, A., Okada, T., and Mori, K. (2001). *Cell* **107**, 881–891.
37. Shen, X., Ellis, R. E., Lee, K., Liu, C. Y., Yang, K., Solomon, A., Yoshida, H., Morimoto, R., Kurnit, D. M., Mori, K., and Kaufman, R. J. (2001). *Cell* **107**, 893–903.
38. Calfon, M., Zeng, H., Urano, F., Till, J. H., Hubbard, S. R., Harding, H. P., Clark, S. G., and Ron, D. (2002). *Nature* **415**, 92–96.
39. Yoshida, H., Haze, K., Yanagi, H., Yura, T., and Mori, K. (1998). *J. Biol. Chem.* **273**, 33741–33749.
40. Lum, L. S. Y., Sultzman, L. A., Kaufman, R. J., Linzer, D. H., and Wu, B. J. (1990). *Mol. Cell Biol.* **10**, 6709–6717.
41. Haze, K., Yoshida, H., Yanagi, H., Yura, T., and Mori, K. (1999). *Mol. Biol. Cell* **10**, 3787–3799.
42. Yoshida, H., Okada, T., Haze, K., Yanagi, H., Yura, T., Negishi, M., and Mori, K. (2000). *Mol. Cell Biol.* **20**, 6755–6767.
43. Li, M., Baumeister, P., Roy, B., Phan, T., Foti, D., Luo, S., and Lee, A. S. (2000). *Mol. Cell Biol.* **20**, 5096–5106.
44. Yoshida, H., Okada, T., Haze, K., Yanagi, H., Yura, T., Negishi, M., and Mori, K. (2001). *Mol. Cell Biol.* **21**, 1239–1248.
45. Wang, Y., Shen, J., Arenzana, N., Tirasophon, W., Kaufman, R. J., and Prywes, R. (2000). *J. Biol. Chem.* **275**, 27013–27020.
46. Clauss, I. M., Chu, M., Zhao, J. L., and Glimcher, L. H. (1996). *Nucleic Acids Res.* **24**, 1855–1864.
47. Ye, J., Rawson, R. B., Komuro, R., Chen, X., Dave, U. P., Prywes, R., Brown, M. S., and Goldstein, J. L. (2000). *Mol. Cell* **6**, 1355–1364.
48. Chen, X., Shen, J., and Prywes, R. (2002). *J. Biol. Chem.* **277**, 13045–13052.
49. Nohturfft, A., Yabe, D., Goldstein, J. L., Brown, M. S., and Espenshade, P. J. (2000). *Cell* **102**, 315–323.
50. Shi, Y., Vattem, K. M., Sood, R., An, J., Liang, J., Stramm, L., and Wek, R. C. (1998). *Mol. Cell Biol.* **18**, 7499–7509
51. Harding, H. P., Zhang, Y., and Ron, D. (1999). *Nature* **397**, 271–274
52. Hinnebusch, A. (2000). In Hershey, J. W. B., Matthews, M. B., and Sonenberg, N., Eds., *Translational Control*, pp.199–244. Cold Spring Harbor Press, Cold Spring Harbor, New York.
53. Scheuner, D., Song, B., McEwen, E., Lui, C., Laybutt, R., Gillespie, P., Saunders, T., Bonner-Weir, S., and Kaufman, R. J. (2001). *Mol. Cell* **7**, 1165–1176.
54. Harding, H. P., Zhang, Y., Bertolotti, A., Zeng, H., and Ron, D. (2000). *Mol. Cell* **5**, 897–904.
55. Wertheimer, E., Sasson, S., Cerasi, E., and Ben-Neriah, Y. (1991). *Proc. Natl. Acad. Sci. USA* **88**, 2525–2529.
56. Harding, H. P., Novoa, I., Zhang, Y., Zeng, H., Wek, R., Schapira, M., and Ron, D. (2000). *Mol. Cell* **6**, 1099–1108.
57. McKnight, S. L., Lane, M. D., and Gluecksohn-Waelsch, S. (1989). *Genes Dev.* **3**, 2021–2024.
58. Shamu, C. E. and Walter, P. (1996). *EMBO J.* **15 (12)**, 3028–3039.
59. Welihinda, A. A. and Kaufman, R. J. (1996). *J. Biol. Chem.* **271**, 18181–18187.
60. Liu, C. Y., Schröder, M., and Kaufman, R. J. (2000). *J. Biol. Chem.* **275**, 24881–24885.
61. Liu, E. S., Ou, J. H., and Lee, A. S. (1992). *J. Biol. Chem.* **267**, 7128–7133.
62. Dorner, A. J., Wasley, L. C., and Kaufman, R. J. (1992). *EMBO J.* **11**, 1563–1571.
63. Morris, J. A., Dorner, A. J., Edwards, C. A., Hendershot, L. M., and Kaufman, R. J. (1997). *J. Biol. Chem.* **272**, 4327–4334.
64. Bertolotti, A., Zhang, Y., Hendershot, L. M., Harding, H. P., and Ron, D. (2000). *Nat. Cell Biol.* **2**, 326–332.
65. Shen, J., Chen, X., Hendershot, L. M., and Prywes, R. (2002). *Dev. Cell* **3**, 99–111.
66. Masaki, T., Yoshida, M., and Noguchi, S. (1999). *Biochem. Biophys. Res. Commun.* **261**, 350–356.
67. Reimold, A. M., Etkin, A., Clauss, I., Perkins, A., Friend, D. S., Zhang, J., Horton, H. F., Scott, A., Orkin, S. H., Byrne, M. C., Grusby, M. J., and Glimcher, L. H. (2000). *Genes Dev.* **14**, 152–157.
68. Reimold, A. M., Iwakoshi, N. N., Manis, J., Vallabhajosyula, P., Szomolanyi-Tsuda, E., Gravallese, E. M., Friend, D., Grusby, M. J., Alt, F., and Glimcher, L. H. (2001). *Nature* **412**, 300–307.
69. Delepine, M., Nicolino, M., Barrett, T., Golamaully, M., Mark Lathrop, G., and Julier, C. (2000). *Nat. Genet.* **25**, 406–409.
70. Harding, H. P., Zeng, H., Zhang, Y., Jungries, R., Chung, P., Plesken, H., Sabatini, D. D., and Ron, D. (2001). *Mol. Cell* **7** (6), 1153–1163.
71. Itoh, N. and Okamoto, H. (1980). *Nature* **283**, 100–102.
72. Ling, Z., Kiekens, R., Mahler, T., Schuit, F. C., Pipeleers-Marichal, M., Sener, A., Kloppel, G., Malaisse, W. J., and Pipeleers, D. G. (1996). *Diabetes* **45**, 1774–1782.
73. Detimary, P., Dejonghe, S., Ling, Z., Pipeleers, D., Schuit, F., and Henquin, J. C. (1998). *J. Biol. Chem.* **273**, 33905–33908.
74. Schuit, F. C., Huypens, P., Heimberg, H., and Pipeleers, D. G. (2001). *Diabetes* **50**, 1–11.

CHAPTER 317

Regulation of mRNA Turnover

Perry J. Blackshear[1,2] and Wi S. Lai[1]

[1]*Office of Clinical Research and Laboratory of Signal Transduction, National Institute of Environmental Health Sciences, Research Triangle Park, North Carolina; and*
[2]*Departments of Medicine and Biochemistry, Duke University, Durham, North Carolina*

Introduction

Steady-state levels of macromolecules in cells are regulated by the balance of synthesis and breakdown. Although historically the control of mRNA levels by regulation of transcription has received the most attention, in recent years the critical role of mRNA breakdown in regulating transcript levels has been more widely appreciated. The stabilities of different transcripts can vary by several orders of magnitude under normal physiological conditions. In addition, recently there has been increased awareness of the possibility of minute-to-minute regulation of mRNA stability, just as regulation of transcription by signal transduction processes has received intense scrutiny during the past two decades. The purpose of this chapter is to provide an update on some recent research findings on rapid regulation of mRNA turnover in response to extracellular agonists in vertebrate cells. Although much of our knowledge of eukaryotic mRNA turnover comes from studies in yeast, space precludes a review of those studies, which have been reviewed [1,2]. The reader is also referred to other reviews on mRNA stability in mammalian cells [3,4].

Current Models of mRNA Stability in Vertebrate Cells

It is now thought that mRNA stability can be described for most cellular mRNAs in vertebrates in terms of the *closed-loop* model. In this model, the poly(A) tail attached the 3′ end of most mRNAs is physically linked to the mRNA cap structure, through direct binding contacts between a widely distributed poly(A) binding protein and the eukaryotic initiation factor eIF4G. This physical interaction is apparently also able to promote the translation of mRNAs. Thus, under situations in which the mRNA is stabilized, the closed-loop configuration simultaneously inhibits mRNA decay and promotes mRNA translation, leading to net increases in protein synthesis (assuming constant transcription rates).

The biochemical changes that occur when this complex is destabilized and the rate of mRNA degradation is increased are still poorly understood. In vertebrates, these steps are thought to involve the deadenylation of the fully polyadenylated transcripts, followed by further nucleolytic action of endo- and exonucleases. In yeast, the removal of both the poly(A) tail and the 5′ cap structure are key early events in mRNA breakdown, followed by a major pathway of 5′ to 3′ exonucleolytic destruction of the mRNA body, as well as a minor pathway of 3′ to 5′ degradation. Similarly, the breakdown of vertebrate mRNAs also is thought to begin with the deadenylation of the polyadenylated transcript, apparently through a poly(A)-specific exonuclease referred to as PARN. However, a role for a putative mammalian decapping activity in the next steps of the process is not clear, although such an activity has been described recently in HeLa cell extracts [5]. For most mRNAs in vertebrate cells, the initial deadenylation is thought to be followed by further exo- and endonucleolytic breakdown of the completely or partially deadenylated mRNA body. Very recently, it has been shown that the deadenylated mRNA can be physically linked to a multiprotein intracellular structure known as an *exosome*, in which further breakdown of the deadenylated mRNA body can take place [6,7].

Presence of Instability Elements in Vertebrate mRNAs

It has been known for many years that instability of certain vertebrate mRNAs is conferred by AU-rich elements (ARE) within the 3′ untranslated region (UTR) of mRNAs [8]. Removal or mutation of these elements results in a more stabilized mRNA, whereas insertion of such elements into otherwise stable mRNAs can destabilize them. These instability-inducing AREs were originally classified into three major groups by Xu *et al.* [9]; this classification has been broadened recently by Willusz *et al.* [2]. A recurring element in these classifications is the AUUUA pentamer, which is linked in five tandem copies in examples of class IIA element-containing mRNAs such as those encoded by granulocyte macrophage colony stimulating factor (GM-CSF) and tumor necrosis factor α (TNFα). Progressively fewer of these pentameric elements are found in mRNAs containing class IIB, C, D, and E AREs, whereas class I mRNAs contain a single pentamer plus a U-rich region, and class III mRNAs are U-rich but do not contain the pentameric motif [2].

Possibly because many of the mRNAs containing class II AREs are of intense clinical interest (e.g., GM-CSF, TNFα, cyclooxygenase 2, interleukin 2, interleukin 3, vascular endothelial growth factor, interferon α), the mechanism by which these elements confer instability on an mRNA has received intense scrutiny. From the work of Chen and Shyu [10], a consensus has emerged that these mRNAs are first subjected to processive poly(A) exonuclease action, resulting in the formation and possible accumulation of deadenylated mRNA bodies; the bodies are then further degraded by other cellular nucleases, perhaps in the exosomal compartment as described earlier. How this process is stimulated by the presence of the class II AREs, and how it is otherwise regulated, are not currently known, but are the subject of much current research.

Effects of ARE Binding Proteins on mRNA Turnover

Given the apparent importance of the class II AREs in the enhanced turnover of the mRNAs that contain them, much research has been focused on the identification and functional analysis of ARE binding proteins. A number of such proteins has been identified, and most seem to exert a protective or inhibitory effect on ARE-containing mRNA destruction. Others appear to have minimal effect on mRNA turnover rate, but instead modulate the translation of the target mRNAs. In most cases, it has not been possible to identify precise mechanisms by which the protective mRNA binding proteins exert their effects, but an obvious possibility is the inhibition of "pro-turnover" binding proteins by physical interference.

Our group has identified one such family of pro-turnover ARE binding proteins, the tristetraprolin (TTP) family of CCCH tandem zinc finger proteins. The prototype of this group, TTP, was originally identified by virtue of its immediate early response gene characteristics in a screen for genes whose transcription was rapidly stimulated by insulin [11]. It was identified independently by groups studying genes rapidly turned on by phorbol esters [12,13] and serum [14]. Its role in the turnover of ARE-containing mRNAs was discovered after TTP knock-out mice exhibited physical characteristics of TNF excess, and macrophages derived from them secreted excess TNF, due in part to increased stability of the TNF mRNA [15–17]. TTP was then identified as a class II ARE binding protein, able to stimulate the destruction of this mRNA by unknown means [17]. Later studies showed a similar effect on the turnover of GM-CSF mRNA, and it was suggested that increased secretion of this factor in the TTP knock-out mice might account for some of the myeloid hyperplasia characteristic of the TTP deficiency syndrome [18,19]. The GM-CSF experiments also demonstrated that the primary effect of TTP deficiency was to inhibit deadenylation of this mRNA, leading to the hypothesis that the primary effect of TTP, after ARE binding, was in some way to promote the deadenylation of ARE-containing mRNAs.

The two other known members of the TTP protein family, now known as Zfp36L1 and Zfp36L2 in mice, are able to exert similar effects to TTP in cell transfection studies [20], but their physiological mRNA targets are still not known. A fourth family member, C3H-4, has been identified as an apparently purely maternal gene product in *Xenopus* and fish oocytes and eggs [21], and it is interesting to speculate that it plays a role in mRNA deadenylation during early development.

Regulation of TTP Activity in Cells

It was apparent from the very earliest studies of TTP induction in response to insulin, growth factors, cytokines, and so on, that cellular levels of this mRNA, and presumably protein, could be regulated rapidly and dramatically through stimulated transcription. The very transient nature of the TTP mRNA induction curve indicates a rapid shut-off of transcription within an hour or so of its stimulation, coupled with a very unstable mRNA. Although the protein levels in cells appear to be rather low, making antibody detection difficult, it has been possible to confirm in several studies that the changes in protein levels largely parallel those of the mRNA. This rapid and dramatic transcriptional response to mitogens and growth factors represents one major mode of regulation of TTP's cellular activity, that is, regulation of its steady-state protein concentrations.

A second mode of regulation was suggested by early experiments in which TTP's subcellular compartmentalization was regulated by the same mitogens and growth factors that stimulate its transcription [22]. These studies indicated that serum, growth factors, and phorbol esters could rapidly (within 5 min) stimulate TTP's translocation from nucleus to cytosol. More recently, we have shown that the nuclear export of TTP and its two mammalian relatives is mediated through the exportin CRM1, using nuclear export sequences on the amino terminus of TTP and on the carboxyl termini of Zfp36L1 and Zfp36L2 [23]. We also demonstrated that the return to the nucleus was mediated through the tandem zinc finger domain itself, although this intracellular movement did not require that the protein be competent to bind mRNA. Thus, a second mode of regulation of TTP's (and related proteins') cellular activity is change in nuclear to cytosol localization. Very little is known about how these processes are regulated, but the rapid nature of the nuclear export in response to growth factors suggests the involvement of phosphorylation or dephosphorylation.

A third mode of regulation was suggested by the early recognition that TTP is phosphorylated in cells in response to many of the same agonists that stimulate its rapid transcription on nuclear to cytoplasmic translocation [24]. We found that recombinant TTP could be phosphorylated in a cell-fee assay by MAP kinase, and that at least one of the serines phosphorylated by MAP kinase in the cell-free assay was also phosphorylated in intact cells in response to agonists that activate MAP kinase; however, replacement of this serine by alanine did not affect the ability of the stimulatory agonists to promote TTP's nuclear to cytoplasmic translocation. It is now clear that many other phosphorylated residues exist on TTP in intact cells, and we are attempting to catalog these and determine which are modified in response to stimulation of cells, and how these modifications might affect TTP's translocation and RNA binding activity.

We recently analyzed the last point as part of a study evaluating the possibility that TTP might be modified in response to another MAP kinase family member, the p38 kinase [25]. Although previous data on the relationship between p38 kinase activation and TNF secretion are somewhat contradictory, most authors have found that the use of relatively specific p38 inhibitors inhibits lipopolysaccharide (LPS)-stimulated TNF secretion from macrophages. A recent example of this demonstrated that, in freshly isolated human mononuclear cells, the inhibitor SB202190 at 3 μM could inhibit LPS-stimulated TNF secretion by approximately 90%, and that this was associated with dramatic destabilization of TNF mRNA [26]. These and other data raised the interesting possibility that the p38 effect on TNF mRNA stability might be mediated in part by changes in the phosphorylation status of TTP. We first demonstrated that TTP-deficient macrophages were relatively refractory to the same p38 inhibitors that were capable of inhibiting TNF secretion from wild-type littermate macrophages. In addition, recombinant TTP could be phosphorylated by p38 in a cell-free system, and its phosphorylation in response to LPS in macrophages could also be inhibited by the p38 inhibitors. These data led us to propose a scheme in which phosphorylation of TTP by p38 kinase would result in a decrease in its affinity for the TNF mRNA, thus allowing for increased TNF mRNA stability and cellular concentrations and for commensurate increases in TNF secretion. This would be reversed by the p38 inhibitors, which would have little influence on the TTP-deficient cells because of loss of this critical target. We attempted to validate this scheme by performing "global" dephosphorylation of TTP expressed in 293 cells by alkaline phosphatase, and then comparing the binding affinities of phosphorylated and dephosphorylated TTP in an RNA gel shift assay. In keeping with our proposed model, the TTP phosphorylated in 293 cells had considerably less affinity for its ARE target than the dephosphorylated form. Thus, although the phosphorylation sites responsible for this change in RNA binding affinity are not known, it appears that TTP activity in cells is likely to be regulated on a minute-to-minute basis by its phosphorylation status. This third mode of TTP regulation is also likely to affect the subcellular localization of the protein and, thus, control its cytosolic concentrations and likely activity indirectly.

Conclusion

It has become clear from numerous studies that mRNA turnover should not be considered a "default" pathway reversing the more regulated process of transcription, but instead a complex set of cellular processes regulated independently. Because many of the ARE-containing mRNAs encode clinically important cytokines, it seems likely that the current intensive study of the regulated turnover of these mRNAs will lead to new insights into the cellular processes of deadenylation, exosomal decay of the mRNA body, relationship of the mRNA cap to its turnover, importance of other *cis*- and *trans*-acting elements, and so on. The molecular details of these processes should be understood more completely in the coming years. In addition, it is possible that increased understanding of these processes may lead to the development of new therapies for diseases in which the goal is modulation of circulating cytokine levels.

References

1. Hilleren, P. and Parker, R. (1999). Mechanisms of RNA surveillance in eukaryotes. *Annu. Rev. Genet.* **33**, 229–260.
2. Wilusz, C. J., Wormington, M., and Peltz, S. W. (2001). The cap-to-tail guide to mRNA turnover. *Nat. Rev. Mol. Cell Biol.* **2**, 237–246.
3. Ross, J. (1995). mRNA stability in mammalian cells. *Microbiology Reviews* **59**, 423–250.
4. Guhaniyogi, J. and Brewer, G. (2001). Regulation of mRNA stability in mammalian cells. *Gene* **265**, 11–23.
5. Gao, M., Wilusz, C. J., Peltz, S. W., and Wilusz, J. (2001). A novel mRNA-decapping activity in HeLa cytoplasmic extracts is regulated by AU-rich elements. *EMBO J.* **20**, 1134–1143.
6. Chen, C. Y., Gherzi, R., Ong, S. E., Chan, E. L., Raijmakers, R., Pruijn, G. J., Stoecklin, G., Moroni, C., Mann, M., and Karin, M. (2001). AU binding proteins recruit the exosome to degrade ARE-containing mRNAs. *Cell* **107**, 451–464.
7. Mukherjee, D., Gao, M., O'Connor, J. P., Raijmakers, R., Pruijn, G., Lutz, C. S., and Wilusz, J. (2002). The mammalian exosome mediates the efficient degradation of mRNAs that contain AU-rich elements. *EMBO J.* **21**, 165–174.
8. Shaw, G. and Kamen, R. (1986). A conserved AU sequence from the 3′-untranslated region of GM-CSF mRNA mediates selective mRNA degradation. *Cell* **46**, 659–667.
9. Xu, N., Chen, C. Y., and Shyu, A. B. (1997). Modulation of the fate of cytoplasmic mRNA by AU-rich elements: Key sequence features controlling mRNA deadenylation and decay. *Mol. Cell. Biol.* **17**, 4611–4621.
10. Chen, C. Y. and Shyu, A. B. (1995). AU-rich elements: characterization and importance in mRNA degradation. *Trends Biochem. Sci.* **20**, 465–470.
11. Lai, W. S., Stumpo, D. J., and Blackshear, P. J. (1990). Rapid insulin-stimulated accumulation of an mRNA encoding a proline-rich protein. *J. Biol. Chem.* **265**, 16556–16563.
12. Varnum, B. C., Lim, R. W., Sukhatme, V. P., and Herschman, H. R. (1989). Nucleotide sequence of a cDNA encoding TIS11, a message induced in Swiss 3T3 cells by the tumor promoter tetradecanoyl phorbol acetate. *Oncogene* **4**, 119–120.

13. Ma, Q. and Herschman, H. R. (1991). A corrected sequence for the predicted protein from the mitogen-inducible TIS11 primary response gene. *Oncogene* **6**, 1277–1278.
14. DuBois, R. N., McLane, M. W., Ryder, K., Lau, L. F., and Nathans, D. (1990). A growth factor-inducible nuclear protein with a novel cysteine/histidine repetitive sequence. *J. Biol. Chem.* **265**, 19185–19191
15. Taylor, G. A., Carballo, E., Lee, D. M., Lai, W. S., Thompson, M. J., Patel, D. D., Schenkman, D. I., Gilkeson, G. S., Broxmeyer, H. E., Haynes, B. F., and Blackshear P. J. (1996). A pathogenic role for tumor necrosis factor α in the syndrome of cachexia, arthritis and autoimmunity resulting from tristetraprolin (TTP) deficiency. *Immunity* **4**, 445–454.
16. Carballo, E., Gilkeson, G. S., and Blackshear, P. J. (1997). Bone marrow transplantation reproduces the TTP-deficiency syndrome in *RAG-2*$^{(-/-)}$ mice: Evidence that monocyte/macrophage progenitors may be responsible for TNFα overproduction. *J. Clin. Invest.* **100**, 986–995.
17. Carballo, E., Lai, W. S., and Blackshear, P. J. (1998). Feedback inhibition of macrophage tumor necrosis factor α production by tristetraprolin. *Science* **281**, 1001–1005.
18. Carballo, E., Lai, W. S., and Blackshear, P. J. (2000). Evidence that tristetraprolin (TTP) is a physiological regulator of granulocyte-macrophage colony-stimulating factor (GM-CSF) mRNA deadenylation and stability. *Blood* **95**, 1891–1899.
19. Carballo, E. and Blackshear, P. J. (2001). Roles of tumor necrosis factor α receptor subtypes in the pathogenesis of the tristetraprolin-deficiency syndrome. *Blood* **98**, 2389–2395.
20. Lai, W. S., Carballo, E., Thorn, J. M., Kennington, E. A., and Blackshear, P. J. (2000). Interactions of CCCH zinc finger proteins with mRNA. Binding of tristetraprolin-related zinc finger proteins to AU-rich elements and destabilization of mRNA. *J. Biol. Chem.,* **275**, 17827–19837.
21. De, J., Lai, W. S., Thorn, J. M., Goldsworthy, S. M., Liu, X., Blackwell, T. K., and Blackshear, P. J. (1999). Identification of four CCCH zinc finger proteins in *Xenopus,* including a novel vertebrate protein with four zinc fingers and severely restricted expression. *Gene* **228**, 133–145.
22. Taylor, G. A., Thompson, M. J., Lai, W. S., and Blackshear, P. J. (1996). Mitogens stimulate the rapid nuclear to cytosolic translocation of tristetraprolin, a potential zinc-finger transcription factor. *Mol. Endocrinol.* **10**, 140–146.
23. Phillips, R. S. and Blackshear, P. J. (2002). Members of the tristetraprolin family of tandem CCCH zinc finger proteins exhibit CRM1-dependent nucleocytoplasmic shuttling. *J. Biol. Chem.* **277**, 11606–11613.
24. Taylor, G. A., Thompson, M. J., Lai, W. S., and Blackshear, P. J. (1995). Phosphorylation of tristetraprolin (TTP), a potential zinc finger transcription factor, by mitogen stimulation in intact cells and by mitogen-activated protein kinase in vitro. *J. Biol. Chem.* **270**, 13341–13347.
25. Carballo, E., Cao, H., Lai, W. S., Kennington, E. A., Campbell, D., and Blackshear, P. J. (2001). Decreased sensitivity of tristetraprolin-deficient cells to p38 inhibitors suggests the involvement of tristetraprolin in the p38 signaling pathway. *J. Biol. Chem.* **276**, 42580–42587
26. Wang, S. W., Pawlowski, J., Wathen, S. T., Kinney, S. D., Lichenstein, H. S., and Manthey, C. L. (1999). Cytokine mRNA decay is accelerated by an inhibitor of p38 mitogen-activated protein kinase. *Inflamm Res.* **48**, 533–538.

CHAPTER 318

CPEB-Mediated Translation in Early Vertebrate Development

Joel D. Richter
Program in Molecular Medicine,
University of Massachusetts Medical School,
Worcester, Massachusetts

Introduction

How a single fertilized egg gives rise to the billions of specialized cells of the adult is the most basic question of developmental biology. While the transcriptional activity of certain master control genes sends cells along certain lineages that culminate in adult tissue differentiation, at very early times of development, especially during germ cell formation, translational control is particularly influential. In female germ cells, several mRNAs that are synthesized and stored during the long period of oogenesis are "masked" or repressed. These mRNAs are subsequently translated in response to exogenous cues such as hormonal stimulation during meiosis (see later discussion) or fertilization. The proteins encoded by masked mRNA are often regulatory, and they dictate the timing of cell division, establish the germ layers, and specify the body axis. Because the translation of maternal mRNAs in vertebrates has been analyzed almost exclusively in the frog *Xenopus laevis* and the mouse, it is worthwhile considering certain key steps of oogenesis in these species.

Soon after their arrival at the genital ridge, the mitotically dividing oocytes enter meiosis and progress through the end of prophase I. During this period, they form the synaptonemal complexes that are required for genetic recombination and synthesize and store mRNA, tRNA, ribosomes, and presumably all the other factors that are necessary for protein synthesis. During their growth phase, the oocytes are arrested in diplotene, the last stage of prophase I, which in many respects resembles G_2 of the mitotic cell cycle. They contain pre-MPF (M-phase promoting factor), an inactive form of the kinase cdc2/cyclin B heterodimer, which when subsequently activated by the MAP kinase cascade, stimulates the oocytes to enter M phase (metaphase I). At the peak of M phase, cyclin B is destroyed, an event that must precede the oocyte's transition into anaphase (AI), which is followed by a second round of MPF activation and entry into metaphase II (MII) [1]. At this point, the cytostatic factor (CSF) is activated, which prevents the oocytes from progressing through the cell cycle until they are fertilized [2–4].

Although many of these events take place in all vertebrate oocytes, some key differences are seen among the animal groups. For example, in *Xenopus*, progesterone, acting through a surface-associated receptor, stimulates reentry into the meiotic divisions, commonly referred to as oocyte maturation. In the mouse, it is the oocyte's liberation from the follicular environment that stimulates maturation. Another important difference in oocyte maturation between these two animals is that only in *Xenopus* is protein synthesis necessary to stimulate maturation. In this species, the MAP kinase cascade that culminates in MPF activation is initiated by Mos, a MAP kinase kinase kinase. Oocytes have no Mos protein, but instead contain dormant mos mRNA that must undergo translational activation for maturation to occur. In the mouse as in *Xenopus*, *de novo* Mos synthesis is required for the MAP kinase cascade [5] and mos mRNA is also under translational control [6]. However, the function of Mos in the mouse is not to stimulate maturation, but to arrest the cell cycle at the end of MII, an activity that is also conserved in *Xenopus* (i.e., CSF, *vida supra*). Indeed, the preponderance of ovarian cysts and tumors in female mos knock-out mice is due to the fact that unovulated oocytes undergo parthenogenetic activation and subsequently exhibit unrestricted cell division [7,8].

Mechanism of Translational Control

In *Xenopus* and mouse oocytes, mos and several others mRNAs such as those encoding the cyclins A1, A2, and B2, are under translational control by cytoplasmic polyadenylation. That is, these mRNAs are dormant in oocytes and are appended with short poly(A) tails, usually ~20–40 nucleotides. On the induction of oocyte maturation, the tails on these messages are elongated, up to ~100–150 nucleotides, and translation ensues. During the past few years, progress has been significant in delineating the steps of cytoplasmic polyadenylation and how this process leads to translational activation. Several dormant mRNAs contain cytoplasmic polyadenylation elements (CPEs) within their 3′ untranslated regions (UTRs). These CPEs, which have the general sequence of UUUUUAU, usually reside within about 100 nucleotides 5′ of the nuclear pre-mRNA cleavage and polyadenylation hexanucleotide AAUAAA. Both of these elements are necessary for cytoplasmic polyadenylation. The CPE is bound by CPEB, an RNA recognition motif (RRM) and zinc finger type of RNA-binding protein [9–11], and the AAUAAA is bound by a cytoplasmic version of cleavage and polyadenylation specificity factor (CPSF) [12,13], a complex of at least three proteins (see later discussion). The initiation of polyadenylation takes place when CPEB serine 174 is phosphorylated by the kinase Aurora (also known as Eg2 or IAK1; for nomenclature, see [14]) [15], which itself is activated soon after the maturation process is initiated [16]. The function of this phosphorylation event is to induce CPEB to bind, with enhanced avidity, CPSF, and possibly stabilize the latter factor's interaction with the AAUAAA [17]. However, note that Dickson *et al.* [18], while confirming the CPEB–CPSF interaction observed no enhanced interaction between these factors when CPEB was phosphorylated (see [19] for review of cytoplasmic polyadenylation). Although the precise function of CPSF in cytoplasmic polyadenylation has not been demonstrated, by analogy with nuclear pre-mRNA polyadenylation, it is to bring poly(A) polymerase (PAP) to the end of the mRNA, where it catalyzes poly(A) addition [18].

Although the basic cytoplasmic polyadenylation reaction involves Aurora, CPEB, CPSF, and PAP, a number of important questions remain regarding this process. First, nuclear CPSF consists of four subunits with molecular sizes of 160, 100, 70, and 30 kDa. At least in *Xenopus* oocytes, the 70 species is missing from the cytoplasm [18]. Whether the activity of this protein is dispensable for CPSF activity or whether it is functionally replaced by a cytoplasmic form of the protein is not known. Second, it has been reported that with some CPE-containing mRNAs, mammalian nuclear CPSF can recapitulate cytoplasmic polyadenylation *in vitro* without a requirement for CPEB [12,18]. However, given that the components of cytoplasmic oocyte CPSF are not known in their entirety, an extrapolation of the *in vitro* activities of nuclear mammalian CPSF to cytoplasmic *Xenopus* CPSF may be open to argument. Clearly, it is important to isolate and characterize cytoplasmic CPSF. Third, although PAP appears to have catalytic activity both before and after maturation [18], the observation that cdc2-mediated phosphorylation inactivates the enzyme during M phase seems paradoxical. That is, as cells enter mitosis (maturation in oocytes), cdc2 (MPF) becomes active and phosphorylates PAP at multiple sites, particularly in the carboxy terminus, which, at least in somatic cells, inactivates the enzyme [20–22]. Given that cytoplasmic polyadenylation is most robust as oocytes mature [23,24], it seems plausible, if not likely, that another PAP that is not inactivated at this time would also be present in cells. Indeed, oocytes do contain a PAP that lacks a large portion of the carboxy terminus and, hence, lacks the major cdc2 phosphorylation sites [25]. In addition, another PAP that is overexpressed in certain mammalian tumors apparently is not phosphorylated M phase and, thus, also has the potential of catalyzing polyadenylation at M phase [26].

CPEB and Early Development

In the four metazoans in which it has been examined, CPEB has been shown to be essential for early development. In *Drosophila*, various mutations in the CPEB (Orb) gene lead to arrested oogenesis (the most severe allele) or defects in embryogenesis (weaker alleles) [27,28]. While Orb has clearly been demonstrated to mediate mRNA localization, it also controls cytoplasmic polyadenylation of several mRNAs, including its own [29–31]. The nematode *Caenorhabditis elegans* has four CPEB genes, and RNAi injection experiments have demonstrated that one of them is involved in spermatogenesis [32]. However, no target mRNAs have been identified, and the molecular functions of the CPEB proteins in worms are not known.

In vertebrates, the essential nature of CPEB and cytoplasmic polyadenylation for development was first demonstrated in *Xenopus*, where the injection of CPEB antibody into oocytes was shown to abrogate meiotic maturation [33]. This block occurred because mos mRNA was not polyadenylated or translated, thus preventing the MAP kinase cascade and the activation of MPF [33,34]. CPEB is also important for cell cycle progression in *Xenopus* embryos. In embryonic cells, as in all cells, progression into M phase requires the activation of MPF, whereas the subsequent destruction of cyclin B and the inactivation of MPF is necessary to transition into interphase [35]. However, at least in embryonic cells, not only is regulated cyclin B mRNA translation important for cell cycle progression, so too is regulated cyclin B mRNA localization. That is, in the blastomeres of the developing embryo, CPEB and maskin, as well as cyclin mRNA and protein are all associated with mitotic spindles and centrosomes [36]. The injection of neutralizing antibody for both CPEB and maskin blocks cell division, as does cordycepin (3′-deoxyadenosine), an inhibitor of polyadenylation. *In vitro* binding experiments demonstrated that CPEB is a microtubule-binding protein, which may explain its association with spindles. Perhaps most importantly, the injection of embryos with a deletion mutant CPEB protein that cannot interact with microtubules, although having little effect on cyclin synthesis, caused cyclin mRNA to dissociate from spines. The result of this dissociation was the cessation of cell division.

The first interpretation of these results is that the dominant negative CPEB protein, because it retained the capacity to bind RNA, outcompeted endogenous CPEB for binding to cyclin mRNA, but because it could not also bind microtubules, it diffused away from the spindles. The second interpretation is that cyclin mRNA translation was required locally, on or near spindles, where it mediates functions related to cell division [36]. These results would also suggest that the polyadenylation of cyclin mRNA is regulated during the cell cycle, and that this regulation is necessary for cell division. Such a hypothesis has recently been tested by Groisman *et al.* [37], who have prepared cell extracts capable of undergoing cell cycle progression. These investigators found that cyclin mRNA polyadenylation increased in the S-phase to M-phase transition and that if polyadenylation was blocked, cyclin was not synthesized nor did the extract enter M phase. Thus, it appears that at least in embryonic cells, cyclin mRNA translation is essential for cell cycle progression.

Finally, a CPEB knock-out (KO) mouse has been generated, and the results firmly demonstrate the necessity of this protein for meiotic progression [38]. Adult CPEB KO females were sterile and devoid of ovaries. Female KO embryos at 18.5 days of development (E18.5) had ovaries, but for the most part these did not contain oocytes. E16.5 embryos did have ovaries with oocytes, but the oocyte nuclei stained aberrantly with hematoxylin and with antibody directed against GCNA1, a germ cell nuclear antigen of unknown function. The staining showed the chromatin to be diffuse and/or fragmented. At 16.5 days of development, the oocytes are normally in pachytene of meiosis I a time of synaptonemal complex (SC) formation. The SC, which facilitates recombination, is comprised of several proteins, but two of them, SCP1 and SCP3, are encoded by CPE-containing mRNAs. In wild-type animals, CPEB was found to be associated with the mRNAs *in vivo*. In the CPEB KO animals, SCP1 and SCP3 mRNAs were neither polyadenylated nor translated, and thus the SC was not formed. Without the SC, there is no recombination, and the oocytes and subsequently the ovary degenerate. CPEB is therefore necessary for two stages of meiosis, at pachytene and during the maturation.

Conclusions

While this brief review has focused on CPEB-induced polyadenylation as one mechanism of translational control in early development, it is by no means the only mechanism. For example, in *Xenopus* oocytes, histone mRNAs, unlike their somatic counterparts, are polyadenylated and mostly dormant. Following the oocytes' reentry into the meiotic divisions, the core histone mRNAs are activated in a poly(A)-independent manner. A stem-loop in the 3′UTRs of histone mRNAs is bound by the stem-loop binding protein (SLBP), which, in a manner yet to be deciphered, facilitates translational repression and activation [39]. In addition, during spermatogenesis in the mouse, the protamine mRNAs are also translationally regulated in a poly(A)-independent manner by other 3′UTR binding proteins [40]. Finally, it is worth noting that the mechanisms of translational control in early development could have a number of other important consequences. Consider that CPEB-mediated polyadenylation also appears to control translation in the mammalian brain, where it could influence synaptic plasticity and long-term memory storage [41,42]. Thus, nature reuses translational control mechanisms in multiple tissues, although this should come as no surprise to the 21st century investigator.

References

1. Nebreda, A. R. and Ferby, I. (2000). Regulation of the meiotic cell cycle in oocytes. *Curr. Opin. Cell Biol.* **12**, 666–675.
2. Sagata, N., Watanabe, N., Vande Woude, G. F., and Ikawa, Y. (1989). The c-mos proto-oncogene product is a cytostatic factor responsible for meiotic arrest in vertebrate eggs. *Nature* **342**, 512–518.
3. Haccard, O., Sarcevic, B., Lewellyn, A., Hartley, R., Roy, L., Izumi, T., Erikson, E., and Maller, J. L. (1993). Induction of metaphase arrest in cleaving *Xenopus* embryos by MAP kinase. *Science* **262**, 1262–1265.
4. Schwab, M. S., Roberts, B. T., Gross, S. D., Tunquist, B. J., Taieb, F. E., Lewellyn, A. L., and Maller, J. L. (2001). Bub1 is activated by the protein kinase p90(Rsk) during *Xenopus* oocyte maturation. *Curr. Biol.* **11**, 141–150.
5. Verlhac, M. H., Kubiak, J. Z., Weber, M., Geraud, G., Colledge, W. H., Evans, M. J., and Maro, B. (1996). Mos is required for MAP kinase activation and is involved in microtubule organization during meiotic maturation in the mouse. *Development* **122**, 815–822.
6. Gebauer, F., Xu, W., Cooper, G. M., and Richter, J. D. (1994). Translational control by cytoplasmic polyadenylation of c-mos mRNA is necessary for oocyte maturation in the mouse. *EMBO J.* **13**, 5712–5720.
7. Hashimoto, N., Watanabe, N., Furuta, Y., Tamemoto, H., Sagata, N., Yokoyama, M., Okazaki, K., Nagayoshi, M., Takeda, N., and Ikawa, Y. (1994). Parthenogenetic activation of oocytes in c-mos-deficient mice. *Nature* **370**, 68–71.
8. Colledge, W. H., Carlton, M. B., Udy, G. B., and Evans, M. J. (1994). Disruption of c-mos causes parthenogenetic development of unfertilized mouse eggs. *Nature* **370**, 65–68.
9. Paris, J., Swenson, K., Piwnica-Worms, H., and Richter, J. D. (1991). Maturation-specific polyadenylation: *in vitro* activation by p34cdc2 and phosphorylation of a 58-kD CPE-binding protein. *Genes Dev.* **5**, 1697–1708.
10. Hake, L. E. and Richter, J. D. (1994). CPEB is a specificity factor that mediates cytoplasmic polyadenylation during *Xenopus* oocyte maturation. *Cell* **79**, 617–627.
11. Hake, L. E., Mendez, R., and Richter, J. D. (1998). Specificity of RNA binding by CPEB: Requirement for RNA recognition motifs and a novel zinc finger. *Mol. Cell. Biol.* **18**, 685–693.
12. Bilger, A., Fox. C. A., Wahle, E., and Wickens, M. (1994). Nuclear polyadenylation factors recognize cytoplasmic polyadenylation elements. *Genes Dev.* **8**, 1106–1116.
13. Dickson, K. S., Bilger, A., Ballantyne, S., and Wickens, M. P. (1999). The cleavage and polyadenylation specificity factor in *Xenopus laevis* oocytes is a cytoplasmic factor involved in regulated polyadenylation. *Mol. Cell. Biol.* **19**, 5707–5717.
14. Nigg, E. A. (2001). Mitotic kinases as regulators of cell division and its checkpoints. *Nat. Rev. Mol. Cell. Biol.* **2**, 21–32.
15. Mendez, R., Hake, L. E., Andresson, T., Littlepage, L. E., Ruderman, J. V., and Richter, J. D. (2000). Phosphorylation of CPE binding factor by Eg2 regulates translation of c-mos mRNA. *Nature* **404**, 302–307.

16. Andresson, T. and Ruderman, J. V. (1998). The kinase Eg2 is a component of the *Xenopus* oocyte progesterone-activated signaling pathway. *EMBO J.* **17**, 5627–5637.
17. Mendez, R., Murthy, K. G., Ryan, K., Manley, J. L., and Richter, J. D. (2000). Phosphorylation of CPEB by Eg2 mediates the recruitment of CPSF into an active cytoplasmic polyadenylation complex. *Mol. Cell* **6**, 1253–1259.
18. Dickson, K. S., Thompson, S. R., Gray, N. K., and Wickens, M. (2001). Poly(A) polymerase and the regulation of cytoplasmic polyadenylation. *J. Biol. Chem.* **276**, 41810–41816.
19. Mendez, R. and Richter, J. D. (2001). Translational control by CPEB: A means to the end. *Nat. Rev. Mol. Cell. Biol.* **2**, 521–529.
20. Colgan, D. F., Murthy, K. G., Zhao, W., Prives, C., and Manley, J. L. (1998). Inhibition of poly(A) polymerase requires p34cdc2/cyclin B phosphorylation of multiple consensus and non-consensus sites. *EMBO J.* **17**, 1053–1062.
21. Colgan, D. F. and Manley, J. L. (1997). Mechanism and regulation of mRNA polyadenylation. *Genes Dev.* **11**, 2755–2766.
22. Colgan, D. F., Murthy, K. G., Prives, C., and Manley, J. L. (1996). Cell-cycle related regulation of poly(A) polymerase by phosphorylation. *Nature* **384**, 282–285.
23. McGrew, L. L., Dworkin-Rastl, E., Dworkin, M. B., and Richter, J. D. (1989). Poly(A) elongation during *Xenopus* oocyte maturation is required for translational recruitment and is mediated by a short sequence element. *Genes Dev.* **3**, 803–815.
24. Fox, C. A., Sheets, M. D., and Wickens, M. P. (1989). Poly(A) addition during maturation of frog oocytes: distinct nuclear and cytoplasmic activities and regulation by the sequence UUUUUAU. *Genes Dev.* **3**, 2151–2162.
25. Gebauer, F. and Richter, J. D. (1995). Cloning and characterization of a *Xenopus* poly(A) polymerase. *Mol. Cell. Biol.* **15**, 3460–3466.
26. Topalian, S. L., Kaneko, S., Gonzales, M. I., Bond, G. L., Ward, Y., and Manley, J. L. (2001). Identification and functional characterization of neo-poly(A) polymerase, an RNA processing enzyme overexpressed in human tumors. *Mol. Cell. Biol.* **21**, 5614–5623.
27. Lantz, V., Chang, J. S., Horabin, J. I., Bopp, D., and Schedl, P. (1994). The *Drosophila* orb RNA-binding protein is required for the formation of the egg chamber and establishment of polarity. *Genes Dev.* **8**, 598–613.
28. Christerson, L. B. and McKearin, D. M. (1994). orb is required for anteroposterior and dorsoventral patterning during *Drosophila* oogenesis. *Genes Dev.* **8**, 614–628.
29. Chang, J. S., Tan, L., and Schedl, P. (1999). The *Drosophila* CPEB homolog, orb, is required for oskar protein expression in oocytes. *Dev. Biol.* **215**, 91–106.
30. Chang, J. S., Tan, L., Wolf, M. R., and Schedl, P. (2001). Functioning of the *Drosophila* orb gene in gurken mRNA localization and translation. *Development* **128**, 3169–3177.
31. Tan, L., Chang, J. S., Costa, A., and Schedl, P. (2001). An autoregulatory feedback loop directs the localized expression of the *Drosophila* CPEB protein Orb in the developing oocyte. *Development* **128**, 1159–1169.
32. Luitjens, C., Gallegos, M., Kraemer, B., Kimble, J., and Wickens, M. (2000). CPEB proteins control two key steps in spermatogenesis in *C. elegans. Genes Dev.* **14**, 2596–2609.
33. Stebbins-Boaz, B., Hake, L. E., and Richter, J. D. (1996). CPEB controls the cytoplasmic polyadenylation of cyclin, Cdk2 and c-mos mRNAs and is necessary for oocyte maturation in *Xenopus. EMBO J.* **15**, 2582–2592.
34. de Moor, C. H. and Richter, J. D. (1997). The Mos pathway regulates cytoplasmic polyadenylation in *Xenopus* oocytes. *Mol. Cell. Biol.* **17**, 6419–6426.
35. King, R. W., Deshaies, R. J., Peters, J. M., and Kirschner, M. W. (1996). How proteolysis drives the cell cycle. *Science* **274**, 1652–1659.
36. Groisman, I., Huang, Y. S., Mendez, R., Cao, Q., Theurkauf, W., and Richter, J. D. (2000). CPEB, maskin, and cyclin B1 mRNA at the mitotic apparatus: Implications for local translational control of cell division. *Cell* **103**, 435–447.
37. Grosiman, I., Jung, M., Sarkissian, M., and Richter, J. D. (2002). Translational control of the embryonic cell cycle (submitted).
38. Tay, J. and Richter, J. D. (2001). Germ cell differentiation and synaptonemal complex formation are disrupted in CPEB knockout mice. *Dev. Cell* **1**, 201–213.
39. Wang, Z. F., Ingledue, T. C., Dominski, Z., Sanchez, R., and Marzluff, W. F. (1999). Two *Xenopus* proteins that bind the 3′ end of histone mRNA: implications for translational control of histone synthesis during oogenesis. *Mol. Cell. Biol.* **19**, 835–845.
40. Zhong, J., Peters, A. H., Lee, K., and Braun, R. E. (1999). A double-stranded RNA binding protein required for activation of repressed messages in mammalian germ cells. *Nat. Genet.* **22**, 171–174.
41. Wu, L., Wells, D., Tay, J., Mendis, D., Abbott, M. A., Barnitt, A., Quinlan, E., Heynen, A., Fallon, J. R., and Richter, J. D. (1998). CPEB-mediated cytoplasmic polyadenylation and the regulation of experience-dependent translation of alpha-CaMKII mRNA at synapses. *Neuron* **21**, 1129–1139.
42. Wells, D. G., Dong, X., Quinlan, E. M., Huang, Y. S., Bear, M. F., Richter, J. D., and Fallon, J. R. (2001). A role for the cytoplasmic polyadenylation element in NMDA receptor-regulated mRNA translation in neurons. *J. Neurosci.* **21**, 9541–9548.

CHAPTER 319

Translational Control in Invertebrate Development

Paul Lasko
Department of Biology, McGill University, Montréal, Québec, Canada

Introduction

Invertebrate genetic model organisms, most notably *Drosophila*, have been instrumental in efforts to understand the role of translational control in regulating gene expression that underlies development. I will summarize some of the mechanisms of translational control that are emerging from this work. Space constraints do not permit this review to be an exhaustive one, yet this article should provide the reader with the means to gain an overview of the field, and to enter the relevant literature in a focused way.

Translational control is a critical mechanism for establishing pattern in the *Drosophila* oocyte, and the axes that are set down in the developing egg define the positional information of the embryo and of the adult fly [1–2]. Whereas the anterior determinant, Bicoid, is deployed mostly through asymmetric localization of its mRNA, many of the proteins required for posterior and germ cell specification are deployed both through localization of their mRNAs and position-dependent translational regulation. These mRNAs accumulate preferentially in a specialized cytoplasm, called *pole plasm*, at the posterior pole of the egg. Their translation is repressed throughout most of the oocyte, but is active within the pole plasm.

Translational Control Targets Oskar to the Pole Plasm

One gene under elaborate translational control is *oskar* (*osk*). The *osk* gene is required for specification of posterior and of the germ line, and *osk* mRNA accumulates in the pole plasm [3]. Unlocalized *osk* is translationally silent, but localized *osk* is translated from two start sites: constitutively at a low level from the more 5′ site to produce an isoform called Long Osk, and from an internal initiation codon to produce a second isoform called Short Osk. Short Osk is sufficient to fully rescue the phenotype of an *osk* mutant, and studies of the translational control of *osk* have mostly concentrated on Short Osk [4].

Translational repression of unlocalized *osk* depends on sequences within its 3′ untranslated region (UTR) and on several proteins and gene activities (Fig. 1). Bruno (Bru), an RRM-type RNA-binding protein, is a translational repressor that binds to three regions of the *osk* 3′UTR called Bru response elements (BREs). Flies bearing an *osk* transgene mutated for the BREs (*oskBRE*⁻) translate *osk* precociously, and BREs are able to confer translational repression on a heterologous mRNA [5]. Bru also represses *osk* translation in extracts in a BRE-dependent manner [6,7]. A second protein, Apontic (Apt), interacts with Bru both in the two-hybrid system and biochemically and is believed to be an essential cofactor [8]. Another protein, p50, cross-links to both the 5′ and 3′ ends of the *osk* mRNA [9]. Transgenes carrying a mutation that specifically reduces p50 binding show precocious *osk* translation, suggesting that p50 is also required for translational repression of *osk*. Similar genetic evidence implicates Bicaudal-C (Bic-C), a KH domain protein, as another translational repressor of *osk* [10].

Activation of *osk* translation also involves several factors. A dsRBD-type RNA-binding protein, Staufen (Stau), is required for translational activation of *osk*, and *oskBRE*⁻, indicating that Stau does not operate simply as a derepressor [5]. Stau is also required for *osk* localization, and both localization and translational regulation of *osk* depend on the region of Stau containing the dsRBDs. However, the precise domains

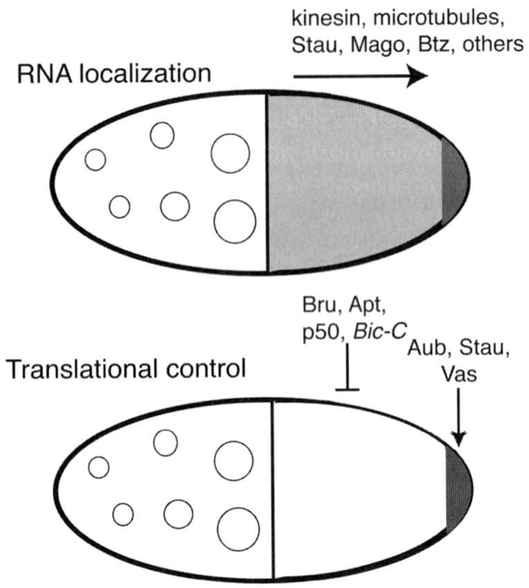

Figure 1 RNA localization and translational control work in concert to restrict the distribution of Oskar (Osk) protein to the pole plasm. *osk* RNA is transcribed in the nurse cells and transported through the ring canals into the early oocyte. Later, *osk* RNA is localized at the posterior pole in a manner that depends on microtubules and several factors, but this process does not result in an absolute restriction of the RNA to the pole plasm. Light gray indicates a low concentration of *osk* RNA, while dark gray represents the high concentration of *osk* RNA in the pole plasm. Translational regulation ensures that protein is not produced from the unlocalized component of *osk* RNA, while the *osk* RNA that is present in the pole plasm is translationally active. Osk protein is depicted in brown.

necessary for localization and translational regulation are distinct [11]. Localization of *osk* mRNA to the pole plasm is probably brought about through kinesin I-mediated microtubule-directed transport, and also requires at least two other proteins, Mago Nashi and Barentsz, which are conserved in evolution but have no previously identified functional domains [12,13].

A key activator of *osk* translation is the Aubergine (Aub) protein, which is related to, but is not the same as, translation initiation factor eIF2C [14]. Aub is a component of polar granules, but, perhaps surprisingly, its ability to activate *osk* translation does not depend on its localization to the pole plasm. Aub cannot work simply by alleviating Bru-mediated repression, because translation of *oskBRE*⁻ remains sensitive to *aub* function. It has been suggested [14] that tenacious binding of localization factors to specific RNAs might itself repress translation. This idea is supported by recent work on ME31B, a DEAD-box helicase that colocalizes in RNPs with several RNAs that localize to the oocyte, and which represses their translation in nurse cells [15]. Aub may then function to alleviate this sort of translational repression.

The RNA helicase Vasa (Vas), which is related to translation initiation factor eIF4A, has also been implicated as regulating *osk* translation [4], but a careful analysis of the temporal sequence of events at the posterior pole of the oocyte indicates that the requirement for Vas is at a later stage of oogenesis than the requirement for Aub [14]. Vas interacts with the essential translation factor eIF5B, originally called dIF2, suggesting that it may activate translation at the step of ribosomal subunit joining [16,17].

Translational Control Targets Nanos to the Pole Plasm

Translational control targets Nanos to the pole plasm, and Nanos establishes the Hunchback gradient by translational control. *nos* mRNA accumulates in the pole plasm but is also present throughout the embryo; only 4% is posteriorly localized and only that localized fraction is translationally active [18]. A region of 90 nucleotides of the *nos* 3′UTR, termed the translational control element (TCE), is predicted to form two extended stem-loops, and mutations abolishing these structures render the element inactive. The loop portion of stem-loop II corresponds to the binding site for the translational repressor Smaug (Smg) [23,24], a protein lacking any previously identified RNA-binding motifs. Smg appears to be the key translational repressor of unlocalized *nos* RNA. Analysis of *nos* in polyribosomal profiles suggests that translational repression of unlocalized *nos* occurs after initiation [25].

Vas has been implicated in activating *nos* translation at the posterior pole, because *nos* transgenes containing either the TCE or the entire 3′UTR of *nos* are not translated in *vas* mutants [20]. When the TCE is deleted, translation no longer depends on *vas*, implying that the role of Vas in activating *nos* translation involves overcoming Smg-mediated repression [19]. Because Smg and Osk interact directly [23], as do Vas and Osk [26], a complex of pole plasm components including Osk and Vas may deactivate Smg on *nos* transcripts that enter the posterior region, leading to derepression of translation.

Nos generates an anterior-to-posterior gradient of the Hb transcription factor by repressing translation of the maternal *hb* transcript in the posterior part of the embryo. To do this, Nos forms a complex with Pumilio (Pum) and the Nos-response element (NRE) sequences within the *hb* 3′UTR [27]. Translational repression of *hb* involves deadenylation promoted by Nos and Pum [28], but can occur in the absence of a poly(A) tail [29]. *hb* repression also requires an association of Nos, Pum, and the NRE with the NHL protein Brain Tumor (Brat) [30]. Nos and Pum have also been implicated in repression of *cyclin B* translation in pole cells [31], and Pum binds to NRE sequences in the maternal *cyclin B* 3′UTR [30]. This function of Nos and Pum, however, is not mediated through Brat, suggesting that the ternary complex of Nos, Pum, and NRE sequences can interact with different cofactors. Proteins related to Pum are very widely conserved in evolution and may have a fundamental role in promoting proliferation of stem cells [32].

Translational Control in the *Drosophila* Nervous System

Although the developing oocyte and preblastoderm embryo are the developmental stages during which translational control is believed to have its most important roles in regulating gene expression, a number of recent studies have implicated translational control mechanisms in later development as well. This is most notable in both the developing and the mature nervous system. For example, mechanosensory bristle development in *Drosophila* requires four successive asymmetric cell divisions, and in the first asymmetric division, one daughter cell (called IIa) gives rise to non-neuronal progeny and the other (called IIb) gives rise to neuronal progeny [33]. A transcriptional repressor, Tramtrack69 (Ttk69), is present only in IIa and is required for specification of the non-neuronal cells, yet Ttk69 mRNA is present at similar levels in both the IIa and IIb cells. This asymmetry of Ttk69 protein distribution is brought about through translational repression of *Ttk69* RNA in IIb by an interaction of the Musashi (Msi) protein with specific elements in its 3′UTR [34]. Msi itself is present in both the IIa and IIb cells, but its activity is inhibited in IIa by Notch activity, indicating that this important intercellular communication pathway is capable of influencing gene expression at the level of translation.

Translational control has also been implicated in synaptic plasticity in the mature nervous system. Aggregates containing essential translation components such as eukaryotic initiation factor 4E (eIF4E), poly(A) binding protein (PABP), and the glutamate receptor subunit II mRNA (GluR-II) accumulate in subsynaptic compartments of the larval neuromuscular junctions of *Drosophila* [35]. Alteration of the dosage of genes encoding these translation components is accompanied by changes in the number of these aggregates and the size of junctions, and with changes in the levels of synaptic proteins such as GluR-II. These results were interpreted as suggesting that localized regulation of translation at the synapse might underlie long-term alternations of neuronal function and connectivity.

Role for Translational Control in Regulating Growth

A critical signaling pathway for regulating growth is that which is activated by insulin and mediated through phosphoinositol 3-kinase (PI3K) and the kinase target of rapamycin (TOR) [36,37]. Downstream targets of insulin include factors that regulate translational activity, including ribosomal protein S6 kinase (S6K) [38] and the eIF4E inhibitor 4E-BP [39]. Genetic manipulations in *Drosophila* that alter the activities of S6K and 4E-BP result in alterations to cell and organismal size [38,39], as does a mutation that blocks phosphorylation of the cap binding protein eIF4E [40]. Thus, the general translation factors may be important targets for growth control signaling pathways.

Translational Repression through MicroRNAs

Two *C. elegans* genes that were discovered on the basis of mutations, *lin-4* and *let-7*, were found to encode very short RNAs of approximately 22 nucleotides that have an antisense sequence to a particular target RNA (lin-14 and let-41, respectively) [41–43]. A recent systematic study [44] has shown that at least 17, and perhaps hundreds, of similar genes encoding microRNAs exist in the *C. elegans* genome, and that numerous such genes are predicted in the *Drosophila* and human genomes as well. Thus microRNA-mediated translational repression could prove to be a widespread mechanism of regulating gene expression.

References

1. Cooperstock, R. L. and Lipshitz, H. D. (2000). RNA localization and translational regulation during axis specification in the *Drosophila* oocyte. *Int. Rev. Cytol.* **203**, 541–566.
2. Johnstone, O. and Lasko, P. (2001). Translational regulation and RNA localization in *Drosophila* oocytes and embryos. *Ann. Rev. Genet.* **35**, 365–406.
3. Ephrussi, A., Dickinson, L. K., and Lehmann, R. (1991). *oskar* Organizes the germ plasm and directs localization of the posterior determinant *nanos*. *Cell* **66**, 37–50.
4. Markussen, F.-H., Michon, A. M., Breitwieser, W., and Ephrussi A. (1995). Translational control of *oskar* generates short OSK, the isoform that induces pole plasm assembly. *Development* **121**, 3723–3732.
5. Kim-Ha, J., Kerr, K., and Macdonald, P. M. (1995). Translational regulation of *oskar* mRNA by Bruno, an ovarian RNA-binding protein, is essential. *Cell* **81**, 403–412.
6. Castagnetti, S., Hentze, M. W., Ephrussi, A., and Gebauer, F. (2000). Control of *oskar* mRNA by Bruno in a novel cell-free system from *Drosophila* ovaries. *Development* **127**, 1063–1068.
7. Lie, Y. S. and Macdonald, P. M. (1999). Translational regulation of oskar mRNA occurs independent of the cap and poly(A) tail in *Drosophila* ovarian extracts. *Development* **126**, 4989–4996.
8. Lie, Y. S. and Macdonald, P. M. (1999). Apontic binds the translational repressor Bruno and is implicated in regulation of *oskar* mRNA translation. *Development* **126**, 1129–1138.
9. Gunkel, N., Yano, T., Markussen, F.-H., Olsen, L. C., and Ephrussi, A. (1998). Localization-dependent translation requires a functional interaction between the 5′ and 3′ ends of *oskar* mRNA. *Genes Dev.* **12**, 1652–1664.
10. Saffman, E. E., Styhler, S., Rother, K., Li, W., Richard, S., and Lasko, P. (1998). Premature translation of *oskar* in oocytes lacking the RNA-binding protein Bicaudal-C. *Mol. Cell. Biol.* **18**, 4855–4862.
11. Micklem, D. R., Adams, J., Grünert, S., and St Johnston, D. (2000). Distinct roles of two conserved Staufen domains in *oskar* mRNA localization and translation. *EMBO J.* **19**, 1366–1377.
12. Micklem, D. R., DasGupta, H., Elliott, H., Gergely, F., Davidson, C., Brand, A., González-Reyes, A., and St Johnston, D. (1997). The *mago nashi* gene is required for the polarisation of the oocyte and the formation of perpendicular axes in *Drosophila*. *Curr. Biol.* **7**, 468–478.
13. van Eeden, F. J. M., Palacios, I. M., Petronczki, M., Weston, M. J. D., and St Johnston, D. (2001). Barentsz is essential for the posterior localization of *oskar* mRNA and colocalizes with it to the posterior pole. *J. Cell Biol.* **154**, 511–523.
14. Harris, A. N. and Macdonald, P. M. (2001). Aubergine Encodes a *Drosophila* polar granule component required for pole cell formation and related to eIF2C. *Development* **128**, 2823–2832.
15. Nakamura, A., Amikura, R., Hanyu, K., and Kobayashi, S. (2001). Me31B silences translation of oocyte-localizing RNAs through the formation of cytoplasmic RNP complex during *Drosophila* oogenesis. *Development* **128**, 3233–3242.
16. Carrera, P., Johnstone, O., Nakamura, A., Casanova, J., Jäckle, H., and Lasko, P. (2000). VASA mediates translation through interaction with a *Drosophila* yIF2 homolog. *Mol. Cell* **5**, 181–187.
17. Pestova, T. V., Lomakin, I. B., Lee, J. H., Choi, S. K., Dever, T. E., and Hellen, C. U. (2000). The joining of ribosomal subunits in eukaryotes requires eIF5B. *Nature* **403**, 332–335.

18. Bergsten, S. E. and Gavis, E. R. (1999). Role for mRNA localization in translational activation but not spatial restriction of *nanos* RNA. *Development* **126**, 659–669.
19. Dahunakar, A. and Wharton, R. P. (1996). The Nanos gradient in *Drosophila* embryos is generated by translational regulation. *Genes Dev.* **10**, 2610–2620.
20. Gavis, E. R., Lunsford, L., Bergsten, S. E., and Lehmann, R. (1996). A conserved 90 nucleotide element mediates translational repression of *nanos* RNA. *Development* **122**, 2791–2800.
21. Smibert, C. A., Wilson, J. E., Kerr, K., and Macdonald, P. M. (1996). Smaug protein represses translation of unlocalized *nanos* mRNA in the *Drosophila* embryo. *Genes Dev.* **10**, 2600–2609.
22. Crucs, S., Chatterjee, S., and Gavis, E. R. (2000). Overlapping but distinct RNA elements control repression and activation of *nanos* translation. *Mol. Cell* **5**, 457–467.
23. Dahunakar, A., Walker, J. A., and Wharton, R. P. (1999). Smaug, a novel RNA-binding protein that operates a translational switch in *Drosophila. Mol. Cell* **4**, 209–218.
24. Smibert, C. A., Lie, Y. S., Shillinglaw, W., Henzel, W. J., and Macdonald, P. M. (1999). Smaug, a novel and conserved protein, contributes to repression of *nanos* mRNA translation *in vitro. RNA* **5**, 1535–1547.
25. Clark, I. E., Wyckoff, D., and Gavis, E. R. (2000). Synthesis of the posterior determinant Nanos is spatially restricted by a novel co-translational regulatory mechanism. *Curr. Biol.* **10**, 1311–1314.
26. Breitwieser, W., Markussen, F.-H., Horstmann, H., and Ephrussi, A. (1996). Oskar protein interaction with Vasa represents an essential step in polar granule assembly. *Genes Dev.* **10**, 2179–2188.
27. Sonoda, J. and Wharton, R. P. (1999). Recruitment of Nanos to *hunchback* mRNA by Pumilio. *Genes Dev.* **13**, 2704–2712.
28. Wreden, C., Verrotti, A. C., Schisa, J. A., Lieberfarb, M. E., and Strickland, S. (1997). Nanos and Pumilio establish embryonic polarity in *Drosophila* by promoting posterior deadenylation of hunchback mRNA. *Development* **124**, 3015–3023.
29. Chagnovich, D. and Lehmann, R. (2001). Poly(A)-independent regulation of maternal hunchback translation in the *Drosophila* embryo. *Proc. Natl. Acad. Sci. USA* **98**, 11359–11364.
30. Sonoda, J. and Wharton, R. P. (2001). *Drosophila* Brain Tumor is a translational repressor. *Genes Dev.* **15**, 762–773.
31. Asaoka-Taguchi, M., Yamada, M., Nakamura, A., Hanyu, K., and Kobayashi, S. (1999). Maternal Pumilio acts together with Nanos in germline development in *Drosophila* embryos. *Nat. Cell Biol.* **1**, 431–437.
32. Wickens, M., Bernstein, D. S., Kimble, J., and Parker, R. (2002). A PUF family portrait: 3′ UTR regulation as a way of life. *Trends Genet.* **18**, 150–157.
33. Jan, Y. N. and Jan, L. Y. (1998). Asymmetric cell division. *Nature* **392**, 775–778.
34. Okabe, M., Imai, T., Kurusu, M., Hiromi, Y., and Okano, H. (2001). Translational repression determines neuronal potential in *Drosophila* asymmetric cell division. *Nature* **411**, 94–98.
35. Sigrist, S. J., Thiel, P. R., Reiff, D. F., Lachance, P. E. D., Lasko, P., and Schuster, C. M. (2000). Postsynaptic translation affects the efficacy and morphology of neuromuscular junctions. *Nature* **405**, 1062–1065.
36. Böhni, R., Riesgo-Escovar, J., Oldham, S., Brogiolo, W., Stocker, H., Andruss, B. F., Beckingham, K., and Hafen, E. (1999). Autonomous control of cell and organ size by CHICO, a *Drosophila* homolog of vertebrate IRS1–4. *Cell* **97**, 865–875.
37. Britton, J. S., Lockwood, W. K., Li, L., Cohen, S. M., and Edgar, B. A. (2002). *Drosophila*'s insulin/PI3-kinase pathway coordinates cellular metabolism with nutritional conditions. *Dev. Cell* **2**, 239–249.
38. Montagne, J., Stewart, M. J., Stocker, H., Hafen, E., Kozma, S. C., and Thomas, G. (1999). *Drosophila* S6 kinase: A regulator of cell size. *Science* **285**, 2126–2129.
39. Miron, M., Verdú, J., Lachance, P. E. D., Birnbaum, M. J., Lasko, P. F., and Sonenberg, N. (2001). The translational inhibitor 4E-BP is an effector of PI(3)K/Akt signaling in *Drosophila. Nat. Cell Biol.* **3**, 596–601.
40. Lachance, P. E. D., Miron, M., Raught, B., Sonenberg, N., and Lasko, P. (2002). Phosphorylation of eukaryotic translation initiation factor 4E is critical for growth. *Mol. Cell. Biol.* **22**, 1656–1663.
41. Lee, R. C., Feinbaum, R. L., and Ambros, V. (1993). The *C. elegans* heterochronic gene lin-4 encodes small RNAs with antisense complementarity to lin-14. *Cell* **75**, 843–854.
42. Wightman, B., Ha, I., and Ruvkun, G. (1993). Posttranscriptional regulation of the heterochronic gene lin-14 by lin-4 mediates temporal pattern formation in *C. elegans. Cell* **75**, 855–862.
43. Slack, F. J., Basson, M., Liu, Z., Ambros, V., Horvitz, H. R., and Ruvkun, G. (2000). The lin-41 RBCC gene acts in the *C. elegans* heterochronic pathway between the *let-7* regulatory RNA and the LIN-29 transcription factor. *Mol. Cell* **5**, 659–669.
44. Lee, R. C. and Ambros, V. (2001). An extensive class of small RNAs in *Caenorhabditis elegans. Science* **294**, 862–864.

CHAPTER 320

332

Role of Alternative Splicing During the Cell Cycle and Programmed Cell Death

Chanseok Shin and James L. Manley
Department of Biological Sciences, Columbia University, New York, New York

Introduction

Alternative splicing plays an important role in gene expression in metazoan organisms. By the mid-1980s, it was estimated that 5% of genes in higher eukaryotes might be subject to alternative splicing [1]. However, recent genome-wide analyses involving computational identification of alternative splicing frequencies indicate that 35–60% of human genes give rise to at least one alternatively spliced form [2–6]. As an extreme example, the *Drosophila* DSCAM gene has the potential of generating more than 38,000 isoforms by alternative splicing [7]. These findings suggest a significant role for alternative splicing in both the generation of protein diversity and also the control of gene expression.

Pre-mRNA splicing occurs in a macromolecular complex, the spliceosome, which consists of small nuclear ribonucleoprotein particles (snRNPs) and a large number of non-snRNP protein factors [8]. Regulation of pre-mRNA splicing provides the main posttranscriptional control of gene expression in higher eukaryotic cells [9]. A combination of genetic and biochemical approaches has allowed progress in identifying both *cis*-acting elements on the pre-mRNA and *trans*-acting protein factors. Among the *cis*-acting elements, exon splicing enhancers stimulate splicing of adjacent upstream introns. SR proteins, a group of *trans*-acting protein factors, play key roles in the activation of splicing enhancers [10,11]. Negative regulatory sequences also exist, and seem to interact with hnRNP proteins as well as SR proteins, and also participate in splicing regulation. Therefore, splicing control can be achieved by coordination between positive and negative regulatory elements in the pre-mRNA and the variation or activity of individual SR proteins in cells [12]. Phosphorylation of splicing factors has been demonstrated to influence pre-mRNA splicing and thus likely also contributes to splicing regulation [13]. In this chapter, we summarize some of the progress made in the past years on pre-mRNA splicing regulation, focusing on genes involved in programmed cell death (apoptosis) and cell cycle control.

Apoptosis and Splicing

A remarkable feature of genes encoding proteins that function in the apoptotic, or programmed cell death (PCD), pathway is that many of them are transcribed into mRNA precursors that can be alternatively spliced to produce proteins with opposite effects on programmed cell death, either facilitating or preventing apoptosis (see Table I for a partial list of such genes; also [14]). The fact that such splicing patterns are commonplace strongly suggests that splicing control plays a significant, physiologically relevant role in affecting programmed cell death, at least in some cases. Although conclusive evidence establishing this is not yet available, a number of studies point in this direction, and they suggest that changes in the activity of splicing factors, such as SR proteins, may be involved.

Table I Alternatively Spliced Products with Opposite Effects on Programmed Cell Death

Receptors	Fas (Apo-1, CD95)
Bcl-2	Bcl-2 Bcl-x Bcl-w Bax Bim Mcl-1
Ced-4	Ced-4(Apaf)
Caspases	Caspase-1 (ICE) Caspase-2 (Ich-1) Caspase-3 Caspase-4 Caspase-6 (Mch-2) Caspase-7 (Mch-3, ICE-LAP3, CMH-1) Caspase-10 (Mch-4)

Evidence that inactivation of an SR protein, ASF/SF2, can lead to apoptosis has recently been described. Specifically, a genetically engineered chicken DT40 B-cell line was constructed such that expression of ASF/SF2 could be strongly repressed by tetracycline [15]. Following depletion of ASF/SF2, the cells stop growing and undergo cell death via an apoptotic pathway ([16]; J. Wang, X. Li, and J. L. Manley, in preparation). Although the exact mechanism by which ASF/SF2 depletion leads to PCD is unknown, it may involve alteration of the splicing pattern of at least one transcript encoding an apoptotic regulator: Bcl-X. Production of mRNA encoding the anti-apoptotic Bcl-X_L is reduced while accumulation of the pro-apoptotic Bcl-X_S isoform increases. Although the changes in splicing are modest, and Bcl-X_L remains the predominant isoform, this change may be sufficient to at least contribute to, or facilitate, PCD in ASF/SF2-depleted cells. Consistent with this, overexpression of a Bcl-X_L cDNA significantly delays onset of cell death following ASF/SF2 depletion (J. Wang, X. Li, and J. L. Manley, in preparation).

The above-described experiments argue that changes in the concentration of an SR protein can alter splicing of apoptotic regulators and contribute to PCD. But do similar situations occur naturally in response to apoptotic stimuli? Several studies indicate that significant changes in SR protein accumulation and/or phosphorylation can occur during apoptosis, for example, induced by Fas ligand, γ-irradiation or UV light [17–19]. These experiments provided evidence that SR proteins become hyperphosphorylation or hypo-hyperphosphorylated in response to these various stimuli and can then be found in complexes, with U1 snRNPs or even snoRNPs, that did not exist prior to apoptosis induction. These changes could have significant affects on SR protein activity. For example, hyperphosphorylation and hypophosphorylation are both known to reduce SR protein activity in *in vitro* splicing assays [20] and hyperphosphorylation can also enhance or stabilize interactions with U1 snRNP [21]. Thus either changes in hyperphosphorylation *per se* or the formation of stable complexes could in effect reduce the concentration of available, active SR proteins, leading to changes in alternative splicing, such as those described earlier for Bcl-X, thereby inducing or enhancing apoptosis.

Another example of slicing regulation involves splicing of the pre-mRNA encoding the Fas receptor and the RNA dividing protein TIA-1 [22]. Alternative splicing of the Fas receptor pre-mRNA results in a membrane-bound form, which promotes PCD, or a soluble form, which inhibits apoptosis. TIA-1 appears to promote apoptosis by activating production of membrane-bound isoform of the Fas receptor. TIA-1 was found to bind to uridine-rich sequences adjacent to the 5′ splice site of exon 5 of the Fas pre-mRNA and recruit U1snRNP to the 5′ splice site, activate exon 6 inclusion, and thereby promote accumulation of the membrane-bound Fas receptor. Depletion of TIA-1 in a fibroblast cell line or deletion of the poly(U) tract adjacent to the 5′ splice site leads to skipping of exon 6, and TIA-1 overexpression increases exon 6 inclusion. Although the physiological relevance of these observations has not been confirmed yet, TIA-1 might promote apoptosis by activating a weak 5′ splice site in the Fas pre-mRNA. This mechanism for regulating alternative splicing by TIA-1 has been shown in other cases, such as the fibroblast growth factor receptor-2 pre-mRNA [23], male-specific lethal-2 pre-mRNA in *Drosophila* [24], and the meiosis-specific MER2 pre-mRNA gene in *Saccharomyces cerevisiae*, by Nam8p, the yeast homolog of TIA-1 [25].

Cell Cycle and Splicing Regulation

Transitions through the cell cycle are highly regulated by the activities of cyclin-dependent kinases (CDKs). Cyclins, which associate with and regulate the CDKs, accumulate only at specific times during the cell cycle. In addition to this temporal control, which involves ubiquitin-mediated proteolysis of the cyclin, CDK activity can be regulated by phosphorylation, and physical interaction with CDK inhibitor proteins (reviewed in [26]).

Links between pre-mRNA splicing and the cell cycle are likely, but have not been fully investigated. In yeast, both *Schizosaccharomyces pombe* and *S. cerevisiae*, several genes initially isolated by virtue of cell cycle phenotypes have been found to encode proteins involved in splicing, and mutations in some known splicing factor genes display cell cycle phenotypes (reviewed in [27]). But little is known about the functional significance of these genetic interactions.

One aspect of cell cycle control is the silencing of gene expression that occurs during mitosis in metazoan cells. Recent years have provided insights into how this occurs and into the underlying mechanisms involved, which, for example, involve phosphorylation of key components of the transcription [28–30] and polyadenylation [31] machineries, and dephosphorylation of a translation initiation factor required for cap-dependent translation [32]. In contrast to all these processes, it is unclear what happens to the splicing machinery during M phase. Our recent studies [32a] have provided insights into this issue. We first addressed the question of whether splicing is inhibited in M-phase cells, and found that it was indeed blocked in whole-cell extracts prepared from nocadazol arrested cells. Intriguingly, though, significant inhibition was only detected in high-salt extracts, suggesting that a mitosis-specific factor associated with insoluble cellular structures is required for inhibition. A candidate is a novel SR protein, SRp38, that we have recently characterized. SRp38 is unlike other SR proteins in that it is unable to activate splicing in *in vitro* assays (reviewed in [10]). Rather, SRp38 seems to be a dedicated splicing repressor and, importantly, its activity is strongly activated by dephosphorylation. Remarkably, SRp38 becomes significantly dephosphorylated during M phase, and is extracted from cell extracts only at the same high-salt concentrations required to detect M-phase-specific repression. These studies have shown that splicing, like other steps in gene expression, is repressed during mitosis. Analogous to the mechanism involved in translational inhibition [32], splicing inhibition appears to involve dephosphorylation, specifically of the SR protein SRp38.

Recently, two cdk-related kinases have been described that may function in splicing. Ko *et al.* [33] described a novel protein kinase as *C*dc2-*r*elated *k*inase with an arginine/serine rich (*RS*) domain (CrkRS). CrkRS has extensive proline-rich regions that match the consensus for SH3 and WW domain binding sites and, most significantly, an N-terminal RS domain. RS domains, as exemplified in SR proteins, are protein–protein interactions characteristic of splicing factors. CrkRS colocalizes in nuclear speckles with other splicing factors, and immunoprecipitates of CrkRS were found to phosphorylate CrkRS itself, exogenously added ASF/SF2, and the GST-tagged C-terminal domain of the RNA polymerase II largest subunit (CTD). CrkRS is hyperphosphorylated in mitosis and hypophosphorylated in interphase. The function of CrkRS is currently unknown, but it has the potential to function in linking transcription and splicing with the cell cycle.

In keeping with the existence of an RS domain-containing cdk, a recent human database search has identified an RS domain-containing cyclin, called CYCLIN L [34]. Recently, the mouse homolog of CYCLIN L, called ania-6, has been described. Intriguingly, ania-6 is rapidly induced in the adult striatum by cocaine or direct dopamine stimulation [35]. Different types of neurotransmitter stimulation cause selective induction of distinct ania-6 isoforms (Ania-6a^{60} and Ania-6a^{25}), through alternative splicing. The longer protein, Ania-6a^{60}, contains a C-terminal RS domain, whereas the shorter isoform, Ania-6a^{25}, does not. Ania-6a^{60} is localized to nuclear speckles, whereas Ania-6a^{25} is not localized to the nucleus. Perhaps significantly, ania-6 sequences are more similar to invertebrate homologs of cyclins than to other mammalian cyclins, and closer to cyclins associated with RNA pol II (such as cyclin K or T) than to cyclins directly involved in cell cycle control (such as cyclin A1). Immunoprecipitation studies with FLAG-Ania-6a^{60} indicate that the protein is associated specifically with the hyperphosphorylated form of RNA polymerase (RNAP IIO), the splicing factor SC35, and the p110 PITSLRE cdk. The idea that RNAP II, and specifically the CTD, participates in mRNA processing is now well accepted (reviewed in [36]) and RNAP IIO has been shown to stimulate pre-mRNA splicing *in vitro* [37]. It seems possible that Ania-6a^{60}, or CYCLIN L, as well as crKRS, play roles in coupling transcription with splicing. How, and if, this coupling functions in regulating expression of specific genes, and whether such controls might be cell cycle related, are interesting questions for the future.

Acknowledgments

We are grateful to X. Li for preparing the table. Work from the authors' lab was supported by NIH grant R37 GM48259.

References

1. Sharp, P. A. (1994). Split genes and RNA splicing. *Cell* **77**, 805–815.
2. Mironov, A. A., Fickett, J. W., and Gelfand, M. S. (1999). Frequent alternative splicing of human genes. *Genome Res.* **9**, 1288–1293.
3. Croft, L., Schandorff, S., Clark, F., Burrage, K., Arctander, P., and Mattick, J. S. (2000). ISIS, the intron information system, reveals the high frequency of alternative splicing in the human genome. *Nat. Genet.* **24**, 340–341.
4. Kan, Z., Rouchka, E. C., Gish, W. R., and States, D. J. (2001). Gene structure prediction and alternative splicing analysis using genomically aligned ESTs. *Genome Res.* **11**, 889–900.
5. Modrek, B., Resch, A., Grasso, C., and Lee, C. (2001). Genome-wide detection of alternative splicing in expressed sequences of human genes. *Nucleic Acids Res.* **29**, 2850–2859.
6. Brett, D., Pospisil, H., Valcarcel, J., Reich, J., and Bork, P. (2002). Alternative splicing and genome complexity. *Nat. Genet.* **30**, 29–30.
7. Schmucker, D., Clemens, J. C., Shu, H., Worby, C. A., Xiao, J., Muda, M., Dixon, J. E., and Zipursky S. L. (2000). *Drosophila* Dscam is an axon guidance receptor exhibiting extraordinary molecular diversity. *Cell* **101**, 671–684.
8. Burge, C. B., Tuschl, T., and Sharp, P. A. (1999). Splicing of precursors to mRNAs by the splicrosomes. In *RNA World II*, pp. 525–560. Cold Spring Harbor Laboratory Press, Cold Spring Harbor, New York.
9. Grabowski, P. J. and Black, D. L. (2001). Alternative RNA splicing in the nervous system. *Prog. Neurobiol.* **65**, 289–308.
10. Manley, J. L. and Tacke, R. (1996). SR proteins and splicing control. *Genes Dev.* **10**, 1569–1579.
11. Graveley, B. R. (2000). Sorting out the complexity of SR protein functions. RNA **6**, 1197–1211.
12. Smith, C. W. J. and Valcárcel, J. (2000). Alternative pre-mRNA splicing: The logic of combinatorial control. *TIBS* **25**, 381–388.
13. Misteli, T. (1999). RNA splicing: What has phosphorylation got to do with it? *Curr. Biol.* **9**, R198-R200.
14. Jiang, Z-H. and Wu, J. Y. (1999). Alternative splicing and programmed cell death. *Proc. Soc. Exp. Biol. Med.* **220**, 64–72.
15. Wang, J., Takagaki Y., and Manley J. L. (1996). Targeted disruption of an essential vertebrate gene: ASF/SF2 is required for cell viability. *Genes Dev.* **10**, 2588–2599.
16. Wang, J Xiao, S. H., and Manley J. L. (1998). Genetic analysis of the SR protein ASF/SF2: Interchangeability of RS domains and negative control of splicing. *Genes Dev.* **12**, 2222–2233.
17. Utz, P. J., Hottelet, M., van Venrooij W. J., and Anderson, P. (1998). Association of phosphorylated serine/arginine (SR) splicing factors with the U1-small ribonucleoprotein (snRNP) autoantigen complex accompanies apoptotic cell death. *J. Exp. Med.* **187**, 547–560.

18. Overzet, K., Gensler T. J., Kim, S. J., Geiger, M. E., Venrooij, W. J., Pollard, K. M., Anderson, P., and Utz, P. J. (2000). Small nuclear RNP scleroderma autoantigens associate with phosphorylated serine/arginine splicing factors during apoptosis. *Arthr. Rheum.* **43**, 1327–1336.
19. Chalfant, C. E., Ogretmen, B., Galadari, S., Kroesen, B. J., Pettus, B. J., Hannun, Y. A. (2001). FAS activation induces dephosphorylation of SR proteins; dependence on the *de novo* generation of ceramide and activation of protein phosphatase 1. *J. Biol. Chem.* **276**, 44848–44855.
20. Prasad, J., Colwill, K., Pawson, T., and Manley, J. L. (1999). The protein kinase Clk/Sty directly modulates SR protein activity: Both hyper- and hypophosphorylation inhibit splicing. *Mol. Cell. Biol.* **19**, 6991–7000.
21. Xiao, S-H. and Manley, J. L. (1997). Phosphorylation of the ASF/SF2 RS domain affects both protein-protein and protein-RNA interactions and is necessary for splicing. *Genes Dev.* **11**, 334–344.
22. Forch, P., Puig, O., Kedersha, N., Martinez, C., Granneman, S., Seraphin, B., Anderson, P., and Valcarcel, J. (2000). The apoptosis-promoting factor TIA-1 is a regulator of alternative pre-mRNA splicing. *Mol. Cell.* **6**, 1089–1098.
23. Del Gatto-Konczak, F., Bourgeois, C. F., Le Guiner, C., Kister, L., Gesnel, M. C., Stevenin, J., and Breathnach, R. (2000). The RNA-binding protein TIA-1 is a novel mammalian splicing regulator acting through intron sequences adjacent to a 5′ splice site. *Mol. Cell Biol.* **20**, 6287–6299.
24. Forch, P., Merendino, L., Martinez, C., and Valcarcel, J. (2001). Modulation of msl-2 5′ splice site recognition by Sex-lethal. *RNA* **7**, 1185–1191.
25. Púig, O., Gottschalk, A., Fabrizio, P., and Seraphin, B. (1999). Interaction of the U1 snRNP with nonconserved intronic sequences affects 5′ splice site selection. *Genes Dev.* **13**, 569–580.
26. Nigg, E. A. (1995). Cyclin-dependent protein kinases: Key regulators of the eukaryotic cell cycle. *Bioessays* **17**, 471–480.
27. Burns, C. G. and Gould, K. L. (1999). Connection between pre-mRNA processing and regulation of the eukaryotic cell cycle, in S. L. Chew, Ed., *Posttranscriptional Regulation of Gene Expression and Its Importance to the Endocrine System,* Vol. 25, pp. 59–82. Karger, Basel, Switzerland.
28. Segil, N., Guermah, M., Hoffmann, A., Roeder, R. G., and Heintz, N. (1996). Mitotic regulation of TFIID: Inhibition of activator-dependent transcription and changes in subcellular localization. *Genes Dev.* **10**, 2389–2400.
29. Akoulitchev, S. and Reinberg, D. (1998). The molecular mechanism of mitotic inhibition of TFIIH is mediated by phosphorylation of CDK7. *Genes Dev.* **12**, 3541–3550.
30. Long, J. J., Leresche, A., Kriwacki, R. W., and Gottesfeld, J. M. (1998). Repression of TFIIH transcriptional activity and TFIIH-associated cdk7 kinase activity at mitosis. *Mol. Cell Biol.* **18**, 1467–1476.
31. Colgan, D. F., Murthy, K. G., Prives, C., and Manley, J. L. (1996). Cell-cycle related regulation of poly(A) polymerase by phosphorylation. *Nature* **384**, 282–285.
32. Pyronnet, S., Dostie, J., and Sonenberg, N. (2001). Suppression of cap-dependent translation in mitosis. *Genes Dev.* **15**, 2083–2093.
32a. Shin, C. and Manley, J. L. (2002). The SR protein SRp38 represses splicing in M phase cells. *Cell* **111**, 407–417.
33. Ko, T. K., Kelly, E., and Pines, J. (2001). CrkRS: A novel conserved Cdc2-related protein kinase that colocalizes with SC35 speckles. *J. Cell. Sci.* **114**, 2591–2603.
34. Boucher, L., Ouzounis, C. A., Enright, A. J., and Blencowe, B. J. (2001). A genome-wide survey of RS domain proteins. *RNA* **7**, 1693–1701.
35. Berke, J. D., Sgambato, V., Zhu, P. P., Lavoie, B., Vincent, M., Krause, M., and Hyman, S. E. (2001). Dopamine and glutamate induce distinct striatal splice forms of Ania-6, an RNA polymerase II-associated cyclin. *Neuron* **32**, 277–287.
36. Hirose, Y. and Manley, J. L. (2000). RNA polymerase II and the integration of nuclear events. *Genes Dev.* **14**, 1415–1429.
37. Hirose, Y., Tacke, R., and Manley, J. L. (1999). Phosphorylated RNA polymerase II stimulates pre-mRNA splicing. *Genes Dev.* **13**, 1234–1239.

CHAPTER 321

NF90 Family of Double-Stranded RNA-Binding Proteins: Regulators of Viral and Cellular Function

Trevor W. Reichman and Michael B. Mathews
*Department of Biochemistry and Molecular Biology,
New Jersey Medical School and the Graduate School of Biomedical Sciences,
University of Medicine and Dentistry of New Jersey,
Newark, New Jersey*

Summary

Double-stranded RNA-binding proteins in the nuclear factor 90 (NF90) family participate in the regulation of gene expression at both transcriptional and translational levels. Well-known members of the family are NF90 itself and NF110, which differ from one another at their C termini and are encoded by the ILF3 gene. We propose a uniform nomenclature for these proteins, many of which have been given alternative names. Related proteins, SPNR/p74 and ZFR, are encoded by distinct genes. The NF90 family proteins contain two tandem copies of the double-stranded RNA-binding motif, flanked by regulatory regions. Present in humans as well as other mammals and *Xenopus*, NF90 family members associate with nucleic acids, both DNA and RNA, and with several proteins including nuclear factor 45 (NF45) with which they copurify. Complexes have been detected containing NF45 as well as other proteins involved in signal transduction, DNA repair, transcription, splicing, and translation. NF110 is more effective than NF90 in activating gene expression at the transcriptional level, and NF45 exerts a modulatory function. In its translational role, NF90 interacts directly with mRNAs as well as with the protein kinase PKR, which phosphorylates initiation factor eIF2. Members of the NF90 family appear to coordinate gene expression by serving as flexible adaptors that integrate cellular signals via recognition of both proteins and structured nucleic acids.

Introduction

Highly structured and double-stranded RNAs (dsRNAs) transmit a multiplicity of signals within cells. In many cases, the signal is transduced by dsRNA-binding proteins containing a motif known as the dsRNA-binding motif (dsRBM, [1]). Although more than 26 dsRBM-containing proteins have been identified in the human genome, the cellular functions of many of these proteins have not yet been fully deciphered [2]. One such group of dsRBM proteins is the NF90 family. NF90 was originally isolated with its heteromeric partner NF45 as a complex that binds to the antigen recognition response element (ARRE-2) present in the interleukin 2 (IL-2) promoter [3,4]. Independently, partial cDNAs were cloned as M-phase phosphorylated proteins [5] and homologs named 4F.1 and 4F.2 were characterized as dsRNA-binding factors present in *Xenopus* extracts [6]. Over the years, several NF90-related proteins have been found in different species, using various techniques, and were named according to the screen through which they were identified. Table I lists the NF90 homologs reported, indicating how they were isolated, the names assigned, and the functions putatively ascribed. An underlying theme that has emerged implicates the proteins as regulators of both viral and cellular gene expression pathways, particularly at the transcriptional and translational levels.

Table I Isolation of NF90 Homologues

Name	Species	Source	Identification	Putative function	Reference
NF90	Human	Jurkat cells	ARRE-2 affinity chromatography	Regulator of activated transcription in T-cells	Kao, P.N., *et al.*, 1994
NF90	Human	293 cells	Adenovirus VA RNA_{II}-specific interacting protein	None assigned	Liao, H.-J., *et al.*, 1998
NF90	Human	Placenta	Component of DNA-PK complex	Facilitated interaction of DNA-PK/Ku complex with dsDNA	Ting, N.S.Y., *et al.*, 1998
NF90	Human	K562 cells	Autoantibodies from pristane treated mice	None assigned	Satoh, M., *et al.*, 1999
NF90	Human	HeLa cells	Interacts with encapsidation signal ε of hepatitis B virus (HBV) RNA	Facilitates binding of HBV polymerase to RNA	Shin, H.J., *et al.*, 2002
NF90	Bovine	Brain	DHS-22 binding protein	Repressor of HLA-DRα gene expression	Sakamoto, S., *et al.*, 1999
MPP4	Human	HeLa and MOLT4	MPM2 reactive protein	None assigned	Matsumoto-Taniura, N., *et al.*, 1996
DRBP76	Human	HeLa cells	Poly (I:C) chromatography and PKR-interacting protein in yeast two hybrid	Substrate of PKR	Patel, R.C., *et al.*, 1999
DRBP76 DRBP76α DRBP76δ ILF3	Human	SK-MeL-28 melanoma cells	Degenerate primers to Rel similarity domain (RSD) of NFAT	Regulator of melanotransferrin gene	Duchange, N., *et al.*, 2000
NFAR1/2	Human	Jurkat cells	PKR-interacting protein in yeast two hybrid	Enhancer of SV40 transcription in mammalian cells	Saunders, L. R., *et al.*, 2001
TCP80	Human	Liver	Interacts with GCase mRNA	Repressor of translation of GCase and other cellular RNAs	Xu, Y.-H. and Grabowski, G.A., 1999
ILF3	Mouse	Male germ cells	Library screen using Prml 3′UTR	RNA metabolism	Buaas, F.W., *et al.*, 1999
ILF3	Mouse	10d embryo	Exportin-5-interacting protein in yeast two-hybrid	None assigned	Brownawell, A.M. and Macara, I.G., 2002
ILF3	Rat	FAO cells	PRMT1-interacting protein in yeast two-hybrid	Substrate and activator of PRMT1	Tang, J., *et al.*, 2000
4F.1/4F.2	Xenopus	Ovary	Library screen using dsRNA	None assigned	Bass, B.A., *et al.*, 1994
CBTF122/ CBTF98	Xenopus	Oocyte extract	Component of CCAAT box binding factor (CBTF)	Regulator of GATA-2 transcription	Orford, R.L., *et al.*, 1998
SPNR	Mouse	Male germ cells	Library screen using Prml 3′UTR	Spermatid RNA metabolism	Schumacher, J. M., *et al.*, 1995
p74	Rat	Smooth muscle	RNA probe	None assigned	Coolidge, C.J., and Patton, J.G., 2000
ZFR	Mouse	Male germ cells	Library screen using Prml 3′UTR	None assigned	Meagher, M. J., *et al.*, 1999

ARRE-2 = Antigen Recognition Response Element-2
CBTF = CCAAT Box Transcription Factor
DRBP76 = Double-stranded RNA Binding Protein 76 kDa
GCase = Acid β-Glucosidase
ILF3 = Interleukin Enhancer Binding Factor 3
MPM2 = Monoclonal antibody reactive to a set of related M-phase phosphorylation sites
MPP4 = M-Phase Phosphoprotein 4
NFAR = Nuclear Factor Associated with dsRNA
SPNR = Spermatid Perinuclear RNA-binding Protein
TCP80 = Translational Control Protein 80 kDa

Members of the NF90 Protein Family

Homologs have been identified in *Xenopus* and three mammals, including humans, and their relationships are summarized in Table II. NF90 is the prototype of a class of human dsRNA-binding proteins generated by the alternate splicing of transcripts from the interleukin enhancer binding factor 3 (ILF3) gene locus located at 19p13 [7,8]. Two regions of the gene undergo alternate splicing events, resulting in the production of at least six unique predicted protein products. Two major species with apparent molecular weights of 90 and 110 kDa are detected by Western blotting. An alternate splicing event with major consequences occurs at the C terminus of the protein coding sequence following exon 17. The 90-kDa species, NF90 (also known as DRBP76 and NFAR1), results from the splicing of exon 17 to exon 18. Two additional cDNAs, DRBP76α and DRBP76δ, are generated via the use of alternate 3′ acceptor sites. These three species have been identified at the mRNA level. Although the corresponding protein products have yet to be conclusively identified *in vivo*, the DRBP76, DRBP76α, and DRBP76δ proteins are predicted to have 702, 694, and 690 amino acids, respectively, and masses of 75–76 kDa [8–10].

Nuclear factor 110 (NF110, also called ILF3 and NFAR2) is generated by the splicing of exon 17 to exons 19, 20, and 21. The resulting slower migrating protein (~110 kDa) is predicted to have 894 amino acids and a mass of 96 kDa [8,9]. Peptides from the unique C terminus of NF110 have been detected [11]. Two mRNA species encoding the NF110 protein were identified in melanoma cells, differing in the length of the 3′ untranslated regions (UTRs) possibly because of alternative polyadenylation site usage [8]. The consequence of the presence or absence of the extended 3′UTR has not been described.

A second alternate splicing event occurs at the exon 13–14 junction and stems from the usage of two different 3′ acceptor sites. Exon 14a contains an additional 12 bp, which leads to the addition of four amino acids (NVKQ) between the two dsRBMs, giving rise to NF90b and NF110b. The corresponding isoforms from which the insert is absent are termed NF90a and NF110a. The 4-amino-acid insert has been identified in both NF90 and NF110 at the mRNA level and in NF90 at the protein level [8,11]. The presence of the insert increases the ability of NF110 to stimulate gene expression in a transfection assay [11,12].

The originally isolated cDNA encoding NF90, referred to as NF90c, appears to contain an additional two nucleotides (TC) after position 1798 of the NF110 mRNA, leading to a frame shift that results in premature termination of the protein. The origin of this variant cDNA, whether it is an artifact of cloning or a bona fide naturally occurring mutant or variant form, remains to be determined. TCP80 is another unique NF90-related cDNA, isolated from a human liver cDNA library [13]. The precise origin of TCP80 by alternative splicing is unknown but it appears to result from the usage of a unique 5′ acceptor site in exon 17 and a novel 3′ acceptor site in exon 21 (Genbank entry AF202445). The expression of this protein isoform in mammalian cells also remains to be verified.

Two additional homologs of NF90, SPNR/p74 and ZFR, are encoded by distinct gene loci and are not alternative splice products of the ILF3 gene. SPNR/p74 was picked up in biochemical screens for RNA-interacting proteins [14,15]. Unlike NF90, which is predominantly nuclear, SPNR binds to microtubules and localizes to the manchette, a spermatid-specific microtubular array [15,16]. SPNR and NF90 do not serve redundant functions in mice, since $SPNR^{-/-}$ mice have neurological, spermatogenic, and sperm morphological abnormalities [17]. ZFR contains three C_2H_2 zinc fingers in its N terminus, followed by a C terminus that is highly homologous to the N terminus of NF90 (amino acids 1–316). The cDNA encoding this protein was identified in an expression library by screening with the protamine 1 (Prm1) 3′UTR [18]. Although the function of ZFR is unknown, it is required for the embryonic development of mice [19].

Table II Comparison of NF90 Homologues from Four Different Species

Species	NF90a/b	NF110a/b	Unique cDNAs
Homo sapiens	DRBP76a DRBP76b NFAR1	ILF3a ILF3b NFAR2	NF90c TCP80 DRBP76α DRBP76δ
Mus musculus	*NI*	ILF3	
Rattus norvegicus	*NI*	ILF3	
Xenopus laevis	4F.1 CBTF[98]	4F.2 CBTF[122]	

NI=cDNA not identified
a or b suffix indicates absence or presence of the NVKQ insertion
For references, see text.

Domain Structure of NF90 Family Proteins

Inspection of the primary sequence of NF90 reveals homology with other proteins and the existence of several known motifs (Fig. 1). Prominent are the two type-A dsRBMs present in the center of the protein between amino acids 404–465 and 526–592. An additional short nucleic acid binding motif, the RGG motif, lies between residues 640 and 659 in NF90, NF110 and their homologs except for NF90c ZFR, and p74/SPNR. This motif was previously identified in tandem with another nucleic acid binding domain in nucleolin, the RNA recognition motif (RRM), with which it worked in concert to enhance nucleic acid binding activity [20]. When expressed in isolation, the RGG domain displays avid affinity for single-stranded (ss)RNA and RNA structures such as G-quartets [21,22]. This region also includes several asymmetric arginine dimethylation consensus sites (RxR and RGG), known targets for protein arginine methyltransferases [23]. A functional bipartite nuclear localization signal is located from amino acids 370 to 394, just upstream of the dsRBMs, and is required for strong nuclear localization of NF90 [24].

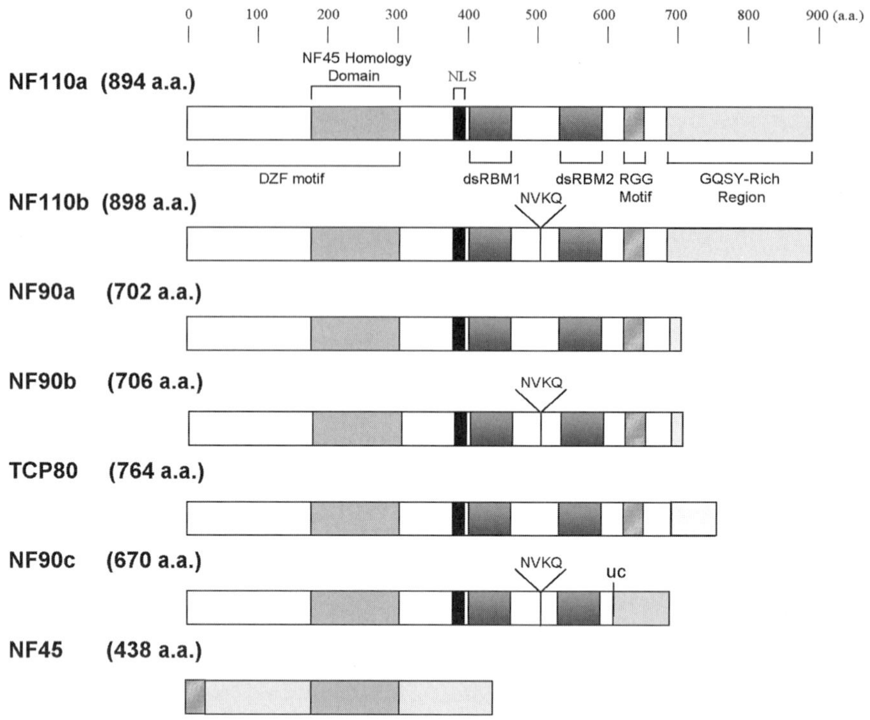

Figure 1 Domain structure of the NF90 homologs. Schematic representation of the proteins encoded by the human ILF3 gene compared with NF45, the heteromeric partner of NF90. NVKQ is the tetrapeptide insertion present between the dsRBMs in the b isoforms, and uc indicates the two bases inserted in the NF110b RNA, which encodes the NF90c protein. DRBP76α and DRBP76δ are not depicted but are identical to NF90 except for the remote C-terminal region. a.a. = amino acid.

NF90 shares extensive homology with its heterodimeric partner NF45. This motif, putatively titled the NF90/NF45 homology region, is located from amino acids 136 to 337. A similar but more extended motif, the DZF motif (acronym for domain in DSRM or ZnF C2H2 domain containing proteins), encompasses the entire N terminus of NF90. This motif is also found in the NF90-related protein, ZFR [18]. In NF90, the NF90/NF45 homology region is the binding site for NF45 and is thought to play a regulatory role by inhibiting the transcriptional activation function of NF90's C terminus. Binding of NF45 to NF90 relieves the inhibitory effect of this domain on the C terminus of the protein, an effect that can be mimicked by deletion of the N terminus from NF90 [24]. It is tempting to speculate that the DZF motif plays a similar regulatory role in ZFR, and that NF45 acts as a general regulator of proteins that contain this motif.

The NF110 isoforms have an additional motif encoded by exons 19, 20, and 21. Their alternatively spliced C-terminal region contains a motif rich in glycine, glutamine, serine, and tyrosine, which we have provisionally titled the GQSY domain. This domain shares limited homology with regions found in other RNA-binding proteins including RNA helicase A (RHA), hnRNP A, and the EWS (Ewing sarcoma) family of proteins that includes EWS, TLS/FUS, and the TATA box protein (TBP)-associated factor $TAF_{II}68$. The GQSY domains of other proteins have been reported to have both nucleic acid binding [25] and protein–protein interaction properties [26]. In NF110, this domain modulates dsRNA binding and increases the protein's ability to enhance gene expression in transient expression assays [9,12]. As discussed later, NF110's GQSY region was found to interact with TLS/FUS and SMN in a yeast two-hybrid assay [9].

Proteins That Interact with NF90

Association of NF90 family members with several different proteins and protein complexes has been reported, suggestive of participation in various cellular processes (Fig. 2). The *Xenopus* orthologs of NF90 and NF110, $CBTF^{98}$ and $CBTF^{122}$ (also known as 4F.1 and 4F.2), were isolated as part of a CCAAT box transcription factor complex that binds to the CCAAT box element present in the GATA-2 promoter. In the early stages of *Xenopus* development, this protein complex is localized in the cytoplasm, anchored to a large pool of untranslated maternal mRNAs via the dsRBMs of CBTF. Prior to the midblastula transition (MBT), the cytoplasmic RNA is degraded and the complex translocates to the nucleus, where it presumably activates transcription of its target genes [27,28].

NF90, NF110, and p74 all interact with and act as substrates for PKR, a critical regulator of translation in mammalian cells [9,10,14,29,30]. PKR's translational function is exerted via the phosphorylation of translation initiation factor eIF2,

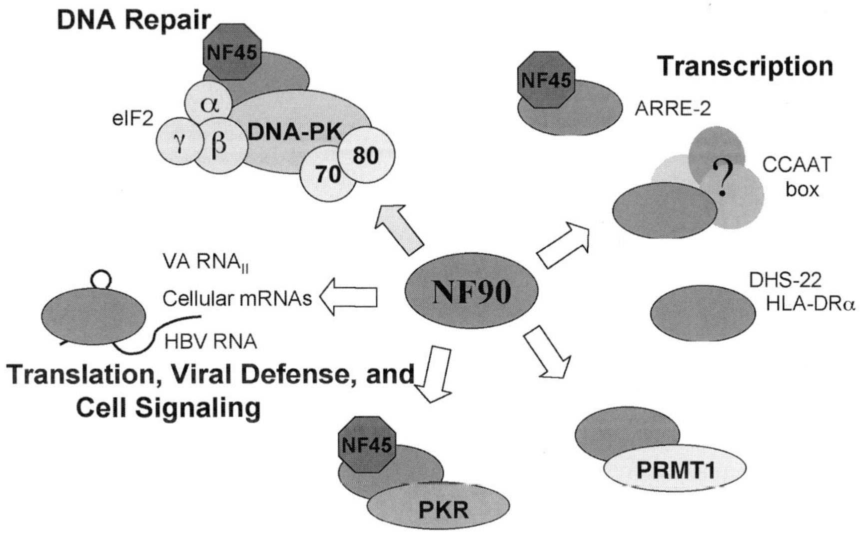

Figure 2 NF90 family proteins: functions and complexes. For details, see text.

resulting in the inhibition of protein synthesis. In an *in vitro* assay, NF90c blocked the activity of PKR, presumably by sequestering dsRNA [30]. NF110 also interacts with and is a substrate for the protein arginine methyltransferase 1 PRMT1. Furthermore, NF110 stimulated the activity of PRMT1 in an *in vitro* methylation assay [23]. NF110 was also isolated as an exportin 5-interacting protein using a yeast two-hybrid screen. Exportin 5 is a karyopherin that functions to transport dsRBM proteins from the nucleus to cytoplasm. Interestingly, the NF110:exportin-5 interaction is inhibited by dsRNA [31].

Two NF110-specific interacting proteins, SMN and TLS, were isolated in a yeast two-hybrid screen conducted with the unique C terminus of NF110 [9]. SMN is an essential component of a complex required for the biogenesis and function of several ribonucleoprotein (RNP) complexes [32–38]. The N terminus of TLS was originally identified as part of a fusion protein resulting from chromosomal translocations in several cancers [39]. Subsequently, TLS itself was found to function in splicing and transcription as well as other cellular processes [40–42]. The interaction of NF110 with TLS and the immunoprecipitation of a spliceosome complex with NF90-specific antibodies suggest a possible role for NF110 in mRNA splicing [9].

NF90 and NF45 copurified in a large complex from placenta that contains the translation factor eIF2, the double-stranded DNA-dependent protein kinase DNA-PK, and the DNA-binding proteins Ku70 and Ku80 [43]. In this complex, NF90/NF45 stabilized the association of Ku/DNA-PK with dsDNA, possibly implying a role for these proteins in DNA repair and perhaps suggesting that they form a link between DNA repair and translation. NF90 and NF45 were also recently identified as components of a complex that binds to the encapsidation signal ε of hepatitis B virus (HBV) RNA and facilitates binding of HBV polymerase to the RNA [44]. This finding extends the range of activities with which NF90 family members are associated to include viral replication, but the functional significance of these diverse interactions remains to be determined.

Nucleic Acid Binding Properties of NF90

Not surprisingly, the nucleic acid binding properties of NF90 and its homologs have been studied extensively. The proteins bind to both DNA and RNA although some of the interactions, especially with DNA, may be indirect. As mentioned earlier, NF90 and its *Xenopus* homologs CBTF^{122} and CBTF^{98} were originally isolated by virtue of their affinity for the enhancer elements ARRE-2 and CCAAT, respectively [3,28]. Furthermore, the CCAAT box DNA-binding factor present in the complex was identified by an expression library screen and determined to be CBTF^{122} [28]. NF90 and NF110 were also identified as factors binding to a unique sequence (DHS-22) present in the DNase I hypersensitive site II in HLA-DRα [45]. Complexes containing NF90, NF110, and NF45 were affinity purified from mammalian cell extracts using ssDNA cellulose and dsDNA cellulose [46]. In many cases, however, the ability of individual purified recombinant proteins to bind directly to dsDNA has yet to be demonstrated, leading to the suggestion that post-transcriptional modification may be required and/or that NF90/NF110 binds DNA via cooperation or interaction with other dsDNA-binding proteins.

In contrast to the absence of a characteristic dsDNA-binding domain, the two dsRBMs of NF90 and its homologs interact tightly with dsRNA and the synthetic dsRNA-like polymer poly(I:C) [6,9,10,14,29,30]. In addition, NF90 binds with somewhat lesser affinity to highly structured ssRNAs such as VA RNA$_{II}$, RNA-DNA hybrids, and ssRNA at high concentrations [6,30,47]. The rat homolog p74 also interacts with additional nucleic acid substrates, having high affinity for poly(dI:dC) and poly(G) and weaker affinity for poly(U) and poly(C). A p74 deletion mutant lacking its two dsRBMs bound almost equally as well to poly(dI:dC), suggesting that the dsDNA-binding domain resides in distinct sequences outside the dsRBM region [14]. It has been inferred from transient expression and RNA-binding assays that the C-terminal RGG and GQSY domains modulate the affinity and specificity of the dsRBMs for dsRNA[12].

Functions of NF90 Homologs

A broad spectrum of functions has been imputed to members of the NF90 family (Table I, Fig. 2) based on their interactions with nucleic acids and other proteins and on their presence in complexes, but the most direct experimental evidence relates to roles in transcription and translation.

NF90 and NF45 were first identified as regulators of the IL-2 promoter. Using polyclonal antibodies specific for NF90 or NF45, this complex was found to be essential for activated transcription from the ARRE-2 enhancer element in an *in vitro* transcription assay using an extract from activated Jurkat T cells [3]. In a transient expression system, NF90 and NF110 function as both positive and negative regulators of gene expression of several cellular and viral promoters in mammalian cells, including the proliferating cell nuclear antigen (PCNA) and the HIV-LTR promoters [9,11,12]. NF90 and NF110 also bind to a specific regulatory sequence in the HLA-DRα gene where they are hypothesized to exert a negative effect on gene expression. Interestingly, NF45 did not copurify as a cofactor in this case, implying that it regulates the activation but not the inhibitory function of NF90 [24,45].

The TCP80 isoform of NF90 acts as a negative regulator of translation in cell-free extracts. Originally isolated as a cytoplasmic protein that binds to the acid β-glucosidase (GCase) mRNA, TCP80 was further characterized as a repressor of translation *in vitro* of several mRNAs including aldolase B, complement protein 8 γ-subunit, and fibronectin receptor β1. This function is dependent on the presence of the N terminus of TCP80 and certain structured regions present in the inhibited mRNAs [13]. TCP80 inhibited the association of GCase mRNA with polysomes with little effect on mRNA translation following initiation [13,48].

Consistent with its role as a regulator of translation, NF90 inhibits the activity of PKR *in vitro* [30]. In *Saccharomyces cerevisiae*, however, NF90 appears to function like PACT, another dsRNA-binding protein [49], causing a slow growth phenotype in the presence of PKR. NF90 also rescued the growth of yeast under conditions of amino acid starvation, an effect that is dependent on PKR activation [50]. Furthermore, NF90 inhibited the translation of both globin mRNA and luciferase mRNA in a cell-free translation system [50]. Using a similar yeast assay, NF45 inhibited the function of PKR and rescued yeast from a slow-growth phenotype. Evidently NF90 and NF45 can function independently of each other and can exert different functions depending on the circumstances.

Cellular Regulation of NF90 and NF45

Despite extensive structural and functional analysis of both NF90 and NF45, little is known about how these proteins are regulated. To date, both NF90 and NF110 have been detected at varying levels via Western blot in every cell line tested, and the two isoforms are always coexpressed. A survey of human tissues revealed that the proteins are expressed at high levels in the testes, ovaries, thymus, and kidney [9]. Similar results were obtained in mice except that no expression was detected in the kidney, but weak expression was detected in the spleen [51]. The association of NF90 with cell growth is further confirmed by the overexpression of NF90 mRNA in nasopharyngeal carcinoma cells [52] and NF90 protein levels in transformed as compared to untransformed fibroblasts [50]. In contrast, THP cells induced to arrest and differentiate using 12-*O*-tetradecanophorbol-13-acetate (TPA) had decreased levels of NF90 mRNA levels as compared to the controls [45]. Examination of the protein levels of NF90 through the cell cycle revealed subtle changes in protein expression with NF90 expression peaking at the G_2/M-phase boundary concomitant with phosphorylation of the protein [5,50].

Even less is known about the expression of the putative regulatory factor of NF90, NF45. The tissue distribution of NF45 has not been reported, but like NF90, NF45 has been found in every cell line examined [4,50,53]. NF45 mRNA is overexpressed in cervical cancers when compared to normal tissue, further supporting a role for NF90/NF45 in regulating genes required for cell growth and proliferation [54]. In contrast, exposure of HL60 cells to X-rays (5 Gy) strongly down-regulated NF45 [55]. Similarly, treatment of bronchial epithelial cells with triptolide, a potent immunosuppressant and anti-inflammatory drug, inhibited TPA-induced expression of NF45 [56]. Examination of the NF90 and NF45 protein levels in various cell types revealed that the levels of the two proteins and the ratios between them vary considerably [50]. These data suggest that the ratio of expression between the two proteins could act to regulate their function; in addition, as suggested by the yeast data discussed earlier [50], these proteins can not only function together in a complex, but may also function separately.

It is likely that NF90 function is regulated by posttranslational modification. As mentioned, NF90 can be phosphorylated and methylated *in vitro* by two known protein modifiers, PKR and PRMT1, respectively [9,23,30]. The PKR phosphorylation site of NF90 maps within the C-terminal portion of the protein containing its two dsRBMs [30]. A close NF90 homolog, MPP4, was identified by virtue of its interaction with MPM2, a monoclonal antibody that recognizes a set of M-phase phosphorylation sites. NF90/ NF110 is phosphorylated specifically during the M-phase of the cell cycle, and homology searches identified five potential MPM2 phosphorylation sites [5]. Seven putative CKII and 10 putative PKC phosphorylation sites have been identified in TCP80 [13]. The consequences of these modifications have yet to be determined, but their existence supports a role for NF90 and its homologs in regulating cell cycle and proliferation.

Conclusions

The NF90 family is a group of dsRNA-binding proteins shown to regulate both transcription and translation in vertebrate cells. They are also implicated, but less firmly, in a diverse set of viral and cellular functions ranging from replication and DNA repair to splicing. Members of this family have been identified in several organisms using various techniques. They are found in a variety of protein complexes, most notably with NF45, and bind to nucleic acids including dsRNA, dsDNA, and structured RNAs including adenovirus VA RNA_{II}, HBV RNA, and some mRNAs. Conceivably they serve as flexible adaptors that recognize classes of both nucleic acids and proteins and modulate macromolecular synthesis at several levels. Further identification and characterization of novel RNA and protein partners should create a more detailed blueprint of the cellular and viral processes regulated by the NF90 family and will provide insight into their roles in normal and abnormal cellular processes.

Acknowledgments

Work in the authors' laboratory is supported by grant R01 AI34552 from the National Institutes of Health to M. B. M. T. W. R was the recipient of predoctoral fellowship 12013CCRS0 from the New Jersey Commission for Cancer Research.

References

1. Fierro-Monti, I. and Mathews, M. B. (2000). Proteins binding duplexed RNA: One motif, multiple functions. *Trends. Biochem. Sci.* **25**, 241–246.
2. Tian, B. and Mathews, M. B. (2003). The occurrence and functions of the double-stranded RNA binding motif: A genome-wide survey *Prog. N. A. Res. Mol. Bio.* **74**, 123–158.
3. Corthésy. B. and Kao, P. N. (1994). Purification by DNA affinity chromatography of two polypeptides that contact the NF-AT DNA binding site in the interleukin 2 promoter. *J. Biol. Chem.* **269**, 20682–20690.
4. Kao, P. N., Chen, L., Brock, G., Ng, J., Kenny, J., Smith, A. J., and Corthésy, B. (1994). Cloning and expression of cyclosporin A- and FK506-sensitive nuclear factor of activated T-Cells: NF45 and NF90. *J. Biol. Chem.* **269**, 20691–20699.
5. Matsumoto-Taniura, N., Pirollet, F., Monroe, R., Gerace, L., and Westendorf, J. M. (1996). Identification of novel M phase phosphoproteins by expression cloning. *Mol. Biol. Cell.* **7**, 1455–1469.
6. Bass, B. L., Hurst, S. R., and Singer, J. D. (1994). Binding properties of newly identified Xenopus proteins containing dsRNA-binding motifs. *Curr. Biol.* **4**, 301–316.
7. Duchange, N., Pidoux, J., Camus, E., and Sauvaget, D. (2000). Alternative splicing in the human interleukin enhancer binding factor 3 (ILF3) gene. *Gene* **261**, 345–353.
8. Saunders, L. R., Perkins, D. J., Balachandran, S., Michaels, R., Ford, R., Mayeda, A., and Barber, G. N. (2001). Characterization of two evolutionarily conserved, alternatively spliced nuclear phosphoproteins, NFAR-1 and -2, that function in mRNA processing and interact with the double-stranded RNA-dependent protein kinase, PKR. *J. Biol. Chem.* **276**, 32300–32312.
9. Saunders, L. R., Jurecic, V., and Barber, G. N. (2001). The 90- and 110-kDa human NFAR proteins are translated from two differentially spliced mRNAs encoded on chromosome 19p13. *Genomics* **71**, 256–259.
10. Patel, R. C., Vestal, D. J., Xu, Z., Bandyopadhyay, S., Guo, W., Erme, S. M., Williams, B. R., and Sen, G. C. (1999). DRBP76, a double-stranded RNA-binding protein, is phosphorylated by the interferon-induced protein kinase, PKR. *J. Biol. Chem.* **274**, 20432–20437.
11. Reichman, T. W., Parrott, A., Fierro-Monti, I., Caron, D. J., Kao, P. N., Lee, C. G., Li, H., and Matthews, M. B. (In Press).
12. Reichman, T. W. and Mathews, M. B. (2003). RNA binding and intramolecular interactions modulate the regulation of gene expression by nuclear factor 110. *RNA* **9**, 543–554.
13. Xu, Y. H., and Grabowski, G. A. (1999). Molecular cloning and characterization of a translational inhibitory protein that binds to coding sequences of human acid beta-glucosidase and other mRNAs. *Mol. Genet. Metab.* **68**, 441–454.
14. Coolidge, C. J. and Patton, J. G. (2000). A new double-stranded RNA-binding protein that interacts with PKR. *Nucleic Acids Res.* **28**, 1407–1417.
15. Schumacher, J. M., Lee, K., Edelhoff, S., and Braun, R. E. (1995). Spnr, a murine RNA-binding protein that is localized to cytoplasmic microtubules. *J. Cell Biol.* **129**, 1023–1032.
16. Schumacher, J. M., Artzt, K., and Braun, R. E. (1998). Spermatid perinuclear ribonucleic acid-binding protein binds microtubules in vitro and associates with abnormal manchettes in vivo in mice. *Biol. Reprod.* **59**, 69–76.
17. Pires-daSilva, A., Nayernia, K., Engel, W., Torres, M., Stoykova, A., Chowdhury, K., and Gruss, P. (2001). Mice deficient for spermatid perinuclear RNA-binding protein show neurologic, spermatogenic, and sperm morphological abnormalities. *Dev. Biol.* **233**, 319–328.
18. Meagher, M. J., Schumacher, J. M., Lee, K., Holdcraft, R. W., Edelhoff, S., Disteche, C., and Braun, R. E. (1999). Identification of ZFR, an ancient and highly conserved murine chromosome-associated zinc finger protein. *Gene* **228**, 197–211.
19. Meagher, M. J. and Braun, R. E. (2001). Requirement for the murine zinc finger protein ZFR in perigastrulation growth and survival. *Mol. Cell. Biol.* **21**, 2880–2890.
20. Ghisolfi, L., Kharrat, A., Joseph, G., Amalric, F., and Erard, M. (1992). Concerted activities of the RNA recognition and the glycine-rich C-terminal domains of nucleolin are required for efficient complex formation with pre-ribosomal RNA. *Eur. J. Biochem.* **209**, 541–548.
21. Darnell, J. C., Jensen, K. B., Jin, P., Brown, V., Warren, S. T., and Darnell, R. B. (2001). Fragile X mental retardation protein targets G quartet mRNAs important for neuronal function. *Cell* **107**, 489–499.
22. Ghisolfi, L., Joseph, G., Amalric, F., and Erard, M. (1992). The glycine-rich domain of nucleolin has an unusual supersecondary structure responsible for its RNA-helix-destabilizing properties. *J. Biol. Chem.* **267**, 2955–2959.
23. Tang, J., Kao, P. N., and Herschman, H. R. (2000). Protein-arginine methyltransferase I, the predominant protein-arginine methyltransferase in cells, interacts with and is regulated by interleukin enhancer-binding factor 3. *J. Biol. Chem.* **275**, 19866–19876.
24. Reichman, T. W., Muniz, L. C., and Mathews, M. B. (2002). The RNA binding protein nuclear factor 90 functions as both a positive and negative regulator of gene expression in mammalian cells. *Mol. Cell. Biol.* **22**, 343–356.
25. Zhang, S. and Grosse, F. (1997). Domain structure of human nuclear DNA helicase II (RNA helicase A). *J. Biol. Chem.* **272**, 11487–11494.
26. Bertolotti, A., Melot, T., Acker, J., Vigneron, M., Delattre, O., and Tora, L. (1998). EWS, but not EWS-FLI-1, is associated with both TFIID and RNA polymerase II: Interactions between two members of the TET family, EWS and hTAFII68, and subunits of TFIID and RNA polymerase II complexes. *Mol. Cell. Biol.* **18**, 1489–1497.
27. Brzostowski, J., Robinson, C., Orford, R., Elgar, S., Scarlett, G., Peterkin, T., Malartre, M., Kneale, G., Wormington, M., and Guille, M. (2000). RNA-dependent cytoplasmic anchoring of a transcription factor subunit during Xenopus development. *EMBO J.* **19**, 3683–3693.

28. Orford, R. L., Robinson, C., Haydon, J. M, Patient, R. K, and Guille, M. J. (1998). The maternal CCAAT box transcription factor which controls GATA-2 expression is novel and developmentally regulated and contains a double-stranded-RNA-binding subunit. *Mol. Cell. Biol.* **18**, 5557–5566.
29. Langland, J. O., Kao, P. N., and Jacobs, B. L. (1999). Nuclear factor-90 of activated T-cells: A double-stranded RNA-binding protein and substrate for the double-stranded RNA-dependent protein kinase, PKR. *Biochemistry* **38**, 6361–6368.
30. Parker, L. M., Fierro-Monti, I., and Mathews, M. B. (2001). Nuclear factor 90 is a substrate and regulator of the eukaryotic initiation factor 2 kinase double-stranded rna-activated protein kinase. *J. Biol. Chem.* **276**, 32522–32530.
31. Brownawell, A. M. and Macara, I. G. (2002). Exportin-5, a novel karyopherin, mediates nuclear export of double-stranded RNA binding proteins. *J. Cell Biol.* **156**, 53–64.
32. Fischer, U., Liu, Q., and Dreyfuss, G. (1997). The SMN-SIP1 complex has an essential role in spliceosomal snRNP biogenesis. *Cell* **90**, 1023–1029.
33. Jones, K. W., Gorzynski, K., Hales, C. M., Fischer, U., Badbanchi, F., Terns, R. M., and Terns, M. P. (2001). Direct interaction of the spinal muscular atrophy disease protein SMN with the small nucleolar RNA-associated protein fibrillarin. J. *Biol. Chem.* **276**, 38645–38651.
34. Meister, G., Buhler, D., Pillai, R., Lottspeich, F., and Fischer, U. (2001). A multiprotein complex mediates the ATP-dependent assembly of spliceosomal U snRNPs. *Nat. Cell. Biol.* **3**, 945–949.
35. Mourelatos, Z., Abel, L., Yong, J., Kataoka, N., and Dreyfuss, G. (2001). SMN interacts with a novel family of hnRNP and spliceosomal proteins. *EMBO J.* **20**, 5443–5452.
36. Pellizzoni, L., Baccon, J., Charroux, B., and Dreyfuss, G. (2001). The survival of motor neurons (SMN) protein interacts with the snoRNP proteins fibrillarin and GAR1. *Curr. Biol.* **11**, 1079–1088.
37. Pellizzoni, L., Charroux, B., Rappsilber, J., Mann, M., and Dreyfuss, G. (2001). A functional interaction between the survival motor neuron complex and RNA polymerase II. *J. Cell. Biol.* **152**, 75–85.
38. Pellizzoni, L., Kataoka, N., Charroux, B., and Dreyfuss, G. (1998). A novel function for SMN, the spinal muscular atrophy disease gene product, in pre-mRNA splicing. *Cell* **95**, 615–624.
39. Ron, D. (1997). TLS-CHOP and the role of RNA-binding proteins in oncogenic transformation. *Curr. Top. Microbiol. Immunol.* **220**, 131–142.
40. Hallier, M., Lerga, A., Barnache, S., Tavitian, A., and Moreau-Gachelin, F. (1998). The transcription factor Spi-1/PU.1 interacts with the potential splicing factor TLS. *J. Biol. Chem.* **273**, 4838–4842.
41. Uranishi, H., Tetsuka, T., Yamashita, M., Asamitsu, K., Shimizu, M., Itoh, M., and Okamoto, T. (2001). Involvement of the pro-oncoprotein TLS (translocated in liposarcoma) in nuclear factor-kappa B p65-mediated transcription as a coactivator. *J. Biol. Chem.* **276**, 13395–13401.
42. Yang, L., Embree, L. J., Tsai, S., and Hickstein, D. D. (1998). Oncoprotein TLS interacts with serine-arginine proteins involved in RNA splicing. *J. Biol. Chem.* **273**, 27761–27764.
43. Ting, N. S., Kao, P. N., Chan, D. W., Lintott, L. G., and Lees-Miller, S. P. (1998). DNA-dependent protein kinase interacts with antigen receptor response element binding proteins NF90 and NF45. *J. Biol. Chem.* **273**, 2136–2145.
44. Shin, H. J., Kim, S. S., Cho, Y. H., Lee, S. G., and Rho, H. M. (2002). Host cell proteins binding to the encapsidation signal epsilon in hepatitis B virus RNA. *Arch. Virol.* **147**, 471–491.
45. Sakamoto, S., Morisawa, K., Ota, K., Nie, J., and Taniguchi, T. (1999). A binding protein to the DNase I hypersensitive site II in HLA-DR alpha gene was identified as NF90. *Biochemistry* **38**, 3355–3361.
46. Satoh, M., Shaheen, V. M., Kao, P. N., Okano, T., Shaw, M., Yoshida, H., Richards, H. B., and Reeves, W. H. (1999). Autoantibodies define a family of proteins with conserved double-stranded RNA-binding domains as well as DNA binding activity. *J. Biol. Chem.* **274**, 34598–34604.
47. Liao, H. J., Kobayashi, R., and Mathews, M. B. (1998). Activities of adenovirus virus-associated RNAs: Purification and characterization of RNA binding proteins. *Proc. Natl. Acad. Sci. USA* **95**, 8514–8519.
48. Xu, Y. H., Busald, C., and Grabowski, G. A. (2000). Reconstitution of TCP80/NF90 translation inhibition activity in insect cells. *Mol. Genet. Metab.* **70**, 106–115.
49. Patel, R. C. and Sen, G. C. (1998). PACT, a protein activator of the interferon-induced protein kinase, PKR. *EMBO J.* **17**, 4379–4390.
50. Parker, L. M., Fierro-Monti, I., Reichman, T. W., Gunnery, S., and Mathews, M. B. (2001). Double-stranded RNA-binding Proteins and the Control of Protein Synthesis and Cell Growth. Cold Spring Harbor Symposia on Quantitative Biology **66**, 485–497.
51. Buaas, F. W., Lee, K., Edelhoff, S., Disteche, C., and Braun, R. E. (1999). Cloning and characterization of the mouse interleukin enhancer binding factor 3 (Ilf3) homolog in a screen for RNA binding proteins. *Mamm. Genome* **10**, 451–456.
52. Fung, L. F., Lo, A. K., Yuen, P.W., Liu, Y., Wang, X. H., and Tsao, S. W. (2000). Differential gene expression in nasopharyngeal carcinoma cells. *Life Sci.* **67**, 923–936.
53. Aoki, Y., Zhao, G., Qiu, D., Shi, L., and Kao, P. N. (1998). CsA-sensitive purine-box transcriptional regulator in bronchial epithelial cells contains NF45, NF90, and Ku. *Am. J. Physiol.* **275**, L1164–1172.
54. Shim, C., Zhang, W., Rhee, C. H., and Lee, J. H. (1998). Profiling of differentially expressed genes in human primary cervical cancer by complementary DNA expression array. *Clin. Cancer Res.* **4**, 3045–3050.
55. Balcer-Kubiczek, E. K., Zhang, X. F, Han, L. H., Harrison, G. H., Davis, C. C., Zhou, X. J., Ioffe, V., McCready, W. A, Abraham, J. M, and Meltzer, S. J. (1998). BIGEL analysis of gene expression in HL60 cells exposed to X rays or 60 Hz magnetic fields. *Radiat. Res.* **150**, 663–672.
56. Zhao, G., Vaszar, L. T, Qiu, D., Shi, L., and Kao, P. N. (2000). Anti-inflammatory effects of triptolide in human bronchial epithelial cells. *Am. J. Physiol. Lung Cell Mol. Physiol.* **279**, L958–966.

CHAPTER 322

Signaling Pathways That Mediate Translational Control of Ribosome Recruitment to mRNA

Nahum Sonenberg and Emmanuel Petroulakis
Department of Biochemistry and McGill Cancer Research Centre, McGill University, Montréal, Québec, Canada

Introduction

Alterations in mRNA translation rates in eukaryotes accompany cell proliferation, differentiation, and growth. Translation initiation is the major target of translational control [1,2]. Specifically, the step of recruitment of ribosomes to the mRNA is tightly regulated. This control is exerted primarily by phosphorylation of initiation factors. Some of the signaling pathways that mediate phosphorylation of initiation factors have been extensively studied and are described in this chapter.

Nuclear transcribed eukaryotic mRNAs possess at their 5′ termini the structure m^{7}GpppX (where X is any nucleotide), termed the *cap* [3,4]. The cap structure strongly facilitates the translation of a large majority of mRNAs [4]. However, some viral mRNAs (such as those of picornaviruses) [5] and a subset of cellular mRNAs translate by a cap-independent mechanism [6]. Cap-dependent mRNA translation initiation is dependent on eIF4E, the cap binding protein, and requires the participation of a set of initiation factors (the eIF4 family), which facilitate the recruitment of the ribosome to the 5′ end of the mRNA [1,4,7]. Cap-dependent translation is counteracted by members of a family of translational inhibitors (the eIF4E binding proteins; 4E-BPs). Many of the initiation factors that promote ribosome recruitment to the mRNA and the repressors of this process are phosphoproteins. The phosphorylation status of these proteins is modulated by extracellular stimuli and environmental stress [1,4]. We describe the current model for cap-dependent, initiation factor-assisted assembly of a ribosome–mRNA complex and how this activity is regulated by translational repressors via specific signaling pathways in mammals. Phosphorylation of initiation factors is also important in regulation of mRNA recruitment in plants, but the signaling pathways have not been elucidated [8]. It is also unclear if and how phosphorylation of initiation factors plays a role in mRNA recruitment to ribosomes in yeast [9].

eIF4F Complex Formation

Affinity chromatography using m^{7}GDP-agarose was used to identify and purify eIF4E, the cap-binding protein [10,11]. Subsequently, it was discovered that eIF4E functions as a complex, termed eIF4F, with two other polypeptides: eIF4A and eIF4G [12]. eIF4A is an RNA helicase, which in combination with two RNA-binding proteins, eIF4B and eIF4H, and accompanied by the hydrolysis of ATP, is believed to unwind the mRNA 5′UTR secondary structure [1,13]. eIF4G serves as a modular scaffolding protein for several proteins, including eIF4A, eIF4E, and eIF3 (a ribosome-associated initiation factor). The interaction between eIF4G and eIF3 is required for the recruitment of the 40S ribosomal subunit with its associated factors (Fig. 1). eIF4G also binds to PABP, an interaction that brings about the circularization of the mRNA [14]. eIF4G also possesses a C-terminal docking site for Mnk kinase, which phosphorylates eIF4E at Ser209 [15].

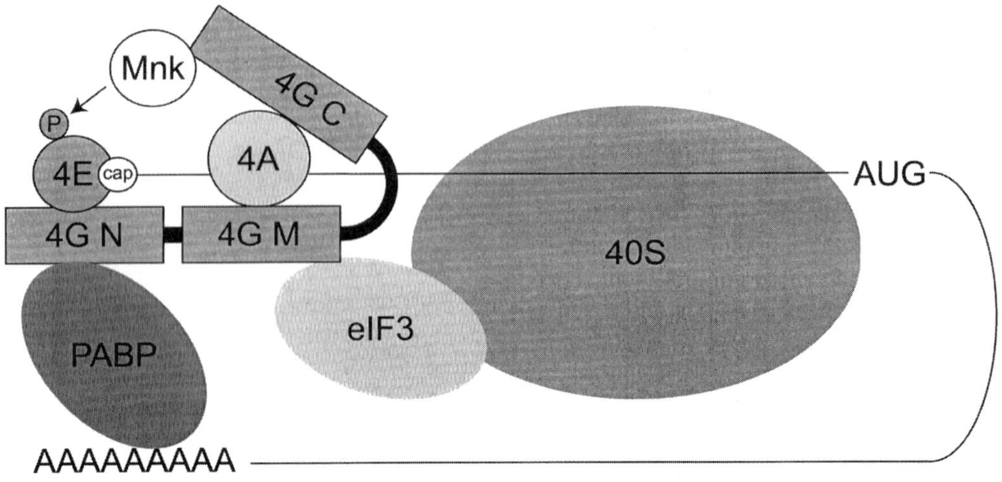

Figure 1 Recruitment of the 40S ribosome to the 5′ end of mRNAs by initiation factors. The eIF4F complex (comprised of eIF4E, eIF4G, and eIF4A) coordinates the recruitment of the 40S ribosomal subunit to the 5′end of eukaryotic mRNAs. The three modular domains of eIF4G (N-terminal domain, 4G N; middle domain, 4G M; C-terminal domain, 4G C) are docking sites for PABP, eIF4E, eIF4A, eIF3, and Mnk1. eIF3 functions to bridge between eIF4G and the 40S ribosomal subunit. eIF4E (indicated as 4E) binds the 5′ cap (m^7G) structure of mRNA. Poly(A)-binding protein (PABP) interacts with the N-terminal domain of eIF4G [62], to mediate circularization of the mRNAs [14]. (Adapted from [61].)

eIF4E Phosphorylation occurs in response to a variety of extracellular stimuli including hormones (such as insulin), growth factors (such as epidermal growth factor), cytokines (tumor necrosis factor α), and mitogens (phorbol ester), and in general positively correlates with cell growth and enhanced translation rates [4]. However, stress induced by arsenite or anisomycin, which inhibit translation, also results in phosphorylation of eIF4E [16]. Both extracellular stimuli and stress activate the Ser/Thr kinase Mnk1 (Mnk2, a homolog of Mnk1 is constitutively active [17]). Mnk-mediated phosphorylation of eIF4E is effected by two pathways: the PD98059-sensitive Ras/Raf/MEK/ERK kinase cascade and the SB203580-sensitive stress-activated p38MAPK signaling pathway (Fig. 2) [16,18]. There is no consensus as to whether phosphorylation of eIF4E stimulates translation, and the mechanism of such stimulation is also controversial (for a review, see [19]). Although some evidence suggests that phosphorylation of eIF4E is not required for general translation *in vitro* [20] or *in vivo* [21], it is noteworthy that flies, which express a mutant eIF4E, Ser209Ala, develop slowly and are smaller than their wild-type counterparts, demonstrating that phosphorylation of Ser209 is biologically significant [22]. In addition to Mnk, PKC (α, β, γ) also phosphorylates eIF4E on Ser209 *in vitro* [23], and in certain circumstances Ser209 is phosphorylated in a PKC-dependent manner *in vivo* [24,25]. Thus, PKC may also play a role in the phosphorylation of Ser209.

Based on the X-ray crystal structure of eIF4E, it was suggested that phosphorylation of Ser209 results in the formation of a salt bridge with Lys159, thus forming a clamp over the mRNA to stabilize its interaction with eIF4E [26]. However, two recent papers [27,28], reported that phosphorylation of Ser209 causes a two to fourfold reduction in the affinity of eIF4E for the cap. Thus, the molecular mechanism by which eIF4E phosphorylation affects its function is unclear.

Repressors of Cap-Dependent Translation

The formation of the eIF4F complex is inhibited by a family of proteins, termed eIF4E binding proteins (4E-BPs) [1]. The 4E-BP family is comprised of three members, 4E-BP1, 4E-BP2, and 4E-BP3, all of which inhibit specifically cap-dependent translation [29,30]. 4E-BPs are also known as PHAS (*p*hosphorylated *h*eat and *a*cid *s*table proteins) [31]. 4E-BPs specifically inhibit cap-dependent translation by binding to eIF4E to impede the eIF4E–eIF4G interaction and, consequently, the assembly of the eIF4F complex [26]. The consensus sequence shared among the 4E-BPs and eIF4Gs, YXXXXLφ (where X is any amino acid and φ is a hydrophobic residue), directly interacts with eIF4E. Biochemical and structural studies demonstrated that the 4E-BPs inhibit the eIF4E–eIF4G interaction by a molecular mimicry mechanism [26]. Indeed, peptides harboring the eIF4E binding motif inhibit cap-dependent translation *in vitro* [26,32].

Modulation of 4E-BP Phosphorylation by FRAP/mTOR

Binding of the 4E-BPs to eIF4E is regulated by phosphorylation (Fig. 2). Hypophosphorylated 4E-BPs bind tightly to eIF4E, resulting in translational inhibition, whereas hyperphosphorylation of 4E-BPs engenders their release from eIF4E [1,29]. Increased phosphorylation of 4E-BPs occurs in response to treatments with hormones, growth factors, mitogens, cytokines, and G-protein-coupled receptor agonists [1].

The PI3K-Akt-FRAP/mTOR signaling pathway plays a critical role in the phosphorylation of 4E-BP1 [33]. PI3Ks regulate a large number of cellular processes including

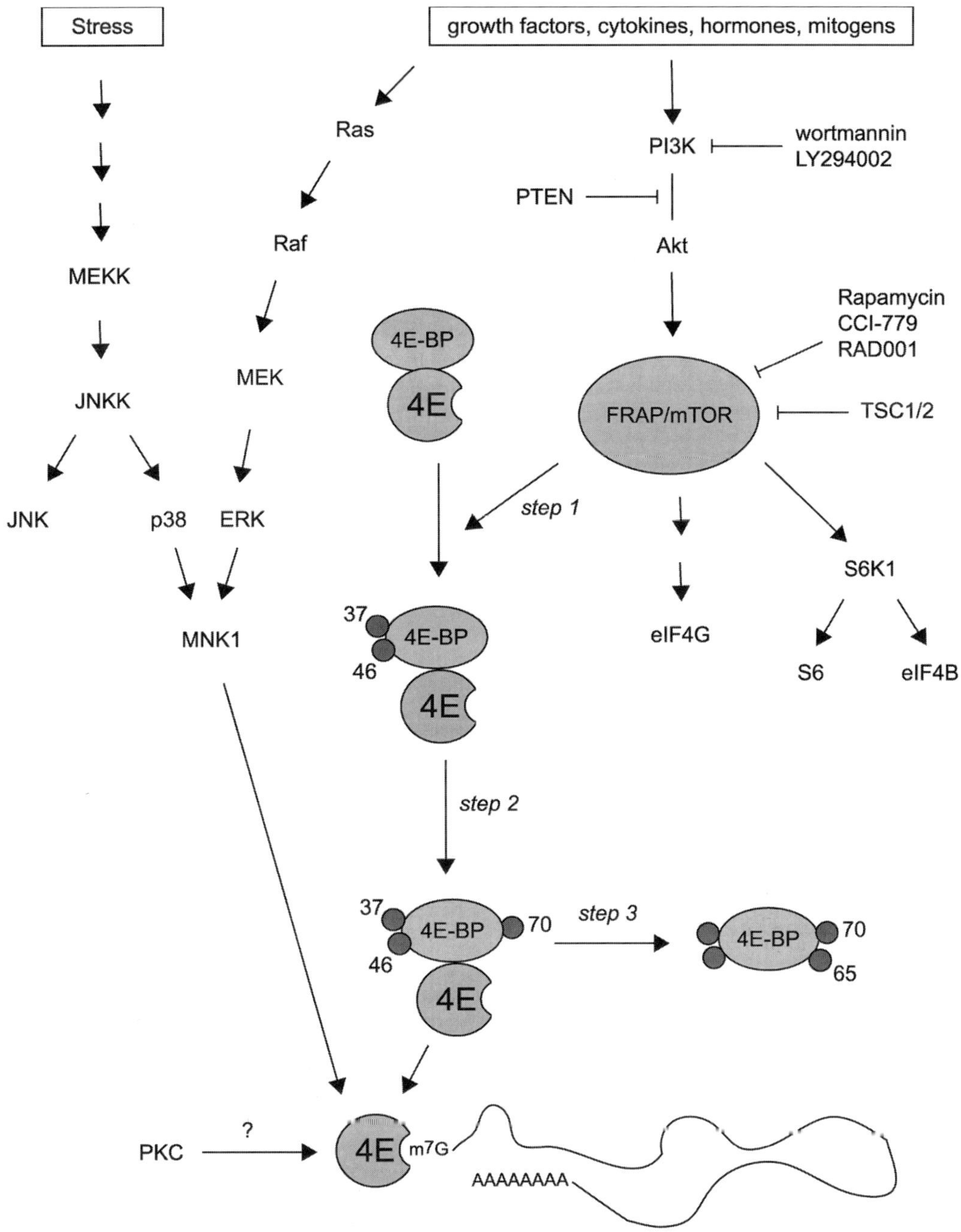

Figure 2 Signaling pathways leading to eIF4E and 4E-BP phosphorylation. Stimuli (i.e., growth factors) activate the ras-signaling and the PI3K-Akt-FRAP/mTOR signaling pathways. 4E-BP1 is phosphorylated through a multistep mechanism. FRAP/mTOR signaling mediates phosphorylation at Thr37 and Thr46 prior to phosphorylation of Ser65 and Thr70. Hyperphosphorylated 4E-BP1 causes its release from eIF4E, allowing for eIF4E to bind to cap (m^7G)-bearing mRNAs. Mutations in PTEN and TSC1/2 contribute to elevated FRAP/mTOR activity, increased levels of unbound eIF4F, and enhanced translation of a subset of mRNAs that harbor extensive secondary structure. FRAP/mTOR activity is inhibited by rapamycin and its derivatives (CCI-779, RAD-001) [63]. The stress-activated and the ras-signaling pathways converge on Mnk1, which phosphorylates eIF4E. PKC may also phosphorylate Ser209.

proliferation, apoptosis, and protein synthesis [34]. PI3K phosphorylates phosphatidylinositol-(4,5)-biphosphate to generate phosphatidylinositol-(3,4,5)-triphosphate. This reaction is negatively regulated by PTEN, a lipid phosphatase, which exhibits tumor activity and is mutated in many cancers [35]. PI3K-generated lipid products bind the pleckstrin homology (PH) domain of the Akt/PKB Ser/Thr protein kinases and target them to the plasma membrane [36]. Overexpression of the p110α PI3K catalytic subunit or Akt/PKB enhances 4E-BP1 phosphorylation [37], providing evidence that 4E-BP1 is a target of PI3K signaling.

FRAP/mTOR (*F*K506 *r*apamycin-*a*ssociated *p*rotein/ *m*ammalian <u>t</u>arget *o*f *r*apamycin) is the mammalian homolog of the yeast TOR proteins, TOR1 and TOR2 [38] (also see chapters by Schmelzle *et al.* and Radimerski and Thomas in this volume). FRAP/mTOR functions as a major integrator of a variety of physiological states, such as nutrient availability (i.e., amino acids, glucose), extracellular stimuli (growth factors, hormones and cytokines), and energy levels (ATP), to effect cell cycle progression, cell growth, cell size, translation, and protein turnover. Importantly, FRAP/mTOR activity is inhibited by a highly specific drug, rapamycin, a bacterial macrolide that forms a gain-of-function complex with FKBP12 (FK506-binding protein). This complex binds with high affinity to FRAP/mTOR to inhibit its kinase activity toward 4E-BPs and p70 S6 kinase [39].

FRAP/mTOR directly phosphorylates 4E-BP1, because a rapamycin-resistant form of FRAP/mTOR confers rapamycin resistance to 4E-BP1 phosphorylation [40–42]. The phosphorylation sites in human 4E-BP1 are Thr37, Thr46, Ser65, Thr70, Ser83, and Ser112. The phosphorylation of these sites except for Ser83 (equivalent to Ser82 in rat 4E-BP1) [43] and Ser112 is dependent on FRAP/mTOR [33,41]. Ser112, which is present only in 4E-BP1, is phosphorylated by ataxia-telangiectasia mutated (ATM) protein kinase [44] and was reported to be a major insulin-dependent phosphorylation site in adipocytes (equivalent to Ser111 in rat 4E-BP1) [45]. Phosphorylation of 4E-BP1 occurs in a hierarchical, multistep process [33,46]. Phosphorylation of Thr37 and Thr46 is necessary for subsequent phosphorylation at Ser65 and Thr70 [46], because substitution of Thr37 and/or Thr46 for alanine residue(s) prevents phosphorylation of Ser65 and Ser70 [46]. In contrast, substitution of serine for glutamic acid at amino acid residues Thr37 and Thr46 partially restores phosphorylation at Ser65 and Ser70. Thus, Thr37 and Thr46 are phosphorylated first. This primes the phosphorylation of Ser70, followed by the phosphorylation of Ser65 [46]. Thr37 and Thr46 are phosphorylated to a considerable extent even in the absence of serum stimulation and in the presence of rapamycin [33]. In comparison, Ser65 and Thr70 are rapidly dephosphorylated in serum-deprived conditions or in the presence of rapamycin [33]. It was proposed that a protein phosphatase 2A (PP2A) is activated by rapamycin treatment and dephosphorylates Ser65 and Thr70 [47]. The proposed model suggests that FRAP/mTOR also phosphorylates PP2A in order to inhibit dephosphorylation of 4E-BP1 (in addition to p70 S6 kinase) [47].

Phosphorylation of eIF4G and eIF4B

Phosphorylation of eIF4G and eIF4B is also rapamycin sensitive, implicating FRAP/mTOR in their phosphorylation. eIF4G becomes phosphorylated on several residues in response to a diverse array of extracellular stimuli. Phosphorylation of eIF4G positively correlates with translational efficiency, and similar to 4E-BP1, a two-step mechanism for eIF4G phosphorylation has been suggested [48]. Two of the phosphorylation sites were mapped to the linker region that separates the middle and the C-terminal eIF4A binding domains (see Fig. 1) [49]. It was suggested that a single eIF4A molecule contacts both the middle domain and the C-terminal domains [48,49]. The linker region could then serve as a hinge, and its phosphorylation might affect the interaction of eIF4A with eIF4G and, in turn, the activity of eIF4F. eIF4B is phosphorylated on two sites (Ser406 and Ser422) by p70 S6 kinase *in vitro*. These sites are highly likely to be phosphorylated *in vivo* by p70 S6 kinase, because their phosphorylation is inhibited by rapamycin. Strong evidence that phosphorylation of eIF4B on these sites is downstream of S6K1 was obtained by expression of a rapamycin-resistant form of S6K1 (J. W. B. Hershey, personal communication). Under these conditions, serum-stimulated phosphorylation is rendered rapamycin resistant. As is the case for eIF4G, phosphorylation of eIF4B positively correlates with enhanced translation [50]. However, it is not known how the phosphorylation of either eIF4G or eIF4B affects their biochemical activities.

Control of Cell Growth and Proliferation by eIF4E: Link to Cancer

eIF4E is a rate-limiting factor for mRNA translation. Overexpression of eIF4E induces malignant transformation of NIH 3T3 and primary rodent cells [51,52] through mechanisms that are thought to selectively enhance the translation of a subset of mRNAs [51,53]. Poorly translated mRNAs generally harbor long and structured (G/C rich) 5′UTRs. In eIF4E-transformed cells, the translation of such mRNAs is enhanced [53], probably as a result of the increased formation of the eIF4F complex that effects mRNA unwinding activity. Consistent with the transformation activity of eIF4E, poorly translated mRNAs usually encode proteins that regulate cell growth, differentiation, and development [54]. Also consistent with the role of eIF4E in cellular transformation, overexpression of 4E-BP1 or 4E-BP2 in eIF4E-transformed cells, or ras- or src-transformed cells, partially reverted the transformed phenotype [55]. eIF4E is overexpressed in many tumors and transformed cell lines [56,57]. It is thus paramount that rapamycin, which inhibits eIF4E activity, is showing promising results as an anticancer drug [58]. It is a tantalizing possibility that rapamycin acts as an anticancer drug by inhibiting eIF4E activity. Consistent with this scenario, it is noteworthy that mutations in the tumor suppressor PTEN result in elevated Akt inputs to FRAP/mTOR signaling [59]. Also, mutations in the tuberous sclerosis complex (TSC1/2), which is a tumor suppressor and inhibits FRAP/mTOR, lead to benign tumors (hamartomas) [60]. Such mutations cause an increase in 4E-BP phosphorylation [60].

Conclusions

FRAP/mTOR plays an important role in the phosphorylation of the 4E-BPs and consequently in regulation of mRNA recruitment to ribosomes. The phosphorylation status of the 4E-BPs is an important readout of FRAP/mTOR activity that directly correlates with eIF4F assembly and activity.

FRAP/mTOR integrates inputs from many diverse extracellular stimuli, intracellular physiological cues, and environmental stress to affect either an anabolic or catabolic output. Understanding the mechanisms that control 4E-BP phosphorylation should shed light on the involvement of 4E-BPs in physiological processes and in disease.

References

1. Raught, B., Gingras, A.-C., and Sonenberg N. (2000). Regulation of ribosomal recruitment in eukaryotes, in Sonenberg, N., Hershey, J. W. B., and Mathews, M. B., Eds., *Translational Control of Gene Expression*, pp. 245–293. Cold Spring Harbor Laboratory Press, Cold Spring Harbor, New York.
2. Mathews, M. B., Sonenberg, N., and Hershey, J. W. B. (2000). Origins and principles of translational control, in Sonenberg, N., Hershey, J. W. B., and Mathews, M. B., Eds., *Translational Control of Gene Expression*, pp. 1–31. Cold Spring Harbor Laboratory Press, Cold Spring Harbor, New York.
3. Shatkin, A. J. (1976). Capping of eucaryotic mRNAs. *Cell* **9**(4, Pt 2), 645–653.
4. Gingras, A. C., Raught, B., and Sonenberg, N. (1999). eIF4 initiation factors: effectors of mRNA recruitment to ribosomes and regulators of translation. *Annu. Rev. Biochem.* 1999. **68**, 913–963.
5. Pelletier, J. and Sonenberg, N. (1988). Internal initiation of translation of eukaryotic mRNA directed by a sequence derived from poliovirus RNA. *Nature* **334**(6180), 320–325.
6. Hellen, C. U. and Sarnow, P. (2001). Internal ribosome entry sites in eukaryotic mRNA molecules. *Genes Dev.* **15**(13), 1593–1612.
7. Hershey, J. and Merrick, W. (2000). Pathway and mechanism of initiation of protein synthesis, in Sonenberg, N., Hershey, J. W. B., and Mathews, M. B., Eds., *Translational Control of Gene Expression*, pp. 33–88. Cold Spring Harbor Laboratory Press, Cold Spring Harbor, New York.
8. Gallie, D. R. (2002). Protein–protein interactions required during translation. *Plant Mol. Biol.* **50**(6), 949–970.
9. McCarthy, J. E. (1998). Posttranscriptional control of gene expression in yeast. *Microbiol. Mol. Biol. Rev.* **62**(4), 1492–1553.
10. Sonenberg, N. *et al.* (1978). A polypeptide in eukaryotic initiation factors that crosslinks specifically to the 5′-terminal cap in mRNA. *Proc. Natl. Acad. Sci. USA* **75**(10), 4843–4847.
11. Sonenberg, N. *et al.* (1979). Eukaryotic mRNA cap binding protein: purification by affinity chromatography on Sepharose-coupled m7GDP. *Proc. Natl. Acad. Sci. USA* **76**(9), 4345–4349.
12. Grifo, J. A. *et al.* (1983). New initiation factor activity required for globin mRNA translation. *J. Biol. Chem.* **258**(9), 5804–5810.
13. Rogers, G. W., Jr., Komar, A. A., and Merrick, W. C. (2002). eIF4A: the godfather of the DEAD box helicases. *Prog. Nucleic Acid Res. Mol. Biol.* **72**, 307–331.
14. Wells, S. E. *et al.* (1998). Circularization of mRNA by eukaryotic translation initiation factors. *Mol. Cell* **2**(1), 135–140.
15. Pyronnet, S. *et al.* (1999). Human eukaryotic translation initiation factor 4G (eIF4G) recruits mnk1 to phosphorylate eIF4E. *EMBO J.* **18**(1), 270–279.
16. Morley, S. J. and McKendrick, L. (1997). Involvement of stress-activated protein kinase and p38/RK mitogen-activated protein kinase signaling pathways in the enhanced phosphorylation of initiation factor 4E in NIH 3T3 cells. *J. Biol. Chem.* **272**(28), 17887–17893.
17. Scheper, G. C. *et al.* (2001). The mitogen-activated protein kinase signal-integrating kinase Mnk2 is a eukaryotic initiation factor 4E kinase with high levels of basal activity in mammalian cells. *Mol. Cell. Biol.* **21**(3), 743–754.
18. Wang, X. *et al.* (1998). The phosphorylation of eukaryotic initiation factor eIF4E in response to phorbol esters, cell stresses, and cytokines is mediated by distinct MAP kinase pathways. *J. Biol. Chem.* **273**(16), 9373–9377.
19. Scheper, G. C. and Proud, C. G. (2002). Does phosphorylation of the cap-binding protein eIF4E play a role in translation initiation? *Eur. J. Biochem.* **269**(22), 5350–5359.
20. Svitkin, Y. V. *et al.* (1996). General RNA binding proteins render translation cap dependent. *EMBO J.* **15**(24), 7147–7155.
21. Morley, S. J. and Naegele, S. (2002). Phosphorylation of eukaryotic initiation factor (eIF) 4E is not required for de novo protein synthesis following recovery from hypertonic stress in human kidney cells. *J. Biol. Chem.* **277**(36), 32855–32859.
22. Lachance, P. E. *et al.* (2002). Phosphorylation of eukaryotic translation initiation factor 4E is critical for growth. *Mol. Cell. Biol.* **22**(6), 1656–1663.
23. Flynn, A. and Proud, G. (1996). Insulin-stimulated phosphorylation of initiation factor 4E is mediated by the MAP kinase pathway. *FEBS Lett.* **389**(2), 162–166.
24. Morley, S. J. and Traugh, J. A. (1990). Differential stimulation of phosphorylation of initiation factors eIF-4F, eIF-4B, eIF-3, and ribosomal protein S6 by insulin and phorbol esters. *J. Biol. Chem.* **265**(18), 10611–10616.
25. Rao, G. N. *et al.* (1994). Angiotensin II induces phosphorylation of eukaryotic protein synthesis initiation factor 4E in vascular smooth muscle cells. *J. Biol. Chem.* **269**(10), 7180–7184.
26. Marcotrigiano, J. *et al.* (1999). Cap-dependent translation initiation in eukaryotes is regulated by a molecular mimic of eIF4G. *Mol. Cell* **3**(6), 707–716.
27. Scheper, G. C. *et al.* (2002). Phosphorylation of eukaryotic initiation factor 4E markedly reduces its affinity for capped mRNA. *J. Biol. Chem.* **277**(5), 3303–3309.
28. Zuberek, J. A. W.-C. *et al.* (2003). Phosphorylation of eIF4E attenuates its interaction with mRNA cap analogs by electrostatic repulsion; intein-mediated protein ligation strategy to obtain phosphorylated protein. RNA (in press).
29. Pause, A. *et al.* (1994). Insulin-dependent stimulation of protein synthesis by phosphorylation of a regulator of 5′-cap function [see comments]. *Nature* **371**(6500), 762–767.
30. Poulin, F. *et al.* (1998). 4E-BP3, a new member of the eukaryotic initiation factor 4E-binding protein family. *J. Biol. Chem.* **273**(22), 14002–14007.
31. Hu, C. *et al.* (1994). Molecular cloning and tissue distribution of PHAS-I, an intracellular target for insulin and growth factors. *Proc. Natl. Acad. Sci. USA* **91**(9), 3730–3734.
32. Fletcher, C. M. *et al.* (1998). 4E binding proteins inhibit the translation factor eIF4E without folded structure. *Biochemistry* **37**(1), 9–15.
33. Gingras, A. C., Raught, B., and Sonenberg, N. (2001). Regulation of translation initiation by FRAP/mTOR. *Genes Dev.* **15**(7), 807–826.
34. Fruman, D. A., Meyers, R. E., and Cantley, L. C. (1998). Phosphoinositide kinases. *Annu. Rev. Biochem.* **67**, 481–507.
35. Di Cristofano, A. and Pandolfi, P. P. (2000). The multiple roles of PTEN in tumor suppression. *Cell* **100**(4), 387–390.
36. Datta, S. R., Brunet, A., and Greenberg, M. E. (1999). Cellular survival: a play in three Akts. *Genes Dev.* **13**(22), 2905–2927.
37. Gingras, A. C. *et al.* (1998). 4E-BP1, a repressor of mRNA translation, is phosphorylated and inactivated by the Akt(PKB) signaling pathway. *Genes Dev.* **12**(4), 502–513.
38. Heitman, J., Movva, N. R., and Hall, M. N. (1991). Targets for cell cycle arrest by the immunosuppressant rapamycin in yeast. *Science* **253**(5022), 905–909.
39. Crespo, J. L. and Hall, M. N. (2002). Elucidating TOR Signaling and Rapamycin Action: Lessons from Saccharomyces cerevisiae. *Microbiol. Mol. Biol. Rev.* **66**(4), 579–591.
40. von Manteuffel, S. R. *et al.* (1997). The insulin-induced signaling pathway leading to S6 and initiation factor 4E binding protein 1 phosphorylation bifurcates at a rapamycin-sensitive point immediately upstream of p70s6k. *Mol. Cell. Biol.* **17**(9), 5426–5436.
41. Brunn, G. J. *et al.* (1997). Phosphorylation of the translational repressor PHAS-I by the mammalian target of rapamycin. *Science* **277**(5322), 99–101.
42. Hara, K. *et al.* (1997). Regulation of eIF-4E BP1 phosphorylation by mTOR. *J. Biol. Chem.* **272**(42), 26457–26463.

43. Fadden, P., Haystead, T. A., and Lawrence, J. C., Jr. (1997). Identification of phosphorylation sites in the translational regulator, PHAS-I, that are controlled by insulin and rapamycin in rat adipocytes. *J. Biol. Chem.* **272**(15), 10240–10247.
44. Yang, D. Q. and Kastan, M. B. (2000). Participation of ATM in insulin signaling through phosphorylation of eIF-4E-binding protein 1. *Nat. Cell Biol.* **2**(12), 893–908.
45. Heesom, K. J. *et al.* (1998). Insulin-stimulated kinase from rat fat cells that phosphorylates initiation factor 4E-binding protein 1 on the rapamycin-insensitive site (serine-111). *Biochem. J.* **336** (Pt 1), 39–48.
46. Gingras, A. C. *et al.* (1999). Regulation of 4E-BP1 phosphorylation: a novel two-step mechanism. *Genes Dev.* **13**(11), 1422–1437.
47. Peterson, R. T. *et al.* (1999). Protein phosphatase 2A interacts with the 70-kDa S6 kinase and is activated by inhibition of FKBP12-rapamycinassociated protein. *Proc. Natl. Acad. Sci. USA* **96**(8), 4438–4442.
48. Raught, B. *et al.* (2000). Serum-stimulated, rapamycin-sensitive phosphorylation sites in the eukaryotic translation initiation factor 4GI. *EMBO J.* **19**(3), 434–444.
49. Imataka, H. and Sonenberg, N. (1997). Human eukaryotic translation initiation factor 4G (eIF4G) possesses two separate and independent binding sites for eIF4A. *Mol. Cell. Biol.* **17**(12), 6940–6947.
50. Duncan, R. and Hershey, J. W. (1985). Regulation of initiation factors during translational repression caused by serum depletion. Covalent modification. *J. Biol. Chem.* **260**(9), 5493–5497.
51. Lazaris-Karatzas, A., Montine, K. S., and Sonenberg, N. (1990). Malignant transformation by a eukaryotic initiation factor subunit that binds to mRNA 5′ cap. *Nature* **345**(6275), 544–547.
52. Lazaris-Karatzas, A. and Sonenberg, N. (1992). The mRNA 5′ cap-binding protein, eIF-4E, cooperates with v-myc or E1A in the transformation of primary rodent fibroblasts. *Mol. Cell. Biol.* **12**(3), 1234–1238.
53. Koromilas, A. E., Lazaris-Karatzas, A., and Sonenberg, N. (1992). mRNAs containing extensive secondary structure in their 5′ non-coding region translate efficiently in cells overexpressing initiation factor eIF-4E. *EMBO J.* **11**(11), 4153–4158.
54. van der Velden, A. W. and Thomas, A. A. (1999). The role of the 5′ untranslated region of an mRNA in translation regulation during development. *Int. J. Biochem. Cell Biol.* **31**(1), 87–106.
55. Rousseau, D. *et al.* (1996). The eIF4E-binding proteins 1 and 2 are negative regulators of cell growth. *Oncogene* **13**(11), 2415–2420.
56. De Benedetti, A. and Harris, A. L. (1999). eIF4E expression in tumors: its possible role in progression of malignancies. *Int. J. Biochem. Cell Biol.* **31**(1), 59–72.
57. Hershey, J. and Miyamoto, S. (2000). Translational control and cancer, in Sonenberg, N., Hershey, J. W. B., and Mathews, M. B., Eds., *Translational Control of Gene Expression,* pp. 637–654. Cold Spring Harbor Laboratory Press, Cold Spring Harbor, New York.
58. Hidalgo, M. and Rowinsky, E. K. (2000). The rapamycin-sensitive signal transduction pathway as a target for cancer therapy. *Oncogene* **19**(56), 6680–6686.
59. Podsypanina, K. *et al.* (2001). An inhibitor of mTOR reduces neoplasia and normalizes p70/S6 kinase activity in Pten$^{+/-}$ mice. *Proc. Natl. Acad. Sci. USA* **98**(18), 10320–10325.
60. Tee, A. R. *et al.* (2002). Tuberous sclerosis complex-1 and -2 gene products function together to inhibit mammalian target of rapamycin (mTOR)-mediated downstream signaling. *Proc. Natl. Acad. Sci. USA* **99**(21), 13571–13576.
61. Pyronnet, S. and Sonenberg, N. (2001). Cell-cycle-dependent translational control. *Curr. Opin. Genet. Dev.* **11**(1), 13–18.
62. Imataka, H., Gradi, A., and Sonenberg, N. (1998). A newly identified N-terminal amino acid sequence of human eIF4G binds poly(A)-binding protein and functions in poly(A)-dependent translation. *EMBO J.* **17**(24), 7480–7489.
63. Huang, S. and Houghton, P. J. (2002). Inhibitors of mammalian target of rapamycin as novel antitumor agents: from bench to clinic. *Curr. Opin. Investig. Drugs* **3**(2), 295–304.

PART IV

Events in Intracellular Compartments

Marilyn Farquhar, Editor

Introduction

Marilyn Farquhar

Overview of Signaling from Intracellular Compartments

Over the last 10 years research in signaling has shifted from an exclusive consideration of events that occur at the cell membrane to focus on those events that occur inside the cell. Among the startling revelations of current research in this field has been the realization that signaling can actually be initiated from intracellular compartments. Classically signaling was assumed to initiated only by cell surface receptors and transduced across the cell membrane to reach intracellular compartments. Many of the other Parts of this book deal with the intracellular consequences of signaling initiated from the outside. Part IV focuses on events initiated or taking place exclusively from the inside of the cell.

The articles in Part IV describe what is known at present about many of the best-studied signaling pathways that are initiated or propagated from intracellular compartments. The information summarized also reveals the interplay between signaling and protein trafficking in that often trafficking of signaling molecules is necessary to effect signaling, and protein targeting and trafficking are typically regulated by classical signaling molecules such as GTPases and phosphoinositides.

The best understood signaling pathways that are initiated from intracellular compartments are those from the endoplasmic reticulum and mitochondria. Two fascinating and well-worked out pathways for signaling from the endoplasmic reticulum are described in the chapters by Espenshade, Goldstein and Brown and by David Ron. In the former the authors summarize their pioneering work demonstrating that the ER contains a sterol sensor SCAP in complex with SREBP, a membrane bound transcription factor. When cells are depleted of sterols, SCAP escorts SREBP to the Golgi where SREBP undergoes proteolytic cleavage liberating a fragment of SREBP that acts as a transcription factor and activates genes involved in cholesterol and fatty acid biosynthesis. This work represents a novel paradigm and provides an intriguing example of how interacting components of intracellular signaling networks located in different intracellular compartments are brought together by regulated trafficking. The chapter by David Ron describes the ER stress response or unfolded protein response—i.e., the pathway initiated by the presence of unfolded proteins in the ER. When cells are stressed a pattern of response similar to that for sterols is initiated involving a putative sensor, proteolysis and generation of a transcriptional regulator that also modifies gene expression and up-regulates stress genes and down-regulates others.

Signals can also be initiated from mitochondria as described in the chapters by Liu and Butow and by Pagliari, Pinkoski and Green. The former reviews how the changes in the functional state of mitochondria elicit specific responses mediated via for example, by increased cytosolic Ca2+ and NAD+, leading to effects on transcription and proliferation of mitochondria. The Chapter by Pagliari *et al.* describes signals for apoptosis or cell death that can be initiated either from mitochondria in response to stress or from the cell membrane. Both involve activation of specific enzymes, the executioner caspases that cleave specific proteins leading to DNA fragmentation and cell death.

Many signaling events, particularly stimulation of growth factor receptors, lead to mitosis and cell proliferation. The article by Clare McGowen provides a comprehensive description of signaling events that initiate the cell cycle, which is controlled by cyclin dependent kinases that act in series to coordinate duplication and segregation of cell components during mitosis.

Another basic tenet that has come to light in the last 10 years has been the realization that signaling and trafficking of cell surface receptors are irrevocably intertwined. Most if not all cell surface receptors must traffic to the cell interior for many of the consequences of their activation to be signaled. Moreover, receptor fate is determined within endosomes where the decision is made that they be either recycled to act again or degraded (down-regulated). The chapter by Benovic and Keen summarizes current information indicating that growth factor and G protein coupled receptors (GPCR) must be taken up and transported to endosomes for effective signaling to take place. The complementary chapter by Difiore and Scita clarifies the interplay between endocytosis and the cytoskeletal system—e.g., actin, myosin VI and dynamin, and the importance of small GTPases of the rho superfamily in the regulation of endocytosis. This chapter brings out the widespread importance of signaling through small GTPases in regulation of the cytoskeleton as well as endocytic trafficking.

The chapter by Eiden provides a more classical description stimulus-secretion coupling which is outside to inside signaling where the stimulus is provided by a ligand or secretagogue and the response is release of secretion granules from endocrine or exocrine organs or synaptic vesicles from neurons or neurosecretory cells. The emphasis is on the role of Ca2+ and cAMP in stimulating exocytosis and the role of snare complexes in recognition and fusion. The chapter by Prudovsky *et al.* describes the generation and release of signaling molecules such as FGF1 that lack classical signal sequences and appear to be transported independent of the ER-Golgi pathway by mechanisms that remain unclear.

Many signaling events affect the expression of specific genes, which requires trafficking of signaling regulators such as transcription factors and cyclins between the nucleus and cytoplsm. The chapter by Cingolari and Gerace describes current knowledge of the molecular mechanisms involved in trafficking of molecules into and out of the nucleus which involves targeting signals, receptors and signaling molecules, as in the case of other protein targeting events. Here, once again, another small GTPase—ran–plays a key role.

Finally, it should be stated, that it seems likely that what we know at present about signaling from intracellular compartments represents only the tip of the iceberg, and it is probable that cells can respond to changes in the environment of many if not all cell organelles. It seems intuitive that the cell must be able to respond and adapt to changes in key metabolites as it responds to changes in cholesterol composition. Although not dealt with here, it is clear that signals are generated via peroxisomes in response to changes in lipid metabolism that lead to peroxisomal proliferation. It seems likely that signaling events can also be initiated from the Golgi apparatus, given its role in trafficking, processing and packaging of macromolecules for transport along the exocytic pathway, and the fact that abundant signaling molecules such as small GTPases, heterotrimeric G proteins, protein kinases, and phosphoinositides, among others, are found there. It follows that many other signaling pathways initiated from inside the cell remain to be discovered, and many gaps in our knowledge remain to be filled in.

CHAPTER 323

SREBPs: Gene Regulation through Controlled Protein Trafficking

Peter J. Espenshade, Joseph L. Goldstein, and Michael S. Brown

Department of Molecular Genetics, University of Texas Southwestern Medical Center at Dallas, Dallas, Texas

Introduction

Cholesterol is an essential component of mammalian cell membranes. Sufficient levels of cellular cholesterol are required for the integrity and impermeability of the plasma membrane, for the proper assembly of cell surface lipid rafts and caveolae, and for the posttranslational modification of at least one protein, the morphogen Hedgehog [1,2]. However, too much unesterified cholesterol is toxic to cells. Thus, levels of intracellular cholesterol must be tightly regulated [3]. Here, the cell faces a fundamental problem: How does a cell measure the concentration of insoluble, membrane-embedded cholesterol and then appropriately adjust its levels? This chapter outlines the molecular mechanism that cells utilize to maintain proper cholesterol homeostasis.

Mammalian cells obtain cholesterol from two sources. Cholesterol can be synthesized *de novo* from acetyl-CoA or taken up in the form of lipoprotein particles by the low density lipoprotein (LDL) receptor [3]. Cells maintain cholesterol homeostasis through a feedback regulatory mechanism that acts at the level of transcription to coordinately control the input of sterols from these two sources [4]. When cells are depleted of sterols, transcription of genes required for the uptake and synthesis of cholesterol, such as the LDL receptor and 3-hydroxy-3-methylglutaryl coenzyme A (HMG CoA) reductase, increases. Conversely, when intracellular cholesterol levels are high, transcription of these target genes decreases, and sterol levels fall. Analysis of the LDL receptor gene identified a *cis*-acting, positive regulatory sequence in the promoter that mediates transcription in the absence of sterols and is silenced by sterols [5]. This sterol-regulatory element (SRE) was used as bait in the purification and subsequent cDNA cloning of a family of membrane-bound transcription factors, designated sterol regulatory element-binding proteins (SREBPs) [5,6].

SREBPs: Membrane-Bound Transcription Factors

SREBPs transmit information to the nucleus about the sterol content of membranes [4]. This information is generated through the process of regulated intramembrane proteolysis (Rip) [7]. Newly synthesized SREBPs are inserted into the membranes of the ER and nuclear envelope in a hairpin orientation such that the NH_2 and COOH termini project into the cytosol (Fig. 1). These termini are separated by two transmembrane segments that surround a short ~30-amino-acid lumenal loop. The NH_2-terminal domain of ~480 amino acids is a transcription factor of the basic helix–loop–helix leucine zipper family, whereas the COOH-terminal domain of ~580 residues performs a regulatory function. Three SREBP proteins, encoded by two genes, are present in humans, hamsters, and mice [8]. *SREBP1* encodes two isoforms (SREBP-1a and SREBP-1c) through the use of alternate promoters that generate different first exons that are spliced to a common second exon. *SREBP-2* encodes a single protein. SREBP-2 preferentially activates genes involved in cholesterol biosynthesis, whereas SREBP-1 activates genes required for fatty acid synthesis. To date, SREBPs have been shown to directly activate more than

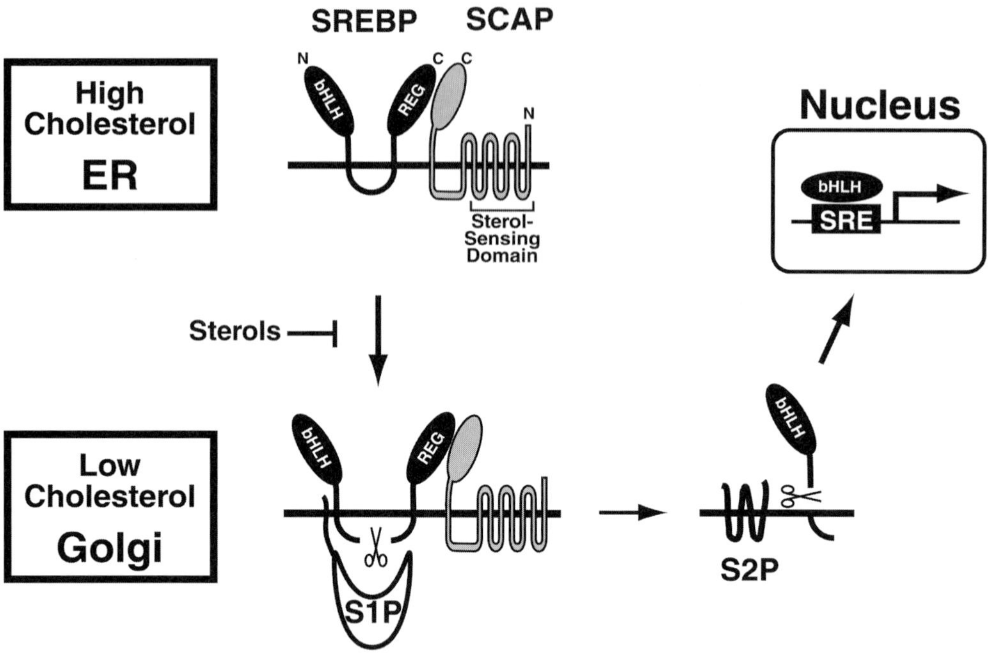

Figure 1 Sterol-regulated transport of SCAP/SREBP complex from ER to Golgi. SCAP/SREBP complexes localize to the ER in the presence of sterols. When cells are depleted of sterols, SCAP escorts SREBP from the ER to the Golgi where SREBP is cleaved at Site-1 by S1P. Then, S2P cuts at a site within the first transmembrane segment. This cleavage by S2P releases the transcription factor from the membrane and allows it to enter the nucleus where it activates transcription of target genes.

30 genes involved in the synthesis and uptake of cholesterol, fatty acids, triglycerides, and phospholipids (see [9] for a detailed review).

When cells are depleted of sterols, SREBPs are activated by two sequential proteolytic cleavage events that release the NH_2-terminal transcription factor from the membrane, allowing it to enter the nucleus and activate transcription of target genes (Fig. 1). The first cleavage occurs in the luminal loop of SREBP at Site-1 and is catalyzed by a transmembrane protease, called Site-1 protease (S1P) [10]. Cleavage by this 1052-amino-acid, subtilisin-related protease occurs after the consensus sequence RXXL and separates the molecule into two halves [11]. S1P is not restricted to proteins involved in cholesterol homeostasis. S1P cleaves and activates ATF6, a transcription factor involved in the endoplasmic reticulum (ER) stress response [12], and it also participates in the processing of Lassa virus glycoprotein precursor GP-C, a step required for production of infectious virus [13].

Following cleavage by S1P, the NH_2-terminal domain of SREBP remains bound to the membrane until a 519-amino-acid, membrane-bound zinc metalloprotease, the Site-2 protease (S2P), cleaves SREBP within the first transmembrane segment [14,15]. The liberated transcription factor now enters the nucleus and activates target gene transcription. Sterols control the activation of SREBPs by regulating cleavage at Site-1. When cholesterol accumulates, cleavage of SREBP by S1P is blocked. Cleavage by S2P is not regulated by sterols, but requires the prior cleavage by S1P [16].

SCAP: Sterol Sensor and Escorter of SREBP from ER to Golgi

Cleavage of SREBP at Site-1 requires a 1276-amino-acid, polytopic membrane protein called SREBP cleavage-activating protein (SCAP) [17]. SCAP is divided into two domains. The NH_2-terminal's 730 amino acids contain eight transmembrane segments that attach SCAP to membranes of the ER and nuclear envelope [18]. The COOH-terminal's 546 amino acids contain five copies of the WD-40 motif, which form β-propeller structures that mediate protein–protein interactions [19]. This domain of SCAP forms a tight complex with the COOH-terminal regulatory domain of SREBP, and this interaction is essential for cleavage of SREBP by S1P (Fig. 1) [20,21]. Chinese hamster ovary (CHO) cells lacking SCAP fail to process SREBP at Site-1 and show reduced levels of SREBP precursor protein, indicating that SCAP is also required for stability of SREBP [22].

SCAP functions as the sterol sensor in the SREBP pathway. The sterol-sensing activity localizes to transmembrane segments 2–6 (~170 amino acids) in the NH_2 terminus of the protein (Fig. 1). Mutant CHO cells that contain single amino acid substitutions in this sterol-sensing domain of SCAP (Y298C and D443N) fail to sense sterols and continue to process SREBPs even in the presence of high levels of sterols [17,23].

Genetic, biochemical, and live-cell microscopy studies have established the following mechanism for sterol-mediated regulation of the processing of SREBP (Fig. 1). In the absence of sterols, SCAP escorts SREBP from the ER to the Golgi where S1P cleaves SREBP at Site-1 and initiates release of the NH_2-terminal transcription factor. Following cleavage of SREBP, SCAP recycles to the ER [24]. In the presence of sterols, the SREBP/SCAP complex remains in the ER and is compartmentally separated from active S1P. The initial evidence for this model emerged from studies of the glycosylation state of SCAP. In wild-type CHO cells cultured in the presence of sterols, the *N*-linked carbohydrate chains of SCAP are sensitive to digestion with endoglycosidase H, indicating that the protein resides in the ER. However, when cells are depleted of sterols, *N*-linked carbohydrates on SCAP become resistant to endoglycosidase H treatment, suggesting that SCAP travels to the Golgi, but only in the absence of sterols [23].

The cloning and characterization of S1P provided further support for this model (Fig. 1). S1P is synthesized as an inactive zymogen (S1P-A). Autocatalytic processing of S1P at two sites in the NH_2 terminus of the protein leads to the removal of an inhibitory prosegment and production of the active form of the enzyme (S1P-C) [25]. Glycosylation and immunolocalization studies demonstrated that S1P-C resides in the Golgi. Therefore, in order to be processed at Site-1, SREBP must move from the ER to the Golgi. SCAP is the key to this movement.

Additional evidence for the role of SCAP in the trafficking of SREBP comes from experiments using SCAP-deficient mutant CHO cells (SRD-13A cells) that possess two copies of a nonfunctional SCAP gene [22]. These cells fail to process SREBPs, and they are therefore unable to synthesize cholesterol. Relocalization of active S1P from the Golgi to the ER bypasses the requirement of SCAP for processing of SREBP in SCAP-deficient SRD-13A cells [26]. This relocalization can be accomplished experimentally in two ways: (1) by treatment of SRD-13A cells with the fungal metabolite, brefeldin A, which results in fusion of the Golgi with the ER; and (2) by transfection of SRD-13A cells with a cDNA encoding a fusion protein of S1P (lacking its membrane-spanning segment) attached to the ER retrieval sequence, KDEL. In both of these situations, cleavage of SREBP occurs constitutively and is not inhibited by sterols [26]. Considered together, these results demonstrate that SCAP functions to escort SREBP from the ER to the Golgi and that sterols act on SCAP to block this transport.

The regulated transport model for SREBP processing is supported by light and electron microscopy studies of CHO/GFP-SCAP cells, a line of CHO cells that stably expresses a fusion protein consisting of the green fluorescent protein (GFP) covalently attached to the NH_2 terminus of SCAP [27]. In the presence of sterols, the GFP-SCAP/SREBP complex localizes to the ER. When CHO/GFP-SCAP cells are depleted of sterols, GFP-SCAP/SREBP moves from the ER to the Golgi by way of the ER–Golgi intermediate compartment (ERGIC) [27,28]. Time-lapse imaging of living CHO/GFP-SCAP cells reveals that on acute depletion of sterols, GFP-SCAP initially accumulates in the ERGIC and then the Golgi [28]. A movie of the ER to Golgi transport of GFP-SCAP can be found at http://www.cell.com/cgi/content/full/102/3/315/DC1. These data suggest that sterols block an early step in the ER to Golgi transport of SCAP.

Sterols Control Sorting of SCAP/SREBP into ER Vesicles

Dissection of the sterol-mediated regulation of ER-to-Golgi transport of SCAP/SREBP necessitated the development of assays that reconstitute this event *in vitro*. Studies in the yeast *Saccharomyces cerevisiae* by Wuestehube and Schekman [29] and in mammalian cells by Rowe and colleagues [30] led to the development of an *in vitro* assay that reconstitutes vesicle formation from the ER using isolated ER membranes and cytosol. These assays were used as a starting point for the development of an *in vitro* reaction that reconstitutes the sterol-regulated formation of SCAP-containing vesicles [28]. For these experiments, Nohturfft *et al.* [27] utilized CHO/VSVG-T7 cells, a cloned line of CHO-K1 cells that stably expresses a temperature-sensitive mutant (tsO45) of vesicular stomatitis virus G protein (VSVG) containing a cytoplasmic, COOH-terminal T7 epitope tag (VSVG-T7). VSVG, a type I transmembrane protein, is a classic marker of the mammalian secretory pathway. Incubation of CHO/VSVG-T7 cells at the nonpermissive temperature (40°C) results in accumulation of VSVG-T7 in the ER [31]. VSVG-T7 subsequently moves to the Golgi when the temperature is lowered to 32°C *in vivo*. When ER membranes are prepared from CHO/ VSVG-T7 cells grown in the absence of sterols, SCAP enters ER vesicles in a reaction that requires ER membranes, rat liver cytosol, ATP, and GTP. When ER membranes are prepared from cells grown in the presence of sterols, vesicles continue to form, as indicated by their content of VSVG-T7. However, SCAP is no longer incorporated into these vesicles. Immunoisolation experiments demonstrate that sterols act by regulating sorting of SCAP into a common class of ER vesicles that contain VSVG, rather than regulating the formation of a unique class of vesicle, specific for SCAP [28].

Recent data indicate that SCAP leaves the ER in COPII-coated vesicles, the same vesicles that carry VSVG and other secretory and membrane-bound proteins destined for the Golgi [28]. Five cytosolic proteins (Sar1, Sec23/24, and Sec13/31, collectively called COPII) can substitute for cytosol in the *in vitro* vesicle formation reaction. Detailed biochemical studies in yeast and mammalian cells have elucidated the mechanism of ER vesicle formation [32]. COPII vesicle formation is initiated by the recruitment of Sar1, a small GTP-binding protein, to the ER membrane through an interaction with a transmembrane, guanine nucleotide exchange protein called Sec12. GTP-bound Sar1 then recruits two heterodimeric complexes, Sec23/24 and Sec13/31, to the membrane. Binding of COPII to the ER membrane forms a protein coat and drives vesicle budding. Hydrolysis of GTP by Sar1 destabilizes the vesicle coat and permits fusion with distal compartments. The Sec23/24 heterodimer is believed to function in cargo selection, while Sec13/31 stimulates the

GTPase activating protein (GAP) activity of the Sec23/24 complex [32].

Characterization of SCAP vesicles produced in the aforementioned *in vitro* vesicle formation assay revealed that SCAP exits the ER in COPII-coated vesicles [28]. But how do sterols control the packaging of SCAP into COPII vesicles? Complexes of Sar1, Sec23/24, and cargo molecules, such as VSVG, can be isolated from ER microsomes that have been incubated with purified COPII proteins and then solubilized with digitonin [33,34]. Using a similar protocol, Espenshade *et al.* [28] showed that SCAP forms a ternary complex with mammalian Sar1 and Sec23/24 and that the formation of this complex is regulated by sterols. Complex formation occurs when ER membranes are prepared from cells cultured in the absence, but not the presence, of sterols [28]. These data indicate that sterols control the packaging of SREBP/SCAP into ER vesicles by regulating the ability of SCAP to interact with COPII vesicle coat proteins (Fig. 2). In the absence of sterols, SCAP/SREBP binds to COPII, enters vesicles, and then moves to the Golgi for proteolytic processing of SREBPs. In the presence of sterols, SCAP/SREBP is unable to interact with the COPII proteins and remains in the ER.

ER Retention of SCAP/SREBP

Clues to the mechanism by which SCAP/SREBP is retained in the ER come from experiments that examine the effects of overexpressing the sterol-sensing domain of SCAP [35]. In the presence of sterols, SCAP/SREBP is retained in the ER. However, when a truncated form of SCAP containing the sterol-sensing domain, transmembrane domains 1–6 (TM1–6), is overexpressed, SCAP/SREBP exits the ER, and SREBP is cleaved even in the presence of sterols. Importantly, overexpression of SCAP(TM1–6) carrying the Y298C mutation that confers sterol resistance has no effect on cleavage of SREBP [35]. These data suggest that in the presence of sterols SCAP binds to a putative retention protein. Overexpression of SCAP(TM1–6) competes for binding of full-length SCAP to this putative retention protein and allows SCAP/SREBP to exit the ER. It is unclear whether the Y298C mutation disrupts the ability of SCAP to sense sterols directly, or whether it impairs its interaction with the putative retention protein (or both).

Current data support the following model, shown schematically in Fig. 2. In the presence of sterols, SCAP/SREBP binds to a retention protein in the ER (protein X in Fig. 2), which prevents its interaction with COPII proteins. This failure to interact with COPII could result from either a conformational change in SCAP itself or from the sequestration of the SCAP/SREBP complex in a domain of the ER that has low vesicle formation activity. In the absence of sterols, SCAP/SREBP no longer interacts with the retention protein and is available to bind to COPII, enters vesicles; it is also transported to the Golgi where SREBP is cleaved sequentially by S1P and S2P.

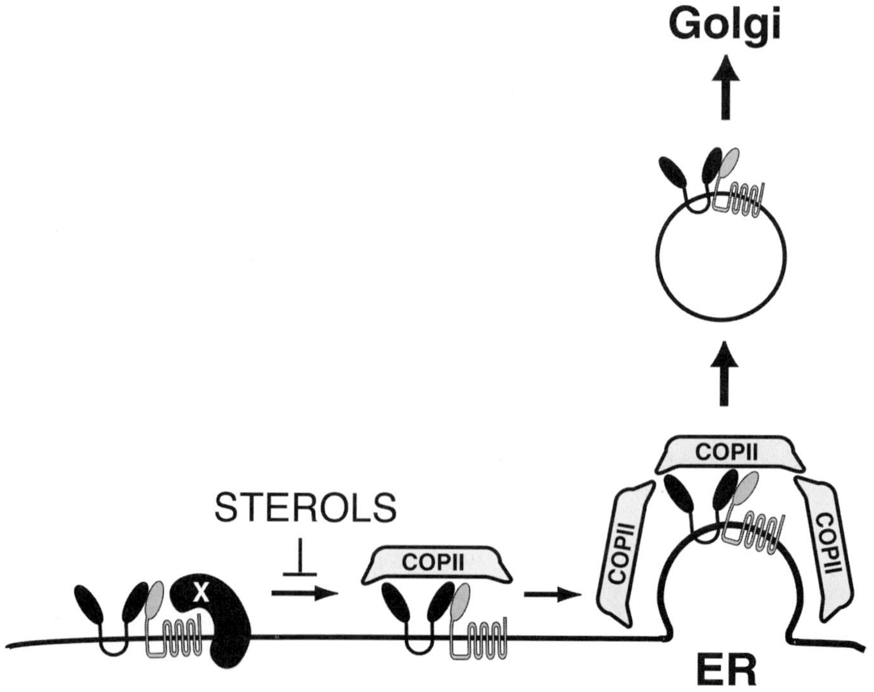

Figure 2 Model for ER retention of SREBP/SCAP complex. In the presence of sterols, SCAP binds to a putative ER retention protein (labeled X). Binding of SCAP to protein X prevents interaction of SREBP/SCAP with COPII proteins. When cells are depleted of sterols, SREBP/SCAP binds to one of the components of COPII, allowing the SREBP/SCAP complex to be packaged into COPII-coated vesicles. SREBP/SCAP is then transported to the Golgi where SREBP is cleaved sequentially by S1P and S2P.

Conclusions

A complete understanding of the mechanism by which sterols regulate the sorting of SCAP/SREBP into ER vesicles awaits the identification of the putative ER retention protein. Detailed structural studies of SCAP will also be required to learn how SCAP senses sterols. Does SCAP directly recognize cholesterol (or an hydroxylated derivative of cholesterol)? Or, does SCAP monitor a more general property of membranes when the local environment surrounding SCAP is perturbed by sterols [36]? Several other proteins that function in various aspects of cholesterol homeostasis contain sterol-sensing domains, including the rate-controlling enzyme in cholesterol synthesis; HMG CoA reductase; the Hedgehog receptor, Patched; and the Niemann-Pick Type C disease gene, NPC1 [4,37]. It remains to be determined whether these other sterol-sensing domains function in a manner similar to the sterol-sensing domain of SCAP. Much has been learned, but more remains to be discovered about this fundamental control mechanism for cellular cholesterol homeostasis.

Acknowledgments

This work was supported by research grants from the National Institutes of Health (HL20948) and the Perot Family Foundation. P. J. E. is the recipient of a Burroughs Wellcome Fund Career Award in the Biomedical Sciences.

References

1. Anderson, R. G. W. (1998). The caveolae membrane system. *Annu. Rev. Biochem.* **67**, 199–225.
2. Mann, R. K. and Beachy, P. A. (2000). Cholesterol modification of proteins. *Biochim. Biophys. Acta* **1529**, 188–202.
3. Brown, M. S. and Goldstein, J. L. (1986). A receptor-mediated pathway for cholesterol homeostasis. *Science* **232**, 34–47.
4. Brown, M. S. and Goldstein, J. L. (1999). A proteolytic pathway that controls the cholesterol content of membranes, cells, and blood. *Proc. Natl. Acad. Sci. USA* **96**, 11041–11048.
5. Wang, X., Briggs, M. R., Hua, X., Yokoyama, C., Goldstein, J. L., and Brown, M. S. (1993). Nuclear protein that binds sterol regulatory element of LDL receptor promoter: II. Purification and characterization. *J. Biol. Chem.* **268**, 14497–14504.
6. Yokoyama, C., Wang, X., Briggs, M. R., Admon, A., Wu, J., Hua, X., Goldstein, J. L., and Brown, M. S. (1993). SREBP-1, a basic helix–loop–helix leucine zipper protein that controls transcription of the LDL receptor gene. *Cell* **75**, 187–197.
7. Brown, M. S., Ye, J., Rawson, R. B., and Goldstein, J. L. (2000). Regulated intramembrane proteolysis: A control mechanism conserved from bacteria to humans. *Cell* **100**, 391–398.
8. Brown, M. S. and Goldstein, J. L. (1997). The SREBP pathway: Regulation of cholesterol metabolism by proteolysis of a membrane-bound transcription factor. *Cell* b 331–340.
9. Horton, J. D., Goldstein, J. L., and Brown, M. S. (2002). SREBPs: Activators of the complete program of cholesterol and fatty acid synthesis in liver. *J. Clin. Invest.* (in press).
10. Sakai, J., Rawson, R. B., Espenshade, P. J., Cheng, D., Seegmiller, A. C., Goldstein, J. L., and Brown, M. S. (1998). Molecular identification of the sterol-regulated luminal protease that cleaves SREBPs and controls lipid composition of animal cells. *Mol. Cell* **2**, 505–514.
11. Duncan, E. A., Brown, M. S., Goldstein, J. L., and Sakai, J. (1997). Cleavage site for sterol-regulated protease localized to a Leu-Ser bond in the lumenal loop of sterol regulatory element-binding protein-2. *J. Biol. Chem.* **272**, 12778–12785.
12. Ye, J., Rawson, R. B., Komuro, R., Dave, U. P., Prywes, R., Brown, M. S., and Goldstein, J. L. (2000). ER stress induces cleavage of membrane-bound ATF6 by the same proteases that process SREBPs. *Mol. Cell* **6**, 1355–1364.
13. Lenz, O., ter Meulen, J., Klenk, H. D., Seidah, N. G., and Garten, W. (2001). The Lassa virus glycoprotein precursor GP-C is proteolytically processed by subtilase SKI-1/S1P. *Proc. Natl. Acad. Sci. USA* **98**, 12701–12705.
14. Rawson, R. B., Zelenski, N. G., Nijhawan, D., Ye, J., Sakai, J., Hasan, M. T., Chang, T. Y., Brown, M. S., and Goldstein, J. L. (1997). Complementation cloning of S2P, a gene encoding a putative metalloprotease required for intramembrane cleavage of SREBPs. *Mol. Cell* **1**, 47–57.
15. Duncan, E. A., Dave, U. P., Sakai, J., Goldstein, J. L., and Brown, M. S. (1998). Second-site cleavage in sterol regulatory element-binding protein occurs at transmembrane junction as determined by cysteine panning. *J. Biol. Chem.* **273**, 17801–17809.
16. Sakai, J., Duncan, E. A., Rawson, R. B., Hua, X., Brown, M. S., and Goldstein, J. L. (1996). Sterol-regulated release of SREBP-2 from cell membranes requires two sequential cleavages, one within a transmembrane segment. *Cell* **85**, 1037–1046.
17. Hua, X., Nohturfft, A., Goldstein, J. L., and Brown, M. S. (1996). Sterol resistance in CHO cells traced to point mutation in SREBP cleavage-activating protein. *Cell* **87**, 415–426.
18. Nohturfft, A., Brown, M. S., and Goldstein, J. L. (1998). Topology of SREBP cleavage-activating protein, a polytopic membrane protein with a sterol-sensing domain. *J. Biol. Chem.* **273**, 17243–17250.
19. Smith, T. F., Gaitatzes, C., Saxena, K., and Neer, E. J. (1999). The WD repeat: A common architecture for diverse functions. *Trends Biochem. Sci.* **24**, 181–185.
20. Sakai, J., Nohturfft, A., Cheng, D., Ho, Y. K., Brown, M. S., and Goldstein, J. L. (1997). Identification of complexes between the COOH-terminal domains of sterol regulatory element-binding proteins (SREBPs) and SREBP cleavage-activating protein. *J. Biol. Chem.* **272**, 20213–20221.
21. Sakai, J., Nohturfft, A., Goldstein, J. L., and Brown, M. S. (1998). Cleavage of sterol regulatory element-binding proteins (SREBPs) at Site-1 requires interaction with SREBP cleavage-activating protein. Evidence from *in vivo* competition studies. *J. Biol. Chem.* **273**, 5785–5793.
22. Rawson, R. B., DeBose-Boyd, R., Goldstein, J. L., and Brown, M. S. (1999). Failure to cleave sterol regulatory element-binding proteins (SREBPs) causes cholesterol auxotrophy in Chinese hamster ovary cells with genetic absence of SREBP cleavage-activating protein. *J. Biol. Chem.* **274**, 28549–28556.
23. Nohturfft, A., Brown, M. S., and Goldstein, J. L. (1998). Sterols regulate processing of carbohydrate chains of wild-type SREBP cleavage-activating protein (SCAP), but not sterol-resistant mutants Y298C or D443N. *Proc. Natl. Acad. Sci. USA* **95**, 12848–12853.
24. Nohturfft, A., DeBose-Boyd, R. A., Scheek, S., Goldstein, J. L., and Brown, M. S. (1999). Sterols regulate cycling of SREBP cleavage-activating protein (SCAP) between endoplasmic reticulum and Golgi. *Proc. Natl. Acad. Sci. USA* **96**, 11235–11240.
25. Espenshade, P. J., Cheng, D., Goldstein, J. L., and Brown, M. S. (1999). Autocatalytic processing of site-1 protease removes propeptide and permits cleavage of sterol regulatory element-binding proteins. *J. Biol. Chem.* **274**, 22795–22804.
26. DeBose-Boyd, R. A., Brown, M. S., Li, W. P., Nohturfft, A., Goldstein, J. L., and Espenshade, P. J. (1999). Transport-dependent proteolysis of SREBP: Relocation of site-1 protease from Golgi to ER obviates the need for SREBP transport to Golgi. *Cell* **99**, 703–712.
27. Nohturfft, A., Yabe, D., Goldstein, J. L., Brown, M. S., and Espenshade, P. J. (2000). Regulated step in cholesterol feedback localized to budding of SCAP from ER membranes. *Cell* **102**, 315–323.
28. Espenshade, P., Li, W. P., and Yabe, D. Sterols control sorting of SCAP into COPII vesicles at the ER (submitted).

29. Wuestehube, L. J. and Schekman, R. W. (1992). Reconstitution of transport from endoplasmic reticulum to Golgi complex using endoplasmic reticulum-enriched membrane fraction from yeast. *Methods Enzymol.* **219**, 124–136.
30. Rowe, T., Aridor, M., McCaffery, J. M., Plutner, H., Nuoffer, C., and Balch, W. E. (1996). COPII vesicles derived from mammalian endoplasmic reticulum microsomes recruit COPI. *J. Cell Biol.* **135**, 895–911.
31. Bergmann, J. E. (1989). Using temperature-sensitive mutants of VSV to study membrane protein biogenesis. *Methods Cell Biol.* **32**, 85–110.
32. Antonny, B. and Schekman, R. (2001). ER export: public transportation by the COPII coach. *Curr. Opin. Cell Biol.* **13**, 438–443.
33. Kuehn, M. J., Herrmann, J. M., and Schekman, R. (1998). COPII–cargo interactions direct protein sorting into ER-derived transport vesicles. *Nature* **391**, 187–190.
34. Aridor, M., Weissman, J., Bannykh, S., Nuoffer, C., and Balch, W. E. (1998). Cargo selection by the COPII budding machinery during export from the ER. *J. Cell Biol.* **141**, 61–70.
35. Yang, T., Goldstein, J. L., and Brown, M. S. (2000). Overexpression of membrane domain of SCAP prevents sterols from inhibiting SCAP. SREBP exit from endoplasmic reticulum. *J. Biol. Chem.* **275**, 29881–29886.
36. Dobrosotskaya, I., Seegmiller, A. C., Brown, M. S., Goldstein, J. L., and Rawson, R. B. (2002). Regulation of SREBP processing and membrane lipid production by phosphatidylethanolamine in *Drosophila*. *Science* In Press.
37. Osborne, T. F. and Rosenfeld, J. M. (1998). Related membrane domains in proteins of sterol sensing and cell signaling provide a glimpse of treasures still buried within the dynamic realm of intracellular metabolic regulation. *Curr. Opin. Lipidol.* **9**, 137–140.

CHAPTER 324

Endoplasmic Reticulum Stress Responses

David Ron
Skirball Institute of Biomolecular Medicine,
New York University School of Medicine,
New York, New York

Introduction

Eukaryotic cells must adapt to wide variation in the load of client proteins that their endoplasmic reticulum (ER) must process. They do so by modulating the synthesis of client proteins and the expression of genes whose products determine the folding capacity of the organelle. The signaling pathways activated to effect these adaptations are part of the unfolded protein response (UPR), which is similar in many ways to the cytoplasmic heat-shock response. In mammalian cells, signaling is initiated by three known ER resident transmembrane proteins (IRE1, ATF6, and PERK) that can sense the stress caused by imbalance between capacity and demand in the ER. IRE1 and PERK are transmembrane protein kinases whose activity is repressed by free chaperones in the ER lumen.

During ER stress, as free chaperones are recruited to serve client proteins, these transducers are activated, resulting in downstream signaling. IRE1 effector functions include endonucleolytic processing of a substrate mRNA (HAC1 in yeast and XBP-1 in metazoans) and activation of JUN N-terminal kinase, resulting in activated gene expression. PERK phosphorylates the translation initiation factor eIF2 and mediates both translational repression of client protein biosynthesis and activation of specific mRNA translation. The latter culminates in increased expression of the transcription factor ATF4 and activates a gene expression program that is also accessed by other forms of stress and is referred to as the *integrated stress response.* ATF6 activation is mediated by proteolytic processing and liberation of an active N-terminal protein fragment that serves as a transcription factor for activating gene expression in the UPR.

ER Stress Defined

Proteins destined for secretion and membrane insertion are synthesized on ribosomes associated with the endoplasmic reticulum. Such client proteins of the ER are translocated into the organelle's lumen where they undergo posttranslational modifications specific to that environment. Chaperone proteins specific to the ER assist in the folding and posttranslational processing of the client proteins, and quality-control mechanisms operating within the organelle clear properly folded client proteins for transport to the Golgi apparatus or earmark others for degradation by the ER associated degradation apparatus (ERAD).

The ER is thus a dynamic cellular compartment whose activity level varies among different cell types and between physiological states. In complex metazoans it is most developed in cells devoted to secretion, such as those of the exocrine and endocrine glands, immunoglobulin-secreting plasma cells, and cells with extensive membranous processes such as neurons and myelin-producing glial cells. In other words, the quantity and intrinsic properties of client proteins synthesized in response to developmental programs and physiological states set the level of *demand* made of the ER. The ability of the ER to *supply* this demand is impacted by the size of the organelle, by the complement of chaperones and other intrinsic protein and lipid components at its disposal, and by the physiological state of the cell. The latter is particularly important in pathological states associated with nutrient limitation, hypoxia, or toxins that degrade the ER's ability to process client proteins (reviewed in [1–3]).

Specific signaling pathways are activated by an imbalance between the demand on the ER and the ability of the organelle to meet the demand, so-called "ER stress." Activation of these pathways occurs under both pathological conditions and in response to normal physiological demands. However, much of what we have learned about them is derived from experiments in which cells were treated with toxins, for example, tunicamycin, an inhibitor of N-linked glycosylation. Because such toxins cause ER stress by interfering with protein folding, the signaling pathways that they activate came to be known as the *unfolded protein response* (UPR) [4,5]. This is a somewhat misleading term; because the pathways in question appear to be activated by the threat of malfolded proteins and do not necessarily require the accumulation of measurable amounts of malfolded proteins.

The UPR in Yeast

The UPR was first studied in yeast where treatment with tunicamycin or mutations that affect protein folding in the ER were noted to activate transcription of the *KAR2* gene encoding the yeast homolog of the mammalian ER chaperone BiP [6,7]. Genetic analysis of signaling from the yeast ER to the *KAR2* promoter unveiled a truly novel pathway. Signaling in response to ER stress is initiated by a type 1 ER resident transmembrane protein encoded by the *IRE1* gene [8,9]. The N-terminal lumenal domain of Ire1p responds to the ER stress signal and the signal is transmitted to the C-terminal, cytoplasmic effector domain in which two distinct activities reside: a protein kinase activity for which the only known substrate is Ire1p itself [8,9] and, remarkably, an endonuclease activity that removes an intron from the *HAC1* mRNA [10–12]. Ire1p activation thus entails transautophosphorylation that unmasks the endonucleolytic activity [13]. The cleaved ends of the *HAC1* mRNA are joined by tRNA ligase to generate an activated, spliced form of the *HAC1* mRNA [14]. The unspliced *HAC1* mRNA is poorly translated, whereas the protein encoded by the spliced form is well expressed. Thus, noncanonical splicing controls expression of Hac1p, a transcription factor that activates downstream target genes of the yeast UPR [15,11].

In yeast the IRE1>HAC1 pathway is a simple linear one that is required for activation of all known target genes of the UPR. These have been the subjects of detailed analysis by expression microarrays. Surprisingly, the genes induced by the UPR in yeast include not only the expected ER chaperones but also components of the apparatus for degrading ER proteins, the machinery required for membrane biosynthesis and non-ER components of the secretory apparatus [16,17]. The yeast UPR, therefore, appears to up-regulate the capacity of the entire endomembrane system to deal with client proteins. In this context the UPR should be viewed as a pathway concerned both with adapting to stress in the short term and with growth and anabolism [18]. It is also clear, however, that the ability to up-regulate target genes of the UPR is an important adaptation, because yeast defective in this function are hypersensitive to ER stress [16,17,19].

IRE1 proteins are conserved in higher eukaryotes [20,21] and their mechanism of action appears to be conserved too (see later discussion), however their preeminent role in regulating all aspects of the UPR has been superseded by new components that are lacking in yeast. One important difference between higher eukaryotes and yeast is that the former rapidly and reversibly attenuate protein biosynthesis in response to ER stress. This adaptation decreases the quantity of client proteins that are delivered to the stressed ER and reduces the demand on the organelle.

The UPR is Metazoans

Translation control by ER stress is mediated by PERK (also known as PEK or EIF2AK3), a type 1 transmembrane ER resident protein kinase related to IRE1 in its stress-sensing lumenal domain. The C-terminal, cytoplasmic effector domain of PERK is a protein kinase that phosphorylates the alpha subunit of translation initiation factor 2 [22]. Phosphorylation of eIF2α inhibits nucleotide exchange on the eIF2 complex and blocks the initiation step of translation. An important component of the regulation of PERK and IRE1 activity depends on an interaction between their related N-terminal lumenal domains and the ER chaperone BiP. In unstressed cells, dispensable BiP protein is available to bind the lumenal domains of PERK and IRE1. BiP binding maintains these proteins in a monomeric, inactive state. When the folding reserve of the ER is challenged, either by increased demand or reduced function, levels of dispensable BiP decrease and PERK-BiP and IRE1-BiP complexes dissociate. Freed from the inhibitory effects of BiP binding, PERK and IRE1 oligomerize, transautophosphorylate, and initiate downstream signaling [23,24].

Cells lacking PERK are unable to reduce translation when experiencing ER stress and consequently are exquisitely sensitive to agents and conditions that cause such stress [25]. Humans and mice with *PERK* mutations develop early onset diabetes mellitus caused by destruction of the insulin-producing pancreatic β cells [26,27]. Protein synthesis in β cells is strongly activated by plasma glucose; PERK modulates this increase, ensuring that the quantity of proteins translocated into the ER does not exceed the capacity of the organelle to fold them. In $PERK^{-/-}$ islets one finds cells with distended ER, and the presence of these correlates with high rates of programmed cell death in the mutant islets. It is clear, therefore, that PERK signaling is activated under normal physiological circumstances, notably in professional secretory cells.

The hypothesis whereby loss of translational control leads to stuffing of the ER with client proteins goes a long way toward explaining the phenotype of the *PERK* mutation. However, this simple hypothesis does not take into account the role of PERK in regulating gene expression in the UPR. eIF2α phosphorylation decreases the efficiency of translation of most mRNAs; however, some mRNAs are translated more efficiently in such circumstances. The mechanism for this has been worked out in great detail in the case of the yeast *GCN4* mRNA, which encodes a transcription factor that activates genes involved in amino acid biosynthesis [28]. Translation of the mammalian *ATF4* mRNA is similarly up-regulated by eIF2α phosphorylation [29]. Thus, ATF4 translation during ER stress and up-regulation of its downstream target genes require PERK. Because this signaling pathway is activated both by ER stress and other conditions associated with eIF2α phosphorylation (e.g., amino acid starvation, arsenite treatment), we propose referring to it as the *integrated stress response* (ISR). Preliminary results suggest that the ISR may contribute substantially to cell survival during ER stress and to developmental programs [30].

GADD34 is an interesting target gene of PERK (and the ISR). The encoded protein binds the catalytic subunit of protein phosphatase 1 and recruits it to dephosphorylate eIF2α. Thus, the integrated stress response induces a negative feedback loop that serves to terminate the response [31]. It seems likely that eIF2α dephosphorylation plays a role in the recovery of protein biosynthesis, which, in turn, is required for the translation of mRNAs that are up-regulated as part of the gene expression program activated during the UPR (and other stresses associated with an ISR). According to this model, the first line of defense against ER stress in complex metazoans is translational repression. It is followed, soon thereafter, by remodeling of the internal milieu; the latter process requires new protein synthesis that is favored by GADD34.

Higher eukaryotes possess a third signaling pathway from the ER to the nucleus (Fig. 1), which is mediated in mammals by two ATF6 genes, α and β, encoding very similar type 2 transmembrane ER resident proteins. The N-terminal cytoplasmic portion of these proteins is a bZIP transcription factor that binds and activates ER stress response elements (ERSEs) in the promoter of such classic UPR target genes as BiP and CHOP [32–34]. In unstressed cells ATF6 is maintained in an inactive state by being tethered to the ER membrane. In response to ER stress, the protein is proteolytically cleaved, generating a free N-terminal fragment that migrates to the nucleus and activates transcription through ERSEs [35,33]. This mode of activation resembles that of another ER-tethered bZIP protein, the sterol-regulated transcription factor SREBP. This is more than a superficial resemblance, because the proteases that process and activate SREBP were found to be essential also for the activation of ATF6 [36]. These proteases are found in the Golgi apparatus. It is likely, therefore, that ATF6 activity is controlled by ER retention, and migration of ATF6 to the Golgi apparatus during ER stress likely leads to processing and liberation of the active transcription factor.

The comprehensive analysis of the yeast UPR, carried out by expression microarrays, has not yet been extended to mammalian cells. Consequently, we do not have detailed information on how the different signaling components of the UPR interact to control gene expression in higher eukaryotes. However, based on even the limited information available, it is clear that the mammalian UPR has diverged considerably from that of the yeast. The ATF6 pathway appears to have assumed a major role in controlling chaperone gene expression. This is revealed by the severe defect in *BiP*

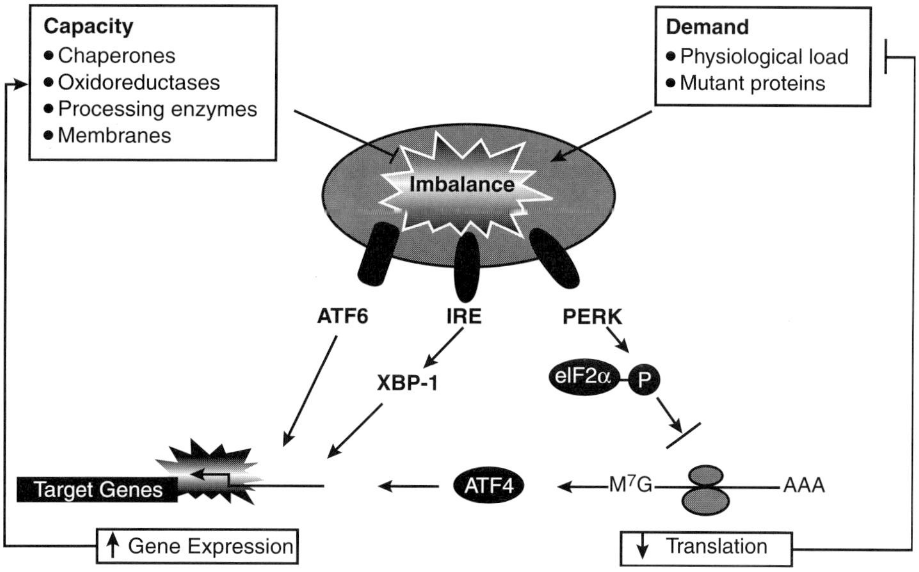

Figure 1 Depiction of the major aspects of the endoplasmic reticulum stress response and its mediators in mammalian cells. An imbalance between the load of client proteins imposed on the ER (demand) and capacity of the ER to fold client proteins is sensed by the three ER stress transducers: ATF6, IRE1, and PERK. These coordinate two functional responses: activation of the expression of genes whose products increase ER folding capacity and inhibition of protein synthesis that attenuated demand on the ER. IRE1 activity is transduced though XBP-1 to activate gene expression, whereas ATF6 directly mediates ER to nuclear signaling. PERK signaling is mediated by eIF2α phosphorylation modifying translation initiation rates. The latter results in a general decrease in demand on the ER and at the same time activates the translation of transcription factors such as ATF4 that contribute to activated gene expression.

expression manifested by cells that lack ATF6-processing protease activity [36]. PERK and the integrated stress response also contribute significantly to gene expression in the UPR [29,30]. Some target genes of the UPR, such as *GADD34* and *CHOP*, are highly dependent on the ISR, whereas other genes, such as *BiP,* are less dependent on PERK signaling and the ISR. IRE1 genes, of which mammals have two, appear to have been diverted from exclusive control of a broad set of UPR genes (as is the case in yeast) toward certain specific cellular functions.

We have found that mouse cells lacking *IRE1* are virtually unimpaired in activating most target genes of the UPR. Furthermore, such mutant cells are not manifestly hypersensitive to pharmacological manipulations that cause ER stress. One measure of the divergence of IRE1 function from yeast to higher eukaryotes is the finding that mammalian IRE1 activates stress-activated protein kinases during ER stress. It does so by recruiting TRAF2 to the ER membrane [37]. The physiological significance of this signaling pathway is not fully understood, but it may link ER stress and programmed cell death. The association between ER stress and programmed cell death is a poorly understood but potentially important subject, particularly in light of recent findings suggesting that ER stress may accompany common pathological conditions [38]. Several signaling pathways leading to CHOP [39], JNK [37], CASPASE-12 [40], and cABL [41] activation may be involved, but the manner in which these are integrated into the UPR remains to be studied in greater detail.

Further clues to the role of mammalian IRE1 have been provided by the identification of a direct downstream target mRNA (analogous to yeast *HAC1*). The *XBP-1* mRNA is processed by mammalian IRE1 in an ER-stress-dependent manner and IRE1 thus controls the expression of this important transcription factor [42–44]. The target genes of XBP-1 have not been identified, but analysis of *XBP-1* knock-out mice reveals its essential role in development of B cells to plasma cells [45]. XBP-1 mediated signaling presumably up-regulates the capacity of the endomembrane system to deal with the load of immunoglobulins that must be secreted by the developing B cell. The finding that XBP-1 is controlled by IRE1 suggests that in mammalian cells the IRE1 genes have specialized to link the load on the ER to certain long-term "developmental" adaptations of the ER. This hypothesis is further supported by the observation that both XBP-1 and IRE1 are enriched in cells that specialize in secretion.

Conclusion

In summary, the mammalian UPR consists of rapid and transient translational repression, which is completely PERK dependent. This short-term adaptation is supplanted by up-regulation of genes that enhance the capacity of ER to deal with increased load. All three known signaling components of the UPR (PERK, IRE1, and ATF6) participate in this latter adaptation. Additional signaling downstream of these and other ER stress responders effects important and poorly understood life and death decisions.

References

1. Lee, A. S. (1992). Mammalian stress response: Induction of the glucose-regulated protein family. *Curr. Opin. Cell Biol.* **4**, 267–273.
2. Aridor, M. and Balch, W. E. (1999). Integration of endoplasmic reticulum signaling in health and disease. *Nat. Med.* **5**, 745–751.
3. Kaufman, R. J. (1999). Stress signaling from the lumen of the endoplasmic reticulum: Coordination of gene transcriptional and translational controls. *Genes Dev.* **13**, 1211–1233.
4. Kozutsumi, Y., Segal, M., Normington, K., Gething, M. J., and Sambrook, J. (1988). The presence of malfolded proteins in the endoplasmic reticulum signals the induction of glucose-regulated proteins. *Nature* **332**, 462–464.
5. Gething, M. J. and Sambrook, J. (1992). Protein folding in the cell. *Nature* **355**, 33–45.
6. Normington, K., Kohno, K., Kozutsumi, Y., Gething, M. J., and Sambrook, J. (1989). S. cerevisiae encodes an essential protein homologous in sequence and function to mammalian BiP. *Cell* **57**, 1223–1236.
7. Kohno, K., Normington, K., Sambrook, J., Gething, M. J., and Mori, K. (1993). The promoter region of the yeast KAR2 (BiP) gene contains a regulatory domain that responds to the presence of unfolded proteins in the endoplasmic reticulum. *Mol. Cell Biol.* **13**, 877–890.
8. Cox, J. S., Shamu, C. E., and Walter, P. (1993). Transcriptional induction of genes encoding endoplasmic reticulum resident proteins requires a transmembrane protein kinase. *Cell* **73**, 1197–1206.
9. Mori, K., Ma, W., Gething, M. J., and Sambrook, J. (1993). A transmembrane protein with a cdc2+/CDC28-related kinase activity is required for signaling from the ER to the nucleus. *Cell* **74**, 743–756.
10. Cox, J. S. and Walter, P. (1996). A novel mechanism for regulating activity of a transcription factor that controls the unfolded protein response. *Cell* **87**, 391–404.
11. Kawahara, T., Yanagi, H., Yura, T., and Mori, K. (1997). Endoplasmic reticulum stress-induced mRNA splicing permits synthesis of transcription factor Hac1p/Ern4p that activates the unfolded protein response. *Mol. Biol. Cell* **8**, 1845–1862.
12. Sidrauski, C. and Walter, P. (1997). The transmembrane kinase Ire1p is a site-specific endonuclease that initiates mRNA splicing in the unfolded protein response. *Cell* **90**, 1031–1039.
13. Shamu, C. E. and Walter, P. (1996). Oligomerization and phosphorylation of the Ire1p kinase during intracellular signaling from the endoplasmic reticulum to the nucleus. *Embo J* **15**, 3028–3039.
14. Sidrauski, C., Cox, J. S., and Walter, P. (1996). tRNA ligase is required for regulated mRNA splicing in the unfolded protein response. *Cell* **87**, 405–413.
15. Chapman, R. E. and Walter, P. (1997). Translational attenuation mediated by an mRNA intron. *Curr. Biol.* **7**, 850–859.
16. Casagrande, R., Stern, P., Diehn, M., Shamu, C., Osario, M., Zuniga, M., Brown, P. O., and Ploegh, H. (2000). Degradation of proteins from the ER of S. cerevisiae requires an intact unfolded protein response pathway. *Mol. Cell* **5**, 729–735.
17. Travers, K. J., Patil, C. K., Wodicka, L., Lockhart, D. J., Weissman, J. S., and Walter, P. (2000). Functional and genomic analyses reveal an essential coordination between the unfolded protein response and ER-associated degradation. *Cell* **101**, 249–258.
18. Cox, J. S., Chapman, R. E., and Walter, P. (1997). The unfolded protein response coordinates the production of endoplasmic reticulum protein and endoplasmic reticulum membrane. *Mol. Biol. Cell* **8**, 1805–1814.
19. Ng, D. T., Spear, E. D., and Walter, P. (2000). The unfolded protein response regulates multiple aspects of secretory and membrane protein biogenesis and endoplasmic reticulum quality control. *J. Cell Biol.* **150**, 77–88.
20. Tirasophon, W., Welihinda, A. A., and Kaufman, R. J. (1998). A stress response pathway from the endoplasmic reticulum to the nucleus

requires a novel bifunctional protein kinase/endoribonuclease (Ire1p) in mammalian cells. *Genes Dev.* **12**, 1812–1824.

21. Wang, X. Z., Harding, H. P., Zhang, Y., Jolicoeur, E. M., Kuroda, M., and Ron, D. (1998). Cloning of mammalian Ire1 reveals diversity in the ER stress responses. *EMBO J* **17**, 5708–5717.
22. Harding, H., Zhang, Y., and Ron, D. (1999). Translation and protein folding are coupled by an endoplasmic reticulum resident kinase. *Nature* **397**, 271–274.
23. Bertolotti, A., Zhang, Y., Hendershot, L., Harding, H., and Ron, D. (2000). Dynamic interaction of BiP and the ER stress transducers in the unfolded protein response. *Nature Cell Biology* **2**, 326–332.
24. Liu, C. Y., Schroder, M., and Kaufman, R. J. (2000). Ligand-independent dimerization activates the stress-response kinases IRE1 and PERK in the lumen of the endoplasmic reticulum. *J Biol Chem* **275**, 17680–17687.
25. Harding, H., Zhang, Y., Bertolotti, A., Zeng, H., and Ron, D. (2000). *Perk* is essential for translational regulation and cell survival during the unfolded protein response. *Mol. Cell* **5**, 897–904.
26. Delepine, M., Nicolino, M., Barrett, T., Golamaully, M., Lathrop, G. M., and Julier, C. (2000). EIF2AK3, encoding translation initiation factor 2-alpha kinase 3, is mutated in patients with Wolcott-Rallison syndrome. *Nat. Genet.* **25**, 406–409.
27. Harding, H., Zeng, H., Zhang, Y., Jungreis, R., Chung, P., Plesken, H., Sabatini, D., and Ron, D. (2001). Diabetes Mellitus and exocrine pancreatic dysfunction in Perk$^{-/-}$ mice reveals a role for translational control in survival of secretory cells. *Mol. Cell* **7**, 1153–1163.
28. Hinnebusch, A. G. (1997). Translational regulation of yeast GCN4. A window on factors that control initiator-tRNA binding to the ribosome. *J. Biol. Chem.* **272**, 21661–21664.
29. Harding, H., Novoa, I., Zhang, Y., Zeng, H., Wek, R. C., Schapira, M., and Ron, D. (2000). Regulated translation initiation controls stress-induced gene expression in mammalian cells. *Mol. Cell* **6**, 1099–1108.
30. Scheuner, D., Song, B., McEwen, E., Gillespie, P., Saunders, T., Bonner-Weir, S., and Kaufman, R. J. (2001). Translational control is required for the unfolded protein response and in-vivo glucose homeostasis. *Mol. Cell* **7**, 1165–1176.
31. Novoa, I., Zeng, H., Harding, H., and Ron, D. (2001). Feedback inhibition of the unfolded protein response by GADD34-mediated dephosphorylation of eIF2α. *J. Cell Biol.* **153**, 1011–1022.
32. Yoshida, H., Haze, K., Yanagi, H., Yura, T., and Mori, K. (1998). Identification of the cis-acting endoplasmic reticulum stress response element responsible for transcriptional induction of mammalian glucose-regulated proteins. Involvement of basic leucine zipper transcription factors. *J. Biol. Chem.* **273**, 33741–33749.
33. Yoshida, H., Okada, T., Haze, K., Yanagi, H., Yura, T., Negishi, M., and Mori, K. (2000). ATF6 activated by proteolysis binds in the presence of NF-Y (CBF) directly to the *cis*-acting element responsible for the mammalian unfolded protein response. *Mol. Cell Biol.* **20**, 6755–6767.
34. Yoshida, H., Okada, T., Haze, K., Yanagi, H., Yura, T., Negishi, M., and Mori, K. (2001). Endoplasmic reticulum stress-induced formation of transcription factor complex ERSF including NF-Y (CBF) and activating transcription factors 6alpha and 6beta that activates the mammalian unfolded protein response. *Mol. Cell Biol.* **21**, 1239–1248.
35. Haze, K., Yoshida, H., Yanagi, H., Yura, T., and Mori, K. (1999). Mammalian transcription factor ATF6 is synthesized as a transmembrane protein and activated by proteolysis in response to endoplasmic reticulum stress. *Mol. Biol. Cell* **10**, 3787–3799.
36. Ye, J., Rawson, R. B., Komuro, R., Chen, X., Dave, U. P., Prywes, R., Brown, M. S., and Goldstein, J. L. (2000). ER stress induces cleavage of membrane-bound ATF6 by the same proteases that process SREBPs. *Mol. Cell* **6**, 1355–1364.
37. Urano, F., Wang, X., Bertolotti, A., Zhang, Y., Chung, P., Harding, H., and Ron, D. (2000). Coupling of stress in the endoplasmic reticulum to activation of JNK protein kinases by transmembrane protein kinase IRE1. *Science* **287**, 664–666.
38. Lee, A. S. (2001). The glucose-regulated proteins: Stress induction and clinical applications. *Trends Biochem. Sci.* **26**, 504–510.
39. Zinszner, H., Kuroda, M., Wang, X., Batchvarova, N., Lightfoot, R. T., Remotti, H., Stevens, J. L., and Ron, D. (1998). CHOP is implicated in programmed cell death in response to impaired function of the endoplasmic reticulum. *Genes Dev.* **12**, 982–995.
40. Nakagawa, T., Zhu, H., Morishima, N., Li, E., Xu, J., Yankner, B. A., and Yuan, J. (2000). Caspase-12 mediates endoplasmic-reticulum-specific apoptosis and cytotoxicity by amyloid-beta. *Nature* **403**, 98–103.
41. Ito, Y., Pandey, P., Mishra, N., Kumar, S., Narula, N., Kharbanda, S., Saxena, S., and Kufe, D. (2001). Targeting of the c-Abl tyrosine kinase to mitochondria in endoplasmic reticulum stress-induced apoptosis. *Mol. Cell Biol.* **21**, 6233–6242.
42. Shen, X., Ellis, R. E., Lee, K., Liu, C.-Y., Yang, K., Solomon, A., Yoshida, H., Morimoto, R., Kurnit, D. M., Mori, K., and Kaufman, R. J. (2001). Complementary signaling pathways regulate the unfolded protein response and are required for *C. elegans* development. *Cell* **107**, 893–903.
43. Yoshida, H., Matsui, T., Yamamoto, A., Okada, T., and Mori, K. (2001). XBP1 mRNA is induced by ATF6 and spliced by IRE1 in response to ER stress to produce a highly active transcription factor. *Cell* **107**, 881–891.
44. Calfon, M., Zeng, H., Urano, F., Till, J. H., Hubbard, S. R., Harding, H. P., Clark, S. G., and Ron, D. (2002). IRE1 couples endoplasmic reticulum load to secretory capacity by processing the *XBP-1* mRNA. *Nature* **415**, 92–96.
45. Reimold, A. M., Iwakoshi, N. N., Manis, J., Vallabhajosyula, P., Szomolanyi-Tsuda, E., Gravallese, E. M., Friend, D., Grusby, M. J., Alt, F., and Glimcher, L. H. (2001). Plasma cell differentiation requires the transcription factor XBP-1. *Nature* **412**, 300–307.

CHAPTER 325

Signaling Pathways from Mitochondria to the Nucleus

Zhengchang Liu and Ronald A. Butow
Department of Molecular Biology,
University of Texas Southwestern Medical Center,
Dallas, Texas

Introduction

Mitochondrial research has entered a new era, which recognizes that mitochondria are key players in intracellular signaling. Cells monitor and respond to changes in the functional state of their mitochondria, resulting in responses as severe as apoptosis and as subtle as changes in nuclear gene expression. The general term for this mitochondria-to-nucleus signaling is *retrograde regulation*. In animal cells, compromises in mitochondrial function often lead to increases in mitochondrial biogenesis, with attendant increases in mitochondrial DNA levels and mitochondrial gene expression. In many instances, changes also occur in the expression of nonmitochondrial components associated with mitochondrial dysfunctions. In animal cells, many of these changes can be attributed to increased cytosolic Ca^{2+}, derived in part from mitochondrial stores. Mitochondria-derived NAD^+ may also be important in signaling the retrograde response. In yeast, both positive and negative regulatory factors, particularly the *RTG* genes, have been identified that control retrograde regulation. Transcriptome analyses reveal that the retrograde regulation is both a stress response and a mechanism to overcome metabolic defects imposed by dysfunctional mitochondria. Many of these effects are signaled by intra- and extracellular glutamate levels. In both yeast and filamentous fungi, the retrograde response is intimately tied to aging and cellular senescence. Altogether, retrograde regulation adapts cells to increased energy demands, and links mitochondrial function to fundamental processes of nutrient sensing, carbohydrate and nitrogen metabolism, stress, and aging.

Milestones in Mitochondrial Research

The history of mitochondrial research can be divided conveniently into three areas of accomplishment, each revealing unique and fascinating intricacies of this eukaryotic organelle. Spanning a period of roughly 50 years, these studies broadly include (1) structure–function, (2) biogenesis, and (3) intracellular signaling. First was the elucidation of the primary function of mitochondria—the synthesis of ATP via oxidative phosphorylation—culminating with the recognition that the structure of mitochondria, the ATP synthase complex, and the organization of components of the electron transport chain within the inner mitochondrial membrane were integral features of the coupling of electron transport to ATP synthesis through an electrochemical gradient.

The second area followed the discovery of mitochondrial DNA (mtDNA) and its role in mitochondrial biogenesis. This led immediately to the general concept of nuclear-mitochondrial cross-talk and the idea that there is coordinate regulation between the nuclear and mitochondrial genomes. These findings also provided a basis for understanding mitochondrial diseases and offered a new paradigm for studies in molecular evolution using mtDNA as an evolutionary clock. Control of the replication, transmission, and expression of the mitochondrial genome is determined largely by nuclear gene products that function either directly within mitochondria, or indirectly through regulation of the expression, targeting, stability, and so on of mitochondrial proteins. These studies culminated with the detailed dissection of the mechanisms of targeting and import of proteins into mitochondria.

The third and most current area of mitochondrial research relates to the unexpected finding that mitochondria are players in intracellular signaling. Changes in the functional state of mitochondria, often subtle, are now known to elicit a wide range of cellular responses: apoptosis, aging and cellular senescence, tumor progression, changes in gene expression leading to metabolic reorganization, and patterns of differentiation, to name a few. Here we review some examples of mitochondrial signaling in animal cells and in fungi, and we also highlight the molecular events that trigger and transduce mitochondrial signals to effect changes in nuclear gene expression. Some aspects of these signaling pathways have been discussed recently [1]. Collectively, signaling from mitochondria to the nucleus is called *retrograde regulation* to distinguish it from nuclear control over mitochondrial biogenesis and mitochondrial activities.

Mitochondrial Signaling

Responses in Animals

Stimulation of Mitochondrial Biogenesis

Different cellular and environmental cues can lead to various tissue-specific increases in mitochondrial biogenesis. Stimulation of skeletal muscle contractility, for example, leads to a dramatic muscle-specific proliferation of mitochondria because of an increased demand for energy [2]. Heat production in brown fat tissue in adaptive thermogenesis is the result of both an increase in mitochondrial mass as well as an increase in uncoupled respiratory activity due to specific expression of uncoupler proteins [3,4]. The regulatory factors that control these responses include the transcription factors NRF-1 and NRF-2, which regulate the expression of a number of nuclear genes encoding mitochondrial proteins [5], and additional activities of the transcription factor PPARγ and its coactivator, PGC-1 [3,4]. Because these proteins also regulate the expression of nuclear genes that control the copy number and expression of mtDNA, they are key components of the interorganelle cross-talk pathway that coordinately regulates expression of the nuclear and mitochondrial genomes.

In cells with dysfunctional mitochondria, increased expression of one or more components of the mitochondrial bioenergetic apparatus, or increased mitochondrial biogenesis, is easily interpretable as an attempt by the cell to overcome the mitochondrial defects. Responses resulting in down-regulation of gene expression, however, are sometimes more difficult to understand. One such example is the decreased steady-state levels of the ADP/ATP translocator mRNA observed in HeLa cells treated with the mitochondrial protein synthesis inhibitor, chloramphenicol, or with the uncoupler, dinitrophenol [6]. This finding contrasts with an earlier observation by the same group that, in ρ° derivatives of human 143B cells (cells that lack mtDNA), expression of an isoform of the ADP/ATP translocator increased [7]. The difference might possibly be related to acute versus chronic failure of mitochondrial function.

In tissues of patients with pathogenic mtDNA mutations, increased expression is seen of nuclear genes encoding components of the oxidative phosphorylation apparatus, including the heart-muscle isoform of adenine nucleotide translocator (ANT1), the ATP synthase β-subunit, and ancillary genes including hexokinase I and glycogen phosphorylase [8]. Additionally, an increase is seen in the expression of mtDNA transcripts, although the amount of mtDNA was unchanged. This contrasts with findings in *Drosophila* that mtDNA copy numbers increase in cells with mtDNA deletion mutations as a mechanism to compensate for reduced expression of genes within the deleted mtDNA [9,10].

In cells with intact, wild-type levels of mtDNA, a retrograde response can often be initiated by perturbation of a specific component or pathway within mitochondria. For example, extensive proliferation of mitochondria is seen in skeletal muscle of mice lacking the heart-muscle ANT1 [11]. Limited transcriptional profiling of skeletal muscle of such ANT1-deficient mice revealed some 17 genes that were up-regulated, including mitochondrial tRNA and rRNA genes, and mitochondrial and nuclear genes encoding components of the oxidative phosphorylation apparatus [12].

An important feature of retrograde regulation is that the response is not restricted to changes in expression of nuclear genes encoding mitochondrial components (see later section). For example, in ρ° chicken cells, an increase is seen in expression of genes encoding elongation factor 1β [13], β-actin, and v-Myc [14] and a decrease in expression of *MEK2* (encoding a dual specificity kinase), the latter having been observed both in ρ° cells and in ρ^{+} cells treated with drugs that block the expression of mtDNA genes [15].

Signals

Ca^{2+} What are the signals that originate from dysfunctional mitochondria in retrograde response pathways, and how are those signals transduced to effect such dramatic changes in nuclear gene expression? It is likely that, depending on the cell type and the kind of mitochondrial lesion, the retrograde response will be initiated by different, but not necessarily exclusive, signals. In the case of the apoptotic response, the release of cytochrome c and other pro-apoptotic factors from mitochondria of many different cell types represents one pathway for the initiation of cell death (see [16] for a recent review).

For less severe responses, specific signals or effectors arise as a consequence of changes in the functional state of mitochondria. One of these effectors is Ca^{2+}. Mitochondria of animal cells transport Ca^{2+} very efficiently, and are thus likely to be important in regulating the distribution of intracellular Ca^{2+} [17]. Given that so many pathways are regulated through Ca^{2+} signaling, it is perhaps not surprising that alterations in the mitochondrial function of animal cells can elicit such profound cellular responses.

Lowering the mtDNA copy number in mouse C2C12 myocytes, or treatment of those cells with mitochondrial inhibitors, results in changes in the expression of a number of cellular and regulatory factors [18]. There is increased expression of the sarcoplasmic reticular ryanodine receptor-1, calcineurin, calcineurin-dependent NFATc, c-Jun N-terminal kinase (JNK)-dependent ATF2 (activated transcription factor 2), and subunit Vb of cytochrome oxidase, and decreased expression of NFκB. These changes can be attributed to increased cytosolic Ca^{2+} levels arising, in part, because of increased Ca^{2+} efflux from defective mitochondria. Treatment of rat PC12 cells with uncouplers similarly increases intracellular Ca^{2+} levels, resulting in activation of the mitogen-activated kinases, ERK1 and ERK2 [19]. These studies clearly establish that mitochondrial stress can disrupt Ca^{2+} homeostasis and either initiate changes in nuclear gene expression or affect signaling pathways directly.

Ca^{2+}-related mitochondria-to-nucleus signaling induces invasive phenotypes in otherwise noninvasive C2H12 myocytes and in human pulmonary carcinoma cells [20]. Expression of cathepsin L and TGFβ, two tumor marker genes, increases in C2H12 cells depleted of their mtDNA or when treated with an uncoupler of oxidative phosphorylation. Increased cathepsin L gene expression in mitochondria-stressed cells results partly from the activation of calcium-dependent protein kinase C. Together, these results suggest a novel connection between retrograde signaling and tumor progression and metastasis. Because of a less efficient repair system, it has long been recognized that mutations are fixed at a higher rate in mtDNA compared to the nuclear genome. This observation has led to some speculation that the accumulation of mtDNA mutations in tumor cells may underlie tumorigenesis by providing some selective advantage to cells, leading eventually to cell populations that contain only mutant mitochondrial genomes (i.e., they are homoplasmic for their mtDNA) [21]. Although it remains possible that some mtDNA mutations could influence tumor progression through a retrograde response, a recent study suggests that simple random segregation of mtDNA in a rapidly dividing cell population, as in tumors, could lead to homoplasmy without any selection [22].

Arnould *et al.* [23] used similar strategies of mtDNA depletion and treatment of cells with mitochondrial poisons to show that mitochondrial dysfunction causes the activation of the cAMP responsive element binding protein (CREB), with attendant activation of CREB target gene expression. CREB is activated by phosphorylation at serine 133 by calcium/calmodulin kinase IV (CaMKIV). The increased cytosolic Ca^{2+} due to mitochondrial dysfunction disrupts the interaction of CaMKIV with protein phosphatase 2A, leading to the activation of the kinase. Among the genes affected by CREB phosphorylation in mtDNA depleted cells is p53. Phosphorylated CREB interacts with p53, affecting p53's transcriptional activity. One consequence is an increased expression of p21, a cyclin-dependent kinase inhibitor, which results in an inhibition of the entry of cells into S phase. These findings suggest that the slower growth rate of cells with mitochondrial defects may not be due solely to compromises in energy metabolism.

Differentiating between a "true" mitochondrial signaling pathway versus secondary responses resulting from decreased ATP supplies because of mitochondrial defects can be discriminated by the specificity of the response. If, for example, cell viability is compromised, many of the molecular indicators of cellular robustness will change; such changes would thus be considered secondary to the mitochondrial dysfunction. This issue was clearly addressed by Rochard *et al.* [24], who observed that treatment of avian myoblast cells in culture with mitochondrial poisons inhibited myogenic development without affecting cell viability or ATP levels. The specificity of the response was evident from the observation that these treatments down-regulated the expression of the muscle-specific regulatory protein, myogenin. Moreover, stimulation of mitochondrial activity enhanced myoblast differentiation, suggesting that some subset of nuclear genes involved in myoblast differentiation is likely to be affected by the functional state of their mitochondria.

NAD+ Besides its role in oxidation-reduction, NAD^+ has been implicated in a variety of signaling pathways related particularly to both mono- and polyADP ribosylation of proteins and to the formation of NAD^+ derivatives, such as cyclic ADP-ribose (cADPR) and nicotinic acid adenine dinucleotide phosphate (NAADP). These derivatives are formed by NADase activities (reviewed in [25]), and both cADPR and NAADP are effective agents for the mobilization of Ca^{2+} from the endoplasmic reticulum (reviewed in [26]).

Di Lisa and Ziegler [25] proposed a model in which mitochondrial alterations could lead to the release of mitochondrial NAD^+, providing additional substrate for nuclear polyADP ribosylation catalyzed by poly(ADP-ribose) polymerase. That enzyme responds to DNA damage—with its attendant changes in nuclear gene expression—and could be linked to the functional state of mitochondria via the generation of reactive oxygen species (ROS) by mitochondrial electron transport activity. Release of mitochondrial NAD^+ would also contribute substrate for cADPR formation catalyzed by an outer mitochondrial membrane NADase activity.

Yeast

THE *RTG* PATHWAY

The most extensive mechanistic information on retrograde regulation has come from studies of *CIT2* gene expression in the budding yeast *Saccharomyces cerevisiae*. *CIT2* encodes a glyoxylate cycle isoform of citrate synthase, one of three citrate synthase genes in yeast. In respiratory competent cells with fully functional mitochondria, *CIT2* expression is low, whereas in cells with dysfunctional mitochondria, *CIT2* transcription, as well as the level of citrate synthase enzyme activity, is dramatically elevated [27–29]. As discussed in a later section, the increases in *CIT2* expression and activity reflect the activation of pathways that resupply intermediates to the TCA cycle when they become severely depleted in cells with dysfunctional mitochondria. Three positive regulatory factors, Rtg1p, Rtg2p, and Rtg3p [29,30], and four negative regulatory factors, Mks1p [31,32], Lst8p [33], and two redundant 14-3-3 proteins encoded by *BMH1* and *BMH2* (Z. Liu, 2002) function in this retrograde pathway. As we will see, the apparent complexity of retrograde regulation in yeast reflects, in part, the intersection of mitochondria-to-nucleus signaling with nutrient sensing, carbohydrate and nitrogen metabolism, and stress.

Rtg1p and Rtg3p are basic helix–loop–helix leucine zipper (bHLH/Zip) transcription factors that bind as a heterodimer to activate transcription at a novel DNA target site, GTCAC, called an R box [30]. This site differs from the canonical E-box recognition site, CANNTG, used by most bHLH proteins. Rtg3p has both C- and N-terminal transactivation domains [34], but only the N-terminal region of the protein is critical for regulated target gene expression [35]. In addition, Rtg3p contains an AD1 (activation domain 1) region at its N terminus [36]. AD1 is a prototypical activation domain conserved from yeast to mammals and is restricted to a subset of bHLH transcription factors. The AD1 domain of Rtg3p is required for its interaction with the nuclear histone acetyltransferase complex, SAGA, to activate transcription. Although the requirement of the AD1 domain in Rtg3p for regulated *CIT2* expression has not been tested directly, mutants lacking components of the SAGA complex show about a 50% reduction in *CIT2* expression [37]. This suggests that the AD1domain in Rtg3p may not be essential for transcriptional activation of target genes. Compared with other bHLH transcription factors, Rtg1p has a truncated basic domain and lacks any transactivation domains. Rtg1p's main function may be to facilitate Rtg3p binding to R boxes.

Rtg2p is a cytoplasmic protein with an amino terminal ATP binding domain similar to the hsp70/actin/sugar kinase superfamily of ATP binding proteins [38,39]. Rtg2p acts as a proximal sensor of mitochondrial dysfunction, transducing signals to the Rtg1p–Rtg3p complex. When the retrograde response is off, Rtg1p and Rtg3p are complexed together in the cytoplasm and Rtg3p is phosphorylated at multiple sites within the protein [35]. When the retrograde response is activated, Rtg3p becomes partially dephosphorylated and accumulates in the nucleus (along with Rtg1p) by processes that require Rtg2p. In *rtg2* mutant cells, Rtg1p and Rtg3p remain in the cytoplasm and Rtg3p is hyperphosphorylated. In addition to its transcriptional function, Rtg1p is also required to sequester Rtg3p in the cytoplasm when the retrograde response is off: Inactivation of *RTG1* results in constitutive nuclear localization of Rtg3p even in the absence of Rtg2p. Rtg3p, but not Rtg1p, contains a nuclear localization signal (NLS) located within its basic domain. Whether retention of Rtg3p in the cytoplasm is due to occlusion of its NLS or to more complicated mechanisms has not been established. Most, if not all, of the phosphorylation sites of Rtg3p are in the N-terminal region of the protein, upstream of the bHLH motif. Mutation of a number of these potential phosphorylation sites leads to constitutive nuclear localization of Rtg3p, but activation of its transcriptional function nevertheless requires additional, but unknown, retrograde signals (T. Sekito, 2002).

Among the genes regulated by the Rtg proteins are *CIT1*, *ACO1*, *IDH1*, and *IDH2* [40]. Those genes encode the enzymes catalyzing the first three steps in the TCA cycle that lead to the formation of α-ketoglutarate. In cells with robust mitochondrial activity, their expression is under the control of the HAP transcription complex [41]. In cells with reduced or compromised mitochondrial function, however, expression of those TCA cycle genes becomes largely dependent on the *RTG* genes [40]. This switch to *RTG* control ensures that sufficient α-ketoglutarate, the direct precursor of glutamate, is made to maintain glutamate supplies in cells with dysfunctional mitochondria. This conclusion is consistent with the observation that (1) none of the TCA cycle genes downstream of α-ketoglutarate is regulated by the *RTG* genes, (2) glutamate is a potent repressor of *RTG*-dependent gene expression, and (3) strains harboring *rtg* mutations are glutamate auxotrophs. Thus, a key function of the retrograde response and *RTG*-dependent gene expression is glutamate homeostasis.

Lst8p is an essential 7 WD-repeat protein that has been implicated in the targeting of the general amino acid permease, Gap1p, from the late-Golgi to the plasma membrane [42]. As a class, WD-repeat proteins are involved in diverse cellular activities through their interactions with other proteins. WD-repeat proteins adopt a seven-bladed β-propeller fold, with each repeating unit consisting of a four-stranded antiparallel β-sheet [43]. WD-repeat proteins can likely use different blades of the β-propeller to interact with different proteins and, thus, regulate different pathways [44]. In this connection, analysis of different Lst8p mutants shows that the protein negatively regulates *RTG*-dependent gene expression at two sites, one downstream of Rtg2p, perhaps at the level of the Rtg1p–Rtg3p complex, and the other upstream of Rtg2p by regulating the targeting of an amino acid sensor, Ssy1p, to the plasma membrane. Ssy1p signaling is coupled to the cytoplasmic proteins Ptr3p and Ssy5p (reviewed in [45]). External glutamate represses *RTG*-dependent gene expression via the Ssy1p/Ptr3p/Ssy5 pathway [33]. This same signal transduction pathway is also responsible for the induction of amino acid permeases [45]. It remains to be determined how these pathways of glutamate sensing, retrograde regulation, and amino acid permease induction and targeting are integrated and where the inductive and repressive components of the pathway branch off.

Mks1p has been reported to be involved in multiple pathways: Ras-cAMP signaling [46], negative regulation of lysine biosynthesis [47], positive regulation of nitrogen catabolism [48], and the generation of the yeast prion [URE3] [49]. [URE3] is a non-Mendelian determinant of Ure2p, a negative regulator of the nitrogen catabolite repression (NCR) pathway, with properties fully consistent with it being an infectious prion. The NCR pathway is activated by release of the GATA transcriptional factor, Gln3p, from a cytoplasmic complex with Ure2p, allowing Gln3 to translocate to the nucleus to activate expression of genes involved in the utilization of nonpreferable nitrogen sources such as urea [50].

Mks1p is a phosphoprotein [31,32] whose phosphorylation state mirrors that of Rtg3p [31]. Negative regulation of the retrograde pathway by Mks1p helps explain the proposed roles for the protein as a negative regulator of lysine biosynthesis and as a positive regulator of [URE3] prion formation. Inactivation of *MKS1* results in the constitutive expression of *RTG* target genes. As a consequence, the levels of α-ketoglutarate, a precursor to lysine, are high, resulting in an increase the intracellular lysine pool. The apparent requirement of Mks1p in [URE3] prion formation is also linked to the retrograde pathway. Inactivation of *rtg2* or *rtg3* results in a 20- to 30-fold increase in the frequency of cells harboring the [URE3] prion [31]. In *rtg2 mks1* double mutant cells, in which *RTG* target gene expression is high (because inactivation of *MKS1* bypasses the requirement for Rtg2p), the frequency of [URE3] formation is suppressed to a very low level. However, in *rtg3 mks1* double mutants, in which *RTG*-target gene expression is blocked because of the transcriptional requirement for Rtg3p, the frequency of [URE3] prion formation is high. In other words, [URE3] prion formation correlates with the presence or absence of *RTG*-dependent gene expression, rather than with the presence or absence of Mks1p. Is the regulation of [URE3] formation by the *RTG* pathway related to glutamate homeostasis? Probably yes. In addition to repressing *RTG*-dependent gene expression, glutamate can also completely suppress [URE3] formation [31]. The *RTG* pathway, and glutamate homeostasis may, therefore, be intimately linked to one way that cells could regulate the sensing of the quality of nitrogen sources—by the conversion of the active negative regulator of the NCR pathway, Ure2p, to its inactive prion form, [URE3].

One recent report suggests that the *RTG* genes play no role in [URE3] generation [51]. The discrepancy might lie in whether glutamate is included in the growth medium in assays of [URE3] prion formation. That assay is based on the ability of cells to take up ureidosuccinic acid (USA) when the NCR pathway is activated. Severe glutamate starvation of an *rtg2* mutant strain would activate the NCR pathway by releasing Gln3p from Ure2p. One of Gln3p's targets is *DAL5*, which encodes an allantoate permease that enables cells to take up USA. Because *rtg2* mutant cells are leaky glutamate auxotrophs, they would thus appear as being USA$^+$ in medium lacking glutamate. Sekito *et al.* [31] supplemented the medium with small amounts of glutamate to allow *rtg* mutant cells to grow robustly if Ure2p were inactivated. In that case, all of the cells that were USA$^+$ were [URE3].

The connection between *RTG*-dependent gene expression and nitrogen metabolism was clarified by the finding that the RTG pathway is also regulated by target of rapamycin (TOR) signaling [52,53]. TOR signaling is mediated by two TOR kinases, Tor1p and Tor2p [54]. Rapamycin, an immunosuppressant that binds to FKBP12, a prolyl isomerase, inactivates the kinase activity of Tor1p and Tor2p. In the presence of rapamycin TOR signaling affects multiple cellular processes including activation of the NCR pathway and the expression of stress-related genes in response to carbon limitation [50,54–56]. When rapamycin is added to yeast cells, or when cells are grown on poor nitrogen sources, *RTG*-dependent gene expression is activated by processes that are, for the most part, similar to their activation in cells with dysfunctional mitochondria [35,53].

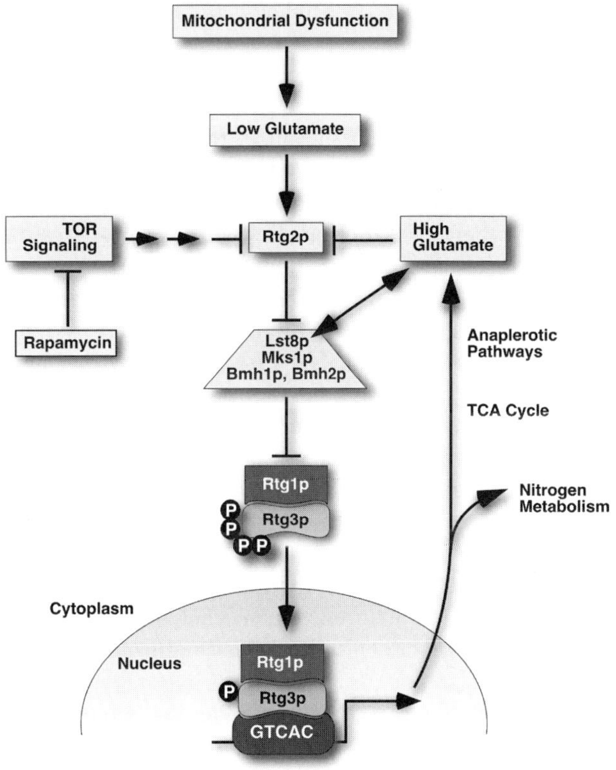

Figure 1 Regulation of *RTG*-dependent gene expression. Translocation of Rtg1p–Rtg3p from the cytoplasm to the nucleus is shown to be regulated by the levels of glutamate and by rapamycin, an inhibitor of the TOR pathway. The negative regulator of the *RTG* pathway, Lst8p, acts both downstream and upstream of Rtg2p. The latter site of action is through the Ssy1p–Ptr3p–Ssy5p signal transduction cascade that responds to external glutamate (see text for details). Inhibition of TOR signaling by rapamycin also activates the RTG pathway through Rtg2p [53], but for reasons that are not yet clear, the relation between nuclear localization of Rtg3p and Rtg3p phosphorylation is opposite to that observed for activation of the RTG pathway by mitochondrial dysfunction or limiting glutamate.

Finally, *BMH1* and *BMH2* also negatively regulate the *RTG* pathway downstream of Rtg2 (Z. Liu, 2002). This is consistent with a recent report suggesting that 14-3-3 proteins are required to keep Rtg3p in an inactive state [57]. Figure 1 summarizes the essential features of the control of *RTG*-dependent gene expression.

ADDITIONAL PATHWAYS

Pleiotropic drug resistance (PDR) is often due to the overexpression of genes encoding ATP-binding cassette (ABC) transporters. In yeast, dysfunctional mitochondria results in the induction of members of this family of PDR genes, such as *PDR5* [58]. The induction of *PDR5* appears to be routed through *RTG*-dependent and *RTG*-independent pathways. Which pathway is chosen may depend on the nature of the mitochondrial lesion [58]. Expression of the *PDR* genes is controlled by two zinc finger transcription factors, Pdr1p and Pdr3p [59]. Enhanced expression of *PDR5* in cells with dysfunctional mitochondria is the result of both increased expression of *PDR3* and posttranslational autoregulation of Pdr3p [58]. In ρ° cells, about one-half of the increased level of *PDR* expression is dependent on the *RTG* pathway. In cells with a mutation of *OXA1*, a gene encoding an inner mitochondrial membrane protein required for cytochrome oxidase and ATPase assembly [60], the up-regulation of *PDR* genes is also observed, but in this case, induction is independent of the *RTG* genes [58]. These findings also suggest that more than one retrograde pathway can affect the same target genes in response to mitochondrial dysfunctions. The *RTG*-independent components of *PDR* expression in ρ° and *oxa1* responses can be traced to the reduction or loss of the F_0 subunit of the mitochondrial ATP synthase complex [61]. How the state of the F_0 subunit is transduced to changes in gene expression is not clear.

It has long been recognized that the functional state of mitochondria is affected by carbon source or growth conditions. In an hypoxic environment, for instance, cells repress the expression of aerobic genes, such as those encoding subunits of the cytochrome oxidase complex; at the same time, they induce the expression of hypoxic genes required for the survival of cells in a low-oxygen environment. In some, but not all cases, a functional mitochondrial electron transport chain is necessary for the induction of hypoxic genes [62]. Repression of cytochrome oxidase subunit (*COX*) genes also occurs in ρ° cells grown aerobically [63,64]; however, under hypoxic growth conditions, expression of *COX* genes in ρ° cells is only about 20% of that in hypoxic ρ^+ cells [63]. These observations suggest that, like the induction of *PDR* genes in ρ° cells, some specific mtDNA-encoded component, perhaps a subunit of the ATP synthase complex, is required to maintain a level of expression of *COX* genes independent of whether the cells have a functional electron transport chain.

As a protection against oxidative damage, cells induce the expression of genes that detoxify ROS. Cytosolic thioredoxin peroxidase I, a thiol-dependent peroxidase encoded by the *TSA1* gene, is an enzyme induced in cells that have been oxidatively stressed, for example, by exposure to high concentrations of H_2O_2 [65]. A role for retrograde signaling in this regulation was noted by the observation that in ρ° cells or in ρ^+ cells treated with the respiratory chain inhibitor, antimycin A, there was a greater induction of *TSA1* expression by H_2O_2 treatment than in ρ^+ cells alone [65]. Induction may be in response to increased ROS production in cells with blocked electron transport activity, and is consistent with the finding that under derepressed growth conditions, inactivation of *TSA1* leads to greater sensitivity to oxidative stress in cells with compromised mitochondrial function.

Genome-Wide Transcriptional Responses to Mitochondrial Dysfunction

Some insights into the scope of the retrograde response have been provided from microarray experiments in yeast by Traven *et al.* [66] and Epstein *et al.* [64] who determined the global changes in gene expression associated with mitochondrial dysfunction in ρ° cells. In addition, Epstein *et al.* [64] evaluated the role of the *RTG* genes in the ρ° responses, and the effects of treatment of respiratory competent cells with three inhibitors of oxidative phosphorylation, antimycin A, an uncoupler (carbonyl cyanide *m*-chlorophenylhydrazone [CCCP]), and an inhibitor of the ATP synthase complex (oligomycin). In comparing the data sets from these studies, it is important to note that different carbon sources were used to grow cells for transcription profiling. Traven *et al.* [66] used glucose, a repressing carbon source, whereas Epstein *et al.* [64] used raffinose, a fermentable but nonrepressing carbon source. In the experiments of Traven *et al.* [66], genes encoding mitochondrial proteins involved in respiratory function and mitochondrial biogenesis were up-regulated in ρ° cells, whereas Epstein *et al.* [64] observed that many of those genes were down-regulated in ρ° petites. One plausible explanation for these differences may be related to the onset of the diauxic shift, in which genes involved in aerobic metabolism become derepressed as glucose is exhausted from the medium [67]; these effects would not be superimposed on a comparison of cells grown on nonrepressing, but fermentable carbon sources. This interpretation is also consistent with the up-regulation of stress response genes and genes involved in the synthesis of reserve carbohydrates in ρ° petites grown on glucose medium [66]. Such responses are typical of yeast cells that begin to deplete glucose from the medium and whose growth rate slows down as they approach stationary phase. Particularly striking was the up-regulation of genes encoding many of the reactions that function to replenish supplies of acetyl-CoA and oxaloacetate to the TCA cycle (i.e., anaplerotic reactions), as well as small molecule transporters, a number of which function in the mobilization of metabolic intermediates of these anaplerotic pathways. In cells grown under derepressing conditions a major source of acetyl-CoA is from peroxisomal β-oxidation of fatty acids. Not surprisingly, respiratory deficient cells dramatically up-regulate peroxisome biogenesis—an outcome that was predicted from the expression profiling data [64].

What is the rationale for these metabolic changes in ρ° petites? A consequence of respiratory deficiency is that the TCA cycle no longer operates as a cycle because of the failure to oxidize succinate. Thus, oxaloacetate, together with its condensation partner, acetyl-CoA, must be provided by other pathways to maintain sufficient amounts of α-ketoglutarate for glutamate biosynthesis. Moreover, the regulation of expression of genes encoding the first three steps of the TCA cycle from which α-ketoglutarate is produced switches from *HAP* control to control by the *RTG*-genes in cells with reduced or compromised mitochondrial function [40]. Loss of respiratory function can also result in the up-regulation of those TCA cycle genes upstream of α-ketoglutarate [66,68].

Transcription profiling reveals that not all genes whose expression changes in respiratory-deficient cells are dependent on the *RTG* genes [64], further supporting the idea of multiple pathways in the retrograde response. Moreover, different mitochondrial lesions clearly elicit different patterns of gene expression as seen from a comparison of the expression profiles of genes up-regulated in respiratory competent cells treated with the different inhibitors of oxidative phosphorylation (Fig. 2). When compared with

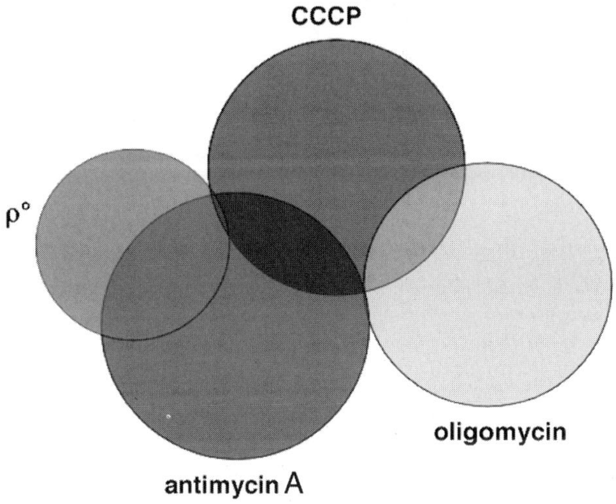

Figure 2 Different mitochondrial inhibitors elicit different genome-wide responses in gene expression. Shown is a Venn diagram of genes that are up-regulated in ρ^+ cells grown in raffinose medium in the presence of the respiratory chain inhibitor, antimycin A, the uncoupler, CCCP, or the ATP synthase inhibitor, oligomycin. These data are compared with a subset of up-regulated genes in a ρ° petite. The diameters of the circles are roughly proportional to the number of up-regulated genes, with the ρ° result set at 42 genes. (Data taken from. [64].)

the response in ρ° cells, only treatment of cells with antimycin A showed a significant overlap in the pattern of gene expression. These findings suggest that in ρ° cells, the predominant observed responses in gene expression are due to respiratory deficiency rather than to the lack of ATP synthesis from oxidative phosphorylation.

Aging and Retrograde Regulation

Aging is a complex phenomenon whose molecular basis is poorly understood. A detailed discussion of aging is well beyond the scope of this review, and so we shall only emphasize the role of mitochondrial function and the retrograde response in aging and cellular senescence. Declining mitochondrial function has often been associated with aging, but, as is the case with most aging-related events, distinguishing between cause and effect has proved difficult. In the filamentous fungi *Podospora anserina* and *Neurospora crassa*, mitochondrial respiratory function and rearrangements of mtDNA eliminating or inactivating essential respiratory genes are intimately related to senescence in these organisms. Senescence in *Podospora anserina* has been associated with the accumulation of senDNAs—circular oligomers of regions of wild-type mtDNA—in which most of the mtDNA coding sequences have been eliminated [69]. Recent studies have now established a direct link between mitochondrial electron transport function, particularly cytochrome oxidase activity, and increased longevity in *P. anserina*. Specific inactivation of nuclear or mtDNA-encoded *COX* genes leads to increased life span [70]. Similarly, life span can be extended by inactivation of the nuclear gene GRISEA [71,72]. GRISEA encodes a transcription factor involved in Cu^{2+} homeostasis. Because Cu^{2+} is an essential component of cytochrome oxidase, inactivation of GRISEA leads to cytochrome oxidase deficiency. In both of these examples, we see an induction of a cyanide-sensitive alternate oxidase (Aox) activity, which shunts electrons directly to oxygen upstream of complex III. It is believed that the presence of Aox induced by retrograde signaling depresses the production of ROS, which otherwise would be high in cytochrome oxidase-deficient mitochondria lacking Aox.

Replicative aging in *S. cerevisiae* is determined by the number of daughter cells that are produced from an individual yeast cell. A relation between the functional state of mitochondria, retrograde signaling, and longevity has become apparent from the finding that some ρ° petite strains show a life span extension that is dependent on the *RTG* genes [73]. Moreover, the correlation between these pathways of life span extension and the *RTG* retrograde pathway was strengthened by the finding that a ρ° petite strain that did not show extended life span had a weak or nonexistent retrograde response as measured by *CIT2* expression. Because of the complete absence of a mitochondrial electron transport chain, life span extension in petite cells, when observed, may be due to decreased production of ROS, to a general increase stress resistance, or to both. We have noted in an earlier section that in one study [66], several stress response genes were reported to be up-regulated in ρ° petites.

The most effective, documented strategy for environmental manipulation of life span is caloric restriction. Restricting calorie intake while maintaining adequate nutrition has been shown to prolong life span in organisms ranging from yeast to mammals. In yeast, glucose limitation results in a clear-cut life span extension, but unlike life span extension in petite cells, life span extension by caloric restriction does not appear to be dependent on the *RTG* genes [74]. On the contrary, inactivation of *RTG*-dependent gene expression in calorically restricted yeast cells has an additive effect on life span extension. These finding led Jiang *et al.* [74] to suggest that life span extension by retrograde regulation and by caloric restriction are separate and exclusive pathways. Just how caloric restriction extends longevity is an issue of some controversy, but it is likely to involve changes in metabolism and cellular redox states. Whether mitochondria play a role here is uncertain. In any case, it is clear that there are *RTG*-independent pathways of mitochondrial retrograde signaling, and it remains to be seen whether any of those pathways participate in life span extension imposed by caloric restriction.

In *Caenorhabditis elegans*, inactivation of an insulin-like signaling pathway and mitochondrial dysfunction resulting from decreased levels of coenzyme Q (CoQ, a component of the electron transport chain), imposed either by diet [75] or by a mutation in CoQ biosynthesis [76,77], result in increased life span. These pathways have been suggested to effect life span extension by an increase in ROS scavenging or production. Because mitochondrial function is affected in both cases, it is plausible that retrograde signaling may also contribute to the control of life span extension in worms.

Conclusions

Mitochondria, once acquired by cells as a free-living organisms, retained, on the one hand, a degree of semiautonomy and, on the other hand, adopted retrograde signaling to alert their "cellular host" to changes in the state of their mitochondria. The responses are usually seen as altered patterns of nuclear gene expression. In some cases, global mitochondrial biogenesis is affected; in other cases, specific changes are seen in the expression of some mitochondrial or nonmitochondrial components. In most instances, retrograde regulation can be rationalized as attempts by cells to make accommodations to their altered or defective mitochondria. Retrograde regulation is likely to play a role in diverse cellular processes, including metabolism, environmental sensing, development, aging, and disease. Some signaling molecules and regulatory factors have been identified, and others are certain to emerge as new retrograde pathways are uncovered.

Acknowledgments

We thank members of the Butow laboratory for suggestions and comments on the manuscript. Work cited from the authors laboratory was supported by grants from the NIH and The Robert A. Welch Foundation.

References

1. Garese, R. and Vallejo, C. G. (2001). Animal mitochondrial biogenesis and function: A regulatory cross-talk between two genomes. *Gene* **263**, 1–16.
2. Williams, R. S., Salmons, S., Newsholme, E. A., Kaufman, R. E., and Mellor, J. (1986). Regulation of nuclear and mitochondrial gene expression by contractile activity in skeletal muscle. *J. Biol. Chem.* **261**, 376–380.
3. Puigserver, P., Wu, Z., Park, C. W., Graves, R., Wright, M., and Spiegelman, B. M. (1998). A cold-inducible coactivator of nuclear receptors linked to adaptive thermogenesis. *Cell* **92**, 829–839.
4. Wu, Z., Puigserver, P., Andersson, U., Zhang, C., Adelmant, G., Mootha, V., Troy, A., Cinti, S., Lowell, B., Scarpulla, R. C., and Spiegelman, B. M. (1999). Mechanisms controlling mitochondrial biogenesis and respiration through the thermogenic coactivator PGC-1. *Cell* **98**, 115–124.
5. Scarpulla, R. C. (1997). Nuclear control of respiratory chain expression in mammalian cells. *J. Bioenerg. Biomem.* **29**, 109–119.
6. Lunardi, J. and Attardi, G. (1991). Differential regulation of expression of the multiple ADP/ATP translocase genes in human cells. *J. Biol. Chem.* **266**, 16534–16540.
7. King, M. P. and Attardi, G. (1989). Human cells lacking mtDNA: Repopulation with exogenous mitochondria by complementation. *Science* **246**, 500–503.
8. Heddi, A., Stepien, G., Benke, P. J., and Wallace, D. C. (1999). Coordinate induction of energy gene expression in tissues of mitochondrial disease patients. *J. Biol. Chem.* **274**, 22968–22976.
9. Beziat, F., Touraille, S., Debise, R., Morel, F., Petit, N., Lecher, P., and Alziari, S. (1997). Biochemical and molecular consequences of massive mitochondrial gene loss in different tissues of a mutant strain of *Drosophila* subobscura. *J. Biol. Chem.* **272**, 22583–22590.
10. Beziat, F., Morel, F., Volz-Lingenhol, A., Saint Paul, N., and Alziari, S. (1993). Mitochondrial genome expression in a mutant strain of *D. subobscura*, an animal model for large scale mtDNA deletion. *Nucleic Acids Res.* **21**, 387–392.
11. Graham, B. H., Waymire, K. G., Cottrell, B., Trounce, I. A., MacGregor, G. R., and Wallace, D. C. (1997). A mouse model for mitochondrial myopathy and cardiomyopathy resulting from a deficiency in the heart/muscle isoform of the adenine nucleotide translocator. *Nat. Genet.* **16**, 226–234.
12. Murdock, D. G., Boone, B. E., Esposito, L. A., and Wallace, D. C. (1999). Up-regulation of nuclear and mitochondrial genes in the skeletal muscle of mice lacking the heart/muscle isoform of the adenine nucleotide translocator. *J. Biol. Chem.* **274**, 14429–14433.
13. Wang, H., Parent, M., and Morais, R. (1994). Cloning and characterization of a cDNA encoding elongation factor 1 alpha from chicken cells devoid of mitochondrial DNA. *Gene* **140**, 155–161.
14. Wang, H. and Morais, R. (1997). Up-regulation of nuclear genes in response to inhibition of mitochondrial DNA expression in chicken cells. *Biochim. Biophys. Acta* **1352**, 325–334.
15. Wang, H., Meury, L., and Morais, R. (1997). Cloning and characterization of cDNAs encoding chicken mitogen-activated protein kinase kinase type 2, MEK2: Downregulation of MEK2 in response to inhibition of mitochondrial DNA expression. *Biochemistry* **36**, 15371–15380.
16. Wang, X. (2001). The expanding role of mitochondria in apoptosis. *Genes Dev.* **15**, 2922–2933.
17. Carfoli, E. (2002). Calcium signaling: A tale for all seasons. *Proc. Natl. Acad. Sci. USA* **99**, 1115–1122.
18. Biswas, G., Adebanjo, O. A., Freedman, B. D., Anandatheerthavarada, H. K., Vijayasarathy, C., Zaidi, M., Kotlikoff, M., and Avadhani, N. G. (1999). Retrograde Ca2+ signaling in C2C12 skeletal myocytes in response to mitochondrial genetic and metabolic stress: A novel mode of inter-organelle crosstalk. *EMBO J.* **18**, 522–533.
19. Luo, Y., Bond, J. D., and Ingram, V. M. (1997). Compromised mitochondrial function leads to increased cytosolic calcium and to activation of MAP kinases. *Proc. Natl. Acad. Sci. USA* **94**, 9705–9710.
20. Amuthan, G., Biswas, G., Zhang, S. Y., Klein-Szanto, A., Vijayasarathy, C., and Avadhani, N. G. (2001). Mitochondria-to-nucleus stress signaling induces phenotypic changes, tumor progression and cell invasion. *EMBO J.* **20**, 1910–1920.
21. Augenlicht, L. H. and Heerdt, B. G. (2001). Mitochondria: integrators in tumorigenesis? *Nat. Genet.* **28**, 104–105.
22. Coller, H. A., Khrapko, K., Bodyak, N. D., Nekhaeva, E., Herrero-Jimenez, P., and Thilly, W. G. (2001). High frequency of homoplasmic mitochondrial DNA mutations in human tumors can be explained without selection. *Nat. Genet.* **28**, 147–150.
23. Arnould, T., Vankoningsloo, S., Renard, P., Houbion, A., Ninane, N., Demazy, C., Remacle, J., and Raes, M. (2002). CREB activation induced by mitochondrial dysfunction is a new signaling pathway that impairs cell proliferation. *EMBO. J.* **21**, 53–63.
24. Rochard, P., Rodier, A., Casas, F., Cassar-Malek, I., Marchal-Victorion, S., Daury, L., Wrutniak, C., and Cabello, G. (2000). Mitochondrial activity is involved in the regulation of myoblast differentiation through myogenin expression and activity of myogenic factors. *J. Biol. Chem.* **275**, 2733–2744.
25. Di Lisa, F. and Ziegler, M. (2001). Pathophysiological relevance of mitochondria in NAD(+) metabolism. *FEBS Lett.* **492**, 4–8.
26. Lee, H. C. (1997). Mechanisms of calcium signaling by cyclic ADP-ribose and NAADP. *Physiol. Rev.* **77**, 1133–1164.
27. Liao, X. S., Small, W. C., Srere, P. A., and Butow, R. A. (1991). Intramitochondrial functions regulate nonmitochondrial citrate synthase (CIT2) expression in *Saccharomyces cerevisiae. Mol. Cell. Biol.* **11**, 38–46.
28. Small, W. C., Brodeur, R. D., Sandor, A., Fedorova, N., Li, G., Butow, R. A., and Srere, P. A. (1995). Enzymatic and metabolic studies on retrograde regulation mutants in yeast. *Biochemistry* **16**, 5569–5576.
29. Liao, X. and Butow, R. A. (1993). *RTG1* and *RTG2:* Two yeast genes required for a novel path of communication from mitochondria to the nucleus. *Cell* **72**, 61–71.
30. Jia, Y., Rothermel, B., Thornton, J., and Butow, R. A. (1997). A basic helix–loop–helix zipper transcription complex functions in a signaling pathway from mitochondria to the nucleus. *Mol. Cell. Biol.* **17**, 1110–1117.
31. Sekito, T., Liu, Z., Thornton, J., and Butow, R. A. (2002). RTG-dependent mitochondria-to-nucleus signaling is regulated by MKS1

and is linked to the formation of the yeast prion [URE3]. *Mol. Biol. Cell*, in press.

32. Chen, C.-Y. and Powers, T. (2002). *MKS1* in concert with TOR signaling negatively regulates *RTG* target gene expression in *S. cerevisiae*. *Curr. Biol.* **157**, 389–395.
33. Liu, Z., Sekito, T., Epstein, C. B., and Butow, R. A. (2002). *RTG*-dependent mitochondria to nucleus signaling is negatively regulated by the seven WD-repeat protein Lst8p. *EMBO J.* **20**, 7209–7219.
34. Rothermel, B., Thornton, J., and Butow, R. A. (1997). Rtg3p, a basic helix–loop–helix/leucine zipper protein functions in mitochondrial-induced changes in gene expression, contains independent activation domains. *J. Biol. Chem.* **272**, 19801–19807.
35. Sekito, T., Thornton, J., and Butow, R. A. (2000). Mitochondria-to-nuclear signaling is regulated by the subcellular localization of the transcription factors Rtg1p and Rtg3p. *Mol. Biol. Cell* **11**, 2103–2115.
36. Massari, M. E., Grant, P. A., Pray-Grant, M. G., Berger, S. L., Workman, J. L., and Murre, C. (1999). A conserved motif present in a class of helix–loop–helix proteins activates transcription by direct recruitment of the SAGA complex. *Mol. Cell* **4**, 63–73.
37. Holstege, F. C., Jennings, E. G., Wyrick, J. J., Lee, T. I., Hengartner, C. J., Green, M. R., Golub, T. R., Lander, E. S., and Young, R. A. (1998). Dissecting the regulatory circuitry of a eukaryotic genome. *Cell* **95**, 717–728.
38. Koonin, E. V. (1994). Yeast protein controlling inter-organelle communication is related to bacterial phosphatases containing the Hsp70-type ATP-binding domain. *Trends Biochem. Sci.* **19**, 156–157.
39. Bork, P., Sander, C., and Valencia, A. (1992). An ATPase domain common to prokaryotic cell cycle proteins, sugar kinases, actin and hsp70 heat shock proteins. *Proc. Natl. Acad. Sci. USA* **89**, 7290–7294.
40. Liu, Z. and Butow, R. A. (1999). A transcriptional switch in the expression of yeast tricarboxylic acid cycle genes in response to a reduction or loss of respiratory function. *Mol. Cell. Biol.* **19**, 6720–6728.
41. Rosenkrantz, M., Kell, C. S., Pennell, E. A., and Devenish, L. J. (1994). The HAP2,3,4 transcriptional activator is required for derepression of the yeast citrate synthase gene, *CIT1*. *Mol. Microbiol.* **13**, 119–131.
42. Roberg, K. J., Bickel, S., Rowley, N., and Kaiser, C. A. (1997). Control of amino acid permease sorting in the late secretory pathway of *Saccharomyces cerevisiae* by *SEC13, LST4, LST7* and *LST8*. *Genetics* **147**, 1569–1584.
43. Wall, M. A., Coleman, D. E., Lee, E., Iniguez-Lluhi, J. A., Posner, B. A., Gilman, A. G., and Sprang, S. R. (1995). The structure of the G protein heterotrimer Gi alpha 1 beta 1 gamma 2. *Cell* **83**, 1047–1058.
44. Smith, T. F., Gaitatzes, C., Saxena, K., and Neer, E. J. (1999). The WD repeat: A common architecture for diverse functions. *Trends Biochem. Sci.* **24**, 181–185.
45. Forsberg, H., Gilstring, C. F., Zargari, A., Martinez, P., and Ljungdahl, P. O. (2001). The role of the yeast plasma membrane SPS nutrient sensor in the metabolic response to extracellular amino acids. *Mol. Microbiol.* **42**, 215–228.
46. Matsuura, A. and Anraku, Y. (1993). Characterization of the MKS1 gene, a new negative regulator of the Ras-cyclic AMP pathway in *Saccharomyces cerevisiae*. *Mol. Gen. Genet.* **238**, 6–16.
47. Feller, A., Ramos, F., Pierard, A., and Dubois, E. (1997). Lys80p of *Saccharomyces cerevisiae*, previously proposed as a specific repressor of LYS genes, is a pleiotropic regulatory factor identical to Mks1p. *Yeast* **13**, 1337–1346.
48. Edskes, H. K., Hanover, J. A., and Wickner, R. B. (1999). Mks1p is a regulator of nitrogen catabolism upstream of Ure2p in *Saccharomyces cerevisiae*. *Genetics* **153**, 585–594.
49. Edskes, H. K. and Wickner, R. B. (2000). A protein required for prion generation: [URE3] induction requires the Ras-regulated Mks1 protein. *Proc. Natl. Acad. Sci. USA* **97**, 6625–6629.
50. Beck, T. and Hall, M. N. (1999). The TOR signaling pathway controls nuclear localization of nutrient-regulated transcription factors. *Nature* **402**, 689–692.
51. Pierce, M. M., Maddelein, M. L., Roberts, B. T., and Wickner, R. B. (2001). A novel Rtg2p activity regulates nitrogen catabolism in yeast. *Proc. Natl. Acad. Sci. USA* **98**, 13213–13218.
52. Shamji, A. F., Kuruvilla, F. G., and Schreiber, S. L. (2000). Partitioning the transcriptional program induced by rapamycin among the effectors of the TOR proteins. *Current Biol.* **10**, 1574–1581.
53. Komeili, A., Wedaman, K. P., O'Shea, E. K., and Powers, T. (2000). Mechanism of metabolic control: target of rapamycin signaling links nitrogen quality to the activity of the Rtg1 and Rtg3 transcription factors. *J. Cell Biol.* **151**, 863–878.
54. Barbet, N. C., Schneider, U., Helliwell, S. B., Stansfield, I., Tuite, M. F., and Hall, M. N. (1996). TOR controls translation initiation and early G1 progression in yeast. *Mol. Biol. Cell* **7**, 25–42.
55. Cardenas, M. E., Cutler, N. S., Lorenz, M. C., Di Como, C. J., and Heitman, J. (1999). The TOR signaling cascade regulates gene expression in response to nutrients. *Genes Dev.* **13**, 3271–3279.
56. Hardwick, J. S., Kuruvilla, F. G., Tong, J. K., Shamji, A. F., and Schreiber, S. L. (1999). Rapamycin-modulated transcription defines the subset of nutrient-sensitive signaling pathways directly controlled by the Tor proteins. *Proc. Natl. Acad. Sci. USA* **96**, 14866–14870.
57. van Heusden, G. P. and Steensma, H. Y. (2001). 14–3–3 Proteins are essential for regulation of RTG3-dependent transcription in *Saccharomyces cerevisiae*. *Yeast* **18**, 1479–1491.
58. Hallstrom, T. C. and Moye-Rowley, S. (2000). Multiple signals from dysfunctional mitochondria activate the pleiotropic drug resistance pathway in *Saccharomyces cerevisiae*. *J. Biol. Chem.* **275**, 34347–7356.
59. Delaveau, T., Delahodde, A., Carvajal, E., Subik, J., and Jacq, C. (1994). PDR3, a new yeast regulatory gene, is homologous to PDR1 and controls the multidrug resistance phenomenon. *Mol. Gen. Genet.* **244**, 501–511.
60. Bonnefoy, N., Chalvet, F., Hamel, P., Slonimski, P. P., and Dujardin, G. (1994). OXA1, a Saccharomyces cerevisiae nuclear gene whose sequence is conserved from prokaryotes to eukaryotes controls cytochrome oxidase biogenesis. *J. Mol. Biol.* **239**, 201–212.
61. Zhang, X. and Moye-Rowley, W. S. (2001). Saccharomyces cerevisiae multidrug resistance gene expression inversely correlates with the status of the F(0) component of the mitochondrial ATPase. *J. Biol. Chem.* **276**, 47844–47852.
62. Kwast, K. E., Burke, P. V., Staahl, B. T., and Poyton, R. O. (1999). Oxygen sensing in yeast: evidence for the involvement of the respiratory chain in regulating the transcription of a subset of hypoxic genes. *Proc. Natl. Acad. Sci. USA* **96**, 5446–5451.
63. Dagsgaard, C., Taylor, L. E., O'Brien, K. M., and Poyton, R. O. (2001). Effects of anoxia and the mitochondrion on expression of aerobic nuclear *COX* genes in yeast. Evidence for a signaling pathway from the mitochondrial genome to the nucleus. *J. Biol. Chem.* **276**, 7593–7601.
64. Epstein, C. B., Waddle, J. A., Hale IV, W., Davé, V., Thornton, J., Macatee, T. L., Garner, H. R., and Butow, H. R. (2001). Genome-wide responses to mitochondrial dysfunctions. *Mol. Biol. Cell* **12**, 297–308.
65. Demasi, A. P., Pereira, G. A., and Netto, L. E. (2001). Cytosolic thioredoxin peroxidase I is essential for the antioxidant defense of yeast with dysfunctional mitochondria. *FEBS Lett.* **509**, 430–434.
66. Traven, A., Wong, J. M., Xu, D., Sopta, M., and Ingles, C. J. (2001). Interorganellar communication. Altered nuclear gene expression profiles in a yeast mitochondrial DNA mutant. *J. Biol. Chem.* **276**, 4020–4027.
67. DeRisi, J. L., Iyer, V. R., and Brown, P. O. (1997). Exploring the metabolic and genetic control of gene expression on a genomic scale. *Science* **278**, 680–686.
68. Hughes, T. R., Marton, M. J., Jones, A. R., Roberts, C. J., Stoughton, R., Armour, C. D., Bennett, H. A., Coffey, E., Dai, H., He, Y. D., Kidd, M. J., King, A. M., Meyer, M. R., Slade, D., Lum, P. Y., Stepaniants, S. B., Shoemaker, D. D., Gachotte, D., Chakraburtty, K., Simon, J., Bard, M., and Friend, S. H. (2000). Functional discovery via a compendium of expression profiles. *Cell* **102**, 109–126.
69. Dujon, B. and Belcour, L. (1989). In Berg, D. E., and Howe, M. M., Eds., Mitochondrial DNA instabilities and rearrangements in yeasts and fungi. *Mobile DNA*. American Society for Microbiology, Washington, D.C.
70. Begel, O., Boulay, J., Albert, B., Dufour, E., and Sainsard-Chanet, A. (1999). Mitochondrial group II introns, cytochrome c oxidase, and senescence in *Podospora anserina*. *Mol. Cell. Biol.* **19**, 4093–4100.

71. Borghouts, C., Kimpel, E., and Osiewacz, H. D. (1997). Mitochondrial DNA rearrangements of *Podospora anserina* are under the control of the nuclear gene grisea. *Proc. Natl. Acad. Sci. USA* **94**, 10768–10773.
72. Borghouts, C. and Osiewacz, H. D. (1998). GRISEA, a copper-modulated transcription factor from *Podospora anserina* involved in senescence and morphogenesis, is an ortholog of MAC1 in *Saccharomyces cerevisiae. Mol. Gen. Genet.* **260**, 492–502.
73. Kirchman, P. A., Kim, S., Lai, C. Y., and Jazwinski, S. M. (1999). Interorganelle signaling is a determinant of longevity in *Saccharomyces cerevisiae. Genetics* **152**, 179–190.
74. Jiang, J. C., Jaruga, E., Repnevskaya, M. V., and Jazwinski, S. M. (2000). An intervention resembling caloric restriction prolongs life span and retards aging in yeast. *FASEB J.* **14**, 2135–2137.
75. Larsen, P. L. and Clarke, C. F. (2002). Extension of life-span in *Caenorhabditis elegans* by a diet lacking coenzyme Q. *Science* **295**, 120–123.
76. Felkai, S., Ewbank, J. J., Lemieux, J., Labbe, J. C., Brown, G. G., and Hekimi, S. (1999). CLK-1 controls respiration, behavior and aging in the nematode *Caenorhabditis elegans. EMBO J.* **18**, 1783–1792.
77. Lakowski, B. and Hekimi, S. (1996). Determination of life-span in *Caenorhabditis elegans* by four clock genes. *Science* **272**, 1010–1013.

CHAPTER 326

Signaling During Exocytosis

Lee E. Eiden
Section on Molecular Neuroscience,
Laboratory of Cellular and Molecular Regulation,
National Institute of Mental Health Intramural Research Program,
Bethesda, Maryland

Introduction

Metazoan cells communicate with one another by secreting informational molecules. Regulated secretion starts, or stops, in response to an extracellular stimulus. In regulated exocytotic secretion, secretory vesicles fuse with the plasma membrane, and extrude their neurotransmitter or hormone contents, after cell stimulation by a secretagogue.

Many types of cells undergo regulated, exocytotic secretion (Table I). Secretion from other cell types is controlled by signaling that stimulates biosynthesis of secreted products, with subsequent constitutive, rather than regulated, secretion (e.g., aldosterone and cortisol secretion from the adrenal cortex, thyroid hormone secretion from thyroid follicular cells, cytokine secretion from T lymphocytes and macrophages, immunoglobulin secretion from B lymphocytes, protein secretion from hepatocytes, androgen and estrogen secretion from the gonads, mucopolysaccharide secretion from lung epithelial cells, bile salt secretion from the liver and gall bladder, and melatonin secretion from pinealocytes). The distinction between exocytotic and nonexocytotic secretion emphasizes the secretory vesicle as a "macromolecular machine" [1] that is an important locus of regulation in cell signaling. Consideration of regulated exocytotic secretion organizes a great deal of information about neuronal, endocrine, exocrine, and immune/inflammatory homeostasis. Solving the biophysical problem of passing packets of informational molecules across cell membranes in a defined period of time, in response to multiple specific extracellular signals, is the special bioengineering triumph of exocytotic secretion.

Some general statements can be made about exocytotic secretion. Endocrine and exocrine exocytosis is mainly "slow" (milliseconds to seconds) and occurs from large dense-core vesicles (LDCVs) or secretory granules (SGs). Neuronal secretion is either fast (release from small synaptic vesicles, or SSVs, within microseconds) or slow (release from LDCVs, as in endocrine cells). Neuroendocrine secretion may be acute (episodic or pulsatile, over minutes) or sustained (over hours). The rate of secretory vesicle exocytosis can be modulated at multiple steps called *docking, priming*, and *fusion*. Fast/acute secretion is calcium dependent, while slow/sustained secretion may involve both calcium and cyclic AMP (cAMP). Exocytosis of a secretory vesicle may deliver all of its contents to the exterior of the cell ("all-or-none") or only a fraction ("kiss-and-run"). Each mode of secretion has unique kinetic features and unique problems of membrane protein retrieval and reutilization. Regulated secretion can be modulated at the level of organellogenesis—the number of secretory vesicles in a particular type of secretory cell. The quality and quantity of the vesicle's content can be further regulated via vesicular transporter trafficking and function, as well as prohormone synthesis, sorting, and processing. The probability of vesicular secretion in response to a given stimulus may vary as a function of the previous stimulatory experience of the cell—this is secretory plasticity.

These factors may account for variation in regulated secretion from different cell types, secretion of different substances from the same cell, and even secretion of the same substance from the same cell in response to different secretagogues. Thus, SSV and LDCV compartments within the same cell behave differently; one secretagogue can interact with multiple receptors to stimulate different modes of regulated secretion at different times; secretion can be augmented or depressed during episodes of continuous stimulation. Differences between fast and slow neurotransmission, SSVs and LDCVs, acute and sustained secretion, and adaptation during continuous secretion have presumably evolved to allow neuronal, endocrine, exocrine, and inflammatory secretory

Table I Some Secretagogues for Regulated Exocytosis and Their Targets

Secretagogue	Source	Target	Secreted material and function
ACh and PACAP	Coreleased from splanchnic nerve at adrenomedullary synapse	Adrenal chromaffin cells	Epinephrine and norepinephrine; hormonal cardiovascular and metabolic regulation
Gastrin and PACAP	Endocrine cells	Enterochromaffin-like cells (ECL cells)	Histamine; gastric acid secretion
Various hormones and paracrine factors	gastrointestinal epithelium	Enterochromaffin cells (EC cells) of gut	Serotonin; to supply 5-HT to platelets and mast cells of the gut
IgE	Circulation/tissue	Mast cells	Histamine; inflammation
Neurotransmitters; action potentials	Presynaptic neurons; nerve terminal depolarization	Neurons (postsynaptic)	Excitatory neurotransmitters (e.g., glutamate) Inhibitory neurotransmitters (e.g., GABA); neurotransmission
Calcium	Systemic	Parathyroid gland	PTH; control of calcium, mineral, and bone metabolism
ACh and cholecystokinin	Parasympathetic nerve terminals; paracrine secretion	Pancreatic acinar cells (exocrine)	Zymogens; amylase secretion into gut to control digestion
Glucose; epinephrine; PST; ACh; galanin and NE	Serum; adrenal medulla; adrenal medulla; islets; parasympathetic nerves; sympathetic nerves	Pancreatic islet cells (endocrine)	Insulin and glucagon; glucose uptake and utilization
Epinephrine	Circulation	Parotid gland	Amylase; regulation of salivation/digestion
Hypophysiotropic hormones (CRH; LH-RH; GH-RH; TRH; DA; somatostatin)	Hypothalamus	Pituitary, anterior	ACTH (+); LH/FSH (+); GH (+); TSH (+); prolactin (−); growth hormone (−). Endocrine regulation of reproduction, growth and metabolism
Sodium (hyperosmolarity)	Systemic	Pituitary, posterior	Oxytocin and vasopressin; salt and fluid retention
Thrombin, collagen, epinephrine, ADP	Circulation, blood vessels	Platelets (dense granules)	Serotonin, ADP & Ca^{2+}; hemostasis (platelet aggregation and vasodilatation, formed element adherence)
Calcium	Systemic	Thyroid C cells	Calcitonin; calcium metabolism

Abbreviations: ACh, acetylcholine; ACTH, adrenocorticotropin hormone; CRH, corticotropin releasing hormone; DA, dopamine; FSH, follicle-stimulating hormone; GABA, gamma-aminobutyric acid; GH, growth hormone; GH-RH, GH releasing hormone; LH, luteinizing hormone; PACAP, pituitary adenylate cyclase-activating polypeptide PST, pancreastatin; PTH, parathyroid hormone; TRH, thyrotropin releasing hormone; TSH, thyroid stimulating hormone

cells to perform homeostasis-by-secretion in response to myriad unique physiological circumstances.

Functional, Morphological, and Historical Aspects of Exocytosis and Stimulus-Secretion Coupling

Bayliss and Starling [2] discovered secretin, and articulated the general concept of regulated hormone secretion, in 1902. The modern understanding of exocytotic secretion, combining its morphological, biochemical, and cell physiological features, evolved during the latter half of the last century. In the 1950s, Hillarp [3] characterized the adrenaline-containing secretory granule, hypothesizing that it was the morphological substrate for catecholamine hormone secretion from the adrenal medulla. Adrenaline was found to be coreleased from the adrenal medulla, upon nerve stimulation, with chromogranins, large proteins also stored in secretory granules. Secretion had to occur via exocytosis to allow passage of these molecules into the circulation in the same ratios as found in the granules [4,5]. Morphological and biochemical evidence for secretory granules and exocytotic hormone release in the adrenal medulla, the exocrine pancreas, from mast cells, and from other secretory tissues followed [6].

Douglas and Rubin [7] showed that acetylcholine-stimulated catecholamine release from the perfused adrenal medulla required calcium influx into chromaffin cells. Douglas proposed the term *stimulus-secretion coupling* for this calcium-dependent process, analogous to stimulus-contraction coupling in myocytes [6]. In stimulus-contraction coupling, acetylcholine release from nerve terminals stimulates muscle cell depolarization, leading to calcium release from the endoplasmic reticulum, triggering muscle contraction. In stimulus-secretion coupling, by contrast, acetylcholine release from nerve terminals stimulates chromaffin cell depolarization, leading to calcium influx and triggering exocytosis. Cytosolic calcium elevation driving exocytosis in most neuronal and endocrine cells is supplied primarily by influx of calcium through voltage-sensitive channels in the plasma membrane. However, the importance of calcium release into the cytosol from within the cell (calcium mobilization) in fine-tuning and in some cases primarily driving exocytotic secretion has been increasingly appreciated in the last decade or so, in neuronal, endocrine, and especially exocrine secretory cells [8,9].

Endocrine and exocrine cells contain relatively large (200 to >500 nm) secretory vesicles called *secretory granules*. Neurons and neuroendocrine cells contain LDCVs 100–200 nm in diameter that store peptides and biogenic amines, and SSVs 40–70 nm in diameter clustered near synapses in nervous tissue and at neuromuscular junctions. Small (40 nm), clear SSVs were first isolated from the electric organ of Torpedo, and shown to contain acetylcholine [10]. Katz, Miledi, and others postulated that these vesicles contained the *quanta* released on invasion of the nerve terminal by a depolarizing action potential, resulting in the miniature end-plate potential (mepp) [11]. Miledi demonstrated in 1973 that calcium entry accompanying the action potential was necessary and sufficient to trigger the release of the acetylcholine quanta at the neuromuscular junction [12,13]. The importance of amine neurotransmitter accumulation into SSVs and LDCVs for neurotransmission in the peripheral nervous system has been appreciated for at least 30 years on pharmacological grounds: The highly specific inhibitor of acetylcholine uptake into cholinergic SSVs, vesamicol, and the highly specific inhibitor of catecholamine uptake into LDCVs, reserpine, are potent inhibitors of cholinergic and adrenergic neurotransmission, respectively, *in vivo* [14,15]. The vesicle hypothesis of neurotransmission [11] has been reconfirmed by modern genetic techniques. Impairment of the gene encoding the protein that transports acetylcholine into vesicles disrupts cholinergic neurotransmission in the nematode *Caenorhabditis elegans*, even though intracellular levels of acetylcholine are increased in the absence of its vesicular accumulation [16]. Similar genetic findings in *C. elegans* support the critical role of vesicular transporters for GABAergic and biogenic amine neurotransmission [17,18].

Several technical and conceptual advances in the last two decades have greatly accelerated progress in understanding signaling during exocytosis. Development of real-time assays for exocytosis have been critical. These include measurement of capacitance changes in whole patch-clamped cells due to plasma membrane addition by vesicle fusion [19]; evanescent wave energy field changes to probe vesicle fusion at the plasma membrane [20,21]; electrophysiological monitoring of single calcium and other ion channels during addition of exogenous compounds to the inside or outside of the cell [22]; instantaneous release, and measurement of, calcium within the cell using caged calcium reagents and low-affinity calcium probes [23]; electrochemical measurement of neurotransmitter release from single vesicles [24]; and measurement of membrane fusion events using lipophilic dyes [25]. Reverse genetics in mammals and high-throughput forward genetics in *C. elegans* and *Drosophila* have also contributed to rapid progress by focusing on the proteins and phenotypes of regulated exocytosis [26]. Added to the classical availability of giant axons and their synapses and the ability to pipet cell-membrane impermeable substances into them during exocytosis [27], and further development of cell-free models for exocytosis [28,29], these new approaches have reconciled and integrated physiological, electrophysiological, biochemical, and molecular biological observations and experiments on the real-time, *in vivo* platform on which exocytosis occurs. The salient features of signaling during regulated exocytosis uncovered by these multiple approaches are summarized here.

Secretion Begins with Secretagogues

Secretagogues are molecules released by other cells into the circulation, a synapse, or the interstitial space that interact with target cell-surface receptors to initiate signaling, leading to exocytosis. In neuronal secretion, the action potential itself is a surrogate secretagogue. It invades and depolarizes the nerve terminal, triggering secretion in a frequency-encoded manner. The action potential is in turn a direct result of depolarizations caused by neurotransmitters released onto the dendrites and soma of the neuron. These neurotransmitters are the effective secretagogues whose messages are encoded by the action potential, transmitted through the axon, and decoded at the nerve terminal. Experimentally, elevated extracellular potassium mimics action potential-generated depolarizations of nerve terminals, through passive depolarization of the cell and opening of voltage-sensitive calcium channels. Depolarization by elevated extracellular potassium, however, even when tightly temporally controlled, only approximates the actual alternating amplitude- and frequency-modulated regulation of exocytosis that happens at synapse-to-axon-to-nerve terminal trios of the central and peripheral nervous systems. A brief illustrative table of physiological secretagogues, their target cells, and secreted products are listed in Table I.

Secretagogues Act at Target Cell Receptors

How do secretagogues act as first messengers to cause regulated secretion from neurons and endocrine and exocrine cells? First, the type of receptor at which the secretagogue acts determines whether secretion will be fast or slow. *Ionotropic* receptors are ion channels whose permeability is altered, or *gated*, by ligand binding. The ionotropic *N*-methyl-D-aspartate and kainic acid receptors allow increased calcium influx upon binding of the excitatory amino acid transmitter glutamate. The nicotinic receptor gates sodium influx upon binding the neurotransmitter acetylcholine, causing subsequent cell depolarization and calcium influx via voltage-gated channels. Ionotropic receptors mediate fast neurotransmission at neuronal synapses (and rapid release of hormones from neuroendocrine and exocrine cells) because their action is immediate and does not depend on the accumulation of a second messenger via intermediary metabolic events. *Metabotropic* receptors show enhanced coupling to intracellular proteins, when liganded to their first messenger, that results in enzyme activation in the target cell. G-protein-coupled receptors (GPCRs), for example, couple to one of four types of G-proteins that mediate stimulation of adenylate cyclase, stimulation of phospholipase C, inhibition of adenylate cyclase, or stimulation of GTP binding proteins of the Ras superfamily [30]. Activation of metabotropic receptors can result in calcium influx just as stimulation of ionotropic receptors does; the difference is that ionotropic receptors mediate increased calcium influx in under a millisecond, whereas metabotropic receptors require tens to hundreds of milliseconds.

Signal propagation mediated by exocytosis through neuronal synapses can occur by either slow or fast transmission. Glutamate in the central nervous system, or acetylcholine at the neuromuscular junction, act via fast transmission. An action potential causes calcium influx, triggering the release of neurotransmitter (glutamate or ACh) from SSVs, and their interaction with ionotropic (NMDA or nicotinic) receptors will result in cation (calcium or sodium) influx into the target cell (neuron or muscle cell), generating a depolarizing signal, or excitatory postsynaptic potential (*epsp*), in less than a millisecond. Dopamine in the striatum acts via slow transmission. Action potentials cause the rapid release of dopamine from SSVs in dopaminergic nerve terminals, but the released dopamine acts on postsynaptic receptors to increase intracellular cAMP, which alters the excitability of the target cell over a period of milliseconds to minutes, or even hours, and affects additional properties of the postsynaptic cell such as cell-specific gene transcription [31]. The fact of slow or fast transmission is not a property of ligand or receptor only, but of the ligand-receptor dyad. ACh released onto cells with muscarinic rather than nicotinic receptors elicits slow responses; and glutamate released onto cells that express metabotropic (mGluR) receptors mediates slow synaptic responses. Mixed fast and slow neurotransmission occurs at the synapse between preganglionic neurons of the splanchnic nerve and chromaffin cells of the adrenal medulla. At high stimulation frequencies, the splanchnic nerve secretes acetylcholine from SSVs, which causes catecholamine release via a rapid, and rapidly desensitizing, action of acetylcholine at nicotinic receptors on chromaffin cells [32]. Acetylcholine can also act at muscarinic chromaffin cell receptors, permitting metabotropic modulation of calcium levels via stimulation of calcium release from intracellular stores mediated by inositol trisphosphate. Stimulation at low frequency causes sustained release of catecholamines from chromaffin cells via release of a peptide transmitter from the splanchnic nerve [33]. The peptide transmitter at the adrenomedullary synapse is PACAP [34]. Presumably the combination of slow and fast transmission, and ionotropic and metabotropic receptors on chromaffin cells, allows fine-tuning of secretory responses at the adrenomedullary synapse. Many central and peripheral neurons contain neuropeptides like PACAP which are released from LDCVs, and classical transmitters like ACh, which are released from SSVs [35,36].

Calcium and Cyclic AMP: The Two Main Second Messengers for Secretion

Exocytotic secretion from neuronal, endocrine, and exocrine cells is a result of elevation of cytosolic calcium or cAMP, or both, following secretagogue receptor stimulation. Calcium influx is the dominant mode of regulation in most secretory cells, intracellular calcium mobilization is more critical in others, and cAMP controls secretion in a few prominent secretory cell systems. In some cases, exocytotic secretion is regulated negatively rather than positively (see later section), and cAMP and calcium signals are attenuated or their effects antagonized.

The dependence on increased intracellular calcium for exocytotic secretion was worked out in chromaffin, mast, and posterior hypophysial cells by several laboratories using various mainly pharmacological maneuvers and observing secretion biochemically or morphologically [6], and in neurons by pipetting calcium ions into the squid giant synapse and observing secretion electrophysiologically [12]. The ability to bypass secretagogue stimulation by increasing cytosolic calcium concentration alone and to "clamp" secretion even in the presence of secretagogue (or electrical stimulation) by removal and/or chelation of calcium have been the sine qua non for calcium as the prime regulator of exocytosis [37]. Strictly speaking, calcium is often the third rather than the second messenger in exocytotic secretion. Thus, glutamate directly opens a calcium channel to allow cytosolic calcium to rise to levels sufficient to trigger exocytosis [38], but acetylcholine gates a channel for sodium, which causes cell depolarization, after which voltage-dependent calcium channels of several types are opened to allow calcium influx [39].

Metabotropic receptor stimulation is responsible for elevation of cytosolic calcium not through opening of calcium channels in the plasma membrane, but via stimulation of other second messengers, such as inositol trisphosphate (IP3), that act together with calcium on receptors on the endoplasmic reticulum (ER) to elevate cytosolic calcium levels. Secretion of insulin from the ß cells of the pancreas by the antidiabetic sulfonylureas can be triggered by calcium influx through L-type voltage-sensitive channels, via inhibition of an ATP-operated potassium channel, blockade of potassium efflux, and subsequent depolarization of the cell [40]. It is not yet clear if depolarization alone can be a second messenger in the absence of calcium entry or mobilization, for example, via sodium influx, changes in the equilibrium potential at the cell membrane, or physical reorganization of lipids and proteolipids at the cell surface. The pore-forming capabilities of

Vo ATPase subunits, found in both the plasma membrane and within the high-molecular-weight vacuolar ATPase of all secretory vesicles, suggest a potential mechanism for calcium-independent vesicular fusion [41].

In neurons and chromaffin cells, physiologically relevant secretion appears to be regulated wholly via calcium as a second messenger (as defined earlier). In other cell types, calcium and cAMP are both required, either in series, or in parallel, for full secretagogue action. Lactotrophs secrete prolactin in response to first messengers that increase intracellular cAMP and their action of forskolin is blocked by verapamil, suggesting that cAMP mediates calcium channel opening resulting in secretion in these cells [42–46]. Secretion from gonadotrophs by luteinizing hormone-releasing hormone requires calcium, supplied via intracellular mobilization following elevation of intracellular IP3 [47], and has also been suggested to require interaction of calcium mobilized from intracellular stores, with activation of PKC and elevation of cAMP [48].

In some secretory systems cAMP is the dominant signaling molecule mediating secretagogue-stimulated exocytosis. CRH-stimulated secretion of ACTH from corticotrophs shows only a partial sensitivity to blockade of calcium influx, and is mimicked by elevation of intracellular cAMP [46]. The calcium influx-dependent component of CRH-induced ACTH release is thought to be mediated by cAMP-dependent protein kinase (PKA), via inhibition of outwardly direct potassium conductance and subsequent cell depolarization [49]. CRH also mediates long-term secretion of catecholamines and enkephalin peptides from adrenal chromaffin cells, a process mimicked by the adenylate cyclase activator forskolin [50,51].

In contrast to regulation of ACTH secretion by CRH in primary corticotrophs, CRH acts to enhance ACTH secretion in AtT20 cells at multiple loci, including cAMP-dependent calcium mobilization independent of effects on potassium conductances [52]. Differences between CRH-regulated ACTH secretion in primary corticotrophs and the corticotroph like AtT20 tumor cell line are illustrative of important differences between primary cells and tumor cell lines with respect to signaling mechanisms underlying regulated hormone secretion. The mechanisms of secretory regulation characterized in tumor cell lines may illuminate, but should not be confused with, those occurring in primary secretory cells, especially since malignant transformation is likely accompanied by profound changes in the regulation of calcium influx, efflux, and buffering that may also impact on exocytotic secretion. There is also evidence that regulated secretory behavior in isolated pituitary cells may differ from that in the perfused pituitary, in part due to the modulatory influence of factors secreted from the folliculostellate cells of the intact gland [53].

The insulin-secreting ß cell of the pancreatic islets is a prototype secretory cell in which calcium is critical for exocytosis, yet cAMP and phosphorylation via PKA play an important role in augmentation of secretion [54]. The primary insulin secretagogue is plasma glucose. When glucose levels are sufficiently high, elevation of intracellular ATP leads to the closure of an outwardly directed, ATP-regulated potassium channel, causing cell depolarization, opening of voltage-sensitive calcium channels, and exocytotic release of insulin [55–57]. Due to its importance in regulating fuel availability and consumption, insulin secretion is also modulated by a large number of peptides and neurotransmitters, including PACAP, acetylcholine, epinephrine, norepinephrine, arginine, VIP, glucose-independent insulinotropic polypeptide (GIP), and glucagon-like peptide 1 (GLP-1) [58]. Peptides can modulate both the fast and slow phases of insulin secretion. These phases describe secretion from the readily releasable pool and mobilization of the reserve pool of secretory granules (distinct from the fast and slow neurotransmission described earlier). Peptide modulation of insulin secretion may involve cAMP/PKA-dependent activation of both phosphatases and other kinases [54], modulation of calcium mobilization via stimulation of second messengers including IP3, and activation of PKC, following elevation of diacylglycerol (DAG) and IP3/Ca^{2+} [58a].

Some secretory cells are either not dependent on calcium for secretion or require it as a permissive rather than a regulatory factor. A prototype is amylase secretion from the parotid gland. Here, cAMP in the absence of extracellular calcium appears to be sufficient to trigger exocytosis [59–61], albeit calcium mobilization from intracellular stores has not been ruled out as a contributor to secretion in these cells.

A recurring theme of coregulation of exocytosis by calcium and cAMP is that the mode of regulation is highly dependent on the kinetics of secretion and, therefore, the rate-limiting step in secretion (see later section). Thus, parotid secretion, characterized by a dominant role for cAMP, is a very slow process (less than one granule per second per cell). Neuronal secretion, characterized by a greater dependence on calcium, is very fast (more than 100 vesicles per millisecond per cell) [62].

Cell permeabilization has been employed to demonstrate a role for GTP in triggering of exocytotic secretion, notably in mast cells [63] and chromaffin cells [64]. Stimulation by GTP can be mimicked by nonhydrolyzable GTP analogs, suggesting the involvement of Ras- or G-protein-type GTP-binding proteins in exocytosis. It is not yet clear, however, what GTP-binding proteins are involved, and whether they are acting in parallel or in series to calcium-mediated secretion elicited by physiological stimulators (e.g., IgG in mast cell secretion). The finding that Rab-like GTP-binding proteins are enriched on secretory vesicle membranes has not yet been developed into a coherent picture of the involvement of GTP-binding protein signaling in exocytosis, although Ras-like GTP-binding proteins are known to be important in shuttling of nonsecretory vesicles within the cell, and Rab3A knock-out mice demonstrate a neurotransmission-defective phenotype [26]. G-proteins may be important in both the priming and docking phases of secretory vesicle preparation for the final fusion event of exocytosis (see later section), and may act by coordinating the interactions of proteins and phospholipids involved in secretion [65,66]. G-protein involvement in signaling during exocytosis awaits the identification of second and third messengers that convey secretagogue signals to them from the cell surface, and of vesicle-associated guanine nucleotide exchange factors, or GEFs, that regulate GTP hydrolysis and therefore the activities of G-proteins involved in exocytosis.

Calcium and the Regulation of Exocytosis

Calcium is important in virtually all cell functions: in exocytosis, calcium is probably involved in the formation of the secretory granule, its guidance to the plasma membrane, delay of fusion prior to a signal for secretion, and the process of exocytosis itself. Because of its primary regulatory role, calcium is extremely well buffered in secretory cells. Calcium is continually pumped out of the cell by a Ca^{2+}ATPase/transporter. It is sequestered within the endoplasmic reticulum and in mitochondria. Cytosolic calcium is normally about 100 n*M*. When it rises above this concentration (globally to about 1 μ*M* and locally to concentrations as high as 250 μ*M*) [9], it does so only briefly, due to multiple, tightly regulated mechanisms for cytoplasmic calcium clearance, calcium buffering, and inhibition of calcium influx and mobilization. These are summarized in detail in other contributions to this Handbook. The features of the regulation of calcium flux that are important in the context of signaling to exocytosis are, first, that virtually all modes of increasing cytosolic calcium, from calcium influx through voltage- and receptor-gated channels, to increased calcium release from endoplasmic reticulum through IP3, ryanodine (RY), cyclic ADP-ribose (cADPR), and nicotinic acid adenine dinucleotide phosphate (NAADP) receptors, can be regulated by first messengers/secretagogues from outside the cell; and, second, that these modes of elevating cytosolic calcium are highly interactive in the regulation of exocytotic secretion by secretagogues both singly and in physiologically relevant combinations [8,67,68].

Calcium influx is regulated by ionotropic secretagogues in two ways. The first is the direct admission of calcium to the cell through ionotropic receptors. The most prominent of these in the central nervous system are the glutamate-gated calcium channels, including the NMDA and AMPA/kainate receptors. These receptors are mainly postsynaptic, so their role in regulation of calcium flux leading to exocytosis at the nerve terminal has not been extensively explored [69,70]. Glutamate has been implicated in modulation of its own release from hippocampal synaptosomes via autoreceptors of the kainate type [71] and may inhibit secretion presynaptically by depolarizing the nerve terminal and decreasing the efficiency of action potential invasion and rapid calcium influx [72]. Acetylcholine is thought to facilitate secretion in the central nervous system by increasing calcium influx (through α_7 subunit-containing presynaptic nicotinic receptors) at presynaptic sites [73]. The chemical neuroanatomy of the striatum allows the unambiguous conclusion that ionotropic receptor regulation of exocytosis in that region is truly presynaptic and heterotypic [74]. Other examples of presynaptic regulation of neuronal exocytosis might represent heterotypic presynaptic modulation of transmission, as in the well-studied Aplysia system, or autoreceptor action, as for dopamine receptors in the striatum.

The second major mode of calcium influx regulation via secretagogue stimulation is indirect, via opening of voltage-sensitive calcium channels (VSCCs), normally closed at resting potential (−70 mV in neurons) upon cellular depolarization. In neurons, depolarization results from invasion of the nerve terminal by an action potential propagated from the cell soma via the axon. In other excitable cells, the action potential is generated directly by ligand-gated ionotropic receptors that allow sodium influx, such as the nicotinic receptor of the chromaffin cell. Generation of an action potential is capable of opening, at least transiently, all of the several types of VSCCs, and several of these are capable of admitting calcium even without complete depolarization. L-[long-lasting; dihydropyridine (DHP)-sensitive] channels are ubiquitous and are the workhorses of skeletal muscle and heart. P (Purkinje)- and N (neuronal)-type VSCCs are DHP resistant, were first identified in Purkinje cells and elsewhere in brain, and are specifically inhibited by snail venoms [75]. L-type channels are found mainly on dendrites and cell bodies of neurons, while non-L-type channels are enriched at nerve terminals [76]. Consistent with their predominant regional localizations, L-type channels are important for regulation of postsynaptic events in neuronal cells [77], whereas non-L-type channels, in various combinations, are responsible for the regulation of calcium flux in response to depolarization that mediates neurotransmitter release from nerve terminals [75]. L-type channels generally control exocytosis in endocrine tissues [78,79].

An interesting and novel third mode of calcium influx regulation involves the modulation of the voltage dependence of a voltage-sensitive general cation channel by cyclic AMP elevation, such that its probability to be open at resting potentials is significantly increased [80]. Norepinephrine and PACAP regulate pinealocyte calcium influx in this manner. It remains to be seen if secretagogues that cause elevation of cAMP in other tissues employ slow depolarization through high-conductance general cation channels to regulate calcium influx and exocytosis.

Finally, depolarization leading to calcium influx has been reported to be stimulated by at least one metabotropic receptor, a variant of the PAC1 receptor for the neuropeptide PACAP [81]. PACAP clearly stimulates calcium influx into neurons and chromaffin cells via voltage-dependent L- and non-L-type channels to stimulate exocytosis [82–85]. The effects of PACAP on calcium influx do not, as in the corticotroph, appear to be mediated by cAMP-dependent modulation of cellular potassium conductance. How this metabotropic receptor, which is not known to directly gate any plasma membrane cation or anion channel, causes the depolarization required for VSCC opening leading to exocytosis is currently a mystery. In sympathetic neurons, PACAP stimulation of depolarization may involve a combination of sodium influx, inhibition of potassium efflux, and IP3-mediated calcium release into the cytoplasm from intracellular stores [86], although the evidence for this as a general mode of regulation of depolarization and calcium influx by PACAP is not compelling at this time.

In neurons, calcium channels are positioned close to sites at which SSVs are docked. Colocalization to this nanodomain presumably allows calcium levels above 100 μM, necessary for triggering synaptic vesicle exocytosis to be attained. By contrast, depolarization-stimulated exocytosis in neuroendocrine cells can be mimicked by intracellular calcium levels of less than 5 μM, as is also the case for secretion induced by release of caged calcium in other endocrine and in exocrine cells. Augustine [87] has succinctly summarized the concept, and proofs for, the involvement of nanodomain high calcium for fast SSV exocytosis, microdomain intermediate calcium levels for moderately fast transmission via SSV exocytosis, and low (2–20 times resting levels) "radial shell" intracellular calcium levels for neuropeptide release from LDCVs in neurons, and hormone release from the SGs of neuroendocrine cells.

As summarized earlier, ionotropic receptor activation can increase calcium influx both directly through gating of calcium influx and indirectly through cell depolarization via increased sodium conductance. Secretagogues acting via metabotropic receptors can also trigger calcium influx through voltage-sensitive channels via elevation of cyclic AMP, which in some cells, like the corticotroph, causes depolarization via inhibition of outwardly direct potassium conductance [49]. More commonly, secretagogue activation of metabotropic receptors causes an increase in cytosolic calcium via release from intracellular stores. Although influx of extracellular calcium is the most common mode of increasing cytosolic calcium for exocytosis, the role of calcium release from within the cell has become increasingly appreciated as an important one in a number of secretory cells. Douglas [88] had already noted in 1975 that calcium required for mast cell degranulation could also be supplied from intracellular stores as well as from outside the cell.

Mitochondria are the major depot for cellular calcium, and they are critical for buffering of cytosolic calcium, but are not a major source of regulated calcium release for exocytosis. Rather, calcium within the ER is released by second messenger stimulation of two or more specific ligand-operated calcium channels on the ER membrane. The generation of intracellular IP3 occurs when Gq-coupled receptors, like those for PACAP and acetylcholine, activate phospholipase C, and cleavage of phosphatidylinositol bisphosphate, causing the liberation of IP3 into the cytosol and leaving DAG within the membrane. IP3 interacts with the ryanodine receptor (RYR), so named because it contains a binding site for the alkaloid ryanodine, causing release of calcium into the cytosol. There is direct evidence that metabotropic secretagogues, such as cholecystokinin (CCK) acting on pancreatic acinar cells, cause exocytosis via IP3-mediated calcium mobilization independently of calcium influx from outside the cell [68].

It is unlikely that release from intracellular stores can supply the concentration of calcium—greater than 100 μM—needed for SSV exocytosis/fast neurotransmission. Levels of cytosolic calcium required for LDCV exocytosis are much lower [62], yet in several cases examined in detail, it appears that modulation of cytosolic calcium from intracellular stores requires the action of multiple ligands, all presumably elevated or liberated by first-messenger secretagogues, to trigger secretion. Two of these, in addition to IP3, are nicotinic acid adenine dinucleotide phosphate (NAADP) [89] and cyclic ADP-ribose (cADPR) [8,90,91]. Cancela and coworkers [68] examined this issue in pancreatic acinar cells using capacitance measurements of exocytosis and intracellular injection of second messengers cADPR, NAADP, and IP3, in conjunction with the secretagogues ACh or CCK. Local calcium spikes insufficient to cause exocytosis elicited by threshold concentrations of ACh or CCK were converted to global sustained calcium elevation in the cell sufficient to cause exocytosis by potentiation with IP3 in the case of CCK, and NAADP or cADPR in the case of ACh, respectively, indicating that CCK mobilizes calcium through NAADP and cADPR, and ACh through IP3, and suggesting a mechanism for combinatorial action of secretagogue regulation of zymogen secretion from the exocrine pancreas [68]. Yule and colleagues [92] have also suggested that sphingolipid (Sp)-activated calcium egress from the ER is involved in the control of zymogen granule exocytosis in the pancreatic acinar cell, thus implying the involvement of four types of calcium channel regulation in the mobilization of calcium from intracellular stores. How does integration of these multiple signaling mechanisms occur? Berridge *et al.* [9] have suggested that calcium initiated calcium release (CIRC) links the individual calcium spikes that result from opening of individual channels by second and third messengers to the building up of a sustained calcium transient, or *global* calcium increase, within the secretory cell. The coregulation of calcium release from intracellular stores by calcium itself, via CIRC, along with the second/third messengers IP3, cyclic ADP ribose (cADPR), NAADP, and Sp provides for instantaneous control of secretion. This type of calcium regulation of secretion is integrated over time, and based on cell sampling of the extracellular concentration of secretagogues the immediate prior history of secretagogue stimulation, and the ability to respond in an exquisitely graded fashion, to the secretory requirements of the organism as communicated by first-messenger/secretagogue signaling to the target secretory cell. In this sense, exocrine and endocrine secretion that is regulated primarily by calcium release from intracellular stores performs in slow motion the highly integrated processing of multiple inputs that is "hard wired" into neurons by their multiple synaptic contacts.

Secretory granules themselves have high concentrations of calcium, and may be involved in supplying calcium required for exocytosis [93]. Yoo and coworkers [94,95] have reported that IP3 receptors are present on bovine chromaffin granules and that IP3 can release calcium from them *in vitro*.

Gerasimenko and coworkers [96] have reported calcium release from zymogen granules of pancreatic acinar cells mediated through IP3 receptor activation, although these results have also been attributed to ER microsomes present in some zymogen granule preparations [97]. The vesicular membrane protein SV2, structurally related to the amine transporters VMAT and VAChT but without a currently assigned synaptic function, has also been implicated in calcium regulation during exocytosis by SSVs [98].

Exocytosis and SNAREs

In this section, we discuss exocytosis and SNAREs their relative importance in different secretory systems; and regulation of the function of core components of the fusion process during exocytosis.

Vesicle shuttling underlies myriad transport processes between membrane compartments within eukaryotic cells. One of the systems for membrane-bound cargo transport between various membrane compartments within the cell is the presence of receptors (SNAREs) on complementary membranes that bind to each other to form an extremely stable complex, driving fusion of the partner membranes and transfer of membrane-bound cargo from vesicles that have budded from a donor compartment, to a recipient membrane compartment. This process is followed by resegregation, mediated by dynamin and other proteins, of the donor vesicle, minus either its membrane or intravesicular cargo, from the recipient membranous compartment. The overall process of fusion and dissociation of the SNARE complex requires recruitment of soluble *N*-ethylmaleimide sensitive factor (NSF) and its attachment proteins (SNAPs), in a process dependent on ATP hydrolysis. In 1993, a landmark paper described the use of a SNAP affinity column to capture SNAREs from brain tissue extracts [99]. A triad of proteins was identified—synaptobrevin/VAMP, syntaxin, and SNAP-25 (for *Synaptosome-associated protein of Mr 25 kDa*)—that could form a 20S fusion particle with NSF and SNAPs. Synaptobrevin/VAMP had previously been identified as a component of small synaptic vesicles, and was therefore postulated to act as the *v-SNARE* or vesicle marker for fusion, and syntaxin and SNAP-25 were postulated to act as the *t-SNAREs* or target membrane receptors, the target in this case being the plasma membrane [100]. A subsequent report indicated that the SNARE complex (synaptobrevin, syntaxin, and SNAP-25) formed remarkably stably in the absence of NSF/SNAPs and could bind a fourth protein, called synaptotagmin, which competes with alpha-SNAP for binding to the SNARE complex. It was postulated that synaptotagmin might function as a "clamp," preventing NSF/SNAP-assisted secretory vesicle fusion until synaptotagmin's regulated removal from the SNARE complex, and replacement with alpha-SNAP [101]. Disruption of depolarization-induced secretion from LDCVs in PC12 cells by microinjection of synaptotagmin antibodies, or soluble fragments of synaptotagmin without a vesicle membrane attachment site, provided direct evidence of a role for synaptotagmin in regulated exocytosis [102]. The next section includes a discussion of the specific role of synaptotagmin as a calcium sensor in exocytosis.

A requirement for SNARE complex formation in regulated exocytosis from a variety of secretory cells has been demonstrated with the aid of toxins that specifically cleave each of the three individual SNAREs and impair their ability to participate in SNARE complex formation [103]. The clostridial botulinum and tetanus toxins (BoNT and TeTx, respectively) are zinc endopeptidases that disrupt neurotransmission by cleaving synaptobrevin/VAMP (BoNT-F, -G and TeTx) [104], SNAP-25 (BoNT-A, -E, and -C), and syntaxin (BoNT-C1). Cells can be permeabilized sufficiently to allow toxin entry without disrupting the exocytotic machinery, most commonly through the use of other bacterial toxins or antibiotics that form controlled-sized membrane pores [29,105], by electropermeabilization [106], or by capacitance measurements of exocytosis during cell perfusion with toxin. This has allowed facile experiments to demonstrate the involvement of the SNARE complex in a variety of secretory systems. The failure to block exocytosis with certain toxins in specialized secretory cells has pointed the way toward the identification of other SNARE-related proteins that take the place of the syntaxin 1 and 2 isoforms in the SNARE complex, to impart variant exocytotic regulatory characteristics to those cells.

Restoration of calcium-dependent secretion in permeabilized cells in which the soluble SNAPs and NSF were allowed to diffuse out, by reintroduction of these factors, has established the importance of these components of the fusion complex in exocytosis in several neuroendocrine cell types [107]. The involvement of synaptobrevin/VAMP in exocytotic release during neurotransmission has been demonstrated by microinjection of BoNT and TeTx, which hydrolyze synaptobrevin-2, into Aplysia neurons, which contain synaptobrevin-2, and rescue from blockade by co-injection of synaptobrevin-1-derived peptides resistant to toxin cleavage [104]. Similarly, BoNT-D and -C1 light chains cleaved synaptobrevin-2/VAMP-2 and blocked calcium-stimulated LDCV exocytosis in chromaffin cells [64]. Exocytosis is blocked in permeabilized pancreatic ß cells with TeTx and BoNTs, implicating both VAMP-2/cellubrevin and SNAP-25 in insulin release from secretory granules [108–110]. BoNT C1 was reported to block 95% of insulin release induced by KCl, implicating syntaxin in secretion, but only 25% of glucose-induced insulin release [111], suggesting that a related, relatively toxin-resistant protein may carry out some proportion of exocytotic secretion of insulin, and reinforcing the notion that multiple vesicle pools may make up what in aggregate is called the readily releasable pool (RRP) which comprises about 1% of the secretory granules of the normal ß cell [58]. Insulin secretion by KCl in the HIT-15, but not the RINm5F insulinoma cell line was blocked by BoNT-F [112]. The RINm5F cell line, however, is impaired in glucose-mediated insulin release and has only 1% of normal cellular levels of insulin compared to ß cells of the pancreatic islets [113], reinforcing the need for caution in applying conclusions drawn from tumor cell lines to the physiological regulation of secretion from specific tissues *in vivo*. Calcium-dependent exocytosis from neutrophils requires VAMP-2 based on its blockade by clostridial toxin treatment [114]. Zymogen granule exocytosis from the exocrine pancreas also requires synaptobrevin-2/VAMP-2, based on clostridial toxin blockade of secretion [115]. VAMP-2 is required for amylase secretion from parotid acinar cells, despite the fact that exocytosis in these cells is controlled predominantly by cyclic AMP rather than calcium and is rather slow even by exocrine secretory standards [116]. Based in part on differential clostridial toxin sensitivity, and on the absence of syntaxin I or II expression in parotid, Fujita-Yoshigaki and colleagues [61,117] have proposed that syntaxin-4 rather than syntaxins-1 or -2, and SNAP-23 rather than SNAP-25 participate with synaptobrevin/VAMP in the SNARE complex for parotid amylase secretion, and have also put forward the intriguing speculation that this SNARE complex may exclude synaptotagmin so that other components, such as Rab3A, can participate in the fusion complex under the control of cAMP/PKA-dependent phosphorylation rather than calcium. Consistent with this hypothesis, syntaxin-4, but not the neuronal isoform syntaxin-1, can be phosphorylated by PKA, and phosphorylation affects its binding to SNAP-23 *in vitro* [118].

Cell permeabilization techniques have also been employed in "run-down" experiments, in which permeabilized cells gradually lose the ability to perform exocytosis, to add back ions, factors, and proteins that restore secretion. Experiments of this type have led to the identification of early and late phases of exocytosis, which may differ in their contributions to the overall rate of secretion in different tissues and may have distinct biochemical requirements as well. Three stages of exocytosis have been defined in secretory cells based on permeabilized cell/reconstitution experiments. They are docking, priming, and fusion [119,120]. Identification of these steps has been useful in assigning spatiotemporal "regulatory domains" not only to calcium and cyclic AMP, but to their sensors in secretory cells (see next section). Operationally, docking defines a step wherein vesicles are associated with the plasma membrane in a way that is not disrupted when cells are ruptured and undocked vesicles are washed free of the plasma membrane [121]. Docking may involve the transfer of vesicles, both SSVs and LDCVs, from a cytoskeletal matrix that either forms a barrier to free diffusion to the plasma membrane for these organelles, or to which they are tethered and must be freed prior to diffusion to the plasma membrane. The roles of kinesin, synapsin, cyclic AMP, PKA, and ATP in this step are considered in the following section.

Priming is an ATP-dependent step that, at the present time, has no morphological substrate to distinguish it from fusion, except that the exocytosis of docked vesicles requires both ATP and calcium, and that the ATP-dependent step precedes the calcium-dependent one [122]. The ATP dependence of priming suggests that NSF and SNAPs could act at this step, positioning the SNARE complex for fusion triggered by elevation of calcium in the neighborhood of the primed vesicle or granule [119], followed by addition of synaptotagmin to the SNARE complex upon elevation of calcium, which increases synaptotagmin/syntaxin affinity about 100-fold [123].

In considering the roles of docking, priming, and fusion and their regulation by calcium and cyclic AMP, it is useful to note briefly the kinetics of secretion in neuronal, endocrine, and exocrine systems, and in immune/inflammatory exocytotic secretion as well. Secretion of amylase from the rather large (>1000 nm) granules of the parotid gland acinar cells is relatively slow—it occurs only by about 10 sec after application of a cAMP-dependent secretagogue such as epinephrine or isoproterenol. Insulin exocytosis from the >200 nm-diameter ß-cell secretory granules also requires a few seconds, whereas secretion from neuronal or neuroendocrine LDCVs, 100–200 nm in size, takes between 1.3 [124] and 40 msec [125]. Exocytosis from neuronal SSVs, about 40 nm in diameter, occurs a few hundred microseconds after elevation of nerve terminal calcium levels above ~25 μM [126]. While it is axiomatic that exocytotic fusion, even in parotid cells, requires or is augmented by calcium, that requirement may be progressively less dominant from a regulatory point of view, as secretion moves along the continuum from (neuronal) very fast, all-or-none secretion of vesicles poised or accumulated in a docked and primed state followed by a refractory state during which replenishment of the releasable pool occurs, to very slow secretion during which vesicles docking, priming, and fusing populate more equivalent proportions of the overall vesicle complement of the cell.

Calcium and cAMP Sensors for Exocytosis

Synaptotagmin has already been alluded to as the primary calcium sensor in neurons, and likely in neuroendocrine cells as well. Recently, analysis of neurotransmission in synaptotagmin-1 knock-out mice, flies, and worms has confirmed its importance in neurotransmission, albeit the full analysis of the role of each of the 12 synaptotagmin isoforms in neuronal and non-neuronal SSV, LDCV, and SG exocytosis remains to be completed in conditional knock-out animals [26,127]. Synaptotagmin-1 was identified in the central nervous system as the major low-affinity sensor for synchronous SSV release based on electrophysiological criteria; and, in fact, asynchronous secretion, elicited by hypertonic medium or alpha latrotoxin, was unaffected. Leaving aside the issue of whether synaptotagmin-1 is a "clamp" whose inhibition of synchronous release is released by calcium, or a calcium-stimulated effector of secretion, its relatively low affinity for calcium accords with estimates, based on caged-calcium release, of a requirement for greater than 20 μM calcium, presumably attainable only in "nanodomains" near voltage-sensitive calcium channels following action potential invasion of the nerve terminal, for neurotransmission via SSV exocytosis in neurons, and consistent with the observation that synaptotagmin/syntaxin affinity is sharply increased only at calcium concentrations above 100 μM [123]. What of calcium-dependent secretion that occurs from LDCVs in neurons, or LDCVs and SGs of neuroendocrine and exocrine cells, that can be triggered by calcium concentrations considerably lower (300–1000 n*M*?), or even SSV secretion from certain nerve terminals that can be triggered by calcium concentrations below the low-affinity calcium binding constant for synaptotagmin-1 [87]? Several possibilities exist, and include (1) altered calcium binding affinity for synaptotagmin due to accessory protein binding in non-neuronal cells, (2) different calcium sensitivities for various synaptotagmin isoforms in cells in which these are expressed and mediate exocytosis [128–130], and (3) calcium sensors other than synaptotagmin in non-neuronal cells, and for LDCVs in neurons.

Dodge and Rahamimoff [131] defined a complex, cooperative, and nonlinear requirement for calcium in exocytosis that appears to apply to a wide variety of excitable, exocytotic cells. Thus, exocytosis might require multiple calcium-sensitive proteins, as well as multiple cooperative calcium binding sites on individual proteins. To act as a sensor for calcium triggering exocytosis, a calcium-binding protein would need to bind calcium with an affinity less than calcium channels in resting cells, depend for its efficacy in promoting exocytosis on that binding event, and possess an affinity compounded of on and off rates for calcium consistent with the kinetics of stimulus-secretion coupling. Synaptotagmin itself could potentially act by binding and enabling fusion/prefusion with different binding partners at different calcium concentrations in a stepwise fashion, the last stage of which is the low-affinity one, which is the rate-limiting step for primed, and possibly "prefused," SSVs. This could account for the requirement for four calcium ions in SSV exocytosis, corresponding to several calcium-dependent synaptotagmin interactions, perhaps even including synaptotagmin dimerization, and less than four calcium ions in other secretory systems involving SGs or LDCVs [125].

Several other calcium-binding proteins have been investigated as calcium sensors for exocytosis in a variety of secretory cells. The calcium-activated protein for secretion (CAPS) is an ubiquitous cytosolic protein that nevertheless binds LDCVs, but not SSVs, in a highly specific and calcium-dependent fashion [132]. Its ablation in *Drosophila* demonstrates that it is essential for LDCV exocytosis [133]. CAPS is present in all secretory cells that exhibit calcium-regulated secretion [134] and, interestingly, is absent from the parotid gland [61]. A puzzling aspect of CAPS involvement in LDCV exocytosis is its relatively low affinity (about 250 μ*M*) for calcium [132], since the calcium threshold for exocytosis in, for example, chromaffin and PC12 cells is less than 10 μM [135]. However, calcium affinities *in vitro* and *in vivo* may easily be discrepant, given both the possibility for cooperativity with other calcium-binding proteins in the exocytotic process, and the need to define very carefully the pool of docked vesicles contributing to secretion in a given experimental paradigm, only a small fraction of which is "readily releasable" [58,136].

The neuronal calcium sensor (NCS) or frequenin family of calcium-binding proteins is made up of candidates for transducing calcium signals during exocytosis that are just beginning to be studied in detail. Overexpression of NCS-1 in AtT-20 cells, for example, has complex effects, blunting secretagogue stimulation of ACTH release in intact cells, yet enhancing calcium- and GTP-gamma-S-induced secretion from permeabilized cells [137] and, thus, possibly representing the link between GTP-binding proteins and calcium in the regulation of exocytosis by nonhydrolyzable GTP analogs reported in mast cells and AtT-20 cells. Frequenin's involvement in exocytosis at the neuromuscular junction appears to be specific for calcium influx through N-type VSCCs, suggesting that this calcium-binding protein may regulate calcium at very specific sites of exocytosis corresponding to calcium channel-vesicle nanodomains [138]. Coscaffolding of frequenin with N-type VSCCs within nerve terminals may explain the observation that both NCS and frequenin demonstrate inhibition of exocytosis at high concentrations, a hallmark of effects on cellular processes upon introduction of high concentrations of scaffolding and scaffolded proteins into many cell types [139].

Cyclic AMP enhances secretion in almost all cell types studied (see [141] for a list and the earlier section on calcium and cyclic AMP). The cAMP sensor is almost universally found, or assumed to be, PKA, due to blockade by PKA inhibitors such as H89. What are the targets for phosphorylation by PKA and, therefore, the effective "cAMP sensors" for exocytosis? One of them may be the cysteine string protein (CSP), a synaptic vesicle protein required for neurotransmission, and first identified by Zinsmaier and Benzer in 1994 [140]. Evans and colleagues [141] have demonstrated that CSP is phosphorylated in neuronal and neuroendocrine cells, is a substrate for PKA-catalyzed phosphorylation *in vitro*, and modulates secretion from chromaffin cells when overexpressed as the wild-type, but not the serine-10-mutated (nonphosphorylatable) form. Secretion from the parotid gland by acetylcholine is likewise potentiated by cAMP, and this may be due to PKA-dependent phosphorylation of the IP3 receptor, resulting in greatly potentiated IP3-dependent intracellular calcium release in response to muscarinic receptor activation [142]. PKA-dependent IP3 receptor phosphorylation is reported to inhibit localized calcium release in pancreatic acinar cells [143]. The finding that PKA may be localized to sites of calcium mobilization, perhaps by specific anchoring proteins or AKAPs [144] suggests that both potentiation and inhibition of secretion may be produced, depending on the specific secretagogue. It has been suggested that the role of PKA-dependent phosphorylation is to regulate intracellular calcium, and therefore exocytosis, by altering the calcium signal patterning of a given secretagogue, rather than simply inhibiting or augmenting calcium fluxes [142,143,143a]. Virtually all members of the SNARE/NSF/SNAP complex contain phosphorylation sites for casein kinase II, CaM kinase II, and PKA, and are phosphorylated by them. Of particular interest in regulation of exocytosis is that the affinity of alpha-SNAP phosphorylated by PKA for the core docking/fusion complex is 10 times weaker than dephospho-alpha-SNAP [145].

Protein phosphatases inhibit, and protein kinases enhance, insulin secretion in mouse pancreatic ß cells [54].

A second sensor for cAMP, the cAMP-dependent guanine nucleotide exchange factors (or Epacs) I and II, have recently been identified [146,147]. cAMP-GEF-II is present in the central nervous system and the adrenal medulla, though not in PC12 cells. Its introduction into PC12 cells, however, results in enhancement of forskolin-stimulated secretion [148] and, therefore, its involvement in cAMP-augmented, or cAMP-driven, exocytosis in other secretory cells, including those of the central nervous system, is an open question.

Thus, an emergent picture of secretagogue regulation of exocytosis is a primary role for calcium in exocytosis in most systems, with an augmentation of secretion by cAMP mediated through PKA phosphorylation of a wide variety of substrates directly involved in exocytosis. A possibility not generally addressed is that first messengers without intrinsic secretagogic activity of their own might well have important modulatory influences on primary secretagogue action, most likely through elevation (or suppression) of cAMP levels in the cell.

Role of Signal Summation in Regulated Exocytosis

Besides the immediate triggering of exocytosis itself, calcium and cyclic AMP (and other second messengers) control secretion in a more temporally extended manner by stimulation of cAMP- and calcium-activated proteins that possess intrinsic, sophisticated means of recording the cellular exocytotic history over time, and tailoring the response to immediate stimuli based on past experience. Calmodulin kinase II (CaMKII) is a prototypical regulatory molecule involved in exocytosis regulation. CaMKII possesses a "molecular memory" for calcium such that its stimulation by calcium in the presence of calmodulin results in a greater than 10-fold discrimination between calcium pulsed at 1, 2, and 4 Hz [149]. Consistent with its regulatory properties, CaMKII appears to play a role in long-term facilitation of exocytotic release in several cell types, via refilling of the docked/primed vesicle pool. In pancreatic ß cells, for example, acetylcholine acting through muscarinic receptors, stimulates late-phase exocytosis several fold, via activation of CaMKII, and this effect appears to be at the level of mobilization of vesicles into the RRP [150]. There are several known substrates for CaMKII relevant to the regulation of exocytosis in secretory cells, including synaptotagmin [151] and synaptobrevin/VAMP [145]. The best documented role for CaMKII in modulation of exocytosis is via phosphorylation of the synaptic vesicle-associated protein synapsin, resulting in vesicle untethering from cytoskeletal elements and mobilization into the "docked" pool [152]. The role for synapsin in exocytosis has been documented in neurons and for SSVs [152,153]. It may obtain as well for some endocrine cells, although many do not express synapsins.

It has been reported that both depolarization-induced secretion of gonadotropin-releasing hormone (GnRH) from the median eminence of the hypothalamus and carbachol or potassium-induced release of catecholamines from PC12 cells is dependent on CaMKII activation, based on blockade by the CaMKII inhibitors KN-62 or KN-93 [154,155]. However, inhibition by the KN compounds was less than 50% in these experiments, and confirmation of involvement of CaMKII in secretion in these systems by nonpharmacological means, in the absence of quantitative controls for the well-documented nonspecific effects of KN-62 and KN-93 directly on calcium influx through VSCCs [156], remains to be determined. Involvement of CaMKII in insulin secretion from ß cells has, however, been deduced from inhibition by KN-62, with no effect of KN-04, a compound that shares the calcium influx inhibition, but not CaMKII inhibitory properties of KN-62 [157].

Calcium signaling may also be responsible for the activation of tyrosine kinases that ultimately affect regulated secretion [158], although the molecular steps intervening between elevation of intracellular calcium and activation of, for example, MAP kinases in bovine chromaffin cells, remain unidentified.

Synaptic facilitation, depression, and other forms of synaptic plasticity that involve altered probabilities of secretion for a given action potential stimulus are mediated by protein kinase stimulation by both calcium and cyclic AMP [159]. The precise cellular loci of such long-term effects on exocytosis remain to be determined, but are likely to involve, at least in part, mobilization of the vesicular pool into the RRP or a pool from which vesicles may be recruited into the readily releasable one. A recent report provides evidence that calcium and cyclic AMP antagonistically regulate SNAP-25 phosphorylation through calcineurin and PKA activation, respectively, thus indirectly affecting the kinetics of vesicle movement into the RRP in chromaffin cells [160]. Long-term facilitation in Aplysia sensory neurons, as well as presynaptic components of long-term potentiation in cerebellum and hippocampus in mammals, may represent in part such temporally delayed regulatory effects on the efficacy of stimulus-secretion coupling [161–163].

Role of PKC and Other PMA Targets in Regulated Secretion

Protein kinase C has been invoked as a regulator of the secretion of histamine from mast cells, catecholamines from chromaffin cells [164], neurotransmitters from nerve terminal SSVs [165], amylase from parotid gland and exocrine pancreas [166], insulin from ß cells, and hormones from the anterior pituitary [167]. Several of these experiments have involved implication of PKC through application of phorbol myristate acetate (PMA), on the assumption that this is the major, if not the sole, target for PMA in secretory tissue. Recently, Rhee and coworkers [168] have reported that the synaptic protein Munc13 is a target for PMA and is the substrate through which PMA and DAG augment synaptosomal neurotransmitter release. Thus, while multiple substrates for PKC, including SNARE complex proteins and cation channels exist, a thorough reassessment of the roles of Munc13, and PKC, in exocytosis can be anticipated. Conventional protein kinases C (PKCc's) possess unique regulatory properties: the C1 (DAG-binding) and C2 (calcium-binding) domains of PKCc's allow this kinase, like calmodulin kinase, to decode frequency-dependent calcium signaling and, in addition, to prolong it when concomitant signaling to phospholipase activation occurs [169]. PKC is implicated in long-term regulation of neurotransmission in the central nervous system through increase in the size of the readily releasable vesicle pool, albeit the extracellular messenger(s) that signal to PKC to cause this event are not identified [165]. The unique response properties and effector functions of PKC and Munc13 make it likely that these two DAG/Ca^{2+} targets function in parallel with each other, and with wholly calcium-dependent processes, in regulating exocytosis.

Negative Regulation of Secretion

Some secretory systems, though dependent on calcium influx or cyclic AMP elevation to trigger secretion, tend to be tonically active *in vivo*, with regulatory function exerted by inhibition of secretion. These systems are highly illustrative both for understanding the triggers of exocytosis and the role of intracellular signaling in negative modulation. Secretion from the prolactin-secreting cells of the anterior pituitary, for example, is vastly increased by removal of the hypothalamic input to the gland [170], raising two questions: How can regulated prolactin secretion also be constitutive and how do factors from the hypothalamus down-modulate this "default-stimulated" system? Lactotrophs exhibit spontaneous action potentials that may drive calcium influx and mobilization in the absence of additional signals, but can be down-regulated by first messengers that act as "antisecretagogues" or inhibitory factors [171]. Dopamine released from the hypothalamus acts through D2 receptors to inhibit prolactin secretion independently of down-regulation of cAMP [172]. Likewise, somatostatin is a critical down-regulator of growth hormone secretion stimulated by GH-RH, ACTH secretion stimulated by CRF, glucagon secretion from α cells of the exocrine pancreas, and neurotransmitter release in the central nervous system. In each case, somatostatin's actions appear to be exerted via multiple mechanisms involving somatostatin receptor/Gi coupling, including restoration of potassium-mediated hyperpolarization by blockade of secretagogue-stimulated inhibition of these conductances via K-channel phosphorylation, activation of calcineurin leading to secretory vesicle "depriming," and as-yet uncharacterized pleiotropic effects of decreasing intracellular levels of cAMP as well as calcium influx [173–175].

Presynaptic inhibition by the action at nerve terminal autoreceptors of newly released transmitters is a critical negative regulatory mechanism in the nervous system. Catecholamines released from nerve terminals interact with presynaptic α_2 receptors that inhibit further secretion and function as a brake on oversecretion [176,177]. The α_2 agonists are therefore important therapeutic agents for hypertension [178]. Presynaptic inhibition is also important for autoreceptors in striatum that limit the physiological release of dopamine controlling extrapyramidal function [176,179,180]. ATP costored with neurotransmitters in secretory vesicles from a variety of neuroendocrine secretory cells is metabolized to adenosine upon release and can inhibit further exocytosis via activation of adenosine receptors at the nerve terminals or endocrine cells from which release occurs [181].

Upstream Regulation of Secretion

Vesicular organellogenesis and trafficking, hormone packaging, sorting and processing, and vesicular filling with small-molecule transmitters via specific transporters are all steps that precede exocytosis and are critical for delivering to the plasma membrane both the vesicle and its informational contents prior to exocytosis. Some aspects of these processes are constitutive, at least with respect to secretagogue signaling and exocytosis, but, surprisingly, some steps are highly dynamic and may be regulated by the same signals that control exocytosis itself.

Exocrine and endocrine secretory granules, and neuroendocrine and neuronal LDCVs begin life as immature secretory granules formed in the trans-Golgi network (TGN) [182], as first described in detail for the prolactin-containing secretory granule of the pituitary lactotroph [183]. Various mechanisms for sorting of secretory proteins into immature secretory granules (ISGs) have been proposed, including cation-assisted secretory protein aggregation and precipitation into membrane-bound nascently budding structures that become the secretory protein-containing ISGs [184,185]. Chromogranin A has recently been demonstrated as a master switch regulating LDCV granulogenesis in PC12 cells [186]. Prolactin may act similarly to promote self-aggregation and granulogenesis in lactotrophs [187,188]. The regulation of biosynthesis and degradation of these proteins is therefore important not only because they are hormones or precursors for hormones themselves, but because their abundance in the TGN will determine the extent and rate of secretory granule formation.

The content of secretory vesicles is determined, for peptide hormones, via copackaging of processing enzymes that themselves may function as sorting chaperones [189], and by the rate of production of mRNAs, encoding them (see next section). The neurotransmitter content of LDCVs and SSVs is determined by the presence of vesicular transporters that drive the uptake of amine and amino acid transmitters into the vesicle against a concentration gradient, providing enrichment of up to 10,000-fold [190,191]. Surprisingly, neurotransmitter uptake is a rate-limited step: Vesicular amine content can be increased by doubling and decreased by halving the amount of vesicular transporter expressed *in vivo* [192–194]. Furthermore, the phosphorylation state of amine transporters can regulate their trafficking and, therefore, availability to a specific vesicle type [195]. Amine loading into vesicles may be regulated by transporter phosphorylation [191,195–198]. Finally, the extent of filling of secretory granules in catecholamine-containing secretory cells may directly affect the degree of processing of cosecreted peptide hormones [199], most likely due to inhibition of proteolytic processing enzymes of the secretory granule by high concentrations of catecholamines [200]. Thus, a complex regulatory web connects exocytosis from the onset of granulogenesis until the presentation of vesicular contents to the extracellular milieu upon vesicle exocytosis.

Far Upstream Regulation of Secretion

Secretory cells participate in organismic homeostasis via regulated exocytosis of informational molecules, and they also possess homeostatic mechanisms for reconciling organismic demand and conservation of cellular resources to provide secretory product over time. *Stimulus-secretion-synthesis coupling* is a term coined to describe the concomitant stimulation of secretion and secretory hormone biosynthesis, so that compensatory biosynthesis enhancement to replete secretory stores can occur during episodes of prolonged secretion [201]. Stimulus-secretion-synthesis coupling may occur in one of two ways. First, a single second messenger produced by a secretagogue may affect both secretion and synthesis. Secondly, dual second messenger signaling initiated by a single first messenger may separately regulate secretion and synthesis. Various types of regulation occur in different types of secretory cells. In corticotrophs, isoproterenol stimulates both release of ACTH and POMC mRNA up-regulation, but increasing intracellular calcium stimulates release without affecting biosynthesis [202]. In chromaffin cells, acetylcholine couples secretion and synthesis via calcium [201], histamine couples secretion and synthesis via both calcium mobilization and influx [203–205], and PACAP couples secretion and synthesis via cAMP and calcium acting both synergistically and independently [83,85].

Conclusions and Future Outlook for Signaling in Exocytosis

This chapter has covered stimulus-secretion coupling as it applies to secretagogue stimulation of calcium influx and cyclic AMP elevation leading to calcium sensor activation, protein phosphorylation, and ultimately secretory vesicle fusion with the plasma membrane. The modes of secretagogue signaling have been discussed. These include depolarization via sodium channel activation leading to VSCC opening and calcium influx; modulation of outwardly directed potassium channels leading to altered calcium influx; direct first-messenger stimulation of VSCC opening; cyclic AMP stimulation of secretory granule exocytosis via interactions with components of the calcium pathway; and cooperative roles of intracellular calcium mobilization and calcium influx in exocytosis. Secretion from SSVs ("fast" neurotransmission) and LDCVs ("slow" regulated secretion) and their separate modes of regulation were discussed. Regulation of exocytosis from calcium influx through secretory vesicle docking, priming, and membrane fusion was described, with SG, LDCV, and SSV exocytosis as special cases of a general process of regulated secretory vesicle fusion, which can be distinguished kinetically, and by differences in the rate-limiting step for exocytosis for each type of vesicle. Secretagogue signaling during exocytosis may include modification of protein components of the "fusion machinery" to modulate their phase associations as well as interactions with other proteins. The role of signaling via PKA- and PKC-mediated phosphorylation was characterized as functionally important in physiological secretion and as affecting virtually all aspects of exocytosis from trafficking and filling to mobilization/docking, priming, and fusion of the vesicles. An additional area not obviously a subject for signaling during exocytosis is the regulation of granulogenesis/vesiculogenesis, including proteins that control aggregation, sorting, hormone processing, and small-molecule vesicular filling prior to and during secretion. Stimulus-secretion-synthesis coupling was discussed to illustrate that secretogogue-based regulation involves control of the export economy of the secretory cell, including both the quantity and quality of the neurosecretory quantum of the SSV, the endocrine hormone cocktail of the LDCV, and the zymogen mixture of the exocrine secretory granule.

In the future, secretion is likely to be viewed within a more unitary framework for neuronal, endocrine, and exocrine secretion, but with a better molecular understanding of the diversity among cell types and various types of exocytosed cargo. Temporal aspects of secretion will receive additional attention, both from the standpoint of speed of the exocytotic events (from less than a millisecond to several seconds) and their persistence (from seconds to hours and days). Various types of neuronal plasticity, such as long- and short-term potentiation, and long-term depression, will be increasingly viewed in the context of ultra-slow components of the process of exocytosis itself, resulting from increased or decreased speed of docking and priming events that are initially not relevant in the time frame of a single exocytotic event, but become rate limiting and significant when encoding the neuronal experience of past strong, weak, and coincident stimuli that underlies learning and memory. Frequency and amplitude modulation of secretion will be increasingly understood in terms of the conversion of calcium spikes with local cellular influences that sum to global calcium changes much as dendritic and somal depolarizations sum to action potentials in neurons. Many regulatory molecules involved in exocytosis act only after recruitment from one membrane compartment to another, or from cytosol to vesicle or plasma membrane, in a transfer that depends on local and specific alterations in membrane lipid composition. Regulation of membrane lipid, via raft formation, inositol removal and addition, and release and addition of lipid acyl moieties from the lipid glycerol backbone have been touched on here only in passing: Their regulation by G-protein coupling, cAMP, and calcium is intimately involved in both vesicular trafficking and cell signaling and creates an important functional interface between these two processes that is beginning to be explored in detail [65,206].

Molecular details of the roles of calcium and cyclic AMP sensors will be learned about in far greater detail by using conditional knock-outs in worms, flies, and mice, including, in particular, the roles of synaptotagmin and other SNARE complex component isoforms, with functions subtly but importantly specialized for performance and regulation in neuroendocrine, endocrine, and exocrine cells [207–210]. The precise order of ATP-, cAMP-, and low-, medium- and high-affinity Ca^{2+}-dependent steps in exocytosis will be determined with greater precision through the use of real-time evanescent wave and other techniques for localizing vesicles with respect to their proximity to the plasma membrane, and through real-time reportage on calcium and cAMP concentrations near a specific protein component of exocytosis using fluorescence resonance energy transfer and other biophysical techniques in conjunction with recombinant reporter molecules. With a greater variety of biochemical, biophysical, pharmacological, and genetic tools, a greater emphasis can be placed on studying secretion from primary cells in intact tissues or *in vivo*, to determine the physiological and pathophysiological conditions under which various secretagogues dominate or even are rate limiting for secretion, for example, before and after meals for secretion from parotid, pancreas, and gut; during processing of sensory information and habit formation in the nervous system; and during metabolic, psychogenic, or physical stress in the neuroendocrine and immunoinflammatory systems. These details will be critical for a mechanistic understanding that will lead to pharmacological intervention in regulated steps of exocytosis to correct malfunctions in fast and slow neurotransmission, in endocrine and exocrine secretion, and in myesthetic and dystrophic, diabetic, and gastrointestinal disease. Regulation of slow transmission, in particular, is a currently underexploited avenue of research into the modulation of brain function in biogenic amine-related disorders including schizophrenia, attention deficit disorder, and depression.

References

1. Avery, J., Ellis D. J., Lang T. *et al.* (2000). A cell-free system for regulated exocytosis in PC12 cells. *J. Cell Biol.* **148**, 317–324.
2. Bayliss, W. M. and Starling, E. H. (1902). On the causation of the so-called 'peripheral reflex secretion' of the pancreas. *Proc. Roy. Soc. Lond.* **69**, 352–353.
3. Hillarp, N.-A. (1958). Isolation and some biochemical properties of the catechol amine granules in the cow adrenal medulla. *Acta Physiol. Scand.* **43**, 82–96.
4. Blaschko, H., Comline, R. S., Schneider, F. H. *et al.* (1967). Secretion of a chromaffin granule protein, chromogranin, from the adrenal gland after splanchnic stimulation. *Nature* **215**, 58–59.
5. Schneider, F. H., Smith, A. D., and Winkler, H. (1967). Secretion from the adrenal medulla: biochemical evidence for exocytosis. *Br. J. Pharmac. Chemother.* **31**, 94–104.
6. Douglas, W. W. (1968). Stimulus-secretion coupling: The concept and clues from chromaffin and other cells. *Br. J. Pharmacol.* **34**, 451–474.
7. Douglas, W. W. and Rubin, R. P. (1961). The role of calcium in the secretory response of the adrenal medulla to acetylcholine. *J. Physiol.* **159**, 40–57.
8. Petersen, O. H. and Cancela, J. M. (1999). New Ca2+-releasing messengers: are they important in the nervous system? *Trends Neurosci.* **22**, 488–495.
9. Berridge, M. J., Lipp, P., and Bootman, M. D. (2000). The versatility and universality of calcium signalling. *Nat. Rev. Mol. Cell. Biol.* **1**, 11–21.
10. Whittaker, V. P., Ed. (1988). The cholinergic synapse. Springer-Verlag, Berlin.
11. Ceccarelli, B. and Hurlbut, W. P. (1980). Vesicle hypothesis of the release of quanta of acetylcholine. *Physiol. Rev.* **60**, 396–441.
12. Miledi, R. (1973). Transmitter release induced by injection of calcium ions into nerve terminals. *Proc. Roy. Soc. Lond. B.* **183**, 421–425.
13. Jeng, J. M. (2002). Ricardo Miledi and the calcium hypothesis of neurotransmitter release. *Nat. Rev. Neurosci.* **3**, 71–76.
14. Parsons, S. M., Prior, C., and Marshall, I. G. (1993). Acetylcholine transport, storage, and release. *Int. Rev. Neurobiol.* **35**, 279–390.
15. Stitzel, R. E. (1977). The biological fate of reserpine. *Pharmacol. Rev.* **28**, 179–205.
16. Alfonso, A., Grundahl, K., Duerr, J. S., *et al.* (1993). The *Caenorhabditis elegans* unc-17 gene: a putative vesicular acetylcholine transporter. *Science* **261**, 617–619.
17. McIntire, S. L., Reimer, R. J., Schuske, K. *et al.* (1997). Identification and characterization of the vesicular GABA transporter. *Nature* **389**, 870–876.
18. Duerr, J. S., Frisby, D. L., Gaskin, J. *et al.* (1999). The cat-1 gene of *Caenorhabditis elegans* encodes a vesicular monoamine transporter required for specific monoamine-dependent behaviors. *J. Neurosci.* **19**, 72–84.
19. Lindau, M. and Neher, E. (1988). Patch-clamp techniques for time-resolved capacitance measurements in single cells. *Pflügers Arch.* **411**, 137–146.
20. Lang, T., Wacker, I., Steyer, J. *et al.* (1997). Ca2+-triggered peptide secretion in single cells imaged with green fluorescent protein and evanescent-wave microscopy. *Neuron* **18**, 857–863.
21. Oheim, M., Loerke, D., Chow, R. H. *et al.* (1999). Evanescent-wave microscopy: a new tool to gain insight into the control of transmitter release. *Phil. Trans. Roy. Soc. Lond. B Biol. Sci.* **354**, 307–318.
22. Hamill, O. P., Marty, A., Neher, E. *et al.* (1981). Improved patch-clamp techniques for high-resolution current recording from cells and cell-free membrane patches. *Pfluegers Arch.* **391**, 85–100.
23. Delaney, K. R. and Zucker, R. S. (1990). Calcium released by photolysis of DM-nitrophen stimulates transmitter release at squid giant synapse. *J. Physiol.* **426**, 473–498.
24. Robinson, I. M., Finnegan, J. M., Monck, J. R. *et al.* (1995). Colocalization of calcium entry and exocytotic release sites in adrenal chromaffin cells. *Proc. Natl. Acad. Sci. USA* **92**, 2474–2478.
25. Cousin, M. A. and Robinson, P. J. (1999). Mechanisms of synaptic vesicle recycling illuminated by fluorescent dyes. *J. Neurochem.* **73**, 2227–2239.
26. Fernández-Chacón, R. and Südhof, T. C. (1999). Genetics of synaptic vesicle function: Toward the complete functional anatomy of an organelle. *Annu. Rev. Physiol.* **61**, 753–776.
27. Llinas, R. R. (1999). The Squid Gia Synapse: Model for Chemical Snaptic Transmitter Oxford University Press, Oxford.
28. Martin, T. F. and Kowalchyk, J. A. (1997). Docked secretory vesicles undergo Ca2+-activated exocytosis in a cell-free system. *J. Biol. Chem.* **272**, 14447–14453.
29. Avery, J., Jahn, R., and Edwardson, J. M. (1999). Reconstitution of regulated exocytosis in cell-free systems: a critical appraisal. *Annu. Rev. Physiol.* **61**, 777–807.
30. Neves, S. R., Ram, P. T., and Iyengar, R. (2002). G protein pathways. *Science* **296**, 1636–1639.
31. Greengard, P. (2001). The neurobiology of slow synaptic transmission. *Science* **294**, 1024–1030.
32. Wakade, A. R. (1981). Studies on secretion of catecholamines evoked by acetylcholine or transmural stimulation of the rat adrenal gland. *J. Physiol.* **313**, 463–480.

33. Wakade, A. R. (1988). Non-cholinergic transmitter(s) maintains secretion of catecholamines from rat adrenal medulla for several hours of continuous stimulation of splanchnic neurons. *J. Neurochem.* **50**, 1302–1308.
34. Hamelink, C., Tjurmina, O., Damadzic, R. *et al.* (2002). Pituitary adenylate cyclase activating polypeptide is a sympathoadrenal neurotransmitter involved in catecholamine regulation and glucohomeostasis. *Proc. Natl. Acad. Sci. USA* **99**, 461–466.
35. Lindh, B. and Hökfelt, T. (1990). Structural and functional aspects of acetylcholine peptide coexistence in the autonomic nervous system. *Prog. Brain Res.* **84**, 175–191.
36. Agoston, D. V., Conlon, J. M., and Whittaker, V. P. (1988). Selective depletion of the acetylcholine and vasoactive intestinal polypeptide of the guinea-pig myenteric plexus by differential mobilization of distinct transmitter pools. *Exp. Brain Res.* **72**, 535–542.
37. Zucker, R. S. (1996). Exocytosis: A molecular and physiological perspective. *Neuron* **17**, 1049–1055.
38. Hollmann, M. and Heinemann S. (1994). Cloned glutamate receptors. *Ann. Rev. Neurosci.* **17**, 31–108.
39. Livett, B. G. (1984). In Cantin, M., Ed. *The Secretory Process in Adrenal Medullary Cells*, pp. 304–358. Karger, Basel.
40. Eliasson, L., Renstrom, E., Ammala, C. *et al.* (1996). PKC-dependent stimulation of exocytosis by sulfonylureas in pancreatic beta cells. *Science* **271**, 813–815.
41. Peters, C., Bayer, M. J., Bühler, S. *et al.* (2001). Trans-complex formation by proteolipid channels in the terminal phase of membrane fusion. *Nature* **409**, 581–588.
42. Dannies, P. S., Gautvik, K. M., and Tashjian, Jr., A. H. (1976). A possible role of cyclic AMP in mediating the effects of thyrotropin-releasing hormone on prolactin release and on prolactin and growth hormone synthesis in pituitary cells in culture. *Endocrinology* **98**, 1147–1159.
43. Swennen, L., Baes, M., and Denef, C. (1985). Beta-adrenergic stimulation of adenosine-3',5'-monophosphate (c-AMP) accumulation and of prolactin and growth hormone secretion in rat anterior pituitary cell cultures. *Neuroendocrinology* **40**, 72–77.
44. Swennen, L., Baes, M., Schramme, C. *et al.* (1985). Beta-adrenergic stimulation of adenosine-3',5'-monophosphate (c-AMP) in primary cultures of rat anterior pituitary cell populations separated by unit gravity sedimentation. Relationship to growth hormone and prolactin release and to nonsecreting cells. *Neuroendocrinology* **40**, 78–83.
45. Waschek, J. A., Dave, J. R., Eskay, R. L. *et al.* (1987). Barium distinguishes separate calcium targets for synthesis and secretion of peptides in neuroendocrine cells. *Biochem. Biophys. Res. Commun.* **146**, 495–501.
46. Dave, J. R., Eiden, L. E., Lozovsky, D. *et al.* (1987). Calcium-independent and calcium-dependent mechanisms regulate corticotropin-releasing factor stimulated proopiomelanocortin peptide secretion and messenger ribonucleic acid production. *Endocrinology* **120**, 305–310.
47. Tse, F. W., Tse, A., Hille, B. *et al.* (1997). Local Ca2+ release from internal stores controls exocytosis in pituitary gonadotrophs. *Neuron* **18**, 121–132.
48. Macrae, M. B., Davidson, J. S., Millar, R. P. *et al.* (1990). Cyclic AMP stimulates luteinizing-hormone (lutropin) exocytosis in permeabilized sheep anterior-pituitary cells. Synergism with protein kinase C and calcium. *Biochem. J.* **271**, 635–639.
49. Lee, A. K. and Tse, A. (1997). Mechanism underlying corticotropin-releasing hormone (CRH) triggered cytosolic Ca2+ rise in identified rat corticotrophs. *J. Physiol.* **504**, 367–378.
50. Udelsman, R., Harwood, J. P., Millan, M. A. *et al.* (1986). Functional corticotropin releasing factor receptors in the primate peripheral sympathetic nervous system. *Nature* **319**, 147–150.
51. Eiden, L. E., Anouar, Y., Hsu, C-M. *et al.* (1998). Transcription regulation coupled to calcium and protein kinase signaling systems through TRE- and CRE-like sequences in neuropeptide genes. *Adv. Pharmacol.* **42**, 264–269.
52. Luini, A., Lewis, D., Guild, S. *et al.* (1985). Hormone secretagogues increase cytosolic calcium by increasing cAMP in corticotropin-secreting cells. *Proc. Natl. Acad. Sci. USA* **82**, 8034–8038.
53. Stojilkovic, S. S. (2001). A novel view of the function of pituitary folliculo-stellate cell network. *Tr. Endocrinol. Metab.* **12**, 378–380.
54. Ammala, C., Eliasson, L., Bokvist, K. *et al.* (1994). Activation of protein kinases and inhibition of protein phosphatases play a central role in the regulation of exocytosis in mouse pancreatic ß cells. *Proc. Natl. Acad. Sci. USA* **91**, 4343–4347.
55. Ashcroft, F. M., Harrison, D. E., and Ashcroft S. J. H. (1984). Glucose induces closure of single potassium channels in isolated rat pancreatic ß-cells. *Nature* **312**, 446–448.
56. Cook, D. L. and Hales, C. N. (1984). Intracellular ATP directly blocks K+ channels in pancreatic β-cells. *Nature* **311**, 271–273.
57. Wollheim, C. B., Lang, J., and Regazzi, R. (1996). The exocytotic process of insulin secretion and its regulation by Ca2+ and G-proteins. *Diabetes Rev.* **4**, 276–297.
58. Bratanova-Tochkova, T. K., Cheng, H., Daniel, S. *et al.* (2002). Triggering and augmentation mechanisms, granule pools, and biphasic insulin secretion. *Diabetes* **51**, S83–S90.
58a. Hille, B., Billiard, J., Babu D. F., *et al.* (1999). Stimulation of exocytosis with a calcium signal. *J. Physiol.* **52**, 23–31.
59. Takuma, T. and Ichida, T. (1986). Does cyclic AMP mobilize Ca2+ for amylase secretion from rat parotid cells? *Biochim. Biophys. Acta.* **887**, 113–117.
60. Takuma, T. and Ichida, T. (1994). Evidence for the involvement of protein phosphorylation in cyclic AMP-mediated amylase exocytosis from parotid acinar cells. *FEBS Lett.* **340**, 29–33.
61. Fujita-Yoshigaki, J. (1998). Divergence and convergence in regulated exocytosis: The characteristics of cAMP-dependent enzyme secretion of parotid salivary acinar cells. *Cell. Signal.* **10**, 371–375.
62. Kasai, H. (1999). Comparative biology of Ca2+-dependent exocytosis: implications of kinetic diversity for secretory function. *TINS.* **22**, 88–43.
63. Gomperts, B. D. (1983). Involvement of guanine nucleotide-binding protein in the gating of Ca2+ by receptors. *Nature* **306**, 64–66.
64. Glenn, D. E. and Burgoyne, R. D. (1996). Botulinum neurotoxin light chains inhibit both Ca(2+)-induced and GTP analogue-induced catecholamine release from permeabilised adrenal chromaffin cells. *FEBS Lett.* **386**, 137–140.
65. Martin, T. F. (1997). Phosphoinositides as spatial regulators of membrane traffic. *Curr. Opin. Neurobiol.* **7**, 331–338.
66. Vitale, N., Gasman, S., Caumont, A. S. *et al.* (2000). Insight in the exocytotic process in chromaffin cells: regulation by trimeric and monomeric G proteins. *Biochimie* **82**, 365–373.
67. Tse, F. W. and Tse, A. (1999). Regulation of exocytosis via release of Ca(2+) from intracellular stores. *Bioessays* **21**, 861–865.
68. Cancela, J. M., Van Coppenolle, F., Galione, A. *et al.* (2002). Transformation of local Ca2+ spikes to global Ca2+ transients: the combinatorial roles of multiple Ca2+ releasing messengers. *EMBO J.* **21**, 909–919.
69. Sheng, M. and Pak, D. T. S. (2000). Ligand-gated ion channel interactions with cytoskeletal and signaling proteins. *Annu. Rev. Physiol.* **62**, 755–778.
70. Magee, J. and Johnston, D. (1997). A synaptically controlled, associative signal for Hebbian plasticity in hippocampal neurons. *Science* **275**, 209–213.
71. Chittajallu, R., Vignes, M., Dev, K. K. *et al.* (1996). Regulation of glutamate release by presynaptic kainate receptors in the hippocampus. *Nature* **379**, 78–81.
72. Miller, R. J. (1998). Presynaptic receptors. *Annu. Rev. Pharmacol. Toxicol.* **38**, 201–227.
73. Girod, R. and Role, L. W. (2001). Long-lasting enhancement of glutamatergic synaptic transmission by acetylcholine contrasts with response adaptation after exposure to low-level nicotine. *J. Neurosci.* **21**, 5182–5190.
74. Charany, A., Deswce, J. M., Godeheu, G. *et al.* (1994). Presynaptic control of dopamine synthesis and release by excitatory amino acids in rat striatal synaptosomes. *Neurochem. Int.* **25**, 145–154.
75. Dunlap, K., Luebke, J. I., and Turner, T. J. (1995). Exocytotic Ca2+ channels in mammalian central neurons. *TINS* **18**, 89–98.
76. Miller, R. J. (1987). Multiple calcium channels and neuronal function. *Science* **235**, 46–52.
77. Bito, H., Deisseroth, K., and Tsien, R. W. (1997). Ca2+-dependent regulation in neuronal gene expression. *Curr. Opin. Neurobiol.* **7**, 419–429.

78. Garcia, A. G., Sala, F., Reig, J. A. *et al.* (1984). Dihydropyridine BAY-K-8644 activates chromaffin cell calcium channels. *Nature* **309**, 69–71.
79. Boyd, A. E. (1992). *J. Cell. Biochem.* **48**, 235–241.
80. Darvish, N. and Russell, J. T. (1998). Neurotransmitter-induced novel modulation of a nonselective cation channel by a cAMP-dependent mechanisms in rat pineal cells. *J. Neurophysiol.* **79**, 2546–2556.
81. Chatterjee, T. K., Sharma, R. V., and Fisher, R. A. (1996). Molecular cloning of a novel variant of the pituitary adenylate cyclase-activating polypeptide (PACAP) receptor that stimulates calcium influx by activation of L-type calcium channels. *J. Biol. Chem.* **271**, 32226–32232.
82. O'Farrell, M. and Marley, P. D. (1997). Multiple calcium channels are required for pituitary adenylate cyclase-activating polypeptide-induced catecholamine secretion from bovine cultured adrenal chromaffin cells. *Naunyn Schmiedebergs Arch. Pharmacol.* **356**, 536–542.
83. Hahm, S. H., Hsu, C.-M., and Eiden, L. E. (1998). PACAP activates calcium influx-dependent and -independent pathways to couple met-enkephalin secretion and biosynthesis in chromaffin cells. *J. Mol. Neurosci.* **11**, 1–15.
84. Taupenot, L., Mahata, M., Mahata, S. K. *et al.* (1999). Time-dependent effects of the neuropeptide PACAP on catecholamine secretion. Stimulation and desensitization. *Hypertension* **34**, 1152–1162.
85. Hamelink, C., Lee, H.-W., Grimaldi, M. *et al.* (2002). Coincident elevation of cyclic AMP and calcium influx by PACAP-27 synergistically regulates VIP gene transcription through a novel PKA-independent signaling pathway. *J. Neurosci.* in press.
86. Beaudet, M. M., Parsons, R. L., Braas, K. M. *et al.* (2000). Mechanisms mediating pituitary adenylate cyclase-activating polypeptide depolarization of rat sympathetic neurons. *J. Neurosci.* **20**, 7353–7361.
87. Augustine, G. J. (2001). How does calcium trigger neurotransmitter release? *Curr. Opin. Neurobiol.* **11**, 320–326.
88. Douglas, W. W. (1975). In Carafoli, E. *et al.* Eds., *Calcium Transport in Contraction and Secretion,* pp. 167–174. North-Holland/American Elsevier, Amsterdam.
89. Cancela, J. M., Churchill, G. C., and Galione, A. (1999). Coordination of agonist-induced Ca2+-signalling patterns by NAADP in pancreatic acinar cells. *Nature* **398**, 74–76.
90. Lee, H. C. (2000). *Science STKE.* 2000.
91. Lee, H. C. (2001). Physiological functions of cyclic ADP-ribose and NAADP as calcium messengers. *Annu. Rev. Pharmacol. Toxicol.* **41**, 317–345.
92. Yule, D. I., Wu, D., Essington, T. E. *et al.* (1993). Sphingosine metabolism induces Ca2+ oscillations in rat pancreatic acinar cells. *J. Biol. Chem.* **268**, 12353–12358.
93. Winkler, H. and Westhead, E. (1980). The molecular organization of adrenal chromaffin granules. *Neuroscience* **5**, 1803–1823.
94. Yoo, S. H. and Albanesi, J. P. (1990). Inositol 1,4,5-trisphosphate-triggered Ca2+ release from bovine adrenal medullary secretory vesicles. *J. Biol. Chem.* **265**, 13446–13448.
95. Yoo, S. H., Oh, Y. S., Kang, M. K. *et al.* (2001). Localization of three types of the inositol 1,4,5-trisphosphate receptor/Ca(2+) channel in the secretory granules and coupling with the Ca(2+) storage proteins chromogranins A and B. *J. Biol. Chem.* **276**, 45806–45812.
96. Gerasimenko, O. V., Gerasimenko, J. V., Beian, P. V. *et al.* (1996). Inositol trisphosphate and cyclic ADP-ribose-mediated release of Ca2+ from single isolated pancreatic zymogen granules. *Cell* **84**, 473–480.
97. Yule, D. I., Ernst, S. A., Ohnishi, H. *et al.* (1997). Evidence that zymogen granules are not a physiologically relevant calcium pool. Defining the distribution of inositol 1,4,5-trisphosphate receptors in pancreatic acinar cells. *J. Biol. Chem.* **272**, 9093–9098.
98. Brose, N. and Rosenmund, C. (1999). SV2: SVeeping up excess Ca2+ or TranSVorming presynaptic Ca2+ sensors? *Neuron* **24**, 766–768.
99. Söllner, T., Whiteheart, S. W., Brunner, M. *et al.* (1993). SNAP receptors implicated in vesicle targeting and fusion. *Nature* **362**, 318–324.
100. Bennett, M. K. and Scheller, R. H. (1993). The molecular machinery for secretion is conserved from yeast to neurons. *Proc. Natl. Acad. Sci. USA* **90**, 2559–2563.
101. Söllner, T., Bennett, M. K., Whiteheart, S. W. *et al.* (1993). A protein assembly-disassembly pathway *in vitro* that may correspond to sequential steps of synaptic vesicle docking, activation and fusion. *Cell* **75**, 409–418.
102. Elferink, L. A., Peterson, M. R., and Scheller, R. H. (1993). A role for synaptotagmin (p65) in regulated exocytosis. *Cell* **72**, 153–159.
103. Ahnert-Hilger, G. and Bigalke, H. (1995). Molecular aspects of tetanus and botulinum neurotoxin poisoning. *Prog. Neurobiol.* **46**, 83–96.
104. Schiavo, G., Benfenati, F., Poulain, B. *et al.* (1992). Tetanus and botulinum-B neurotoxins block neurotransmitter release by proteolytic cleavage of synaptobrevin. *Nature* **359**, 832–835.
105. Schiavo, G. and van der Goot, F. G. (2001). The bacterial toxin toolkit. *Nat. Rev. Mol. Cell Biol.* **2**, 530–537.
106. Baker, P. F. and Knight, D. E. (1978). Calcium-dependent exocytosis in bovine adrenal medullary cells with leaky plasma membranes. *Nature* **276**, 620–622.
107. Kiraly-Borri, C. E., Morgan, A., Burgoyne, R. D. *et al.* (1996). Soluble N-ethylmaleimide-sensitive-factor attachment protein and N-ethylmaleimide-insensitive factors are required for Ca2+-stimulated exocytosis of insulin. *Biochem. J.* **314**, 199–203.
108. Sadoul, K., Lang, J., Montecucco, C. *et al.* (1995). SNAP-25 is expressed in islets of Langerhans and is involved in insulin release. *J. Cell Biol.* **128**, 1019–1028.
109. Regazzi, R., Wollheim, C. B., Lang, J. *et al.* (1995). VAMP-2 and cellulbrevin are expressed in pancreatic beta-cells and are essential for Ca(2+)-but not for GTP gamma S-induced insulin secretion. *EMBO J.* **14**, 2723–2730.
110. Wheeler, M. B., Sheu, L., Ghai, M. *et al.* (1996). Characterization of SNARE protein expression in beta cell lines and pancreatic islets. *Endocrinology* **137**, 1340–1348.
111. Lang, J., Zhang, H., Vaidyanathan, V.-V. *et al.* (1997). Transient expression of botulinum neurotoxin C1 light chain differentially inhibits calcium and glucose induced insulin secretion in clonal ß-cells. *FEBS Lett.* **419**, 13–17.
112. Boyd, R. S., Duggan, M. J., Shone, C. C. *et al.* (1995). The effect of botulinum neurotoxins on the release of insulin from the insulinoma cell lines HIT-15 and RINm5F. *J. Biol. Chem.* **270**, 18216–18218.
113. Praz, G. A., Halban, P. A., Wollheim, C. B. *et al.* (1983). Regulation of immunoreactive-insulin release from a rat cell line (RINm5F). *Biochem. J.* **210**, 345–352.
114. Brumell, J. H., Volchuk, A., Sengelov, H. *et al.* (1995). Subcellular distribution of docking/fusion proteins in neutrophils, secretory cells with multiple exocytotic compartments. *J. Immunol.* **155**, 5750–5790.
115. Gaisano, H. Y., Sheu, L., Foskett, J. K. *et al.* (1994). Tetanus toxin light chain cleaves a vesicle-associated membrane protein (VAMP) isoform 2 in rat pancreatic zymogen granules and inhibits enzyme secretion. *J. Biol. Chem.* **269**, 17062–17066.
116. Fujita-Yoshigaki, J., Dohke, Y., Hara-Yokoyama, M. *et al.* (1996). Vesicle-associated membrane protein 2 is essential for cAMP-regulated exocytosis in rat parotid acinar cells. The inhibition of cAMP-dependent amylase release by botulinum neurotoxin B. *J. Biol. Chem.* **271**, 13130–13134.
117. Fujita-Yoshigaki, J., Dohke, Y., Hara-Yokoyama, M. *et al.* (1999). Presence of a complex containing vesicle-associated membrane protein 2 in rat parotid acinar cells and its disassembly upon activation of cAMP-dependent protein kinase. *J. Biol. Chem.* **274**, 23642–23646.
118. Foster, L. J., Yeung, B., Mohtashami, M. *et al.* (1998). Binary interactions of the SNARE proteins syntaxin-4, SNAP23, and VAMP-2 and their regulation by phosphorylation. *Biochemistry* **37**, 11089–11096.
119. Augustine, G. J., Burns, M. E., DeBello, W. M. *et al.* (1996). Exocytosis: proteins and perturbations. *Annu. Rev. Pharmacol. Toxicol.* **36**, 659–701.
120. Jahn, R. and Südhof, T. C. (1999). Membrane fusion and exocytosis. *Annu. Rev. Biochem.* **68**, 863–911.
121. Hay, J. C. and Martin, T. F. (1992). Resolution of regulated secretion into sequential MGATP-dependent and calcium-dependent stages mediated by distinct cytosolic proteins. *J. Cell Biol.* **119**, 139–151.
122. Holz, R. W., Bittner, M. A., Peppers, S. C. *et al.* (1989). MgATP-independent and MgATP-dependent exocytosis. *J. Biol. Chem.* **264**, 5412–5419.

123. Chapman, E. R., Hanson, P. I., Ahn, S. *et al.* (1995). Ca2+ regulates the interaction between synaptotagmin and syntaxin I. *J. Biol. Chem.* **270**, 23667–23671.

124. Bruns, D. and Jahn, R. (1995). Real-time measurement of transmitter release from single synaptic vesicles. *Nature* **377**, 62–65.

125. Thomas, P., Wong, J. G., Lee, A. K. *et al.* (1993). A low affinity Ca2+ receptor controls the final steps in peptide secretion from pituitary melanotrophs. *Neuron* **11**, 93–104.

126. von Gersdorff, H. and Matthews, G. (1994). Dynamics of synaptic vesicle fusion and membrane retrieval in synaptic terminals. *Nature* **367**, 735–739.

127. Geppert, M., Goda, Y., Hammer, R. E. *et al.* (1994). Synaptotagmin I: a major Ca2+ sensor for transmitter release at a central synapse. *Cell* **79**, 717–727.

128. Jacobsson, G., Bean, A. J., Scheller, R. H. *et al.* (1994). Identification of synaptic proteins and their isoform mRNAs in compartments of pancreatic endocrine cells. *Proc. Natl. Acad. Sci. USA* **91**, 12487–12491.

129. Martinez, I., Chakrabarati, S., Hellevik, T. *et al.* (2000). Synaptotagmin VII regulates Ca(2+)-dependent exocytosis of lysosomes in fibroblasts. *J. Cell Biol.* **148**, 1141–1149.

130. Fukuda, M., Kowalchyk, J. A., Zhang, X. *et al.* (2002). Synaptotagmin IX regulates Ca2+-dependent secretion in PC12 cells. *J. Biol. Chem.* **277**, 4601–4604.

131. Dodge, F. A. J. and Rahamimoff, R. (1967). Co-operative action of calcium ions in transmitter release at the neuromuscular junction. *J. Physiol.* **193**, 419–432.

132. Ann, K., Kowalchyk, J. A., Loyet, K. M. *et al.* (1997). Novel Ca2+-binding protein (CAPS) related to UNC-31 required for Ca2+-activated exocytosis. *J. Biol. Chem.* **272**, 19637–19640.

133. Renden, R., Berwin, B., Davis, W. *et al.* (2001). *Drosophila* CAPS is an essential gene that regulates dense-core vesicle release and synaptic vesicle fusion. *Neuron* **31**, 421–437.

134. Walent, J. H., Porter, B. W., and Martin, T. F. (1992). A novel 145 kd brain cytosolic protein reconstitutes Ca(2+)-regulated secretion in permeable neuroendocrine cells. *Cell* **70**, 765–775.

135. Kishimoto, T., Liu, T-T., Ninomiya, Y. *et al.* (2001). Ion selectivities of the Ca2+ sensors for exocytosis in rat phaeochromocytoma cells. *J. Physiol.* **555**, 627–637.

136. Parsons, T. D., Coorssen, J. R., Horstmann, H. *et al.* (1995). Docked granules, the exocytic burst, and the need for ATP hydrolysis in endocrine cells. *Neuron* **15**, 1085–1096.

137. Guild, S., Murray, A. T., Wilson, M. L. *et al.* (2001). Over-expression of NCS-1 in AtT-20 cells affects ACTH secretion and storage. *Mol. Cell. Endocrinol.* **184**, 51–63.

138. Wang, C. Y., Yang, F., He, X. *et al.* (2001). Ca(2+) binding protein frequenin mediates GDNF-induced potentiation of Ca(2+) channels and transmitter release. *Neuron* **32**, 99–112.

139. Burack, W. R. and Shaw, A. S. (2000). Signal transduction: hanging on a scaffold. *Curr. Opin. Cell Biol.* **12**, 211–216.

140. Zinsmaier, K. E., Ebele, K. K., Buchner, E. *et al.* (1994). Paralysis and early death of cystein string protein mutants of *Drosophila.* *Science* **263**, 977–980.

141. Evans, G. J. O., Wilkinson, M. C., Graham, M. E. *et al.* (2001). Phosphorylation of cysteine string protein by protein kinase A. Implications for the modulation of exocytosis. *J. Biol. Chem.* **276**, 47877–47885.

142. Bruce, J. I. E., Shuttlesworth, T. J., Giovannucci, D. R. *et al.* (2002). Phosphorylation of inositol 1,4,5-trisphosphate receptors in parotid acinar cells. A mechanism for the synergistic effects of cAMP on Ca2+ signalling. *J. Biol. Chem.* **277**, 1340–1348.

143. Giovannucci, D. R., Groblewski, G. E., Sneyd, J. *et al.* (2000). Targeted phosphorylation of inositol 1,4,5-trisphosphate receptors selectively inhibits localized Ca2+ release and shapes oscillatory Ca2+ signals. *J. Biol. Chem.* **275**, 33704–33711.

143a. Straub, S. V., Giovanucci, D. R., Bruce, J. I. *et al.* (2002). A role for phosphorylation of inosital 1,4,5-trisphosphate receptors in defining calcium signals induced by peptide agonists in pancreatic acinar cells. *J. Biol. Chem.* **277**, 31949–31956.

144. Dell'Acqua, M. L. and Scott, J. D. (1997). Protein kinase A anchoring. *J. Biol. Chem.* **272**, 12881–12884.

145. Hirling, H. and Scheller, R. H. (1996). Phosphorylation of synaptic vesicle proteins: Modulation of the alphaSNAP interaction with the core complex. *Proc. Natl. Acad. Sci. USA* **93**, 11945–11949.

146. de Rooij, J., Zwartkruis, F. J. T., Verheijen, M. H. G. *et al.* (1998). Epac is a Rap1 guanine-nucleotide-exchange factor directly activated by cyclic AMP. *Nature* **396**, 474–477.

147. Kawasaki, H., Springett, G. M., Mochizuki, N. *et al.* (1998). A family of cAMP-binding proteins that directly activate Rap1. *Science* **282**, 2275–2279.

148. Ozaki, N., Shibasaki, T., Kashima, Y. *et al.* (2000). cAMP-GEFII is a direct target of cAMP in regulated exocytosis. *Nature Cell Biol.* **2**, 805–811.

149. De Koninck, P. and Schulman, H. (1998). Sensitivity of CaM kinase II to the frequency of Ca2+ oscillations. *Science* **279**, 227–230.

150. Gromada, J., Hoy, M., Renström, E. *et al.* (1999). CaM kinase II-dependent mobilization of secretory granules underlies acetylcholine-induced stimulation of exocytosis in mouse pancreatic B-cells. *J. Physiol.* **518**, 745–759.

151. Hilfiker, S., Pieribone, V. A., Nordstedt, C. *et al.* (1999). Regulation of synaptotagmin I phosphorylation by multiple protein kinases. *J. Neurochem.* **73**, 921–932.

152. Llinas, R., Gruner, J. A., Sugimori, M. *et al.* (1991). Regulation by synapsin I and Ca(2+)-calmodulin-dependent protein kinase II of the transmitter release in squid giant synapse. *J. Physiol.* **436**, 257–282.

153. Nichols, R. A., Chilcote, T. J., Czernik, A. J. *et al.* (1992). Synapsin I regulates glutamate release from rat brain synaptosomes. *J. Neurochem.* **58**, 783–785.

154. Schweitzer, E. S., Sanderson, M. J., and Wasterlain, C. G. (1995). Inhibition of regulated catecholamine secretion from PC12 cells by the Ca2+/calmodulin kinase II inhibitor KN-62. *J. Cell Sci.* **108**, 2619–2628.

155. Waters, W. W., Chen, P. L., McArthur, N. H. *et al.* (1998). Calcium/calmodulin-dependent protein kinase II involvement in release of gonadotropin-releasing hormone. *Neuroendocrinology* **67**, 145–152.

156. Maurer, J. A., Wenger, B. W., and McKay, D. B. (1996). Effects of protein kinase inhibitors on morphology and function of cultured bovine adrenal chromaffin cells: KN-62 inhibits secretory function by blocking stimulated Ca2+ entry. *J. Neurochem.* **66**, 105–113.

157. Wenham, R. M., Landt, M., Walters, S. M. *et al.* (1992). Inhibition of insulin secretion by KN-62, a specific inhibitor of the multifunctional Ca2+/calmodulin dependent protein kinase II. *Biochem. Biophys. Res. Commun.* **189**, 128–133.

158. Cox, M. E., Ely, C. M., Catling, A. D. *et al.* (1996). Tyrosine kinases are required for catecholamine secretion and mitogen-activated protein kinase activation in bovine adrenal chromaffin cells. *J. Neurochem.* **66**, 1103–1112.

159. Schneggenburger, R., Sakaba, T., and Neher, E. (2002). Vesicle pools and short-term synaptic depression: lessons from a large synapse. *Trends Neurosci.* **25**, 206–212.

160. Nagy, G., Matti, U., Binz, T. *et al.* (2002). Phosphorylation of SNAP-25 by protein kinase A regulates the size of the releasable vesicle pools in adrenal chromaffin cells. *Soc. Neurosci. Abstr.* **520.4**.

161. Salin, P. A., Malenka, R. C., and Nicoll, R. A. (1996). Cyclic AMP mediates a presynaptic form of LTP at cerebellar parallel fiber synapses. *Neuron* **16**, 797–803.

162. Wong, S. T., Athos, J., Figueroa, X. A. *et al.* (1999). Calcium-stimulated adenylyl cyclase activity is critical for hippocampus-dependent long-term memory and late-phase LTP. *Neuron* **23**, 787–798.

163. Kandel, E. R. (2001). The molecular biology of memory storage: a dialogue between genes and synapses. *Science* **294**, 1030–1038.

164. Holz, R. W. (1988). Control of exocytosis from adrenal chromaffin cells. *Cell. Mol. Neurobiol.* **8**, 259–268.

165. Stevens, C. F., and Sullivan, J. M. (1998). Regulation of the readily releasable vesicle pool by protein kinase C. *Neuron* **21**, 885–893.

166. Suwandito, L., Leung, Y. K., Firmansyah, A. *et al.* (1989). Phorbol ester potentiates VIP-stimulated amylase release in rat pancreatic acini. *Pancreas* **4**, 459–463.

167. Nishizuka, Y. (1986). Studies and perspectives of protein kinase C. *Science* **233**, 305–312.
168. Rhee, J.-S., Betz, A., Pyott, S. *et al.* (2002). ß Phorbol ester- and diacylglycerol-induced augmentation of transmitter release is mediated by Munc13s and not by PKCs. *Cell* **108**, 121–133.
169. Oancea, E. and Meyer, T. (1998). Protein kinase C as a molecular machine for decoding calcium and diacylglycerol signals. *Cell* **95**, 307–318.
170. McCann, S. M., Lumpkin, M. D., Mizunuma, H. *et al.* (1984). Peptidergic and dopaminergic control of prolactin release. *Ends in hemoscience* 127–131.
171. Israel, J. M., Kirk, C., and Vincent, J. D. (1987). Electrophysiological responses to dopamine of rat hypophysial cells in lactotroph-enriched primary cultures. *J. Physiol.* **390**, 1–22.
172. Delbeke, D., Scammell, J. G., Martinez-Campos, A. *et al.* (1986). Dopamine inhibits prolactin release when cyclic adenosine 3',5'-monophosphate levels are elevated. *Endocrinology* **118**, 1271–1277.
173. Pittman, Q. J. and Siggins, G. R. (1981). Somatostatin hyperpolarizes hippocampal pyramidal cells *in vitro*. *Brain Res.* **221**, 402–408.
174. Luini, A., Lewis, D., Guild, S. *et al.* (1986). Somatostatin, an inhibitor of ACTH secretion, decreases cytosolic free calcium and voltage-dependent calcium current in a pituitary cell line. *J. Neurosci.* **6**, 3128–3132.
175. Gromada, J., Hoy, M., Bschard, K. *et al.* (2001). Somatostatin inhibits exocytosis in rat pancreatic alpha-cells by Gi2-dependent activation of calcineurin and depriming of secretory granules. *J. Physiol.* **535**, 519–532.
176. Starke, K., Gothert, M., and Kilbinger, H. (1989). Modulation of neurotransmitter release by presynaptic autoreceptors. *Physiol. Rev.* **69**, 864–989.
177. Langer, S. Z. and Schoemaker, H. (1989). Alpha-adrenoceptor subtypes in blood vessels: physiology and pharmacology. *Clin. Exp. Hypertens. A.* **11**(Suppl. 1), 21–30.
178. Langer, S. Z., Duval, N., and Massingham, R. (1985). Pharmacologic and therapeutic significance of alpha-adrenoceptor subtypes. *J. Cardiovasc. Pharmacol.* **7**, S1–S8.
179. Arbilla, S., Nowak, J. Z., and Langer, S. Z. (1985). Rapid desensitization of presynaptic dopamine autoreceptors during exposure to exogenous dopamine. *Brain Res.* **337**, 11–17.
180. Lee, T. H., Gee, K. R., Davidson, C. *et al.* (2002). Direct, real-time assessment of dopamine release autoinhibition in the rat caudate-putamen. *Neuroscience* **112**, 647–654.
181. Redman, R. S. and Silinsky, E. M. (1994). ATP released together with acetylcholine as the mediator of neuromuscular depression at frog motor nerve endings. *J. Physiol.* **477**, 117–127.
182. Tooze, S. A. (1998). Biogenesis of secretory granules in the trans-Golgi network of neuroendocrine and endocrine cells. *Biochim. Biophys. Acta.* **1404**, 231–244.
183. Farquhar, M. G., Reid, J. J., and Daniell, L. W. (1978). Intracellular transport and packaging of prolactin: a quantitative electron microscope autoradiographic study of mammotrophs dissociated from rat pituitaries. *Endocrinology* **102**, 296–311.
184. Iacangelo, A. L. and Eiden, L. E. (1995). Chromogranin A: Current status as a precursor for bioactive peptides and a granulogenic/sorting factor in the regulated secretory pathway. *Regul. Peptides.* **8**, 65–88.
185. Gorr, S.-U., Jain, R. K., Kuehn, U. *et al.* (2001). Comparative sorting of neuroendocrine secretory proteins: a search for common ground in a mosaic of sorting models and mechanisms. *Mol. Cell. Endocrinol.* **172**, 1–6.
186. Kim, T., Tao-Cheng, J.-H., Eiden, L. E. *et al.* (2001). Chromogranin A, an "on/off" switch controlling dense-core secretory granule biogenesis. *Cell* **106**, 1–13.
187. Sun, Z., Lee, M. S., Rhee, H. K. *et al.* (1997). Inefficient secretion of human H27A-prolactin, a mutant that does not bind Zn2+. *Mol. Endocrinol.* **11**, 1544–1551.
188. Sankoorikal, B. J., Zhu, Y. L., Hodsdon, M. E. *et al.* (2002). Aggregation of human wild-type and H27A-prolactin in cells and in solution: roles of Zn2+, Cu2+, and pH. *Endocrinology* **143**, 1302–1309.
189. Zhang, C. F., Snell, C. R., and Loh, Y. P. (1999). Identification of a novel prohormone sorting signal-binding site on carboxypeptidase E, a regulated secretory pathway-sorting receptor. *Mol. Endocrinol.* **13**, 527–536.
190. Edwards, R. H. (1992). The transport of neurotransmitters into synaptic vesicles. *Curr. Opin. Neurobiol.* **2**, 586–594.
191. Eiden, L. E. (2000). The vesicular neurotransmitter transporters: current perspectives and future prospects. *FASEB J.* **14**, 2396–2400.
192. Fon, E. A., Pothos, E. N., Sun, B.-C. *et al.* (1997). Vesicular transport regulates monoamine storage and release but is not essential for amphetamine action. *Neuron* **19**, 1271–1283.
193. Song, H., Ming, G., Fon, E. *et al.* (1997). Expression of a putative vesicular acetylcholine transporter facilitates quantal transmitter packaging. *Neuron* **18**, 815–826.
194. Wang, Y.-M., Gainetdinov, R. R., Fumagalli, F. *et al.* (1997). Knockout of the vesicular monoamine transporter 2 gene results in neonatal death and supersensitivity to cocaine and amphetamine. *Neuron* **19**, 1285–1296.
195. Liu, Y., Krantz, D. E., Waites, C. *et al.* (1999). Membrane trafficking of neurotransmitter transporters in the regulation of synaptic transmission. *Trends Cell Biol.* **9**, 356–363.
196. Nakanishi, N., Onozawa, S., Matsumoto, R. *et al.* (1995). Effects of protein kinase inhibitors and protein phosphatase inhibitors on cyclic AMP-dependent down-regulation of vesicular monoamine transport in pheochromocytoma PC12 cells. *FEBS Lett.* **368**, 411–414.
197. Nakanishi, N., Onozawa, S., Matsumoto, R. *et al.* (1995). Cyclic AMP-dependent modulation of vesicular monoamine transport in pheochromocytoma cells. *J. Neurochem.* **64**, 600–607.
198. Krantz, D. E., Peter, D., Liu, Y. *et al.* (1997). Phosphorylation of a vesicular monoamine transporter by casein kinase II. *J. Biol. Chem.* **272**, 6752–6759.
199. Hussain, I., Bate, G. W., Henry, J. *et al.* (1999). Modulation of gastrin processing by vesicular monoamine transporter type 1 (VMAT1) in rat gastrin cells. *J. Physiol.* **517**, 495–505.
200. Wolkersdorfer, M., Laslop, A., Lazure, C. *et al.* (1996). Processing of chromogranins in chromaffin cell culture: effects of reserpine and alpha-methyl-p-tyrosine. *Biochem. J.* **316**, 953–958.
201. Eiden, L. E., Giraud, P., Dave, J. *et al.* (1984). Nicotinic receptor stimulation activates both enkephalin release and biosynthesis in adrenal chromaffin cells. *Nature* **312**, 661–663.
202. Dave, J. R., Eiden, L. E., Lozovsky. D. *et al.* (1987). Differential role of calcium in stimulus-secretion-synthesis coupling in lactotrophs and corticotrophs of rat anterior pituitary. *Ann. N. Y. Acad. Sci.* **493**, 577–580.
203. Bauer, J. W., Kirchmair, R., Egger, C. *et al.* (1993). Histamine induces a gene-specific synthesis regulation of secretogranin II but not of chromogranin A and B in chromaffin cells in a calcium-dependent manner. *J. Biol. Chem.* **268**, 1586–1589.
204. Bommer, M., Liebisch, D., Kley, N. *et al.* (1987). Histamine affects release and biosynthesis of opioid peptides primarily via H1-receptors in bovine chromaffin cells. *J. Neurochem.* **49**, 1688–1696.
205. Donald, A. N., Wallace, D. J., McKenzie, S. *et al.* (2002). Phospholipase C-mediated signalling is not required for histamine-induced catecholamine secretion from bovine chromaffin cells. *J. Neurochem.* **81**, 1116–1129.
206. Martin, T. F., Loyet, K. M., Barry, V. A. *et al.* (1997). The role of PtdIns(4,5)P2 in exocytotic membrane fusion. *Biochem. Soc. Trans.* **25**, 1137–1141.
207. Kavalali, E. T. (2002). SNARE interactions in membrane trafficking: a perspective from mammalian central synapses. *Bioessays* **24**, 926–936.
208. el Far, O., O'Connor, V., Dresbach, T. *et al.* (1998). Protein interactions implicated in neurotransmitter release. *J. Physiol. Paris* **92**, 129–133.
209. Chapman, E. R. (2002). Synaptotagmin: A Ca^{2+} Sensor that triggers exocytosis? *Nat. Revs. Mol. Cell Biol.*, **3**, 1–11.
210. Sollner, T. H. (2003). Regulated exocytosis and SNARE function. *Mol. Membr. Biol.* **20**, 209–220.

CHAPTER 327

Nonclassical Pathways of Protein Export

Igor Prudovsky, Anna Mandinova, Cinzia Bagala, Raffaella Soldi, Stephen Bellum, Chiara Battelli, Irene Graziani, and Thomas Maciag

Center for Molecular Medicine, Maine Medical Center Research Institute, Scarborough, Maine

Introduction

The majority of proteins that are destined to be secreted are usually translated with an NH_2-terminal cleavable signal peptide sequence. This structure is rich in hydrophobic amino acid residues that enable the polypeptide to dock and insert into the lumen of the endoplasmic reticulum for its further transport to the Golgi apparatus and eventual secretion as an intravesicular polypeptide [1]. However, there are groups of extracellular proteins that neither display a signal peptide sequence in their primary structure nor utilize the classical endoplasmic reticulum (ER)–Golgi pathway for their release. In this review, we focus on the mechanisms of stress-induced release of two signaling proteins belonging to this group, fibroblast growth factor 1 (FGF-1) and interleukin 1α (IL-1α). However, before discussing the export of these polypeptides, we briefly review the literature concerning the nonclassical release of other signal peptide-less proteins.

One of the earliest studies of nonclassical protein release was devoted to secretory transglutaminase (TG), an enzyme that cross-links extracellular polypeptides and is released into the extracellular space by cells of the prostate [2,3]. Aumuller and colleagues [3] extensively investigated TG release using immunoelectron microscopy and observed that TG appeared to be synthesized on non-ER-associated ribosomes and was not found in the lumen of the ER and Golgi structures but rather accumulated in cytoplasmic blebs protruding from the apical surface of the secretory cells. Interestingly, the detachment of these apical blebs followed by the release of their contents has long been known as *apocrine secretion* [4]. Although the expression and release of TG is dependent on stimulation by androgens, the regulatory mechanism of TG release has not been investigated, and the mechanism of TG transport to the apical cell membrane as well as the detachment of apical blebs remains obscure. Likewise, the human immunodeficiency virus 1 TAT protein [5] and sphingosine kinase (SK), an enzyme involved in the biosynthesis of shingosine-1-phosphate [6], galectins [78], annexin1 [79], and some other signal peptide-less polypeptides [80] are present in the extracellular compartment. The release of SK and TAT into the extracellular compartment occurs independent of the function of the ER–Golgi apparatus yet their intracellular traffic and export activity has not yet been studied in detail [5,6]. Thioredoxin and S100 proteins are also known to be released by a nonclassical mechanism and these are discussed later in this review.

Fibroblast Growth Factor Export Pathways

FGF-1 and FGF-2 are the signal peptide-less prototypic members of the FGF gene family and they are involved in mesodermal and neuroectodermal induction, angiogenesis, chondrogenesis, neurogenesis, wound healing, and tumor formation [7]. In contrast to the export of SK, TAT, and TG, the release of the FGF prototypes has been studied in some detail. Although the transcripts encoding the FGF prototypes are expressed in all tissues and cells [7], expression of the FGF prototype polypeptides is most evident in mononuclear

..s especially within macrophages [8,9]. The release of the FGF prototypes into the extracellular compartment is a necessary precondition for their biological activity because their association with the glycosylaminoglycan, heparan sulfate, and their high-affinity tyrosine kinase receptors are requisite for cellular signaling [7,10]. However, the absence of a classical signal peptide sequence within the structures of the FGF prototypes excludes a possibility of their transport through the ER–Golgi pathway. Indeed, *in vitro*, cells transfected with FGF-1 do not release it under normal cell culture conditions, and the addition of a classical signal peptide sequence at the NH_2 terminus of FGF-1 results in the generation of an oncogenic form of FGF-1 as a consequence of an extracellular autocrine signal established by its secretion through the ER–Golgi compartments [11]. FGF-2 also exhibits similar structural and functional behavior [12], yet the majority of the members of the FGF gene family have been characterized as oncogenes as a result of the presence of a functional signal peptide sequence that enables them to function in an autocrine and paracrine manner [10,13].

The mechanisms responsible for the regulation of the release of the FGF prototypes have diverged because we know that FGF-2 is released in a constitutive manner [14–17], whereas the release of FGF-1 utilizes a stress-inducible pathway [18]. Interestingly, the invertebrate FGF homolog genes encode FGFs containing a signal peptide sequence [13,19] and it is not known where or why the vertebrate FGF prototypes lost their signal peptide structures. One can speculate that as evolution selected for species with extended life spans, the maintenance of organ and tissue physiology required the function of nononcogenic forms of the FGFs, and this may explain the divergent nature of the pathways utilized to export the FGF prototypes. However, a common feature shared by the FGF prototypes is their independence of the function of the ER–Golgi apparatus since the release of FGF-1 and FGF-2 is insensitive to treatment with brefeldin A [14,15,18].

Although the export of FGF-2 is constitutive, the mechanism of FGF-2 release involves the function of the Na^+/K^+-ATPase transporter and thus may involve changes in membrane potential for the export of FGF-2 as a functional mitogen [20,21]. In contrast, however, the export of FGF-1 is inducible and involves the function of intracellular Cu^{2+} to assemble a multiprotein complex that enables the release of FGF-1 as a biologically inactive Cys30 FGF-1 homodimer [22]. A variety of cellular stresses have been implicated as the initiation signal for FGF-1 export and these include heat shock, anoxia, hypoxia, irradiation, serum deprivation, and oxidized LDL [7,18,23–25]. Because many of these stimuli are involved in establishing an inflammatory response and FGF-1 expression is exaggerated in mononuclear cells [9], it is likely that FGF-1 is delivered to inflammatory sites for tissue/organ repair by mononuclear cells including macrophages [9]. Because this would involve the function of pro-inflammatory cytokines as signals for this recruitment process [26], it is intriguing to speculate that members of the IL-1 gene family, which also lack a classical signal peptide sequence [27], may be involved in mediating this process. However, we should emphasize that the genes encoding the FGF prototypes are not transcriptionally responsive to environmental stress (S. Friedman and T. Maciag, unpublished results, 1993) and thus the export pathways utilized by the FGF prototypes are regulated primarily by either translational or posttranslational events.

The appearance of FGF-1 in the extracellular compartment in response to cellular stress as a Cys30 FGF-1 homodimer is interesting since the dimeric form of FGF-1 can be formed under cell-free conditions by Cu^{2+} oxidation [28]. Yet FGF-2, its close structural homolog, is unable to form a disulfide-linked homodimer even though two of three cysteine residues are conserved between the FGF prototypes [29,30]. It is also interesting that the nonfunctional FGF-1 homodimer exhibits a significantly reduced affinity for immobilized heparin [28]. Because heparin affinity is a hallmark of the proteins encoded by the FGF gene family and interactions with heparin are known to protect FGFs from proteolytic inactivation [31], the failure to achieve an extracellular reducing environment in response to cellular stress would ensure proteolytic inactivation of the latent FGF-1 homodimer and prevent FGF-1 from being functional at inopportune situations. It is of interest to note that the thioredoxin gene also encodes a signal peptide-less redox regulatory polypeptide that is released from dendritic cells following the interaction with antigen-recognizing T lymphocytes and provides a reducing microenvironment in the extracellular compartment [32]. However, although it is not known whether thioredoxin functions to regulate the redox potential of the latent FGF-1 homodimer, it is clear that an extracellular reducing agent is required to activate the latent FGF-1 homodimer as a heparin-binding polypeptide mitogen.

The Export of FGF-1 as a Multiprotein Complex

Early observations that bovine neural tissues contain high-molecular-weight noncovalent FGF-1 complexes [33] facilitated the identification and characterization of polypeptides that are involved in the export of FGF-1 [34–36]. Indeed, the identities of two of the four polypeptides associated with the high-molecular-weight form of brain-derived FGF-1 were determined to be synaptotagmin (Syt)1 and S100A13 on the basis of microsequencing and immunoblot analysis [36,37]. Syt1 is a 65-kDa transmembrane protein involved in the regulation of exocytotic vesicle traffic, and structural information has implicated the function of the p40 extravesicular fragment of p65 Syt1 as the FGF-1-associated chaperone [38]. The p40 form of Syt1 appears to be the result of proteolytic cleavage at an extravesicular hypersensitive proteolytic cleavage site near the transmembrane domain of p65 Syt1 [39]. In addition, the p40 form of Syt1 contains a single cysteine residue, as well as two Ca^{2+}-binding C2 domains [39]. In contrast, S100A13 is a member of the S100 gene family characterized as EF-hand-mediated Ca^{2+}-binding proteins implicated in a variety of calmodulin- and F-actin-dependent cytoskeletal processes [40]. However, unlike the majority of the members of the S100 gene family, the structure of S100A13 is novel because it lacks cysteine residues and contains a basic residue-rich domain positioned at its carboxy-terminus [41,42]. The function of both Syt1 and S100A13 in the regulation of the release of FGF-1 in response to cellular stress could be demonstrated by the expression of dominant-negative mutants of either Syt1 or S100A13 that are able to attenuate FGF-1 export [35,43]. However, the Ca^{2+}-binding properties of p40 Syt1 and S100A13 are not likely to be involved in the stress-induced release of FGF-1 because the Ca^{2+} ionophore does not impair or interfere with the release of either FGF-1, S100A13, or p40 Syt1 into the extracellular compartment [35]. Interestingly, a novel feature of both p40 Syt1 and S100A13 is their ability to be constitutively released from cells under normal cell culture conditions [35,43] in spite of the absence of a classical signal peptide sequence in S100A13 and the fact that the signal peptide sequence in p65 Syt1 remains structurally intact within the intravesicular domain (p25) of the transmembrane polypeptide. Although the mechanisms(s) utilized by p40 Syt1 and S100A13 for constitutive export is not known, the stress-induced presence of p40 Syt1 and S100A13 in the extracellular compartment with FGF-1 implies that these intracellular proteins may indeed assemble into a multiprotein complex prior to export (Fig. 1) and that formation of p40 Syt1 from p65 Syt1 may involve the function of an intracellular protease that cleaves p65 Syt1 at its hyperproteolytic site within the extravesicular domain.

Evidence for the self-assembly of the FGF-1:p40 Syt1:S100A13 complex was obtained from cell-free studies utilizing the recombinant forms of these polypeptides. Interestingly, the self-assembly of this multiprotein complex requires the oxidative capacity of Cu^{2+}, and it has been possible to define the molar ratio of the Cu^{2+}-dependent multiprotein complex containing p40 Syt1:FGF-1:S100A13 as 1:2:2, respectively [22]. Additional studies utilizing *in vitro* methods have confirmed this stoichiometry and it appears that Cu^{2+} is able to facilitate the formation of a heterotetramer FGF-1:S100A13 complex containing p40 Syt1, which is assembled and released in response to cellular stress [22].

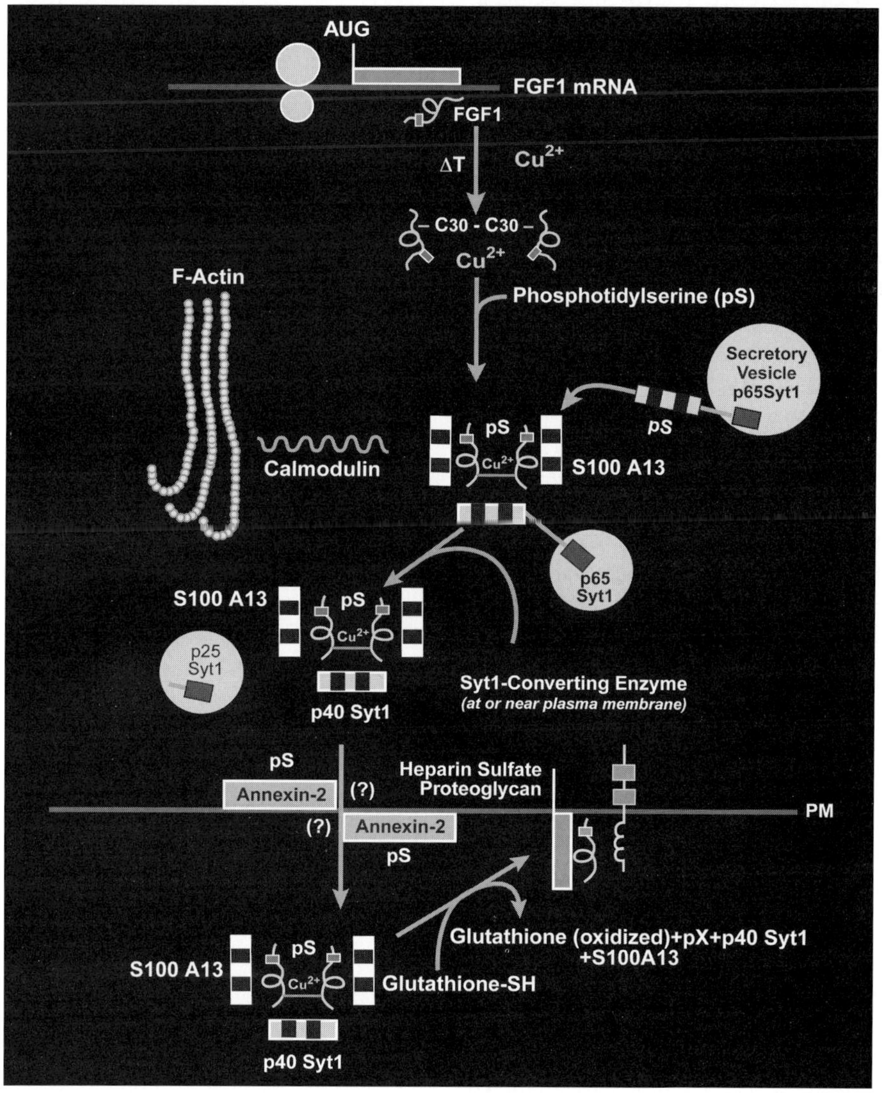

Figure 1 FGF-1 release pathway.

The observations that FGF-1, p40 Syt1, and S100A13 are Cu^{2+}-binding proteins [22,28] and that the Cu^{2+} chelator, tetrathiomolybdate [44], is able to repress the stress-induced release of these polypeptides *in vitro* [22] adds additional credence to the involvement of intracellular Cu^{2+} oxidation in the release of FGF-1.

Interestingly, the overexpression of S100A13 completely repressed the sensitivity of FGF-1 release to both cycloheximide and actinomycin D [43], and this suggests that the Cu^{2+}-dependent assembly mechanism does not require the transcriptional activation and translation of a stress-induced chaperone to facilitate the intracellular assembly of this multiprotein complex. Note that while some members of the S100 gene family are stress-induced genes, neither Syt1 nor S100A13 is transcriptionally activated by cellular stress (T. Lavallee, F. Tarantini, and T. Maciag, unpublished results, 1998). Thus, it is likely that the mechanism utilized by cellular stress to assemble the FGF-1 release complex may involve alterations in the distribution of intracellular Cu^{2+} in order for the divalent oxidant to access S100A13 for the formation of the FGF-1 homodimer. Indeed, recent kinetic data obtained using confocal immunofluorescence microscopy suggest that the intracellular distribution of S100A13, p40 Syt1, and FGF-1 from the cytosol to the inner surface of the plasma membrane in response to cellular stress occurs prior to the assembly of the multiprotein complex [45].

The observation that the assembly of the multiprotein complex follows the intracellular trafficking of the individual components to the inner surface of the plasma membrane [45] suggests a potential role for the cytoskeleton in this process. The first evidence for a role of F-actin filaments in this process came as a result of studies with amlexanox, an anti-inflammatory agent [36,46]. Oyama *et al.* [47], using amlexanox affinity chromatography, purified S100A13 complexed with annexin 2 as an amlexanox-binding complex from lung extracts. Because S100A13 expression is important for the release of FGF-1, the ability of amlexanox to inhibit the release of FGF-1, p40 Syt1, and S100A13 in response to cellular stress was consistent with the ability of amlexanox to associate with S100A13 [36,43]. Interestingly, amlexanox was also able to stimulate the activation of Src and, as a result, collapse the F-actin cytoskeleton [46]. The ability of a dominant-negative Src construct to inhibit the stress-induced release of FGF-1 [46] and the ability of amlexanox to inhibit the translocation of FGF-1 to the inner surface of the plasma membrane [45] is consistent with a role of the F-actin cytoskeleton in this process. However, neither FGF-1, S100A13, nor p40 Syt1 appears to be phosphorylated during either the assembly or release process (T. Lavallee, F. Tarantini, M. Landriscina, and T. Maciag, unpublished results, 1998, 2000) suggesting that the role of Src in FGF-1 export is involved in the maintenance of cytoskeletal integrity. Thus it is likely that the intracellular trafficking of the individual components of the multiprotein complex may utilize the F-actin cytoskeleton for transport to the inner surface of the plasma membrane where the FGF-1:p40 Syt1:S100A13 complex is assembled for export into the extracellular compartment.

The role of Syt1 in the stress-induced release of FGF-1 requires the function of an intracellular protease to convert the transmembrane form of p65 Syt1 to soluble intracellular p40 Syt1. Although the identity of this protease is not known, it is probable that the enzymatic activity of this converting enzyme is regulated by cellular stress. However, it is unlikely that this regulation occurs at the transcriptional level since the Cu^{2+}-dependent assembly of the multiprotein complex prior to export is independent of new transcription and translation [22,43]. It is interesting to note that thrombin induces annexin 2 flipping from the inner surface to the outer surface of the plasma membrane [81] where it functions as the receptor for the fibrinolytic enzymes plasminogen and tissue plasminogen activator [48]. Because S100A13 is able to associate with annexin 2 [47], it is intriguing to speculate that intracellular annexin 2 may not only be the site of residence for the p65 Syt1-converting enzyme but may also be the site of assembly for the multiprotein complex prior to export.

Interleukin 1 Export Pathways

The IL-1 gene family is presently comprised of eight members that are known to function as extracellular receptor-dependent pro-inflammatory cytokines [27,49]. However, the IL-1 gene family contains only a single member with a classical signal peptide sequence, the IL-1 receptor antagonist [27,49]. The majority of the members of the IL-1 gene family are translated as precursor structures and the IL-1 gene family prototypes, IL-1α and IL-1β, require proteolytic modification by either calpain (IL-1α) or the IL-1-converting enzyme (IL-1β) to generate the mature forms of these cytokines [27]. Interestingly, although a primary sequence similarity exists between the FGF and IL-1 prototypes, the crystallographic structures of the IL-1 gene family prototypes are remarkably similar to the crystallographic structures exhibited by the FGF prototypes [50–52] and this structural feature implies that there may be other similarities between the FGF and IL-1 gene families as well. Further, the differences observed between the FGF-1 (stress-induced) and FGF-2 (constitutive) release pathways may be conserved between the members of the IL-1 gene family. Indeed, with respect to nonclassical release, the release of IL-1α, like FGF-1, is regulated by cellular stress and utilizes a pathway similar yet distinct from FGF-1 [53]. The release of IL-1α requires the proteolytic activity of calpain since the mature form of IL-1α but not the precursor form of IL-1α is released in response to cellular stress [53]. The release of mature IL-1α exhibits identical pharmacologic and kinetic properties as those described for FGF-1 [53] and, like the FGF-1 release pathway, IL-1α export requires the function of S100A13 and intracellular Cu^{2+} [82]. Unlike FGF-1, mature IL-1α is released as a functional cytokine and does not contain cysteine residues within its structure that are conserved among IL-1α structures from

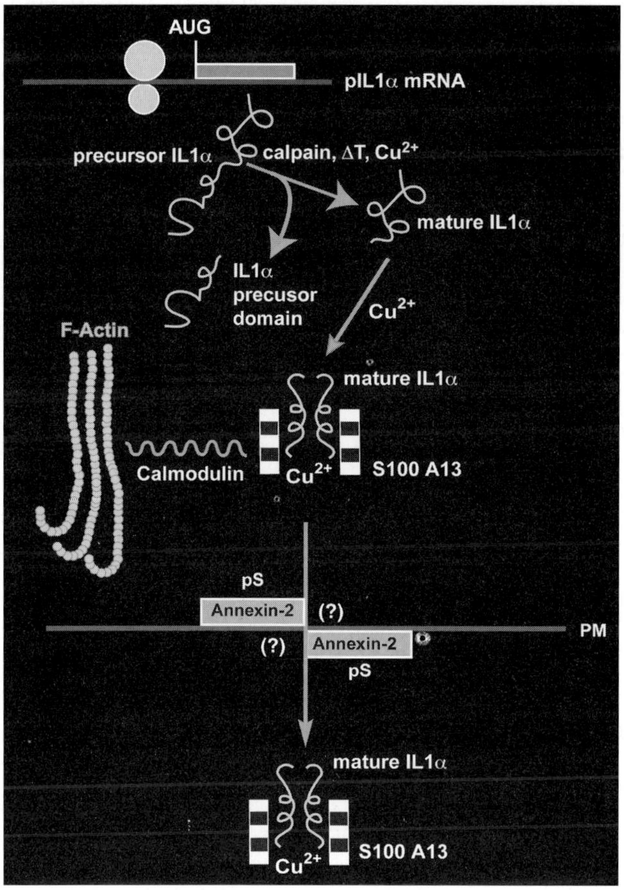

Figure 2 IL-1α release pathway.

other species [54,55]. Thus it is likely that the function of S100A13 and Cu^{2+} is not to establish an intracellular IL-1α:S100A13 heterotetramer by the oxidation of a disulfide bond between two monomeric forms of IL-1α, but rather to form a noncovalent heterotetramer (Fig. 2). The observation that S100A13 can enable the release of a cysteine-free mutant of FGF-1 as a noncovalent heterotetramer [43] is consistent with this premise. However, unlike the stress-induced FGF-1 release pathway, the release of the mature form of IL-1α does not require Syt1 because expression of a dominant-negative mutant of Syt1, which represses FGF-1 export, does not attenuate the release of IL-1α under similar conditions [53]. Thus, if the function of the Ca^{2+}-binding C2 domains in Syt1 is important for the release of both FGF-1 and IL-1α, another intracellular polypeptide containing a C2 domain may be involved. Last, the expression of the precursor but not the mature form of IL-1α represses the release of FGF-1 in response to cellular stress [53] and this observation further argues in favor of the convergent nature of the FGF-1 and IL-1α release pathways.

Because of the convergence between the FGF-1 and IL-1α release pathways, it is possible that other members of the IL-1 gene family may be convergent with the FGF-2 release pathway. However, because the release of the mature form of IL-1β is enhanced by cellular stress including temperature [56] and lipopolysaccharides [57], it is unlikely that the mechanism of IL-1β release will have significantly diverged from the FGF-1 and IL-1α release pathways. Although there is no information about the potential role of S100A13 and/or Cu^{2+} ions on the regulation of IL-1β release, it is noteworthy that IL-1β, like FGF-1 and IL-1α, is a Cu^{2+}-binding protein (S. Bellum, A. Mandinova, and T. Maciag, unpublished results, 2002), yet unlike the release of FGF-1 [58] and IL-1α (F. Tarantini and T. Maciag, unpublished results, 1999), IL-1β export is sensitive to methylamine, an inhibitor of exocytosis [56]. Recent evidence has suggested that IL-1β is contained within intracellular vesicles that express markers of the lysosome but not ER–Golgi, and lysosomotropic agents such as chloroquine and NH_4C1 inhibit the export of IL-1β [57]. A similar inhibitory effect was also observed with sulfonylurea glibenclamide (SG), an inhibitor of ABC1, a mammalian ATP-binding cassette (ABC) translocator [59] and it is known that the nonclassical export pathway employed in unicellular organisms utilizes ABC transporters [60]. However, SG was unable to repress the release of FGF-1 and IL-1α (F. Tarantini and T. Maciag, unpublished results, 2000). Thus, comparative analysis of the IL-1α and IL-1β export pathways suggests that, like the FGF-1 and FGF-2 export pathways, these pathways may have also diverged with the IL-1β export pathway dependent on translocation into the lumen of lysosomes and subsequent lysosomal fusion with the cell membrane possibly using classical exocytotic principles.

Acidic Phospholipids and the Molten Globule Hypothesis

The mechanism utilized by these polypeptides for translocation through the plasma membrane is not known. However, there are a few properties of these proteins that may be consistent with their ability to traverse lipid bilayers. Phosphatidylserine (pS) is an acidic phospholipid that is known to flip from the inner to the outer surface of plasma membranes in response to cellular stress. Although pS flipping is an important component of the intrinsic coagulation system [61] and is widely used in its exaggerated form as evidence for cellular apoptotic behavior as a result of annexin 5-binding [62], all of the known and assumed components of the FGF-1 and IL-1α release pathway are pS-binding proteins. Indeed, IL-1α [63], FGF-1 [64], the S100 gene family [41], p40 Syt1 [39], and annexin 2 [65] are able to associate with pS under cell-free conditions. Although the flipping of pS is known to be ATP dependent [61], as is the release of FGF-1 [35] and IL-1α [53] in response to cellular stress, the molecular basis for pS translocation has not been firmly established [61]. However, FGF-1 [66–68] and IL-1α [82] are known to permeabilize acidic phospholipid-rich vesicles in cell-free systems and are endowed with a *molten globule* character [66–68]. Although the physical protein biochemistry of the molten globule is in its earliest stage of investigation, the term defines the state of a protein that is able to display uncooperative unfolded states or transitions in which the structure of the protein attains multiple conformational states with high secondary but low tertiary structure [69].

As a result of these changes, the partially unfolded conformations are able to achieve low solubility in aqueous environments, resulting in the association with lipid bilayers that they are able to transverse [70]. Although molten globule states are usually formed at low pH and physiological temperatures, the molten globule character of FGF-1 begins to be exaggerated at temperatures between 37° and 42°C and this is facilitated by the presence of acidic phospholipids in the lipid bilayer [66–68].

Thus it is possible that the combination of pS affinity, pS-flipping, and molten globule character may enable FGF-1 to traverse the lipid bilayer. Whether or not S100A13 and p40 Syt1 exhibit similar biophysical characteristics is not known, but it should be emphasized that cells stably transfected with these genes do exhibit constitutive export of both S100A13 and p40 Syt1 into the extracellular compartment at 37°C [35,43]. Interestingly, the S100A13 and p40 Syt1 expression in a FGF-1 background represses their constitutive release of S100A13 but not p40 Syt1, suggesting that S100A13 but not the extravesicular domain of Syt1 does interact with FGF-1 under stress-free conditions [43]. Indeed, if S100A13 and/or p40 Syt1 do exhibit molten globule characters, it is possible that their presence with FGF-1 in the multiprotein complex may facilitate the formation of multiple conformational states and cooperate to exaggerate the low solubility of FGF-1 in an aqueous environment. Because crystallographic data demonstrate that the cysteine residue at position 30 in FGF-1 is buried [50], and S100A13 is able to expose this cysteine residue to solvent for Cu^{2+} oxidation [22,43], these data are consistent with this hypothesis because they suggest that S100A13 is able to alter the conformation of FGF-1.

The Potential Pathophysiological Implications of Nonclassical Release

Although the biological significance of nonclassical release mechanisms is not well understood, some intriguing observations are noteworthy. Interestingly, the pro-inflammatory and angiogenic character of Cu^{2+} is well appreciated [44,71]; however, it is only recently that efficient and specific Cu^{2+} chelators have been utilized in the clinical management of human disease. Indeed, in two phase 1 clinical trials with stage 4 human cancer patients [72,73] including a study limited to head and neck tumors, which are extremely difficult to clinically manage, the Cu^{2+} chelator tetrathiomolybdate (TTM) has been shown to be a valuable addition to the arsenal of antitumor agents as a palliative treatment for the clinical management of solid human tumor growth. Initially described for the treatment of Wilson's disease [74], a disease of Cu^{2+} excess in man, TTM has been documented as an anti-angiogenic effector both in preclinical [83] and clinical [72] settings.

Although the mechanism utilized by TTM is not fully understood, it is possible that the ability of TTM to inhibit the release of the pro-inflammatory cytokine IL-1α from the tumor microvasculature could prevent the tumor from recruiting FGF-1-rich mononuclear cells [9,75], which would significantly reduce the tumor angiogenic response. Indeed, recent experiments on knockout mice demonstrated that IL1 is required for tumer angiogenesis [84]. Further, TTM could also serve to impair the release of FGF-1 from the tumor itself or its surrounding microenvironment to further impair development of the tumor microvasculature. If this hypothesis is indeed accurate, there are other human pathologies in which inflammatory angiogenesis is thought to play an important role including rheumatoid arthritis [76], Alzheimer disease [86], psoriasis [87], and restenosis [77]. Thus, it may be potentially quite interesting to determine whether these nonclassical export pathways do contribute to the development of human pathology.

References

1. Blobel, G. (1995). Unidirectional and bidirectional protein traffic across membranes. *Cold Spring Harb. Symp. Quant. Biol.* **60**, 1–10.
2. Seitz, J., Keppler, C., Rausch, U., and Aumuller, G. (1990). Immunohistochemistry of secretory transglutaminase from rodent prostate. *Histochemistry* **93**, 525–530.
3. Aumuller, G., Wilhelm, B., and Seitz, J. (1999). Apocrine secretion—fact or artifact? *Anat. Anz.* **181**, 437–446.
4. Schiefferdecke, P. (1922). Die Hautdrusen des Menschen und der Saugetiere, ihre biologische und rassenanatomische Bedeutung, sowie die Muscularis sexualis. Stuttgart, E. Schweizerbart'sche Verlagsbuchhandlung.
5. Chang. H. C., Samaniego, F., Nair, B. C., Buonaguro, L., and Ensoli, B. (1997). HIV-1 Tat protein exits from cells via a leaderless secretory pathway and binds to extracellular matrix-associated heparan sulfate proteoglycans through its basic region. *AIDS* **11**, 1421–1431.
6. Ancellin, N., Colmont, C., Su, J., Li, Q., Mittereder, N., Chae, S. S., Stefansson, S., Liau, G., and Hla, T. (2002). Extracellular export of sphingosine kinase-1 enzyme. Sphingosine 1-phosphate generation and the induction of angiogenic vascular maturation. *J. Biol. Chem.* **277**, 6667–6675.
7. Friesel, R. and Maciag, T. (1999). Fibroblast growth factor prototype release and fibroblast growth factor receptor signaling. *Thromb. Haemost.* **82**, 748–754.
8. Sano, H., Forough, R., Maier, J. A., Case, J. P., Jackson, A., Engleka, K., Maciag, T., and Wilder, R. L. (1990). Detection of high levels of heparin binding growth factor-1 (acidic fibroblast growth factor) in inflammatory arthritic joints. *J. Cell Biol.* **110**, 1417–1426.
9. Kuwabara, K., Ogawa, S., Matsumoto, M., Koga, S., Clauss, M., Pinsky, D. J., Lyn, P., Leavy, J., Witte, L., Joseph-Silverstein, J. *et al.* (1995). Hypoxia-mediated induction of acidic/basic fibroblast growth factor and platelet-derived growth factor in mononuclear phagocytes stimulates growth of hypoxic endothelial cells. *Proc. Natl. Acad. Sci. USA* **92**, 4606–4610.
10. Szebenyi, G. and Fallon, J. F. (1999). Fibroblast growth factors as multifunctional signaling factors. *Int. Rev. Cytol.* **185**, 45–106.
11. Forough, R., Xi, Z., MacPhee, M., Friedman, S., Engleka, K. A., Sayers, T., Wiltrout, R. H., and Maciag, T. (1993). Differential transforming abilities of non-secreted and secreted forms of human fibroblast growth factor-1. *J. Biol. Chem.* **268**, 2960–2968.
12. Gately, S., Tsanaclis, A. M., Takano, S., Klagsbrun, M., and Brem, S. (1995). Cells transfected with the basic fibroblast growth factor gene fused to a signal sequence are invasive *in vitro* and in situ in the brain. *Neurosurgery* **36**, 780–788.
13. Coulier, F., Pontarotti, P., Roubin, R., Hartung, H., Goldfarb, M., and Birnbaum, D. (1997). Of worms and men: An evolutionary perspective on the fibroblast growth factor (FGF) and FGF receptor families. *J. Mol. Evol.* **44**, 43–56.

14. Mignatti, P., Morimoto, T., and Rifkin, D. B. (1992). Basic fibroblast growth factor, a protein devoid of secretory signal sequence, is released by cells via a pathway independent of the endoplasmic reticulum-Golgi complex. *J. Cell Physiol.* **151**, 81–93.
15. Florkiewicz, R. Z., Majack, R. A., Buechler, R. D., and Florkiewicz, E. (1995). Quantitative export of FGF-2 occurs through an alternative, energy-dependent, non-ER/Golgi pathway. *J. Cell Physiol.* **162**, 388–399.
16. Trudel, C., Faure-Desire, V., Florkiewicz, R. Z., and Baird, A. (2000). Translocation of FGF2 to the cell surface without release into conditioned media. *J. Cell Physiol.* **185**, 260–268.
17. Hannon, K., Kudla, A. J., McAvoy, M. J., Clase, K. L., and Olwin, B. B. (1996). Differentially expressed fibroblast growth factors regulate skeletal muscle development through autocrine and paracrine mechanisms. *J. Cell Biol.* **132**, 1151–1159.
18. Jackson, A., Friedman, S., Zhan, X., Engleka, K. A., Forough, R., and Maciag, T. (1992). Heat shock induces the release of fibroblast growth factor 1 from NIH 3T3 cells. *Proc. Natl. Acad. Sci. USA* **89**, 10691–10695.
19. Sutherland, D., Samakovlis, C., and Krasnow, M. A. (1996). Branchless encodes a *Drosophila* FGF homolog that controls tracheal cell migration and the pattern of branching. *Cell* **87**, 1091–1101.
20. Florkiewicz, R. Z., Anchin, J., and Baird, A. (1998). The inhibition of fibroblast growth factor-2 export by cardenolides implies a novel function for the catalytic subunit of Na+,K+ -ATPase. *J. Biol. Chem.* **273**, 544–551.
21. Dahl, J. P., Binda, A., Canfield, V. A., and Levenson, R. (2000). Participation of Na,K-ATPase in FGF-2 secretion: Rescue of ouabain-inhibitable FGF-2 secretion by ouabain-resistant Na,K-ATPase alpha subunits. *Biochemistry* **39**, 14877–14883.
22. Landriscina, M., Bagala, C., Mandinova, A., Soldi, R., Micucci, I., Bellum, S., Prudovsky, I., and Maciag, T. (2001). Copper induces the assembly of a multiprotein aggregate implicated in the release of fibroblast growth factor 1 in response to stress. *J. Biol. Chem.* **276**, 25549–25557.
23. Shin, J. T., Opalenik, S. R., Wehby, J. N., Mahesh, V. K., Jackson, A., Tarantini, F., Maciag, T., Thompson, J. A. (1996). Serum-starvation induces the extracellular appearance of FGF-1. *Biochim. Biophys. Acta.* **1312**, 27–38.
24. Ananyeva, N. M., Tjurmin, A. V., Berliner, J. A., Chisolm, G. M., Liau, G., Winkles, J. A., and Haudenschild, C. C. (1997). Oxidized LDL mediates the release of fibroblast growth factor-1. *Arterioscler. Thromb. Vasc. Biol.* **17**, 445–453.
25. Mouta Carreira, C., Landriscina, M., Bellum, S., Prudovsky, I., and Maciag, T. (2001). The comparative release of FGF1 by hypoxia and temperature stress. *Growth Factors* **18**, 277–285.
26. Vaday, G. G., Franitza, S., Schor, H., Hecht, I., Brill, A., Cahalon, L., Hershkoviz, R., and Lider, O. (2001). Combinatorial signals by inflammatory cytokines and chemokines mediate leukocyte interactions with extracellular matrix. *J. Leukoc. Biol.* **69**, 885–892.
27. Dinarello, C. A. (1998). Interleukin-1, interleukin-1 receptors and interleukin-1 receptor antagonist. *Int. Rev. Immunol.* **16**, 457–499.
28. Engleka, K. A. and Maciag, T. (1992). Inactivation of human fibroblast growth factor-1 (FGF-1) activity by interaction with copper ions involves FGF-1 dimer formation induced by copper-catalyzed oxidation. *J. Biol. Chem.* **267**, 11307–11315.
29. Jaye, M., Howk, R., Burgess, W., Ricca, G. A., Chiu, I. M., Ravera, M. W., O'Brien, S. J., Modi, W. S., Maciag, T., and Drohan, W. N. (1986). Human endothelial cell growth factor: Cloning, nucleotide sequence, and chromosome localization. *Science* **233**, 541–545.
30. Abraham, J. A., Whang, J. L., Tumolo, A., Mergia, A., and Fiddes, J. C. (1986). Human basic fibroblast growth factor: Nucleotide sequence, genomic organization, and expression in mammalian cells. *Cold Spring Harb. Symp. Quant. Biol.* **51**, 657–668.
31. Rosengart, T. K., Johnson, W. V., Friesel, R., Clark, R., and Maciag, T. (1988). Heparin protects heparin-binding growth factor-I from proteolytic inactivation *in vitro*. *Biochem. Biophys. Res. Commun.* **152**, 432–440.
32. Angelini, G., Gardella, S., Ard, M., Ciriolo, M. R., Filomeni, G., Di Trapani, G., Clarke, F., Sitia, R., Rubartelli, A. (2002). Antigen-presenting dendritic cells provide the reducing extracellular microenvironment required for T lymphocyte activation. *Proc. Natl. Acad. Sci. USA*. **99**, 1491–1496.
33. Maciag, T., Hoove, G. A., and Weinstein, R. (1982). High and low molecular weight forms of endothelial cell growth factor. *J. Biol. Chem.* **257**, 5333–5336.
34. Tarantini, F., LaVallee, T., Jackson, A., Gamble, S., Carreira, C. M., Garfinkel, S., Burgess ,W. H., and Maciag. T. (1998). The extravesicular domain of synaptotagmin-1 is released with the latent fibroblast growth factor-1 homodimer in response to heat shock. *J. Biol. Chem.* **273**, 22209–22216.
35. LaVallee, T. M., Tarantini, F., Gamble, S., Carreira, C. M., Jackson, A., and Maciag, T. (1998). Synaptotagmin-1 is required for fibroblast growth factor-1 release. *J. Biol. Chem.* **273**, 22217–22223.
36. Mouta Carreira, C. M., LaVallee, T. M., Tarantini, F., Jackson, A., Lathrop, J. T., Hampton, B., Burgess, W. H., and Maciag, T. (1998). S100A13 is involved in the regulation of fibroblast growth factor-1 and p40 synaptotagmin-1 release *in vitro*. *J. Biol. Chem.* **273**, 22224–22231.
37. Burgess, W. H., Mehlman , T., Friesel, R., Johnson, W. V., and Maciag, T. (1985). Multiple forms of endothelial cell growth factor. Rapid isolation and biological and chemical characterization. *J. Biol. Chem.* **260**, 11389–11392.
38. Sudhof, T. C. and Rizo, J. (1996). Synaptotagmins: C2-domain proteins that regulate membrane traffic. *Neuron* **17**, 379–388.
39. Marqueze, B., Berton, F., and Seagar, M. (2000). Synaptotagmins in membrane traffic: Which vesicles do the tagmins tag? *Biochimie* **82**, 409–420.
40. Schafer, B. W. and Heizmann, C. W. (1996). The S100 family of EF-hand calcium-binding proteins: Functions and pathology. *Trends Biochem. Sci.* **21**, 134–140.
41. Heizmann, C. W. and Cox, J. A. (1998). New perspectives on S100 proteins: A multi-functional Ca(2+)-, Zn(2+)- and Cu(2+)-binding protein family. *Biometals* **11**, 383–397.
42. Wicki, R., Schafer, B. W., Erne, P., and Heizmann, C. W. (1996). Characterization of the human and mouse cDNAs coding for S100A13, a new member of the S100 protein family. *Biochem. Biophys. Res. Commun.* **227**, 594–599.
43. Landriscina, M., Soldi, R., Bagala, C., Micucci, I., Bellum, S., Tarantini, F., Prudovsky, I., and Maciag, T. (2001). S100A13 participates in the release of fibroblast growth factor 1 in response to heat shock *in vitro*. *J. Biol. Chem.* **276**, 22544–22552.
44. Brewer, G. J. (2001). Copper control as an antiangiogenic anticancer therapy: Lessons from treating Wilson's disease. *Exp. Biol. Med. (Maywood)*. **226**, 665–673.
45. Prudovsky, I., Bagala, C., Tarantini, F., Mandinova, A., Soldi, R., Bellum, S., and Maciag, T. (2002). The intracellular translocation of the components of the FGF1 release complex precedes their assembly prior to export. *J. Cell Biol.* in press.
46. Landriscina, M., Prudovsky, I., Carreira, C. M., Soldi, R., Tarantini, F., and Maciag, T. (2000). Amlexanox reversibly inhibits cell migration and proliferation and induces the Src-dependent disassembly of actin stress fibers *in vitro*. *J. Biol. Chem.* **275**, 32753–32762.
47. Oyama, Y., Shishibori, T., Yamashita, K., Naya, T., Nakagiri, S., Maeta, H., and Kobayashi, R. (1997). Two distinct anti-allergic drugs, amlexanox and cromolyn, bind to the same kinds of calcium binding proteins, except calmodulin, in bovine lung extract. *Biochem. Biophys. Res. Commun.* **240**, 341–347.
48. Kim, J. and Hajja, K. A. (2002). Annexin II: A plasminogen-plasminogen activator co-receptor. *Front. Biosci.* **7**, D341–D348.
49. Stylianou, E. and Saklatvala, J. (1998). Interleukin-1. *Int. J. Biochem. Cell Biol.* **30**, 1075–1079.
50. Zhu, X., Komiya, H., Chirino, A., Faham, S., Fox, G. M., Arakawa, T., Hsu, B. T., and Rees, D. C. (1991). Three-dimensional structures of acidic and basic fibroblast growth factors. *Science* **251**, 90–93.
51. Graves, B. J., Hatada, M. H., Hendrickson, W. A., Miller, J. K., Madison, V. S., and Satow, Y. (1990). Structure of interleukin 1 alpha at 2.7-A resolution. *Biochemistry* **29**, 2679–2684.

52. Treharne, A. C., Ohlendorf, D. H., Weber, P. C., Wendoloski, J. J., and Salemme, F. R. (1990). X-ray structural studies of the cytokine interleukin 1-beta. *Prog. Clin. Biol. Res.* **349**, 309–319.
53. Tarantini, F., Micucci, I., Bellum, S., Landriscina, M., Garfinkel, S., Prudovsky, I., and Maciag, T. (2001). The precursor but not the mature form of IL1alpha blocks the release of FGF1 in response to heat shock. *J. Biol. Chem.* **276**, 5147–5151.
54. Furutani, Y., Notake, M., Fukui, T., Ohue, M., Nomura, H., Yamada, M., and Nakamura, S. (1986). Complete nucleotide sequence of the gene for human interleukin 1 alpha. *Nucleic Acids Res.* **14**, 3167–3179.
55. Lomedico, P. T., Gubler, U., Hellmann, C. P., Dukovich, M., Giri, J. G., Pan, Y. C., Collier, K., Semionow, R., Chua, A. O., and Mizel, S. B. (1984). Cloning and expression of murine interleukin-1 cDNA in Escherichia coli. *Nature* **312**, 458–462.
56. Rubartelli, A., Cozzolino, F., Talio, M., and Sitia, R. (1990). A novel secretory pathway for interleukin-1 beta, a protein lacking a signal sequence. *EMBO J.* **9**, 1503–1510.
57. Andrei, C., Dazzi, C., Lotti, L., Torrisi, M. R., Chimini, G., and Rubartelli, A. (1999). The secretory route of the leaderless protein interleukin 1beta involves exocytosis of endolysosome-related vesicles. *Mol. Biol. Cell.* **10**, 1463–1475.
58. Jackson, A., Tarantini, F., Gamble, S., Friedman, S., and Maciag, T. (1995). The release of fibroblast growth factor-1 from NIH 3T3 cells in response to temperature involves the function of cysteine residues. *J. Biol. Chem.* **270**, 33–36.
59. Becq, F., Hamon, Y., Bajetto, A., Gola, M., Verrier, B., and Chimini, G. (1997). ABC1, an ATP binding cassette transporter required for phagocytosis of apoptotic cells, generates a regulated anion flux after expression in Xenopus laevis oocytes. *J. Biol. Chem.* **272**, 2695–2699.
60. Kuchler, K. E. R. (1997). Unusual protein secretion and translocation pathways in yeast: Implication of ABC transporters, in Kuchler, K., Ed., *From Bacteria to Man.* Landes BioScience, Austin, TX.
61. Sims, P. J. and Wiedmer, T. (2001). Unraveling the mysteries of phospholipid scrambling. *Thromb. Haemost.* **86**, 266–275.
62. van Engeland, M., Ramaekers, F. C., Schutte, B., and Reutelingsperger, C. P. (1996). A novel assay to measure loss of plasma membrane asymmetry during apoptosis of adherent cells in culture. *Cytometry* **24**, 131–139.
63. Kobayashi, Y., Oppenheim, J. J., and Matsushima, K. (1990). Calcium-dependent binding of phosphorylated human pre interleukin 1 alpha to phospholipids. *J. Biochem. (Tokyo)* **107**, 666–670.
64. Tarantini, F., Gamble, S., Jackson, A., and Maciag, T. (1995). The cysteine residue responsible for the release of fibroblast growth factor-1 residues in a domain independent of the domain for phosphatidylserine binding. *J. Biol. Chem.* **270**, 29039–29042.
65. Jost, M., Weber, K., and Gerke, V. (1994). Annexin II contains two types of Ca(2+)-binding sites. *Biochem. J.* **298** Pt 3, 553–559.
66. Mach, H. and Middaugh, C. R. (1995). Interaction of partially structured states of acidic fibroblast growth factor with phospholipid membranes. *Biochemistry* **34**, 9913–9920.
67. Samuel, D., Kumar, T. K., Srimathi, T., Hsieh, H., and Yu, C. (2000). Identification and characterization of an equilibrium intermediate in the unfolding pathway of an all beta-barrel protein. *J. Biol. Chem.* **275**, 34968–34975.
68. Sanz, J. M., Jimenez, M. A., and Gimenez-Gallego, G. (2002). Hints of nonhierarchical folding of acidic fibroblast growth factor. *Biochemistry* **41**, 1923–1933.
69. Ptitsyn, O. B. (1995). Molten globule and protein folding. *Adv. Protein Chem.* **47**, 83–229.
70. Bychkova, V. E., Pain, R. H., and Ptitsyn, O. B. (1988). The 'molten globule' state is involved in the translocation of proteins across membranes? *FEBS Lett.* **238**, 231–234.
71. Zoli, A., Altomonte, L., Caricchio, R., Galossi, A., Mirone, L., Ruffini, M. P., and Magaro, M. (1998). Serum zinc and copper in active rheumatoid arthritis: Correlation with interleukin 1 beta and tumour necrosis factor alpha. *Clin. Rheumatol.* **17**, 378–382.
72. Brewer, G. J., Dick, R. D., Grover, D. K., LeClaire, V., Tseng, M., Wicha, M., Pienta, K., Redman, B. G., Jahan, T., Sondak, V. K., Strawderman, M., LeCarpentier, G., and Merajver, S. D. (2000). Treatment of metastatic cancer with tetrathiomolybdate, an anticopper, antiangiogenic agent: Phase I study. *Clin. Cancer Res.* **6**, 1–10.
73. Cox, C., Teknos, T. N., Barrios, M., Brewer, G. J., Dick, R. D., and Merajver, S. D. (2001). The role of copper suppression as an antiangiogenic strategy in head and neck squamous cell carcinoma. *Laryngoscope* **111**, 696–701.
74. Brewer, G. J., Johnson, V. D., Dick, R. D., Hedera, P., Fink, J. K., and Kluin, K. J. (2000). Treatment of Wilson's disease with zinc. XVII: Treatment during pregnancy. *Hepatology* **31**, 364–370.
75. Sano, H., Hla, T., Maier, J. A., Crofford, L. J., Case, J. P., Maciag, T., and Wilder, R. L. (1992). *In vivo* cyclooxygenase expression in synovial tissues of patients with rheumatoid arthritis and osteoarthritis and rats with adjuvant and streptococcal cell wall arthritis. *J. Clin. Invest.* **89**, 97–108.
76. Walsh, D. A. and Pearson, C. I. (2001). Angiogenesis in the pathogenesis of inflammatory joint and lung diseases. *Arthritis Res.* **3**, 147–153.
77. Wang, X., Romanic, A. M., Yue, T. L., Feuerstein, G. Z., and Ohlstein, E. H. (2000). Expression of interleukin-1beta, interleukin-1 receptor, and interleukin-1 receptor antagonist mRNA in rat carotid artery after balloon angioplasty. *Biochem. Biophys. Res. Commun.* **271**, 138–143.
78. Hughes, R. C. (1999). Secretion of the galectin family of mammalian carbohydrate-binding proteins. *Biochim. Biophys. Acta.* **1473**, 172–185.
79. Chapman, L. P., Epton, M. J., Buckingham, J. C., Morris, J. F., and Christian, H. C. (2003). Evidence for a role of the adenosine 5′-triphosphate-binding cassette transporter A1 in the externalization of annexin 1 from pituitary folliculo-stellate cells. *Endocrinology* **144**, 1062–1073.
80. Nickel, W. (2003). The mystery of nonclassical protein secretion. A current view on cargo proteins and potential export routes. *Eur. J. Biochem.* **270**, 2109–2119.
81. Peterson, E., Sutherland, M., Nesheim, M., and Pryzdial, E. (2003). Thrombin induces endothelial cell-surface exposure of the plasminogen receptor annexin 2. *J. Cell Sci.* **116**, 2399–2408.
82. Mandinova, A., Soldi, R., Graziani, I., Bagala, C., Bellum, S., Landriscina, M., Tarantini, F., Prudovsky, I., Maciag, T. (2003). S100A13 Mediates the Copper-Dependent Stress-Induced Release of IL1 alpha from Both Human U937 and Murine NIH 2T2 Cells. *J. Cell Science* **116**, 2687–2696.
83. Pan, Q., Kleer, C. G., van Golen, K. L., Irani, J., Bottema, K. M., Bias, C., De Carvalho, M., Mesri, E. A., Robins, D. M., Dick, R. D., Brewer, G., and Merajver, S. (2002). Copper deficiency induced by tetrathiomolybdate suppresses tumor growth and angiogenesis. *Cancer Res.* **62**, 4854–4859.
84. Voronov, E., Shouval, D. S., Krelin, Y., Cagnano, E., Benharroch, D., Iwakura, Y., Dinarello, C. A., and Apte, R. N. (2003). IL-1 is required for tumor invasiveness and angiogenesis. *Proc. Natl. Acad. Sci. U.S.A.* **100**, 2645–2650.
85. Mandinov, L., Mandinova, A., Kyurkchiev, S., Kehayov, I., Kolev, V., Soldi, R., Bagala, C., De Muinck, E. D., Lindner, V. *et al.* (2003). Copper chelation represses the vascular response to injury. *Proc. Natl. Acad. Sci. U.S.A.* **100**, 6700–6705.
86. Cherney, R. A., Atwood, C. S., Xilinas, M. E., Gray, D. N., Jones, W. D., McLean, C. A., Barnham, K. J., Volitakis, I., Fraser, F. W., Kim, Y., Huang, X., Goldstein, L. E., Moir, R. D., Lim, J. T., Beyruether, K., Zheng, H., Tanzi, R. E., Masters, C. L., and Bush, A. I. (2001). Treatment with a copper-zinc chelator markedly and rapidly inhibits beta-amyloid accumulation in Alzheimer's disease transgenic mice. *Neuron* **30**, 665–676.
87. Creamer, J. D. and Barker, J. N. (1995). Vascular proliferation and angiogenic factors in psoriasis. *Clin. Exp. Dermatol.* **20**, 6–9.

CHAPTER 328

Regulation of Cell Cycle Progression

Clare H. McGowan
Department of Molecular Biology and Department of Cell Biology, The Scripps Research Institute, La Jolla, California

Introduction

The fundamental purpose of the mitotic cycle is the generation of genetically identical daughter cells. Because growth, DNA replication, and organelle duplication are physically independent processes, successful division is dependent on ensuring that the duplication and segregation of cellular components is tightly coordinated. Coordination is achieved by placing the execution of all cell cycle events under the control of a family of highly regulated protein kinases. These kinases are known as cyclin-dependent kinases (Cdks) and each phase of the cell cycle is defined by the activation and inactivation of distinct members of the family. The general principles by which the sequential activation and inactivation of Cdks are achieved are as follows: (1) Processes that initiate Cdk activation also lead to the down-regulation or destruction of the activity. (2) The activity of one Cdk-cyclin sets up the conditions needed for the activation of the next. (3) The destruction of cyclins ensures a unidirectional cell cycle. (4) The inhibition of assembled Cdk-cyclin complexes, either by phosphorylation or by the binding of inhibitory protein, allows the accumulation of inactive complex. (5) When cellular or environmental conditions are not favorable, checkpoint signals delay the activation of performed Cdk-cyclin complexes beyond completion of previous cell cycle events.

Being There: Cyclins Define Cell Cycle Phase

As the name implies, Cdks are protein kinases that need to bind a cyclin subunit to be active. Cyclins are synthesized and degraded in a highly coordinated process as cells progress through the cell cycle, thus, the most fundamental level of control exercised over the activity of Cdks is the periodic presence and absence of the cyclin subunit. When resting (quiescent) cells are stimulated to enter the cell cycle, the first cyclin to be expressed is cyclin D. Expression of cyclin D is dependent on sustained and coordinate signaling of growth factors, through receptor tyrosine kinases, and of extracellular matrix components through integrins (reviewed in [1]). Cyclin D expression is also increased by increased mRNA translation and stability in a phosphoinositol-3-kinase-dependent pathway (reviewed in [2]). There are three D-type cyclins in human cells, with D1 being the most ubiquitously expressed [3]. Cyclin D assembles with Cdk4 and Cdk6; the complex enters the nucleus where it is phosphorylated by a Cdk-activating kinase (Cak, see later section) to form an active kinase that phosphorylates the retinoblastoma protein (Rb) and the related pocket proteins p107 and p130 [4]. Cdk4 and Cdk6 are coexpressed in many cell types and are thought to have partially, but not completely, overlapping functions [5]. Both Cdk4 and Cdk6 are constitutively expressed throughout the cell cycle, because the monomeric kinase subunits are relatively unstable and the abundance of cyclin D is limiting for the stabilization and activation of Cdk4 and Cdk6.

The same mitogenic signals that induce cyclin D expression also induce expression of a second cyclin, cyclin E, and of two inhibitors of cyclin-dependent kinases, $p21^{cip1}$ and $p27^{kip1}$. The induction of Cdk inhibitory proteins as cells enter a new round of growth and division seems counterproductive. However, p21 and p27 are actually required for the efficient assembly and nuclear import of cyclin D/Cdk4 complexes. p21 and p27 bind cyclin D/Cdk4 without inhibiting kinase activity [6,7]. By contrast, they are effective inhibitors of cyclin E/Cdk2 activity, thus, the presence of these proteins early in G1 promotes the formation of active cyclin D/Cdk4 complexes at the same time that it delays the activation of cyclin E/Cdk2 complexes.

Cyclin D/Cdk4 plays two distinct roles in promoting cell cycle progression. One function stems from the ability to phosphorylate Rb and other pocket proteins, p107 and p130. Unphosphorylated pocket proteins repress transcription by inhibiting the E2F family of transcription factors and by recruiting transcriptional repressors, such as histone deacetylase and the chromosomal remodeling complex SWI/SNF, to the promoters of specific G_1/S-phase genes. By phosphorylating Rb and the other pocket proteins cyclin D/Cdk4 inhibits the ability of Rb to repress transcription. Free E2F promotes transcription from a number of promoters including those of cyclin E, cyclin A, and the several proteins needed for S phase (reviewed in [8]). The second role cyclin D plays in promoting cell cycle progression is noncatalytic and depends on the ability of cyclin D/Cdk4 to bind p27. As cyclin D expression increases in G_1 the increasing abundance of cyclin D/Cdk4 provides a binding site for p27. p27 Is recruited away from cyclin E/Cdk2, and the amount of cyclin E/Cdk2 relieved from the inhibitory effect of p27 increases. Hence, the continued expression of cyclin D indirectly stimulates the activation of cyclin E/Cdk2, and the transcriptional program initiated by mitogenic stimuli drives two temporally distinct waves of Cdk activation (Fig. 1).

Cyclin E/Cdk2 collaborates with cyclin D-Cdk4/6 in phosphorylating and inactivating Rb and the pocket proteins. This dual phosphorylation is required for the full activation of the S-phase transcriptional program [9]. The shift in cyclin D/Cdk4/6 catalyzed phosphorylation of Rb to that catalyzed by cyclin E/Cdk2 correlates with the transition to a state in which further cell cycle progression will occur even if extracellular growth factors are removed. Once this stage of the cell cycle is reached, the signals to continue with the cycle all come from within the cell. Internal or external events can delay, or prevent cell cycle progression. However, if no signal to stop is received, the initiated cycle of growth, replication, segregation, and division will continue without further extracellular input. If extracellular conditions remain favorable, mitogenic signaling maintains cyclin D expression through the cell cycle, and on completion of mitosis the new daughter cells will be ready to initiate a new round of cell division immediately. If, on the other hand, mitogenic stimuli are not maintained, cyclin D is phosphorylated in a glycogen synthase kinase 3 (GSK3)-dependent process and is targeted for destruction by the proteosome (reviewed in [10]). Progression into another round of replication is thus dependent on sustained, favorable extracellular signals.

The activation of cyclin E/Cdk2 leads to the initiation of DNA replication, and a number of substrates required to carry out the task of duplicating and segregating cellular contents

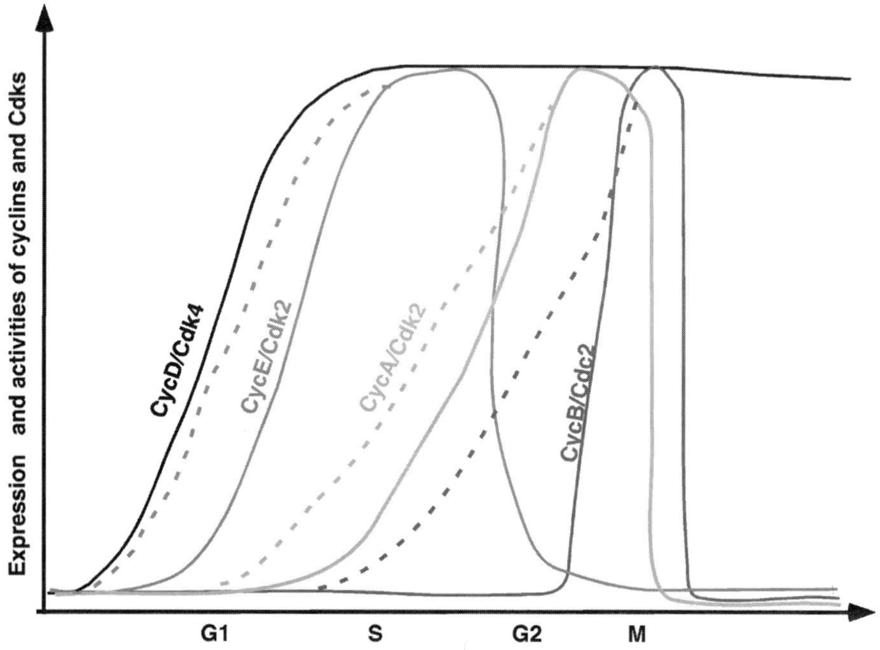

Figure 1 Expression pattern of cyclins and Cdk activities during the cell cycle. Expression of the specified cyclin is indicated by the dashed lines; activity of the specified cyclin/Cdk is indicated by the solid line. Sustained expression of cyclin D is dependent on favorable extracellular conditions as explained in the text. Note that the delay between expression of cyclin E and activation of cyclin E/Cdk2 is primarily due to inhibition of kinase activity by p27. The delay between expression of cyclin B and activation of cyclin B/Cdc2 is primarily due to inhibitory phosphorylation. Note that cyclin D also binds to Cdk4 and that cyclin A also binds to Cdc2.

have been identified. Specifically, cyclin E/Cdk2 phosphorylates components of the centrosome, initiating the process of centrosome duplication and, hence, segregation of the replicated sister chromatids [11]. It is likely that more cyclin E/Cdk2 substrates exist [12,13]. Cyclin E/Cdk2 also phosphorylates the Cdk inhibitor p27. This converts p27 to a form that is recognized by an ubiquitin ligase and is targeted for destruction by the proteosome (see later section) [14]. Thus, activation of cyclin E/Cdk2 triggers the destruction of its negative regulator. Cyclin E/Cdk2 autophosphorylates on cyclin E, targeting the cyclin for degradation [15,16]. This feature of priming an inhibitor for degradation and priming the cyclin subunit for degradation renders the kinase complex both self-activating and self-limiting.

Cyclin A is expressed soon after cyclin E at the G_1/S boundary. Both cyclin E/Cdk2 and cyclin A/Cdk2 activities are essential for the initiation and completion of DNA replication [17–19] and for ensuring that replication occurs only once in each cell cycle. Our understanding of the mechanism by which cells ensure that DNA is replicated once and only once in each cycle derives from experiments in many species, and the exact details may vary among organisms. Nevertheless, the principles by which it is achieved are at least partly conserved. The initiation of DNA replication requires the assembly of replication machinery complexes on DNA. One part of the machinery, the origin recognition complex, is present throughout the cell cycle, whereas other proteins, including the polymerase loading factors Cdc6 (Cdc18), Cdt1, and the MCM proteins, are loaded stepwise in G_1. The loading of these proteins and the activation of replication are under the control of cyclin E/Cdk2 and the S-phase activated kinase Cdc7 [20].

Once replication has initiated Cdc6 (Cdc18) and other essential replication cofactors, including the MCM protein, are phosphorylated and inactivated (reviewed in [21,22]). Inactivating phosphorylation triggers the degradation of some proteins, and it prevents access to the nucleus or to chromatin of others. The prereplicative state, that is, that which allows loading of the replication machinery on to chromatin, cannot be reestablished until division is complete and Cdk activity drops following mitosis [23]. In addition to this, metazoan cells express Geminin, an inhibitor of origin firing in S phase. Expression begins in S phase, and the protein inhibits the initiation of DNA replication until the metaphase-to-anaphase transition when it is degraded by the anaphase promoting complex (APC) (reviewed in [24]). The APC is a multiprotein complex that ubiquitinates several mitotic protein and then targets them for degradation by the 26S proteosome (see later section). In addition to its role in controlling replication, cyclinA/Cdk2 also promotes the efficient execution of S phase by increasing transcription of histone and other genes needed to accommodate replication [10].

After growing and duplicating its contents, the cell is faced with the challenge of dividing itself into two viable daughter cells. This is achieved in mitosis, which begins at prophase, with the condensation of chromosomes and the formation of a mitotic spindle. In prometaphase, the spindle microtubules attach to the two sister chromatids via kinetochores, and the nuclear envelope breaks down. Once everything is correctly aligned, the spindle pulls the sister chromatids apart in anaphase. Following separation of the replicated genomes, the spindle is disassembled, the chromatids decondense, and the nuclear envelope begins to reform, in telophase. Finally, at cytokinesis, the formation of the two daughter cells is completed by the partitioning of the cytoplasm. These dramatic morphological changes are under the control Cdc2 (Cdk1) in association with cyclin A and cyclin B. Expression of cyclin B lags behind that of cyclin A, rising late in S phase and remaining high throughout G_2 and mitosis. The gradual increase in cyclin B protein is not reflected in gradually increased kinase (Fig. 1) because the complex is largely maintained in a phosphorylated inactive form until the G_2/M boundary. Dephosphorylation and activation of cyclin B/Cdc2 correlates very closely with the morphological changes that accompany mitosis. Nuclear lamins, nucleolar proteins, centrosomal proteins, and Eg5 (a kinesin relates motor) have all been described as cyclin B/Cdc2 substrates [10]. Cdc2 in association with a second cyclin B subunit, cyclinB2, which localizes predominantly to the endoplasmic reticulum, may play a role in dispersing the Golgi apparatus so that cytokinesis will provide the two daughters with sufficient components to rebuild the secretory apparatus [25]. Cyclin B/Cdc2 also phosphorylates and activates at least two other kinases in mitosis, polo-like kinase (Plk) and the Aurora-related kinases. Plk and Aurora have roles in centrosome separation, spindle assembly, and chromosome segregation [26,27]. As an activator of Cdc25C, the phosphatase that activates Cdc2/ CycB, Plk1 is thought to play a role in the feedback loop to ensure that cyclinB/Cdc2 is rapidly and fully activated in mitosis (Fig. 2) [26]. Cyclin B/Cdc2 also stimulates its own destruction by phosphorylating and activating components of the APC.

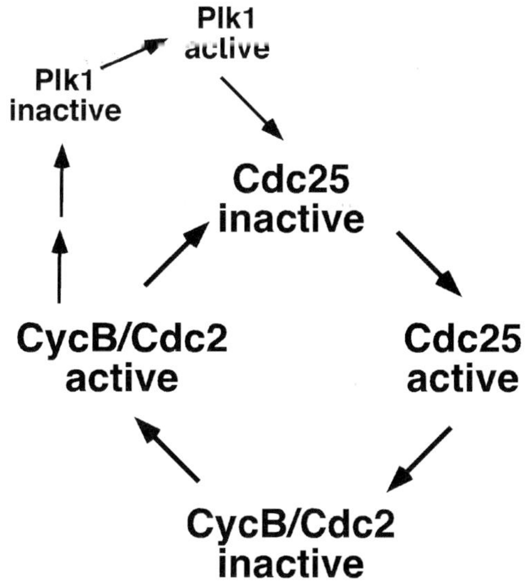

Figure 2 Activation of a small amount of cyclin B/Cdc2 leads to the activation of Cdc25, which promotes the activation of more cyclin B/Cdc2. Stimulation of Cdc25 by Plk1, which is activated in mitosis, could reinforce the positive feedback loop.

Signals to Slow Processes: Regulation of Cdks by Inhibitory Proteins

An important mechanism for negatively regulating Cdk/cyclin is the interaction with small inhibitory proteins. Two distinct families of Cdk inhibitors (CKIs) have been described in mammals. The Ink4 family (p15, p16, p18, and p19), named for the ability to inhibit Cdk4, specifically inhibits the cyclin D-containing Cdks, and the Cip/Kip family ($p21^{cip1}$, $p27^{kip1}$ and $p57^{kip2}$) strongly inhibits Cdk2-containing complexes [4].

The Ink family of CKIs bind to monomeric Cdk4 and Cdk6. The inhibitors bind to the same region of the Cdk as cyclin D, thus, high Ink4 expression blocks formation of the active cyclinD/Cdk4 complex. Because Ink4 prevents cyclin D/Cdk4 complex formation it prevents binding of p27 to cyclinD/Cdk4. Increased Ink4 expression therefore increases the pool of p27 available to bind to and inhibit cyclin E/Cdk2. Hence, up-regulation of Ink4 directly inhibits cyclin D/Cdk4 activity, and indirectly inhibits cyclin E/Cdk2 activity [4]. The regulation of distinct Ink4 family members is complex and not fully elucidated. Nevertheless, the well-characterized pathway by which the cytokine, TGFβ, induces Ink4 elegantly illustrates one mechanism by which signals from outside the cell delay progression if extracellular conditions are not favorable for division [28].

The Cip/Kip family of inhibitors binds to preformed cyclin/Cdk complex blocking substrate access. The first of these inhibitors to be characterized, p21, is up-regulated by p53 in response to DNA damage. The p21 inhibitor predominantly interacts with and inhibits cyclin E/Cdk2 and cyclin A/Cdk2, thereby it delays entry into S phase. Also, p21 was identified as a protein that is up-regulated during the transition to replicative senescence, that is, in cells that remain metabolically intact but do not to under go further rounds of division [31]. The role p21 plays in adaptation to replicative senescence may be comparable with the role it plays in permanently arresting cells that have suffered irreparable DNA damage.

Cdks Are Positively and Negatively Regulated by Phosphorylation

The association of kinase and cyclin subunits results in a partially activate kinase complex. Full activation requires phosphorylation of a threonine residue within the activation or T-loop region of the kinase. Crystallographic studies and biochemical analyses support a model in which cyclin binding to Cdks induces a conformational change in the kinase, exposing the T-loop threonine and making it more accessible to the activating kinase [32]. Phosphorylation of the T-loop induces a second conformational shift that moves the T-loop away from the ATP binding region and thus facilitates catalysis.

The kinases responsible for catalyzing the activating threonine phosphorylation are known as Cak, for Cdk activating kinases. In mammals, Cak itself is a Cdk comprised of a Cdk subunit, Cdk7, and a cyclin subunit, cyclin H [33]. Cdk7 exists in a ternary complex of cyclin H/Cdk7 and a ring finger protein Mat1. Mat1 stabilizes the cyclin H/Cdk7 complex and the kinase complex is active in the absence of T-loop phosphorylation [34]. However, the finding that cyclin A/Cdk2 and cyclin B/Cdc2 activate cyclin H/Cdk7 by phosphorylating residues within the T-loop implies that an intriguing, reciprocal Cdk activation loop may be operational [35]. The activity of cyclin H/Cdk7 is constitutive across the cell cycle and there is little evidence that modulation of CAK activity regulates cell cycle progression. Cyclin H/Cdk7 has a second function in transcription-coupled repair. A pool cyclin H/Cdk7 associates with the transcription machinery and phosphorylates the carboxyl-terminal domain (CTD) of RNA polymerase II, promoting the transition from the initiation-competent form to the elongation-competent form of the complex [36]. The dual functions of cyclin H/Cdk7 in basal transcription and in cell cycle progression may help to coordinate these two processes.

Phosphorylation of Cdks also negatively regulates kinase activity. The activity of Cdc2/cyclin B is perhaps the best characterized example of regulation by inhibitory phosphorylation; however, the activities of cyclin A/Cdk2, cyclin E/Cdk2, and CylinD/Cdk4 are also, at least partially, controlled this way [37,38]. Inhibitory phosphorylation occurs near the amino terminus in the ATP binding cleft of the kinase subunit, at sites corresponding to threonine 14 and tyrosine 15 in human Cdc2. Wee1 is a nuclear kinase that shows a preference for phosphorylating tyrosine 15. Whereas Myt1, a cytoplasmic protein that associates with the membranes of the endoplasmic reticulum and Golgi, shows a preference for phosphorylating threonine 14. The distinct location of these two kinases suggests a mechanism by which Cdc2 can be negatively regulated in response to both cytoplasmic and nuclear conditions. Wee1 and Myt1 are active throughout interphase, and coincident with the activation of cyclin B/Cdc2 in mitosis, both kinases are inactivated [39–41]. Two kinases have been reported to inactivate Myt1 in maturing oocytes [42,43], however the exact mechanism of inactivation of Myt1 and Wee1 in mitotic cells is not known [41].

The inhibitory phosphates on Cdks are removed by the Cdc25 family of dual-specificity phosphatases. Three forms of Cdc25 are expressed in mammalian cells. Cdc25A acts early in the cell cycle to promote the dephosphorylation and activation of cyclin E/Cdk2. Expression of the predominantly cytoplasmic Cdc25B rises in S phase and remains high in G_2. The abundance of Cdc25C is constant; however, its activity increase in mitosis consistent with a function in the G_2/M transition (reviewed in [44]). Curiously the Cdc25 proteins are themselves substrates of the Cdks: Cdc25A is phosphorylated and activated by cyclin E/Cdk2, and Cdc25C is activated by phosphorylation by cyclin B/Cdc2 [45,46]. The activation of Cdc25 by the kinases that it activates led to the idea of a positive feedback loop that stimulates the rapid and full activation of Cdc2 as shown in Fig. 2. In the case of the G_2/M transition, the feedback loop may require the activity of a non-Cdk kinase, Plk1, which phosphorylates and activates Cdc25 and is itself activated in mitosis [47,48].

Antisense and microinjection of antibodies suggest that both Cdc25B and Cdc25C promote the G_2/M transition [49–51]. Surprisingly, disruption of the mouse genes encoding Cdc25B or Cdc25C is not associated with defects in proliferative rate, cell cycle profile, or checkpoint responses [52,53]. Thus, despite differences in timing of expression and in subcellular location, it is possible that Cdc25B compensates for the loss of Cdc25C and vice versa. The phenotype of the double disruptant is needed to validate the importance of Cdk dephosphorylation in the regulation of cell cycle progression and checkpoint control.

Degradation: The Importance of Being Absent

Each cell cycle phase is defined by the form of the cyclin/Cdk complex that is present and active. It is therefore important that cyclin/Cdks be inactivated and destroyed in a timely manner. Ubiquitin-mediated degradation is the main pathway for ridding cells of cell cycle proteins that have executed their function. The addition of ubiquitin residues to substrate proteins is the signal for degradation by the 26S proteasome, a multisubunit cellular complex that specializes in the unfolding and proteolysis of ubiquitin-tagged proteins. Traffic of cell cycle proteins into the 26S proteosome is controlled by the activity of two structurally and functionally similar complexes, the SCF (Skp-Cullin F box) and APC. The SCF E3 ubiquitin ligase is active in G_1, through S phase, and into G_2. The APC is also an E3 ubiquitin ligase; it is activated in mitosis and remains active in G_1. The specificity of these ubiquitin ligases is controlled by substrate recognition proteins. In the case of SCF a large number of substrate recognition proteins, characterized by the presence of F boxes, leucine-rich repeats, or WD-40 repeats, are known to exist [54]. In the case of the APC, two substrate-specific interacting proteins, Cdc20 and Cdh1, are known, but more may exist [23]. Specificity and appropriate temporal degradation are achieved by a combination of the fact that Cdk catalyzed phosphorylation of target proteins is required for recognition by F-box proteins, and that components of APC are regulated by Cdk-dependent phosphorylation.

As discussed earlier, the degradation of p27 and cyclin E is initiated when these proteins are phosphorylated by active cyclin E/Cdk2. In a similar manner, the transcriptional activator E2F is inactivated by cyclinA/Cdk2-dependent phosphorylation and degradation by the SCF [55]. In both cases degradation is initiated when the protein has been phosphorylated and thus degradation is under the direct control of Cdk activity. Compound mechanisms regulate the traffic of proteins through the APC. The rise in expression of the substrate recognition protein, Cdc20, in mitosis initiates degradation of APC/Cdc20 substrates. Phosphorylation of Cdc20 and core components of the APC by cyclinB/Cdc2 stimulates ubiquitin ligase activity.

A major role of the APC/Cdc20 complex is to promote the degradation of proteins that maintain sister chromatid cohesion and, thus, to allow the metaphase-to-anaphase transition. If all of the chromosomes are not correctly aligned, the presence of an unattached kinetochore generates a signal that inhibits APC/Cdc20. Activated Mad2 and BubR1 have been shown to independently inhibit the metaphase-to-anaphase transition by inhibiting APC function [56,57]. By inhibiting the APC/Cdc20, the spindle checkpoint prevents sister chromatid separation until all chromatids are properly attached to the spindle microtubules that will pull the two sister chromatids to opposite sides of the cell. Expression of the second APC specificity factor, Cdh1, rises in S phase and persists through G_1. However, APC/Cdh1 is inhibited by high cyclinB/Cdc2 activity, thus degradation of APC/Cdh1 substrates cannot occur until after degradation of cyclin B. Once APC/Cdh1 is activated, proteins that prevent anaphase including remaining mitotic cyclins are destroyed. The inhibition of Cdh1 by mitotic Cdks prevents the onset of anaphase if metaphase is delayed. APC/Cdh1 remains active into G_1 ensuring that components of the previous mitosis are destroyed before initiation of the next cycle. The APC also catalyzes the destruction of Cdc20, thereby preventing APC activity until Cdc20 is synthesized again late in the next cycle (Fig. 3).

Location, Location, Location

The substrates of Cdks are predominantly nuclear; thus, the presence or absence of active Cdks to the nucleus is a potentially important mechanism of regulation. Just as the cyclin subunit controls the temporal activation of Cdks, there is evidence that cyclins also control the subcellular location of cyclin/Cdk complexes (reviewed in [58]). For example, monomeric Cdc2 appears to be pancellular; however, as

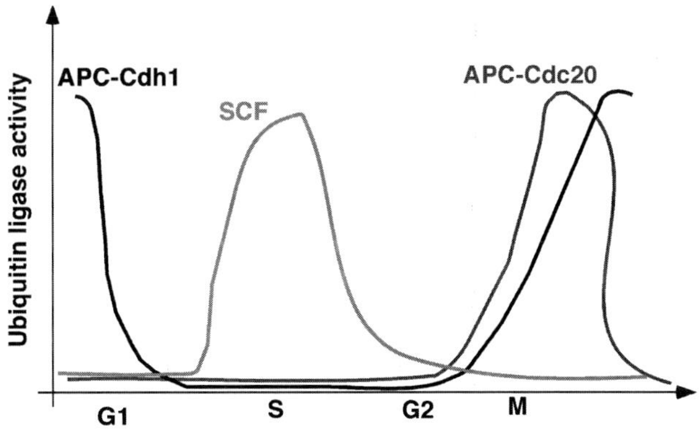

Figure 3 Activity profiles of the ubiquitin ligase complexes that tag cell cycle proteins for degradation. Degradation of cyclins and other cell cycle regulators is required for progression. Degradation of SCF targets is maximal when Cdk2 complexes are available to phosphorylate target proteins. The activity of APC/Cdc20 increases with the abundance of Cdc20 and is stimulated in mitosis by cyclin B/Cdc2 phosphorylation. The APC/Cdh1 degrades both cyclin B and Cdc20, allowing the completion of anaphase and entry into a new G1.

cyclin B expression increases, the complex accumulates in the cytoplasm, predominantly in a perinuclear region. At the start of mitosis, cyclin B/Cdc2 translocates to the nucleus, just prior to nuclear envelope breakdown. The cytoplasmic location of cyclin B/Cdc2 is due both to active retention in the cytoplasm and to active transport out of the nucleus [59]. The signals that mediate cytoplasmic retention and nuclear export are clustered in the amino-terminal region of the protein. Phosphorylation of sites within this region alters the balance of import–export such that during prophase, the bulk of cyclin B/Cdc2 becomes nuclear [59]. The coincident activation and nuclear import of cyclin B/Cdc2 is thought to contribute to the rapid, switch from G_2 into mitosis [60]. However, expression of constitutively nuclear cyclin B mutant is not sufficient to induce premature mitosis, which suggests that regulated localization is perhaps one of the redundant mechanisms used by higher eukaryotes to regulate cell cycle progression [59,61].

An important issue of spatial regulation is how Cdks coordinate nuclear and cytoplasmic functions. In part, the answer lies in the realization that proteins are much less static than immunofluorescence images might suggest. For instance, CycE and CycA complexes appear to be predominantly nuclear when examined in cells that have been fixed. However, live cell analysis shows that they constantly shuttle between the cytoplasm and nucleus [62]. The shuttling of these complexes allows them to phosphorylate both nuclear and cytoplasmic substrates. Thus, for example, cyclin E/Cdk2 phosphorylates nuclear proteins at sites of ongoing replication and cytoplasmic proteins in the replicating centrosome. Shuttling presumably also allows access to regulators if they are present in different compartments.

Checkpoint Signaling

Once it has passed the mitogen-dependent stage, the cell cycle is an autonomous signaling system in which the completion of each step promotes progression through the next. Despite this constant drive to move forward, a problem in any of the subpathways of the cell cycle can generate a signal that stops progression in all other subpathways. The mechanisms that inhibit cell cycle progression in response to intracellular problems are known as *checkpoints.* Checkpoint responses may, in fact, be activated in every unperturbed cell cycle; however, they are most easily seen and best understood in the context of cells that have been subjected to unusual levels of damage or unanticipated nucleotide deprivation. The concept of checkpoint signaling and how they integrate with cell cycle regulation has been reviewed [21,63].

If the genome is damaged, cell cycle progression is delayed until the damage is repaired. In addition to delaying cell cycle progression, the DNA damage response must also activate relevant repair pathways. The most appropriate repair pathway depends not only on the nature of the damage but also on the stage of the cell cycle. Thus, for instance, a double-strand break that occurs late in S phase or in G_2 can be repaired by the process of homologous recombination, which requires the presence of a sister chromatid as a template for repair. If, however, a double-strand break occurs in G_1 the process of end joining is likely to be the only viable repair pathway available [64]. Other repair processes that are active throughout the cell cycle [65] are up-regulated when elevated rates of damage are incurred.

As far as we know all DNA damage checkpoints function through two protein kinases, ATM and ATR [66]. Arrest in the G_1 phase of the cell cycle is mainly mediated by the p53 tumor suppressor, a transcription factor that induces cell cycle arrest, damage repair, and apoptosis [67]. ATM and ATR activate the p53 protein in at least three ways (Fig. 4). First, both kinases directly phosphorylate p53 on serine 15, a site that increases stability and transcriptional activity [68–70]. Second, ATM phosphorylates Mdm2, an ubiquitin ligase that tags p53 for degradation. Phosphorylation of Mdm2 decreases its ability to promote the degradation of p53 [71]. Third, ATM phosphorylates and activates another protein kinase Cds1 (Chk2) that phosphorylates p53 on serine 20, disrupting the association with Mdm2 [72,73]. Stabilized and activated p53 drives the transcription of several genes, including the Cdk inhibitor p21. The induction of p21 in response to DNA damage is the major mechanism enforcing arrest in G_1 in response to DNA damage or nucleotide pool perturbation [29,30]. In addition to the role p53 plays in the G_1 checkpoint, p53-mediated transcription of p21 and a 14-3-3 protein, 14-3-3σ, is required for maintaining a long-term G_2 arrest [74]. The 14-3-3 proteins are a family of ubiquitously and highly expressed phosphobinding proteins (reviewed in [75]). The 14-3-3σ protein is thought to prevent mitosis by retaining cyclin B/Cdc2 in the cytoplasm [76].

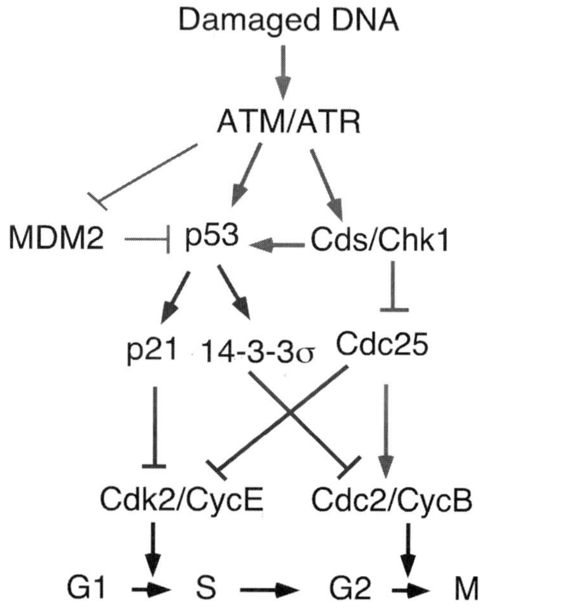

Figure 4 The DNA damage checkpoints inhibit Cdk activity. Phosphorylation and degradation events that are expected to function rapidly following DNA damage are indicated by gray lines and arrows. Processes that require transcription, and are required for the long-term maintenance of the arrest are indicated in black.

In mammalian cells, the initial G_2-damage-induced checkpoint works by maintaining the inhibitory phosphorylation of cyclinB/Cdc2. Cdc2 dephosphorylation and activation can be catalyzed *in vivo* by the two isoforms of Cdc25, Cdc25B and Cdc25C [49]. The checkpoint kinases Chk1 and Cds1 (Chk2) phosphorylate Cdc25C on serine 216, a site that induces 14-3-3 binding [77,78]. This site is constitutively phosphorylated throughout interphase, however, it is not phosphorylated in the hyperactive mitotic form of Cdc25C. Phosphorylation of serine 216 by the checkpoint kinases is proposed to reduce Cdc25 function by sequestrating Cdc25 away from its substrate or by preventing its nuclear accumulation [78]. Expressing a mutant Cdc25C that cannot bind 14-3-3 has a modest effect on the G_2 checkpoint [78]; however, expression of a constitutively nuclear mutant of Cdc25 is not sufficient to overcome checkpoint regulation [79]. Thus, additional sites on Cdc25, or other factors may contribute to the G_2 damage checkpoint. The phosphorylation of Cdc25C by Chk1 and Cds1 on sites other than S216 that inhibit phosphatase activity directly may be required to enforce the G_2/M checkpoint [80]. In addition, the possibility that the DNA damage checkpoint additionally up-regulates inhibitory kinases, such as Wee1 or Myt1, has not been discounted in mammalian cells.

Chk1 and Cds1 also play a role in delaying progression through G_1 and in slowing DNA replication. In this case, the checkpoint kinases phosphorylate Cdc25A on a site that targets the protein for ubiquitin-mediated degradation (reviewed in [21]). Thus, it appears that the down-regulation of Cdc25 by multiple overlapping mechanisms is a common theme in preventing Cdk activation following DNA damage. The degradation of Cdc25A presumably acts synergistically with p21 to inhibit cyclin E/Cdk2 and hence delays the G_1/S transition and progression through S phase.

An important factor in checkpoint control is the timescale on which they operate. Checkpoint mechanisms that act through phosphorylation or degradation of key regulators might be expected to operate very rapidly in response to damage and be capable of delaying the onset of the next cell cycle step right up until the moment of initiation. On the other hand, the transcriptional induction of cell cycle inhibitor proteins is a process that functions over the course of hours rather than minutes. It is becoming increasingly clear that quick acting and sustained checkpoint responses are required to effectively inhibit the activity of the cyclin/Cdks that drive cell cycle progression.

References

1. Roovers, K., and Assoian, R. K. (2000). Integrating the MAP kinase signal into the G1 phase cell cycle machinery. *Bioessays* **22**(9), 818–826.
2. Ekholm, S. V. and Reed, S. I. (2000). Regulation of G(1) cyclin-dependent kinases in the mammalian cell cycle. *Curr. Opin. Cell Biol.* **12**(6), 676–684.
3. Sherr, C. J. (1993). Mammalian G1 cyclins. *Cell* **73**(6), 1059–1065.
4. Sherr, C. J. and Roberts, J. M. (1999). CDK inhibitors: Positive and negative regulators of G1-phase progression. *Genes Dev.* **13**(12), 1501–1512.
5. Tsutsui, T., Hesabi, B., Moons, D. S., Pandolfi, P. P., Hansel, K. S., Koff, A., and Kiyokawa, H. (1999). Targeted disruption of CDK4 delays cell cycle entry with enhanced p27(Kip1) activity. *Mol. Cell. Biol.* **19**(10), 7011–7019.
6. Cheng, M., Olivier, P., Diehl, J. A., Fero, M., Roussel, M. F., Roberts, J. M., and Sherr, C. J. (1999). The p21(Cip1) and p27(Kip1) CDK 'inhibitors' are essential activators of cyclin D-dependent kinases in murine fibroblasts. *EMBO J.* **18**(6), 1571–1583.
7. LaBaer, J., Garrett, M. D., Stevenson, L. F., Slingerland, J. M., Sandhu, C., Chou, H. S., Fattaey, A., and Harlow, E. (1997). New functional activities for the p21 family of CDK inhibitors. *Genes Dev.* **11**(7), 847–862.
8. Harbour, J. W. and Dean, D. C. (2000). The Rb/E2F pathway: Expanding roles and emerging paradigms. *Genes Dev.* **14**(19), 2393–2409.
9. Harbour, J. W., Luo, R. X., Dei Santi, A., Postigo, A. A., and Dean, D. C. (1999). Cdk phosphorylation triggers sequential intramolecular interactions that progressively block Rb functions as cells move through G1. *Cell* **98**(6), 859–869.
10. Obaya, A. J. and Sedivy, J. M. (2002). Regulation of cyclin-Cdk activity in mammalian cells. *Cell Mol. Life Sci.* **59**(1), 126–142.
11. Hinchcliffe, E. H. and Sluder, G. (2001). "It takes two to tango": Understanding how centrosome duplication is regulated throughout the cell cycle. *Genes Dev.* **15**(10), 1167–1181.
12. Zhao, J., Dynlacht, B., Imai, T., Hori, T., and Harlow, E. (1998). Expression of NPAT, a novel substrate of cyclin E/CDK2, promotes S-phase entry. *Genes Dev.* **12**(4), 456–461.
13. Hall, C., Nelson, D. M., Ye, X., Baker, K., DeCaprio, J. A., Seeholzer, S., Lipinski, M., and Adams, P. D. (2001). HIRA, the human homologue of yeast Hir1p and Hir2p, is a novel cyclin-cdk2 substrate whose expression blocks S-phase progression. *Mol. Cell. Biol.* **21**(5), 1854–1865.
14. Sheaff, R. J., Groudine, M., Gordon, M., Roberts, J. M., and Clurman, B. E. (1997). Cyclin E/CDK2 is a regulator of p27Kip1. *Genes Dev.* **11**(11), 1464–1478.
15. Clurman, B. E., Sheaff, R. J., Thress, K., Groudine, M., and Roberts, J. M. (1996). Turnover of cyclin E by the ubiquitin-proteasome pathway is regulated by cdk2 binding and cyclin phosphorylation. *Genes Dev.* **10**(16), 1979–1990.
16. Won, K.-A. and Reed, S. (1996). Activation of Cyclin E/CDK2 is coupled to site-specific autophosphorylation and ubiquitin-dependent degradation of cyclin E. *EMBO J.* **16**(16), 4182–4193.
17. Pagano, M., Pepperkok, R., Verde, F., Ansorge, W., and Draetta, G. (1992). Cyclin A is required at two points in the human cell cycle. *EMBO J.* **11**, 961–971.
18. Girard, F., Strausfeld, U., Fernandez, A., and Lamb, N. J. C. (1991). Cyclin A is required for the onset of DNA replication in mammalian fibroblasts. *Cell* **67**, 1169–1179.
19. Ohtsubo, M., Theodoras, A. M., Schumacher, J., Roberts, J. M., and Pagano, M. (1995). Human cyclin E, a nuclear protein essential for the G1-to-S phase transition. *Mol. Cell. Biol.* **15**(5), 2612–2624.
20. Zou, L. and Stillman, B. (2000). Assembly of a complex containing Cdc45p, replication protein A, and Mcm2p at replication origins controlled by S-phase cyclin-dependent kinases and Cdc7p-Dbf4p kinase. *Mol. Cell. Biol.* **20**(9), 3086–3096.
21. Bartek, J. and Lukas, J. (2001). Mammalian G1- and S-phase checkpoints in response to DNA damage. *Curr. Opin. Cell Biol.* **13**(6), 738–747.
22. Kelly, T. J. and Brown, G. W. (2000). Regulation of chromosome replication. *Annu. Rev. Biochem.* **69**, 829–880.
23. Zachariae, W. and Nasmyth, K. (1999). Whose end is destruction: Cell division and the anaphase-promoting complex. *Genes Dev.* **13**(16), 2039–2058.
24. Diffley, J. F. (2001). DNA replication: Building the perfect switch. *Curr. Biol.* **11**(9), R367–370.
25. Draviam, V. M., Orrechia, S., Lowe, M., Pardi, R., and Pines, J. (2001). The localization of human cyclins B1 and B2 determines CDK1 substrate specificity and neither enzyme requires MEK to disassemble the Golgi apparatus. *J. Cell Biol.* **152**(5), 945–958.
26. Nigg, E. A. (1998). Polo-like kinases: Positive regulators of cell division from start to finish. *Curr. Opin. Cell Biol.* **10**(6), 776–783.

27. Bischoff, J. R. and Plowman, G. D. (1999). The Aurora/Ipl1p kinase family: Regulators of chromosome segregation and cytokinesis. *Trends Cell Biol.* **9**(11), 454–459.
28. Massague, J. and Wotton, D. (2000). Transcriptional control by the TGF-beta/Smad signaling system. *EMBO J.* **19**(8), 1745–1754.
29. Brugarolas, J., Chandrasekaran, C., Gordon, J. I., Beach, D., Jacks, T., and Hannon, G. J. (1995). Radiation-induced cell cycle arrest compromised by p21 deficiency. Nature **377**(6549), 552–557.
30. Deng, C., Zhang, P., Harper, J. W., Elledge, S. J., and Leder, P. (1995). Mice lacking p21CIP1/WAF1 undergo normal development, but are defective in G1 checkpoint control. *Cell* **82**(4), 675–684.
31. Noda, A., Ning, Y., Venable, S. F., Pereira-Smith, O. M., and Smith, J. R. (1994). Cloning of senescent cell-derived inhibitors of DNA synthesis using an expression screen. *Exp. Cell Res.* **211**(1), 90–98.
32. Johnson, L. N., Noble, M. E., and Owen, D. J. (1996). Active and inactive protein kinases: Structural basis for regulation. *Cell* **85**(2), 149–158.
33. Kaldis, P. (1999). The cdk-activating kinase (CAK): From yeast to mammals. *Cell Mol. Life Sci.* **55**(2), 284–296.
34. Devault, A., Martinez, A. M., Fesquet, D., Labbe, J. C., Morin, N., Tassan, J. P., Nigg, E. A., Cavadore, J. C., and Doree, M. (1995). MAT1 ("menage a trois"): A new RING finger protein subunit stabilizing cyclin H/Cdk7 complexes in starfish and *Xenopus* CAK. *EMBO J.* **14**(20), 5027–5036.
35. Garrett, S., Barton, W. A., Knights, R., Jin, P., Morgan, D. O., and Fisher, R. P. (2001). Reciprocal activation by cyclin-dependent kinases 2 and 7 is directed by substrate specificity determinants outside the T loop. *Mol. Cell. Biol.* **21**(1), 88–99.
36. Nigg, E. A. (1996). Cyclin-dependent kinase 7: At the cross-roads of transcription, DNA repair and cell cycle control? *Curr. Opin. Cell Biol.* **8**(3), 312–317.
37. Terada, Y., Tatsuka, M., Jinno, S., and Okayama, H. (1995). Requirement for tyrosine phosphorylation pf Cdk4 in G1 arrest induced by ultraviolet irradiation. *Nature* **376**, 358–362.
38. Dulic, V., Lees, E., and Reed, S. I. (1992). Association of human cyclin E with a periodic G_1-S phase protein kinase. *Science* **257**, 1958–1961.
39. McGowan, C. H., and Russell, P. (1995). Cell cycle regulation of human WEE1. *EMBO J.* **14**(10), 2166–2175.
40. Watanabe, N., Broome, M., and Hunter, T. (1995). Regulation of the human WEE1Hu CDK tyrosine 15-kinase during the cell cycle. *EMBO J.* **14**(9), 1878–1891.
41. Booher, R. N., Holman, P. S., and Fattaey, A. (1997). Human Myt1 is a cell cycle-regulated kinase that inhibits Cdc2 but not Cdk2 activity. *J. Biol. Chem.* **272**(35), 22300–22306.
42. Okumura, E., Fukuhara, T., Yoshida, H., Hanada Si, S., Kozutsumi, R., Mori, M., Tachibana, K., and Kishimoto, T. (2002). Akt inhibits Myt1 in the signalling pathway that leads to meiotic G2/M-phase transition. *Nat. Cell Biol.* **4**(2), 111–116.
43. Palmer, A., Gavin, A. C., and Nebreda, A. R. (1998). A link between MAP kinase and p34(cdc2)/cyclin B during oocyte maturation: p90(rsk) phosphorylates and inactivates the p34(cdc2) inhibitory kinase Myt1. *EMBO J.* **17**(17), 5037–5047.
44. Nilsson, I. and Hoffmann, I. (2000). Cell cycle regulation by the Cdc25 phosphatase family. *Prog. Cell Cycle Res.* **4**, 107–114.
45. Hoffmann, I., Draetta, G., and Karsenti, E. (1994). Activation of the phosphatase activity of human cdc25A by a cdk2-cyclin E dependent phosphorylation at the G1/S transition. *EMBO J.* **13**(18), 4302–4310.
46. Hoffmann, I., Clarke, P. R., Marcote, M. J., Karsenti, E., and Draetta, G. (1993). Phosphorylation and activation of human cdc25-C by cdc2-cyclin B and its involvement in the self-amplification of MPF at mitosis. *EMBO J.* **12**, 53–63.
47. Kumagai, A. and Dunphy, W. G. (1996). Purification and molecular cloning of Plx1, a Cdc25-regulatory kinase from *Xenopus* egg extracts. *Science* **273**, 1377–1380.
48. Qian, Y. W., Erikson, E., Taieb, F. E., and Maller, J. L. (2001). The polo-like kinase Plx1 is required for activation of the phosphatase Cdc25C and cyclin B/Cdc2 in *Xenopus* oocytes. *Mol. Biol. Cell* **12**(6), 1791–1799.
49. Karlsson, C., Katich, S., Hagting, A., Hoffmann, I., and Pines, J. (1999). Cdc25B and Cdc25C differ markedly in their properties as initiators of mitosis. *J. Cell Biol.* **146**(3), 573–584.
50. Millar, J. B. A., Blevitt, J., Gerace, L., Sadhu, K., Featherstone, C., and Russell, P. (1991). $p55^{CDC25}$ is a nuclear protein required for the initiation of mitosis in human cells. *Proc. Natl. Acad. Sci. USA* **88**, 10500–10504.
51. Lammer, C., Wagerer, S., Saffrich, R., Mertens, D., Ansorge, W., and Hoffmann, I. (1998). The cdc25B phosphatase is essential for the G2/M phase transition in human cells. *J. Cell Sci.* **111**(Pt 16), 2445–2453.
52. Chen, M. S., Hurov, J., White, L. S., Woodford-Thomas, T., and Piwnica-Worms, H. (2001). Absence of apparent phenotype in mice lacking Cdc25C protein phosphatase. *Mol. Cell. Biol.* **21**(12), 3853–3861.
53. Lincoln, A. J., Wickramasinghe, D., Stein, P., Schultz, R. M., Palko, M. E., De Miguel, M. P., Tessarollo, L., and Donovan, P. J. (2002). Cdc25b phosphatase is required for resumption of meiosis during oocyte maturation. *Nat. Genet.* **30**(4), 446–449.
54. Kipreos, E. T. and Pagano, M. (2000). The F-box protein family. *Genome Biol.* **1**(5).
55. Harper, J. W. and Elledge, S. J. (1999). Skipping into the E2F1-destruction pathway. *Nat. Cell Biol.* **1**(1), E5–E7.
56. Shah, J. V. and Cleveland, D. W. (2000). Waiting for anaphase: Mad2 and the spindle assembly checkpoint. *Cell* **103**(7), 997–1000.
57. Gillett, E. S. and Sorger, P. K. (2001). Tracing the pathway of spindle assembly checkpoint signaling. *Dev. Cell* **1**(2), 162–164.
58. Pines, J. (1999). Four-dimensional control of the cell cycle. *Nat. Cell Biol.* **1**(3), E73–E79.
59. Hagting, A., Jackman, M., Simpson, K., and Pines, J. (1999). Translocation of cyclin B1 to the nucleus at prophase requires a phosphorylation-dependent nuclear import signal. *Curr. Biol.* **9**(13), 680–689.
60. Ferrell, J. E., Jr. (1998). How regulated protein translocation can produce switch-like responses. *Trends Biochem. Sci.* **23**(12), 461–465.
61. Jin, P., Hardy, S., and Morgan, D. O. (1998). Nuclear localization of cyclin B1 controls mitotic entry after DNA damage. *J. Cell Biol.* **141**(4), 875–885.
62. Jackman, M., Kubota, Y., den Elzen, N., Hagting, A., and Pines, J. (2002). Cyclin A- and cyclin E-Cdk complexes shuttle between the nucleus and the cytoplasm. *Mol. Biol. Cell* **13**(3), 1030–1045.
63. Zhou, B. B. and Elledge, S. J. (2000). The DNA damage response: Putting checkpoints in perspective. *Nature* **408**(6811), 433–439.
64. Karran, P. (2000). DNA double strand break repair in mammalian cells. *Curr. Opin. Genet. Dev.* **10**(2), 144–150.
65. Hoeijmakers, J. H. (2001). Genome maintenance mechanisms for preventing cancer. *Nature* **411**(6835), 366–374.
66. Abraham, R. T. (2001). Cell cycle checkpoint signaling through the ATM and ATR kinases. *Genes Dev.* **15**(17), 2177–2196.
67. Wahl, G. M., Linke, S. P., Paulson, T. G., and Huang, L. C. (1997). Maintaining genetic stability through TP53 mediated checkpoint control. *Cancer Surv.* **29**, 183–219.
68. Banin, S., Moyal, L., Shieh, S., Taya, Y., Anderson, C. W., Chessa, L., Smorodinsky, N. I., Prives, C., Reiss, Y., Shiloh, Y., and Ziv, Y. (1998). Enhanced phosphorylation of p53 by ATM in response to DNA damage. *Science* **281**(5383), 1674–1677.
69. Canman, C. E., Lim, D. S., Cimprich, K. A., Taya, Y., Tamai, K., Sakaguchi, K., Appella, E., Kastan, M. B., and Siliciano, J. D. (1998). Activation of the ATM kinase by ionizing radiation and phosphorylation of p53. *Science* **281**(5383), 1677–1679.
70. Tibbetts, R. S., Brumbaugh, K. M., Williams, J. M., Sarkaria, J. N., Cliby, W. A., Shieh, S. Y., Taya, Y., Prives, C., and Abraham, R. T. (1999). A role for ATR in the DNA damage-induced phosphorylation of p53. *Genes Dev.* **13**(2), 152–157.
71. Maya, R., Balass, M., Kim, S. T., Shkedy, D., Leal, J. F., Shifman, O., Moas, M., Buschmann, T., Ronai, Z., Shiloh, Y., Kastan, M. B., Katzir, E., and Oren, M. (2001). ATM-dependent phosphorylation of Mdm2 on serine 395: Role in p53 activation by DNA damage. *Genes Dev.* **15**(9), 1067–1077.

72. Chehab, N. H., Malikzay, A., Appel, M., and Halazonetis, T. D. (2000). Chk2/hCds1 functions as a DNA damage checkpoint in G(1) by stabilizing p53. *Genes Dev.* **14**(3), 278–288.
73. Hirao, A., Kong, Y. Y., Matsuoka, S., Wakeham, A., Ruland, J., Yoshida, H., Liu, D., Elledge, S. J., and Mak, T. W. (2000). DNA damage-induced activation of p53 by the checkpoint kinase Chk2. *Science* **287**(5459), 1824–1827.
74. Bunz, F., Dutriaux, A., Lengauer, C., Waldman, T., Zhou, S., Brown, J. P., Sedivy, J. M., Kinzler, K. W., and Vogelstein, B. (1998). Requirement for p53 and p21 to sustain G2 arrest after DNA damage. *Science* **282**(5393), 1497–1501.
75. van Hemert, M. J., Steensma, H. Y., and van Heusden, G. P. (2001). 14–3–3 proteins: Key regulators of cell division, signalling and apoptosis. *Bioessays* **23**(10), 936–946.
76. Chan, T. A., Hermeking, H., Lengauer, C., Kinzler, K. W., and Vogelstein, B. (1999). 14–3–3Sigma is required to prevent mitotic catastrophe after DNA damage. *Nature* **401**(6753), 616–620.
77. Sanchez, Y., Wong, C., Thoma, R. S., Richman, R., Wu, Z., Piwnica-Worms, H., and Elledge, S. J. (1997). Conservation of the Chk1 checkpoint pathway in mammals: Linkage of DNA damage to Cdk regulation through cdc25. *Science* **277**, 1497–1501.
78. Peng, C.-Y., Graves, P. R., Thoma, R. S., Wu, Z., Shaw, A. S., and Piwnica-Worms, H. (1997). Mitotic and G2 checkpoint control: Regulation of 14–3–3 protein binding by phosphorylation of Cdc25C on serine-216. *Science* **277**, 1501–1505.
79. Graves, P. R., Lovly, C. M., Uy, G. L., and Piwnica-Worms, H. (2001). Localization of human Cdc25C is regulated both by nuclear export and 14–3–3 protein binding. *Oncogene* **20**(15), 1839–1851.
80. Blasina, A., de Weyer, I. V., Laus, M. C., Luyten, W., Parker, A. E., and McGowan, C. H. (1999). A human homologue of the checkpoint kinase cds1 directly inhibits cdc25 phosphatase. *Curr. Biol.* **9**(1), 1–10.

Chapter 329

Endocytosis and Cytoskeleton

Pier Paolo Di Fiore[1,2,3] and Giorgio Scita[1]

[1]*Department of Experimental Oncology, Istituto Europeo di Oncologia, Milan, Italy*
[2]*Medical School, University of Milan, Milan Italy*
[3]*FIRC Institute for Molecular Oncology, Milan, Italy*

Introduction

Signals emanating from membrane receptors regulate a number of fundamental cellular events, thereby dictating whether a cell will differentiate or proliferate, die or survive, rest or migrate. This pleiotropism is achieved through the regulation of a heterogeneous set of signaling molecules, which are organized in complex networks characterized by extensive cross-talk, integration between different linear pathways, and the formation of large signaling units. Additional complexity is brought about by coordinated cellular processes such as receptor internalization, membrane trafficking, and compartmentalization, which govern the duration and the intensity of signals. These latter processes are affected, and at least in part regulated, by the overall architecture of a cell, which, in turn, is dramatically remodeled following membrane receptor stimulation. For these reasons, increasing efforts have been directed toward the identification of the molecular machinery allowing such interplay. Exemplar are the GTPases, such as those belonging to the small GTPase family and the dynamins. The former have emerged as master regulators of several biological processes. They are frequently activated by upstream signals, originating from membrane receptors, and further propagate these signals to downstream effectors. GTPases of the Rho subfamily have been implicated at various steps of the regulation of actin cytoskeleton [1], whereas GTPases of the Rab subfamily govern several aspects of membrane traffic and vesicle dynamics [2]. Evidence suggests integration of these regulatory networks in which Rho GTPases also control endocytosis, and Rab GTPases (in particular, Rab5) participate in the regulation of actin dynamics.

Dynamins, on the other hand, are key endocytic molecules, and act at the step of vesicle fission from the membrane. Recently, however, dynamins have also been implicated in processes as diverse as apoptosis and actin cytoskeleton dynamics, thus emerging as key candidates for integrating multiple signaling pathways.

In this chapter, we cover the emerging evidence supporting the involvement of actin dynamics in endocytosis. We also briefly discuss possible roles for microtubules during the initial and later phases of the endocytic process. Finally, we describe the known molecular circuitries that link and integrate endocytosis and actin dynamics.

Actin Dynamics and Endocytosis

Actin Cytoskeleton and Endocytosis in Yeast

Endocytosis of membrane receptors is governed by a complex structural machinery [3,4], which is in turn controlled by a wealth of accessory regulatory proteins [5,6]. Links between endocytosis and the actin cytoskeleton have been revealed by a number of studies in yeast A variety of the so-called "end" or "dim" mutations, isolated in screening for endocytic defects, also disrupt the actin cytoskeleton (for reviews, see [7–9]). In contrast, many temperature-sensitive mutations in actin and actin-binding proteins display endocytic defects. Notably, a collection of isogenic mutants of the actin gene contains multiple alleles that are defective in endocytosis. Most of these mutations cluster in a region identified as the binding site for fimbrim, an actin filament-binding protein, suggesting that the integrity of actin filaments is required for endocytosis. Additionally, drugs largely used in mammalian systems to disrupt actin dynamics also affect endocytic processes in yeast [8,9]. Moreover, a role in endocytosis for the actin-regulating Arp2/3 complex is emerging. The Arp2/3 complex, which is composed of seven subunits, is endowed with the property of nucleating actin filaments and is responsible for the formation of a branched actin network. A variety of activators of Arp2/3, including Las17/Bee1p, Abp1, and the type I myosin Myo3/5p (reviewed in [9]), have been reported to affect endocytosis to various extents. On the other hand, the yeast Eps15 homology domain (EH)-containing protein Pan1p, which is part of the EH domain complex involved in endocytosis [10], can bind and activate the Arp2/3 actin nucleating function [11]. Interestingly, a mutant of Pan1p has recently being described, in which the endocytic defects are separable from those on actin dynamics [11], ruling out the possibility that the effects on endocytosis are the results of the disruption of the actin cytoskeleton.

A certain caution must be exerted when generalizing these observations to mammalian cells, because some important differences exist in the endocytic process between yeasts and mammals. The most striking difference is that no dynamin-like protein seems to participate in the uptake step in yeast. Similarly, clathrin, which is a fundamental component for the formation of a coated pit in mammals, is only "mildly" required in yeasts, because it is dispensable for cell growth, protein secretion, or receptor endocytosis [7].

Actin Dynamics at Various Steps of the Endocytic Process in Mammals

Links between actin cytoskeletal organization and endocytosis in mammalian cells also exist. For example, drugs or proteins that disrupt the actin cytoskeleton by perturbing actin treadmilling, also display a variety of effects on early steps of receptor-mediated endocytosis (reviewed in [12]). These studies have converged to establish a model in which *de novo* actin polymerization (or actin remodeling) is required in endocytosis, while preexisting actin filaments are not. Actin remodeling might participate in the early phases of endocytosis in various ways (Fig. 1). In the case of the internalization of the B-cell receptor, for example, actin polymerization appears to define the topology of endocytosis by restricting the localizations at which receptor internalization can take place [13].

A complex role of actin dynamics in the process of SV-40 internalization has been recently uncovered. SV-40 is endocytosed through caveolae instead of clathrin-coated pits. Entry of viral particles induces a transient breakdown of stress fibers [14]. Actin is then recruited, at the entry site, in patches, which then serve as actin nucleation foci for "comet tail" formation (see later discussion). Prevention of stress fiber breakdown by the actin filament stabilizing drug jasplakinolide reduces virus internalization and infection by inhibiting the dissociation of actin stress fibers, with a consequent block in the formation of actin patches and in the polymerization of actin tails. Conversely, treatment with latrunculin A, an actin monomer-sequestering agent, reduces viral internalization by preventing the formation of actin tails, but not of actin patches, thereby inhibiting viral passage deeper into the cytoplasm [14]. Interestingly, dynamin, like actin, is also recruited to virus-containing

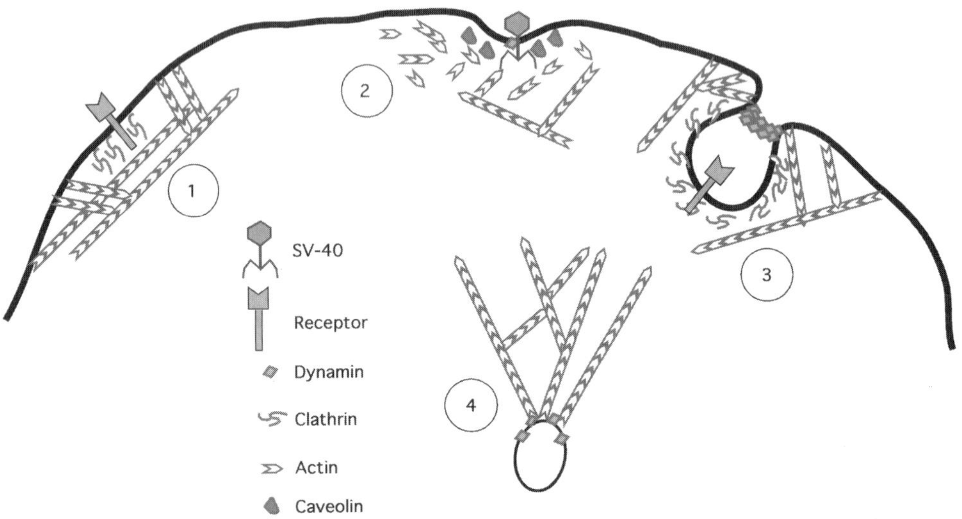

Figure 1 Actin dynamics at various steps of the endocytic route. The illustration shows the potential steps in the endocytic process that might be controlled by actin cytoskeleton remodeling. (1) Actin cytoarchitecture might impose topological constraints on the lateral mobility of the endocytic machinery, resulting in the determination of sites (hot spots) on the plasma membrane where endocytosis take place, as in the case of B-cell receptor internalization [13]. (2) Endocytosis of SV-40 occurs through caveolae. This requires the depolymerization of preexisting actin filaments [14]. (3) Actin might facilitate the fission of vesicles from the plasma membrane as indicated by the increasing molecular connections between the activity of dynamin and dynamin-associated proteins and the actin polymerizing machinery. (4) Actin might participate in later endocytic events by propelling endocytic vesicles in their cytosolic routes. Actin tails have been seen in association with newly formed endocytic vesicles, "rocketing endosomes," with Golgi-derived clathrin-coated vesicles, and with SV-40-containing, caveolae-enriched vesicles.

caveolae and interference with its function reduces virus internalization. Thus, both actin and dynamin are important for the closure of the caveolae by membrane fission and for the movement of vesicles deep inside the cytoplasm.

The participation of actin in the pinching off of vesicles from the plasma membrane (Fig. 1) is further supported by increasing molecular connections between the activity of dynamin and the actin polymerizing machinery (see later discussion) and by studies of myosin VI, the first motor protein implicated in clathrin-mediated endocytosis. Although the mechanisms whereby myosin VI participates in the endocytic process remain to be investigated, myosin VI's ability to move toward the pointed end of actin filaments suggests the intriguing possibility that it may be involved in pulling away the newly formed vesicle along actin filaments, which extends their plus end toward the plasma membrane, leaving their minus ends buried in the cytoplasm (Fig. 1) (see [12] for review).

In addition to this possible role in the early steps of endocytosis, actin might also participate in later events by propelling endocytic vesicles in their cytosolic routes. Actin tails have been seen in association with newly formed endocytic vesicles, "rocketing endosomes," and with Golgi-derived clathrin-coated vesicles [15,16]. These actin comet-like tails are similar to those that drive the movement of intracellular parasites, such as *Listeria* and *Shigella* [17]. This process has been extensively characterized and it requires the recruitment of the basic machinery of actin polymerization, which is in turn regulated by its association with the GTPases Cdc42 and to phosphatidylinositol 4,5 phosphates (PIP2) [18]. Interestingly, PIP2 was also shown to regulate several steps of the formation of the clathrin coat, suggesting that the localized production of this phospholipid, required for the dynamic assembly of the clathrin coat, may also stimulate actin polymerization and drive the movement of vesicles derived from membranes [18].

Role of Microtubule Cytoskeleton in Receptor Endocytosis

Microtubules are long filaments that are involved both in maintaining a precise spatial organization of intracellular organelles and in the transport of cargo between organelles [19]. Movement of some cargo-laden vesicles can be driven by diffusion. However, especially in highly polarized cells, it predominantly occurs along microtubular "tracks" powered by molecular motor proteins like kinesins, dyneins, and myosins. This ensures that vesicles are transported efficiently (at a speed of around 1 μm/s) and for a relatively long distance. In nonpolarized cells, the minus ends of microtubules are located toward the cell center near the centrosome, and the plus ends point radially to the cell periphery. In highly polarized cells, the plus ends of microtubules are localized basolaterally, whereas the minus ends are at the apical side. Movement along the microtubular tracks occurs in both directions. Kinesins are the motor proteins responsible for the movement toward the fast growing plus ends, whereas dyneins are involved in the movement toward the minus ends [20].

The role of microtubules in the endocytic process has been established mainly with the use of microtubule depolymerizing drugs, such as nocodazole (for a review, see [21]). Although the large majority of these studies showed that the microtubule cytoskeleton is not essential for the first step of clathrin-mediated endocytosis, examples are seen in which nocodazole treatment impairs adsorptive and fluid phase endocytosis. Moreover, in cells kept in suspension, but not in cells grown on substrates, nocodazole was reported to reduce by 40% the initial rate of transferrin receptor internalization (for a review, see [21]). This suggests that the overall structural organization of a cell might affect microtubule organization and dynamics, leading to alterations in membrane plasticity, and consequently affecting the ability of a receptor to be recruited to the invaginating clathrin-coated pit.

A much broader consensus has been reached on the requirement of microtubule dynamics at later steps of the endocytic route. For example, the transport of receptors from early to late endosome, their delivery to lysosomes, and their recycling back to the plasma membrane are all severely impaired in nocodazole-treated cells [21]. This is mirrored by the finding that cytoplasmic dynein, in conjunction with the dynactin complex that mediates the association of cargo structures with microtubules, facilitates the inward movement of endocytic vesicles from early to late endosome (for a review, see [22]). It remains unclear, however, to which extent alterations in microtubule dynamics may affect the overall biological activity of membrane receptors, and whether signaling emanating form membrane receptors may in turn modulate microtubule-based vesicle movements. A possible connection is represented by the small GTPase Rab5, which regulates endosome interaction with the microtubule network [23]. *In vitro*, Rab5 was shown to stimulate the association of early endosomes with microtubules and their motility toward the minus ends. *In vivo* Rab5-positive endosomes were shown to move along microtubule tracks. Moreover, the Rab5 effector Vps34, a phosphoinositide-3 kinase, is required for this function, indicating that GTP-loading of Rab5 mediates this effect. Because Rab5 activity is directly controlled by RTKs (see later discussion), it is intriguing to speculate that signaling emanating from activated receptors may also coordinate microtubule-dependent late steps in the endocytic route.

Physical and Functional Interactions of Dynamin and Dynamin-Interacting Proteins with the Actin Cytoskeleton

Dynamin and Actin Cytoskeleton

Dynamin is a GTPase that plays a critical role in endocytosis, by determining the fission of vesicles from the plasma membrane. A number of recent observations, obtained in neuronal cells, functionally linked dynamin to the actin cytoskeleton (reviewed in [24,25]). In addition, expression of a dominant-negative mutant of dynamin-1, in HeLa cells, causes redistribution of actin stress fibers to the cell cortex [26]. Further links were obtained with studies of podosomes and actin comet tails. In the former case, a dynamin isoform, dynamin-2aa, was found to colocalize with filamentous actin at membrane ruffles [27] and at podosomes. A temperature-sensitive mutant of dynamin-2aa disrupted podosome formation [28]. Interestingly, another dynamin mutant (K44A), which is impaired in GTP hydrolysis and potently inhibits endocytosis, only delayed actin dynamics at podosomes, but did not prevent their formation [28]. This latter result suggests that the interference of dynamin with actin remodeling is not due to secondary effects of an endocytic block, but rather to a direct functional role on the actin cytoskeleton.

Two recent reports identified dynamin 1 and 2 as components of actin comet tails generated by *Listeria*, an intracellular pathogen known to utilize an actin tail for movement within the cytoplasm, and by type I PIP kinase. An inactive (K44A) dynamin mutant significantly reduced comet number, length, velocity, and efficiency of movement, suggesting that dynamin is part of a protein network that controls nucleation of actin from the membranes [29,30].

A further intriguing connection emerged from the functional characterization of the putative tumor suppressor protein nm23H1 (Fig. 2). The nm23 protein belongs to a set of structurally conserved nucleoside diphosphatase kinases (NDKs), a family of enzymes that synthesizes triphosphates from their respective nucleotide diphosphates [31]. In *Drosophila*, nm23, regulates synaptic vesicle internalization at a stage where the function of the dynamin GTPase activity is required [32], suggesting a model whereby the NDK activity of nm23 is required locally to increase the concentration of GTP, in order to facilitate the loading of dynamin, which has an unusually low affinity for GTP. Thus, interference with the function of nm23 should impair receptor internalization, resulting in prolonged exposure of the receptor in the plasma membrane and sustained signaling, a possibility compatible with its postulated metastasis suppressor role. This effect may be further strengthened through an additional activity of nm23H1, which also inhibits Tiam-1-induced production of GTP-bound Rac-1 [33]. Tiam-1 is a Rac-specific GEF originally identified as an invasion and metastasis-inducing gene [34]. Nm23h1 interacts directly with Tiam-1, and blocks its GEF activity *in vivo* [33]. Thus, given the inhibitory effects exerted by GTP-loaded Rac in receptor internalization (see later section), the lack of nm23h1 may result in enhanced Rac activation, contributing to the prolonged permanence of receptors on the plasma membrane. Concomitantly, Rac activation may enhance cell migration, thereby further contributing to the onset of the phenotypic features typical of metastatic cells.

Dynamin-Interacting Proteins and Actin Cytoskeleton

Clues as to the molecular mechanisms through which actin dynamics might influence the function of dynamin are emerging from studies of dynamin-interacting proteins (Fig. 2). Syndapin, for example, binds to numerous proteins involved in the endocytic process (dynamin, synaptojanins, and synapsins) and in the regulation of actin dynamics (N-Wasp, Sos-1). Accordingly, syndapin affects both endocytosis and actin cytoskeleton by a mechanism that coordinates vesicle fission and actin nucleation, thereby triggering a burst of actin polymerization around the vesicle neck, which favors detachment from the plasma membrane and propulsion into the cytoplasm.

Amphiphysin I and II represent another class of dynamin-binding proteins with a dual involvement in endocytosis and actin cytoskeleton. Amphiphysins are thought to target dynamin to the plasma membrane (reviewed in [35]). In addition, expression of their SH3 domains reduces receptor endocytosis, probably by preventing the formation of the dynamin ring [35]. Treatment of primary hippocampal neurons with amphyphysin I antisense oligonucleotides inhibits neurite outgrowth, an effect similar to that exerted by suppression of dynamin expression in the same system [36]. Although the mechanism remains to be determined, it is tempting to speculate that suppression of neurite outgrowth might reflect a role of amphiphysin in actin cytoskeleton, possibly exerted through cooperation with the amphiphysin-interacting protein synaptojanin, an inositol-5 phosphatase that cleaves phosphoinositides, which are in turn able to regulate various aspects of actin remodeling [37].

A final example is provided by Abp1, which binds to F-actin and it is associated to dynamic actin structures in lamellipodia and filopodia, in a fashion controlled by the Rac1 GTPase [38]. Abp1 binds to dynamin both in vitro and in vivo, via its SH3 domain. Ectopic expression of the Abp1-SH3 domain leads to a reduction in receptor-mediated endocytosis, an effect reverted by the simultaneous overexpression of dynamin. Mammalian Abp1 is also enriched at sites where both dynamin and actin are enriched [38]. This supports the possibility that it may serve as a link between the endocytic machinery, via its SH3 domain, and the actin cytoskeleton, through its N-terminal F-actin binding region. The recent discovery that Abp1 modulates actin polymerization through its binding and regulation of the activity of the Arp2/3 actin nucleation complex [39] further strengthens the possibility that actin dynamics may play an active role in driving the completion of clathrin coated vesicle budding.

Integration of Signals in Endocytosis and Actin Dynamics by Small GTPases

Rho GTPases in Actin Cytoskeleton Remodeling and Endocytosis

Actin cytoskeleton remodeling induced by membrane receptors requires the sequential activation of small GTPases of the Rho subfamily, which are in turn connected through a complex network of interactors-effectors. The function of several of these proteins is also required in

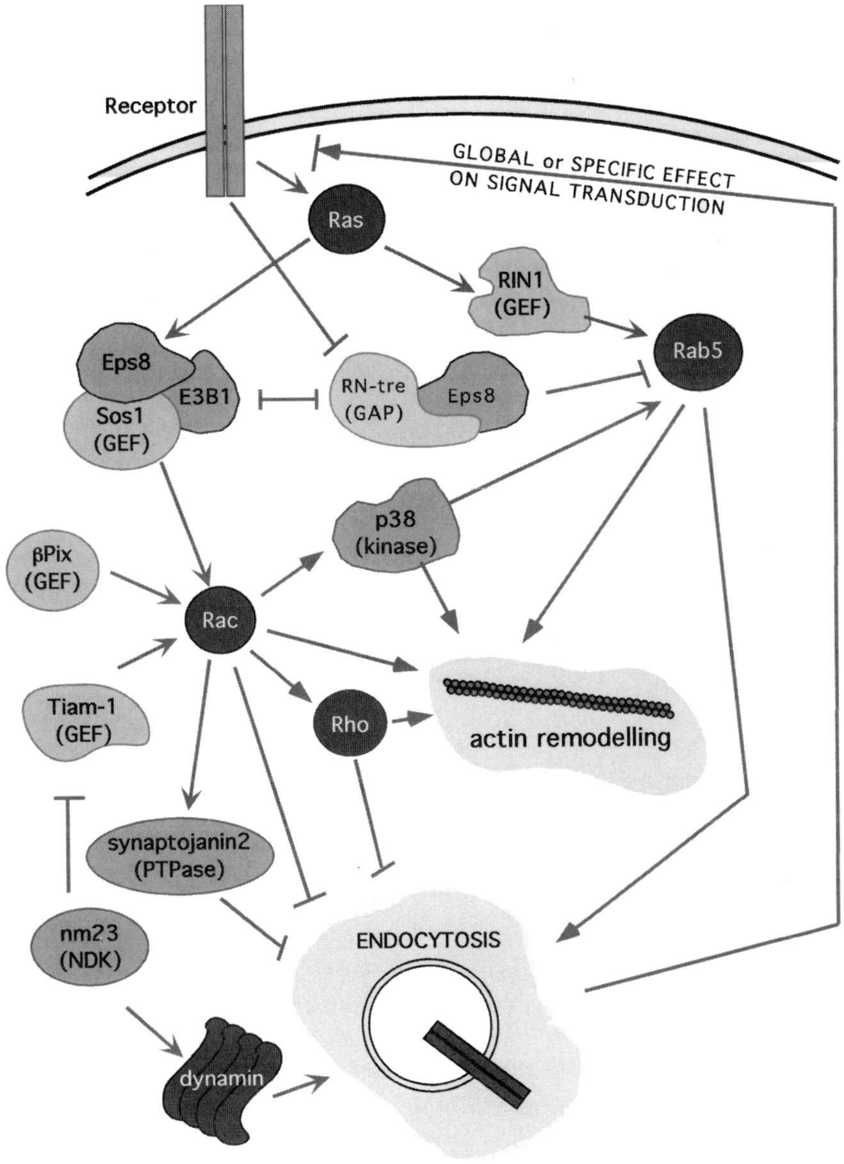

Figure 2 Molecular circuitries connecting actin cytoskeleton and endocytosis. This illustration demonstrates how signaling pathways leading to actin cytoskeleton remodeling and endocytosis are connected and integrated. The scheme is intended to represent a combination of biochemical, genetic, and biological evidence, and it is not exhaustive of all reported interactions. (From [12], Fig. 5, p. 47.)

endocytosis, suggesting that they could represent molecular bridges between cytoskeleton reorganization and vesicular trafficking. Initial indications for a role of Rho GTPases in membrane trafficking derived from the analysis of their intracellular localization. Rho B localizes to late endosomes/lysosomes [40], whereas Cdc42 is associated, in part, with the Golgi [41]. Furthermore, biological studies demonstrated how receptor-mediated endocytosis is inhibited by constitutively active mutants of Rho A and Rac1 [42]. Similarly, activated Rac1 inhibited apical and basolateral endocytosis in polarized epithelial cells [43], while Cdc42 controlled secretory and endocytic transport to the basolateral membrane [44]. This might lead to important consequences. For example, activation of Rho-GTPases by membrane receptors can be predicted to inhibit both receptor internalization and the ensuing inactivation by degradation, thus allowing the sustained and prolonged signaling required for cell proliferation. In addition, increasing evidence is also supporting a requirement for the endocytic process in order to allow full activation of certain signaling pathways, such as those leading to activation of ERKs (reviewed in [45]). Thus delayed endocytosis, obtained through signaling components such as Rho-GTPases, might serve the purpose of delaying the activation of certain signaling, while permitting others, thereby controlling the temporal progression of signaling (Figs. 1 and 2).

How actin dynamics plays a role in this remains to be elucidated. The molecular mechanism of the action of Rho GTPases in endocytosis is not clear. On the one hand, it is possible that the influence of these GTPases on endocytic process is simply a "by-product" (albeit biologically relevant) of their role in actin cytoskeleton remodeling. On the other, a more direct involvement in the molecular machinery of endocytosis is likely, as witnessed by the observation that Rac1 interacts directly with synaptojanin 2, a polyphosphoinositide phosphatase implicated in the uncoating of endocytic clathrin-coated vesicles [46]. Rac may therefore contribute to the regulation of endocytosis with a dual mechanism: directly, by targeting synaptojanin 2 to the plasma membrane, and indirectly, by controlling actin polymerization. The coordination of these events might dictate the final output and contribute to the regulation of the duration and of the intensity of signals (Fig. 2). Another intriguing example supporting a more complex role of GTPase is provided by the diverse effect exerted by RhoA and RhoB on membrane receptor endocytosis. These GTPases are highly related and display virtual identity at the level of their effector-binding region, suggesting that they may interact with the same subset of downstream molecules. Their intracellular localization is, however, remarkably different. RhoA is mainly cytosolic and relocalizes to the plasma membrane upon RTK activation; RhoB is strictly localized to the cytosolic face of endocytic vesicle [40,47]. This is mirrored by the differential effects that RhoA and RhoB exert on endocytosis. An activated mutant of RhoA inhibited constitutive transferring-receptor internalization [42]; however, it did not affect ligand-induced EGFR-endocytosis, at least in nonpolarized cells [47]. This suggests that RhoA-mediated interference on endocytosis is unlikely due to a general perturbation of the actin cytoskeleton. The identification of RhoA effectors capable of mediating its endocytic function(s) will, however, be required to provide a molecular understanding of this putative endocytic-specific function. RhoB, on the other hand, probably acting through its target PRK1 (serine/threonine kinase) retarded the traffic of EGFR from endosome to a prelysosomal compartment [47], where receptors are degraded. Interestingly, RhoB has been recently shown to be activated by EGF stimulation through the guanine nucleotide exchange factor, Vav2 [48]. This suggests the possibility of a positive feedback loop whereby signals triggered by activated RTK also regulate its trafficking, thus determining the duration and intensity of the biological response. Finally, RhoB transcription was shown to be rapidly induced following EGF stimulation [49], indicating that RhoB may be involved in mediating both acute and adaptive response to EGF stimulation.

Linking Rab5 and Rho GTPase Signaling Pathways

The recognized function of Rab5 is to regulate endosome fusion [2]. However, an important role for this small GTPase in endocytosis is also emerging [12]. Studies of the Rab5:GDI (guanine-nucleotide dissociation inhibitor) complex demonstrated its indispensability for receptor sequestration into clathrin-coated pits [50]. GDI is able to extract from membranes the inactive GDP-bound form of Rab proteins [51] and to deliver them to their membrane site(s) of action where nucleotide exchange occurs. In the case of Rab5, it has been proposed that GDI delivers it to the forming pit after coat assembly, to promote efficient sequestration of ligand by an as yet unknown mechanism [50]. Rab5 has also been implicated in actin cytoskeleton remodeling. A constitutively activated form of Rab5 (Rab5Q79L) induces lamellipodia formation in NIH3T3-A14 cells [52] and enhanced cell migration.

Not surprisingly, the dual function of Rab5 and Rho-GTPases in the control of actin cytoskeleton and endocytosis is mirrored by common levels of regulation and actual "integration" of the two signaling systems (Fig. 2), mediated by the small GTPase Ras. One of the best studied functions of active Ras is to activate actin remodeling via physical interaction with phosphatidylinositol 3-kinase (PI3-K) with ensuing activation of Rac [53]. Interestingly, RTKs activate Rab5 via Ras [54]. Such activation seems to be due to RIN1 [20], which possesses Rab5-specific GEF activity [56]. Binding of RIN1 to Ras-GTP potentiates its GEF activity toward Rab5 *in vitro*, presumably increasing the levels of Rab5-GTP *in vivo*. Consistent with the biochemical data, it was shown that overexpression of RIN1 stimulates internalization of the EGFR, whereas its dominant-negative mutant inhibited it [56]. Because overexpression of RIN1 also increases Rab5-dependent endosome fusion, it is reasonable to assume that the effects of RIN1 on endocytosis are due to activation of Rab5. Thus signals emanating from Ras-GTP coordinately activate Rab5 and Rac (Fig. 2).

Similar levels of coordination apply to the deactivation phase of these GTPases, which is controlled by GAPs (Fig. 2). In the case of Rab5 also, this step appears to be under the influence of RTK-originated signals, as demonstrated by studies of RN-tre, a specific Rab5 GAP. RN-tre, whose function is negatively modulated by EGF treatment, inhibits endocytosis by keeping Rab5 inactive [57]. At least in the case of EGFR internalization, RN-tre requires binding to Eps8 to attenuate endocytosis [57]. The best characterized function of Eps8 is to activate Rac by stimulating the Rac-GEF activity of Sos-1, a dual GEF displaying both Ras-GEF and Rac-GEF activities [58]. Thus, Eps8 represents yet another intersection of pathways regulating actin remodeling (through Sos-1) and endocytosis (through RN-tre).

Further coordination is exerted at the level of GDI (Fig. 2). It has recently been demonstrated that the ability of GDI to extract Rab5 from membranes is regulated via its phosphorylation. The kinase responsible for GDI phosphorylation is the stress-induced MAP kinase p38 [59]. It stimulates the formation of the Rab5:GDI complex, which is limiting in steady-state conditions, thereby accelerating endocytosis [59]. However, it has recently been found that the p38 MAP kinase is also necessary for the formation of membrane ruffles induced by βPix, another GEF protein for Cdc42/Rac [60]. Overexpression of βPix induces both ruffle formation and increased activity of p38 MAP kinase, which are abrogated by specific inhibitors of p38, strongly suggesting that the activity of p38 is required for the βPix-induced ruffling [60]. Thus, p38 might positively modulate both the activity of the Rab5 cycle and the cascade that from activated Cdc42/Rac leads to actin reorganization (Fig. 2).

Conclusions

Following biological and genetic evidence of a functional link between endocytosis and actin dynamics, the identification of a number of proteins that connect the two pathways is starting to provide the necessary molecular knowledge to define how these cellular functions are integrated. Small GTPases, through their action as molecular switches, seem to act as the master players in this scenario. Given the research efforts and rate of discoveries, it is reasonable to predict that it will not be long before a final and coherent picture is obtained.

References

1. Hall, A. (1998). Rho GTPases and the actin cytoskeleton. *Science* **279**, 509–514.
2. Zerial, M. and McBride, H. (2001). Rab proteins as membrane organizers. *Nat. Rev. Mol. Cell Biol.* **2**,107–117.
3. Marsh, M. and McMahon, H. T. (1999). The structural era of endocytosis. *Science* **285**, 215–220.
4. Owen, D. J. and Luzio, J. P. (2000). Structural insights into clathrin-mediated endocytosis. *Curr. Opin. Cell Biol.*, 467–474.
5. Robinson, M. S. and Bonifacino, J. S. (2001). Adaptor-related proteins. *Curr. Opin. Cell Biol.* **13**, 444–453.
6. Schmid, S. L. (1997). Clathrin-coated vesicle formation and protein sorting: An integrated process. *Annu. Rev. Biochem.* **66**, 511–548.
7. Geli, M. I. and Riezman, H. (1998). Endocytic internalization in yeast and animal cells: Similar and different. *J. Cell Sci.* **111**, 1031–1037.
8. Munn, A. L. (2001). Molecular requirements for the internalisation step of endocytosis: Insights from yeast. *Biochim. Biophys. Acta* **1535**, 236–257.
9. Shaw, J. D., Cummings, K. B., Huyer, G., Michaelis, S., and Wendland, B. (2001). Yeast as a model system for studying endocytosis. *Exp. Cell Res.* 271:1 1–9.
10. Wendland, B., and Emr, S. D. (1998). Pan1p, yeast eps15, functions as a multivalent adaptor that coordinates protein-protein interactions essential for endocytosis. *J. Cell Biol.* **141**, 171–184.
11. Duncan, M. C., Cope, M. J., Goode, B. L., Wendland, B., and Drubin, D. G. (2001). Yeast Eps15-like endocytic protein, Pan1p, activates the Arp2/3 complex. *Nat. Cell Biol.* **3**, 687–690.
12. Lanzetti, L., Di Fiore, P. P., and Scita, G. (2001). Pathways linking endocytosis and actin cytoskeleton in mammalian cells. *Exp. Cell Res.* **271**, 45–56.
13. Brown, B. K. and Song, W. (2001). The actin cytoskeleton is required for the trafficking of the b cell antigen receptor to the late endosomes. *Traffic* **2**, 414–427.
14. Pelkmans, L., Puntener, D., and Helenius, A. (2002). Local actin polymerization and dynamin recruitment in SV40-induced internalization of caveolae. *Science* **296**, 535–539.
15. Frischknecht, F., Cudmore, S., Moreau, V., Reckmann, I., Rottger, S., and Way, M. (1999). Tyrosine phosphorylation is required for actin-based motility of vaccinia but not Listeria or Shigella. *Curr. Biol.* **9**, 89–92.
16. Taunton, J. (2001). Actin filament nucleation by endosomes, lysosomes and secretory vesicles. *Curr. Opin. Cell Biol.* **13**, 85–91.
17. Frischknecht, F. and Way, M. (2001). Surfing pathogens and the lessons learned for actin polymerization. *Trends Cell Biol.* **11**, 30–38.
18. Sechi, A. S. and Wehland, J. (2000). The actin cytoskeleton and plasma membrane connection: PtdIns(4,5)P(2) influences cytoskeletal protein activity at the plasma membrane. *J. Cell Sci.* **113**, 3685–3695.
19. Bloom, G. S. and Goldstein, L. S. (1998). Cruising along microtubule highways: How membranes move through the secretory pathway. *J. Cell Biol.* **140**, 1277–1280.
20. Hirokawa, N. (1998). Kinesin and dynein superfamily proteins and the mechanism of organelle transport. *Science* **279**, 519–526.
21. Apodaca, G. (2001). Endocytic traffic in polarized epithelial cells: Role of the actin and microtubule cytoskeleton. *Traffic* **2**, 149–159.
22. Holleran, E. A., Karki, S., and Holzbaur, E. L. (1998). The role of the dynactin complex in intracellular motility. *Int. Rev. Cytol.* **182**, 69–109.
23. Nielsen, E., Severin, F., Backer, J. M., Hyman, A. A., and Zerial, M. (1999). Rab5 regulates motility of early endosomes on microtubules. *Nat. Cell Biol.* **1**, 376–382.
24. McNiven, M. A., Cao, H., Pitts, K. R., and Yoon, Y. (2000). The dynamin family of mechanoenzymes: Pinching in new places. *Trends Biochem. Sci.* **25**, 115–120.
25. Sever, S., Damke, H., and Schmid, S. L. (2000). Garrotes, springs, ratchets, and whips: Putting dynamin models to the test. *Traffic* **1**, 385–392.
26. Damke, H., Baba, T., Warnock, D. E., and Schmid, S. L. (1994). Induction of mutant dynamin specifically blocks endocytic coated vesicle formation. *J. Cell Biol.* **127**, 915–934.
27. Cao, H., Garcia, F., and McNiven, M. A. (1998). Differential distribution of dynamin isoforms in mammalian cells. *Mol. Biol. Cell.* **9**, 2595–2609.
28. Ochoa, G. C., Slepnev, V. I., Neff, L., Ringstad, N., Takei, K., Daniell, L., Kim, W., Cao, H., McNiven, M., Baron, R., and De Camilli, P. (2000). A functional link between dynamin and the actin cytoskeleton at podosomes. *J. Cell Biol.* **150**, 377–389.
29. Lee, E. and De Camilli, P. (2002). Dynamin at actin tails. *Proc. Natl. Acad. Sci. USA* **99**, 161–166.
30. Orth, J. D., Krueger, E. W., Cao, H., and McNiven, M. A. (2002). The large GTPase dynamin regulates actin comet formation and movement in living cells. *Proc. Natl. Acad. Sci. USA* **99**, 167–172.
31. Wagner, P. D. and Vu, N. D. (1995). Phosphorylation of ATP-citrate lyase by nucleoside diphosphate kinase. *J. Biol. Chem.* **270**, 21758–21764.
32. Krishnan, K. S., Rikhy, R., Rao, S., Shivalkar, M., Mosko, M., Narayanan, R., Etter, P., Estes, P. S., and Ramaswami, M. (2001). Nucleoside diphosphate kinase, a source of GTP, is required for dynamin-dependent synaptic vesicle recycling. *Neuron* **30**, 197–210.
33. Otsuki, Y., Tanaka, M., Yoshii, S., Kawazoe, N., Nakaya, K., and Sugimura, H. (2001). Tumor metastasis suppressor nm23H1 regulates Rac1 GTPase by interaction with Tiam1. *Proc. Natl. Acad. Sci. USA* **98**, 4385–4390.
34. Habets, G. G., Scholtes, E. H., Zuydgeest, D., van der Kammen, R. A., Stam, J. C., Berns, A., and Collard, J. G. (1994). Identification of an invasion-inducing gene, Tiam-1, that encodes a protein with homology to GDP-GTP exchangers for Rho-like proteins. *Cell* **77**, 537–549.
35. Wigge, P. and McMahon, H. T. (1998). The amphiphysin family of proteins and their role in endocytosis at the synapse. *Trends Neurosci.* **21**, 339–344.
36. Mundigl, O., Ochoa, G. C., David, C., Slepnev, V. I., Kabanov, A., and De Camilli, P. (1998). Amphiphysin I antisense oligonucleotides inhibit neurite outgrowth in cultured hippocampal neurons. *J. Neurosci.* **18**, 93–103.
37. Micheva, K. D., Kay, B. K., and McPherson, P. S. (1997). Synaptojanin forms two separate complexes in the nerve terminal. Interactions with endophilin and amphiphysin. *J Biol Chem.* **272**, 27239–27245.
38. Kessels, M. M., Engqvist-Goldstein, A. E., and Drubin, D. G. (2000). Association of mouse actin-binding protein 1 (mAbp1/SH3P7), an Src kinase target, with dynamic regions of the cortical actin cytoskeleton in response to Rac1 activation. *Mol. Biol. Cell* **11**, 393–412.

39. Goode, B. L., Rodal, A. A., Barnes, G., and Drubin, D. G. (2001). Activation of the Arp2/3 complex by the actin filament binding protein Abp1p. *J. Cell Biol.* **153**, 627–634.
40. Adamson, P., Paterson, H. F., and Hall, A. (1992). Intracellular localization of the P21rho proteins. *J. Cell Biol.* **119**, 617–627.
41. Erickson, J. W., Zhang, C., Kahn, R. A., Evans, T., and Cerione, R. A. (1996). Mammalian Cdc42 is a brefeldin A-sensitive component of the Golgi apparatus. *J. Biol. Chem.* **271**, 26850–26854.
42. Lamaze, C., Chuang, T. H., Terlecky, L. J., Bokoch, G. M., and Schmid, S. L. (1996). Regulation of receptor-mediated endocytosis by Rho and Rac. *Nature* **382**, 177–179.
43. Jou, T. S., Leung, S. M., Fung, L. M., Ruiz, W. G., Nelson, W. J., and Apodaca, G. (2000). Selective alterations in biosynthetic and endocytic protein traffic in Madin-Darby canine kidney epithelial cells expressing mutants of the small GTPase Rac1. *Mol. Biol. Cell* **11**, 1287–304.
44. Kroschewski, R., Hall, A., and Mellman, I. (1999). Cdc42 controls secretory and endocytic transport to the basolateral plasma membrane of MDCK cells. *Nat. Cell Biol.* **1**, 8–13.
45. Di Fiore, P. P. and De Camilli, P. (2001). Endocytosis and signaling: An inseparable partnership. *Cell* **106**, 1–4.
46. Malecz, N., McCabe, P. C., Spaargaren, C., Qiu, R., Chuang, Y., and Symons, M. (2000). Synaptojanin 2, a novel Rac1 effector that regulates clathrin-mediated endocytosis. *Curr. Biol.* **10**, 1383–1386.
47. Gampel, A., Parker, P. J., and Mellor, H. (1999). Regulation of epidermal growth factor receptor traffic by the small GTPase rhoB. *Curr. Biol.* **9**, 955–958.
48. Gampel, A. and Mellor, H. (2002). Small interfering RNAs as a tool to assign Rho GTPase exchange-factor function *in vivo*. 393–398.
49. Zalcman, G., Closson, V., Linares-Cruz, G., Lerebours, F., Honore, N., Tavitian, A., and Olofsson, B. (1995). Regulation of Ras-related RhoB protein expression during the cell cycle. Oncogene **10**, 1935–1945.
50. Novick, P. and Zerial, M. (1997). The diversity of Rab proteins in vesicle transport. *Curr. Opin. Cell Biol.* **9**, 496–504.
51. McLauchlan, H., Newell, J., Morrice, N., Osborne, A., West, M., and Smythe, E. (1998). A novel role for Rab5-GDI in ligand sequestration into clathrin-coated pits. *Curr. Biol.* **8**, 34–45.
52. Spaargaren, M. and Bos, J. L. (1999). Rab5 induces Rac-independent lamellipodia formation and cell migration. *Mol. Biol. Cell* **10**, 3239–3250.
53. Scita, G., Tenca, P., Frittoli, E., Tocchetti, A., Innocenti, M., Giardina, G., and Di Fiore, P. P. (2000). Signaling from Ras to Rac and beyond: Not just a matter of GEFs. *EMBO J.* **19**, 2393–2398.
54. Barbieri, M. A., Roberts, R. L., Gumusboga, A., Highfield, H., Alvarez-Dominguez, C., Wells, A., and Stahl, P. D. (2000). Epidermal growth factor and membrane trafficking. EGF receptor activation of endocytosis requires Rab5a. *J. Cell Biol.* **151**, 539–550.
55. Han, L. and Colicelli, J. (1995). A human protein selected for interference with Ras function interacts directly with Ras and competes with Raf1. *Mol. Cell Biol.* **15**, 31318–31323.
56. Tall, G. G., Barbieri, M. A., Stahl, P. D., and Horazdovsky, B. F. (2001). Ras-activated endocytosis is mediated by the Rab5 guanine nucleotide exchange activity of RIN1. *Dev. Cell* **1**, 173–182.
57. Lanzetti, L., Rybin, V., Malabarba, M. G. Christoforidis, S., Scita, G., Zerial, M., and Di Fiore, P. P. (2000). The Eps8 protein coordinates EGF receptor signalling through Rac and trafficking through Rab5. *Nature* **408**, 374–377.
58. Scita, G., Nordstrom, J., Carbone, R., Tenca, P., Giardina, G., Gutkind, S., Bjarnegard, M., Betsholtz, C., and Di Fiore, P. P. (1999). EPS8 and E3B1 transduce signals from Ras to Rac. *Nature* **401**, 290–293.
59. Cavalli, V., Vilbois, F., Corti, M., Marcote, M. J., Tamura, K., Karin, M., Arkinstall, S., and Gruenberg, J. (2001). The stress-induced MAP kinase p38 regulates endocytic trafficking via the GDI:Rab5 complex. *Mol. Cell* **7**, 421–432.
60. Lee, S. H., Eom, M., Lee, S. J., Kim, S., Park, H. J., and Park, D. (2001). Beta-Pix-enhanced p38 Activation by Cdc42/Rac/PAK/MKK3/6-mediated Pathway. *J. Biol. Chem.* **276**, 25066–25072.

Molecular Basis for Nucleocytoplasmic Transport

Gino Cingolani and Larry Gerace
Department of Cell Biology, The Scripps Research Institute, La Jolla, California

Introduction

Transport of proteins and RNAs between the cytoplasm and the nucleus in eukaryotic cells proceeds through the nuclear pore complex (NPC), a massive proteinaceous structure that spans the nuclear envelope (Fig. 1; reviewed in [1–5]). During the past 15 years, the protein components of the NPC have been completely cataloged by a combination of traditional approaches and proteomic analysis. Moreover, permeabilized cell transport assays developed in higher eukaryotes [6], in combination with *in vivo* and genetic approaches in *Saccharomyces cerevisiae* and other organisms, have led to the identification and characterization of major factors and steps in protein and RNA translocation through the NPC. Finally, X-ray crystallography [7,8] has started to reveal the structures of individual components of the transport machinery at an atomic level, providing complementary information for understanding transport mechanisms.

Macromolecular transport through the NPC is a complex multistep process that involves the movement of cargoes over a distance of ~200 nm. The overall process of nuclear transport is energy dependent. Because individual NPCs are simultaneously engaged in nuclear import and export, one of the key questions about the mechanics is how the NPC achieves selectivity of transport in the face of enormous cargo flux in the import and export directions. Moreover, the transport of specific cargoes is usually highly directional, often occurring against a concentration gradient. Four major determinants of nuclear transport have been described: (1) a transport signal in protein or RNA cargoes, (2) nucleocytoplasmic shuttling receptors that recognize transport signals and carry cargoes through the NPC, (3) the GTPase Ran, and (4) specific proteins of the NPC (nucleoporins) that facilitate transport. In this review, we discuss these four transport determinants, with emphasis on the underlying molecular and structural mechanisms. As discussed later, the majority of nuclear transport receptors are part of a receptor superfamily related to importin/karyopherin β [5]. This family contains 14 members in *S. cerevisiae* and is likely to be larger in higher eukaryotes.

Transport Signals

Nuclear transport signals direct the passage of cargoes through the NPC by specifying binding to specific transport receptors or adaptors, which dictate the directionality of nuclear import and export [11]. Transport signals in proteins most commonly consist of short amino acid stretches, such as the classical basic amino acid-rich nuclear localization signal (NLS) or the leucine-rich nuclear export signal (NES). Some RNAs, such as tRNAs and type D retroviral mRNAs, contain nuclear transport signals within specific nucleotide stretches.

Nuclear Import Signals

Import signals can be recognized directly by an import receptor or by a specific adaptor that in turn binds to the import receptor. The best studied import signals are recognized by the receptor importin β. Two known classes of signal associate with importin β (Fig. 2A): *classical* NLSs, which are recognized by the importin β-binding adaptor

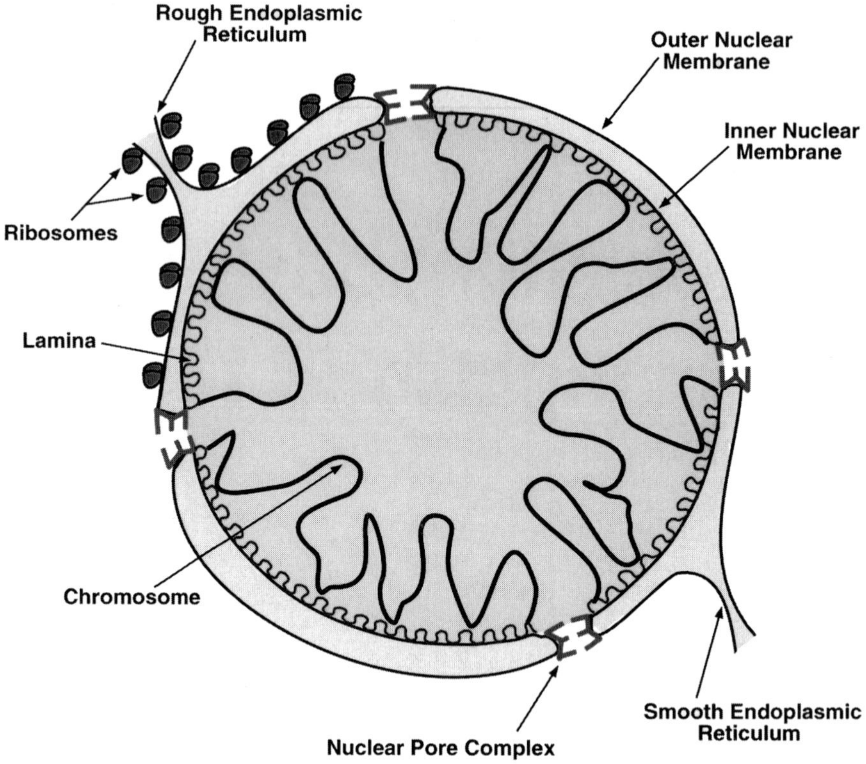

Figure 1 Organization of the nuclear envelope. The nuclear envelope, which separates the nucleoplasm from the cytoplasm, consists of a double membrane that is continuous with the endoplasmic reticulum. The inner and outer nuclear membranes are joined at nuclear pore complexes, which provide the only known transport route across the nuclear envelope. A filamentous network of proteins, known as the nuclear lamina, lines the inner nuclear membrane and serves to anchor chromosomes to the nuclear periphery.

importin α, and *nonclassical* NLSs, which associate directly with importin β. Classical NLSs (cNLS) can be monopartite, as originally discovered in the large T antigen of SV40, or bipartite, as in nucleoplasmin. A classical monopartite NLS consists of approximately seven residues, four or five of which are lysines (or arginines). In contrast, bipartite NLSs contain two discrete basic amino acid stretches separated by approximately 10 amino acids (Fig. 2A). Atomic structures have been determined for several NLS peptides in complex with the adaptor importin α. Both monopartite and bipartite NLSs display an extended conformation (Fig. 2B) [10,11], with the conserved side chains of Lys and Arg engaging in specific electrostatic interactions with the acidic surface of the importin α adaptor. Despite the shortness of the signal, such as in the case of monopartite cNLSs, the binding specificity and affinity for the adaptor can be very high. In the case of the SV40-NLS, the dissociation constant for the adaptor importin α is approximately $K_d \sim 10\,nM$ [12].

In contrast to classical NLSs, which associate with importin β via the adaptor importin α, adaptor-independent NLSs are recognized directly by importin β. Two subclasses of adaptor-independent NLSs for importin β have been described thus far: those similar to the IBB domain and those distinct from IBB. The IBB domain was originally described as the *i*mportin β-*b*inding domain of importin α, which mediates the heterodimerization of importin α/β during nuclear import. It consists of approximately 40 amino acids arranged as a long α-helix after binding to the receptor importin β (Figs. 2B and 3B) [13], but it is unfolded in the unbound state [14,15]. An IBB domain also has been found in the adaptors snurportin [1,2] and XRIPα [16]. Several other cargoes like the parathyroid hormone-related protein (PTHrP) [17] and ribosomal protein L5 [18] (Fig. 2A) bind importin β in the absence of an adaptor, but apparently utilize a cargo binding site that is distinct from the region recognized by the IBB domain and thus are expected to have unique properties. Intriguingly, the nonclassical NLS of ribosomal protein L25 (also known as the BIB domain) can be recognized by at least four receptors of the importin β superfamily, namely, importin β-1, RanBP7 (importin 7), importin 3, and importin 4, which can promote, *in vitro*, efficient transport of the cargo into the nucleus [18]. This finding suggests that the recognition of a transport signal by a receptor is not absolute and multiple receptors may cooperate to import the same cargo.

A second well-characterized signal, the so-called "M9" domain (Fig. 2A), is recognized by transportin (karyopherin β-2), a member of the importin β superfamily of import receptors. The M9 signal was originally described in the mRNA binding protein hnRNP A1 and is a 38-residue-long sequence poor in basic residues and rich in glycines and serines. Unfortunately, the binding site and molecular details

A

IMPORT SIGNALS	
cNLS–monopartite	126–PKKKRKV–132
cNLS–bipartite	150–KRPAAIKKAGQAKKKK–170
IBB–domain	11–AARLHRFKNKGKDSTEMRRRRIEVNVELRKAKKDDQMLKRRNVS–54
PTHrP–(66–94)	66–RYLTQETNKVETYKEQPLKTPGKKKKGKP–94
L23a–BIB–domain	32–HSHKKKKIRTSPTFRRPKTLRLRRQPKYPRKSAPRRNKLDHY–74
M9	268–NQSSNFGPMKGGNFGGRSSGPYGGGGQYFAKPRNQGGY–305
EXPORT SIGNALS	
Rev	73–LQLPPLERLTLD–84

B

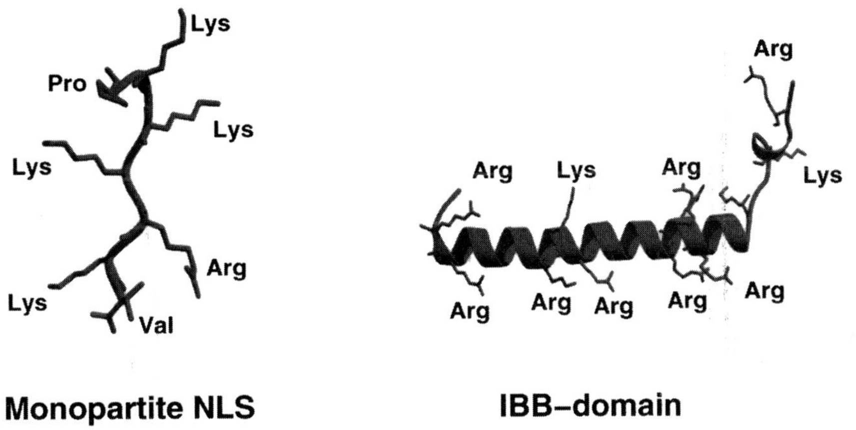

Figure 2 Transport Signals. (A) Signals for nuclear import and export. (B) Structure of classical monopartite SV-40 NLS [10] and IBB domain [13].

of the interaction between the M9 peptide and transportin have not yet been elucidated.

Nuclear Export Signals

NESs are less well characterized than NLSs and lack a strong unique consensus sequence. NESs were originally identified in the inhibitor of the cyclic AMP-dependent protein kinase (PKI) and the HIV Rev protein and subsequently were found in several other cellular and viral proteins [1–5]. All of these originally defined NESs, which are recognized by a single member of the importin β superfamily termed Crm1, are rich in leucine and other hydrophobic residues (Fig. 2A), although both the relative spacing among leucine residues and number of leucines can vary from protein to protein. Certain NESs of this class do not contain multiple leucines, as with the influenza NES protein or the NFAT transcription factor [19], suggesting that it is not the leucine *per se* that confers the molecular recognition, but rather the presence of hydrophobic residues distributed within a short stretch of amino acids.

Transport Receptors

Nuclear transport receptors are soluble proteins that recognize a transport cargo via a direct or indirect interaction with its transport signal and, consequently, are able to carry the cargo through the NPC by virtue of specific interactions with nucleoporins. These transport receptors can be classified into two major classes: members of the importin β superfamily, and receptors unrelated to importin β.

Importin β Superfamily

Receptors of the importin β superfamily have been identified on the basis of their similarity to importin β at the primary structure level [5], and they may have a common evolutionary origin. The members of this superfamily share a number of interesting biochemical properties, including a predominance of α-helical structure, which is present in a tandem array of HEAT repeats (Fig. 3B), a preponderance of acidic over basic amino acid residues, an N-terminal RanGTP

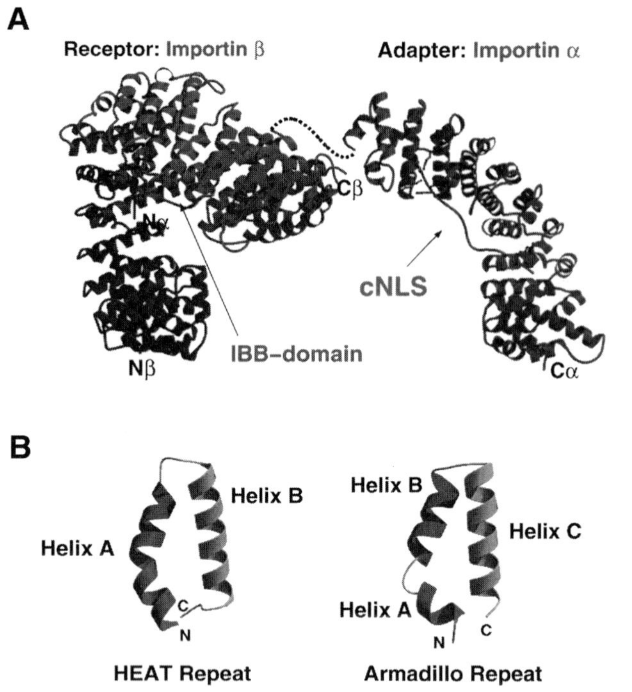

Figure 3 Model for importin α/β heterodimer:cargo complex. (A) The model was built from the structure of importinβ:IBB domain complex [13] (left) and importin α:bipartite NLS complex [10] (right), respectively. The angle between the two proteins could vary according the size of the cargo bound to importin α adaptor. (B) Characteristic HEAT and Armadillo motifs found in importin β and importin α, respectively.

binding domain, and nucleoporin binding sites. Even though the sequence similarity among different members of the family is often difficult to detect, it is possible that all receptors of the importin β superfamily may exhibit a very similar three-dimensional structure.

Import Receptors

Importin β is the best characterized nuclear transport receptor. In complex with the adaptor importin α (Fig. 3A), it mediates nuclear accumulation of cargoes bearing classical NLSs. Importin β presents a characteristic modular structure consisting of 19 tandemly repeated HEAT motifs (Fig. 3B). Each HEAT repeat is formed by two α-helices (helices A and B) connected by a turn. When packed in importin β, the HEAT repeats are arranged to form a superhelix of helices [13]. A-helices are located on the exterior of the protein and form a convex (outer) surface where nucleoporin binding sites are located (Fig. 4A). In contrast, B-helices are arranged inside the protein to create a concave (inner) face, which is involved in binding to cargoes and RanGTP. The IBB domain of importin α interacts with HEAT repeats 7–19 of importin β, and folds as a long α-helix (residues 23–51) followed by a short N-terminal coiled moiety (residues 11–21). The protein–peptide interaction is largely electrostatic in nature and contains highly specific cation-π interactions between critically conserved tryptophan residues of importin β and conserved Lys and Arg residues of the IBB domain.

The dissociation of the importin α/β heterodimer and the subsequent release of the NLS cargo into the nucleoplasm is mediated by the small GTPase Ran (see later section), which also plays a critical role in releasing the transport receptor from nucleoporin binding sites. RanGTP binds to the inner surface of importin β between HEAT repeats 1 and 8, in a region that overlaps minimally with the IBB binding domain [20]. It is plausible that Ran induces dramatic conformational changes in importin β that allow the simultaneous release of cargo and nucleoporins. This model is supported by the structure of importin β-2 (karyopherin β-2, transportin) complexed with RanGppNHp [21]. Transportin is an importin β superfamily protein involved in the import of hnRNP A1 into the nucleus. Similar to importin β, it is a fully helical protein and consists of tandemly repeated HEAT motifs. In complex with RanGppNHp, transportin displays a rather open conformation consisting of two perpendicular arches. RanGppNHp binds to the N-terminal arch, whereas the cargo is predicted to bind to the C-terminal arch.

Several other import receptors have been described [22–24], and the importin β superfamily is likely to grow in the next several years. Based on the presence of tandemly repeated HEAT repeats, it has been predicted that the molecular interactions of these proteins are similar and strongly dependent on the surface polarization between the convex and concave surfaces. The need for adaptor molecules appears as a specific feature of importin β, rather than as a general requirement for all transport receptors.

Transport Adaptors

As previously mentioned, importin β binds classical NLS-bearing cargoes via interaction with the adaptor importin α. The crystal structures of both murine [10] and yeast [11] importin α reveal an elongated banana-like structure (Fig. 3A) consisting of 10 and 11 Armadillo (Arm) repeats, respectively. Each Armadillo motif (Fig. 3B) is formed by three α-helices connected by a loop, having high structural resemblance to the HEAT repeats found in importin β. The NLS binding domain of importin α consists of an acidic binding groove formed by tandemly repeated Arm repeats. Monopartite NLSs can bind to two distinct binding sites in this cavity, a major one located within Arm repeats 2–3, and a minor within Arm repeats 6–7 [10]. As expected, a bipartite NLS peptide spans both cargo-binding sites and thus adopts an extended conformation [11]. Arrays of conserved Trp residues in importin α are fundamentally important for sandwiching the aliphatic portion of conserved arginine and lysine side chains of the NLS. In the murine importin α structure, a portion of the IBB domain occupies the NLS binding site, suggesting a mechanism for auto-inhibition of importin α in absence of a bound NLS [15].

Interestingly, six distinct isoforms of importin α are found in higher eukaryotic cells, which show characteristic tissue-specific expression patterns and altered affinities for distinct NLS cargoes [25]. The functional specialization of the adaptor importin α contributes to an increase in the

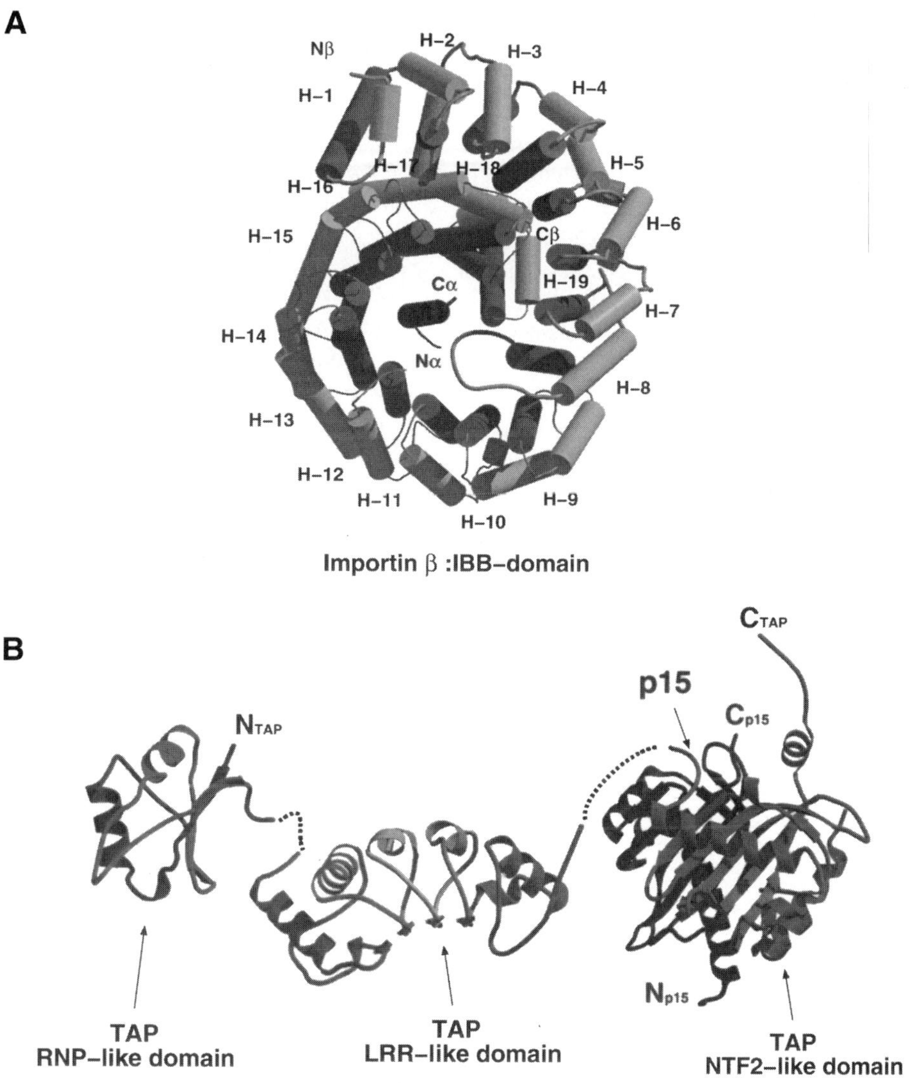

Figure 4 Ribbon representations of nuclear transport receptors. (A) Importin β (1–876) is bound to the IBB domain of importin α. The A-helices of HEAT repeats are located outside the protein, whereas B-helices are arrayed inside. The IBB domain is shown as a rod intimately bound to B-helices of HEAT repeats 7–19. (B) Modular organization of TAP. The model for full-length TAP was created using the structure of TAP fragments reported in [31,34].

range of signals that can be transported by importin β. Furthermore, the definition of receptor and adaptor is not always strict. The transport receptor RanBP7 (Importin β-7), which is involved in nuclear import of ribosomal proteins, also can act as an adaptor for importin β in the import pathway for histone H1 [1].

Export Receptors

A number of export receptors of the importin β superfamily have been described in higher eukaryotes. Although some of these proteins have been extensively characterized by functional and biochemical approaches, no three-dimensional structural information on these proteins has yet been obtained. Export receptors of the importin β superfamily have been described for various cargoes [1–5,26]: tRNAs, dsRNA binding protein, importin α and various proteins bearing leucine-rich NES.

Whereas RanGTP triggers the dissociation of cargo from import receptors of the importin β superfamily, RanGTP has the opposite effect on export receptors of this family, because the binding of RanGTP cooperatively enhances the affinity of the export receptor for the cargo [1–5]. It is possible that export receptors may fully expose the cargo-binding site after binding to RanGTP. Unloading of the cargo in the cytoplasm is mediated by the combined action of RanBP1/RanBP2 (see later discussion) and RanGAP1, which in turn trigger hydrolysis of GTP, disassembly of Ran from the export complex, and subsequent release of the cargo.

Crm1 is the export receptor for cargoes with leucine-rich NESs. Crm1 was originally identified as a protein essential for maintaining chromosome structure in the fission yeast *Schizosaccharomyces pombe*. The interaction between Crm1 and at least some cargoes is stabilized by Ran-binding protein 3 (RanBP3), which binds directly to Crm1 and enhances the affinity for both RanGTP and cargo [27]. Therefore, in some cases the actual export complex moving through the NPC out of the nucleus into the cytoplasm is a quaternary complex containing Crm1, RanBP3, RanGTP, and cargo. Crm1 is also the specific export receptor of snurportin [1], an import adaptor involved in import of m3G-capped U snRNPs. In this case, the interaction with the cargo is not mediated by a short peptide like a leucine-rich NES, but rather by a large domain of the cargo. Moreover, Crm1 binds snurportin with approximately 50-fold higher affinity than the Rev protein and a 5000-fold stronger affinity than the minimum Rev-NES.

Interestingly, the compound leptomycin B (LMB) can block both *in vivo* and *in vitro* Crm1-dependent nuclear export of NES cargoes. The inhibition is caused by the selective alkylation of cysteine 529 [28] of Crm1 by the α, β-unsaturated δ-lactone group of LMB.

Transport Receptors Distinct from the Importin β Superfamily

TAP is part of a larger family of conserved proteins (known as the NXF family) unrelated to importin β [29], which exhibits the two essential properties of an mRNA export receptor: binding to mRNA and to nucleoporins. In the case of human TAP it has been shown that it acts as an export receptor of the *cis*-acting constitutive transport element (CTE) of simian type D retrovirus mRNA and as general mRNA export receptor [30]. Whereas TAP directly binds to the CTE, the binding of TAP to cellular mRNAs is promoted by other unidentified cellular proteins.

Human TAP has a modular structure and consists of at least four structural domains (Fig. 4B) [27]. The minimal CTE-binding fragment has been mapped in the N-terminal domain of the export factor and consists of a ribonucleoprotein fold-like domain (known as RNP or RBD) and a leucine-rich repeat domain (LRR) [31], similar to those originally observed in the ribonuclease inhibitor structure [32]. Both the RNP and LRR domains are required for efficient binding to the CTE element of simian type D retroviral mRNA. In contrast, the mRNA export activity of TAP is directed by the LRR domain and additional cellular proteins such as p15 (NXT1), a transport factor that shares significant homology with NTF2 [33].

A recent crystal structure of the C-terminal half of TAP in complex with the p15 (Fig. 4B) [34] has revealed that the C-terminal domain of TAP contains an NTF-2 like domain, which heterodimerizes with p15, to closely resemble the NTF2 homodimer [35]. NTF2 is a small transport factor involved in the nuclear import of Ran and known to move through the NPC by interaction with nucleoporins. The NTF2-like domain of TAP, but not that of p15, interacts specifically with Phe-Gly (FG)-rich nucleoporins [36], providing a scaffold for the interaction with the NPC components and providing an explanation for the translocation-promoting properties of TAP. In contrast to receptors of the importin β superfamily, TAP does not seem to bind Ran directly in either nucleotide state. However, the TAP-interacting protein p15 can associate with the GTP-bound state of Ran, and this may influence export complex disassembly.

The Small GTPase Ran

Ran Structure

Ran is a small GTPase (~25 kDa) of the Ras superfamily that plays a key role in nucleocytoplasmic transport [1–5,7,8]. Ran contains the characteristic G-domain fold of other GTPases (Fig. 5A), consisting of a six-stranded β-sheet surrounded by five helices. The overall structure of Ran closely resembles other guanine nucleotide-binding proteins involved in several cellular processes such as Ras, Rho, Arf, and Sar1 [37].

Similar to other proteins of the Ras superfamily, Ran undergoes dramatic conformational changes upon hydrolysis of the nucleotide in two regions of the protein that are referred to as switch regions I and II (Fig. 5A). However, unique among all Ras superfamily GTPases, Ran also contains a long C-terminal extension, which consists of a linker without secondary structure followed by a 16-residue-long helix (also named switch III), situated opposite the switch I region. The end of this helix contains a highly acidic C-terminal DEDDDL motif (residues 211–216), whose charge is conserved in Ran for all species examined.

In the GDP-bound state, both switch regions I and II are in a more open conformation and extend far from the nucleotide, whereas the C terminus is packed as an α-helix against the G domain. In contrast, in the GTP-bound state, switch regions I and II close up to interact with the γ-phosphate of the GTP, and the carboxy-terminal tail is flipped away from the G domain (Fig. 5A). In this structural conformation, the binding of Ran to importin β superfamily receptors is about 10^4-fold higher than in the GDP-bound state [37].

Cellular Asymmetry of Ran

A fundamental feature of Ran as GTPase is the intrinsic low rate of GTP hydrolysis in the absence of effectors. As a direct consequence of this weak activity, the cellular distribution of Ran in the GTP- or GDP-bound state strongly reflects the cellular distribution of Ran effectors. The GTP-bound state of Ran occurs predominantly in the nucleoplasm of eukaryotic cells [38], where the guanine nucleotide exchange factor RCC1 is localized and largely bound to chromatin. Conversely, the GDP-bound form of Ran predominates in the cytoplasm, where RanGAP and Ran binding proteins (RanBP) 1 and 2, which stimulate the GTP hydrolysis activity of Ran, are localized. Of note is that the soluble pools of both RanGAP1 and RanBP1 are concentrated in the cytoplasm at steady state due to the presence of a strong NES in each protein. In higher eukaryotes, a major portion of RanGAP1 also is anchored to the cytoplasmic face of the

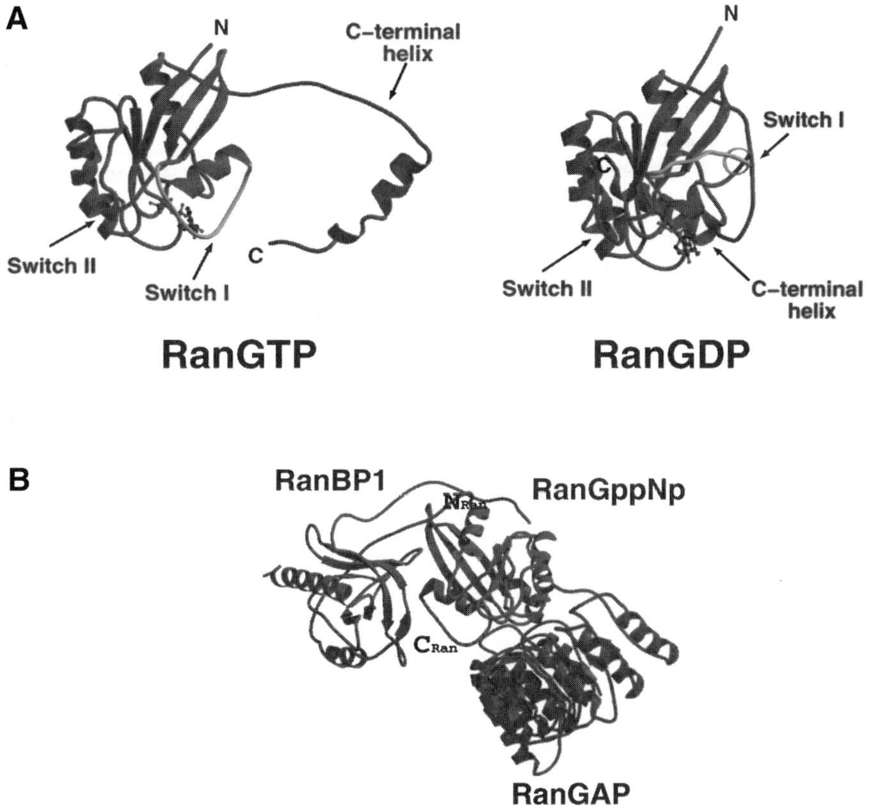

Figure 5 The small GTPase Ran. (A) Conformational changes taking place in Ran in going from a GTP to a GDP-bound state. Switch I and II are indicated by arrows. (B) Structure of RanGppNp complexed to RanBP1 and RanGAP1 (38). The C-terminal arm of Ran, also known as switch III, is flipped away from the G domain and embraces RanBP1.

NPC by the small ubiquitin-related modifier SUMO-1 [1–5]. Similarly, four RanBP1-like modules (termed RanBDs 1–4) are found in the gigantic nucleoporin RanBP2, which localizes at the cytoplasmic face of the NPC. The pool of RanGAP1 and RanBDs localized at the cytoplasmic face of the NPC presumably is involved in the disassembly of export complexes [1–5].

The detailed molecular mechanism of the GTPase stimulation of Ran by the combined action of RanGAP1 and RanBP1 has been revealed by the three-dimensional structure of a Ran–RanBP1–RanGAP ternary complex (Fig. 5B) [39]. RanBP1 does not participate directly in the RanGAP1-stimulated hydrolysis of the GTP by Ran. In fact, the carboxy terminus of Ran wraps around RanBD1 and contacts a basic patch on RanBD1 via its acidic C terminus. This molecular opening of RanGTP facilitates GTP hydrolysis by the GTPase-activating protein RanGAP. However, the molecular action of RanGAP1 is different from all of the GTPase-activating proteins (GAPs) previously characterized, which increases the rate of GTP hydrolysis by insertion of a so-called "arginine finger" into the phosphate-binding pocket. RanGAP1 stimulates the rate of Ran's GTP hydrolysis by stabilizing and positioning the catalytically essential residue glutamine-69 of Ran. In other words, RanGAP1 offers an overall structural scaffold to Ran, but does not intervene directly in the enzymatic mechanisms of catalysis.

Nuclear Import of Ran

In addition to the asymmetric distribution of RanGTP, which is more abundant in the nucleoplasm than in the cytoplasm, much of the total cellular pool of Ran is localized to the nucleoplasm. Similar to other several small proteins, Ran (~25 kDa) could, in principle, diffuse through the nuclear pore complex (see later discussion). Nevertheless, nuclear accumulation of Ran appears to be regulated by the soluble transport factor NTF2 (nuclear transport factor 2, also known as p10), which facilitates the diffusion of Ran by engaging in simultaneous interactions with nucleoporins and RanGDP. In the cell, NTF2 exists as a homodimer, where each monomer is formed by six β-sheets and two α-helices [35]. The tight structure of the NTF2 homodimer opens at one end to form a distinct hydrophobic cavity, forming a potential binding site for certain hydrophobic nucleoporins rich in the FG motif. RanGDP [40] binds the opposite sides of NTF2 homodimer structure, suggesting that NTF2 facilitates the movement of Ran through the pore simply by offering simultaneous binding to nucleoporins and Ran itself. The specificity of NTF2 for RanGDP is also confirmed by the crystal structure of the NTF2–RanGDP complex [40], where switch region II of Ran interacts largely with NTF2. The conformational changes of this region caused by the presence of GTP would sterically impair the binding to NTF2.

Nuclear Pore Complex

The NPC is a large proteinaceous structure spanning the nuclear envelope and providing a passageway for molecular traffic between the cytoplasm and the nucleoplasm [41–44]. The NPC ranges in size from ~125 MDa in *Xenopus* oocyte to ~60 MDa in *S. cerevisiae*. The architecture of the NPC has been extensively investigated by electron microscopy, which reveals a strong eightfold symmetry about the nuclear–cytoplasmic axis (Fig. 6). Three morphologically distinct regions of the vertebrate NPC can be distinguished: (1) the central ring-spoke assembly of 120 × 80 nm that is roughly symmetrical about the pore midplane and that surrounds a central transport channel, (2) flexible ~50-nm-long fibrils emanating into the cytoplasm from the cytoplasmic ring, and (3) a nuclear basket consisting of fibrils joined at their distal ends, which protrudes ~50–100 nm into the nucleoplasm from the nuclear ring. The central channel provides the primary site for restriction of diffusional movement through the NPC, and it is the region where transport receptors move through the NPC. Its exact composition is unknown, and often it appears to contain polymorphic material trapped in the act of being transported.

Based on its molecular mass and dimensions, the NPC originally was thought to contain as many as 100 different proteins (nucleoporins). However, in yeast [45] and mammalian cells [46] recent proteomic analysis has shown that the NPC consists of only about 30 different polypeptides, each of which is repeated multiple times.

Subcomplexes of nucleoporins that are moderately stable to chemical extraction have been biochemically defined including the Nup62 complex [47,48] and the Nup84 complex [49] in metazoans and yeast, respectively. The Nup62 complex contains four proteins, Nup62, Nup58, Nup54, and Nup45, forming a donut-shaped particle that presumably contains one copy of each of the four subunits. The Nup62 complex has been shown to localize on both sides of the NPC near the central channel and appears to be critically involved in nuclear import [50,51].

The existence of subcomplexes of nucleoporins suggests that NPC architecture may be assembled in a hierarchical way. Moreover, it appears that only two membrane-spanning nucleoporins in higher eukaryotes are involved in NPC anchoring to the nuclear membrane. Thus, most NPC organization appears to be based on extensive protein–protein interactions between subcomplexes. These probably remain partially assembled during mitosis in higher eukaryotes, when the NPC disassembles. The molecular mechanisms of NPC assembly and disassembly are not understood, but may be controlled by phosphorylation of nucleoporins, at least in part [41–44].

Several vertebrate nucleoporins are posttranslationally modified with O-linked *N*-acetylglucosamine and directly participate in nuclear protein import. All O-linked glycoproteins contain multiple repeats of the FG motif, which are often integral parts of larger motifs such as FxFG or GLGF. A large body of evidence suggests that these proteins function as binding sites for nuclear transport receptors [1–5,41–44]. FG-rich nucleoporins are also found in yeast and are localized on both cytoplasmic and nuclear faces of the pore complex. In virtue of the eightfold rotational symmetry of the pore and the repetition of FG motifs within the same nucleoporin,

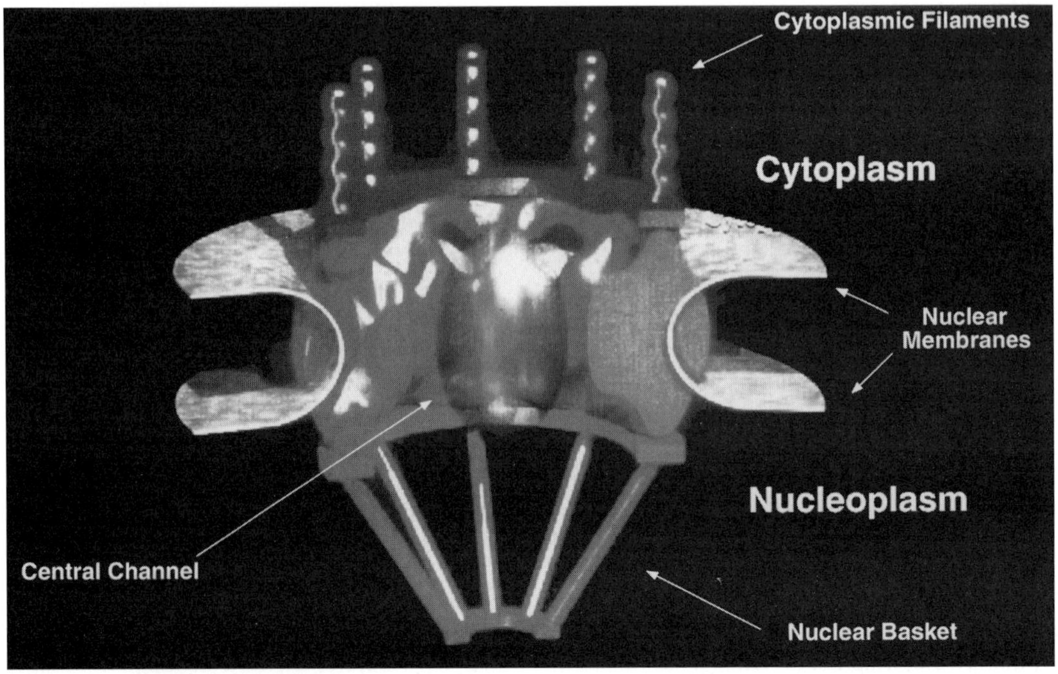

Figure 6 The architecture of the nuclear pore complex in 3D. The volume inside the pore is the location of the central channel, which represents the major diffusional restriction for passage of macromolecules. (Courtesy of Dr. U. Aebi.)

the effective FG-motif concentration in the NPC may be very high.

Mechanism of Transport

Passive and Facilitated Diffusion through the NPC

The NPC provides a selective barrier that restricts the passive diffusion of most proteins and simultaneously acts as an efficient conduit for receptor-mediated translocation of small and large cargoes. Passive diffusion through the NPC has been extensively investigated *in vivo* by cell injection of inert tracers, which suggests that the NPC has an ~10-nm-diameter diffusional channel [41–44]. Passive diffusion through the NPC does not occur at a significant rate for most molecules larger than 20–40 kDa, restricting these molecules to receptor-mediated transport. However, even cargoes that are smaller than 40 kDa often use transport receptors to accelerate their transport rate, which could promote vital processes like DNA replication and/or transcription. In addition to facilitating their rate of transport, transport receptors of the importin β superfamily can serve as chaperones to keep small basic proteins from aggregating [24].

The NPC can support an extraordinary rate of receptor-mediated transport. In an *in vitro* assay where the nuclear import of cargo-free transportin was studied in the absence of competing receptors, a maximum of ~700 transport events NPC^{-1} sec^{-1} was measured [52]. The movement of cargoes through the NPC can occur against a concentration gradient, which results in active accumulation of substrates in either cytoplasm of nucleoplasm. This transport of cargo up a concentration gradient is supported by GTP hydrolysis by Ran, which maintains the nuclear–cytoplasmic RanGTP gradient.

Interestingly, the requirement for GTP hydrolysis by Ran, which is a necessary feature of transport through the NPC *in vivo*, does not appear to be an obligatory step in NPC transit of certain cargoes *in vitro* under conditions in which receptor recycling is not required. Both Crm1-mediated export and importin β import of small protein cargoes [1–5] can proceed in the presence of nonhydrolyzable analogs of GTP. Furthermore, transportin-mediated protein import required neither Ran nor GTP [1–5]. These findings provided strong support for the model that transport receptors can move through the NPC by a facilitated diffusion mechanism, involved repeated binding and dissociation from a series of nucleoporins (see later discussion). However, in contrast to studies with the small cargoes, both Ran and GTP hydrolysis are required for efficient importin β and transportin-mediated import of certain large cargoes [53]. In this case, the presence of RanGTP could promote the facilitated diffusion process by increasing the off-rate of receptor–cargo complexes from nucleoporins [53]. In addition to size, factors such as hydrophobicity [54] and charge of the cargo also can play a role in dictating movement though the NPC. Moreover, different structural conformations of a specific transport receptor when bound to different cargoes may influence the affinity for nucleoporins, which in turn could affect the RanGTP requirement.

Interaction between Transport Receptors and FG-Rich Nucleoporins

Both yeast and higher eukaryotic nucleoporins contain multiple dispersed repeats (up to 30 or more) of the Phe-Gly motif, which provide binding sites for transport receptors of the importin β superfamily as well as for the structurally unrelated TAP. RanGTP is usually required to induce the dissociation of the receptor from nucleoporins during transport when the affinity of the former is substantial. However, even for receptors like transportin that have only a low affinity for nucleoporins, RanGTP is required for transport of large cargoes [55].

Two structural analyses of transport receptors bound to FG-rich peptides were recently reported. In one case, the structure of an N-terminal fragment of human importin β (1–442) was solved in complex with a yeast nucleoporin fragment of Nsp1p containing 5 FxFG repeats [55]. This structure reveals two major FxFG binding sites located on the outer convex surface of the importin β, between HEAT repeats 5–4 and 7–8. The interaction between the transport receptor importin β and the FG peptide is primarily hydrophobic and involves almost exclusively the two Phes of the FxFG core, which are buried in a hydrophobic pocket generated by the A-helices of HEAT repeats 5–6.

In a second case, the structure of a C-terminal NTF2-like domain of the export receptor TAP:p15 heterodimer [34] was solved in complex with a FG-bearing peptide derived from the nucleoporin Nup214. In this case, the interaction is driven by a single Phe residue of the FG peptide, buried inside a highly hydrophobic pocket of TAP. The binding pockets of importin β and TAP interacting with the FG motif are structurally different and arise from distinct classes of transport receptors. Nonetheless, the FG nucleoporin repeat assumes a similar conformation upon binding to the two receptors, mostly likely due to the structural flexibility of the glycine residue and the ability of the phenylalanine residue to insert into hydrophobic pockets. It is possible that FG binding sites on nucleoporins may act synergistically, thus strengthening the affinity of a specific FG-rich nucleoporin for a certain receptor. However, it does not appear that the number of FG repeats in a nucleoporin correlates directly to the affinity for a certain transport receptor, such as importin β. The critical parameter is likely to be in the accessibility of an FG sequence rather than in the number of FG repeats.

Models for Translocation

Cargo movement through the NPC (over a distance of ~200 nm) is thought to occur by the sequential binding and release of the receptor–cargo complex to a series of nucleoporins that extend across the NPC [1–5]. This model raises the problem of how directionality of the movement is achieved [56]. It is clear that one of the key determinants of transport directionality is the compartmentalization of RanGTP and its effectors (see earlier discussion). RanGTP promotes the disassembly of import complexes when they reach the nucleus and also promotes the assembly of export complexes. Moreover, RanGAP and RanBP1/RanBP2 promote the disassembly of export complexes once they reach the cytoplasm.

An additional mechanism to promote transport directionality could involve a progressively higher affinity of a transport receptor for the nucleoporins along its transport pathway. This could promote vectorial transport by guiding the receptor to a terminal binding site in the pathway. To investigate this model, the affinity of the interaction between the transport receptor importin β and several bacterially expressed FG-containing nucleoporins has been analyzed by solid phase binding assays using recombinant nucleoporins [51]. The observation was made that the affinity of importin β for nucleoporins increases progressively in a cytoplasmic to nuclear direction, from a K_d of ~200 n*M* for a nucleoporin in the cytoplasmic fibrils (Nup358) to a K_d of ~10 n*M* for nucleoporin of the nuclear basket (Nup153). It is not known whether there is a similar affinity gradient for other importins, but there may be an affinity gradient in the reverse direction for Crm1 export complexes [57].

As discussed earlier, the lack of an energy requirement for the NPC transit of certain receptor–cargo complexes *in vitro* suggests that receptors can move through the NPC by facilitated diffusion. However this does not explain how the NPC selectively transports only receptor-bound macromolecules, while retaining a diffusion barrier for macromolecules lacking transport signals. To explain the movement of a receptor–cargo complex through the diffusionally restricted central channel of the NPC, Rout *et al.* [45] proposed a Brownian affinity gate translocation model, which suggests that the NPC is not physically gated, but is instead screened by a high concentration of flexible NPC filaments as a result of their Brownian motion. Transport receptors would be concentrated within this FG mesh in virtue of their affinity for FG repeats, in turn, strongly increasing the probability of entering and diffusing through the pore.

In contrast, Ribbeck and Görlich [52] have proposed that the NPC contains a central plug that constitutes a hydrophobic phase in which transport receptors can partition. This hydrophobic phase is proposed to be formed by an internal FG-rich nucleoporin meshwork that is held together by the mutual attraction between FG repeats. Accordingly, the network of cross-linked nucleoporins would give rise to a molecular sieve, allowing only the passage of molecules smaller than the mesh size, and restricting the movement of larger molecules. Transport receptors could "solubilize" the hydrophobic nucleoporin phase by engaging in interactions with FG repeats in nucleoporins, thereby allowing the partitioning of the receptor–cargo complex in the hydrophobic phase of the NPC and permitting translocation through the pore. This model assumes low-affinity interactions among nucleoporins and between nucleoporins and receptors and the physical existence of a hydrophobic nucleoporin phase in the central channel.

Future Directions

The mechanism of nucleocytoplasmic transport is a key problem in cell biology. Because many of the signals, soluble factors, and nucleoporins involved in nuclear transport have now been described, research is entering a new phase that is directed at obtaining a detailed mechanistic understanding of the transport process. To achieve this goal, we need to understand the three-dimensional structure of the NPC in much greater detail. This is technically very challenging, because the NPC has a massive and inherently very flexible architecture. It is likely that analysis of NPC substructures by a combination of X-ray crystallography and three-dimensional reconstruction based on electron microscopy will be needed to achieve this objective. It will be especially important to understand the organization of the central channel, because this is the major site of transport selectivity. In parallel, it will be important to develop new biophysical approaches and models for understanding the individual steps involved in transport of the receptor–cargo complexes between specific nucleoporins. Closely linked to the question of NPC architecture is the issue of how the NPC is assembled, both during interphase and at the end of mitosis. These questions can feasibly be addressed with a combination of genetic studies and *in vivo* and *in vitro* nuclear assembly analyses. Lastly, the nuclear transport machinery is clearly regulated in different functional states of the cells. Understanding how NPC regulation occurs in response to cell signaling and understanding how nuclear transport is integrated with other cellular pathways are major goals for the future.

References

1. Gorlich, D. and Kutay, U. (1999). Transport between the cell nucleus and the cytoplasm. *Annu. Rev. Cell Dev. Biol.* **15**, 607–660.
2. Kuersten, S., Ohno, M., and Mattaj, I. W. (2001). Nucleocytoplasmic transport: Ran, beta and beyond. *Trends Cell Biol.* **12**, 497–503.
3. Macara, I. G. (2001). Transport into and out of the nucleus. *Microbiol. Mol. Biol. Rev.* **65**, 570–594.
4. Bayliss, R., Corbett, A. H., and Stewart, M. (2000). The molecular mechanism of transport of macromolecules through nuclear pore complexes. *Traffic* **1**, 448–456.
5. Ohno, M., Fornerod, M., and Mattaj, I. W. (1998). Nucleocytoplasmic transport: The last 200 nanometers. *Cell* **92**, 327–336.
6. Adam, S. A., Sterne-Marr, R., and Gerace, L. (1992). Nuclear protein import using digitonin-permeabilized cells. *Methods Enzymol.* **219**, 97–110.
7. Conti, E. and Izaurralde, E. (2001). Nucleocytoplasmic transport enters the atomic age. *Curr. Opin. Cell Biol.* **13**, 310–319.
8. Chook, Y. M. and Blobel, G. (2001). Karyopherins and nuclear import. *Curr. Opin. Struct. Biol.* **11**, 703–715.
9. Jans, D. A., Xiao, C. Y., and Lam, M. H. (2000). Nuclear targeting signal recognition: A key control point in nuclear transport? *Bioessays* **22**, 532–544.
10. Conti, E., Uy, M., Leighton, L., Blobel G., and Kuriyan, J. (1998). Crystallographic analysis of the recognition of a nuclear localization signal by the nuclear import factor karyopherin alpha. *Cell* **94**, 193–204.
11. Conti, E. and Kuriyan, J. (2000). Crystallographic analysis of the specific yet versatile recognition of distinct nuclear localization signals by karyopherin alpha. *Structure Fold. Des.* **8**, 329–338.

12. Hodel, M. R., Corbett, A. H., and Hodel, A. E. (2001). Dissection of a nuclear localization signal. *J. Biol. Chem.* **276**, 1317–1325.
13. Cingolani, G., Petosa, C., Weis, K., and Muller, C. W. (1999). Structure of importin-beta bound to the IBB domain of importin-alpha. *Nature* **399**, 221–229.
14. Cingolani, G., Lashuel, H. A., Gerace, L., and Muller, C. W. (2000). Nuclear import factors importin alpha and importin beta undergo mutually induced conformational changes upon association. *FEBs Lett.* **484**, 291–298.
15. Kobe, B. (1999). Autoinhibition by an internal nuclear localization signal revealed by the crystal structure of mammalian importin alpha. *Nat. Struct. Biol.* **6**, 388–397.
16. Jullien, D., Gorlich, D., Laemmli, U. K., and Adachi, Y. (1999). Nuclear import of RPA in Xenopus egg extracts requires a novel protein XRIPalpha but not importin alpha. *EMBO J.* **18**, 4348–4358.
17. Lam, M. H., Briggs, L. J., Hu, W., Martin, T. J., Gillespie, M. T., and Jans, D. A. (1999). Importin beta recognizes parathyroid hormone-related protein with high affinity and mediates its nuclear import in the absence of importin alpha. *J. Biol. Chem.* **274**, 7391–7398.
18. Jakel, S. and Gorlich, D. (1998). Importin beta, Transportin, RanBP5 and RanBP7 mediate nuclear import of ribosomal proteins in mammalian cells. *EMBO J.* **17**, 4491–4502.
19. Klemm, J. D., Beals, C. R., and Crabtree, G. R. (1997). Rapid targeting of nuclear proteins to the cytoplasm. *Curr. Biol.* **7**, 638–644.
20. Vetter, I. R., Arndt, A., Kutay, U., Gorlich, D., and Wittinghofer, A. (1999). Structural view of the Ran-Importin beta interaction at 2.3 A resolution. *Cell* **97**, 635–646.
21. Chook, Y. M. and Blobel, G. (1999). Structure of the nuclear transport complex karyopherin-beta2-Ran x GppNHp. *Nature* **399**, 230–236.
22. Plafker, S. M. and Macara, I. G. (2000). Importin-11, a nuclear import receptor for the ubiquitin-conjugating enzyme, UbcM2. *EMBO J.* **19**, 5502–5513.
23. Mingot, J. M., Kostka, S., Kraft, R., Hartmann, E., and Gorlich, D. (2001). Importin 13: A novel mediator of nuclear import and export. *EMBO J.* **20**, 3685–3694.
24. Jakel, S., Mingot, J. M., Schwarzmaier, P., Hartmann, E., and Gorlich, D. (2002). Importins fulfills a dual function as nuclear import receptors and cytoplasmic chaperones for exposed basic domains. *EMBO J.* **21**, 377–386.
25. Kohler, M., Speck, C., Christiansen, M., Bischoff, F. R., Prehn, S., Haller, H., Gorlich, D., and Hartmann, E. (1999). Evidence for distinct substrate specificities of importin alpha family members in nuclear protein import. *Mol. Cell Biol.* **19**, 7782–7791.
26. Brownawell, A. M. and Macara, I. G. (2002). Exportin-5, a novel karyopherin, mediates nuclear export of double-stranded RNA binding proteins. *J. Cell Biol.* **156**, 53–64.
27. Lindsay, M. E., Holaska, J. M., Welch, K., Paschal, B. M., and Macara, I. G. (2001). Ran-binding protein 3 is a cofactor for Crm1-mediated nuclear protein export. *J. Cell Biol.* **153**, 1391–1402.
28. Kudo, N., Matsumori, N., Taoka, H., Fujiwara, D., Schreiner, E. P., Wolff, B., Yoshida, M., and Horinouchi, S. (1999). Leptomycin B inactivates CRM1/exportin 1 by covalent modification at a cysteine residue in the central conserved region. *Proc. Natl. Acad. Sci. USA* **96**, 9112–9117.
29. Herold, A., Suyama, M., Rodrigues, J. P., Braun, I. C., Kutay, U., Carmo-Fonseca, M., Bork, P., and Izaurralde, E. (2000). TAP (NXF1) belongs to a multigene family of putative RNA export factors with a conserved modular architecture. *Mol. Cell Biol.* **20**, 8996–9008.
30. Gruter, P., Tabernero, C., von Kobbe, C., Schmitt, C., Saavedra, C., Bachi, A., Wilm, M., Felber, B. K., and Izaurralde, E. (1998). TAP, the human homolog of Mex67p, mediates CTE-dependent RNA export from the nucleus. *Mol. Cell.* **1**, 649–659.
31. Liker, E., Fernandez, E., Izaurralde, E., and Conti, E. (2000). The structure of the mRNA export factor TAP reveals a cis arrangement of a non-canonical RNP domain and an LRR domain. *EMBO J.* **19**, 5587–5598.
32. Kobe, B. and Kajava, A. V. (2001). The leucine-rich repeat as a protein recognition motif. *Curr. Opin. Struct. Biol.* **11**, 725–732.
33. Black, B. E., Levesque, L., Holaska, J. M., Wood, T. C., and Paschal, B. M. (1999). Identification of an NTF2-related factor that binds Ran-GTP and regulates nuclear protein export. *Mol. Cell Biol.* **19**, 8616–8624.
34. Fribourg, S., Braun, I. C., Izaurralde, E., and Conti, E. (2001). Structural basis for the recognition of a nucleoporin FG repeat by the NTF2-like domain of the TAP/p15 mRNA nuclear export factor. *Mol. Cell.* **8**, 645–656.
35. Bullock, T. L., Clarkson, W. D., Kent, H. M., and Stewart, M. (1996). The 1.6 angstroms resolution crystal structure of nuclear transport factor 2 (NTF2). *J. Mol. Biol.* **260**, 422–431.
36. Braun, I. C., Herold, A., Rode, M., Conti, E., and Izaurralde E. (2001). Overexpression of TAP/p15 heterodimers bypasses nuclear retention and stimulates nuclear mRNA export. *J. Biol. Chem.* **276**, 20536–2043.
37. Vetter, I. R. and Wittinghofer, A. (2002). The guanine nucleotide-binding switch in three dimensions. *Science* **294**, 1299–1304.
38. Kalab, P., Weis, K., and Heald, R. (2002). Visualization of a Ran-GTP gradient in interphase and mitotic Xenopus egg extracts. *Science* **295**, 2452–2456.
39. Seewald, M. J., Korner, C., Wittinghofer, A., and Vetter, I. R. RanGAP mediates GTP hydrolysis without an arginine finger. *Nature* **415**, 662–666.
40. Stewart, M., Kent, H. M., and McCoy, A. J. (1998). Structural basis for molecular recognition between nuclear transport factor 2 (NTF2) and the GDP-bound form of the Ras-family GTPase Ran. *J. Mol. Biol.* **277**, 635–646.
41. Vasu, S. K. and Forbes, D. J. (2001). Nuclear pores and nuclear assembly. *Curr. Opin. Cell Biol.* **13**, 363–375.
42. Fahrenkrog, B., Stoffler, D., and Aebi, U. (2001). Nuclear pore complex architecture and functional dynamics. *Curr. Top. Microbiol. Immunol.* **259**, 95–117.
43. Stoffler, D., Fahrenkrog, B., and Aebi, U. (1999). The nuclear pore complex: From molecular architecture to functional dynamics. *Curr. Opin. Cell Biol.* **11**, 391–401.
44. Ryan, K. J. and Wente, S. R. (2000). The nuclear pore complex: A protein machine bridging the nucleus and cytoplasm. *Curr. Opin. Cell Biol.* **12**, 361–371.
45. Rout, M. P., Aitchison, J. D., Suprapto, A., Hjertaas, K., Zhao, Y., and Chait, B. T. (2000). The yeast nuclear pore complex: Composition, architecture, and transport mechanism. *J. Cell Biol.* **148**, 635–651.
46. Cronshaw, J. M., Krutchinsky, A. N., Zhang, W., Chait, B. T., and Matunis, M. J. (2002). Proteomic analysis of the mammalian nuclear pore complex. *J Cell Biol.* **158**, 915–927.
47. Guan, T., Muller, S., Klier, G., Pante, N., Blevitt, J. M., Haner, M., Paschal, B., Aebi, U., and Gerace, L. (1995). Structural analysis of the p62 complex, an assembly of O-linked glycoproteins that localizes near the central gated channel of the nuclear pore complex. *Mol. Biol. Cell.* **6**, 1591–1603.
48. Hu, T., Guan, T., and Gerace, L. (1996). Molecular and functional characterization of the p62 complex, an assembly of nuclear pore complex glycoproteins. *J. Cell Biol.* **134**, 589–601.
49. Lutzmann, M., Kunze, R., Buerer, A., Aebi, U., and Hurt, E. (2002). Modular self-assembly of a Y-shaped multiprotein complex from seven nucleoporins. *EMBO J.* **21**, 387–397.
50. Meier, E., Miller, B. R., and Forbes, D. J. (1995). Nuclear pore complex assembly studied with a biochemical assay for annulate lamellae formation. *J. Cell Biol.* **129**, 1459–1472.
51. Ben-Efraim, I. and Gerace, L. (2001). Gradient of increasing affinity of importin beta for nucleoporins along the pathway of nuclear import. *J. Cell Biol.* **152**, 411–417.
52. Ribbeck, K. and Gorlich, D. (2001). Kinetic analysis of translocation through nuclear pore complexes. *EMBO J.* **20**, 1320–1330.
53. Lyman, S. K., Guan, T., Bednenko, J., Wodrich, H., and Gerace, L. (2002). Influence of cargo size on Ran and energy requirements for nuclear protein import. *J. Cell Biol.* **159**, 55–67.

54. Ribbeck, K. and Gorlich, D. (2002). The permeability barrier of nuclear pore complexes appears to operate via hydrophobic exclusion. *EMBO J.* **21**, 2664–2671.
55. Bayliss, R., Littlewood, T., and Stewart, M. (2000). Structural basis for the interaction between FxFG nucleoporin repeats and importin-beta in nuclear trafficking. *Cell* **102**, 99–108.
56. Rabut, G. and Ellenberg, J. (2001). Nucleocytoplasmic transport: Diffusion channel or phase transition? *Curr. Biol.* **11**, 551–554.
57. Kehlenbach, R. H., Dickmanns, A., Kehlenbach, A., Guan, T., and Gerace, L. (1999). A role for RanBP1 in the release of CRM1 from the nuclear pore complex in a terminal step of nuclear export. *J. Cell Biol.* **145**, 645–657.

CHAPTER 331

Apostosis Signaling: A Means to an End

Lisa J. Pagliari, Michael J. Pinkoski, and Douglas R. Green
Division of Cellular Immunology, La Jolla Institute for Allergy and Immunology, San Diego, California

Introduction

Programmed cell death is a fundamental biological process of all multicellular organisms and plays important roles in tissue homeostasis, host defense, development, metamorphosis, and morphogenesis. In animals, programmed cell death occurs via apoptosis, a morphologically defined form of cell death that has a number of biochemical features. In addition to its physiological roles, apoptosis contributes to several pathological conditions, such as cancer, AIDS, aging, and cardiovascular, neurodegenerative, and autoimmune diseases. The mechanisms that govern a cellular decision to live or die are complex and tightly regulated by a plethora of molecules with distinct roles in the signaling process. Generally, cell death occurs after an initial apoptotic signal spurs a cascade of subsequent events from which a cell cannot recover.

Apoptosis is defined by its morphological features of membrane blebbing, cellular shrinkage, and chromosomal condensation. Endonuclease activation that results in a characteristic 200-bp nucleosomal DNA ladder is a common feature of apoptosis, although not definitive [1]. Intense research has indicated that the cellular events leading to apoptosis are complex and varied, often depending on the cell type and stimulus utilized. However, several aspects of these death pathways are common among various stimuli and cell types.

The End of the Road

It is the cleavage of key cellular substrates and not a general proteolytic digestion that orchestrates the morphological and biochemical changes that characterize cell death by apoptosis. The degradative phase of apoptosis is mediated by a highly conserved family of cysteine proteases (caspases) that cleave specific proteins, including other caspases, at the C-terminal end of aspartic acid residues [2]. Figure 1 classifies caspases 1–10 and depicts several structural motifs. Most, if not all, caspases exist intracellularly as inactive zymogens. Following an apoptotic signal, the inactive zymogen is cleaved between what will be the large and small subunits of the mature enzyme to generate an active caspase [3,4]. Caspases-3, -6, and -7 execute the degradative events and are, therefore, classified as effector caspases [2]. The inactive proforms of these appear to preexist as dimers, which can only be activated by proteolytic cleavage to create a mature executioner caspase with two active sites [5]. The activation of an executioner caspase by proteolytic cleavage is illustrated in Fig. 2.

The active executioner caspases then cleave key substrates in the cell to promote apoptosis. For example, a complex of a nuclease and its chaperone/inhibitor is activated when caspase-3 cleaves the inhibitor of CAD (iCAD), releasing the nuclease CAD (caspase-activated DNase), to cleave DNA [6]. Activated caspases also promote the blebbing of a dying cell through the cleavage and activation of several molecules, including gelsolin, p21-activated kinase, and ROCK-1 [7,8]. Perhaps most importantly, caspases promote the exposure of phagocytosis markers on the surface of a dying cell, such as the externalization of phosphatidylserine [9], which binds to specific receptors on phagocytic cells. However, exactly how caspases induce phosphatidylserine externalization remains unknown. Active caspases are therefore crucial for this specialized form of cell death that ensures that the dying cells will be packed and

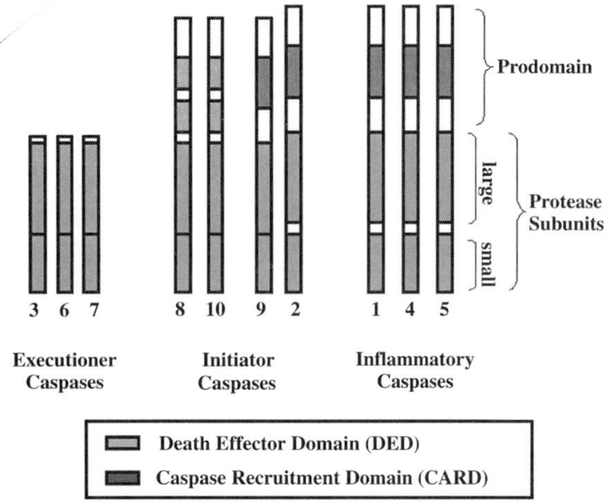

Figure 1 Caspase classification. Caspases 1–10 can be divided into three categories based on function and structural composition. Executioner caspases-3, -6, and -7 contain the large and small protease subunits and a small prodomain. In contrast, initiator caspases, such as caspases-8, -9, and -10, possess large prodomains consisting of protein–protein interaction motifs, including death effector domains (DEDs) and caspase recruitment domains (CARDs). Caspases-1, -4, and -5 also contain CARDs within large prodomains, but are termed inflammatory caspases for their role in inflammation. Caspases are cleaved after aspartic acid residues that are located between the prodomain and the protease subunits and the large and small protease subunits.

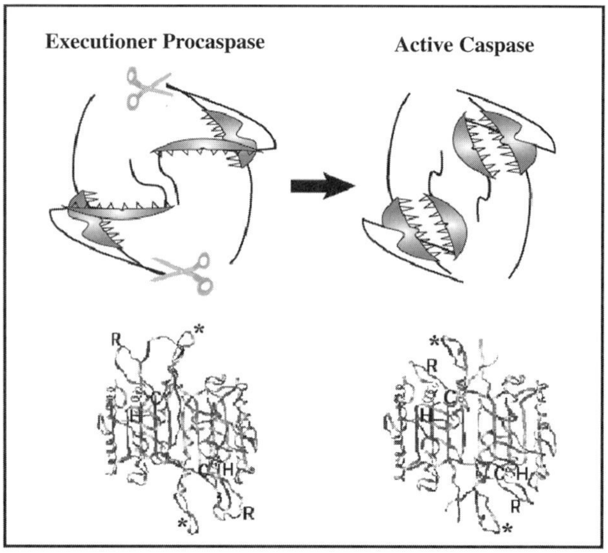

Figure 2 Activation of executioner caspases. Executioner caspases exist as inactive dimers that are activated following cleavage between the small and large subunits of each monomer. Although separated, the large and small subunits remain intimately associated and both contribute residues to the active site that are necessary for substrate binding and proteolysis. The drawing depicts where cleavage between the subunits occurs and the resultant conformational change. The crystal structure of an executioner caspase (caspase-7) shows the formation of the active sites (represented by asterisks) that occurs following cleavage. The amino acids critical for the active site (C, H) and for the aspartate specificity (R) of the protease are indicated.

marked for clearance by professional phagocytes, including macrophages and surrounding epithelial cells, to avoid the induction of an inflammatory response by released intracellular molecules.

Although inhibition of caspase activity prevents many phenotypic characteristics of apoptosis, such as DNA fragmentation, cell death may still ensue in the absence of caspase activation [10–12]. Caspase-independent cell death proceeds following the loss of mitochondrial function and may involve the release of alternative death-inducing molecules, suggesting that mitochondrial damage may signify a "point of no return" for dying cells [10–12].

A family of naturally occurring inhibitors of apoptosis proteins (IAPs) blocks the activity of caspases and may target active caspases for degradation, thereby averting apoptosis at critical commitment steps of the pathway [13]. IAP orthologs have been identified in several species, including yeast, nematodes, and flies. The mammalian family members c-IAP1 and 2 function to inhibit the activation of caspase-8, an initiator caspase responsible for the cleavage and activation of downstream effector caspases. X-linked IAP (XIAP) inhibits both effector caspase-3 and caspase-9, another initiator caspase (Fig. 3). Whereas certain family members appear to directly bind specific caspases (namely, caspases-3, -7, and -9), the mechanism by which IAPs inhibit caspase enzymatic activity has yet to be clarified. Some of the IAPs (such as yeast Bir-1 and possibly mammalian survivin) function in cell cycle regulation rather than by controlling caspases or apoptosis [14].

The initiator caspases bear long prodomains containing one or more protein interaction motifs (Fig. 1) that solicit binding by adaptor molecules and subsequent activation by dimerization [2]. The process through which initiator caspases are activated often characterizes the pathways that lead to apoptosis. There are two general types of initiator caspase activation that are best understood: one is activation by receptor ligation and another occurs following the release of mitochondrial intermembrane proteins into the cytosol. The initiator caspases that are predominantly involved in these two pathways are caspase-8 and -9, respectively.

Caspase-8 Activation via Death Receptors

The death receptors are a subset of the tumor necrosis factor receptor (TNFR) superfamily and include Fas (CD95), TNFR1, TRAIL receptors-1 and -2, and death receptor-3 (DR3). These relay extracellular apoptotic signals through binding of specific ligands to trimeric receptors [15]. Death domains (DDs) found in the cytoplasmic portion of the death receptor interact with DDs on adaptor molecules, namely, Fas-associated death domain (FADD) or TNFR-associated death domain (TRADD), through homotypic interactions (Fig. 3). FADD also associates with TRADD by DD interactions, providing a common scaffold for death receptor-induced caspase activation. Engagement of DDs on FADD exposes another domain termed the death effector domain (DED) that interacts with DEDs present in the prodomains

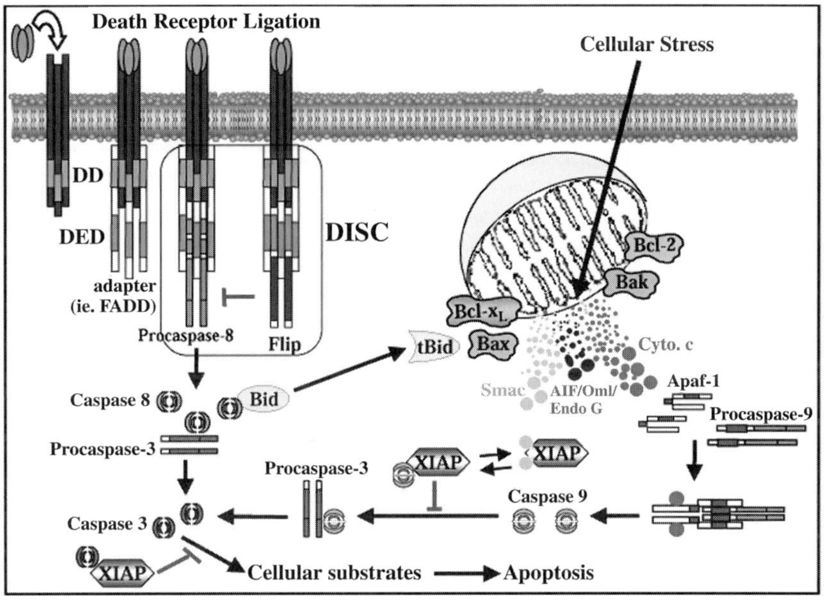

Figure 3 Apoptotic signaling from two distinct pathways. The death receptor (i.e., Fas) pathway involves ligand binding, recruitment of adaptor proteins (i.e., FADD) through death domain (DD) interactions, binding of procaspase-8 through death effector domain (DED) interactions, and proteolytic activation of the initiator caspase-8. c-Flip expression blocks caspase-8 activation by inhibiting binding with FADD at the DISC. Caspase-8 may directly activate executioner caspase-3 or may cleave Bid to target the mitochondria. The mitochondria are regulated by the pro-apoptotic (Bid, Bax, Bak) and anti-apoptotic (Bcl-2, Bcl-x_L) Bcl-2 family proteins, and disruption of the mitochondrial membrane results in the release of cytochrome c, Smac/Diablo, AIF, Omi/Htra2, and Endo G. The binding of cytochrome c to Apaf-1 and recruitment of procaspase-9 forms the apoptosome, which allows aggregation and activation of caspase-9. Caspase-9 then activates caspase-3 (and caspase-7) to cleave key cellular substrates and dismantle the dying cell. Proteolytic activity of caspase-9 or -3 may be inhibited by XIAP and cytosolic Smac reverses this inhibition upon binding XIAP.

of procaspases-8 and -10. Although distinct in sequence, the structures of the DD and DED are very similar, forming a so-called "death fold." The result is the formation of a DISC (death-inducing signaling complex), which leads to rapid aggregation and activation of caspase-8 or -10 [16]. When two procaspase-8 molecules are brought together, each cleaves after the aspartic acid residues in the other's chain, separating the large and small subunits, removing the prodomain, and forming a mature and fully active caspase with two active sites [17]. It is likely that caspase-10, a similar caspase, is activated in this way as well [18]. The mature initiator caspases can now cleave and activate the executioner caspases to promote apoptosis.

The regulation of the DISC is complex and not fully understood. A naturally occurring inhibitor of DISC formation, termed c-Flip (also known as Casper, Clarp, Flame-1, I-Flice, Cash, Ursurpin, and Mrit), impedes the binding of procaspase-8 to FADD, thereby preventing apoptosis induced by death receptors [19] (Fig. 3). Decreased Flip expression sensitizes cells to Fas and TNFα-induced apoptosis; however, the precise mechanism by which this occurs remains to be elucidated [20–22]. Evidence suggests that Flip is a caspase regulator that may also function to promote caspase activation by facilitating caspase aggregation under some circumstances [23]. It appears that the level of Flip expression, relative to other members of the DISC, may determine its role in apoptosis.

Cells that are resistant to apoptosis induced by the death ligands TNFα, Fas ligand, or TRAIL can often be sensitized by inhibiting macromolecular synthesis (e.g., by addition of actinomycin D or cycloheximide), indicating that an active pathway of protection exists. In the case of TNFR1, the receptor induces both an apoptotic signal through TRADD and a survival signal through the activation of the transcription factor nuclear factor κB (NFκB) [24]. Although the inhibition of NFκB often sensitizes cells to death by TNFα, exactly how NFκB activation blocks TNFα-induced apoptosis remains unclear. The ability of NFκB to block TNFR1 signaling may be in part through the expression of cIAP1 and cIAP2, which bind to TNFR1 adaptor molecules [25]. Moreover, NFκB has recently been implicated in the regulation of c-Jun N-terminal kinase (JNK) activity, which may play a role in TNFα-induced apoptosis [26,27]. Although NFκB can interfere with apoptosis in some cases, in others, such as Fas-induced apoptosis [28], it does not. There are also instances in which NFκB activation actually enhances apoptosis, indicating that the response to NFκB activation may be stimulus specific [29]. The balance between signals for survival versus death appears to be crucial for determining a cell's ultimate fate.

Cells can be loosely categorized by the apoptotic pathway employed following death receptor ligation [30] (Fig. 3). In short, ligation of a death receptor triggers apoptosis through activation of initiator caspases-8 or -10, which can cleave and activate executioner caspases that orchestrate the death of the cell. Alternatively, activated caspase-8 may cleave and activate the pro-apoptotic Bcl-2 family protein Bid [31,32], which induces mitochondrial membrane permeabilization and the release of apoptotic factors from mitochondria [33] (discussed later). Certain cell types, termed type 1 cells, generate effective DISC formation and rapid, abundant caspase-8 cleavage that can result in direct activation of caspase-3. In contrast, type 2 cells form small amounts of DISC with only slight caspase-8 activation, necessitating an amplification of the apoptotic signal through Bid cleavage and disruption of the mitochondrial outer membrane [30].

Mitochondria and the Activation of Caspase-9

The activation of the second class of initiator caspase, caspase-9, represents a fundamentally different pathway from that involving caspase-8. Procaspase-9 also has a large prodomain containing a protein interaction domain called a CARD (caspase recruitment domain) (Fig. 1) that structurally resembles DD and DED and also forms a death fold [34]. During its activation, the prodomain of caspase-9 is not removed. Unlike the other caspases discussed, cleavage between the large and small subunits is not necessary for activation of procaspase-9 [35]. Instead, procaspase-9 appears to be activated by binding to an adaptor, Apaf-1 (apoptotic protease activating factor-1) [36].

Inactive Apaf-1 is present as a monomer in the cytosol. Cytochrome c released from mitochondria binds the inert Apaf-1 and promotes Apaf-1 oligomerization and activation [37]. This is illustrated in Fig. 4. In its active configuration, Apaf-1 forms a complex of seven Apaf-1 molecules with exposed CARDs in the central "hub." The CARDs of Apaf-1 bind the CARDs within the caspase-9 prodomains, allowing for aggregation and activation of the protease, which must remain associated to be active [38]. This complex has been termed the *apoptosome* and its formation leads to caspase-9 multimerization and a conformational change that permits only one of the two active sites in the caspase-9 tetramer to be active at a time [39]. Evidently, the conformational change necessary for the formation of a specificity-determining groove pulls the other active site out of alignment (Fig. 4). It is proposed that the binding of Apaf-1 to the prodomain of caspase-9 allows this conformational change to occur. Once caspase-9 is activated, it cleaves and activates executioner caspase-3 to induce the degradative events of apoptosis.

The release of cytochrome c from mitochondria is a critical event for apoptosome formation [36]. Cellular stresses, such as DNA damage, cytoskeletal damage, and metabolic disruption, induce pro-apoptotic Bcl-2 family molecules (discussed later) to compromise the barrier functions of the mitochondrial outer membrane, ultimately leading to the dissipation of the inner transmembrane potential and release

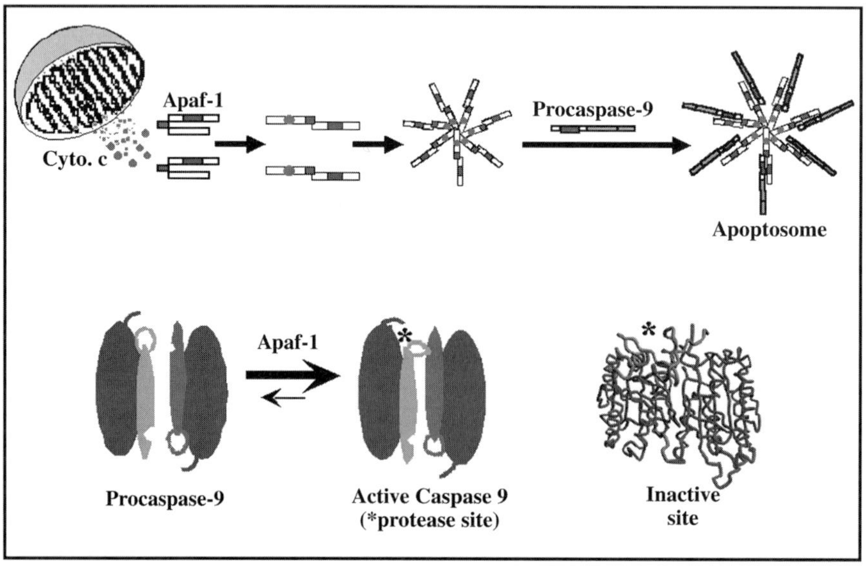

Figure 4 Apoptosome formation and activation of caspase-9. Cytochrome c (cyto. c) released from mitochondria binds to inactive Apaf-1 and induces a conformational change and oligomerization of seven Apaf-1 molecules. This assembly exposes the caspase recruitment domain (CARD) of Apaf-1 and allows interaction of Apaf-1 CARD with the CARD of procaspase-9. The resultant formation is termed an apoptosome. Within the apoptosome, bound procaspase-9 is cleaved between the large and small subunits, which remain associated and form a proteolytic active site. Multimerization of caspase-9 induces a conformational change that allows for only one proteolytic site to be active at a time (identified by the asterisk).

of cytochrome c [40]. Normally found in the intermembrane space of the mitochondria, cytochrome c must be released into the cytosol to induce apoptosome formation. Embryonic cells lacking cytochrome c did not undergo apoptosis in response to several inducers, demonstrating the importance for cytochrome c in the mitochondrial death pathway [41]. Cells from mice deficient in either Apaf-1 or caspase-9 display similar phenotypes (resistance to a variety of apoptotic stimuli) to cells lacking cytochrome c, suggesting that these three molecules serve a cooperative function in this pathway of apoptosis [42,43].

In addition to the release of cytochrome c, other mitochondrial proteins, including, Smac (second mitochondria-derived activator of caspase)/Diablo (direct IAP binding protein with low pI), Endo G (endonuclease G), Omi/Htra2, and AIF (apoptosis-inducing factor), are expelled into the cytosol following mitochondrial membrane disruption and potentially play a role in apoptosis. Because the function of cytochrome c in apoptosis appears to be fundamentally distinct from its role in electron transport, it is possible that several of the basic mitochondrial functions necessary for cell survival may also signal apoptosis when mitochondria are disrupted. Thus, various molecules associated with mitochondria may serve dual roles to maintain cellular function as well as to induce death, depending on cellular conditions. For instance, AIF, which possesses redox activity in the mitochondria, translocates to the nucleus and may promote DNA damage and cell death via a caspase-independent mechanism [44,45]. Endo G, an endonuclease potentially involved in mitochondrial replication, is released from mitochondria and has been suggested to mediate nuclear DNA damage in the absence of caspases [46]. Similarly, Omi/HtrA2, a mitochondrial serine protease, may promote caspase-independent cell death via its protease activity (distinct from that of caspases) when released into the cytosol [47]. These proteins, along with cytochrome c, are conserved from mammals to yeast and apparently retain a potential to kill cells independent of their role in mitochondria function.

While some of the death-inducing factors released from the mitochondria contribute to caspase-independent cell death, others promote caspase activation to enhance apoptosis. Both Smac/Diablo [48,49] and Omi/HtrA2 [47,50–52] can block IAP-mediated caspase inhibition by binding XIAP via an AVP(I/S) sequence at the N terminus and disrupting the association of caspase-9 with XIAP (Fig. 3). Thus, these mitochondrial proteins function in the cytosol to negate IAPs' anti-apoptotic function. Moreover, Smac/Diablo may serve an additional pro-apoptotic role that is independent from its ability to bind IAPs [53].

Mitochondrial Outer Membrane Permeabilization

The coincidental release of proteins from the mitochondrial intermembrane space during apoptosis suggests the occurrence of a mitochondrial outer membrane permeabilization. Although the onset of cytochrome c release may vary depending on the initial apoptotic stimulus utilized, once release occurs, it is rapid and complete [54]. In most instances, the molecules discussed earlier (e.g., Smac/Diablo, AIF) are also expelled into the cytosol with cytochrome c; however, the mechanism for this event remains unclear and it is possible that selective release can occur.

Essentially three models (reviewed in [55]) account for outer membrane permeabilization during apoptosis: (1) the pore-forming model in which permeabilization is mediated by a change in the outer membrane, permitting protein release without involving the inner membrane [33]; (2) the permeability transition model, in which an opening of the adenine nucleotide transporter (ANT) in the inner membrane causes matrix swelling, leading to outer membrane disruption [56]; and (3) the voltage-dependent anion channel (VDAC) closure model, in which metabolic signals trigger a closure of VDAC in the outer membrane, resulting in inner membrane perturbations that cause matrix swelling and outer membrane disruption [57]. Models 2 and 3 propose an active role for mitochondria in apoptosis, whereas model 1 defines this organelle as a repository of apoptogenic factors that are released by the actions of mediators on the outer membrane (discussed later). However, the function of the electron transport chain, which is essential for most mitochondrial functions including ATP generation, is lost following permeabilization, and this loss is greatly facilitated by caspase activation [54,58]. Mitochondrial dysfunction may have a fundamental role in apoptotic cell death, but the extent to which the loss of mitochondrial function versus the activity of death-promoting proteins accounts for the death of cells following mitochondrial outer membrane permeabilization is currently unresolved.

The Bcl-2 Family

Although the precise nature of mitochondrial membrane permeabilization remains elusive, some of the factors controlling it are beginning to be understood. A group of related molecules known as the Bcl-2 family proteins can target the mitochondria and regulate membrane permeabilization [59]. Bcl-2-related proteins are categorized by their ability to either induce or inhibit apoptosis. Apoptotic inducers (Bax, Bad, Bak, Bik, Bok, Bim, Bip, Bid, Diva, Hrk, or Blk) and protectors (Bcl-x_L, Mcl-1, A1, and Bcl-w) that share homologous regions (BH domains) with anti-apoptotic Bcl-2 have been identified (Fig. 5) [60]. Anti-apoptotic members Bcl-2 and Bcl-x_L prevent mitochondrial outer membrane permeabilization, whereas pro-apoptotic members Bax and Bak produce it [59]. Thus, the balance of anti-apoptotic and pro-apoptotic Bcl-2 family members may be responsible for maintaining membrane integrity.

The pro-apoptotic Bcl-2 family members have been further divided into a subset termed *BH3-only* proteins (Bid, Bim, Bik, Blk, Bmf, Bad, Hrk, BNIP3, Puma, Noxa) that contain only the BH3 of the four BH domains and may function as "sensors" to assess the status of death or survival signals.

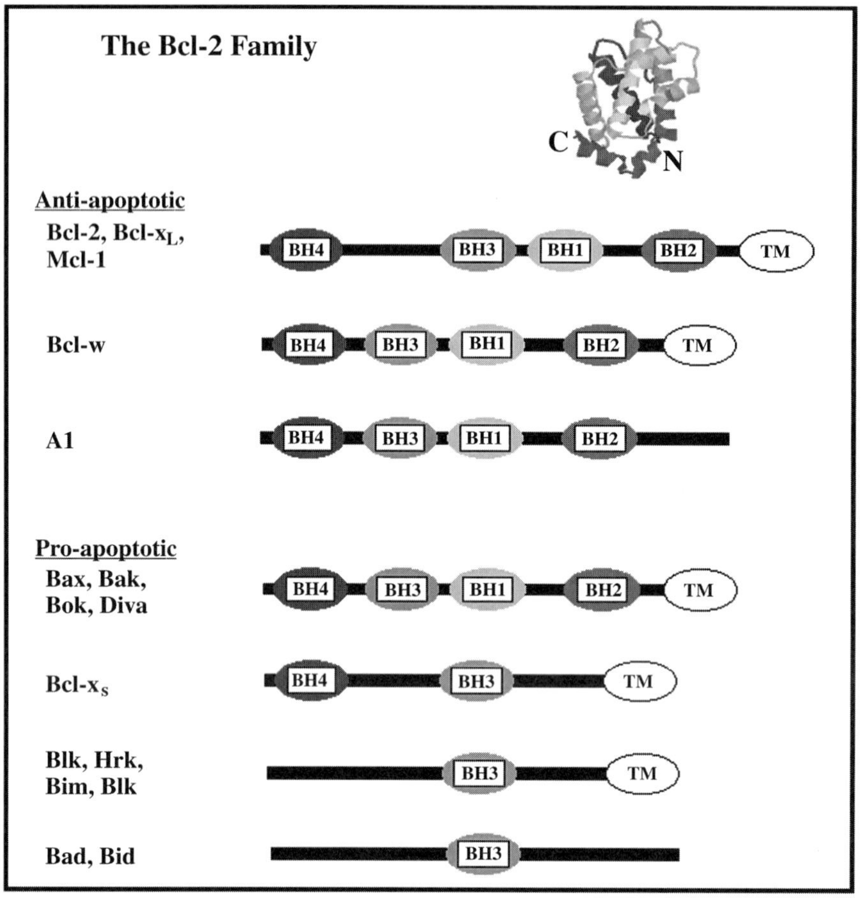

Figure 5 The Bcl-2 family. The proteins belonging to the Bcl-2 family are related based on the presence of Bcl-2 homology (BH) domains, but are categorized by function. The family members are divided based on their ability to inhibit (anti-apoptotic) or induce (pro-apoptotic) apoptosis. The Bcl-2 family is further sorted by the presence or absence of the four BH domains. For instance, the BH3-only proteins are a subdivision of the pro-apoptotic molecules that contain only the BH3 domain. Several of the family members also contain a hydrophobic transmembrane (TM) domain that allows for membrane insertion and localization to the mitochondria. The presence of common BH domains allows for oligomerization of family members, including binding between pro- and anti-apoptotic proteins, which may determine the effect on mitochondrial integrity.

For example, Bid "senses" protease activation and becomes activated itself through proteolytic cleavage by caspases [61], the cytotoxic lymphocyte protease granzyme B [62], and lysosomal proteases [63]. Bim [64] and Bmf [65] appear to "sense" cytoskeleton status, and Bad "senses" the status of growth factor receptor signaling. Noxa [66] and Puma [67,68] "sense" the activation of the pro-apoptotic transcription factor p53 through a p53-dependent increase in their expression. Although they may not directly induce mitochondrial outer membrane effects, the BH3-only proteins interact with other pro-apoptotic Bcl-2 family members, namely, Bax and Bak, to induce mitochondrial membrane permeabilization [69]. Bax and Bak have three of the four BH domains and are called *multidomain* or *BH-123* proteins. In this scenario, the BH3-only molecules sense a particular cellular apoptotic signal and direct Bax and Bak to disrupt the mitochondria. Isolated mitochondria from cells lacking both Bax and Bak (but not single knock-outs) are strikingly resistant to the induction of mitochondrial permeabilization and this pathway of apoptosis [70], suggesting that these two proteins are intimately, perhaps directly, involved in the mechanism of membrane permeabilization.

The ability of Bax and Bak to induce permeabilization may depend on oligomerization. Bax and Bak, which are highly homologous, exist as inactive monomers that can be induced to oligomerize by BH3-only proteins [71,72]. Oligomerization correlates with induction of membrane permeabilization; however, the precise role of oligomerization has not been directly demonstrated. It is possible that oligomerized pro-apoptotic Bcl-2 proteins form a pore or alter the lipids of the mitochondrial outer membrane to induce permeabilization. In support of this idea, Bcl-2, Bcl-x$_L$, Bax, and Bid contain similar structures that resemble the pore-forming chain of some bacterial toxins and have been shown to have weak channel-forming activity for small ions through lipid membranes [55]. It has yet to be determined if these molecules form pores on their own or through interactions with other proteins in the outer membrane, such as VDAC [55,73].

Furthermore, the anti-apoptotic Bcl-2 proteins may act by inhibiting oligomerization and possible pore formation by the pro-apoptotic Bcl-2 proteins [74]. However, whether or not Bcl-2 family members directly regulate apoptosis in this manner remains controversial.

Cell Cycle versus Apoptosis

To maintain homeostasis, cell death and proliferation must be precisely balanced, and communication between these two distinct signaling pathways is critical. Several proteins that engage the cell cycle, including c-Myc and E2F1, sensitize cells to apoptosis by increasing susceptibility to mitochondrial membrane permeabilization. In response to apoptotic stimuli, such as growth factor withdrawal or DNA damaging agents, c-Myc activation stimulates a caspase-independent release of cytochrome c from mitochondria [75]. Furthermore, activation of c-Myc may provoke Fas-mediated apoptosis, possibly through a c-Myc-dependent up-regulation of Fas ligand [76,77] and/or sensitization to death receptor signaling [78].

Anti-apoptotic Bcl-2 family members may function not only to protect mitochondria from membrane permeabilization, but also to repress cell cycle entry. The ability of Bcl-2 to prevent or delay entry into G_1 appears to be independent of its role in maintaining mitochondria integrity, as point mutations in Bcl-2 have been described that eliminate the cell cycle-inhibitory activity without affecting the anti-apoptotic activity [79]. Therefore, while entry into the cell cycle does not directly induce apoptosis, these two processes are linked and a variety of proteins have coevolved to play distinct roles in each pathway. The connection between induction of the cell cycle and cell death, termed *antagonistic pleiotropy* [80], is a fundamental mechanism to prevent cancer and may be "hard wired" in animals in which cell proliferation is required for tissue homeostasis. Nevertheless, this does not mean that entry into the cell cycle is a requirement for apoptosis, nor that blockade of the cell cycle will necessarily promote (or inhibit) cell death via apoptosis. These are distinct but evolutionarily linked cellular processes.

Conclusions

When properly regulated, apoptosis is essential for maintaining homeostasis of several cellular systems, including development, tissue turnover, immune regulation, and control of oncogenesis. However, inappropriate cell death influences several disease states, because too much (i.e., Alzheimer's disease) or too little (i.e., cancer, autoimmunity) apoptosis creates an adverse imbalance. Considering the vast number of molecules involved in orchestrating an apoptotic response, defining the precise mechanisms and targets that may contribute to apoptosis in a specific setting is proving to be extremely challenging. In addition, many apoptotic signaling pathways are stimulus and cell-type specific, further complicating the process. Novel proteins and pathways are being discovered at a fervent pace and although tremendous gains have been made in elucidating the mechanisms of apoptosis, this is a highly complex process that remains only partially understood.

References

1. Wyllie, A. H., Kerr, J. F., and Currie, A. R. (1980). Cell death: The significance of apoptosis. *Int. Rev. Cytol.* **68**, 251–306.
2. Thornberry, N. A. and Lazebnik, Y. (1998). Caspases: Enemies within. *Science* **281**, 1312–1316.
3. Wilson, K. P., Black, J. A., Thomson, J. A., Kim, E. E., Griffith, J. P., Navia, M. A., Murcko, M. A., Chambers, S. P., Aldape, R. A., Raybuck, S. A. *et al.* (1994). Structure and mechanism of interleukin-1 beta converting enzyme. *Nature* **370**, 270–5.
4. Walker, N. P., Talanian, R. V., Brady, K. D., Dang, L. C., Bump, N. J., Ferenz, C. R., Franklin, S., Ghayur, T., Hackett, M. C., Hammill, L. D. *et al.* (1994). Crystal structure of the cysteine protease interleukin-1 beta-converting enzyme: A (p20/p10)2 homodimer. *Cell* **78**, 343–52.
5. Shi, Y. (2002). Mechanisms of caspase activation and inhibition during apoptosis. *Mol Cell* **9**, 459–70.
6. Enari, M., Sakahira, H., Yokoyama, H., Okawa, K., Iwamatsu, A., and Nagata, S. (1998). A caspase-activated DNase that degrades DNA during apoptosis, and its inhibitor ICAD. *Nature* **391**, 43–50.
7. Stroh, C. and Schulze-Osthoff, K. (1998). Death by a thousand cuts: An ever increasing list of caspase substrates. *Cell Death Differ.* **5**, 997–1000.
8. Coleman, M. L., Sahai, E. A., Yeo, M., Bosch, M., Dewar, A., and Olson, M. F. (2001). Membrane blebbing during apoptosis results from caspase-mediated activation of ROCK I. *Nat. Cell Biol.* **3**, 339–45.
9. Fadok, V. A., de Cathelineau, A., Daleke, D. L., Henson, P. M., and Bratton, D. L. (2001). Loss of phospholipid asymmetry and surface exposure of phosphatidylserine is required for phagocytosis of apoptotic cells by macrophages and fibroblasts. *J. Biol. Chem.* **276**, 1071–7.
10. Sun, X. M., MacFarlane, M., Zhuang, J., Wolf, B. B., Green, D. R., and Cohen, G. M. (1999). Distinct caspase cascades are initiated in receptor-mediated and chemical-induced apoptosis. *J. Biol. Chem.* **274**, 5053–60.
11. Xiang, J., Chao, D. T., and Korsmeyer, S. J. (1996). BAX-induced cell death may not require interleukin 1β-converting enzyme-like proteases. *Proc. Natl. Acad. Sci. USA* **93**, 14559–14563.
12. McCarthy, N. J., Whyte, M. K., Gilbert, C. S., and Evan, G. I. (1997). Inhibition of Ced-3/ICE-related proteases does not prevent cell death induced by oncogenes, DNA damage, or the Bcl-2 homologue Bak. *J. Cell Biol.* **136**, 215–27.
13. Salvesen, G. S. and Duckett, C. S. (2002). IAP proteins: Blocking the road to death's door. *Nat. Rev. Mol. Cell Biol.* **3**, 401–10.
14. Silke, J. and Vaux, D. L. (2001). Two kinds of BIR-containing protein-inhibitors of apoptosis, or required for mitosis. *J. Cell Sci.* **114**, 1821–7.
15. Locksley, R. M., Killeen, N., and Lenardo, M. J. (2001). The TNF and TNF receptor superfamilies: Integrating mammalian biology. *Cell* **104**, 487–501.
16. Green, D. R. (1998). Apoptotic pathways: The roads to ruin. *Cell* **94**, 695–698.
17. Muzio, M., Stockwell, B. R., Stennicke, H. R., Salvesen, G. S., and Dixit, V. M. (1998). An induced proximity model for caspase-8 activation. *J. Biol. Chem.* **273**, 2926–30.
18. Wang, J., Chun, H. J., Wong, W., Spencer, D. M., and Lenardo, M. J. (2001). Caspase-10 is an initiator caspase in death receptor signaling. *Proc. Natl. Acad. Sci. USA* **98**, 13884–8.
19. Irmler, M., Thome, M., Hahne, M., Schneider, P., Hofmann, K., Steiner, V., Bodmer, J.-L., Schroter, M., Burns, K., Mattmann, C., Rimoldi, D., French, L. E., and Tschopp, J. (1997). Inhibition of death receptor signals by cellular FLIP. *Nature* **388**, 190–195.

20. Perlman, H., Pagliari, L. J., Georganas, C., Mano, T., Walsh, K., and Pope, R. M. (1999). FLICE-inhibitory protein expression during macrophage differentiation confers resistance to fas-mediated apoptosis. *J. Exp. Med.* **190**, 1679–88.
21. Yeh, W. C., Itie, A., Elia, A. J., Ng, M., Shu, H. B., Wakeham, A., Mirtsos, C., Suzuki, N., Bonnard, M., Goeddel, D. V., and Mak, T. W. (2000). Requirement for Casper (c-FLIP) in regulation of death receptor-induced apoptosis and embryonic development. *Immunity* **12**, 633–42.
22. Scaffidi, C., Schmitz, I., Krammer, P. H., and Peter, M. E. (1999). The role of c-FLIP in modulation of CD95-induced apoptosis. *J. Biol. Chem.* **274**, 1541–1548.
23. Chang, D. W., Xing, Z., Pan, Y., Algeciras-Schimnich, A., Barnhart, B. C., Yaish-Ohad, S., Peter, M. E., and Yang, X. (2002). c-FLIP(L) is a dual function regulator for caspase-8 activation and CD95-mediated apoptosis. *EMBO J.* **21**, 3704–3714.
24. Chen, G. and Goeddel, D. V. (2002). TNF-R1 signaling: A beautiful pathway. *Science* **296**, 1634–5.
25. Wang, C.-Y., Mayo, M. W., Korneluk, R. G., Goeddel, D. V., and Baldwin A. S. Jr. (1998). NF-κB antiapoptosis: Induction of TRAF1 and TRAF2 and c-IAP1 and c-IAP2 to suppress caspase-8 activation. *Science* **281**, 1680–1683.
26. Tang, G., Minemoto, Y., Dibling, B., Purcell, N. H., Li, Z., Karin, M., and Lin, A. (2001). Inhibition of JNK activation through NF-kappaB target genes. *Nature* **414**, 313–7.
27. De Smaele, E., Zazzeroni, F., Papa, S., Nguyen, D. U., Jin, R., Jones, J., Cong, R., and Franzoso, G. (2001). Induction of gadd45beta by NF-kappaB downregulates pro-apoptotic JNK signalling. *Nature* **414**, 308–13.
28. Kasibhatla, S., Brunner, T., Genestier, L., Echeverri, F., Mahboubi, A., and Green, D. R. (1998). DNA damaging agents induce expression of Fas ligand and subsequent apoptosis in T lymphocytes via the activation of NF-kappa B and AP-1. *Mol. Cell* **1**, 543–51.
29. Ryan, K. M., Ernst, M. K., Rice, N. R., and Vousden, K. H. (2000). Role of NF-kappaB in p53-mediated programmed cell death. *Nature* **404**, 892–7.
30. Scaffidi, C., Fulda, S., Srinivasan, A., Friesen, C., Li, F., Tomaselli, K. J., Debatin, K. M., Krammer, P. H., and Peter, M. E. (1998). Two CD95 (APO-1/Fas) signaling pathways. *EMBO J.* **17**, 1675–1687.
31. Li, H., Zhu, H., Xu, C., and Yuan, J. (1998). Cleavage of BID by caspase 8 mediates the mitochondrial damage in the Fas pathway of apoptosis. *Cell* **94**, 491–501.
32. Luo, X., Budihardjo, I., Zou, H., Slaughter, C., and Wang, X. (1998). Bid, a Bcl-2 interacting protein mediates cytochrome c release from mitochondria in response to activation of cell surface death receptors. *Cell* **94**, 481–490.
33. Wei, M. C., Zong, W. X., Cheng, E. H., Lindsten, T., Panoutsakopoulou, V., Ross, A. J., Roth, K. A., MacGregor, G. R., Thompson, C. B., and Korsmeyer, S. J. (2001). Proapoptotic BAX and BAK: A requisite gateway to mitochondrial dysfunction and death. *Science* **292**, 727–30.
34. Hofmann, K., Bucher, P., and Tschopp, J. (1997). The CARD domain: A new apoptotic signalling motif. *Trends Biochem. Sci.* **22**, 155–6.
35. Stennicke, H. R., Deveraux, Q. L., Humke, E. W., Reed, J. C., Dixit, V. M., and Salvesen, G. S. (1999). Caspase-9 can be activated without proteolytic processing. *J. Biol. Chem.* **274**, 8359–62.
36. Li, P., Nijhawan, D., Budihardjo, I., Srinivasula, S. M., Ahmad, M., Alnemri, E. S., and Wang, X. (1997). Cytochrome c and dATP-dependent formation of Apaf-1/caspase-9 complex initiates an apoptotic protease cascade. *Cell* **91**, 479–489.
37. Zou, H., Henzel, W. J., Liu, X., Lutschg, A., and Wang, X. (1997). Apaf-1, a human protein homologous to C. elegans CED-4, participates in cytochrome c-dependent activation of caspase-3. *Cell* **90**, 405–413.
38. Shiozaki, E. N., Chai, J., and Shi, Y. (2002). Oligomerization and activation of caspase-9, induced by Apaf-1 CARD. *Proc. Natl. Acad. Sci. USA* **99**, 4197–202.
39. Renatus, M., Stennicke, H. R., Scott, F. L., Liddington, R. C., and Salvesen, G. S. (2001). Dimer formation drives the activation of the cell death protease caspase 9. *Proc. Natl. Acad. Sci. USA* **98**, 14250–5.
40. Wang, X. (2001). The expanding role of mitochondria in apoptosis. *Genes Dev.* **15**, 2922–33.
41. Li, K., Li, Y., Shelton, J. M., Richardson, J. A., Spencer, E., Chen, Z. J., Wang, X., and Williams, R. S. (2000). Cytochrome c deficiency causes embryonic lethality and attenuates stress-induced apoptosis. *Cell* **101**, 389–99.
42. Kuida, K., Haydar, T. F., Kuan, C., Gu, Y., Taya, C., Karasuyama, H., Su, M. S., Rakic, P., and Flavell, R. A. (1998). Reduced apoptosis and cytochrome c-mediated caspase activation in mice lacking caspase 9. *Cell* **94**, 325–337.
43. Yoshida, H., Kong, Y. Y., Yoshida, R., Elia, A. J., Hakem, A., Hakem, R., Penninger, J. M., and Mak, T. W. (1998). Apafl is required for mitochondrial pathways of apoptosis and brain development. *Cell* **94**, 739–50.
44. Miramar, M. D., Costantini, P., Ravagnan, L., Saraiva, L. M., Haouzi, D., Brothers, G., Penninger, J. M., Peleato, M. L., Kroemer, G., and Susin, S. A. (2001). NADH oxidase activity of mitochondrial apoptosis-inducing factor. *J. Biol. Chem.* **276**, 16391–8.
45. Susin, S. A., Lorenzo, H. K., Zamzami, N., Marzo, I., Brenner, C., Larochette, N., Prevost, M. C., Alzari, P. M., and Kroemer, G. (1999). Mitochondrial release of caspase-2 and -9 during the apoptotic process. *J. Exp. Med.* **189**, 381–393
46. Li, L. Y., Luo, X., and Wang, X. (2001). Endonuclease G is an apoptotic DNase when released from mitochondria. *Nature* **412**, 95–9.
47. Hegde, R., Srinivasula, S. M., Zhang, Z., Wassell, R., Mukattash, R., Cilenti, L., DuBois, G., Lazebnik, Y., Zervos, A. S., Fernandes-Alnemri, T., and Alnemri, E. S. (2002). Identification of Omi/HtrA2 as a mitochondrial apoptotic serine protease that disrupts inhibitor of apoptosis protein-caspase interaction. *J. Biol. Chem.* **277**, 432–8.
48. Du, C., Fang, M., Li, Y., Li, L., and Wang, X. (2000). Smac, a mitochondrial protein that promotes cytochrome c-dependent caspase activation by eliminating IAP inhibition. *Cell* **102**, 33–42.
49. Verhagen, A. M., Ekert, P. G., Pakusch, M., Silke, J., Connolly, L. M., Reid, G. E., Moritz, R. L., Simpson, R. J., and Vaux, D. L. (2000). Identification of DIABLO, a mammalian protein that promotes apoptosis by binding to and antagonizing IAP proteins. *Cell* **102**, 43–53.
50. Suzuki, Y., Imai, Y., Nakayama, H., Takahashi, K., Takio, K., and Takahashi, R. (2001). A serine protease, HtrA2, is released from the mitochondria and interacts with XIAP, inducing cell death. *Mol. Cell* **8**, 613–21.
51. Martins, L. M., Iaccarino, I., Tenev, T., Gschmeissner, S., Totty, N. F., Lemoine, N. R., Savopoulos, J., Gray, C. W., Creasy, C. L., Dingwall, C., and Downward, J. (2002). The serine protease Omi/HtrA2 regulates apoptosis by binding XIAP through a reaper-like motif. *J. Biol. Chem.* **277**, 439–44.
52. Verhagen, A. M., Silke, J., Ekert, P. G., Pakusch, M., Kaufmann, H., Connolly, L. M., Day, C. L., Tikoo, A., Burke, R., Wrobel, C., Moritz, R. L., Simpson, R. J., and Vaux, D. L. (2002). HtrA2 promotes cell death through its serine protease activity and its ability to antagonize inhibitor of apoptosis proteins. *J. Biol. Chem.* **277**, 445–54.
53. Roberts, D. L., Merrison, W., MacFarlane, M., and Cohen, G. M. (2001). The inhibitor of apoptosis protein-binding domain of Smac is not essential for its proapoptotic activity. *J. Cell Biol.* **153**, 221–8.
54. Goldstein, J. C., Waterhouse, N. J., Juin, P., Evan, G. I., and Green, D. R. (2000). The coordinate release of cytochrome c during apoptosis is rapid, complete and kinetically invariant. *Nat. Cell Biol.* **2**, 156–62.
55. Martinou, J. C. and Green, D. R. (2001). Breaking the mitochondrial barrier. *Nat. Rev. Mol. Cell Biol.* **2**, 63–7.
56. Zamzami, N. and Kroemer, G. (2001). The mitochondrion in apoptosis: How Pandora's box opens. *Nat. Rev. Mol. Cell Biol.* **2**, 67–71.
57. Vander Heiden, M. G. and Thompson, C. B. (1999). Bcl-2 proteins: Regulators of apoptosis or of mitochondrial homeostasis? *Nat. Cell Biol.* **1**, E209–16.
58. Waterhouse, N. J., Goldstein, J. C., von Ahsen, O., Schuler, M., Newmeyer, D. D., and Green, D. R. (2001). Cytochrome c maintains

mitochondrial transmembrane potential and ATP generation after outer mitochondrial membrane permeabilization during the apoptotic process. *J. Cell Biol.* **153**, 319–28.
59. Harris, M. H. and Thompson, C. B. (2000). The role of the Bcl-2 family in the regulation of outer mitochondrial membrane permeability. *Cell Death Differ.* **7**, 1182–91.
60. Adams, J. M. and Cory, S. (1998). The Bcl-2 protein family: Arbiters of cell survival. *Science* **281**, 1322–1326.
61. Budihardjo, I., Oliver, H., Lutter, M., Luo, X., and Wang, X. (1999). Biochemical pathways of caspase activation during apoptosis. *Annu. Rev. Cell Dev. Biol.* **15**, 269–90.
62. Pinkoski, M. J., Waterhouse, N. J., Heibein, J. A., Wolf, B. B., Kuwana, T., Goldstein, J. C., Newmeyer, D. D., Bleackley, R. C., and Green, D. R. (2001). Granzyme B-mediated apoptosis proceeds predominantly through a Bcl-2-inhibitable mitochondrial pathway. *J. Biol. Chem.* **276**, 12060–7.
63. Stoka, V., Turk, B., Schendel, S. L., Kim, T. H., Cirman, T., Snipas, S. J., Ellerby, L. M., Bredesen, D., Freeze, H., Abrahamson, M., Bromme, D., Krajewski, S., Reed, J. C., Yin, X. M., Turk, V., and Salvesen, G. S. (2001). Lysosomal protease pathways to apoptosis. Cleavage of bid, not pro-caspases, is the most likely route. *J. Biol. Chem.* **276**, 3149–57.
64. Puthalakath, H., Huang, D. C., O'Reilly, L. A., King, S. M., and Strasser, A. (1999). The proapoptotic activity of the Bcl-2 family member Bim is regulated by interaction with the dynein motor complex. *Mol. Cell* **3**, 287–96.
65. Puthalakath, H., Villunger, A., O'Reilly, L. A., Beaumont, J. G., Coultas, L., Cheney, R. E., Huang, D. C., and Strasser, A. (2001). Bmf: A proapoptotic BH3-only protein regulated by interaction with the myosin V actin motor complex, activated by anoikis. *Science* **293**, 1829–32.
66. Oda, E., Ohki, R., Murasawa, H., Nemoto, J., Shibue, T., Yamashita, T., Tokino, T., Taniguchi, T., and Tanaka, N. (2000). Noxa, a BH3-only member of the Bcl-2 family and candidate mediator of p53-induced apoptosis. *Science* **288**, 1053–8.
67. Nakano, K. and Vousden, K. H. (2001). PUMA, a novel proapoptotic gene, is induced by p53. *Mol. Cell* **7**, 683–94.
68. Yu, J., Zhang, L., Hwang, P. M., Kinzler, K. W., and Vogelstein, B. (2001). PUMA induces the rapid apoptosis of colorectal cancer cells. *Mol. Cell* **7**, 673–82.
69. Lutz, R. J. (2000). Role of the BH3 (Bcl-2 homology 3) domain in the regulation of apoptosis and Bcl-2-related proteins. *Biochem. Soc. Trans.* **28**, 51–6.
70. Lindsten, T., Ross, A. J., King, A., Zong, W. X., Rathmell, J. C., Shiels, H. A., Ulrich, E., Waymire, K. G., Mahar, P., Frauwirth, K., Chen, Y., Wei, M., Eng, V. M., Adelman, D. M., Simon, M. C., Ma, A., Golden, J. A., Evan, G., Korsmeyer, S. J., MacGregor, G. R., and Thompson, C. B. (2000). The combined functions of proapoptotic Bcl-2 family members bak and bax are essential for normal development of multiple tissues. *Mol. Cell* **6**, 1389–99.
71. Eskes, R., Desagher, S., Antonsson, B., and Martinou, J. C. (2000). Bid induces the oligomerization and insertion of Bax into the outer mitochondrial membrane. *Mol. Cell Biol.* **20**, 929–35.
72. Wei, M. C., Lindsten, T., Mootha, V. K., Weiler, S., Gross, A., Ashiya, M., Thompson, C. B., and Korsmeyer, S. J. (2000). tBID, a membrane-targeted death ligand, oligomerizes BAK to release cytochrome c. *Genes Dev.* **14**, 2060–71.
73. Shimizu, S., Matsuoka, Y., Shinohara, Y., Yoneda, Y., and Tsujimoto, Y. (2001). Essential role of voltage-dependent anion channel in various forms of apoptosis in mammalian cells. *J. Cell Biol.* **152**, 237–50.
74. Cheng, E. H., Wei, M. C., Weiler, S., Flavell, R. A., Mak, T. W., Lindsten, T., and Korsmeyer, S. J. (2001). BCL-2, BCL-X(L) sequester BH3 domain-only molecules preventing BAX- and BAK-mediated mitochondrial apoptosis. *Mol. Cell* **8**, 705–11.
75. Juin, P., Hueber, A. O., Littlewood, T., and Evan, G. (1999). c-Myc-induced sensitization to apoptosis is mediated through cytochrome c release. *Genes Dev.* **13**, 1367–81.
76. Kasibhatla, S., Beere, H. M., Brunner, T., Echeverri, F., and Green, D. R. (2000). A "non-canonical" DNA-binding element mediates the response of the Fas-ligand promoter to c-Myc. *Curr. Biol.* **10**, 1205–8.
77. Brunner, T., Kasibhatla, S., Pinkoski, M. J., Frutschi, C., Yoo, N. J., Echeverri, F., Mahboubi, A., and Green, D. R. (2000). Expression of Fas ligand in activated T cells is regulated by c-Myc. *J. Biol. Chem.* **275**, 9767–72.
78. Hueber, A. O., Zornig, M., Lyon, D., Suda, T., Nagata, S., and Evan, G. I. (1997). Requirement for the CD95 receptor-ligand pathway in c-Myc-induced apoptosis. *Science* **278**, 1305–9.
79. Huang, D. C. S., O'Reilly, L. A., Strasser, A., and Cory, S. (1997). The anti-apoptotic function of Bcl-2 can be genetically separated from its inhibitory effect on cell cycle entry. *EMBO J.* **16**, 4628–4638
80. Green, D. R. and Evan, G. I. (2002). A matter of life and death. *Cancer Cell* **1**, 19–30.

CHAPTER 332

Signaling Down the Endocytic Pathway

Jeffrey L. Benovic and James H. Keen
Cell Biology and Signaling Program, Kimmel Cancer Center, Thomas Jefferson University, Philadelphia, Pennsylvania

Introduction

Intercellular signaling has long been recognized to be initiated at the cell surface, and receptor clearance from the plasma membrane by internalization mechanisms has correspondingly been associated with termination of signaling. Recognition of the existence of multiple pathways for internalization as well as the complexity of intracellular itineraries that ligands and receptors can utilize clearly indicates that cellular signaling has not only temporal but also spatial properties. As ligand–receptor complexes proceed down the endocytic pathway, they encounter unique components and environments at both the cytoplasmic and lumenal interfaces. Relevant "geographical" effects are only recently beginning to come into focus, as potential intracellular sites of action are revealed. Several reviews of the interface between signaling and endocytosis that the reader may wish to consult have appeared [1–3]; the focus here is on aspects of signaling following internalization of receptor tyrosine kinases (RTKs) and G-protein-coupled receptors (GPCRs).

RTK Signaling from the Cell Surface

Receptor tyrosine kinases have been investigated for more than 25 years. Ligand binding to epidermal growth factor receptors (EGFR), a particularly well-studied system, induces dimerization and autophosphorylation, followed by recruitment of multiple adaptor proteins and activation of several signaling systems at the cell surface including Ras, Jak/Stat, and PLCγ pathways (reviewed in [1,4,5]). Sustained occupancy of cell surface EGFRs, achieved either through continuous supply of external ligand or through inhibition of removal of surface ligand–receptor complexes, is known to be required for mitogenesis. Thus, while the wild-type form of the EGF receptor (EGFR) is rapidly internalized and is not oncogenic, a mutant form that is not endocytosed exhibits a more pronounced transformation phenotype in the same setting [6]. Another RTK, the insulin receptor, exhibits a corollary phenomenon. When insulin receptor endocytosis is blocked, certain activities initiated at the cell surface such as glucose and amino acid transport are increased [7].

RTK Signaling from Endocytic Compartments

Although the observations just mentioned establish that signaling occurs on the cell surface, they do not exclude the possibility of signaling from internal compartments, and the two possibilities are not mutually exclusive. Indeed, there have been substantive indications that signaling from internal compartments does occur. One line of investigation has revealed that signaling is attenuated or altered when internalization is blocked by inhibitors of endocytosis. For example, upon expression of a mutant form of dynamin that blocks endocytosis, EGF binding induces Shc recruitment to the stimulated receptor but MAP kinase activation is blocked [8]. Indications of the potential complexity of interpreting these experimental designs are the observations that Ras, Raf, and MEK1 are actually phosphorylated under these conditions, suggesting that the block in MAP kinase activation may actually result from a failure to internalize MEK1 [9], and that high-affinity EGFRs disappear from the cell surface [10]. Yet an additional

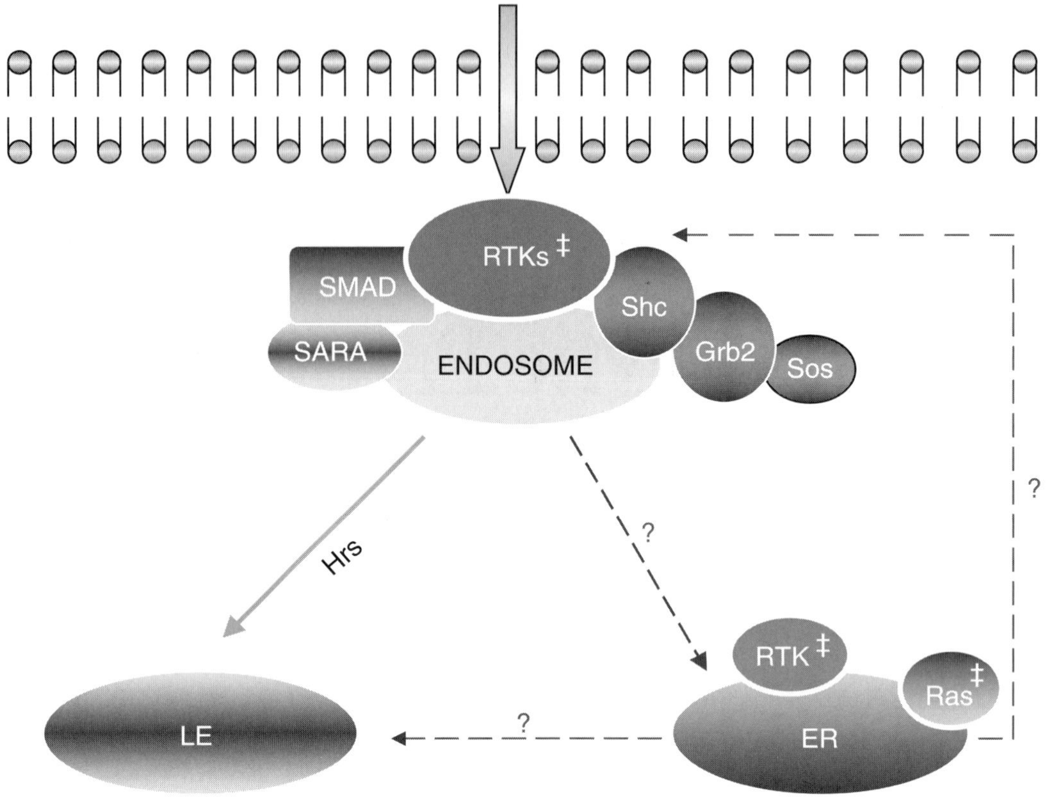

Figure 1 Receptor tyrosine kinase signaling pathway components in internal compartments.

Activated (phosphorylated) receptor tyrosine kinases (RTKs‡) and other functional signaling molecules, some of which are shown here, can be detected in endosomal compartments. Degradation of internalized receptors occurs in multivesicular bodies of the late endosomal (LE) system in which the RTK substrate Hrs plays a major role. There is also evidence for the presence of activated receptors and signaling molecules such as Ras on the endoplasmic reticulum (ER) though their precise origin, function and fate remains uncertain. See text for details.

wrinkle is suggested by the observation that agonist-dependent trafficking of Ras and Raf1 may be required for activation of the MAP kinase cascade [11]. Finally, inhibition of endocytosis has also been reported to block delivery of the RTK (EGFR/PDGFR) activated transcription factor Stat3 to the nucleus [12]. Colocalization of Stat3 with AP-2 decorated vesicles (usually restricted to the plasma membrane) reported in this study is surprising, but the possibility that transcription factors use the endocytic pathway for inward vectorial transport is intriguing. A related observation is the report that EGFR itself can function as a coactivator or transcription factor; this is of particular significance because the receptor is frequently found in the nucleus in various cell lines and tissues [13].

Another line of study that has implicated signaling from cytoplasmic compartments following RTK activation is the observation that activated EGFRs are present on internal compartments [14]. Furthermore, biochemical preparations derived predominantly from endosome-enriched fractions had markedly different phosphotyrosine patterns than those derived from plasma membrane preparations [15]. This has subsequently been confirmed and extended by morphological studies at both the light and ultrastructural levels that visualize specific localization of phosphospecific antibodies [16,17]. Signaling from internal compartments might also be expected to require the presence of a known cast of adaptor molecules (Fig. 1). Indeed, endosome preparations from stimulated cells have been shown to contain Grb2, SOS, and Shc in association with activated EGFR [18] and more recent reports have identified other signaling molecules including Src [19], MEK, and Raf [20].

Though the interpretation of these kinds of biochemical experiments is necessarily dependent on the purity of the endosomal preparations—no minor hurdle given the pleomorphic nature of the endosomal compartment—they have found support in two additional lines of study. There is evidence that the internal pool of EGFR is intrinsically competent for signal initiation because activation of MAP kinase [21], PLCγ [22], and Ras [23] could be demonstrated in broken cell lysates. Elegant fluorescence resonance energy transfer (FRET) experiments utilizing tagged Ras and Ras binding domain proteins provide evidence for pools of activated Ras on the endoplasmic reticulum and Golgi in response to serum activation [24]; while endocytosis of activated receptors did not seem to be required in this system, its role in activating endogenous levels of Ras remains a possibility. And in an exogenous expression approach in live cells, Jiang and Sorkin [25] have shown that Grb2, Shc, and Ras can be visualized in pairwise combination accumulating on endosomes following EGF stimulation. Finally, Haj and colleagues [26] have demonstrated that phosphorylated and endocytosed EGFR and PDGF receptors interact on the endoplasmic reticulum with the protein tyrosine phosphatase PTP1B. Collectively, these results are provocative in demonstrating that the machinery for implementing and terminating signaling is present on internal compartments.

Accessibility to appropriate internal substrates for signaling might be expected to be subject to tight regulation. For example, activated EGFRs can be shown to persist after internalization (consistent with the distinct phosphotyrosine patterns noted earlier), as do tyrosine-phosphorylated and activated forms of its downstream effector phospholipase Cγ; however phosphatidylinositol 4,5-bisphosphate (PIP_2) hydrolysis appears to occur only at the plasma membrane [22]. Using fluorescent reporters, evidence for phosphotyrosine production at internal sites could be visualized, but PIP_2 localization was restricted to the plasma membrane localization [27], suggesting that accessibility to appropriate downstream substrates for signal generation is tightly regulated in the intact cell.

Signaling unique to internal compartments might also be expected to involve additional specialized components not present on the cell surface. Recent findings indicate that the protein Hrs (hepatocyte growth factor regulated tyrosine kinase substrate) fulfills this role [28]. Hrs is a tyrosine kinase substrate for several RTKs [29] that is phosphorylated only after endocytosis [30]. It contains a phosphatidylinositol 3-phosphate (PI3P) binding FYVE domain that contributes to its localization to early endosomes [29], and it is the mammalian homolog of the Class E yeast Vps27p, which regulates sorting of proteins to vacuole. Interestingly, an *hrs* null mutant in mice exhibited enlarged early endosomes [31]. More recent studies of *Drosophila* larvae expressing mutant Hrs reveal an impairment in the formation of inwardly budded vesicles in endosomes that are thought to give rise to multivesicular bodies (MVBs), missing in these cells [32]. Although cells expressing this mutant *hrs* possess the ability to internalize EGFR, they fail to degrade it or other RTKs. Consequently, the cells exhibit increased levels of activated EGFR, which are paralleled by increased levels of MAP kinase activation and an expansion of cells whose proliferation in the larvae is controlled by EGFR signaling. Thus Hrs appears to be a "housekeeping" component critical for EGFR degradation, and its absence provides evidence that functional signaling can emanate from internal compartments under physiological conditions. These results then provide compelling evidence that specific components and associated mechanisms are found in internal cellular compartments to regulate signaling, that these factors are under tight regulation, and that their inactivation results in aberrant signaling.

Although less extensively investigated than the EGFR, there is also recent support for critical steps in intracellular signaling following activation of receptors belonging to the transforming growth factor β (TGF-β) superfamily [33]. These homodimeric and heterodimeric receptor complexes phosphorylate a family of SMAD proteins, which in combination with a cell-specific complement of DNA binding proteins yields the range of signaling possibilities characteristic of this ubiquitous signaling system [34]. For some of these receptors, phosphorylation of the SMAD proteins is regulated by the SARA (SMAD anchor for receptor activation) protein. Interestingly, SARA contains a FYVE domain that binds tightly to PI3P [35], and has been localized to early endosomes [36]. Although still controversial [37], it seems likely that TGF-β receptors can encounter and activate SMAD proteins on endosomes as the receptors are known to undergo rapid endocytosis [38], and mutant SARA proteins lacking the FYVE domain have been reported to inhibit downstream signaling [36].

GPCR Signaling Paradigms and Desensitization

The superfamily of G-protein-coupled receptors mediates intracellular signaling to a diverse array of stimuli including light, odorants, neurotransmitters, lipids, and various peptides and proteins (reviewed in [39]). GPCR activation promotes interaction with heterotrimeric GTP binding proteins (G-proteins) resulting in GDP/GTP exchange and subsequent dissociation of the α-GTP and βγ subunits. These subunits function to regulate the activity of multiple effector proteins including adenylyl cyclases, phospholipases, phosphodiesterases, ion channels, and phosphatidylinositol 3-kinase (PI3K). Thus, GPCRs regulate numerous biological functions including sensory perception, neurotransmission, chemotaxis, development, cellular proliferation, differentiation, and survival.

Although GPCR activation of heterotrimeric G-proteins was initially identified some 30 years ago, more recent studies have revealed that GPCRs can also interact with a myriad of additional proteins (reviewed in [40,41]). These include various protein kinases and a number of adaptor and scaffolding proteins. Here we briefly discuss those interactions that have been implicated in GPCR signaling and endocytosis. Although GPCR activation initiates signaling via heterotrimeric G-proteins, many activated GPCRs also interact with two additional protein families, G-protein-coupled receptor kinases (GRKs) and arrestins (reviewed in [42,43]). GRKs specifically bind to activated GPCRs, leading to increased receptor phosphorylation and enhanced arrestin binding. Arrestins have been implicated in the process of receptor desensitization, that is, the reduced ability of a receptor to respond to hormones following prolonged stimulation. Indeed, GRKs and arrestins were first identified in rod cells and implicated in quenching phototransduction. Arrestins bind to activated-phosphorylated GPCRs and thereby attenuate further G-protein interaction and signaling. Although a role for arrestins in GPCR desensitization is well established, recent studies reveal that the two nonvisual mammalian arrestins (termed arrestin-2 or β-arrestin-1 and arrestin-3 or β-arrestin-2) also function in regulating various aspects of GPCR trafficking and signaling (reviewed in [44]). These functions of arrestins are discussed in more detail later.

Additional interactions that may contribute to GPCR signaling and provide a potential link with endocytosis are more receptor specific and have been most extensively characterized for the β-adrenergic receptors (βARs) (reviewed in [41]). For example, β_1AR interaction with endophilins enhances agonist-promoted internalization of the receptor [45], whereas interaction with PSD-95 attenuates receptor internalization and facilitates interaction with *N*-methyl-D-aspartate receptors [46]. Interestingly, GRK5 phosphorylation of the β_1AR dramatically reduces PSD-95 interaction, suggesting that this association is also regulated by the activation state of the receptor [46]. The β_2AR interacts with the PDZ domain-containing protein Na^+/H^+ exchanger regulatory factor (NHERF) via a C-terminal PDZ binding domain [47]. NHERF interaction with the β_2AR positively regulates Na^+/H^+ exchange but also appears to regulate the sorting of internalized β_2ARs [48]. Interestingly, interaction of the β_2AR and NHERF appears to be negatively regulated by GRK5 phosphorylation of the β_2AR, again providing a link with the activation state of the receptor [48]. *N*-Ethylmaleimide-sensitive factor (NSF) also interacts with the C-terminal region of the β_2AR and facilitates receptor internalization and recycling [49]. In addition, the scaffolding protein gravin (also termed AKAP250) interacts with the β_2AR and has been implicated in regulating receptor resensitization and internalization [50,51]. Thus, multiple proteins function to dynamically regulate GPCR signaling and trafficking.

Control of RTK and GPCR Trafficking Leading to Degradation

Although multiple mechanisms contribute to receptor trafficking, most GPCRs endocytose via clathrin-coated pits (Fig. 2). As discussed previously, GPCR activation often results in GRK-mediated phosphorylation, arrestin binding, and desensitization. Several years ago, two reports suggested a role for arrestins in GPCR trafficking [52,53]. Initial mechanistic insight into this process revealed that nonvisual arrestins specifically interact with the clathrin heavy chain via a LϕXϕE motif (where ϕ is a bulky aliphatic residue) found within a C-terminal insert unique to arrestin-2 and -3 [54]. The interaction of arrestin and clathrin appears to play a critical role in mediating GPCR trafficking [54]. Additional arrestin interactions that have been implicated in GPCR trafficking include phosphoinositide [55], the adaptor protein AP2 [56,57], NSF [58], and ADP-ribosylation factor 6 and its exchange factor ARNO [59]. While detailed mechanistic insight into how these various interactions are coordinated is lacking, it is evident that arrestins play an important role in mediating GPCR trafficking.

For many GPCRs, prolonged activation results in down-regulation, a process that involves increased receptor degradation and an overall reduction in receptor levels (reviewed in [60]). Although multiple mechanisms contribute to receptor down-regulation, a number of studies have suggested

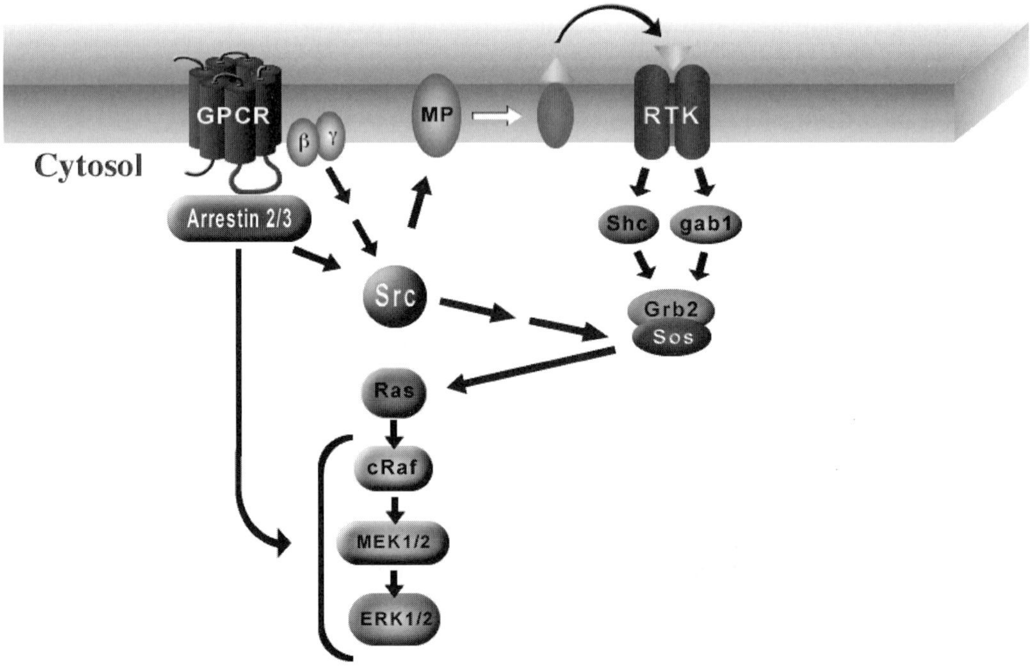

Figure 2 G protein-coupled receptor activation of MAP kinases.
Multiple pathways may provide a link between GPCR activation of MAP kinases and endocytic processes. These include the ability of arrestins to function as scaffolding proteins for various MAP kinases including Raf and ERK1/2 thereby providing a direct connection between GPCRs and MAP kinase localization and activation. Arrestins also function to link GPCRs with activation of Src and Src-like kinases providing a potential mechanism to connect endocytic processes, via the phosphorylation of dynamin, and stimulation of MAP kinases, via the activation of Ras guanine nucleotide exchange factors. GPCRs can also activate MAP kinases via their ability to transactivate receptor tyrosine kinases (RTKs) such as the epidermal growth factor receptor which in some cases involves the activation of metalloproteinases (MP) that result in the release of RTK ligands.

that GPCR endocytosis and subsequent sorting to lysosomes play an important role in this process [61–64]. Interestingly, two recent studies suggest that receptor ubiquitination contributes to the sorting and degradation of mammalian GPCRs [64,65]. Interesting correlates to these findings are the mating factor receptors in *Saccharomyces cerevisiae*. Previous studies have revealed that the mating factor receptors undergo agonist-dependent phosphorylation, ubiquitination, endocytosis, and degradation (reviewed in [66]). Although yeast do not contain arrestins, significant overlap may occur in the mechanisms used to sort and degrade mammalian and *S. cerevisiae* GPCRs.

Molecular mechanisms utilizing receptor ubiquitination have also recently been shown to play a role in termination of signaling and down-regulation of RTKs. In the case of the EGFR, recent results have shown that the E3 ubiquitin ligase c-Cbl can bind the receptor through its RING domain and induce receptor ubiquitination, thereby targeting the receptor for intracellular degradation [67,68]. A virally coded form of v-Cbl that lacks the RING domain does not down-regulate the receptor, resulting in increased signaling [69,70]. Furthermore, a role for c-Cbl in inducing rapid internalization as well as shunting internalized receptors for degradation has also recently been reported. Cbl was shown to associate with endophilin through CIN85 to induce rapid internalization of EGFR/c-met receptors; inhibition of this association blocked internalization and again enhanced biological signaling [71,72]. Collectively, these observations point to ubiquitination of the type I RTKs as critical in their down-regulation and, therefore, in overall decrease in their signal generation.

GPCR Activation of MAP Kinases

It is well established that many GPCRs can activate mitogen-activated protein kinases (MAPKs), including members of the extracellular signal-regulated kinase (ERK), c-Jun NH_2-terminal kinase (JNK), and p38 families (reviewed in [73]). Although multiple mechanisms mediate GPCR activation of MAPKs, three pathways in particular have been extensively characterized: (1) transactivation of RTKs, (2) formation of focal adhesion complexes, and (3) direct scaffolding of signaling proteins. These pathways are reviewed in [74] so we primarily discuss them here from the perspective of the potential link between signaling and endocytosis.

RTK transactivation by GPCRs was initially identified by Daub and coworkers [75,76] who showed that Gq and Gi-coupled receptors can activate ERK via a pathway that is dependent on activated EGFR, SHC, and Src. These findings were extended by Luttrell and colleagues [77] who implicated a role for G-protein $\beta\gamma$ subunits in EGFR transactivation. The mechanism of RTK transactivation has not been completely elucidated although recent studies suggest that GPCR activation results in activation of a metalloproteinase that in turn leads to the release of heparin binding EGF (HB-EGF) [78]. HB-EGF can then act in either an autocrine or paracrine fashion to activate cell surface EGFRs. Thus, GPCR activation of MAPKs that occurs via RTK transactivation might be expected to show the same correlation with endocytosis as outlined earlier for RTKs [79].

The other major pathway that may provide a link between GPCR signaling and endocytosis involves the ability of GPCRs to function as scaffolding proteins. In several instances this may involve a direct interaction of a GPCR with a signaling molecule such as Src interaction with the β_3-adrenergic receptor [80]. However, arrestins also appear to interact directly with Src family members, thus providing a broader link between GPCRs, Src, and MAPKs since most GPCRs bind arrestins. Initial studies in this area identified formation of an agonist-promoted complex containing the β_2AR, arrestin-2, and c-Src [81]. Inhibition of complex formation by mutation of proline-rich regions in arrestin-2 resulted in a reduction in β_2AR activation of ERK1/2. Activation of the chemokine receptor CXCR1 results in formation of an arrestin-2 complex with the Src family kinases Hck and c-Fgr [82], whereas formation of an endothelin-1 receptor complex with arrestin-2 and Yes mediates endothelin promoted GLUT4 translocation [83]. Because Src has been implicated in dynamin phosphorylation and endocytosis [84], formation of GPCR/arrestin/Src complexes may provide an important link between signaling and endocytosis.

Several recent studies also suggest an important role for arrestins in linking GPCRs, MAPK activation, and endocytosis. As discussed previously, nonvisual arrestins bind to activated phosphorylated GPCRs and target the receptors for clathrin-coated pit mediated endocytosis. Surprisingly, arrestins also appear to function as direct scaffolding proteins for MAPKs. DeFea and coworkers [85] initially identified formation of a complex containing a GPCR (proteinase-activated receptor 2 or PAR2), an arrestin (arrestin-2), a MAPK kinase kinase (Raf-1), and a MAPK (ERK1/2). Similar studies identified an agonist-promoted complex containing the neurokinin-1 receptor, arrestin-2, Src, and ERK1/2 [86]. Luttrell and coworkers [87] also observed such complexes containing, for example, angiotensin II type 1a receptors, arrestin-3, Raf-1, the MAPK kinase MEK1, and ERK2 [87]. In addition, McDonald and colleagues [88] also found that arrestin-3 can form a specific complex with the MAPK kinase kinase ASK1 and JNK3. Two interesting aspects of these studies are that formation of these complexes appears to be enhanced by receptor activation and that the resulting receptor/arrestin/MAPK complex appears to be endocytosed. Although a direct requirement for GPCR endocytosis in mediating MAPK activation remains controversial (reviewed in [74]), this may provide a mechanism for endocytosing and retaining an active signaling complex in the cytoplasm, thereby preventing nuclear translocation of the MAPK and potentially altering substrate specificity.

Endocytic Signaling in Developmental Systems

Vesicular trafficking and signaling on the "supracellular" level has also recently been recognized as playing a critical role in establishing and regulating morphogen gradients (reviewed in [89–91]). Morphogens have been defined as secreted molecules that convey signaling information that affects cell fate determination in developing systems, often in a concentration-dependent manner. Originally considered to be spread from cellular sites of synthesis to target cells by simple diffusion, it is now recognized that internalization and intracellular trafficking pathways in the peripheral cells play critical roles in regulating local morphogen concentrations along developmental axes. For example, establishment and maintenance of concentration gradients of transcriptional activators such as decapentaplegic (Dpp), Wingless, and Hedgehog in the fly embryo are controlled by endocytosis and degradation of morphogen. Somewhat more remarkably, the extent of spread of some morphogens is actually amplified by internalization and transcytosis across target cells, apparently establishing longer range gradients than would be effectively established by diffusion alone. This then raises the possibility that the balance between recycling and degradation will control not only intracellular signaling but also the spatial range and vectoriality of cell fate determination, an area of active investigation [92].

Signaling between Neuronal Cell Body and Terminal

Neurons represent an extreme spatial case in which cellular signaling initiated at the terminal region may culminate in biological effects at the cell body, often millimeters to tens of centimeters away. Indeed neurotrophins such as nerve growth factor (NGF), which bind to TrkA receptors at the terminal region, elicit either cell survival or differentiation effects mediated by nuclear events. An intriguing "signaling endosome" hypothesis [93] has been put forward to explain this signal transmission event, invoking internalization of neurotrophin–Trk complexes, and their retrograde transport down the axon to the cell body. Considerable evidence supports the transport of activated receptors (reviewed in [94]). Vesicle-bound forms of activated phosphorylated TrkA receptors can be detected throughout the axon of stimulated neurons [95], and receptor internalization is required for neuronal differentiation though not for Akt-mediated survival [96]. Activated TrkA receptors, NGF, and components of the Ras-MAP kinase signaling pathway as well as clathrin and AP-2 are recovered in clathrin-coated vesicle fractions prepared from stimulated neurons, but not the early endosomal marker EEA1 [97]. This may suggest that in neurons signaling components have been recruited to coated membranes as well as endosomes or that neuronal signaling endosomes have unique characteristics. Dissection of the role of the retrograde transport of hormone–receptor complexes, as well as nonvesicular transport, which may also contribute to extremely rapid signal transmission [95,98], promises to be an exciting task.

Acknowledgment

We are grateful to Dr. Yanqui Zhao for assistance with the text and figures.

References

1. Carpenter, G. (2000). The EGF receptor: A nexus for trafficking and signaling. *Bioessays,* **22**, 697–707.
2. Clague, M. J. and Urbe, S. (2001). The interface of receptor trafficking and signalling. *J. Cell Sci.* **114**, 3075–3081.
3. Leof, E. B. (2000). Growth factor receptor signalling: Location, location, location. *Trends Cell Biol.* **10**, 343–348.
4. Zwick, E. *et al.* (1999). The EGF receptor as central transducer of heterologous signalling systems. *Trends Pharmacol. Sci.* **20**, 408–412.
5. Geschwind, A. *et al.* (2001). Cell communication networks: Epidermal growth factor receptor transactivation as the paradigm for interreceptor signal transmission. *Oncogene* **20**, 1594–1600.
6. Wells, A. *et al.* (1990). Ligand-induced transformation by a noninternalizing epidermal growth factor receptor. *Science* **247**, 962–964.
7. Ceresa, B. P. *et al.* (1998). Inhibition of clathrin-mediated endocytosis selectively attenuates specific insulin receptor signal transduction pathways. *Mol. Cell Biol.* **18**, 3862–3870.
8. Vieira, A. V., Lamaze, C., and Schmid, S. L., (1996). Control of EGF receptor signaling by clathrin-mediated endocytosis. *Science* **274**, 2086–2089.
9. Kranenburg, O., Verlaan, I., and Moolenaar, W. H. (1999). Dynamin is required for the activation of mitogen-activated protein (MAP) kinase by MAP kinase kinase. *J. Biol. Chem.* **274**, 35301–35304.
10. Ringerike, T. *et al.* (1998). High-affinity binding of epidermal growth factor (EGF) to EGF receptor is disrupted by overexpression of mutant dynamin (K44A). *J. Biol. Chem.* **273**, 16639–16642.
11. Rizzo, M. A. *et al.* (2001). Agonist-dependent traffic of raft-associated Ras and Raf-1 is required for activation of the mitogen-activated protein kinase cascade. *J. Biol. Chem.* **276**, 34928–34933.
12. Bild, A. H., Turkson, J., and Jove, R. (2002). Cytoplasmic transport of Stat3 by receptor-mediated endocytosis. *EMBO J.* **21**, 3255–3263.
13. Lin, S. Y. *et al.* (2001). Nuclear localization of EGF receptor and its potential new role as a transcription factor. *Nat. Cell Biol.* **3**, 802–808.
14. Lai, W. H. *et al.* (1989). Ligand-mediated autophosphorylation activity of the epidermal growth factor receptor during internalization. *J. Cell Biol.* **109**, 2751–2760.
15. Nesterov, A. *et al.* (1994). Phosphorylation of the epidermal growth factor receptor during internalization in A-431 cells. *Arch. Biochem. Biophys.* **313**, 351–359.
16. Emlet, D. R. *et al.* (1997). Subsets of epidermal growth factor receptors during activation and endocytosis. *J. Biol. Chem.* **272**, 4079–4086.
17. Oksvold, M. P. *et al.* (2000). Immunocytochemical localization of Shc and activated EGF receptor in early endosomes after EGF stimulation of HeLa cells. *J. Histochem. Cytochem.* **48**, 21–33.
18. Di Guglielmo, G. M. *et al.* (1994). Compartmentalization of SHC, GRB2 and mSOS, and hyperphosphorylation of Raf-1 by EGF but not insulin in liver parenchyma. *EMBO J.* **13**, 4269–4277.
19. Kaplan, K. B. *et al.* (1992). Association of p60c-src with endosomal membranes in mammalian fibroblasts. *J. Cell Biol.* **118**, 321–333.
20. Pol, A., Calvo, M., and Enrich, C. (1998). Isolated endosomes from quiescent rat liver contain the signal transduction machinery. Differential distribution of activated Raf-1 and Mek in the endocytic compartment. *FEBS Lett.* **441**, 34–38.
21. Xue, L. and Lucocq, J. (1998). ERK2 signalling from internalised epidermal growth factor receptor in broken A431 cells. *Cell Signal* **10**, 339–348.
22. Haugh, J. M. *et al.* (1999). Effect of epidermal growth factor receptor internalization on regulation of the phospholipase C-gamma1 signaling pathway. *J. Biol. Chem.* **274**, 8958–8965.
23. Haugh, J. M. *et al.* (1999). Internalized epidermal growth factor receptors participate in the activation of p21(ras) in fibroblasts. *J. Biol. Chem.* **274**, 34350–34360.
24. Chiu, V. K. *et al.* (2002). Ras signalling on the endoplasmic reticulum and the Golgi. *Nat. Cell Biol.* **4**, 343–350.
25. Jiang, X. and Sorkin, A. (2002). Coordinated traffic of Grb2 and Ras during epidermal growth factor receptor endocytosis visualized in living cells. *Mol. Biol. Cell* **13**, 1522–1535.

26. Haj, F. G. *et al.* (2002). Imaging sites of receptor dephosphorylation by PTP1B on the surface of the endoplasmic reticulum. *Science* **295**, 1708–1711.
27. Haugh, J. M. and Meyer, T. (2002). Active EGF receptors have limited access to PtdIns(4,5)P(2) in endosomes: Implications for phospholipase C and PI 3-kinase signaling. *J. Cell Sci.* **115**, 303–310.
28. Raiborg, C. *et al.* (2001). Function of Hrs in endocytic trafficking and signalling. *Biochem. Soc. Trans.* **29**, 472–475.
29. Komada, M. *et al.* (1997). Hrs, a tyrosine kinase substrate with a conserved double zinc finger domain, is localized to the cytoplasmic surface of early endosomes. *J. Biol. Chem.* **272**, 20538–20544.
30. Urbe, S. *et al.* (2000). Endosomal localization and receptor dynamics determine tyrosine phosphorylation of hepatocyte growth factor-regulated tyrosine kinase substrate. *Mol. Cell Biol.* **20**, 7685–7692.
31. Komada, M. and Soriano, P. (1999). Hrs, a FYVE finger protein localized to early endosomes, is implicated in vesicular traffic and required for ventral folding morphogenesis. *Genes Dev.* **13**, 1475–1485.
32. Lloyd, T. E. *et al.* (2002). Hrs regulates endosome membrane invagination and tyrosine kinase receptor signaling in *Drosophila*. *Cell* **108**, 261–269.
33. Penheiter, S. G. *et al.* (2002). Internalization-dependent and -independent requirements for transforming growth factor beta receptor signaling via the Smad pathway. *Mol. Cell Biol.* **22**, 4750–4759.
34. Attisano, L. and Wrana, J. L. (2002). Signal transduction by the TGF-beta superfamily. *Science* **296**, 1646–1647.
35. Tsukazaki, T. *et al.* (1998). SARA, a FYVE domain protein that recruits Smad2 to the TGFbeta receptor. *Cell* **95**, 779–791.
36. Panopoulou, E. *et al.* (2002). Early endosomal regulation of Smad-dependent signaling in endothelial cells. *J. Biol. Chem.* **277**, 18046–18052.
37. Lu, Z. *et al.* (2002). TGFb activates Smad2 in the absence of receptor endocytosis. *J. Biol. Chem.* **28**, 28.
38. Massague, J. and Kelly, B. (1986). Internalization of transforming growth factor-beta and its receptor in BALB/c 3T3 fibroblasts. *J. Cell Physiol.* **128**, 216–222.
39. Marinissen, M. J. and Gutkind, J. S. (2001). G-protein-coupled receptors and signaling networks: Emerging paradigms. *Trends Pharmacol. Sci.* **22**, 368–376.
40. Hall, R. A., Premont, R. T., and Lefkowitz, R. J. (1999). Heptahelical receptor signaling: Beyond the G protein paradigm. *J. Cell Biol.* **145**, 927–932.
41. Brady, A. E. and Limbird, L. E. (2002). G protein-coupled receptor interacting proteins: Emerging roles in localization and signal transduction. *Cell Signal.* **14**, 297–309.
42. Krupnick, J. G. and Benovic, J. L. (1998). The role of receptor kinases and arrestins in G protein-coupled receptor regulation. *Annu. Rev. Pharmacol. Toxicol.* **38**, 289–319.
43. Ferguson, S. S. (2001). Evolving concepts in G protein-coupled receptor endocytosis: The role in receptor desensitization and signaling. *Pharmacol. Rev.* **53**, 1–24.
44. Perry, S. J. and Lefkowitz, R. J. (2002). Arresting developments in heptahelical receptor signaling and regulation. *Trends Cell Biol.* **12**, 130–138.
45. Tang, Y. *et al.* (1999). Identification of the endophilins (SH3p4/p8/p13) as novel binding partners for the beta1-adrenergic receptor. *Proc. Natl. Acad. Sci. USA* **96**, 12559–12564.
46. Hu, L. A. *et al.* (2002). G protein-coupled receptor kinase 5 regulates beta 1-adrenergic receptor association with PSD-95. *J. Biol. Chem.* **277**, 1607–1613.
47. Hall, R. A. *et al.* (1998). The beta2-adrenergic receptor interacts with the Na+/H+-exchanger regulatory factor to control Na+/H+ exchange. *Nature* **392**, 626–630.
48. Cao, T. T. *et al.* (1999). A kinase-regulated PDZ-domain interaction controls endocytic sorting of the beta2-adrenergic receptor. *Nature* **401**, 286–290.
49. Cong, M. *et al.* (2001). Binding of the beta2 adrenergic receptor to N-ethylmaleimide-sensitive factor regulates receptor recycling. *J. Biol. Chem.* **276**, 45145–45152.
50. Shih, M. *et al.* (1999). Dynamic complexes of beta2-adrenergic receptors with protein kinases and phosphatases and the role of gravin. *J. Biol. Chem.* **274**, 1588–1595.
51. Lin, F., Wang, H., and Malbon, C. C. (2000). Gravin-mediated formation of signaling complexes in beta 2-adrenergic receptor desensitization and resensitization. *J. Biol. Chem.* **275**, 19025–19034.
52. Ferguson, S. S. *et al.* (1996). Role of beta-arrestin in mediating agonist-promoted G protein-coupled receptor internalization. *Science* **271**, 363–366.
53. Goodman, Jr., O. B. *et al.* (1996). Beta-arrestin acts as a clathrin adaptor in endocytosis of the beta2-adrenergic receptor. *Nature* **383**, 447–450.
54. Krupnick, J. G. *et al.* (1997). Arrestin/clathrin interaction. Localization of the clathrin binding domain of nonvisual arrestins to the carboxy terminus. *J. Biol. Chem.* **272**, 15011–15016.
55. Gaidarov, I. *et al.* (1999). Arrestin function in G protein-coupled receptor endocytosis requires phosphoinositide binding. *EMBO J.* **18**, 871–881.
56. Laporte, S. A. *et al.* (1999). The beta2-adrenergic receptor/betaarrestin complex recruits the clathrin adaptor AP-2 during endocytosis. *Proc. Natl. Acad. Sci. USA* **96**, 3712–3717.
57. Laporte, S. A. *et al.* (2000). The interaction of beta-arrestin with the AP-2 adaptor is required for the clustering of beta 2-adrenergic receptor into clathrin-coated pits. *J. Biol. Chem.* **275**, 23120–23126.
58. McDonald, P. H. *et al.* (1999). Identification of NSF as a beta-arrestin1-binding protein. Implications for beta2-adrenergic receptor regulation. *J. Biol. Chem.* **274**, 10677–10680.
59. Claing, A. *et al.* (2001). beta-Arrestin-mediated ADP-ribosylation factor 6 activation and beta 2-adrenergic receptor endocytosis. *J. Biol. Chem.* **276**, 42509–42513.
60. Tsao, P. and von Zastrow, M. (2000). Downregulation of G protein-coupled receptors. *Curr. Opin. Neurobiol.* **10**, 365–369.
61. Gagnon, A. W., Kallal, L., and Benovic, J. L. (1998). Role of clathrin-mediated endocytosis in agonist-induced down-regulation of the beta2-adrenergic receptor. *J. Biol. Chem.* **273**, 6976–6981.
62. Trejo, J., Hammes, S. R., and Coughlin S. R. (1998). Termination of signaling by protease-activated receptor-1 is linked to lysosomal sorting. *Proc. Natl. Acad. Sci. USA* **95**, 13698–13702.
63. Tsao, P. I. and von Zastrow, M. (2000). Type-specific sorting of G protein-coupled receptors after endocytosis. *J. Biol. Chem.* **275**, 11130–11140.
64. Marchese, A. and Benovic, J. L. (2001). Agonist-promoted ubiquitination of the G protein-coupled receptor CXCR4 mediates lysosomal sorting. *J. Biol. Chem.* **276**, 45509–45512.
65. Shenoy, S. K. *et al.* (2001). Regulation of receptor fate by ubiquitination of activated beta 2-adrenergic receptor and beta-arrestin. *Science* **294**, 1307–1313.
66. Hicke, L. (1999). Gettin' down with ubiquitin: Turning off cell-surface receptors, transporters and channels. *Trends Cell Biol.* **9**, 107–112.
67. Levkowitz, G. *et al.* (1999). Ubiquitin ligase activity and tyrosine phosphorylation underlie suppression of growth factor signaling by c-Cbl/Sli-1. *Mol. Cell* **4**, 1029–1040.
68. Joazeiro, C. A. *et al.* (1999). The tyrosine kinase negative regulator c-Cbl as a RING-type, E2-dependent ubiquitin-protein ligase. *Science* **286**, 309–312.
69. Langdon, W. Y. *et al.* (1989). v-cbl, an oncogene from a dual-recombinant murine retrovirus that induces early B-lineage lymphomas. *Proc. Natl. Acad. Sci. USA* **86**, 1168–1172.
70. Lill, N. L. *et al.* (2000). The evolutionarily conserved N-terminal region of Cbl is sufficient to enhance down-regulation of the epidermal growth factor receptor. *J. Biol. Chem.* **275**, 367–377.
71. Petrelli, A. *et al.* (2002). The endophilin-CIN85-Cbl complex mediates ligand-dependent downregulation of c-Met. *Nature* **416**, 187–190.
72. Soubeyran, P. *et al.* (2002). Cbl-CIN85-endophilin complex mediates ligand-induced downregulation of EGF receptors. *Nature* **416**, 183–187.
73. Gutkind, J. S. (2000). Regulation of mitogen-activated protein kinase signaling networks by G protein-coupled receptors. *Sci. STKE* **2000**, RE1.

74. Pierce, K. L., Luttrell, L. M., and Lefkowitz, R. J. (2001). New mechanisms in heptahelical receptor signaling to mitogen activated protein kinase cascades. *Oncogene* **20**, 1532–1539.
75. Daub, H. *et al.* (1996). Role of transactivation of the EGF receptor in signalling by G-protein-coupled receptors. *Nature* **379**, 557–560.
76. Daub, H. *et al.* (1997). Signal characteristics of G protein-transactivated EGF receptor. *EMBO J.* **16**, 7032–7044.
77. Luttrell, L. M. *et al.* (1997). Gbetagamma subunits mediate Src-dependent phosphorylation of the epidermal growth factor receptor. A scaffold for G protein-coupled receptor-mediated Ras activation. *J. Biol. Chem.* **272**, 4637–4644.
78. Prenzel, N. *et al.* (1999). EGF receptor transactivation by G-protein-coupled receptors requires metalloproteinase cleavage of proHB-EGF. *Nature* **402**, 884–888.
79. Damke, H. *et al.* (1994). Induction of mutant dynamin specifically blocks endocytic coated vesicle formation. *J. Cell Biol.* **127**, 915–934.
80. Cao, W. *et al.* (2000). Direct binding of activated c-Src to the beta 3-adrenergic receptor is required for MAP kinase activation. *J. Biol. Chem.* **275**, 38131–38134.
81. Luttrell, L. M. *et al.* (1999). Beta-arrestin-dependent formation of beta2 adrenergic receptor-Src protein kinase complexes. *Science* **283**, 655–661.
82. Barlic, J. *et al.* (2000). Regulation of tyrosine kinase activation and granule release through beta-arrestin by CXCRI. *Nat. Immunol.* **1**, 227–233.
83. Imamura, T. *et al.* (2001). beta-Arrestin-mediated recruitment of the Src family kinase Yes mediates endothelin-1-stimulated glucose transport. *J. Biol. Chem.* **276**, 43663–43667.
84. Ahn, S. *et al.* (1999). Src-mediated tyrosine phosphorylation of dynamin is required for beta2-adrenergic receptor internalization and mitogen-activated protein kinase signaling. *J. Biol. Chem.* **274**, 1185–1188.
85. DeFea, K. A. *et al.* (2000). beta-Arrestin-dependent endocytosis of proteinase-activated receptor 2 is required for intracellular targeting of activated ERK1/2. *J. Cell Biol.* **148**, 1267–1281.
86. DeFea, K. A. *et al.* (2000). The proliferative and antiapoptotic effects of substance P are facilitated by formation of a beta-arrestin-dependent scaffolding complex. *Proc. Natl. Acad. Sci. USA* **97**, 11086–11091.
87. Luttrell, L. M. *et al.* (2001). Activation and targeting of extracellular signal-regulated kinases by beta-arrestin scaffolds. *Proc. Natl. Acad. Sci. USA* **98**, 2449–2454.
88. McDonald, P. H. *et al.* (2000). Beta-arrestin 2: A receptor-regulated MAPK scaffold for the activation of JNK3. *Science* **290**, 1574–1577.
89. Seto, E. S., Bellen, H. J., and Lloyd, T. E. (2002). When cell biology meets development: Endocytic regulation of signaling pathways. *Genes Dev.* **16**, 1314–1336.
90. Cadigan, K. M. (2002). Regulating morphogen gradients in the *Drosophila* wing. *Semin. Cell Dev. Biol.* **13**, 83–90.
91. Entchev, E. V. and Gonzalez-Gaitan, M. A. (2002). Morphogen gradient formation and vesicular trafficking. *Traffic* **3**, 98–109.
92. Entchev, E. V., Schwabedissen, A., and Gonzalez-Gaitan, M. (2000). Gradient formation of the TGF-beta homolog Dpp. *Cell* **103**, 981–991.
93. Grimes, M. L. *et al.* (1996). Endocytosis of activated TrkA: Evidence that nerve growth factor induces formation of signaling endosomes. *J. Neurosci.* **16**, 7950–7964.
94. Ginty, D. D. and Segal, R. A. (2002). Retrograde neurotrophin signaling: Trk-ing along the axon. *Curr. Opin. Neurobiol.* **12**, 268–274.
95. Bhattacharyya, A. *et al.* (1997). Trk receptors function as rapid retrograde signal carriers in the adult nervous system. *J. Neurosci.* **17**, 7007–7016.
96. Zhang, Y. *et al.* (2000). Cell surface Trk receptors mediate NGF-induced survival while internalized receptors regulate NGF-induced differentiation. *J. Neurosci.* **20**, 5671–5678.
97. Howe, C. L. *et al.* (2001). NGF signaling from clathrin-coated vesicles: Evidence that signaling endosomes serve as a platform for the Ras-MAPK pathway. *Neuron* **32**, 801–814.
98. MacInnis, B. L. and Campenot, R. B. (2002). Retrograde support of neuronal survival without retrograde transport of nerve growth factor. *Science* **295**, 1536–1539.

PART V

Cell–Cell and Cell–Matrix Interactions

E. Brad Thompson, Editor

Part V

Introduction

E. Brad Thompson

In Part V of the Handbook we provide overviews that integrate the *cis*-signaling systems involved in the regulation of many major organs or tissue systems. For the most part, these chapters will not attempt to repeat the information concerning the pathway details that have already been covered in previous sections. Instead, the theme here is to provide overviews of the multiple signaling molecules and pathways that must be integrated at the tissue and organ level in order for cellular homeostasis to be maintained. This part begins with an overview of cell–cell and cell–matrix interactions that introduces readers unfamiliar with the concepts to the various types of ways in which signals are delivered. These include endocrine, paracrine, autocrine, and intracrine signal systems. These various ways in which signaling molecules are presented allow for control of the signals in a three-dimensional, geographic sense. The use of some signals are restricted locally to individual cells or nearby cells, while others are delivered from one set of cells to others at varying degrees of distance. Following this introductory discussion, individual tissues and systems are considered in the succeeding chapters.

General overviews of the various major signals to lung, bone, prostate, breast, gastrointestinal tract, B-lymphoid cells and T-lymphoid cells, pancreas, testis and ovary, and the hematopoietic system are presented. Two aspects of signaling in the nervous system are covered. Neurotrophic factors and how their responses are integrated are discussed in the chapter by Perez-Polo. A recently discovered aspect of nervous system signaling, namely, retrograde signaling by way of dorsal root reflexes, is dealt with in the chapter by Willis. Among the many signaling systems that play on the liver, the stress response and its variation in the young versus the aged animals is discussed in the chapter by Papaconstantinou. The cardiovascular system is represented by two chapters: one dealing with the signaling pathways controlling cardiogenesis, and the second with the regulation of angiogenesis in the postnatal state. The latter obviously has relevance not only to normal development, but also to malignancy, since every growing tissue must be supplied with blood in order to continue to grow.

In most of these discussions, the reader will find that the focus of recent research has been on the discovery of increasing numbers of signaling molecules and how they play upon each tissue or system. Much remains to be done to understand how, in any particular tissue, all the known signals create an integrated response. How the multiple signaling systems interact to achieve the overall homeostatic regulation of an organ has not yet been determined in most cases. Our state of knowledge is one in which we are uncovering the regulatory molecules and the pathways to which they individually act. In very recent years, we have begun to discover the interactions among those pathways, but exactly what the interplay of pathways is that results in the normal regulation of the cell or a tissue is not yet understood, in most cases. This next order of understanding will doubtless be the pursuit of the coming generation of scientists, but it must be based on knowledge of the entire repertoire of signaling molecules that are relevant to that tissue, the pathways through which they act and precise knowledge of how those pathways govern cell behavior molecularly, as well as precise knowledge of the interpathway interactions. A combination of cell biology, biochemistry, molecular biology, and mathematics will ultimately, in the opinion of this editor, be required to solve this large and intricate problem.

Overview of Cell–Cell and Cell–Matrix Interactions

E. Brad Thompson[1] and Ralph A. Bradshaw[2]

[1]*Department of Human Biological Chemistry and Genetics, The University of Texas Medical Branch, Galveston, and*
[2]*Departments of Physiology and Biophysics, Anatomy and Neurobiology, College of Medicine, University of California, Irvine.*

Cell-signaling pathways are not simply linear, but in fact form extensive interactive networks. Indeed it is the overlapping and interconnecting nature of these that provides the distinctive features distinguishing many of the response properties of specific tissues and organs. The complexity of these networks will require a great deal of research before their organization is understood in detail, but some generalities are beginning to emerge [1]. The advent of the techniques of genomics, including microchip arrays of genes and proteomics, will stimulate much more rapid development of understanding of pathway interactions as we see how a given signal reverberates through the tissue and cellular systems. Indeed, a considerable amount is already known, and in many systems the overall patterns are beginning to be made clear. Completing this knowledge will undoubtedly be the goal of a great deal of research in the near future. This section describes how many major tissue and organ systems work as they consider the many signals that play upon them.

As one reads entries in this section, it may be necessary for the uninitiated peruser to refer to earlier entries in other sections, which explain in detail how particular signal pathways function, as this section deals only with the identification of those pathways that are important in the control of organ and tissue function and not the iteration of how each pathway works. However, one concept not covered previously needs to be introduced, i.e. the idea that chemical signals may be provided locally or regionally in tissues by a group of mechanisms that have become known as paracrine, autocrine, intracrine and juxtacrine interactions. These constitute means for regulating tissue-specific signal responses–by providing the needed signals only on a local basis.

These concepts were born from the field of endocrinology. Traditionally this discipline held that certain specific organs produce and secrete particular signaling molecules into the blood stream, which delivered them elsewhere in the body to carry out their signaling activity. A classic example is the production of insulin by the beta cells of the pancreas, with the hormone transferred systemically to many other tissues, where it activates its receptors thus affecting glucose metabolism, among other responses. Over the last 10 to 20 years, it has become clear that, in addition to this classic **endocrine** notion, signaling molecules are also produced to function more locally. That is to say, although they may not enter the general circulation and consequently act only locally, they nevertheless work by binding receptors–either on the surface or within cells–and set off the same types of signal transduction pathways, as do traditional hormones. In fact, a general understanding has evolved that signal transduction pathways that usually have been addressed separately because of their inclusion in a particular discipline, e.g. endocrinology as contrasted with immunology, behave in much the same ways. A great unanimity in general mechanisms is seen as signals are transmitted between cells, whether they be signals from one immune cell to another or signals from a classic endocrine target organ to another tissue. In this sense, the same types of chemical and physical behaviors ultimately carry out cell signaling universally. The specific types of mechanisms mentioned above, in addition to endocrine, are defined according to the degree of localization of effect, but the signals generated carry out their functions by the same sorts of receptor-transduced pathways, as do the ubiquitous signaling molecules.

In addition to the classical endocrine mechanism, localized signaling mechanisms can be conveniently sub-grouped into four types. **Paracrine** interactions induce signaling activities that occur from cell-to-cell within a given tissue or organ, rather than through the general circulation. This takes place as locally produced hormones or other small signaling molecules exit their cell of origin, and then by diffusion or local circulation act only regionally on other cells of a different type within that tissue. This has been found to be important in many organs and is a field of investigation that continues to develop rapidly. The local concentrations of paracrine signals can be quite high compared to the circulating levels, and thus can trigger effects by acting on low affinity receptors or by supplying sufficient local signal to bind to high affinity receptors even when the circulating level of a molecule that produces a similar signal is too low to do so. Paracrine signaling molecules sometimes are very rapidly metabolized locally to further limit the physical extent of their action. Examples are the prostaglandins and nitric oxide. Other, longer-acting signaling molecules also are employed in a paracrine fashion. Limiting their access to the blood supply and/or the total amounts produced within a tissue, so that local receptors can bind them before the general circulation is reached, keeps them acting in a paracrine fashion. Sometimes paracrine signaling molecules are moved by local physical means, so as to provide greater or lesser regional concentrations. An example is found in the development of the heart (see Chapter on Cardiogenesis). Many growth factors function by paracrine mechanisms.

The term **autocrine** refers to entities that are released from a cell and bind to receptors on that same cell, thereby activating it. This sort of self-stimulation occurs in carefully timed phases during normal embryonic development and tissue differentiation. It is also used in inflammation and wound healing. Such localized signals help direct the concentration of appropriate cells at the wound or inflamed tissue. In addition to their importance in normal tissues, these localized signaling systems have been discovered to be quite important in understanding the autonomy achieved by cancer cells. Quite often, one of the contributing mechanisms by which a malignant cell population escapes the normal control mechanisms for regulated growth is by producing autocrine, paracrine, and other localized types of signals that stimulate cell division or other activities that favor survival and expansion of the cancer cell population.

It is interesting to note that historically the underlying concepts inherent in these mechanisms were appreciated, although not at the molecular level, as early as 1775. De Bordeu, and later Brown-Sequard in 1891 [2], proposed that every cell, not just tissues and organs, actively secreted into the circulation substances that influenced other tissues. The focus on the role of endocrine glands (thyroid, pituitary, adrenal, pancreas, etc.) in providing these 'internal secretions' obscured the existence of the autocrine and paracrine messengers (as did the emphasis on the circulation) for some time, but the appreciation that all cells can and do actively secrete regulatory elements was an essential concept that was clearly recognized (but alas not considered to any great extent until relatively recently) over 200 years ago.

Both normal and pathological conditions can use the same hormone for autocrine and paracrine interactions. Autocrine regulation of a phase of keratinocyte development by NGF is well established [3], while the same factor generally acts as a paracrine regulator of sympathetic and selected sensory neurons [4]. At the same time, many tumors progress by autocrine stimulation by any one of several mitogenic factors, such EGF and FGF. These same substances also participate in numerous normal tissue situations using paracrine mechanisms (See entries on these factors in earlier sections).

Two additional mechanisms have been proposed that further extend cell signaling beyond the action of circulating messengers. These are **juxtacrine** signals, in which the signaling entity (receptor ligand) is not soluble but is membrane-bound on one cell and is delivered by cell-cell physical approximation to the cell bearing the receptor (usually but not necessarily different in type from the target cell) and **intracrine signals**, in which both receptor and ligand are expressed intracellularly and signals are generated without external stimuli. The former mechanism is now well established and is exemplified by a number of systems, such as the notch receptor [5] and the tyrosine kinase receptor family, Eph [6]. There is less compelling evidence for intracrine mechanisms, although there are a number of growth factors, e.g. FGF1 and 2 and interleukin 1, that are not exported in the usual manner via the endoplasmic reticulum and that clearly exhibit both extracellular and intracellular concentrations of factor (see chapter by Maciag). Thus the intracellular material could be appropriate for such signaling.

Throughout the chapters of this section, the reader will find multiple applications of these concepts as they are used in physiologically relevant systems.

References

1. Barolo, S. and Posakony, J. W. (2002). Three habits of highly effective signaling pathways: principles of transcriptional control by developmental cell signaling. *Genes & Development.* **16**, 1167–1181.
2. Brown-Sequard, C. E. and D'Absonval, A. (1891). *Compt. rend Soc. biol.* **5s**, 248.
3. Di_Marco, E., Mathor, M., Bondanza, S., Cutuli, N., Marchisio, P. C., Cancedda, R., and De_Luca, M. (1993). Nerve growth factor binds to normal human keratinocytes through high and low affinity receptors and stimulates their growth by a novel autocrine loop. *The Journal of Biological Chemistry* **268**, 22838–22846.
4. Huang, E. J. and Reichardt, L. F. (2001). Neurotrophins: Roles in Neuronal Development and Function. *Annual Reviews of Neuroscience* **24**, 677–736.
5. Weinmaster, G. (2000). Notch signal transduction: A real rip and more. *Current Opinion in Genetics & Development* **10**, 363–369.
6. Zisch, A. H. and Pasquale, E. B. (1997). The Eph family: A multitude of receptors that mediate cell recognition signals. *Cell and Tissue Research* **290**, 217–226.

Chapter 334

Angiogenesis: Cellular and Molecular Aspects of Postnatal Vessel Formation

Carla Mouta, Lucy Liaw, and Thomas Maciag
Center for Molecular Medicine,
Maine Medical Center Research Institute, Scarborough, Maine

Introduction

Angiogenesis is not the process by which vessels originate *de novo* in the embryo nor does it necessarily imply the use of stem or precursor cells in the adult. Those processes would be more typically referred to as vasculogenesis, and are not the subject of this review. In other words, angiogenesis does not mean the inceptive making of a vessel from the assembly of all its individual components. Rather, angiogenesis is the growth of a vessel from a preexisting one by mechanisms that include sprouting, branching, and intussusception. Historically, the term *angiogenesis* was used indiscriminately to describe the growth of both blood and lymph vessels. However, recent observations suggest that these two vessel types are formed and grow by different mechanisms. Indeed, hemangiogenesis (blood vessels) and lymphangiogenesis (lymphatics) are important multistep biological processes integral to both normal mammalian physiology (e.g., wound healing, adaptation of exercised muscles, development, maturation and endocrine function of the female reproductive tract) and disease (e.g., lymphedema, collateral vessel development in ischemic hearts and limbs, diabetic retinopathy, age-related macular degeneration, restenosis, rheumatoid arthritis, and tumor growth and metastasis).

The study of the angiogenic mechanisms can be divided into two disciplines: mechanisms responsible for the induction of vessel formation and mechanisms responsible for its inhibition. Because the latter have been quite controversial, our focus will be on the former. Here we present a basic analysis of the overriding principles of postnatal angiogenesis. Clearly, a comprehensive description of the method is outside the scope of this review. For a dissection of the intra- and intercellular signaling pathways involved in angiogenesis and vasculogenesis, the reader is referred to the accompanying work of Holash, Alitalo, and Donoghue (Part H) and various recent reviews [1–4].

Initiators of Angiogenesis: Cellular, Metabolic, and Mechanical

Apart from the self-directed and abnormal growth of vascular malformations and vascular tumors (e.g., hemangiomas and lymphangiomas) [5], postnatal angiogenesis and lymphangiogenesis occur only upon local tissue demands. For instance, blood vessels can either grow or degenerate according to tissue-specific needs for oxygen [6], glucose [7], and alterations in pH [8]. Blood rheology, as well as mechanical stress around a vessel, also controls angiogenesis [9]. Much less is known about the needs that lymph vessels fulfill and what specific tissue demands govern their morphogenesis [10,11]. However, the contribution of lymphatics to local fluid balance in tissues, to the drainage of metabolic products, as well as to immune surveillance suggests that parameters as diverse as hydrostatic pressure [12] and inflammatory stromal reactions [13] are likely triggers for lymphangiogenesis [11]. Other driving forces may include long-term alterations of extracellular pH and increases in cell death. Consequently, there are multiple factors that influence the growth of any particular vessel and perhaps an equally large set of mechanisms by which this growth proceeds. Notably, "angiogenesis is not angiogenesis is not angiogenesis" [14] because vessel growth occurs through different processes in tissue repair (wound healing [14], transplantation), chronic inflammation (psoriasis [15], rheumatoid arthritis [16], corneal neovascularization [17]), and tumorigenesis [4,18].

Vessel-Specific Requirements in Angiogenesis

Not only can different mechanisms be utilized in the making of a blood or lymph-carrying vessel, the complexity of those increases with the number of layers that comprise that vessel (intima, media, and adventitia). In vertebrates, functional vessels are more than endothelial-lined tubes. Vessels must adapt, accommodate, and respond to the temperature, pH, rheological, and mechanical parameters of both the surrounding tissue microenvironment and the fluid they carry [10,19]. That is how the blood vasculature most effectively utilizes its capillaries, sinusoids, venules, veins, and arteries [20] and how the lymph vasculature capitalizes on its network of initial lymphatics, collector lymphatics, ducts, and trunks [21]. Each branch serves a specific purpose, and each one of the latter is accomplished by a diverse set of cellular and extracellular interactions. Specific endothelial cell phenotypes (e.g., continuous, fenestrated, discontinuous, lymphatic valves, and anchoring filaments) are accompanied by the presence or absence of a particular mural cell [e.g., pericytes, smooth muscle cells (SMCs), podocytes, astrocytes] [22] and tailored perivascular matrices (e.g., variable collagen and laminin subtypes within the basement membrane). Certainly, *in vitro*, endothelial cells (ECs) appear to contain all of the necessary ingredients for forming a tubelike structure within a given provisional matrix [23,24]. But how functional are these tubes? In other words, how effective would these vessel-like structures be with regard to fluid uptake and transport *in vivo*? Recent studies have suggested that further components need to be incorporated when these vessel-like tubes are implanted *in vivo* [25]. Thus, while some of these questions are only now beginning to be elucidated, the answers will be critical to the development of successful cell-based strategies [26] and better biomaterials [27] for tissue repair.

Cellular and Soluble Regulators

The complete set of the cellular and molecular players involved in the production of each layer of a functional vessel is unknown, even for the most rudimentary of the capillaries. In hemangiogenesis, the preexisting vessel itself provides some but not all components and instructions for the formation of its new sprouts and branches (Table I). For instance, on one hand are the endothelial and mural cells (e.g., pericytes and SMCs) that migrate and multiply on extravasated clotted plasma (provisional matrix of fibrin and fibronectin) [24]. However, also frequently utilized are fibroblasts (which release angiogenic cytokines, as well as chemokines that attract inflammatory cells) [28] and a variety of subtypes of leukocytes [29]. The latter are important sources of other chemokines and proangiogenic factors [matrix metalloproteinases (MMPs), interleukin 1 (IL-1), IL-8, fibroblast growth factors (FGFs), interferons (IFNs), and others], in particular for angiogenesis occurring during wound healing and repair [3,30]. It is likely that the growth of blood and lymph vessels utilizes some common principles, but distinct players and mechanisms probably exist as well. For instance, lymphatic endothelial cells (LECs) seem to prefer VEGF-C to VEGF-A and indeed preferably express its major receptor Flt4/VEGFR3, a member of the VEGFR family [31]. However, although VEGF-C overexpression can lead to lymphatic hyperplasia [32], the factors involved in the promotion of lymphatic sprouting remain unknown. Furthermore, another member of the VEGFR family, neuropilin 2 (NRP-2), is expressed in visceral but not cutaneous lymphatics in the adult mouse [33]. A similar organ-specific modulation of lymphatic formation has been recently reported for angiopoietin 2 (Ang2) [34]. These observations further support the premise that there will be tissue-specific regulators of lymphangiogenesis, much like there are tissue-specific mitogens for blood vessel endothelial cells such as EG-VEGF [35]. Unfortunately, much less is known about embryonic and adult lymph vessel formation than about hemangiogenesis. Thus, we are limited in our description to some instances in which clear differences between these processes have been observed.

Coordination of Angiogenesis by Cellular and Molecular Interactions

During angiogenesis, vascular cells break up and establish new contacts (gap, tight, and adherens junctions), including contact with the surrounding matrix [36–38]. Moreover, angiogenic cells continuously lay down and degrade the extracellular matrix (ECM) to facilitate their migration and proliferation. Thus, angiogenesis likely results from the simultaneous and harmonized occurrence of multiple cellular and molecular processes. Altogether, a functional vessel may be obtained through the following temporal sequence of events: An initial angiogenic stimulus triggers cell activation, which is followed by cell migration, division, and alignment, vessel pruning and maturation, and phenotypic and organotypic differentiation [19] (Fig. 1). According to this process, the onset of hemangiogenesis is the opening of intercellular junctions in the endothelial lining, allowing the leakage of plasma into the subendothelial space. Thus, one of its possibly

Table I Cell–Cell, Cell–Matrix Interactions and Their Molecular Mediators in Angiogenesis

	Cell activation & local remodeling	Cell migration & vessel sprouting	Cell division	Lumen formation & maturation	Differentiation (Phenotypic & organotypic)
Adventitial cells	Protease production ECM degradation Release of growth factors (VEGF, FGF1 and 2, IGF1)	Migration stimuli S1P from aggregated platelets and mast cells is chemotactic for EC and inhibits PDGF-induced migration of SMC	Mitogen Production VEGFs, FGFs, TGFβ, IL1s, PDGF, and IGF1	Cell differentiation/survival S1P regulates cell differentiation via cadherins/adherens junction assembly; it can promote cell survival from apoptosis due to anoikis	EC-Inflammatory cells Different vascular beds acquire specific subsets of homing receptors for blood cells (selectins)
Leukocytes fibroblasts	TSP-1 stimulates the release of growth factors by fibroblasts				
Mural cells	Disruption of EC-mural cell interactions and underlying matrix	Cell-Matrix remodeling Integrin interaction with FN, laminin, vitronectin, osteopontin, HA, and others facilitates vessel sprouting	SMC proliferation regulated by ECM components (elastin and others)	Investment Ang1 produced by mesenchymal cells tightens EC investment by mural cells, which then inhibit EC migration and proliferation via TGF-β	Vascular-bed-specific mural cells brain astrocytes kidney podocytes venular pericytes arterial SMC Provide survival and vasomotor agents, growth factors and cytokines
Pericytes smooth muscle cells	Ang2	PDGF-BB induces migration and proliferation on pericytes	FGFs and PDGFs recruit mural cells to nascent vessel sprouts		Mesenchymal cell differentiation and ECM deposition TF, TGFβ endoglin, dHAND
Lumenal cells	Adhesion molecule redistribution PECAM, VE-Cadherin, endoglin	Cell-Matrix remodeling Disruption of EC-EC contacts (PECAM-1 VE-cadherin, ephrins) VEGF-E, PlGF, modulate degradation of ECM by MMPs, uPA	Mitogens FGFs TNF VEGFs TGF PDGF-BB (mural cell growth)	Lumen Formation VEGF121 & 165 combined with Ang1 (↑) but VEGF189, TSP-1 (↓)	Junctional communication Increase in tight, gap, adherens junctions (cadherins, connexins) Individual Ephrin bHLH factors determine vessel specification
Endothelial cells	Cell invasion Degradation of ECM by uPA, tPA. Activation of plasmin and MMPs TNFα, LPS, and IL1s induce EphA1 in EC	Cell Movement, Sprouting Facilitated by EphA1 activation, adhesion to FN and Tiam1 redistribution in adherens junctions	Tissue-specific mitogens for EC (e.g. EG-VEGF)	Branching, Tube Formation Integrins, Notch:Delta/Jagged FGFs, signals influence branching, S1P:Edg system is involved in tube-formation	

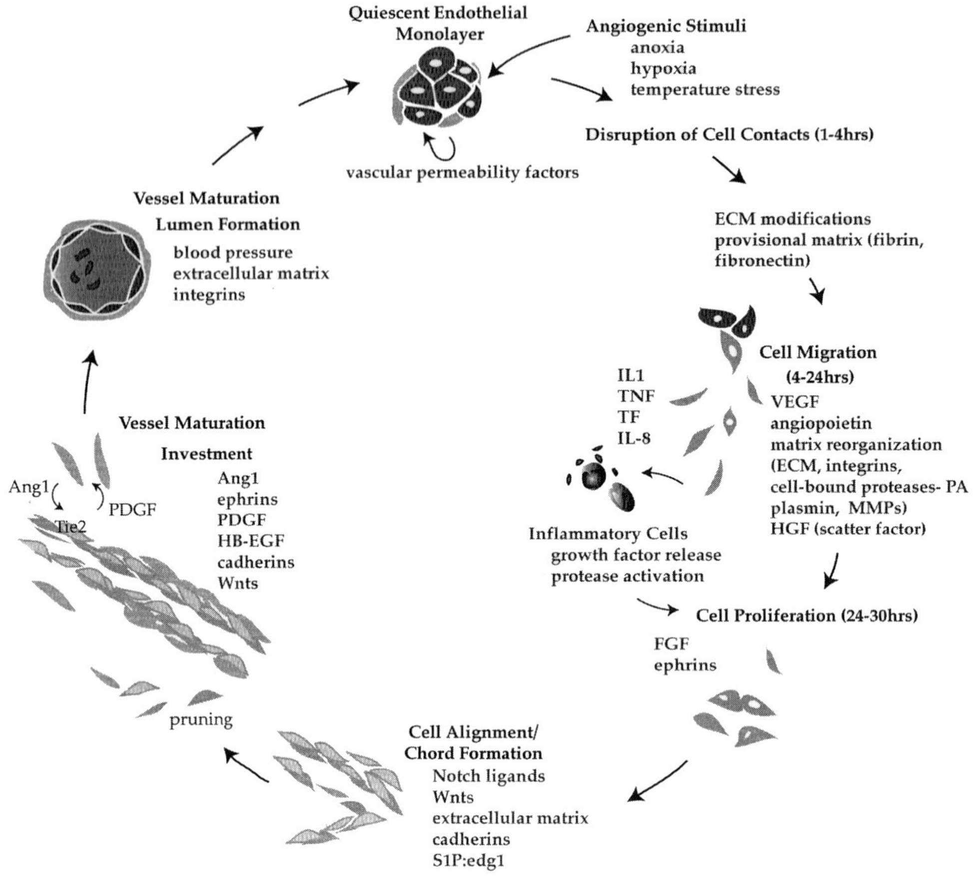

Figure 1 Regulation of the angiogenic cycle. The major steps involved in angiogenesis are receipt of an angiogenic stimulus, disruption of endothelial cell contacts and cell migration, cell proliferation, cell alignment and tubule formation, and the maturation of vascular structures into vessels by investment with mural cells and formation of a lumen continuous with the circulation. Factors that control the various steps are listed.

earliest modulators is vascular permeability factor (VPF) [39,40], the prototype member of the VEGF family. VPF, which may have variable cell sources including fibroblasts [41] and platelets [42], mediates EC dilation [43], permeability, and migration [44] both through the stimulation of nitric oxide (NO) production and as a result of NO accumulation [45]. Alternative vasodilators useful in angiogenesis are prostaglandins, lipid products of arachidonic acid metabolism by cyclooxygenases (COX-1 and COX-2) [46,47] that regulate multiple steps in angiogenesis.

Resident macrophages, polymorphonuclear leukocytes (PMNs), and mast cells can be attracted to sites in need of new vessels by proinflammatory growth factors and by cytokines and chemokines such as IP-10 and PF4 [29,30]. Once stimulated under angiogenesis-promoting conditions such as hypoxia and high lactate concentrations, these cells release IL-1s, TNFs, FGFs, and proteolytic enzymes (elastase, cathepsins, and proteinase-3) that facilitate ECM degradation and activation of MMPs [48]. In addition, they also induce the ECs, fibroblasts, and keratinocytes to release another set of proteases that further degrade the basal lamina, facilitating EC migration and growth toward the chemotactic source [49]. Consequently, there is an overall loosening of the original cell–cell and cell–matrix contacts during the earliest angiogenic events. These processes involve a large group of proteins of the integrin, selectin, and cadherin families, as well as members of the immunoglobin-like gene family (ICAM, VCAM, PECAM) [50]. For instance, the loosening affects PECAM-1 in that it becomes redistributed out of EC–EC junctions in response to VEGF and TNF-α [51]. As more PECAM-1 becomes available for interactions with integrins, EC migration is triggered. Notwithstanding, proteoglycans [52], as well as CD44 [53], LYVE-1 [11,54], and other receptors for hyaluronan, which is the most abundant sugar responsible for hydration of the ECM, may also facilitate cell migration. One of these molecules, LYVE-1, is concentrated in lymphatics and a subset of blood vessels (liver sinusoids) [11]. Thus, LYVE-1 may participate in hemangiogenesis and lymphangiogenesis as well [54]. Finally, proinflammatory molecules such as TNF, IL-1, and LPS as well as VEGF induce the expression of certain ephrins. It is thought that the interaction of ephrins with specific Eph receptors [3,38] on adjacent ECs promotes their sprouting, migration, and capillary tube formation [3,38]. Similar events may also be mediated by the Notch ligands, Jagged and Delta [55].

An adult human has approximately two pounds of ECs [56], primarily in a resting or quiescent state (G_0 phase, "outside" of the cell cycle). Notably, the turnover rates of the endothelia are of the order of hundreds of days (5000 days for the retinal blood vessels), and during normal conditions less than 0.01% of the ECs are "in cycle" [4]. Thus, after their initial activation and migration, EC proliferation upon exposure to mitogens (such as FGFs and PDGFs) begins by reentry into the G_1 phase of the cell cycle [57,58]. However, commitment to division may actually require the continuous presence of potent mitogenic factors such as FGFs throughout most of the remaining phases [59]. EC mitogens activate the cell cycle via Rb phosphorylation as well as cyclin production and activation, but progression is also influenced by cues from EC contacts with the ECM. For instance, integrin-mediated signals from the matrix regulate the levels of p21, p27, and p53 [60]. It is anticipated that the recent development of methods allowing LEC identification *in vivo* [61], in addition to permitting establishment of LEC cultures *in vitro* [62], will provide a more clear insight into the mechanisms of LEC quiescence and proliferation. Indirectly, the inflammatory cytokines IL-1 and TNF may also induce LEC proliferation by promoting the expression of VEGF-C by fibroblasts and ECs [63]. As such, it has been proposed that lymphatics are activated in inflammation [11,63] and that the resulting lymphangiogenic response plays a role in this process by controlling the composition and interstitial fluid pressure, as well as by facilitating trafficking of immune cells [13,63].

For sprouting of new vessels, contacts between ECs (tight- and adherens-type junctions) have to be loosened, and new contacts established later [64,65]. During these processes, ECM deposition and production by the tissue stroma change remarkably, resulting in a transient provisional matrix. In blood vessels, this matrix serves both as a signal for EC adhesion, survival, and migration and as a regulator of growth factor activity. An early event in the establishment of the provisional matrix is the deposition of fibrin/fibrinogen, fibronectin (FN), and vitronectin (VN) partly due to increased vascular leakage [66]. Proteins including thrombospondin 1 (TSP-1), TSP-2, osteopontin (OPN), SPARC, tenascin-C [67], and del1 [68] are components of this provisional matrix and are postulated to participate in regulating EC behavior and hemangiogenic events. Recent *in vivo* genetic approaches using transgenes or null mutant animals have demonstrated that TSP-1 and -2 are context-dependent natural inhibitors of angiogenesis [69,70], whereas OPN promotes angiogenesis [71]. The combination of adhesive glycoproteins and those with antiadhesive properties illustrates that cells need to constantly adapt when undergoing remodeling events [72]. The observation that many of the components of the "angiogenic matrix" interact with alpha v-containing integrins is consistent with the importance of $\alpha v\beta 3$ and $\alpha v\beta 5$ in the angiogenic process [73]. Indeed, although many integrins are expressed in quiescent endothelium *in vivo* and interact with normal collagens and laminin in the basement membrane, EC integrin profiles change during angiogenesis. In particular, both $\alpha v\beta 3$ and $\alpha 5\beta 1$ are increased, and that correlates to their activities as receptors for multiple components of the provisional matrix [65]. Finally, continued remodeling of the cell environment is accomplished by modification/refinement of the provisional matrix by proteases. For instance, MMPs not only degrade proteins, but also modify the activity of matrix proteins as well as cellular receptors to provide a matrix suitable for cellular migration and proliferation [74].

Furthermore, most but not all vessels are invested with a SMC-like layer that provides them contractility, stability, and survival factors [75,76]. Notable exceptions are some blood sinusoids and the initial lymphatics [77]. The latter nevertheless appear to have evolved anchoring filaments that protrude from the LEC and establish close contacts with ECM that also involve integrins localized to focal adhesions [78]. These structures are not apparent in lymphatic ducts and trunks, which in turn are invested by mural cells. Consequently, in most situations, as an EC begins to migrate and divide, it does loosen its association with the eventual pericyte or SMC involving it. The investment of the endothelium by mural cells is largely regulated by Ang-1 and Ang-2/Tie receptor complexes at the EC surface and involves both tight and gap junctions, as well as adhesion plaques [64]. These interactions influence the EC proliferative state, and later help in EC survival during vessel pruning and maturation at a time in which the levels of proangiogenic growth factors (VEGFs, FGFs) and other molecules (Ang2) become lower than those of endogenous antiangiogenic molecules [34] (e.g., TIMPs, fragments of several ECM molecules, TSP-1, IL-4). SMC and pericytes also provide several useful proteases (e.g., plasmin, urokinase, tPA, MMPs) that may release ECM-stored growth factors (FGF and VEGF) and other molecules that participate in the angiogenesis process [22].

A primitive vessel thus assembles along the concentration gradient of angiogenic factors (mitogens, proteases, and others), laying the foundation for a new branch of the vasculature that is still only barely functional. As VEGF, FGFs, and Ang2 levels begin to fall (for instance, as a result of improved tissue oxygenation), Ang1 produced by mesenchymal cells activates the Tie2 receptor on EC, and this in turn leads to the production and release of a recruitment signal for pericytes (PDGF-BB) and SMCs (PDFG-AA, HB-EGF) [1]. Once these cells arrive and contact the endothelium, TGF-ß may be activated, inhibiting EC proliferation, altering integrin expression profiles, and stimulating matrix deposition [22]. Similarly, it is likely that ECs either secrete or express surface molecules that contribute to pericyte and SMC quiescence [79]. Vessel integrity is thus strengthened; on the other hand, some vessels get pruned, as is the case of premature blood vessels that fail to become properly invested by mural cells [76]. Thus, an intriguing question is how do initial lymphatics manage to survive and function without mural cells? Is their role somehow served by the anchoring filaments instead? As blood vessels mature, they also establish a different set of interactions with the newly deposited ECM (e.g., a continuous layer of laminin and type IV collagen surrounded by the interstitial collagens type I and III) that replaces the provisional matrix (e.g., a fibrillar network of fibronectin and type V collagen with patchy deposits of laminin and type IV collagen). These EC–ECM contacts are important in lumen formation, vessel elongation, and acquisition of a vessel-specific EC differentiated phenotype. Furthermore, proper sprouting, branching, and lumen formation of vessels seem also to utilize helix–loop–helix (HLH)-containing Id (inhibitors of differentiation) proteins. This is partly possible because Id proteins can regulate the function of various transcription factors in EC, thus altering their production of integrins, MMPs, and other proteins utilized for EC–ECM interactions and remodeling [80].

Finally, the EC–mural cell interaction is important for effective function of the new vessel—and not just to its growth control and maintenance [22,81]. For instance, vessel tone is controlled by the action of EC-derived factors (NO, endothelin-1) on adjacent SMCs. Astrocytes contribute to the establishment of the blood–brain barrier and blood–retinal barrier in neuronal vessels by fostering the formation of tight junctions between ECs. Finally, type II alveolar epithelial cells play a critical role in the development of a correct vasculature in pulmonary alveoli, probably through VEGF-A [88]. The progression from an actively migratory and proliferative phenotype to quiescence is further accompanied by the acquisition of a more differentiated phenotype for each of the cellular vessel components. For instance, the endothelium itself can have microvalves such as in the lymphatics [82] and can be continuous, discontinuous, fenestrated, or a combination of the above according to the specific needs of each vascular bed [20]. Constitutive expression of VEGF/VEGFR complexes in adult tissues (choroid plexus and kidney glomeruli) suggests that this system can promote organotypic EC differentiation and permeability of fenestrated endothelium [20]. Thus, the variability of local organ and tissue microenvironments (e.g., biochemical, mechanical, and biophysical environments, presence or absence of inflammatory stimuli) expectedly requires a large array of subtypes of EC, each one with its unique differentiated phenotype to serve specific cellular and matrix environments. In other words, the exact properties of an EC are not easily predicted from the vessel from which it derives nor are they immutable. Some of the molecules that determine the fate and identity of specific EC have been recently discovered (Notch/Jagged and Delta [55,83] ephrins/EphRs [38], Frizzleds/Wnts [84], and Prox 1 [10,85]), but most are likely still to be determined. Whether these novel ligand-receptor signaling pathways integrate into the activated receptor tyrosine kinase-mediated signaling pathways is also currently not known. However, it is anticipated that they will indeed do so to direct the fate of an individual vessel in response to microenvironmental cues.

Therapeutic Implications

It is becoming increasingly evident that inflammation may be the common denominator to most post-natal and pathological microenvironments in which blood or lymph vessel growth occurs. Less clear, however, is how the multitude of cellular and molecular players that are present in inflammatory settings act in concert to induce the timely proliferation and regression of these vessels. Because of their multiple sources and target cells, the IL-1 prototypes, their structurally-related cytokines, the FGF1 prototypes (86), and the VEGFs are possibly some of the most crucial elements of the pro-inflammatory angiogenic response (Figure 2). Indeed, their relevance as therapeutic targets is highlighted by the recent observation that TTM, a specific chelator of physiological copper, can exert potent anti-angiogenic effects largely by complex biochemical mechanisms that involve sequestration of intracellular IL-1α and FGF1 thereby effectively reducing the extracellular pool of these proteins (87). Moreover, the very recent discovery that mice which are genetically null for the IL1 prototypes exhibit significantly impaired tumor invasiveness and angiogenesis (88) reinforces the role of these pro-inflammatory and angiogenic signals in the regulation of the microvasculature in response to inflammatory challenges. These results offer new and exciting venues for the treatment of cancer, atherosclerosis, lupus, psoriasis and a plethora of other autoimmune disorders associated with abnormal vessel proliferation.

Acknowledgments

The authors thank Norma Albrecht and Gloria Ledoux for expert administrative assistance. This work was supported in part by NIH grants AG 98503 and HL 32348 to T. M.

References

1. Conway, E. M., Collen, D., and Carmeliet, P. (2001). Molecular mechanisms of blood vessel growth. *Cardiovasc. Res.* **49**, 507–521.
2. Karkkainen, M. J. and Alitalo, K. (2002). Lymphatic endothelial regulation, lymphoedema, and lymph node metastasis. *Semin. Cell Dev. Biol.* **13**, 9–18.
3. Carmeliet, P. and Jain, R. K. (2000). Angiogenesis in cancer and other diseases. *Nature* **407**, 249–257.
4. Folkman, J. (2000). Tumor angiogenesis. In Holland, III, J. F. F. E., Bast, Jr., R. C., Kufe, D. W., Pollock, R. E., and Weichselbaum, R. R., Eds., *Cancer Medicine*, 5th ed., Vol. 5., pp. 132–152. Decker, Ontario.
5. Cohen, Jr., M. M. (2002). Vasculogenesis, angiogenesis, hemangiomas, and vascular malformations. *Am. J. Med. Genet.* **108**, 265–274.
6. Berra, E., Milanini, J., Richard, D. E., Le Gall, M., Vinals, F., Gothie, E., Roux, D., Pages, G., and Pouyssegur, J. (2000). Signaling angiogenesis via p42/p44 MAP kinase and hypoxia. *Biochem. Pharmacol.* **60**, 1171–1178.
7. Satake, S., Kuzuya, M., Miura, H., Asai, T., Ramos, M. A., Muraguchi, M., Ohmoto, Y., and Iguchi, A. (1998). Up-regulation of vascular endothelial growth factor in response to glucose deprivation. *Biol. Cell* **90**, 161–168.
8. Fukumura, D., Xu, L., Chen, Y., Gohongi, T., Seed, B., and Jain, R. K. (2001). Hypoxia and acidosis independently up-regulate vascular endothelial growth factor transcription in brain tumors *in vivo. Cancer Res.* **61**, 6020–6024.
9. Koike, C., McKee, T. D., Pluen, A., Ramanujan, S., Burton, K., Munn, L. L., Boucher, Y., and Jain, R. K. (2002). Solid stress facilitates spheroid formation: Potential involvement of hyaluronan. *Br. J. Cancer* **86**, 947–953.

10. Oliver, G. and Detmar, M. (2002). The rediscovery of the lymphatic system: Old and new insights into the development and biological function of the lymphatic vasculature. *Genes Dev.* **16**, 773–783.
11. Mouta Carreira, C. M., Nasser, S. M., di Tomaso, E., Padera, T. P., Boucher, Y., Tomarev, S. I., and Jain, R. K. (2001). LYVE-1 is not restricted to the lymph vessels: Expression in normal liver blood sinusoids and down-regulation in human liver cancer and cirrhosis. *Cancer Res.* **61**, 8079–8084.
12. Jain, R. K. and Fenton, B. T. (2002). Intratumoral lymphatic vessels: A case of mistaken identity or malfunction? *J. Natl. Cancer Inst.* **94**, 417–421.
13. Schoppmann, S. F., Horvat, R., and Birner, P. (2002). Lymphatic vessels and lymphangiogenesis in female cancer: Mechanisms, clinical impact and possible implications for anti-lymphangiogenic therapies (Review). *Oncol. Rep.* **9**, 455–460.
14. Pettersson, A., Nagy, J. A., Brown, L. F., Sundberg, C., Morgan, E., Jungles, S., Carter, R., Krieger, J. E., Manseau, E. J., Harvey, V. S., Eckelhoefer, I. A., Feng, D., Dvorak, A. M., Mulligan, R. C., and Dvorak, H. F. (2000). Heterogeneity of the angiogenic response induced in different normal adult tissues by vascular permeability factor/vascular endothelial growth factor. *Lab. Invest.* **80**, 99–115.
15. Nickoloff, B. J. (2000). Characterization of lymphocyte-dependent angiogenesis using a SCID mouse: Human skin model of psoriasis. *J. Invest. Dermatol. Symp. Proc.* **5**, 67–73.
16. Wernert, N., Justen, H. P., Rothe, M., Behrens, P., Dreschers, S., Neuhaus, T., Florin, A., Sachinidis, A., Vetter, H., and Ko, Y. (2002). The Ets 1 transcription factor is upregulated during inflammatory angiogenesis in rheumatoid arthritis. *J. Mol. Med.* **80**, 258–266.
17. Lai, C. M., Brankov, M., Zaknich, T., Lai, Y. K., Shen, W. Y., Constable, I. J., Kovesdi, I., and Rakoczy, P. E. (2001). Inhibition of angiogenesis by adenovirus-mediated sFlt-1 expression in a rat model of corneal neovascularization. *Hum. Gene Ther.* **12**, 1299–1310.
18. Rak, J., Yu, J. L., Klement, G., and Kerbel, R. S. (2000). Oncogenes and angiogenesis: Signaling three-dimensional tumor growth. *J. Invest. Dermatol. Symp. Proc.* **5**, 24–33.
19. Augustin, H. G. (2001). Tubes, branches, and pillars: The many ways of forming a new vasculature. *Circ. Res.* **89**, 645–647.
20. Stevens, T., Rosenberg, R., Aird, W., Quertermous, T., Johnson, F. L., Garcia, J. G., Hebbel, R. P., Tuder, R. M., and Garfinkel, S. (2001). NHLBI workshop report: Endothelial cell phenotypes in heart, lung, and blood diseases. *Am. J. Physiol. Cell Physiol.* **281**, C1422–C1433.
21. Swartz, M. A. (2001). The physiology of the lymphatic system. *Adv. Drug Deliv. Rev.* **50**, 3–20.
22. Hirschi, K. K. and D'Amore, P. A. (1997). Control of angiogenesis by the pericyte: Molecular mechanisms and significance. *Exs.* **79**, 419–428.
23. Maciag, T., Kadish, J., Wilkins, L., Stemerman, M. B., and Weinstein, R. (1982). Organizational behavior of human umbilical vein endothelial cells. *J. Cell Biol.* **94**, 511–520.
24. Dvorak, H. F., Nagy, J. A., Feng, D., Brown, L. F., and Dvorak, A. M. (1999). Vascular permeability factor/vascular endothelial growth factor and the significance of microvascular hyperpermeability in angiogenesis. *Curr. Top. Microbiol. Immunol.* **237**, 97–132
25. Schechner, J. S., Nath, A. K., Zheng, L., Kluger, M. S., Hughes, C. C., Sierra-Honigmann, M. R., Lorber, M. I., Tellides, G., Kashgarian, M., Bothwell, A. L., and Pober, J. S. (2000). *In vivo* formation of complex microvessels lined by human endothelial cells in an immunodeficient mouse. *Proc. Natl. Acad. Sci. USA* **97**, 9191–9196.
26. Luttun, A., Carmeliet, G., and Carmeliet, P. (2002). Vascular progenitors: From biology to treatment. *Trends Cardiovasc. Med.* **12**, 88–96.
27. Kidd, K. R., Nagle, R. B., and Williams, S. K. (2002). Angiogenesis and neovascularization associated with extracellular matrix-modified porous implants. *J. Biomed. Mater. Res.* **59**, 366–377.
28. Zhao, L. and Eghbali-Webb, M. (2001). Release of pro- and anti-angiogenic factors by human cardiac fibroblasts: effects on DNA synthesis and protection under hypoxia in human endothelial cells. *Biochim. Biophys. Acta* **1538**, 273–282.
29. Yasuda, M., Shimizu, S., Tokuyama, S., Watanabe, T., Kiuchi, Y., and Yamamoto, T. (2000). A novel effect of polymorphonuclear leukocytes in the facilitation of angiogenesis. *Life Sci.* **66**, 2113–2121.
30. Lingen, M. W. (2001). Role of leukocytes and endothelial cells in the development of angiogenesis in inflammation and wound healing. *Arch. Pathol. Lab. Med.* **125**, 67–71.
31. Kaipainen, A., Korhonen, J., Mustonen, T., van Hinsbergh, V. W., Fang, G. H., Dumont, D., Breitman, M., and Alitalo, K. (1995). Expression of the fms-like tyrosine kinase 4 gene becomes restricted to lymphatic endothelium during development. *Proc. Natl. Acad. Sci. USA* **92**, 3566–3570.
32. Jeltsch, M., Kaipainen, A., Joukov, V., Meng, X., Lakso, M., Rauvala, H., Swartz, M., Fukumura, D., Jain, R. K., and Alitalo, K. (1997). Hyperplasia of lymphatic vessels in VEGF-C transgenic mice. *Science* **276**, 1423–1425.
33. Karkkainen, M. J., Saaristo, A., Jussila, L., Karila, K. A., Lawrence, E. C., Pajusola, K., Bueler, H., Eichmann, A., Kauppinen, R., Kettunen, M. I., Yla-Herttuala, S., Finegold, D. N., Ferrell, R. E., and Alitalo, K. (2001). A model for gene therapy of human hereditary lymphedema. *Proc. Natl. Acad. Sci. USA* **98**, 12677–12682.
34. Jones, N., Iljin, K., Dumont, D. J., and Alitalo, K. (2001). Tie receptors: New modulators of angiogenic and lymphangiogenic responses. *Nat. Rev. Mol. Cell Biol.* **2**, 257–267.
35. LeCouter, J. and Ferrara, N. (2002). EG-VEGF and the concept of tissue-specific angiogenic growth factors. *Semin. Cell Dev. Biol.* **13**, 3–8.
36. Bazzoni, G., Dejana, E., and Lampugnani, M. G. (1999). Endothelial adhesion molecules in the development of the vascular tree: The garden of forking paths. *Curr. Opin. Cell Biol.* **11**, 573–581.
37. Carmeliet, P. and Collen, D. (2000). Molecular basis of angiogenesis. Role of VEGF and VE-cadherin. *Ann. NY Acad. Sci.* **902**, 249–262; discussion 262–244.
38. Cheng, N., Brantley, D. M., and Chen, J. (2002). The ephrins and Eph receptors in angiogenesis. *Cytokine Growth Factor Rev.* **13**, 75–85.
39. Senger, D. R., Galli, S. J., Dvorak, A. M., Perruzzi, C. A., Harvey, V. S., and Dvorak, H. F. (1983). Tumor cells secrete a vascular permeability factor that promotes accumulation of ascites fluid. *Science* **219**, 983–985.
40. Ferrara, N. and Henzel, W. J. (1989). Pituitary follicular cells secrete a novel heparin-binding growth factor specific for vascular endothelial cells. *Biochem. Biophys. Res. Commun.* **161**, 851–858.
41. Fukumura, D., Xavier, R., Sugiura, T., Chen, Y., Park, E. C., Lu, N., Selig, M., Nielsen, G., Taksir, T., Jain, R. K., and Seed, B. (1998). Tumor induction of VEGF promoter activity in stromal cells. *Cell* **94**, 715–725.
42. Wartiovaara, U., Salven, P., Mikkola, H., Lassila, R., Kaukonen, J., Joukov, V., Orpana, A., Ristimaki, A., Heikinheimo, M., Joensuu, H., Alitalo, K., and Palotie, A. (1998). Peripheral blood platelets express VEGF-C and VEGF which are released during platelet activation. *Thromb. Haemost.* **80**, 171–175.
43. Feng, D., Nagy, J. A., Pyne, K., Hammel, I., Dvorak, H. F., and Dvorak, A. M. (1999). Pathways of macromolecular extravasation across microvascular endothelium in response to VPF/VEGF and other vasoactive mediators. *Microcirculation* **6**, 23–44.
44. Kumar, R., Yoneda, J., Bucana, C. D., and Fidler, I. J. (1998). Regulation of distinct steps of angiogenesis by different angiogenic molecules. *Int. J. Oncol.* **12**, 749–757.
45. Lal, B. K., Varma, S., Pappas, P. J., Hobson, II, R. W., and Duran, W. N. (2001). VEGF increases permeability of the endothelial cell monolayer by activation of PKB/akt, endothelial nitric-oxide synthase, and MAP kinase pathways. *Microvasc. Res.* **62**, 252–262.
46. Hla, T., Ristimaki, A., Appleby, S., and Barriocanal, J. G. (1993). Cyclooxygenase gene expression in inflammation and angiogenesis. *Ann. NY Acad. Sci.* **696**, 197–204.
47. Jones, M. K., Szabo, I. L., Kawanaka, H., Husain, S. S., and Tarnawski, A. S. (2002). von Hippel Lindau tumor suppressor and HIF-1alpha: New targets of NSAIDs inhibition of hypoxia-induced angiogenesis. *FASEB J.* **16**, 264–266.
48. Polverini, P. J. (1996). How the extracellular matrix and macrophages contribute to angiogenesis-dependent diseases. *Eur. J. Cancer* **32A**, 2430–2437.
49. Haas, T. L. and Madri, J. A. (1999). Extracellular matrix-driven matrix metalloproteinase production in endothelial cells: implications for angiogenesis. *Trends Cardiovasc. Med.* **9**, 70–77.

50. Brooks, P. C. (1996). Cell adhesion molecules in angiogenesis. *Cancer Metastasis. Rev.* **15**, 187–194.
51. Romer, L. H., McLean, N. V., Yan, H. C., Daise, M., Sun, J., and DeLisser, H. M. (1995). IFN-gamma and TNF-alpha induce redistribution of PECAM-1 (CD31) on human endothelial cells. *J. Immunol.* **154**, 6582–6592.
52. Iozzo, R. V. and San Antonio, J. D. (2001). Heparan sulfate proteoglycans: heavy hitters in the angiogenesis arena. *J. Clin. Invest.* **108**, 349–355.
53. Savani, R. C., Cao, G., Pooler, P. M., Zaman, A., Zhou, Z., and DeLisser, H. M. (2001). Differential involvement of the hyaluronan (HA) receptors CD44 and receptor for HA-mediated motility in endothelial cell function and angiogenesis. *J. Biol. Chem.* **276**, 36770–36778.
54. Banerji, S., Ni, J., Wang, S. X., Clasper, S., Su, J., Tammi, R., Jones, M., and Jackson, D. G. (1999). LYVE-1, a new homologue of the CD44 glycoprotein, is a lymph-specific receptor for hyaluronan. *J. Cell Biol.* **144**, 789–801.
55. Zimrin, A. B., Pepper, M. S., McMahon, G. A., Nguyen, F., Montesano, R., and Maciag, T. (1996). An antisense oligonucleotide to the notch ligand jagged enhances fibroblast growth factor-induced angiogenesis *in vitro*. *J. Biol. Chem.* **271**, 32499–32502.
56. Cines, D. B., Pollak, E. S., Buck, C. A., Loscalzo, J., Zimmerman, G. A., McEver, R. P., Pober, J. S., Wick, T. M., Konkle, B. A., Schwartz, B. S., Barnathan, E. S., McCrae, K. R., Hug, B. A., Schmidt, A. M., and Stern, D. M. (1998). Endothelial cells in physiology and in the pathophysiology of vascular disorders. *Blood* **91**, 3527–3561.
57. Patan, S. (2000). Vasculogenesis and angiogenesis as mechanisms of vascular network formation, growth and remodeling. *J. Neurooncol.* **50**, 1–15.
58. Ingber, D. E., Prusty, D., Sun, Z., Betensky, H., and Wang, N. (1995). Cell shape, cytoskeletal mechanics, and cell cycle control in angiogenesis. *J. Biomech.* **28**, 1471–1484.
59. Zhan, X., Hu, X., Friesel, R., and Maciag, T. (1993). Long term growth factor exposure and differential tyrosine phosphorylation are required for DNA synthesis in BALB/c 3T3 cells. *J. Biol. Chem.* **268**, 9611–9620.
60. Schwartz, M. A. and Assoian, R. K. (2001). Integrins and cell proliferation: Regulation of cyclin-dependent kinases via cytoplasmic signaling pathways. *J. Cell Sci.* **114**, 2553–2560.
61. Sleeman, J. P., Krishnan, J., Kirkin, V., and Baumann, P. (2001). Markers for the lymphatic endothelium: In search of the holy grail? *Microsc. Res. Tech.* **55**, 61–69.
62. Makinen, T., Veikkola, T., Mustjoki, S., Karpanen, T., Catimel, B., Nice, E. C., Wise, L., Mercer, A., Kowalski, H., Kerjaschki, D., Stacker, S. A., Achen, M. G., and Alitalo, K. (2001). Isolated lymphatic endothelial cells transduce growth, survival and migratory signals via the VEGF-C/D receptor VEGFR-3. *EMBO J.* **20**, 4762–4773.
63. Narko, K., Enholm, B., Makinen, T., and Ristimaki, A. (1999). Effect of inflammatory cytokines on the expression of the vascular endothelial growth factor-C. *Int. J. Exp. Pathol.* **80**, 109–112.
64. Darland, D. C. and D'Amore, P. A. (2001). Cell-cell interactions in vascular development. *Curr. Top. Dev. Biol.* **52**, 107–149.
65. Eliceiri, B. P. and Cheresh, D. A. (2001). Adhesion events in angiogenesis. *Curr. Opin. Cell Biol.* **13**, 563–568.
66. van Hinsbergh, V. W., Collen, A., and Koolwijk, P. (2001). Role of fibrin matrix in angiogenesis. *Ann. NY Acad. Sci.* **936**, 426–437.
67. Talts, J. F., Wirl, G., Dictor, M., Muller, W. J., and Fassler, R. (1999). Tenascin-C modulates tumor stroma and monocyte/macrophage recruitment but not tumor growth or metastasis in a mouse strain with spontaneous mammary cancer. *J. Cell Sci.* **112**, 1855–1864.
68. Penta, K., Varner, J. A., Liaw, L., Hidai, C., Schatzman, R., and Quertermous, T. (1999). Del1 induces integrin signaling and angiogenesis by ligation of alphaVbeta3. *J. Biol. Chem.* **274**, 11101–11109.
69. Lawler, J., Miao, W. M., Duquette, M., Bouck, N., Bronson, R. T., and Hynes, R. O. (2001). Thrombospondin-1 gene expression affects survival and tumor spectrum of p53-deficient mice. *Am. J. Pathol.* **159**, 1949–1956.
70. Kyriakides, T. R., Zhu, Y. H., Yang, Z., Huynh, G., and Bornstein, P. (2001). Altered extracellular matrix remodeling and angiogenesis in sponge granulomas of thrombospondin 2-null mice. *Am. J. Pathol.* **159**, 1255–1262.
71. Yumoto, K., Ishijima, M., Rittling, S. R., Tsuji, K., Tsuchiya, Y., Kon, S., Nifuji, A., Uede, T., Denhardt, D. T., and Noda, M. (2002). Osteopontin deficiency protects joints against destruction in anti-type II collagen antibody-induced arthritis in mice. *Proc. Natl. Acad. Sci. USA* **99**, 4556–4561.
72. Murphy-Ullrich, J. E. (2001). The de-adhesive activity of matricellular proteins: Is intermediate cell adhesion an adaptive state? *J. Clin. Invest.* **107**, 785–790.
73. Stromblad, S., Fotedar, A., Brickner, H., Theesfeld, C., Aguilar de Diaz, E., Friedlander, M., and Cheresh, D. A. (2002). Loss of p53 compensates for alpha v-integrin function in retinal neovascularization. *J. Biol. Chem.* **277**, 13371–13374.
74. Chang, C. and Werb, Z. (2001). The many faces of metalloproteases: cell growth, invasion, angiogenesis and metastasis. *Trends Cell Biol.* **11**, S37–S43.
75. Benjamin, L. E. (2000). The controls of microvascular survival. *Cancer Metastasis Rev.* **19**, 75–81.
76. Benjamin, L. E., Hemo, I., and Keshet, E. (1998). A plasticity window for blood vessel remodelling is defined by pericyte coverage of the preformed endothelial network and is regulated by PDGF-B and VEGF. *Development* **125**, 1591–1598.
77. Swartz, M. A. and Skobe, M. (2001). Lymphatic function, lymphangiogenesis, and cancer metastasis. *Microsc. Res. Tech.* **55**, 92–99.
78. Gerli, R., Solito, R., Weber, E., and Agliano, M. (2000). Specific adhesion molecules bind anchoring filaments and endothelial cells in human skin initial lymphatics. *Lymphology* **33**, 148–157.
79. Chakravarthy, U. and Gardiner, T. A. (1999). Endothelium-derived agents in pericyte function/dysfunction. *Prog. Retin. Eye Res.* **18**, 511–527.
80. Lyden, D., Hattori, K., Dias, S., Costa, C., Blaikie, P., Butros, L., Chadburn, A., Heissig, B., Marks, W., Witte, L., Wu, Y., Hicklin, D., Zhu, Z., Hackett, N. R., Crystal, R. G., Moore, M. A., Hajjar, K. A., Manova, K., Benezra, R., and Rafii, S. (2001). Impaired recruitment of bone-marrow-derived endothelial and hematopoietic precursor cells blocks tumor angiogenesis and growth. *Nat. Med.* **7**, 1194–1201.
81. Nicosia, R. F. and Villaschi, S. (1999). Autoregulation of angiogenesis by cells of the vessel wall. *Int. Rev. Cytol.* **185**, 1–43.
82. Trzewik, J., Mallipattu, S. K., Artmann, G. M., Delano, F. A., and Schmid-Schonbein, G. W. (2001). Evidence for a second valve system in lymphatics: endothelial microvalves. *FASEB J.* **15**, 1711–1717.
83. Lindner, V., Booth, C., Prudovsky, I., Small, D., Maciag, T., and Liaw, L. (2001). Members of the Jagged/Notch gene families are expressed in injured arteries and regulate cell phenotype via alterations in cell matrix and cell-cell interaction. *Am. J. Pathol.* **159**, 875–883.
84. Ishikawa, T., Tamai, Y., Zorn, A. M., Yoshida, H., Seldin, M. F., Nishikawa, S., and Taketo, M. M. (2001). Mouse Wnt receptor gene Fzd5 is essential for yolk sac and placental angiogenesis. *Development* **128**, 25–33.
85. Wigle, J. T., Harvey, N., Detmar, M., Lagutina, I., Grosveld, G., Gunn, M. D., Jackson, D. G., and Oliver, G. (2002). An essential role for Prox1 in the induction of the lymphatic endothelial cell phenotype. *EMBO J.* **21**, 1505–1513.
86. Prudovsky, I., Mandinova, A., Bagala, C., Soldi, R., Bellum, S., Battelli, C., Graziani, I., Maciag, T., Non-Classical Pathways of Protein Export. In *Handbook of Cellular Signaling*, Edited by Bradshaw, R. and Dennis, E, Academic Press, 2002, In Press.
87. Mandinov, L., Mandinova, A., Kyurkchiev, S., Kehayov, I., Kolev, V., Soldi, R., Bagala, C., De Muinck, E. D., Lindner, V. *et al.* (2003). Copper chelation represses the vascular response to injury. *Proc. Natl. Acad. Sci. U.S.A.* **100**, 6700–6705.
88. Voronov, E., Shouval, D. S., Krelin, Y., Cagnano, E., Benharroch, D., Iwakura, Y., Dinarello, C. A. and Apte, R. N. (2003). IL-1 is required for tumor invasiveness and angiogenesis. *Proc. Natl. Acad. Sci. U.S.A.* **100**, 2645–2650.

CHAPTER 335

Signaling Pathways Involved in Cardiogenesis

Deepak Srivastava
Departments of Pediatrics and Molecular Biology, University of Texas Southwestern Medical Center, Dallas, Texas

Introduction

The heart is a complex organ that is derived from multiple cell types and requires extensive cell–cell signaling events that are often guided by specialized forms of extracellular matrices. As the earliest organ to form in an embryo, the heart must be functional well before it has the opportunity to shape itself into a mature organ. The combination of multifarious morphogenetic events necessary for cardiogenesis and the superimposed hemodynamic influences may contribute to the exquisite sensitivity of the heart to perturbations. This phenomenon is reflected in the estimated 10% incidence of severe cardiac malformations observed in early miscarriages and in the nearly 1% of live births affected by cardiac developmental defects.

The recognition that cardiac genetic pathways and morphogenetic steps are highly conserved across vastly diverse species from flies to man has resulted in a rapid growth of knowledge from studies in more tractable and accessible biological models than *Homo sapiens*. The fruit fly (*Drosophila*) has been a source of discovery for genes involved in early cardiac determination events in spite of its relatively simple cardiovascular system composed of a linear heart tube that pumps hemolymph through an open circulatory system. The reverse genetics available in flies are also possible in zebrafish, which have several distinct advantages. Zebrafish are vertebrates with a more complex two-chambered heart that is not necessary for survival during the period of cardiac development. Chick and mouse embryos have four-chambered hearts similar to humans, but the elegant genetics available in the latter have made it a particularly rich system. In a simplified view, it appears that higher organisms have retained the morphologic steps utilized by lower organisms and have built complexity into the heart in a modular fashion [1].

Here, some of the critical signaling events necessary for normal cardiogenesis are reviewed to provide a framework for considering the intricate interactions necessary between distinct cell types. Multiple signaling pathways often converge on individual events to establish domains or cell fates within the embryo in a combinatorial fashion. In addition, common pathways are often reemployed in various regions of the heart, although their ultimate morphogenetic effects are vastly different based on activation of unique transcriptional programs. Therefore, we consider the combined signaling events that contribute to specific steps of cardiogenesis using knowledge gained from multiple species.

Cardiomyocyte and Heart Tube Formation

Soon after gastrulation (about embryonic day 20 in humans), progenitor cells within the anterior lateral plate mesoderm become committed to a cardiogenic fate in response to an inducing signal thought to emanate from the adjacent endoderm [2,3]. The complete set of signaling molecule(s) responsible for cardiogenic commitment remains to be identified, although several pathways have now been implicated in defining the cardiogenic domain and in cardiac cell fate determination. In flies, a member of the transforming growth factor ß (TGF-ß) family, *dpp*, is essential for the initial determination of a cardioblast [4]. In chick and frog, dpp-like members of the TGF-ß family, including bone morphogenetic proteins (Bmps) -2, -4, and -7 are secreted from the endoderm in the anterior half of the embryo and act through Bmp receptors in the adjacent mesoderm to activate cardiac gene expression [3,5]. In flies or vertebrates, dpp or Bmp signaling is mediated by transcription factors of the Smad family that directly interact with cis elements in the enhancer of *tinman* or *Nkx2.5*, respectively [6,7]. Mutation of *tinman*, a homeodomain-containing transcription factor, in flies results in an organism similar to the *Wizard of Oz* character who is lacking a heart [8]. In contrast to the requirement of *tinman* for heart formation in flies, its mammalian ortholog, *Nkx2.5*, is not essential for specification of the cardiac lineage in mice, suggesting that other genes may share related functions with *Nkx2.5* or that cardiogenesis in flies and vertebrates differs with respect to its dependence on this family of homeobox genes [9,10]. Tinman and Nkx2.5 cooperate with zinc-finger transcription factors of the GATA family to activate cardiac gene expression [11].

Only certain spatial domains of the embryo are competent to respond to Bmp signals in a cardiogenic fashion, suggesting that other factors are also important. The expression of fibroblast growth factors such as Fgf4 and Fgf8 overlaps with Bmp in the cardiogenic field and the two appear to cooperate to promote cardiac cell fate determination [12]. The recent identification of Wnt signals emanating from the midline of the embryo provides insight into the cues that pattern the anterior domain, which is competent to respond to Bmp and Fgf signaling. Wnt molecules diffuse laterally from the midline notochord and inhibit cardiac gene expression [13,14]. An inhibitor of the Wnt receptor, frizzled, is expressed specifically in the precardiac mesoderm and is capable of promoting cardiac cell fate in frogs and chicks. Interestingly, the fly ortholog of Wnt, *wingless*, activates *dpp* signaling and heart formation, thus functioning opposite of Wnt signaling in vertebrates [15]. Nevertheless, a combination of Bmp, Fgf, and Wnt signaling is important in establishing the initial cardiac precursors, although it is clear that yet unknown factors are also involved.

Soon after their specification, cardiac muscle cells converge along the ventral midline of the embryo to form a beating linear heart tube composed of distinct myocardial and endocardial layers separated by a rich extracellular matrix. Mutations of GATA proteins in mice and zebrafish have demonstrated a critical role for this family of transcription factors in midline fusion of the heart tube [16–18]. In an example of the power of zebrafish genetics, a gene responsible for a cardiac bifida phenotype, *miles apart*, has recently been identified [19]. This gene encodes a novel sphingosine 1-phosphate receptor that may be mediating a midline signal to attract cardiomyocytes from the lateral aspect of the early embryo; however, the mechanisms of cell movement and fusion are yet to be determined.

As the straight heart tube takes shape, four distinct tubular segments form in a temporal sequence along the anterior-posterior (AP) axis. The primitive right and left ventricles are the first to be distinguished, followed by the atrioventricular canal segment. The sinoatrial segment forms most caudally and has distinct left–right asymmetry, with the right and left limbs of this segment later contributing to the right and left atria, respectively. The conotruncus is the last segment to form and lies in the most anterior portion of the heart tube. This segment appears to arise from cells that respond to Fgf10 signals and migrate from the pharyngeal mesoderm to populate the heart tube after its initial formation [20]. As the heart tube loops to the right, the cardiac chambers begin to become distinguished morphologically and adopt their left–right orientation (Fig. 1).

Cardiac Looping and Left–Right Asymmetry

The pathways that control the direction of cardiac looping along the left–right axis have recently been elucidated (reviewed in [21]) (Fig. 2). The heart is the first organ to break the bilateral symmetry present in the early embryo and the rightward direction of its looping reflects a more global establishment of left–right (LR) asymmetry that affects the lungs, liver, spleen, and gut. Defects in establishment of LR asymmetry in humans are associated with a wide range of cardiac alignment defects, suggesting that pathways regulating LR asymmetry dramatically affect cardiac development.

A cascade of signaling molecules regulating the establishment of embryonic LR asymmetry has been revealed from recent studies of chick embryonic development. Before the formation of organs in the developing embryo, asymmetric expression of the morphogen Sonic hedgehog (Shh) on the left side of Hensen's node leads to left lateral mesoderm expression of nodal and lefty, members of the transforming growth factor ß (TGF-ß) family [22]. Transfer of this signal from the node to the lateral mesoderm is mediated by the secreted molecule, caronte. Caronte inhibits Bmp on the left side, relieving Bmp-mediated repression of nodal in the left lateral plate mesoderm [23]. Left-sided expression of nodal induces rightward looping of the midline heart tube. Fibroblast growth factor and activin receptor-mediated pathways suppress caronte expression on the right side and the resulting activity of Bmp signaling results in suppression of right-sided nodal expression. Conversely, the snail-related (cSnR-1) zinc finger transcription factor is expressed in the right lateral mesoderm and is repressed by Shh on the left [24]. The above signaling pathways are active in the lateral plate mesoderm, but not in the heart or other organs that actually display LR asymmetry. Ultimately, the nodal-dependent pathways result in expression of a homoedomain protein, Pitx2, on the left side of visceral organs and repression of Pitx2 on the right [25]. Pitx2 appears to be the major factor that interprets the LR signaling cascade at the organ level. Asymmetric expression of Pitx2 is sufficient for establishing the LR asymmetry of the heart, lungs, and gut [26].

The mechanisms that control directionality of cardiac looping have also been explored by genetic analysis of mouse mutants with abnormalities in left–right asymmetry.

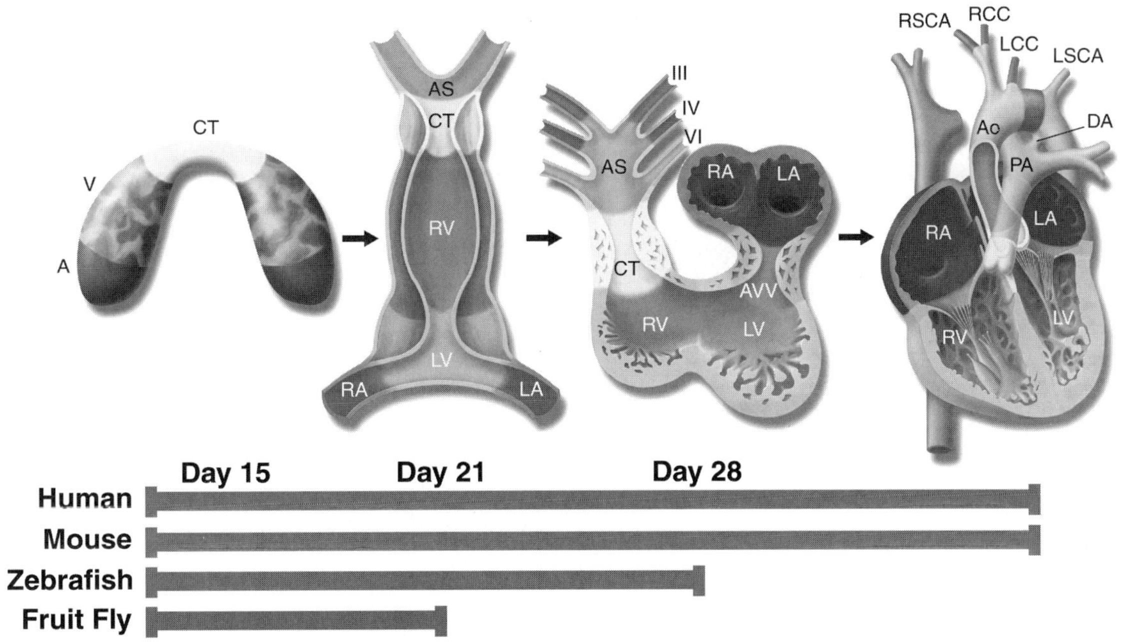

Figure 1 Schematic of cardiac morphogenesis. Illustrations depict cardiac development of morphologically related regions, seen from a ventral view. Cardiogenic precursors form a crescent (leftmost panel) that is specified to form specific segments of the linear heart tube, which is patterned along the AP axis to form the various regions and chambers of the looped and mature heart. Each cardiac chamber balloons from the outer curvature of the looped heart tube in a segmental fashion. Neural crest cells populate the bilaterally symmetric aortic arch arteries (III, IV, and VI) and aortic sac (AS) that together contribute to specific segments of the mature aortic arch. Mesenchymal cells form the cardiac valves from the conotruncal (CT) and atrioventricular valve (AVV) segments. Corresponding days of human embryonic development are indicated. RV, right ventricle; LV, left ventricle; RA, right atrium; LA, left atrium; PA, pulmonary artery; Ao, aorta; DA, ductus arteriosus; RSCA, right subclavian artery; RCC, right common carotid; LCC, left common carotid; LSCA, left subclavian artery. [Reproduced with permission from Srivastava and Olson, *Nature*, **407**, 221–226 (2000)].

Mice homozygous for mutation in the *left–right dynein* gene (*iv/iv*) display randomization of left–right orientation of the heart and viscera and have bilaterally symmetric, absent, or randomized expression of nodal and Pitx2 [27,28]. Nodal and Pitx2 are expressed along the right lateral mesoderm rather than the left, displaying complete reversal of the LR signals and have bilaterally symmetric, absent, or randomized nodal and Pitx2 expression. In contrast, in the situs inversus (*inv*) mouse [29], which has nearly 100% reversal of left–right asymmetry, nodal and Pitx2 are expressed along the right lateral mesoderm rather than the left, displaying complete reversal of the LR signals. *Pitx2* mutant mice have abnormal LR asymmetry of the lungs and a low penetrance of reversed cardiac looping, similar to *Shh* and *Fgf8* mutant mice [30,31]. Oddly, the initial LR asymmetry and roles of Fgf and Shh are opposite in mice and chicks, however the left–right sidedness of later events involving nodal and Pitx2 are conserved [32].

While the necessity of LR asymmetric gene expression is intuitive, how the initial asymmetry of molecules is established remains in question. Initial clues came from studies of immotile cilia syndrome, also known as Kartagener's syndrome, in which individuals had situs inversus totalis, with mirror-image reversal of all organs. It was recently found that, prior to organ formation, Hensen's node contains ciliary processes that beat in a vortical fashion, pushing morphogens to the left side of the embryo [33]; concurrent establishment of a midline barrier, possibly by *lefty* gene expression along the left midline, may be responsible for subsequent asymmetric gene expression. Mice lacking ciliary movement in the node display abnormal LR patterning, consistent with this model.

Patterning of the Developing Heart Tube

Numerous transcription factors are expressed in a chamber-specific fashion, providing a possible mechanism to explain how distinct segments of the heart adopt their respective fates. Two related basic helix–loop–helix (bHLH) transcription factors, *dHAND* and eHAND, are expressed predominantly in the primitive right and left ventricle segments, respectively, during mouse heart development [34,35]. Deletion of *dHAND* in mice results in hypoplasia of the right ventricular segment from a cell survival defect. Mice lacking eHAND die early from placental defects precluding detailed analysis of its role in left ventricular development [36]. eHAND is down-regulated in Nkx2.5-deficient mice, which fail to precisely segment the heart tube and die around the stage of cardiac looping [37]. Disruption of both *dHAND* and *Nkx2.5* in combination results in the absence of the right and left ventricle, suggesting that the combined function of dHAND and Nkx2.5, possibly through their regulation

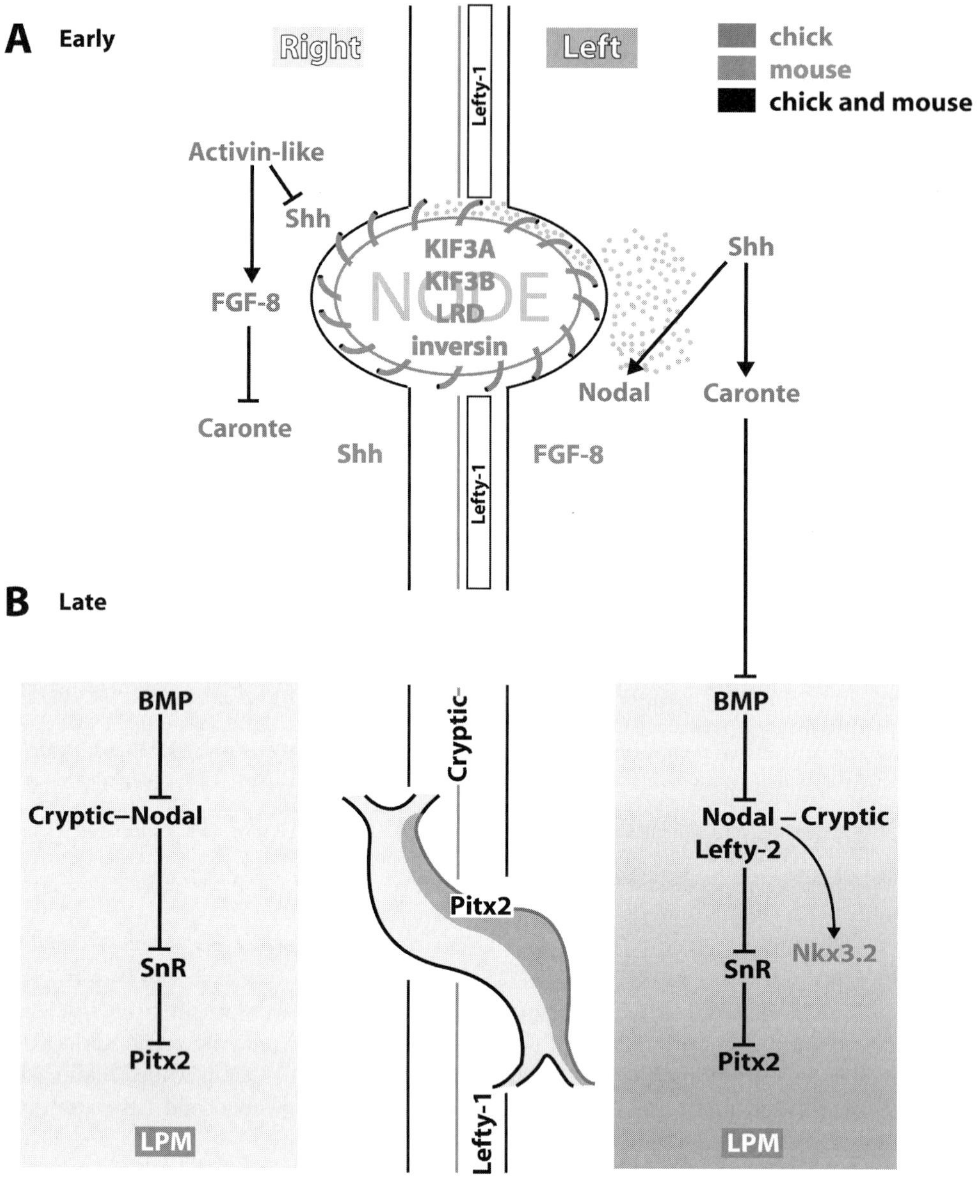

Figure 2 Cascades regulating LR asymmetry. Early asymmetrical gene expression around the node (A) results in activation or repression of sonic hedgehog (Shh) or fibroblast growth factor (Fgf)-8 dependent pathways on the right or left (ventral view). Early roles of Shh and Fgf8 are reversed in mouse and chick. Leftward flow of morphogens by nodal cilia establishes the asymmetric gradient around the node in mice. Expression of Lefty-1 near the midline may serve as a barrier to maintain left-sided asymmetry of morphogens. At later stages of organogenesis, LR asymmetric information at the node is transferred to the lateral plate mesoderm (LPM) by Caronte. Caronte relieves Bmp inhibition on the left, initiating a cascade of events culminating in expression of Pitx2 in the left LPM and in the left side of the heart tube (B). Consequently, a "leftness" signal appears to be actively propagated to overcome a default "rightness" program. [Reproduced with permission from Kathiriya and Srivastava, *Am. J. Med. Genet.* 97:271–279 (2001).]

of eHAND, is necessary for ventricular formation [38]. In zebrafish, which have a single ventricle, only one *HAND* gene has been identified (*dHAND*), disruption of which results in lack of a ventricular segment of the heart, similar to that seen in the absence of *dHAND* and *Nkx2.5* [39]. Expression of the ventricular-specific homeobox gene of the Iroquois family, *Irx4*, is dependent on both dHAND and Nkx2.5 [38,40], and misexpression of Irx4 in the atria is sufficient to activate ventricle-specific gene expression [41]. These findings suggest that HAND and Nkx2.5 proteins may cooperate in early ventricle-specific decisions. A cardiac-specific chromatin remodeling protein, Bop, is also essential for normal ventricular development and is required for activation of *dHAND* gene expression [42]. Deletion of one of the four MEF2 factors in mice, MEF2C, which also may be involved in chromatin reorganizing events, results in hypoplasia of the right and left ventricles as well, but not of the atria [43]. The chamber-specific role of MEF2C, in spite of its homogenous expression in the heart, suggests that MEF2C might be a necessary cofactor for one or more of the other ventricular-restricted regulatory proteins.

How the segmental pattern of gene expression is established remains unclear; however, retinoid signaling has been implicated in atrial specification and positioning of the atrioventricular (AV) border along the AP axis of the heart tube [44]. The recent discovery of a novel class of hairy-related transcription factors (HRT1, HRT2, HRT3) may also provide some insight [45]. Hairy proteins often function downstream of the transmembrane receptor Notch in establishing boundaries of gene expression. Interestingly, HRT1 and HRT2 are expressed in a complementary fashion in the atria and ventricles, respectively. How the many transcription factors function in a coordinated manner to regulate chamber specification and differentiation remains to be determined.

Myocardial Growth

Mutations of a wide variety of genes in mice result in hypoplasia of the muscular wall of the heart. Mice homozygous for a null mutation in the *retinoid X receptor-alpha* (*RXRα*) gene display ventricular chamber hypoplasia and have a defect in compaction of the myocardium, although this may not be a cell autonomous effect [46]. Signaling between the endocardium and the myocardium also appears to be important for ventricular growth. Neuregulin growth factors are expressed in the endocardium and are required for the development of trabeculae, the finger-like projections of the ventricular myocardium [47]. In mice deficient in neuregulin or its receptors, erbB2 and erbB4, the ventricular trabeculae fail to form, possibly as a result of decreased endocardial signals. Similar defects in ventricular trabeculation have been observed in mice lacking angiogenic factors that are also expressed in the endocardium [48,49].

Cardiac Valve Formation

Appropriate placement and function of cardiac valves is essential for chamber septation and for unidirectional flow of blood through the heart. During early heart tube formation, "cushions" of extracellular matrix between the endocardium and myocardium presage valve formation at each end of the heart tube. Reciprocal signaling, mediated in part by TGF-ß family members, between the myocardium and endocardium in the cushion region induces a transformation of endocardial cells into mesenchymal cells that migrate into the cushion extracellular matrix (ECM). These mesenchymal cells differentiate into the fibrous tissue of the valves and are involved in septation of the common atrioventricular canal into right and left sided orifices.

The Smad proteins are intracellular transcriptional mediators of signaling initiated by TGF-β ligands. Smad6 is specifically expressed in the atrioventricular cushions and outflow tract during cardiogenesis and is a negative regulator of TGF-β signaling. Targeted disruption of *Smad6* in mice results in thickened and gelatinous atrioventricular and semilunar valves, similar to those observed in human disease [50]. In addition to *Smad6*, there are likely other genes in the TGF-β signaling pathway that, when mutated, result in the formation of hyperplastic valves that may be a result of excessive transformation of endocardial cells.

In mouse models, the absence of *PTPN11*, which encodes the protein tyrosine phosphatase Shp-2, has been shown to result in dysplastic cardiac valves by its involvement in a signaling pathway mediated by epidermal growth factor receptor [51]. The importance of *PTPN11* in congenital heart disease was shown by the identification of point mutations in *PTPN11* in patients with Noonan syndrome, whose phenotype commonly includes pulmonic valve thickening [52].

Nuclear factor of activated T cells-c (NF-ATc) is a transcription factor that is needed for cytokine gene expression in activated lymphocytes. It is controlled by a calcium-regulated phosphatase, calcineurin. In the heart, NF-ATc expression is restricted to the endocardium. Unlike the factors described earlier, mice lacking NF-ATc fail to form aortic or pulmonary valves, suggesting that this factor is necessary for formation of the semilunar valves [53,54]. The mechanisms through which NF-Atc regulates valve formation remain unknown.

Cardiac Outflow Tract and Aortic Arch Development

The cardiac outflow tract (conotruncus) and aortic arch undergo extensive and rather complex morphogenetic changes. The cardiac outflow is initially a single vessel, the truncus arteriosus, that becomes septated by mesenchymal cells into the aorta and pulmonary arteries. Six bilaterally symmetric vessels known as aortic arch arteries arise sequentially along the AP axis and undergo extensive remodeling to ultimately form distinct regions of the mature aortic arch and proximal pulmonary arteries (Fig. 3). In particular, the left fourth aortic arch artery forms the transverse aortic arch between the left common carotid and left subclavian arteries, while the sixth arch artery contributes to the proximal pulmonary artery and the ductus arteriosus.

A unique population of cells along the crest of the neural folds (neural crest cells) migrates away from the neural

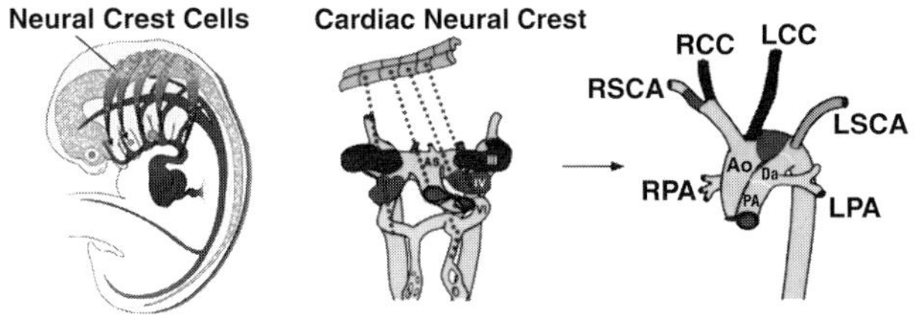

Figure 3 Migration and contribution of the cardiac neural crest. Neural crest cells arise from the crest of the neural folds between the otic placode and the third somite. Subsequent migration of the neural crest into the aortic arch arteries and cardiac outflow tract is required for arch and conotruncal development.

folds and retains the ability to differentiate into multiple cell types; therefore, these cells are pluripotent. Their migratory path and ultimate cell fates are dependent on their relative position of origin along the AP axis and their interactions with nearby matrices and adjacent epithelial cells. Neural crest cells differentiate and contribute to diverse embryonic structures, including the cranial ganglia, peripheral nervous system, adrenal glands, and melanocytes. Cranial neural crest cells migrate through the developing pharyngeal arches and populate the mesenchyme of each of the pharyngeal and aortic arch arteries, the conotruncus, and the conotruncal septum [55]. Because of their migratory path, this segment of the neural crest is often referred to as the cardiac neural crest.

Insight into the genes that regulate cardiac neural crest development has come from studies in other vertebrate models. Mice lacking the 21-amino-acid signaling peptide, endothelin-1 (ET-1), or its G-protein-coupled receptor, ET_A, show postmigratory cardiac neural crest defects, cleft palate, and other craniofacial anomalies reminiscent of DiGeorge syndrome in humans [56,57]. dHAND and eHAND, normally expressed in the neural crest-derived pharyngeal and aortic arches, are down-regulated in these structures in *ET-1*- and ET_A-deficient mice, suggesting that they are regulated by ET-1 signaling [58]. Analysis of the upstream regulatory region of *dHAND* revealed that ET-1 signals are mediated by D1x6, which directly activates dHAND expression [59]. This pathway has been developed further by the observation that neuropilin-1, a semaphorin and VEGF receptor, is down-regulated in *dHAND* mutants [60]. Targeted mutation of *neuropilin-1* or its ligand, semaphorinC, results in a phenotype similar to that of *ET-1* mutants, suggesting that ET-1, dHAND, and neuropilin-1 may function in a common pathway regulating neural crest development [61,62].

The Notch signaling pathway is involved in cell fate and differentiation decisions throughout the embryo, but has only recently been implicated in cardiovascular development. Alagille syndrome is an autosomal dominant disorder characterized by biliary atresia and defects involving the cardiac outflow tract. This syndrome is caused by mutations in *JAGGED-1*, a ligand for the Notch receptor [63,64]. Isolated outflow tract defects have also been associated with *JAGGED-1* mutations [65]. The intracellular events downstream of Notch signaling are mediated by a family of bHLH transcriptional repressors known as Hairy proteins. In mammals, the hairy-related transcription factors, HRT1 and HRT2, are expressed in the cardiac outflow tract and are activated by Notch signaling [45], raising the possibility that they may mediate the effects of JAGGED-1. The zebrafish ortholog of HRT2, *gridlock*, is activated by Notch signaling, similar to its mammalian counterpart, and is necessary and sufficient for the initial parsing of hemangioblasts into arterial rather than venous endothelial cells in the fish [66]. Hypomorphic mutations of *gridlock* result in narrowing of a specific region of the aorta, a defect commonly observed in humans, possibly because of an insufficient number of arterial endothelial cells in specific vascular locations [67].

Conclusions

The steps of cardiogenesis described here illustrate some of the signaling networks necessary for multiple cell types to communicate with one another in order to form a functioning organ. Reciprocal interactions between cell layers function to guide cells in the correct temporospatial pattern and ultimately to adopt specific cell fates and achieve terminal differentiation. Disruption of such signaling events often underlies pathologic development of the heart that manifests as congenital heart disease. Because fetal gene programs are often reactivated in the adult diseased heart with negative consequences, it is possible that inhibition or activation of specific signaling pathways involved in cardiogenesis may prove to have therapeutic value, even in late-onset heart disease.

References

1. Srivastava, D. and Olson, E. N. (2000). A genetic blueprint for cardiac development. *Nature* **407**, 221–226
2. Schultheiss, T. M., Xydas, S., and Lassar, A. B. (1995). Induction of avian cardiac myogenesis by anterior endoderm. *Development* **121**, 4203–4214.
3. Schultheiss, T. M., Burch, J. B., and Lassar, A. B. (1997). A role for bone morphogenetic proteins in the induction of cardiac myogenesis. *Genes Dev* **11**, 451–462.
4. Frasch, M. (1995). Induction of visceral and cardiac mesoderm by ectodermal Dpp in the early Drosophila embryo. *Nature* **374**, 464–467.
5. Andree, B., Duprez, D., Vorbusch, B., Arnold, H. H., and Brand, T. (1998). BMP-2 induces ectopic expression of cardiac lineage markers and interferes with somite formation in chicken embryos. *Mech. Dev.* **70**, 119–131.
6. Xu, X., Yin, Z., Hudson, J. B., Ferguson, E. L., and Frasch, M. (1998). Smad proteins act in combination with synergistic and antagonistic regulators to target Dpp responses to the Drosophila mesoderm. *Genes Dev.* **12**, 2354–2370.
7. Lien C. L., McAnally J., Richardson J. A., and Olson E. N. (2002). Cardiac-specific activity of an Nkx2-5 enhancer requires an evolutionarily conserved Smad binding site. *Dev Biol.* **244**(2), 257–266.
8. Bodmer, R. (1993). The gene tinman is required for specification of the heart and visceral muscles in Drosophila. *Development* **118**, 719–729.
9. Lyons, I. *et al.* (1995). Myogenic and morphogenetic defects in the heart tubes of murine embryos lacking the homeo box gene Nkx2–5. *Genes Dev.* **9**, 1654–1666.
10. Tanaka, M., Chen, Z., Bartunkova, S., Yamasaki, N., and Izumo, S. (1999). The cardiac homeobox gene Csx/Nkx2.5 lies genetically upstream of multiple genes essential for heart development. *Development* **126**, 1269–1280.
11. Durocher, D., Charron, F., Warren, R., Schwartz, R. J., and Nemer, M. (1997). The cardiac transcription factors Nkx2–5 and GATA-4 are mutual cofactors. *EMBO J.* **16**, 5687–5696.
12. Alsan, B. H. and Schultheiss, T. M. (2000). Regulation of avian cardiogenesis by Fgf8 signaling. *Development* **129**, 1935–1943.
13. Marvin, M. J., Di Rocco, G., Gardiner, A., Bush, S. M., and Lassar, A. B. (2001). Inhibition of Wnt activity induces heart formation from posterior mesoderm. *Genes Dev.* **15**, 316–327.
14. Schneider, V. A. and Mercola, M. (2001). Wnt antagonism initiates cardiogenesis in *Xenopus laevis. Genes Dev.* **15**, 304–315.
15. Park, M., Wu, X., Golden, K., Axelrod, J. D., and Bodmer, R. (1996). The wingless signaling pathway is directly involved in Drosophila heart development. *Dev. Biol.* **177**, 104–116.
16. Molkentin, J. D., Lin, Q., Duncan, S. A., and Olson, E. N. (1997). Requirement of the transcription factor GATA4 for heart tube formation and ventral morphogenesis. *Genes Dev.* **11**, 1061–1072.

17. Kuo, C. T. *et al.* (1997). GATA4 transcription factor is required for ventral morphogenesis and heart tube formation. *Genes Dev.* **11**, 1048–1060.
18. Reiter, J. F. *et al.* (1999). Gata5 is required for the development of the heart and endoderm in zebrafish. *Genes Dev.* **13**, 2983–2995.
19. Walsh, E. C. and Stainier, D. Y. (2001). UDP-glucose dehydrogenase required for cardiac valve formation in zebrafish. *Science* **293**, 1670–1673.
20. Kelly, R. G., Brown, N. A., and Buckingham, M. E. (2001). The arterial pole of the mouse heart forms from Fgf10-expressing cells in pharyngeal mesoderm. *Dev. Cell* (3), 435–440.
21. Kathiriya, I. S. a. S., D. (2001). Left–right symmetry and cardiac looping: Implications for cardiac development and congenital heart disease. *Amer. J. Med. Gen.* **97**, 271–279.
22. Levin, M., Johnson, R. L., Stern, C. D., Kuehn, M., and Tabin, C. (1995). A molecular pathway determining left–right asymmetry in chick embryogenesis. *Cell* **82**, 803–814.
23. Rodriguez Esteban, C., Capdevila, J., Economides, A. N., Pascual, J., Ortiz, A., and Izpisua Belmonte, J. C. (1999). The novel Cer-like protein Caronte mediates the establishment of embryonic left–right asymmetry. *Nature* **401**, 243–251.
24. Isaac, A., Sargent, M. G., and Cooke, J. (1997). Control of vertebrate left–right asymmetry by a snail-related zinc finger gene. *Science* **275**, 1301–1304.
25. Piedra, M. E., Icardo, J. M., Albajar, M., Rodriguez-Rey, J. C., and Ros, M. A. (1998). Pitx2 participates in the late phase of the pathway controlling left–right asymmetry. *Cell* **94**, 319–324.
26. Logan, M., Pagan-Westphal, S. M., Smith, D. M., Paganessi, L., and Tabin, C. J. (1998). The transcription factor Pitx2 mediates situs-specific morphogenesis in response to left–right asymmetric signals. *Cell* **94**, 307–317.
27. Supp, D. M., Witte, D. P., Potter, S. S., and Brueckner, M. (1997). Mutation of an axonemal dynein affects left–right asymmetry in inversus viscerum mice. *Nature* **389**, 963–966.
28. Supp, D. M. *et al.* (1999). Targeted deletion of the ATP binding domain of left–right dynein confirms its role in specifying development of left–right asymmetries. *Development* **126**, 5495–5504.
29. Yokoyama, T., Copeland, N. G., Jenkins, N. A., Montgomery, C. A., Elder, F. F., and P. A. O. (1993). Reversal of left–right asymmetry: a situs inversus mutation. *Science* **260**, 679–682.
30. Lin, C. R. *et al.* (1999). Pitx2 regulates lung asymmetry, cardiac positioning and pituitary and tooth morphogenesis. *Nature* **401**, 279–282.
31. Campione, M. *et al.* (1999). The homeobox gene Pitx2: Mediator of asymmetric left–right signaling in vertebrate heart and gut looping. *Development* **126**, 1225–1234.
32. Capdevila, J., Vogan, K. J., Tabin, C. J., and Izpisua Belmonte, J. C. (2000). Mechanisms of left–right determination in vertebrates. *Cell* **101**, 9–21.
33. Nonaka, S. *et al.* (1998). Randomization of left–right asymmetry due to loss of nodal cilia generating leftward flow of extraembryonic fluid in mice lacking KIF3B motor protein. *Cell* **95**, 829–837.
34. Srivastava, D., Cserjesi, P., and Olson, E. N. (1995). A subclass of bHLH proteins required for cardiac morphogenesis. *Science* **270**, 1995–1999.
35. Srivastava, D. *et al.* (1997). Regulation of cardiac mesodermal and neural crest development by the bHLH transcription factor, dHAND. *Nat. Genet.* **16**, 154–160.
36. Firulli, A. B., McFadden, D. G., Lin, Q., Srivastava, D., and Olson, E. N. (1998). Heart and extra-embryonic mesodermal defects in mouse embryos lacking the bHLH transcription factor Hand1. *Nat. Genet.* **18**, 266–270.
37. Biben, C. a. H., R. P. (1997). Homeodomain factor Nkx2–5 controls left/right asymmetric expression of bHLH gene eHAND during murine heart development. *Genes Dev.* **11**, 1357–1369.
38. Yamagishi, H., Yamagishi, C., Harvey, R. P., Nakagawa, O., Olson, E. N., and Srivastava, D. (2001). Combinatorial activities of Nkx2.5 and dHAND is essential for cardiac ventricle formation. *Dev. Biol.* **239**, 190–203.
39. Yelon, D. *et al.* (2000). The bHLH transcription factor hand2 plays parallel roles in zebrafish heart and pectoral fin development. *Development* **127**, 2573–2582.
40. Bruneau, B. G. *et al.* (2000). Cardiac expression of the ventricle-specific homeobox gene Irx4 is modulated by Nkx2–5 and dHand. *Dev. Biol.* **217**, 266–277.
41. Bao, Z. Z., Bruneau, B. G., Seidman, J. G., Seidman, C. E., and Cepko, C. L. (1999). Regulation of chamber-specific gene expression in the developing heart by Irx4. *Science* **283**, 1161–1164.
42. Gottlieb, P., Pierce, S. A., Sims, R. J., Yamagishi, H., Weihe, E. K., Harriss, J. V., Maika, S. D., Kuziel, W. A., King, H. L., Olson, E. N., Nakagawa, O., and Srivastava, D. (2002). Bop encodes a muscle-restricted MYND and SET domain-containing protein essential for cardiac differentiation and morphogenesis. *Nat. Genet.*
43. Lin, Q., Schwarz, J., Bucana, C., and Olson, E. N. (1997). Control of mouse cardiac morphogenesis and myogenesis by transcription factor MEF2C. *Science* **276**, 1404–1407.
44. Dyson, E. *et al.* (1995). Atrial-like phenotype is associated with embryonic ventricular failure in retinoid X receptor alpha –/– mice. *Proc. Natl. Acad. Sci. USA* **92**, 7386–7390.
45. Nakagawa, O., Nakagawa, M., Richardson, J., Olson, E. N., and Srivastava, D. (1999). HRT1, HRT2 and HRT3: A new family of bHLH transcription factors marking specific cardiac, somitic and branchial arch segments. *Developmental Biology* **216**, 72–84.
46. Tran, C. M. and Sucov, H. M. (1998). The RXRalpha gene functions in a non-cell-autonomous manner during mouse cardiac morphogenesis. *Development* **25** (10), 1951–1956.
47. Lee, K. F. *et al.* (1995). Requirement for neuregulin receptor erbB2 in neural and cardiac development. *Nature* **378**, 394–348.
48. Carmeliet, P. *et al.* (1996). Abnormal blood vessel development and lethality in embryos lacking a single VEGF allele. *Nature* **380**, 435–439.
49. Suri, C. *et al.* (1996). Requisite role of angiopoietin-1, a ligand for the TIE2 receptor, during embryonic angiogenesis. *Cell* **87**, 1171–1180.
50. Galvin, K. M. *et al.* (2000). A role for smad6 in development and homeostasis of the cardiovascular system. *Nat. Genet.* **24**, 171–174.
51. Chen, B. *et al.* (2000). Mice mutant for Egfr and Shp2 have defective cardiac semilunar valvulogenesis. *Nat. Genet.* **24**, 296–299.
52. Tartaglia, M. *et al.* (2001). Mutations in PTPN11, encoding the protein tyrosine phosphatase SHP-2, cause Noonan syndrome. *Nat. Genet.* **29**, 465–468.
53. de la Pompa, J. L. *et al.* (1998). Role of the NF-ATc transcription factor in morphogenesis of cardiac valves and septum. *Nature* **392**, 182–186.
54. Ranger, A. M. *et al.* (1998). The transcription factor NF-ATc is essential for cardiac valve formation. *Nature* **392**, 186–190.
55. Kirby, M. L. and Waldo, K. L. (1995). Neural crest and cardiovascular patterning. *Circ. Res.* **77**, 211–215.
56. Kurihara, Y. *et al.* (1995). Aortic arch malformations and ventricular septal defect in mice deficient in endothelin-1. *J. Clin. Invest.* **96**, 293–300.
57. Clouthier, D. E. *et al.* (1998). Cranial and cardiac neural crest defects in endothelin-A receptor-deficient mice. *Development* **125**, 813–824.
58. Thomas, T. *et al.* (1998). A signaling cascade involving endothelin-1, dHAND and msx1 regulates development of neural-crest-derived branchial arch mesenchyme. *Development* **125**, 3005–3014.
59. Charite, J., McFadden, D. G., Merlo, G., Levi, G., Clouthier, D. E., Yanagisawa, M., Richardson, J. A., and Olson, E. N. (2001). Role of Dlx6 in regulation of an endothelin-1-dependent, dHAND branchial arch enhancer. *Genes Dev.* **15**(22), 3039–3049.
60. Yamagishi, H., Olson, E. N., and Srivastava, D. (2000). The bHLH transcription factor, dHAND, is required for vascular development. *J. Clin. Invest.* **105**, 261–270.
61. Kawasaki, T. *et al.* (1999). A requirement for neuropilin-1 in embryonic vessel formation. *Development* **126**, 4895–4902.
62. Feiner, L., Webber, A. L., Brown, C. B., Lu, M. M., Jia, L., Feinstein, P., Mombaerts, P., Epstein, J. A., and Raper, J. A. (2001). Targeted disruption of semaphorin 3C leads to persistent truncus arteriosus and aortic arch interruption. *Development* **128**(16), 3061–3070.

63. Li, L. *et al.* (1997). Alagille syndrome is caused by mutations in human Jagged1, which encodes a ligand for Notch1. *Nat. Genet.* **16**, 243–251.
64. Oda, T., Elkahloun, A. G., and Pike, B. L. (1997). Mutations in the human Jagged1 gene are responsible for Alagille syndrome. *Nat. Genet.* **16**, 235–242.
65. Krantz, I. D. *et al.* (1999). Jagged1 mutations in patients ascertained with isolated congenital heart defects. *Am J. Med Genet* **84**, 56–60.
66. Zhong, T. P., Childs, S., Leu, J. P., and Fishman, M. C. (2001). Gridlock signalling pathway fashions the first embryonic artery. *Nature* **414**, 216–220.
67. Zhong, T. P., Rosenberg, M., Mohideen, M. A., Weinstein, B., and Fishman, M. C. (2000). gridlock, An HLH gene required for assembly of the aorta in zebrafish. *Science* **287**, 1820–1824.

CHAPTER 336

Development and Regulatory Signaling in the Pancreas

Murray Korc
Division of Endocrinology, Diabetes and Metabolism,
University of California at Irvine,
Irvine, California

Introduction

The word *pancreas* derives from the Greek words *pan* (meaning "all") and *kreos* (meaning "flesh"), because the ancients believed that the pancreas was a pillow of flesh that was destined to provide the stomach with a comfortable cushion. In fact, the mammalian pancreas is a complex organ that consists of acinar cells that synthesize and secrete digestive enzymes, duct cells that produce bicarbonate rich fluid, and islets of hormone-secreting cells that are dispersed throughout the exocrine tissue. These endocrine islets were first described by Langerhans in 1869, who initially believed them to be intrapancreatic lymph nodes [1]. The concept that these islets may produce insulin arose following the discovery that pancreatectomized dogs develop diabetes mellitus [2–4]. Eventually, an active pancreatic extract that was rich in islet-cell-derived insulin was used to treat Leonard Thompson, the first human to receive live-saving insulin therapy [5]. This chapter delineates some of the aspects of pancreatic ontogeny and describes unique anatomic and cellular interactions that allow the pancreas to perform its complex functions.

Ontogeny of the Pancreas

The embryonic pancreas arises from endodermal cells in the duodenal region of the gut [6]. Commitment to pancreatic fate occurs at embryonic day 8.5 (E8.5) in the mouse, and dorsal and ventral pancreatic buds become evident on day E9 [7]. Pancreatic fate is dictated by the expression in these cells of the HOX-like homeoprotein pancreas duodenum homeobox-1 (PDX-1), a well-characterized transcription factor [8]. During organogenesis, PDX-1 is expressed in both the exocrine and endocrine components of the pancreas [8,9]. However, in the adult pancreas its expression is restricted to the insulin-containing β cells and the glucagon-containing α cells [9,10]. The hedgehog family of proteins also modulates pancreatic development, but in a negative manner. Thus, expression of *Sonic hedgehog* (*shh*) dictates intestinal development, and prevention of Shh expression, effected in part by fibroblast growth factor 2 (FGF-2), is absolutely essential for pancreatic development [11].

Studies of pancreas development in the mouse using classical approaches and gene knock-out strategies have revealed that there is a network of signaling molecules and transcription factors that together dictate the ultimate fate of pancreatic progenitor cells. For example, mice deficient for the type IIB activin receptor (ActRIIB) exhibit pancreatic hypoplasia and marked hyposplenism [12]. The pancreatic hypoplasia is due to loss of activin signaling, which acts like FGF-2 to suppress Shh expression [12]. Mice that are heterozygous for Smad2 deficiency also exhibit pancreatic endocrine defects [7], indicating that both activin and transforming growth factor β (TGF-β) pathways are important in islet cell development. Conversely, in mice lacking the transcription factor p48, there is a complete absence of exocrine pancreas development, whereas hormones expressing islet cells survive and are found in the spleen [13].

Inhibition of Notch signaling also blocks acinar cell formation and promotes islet cell formation [14]. Activation of the Notch receptor leads to enhanced transcription of hairy/enhancer-of-split genes (*Hes*), which encodes basic helix–loop–helix (bHLH) transcription factors that repress neurogenin expression [7,15]. *Ngn3*, a member of the neurogenin family, is expressed at high levels in both α and β cells, and *Hes1*-deficient mice exhibit acinar cell apoptosis and hypoplasia as well as precocious development of endocrine cells [16]. These observations point to a common precursor cell that gives rise to mature pancreatic acinar and endocrine cells. The hypothesis of a common progenitor cell in the endocrine pancreas is supported by the observation that islet cell tumors may give rise to cloned cell lines that express several islet cell hormones [17].

Pancreatic Islet–Acinar Interactions

The pancreatic endocrine cells are dispersed throughout the exocrine pancreas, which makes up approximately 95% of the total pancreatic tissue mass. Of the four major types of pancreatic endocrine cells, the β cells are the most important. In these cells, insulin is packaged into heterogeneous granules with electron-dense cores [18]. In contrast, the pancreatic polypeptide (PP)-containing cells have oblong granules that are markedly electron dense; the α cells exhibit uniform electron-dense granules that contain glucagon; the δ cells are small, somatostatin-containing cells that have a dendritic shape [18]; and the acinar cells store their digestive proenzymes in zymogen granules. Islet cells may modulate each other's actions as a result of intra-islet cell-to-cell communications through gap junctions [19]. However, the α, δ, and PP-containing cells are located at the periphery of the islets, and they release their hormones into venules that are located on the islet surface, thereby mostly avoiding the centrally located β cells [20]. In contrast, within the islet microcirculation some of the insulin is carried to the α, δ, and PP cells, thereby allowing for direct effects by this hormone on the islet cells [21]. Thus, the network of signaling molecules and transcription factors that dictates pancreatic development gives rise to cells that exhibit a high degree of specialization and an anatomic organization that allows for complex and highly regulated cell-to-cell interactions.

Although the endocrine islets constitute approximately 2% of the pancreatic volume, they receive approximately 20% of the intrapancreatic blood flow [22,23]. The venous effluent from the larger islets appears to bypass the surrounding exocrine tissue, whereas the effluent from smaller but more numerous islets forms an intrapancreatic portal circulation that passes through the exocrine tissue [20]. This anatomic arrangement has given rise to the hypothesis that islet cell hormones may participate in the regulation of pancreatic exocrine function [24]. Indeed, the pancreatic acinar cell has specific high-affinity receptors for insulin and somatostatin [25,26], and insulin deficiency is associated with enhanced cholecystokinin (CCK) binding and decreased epidermal growth factor (EGF) binding [27,28].

Several additional lines of evidence support a role for insulin in regulating pancreatic exocrine function. Insulin directly enhances pancreatic acinar cell glucose oxidation and transport, protein synthesis, and the secretory effects of the gastrointestinal hormone CCK, a calcium-mobilizing secretagogue [29–35]. Insulin markedly increases pancreatic amylase mRNA levels in insulin-deficient rats, but only slightly alters parotid amylase mRNA levels [36]. Pancreatic acini isolated from insulin-deficient rats exhibit attenuated increases in cytosolic free calcium and inositol phosphate levels [37,38]. Pancreatic duct cell function is also altered in states of insulin deficiency, as evidenced by the observations that patients with poorly controlled, long-standing type I diabetes mellitus exhibit a decreased ability to secrete bicarbonate-rich fluid and pancreatic digestive enzymes in response to pancreatic secretagogues [39–41], and that insulin increases the stimulatory effects of secretin on pancreatic juice secretion in the perfused rat pancreas [42]. Thus, the propinquity of the islet and exocrine cells in the pancreas allows for unique cellular interactions.

Cell–Cell and Matrix Interactions in the Endocrine Pancreas

Several growth factors have an important role in epithelial mesenchymal interactions during pancreatic development and may also exert a role in the adult pancreas. In addition to FGF-2 and activin, keratinocyte growth factor, or FGF-7, has been implicated in this process. FGFs act by binding to the protein products of four distinct genes encoding high-affinity FGF receptors designated as FGFR-1, FGFR-2, FGFR-3, and FGFR-4 [43]. These receptors possess an extracellular cytoplasmic ligand-binding domain that has three immunoglobulin-like (Ig-like) regions, a hydrophobic transmembrane domain, and a discontinuous intracellular tyrosine kinase domain exhibiting a short intervening sequence [43]. The presence of an intron–exon boundary in the third Ig-like loop (domain III) allows for the generation of two alternative carboxyl-terminal domains (IIIb or IIIc) of FGFR-1, -2, and -3. Expression of domain III isoforms is often restricted to specific cell types. For example, the IIIb splice form of FGFR-1 is generally found in epithelial cell types, whereas the IIIc splice form is often restricted to mesenchymal cell types [43]. This domain confers ligand-binding specificity, as underscored by the observation that the IIIb splice form of FGFR-2 is activated by FGF-7 but not by FGF-2 [43]. FGF-7 stimulates the proliferation of pancreatic epithelial cells that are the precursors of endocrine cells, and its removal leads to their differentiation into endocrine cells [44]. Furthermore, mice deficient for the FGFR-2 IIIb receptor exhibit abnormal pancreatic development [44], indicating that this receptor isoform and FGF-7 participate in important epithelial–mesenchymal interactions in the pancreas. Similarly, FGF-10 is expressed in the mesenchyme that is adjacent to the early pancreatic buds, and in mice that are deficient for FGF-10 there is an arrest in branching of the PDX-1 positive epithelium that is destined to become the pancreas [45]. PDX-1 positive cells reappear in culture in the presence of exogenous FGF-10, confirming the importance of this growth factor in pancreatic organogenesis and in dictating the size of the pool of cells that are destined to become the pancreas [45]. EGF also enhances the proliferation of these cells [46], and all of these growth factors may, under certain circumstances, have the potential to expand the pool of endocrine cells in the adult pancreas.

Another important component of cell–matrix interactions in the pancreas is dependent on signaling pathways activated by the integrin family of receptors and their ligands. Integrins $\alpha v\beta 3$ and $\alpha v\beta 5$, and their major ligands vitronectin and fibronectin, are expressed in islet progenitor cells and have been implicated in islet cell migration and morphogenesis [47]. Both $\alpha v\beta 3$ and $\alpha v\beta 5$ mediate islet cell migration in culture, and both appear to contribute to the structural integrity of mature islets [47]. Furthermore, the survival and function of islets in culture is enhanced in the presence of extracellular matrix (ECM) [48,49]. These observations indicate that both islet differentiation and the maintenance of islet cell-differentiated functions are dependent on interactions with various components of the ECM.

Although exocrine and endocrine cells arise from a common population of progenitor endoderm cells [50], the endocrine cells are able to migrate toward the mesenchyme, attract endothelial cells, and eventually form the highly vascularized islets [6]. Integrins $\alpha v\beta 3$, which are expressed in invasive and proliferating endothelial cells [51], may have a role in islet vascularization. In addition, rodent and human islets express relatively high levels of vascular endothelial growth factor (VEGF), and pancreatic endothelial cells express the two high-affinity receptors VEGFR1 and VEGFR2 [52–54]. Inasmuch as VEGF is mitogenic toward endothelial cells [55], it is possible that islet-derived VEGF may recruit endothelial cells and induce their proliferation during development. Furthermore, the appearance of insulin-expressing cells during islet development requires contact with blood vessel endothelium [56] and both VEGF and its receptors in the islet cells in adult humans [54]. Together, these observations point to the existence of complex regulatory interactions between pancreatic endocrine and vascular cells during organogenesis and in the mature pancreas, and they raise the possibility that VEGF-dependent pathways participate in the regulation of differentiated islet cell function.

Matrix and Cell–Cell Interactions in the Exocrine Pancreas

The preponderant cell type in the exocrine pancreas is the acinar cell. It has specific, high-affinity receptors for numerous agonists, including CCK, which acts via calcium, and vasoactive intestinal polypeptide (VIP), which acts via cAMP [57–59]. CCK mobilizes calcium from intracellular stores, and the resulting calcium waves appear to spread from acinar cell to acinar cell, partly as a consequence of open gap junctions that allow for this type of cell-to-cell communication [57].

Two lines of evidence indicate that pancreatic acinar cell functions can be modulated by interactions with the ECM. First, when pancreatic acini are placed in culture, they dedifferentiate into ductal-like cells [60], implying that maintenance of a differentiated acinar-like state is dependent on interactions with the pancreatic ECM. Second, the pancreatic acinar cell possesses a $\beta 1$ integrin receptor, and ligation of this receptor with an anti-$\beta 1$ integrin antibody leads to tyrosine phosphorylation phospholipase C-$\gamma 1$, translocation of protein kinase Cα to the cell membrane, and a rise in cytosolic free calcium levels, pointing to a direct effect by the ECM on acinar cell differentiated functions [61,62].

The exocrine pancreas also has a defined but small stromal compartment that consists of connective tissue and fibroblasts surrounding the interlobular ducts, a small number of pancreatic stellate cells, vascular endothelial cells, and thin stromal partitions that separate adjacent pancreatic lobules [63,64]. In pathological conditions such as chronic pancreatitis, there is a marked expansion of this stromal compartment, with extensive deposition of fibronectin, laminin, and collagen types I, III, and IV [65]. This fibrotic reaction occurs, in part, as a consequence of excessive production of growth factors such as TGF-α, TGF-β, connective tissue growth factor, FGFs, and platelet-derived growth factor [66–70]. In addition, enhanced fibroblast proliferation and activation of pancreatic stellate cells are seen [64]. As a consequence, the pancreas exhibits regions of acinar cell degeneration and apoptosis and ductal cell proliferation, in association with pancreatic exocrine and endocrine dysfunctions [71,72], underscoring the importance of maintaining normal epithelial mesenchymal interactions in the adult pancreas.

Conclusions

Pancreatic function is regulated by complex interactions that include a multiplicity of extracellular signals from hormones, neurotransmitters, and nutrients. Proper coordination of the exocrine and endocrine components of the pancreas ensures the timely digestion of nutrients in the gut lumen and the subsequent assimilation of the absorbed components by target tissues through the actions of islet hormones. Multiple endocrine–exocrine–ECM interactions are necessary to maintain these differentiated functions. In addition, the dispersal of the islets throughout the exocrine tissue, the rich vascularization of the islets, and the existence of an intrapancreatic portal circulation translate into unique cell–cell interactions that are yet to be fully understood.

References

1. Langerhans, P. (1869). Contributions to the microscopic anatomy of the pancreas. MD Thesis, Berlin.
2. Merling, von J. and Minkowski, O. (1890). Diabetes mellitus nach pancreas extirpation. *Arch. Exp. Pathol. Pharmakol.* **26**, 371–387.
3. Laguesse, E. (1893). Sur la formation des flots de Langerhans dans le pancreas. *C. R. Soc. Biol. (Paris)* **14**, 819–820.

4. Opie, E. L. (1990). The relation of diabetes mellitus to lesions of the pancreas. Hyaline degeneration of the islands of Langerhans. *J. Exp. Med.* **5**, 527–540.
5. Bliss, M. (1982). *The Discovery of Insulin.* University of Chicago Press, Chicago.
6. Slack, J. M. (1995). Developmental biology of the pancreas. *Development* **121**, 1569–1580.
7. Kim, S. K. and Hebrok, M. (2001). Intercellular signals regulating pancreas development and function. *Genes Dev.* **15**, 111–127.
8. Hui, H. and Perfetti, R. (2002). Pancreas duodenum homeobox-1 regulates pancreas development during embryogenesis and islet cell function in adulthood. *Eur. J. Endocrinol.* **146**, 129–141.
9. Sander, M. and German, M. S. (1997). The beta cell transcription factors and development of the pancreas. *J. Mol. Med.* **75**, 327–340.
10. Edlund, H. (2001). Factors controlling pancreatic cell differentiation and function. *Diabetologia* **44**, 1071–1079.
11. Hebrok, M., Kim, S. K., St. Jacques, B., McMahon, A. P., and Melton, D. A. (2000). Regulation of pancreas development by hedgehog signaling. *Development* **127**, 4905–4913.
12. Kim, S. K., Hebrok, M., Li, E., Oh, S. P., Schrewe, H., Harmon, E. B., Lee, J. S., and Melton, D. A. (2000). Activin receptor patterning of foregut organogenesis. *Genes Dev.* **14**, 1866–1871.
13. Krapp, A., Knofler, M., Ledermann, B., Burki, K., Berney, C., Zoerkler, N., Hagenbuchle, O., and Wellauer, P. K. (1998). The bHLH protein PTF1-p48 is essential for the formation of the exocrine and the correct spatial organization of the endocrine pancreas. *Genes Dev.* **12**, 3752–3763.
14. Apelqvist, A., Li, H., Sommer, L., Beatus, P., Anderson, D. J., Honjo, T., Hrabe de Angelis, M., Lendahl, U., and Edlund, H. (1999). Notch signalling controls pancreatic cell differentiation. *Nature* **400**, 877–881.
15. Jensen, J., Heller, R.. S., Funder-Nielsen, T., Pedersen, E. E., Lindsell, C., Weinmaster, G., Madsen, O. D., and Serup, P. (2000). Independent development of pancreatic alpha- and beta-cells from neurogenin3-expressing precursors: a role for the notch pathway in repression of premature differentiation. *Diabetes* **49**, 163–176.
16. Jensen, J., Pedersen, E. E., Galante, P., Hald, J., Heller, R. S., Ishibashi, M., Kageyama, R., Guillemot, F., Serup, P., and Madsen, O. D. (2000). Control of endodermal endocrine development by Hes-1. *Nat. Genet.* **24**, 36–44.
17. Philippe. J, Chick W. L., and Habener J. F. (1987). Multipotential phenotypic expression of genes encoding peptide hormones in rat insulinoma cell lines. *J. Clin. Invest.* **79**, 351–358.
18. Falkmer, S. and Ostberg, Y. (1977). Comparative morphology of pancreatic islets in animals. In Volk, B. W., and Wellman, K. F., Eds., *The Diabetic Pancreas*, pp. 15–60. Plenum Press, New York.
19. Meda, P., Perrelet, A., and Orci, L. (1984). Gap junctions and cell to cell coupling in endocrine glands. *Mod. Cell. Biol.* **3**, 131–196.
20. Bonner-Weir, S. and Orci, L. (1982). New perspectives on the microvasculature of the islets of Langerhans in the rat. *Diabetes* **31**, 883–889.
21. Bonner-Weir, S. (1988). Morphological evidence for pancreatic polarity of B-cell within the islets of Langerhans. *Diabetes* **37**, 616–621.
22. Lifson, N., Kramlinger, K. G., Mayrand, R. R., and Linder, E. J. (1980). Blood flow to the rabbit pancreas with specific reference to the islets of Langerhans. *Gastroenterology* **79**, 466–473.
23. Lifson, N., Lassa, C. V., and Dixit, P. K. (1985). Relation between blood flow and morphology in islet organ of rat pancreas. *Am. J. Physiol.* **249**, E43–E48.
24. Henderson, J. R., Daniel, P. M., and Frasier, P. A. (1981). The pancreas as a single organ: The influence of the endocrine upon the exocrine part of the gland. *Gut* **22**, 158–167.
25. Korc, M., Sankaran, H., Wong, K. Y., Williams J. A., and Goldfine, I. D. (1978). Insulin receptors in isolated mouse pancreatic acini. *Biochem. Biophys. Res. Commun.* **84**, 293–299.
26. Garry, D. J., Garry, M. G., Williams, J. A., Mahoney, W. C., and Sorenson, R. L. (1989). Effects of islet hormones on amylase secretion and localization of somatostatin binding sites. *Am. J. Physiol.* **256**, G897–904.
27. Otsuki, M., Goldfine, I. D., and Williams, J. A. (1984). Diabetes in the rat is associated with a reversible postreceptor defect in cholecystokinin action. *Gastroenterology* **87**, 882–887.
28. Korc, M., Matrisian, L. M., Nakamura, R., and Magun, B. E. (1984). Epidermal growth factor binding is altered in pancreatic acini from diabetic rats. *Life Sci.* **35**, 2049–2055.
29. Williams, J. A., Bailey, A. C., Preissler, M., and Goldfine, I. D. (1982). Insulin regulation of sugar transport in isolated pancreatic acini from diabetic mice. *Diabetes* **31**, 674–682.
30. Danielsson, A. and Sehlin, J. (1974). Transport and Oxidation of amino acids and glucose in the isolated exocrine mouse pancreas: Effects of insulin and pancreozymin. *Acta Physiol. Scand.* **91**, 557–565.
31. Korc, M., Iwamoto, Y., Sankaran, H., Williams, J. A., and Goldfine, I. D. (1981). Insulin action in pancreatic acini from streptozotocin-treated rats. I. Stimulation of protein synthesis. *Am. J. Physiol.* **240**, G56–G62.
32. Okabayashi, Y., Moessner, J., Logsdon, C. D., Goldfine, I. D., and Williams, J. A. (1987). Insulin and other stimulants have nonparallel translational effects on protein synthesis. *Diabetes* **36**, 1054–1060.
33. Saito, A., Williams, J. A., and Kanno, T. (1980). Potentiation of cholecystokinin-induced exocrine secretion by both exogenous and endogenous insulin in isolated and perfused rat pancreata. *J. Clin. Invest.* **65**, 777–782.
34. Matsushita, K., Okabayashi, Y., Koide, M., Hasegawa, H., Otsuki, M., and Kasuga, M. (1994). Potentiating effect of insulin on exocrine secretory function in isolated rat pancreatic acini. *Gastroenterology* **106**, 200–206.
35. Otsuki, M. and Williams J. A. (1983). Direct modulation of pancreatic CCK receptors and enzyme secretion by insulin in isolated pancreatic acini from diabetic rats. *Diabetes* **32**, 241–246.
36. Korc, M., Owerbach, D., Quinto, C., and Rutter, W. J. (1981). Pancreatic islet-acinar cell interaction: Amylase messenger RNA levels are determined by insulin. *Science* **213**, 351–352.
37. Korc, M. and Schoni, M. L. (1988). Quin 2 and manganese define multiple alternations in cellular calcium homeostasis in diabetic rat pancreas. *Diabetes* **37**, 13–20.
38. Chandrasekar, B. and Korc, M. (1991). Alteration of cholecystokinin-mediated phosphatidylinositol hydrolysis in pancreatic acini from insulin-deficient rats. *Diabetes* **40**, 1282–1291.
39. Vacca, J. B., Henke, W. H., and Knight, W. A., Jr. (1964). The exocrine pancreas in diabetes mellitus. *Ann. Intern. Med.* **61**, 242–247.
40. Frier, B. M., Saunders, J. H. B., Wormsley, K. G., and Bouchier, I. A. D. (1976). Exocrine pancreatic function in juvenile-onset diabetes mellitus. *Gut* **17**, 685–691.
41. Frier, B. M., Faber, O. K., Binder, C., and Elliot, H. L. (1978). The effect of residual insulin secretion in juvenile onset diabetes mellitus. *Diabetologia* **14**, 301–304.
42. Hasegawa, H., Okabayashi, Y., Koide, M., Kido, Y., Okutani, T., Matsushita, K., Otsuki, M., and Kasuga, M. (1993). Effect of islet hormones on secretin-stimulated exocrine secretion in isolated perfused rat pancreas. *Dig. Dis. Sci.* **38**, 1278–1283.
43. Ornitz, D. M., Xu, J., Colvin, J. S., McEwen, D. G., MacArthur, C. A., Coulier, F., Gao, G., and Goldfarb, M. (1996). Receptor specificity of the fibroblast growth factor family. *J. Biol. Chem.* **271**, 15292–15297.
44. Elghazi, L., Cras-Meneur, C., Czernichow, P., and Scharfmann, R. (2002). Role for FGFR2IIIb-mediated signals in controlling pancreatic endocrine progenitor cell proliferation. *Proc. Natl. Acad. Sci. USA* **99**, 3884–3889.
45. Bhushan, A., Itoh, N., Kato, S., Thiery, J. P., Czernichow, P., Bellusci, S., and Scharfmann, R. (2001). Fgf10 is essential for maintaining the proliferative capacity of epithelial progenitor cells during early pancreatic organogenesis. *Development* **128**, 5109–5117.
46. Cras-Meneur, C., Elghazi, L., Czernichow, P., and Scharfmann, R. (2001). Epidermal growth factor increases undifferentiated pancreatic embryonic cells in vitro: a balance between proliferation and differentiation. *Diabetes* **50**, 1571–1579.
47. Cirulli, V., Beattie, G. M., Klier, G., Ellisman, M., Ricordi, C., Quaranta, V., Frasier, F., Ishii, J. K., Hayek, A., and Salomon, D. R. (2000). Expression and function of alpha(v)beta(3) and alpha(v)beta(5)

integrins in the developing pancreas: roles in the adhesion and migration of putative endocrine progenitor cells. *J. Cell Biol.* **150**, 1445–1460.

48. Wang, R. N. and Rosenberg, L. (1999). Maintenance of beta-cell function and survival following islet isolation requires re-establishment of the islet-matrix relationship. *J. Endocrinol.* **163**, 181–190.
49. Bosco, D., Meda, P., Halban, P. A., and Rouiller, D. G. (2000). Importance of cell–matrix interactions in rat islet beta-cell secretion *in vitro:* Role of alpha6beta1 integrin. *Diabetes* **49**, 233–243.
50. Gittes, G. K., Galante, P. E., Hanahan, D., Rutter, W. J., and Debase, H. T. (1996). Lineage-specific morphogenesis in the developing pancreas: Role of mesenchymal factors. *Development* **122**, 439–447.
51. Clark, R. A., Tonnesen, M. G., Gailit, J., and Cheresh, D. A. (1996). Transient functional expression of alphaVbeta 3 on vascular cells during wound repair. *Am. J. Pathol.* **148**, 1407–1421.
52. Christofori, G., Naik, P., and Hanahan, D. (1995). Vascular endothelial growth factor and its receptors, flt-1 and flk-1, are expressed in normal pancreatic islets and throughout islet cell tumorigenesis. *Mol. Endocrinol.* **9**, 1760–1770.
53. Vasir, B., Jonas, J. C., Steil, G. M., Hollister-Lock, J., Hasenkamp, W., Sharma, A., Bonner-Weir, S., and Weir, G. C. (2001). Gene expression of VEGF and its receptors Flk-1/KDR and Flt-1 in cultured and transplanted rat islets. *Transplantation* **71**, 924–935.
54. Itakura, J., Ishiwata, T., Shen, B., Kornmann, M., and Korc, M. (2000). Concomitant over-expression of vascular endothelial growth factor and its receptors in pancreatic cancer. *Int. J. Cancer* **85**, 27–34.
55. Ferrara, N. (1999). Molecular and biological properties of vascular endothelial growth factor. *J. Mol. Med.* **77**, 527–543.
56. Lammert, E., Cleaver, O., and Melton, D. (2001). Induction of pancreatic differentiation by signals from blood vessels. *Science* **294**, 564–567.
57. Williams, J. A. (2001). Intracellular signaling mechanisms activated by cholecystokinin-regulating synthesis and secretion of digestive enzymes in pancreatic acinar cells. *Annu. Rev. Physiol.* **63**, 77–97.
58. Fitzsimmons, T. J., Gukovsky, I., McRoberts, J. A., Rodriguez, E., Lai, F. A., and Pandol, S. J. (2000). Multiple isoforms of the ryanodine receptor are expressed in rat pancreatic acinar cells. *Biochem. J.* **351**, 265–271.
59. Ito, T., Hou, W., Katsuno, T., Igarashi, H., Pradhan, T. K., Mantey, S. A., Coy, D. H., and Jensen, R. T. (2000). Rat and guinea pig pancreatic acini possess both VIP(1) and VIP(2) receptors, which mediate enzyme secretion. *Am. J. Physiol.* **278**, G64–74.
60. Logsdon. C. D. and Williams. J. A. (1986). Pancreatic acinar cells in monolayer culture: Direct trophic effects of caerulein *in vitro. Am. J. Physiol.* **250**, G440–447.
61. Wrenn, R. W. and Herman, L. E. (1995). Integrin-linked tyrosine phosphorylation increases membrane association of protein kinase C alpha in pancreatic acinar cells. *Biochem. Biophys. Res. Commun.* **208**, 978–984.
62. Wrenn, R. W., Creazzo, T. L., and Herman, L. E. (1996). Beta 1 integrin ligation stimulates tyrosine phosphorylation of phospholipase C gamma 1 and elevates intracellular Ca2+ in pancreatic acinar cells. *Biochem. Biophys. Res. Commun.* **226**, 876–882.
63. Kern, H. (1993). Fine structure of the human pancreas. In Scheele, G., Ed., *The Pancreas*, pp. 9–19. Raven Press, New York.
64. Luttenberger, T., Schmid-Kotsas, A., Menke, A., Siech, M., Beger, H., Adler, G., Grunert, A., and Bachem, M. G. (2000). Platelet-derived growth factors stimulate proliferation and extracellular matrix synthesis of pancreatic stellate cells: Implications in pathogenesis of pancreas fibrosis. *Lab. Invest.* **80**, 47–55.
65. Gress, T. M., Menke, A., Bachem, M., Muller-Pillasch, F., Ellenrieder, V., Weidenbach, H., Wagner, M., and Adler, G. (1998). Role of extracellular matrix in pancreatic diseases. *Digestion* **59**, 625–367.
66. Korc, M., Friess, H., Yamanaka, Y., Kobrin, M. S., Büchler, M., and Beger, H. G. (1994). Chronic pancreatitis is associated with increased concentrations of epidermal growth factor receptor, transforming growth factor alpha, and phospholipase C gamma. *Gut* **35**, 1468–1473.
67. Muller-Pillasch, F., Menke, A., Yamaguchi, H., Elsasser, H. P., Bachem, M., Adler, G., and Gress, T. M. (1999). TGF beta and the extracellular matrix in pancreatitis. *Hepatogastroenterology* **46**, 2751–2756.
68. di Mola, F. F., Friess, H., Martignoni, M. E., Di Sebastiano, P., Zimmermann, A., Innocenti, P., Graber, H., Gold, L. I., Korc, M., and Büchler, M. W. (1999). Connective tissue growth factor is a regulator for fibrosis in human chronic pancreatitis. *Ann. Surg.* **230**, 63–71.
69. Kornmann, M., Beger, H. G., and Korc, M. (1998). Role of fibroblast growth factors and their receptors in pancreatic cancer and chronic pancreatitis. *Pancreas* **17**, 169–175.
70. Ebert, M., Kasper, H. U., Hernberg, S., Friess, H., Büchler, M. W., Roessner, A., Korc, M., and Malfertheiner, P. (1998). Overexpression of platelet-derived growth factor (PDGF) B chain and type beta PDGF receptor in human chronic pancreatitis. *Dig. Dis. Sci.* **43**, 567–574.
71. Bockman, D. E. (1995). Toward understanding pancreatic disease: From architecture to cell signaling. *Pancreas* **11**, 324–329.
72. Lankisch, P. G., Lohr-Happe, A., Otto, J., and Creutzfeldt, W. (1993). Natural course in chronic pancreatitis. Pain, exocrine and endocrine pancreatic insufficiency and prognosis of the disease. *Digestion* **54**, 148–155.

CHAPTER 337

Tropic Effects of Gut Hormones in the Gastrointestinal Tract

B. Mark Evers and Robert P. Thomas
Department of Surgery, The University of Texas Medical Branch, Galveston, Texas

Introduction

The mucosa of the gastrointestinal (GI) tract is a complex and constantly renewing tissue that is characterized by rapid proliferation, differentiation, and subsequent apoptosis, followed by extrusion into the GI lumen. These events occur as GI luminal epithelial cells ascend the vertical axis of the microfolded crypts lining the GI tract [1,2]. This process normally takes 3–8 days, depending on the species and the location along the GI tract [3,4]. Numerous factors can contribute to growth of the GI mucosa. This chapter focuses specifically on the effects of GI hormones on the proliferation and repair of non-neoplastic tissues and on the receptors and signaling pathways that transmit signals from the cell surface to the nucleus.

By definition, any agent that stimulates growth can be considered a growth factor; however, these growth-stimulating agents are usually divided into those produced by normal cells and thought to act locally to control proliferation and hormones that are thought to act at a distance. Peptide growth factors that act locally include members of the epidermal growth factor (EGF) family, the transforming growth factor β (TGF-β) family, the insulin-like growth factor (IGF) family, the fibroblast growth factor (FGF) family, the trefoil factor (TFF) family, the colony-stimulating factor (CSF) family, and a few other unrelated regulatory peptides, such as hepatocyte growth factor (HGF), platelet-derived growth factor (PDGF), various interleukins, interferons, and tumor necrosis factor-related proteins [5].

Various gut hormones can regulate growth of the GI mucosa, usually through an endocrine effect although, on occasion, an autocrine or paracrine mechanism has been postulated for the proliferative effects of these trophic peptides [6,7]. The gut peptides that have been best described in their role as stimulating mucosal proliferation of the stomach, small bowel, or colon include gastrin, bombesin (BBS)/gastrin-releasing peptide (GRP), neurotensin (NT), glucagon-like peptide-2 (GLP-2), and peptide YY (PYY) [1]. These hormones are secreted by endocrine cells that are widely distributed throughout the GI mucosa and pancreas. In addition to mucosal proliferation, these gut peptides control many other functions in the GI tract, including regulation of secretion, motility, absorption, and digestion (Table I).

Tropic Effects of Gut Peptides in the Stomach, Small Bowel, and Colon

Stomach

Gastrin is the GI hormone that has been best characterized for its trophic effects in the stomach. Gastrin stimulates acid secretion from gastric parietal cells and is the single most important trophic hormone for the gastric mucosa. The trophic effect of gastrin on gastric mucosa was initially demonstrated with the synthetic gastrin analog, pentagastrin, which, when given to rats, stimulated protein synthesis and parietal cell mass [8,9]. These results were further confirmed using the natural amidated gastrins, G17 and G34, with most pronounced effects noted in the oxyntic acid-secreting mucosa and enterochromaffin-like cells. Resection of the gastric antrum, which removes endogenous gastrin, results in gastric mucosal atrophy; this atrophy can be prevented by administration of exogenous gastrin [10].

Table I Gut Hormones Contributing to GI Mucosal Growth

Hormone	Location	Primary effects
Gastrin	Antrum, duodenum (G cells)	- Stimulates gastric acid and pepsinogen secretion - Stimulates gastric mucosal growth
Gastrin releasing peptide (GRP) (mammalian equivalent of bombesin [BBS])	Small bowel	- Stimulates release of all GI hormones - Stimulates GI secretion and motility - Stimulates gastric acid secretion and release of antral gastrin - Stimulates growth of intestinal mucosa and pancreas
Neurotensin (NT)	Small bowel (N cells)	- Stimulates pancreatic water and bicarbonate secretion - Inhibits gastric secretion - Stimulates growth of small and large bowel mucosa
Glucagon-like peptide-2 (GLP-2)	Small bowel (L cells)	- Potent enterotrophic factor
Peptide YY (PYY)	Distal small bowel, colon	- Inhibits gastric and pancreatic secretion - Inhibits gallbladder contraction - Stimulates intestinal growth?

Further confirmation of the effects of gastrin on gastric mucosal growth is provided by transgenic mice that either overexpress gastrin or are gastrin deficient. In mice overexpressing either unprocessed gastrin or the amidated gastrins (G17 and G34), a marked thickening is seen in the oxyntic mucosa with increased BrdU labeling, representing an 85% increase in cells undergoing proliferation [11,12]. In gastrin-deficient mice, a 35% decrease in parietal cell mass with no decrease in basal fundic proliferation rate is noted when compared with wild-type control animals [13,14].

Another peptide that has been shown to stimulate gastric mucosal proliferation is BBS/GRP, which stimulates pancreatic, gastric, and intestinal secretion, gut motility, smooth muscle contraction, and release of all gut hormones [15]. In addition, these peptides can stimulate growth of GI mucosa and pancreas. BBS stimulates gastric weight, fundic and antral mucosal height, and density of parietal cells in neonatal rats compared with saline-treated controls [16,17]. These results were confirmed in adult rats given BBS for 7 days, demonstrating increased weight, increased RNA, and DNA content of the oxyntic mucosa of the stomach and the duodenal mucosa; the inhibitory hormone, somatostatin, attenuated the proliferative effect of BBS [18]. In another study, the BBS receptor antagonist, RC-3095, prevented the proliferative effect of BBS on the gastric mucosa, thus providing evidence that BBS/GRP stimulates growth of stomach and duodenum, predominantly due to a direct effect of hormone stimulation and not secondary to release of other gut hormones [19].

Small Intestine

The intestinal hormones that have been shown to stimulate growth of small intestine mucosa include NT, BBS/GRP, PYY, and GLP-2. Wood *et al.* [20] first noted that NT stimulated the small bowel mucosa of rats fed a normal diet. Investigators in our laboratory have shown that administration of NT prevents gut mucosal atrophy induced by feeding rats an elemental diet and stimulates mucosal growth in defunctionalized, self-emptying jejunoileal loops or isolated small bowel loops termed *Thiry-Vella fistulas* (TVFs), thus supporting a direct role for NT in the stimulation of gut mucosal growth [21,22]. Vagianos *et al.* [23] reported that NT restores gut mucosal integrity in rats and prevents the translocation of indigenous bacteria after radiation-induced mucosal injury. Furthermore, Izukura *et al.* [24] and de Miguel *et al.* [25] demonstrated, in separate studies, that administration of NT can augment the normal adaptive hyperplasia of gut mucosa that is associated with a massive small bowel resection.

BBS also stimulates growth of the small bowel mucosa. Administration of BBS effectively prevented mucosal atrophy associated with feeding rats a liquid elemental diet [26]. Furthermore, BBS was noted to increase mucosal weight, DNA, and protein content in both jejunal and ileal TVFs compared to control animals, suggesting that the effects of BBS were directly mediated as opposed to indirect effects of stimulation of luminal pancreatic or biliary secretion [27]. In addition to its effects on gut mucosal growth, BBS exhibits protective effects on the gut after injury [28]. Using a lethal enterocolitis model in rats induced by the chemotherapeutic agent methotrexate (MTX), BBS enhanced gut mucosal growth and significantly inhibited mortality. A beneficial effect of BBS on survival was noted when BBS was given prior to or at the same time as MTX, which suggested that BBS may act through additional mechanisms other than gut mucosal growth alone. One possibility is that BBS may produce its beneficial effects through enhancement of the immune system, a known action of BBS.

Although the data are somewhat controversial, the gut peptide PYY has likewise been shown to produce a trophic effect in small bowel mucosa of both rat and mouse [29]. These effects were noted at relatively high dosages. Similarly, Chance *et al.* [30] found that PYY treatment in Sprague-Dawley rats given total parenteral nutrition (TPN) produced significant increases in jejunal, ileal, and colonic protein contents.

A trophic effect for glucagon-derived peptides in the intestinal mucosa has been postulated since the description of a glucagon-secreting tumor of the kidney associated with small bowel mucosal hypertrophy. Drucker *et al.* [31] was the first to demonstrate that the intestinal trophic factor was GLP-2, which produced a 50% increase in small bowel weight and a significant increase in mucosal thickness. Similarly, Ghatei *et al.* [32] demonstrated prominent trophic effects of GLP-2 in Wistar rats, and Litvak *et al.* [33] demonstrated that GLP-2 significantly increased the weight of jejunum, ileum, and colon of athymic nude mice compared to control mice. In addition to the effects of GLP-2 on normal mucosa, the effects of this agent during periods of gut injury or atrophy have also been assessed. Mice treated with indomethacin developed small bowel enteritis associated with significant mortality at 48–72 hr after administration; treatment with human [Gly^2]-GLP-2, either before, during, or after indomethacin administration, resulted in reduced mortality and decreased mucosal injury [34]. The protective effects were attributed to the significantly increased crypt cell proliferation and decreased crypt compartment apoptosis. The effect of GLP-2 on chemotherapy-induced intestinal mucositis has also been assessed. Pretreatment of mice with human [Gly^2]-GLP-2 before administration of the topoisomerase inhibitor, irinotecan, resulted in reduced bacterial translocation, intestinal damage, and mortality [35]. Histological and biochemical analyses revealed significant reductions in crypt compartment apoptosis and reduced caspase-8 activation. Consistent with these reports, Tavakkolizadeh *et al.* [36] noted decreased intestinal damage in rats given GLP-2 in combination with the chemotherapeutic agent 5-fluorouracil. Finally, repeated cyclical administration of human [Gly^2]-GLP-2 resulted in significantly decreased mortality in groups of Balb/c mice given irinotecan.

Colon

Colonic mucosal growth may be affected by the gut peptides gastrin, BBS/GRP, NT, and GLP-2. Earlier reports suggested a role for amidated gastrin (i.e., G17 and G34) as trophic factors in the colon [37]. Recent studies now suggest that glycine-extended progastrin (G-Gly) may be the responsible agent producing the effects noted with gastrin administration. These findings have sparked renewed interest in a role for gastrin precursor products in colonic growth. Koh *et al.* [38] generated mice that overexpressed progastrin truncated at glycine-72. These mice demonstrate elevated serum and mucosal levels of G-Gly compared with wild-type mice. Mice overexpressing G-Gly displayed a 43% increase in colonic mucosal thickness and a 41% increase in the percentage of goblet cells per crypt. Furthermore, administration of G-Gly to gastrin-deficient mice resulted in a 10% increase in colonic mucosal thickness and an 81% increase in colonic proliferation as measured by BrdU incorporation.

Although the small bowel is significantly more sensitive to the effects of GLP-2, studies have shown that GLP-2 and analogs can stimulate the growth of colonic mucosa. Litvak *et al.* [33] demonstrated the trophic effect of GLP-2 on the colonic mucosa of athymic nude mice. Drucker *et al.* [39] demonstrated an increase in colonic growth using dipeptidyl peptidase IV-resistant GLP-2 analog, human [Gly^2]-GLP-2, in 6-week-old female mice. A significant increase in large bowel mass was detected in mice treated with this analog for 10 days. Furthermore, the combination of this agent with either IGF-1 or an IGF-1 analog produced a greater increase in large bowel mass than mice treated with [Gly^2]-GLP-2 alone. Administration of GLP-2 increased colonic weight in Wistar rats with atrophic colonic mucosa induced by TPN administration and reduced colonic mucosal injury in a dextran sulfate-induced colitis model [32].

Other intestinal hormones may play a contributory role in colonic proliferation; however, the effects are relatively minimal. For example, BBS administered three times a day for 7 days stimulated rat colonic mucosal growth [40] and, moreover, administration of BBS orally during the neonatal period stimulated colonic growth [17]. Investigators in our laboratory have shown that, in rats given an elemental diet, the proliferative effect of BBS was confined to the proximal colon [41]. Colonic proliferation is likewise noted with NT administration; however, the effects of NT on the colon are much less pronounced than in the small bowel [42]. NT-induced colonic proliferation appears to be dependent on age, with hyperplasia noted in the colon of young rats given NT, whereas NT significantly increased hypertrophy in aged rats. Similarly, PYY has been shown to have a modest effect on growth of the colonic mucosa [29].

GI Hormone Receptors and Signal Transduction Pathways

GI hormone-stimulated signal transduction occurs with the binding of hormones to their cognate cell surface receptors, which are G-protein-coupled receptors (GPCRs) [43]. These receptors have the typical structural features of G-protein-binding seven-transmembrane receptors, which can regulate a number of physiological processes, including proliferation, growth, and development. It was originally thought that in order for GPCR signaling to occur, specific interactions between the GI hormone and the receptor were necessary to produce conformational changes in the receptor and stimulate intercellular signal transduction pathways. However, recent studies suggest a more complex regulation of the GPCRs through (1) dimerization with themselves and other receptors, (2) activation of differing G-proteins, (3) internalization and desensitization, and (4) ability to change in conformation and interactions with empty or inactive receptors [44].

The seven-transmembrane-spanning α-helical domains function as ligand-regulated, guanine nucleotide exchange factors for the intercellular heterotrimeric G-proteins [43]. Heterotrimeric G-proteins are composed of the products of three gene families encoding α, β, and γ subunits.

The agonist-activated GPCR catalyzes the exchange of GTP for GDP bound to the Gα subunit, as well as the dissociation of GTP-Gα from its cognate Gβγ dimer. The activated GTP-Gα and Gβγ subunits, in turn, regulate the activity of various intercellular effector proteins, such as phospholipases, adenyl cyclases, protein kinases, membrane ion channels, and members of the Ras family of GTP-binding proteins. In addition, based on structural similarities, the 20 identified Gα subunits have been divided into four subfamilies: (1) the cholera toxin-sensitive (α) subunits that stimulate adenyl cyclase and increase cyclic AMP levels, (2) the pertussin toxin-sensitive ($\alpha_{i/o}$) subunits that inhibit adenyl cyclase activity, (3) the pertussin toxin-sensitive ($\alpha_{q/11/14}$) subunits that stimulate membrane phospholipases, and (4) the ($\alpha_{12/13}$) subfamily that links GPCR to the Ras-related GTP binding protein, Rho [43]. Additionally, twelve Gγ and six Gβ subunits have been identified. These βγ dimers have been linked to the signaling molecules, phosphatidylinositol 3-kinase (PI3K), and select forms of adenyl cyclase and receptor kinases.

Among the multiple intercellular signaling pathways that mediate the proliferative effects of GPCRs, a family of related serine-threonine kinases, collectively known as the mitogen-activated protein kinases (MAPKs), appear to play a central role [45,46]. Hormones act as ligands to eventually activate p42 and p44 ERKs, which occurs through the involvement of a complex interplay of several known nonreceptor kinases and receptor kinases. The ability of tyrosine kinase inhibitors to reduce the activation of MAPK by GPCR and the rapid tyrosine phosphorylation of Shc (*src* homology and collagen) following GPCR stimulation with the consequent formation of Shc-Grb2 (growth factor receptor bound 2) complexes provides evidence that tyrosine kinases link GPCRs to the Ras-MAPK pathway [47,48]. Additionally, GPCRs link to the Jun-N terminal kinase (JNK), p38, MAPK, and the big mitogen-activated kinase-1 (BMK-1) or ERK5 pathways [43].

The molecular mechanisms though which GPCRs transduce signals are complex and likely involve multiple signaling pathways. In addition, the signaling pathways are likely cell specific, which may explain the diverse physiological functions controlled by gut hormones, ranging from regulation of secretion, motility, and, in some instances, growth, depending on the target tissue.

Signaling Pathways Mediating the Effects of Intestinal Peptides

Figure 1 summarizes how signaling pathways mediate the effects of intestinal peptides. Once a trophic GI peptide binds its seven-transmembrane GPCR, signal transduction pathways are activated that ultimately can lead to cell proliferation depending on cell type [43,49–52]. A number of pathways and proteins have been identified that are stimulated by the trophic gut peptides. For the most part, these pathways have been identified using neoplastic cells that possess the receptor for the trophic gut hormone.

Pathways Involving Phospholipase C, Phosphatidylinositol Activation, Calcium Mobilization, and Protein Kinase C

An early event associated with binding of trophic peptides to its receptor is activation of the phospholipase C (PLC) signal transduction pathway. For example, gastrin stimulates PLC in a number of cell types, including gastric parietal

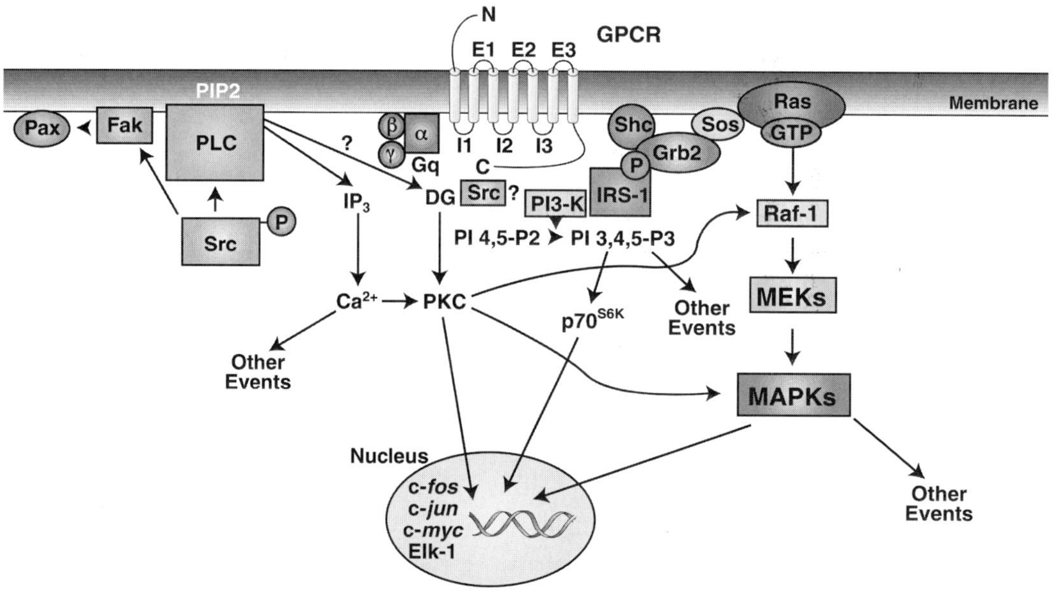

Figure 1 Summary of pathways that can mediate the trophic effect of gastrin. Although these pathways have been best described for gastrin, other gut peptides can interact with their specific G-protein-coupled receptor (GPCR) to stimulate similar pathways in receptor-positive cells. See text for details and abbreviations. [Adapted from Yassin, R. R., *et al.* (1999). Signaling pathways mediating gastrin's growth-promoting effects. *Peptides* **20**, 885–898.]

cells and colonic epithelial cells, NIH-3T3 fibroblasts that express the gastrin receptor (CCK-B/gastrin), and various neoplastic cell lines (reviewed in [53,54]). The activation of PLC, which may involve coupling of the CCK-B/gastrin receptor with selected members of the G-protein superfamily, induces the breakdown of membrane phosphatidylinositol 4,5-bisphosphate (PIP_2) leading to the formation of two second messengers, IP_3 and 1,2-diacylglycerol (DAG). IP_3 binds to its intracellular receptor and triggers the release of calcium from internal stores. Gastrin stimulates inositol phosphate production in a variety of cell types, leading to IP_3 and DAG formation, protein kinase C (PKC) activation, and intracellular calcium mobilization [53,54]. These effects can be blocked by CCK-B/gastrin receptor antagonists. In addition to gastrin, other peptides, such as NT, have been shown to stimulate the PKC pathway, IP_3 turnover, and calcium mobilization in a number of cell types, including the colon cancer cell lines, HT29 and KM20, as well as the pancreatic cancer cell line, MIA PaCa-2 [55–58].

The trophic GI hormones also activate downstream PKC isoenzymes. Gastrin activates classic calcium- and phospholipid-dependent PKCs as demonstrated by translocation of the cytosolic activity of PKC to the membrane compartment of rat colonic epithelial cells [59]. The isoforms, PKC-α and -β, are responsive to gastrin treatment and mobilize from the cytosol with treatment [60]. In addition, recent results demonstrate that gastrin induces the novel protein kinase D (PKD, also known as PKC-μ), which has distinct structural and enzymological properties from the PKCs [61], in Rat-1 cells transfected with the human CCK-B/gastrin receptor [62]. Similarly, Guha *et al.* [63] recently reported that NT induced a rapid activation of PKD, which was linked to the mitogenic effect of NT in pancreatic cancer cells.

Tyrosine Kinases, Tyrosine Phosphorylation of Focal Adhesion Kinase, Paxillin, and CRK-Associated Substrate

Through pairing with nonreceptor tyrosine kinases, GPCRs utilize the tyrosine kinase pathway to stimulate cell growth [64]. Gastrin stimulates tyrosine kinase activity and tyrosine phosphorylation of a 57-kDa membrane protein in rat colonic mucosal cells and also membrane-associated protein tyrosine kinase and tyrosine phosphorylation of endogenous proteins in the IEC-6 intestinal cell line [65–67]. In addition, gastrin results in tyrosine phosphorylation of 62- and 54-kDa Src-like proteins in IEC-6 cells and $pp60^{c\text{-}src}$ kinase in rat colonic epithelial cells, which leads to tyrosine phosphorylation and activation of PLCγ1 [67,68].

Growth factor receptors with intrinsic kinase activity and those that signal through G-proteins can promote tyrosine phosphorylation of the adaptor protein Shc, which links activated growth factor receptors to the Ras signaling pathway, and its subsequent association with the Grb2-SOS (growth factor receptor binding protein-2/SON of sevenless) complex [69]. Gastrin promotes a rapid and transient increase in tyrosine phosphorylation of the Shc proteins and association with Grb2-SOS, leading to activation of MAPKs in the human gastric cancer cell line, AGS-B, which expresses the human CCK-B/gastrin receptor [70].

The involvement of PI3K, as well as other SH2 anchoring proteins, in gastrin's mitogenic pathway has been reported. Gastrin induces tyrosine phosphorylation of the insulin receptor substrate 1 (IRS-1) and association with the 85-kDa subunit of PI3K [71]. Gastrin also stimulates the association of phosphorylated IRS-1 with the adaptor Grb2, indicating that tyrosine phosphorylation of IRS-1 may be a mechanism whereby gastrin activates PI3K and other adapters [71]. Therefore, mobilization of the adapter proteins IRS-1, Shc, Grb2, and SOS could serve to link the CCK-B/gastrin receptor to the Ras-MAPK cascade, ultimately leading to transcriptional regulation [53].

Gastrin treatment results in the tyrosine phosphorylation of various proteins [67,72]. For example, gastrin promotes tyrosine phosphorylation of focal adhesion kinase ($p125^{fak}$), a tyrosine kinase that localizes at focal adhesions and is important for cell adhesion and transformation [72]. The focal adhesion proteins paxillin and CAS are potential downstream targets for $p125^{fak}$ and function as adaptor proteins. Gastrin induces phosphorylation of $p125^{fak}$, CAS, and paxillin in Rat-1 and NIH-3T3 cells transfected with the CCK-B/gastrin receptor [72,73].

MAPK Pathway

The MAPKs are a family of highly conserved serine-threonine kinases that are activated by a variety of extracellular signals and relay mitogenic signals to the nucleus [74]. The activation of the MAPK pathway by gastrin has been demonstrated in a variety of cell types, both containing the endogenous CCK-B/gastrin receptor and cells stably transfected with this receptor [69,70,72]. Gastrin also stimulates the serine-threonine kinase Raf-1, the cellular homolog of the Raf oncogene and an upstream modulator of the MAPKs [70,72]. The mechanism for the effect of these hormones on MAPK activation is dependent on cell type. For example, in the AR4-2J pancreatic cell line, gastrin activated ERK, which was attenuated by treatment with agents that interfere with calcium mobilization or PKC activation [75]. In other cells, the activation of MAPK by gastrin does not involve PKC. In Chinese hamster ovary (CHO) cells transfected with the CCK-B/gastrin receptor, gastrin induces ERK activation partly through the Src and PI3K pathways and partly through PKC [76]. Conversely, gastrin-stimulated Raf-1 and ERK activation in Rat-1 fibroblasts stably transfected with the CCK-B/gastrin receptor occurs independent of PKC [72]. Likewise, PYY stimulates MAPKs through binding of the Y1 receptor [77]. The stimulation of MAPK can occur via multiple and diverse pathways, but in the case of PYY, it has been shown that PKC plays a major role in the signaling pathway between the Y1 receptor and MAPK, acting between the EGFR and MAPK. In CHO cells transfected with the human Y1 receptor, both PKC and Ras are needed for activation of MAPK [78]. Specifically, only PKC-ε, an isoform that has been specifically linked to mitogenic effects in gut epithelium [79], was activated by PYY in the IEC-6 intestinal cell line. NT has been shown to stimulate ERK and JNK activity in various neoplastic cell types containing endogenous NT receptors, ultimately leading to transcription factor activation [57,58]. Therefore, activation of MAPKs appears to be an important mechanism for the mitogenic effects of intestinal hormones; this activation can occur by a variety of signal transduction mechanisms and depends on cell context as to which pathway is active in which cell type.

Downstream Transcription Factors

Ultimately, stimulation of various signaling pathways, such as PKC, the MAPKs, or ribosomal S6 kinase ($p70^{S6K}$), can lead to activation of downstream transcription factors. Gastrin stimulates the expression of early response genes, including c-*fos* and c-*jun* in AR4-2J cells, and c-*fos* and c-*myc* in NIH-3T3 cells transfected with the CCK-B/gastrin receptor [73,80]. Gastrin induced ERK-mediated phosphorylation and activation of these transcription factors was prevented by pharmacologic inhibition of PKC [81], suggesting an important role for the PKCs and ERKs in the activation of the AP-1 transcription factors. In addition, NT and BBS have been shown to stimulate expression of c-*jun* and c-*fos* in cell types possessing endogenous receptors for these peptides [57,58,82]. It is likely that the activation of these, as well as other transcription factors, such as Elk-1, ultimately plays a major role in the mitogenic response of these hormones.

Conclusions

The growth of GI mucosa is modulated by multiple factors, including intraluminal nutrients and the local release of growth factors and various GI hormones. These hormones bind to their specific receptors, stimulating many signal transduction pathways, which ultimately lead to the mitogenic effects of these gut peptides. In this regard, the GI hormones have been suggested as potential therapeutic agents in disease states in the non-neoplastic GI tract related to gut disuse or atrophy, mucosal ulcers, or inflammatory conditions. For example, the trophic peptides, NT, BBS/GRP, and GLP-2, can augment or maintain GI mucosal growth during periods of gut disuse or atrophy [21,22,27,34]. In addition, these peptides enhance adaptive hyperplasia associated with massive intestinal resection [24,25]. In a limited clinical trial, administration of GLP-2 improved intestinal energy absorption, decreased energy excretion, increased body weight and lean body mass, and enhanced urinary creatinine excretion in patients with short bowel syndrome [83]. In addition, GI hormones may play a role in preventing the severe sequelae of chemotherapeutic agents on the intestinal mucosa. Both BBS and GLP-2 have been shown to prevent the severe mucosal inflammation associated with various chemotherapeutic agents [28,35,36]. In the future, it will be important to further define the signaling pathways regulated by the trophic intestinal peptides so that more effective agents can be developed that can take advantage of the gut-specific effects of these hormones on the proliferation and maintenance of the GI mucosa.

Acknowledgments

The authors thank Liz Cook for manuscript preparation and Karen Martin for assistance with preparation of the figures.

References

1. Babyatsky, M. W. and Podolsky, D. K. (1999). Growth and development of the gastrointestinal tract. In Yamada, T. (Ed.), *Textbook of Gastroenterology,* Vol. 1, 3rd ed., pp. 547–584. Lippincott Williams & Wilkins Publishers, Philadelphia.
2. Cheng, H. and Leblond, C. P. (1974). Origin, differentiation and renewal of the four main epithelial cell types in the mouse small intestine. I. Columnar cell. *Am. J. Anat.* **141**(4), 461–479.
3. Pritchard, D. M. and Watson, A. J. (1996). Apoptosis and gastrointestinal pharmacology. *Pharmacol. Ther.* **72**(2), 149–169.
4. Potten, C. S. (1997). Epithelial cell growth and differentiation. II. Intestinal apoptosis. *Am. J. Physiol.* **273**(2 Pt 1), G253–G257.
5. Dignass, A. U. and Sturm, A. (2001). Peptide growth factors in the intestine. *Eur. J. Gastroenterol. Hepatol.* **13**(7), 763–770.
6. Walsh, J. H. (1994). Gastrointestinal hormones. In Johnson, L. R., Alpers, D. H., Christensen, J., Jacobson, E. D., and Walsh, J. H. (Eds.), *Physiology of the Gastrointestinal Tract,* 3rd ed., pp. 1–128. Raven Press, New York.
7. Larsson, L. I. (1980). Gastrointestinal cells producing endocrine, neurocrine and paracrine messengers. *Clin. Gastroenterol.* **9**(3), 485–416.
8. Johnson, L. R., Aures, D., and Hakanson, R. (1969). Effect of gastrin on the in vivo incorporation of 14C-leucine into protein of the digestive tract. *Proc. Soc. Exp. Biol. Med.* **132**(3), 996–998.
9. Crean, G. P., Marshall, M. W., and Rumsey, R. D. (1969). Parietal cell hyperplasia induced by the administration of pentagastrin (ICI 50,123) to rats. *Gastroenterology* **57**(2), 147–155.
10. Johnson, L. R. (1976). The trophic action of gastrointestinal hormones. *Gastroenterology* **70**(2), 278–288.
11. Wang, T. C., Koh, T. J., Varro, A., Cahill, R. J., Dangler, C. A., Fox, J. G., and Dockray, G. J. (1996). Processing and proliferative effects of human progastrin in transgenic mice. *J. Clin. Invest.* **98**(8), 1918–1929.
12. Wang, T. C., Dangler, C. A., Chen, D., Goldenring, J. R., Koh, T., Raychowdhury, R., Coffey, R. J., Ito, S., Varro, A., Dockray, G. J., and Fox, J. G. (2000). Synergistic interaction between hypergastrinemia and *Helicobacter* infection in a mouse model of gastric cancer. *Gastroenterology* **118**(1), 36–347.
13. Friis-Hansen, L., Sundler, F., Li, Y., Gillespie, P. J., Saunders, T. L., Greenson, J. K., Owyang, C., Rehfeld, J. F., and Samuelson, L. C. (1998). Impaired gastric acid secretion in gastrin-deficient mice. *Am. J. Physiol.* **274**(3 Pt 1), G561–G568.
14. Nagata, A., Ito, M., Iwata, N., Kuno, J., Takano, H., Minowa, O., Chihara, K., Matsui, T., and Noda, T. (1996). G protein-coupled cholecystokinin-B/gastrin receptors are responsible for physiological cell growth of the stomach mucosa in vivo. *Proc. Natl. Acad. Sci. USA* **93**(21), 11825–11830.
15. Greeley, G. H., Jr. and Newman, J. (1987). Enteric bombesin-like peptides. In Thompson, J. C., Greeley, G. H., Jr., Rayford, P. L., and Townsend, C. M. Jr., (Eds.), *Gastrointestinal Endocrinology,* pp. 322–329. McGraw-Hill, New York.
16. Puccio, F. and Lehy, T. (1989). Bombesin ingestion stimulates epithelial digestive cell proliferation in suckling rats. *Am. J. Physiol.* **256** (2 Pt 1), G328–G334.

17. Lehy, T., Puccio, F., Chariot, J., and Labeille, D. (1986). Stimulating effect of bombesin on the growth of gastrointestinal tract and pancreas in suckling rats. *Gastroenterology* **90**(6), 1942–1949.
18. Dembinski, A., Konturek, P. K., and Konturek, S. J. (1990). Role of gastrin and cholecystokinin in the growth-promoting action of bombesin on the gastroduodenal mucosa and the pancreas. *Regul. Pept.* **27**(3), 343–354.
19. Dembinski, A., Warzecha, Z., Konturek, S. J., Banas, M., Cai, R. Z., and Schally, A. V. (1991). The effect of antagonist of receptors for gastrin, cholecystokinin and bombesin on growth of gastroduodenal mucosa and pancreas. *J. Physiol. Pharmacol.* **42**(3), 263–277.
20. Wood, J. G., Hoang, H. D., Bussjaeger, L. J., and Solomon, T. E. (1988). Neurotensin stimulates growth of small intestine in rats. *Am. J. Physiol.* **255**(6 Pt 1), G813–G817.
21. Evers, B. M., Izukura, M., Townsend, C. M., Jr., Uchida, T., and Thompson, J. C. (1992). Neurotensin prevents intestinal mucosal hypoplasia in rats fed an elemental diet. *Dig. Dis. Sci.* **37**(3), 426–431.
22. Chung, D. H., Evers, B. M., Shimoda, I., Townsend, C. M., Jr., Rajaraman, S., and Thompson, J. C. (1992). Effect of neurotensin on gut mucosal growth in rats with jejunal and ileal Thiry-Vella fistulas. *Gastroenterology* **103**(4), 1254–1259.
23. Vagianos, C., Karatzas, T., Scopa, C. D., Panagopoulos, C., Tsoni, I., Spiliopoulou, I., and Kalfarentzos, F. (1992). Neurotensin reduces microbial translocation and improves intestinal mucosa integrity after abdominal radiation. *Eur. Surg. Res.* **24**(2), 77–83.
24. Izukura, M., Evers, B. M., Parekh, D., Yoshinaga, K., Uchida, T., Townsend, C. M., Jr., and Thompson, J. C. (1992). Neurotensin augments intestinal regeneration after small bowel resection in rats. *Ann. Surg.* **215**(5), 520–527.
25. de Miguel, E., Gomez de Segura, I. A., Bonet, H., Rodriguez Montes, J. A., and Mata, A. (1994). Trophic effects of neurotensin in massive bowel resection in the rat. *Dig. Dis. Sci.* **39**(1), 59–64.
26. Evers, B. M., Izukura, M., Townsend, C. M., Jr., Uchida, T., and Thompson, J. C. (1990). Differential effects of gut hormones on pancreatic and intestinal growth during administration of an elemental diet. *Ann. Surg.* **211**(5), 630–638.
27. Chu, K. U., Higashide, S., Evers, B. M., Ishizuka, J., Townsend, C. M., Jr., and Thompson, J. C. (1995). Bombesin stimulates mucosal growth in jejunal and ileal Thiry-Vella fistulas. *Ann. Surg.* **221**(5), 602–611.
28. Chu, K. U., Higashide, S., Evers, B. M., Rajaraman, S., Ishizuka, J., Townsend, C. M., Jr., and Thompson, J. C. (1994). Bombesin improves survival from methotrexate-induced enterocolitis. *Ann. Surg.* **220**(4), 570–577.
29. Gomez, G., Zhang, T., Rajaraman, S., Thakore, K. N., Yanaihara, N., Townsend, C. M., Jr., Thompson, J. C., and Greeley, G. H. (1995). Intestinal peptide YY: Ontogeny of gene expression in rat bowel and trophic actions on rat and mouse bowel. *Am. J. Physiol.* **268**(1 Pt 1), G71–G81.
30. Chance, W. T., Zhang, X., Balasubramaniam, A., and Fischer, J. E. (1996). Preservation of intestine protein by peptide YY during total parenteral nutrition. *Life Sci.* **58**(21), 1785–1794.
31. Drucker, D. J., Erlich, P., Asa, S. L., and Brubaker, P. L. (1996). Induction of intestinal epithelial proliferation by glucagon-like peptide 2. *Proc. Natl. Acad. Sci. USA* **93**(15), 7911–7916.
32. Ghatei, M. A., Goodlad, R. A., Taheri, S., Mandir, N., Brynes, A. E., Jordinson, M., and Bloom, S. R. (2001). Proglucagon-derived peptides in intestinal epithelial proliferation: glucagon-like peptide-2 is a major mediator of intestinal epithelial proliferation in rats. *Dig. Dis. Sci.* **46**(6), 1255–1263.
33. Litvak, D. A., Hellmich, M. R., Evers, B. M., Banker, N. A., and Townsend, C. M., Jr. (1998). Glucagon-like peptide 2 is a potent growth factor for small intestine and colon. *J. Gastrointest. Surg.* **2**(2), 146–150.
34. Boushey, R. P., Yusta, B., and Drucker, D. J. (1999). Glucagon-like peptide 2 decreases mortality and reduces the severity of indomethacin-induced murine enteritis. *Am. J. Physiol.* **277**(5 Pt 1), E937–E947.
35. Boushey, R. P., Yusta, B., and Drucker, D. J. (2001). Glucagon-like peptide (GLP)-2 reduces chemotherapy-associated mortality and enhances cell survival in cells expressing a transfected GLP-2 receptor. *Cancer Res.* **61**(2), 687–693.
36. Tavakkolizadeh, A., Shen, R., Abraham, P., Kormi, N., Seifert, P., Edelman, E. R., Jacobs, D. O., Zinner, M. J., Ashley, S. W., and Whang, E. E. (2000). Glucagon-like peptide 2: a new treatment for chemotherapy-induced enteritis. *J. Surg. Res.* **91**(1), 77–82.
37. Johnson, L. R. (1977). New aspects of the trophic action of gastrointestinal hormones. *Gastroenterology* **72**(4 Pt.2), 788–792.
38. Koh, T. J., Dockray, G. J., Varro, A., Cahill, R. J., Dangler, C. A., Fox, J. G., and Wang, T. C. (1999). Overexpression of glycine-extended gastrin in transgenic mice results in increased colonic proliferation. *J. Clin. Invest.* **103**(8), 1119–1126.
39. Drucker, D. J., DeForest, L., and Brubaker, P. L. (1997). Intestinal response to growth factors administered alone or in combination with human [Gly2]glucagon-like peptide 2. *Am. J. Physiol.* **273**(6 Pt 1), G1252–G1262.
40. Johnson, L. R. and Guthrie, P. D. (1983). Regulation of antral gastrin content. *Am. J. Physiol.* **245**(6), G725–G779.
41. Chu, K. U., Evers, B. M., Ishizuka, J., Townsend, C. M., Jr., and Thompson, J. C. (1995). Role of bombesin on gut mucosal growth. *Ann. Surg.* **222**(1), 94–100.
42. Evers, B. M., Izukura, M., Chung, D. H., Parekh, D., Yoshinaga, K., Greeley, G. H., Jr., Uchida, T., Townsend, C. M., Jr., and Thompson, J. C. (1992). Neurotensin stimulates growth of colonic mucosa in young and aged rats. *Gastroenterology* **103**(1), 86–91.
43. Marinissen, M. J. and Gutkind, J. S. (2001). G-protein-coupled receptors and signaling networks: emerging paradigms. *Trends Pharmacol. Sci.* **22**(7), 368–376.
44. Wilkie, T. M. (2001). Treasures throughout the life-cycle of G-protein-coupled receptors. *Trends Pharmacol. Sci.* **22**, 396–397.
45. Davis, R. J. (1995). Transcriptional regulation by MAP kinases. *Mol. Reprod. Dev.* **42**(4), 459–467.
46. Gutkind, J. S. (1998). The pathways connecting G protein-coupled receptors to the nucleus through divergent mitogen-activated protein kinase cascades. *J. Biol. Chem.* **273**(4), 1839–1842.
47. Hordijk, P. L., Verlaan, I., van Corven, E. J., and Moolenaar, W. H. (1994). Protein tyrosine phosphorylation induced by lysophosphatidic acid in Rat-1 fibroblasts. Evidence that phosphorylation of map kinase is mediated by the Gi-p21ras pathway. *J. Biol. Chem.* **269**(1), 645–651.
48. van Biesen, T., Hawes, B. E., Luttrell, D. K., Krueger, K. M., Touhara, K., Porfiri, E., Sakaue, M., Luttrell, L. M., and Lefkowitz, R. J. (1995). Receptor-tyrosine-kinase- and G beta gamma-mediated MAP kinase activation by a common signalling pathway. *Nature* **376**(6543), 781–784.
49. Ji, T. H., Grossmann, M., and Ji, I. (1998). G protein-coupled receptors. I. Diversity of receptor-ligand interactions. *J. Biol. Chem.* **273**(28), 17299–17302.
50. Wank, S. A. (1998). G protein-coupled receptors in gastrointestinal physiology. I. CCK receptors: an exemplary family. *Am. J. Physiol.* **274**(4 Pt 1), G607–G613.
51. Rozengurt, E. (1986). Early signals in the mitogenic response. *Science* **234**(4773), 161–166.
52. Rozengurt, E. (1998). Signal transduction pathways in the mitogenic response to G protein-coupled neuropeptide receptor agonists. *J. Cell Physiol.* **177**(4), 507–517.
53. Yassin, R. R. (1999). Signaling pathways mediating gastrin's growth-promoting effects. *Peptides* **20**(7), 885–898.
54. Rozengurt, E. and Walsh, J. H. (2001). Gastrin, CCK, signaling, and cancer. *Annu. Rev. Physiol.* **63**, 49–76.
55. Slogoff, M. and Evers, B. M. (2003). Neurotensin. In Henry, H. L. Norman, A. W., *et al.* (Eds.), *Encyclopedia of Hormones* Vol. 3, pp. 45–53. Academic Press, San Diego.
56. Vincent, J. P., Mazella, J., and Kitabgi, P. (1999). Neurotensin and neurotensin receptors. *Trends Pharmacol. Sci.* **20**(7), 302–309.
57. Ehlers, R. A., Zhang, Y., Hellmich, M. R., and Evers, B. M. (2000). Neurotensin-mediated activation of MAPK pathways and AP-1 binding in the human pancreatic cancer cell line, MIA PaCa-2. *Biochem. Biophys. Res. Commun.* **269**(3), 704–708.

58. Ehlers, R. A., 2nd, Bonnor, R. M., Wang, X., Hellmich, M. R., and Evers, B. M. (1998). Signal transduction mechanisms in neurotensin-mediated cellular regulation. *Surgery* **124**(2), 239–247.
59. Yassin, R. R., Clearfield, H. R., and Little, K. M. (1993). Gastrin's trophic effect in the colon: identification of a signaling pathway mediated by protein kinase C. *Peptides* **14**(6), 1119–1124.
60. Yassin, R. R. and Little, K. M. (1995). Early signalling mechanism in colonic epithelial cell response to gastrin. *Biochem. J.* **311** (Pt 3), 945–950.
61. Valverde, A. M., Sinnett-Smith, J., Van Lint, J., and Rozengurt, E. (1994). Molecular cloning and characterization of protein kinase D: A target for diacylglycerol and phorbol esters with a distinctive catalytic domain. *Proc. Natl. Acad. Sci. USA* **91**(18), 8572–8576.
62. Chiu, T. T., Duque, J., and Rozengurt, E. (1999). Protein kinase D (PKD) activation is a novel early event in CCKB/gastrin receptor signaling. *Gastroenterology* **116**, A597.
63. Guha, S., Rey, O., and Rozengurt, E. (2002). Neurotensin induces protein kinase C-dependent protein kinase D activation and DNA synthesis in human pancreatic carcinoma cell line PANC-1. *Cancer Res.* **62**(6), 1632–1640.
64. Malarkey, K., Belham, C. M., Paul, A., Graham, A., McLees, A., Scott, P. H., and Plevin, R. (1995). The regulation of tyrosine kinase signalling pathways by growth factor and G-protein-coupled receptors. *Biochem. J.* **309**(Pt 2), 361–375.
65. Majumdar, A. P. (1990). Role of tyrosine kinases in gastrin induction of ornithine decarboxylase in colonic mucosa. *Am. J. Physiol.* **259** (4 Pt 1), G626–G630.
66. Malecka-Panas, E., Tureaud, J., and Majumdar, A. P. (1996). Gastrin activates tyrosine kinase and phospholipase C in isolated rat colonocytes. *Acta Biochim. Pol.* **43**(3), 539–546.
67. Singh, P., Narayan, S., and Adiga, R. B. (1994). Phosphorylation of pp62 and pp54 src-like proteins in a rat intestinal cell line in response to gastrin. *Am. J. Physiol.* **267**(2 Pt 1), G235–G244.
68. Yassin, R. R. and Abrams, J. T. (1998). Gastrin induces IP3 formation through phospholipase C gamma 1 and $pp60^{c\text{-}src}$ kinase. *Peptides* **19**(1), 47–55.
69. Seva, C., Kowalski-Chauvel, A., Blanchet, J. S., Vaysse, N., and Pradayrol, L. (1996). Gastrin induces tyrosine phosphorylation of Shc proteins and their association with the Grb2/Sos complex. *FEBS Lett.* **378**(1), 74–78.
70. Hocker, M., Henihan, R. J., Rosewicz, S., Riecken, E. O., Zhang, Z., Koh, T. J., and Wang, T. C. (1997). Gastrin and phorbol 12-myristate 13-acetate regulate the human histidine decarboxylase promoter through Raf-dependent activation of extracellular signal-regulated kinase-related signaling pathways in gastric cancer cells. *J. Biol. Chem.* **272**(43), 27015–27024.
71. Kowalski-Chauvel, A., Pradayrol, L., Vaysse, N., and Seva, C. (1996). Gastrin stimulates tyrosine phosphorylation of insulin receptor substrate 1 and its association with Grb2 and the phosphatidylinositol 3-kinase. *J. Biol. Chem.* **271**(42), 26356–26361.
72. Seufferlein, T., Withers, D. J., Broad, S., Herget, T., Walsh, J. H., and Rozengurt, E. (1995). The human CCKB/gastrin receptor transfected into rat1 fibroblasts mediates activation of MAP kinase, p74raf-1 kinase, and mitogenesis. *Cell Growth Differ.* **6**(4), 383–393.
73. Taniguchi, T., Matsui, T., Ito, M., Murayama, T., Tsukamoto, T., Katakami, Y., Chiba, T., and Chihara, K. (1994). Cholecystokinin-B/gastrin receptor signaling pathway involves tyrosine phosphorylations of p125FAK and p42MAP. *Oncogene* **9**(3), 861–867.
74. Widmann, C., Gibson, S., Jarpe, M. B., and Johnson, G. L. (1999). Mitogen-activated protein kinase: conservation of a three-kinase module from yeast to human. *Physiol. Rev.* **79**(1), 143–180.
75. Daulhac, L., Kowalski-Chauvel, A., Pradayrol, L., Vaysse, N., and Seva, C. (1997). Ca^{2+} and protein kinase C-dependent mechanisms involved in gastrin-induced Shc/Grb2 complex formation and P44-mitogen-activated protein kinase activation. *Biochem. J.* **325** (Pt 2), 383–389.
76. Daulhac, L., Kowalski-Chauvel, A., Pradayrol, L., Vaysse, N., and Seva, C. (1999). Src-family tyrosine kinases in activation of ERK-1 and p85/p110-phosphatidylinositol 3-kinase by G/CCKB receptors. *J. Biol. Chem.* **274**(29), 20657–20663.
77. Mannon, P. J. (2002). Peptide YY as a growth factor for intestinal epithelium. *Peptides* **23**(2), 383–388.
78. Mannon, P. J. and Raymond, J. R. (1998). The neuropeptide Y/peptide YY Y1 receptor is coupled to MAP kinase via PKC and Ras in CHO cells. *Biochem. Biophys. Res. Commun.* **246**(1), 91–94.
79. Perletti, G. P., Folini, M., Lin, H. C., Mischak, H., Piccinini, F., and Tashjian, A. H., Jr. (1996). Overexpression of protein kinase C epsilon is oncogenic in rat colonic epithelial cells. *Oncogene* **12**(4), 847–854.
80. Todisco, A., Takeuchi, Y., Seva, C., Dickinson, C. J., and Yamada, T. (1995). Gastrin and glycine-extended progastrin processing intermediates induce different programs of early gene activation. *J. Biol. Chem.* **270**(47), 28337–28341.
81. Stepan, V. M., Tatewaki, M., Matsushima, M., Dickinson, C. J., del Valle, J., and Todisco, A. (1999). Gastrin induces c-*fos* gene transcription via multiple signaling pathways. *Am. J. Physiol.* **276**(2 Pt 1), G415–G424.
82. Kim, H. J., Evers, B. M., Guo, Y., Banker, N. A., Hellmich, M. R., and Townsend, C. M., Jr. (1996). Bombesin-mediated AP-1 activation in a human gastric cancer (SIIA). *Surgery* **120**(2), 130–137.
83. Jeppesen, P. B., Hartmann, B., Thulesen, J., Graff, J., Lohmann, J., Hansen, B. S., Tofteng, F., Poulsen, S. S., Madsen, J. L., Holst, J. J., and Mortensen, P. B. (2001). Glucagon-like peptide 2 improves nutrient absorption and nutritional status in short-bowel patients with no colon. *Gastroenterology* **120**(4), 806–815.

CHAPTER 338

Integrated Response to Neurotrophic Factors

J. Regino Perez-Polo
University of Texas Medical Branch, Galveston, Texas

Introduction

Neurotrophic growth factors are proteins responsible for cell survival, proliferation, migration, neurite elongation, neurotransmitter expression, and synaptic maturation in the nervous system. The cellular targets of neurotrophic factors are not restricted to neurons because they act on both neurons and glia (for example, astrocytes, oligodendroglia, microglia, and Schwann cells). The mechanisms of action of neurotrophic factors are probably best understood as they apply to effects on cell survival, the focus of this chapter.

Neural Cell Death

Programmed cell death during development of the nervous system, also called *apoptosis*, is a tightly controlled, highly ordered, physiological form of cell death wherein a cell actively commits to apoptosis if it fails to garner sufficient trophic factor activation of its endogenous programs regulating metabolism [1]. The distinction is rarely made between the original term of apoptosis and the mechanism of programmed or delayed cell death, and these terms are used interchangeably. Apoptosis can be triggered at the level of the individual neuron or glia by a wide variety of ligand/receptor-mediated stimuli during embryonic development [2]. Thus, various ligands, or a combination of ligands, can set off signals that culminate in apoptotic cell death, appropriate ongoing physiological function, or proliferative expansion. In addition, neural cells (neurons or glia) can also commit to cell death, exhibiting apoptotic features as part of a spectrum of responses to trauma and disease [3]. These responses have a common component of oxidative stress, and the resultant cell death is also referred to as delayed cell death [4]. Although these forms of cell death can have different properties, they all commit a neural cell to death with some common cellular and molecular mechanisms. Here, we refer to these forms of cell death by the name of *apoptosis* in contrast to another form of cell death, *necrosis*, which results from the abrupt impairment of energy metabolism associated with sudden trauma. While the morphological hallmarks of apoptosis include cell shrinkage and membrane blebbing, nuclear condensation and DNA fragmentation with a ladder-like electrophoretic appearance [4], necrosis is a pathological form of cell death that results from massive cellular injury and a resultant prompt and robust energy depletion that is characterized by swelling of the cytoplasm, organelle dissolution, and rupture of the plasma membrane followed by inflammation [2–4]. Apoptosis also displays characteristic molecular changes that include increases in intracellular calcium ion fluxes, increased activation of caspase enzymes, and shifts in the levels and the intracellular occupancy of the Bcl-2 family of proteins [5–8].

During the last two decades, the morphologic appearances of dying cells and their relationship to specific metabolic pathways responsible for cell death have been the focus of extensive studies and discussions. The literature regarding the roles of apoptosis and necrosis in brain trauma is complex and sometimes contradictory, with morphological studies often giving results that differ from biochemical studies (see [3] for a review). Furthermore, when the adult nervous system undergoes injury, the dying neural cells typically first display necrosis and only later display versions of apoptosis with properties that differ from that observed in either

...urbed developmental cell death or the cell death that ...ults from injury to a developing nervous system [3]. For example, in adult animal models of injury, both necrosis and apoptosis have been reported following focal cerebral ischemia in rats and mice [9–13] or after global cerebral ischemia [14,15], and in both instances significant delayed neuronal death occurs in the same regions of the hippocampus. Others have stated that, based on electron microscopic morphological criteria, no neuronal apoptosis occurs after global cerebral ischemia in adult animals [16–18]. Recently, it has been suggested that neurons exhibit a spectrum of morphological and biochemical changes indicative of cell death, with apoptosis and necrosis at opposite ends of the spectrum of observed changes [3,8,17,19–21]. Thus, the specific morphological features seen in a given cell may reflect a variety of factors that influence the cellular events that follow injury, including cellular phenotype and developmental stage, the severity of the insult, the survival time after insult, and the history and aftermath of previous insults.

In immature animals experiencing trauma early in development, the light microscopic appearance of dying neurons is different from that observed in adults. For example, the term *karyorrhexis* has been used in immature brains to describe a nucleus that in the early stages after ischemic injury breaks up into relatively large, darkly basophilic, rounded clumps of chromatin. In the human fetal and neonatal brain, Scott and Hegyi [22] found the presence of karyorrhexic neurons in the basis pontis, an area known to be susceptible to selective neuronal death (*pontosubicular necrosis*) and suggested that karyorrhexis has features common to both necrosis and apoptosis. Other investigators have suggested that both necrosis and apoptosis occur after perinatal ischemia, or any form of excitotoxic injury, in the immature rat [23,24] and that the type of cell death observed varies among brain regions [17,25]; the severity of injury [26,27], the specific stage of development of the animal, and the individual cell phenotype studied [28]. Other factors that may determine whether a necrotic or apoptotic outcome results from trauma to the nervous system are interactions among different neuronal phenotypes—oligodendrocytes, astrocytes, and microglia—together with the different trophic and hormonal activity levels present at the time of injury.

In summary, responses to trauma display a spectrum of responses ranging from cellular commitment to necrosis to apoptosis, or, of more significance, to recovery events that result in a complete return to physiological function [8]. Furthermore, there is evidence for overlap in the features of these different end points that depends on neuronal development and the nature of injury. Here, we focus on apoptosis and apoptotic cell death to reflect the fact that the phenotype associated with cell death during development of the nervous system is the best understood in terms of its regulation by epigenetic factors (neurotrophic factors) whose activity is dependent on cell–cell interactions [29]. By contrast, trauma-induced delayed cell death typically displays morphological and molecular features in common with apoptosis and dissimilar from necrosis [2]; the broad spectrum of cell-type-specific features in both kinds of cell death outcomes has not been fully characterized [3].

The Neurotrophic Hypothesis

The strategy for establishing the final neural net that allows a nervous system to process sensory signals, establish memory engrams, and generate responses ranging from the endocrine and motor to the sensation of self-awareness depends in some fashion on the principle of the *neurotrophic hypothesis*, perhaps best illustrated during the early neuronal development of the sympathetic nervous system. Early in development there is an overproduction of precursor sympathetic neurons, sympathicoblasts, that when exposed to the nerve growth factor protein (NGF), in the absence of glucocorticoids, extend neurites oriented to the NGF source (Fig. 1A). Over time a decrease is seen in the ambient levels of NGF even as the extended growth cones contact target tissues that synthesize and release the reduced amounts of NGF. Synaptic contact

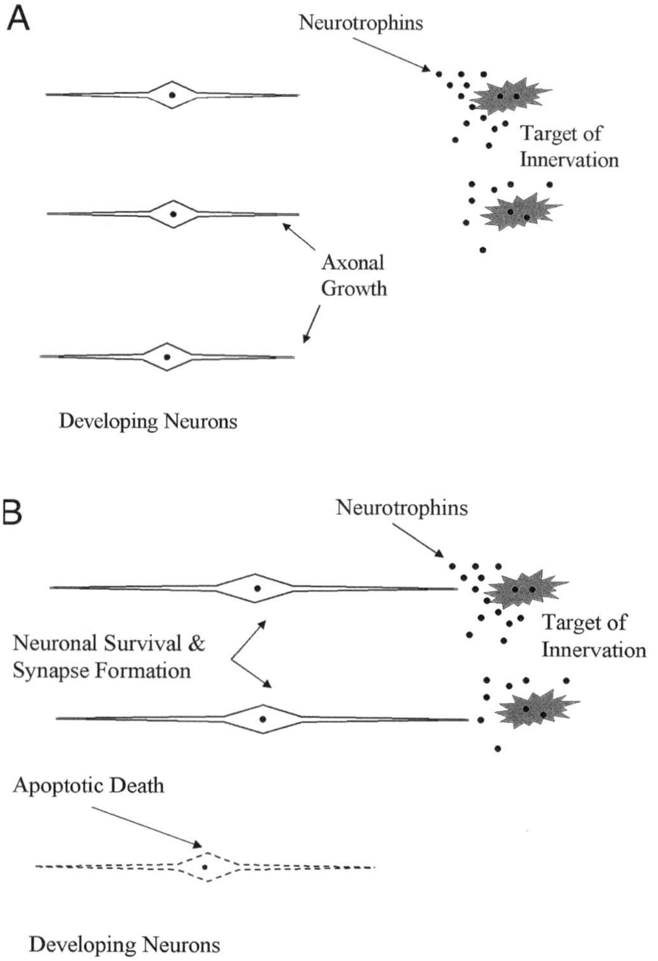

Figure 1 Neurotrophic hypothesis. (A) Target tissues release neurotrophic factors that guide axons. (B) Neurons that reach targets form synapses with targets and compete successfully for neurotrophic factors, whereas neurons that do not make successful contact experience neurite pruning and cell death with apoptotic properties.

between the neuronal growth cones that display NGF receptors takes place and binding of nearby NGF molecules to these receptors occurs, followed by internalization and the eventual retrograde transport to the neuronal soma of both encapsulated NGF-receptor complexes and activated NGF receptors [30]. Those sympathicoblasts whose axons fail to reach a target become growth factor deprived, experience neurite pruning, and perish via apoptosis, the fate of half of all sympathetic neurons during development (Fig. 1B). This is not a unique event because the fate of neuronal apoptosis during development is common to almost all neuronal populations [29].

Experiments that support the neurotrophic hypothesis have shown that an overproduction of neurons early in development is followed by a significant decrease in their number as maturity is reached; that the synthesis of neurotrophic substances takes placc at tissues that are targets of innervation, but not the neurons that innervate them; and that cognate high-affinity neurotrophic receptors are present in the innervating neurons but not in their target tissues. Furthermore, reducing the levels of the neurotrophic factors or of their cognate receptors or interrupting the retrograde transport of neurotrophin results in the death of the innervating neurons, an event that can be abrogated by the external application of the appropriate growth factors to the innervating neurons [32–36].

Neurotrophins

Although a large number of growth factor proteins have been shown to display neurotrophic activity, the best characterized are the neurotrophins. During the last half-century four neurotrophins (NGF, BDNF, NT-3, and NT-4/5) were isolated [37–52] and shown to display a 50–60% sequence homology among themselves [52] and 25% sequence homology to proinsulin [42]. The neurotrophins are a family of related protein growth factors that share certain structural features with other signal proteins belonging to the cysteine knot superfamily [53]. The predominant feature of the cysteine knot family of proteins consists of ring structures made up of intracellular disulfide bridges that form a tightly packed "cysteine knot" that allows for homodimers with extensive surfaces in contact between antiparallel disposed monomers and a strong noncovalent bond between dimer members ($K_d \sim 10^{-13}\ M$ for NGF). Neurotrophins are typically very basic small proteins (pI ~ 9–10; MW ~ 12–14 kDa). For one such neurotrophin, NGF, the biologically active dimer (β-NGF) can be isolated as part of a hetero hexamer made up of the β-NGF dimer and two other protein dimer kallikreins, one active, γ-NGF, and one inactive, α-NGF, which together with two zinc atoms form a stable equilibrium complex ($\alpha_2\beta_2\gamma_2Zn_2$) called 7S NGF based on its equilibrium sedimentation constant [52].

Murine 7S NGF is a stable multimer within a pH range of $5 \leq pH \leq 9$ and at concentrations consistent with its dissociation equilibrium constant ($K_d = 10^{-9}\ M$) [52]. Dissociation and association of the 7S complex serves regulatory functions given that, whereas cross-linked β-NGF is biologically active, the cross-linked 7S multimer is not active and the equilibrium dissociation constant for 7S NGF is in the same range as the binding equilibrium constant of neurotrophins for the $p75^{NTR}$ receptor [53]. The α-NGF subunit stabilizes the 7S NGF complex, but has no other known biological function [54] and the γ-NGF subunits are arginine-specific esteropeptidases of the serine family that process the pro-β-NGF precursor yielding the mature NGF form [55]. The significant levels of neurotrophin precursor forms present throughout adulthood may serve a complex regulatory function given the reported widespread presence of proneurotrophins and the selective cleavage of pro-NGF by γ-NGF and of pro-BDNF by the matrix metalloproteimerase MMP-7 but not MMP-2 or MMP-3 [56]. The nature and processing of NT-3 and NT-4/5 are not known.

It has been proposed that the relatively high ambient levels of proneurotrophins, compared to their mature forms, suggest a regulatory role for proneurotrophin [56]. For example, the NGF $p75^{NTR}$ receptor has been shown to exhibit preferential binding to proNGF as compared to the mature β-NGF dimer, with likely proapoptotic consequences [56]. Given that $p75^{NTR}$ itself enhances the affinity of the other NGF receptor, TrkA, for NGF and decreases the affinity of TrkA for NT-3, it is likely that proneurotrophin processing and binding of precursor forms to $p75^{NTR}$, together with environmental effects of pH and neurotrophin concentration on neurotrophin multimer dissociation, such as 7S NGF, provide a delicately balanced set of interactions that can selectively affect cell survival in response to both developmental cues and environmental changes during development or in the aftermath of stress-related events.

Although the large proportion of neurons affected by the neurotrophic hypotheses suggests that this is a dominant mechanism for the appropriate matching of neuronal numbers and connections and targets to be innervated, it also serves as a mechanism to correct inappropriate or aberrant connections and the culling of neuronal populations that serve transient functions during narrow developmental windows [57–60]. Given that regulation of the synthesis of neurotrophic factors and their receptors is itself regulated by steroids and cytokine and chemokine signaling molecules, it is not surprising that neurotrophic factors play a role in the establishment of sexual dimorphism in the central nervous system [61,62]. Thus coexpression of steroids and neurotrophins in the central nervous system is consistent with the regulation of neurotrophin synthesis by steroids both *in vitro* and *in vivo* [61,62]. For example, estrogen stimulates BDNF and TrkA synthesis and dexamethasone treatment inhibits $p75^{NTR}$ expression [63]. Cytokines have also been shown to stimulate NGF synthesis by astroglial cells in the central nervous system [64]. Given the role of steroids and cytokines/chemokines in the stress response signaling pathways, it is not surprising that the neurotrophin trophic factors attenuate oxidative-stress-induced cell death via stimulation of antioxidant enzymatic conversion of ROS to nontoxic molecules, inhibition of DNA repair energy depletion, and stimulation of anti-apoptotic activity such as the synthesis and translocation of members of the Bcl-2 family of proteins [65–70].

The neurotrophic hypothesis is dependent on the retrograde transport of neurotrophic factors from nerve terminals with access to growth factor secreting tissues. There is also evidence for the anterograde transport of growth factors (principally NT-3 and BDNF; [71–73]) in the avian and mammalian nervous system. While retrograde transport of neurotrophins, a basic element of the trophic hypothesis, has been best documented for NGF, the evidence for anterograde transport is best demonstrated for BDNF, in the context of a response to increased neurotransmitter signaling [72,73]. It is tempting to suggest that although NGF is principally involved in cell survival outcomes, BDNF plays a role in synaptic plasticity in the adult nervous system. Such speculation would be consistent with the temporal pattern of expression of the neurotrophins and neuronal dependence on the different neurotrophin molecules during development with NT-3 and NT-4/5 playing a major role earlier in development at a time when cellular proliferation, migration, and differentiation are central, and BDNF and NGF appearing later in development as both neurite growth and synaptic connectivity patterns are established.

The evidence for an interaction between neuronal activity and neurotrophin action is not limited to BDNF. Both limbic seizures and physical exercise have been shown to stimulate cortical and hippocampal NGF levels [74–76]. Whether the mechanisms linking physiological activity and neurotrophin levels and effects are the same as those triggered by injury is not known. In part, this follows from the ubiquitous nature of the glutamate-ROS-cytokine inflammatory cascade triggered by all trauma and the close linkage between neurotransmitter action (glutaminergic, cholinergic, dopaminergic and serotonergic) and oxidative stress in the nervous system. For example, this is especially true of brain regions involved in memory and cognition where synaptic and functional plasticity occur via neurotransmitter systems whose robust action is associated with significant oxidative stress. Thus, exaggerated plasticity results in toxicity; although the quantitative determination of the correlation between neurotransmitter action and toxicity has not been determined in terms of specific molecular pathways.

Neurotrophin Receptors

The initial event in the regulation of cell death by neurotrophins during development is the binding of a neurotrophin ligand to a neurotrophin receptor, and for the neurotrophin factors (NGF, BDNF, NT-3, NT-4/5; [77]) there are two categories of receptors: $p75^{NTR}$ and the Trk receptors. The structure–function features of the actions of these ligands with their cognate receptors ($p75^{NTR}$, TrkA, TrkB, TrkC) is fairly well understood, although the details of the machinery activated may be less so [77,78]. The $p75^{NTR}$ receptor belongs to the tumor necrosis factor (TNF), fas antigen receptor (Fas) family of receptors, all of which contain an intracellular death domain. The $p75^{NTR}$ is a relatively small receptor that is able to trigger ceramide signaling and NFκB activation [79–82]. The ability of $p75^{NTR}$ to mediate both the cell survival and cell death of neurons and glia depends on its interaction with other neurotrophin receptors [53,83–90]. This is because $p75^{NTR}$ is a relatively low-affinity ($K_d \sim 10^{-9} M$) receptor typically present at 25- to 40-fold higher levels than the Trk high-affinity ($K_d \sim 10^{-11} M$) neurotrophin receptors. The three Trk receptors ($p140^{TrkA}$, $p145^{TrkB}$ and $p145^{TrkC}$) belong to the tyrosine kinase family of receptors that share an amino acid sequence homology with the tropomyosin receptor kinase. They are commonly called TrkA, TrkB, and TrkC. The Trk receptors display differential binding to the different neurotrophins [53,89]. Whereas the $p75^{NTR}$ receptors appear to bind equally to any of the neurotrophins, the Trk receptors do show some specificity with TrkA preferentially binding to NGF, TrkB preferentially binding to BDNF and NT-4/5, and TrkC preferentially binding to NT-3 (Fig. 2) [89]. As with other tyrosine kinase receptors, there are nonfunctional truncated TrkB and TrkC receptors, which may serve to modulate neurotrophin levels extracellularly, although this has not been unequivocally demonstrated [90]. In addition, although the $p75^{NTR}$ receptor is supposed to bind any of the neurotrophin ligands, there are reports of selective activation of ceramide signaling only in the presence of NGF, and not the other neurotrophins, which suggests that NGF signaling is the more selective and specific of the Trk neurotrophin interactions. Not surprisingly, BDNF and NT-3 Trk receptors are widely distributed in the nervous system compared to the more restricted distribution of TrkA receptors.

Studies using neurotrophin and neurotrophin receptor null mice would suggest that abolition of the TrkA gene disturbs sensory neurons in the periphery to a greater extent than the TrkA bearing neurons in the central nervous system (basal forebrain cholinergic neurons and striatal cholinergic interneurons) [91,92]. It is also interesting that both TrkB and TrkC are widely distributed in the central and peripheral nervous system and that null mice lacking individual neurotrophin or neurotrophin receptor genes have phenotypes in

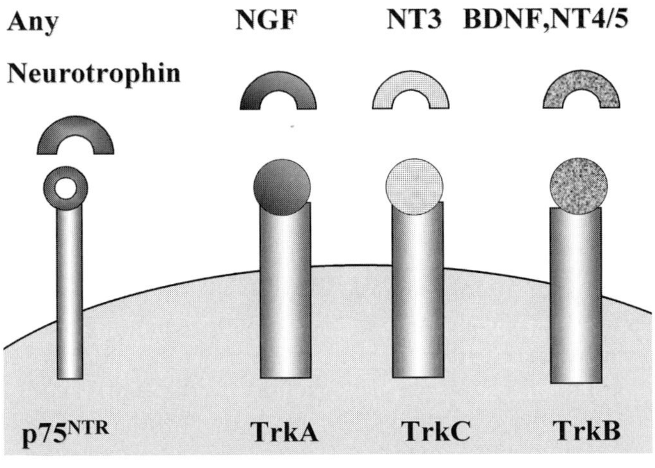

Figure 2 Neurotrophins and neurotrophin receptors.

which it appears that the absence of one growth factor or its receptor is compensated by the expression of others [93–96]. Thus, although there are specific developmental stages for the maximal expression of the different neurotrophins, null mice studies would suggest that there is also a great deal of compensation for the absence of a neurotrophin at a specific developmental stage. This is in agreement with more recent gene network descriptions that would suggest that specific physiological functions dependent on epigenetic growth factors are dependent on the actions of clusters of growth factors working in a complementary, but also redundant, fashion [91,92].

Nevertheless, cell death regulation by neurotrophins can be very selective. For example, in the central nervous system, NGF is synthesized and released by glutaminergic neurons in the hippocampus, where it is taken up by axons projecting from the basal forebrain cholinergic neurons, which express $p75^{NTR}$ receptors [53,83–90,98–104]. TrkA is not exclusively expressed by the cholinergic neurons of the basal forebrain; TrkB is also expressed there in association with minor responses to BDNF [97]. The basal forebrain cholinergic neuron $p75^{NTR}$ receptors can also bind and internalize the well-characterized monoclonal antibody to $p75^{NTR}$, 192 IgG. When 192 IgG is cross-linked via a disulfide bond to the ribosomal inactivating protein, saporin, the immunotoxin 192 IgG–saporin complex is itself internalized into cholinergic basal forebrain neurons [99–104]. Thus, after an i.c.v. injection of 192 IgG–saporin, the immunotoxins are internalized at the terminals of $p75^{NTR}$-bearing cholinergic basal forebrain neurons, retrogradely transported, and accumulated in the cell bodies of cholinergic basal forebrain neurons where they cause cell death. Treatment with intraventricular administration of immunotoxin produces selective and dose-dependent cell death of up to 90% of all $p75^{NTR}$-bearing cholinergic basal forebrain neurons with substantial reductions in the activities of the acetylcholine degrading and synthesizing enzymes, acetyl cholinesterase and choline acetyltransferase, in the rat basal forebrain and its neocortical and hippocampal afferents [99–104]. These cell losses and cholinergic deficits accompany cognitive deficits.

Neurotrophin Signaling Pathways

The neurotrophic hypothesis states that the competition for target-derived trophic factors provides a mechanism for the innervation of target tissues during the development of the nervous system. The properties of neurons that perish apoptotically during development, the transient or permanent nature of their dependence on those trophic growth factors at the termination of the developmental sequelae, or the role of these growth factors in adult neuronal function or responses to injury can vary greatly. However, in spite of these variations, certain common features are worth examination.

Two triggers bring about the death of a neuron. First, we have the intrinsic signaling pathways, which act via binding of neurotrophins or cytokines (for example, NGF or TNF-α) to cognate receptors, such as $p75^{NTR}$, that can irreversibly commit a cell to apoptosis [84,98,105–110]. Second, we have the signaling pathways, which result from changes in membrane transport, energy metabolism, and cellular redox and pH that can directly activate intracellular signaling cascades committing a neuron to apoptosis.

The nature and magnitude of Trk and $p75^{NTR}$ receptor activity are modulated by interactions between these two receptor families [84]. For example, although $p75^{NTR}$ can act as an apoptotic stimulus and Trks typically mediate anti-apoptotic outcomes, $p75^{NTR}$ can also enhance neurotrophin binding to Trk receptors by increasing the local neurotrophin concentration in the environment of a Trk receptor, a function that is dependent on the ratio of $p75^{NTR}$ to Trk receptors present on a particular cellular membrane. There is also evidence for a direct $p75^{NTR}$-mediated conformational change in TrkA that augments binding and activity. The antagonistic actions of $p75^{NTR}$ and the Trk can serve two functions: (1) In the absence of growth factor in the environment of a neural cell early in development, cells die promptly and when neurotrophins become available, high-affinity binding to Trk receptors abrogates the $p75^{NTR}$ apoptotic effect; and (2) once a neuron extends neurites, if it reaches an inappropriate target that releases the "wrong" neurotrophin, its weak binding to the inappropriate Trk will not able to overcome $p75^{NTR}$ proapoptotic signaling. In this way wrong "connections" are abolished. Thus, having two antagonistic signaling systems for the regulation of cell survival allows for developmental adjustments of neuronal number to target size and the elimination of errors in innervation. Experimental evidence for these two developmental regulatory phenomena are present for both peripheral and central neurons [106–108] and glial cells [109,110] that bind NGF, BDNF, and NT-3 [105].

Although the intracellular signaling pathways triggered by neurotrophin binding to $p75^{NTR}$ and Trk receptors differ significantly, consistent with the different outcomes of their activity, there are also common signaling elements acting in both instances (Figs. 3 and 4). The signal transduction pathway activated by $p75^{NTR}$ consists of three major interactive pathways. $p75^{NTR}$ stimulates sphingolipase-mediated hydrolysis of sphingomyelin to ceramide, which will alter membrane fluidity and hence Trk-Trk association, known to markedly stimulate Trk autophosphorylation [111,112]. Increased ceramide levels can also directly stimulate activation of the transcription factor NFκB proapoptotic pathways directly [80,109,110]. Neurotrophin binding to $p75^{NTR}$ can also stimulate NFκB via TNFR-associated factor (TRAF) 6, as part of a complex pathway regulating cell cycle progression with anti-apoptotic properties that is likely to stimulate transcription of Bcl-2 and Bcl-x in early development and Bcl-x in the adult central nervous system [70,80–82,113,114]. A third apoptotic pathway that is stimulated by $p75^{NTR}$ is the mitogen-activated protein kinase (MAPK) kinase (MEKK) $\rightarrow$ Jun amino-terminal kinase (JNK) $\rightarrow$ c-Jun $\rightarrow$ p53 $\rightarrow$ Bax pathway [115–118].

The Trk receptors mediate a number of other neural functions in addition to their well-known survival effects.

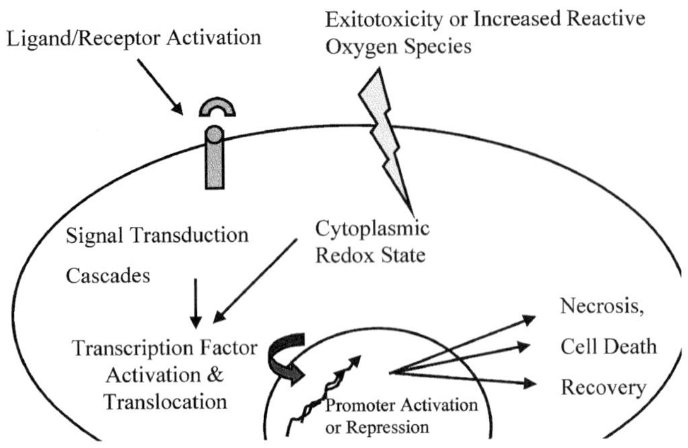

Figure 3 Schematic of activation of stress response signaling mechanisms that determine cellular outcomes: necrosis, apoptosis, recovery.

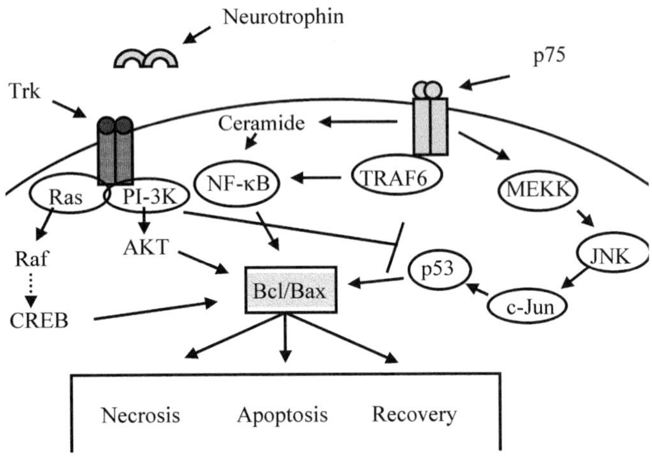

Figure 4 Regulation of cellular commitment to cell death or recovery by neurotrophin receptor activation.

These include regulation of neurite elongation, synapse maturation, neurotransmitter phenotype, and activity levels. Although the specific pathways involved in some of these developmental and plasticity-related functions are not known, it is known that the Ras → Raf → MEK → MAPK pathway, which also is a determinant of cellular commitment to apoptosis, is also likely to be involved. The Ras pathway, together with the PI3K → AKT pathway also promote cell survival by stimulating Bcl-2, Bcl-x while inhibiting p53 effects on Bax expression [116–124]. Furthermore, synergistic and antagonistic effects are acting on Bcl/Bax expression via phosphorylation cascades involving several of the pathways affected by neurotrophin occupancy of $p75^{NTR}$ and Trks in addition to transcriptional regulation via the AP-1 (Fos/Jun family) and NFκB transcription factors. To summarize, there are both antagonistic and mutually interactive effects at the level of proneurotrophin to mature neurotrophin processing, generation of truncated versus full-length Trk receptors, differential triggering of receptor-associated phosphorylation cascades, and AP-1 and NFκB differential transcription factor activation and repression of specific genes that orchestrate the balance of apoptotic versus anti-apoptotic determinants and shift the cell from a necrotic to an apoptotic outcome or from an apoptotic outcome to unimpeded survival and function.

The common signaling elements underlying the events that can bring about apoptosis act via reactive oxygen and nitrogen molecular species and the phosphorylation of proteins as activating or deactivating events. For example, NO generation in the glutaminergic system, or oxygen radical generation in the dopaminergic system, are triggers for cascades whose end point can be cell death and neurological dysfunction after injury or ischemia [4,6]. Furthermore, both NO and H_2O_2 participate in cross-reacting signaling pathways that yield the very reactive hydroxyl and peroxynitrite radicals that, while required for learning and memory processing in brain, are also toxic to neurons and oligodendroglia and over time, or when exacerbated after acute trauma, interfere with proper signal processing by neurons. Downstream from the cytoplasmic signaling cascades responsible for cellular regulation of apoptosis [105], the AP-1 and NFκB transcription factors determine neuronal death outcomes during development or after trauma.

Transcriptional Regulation

The final outcomes of neurotrophin action in terms of cell death versus survival, or aspects of neuronal growth or regeneration, and synaptic plasticity result from the balanced integration of signaling cascades triggered by ligand binding to cell-surface receptors and the subsequent signal transfer to transcription factor assemblies, themselves responsive to the cytoplasmic redox milieu [5,105]. In turn, the selective activation of transcription factors determines the activation or repression of hosts of genes that determine physiological outcomes. Although the number of transcription factor families that participate in these orchestrated events is both large and ever growing [125], we discuss here just two transcription factors that regulate cell death commitment: activator protein-1 (AP-1) and nuclear factor kappa B (NFκB). The first step in AP-1 activation is via prompt *de novo* synthesis of AP-1 proteins and their nuclear translocation and phosphorylation in response to redox changes that activate or repress transcription. Direct activation of NFκB results from the phosphorylation of inactive cytoplasmic protein complexes that permit nuclear translocation and binding to DNA consensus sequences on gene promoters.

AP-1

Members of the AP-1 transcription factor family have been shown to play an important role in cell proliferation, differentiation, and survival [126]. The AP-1 complex recognizes a set of specific DNA sequences (TREs) present in many enhancer or promoter regions of genes [126]. The AP-1 Jun proteins have two functional domains that bind to a DNA consensus

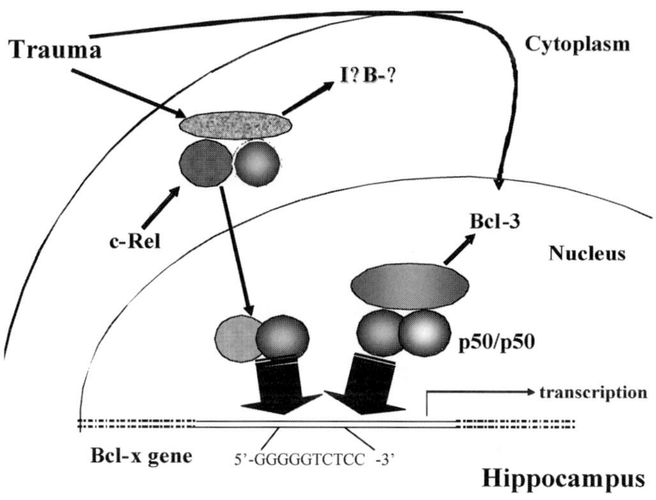

Figure 5 The AP-1 proteins.

binding sequence CCCAGGC and to Fos proteins, respectively. The AP-1 contains a leucine zipper. Cellular AP-1 typically consists of dimeric complexes made up of different Fos and Jun protein family members (Fig. 5) made up of Jun protein homodimers or Fos/Jun heterodimers. The Fos family of proteins consists of c-Fos, Fos B, Fra 1, and Fra 2. The Jun family of proteins consists of c-Jun, Jun B, and Jun D.

Changes in the relative levels of the different members of the AP-1 family or their state of phosphorylation serve to provide specific outcomes in terms of cellular commitment to modes of cell death versus cellular proliferation, for example. In some cell types, repression of AP-1 activity has been shown to trigger apoptosis. For example, both dexamethasone and retinoic acid repress AP-1 function and induce apoptosis in lymphocytes, and E1A, which represses AP-1 transaction activity in transfected cells, also induces apoptosis in several cell types [127]. In other instances, induction of the AP-1 c-Jun protein appears to cause apoptosis. It has been shown that overexpression of c-fos and c-jun is required for apoptosis in a murine cytokine-dependent lymphoblastoid cell line after growth factor deprivation [128]. Of relevance here is that it has been shown that the c-Jun protein plays a necessary role in the death of sympathetic neurons after NGF withdrawal and that AP-1 activity is necessary for apoptosis under these conditions [129]. However, in a rat pheochromocytoma cell line with some sympathetic cell properties (PC12), treatment with NGF induces differentiation into sympathetic-like neurons and prompt and transient expression of c-fos and c-jun mRNA [130] with increased AP-1 DNA binding activity [131]. A more thorough examination of AP-1 DNA binding activity, and the composition of AP-1 and AP-1-mediated transcription during NGF-mediated rescue in serum-deprived PC12 cells [132] and also the hippocampi of rats exposed to oxidative stress [133], showed that although there is no significant change in the overall composition of AP-1 in PC12 cells after serum deprivation, protection with NGF treatment significantly increased the activity levels of participation of Fra 2 and Jun B in AP-1 DNA binding, as determined by immunodepletion/supershift assays [132]. Thus, although transcription of some common genes may take place under both serumless and NGF treatment conditions, a unique set of genes is transcribed after exposure to NGF via increased levels of Fra 2 and Jun B that contribute to PC12 cell differentiation and to NGF-mediated anti-apoptotic outcomes [134,135]. Thus, the differences in the AP-1 dimer proteins induced or activated by phosphorylation of c-Jun can alter cell death outcomes.

NFκB Transcription Factor

The NFκB family of transcription factors regulates genes mediating inflammation, responses to infection, oxidative stress, and the aftermath of ambient necrotic events [136]. NFκB is made up of five structurally related protein subunits: p50, p52, p65/RelA, c-Rel, and RelB, which form a variety of homo- and heterodimers in multiple tissues [70]. Unlike p65, RelB, and c-Rel, the p50 and p52 subunits result from proteolytic cleavage of their p105 and p100 precursor molecules, respectively. The precursor forms include in their structure C-terminal IκB-like ankyrin repeats that inhibit p50 and p52 activity by virtue of their high affinity for these two proteins. The Bcl-3 proteins (not related to the Bcl-2 family) are nuclear proteins with sequence homology to the IκB inhibitor proteins that can "bind away" p50 or p52 homodimers from promoter NFκB binding sites in nuclear DNA and allow p65/p50 or cRel/p50 activating heterodimers to bind the unoccupied NFκB binding sites and thus stimulate gene-specific transcription [137]. All five NFκB/Rel proteins share a 300-amino-acid region, the Rel homology domain, which is the structural basis for dimerization, DNA binding, and nuclear localization [138]. The NFκB proteins p65/RelA, c-Rel, and RelB subunits contain a C-terminal acidic activation domain not present in the p50 and p52 subunits. Thus, heterodimers containing p52 or p50 combined with p65, c-Rel, or RelB are capable of activating transcription [138], whereas p52/49 and p50/p50 are generally considered transcriptional repressors, although these homodimers can participate in promoter activation through interaction with other non-Rel proteins such as the IκB family member Bcl-3 [137]. Despite this diversity, the active heterodimer p50/p65 is generally referred to as NFκB. NFκB activation can be measured by determining levels of cytoplasmic IκB; nuclear levels of p65, c-Rel, or RelB; nuclear levels of Bcl-3; or binding of nuclear extracts to oligonucleotides bearing the different NFκB binding consensus DNA sequences present in specific gene promoters of interest identified by immunodepletion/supershift assays (Fig. 6).

Role of IκB

One major regulatory influence on NFκB activity is its intracellular localization. Inactive NFκB homo- and

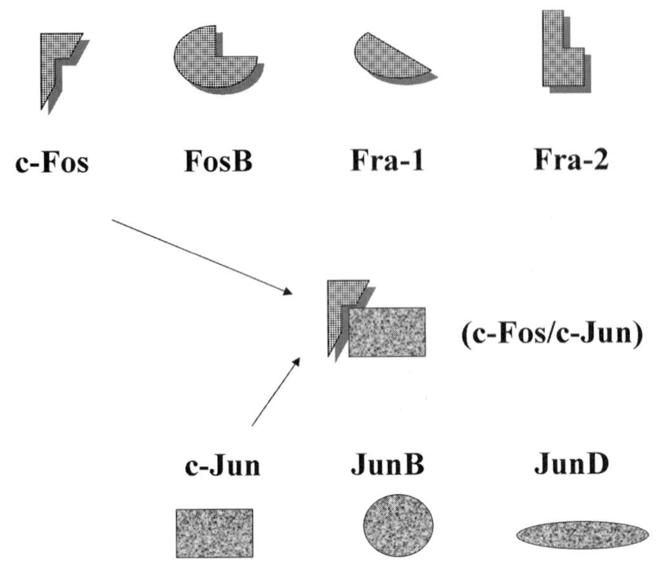

Figure 6 NFκB proteins modify the activate gene transcription (two separate functions). The IκB proteins modify the activity of the NFκB protein dimers. For example, Bcl-3 can bind to p50/p50 homodimers and temove then from promoter sites, facilitating the binding of activating NFκB heterodimers p65/p50 or c-Rel/p50.

heterodimers are sequestered in the cytoplasm of cells complexed with one of several distinct inhibitory subunits, including IκB-α [139]. IκBs contain six copies of an ankyrin repeat structural motif that are required for protein–protein interactions with NFκB dimers [140]. IκB-γ is the C-terminal domain of the 105-kDa precursor protein of the p50 subunit and binds p50 monomers via its seven ankyrin repeats [141]. IκB-γ can also exist as an independent 70-kDa form that results from transcripts using an alternate, internal promoter within the p50/p105 gene [142]. An additional level of regulation is imparted to the NFκB system since different IκB molecules show preference for inhibiting different subunit combinations, whereas IκB-α and IκB-β sequester p65/p50, p65/p52, c-Rel/p50, or c-Rel/p52 homodimers and heterodimers. IκB-ε complexes only with dimers containing p65 and c-Rel [143].

Stimuli that activate NFκB are generally either pathogens (bacterial lipopolysaccharide, viruses, hydrogen peroxide, UV light), or cytokines (tumor necrosis factor, interleukin 1 and 2), and signals associated with these stimuli, which result in the activation of IκB kinase (IKK), a large, multisubunit complex containing two catalytic subunits (IKK-α and IKK-β) and a regulatory subunit (IKK-γ) [144]. IκB molecules are subject to phosphorylation by IKK at two serine residues (serines 32 and 36 in human IκB-α) in their N-terminal domain. These are critical for making the IκB molecule a substrate for ubiquitination at lysine residues [145]. Once ubiquitinated by an ubiquitin ligase [146], the IκB molecule is degraded by 26S proteasomes, releasing NFκB, which is active because the loss of the IκB molecule from the NFκB dimer exposes nuclear localization signals that allow NFκB translocation into the nucleus via interactions of Rel homology domain nuclear localization sequences with importin-α/importin-β heterodimers at the nuclear pore [147,148]. NFκB dimers bind target gene regulatory regions through a wide variety of binding sites that generally match a 5′-GGGRNTY(C/T)C-3′ consensus (R = A or G, Y = C or T, N = any nucleotide). NFκB achieves target gene specificity in part through preferential binding of different subunit combinations to the different DNA consensus binging sequences [70,113,149,150]. Although the inhibition of total NFκB binding to DNA consensus sequences present in the IgG-κB promoter has been shown to increase apoptotic cell death [81,82], the consequences of more selective interventions in the interactions between specific NFκB protein dimers and gene-specific DNA consensus sequences present in the promoters of genes whose expression is known to be altered by oxidative stress and to have significant effects on apoptotic outcomes have not been determined. Based on the outcomes observed for changes in $p75^{NTR}$ versus Trks or intracellular phosphorylation cascades (Figs. 2, 3, and 4), it is likely that balanced counteracting transcriptional events are regulated by a complex symphonic array of NFκB homo- and heterodimers with the eventual determination of life versus death at the cellular level as the outcome.

Conclusions

The aggregate of these observations is consistent with the hypotheses that cellular events that take place when cellular homeostatic set points are shifted as part of the developmental process or in response to abrupt trauma or disease trigger signaling pathways with quite opposite end points. Consequently, eventual physiological outcomes will depend on delicate interactions at the level of receptors, receptor-associated pathways of activation, and differential transcriptional regulation by transcription factor dimers binding to similar but different promoter sites.

Neurotrophic regulation of neuronal fate during phenotypic changes associated with the development of the nervous system or during neuronal recovery after injury or disease rely on a complex orchestration of signaling pathways and gene activating events balanced by virtue of shared molecular components and of binding events at the cell surface, in the cytoplasm, or at gene promoter sites that are similar but different enough to selectively promote or prevent cell death.

References

1. Burek, M. J. and Oppenheim, R. (1999). In Koliatsos, V. E., and Ratan, R. R., Eds., *Cell Death and Diseases of the Nervous System*, pp. 145–180. Humana Press, Totowa, NJ.
2. Kerr, J., Wyllie, A., and Currie, A. (1972). Apoptosis: A basic biological phenomenon with wide-ranging implications in tissue kinetics. *Br. J. Cancer* **26**, 239–257.
3. Chu, D., Qiu, J. X., Grafe, M., Fabian, R., Kent, T. A., Rassin, D., Nesic, O., Werrbach-Perez, K., and Perez-Polo, R. (2002). Delayed cell death signaling in traumatized central nervous system: hypoxia. *Neurochem. Res.* **27**, 97–106.

4. Thompson, C. (1995). Apoptosis in the pathogenesis and treatment of disease. *Science* **267**, 1456–1462.
5. Tong, L., Toliver-Kinsky, T., Taglialatela, G., and Perez-Polo, J. (1998). Signal transduction in neuronal death. *J. Neurochem.* **71**, 447–459.
6. Dykens, J. A. (1999). In Koliatsos, V. E., and Ratan, R. R., Eds., *Cell Death and Diseases of the Nervous System*, pp. 145–180. Humana Press, Totowa, NJ.
7. Davies, A. M. (1995). The Bcl-2 family of proteins, and the regulation of neuronal survival. *Trends Neurosci.* **18**, 355–358.
8. Cole, K. K. and Perez-Polo, J. R. Poly(ADP-ribose) polymerase inhibition prevents both apoptotic-like delayed neuronal death and necrosis after H_2O_2 injury. *J. Neurochem.* (in press).
9. Charriaut-Marlangue, C., Margaill, I., Represa, A., Popovici, T., Plotkine, M., and Ben-Ari, Y. (1996). Apoptosis and necrosis after reversible focal ischemia: An *in situ* DNA fragmentation analysis. *J. Cereb. Blood Flow Metab.* **16**, 186–194.
10. Chen, J., Jin, K., Chen, M., Pei, W., Kawaguchi, K., Greenberg, D. A., and Simon, R. P. (1997). Early detection of DNA strand breaks in the brain after transient focal ischemia: Implications for the role of DNA damage in apoptosis and neuronal cell death. *J. Neurochem.* **69**, 232–245.
11. Endres, M., Namura, S., Shimizu-Sasamata, M., Waeber, C., Zhang, L., Gomez-Isla, T., Hyman, B. T., and Moskowitz, M. A. (1998). Attenuation of delayed neuronal death after mild focal ischemic in mice by inhibition of the caspase family. *J. Cereb. Blood Flow Metab.* **18**, 238–247.
12. Matsushita, K., Matsuyama, T., Kitagawa, K., Matsumoto, M., Yanagihara, T., and Sugita, M. (1998). Alterations of Bcl-2 family proteins precede cytoskeletal proteolysis in the penumbra, but not in infarct centres following focal cerebral ischemia in mice. *J. Neurosci.* **83**, 439–448.
13. Guégan, C. and Sola, B. (2000). Early and sequential recruitment of apoptotic effectors after focal permanent ischemic in mice. *Brain Res.* **856**, 93–100.
14. Nitatori, T., Sato, N., Waguri, S., Karasawa, Y., Araki, H., Shibanai, K., Kominami, E., and Uchiyama, Y. (1995). Delayed neuronal death in the CA1 pyramidal cell layer of the gerbil hippocampus following transient ischemia is apoptosis. *J. Neurosci.* **15**, 1001–1011.
15. Ni, B., Wu, X., Su, Y., Stephenson, D., Smalstig, E. B., Clemens, J., and Paul, S. M. (1998). Transient global forebrain ischemia induces a prolonged expression of the caspase-3 mRNA in rat hippocampal CA1 pyramidal neurons. *J. Cereb. Blood Flow Metab.* **18**, 248–256.
16. Martin, L. J., Al-Abdulla, N. A., Brambrink, A. M., Kirsch, J. R., Sieber, F. E., and Portera-Cailliau, C. (1998). Neurodegeneration in excitotoxicity, global cerebral ischemia, and target deprivation: A perspective on the contributions of apoptosis and necrosis. *Brain Res. Bull.* **46**, 281–309.
17. Martin, L. J., Brambrink, A. M., Price, A. C., Kaiser, A., Agnew, D. M., Ichord, R. N., and Traystman, R. J. (2000). Neuronal death in newborn striatum after hypoxia-ischemia is necrosis and evolves with oxidative stress. *Neurobiol. Dis.* **7**, 169–191.
18. Colbourne, F., Sutherland, G. R., and Auer, R. N. (1999). Electron microscopic evidence against apoptosis as the mechanism of Neuronal death in global ischemia. *J. Neurosci.* **19**, 4200–4210.
19. Leist, M. and Nicotera, P. (1998). Apoptosis, excitotoxicity, and neuropathology. *Exp. Cell Res.* **239**, 183–201.
20. Pieper, A. A., Verma, A., Zhang, J., and Snyder, S. H. (1999). Poly (ADP-ribose) polymerase, nitric oxide and cell death. *Trends Pharmacol. Sci.* **20**, 171–181.
21. Formigli, L., Papucci, L., Tanni, A., Schiavone, N., Tempestini, A., Orlandini, G. E., Capaccioloi, S., and Zecchi Orlandini, S. (2000). Aponecrosis: Morphological and biochemical exploration of a syncretic process of cell death sharing apoptosis and necrosis. *J. Cell Physiol.* **182**, 41–49.
22. Scott, R. J. and Hegyi, L. (1997). Cell death in perinatal hypoxic-ischaemic brain injury. *Neuropath. Appl. Neurobiol.* **23**, 307–314.
23. Ferrer, I., Tortosa, A., Macaya, A., Sierra, A., Moreno, D., Munell, F., Blanco, R., and Squier, W. (1994). Evidence of nuclear DNA fragmentation following hypoxia-ischemia in the infant rat brain, and transient forebrain ischemia in the adult gerbil. *Brain Pathol.* **4**, 115–122.
24. Portera-Cailliau, C, Price, D. L., and Martin, L. J. (1997). Excitotoxic neuronal death in the immature brain is an apoptosis-necrosis morphological continuum. *J. Comp. Neurol.* **378**, 70–87.
25. Northington, F. J., Ferriero, D. M., Flock, D. L., and Martin, L. J. (2001). Delayed neurodegeneration in neonatal rat thalamus after hypoxia-ischemia is apoptosis. *J. Neurosci.* **27**, 1931–1938.
26. Beilharz, E. J., Williams, C. E., Dragunow, M., Sirimanne, E. S., and Gluckman, P. D. (1995). Mechanisms of delayed cell death following hypoxic-ischemic injury in the immature rat: Evidence for apoptosis during selective neuronal loss. *Brain Res. Mol. Brain Res.* **29**, 1–14.
27. Walton, M., Connor, B., Lawlor, P., Young, D., Sirimanne, E., Gluckman, P., Cole, G., and Dragunow, M. (1999). Neuronal death and survival in two models of hypoxic-ischemic brain damage. *Brain Res. Brain Res. Rev.* **29**, 137–168.
28. Sidhu, R. S., Tuor, U. I., and Del Bigio, M. R. (1997). Nuclear condensation and fragmentation following cerebral hypoxia-ischemia occurs more frequently in immature than older rats. *Neurosci. Lett.* **223**, 129–132.
29. Oppenheim, R. W. (1991). Cell death during development of the nervous system. *Annu. Rev. Neurosci.* **14**, 453–501.
30. Miller, F. D. and Kaplan, D. R. (2002). TRK makes the retrograde. *Science* **295**, 1471–1473.
31. Davies, A. M., Bandtlow, C., Heumann, R., Korsching, S., Rohrer, H., and Thoenen, H. (1987). Timing and site of nerve growth factor synthesis in developing skin in relation to innervation and expression of the receptor. *Nature* **326**, 353–358.
32. Smeyne, R. J., Klein, R., Schnapp, A., Long, L. K., Brynat, S., Lewin, A., Lira, S. A., and Barbacid, M. (1994). Severe sensory and sympathetic neuropathy in mice carrying a disrupted trk/NGF receptor gene. *Nature* **365**, 246–249.
33. Johnson, E. M., Gorin, P. M., Brandeis, L. D., and Pearson, J. (1980). Dorsal root ganglion neurons are destroyed by exposure *in utero* to maternal antibody to nerve growth factor. *Science* **210**, 916–918.
34. Levi-Montalcini, L. (1987). The nerve growth factor 35 years later. *Science* **237**, 1154.
35. Johnson, E. M. (1978). Destruction of the sympathetic nervous system in neonatal rats and hamsters by vinblastine: Prevention by concomitant administration of nerve growth factor. *Brain Res.* **141**, 105–118.
36. Oppenehim, R. W., Maderdrut, J. L., and Wells, D. J. (1982). Cell death of motoneurons in the chick embryo spinal cord. VI. Reduction of naturally occurring cell death in the thoracolumbar column of terni by nerve growth factor. *J. Comp. Neurol.* **210**, 174–189.
37. Levi-Montalcini, R. and Hamburger, V. (1953). A diffusible agent of mouse sarcoma, producing hyperplasia of sympathetic ganglia and hyperneurotization of viscera in the chick embryo. *J. Exp. Zool.* **123**, 233–287.
38. Cohen, S. (1959). Purification and metabolic effects of a nerve growth-promoting protein from snake venom. *J. Biol. Chem.* **234**, 1129–1137.
39. Bocchini, V. and Angeletti, P. U. (1969). The nerve growth factor: Purification as a 30,000-molecular-weight protein. *Proc. Natl. Acad. Sci. USA* **64**, 787–794.
40. Angeletti, R. H. and Bradshaw, R. A. (1971). Nerve growth factor from mouse submaxillary gland: Amino acid sequence. *Proc. Natl. Acad. Sci. USA* **68**, 2417–2420.
41. Perez-Polo, J. R., de Jong, W. W., Strauss, D., and Shooter, E. M. (1972). The physical and biological properties of 7S and beta-NGF from the mouse submaxillary gland in functional and structural proteins of the nervous system. *Adv. Exp. Med. Biol.* **32**, 91.
42. Barde, Y.-A., Edgard, D., and Thoenen, H. (1982). Purification of a new neurotrphic factor from mammalian brain. *EMBO J.* **1**, 549–553.
43. Ernfors, P., Ibez, C. F., Ebendal, T., Olson, L., and Persson, H. (1990). Molecular cloning and neurotrophic activities of a protein with structural similarities to nerve growth factor: Developmental and topographical expression in the brain. *Proc. Natl. Acad. Sci. USA* **87**, 5454–5458.

44. Hon, A., Liebrock, J., Bailey, K., and Barde, Y-A. (1990). Identification and characterization of a novel member of the nerve growth factor/brain-derived neurotrphic factor family. *Nature* **344**, 339–341.
45. Kaisho, Y., Yoshimura, K., and Nakaham, K. (1990). Cloning and expression of a cDNA encoding a novel human neurotrophic factor. *FEBS* **266**, 187–191.
46. Maisonpierre, P. C., Belluscio, L., Squinto, S., Ip, N. Y., Furth, M. E., Lindsay, R. M., and Yancopoulos, G. D. (1990). Neurotrophin-3: A neurotrophic favor related to NGF and BDNF. *Science* **247**, 1446–1451.
47. Rosenthal, A., Goeddel, D. V., Nguyen, T., Lewis, M., Shih, A., Laramee, G. R., Nikolics, K., and Winslow, J. W. (1990). Primary structure and biological activity of a novel human neurotrophic factor. *Neuron* **4**, 767–773.
48. Berkemeier, L. R., Winslow, J. W., Kaplan, D. R., Nikolics, K., Goeddel, D. V., and Rosenthal, A. (1991). Neurotrpohin-5: A novel neurotrophic factor that activates trk and trkB. *Neuron* **7**, 857–866.
49. Hallböök, F., Ibáñez, C. F., and Persson, H. (1991). Evolutionary studies of the nerve growth factor family reveal a novel member abundantly expressed in *Xenopus* ovary. *Neuron* **6**, 845–858.
50. Ip, N. Y., Ibáñez, C. F., Nye, S. H., McClain, J., Jones, P. F., Gies, D. R., Belluscio, L., Le Beau, M. M., Espinosa, III, R., Squinto, S. P., Persson, H., and Yancopoulos, G. D. (1992). Mammalian neurotrophin-4: Structure, chromosomal localization, tissue distribution, and receptor specificity. *Proc. Natl. Acad. Sci. USA* **889**, 3060–3064.
51. Ibáñez, C. F. (1995). Neurotrophic factors: From structure–function studies to designing effective therapeutics. *Trends Biotechnol.* **13**, 217–227.
52. Shooter, E. M. (2001). Early days of the nerve growth factor proteins. *Ann. Rev. Neurosci.* **24**, 601–629.
53. Chao, M. V. and Bothwell, M. (2002). Neurotrophins: To cleave or not to cleave. *Neuron* **33**, 9–12.
54. Thomas, K. A., Baglan, N. C., and Bradshaw, R. A. (1981). The amino acid sequence of the γ-subunit of mouse submaxillary gland 7S nerve growth factor. *J. Biol. Chem.* **256**, 9156–9166.
55. Thomas, K. A., Silverman, R. E., Jeng, I., Baglan, N. C., and Bradshaw, R. A. (1981). Electrophoretic heterogeneity and polypeptide chain structure of the γ-subunit of mouse submaxillary 7S nerve growth factor. *J. Biol. Chem.* **256**, 9147–9153.
56. Lee, R., Kermani, P., Teng, K. K., and Hempstead, B. L. (2001). Regulation of cell survival by secreted proneurotrophins. *Science* **294**, 1945–1948.
57. Forger, N. G., Roberts, S., Wong, V., and Breedlove, S. M. (1993). Ciliary neurotrophic factor maintains motoneurons and their target muscles in developing rats. *J. Neurosci.* **13**, 4720–4726.
58. Chalupa, L. M. and Dreher, B. (1991). High precision systems require high precision blueprints: A new view regarding the formation of connections in the mammalian visual system. *J. Cognit. Neurosci.* **3**, 209–219.
59. O'Leary, D. D. M., Fawcett, J. W., and Cowan, W. M. (1986). Topographic targeting errors in the retino-collicular projections and their elimination by selective ganglion cell death. *J. Neurosci.* **6**, 3692–3705.
60. Catsicas, S., Thanos, S., and Clarke, P. G. H. (1987). Major role for neuronal death during brain development refinement of topographical connections. *Proc. Natl. Acad. Sci. USA* **84**, 8165–8168.
61. Breedlove, S. M. (1992). Sexual dimorphism in the vertebrate nervous system. *J. Neurosci.* **12**, 4133–4142.
62. Perez-Polo, J. R., Hall, K., Livingston, K., and Westlund, K. (1977). Steroid induction of nerve growth factor synthesis in cell culture. *Life Sci.* **21**, 1535–1544.
63. Solum, D. T. and Handa, R. J. (2002). Estrogen regulates the development of brain-derived neurotrophic factor mRNA and protein in the rat hippocampus. *J. Neurosci.* **22**, 2650–2659.
64. Wu, V. W., Nishiyama, N., and Schwartz, J. P. (1998). A culture model of reactive astrocytes: Increased nerve growth factor synthesis and reexpression of cytokine responsiveness. *J. Neurochem.* **71**, 749–756.
65. Coyle, J. T. and Puttfarcken, P. (1993). Oxidative stress, glutamate and neurodegenerative disorders. *Science* **262**, 689–695.
66. Jackson, G. R., Apffel, L., Werrbach-Perez, K., and Perez-Polo, J. R. (1990). Role of nerve growth factor in oxidant-anti-oxidant balance and neuronal injury. I. Stimulation of hydrogen peroxide resistance. *J. Neurosci. Res.* **25**, 360–368.
67. Pan, Z. and Perez-Polo, R. (1993). Role of nerve growth factor in oxidant homeostasis: Glutathione metabolism. *J. Neurochem.* **61**, 1713–1721.
68. Sampath, D., Jackson, G. R., Werrbach-Perez, K., and Perez-Polo, J. R. (1994). Effects of nerve growth factor on glutathione peroxidase and catalase in PC12 cells. *J. Neurochem.* **62**, 2476–2479.
69. Spina, M. B., Squinto, S. P., Miller, J., Lindasy, R. M., and Hyman, C. (1992). Brain-derived neurotrophic factor protects dopamine neurons against 6-hydroxydopamine and N-methyl-4-phenyl-piridium ion toxicity: Involvement of the glutathione system. *J. Neurochem.* **59**, 99–106.
70. Qiu, J.-X., Grafe, M. R., Schmura, S. M., Glasgow, J., Kent, T. A., Rassin, D. K., and Perez-Polo, J. R. (2001). Differential NF-κB regulation of bcl-x gene expression in hippocampus and basal forebrain in response to hypoxia. *J. Neurosci. Res.* **64**, 223–234.
71. von Bartheld, C. S., Byers, M. R., Williams, R., and Bothwell, M. (1996). Anterograde transport of neurotrophins and axodendritic transfer in the developing visual system. *Nature* **379**, 830–833.
72. Smith, M. A., Zhang, L.-X., Lyon, W. E., and Mamounas, L. A. (1997). Anterograde transport of endogenous brain-derived neurotrphic factor in hippocampal mossy fibers. *Neuro Report* **8**, 1829–1834.
73. Tao, H. W. and Poo, M. (2001). Retrograde signaling at central synapses. *Proc. Natl. Acad. Sci. USA* **98**, 11009–11015.
74. Gall, C. M. and Isackson, P. J. (1989). Limbic seizures increase production of messenger RNA for nerve growth factor. *Science* **245**, 758–761.
75. Ernfors, P., Bengzon, J., Kokaia, Z., Persson, H., and Lindvall, O. (1991). Increased levels of messenger RNAs for neurotrophic factors in the brain during kindling epileptogenesis. *Neuron* **7**, 165–176.
76. Neeper, S. A., Gomez-Pinilla, F., Choi, J., and Cotman, C. W. (1996). Physical activity increases mRNA for brain-derived neurotrophic factor and nerve growth factor in rat brain. *Brain Res.* **726**, 49–56.
77. Bibel, M., Hoppe, E., and Barde, Y. A. (1999). Biochemical and functional interactions between the neurotrophin receptors trk and p75NTR. *EMBO J.* **18**, 616–622.
78. Meakin, S. O. and Shooter, E. M. (1992). The nerve growth factor family of receptors. *Trends Neurosci.* **15**, 323–331.
79. Dobrowsky, R. T., Werner, M. H., Castellino, A. M., Chao, M. V., and Hannum, Y. A. (1994). Activation of the sphingomyelin cycle through the low-affinity neurotrophin receptor. *Science* **265**, 1596–1599.
80. Carter, B. D, Kaltschmidt, C., Kaltschmide, B., Offenhäuser, N., Böhm-Matthaei, R., Baeuerle, P. A., and Barde, Y.-A. (1996). Selective activation of NF-kappaB by nerve growth factor through the neurotrpohin receptor p75. *Science* **272**, 542–545.
81. Taglialatela, G., Robinson, R., and Perez-Polo, J. R. (1997). Inhibition of nuclear factor kappa B (NF-kB) activity induces NGF-resistant apoptosis in PC12 cells. *J. Neurosci. Res.* **47**, 155–162.
82. Taglialatela, G., Kauffman, J. A., Trevino, A., and Perez-Polo, J. R. (1998). Central nervous system DNA fragmentation induced by the inhibition of nuclear factor Kappa B. *Neuro Report* **9**, 489–493.
83. Taglialatela, G., Hibbert, C. J., Werrbach-Perez, K., and Perez-Polo, J. R. (1996). Suppression of $p140^{trkA}$ does not abolish nerve growth factor rescue of apoptotic PC12 cells and enables brain-derived neurotrophic factor to promote cell survival. *J. Neurochem.* **66**, 1826–1835.
84. Hirata, H., Hibasami, H., Yoshida, T., Ogawa, M., Matsumoto, M., Morita, A., and Uchida, A. (2001). Nerve growth factor signaling of p75 induces differentiation and ceramide-mediated apoptosis in Schwann cells cultured from degenerating nerves. *GLIA* **36**, 245–258.
85. Dechant, G. (2001). Molecular interactions between neurotrophin receptors. *Cell Tissue Res.* **305**, 229–238.
86. Miller, F. D. and Kaplan, D. R. (2001). On Trk for retrograde signaling. *Neuron* **32**, 767–770.

87. Radeke, M. J., Misko, T. P., Hsu, C., Herzenberg, L. A., and Shooter, E. M. (1987). Gene transfer and molecular cloning of the rat nerve growth factor receptor. *Nature* **325**, 593–597.
88. Rodriguez-Tebar, A., Dechant, G., and Barde, Y.-A. (1990). Binding of brain-derived neurotrophic factor to the nerve growth factor receptor. *Neuron* **4**, 487–492.
89. McDonald, N. Q. and Chao, M. V. (1995). Structural determinants of neurotrophin action. *J. Biol. Chem.* **270**, 19669–19672.
90. Middlemas, D. S., Lindberg, R. A., and Hunter, T. (1991). trkB, a neural receptor protein-tyrosine kinase: Evidence for a full-length and two truncated receptors. *Mol. Cell Biol.* **11**, 143–153.
91. Smeyne, R. J., Klein, R., Schnapp, A., Long, L. K., Brynat, S., Lewin, A., Lira, S. A., and Barbacid, M. (1994). Severe sensory and sympathetic neuropathy in mice carrying a disrupted trk/NGF receptor gene. *Nature* **365**, 246–249.
92. Crowley, C., Spencer, S. D., Nishimura, M. C., Chen, K. S., Pitts-Meek, S., Armanini, M. P., Ling, L. H., McMahon, S. B., Shelton, D. L., Levinson, A. D., and Phillips, H. S. (1994). Mice lacking nerve growth factor display perinatal loss of sensory and sympathetic neurons yet develop basal forebrain cholinergic neurons. *Cell* **76**, 1001–1011.
93. Jones, K. R., Fariñas, I., Backus, C., and Reichardt, L. F. (1994). Targeted disruption of the BDNF gene perturbs brain and sensory neuron development but not motor neuron development. *Cell* **76**, 989–999.
94. Conover, J. C., Erickson, J. T., Katz, D. M., Bianchi, L. M., Poueymirou, W. T., McClain, J., Pan, L., Helgren, M., Ip, N. Y., Boland, P., Friedman, B., Weigand, S., Vejsada, R., Kato, A. C., DeChiara, T. M., and Yancopoulos, G. D. (1995). Neuronal deficits, not involving motor neurons, in mice lacking BDNF and/or NT4. *Nature* **375**, 235–238.
95. Ernfors, P., Lee, K. F., and Jaenisch, R. (1994). Mice lacking brain-derived neurotrophic factor develop with sensory deficits. *Nature* **368**, 147–150.
96. Ernfors, P., Lee, K. F, Kucera, J., and Jaenisch, R. (1994). Lack of neurotrophin-3 leads to deficiencies in the peripheral nervous system and loss of limb proprioceptive afferents. *Cell* **77**, 503–512.
97. Koliatsos, V. E., Price, D. L., Gouras, G. K., Cayouette, M. H., Burton, L. E., and Winslow, J. W. (1994). Highly selective effects of nerve growth factor, brain-derived neurotrphic factor, and neurotrophin-3 on intact and injured basal forebrain magnocellular neurons. *J. Comp. Neurol.* **343**, 247–262.
98. Naumann, T., Casademunt, E., Hollerbach, E., Hofmann, J., Dechant, G., Frotscher, M., and Barde, Y.-A. (2002). Complete deletion of the neurotrophin receptor $p75^{NTR}$ lead to long-lasting increases in the number of basal forebrain cholinergic neurons. *J. Neurosci.* **22**, 2409–2418.
99. Rossner, S., Schliebs, R., Perez-Polo, J. R., Wiley, R.G., and Bigl, V. (1995). Differential changes in cholinergic markers from selected brain regions after specific immunolesion of rat cholinergic basal forebrain system. *J. Neurosci. Res.* **40**, 31–43.
100. Yu, J., Wiley, R. G., and Perez-Polo, J. R. (1996). Altered NGF protein levels in different brain areas after immunolesion. *J. Neurosci. Res.* **43**, 213–223.
101. Yu, J., Pizzo, D. P., Hutton, L. A., and Perez-Polo, J. R. (1995). Role of the cholinergic system in the regulation of neurotrophin synthesis. *Brain Res.* **705**, 247–254.
102. Rossner, S., Schliebs, R., Härtig, W., Perez-Polo, J. R., and Bigl, V. (1997). Selective induction of c-Jun and NGF in reactive astrocytes after cholinergic degenerations in rat basal forebrain. *Neuro Report* **8**, 2199–2202.
103. Rossner, S., Woertwein, G., Gu, Z., Yu, J., Schliebs, R., Bigl, V., and Perez-Polo, J. R. (1997). Cholinergic control of nerve growth factor in adult rats: Evidence from cortical cholinergic deafferentation and chronic drug treatment. *J. Neurochem.* **69**, 947–953.
104. Gu, Z., Yu, J., and Perez-Polo, J. R. (1998). Long term changes in brain cholinergic markers and NGF levels after partial immunolesion. *Brain Res.* **801**, 190–197.
105. Miller, F. D. and Kaplan, D. R. (2001). Neurotrophin signaling pathways regulating neuronal apoptosis. *Cell Mol. Life Sci.* **58**, 1045–1053.
106. Bamji, S. X., Majdan, M., Pozniak, C. D., Belliveau, D. J., Aloyz, R., Kohn, J. *et al.* (1998). The p75 neurotrophin receptor mediates neuronal apoptosis and is essential for naturally occurring sympathetic neurons death. *J. Cell Biol.* **140**, 911–923.
107. Davey, F. and Davies, A. M. (1998). TrkB signaling inhibits p75-mediated apoptosis induced by nerve growth factor in embryonic proprioceptive neurons. *Curr. Biol.* **8**, 915–918.
108. Friedman, W. (2000). Neurotrophins induce death of hippocampal neurons via the p75 receptor. *J. Neurosci.* **20**, 6340–6346.
109. Casaccia-Bonnefil, P., Carter, B. C., Dobrowsky, R. T., and Chao, M. V. (1996). Death of oligodendrocytes mediated by the interaction of nerve growth factor with its receptor p75. *Nature* **383**, 716–719.
110. Yoon, S. O., Carter, B. D., Casaccia-Bonnefil, P., and Chao, M. V. (1998). Competitive signaling between TrkA and p75 nerve growth factor receptors determined cell survival. *J. Neurosci.* **18**, 3273–3281.
111. Jing, S. Q., Tapley, P., and Barbacid, M. (1992). Nerve growth factor mediates signal transduction through trk homodimer receptors. *Neuron* **9**, 1067–1079.
112. Kaplan, D. R. and Stephens, R. M. (1994). Neurotrophin signal transduction by the Trk receptor. *J. Neurobiol.* **25**, 1404–1417.
113. Glasgow, J. N., Wood, T., and Perez-Polo, J. R. (2000). Identification and characterization of NF-κB binding sites in the murine bcl-x promoter. *J. Neurochem.* **75**, 1377–1389.
114. Macdonald, N. J., Perez-Polo, J. R., Bennett, A. D., and Taglialatela, G. (1999). NGF-resistant PC12 cell death induced by arachidonic acid is accompanied by a decrease of active PKC ζ and nuclear factor Kappa B. *J. Neurosci. Res.* **57**, 219–226.
115. Khursigara, G., Orlinick, J. R., and Chao, M. V. (1999). Association of p75 neurotrophin receptor with TRAF6. *J. Biol. Chem.* **274**, 2597–2600.
116. Aloyz, R. S., Bamji, S. X., Pozniak, C. D., Toma, J. G., Atwal, J., Kaplan, D. R., and Miller, F. D. (1998). P53 is essential for developmental neuron death as regulated by the TrkA and p75 neurotrophin receptors. *J. Cell Biol.* **143**, 1691–1703.
117. Yang, D. D., Kuan, C., Whitmarsh, A. J., Rincon, M., Zheng, T. S., Davis, R. J., Rakic, P., and Flavell, R. A. (1997). Absence of excitotoxicity-induced apoptosis in the hippocampus of mice lacking the Jnk3 gene. *Nature* **389**, 865–870.
118. Morrison, R. S., Wenzel, H. J., Kinoshita, Y., Robbins, C. A., Donehower, L. A., and Schwartzkroin, P. A. (1996). Loss of the p53 tumor suppressor gene protects neurons from kainate-induced cell death. *J. Neurosci.* **16**, 1337–1345.
119. Vaillant, A. R., Mazzoni, I., Tudan, C., Boudreau, M., Kaplan, D. R., and Miller, F. D. (1999). Depolarization and neurotrophins converge on the phosphatidylinositol 3-kinase-Akt pathway to synergistically regulate neuronal survival. *J. Cell Biol.* **146**, 955–966.
120. Hetman, M., Kanning, K., Cavanaugh, J. E., and Xia, Z. (1999). Neuroprotection by brain-derived neurotrophic factor is mediated by extracellular signal-regulated kinase and phosphatidylinositol 3-kinase. *J. Biol. Chem.* **274**, 22569–22580.
121. Blair, L. A., Bence-Hanulec, K. K., Mehta, S., Franke, T., Kaplan, D., and Marshall, J. (1999). Akt-dependent potentiation of L channels by insulin-like growth factor-1 is required for Neuronal survival. *J. Neurosci.* **19**, 1940–1951.
122. Vanhaesebroeck, B. and Alessi, D. R. (2000). The PI3K-PDk1 connection: More than just a road to PKB. *Biochem. J.* **346**, 561–576.
123. Michaelidis, T. M., Sendtner, M., Cooper, J. D., Airaksinen, M. S., Holtman, B., Meyer, M., and Thoenen, H. (1996). Inactivation of bcl-2 results in progressive degeneration of motoneurons, sympathetic and sensory Neurons during early postnatal development. *Neuron* **17**, 75–89.
124. Liu, Y.-Z., Boxer, L. M., and Latchman, D. S. (1999). Activation of the Bcl-2 promoter by nerve growth factor is mediated by the p42/p44 MAPK cascade. *Nucleic Acids Res.* **27**, 2086–2090.

125. Brivanlou, A. H. and Darnell, J. E. (2002). Signal transduction and the control of gene expression. *Science* **295**, 813–818.
126. Angel, P. and Karin, M. (1991). The role of Jun, Fos and the AP-1 complex in cell proliferation and transformation. *Biochim. Biophys. Acta* **1072**, 129–157.
127. Hagmeyer, B. M., König, I. H., Offring, R., Zantema, A. J., Herrlich, P., and Angel, P. (1993). Adenovirus E1A negatively and positively modulates transcription of AP-1 dependent genes by dicer-specific regulation of the DNA binding and transactivation activities of Jun. *EMBO J.* **12**, 3559–3572.
128. Colotta, F., Polentarutti, N., Sironi, M., and Mantovani, A. (1992). Expression and involvement of c-fos and c-jun protooncogenes in programmed cell death induced by growth factor deprivation in lymphoid cell lines. *J. Biol. Chem.* **267**, 18278–18283.
129. Ham, J., Babij, C., Whitfield, J., Pfarr, C. M., Lallemand, D., Yaniv, M., and Rubin, L. L. (1995). A c-Jun dominant negative mutant protects sympathetic neurons against programmed cell death. *Neuron* **14**, 927–939.
130. Batistatou, A., Volont'e, C., and Greene, L. A. (1992). Nerve growth factor employs multiple pathways to induce primary response genes in PC12 cells. *Mol. Biol. Cell* **3**, 363–371.
131. Visvader, J., Sassone, M. A., and Verma, I. M. (1988). Two adjacent promoter elements mediate nerve growth factor activation of the c-fos gene and bind distinct nuclear complexes. *Proc. Natl. Acad. Sci. USA* **85**, 9474–9478.
132. Tong, L, Werrbach-Perez-Polo, K., and Perez-Polo, J. R. (1999). Prolonged activation of transcription factor AP-1 during NGF-mediated rescue in apoptotic cell death in PC12 cells. *Neurochem. Res.* **24**, 1431–1441.
133. Tong, L., Toliver-Kinsky, T., Rassin, D., Werrbach-Perez, K., and Perez-Polo, J. R. (2002). Hyperoxia increases AP-1 DNA binding in rat brain. *Neurochem. Res.* (in press).
134. Schlingensiepen, K. H., Schlingensiepen, R., Kunst, M., Klinger, I., Gerdes, W., Seifert, W., and Brysch, W. (1993). Opposite functions of jun-B and c-jun in growth regulation and neuronal differentiation. *Dev. Genet.* **14**, 305–312.
135. Minden, A., Lin, A., McMahon, M., Lange-Carter, C., Derijard B., Davis, R. J., Johnson, G. L., and Karin, M. (1995). Differential activation of ERK and JNK mitogen-activated protein kinases by Raf-1 and MEKK. *Science* **266**, 1719–1723.
136. Beg, A. and Baltimore, D. (1996). An essential role for NF-kappa B in preventing TNF-alpha-induced cell death. *Science* **274**, 782–784.
137. Bours, V., Franzoso, G., Azarenko, V., Park, S., Kanno, T., Brown, K., and Siebenlist, U. (1993). The oncoprotein Bcl-3 directly transactivates through kappa B motifs via association with DNA-binding p50B homodimers. *Cell* **72**, 729–739.
138. Liou, H.-C. and Baltimore, D. (1993). Regulation of the NF-kappa B/rel transcription factor and I kappa B inhibitor system. *Curr. Opin. Cell Biol.* **5**, 477–487.
139. Ghosh, S. and Baltimore, D. (1990). Activation *in vitro* of NF-kappa B by phosphorylation of its inhibitor, I kappa B. *Nature* **344**, 678–682.
140. Beg, A., Ruben, S., Scheinman, R., Haskill, S., Rosen, C., and Baldwin, A. J. (1992). I kappa B interacts with the nuclear localization sequences of the subunits of NF-kappa B: A mechanism for cytoplasmic retention. *Genes Dev.* **6**, 1899–1913.
141. Rice, N., MacKichen, M., and Israel, A. (1992). The precursor to NF-kappa B p50 has I kappa B-like functions. *Cell* **71**, 243–253.
142. Inoue, J., Kerr, L., Kakizuka, A., and Verma, I. (1992). I kappa B gamma, a 70 Kd protein identical to the c-terminal half of p110 NF-kappa B: A new member of the I kappa B family. *Cell* **68**, 1109–1120.
143. Whiteside, S., Epinat, J., Rice, N., and Israel, A. (1997). I kappa B epsilon, a novel member of the I kappa B family, controls RelA and cRel activity. *EMBO J.* **16**, 1413–1426.
144. Zandi, E., Chen, Y., and Karin, M. (1998). Direct phosphorylation of I kappa B by IKK-alpha and IKK-beta: Discrimination between free and NF-kappa B bound substrate. *Science* **281**, 1360–1363.
145. Beg, A., Finco, T., Nantermet, P., and Jr, B. A. (1993). Tumor necrosis factor and interleukin-1 lead to phosphorylation and loss of I kappa B alpha: A mechanism for NF-kappa B activation. *Mol. Cell Biol.* **13**, 3310–3310.
146. Yaron, A., Hatzubai, A., Davis, M., Lavon, I., Amit, S., Manning, A., Anderson, J., Mann, M., Mercurio, F., and Ben-Neriah, Y. (1998). Identification of the receptor component of the I kappa B alpha-ubiquitin ligase. *Nature* **396**, 590–594.
147. Didonato, J., Mercurio, F., and Karin, M. (1995). Phosphorylation of I kappa B alpha precedes but is not sufficient for its dissociation from NF-kappa B. *Mol. Cell Biol.* **15**, 1302–1311.
148. Torgerson, T., Colosia, A., Donahue, J., Lin, Y.-Z., and Hawiger, J. (1998). Regulation of NF-kappa B, AP-1, NFAT, and STAT1 nuclear transport in T lymphocytes by noninvasive delivery of peptide carrying the nuclear localization sequence of NF-kappa B p50. *J. Immunol.* **161**, 6084–6092.
149. Perkins, N., Schmidt, R., Duckett, C., Leung, K., Rice, N., and Nabel, G. (1992). Distinct combination of NF-kappa B subunits determine the specificity of transcriptional activation. *Proc. Natl. Acad. Sci. USA* **89**, 1529–1533.
150. Toliver-Kinsky, T., Wood, T., and Perez-Polo, J. R. (2000). Nuclear factor kappa B/p52 is a negative regulatory factor in nerve growth factor-induced choline acetyltransferase promoter activity in PC12 cells. *J. Neurochem.* **75**, 2241–2251.

Chapter 339

Cell–Cell and Cell–Matrix Interactions in Bone

L. F. Bonewald

University of Missouri at Kansas City, School of Dentistry, Kansas City, Missouri

Introduction

Far from being the static, hard skeleton hanging in the anatomy classroom, the skeleton within the body is dynamic and constantly responding to internal and external forces. The internal forces include cytokines, growth factors, and hormones, and the external force is response to muscle and to strain placed on the skeleton. In fact, the adult skeleton undergoes greater remodeling than other organs in the body. Several hormones have been shown to play important roles in the skeleton such as the estrogens and androgens, parathyroid hormone (PTH), and 1,25-dihydroxyvitamin vitamin D_3 (1,25-D_3). The three major bone cell types, the osteoclasts, osteoblasts, and osteocytes, are in constant communication with each other and with cells of the immune and hemopoietic systems. Not only are bone cells in constant communication with cells of other systems but also with the extracellular matrix (ECM) composed of osteoid, nonmineralized bone tissue and the mineralized bone matrix. Although previously viewed as mainly a support structure for bone cells, it is now clear that the bone ECM controls and directs bone cell function.

Diseases of Bone

Manifestation of bone disease is usually later or slower than manifestation of disease in other organs. For example, bone cancer is usually discovered after manifestation in other tissues such as breast or lung. Bone cancer such as osteosarcoma usually does not present until fracture or pain occurs, often after the cancer is fully entranced and difficult, if not impossible, to cure. Another example is osteoporosis in which bone loss can occur over decades before being identified and treated. At present, the only hope of treatment for many bone malformations is surgery. Therefore, a greater understanding of normal bone function and pathology is required for the design of preventive therapy and for treatment of disease.

Osteoporosis

Osteoporosis has become a major medical problem as the world population ages. Bone strength is reduced in the postmenopausal female and in both sexes with aging. Bone strength is a function of size, connectivity of trabecular structures, level of remodeling, and the intrinsic strength of the bone itself. Osteoporosis is defined as "the condition of generalized skeletal fragility in which bone strength is sufficiently weak that fractures occur with minimal trauma, often no more than is applied by routine daily activity" [1]. *Primary osteoporosis* is a disorder of postmenopausal women and of older men and women. *Secondary osteoporosis* occurs due to clinical disorders such as endocrinopathies and genetic diseases or to drugs such as in glucocorticoid-induced bone loss. As the U.S. population ages, osteoporosis takes a greater and greater toll in terms of both suffering and economic cost. Each year osteoporosis is the underlying basis for 1.5 million fractures. These cause not only pain and morbidity, but also diminish the quality of life for these individuals, as they lose their independence. Hip fracture results in up to 24% mortality, 25% of hip fracture patients require long-term care, and only a third regain their prefracture level of independence [2]. Treatment for osteoporosis includes hormone therapy and the use of bone resorption inhibitors such as calcitonin and the bisphosphonates. First-, second-, and third-generation bisphosphonates have been developed based on when the drug became available, its structure, and potency. These are not only used for the treatment of osteoporosis but also to prevent bone loss due to glucocorticoid use and Paget's disease. These compounds are useful not only to treat bone loss but also to treat and reduce bone metastasis in metastatic breast cancer and multiple myeloma [3].

Skeletal Malignancies

It has become clear that tumor cells use disruption or enhancement of normal cell–cell and cell–ECM interactions to enhance their own growth and metastasis. A prime example is multiple myeloma in which the tumor cells express an integrin complex, VLA4, that allows it to home to bone marrow [4,5]. Myeloma is characterized by extensive bone destruction; therefore, efforts are under way to prevent metastasis to bone through the use of agents that block myeloma–bone ECM interactions. Breast, lung, and prostate cancer preferentially metastasize to bone [6]. Factors that may play a role in osteoblastic bone metastasis include fibroblast growth factors (FGFs), transforming growth factor β (TGF-β), platelet-derived growth factor (PDGF), and, more recently, endothelin-1 [7]. Systemic syndromes can be associated with such cancers; these include leukocytosis and hypercalcemia. A well-studied factor clearly associated with hypercalcemia is parathyroid hormone-related peptide (PTHrp) that is normally produced by keratinocytes, uterus, placenta, and mammary tissue. This factor mimics PTH action by binding to the same receptor. Other factors implicated in osteolytic bone loss due to malignancy include interleukin (IL)-1, IL-6, tumor necrosis factor α (TNF-α), RANKL, and MIP-1α [8]. These factors are discussed in more detail later.

Bone Malformations and Genetic Defects

A baby with craniofacial defects is born every hour, and 75% of all malformations seen at birth fall under this category. Cleft lip/palate occurs in 1 out of 500 live births, and the cost of treating one of these children over a lifetime is more than $100,000. Osteogenesis imperfecta is a heritable disease of bone characterized by recurring bone fractures. It is caused by mutations affecting the structure of the type 1 collagen molecule and is the most common single gene defect causing bone disease [9]. Mutations in type 1 collagen can cause moderate disease or can be lethal during the perinatal period.

As bone growth proceeds through the growth plate in long bones, defects or mutations in a number of factors essential in this process lead to chondrogenic dysplasias. Jansen's metaphyseal chondrodysplasia and Blomstrand's lethal chondrodysplasia are due to mutations in the PTH receptor [10]. The mutations responsible for other genetic disorders have not yet been identified, such as those responsible for Paget's disease of bone, a common disorder with variable clinical presentations [11]. Complications of this disease include deformations of skull, face, and lower extremities, pain, degenerative arthritis, hearing loss, hypercalcemia, and hyperuricemia. It has been proposed that individuals with this disease are predisposed to the effects of Paramyxoviridae viruses on their osteoclasts, however, this is still controversial [12].

Bone Cells and Their Functions

Osteoclasts

The sole function of the osteoclast is to resorb bone. The mature osteoclast is described histologically as a multinucleated, tartrate-resistant acid phosphatase (TRAP) positive cell. However, macrophage polykaryons can have these same characteristics, so the "gold standard" for identifying an osteoclast is the formation of resorption lacunae or "pits" on a mineralized surface. Other characteristics of the osteoclast include the expression of calcitonin receptors, enzymes such as cathepsin K, and matrix metalloproteinase-9 (MMP-9), which play a role in matrix degradation, and the vacuolar proton pump for the transport of protons to the resorption lacunae. For the osteoclast to resorb, it must form a sealing zone around the periphery of its attached area to concentrate its secreted proteases and protons into a limited area. Underneath the cell, a ruffled border is formed and in this region the pH is reduced to approximately 2–3, which enhances the degradation of mineralized matrix. [13].

Osteoclast precursors are derived from hematopoietic precursors, the same stem cells that become granulocytes and monocytes/macrophages. Cell lines such as RAW 267.4 and MOPC-5 are available that represent osteoclast precursors, because these cells can form TRAP positive multinucleated cells that resorb bone [14]. (However, one must be careful when interpreting results from these cell lines because it appears that only a portion of these cells can fully differentiate into osteoclasts.) It has been well known for the last 10–15 years in the bone field that osteoclast precursors require supporting cells for osteoclast formation. The importance of macrophage colony-stimulating factor (M-CSF) as a supporter of proliferation of osteoclast precursors has been determined [15]. Critical factors and cell surface molecules involved in osteoclast formation have only recently been elucidated with the discovery of RANK ligand (RANKL) and osteoprotegerin (OPG) [16,17]. The osteoclast precursor expresses a receptor known as RANK (receptor activator of NFκB) that signals through the NFκB pathway. The binding of the cell membrane-bound ligand, RANKL, activates RANK receptor. However, a soluble factor, OPG, acting as a "decoy" receptor can bind to RANKL, preventing osteoclast formation. The expression of RANKL on the surface of supporting cells occurs when these cells are exposed to bone resorbing cytokines, hormones, and factors such as IL-1, IL-6, IL-11, PTH, PTHrp, oncostatin M, leukemia inhibitory factor, prostaglandin E_2, or 1,25-D_3 [18]. These factors up-regulate RANKL to a level capable of overcoming the effects of circulating OPG, thereby resulting in osteoclast formation. Efforts to generate osteoclasts without supporting cells have only recently been accomplished *in vitro* by using an artificial, soluble form of RANKL [19].

Osteoblasts

The formation of bone matrix on bone-forming surfaces has been well studied. The osteoblast is derived from an as yet to be identified precursor stem cell of mesenchymal origin and in cell culture behaves similarly to fibroblasts, except for its ability to form mineralized matrix. The osteoblast undergoes three major phenotypically identifiable stages of differentiation that Pockwinse and coworkers [20] have characterized as proliferation, matrix production, and maturation in which mineralization occurs. During the proliferation phase, high-level expression of c-fos and histone H4 occurs, and during matrix production, transforming growth factor β type 1 (TGF-β1) and type 1 collagen. During the maturation phase these proteins decrease and the expression of alkaline phosphatase, osteopontin, and core binding factor 1 (Cbfa1) increases. During the mineralization phase, these previous proteins decline in expression, whereas proteins such as osteocalcin increase with increases in mineralized bone formation. Osteoblast cells in each of these phases are often described as early preosteoblasts, proliferating osteoblasts, mature osteoblasts, and finally preosteocytes/osteocytes within the mineralized matrix.

Osteocytes

The osteocyte is the bone cell that we know least about and, therefore, the function of this cell is the focus of numerous hypotheses. These potential functions include the capacity to regulate calcium homeostasis, to respond to mechanical strain, and to send signals of bone formation or bone resorption to the bone surface. These functions are proposed to be accomplished through gap junctions, through the secretion of factors, through glutamate receptors and through the direct dendritic contact with cells on the bone surface [21].

Osteocytes are terminally differentiated osteoblasts, making up the majority, more than 90%, of all bone cells. During osteocyte ontogeny, the matrix producing osteoblast becomes either a lining cell or a preosteocyte embedded in the newly formed osteoid. These preosteocytes produce factors that locally inhibit mineralization (such as osteocalcin) and form a lacunae around the main body of the osteocyte and canaliculi around the dendritic processes [22,23]. A mature osteocyte is defined as a cell surrounded by mineralized bone and is described as a stellate or star-shaped cell with a large number of slender, cytoplasmic processes radiating in all directions but generally perpendicular to the bone surface. Osteocytes first attracted the attention of electron microscopists because of their extensive networks within the mineralized bone matrix that connect the embedded osteocytes to form an extensive syncytium that also connects with cells on the surface of bone. Mature osteocytes are most likely coupled by GAP junctions and appear to be linked to lining cells by the same connections [24]. Osteocytes are attached to mineralized matrix via integrins [25]. Osteocytes have been shown to die by apoptosis due to lack of estrogen or high levels of glucocorticoids [26,27], suggesting that osteocyte viability is important for bone integrity. These properties support the hypothesis that osteocytes function as a network of mechanosensory cells in bone that can mediate their effects through their extensive communication network.

Mechanical Strain

Julius Wolff in 1892 was the first to suggest that bone accommodates or responds to strain. To paraphrase Wolff's law, the law of bone remodeling, alteration of internal and external architecture occurs as a consequence of the stressing of bone. In general, athletes such a wrestlers and chronic exercisers such as tennis players have higher bone mineral density and mass than matched, nonexercising controls. Astronauts subjected to long periods of weightlessness during space flight lose bone. The cells of bone with the potential for sensing mechanical strain and translating these forces into biochemical signals include bone lining cells, osteoblasts, and osteocytes.

The skeleton adapts to mechanical usage. When the skeleton is not used, as in immobilization, bone is lost. During growth the skeleton is in "mild overload," which results in bone modeling and resulting new bone formation. When growth ceases, muscle strength is no longer in overload and bone strain is therefore reduced to the adapted level of strain. The estimated levels of microstrain for each "window" or level of strain have been determined by *in vivo* animal and human experiments. At less than 100 υE, only resorption occurs, and this has been called the "disuse window." The "adapted window" is where remodeling (resorption followed by formation) occurs, between 100 and 1000 υE. This adapted window can be raised or lowered depending on the hormonal environment. The "mild overload window" is where modeling (formation only) occurs between 1000 and 3000 υE and this only occurs in growing animals. The "pathologic overload window" is where microdamage (resorption followed by formation) occurs at strains greater than 3000 υE. Microdamage can occur in race horses and recruits during basic training. Microdamage due to pathological overload results in rapid resorption followed by formation at areas of microdamage. Fracture strain is approximately 25,000. [28].

It has been proposed that bone possesses a *mechanostat*, a mechanism whereby bone can reset its response to particular levels of strain [29,30]. Hormones have been proposed to lower the mechanostat, that is, addition of hormones such as estrogen, PTH, or 1,25-D_3 can lower the magnitude of strain necessary to induce a response [31–33]. For example, if bone normally responds to 2000 υE with an increase in modeling, in the presence of hormone, the same response would occur to lower strain levels, say, 1000 υE. However, it has been shown that the skeleton cannot respond optimally to mechanical strain without the presence of estrogen; therefore, estrogen is essential for the normal response of bone and not for just resetting the mechanostat [34]. Therefore, for normal skeletal growth and maintenance, both biochemical signals and mechanical strain are essential. For the skeleton to optimally respond to mechanical strain, hormones must be present and, conversely, bone cannot develop normally in the absence of strain.

Hormones Responsible for Bone Development, Growth, and Maintenance

A number of hormones play a role in the maintenance of bone, including estrogen, progesterone, aldosterone, androgens, vitamin A, and glucocorticoids. These hormones' actions are mediated by hormone-activated transcription factors belonging to the superfamily of ligand-dependent nuclear receptors. These nuclear receptors, upon binding to ligand, can form homodimers or heterodimers at specific DNA-binding sites. They also interact with a wide array of other transcription factors, as well as general and specific coregulatory proteins. The complexity of these interactions leads to intrinsic specificity of gene regulation, but the exact determinants of specificity are still under study.

Estrogen

Estrogen is clearly the major sex hormone affecting growth, remodeling, and homeostasis of the skeleton. Reduction in estrogen levels that occurs with menopause or through ovariectomy can result in bone loss. However, estrogen also plays a role in skeletal integrity in the male. Strong support for this comes from the clinical reports of two human males, one with a mutation in the alpha isomer of the estrogen receptor (ER-α) gene, causing partial estrogen resistance, and another individual with aromatase p450 deficiency causing complete estrogen deficiency. They both showed continued longitudinal bone growth due to delayed epiphyseal growth plate ossification and osteopenia [35,36]. Therefore, there is increased clinical interest in the use of selective estrogen receptor modulators (SERMS) in the treatment of bone loss. These synthetic ligands appear to have greater specificity for estrogen receptor in bone and, therefore, the undesired side effects of estrogen in other tissues is avoided while bone mass is maintained.

Estrogen appears to have direct effects on osteoblasts, osteocytes, and osteoclasts. Estrogen is a viability factor for osteoblasts and osteocytes, but appears to induce apoptosis of osteoclasts [37,38]. Estrogen also appears to down-regulate the production of several factors that play a role in bone resorption such as IL-1, TNF, and IL-6 [39]. Therefore, considerable interest has been directed to inhibitors of these cytokines such as their receptor antagonists.

There are two forms of the ER, designated α and β, and both forms are found in bone. The phenotypes of mice lacking either or both of these receptors are complicated and show sexual dimorphism. One receptor isoform can partially compensate for the other. Also, it appears that estrogen can have opposite effects on mice compared to humans, especially on longitudinal bone growth where estrogen enhances long bone growth in rodents but causes epiphyseal closure in humans. Homozygous deletion of ER-α results in reduced cortical bone formation and density in both male and female mice [40]. Female mice still lose bone with ovariectomy and estrogen responsiveness is reduced. Studies with female estrogen receptor β (ER-β) knock-out (–/–) mice indicate that ER-β is involved in the regulation of trabecular bone during adulthood by suppressing bone resorption, whereas this is not the case for male mice where no effect is seen [41]. Both sexes exhibit delayed growth plate closure. Mice with both ER isoforms deleted generate a similar skeleton in the male as the ER-α knock-out, but in contrast, females exhibit a more pronounced phenotype with reduced cortical thickness and trabecular bone density [42]. It is clear that ER-α and ER-β perform different functions in cortical and trabecular bone and that these functions differ between the sexes.

1α,25-Dihydroxyvitamin D_3

Vitamin D is well known to prevent rickets; however, the compounds ergocalciferol (vitamin D_2) and cholecalciferol (vitamin D_3), are really prohormones that are converted to the biologically active form, 1α,25$(OH)_2$vitamin D_3, which functions in a manner analogous to that of the steroid hormones [43]. Metabolism of the precursors occurs first in the liver and is completed in the kidney. The final active form then acts on the major target organs, bone, intestine and kidney. A vitamin D binding protein (DBP) transports the hormone in the circulation for delivery to cells, where the hormone is freed from its binding protein for binding to the vitamin D receptor (VDR), a canonical nuclear receptor. However, there is mounting evidence that 1,25-D_3 can also signal through membrane receptors [44, 45] and can have nongenomic effects [46]. These rapid responses, such as enhanced transport of Ca^{2+} and activation of protein kinase C and phospholipase A2, have been attributed to the interaction of the hormone with a membrane receptor and not with the canonical VDR. This membrane receptor remains to be completely characterized.

Hereditary vitamin D-resistant rickets is a rare autosomal recessive disease in which patients exhibit defective bone mineralization and hypocalcemia, due to decreased intestinal calcium absorption. These patients are unresponsive to 1,25-D_3 due to mutations in the VDR. Targeted disruption of this receptor in mice results in animals that appear normal at birth, but after weaning show growth retardation and alopecia [16,47]. These mice are also infertile, which suggests a role for vitamin D for gonadal function. Much attention has focused on 1α,25$(OH)_2$vitamin D_3, the major metabolite of vitamin D. The second metabolite 24R,25$(OH)_2D_3$, may also be important. Deletion of the hydroxylase necessary for the generation of this metabolite results in mice that show poor viability and bones with an accumulation of unmineralized matrix [48].

Parathyroid Hormone and PTH-Related Protein

PTH is responsible for calcium homeostasis in the body by its direct actions on bone, that of calcium release from bone, and enhanced calcium reabsorption from kidney, and indirectly by actions on the gastrointestinal tract, that of conversion of $25(OH)D_3$ to $1,25(OH)_2D_3$. The principal form is intact PTH (1-84) however, there are several circulating cleavage forms, the functions of which are not clear [49]. It is clear that both the intact and amino terminal forms of PTH bind to the PTH type 1 receptor, a G-protein-coupled, seven-transmembrane receptor that signals through cAMP and potentially also through PKC and calcium activation. Mutations in this receptor result in various forms of chondrodysplasias in humans that resemble the disorganized growth plate phenotype of mice lacking this receptor [50]. Whereas both PTH and PTHrp bind to the PTH type 1 receptor, a second receptor, PTH type 2, has been identified that responds to PTH, but not to PTHrp [51]. Recently, it has been clearly demonstrated that receptors also exist that are specific for carboxy terminal fragments of PTH [52].

PTH has been shown to have both anabolic and catabolic effects on bone *in vivo*, to stimulate activation of osteoclasts, and to either stimulate or inhibit osteoblast proliferation and matrix production *in vitro*. These effects of PTH have generated considerable debate about the role of PTH in bone remodeling. It appears that continuous PTH is responsible for resorption, whereas intermittent PTH can induce new bone formation [53]. Studies suggest that continuous exposure of cells results in down-regulation of PTH receptor. Injections of either PTH or PTHrp into humans and animals can result in new bone formation, and intermittent application can stimulate mineralization of osteoblasts *in vitro* [54,55].

PTHrp was first identified as the factor responsible for causing humoral hypercalcemia of malignancy, but soon afterward was identified in many normal tissues. Nonetheless, it was undetectable in the normal circulation [56]. This molecule shares homology with PTH at the amino acid level with 8 of the first 13 amino acids being identical and three-dimensional homology in regions 13–34. PTHrp mRNA is alternatively spliced to yield three isoforms of 139, 141, and 173 amino acids. PTHrp is required for development of cartilage, morphogenesis of mammary gland, and tooth eruption as demonstrated with tissue-specific rescue of in PTHrp null mice that normally die at birth [57]. PTHrp also plays a role in calcium transfer across the placenta and appears to play a role in smooth muscle contractility. Generally, bone cells appear to respond in a similar manner to PTHrp as to PTH.

Growth and Transcription Factors Responsible for Bone Development and Growth

Bone formation proceeds through two ossification processes: endochondral ossification and intramembranous ossification. The former process involves a cartilage intermediate and occurs during most of the skeletal ossification, postnatal growth, bone remodeling, and fracture repair processes. The second process, by which bones form from mesenchymal condensations without a cartilage intermediate, only occurs in some craniofacial bones. The process of chondrogenesis and osteogenesis is tightly regulated at specific times and sites. Several transcription factors are important in this process such as Indian hedgehog and Sonic hedgehog, and the FGFs and their receptors (see later discussion). Indian hedgehog, expressed in prehypertrophic and hypertrophic chondrocytes [58], couples chondrogenesis to osteogenesis through PTHrP-dependent and PTHrP-independent pathways [59,60]. The PTH signaling pathway also plays a critical role in growth plate development (see later discussion).

Core Binding Factor 1, a Master Gene for Bone

Cbfa1, also known as Pebp2a1, Aml3, and Runx2, was originally thought to be T-cell specific [61]. However, direct evidence that Cbfa1 is essential in bone and tooth development comes from Cbfa1 gene knock-out experiments [62]. In these mice, there is a total absence of bone, as well as arrested tooth development. The membranous bones of the skull are replaced by fibrous tissue and endochondral bone does not replace the cartilaginous skeleton. Heterozygous Cbfa1 mice express a phenotype that is similar to the clinical manifestations of cleidocranial dysplasia, in which functional mutations of the Cbfa1 gene have been identified. Cleidocranial dysplasia is characterized by hypoplasia/aplasia of the clavicles, patent fontanelles, supernumerary teeth, short stature, and changes in skeletal patterning and growth. Cbfa1 is the earliest and most specific marker of osteogenesis identified to date [63]. Several homeodomain transcription factors, such as Msx2, Dlx5, Bapx1, and Hoxa-2, have been suggested to regulate Cbfa1 expression.

A second transcription factor that is required for osteoblast differentiation during development that acts downstream of Cbfa1 is osterix (Osx) [64]. Mice lacking Osx have a similar phenotype to those lacking Cbfa1. Target genes for Cbfa1 and Osx include osteocalcin, collagen type 1, collagenase 3, TGF-β type 2 receptor, and other genes necessary for osteoblast function. Although Cbfa1 mRNA do not correlate with target gene regulation, phosphorylation of Cbfa1 protein does [65].

Transforming Growth Factor β

TGF-β1 is the prototype and the founding molecule for the TGF-β superfamily [21]. This family has grown to include more than 40 members including the TGF-β isoforms, the activins and inhibins, Mullerian inhibitory substance, growth differentiation factors (GDFs), and an ever-increasing number of bone morphogenetic proteins (BMPs). Members of this superfamily appear to mediate many key events in growth and development evolutionarily maintained from fruit flies to mammals. The actions of these proteins appear to be mediated through structurally similar serine/threonine kinase transmembrane receptors.

Members of the TGF-β superfamily bind to two distinct forms of serine/threonine kinase receptors, called type I and type II [66]. The constitutively active type II receptor initially binds to active TGF-β and upon binding subsequently associates with the type I receptor and phosphorylates it. The direct substrates for the phosphorylated type I receptor appear to be Smad-2 and Smad-3, also known as receptor-activated Smads (R-Smads), whereas negative regulators of the type I receptor include Smad-6 and Smad-7, the inhibitory Smads (I-Smads) [67]. The discovery and naming of the Smads originated from studies of the *dpp* signaling pathway in *Drosophila* [68]. Smad-2 or Smad-3 heterozygous mutant mice are viable, but the compound heterozygous Smad-2/Smad-3 mutant is lethal, suggesting a gene dosage effect and that the relative expression level of Smad-2 and-3 in the cell may influence the nature of the TGF-β response. Loss of Smad-3 results in a lower bone formation rate and osteopenia in mice [69]. Smad-4, also called a common mediator (Co-Smad), appears to bring the cytoplasmic Smad-2 and Smad-3 into the nucleus where together they can regulate transcription of target genes. Smad-4 was found to be homologous to a gene deleted in pancreatic carcinomas called *deleted in pancreatic cancer-4* or DPC-4 [70]. Smad-4 is not always required for TGF-β signaling, since a number of Smad-4 independent TGF-β responses have been identified [67]. These include Jun N-terminal kinase (JNK) and extracellular signal-related kinases (ERK) mitogen-activated protein (MAP) kinase pathways. The potential exists for other Co-Smads to be identified.

The major function of TGF-β in bone is as an induction of matrix formation. Mice lacking specific isoforms of TGF-β have boney defects. Injections of TGF-β can induce new bone formation or prevent bone loss, but inappropriate expression of TGF-β or its receptor can lead to bone loss. Even though TGF-β stimulates osteoid production, it actually inhibits mineralization of osteoid. Therefore, this factor must be activated in a specific time and tissue and then inactivated for normal bone remodeling to occur. TGF-β can enhance either bone formation or bone resorption depending on the assay system and the presence of other factors. TGF-β was proposed to be a "coupling" factor, coupling bone resorption with bone formation as this factor is released from the bone matrix where it is stored by resorbing osteoclasts [21,71].

TGF-β is well known among growth factors for its potent and widespread actions. Almost every cell in the body has been shown to make some form of TGF-β and almost every cell expresses receptors for TGF-β. The largest source of TGF-β in the body is bone. This growth factor must be tightly regulated to prevent disease. Appropriately, the mechanisms of regulation of TGF-β are extensive and complex. One unique set of regulatory mechanisms centers around the fact that TGF-β is produced in a latent form that must be activated to produce biologically active TGF-β. The mechanisms of regulation not only include regulation of the latency of the molecule, but the production of different latent forms, such as the small and large latent complexes, TGF-β targeting to matrix for storage or to cells for activation, and the various means of activation. The extracellular matrix protein, latent TGF-β binding protein (LTBP-1), appears to play a major role in the regulation of TGF-β (see later discussion).

Bone Morphogenetic Factors

Unlike the TGF-βs that can only induce bone when injected in proximity to existing bone, the BMPs can induce new bone formation when injected into muscle. Urist [72] was the first to describe bone regenerative capacity of bone extracts, but Celeste and coworkers [73] were the first to identify the factors responsible through the use of peptide sequences from these mixtures and then cloning the resulting recombinant DNA for *in vivo* studies to provide the first identification of these factors. The BMPs are more closely related to proteins involved in differentiation during embryogenesis than they are to the TGF-βs. In fact, although it is clear that these factors are important or essential for development, it is not clear if these factors play an important role in the adult skeleton. This will not be known until time- and tissue-specific null mice are generated.

At present it is known that deletion of BMP2 or BMP4 is embryonic lethal, whereas deletion of other BMPs is not so dramatic. Deletion of BMP7 results in mice with mild limb skeleton abnormalities, BMP6$^{-/-}$ mice appear normal, and BMP5$^{-/-}$ mice exhibit the short ear phenotype. These results suggest that BMPs in some cases can compensate for deletion of one member. BMP3$^{-/-}$ mice are normal and, in fact, have increased bone density that may explain why injection of recombinant BMP3 has never induced bone formation. The growth and differentiation factors (GDFs) are also members of the BMP family. GDF5$^{-/-}$ mice exhibit brachypodism, reduction in number of digits, and misshapen bones. Deletion of GFD11 leads to defects in skeletal patterning and palate abnormalities. Even deletion of the some of the receptors for BMPs (see later discussion) do not result in severe or lethal phenotypes, suggesting that the receptors can compensate for one another. Deletion of either the BMPR1B receptor or the ActRIIA receptor is not severe, whereas the deletion of the BMPR1A is embryonic lethal [74]. Negative regulators of the BMPs include the following: noggin for BMPs 2, 4, 7 and GDF 5 and 6; chordin for BMPs 2, 4, and 7; follistatin for BMPs 2, 4, 7, 11; and gremlin for BMP 2, 4, and GDF 5.

Like the TGF-βs, the BMPs signal through type I and type II receptors, however, BMPs can signal through type II receptors alone, yet this signal is enhanced when both receptors are engaged [75]. Seven type I receptors have been identified, called *activin receptor-like kinases* (ALKs), and three type II receptors. Members of the BMP family bind with different affinities to the type I and type II receptors adding to redundancy and complexity in signaling. BMP receptors also signal through the Smads. The R-Smads for BMPs include Smad-1, -5, and -8, (for TGF-β they are 2 and 3 as described above). Smad-4 is the only ad that is shared by both the BMPs and TGF-β. I-Smads are Smad-6 and -7. Smad-5-deficient mice have defects in angiogenesis and Smad-6 mice exhibit cardiac defects. Transcriptional corepressors of the Smads include TGIF, c-Ski, and SnoN. Target genes of the BMPs such as Tlx-2, a homeobox gene related to human HoxII, Dad (Daughters against Dpp), and Id gene products, are generally responsible for patterning and development.

Recombinant BMPs 2, 4, and 7 are being used for clinical studies to induce fracture repair, augmentation of alveolar bone, and for gene therapy. To date, none of these applications has been approved for general application, probably due to the difficulties in determining ideal doses, times for administration, and ideal carriers. However, the potential still exists for therapeutic treatment.

Insulin-Like Growth Factors

The IGFs were shown to be the mediators of the effects of growth hormone. IGF I and II are 7-kDa proteins that share homology with proinsulin. Originally it was found that IGF is made by the liver, but it has also been shown that osteoblasts produce this growth factor. Bone is the major storage organ for IGFs, and IGF II is the most abundant of all the growth factors stored in the skeleton [76]. Factors such as PTH, estrogen, prostaglandin E2, and BMP-2 will increase IGF expression in osteoblasts.

Animals with targeted overexpression of IGF in bone using the osteocalcin promoter have greater bone mineral density [77]. IGF-I-deficient neonates have a marked increase in death rate compared to IGF-II-deficient animals that have normal survival rates. Mice lacking the receptor die at birth. Regardless of whether ligand or receptor are deleted, the pups express normal morphogenesis [71]. This suggests that the major function of the IGFs are growth, not morphogenesis, as is true for other growth factors such as the BMPs and FGFs.

The actions of the IGFs appear to be tightly regulated by the IGF binding proteins. These are found in serum and in bone matrix. Six have been cloned and characterized. They bind with high affinity to the IGFs, preventing their interaction with the IGF receptor. IGFBP-1 can inhibit or enhance IGF action dependent on its phosphorylation state and may be responsible for suppression of bone formation in malnourished individuals. IGFPB-2 is the major binding protein secreted by osteoblasts. IGF-PB3 has both inhibitory and stimulatory activity depending on location within the cell. IGF-BP5 is not normally in the circulation, but is preferentially found in the bone matrix where it appears to be protected from proteases and appears to potentiate IGF activity. IGF-BP6 has a selective affinity for IGF II over IGF I. [71,78]. The binding proteins can be degraded by specific and nonspecific proteases, thereby adding another level of regulation of IGF activity. For example, cathepsin D will degrade IGFBPs 1–5, whereas pregnancy-associated plasma protein-A will specifically proteolyze IGFBP-4.

The IGFs appear to stimulate new bone formation *in vivo*, with little or no preliminary resorption phase. In animal models, IGF I can enhance longitudinal growth, bone formation, and bone mass in various but not all models. For example, growth can be restored in hypophysectomized rats, but little effect is seen in normal rats. It is clear that recombinant IGF I can enhance trabecular and cortical bone mineral density in humans with an impaired growth hormone–IGF axis, but such data are not available for normal or older adults. Therefore, at this time, use of IGF can only be justified in specific conditions such as growth hormone-resistant short stature [79].

Fibroblast Growth Factors

The FGFs were first so named for their ability to stimulate the growth of 3T3 fibroblasts. The FGF family has grown to include at least 23 genes [71,80]. The first two FGFs were called acidic and basic FGF, based on their isoelectric points, but have since been renamed FGF-1 and FGF-2. In addition to promoting cell growth, these factors can induce a mitogenic response, stimulate cell migration, angiogenesis, vasculogenesis, transformation, morphogenesis, wound healing, and tissue repair. FGF-2 and -3 are distinguished from all other growth factors by a novel translation initiation mechanism. Four high-molecular-weight isoforms of FGF-2 are initiated with an unconventional CUG translation codon, whereas a smaller 18-kDa isoform is initiated by the classical AUG codon. Another interesting feature of some of the FGFs, like IL-1, is their ability to be nonclassically secreted even though they do not contain hydrophobic signal peptide sequences. On the cell surface, the FGFs interact with at least three types of molecules, including four high-affinity signaling receptors (FGFRs 1–4), low-affinity receptors such as perlecan and syndecan that potentiate ligand/receptor interactions, and cysteine-rich nonsignaling receptors that may function to antagonize and remove ligand.

FGFs clearly stimulate new bone formation; however, injections of FGF cause serious side effects such as acute hypotension. Mutations in FGF receptors result in a number of human dysmorphic (dwarfism) syndromes such as achondroplasia, thanatophoric dysplasia, Jackson-Weiss syndrome, and Pfeiffer syndrome. The clearest indications that FGFs are important in bone development is revealed through the bone phenotypes of null mice lacking the FGF receptors, rather than mice lacking a particular FGF as compensation among this family of factors appears to occur. Disruption of FGF-2, however, results in decreased bone mass and bone formation [81]. Mice expressing activated *FGFR3* mutants reproduce the dwarfism phenotype of the chondroplasias and show a marked decrease in the proliferation rate of the columnar proliferating chondrocytes and a decrease in size of the zone of hypertrophic chondrocytes [82,83,84]. Thus, a normal function of FGF signaling in chondrocytes is to inhibit chondrocyte proliferation.

FGFs may have additional effects on the skeleton. The newest member of the family, FGF-23, appears to play a key role in hypophosphatemic disorders [85]. This FGF is produced by tumors that cause osteomalacia, and when injected into mice causes hypophosphatemic rickets. In all tumors causing hypophosphatemic osteomalacia, mutations around Ser180 have been identified, resulting in a noncleaved 32-kDa protein. In a condition of autosomal dominant hypophosphatemic rickets, FGF-23 is also not cleaved. It is presumed that under normal conditions, FGF-23 is cleaved at residue Ser180 and that these mutations may cause a gain of function for FGF-23. The mechanisms for the role of this FGF in hypophosphatemia are under intense investigation.

Low-Density Lipoprotein Receptor-Related Protein 5 as a High Bone Mass Gene

Recently, a mutation in the extracellular domain of the low-density lipoprotein receptor-related protein 5 (Lrp5) gene was shown to result in extremely high bone mass in a human cohort. These individuals essentially never break their bones and have no other clinical features, suggesting solely positive effects of this mutation [86]. However, mutations in the intracellular domain of this receptor result in a condition called osteoporosis pseudoglioma syndrome of juvenile onset. These individuals have osteoporosis and exhibit progressive blindness [87]. The extracellular mutation appears to result in constitutive activation and the intracellular mutation results in a loss of function. Surprisingly, these studies show the importance of Lrp5 in regulation of bone mass. This was unexpected because this protein is ubiquitously expressed and had only been associated with lipoproteins and liver function.

Lrp5 is a coreceptor with the seven-transmembrane receptor, frizzled, in the canonical Wnt signaling pathway [88]. The potential role of Wnt signaling in maintenance of bone density will be an exciting area of study, sure to involve other members of this pathway such as β-catenin, glycogen synthase kinase 3B, dishevelled, dickkopt-1, axin, and targets of this pathway such as members of the t-cell factor/lymphocyte enhancer factor family such as Cox-2, c-jun, and connexin 43 [89,90]. This gene may regulate bone mass during development and also may be responsive to mechanical strain. This is a novel area of investigation, because the role of Lrp5 in regulation of bone remodeling and signaling pathways is speculative at this time.

Bone Extracellular Matrix

The ECM and the proteins it contains have not received the same attention as other areas of bone biology such as cytokines, receptors, cell signaling, and transcription factors. This is partially due to the difficulty in determining the potential functions of large, extensively modified ECM proteins. Recently, more attention has focused on the extracellular components of bone due to the advent of new technologies. Transgenic animals and null mice have greatly assisted in determining the functions of these proteins [91]. Knock-outs of the ECM proteins lead to various bone defects, such as thickened bones in osteocalcin null mice, and an osteoporosis-like phenotype in biglycan null mice. On the other hand, no bone phenotype was observed in the decorin or osteonectin null mice [91]. Occasionally deleting a specific ECM protein gene, such as that for fibronectin, can result in embryonic lethality that, unfortunately, does not give information concerning the specific function of the matrix protein. In these cases, the phenotype of a transgenic animal or a deletion heterozygote can be more informative.

Although the major ECM protein in bone is collagen type 1, there are numerous noncollagenous proteins. These include proteoglycans such as decorin and biglycan that are characterized by glycosaminoglycans attached to core protein, chondroitin sulfate proteoglycans such as aggrecan and versican, glycoproteins such as osteonectin, vitronectin and thrompospondins, proteins containing γ carboxy glutamic acid such as matrix Gla protein and osteocalcin, and a group of proteins known as the SIBLINGS for small integrin-binding ligands with *N*-linked glycosylation [92]. Members of the SIBLINGS include osteopontin, bone sialoprotein, dentin matrix protein-1 (DMP-1), dentin sialophosphoprotein, and MEPE. Bone proteins are proposed to have a role in the mineralization process. Whereas deletion of osteocalcin, osteonectin, and bone sialoprotein has not resulted in significant changes in the bone phenotype, deletion of DMP-1 has dramatic effects on the growth plate, and the mice display a chondrodysplastic phenotype and dwarfism [93]. DMP-1 may also be important in osteocyte function [94,95].

Latent TGF-β Binding Proteins

Latent TGF-β binding proteins (LTBPs) appear to be an important mechanism whereby TGF-β is controlled. To date, four LTBP genes have been isolated (LTBPs 1–4) containing cysteine and EGF-like repeating domains. LTBPs are highly homologous to fibrillins 1 and 2, major constituents of connective tissue microfibrils. LTBP 2 does not bind latent TGF-β and therefore may be more homologous than LTBP 1, 3, and 4 to the fibrillins. The third eight-cysteine repeat in LTBP-1 forms a covalent disulfide bond with the TGF-β1 precursor or "latency associated peptide." The major isoform of TGF-β stored in bone matrix is TGF-β1 (80–90%) as part of a latent complex containing LTBP-1.

LTBP-1 does not confer latency to the TGF-β complex, but has other unique functions. The latent TGF-β complex produced by matrix-forming osteoblasts is targeted by LTBP-1 to fibrillar structures known as microfibrils in bone ECM [92]. Although LTBP-1 covalently associates with small latent TGF-β1, it is also produced by osteoblasts in a free form (80% of total) not associated with latent TGF-β1. This molar excess suggests a function separate and distinct from its association with TGF-β. Many ECM proteins contain EGF-like repeats that mediate protein–protein interactions, suggesting that LTBP-1 may have similar functions [21].

Deletion of LTBP 2 is lethal, whereas deletion of LTBP 3 results in mice with ossification of synchondroses at 2 weeks that normally do not ossify [97]. At 6 and 9 months, these animals develop osteosclerosis and osteoarthritis. To date, the LTBP-1 gene has not been successfully deleted.

Microfibrils

Recently the role of components of microfibrils in bone development has attracted attention. The components of microfibrils include fibronectin, fibrillins 1 and 2, elastin, microfibril-associated glycoprotein (MAGPs) 1 and 2, fibulin, and others [98,99]. Mice lacking the genes for the microfibril proteins often have a more dramatic bone phenotype than mice lacking genes for many of the bone-specific matrix proteins. For example, deletion of the bone-specific gene osteocalcin results in a modest bone phenotype, whereas deletion of LTBP 3 appears to have a more dramatic effect on bone [100].

The deletion of fibronectin, a major component of ECM and microfibrils, is a neonatal lethal and there are no known human mutations of this protein [96,97]. The fibrillin 1 knock-out does not appear to have a bone phenotype, whereas fragments of overexpressed fibrillin 1 result in mice with overgrowth of ribs and long bones [98]. It was actually suggested that microfibrils control bone growth in a negative fashion based on these results. The fibrillin 2 null mouse has a bone phenotype with contracture at birth that resolves with age, rear joints that do not flex, and fusion of three toes in the hind limbs into one phalange. The elastin null mouse dies at 4.5 days postpartum due to arterial obstruction but the heterozygote shows many features of supervalvolar aortic stenosis in humans [99]. Proteins that are components of microfibrils could function by physical influence through alteration of the physical properties of matrix, through indirect signaling by way of retaining and releasing growth factors such as TGF-β, and through presentation and binding of protein to receptors or signaling molecules on cell surfaces.

Components of microfibrils have strong protein–protein interactions and protein–cell interactions. Cells have been shown to bind to fibrillin through integrins [100]. MAGP-2 has an RGD sequence motif that modulates cell-to-microfibril interactions and binds to $\alpha v\beta 3$ integrins as does fibrillin-1 [101]. Fibulins all have long EGF repeats and bind to nidogen, aggrecan, versican, fibronectin, endostatin, collagen IV, laminin $\alpha 2$, and perlecan [102,103]. The matrix glycoproteins laminin, perlecan, tenascin, and nidogen all have calcium-binding EGF-like repeats. These calcium-binding repeats are necessary for stabilizing their tightly folded structures. Several growth factors have also been shown to bind to components of microfibrils, for example, connective tissue growth factor binds to fibrillin.

Matrix Metalloproteinases

An intricate balance between deposition and breakdown of ECM is critical for growth and development of bone, and significant progress has been made in understanding the roles of MMPs in the balance between osteoblasts and osteoclasts [104]. MMPs belong to a family of zinc- and calcium-dependent endopeptidases that catalyze the proteolysis of components of ECM at neutral pH. Each member has specificity for a particular subset of ECM components. The most important members are MMP-2 and MMP-9. Martignetti and colleagues [105] found that a human disease with an enhanced degradation of ECM, which features osteolytic lesions in facial bones, arthritis, and subcutaneous nodules, is due to the lack of a single proteolytic enzyme, MMP-2. Similarly, deficiency of mouse MT1-MMP, which activates MMP-2, results in a decrease of collagen breakdown by osteoblasts, a decrease in bone formation, and an increase in the number of osteoclasts [76,106]. MMP-13 is predominantly expressed in the skeleton, and null mice have elongated growth plates and reduced bone resorption, suggesting that MMP13 directly or indirectly inhibits chondrocyte growth and stimulates osteoclastogenesis [107] while overexpression of MMP-13 leads to osteoarthritis [108].

Conclusions

This chapter merely touches on many important areas in bone research, as room is not available for review of others topics such as the bone resorbing cytokines including IL-1, -6, and -11, tumor necrosis factors, and regulatory factors such as arachidonic acid metabolites including the prostaglandins, nitric oxide and its regulatory enzymes, and so on. Although many of the factors involved in regulation of bone function are similar to those in other organs, a level of complexity is added due to the mineralized nature of bone. Hematopoietic and immune cells have been well characterized because they are relatively easy to obtain, but this is not the case for bone cells, especially for osteocytes embedded in bone. Investigators in the bone field have often referred back to the areas of hematology, immunology, and development to understand the potential role of factors in bone. Determining the function of matrix proteins in bone has relied heavily on studies in other tissues, such as cartilage, skin, and other connective tissues. However, bone biologists are not able to rely on studies in other organs with regard to mineralization, the unique feature of the skeleton.

In summary, bone is a storehouse of factors ready to be released during resorption that can modify the bone coupling process or provide circulating growth factors. A number of transcription factors have been identified that are specific for bone induction and development. Clearly these growth factors and transcription factors are regulated by a number of circulating hormones such as parathyroid hormone, estrogen, and $1,25(OH)_2$ vitamin D_3. Another layer of complexity is added as bone is also regulated by mechanical strain. Understanding the normal physiology of bone and its diseases should lead to prevention and treatment of disease, acceleration and initiation of repair, and treatment or reversal of abnormal development.

References

1. Marcus, R. and Majumder, S. (2001). The nature of osteoporosis. In Marcus, R., Feldman, D., and Kelsey, J., Eds., *Osteoporosis,* Vol. 2, pp. 3–1. Academic Press, San Diego.
2. National Osteoporosis Foundation (2002). America's bone health: The state of osteoporosis and low bone mass in our ation. *Fighting Osteoporosis & Promoting Bone Health.*

3. Brown, D. L. and Robbins, R. (1999). Developments in the therapeutic applications of bisphosphonates. *J. Clin. Pharmacol.* **39**, 651–660.
4. Michigami, T., Shimizu, N., Williams, P. J., Niewolna, M., Dallas, S. L., Mundy, G. R., and Yoneda, T. (2000). Cell–cell contact between marrow stromal cells and myeloma cells via VCAM-1 and alpha4beta1 integrin enhances production of osteoclast stimulating activity. *Blood* **96**, 1953–1960.
5. Teoh, G. and Anderson K. C. (1997). Interaction of tumor and host cells with adhesion and extracellular matrix molecules in the development of multiple myeloma. *Hematol. Oncol. Clinics North Am.* **11**, 27–42.
6. Mundy, G. R. and Martin T. J. (1993). Pathophysiology of skeletal complications of cancer. In Mundy, G., and Martin, T. J., Eds., *Physiology and Pharmacology of Bones: Handbook of Experimental Pharmacology*, Vol. 18, pp. 642–647. Springer-Verlag, Berlin Heidelberg, Germany.
7. Yin, J. J., Grubbs, B. G., Cui, Y., Wu-Wong, J. R., Wessale, J., Padley, R. J., and Guise, T. A. (2000). Endothelin A receptor blockade inhibits osteoblastic metastases. *J. Bone Miner. Res.* **15**, 1254.
8. Mundy, G. R., Toshiyuki, Y., Guise, T. A., and Oyajobi, B. (2002). Local factors in skeletal malignancy. In Bilezikian, J. P., Raisz, L. G., and Rodan, G. A., Eds., *Principles of Bone Biology*, Vol. 2, pp. 1093–1104, Academic Press, San Diego.
9. Rowe, D. W. (2002). Osteogenesis imperfecta. In Bilezikian, J. P., Raisz, L. G., and Rodan, G. A., Eds., *Principles of Bone Biology*, Vol. 2, pp. 1177–1193, Academic Press, San Diego.
10. Juppner, H. S. E. and Silve, C. (2002). Jansen's metaphyseal chondrodysplasia and Blomstrand's lethal chondrodysplasia: two genetic disorders caused by PTH/PTHrP receptor mutation. In Bilezikian, J. P., Raisz, L. G., and Rodan, G. A., Eds., *Principles of Bone Biology*, Vol. 2, pp. 1117–1135, Academic Press, San Diego.
11. Singer, F. R. and Krane, S. M. (1998). Paget's disease of bone. In Avioli, L. V., Krane, S. M., Eds., *Metabolic Bone Disease*, pp. 546–615. Saunders, Philadelphia.
12. Kurihara, N., Reddy, S. V., Menaa, C., and Rodman, G. D. (2000). Osteoclasts formed by normal human bone marrow cells transduced with the measles virus nucleocapsid gene express a pagetic phenotype. *J. Clin. Invest.* **105**, 607–614.
13. Roodman, G. D. (1996). Advances in bone biology: The osteoclast. *Endocrinol. Rev.* **17**, 308–332.
14. Chen, W. and Li, Y. P. (1998). Generation of mouse osteoclastogenic cell lines immortalized with SV40 large T antigen. *J. Bone Miner. Res.* **13**, 1112–1123.
15. Kodama, H., Nose, M., Niida, S., and Yamasaki, A. (1991). Essential role of macrophage colony-stimulating factor in the osteoclast differentiation supported by stromal cells. *J. Exp. Med.* **173**, 1291–1294.
16. Yoshizawa, T., Handa, Y., Uematsu, Y., Takeda, S., Sekine, K., Yoshihara, Y., Kawakami, T., Arioka, K., Sato, H., Uchiyama, Y., Masushige, S., Fukamizu, A., Matsumoto, T., and Kato, S. (1997). Mice lacking the vitamin D receptor exhibit impaired bone formation, uterine hypoplasia and growth retardation after weaning. *Nat. Genet.* **16**, 391–396.
17. Simonet, W. S., Lacey, D. L., Dunstan, C. R., Kelley, M., Chang, M. S., Luthy, R., Nguyen, H. Q., Wooden, S., Bennett, L., Boone, T., Shimamoto, G., DeRose, M., Elliott, R., Colombero, A., Tan, H. L., Trail, G., Sullivan, J., Davy, E., Bucay, N., Renshaw-Gegg, L., Hughes, T. M., Hill, D., Pattison, W., Campbell, P., and Boyle, W. J. (1997). Osteoprotegerin: A novel secreted protein involved in the regulation of bone density. *Cell* **89**, 309–319.
18. Aubin, J. E. and Bonnelye, E. (2000). Osteoprotegerin and its ligand: A new paradigm for regulation of osteoclastogenesis and bone resorption. *Medscape Womens Health* **5**, 5.
19. Quinn, J. M., Elliott, J., Gillespie, M. T., and Martin, T. J. (1998). A combination of osteoclast differentiation factor and macrophage-colony stimulating factor is sufficient for both human and mouse osteoclast formation in vitro. *Endocrinology* **139**, 4424–4427.
20. Pockwinse, S. M., Wilming, L. G., Conlon, D. M., Stein, G. S., and Lian, J. B. (1992). Expression of cell growth and bone specific genes at single cell resolution during development of bone tissue-like organization in primary osteoblast cultures. *J. Cell Biochem.* **49**, 310–323.
21. Bonewald, L. F. (2002). Osteocytes: A proposed multifunctional bone cell. *J. Musculoskeletal Neuronal Interact.* **2**, 239–241.
22. Marotti, G., Cane, V., Palazzini, S., and Palumbo, C. (1990). Structure–function relationships in the osteocyte. *Ital. J. Min. Electrolyte Metab.* **4**, 93–106.
23. Marotti, G. (1996). The structure of bone tissues and the cellular control of their deposition. *Ital. J. Anat. Embryol.* **101**, 25–79.
24. Doty, S. B. (1981). Morphological evidence of gap junctions between bone cells. *Calcified Tissue Int.* **33**, 509–512.
25. van der Plas, A., Aarden, E. M., Feijen, J. H., de Boer, A. H., Wiltink, A., Alblas, M. J., de Leij, L., and Nijweide, P. J. (1994). Characteristics and properties of osteocytes in culture. *J. Bone Miner. Res.* **9**, 1697–1704.
26. Weinstein, R. S., Jilka, R. L., Parfitt, A. M., and Manolagas, S. C. (1998). Inhibition of osteoblastogenesis and promotion of apoptosis of osteoblasts and osteocytes by glucocorticoids. Potential mechanisms of their deleterious effects on bone. *J. Clin. Invest.* **102**, 274–282.
27. Tomkinson, A., Reeve, J., Shaw, R. W., and Noble, B. S. (1997). The death of osteocytes via apoptosis accompanies estrogen withdrawal in human bone. *J. Clin. Endocrinol. Metabol.* **82**, 3128–3135.
28. Frost, H. M. (1992). Perspectives: Bone's mechanical usage windows. *Bone Miner.* **19**, 257–271.
29. Frost, H. M. (1987). Bone "mass" and the "mechanostat": A proposal. *Anat. Rec.* **219**, 1–9.
30. Martin, R. B. and Burr, D. B. (1989). *Structure, Function and Adaptation of Compact Bone.* Raven Press, New York.
31. Turner, C. H., Riggs, B. L., and Spelsberg, T. C. (1994). Skeletal effects of estrogen. *Endocrinol. Rev.* **15**, 275–300.
32. Cheng, M. Z., Zaman, G., and Lanyon, L. E. (1994). Estrogen enhances the stimulation of bone collagen synthesis by loading and exogenous prostacyclin, but not prostaglandin E2, in organ cultures of rat ulnae. *J. Bone Miner. Res.* **9**, 805–816.
33. Cheng, M. Z., Zaman, G., Rawlinson, S. C., Suswillo, R. F., and Lanyon, L. E. (1996). Mechanical loading and sex hormone interactions in organ cultures of rat ulna. *J. Bone Miner. Res.* **11**, 502–511.
34. Damien, E., Price, J. S., and Lanyon, L. E. (2000). Mechanical strain stimulates osteoblast proliferation through the estrogen receptor in males as well as females. *J. Bone Miner. Res.* **15**, 2169–2177.
35. Carani, C., Qin, K., Simoni, M., Faustini-Fustini, M., Serpente, S., Boyd, J., Korach, K. S., and Simpson, E. R. (1997). Effect of testosterone and estradiol in a man with aromatase deficiency. *N. Engl. J. Med.* **337**, 91–95.
36. Smith, E. P., Boyd, J., Frank, G. R., Takahashi, H., Cohen, R. M., Specker, B., Williqms, T. C., Lubahn, D. B., and Korach, K. S. (1994). Estrogen resistance caused by a mutation in the estrogen receptor gene in a man. *N. Engl. J. Med.* **331**, 1056–1061.
37. Hughes, D. E., Dai, A., Tiffee, J. C., Li, H. H., Mundy, G. R., and Boyce, B. F. (1996). Estrogen promotes apoptosis of murine osteoclasts mediated by TGF-beta. *Nat. Med.* **2**, 1132–1136.
38. Kousteni, S., Bellido, T., Plotkin, L. I., O'Brien, C. A., Bodenner, D. L., Han, L., Han, K., DiGregorio, G. B., Katzenellenbogen, J. A., Katzenellenbogen, B. S., Roberson, P. K., Weinstein, R. S., Jilka, R. L., and Manolagas, S. C. (2001). Nongenotropic, sex-nonspecific signaling through the estrogen or androgen receptors: Dissociation from transcriptional activity. *Cell* **104**, 719–730.
39. Pacifici, R. (1996). Estrogen, cytokines and pathogenesis of postmenopause osteoporosis. *J. Bone Miner. Res.* **11**, 1043–1051.
40. Korach, K. S., Taki, M., and Kimbro, K. S. (1997). *The Effects of Estrogen Receptor Gene Disruption on Bone.* Kluwer Academic and Fondazione Giovanni Lorenzini.
41. Vidal, O., Lindberg, M. K., Hollberg, K., Baylink, D. J., Andersson, G., Lubahn, D. B., Mohan, S., Gustafsson, J. A., and Ohlsson, C. (2000). Estrogen receptor specificity in the regulation of skeletal growth and maturation in male mice. *Proc. Natl. Acad. Sci. USA*, **97**, 5474–5479.
42. Sims, N. A., Dupont, S., Resche-Rogon, M., Clement-Lacroix, P., Bouali, Y., DaPonte, F., Galien, R., Gaillard-Kelly, M., and Baron, R. (2000). *In vivo* analysis of male and female estrogen receptor a, b, and double knockouts reveals a dual role for ER beta in bone remodelling. *J. Bone Miner. Res.* **15**, S160.

43. Norman, A. W. (1996). Vitamin D. In Ziegler, E. E., and Filer, L. J., Eds., *Present Knowledge in Nutrition (PKN7).* International Life Sciences Institute, Washington.
44. Nemere, I., Yoshimoto, Y., and Norman, A. W. (1984). Studies on the model of action of calciferol. LIV. Calcium transport in perfused duodena from normal chicks: Enhancement with 14 minutes of exposure to 1 alpha, 25-dihydroxvitamin D3. *Endocrinology* **115**, 1476–1483.
45. Pedrozo, H. A., Schwartz, Z., Rimes, S., Sylvia, V. L., Nemere, I., Posner, G. H., Dean, D. D., and Boyan, B. D. (1999). Physiological importance of the 1, 25(OH)2D3 membrane receptor and evidence for a membrane receptor specific for 24, 25(OH)2D3. *J. Bone Miner. Res.* **14**, 856–876.
46. Boyan, B. B., Swartz, Z., Snyder, S. P., Dean, D. D., Yang, F., Twardzik, D., and Bonewald, L. F. (1994). Latent transforming growth factor beta is produced by chondrocytes and activated by extracellular matrix vesicles upon exposure to 1, 25(OH)2D3. *J. Biol. Chem.* **269**, 28374–28381.
47. Li, Y. C., Pirro, A. E., Amling, M., Delling, G., Baroni, R., and Demay, M. B. (1997). Targeted ablation of the vitamin D receptor: An animal model of vitamin D-dependent rickets type II with alopecia. *Proc. Natl. Acad. Sci. USA* **94**, 9831–9835.
48. St. Arnaud, R., Arabian, A., Travers, R., and Glorieux, F. H. (1997). Abnormal intramembranous ossification in mice deficient for the vitamin D 24-hydroxylase. In Norman, A. W., Bouillon, R., Thomasset, M., Eds., *Vitamin D: Chemistry, Biology and Clinical Application of the Steroid Hormone,* pp. 635–644. University of California Press.
49. Hock, J. M., Fitzpatric, L. A., and Bilezikian, J. P. (2002). Actions of parathyroid hormone. In Bilezikian, J. P., Raisz, L. G., and Rodan, G. A., Eds., *Principles of Bone Biology,* Vol. 1, pp. 463–481, Academic Press, San Diego.
50. Lanske, B. (1999). Ablation of the PTHrP gene or the PTH/PTHrP receptor gene leads to distinct abnormalities in bone development. *J. Clin. Invest.* **104**, 399–407.
51. Juppner, H. (1996). Receptors for parathyroid hormone and parathyroid hormone-related peptide: from molecular cloning to definition of diseases. *Curr. Opin. Nephrol. Hypertension* **5**, 300–306.
52. Divieti, P. (2001). Receptors for the carboxyl-terminal region of pth(1–84) are highly expressed in osteocytic cells. *Endocrinology* **142**, 916–925.
53. Schaefer, F. (2000). Pulsatile parathyroid hormone secretion in health and disease. *Novartis Foundation Symp.* **227**, 225–239; and Discussion, 239–243.
54. Bauer, E., Aub, J., and Algright, F. (1929). Studies of calcium and phosphorus metabolism. V. A study of the bone trabeculae as a readily available reserve supply of calcium. *J. Exp. Med.,* **49**, 145–162.
55. Ishizuya, T., Yokose, S., Hori, M., Noda, T., Suda, T. Yoshiki, S., and Yamaguchi, A. (1997). Parathyroid hormone exerts disparate effect on osteoblast differentiation depending on exposure time in rat osteoblastic cells. *J. Clin. Invest.* **99**, 2961–2970.
56. Philbrick, W. M. (1996). Defining the roles of parathyroid hormone-related protein in normal physiology. *Physiol. Rev.* **76**, 127–173.
57. Karaplis, A. C., Luz, A., Glowacki, J., Bronson, R. T., Tybulewicz, V. L., Kronenberg, H. M., and Mulligan, R. C. (1994). Lethal skeletal dysplasia from targeted disruption of the parathyroid hormone-related peptide gene. *Genes Dev.* **8**, 277–289.
58. Bitgood, M. J. and McMahon, A. P. (1995). Hedgehog and Bmp genes are coexpressed at many diverse sites of cell–cell interaction in the mouse embryo. *Dev. Biol.* **172**, 126–138.
59. Karp, S. J., Schipani, E., St-Jacques, B., Hunzelman, J., Kronenberg, H., and McMahon, A. P. (2000). Indian hedgehog coordinates endochondral bone growth and morphogenesis via parathyroid hormone related-protein-dependent and -independent pathways. *Development* **127**, 543–548.
60. Chung, U. I., Schipani, E., McMahon, A. P., and Kronenberg, H. M. (2001). Indian hedgehog couples chondrogenesis to osteogenesis in endochondral bone development. *J. Clin. Invest.* **107**, 295–304.
61. Ogawa, E., Maruyama, M., Kagoshima, H., Inuzuka, M., Lu, J., Satake, M., Shigesada, K., and Ito, Y. (1993). PEBP2/PEA2 represents a family of transcription factors homologous to the products of the *Drosophila* runt gene and the human AML1 gene. *Proc. Natl. Acad. Sci. USA* **90**, 6859–6863.
62. Ducy, P., Zhang, R., Geoffroy, V., Ridall, A. L., and Karsenty, G. (1997). Osf2/Cbfa1: a transcriptional activator of osteoblast differentiation. *Cell* **89**, 747–754.
63. Karsenty, G. (2000). How many factors are required to remodel bone? *Nat. Med.* **6**, 970–971.
64. Nakashima, K., Zhou, X., Kunkel, G., Zhang, Z., Deng, J. M., Behringer, R. R., and de Crombrugghe, B. (2002). The novel zinc finger-containing transcription factor osterix is required for osteoblast differentiation and bone formation. *Cell* **108**, 17–29.
65. Xiao, G., Jiang, D., Thomas, P., Benson, M. D., Guan, K., Karsenty, G., and Franceschi, R. T. (2000). MAPK pathways activate and phosphorylate the osteoblast-specific transcription factor, Cbfa1. *J. Biol. Chem.* **275**, 4453–4459.
66. Heldin, C. H., Miyazono, K., and ten Dijke, P. (1997). TGF-beta signalling from cell membrane to nucleus through SMAD proteins. *Nature* **390**, 465–471.
67. de Caestecker, M. P., Piek, E., and Roberts, A. B. (2000). Role of transforming growth factor-beta signaling in cancer. *J. Natl. Cancer Inst.* **92**, 1388–1402.
68. Massague, J. (1998). TGF-beta signal transduction. *Annu. Rev. Biochem.* **67**, 753–791.
69. Borton, A. J., Frederick, J. P., Datto, M. B., Wang, X. F., and Weinstein, R. S. (2001). The loss of Smad3 results in a lower rate of bone formation and osteopenia through dysregulation of osteoblast differentiation and apoptosis. *J. Bone Miner. Res.* **16**, 1754–1764.
70. Hahn, S. A. and Schmiegel, W. H. (1998). Recent discoveries in cancer genetics of exocrine pancreatic neoplasia. *Digestion* **59**, 493–501.
71. Bonewald, L. F. and Dallas, S. L. (1998). The role of growth factors in bone formation. In *Advances in Oral Biology,* Vol. 5B, pp. 591–613. JAI Press, New York, New York.
72. Urist, M. R. (1965). Bone: Formation by autoinduction. *Science* **150**, 893–899.
73. Celeste, A. J., Iannazzi, J. A., Taylor, R. C., Hewick, R. M., Rosen, V., Wang, E. A., and Wozney, J. M. (1990). Identification of transforming growth factor beta family members present in bone-inductive protein purified from bovine bone. *Proc. Natl. Acad. Sci. USA* **87**, 9843–9847.
74. Rosen, V. and Wozney, J. M. (2002). Bone morphogenetic proteins. In Bilezikian, J. P., Raisz, L. G., and Rodan, G. A., Eds., *Principles of Bone Biology,* Vol. 2, pp. 919–928, Academic Press, San Diego.
75. Kohei, M. (2002). Bone morphogenetic protein receptor and actions. In Bilezikian, J. P., Raisz, L. G., and Rodan, G. A., Eds., *Principles of Bone Biology,* Vol. 2, pp. 929–942, Academic Press, San Diego.
76. Mohan, S., Linkhart, T. A., Jennings, J. C., and Baylink, D. J. (1987). Identifaction and quantification of four distinct growth factors stored in human bone matrix. *J. Bone Miner. Res.* **2**, 44–47.
77. Zhao, G., Monier-Faugere, M. C., Langub, M. C., Geng, Z., Nakayama, T., Pike, J. W., Chernausek, S. D., Rosen, C. J., Donahue, L. R., Malluche, H. H., Fagin, J. A., and Clemens, T. L. (2000). Targeted overexpression of insulin-like growth factor I to osteoblasts of transgenic mice: increased trabecular bone volume without increased osteoblast proliferation. *Endocrinology* **141**, 2674–2682.
78. Conover, C. A. and Rosen, C. (2002). The role of insulin-like growth factors and binding proteins in bone cell biology. In Bilezikian, J. P., Raisz, L. G., and Rodan, G. A., Eds., *Principles of Bone Biology,* Vol. 2, pp. 801–815, Academic Press, San Diego.
79. Wuster, C. and Rosen, C. (2001). Growth hormone, insulin-like growth factors: potential applications and limitations in the management of osteoporosis. In Marcus, R., Feldman, D., and Kelsey, J., Eds., *Osteoporosis,* Vol. 2, pp. 747–767. Academic Press, San Diego.
80. Hurley, M. M., Marie, P. J., and Florkiewicz, R. Z. (2002). Fibroblast growth factor (FGF) and FGF receptor families in bone. In Bilezikian, J. P., Raisz, L. G., and Rodan, G. A., Eds., *Principles of Bone Biology,* Vol. 1, pp. 825–851, Academic Press, San Diego.
81. Montero, A., Okada, Y., Tomita, M., Ito, M., Tsurukami, H., Nakamura, T., Doetschman, T., Coffin, J. D., and Hurley, M. M. (2000). Disruption of

the fibroblast growth factor-2 gene results in decreased bone mass and bone formation. *J. Clin. Invest.* **105**, 1085–1093.

82. Iwata, T., Li, C. L., Deng, C. X., Francomano, C. A. (2001). Highly activated Fgfr3 with the K644M mutation causes prolonged survival in severe dwarf mice. *Hum. Mol. Genet.* **10**, 1255–64.
83. Naski, M. C., Ornitz, D. M. (1998). FGF signaling in skeletal development. *Front. Biosci.* **3**, D781–94.
84. Chen, L., Li, C., Qiao, W., Xu, X., Deng, C. (2001). A Ser (365)->Cys mutation of fibroblast growth factor receptor 3 in mouse down regulates Ihh/PTHrP signals and causes severe achondroplasia. *Hum. Mol. Genet.* **10**, 457–65.
85. Thakker, R. V. (2001). Hereditary hypophosphataemic rickets: Role for a fibroblast growth factor, FGF23. IBMS BoneKEy. Website.
86. Little, R. D., Carulli, J. P., Del Mastro, R. G., Dupuis, J., Osborne, M., Folz, C., Manning, S. P., Swain, P. M., Zhao, S. C., Eustace, B., Lappe, M. M., Spitzer, L., Zweier, S., Braunschweiger, K., Benchekroun, Y., Hu, X., Adair, R., Chee, L., FitzGerald, M. G., Tulig, C., Caruso, A., Tzellas, N., Bawa, A., Franklin, B., McGuire, S., Nogues, X., Gong, G., Allen, K. M., Anisowicz, A., Morales, A. J., Lomedico, P. T., Recker, S. M., Van Eerdewegh, P., Recker, R. R., and Johnson, M. L. (2002). A mutation in the LDL receptor-related protein 5 gene results in the autosomal dominant high-bone-mass trait. *Am. J. Hum. Genet.* **70**, 11–19.
87. Gong, Y., Vikkula, M., Boon, L., Liu, J., Beighton, P., Ramesar, R., Peltonen, L., Somer, H., Hirose, T., Dallapiccola, B., De Paepe, A., Swoboda, W., Zabel, B., Superti-Furga, A., Steinmann, B., Brunner, H. G., Jans, A., Boles, R. G., Adkins, W., van den Boogaard, M. J., Olsen, B. R., and Warman, M. L. (1996). Osteoporosis-pseudoglioma syndrome, a disorder affecting skeletal strength and vision, is assigned to chromosome region 11q12–13. *Am. J. Hum. Genet.* **59**, 146–151.
88. Wehrli, M., Dougan, S. T., Caldwell, K., O'Keefe, L., Schwartz, S., Vaizel-Ohayon, D., Schejter, E., Tomlinson, A., and DiNardo, S. (2000). Arrow encodes an LDL-receptor-related protein essential for Wingless signalling. *Nature* **407**, 527–530.
89. Mann, B., Gelos, M., Siedow, A., Hanski, M. L., Gratchev, A., Ilyas, M., Bodmer, W. F., Moyer, M. P., Riecken, E. O., Buhr, H. J., and Hanski, C. (1999). Target genes of beta-catenin-T cell-factor/lymphoid-enhancer-factor signaling in human colorectal carcinomas. *Proc. Natl. Acad. Sci. USA* **96**, 1603–1608.
90. Mao, J., Wang, J., Liu, B., Pan, W., Farr, G. H., 3rd, Flynn, C., Yuan, H., Takada, S., Kimelman, D., Li, L., and Wu, D. (2001). Low-density lipoprotein receptor-related protein-5 binds to Axin and regulates the canonical Wnt signaling pathway. *Mol. Cell* **7**, 801–809.
91. Lian, J. B., Stein, G. S., Canalis, E., Robey, P., and Boskey, A. L. (1999). Osteoblast lineage cells, growth factors, matrix proteins, and the mineralization process. In *Primer on the Metabolic Bone Diseases and Disorders of Mineral Metabolism: Bone Formation,* pp. 14–19. Lippincott, Williams & Wilkins. Philadelphia.
92. Robey, P. G. (2002). Bone matrix proteoglycans and glycoproteins. In Bilezikian, J. P., Raisz, L. G., and Rodan, G. A., Eds., *Principles of Bone Biology,* Vol. 1, pp. 225–237, Academic Press, San Diego.
93. Feng, J. Q., Ye, L., Huang, H., Lu, Y., Zhang, J., Li, G., Dallas, S., Harris, S., Bonewald, L., and Mishina, Y. (2002). DMP-1 deficient mice develop dwarfism, chondrodysplasia, and disorganized bone remodeling and mineralization during postnatal development. *J. Bone Miner. Res.* **17**, S127 Abstract.
94. Toyosawa, S., Shintani, S., Fujiwara, T., Ooshima, T., Sato, A., Ijuhin, N., and Komori, T. (2001). Dentin matrix protein 1 is predominantly expressed in chicken and rat osteocytes but not in osteoblasts. *J. Bone Miner. Res.* **16**, 2017–2026.
95. Gluhak-Heinrich, J., Bonewald, L., Feng, J. Q., MacDougall, M., Harris, S. E., and Pavlin, D. (2003). Mechanical loading stimulates dentin matrix protein 1 (DMP1) expression in osteocytes *in vivo. J. Bone Miner. Res.* **18**, 807–817.
96. Dallas, S. L., Keene, D. R., Bruder, S. P., Saharinen, J., Sakai, L. Y., Mundy, G. R., and Bonewald, L. F. (2000). Role of the latent transforming growth factor beta binding protein 1 in fibrillin-containing microfibrils in bone cells in vitro and *in vivo. J. Bone Miner. Res.* **15**, 68–81.
97. Dabovic, B., Chen, Y., Colarossi, C., Óbata, H., Zambuto, L., Perle, M. A., Rifkin, D. B. (2002). Bone abnormalities in latent TGFβ binding protein (LTBP)-3 null mice indicate a role for LTBP-3 in modulating TGFβ availability. *J. Cell Biol.* **156**, 227–32.
98. Christiano, A. M. and Uitto, J. (1994). Molecular pathology of the elastic fibers. *J. Invest. Dermatol.* **103**, 53S–57S.
99. Rosenbloom, J., Abrams, W. R., and Mecham, R. (1993). Extracellular matrix 4: The elastic fiber. *FASEB J.* **7**, 1208–1218.
100. Dabovic, B., Chen, Y., Colarossi, C., Obata, H., Zambuto, L., Perle, M. A., and Rifkin, D. B. (2002). Bone abnormalities in latent TGF-beta binding protein (Ltbp)-3-null mice indicate a role for Ltbp-3 in modulating TGF-beta bioavailability. *J. Cell Biol.* **156**, 227–232.
101. George, E. L., Baldwin, H. S., and Hynes, R. O. (1997). Fibronectins are essential for heart and blood vessel morphogenesis but are dispensable for initial specification of precursor cells. *Blood* **90**, 3073–3081.
102. George, E. L., Georges-Labouesse, E. N., Patel-King, R. S., Rayburn, H., and Hynes, R. O. (1993). Defects in mesoderm, neural tube and vascular development in mouse embryos lacking fibronectin. *Development* **119**, 1079–1091.
103. Pereira, L., Lee, S. Y., Gayraud, B., Andrikopoulos, K., Shapiro, S. D., Bunton, T., Biery, N. J., Dietz, H. C., Sakai, L. Y., and Ramirez, F. (1999). Pathogenetic sequence for aneurysm revealed in mice underexpressing fibrillin-1. *Proc. Natl. Acad. Sci. USA* **96**, 3819–3823.
104. Dietz, H. C. and Mecham, R. P. (2000). Mouse models of genetic diseases resulting from mutations in elastic fiber proteins. *Matrix Biol.* **19**, 481–488.
105. Pfaff, M., Reinhardt, D. P., Sakai, L. Y., and Timpl, R. (1996). Cell adhesion and integrin binding to recombinant human fibrillin-1. *FEBS Lett.* **384**, 247–250.
106. Gibson, M. A., Leavesley, D. I., and Ashman, L. K. (1999). Microfibril-associated glycoprotein-2 specifically interacts with a range of bovine and human cell types via alphaVbeta3 integrin. *J. Biol. Chem.* **274**, 13060–13065.
107. Pan, T. C., Kluge, M., Zhang, R. Z., Mayer, U., Timpl, R., and Chu, M. L. (1993). Sequence of extracellular mouse protein BM-90/fibulin and its calcium-dependent binding to other basement-membrane ligands. *Eur. J. Biochem.* **215**, 733–740.
108. Balbona, K., Tran, H., Godyna, S., Ingham, K. C., Strickland, D. K., and Argraves, W. S. (1992). Fibulin binds to itself and to the carboxyl-terminal heparin-binding region of fibronectin. *J. Biol. Chem.* **267**, 20120–20125.
109. Vu, T. H. (2001). Don't mess with the matrix. *Nat. Genet.* **28**, 202–203.
110. Martignetti, J. A., Aqeel, A. A., Sewairi, W. A., Boumah, C. E., Kambouris, M., Mayouf, S. A., Sheth, K. V., Eid, W. A., Dowling, O., Harris, J., Glucksman, M. J., Bahabri, S., Meyer, B. F., and Desnick, R. J. (2001). Mutation of the matrix metalloproteinase 2 gene (MMP2) causes a multicentric osteolysis and arthritis syndrome. *Nat. Genet.* **28**, 261–265.
111. Holmbeck, K., Bianco, P., Caterina, J., Yamada, S., Kromer, M., Kuznetsov, S. A., Mankani, M., Robey, P. G., Poole, A. R., Pidoux, I., Ward, J. M., and Birkedal-Hansen, H. (1999). MT1-MMP-deficient mice develop dwarfism, osteopenia, arthritis, and connective tissue disease due to inadequate collagen turnover. *Cell* **99**, 81–92.
112. Krane, S. M. (2002), *Meeting Report from the Frontiers of Skeletal Biology: Ninth Workshop on Cell Biology of Bone and Cartilage in Health and Disease,* Davos, Switzerland.
113. Neuhold, L. A., Killar, L., Zhao, W., Sung, M. L., Warner, L., Kulik, J., Turner, J., Wu, W., Billinghurst, C., Meijers, T., Poole, A. R., Babij, P., and DeGennaro, L. J. (2001). Postnatal expression in hyaline cartilage of constitutively active human collagenase-3 (MMP-13) induces osteoarthritis in mice. *J. Clin. Invest.* **107**, 35–44.

CHAPTER 340

Cell-to-Cell Interactions in Lung

Joseph L. Alcorn
Department of Pediatrics, University of Texas Health Science Center, Houston, Texas

Introduction

The lung is a remarkably complex organ that requires an intricate interaction of the processes of branching morphogenesis and cell proliferation for organ development in an intrauterine environment and for the function of gaseous exchange upon transition to an extrauterine environment [1–3]. Once developed, the lung processes 10,000 liters of air per day, exposing it to numerous pathogenic and noxious environmental insults, thus requiring programs for repair responses to these insults. The complexity of the organ is demonstrated by the fact that more than 40 different cell types with respiratory and nonrespiratory functions are present in the lung; and these cell types can be divided into several classes: type I epithelial cells, type II epithelial cells, endothelial cells, interstitial or mesenchymal cells, neuroendocrine cells, and alveolar macrophages.

The number of different pulmonary cell types and their individual functions require precise and timely communication in order to embark on a developmental program in the fetus and to maintain the integrity of lung function in the face of damage and injury. Cell–cell communication and cell–matrix interactions play an important role in aspects of both lung development and repair [4]; it is believed that repair following lung injury recapitulates some of the pathways and processes mediating lung formation. This chapter deals with various soluble cell-to-cell signals that are involved in the processes of lung development and lung repair. However, due to the complexity of the lung and space limitations, the primary focus of this chapter is on cell-to-cell interactions that involve epithelial cells of the lung.

Lung Organogenesis and Development

Organogenesis and development of the lung begins at gestational age day 26 in the human embryo as a ventral out-pouching of the foregut that results in a tracheal bud. Lung development continues in distinct stages from this prenatal period until the postnatal period in early childhood [5]. During the embryonic period (1–7 weeks), lung lobe formation is established and subsegmental branching of the bronchial tree occurs, which seems to be determined by interactions with the mesenchyme. At this time, vascular connections to the pulmonary system are established. In the pseudoglandular stage (5–17 weeks), formation of the bronchiolar tree is completed. Formation of the upper (proximal) airway, development to the alveolar ducts, and innervation of vasculature takes place during this stage as well. The transition to the canalicular stage (16–26 weeks) marks the point at which the lung parenchyma is canalized by capillaries. The epithelial cells that line the air ducts start differentiating into type I and type II cells and formation of a thin air–blood barrier are other hallmarks of this stage.

The final prenatal stage, the saccular (24 weeks to term), encompasses the time in which the peripheral or distal airways form the typical terminal clusters of widened air spaces called *saccules*. During this time, the interstitial tissue between the air spaces is decreased and the capillary networks get closer together. Postnatally, lung development entails changes in which the size of the lung is increased and the ability to perform efficient gaseous exchange is enhanced. The alveolar stage (up to 18 months) is the time when the number of alveolar sacs is increased. While alveolarization proceeds, the stage of microvascular maturation (about 6 months to 3 years) becomes more pronounced. During this time, the capillary networks of the lung are established.

Soluble Factors of Cell-to-Cell Interactions Involved in Lung Cell Differentiation

During all stages, the various types of epithelial cells in the respiratory tract are derived from common progenitors and are generated through cell–cell, matrix–cell, and autocrine–paracrine pathways: The mesenchyme produces growth factors and patterning information that determines how epithelial cells will proliferate and differentiate. The soluble factors listed next are those believed to be involved in these interactions.

Fibroblast Growth Factor

Fibroblast growth factor (FGF) signaling is critical for the regulation of lung development and epithelial cell differentiation. FGF receptor-2 (FGFR-2) is preferentially expressed in epithelial cells during development, and ablation of FGFR-2 signaling results in a nearly complete lack of lung formation in the mouse due to the failure of branching morphogenesis [6]. FGF-3, FGF-7 (keratinocyte growth factor, KGF), and mesenchymal-derived FGF-10 bind to FGFR-2 [7,8]. A lack of lung development is observed in FGF-10-deficient mice [9]. In addition to being critical for lung formation, FGF is also involved in epithelial cell proliferation and differentiation in postnatal lung. KGF and FGF-10 increase epithelial cell proliferation [10]. KGF regulates alveolar type II cell differentiation in adult lung [11], and expression of KGF during the pseudoglandular stage of development disrupted normal pulmonary branching morphogenesis, demonstrating the importance of correct temporal expression of members of the FGF family in lung development [12].

Epidermal Growth Factor and Transforming Growth Factor α

Epidermal growth factor (EGF) has been shown to accelerate epithelial cell development when administered *in vivo* [13]. Both transforming growth factor α (TGF-α) and EGF act through the EGF receptor (EGFR), and TGF-α, EGF, and EGFR colocalize to epithelial and smooth muscle cells of bronchioles and bronchi and in epithelial cells of saccules of rat lung [14]. Epithelial cells cultured from late-gestation fetal rat lung express TGF-α and EGFR, whereas cultured fibroblasts express EGFR mRNA, but no detectable TGF-α mRNA. These results suggest that TGF-α produced by fetal lung epithelial cells might act through an autocrine or paracrine mechanism with epithelial and mesenchymal cells. In addition, colocalization of TGF-α and EGF suggests that these agents might act in parallel in lung development [15]. EGFR-deficient mice show evidence of lung immaturity due to impaired branching and deficient alveolization and septation as well as evidence of type II cell immaturity [14].

Platelet-Derived Growth Factor

The platelet-derived growth factor (PDGF) family acts through two tyrosine kinase receptors, PDGFR α and β, and three of the ligands (PDGF-A, -B, and -C) bind to PDGFR-α with high affinity. Although PDGF-A and PDGF-B can be found in the lung, expression of PDGFR-α and PDGFR-β are mainly restricted to lung mesenchyme and PDGF-A expression is restricted to developing lung epithelium [16]. Early data suggested that PDGF-BB stimulation of both receptors leads to lung growth, whereas PDGF-AA stimulation of PDGFR-α induces transduction pathways that lead lung branching [17]. However, mice deficient in the PDGF-A gene have evidence of branching morphogenesis but lack lung alveolar smooth muscle cells, exhibit reduced deposition of elastin fibers in the lung parenchyma, and develop lung emphysema due to complete failure of alveogenesis. Failure of postnatal alveogenesis in these mice is apparently due to a prenatal block in the distal spreading of PDGFR-α+ cells along the tubular lung epithelium during the canalicular stage of lung development [16].

Studies with mice deficient in the PDGFR-α gene have demonstrated its role in the recruitment of smooth muscle cells to the alveolar sacs and their further compartmentalization into alveoli [18]. These data suggest that although PDGF-AA signaling through PDGFR-α may have direct or indirect roles in overall lung growth, it does not specifically control early branching of the lung epithelium. Transgenic mice expressing PDGF-A in the distal lung epithelium were shown to have increased mesenchymal cells and acinar buds and decreased bronchioles and dilated air spaces. These results indicate that PDGF-A is a potent growth factor for mesenchymal cells in the developing lung and that the downregulation of PDGF-A expression that normally occurs in the lung during late gestation is required for transition from the canalicular to the saccular stage of lung development [19].

Insulin-Like Growth Factors

Insulin-like growth factors (IGFs) are mitogens that share structural homology with proinsulin and are involved in the growth and development of many organs, including lung. The regulation of IGF action depends not only on the expression of IGFs and IGF receptors, but also on the modulation of IGF activity by IGF-binding proteins (IGFBPs). In human lung, IGF I and predominantly IGF II mRNAs are expressed throughout gestation with decreasing expression of both IGFs after the canalicular phase of gestation. They are mainly detected in the mesodermal-derived components of the respiratory tract, especially in the undifferentiated mesenchyme of the lung buds before and during the canalicular phase of gestation. Expression of the IGFs is weak in the epithelium during the first stages of growth and progressively decreases on maturation. In the endothelium, expression of IGF II, but not IGF I, is consistent throughout gestation. The levels of expression and the cellular distribution of the IGF receptors, IGFR-1 and IGFR-2, are very complex and vary with gestational age [20,21]. IGF and IGFBP-2 mRNAs were expressed both in mesenchymal and epithelial cells. Type I IGF receptor transcripts are present throughout the developing lung, with the exception of the epithelial cells of the bronchi later in gestation. IGFBP-2, -4, and -5 genes are differentially regulated during embryonic development and suggest that each may have a discrete function, and a possible role for these IGFBPs is to participate in the regulation of cell-specific IGF responses during mouse lung development [22]. The IGFs may act on lung epithelial cell proliferation in both autocrine and paracrine ways, and may also stimulate the maturation of the connective tissue [23].

Hepatocyte Growth Factor

Hepatocyte growth factor (HGF) is a mesenchyme-derived heparin-binding growth factor that has mitogenic, motogenic, and morphogenic activities on lung epithelial cells and may be a mediator of epithelial–mesenchymal interaction during lung organogenesis and regeneration following lung injury [24]. HGF is expressed in mesenchymal fibroblasts and the c-met/HGF receptor is expressed in epithelium in the developing lung. HGF apparently stimulates branching morphogenesis of the fetal lung, and appears to work in collaboration with the FGF family in epithelial branching [25]. On the other hand, mice deficient in the HGF gene have been shown to have lung development that is identical to wild-type mice, suggesting that HGF may not be of primary importance in lung development [26,27].

Vascular Endothelial Growth Factor

Vascular endothelial growth factor (VEGF) is a potent endothelial cell mitogen and is abundant in distal epithelium of midtrimester human fetal lung [28]. KDR, a major VEGF-specific receptor, can be localized to distal lung epithelial cells of human fetal lung tissue, suggesting a possible autocrine or paracrine regulatory role for VEGF in pulmonary epithelial cell growth and differentiation [28]. Expression of VEGF in the developing respiratory epithelium of transgenic mice increases growth of the pulmonary blood vessels, disrupts branching morphogenesis of the lung, and inhibits type I cell differentiation [29]. Addition of exogenous VEGF to human fetal lung explants causes proliferation of distal airway epithelial cells and expression of type II cell-specific markers [28], indicating that VEGF may be an important autocrine growth factor for distal airway epithelial type II cells in the developing human lung.

Fibroblast-Pneumonocyte Factor

Fibroblast-pneuomonocyte factor (FPF) was first discovered as an organ-specific differentiation factor produced by glucocorticoid-treated fetal lung fibroblasts that enhances the differentiation of the alveolar type II epithelial cells [30]. The identification and characterization of the agent has been elusive, but it is thought to be very stable with a molecular mass of 5–15 kDa [31]. FPF production is positively regulated by glucocorticoids and negatively regulated by dihydrotestosterone, which interferes with progression of lung development by delaying the appearance of FPF production by the fibroblast, and TGF-β, which appears to interfere with type II epithelial cell development such that the cell cannot respond to FPF [32].

Bombesin-Like Peptides

Bombesin-like peptide (BLP) or gastrin-releasing peptide (GRP) is produced by pulmonary neuroendocrine cells (PNEC). The function of GRP as a growth factor involves lung morphogenesis, but the precise mechanism and the target cells have not been defined. GRP-R mRNA was localized mainly in the distal airway epithelial tubes and surrounding mesenchyme, whereas proximal airways showed decreased signal. Postnatal lung, a strong signal for GRP-R mRNA, was localized in the airway epithelial cells. PNEC in lung expressed mRNA and peptide. The expression of GRP-R in mammalian lung is developmentally regulated, peaking both spatially and temporally during the phase of rapid airway growth and differentiation. Both epithelial and mesenchymal components express GRP-R consistent with paracrine mechanism for GRP activity during lung morphogenesis [33]. BLP stimulates type II cell differentiation, neuroendocrine differentiation, and cell proliferation in fetal lung [34,35].

Transforming Growth Factor β

Unlike the aforementioned factors, TGF-β has been shown to inhibit rather than activate branching morphogenesis fetal lung through its receptor: the TGF-β type II receptor. TGF-β is believed to promote expression of extracellular matrix components around the airways, prevent budding of branches, and suppress epithelial cell proliferation [1,3]. TGF-β mRNA is confined to fibroblasts around epithelial ducts and to smooth muscle cells in the lung [36]. It is not known whether the function of TGF-β resides in lung epithelium or mesenchyme, but it is believed that TGF-β signaling inhibits lung morphogenesis by hampering the inductive and permissive effects of other growth factors such as EGF, TGF-α, and PDGF-AA [37]. Addition of TGF-β to explant cultures of fetal mouse lungs inhibits branching morphogenesis, and its effect could be reversed by washing out the compound [38]. In the development of glandular organs, it has been suggested that activins, dimeric proteins that are members of the TGF-β superfamily, are key growth and differentiation factors in the process of branching morphogenesis [39]. Although activin A has been shown to be expressed by and stimulate pulmonary fibroblast proliferation and differentiation [40], it seems that TGF-β plays a more important role in branching morphogenesis.

Bone Morphogenic Protein

Bone morphogenic proteins (BMPs) are members of the TGF-β family that act to suppress the proliferative effects of the various growth factors in the lung. Distal expression of BMP4 has profound effects on cell proliferation and cell fate in the lung bud [41]. Interestingly, expression of BMP4 is controlled by FGF-10. BMP4 is expressed in the epithelial cells in the tips of growing lung buds. Inhibiting BMP4 signaling in the developing lung results in a severe reduction in distal epithelial cell types and a simultaneous increase in proximal cell types [42].

Soluble Factors of Cell-to-Cell Interactions Involved in Lung Injury

Hyperoxia or exposure to cytotoxic reactive oxygen species results in the death of endothelial cells followed by type I epithelial cells, then type II cells. Although we do not know how endothelial cells are replaced, it is clear that an important process in the repair of lung following damage is the proliferation and transdifferentiation of cuboidal alveolar type II cells into a flattened type I cell. Normal repair requires interaction between fibroblasts and epithelium in which there is balanced proliferation and differentiation of both epithelial and mesenchymal cells. To this end, the lung performs the processes of repair by the use of many of the signals that are involved in lung development and lung cell differentiation. It has been shown that fibroblasts exposed to 100% O_2 produce unidentified growth factors for type II cells [43]. Expression of KGF, FGF, and IGF-1 from the mesenchyme increases in lung on exposure to high levels of O_2 [44,45], while TGF-β, a growth inhibitory signal, and parathyroid hormone (PTrH), which inhibits type II cell proliferation, are decreased [46,47]. Recent studies have demonstrated a strong expression of activin in repair processes in the lung with roles in epithelial differentiation, fibroblast proliferation, and expression of matrix molecules by these cells [48]. BLP has been shown to stimulate type II cell differentiation and cell proliferation in response to lung injury [34,35]. Apart from its role in lung development, KGF has been implicated in protection from hyperoxic damage [49]. FGF-3 appears to have a role in repair following lung injury in adult lung [50]. Low levels of FGF-3 expression resulted in massive free alveolar macrophage infiltration, indicating that FGF may influence cytokine and chemokine levels involved in host defense mechanisms. On the other hand, high levels of FGF-3 expression resulted in diffuse alveolar type II cell hyperplasia. HGF is a potent mitogen for airway epithelial cells and alveolar epithelial cells, and may act as a pulmotrophic factor responsible for airway and alveolar regeneration during lung regeneration after acute lung injury. Expression of HGF is rapidly induced in the lung after acute lung injury in experimental animals and that HGF levels are elevated in blood of patients with lung diseases [51]. VEGF is a specific mitogen for endothelial cells that is expressed by alveolar epithelial cells. Hyperoxic injury decreases neonatal lung VEGF mRNA and protein, which may be a contributory mechanism of impaired postnatal microvascular development in oxygen injury [52].

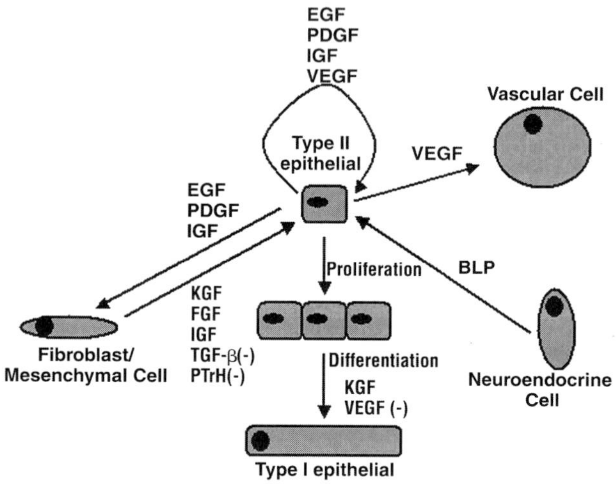

Figure 1 Summary of the cell-to-cell interactions involved in the processes of proliferation and differentiation of lung cells.

Conclusions

The lung is a complex organ and the various forms of interaction between the 40 or so cell types cannot be described fully in the space allotted. The focus of this chapter has been on the soluble factors that play a role in cell-to-cell communication during development and repair of the lung, particularly as they pertain to interactions between epithelial cells and other cell types in the lung. It has often been said that a picture is worth a thousand words, so in the interest of space, Fig. 1 has been included to summarize the interactions between the signals produced and signals received by the various cells, resulting in the proliferation and differentiation of lung cells. Although the figure does not fully address all of the interactions in the lung, it should give the reader a sense of the balance that is required for the development of the lung and to maintain integrity of its function in the event of damage postnally.

References

1. Cardoso, W. V. (2001). Molecular regulation of lung development. *Annu. Rev. Physiol.* **63**, 471–494.
2. Cardoso, W. V. (2000). Lung morphogenesis revisited: old facts, current ideas. *Dev. Dyn.* **219**, 121–130.
3. Warburton, D., Schwarz, M., Tefft, D., Flores-Delgado, G., Anderson, K. D., and Cardoso, W. V. (2000). The molecular basis of lung morphogenesis. *Mech. Dev.* **92**, 55–81.
4. Desai, T. J. and Cardoso, W. V. (2002). Growth factors in lung development and disease: friends or foe? *Respir. Res.* **3**, 2.
5. Burri, P. H. (1997). Structural aspects of prenatal and postnatal development and growth of the lung. In McDonald, J. A., Ed., *Lung Growth and Development,* pp. 1–35. Marcel Dekker, New York.
6. Peters, K., Werner, S., Liao, X., Wert, S., Whitsett, J., and Williams, L. (1994). Targeted expression of a dominant negative FGF receptor blocks branching morphogenesis and epithelial differentiation of the mouse lung. *EMBO J.* **13**, 3296–3301.
7. Lizarraga, G., Ferrari, D., Kalinowski, M., Ohuchi, H., Noji, S., Kosher, R. A., and Dealy, C. N. (1999). FGFR2 signaling in normal and limbless chick limb buds. *Dev. Genet.* **25**, 331–338.
8. Kato, S. and Sekine, K. (1999). FGF-FGFR signaling in vertebrate organogenesis. *Cell Mol. Biol.* **45**, 631–638.
9. Min, H., Danilenko, D. M., Scully, S. A., Bolon, B., Ring, B. D., Tarpley, J. E., DeRose, M., and Simonet, W. S. (1998). Fgf-10 is required for both limb and lung development and exhibits striking functional similarity to *Drosophila* branchless. *Genes Dev.* **12**, 3156–3161.
10. Ulich, T. R., Yi, E. S., Cardiff, R., Yin, S., Bikhazi, N., Biltz, R., Morris, C. F., and Pierce, G. F. (1994). Keratinocyte growth factor is a growth factor for mammary epithelium *in vivo.* The mammary

epithelium of lactating rats is resistant to the proliferative action of keratinocyte growth factor. *Am. J. Pathol.* **144**, 862–868.
11. Shannon, J. M., Gebb, S. A., and Nielsen, L. D. (1999). Induction of alveolar type II cell differentiation in embryonic tracheal epithelium in mesenchyme-free culture. *Development* **126**, 1675–1688.
12. Simonet, W. S., DeRose, M. L., Bucay, N., Nguyen, H. Q., Wert, S. E., Zhou, L., Ulich, T. R., Thomason, A., Danilenko, D. M., and Whitsett, J. A. (1995). Pulmonary malformation in transgenic mice expressing human keratinocyte growth factor in the lung. *Proc. Natl. Acad. Sci. USA* **92**, 12461–12465.
13. Catterton, W. Z., Escobedo, M. B., Sexson, W. R., Gray, M. E., Sundell, H. W., and Stahlman, M. T. (1979). Effect of epidermal growth factor on lung maturation in fetal rabbits. *Pediatr. Res.* **13**, 104–108.
14. Miettinen, P. J., Warburton, D., Bu, D., Zhao, J. S., Berger, J. E., Minoo, P., Koivisto, T., Allen, L., Dobbs, L., Werb, Z., and Derynck, R. (1997). Impaired lung branching morphogenesis in the absence of functional EGF receptor. *Dev. Biol.* **186**, 224–236.
15. Strandjord, T. P., Clark, J. G., and Madtes, D. K. (1994). Expression of TGF-alpha, EGF, and EGF receptor in fetal rat lung. *Am. J. Physiol.* **267**, L384–389.
16. Lindahl, P., Karlsson, L., Hellstrom, M., Gebre-Medhin, S., Willetts, K., Heath, J. K., and Betsholtz, C. (1997). Alveogenesis failure in PDGF-A-deficient mice is coupled to lack of distal spreading of alveolar smooth muscle cell progenitors during lung development. *Development* **124**, 3943–3953.
17. Souza, P., Tanswell, A. K., and Post, M. (1996). Different roles for PDGF-alpha and -beta receptors in embryonic lung development. *Am. J. Respir. Cell Mol. Biol.* **15**, 551–562.
18. Bostrom, H., Gritli-Linde, A., and Betsholtz, C. (2002). PDGF-A/PDGF alpha-receptor signaling is required for lung growth and the formation of alveoli but not for early lung branching morphogenesis. *Dev. Dyn.* **223**, 155–162.
19. Li, J. and Hoyle, G. W. (2001). Overexpression of PDGF-A in the lung epithelium of transgenic mice produces a lethal phenotype associated with hyperplasia of mesenchymal cells. *Dev. Biol.* **239**, 338–349.
20. Maitre, B., Clement, A., Williams, M. C., and Brody, J. S. (1995). Expression of insulin-like growth factor receptors 1 and 2 in the developing lung and their relation to epithelial cell differentiation. *Am. J. Respir. Cell Mol. Biol.* **13**, 262–270.
21. Moats-Staats, B. M., Price, W. A., Xu, L., Jarvis, H. W., and Stiles, A. D. (1995). Regulation of the insulin-like growth factor system during normal rat lung development. *Am. J. Respir. Cell Mol. Biol.* **12**, 56–64.
22. Schuller, A. G., van Neck, J. W., Beukenholdt, R. W., Zwarthoff, E. C., and Drop, S. L. (1995). IGF, type I IGF receptor and IGF-binding protein mRNA expression in the developing mouse lung. *J. Mol. Endocrinol.* **14**, 349–355.
23. Lallemand, A. V., Ruocco, S. M., Joly, P. M., and Gaillard, D. A. (1995). *In vivo* localization of the insulin-like growth factors I and II (IGF I and IGF II) gene expression during human lung development. *Int. J. Dev. Biol.* **39**, 529–537.
24. Mason, R. J. (2002). Hepatocyte growth factor: The key to alveolar septation? *Am. J. Respir. Cell Mol. Biol.* **26**, 517–520.
25. Ohmichi, H., Koshimizu, U., Matsumoto, K., and Nakamura, T. (1998). Hepatocyte growth factor (HGF) acts as a mesenchyme-derived morphogenic factor during fetal lung development. *Development* **125**, 1315–1324.
26. Schmidt, C., Bladt, F., Goedecke, S., Brinkmann, V., Zschiesche, W., Sharpe, M., Gherardi, E., and Birchmeier, C. (1995). Scatter factor/hepatocyte growth factor is essential for liver development. *Nature* **373**, 699–702.
27. Uehara, Y., Minowa, O., Mori, C., Shiota, K., Kuno, J., Noda, T., and Kitamura, N. (1995). Placental defect and embryonic lethality in mice lacking hepatocyte growth factor/scatter factor. *Nature* **373**, 702–705.
28. Brown, K. R., England, K. M., Goss, K. L., Snyder, J. M., and Acarregui, M. J. (2001). VEGF induces airway epithelial cell proliferation in human fetal lung in vitro. *Am. J. Physiol. Lung Cell Mol. Physiol.* **281**, L1001–1010.
29. Zeng, X., Wert, S. E., Federici, R., Peters, K. G., and Whitsett, J. A. (1998). VEGF enhances pulmonary vasculogenesis and disrupts lung morphogenesis in vivo. *Dev. Dyn.* **211**, 215–227.
30. Smith, B. T. and Post, M. (1989). Fibroblast-pneumonocyte factor. *Am. J. Physiol.* **257**, L174–178.
31. Floros, J., Post, M., and Smith, B. T. (1985). Glucocorticoids affect the synthesis of pulmonary fibroblast-pneumonocyte factor at a pretranslational level. *J. Biol. Chem.* **260**, 2265–2267.
32. Nielsen, H. C., Kellogg, C. K., and Doyle, C. A. (1992). Development of fibroblast-type-II cell communications in fetal rabbit lung organ culture. *Biochim. Biophys. Acta* **1175**, 95–99.
33. Wang, D., Yeger, H., and Cutz, E. (1996). Expression of gastrin-releasing peptide receptor gene in developing lung. *Am. J. Respir. Cell Mol. Biol.* **14**, 409–416.
34. Emanuel, R. L., Torday, J. S., Mu, Q., Asokananthan, N., Sikorski, K. A., and Sunday, M. E. (1999). Bombesin-like peptides and receptors in normal fetal baboon lung: Roles in lung growth and maturation. *Am. J. Physiol.* **277**, L1003–1017.
35. Sunday, M. E., Hua, J., Dai, H. B., Nusrat, A., and Torday, J. S. (1990). Bombesin increases fetal lung growth and maturation *in utero* and in organ culture. *Am. J. Respir. Cell Mol. Biol.* **3**, 199–205.
36. Pelton, R. W., Johnson, M. D., Perkett, E. A., Gold, L. I., and Moses, H. L. (1991). Expression of transforming growth factor-beta 1, -beta 2, and -beta 3 mRNA and protein in the murine lung. *Am. J. Respir. Cell Mol. Biol.* **5**, 522–530.
37. Zhao, J., Sime, P. J., Bringas, P., Jr., Gauldie, J., and Warburton, D. (1998). Epithelium-specific adenoviral transfer of a dominant-negative mutant TGF-beta type II receptor stimulates embryonic lung branching morphogenesis in culture and potentiates EGF and PDGF-AA. *Mech. Dev.* **72**, 89–100.
38. Serra, R., Pelton, R. W., and Moses, H. L. (1994). TGF beta 1 inhibits branching morphogenesis and N-myc expression in lung bud organ cultures. *Development* **120**, 2153–2161.
39. Ball, E. M. and Risbridger, G. P. (2001). Activins as regulators of branching morphogenesis. *Dev. Biol.* **238**, 1–12.
40. Ohga, E., Matsuse, T., Teramoto, S., and Ouchi, Y. (2000). Activin receptors are expressed on human lung fibroblast and activin A facilitates fibroblast-mediated collagen gel contraction. *Life Sci.* **66**, 1603–1613.
41. Weaver, M., Dunn, N. R., and Hogan, B. L. (2000). Bmp4 and Fgf10 play opposing roles during lung bud morphogenesis. *Development* **127**, 2695–2704.
42. Weaver, M., Yingling, J. M., Dunn, N. R., Bellusci, S., and Hogan, B. L. (1999). Bmp signaling regulates proximal-distal differentiation of endoderm in mouse lung development. *Development* **126**, 4005–4015.
43. Everett, M. M., King, R. J., Jones, M. B., and Martin, H. M. (1990). Lung fibroblasts from animals breathing 100% oxygen produce growth factors for alveolar type II cells. *Am. J. Physiol.* **259**, L247–254.
44. Buch, S., Han, R. N., Liu, J., Moore, A., Edelson, J. D., Freeman, B. A., Post, M., and Tanswell, A. K. (1995). Basic fibroblast growth factor and growth factor receptor gene expression in 85% O_2-exposed rat lung. *Am. J. Physiol.* **268**, L455–464.
45. Charafeddine, L., D'Angio, C. T., Richards, J. L., Stripp, B. R., Finkelstein, J. N., Orlowski, C. C., LoMonaco, M. B., Paxhia, A., and Ryan, R. M. (1999). Hyperoxia increases keratinocyte growth factor mRNA expression in neonatal rabbit lung. *Am. J. Physiol.* **276**, L105–113.
46. Hastings, R. H., Ryan, R. M., D'Angio, C. T., Holm, B. A., Patel, A., Quintana, R., Biederman, E., Burton, D. W., and Deftos, L. J. (2002). Parathyroid hormone-related protein response to hyperoxic lung injury. *Am. J. Physiol. Lung Cell Mol. Physiol.* **282**, L1198–1208.
47. Moore, A. M., Buch, S., Han, R. N., Freeman, B. A., Post, M., and Tanswell, A. K. (1995). Altered expression of type I collagen, TGF-beta 1, and related genes in rat lung exposed to 85% O2. *Am. J. Physiol.* **268**, L78–84.
48. Hubner, G., Alzheimer, C., and Werner, S. (1999). Activin: A novel player in tissue repair processes. *Histol. Histopathol.* **14**, 295–304.

49. Barazzone, C., Donati, Y. R., Rochat, A. F., Vesin, C., Kan, C. D., Pache, J. C., and Piguet, P. F. (1999). Keratinocyte growth factor protects alveolar epithelium and endothelium from oxygen-induced injury in mice. *Am. J. Pathol.* **154**, 1479–1487.
50. Zhao, B., Chua, S. S., Burcin, M. M., Reynolds, S. D., Stripp, B. R., Edwards, R. A., Finegold, M. J., Tsai, S. Y., and DeMayo, F. J. (2001). Phenotypic consequences of lung-specific inducible expression of FGF-3. *Proc. Natl. Acad. Sci. USA* **98**, 5898–5903.
51. Ohmichi, H., Matsumoto, K., and Nakamura, T. (1996). *In vivo* mitogenic action of HGF on lung epithelial cells: Pulmotrophic role in lung regeneration. *Am. J. Physiol.* **270**, L1031–1039.
52. Maniscalco, W. M., Watkins, R. H., D'Angio, C. T., and Ryan, R. M. (1997). Hyperoxic injury decreases alveolar epithelial cell expression of vascular endothelial growth factor (VEGF) in neonatal rabbit lung. *Am. J. Respir. Cell Mol. Biol.* **16**, 557–567.

CHAPTER 341

Mechanisms of Stress Response Signaling and Recovery in the Liver of Young versus Aged Mice: The p38 MAPK and SOCS Families of Regulatory Proteins

John Papaconstantinou
University of Texas Medical Branch,
Department of Human Biological Chemistry and Genetics,
Galveston, Texas

Introduction

Biological signaling is a highly specific process that regulates, by activation or repression, gene expression in response to environmental factors, growth factors, and hormones. Signaling processes involve a series of protein–protein interactions that comprise pathways that target the activities of specific genes. Furthermore, biological signaling also involves protein–protein interactions that attenuate or turn off the signals in the process of recovery. This chapter discusses the molecular mechanisms of the protein–protein interactions of two major hepatic signaling pathways: (1) the p38 MAPK stress response pathway that involves the activation of stress response genes and (2) the mechanisms of turning off cytokine signaling pathway(s) by the SOCS family of negative regulators. The p38 MAPK proteins contain docking domains that facilitate (1) activation by upstream signaling proteins, (2) translocation to the nucleus, and (3) activation of stress response transcription factors that regulate the expression of stress response genes. The SOCS family of negative regulators interacts with various protein components of cytoplasmic domains of membrane receptors, thereby preventing the transduction of signals to downstream activator proteins. The characteristics of the proteins that interact to transduce and to turn off these signals as they pertain to the p38 MAPK and SOCS families of proteins are discussed.

The p38 MAPK Pathway in Stress Response Signaling

Biological signaling is a highly specific process that regulates gene expression in response to changes in intrinsic and extrinsic factors. Any alteration in these biological processes affects organismal function and survival. Thus, the status of biological signaling contributes to the overall biochemical phenotype of tissues at all stages of growth and development of aging. Hepatic signaling is, therefore, an important factor in this tissue-specific response to various

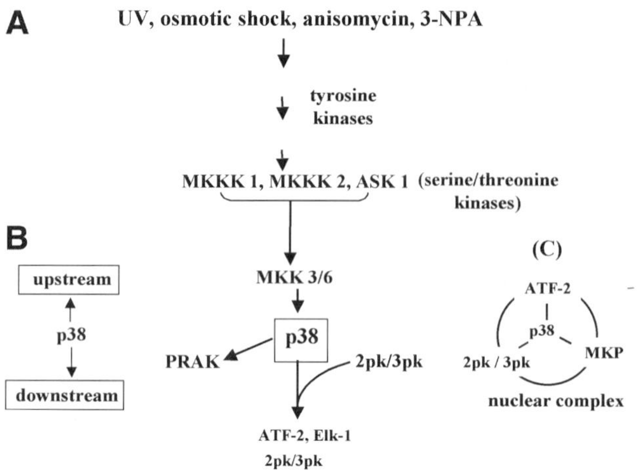

Figure 1 A map of the serine/threonine kinases of the p38 MAPK stress response pathway. The individual upstream tyrosine kinases are not included in this map. (A) Environmental factors (UV, osmotic shock, anisomycin, and 3-nitropropionic acid) activate stress response genes via the p38 MAPK signaling pathway. p38 is depicted as a central component of the pathway because of its role in the translocation of signals to the nucleus where transcription factors and gene activation are carried out. (B) The upstream activators of the p38 MAPK pathway are MAP kinase kinase kinases (MKKKs 1, 2, and ASK1) and the MAP kinase kinases (MKKs 3 and 6). (C) A nuclear complex of p38 MAPK–MAPKAPK-2/3–MAPK–phosphatase. The complex represents a potential mechanism for the activation of transcription factors (ATF-2) by p38 MAPK and 2pk/3pk. (See [105] for recent data on activation of ATF-2 by p38 MAPK pathway.)

factors including growth factors, hormones, intermediates of metabolism, and environmental factors. In particular, members of the mitogen-activated protein kinase (MAPK)/stress-activated protein kinase (SAPK/JNK) families of signaling proteins are activated by a wide variety of extracellular signals (Fig. 1).

The family has three major subgroups: ERKs (extracellular regulators of kinases), p38, and SAPK/JNK. ERKs are activated by growth factors (extracellular matrix interactions with growth factors) and phorbol esters, whereas p38 and SAPK/JNK are activated by such extracellular stresses as UV irradiation, osmotic stress, and inflammatory cytokines. Activation of MAPK signaling leads to the regulation of gene clusters that mediate complex biological responses such as inflammation, cell proliferation, differentiation, and apoptosis [1–3]. The proteins in the MAPK pathways comprise a cascade of multiple tyrosine and serine/threonine kinases and phosphatases; the signals are transduced by protein kinase phosphorylation/dephosphorylation events that mediate docking and enhance specific serine/threonine kinase/phosphatase activities that attenuate the signaling. Although these molecular interactions are both efficient and specific, they also exhibit cross-talk among the MAPK cascades or with other signaling pathways, thereby mediating complex biological processes. Numerous reviews have been published on the structure and function of these signaling proteins [1–3]. In this chapter, we focus first on the serine/threonine kinase components of the p38 MAPK stress response signaling pathway, its mechanisms of signal transduction, and its relevance to liver function and, secondly, the attenuation of hepatic stress signaling pathways.

The p38 MAPK Stress Response Pathway

p38 MAPK is a focal point of interactions of the serine/threonine kinases that transduce stress signals via upstream cytoplasmic activator proteins to their downstream substrates or nuclear targets, that is, the stress response genes (Fig. 1). The pathway in Fig. 1 shows that the transduction of biological signals involves interactions of upstream tyrosine kinases with downstream serine/threonine kinases thereby linking external environmental signals to gene targets. Furthermore, the p38 MAPK isoforms, and the SAPK/JNK isoforms are major nucleocytoplasmic trafficking proteins of these pathways and are, therefore, the crossroads where the interactions of these isoforms with upstream activators and downstream targets (genes) deliver the biological responses to environmental challenges (Fig. 1). Thus, the protein–protein interactions of the MAPK pathways are the major mechanism for controlling the activation of transcription factors and their targeted genes in response to extrinsic (environmental) and intrinsic (biological) factors.

The p38 MAPK Stress Signaling Pathway Specificities and Mechanisms of Protein–Protein Interactions

p38α Was first isolated as a 38-kDa protein rapidly phosphorylated on Tyr182 in response to the inflammatory stress caused by bacterial lipopolysaccharide (LPS). Treatment of mammalian cells in culture [4–6] in mammals indicates that the p38 MAPK family consists of α, β, γ, and δ isoforms: p38α and p38β MAPK are 60% identical to p38γ and p38δ, indicating that they represent related but distinct MAPK subgroups [7]. p38α and p38β Are ubiquitously expressed but at significantly different levels in each tissue. Both p38α and p38β are abundant in the liver. Both p38γ and p38δ are also differentially expressed; p38γ is predominant in skeletal muscle [8,9] and p38δ is enriched in lung, kidney, testis, pancreas, and small intestine [10]. The isoforms have distinct biological functions *in vivo* due, in part, to their ability to target (phosphorylate and so activate) specific downstream transcription factors and to be selectively activated by extracellular stimuli via the upstream tyrosine kinases. Importantly, the p38 MAPK kinases serve as a focal point for the transduction of biological signals from the cytoplasm to their nuclear gene targets. Thus, the pathway consists of a cascade of three protein kinase families: MAP kinase kinase kinases (MKKK), MAP kinase kinases (MKK), and MAP kinases, that is, the p38 MAPK proteins (Fig. 1). These proteins are activated in series such that the MKKKs phosphorylate the MKK activation loop serines, which then activate the p38 MAPKs by phosphorylation of Thr180 and Tyr182.

It is through the p38 MAPK that transcription factors are activated on translocation to the nucleus. The p38 MAPKs exhibit differential responses to specific drugs and inflammatory agents [7,11]. Thus, the four p38 isoforms can target genes in response to specific drugs and inflammatory agents. Its diverse functions are seen, for example, in its role in the Fas-mediated apoptosis of endothelial cells of the murine liver sinusoids [12] versus its role as the MAP kinase required for pathogen defense against *Pseudomonas* infection in *Caenorhabditis elegans* [13]. A most recent study has demonstrated that *B. anthracis* lethal factor selectively induces apoptosis by cleaving the amino-terminal domain of MKKs that serves as a docking domain for the activation of p38 MAPK [13a]. This dismantling of the p38 MAPK–MKK interaction is the mechanism by which *B. anthracis* paralyzes the host inmate immunity. Similarly, in HeLa cells, p38α induces apoptosis while p38β promotes cell survival, suggesting both overlapping and distinct physiological roles for the p38 isoforms, but clearly demonstrating their roles in establishing a biochemical phenotype [14]. One subgroup (p38α and p38β) is inhibited by pyridinyl imidazole derivatives, drugs that inhibit the production of proinflammatory cytokines, while the others (p38γ and p38δ) are insensitive to these drugs [15].

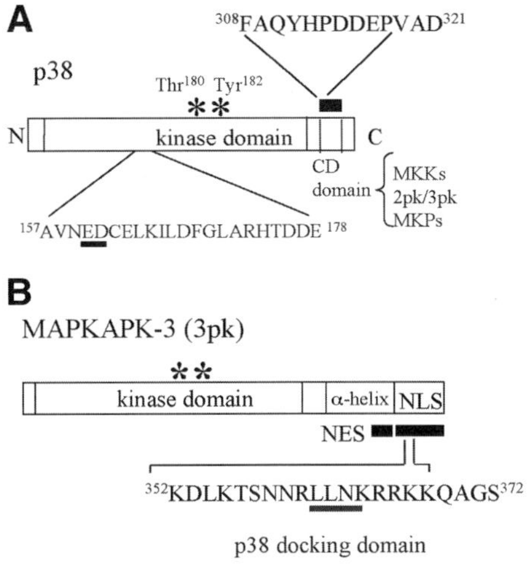

Figure 2 Maps showing the p38 MAPK and 3pk docking sites. (A) The p38α docking sites for upstream activators (MKKs) and downstream substrates (2pk/3pk). (B) A map of the MAPKAPK3 domains. The nuclear localization signal (NLS), nuclear export signal (NES), and p38 docking domains are at the C-terminal end of 3pk.

Transduction of Stress Signals via p38 MAPK Pathway

The activation of p38 MAPK signaling occurs by interaction with one of two specific upstream serine/threonine MAPK kinases, MKK3 and MKK6 (Fig. 1). Targeted disruption of the *mkk3* and *mkk6* genes shows that they have nonredundant functions [15,16]. For example, coexpression of MKK3 with p38β enhances hypertrophy, whereas coexpression with p38α enhances apoptosis [18,19]. This signaling specificity is an example of the generation of specific biological responses by this kinase pathway and emphasizes the importance of how specific complexing of pathway components target and affect specific biological processes.

Although MKK6 is a common activator of p38α, β, γ, and δ, MKK3 activates only p38α, γ, and δ [20]. This exclusive activation of p38β by MKK6 contributes to this pathway's specificity of signal transduction.

The following factors contribute to the specificity of p38 MAPK activation [4,18]: (1) the selective formation of functional complexes between MKKs and the p38 MAPK isoforms, which requires the presence of a MAPK docking site at the N terminus of the MKKs (Fig. 2); and (2) selective recognition of the activation loop (T-loop) or catalytic domain of p38 MAPK isoforms; the T-loop contains the Thr180-Tyr182 residues involved in kinase activation. Together, these provide a mechanism for the selective activation of p38 MAPKs in response to activated MKKs (Fig. 2).

Structural Basis for Protein–Protein Interactions of p38 MAPK and MKK Proteins

Enslen *et al.* (18) showed that interaction with a p38 docking domain in MKK6 contributes to the activation of all four p38 isoforms (Fig. 2). This docking site identified in the N-terminal region of MKK1 consists of highly conserved sequences and has also been identified within the N-terminal region of MKK6. Synthetic peptides based on the primary sequence of the MKK6 docking sites inhibit activation of p38β, indicating that MKK binding is necessary for the activation. MKK3, on the other hand, which lacks this docking site, can activate p38α, p38γ and p38δ, but not p38β; this accounts for the selective activation of p38α but not p38β by MKK3. Differences in the primary sequences of the T-loop of the p38 isoforms also contribute to signaling specificity (discussed later). Thus, selective activation of p38 by MKKs requires multiple molecular determinants present in both kinases, that is, p38 and MKKs.

A conserved docking motif common to interacting substrates, activators, and regulators has been localized to the C-terminal portion of p38 MAPKs and is commonly used for binding to MKKs, nuclear localized MAP-kinase-activated protein kinases (MAPKAPKs; 2pk, 3pk), and MAPK phosphatases (MPK). This common docking (CD) domain is characterized by negatively charged amino acids (MPKs; Fig. 2).

In p38α the CD domain is outside of the active center (Fig. 2). Conceptually, therefore, recognition between p38 MAPKs and interacting proteins involves both the docking interaction domains and the transient enzyme–substrate interaction at the T-loop (Thr180-Tyr182), both of which regulate the efficiency and specificity of the enzymatic (kinase) reactions [4,19–24].

Beside the CD domain, another domain, the ED site, determines docking specificity toward 2pk and 3pk, which are the nuclear enzymes that associate with p38 MAPK and transcription factors [23]. In p38 this site is located at Glu160 and Asp161; exchange of these amino acids with the corresponding residues in ERK2 and vice versa also changes the docking specificity. A 3D molecular model of MAPKs shows the proximity of the CD and the ED domains, which would explain their ability to serve as a common docking groove [25]. It is proposed that every MAPK-interacting molecule may bind to this docking groove and that each residue therein is differentially involved in each docking interaction [26]. This study suggests that on translocation to the nucleus, p38 isoforms interact with 3pk (or 2pk) to form a complex that targets transcription factors (Fig. 3). In addition, the 2pk and 3pk proteins contain trafficking domains that mediate the import as well as the export of a p38 MAPK–2pk complex from the nucleus to the cytoplasm (Figs. 3 and 4) [26].

Activation of the p38 Catalytic Site

The MAPKs are activated by dual phosphorylation of Thr180 and Tyr182 in the T-loop. Whereas p38α is phosphorylated preferentially on Tyr182 by MKKs that lack a MAPK docking site (e.g., MKK3), those with a MAPK docking site (MKK6) phosphorylate p38α MAPK on both Thr180 and Tyr182 residues. Thus, this differential phosphorylation by MKK3 versus MKK6 participates in the specificity of signaling of these upstream activators. It has been suggested that docking may increase the processivity of MAPK phosphorylation, leading to dual phosphorylation and increased activation [26]. Similarly, differential phosphorylation occurs with MKK4 which preferentially phosphorylates the tyrosine residue of the JNK1 catalytic site while MKK7 phosphorylates the threonine residue of this catalytic site.

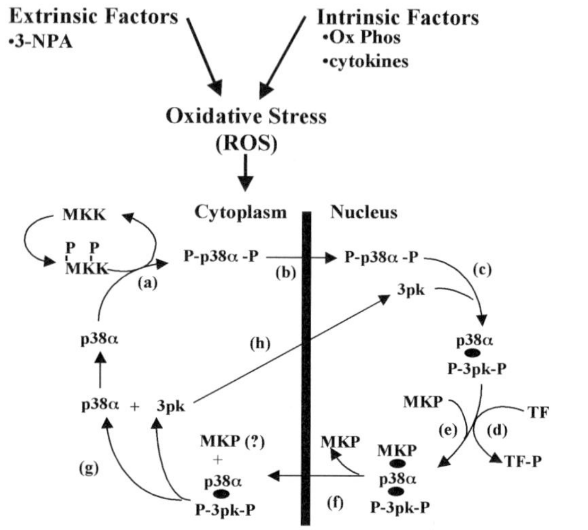

Figure 3 A proposed pathway for the nucleocytoplasmic trafficking of p38α signaling proteins, its upstream activators (MKKs), and downstream substrates (3pk and transcription factors). (a) Activation of p38α in the cytoplasm; (b) translocation to the nucleus; (c) activation of 2pk/3pk; (d) activation of ATF-2 transcription factor (TF); (e) interaction of MAPK-phosphatase (MKP) with the complex and inactivation of signaling; (f) translocation of the complex to the cytoplasm; (g) dissociation of the p38α-3pk-MKP complex; (h) dephosphorylated 3pk is translocated to the nucleus where it reenters the transcription factor activation cycle; p38α is reactivated in the cytoplasm in response to a new stress.

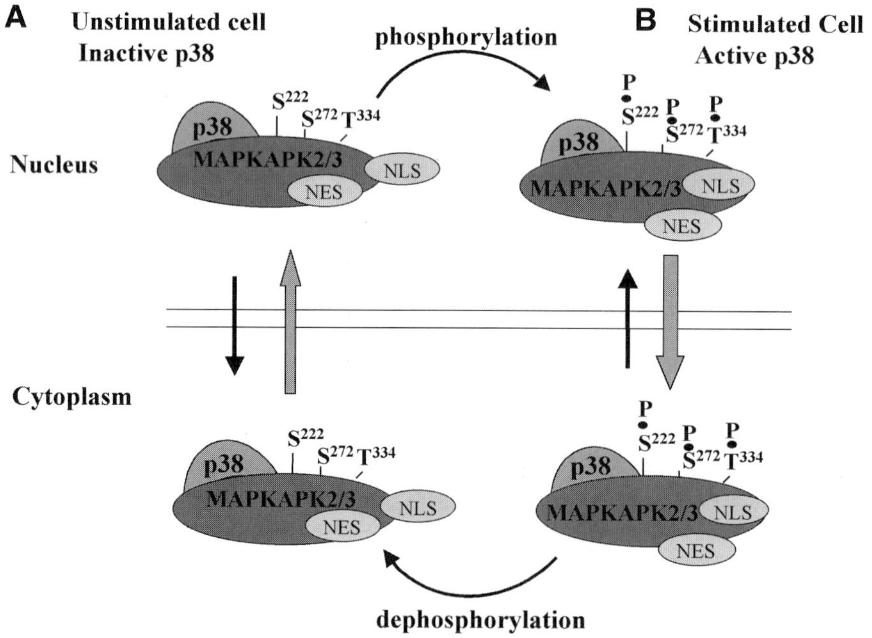

Figure 4 A model of the regulation of nuclear and cytoplasmic localization of a MAPKAPK-2/3–p38 MAPK complex as a mechanism for the activation of p38 MAPK in the nucleus. (A) In the unstimulated cell a MAPKAPK-2/3–p38 MAPK complex is formed in the cytoplasm. The nuclear localization signal (NLS) of MAPKAPK is exposed and mediates the translocation of the complex to the nucleus. (B) In the stimulated cell, the active complex mediates its transcription factor activation. The nuclear export signal (NES) is exposed and the complex is translocated to the cytoplasm. In the cytoplasm of the recovered cell, MAPKAPK-2/3 is dephosphorylated and the NLS is exposed. This results in the translocation of the complex to the nucleus. In the model the p38 MAPK may be phosphorylated in either the cytoplasm or nucleus.

Substrate Specificity of p38 MAPK Isoforms

Binding of the p38 MAPK complex to transcription factors is a critical determinant of kinase specificity. The mammalian MAP kinases (p38 MAPK, JNK, and ERK) require docking sites on transcription factors for correct function. For example, p38α and p38β phosphorylate and activate the muscle-specific transcription factors, MEF2A and MEF2C [27]. These p38 isoforms target transcription factors by a docking domain distinct from the phosphoacceptor motifs that confer responsiveness for p38α and p38β, but not p38γ or p38δ [4,28,29]. Similar binding motifs have been identified in several MAPK substrates, including c-Jun [23], Jun B [30,31], Elk-1 [27,32], NFAT4 [33], and ATF-2 [34,35].

The ability of p38 MAPK to bind to related sequences in both their downstream substrates (transcription factors) and their upstream MKKs enzymes suggests an upstream–downstream competition for binding to the MAPK. This hypothesis predicts that phosphorylation of transcription factors by MAPK can only occur after release of activated MAPK from MKK. However, the complexity of the signaling interactions increases significantly because of the interactions of p38 with the nuclear kinases, 2pk/3pk, which may interact with the transcription factors as well as mediate the nuclear export of p38 (Figs. 3 and 4) [26].

Subcellular Localization and Nucleocytoplasmic Transport of the p38 MAPK Signal

Subcellular localization is an integral part of the functional p38 MAPK signaling pathway (Figs. 3 and 4). Nucleocytoplasmic trafficking facilitates the regulation of transcription factor activity. It is thus important to understand the cellular localization of the p38 MAPK pathway components before and after stimulation. Nuclear substrates of p38 include 2pk, 3pk, MAP-kinase-interacting kinase (Mnk), and a variety of transcription factors, including CHOP, Elk-1, CREB, and Sap-1a. The 2pk and 3pk have both nuclear localization sequences (NLSs) and nuclear export sequences (NESs), and are localized in the nucleus in unstimulated cells, where, on stimulation, they dock with and are phosphorylated by p38 MAPK [16,23,34]. Although this complex phosphorylates (activates) specific transcription factors, it is also exported to the cytoplasm [16,23,34]. For example, phosphorylation of CREB may be mediated by 2pk. Thus, phosphorylation of 2pk and 3pk by p38 MAPK not only activates the kinase so that it can phosphorylate its transcription factor substrates, but it has also been postulated to expose the NES that results in nuclear export of both proteins (Fig. 4) [18,26,36].

The NES and NLS of 2pk and 3pk are located at their C-terminal ends (Fig. 2) [23,36]. In addition, 3pk is localized in the nucleus before osmotic stress and in the cytoplasm upon recovery. The docking of p38 with 3pk and 2pk is essential for their phosphorylation and nucleocytoplasmic export of the complex. Thus, the docking interaction between p38 and 3pk is achieved via direct interaction of the CD domain and the ED site of p38 with the C-terminal portion of 3pk. (Figs. 2 and 4).

Although some data indicate that p38 is activated in the cytoplasm, followed by rapid transport to the nucleus where it phosphorylates its substrates and then is exported to the cytoplasm, there is equally strong evidence that it may also be activated within the nucleus by 2pk or 3pk, followed by translocation of the activated complex to the cytoplasm [26]. Thus, the model in Fig. 4 suggests that a p38–2pk/3pk complex may be formed in the cytoplasm of unstimulated cells and translocated to the nucleus where it may be activated. The fact that there are cytoplasmic substrates (PRAK and Hsp27) that are phosphorylated by p38 in the cytoplasm points to the diversity of p38 MAPK activity for downstream targets. Interestingly, the activation of p38 MAPK by DNA-damaging agents supports the idea that the p38 cascade might be initiated in the nucleus [37]. Activation of p38 in the nucleus, on the other hand, would indicate that MKK3 and MKK6 must be localized there. Ben-Levy *et al.* [26] found that both MKK3 and MKK6 are localized in both the cytoplasm and nucleus. Thus, 2pk and 3pk may serve a dual function, both as effectors of p38 by phosphorylating substrates such as Hsp27 (cytoplasmic) and CREB (nuclear), and as determinants of p38 localization (Figs. 3 and 4).

Feedback Control of MAPK-Regulated Transcription

The dual-specificity MAP kinase phosphatases (MKPs) inactivate MAPKs by targeting their two regulatory phosphorylation sites [38]. There are about nine mammalian MKPs, divided into two groups according to their patterns of transcriptional regulation and subcellular localization [39]. The nuclear MKPs are rapidly and highly inducible by many of the stimuli that activate MAPKs. It is postulated, therefore, that these MKPs play an important role in the feedback control of MAPK signaling in the nucleus [39–41]. Hutter *et al.* [42] showed that MKP-1 interacts with the C-terminal end of nuclear p38 to activate MKP-1 catalytically. This raises the question of whether a 2pk/3pk–p38–MKP complex forms in the nucleus and is exported into the cytoplasm (Fig. 3) and whether this complex is targeted in the process of inactivation of p38 MAPK?

Several cytosolic MKPs can be triggered by direct interaction with MAPK [43,44]. MKP-3 interacts specifically with ERK [19,41]; binding of ERK2 to MKP-3 dramatically enhances the latter's catalytic activity [43]. On the other hand, MKP-4 interacts with all members of the three major MAPK subfamilies to become catalytically activated [43]. These novel mechanisms ensure the tight feedback control of MAPK signaling in the cytosol.

Response of the Aging Liver to Stress Challenges

The free radical theory of aging proposes that endogenous oxygen radicals generated by cells are a basic cause of progressive age-associated declines in cell and tissue function, and that oxidative stress generated by extrinsic or environmental factors accelerates this decline (Fig. 3) [45–50]. Identification of mitochondria and peroxisomes as major sources of endogenous reactive oxygen species (ROS) and of superoxide dismutases as enzymes that remove superoxide anions supports this theory. More precisely, the theory postulates that accumulated oxidative damage to DNA, RNA, proteins, and lipids leads to the age-associated decline in tissue function and development of the age-associated biochemical phenotype [49–54]. In fact, the accumulation of ROS-damaged macromolecules, for example, nuclear and mtDNA, and carbonylated proteins [55,56] suggests that oxidative damage of macromolecules may be a major contributing factor to the decline in specific biological functions. This idea is strongly supported by studies with nematodes [52], *Drosophila* [49,51], and rodents [53,54,57], showing longevity to be associated with resistance to oxidative stress. Thus, damaged mitochondria release more ROS, initiating a vicious cycle of increasing DNA damage, leading to increased ROS production, more DNA damage, and mitochondrial dysfunction [53,55–59]. Based on this hypothesis, it might be expected that (1) the increased level of persistent (long-term) oxidative stress may, through oxidative damage of macromolecules, affect the function of stress signaling pathways; (2) as aging progresses, the constitutive activity of stress-activated signal pathways (e.g., p38 and SAPK/JNK) would increase; and (3) this new level of activity is stabilized—becoming a basic factor in the development of chronic stress in aged tissues.

Mitochondrial Dysfunction as a Source of Oxidative Stress

Phosphorylation of proteins in the p38 MAPK pathway increases dramatically in the aged liver without an increase in pool levels, suggesting that these modifications may mediate functional changes in these stress-signaling proteins [37,60]. Furthermore, p38 MAPK is activated in the young mouse livers by methyl methane sulfonate, a DNA damaging agent, and 3-nitropropionic acid (3-NPA), a potent inhibitor of complex II (succinic dehydrogenase). Both of these agents are generators of oxidative stress (ROS). In the livers of aged mice, however, this response to oxidative stress does not occur. Thus, the increased constitutive levels of phosphorylation, but failure to respond to the oxidative stress, suggests that the structure of these proteins may be altered in the aged tissue [60]. It has been proposed that changes in the p38 MAPK and SAPK/JNK signaling pathway activities may be a primary causative factor in the age-associated alteration of structure and function of transcription factors targeted by these pathways. A summary of these interactions affecting signaling in the liver is shown in Fig. 5 and listed here: (1) "Locked-in" generation of ROS is a major cause of age-associated mitochondrial dysfunction and a basic factor in the age-associated changes in activity of p38 MAPK and SAPK/JNK signaling pathways; (2) similarly, the increased level of ROS may cause nuclear dysfunction; (3) these structural changes alter the activities of transcription factors targeted by these signaling proteins, resulting in the development of symptoms of chronic stress (oxidative stress); (4) p38 MAPK pathway proteins' failure to respond to MMS or 3-NPA-generated oxidative stress in the aged rodent liver is due to structural changes in catalytic and/or regulatory domains of these kinases, decreasing their ability to interact/dock to form complexes essential for normal signal transduction; and (5) similar age-associated characteristics seen in the aging liver are also observed in other tissues such as brain and muscle.

Mitochondrial Dysfunction and the p38 MAPK Stress Signaling Pathway

3-NPA, a suicide inhibitor of complex II (succinic dehydrogenase), increases ROS by interrupting the electron cascade and preventing phosphorylation [61–65]. 3-NPA increases

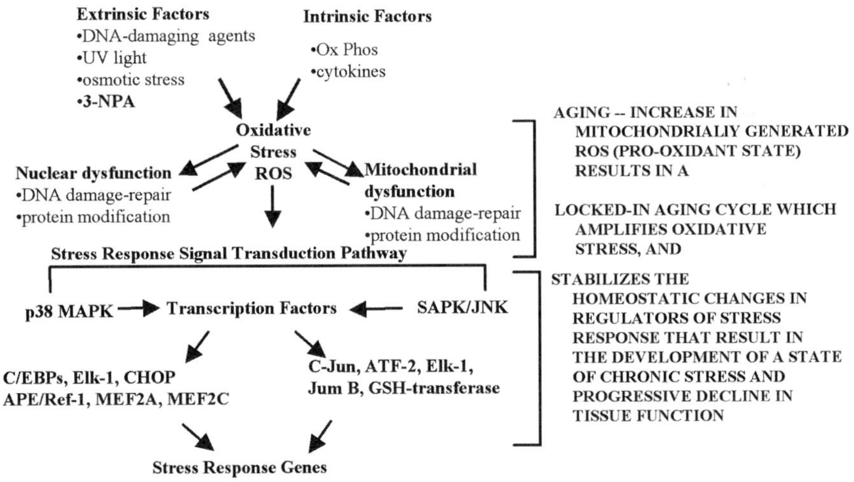

Figure 5 The oxidative chronic stress cycle of aging that depicts the relationship of oxidative stress to mitochondrial dysfunction, activation of stress response signaling pathways, stress response transcription factors, and their targeted genes.

production of free radicals (about twofold) in both mitochondria and microsomes of the liver and brain [68–70].

The mechanism of inactivation of complex II (succinic dehydrogenase) involves the alkylation of an—SH group in the enzyme's catalytic site, located in the 70-kDa subunit [66]. The inactivation mechanism involves a nucleophilic stack by this essential—SH group on the double bond of 3-nitroacrylate (3-NA, the product of SDH oxidation of 3-NPA. This mechanism proposes that an $O^{2.-}$ radical may be the ROS generated by the oxidation–reduction of FAD resulting from the oxidation of 3-NPA [66]. Interestingly, it has been shown that oxidative stress generated by rotenone, an inhibitor of complex I, also activates p38 MAPK [67]. Although these experiments were done with cells in culture, these data demonstrate that ROS generated by dysfunction of components of the mitochondrial electron transport system participate in activation of the p38 MAPK signaling pathway. It has also been proposed that the nitro group of 3-NPA may be reduced by the nitro-reductase system to produce nitro superoxide anions. This initiates lipid peroxidation, leading to cell membrane damage [70]. Thus, 3-NPA can stimulate free-radical production both *in vivo* and *in vitro*. The 3-NPA-mediated increase in ROS (acute oxidative stress) can activate both the p38 MAPK and SAPK/JNK stress response signaling pathways and so induce an inflammatory response [37,60]. These studies suggest a possible linking of mitochondrial dysfunction, that is, generation of ROS to the activation of specific stress response signaling pathway(s) and the targeting of stress response genes. Furthermore, these studies have shown that aging affects the ability of liver cells to respond to this oxidative stress as is indicated by either failure to respond or an increased response (JNKs) to these challenges.

Aging, Oxidative Stress, and Cellular Signaling in the Liver

Both extrinsic and intrinsic signals are transduced to nuclear gene targets by a complex series of protein–protein interactions. These characteristics are determined by 3D protein structures that determine the proximity of specific amino acid clusters and their functional capacity [71,72]. Modifications of these clusters by phosphorylation/dephosphorylation modulate such functions as enzyme activity (activation), docking ability, and intracellular localization. The activities of the stress-activated p38 MAPK and JNK signaling pathways in aged liver suggest that these molecular interactions are altered by and affect the level and efficiency of signal transduction.

SOCS Family of Negative Regulators of Inflammatory Response

Cytokines activate signal transduction processes in response to inflammatory challenges as well as to growth factors and hormones. They initiate their signaling cascades by binding to cell-surface receptors whose cytoplasmic domains activate intracellular signal cascades of the JAK-STAT pathway [73]. In the liver, cytokine receptors (e.g., IL6) are associated with the Janus kinase (JAK) family of tyrosine kinases, for example, TYK2, JAK1, JAK2, and JAK3. Thus, cytokine binding induces receptor aggregation, which then activates JAKs via cross-phosphorylation of tyrosine residues in the cytoplasmic domain of cytokine receptors. This creates recruitment sites for signaling proteins containing Src homology 2 (SH2) or phosphotyrosine binding (PTB) domains. By this mechanism, cytokine stimulation initiates multiple signal transduction cascades, one of which is the *s*ignal *t*ransducers and *a*ctivators of *t*ranscription (Stats). These signaling pathways then target genes whose activities lead to such biological processes as inflammatory response, cell differentiation, proliferation, and apoptosis.

Induction of SOCS Transcription and the Negative Regulation of Cytokine Gene Expression

Cytokine signaling is negatively regulated by the *s*uppressor *o*f *c*ytokine *s*ignaling (SOCS) family of proteins (SOCS1–SOCS7, CIS) [73]. SOCS gene transcription is up-regulated in response to cytokine stimulation, and the corresponding SOCS proteins inhibit the cytokine-induced signaling pathway. Modulation of this signaling occurs by several mechanisms, including (1) inactivation of the JAKs, (2) blocking access of the Stats to receptor binding sites, and (3) ubiquitination of signaling proteins and targeting to degradation via the proteosome.

The molecular mechanism of negative regulation of cytokine signaling by SOCS proteins is shown in Fig. 6. The models show that the cytokine activation signal is initiated by binding the dimerized cell-surface receptor. This stimulates the JAK kinase phosphorylation of the Stat transcription factors. The dimerized Stat-P translocates to the nucleus where it activates SOCS genes. Transcription and translation of the SOCS mRNA produces the SOCS proteins. Inhibition of the cytokine signal is then mediated by several interactions between a specific SOCS protein and the JAK or Stat proteins.

SOCS mRNAs are induced by cytokines and the corresponding SOCS proteins through a classical negative feedback mechanism that attenuates the signaling pathways that stimulated their production (Fig. 6). Furthermore, because the induction of SOCS proteins by one cytokine can inhibit signaling by another, it appears that these proteins might also mediate an overall down-regulation via their ability to cross-talk between cytokine receptors.

Overexpression of CIS, SOCS1, or SOCS3 in cell lines has shown their inhibitory effect on a wide variety of cytokines, hormones, and growth factors. Transgenic mouse models have revealed the ability of SOCS proteins to inhibit cytokine inducible signaling *in vivo*. For example, transgenic mice that express *cis* under the control of the β-actin promoter exhibit a significant down-regulation of GH-induced

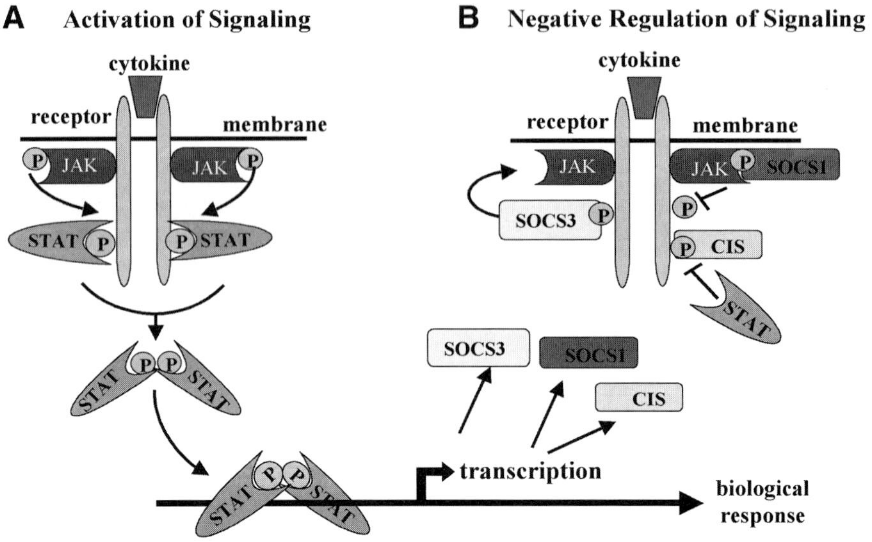

Figure 6 A model showing the cytokine-mediated activation of the JAK/Stat signaling pathway and the SOCS-mediated negative regulation of the JAK/Stat signal. (A) Activation of cytokine signaling is initiated by the dimerization of a cytokine membrane receptor complex. Phosphorylation of JAKs occurs upon dimerization of the receptor. Activation of the JAK/Stat signaling induces SOCS gene expression. (B) The SOCS proteins (CIS, SOCS1, SOCS3) inhibit the signaling pathway by negative feedback. The inhibition by SOCS1 involves binding to the JAKs and inhibition of JAKS catalytic activity. Inhibition by SOCS3 involves binding to the JAKs binding site on the receptor. Inhibition by CIS involves blocking of STAT binding to the receptor.

tyrosine phosphorylation of Stat 5 in the liver [74]. The fact that the phenotype of CIS transgenic mice resembles that of mice lacking Stat5a and/or Stat5b suggests that CIS might contribute to the negative regulation of Stat5 [74].

Physiological Function of SOCS Protein

Understanding of the physiological functions of SOCS genes advanced rapidly upon the successful generation and analysis of several knock-out mice. Although the SOCS1 (*socs1*$^{-/-}$) mutants are normal at birth they exhibit stunted growth and die at 3 weeks of age with a syndrome characterized by severe lymphopenia, activation of T cells, fatty degeneration and necroses of the liver, and macrophage infiltration of major organs [75]. It appears that the neonatal defects of the *socs1*$^{-/-}$ mutants may be due to the overexpression of IFN-γ signaling. More specifically the studies with the mutants suggest that SOCS1 plays a key role in the negative regulation of signaling by IFN-γ. This is supported by the fact that administration of IFN-γ to neonatal wild-type mice induces a pathology similar to that observed for *socs1*$^{-/-}$ mutants. It has been suggested that SOCS1 may regulate I-cell differentiation to prevent the emergence of activated I cells that produce excess IFN-γ, a situation that may have secondary consequences such as the up-regulation of proinflammatory cytokines such as TNF-α.

Mice lacking the *socs2* gene appear normal until 3 weeks after birth after which they grow 30–40% larger than their wild-type littermates. The weight increase is due to a significant increase in visceral organ weight, (not excess fat), carcass weight, long bone length, and body length [75]. Because *socs2*$^{-/-}$ mice are excessively large, it has been proposed that SOCS2 may negatively regulate growth promoting cytokines such as GH and IGF-I. For example, *socs2*$^{-/-}$ mice exhibit abnormalities associated with GH and IGF-I signaling. One outcome of GH signaling is the production of IGF-1; *socs2*$^{-/-}$ mutants exhibit characteristics of deregulation of GH and IGF-I signaling. These include elevated levels of IGF-I in heart, lungs, and spleen and decreased levels of the major urinary proteins (MUPs) [75]. MUPS have been shown to be down-regulated in GH-deficient mice [76,77]. In addition, the regulation of MUPs, which is GH-pulse dependent, is down-regulated in mice that overexpress GH. This is attributed to the disruption of the GH-pulsatile pattern. The decrease in MUPs in *socs2*$^{-/-}$ urine suggests, therefore, that GH signaling might be disregulated in these mice [75]. SOCS2, therefore, is a major component of the SOCS family synthesized in the liver and appears to play an essential physiological role in the regulation of growth, possibly due to its ability to modulate GH and/or IGF-I signaling. Thus, the ability of SOCS2 proteins to regulate GH/IGF-I signaling indicates that it plays a key role in the regulation of multiple pathways of biological processes.

Analysis of *socs3*$^{-/-}$ mice has shown that these mutants, as well as those overexpressing SOCS3, are embryonic lethals. The transgenic mice that overexpress *socs3* lack fetal liver erythropoiesis, whereas the *socs3*$^{-/-}$ embryos exhibit a marked erythropoiesis throughout the embryo [78]. There appears to be some controversy with respect to the characteristics of this mutation since a more recent study describes severe placental dysfunction as the major consequences of this defect [79]. SOCS3 is especially expressed in fetal liver erythroid progenitors during the stage of erythropoiesis in which the erythropoietin (Epo)-dependent expansion of erythroid lineage cells occurs. It has been proposed, therefore, that the *socs3*$^{-/-}$ gene plays an important role in the regulation of fetal liver erythropoiesis and that the mechanism involves modulation of Epo signaling. Thus, SOCS3 deficiency in the *socs3*$^{-/-}$ mutant results in inability to attenuate Epo signaling. SOCS3 is also expressed in low levels in adult tissues; it is inducible by LPS in adult liver and plays a key role in the regulation of leptin in brain.

Molecular Mechanisms of Negative Feedback Regulation of Cytokine Signaling

The major SOCS family members of the liver, CIS, SOCS1, SOCS2, and SOCS3, inhibit cytokine signaling by several different mechanisms (Fig. 6): SOCS1 reacts directly with JAK1, JAK2, and JAK3, and TYK2, thereby inhibiting their catalytic (kinase) activity [80–83]. SOCS3 also inhibits JAK2 activity, but it binds to JAK2 with a lower affinity, suggesting it is a less efficient inhibitor than SOCS1 [84–87]. CIS and SOCS2 do not bind to or inhibit the JAKs [83,84].

The mechanisms by which SOCS1 inhibits JAK activity involve its binding to phosphorylated Tyr1007 of JAK2, which lies within the JAK2 activation loop, thereby regulating JAK activity [83]. The SOCS1 SH2 domain and 24-amino-acid residues immediately N terminal to the SH2 domain are required for high-affinity binding and inhibition of JAK2 activity [82,83]. There is also a kinase inhibitory region (KIR) within the 24-residue region (both SOCS1 and SOCS3) that is required for inhibition of JAK2 activity [83]. It is proposed that the KIR might inhibit JAK activity by acting as a pseudosubstrate that prevents access of legitimate substrates to the JAK catalytic pocket [83].

The negative regulatory activity of CIS occurs via another mechanism that involves the competition of Stats versus SOCS proteins for the phosphotyrosine binding (PTB) sites within the cytoplasmic domain of the cytokine receptor (Fig. 6). This is supported by the observation that both CIS and Stat5b bind to and compete for an overlapping set of phosphotyrosine residues on the GH receptor [88–90]. Thus, CIS inhibits GH signaling and Epo signaling by binding to their receptors [90].

The ability of SOCS3 to inhibit GH-induced JAK2 activation increases when the expression level of GHR increases [89]. Similarly the ability of SOCS3 to inhibit IL-2-induced JAK1 activation is enhanced in the presence of IL-2 [91]. These studies indicate that the inhibition of JAK activity by SOCS3 is augmented when SOCS3 is bound to cytokine receptors.

Recently, Nicholson *et al.* [92] demonstrated that SOCS3 binds with high affinity to a phosphopeptide corresponding to a region surrounding Tyr757 within gp130 of the IL-6 receptor. Mutation of Tyr757 causes decreased ability of SOCS3 to inhibit gp130 signaling. These data suggest that optimal inhibition by SOCS3 occurs when SOCS3 is bound to gp130. A similar mechanism is suggested by the inhibition of the SOCS3 binding site on leptin or Epo receptors, which also interferes with ability of SOCS3 to inhibit signaling from these receptors. Thus, it is proposed that the mechanism of SOCS3 inhibition binding to activated cytokine receptors in the region of the JAKs (Fig. 6). Once localized at the receptor, SOCS3 inhibits JAK activity through its KIR.

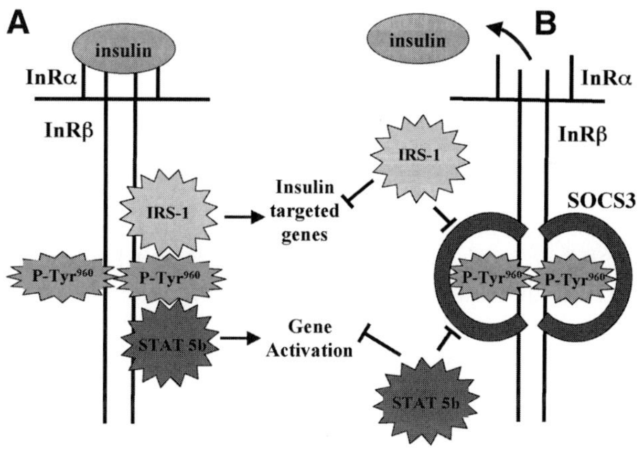

Figure 7 A model of the inhibition of Stat5b and IRS-1 binding to the insulin receptor by SOCS3. (A) The IRS-1, through its PTB domain interacts with the insulin receptor P-Tyr960 and the amino acid residues at the C terminus of the motif containing the P-Tyr. Stat5b interacts with the P-Tyr960 and the amino acid residues at the N terminus of the motif containing the P-Tyr. SOCS3 binds to the cytoplasmic domain of the insulin receptor (InRβ) via its SH2 domain and interaction with P-Tyr960. The binding of SOCS3 to the P-Tyr960 containing motif prevents IRS-1 and Stat5B from binding to the insulin receptor.

Role of SOCS Proteins in Down-Regulation of Insulin Signaling

Insulin induces a transient expression of SOCS3 in liver, skeletal muscle, and white adipose tissue *in vivo*, and in 3T3-L1 cells *in vitro* [92]. Furthermore, SOCS3 antagonizes insulin-induced IRS-1 tyrosine phosphorylation and decreases IRS-1 association with p85 (PI3K). This involves the binding of SOCS3 to Tyr960 of the insulin receptor, thereby preventing Stat5b activation by insulin (Fig. 7).

Thus, the binding of SOCS3 to Tyr960 of InR has multiple regulatory functions, for example, attenuation of STATb targeted genes, attenuation of the downstream activation of IRS-1 tyrosine phosphorylation, and decreases in docking with p85 (PI3K), thereby demonstrating a sequential attenuation of downstream signal transduction.

Analysis of RNA from livers of mice starved overnight and injected with insulin showed that SOCS3 was induced 5.5-fold 1 hr after injection. Expression was sustained for 1 hr and returned to basal levels within 2 hr after injection. Similarly starved mice were injected with TNF-α; SOCS3 was induced in liver within 2 hr after injection and returned to normal at 4 hr. The level of induction of SOCS3 by TNF-α in liver was less than that by insulin.

These studies have shown that SOCS3 interferes with the tyrosine phosphorylation of IRS-1. IRS-1, through its interaction with PI3K, Grb2, and SHP2, controls several major cellular processes regulated by insulin such as glucose uptake, protein and glycogen synthesis, and gene expression. IRS-1 binds to InR using its PTB domain, whereas Stat5b and SOCS3 bind to the InR via their SH2 domains. However, the SH2 domain interacts with P-Tyr960 and amino acids located in the N terminus of the motif containing P-Tyr960, whereas the PTB domain binds P-Tyr960 and is dependent on residues located at the C terminus of the motif containing P-Tyr960. Another difference is that IRS-1 possesses a PH domain that also plays a critical role in insulin-induced IRS-1 phosphorylation in intact cells (Fig. 7) [93].

SOCS3 expression attenuates insulin-induced tyrosine phosphorylation of IRS-1 and its subsequent association with $p85^{PI3K}$. Thus, binding of SOCS3 to Tyr960 prevents the docking of IRS-1 with InR both through its PTB and PH domains. The interference of IRS-1 phosphorylation and p85 docking should attenuate specific biological functions. Through this protein–protein complex, SOCS-3 serves as a potent inhibitor of insulin signaling. It is proposed that SOCS3 expression is increased in several situations associated with insulin resistance.

All reports point toward an important role in IRS-1 phosphorylation on serine residues thought to be responsible for the decrease in insulin signaling [94]. There is a pronounced decrease in SOCS3 expression in (obese) mice lacking TNF-α signaling. Thus, elevated SOCS3 levels (found in obesity) in inflammatory response may be linked to increased TNF-α expression. To explain TNF-α-induced insulin resistance it has been proposed that phosphorylation of serine residues in IRS-1 may be responsible for the decrease in insulin signaling. This is based on the proposal that phosphorylation of serine residues of IRS-1 prevents its docking to form an active signaling complex [95].

The InR mediates its biological effects through tyrosine phosphorylation of several substrates such as IRS, Shc, and Stat5b. This phosphorylation is dependent on a functional coupling between the InR and its substrates. Thus, any molecular event impairing the coupling between the insulin receptor and its substrates should lead to a decrease in the tyrosine phosphorylation of the substrate molecules. These studies have shown that SOCS3 expression reduces insulin-induced DNA binding activity of Stat5b, probably via a competition between Stat5b and SOCS3 for binding to the insulin receptor P-Tyr960. Thus, SOCS3 negatively regulates insulin signaling; increased SOCS3 participates in development of insulin resistance; SOCS3 also regulates Stat5b and its targeted genes.

Role of SOCS Proteins in Down-Regulation of Growth Hormone Signaling

GH resistance is known to develop in certain conditions of severe illness, for example, sepsis and burns. Sublethal doses of LPS have been shown to result in a decreased ability of GH to stimulate Stat5 tyrosine phosphorylation in the liver. This occurs for at least 6 hr after single endotoxin injection [96]. There is no change in Stat5 pool levels in the liver, and the decrease in Stat5 phosphorylation results in decreased GH-mediated gene activation. This is proposed to be the mechanism of the resistance to GH induced by endotoxin or other catabolic states.

It has also been shown that the GH-GHR activation of JAK2 tyrosine kinase by autophosphorylation of tyrosine residues is also decreased in response to LPS treatment [94,97]. In this case, however, there is also an increase in the nonphosphorylated JAK2 pool level, which results from an increase in JAK2 mRNA. Although the mechanism of up-regulation of Jak2 gene transcription, for example, the transcription factors associated with this regulation, is not known the overall mechanism appears to involve both post-translational and transcriptional processes.

The abundance of GHR in the liver is sensitive to such stresses as nutrient regulation (starvation), surgery, sepsis, and so on. GHR levels were not affected by endotoxin 4 hr after treatment at a time when the Stat5 phosphorylation is lowest and JAK2 mRNA level is at its peak. Thus, these changes in Stat5 and JAK2 are independent of GHR abundance in the liver and suggest that resistance to GH occurs in the absence of receptor abundance.

The SOCS proteins are candidates for the endotoxin-induced postreceptor GH resistance. Three possible candidates of the SOCS family that are abundant in the liver are CIS, SOCS2, and SOCS3. These genes respond to cytokines and endotoxins; Northern analyses showed a rapid 4-fold increase in CIS and 10-fold increase in SOCS3 mRNAs but only a small transient increase in SOCS2 after endotoxin treatment. The SOCS3 mRNA remained elevated for at least 6 hr.

The increased SOCS gene mRNAs correlate with endotoxin-induced resistance to GH-stimulated Stat5 and JAK2 tyrosine phosphorylation, suggesting that SOCS proteins mediate these effects on GH signaling. In addition further evidence is provided by SOCS3-mediated inhibition of Spi 2.1 promoter activity induced by GH transactivation. Thus, SOCS3 negatively regulate the GH signaling that activates Spi 2.1. This negative regulation was not seen with CIS.

GH itself induces SOCS3 mRNA in the mouse liver; the resultant SOCS3 protein then functions as a negative feedback loop to attenuate GH signaling. This mechanism is similar to that proposed for the attenuation of cytokine signaling by SOCS proteins. After endotoxin administration, multiple cytokines are elaborated and each may interact with the liver to induce SOCS-family gene expression. Thus, the activation of SOCS genes in the liver may serve as attenuators of GH as well as cytokine signaling via a negative feedback mechanism.

In summary, endotoxin causes the decrease of GH stimulated tyrosine phosphorylation of Stat5 in the liver. This occurs in the absence of any change in Stat5 pool levels and an increase in JAK2 mRNA and protein pool levels. There is also a 50% decrease in the amount of GH stimulated phosphotyrosine per JAK2, which is preceded by the marked

Figure 8 The role of TNF-α in development of resistance to IFN-α therapy. The sequence of biochemical events that leads to resistance to IFN-α therapy in patients with liver disease.

increase in JAK2 mRNA. The GH signaling changes due to LPS-mediated resistance are not due to changes in the abundance of GHR. But SOCS mRNAs (CIS, SOCS2, SOCS3) are rapidly up-regulated by LPS treatment. Thus, the SOCS proteins are involved in LPS-mediated alteration of GH signaling.

Role of SOCS Proteins in the Regulation of Liver Regeneration

Adult animal hepatocytes regenerate following partial hepatectomy in rats. Quiescent hepatocytes rapidly and synchronously reenter the cell cycle and divide until the original mass of liver tissue is restored (~10 days). One of the Stat transcription factors, Stat3, which is associated with cytokine signaling, is activated during the first few hours after PH [97]. For example, Stat3 activation in rat hepatocytes peaks at 3 hr after partial hepatectomy and recovers by 12 hr. The Stat3 activated hepatocytes were initially detected in the periportal zones and then extended to the pericentral zones. This coincides with the transcriptional induction of immediate early growth response genes [96]. This follows the pattern of DNA synthesis [98]. DNA synthesis and Stat3 binding activity, after partial hepatectomy in TNFR-I or IL-6(–/–), are severely impaired. This can be corrected by injection of IL-6 before partial hepatectomy [99–101].

For example, Stat3 is also activated in Kupffer cells as well as sinusoidal endothelial cells prior to activation in hepatocytes. Stat3 is activated rapidly in regenerating or sham-operated livers [101]. Stat1, -4, or -6 is not activated. Thus, SOCS1 and SOCS3 negatively regulate Stat3 signaling in the early phase of liver regeneration.

Role of SOCS and SHP2 in Regulation of Interferon Signaling

Patients having high levels of TNF-α have been found to respond poorly to IFN-α therapy, suggesting that TNF-α may be involved in resistance to IFN-α therapy. The studies by Hong *et al.* [102,103] have shown that TNF-α induces Stat1 protein expression but inhibits IFN-1α activated tyrosine phosphorylation of both Stat1α (91 kDa), and Stat1β, (84 kDa) at Tyr701. These data indicate Stat1 gene expression is activated.

TNF-α inhibition of IFN-α activated Stat1 in the liver occurs 2 hr after injection, which is consistent with new protein synthesis. The role of two protein families [SOCS and PIAS (protein inhibitor of activated Stat)] in the inhibition of IFN-α signaling was examined. The data show that TNF-α specifically induces SOCS3 mRNA expression in the liver; that SOCS3 attenuated IFN-α-induced reporter gene expression. These data suggest that SOCS3 plays a significant role in the TNF-α-mediated inhibition of IFN-α signaling in the liver (Fig. 8).

Protein tyrosine phosphatase (PTP) is another major pathway involved in suppression of IFN-α signaling. Injection of TNF-α into mouse showed that SHP2 is markedly induced, whereas several others were not affected. TNF-α specifically induces SHP2 protein expression in the liver. Coimmunoprecipitation showed that JAK1 and TYK2 coimmunoprecipitate with SHP2 in normal untreated (TNF-α) liver thus demonstrating an *in vivo* complex formation. TNF-α markedly enhances association of SHP2 with JAK1 or TYK2 and increased level of expression of SHP2 protein. These data suggest that induction of SHP2 protein expression and an increase in association of SHP2 with JAKs may contribute to TNF-α mediated inhibition of IFN-α signaling in liver, *in vivo* [103].

Although TNF-α and SOCS3 are components of the TNF-α inhibition of IFN-α signaling in liver *in vivo*, TNF-α does not directly induce SOCS3 expression or inhibit IFN-α signaling. TNF-α does stimulate IL-6 gene expression and it is the IL-6 that markedly stimulates SOCS3 expression in primary rat hepatocytes. These studies have shown that IL-6 is involved in the IFN-α induction of Stat1 protein expression but not in the induction of either SOCS3 or SHP2 in liver. The failure of TNF-α to induce SOCS3 or SHP2 was shown using IL-6(–/–) mutant. TNF-α can induce SOCS3 and SHP2 in IL-6(–/–) mutant livers.

CCl_4 was used to induce liver injury and inflammation to examine whether IFN-α is inhibited by the TNF-α mechanism. CCl_4 markedly induces TNF-α and attenuates IFN-α activated Stat1 in the liver. The Stat1 protein level goes up. Using TNFR1 and TNFR2 knock-out mice, it was shown that CCl_4 attenuated IFN-α-activated Stat1 in wild-type mice but not in knock-out mice. These studies indicate that activation of TNFR1 by TNF-α is involved in liver injury-mediated suppression of IFN-α-activated Stat1 in the liver.

In summary, (1) TNF-α, which is elevated in serum of patients with viral hepatitis (Fig. 8), inhibits IFN-α signaling in the liver *in vivo*; (2) TNF-α is involved in resistance to IFN-α therapy; (3) both SOCS3 and SHP2 are involved in TNF-α-mediated inhibition of IFN-α signaling in liver; (4) three families of proteins are implicated in the down-regulation of the JAK/STAT signaling: SOCS, PIAS, and PTP; (5) TNF-α induces SOCS 3 and SHP2, but not other SOCS and PTPs; and (6) induction of SOCS3 and SHP2 participate in TNF-α-mediated inhibition of IFN-α signaling.

Role of SOCS in Hepatocellular Carcinoma

Abnormalities of the JAK/STAT pathway have been reported in association with cancer; activation of JAK/STAT expression plays a significant role in oncogenesis [104]. Aberrant methylation of promoter regions, which silence transcription, has been shown to be associated with cancer. Using restriction landmark genomic scanning analysis of hepatocellular carcinoma (HCC), Yoshikawa *et al.* [104] identified the SOCS1 gene sequences in the DNA fragment and determined that the *NotI* site was methylated in the HCC. These studies showed that methylation silenced SOCS1 and the growth-suppression activity of SOCS1 protein. Examination of a variety of HCC cell lines showed that the expression status of SOCS1 correlates with the dense methylation of CpG islands in the promoter. Analyses also showed that the matrix association of SOCS1 is altered in association with methylation status and that the matrix attachment was reduced in the SOCS1 methylated HCC cells.

Because SOCS1 inhibits JAK2 tyrosine phosphorylation, there should be functional consequences of the silencing of SOCS1 by methylation. The cells showing silenced SOCS1 also exhibited phosphorylated JAK2. Furthermore, because phosphorylation of Stat3 is in the JAK/STAT pathway, the methylated SOCS1 should result in constitutive Stat3 phosphorylation. The data show that SOCS1 inactivated-JAK2 phosphorylated cells exhibit constitutive phosphorylation of Stat3. Thus, SOCS1 inactivation leads to a basal level of activation of JAK2/Stat3 pathway.

Restoration of SOCS1 in silenced HCC cells selectively suppressed the growth of these cells by inducing apoptosis. Thus, SOCS1 suppresses anchorage-independent growth as well as growth in a monolayer. The following model is proposed: (1) In normal cells, SOCS1 activation blocks JAK activation and terminates cytokine signaling due to stress and growth factor; (2) in cancer cells, SOCS1 is silenced by methylation and is unable to terminate these signaling pathways, making JAK activation constitutive; (3) thus, SOCS-1 silenced cells have unopposed growth stimulation by various cytokines, growth factors, and hormones; and (4) SOCS1 normally functions to suppress hepatocyte growth.

Conclusions

Biological signaling is the highly specific process that enables cells and tissues to regulate their responses to their environment, both intrinsic and extrinsic. This chapter focused on two major signaling pathways: the p38 MAP kinase pathway, which responds to mitogenic and stress activating signaling molecules, and the SOCs family, which "turns off" or mediates the recovery from these signaling stimulations. Although the p38 MAPK stress response pathway was first discovered in response to an inflammatory challenge by bacterial endotoxin (LPS), its diverse functions have been rapidly identified, thus making it a major and critical pathway that regulates numerous biological processes. This is seen in its regulation of differentiation, apoptosis, and immune functions.

The recent demonstration that the mechanism of cellular destruction by anthrax lethal factor involves blocking the p38 pathway adds to the importance of this pathway. Most certainly, there are other equally important signaling pathways that play a key role in the response to stress factors, mitogens, hormones, and so on, such as the ERK and SAPK/JNK pathways and the hormone activated pathways such as insulin/IGF-1 and GH pathways. All of these pathways share several mechanisms of signal transduction: protein–protein interactions (docking), protein modifications (phosphorylation), and intracellular trafficking (nucleocytoplasmic, mitochondrial). Thus, in our discussion of the mechanisms of p38 MAPK and SOCs pathways, our purpose was to familiarize the reader with these mechanisms that occur in most signaling processes. Furthermore, our focus on how aging affects the function of the p38 MAP kinase pathway is meant to demonstrate that in addition to environmental and intrinsic factors, signaling pathways also play a key regulating role in biological processes during the entire life cycle, that is, spanning the embryonic young adult and aging phases of life.

Although signaling is a critical factor in the activation of biological processes, an important part of the signaling process is "turning the signals off." Thus, by describing the negative feedback regulation of signaling by members of the SOCs gene family we demonstrate the basic mechanisms required for this regulation. Again, we see that protein–protein interactions, protein modifications, and intracellular trafficking are important processes in the mechanism. Thus, the regulation of biological signaling, both "in" and "off" plays a key role in establishing tissue homeostasis. It is important therefore, to understand these mechanisms in order to understand the biological processes of growth, development, and aging as well as their role, in many cases, in dysregulation in pathological circumstances.

References

1. Hunter, T. (2000). Signaling—2000 and beyond. *Cell* **100**, 113–127.
2. Schaeffer, H. J. and Weber, M. J. (1999). Minireview: Mitogen activated protein kinases: Specific messages from ubiquitous messengers. *Mol. Cell. Biol.* **19**, 2435–2444.
3. Pawson, T. and Nash, P. (2000). Protein–protein interactions define specificity in signal transduction. *Genes Dev.* **14**, 1027–1047.
4. Ono, K. and Han, J. (2000). The p38 signal pathway activation and function. *Cell. Signaling* **12**, 1–13.
5. Raingeaud, J. Gupta, S., Rogers, J. S., Dickens, M., Han, J., Ulevich, R. J., and Davis, R. J. (1995). Pro-inflammatory cytokines and environmental stress cause p38 mitogen-activated protein kinase activation by dual phosphorylation on tyrosine and threonine. *J. Biol. Chem.* **270**, 7420–7426.
6. Han, J., Lee, J. D., Bibbs, L., and Ulevitch, R. J. (1994). A map kinase-targeted by endoctoxins and hyperosmolarity in mammalian cells. *Science* **265**, 808–811.
7. Cohen, P. (1997). The search for physiological substrates of MAP and SAP kinases in mammalian cells. *Trends Cell Biol.* **7**, 353–361.
8. Lechner, C., Zahalka, M. A., Giot, F-F., Moler, N. P. H., and Ullrich, A. (1996). ERK6, a mitogen-activated protein kinase involved in G2C12 myoblast differentiation. *Proc. Natl. Acad. Sci USA* **93**, 4355–4359.

9. Li, Z., Jiang, Y., Ulevitch, R. J., and Han, J. (1996). The primary structure of p38gamma: A new member of p38 group of MAP kinases. *Biochem. Biophys. Res. Comm.* **228**, 334–340.
10. Jiang, Y., Gram, H., Zhao, M., New, L., Gu, J., Feng, L., Padova, F., Ulevitch, R., and Han, J. (1997). Characterization of the structure and function of the fourth member of p38 group mitogen-activated Protein Kinases, p38δ. *J. Biol Chem.* **333**, 30122–30128.
11. Conrad, P. W., Rust, R. T., Han, J., Milkorn, D. E., and Beitner-Johnson, D. (1999). Selective activation of p38α and p38β by anoxia. Role in regulation of cyclin D1 by hypoxia in PC12 cells. *J. Biol. Chem.* **274**, 23570–23576.
12. Cardier, J. E. and Erickson-Miller, C. L. (2002). Fas (CD95)-and tumor necrosis factor mediated apoptosis in liver endothelial cells; role of caspase-3 and the p38 MAPK. *Mricrovasc. Res.* **63**, 10–18.
13. Kim, D. H., Feinbaum, R., Alloing, G., Emerson, F. E., Garsin, D. A., Inoue, H., Tanaka-Hino, M., Hisamoto, N., Matsumoto, K., Tan, V.-W., and Ausubel, F. M. (2002). A conserved p38 MAP kinase pathway in *Caenorhabditis elegans* innate immunity. *Science* **297**, 623–626.
13a. Park, J. M., Greten, F. R., Li, 2.-W., and Karin, M. (2002). Macrophage apoptosis by anthrax lethal factor through p38 MAP kinase inhibition. *Science* **297**, 2048–2051.
14. Nemoto, S., Xiang, J., Huang, S., and Lin, A. (1998). Induction of apoptosis by SB202190 through inhibition of p382 mitogen-activated protein kinase. *J. Biol. Chem.* **273**, 16415–16420.
15. Lee, J. C., Kassis, S., Kumar, S., Badger, A., and Adams, J. L. (1999). p38-mitogen activated protein kinase inhibitors–mechanisms and therapeutic potentials. *Pharmacol. Ther.* **82**, 389–397.
16. Lu, H. T., Yang, D. D., Wysk, M., Gatti, E., Mellman, I., Davis, R. J., and Flavell, R. A. (1999). Defective IL-12 production in mitogen-activated protein (MAP) kinase 3 (Mkk3)-deficient mice. *EMBO J.* **18**, 1845–1857.
17. Wysk, M., Yang, D. D., Lu, H. T., Flavell, R. A., and Davis, R. J. (1999). Requirement of mitogen-activated protein kinase kinase 3 (MKK3) for tumor necrosis factor-induced cytokine expression. *Proc. Natl. Acad. Sci. USA* **96**, 3763–3768.
18. Enslen, H., Raingeaud, J., and Davis, R. J. (1998). Selective activation of p38 mitogen-activated protein (MAP) kinase isoforms by the MAP kinase kinases MKK3 and MKK6. *J. Biol. Chem.* **273**, 1741–1748.
19. Wang, Y. U., Huang, S., Sah, V. P., Ross, J. Jr., Brown, J. H., Han, J., and Chien, K. R. (1998). Cardiac muscle cell hypertrophy and apoptosis induced by distinct members of the p38 mitogen-activated protein kinase family. *J. Biol. Chem.* **273**, 2161–2168.
20. Yang, S. H., Whitemarsh, A. J., Davis, R. J., and Sharrocks, A. D. (1998). Differential targeting of MAP kinases to the ETS-domain transcription factor E1k-1. *EMBO J.* **17**, 1740–1749.
21. Muda, M., Theodosiou, A., Gillieron, C., Smith, A., Chabert, C., Camps, M., Boschert, U., Rodrigues, N., Davies, K., Ashworth, A., and Arkinstall, S. (1998). The mitogen-activated protein kinase phosphatase-3 N-terminal noncatalytic region is responsible for tight substrate binding and enzymatic specificity. *J. Biol. Chem.* **273**, 9323–9329.
22. Xia, Y. and Karin, M (1998). JNKK1 organizes a MAPK module through specific and sequential interactions with upstream and downstream components mediated by its amino terminal extension. *Genes Dev.* **12**, 3369–3381.
23. Gavin, A. C. and Nebreda, A. R. (1999). A MAP kinase docking site is required for phosphorylation and activation of p90RSK/MAPKAPK-1. *Curr. Biol.* **9**, 281–284.
24. Smith, J. A., Poteit-Smith, C. E., Malarkey, K., and Sturgill, T. W. (1999). Identification of an extracellular signal-related kinase (ERK) docking site in ribosomal S6 kinase, a sequence critical for activation by ERK *in vivo*. *J. Biol. Chem.* **274**, 2893–2898.
25. Tanoue, T., Ryota, M., Adachi, M., and Nichida, E., (2001). Identification of docking groove on ERK and p38 MAP kinases that regulate the specificity of docking interactions. *EMBO J.* **20**, 466–479.
26. Ben-Levy, R., Hooper, S., Wilson, R., Paterson, H. F., and Marshall, C. J. (1998). Nuclear export of the stress-activated protein kinase p38 mediated by its substrate MAPKAP kinase-2. *Curr. Biol.* **8**, 1049–1057.
27. Yang, S.-H., Galanis, A., and Sharrocks, A. D. (1999). Targeting of p38 mitogen-activated protein kinase to MEF2 transcription factors. *Mol. Cell Biol.* **19**, 4028–4038; see also Zhao, M. Liguo, N., Kravchenko, V. V., Kato, Y., Gram, H., diPadova, F., Olson, E. N., Ulevitch, R. J., and Han, J. (1999). Regulation of the MEF2 family of transcription families by p38. *Mol. Cell. Biol.* **19**, 21–30.
28. Zhao, M., Liguo, N., Kravchenko, V. V., Kato, Y., Gram, H., diPadova, F., Olson, E. N., Ulevitch, R. J., and Han, J. (1999). Regulation of the MEF2 family of transcription families by p38. *Mol. Cell Biol.* **19**, 21–30.
29. Hibi, M., Lin, A., Smeal, T., Minden, A., and Karin, M. (1993). Identification of an oncoprotein- and UV-responsive protein kinase that binds and potentiates the c-Jun activation domain. *Genes Dev.* **7**, 2135–2148.
30. Kallunki, T., Deng, T., Hibi, M., and Karin, M. (1996). c-Jun recruits JNK to phosphorylate K dimerization partners via specific docking interactions. *Cell* **87**, 929–939.
31. Gupta, S., Barrett, T., Whitemarsh, A. J., Cavanaugh, J., Sluss, H. K., Derijard, B., and Davis, R. J. (1996). Selective interaction of JNK protein kinase isoforms with transcription factors. *EMBO J.* **15**, 2760–2770.
32. Hibi, M., Lin, A., Smeal, T., Minden, A., and Karin, M. (1993). Identification of an oncoprotein- and UV-responsive protein kinase that binds and potentiates the c-Jun activation domain. *Genes Dev.* **7**, 2135–2148.
33. Chow, C. W., Rincon, M., Cavanagh, J., Dickens, M., and Davis, R. J. (1997). Nuclear accumulation of NFAT4 opposed by JNK Signal transduction pathway. *Science* **278**, 1638–1641.
34. Gupta, S., Campbell, D., Derijard, B., and Davis, R. J. (1995). Transcription factor ATF2 regulation by the JNK signal transduction pathway. *Science* **267**, 389–393.
35. Livingstone, C., Patel, G., and Jones, N. (1995). ATF-2 contains a phosphorylation-dependent transcriptional activation domain, *EMBO J.* **14**, 1785–1797.
36. Engel, K., Schultz, H., Martin, F., Kotlyarov, A., Plath, K., Hahn, M. *et al.* (1995). Constitutive activation of mitogen-activated protein kinase 2 by mutation of phosphorylation sites and an A-helix motif. *J. Biol. Chem.* **270**, 27213–27221.
37. Suh, Y. (2001). Age-specific changes in expression, activity, and activation of the c-Jun NH(2)-terminal kinase and p38 mitogen-activated protein kinases by methyl methanesulfonate in rats. *Mech. Age. Dev.* **122**, 1797–1811.
38. Sun, H., Charles, C. H., Lau, L. F., and Tonks, N. K. (1993). MAKP-1 (3CH134) an immediate early gene product, is a dual specificity phosphatase that dephosphorylates MAP kinase *in vivo*. *Cell* **75**, 487–493.
39. Keyse, S. M. (2000). Protein phosphatases and the regulation of mitogen-activated protein kinase signaling. *Curr. Opin. Cell Biol.* **12**, 186–192.
40. Rohan, P. J., Davis, P., Moskaluk, C. A., Kearns, M., Krutzsch, H., Siebenlist, U., and Kelly, K. (1993). PAC-1 a mitogen-induced nuclear protein tyrosine phosphatase. *Science* **259**, 1763–1766.
41. Liu, Y., Gorospe, M., Yang, C., and Holbrook, N. J. (1995). Role of mitogen activated protein kinase phosphatase during the cellular response to genotoxic stress. Inhibition of c-Jun N-terminal kinase activity and AP-1 dependent gene activation. *J. Biol. Chem.* **270**, 8377–8380.
42. Hutter, D., Chen, P., Barnes, J., and Liu, Y. (2000). Catalytic activation of mitogen-activated protein (MAP) kinase phosphatase-1 by binding to p38 MAP kinase: critical role of the p38 C-terminal domain in its negative regulation. *Biochemistry* **352**, 155–163.
43. Camps, M., Nichols, A., Gillieron, C., Antonsson, B., Muda, M., Chabert, C., Boschert, U., and Arkinstall, S. (1998). Catalytic activation of the phosphatase MKP-3 by ERK-2 mitogen activated protein kinase. *Science* **280**, 1262–1265.
44. Dowd, S., Sneddon, A. A., and Keyse, S. M. (1998). Isolation of the human genes encoding the Pyst1 and Pyst2 as a cytosolic dual-specificity MAP kinase phosphatase and its catalytic activation by both MAP and SAP kinase phosphatase and its catalytic activation by both MAP and SAP kinases. *J. Cell Sci.* **111**, 3389–3399.

45. Harman, D. (1956). Aging: A theory based on free radical chemistry. *Gerontology* **11**, 298–300.
46. Stadtman, E. R. (1992). Protein oxidation and aging. *Science* **257**, 1220–1224.
47. Ames, B. N., Shigenaga, M. K., and Hagen, T. M. (1993). Oxidants, antioxidants, and the degenerative diseases of aging. *Proc. Natl. Acad. Sci. USA* **90**, 7915–7922.
48. Papaconstantinou, J. (1994). Unifying model of the programmed (intrinsic) and stochastic (extrinsic) theories of aging: The stress response genes, signal transduction-redox pathways and aging. *Ann. NY Acad. Sci.* **719**, 195–211.
49. Sohal, R. S. and Orr, W. C. (1995). In Esser, K. and Martin, G. M., Eds., *Molecular Aspects of Aging,* pp. 109–127. Wiley, New York.
50. Hamilton, M. L., Remmen, H. V., Drake, J. A., Yang, H., Guo, Z. M., Kewitt, K., Walter, C. A., and Richardson, A. (2001). Does oxidative damage to DNA increase with age? *Proc. Natl. Acad. Sci USA* **98**, 10469–10474.
51. Sohal, R. S. and Weindruch, R. (1996). Oxidative stress, caloric restriction, and aging, *Science* **273**, 59–63.
52. Lithgow, G. I., White, T. M., Melov, S., and Johnson, T. E. (1995). Thermotolerance and extended life-span conferred by single-gene mutations and induced by thermal stress. *Proc. Natl. Acad. Sci. USA* **92**, 7540–7544.
53. Martin, G. M., Austad, S. N., and Johnson, T. E. (1996). Genetic analysis of aging: role of oxidative damage and environmental stresses. *Nat. Genet.* **13**, 25–34.
54. Finkel, T. and Holbrook, N. J. (2000). Oxidants, oxidative stress and biology of aging. *Nature* **408**, 239–247.
55. Yan, L.-J., Levine, R. L., and Sohal, R. S. (1997). Oxidative damage during aging targets mitochondrial aconitase. *Proc. Natl. Acad. Sci. USA* **94**, 11168–11172.
56. Goto, S., Nakamura, A., Radak, Z., Nakamoto, H., Takahashi, R., Yasuda, K., Sakurai, Y., and Ishi, N. (1999). Carbonylated proteins in aging and exercise: immunoblot approaches. *Mech. Age. Dev.* **107**, 245–253.
57. Beckman, K. B. and Ames, B. N. (1998). The free radical theory of aging matures. *Physiol. Rev.* **78**, 547–581.
58. Golden, T. R. and Melov, S. (2001). Mitochondrial DNA mutations, oxidative stress, and aging. *Mech. Aging Dev.* **122**, 1577–1589.
59. Esposito, L. A., Melov. S., Panov. A., Cottrell, B. A., and Wallace, D. C. (1999). Mitochondrial disease in mouse results in increased oxidative stress. *Proc. Natl. Acad. Sci. USA* **96**, 4829–4825.
60. Hsieh, C.-C. and Papaconstantinou (2002). The effect of aging on p38 signaling pathway activity in the mouse liver and in response to ROS generated by 3-nitropropionic acid. *Mech. Age Dev.* (in press).
61. Hamilton, B. C. and Gould D. H. (1987). Correlation of morphologic brain lesions with physiologic alterations and blood-brain barrier impairment in 3-nitropropionic acid toxicity in rats. *Acta Neuropathol.* **74**, 67–74.
62. Gould, D. H., Wilson, M. P., and Hamar, D. W. (1985). Brain enzyme and clinical alterations induced in rats and mice by nitroaliphatic toxicants. *Toxic. Lett.* **27**, 83–89.
63. Coles, C. J., Edmondson, D. E., and Singer, T. P. (1979). Inactivation of succinate dehydrogenase by 3-nitropropionate. *J. Biol. Chem* **254**, 5161–5167.
64. Hamilton, B. F. and Gould D. H. (1987). Nature and distribution of brain lesions in the rat intoxicated with 3-nitropropionic acid: a type of hypoxic (energy deficient) brain damage. *Acta Neuropathol. (Berlin)* **74**, 67–74.
65. Ludolph, A. C., Seelig, M., Ludolph, A. G., Sabri, M., Spencer, P. S. *et al.* (1992). ATP deficits and neuronal degeneration induced by 3-nitropropionic acid. *Ann. NY Acad. Sci.* **648**, 300–302.
66. Coles, C. J., Edmondson, D. E., and Singer, T. P. (1979). Inactivation of succinic dehydrogenase by 3-nitropropionate. *J. Biol. Chem.* **254**, 5161–5167.
67. Kulisz, A., Chen, N., Chandel, N. S., Shao, Z., and Schumaker, P. T. (2002). Mitochondrial ROS initiate phosphorylation of p38 MAP kinase during hypoxia in cardiomyocytes. *Am J. Physiol. Lung Mol. Physiol.* **282**, L1324–L1329.
68. Bossi, S. R., Simpson, J. R., and Isacson, O. (1993). Age dependence of striatal neuronal death caused by mitochondrial dysfunction. *Neuro Report* **4**, 73–76.
69. Fu, Y.-T., He, F.-S., Zhang, S.-L., and Zhang, J.-S. (1995). Lipid peroxidation in rats intoxicated with 3-nitropropionic acid. *Toxicology* **33**, 327–331.
70. Kappus, H. (1986). Overview of enzyme systems involved in bioreduction of drugs and in redox cycling. *Biochem. Pharmac.* **35**, 1–6.
71. Wang, Z., Harkins, P. C., Ulevitch, R. J., Han, J., Cobb, M. H., and Goldsmith, E. J. (1997). The structure of mitogen activated protein kinase p38 at 2.1 A° resolution. *Proc. Natl. Acad. Sci. USA* **94**, 2327–2332.
72. Wilson, K. P., Fitzgibbon, M. J., Caron, P. R., Griffith, J. P., Chen, W., McCaffrey, P. A., Chambers, S. P., and Su, M. S.-S. (1996). Crystal structure of p38 mitogen-activated protein kinase. *J. Biol. Chem.* **271**, 27696–27700.
73. Krebs, D. L. and Hilton, D. J. (2001). SOCS proteins: Negative regulators of cytokine signaling. *Stem Cells* **19**, 378–387.
74. Mastsumoto, A., Seki, Y., Kubo, M., Ohtsuka, S., Suzuki, A., Hayashi, I., Tsuji, K., Nakahata, T., Okabe, M., Yamada, S., and Yoshimura, A. (1999). Suppression of STAT 5 functions in liver, mammary glands, and T cells in cytokine-inducible SH2-containing protein 1 transgenic mice. *Mol. Cell. Biol.* **19**, 6396–6407.
75. Metcalf, D., Greenhalgh, C. J., Viney, E., Willson, T. A., Starr, R., Nicola, N. A., Hilton, D. J., and Alexander, W. S. (2000). Gigantism in mice lacking suppressor of cytokine signaling-2. *Nature* **405**, 1069–1073.
76. Dozmorov, I., Galecki, A., Chang, Y., Krzesicki, R., Vergara, M., and Miller, R. Q. (2002). Gene expression profile of long-lived snell dwarf mice. *J. Gerontol.* **57A**, B99–B108.
77. Knopf, J. L., Gallager, J. F., and Held, W. A. (1983). Differential, multihormonal regulation of the mouse major urinary protein gene family in the liver. *Mol. Cell. Biol.* **12**, 2232–2240.
78. Marine, J. C., McKay, C., Wang, D., Topham D. J., Parganas, E., Nakajima, H., Pendeville, H., Yasukawa, H., Sasaki, A., Yoshimura A., and Ihle, J. N. (1999). SOCS3 is essential in the regulation of fetal liver erythropoiesis. *Cell* **98**, 617–627.
79. Roberts, A. W., Robb, L., Rakar, S., Hartley, L., Cluse, L., Nicola, A. A., Metcalf, D., Hilton, D. J., and Alexander, W. S. (2001). Placental defects and embryonic lethality in mice lacking suppressor of cytokine signaling 3. *Proc. Natl. Acad. Sci. USA* **98**, 9324–9329.
80. Naka, T., Narazaki, M., Hirata, M., Matsumoto, T., Minamoto, S., Aono, A., Nishimoto, N., Kajita, T., Taga, T., Yoshizaki, K., Akira, S., and Kishimoto, T. (1997). Structure and function of a new STAT-induced STAT inhibitor. *Nature* **387**, 924–929.
81. Endo, T. A., Masuhara, M., Yokouchi, M., Suzuki, T., Sakamoto, H., Mitsui, K., Matsumoto, A., Tanimura, S., Ohtsubo, M., Misana, H., Miyazaki, T., Leonor, N., Taniguchi, T., Fujita, T., Kanakura, Y., Komiya, S., and Yoshimura, A. (1997). A new protein containing an SH2 domain that inhibits JAK kinases. *Nature* **387**, 921–924
82. Nicholson, S. E., Wilson, T. A., Farley, A., Starr, R., Zhang, J. G., Baca, M., Alexander, W. S., Metcalf, D., and Nicola, N. A. (1999). Mutational analysis of the SOCS proteins suggest a dual domain requirement but distinct mechanisms for inhibition of LIF and IL-6 signal transduction. *EMBO J.* **18**, 375–385.
83. Yasukawa, H., Misawa, H., Sakamoto, H., Masuhara, M., Sasaki, A., Wakioka, T., Ohtsuka, S., Imaizumi, T., Matsuda, T., Ihle, J. N., and Yoshimura, A. (1999). The JAK binding protein JAB inhibits Janus tyrosine kinase activity through binding in the activation loop. *EMBO J.* **18**, 1309–1320.
84. Pezet, A., Favre, H., Kelley, P. A., and Edery, M. (1999). Inhibition and restoration of prolactin signal transduction by suppressors of cytokine signaling. *J. Biol. Chem.* **274**, 24497–24502.
85. Mashuhara, M., Sakamoto, H., Matsumoto, A., Suzuki, R., Yasukawa, H., Mitsui, K., Wakioka, T., Tanimura, S., Sasaki, A., Misawa, H., Yokouchi, M., Ohtsubo, M., and Yoshimura, A. (1997). Cloning and characterization of novel CIS family genes. *Biochem. Biophys. Res. Commun.* **239**, 439–446.
86. Suzuki, R., Sakamoto, H., Yasukawa, H., Masuhara, M., Wahioka, T., Sasaki, A., Yuge, K., Komiya, S., Inoue, A., and Yoshimura, A. (1998).

CIS3 and JAB have different regulatory roles in interleukin-6 mediated differentiation and Stat3 activation in NM1 leukemia cells. *Oncogene* **17**, 2271–2278.

87. Sasaki, A., Yasukawa, H., Suzuki, A., Kamizono, S., Syoda, T., Kinijyo, I., Sasaki, M., Johnston, J. A., and Yoshimura, A. (1999). Cytokine inducible SH2 protein-3 (CIS3/SOCS3) inhibits Janus tyrosine kinase by binding through the N-terminal kinase inhibitory region as well as SH2 domain. *Genes Cells* **4**, 339–351.
88. Ram, P. A. and Waxman, D. J. (1999). SOCS/CIS protein inhibition of growth hormone-stimulated STAT 5 signaling by multiple mechanisms. *J. Biol. Chem.* **274**, 35553–35556.
89. Ram P. A. and Waxman, D. J. (2000). Role of the cytokine-inducible SH2 protein CIS in desensitization of STAT 5b signaling by continuous growth hormone. *J. Biol. Chem.* **275**, 39487–39496.
90. Hansen, J. A., Lindberg, K., Hilton, D. J., Nielsen, J. H., and Billestrup, N. (1999). Mechanism of inhibition of growth hormone receptor signaling by suppressor of cytokine signaling proteins. *Mol. Endocrinol.* **13**, 1832–1843.
91. Verdier, F., Cretien, S., Muller, O., Varlet, P., Yoshimura, A., Gisselbrecht, S., Lacombe, C., and Mayeux, P. (1998). Proteosomes regulate erythropoietin receptor and signal transducer and activator of transcription 5 (STAT 5) activation. Possible involvement of the ubiquitinated Cis protein. *J. Biol. Chem.* **273**, 28185–28190.
92. Cohney, S. J., Sanden, D., Cacalano, N. A., Yoshimura, A., Mui, A., Migone, T. S., and Johnston, J. A. (1999). SOCS3 is phosphorylated in response to interleukin-2 and suppresses STAT 5 phosphorylation and lymphocyte proliferation. *Mol. Cell. Biol.* **19**, 4980–4988.
93. Emanuelli, B., Peraldi, P., Filloux, C., Chavey, C., Freidinger, K., Hilton, D. J., Hotamisliogil, H. B., and Van Obberghen, E. (2001). SOCS-3 inhibits insulin signaling and is up regulated in response to tumor necrosis factor-α in the adipose tissue of obese mice. *J. Biol. Chem.* **276**, 47944–47949.
94. Paz, K., Hemi, R., LeRoith, D., Karasik, A., Elhanany, E., Kanety, H., and Ziek, Y. (1997). A molecular basis for insulin resistance. Elevated serine/threonine phosphorylation of IRS-1 and IRS-2 inhibits their binding to the juxtamembrane region of the insulin receptor and impairs their ability to undergo insulin-induced tyrosine phsophorylation. *J. Biol. Chem.* **272**, 29911–29918.
95. Yenush, L., Makati, K. J., Smith-Hall, J., Ishibasi, O., Myers, M. G. Jr., and White, M. F. (1996). The plekstrin homology domain is the principal link between the insulin receptor and IRS-1. *J. Biol Chem.* **271**, 24300–24306.
96. Sakuda, S., Tamura, S., Yamada, A., Miyagawa, J-i., Yamamoto, K., Kiso, S., Ito, N., Imanaka, K., Wada, A., Naka, T., Kishimoto, T., Kawata, S., and Matsuzawa, Y. (2002). Activation of signal transducer and activator transcription 3 and expression of suppressor of cytokine signal 1 during liver regeneration in rats. *J. Hepatol.* **36**, 378–384.
97. Mao, Y., Ling, P.-R. Fitzgibbon, T. P., McCowen, K. C., Frick, G. P., Bistrian, B. R., and Smith, R. J. (1999). Endotoxin induced inhibition of growth hormone receptor signaling in rat liver, *in vivo. Endocrinology* **140**, 5505–5515.
98. Cressman, D. E., Greenbaum, L. E., Haber, B. A., and Taub, R. (1995). Rapid activation of the STAT3 transcription complex in liver regeneration. *Hepatology* **21**, 1443–1449.
99. Grisham, J. W. (1962). A morphologic study of deoxyribonucleic acid synthesis and cell proliferation in regenerating rat liver; Autoradiography with thymidine-H^3. *Cancer Res.* **22**, 842–849.
100. Cressman, D. E., Greenbaum, L. E., DeAngelis, R. A., Ciliberto, G., Furthe, E. E., Poli, V. *et al.* (1996). Liver failure and defective hepatocyte regeneration in interleukin-6-deficient mice. *Science* **274**, 1379–1383.
101. Yamada, Y., Kirillova, I., Peschon, J., and Fausto, N. (1997). Initiation of liver growth by tumor necrosis factor: Deficient liver regeneration in mice lacking type 1 tumor necrosis factor receptor. *Proc. Natl. Acad. Sci. USA* **94**, 1441–1446.
102. Heim, M. H., Gamboni, G., Beglinger, C., and Gyr, C. (1997). Specific activation of AP-1 but not Stat3 in regenerating liver in mice. *Eur. J. Clin. Invest.* **27**, 948–955.
103. Hong, F., Nguyen, V.-A., and Gao, B. (2001). Tumor necrosis factor α attenuates interferon α signaling in the liver: involvement of SOCS3 and SHP2 and implication in resistance to interferon therapy. *FASEB J.* **15**, 1595–1597.
104. Yoshikawa, H., Matsubara, K., Qian, G.-S., Jackson, P., Gropman, J. D., Manning, J. E., Harris, C.-C., and Herman J. G. (2001). SOCS1, a negative regulator of the JAK/STAT pathway, is silenced by methylation in human hepatocellular carcinoma and shows growth-suppression activity. *Nat. Genet.* **28**, 29–35.
105. Ouwens, D. M., deRuiter, N. D., van der Zon, G. C. M., Carter, A. P., Schouten, J., van der Burgt, C., Kooistra, K., Bos, J. L., Maassen, J. A., and Van Dam, H. (2002). Growth factors can activate ATF-2 via a two-step mechanism: phosphorylation of Thr^{71} through the Ras-MEK-ERK pathway and of Thr^{60} through RalGDS-Src-p38. *EMBO J.* **21**, 3782–3793.

CHAPTER 342

Cell–Cell Signaling in the Testis and Ovary

Michael K. Skinner
Center for Reproductive Biology,
School of Molecular Biosciences,
Washington State University,
Pullman, Washington

Introduction

The evolution of multicellular organisms was facilitated by the ability of different cells to communicate and interact. This cell–cell signaling generates a higher order functional state than that possible with individual cell types. Cell–cell interactions have become an essential requirement for the physiology of any organ or tissue and are critical in the regulation of any cell's biology. For this reason, elaborate networks of cell–cell interactions have evolved to control the development and maintenance of tissue functions. The focus of the current chapter is on the regulatory signals that mediate cell–cell interactions in the testis and ovary.

Several previous reviews have discussed the cell–cell interactions in the testis [1,2] and ovary [3,4]. These include a focus on secretory products of the various cell types and actions of individual regulatory molecules. The current chapters briefly discusses the advances in cell–cell signaling in these organs.

Many different types of cell–cell interactions are required for the control of tissue physiology and cellular functions. These have been previously categorized into regulatory, nutritional, and environmental classifications [1]. Regulatory interactions are generally mediated by extracellular factors that through receptor-mediated events cause a signaling event to modulate cell functions. Nutritional interactions generally involve the transport of nutritional substances, energy metabolites, or metabolic substrates between cells. Environmental interactions involve extracellular environmental factors that affect cell contacts and cytoarchitecture. The focus of this chapter is primarily on regulatory-type interactions that involve a receptor-mediated signaling event. It is this type of cellular signaling that regulates a cell's function on a molecular level. The factors involved are generally paracrine and autocrine agents such as growth factors and cytokines.

Both the testis and ovary are endocrine organs. Endocrine hormones from the pituitary [i.e., gonadotropins, follicle-stimulating hormone (FSH), and luteinizing hormone (LH)] act on various cell types to influence cellular functions and cell–cell interactions. The influence these endocrine hormones have on cell–cell signaling events is briefly reviewed. The testis and ovary are also sites for the production of hormones. These gonadal hormones have an endocrine role in regulating a wide variety of tissues in the body, but can also act in a paracrine manner within the gonads to influence cell–cell signaling and cellular functions. Again the role these gonadal steroids and peptide hormones play in the regulation of cell–cell signaling within the gonad are discussed in this chapter.

Cell–Cell Signaling in the Testis

Testis Cell Biology

The adult testis is a complex organ that is composed of seminiferous tubules that are enclosed by a surrounding interstitium. The seminiferous tubules are the site of spermatogenesis where germ cells develop into spermatozoa in close interaction with Sertoli cells (Fig. 1). The Sertoli cell

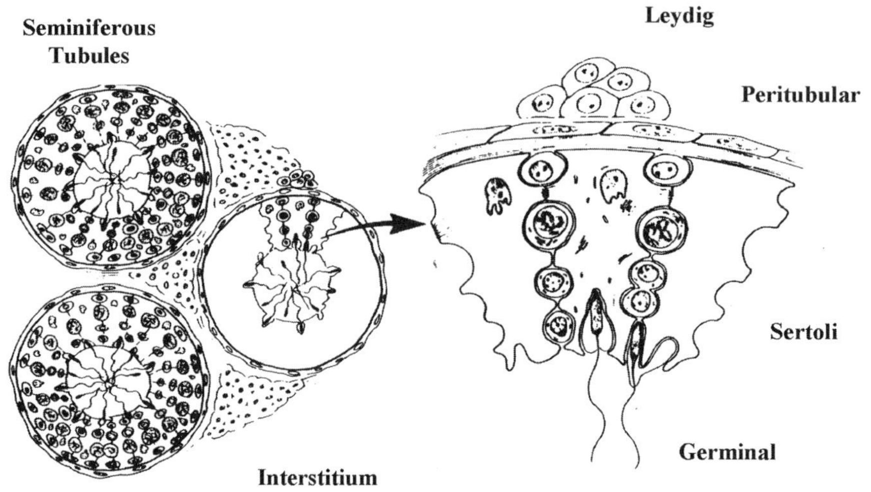

Figure 1 Testis cell biology.

is an important testicular somatic cell that controls the germ cell environment by the secretion and transport of nutrients and regulatory factors. The Sertoli cells [5] form the basal and apical surface of the seminiferous tubule and provide the cytoarchitectural arrangements for the developing germ cells [6]. Tight junctional complexes between the Sertoli cells contribute to the maintenance of a blood–testis barrier [7] and create a unique environment within the tubule [8]. The structure of the Sertoli cell has been reviewed by several investigators (for a review, see [6]), and a three-dimensional reconstruction has increased appreciation for the complexity of the structural relationships between cells within the seminiferous tubule [9]. The biochemical analysis of the Sertoli cell has primarily focused on an examination of the components synthesized and secreted by the cell. The list of products includes steroids such as estradiol [10], metabolites such as lactate [11], and various proteins such as plasminogen activator [12], testicular transferrin [13], testicular ceruloplasmin [14], inhibin, and others (for a review, see [2]). The majority of the secretory products are hormonally regulated and provide useful markers of Sertoli cell differentiation.

Surrounding the basal surface of the Sertoli cells is a layer of peritubular myoid cells (Fig. 1) that function in contraction of the tubule. The peritubular cells surround and form the exterior wall of the seminiferous tubule. Peritubular cells are mesenchymally derived cells that secrete fibronectin [15] and several extracellular matrix components [16]. Both the peritubular and the Sertoli cells form the basement membrane surrounding the seminiferous tubule and their interactions are important in germ cell development.

The interstitial space around the seminiferous tubules contains another somatic cell type, the Leydig cell (Fig. 1), which is responsible for testosterone production. Leydig cells have a major influence on spermatogenesis through the actions of testosterone on both the seminiferous tubule and the pituitary. Although the Leydig cell has numerous secretory products [1], testosterone is the most significant secretory product of the cells. Thus, interaction of all three somatic cells—Sertoli, peritubular, and Leydig—are important for regulation of normal spermatogenic function in the testis (for a review, see [1]).

Testis Development

The process of fetal testis formation occurs late in embryonic development (embryonic day 13 where plug date = E0 (E13) in the rat) and is initiated by migration of primordial germ cells, first from the yolk sac to the hindgut and then from the hindgut to the genital ridge. The first phase of migration is proposed to occur through a mechanism in which transient interactions between fibronectin molecules on the extracellular matrix and corresponding receptors on the primordial germ cells cause movement of the germ cells. The second migration is thought to occur by the release of chemoattractant factors from the genital ridge. Kit ligand and its receptor c-kit appear to be involved first in the migration to the genital ridge and later in the proliferation of germ cells after colonization of the genital ridge. Expression of kit ligand has been localized to cells along the migratory pathway, and c-kit is expressed by primordial germ cells at this time in development (for a review, see [17]). After migration, germ cell differentiation in the gonad is dependent on locally produced factors such as prostaglandins [18] and the induction of specific transcription factors [19]. It is a complex network of cellular interactions that control testis and germ cell development.

The gonad is bipotential after germ cell migration and can be distinguished morphologically from the adjoining mesonephros (E12 in rat), but cannot be identified as an ovary or a testis. A variety of genes such as SRY, SOX-9, SF1, and DMRT1 are involved in the transcriptional induction of sex determination and testis development [20–27]. Two morphological events occur early on E13 to alter the bipotential gonad. First, Sertoli cells, which are proposed to be the first cell in the testis to differentiate, aggregate around primordial germ cells [28,29]. Secondly, migration of mesenchymal cells occurs from the adjoining mesonephros into the developing gonad to surround the Sertoli cell–germinal cell aggregates. The migrating population of cells has been speculated to be preperitubular cells [30–32]. The mechanism for this migration is unknown, but a signal from the testis is proposed to occur and cause cell migration. This is postulated due to the observation that ovarian mesonephros can also be stimulated to initiate cell migration after close interaction with a developing testis [33]. In addition, using an organ culture system in which mesonephros and embryonic testis were separated by an embryonic ovary, mesonephros cells migrated through the ovary to the testis [30]. Therefore, during early testis development Sertoli–peritubular cell interactions may allow for cord formation to occur. The cords develop neonatally into seminiferous cords and at the onset of puberty develop into the seminiferous tubules. Sertoli cells have been postulated to originate from stem cells in the coelomic epithelium at an early stage in gonadal development. Other cells that may potentially originate from the coelomic epithelium are interstitial or Leydig cells [34].

Seminiferous cords, precursors of adult seminiferous tubules, form as the Sertoli cell–primordial germ cell aggregates become more organized and are fully surrounded by mesenchymal cells. The formation of the seminiferous cords (E14 in rat) is a critical event in the morphogenesis of the testis since this is the first indication of male sex differentiation [26]. During the process of cord formation, Sertoli cells undergo a number of morphological changes including a change in expression of mesenchymal to epithelial cell markers (vimentin to cytokeratin; [35]), a change in expression of cytokeratin 19 to cytokeratin 18 (cytokeratin 21 expressed in ovary; [36]), and expression of Müllerian inhibiting substance (MIS), which inhibits the development of the Mullerian duct, the precursor of the female uterus, cervix, fallopian tubes, and upper vagina [37–39].

Outside of the seminiferous cords, the peritubular layer of cells becomes identifiable from the interstitium or Leydig cells at E15 [38] and 3βHSD production is detected after E15 [39]. Leydig cells have been hypothesized to differentiate after cord formation and Sertoli cell differentiation is completed [40,41]. This is important because the production of testosterone and androgens by the Leydig cells has been demonstrated to stabilize the Wolffian duct derivatives, hence allowing normal male duct development [42,43]. Therefore, appropriate differentiation of somatic cell types in the testis around the time of cord formation is crucial not only to the normal development of the testis, but also for the continued presence of the Wolffian duct and normal male reproductive tract development.

Testis Cell–Cell Interactions

Table I outlines a number of the factors produced locally in the testis that mediate cell–cell signaling events in the control of spermatogenesis and testis function. Several reviews address the topic of cell–cell interactions in the testis and the control of spermatogenesis [1,2,44,45]. Recent observations are cited next.

Transforming growth factor α (TGF-α) is an epidermal growth factor (EGF) superfamily member and is produced by Sertoli, peritubular, and Leydig cells. TGF-α can act as a growth stimulator on all major cell types in the testis [46–48]. As with TGF-α, transforming growth factor β (TGF-β) is also produced by Sertoli, peritubular, and Leydig cells and can act on all the major cells in the testis [49–53]. In contrast, TGF-β primarily acts as a growth inhibitor and can stimulate a variety of differentiated functions. Another example of a factor that is produced by all of the somatic cells and acts on all major cell types in the testis is insulin-like growth factor 1 (IGF-1) [54–56]. IGF-1 plays a general role in regulation of the growth cycle and homeostasis of the testis. A related family member, IGF-2, mediates paracrine interactions between Sertoli cells and germ cells [57]. These are examples of regulatory factors that mediate cell–cell signaling events between the majority of the cell types in the testis.

Several interleukins (IL-1α, IL-1β, IL-6) are produced in the testis by Sertoli cells and Leydig cells. These interleukins can regulate Sertoli, Leydig, and germ cell growth and differentiated functions [58–65]. Although further analysis is needed, interleukins appear to mediate primarily Sertoli–germ cell and Leydig–Sertoli cell interactions, as well as autocrine roles for these factors.

Several hormonal factors produced in the testis also act locally within the testis as paracrine factors. An example is inhibin and its related peptide activin [66–68]. Inhibin is primarily produced by Sertoli cells and can act on germ cells and Leydig cells. Further investigation of the actions of inhibin and related compounds within the testis is needed. Another major endocrine factor produced in the testis is testosterone by Leydig cells that can in turn act on Sertoli, peritubular, and Leydig cells [69]. Androgens have a major role in the maintenance of testis function by inducing cellular differentiated functions.

Fibroblast growth factor (FGF) family members have been shown to be expressed in the testis and regulate the growth and differentiation of a variety of cells [70–75]. FGF receptors are predominant in germ cells and Leydig cells, but are also present in the other cell types [71]. FGF-14 has recently been shown to be expressed in spermatocytes and may influence adjacent Sertoli or peritubular cells. FGF-9 null mutants also suggest a role in early testis development, but remain to be investigated in the adult [74]. Basic FGF is produced by Sertoli cells and can also act on the other cells [75]. The variety of FGF ligands and receptors role in testis function remains to be elucidated.

Platelet-derived growth factor (PDGF) has been shown to be produced by Sertoli cells and influence peritubular cells and Leydig cells [76–79]. Although PDGF in the adult may also be produced by the Leydig cell [76], it appears to be a factor produced within the seminiferous tubules that acts on adjacent peritubular cells and Leydig cells. Another factor that is only produced by Sertoli cells is stem cell factor (SCF)/kit ligand (KL), which has a direct role in regulating spermatogonial cell proliferation [80–83]. Mutations in SCF/KL block the process of spermatogenesis. This is one of the better examples of a somatic–germ cell interaction.

Leukemia inhibitory factor (LIF) is a pleiotropic cytokine that influences stem cell growth and survival. LIF is

Table I Cell–Cell Signaling Factors in the Testis

Signaling factor	Site production	Site action	Functions	Ref
Transforming growth factor α (TGFα)	Sertoli Peritubular Leydig	Sertoli Peritubular Leydig Germ	Growth stimulation	46–48
Transforming growth factor β (TGFβ)	Sertoli Peritubular Leydig	Sertoli Peritubular Leydig Germ	Growth inhibition Differentiation stimulation	49–53
Insulin-like growth factor (IGF1)	Sertoli Peritubular Leydig	Sertoli Peritubular Leydig Germ	Homeostasis and DNA synthesis	54–56
Interleukin-s	Sertoli Leydig	Sertoli Leydig Germ	Growth regulation Cellular differentiation	58–65
Inhibin	Sertoli	Germ Leydig	Cellular differentiation	66–68
Androgen	Leydig	Sertoli Peributular Leydig	Cellular differentiation	69
Fibroblast growth factors	Sertoli Germ Leydig	Germ Peritubular Sertoli Leydig	Growth stimulation	70–75
Platelet derived growth factor (PDGF)	Sertoli	Peritubular Leydig	Growth stimulation Cellular differentiation	76–79
Stem cell factor/Kit ligand (SCF/KL)	Sertoli	Germ	Growth stimulation	80–83
Leukemia inhibitory factor (LIF)	Peritubular Sertoli Leydig	Germ	Growth stimulation Cell survival	84–85
Tumor necrosis factors	Germ Leydig	Sertoli Germ	Cellular apoptosis Cellular differentiation	86–88
Hepatocyte growth factor (HGF)	Peritubular	Leydig Peritubular Sertoli	Growth stimulation Tubule formation	89–91
Neurotropins	Germ Sertoli	Sertoli Peritubular	Growth stimulation Cell migration Cellular differentiation	1,92

predominantly produced by peritubular cells, but also by Sertoli cells and Leydig cells [84]. Although LIF has been shown to influence germ cell growth and survival [85], other functions remain to be elucidated.

Tumor necrosis factors (TNF-α) and related ligands (TRAIL) are produced in the testis by germ cells and Leydig cells. Both TNF and TRAIL have a role in regulating germ cells and Sertoli cells [86–88]. The function of these regulatory factors for the germ cells may be more for apoptosis regulation, unlike for Sertoli cells, which may be more for cellular differentiated functions.

Hepatocyte growth factor (HGF) is generally a mesenchymal-derived factor that acts on adjacent epithelial cells. HGF was found to be expressed by the mesenchymal-derived peritubular cells, and its receptor (c-met) was found on both Sertoli cells and Leydig cells [89–91]. Interestingly, c-met was also found in the peritubular cells. HGF may also have a role in seminiferous tubule formation [91].

Several neurotropins have been shown to be expressed in the testis. Nerve growth factor (NGF) is produced by the germ cells in the adult and appears to act on the Sertoli cells [1]. In embryonic development neurotropin-3 is expressed by Sertoli cells and acts on the migrating mesonephros cells to promote seminiferous cord formation [92]. Further investigations are needed to elucidate the roles of these and other neurotropins in the testis.

Additional factors are anticipated to be identified and have critical roles in testis development. Recently identified factors such as erythropoietin expression in Sertoli cells and peritubular cells [93], or interferon-gamma actions on

Sertoli cells [94], or relaxin-like factor (RLF) expression by Leydig cells [95] are all likely to have roles in cell–cell signaling in the testis. These and other factors such as PModS [96] need to be further investigated to determine their roles in testis cell biology. Clearly, a complex network of cell–cell signaling events and factors regulates testis function and spermatogenesis.

Cell–Cell Signaling in the Ovary

Ovarian Cell Biology

The ability of somatic cells in the gonad to control and maintain the process of gametogenesis is an essential requirement for reproduction. The basic functional unit in the ovary is the ovarian follicle, which is composed of somatic cells and the developing oocyte (Fig. 2). The two primary somatic cell types in the ovarian follicle are the theca cells and granulosa cells. These two somatic cell types are the site of action and synthesis of a number of hormones that promote complex regulation of follicular development. The proliferation of these two cell types is in part responsible for the growth of the ovarian follicle. The elucidation of factors that control ovarian somatic cell growth and development is critical to an understanding of ovarian physiology.

Granulosa cells are the primary cell type in the ovary that provides the physical support and microenvironment required for the developing oocyte (Fig. 2). Granulosa cells are an actively differentiating cell with several distinct populations. Alteration in cellular differentiation is required during folliculogenesis from a primordial stage of development through ovulation to a luteal stage of development. Regulation of granulosa cell cytodifferentiation requires the actions of a number of hormones and growth factors. Specific receptors have been demonstrated on granulosa cells for the gonadotropins FSH [97] and LH [98]. In addition, receptors have been found for factors such as EGF [99,100], insulin-like growth factor [101], and anti-Mullerian hormone [102]. The actions of these hormones and growth factors on granulosa cells vary with the functional marker being examined and the stage of differentiation. The biosynthesis of two important ovarian steroids, estradiol and progesterone, is a primary function of the granulosa cells in species such as the bovine, human, and rodent. Estrogen biosynthesis is controlled by the enzyme aromatase, which requires androgen as a substrate. Progesterone is synthesized from cholesterol by a series of steroidogenic enzymes. As the follicle develops, granulosa cells differentiate and estrogen biosynthesis increases. FSH promotes this follicular development via the actions of cAMP. As the follicle reaches stages before ovulation, the granulosa cells develop an increased capacity to synthesize and secrete progestins under the control of LH. In contrast, the early follicle stage (e.g., primordial) granulosa appear hormone independent and are nonsteroidogenic.

Another important cell type in the ovary is the ovarian theca cell (Fig. 2). These are differentiated stromal cells that surround the follicle and have also been termed theca interstitial cells [103]. The inner layer of cells, the theca interna, has a basement membrane separating it from the outermost layer of mural granulosa cells. One of the major functions of theca cells in species such as the bovine, human, and rodent is the secretion of androgens [104]. Theca cells respond to LH by increasing the production of androgens from cholesterol [105] (Fig. 2). Theca cells also produce progestins under gonadotropin control [106–109]. Other secretory products of theca cells have not been thoroughly investigated. At the primordial stage no theca cells are present; however, during transition to the primary stage, theca cells (i.e., precursor cells) are recruited to the follicle [3].

Follicle Growth and Differentiation (Folliculogenesis)

The control of ovarian follicle development is complex and involves multiple waves of growth [110]. In both the human and bovine ovary, two or three waves of follicles are

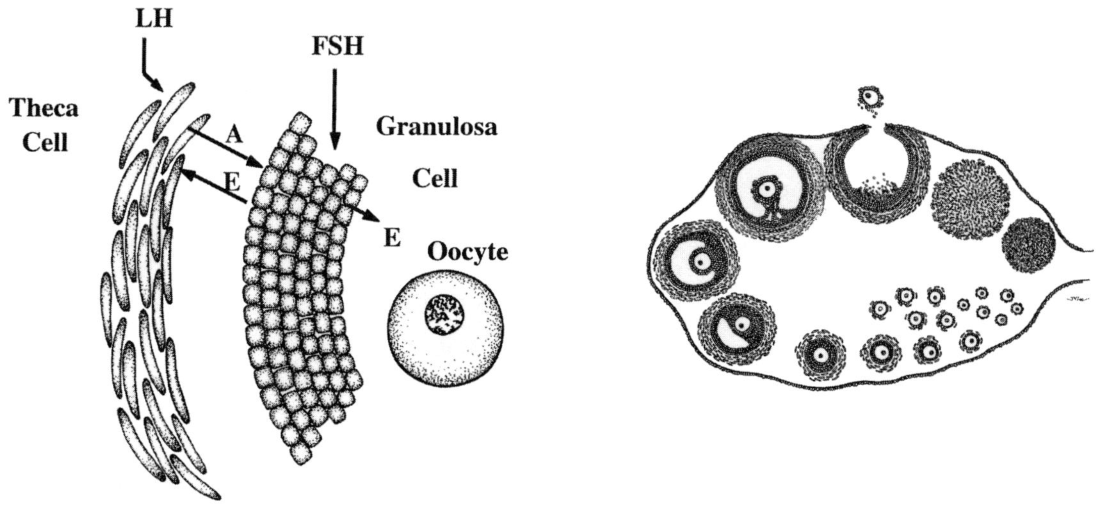

Figure 2 Ovary cell biology.

initiated to develop in a single ovarian cycle [111,112]. For both of these species, follicles expand from several millimeters to up to 2 cm during this process. A combination of granulosa cell growth, theca cell growth, and antrum formation results in the expansion of the ovarian follicle. Although rapid stimulation of cell growth is required for the ovulatory follicle to develop, the vast majority of follicles undergo atresia in which cell growth is arrested at various stages of follicle development. Hormones such as estrogen and FSH have been shown to promote follicle cell growth *in vivo*, however, these hormones alone do not stimulate growth of ovarian cells *in vitro* [113]. The possibility that these hormones may act indirectly through the local production of growth factors is proposed for later stages of development. Therefore, the regulation of ovarian cell growth is a complex process that requires an array of externally and locally derived regulatory agents. Interactions between theca cells, granulosa cells, and oocytes are required for follicular maturation [114]. The individual processes such as dominant follicle selection [115] and follicle cell apoptosis/atresia [116,117] also require integrated cell–cell interactions. A variety of specific growth factors produced in the follicle appear to mediate many of these cellular interactions in later stages of follicle development.

Ovarian Cell–Cell Interactions

Table II outlines a number of the factors produced locally in the ovary that mediate cell–cell signaling events in the control of follicle development and ovarian function. Several reviews address the topic of cell–cell interactions in the ovary and the control of follicle development [3,4,118,119]. Recent observations are cited next.

Table II Cell–Cell Signaling Factors in the Ovary

Signaling factor	Site production	Site action	Functions	Ref
Transforming growth factor α (TGFα)	Theca	Granulosa Theca	Growth stimulation	120–127
Transforming growth factor β (TGFβ)	Theca Granulosa	Granulosa Theca	Growth inhibition Cellular differentiation	128–132
Hepatocyte growth factor (HGF)	Theca	Granulosa	Growth stimulation	133–136
Keratocyte growth factor (KGF)	Theca	Granulosa	Growth stimulation	137–139
Colony stimulating factor (CSF)	Theca	Granulosa Theca	Growth regulation	140–141
Tumor necrosis factor (TNF)	Granulosa Theca Oocyte	Oocyte Granulosa Theca	Apoptosis Growth regulation	142–145
Fas ligand	Granulosa Theca Oocyte	Oocyte Granulosa Theca	Apoptosis	146–149
Nerve growth factor (NGF)	Theca	Granulosa Theca	Growth stimulation Ovulation	150
Fibroblast growth factor (bFGF)	Granulosa Theca Oocyte	Granulosa Theca	Growth stimulation	152–154
Growth differentiation factor -9 (GDF-9)	Oocyte	Granulosa Theca	Cellular differentiation	155–159
Bone morphogenic poteins (BMP)	Oocyte Theca	Granulosa Theca	Cellular differentiation	160–166
Kit Ligand/Stem cell factor (KL)	Granulosa	Oocyte Theca	Growth stimulation	167–171
Leukemia inhbitory factor (LIF)	Granulosa	Oocyte Theca	Growth stimulation Cellular differentiation	172–173
Vascular endothelial factor (VEGF)	Theca Granulosa	Edothelium Granulosa	Angiogenesis	174–178
Interleukins	Granulosa Theca	Granulosa Theca	Cellular differentiation	179–182
Insulin-Like growth factor (IGF-1)	Granulosa Theca	Oocyte Granulosa Theca	Growth stimulation Cellular differentiation	183–185
Inhibin	Granulosa	Oocyte Theca Granulosa	Cellular differentiation	186–187

TGF-α has been shown to be produced by theca cells [120–123] and to influence the growth of both theca and granulosa cells [120,124]. Several *in vivo* experiments have shown that TGF-α can influence follicle development [125,126]. TGF-α appears to be important for follicle development and involves theca cell–granulosa cell interactions. TGF-α has also been localized to isolated granulosa cells but appears predominantly in theca cells [127]. The primary function of TGF-α is growth stimulation. TGF-β is also predominately produced by theca cells [128], but is produced by isolated granulosa cells in selected follicle stages [129]. TGF-α and TGF-β differentially regulate granulosa and theca cell differentiated functions and growth [130–132]. Although TGF-β inhibits TGF-α growth stimulation, TGF-β also can influence cell functions.

HGF is produced by theca cells and acts on granulosa cells to promote cell proliferation and function [133, 134]. This is an excellent example of the role HGF plays in mediating mesenchymal–epithelial interactions in tissues. Interestingly, SCF/KL produced by the granulosa cells can feed back on the theca to regulate HGF production [135,136]. In a similar manner, keratinocyte growth factor (KGF) is produced by theca cells and acts on granulosa cells to regulate cell growth [137,138]. KGF is also expressed in the corpus luteum [139]. SCF/KL was found to also stimulate KGF expression by theca [135]. These factors reflect the importance of the theca cell in the regulation of follicle growth.

Granulocyte-macrophage colony-stimulating factor (GM-CSF) was found to primarily be expressed by theca cells and most cells in the ovary [140,141]. The GM-CSF can influence granulosa cell growth and function. Null mice had abnormal follicle development that suggested effects on the local cell–cell interactions [141].

Apoptosis is an essential aspect of follicle development and ovarian function. The vast majority of follicles undergo atresia and apoptosis. TNF has been shown to be produced by most cell types in the ovary associated with apoptosis [142–145]. TNF can act on all cell types and induce apoptosis and growth regulation. Another death ligand that binds death receptors to induce apoptosis is Fas ligand. Fas is also produced by all the cells associated with apoptosis and acts to promote apoptosis in the atretic follicles [146–149]. These signaling molecules are essential for ovarian function in the promotion of follicle atresia during folliculogenesis.

Nerve growth factor (NGF) was found to be expressed by theca cells and act on theca and granulosa cells [150]. The localization and actions of NGF suggest a potentially important role at the time of ovulation [150]. Other neurotropins (e.g., NT-4) are also expressed at various stages of ovary development [151] and require further investigation.

Basic fibroblast growth factor (bFGF) has been shown to be expressed by granulosa cells and to a lesser extent by theca cells [152]. Basic FGF can regulate both granulosa cell and theca cell growth and differentiated functions [153,154]. During follicle development the expression of bFGF changes, being initially in the oocyte at the primordial stage and then in the granulosa in the primary stage [3]. The role of other FGF family members has not been rigorously addressed.

Growth differentiation factor 9 (GDF-9) is a member of the TGF-β superfamily and is specifically localized to the oocyte. GDF-9 can act on both granulosa cells and theca cells to regulate steroidogenesis and differentiated functions [155–159]. The actions of GDF-9 are follicle stage specific and it appears to be expressed in a variety of species. GDF-9 may regulate the expression of other paracrine factors such as SCF/KL in the developing follicle [159]. This is one of the few oocyte-specific products identified to be involved in cell–cell signaling in the ovary.

Another factor specifically expressed in the oocyte that appears to regulate granulosa cell function is bone morphogenic protein 15 (BMP-15) [160,161]. BMP-15 and GDF-9 may act synergistically during follicle development. Other BMP family members include BMP-4 and -7, which are primarily localized in the theca cells and appear to act on the granulosa cells [162]; BMP-2, which acts on granulosa cells [163]; and BMP-6, which is also expressed in the oocyte and acts on the granulosa cells [164]. The BMP family of growth factors are TGF-β superfamily members and appear to be critical to follicle development [165,166].

SCF/KL is produced by the granulosa cell and acts on the oocyte and theca cells [167–171]. The null mutant suggests a critical role in oocyte viability and recruitment of primordial follicles. In addition to the role in granulosa–oocyte interactions, granulosa KL also influences theca cell function and development [170]. Oocytes appear to have a regulatory role in influencing the expression of KL by granulosa cells [171]. As found in the testis, this is a critical somatic–germ cell interaction. Another factor found to be expressed by granulosa cells that regulates oocytes is LIF [172,173]. LIF is also produced by stromal cells in the ovary. This action of LIF in mediating granulosa–oocyte interactions is supported by levels of LIF that increase in follicular fluid as the follicle develops [172,173].

Vascular endothelial growth factor (VEGF) has a critical role in angiogenesis. This process is important for developing follicles past the primary stage of development. VEGF is primarily expressed in theca cells and to a reduced level by granulosa cells [174–178]. VEGF has a major role in acting on endothelial cells to promote angiogenesis, but also can influence granulosa cell functions [175]. This cell–cell signaling event controlled by VEGF is critical for follicle development.

Cytokines as seen with the testis also influence ovary function. Interleukins 1, 6, and 8 have all been shown to regulate follicle development. IL-1 is expressed by the granulosa and affects granulosa function [179,180]. IL-8 is primarily expressed by the theca and to a lesser extent by granulosa and influences cellular function [181]. IL-6 is also expressed by granulosa cells an acts on various cells, including granulosa [182]. Further investigation of the specific roles of these and other member of the interleukin family is needed.

IGF-1 also has a role in the ovarian follicle [183]. IGF-1 is expressed by granulosa and theca cells and acts on the oocyte, granulosa, and theca cells [183–185]. Mice with null mutations in IGF-1 have impaired follicle development [184]. Other members of the family IGF-2 and the IGF-binding proteins also have a critical role in follicle development [185].

Inhibin also has a paracrine role in the developing follicle. Inhibin is primarily produced by the granulosa cells and acts on the oocyte, theca, and granulosa cells [186,187]. Related family members such as the activins are also anticipated to have similar roles. This is distinct from the roles these factors have in the endocrine system.

Additional signaling factors are anticipated to be essential for ovarian function and follicle development. One example is anti-Müllerian hormone (AMH), which is expressed by the granulosa cells, but specific biological function remains to be determined [188]. AMH may have a role as a negative regulator of oocyte viability or of primordial follicle development. Local steroid production is also expected to influence the network of local cell–cell signaling events. This includes both androgen and estrogen production [189].

Conclusions

The preceding descriptive discussion of cell–cell signaling in the testis and ovary demonstrate a growing complexity in the networks of cellular interactions and factors. It is anticipated that some of these factors will have compensatory roles to ensure growth and differentiation of the tissues. The list of factors provided is likely to be only partially complete and more will be added as investigation of cell–cell interactions in the gonads expands. Currently, we are primarily in the identification of the site of production and action research phase. The functions of some individual factors are also being analyzed. However, the next research phase of cell–cell signaling will involve a more systems biology type of approach that should tie together all potential interactions and provide more insight into the regulation of testis and ovary function.

The specific cell–cell signaling events identified are shown in most cases to change during development. The requirements and physiology of the embryonic testis and ovary are very different from the adult. Another research area to expand is the elucidation of cell–cell signaling at these different stages of development.

A comparison of the cell–cell signaling events between the testis and ovary is very useful. Some signaling events are the same, for example, the role SCF has in mediating direct somatic–germ cell interaction or the role HGF and KGF play in mesenchymal–epithelial cell interactions. A direct correlation among the cell–cell interactions of the testis and ovary will be invaluable in elucidating the system biology approach to understand gonadal function.

Elucidation of cell–cell signaling events is required for the future development of therapeutic agents to control fertility and treat reproductive diseases. By understanding the signaling events involved in testis and ovarian function, basic information is provided to design more effective therapeutics. Significant advances are anticipated to be in the area of contraceptive and fertility agent development and treatment of diseases such as polycystic ovarian disease or premature ovarian failure. Although an understanding of the intracellular signaling events is essential for understanding how a factor acts, the elucidation of the network of extracellular signaling molecules that regulates a cells function is essential to our understanding of how a tissue or organism functions.

References

1. Skinner, M. K. (1991). Cell–cell interactions in the testis. *Endocr. Rev.* **12**, 45–77.
2. Griswold, M. D. (1988). Protein secretions of Sertoli cells. *Intl. Rev. Cytol.* **110**, 133–156.
3. Nilsson, E. E. and Skinner, M. K. (2001). Cellular interactions that control primordial follicle development. *J. Soc. Gynecol. Investig.* **8**, S17–S18.
4. Hirshfield, A. N. (1991). Development of follicles in the mammalian ovary. *Intl. Rev. Cytol.* **124**, 43–101.
5. Sertoli, E. (1865). On the existence of special branched cells in the seminiferous tubule of the human testes. *Morgangni* **7**, 31–39.
6. Fawcett, D. W. (1975). The ultrastructure and functions of the Sertoli cell. In Greep, R. O., and Hamilton, E. W., Eds., *Handbook of Physiology,* Vol. V, pp. 22–55. American Physiology Society, Washington, DC.
7. Setchell, B. P. and Waites, G. M. H. (1975). The blood-testis barrier. In Greep, R. O., and Hamilton, E. W., Eds., *Handbook of Physiology,* Vol. V, pp. 143–172. American Physiology Society, Washington, DC.
8. Waites, G. M. H. and Gladwell, R. T. (1982). Physiological significance of fluid secretion in the testis and blood–testis barrier. *Physiol. Rev.* **62**, 624–671.
9. Russell, L. D., Tallon-Doran, M., Weber, J. E., Wong, V., and Peterson, R. N. (1983). Three-dimensional reconstruction of a rat stage V Sertoli cell: III. A study of specific cellular relationships. *Am. J. Anat.* **167**, 181–192.
10. Dorrington, J. H., Fritz, I. B., and Armstrong, D. T. (1978). Control of testicular estrogen synthesis. *Biol. Reprod.* **18**, 55–65.
11. Robinson, R. and Fritz, I. B. (1981). Metabolism of glucose by Sertoli cells in culture. *Biol. Reprod.* **24**, 1032–1041.
12. Lacroix, M., Smith, F. E., and Fritz, I. B. (1977). Secretion of plasminogen activator by Sertoli cell enriched cultures. *Mol. Cell Endocrinol.* **9**, 227–236.
13. Skinner, M. K. and Griswold, M. D. (1980). Sertoli cells synthesize and secrete transferrin-like protein. *J. Biol. Chem.* **255**, 9523–9525.
14. Skinner, M. K., and Griswold, M. D. (1983). Sertoli cells synthesize and secrete a ceruloplasmin-like protein. *Biol. Reprod.* **28**, 1225–1229.
15. Tung, P. S., Skinner, M. K., and Fritz, I. B. (1984). Fibronectin synthesis is a marker for peritubular cell contaminants in Sertoli cell-enriched cultures. *Biol. Reprod.* **30**, 199–121.
16. Skinner, M. K., Tung, P. S., and Fritz, I. B. (1985). Cooperativity between Sertoli cells and testicular peritubular cells in the production and deposition of extracellular matrix components. *J. Cell Biol.* **100**, 1941–1947.
17. Kierszenbaum, A. L. (1994). Mammalian spermatogenesis *in vivo* and *in vitro:* A partnership of spermatogenic and somatic cell lineages. *Endocr. Rev.* **15**,116–134.
18. Adams, I. R. and McLaren, A. (2002). Sexually dimorphic development of mouse primordial germ cells: switching from oogenesis to spermatogenesis. *Development* **129**, 1155–1164.
19. Takaski, N., Rankin, T., and Dean, J. (2001). Normal gonadal development in mice lacking GPBOX, a homeobox protein expressed in germ cells at the onset of dimorphism. *Mol. Cell Biol.* **21**, 8197–8202.
20. Vaillant, S., Magre, S., Dorizzi, M., Pieau, C., and Richard-Mercier, N. (2001). Expression of AMH, SF1 and SOX9 in gonads of genetic

female chicken reversal induced by an aromatase inhibitor. *Dev. Dyn.* **21**, 228–237.

21. Ikeda, Y., Takeda, Y., Shikayama, T., Mukai, T., Hisano, S., and Morohashi, K. I. (2001). Comparative localization of Dax-1 and Ad4BP/SF-1 during development of hypothalamic-pituitary-gonadal axis suggests their closely related and distinct functions. *Dev. Dyn.* **220**, 363–376.
22. Clinton, M. and Haines, L. C. (2001). An overview of factors influencing sex determination and gonadal development in birds. *EXS* **91**, 97–115.
23. Parker, K. L., Schimmer, B. P., and Schedl, A. (2001). Genes essential for early events in gonadal development. *EXS* **91**, 11–24.
24. Ostrer, H. (2000). Sexual differentiation. *Semin. Reprod. Med.* **18**, 41–49.
25. Drews, U. (2000). Local mechanisms in sex specific morphogenesis. *Cytogenet. Cell Genet.* **91**, 72–80.
26. McLaren, A. (2000). Germ and somatic cell lineages in the developing gonad. *Mol. Cell Endocrinol.* **163**, 3–9.
27. Raymond, C. S., Murphy, M. W., O'Sullivan, M. G., Bardwell, V. J., and Zarkower, D. (2000). Dmrt 1, a gene related to worm and fly sexual regulators, is required for mammalian testis differentiation. *Genes Dev.* **14**, 2587–2595.
28. Jost, A., Magre, S., and Agelopoulou, R. (1981). Early stages of testicular differentiation in the rat. *Hum. Genet.* **58**, 59–63.
29. Magre, S. and Jost, A. (1980). The initial phases of testicular organogenesis in the rat. An electron microscopy study. *Arch. Anat. Microsc. Morphol. Exp.* **69**, 297–318.
30. Buehr, M., Gu, S., and McLaren, A. (1993). Mesonephric contribution to testis differentiation in the fetal mouse. *Development* **117**, 273–281; see also Erratum (1993). *Development* **118**, following 1384.
31. Merchant-Larios, H., Moreno-Mendoza, N., and Buehr, M. (1993). The role of the mesonephros in cell differentiation and morphogenesis of the mouse fetal testis. *Int. J. Dev. Biol.* **37**, 407–415.
32. Ricci, G., Catizone, A., Innocenzi, A., and Galdieri, M. (1999). Hepatocyte growth factor (HGF) receptor expression and role of HGF during embryonic mouse testis development. *Dev. Biol.* **216**, 340–347.
33. McLaren, A. (1991). Development of the mammalian gonad: The fate of the supporting cell lineage. *Bioassays* **13**, 151–156.
34. Karl, J., and Blanche, C. (1998). Sertoli cells of the mouse testis originate from the coelomic epithelium. *Dev. Biol.* **203**, 323–333.
35. Frojdman, K., Paranko, J., Virtanen, I., and Pelliniemi, L. J. (1992). Intermediate filaments and epithelial differentiation of male rat embryonic gonad. *Differentiation* **50**, 113–123.
36. Fridmacher, V., Le Bert, M., Buillou, F., and Magre, S. (1995). Switch in the expression of the K19/K18 keratin genes as a very early evidence of testicular differentiation in the rat. *Mech. Dev.* **32**, 199–207.
37. Blanchard, M. G., and Josso, N. (1974). Source of the anti-Mullerian hormone synthesized by the fetal testis: Mullerian-inhibiting activity of fetal bovine Sertoli cells in tissue culture. *Pediatr. Res.* **8**, 968–971.
38. Greene, R. R. (1944). Embryology of sexual structure and hermaphroditism. *J. Clin. Endocrinol.* **4**, 335–348.
39. Magre, S. and Jost, A. (1991). Sertoli cells and testicular differentiation in the rat fetus. *J. Electron Microsc. Tech.* **19**, 172–188.
40. Greco, T. L. and Payne, A. H. (1994). Ontogeny of expression of the genes for steroidogenic enzymes P450 side-chain cleavage, 3B-hydroxysteroid dehydrogenase, P45017 ∝ hydroxylase/C 17–20 lyase, and P450 aromatase in fetal mouse gonads. *Endocrinology* **135**, 262–268.
41. Byskov, A. G. (1986). Differentiation of mammalian embryonic gonad. *Physiol. Rev.* **66**, 71–117.
42. Orth, J. M., Weisz, J., Ward, O. B., and Ward, I. L. (1983). Environmental stress alters the developmental pattern of delta 5–3 beta-hydroxysteroid dehydrogenase activity in Leydig cells of fetal rats: A quantitative cytochemical study. *Biol. Reprod.* **28**, 625–631.
43. Bloch, E., Lew, M., and Klein, M. (1971). Studies on the inhibition of fetal androgen formation. Inhibition of testosterone synthesis in rat and rabbit fetal testes with observations on reproductive tract development. *Endocrinology* **89**, 16–31.
44. Weinbauer, F. G. and Wessels, J. (1999). Paracrine control of spermatogenesis. *Andrologia* **31**, 249–262.
45. Roser, J. F. (2001). Endocrine and paracrine control of sperm production in stallions. *Animal Reprod. Sci.* **68**, 139–151.
46. Levine, E., Cupp, A. S., Miyashiro, L., and Skinner, M. K. (2000). Role of transforming growth factor-alpha and the epidermal growth factor receptor in embryonic rat testis development. *Biol. Reprod.* **62**, 477–490.
47. Petersen, C., Boitani, C., Froysa, B., and Soder, O. (2001). Transforming growth factor-alpha stimulates proliferation of rat Sertoli cells. *Mol. Cell Endocrinol.* **181**, 221–227.
48. Mendis-Handagama, S. M. and Ariyaratne, H. B. (2001). Differentiation of the adult Leydig cell population in the postnatal testis. *Biol. Reprod.* **65**, 660–671.
49. Avallet, O., Gomez, E., Vigier, M., Jegou, B., and Saez, J. M. (1997). Sertoli cell-germ cell interaction and TGF beta 1 expression and secretion *in vitro*. *Biochem. Biophys. Res. Commun.* **238**, 905–909.
50. Lui, W. Y., Lee, W. M., and Cheng, C. Y. (2001). Transforming growth factor-beta3 perturbs the inter-Sertoli tight junction permeability barrier *in vitro* possibly mediated via its effects on occludin, zonula occludens-1, and claudin-11. *Endocrinology* **142**, 1865–1877.
51. Olaso, R., Pairault, C., and Habert, R. (1998). Expression of type I and II receptors for transforming growth factor beta in the adult rat testis. *Histochem. Cell Biol.* **110**, 613–618.
52. Wang, R. A., and Zhao, G. Q. (1999). Transforming growth factor beta signal transducer Smad2 is expressed in mouse meiotic germ cells, Sertoli cells, and Leydig cells during spermatogenesis. *Biol. Reprod.* **61**, 999–1004.
53. Konrad, L., Albrecht, M., Renneberg, H., and Aumuller, G. (2000). Transforming growth factor-beta2 mediates mesenchymal-epithelial interactions of testicular somatic cells. *Endocrinology* **141**, 3679–3686.
54. Rouiller-Fabre, V., Lecref, L., Gautier, C., Saez, J. M., and Habert, R. (1998). Expression and effect of insulin-like growth factor I on rat fetal Leydig cell function and differentiation. *Endocrinology* **139**, 2926–2934.
55. Le Roy, C., Lejeune, H., Chuzel, F., Saez, J. M., and Langlois, D. (1999). Autocrine regulation of Leydig cell differentiated functions by insulin-like growth factor I and transforming growth factor beta. *J. Steroid Biochem. Mol. Biol.* **69**, 379–384.
56. Santos, R. L., Silva, C. M., Ribeiro, A. F., Vasconcelow, A. C., Pesquero, J. L., Coelho, S. G., Serakides, R., and Reis, S. R. (1999). Effect of growth hormone and induced IGF-1 release on germ cell population and apoptosis in the bovine testis. *Theriogenology* **51**, 975–984.
57. Tsuruta, J. K., Eddy, E. M., and O'Brien, D. A. (2000). Insulin-like growth factor-II/cation-independent mannose 6-phosphate receptor mediates paracrine interactions during spermatogonial development. *Biol. Reprod.* **63**, 1006–1013.
58. Soder, O., Sultana, T., Jonsson, C., Wahlgren, A., Petersen, C., and Holst, M. (2000). The interleukin-1 system in the testis. *Andrologia* **32**, 52–55.
59. Huleihel, M. and Lunenfeld, E. (2002). Involvement of intratesticular IL-1 system in the regulation of Sertoli cell functions. *Mol. Cell Endocrinol.* **187**, 125–132.
60. Stephan, J. P., Syed, V., and Jegou, B. (1997). Regulation of Sertoli cell IL-1 and IL-6 production *in vitro*. *Mol. Cell Endocrinol.* **134**, 109–118.
61. Nehar, D., Mauduit, C., Bousouar, F., and Benahmed, M. (1998). Interleukin 1 alpha stimulates lactate dehydrogenase A expression and lactate production in cultured porcine Sertoli cells. *Biol. Reprod.* **59**, 1425–1432.
62. Zeyse, D., Lunenfeld, E., Beck, M., Prinsloo, I., and Huleihel, M. (2000). Interleukin-1 receptor antagonist is produced by Sertoli cells *in vitro*. *Endocrinology* **141**, 1521–1527.
63. Jenab, S. and Morris, P. L. (2000). Interleukin-6 regulation of kappa opioid receptor gene expression in primary Sertoli cells. *Endocrine* **13**, 11–15.
64. Meroni, S. B., Suburo, A. M., and Cigorraga, S. B. (2000). Interleukin-1beta regulates nitric oxide production and gamma-glutamyl transpeptidase activity in Sertoli cells. *J. Androl.* **21**, 855–161.
65. Petersen, C., Boitani, C., Froysa, B., and Soder, O. (2002). Interleukin-1 is a potent growth factor for immature rat Sertoli cells. *Mol. Cell Endocrinol.* **186**, 37–47.

66. De Kretser, D. M., Meinhardt, A., Meehan, T., Phillips, D. J., O'Bryan, M. K., and Loveland, K. A. (2000). The roles of inhibin and related peptides in gonadal function. *Mol. Cell Endocrinol.* **161**, 43–46.
67. Risbridger, G. P. and Cancilla, B. (2000). Role of activins in the male reproductive tract. *Rev. Reprod.* **5**, 99–104.
68. Ethier, J. F. and Findlay, J. K. (2001). Roles of activin and its signal transduction mechanisms in reproductive tissues. *Reproduction* **121**, 667–675.
69. Schlatt, S., Meinhardt, A., and Nieschlag, E. (1997). Paracrine regulation of cellular interacts in the testis: Factors in search of a function. *Eur. J. Endocrinol.* **137**, 107–117.
70. Cancilla, B., Davies, A., Ford-Perriss, M., and Risbridger, G. P. (2000). Discrete cell-and stage-specific localisation of fibroblast growth factors and receptor expression during testis development. *J. Endocrinol* **164**, 149–159.
71. Cancilla, B. and Risbridger, G. P. (1998). Differential localization of fibroblast growth factor receptor-1, -2, -3, and -4 in fetal, immature, and adult rat testes. *Biol. Reprod.* **58**, 1138–1145.
72. Yamamoto, S., Mikami, T., Konishi, M., and Itoh, N. (2000). Stage-specific expression of a novel isoform mouse FGF-14 (FHF-4) in spermatocytes. *Biochim. Biophys. Acta* **1490**, 121–124.
73. Yamamoto, H., Ochiya, T., Takahama, Y., Ishii, Y., Osumi, N., Sakamoto, H., and Terada, M. (2000). Detection of spatial localization of Hst-1/Fgf-4 gene expression in brain and testis from adult mice. *Oncogene* **19**, 3805–3810.
74. Colvin, J. S., Green, R. P., Schmahl, J., Capel, B., and Ornitz, D. M. (2001). Male-to-female sex reversal in mice lacking fibroblast growth factor 9. *Cell* **104**, 875–889.
75. Schteingart, H. F., Meroni, S. B., Canepa, D. F., Pellizzari, E. H., and Cigorraga, S. B. (1999). Effects of basic fibroblast growth factor and nerve growth factor on lactate production, γ-glutamyl transpeptidase and aromatase activities in cultured Sertoli cells. *Eur. J. Endocrinol.* **141**, 539–545.
76. Mariani, S., Basciani, S., Arizzi, M., Spera, G., and Gnessi, L. (2002). PDGF and the testis. *Trends Endocrinol. Metab.* **13**, 11–17.
77. Gnessi, L., Basciani, S., Mariani, S., Arizzi, M., Spera, G., Wang, C., Bondjers, C., Karlsson, L., and Betsholtz, C. (2000). Leydig cell loss and spermatogenic arrest in platelet-derived growth factor (PDGF)-A-deficient mice. *J. Cell Biol.* **149**, 1019–1026.
78. Chiarenza, C., Filippini, A., Tripiciano, A., Beccari, E., and Palombi, F. (2000). Platelet-derived growth factor-BB stimulates hypertrophy of peritubular smooth muscle cells from rat testis in primary cultures. *Endocrinology* **141**, 2971–2981.
79. Basciani, S., Mariani, S., Arizi, M., Ulisse, S., Rucci, N., Jannini, E. A., Rocca, C. D., Manicone, A., Carani, C., Spera, G., and Gnessi, L. (2002). Expression of platelet-derived growth factor-A (PDGF-A), PDGF-B, and PDGF Receptor-alpha and -beta during human testicular development and disease. *J. Clin. Endocrinol. Metab.* **87**, 2310–2319.
80. Loveland, K. L. and Schlatt, S. (1997). Stem cell factor and c-kit in the mammalian testis: Lessons originating from Mother Nature's gene knockouts. *J. Endocrinol.* **153**, 337–344.
81. Mauduit, C., Hamamah, S., and Benahmed, M. (1999). Stem cell factor/c-kit system in spermatogenesis. *Hum. Reprod. Update* **5**, 535–545.
82. Rossi, P., Sette, C., Dolci, S., and Geremia, R. (2000). Role of c-kit in mammalian spermatogenesis. *J. Endocrinol. Invest.* **23**, 609–615.
83. Vincent, S., Segretain, D., Nishikawa, S., Nishikawa, S. I., Sage, J., Cuzin, F., and Rassoulzadegan, M. (1998). Stage-specific expression of the kit receptor and its ligand (KL) during male gametogenesis in the mouse: A Kit-KL interaction critical for meiosis. *Development* **125**, 4585–4593.
84. Piquest-Pellorce, C., Dorval-Coiffec, I., Pham, M. D., and Jegou, B. (2000). Leukemia inhibitory factor expression and regulation within the testis. *Endocrinology* **141**, 1136–1141.
85. Hara, T., Tamura, K., de Miguel, M. P., Mukouyama, Y. S., Kim, H. J., Kogo, H., Donovan, P. J., and Miyajima, A. (1998). Distinct roles of oncostatin M and leukemia inhibitory factor in the development of primordial germ cells and Sertoli cells in mice. *Dev. Biol.* **201**, 144–153.
86. Boussouar, F., Grataroli, R., Ji, J., and Benahmed, M. (1999). Tumor necrosis factor-alpha stimulates lactate dehydrogenase A expression in porcine cultured Sertoli cells: mechanisms of action. *Endocrinology* **140**, 3054–3062.
87. Riera, M. F., Meroni, S. B., Gomez, G. E., Schteingart, H. F., Pellizzari, E. H., and Cigorraga, S. B. (2001). Regulation of lactate production by FSH, iL1beta, and TNF alpha in rat Sertoli cells. *Gen. Comp. Endocrinol.* **122**, 88–97.
88. Grataroli, R., Vindrieux, D., Gougeon, A., and Benahmed, M. (2002). Expression of tumor necrosis factor-alpha-related apoptosis-inducing ligand and its receptors in rat testis during development. *Biol. Reprod.* **66**, 1707–1715.
89. Catizone, A., Riccci, G., Arista, V., Innocenzi, A., and Galdieri, M. (1999). Hepatocyte growth factor and c-MET are expressed in rat prepuberal testis. *Endocrinology* **140**, 3106–3113.
90. Catizone, A., Ricci, G., and Galdieri, M. (2001). Expression and functional role of hepatocyte growth factor receptor (C-MET) during postnatal rat testis development. *Endocrinology* **142**, 1828–1834.
91. van der Wee, K. and Hofmann, M. C. (1999). An *in vitro* tubule assay identifies HGF as a morphogen for the formation of seminiferous tubules in the postnatal mouse testis. *Exp. Cell Res.* **252**, 175–185.
92. Cupp, A. S., Kim, G. H., and Skinner, M. K. (2000). Expression and action of neurotropin-3 and nerve growth factor in embryonic and early postnatal rat testis development. *Biol. Reprod.* **63**, 1617–1628.
93. Magnanti, M., Gandini, O., Giulianai, L., Gazzaniga, P., Marti, H. H., Gradilone, A., Frati, L., Agliano, A. M., and Gassmann, M. (2001). Erythropoietin expression in primary rat Sertoli and peritubular myoid cells. *Blood* **98**, 2872–2874.
94. Kanzaki, M. and Morris, P. L. (1998). Identification and regulation of testicular interferon-gamma (IFNgamma) receptor subunits: IFNgamma enhances interferon regulatory factor-1 and interleukin-1 beta converting enzyme expression. *Endocrinology* **139**, 2636–2644.
95. Ivell, R. (1997). Biology of the relaxin-like factor (RLF). *Rev. Reprod.* **2**, 133–138.
96. Verhoeven, G., Hoeben, E., and De Gendt, K. (2000). Peritubular cell-Sertoli cell interactions: Factors involved in PModS activity. *Andrologia* **32**, 41–64.
97. Midgley, A. J. R. (1973). Autoradiographic analysis of gonadotropin binding to rat ovarian tissue sections. *Adv. Exp. Med. Biol.* **36**, 365–378.
98. Richards, J. S. and Midgley, A. R. (1976). Protein hormone action: A key to understanding ovarian follicular and luteal cell development. *Biol. Reprod.* **14**, 82–94.
99. Vlodavsky, I., Brown, K. D., and Gospodarowicz, D. (1978). A comparison of the binding of epidermal growth factor to cultured granulosa and luteal cells. *J. Biol. Chem.* **253**, 3744–3750.
100. Wandji, S. A., Pelletier, G., and Sirard, M. A. (1992). Ontogeny and cellular localization of 125I-labeled basic fibroblast growth factor and 125I labeled epidermal growth factor binding sites in ovaries from bovine fetuses and neonatal calves. *Biol. Reprod.* **47**, 807–813.
101. Adashi, E. Y. (1998). The IGF family and folliculogenesis. *J. Reprod. Immunol.* **39**, 13–19.
102. Peng, C., Ohno, T., Khorasheh, S., and Leung, P. C. (1996). Activin and follistatin as local regulators in the human ovary. *Biol. Signals* **5**, 81–89.
103. Erickson, G. F. (1983). Primary cultures of ovarian cells in serum-free medium as models of hormone-dependent differentiation. *Mol. Cell Endocrinol.* **29**, 21–49.
104. Fortune, J. E., and Armstrong, D. T. (1977). Androgen production by theca and granulosa isolated from proestrous rat follicles. *Endocrinology* **100**, 1341–1347.
105. Erickson, G. F., and Ryan, K. J. (1976). Stimulation of testosterone production in isolated rabbit thecal tissue by LH/FSH, dibutyryl cyclic AMP, PGE2alpha, and PGE2. *Endocrinology* **99**, 452–458.
106. Haney, A. F. and Schomberg, D. W. (1981). Estrogen and progesterone production by developing porcine follicles *in vitro:* Evidence for estrogen formation by theca. *Endocrinology* **109**, 971–977.
107. Evans, G., Dobias, M., King, G. J., and Armstrong, D. T. (1981). Estrogen, androgen, and progesterone biosynthesis by theca and granulosa of preovulatory follicles in the pig. *Biol. Reprod.* **25**, 673–682.
108. Channing, C. P. (1980). Progesterone and estrogen secretion by cultured monkey ovarian cell types: Influences of follicular size,

serum luteinizing hormone levels, and follicular fluid estrogen levels. *Endocrinology* **107**, 342–352.

109. McNatty, K. P., Makris, A., DeGrazia, C., Osathanondh, R., and Ryan, K. J. (1979). The production of progesterone, androgens, and estrogens by granulosa cells, thecal tissue, and stromal tissue from human ovaries *in vitro*. *J. Clin. Endocrinol. Metab.* **49**, 687–699.
110. Fortune, J. E. (1994). Ovarian follicular growth and development in mammals. *Biol. Reprod.* **50**, 225–232.
111. Sirois, J. and Fortune, J. E. (1988). Ovarian follicular dynamics during the estrous cycle in heifers monitored by real-time ultrasonography. *Biol. Reprod.* **39**, 308–317.
112. Fortune, J. E. (1994). Ovarian follicular growth and development in mammals. *Biol. Reprod.* **50**, 225–232.
113. Hsueh, A. J., McGee, E. A., Hayashi, M., and Hsu, S. Y. (2000). Hormonal regulation of early follicle development in the rat ovary. *Mol. Cell Endocrinol.* **163**, 95–100.
114. Yada, H., Hosokawa, K., Tajima, K., Hasegawa, Y., and Kotsuji, F. (1999). Role of ovarian theca and granulosa cell interaction in hormone production and cell growth during the bovine follicular maturation process. *Biol. Reprod.* **61**, 1480–1486.
115. Baker, S. J. and Spears, N. (1999). The role of intra-ovarian interactions in the regulation of follicle dominance. *Hum. Reprod. Update* **5**, 153–165.
116. Hsueh, A. J., Eisenhauer, K., Chun, S. Y., Hsu, S. Y., and Billig, H. (1996). Gonadal cell apoptosis. *Recent Prog. Horm. Res.* **51**, 433–455.
117. Chun, S. Y., Eisenhauer, K. M., Minami, S., and Hsueh, A. J. (1996). Growth factors in ovarian follicle atresia. *Semin. Reprod. Endocrinol.* **14**, 197–202.
118. Erickson, G. F. and Shimasaki, S. (2001). The physiology of folliculogenesis: The role of novel growth factors. *Fertil. Steril.* **76**, 943–949.
119. Einspanier, R., Lauer, B., Gabler, C., Kamhuber, M., and Schams, D. (1997). Egg-cumulus-oviduct interactions and fertilization. *Adv. Exp. Med. Biol.* **424**, 279–289.
120. Skinner, M. K. and Coffey, R. J. (1988). Regulation of ovarian cell growth through the local production of transforming growth factor-alpha by theca cells. *Endocrinology* **123**, 2632–2638.
121. Derynck, R. (1986). Transforming growth factor-alpha: Structure and biological activities. *J. Cell Biochem.* **32**, 293–304.
122. Kudlow, J. E., Kobrin, M. S., Purchio, A. F., Twardzik, D. R., Hernandez, E. R., Asa, S. L., and Adashi, E. Y. (1987). Ovarian transforming growth factor-alpha gene expression: Immunohistochemical localization to the theca-interstitial cells. *Endocrinology* **121**, 1577–1579.
123. Lobb, D., Dobrin, M., Kudlow, J., and Dorrington, J. (1989). Transforming growth factor-alpha in the adult bovine ovary: identification in growing ovarian follicles. *Biol. Reprod.* **40**, 1087–1093.
124. Skinner, M. K. (1989). Transforming growth factor production and actin in the ovarian follicle: Theca cell and granulosa cell interactions. In Hirshfeld, A., Ed., *Growth Factors and the Ovary,* p. 141. Plenum Press, New York.
125. Ma, Y. J., Dissen, G. A., Merlino, G., Coquelin, A., and Ojeda, S. R. (1994). Overexpression of a human transforming growth factor-alpha (TGF alpha) transgene reveals a dual antagonistic role of TGF alpha in female sexual development. *Endocrinology* **135**, 1392–1400.
126. Campbell, B. K., Gordon, B. M., and Scaramuzzi, R. J. (1994). The effect of ovarian arterial infusion of transforming growth factor alpha on ovarian follicle populations and ovarian hormone secretion in ewes with an autotransplanted ovary. *J. Endocrinol.* **143**, 13–24.
127. Qu, J., Nisolle, M., and Donnez, J. (2000). Expression of transforming growth factor-alpha, epidermal growth factor, and epidermal growth factor receptor in follicles of human ovarian tissue before and after cryopreservation. *Fertil. Steril.* **74**, 113–121.
128. Skinner, M. K., Keski-Oja, J., Osteen, K., and Moses, H. L. (1987). Ovarian thecal cells produce transforming growth factor-beta which can regulate granulosa cell growth. *Endocrinology* **121**, 786–792.
129. Christopher, B. (2000). Immunolocalization of transforming growth factor-beta1 during follicular development and atresia in the mouse ovary. *Endocr. J.* **47**, 475–480.
130. Roberts, A. and Skinner, M. K. K. (1991). Transforming growth factor-alpha and -beta differentially regulate growth and steroidogenesis of bovine thecal cells during antral follicle development. *Endocrinology* **129**, 2041–2048.
131. Feng, P., Catt, K. J., and Knecht, M. (1986). Transforming growth factor beta regulates the inhibitory actions of epidermal growth factor during granulosa cell differentiation. *J. Biol. Chem.* **261**, 14167–14170.
132. Ying, S. Y., Becker, A., Ling, N., Ueno, N., and Guillemin, R. (1986). Inhibin and beta type transforming growth factor (TGF beta) have opposite modulating effects on the follicle stimulating hormone (FH)-induced aromatase activity of cultured rat granulosa cells. *Biochem. Biophys. Res. Comm.* **136**, 969–975.
133. Parrott, J. A. and Skinner, M. K. (1998). Developmental and hormonal regulation of hepatocyte growth factor expression and action in the bovine ovarian follicle. *Biol. Reprod.* **59**, 553–560.
134. Zachow, R. J., Ramski, B. E., and Lee, H. (2000). Modulation of estrogen production and 17beta-hydroxysteroid dehydrogenase-type 1, cytochrome P450 aromatase, c-met, and protein kinase Balpha messenger ribonucleic acid content in rat ovarian granulosa cells by hepatocyte growth factor and follicle-stimulating hormone. *Biol. Reprod.* **62**, 1851–1857.
135. Parrott, J. A. and Skinner, M. K. (1998). Thecal cell-granulosa cell interactions involve a positive feedback loop among keratinocyte growth factor, hepatocyte growth factor, and Kit ligand during ovarian follicular development. *Endocrinology* **139**, 2240–2245.
136. Ito, M., Harada, T., Tanikawa, M. M., Fujii, A., Shiota, G., and Terakawa, N. (2001). Hepatocyte growth factor and stem cell factor involvement in paracrine interplays of theca and granulosa cells in the human ovary. *Fertil. Steril.* **75**, 973–979.
137. Parrott, J. A. and Skinner, M. K. (1998). Developmental and hormonal regulation of keratinocyte growth factor expression and action in the ovarian follicle. *Endocrinology* **139**, 228–235.
138. Osuga, Y., Koga, K., Tsutsumi, O., Yano, T., Kugu, K., Momoeda, M., Okagaki, R., Suenaga, A., Fujiwara, T., Fujimoto, A., Matsumi, H., Hiroi, H., and Tasketani, Y. (2001). Evidence for the presence of keratinocyte growth factor (KGF) in human ovarian follicles. *Endocr. J.* **48**, 161–166.
139. Salli, U., Bartol, F. F., Wiley, A. A., Tarleton, B. J., and Braden, T. D. (1998). Keratinocyte growth factor expression by the bovine corpus luteum. *Biol. Reprod.* **59**, 77–83.
140. Tamura, K., Tamura, H., Kumasaka, K., Miyajima, A., Suga, T., and Kogo, H. (1998). Ovarian immune cells express granulocyte-macrophage colony-stimulating factor (GM-CSF) during follicular growth and luteinization in gonadotropin-primed immature rodents. *Mol. Cell Endocrinol.* **142**, 153–163.
141. Gilchrist, R. B., Rowe, D. B., Ritter, L. J., Roberston, S. A., Norman, R. J., and Armstrong, D. T. (2000). Effect of granulocyte-macrophage colony-stimulating factor deficiency on ovarian follicular cell function. *J. Reprod. Fertil.* **120**, 283–292.
142. Morrison, L. J. and Marcinkiewicz, J. L. (2002). Tumor necrosis factor alpha enhances oocyte/follicle apoptosis in the neonatal rat ovary. *Biol. Reprod.* **66**, 450–457.
143. Prabge-Kiel, J., Kreutzkamm, C., Wehrenberg, U., and Rune, G. M. (2001). Role of tumor necrosis factor in preovulatory follicles of swine. *Biol. Reprod.* **65**, 928–935.
144. Spicer, L. J. (2001). Receptors for insulin-like growth factor-I and tumor necrosis factor-alpha are hormonally regulated in bovine granulosa and thecal cells. *Anim. Reprod. Sci.* **67**, 45–58.
145. Spaczynski, R. Z., Arici, A., and Duleba, A. J. (1999). Tumor necrosis factor-alpha stimulates proliferation of rat ovarian theca-interstitial cells. *Biol. Reprod.* **61**, 993–998.
146. Bridgham, J. T. and Johnson, A. L. (2001). Expression and regulation of Fas antigen and tumor necrosis factor receptor type I in hen granulosa cells. *Biol. Reprod.* **65**, 733–739.
147. Quirk, S. M., Porter, D. A., Huber, S. C., and Cowan, R. G. (1998). Potentiation of Fas-mediated apoptosis of murine granulosa cells by interferon-gamma, tumor necrosis factor-alpha, and cycloheximide. *Endocrinology* **139**, 4860–4869.

148. Vickers, S. L., Cowan, R. G., Harman, R. M., Porter, D. A., and Quirk, S. M. (2000). Expression and activity of the Fas antigen in bovine ovarian follicle cells. *Biol. Reprod.* **62**, 54–61.
149. Quirk, S. M., Harman, R. M., and Cowan, R. G. (2000). Regulation of Fas antigen (Fas, CD95)-mediated apoptosis of bovine granulosa cells by serum and growth factors. *Biol. Reprod.* **63**, 1278–1284.
150. Dissen, G. A., Parrott, J. A., Skinner, M. K., Hill, D. F., Costa, M. E., and Ojeda, S. R. (2000). Direct effects of nerve growth factor on thecal cells from antral ovarian follicles. *Endocrinology* **141**, 4736–4750.
151. Anderson, R. A., Robinson, L. L., Brooks, J., and Spears, N. (2002). Neurotropins and their receptors are expressed in the human fetal ovary. *J. Clin. Endocrinol. Metab.* **87**, 890–897.
152. Yamamoto, S., Konishis, I., Nanbu, K., Komatsu, T., Mandai, M., Kuroda, H., Matsushita, K., and Mori, T. (1997). Immunohistochemical localization of basic fibroblast growth factor (bFGF) during folliculogenesis in the human ovary. *Gynecol. Endocrinol.* **11**, 223–230.
153. Peluso, J. J., Pappalardo, A., and Fernandez, G. (2001). Basic fibroblast growth factor maintains calcium homeostasis and granulosa cell viability by stimulating calcium efflux via a PKC delta-dependent pathway. *Endocrinology* **142**, 4203–4211.
154. Puscheck, E. E., Patel, Y., and Rappolee, D. A. (1997). Fibroblast growth factor receptor (FGFR)-4, but not FGFR-3 is expressed in the pregnant ovary. *Mol. Cell Endocrinol.* **132**, 169–176.
155. Elvin, J. A., Clark, A. T., Wang, P., Wolfman, N. M., and Matzuk, M. M. (1999). Paracrine actions of growth differentiation factor-9 in the mammalian ovary. *Mol. Endocrinol.* **13**, 1035–1048.
156. Solovyeva, E. V., Hayashi, M., Margi, K., Barkats, C., Klein, C., Amsterdam, A., Hsueh, A. J., and Tsafriri, A. (2000). Growth differentiation factor-9 stimulates rat theca-interstitial cell androgen biosynthesis. *Biol. Reprod.* **63**, 1214–1218.
157. Vit, U. A. and Hsueh, A. J. (2002). Stage-dependent role of growth differentiation factor-9 in ovarian follicle development. *Mol. Cell Endocrinol.* **186**, 211–217.
158. Fitzpatrick, S. L., Sindoni, D. M., Shughrue, P. J., Lane, M. V., Merchenthaler, I. J., and Frail, D. E. (1998). Expression of growth differentiation factor-9 messenger ribonucleic acid in ovarian and nonovarian rodent and human tissues. *Endocrinology* **139**, 2571–2580.
159. Joyce, I. M., Clark, A. T., Pendola, F. L., and Eppig, J. J. (2000). Comparison of recombinant growth differentiation factor-9 and oocyte regulation of KIT ligand messenger ribonucleic acid expression mouse ovarian follicles. *Biol. Reprod.* **63**, 1669–1675.
160. Otsuka, F., Yamamoto, S., Erickson, G. F., and Shimasaki, S. (2001). Bone morphogenetic protein-15 inhibits follicle-stimulating hormone (FSH) action by suppressing FSH receptor expression. *J. Biol. Chem.* **276**, 11387–11392.
161. Yan, C., Wang, P., DeMayo, J., DeMayo, F. J., Elvin, J. A., Carino, C., Prasad, S. V., Skinner, S. S., Dunbar, B. S., Dube, J. L., Celeste, A. J., and Matzuk, M. M. (2001). Synergistic roles of bone morphogenetic protein 15 and growth differentiation factor 9 in ovarian function. *Mol. Endocrinol.* **15**, 854–866.
162. Lee, W. S., Otsuka, F., Moore, R. K., and Shimasaki, S. (2001). Effect of bone morphogenetic protein-7 on folliculogenesis and ovulation in the rat. *Biol. Reprod.* **65**, 994–999.
163. Souza, C. J., Campbell, B. K., McNeilly, A. S., and Baird, D. (2002). Effect of bone morphogenetic protein 2 (BMP2) on oestradiol and inhibin A production by sheep granulosa cells, and localization of BMP receptors in the ovary by immunohistochemistry. *Reproduction* **123**, 363–369.
164. Otsuka, F., Moore, R. K., and Shimasaki, S. (2001). Biological function and cellular mechanism of bone morphogenetic protein-6 in the ovary. *J. Biol. Chem.* **276**, 32889–32895.
165. Shimasaki, S., Zachow, R. J., Li, D., Kim, H., Iemura, S., Ueno, N., Sampath, K., Chang, R. J., and Erickson, G. F. (1999). A functional bone morphogenetic protein system in the ovary. *Proc. Natl. Acad. Sci. USA* **96**, 7282–7287.
166. Elvin, J. A., Yan, C., and Matzuk, M. M. (2000). Oocyte-expressed TGF-beta superfamily members in female infertility. *Mol. Cell Endocrinol.* **159**, 1–5.
167. Driancourt, M. A., Reynaud, K., Cortvrindt, R., and Smitz, J. (2000). Roles of KIT and KIT LIGAND in ovarian function. *Rev. Reprod.* **5**, 143–152.
168. Klinger, F. G. and De Felici, M. (2002). *In vitro* development of growing oocytes from fetal mouse oocytes: stage-specific regulation by stem cell factor and granulosa cells. *Dev. Biol.* **244**, 85–95.
169. Sette, C., Dolci, S., Geremia, R., and Rossi, P. (2000). The role of stem cell factor and of alternative c-kit gene products in the establishment, maintenance and function of germ cells. *Int. J. Dev. Biol.* **44**, 599–608.
170. Parrott, J. A. and Skinner, M. K. (2000). Kit ligand actions on ovarian stromal cells: Effects on theca cell recruitment and steroid production. *Mol. Reprod. Dev.* **55**, 55–64.
171. Joyce, I. M., Pendola, F. L., Wigglesworth, K., and Eppig, J. J. (1999). Oocyte regulation of kit ligand expression in mouse ovarian follicles. *Dev. Biol.* **214**, 342–353.
172. Arici, A., Oral, E., Bahtiyar, O., Engin, O., Seli, E., and Jones, E. E. (1997). Leukaemia inhibitory factor expression in human follicular fluid and ovarian cells. *Hum. Reprod.* **12**, 1233–1239.
173. Coskun, S., Uzumcu, M., Jaroudi, K., Hollanders, J. M., Parhar, R. S., and al-Sedairy, S. T. (1998). Presence of leukemia inhibitory factor and interleukin-12 in human follicular fluid during follicular growth. *Am J. Reprod. Immunol.* **40**, 13–18.
174. Yamamoto, S., Konishi, I., Tsuruta, Y., Nanbu, K., Mandai, M., Kuroda, H., Matsushita, K., Hamid, A. A., Yura, Y., and Mori, T. (1997). Expression of vascular endothelial growth factor (VEGF) during folliculogenesis and corpus luteum formation in the human ovary. *Gynecol. Endocrinol.* **11**, 371–381.
175. Hazzard, T. M., Molskness, T. A., Chaffin, C. L., and Stouffer, R. L. (1999). Vascular endothelial growth factor (VEGF) and angiopoietin regulation by gonadotrophin and steroids in macaque granulosa cells during the peri-ovulatory interval. *Mol. Hum. Reprod.* **5**, 1115–1121.
176. Berisha, B., Schams, D., Kosmann, M., Amselkgruber, W., and Einspanier, R. (2000). Expression and localisation of vascular endothelial growth factor and basic fibroblast growth factor during the final growth of bovine ovarian follicles. *J. Endocrinol.* **167**, 371–382.
177. Garrido, N., Albert, C., Krussel, J. S., O'Connor, J. E., Remohi, J., Simon, C., and Pellicer, A. (2001). Expression, production, and secretion of vascular endothelial growth factor and interleukin-6 by granulosa cells is comparable in women with and without endometriosis. *Fertil. Steril.* **76**, 568–575.
178. Barboni, B., Turriani, M., Galeati, G., Spinaci, M., Bacci, M. L., Forni, M., and Mattioli, M. (2000). Vascular endothelial growth factor production in growing pig antral follicles. *Biol. Reprod.* **63**, 858–864.
179. Buscher, U., Chen, F. C., Kentenich, H., and Schmiady, H. (1999). Cytokines I the follicular fluid of stimulated and non-stimulated human ovaries; is ovulation a suppressed inflammatory reaction? *Hum. Reprod.* **14**, 162–166.
180. Ghersevich, S., Isomaa, V., and Vihko, P. (2001). Cytokine regulation of the expression of estrogenic biosynthetic enzymes in cultured rat granulosa cells. *Mol. Cell Endocrinol.* **172**, 21–30.
181. Runesson, E., Ivarsson, K., Janson, P. O., and Brannstrom, M. (2000). Gonadotropin- and cytokine-regulated expression of the chemokine interleukin 8 in the human preovulatory follicle of the menstrual cycle. *J. Clin. Endocrinol. Metab.* **85**, 4387–4395.
182. Salmassi, A., Lu, S., Hedderich, J., Oettinghaus, C., Jonat, W., and Mettler, L. (2001). Interaction of interleukin-6 on human granulosa cell steroid secretion. *J. Endocrinol.* **170**, 471–478.
183. Monget, P. and Bondy, C. (2000). Importance of the IGF system in early folliculogenesis. *Mol. Cell Endocrinol.* **163**, 89–93.
184. Kadakia, R., Arraztoa, J. A., Bondy, C., and Zhou, J. (2001). Granulosa cell proliferation is impaired in the Igf1 null ovary. *Growth Horm. IGF Res.* **11**, 220–224.

185. Giudice, L. C. (2001). Insulin-like growth factor family in Graafian follicle development and function. *J. Soc. Gynecol. Investig.* **8**, S26–S29.
186. Lanuza, G. M., Groome, N. P., Baranao, J. L., and Campo, S. (1999). Dimeric inhibin A and B production are differentially regulated by hormones and local factors in rat granulosa cells. *Endocrinology* **140**, 2549–2554.
187. Findlay, J. K., Drummond, A. E., Dyson, M., Baillie, A. J., Roberston, D. M., and Ethier, J. F. (2001). Production and actions of inhibin and activin during folliculogenesis in the rat. *Mol. Cell Endocrinol.* **180**, 139–144.
188. Josso, N., di Clemente, N., and Gouedard, L. (2001). Anti-Mullerian hormone and its receptors. *Mol. Cell Endocrinol.* **179**, 25–32.

CHAPTER 343

T Lymphocytes

Rolf König and Wenhong Zhou
Department of Microbiology and Immunology,
Sealy Center for Molecular Science,
The University of Texas Medical Branch,
Galveston, Texas

Introduction

The immune system provides a highly sophisticated surveillance mechanism to detect diverse antigens and protect the host organism from invading pathogens and altered cells (e.g., virus-infected and tumor cells). Adaptive immune responses depend on the recognition of antigen by specific antigen receptors that are expressed on the surface of T and B lymphocytes. To initiate effector mechanisms, binding of the antigen must induce intracellular signaling cascades that activate the lymphocytes and promote their differentiation to an effector cell type appropriate for the particular antigenic challenge. Importantly, regulatory mechanisms must also be present to safeguard against inadvertent self-reactivity, which could lead to autoimmunity, and to terminate immune responses, thereby avoiding overexposure of the organism to toxic effectors (e.g., cytotoxic T cells, cytokines).

T lymphocytes are derived from the lymphoid lineage of hematopoietic stem cells. T-cell progenitors enter the thymus where they develop into mature T lymphocytes. During thymic development, the immature thymocytes undergo rearrangement of first the β and then the α T-cell receptor (TCR) genes [1]. This process increases the diversity of available TCRs and ensures that each T cell expresses only a single type of TCR. Only those thymocytes that have successfully completed TCR gene rearrangement will be allowed to survive; unsuccessful rearrangement leads to programmed cell death. Following rearrangement, the functionality of the maturing thymocytes is tested by interactions with thymic antigen-presenting cells (APCs).

To be functional, the thymocytes' TCR must be able to engage epitopes formed by short peptides bound to molecules encoded by the major histocompatibility complex (MHC). Because of the large isotypic and allelic variability of MHC molecules within each species, not all recombination events of TCR genes lead to matches with a peptide–MHC complex potentially present in the individual organism. Therefore, a positive selection event is required. Two classes of MHC molecules select T cells with different functions. MHC class I molecules bind peptides derived from proteins synthesized by the presenting cell and proteolytically processed by the cell's proteosome. These peptides can combine with nascent MHC class I molecules in the endoplasmic reticulum. MHC class II molecules bind peptides derived from extracellular, endocytosed proteins that are processed in lysosomes. Nascent and recycling MHC class II molecules bind these peptides while trafficking through lysosomes [2].

Positive selection of thymocytes requires that the interaction between the TCR and peptide–MHC complex induces a signal of sufficient strength and duration. Thymocytes that do not receive a TCR-mediated signal during positive selection die. Positive selection occurs on thymic stromal epithelial cells, which do not present a full complement of all of the possible antigens that the T cell might encounter in its lifetime. Thus, selected T cells have partial self-reactivity and the potential to recognize antigens that will only be encountered later in life. Maturing thymocytes express both CD4 and CD8 coreceptors. Following positive selection, one coreceptor gene is silenced. Recent experimental results suggest that the duration of the selection signal determines lineage commitment with a short duration leading to differentiation to $CD8^+$, and a long duration favoring differentiation to $CD4^+$ thymocytes [3,4].

A separate selection step is a negative test to eliminate thymocytes that overtly respond to self-antigens. Estimates suggest that of all thymocytes that mature up to the double-positive ($CD4^+CD8^+$) stage, only about 3% mature and emigrate from the thymus to peripheral lymphoid tissues (e.g., blood, lymph nodes, and spleen). Negative selection is mediated by thymic stromal cells or APCs (mainly dendritic cells) that migrate from peripheral organs into the thymus. These cells provide a large battery, but not a complete complement, of different peptides derived from self and foreign proteins. Thymocytes that vigorously respond to the stimuli presented during negative selection will undergo apoptosis [5,6].

This chapter provides a broad overview of T-cell signaling, with emphasis on the signaling molecules and pathways important for antigen-induced signal transduction and elicitation of effector functions. General principles are discussed, and open questions and areas of current research are also highlighted.

Signaling Receptors in T Cells Form Dynamic Macromolecular Signaling Complexes

Antigen-Specific T-Cell Receptors

The antigen-specific T-cell receptor is composed of multiple polypeptides. Following thymic selection, peripheral T lymphocytes express a TCR that is composed of two transmembrane protein chains linked by a disulfide bridge. Most mature T cells express a recombined α TCR gene combined with a recombined β TCR chain, but some T cells, especially those in mucosal tissues, express a γδ TCR. The αβ TCR is the antigen-recognition unit of the T cell, but it does not have intrinsic signaling capability. The ability to transduce intracellular signals is conveyed to the TCR via its mandatory and constitutive association with a multiprotein structure, termed the CD3-ς2 complex. Neither of the individual components of the TCR/CD3-ς2 signaling machine can be transported to the cell surface without the full assembly of the complex [7].

The CD3 complex is composed of four transmembrane polypeptide chains, a γε and a δε heterodimer. These proteins have very short extracellular domains and each intracellular domain contains a conserved protein tyrosine kinase (PTK) recognition motif, termed an *immunoreceptor tyrosine-based activation motif* (ITAM). The associated, disulfide-linked ς-dimer, sometimes substituted with a ς-η heterodimer, contains three ITAMs per protein chain. ITAMs are substrates for *src* family PTKs, and their phosphorylation is a determining initiation event for T-cell signaling [8].

TCR Engagement and the Formation of Signalosomes

The binding of ligands to the TCR triggers the activation of receptor-associated *src* family PTKs, such as $p56^{lck}$ and $p59^{fyn}$, leading to the rapid tyrosine phosphorylation of numerous proteins. The phosphorylation of ITAMs located in the cytoplasmic tails of CD3 and $ς_2$ generates binding sites for proteins bearing Src homology 2 (SH2) domains, such as the cytosolic *syk* family PTK *ς-associated protein of 70 kDa* (ZAP-70) [8]. Recruitment of ZAP-70 allows enhanced activation of that kinase. ZAP-70 in turn phosphorylates components of distinct downstream signaling pathways [9–11]. Thus, T-cell activation depends on activation of both *src* family kinases and ZAP-70.

Before we follow the signal transduction cascade further, it is worthwhile to briefly discuss the complex temporal and spatial arrangement of signaling complexes and networks activated by TCR engagement. Slightly different compositions of these signaling machines, also termed *signalosomes* [12], can induce different second messenger signals and lead to drastically diverse cellular responses. In addition, the dynamic assembly and disassembly of signalosomes is likely a major factor in regulating signal transduction networks. TCR signalosomes consist of transmembrane receptors, protein kinases, phosphatases, and their substrates, all of which are organized into signaling machines by anchoring, adapter, and scaffolding proteins. Signalosomes connect events on the plasma membrane to distal signaling cascades, which ultimately modulate T-cell biology. Several protein adapters, in particular, linker of activated T cells (LAT), act as central switches that translate the quality, quantity, and duration of signals into the correct activation of specific downstream pathways [8].

Formation of signalosomes is aided by compartmentalization of the plasma membrane into detergent-insoluble, sphingolipid/cholesterol-enriched microdomains, which promote the recruitment of signal transduction molecules to the TCR-signaling machine upon TCR engagement [13,14] (Fig. 1). These areas of the T-cell surface are also known as lipid "rafts." Palmitoylation constitutively embeds several components, such as Lck, Fyn, and LAT, into these lipid microdomains, whereas others, such as ZAP-70, relocalize into rafts upon TCR engagement [12,13].

Returning to the events following TCR engagement, it is now clear that the relocalization of signalosomes to receptor-associated scaffolds is crucial for effective signal transduction [12,15,16]. Adapter proteins with SH2 domains bind to the phosphorylated ς chain. Among these proteins is the Src homology 2 protein of beta cells (Shb), which recruits LAT via its central, phosphotyrosine-binding (PTB) domain-like motif [17,18]. LAT contains nine tyrosine phosphorylation sites and is a substrate for ZAP-70 [18]. Tyrosine phosphorylation of LAT leads to the recruitment of additional signaling molecules with SH2 motifs, including the adapters growth factor receptor-bound protein 2 (Grb2) and Gads, as well as the phospholipase Cγ1 (PLCγ1), and the p85 subunit of phosphatidylinositol 3-kinase (PI3K) [18–21] (Fig. 2).

Gads is associated with the SH2 domain-containing leukocyte protein of 76 kDa (SLP-76), and together they form a macromolecular scaffold, which stabilizes the interaction of PLCγ1 with the TCR signalosome [21,22]. SLP-76 is essential in the activation of PLCγ1 and downstream signaling [23]. PLCγ1 is tyrosine phosphorylated in antigen-activated T cells, an event required for its activation. Active PLCγ1 hydrolyzes phosphotidylinositol biphosphate (PIP_2),

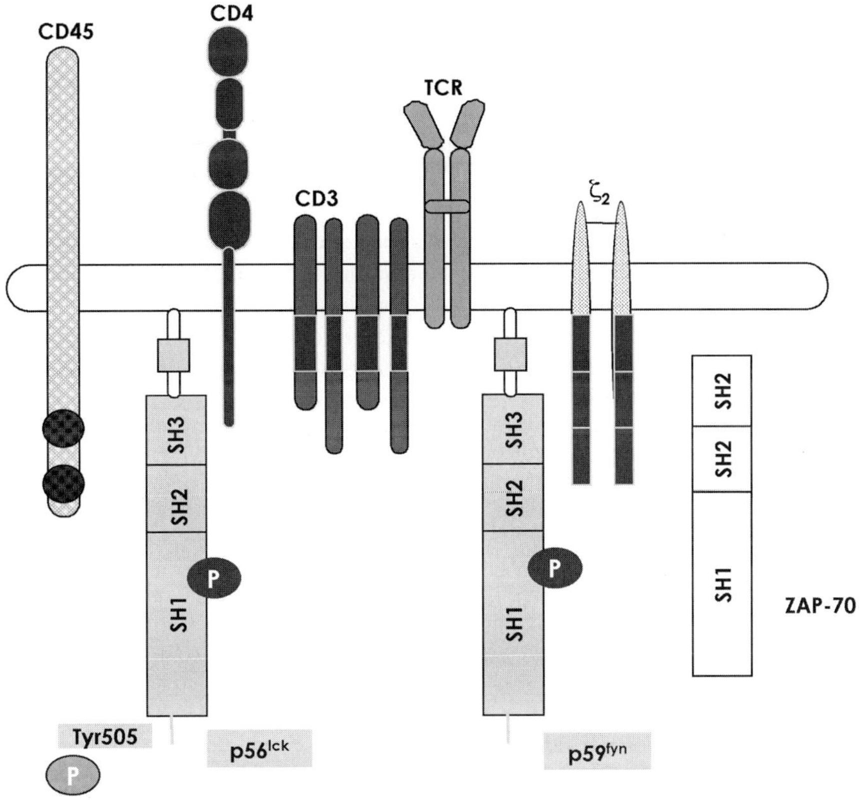

Figure 1 The TCR and its signaling machinery. The initial signalosome consists of the TCR, the associated CD3/ς_2 complex, the coreceptor (CD4 or CD8), the phosphatase CD45, and the *srk* kinases p56^{lck} and p59^{fyn}. The *syk* kinase ZAP-70 is recruited following phosphorylation of tyrosines located in ITAMs of CD3 and ς_2 by srk kinases.

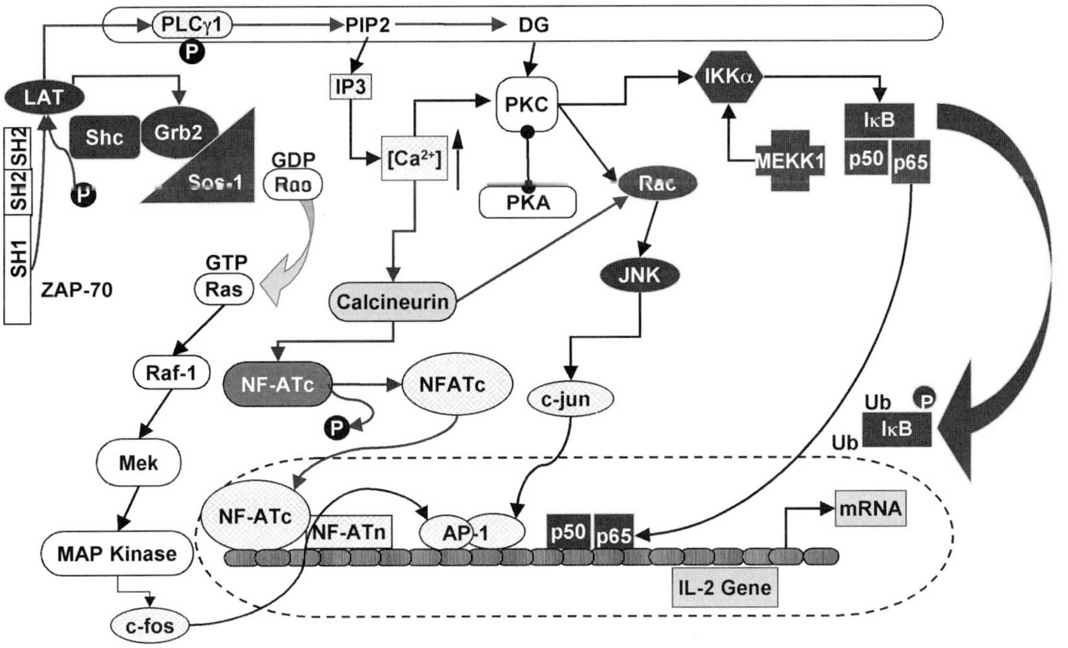

Figure 2 The signaling pathways induced by stimulation of the TCR/CD3/ς_2 complex mediate translocation of NFAT, AP-1, and NFκB transcription factors to the nucleus, leading to IL-2 gene expression. See text for details.

producing diacylglycerol (DAG) and 1,4,5-inositol triphosphate (IP_3). DAG in turn activates the serine/threonine kinase family of protein kinase C (PKC), while IP_3 induces calcium (Ca^{2+}) mobilization in the cytosol. Thus, ZAP-70 amplifies the TCR signal by specifically phosphorylating downstream components such as LAT [18,20] and PLCγ1 [23].

In this way the signalosome expands in molecular complexity and amplifies the TCR-initiated signal. Importantly, LAT can also bind proteins that negatively regulate TCR signaling. The SH2 domain-containing hematopoietic phosphotyrosine phosphatase, SHP-1, associates with LAT upon TCR stimulation [24,25] and prevents further phosphorylation of the adapter by ZAP-70, suggesting a potential conversion from an "activating" to an "inhibiting" signalosome. Similarly, the C-terminal *src* kinase (Csk) relocalizes to rafts by docking to the transmembrane adapter Csk-binding protein (Cbp), also known as phosphoprotein associated with glycosphingolipid-enriched microdomains (PAG) [26]. In rafts, Csk inhibits *src* family PTKs by phosphorylating their regulatory tyrosines and, thus, blocking TCR-mediated signal transduction [27].

Coreceptor and Costimulatory Proteins Modulate T-Cell Signaling Pathways

Before discussing the second messenger signals induced by TCR engagement, we need to introduce two sets of T-cell surface proteins that crucially modulate TCR/CD3-ς_2-mediated signals: the coreceptors and the costimulators.

Coreceptors, TCRs, and Formation of Signalosomes

Coreceptors are associated with the TCR/CD3-ς2 complex upon T-cell activation. Their presence in the TCR multicomponent signaling machine amplifies or modulates the activation signal. Often, their presence is absolutely required, but not sufficient, for productive signaling, that is, signaling that results in cell cycle progression and effector functions.

CD4 AND CD8

CD4 and CD8 are membrane glycoproteins associated with MHC class II and class I restriction, respectively. In mature T cells, their expression is mutually exclusive. In general, $CD4^+$ T helper (Th) cells respond to antigen presented by MHC class II molecules, and $CD8^+$ cytotoxic T (Tc) cells respond to MHC class I-presented antigens. CD4 and CD8 have been thought of as molecules that enhance the stability of the tripartite complex between TCR, antigen, and MHC. However, in recent years, it has become clear that their function is more complex and dynamic. For example, TCR antigen–MHC interactions initiate the formation of a specialized junction between T cells and APCs, the immunological synapse. Stimulation- and cytoskeleton-dependent processes cluster TCR/CD3 complexes in the center of the synapse, also termed the *central zone of the supramolecular activation cluster* (cSMAC), whereas adhesion molecules such as LFA-1 form a ring surrounding the central area called the *peripheral SMAC* (pSMAC) [28,29]. After stimulation of the T cell, CD4 coreceptors are rapidly recruited into the cSMAC, but migrate toward the periphery within a few minutes, whereas TCR/CD3 complexes stabilize within the central area [30]. Both CD4 and CD8 associate with the PTK $p56^{lck}$, and the efficient transport of $p56^{lck}$ into the cSMAC is a major function of these coreceptors [31]. In addition, coreceptor interactions with MHC molecules regulate peripheral T-cell homeostasis and the survival of naïve T cells in the absence of antigenic stimulation [32,33]. The recent discovery that CD4 can induce signals independent of the TCR suggests complex regulatory effects of coreceptors on T-cell function [70].

CD5

The CD5 lymphocyte glycoprotein is expressed on thymocytes and all mature T cells. CD5 can act as a costimulatory molecule for resting T cells by augmenting CD3-mediated signaling [34]. In mature, peripheral T cells, CD5 is present in lipid rafts of the T-cell surface, where it promotes CD3 redistribution into rafts, thus markedly up-regulating ZAP-70 and LAT activation and also Ca^{2+} influx [35]. However, CD5 is also constitutively associated with SHP-1, an interaction that increases upon TCR stimulation and negatively regulates TCR-mediated activation [36]. The differential modulatory properties of CD5 depend on the context of lymphocyte subset and their differentiation stage. A recent review discusses contradictory and complementary reports of CD5-mediated molecular intracellular signaling events [37].

CD45

CD45 is a membrane-bound tyrosine phosphatase present on hematopoietic cells. Multiple exons and differential glycosylation allow the expression of different isoforms in cell- and development-specific fashion. These different CD45 isoforms can be distinguished by monoclonal antibodies specific for splicing- and glycosylation-dependent epitopes [38]. In peripheral human $CD4^+$ T cells, the naïve subset, that is, the cells that have not been stimulated by antigen after thymic selection, expresses the high-molecular-weight form, CD45RA, whereas activated and memory $CD4^+$ T cells express the low-molecular-weight form, CD45RO [38]. Importantly, TCR-dependent intracellular signaling events differ in relation to CD45 isoform expression [39]. Another distinguishing feature of CD45 is its distribution in activated T cells: Whereas the CD4/CD8 coreceptors migrate into the cSMAC before relocating to the pSMAC, CD45 is completely excluded from the immunological synapse [40].

MULTIPLE FUNCTIONS OF CORECEPTORS

Views on coreceptor function have evolved with available technologies. The traditional definition identifies coreceptors as proteins that associate with the TCR upon T-cell stimulation, and stabilize the TCR's interactions with its ligands. However, the ability to identify PTKs and to measure their activities has demonstrated associations between p56^{lck} and CD4 or CD8, inspiring the realization that these coreceptors enhance and regulate TCR functions by transporting Lck into the TCR/CD3-ς_2 complex. Similarly, the recently identified coreceptor CD160/BY55, which is expressed by most intestinal intraepithelial lymphocytes and by a minor subset of circulating lymphocytes including NK, γδ TCR, and cytotoxic effector CD8$^+$ T lymphocytes, associates with Lck and tyrosine-phosphorylated ς_2 upon TCR/CD3 cell activation [41]. In addition, CD5 is associated with SHP-1 upon T-cell activation [36], and CD45 contains intrinsic phosphatase activity. Thus, physiological responses to TCR engagement at the T-cell–APC contact site are the result of localized and finely tuned alterations in the balance between cellular kinases and phosphatases [40].

The ability of coreceptors to transduce signals independent of TCR stimulation suggests that they are not merely scaffolding or transport proteins, but rather exert complex regulatory effects on T-cell activation [70]. Functionally, the distinction between coreceptors and costimulators blurs and the distinguishing feature of coreceptors may be their temporal association with the TCR/CD3-ς_2 complex.

Costimulatory Receptors and TCR-Mediated Activation

The concept of T-cell costimulation was born of the necessity to explain the phenomenon of self-tolerance within the concept of Burnet's clonal selection theory. Because not all self-proteins traffic to the thymus or are expressed at all stages of development, central tolerance induced by negative selection of autoreactive thymocytes in the thymus cannot fully explain the unresponsiveness of mature T cells to peripheral self-antigens. Therefore, several investigators introduced the two-signal model of activation [42–44]. The model postulates that any naïve lymphocyte stimulated via engagement of only the antigen receptor will enter a state of anergy, which is characterized by unresponsiveness to future antigen-mediated stimulation. Only if a second, costimulatory signal is given during antigen stimulation is the lymphocyte fully activated to display effector functions and to proliferate. Costimulatory signals for B cells and CD8$^+$ cytotoxic T cells are mainly provided by activated CD4$^+$ Th cells, whereas a variety of surface molecules expressed on hematopoietic APCs (e.g., dendritic cells, activated macrophages, activated B cells) can induce costimulation in Th cells.

Th-cell costimulatory receptors do not have to associate with the TCR/CD3 complex to exert their function. They can transduce signals independent of the TCR/CD3. Originally, they were thought to only induce cellular effector functions in combination with TCR-transduced signals. This dogma was recently challenged as a result of the realization that CD28, one of the most important costimulatory receptors for T-cell activation, can up-regulate cytokine gene transcription independently of TCR stimulation [45]. Some of the coreceptors discussed in the previous chapter also have costimulatory functions.

CD28

Interactions of CD28 with its ligand B7 on APCs activate the CD28-responsive element (CD28RE), contained within the interleukin 2 (IL-2) gene promoter, and thus promote induction of IL-2 gene expression. Interaction of CD28 with three intracellular proteins—PI3K, the T-cell-specific Tec-family kinase Itk, and the complex between Grb2 and the guanine nucleotide exchange protein "son of sevenless" (SOS)–activates the MAPK/ERK kinase kinase (MEKK1), which contributes to full activation of the CD28RE and regulates activation of the transcription factors NFκB and AP-1 [46]. The MEKKs are serine/threonine kinases that are upstream regulators of MAPKs (Fig. 2).

The CD28 signal also amplifies activation of PLCγ1 and mobilization of Ca^{2+}. Further, CD28 engagement activates p56^{lck} molecules that are phosphorylated at the regulatory tyrosine residue 505, generally considered to be an inactive form of p56^{lck} [31]. Thus CD28 amplifies signals from the TCR that would otherwise be too weak for T-cell activation. Interestingly, the PTKs p56^{lck} and p59^{fyn} phosphorylate CD28, and Itk binding to CD28 is dependent on the presence of p56^{lck}. Thus, p56^{lck} is likely to be a central switch in T-cell activation, with the dual function of regulating CD28-mediated costimulation as well as TCR/CD3/CD4 signaling.

Because CD28 provides cosignals in T-cell responses, a key question is whether the CD28 operates exclusively via TCR/CD3-ς_2 or also operates as an independent signaling unit. Indeed, CD28 can cooperate with Vav and the SLP-76 adapter to up-regulate IL-2 and -4 transcription independently of TCR engagement [45,47]. In addition, CD28 stimulation alone activates the p38 alpha MAPK [48].

Importantly, engagement of the TCR alone may result in an anergic state or T-cell deletion, both of which can induce tolerance to antigen stimulation. Insight into the regulation of CD28 dependency comes from genetic experiments. T cells that are deficient in the adapter molecule Cbl-b do not require CD28 engagement for IL-2 production. Also, whereas B cells responding to T-cell-dependent antigens cannot undergo isotype switching from IgM to IgG in CD28-deficient mice, T-cell help is fully restored in CD28/Cbl-b double-deficient mice [49]. The function of Cbl-b is to selectively suppress TCR-mediated Vav activation, thus rendering T cells dependent on CD28 costimulation [49].

CD40L

The interaction of the tumor necrosis factor (TNF) family member, CD40L, on activated T cells with its receptor, the TNF receptor family member, CD40, which is expressed on macrophages, dendritic cells, and activated B cells, provides a strong signal for IL-12 production [50]. An important aspect of CD40-CD40L signaling is its synergistic relationship with the CD28-B7 signal. CD40L cell surface expression is up-regulated by CD28 signaling. The subsequent interaction between CD40L and CD40 induces B7 up-regulation on APCs, enhancing the costimulatory activity of macrophages, dendritic cells, and B cells [51].

Intracellular Signaling Pathways Induced by Antigen Stimulation of T Cells

Calcium Mobilization

The TCR-induced signal transduction leads to activation of PLCγ1, which hydrolyzes PIP_2 to DAG and IP_3. Binding of IP_3 to its receptor in the endoplasmic reticulum membrane induces the release of Ca^{2+} into the cytosol. The subsequent increase in intracellular free Ca^{2+} opens Ca^{2+}-regulated Ca^{2+} channels in the plasma membrane, inducing additional Ca^{2+} influx. Intracellular free Ca^{2+} acts as an essential second messenger for T-cell activation. Its regulatory effects on T-cell activation are mediated via calmodulin, a Ca^{2+}-binding protein expressed in all eukaryotic cells. Effective T-cell activation leading to IL-2 secretion requires that intracellular Ca^{2+} levels be elevated for a period of 1–2 hr. Sustained Ca^{2+} signaling is required for maintaining the transcription factor nuclear factor of activated T cells (NFAT) in the nucleus in an active form [52,53]. NFAT is a key transcriptional regulator of the IL-2 gene and other cytokine genes.

Ca^{2+} signaling is required for various lymphocyte activities, for example, cell mobility, change of cytoskeletal structure, cell death, differentiation, and activation. Thus, a single second messenger can elicit multiple cellular responses. The type of response induced may depend on the amplitude, duration, and temporal fluctuations of Ca^{2+} mobilization. For example, activation of NFκB is induced by high levels of Ca^{2+} because of this transcription factor's low Ca^{2+} sensitivity. In contrast, long-lasting, low levels of Ca^{2+} selectively activate NFAT, because NFAT is highly sensitive to Ca^{2+}, but is rapidly inactivated after Ca^{2+} removal [54].

PKC

Release of DAG stimulates PKC, a family of serine/threonine kinases. In T cells, PKC isoforms α, β, η, δ, ζ, ε, and θ are expressed. One of the major functions of PKC is to induce MAPKs. PKCα directly phosphorylates and activates Raf-1, another serine/threonine kinase. Activation of Raf-1 triggers a protein kinase cascade by directly phosphorylating MAPK kinase. Activation of PKC also mediates the rapid accumulation of the active, GTP-bound form of $p21^{ras}$.

Transcription Factors: Activation and Gene Expression

The multiple signaling pathways originating from T-cell surface molecules initiate the expression of genes responsible for proliferation and immune functions in a cooperative manner. The most extensively studied example is the induction of the IL-2 gene. The IL-2 gene promoter contains at least seven distinct binding sites for transcription factors. Therefore, maximal transcription demands the simultaneous presence of all factors, among which AP-1, NFAT, and NFκB are the best characterized (Fig. 2).

AP-1 is a heterodimer of c-Fos and c-Jun. Maximum activation of AP-1 requires *de novo* synthesis of c-Jun and c-Fos, and phosphorylation by MAPKs of the activation domains of both proteins, leading to translocation into the nucleus.

Activation of NFAT requires dephosphorylation by the serine/threonine protein phosphatase, calcineurin, followed by translocation into the nucleus. In the nucleus, NFAT cooperates with AP-1 in gene transactivation, resulting in a 20-fold increase in the stability of NFAT/AP-1/DNA complexes as compared with NFAT/DNA complexes.

NFκB is a homodimer or heterodimer of a family of structurally related proteins. Each member of this family contains a conserved N-terminal Rel-homology domain (RHD), which mediates dimerization and binding to DNA. The RHD contains a nuclear localization sequence that promotes NFκB translocation to the nucleus following release of NFκB from IκB. In its inactive form, NFκB is sequestered in the cytosol by noncovalent interactions with the inhibitory protein, IκB, which masks the nuclear translocation signal. Phosphorylation of IκB by IκB kinase targets IκB for destruction by the ubiquitin-protease system. NFκB induces IL-2 gene expression, and in a negative feedback loop, promotes transcription of the IκB gene. Importantly, PKCθ associates with IκB kinase to mediate activation of NFκB following TCR/CD28 stimulation [55,56].

Role of Cyclic AMP in Th-Cell Activation

cAMP, Adenylyl Cyclases, and Phosphodiesterases

Cyclic AMP (cAMP) is an intracellular second messenger to a wide variety of hormones and neurotransmitters. In T cells, elevated cAMP levels antagonize activation by inhibiting T-cell proliferation [57,58] and by suppressing the production of IL-2 and IFN-γ [59]. TCR signaling alone (in the absence of CD28 costimulation) elevates cAMP levels and adenylyl cyclase (AC) activity [60,61]. The regulation of cAMP during T-cell activation is also mediated via other T-cell surface molecules, such as CD28 and CD44 [62,63], as well as via the coreceptor CD4 [70].

In T cells, the cAMP level is controlled by two types of enzymes: ACs and phosphodiesterases (PDEs). ACs catalyze cAMP production from ATP, whereas PDEs control the rate of cAMP degradation to AMP. Members of the PDE families 1, 3, 4, and 7 are expressed in T cells. Different mechanisms control the activity of the different families of PDEs: Ca^{2+}/calmodulin stimulates PDE1, cyclic GMP inhibits PDE 3, p70S6 kinase and the MAPK pathway can activate PDE 4, and CD28-mediated signals activate PDE 7 [62]. Importantly, TCR/CD28 stimulation of human T cells transiently up-regulates AC and PDE activities with different kinetics for different PDE isozymes [64]. Thus, an initial

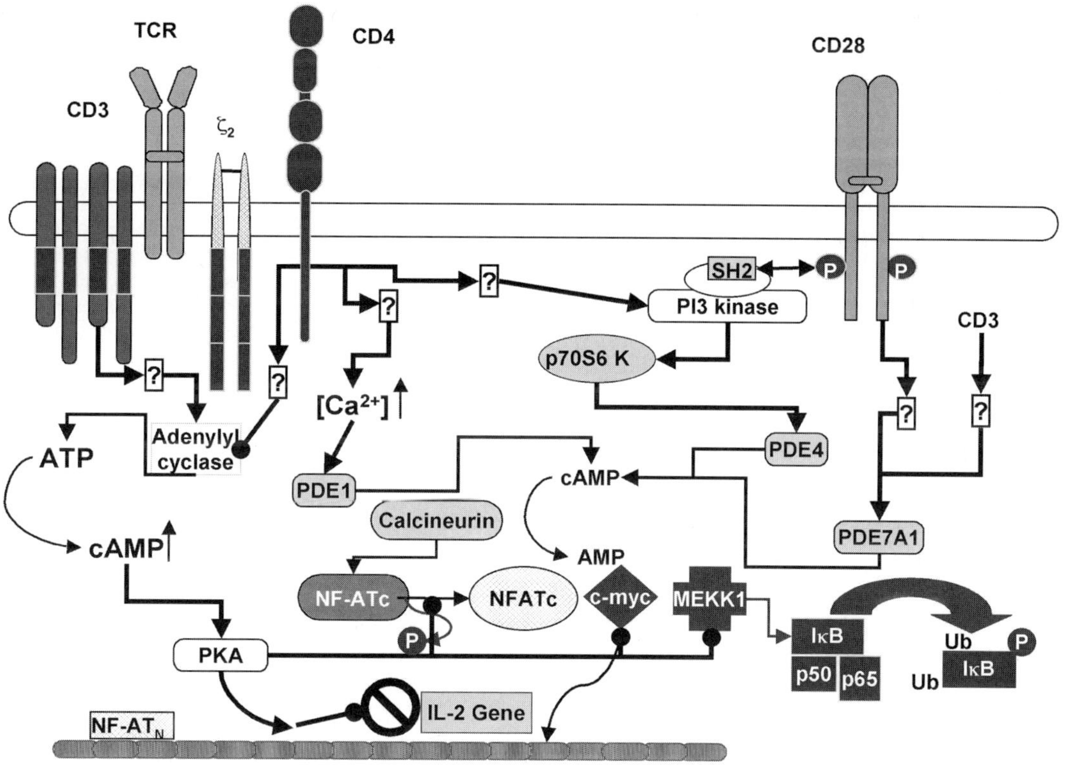

Figure 3 Cyclic AMP exerts multiple regulatory effects on T cell activation and is itself regulated by multiple signaling pathways that activate or block adenylate cyclase and activate phosphodiesterases. Stimulatory effects are indicated by a line ending in an arrow, whereas inhibitory effects are depicted by a line ending in a filled circle. The overall effect of PKA I activation is inhibition of IL-2 gene expression.

increase, followed by a rapid decrease, in intracellular cAMP is required for T-cell activation, suggesting a precise kinetic regulation of cAMP production and degradation (Fig. 3).

cAMP-DEPENDENT KINASE

The cAMP-dependent protein kinase (PKA) is the principal intracellular cAMP receptor. In the absence of cAMP, PKA is an enzymatically inactive, tetrameric holoenzyme, consisting of two catalytic (C) subunits and two regulatory (R) subunits. The cooperative binding of four cAMP molecules to two sites on each R subunit induces dissociation into dimeric R and two monomers of C subunits. Once freed from the R subunits, the C subunits display serine/threonine kinase activity.

PKA I, but not PKA II, mediates the inhibitory role of cAMP on T-cell proliferation induced by TCR signaling [65]. PKA I activates Csk to inhibit Lck activity [27]. It also phosphorylates Ser-43 of Raf-1 to block the MAP kinase pathway [66]. In the nucleus, activation of PKA prevents stable protein–DNA interactions at the NFκB, NFAT, and AP-1 binding sites of the IL-2 enhancer [67] (Fig. 3). In addition, PKA I activity also inhibits cyclin D3 expression and induces the cyclin-dependent kinase inhibitor p27*kip1* [68]. For T cells to enter the S phase of the cell cycle, D-type cylins, including cyclin D3, are synthesized during the G_1 phase [69]. These cyclins can bind to cyclin-dependent kinase (Cdk) and form an active kinase complex that phosphorylates and inactivates retinoblastoma protein (pRb). Inactivation of pRb then allows cells to pass through the late G_1-phase restriction point and enter the S phase. However, cyclin D/Cdk complexes can associate with the Cdk inhibitor p27*kip1*, thus be rendered inactive. Therefore, in addition to induction of cyclin D, down-regulation of p27*kip1* is required for the initiation of T-cell proliferation. Hence, inhibition of cyclin D3 expression and induction of p27*kip1* by PKA I both block T-cell cycle progression.

Conclusions

Signal transduction research in T lymphocytes has focused on identifying the receptors and intracellular proteins affecting specific signaling pathways. It has become clear that signaling machines in T cells differ in composition depending on the extracellular signal received and the requirements of the stimulated cell. The composition of an established signalosome could also change over time in order to fine-tune or alter effector signaling pathways. In the future, it will be important to determine the kinetics of assembly, the dynamics of composition, and the compartmentalization for all signalosomes induced in T cells, as well as to define the interactions between different signaling pathways. Future research will utilize functional genomics approaches to characterize the *signalome*—the entire complement of a T cell's signaling molecules and interactions in a temporal and spatial context.

Acknowledgments

We thank Ms. Jennifer Junemann for help with the figures and Ms. Mardelle Susman for editorial assistance. We apologize to all of the colleagues whose valuable contributions to the scientific literature could not be adequately cited because of space limitations.

References

1. Khor, B. and Sleckman, B. P. (2002). Allelic exclusion at the TCRbeta locus. *Curr. Opin. Immunol.* **14**, 230–234.
2. Germain, R. N. (1994). MHC-dependent antigen processing and peptide presentation: Providing ligands for T lymphocyte activation. *Cell* **76**, 287–299.
3. Germain, R. N. (2002). T-cell development and the CD4-CD8 lineage decision. *Nat. Rev. Immunol.* **2**, 309–322.
4. Singer, A. (2002). New perspectives on a developmental dilemma: The kinetic signaling model and the importance of signal duration for the CD4/CD8 lineage decision. *Curr. Opin. Immunol.* **14**, 207–215.
5. Robey, E. and Fowlkes, B. J. (1994). Selective events in T cell development. *Annu. Rev. Immunol.* **12**, 675–705.
6. Williams, C. B., Engle, D. L., Kersh, G. J., Michael White J., and Allen, P. M. (1999). A kinetic threshold between negative and positive selection based on the longevity of the T cell receptor-ligand complex. *J. Exp. Med.* **189**, 1531–1544.
7. Sun, Z. Y. J, Kim, H. S., Wagner, G., and Reinherz, E. L. (2001). Mechanisms contributing to T cell receptor signaling and assembly revealed by the solution structure of an ectodomain fragment of the CD3 epsilon gamma heterodimer. *Cell* **105**, 913–923.
8. Germain, R. N. (2001). The T cell receptor for antigen: Signaling and ligand discrimination. *J. Biol. Chem.* **276**, 35223–35226.
9. Elder, M. E., Skoda-Smith, S. , Kadlecek, T. A., Wang, F., Wu, J., and Weiss, A. (2001). Distinct T cell developmental consequences in humans and mice expressing identical mutations in the DLAARN motif of ZAP-70. *J. Immunol.* **166**, 656–661.
10. Gong, Q., Jin, X., Akk, A. M., Foger, N., White, M., Gong, G., Wardenburg, J. B., and Chan, A. C. (2001). Requirement for tyrosine residues 315 and 319 within zeta chain-associated protein 70 for T cell development. *J. Exp. Med.* **194**, 507–518.
11. Magnan, A., Di Bartolo, V., Mura, A. M., Boyer, C., Richelme, M., Lin, Y. L., Roure, A., Gillet, A., Arrieumerlou, C., Acuto, O., *et al.* (2001). T cell development and T cell responses in mice with mutations affecting tyrosines 292 or 315 of the ZAP-70 protein tyrosine kinase. *J. Exp. Med.* **194**, 491–505.
12. Werlen, G., Hausmann, B., and Palmer, E. (2000). A motif in the alphabeta T-cell receptor controls positive selection by modulating ERK activity. *Nature* **406**, 422–426.
13. Harder, T. and Kuhn, M. (2001). Immunoisolation of TCR signaling complexes from Jurkat T leukemic cells. *Sci STKE.* **2001**, PL1.
14. Leitenberg, D., Balamuth, F., and Bottomly, K. (2001). Changes in the T cell receptor macromolecular signaling complex and membrane microdomains during T cell development and activation. *Semin Immunol.* **13**, 129–138.
15. Delgado, P., Fernandez, E., Dave, V., Kappes, D., and Alarcon, B. (2000). CD3delta couples T-cell receptor signalling to ERK activation and thymocyte positive selection. *Nature* **406**, 426–430.
16. Harder, T. and Kuhn, M. (2000). Selective accumulation of raft-associated membrane protein LAT in T cell receptor signaling assemblies. *J. Cell Biol.* **151**, 199–208.
17. Welsh, M., Songyang, Z., Frantz, J. D., Trub, T., Reedquist, K. A., Karlsson, T., Miyazaki, M., Cantley, L. C., and Shoelson, S. E. (1998). Stimulation through the T cell receptor leads to interactions between SHB and several signaling proteins. *Oncogene* **16**, 891–901.
18. Zhang, W., Sloan-Lancaster, J., Kitchen, J., Trible, R. P., and Samelson, L. E. (1998). LAT: The ZAP-70 tyrosine kinase substrate that links T cell receptor to cellular activation. *Cell* **92**, 83–92.
19. Sommers, C. L., Menon, R. K., Grinberg, A., Zhang, W., Samelson, L. E., and Love, P. E. (2001). Knock-in mutation of the distal four tyrosines of linker for activation of T cells blocks murine T cell development. *J. Exp. Med.* **194**, 135–142.
20. Paz, P. E., Wang, S. J., Clarke, H., Lu, X. B., Stokoe, D., and Abo, A. (2001). Mapping the Zap-70 phosphorylation sites on LAT (linker for activation of T cells) required for recruitment and activation of signalling proteins in T cells. *Biochem. J.* **356**, 461–471.
21. Yablonski, D. and Weiss, A. (2001). Mechanisms of signaling by the hematopoietic-specific adaptor proteins, SLP-76 and LAT and their B cell counterpart, BLNK/SLP-65. *Adv. Immunol.* **79**, 93–128.
22. Yoder, J., Pham, C., Iizuka, Y. M., Kanagawa, O., Liu, S. K., McGlade, J., and Cheng, A. M. (2001). Requirement for the SLP-76 adaptor GADS in T cell development. *Science* **291**, 1987–1991.
23. Yablonski, D., Kadlecek,T., and Weiss, A. (2001). Identification of a phospholipase C-gamma1 (PLC-gamma1) SH3 domain-binding site in SLP-76 required for T-cell receptor-mediated activation of PLC-gamma1 and NFAT. *Mol. Cell Biol.* **21**, 4208–4218.
24. Kosugi, A., Sakakura, J., Yasuda, K., Ogata, M., and Hamaoka, T. (2001). Involvement of SHP-1 tyrosine phosphatase in TCR-mediated signaling pathways in lipid rafts. *Immunity* **14**, 669–680.
25. Su, M. W., Yu, C. L., Burakoff, S. J., and Jin, Y. J. (2001). Targeting Src homology 2 domain-containing tyrosine phosphatase (SHP-1) into lipid rafts inhibits CD3-induced T cell activation. *J. Immunol.* **166**, 3975–3982.
26. Brdicka, T., Pavlistova, D., Leo, A., Bruyns, E., Korinek, V., Angelisova, P., Scherer, J., Shevchenko, A., Hilgert, I., Cerny, J., *et al.* (2000). Phosphoprotein associated with glycosphingolipid-enriched microdomains (PAG), a novel ubiquitously expressed transmembrane adaptor protein, binds the protein tyrosine kinase csk and is involved in regulation of T cell activation. *J. Exp. Med.* **191**, 1591–1604.
27. Vang, T., Torgersen, K. M., Sundvold, V., Saxena, M., Levy, F. O., Skalhegg, B. S., Hansson, V., Mustelin, T., and Tasken, K. (2001). Activation of the COOH-terminal Src kinase (Csk) by cAMP-dependent protein kinase inhibits signaling through the T cell receptor. *J. Exp. Med.* **193**, 497–507.
28. Grakoui, A., Bromley, S. K., Sumen, C., Davis, M. M., Shaw, A. S., Allen, P. M., and Dustin, M. L. (1999). The immunological synapse: A molecular machine controlling T cell activation. *Science* **285**, 221–227.
29. Delon, J. and Germain, R. N. (2000). Information transfer at the immunological synapse. *Curr. Biol.* **10**, R923–933.
30. Krummel, M. F., Sjaastad, M. D., Wulfing, C., and Davis, M. M. (2000). Differential clustering of CD4 and CD3zeta during T cell recognition. *Science* **289**, 1349–1352.
31. Holdorf, A. D., Lee, K. H., Burack, W. R., Allen, P. M., and Shaw, A. S. (2002). Regulation of Lck activity by CD4 and CD28 in the immunological synapse. *Nat. Immunol.* **3**, 259–264.
32. König, R. (2002). Interactions between MHC molecules and co-receptors of the TCR. *Curr. Opin. Immunol.* **14**, 75–83.
33. König, R., Shen, X., Maroto, R., and Denning, T. L. (2002). The role of CD4 in regulating homeostasis of T helper cells. *Immunol Res.* **25**, 115–130.
34. Berney, S. M., Schaan, T., Wolf, R. E., Kimpel, D. L., van der Heyde, H., and Atkinson, T. P. (2001). CD5 (OKT1) augments CD3-mediated intracellular signaling events in human T lymphocytes. *Inflammation* **25**, 215–221.
35. Yashiro-Ohtani, Y., Zhou, X. Y., Toyo-Oka, K., Tai, X. G., Park, C. S., Hamaoka, T., Abe, R., Miyake, K., and Fujiwara, H. (2000). Non-CD28 costimulatory molecules present in T cell rafts induce T cell costimulation by enhancing the association of TCR with rafts. *J. Immunol.* **164**, 1251–1259.
36. Perez-Villar, J. J., Whitney, G. S., Bowen, M. A., Hewgill, D. H., Aruffo, A. A., and Kanner, S. B. (1999). CD5 negatively regulates the T-cell antigen receptor signal transduction pathway: involvement of SH2-containing phosphotyrosine phosphatase SHP-1. *Mol. Cell Biol.* **19**, 2903–2912.

37. Lozano, F., Simarro, M., Calvo, J., Vila, J. M., Padilla, O., Bowen, M. A., and Campbell, K. S. (2000). CD5 signal transduction: Positive or negative modulation of antigen receptor signaling. *Crit. Rev. Immunol.* **20**, 347–358.
38. Trowbridge, I. S. and Thomas, M. L. (1994). CD45: An emerging role as a protein tyrosine phosphatase required for lymphocyte activation and development. *Annu. Rev. Immunol.* **12**, 85–116.
39. Dornan, S., Sebestyen, Z., Gamble, J., Nagy, P., Bodnar, A., Alldridge, L., Doe, S., Holmes, N., Goff, L. K., Beverley, P., *et al.* (2002). Differential association of CD45 isoforms with CD4 and CD8 regulates the actions of specific pools of p56lck tyrosine kinase in T cell antigen receptor signal transduction. *J. Biol. Chem.* **277**, 1912–1918.
40. Leupin, O., Zaru, R., Laroche, T., Muller S., and Valitutti, S. (2000). Exclusion of CD45 from the T-cell receptor signaling area in antigen-stimulated T lymphocytes. *Curr. Biol.* **10**, 277–280.
41. Nikolova, M., Marie-Cardine, A., Boumsell, L., and Bensussan, A. (2002). BY55/CD160 acts as a co-receptor in TCR signal transduction of a human circulating cytotoxic effector T lymphocyte subset lacking CD28 expression. *Int. Immunol.* **14**, 445–451.
42. Bretscher, P. and Cohn, M. (1970). A theory of self-nonself discrimination. *Science.* **169**, 1042–1049.
43. Lafferty, K. J. and Cunningham, A. J. (1975). A new analysis of allogeneic interactions. *Aust. J. Exp. Biol. Med. Sci.* **53**, 27–42.
44. Jenkins, M. K., Pardoll, D. M., Mizuguchi, J., Quill, H., and Schwartz, R. H. (1987). T-cell unresponsiveness in vivo and *in vitro:* Fine specificity of induction and molecular characterization of the unresponsive state. *Immunol. Rev.* **95**, 113–135.
45. Raab, M., Pfister, S., and Rudd, C. E. (2001). CD28 signaling via VAV/SLP-76 adaptors: regulation of cytokine transcription independent of TCR ligation. *Immunity* **15**, 921–933.
46. Marinari, B., Costanzo, A., Viola, A., Michel, F., Mangino, G., Acuto, O., Levrero, M., Piccolella, E., and Tuosto, L. (2002). Vav cooperates with CD28 to induce NF-kappaB activation via a pathway involving Rac-1 and mitogen-activated kinase kinase 1. *Eur. J. Immunol.* **32**, 447–456.
47. Herndon, T. M., Shan, X. C., Tsokos, G. C., and Wange, R. L. (2001). ZAP-70 and SLP-76 regulate protein kinase C-theta and NF-kappa B activation in response to engagement of CD3 and CD28. *J. Immunol.* **166**, 5654–5664.
48. Schafer, P. H., Wadsworth, S. A., Wang, L., and Siekierka, J. J. (1999). p38 alpha mitogen-activated protein kinase is activated by CD28-mediated signaling and is required for IL-4 production by human CD4+CD45RO+ T cells and Th2 effector cells. *J. Immunol.* **162**, 7110–7119.
49. Chiang, Y. J., Kole, H. K., Brown, K., Naramura, M., Fukuhara, S., Hu, R. J., Jang, I. K., Gutkind, J. S., Shevach, E., and Gu, H. (2000). Cbl-b regulates the CD28 dependence of T-cell activation. *Nature* **403**, 216–220.
50. Bullens, D. M., Kasran, A., Thielemans, K., Bakkus, M., and Ceuppens, J. L. (2001). CD40L-induced IL-12 production is further enhanced by the Th2 cytokines IL-4 and IL-13. *Scand. J. Immunol.* **53**, 455–463.
51. Grewal, I. S. and Flavell, R. A. (1996). The role of CD40 ligand in costimulation and T-cell activation. *Immunol. Rev.* **153**, 85–106.
52. Loh, C., Carew, J. A., Kim, J., Hogan, P. G., and Rao, A. (1996). T-cell receptor stimulation elicits an early phase of activation and a later phase of deactivation of the transcription factor NFAT1. *Mol. Cell Biol.* **16**, 3945–3954.
53. Timmerman, L. A., Clipstone, N. A., Ho, S. N., Northrop, J. P., and Crabtree, G. R. (1996). Rapid shuttling of NF-AT in discrimination of Ca2+ signals and immunosuppression. *Nature* **383**, 837–840.
54. Dolmetsch, R. E., Lewis, R. S., Goodnow, C. C., and Healy, J. I. (1997). Differential activation of transcription factors induced by Ca2+ response amplitude and duration. *Nature* **386**, 855–858.
55. Khoshnan, A., Bae, D., Tindell, C. A., and Nel, A. E. (2000). The physical association of protein kinase C theta with a lipid raft-associated inhibitor of kappa B factor kinase (IKK) complex plays a role in the activation of the NF-kappa B cascade by TCR and CD28. *J. Immunol.* **165**, 6933–6940.
56. Coudronniere, N., Villalba, M., Englund, N., and Altman, A. (2000). NF-kappa B activation induced by T cell receptor/CD28 costimulation is mediated by protein kinase C-theta. *Proc. Natl. Acad. Sci. USA* **97**, 3394–3399.
57. Baker, P. E., Fahey, J. V., and Munck, A. (1981). Prostaglandin inhibition of T-cell proliferation is mediated at two levels. *Cell Immunol.* **61**, 52–61.
58. Lingk, D. S., Chan, M. A., and Gelfand, E. W. (1990). Increased cyclic adenosine monophosphate levels block progression but not initiation of human T cell proliferation. *J. Immunol.* **145**, 449–455.
59. Novak, T. J. and Rothenberg, E. V. (1990). cAMP inhibits induction of interleukin 2 but not of interleukin 4 in T cells. *Proc. Natl. Acad. Sci. USA* **87**, 9353–9357.
60. Bihoreau, C., Heurtier, A., Enjalbert, A., Corvaia, N., Bensussan, A., Degos, L., and Kordon, C. (1991). Activation of the CD3/T cell receptor (TcR) complex or of protein kinase C potentiate adenylyl cyclase stimulation in a tumoral T cell line: Involvement of two distinct intracellular pathways. *Eur. J. Immunol.* **21**, 2877–2882.
61. Feuerstein, N., Firestein, R., Aiyar, N., He, X., Murasko, D., and Cristofalo,V. (1996). Late induction of CREB/ATF binding and a concomitant increase in cAMP levels in T and B lymphocytes stimulated via the antigen receptor. *J. Immunol.* **156**, 4582–4593.
62. Li, L., Yee, C., and Beavo, J. A. (1999). CD3- and CD28-dependent induction of PDE7 required for T cell activation. *Science* **283**, 848–851.
63. Rothman, B. L., Kennure, N., Kelley, K. A., Katz, M., and Aune, T. M. (1993). Elevation of intracellular cAMP in human T lymphocytes by an anti-CD44 mAb. *J. Immunol.* **151**, 6036–6042.
64. Kanda, N. and Watanabe, S. (2001). Regulatory roles of adenylate cyclase and cyclic nucleotide phosphodiesterases 1 and 4 in interleukin-13 production by activated human T cells. *Biochem. Pharmacol.* **62**, 495–507.
65. Aukrust, P., Aandahl, E. M., Skalhegg, B. S., Nordoy, I., Hansson, V., Tasken, K., Froland, S. S., and Muller, F. (1999). Increased activation of protein kinase A type I contributes to the T cell deficiency in common variable immunodeficiency. *J. Immunol.* **162**, 1178–1185.
66. Ramstad, C., Sundvold, V., Johansen, H. K., and Lea, T. (2000). cAMP-dependent protein kinase (PKA) inhibits T cell activation by phosphorylating ser-43 of raf-1 in the MAPK/ERK pathway. *Cell Signal.* **12**, 557–563.
67. Chen, D. and Rothenberg, E. V. (1994). Interleukin 2 transcription factors as molecular targets of cAMP inhibition: delayed inhibition kinetics and combinatorial transcription roles. *J. Exp. Med.* **179**, 931–942.
68. van Oirschot, B. A., Stahl, M., Lens, S. M., and Medema, R. H. (2001). Protein kinase A regulates expression of p27(kip1) and cyclin D3 to suppress proliferation of leukemic T cell lines. *J. Biol. Chem.* **276**, 33854–33860.
69. Boonen, G. J., van Dijk, A. M., Verdonck, L. F., van Lier, R. A., Rijksen, G., and Medema, R. H. (1999). CD28 induces cell cycle progression by IL-2-independent down-regulation of p27kip1 expression in human peripheral T lymphocytes. *Eur. J. Immunol.* **29**, 789–798.
70. Zhou, W. and König, R. (2003) T cell receptor-independent CD4 signaling: CD4-MHC class II interactions regulate intracellular calcium and cyclic AMP. Cell. Signal. 15:751–762.

CHAPTER 344

Signal Transduction via the B-Cell Antigen Receptor: A Crucial Regulator of B-Cell Biology

Louis B. Justement
Division of Developmental and Clinical Immunology,
Department of Microbiology,
University of Alabama at Birmingham,
Birmingham, Alabama

Introduction

B lymphocytes express antigen receptors (BCR) on their surface that impart the ability to detect foreign antigens. Binding of antigen to the BCR initiates a signal transduction cascade that in turn leads to cellular activation, proliferation, and ultimately differentiation into antibody-secreting plasma cells. Additionally, the BCR plays an important role in internalization of antigen resulting in its processing and presentation in the context of MHC class II to helper T cells, which promote clonal expansion and differentiation of the antigen-specific B cells [1,2]. Although signal transduction via the BCR is central to the generation of a productive humoral immune response, its role in regulating B-cell biology is more complex and multifaceted. Indeed the BCR is involved in determining the fate of the B cell throughout its development and differentiation [3–5]. During B-cell development in the bone marrow, the pre-BCR is crucial for transducing signals that ensure the formation and expression of a functional, mature BCR on the surface of the B cell. Additionally, the pre-BCR is involved in regulating allelic exclusion to ensure that only BCRs with a single antigenic specificity are expressed on the surface of any given B cell. Once a mature BCR is expressed on the surface of an immature B cell, signals delivered through it can either positively select those cells that will ultimately enter into the periphery to patrol for foreign antigen, or negatively select those cells that recognize self-antigens with a high affinity. Negative selection mediated by signals delivered via the BCR results in editing of the BCR to change its specificity or in deletion of the self-reactive clone [6–8]. Thus the BCR is directly involved in determining the fate of the cell presumably by virtue of its ability to transduce signals that vary both quantitatively as well as qualitatively. Additionally, the response of the B cell to such signals is likely to be regulated in a developmental manner [9,10]. In conclusion, the BCR serves a central role in regulating B-cell biology by virtue of its ability to selectively transduce signals in response to antigen binding. The discussion that follows deals primarily with the basic signal transduction pathways that are activated in response to binding of antigen to BCRs expressed on mature, quiescent B cells.

Initiation of Signal Transduction through the BCR

The BCR complex is comprised of an antigen recognition structure, membrane immunoglobulin (mIg), and an associated signal transducing heterodimer. Membrane Ig consists of two heavy chains and two light chains that are disulfide bonded to one another to form the mature antigen recognition structure. Membrane Ig is noncovalently associated with one transmembrane heterodimer consisting of disulfide-linked Igα (CD79a) and Igβ (CD79b) polypeptides [11,12]. The Igα/β heterodimer functions both as a transporter and a signal transducing structure [12]. Initiation of signal transduction through the BCR was originally thought to occur in response to BCR cross-linking mediated by binding of bivalent or multivalent antigen to the bivalent mIg molecule. This was thought to induce BCR dimerization and/or multimerization, which in turn facilitated the ability of associated Src family protein tyrosine kinases (PTK) (e.g., Lyn, Fyn) to phosphorylate tyrosine residues in the cytoplasmic domains of Igα/β [13]. Although the BCR in resting B cells is constitutively associated with Src family PTKs at a low stoichiometry, the net level of Igα/β phosphorylation is presumably minimal due to the fact that individual receptor complexes are distributed throughout the membrane and the phosphorylation of Igα/β is counterbalanced by one or more protein tyrosine phosphatases (PTP). In contrast, BCR aggregation presumably favors a net increase in the total level of protein tyrosine phosphorylation associated with the activation complex due to the physical colocalization of Src family PTKs and their substrates (i.e., Igα/β) [13].

More recently, however, it has been proposed that the BCR exists in the membrane of unstimulated B cells in preformed oligomers and that initiation of signaling occurs when antigen binds to one or more mIg molecules within an oligomer, thereby causing a conformational change or reorganization of the individual subunits that promotes signal transduction [11,12,14,15]. In this regard, the transmembrane region of mIg heavy chains is thought to assume an α-helical conformation in the membrane in which one face of the α-helix is comprised of highly conserved hydrophilic amino acid residues that mediate the interaction with the Igα/β heterodimer [14]. The other face of the α-helix is specific to each of the mIg isotypes and contains several hydrophilic amino acid residues with large polar side groups. Because such hydrogen-binding, hydrophilic moieties exhibit a strong bias against being located within a lipid membrane, it is possible that this face of the helix constitutes an interaction surface that is involved in BCR oligomerization [14]. Although the molecular composition of the resting BCR oligomers has yet to be elucidated, it has been hypothesized that the oligomers are associated with the PTK Syk [16]. This PTK presumably promotes phosphorylation of Igα/β heterodimers within the oligomer, which in turn facilitates binding of Syk to phosphotyrosine residues via its tandem Src homology 2 (SH2) domains. It is further hypothesized that a low basal level of tyrosine phosphorylation is maintained by the action of a PTP called SHP-1, which acts to constitutively down-regulate Syk kinase activity, and perhaps to dephosphorylate Igα/β [13]. Binding of antigen to the complex is thought to cause a conformational change in the resting BCR oligomer such that SHP-1 activity is attenuated, possibly by causing the PTP to become physically dissociated from the BCR complex. Dissociation of SHP-1 would presumably favor a net increase in tyrosine phosphorylation of Igα/β leading to initiation of signal transduction and cellular activation. It has been reported that SHP-1 constitutively interacts with the BCR isolated from resting B cells and is induced to dissociate from the BCR upon activation in agreement with this proposed mechanism [17].

Another recent finding is that binding of antigen causes a rapid translocation of the BCR to glycosphingolipid-enriched microdomains (GEMs) within the plasma membrane. These microdomains are enriched for Src family PTKs including Lyn and it has been hypothesized that translocation of the BCR into GEMs physically localizes the complex with Src PTKs, thereby promoting tyrosine phosphorylation of Igα/β by Lyn [18–20]. Interestingly, translocation of the BCR into GEMs appears to constitute a novel step that precedes signal transduction because translocation does not require the Igα/β heterodimer and is resistant to blockade by PTK inhibitors [21]. Thus, it is possible that binding of antigen to BCR oligomers on resting B cells leads to a reorganization/conformational change in the oligomer that promotes its translocation to GEMs. As stated, localization of the BCR to GEMs promotes Lyn-mediated phosphorylation of Igα/β, as well as phosphorylation of Syk, resulting in potentiation of its catalytic activity. Simultaneously, translocation of the BCR to GEMs may lead to dissociation of SHP-1, thereby causing a net increase in tyrosine phosphorylation of Igα/β and initiation of signal transduction. Alternatively, it is possible that Lyn activity is relatively resistant to the inhibitory action of SHP-1, in which case colocalization of Lyn with the BCR may be sufficient to favor a net increase in Igα/β phosphorylation despite the presence of SHP-1 [14].

Regardless of whether antigen binding promotes BCR aggregation or reorganization of preexisting oligomers, it is clear that the BCR translocates to GEMs where there is a net increase in the tyrosine phosphorylation of Igα and Igβ (Fig. 1). Both of these polypeptides contain immunoreceptor tyrosine-based activation motifs (ITAMs) within their cytoplasmic domains that function as docking sites for SH2 domain-containing proteins including Syk, Shc, and BLNK [13]. The net increase in tyrosine phosphorylation of the Igα/β ITAMs promotes the formation of multimolecular complexes that are targeted to the BCR and are critical for propagation of signal transduction. The formation of these complexes enables the BCR to activate several signaling pathways that are distinct yet interrelated, including the phospholipase C (PLCγ) pathway, the phosphatidylinositol 3-kinase (PI3K) pathway, and signaling processes that are controlled by the Ras, Rac1, and Rap1 GTPases (for reviews, see [22–26]).

Propagation of Signal Transduction via the BCR

Numerous studies have demonstrated that ligand binding to the BCR results in inducible tyrosine phosphorylation and activation of the PTK Syk [27–29]. The critical role that Syk serves during BCR-mediated signaling has been demonstrated by the analysis of Syk-deficient B cells in which

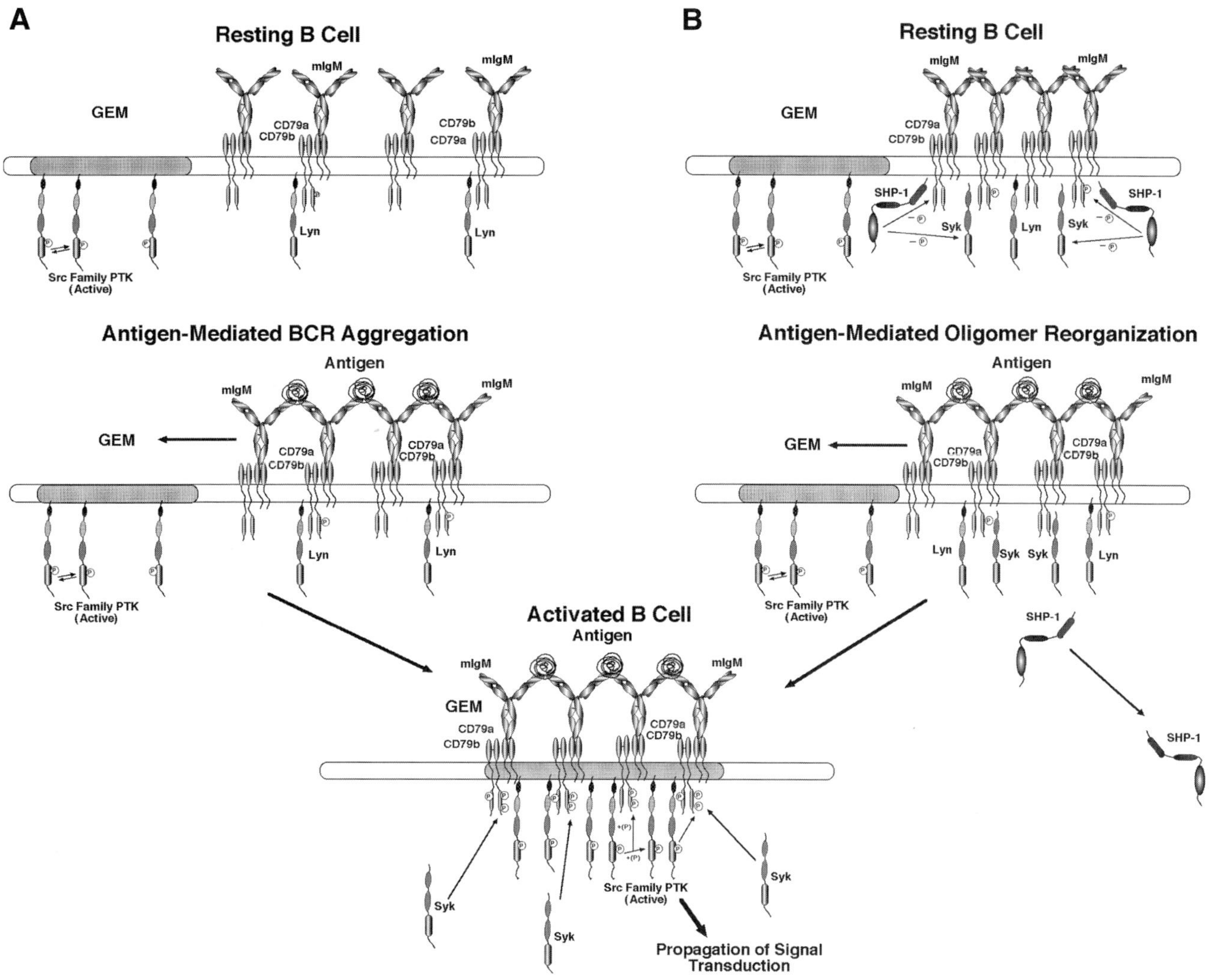

Figure 1 Initiation of signal transduction via the BCR. (A) Initiation of signal transduction in response to antigen-mediated aggregation of BCR complexes. In the resting B cell the BCR exists as monomeric units in the plasma membrane. The net level of tyrosine phosphorylation associated with the BCR is maintained at a low level by the physical dispersal of the BCR complexes throughout the membrane, by the low stoichiometry of interaction with Src family PTKs, and by the action of PTPs (e.g., CD45 or SHP-1), which dephosphorylate the BCR complex. Ligation of the BCR by antigen induces aggregation of the individual BCRs, which then promotes their translocation into GEMs in a signaling-independent manner. Once the aggregated BCRs enter the GEM, they are colocalized with active Lyn, which phosphorylates CD79a/b (Igα/β). This promotes recruitment of Syk and the initiation of signal transduction. (B) Initiation of signaling via antigen-mediated reorganization of preexisting BCR oligomers. In this model, the BCRs on resting B cells exist in preformed oligomers that contain Syk and the PTP SHP-1. Low basal levels of phosphorylation are presumably maintained by SHP-1, which inhibits Syk activation and may dephosphorylate CD79a/b. Antigen binding causes a reorganization of the BCR oligomer that promotes translocation to GEMs, as well as the dissociation of SHP-1 from the complex. Once the BCRs enter the GEM, initiation of signaling can proceed.

abrogation of Syk expression results in the loss of IP_3 generation and Ca^{2+} mobilization [30]. Moreover, studies using bone marrow cells derived from Syk-deficient mice have confirmed that Syk is essential for proper signal transduction via the BCR based on the inability of these cells to develop normally into mature B cells in chimeric mice [31,32]. Recruitment of Syk to the BCR complex is mediated by tyrosine phosphorylation of the ITAMs in Igα and Igβ, creating docking sites that are recognized by the dual SH2 domains of Syk [33–35]. Efficient recruitment of Syk requires that both conserved tyrosines within a given ITAM be present and that they be phosphorylated. Additionally, both SH2 domains of Syk must be present and functional in order to observe optimal association of the kinase with Igα and Igβ and its subsequent activation through the physical interaction with these polypeptides [13]. Binding of Syk to the ITAMs of Igα and Igβ may also play a crucial role in its activation by focusing Syk to the BCR activation complex within GEMs, which effectively brings Syk into proximity with Lyn. Studies support the conclusion that Lyn phosphorylates key tyrosine residues on Syk, resulting in potentiation of its kinase activity [13].

The ability of Syk to phosphorylate downstream signaling effector proteins is dependent on the recruitment of an adaptor protein called BLNK (SLP65, BASH) to the BCR activation complex [25,26,36]. Recent studies have determined that BLNK is recruited to the BCR complex by virtue of its ability to bind to phosphorylated tyrosine residues in the cytoplasmic tail of Igα that lie outside the ITAM [37,38]. It has been shown that mutation of tyrosines 176 and 204, which flank the ITAM, abrogates BLNK-dependent signaling [38]. Moreover, it was shown that the SH2 domain of BLNK binds directly to tyrosine 204 of Igα [37,38]. Thus it appears that Syk and BLNK are colocalized to the BCR complex through their interaction with Igα. This facilitates the ability of Syk to phosphorylate BLNK, which in turn generates phosphotyrosine-dependent binding sites for recruitment of additional signal transducing proteins [16,39]. The formation of the BCR:Syk:BLNK complex constitutes the basic unit that is required for propagation of signal transduction leading to the activation of several interrelated downstream pathways (Fig. 2).

Activation of PLCγ2-Dependent Signaling

The initial activation of PTKs is responsible for mediating the production of second messengers that subsequently regulate intermediate signaling processes leading to transcription factor activation. Studies have elegantly demonstrated that PLCγ2 is activated in response to BCR ligation and is responsible for the production of diacylglycerol (DAG) and inositol 1,4,5-trisphosphate (IP_3) [40–42]. These in turn promote PKC activation and mobilization of Ca^{2+}, respectively [22–26]. It has been demonstrated using a non-lymphoid cell reconstitution system that expression of Igα/β, Syk, and PLCγ2 is not sufficient to mediate activation of PLCγ2 and Ca^{2+} mobilization in response to BCR ligation, suggesting that one or more key components were missing [43]. Subsequent work has clearly shown that PLCγ2 is recruited to tyrosine phosphorylated BLNK via its tandem SH2 domains. In cells that lack BLNK, PLCγ2 is not observed to translocate from the cytoplasm to GEMs and exhibits decreased activation [36]. Additionally, mutation of the amino-terminal SH2 domain of PLCγ2 blocks its binding to phosphorylated BLNK and blocks localization to GEMs [44]. Therefore, it is apparent that BLNK provides a scaffold to localize PLCγ2 to the BCR activation complex within GEMs where it can be activated.

Based on numerous findings, it has been concluded that Syk and the PTK Btk act in concert to regulate the phosphorylation and activity of PLCγ2 [45–47]. In addition to the Src family PTKs and Syk, Btk is inducibly activated in response to BCR ligation and plays an important role in signal transduction [46,48]. Like the Src family PTKs, Btk contains contiguous SH3, SH2, and SH1 domains, although it does not possess a carboxyl-terminal negative regulatory tyrosine residue or a myristylation site. In addition to the SH domains, Btk contains an amino-terminal pleckstrin homology (PH) domain and an adjacent proline- and cysteine-rich Tec homology (TH) domain [46]. It has been shown that the SH2 domain of Btk is required for PLCγ2 phosphorylation and activation and that it exhibits restricted binding specificity for tyrosine phosphorylated BLNK. Thus, PLCγ2 and Btk are colocalized to the BCR activation complex by virtue of their interaction with BLNK.

Studies in the mouse and human have clearly demonstrated that Btk expression is essential for BCR-mediated Ca^{2+} mobilization [49,50]. Moreover, expression of Btk has been shown to restore calcium signaling in Btk-deficient XLA B cells. Although expression of Syk is required for normal mobilization of Ca^{2+} in B cells, overexpression of this PTK was not observed to restore calcium signaling in XLA B cells, demonstrating that it cannot compensate for the loss of Btk. This finding indicates that these PTKs must act in concert in a nonredundant manner to regulate PLCγ2 function [36,47]. The ability of Btk to reconstitute calcium signaling is dependent on its catalytic function and the Btk activation loop tyrosine (Tyr551), which is phosphorylated by Syk. Mutation of the Syk transphosphorylation site or the ATP-binding site abrogates Btk-dependent phosphorylation of PLCγ2 [45,47]. Recent studies have demonstrated that specific tyrosine residues in PLCγ2 (tyrosines 753, 759, 1197, and 1217) are direct targets for Btk and that their phosphorylation is essential for optimal activation of its catalytic function [51,52]. In conclusion, it appears that Syk is required for PLCγ2 activation by virtue of its role in phosphorylating BLNK to generate docking sites for Btk and PLCγ2, and through its role in phosphorylating Tyr551 on Btk leading to its catalytic activation. Btk appears to be specifically involved in phosphorylating multiple tyrosine residues on PLCγ2 that are required for optimal catalytic activation, leading to hydrolysis of phosphatidylinositol 4,5-bisphosphate to produce IP_3 and DAG.

The production of IP_3 in response to activation of PLCγ2 results in mobilization of Ca^{2+} from the endoplasmic reticulum (ER) [53]. IP_3-mediated release of Ca^{2+} involves IP_3 receptors (IP_3R) located in the ER, which form heterotetrameric channels. It has been shown that the nature of the calcium mobilization response in B cells may be controlled by the combinatorial function of three IP_3R receptor subtypes [54]. Additional evidence suggests that the function of IP_3Rs may be regulated by phosphorylation, as well as in response to binding of IP_3. A B-cell restricted scaffold protein with ankyrin repeats (BANK) has recently been cloned that is inducibly phosphorylated on tyrosine residues in response to BCR ligation [55]. Overexpression of BANK results in enhanced BCR-induced calcium mobilization. Interestingly, BANK has been shown to interact with Lyn and IP_3Rs, suggesting that this adaptor functions to bring Lyn and IP_3Rs into proximity, thereby promoting tyrosine phosphorylation of IP_3Rs by Lyn [55]. This may play a key role in potentiating calcium mobilization from the ER because it has been shown that tyrosine phosphorylation of IP_3Rs up-regulates their channel activity. The release of Ca^{2+} from intracellular stores has been shown to promote Ca^{2+} influx across the plasma membrane, presumably through Ca^{2+} release activated Ca^{2+} channels (CRAC) [56,57]. CRAC channel activation is mediated by the accumulation of a poorly defined calcium influx factor(s) that is produced in response to depletion of Ca^{2+} from intracellular stores.

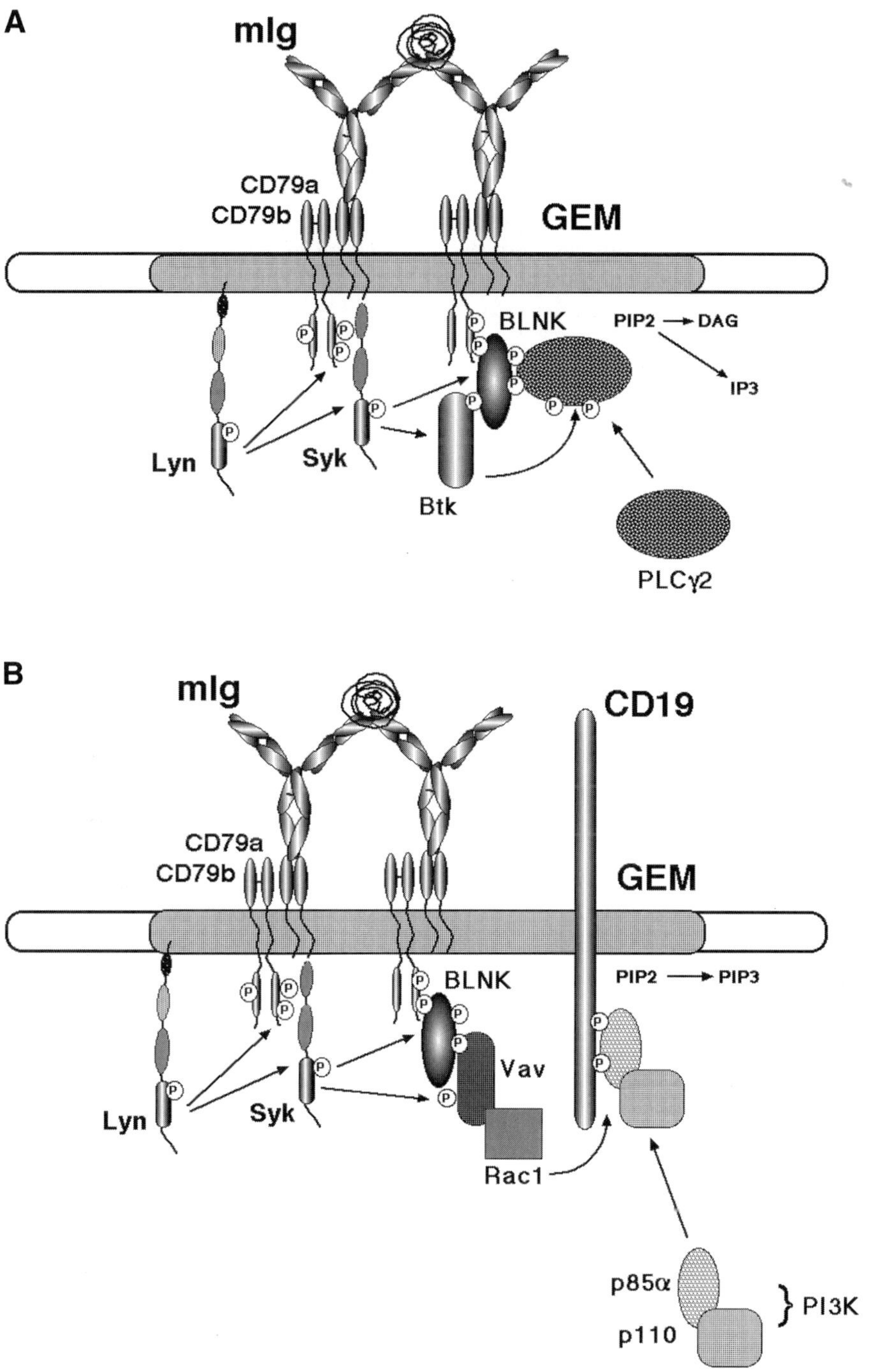

Figure 2 Propagation of signal transduction via the BCR. (A) Activation of PLCγ2-dependent signal transduction. Translocation of the BCR to GEMs facilitates phosphorylation of CD79a/b by Lyn, which in turn promotes recruitment of Syk and BLNK. Syk and BLNK bind to phosphorylated CD79a after which Syk is activated by Lyn-mediated phosphorylation. Syk then phosphorylates BLNK generating phosphotyrosine motifs that recruit Btk and PLCγ2, which bind to BLNK via their SH2 domains. Btk is further activated by Syk and in turn phosphorylates PLCγ2 resulting in its catalytic activation. PLCγ2 then hydrolyzes PIP_2 to produce DAG and IP_3, which mediate PKC activation and Ca^{2+} mobilization, respectively. (B) Activation of PI3K-dependent signal transduction. CD19 is inducibly phosphorylated on tyrosine in response to BCR ligation, enabling it to recruit PI3K to the membrane. Phosphorylated BLNK associated with the BCR complex recruits Vav, which binds to BLNK via its SH2 domain. Vav is phosphorylated and activated by Syk and then activates Rac1. Activated Rac1 binds to PI3K and potentiates its catalytic activity. Active PI3K phosphorylates PIP_2 to form PIP_3. PIP_3 recruits numerous PH domain-containing signaling proteins to the membrane including Btk, PLCγ2, Vav, and the PDK1/Akt kinases.

One of the most important downstream targets of the calcium signaling pathway in B cells is the transcription factor NFAT [58], although increased Ca^{2+} levels in the cell also play a role in the activation of protein kinase C (PKC) family members and in the activation of small molecular weight G-proteins [22–26].

The production of DAG by PLCγ2 leads to the activation of members of the PKC family. Both conventional (PKC-α, β, βII, γ) and novel (PKC-δ, ε, η, θ, μ) members of the PKC family require DAG for their activation, whereas the former require Ca^{2+} as well [59]. Studies have detected PKC-α, β, γ, δ, ε, η, θ, and μ expression in B cells and have also shown that several of these kinases (PKC-α, βII, δ, ε) translocate from the cytoplasm to the membrane in response to BCR ligation [59,60]. It has further been shown that PKC-μ interacts with the BCR and exhibits increased specific activity in response to BCR ligation [61]. The activation of members of the PKC family has been shown to play a role in regulating activation of several transcription factors in B cells [22–26]. PKC activation leads to activation of the mitogen-activated protein (MAP) kinase ERK2 via the classical Ras/Raf-1/Mek/ERK2 pathway [22–26]. ERK2 translocates to the nucleus where it regulates phosphorylation and activation of transcription factors including Elk-1 and members of the Ets family [62,63]. Additionally, it has recently been shown that novel PKC members PKC-θ and δ may play a critical role in B-cell activation by virtue of their ability to regulate the activation of the transcriptional regulator NFκB and the MAP kinase JNK [64].

Activation of Phosphoinositide 3-Kinase-Dependent Signaling

Another major signaling pathway that is activated upon BCR ligation involves the lipid kinase PI3K. Mice lacking the p85 subunit of PI3K exhibit severe defects in B-cell development as well as impaired proliferative responses to stimulation through the BCR [65,66]. Thus it is apparent that PI3K plays a critical role in signaling via the BCR. Recruitment and activation of PI3K to the BCR complex have been shown to occur via the transmembrane protein CD19 [67]. BCR ligation leads to tyrosine phosphorylation of several tyrosine residues in the cytoplasmic domain of CD19. Two of these residues, Tyr482 and Tyr513, have been shown to mediate binding of the p85 subunit of PI3K via its tandem SH2 domains [68]. Because CD19 localizes to GEMs in response to B-cell activation, its interaction with PI3K provides a mechanism to focus this kinase in regions of the cell that are enriched for its substrate phosphoinositide 4,5-bisphosphate (PIP_2) [69]. It is important to note, however, that loss of CD19 expression causes a less severe defect in B-cell development and function than is observed in B cells that lack the p85 subunit of PI3K. This suggests that additional adaptor proteins other than CD19 may mediate PI3K recruitment and activation [25].

Recent studies have identified a protein called B-cell adaptor for phosphoinositide 3-kinase (BCAP) that is inducibly phosphorylated by Syk and Btk in response to BCR ligation [70]. Of the 31 potential tyrosine residues that can be phosphorylated on BCAP, four are contained within consensus motifs for binding to the SH2 domains of the PI3K p85 subunit. Mutation of these tyrosine residues ablates the ability of BCAP to interact with p85 or to restore Akt activation in cells that lack BCAP [70]. Additionally, loss of BCAP expression has been shown to attenuate the recruitment of PI3K to GEMs suggesting that this adaptor plays an important role in targeting PI3K to the BCR activation complex. Nevertheless, loss of BCAP expression does not entirely abrogate the production of phosphatidylinositol 3,4,5-trisphosphate (PIP_3) or the activation of the downstream target Akt [70]. Therefore, it is clear that CD19 as well as other potential adaptors may be able to promote PI3K recruitment to GEMs where it is activated. A candidate for such an adaptor is Gab1, which is inducibly phosphorylated on tyrosine residues in response to BCR ligation mediating its direct interaction with the SH2 domains of PI3K, Shc, and the PTP SHP-2 in a phosphotyrosine-dependent manner [71]. Overexpression of Gab1 in B cells was observed to potentiate BCR-mediated phosphorylation of Akt, which is a PI3K-dependent response. Importantly, it was observed that the pleckstrin homology domain of Gab1 is required for its ability to translocate from the cytoplasm to the plasma membrane and for its ability to potentiate activation of PI3K [72]. PH domains have been shown to bind to PIP_3, which is produced in response to activation of PI3K. Thus it is likely that Gab1 functions as an amplifier of PI3K-dependent signaling due to the fact that its recruitment to the membrane and function require the initial production of PIP_3 by PI3K [72].

Activation of PI3K is mediated by virtue of its ability to bind to tyrosine-phosphorylated proteins such as CD19 and BCAP via the SH2 domains of the p85 subunit. This in turn promotes translocation of PI3K to GEMs, where it is able to access its substrate PIP_2, which is enriched in these microdomains of the plasma membrane. Although PI3K binding to adaptor proteins and translocation is required for activation, it is not sufficient for full activation of PI3K catalytic function. Studies have indicated that Vav family members may play a role in potentiating the activation of PI3K [25]. Generation of B cells lacking Vav3 resulted in significant decreases in PIP_3 production and activation of Akt [73]. These defects could be corrected by expressing Vav3 or Vav2 in Vav3-deficient B cells. However, the guanine nucleotide exchange factor (GEF) mutant of Vav3 was not able to restore PI3K function, indicating that Vav3 may regulate PI3K activity through its target Rac1 [73]. This was indeed found to be the case based on several experimental strategies. In conclusion, Vav appears to be important for potentiating PI3K activity in response to BCR ligation. Nevertheless, it has yet to be formally determined whether all members of the Vav family share the ability to potentiate PI3K activity through activation of Rac1, although it seems likely based on the available experimental evidence.

Activation of PI3K results in the phosphorylation of PIP_2 to form PIP_3, which regulates the activation of numerous downstream signaling proteins that contain PH domains [74]. It has been shown that activation of PLCγ2, Btk, Rac1, and the kinase network including PDK1, Akt, GSK-3, and mTOR is regulated by the ability of these proteins to bind to PIP_3 via their PH domains [22–26]. Presumably, the production of PIP_3 by PI3K plays an important role in promoting recruitment of PH domain-containing proteins such as PLCγ2 and Btk to the membrane where they are colocalized through their interaction with BLNK. Additionally, it is possible that the production of PIP_3 functions to maintain activated PLCγ2 and/or Btk at the membrane once they have dissociated from BLNK, thereby prolonging the signal transduction response. It is now well documented that PI3K-dependent production of PIP_3 plays a crucial role in activation of the PDK1/Akt/GSK-3 kinase cascade, which in turn promotes cell survival [74]. PIP_3-dependent recruitment of PDK1 and Akt to the membrane facilitates the ability of PDK1 to phosphorylate Akt on serine/threonine residues, which in turn leads to activation of Akt [74]. Numerous studies have documented activation of Akt in response to BCR ligation [75–78]. Kinetic studies have demonstrated that Akt transiently translocates to the plasma membrane in B cells where it is activated and then migrates to the cytoplasm and nucleus where it presumably can interact with potential substrates [79]. Akt is a serine/threonine kinase that phosphorylates numerous downstream substrates including Bad and GSK-3. Phosphorylation of the apoptosis-inducing Bad protein creates a binding site for 14-3-3 proteins preventing Bad from binding to Bcl2 and Bcl-X_L thereby releasing them to mediate cell survival [74]. Phosphorylation of GSK-3, which is a serine/threonine kinase, inactivates its catalytic function. GSK-3 is constitutively active in resting cells and phosphorylates numerous proteins including c-myc and cyclin D, maintaining them in an inactive state [74]. GSK-3 has also been shown to phosphorylate NFAT causing a change in its conformation that reveals a nuclear export signal [80]. Thus, inhibition of GSK-3 activity by Akt promotes retention of NFAT in the nucleus. Therefore, phosphorylation of GSK-3 by Akt promotes the activation of proteins that regulate cell cycling and cell survival.

Activation of Small Molecular Weight G-Proteins and Their Signaling Pathways

BCR ligation promotes the activation of small molecular weight G-proteins that in turn regulate signal transduction pathways that control transcription factor activation and cytoskeletal reorganization, which affects cell morphology and motility. In many instances it appears that BLNK plays a role in colocalizing critical signal effector proteins leading to activation of G-proteins. It has been shown that BLNK interacts with the adaptor protein Grb2, which appears to bind constitutively to a proline-rich region on BLNK via one of its SH3 domains [81,82]. The interaction between Grb2 and BLNK can also be potentiated in a tyrosine phosphorylation-dependent manner in which the SH2 domain of Grb2 binds to phosphotyrosine on BLNK. In either case, Grb2 recruits the guanine nucleotide exchange factor (GEF) Sos to the complex. Another potential mechanism whereby Sos is recruited to the membrane involves the formation of a Shc/Grb2/Sos complex in which Shc is inducibly phosphorylated in response to BCR ligation promoting binding of Grb2 via its SH2 domain. It has been proposed that Shc in turn binds via its SH2 domain to tyrosine residues within the cytoplasmic domain of Igα/β [13,22,23]. An alternative mechanism by which the Shc/Grb2/Sos complex may be recruited to the membrane is through binding of Shc to tyrosine phosphorylated Gab1 [71]. Regardless of the specific mechanism by which Sos is recruited to the plasma membrane, its localization leads to direct activation of the small molecular weight G-protein Ras by virtue of its GEF activity. GTP-bound Ras controls the activation of a kinase cascade that culminates in the activation of the MAP kinases ERK1 and ERK2 [22,23,59]. This is mediated by binding of activated Ras to the Raf-1 kinase, which then phosphorylates and activates MEK1 and 2, and these phosphorylate the ERKs. Activated ERK1 and ERK2 translocate to the nucleus where they phosphorylate and regulate the activity of transcription factors including Elk-1 and Sap1a. The Ras/Raf-1/ERK pathway also functions to regulate cell cycle progression by virtue of its ability to up-regulate the expression of cyclins D and E1 while at the same time inhibiting the expression of the cell cycle inhibitor $p27^{Kip1}$ [83–85]. The function of the Ras/Raf-1/ERK pathway is negatively regulated by the Rap1 G-protein, which is inducibly activated in response to BCR ligation via a DAG-dependent mechanism [86]. Rap1 possesses the same effector binding domain as Ras, suggesting that it can compete with Ras for binding to downstream proteins in the Ras/Raf-1/ERK cascade. Indeed Rap1 has been shown to bind to Raf-1, but does not activate it [87]. Thus Rap1 may sequester components of the Ras/Raf-1/ERK pathway thereby blocking activation of ERK1 and 2. Alternatively it has been proposed that Rap1 activates a distinct pathway that may antagonize the function of the Ras pathway [88].

Activation of the Rac1 GTPase is mediated through the GEF activity of Vav [89]. Vav is effectively recruited to the plasma membrane by virtue of its PH domain, which binds to PIP_3, as well as its ability to interact with adaptor proteins via SH2/phosphotyrosine-dependent interactions. It has been shown that Vav is recruited to tyrosine phosphorylated BLNK where it is colocalized with the PTK Syk [36]. Phosphorylation of Vav by Syk potentiates its GEF activity leading to activation of Rac1. Additionally, it has been shown that Vav interacts with CD19, suggesting that this may constitute another mechanism for targeting it to the membrane. Rac1 is important for regulation of receptor-induced actin polymerization and cytoskeletal reorganization [90]. Such processes are likely to play an important role in organization of signaling proteins into effective complexes that promote B-cell activation. Rac1 activation also plays an important role in linking BCR signaling to activation of the downstream MAP kinases JNK and p38 [22,23,59]. Studies have shown that activation of these kinases is abrogated in cells that lack BLNK and that activation cannot be restored by reconstitution of PLCγ2 signaling alone [36]. This indicates that binding of Vav to BLNK and its activation by Syk is crucial for subsequent activation of Rac1 and the JNK and p38 MAP kinases. Activated forms of these MAP kinases translocate to the nucleus where they play an important role in regulating the function of transcription factors through phosphorylation. JNK can activate Elk-1 and Sap1a transcription factors as well as c-Jun. It has been shown that p38 can regulate the activation of ATF-2, Sap1a, CHOP, and MEF2C transcription factors [59].

Conclusions

It is clear that the BCR is able to access several distinct, yet interrelated signaling pathways in response to binding of antigen. The initiation of signaling is dependent on antigen-driven changes in the organization of the BCR complexes expressed on the surface of a quiescent B cell. Although the exact nature of the changes that are elicited in response to antigen binding have yet to be elucidated, it is apparent that perturbation of the BCR complexes, whether they exist as individual monomeric structures or in preformed oligomers, leads to translocation to GEMs. This is critical for promoting a net increase in tyrosine phosphorylation of the BCR-associated Igα/β heterodimer and for recruitment of Syk and BLNK, which form the central initiation complex. Formation of this initiation complex leads to propagation of signaling transduction via pathways that are regulated by activation of PLCγ2, PI3K, PKC, Ras, and Rac1. These pathways in turn lead to activation of numerous transcription factors that regulate gene expression.

The ability of the BCR to control the various biological outcomes associated with B-cell development, selection, and activation is ultimately regulated by both intrinsic as well as extrinsic factors that affect the qualitative and quantitative nature of the overall signal transduced via the BCR. Extrinsic factors include the physical nature of the antigen, the duration of exposure to antigen, and the B cell's previous exposure to that antigen. Intrinsic factors include the developmental state of the B cell, which may affect the response of the cell to a given antigenic signal at the genetic level.

Additionally, it is clear that the maturational and differentiation states of the B cell dramatically alter its response to antigenic challenge. This can occur through changes in the expression of proximal BCR signaling proteins such as PTKs and PTPs, as well as through changes in the expression of other transmembrane receptors that function as coreceptors for the BCR. Examples of these include CD19, CD22, FcγRIIb, and PIR-B. These coreceptors have the ability to provide contextual information to the B cell through their ability to detect extracellular ligands that affect their ability to engage the BCR and to recruit signal transducing proteins that modify the nature of the signal transduced via the BCR [91–94]. Thus it is clear that numerous mechanisms exist for modulating the basic signal transduced via the BCR and this in turn alters the complement of transcriptional regulators that are activated as well as the genes that may be accessible to them. This then dramatically alters the expression of key regulators that determine whether the B cell undergoes apoptosis or mounts an active immune response.

References

1. Wagle, N. M., Cheng, P., Kim, J., Sproul, T. W., Kausch, K. D., and Pierce, S. K. (2000). B-lymphocyte signaling receptors and the control of class-II antigen processing. *Curr. Top. Microbiol. Immunol.* **245**, 101–126.
2. Bishop, G. A. and Hostager, B. S. (2001). B lymphocyte activation by contact-mediated interactions with T lymphocytes. *Curr. Opin. Immunol.* **13**, 278–285.
3. Cariappa, A. and Pillai, S. (2002). Antigen-dependent B-cell development. *Curr. Opin. Immunol.* **14**, 241–249.
4. Kurosaki, T. (2002). Regulation of B cell fates by BCR signaling components. *Curr. Opin. Immunol.* **14**, 341–347.
5. Benschop, R. J. and Cambier, J. C. (1999). B cell development: Signal transduction by antigen receptors and their surrogates. *Curr. Opin. Immunol.* **11**, 143–151.
6. Nemazee, D. (2000). Receptor editing in B cells. *Adv. Immunol.* **74**, 89–126.
7. Glynne, R., Ghandour, G., Rayner, J., Mack, D. H., and Goodnow, C. C. (2000). B-lymphocyte quiescence, tolerance and activation as viewed by global gene expression profiling on microarrays. *Immunol. Rev.* **176**, 216–246.
8. Monroe, J. G. (1997). Molecular mechanisms regulating B-cell negative selection. *Biochem. Soc. Trans.* **25**, 643–647.
9. King, L. B. and Monroe, J. G. (2000). Immunobiology of the immature B cell: Plasticity in the B-cell antigen receptor-induced response fine tunes negative selection. *Immunol. Rev.* **176**, 86–104.
10. Gauld, S. B., Dal Porto, J. M., and Cambier, J. C. (2002). B cell antigen receptor signaling: Roles in Cell development and disease. *Science* **296**, 1641–1642.
11. Schamel, W. W. and Reth, M. (2000). Monomeric and oligomeric complexes of the B cell antigen receptor. *Immunity* **13**, 5–14.
12. Reth, M., Wienands, J., and Schamel, W. W. (2000). An unsolved problem of the clonal selection theory and the model of an oligomeric B-cell antigen receptor. *Immunol. Rev.* **176**, 10–18.
13. Justement, L. B. (2000). Signal transduction via the B-cell antigen receptor: The role of protein tyrosine kinases and protein tyrosine phosphatases. *Curr. Top. Microbiol. Immunol.* **245**, 2–51.
14. Reth, M. (2001). Oligomeric antigen receptors: A new view on signaling for the selection of lymphocytes. *Trends Immunol.* **22**, 356–360.
15. Matsuuchi, L. and Gold, M. R. (2001). New views of BCR structure and organization. *Curr. Opin. Immunol.* **13**, 270–277.
16. Zhang, Y., Wienands, J., Zurn, C., and Reth, M. (1998). Induction of the antigen receptor expression on B lymphocytes results in rapid competence for signaling of SLP-65 and Syk. *EMBO J.* **17**, 7304–7310.
17. Pani, G., Kozlowski, M. Cambier, J. C., Mills, G. B., and Siminovitch, K. A. (1995). Identification of the tyrosine phosphatase PTP1C as a B cell antigen receptor-associated protein involved in the regulation of B cell signaling. *J. Exp. Med.* **181**, 2077–2084.
18. Cheng, P. C., Dykstra, M. L., Mitchell, R. N., and Pierce, S. K. (1999). A role for lipid rafts in B cell antigen receptor signaling and antigen targeting. *J. Exp. Med.* **190**, 1549–1560.
19. Cherukuri, A., Dykstra, M., and Pierce, S. K. (2001). Floating the raft hypothesis: Lipid rafts play a role in immune cell activation. *Immunity* **14**, 657–660.
20. Pierce, S. K. (2002). Lipid rafts and B-cell activation. *Nature Rev. Immunol.* **2**, 96–105.
21. Cheng, P. C., Brown, B. K., Song, W., and Pierce, S. K. (2001). Translocation of the B cell antigen receptor into lipid rafts reveals a novel step in signaling. *J. Immunol.* **166**, 3693–3701.

22. DeFranco, A. L. (1997). The complexity of signaling pathways activated by the BCR. *Curr. Opin. Immunol.* **9**, 296–308.
23. Campbell, K. S. (1999). Signal transduction from the B cell antigen-receptor. *Curr. Opin. Immunol.* **11**, 256–264.
24. Kurosaki, T. (2000). Functional dissection of BCR signaling pathways. *Curr. Opin. Immunol.* **12**, 276–281.
25. Kelly, M. E. and Chan, A. C. (2000). Regulation of B cell function by linker proteins. *Curr. Opin. Immunol.* **12**, 267–275.
26. Kurosaki, T. (2002). Regulation of B-cell signal transduction by adaptor proteins. *Nat. Rev. Immunol.* **2**, 354–363.
27. Hutchcroft, J. E., Harrison, M. L., and Geahlen, R. L. (1991). B lymphocyte activation is accompanied by phosphorylation of a 72-kDa protein-tyrosine kinase. *J. Biol. Chem.* **266**, 14846–14849.
28. Hutchcroft, J. E., Harrison, M. L., and Geahlen, R. L. (1992). Association of the 72-kDa protein tyrosine kinase PRK 72 with the B cell antigen receptor. *J. Biol. Chem.* **267**, 8613–8619.
29. Saouaf, S. J., Mahajan, S., Rowley, R. B., Kut, S. A., Fargnoli, J., Burkhardt, A. L., Tsukada, S., Witte, O. N., and Bolen, J. B. (1994). Temporal differences in the activation of three classes of non-transmembrane protein tyrosine kinases following B-cell antigen receptor surface engagement. *Proc. Natl. Acad. Sci. USA* **91**, 9524–9528.
30. Takata, M., Sabe, H., Hata, A., Inazu, T., Homma, Y., Nukada, T., Yamamura, H., and Kurosaki, T. (1994). Tyrosine linases Lyn and Syk regulate B cell receptor-coupled Ca^{2+} mobilization through distinct pathways. *EMBO J.* **13**, 1341–1349.
31. Turner, M., Mee, J. P., Costello, P. S., Williams, O., Price, A. A., Duddy, L. P., Furlong, M. T., Geahlen, R. L., and Tybulewicz, V. L. J. (1995). Perinatal lethality and blocked B-cell development in mice lacking the tyrosine kinase Syk. *Nature* **378**, 298–302.
32. Cheng, A. M., Rowley, B., Pao, W., Hayday, A., Bolen, J. B., and Pawson, T. (1995). Syk Tyrosine kinase required for mouse viability and B-cell development. *Nature* **378**, 303–306.
33. Rowley, R. B., Burkhardt, A. L., Chao, H.-G., Matsueda, G. R., and Bolen, J. B. (1995). Syk protein-tyrosine kinase is regulated by tyrosine-phosphorylated Igα/Igβ immunoreceptor tyrosine activation motif binding and autophosphorylation. *J. Biol. Chem.* **270**, 11590–11594.
34. Kurosaki, T., Johnson, S. A., Pao, L., Sada, K., Yamamura, H., and Cambier, J. C. (1995). Role of the Syk autophosphorylation site and SH2 domains in B cell antigen receptor signaling. *J. Exp. Med.* **182**, 1815–1823.
35. Turner, M., Schweighoffer, E., Colucci, F., Di Santo, J. P., and Tybulewicz, V. L. (2000). Tyrosine kinase SYK: Essential functions for immunoreceptor signaling. *Immunol. Today* **21**, 148–154.
36. Ishiai, M., Kurosaki, M., Pappu, R., Okawa, K., Ronko, I., Fu, C., Shibata, M., Iwamatsu, A., Chan, A. C., and Kurosaki, T. (1999). BLNK required for coupling Syk to PLCγ2 and Rac1-JNK in B cells. *Immunity* **10**, 117–125.
37. Su, Y.-W., Zhang, Y., Schwelkert, J., Koretsky, G. A., Reth, M., and Wienands, J. (1999). Interaction of SLP adaptors with the SH2 domain of Tec family kinases. *Eur. J. Immunol.* **19**, 3702–3711.
38. Kabak, S., Skaggs, B. J., Gold, M. R., Affolter, M., West, K. L., Foster, M. S., Siemasko, K., Chan, A. C., Aebersold, R., and Clark, M. R. (2002). The direct recruitment of BLNK to immunoglobin α couples the B-cell antigen receptor to distal signaling pathways. *Mol. Cell. Biol.* **22**, 2524–2535.
39. Wienands, J., Schweikert, J., Wollscheid, B., Jumaa, H., Nielsen, P. J., and Reth, M. (1998). A new signaling component in B lymphocytes which requires expression of the antigen receptor for phosphorylation. *J. Exp. Med.* **188**, 791–795.
40. Coggeshall, M. K. (1998). Inhibitory signaling by B cell FcγRIIb. *Curr. Opin. Immunol.* **10**, 306–312.
41. Hempel, W. M., Schatzman, R. C., and DeFranco, A. L. (1992). Tyrosine phosphorylation of phospholipase C-γ2 upon crosslinking of membrane Ig on murine B lymphocytes. *J. Immunol.* **148**, 3021–3027.
42. Carter, R. H., Park, D. J., Rhee, S. G., and Fearon, D. T. (1991). Tyrosine phosphorylation of phospholipase C induced by membrane immunoglobulin crosslinking in B lymphocytes. *Proc. Natl. Acad. Sci. USA* **88**, 2745–2749.
43. Richards, J. D., Gold, M. R., Hourihane, S. L., DeFranco, A. L., and Matsuuchi, L. (1996). Reconstitution of B cell antigen receptor-induced signaling events in a nonlymphoid cell line by expressing the Syk protein-tyrosine kinase. *J. Biol. Chem.* **271**, 6458–6466.
44. Ishiai, M., Sugawara, H., Kurosaki, M., and Kurosaki, T. (1999). Cutting edge: Association of phospholipase C-γ2 Src homology 2 domains with BLNK is critical for B-cell antigen receptor signaling. *J. Immunol.* **163**, 1746–1749.
45. Takata, M. and Kurosaki, T. (1996). A role for Bruton's tyrosine kinase in B cell antigen receptor-mediated activation of phospholipase C-γ2. *J. Exp. Med.* **184**, 31–40.
46. Desiderio, S. (1997). Role of Btk in B cell development and signaling. *Curr. Opin. Immunol.* **9**, 534–540.
47. Fluckiger, A.-C., Li, Z., Kato, R. M., Wahl, M. I., Ochs, H. D., Longnecker, R., Kinet, J.-P., Witte, O. N., Scharenberg, A. M., and Rawlings, D. J. (1998). Btk/Tec kinases regulate sustained increases in intracellular Ca^{2+} following B cell receptor activation. *EMBO J.* **17**, 1973–1985.
48. Satterthwaite, A. B., Cheroutre, H., Khan, W. N., Sideras, P., and Witte, O. N. (1997). Btk dosage determines sensitivity to B cell antigen receptor cross-linking. *Proc. Natl. Acad. Sci. USA* **94**, 13152–13157.
49. Rigley, K. P., Harnett, M. M., Phillips, R. J., and Klaus, G. G. B. (1989). Analysis of signaling via surface immunoglobulin receptors on B cells from CBA/N mice. *Eur. J. Immunol.* **19**, 2081–2086.
50. Wickler, L. S. and Scher, I. (1986). X-linked immune deficiency (xid) of CBA/N mice. *Curr. Top. Microbiol. Immunol.* **124**, 87–101.
51. Rodriguez, R., Matsuda, M., Perisic, O., Bravo, J., Paul, A., Jones, N. P., Light, Y., Swann, K., Williams, R. L., and Katan, M. (2001). Tyrosine residues in phospholipase Cγ2 essential for the enzyme function in B-cell signaling. *J. Biol. Chem.* **276**, 47982–47992.
52. Watanabe, D., Hashimoto, S., Ishiai, M., Matsushita, M., Baba, Y., Kishimoto, T., Kurosaki, T., and Tsukada, S. (2001). Four tyrosine residues in phospholipase C-γ2, identified as Btk-dependent phosphorylation sites, are required for B cell antigen receptor-coupled calcium signaling. *J. Biol. Chem.* **276**, 38595–38601.
53. Berridge, M. J. (2001). The versatility and complexity of calcium signaling. *Novartis Found. Symp.* **239**, 64–67.
54. Miyakawa, T., Maeda, A., Yamazawa, T., Hirose, K., Kurosaki, T., and Iino, M. (1999). Encoding of Ca^{2+} signals by differential expression of IP3 receptor subtypes. *EMBO J.* **18**, 1303–1308.
55. Yokoyama, K., Su, I.-H., Tezuka, T., Yasuda, T., Mikoshiba, K., Tarakhovsky, A., and Yamamoto, T. (2002). BANK regulates BCR-induced calcium mobilization by promoting tyrosine phosphorylation of IP3 receptor. *EMBO J.* **21**, 83–92.
56. Clapham, D. E. (1995). Calcium signaling. *Cell* **80**, 259–268.
57. Lewis, R. S. and Cahalan, M. D. (1995). Potassium and calcium channels in lymphocytes. *Annu. Rev. Immunol.* **13**, 623–653.
58. Crabtree, G. R. and Olson, E. N. (2002). NFAT signaling: Choreographing the social lives of cells. *Cell* **109**, S67–S79.
59. Gold, M. R. (2000). Intermediary signaling effectors coupling the B-cell Receptor to the nucleus. *Curr. Top. Microbiol. Immunol.* **245**, 78–134.
60. Mischak, H., Kolch, W., Goodnight, J., Davidson, W. F., Rapp, U., Rose-John, S., and Mushinski, J. F. (1991). Expression of protein kinase C genes in hemopoietic cells is cell-type- and B cell-differentiation stage specific. *J. Immunol.* **147**, 3981–3987.
61. Sidorenko, S. P., Law, C.-L., Klaus, S. J., Chandran, K. A., Takata, M., Kurosaki, T., and Clark, E. A. (1996). Protein linase C μ (PKCμ) associates with the B cell antigen receptor complex and regulates lymphocyte signaling. *Immunity* **5**, 353–363.
62. Janknecht, R. and Hunter, T. (1997). Convergence of MAP kinase pathways on the ternary complex factor Sapla. *EMBO J.* **16**, 1620–1627.
63. Treisman, R. (1996). Regulation of transcription by MAP kinase cascades. *Curr. Opin. Cell Biol.* **8**, 205–215.
64. Krappmann, D., Patke, A., Heissmeyer, V., and Scheidereit, C. (2001). B-cell receptor- and phorbol ester-induced NF-κB and c-Jun N-terminal kinase activation in B cells requires novel protein kinase C's. *Mol. Cell Biol.* **21**, 6640–6650.

65. Suzuki, H., Terauchi, Y., Fujiwara, M., Aizawa, S., Yazaki, Y., Kadowaki, T., and Koyasu, S. (1999). Xid-like immunodeficiency in mice with disruption of the p85α subunit of phosphoinositide 3-kinase. *Science* **283**, 390–392.
66. Fruman, D. A., Snapper, S. B., Yballe, C. M., Davidson, L., Yu, J. Y., Alt, F. W., and Cantley, L. C. (1999). Impaired B cell development and proliferation in absence of phosphoinositide 3-kinase p85α. *Science* **283**, 393–397.
67. Poe, J. C., Hasegawa, M., and Tedder, T. F. (2001). CD19, CD21, and CD22: Multifaceted response regulators of B lymphocyte signal transduction. *Int. Rev. Immunol.* **20**, 739–762.
68. Tuveson, D. A., Carter, R. H., Soltoff, S. P., and Fearon, D. T. (1993). CD19 of B cells as a surrogate kinase insert region to bind to phosphatidylinositol 3-kinase. *Science* **260**, 986–989.
69. Cherukuri, A., Cheng, P. C., Sohn, H. W., and Pierce, S. K. (2001). The CD19/CD21 complex functions to prolong B cell antigen receptor signaling from lipid rafts. *Immunity* **14**, 169–179.
70. Okada, T., Maeda, A., Iwamatsu, A., Gotoh, K., and Kurosaki, T. (2000). BCAP: The tyrosine kinase substrate that connects B cell receptor to phosphoinositide 3-kinase activation. *Immunity* **13**, 817–827.
71. Ingham, R. J., Holgado-Madrug, M., Siu, C., and Wong, A. J. (1998). The Gab1 protein is a docking site for multiple proteins involved in signaling by the B cell antigen receptor. *J. Biol. Chem.* **273**, 30630–30637.
72. Ingham, R. J., Santos, L., Dang-Lawson, M., Holgado-Madruga, M., Dudek, P., Maroun, C. R., Wong, A. J., Matsuuchi, L., and Gold, M. R. (2001). The Gab1 docking protein links the B cell antigen receptor to the phosphatidylinositol 3-kinase/Akt signaling pathway and to the SHP2 tyrosine phosphatase. *J. Biol. Chem.* **276**, 12257–12265.
73. Inabe, K., Ishiai, M., Scharenberg, A. M., Freshney, N., Downward, J., and Kurosaki, T. (2002). Vav3 modulates B cell receptor responses by regulating phosphoinositide 3-kinase activation. *J. Exp. Med.* **195**, 189–200.
74. Cantley, L. C. (2002). The phosphoinositide 3-kinase pathway. *Science* **296**, 1655–1657.
75. Li, H.-L., Davis, W. W., Whiteman, E. L., Birnbaum, M. J., and Pure, E. (1999). The tyrosine kinases Syk and Lyn exert opposing effects on the activation of protein kinase Akt/PKB in B lymphocytes. *Proc. Natl. Acad. Sci. USA* **96**, 6890–6895.
76. Gold, M. R., Scheid, M. P., Santos, L., Dang-Lawson, M., Roth, R. A., Matsuuchi, L., Duronio, V., and Krebs, D. L. (1999). The B cell antigen receptor activates the Akt (protein kinase B)/glycogen synthase kinase-3 signaling pathway via phosphatidylinositol 3-kinase. *J. Immunol.* **163**, 1894–1905.
77. Pogue, S. L., Kurosaki, T., Bolen, J., and Herbst, R. (2000). B cell antigen receptor-induced activation of Akt promotes B cell survival and is dependent on Syk kinase. *J. Immunol.* **165**, 1300–1306.
78. Craxton, A., Jiang, A., Kurosaki, T., and Clark, E. A. (1999). Syk and Bruton's tyrosine kinase are required for B cell antigen receptor-mediated activation of the kinase Akt. *J. Biol. Chem.* **274**, 30644–30650.
79. Astoul, E., Watton, S., and Cantrell, D. (1999). The dynamics of protein kinase B regulation during B cell antigen receptor engagement. *J. Cell Biol.* **145**, 1511–1520.
80. Beals, C. R., Sheridan, C. M., Turck, C. W., Gardner, P., and Crabtree, G. R. (1997). Nuclear export of NF-ATc enhanced by glycogen synthase kinase-3. *Science* **275**, 1930–1934.
81. Fu, C., Turck, C. W., Kurosaki, T., and Chan, A. C. (1998). BLNK: A central linker protein in B cell activation. *Immunity* **22**, 267–272.
82. Wienands, J., Schweikert, J., Wollscheid, B., Jumaa, H., Nielsen, P. J., and Reth, M. (1998). SLP-65: A new signaling component in B lymphocytes which requires expression of the antigen receptor for phosphorylation. *J. Exp. Med.* **188**, 791–795.
83. Ekholm, S. V. and Reed, S. I. (2000). Regulation of G(1) cyclin-dependent kinases in the mammalian cell cycle. *Curr. Opin. Cell Biol.* **12**, 676–684.
84. Piatelli, M. J., Doughty, C., and Chiles, T. C. (2002). Requirement for a Hsp90 chaperone-dependent MEK1/2-ERK pathway for B cell antigen receptor-induced cyclin D2 expression in mature B lymphocytes. *J. Biol. Chem.* **277**, 12144–12150.
85. Sherr, C. J. and Roberts, J. M. (1999). CDK inhibitors: Positive and negative regulators of G1-phase progression. *Genes Dev.* **13**, 1501–1512.
86. McLeod, S. J., Ingham, R. J., Bos, J. L., Kurosaki, T., and Gold, M. R. (1998). Activation of the Rap1 GTPase by the B cell antigen receptor. *J. Biol. Chem.* **273**, 29218–29223.
87. Zwartkruis, F. J. and Bos, J. L. (1999). Ras and Rap1: Two highly related small GTPases with distinct function. *Exp. Cell Res.* **253**, 157–165.
88. Bos, J. L. (1998). All in the family? New insights and questions regarding interconnectivity of Ras, Rap1, and Ral. *EMBO J.* **17**, 6776–6782.
89. Crespo, P., Schuebel, K. E., Ostrom, A. A., Gutkind, J. S., and Bustelo, X. R. (1997). Phosphotyrosine-dependent activation of Rac-1 GDP/GTP exchange by the Vav Proto-Oncogene Product. *Nature* **385**, 169–172.
90. Kaibuchi, K., Kuroda, S., and Amano, M. (1999). Regulation of the cytoskeleton and cell adhesion by the Rho family GTPases in mammalian cells. *Annu. Rev. Biochem.* **68**, 459–486.
91. Cyster, J. G. and Goodnow, C. C. (1997). Tuning antigen receptor signaling by CD22: Integrating cues from antigens and the microenvironment. *Immunity* **6**, 509–517.
92. Fearon, D. T. and Carroll, M. C. (2000). Regulation of B lymphocyte responses to foreign and self-antigens by the CD19/CD21 complex. *Annu. Rev. Immunol.* **18**, 393–422.
93. Coggeshall, K. M. (2000). Positive and negative signaling in B lymphocytes. *Curr. Top. Microbiol. Immunol.* **245**, 213–260.
94. Takai, T. and Ono, M. (2001). Activating and inhibitory nature of the murine paired immunoglobulin-like receptor family. *Immunol. Rev.* **181**, 215–222.

CHAPTER 345

Signaling Pathways in the Normal and Neoplastic Breast

Danica Ramljak and Robert B. Dickson
Department of Oncology,
Lombardi Cancer Center,
Georgetown University,
Washington, D.C.

Introduction

Deregulated cell and tissue growth is a defining feature of all neoplasms, both benign and malignant. Malignant neoplasms have the capacity to invade normal tissues, to induce the development of a local vasculature, to metastasize and to grow at distant body sites. Deregulated growth occurs as a result of perturbed signal transduction. Signal transduction pathways include, in their broadest sense, all biochemical cellular pathways that modulate or alter cellular behavior or function. Consequently, cancers do not necessarily arise as a result of an increased rate of cellular proliferation. Rather, carcinogenesis is a combination of defects in cell cycle progression (cellular division), immortalization, genomic instability, programmed cell death (apoptosis) cell–cell and cell–substrate adhesion, and angiogenesis. During normal embryonic development and in adult life, signaling needs to be precisely coordinated and integrated at all times, because properly regulated differentiation signals are critical for preventing oncogenesis.

Studies of the normal and neoplastic breast during the last several decades of the twentieth century focused on identification of the mechanisms of action of estrogen and progesterone at the local tissue level, in the normal breast, and early promotion and later progression to malignancy. Breast tissue regulation by these hormones is modulated in a rather complex fashion by autocrine and paracrine growth factors and by a variety of transcription factors (such as coactivators and corepressors), controlling epithelial cellular differentiation, epithelial cell–cell and cell–stromal adhesion. A large body of breast cancer research has been focused on understanding the complex interactions among growth factors, deregulated growth regulatory genes (oncogenes or proto-oncogenes), in mediating or modulating endocrine steroid action in breast cancer.

A second important topic has been the involvement of growth factors in facilitating malignant progression of the disease. One area of research has examined defective tumor host interactions, resulting in aberrant stromal collagen synthesis (desmoplasia), epithelial cell invasion, and vascular infiltration (angiogenesis) to promote distant metastasis. Also, studies have found that certain growth factors may suppress the host immune response to the tumor and may influence a tumor's response to therapy.

In the past few years, significant progress has been made in understanding the roles of different signaling molecules in normal and malignant breast. However, for the purpose of this chapter we focus only on several major molecules and pathways, such as the epidermal growth factor family (EGF), TGF-β family/Smads, and some other families of growth factors that have been shown to be important in breast signaling. In addition, three major proliferation/survival molecules/pathways, Mek/Erk, PI3-K/Akt, and Stats, are discussed. We do not address the roles of cell adhesion molecules or steroid hormones in signaling; other reviews covering these topics are available [1–3].

Signaling Molecules: A Class of Growth Factors

Epidermal Growth Factor Family

Two major classes of structurally and functionally distinct TGF-families were initially characterized by the prototypic transforming growth factor α (TGF-α) and transforming growth factor β (TGF-β). TGF-α is closely homologous to epidermal growth factor (EGF), and both bind to the common EGF receptor (EGFR) [4]. EGFR is tyrosine-kinase that, on activation, stimulates several cellular responses including survival, proliferation, motility, and differentiation [4]. The EGF family members that bind to EGFR are EGF, TGF-α, amphiregulin (AR, a heparin-binding factor), heparing-binding EGF (HbEGF), epiregulin, and β-cellulin [5]. Cripto (CR-1) is another EGF family member with a demonstrated role in embryogenesis and mammary gland development and it has been overexpressed in several human tumors [6,7]. Recently, the nature of its receptor has been identified as an ALK4, a type I serine/threonine kinase receptor for activin. CR-1 binds to cell surface ALK4 that is expressed on mammalian epithelial cells [8]. Also recently, molecules known as *heregulins* (human) and NDFs (*neu* differentiation factor, from rat) were cloned and identified [9,10]. *Neuregulin* is a widely used, common term for both NDFs and heregulins. Unlike most of the other EGF family members that bind directly to EGFR, neuregulins bind to two other EGF family members termed c-Erb-B3 and c-Erb-B4, which are structurally related [11]. Whereas c-Erb-B4 is a tyrosine kinase, c-Erb-B3 is kinase defective. A fourth member, c-Erb-B2 (HER2/neu), binds no secreted ligand and is activated through heterodimeric interaction with other family members. c-Erb-B2 is particularly important as an oncogene in breast cancer [12]. All four members of the EGFR family (Table I), and most of the EGFR-related growth factors, have been shown to play a role in tumor growth and development and in progression of human breast cancer.

Recently, mouse models have provided important corroboration and new insights regarding the role of EGF family members in mammary gland development. The view emerging from recent work is that signaling by EGFR and possibly c-Erb-B2 is critical for ductal outgrowth in pubescent glands, while signaling by c-Erb-B2, c-Erb-B3, and c-Erb-B4 is important for alveolar morphogenesis and lactation. Currently, it is unclear whether EGFR has a significant role in the latter processes. In a recent study only EGFR and c-Erb-B2 proteins appeared to play a role in the developing virgin gland [11], although all four members of the receptor family were expressed and maximally phosphorylated during pregnancy and lactation. Results of several studies indicate that members of the EGF family could act as local mediators of ductal morphogenesis. Analysis of triple null mice lacking EGF, TGF-α, and AR, alone or in various combinations, suggested a physiological role for AR ligand, a very surprising result in light of this ligand's reduced affinity for EGFR [13]. These results from animal models demonstrate a requirement for activation of EGFR by AR in ductal development in the adolescent mammary gland. Elongating ducts of adolescent mice contained high levels of the AR transcript, whereas expression was not detected in surrounding fibroblasts or fat cells, indicating that the epithelium is the principal source of AR in the developing mammary gland [13]. In contrast, EGFR protein and mRNA were detected in all cellular compartments, including epithelial cells, fibroblasts, and adipocytes. Thus, ductal outgrowth could be stimulated by autocrine and/or paracrine activation of EGFR. However, available evidence supports a paracrine model. Furthermore, although the role of EGFR in mammary gland development is obvious, until recently it was less clear whether this receptor also functions in alveolar morphogenesis and lactation. Current evidence supports such a role. Overexpression of TGF-α from WAP-driven transgene induced precocious alveolar development and delayed involution in transgenic mice [14], suggesting that these processes might be normally regulated by EGFR. EGFR levels and EGFR's phosphorylation were also shown to coordinately peak in late pregnancy and lactation [15]. In addition, aberrant epithelial differentiation and impaired lobular expansion were observed in the mammary gland of the AR knock-out mice, a phenotype that was aggravated by additional loss of EGF and TGF-α [13].

Steroid–growth factor interactions have been studied in human mammary tissue only in the context of malignant epithelium, although they are almost certainly crucial as well in the regulation of the normal gland and in the development of cancer. In hormone-responsive human breast cancer cells, estrogen-induced proliferation is accompanied by an increase in growth stimulatory TGF-α, AR, and IGF-II, modulation of IGF-binding proteins, induction of EGF and IGF-I receptors, inhibition of IGF-II and c-Erb-B2 receptors, and inhibition of TGF-β [16].

As mentioned earlier, the role of EGF family members in tumor onset and progression is well established [5]. TGF-α has been shown, both in cell lines *in vitro* and in experimental animal models *in vivo*, to be a positive modulator of cellular transformation. These effects of TGF-α are due to its effects on proliferation, survival, and motility, as well as

Table I Members of the EGFR Family and Their Ligands[a]

Receptor	Ligands
EGFR (c-Erb-B1)	Epidermal growth factor (EGF), transforming growth factor-α, amphiregulin, heparin-binding EGF-like growth factor (Hb EGF), betacellulin (BC), epiregulin (EP)
c-Erb-B2 (HER2)	Not established
c-Erb-B3 (HER3)	Neuregulins, Hb EGF, BC
c-Erb-B4 (HER4)	Neuregulins, Hb EGF, BC

[a]The four ErbB receptors are shown along with their ligands. Note that ErbB2 does not have a direct binding ligand.

modulation of differentiation [17]. Several laboratories, including our own, previously developed and/or utilized mouse models in which the EGFR ligand, TGF-α, was overexpressed in the mammary gland under control of MMTV LTR or WAP promoter [18–20]. The presence of the transgene in the mammary gland caused anomalous development: Alveoli appeared precociously, and postlactational involution was impaired, resulting in persistent epithelial structures, termed *hyperplastic alveolar nodules*. The mice also developed focal mammary tumors with high efficiency and short latency [18], indicating that ErbB signaling leads to neoplastic progression in this tissue. Recently, Humphreys and Hennighausen [19,21], compared TGF-α-induced mammary tumorigenesis in wild-type mice versus those lacking a functional Stat5a gene. They found that the absence of Stat5a delayed hyperplasia and tumor development and, coincidentally, promoted more epithelial regression. These effects were not observed in TGF-α transgenic mice containing Stat5a.

Our own studies and the research of other laboratories have been focused on understanding the cooperation and synergy between TGF-α and c-Myc that is frequently amplified and overexpressed in human breast cancer. c-Myc is a downstream effector of the c-Erb-B2 oncogene. Bitransgenic c-Myc/TGF-α mice developed multiple aggressive mammary tumors with dramatically shorter latency compared to either single transgenic lineage [18]. These results indicate a strong synergy between TGF-α and c-Myc that could reflect the latter's ability to amplify autocrine growth circuits, including those involving EGFR, in cultured cells. In addition, our recent data indicate that there is another possibility involving cooperation between cyclin D1 and TGF-α [22]. It appears that the early up-regulation of cyclin D1 by TGF-α might circumvent the normal ability of c-Myc to repress this cell cycle regulator, perhaps obviating the need for genetic alterations that would otherwise uncouple c-Myc and cyclin D1 during neoplastic progression. Work by Shroeder *et al.* [20] focused on identifying genes that cooperate with TGF-α in mammary tumorigenesis. Based on their results it appears that TGF-α caused up-regulation of the Wnt 3 gene, suggesting that the synergy between EGFR and Frizzled signaling pathways might contribute to neoplastic progression. Both Wnt and ErbB receptors can regulate β-catenin activity leading to cellular disaggregation. Therefore, it is possible to speculate that the observed synergy involves this pathway.

Besides the role of TGF-α in mammary tumorigenesis, other EGFR ligands might have a role in the proliferation process. The closely related factor, EGF, has been shown to act as an oncogene-like molecule when transfected and overexpressed in immortalized rodent fibroblasts [11]. Furthermore, it has been shown that besides TGF-α, both AR and CR-1 may be important in early stages of the onset of breast cancer [15,23–27]. Recently, a new transgenic model of EGFR overexpression in mammary gland has been described [28]. Mammary glands of virgin mice harboring an EGFR transgene, under the control of the MMTV or β-lactoglobulin (BLG) promoters, developed abnormally and displayed epithelial hyperplasias. With multiple pregnancies, dysplasias and tubular adenocarcinomas were also observed. Differentiation of mammary epithelium was perturbed in response to deregulated EGFR, as reflected by fewer alveoli developing in whole mount organ cultures. Similar to EGFR, overexpression of c-Erb-B2 in the transgenic mouse, or in the transgenic mammary gland after retroviral transfer, also leads to pregnancy-induced mammary tumors [11]. Some limited studies have also addressed the function of the c-Erb-B/c-Erb-B4 ligand family and neuregulins in breast cancer [11, 29].

Clearly, signaling by the ErbB network is important to the complex overall regulation of the mammary gland. It appears that all the models described above have firmly established that altered regulation or signaling of ErbB receptors and ligands can facilitate tumor promotion and/or neoplastic progression in the mammary gland. Future issues to be solved will include determination of definitive roles of c-Erb-B3 and c-Erb-B4 in mammary tumorigenesis, exploration of cooperative roles and interaction of EGFR and c-Erb-B2 in carcinogenesis, and the interactions of the c-ErbB family with other oncogenes and hormones in human breast cancers. In addition, it will be of interest to establish the identity of critical intracellular signaling pathways downstream of these receptors in human breast cancer at various steps in malignancy progression. Our own preliminary work and the work of others indicate that both the PI3K and MEK/Erk pathways are two key survival and proliferation signaling pathways, conveying signals downstream of the EGFR and possibly other members of the family. These two pathways are discussed later in this chapter.

Transforming Growth Factor β Family

The TGF-β family consists of several related gene products, each forming 25-kDa homodimeric or heterodimeric species, found in both normal mammary epithelium and in breast cancer. Three membrane-binding proteins interact with this family of growth factors. These were initially termed *receptors* (type I, II, and III), but type III receptors seem to be nonsignaling proteins, whereas the other two types (type I and II) are serine/threonine kinases and have been shown to deliver intracellular signals [30–34]. Four different type II receptors have been cloned, and they each may associate with one of several type II receptors. The function of type II receptors is determined on the basis of which type I receptor is recruited into dimer formation. TGF-β ligands only directly bind to type II receptors.

The epithelial-inhibitory TGF-β family is found in normal and malignant mammary epithelium and human milk [35]. TGF-β clearly has a negative effect on ductal epithelial proliferation and lactation in mouse mammary glands *in vivo* [36]. It is important to note that the production of TGF-β increases as breast cancer progresses; its accumulation appears to be important in the characteristic fibrous desmoplastic stroma of the disease [37], in immune suppression, and in tumor angiogenesis. It seems that, although TGF-β clearly serves a growth-inhibitory role in the normal gland, in which it may be tumor suppressive, overproduction of TGF-β may contribute to aberrant tumor–host interactions in the later progression of breast cancer [35].

In the review by Wakefield and Roberts [38] on TGF-β signaling in mammary gland development and tumorigenesis, several important issues have been summarized. First, ligands of the TGF-β superfamily are unique, in that they signal through transmembrane receptor serine/threonine kinases, rather than the tyrosine kinases of many other well-known growth factors in the breast. The TGF-β receptor complex couples to a signal transduction pathway involving a novel family of proteins, the Smads. On phosphorylation, Smads translocate to the nucleus where they modulate transcriptional responses. However, TGF-β can also activate the mitogen-activated protein kinase (MAPK) 4 pathway, and the different biological responses to TGF-β depend to varying degrees on activation of either or both of these two pathways. The Smad pathway is a nexus for cross-talk with other signal transduction pathways and for modulation by many different interacting proteins. Despite compelling evidence that TGF-β has tumor suppressor activity in the mammary gland, neither TGF-β receptors nor Smads are genetically inactivated in human breast cancer, although receptor expression is reduced. Work by Xie *et al.* [39] provides new information on the role of Smads in human breast cancer. It shows that alterations of Smad signaling in human breast carcinoma are associated with poor outcome. Among 456 cases of human breast carcinoma assembled in tissue microarrays, the majority (92%) expressed Smad2, Smad2P, and Smad4, indicating their ability to proliferate within a microenvironment that contains bioactive TGF-β.

Although some of the roles of TGF-β in the mammary gland signaling are obvious, some of the mechanisms in which TGF-β might play a role in development of human breast cancer are poorly understood. This needs to be addressed in future research initiatives.

Other Families of Growth Factors

In addition to EGFR and TGF-β growth factor families, at least five other growth factor stimulatory molecules are found to play a role in breast cancer. These are summarized in Table II: insulin-like growth factors (type I and II) [40], members of the Wnt (wingless) growth factor receptor family (Wnt-2, Wnt-3, Wnt-4, Wnt5a, and Wnt-7b) [41], platelet-derived growth factors A and B [42], and the fibroblast growth factor (FGF) family [43]. Each of these growth factor classes binds to one or more specific tyrosine kinase-encoding receptors. Vascular endothelial growth factor (VEGF, a member of a different family of tyrosine kinase receptor-binding factors) [44], pleiotrophin (a developmental, neurotropic factor) [45], and hepatocyte growth factor (HGF; also called scatter factor) and its tyrosine kinase encoding receptor, c-Met [46] are all produced by breast cancer. In addition, breast cancer cells also produce the hormone prolactin [47] and mammary-derived growth factor 1 (MDGF-1) [48].

Table II Diverse Group of Growth Factors and Other Molecules Thought to Play Roles in Breast Cancer[a]

Signaling molecules	Cellular response
Insulin-like growth factors (type I and II)	Stimulatory/tumor cells
Wnt growth factor family	Stimulatory/tumor cells
Platelet-derived growth factors A and B	Stimulatory/stromal cells[b]
Fibroblast growth factors	Stimulatory/tumor cells
Vascular endothelial growth factor	Angiogenesis/vascular cells[b]
Hepatocyte growth factor, receptor c-Met	Stimulatory/tumor cells
Prolactin	Stimulatory/tumor cells
Mammary-derived growth factor 1	Stimulatory/tumor cells
Pleiotrophin (developmental neurotropic factor)	Stimulatory/tumor cells[b]

[a] Proposed autocrine and paracrine factors in breast cancer. Tumor cells release variety of growth factors that might play autocrine roles *in vivo* (this is based mostly on their activity *in vitro*). Several of the same factors also are known to play paracrine and endocrine roles as well.

[b] Additional, angiogenic effects of these factors.

Although the roles of some of the molecules just mentioned are well established in breast signaling, much more research is needed to fully understand their role in both mammary gland development and in human breast cancer.

PI3K/Akt, MEK/Erk, and Stats: Major Proliferation/Survival Molecules Downstream of Growth Factor Receptors in Breast

Information is limited on the signaling pathways linking EGFR and other growth factor receptors to the regulation of cellular survival in mouse and human mammary epithelial and carcinoma cells. However, studies in some nonmammary epithelial cells (hepatic carcinoma cells and keratinocytes) have identified two survival pathways downstream of EGFR: phosphatidylinositol 3-kinase (PI3K)/Akt and extracellular signal regulated kinase (Erk1/2). In most cases, the PI3K/Akt pathway delivers the most potent survival signals downstream of EGFR [49,50].

Akt is a serine/threonine kinase, downstream of PI3K, that delivers strong survival signals in many cell types [51–53]. Both growth factors and integrins activate Akt through activation of PDK1 and a putative PDK2 kinase, which subsequently phosphorylate Akt at Thr308 and Ser473, respectively [54]. There are several isoforms of Akt (Akt1, Akt2, and Akt3); each has been shown to be expressed at different levels in various tissues [55]. The targets of Akt in epithelial cells, including mouse and human breast cells, include proteins involved in cell growth, metabolism, and apoptosis. The Akt targets involved in apoptosis include Bad, a proapoptotic member of the Bcl-2 family of proteins [56], caspase-9 [57], and the forkhead transcription factor [58]. Recently, Akt also has been reported to up-regulate the expression of antiapoptotic proteins in lymphoid cells such as Bcl-x_L [59], Bcl-2 [60], and Mcl-1 [61]. Akt activates NFκB in both fibroblasts and epithelial cells [62,63].

Similar to Akt, in most cell types, both growth factors and integrins are capable of activating the MAPK/Erk pathway. Of particular relevance to our studies involving models of human breast cancer is the fact that the MAPK/Erk pathway was previously shown to convey survival signals in response to EGF [64,65].

Recently, it has been shown that the PI3K/Akt and the MAPK/Erk pathways can cooperate in the inhibition of Bad in some cell types [66]. However, the prosurvival targets and the interactions of EGFR-activated PI3K/Akt and MAPK/Erk pathways have not been established in murine mammary epithelial cells (MMECs), human breast epithelial cells, or human carcinomas. Our earlier studies using mouse mammary epithelial cells [67a], established that activation of EGFR can promote survival in association with upregulation of Bcl-x_L a prosurvival family member of the Bcl-2 family. However, the mechanism(s) responsible for up-regulation of Bcl-x_L is unknown.

In our recent studies [67] we have chosen proapoptotic, c-Myc-overexpressing MMECs derived from MMTV-c-Myc transgenic mouse tumors to address these issues. We demonstrated that EGF strongly activates both Akt and Erks over a similar time frame and that this activity is dependent on EGFR kinase activity. The importance of Akt and Erk1/Erk2 in EGF-mediated survival signaling in c-Myc-overexpressing cells was further confirmed through the finding that prolonged inhibition of PI3K and MAPK/MEK activity leads to apoptosis. Apoptosis, induced by inhibition of PI3K activity, was paralleled by down-regulation of protein expression of both Akt and Erk1/Erk2. In addition, overexpression of constitutively activated Akt (Myr-Akt) prevented c-Myc-mediated apoptosis, triggered by the inhibition of the EGFR tyrosine kinase activity. The mechanism through which Akt, and possibly Erk, promote survival in c-Myc-overexpressing cells appears to involve the up-regulation of Bcl-x_L, suggesting that Bcl-x_L might be a novel target of Akt in MMECs (Fig. 1). With the emerging role of activated EGFR family in human breast cancer, it would be important to determine the antiapoptotic targets of EGFR-stimulated Akt and Erk in MMECs and human breast cancer cells. In support of *in vitro* studies, work by Hutchinson *et al.* [68] directly assessed the role of Akt in mammary epithelial development and tumorigenesis, using transgenic mice. The results indicate that activated Akt provides a critical cell survival signal required for tumor progression, without causing metastatic progression.

Another group of very important molecules that plays a role in mammary gland signaling is the Stats, transcription factors that are sequestered in the cytoplasm in an inactive form. Two members of the Stat family of transcription factors play a vital role in mouse mammary gland development. Stat5a was originally described as a regulator of milk protein gene expression and was subsequently shown to be essential for mammary development and lactogenesis. In contrast, Stat3 is an essential mediator of apoptosis and postlactational regression. Other members of the Stat family may have specific, but as yet undemonstrated, functions in mammary development. However, since Stat1 activity is regulated during mammary development in a pattern different from Stats 3 and 5, this factor also may have a functional role. Phosphorylation by activated cytokines and growth factor receptors leads to dimerization and translocation to the nucleus where Stats can activate specific sets of genes [69]. Constitutive activation of Stats has been associated with invasive breast carcinomas and breast cancer cell lines, but not with *in situ* carcinoma or benign lesions. Watson [70] postulates that cross-talk between Stat pathways and the ErbB family and Src may be an important regulator of breast cancer progression.

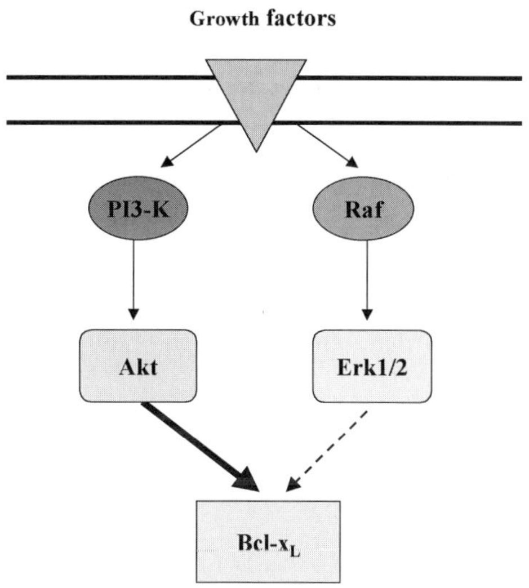

Figure 1 Proposed model of Bcl-x_L regulation by Akt and possibly Erk. Activated Akt up-regulates Bcl-x_L protein by a presently unknown mechanism in mouse mammary epithelial cells overexpressing c-Myc. This might lead to inhibition of apoptosis.

Stats have been shown to have a binding site in the Bcl-x_L promoter region [71]; therefore, when activated Stats might also contribute to cellular survival through up-regulation of this important prosurvival molecule. It has been reported that STATs 1 and 3 expression in primary breast carcinomas correlate with EGFR, HER2, p53, ER, PR, p21/waf1, Bcl-x_L, and Ki-67 expression [72].

Conclusions and Future Prospects

This chapter highlighted the importance of some of the molecules in breast signaling. Clearly, the importance of signaling by the c-ErbB family, TGF-β/Smads, several other growth factors, PI3K/Akt, MEK/Erk, and Stats in mammary gland development and mammary tumorigenesis has been well documented. Despite the fact that the recent insights revealed a significant amount of evidence for their role, mostly by research using *in vitro* human cellular models and mouse models of human breast cancer, it appears that many more questions remain unanswered. Indeed, studies to date have only started to answer some of the fundamental questions about how the signaling molecules mentioned here contribute to physiological and pathological activities in the breast tissue. New and more sophisticated insights will undoubtedly provide crucial information that will help our understanding of the role of these molecules in breast cancer and will help in providing valuable information for future therapies.

References

1. Debies, M. T. and Welch, D. R. (2001). Genetic basis of human breast cancer metastasis. *J. Mammary Gland Biol. Neoplasia* **6**, 441–451.
2. Dickson, R. B. and Stancel, G. M. (2000). Estrogen receptor-mediated processes in normal and cancer cells. *J. Natl. Cancer Inst. Monogr.* **27**, 135–145.
3. Haslam, S. Z. and Woodward, T. L. (2001). Reciprocal regulation of extracellular matrix proteins and ovarian steroid activity in the mammary gland. *Breast Cancer Res.* **3**(6), 365–372.
4. Kim, E. S., Khuri, F. R., and Herbst, R. S. (2001). Epidermal growth factor receptor biology (IMC-C225). *Curr. Opin. Oncol.* **13**, 506–513.
5. Prenzel, N., Zwick, E., Leserer, M., and Ullrich, A. (2000). Tyrosine kinase signaling in breast cancer. Epidermal growth factor receptor: Convergence point for signal integration and diversification. *Breast Cancer Res.,* **2**, 184–190.
6. Adamson, E. D., Minchiotti, G., and Salomon, D. S. (2002). Cripto: A tumor growth factor and more. *J. Cell Physiol.* **190**, 267–278.
7. Salomon, D. S., Bianco, C., Ebert, A. D., Khan, N. I., De Santis, M., Normanno, N., Wechselberger, C., Seno, M., Williams, K., Sanicola, M., Foley, S., Gullick, W. J., and Persico, G. (2000). The EGF-CFC family: Novel epidermal growth factor-related proteins in development and cancer. *Endocr. Relat. Cancer* **7**, 199–226.
8. Bianco, C., Adkins, H. B., Wechselberger, C., Seno, M., Normanno, N., De Luca, A., Sun, Y., Khan, N., Kenney, N., Ebert, A., Williams, K. P., Sanicola, M., and Salomon, D. S. (2002). Cripto-1 activates nodal- and ALK4-dependent and -independent signaling pathways in mammary epithelial cells. *Mol. Cell Biol.* **22**, 2586–2589.
9. Holmes, W. E., Sliwkowski, M. X., Akita, R. W., Henzel, W. J., Lee, J., Park, J. W., Yansura, D., Abadi, N., Raab, H., Lewis G. D., *et al.* (1992). Identification of heregulin, a specific activator of p185erbB2. *Science* **256**, 1205–1210.
10. Peles, E., Bacus, S. S., Koski, R. A., Lu, H. S., Wen, D., Ogden, S. G., Levy, R. B., and Yarden, Y. (1992). Isolation of the neu/HER-2 stimulatory ligand: A 44 kd glycoprotein that induces differentiation of mammary tumor cells. *Cell* **69**, 205–216.
11. Troyer, K. L. and Lee, D. C. (2001). Regulation of mouse mammary gland development and tumorigenesis by the ERBB signaling network. *J. Mammary Gland Biol. Neoplasia* **6**, 7–21.
12. Eccles, S. A. (2001). The role of c-erbB-2/HER2/neu in breast cancer progression and metastasis. *J. Mammary Gland Biol. Neoplasia* **6**, 393–406.
13. Luetteke, N. C., Qiu, T. H., Fenton, S. E., Troyer, K. L., Riedel, R. F., Chang, A., and Lee, D. C. (1999). Targeted inactivation of the EGF and amphiregulin genes reveals distinct roles for EGF receptor ligands in mouse mammary gland development. *Development* **26**, 2739–2750.
14. Sandgren, E. P., Schroeder, J. A., Qui, T. H., Palmiter, R. D., Brinster, R. L., and Lee, D. C. (1995). Inhibition of mammary gland involution is associated with transforming growth factor alpha but not c-myc-induced tumorigenesis in transgenic mice. *Cancer Res.* **55**, 3915–3927.
15. Schroeder, J. A. and Lee, D. C. (1998). Dynamic expression and activation of ERBB receptors in the developing mouse mammary gland. *Cell Growth Differ.* **9**, 451–464.
16. Dickson, R. B. and Lippman, M. E. (2000). Autocrine and paracrine growth factors in the normal and neoplastic breast. In Harris, J. R., Ed., *Diseases of the Breast*, 2nd ed., pp. 303–317. Lippincott Williams & Wilkins, Philadelphia.
17. Amundadottir, L. T. and Leder, P. (1998). Signal transduction pathways activated and required for mammary carcinogenesis in response to specific oncogenes. *Oncogene* **16**, 737–746.
18. Amundadottir, L. T., Nass, S. J., Berchem, G. J., Johnson, M. D., and Dickson, R. B. (1996). Cooperation of TGF-alpha and c-Myc in mouse mammary tumorigenesis: coordinated stimulation of growth and suppression of apoptosis. *Oncogene* **13**, 757–765.
19. Humphreys, R. C. and Hennighausen, L. (2000). Transforming growth factor alpha and mouse models of human breast cancer. *Oncogene* **19**, 1085–1091.
20. Schroeder, J. A., Troyer, K. L., and Lee, D. C. (2000). Cooperative induction of mammary tumorigenesis by TGF-alpha and Wnts. *Oncogene* **19**, 3193–3199.
21. Humphreys, R. C. and Hennighausen, L. (1999). Signal transducer and activator of transcription 5a influences mammary epithelial cell survival and tumorigenesis. *Cell Growth Differ.* **10**, 685–694.
22. Liao, D. J., Natarajan, G., Deming, S. L., Jamerson, M. H., Johnson, M., Chepko, G. and Dickson R. B. (2000). Cell cycle basis for the onset and progression of c-Myc-induced, TGF-alpha-enhanced mouse mammary gland carcinogenesis. *Oncogene* **19**, 1307–1317.
23. Kenney, N. J., Smith, G. H., Johnson, M. D. *et al.* (1997). Cripto-1 activity in the intact and ovariectomized virgin mouse mammary gland. *Pathogenesis* **1**, 57.
24. Kenney, N. J., Smith, G. H., Rosenberg, K., Cutler, M. L., and Dickson, R. (1996). Induction of ductal morphogenesis and lobular hyperplasia by amphiregulin in the mouse mammary gland. *Cell Growth Differ.* **7**, 1769–1781.
25. Amundattottir, L. T., Merlino, G. T., and Dickson, R. B. (1996). Transgenic models of breast cancer. *Breast Cancer Res. Treatment* **39**, 119–135.
26. Schroeder, J. A. and Lee, D. C. (1997). Transgenic mice reveals roles for TGF-alpha and EGF receptor in mammary gland development and neoplasia. *Mammary Gland Biol. Neoplasia* **2**, 119–129.
27. Edwards, P. A. W. (1998). Control of three dimensional growth pattern of mammary epithelium: Role of genes of the WNT and erbB families studied using reconstituted epithelium. In Rudland, P. S., Fernig, D. G., Leinster, S., and Lunt, G. G., Eds., *Mammary Development and Cancer*, pp. 21–26. Portland Press, London.
28. Brandt, R. Eisenbrandt, R., Leenders, F., Zschiesche, W., Binas, B., Juergensen, C., and Theuring, F. (2000). Mammary gland specific hEGF receptor transgene expression induces neoplasia and inhibits differentiation. *Oncogene* **19**, 2129–2137.
29. Srinivasan, R., Poulsom, R., Hurst, H. C., and Gullick, W. J. (1998). Expression of the c-erbB-4/HER4 protein and mRNA in normal human fetal and adult tissues and in a survey of nine solid tumor types. *J. Pathol.* **185**(3), 236–245.
30. Heldin, C. H., Miyazono, K., and Dijke, P. T. (1997). TGF-β signaling from cell membrane to nucleus through SMAD proteins. *Nature* **390**, 465–471.
31. Massague, J. (1992). Receptors for the TGF-beta family. *Cell* **69**, 1067–1070.
32. Ebner, R., Chen, R. H., Shum, L., Lawler, S., Zioncheck, T. F., Lee, A., Lopez, A. R., and Derynck, R. (1993). Cloning of a type I TGF-beta receptor and its effect on TGF-beta binding to the type II receptor. *Science* **260**, 1344–1348.
33. Attisano, L., Carcamo, J., Ventura, F., Weis, F. M., Massague, J, and Wrana, J. L. (1993). Identification of human activin and TGF-beta type I receptors that form heteromeric kinase complexes with type II receptors. *Cell* **75**, 671–680.
34. Liu, X., Yue, J., Frey, R. S., Zhu, Q., and Mulder, K. M. (1998). Transforming growth factor beta signaling through smad1 in human breast cancer cells. *Cancer Res.* **58**, 4752–4757.
35. McCune, B. K., Mullin, B. R., Flanders, K. C., Jaffurs, W. J., Mullen, L. T., and Sporn, M. B. (1992). Localization of transforming growth factor-beta isotypes in lesions of the human breast. *Hum. Pathol.* **23**, 13–20.
36. Robinson, S. D., Silberstein, G. B., Roberts, A. B., Flanders, K. C., and Daniel, C. W. (1991). Regulated expression and growth inhibitory

effects of transforming growth factor-beta isoforms in mouse mammary gland development. *Development* **113**, 867–878.

37. Stampfer, M. R., Yaswen, P., Alhadeff, M., and Hosoda, J. (1993). TGF-beta induction of extracellular matrix associated proteins in normal and transformed human mammary epithelial cells in culture is independent of growth effects. *J. Cell Physiol.* **155**, 210–221.
38. Wakefield, L. M. and Roberts, A. B. (2002). TGF-beta signaling: positive and negative effects on tumorigenesis. *Curr. Opin. Genet. Dev.* **12**, 22–29.
39. Xie, W., Mertens, J. C., Reiss, D. J., Rimm, D. L., Camp, R. L., Haffty, B. G., and Reiss, M. (2002). Alterations of Smad signaling in human breast carcinoma are associated with poor outcome: a tissue microarray study. *Cancer Res.* **62**, 497–505.
40. Ellis, M. J. (1999). The insulin-like growth factor network and breast cancer. In Bowcock, A., Ed., *Breast Cancer: Molecular Genetics, Pathogenesis, and Therapeutics*, p. 121. Humana Press, Totowa, NJ.
41. Bergstein, I. and Brown, A. M. C. (1999). WNT genes and breast cancer. In Bowcock, A., Ed., *Breast Cancer: Molecular Genetics, Pathogenesis, and Therapeutics*, p. 181. Humana Press, Totowa, NJ.
42. Bronzert, D. A., Pantazis, P., Antoniades, H. N., Kasid, A., Davidson, N., Dickson, R. B., and Lippman, M. E. (1987). Synthesis and secretion of platelet-derived growth factor by human breast cancer cell lines. *Proc. Natl. Acad. Sci. USA* **84**, 5763–5767.
43. Kern, F. G. (1999). The role of fibroblast growth factors in breast cancer pathology and progression. In. Bowcock, A., Ed., *Breast Cancer: Molecular Genetics, Pathogenesis, and Therapeutics*, p. 59. Humana Press, Totowa, NJ.
44. Ferrara, N., Houck, K., Jakeman, L., and Leung, D. W. (1992). Molecular and biological properties of the vascular endothelial growth factor family of proteins. *Endocr. Rev.* **13**, 18–32.
45. Wellstein, A., Fang, W. J., Khatri, A., Lu, Y., Swain, S. S., Dickson, R. B., Sasse, J., Riegel, A. T., and Lippman, M. E. (1992). A heparin-binding growth factor secreted from breast cancer cells homologous to a developmentally regulated cytokine. *J. Biol. Chem.* **267**, 2582–2587.
46. Rong, S., Bodescot, M., Blair, D., Dunn, J., Nakamura, T., Mizuno, K., Park, M., Chan, A, Aaronson, S., and Vande Woude, G. F. (1992). Tumorigenicity of the met proto-oncogene and the gene for hepatocyte growth factor. *Mol. Cell. Biol.* **12**, 5152–5158.
47. Clevenger, C. V., Chang, W. P., Ngo, W., Pasha, T. L., Montone, K. T., and Tomaszewski, J. E. (1995). Expression of prolactin and prolactin receptor in human breast carcinoma. Evidence for an autocrine/paracrine loop. *Am. J. Pathol.* **146**, 695–705.
48. Bano, M., Kidwell, W. R., and Dickson, R. B. (1994). MDGF-1: A multifunctional growth factor in human milk and human breast cancer. In Dickson, R. B., and Lippman, M. E., Eds., *Mammary Tumorigenesis and Malignant Progression*, p. 193. Kluwer, Boston.
49. Roberts, R. A., James, N. H., and Cosulich, S. C. (2000). The role of protein kinase B and mitogen-activated protein kinase in epidermal growth factor and tumor necrosis factor alpha-mediated rat hepatocyte survival and apoptosis. *Hepatology* **31**, 420–427.
50. Sibilia, M., Fleischmann, A., Behrens, A., Stingl, L., Carroll, J., Watt, F. M., Schlessinger, J., and Wagner, E. F. (2000). The EGF receptor provides an essential survival signal for SOS-dependent skin tumor development. *Cell* **10**, 211–220.
51. Downward, J. (1998). Mechanisms and consequences of activation of protein kinase B/Akt. *Curr. Opin. Cell Biol.* **10**, 262–267.
52. Kandel, E. S. and Hay, N. (1999). The regulation and activities of the multifunctional serine/threonine kinase Akt/PKB. *Exp. Cell Res.* **253**, 210–229.
53. Datta, S. R., Brunet, A., and Greenberg, M. E. (1999). Cellular survival: A play in three Akts. *Genes Dev.* **13**, 2905–2927.
54. Coffer, P. J., Jin, J., and Woodgett, J. R. (1998). Protein kinase B (c-Akt): A multifunctional mediator of phosphatidylinositol 3-kinase activation. *Biochem. J.* **335**, 1–13.
55. Okano, J., Gaslightwala, I., Birnbaum, M. J., Rustgi, A. K., and Nakagawa, H. (2000). Akt/protein kinase B isoforms are differentially regulated by epidermal growth factor stimulation. *J. Biol. Chem.* **275**, 30934–30942.
56. Datta, S. R., Dudek, H., Tao, X., Masters, S., Fu, H., Gotoh, Y., and Greenberg, M. E. (1997). Akt phosphorylation of BAD couples survival signals to the cell-intrinsic death machinery. *Cell* **9**, 231–241.
57. Cardone, M. H., Roy, N., Stennicke, H. R., Salvesen, G. S., Franke, T. F., Stanbridge, E., Frisch, S., and Reed, J. C. (1998). Regulation of cell death protease caspase-9 by phosphorylation. *Science* **282**,1318–1321.
58. Brunet, A., Bonni, A., Zigmond, M. J., Lin, M. Z., Juo, P., Hu, L. S., Anderson, M. J., Arden, K. C., Blenis, J., and Greenberg, M. E. (1999). Akt promotes cell survival by phosphorylating and inhibiting a Forkhead transcription factor. *Cell* **96**, 857–868.
59. Jones, R. G., Parsons, M., Bonnard, M., Chan, V. S., Yeh, W. C., Woodgett, J. R., and Ohashi, P. S. (2000). Protein kinase B regulates T lymphocyte survival, nuclear factor kappaB activation, and Bcl-X(L) levels *in vivo*. *J. Exp. Med.* **191**, 1721–1734.
60. Pugazhenthi, S., Nesterova, A., Sable, C., Heidenreich, K. A., Boxer, L. M., Heasley, L. E., and Reusch, J. E. (2000). Akt/protein kinase B up-regulates Bcl-2 expression through cAMP-response element-binding protein. *J. Biol. Chem.* **275**, 10761–10766.
61. Kuo, M. L., Chuang, S. E., Lin, M. T., and Yang, S. Y. (2001). The involvement of PI 3-K/Akt-dependent up-regulation of Mcl-1 in the prevention of apoptosis of Hep3B cells by interleukin-6. *Oncogene* **20**, 677–685.
62. Romashkova, J. A. and Makarov, S. S. (1999). NF-kappaB is a target of AKT in anti-apoptotic PDGF signalling. *Nature (Lond.)* **401**, 86–90.
63. Pianetti, S., Arsura, M., Romieu-Mourez, R., Coffey, R. J., and Sonenshein, G. E. (2001). Her-2/neu overexpression induces NF-kappaB via a PI3Kinase/Akt pathway involving calpain-mediated degradation of IkappaB-alpha that can be inhibited by the tumor suppressor PTEN. *Oncogene* **20**, 1287–1299.
64. Moro, L., Venturino, M., Bozzo, C., Silengo, L., Altruda, F., Beguinot, L., Tarone, G., and Defilippi, P. (1998). Integrins induce activation of EGF receptor: Role in MAP kinase induction and adhesion-dependent cell survival. *EMBO J.* **17**, 6622–6632.
65. Walker, F., Kato, A., Gonez, L. J., Hibbs, M. L., Pouliot, N., Levitzki, A., and Burgess. A. W. (1998). Activation of the Ras/mitogen-activated protein kinase pathway by kinase-defective epidermal growth factor receptors results in cell survival but not proliferation. *Mol. Cell Biol.* **18**, 7192–7204.
66. Hayakawa, J., Ohmichi, M., Kurachi, H., Kanda, Y., Hisamoto, K., Nishio, Y., Adachi, K., Tasaka, K., Kanzaki, T., and Murata, Y. (2000). Inhibition of BAD phosphorylation either at serine 112 via extracellular signal-regulated protein kinase cascade or at serine 136 via Akt cascade sensitizes human ovarian cancer cells to cisplatin. *Cancer Res.*, **60**, 5988–5994
67. Ramljak, D., Coticchia, C. M., Nishanian, T. G., Saji, M., Ringel, M. D. Conzen, S. D., and Dickson, R. B. (2003). Epidermal growth factor inhibition of c-Myc-mediated apoptosis through Akt and Erk involves Bcl-x_L upregulation. *Exp. Cell Res.* **287**, 397–410.

67a. Nass, S. J., Li, M., Amundadottio, L. T., Furth, P. A., and Dickson, R. B. (1996). Role for Bc1-xL in the regulation of apoptosis by EGF and TSFβ in c-Mgc-overexpressing mammary epithelial cells. *Bichem. Biophys. Res. Comm.* **277**, 248–256.

68. Hutchinson, J., Jin, J., Cardiff, R. D., Woodgett, J. R., and Muller, W. J. (2001). Activation of Akt (protein kinase B) in mammary epithelium provides a critical cell survival signal required for tumor progression. *Mol. Cell. Biol.* **21**, 2203–2212.
69. Rane, S. G. and Reddy, E. P. (2002). JAKs, STATs and Src kinases in hematopoiesis. *Oncogene* **21**, 3334–3358.
70. Watson, C. J. (2001). Stat transcription factors in mammary gland development and tumorigenesis. *J. Mammary Gland Biol. Neoplasia* **6**, 115–127.
71. de Groot, R. P., Raaijmakers, J. A., Lammers, J. W., and Koenderman, L. (2000). STAT5-dependent cyclin D1 and Bcl-xL expression in Bcr-Abl-transformed cells. *Mol. Cell Biol. Res. Commun.* **3**, 299–305.
72. Berclaz, G., Altermatt, H. J., Rohrbach, V., Siragusa, A., Dreher, E., and Smith, P. D. (2001). EGFR dependent expression of STAT3 (but not STAT1) in breast cancer. *Int. J. Oncol.* **19**, 1155–1160.

CHAPTER 346

Kidney

Elsa Bello-Reuss[1] and William J. Arendshorst[2]

[1]Department of Internal Medicine, Division of Nephrology and Department of Physiology and Biophysics, University of Texas Medical Branch, Galveston, Texas

[2]Department of Cell and Molecular Physiology, University of North Carolina at Chapel Hill, Chapel Hill, North Carolina

Overview of Kidney Function and Cell-to-Cell Interactions

The kidney is a complex organ comprising diverse cell types that work in coordination to perform a broad spectrum of functions, including the maintenance of body fluid and electrolyte balance, pH regulation, secretion of renin and erythropoietin, activation of vitamin D, excretion of numerous drugs and toxins, and regulation of blood pressure. Several of these functions are related and involve the process of urine formation, which takes place in the functional unit of the kidney, the nephron. Each human kidney has about one million nephrons that operate in parallel. The nephron consists of a glomerulus and a tubule arranged in series. The glomerulus is formed by a capillary network (glomerular tuft) extending between the afferent and efferent glomerular arterioles. The capillary loops are separated by intercellular material and surrounded by a layer of specialized epithelial cells (*podocytes*) attached to the basement membrane of the capillaries. The epithelial layer reflects on itself in the vascular pole, forming the Bowman space (or urinary space) that is continued by the lumen of the proximal tubule (Fig. 1). Urine formation is the result of filtration at the glomerulus and reabsorption or secretion at the tubule. The net flux from tubule lumen to blood constitutes reabsorption and the net flux in the opposite direction constitutes secretion. The renal-tubule cells can perform net transport between solutions of very similar or identical composition because they are polarized, that is, different transport proteins are expressed in the apical (lumen-facing) and basolateral (blood-facing) membrane domains, allowing them to carry out directional transport of specific solutes. Water transport is osmotic and occurs in the direction of net solute transport. Further details about renal organization and function can be found in recent texts [1,2].

Glomerular filtration, secretion, and tubule transport are finely regulated processes. The proximity of renal tubules to each other, as well as to capillaries and interstitial cells, facilitates paracrine interactions between different cell types. The glomerular urinary space is in series with the lumen of the renal tubule segments, so that cells in consecutive structures communicate with each other via the luminal fluid. The composition of this fluid can be changed by the rate of filtration (that increases the NaCl load) or by the secretion of signaling molecules into the lumen. Both changes in luminal NaCl concentration and the presence of signaling molecules are sensed by downstream tubule segments. This signaling mode, unique to the kidney, provides functional integration at the level of the single nephron. This theme is one of the central topics in this article.

As in other organs, cell-to-cell communication in the kidney is largely mediated by paracrine and autocrine agents that may act in the extracellular or intracellular compartments, activating signal transduction systems. These complex interactions unify membrane transport processes and urine formation with homeostatic regulation of renal hemodynamics and glomerular filtration rate (GFR) [3–5].

The kidney is divided in two regions: the superficial cortex and the interior medulla (Fig. 1, top panel). The cortex, 70% of the renal parenchyma, contains three classes of cells: *vascular* (large and small arteries, arterioles, capillaries), *epithelial* [proximal convoluted tubule (PCT), loop of Henle (LH) of short-looped cortical nephrons, distal convoluted tubule (DCT), cortical collecting tubule (CCT)], and *interstitial* cells. The medulla comprises 30% of the kidney mass

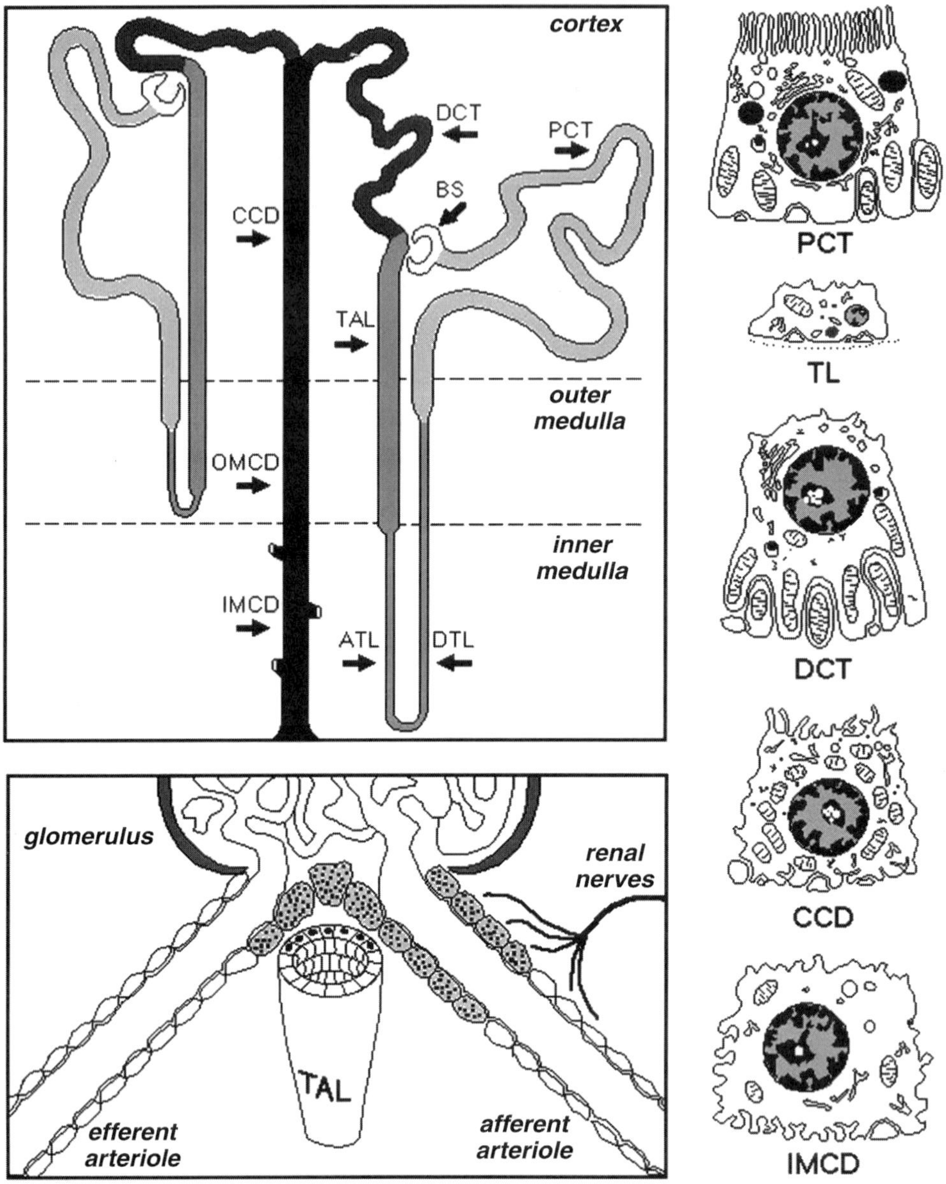

Figure 1 (Top) Scheme of the structure of deep and superficial nephrons. BS, Bowman's space; PCT, proximal convoluted tubule; DTL, descending thin limb of Henle's loop; ATL, ascending thin limb of Henle's loop; TAL, thick ascending limb of Henle's loop; DCT, distal convoluted tubule; CCD, cortical collecting duct; OMCD, outer medullary collecting duct; IMCD, inner medullary collecting duct. Shown on the right are simplified drawings portraying major structural differences between epithelial cells along the nephron. (Bottom) Scheme of the juxtaglomerular apparatus, which consists of vascular smooth muscle cells of afferent and efferent arterioles, juxtaglomerular granular cells at the glomerular end of an afferent arteriole, macula densa cells of the thick ascending limb of Henle's loop (TAL), glomerulus and extraglomerular mesangial cells.

and contains vasa recta capillaries, long loops of Henle of deep (juxtamedullary) nephrons, collecting ducts (CD), and interstitial cells. Thin-descending and thin-ascending LH are present in the inner medulla. The thick ascending limb of Henle's loop (TAL) resides in the outer medulla and cortex.

The vascular resistance of the kidney is lower than those of other organs, explaining the high blood flow of the organ, on average 1200 mL/min (about 25% of cardiac output) or 4 mL/min·g kidney wt. The renal artery and its branches carry 85–90% of the total renal blood flow to the cortex, where the microvasculature consists of two capillary beds arranged in series: the glomerular capillaries (after the afferent arteriole of the glomerulus) and the peritubular capillaries (after the efferent arteriole of the glomerulus). Afferent and efferent arterioles regulate blood flow and glomerular capillary hydrostatic pressure, the latter being a major determinant of the GFR. The transcapillary hydrostatic pressure gradient exceeds the plasma colloid osmotic pressure, thus causing fluid ultrafiltration all along the length of the glomerular capillaries. The glomerular-capillary wall is formed, from blood to Bowman's space, by the endothelial layer (that is fenestrated), the basement membrane, and the epithelial layer (foot processes separated by slits). The basement membrane, that is, the effective barrier of the glomerular-capillary wall, has a low permeability to plasma proteins and other molecules larger than albumin, whereas its permeability to water, ions, and other small solutes is considerably higher than in systemic capillaries. Fluid flows from Bowman's space into the proximal tubule, where solutes and water are reabsorbed by transport across both the cell membranes and the intercellular pathway into the peritubular space and then the capillaries. The plasma colloid osmotic pressure in the peritubular capillaries (elevated because in the glomerulus water is filtered, but not protein) exceeds the transcapillary hydrostatic pressure gradient and thus the reabsorbate readily flows into peritubular capillaries, both in the cortex and in the medulla. The capillaries in the latter are the hairpin-shaped *vasa recta*.

Medullary blood flow is 10–15% of the total renal blood flow. Its control is essential for the operation of the countercurrent multiplication and exchange mechanisms that accumulate solute in the medullary interstitial space providing the driving forces to concentrate or dilute the urine. In addition to upstream control by efferent arterioles, local blood-flow regulation is provided by smooth muscle-like pericytes around descending vasa recta. The medullary blood vessels are more sensitive to vasoactive agents than the cortical resistance vessels. Abnormal regulation of the renal-medullary circulation can lead to salt retention and hypertension [6–8].

Redundant and complementary control mechanisms form the basis of structural–functional relationships that are central to two of the primary functions of the kidney, namely, the maintenance of salt and water balances and the control of blood pressure. In the steady state, the output of fluid and electrolytes (excretion) matches the input (intake plus production and administration), so that the volume and composition of the extracellular fluid remain constant [3,5,9]. In a normal hydration state, the renal tubules reabsorb more than 99% of the glomerular filtrate. The rate of glomerular filtration in adult humans is about 120 mL/min, or 170 L/day. About 70% of the reabsorption occurs in the PT, 15% in the LH, and the rest in the distal nephrons (DCT, CCD, MCD). Figure 1 outlines the nephron segments and the basic structural features of the epithelial cells of each segment. The luminal fluid flows from PT to the descending and ascending loops of Henle, DCT, and CD. The transport processes and permeability properties of each nephron segment are regulated by paracrine factors generated by vascular and epithelial cells. NaCl is actively reabsorbed in all nephron segments, except the thin loops of Henle. The basolateral membranes of all renal-tubule cells express Na^+-K^+/ATPase, or Na^+ pump, a primary-active transport protein that extrudes Na^+ and takes up K^+, establishing chemical and electrical gradients at the luminal membrane that favor passive Na^+ entry. The luminal membranes of the different tubule segments express diverse Na^+ transport proteins [10–19]. The main mechanisms of Na^+ entry across the luminal membrane are: Na^+-H^+ exchange in the PCT (mediated by the antiporter NHE3, sensitive to high concentrations of the drug amiloride), Na^+-K^+-$2Cl^-$ in the TAL (mediated by the symporter NKCC2, sensitive to the diuretic furosemide), Na^+-Cl^- cotransport in the DCT (mediated by the symporter NCC, sensitive to diuretics of the thiazide family), and channel-mediated entry in the principal cells of the CD (via ENaC, the epithelial Na^+ channel, blocked by low concentrations of amiloride). Reabsorption of Na^+ is accompanied by anion (largely Cl^- and HCO_3^-) transport in the same direction and/or transport of another cation (H^+ or K^+) in the opposite direction to Na^+. The transport of Na^+ is coupled to transport of other solutes at the molecular level (examples given earlier) or by changing the driving force for transport of the other solute (i.e., transcellular Na^+ absorption favors paracellular Cl^- absorption).

Vasopressin, secreted by the pituitary, acts on CD principal cells to elicit exocytotic insertion of aquaporin (AQP2) and urea transporter (UT1) in the luminal membrane, thus increasing the permeability to both urea and water. Aldosterone, secreted by the adrenal cortex, also exerts its main action on the principal cells of the CD, stimulating Na^+ reabsorption by genomic and nongenomic mechanisms that enhance the number of Na^+ channels at the luminal membrane and the number of Na^+ pumps at the basolateral membrane. Because of the differences in the salt and water permeabilities among renal-tubule segments, the fractional reabsorptions of salt and water are somewhat different in specific segments. In normal individuals, fractional Na^+ reabsorption is about 67% in the PT, 20% in the LH, 7% in the DCT, 5% in the CD, and 1% in the IMCD.

In summary, the kidney is a highly sophisticated organ serving multiple functions ranging from synthesis and release of the proteolytic enzyme renin to maintenance of extracellular fluid volume and composition. Several of these functions depend critically on cell-to-cell communication. Homeostasis is regulated by multiple systems featuring coordination of hormonal, neural, paracrine, and autocrine signals as they act to control both the renal microcirculation and epithelial transport along the renal tubule.

This chapter focuses on cell-to-cell communication within the kidney. The extrarenal neural-hormonal regulation and other important functions, such as regulation of acid–base and calcium and potassium balances, are not discussed (for recent reviews, see [5,6,9,14,16,20–22]). Signal transduction among different cells is highlighted, at the expense of in-depth discussion of mechanisms underlying effector cell responses. Due to space limitations, not all intrarenal events are discussed. The reader is referred to recent reviews on contractile mechanisms of vascular smooth muscle cells [23–27] and basic transport mechanisms in renal epithelial cells [10–19].

Vascular Endothelial Cells

Endothelial cells [28–30] are in contact with blood elements and subjected to hemodynamic forces such as

Table I Summary of Major Receptors and Signaling Pathways in the Renal Vasculature

Hormone/paracrine/ autocrine agents	Receptor	G-protein	Intermediates/messengers
Vasoconstrictor Agents (direct actions on VSMC)			
Adenosine	P_1-A_1	$G_{\alpha i}$	↓cAMP/PKA
ATP	P_{2X}	—	↓K channel activity/ Depolarization Non-selective cation channel/Ca^{2+}
	P_{2Y}	$G_{\alpha q/11}$	Ca^{2+}/PKC Ca^{2+}/PLA_2/Cyt P450/HETE
Angiotensin II	AT_1	$G_{\alpha q/11}$	Ca^{2+}/PKC Ca^{2+}/PLA_2/Cyt P450/HETE
Catecholamines	α_1 α_2	$G_{\alpha q/11}$ $G_{\alpha i}$	Ca^{2+}/PKC ↓cAMP/PKA
Endothelin	ET_A	$G_{\alpha q/11}$	Ca^{2+}/PKC Ca^{2+}/PLA_2/Cyt P450/HETE
	ET_B	$G_{\alpha q/11}$	Ca^{2+}/PKC Ca^{2+}/PLA2/Cyt P450/HETE
PGE_2	EP_1 EP_3	$G_{\alpha q/11}$ $G_{\alpha i}$	Ca^{2+}/PKC ↓cAMP/PKA
$PGF_{2\alpha}$	FP	$G_{\alpha q/11}$	Ca^{2+}/PKC
Thromboxane	TP	$G_{\alpha q/11}$	Ca^{2+}/PKC
Vasopressin	V_1	$G_{\alpha q/11}$	Ca^{2+}/PKC
20-HETE	*	—	K channel (Ca^{2+} activated)/ depolarization
Vasodilator Agents (direct actions on VSMC)			
Adenosine	P_1-A_2	$G_{\alpha s}$	cAMP/PKA
Catecholamines	β	$G_{\alpha s}$	cAMP/PKA
Dopamine	D_1	$G_{\alpha s}$	cAMP/PKA
PGE_2	EP_4	$G_{\alpha s}$	cAMP/PKA
PGI_2	IP	$G_{\alpha s}$	cAMP/PKA
EDHF	*	—	K channel (Ca^{2+} activated)/ hyperpolarization
11,12-/14,15-EET	*	—	K channel (Ca^{2+} activated)/ hyperpolarization
Nitric oxide	sGC	—	cGMP/PKG cGMP/↓ PDE/↑ cAMP ↓ Cyt P450/↓ 20-HETE/ hyperpolarization
Carbon monoxide	sGC	—	cGMP/PKG
Vasodilator Agents (with preferential actions on Endothelial Cells)			
Endothelin	ET_B	$G_{\alpha q/11}$	Ca^{2+} NOS/NO Ca^{2+} PLA_2/COX/PGs PLA_2/EET
Bradykinin	B_2	$G_{\alpha q/11}$	Ca^{2+} NOS/NO Ca^{2+} PLA_2/COX/PGs PLA_2/EET
Acetylcholine	M	$G_{\alpha q/11}$	Ca^{2+} NOS/NO Ca^{2+} PLA_2/COX/PGs PLA_2/EET
Vasoconstrictor Agents (that also stimulate production of vasodilator substances by Endothelial Cells)			
Adenosine	P_1-A_1	Gαq/11	Ca^{2+} NOS/NO Ca^{2+} PLA_2/COX/PGs PLA_2/Cyt-P450/EET
Angiotensin II	AT_1	$G_{\alpha q/11}$	Ca^{2+} NOS/NO Ca^{2+} PLA_2/COX/PGs PLA_2/Cyt-P450/EET
ATP	P_{2Y}	$G_{\alpha q/11}$	Ca^{2+} NOS/NO Ca^{2+} PLA_2/COX/PGs PLA_2/Cyt-P450/EET
Vasopressin	V_1	$G_{\alpha q/11}$	Ca^{2+} NOS/NO Ca^{2+} PLA_2/COX/PGs PLA2/Cyt-P450/EET

Abbreviations in Table I:
cAMP, cyclic adenosine monophosphate
ATP, adenosine triphosphate
CO, carbon dioxide
COX, cyclooxygenase
cGMP, cyclic guanine monophosphate
Cyt-P450, cytochrome P450 enzyme
EET, Epoxy-eicosatrienoic acid
HETE, Hydroxy-eicosatrienoic acid
HO, heme oxygenase enzyme
NO, nitric oxide
NOS, nitric oxide synthase
PDE, phosphodiesterase
PGs, prostaglandins
PLA_2, phospholipase A_2
SGC, Soluble guanylyl cyclase
*, undetermined
↓, inhibits

hydrostatic pressure and shear stress. The hydrostatic pressure tends to distend the arterioles, causing a contractile myogenic response, whereas the shear stress, depending on the flow velocity, causes release of vasodilator agents including nitric oxide (NO), cyclooxygenase products and EDHP, and vasoconstrictor factors including endothelin-1, cytochrome P450 (Cyt-P450) products, and COX II derivatives. The interplay of these mechanisms determines the basal tone in resistance arteries and allows for rapid adaptations to fluctuations in flow and pressure.

Endothelial cells constitute a barrier whose permeability varies in different organs, largely depending on the presence of fenestrate and the properties of the tight junctions that separate the endothelial cells. Endothelial cells synthesize paracrine agents (gases, fatty acids, and peptides) that act on both nearby vascular smooth muscle cells (VSMC) and epithelial cells.

The release of both relaxing and constricting vasoactive factors modulates the degree of contraction of VSMC (Table I). Vasodilators include NO, PGI_2, and endothelium-derived hyperpolarizing factor (EDHF). The most potent vasoconstrictor is endothelin (ET). In addition, endothelial cells express angiotensin-converting enzyme (ACE), tethered to the blood-facing membrane; this enzyme catalyzes the local production of the vasoconstrictor angiotensin II (Ang II). Under physiological conditions, the actions of dilator agents predominate, whereas constrictor systems tend to prevail in stressful and disease states. The physiological actions of vasodilator agents are more important in counteracting the actions of constrictor systems, for example, angiotensin II and norepinephrine, than in producing vasodilation per se.

Nitric Oxide

Nitric oxide [31–34] synthase (NOS) is activated by hypoxia, shear stress, cell deformation, or vasoactive substances, to produce NO from L-arginine and molecular oxygen. NO is an important mediator of communication between endothelial cells and VSMC, as well as between endothelial cells and tubule cells. Constitutively active endothelial NOS (eNOS) is richly expressed in vessels and IMCD. Hemoglobin rapidly inactivates NO. Vasoactive agents such as Ang II and bradykinin stimulate eNOS, and ATP can stimulate NOS isoenzymes. The activation by these agents is mediated by G-protein-coupled receptors and involves activation of phospholipase C (PLC), increase in cytosolic calcium concentration ($[Ca^{2+}]_i$), and calmodulin-dependent enzyme activation. Thus, many vasoconstrictor-ligand-receptor interactions leading to increased $[Ca^{2+}]_i$ elicit primarily VSMC contraction but also stimulate endothelial production of the vasodilator NO, which limits the degree of constriction. The final balance of these two effects depends on the agonist and the density of receptors/enzymes in endothelial cells and VSMC. Ang II produces net vasoconstriction by virtue of the dominant action of AT_1 receptors on VSMC, whereas bradykinin and ATP cause net vasodilation due to their dominant effects on endothelial cells and NO production.

The low renal vascular resistance, compared to other vascular beds, is explained in part by a high dependence on NO. Inhibition of all three NOS isoforms (type I or neuronal, type II or inducible, and type III or endothelial) reduces renal blood flow (RBF) by about 30%, reduces GFR to a lesser extent, and increases arterial pressure by about 30 mm Hg. Constriction occurs along both afferent and efferent glomerular arterioles, and the glomerular filtration coefficient (hydraulic conductivity per glomerulus = capillary surface area × hydraulic conductivity per unit surface area) is reduced.

Endothelin

Endothelins [35–37] are potent paracrine vasoconstrictors produced by endothelial cells. They are 21-amino-acid peptides and three isoforms have been described, ET-1 being the most prevalent one in humans. Human ET is derived by successive proteolytic steps from preproendothelin (212 amino acids), which is hydrolyzed by a specific endopeptidase to a 37- to 39-amino-acid molecule named prepro-ET or big-ET; ET-1 is formed by the action of endothelial-bound ET converting enzyme on pro-ET.

The stimuli for endothelin secretion are bradykinin, ATP, shear stress, and cytokines. ET is predominantly secreted toward adjacent VSMC and the local effects of endogenous ET are more pronounced than those elicited by the usually low concentrations of circulating ET. Both ET_A and ET_B receptors are found in renal VSMC and both activate $G_{\alpha q/11}$ promoting an increase in $[Ca^{2+}]_i$ and triggering a contraction more pronounced and longer than those elicited by norepinephrine or Ang II, due to higher affinity of ET to its receptor. Only ET_B receptors are expressed in endothelial cells, where their activation increases $[Ca^{2+}]_i$ and stimulates NOS to produce the vasodilator NO; generation of vasodilator prostaglandins may also increase. Whether ET exerts a dilator or a constrictor action on arterioles depends on the relative abundance of ET receptors on the endothelium versus the VSMC. ET is thought to be primarily a vasoconstrictor agent in pathological conditions such as congestive heart failure and chronic renal failure, acting on both afferent and efferent arterioles of the glomerulus.

Arachidonic Acid Metabolites

Eicosanoid production is governed by the availability of the membrane fatty acid arachidonic acid. Its release is mediated by calcium-dependent, cytosolic phospholipase A2 (PLA_2) and the activity of enzymes of the cyclooxygenase (COX), lipoxygenase, and Cyt-P450 families. COX I and II synthesize prostaglandin H_2, from which different eicosanoids (PGE_2, PGI_2, and thromboxane TxA_2) are produced by the action of PGE_2-isomerase, prostacyclin synthase, and thromboxane synthase, respectively. The lipoxygenase family generates leukotrienes. Very little is known about lipoxygenase activity in physiological conditions. Finally, the Cyt-P450 monooxygenase family yields epoxy-eicosatrienoic acids (EETs) and hydroxy-eicosatrienoic acids (HETEs).

ARACHIDONIC ACID-COX METABOLITES

Prostaglandins (PGs) and thromboxanes (TXs) are synthesized by endothelial cells and act on vascular and tubule cells to function as autocrine or paracrine agents, contributing to the regulation of renal hemodynamics, renin release, and salt and water balance. When renin levels are normal, the net effects of COX metabolites [38–40] are vasodilator and natriuretic (i.e., promoting Na^+ excretion), reflecting the predominant production and actions of PGE_2 and/or PGI_2 over $PGF_{2\alpha}$ and TxA_2. In the cortical vasculature, PGI_2 is the major COX metabolite whereas tubule cells, especially IMCD cells, produce primarily PGE_2.

Prostaglandins activate G-protein-coupled receptors. In the renal vasculature, the main isotypes are EP_4 for dilator PGE_2, IP for dilator PGI_2, FP for constrictor $PGF_{2\alpha}$, and TP for constrictor TxA_2. EP_4 and IP receptors signal through $G_{\alpha s}$-proteins and cAMP/protein kinase A (PKA) cascade to relax VSMC. Under resting conditions, vascular endothelial cells primarily produce PGI_2. Descending vasa recta have dilatory IP and EP_4 receptors. TP receptors favor platelet aggregation and are vasoconstrictors, stimulating $G_{\alpha q/11}$proteins to activate PLC. However, PGE_2 can have a vasoconstricting effect via EP1 and EP3 receptors (see Table I).

Vasoconstrictors such as Ang II and ET increase $[Ca^{2+}]_i$ in endothelial cells as well as in VSMC. The $[Ca^{2+}]_i$ increase stimulates PLA_2 to release arachidonic acid, the rate-limiting step in eicosanoid production. Vasodilator COX metabolites (PGE_2 and PGI_2) are important in limiting the degree of vasoconstriction and thus maintaining adequate renal blood flow during low-salt diet and pathophysiological conditions such as heart failure and chronic renal failure. Production of vasodilating prostaglandins is increased by chronic salt restriction, an effect presumably mediated in part by high levels of Ang II and continuous elevation of $[Ca^{2+}]_i$ in VSMC. COX inhibition increases the vasoconstrictor effect of Ang II. The enhanced vasodilator effect of endogenous PGs during the course of a low-sodium diet is due to a combination of Ang II stimulation of PG production and apparent up-regulation of EP_4 receptors. A greater density of EP_4 receptors also may contribute to elevated renin release during extracellular volume contraction.

ARACHIDONIC ACID-CYT-P450 METABOLITES

In renal endothelial cells, VSMCs, and tubule cells, arachidonic acid metabolism by Cyt-P450s generates EETs, diHETEs, and 20-HETE, substances that are autocrine agents and second messengers active in the kidney [29,40–46]. These agents are lipophilic, bind to proteins, and partition into phospholipids. Their actions may be independent of conventional receptors. Large cortical arteries produce 20-HETE via the Cyt-P450-4A hydroxylase gene family. Smaller arterioles produce a combination of vasodilator EETs and constrictor 20-HETE. The metabolites 5,6-EET, 11,12-EET, and 14,15-EET are derived from the P450-2C epoxygenase family. Vasodilator 11,12-14,15-EETs are primarily produced in the renal cortex, whereas 20-HETE is preferentially synthesized in the medulla, especially the TAL.

Endothelium-Dependent Hyperpolarizing Factor

In response to stimulation by bradykinin or acetylcholine, endothelial cells release a vasodilating agent distinct from prostaglandins and NO. Endothelial-derived hyperpolarizing factor (EDHF) [29,45,46] is a diffusible factor that relaxes VSMC by hyperpolarization caused by activation of high-conductance, Ca^{2+}-dependent K^+ channels or by propagation of the endothelial-cell hyperpolarization via myoendothelial gap junctions. Recent evidence implicates 11,12-EET and 14,15-EET as EDHFs, with greater relaxing effects in the microcirculation than in conduit arteries.

Heme Oxygenases/Carbon Monoxide System

Microsomal heme-oxygenase (HO) [47,48] catalyzes the metabolism of heme to CO, biliverdin, and free iron. Two isoforms, HO-1 and HO-2, can be expressed in the kidney. Constitutive HO-2 is ubiquitous, present in the renal vasculature and in almost all nephron segments in both the cortex and medulla. Under basal conditions, inducible HO-1 appears to be at low levels or absent from renal structures. CO contributes to the renal vascular reactivity by acting as a vasodilator agent counteracting the vasoconstriction produced by Ang II, catecholamines, or pressure-induced myogenic tone. CO activates calcium-sensitive K^+ channels and increases the membrane potential of VSMC, therefore reducing Ca^{2+} entry. Little is known about the effects of endogenous CO on tubule transport and renal excretion of salt and water. Oxidative stress and injury, as well as chronic activation of AT_1 receptors by Ang II or V_1 receptors by vasopressin (AVP), induce HO-1 mRNA by Ca^{2+}- and perhaps PKC-dependent mechanisms.

Endothelial Cell Connections: Connexins and Gap Junctions

Electrical signals are transmitted along thin sheets of endothelial cells as well as to the surrounding musculature [49]. Gap junctions (GJs) underlie this communication. GJs are cell-to-cell channels permeable to molecules up to 1000 Da, thus allowing for the rapid exchange of water, ions, and other solutes between adjacent cells. GJs are formed by membrane proteins called connexins (Cx), of which about 20 isoforms appear to exist in humans. Cx37, -40, and -43 (number denotes approximate molecular weight in kilodaltons) are expressed in vascular endothelial cells. Cxs can form homomeric or heteromeric channels with different gating and permeability properties. GJ-mediated intercellular communication is stimulated by hypoxia and inhibited by low intracellular pH. PKC-mediated Cx phosphorylation reduces gap-junctional communication. Myoendothelial GJs render endothelial cells and VSMC into a coordinated functional unit. Little is known about gap junctions and their function in the renal vasculature. Cx40 and Cx43 are the main isoforms in VSMC, and Cx40 predominates in the renal preglomerular vasculature. Gap junctional communication between afferent arterioles from different glomeruli can occur via a common interlobular artery.

Summary

Vascular endothelial cells play an important role in regulating renal vascular resistance as well as ion and water transport by the renal tubules. Endothelial cells produce vasodilator and vasoconstrictor agents whose actions are finely balanced to regulate extracellular fluid homeostasis and blood pressure. Under physiological conditions, endothelium-derived vasodilators such as PGI_2, nitric oxide, carbon monoxide, and EDHF counteract the renal vasoconstrictor effects produced by circulating agents such as vasopressin, and norepinephrine, and local, as well as circulating, angiotensin II. In disease states such as congestive heart failure, chronic renal disease, and hypertension, the vascular endothelium can produce large amounts of constrictor substances such as ET and TxA_2. Our understanding of cell–cell connections between endothelial and VSMC, as well as the paracrine influences of vascular cells on tubular cells, is still quite limited.

Vascular Smooth Muscle Cells

VSMC encircle endothelial cells and contract to regulate the blood-vessel diameter and thereby its resistance to blood flow. Contraction and relaxation of VSMC in interlobular arteries and in afferent and efferent glomerular arterioles are the primary determinants of total renal vascular resistance. In addition to circulating hormones and nerve activity, paracrine/autocrine factors regulate the contractile function of VSMC of arteries, arterioles, and mesangial cells (cells of smooth muscle origin located between the glomerular capillary loops.) Vasomotor tone is regulated by communication between endothelial cells and the contractile VSMC. Just before an afferent arteriole enters a glomerulus, there is a short segment of juxtaglomerular granular cells responsible for the production, storage, and release of renin and Ang II. Renin release depends on signals from the macula densa cells of the TAL, activity of sympathetic nerve terminals, and baroreceptor-like effects of luminal pressure directly on juxtaglomerular granular cells (see Fig. 1, bottom panel).

Vasoconstrictor Mechanisms

The VSMC of renal microvessels respond to local paracrine factors in addition to circulating hormones and sympathetic nerve activity [3,5,23–26]. Contraction of VSMC is regulated by physical and chemical factors, with multiple signal transduction pathways coupling agonist-receptor interactions to $[Ca^{2+}]_i$, followed by activation of the actin-myosin contractile machinery. G-protein-coupled receptors play a prominent role in these processes. Heterotrimeric G-proteins transduce stimulatory or inhibitory signals to the cell interior. Table I shows a list of representative cell-surface receptors found in afferent and efferent arterioles. Ligand binding to the receptor triggers the intracellular signals, inducing changes in vasomotor tone and also causes receptor inactivation by phosphorylation (G-protein-coupled receptor serine/threonine protein kinases), thus limiting the duration of the effect of the agonist. Most of the receptors mediating vasoconstriction couple primarily with $G_{\alpha q/11}$ to activate PLC, form inositol trisphosphate (IP_3) and diacyl glycerol (DAG), and thereby elevate $[Ca^{2+}]_i$ and activate protein kinase C (PKC). Rapid increases in $[Ca^{2+}]_i$ are mediated by mobilization from internal stores and more sustained elevations result from increased Ca^{2+} entry via plasma membrane ion channels. The Ca^{2+} sensitivity of the contractile machinery of VSMC is enhanced by activation of PKC and Rho kinase. A variety of cell-surface receptors for vasoconstrictor agents are phosphorylated and inactivated or desensitized indirectly by PKC-mediated phosphorylations.

Vasodilator Mechanisms: Responses to Endothelium-Derived Agents

EICOSANOIDS

Vasoconstrictor agents may elicit weak vasodilation mediated by $[Ca^{2+}]_i$ stimulation of PLA_2 to form prostanoids and of NOS to produce NO [27,38,39,50–52]. Dilator PGE_2 and PGI_2 activate EP_4 and IP receptors, respectively. Both receptors are coupled to $G_{\alpha s}$-proteins and the cAMP-PKA signaling pathway. Prostanoids and NO exert similar effects on afferent- and efferent-arteriole VSMC. 5,6-EET appears to elicit vasodilation secondary to formation of the COX metabolites PGI_2 and PGE_2. Bradykinin and acetylcholine are thought to produce vasodilation due to their ability to produce endothelial-derived EET and di-HETE (or EDHF) in addition to stimulating COX and NOS.

NITRIC OXIDE AND CARBON MONOXIDE

The gases NO and CO [27,31,33,47,52] reduce vasomotor tone by cGMP-dependent and cGMP-independent mechanisms that ultimately reduce $[Ca^{2+}]_i$ in VSMC. NO and CO rapidly permeate plasma membranes and bind to the heme moiety of soluble (no membrane tether) cytosolic guanylyl cyclase to form a heterodimer of α and β subunits that catalyze production of cGMP.

cGMP/PKG reduce IP3-mediated Ca^{2+} release and stimulate the sarcoplasmic-reticulum Ca^{2+}-ATPase. Both effects lower $[Ca^{2+}]_i$, reducing smooth muscle contraction. PKG also inhibits a cGMP-dependent phosphodiesterase responsible for catabolism of cAMP, thus contributing to the reduction of $[Ca^{2+}]_i$. A third vasodilatory mechanism is NO inhibition of Cyt-P450-mediated production of the vasoconstrictor 20-HETE. In disease states, NO is involved in the generation of vasoconstricting agents such as superoxide radicals and other reactive oxygen species.

Vasoactive Paracrine/Autocrine Agents Produced by Smooth Muscle Cells: Cyt-P450 Metabolites

EETs and HETEs can exert dilator and constrictor actions respectively on the VSMC and, hence, can have opposing effects on tubule transport [40–44]. Vasoconstrictors such as Ang II and ET increase vascular production of 20-HETE that reinforces agonist-induced vasoconstriction by receptor-mediated increases in $[Ca^{2+}]_i$ and activation of PLA_2, with 20-HETE formation via Cyt-P450 ω-hydroxylation of arachidonic acid. 20-HETE produces vasoconstriction due to increased $[Ca^{2+}]_i$ and reduces the open-state probability of a Ca^{2+}-activated high-conductance K^+ channel, causing depolarization and activation of voltage-gated Ca^{2+} channels.

Summary

VSMC contract and relax in response to a host of vasoactive factors to change the diameter of the interlobular arteries and glomerular arterioles, the major renal resistance vessels. The balance of afferent and efferent arteriolar tone determines glomerular capillary pressure and in turn glomerular filtration rate. In a regulated fashion, tubular reabsorption returns between 90% and 99% of the filtered fluid and solutes to peritubular capillaries and vasa recta and eventually the systemic circulation. Most of the constrictor agents such as Ang II, vasopressin, and norepinephrine activate specific G-protein-coupled cell-surface receptors leading to activation of a phospholipase C that elicits IP3 and DAG signals that lead to increased intracellular $[Ca^{2+}]$ and stimulation of PKC. Vasodilator agents such as PGE_2 and PGI_2 activate specific receptors linked to the cAMP/PKA signaling pathway and reductions in intracellular $[Ca^{2+}]$. On the other hand, the vasodilators nitric oxide and carbon monoxide exert their relaxing effects on VSMC through a cGMP/PKC pathway, as well as mechanisms independent of cGMP. Dilator EETs and constrictor HETEs act by stimulating or inhibiting K^+ channels, respectively, to change membrane potential and thus the activity of L-type Ca^{2+} channels.

Tubulovascular Interactions: The Juxtaglomerular Apparatus

The juxtaglomerular apparatus (Fig. 1, bottom panel) [3–5,53–55] is a structural and functional unit in which paracrine signals are transmitted between the macula densa cells (differentiated tubule cells at the end of the TAL), the extraglomerular mesangial cells (interposed between the glomerular arterioles and the macula densa), the arteriolar VSMC, and the renin-producing granular cells in the afferent arteriole.

Renin Release: Juxtaglomerular Granular Cell Signaling

VASCULAR AND NEURAL CONTROL OF RENIN SECRETION

Granular cells at the end of the afferent arteriole synthesize and release renin by a process regulated by local changes in arteriolar hydrostatic pressure and cell stretch (or tension) and β-adrenoceptor stimulation by norepinephrine released from perivascular sympathetic nerve terminals [20,56–60]. Increased afferent arteriolar pressure inhibits renin release, an effect presumably due to stretch of JGA cells causing an increase in Ca^{2+} influx that, in contrast with the effects in most secretory cells, inhibits renin secretion. β-Adrenoceptors activate a $G_{\alpha s}$-protein linked to the cAMP-PKA messenger pathway, which stimulates renin secretion.

Circulating hormones and paracrine factors also influence renin secretion, some of them acting via the cAMP/PKA signaling pathway. Endothelium-derived PGE_2 and PGI_2 and β-adrenergic receptor agonists are potent stimuli of renin secretion. Dopamine stimulates renin release from granular cells via D_1 receptors and cAMP generation. Interactions such as Ang II with AT_1 receptors, AVP with V_1 receptors, or ET with ET_A receptors activate a $G_{\alpha q/11}$-protein, $[Ca^{2+}]_i$ increases, and PKC is activated, causing a decrease in renin secretion. It is not clear whether the key signal is increased $[Ca^{2+}]_i$ by itself, Ca^{2+} activation of Cl^- channels and a fall in $[Cl^-]_i$ per se or via the depolarization associated with Cl^- efflux, or increased PKC activity.

NO has biphasic effects on granular cells. Acute increases are inhibitory, whereas long-term exposure stimulates renin release. In this regard, endothelial NO appears to have a tonic effect, mediated by the ability of cGMP to elevate cAMP by inhibiting phosphodiesterase and thereby cAMP breakdown. Under certain conditions, NO can inhibit renin release via cGMP-dependent protein kinase (PKG) activity. In contrast, NO generated from nNOS in TAL and macula densa cells does not mediate changes in renin secretion in individuals on a low-sodium diet and renin secretion is normally regulated in eNOS-deficient mice.

MACULA DENSA CONTROL OF RENIN RELEASE: ARACHIDONIC ACID METABOLITES

The macula densa participates in the regulation of renin release from juxtaglomerular granular cells [38–41,61]. Renin secretion depends on NaCl delivery to and reabsorption by the macula densa cells at the end of the TAL. Inhibition occurs when solute delivery to this section of the renal tubule is high, and stimulation is associated with low solute delivery. Renin release leads to increased concentration of Ang I and Ang II in the adjacent interstitial compartment as well as in the systemic circulation. This is a regulatory mechanism mediated by macula densa metabolites of inducible COX II and Cyt-P450 enzymatic pathways. The COX II metabolites PGE_2 and PGI_2 stimulate cAMP and PKA, thereby enhancing renin secretion. COX II expression can be induced by either chronic sodium restriction or by inhibition of TAL NaCl reabsorption by furosemide. In contrast, the Cyt-P450 metabolite 20-HETE inhibits renin secretion, presumably by elevating $[Ca^{2+}]_i$ in effector cells.

MACULA DENSA CONTROL OF RENIN RELEASE: PURINERGIC AGENTS

Macula densa cells may signal granular cells to inhibit renin release by secreting adenosine and/or ATP in response to increased sodium delivery [56,62]. The precise mechanism of this effect is not clear. Adenosine stimulates granular cell A_1 and A_2 receptors. A_1-receptor activation inhibits renin secretion via stimulation of PCL, by increasing $[Ca^{2+}]_i$ and reducing cAMP/PKA. As stated earlier, this is an exception in secretory cells. Conversely, A_2 receptors stimulate renin secretion via cAMP/PKA signaling.

Summary

The juxtaglomerular apparatus has unique anatomical and functional properties. Macula densa cells at the end of the thick ascending limb of Henle's loop respond to low lumen [NaCl] by signaling to juxtaglomerular granular cells at the end of the afferent arteriole to increase renin secretion. Other stimuli for renin secretion are high sympathetic nerve activity, low systemic and afferent-arteriole pressure, and increased PGE_2. High rates of NaCl reabsorption by macula densa cells inhibit renin secretion. The signal mediating the functional connection between macula densa cells and afferent-arteriolar granular cells has been elusive. Attractive candidates are adenosine and ATP.

Tubulovascular Interactions: The Juxtaglomerular Apparatus and Tubuloglomerular Feedback

Tubuloglomerular feedback (TGF) [3,53,63], a process mediated by the juxtaglomerular apparatus (Fig. 1, bottom panel), is the direct relationship between the NaCl concentration in the tubule fluid at the macula densa and the capillary pressure and filtration rate of the same nephron. In functional terms, an increase in distal delivery of lumenal NaCl produces contraction of the afferent arteriole and a decrease in glomerular filtration.

The kidney regulates RBF and GFR during changes in arterial pressure by an intrinsic mechanism consisting of adjustments in preglomerular vascular resistance. This is referred to as *autoregulation* and is mediated by two mechanisms. First, a myogenic response intrinsic to the afferent-arteriole VSMC. These cells contract in response to the increase in tension of the wall when stretched by the higher blood pressure. Second, TGF. Primary increases in arterial pressure increase glomerular filtration and fluid delivery to the TAL, where the macula densa cells respond by sending a vasoconstricting signal to the afferent-arteriole VSMC. The nature of the signal mediating TGF is uncertain. Current postulates include ATP, adenosine, and other factors. As NaCl transport by the macula densa cells increases, so does $[Ca^{2+}]_i$, which can stimulate PLA_2 and nNOS and also elicit ATP release from mitochondria. The Cyt-P450 metabolite 20-HETE and NO may also play roles in adjusting preglomerular tone, and extraglomerular mesangial cells may be also involved. These agents may act as either mediators or modulators. For example, NO generated by macula densa nNOS acts as a modulator of TGF responsiveness as it inhibits macula densa NaCl reabsorption at high tubule flow rates, compared to little or no effect when [NaCl] at the macula densa is low. NO is thought to act directly on macula densa cells suppressing release of a constrictor agent, rather than diffusing to the afferent arteriole. Also, NO produced at upstream nephron sites may contribute to inhibition of TGF due to its ability to inhibit NaCl transport by macula densa cells. On the other hand, nNOS deficit in transgenic animals does not affect TGF.

It is clear that Ang II does not mediate TGF. An increase in renin release and Ang II formation occurs when NaCl delivery to the macula densa is low and the preglomerular vessels dilate, not contract. However, Ang II modulates TGF sensitivity and glomerular vascular reactivity by a mechanism that remains elusive. Other vasoconstrictors, such as norepinephrine and ET, have little effect on TGF. An inverse relationship exists between the degree of TGF and chronic salt intake. High renin and Ang II levels are associated with strong TGF during salt restriction. The attenuated TGF during a high-salt diet is attributable to enhanced NO production and low Ang II levels.

The involvement of eicosanoids in TGF is uncertain. COX II is present in macula densa and TAL cells; its expression is enhanced during chronic sodium restriction and COX II metabolites diminish the vasoconstriction accompanying increased NaCl delivery. However, TGF is normal in animals null for COX II or thromboxane receptors.

Purine Nucleotides and Purinoceptors: Role in Tubuloglomerular Feedback

The purinergic agents adenosine and ATP can be released from epithelial cells into the interstitial compartment, providing a metabolic link between epithelial cell NaCl transport and preglomerular resistance [62,64–67]. As load-dependent reabsorption of NaCl by the macula densa increases, so does ATP hydrolysis and hence adenosine production. Adenosine diffuses to the afferent arterioles to activate vascular P_1 (adenosine) receptors and thereby elicit vasoconstriction. ATP can be released by exocytosis from nerve terminals or may exit from endothelial cells, VSMC, and epithelial cells, probably through membrane channels, to activate P_2 (ATP) receptors.

Extracellular adenosine acts on P_1-A_1 purinergic receptors that respond to AMP but not to ADP or ATP. P_1-A_1 receptors couple to $G_{\alpha i}$-proteins and decrease cAMP/PKA activity in VSMC. In epithelial and endothelial cells, P_1-A_1 receptors couple to $G_{\alpha q/11}$, stimulating PLC signaling, increasing both $[Ca^{2+}]_i$ and PKC, and activating Ca^{2+}-dependent eNOS in endothelial cells. Afferent arterioles have more P_1-A_1 receptors than efferent arterioles. Renal P_1-A_2 receptors linked to $G_{\alpha s}$-proteins stimulate adenylyl cyclase to activate the cAMP/PKA pathway. P_1-A_2 receptor stimulation by adenosine causes vasodilation of both afferent and efferent arterioles and a natriuresis without a change in the filtered sodium load. Adenosine has a greater affinity for P_1-A_1 receptors. Therefore, low adenosine concentrations elicit vasoconstriction, whereas high concentrations produce vasodilation. A_1-receptor antagonists attenuate TGF activity, as do mutations of A_1 receptors. In the medulla, adenosine activation of epithelial A_1 receptors appears to be antinatriuretic, compared to A_2-receptor-mediated increase in medullary blood flow and natriuresis. The balance between P_1-A_2 vasodilating receptors and P_1-A_1 vasoconstricting receptors varies with salt intake. P_1 receptor stimulation is more effective in animals maintained on a low-salt diet. In certain pathological conditions, Ang II and adenosine are synergistic, that is, Ang II enhances the vasoconstrictor response to adenosine and vice versa.

ATP can be released from nerve terminals, endothelial cells, VSMC, and epithelial cells and acts locally to produce vasoconstriction. Extracellular ATP and ADP preferentially activate P_2 receptors, which have less affinity for adenosine or AMP. P_2 receptors are present on preglomerular vessels (endothelial cells and VSMC), glomerular mesangial cells, and tubule cells (PCT and CD). The P_{2x} and P_{2y} subtypes are the most common. P_{2x} receptors are ligand-gated ion channels with two membrane-spanning domains. In preglomerular VSMC, P_{2x} receptors depolarize the plasma membrane by inhibiting K^+ channels and triggering Ca^{2+} entry through voltage-gated L-type Ca^{2+} channels. ATP activation of P_{2x} receptors activates Cyt-P450 production of 20-HETE. P_{2y} receptors are classic $G_{\alpha q/11}$-protein-coupled receptors. Endothelial P_{2y} receptors are coupled to $G_{\alpha q/11}$-proteins that activate PLC to mobilize Ca^{2+} from sarcoplasmic reticular stores and stimulate PKC. Ca^{2+}-dependent eNOS and COX I produce the vasodilators NO and PGI_2. Intra-arterial administration of ATP elicits variable responses. The effects of circulating ATP on the renal vasculature involve vasodilator receptors on endothelial cells and vasoconstrictor P_{2x} receptors on VSMC. When the vasodilating component (P_{2y} receptors) is eliminated by inhibition of NO production, then the net response becomes vasoconstriction (P_{2x} receptors). ATP constricts the afferent arteriole considerably more than the efferent arteriole. Intracellular ATP can modulate vascular resistance by regulating ATP-sensitive K^+ channels; high ATP levels lead to VSMC hyperpolarization and vasodilatation.

Intrarenal Angiotensin II Production, Storage, and Actions

Circulating Ang II is formed by the action of angiotensin converting enzyme (ACE) on plasma Ang I in the lungs and the renal vasculature [56–58,66,68,69]. Ang I can be generated from angiotensinogen by renin and other proteolitic enzymes, for example, chymase and catepsin. Recent evidence suggests important formation, storage, and actions of intrarenal Ang II. All necessary substrates and enzymes for Ang II production are present in juxtaglomerular granular cells and in PCT cells. ACE lines endothelial cells and luminal and basolateral membranes of PCT cells. About 20% of circulating inactive Ang I is converted to biologically active Ang II in the kidney. The PCT can produce Ang II from circulating Ang I, concentrates Ang II by AT_1-receptor-mediated uptake, and stores it in the cells. Ang II is secreted into the lumen, where it can act locally or downstream, binding to luminal-membrane receptors in TAL, DCT, and CD. The effects are to increase Na^+, Cl^-, and HCO_3^- reabsorption.

There are two classes of Ang II receptors. AT_1 receptors, the predominant if not exclusive class under normal conditions, are present along the renal vasculature, including juxtaglomerular granular cells, glomeruli and vasa recta, multiple nephron segments (PCT, TAL, DCT, CD), and medullary interstitial cells. AT_1 receptors couple to $G_{\alpha q/11}$-proteins, PLC and the classical IP_3-Ca^{2+} and DAG-PKC pathways to cause vasoconstriction and decrease salt excretion.

Elevated Ang II levels, as seen with chronic salt restriction, constrict both afferent and efferent arterioles, reducing the renal blood flow (RBF) more than GFR. High Ang II concentrations desensitize the blood vessels by reducing the AT_1 receptor density and thus reducing the chronic effects of increased levels of the vasoconstrictor. In contrast, high endogenous Ang II concentration up-regulates tubule AT_1 receptors, favoring Na^+ retention. AT_1 receptors on granular cells exert short-loop feedback inhibition of renin release in association with increased $[Ca^{2+}]_i$ and PKC activation. Stimulation of AT_1 receptors activates the Ca^{2+}-dependent enzymes PLA_2 and NOS in endothelial cells, promoting production of prostanoids and NO. AT_1 receptors stimulate Na^+-H^+ exchange in PCT and TAL. Renal vascular AT_2 receptors are rare in a healthy, adult kidney. They may be up-regulated during conditions of a low-salt diet, with vascular and tubule functions opposite those of the predominant AT_1 receptors.

Vascular Actions of Bradykinin

Intrarenal bradykinin has vascular actions in the cortex and medulla mediated primarily, if not exclusively, by B_2 receptors [3,33,70]. The main action of bradykinin, vasodilatation, is mediated by a predominance of B_2 receptors on endothelial cells, which trigger PLC and increase $[Ca^{2+}]_i$ to stimulate eNOS activity and perhaps release EDHF/EET. A smaller population of constrictor B_2 receptors coupled to $G_{\alpha q/11}$ reside on VSMC.

Summary

Tubuloglomerular feedback is an adaptive mechanism that links the rate of glomerular filtration to the concentration of salt in the tubule fluid at the macula densa. A high [NaCl] at this level causes contraction of the afferent arteriole and a reduction in GFR. This autoregulatory mechanism is intrinsic to the kidney; that is, it does not require neural or humoral agents. The nature of the signal is still controversial. It could well involve adenosine or ATP. A renin-Ang II system exists within the kidney, with substrates and enzymes localized to proximal tubular cells in addition to juxtaglomerular granular cells. The major effects of local and circulating Ang II are vasoconstriction and Na^+ retention, both of which are primarily mediated by AT1 receptors.

Vasculotubular Communication

Receptors in the basolateral membranes of renal-tubule cells convey chemical information from the blood, vascular cells, and interstitial cells. Interactions between extracellular matrix and cell functions are beginning to be understood. In general, vasodilating agents inhibit Na^+ reabsorption by one or more nephron segments. NO, PGE_2/PGI_2, dopamine, and bradykinin conform to this oversimplified notion. Vasoconstrictors such as Ang II and norepinephrine usually stimulate Na^+ reabsorption. Notable exceptions are ET and 20-HETE, both of which are natriuretic. Paracrine agents produced by vascular cells exert actions on tubule cells by receptor-mediated events such as those elicited by agents produced by epithelial cells. (To minimize duplication, paracrine control of carrier proteins is discussed later under the heading of Tubule–Tubule Communication.)

Pressure Natriuresis

The mechanisms by which acute changes in arterial pressure influence tubule transport and Na^+ excretion are incompletely understood [3,6]. Increased vascular hydrostatic pressure may release NO and PGE_2 from endothelial cells. Blood flow in the inner medulla is more sensitive to changes in arterial pressure than cortical blood flow, and increased medullary blood flow is thought to elevate capillary and interstitial hydrostatic pressure and decrease colloid-osmotic pressure. Together, these changes cause a reduction in net salt and water reabsorption along the loops of Henle and collecting ducts.

Tubule–Tubule Communication: Paracrine Agents Released from Epithelial Cells

Paracrine Control of Solute Carrier Proteins

The major nephron segments produce and release autocrine and paracrine factors that modulate transepithelial solute and water transport [10,13,38,39,71,72]. The main transport mechanisms for solutes and water are listed in Table II. The PCT produces dopamine, NO, and Ang II; the TAL synthesizes ET, 20-HETE, NO, and PGE_2; the DCT generates kallikrein 20-HETE, NO, and COX II metabolites; and the CD produces ET, NO, and PGE_2. All renal-epithelial cells consume ATP and produce adenosine and purine nucleotides.

Angiotensin II

Ang II exerts important effects on water and electrolyte transport by indirect and direct mechanisms [15,44,58,73]. Its main indirect effect is the stimulation of secretion of the sodium-retaining hormone aldosterone by the adrenal cortex. Ang II directly stimulates NaCl reabsorption in PCT, TAL, DCT, and CD. In the PCT this effect is mediated by activation of AT_1 receptors expressed on the basolateral membrane. Receptor activation inhibits the cAMP/PKA pathway via $G_{\alpha i}$-protein and stimulates IP3 and PKC formation via $G_{\alpha q/11}$-protein. Ang II is synthesized in the PT and secreted to the lumen, reaching concentrations 100- to 1000-fold higher than in the systemic concentration ($<10^{-12}$ *M*). The effects of Ang II on the PCT are dose dependent: Physiological concentrations stimulate salt and fluid reabsorption, whereas pharmacologically high concentrations are inhibitory. The stimulatory effect in the PCT is mediated by activation of several target transport proteins: In the apical membrane, the Na^+/H^+ exchanger; in the basolateral membrane, the Na^+-K^+/ATPase, the Na^+/HCO_3^- cotransporter, and the K^+ channels all transport proteins that coordinately account for Na^+ reabsorption. The inhibitory effect of high levels of Ang II is mediated by binding to AT_2 receptors, resulting in PLA_2-mediated AA release. In the TAL, low Ang II concentrations ($<10^{-12}$ *M*) stimulate 20-HETE production and inhibit Na^+ reabsorption by inhibiting the apical-membrane K^+ channel, an effect that decreases K^+ recycling and K^+ availability to the Na^+-K^+-$2Cl^-$ cotransporter. $NaHCO_3$ transport in this segment is also inhibited by Ang II (10^{-8} *M*) via a mechanism that can be inhibited by Cyt-P450 blockade and by inhibitors of AA metabolism and by intracellular events that are not yet understood. Higher Ang II concentrations stimulate AT_1 receptors, elevating PLC and PKC activities and stimulating the Na^+-K^+-$2Cl^-$ cotransporter in the apical membrane. There are additional binding sites for Ang II in the DCT and cortical and medullary CD consistent with an action of Ang II in these segments. A significant effect of Ang in the distal segments is demonstrated by studies of

Table II Major Renal Epithelial Proteins Responsible for Transport of NaCl, Water and Urea and Regulated by Hormones and Paracrine/Autocrine Agents

Tubule segment	Apical membrane	Basolateral membrane
Proximal Tubule	Na^+-H^+ exchanger (NHE3) Na^+-glucose cotransporters (SGLT1, 2)	Na^+-K^+-ATPase Na^+-$3HCO_3^-$ cotransporter (NBC1)
Descending Limb of Henle's Loop	None	None
Thick-Ascending Limb of Henle's Loop	Na^+-K^+-$2Cl^-$-cotransporter (NKCC2) K^+ channel (ROMK, Kir 1.1) Na^+-H^+ exchanger (NHE3)	Na^+-K^+-ATPase Cl^- channel (CLC-K2)
Distal Tubule	Na^+-Cl^--cotransporter (NCC) Na^+-H^+ exchanger (NHE) Na^+ channel (ENaC) K^+ channel (ROMK, Kir 1.1) H^+-K^+-ATPase	Na^+-K^+-ATPase Cl^- channel (CLC-K1)
Collecting Duct (Principal cell)	Na^+ channel (ENaC) Aquaporin (AQP2) Urea transporter (UT-1) K^+ channel (ROMK, Kir 1.1)	Na^+-K^+-ATPase
(Intercalated cell)	H^+-ATPase Cl^--HCO_3^- exchanger Na^+-H^+ exchanger	Na^+-K^+-ATPase

knock-out mice for tissue ACE; the mice exhibit a defect in urine concentration that is concomitant with a decrease in transport proteins (UT-A, ClC-K1, NKCC2/BSC1, and AQP1).

In summary, Ang II modulates water and electrolyte transport indirectly, via the stimulation of aldosterone secretion, and directly by stimulating NaCl reabsorption in PCT, TAL, DCT, and CD. In the PCT this activation is mediated by AT_1 receptors that inhibit the cAMP/PKA pathway via $G_{\alpha i}$-proteins and stimulate IP3 and PKC formation via $G_{\alpha q/11}$-protein. In the PCT, physiological concentrations of Ang II stimulate salt and fluid reabsorption, whereas in pharmacological concentrations, high concentrations inhibit; the latter effect is mediated by AT_2 receptors. In the TAL, low concentrations of Ang II stimulate production of 20-HETE and inhibit Na^+ reabsorption by intracellular mechanisms that seem to differ for the different Na^+ transporters in the segment. At high concentrations, Ang II, elevates PLC and PKC activities and stimulates the Na^+-K^+-$2Cl^-$ cotransporter in the apical membrane. Effects in more distal segments are suspected from the existence of Ang II receptors in the cells of these segments and studies in ACE knock-out mice.

Dopamine

Dopamine [74–77] is an intrarenal hormone or paracrine agent that elicits vasodilation and natriuresis. Renal dopamine is generated mainly in the PCT (from L-dopa by the action of aromatic l-amino acid decarboxylase) and secreted across both apical and basolateral membranes, to the lumen and the blood, respectively. Renal nerves are an additional, minor source of dopamine. During salt loading, dopamine production and excretion are increased in parallel with sodium excretion.

Dopamine receptors are classified as D_1-like and D_2-like receptors. Both classes have been identified in the kidney: D_1-like receptors comprise D_1 and D_5. D_2-like receptors comprise D_2, D_3, and D_4. D_1 receptors predominate in the vasculature (arcuate arteries, interlobular arteries, afferent and efferent arterioles), which also expresses D_2-like receptors (D_3 and D_4). Both D_1-like and D_2-like receptors are present in the renal tubule cells. PTC cells express predominantly D_1, but also D_5 and also D_3. The mTAL expresses D_5. The CCD, expresses predominantly D_5 and less D_1, D_3, and D_4. The overall effect of dopamine is inhibition of Na^+ reabsorption. Activation of D_1-like receptors stimulates both the cAMP/PKA and the PLC pathways, with downstream activation of the IP3-Ca^{2+} and DAG-PKC signaling pathways. Both paths inhibit the Na^+-K^+/ATPase. D_2-like receptors inhibit cAMP/PKA production, which stimulates the luminal Na^+-H^+ exchanger and the basolateral Na^+-K^+/ATPase. Thus, D_1-like and D_2-like receptor actions on salt reabsorption oppose each other, but in some instances D_2 receptors enhance D_1 receptor effects, perhaps by shifting the effect of D_2 from inhibition of adenylcyclase to stimulation of phospholipase A_2, resulting in an increase of AA production. Nevertheless, the dominant effect of dopamine is to reduce Na^+ reabsorption. An additional inhibitory effect on the apical-membrane Na^+-H^+ antiporter in the PCT has been reported, which also contributes to the natriuretic effect [76]. In the TAL, D_1 receptors inhibit the Na^+-H^+ exchanger and the basolateral Na^+-K^+/ ATPase, with weaker stimulation of Na^+-K^+-$2Cl^-$ cotransport via the cAMP/PKA pathway. The net effect is a decrease in Na^+ transport due to the predominance of the inhibition of the Na^+ pump. Finally, D_1 receptors may also inhibit Na^+ reabsorption in the CCD, perhaps by antagonizing the stimulatory effects of aldosterone as well as inhibiting the Na^+-K^+/ATPase.

In summary, renal dopamine is generated mainly in the PCT and can exert effects on more distal segments of the nephron. During salt loading, dopamine production and excretion increase in parallel with Na^+ excretion. Activation of D_1-like receptors in the PCT results in stimulation of both cAMP/PKA signaling and PLC activity, with downstream activation of the IP_3-Ca^{2+} and DAG-PKC signaling pathways resulting in inhibition of Na^+-K^+/ATPase. D_1 receptor stimulation also inhibits the Na^+-H^+ exchanger and the Na^+-K^+/ATPase. Inhibition of the Na^+ pump may also account for the natriuretic effect of dopamine in the mTAL and the CD.

Nitric Oxide

The primary action of endogenous NO [31,63,71,78–80] on renal-tubule cells is to inhibit Na^+ reabsorption, causing natriuresis and diuresis. The systemic inhibition of NOS results in Na^+ retention if the resulting hypertension and pressure natriuresis are pharmacologically prevented. The L-arginine/NO system inhibits solute and water reabsorption in most nephron segments, including PCT, TAL, and cortical and medullary CD. Supporting the notion that NO is a regulator operating in physiological conditions, the natriuresis associated with a chronic high-salt diet is accompanied by high levels of NOS activity and increases in NO production in the kidney and excretion in the urine.

All three NOS isoforms (type I or neuronal, type II or inducible, and type III or endothelial) are expressed in tubule cells. Endothelial NOS is present in PCT, TAL, and CD. Inducible NOS is found in PCT, TAL, DCT, and cortical and inner medullary CD. Neuronal NOS is limited to the TAL, macula densa, and CD. Ang II is a potent stimulant of NO production by PCT cells. In turn, NO inhibits the stimulatory action of Ang II on NaCl reabsorption.

A direct effect of NO on PCT transport is controversial, as both stimulation and inhibition of bicarbonate and fluid transport have been reported. However, there is agreement that NO inhibits both the luminal Na^+-H^+ exchanger and the basolateral Na^+-K^+/ATPase. NO may also influence ion transport by reducing cell ATP, an effect of potential importance in hypoxic injury. Very high [NO] inhibits the Na^+-K^+/ATPase via activation of PKC. Segment-dependent differences and perhaps concentration-dependent effects, since inhibition of cGMP analogs and of NO donors require large doses of blockers, may explain some contradictory findings reported in the literature.

NO is produced by eNOS in the TAL, where it inhibits NaCl reabsorption and thereby attenuates TGF at high tubule flow rates. The inhibition of NaCl reabsorption is primarily by decreasing luminal-membrane Na^+-K^+-$2Cl^-$ cotransport, with a secondary inhibitory effect on the Na^+-H^+ exchanger. These effects result from stimulation of soluble guanylyl cyclase to form cGMP and consequent stimulation of phosphodiesterase II, reducing cAMP levels.

NOS isoforms are expressed in DCT, but there is no information on effects of NO on ion transport in this nephron segment. The CD produces NO, which inhibits NaCl reabsorption by principal cells by a mechanism independent of the Na^+-K^+/ATPase, probably involving mobilization of intracellular Ca^{2+} and inhibition of ENaC. At high concentrations, NO also impairs urine acidification by inhibiting the H^+-ATPase in intercalated cells. At low NO concentrations, a stimulatory pathway has been described that is mediated by small increases in cGMP that activate basolateral K^+ channels, hyperpolarizing the CD cell and increasing the driving force for Na^+ entry across the apical membrane. A stimulus of CD-NOS is AVP, acting via V_2 receptors. In turn, NO inhibits AVP-stimulated osmotic water permeability in the medullary CD via decreased cAMP secondary to activation of guanylyl cyclase and cGMP-dependent PKG.

In summary, the L-arginine/NO system inhibits solute and water reabsorption in most nephron segments, including PCT, TAL, and cortical and medullary CD. Ang II is a potent stimulus of NO production by PCT cells, and NO inhibits the stimulatory action of Ang II on NaCl reabsorption. A direct effect of NO on PCT transport is controversial; however, it is likely that NO inhibits both the luminal Na^+-H^+ exchanger and the basolateral Na^+/K^+/ATPase. In the TAL, NO inhibits NaCl transport and may thereby contribute to attenuation of TGF at high tubule-flow rates. The CD also produces NO, which inhibits salt transport in principal cells by a mechanism independent of the Na^+/K^+/ATPase, probably involving mobilization of intracellular Ca^{2+} and inhibition of ENaC.

Endothelin

The highest concentration of ET-1 in the body is in the renal medulla [35–37, 80–82]. CD cells secrete ET-1 across the basolateral membrane. The major stimulant is AVP, acting via V_2 receptors, but production of ET-1 is also elicited by Ang II, adrenaline, insulin, cortisol, IL-1, transforming growth factor-β, low-density lipoproteins, hypoxia, and endothelin itself. ET-1 synthesis is inhibited by atrial natriuretic peptide, NO, and prostacyclins.

The ET receptors ET_A and ET_B (both of them G-protein coupled) are found in the inner medullary collecting tubule cells where ET may exert an autocrine effect. ET provides acute negative feedback control of the AVP-induced osmotic water permeability in medullary CD. However, a chronic increase of interstitial osmolality may inhibit ET production/secretion. TAL and CD have ET_B receptors. ET inhibits NaCl transport in these nephron segments by increasing $[Ca^{2+}]_i$ and stimulating nNOS to produce NO, which in turn inhibits both Na^+/K^+/ATPase and ENaC at downstream sites, that is, the DCT and CD. ET is natriuretic by virtue of its action on ET_B receptors and NO production in the terminal CD. ET_B-receptor-deficient mice develop salt-dependent hypertension that is reversed by amiloride inhibition of ENaC channels in the distal nephron.

In summary, the main source of ET-1 in the renal medulla are the CD cells. ET production is elicited by AVP, Ang II, adrenaline, insulin, cortisol, IL-1, TGF-β, low-density lipoproteins, hypoxia, and endothelin itself. ET synthesis is inhibited by atrial natriuretic peptide, NO, and prostacyclins. ET is natriuretic by virtue of its action on ET_B receptors. TAL and CD have ET_B receptors. ET inhibits NaCl transport in these nephron segments by elevating $[Ca^{2+}]_i$ and nNOS to produce NO that inhibits both Na^+/K^+/ATPase and ENaC.

Eicosanoids

Arachidonic acid metabolites affect renal-tubule transport as well as renal hemodynamics [15,38,39,43]. Tubule segments containing cyclooxygenases (COX) are the cortical TAL, which expresses the inducible COX II, and the medullary TAL and CD, which express the constitutive COX I. Little is known about the lipoxygenase activity along the nephron. Cyt-P450 monoxygenases are present in PCT, TAL, and CD, where both EETs and 20-HETE are produced. The Cyt-P450-2 enzyme family in PCT and CD leads to preferential production of EETs in response to stimulation by Ang II, bradykinin, or ET. Increased levels of NO and CO inhibit Cyt-P450 activity and thus decrease production of EET and 20-HETE. A high-salt diet is associated with increased production of EET by PCT and CD and inhibition of NaCl transport. Agents that signal via cAMP/PKA, such as dopamine and PTH, increase 20-HETE formation in PCT. 20-HETE is the exception to the dictum that vasoconstrictors favor Na^+ retention; it inhibits NaCl reabsorption in both PCT and TAL, by PKC-mediated phosphorylation and inhibition of the α subunit of the Na^+-K^+/ATPase. Ang II stimulates 20-HETE production in the TAL by increasing $[Ca^{2+}]_i$ and stimulating PLA_2 to liberate arachidonic acid. In the TAL, 20-HETE inhibits NaCl reabsorption by acting on the luminal Na^+-K^+-$2Cl^-$ cotransporter and the basolateral Na^+-K^+/ ATPase, opposing the stimulatory action of AT_1 receptors. In addition to being stimulated by paracrine/autocrine factors, PGE_2 and 20-HETE production in TAL are triggered by activation of a luminal Ca^{2+}-sensing receptor that is coupled to $G_{\alpha q/11}$-proteins and induces PLA_2 activation and an increase in $[Ca^{2+}]_i$.

High PGE_2 in the inner medulla, produced locally by epithelial cells and interstitial cells, antagonizes vasopressin-stimulated salt reabsorption in the TAL and water reabsorption in the CD. The major tubule-cell PGE_2 receptors are EP_1 and EP_3, with highest densities found in TAL and CD. Activation of EP_1 receptors inhibits sodium reabsorption through a $G_{\alpha q/11}$-PLC system that increases $[Ca^{2+}]_i$ and stimulates PKC, inhibiting the basolateral Na^+-K^+/ATPase. EP_3 receptors stimulate a $G_{\alpha i}$-protein and reduce cAMP/PKA activity. EP_3 receptors inhibit NaCl reabsorption in the TAL by down-regulating the density of luminal-membrane Na^+-K^+-$2Cl^-$ cotransporters independently of $[Ca^{2+}]_i$, and in CD by inhibiting ENaC. Thought to be of minor functional significance during physiological conditions, FP receptors sensitive to $PGF_{2\alpha}$ predominate along the DCT. They signal through PKC and perhaps also a PKC-independent Rho-mediated pathway.

In summary, the main effects of arachidonic acid metabolites are natriuresis and water diuresis. Tubule segments distal to the PT contain COX, leading to the production of PGE_2. Cyt-P450 monoxygenases are present in PCT, TAL, and CD, where both EETs and 20-HETE are produced. In PCT and CD, preferential production of EETs occurs in response to Ang II, bradykinin, or ET. Increased levels of NO and CO inhibit Cyt-P450 activity and thus decrease production of EET and 20-HETE. In the TAL, 20-HETE inhibits NaCl reabsorption by acting on the luminal Na^+-K^+-$2Cl^-$ cotransporter and the basolateral Na^+-K^+/ATPase. High concentrations of PGE_2 in the inner medulla, produced locally by epithelial cells and interstitial cells, antagonize vasopressin-stimulated salt reabsorption in the TAL and water reabsorption in the CD. Activation of PGE_2 EP_1 receptors inhibits Na^+ reabsorption through a $G_{\alpha q/11}$-PLC mechanism that increases $[Ca^{2+}]_i$ and stimulates PKC, inhibiting the Na^+-K^+/ATPase. EP_3 receptors inhibit NaCl reabsorption in the TAL by reducing the density of luminal Na^+-K^+-$2Cl^-$ transporters independently of $[Ca^{2+}]_i$ and in CD by inhibiting ENaC.

Bradykinin

Bradykinin [70,82] of renal origin is vasodilator and natriuretic. The enzyme kallikrein is produced and released from the distal convoluted tubule and connecting segment by exocytosis at both the apical and basolateral membranes, to the lumen and interstitial fluid, respectively. Secretion at this site makes the enzyme available at the site of substrate (kininogen) production, the principal cells of the CD. Bradykinin produced by the action of kallikrein on kininogen is present in both the lumen of the CD and the interstitial fluid. Renal bradykinin formation is normally low and is increased during sodium restriction and water deprivation. Bradykinin is inactivated by kininase II, the same enzyme as ACE. The B_2 receptors are expressed in mesangial cells, juxtaglomerular granular cells, TAL, cortical and medullary CD, and renomedullary interstitial cells. In the TAL, B_2 receptor activation inhibits NaCl reabsorption. In the CD, luminal bradykinin acts on B_2 receptors and inhibits both NaCl and water transport. The mechanism of natriuresis is the inhibition of ENaC, an effect probably mediated by PGE_2, although NO may have a role by modifying the local metabolism of atrial natriuretic peptide. Although absent during physiological conditions, tubule-cell B_1 receptors are induced during inflammation, primarily in efferent arterioles, PCT, TAL, and DCT.

In summary, bradykinin of renal origin is vasodilator and natriuretic. Bradykinin is produced by the action of kallikrein on kininogen and is present in both the lumen of the CD and the interstitial fluid. Renal bradykinin formation is normally low and increases during sodium restriction and water deprivation. In the CD, luminal bradykinin acts on B_2 receptors to inhibit NaCl and water transport.

Adenosine and ATP

Adenosine and ATP [62–65,67,83–85] are produced by renal-tubule cells and can affect their transport functions by autocrine or paracrine mechanisms. Purinergic receptors (P_1 and P_2) are responsible for these effects. P_1 receptors comprise four subtypes of which A_1 and A_2 are expressed in the collecting duct and A_1 in the thick ascending loop of Henle. When these receptors are activated by adenosine, they couple to different heterotrimeric G-proteins and have effects on Na^+ and water excretion. The P_2 receptors respond predominantly to ATP and can be P_{2x} (Ca^{2+}-permeable nonselective cation channels) and P_{2y} (G-protein-coupled receptor). The A_1 receptor block produces a reduction in Na^+ reabsorption and natriuresis. In cultures of PT cells, A_1 receptor activation increases Na^+-glucose and Na^+-phosphate cotransport at the luminal membrane and also increases $Na^+HCO_3^-$ transport at the basolateral membrane. Adenosine can be formed from cAMP secreted in the proximal tubule by an ectophosphodiesterase. The effects of adenosine are thought to be mediated by inhibition of adenylyl cyclase, because the natriuresis can be inhibited by a block of G_i-proteins with Pertussis toxin. In the TAL, activation of A_1 receptors decreases the transepithelial voltage and inhibits AVP-induced $NaHCO_3$ reabsorption, perhaps by reducing cAMP/PKA activity. These two effects suggest a reduction in salt reabsorption. In the CD, A_1 receptor activation decreases the hydraulic water permeability, increasing excretion of water in the urine. Activation of A_2 receptors also decreases the water permeability of the CD.

The PCT is a rich source of ATP that is secreted to the lumen and travels downstream to regulate transport in more distal nephron segments. Luminal P_{2x} and P_{2y} receptors in PCT and IMCD inhibit Na^+ reabsorption by Ca^{2+}-, PKC-, and arachidonic acid-dependent pathways. Activation of P_{2y} receptors in DCT and CD increases PLC activity and $[Ca^{2+}]_i$, resulting in increased luminal-membrane Cl^- permeability and inhibition of Na^+ reabsorption, at least in part because of the resulting apical-membrane depolarization. As mentioned earlier, adenosine and/or ATP may play important roles in TGF and autoregulation.

In summary, adenosine and ATP are produced by renal-tubule cells and can affect their transport functions by autocrine or paracrine mechanisms. Their effects are mediated by stimulation of the purinergic receptors P_1 and P_2. Adenosine is coupled to different heterotrimeric G-proteins, altering Na^+ and water excretion. The ATP produced in the PCT is secreted to the lumen and regulates transport in more distal nephron segments. Inhibition of Na^+ transport is mediated by both P_2 receptors types.

Gap Junctions

GAP junctions [86–88] are cell-to-cell channels that communicate epithelial and vascular smooth muscle cells (see earlier section). Cx43-formed gap junctions allow permeation of hydrophilic molecules of up to 1 kDa. Functional regulation of Cx by phosphorylation is isoform dependent. In addition, GJ communication is modulated by membrane voltage, $[Ca^{2+}]_i$, and intracellular pH. GJs may coordinate oscillations of $[Ca^{2+}]_i$ and hence participate in intercellular signaling. Cells with constitutive higher sensitivity to a given stimulus may act as pacemakers, initiating a response and signaling neighboring cells by intercellular permeation of ions or second-messenger molecules.

Interstitial Cell–Tubule Communication

Renal interstitial cells occupy the space that surrounds the blood vessels and tubules [38,39,61]. In the renal cortex, the interstitial cells are stellate fibroblasts and lymphocyte-like cells. In the inner medulla, there are lipid-laden cells (located between thin descending limbs of Henle in a ladder-like appearance), cells similar to the fibroblasts and lymphocytes found in the cortex, and SMC-like pericytes that encircle descending vasa recta capillaries and act as sphincters to regulate medullary blood flow.

Cortical Interstitial Cells

Fibroblasts have the potential to differentiate to myofibroblasts containing α-smooth muscle actin and desmin, a process observed in response to inflammatory cytokines and thought to be important in the interstitial fibrosis of kidney diseases [89,90]. In response to hypoxia, cortical fibroblasts close to the outer medullary border produce erythropoietin, a glycopeptide hormone that stimulates erythropoiesis in the bone marrow, as well as the synthesis of hemoglobin. Little is known about the function of lymphocyte-like cells that reside in the cortical interstitium.

Medullary Interstitial Cells

Medullary interstitial cells [61,91–93] produce COX II metabolites, primarily PGE_2, in response to Ang II, AVP, ET, and bradykinin. These cells express an unusually high density of AT_1 and ET_A and ET_B receptors. The common signaling pathway involves $G_{\alpha q/11}$ activation of PLC and IP_3-induced mobilization of $[Ca^{2+}]_i$, which activates PLA_2. Locally produced PGE_2 causes relaxation of vasa recta pericytes and inhibits NaCl reabsorption in the TAL and medullary CD. In addition, PGE_2 can inhibit the ability of AVP to mobilize aquaporin-2 water channels to the luminal membrane and thus reduces the osmotic water permeability in the inner medullary CD.

Medullipins are lipids produced by medullary interstitial cells. An inactive form is released into the systemic circulation, markedly so after surgical relief from renal arterial stenosis. Medullipin becomes activated by a liver Cyt-P450 oxidase and acts as a vasodilating and natriuretic agent. It may also suppress sympathetic nerve activity. The physiological roles of medullipins, if any, are not understood.

In summary, cortical interstitial fibroblasts respond to hypoxia producing erythropoietin, a hormone that stimulates erythropoiesis in the bone marrow. Medullary interstitial cells produce COX II metabolites, primarily PGE_2, in response to Ang II, AVP, ET, and bradykinin. The common signaling pathway involves $G_{\alpha q/11}$ activation of PLC and IP_3-induced mobilization of $[Ca^{2+}]_i$, which activates PLA_2.

Conclusions

The kidney is a complex, highly sophisticated organ containing diverse cells types responsible for specialized functions, both in the renal vasculature and nephron segments. The structure of the kidney and the process of urine formation allow for cell-to-cell influences along single nephron segments that are distant in space. In addition, the proximity of parallel structures permits lateral communication between tubules, capillaries, and interstitial cells. These two kinds of cell-to-cell communication are central for functional integration at the single-nephron and whole-organ levels. Communication is mediated by paracrine and autocrine agents that act extracellularly or intracellularly by turning on signaling systems, thus eventually unifying homeostatic regulation of renal hemodynamics, glomerular filtration rate, tubule-transport processes, and urinary excretion. Endothelial cells regulate renal vascular resistance and reabsorption of salt and water by the renal tubules via a variety of messenger molecules, including nitric oxide, prostaglandins, EETs, 20-HETE, and endothelin. The vascular and epithelial cells regulate their own functions via the production of the autocrine agent 20-HETE.

Tubule cells produce dopamine, purine nucleotides, nitric oxide, prostanoids, and endothelin. These agents act locally to regulate the vasculature and transepithelial transport of salt and water. Despite a rapidly growing understanding of cell–cell signaling in the kidney, the wide variety of autocrine and paracrine systems and their actions on multiple cell types are not completely understood. Interactions with hormones and neural control systems add to this complexity. Current research is largely reductionist. Integrative studies are also needed, in particular, to discern the relative importance, cross-talk, redundancy, and compensatory effects of the various systems under physiological and pathophysiological conditions. A thorough understanding of these systems will be the foundation for rational and effective approaches to the treatment of renal diseases.

Acknowledgments

We thank Michael Goy for assistance in creating Fig. 1.

References

1. Kriz, W. and Kaissling, B. (2000). Structural organization of the mammalian kidney. In Seldin, D. W. and Giebisch, G., Eds., *The Kidney: Physiology and Pathophysiology*, 3rd ed., pp. 1249–1288. Raven Press, New York.
2. Giebisch, G. and Windhager, E. (2003). Organization of the urinary system. In Boron, W. F. and Boulpaep, W. F., Eds., *Medical*

Physiology: A Cellular and Molecular Approach, pp. 737–756. W. B. Saunders, Philadelphia.
3. Arendshorst, W. J. and Navar, L. G. (2001). Renal circulation and glomerular hemodynamics. In Schrir, R. W., Ed., *Diseases of the Kidney*, pp. 59–107. Lippincott Williams & Wilkins, Philadelphia.
4. Navar, L. G. (1998). Integrating multiple paracrine regulators of renal microvascular dynamics. *Am. J. Physiol.* **274**, 433–444.
5. Navar, L. G., Inscho, E. W., Majid, S. A., Imig, J. D., Harrison-Bernard, L. M., and Mitchell, K. D. (1996). Paracrine regulation of the renal microcirculation. *Physiol. Rev.* **76**, 425–536.
6. Cowley, A. W., Jr. (1997). Role of the renal medulla in volume and arterial pressure regulation. *Am. J. Physiol.* **273**, R1–R15.
7. Edwards, A., Silldforff, E. P., and Pallone, T. L. (2000). The renal medullary microcirculation. *Front. Biosci.* **5**, E36–E52.
8. Pallone, T. L., Robertson, C. R., and Jamison, R. L. (1990). Renal medullary microcirculation. *Physiol. Rev.* **70**, 885–920.
9. Hall, J. E., Brands, M. W., and Henegar, J. R. (1999). Angiotensin II and long-term arterial pressure regulation: The overriding dominance of the kidney. *J. Am. Soc. Nephrol.* **10**(Suppl. 12), S258–S265.
10. Andreoli, T. E. (1999). An overview of salt absorption by the nephron. *J. Nephrol.* **12**(Suppl. 2), S3–15.
11. Aronson, P. S. and Giebisch, G. (1997). Mechanisms of chloride transport in the proximal tubule. *Am. J. Physiol.* **273**, F179–F192.
12. Feraille, E. and Doucet, A. (2001). Sodium-potassium-adenosine-triphosphatase-dependent sodium transport in the kidney: hormonal control. *Physiol. Rev.* **81**, 345–418.
13. Knepper, M. A. and Brooks, H. L. (2001). Regulation of the sodium transporters NHE3, NKCC2 and NCC in the kidney. *Curr. Opin. Nephrol. Hypertens.* **10**, 655–659.
14. Giebisch, G. (1998). Renal potassium transport: Mechanisms and regulation. *Am. J. Physiol.* **274**, F817–F833.
15. Good, D. W., George, T., and Wang, D. H. (1999). Angiotensin II inhibits HCO-3 absorption via a cytochrome P-450-dependent pathway in MTAL. *Am. J. Physiol.* **276**, F726–F736.
16. Reeves, W. B., Winters, C. J., and Andreoli, T. E. (2001). Chloride channels in the loop of Henle. *Annu. Rev. Physiol.* **63**, 631–645.
17. Schafer, J. A. (2002). Abnormal regulation of ENaC: Syndromes of salt retention and salt wasting by the collecting duct. *Am. J. Physiol.* **283**, F221–F235.
18. Schafer, J. A., Patlak, C. S., and Andreoli, T. E. (2000). Fluid absorption and active and passive ion flows in the rabbit superficial pars recta. *J. Am. Soc. Nephrol.* **11**, 784–800.
19. Wang, W., Hebert, S. C., and Giebisch, G. (1997). Renal K^+ channels: Structure and function. *Annu. Rev. Physiol.* **59**, 413–436.
20. DiBona, G. F. and Kopp, U. C. (1997). Neural control of renal function [Review]. *Physiol. Rev.* **77**, 75–197.
21. Alpern, R. J. (2000). Renal acidification mechanisms. In Brenner, B. M., Ed., *Brenner and Rector's The Kidney*, 6th ed., pp. 455–519. W. B. Saunders, Philadelphia.
22. Suki, W. N., Lederer, E. D., and Rouse, D. (2000). Renal transport of calcium, magnesium, and phosphate. In Brenner, B. M., Ed., *Brenner and Rector's The Kidney*, 6th ed., pp. 520–574. W. B. Saunders, Philadelphia.
23. Abdel-Latif, A. A. (2001). Cross talk between cyclic nucleotides and polyphosphoinositide hydrolysis, protein kinases, and contraction in smooth muscle. *Exp. Biol. Med. (Maywood)* **226**, 153–163.
24. Horowitz, A., Menice, C. B., Laporte, R., and Morgan, K. G. (1996). Mechanisms of smooth muscle contraction. *Physiol. Rev.* **76**, 967–1003.
25. Somlyo, A. P., Wu, X., Walker, L. A., and Somlyo, A. V. (1999). Pharmacomechanical coupling: The role of calcium, G-proteins, kinases and phosphatases. *Rev. Physiol. Biochem. Pharmacol.* **134**, 201–234.
26. Gollasch, M. and Nelson, M. T. (1997). Voltage-dependent Ca^{2+} channels in arterial smooth muscle cells [Review]. *Kidney Blood Press. Res.* **20**, 355–371.
27. Lincoln, T. M., Dey, N., and Sellak, H. (2001). Invited review: cGMP-dependent protein kinase signaling mechanisms in smooth muscle: From the regulation of tone to gene expression. *J. Appl. Physiol.* **91**, 1421–1430.
28. Ballermann, B. J. (1998). Endothelial cell activation. *Kidney Int.* **53**, 1810–1826.
29. Campbell, W. B. and Gauthier, K. M. (2002). What is new in endothelium-derived hyperpolarizing factors? *Curr. Opin. Nephrol. Hypertens.* **11**, 177–183.
30. Griendling, K. K. and Alexander, R. W. (1996). Endothelial control of the cardiovascular system: Recent advances. *FASEB J.* **10**, 283–292.
31. Gabbai, F. B. and Blantz, R. C. (1999). Role of nitric oxide in renal hemodynamics. *Semin. Nephrol.* **19**, 242–250.
32. Moncada, S. (1997). Nitric oxide in the vasculature: physiology and pathophysiology. *Ann. NY Acad. Sci.* **811**, 60–67.
33. Raij, L. and Baylis, C. (1995). Glomerular actions of nitric oxide [Editorial]. *Kidney Int.* **48**, 20–32.
34. Fleming, I. and Busse, R. (1999). Signal transduction of eNOS activation. *Cardiovasc. Res.* **43**, 532–541.
35. Dussaule, J. C., Boffa, J. J., Tharaux, P. L., Fakhouri, F., Ardaillou, R., and Chatziantoniou, C. (2000). Endothelin, renal diseases, and hypertension. *Adv. Nephrol. Necker Hosp.* **30**, 281–303.
36. Pollock, D. M. (2000). Renal endothelin in hypertension. *Curr. Opin. Nephrol. Hypertens.* **9**, 157–164.
37. Kedzierski, R. M. and Yanagisawa, M. (2001). Endothelin system: The double-edged sword in health and disease. *Annu. Rev. Pharmacol. Toxicol.* **41**, 851–876.
38. Breyer, R. M., Bagdassarian, C. K., Myers, S. A., and Breyer, M. D. (2001). Prostanoid receptors: Subtypes and signaling. *Annu. Rev. Pharmacol. Toxicol.* **41**, 661–690.
39. Breyer, M. D. and Breyer, R. M. (2001). G protein-coupled prostanoid receptors and the kidney. *Annu. Rev. Physiol.* **63**, 579–605.
40. Imig, J. D. (2000). Eicosanoid regulation of the renal vasculature. *Am. J. Physiol.* **279**, F965–F981.
41. Roman, R. J. (2002). P-450 metabolites of arachidonic acid in the control of cardiovascular function. *Physiol. Rev.* **82**, 131–185.
42. Capdevila, J. H., Harris, R. C., and Falck, J. R. (2002). Microsomal cytochrome P450 and eicosanoid metabolism. *Cell Mol. Life Sci.* **59**, 780–789.
43. McGiff, J. C. and Quilley, J. (1999). 20-HETE and the kidney: Resolution of old problems and new beginnings. *Am. J. Phyisol.* **277**, R607–R623.
44. Rahman, M., Wright, J. T., Jr., and Douglas, J. G. (1997). The role of the cytochrome P450-dependent metabolites of arachidonic acid in blood pressure regulation and renal function: A review. *Am. J. Hypertens.* **10**, 356–365.
45. Pratt, P. F., Li, P., Hillard, C. J., Kurian, J., and Campbell, W. B. (2001). Endothelium-independent, ouabain-sensitive relaxation of bovine coronary arteries by EETs. *Am. J. Physiol.* **280**, H1113–H1121.
46. Quilley, J. and McGiff, J. C. (2000). Is EDHF an epoxyeicosatrienoic acid? *Trends Pharmacol. Sci.* **21**, 121–124.
47. Zhang, F., Kaide, J. I., Rodriguez-Mulero, F., Abraham, N. G., and Nasjletti, A. (2001). Vasoregulatory function of the heme-heme oxygenase-carbon monoxide system. *Am. J. Hypertens.* **14**, 62S–67S.
48. Zou, A. P., Billington, H., Su, N., and Cowley, A. W., Jr. (2000). Expression and actions of heme oxygenase in the renal medulla of rats. *Hypertension* **35**, 342–347.
49. Welsh, D. G. and Nelson, M. T. (2000). A case for myoendothelial gap junctions. *Circ. Res.* **87**, 427–428.
50. Lucas, K. A., Pitari, G. M., Kazerounian, S., Ruiz-Stewart, I., Park, J., Schulz, S., Chepenik, K. P., and Waldman, S. A. (2000). Guanylyl cyclases and signaling by cyclic GMP. *Pharmacol. Rev.* **52**, 375–414.
51. Purdy, K. E. and Arendshorst, W. J. (2001). Iloprost inhibits inositol-1,4,5-trisphosphate-mediated calcium mobilization stimulated by angiotensin II in cultured preglomerular vascular smooth muscle cells. *J. Am. Soc. Nephrol.* **12**, 19–28.
52. Walsh, M. P., Kargacin, G. J., Kendrick-Jones, J., and Lincoln, T. M. (1995). Intracellular mechanisms involved in the regulation of vascular smooth muscle tone. *Can. J. Physiol. Pharmacol.* **73**, 565–573.
53. Schnermann, J. (1998). Juxtaglomerular cell complex in the regulation of renal salt excretion. *Am. J. Phyisol.* **274**, R263–R279.

54. Schnermann, J. (1999). Micropuncture analysis of tubuloglomerular feedback regulation in transgenic mice. *J. Am. Soc. Nephrol.* **10**, 2614–2619.
55. Thomson, S. C. and Blantz, R. C. (2000). Ions and signal transduction in the macula densa. *J. Clin. Invest.* **106**, 633–635.
56. Kurtz, A. and Wagner, C. (1999). Cellular control of renin secretion. *J. Exp. Biol.* **202**(Part 3), 219–225.
57. Coffman, T. M. (1998). Gene targeting in physiological investigations: studies of the renin-angiotensin system. *Am. J. Physiol.* **274**, F999–F1005.
58. Navar, L. G., Imig, J. D., Zou, L., and Wang, C. T. (1997). Intrarenal production of angiotensin II. *Semin. Nephrol.* **17**, 412–422.
59. Schnermann, J. and Briggs, J. P. (2000). Function of the juxtaglomerular apparatus: Control of glomerular hemodynamics and renin secretion. In Seldin, D. W. and Giebisch, Eds., *The Kidney: Physiology and Pathophysiology*, 3rd ed., pp. 945–980. Raven Press, New York.
60. Davis, M. J. and Hill, M. A. (1999). Signaling mechanisms underlying the vascular myogenic response. *Physiol. Rev.* **79**, 387–423.
61. Harris, R. C. and Breyer, M. D. (2001). Physiological regulation of cyclooxygenase-2 in the kidney. *Am. J. Physiol.* **281**, F1–F11.
62. Bodin, P. and Burnstock, G. (2001). Purinergic signaling: ATP release. *Neurochem. Res.* **26**, 959–969.
63. Welch, W. J., Wilcox, C. S., and Thomson, S. C. (1999). Nitric oxide and tubuloglomerular feedback. *Semin. Nephrol.* **19**, 251–262.
64. Bailey, M. A., Imbert-Teboul, M., Turner, C., Marsy, S., Srai, K., Burnstock, G., and Unwin, R. J. (2000). Axial distribution and characterization of basolateral P2Y receptors along the rat renal tubule. *Kidney Int.* **58**, 1893–1901.
65. Chan, C. M., Unwin, R. J., Bardini, M., Oglesby, I. B., Ford, A. P., Townsend-Nicholson, A., and Burnstock, G. (1998). Localization of P2X1 purinoceptors by autoradiography and immunohistochemistry in rat kidneys. *Am. J. Physiol.* **274**, F799–F804.
66. Griendling, K. K., Ushio-Fukai, M., Lassegue, B., and Alexander, R. W. (1997). Angiotensin II signaling in vascular smooth muscle. New concepts. *Hypertension* **29**, 366–373.
67. Schwiebert, E. M. (2001). ATP release mechanisms, ATP receptors and purinergic signaling along the nephron. *Clin. Exp. Pharmacol. Physiol.* **28**, 340–350.
68. Arendshorst, W. J., Brannstrom, K., and Ruan, X. (1999). Actions of angiotensin II on the renal microvasculature. *J. Am. Soc. Nephrol.* **10**, S149–S161.
69. Guo, D. F., Sun, Y. L., Hamet, P., and Inagami, T. (2001). The angiotensin II type 1 receptor and receptor-associated proteins. *Cell Res.* **11**, 165–180.
70. Erdos, E. G. (2002). Kinins, the long march—a personal view. *Cardiovasc. Res.* **54**, 485–491.
71. Ortiz, P. A. and Garvin, J. L. (2002). Role of nitric oxide in the regulation of nephron transport. *Am. J. Physiol.* **282**, F777–F784.
72. Laghmani, K., Preisig, P. A., and Alpern, R. J. (2002). The role of endothelin in proximal tubule proton secretion and the adaptation to a chronic metabolic acidosis. *J. Nephrol.* **15**(Suppl. 5), S75–S87.
73. Harris, P. J., Hiranyachattada, S., Antoine, A. M., Walker, L., Reilly, A. M., and Eitle, E. (1996). Regulation of renal tubular sodium transport by angiotensin II and atrial natriuretic factor. *Clin. Exp. Pharmacol. Physiol.* Suppl. 3, S112–S118.
74. Carey, R. M. (2001). Theodore Cooper Lecture: Renal dopamine system: Paracrine regulator of sodium homeostasis and blood pressure. *Hypertension* **38**, 297–302.
75. Jose, P. A., Raymond, J. R., Bates, M. D., Aperia, A., Felder, R. A., and Carey, R. M. (1992). The renal dopamine receptors. *J. Am. Soc. Nephrol.* **2**, 1265–1278.
76. Jose, P. A., Eisner, G. M., and Felder R. A. (2000). Paracrine regulation of renal function by dopamine. In Seldin, D. W. and Giebisch, Eds., *The Kidney: Physiology and Pathophysiology*, 3rd ed., pp. 915–930. Raven Press, New York.
77. Aperia, A. C. (2000). Intrarenal dopamine: a key signal in the interactive regulation of sodium metabolism. *Annu. Rev. Physiol.* **62**, 621–647.
78. Wilcox, C. S. (2000). L-arginine-nitric oxide pathway. In Seldin, D. W. and Giebisch, Eds., *The Kidney: Physiology and Pathophysiology*, 3rd ed., pp. 849–871. Raven Press, New York.
79. Bachmann, S. and Mundel, P. (1994). Nitric oxide in the kidney: Synthesis, localization, and function. *Am. J. Kidney Dis.* **24**, 112–129.
80. Plato, C. F., Pollock, D. M., and Garvin, J. L. (2000). Endothelin inhibits thick ascending limb chloride flux via ET(B) receptor-mediated NO release. *Am. J. Physiol.* **279**, F326–F333.
81. Simonson, M. S. and Dunn, M. J. (1993). Endothelin peptides and the kidney. *Annu. Rev. Physiol.* **55**, 249–265.
82. Kon, V. and Hunley, T. E. (2000). Kinin and endothelin. In Seldin, D. W. and Giebisch, Eds., *The Kidney: Physiology and Pathophysiology*, 3rd ed., pp. 889–903. Raven Press, New York.
83. Schwiebert, E. M. and Bellamkonda K. K. (2001). Extracellular nucleotide signaling along the renal epithelium. *Am. J. Physiol.* **280**, F945–F963.
84. Jackson, E. K. and Dubey, R. K. (2001). Role of extracellular cAMP-adenosine pathway in renal physiology. *Am. J. Physiol.* **281**, F597–F612.
85. North, R. A. (2002). Molecular Physiology of P2X receptors. *Physiol. Rev.* **84**, 1013–1067.
86. Harris, A. L. (2001). Emerging issues of connexin channels: Biophysics fills the gap. *Quart. Revs. Biophys.* **34**, 325–472.
87. Seemes, E. (2000). Components of astrocytic intercellular calcium signaling. *Mol. Neurobiol.* **22**, 167–179.
88. Nicholson, S. M. and Bruzzone, R. (1997). Gap junctions: Getting the message through. *Curr. Biol.* **7**, 340–344.
89. Bieber, E. (2001). Erythropoietin, the biology of erythropoiesis and epoietin alfa. An overview. *J. Reprod. Med.* **46**, 521–530.
90. Donnelly, S. (2001). Why is erythropoietin made in the kidney? The kidney functions as a critmeter. *Am. J. Kidney Dis.* **38**, 415–425.
91. Muirhead, E. E. (1991). The medullipin system of blood pressure control. *Am. J. Hypertens.* **4**, 556S–568S.
92. Cowley, A. W., Jr. (1994). Franz Volhard Lecture. Evolution of the medullipin concept of blood pressure control: A tribute to Eric Muirhead. *J. Hypertens.* Suppl. 12, S25–S34.
93. Agre, P., Nielsen, S., and Knepper, M. A. B. (2000). Aquaporin water channels in mammalian kidney. In Seldin, D. W. and Giebisch, Eds., *The Kidney: Physiology and Pathophysiology*, 3rd ed., pp. 363–377. Raven Press, New York.

CHAPTER 347

Prostate

Jean Closset[1] and Eric Reiter[2]
[1]*Biochimie, Faculté de Médecine, Institut de Pathologie, Université de Liège, Liège, Belgium*
[2]*Station de Physiologie de la Reproduction et des Comportements, INRA/CNRS/Université de Tours, Nouzilly, France*

Introduction

The prostate is an exocrine gland with a heterogeneous epithelial and stromal cell composition. Located below the bladder, it surrounds the urethra. Its development, growth, and function are controlled by hormones. Androgens play a major role in this regulation. In subjects who have congenital 5α-reductase deficiency and who no longer synthesize DHT (dihydrotestosterone), the prostate no longer develops [1]. Androgen depletion induces massive apoptosis of prostatic cells. This cell death affects the epithelial cells more than it does the cells of the fibromuscular stroma [2]. Administration of testosterone to castrated animals causes an increase in the mitotic activity of the prostatic epithelial cells and restores the function of the gland. Castration early in life prevents the development of prostatic cancer and hyperplasia during ageing.

The work of Cunha and coworkers [3] showed that androgens act indirectly via the mesenchyme to ensure the growth and differentiation of the prostate. Paracrine-factor-mediated interactions between the epithelium and stroma are thus of capital importance. Not all relevant paracrine factors have been identified. The known ones include nonandrogenic steroids, peptide hormones, neuropeptides, neurotransmitters, and growth factors. All of them act via their own specific intracellular signaling pathways, but these pathways are tightly interconnected so as to elicit an integrated response ensuring coherent development and functioning of the gland.

It is noteworthy that the nature of androgen-mediated effects changes through the major periods of life: (1) During fetal life, the prostate develops first in a context where androgen levels are low but sufficient to ensure the gland's morphogenesis. (2) During adult life, prostatic growth is slight, although androgen levels are high. Their role is then to ensure the function and maintain the integrity of the prostate. (3) During aging, the prostate continues to grow despite decreasing androgen levels in the bloodstream [4].

Furthermore, androgens alone are not sufficient to control the development and function of the prostate. Pituitary hormones (prolactin, somatotropin, luteinizing hormone) and certain growth factors such as IGF-I act directly via their specific receptors present in the prostate [5,6]. These different factors induce the synthesis of proteins that are important to prostate function. Through their systemic action, these factors might be involved in resistance to androgen-deprivation-induced apoptosis. One-third of all androgen-dependent prostate tumours overexpress the androgen receptor; another third display a receptor that is neither mutated nor amplified; the remaining third have lost the receptor or express a mutated or nonfunctional receptor [7]. These differences in androgen receptivity highlight the diversity and complexity of the mechanisms involved in the uncontrolled growth of this type of cancer. The fact that these tumors become resistant to androgen-deprivation-induced apoptosis shows that within the tumor cells, survival factors also contribute to tumor progression.

If we are to understand the appearance of the various pathologies that affect the prostate in aging men, knowledge is needed about the various signals acting on the prostate and on the signaling pathways controlling the gland's function throughout the different stages of life. These signals and the intracellular mechanisms underlying their action constitute potential new targets for the development of effective prostate cancer therapies. Prostate cancer has become the number two cause of cancer deaths in Western countries, and its impact on health care, combined with that of benign prostatic hypertrophy [BPH], is considerable [8]. This chapter provides an integrated picture of the signals and signaling pathways that control the development of the prostate during fetal and adult life and during aging.

Development of the Prostate During Fetal Life

The prostate is formed by outgrowth of the urogenital sinus epithelium (of endodermal origin) into the urogenital sinus mesenchyme (of mesodermal origin). Its morphogenesis has been remarkably described by Cunha *et al.* [3].

Rodent and human prostates are organized differently. The same applies to the branching morphogenesis of their epithelial ducts during fetal and neonatal development. The rodent prostate is organized in lobes that differ from each other both morphologically and biochemically. The lobes are named according to their anatomical position in the organ: anterior, dorsal, dorsolateral, and ventral [9]. The human prostate does not show this discrete structure. McNeal [10] has subdivided it into three zones: a central zone, a peripheral zone, and a transition zone. At least three cell types can be distinguished in the epithelial compartment of the prostate: luminal, basal, and neuroendocrine cells. The stromal compartment contains fibroblasts, muscle cells, endothelial cells, nerve cells, and blood cells.

Androgens

Androgens are required to initiate prostate development and to ensure the gland's fetal and neonatal growth [11]. The fetal period is the life stage at which prostate development is most intense. Growth and development follow the differentiation of the gonads and the establishment of the pituitary-gonadal regulatory axis. In man, the androgen receptor and 5α reductase (responsible for production of DHT) are present in the mesenchyme from the eighth week postconception. The particular distribution of the androgen receptor may be related to the branching differences observed in the various lobes of the rodent prostate during morphogenesis [12]. Androgen receptors are also found in the epithelium, but their presence there does not seem to be required for development of the gland. The work of Cunha *et al.* [13] has demonstrated that the mesenchyme expresses all of the factors required for prostate morphogenesis. Androgens act via their receptors in the mesenchyme to ensure the growth of the epithelium and differentiation of the stroma. Yet to date, there has been no confirmation of the existence of androgen-dependent mesenchyme-derived factors that alone ensure the fetal development of the prostate.

Data are incomplete and contradictory regarding the nature and mode of action of the paracrine factors mediating the action of androgens on the development of the fetal prostate. According to the types of effects they exert, these factors can be classified either as mitogenic factors, such as fibroblast growth factors (FGFs), epidermal growth factors (EGFs), and insulin-like growth factors (IGFs), or as morphogenic factors such as transforming growth factor β (TGF-β), hepatocyte growth factor (HGF), and protein PS20 [14], capable of inhibiting proliferation in the fetal prostate.

The Main Mitogenic Factors

Members of the FGF family such as FGF-7 (also called KGF for keratinocyte growth factor) and FGF-10 play important roles in prostate development during the fetal period. In the fetal prostate, expression of these two factors does not appear to be regulated by androgens.

FGF-7 is produced by fibroblasts and mesenchymal cells, but these cells do not express the receptor for this growth factor (FGF-R2iii b). Conversely, epithelial cells possess the FGF-7 receptor but do not synthesize FGF-7. FGF-7 exerts mitogenic action on the epithelium *in vitro*. FGF-7 transcripts are abundant during the development of the rat ventral prostate, and large amounts of FGF-7 are produced during periods of active growth [15]. In the rat, furthermore, FGF-7 can replace testosterone for epithelial ductal branching during morphogenesis of the ventral lobe [16]. On the other hand, FGF-7 knock-out mice show unperturbed development of the urogenital tract [17].

FGF-10 is also an important paracrine regulator of prostate development. It appears to act in the early stages of the organ's development [18]. By *in situ* hybridization, FGF-10 transcripts have been observed in the mesenchymal cells of buds undergoing branching morphogenesis of the epithelial ducts. In addition, the FGF-10 protein stimulates the development of ventral prostate explants cultured in serum-free medium. FGF-10 stimulates proliferation of the epithelium but not that of the stroma [18].

Although FGF-7 and FGF-10 show high sequence identity, they are not functionally interchangeable. Deletion of the FGF-10 gene in knock-out mice leads to the loss of several organs, including the prostate [19,20]. FGF-10 thus appears as a mesenchymal factor required for organogenesis of the prostate [21]. It has been hypothesized that FGF-10 might take over the action of FGF-7 in FGF-7 knock-out mice, thus compensating for the deletion.

IGF-I and IGF-II and their receptors have been implicated in fetal development [22]; both IGF-I and -II transcripts are present in fetal tissues [23]. We have very little information about IGF expression, IGF binding proteins, and IGF receptors in the developing prostate. Data on systemic and locally produced IGF-I are equally scant. In the adult, IGF-I and -II are present mainly in the fibromuscular stroma, whereas the type I receptor, which binds IGF-I and -II with different affinities, is present in both compartments. The type I receptor is a receptor tyrosine kinase. Arguments in favor of the involvement of IGFs in fetal development come from xenotransplant experiments [24–26]. When urogenital sinus tissue is transplanted from mice with a knocked-out IGF-I or type I receptor gene into athymic mice, the transplants show delayed growth when compared to grafts of normal fetal tissue. This shows that the IGF-I produced by the host cannot totally compensate for the lack of locally produced IGF-I with regards to the latter's effects on the fetal prostate. IGF-I is thus important for growth of the fetal epithelium, but the latter is able to differentiate in the factor's absence.

Although the epithelium of the fetal prostate is modulated by mesenchymal factors, epithelial signals are important for differentiation and proliferation of the prostatic epithelium. The product of the Sonic hedgehog (Shh) gene is a glycoprotein produced by the epithelium of the urogenital sinus. Its overexpression in cells coincides with an increased testosterone level. The product of the Shh gene is involved in formation of the urethra and epithelial ducts [27].

Also known is the importance of the nervous system in the development of nerves and smooth muscle in fetuses with meningomyelocele. Under these conditions, there is no smooth muscle and the posterior prostate is underdeveloped. Likewise, prostate size is affected by the absence of innervations. These observations are in keeping with some very interesting observations of Kyprianou *et al.* [28] who demonstrated the role of α_1 receptor agonists in the growth of the adult prostate. Data are scant on the secretions of the neuroendocrine cells of the prostate during fetal life. What we do know is that such cells are present in the fetal prostate from the 13th week of gestation, and that they might be associated with the differentiation of the epithelium [29].

The Main Inhibitors of Proliferation

TGF-β and its isoforms (1 to 3) inhibit the proliferation of epithelial cells, alter cell differentiation, and stimulate fibroblast growth. All of the TGF-β isoforms are down-regulated by androgens. TGF-β1 has recently been shown to be capable of altering the nuclear-to-cytoplasmic distribution of androgen receptor in prostatic smooth muscle cells *in vitro* [30]. Remarkably, the effects of TGF-β on the prostate are similar to those observed after neonatal exposure of the gland to estrogens. Expression of the three TGF-β isoforms in the rodent prostate is time and cell type dependent [31,32]. Chang *et al.* [33] have shown that TGF-β1 and the RII receptor are expressed in the smooth muscle cells surrounding the epithelial ducts. TGF-β2 and TGF-β3 are present mainly in differentiating epithelial cells. The RII receptor is expressed weakly in the epithelium but more strongly in the smooth fibromuscular cells; this may reflect the fact that TGF-β-mediated effects in the prostate occur preferentially at the level of muscle cells and that TGF-β has been shown to be capable of inducing smooth muscle markers in stromal cells [34].

Activin and follistatin, which belong to the TGF-β family, have also been implicated in the branching morphogenesis of the fetal prostate. Activin A (a dimer of subunit β) and its binding protein, follistatin, have opposite effects *in vitro* on branching of the prostatic ducts. A down-regulating role of activin A on morphogenesis *in vivo* is imaginable, since activin A has been found in the epithelium and in mesenchymal aggregates at the ductal tips [35].

PS-20 has many characteristics of smooth muscle-derived inhibitor of epithelial cell proliferation. The protein was first isolated from a fetal urogenital sinus mesenchymal cell line (U4-F) and possesses growth inhibitory properties on bladder carcinoma and PC-3 cell proliferation [36]. PS-20 is localized in the smooth muscle compartment of the prostate and its sequence analysis has revealed that the protein is a member of a WAP-type serine proteinase inhibitor [14]. Details of the molecular mechanisms by which PS-20 may exert its effects are currently unknown.

Data on the involvement of EGF and of its homolog TGF-α on prostatic morphogenesis come from experiments conducted on transgenic mice with a knocked-out phospholipase Cγ gene. (Phospholipase Cγ is involved in the pathway used by EGF to regulate cell motility [37].) These mice show markedly delayed development of the prostate. Their prostatic cells are less differentiated and the ducts much less numerous. These results, like those of Xie *et al.* [38] concerning the mammary gland, show that early prostate development depends on intracellular signaling pathways involving PLCγ, known less for its role in proliferation than as a mediator of effects on cell motility.

It has been hypothesized that during aging some latent embryonic potential is reactivated, and that this is responsible for the appearance of cancer of the prostate and benign prostatic hyperplasia in elderly men. In the fetal prostate, signaling by the SSH gene product is required for expression of the gene encoding the NKX3.1 transcription factor. When this gene is deleted in mice, increased proliferation of the prostatic epithelium is observed [39]. NKX3.1 might also play a role in BPH, since Bhatia-Gaur *et al.* [40] have shown that deletion of the NKX3 gene causes epithelial hyperplasia in elderly males. Conditional loss of the NKX3.1 gene in adult mice induces prostatic intraepithelial neoplasia [41]. The gene might thus be a tumor suppressor gene.

Development of the prostate during fetal life also requires the expression of genes involved in controlling angiogenesis. Vascular endothelial growth factor (VEGF) is expressed in the human prostate from the ninth week of life. VEGF expression by cultured fibroblasts is increased by androgens [42], showing that androgen-dependent genes are important in angiogenesis during the early development of the prostate. The extracellular matrix plays an interesting role in angiogenesis. In the matrix, plasminogen and its activators are activated, and this leads to production of powerful inhibitors of angiogenesis. Elfman *et al.* [43] have recently shown that the amino-terminal peptide of the urokinase plasminogen activator inhibits branching morphogenesis of the ventral prostate in mice. This activity is related to that of angiostatin, an active plasminogen fragment obtained by limited proteolytic cleavage of the molecule. Interestingly, angiostatin-like peptides have been produced through the action of PSA, the prostate-specific antigen, a kallikrein-like enzyme [44].

The mesenchyme also produces other inhibitors of epithelial proliferation during prostate development. Bone morphogenetic protein 4 (BMP-4) is expressed from the 14th day of gestation. As the fetus develops, this expression gradually becomes limited to the mesenchyme. *In vitro*, BMP-4 inhibits epithelial proliferation and duct branching in a dose-dependent manner [45].

Last, fucosyl transferase-1 has also been identified as a modulator of epithelial proliferation during fetal development [46] and of branching morphogenesis *in vitro*. Fucosyl transferase is an enzyme present at the surfaces of epithelial cells. Its involvement demonstrates the importance of cell–cell interactions in the fetal development of the prostate.

Estrogens

The developing prostate is also sensitive to the action of estrogens and endocrine disruptors. Two estrogen receptors, α and β, are present in the prostates of rodents and humans alike. In rodents the α receptor is located exclusively on fibroblasts and smooth muscle cells. Expression of the β receptor is low during fetal life, but it increases after birth as epithelial cells differentiate into lumino-epithelial cells. Furthermore, an increasing gradient of β receptor concentration has been observed from the proximal part of the ducts to the distal part. It has been hypothesized that the β receptor might play an important role in epithelial differentiation in the course of prostate development [47,48].

During morphogenesis of the rat or mouse prostate, large doses of endogenous or exogenous estrogens induce permanent changes in prostatic growth. In the mouse, these estrogen-induced effects are dose dependent. High estrogen doses reduce the size of the adult prostate, whereas low doses increase it. These low doses of estrogen also favor a larger number of epithelial ducts and an increased level of androgen receptor [49]. When a rat is treated during the neonatal period with high doses of estrogen, it shows in adult life a highly abnormal ventral prostate, with numerous hyperplasias and dysplasias, despite the delayed prostatic growth observed after treatment [50]. The prostatic lesions observed may result from aberrant cell–cell interactions and defective cell–cell communications [51] and from epithelial differentiation defects involving reduced expression of the β receptor. Chang and coworkers [52] have recently shown that estrogen exposure during the neonatal period in the rat perturbs the TGF-β signaling pathways and thereby blocks $p21^{cip1/WAF1}$-mediated epithelial differentiation and stops proliferation of the fibroblasts surrounding the prostatic ducts.

The Adult Prostate

In the adult prostate, the daily renewal of 1–2% of the cells is compensated by cell death due to apoptosis. This cell homeostasis results from a balance between the proliferation-enhancing effects of certain signals (growth factors, neuropeptides, etc.) on the one hand, and proliferation-inhibiting and apoptosis-inducing effects on the other. Androgens modulate the pathways that stimulate proliferation and they antagonize apoptosis. These proliferative signals act univocally or interfere with other signaling pathways.

Factors Involved in Controlling Proliferation

The main mitogenic signals exerting their effects on the adult prostate are EGF/TGF-α and basic FGF (bFGF). These two growth factors account for 80% of all proliferation-stimulating effects. Several FGF isoforms are also expressed during adult life by the prostatic epithelium, but their role in proliferation and differentiation seems less important than that of bFGF. Epithelial and stromal cells are not the only cells to produce bFGF—so do macrophages and endothelial cells [53]. The mesenchyme is the main target of bFGF *in vivo*. No action of bFGF on the epithelium had been demonstrated *in vivo*, but bFGF does stimulate the proliferation of epithelial cells in culture [54]. The prostate contains high levels of EGF. EGF, heparin-binding EGF, and TGF-α are all secreted by prostatic stromal cells on which they exert their main mitogenic effects. EGF is also secreted by human epithelial cancer cell lines and perhaps by normal epithelial cells, at low levels. The selective response of adult prostatic epithelial cells to the paracrine action of TGF-α rather than to the autocrine action of EGF had led to the suggestion that the EGF receptor might be located on the basolateral surface of the epithelial cells. Tight junctions would prevent EGF from reaching the receptors [37]. The mitogenic effects are mediated by transactivation of the EGF receptor, requiring cleavage of the ligand by extracellular matrix metalloproteinases [55].

In the adult, most of the mitogenic signals occurring in the prostate act through receptor tyrosine kinases and the corresponding transduction pathways. Other signals, released by neuroendocrine cells (such as bombesin, somatostatin, bradykinin) or by the endothelium (such as endothelin-1) exert their proliferation-enhancing activity via G-protein-coupled receptors (GPCRs) and the MAP kinase pathway (and, in particular, via ERK, the extracellular signal regulated kinase). This pathway is an intracellular point of convergence of mitotic signals acting via GPCRs and signals acting via receptor tyrosine kinases. In this context, Putz *et al.* [56] have shown that blocking of the EGFR inhibits not only the action of EGF on prostatic cells, but also that of IGF-I and PKA activators.

The androgen receptor whose expression is constant in the adult prostate undergoes ligand-independent transactivation via phosphorylations induced by receptor tyrosine kinases or serine/threonine kinases and via protein kinases A and C [57]. Culig *et al.* [58] have shown that IGF-I, FGF-7, and EGF activate transcription mediated by androgens and their receptors and that the antiandrogen bicalutamide inhibits this effect. The CREB protein, involved in transducing messages transiting via cyclic AMP and PKA, also intervenes in the activation of transcription by the androgen receptor [59,60].

IGF family factors exert major effects on the regulation of growth in general. IGF-I and IGF-II are synthesized mainly by the liver in response to growth hormone stimulation. They stimulate cell proliferation and differentiation and inhibit apoptosis. Recent interest in the IGF family in relation to the prostate stems from *in vivo* observations. Transgenic mice overexpressing the bovine growth hormone display oversized prostates. The effect is attributed to increased systemic levels of IGF-I [61]. Transgenic mice with knocked-out IGF-I and type I receptor genes have underdeveloped prostates [62]. When rats are treated systemically with IGF-I, the treatment stimulates prostatic growth [6]. Young untreated acromegaly patients have an overdeveloped prostate compared to healthy patients [63].

IGFs are also produced locally by many tissues. In the prostate, they have an autocrine/paracrine action via the type I receptor tyrosine kinase. The type II receptor is not involved in intracellular signaling. Its role would appear to be related to the uptake and degradation of IGF-II. IGF availability in prostatic cells is modulated by carrier proteins (called IGF binding proteins or IGF-BPs), at least seven of which have been well characterized [64]. IGF-BPs affect IGF availability in a cell-type-specific manner. IGF-BP3 binds more than 90% of the circulating IGFs. At prostate level, it can be cleaved *in situ* by prostate-specific antigen (PSA), whose proteolytic activity is kallikrein-like. The effect of this proteolytic action is to make IGF-I available to its type I receptor and to favor IGF-mediated mitogenic effects [65]. Moreover, TGF-β1 is believed to exert its antiproliferative action by stimulating IGF-BP3.

IGF-BP5 is bound to the extracellular matrix. It potentiates the action of IGFs on DNA synthesis. Gregory *et al.* [66] have shown in the xenograft CWR22 model that androgens can regulate the expression of IGF-BP5 but not that of other IGF-BPs. High levels of IGF-BP5 would thus enable prostatic cells to recruit more IGF-I, and this would favor cell proliferation in the presence of androgens. On the other hand, production of IGF-I and its receptor is regulated by androgens: Nickerson *et al.* [67] have shown that they are both overexpressed during androgen-deprivation-induced apoptosis.

In transgenic mice overexpressing bovine growth hormone (GH), the observed increase in prostate weight might also result from the direct action of the hormone on the prostate. Bovine GH has no prolactin activity in rodents and a specific growth hormone receptor is present in the prostate. This receptor mediates effects that are at least partially androgen independent [5]. Growth hormone receptors are also expressed more strongly by LNCAP and PC-3 cells than by normal prostatic cells. This implies that growth hormone might be involved in the development of prostate cancer. This hypothesis is supported by the work of Jungwirth *et al.* [68], who demonstrated *in vivo* a major inhibitory effect of a somatostatin analog on xenografted prostatic tumor cell lines.

In this context, it is well established that prolactin can also affect the proliferation and differentiation of prostatic cells [69] via activation of specific receptors [70]. The prostates of transgenic mice overexpressing prolactin weigh 20 times as much as those of normal animals [61]. These effects are attributed in part to the action of prolactin on testicular steroidogenesis and to its major role in Leydig cell proliferation. Because the prolactin receptor is also present in the prostate, the hormone may also act directly to stimulate transcription of certain genes via activation of Stat5 [71]. Some of these effects may be androgen independent [72]. Prolactin is also produced by the prostate [73] and can thus exert paracrine/autocrine effects within the organ. Untergasser *et al.* [74] have shown that this local network of interactions is not involved in prostatic pathologies. The relative importance of prolactin's endocrine and paracrine/autocrine effects could doubtless best be evaluated in transgenic models. Prolactin and prolactin receptor knock-out mice have been produced [75,76] but the effects of these deletions on the prostate have not yet been assessed.

Of the gonadotropins playing a major role in reproduction, luteinizing hormone (LH) and follicle-stimulating hormone (FSH), only LH has receptors in the prostate [77]. These G-protein-coupled receptors acting via cyclic AMP are functional, since binding of the hormone causes the accumulation of certain mRNAs [78]. Because these effects have been observed on prostatic explants cultured in androgen-free medium, LH might act on the prostate in an androgen-independent manner. In addition, Tao *et al.* [78a] have shown that LH/hCG is expressed in BPH. The study of LH signaling in the prostate thus opens interesting prospects for further investigation.

Factors Involved in Apoptosis

In the adult prostate, androgen deprivation leads to apoptosis becoming a major, macroscopic process. In this phenomenon, the cells concerned no longer proliferate, but remain at stage G_0 or engage in a defective cell cycle. In the presence of androgens, cells enter apoptosis in response to various stimuli such as glucocorticoids, cytokines, and cell surface death signal transmitting receptors [79,80] or to physical factors such as hypothermia and ionizing radiation [81].

The main role of TGF-β1 in the adult prostate is to inhibit the gland's growth by controlling cell proliferation and inducing epithelial cell apoptosis [82]. TGF-β1 also inhibits fibroblast growth. Bretland *et al.* [83] have shown on stromal cells derived from prostatic adenomas (BPH) that the nature of the effect observed depends on the TGF-β concentration used. It is generally accepted that TGF-β plays its inhibitory role via induction of the cyclin-dependent kinase (CDK) inhibitors $P15^{INK}$ and $P21^{CIP}$ [84]. This induction results in cell growth arrest. Induction of $P15^{INK}$ and $P21^{CIP}$ is mediated by Smad2 and Smad3 [85]. In androgen-deprivation-induced prostatic apoptosis, TGF-β plays an inducing role that has been described at length [82,86,87]. Most interestingly, Bruckheimer and Kyprianou [88] have recently shown that dihydrotestosterone, at physiological concentrations, stimulates TGF-β-induced apoptosis. These results provide a molecular basis for androgen priming of prostate cancer cells in order to obtain maximal apoptosis in antiandrogen treatment of the prostate. It is noteworthy in this context that TGF-β1 has been implicated in determining the nuclear or cytoplasmic localization of the androgen receptor [89]. By favoring the receptor's cytoplasmic location, TGF-β might inhibit the cell response to androgens. Moreover, Reddy *et al.* [90] have shown that clusterin interacts with the type II TGF-β receptor at the level of its intracytoplasmic enzymatic domain. Clusterin is the protein most strongly overexpressed in androgen-dependent apoptosis, and its intracellular localization also depends on TGF-β [91]. These results suggest a regulatory role for clusterin in prostatic cell apoptosis.

In androgen-dependent prostate regression, the FAS ligand and its receptor also appear to be involved. The FAS ligand is expressed by epithelial cells, and this expression is increased in the prostate after castration [92]. In mutant mice with an altered FAS antigen, moreover, prostatic tumor transplants regress less after castration than grafts implanted in the wild type [93]. Binding of the FAS ligand to the FAS receptor leads to recruitment of the FADD death domain protein. This results in the activation of caspases-8 and -10 via the RIP protein, and in initiation of apoptosis by these proteins. TNFα also induces prostatic apoptosis by binding to its type I membrane receptor. The complex formed recruits TRADD, which in turn recruits FADD. From this association arises an apoptosis-activating pathway between FAS and TNFα.

The TNF-α-related-apoptosis-inducing ligand (TRAIL, also called AP0-2L) [94,95] and APO-3L are other members of the TNFα family. TRAIL can induce apoptosis of several human prostatic cell lines: ALVA-31, DV-145, PC-3, JCA-1, and TSV-Pr1 [96]. It has been shown that TRAIL can trigger apoptosis of normal prostatic epithelial cells within 6 hr of treatment. Hence, TRAIL might participate in cell homeostasis in the prostatic epithelium [97]. TRAIL is a type II transmembrane protein. It exerts its apoptotic action via the DR-4 and DR-6 receptors [98]. The intensity of this action is modulated by decoy receptors lacking the death domain and conferring resistance to the action of TRAIL. Expression of osteoprotegerin (OPG), a decoy receptor of TRAIL, by prostate cancer cells could confer a survival advantage on these cells [99]. TRAIL-triggered apoptosis, like FAS-triggered apoptosis, involves FADD recruitment and activation of the caspase pathway. Although TRAIL expression is ubiquitous, this factor inhibits tumor growth *in vivo* without causing any toxic effect in rodents or nonhuman primates. This makes it, unlike FAS-L, an interesting molecule for treating many types of tumors [100,101].

Because cell renewal occurs at a low rate in the adult prostate, we can assume that apoptosis is counteracted by the action of various survival factors. The many messages that activate receptor tyrosine kinases or GPCRs in the prostate may induce this type of effect via activation of the protein kinase B/Akt [102] and also via the transcription factor NFχB [102]. The inhibitory action of NFχB on apoptosis is controversial, but it might result from expression of the genes involved in oxidative stress [103]. Other signaling pathways are used to activate resistance to apoptosis. Prolactin delays androgen-dependent apoptosis in cultured explants and it can replace testosterone as an agent maintaining rat prostatic explants in long-term culture [104]. In these explant cultures, prolactin acts on the survival of prostatic epithelial cells by activating Janus kinases and Stat5a/5b [105]. This transduction pathway is prostate specific and might be used specifically to maintain the viability of androgen-deprived prostatic epithelial cells. Other pathways such as the B/AKT kinase pathway have also been proposed as mediating the antiapoptotic effects of PRL [106]. In the prostatic epithelium, PRL receptors are located preferentially on the apical side of the cells. This argues in favor of a paracrine/autocrine action of the hormone in this organ [107]. Prolactin also mediates the inflammatory effects observed in the dorsolateral lobe of Wistar rats treated with estradiol [108] and in Noble rats treated with a mixture of testosterone and estradiol [109]. On the other hand, hyperprolactinemia induced by sulpiride injection causes an increase in size and inflammation of the lateral lobe of the prostate, associated with Bcl-2 overexpression. This overexpression might lead to deficient apoptosis at the level of this lobe [110]. The pituitary gland secretes, depending on the physiological circumstances, a phosphorylated variant of prolactin, which acts as an antagonist at receptor level [111]. A nonhydrolyzable variant of the unphosphorylated molecule has been shown to delay effectively *in vivo* the androgen-independent growth of DU145 cells in nude mice [112]. These results have raised interest in the prolactin signaling pathway as a potential target for new therapies for androgen-independent prostate cancer.

Several publications have shown the importance of Bcl-2 in resistance to apoptosis. Very recently, Banerjee *et al.* [113] showed that Bcl-2 is differentially expressed in the different lobes of the rodent prostate. Its overexpression is age dependent and it correlates with cell survival in the absence of androgens. Bcl-2 and Bcl xL are known to inhibit the proapoptotic effects of medicinal treatments and radiotherapy on various cancers. It has been shown that Bcl-2 protects against apoptosis induced by the FAS ligand and TNF [114]. Likewise, Bcl-2 has recently been shown to inhibit TRAIL-induced apoptosis in three human prostatic cell lines [115]. The link between Bcl-2 and prostatic cancer is well established. It has recently been shown by immunohistochemistry that Bcl-2, Bcl-x2, and Mcl-1 increase as prostate cancer progresses [116]. It is generally accepted that the main site of the antiapoptotic action of Bcl-2 is the mitochondrion [117], but it has been reported that Bcl-2 could be a chaperone interacting with certain procaspases to prevent their activation [118]. Bax, in this context, would exert an apoptosis-inducing effect by binding to Bcl-2-procaspase complexes. This competition between Bax and Bcl-2 would probably explain why the Bax/Bcl-2 ratio is high at the onset of androgen-deprivation-induced apoptosis [119].

It is known that the neuroendocrine cells of the prostate resist androgen-dependent apoptosis and do not express Bcl-2. This gives them a privileged role in progression of androgen-resistant prostatic cancer. It has recently been shown that the neuroendocrine cells of the prostate express survivin [120]. Survivin is a member of the recently discovered apoptosis inhibitor family [117]. Survivin expression in prostatic cancer cells has recently been demonstrated and resistance to apoptosis of these cells could be due in part to inhibitors of apoptosis proteins (IAP) expression. It has recently been shown that three of the most commonly used prostate cancer cell lines (DU-145, 2LNCap, and PC-3) expressed IAPs and survivin [121], suggesting that these inhibitors may make an important contribution to apoptotic resistance in patients with prostate cancer.

P75 neurotrophin receptor (P75^{NTR}) has been suggested to be a candidate tumor suppressor gene in human prostate [122]. Indeed, P75^{NTR} is a low-affinity receptor for NGF that is expressed to varying degrees in epithelial cells [123–125]. In these cells, P75^{NTR} inhibits cell proliferation and promotes apoptosis [126]. Loss of expression of $P75^{NTR}$ in prostate cancer cells appears to block a potential programmed cell death pathway facilitating cell survival [122].

Because expression of 75^{NTR} is lost in prostate cancer, NGF-mediated growth of cancer cells occurs via the high-affinity Trk receptor (Tropomyosin receptor kinase), which is expressed in nonmetastatic prostate cancer and in cancer cell lines derived from metastases [127]. Although the Trk receptor was first identified as a colon cancer oncogene, Trk mutations within the human prostate cancer have never been found [128]. Thus Trk signaling pathways might be important in prostate cancer development.

The autonomous nervous system, which ensures the maintenance of the gland's structure and controls its secretions [129], has recently been implicated in the regulation of apoptosis in the prostate. Kyprianou *et al.* [28] have shown that α_1 adrenergic receptors are involved in controlling apoptosis and its induction by α_1 antagonists. These authors have also demonstrated a proapoptotic effect of Doxazosin and Terazosin on prostatic cells. These results suggest that α adrenergic antagonists have therapeutic potential in the treatment of BPH and cancer of the prostate.

The Prostate during Aging

Steroid Hormones during Aging

During aging, the prostate can be affected by two major pathologies: BPH and prostate cancer. Among the various species, only man spontaneously develops prostate cancer in old age. BPH and cancer develop from two different zones of the human prostate: the transition zone for BPH and the peripheral zone for prostate cancer. BPH is due mainly to proliferation of the fibromuscular stroma, whereas prostatic cancer develops in the epithelium. Causal relationships between these two diseases have not been clearly established. The development of prostatic cancer only rarely results from a malignant evolution of BPH. The cellular and biochemical mechanisms involved in initiating these two pathologies are different. Men as young as 30 may show intraprostatic tumor foci, whereas the first signs of BPH often appear after the age of 40. Yet these two diseases progress in an identical steroid context characterized by slowly decreasing bloodstream androgen levels and increasing estrogen levels.

Krieg *et al.* [130] have given a good description of the androgen context of the prostate during aging. They have shown that the accumulation of estrogens in the gland is an age-dependent phenomenon. Both 17β-estradiol and estrone increase in the stroma with age, whereas their concentrations in the epithelium remain constant. The dihydrotestosterone level varies in opposite fashion, decreasing in the epithelium and remaining constant in the stroma. What's more, it has been shown that in the transition zone of the human prostate, that is, the tissue where BPH preferentially develops, the estrogen-to-androgen ratio increases with age [131]. Given the action of androgens and estrogens on the development of the fibromuscular stroma, the role of these steroids seems fundamental in the development of prostatic pathologies [132].

During aging, expression of the androgen receptor remains constant in the normal prostate and in BPH [133]. Both the androgen receptor and 5α reductase are expressed mainly in the stroma. The epithelium also expresses the receptor but much less strongly. In the hyperplastic epithelium, the androgen receptor is expressed more abundantly in the luminal cells than in the basal cells. In the stroma, the receptor is present on the smooth muscle cells and fibroblasts. Increased androgen and prolactin levels resulting from knock out of the aromatase gene (ArKO) in transgenic mice lead to prostatic hyperplasia. Yet testosterone supplementation in elderly men seems to have no effect on prostatic growth. In the ArKO mouse model, the absence of estrogens does not lead to any cancerous development of the prostate [134]. These results show that both androgens and estrogens are required for the development of prostate cancer.

Estrogens exert their effects by binding to two types of receptors, Erα and Erβ, which are differentially expressed in the adult prostate. Erα is stromal and Erβ is present in the glandular epithelium of the rat prostate [48]. In the adult rat, there is an increasing gradient of Erβ expression from the proximal part of the ducts to the distal part. Estrogens are produced by peripheral aromatization of testosterone by aromatase. This enzyme is present in the adult prostate but very little is known about its activity during aging. On the other hand, aromatase is present in two androgen-independent human cell lines, PC3 and DU-145 [135], where it might mediate the effects of testosterone. In man, circulating estrogen levels increase with age. This increase is due to increased aromatization in the adrenal gland after the age of 50. The increased estrogen levels inside the prostate may be due to the increased serum levels of these steroids at the andropause. Because of its stromal localization, the Erα has been implicated in the preferential development of the stroma. Likewise, the epithelial localization of Erβ suggests that it might be involved in the development of prostatic cancer. Yet Erβ has also been implicated in BPH as a result of observations on transgenic mice. Erβ knock-out mice develop prostatic hyperplasia late in life [136]. Erβ might thus play a protective role during abnormal growth of the gland by exerting an antiproliferative action. More detailed studies of the prostate, however, failed to corroborate these initial results [137,138]. The reason for such a discrepancy is still unclear. In hyperplasia caused by estrogen treatment during fetal or neonatal life, the level of Erβ is reduced in adult life. It is important to establish the role of the β receptor and its interactions with Erα. In the prostate, the effects mediated by these two receptors appear to be mutually antagonistic. Paech *et al.* [139] have shown that tamoxifen and genistein exert agonistic effects on Erα but antagonistic effects on Erβ.

It is interesting to note that some *in vitro* studies show that estrogens can bind to androgen receptors. This was observed on LNCaP cells, which are androgen sensitive and possess a mutated androgen receptor [140]. It is believed, however, that estrogens act essentially via coactivators, to modulate transcription of androgen-controlled genes. ARA70 is an example of a coactivating protein [141]. It has been shown in DU-145 cells transfected with the gene encoding the androgen receptor that 17β-estradiol binds to this receptor and activates transcription in the presence of ARA70 [142]. This estrogen-mediated transcriptional activation is specific to the androgen receptor and does not require the presence of an estrogen receptor. This result clearly shows the importance of the joint action of estrogens and androgens in regulating gene expression in the prostate. The action of estradiol on the transcription of androgen-dependent genes supports the idea of using antiestrogens like diethylstilbestrol (DES) to treat pathologies of the prostate.

Prostatic cancer evolves from androgen dependence to stages where it requires for its progression only local factors or factors that deregulate the intracellular mechanisms that make the gland's growth androgen dependent. This progression of prostatic cancer usually occurs in man after the age of 50, although prostatic tumor foci may be present in the gland around the age of 30. When cancer is diagnosed at a sufficiently early stage, the androgen receptor is functional and sometimes amplified at both the gene and protein levels. This greater androgen receptivity of prostate tumors may explain the development, growth, and progression of androgen-dependent prostate cancer: The process would be triggered by transcriptional activation of androgen-dependent genes. PSA is the prototype of these androgen-regulated factors and is used in the early diagnosis of prostate cancer. Yet PSA levels remain very high in androgen-resistant prostatic cancer after androgen depletion. There is an apparent paradox here, because it is generally accepted that the androgen receptor is nonfunctional in prostate cancer, once the cancer has become hormone resistant. Yet it has been demonstrated that androgen receptor expression also increases in some androgen-resistant and metastasizing tumors [143,144]. Compensatory sensitivity to androgens has also been proposed to explain progression of androgen-resistant prostate cancer. This concept stems from observations on LNCaP cells cultured in androgen-free medium. Cells appearing in these cultures express three times more androgen receptor than cells cultured in the presence of DHT or testosterone. The overproduced receptor is functional, because it responds to androgen stimulation. The mitogenic action of androgens can also be amplified in androgen-sensitive tumors by mechanisms involving coactivators of steroid hormone receptors [145]. These coactivators are important to the assembly of the RNA polymerase II preinitiation complex to the translation start site. *In vivo* studies suggest that overexpression of these coactivators associated with that of the steroid hormone receptors may play a role in cancer development by enabling the receptors to activate their target genes independently of their ligand.

It has been shown *in vitro* that in addition to ARA70, other factors can up-regulate androgen receptor activity. Brady *et al.* [146] have shown that the Tip 60 protein is a coactivator of androgen-receptor-controlled transcription. The ARA54 factor, whose gene has been cloned by Kang *et al.* [147], is also a protein binding to the androgen receptor and modulating its activity. Some *in vivo* studies have shown that certain coactivators can be overexpressed or mutated in hormone-dependent cancers. In an *in vivo* model where the SRC-1 gene was inactivated (SCR-1 stands for steroid receptor coactivator-1), Xu *et al.* [148] have shown that the product of this gene is involved in the growth of the prostate and other hormone-dependent organs. The P-300/CBP coactivator is mutated in prostate cancer. It may be involved in controlling the expression of P53, whose role in the induction of apoptosis is well known [149].

The cAMP Signaling Pathway

As mentioned earlier, cross-talk between steroid hormone receptors and receptor tyrosine kinases, Ser/Thr kinases, and GPCRs may explain the progression of prostatic pathologies. IGF-I, KGF (FGF-7), and EGF can activate androgen-receptor-mediated transcription [58]. The CREB protein also intervenes in androgen-receptor-mediated effects [150]. On the other hand, it has also been shown that EGF, IGF-I, and cyclic AMP (cAMP) activate the signaling pathway of the α receptor via its AF-1 domain [151]. Estradiol can also cause rapid accumulation of cAMP in the prostate, through action at the level of the plasma membrane [152]. In addition, binding of oestradiol to sex hormone binding globulin (SHBG) can cause an increase in the intracellular level of cAMP. SHBG is the carrier protein for sex steroids in the bloodstream. When not bound to its ligand, SHBG binds to membrane receptors at the surface of prostatic cells. When estradiol binds to SHBG, the receptor–SHBG complex is activated, and this causes an increase in the intracellular level of cAMP and activation of PKA. This leads to increased expression of PSA in prostatic cells, specific to PKA activation. This alternative pathway for estrogen action might be involved in controlling the effects of male and female sex hormones on the prostate. It is noteworthy that SHBG can bind testosterone, DHT, or estradiol, albeit with different affinities. Binding of one of these steroid hormones to SHBG might thus lead to local competition at the level of prostate cells [153].

Classically, high cAMP levels activate protein kinase A (PKA) and the latter in turn can activate several signaling pathways. In the prostate, cAMP notably increases the activity of MAP kinases [154]. Furthermore, Nazareth *et al.* [154] have shown that PKA is responsible for activation of the N-terminal region of the androgen receptor, involved in DNA binding. PKC has also been implicated in androgen receptor modulation. De Ruiter and coworkers [155] have shown in transient transfection experiments performed on Chinese hamster ovary (CHO) cells that PKC acts in synergy with the activated androgen receptor. Activation by PKC, however, fails to induce ligand-independent androgen receptor activation.

Last, an increased cAMP level can also result, *in vivo*, from the action of several neuropeptides or hormones acting via adenylate-cyclase-coupled GPCRs. The corresponding receptors are present in the prostate [5] and some of them are more abundant in BPH and prostate cancer [157,158]. These ligands can act systemically like the pituitary hormones or in paracrine/autocrine fashion like certain ligands produced by endothelial and neuroendocrine cells.

Growth Factor Signaling

Growth factors and their receptors are involved in the gland's normal growth. Several of them are proto-oncogenes; changes in their expression lead to the development of prostate cancer and to its uncontrolled growth. This is the case of EGF and TGF-α, which act via the same receptor and whose expression, according to some investigators, is altered in prostatic cancer [159], although other authors say it is not [160]. EGF is mitogenic toward prostatic epithelial cancer cells, and its receptor is present at their surface both *in vivo* and *in vitro* [161]. EGF and its receptor exert pleiotropic effects on prostatic cancer cells, stimulating their proliferation and increasing their invasiveness.

Another mechanism involves, for example, up-regulation of a receptor. HER-2/Neu, HER-3, and HER-4 are proto-oncogenes belonging to the EGF receptor family. HER-2 is expressed differentially in the normal prostate and in cancerous tissue; HER-2/Neu is not expressed by the normal epithelium but most epithelial cancer cells express it. HER-2/Neu induces proliferation of these cells and increases their invasiveness. It has also been implicated in progression of androgen-dependent cancer, as a coactivator of the androgen signaling pathway. Furthermore, HER-2/Neu-dependent activation of the androgen receptor has been shown to involve the MAP kinase pathway [162]. In the absence of androgens, HER-2/Neu increases the growth and survival of prostatic cancer cells by stimulating the PI3K/AKT pathway. AKT transactivates the androgen receptor by binding to it specifically and phosphorylating its residues Ser213 and Ser791 [163]. These results provide a molecular mechanism that would explain the role of Her-2/Neu in the progression of androgen-independent prostatic cancer, where overexpression and genetic amplification have been reported [164]. Yet other studies indicate that HER-2/Neu overexpression is relatively rare in this type of cancer. Heregulin binds to the HER-3 receptor, an isoform of the EGFR family, and activates HER-2, HER-3, and HER-4 by inducing formation of heterodimers between the different receptors of the family. A naturally secreted soluble form of HER-3 (p-85s HER-3) has been identified and recently been shown to inhibit activation of HER-2, HER-3, and HER-4 [165]. Identification of this natural inhibitor opens up prospects for its use in treating tumors whose growth is heregulin dependent.

Tumor progression is also influenced by TGF-β1, its isoforms, and other, more distant members of the same family, inhibin α, activin, follistatin, and certain BMPs. *In vivo*, TGF-β1 increases tumor growth and metastasis. The mechanism prevailing in this family appears to be down-regulation of the type I and II TGF receptors in tumor cells. At the same time, TGF-β1 is overexpressed in the same tumor cells. The result is a strong antiproliferative effect on the neighboring healthy cells. This autocrine/paracrine mechanism favors tumor progression at the expense of the host's cells. The involvement of this mechanism in tumor growth is confirmed by the observation that cancer cells in which TGF-β RII expression is restored undergo apoptosis when transplanted into a nude mouse. BMP family proteins induce bone formation *in vivo* and have been implicated in the development of prostate cancer bone metastases. Several BMPs are expressed by normal epithelial tissue, including BMP-2, -3, and -4. BMP-6 seems *in vivo* to be most directly involved in advanced prostatic cancer [166]. Its expression correlates with the Gleason score used to stage prostatic tumors [167].

FGF-2 (basic FGF) affects cell proliferation by an autocrine mechanism of cancer cells. It also stimulates angiogenesis and this favors the appearance of metastases. The capacity of FGF-2 to stimulate metastase formation reflects its capacity to regulate turnover of the extracellular matrix by modulating protease expression and collagen and fibronectin synthesis.

Another way to modify the growth of prostatic tumors is illustrated by the action of IGFs. These growth factors bind to seven different binding proteins. Located in the extracellular matrix, these proteins recruit to the matrix both circulating and locally produced IGFs. IGF-BP5 seems to be the only androgen-regulated BP, whence its role in the proliferation of tumor cells in the presence of androgens. However, overexpression of the inhibitory IGF-BP4 has been shown to delay the onset of prostate tumor formation *in vivo* [168] IGF-I is also a powerful mitogen for prostatic tumor cells. *In vitro*, IGF-I stimulates growth of PC3 and DU-145 cells, and an antisense oligonucleotide corresponding to the IGF-I receptor inhibits this growth [169,170]. On the other hand, evidence demonstrates that synergism exists between IGF-I and EGF at low androgen concentrations [171].

Last, growth factors have angiogenic properties. VEGF, a glycoprotein [172], which might be produced by the neuroendocrine or cancer cells, seems to be involved in neovascularization. Hypoxia may be the mechanism controlling VEGF synthesis [173]. It has recently been shown that testosterone itself controls the blood flow to the prostate [174] and that in castration-induced apoptosis, the blood flow first diminishes with the onset of apoptosis. This observation may explain the increase in VEGF enabling reconstruction of the vascular network during testosterone supplementation. Testosterone might also stimulate VEGF synthesis and, hence, the development of metastases.

Conclusions

Prostatic cancer is a heterogeneous disease. This applies both to its causes and to the mechanisms that facilitate its progression. Several molecular changes occur during this progression, giving the cancer cells a selective advantage over normal cells. Although tumor progression occurs in the context of androgen independence, the changes occurring in the tumor are only partially independent from the signaling pathways involving the androgen receptor. There is cross-talk between the intracellular signaling pathways used by the androgen receptor and those used by other steroid hormones or by various growth factors. A therapeutic consequence is that in a patient having undergone total androgen suppression, the expression of certain androgen-regulated genes may be maintained. It is important to identify the genes whose regulation occurs partially via the signaling pathway involving the (possibly mutated) androgen receptor, in order to develop effective therapies for androgen-independent prostatic cancer. It is equally interesting to identify those genes that, in the normal prostate, are repressed by androgens. These genes might encode survival factors enabling cancer cells to elude therapies based on androgen deprivation.

Prostate cancer, like benign prostatic hyperplasia, develops also in a context where androgen levels are decreasing in the bloodstream and estrogen levels are rising. It would certainly be interesting to identify the genes regulated by estrogens via the α and β receptors in the normal and cancerous prostate. At present, treatment of hormone refractory prostatic cancer with the antiestrogen tamoxifen must be considered preliminary [175].

Tumor progression toward androgen-resistant stages is accompanied at the cell level by changes involving the acquisition or overproduction of growth factors and their receptors, deregulation of the local synthesis of binding proteins, and overproduction of angiogenic and survival factors. All of these mechanisms use highly interconnected signaling pathways. By identifying the common parts of these pathways, it should be possible to target simultaneously several mechanisms leading to androgen independence. Various experimental approaches have led to identifying the MAP kinase and ERK pathways as potential targets [176]. Another potentially interesting approach would be to reintroduce apoptosis into prostatic tumor cells by gene therapies targeting the caspase pathway, and particularly caspases 3 and 7 [177].

The use of prodrugs activated by the PSA is certainly another possibility [178], as is the use of peptides with arginine-glycine-aspartic acid [179] structures or nonpeptidic molecules [180]. Prostate cancer control might also involve systemic factors such as prolactin and the various partners of the growth hormone/IGF axis. The use of antagonists of these messengers, of inhibitors of their synthesis, and of enzymes promoting the degradation of their binding proteins might be studied at the level of both endocrine and paracrine/autocrine regulatory mechanisms. Other targets for therapy certainly remain to be discovered.

It is important to realize that progress in this field will depend on learning about cell signaling mechanisms in both the normal and the pathological prostate.

References

1. Imperato-McGinley, J., Guerrero, L., Gautier, T., and Peterson R. E. (1974). Steroid 5alpha-reductase deficiency in man: An inherited form of male pseudohermaphroditism. *Science* **186**, 1213–1215.
2. Isaacs, J. T., Furuya, Y., and Berges, R. (1994). The role of androgen in the regulation of programmed cell death/apoptosis in normal and malignant prostatic tissue. *Semin. Cancer Biol.* **5**, 391–400.
3. Cunha, G. R., Donjacour, A. A., Cooke, P. S., Mee, S., Bigsby, R. M., Higgins, S. J., and Sugimura, Y. (1987). The endocrinology and developmental biology of the prostate. *Endocr. Rev.* **8**, 338–362.
4. Kaufman, J. M. and Vermeulen, A. (1998). Androgen in male senescence. In Nieslag, E. and Behre, Eds., *Testosterone Action, Deficiency Substitution,* pp. 437–471, Springer, Berlin.
5. Reiter, E., Hennuy, B., Bruyninx, M., Cornet, A., Klug, M., McNamara, M., Closset, J., and Hennen G. (1999). Effects of pituitary hormones on the prostate. *Prostate* **38**, 159–165.
6. Torring, N., Vinter-Jensen, L., Pedersen, S. B. *et al.* (1998). Systemic administration of insulin-like growth factor (IGF-I). causes growth of the rat prostate. *J. Urol.* **158**, 222–227.
7. Abate-Shen, C. and Shen, M. M. (2000). Molecular genetics of prostate cancer. *Genes Dev.* **14**, 2410–2434.
8. Greenlee, R. T., Hill-Harmon, M. B., Murray, T., and Thun, M. (2001). Cancer statistics. *CA Cancer J. Clin.* **51**, 15–36.
9. Jesik, C. J., Holland, J. M., and Lee, C. (1982). An anatomic and histologic study of the rat prostate. *Prostate* **3**, 81–97.
10. McNeal, J. E. (1983). The prostate gland: Morphology and pathobiology. *Monogr. Urol.* **4**, pp 3–37.
11. Price, D. and Ortiz, E. (1965). The role of fetal androgens in sex differentiation in mammals. In Urspring, R. L. and Dehaan, H., Eds., *Organogenesis,* pp 629–652. Holt, Rinehart and Winston, New York.
12. Lee, C., Sensibar, J. A., Dudek, S. M., Hiipakka, R. A., and Liao, S. T. (1990). Prostatic ductal system in rats: Regional variation in morphological and functional activities. *Biol. Reprod.* **43**, 1079–1086.
13. Cunha, G. R., Lung, B., and Reese, B. (1980). Glandular epithelial induction by embryonic mesenchyme in adult bladder epithelium of BALB/c mice. *Invest. Urol.* **17**, 302–304.
14. Larsen, M., Ressler, S. J., Lu, B., Gerdes, M. J., McBride, L., Dang, T. D., and Rowley, D. R. (1998). Molecular cloning and expression of ps20 growth inhibitor. A novel WAP-type "four-disulfide core" domain protein expressed in smooth muscle. *J. Biol. Chem.* **273**, 4574–4584.
15. Thomson, A. A., Foster, B. A., and Cunha, G. R. (1997). Analysis of growth factor and receptor mRNA levels during development of the rat seminal vesicle and prostate. *Development* **124**, 2431–2439.
16. Sugimura, Y., Foster, B. A., and Hom, Y. K. (1996). Keratinocyte growth factor can replace testosterone in the ductal branching morphogenesis of the rat ventral prostate. *Int. J. Dev. Biol.* **40**, 941–951.
17. Guo, L., Degenstein, L., and Fuchs, E. (1996). Keratinocyte growth factor is required for hair development but not for wound healing. *Genes Dev.* **10**, 165–175.
18. Thomson, A. A. and Cunha, G. R. (1999). Prostatic growth and development are regulated by FGF10. *Development* **126**, 3693–3701.
19. Min, H., Danilenko, D. M., Scully, S. A., Bolon, B., Ring, B. D., Tarpley, J. E., DeRose, M., and Simonet, W. S. (1998). FGF-10 is required for both limb and lung development and exhibits striking functional similarity to Drosophila branchless. *Genes Dev.* **12**, 3156–3161.
20. Sekine, K., Ohuchi, H., Fujiwara, M., Yamasaki, M., Yoshizawa, T., Sato, T., Yagishita, N., Matsui, D., Koga, Y., Itoh, N., and Kato, S. (1999). FGF10 is essential for limb and lung formation. *Nat. Genet.* **21**, 138–141.
21. Thomson, A. A. (2001). Role of androgens and fibroblast growth factors in prostatic development. *Reproduction* **121**, 187–195.
22. DeChiara, T. M., Efstratiadis, A., and Robertson, E. J. (1990). A growth-deficiency phenotype in heterozygous mice carrying an insulin-like growth factor II gene disrupted by targeting. *Nature* **345**, 78–80.
23. Adamo, M., Lowe, W. L. Jr, LeRoith, D., and Roberts, C. T. Jr. (1989). Insulin-like growth factor I messenger ribonucleic acids with alternative 5′-untranslated regions are differentially expressed during development of the rat. *Endocrinology* **124**, 2737–2744.
24. Hayward, S. W. and Cunha, G. R. (2000). The prostate: Development and physiology. *Radiol. Clin. North Am.* **38**, 1–14.
25. Baker, J., Hardy, M. P., Zhou, J., Bondy, C., Lupu, F., Bellve, A. R., and Efstratiadis, A. (1996). Effects of an Igf1 gene null mutation on mouse reproduction. *Mol. Endocrinol.* **10**, 903–918.

26. Powell-Braxton, L., Hollingshead, P., Giltinan, D., Pitts-Meek, S., and Stewart, T. (1993). Inactivation of the IGF-I gene in mice results in perinatal lethality. *Ann. NY Acad. Sci.* **692**, 300–301.

27. Podlasek, C. A., Barnett, D. H., Clemens, J. Q., Bak, P. M., and Bushman, W. (1999). Prostate development requires Sonic hedgehog expressed by the urogenital sinus epithelium. *Dev. Biol.* **209**, 28–39.

28. Kyprianou, N., Litvak, J. P., Borkowski, A., Alexander, R., and Jacobs, S. C. (1998). Induction of prostate apoptosis by doxazosin in benign prostatic hyperplasia. *J. Urol.* **159**, 1810–1815.

29. Xue, Y., van der Laak, J., Smedts, F., Schoots, C., Verhofstad, A., de la Rosette, J., and Schalken, J. (2000). Neuroendocrine cells during human prostate development: Does neuroendocrine cell density remain constant during fetal as well as postnatal life? *Prostate* **42**, 116–123.

30. Gerdes, M. J., Dang, T. D., Larsen, M., and Rowley, D. R. (1998). Transforming growth factor-beta1 induces nuclear to cytoplasmic distribution of androgen receptor and inhibits androgen response in prostate smooth muscle cells. *Endocrinology* **139**, 3569–3577.

31. Timme, T. L., Truong, L. D., Merz, V. W., Krebs, T., Kadmon, D., Flanders, K. C., Park, S. H., and Thompson, T. C. (1994). Mesenchymal-epithelial interactions and transforming growth factor-beta expression during mouse prostate morphogenesis. *Endocrinology* **134**, 1039–1045.

32. Itoh, N., Patel, U., Cupp, A. S., and Skinner, M. K. (1998). Developmental and hormonal regulation of transforming growth factor-beta1 (TGFβ1), -2, and -3 gene expression in isolated prostatic epithelial and stromal cells: Epidermal growth factor and TGFβ interactions. *Endocrinology* **139**, 1378–1388.

33. Chang, W. Y., Birch, L., Woodham, C., Gold, L. I., and Prins, G. S. (1999). Neonatal estrogen exposure alters the transforming growth factor-beta signaling system in the developing rat prostate and blocks the transient p21(cip1/waf1) expression associated with epithelial differentiation. *Endocrinology* **140**, 2801–2813.

34. Peehl, D. M. and Sellers, R. G. (1997). Induction of smooth muscle cell phenotype in cultured human prostatic stromal cells. *Exp. Cell Res.* **232**, 208–215.

35. Cancilla, B., Jarred, R. A., Wang, H., Mellor, S. L., Cunha, G. R., and Risbridger, G. P. (2001). Regulation of prostate branching morphogenesis by activin A and follistatin. *Dev. Biol.* **237**, 145–158.

36. Rowley, D. R., Dang, T. D., Larsen, M., Gerdes, M. J., McBride, L., and Lu, B. (1995). Purification of a novel protein (ps20) from urogenital sinus mesenchymal cells with growth inhibitory properties *in vitro*. *J. Biol. Chem.* **270**, 22058–22065.

37. Kim, H. G., Kassis, J., Souto, J. C., Turner, T., and Wells, A. (1999). EGF receptor signaling in prostate morphogenesis and tumorigenesis. *Histol. Histopathol.* **14**, 1175–1182.

38. Xie, W., Paterson, A. J., Chin, E., Nabell, L. M., and Kudlow, J. E. (1997). Targeted expression of a dominant negative epidermal growth factor receptor in the mammary gland of transgenic mice inhibits pubertal mammary duct development. *Mol. Endocrinol.* **11**, 1766–1781.

39. Tanaka, M., Komuro, I., Inagaki, H., Jenkins, N. A., Copeland, N. G., and Izumo, S. (2000). Nkx3.1, a murine homolog of Drosophila bagpipe, regulates epithelial ductal branching and proliferation of the prostate and palatine glands. *Dev. Dyn.* **219**, 248–260.

40. Bhatia-Gaur, R, Donjacour, A. A., Sciavolino, P. J., Kim, M., Cunha, G. R., Abate-Shen, C., and Shen, M. M. (1999). Roles for NKX3.1 in prostate development and cancer. *Genes Dev.* **13**, 966–977.

41. Abdulkadir, S. A., Magee, J. A., Peters, T. J., Kaleem, Z., Naughton, C. K., Humphrey, P. A., and Milbrandt, J. (2002). Conditional loss of Nkx3.1 in adult mice induces prostatic intraepithelial neoplasia. *Mol. Cell. Biol.* **22**, 1495–1503.

42. Levine, A. C., Liu, X. H., Greenberg, P. D., Eliashvili, M., Schiff, J. D., Aaronson, S. A., Holland, J. F., and Kirschenbaum, A. (1998). Androgens induce the expression of vascular endothelial growth factor in human fetal prostatic fibroblasts. *Endocrinology*. **139**, 4672–4678.

43. Elfman, F., Bok, R., Conn, M., Shuman, M., and Cunha, G. (2001). Urokinase plasminogen activator amino-terminal peptides inhibit development of the rat ventral prostate. *Differentiation* **69**, 108–120.

44. Heidtmann, H. H., Nettelbeck, D. M., Mingels, A., Jager, R., Welker, H. G., and Kontermann, R. E. (1999). Generation of angiostatin-like fragments from plasminogen by prostate-specific antigen. *Br. J. Cancer* **81**, 1269–1273.

45. Lamm, M. L., Podlasek, C. A., Barnett, D. H., Lee, J., Clemens, J. Q., Hebner, C. M., and Bushman, W. (2001). Mesenchymal factor bone morphogenetic protein 4 restricts ductal budding and branching morphogenesis in the developing prostate. *Dev. Biol.* **232**, 301–314.

46. Marker, P. C., Stephan, J. P., Lee, J., Bald, L., Mather, J. P., and Cunha, G. R. (2001). Fucosyltransferase1 and H-type complex carbohydrates modulate epithelial cell proliferation during prostatic branching morphogenesis. *Dev. Biol.* **233**, 95–108.

47. Prins, G. S., Marmer, M., Woodham, C., Chang, W., Kuiper, G., Gustafsson, J. A., and Birch, L. (1998). Estrogen receptor-beta messenger ribonucleic acid ontogeny in the prostate of normal and neonatally estrogenized rats. *Endocrinology* **139**, 874–883.

48. Chang, W. Y. and Prins, G. S. (1999). Estrogen receptor-beta: Implications for the prostate gland. *Prostate* **40**, 115–124.

49. vom Saal, F. S., Timms, B. G., Montano, M. M., Palanza, P., Thayer, K. A., Nagel, S. C., Dhar, M. D., Ganjam, V. K., Parmigiani, S., and Welshons, W. V. (1997). Prostate enlargement in mice due to fetal exposure to low doses of estradiol or diethylstilbestrol and opposite effects at high doses. *Proc. Natl. Acad. Sci. USA* **94**, 2056–2061.

50. Putz, O., Schwartz, C. B., Kim, S., LeBlanc, G. A., Cooper, R. L., and Prins, G. S. (2001). Neonatal low- and high-dose exposure to estradiol benzoate in the male rat: I. Effects on the prostate gland. *Biol. Reprod.* **65**, 1496–1505.

51. Habermann, H., Chang, W. Y., Birch, L., Mehta, P., and Prins, G. S. (2001). Developmental exposure to estrogens alters epithelial cell adhesion and gap junction proteins in the adult rat prostate. *Endocrinology* **142**, 359–369.

52. Chang, W. Y., Birch, L., Woodham, C., Gold, L. I., and Prins, G. S. (1999). Neonatal estrogen exposure alters the transforming growth factor-beta signaling system in the developing rat prostate and blocks the transient p21(cip1/waf1) expression associated with epithelial differentiation. *Endocrinology* **140**, 2801–2813.

53. Ittman, M. and Mansukhani, A. (1997). Expression of fibroblast growth factors (FGFs), and FGF receptors in human prostate. *J. Urol.* **157**, 351–356.

54. McKeehan, W. L. and Adams, P. S. (1988). Heparin-binding growth factor/prostatropin attenuates inhibition of rat prostate tumor epithelial cell growth by transforming growth factor type beta. *In Vitro Cell Dev. Biol.* **24**, 243–246.

55. Prenzel, N., Zwick, E., Daub, H., Leserer, M., Abraham, R., Wallasch, C., and Ullrich, A. (1999). EGF receptor transactivation by G-protein-coupled receptors requires metalloproteinase cleavage of proHB-EGF. *Nature* **402**, 884–888.

56. Putz, T., Culig, Z., Eder, I. E., Nessler-Menardi, C., Bartsch, G., Grunicke, H., Uberall, F., and Klocker, H. (1999). Epidermal growth factor (EGF) receptor blockade inhibits the action of EGF, insulin-like growth factor I, and a protein kinase A activator on the mitogen-activated protein kinase pathway in prostate cancer cell lines. *Cancer Res.* **59**, 227–233.

57. Luttrell, L. M., Daaka, Y., and Lefkowitz, R. J. (1999). Regulation of tyrosine kinase cascades by G-protein-coupled receptors. *Curr. Opin. Cell Biol.* **11**, 177–183.

58. Culig, Z., Hobisch, A., Cronauer, M. V., Radmayr, C., Trapman, J., Hittmair, A., Bartsch, G., and Klocker, H. (1994). Androgen receptor activation in prostatic tumor cell lines by insulin-like growth factor-I, keratinocyte growth factor, and epidermal growth factor. *Cancer Res.* **54**, 5474–5478.

59. Aarnisalo, P., Palvimo, J. J., and Janne, O. A. (1998). CREB-binding protein in androgen receptor-mediated signaling. *Proc. Natl. Acad. Sci. USA* **95**, 2122–2127.

60. Darne, C., Veyssiere, G., and Jean, C. (1998). Phorbol ester causes ligand-independent activation of the androgen receptor. *Eur. J. Biochem.* **256**, 541–549.

61. Wennbo, H., Kindblom, J., Isaksson, O. G., and Tornell, J. (1997). Transgenic mice overexpressing the prolactin gene develop dramatic enlargement of the prostate gland. *Endocrinology* **138**, 4410–4415.

62. Ruan, W., Powell-Braxton, L., Kopchick, J. J., and Kleinberg, D. L. (1999). Evidence that insulin-like growth factor I and growth hormone are required for prostate gland development. *Endocrinology* **140**, 1984–1989.
63. Colao, A., Marzullo, P., Ferone, D., Spiezia, S., Cerbone, G., Marino, V., Di Sarno, A., Merola, B., and Lombardi, G. (1998). Prostatic hyperplasia: An unknown feature of acromegaly. *J. Clin. Endocrinol. Metab.* **83**, 775–779.
64. Clemmons, D. R. (1997). Insulin-like growth factor binding proteins and their role in controlling IGF actions. Cytokine. *Growth Factor Rev.* **8**, 45–62.
65. Cohen, P., Graves, H. C., Peehl, D. M., Kamarei, M., Giudice, L. C., and Rosenfeld, R. G. (1992). Prostate-specific antigen (PSA) is an insulin-like growth factor binding protein-3 protease found in seminal plasma. *J. Clin. Endocrinol. Metab.* **75**, 1046–1053.
66. Gregory, C. W., Kim, D., Ye, P., D'Ercole, A. J., Pretlow, T. G., Mohler, J. L., and French, F. S. (1999). Androgen receptor up-regulates insulin-like growth factor binding protein-5 (IGFBP-5) expression in a human prostate cancer xenograft. *Endocrinology* **140**, 2372–2381.
67. Nickerson, T., Pollak, M., and Huynh, H. (1998). Castration-induced apoptosis in the rat ventral prostate is associated with increased expression of genes encoding insulin-like growth factor binding proteins 2, 3, 4 and 5. *Endocrinology* **139**, 807–810.
68. Jungwirth, A., Schally, A. V., Pinski, J., Halmos, G., Groot, K., Armatis, P., and Vadillo-Buenfil, M. (1997). Inhibition of in vivo proliferation of androgen-independent prostate cancers by an antagonist of growth hormone-releasing hormone. *Br. J. Cancer* **75**, 1585–1592.
69. Negro-Vilar, A., Saad, W. A., and McCann, S. M. (1977). Evidence for a role of prolactin in prostate and seminal vesicle growth in immature male rats. *Endocrinology* **100**, 729–737.
70. Thompson, S. A., Johnson, M. P., and Brooks, C. L. (1998). Biochemical and immunohistochemical characterization of prolactin binding in rat ventral, lateral, and dorsal prostate lobes. *Prostate* **3**, 45–58.
71. Wakao, H., Gouilleux, F., and Groner, B. (1994). Mammary gland factor (MGF) is a novel member of the cytokine regulated transcription factor gene family and confers the prolactin response. *EMBO J.* **13**, 2182–2191.
72. Costello, L. C. and Franklin, R. B. (1994). Bioenergetic theory of prostate malignancy. *Prostate* **25**, 162–166.
73. Nevalainen, M. T., Valve, E. M., Ingleton, P. M., Nurmi, M., Martikainen, P. M., and Harkonen, P. L. (1997). Prolactin and prolactin receptors are expressed and functioning in human prostate. *J. Clin. Invest.* **99**, 618–627.
74. Untergasser, G., Rumpold, H., Hermann, M., Dirnhofer, S., Jilg, G., and Berger, P. (1999). Proliferative disorders of the aging human prostate: Involvement of protein hormones and their receptors. *Exp. Gerontol.* **34**, 275–287.
75. Ormandy, C. J., Camus, A., Barra, J., Damotte, D., Lucas, B., Buteau, H., Edery, M., Brousse, N., Babinet, C., Binart, N., and Kelly, P. A. (1997). Null mutation of the prolactin receptor gene produces multiple reproductive defects in the mouse. *Genes Dev.* **11**, 167–178.
76. Horseman, N. D., Zhao, W., Montecino-Rodriguez, E., Tanaka, M., Nakashima, K., Engle, S. J., Smith, F., Markoff, E., and Dorshkind, K. (1997). Defective mammopoiesis, but normal hematopoiesis, in mice with a targeted disruption of the prolactin gene. *EMBO J.* **16**, 6926–6935.
77. Reiter, E., McNamara, M., Closset, J., and Hennen, G. (1995). Expression and functionality of luteinizing hormone/chorionic gonadotropin receptor in the rat prostate. *Endocrinology* **136**, 917–923.
78. Reiter, E., Poncin, J., Hennuy, B., Bruyninx, M., Klug, M., Cornet, A., Closset, J., and Hennen, G. (1997). Luteinizing hormone increases the abundance of various transcripts, independently of the androgens, in the rat prostate. *Biochem. Biophys. Res. Commun.* **233**, 108–112.
78a. Tao, Y. X., Lei, Z. M., Woodworth, S. H., and Rao, C. V. (1995). Novel expression of luteinizing hormone/chorionic gonadotropin receptor gene in rat prostates. *Mol. Cell Endocrinol.* **111**, 9–12.
79. Simons, J. W. and Mikhak, B. (1998). Ex-vivo gene therapy using cytokine-transduced tumor vaccines: Molecular and clinical pharmacology. *Semin. Oncol.* **25**, 661–676.
80. Rokhlin, O. W., Bishop, G. A., Hostager, B. S., Waldschmidt, T. J., Sidorenko, S. P., Pavloff, N., Kiefer, M. C., Umansky, S. R., Glover, R. A., and Cohen, M. B. (1997). Fas-mediated apoptosis in human prostatic carcinoma cell lines. *Cancer Res.* **57**, 1758–1768.
81. Li, W. X. and Franklin, W. A. (1998). Radiation- and heat-induced apoptosis in PC-3 prostate cancer cells. *Radiat. Res.* **150**, 190–194.
82. Martikainen, P., Kyprianou, N., and Isaacs, J. T. (1990). Effect of transforming growth factor-β1 on proliferation and death of rat prostatic cells. *Endocrinology* **127**, 2963–2968.
83. Bretland, A. J., Reid, S. V., Chapple, C. R., and Eaton, C. L. (2001). Role of endogenous transforming growth factor beta (TGFβ)1 in prostatic stromal cells. *Prostate* **48**, 297–304.
84. Hannon, G. J. and Beach, D. (1994). p15INK4B is a potential effector of TGF-beta-induced cell cycle arrest. *Nature* **371**, 257–261.
85. Derynck, R., Zhang, Y., and Feng, X. H. (1998). Smads: Transcriptional activators of TGF-β responses. *Cell* **95**, 737–740.
86. Buttyan, R., Olsson, C. A., Pintar, J., Chang, C., Bandyk, M., Ng, P. Y., and Sawczuk, I. S. (1989). Induction of the TRPM-2 gene in cells undergoing programmed death. *Mol. Cell Biol.* **9**, 3473–3481.
87. Kyprianou, N. and Isaacs, J. T. (1988). Identification of a cellular receptor for transforming growth factor-β in rat ventral prostate and its negative regulation by androgens. *Endocrinology* **123**, 2124–2131.
88. Bruckheimer, E. M. and Kyprianou, N. (2001). Dihydrotestosterone enhances transforming growth factor-induced apoptosis in hormone-sensitive prostate cancer cells. *Endocrinology* **142**, 2419–2426.
89. Gerdes, M. J., Dang, T. D., Larsen, M., and Rowley, D. R. (1998). Transforming growth factor-beta1 induces nuclear to cytoplasmic distribution of androgen receptor and inhibits androgen response in prostate smooth muscle cells. *Endocrinology* **139**, 3569–3577.
90. Reddy, K. B., Karode, M. C., Harmony, A. K., and Howe, P. H. (1996). Interaction of transforming growth factor β receptors with apolipoprotein J/clusterin. *Biochemistry* **35**, 309–314.
91. Reddy, K. B., Jin, G., Karode, M. C., Harmony, J. A., and Howe, P. H. (1996). Transforming growth factor beta (TGFβ)-induced nuclear localization of apolipoprotein J/clusterin in epithelial cells. *Biochemistry* **35**, 6157–6163.
92. de la Taille, A., Chen, M. W., Shabsigh, A., Bagiella, E., Kiss, A., and Buttyan, R. (1999). Fas antigen/CD-95 upregulation and activation during castration-induced regression of the rat ventral prostate gland. *Prostate* **40**, 89–96.
93. Suzuki, T., Kurokawa, K., Suzuki, K., Matsumoto, K., and Yamanaka, H. (1995). Histological and immunohistochemical changes after transurethral balloon laser hyperthermia in the canine prostate. *Tohoku J. Exp. Med.* **177**, 39–48.
94. Wiley, S. R., Schooley, K., Smolak, P. J., Din, W. S., Huang, C. P., Nicholl, J. K., Sutherland, G. R., Smith, T. D., Rauch, C., Smith, C. A. *et al.* (1995). Identification and characterization of a new member of the TNF family that induces apoptosis. *Immunity* **3**, 673–682.
95. Pitti, R. M., Marsters, S. A., Ruppert S., Donahue, C. J., Moore, A., and Ashkenazi, A. (1996). Induction of apoptosis by Apo-2 ligand, a new member of the tumor necrosis factor cytokine family. *J. Biol. Chem.* **271**, 12687–12690.
96. Nesterov, A., Lu, X., Johnson, M., Miller, G. J., Ivashchenko, Y., and Kraft, A. S. (2001). Elevated AKT activity protects the prostate cancer cell line LNCaP from TRAIL-induced apoptosis. *J. Biol. Chem.* **276**, 10767–10774.
97. Nesterov, A., Ivashchenko, Y., and Kraft, A. S. (2002). Tumor necrosis factor-related apoptosis-inducing ligand (TRAIL) triggers apoptosis in normal prostate epithelial cells. *Oncogene* **21**, 1135–1140.
98. Griffith, T. S. and Lynch, D. H. (1998). TRAIL: A molecule with multiple receptors and control mechanisms. *Curr. Opin. Immunol.* **10**, 559–563.
99. Holen, I., Croucher, P. I., Hamdy, F. C., and Eaton, C. L. (2002). Osteoprotegerin (OPG) is a survival factor for human prostate cancer cells. *Cancer Res.* **62**, 1619–1623.
100. Walczak, H., Miller, R. E., Ariail, K., Gliniak, B., Griffith, T. S., Kubin, M., Chin, W., Jones, J., Woodward, A., Le, T., Smith, C., Smolak, P., Goodwin, R. G., Rauch, C. T., Schuh, J. C., and

Lynch, D. H. (1999). Tumoricidal activity of tumor necrosis factor-related apoptosis-inducing ligand in vivo. *Nat. Med.* **5**, 157–163.
101. Ashkenazi, A., Pai, R. C., Fong, S., Leung, S., Lawrence, D. A., Marsters, S. A., Blackie, C., Chang, L., McMurtrey, A. E., Hebert, A., DeForge, L., Koumenis, I. L., Lewis, D., Harris, L., Bussiere, J., Koeppen, H., Shahrokh, Z., and Schwall, R. H. (1999). Safety and antitumor activity of recombinant soluble Apo2 ligand. *J. Clin. Invest.* **104**, 155–162.
102. Raj, G. V., Barki-Harrington, L., Kue, P. F., and Daaka, Y. (2002). Guanosine phosphate binding protein coupled receptors in prostate cancer: A review. *J. Urol.* **167**, 1458–1463.
103. Wong, G. H., Elwell, J. H., Oberley, L. W., and Goeddel, D. V. (1989). Manganous superoxide dismutase is essential for cellular resistance to cytotoxicity of tumor necrosis factor. *Cell* **58**, 923–931.
104. Ahonen, T. J., Harkonen, P. L., Laine, J., Rui, H., Martikainen, P. M., and Nevalainen, M. T. (1999). Prolactin is a survival factor for androgen-deprived rat dorsal and lateral prostate epithelium in organ culture. *Endocrinology* **140**, 5412–5421.
105. Ahonen, T. J., Harkonen, P. L., Rui, H., and Nevalainen, M. T. (2002). PRL signal transduction in the epithelial compartment of rat prostate maintained as long-term organ cultures in vitro. *Endocrinology* **143**, 228–238.
106. Al-Sakkaf, K. A., Mooney, L. M., Dobson, P. R., and Brown, B. L. (2000). Possible role for protein kinase B in the anti-apoptotic effect of prolactin in rat Nb2 lymphoma cells. *J. Endocrinol.* **167**, 85–92.
107. Nevalainen, M. T., Valve, E. M., Ingleton, P. M., and Harkonen, P. L. (1996). Expression and hormone regulation of prolactin receptors in rat dorsal and lateral prostate. *Endocrinology* **137**, 3078–3088.
108. Tangbanluekal, L. and Robinette, C. L. (1993). Prolactin mediates estradiol-induced inflammation in the lateral prostate of Wistar rats. *Endocrinology* **132**, 2407–2416.
109. Lane, K. E., Leav, I., Ziar, J., Bridges, R. S., Rand, W. M., and Ho, S. M. (1997). Suppression of testosterone and estradiol-17β-induced dysplasia in the dorsolateral prostate of Noble rats by bromocriptine. *Carcinogenesis* **18**, 1505–1510.
110. Van Coppenolle, F., Slomianny, C., Carpentier, F., Le Bourhis, X., Ahidouch, A., Croix, D., Legrand, G., Dewailly, E., Fournier, S., Cousse, H., Authie, D., Raynaud, J. P., Beauvillain, J. C., Dupouy, J. P., and Prevarskaya, N. (2001). Effects of hyperprolactinemia on rat prostate growth: Evidence of androgen-dependence. *Am. J. Physiol. Endocrinol. Metab.* **280**, E120–E129.
111. Wang, Y. F., Liu, J. W., Mamidi, M., and Walker, A. M. (1996). Identification of the major site of rat prolactin phosphorylation as serine 177. *J. Biol. Chem.* **271**, 2462–2469.
112. Xu, X., Kreye, E., Kuo, C. B., and Walker, A. M. (2001). A molecular mimic of phosphorylated prolactin markedly reduced tumor incidence and size when DU-145 human prostate cancer cells were grown in nude mice. *Cancer Res.* **61**, 6098–6104.
113. Banerjee, P. P., Banerjee, S., and Brown, T. R. (2002). Bcl-2 protein expression correlates with cell survival and androgen independence in rat prostatic lobes. *Endocrinology* **143**, 1825–1832.
114. Boise, L. H. and Thompson, C. B. (1997). Bcl-x(L) can inhibit apoptosis in cells that have undergone Fas-induced protease activation. *Proc. Natl. Acad. Sci. USA* **94**, 3759–3764.
115. Munshi, A., Pappas, G., Honda, T., McDonnell, T. J., Younes, A., Li, Y., and Meyn, R. E. (2001). TRAIL (APO-2L) induces apoptosis in human prostate cancer cells that is inhibitable by Bcl-2. *Oncogene* **20**, 3757–3765.
116. Krajewska, M., Krajewski, S., Epstein, J. I., Shabaik, A., Sauvageot, J., Song, K., Kitada, S., and Reed, J. C. (1996). Immunohistochemical analysis of bcl-2, bax, bcl-X, and mcl-1 expression in prostate cancers. *Am. J. Pathol.* **148**, 1567–1576.
117. Schulze-Osthoff, K., Ferrari, D., Los, M., Wesselborg, S., and Peter, M. E. (1998). Apoptosis signaling by death receptors. *Eur. J. Biochem.* **254**, 439–459.
118. Villa, P., Kaufmann, S. H., and Earnshaw, W. C. (1997). Caspases and caspase inhibitors. *Trends Biochem. Sci.* **22**, 388–393.
119. Perlman, H., Zhang, X., Chen, M. W., Walsh, K., and Buttyan, R. (1999). An elevated bax/bcl-2 ratio corresponds with the onset of prostate epithelial cell apoptosis. *Cell Death Differ.* **6**, 48–54.
120. Xing, N., Qian, J., Bostwick, D., Bergstralh, E., and Young, C. Y. (2001). Neuroendocrine cells in human prostate over-express the anti-apoptosis protein survivin. *Prostate* **48**, 7–15.
121. McEleny, K. R., Watson, R. W., Coffey, R. N., O'Neill, A. J., and Fitzpatrick, J. M. (2002). Inhibitors of apoptosis proteins in prostate cancer cell lines. *Prostate* **51**, 133–140.
122. Perez, M., Regan, T., Pflug, B., Lynch, J., and Djakiew, D. (1997). Loss of low-affinity nerve growth factor receptor during malignant transformation of the human prostate. *Prostate* **30**, 274–279.
123. MacGrogan, D., Saint-Andre, J. P., and Dicou, E. (1992). Expression of nerve growth factor and nerve growth factor receptor genes in human tissues and in prostatic adenocarcinoma cell lines. *J. Neurochem.* **59**, 1381–1391.
124. Paul, A. B., Grant, E. S., and Habib, F. K. (1996). The expression and localisation of beta-nerve growth factor (β-NGF) in benign and malignant human prostate tissue: Relationship to neuroendocrine differentiation. *Br. J. Cancer* **74**, 1990–1996.
125. Graham, C. W., Lynch, J. H., and Djakiew, D. (1992). Distribution of nerve growth factor-like protein and nerve growth factor receptor in human benign prostatic hyperplasia and prostatic adenocarcinoma. *J. Urol.* **147**, 1444–1447.
126. Pflug, B. and Djakiew, D. (1998). Expression of p75NTR in a human prostate epithelial tumor cell line reduces nerve growth factor-induced cell growth by activation of programmed cell death. *Mol. Carcinog.* **23**, 106–114.
127. Angelsen, A., Sandvik, A. K., Syversen, U., Stridsberg, M., and Waldum, H. L. (1998). NGF-β, NE-cells and prostatic cancer cell lines. A study of neuroendocrine expression in the human prostatic cancer cell lines DU-145, PC-3, LNCaP, and TSU-pr1 following stimulation of the nerve growth factor-β. *Scand. J. Urol. Nephrol.* **32**, 7–13.
128. George, D. J., Suzuki, H., Bova, G. S., and Isaacs, J. T. (1998). Mutational analysis of the TrkA gene in prostate cancer. *Prostate* **36**, 172–180.
129. McVary, K. T., McKenna, K. E., and Lee, C. (1998). Prostate innervation. *Prostate* **8**, 2–13.
130. Krieg, M., Nass, R., and Tunn, S. (1993). Effect of aging on endogenous level of 5α-dihydrotestosterone, testosterone, estradiol, and estrone in epithelium and stroma of normal and hyperplastic human prostate. *J. Clin. Endocrinol. Metab.* **77**, 375–381.
131. Shibata, Y., Ito, K., Suzuki, K., Nakano, K., Fukabori, Y., Suzuki, R., Kawabe, Y., Honma, S., and Yamanaka, H. (2000). Changes in the endocrine environment of the human prostate transition zone with aging: Simultaneous quantitative analysis of prostatic sex steroids and comparison with human prostatic histological composition. *Prostate* **42**, 45–55.
132. Brown, T. R. and Lee, C. (2000). Conference summary on prostate growth and aging. *Prostate* **48**, 54–65.
133. Cooke, P., Young, P., and Cunha. (1991). Androgen receptor expression in developing male reproductive organs. *Endocrinology* **128**, 2867–2873.
134. McPherson, S. J., Wang, H., Jones, M. E., Pedersen, J., Iismaa, T. P., Wreford, N., Simpson, E. R., and Risbridger, G. P. (2001). Elevated androgens and prolactin in aromatase-deficient mice cause enlargement, but not malignancy, of the prostate gland. *Endocrinology* **142**, 2458–2467.
135. Negri-Cesi, P., Colciago, A., Poletti, A., and Motta, M. (1999). 5α-reductase isozymes and aromatase are differentially expressed and active in the androgen-independent human prostate cancer cell lines DU-145 and PC3. *Prostate* **41**, 224–232.
136. Krege, T. H., Hodgin, J. B., Couse, J. F., Enmark, E., Warner, M., Mahler, J. F., Sar, M., Korach, K. S., Gustafsson, J. A., and Smithies, O. (1998). Generation and reproductive phenotypes of mice lacking estrogen receptor β. *Proc. Natl. Acad. Sci. USA* **95**, 15677–15682.

137. Couse, J. F., Curtis Hewitt, S., and Korach, K. S. (2000). Receptor null mice reveal contrasting roles for estrogen receptor α and β in reproductive tissues. *J. Steroid Biochem. Mol. Biol.* **74**, 287–296.
138. Dupont, S., Krust, A., Gansmuller, A., Dierich, A., Chambon, P., and Mark, M. (2000). Effect of single and compound knockouts of estrogen receptor α (ERα) and β (ERβ) on mouse reproductive phenotypes. *Development* **127**, 4277–4291.
139. Paech, K., Webb, P., Kuiper, G. G., Nilsson, S., Gustafsson, J., Kushner, P. J., and Scanlan, T. S. (1997). Differential ligand activation of estrogen receptors ERα and ERβ at AP1 sites. *Science* **277**, 1508–1510.
140. Veldscholte, J., Voorhorst-Ogink, M. M., Bolt-de Vries, J., van Rooij, H. C., Trapman, J., and Mulder E. (1990). Unusual specificity of the androgen receptor in the human prostate tumor cell line LNCaP: High affinity for progestogenic and estrogenic steroids. *Biochim. Biophys. Acta* **1052**, 187–194.
141. Yeh, S. and Chang, C. (1996). Cloning and characterization of a specific coactivator, ARA70, for the androgen receptor in human prostate cells. *Proc. Natl. Acad. Sci. USA* **93**, 5517–5521.
142. Yeh, S., Miyamoto, H., Shima, H., and Chang, C. (1998). From estrogen to androgen receptor: A new pathway for sex hormones in prostate. *Proc. Natl. Acad. Sci. USA* **95**, 5527–5532.
143. Culig, Z., Klocker, H., Eberle, J., Kaspar, F., Hobisch, A., Cronauer, M. V., and Bartsch, G. (1994). DNA sequence of the androgen receptor in prostatic tumor cell lines and tissue specimens assessed by means of the polymerase chain reaction. *Prostate* **22**, 11–22.
144. Klotz, L. (2000). Hormone therapy for patients with prostate carcinoma. *Cancer* **88**, 3009–3014.
145. Katzenellenbogen, J. A., M'Malley, B. W., and Katzenellenbogen, B. S. (1996). Tripartite steroid hormone receptor pharmacology: Interaction with multiple effector sites as a basis for the cell- and promoter-specific action of these hormones. *Mol. Endocrinol.* **10**, 119–131.
146. Brady, M. E., Ozanne, D. M., Gaughan, L., Waite, I., Cook, S., Neal, D. E., and Robson, C. N. (1999). Tip60 is a nuclear hormone receptor coactivator. *J. Biol. Chem.* **274**, 17599–17604.
147. Kang, H. Y., Yeh, S., Fujimoto, N., and Chang, C. (1999). Cloning and characterization of human prostate coactivator ARA54, a novel protein that associates with the androgen receptor. *J. Biol. Chem.* **274**, 8570–8576
148. Xu, J., Qiu, Y., DeMayo, F. J., Tsai, S. Y., Tsai, M. J., and O'Malley B. W. (1998). Partial hormone resistance in mice with disruption of the steroid receptor coactivator-1 (SRC-1) gene. *Science* **279**, 1922–1925.
149. Scolnick, D. M., Chehab, N. H., Stavridi, E. S., Caruso, L., Moran, E., Berger, S. L., and Halazonetis, T. D. (1997). CREB-binding protein and p300/CBP-associated factor are transcriptional coactivators of the p53 tumor suppressor protein. *Cancer Res.* **57**, 3693–3696.
150. Aronica, S. M. and Katzenellenbogen, B. S. (1993). Stimulation of estrogen receptor-mediated transcription and alteration in the phosphorylation state of the rat uterine estrogen receptor by estrogen, cyclic adenosine monophosphate, and insulin-like growth factor-I. *Mol. Endocrinol.* **7**, 743–752.
151. Nakhla, A. M., Khan, M. S., Romas, N. P., and Rosner, W. (1994). Estradiol causes the rapid accumulation of cAMP in human prostate. *Proc. Natl. Acad. Sci. USA* **91**, 5402–5405.
152. Ding, V. D. H., Moller, D. E., and Feeney, W. P. (1998). Sex hormone-binding globulin mediates prostate androgen receptor action via a novel signaling pathway. *Endocrinology* **139**, 213–218.
153. Chen, T., Cho, R. W., Stork, P. J., and Weber, M. J. (1999). Elevation of cyclic adenosine 3′,5′-monophosphate potentiates activation of mitogen-activated protein kinase by growth factors in LNCaP prostate cancer cells. *Cancer Res.* **59**, 213–218.
154. Nazareth, L. V. and Weigel, N. L. (1996). Activation of the human androgen receptor through a protein kinase A signaling pathway. *J. Biol. Chem.* **271**, 19900–19907.
155. de Ruiter, P. E., Teuwen, R., Trapman, J., Dijkema, R., and Brinkmann, A. O. (1995). Synergism between androgens and protein kinase-C on androgen-regulated gene expression. *Mol. Cell Endocrinol.* **110**, R1–R6.
156. Bonkhoff, H., Wernert, N., Dhom, G., and Remberger, K. (1991). Relation of endocrine-paracrine cells to cell proliferation in normal, hyperplastic, and neoplastic human prostate. *Prostate* **19**, 91–98.
157. Ben-Josef, E., Yang, S. Y., Ji, T. H., Bidart, J. M., Garde, S. V., Chopra, D. P., Porter, A. T., and Tang, D. G. (1999). Hormone-refractory prostate cancer cells express functional follicle-stimulating hormone receptor (FSHR). *J. Urol.* **161**, 970–976.
158. Davies, P. and Eaton, C. L. (1989). Binding of epidermal growth factor by human normal, hypertrophic, and carcinomatous prostate. *Prostate* **14**, 123–132.
159. Turkeri, L. N., Sakr, W. A., Wykes, S. M., Grignon, D. J., Pontes, J. E., and Macoska, J. A. (1994). Comparative analysis of epidermal growth factor receptor gene expression and protein product in benign, premalignant, and malignant prostate tissue. *Prostate* **25**, 199–205.
160. Traish, A. M. and Wotiz, H. H. (1987). Prostatic epidermal growth factor receptors and their regulation by androgens. *Endocrinology* **121**, 1461–1467.
161. Yeh, S., Lin, H. K., Kang, H. Y., Thin, T. H., Lin, M. F., and Chang, C. (1999). From HER2/Neu signal cascade to androgen receptor and its coactivators: A novel pathway by induction of androgen target genes through MAP kinase in prostate cancer cells. *Proc. Natl. Acad. Sci. USA* **96**, 5458–5463.
162. Wen, Y., Hu, M. C., Makino, K., Spohn, B., Bartholomeusz, G., Yan, D. H., and Hung, M. C. (2000). HER-2/neu promotes androgen-independent survival and growth of prostate cancer cells through the Akt pathway. *Cancer Res.* **60**, 6841–6845.
163. Signoretti, S., Montironi, R., Manola, J., Altimari, A., Tam, C., Bubley, G., Balk, S., Thomas, G., Kaplan, I., Hlatky, L., Hahnfeldt, P., Kantoff, P., and Loda, M. (2000). Her-2-neu expression and progression toward androgen independence in human prostate cancer. *J. Natl. Cancer. Inst.* **92**, 1918–1925.
164. Reese, D. M., Small, E. J., Magrane, G., Waldman, F. M., Chew, K., and Sudilovsky, D. (2001). HER2 protein expression and gene amplification in androgen-independent prostate cancer. *Am. J. Clin. Pathol.* **116**, 234–239.
165. Lee, H., Akita, R. W., Sliwkowski, M. X., and Maihle, N. J. (2001). A naturally occurring secreted human ErbB3 receptor isoform inhibits heregulin-stimulated activation of ErbB2, ErbB3, and ErbB4. *Cancer Res.* **61**, 4467–4473.
166. Barnes, J., Anthony, C. T., Wall, N., and Steiner, M. S. (1995). Bone morphogenetic protein-6 expression in normal and malignant prostate. *World Urol. J.* **13**, 337–343.
167. Damon, S., Maddison, L., Ware, J., and Plymate, S. (1998). Overexpression of an inhibitory insulin-like growth factor binding protein (IGFBP), IGFBP-4, delays onset of prostate tumor formation. *Endocrinology* **139**, 3456–3464.
168. Iwamura, M., Sluss, P. M., Casamento, J. B., and Cockett, A. T. K. (1993). Insulin-like growth factor-I: Action and receptor characterization in human prostate cancer cell lines. *Prostate* **22**, 243–252.
169. Kimura, G., Kasuya, J., Giannini, S., Honda, Y., Mohan, S., Kawachi, M., Akimoto, M., and Fujita-Yamaguchi, Y. (1996). Insulin-like growth factor (IGF) system components in human prostate cancer cell lines: LNCaP, DU-145 and PC-3 cells. *Int. J. Urol.* **3**, 39–46.
170. Kovisto, P., Kononen, J., and Palmberg, C. (1997). Androgen receptor gene amplification: A possible mechanism for androgen deprivation therapy failure in prostate cancer. *Cancer Res.* **57**, 314–319.
171. Ferrara, N. and Davis-Smyth, T. (1997). The biology of vascular endothelial growth factor. *Endocr. Rev.* **18**, 4–25.
172. Minchenko, A., Bauer, T., Salceda, S., and Caro, J. (1994). Hypoxic stimulation of vascular endothelial growth factor expression in vitro and in vivo. *Lab. Invest.* **71**, 374–379.
173. Franck-Lissbrant, L., Haggstrom, S., Damber, J. E., and Bergh, A. (1998). Testosterone stimulates angiogenesis and vascular regrowth in the ventral prostate in castrated adult rats. *Endocrinology* **139**, 451–456.
174. Bergan, R. C., Reed, E., Myers, C. E., Headlee, D., Brawley, O., Cho, H. K., Figg, W. D., Tompkins, A., Linehan, W. M., Kohler, D.,

Steinberg, S. M., and Blagosklonny, M. V. (1999). A phase II study of high-dose tamoxifen in patients with hormone-refractory prostate cancer. *Clin. Cancer Res.* **5**, 2366–2373.

175. Griffiths, K., Cockett, A. T. K., Coffey, D. S., Krieg, M., Lee, C., McKechan, M., Nehl, D. E., Partin, A., Di Sant'Agnese, A., and Schalken, J. (1998). Regulation of prostate growth. *In* Griffiths, D. L., Khoury, S., Cockett, A. T. C., McConnell, J., Chatelain, C., Murphy, G., and Yoshida, E., Eds., *Fourth International Consultation on BPH*, pp 83–125. SCI, Paris.

176. Marcelli, M., Cunningham, G. R., Walkup, M., He, Z., Sturgis, L., Kagan, C., Mannucci, R., Nicoletti, I., Teng, B., and Denner, L. (1999). Signalling pathway activated during apoptosis of the prostate cancer cell line LNCaP: Overexpression of caspase-7 as a new gene therapy strategy for prostate cancer. *Cancer Res.* **59**, 382–390.

177. Denmeade, S. R., Nagy, A., Gao, J., Lilja, H., Schally, A. V., and Isaacs, J. T. (1998). Enzymatic activation of a doxorubicin-peptide prodrug by prostate-specific antigen. *Cancer Res.* **58**, 2537–2540.

178. Broxterman, H. J. and Hoekman, K. (1999). Direct activation of caspases by RGD-peptides may increase drug sensitivity of tumour cells. *Drug Resist. Updat.* **2**, 139–141.

179. Enyedy, I. J., Ling, Y., Nacro, K., Tomita, Y., Wu, X., Cao, Y., Guo, R., Li, B., Zhu, X., Huang, Y., Long, Y. Q., Roller, P. P., Yang, D., and Wang, S. (2001). Discovery of small-molecule inhibitors of Bcl-2 through structure-based computer screening. *J. Med. Chem.* **44**, 4313–4324.

CHAPTER 348

Retrograde Signaling in the Nervous System: Dorsal Root Reflexes

William D. Willis
Department of Anatomy and Neurosciences, University of Texas Medical Branch, Galveston, Texas

Cell-to-Cell Signaling in the Nervous System

Anterograde Signaling

The nervous system is designed for rapid cell-to-cell communication, which allows for quick adjustments in behavior in response to environmental imperatives. Some of the special features of nerve cells that provide for rapid communications between neurons and other nerve cells or effectors, such as muscle or gland cells, include nerve impulse propagation along the axons of neurons and synaptic transmission between neurons and other cells.

The great Spanish neuroanatomist, Santiago Ramón y Cajal, proposed the "law of dynamic polarization" of the nervous system [1]. This "law" states that nerve impulses propagate in dendrites toward the neuronal cell body and in axons away from the cell body. However, Sherrington [2] demonstrated that nerve impulses can propagate in either direction along axons and that it is the synapses that channel information flow in a forward direction (for example, from sensory input to motor output) through the nervous system [3]. Synapses serve as rectifiers. This is generally true whether the synapses are of the chemical or the electrical variety [4,5]. Neural signals are conveyed from a neuron to another cell in one direction across the synapse. Nerve impulses in the postsynaptic cell do not normally have a substantial effect on the presynaptic cell.

Several additional corrections to Cajal's concept are also necessary. Information flow in dendrites is generally in the form of synaptic currents, which affect the membrane potential at the action potential trigger zone in the axon hillock, where a nerve impulse may be triggered. The action potential then conducts distally along the axon, but also backward (antidromically) into the dendrites [6].

Dorsal root ganglion cells, the primary afferent sensory neurons of the spinal cord, do not have dendrites. Instead, they have a bifurcating axon that ends in sense organs in the periphery and in synapses within the spinal cord or caudal medulla on upstream neurons that are involved in motor responses or in the transmission of sensory information to higher levels of the nervous system. The peripheral axon normally conducts nerve impulses evoked in response to sensory stimulation toward the cell body of the dorsal root ganglion cell, and the central axon conducts nerve impulses away from the cell body and toward the synapses in the CNS (Fig. 1, topmost primary afferent neuron).

Retrograde Signaling

In some instances, chemical communication occurs between post- and presynaptic elements of a synapse, although this generally does not result in nerve impulse propagation in the reverse direction to the normal. For example, in the process of long-term potentiation, as studied in slices of the hippocampus, one of the factors that enhances synaptic transmission may be the generation of nitric oxide (NO) in the postsynaptic cell and the retrograde diffusion of the NO

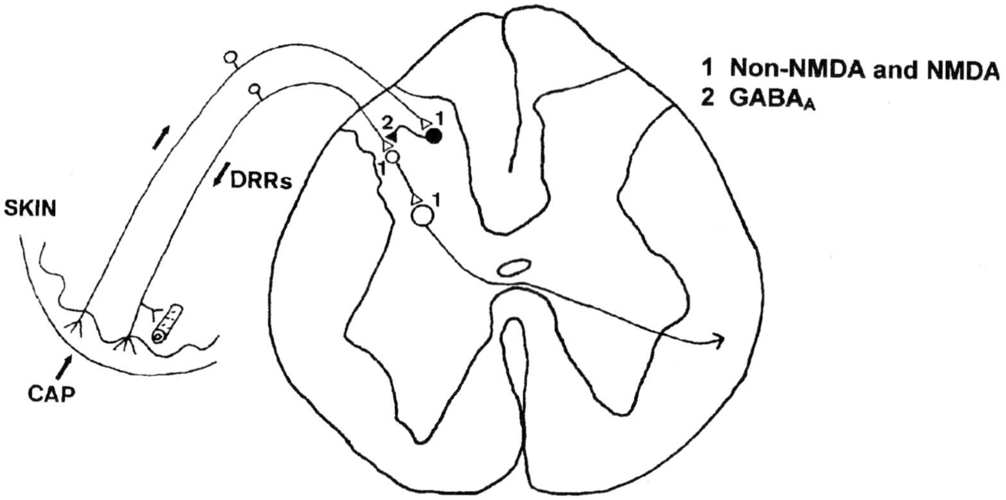

Figure 1 The upper left arrow in the diagram shows the normal direction of nerve impulse conduction in a primary afferent neuron. Conduction is from peripheral receptor endings, in this case in the skin, into the spinal cord. Also shown is a circuit involving a GABAergic inhibitory interneuron (shown in black) that releases GABA onto the synaptic terminals in the dorsal horn of another primary afferent neuron. Propagation of dorsal root reflexes (DRRs) in another primary afferent neuron toward the periphery is indicated by the lower right arrow. The receptor types that are involved in transmission within the dorsal horn circuit include non-NMDA and NMDA glutamate receptors and GABA$_A$ receptors. (Reprinted with permission from [38].)

across the synapse. Activation of intracellular signaling pathways in the presynaptic terminals leads to enhanced neurotransmitter release and stronger synaptic excitation [7].

At the level of neural circuits, there is often an arrangement of neurons that allows a postsynaptic neuron to provide feedback inhibition or excitation within a circuit. For example, activation of motor neurons in the spinal cord can lead not only to contraction of the skeletal muscle fibers supplied by the motor neurons, but also to inhibition of the same and related motor neurons. This inhibition involves the propagation of nerve impulses along recurrent collaterals of the motor axons and activation of inhibitory interneurons (called Renshaw cells after their discoverer), which in turn cause postsynaptic inhibition of the motor neurons on which the axons of the Renshaw cells synapse [4].

Neurogenic Inflammation

Antidromic Vasodilation

The neurotransmitter released at synapses in the spinal cord by most dorsal root ganglion cells is glutamate [8]. However, many small caliber primary afferent fibers, such as nociceptors, also contain and release peptides, such as substance P and/or calcitonin gene-related peptide [9]. Transmitter release is triggered by the invasion of synaptic terminals by nerve impulses, which cause the opening of voltage-gated calcium channels, Ca^{2+} influx into the presynaptic terminals, and exocytosis of synaptic vesicles that contain neurotransmitter [4,8]. Note, however, that this neurotransmitter release mechanism is present not only in the central terminals of dorsal root ganglion cells within the spinal cord but also in the terminals of these sensory neurons in the periphery [10]. This implies that the propagation of nerve impulses in the antidromic direction in sensory neurons will result in the release of transmitter substances in peripheral tissue. What would be the consequence of such an event?

Bayliss [11] stimulated the distal stump of a cut dorsal root electrically so that nerve impulses were conducted toward the periphery. The antidromically propagated action potentials in the sensory axons resulted in vasodilation in the skin. *Antidromic vasodilation* produced in this manner is accompanied by plasma extravasation, which leads to neurogenic edema [12,13]. For antidromic vasodilation and neurogenic edema both to occur, the electrical stimuli have to activate unmyelinated nociceptive sensory axons (C-fibers), although finely myelinated (Aδ) fibers can contribute to antidromic vasodilation [14–16]. Treatment of neonatal rats with capsaicin prevents antidromic vasodilation, indicating that the sensory fibers involved are capsaicin sensitive [13,17]. The vasodilation is attributable chiefly to the release of calcitonin gene-related peptide, although release of substance P probably also contributes, and the neurogenic edema to substance P release [10].

Inflammation

Inflammation is characterized in classical medical terminology by the occurrence of rubor, calor, tumor, and dolor. Inflammation in the skin often depends on an "efferent function" of its "afferent innervation." If inflammation is prevented when the skin is denervated, this type of inflammation is termed *neurogenic inflammation.*

Observations such as those of Bayliss [11] led to the proposal by Thomas Lewis [18] that inflammation of the skin

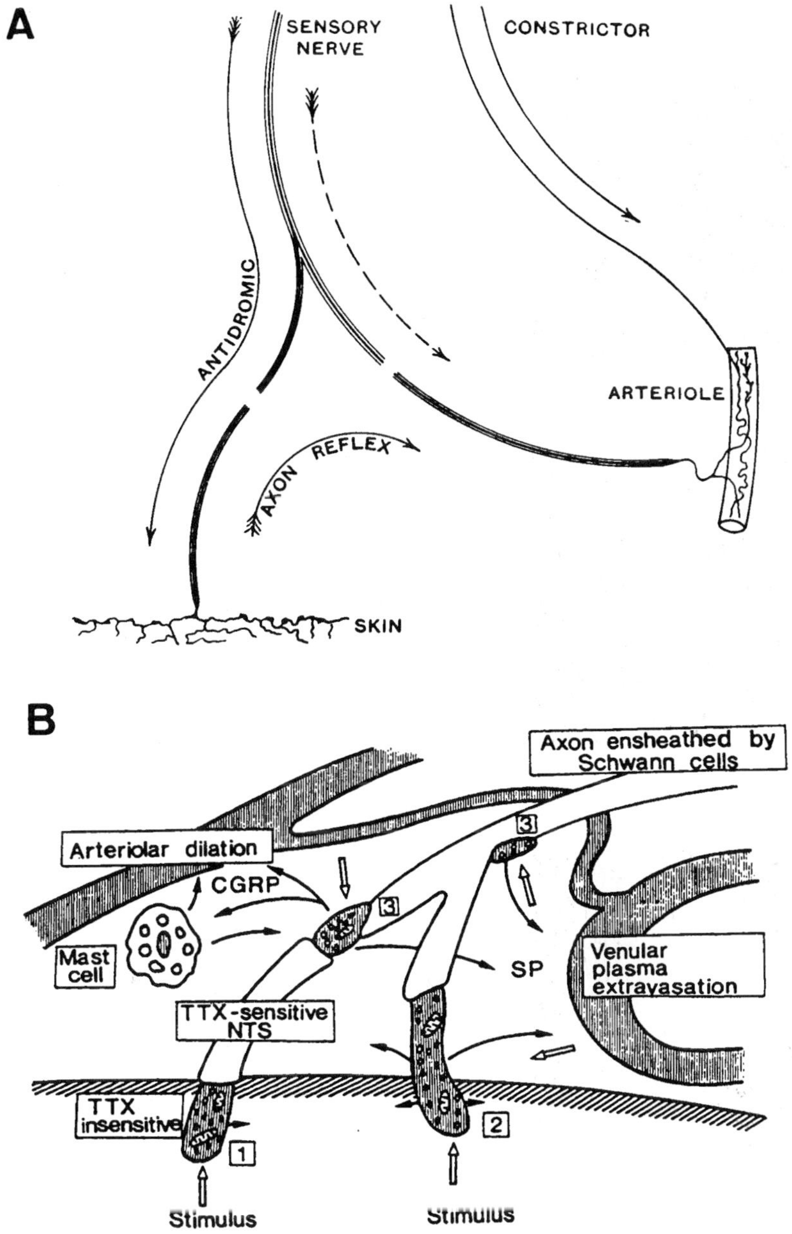

Figure 2 (A) The axon reflex as conceived by Lewis [18]. A sensory axon has terminals in the ski and projects to the spinal cord. However, it gives off a side branch that has endings near an arteriole. Activation of the sensory endings causes a nerve impulse to propagate as an axon reflex through the side branch, where a vasodilator substance is released on the arteriole. Antidromic activation of the same sensory axon by electrical stimulation of the dorsal root through which it enters the spinal cord also causes an axon reflex action on the arteriole. (B) Magnified view of a conceptualization similar to that of (A). A nerve terminal is shown with several sites for the release of the peptides, calcitonin gene-related peptide (CGRP) and substance P (SP). Stimuli applied to the nerve terminals can cause direct release of the peptides without the requirement for action potential generation (i.e., release still occurs in the presence of tetrodotoxin). However, release from the axon away from the terminals depends on propagation of an action potential (axon reflex). The CGRP causes vasodilation (and also stimulates a mast cell), and the SP produces plasma extravasation from a precapillary venule. (Part A reprinted with permission from [17]. Part B reprinted with permission from CRC Press, Boca Raton, Florida.)

is initiated by release of substances from nerve terminals, and that a damaging stimulus could cause not only direct release of substances from the affected nerve terminals but also from branches of these nerve terminals by propagation of an *axon reflex* (Fig. 2A). Although Lewis thought that histamine might be the chemical substance that produced inflammation, as mentioned, it is now believed that calcitonin gene-related peptide is the most important agent producing

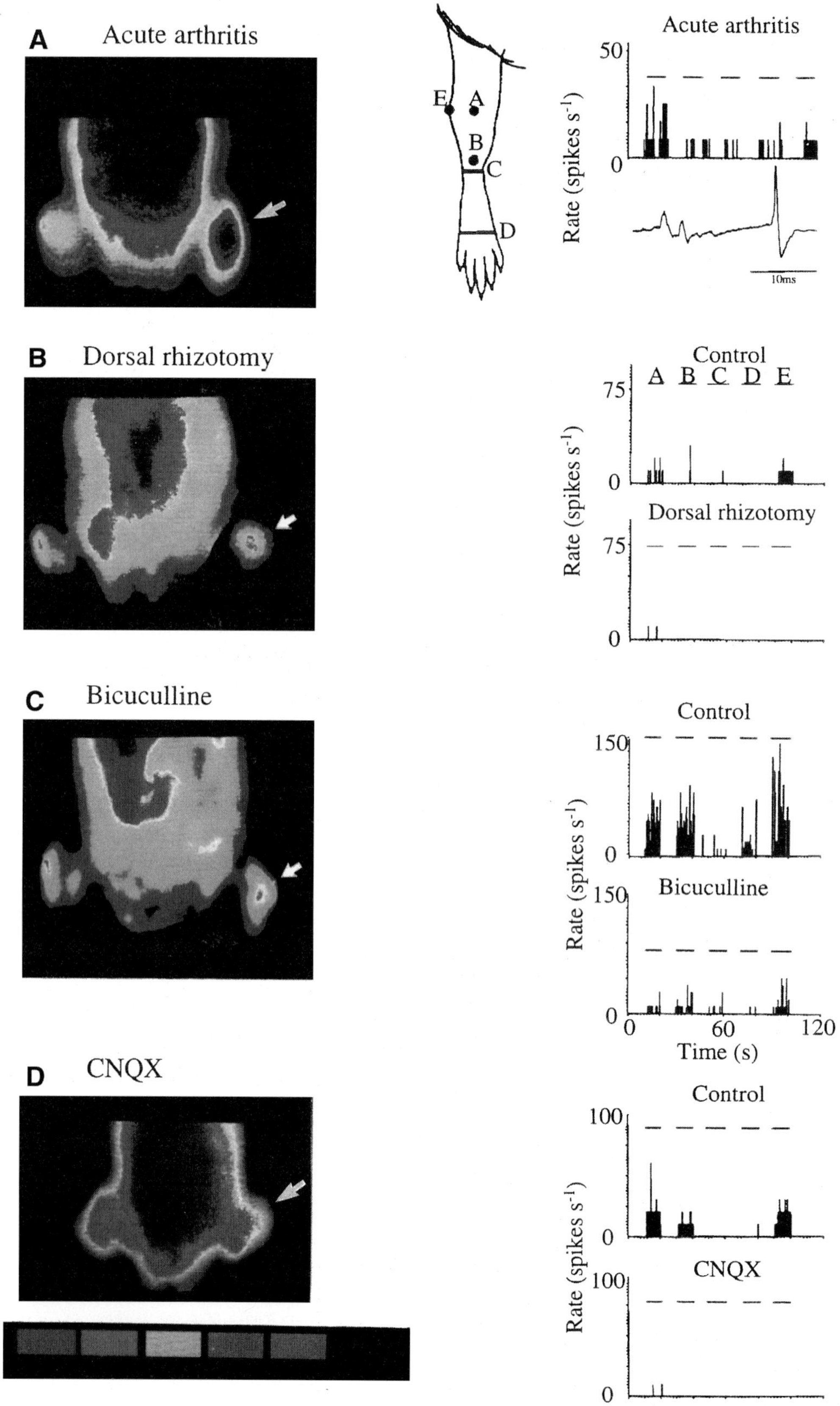

Figure 3 The temperature of the body surfaces of rats was monitored by use of a thermograph. The abdomen and knees of each rat were placed in contact with the thermograph, which displayed the temperature of just these parts of the body surface. An arrow by each thermograph illustrated indicates the knee that had been injected with kaolin and carrageenan to initiate inflammation. The amount of swelling and the temperatures of knee joints on the two sides can be judged by their size and color. The temperature scale at the bottom indicates that the highest

vasodilation and that substance P is responsible for neurogenic edema (Fig. 2B). Other substances that are involved in the progression of inflammation are released by injured cells, including platelets and immune cells, or are synthesized from plasma proteins. Some of these substances include not only histamine, but also serotonin, bradykinin, prostaglandins, and cytokines. A lowered pH and raised temperature are other important factors.

Dorsal Root Reflexes as Retrograde Signals

A form of retrograde signaling that occurs in the spinal cord and brain stem under certain conditions involves the triggering of *dorsal root reflexes.* It is unclear if these represent a physiological or a pathological process. Nevertheless, dorsal root reflexes that occur in certain types of primary afferent neurons can have dramatic peripheral effects that contribute to neurogenic inflammation [19].

Dorsal Root Reflexes and Primary Afferent Depolarization

Dorsal root reflexes are action potentials that are initiated in the axon terminals of dorsal root ganglion cells within the spinal cord. Dorsal root reflexes are triggered by dorsal root potentials [20,21], which are depolarizations that can be large enough to reach threshold in the primary afferent endings [22,23]. The dorsal root reflexes then propagate retrogradely (antidromically) back out the dorsal root and peripheral nerve to the sensory terminals in peripheral tissue (Fig. 1, DRRs) [24,25]. A depolarizing dorsal root potential (or *primary afferent depolarization*) normally results in a form of presynaptic inhibition [4]. However, when dorsal root reflexes occur, enhanced synaptic transmission is seen centrally [26,27], as well as peripheral release of transmitters. Primary afferent depolarization results from the activation of GABAergic interneurons in the spinal cord, and release of GABA causes depolarization of primary afferent terminals through an action on $GABA_A$ receptors (Fig. 1) [28]. This depolarization results from an efflux of Cl^- through Cl^- channels ([29]; also see review by Willis [19]).

Most investigators have assumed that dorsal root reflexes do not occur under normal conditions. However, recently dorsal root reflexes have been recorded from proprioceptive afferents during locomotion under relatively physiological conditions [30]. It was suggested that these nerve impulses might help determine the membrane excitability of the sensory terminals and hence the intensity of sensory discharges during certain phases of the step cycle. However, when dorsal root reflexes occur in fine afferent fibers, especially in C and Aδ nociceptors, peripheral events occur that resemble those that are produced by stimulation of a dorsal root.

Experimental Arthritis

Injection of the irritants kaolin and carrageenan into the knee joint in an experimental animal will result in the development of acute arthritis [31,32]. This arthritis is characterized by swelling of the joint (neurogenic edema) and by an increase in joint temperature (due to vasodilation). As the arthritis develops, dorsal root reflexes can be observed both in joint nerves and in filaments of dorsal root [33–35]. Part of the neurogenic inflammation of the knee joint can be attributed to a direct action of the irritants on the nerve terminals in the joint, which presumably causes the release of active substances from the terminals (see Fig. 2B). However, half of the swelling and part of the elevation in temperature can be prevented when dorsal root reflexes in the joint afferents are blocked by spinal cord administration of either CNQX or bicuculline (Fig. 3) [34,36,37]. Thus, part of the neurogenic inflammation is attributable to dorsal root reflexes.

Acute Inflammation of the Skin by Capsaicin

Similarly, when capsaicin is injected into the skin, a local reaction occurs near the injection site that is probably due to a direct effect of the capsaicin on peripheral endings of cutaneous nerve fibers. However, vasodilation and neurogenic edema also occur in the surrounding area of skin. The vasodilation or "flare" spreads at least 3 cm from the injection site [38]. However, in rats, the peripheral arborization of the terminals of C fibers is only 1 6 mm in diameter [16], and so the extent of the flare cannot be explained on the basis of axon reflexes. Intradermal injection of capsaicin results in powerful dorsal root reflexes that can be demonstrated in fine primary afferent fibers, including both C and Aδ fibers, but not in the large Aβ fibers (Fig. 4) [39].

If the appropriate peripheral nerves or dorsal roots are cut before the capsaicin injection, the flare is blocked (Fig. 5A) [38]. This surgery would prevent the spread of dorsal root reflexes into the nerve endings, but it should not interfere

Figure 3, *cont'd* temperature is signified (counterintuitively) by blue and the lowest by red. Note that there is a large difference between the size and temperature of the inflamed knee and the contralateral knee of an untreated, arthritic rat (A) but that the knees are nearly symmetrical when the dorsal roots were cut prior to the knee injection (B) or when the animals were pretreated by spinal cord administration of bicuculline (C) or CNQX (D). Therefore, cutting the dorsal roots or administering drugs to the spinal cord that block $GABA_A$ or glutamate receptors prevented much of the inflammation of the knee. The drawing of the rat hind limb at the top center shows the sites stimulated to evoke the dorsal root reflexes (A–E) shown in the right panels. In each of these panels, a peristimulus time histogram shows the dorsal root reflexes that resulted from stimulation at sites A–E. The trace below the uppermost panel illustrates dorsal root reflexes recorded from the medial articular nerve on the injected side in an untreated, arthritic rat. The other panels show the dorsal root reflexes recorded before and after dorsal rhizotomy or spinal administration of bicuculline or CNQX. (Reprinted with permission from [40].)

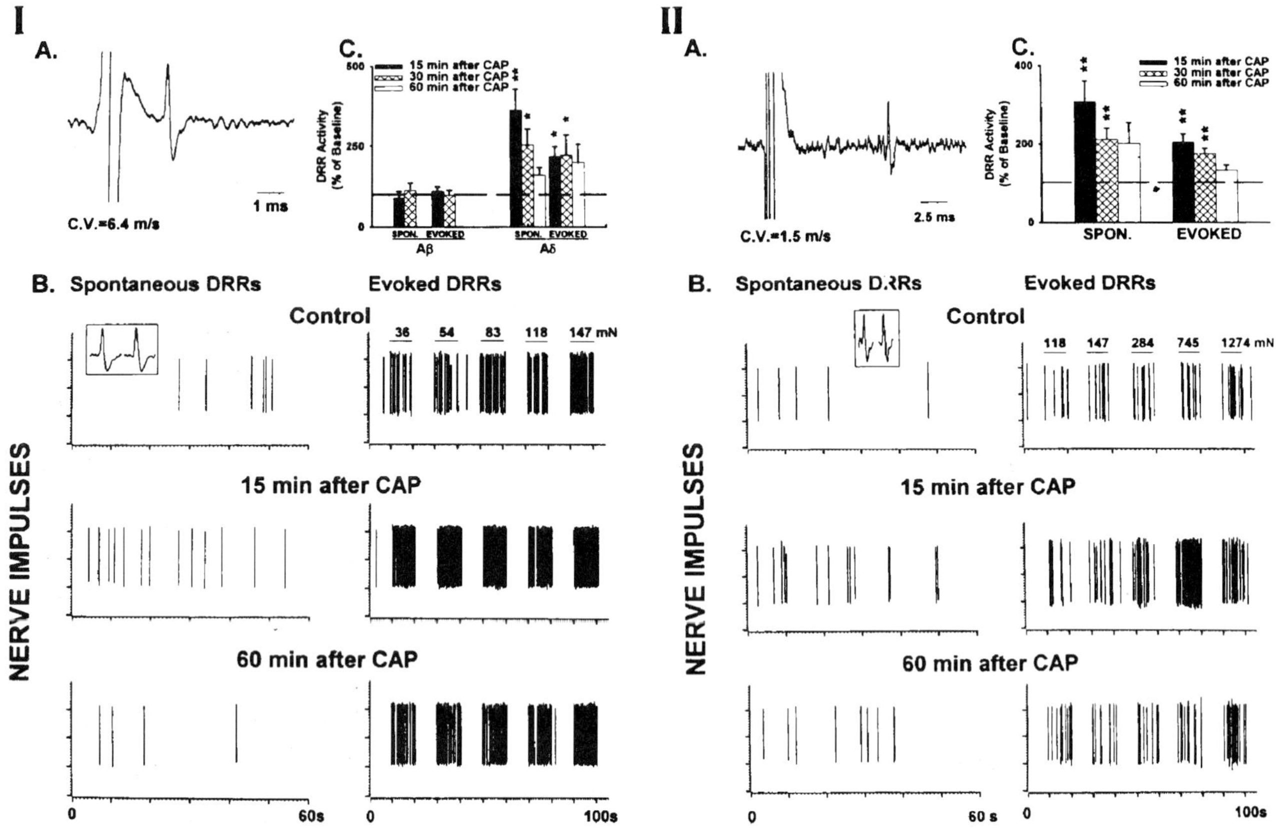

Figure 4 Changes in spontaneous and evoked dorsal root reflex activity before and after intradermal injection of capsaicin. (I) The recordings in (A) and (B) were from an Aδ fiber in a dorsal root filament (cut distally; other filaments of the dorsal root and adjacent dorsal roots were left intact). The conduction velocity of the axon was determined to be 6.4 m/sec based on the latency of the action potential following stimulation of the dorsal rootlet near the spinal cord in (A) and the measured conduction distance. The summary bar graphs in (C) show the increases in both the spontaneous and evoked dorsal root reflexes in Aδ ($n = 11$) and the absence of such changes in Aβ fibers ($n = 6$). (II) The responses of a C fiber recorded in the same way and summary histograms for recordings from C fibers ($n = 10$). (Reprinted with permission from [39].)

with "axon reflexes." The flare can also be prevented by spinal cord administration of antagonist drugs that block $GABA_A$ receptors (bicuculline) or receptors for excitatory amino acids (CNQX, AP7), as shown in Fig. 5B. This can be explained by prevention of primary afferent depolarization by blocking responses of $GABA_A$ receptors and by prevention of the activation of GABAergic interneurons by sensory volleys that enter the spinal cord from the skin and release glutamate to excite these interneurons. The neurogenic edema that occurs following capsaicin injection is also blocked by these surgical and pharmacological interventions, as are the dorsal root reflexes evoked by the injection [38].

Conclusions

Neurons are designed to communicate rapidly with other neurons and with effectors so that behavior can be adjusted quickly. Intercellular communication in the nervous system involves synaptic transmission between individual elements of a neural network. Propagation of information in such a network is generally unidirectional. However, instances are known of retrograde signaling. What may prove to be an important peripheral mechanism is the production of dorsal root reflexes triggered in primary afferent nociceptive neurons during inflammation. Dorsal root reflexes depend on a synaptic process known as primary afferent depolarization, which results from the release of γ-aminobutyric acid onto the terminals of the afferent fibers in the spinal cord and an action on $GABA_A$ receptors. Normally, the depolarization is subthreshold, but when it is large enough to exceed threshold for the afferent endings, nerve impulses are triggered, and these propagate retrogradely to the periphery. Damage causes an increase in the probability of dorsal root reflexes. The peripheral terminals of the nociceptive afferents then release neurotransmitters, including substance P and calcitonin gene-related peptide, which cause vasodilation and neurogenic edema. Experiments in which dorsal root reflexes are blocked by spinal cord administration of drugs that prevent GABAergic transmission interfere with the development of inflammation in the knee joint and in the skin, supporting the hypothesis that the retrogradely transmitted dorsal root reflexes contribute to neurogenic inflammation.

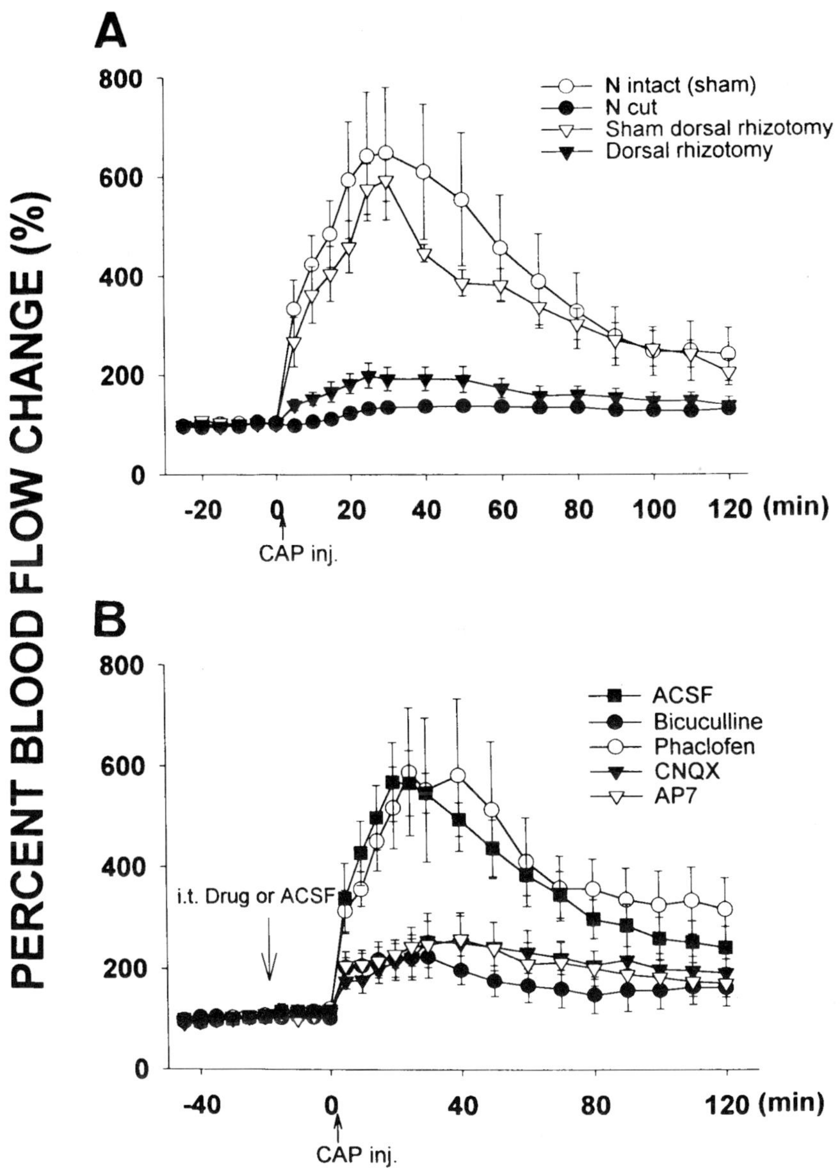

Figure 5 Blood flow changes following intradermal injection of capsaicin. Blood flow was measured by a laser Doppler blood flow probe placed on the plantar skin of the foot 15–20 mm distal to the site of capsaicin injection. The blood flow after capsaicin is presented as a percentage of that recorded before the injection. The experiments in (A) involved surgical procedures, whereas those shown in (B) involved pretreatment with an intrathecal infusion of a drug. Control groups had either sham surgery or intrathecal administration of artificial cerebrospinal fluid (ACSF). (A) shows that cutting the appropriate peripheral nerves or performing dorsal rhizotomies prevented most of the blood flow change induced by capsaicin injection. (B) shows that intrathecal pretreatment with the $GABA_A$ receptor antagonist, bicuculline, the non-NMDA glutamate receptor antagonist, CNQX, or the NMDA receptor antagonist, AP7, reduced the blood flow change, whereas the $GABA_B$ receptors antagonist, phaclofen, did not. (Reprinted with permission from [38].)

References

1. Cajal, S. R. (1909). *Histology of the Nervous System of Man and Vertebrates,* Vol. I. (Trans. by Swanson, N., and Swanson, L. W., from the 1909 French edition). Oxford University Press, New York, 1995.
2. Sherrington, C. S. (1900). The spinal cord. In Schäfer, E. A. Ed., *Textbook of Physiology.* Caxton, London.
3. Berlucchi, G. (1999). Some aspects of the history of the law of dynamic polarization of the neurons. From William James to Sherrington, from Cajal and Van Gehuchten to Golgi. *J. Hist. Neurosci.* **8**, 191–201.
4. Eccles, J. C. (1964). *The Physiology of Synapses.* Academic Press, New York.
5. Nicholls, J. G., Martin, A. R., and Wallace, B. G. (1992). *From Neuron to Brain: A Cellular and Molecular Approach to the Function of the Nervous System,* 3rd ed. Sinauer Associates, Sunderland, MA.
6. Eccles, J. C. (1957). *The Physiology of Nerve Cells.* Johns Hopkins Press, Baltimore, MD.

7. O'Dell, T. J., Hawkins, R. D., Kandel, E. R., and Arancio, O. (1991). Tests of the roles of two diffusible substances in long-term potentiation: evidence for nitric oxide as a possible early retrograde messenger. *Proc. Natl. Acad. Sci. USA* **88**, 11285–11289.
8. Robinson, J. D. (1990). *Mechanisms of Synaptic Transmission: Bridging the Gaps (1890–1990).* Oxford University Press, New York.
9. Willis, W. D. and Coggeshall, R. E. (1991). *Sensory Mechanisms of the Spinal Cord,* 2nd ed. Plenum Press, New York.
10. Geppetti, P. and Holzer, P., Eds. (1996). *Neurogenic Inflammation.* CRC Press, Boca Raton, FL.
11. Bayliss, W. M. (1901). On the origin from the spinal cord of the vasodilator fibres of the hind-limb, and on the nature of these fibres. *J. Physiol.* **26**, 173–209.
12. Jancsó, N., Jancsó-Gábor, A., and Szolcsányi, J. (1967). Direct evidence for neurogenic inflammation and its prevention by denervation and by pretreatment with capsaicin. *Br. J. Pharmacol.* **31**, 138–151.
13. Lembeck, F. and Holzer, P. (1979). Substance P as neurogenic mediator of antidromic vasodilatation and neurogenic plasma extravasation. *Naunyn Schmiedebergs Arch. Pharmacol.* **310**, 175–183.
14. Kenins, P. (1981). Identification of the unmyelinated sensory nerves which evoke plasma extravasation in response to antidromic stimulation. *Neurosci. Lett.* **25**, 137–141.
15. Jänig, W. and Lisney, S. J. W. (1989). Small diameter myelinated afferents produce vasodilatation but not plasma extravasation in rat skin. *J. Physiol.* **415**, 477–486.
16. Bharali, L. A. M. and Lisney, S. J. W. (1992). The relationship between unmyelinated afferent type and neurogenic plasma extravasation in normal and reinnervated rat skin. *Neuroscience* **47**, 703–712.
17. Szolcsányi, J. (1988). Antidromic vasodilatation and neurogenic inflammation. *Agents Actions* **23**, 4–11.
18. Lewis, T. (1927). *The Blood Vessels of the Human Skin and their Responses.* Shaw and Sons, London.
19. Willis, W. D. (1999). Dorsal root potentials and dorsal root reflexes: A double-edged sword. *Exp. Brain Res.* **124**, 395–421.
20. Barron, D. H. and Matthews, B. H. C. (1938). The interpretation of potential changes in the spinal cord. *J. Physiol.* **92**, 276–321.
21. Lloyd, D. P. C. and McIntyre, A. K. (1949). On the origins of dorsal root potentials. *J. Gen. Physiol.* **32**, 409–443.
22. Koketsu, K. (1956). Intracellular potential changes of primary afferent nerve fibers in spinal cords of cats. *J. Neurophysiol.* **19**, 375–392.
23. Eccles, J. C., Kozak, W., and Magni, F. (1961). Dorsal root reflexes of muscle group I afferent fibres. *J. Physiol.* **159**, 128–146.
24. Barron, D. H. and Matthews, B. H. C. (1935). Intermittent conduction in the spinal cord. *J. Physiol.* **85**, 73–103.
25. Barron, D. H. and Matthews, B. H. C. (1935). Recurrent fibers of the dorsal roots. *J. Physiol.* **85**, 104–108.
26. Eccles, R. M. and Willis, W. D. (1962). Presynaptic inhibition of the monosynaptic reflex pathway in kittens. *J. Physiol.* **165**, 403–420.
27. Duchen, M. R. (1986). Excitation of mouse motoneurones by GABA-mediated primary afferent depolarization. *Brain Res.* **379**, 182–187.
28. Eccles, J. C., Schmidt, R. F., and Willis, W. D. (1963). Pharmacological studies on presynaptic inhibition. *J. Physiol.* **168**, 500–530.
29. Gallagher, J. P., Higashi, H., and Nishi, S. (1978). Characterization and ionic basis of GABA-induced depolarizations recorded *in vitro* from cat primary afferent neurones. *J. Physiol.* **275**, 263–282.
30. Roissignol, S., Beloozerova, I., Gossard, J. P., and Dubuc, R. (1998). Presynaptic mechanisms during locomotion. In Rudomin, P., Romo, R., and Mendell, L. M., Eds., *Presynaptic Inhibition and Neural Control,* pp. 385–397. Oxford University Press, New York.
31. Schaible, H. G. and Schmidt, R. F. (1985). Effects of an experimental arthritis on the sensory properties of fine articular afferent units. *J. Neurophysiol.* **54**, 1109–1122.
32. Sluka, K. A. and Westlund, K. N. (1992). An experimental arthritis in rats: Dorsal horn aspartate and glutamate increases. *Neurosci. Lett.* **145**, 141–144.
33. Rees, H., Sluka, K. A., Westlund, K. N., and Willis, W. D. (1994). Do dorsal root reflexes augment peripheral inflammation? *Neuro Report* **5**, 821–824.
34. Rees, H., Sluka, K. A., Westlund, K. N., and Willis, W. D. (1995). The role of glutamate and GABA receptors in the generation of dorsal root reflexes by acute arthritis in the anesthetized rat. *J. Physiol.* **484**, 437–445.
35. Sluka, K. A., Rees, H., Westlund, K. N., and Willis, W. D. (1995). Fiber types contributing to dorsal root reflexes induced by joint inflammation in cats and monkeys. *J. Neurophysiol.* **74**, 981–989.
36. Sluka, K. A. and Westlund, K. N. (1993). Centrally administered non-NMDA but not NMDA receptor antagonists block peripheral knee joint inflammation. *Pain* **55**, 217–225.
37. Sluka, K. A., Willis, W. D., and Westlund, K. N. (1993). Joint inflammation and hyperalgesia are reduced by spinal bicuculline. *Neuro Report* **5**, 109–112.
38. Lin, Q., Wu, J., and Willis, W. D. (1999). Dorsal root reflexes and cutaneous neurogenic inflammation after intradermal injection of capsaicin in rats. *J. Neurophysiol.* **82**, 2602–2611.
39. Lin, Q., Zou, X., and Willis, W. D. (2000). Aδ and C primary afferents convey dorsal root reflexes after intradermal injection of capsaicin in rats. *J. Neurophysiol.* **84**, 2695–2698.
40. Willis, W. D., Sluka, K. A., Rees, H., and Westlund, K. N. (1998). A contribution of dorsal root reflexes to peripheral inflammation. In Rudomin, P., Romo, R., and Mendell, L. M., Eds., *Presynaptic Inhibition and Neural Control,* pp. 407–423. Oxford University Press, New York.

CHAPTER 349

Cytokines and Cytokine Receptors Regulating Cell Survival, Proliferation, and Differentiation in Hematopoiesis

Fiona J. Pixley and E. Richard Stanley

General Aspects of Hematopoiesis

Blood contains red cells, megakaryocytes, lymphocytes, monocytes, and the various types of granulocytes. All mature blood cell types turn over rapidly, requiring active synthesis of large numbers of new cells to maintain a steady state. Differentiated blood cells are all ultimately derived from a small pool of undifferentiated, pluripotent hematopoietic stem cells. This process, involving extensive cell proliferation and differentiation, is known as *Hematopoiesis* and begins in the embryonic yolk sac (primitive hematopoiesis) before maturing into definitive hematopoiesis in the fetal liver and adult bone marrow. Hematopoiesis is regulated by a large number of cytokines that are present in the microenvironment. Specific subsets of these cytokines influence each step in the process. This chapter reviews our current knowledge of the biology of adult hematopoiesis with an emphasis on the cytokines and their target signaling pathways important in steady-state maintenance of nonlymphoid cell lineages. Due to page limitations, only selected papers or comprehensive reviews are referenced.

Till and McCulloch [1] introduced the concept of a hematopoietic stem cell with the capacity to (1) self-replicate, (2) proliferate to produce many cell progeny, and (3) differentiate to generate all the mature blood cell types. The pool of such stem cells, which are normally quiescent or cycling slowly, represents only 10^{-5} of the total nucleated bone marrow cells in the mouse. Upon division, a stem cell gives rise to an indistinguishable daughter cell, and a daughter cell proliferates extensively and differentiates to give rise to common myeloid and lymphoid progenitor cells, which, as they proliferate and differentiate, progressively develop a more restricted capacity for differentiation, eventually giving rise to cells that are capable of forming only one mature blood or lymphoid cell type [2] (Fig. 1).

Many different cytokines in the microenvironment of the bone marrow stimulate the development of cells of different lineages. These growth factors may be circulating or bound to either the surface of their producing cells or to the extracellular matrix. Under the influence of specific cytokines, progenitor cells can also proliferate and differentiate in semisolid culture to form macroscopic colonies of differentiated cells, hence the term *colony-stimulating factor* (CSF) for the responsible growth factor [3]. The analysis of mice with targeted inactivations of the genes encoding most of these cytokines and their receptors has greatly increased our understanding of the biology of hematopoiesis. The dominant role of cytokines in the regulation of hematopoiesis has led to their rapid clinical application in maintaining normal hematopoiesis in the face of a variety of pathological conditions.

Hematopoietic Cytokines

The hematopoietic cytokines are glycoproteins, which are either constitutively present in the circulation (e.g., CSF-1, SCF, FL, G-CSF, EPO, and TPO) or induced to appear in

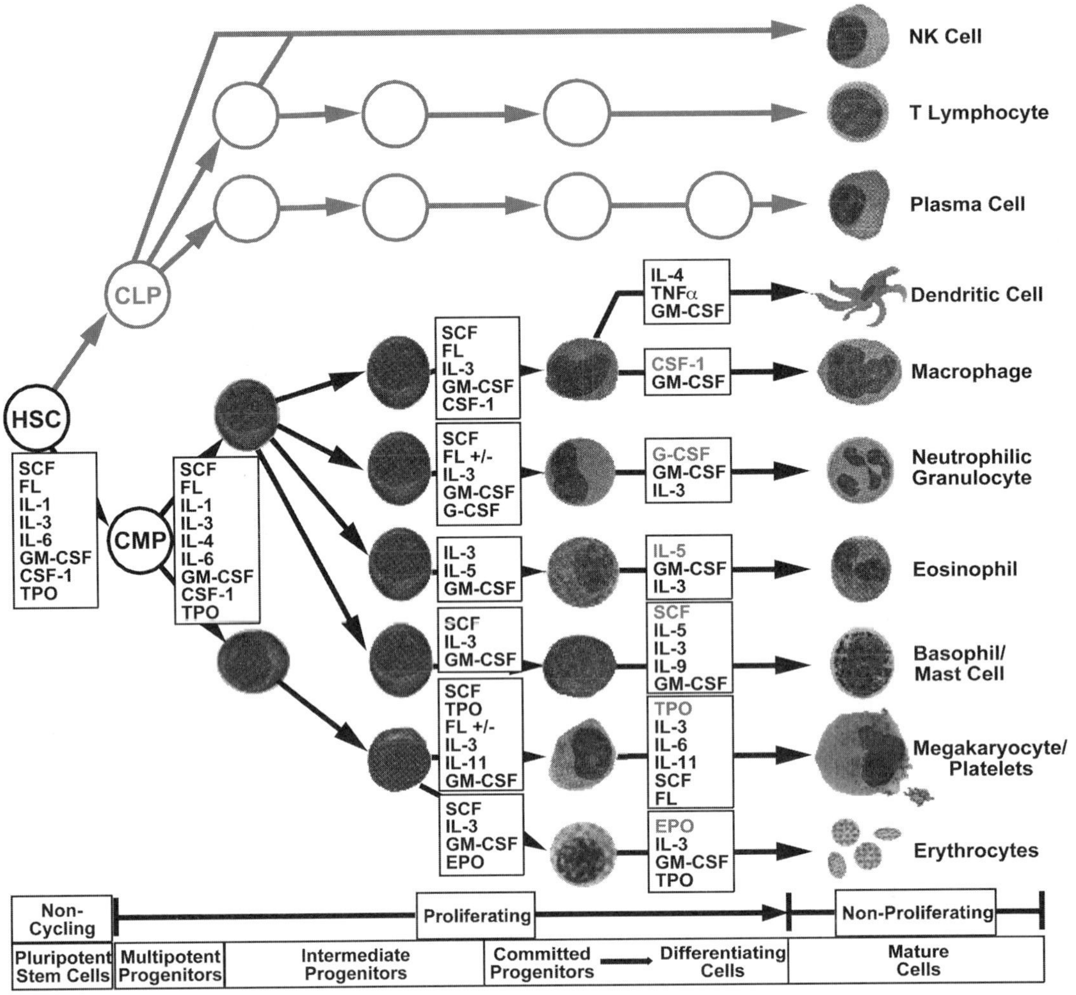

Figure 1 Hematopoietic cells, indicating the points of regulation by hematopoietic cytokines. The primary cytokines regulating proliferation and differentiation of committed progenitors of individual lineages are colored in red. Abbreviations: HSC, hematopoietic stem cell; CLP, common lymphoid progenitor; CMP, common myeloid progenitor; SCF, stem cell factor; FL, flt ligand; IL, interleukin; GM-CSF, granulocyte/macrophage colony-stimulating factor; CSF-1, colony-stimulating factor 1; G-CSF, granulocyte-CSF; TPO, thrombopoietin; EPO, erythropoietin; TNF, tumor necrosis factor. (A color representation of this figure is available on the CD version of the *Handbook of Cell Signaling*.)

response to infection or inflammation (e.g., GM-CSF, IL-3, IL-5, IL-6, and IL-11), and their concentrations may be increased in response to specific triggering conditions such as EPO, which is increased in response to hypoxia and/or anemia (Table I).

Primitive multipotent hematopoietic cells coexpress different lineage-specific cytokine receptors at low levels require a combination of cytokines (e.g., SCF, IL-1, IL-3, IL-6, GM-CSF, and CSF-1) for lineage commitment. As these cells differentiate, they lose receptors for some cytokines (e.g., SCF or IL-3), while increasing expression of receptors for the late-acting cytokines (e.g., CSF-1 or EPO). When they reach the stage of committed progenitor cell, their further proliferation and differentiation is along one particular lineage and is regulated by one or more late-acting cytokines. Within specific lineages, the most primitive cells respond by both proliferating and differentiating (e.g., committed macrophage progenitors → monoblasts → promonocutes → monocytes → macrophages), whereas differentiating, nondividing cells (e.g., peritoneal macrophages) require the cytokines for survival, activation, and function. CSF-1 and GM-CSF, for example, prime mature cells for neutrophil and macrophage activation by other agents [3]. Despite an apparent overlap in target cell specificity of several cytokines, their functions are largely nonredundant as indicated by the distinct hematopoietic phenotypes of cytokine or receptor nullizygous mice (Table I). Transcription factors such as GATA-1 and PU.1, which regulate cytokine receptor expression, are also important in hematopoietic cell commitment.

The "permissive" model of hematopoietic cell regulation by cytokines is one in which the growth factor does not have a role in multipotent progenitor cell commitment but simply allows the survival and proliferation of committed cells. In contrast, the "instructive" model posits that specific cytokines direct multipotent progenitors to become committed to a specific lineage. It is uncertain whether cytokine regulation of differentiation is simply permissive or instructive since there is good evidence for both mechanisms. Both may be

Table I Hematopoietic Cytokines

Cytokine	Sources of cytokine- or cytokine-receptor	Primary Hematopoietic Phenotype deficient mice[a]
SCF	Bone marrow stroma, Fibroblasts, Placenta, Others	Severe macrocytic anemia
FL	Ubiquitous	NK cell and dendritic cell deficiencies
CSF-1	Endothelial cells, Fibroblasts, Uterine epithelium, Macrophages	Osteopetrosis
EPO	Kidney proximal tubular cells, Liver	Severe anemia
G-CSF	Activated bone marrow stroma, Macrophages, Fibroblasts, Endothelial cells	Neutropenia
TPO	Hepatocytes, Endothelial cells Fibroblasts	Thrombocytopenia
IL-3	Activated T cells	Reduced delayed hypersensitivity
GM-CSF	Bone marrow stroma, Activated T cells, Endothelial cells, Fibroblasts, Macrophages	Pulmonary alveolar proteinosis
IL-5	Activated helper T cells	Eosinophil deficiency
IL-2	Activated T cells	Autoimmune disease
IL-15	Ubiquitous, increased by activation	Lymphopenic, deficient in NK, NK-T, CD8+ cells and gδT cells
IL-7	Fetal liver, bone marrow and thymic, stromal cells, lymphoid cells, others	Reduces T and B cells
IL-4	TN2 and NK1.1+ T cells, mast cells, basophils and eosinophils	TN2 deficient
IL-9	TN2 cells, mast cells and eosinophils	Pulmonary mastocytosis and goblet cell hyperplasia
IL-6	Ubiquitous, in response to inflammatory stimuli	Reduced T cells, IgG & IgA responses, impaired neutrophil/macrophage function
IL-11	Ubiquitous, in response to inflammatory stimuli	Embryonic Lethal
LIF	Monocytes, bone marrow stroma	None reported

[a]References for these cytokine and cytokine receptor-deficient mice, except for the IL-9 deficient mouse [17] are listed in [3].

utilized depending on the receptors and commitment steps involved [3].

The phenomenon of synergism between predominantly late-acting, lineage-restricted cytokines, such as CSF-1, EPO, and G-CSF, with predominantly early-acting cytokines such as SCF, in stimulating the proliferation and differentiation of primitive multipotent cells provides a mechanism for coupling the changes in levels of the late-acting cytokine. These late-acting cytokines are tightly regulated by the primary stimuli, to the channeling of multipotent cells into a lineage in order to satisfy the demand for differentiated cells. The mechanisms underlying synergism between cytokines in the regulation of primitive hematopoietic cell proliferation can occur directly at the level of the receptors for the synergizing cytokines or be due to synergistic effects between postreceptor signal transduction pathways [3].

Signaling through Cytokine Receptors

General

Hematopoietic cytokine action on target cells is mediated by specific, high-affinity, cell-surface receptors that signal for progenitor cell survival, proliferation, and differentiation, and mature cell survival and activation. Cytokine receptors can be classified according to the presence of either intrinsic or associated tyrosine kinase activity, the structure of the extracellular domain (ECD), and the requirement for common shared receptor subunits (Fig. 2). The end result of any cytokine binding to its cognate receptor is the phosphorylation of particular receptor intracellular domain (ICD) tyrosine residues that act as binding sites for downstream signaling molecules. Different activated receptor cytoplasmic

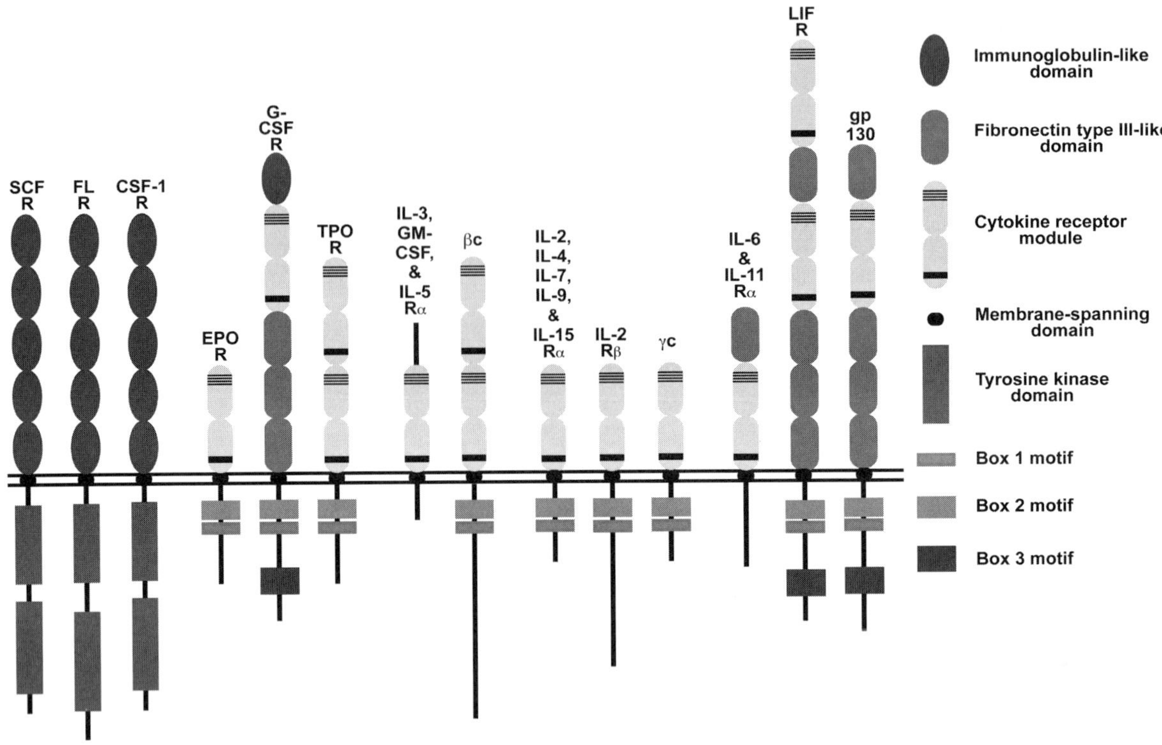

Figure 2 Schematic of selected high-affinity hematopoietic cytokine receptors showing the similarity between those containing intrinsic tyrosine kinase domains (left) and the modular nature of the cytokine receptor family that are preassociated with cytoplasmic tyrosine kinases.

domains often bind a common signaling molecule or family of signaling molecules. Targeted downstream pathways include those regulating gene transcription, protein translation, actin cytoskeletal remodeling, and cell adhesion and motility [3–5].

Tyrosine Kinase Receptors

The three hematopoietic cytokines signaling through tyrosine kinase receptors, stem cell factor (SCF), flt3 ligand (FL), and colony-stimulating factor 1 (CSF-1), are members of a family of homodimeric cytokines that share some sequence and structural similarity. SCF and CSF-1 have been shown to have effects on nonhematopoietic as well as hematopoietic cells.

The SCF, CSF-1, and FL receptors, all members of the PDGF receptor family, possess ECDs comprised of five heavily glycosylated immunoglobulin-like repeats, a transmembrane domain, and an intracellular tyrosine kinase domain that is interrupted by a kinase insert domain. Binding of their cognate bivalent ligands by this class of receptors stabilizes their noncovalent dimerization, permitting receptor activation and trans-tyrosine phosphorylation of one ICD by the other. The receptor phosphotyrosines act as "docking sites" for *src* homology region 2 (SH2) and other phosphotyrosine binding domains of signaling and adaptor proteins that bind the receptor and may themselves become tyrosine phosphorylated (Fig. 3A). Many of the signaling pathways activated by these receptors, including the MAP kinase (MAPK) cascade, the JAK/STAT pathway, Src family members, and PI3-kinase, are shared. Two of the receptors, SCFR and CSF-1R, are encoded by the protooncogenes c-*kit* and c-*fms*, respectively. The v-*kit* and v-*fms* oncogenes are present in feline sarcomatous retroviruses and contain mutations of the normal cellular genes that render their encoded receptors constitutively active in the absence of cytokines.

SCF and the SCF Receptor (SCFR)

SCF influences development in the three different lineages involving pigmentation, hematopoiesis, and fertility. It is widely expressed, as both secreted and cell surface forms, during embryogenesis and in a variety of adult tissues, including bone marrow stromal cells, fibroblasts, and endothelial cells, as well as the yolk sac and placenta. Cells expressing the SCFR are frequently contiguous with SCF-expressing cells and include germ cells, interstitial cells of Cajal in the gut, melanocytes, and early hematopoietic cells. Hematopoetic cell SCFR expression is low in very primitive multipotent progenitors, highest in committed progenitors, and decreases as cells mature. Because pluripotent stems cells do not appear to express the SCFR and can survive *in vitro* in the absence of SCF, it may not act on the earliest hematopoietic precursors. Although SCF has a broad spectrum of activity on hematopoietic cells, it has little activity alone but acts synergistically with many of the hematopoietic cytokines, especially IL-6 and IL-3, to increase the numbers of precursors of most, if not

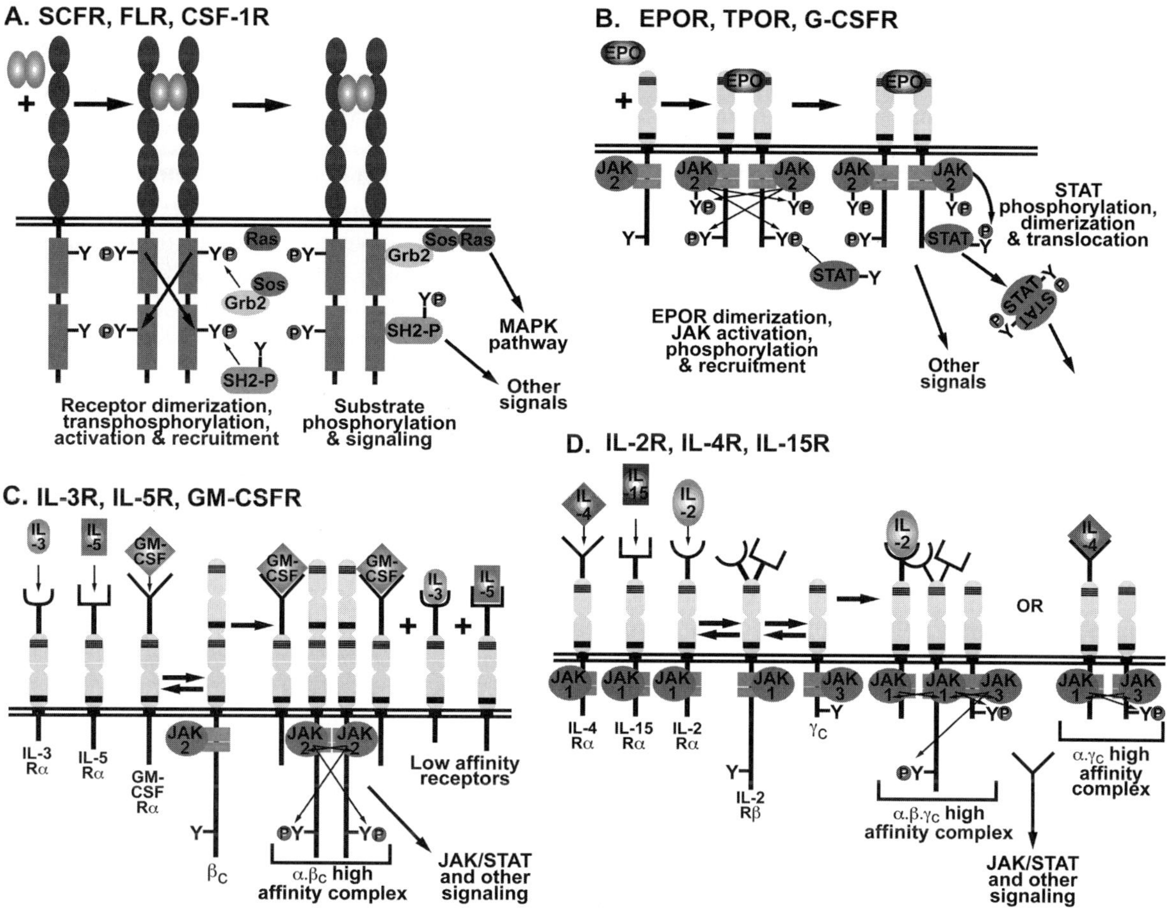

Figure 3 Models for the activation and signaling of hematopoietic growth factor receptors. (A) Homodimeric receptors possessing intrinsic tyrosine kinase domains; SH2-P, SH2 domain- or PTB domain-containing signaling molecules. (B) Homodimeric cytokine family receptors. (C) Multimeric cytokine receptors sharing the common β subunit, βc, illustrating the concept of cross-competition between cytokines for the β subunit that is necessary for the formation of the high-affinity signaling complex. GM-CSF is shown forming the high-affinity complex. (D) Receptors sharing the common γ subunit, γc, form a high-affinity α · γc receptor complex (e.g., IL-4) or a high-affinity α · β· γc receptor complex utilizing a second signaling subunit, IL-2β, (e.g., IL-2, IL-15).

all, lineages. Spontaneously occuring SCF-null *Sl/Sl* mice are embryonic lethals, but a partially functional Sl^d allele allows embryonic survival, yielding severely anemic and mast-cell-deficient mice, indicating that the major effects of SCF are on erythropoiesis and mast cell development. In combination with EPO, SCF enhances the number of erythroid precursor cells generated from primitive multipotent cells and allows precursors to respond to levels of EPO that are too low to elicit a response in the absence of SCF. Mast cells and their progenitors require SCF throughout the differentiation of the lineage from early precursors to mature, primed tissue mast cells. There is also strong evidence of biologic activity for SCF on the megakaryocyte, granulocyte/macrophage, and lymphoid lineages, yet the effects of the absence of SCF on their development is minimal, implying some redundancy in cytokine action on these lineages [6,7].

FL AND FLT3

FL, which occurs largely as a cell-surface, noncovalently associated homodimer, regulates the proliferation of primitive hematopoietic cells. It is widely expressed while hematopoietic expression of its cognate receptor, flt3, is predominantly restricted to the progenitor/stem cell compartment. Like SCF, FL alone cannot stimulate the proliferation of its primitive hematopoietic target cells, but rather synergizes with other hematopoietic cytokines. In contrast to SCF, however, there is little or no effect of FL on erythroid, or megakaryocyte progenitor cells, because flt3-deficient mice have no defects in red cell, megakaryocyte, or platelet production. In combination with GM-CSF, TNF, and IL-4, FL enhances the production of denditric cells, and dendritic cell numbers are reduced in FL-deficient mice [7].

CSF-1 AND THE CSF-1 RECEPTOR (CSF-1R)

CSF-1 regulates the survival, differentiation, and function of cells of the mononuclear phagocytic (monocyte/macrophage) lineage and the function of cells of the female reproductive tract. Mature forms of the disulfide-linked CSF-1 homodimer include a secreted glycoprotein, a secreted proteoglycan, and a cell-surface glycoprotein. CSF-1 is synthesized by a variety of cell types, including fibroblasts, endothelial cells, bone marrow stromal cells, osteoblasts, keratinocytes, astrocytes, myoblasts, and breast and uterine epithelial cells. Circulating CSF-1 is elevated in response to bacterial, viral, and parasitic infections and is primarily cleared by Kupffer cells, so the number of sinusoidally located macrophages determines the concentration of the cytokine responsible for their production, a simple feedback control. CSF-1 homodimer binding by the CSF-1R results in the formation or stabilization and activation of a receptor dimer. After activation and tyrosine phosphorylation of signaling molecules, most of the receptor–ligand complexes are internalized, ubiquitinated, and destroyed.

The osteopetrotic (*Csf-1^{op}/Csf-1^{op}*) mouse, which possesses an inactivating mutation in the CSF-1 gene, exhibits impaired bone resorption associated with a paucity of osteoclasts, no incisors, poor fertility, a lower body weight, a shortened life span, and deficiencies in blood monocytes and macrophages in most tissues, indicating that CSF-1 is the primary regulator of mononuclear phagocytes. Restoration of circulating CSF-1 in newborn *Csf-1^{op}/Csf-1^{op}* mice cures their osteopetrosis, monocytopenia, and some but not all of the tissue macrophage populations, demonstrating additional local regulation by CSF-1. CSF-1 regulates the development of macrophages with trophic and scavenger (i.e., physiological) functions, whereas the development of macrophages involved in inflammatory and immunologic (i.e., pathological) functions such as lymph node and thymic macrophages apparently depends on other cytokines. CSF-1 synergizes with other cytokines (e.g., IL-1, SCF, IL-3, IL-6) to stimulate the proliferation and differentiation of multipotent cells to committed macrophage progenitors that respond to CSF-1 alone [8,9].

Cytokine Signaling by Homodimerization of a Single, Non-Tyrosine Kinase Receptor Polypeptide Chain

Erythropoietin (EPO), granulocyte colony-stimulating factor (G-CSF), and thrombopoietin (TPO) signal by initiating homodimerization of their cognate non-tyrosine kinase receptors, although additional receptor subunits may be involved. These receptors belong to a large family of receptors, the cytokine receptor family, defined by the presence of an ECD cytokine receptor module (Fig. 2). These modules contain conserved short amino acid sequence elements, particularly two pairs of conserved cysteine residues near the amino terminal end of the motif and a Trp-Ser-X-Trp-Ser (WSXWS) sequence, the function of which is not clear, near the transmembrane domain. The EPOR, G-CSFR, and TPOR each contain conserved motifs in their ICDs that mediate constitutive association with members of the cytosolic Janus kinase (JAK) family (Figs. 2 and 3B). The associated JAKs (JAK2 with EPOR and TPOR; JAK1 with G-CSFR) are activated by formation of the ligand–receptor complex, leading to their tyrosine phosphorylation and to tyrosine phosphorylation of the signaling subunit. This creates docking sites for molecules with protein tyrosine binding (PTB) or SH2 domains, and to phosphorylation and activation of these recruited signaling molecules. Recruited molecules include the signal transducers and activators of transcription (STATs), additional JAKs, and other cytosolic tyrosine kinases and SH2 domain-containing protein tyrosine phosphatases. The MAPK pathway and PI3 kinase are also activated in response to these cytokines.

EPO and the EPO Receptor (EPOR)

Epo is the primary regulator of erythropoiesis and is synthesized primarily by the proximal convoluted tubules of the kidney in response to hypoxia. Expression and homodimerization of the EPOR is necessary and sufficient for cell responsiveness to EPO although, to facilitate the synergistic interaction of EPO and SCF, it functionally and physically interacts with the SCFR and possibly also with the IL-3R β chain. EPOR expression, which is correlated with the proliferative and differentiation effects of EPO in erythroid lineage cells, increases from BFU-E to CFU-E and erythroblasts (Fig. 1) before decreasing again during the terminal stages of differentiation. EPO also stimulates the release of maturing normoblasts from the bone marrow and increases the amount of hemoglobin synthesized per erythrocyte [10].

G-CSF and the G-CSF Receptor (G-CSFR)

G-CSF, which is a monomer that shares sequence homology with IL-6, is synthesized by a variety of cell types, including stromal cells, fibroblasts, and endothelial cells, usually in response to inflammatory stimuli such as lipopolysaccharide (LPS), tumor necrosis factor (TNF), and IL-1. The large ECD of its receptor contains one immunoglobulin and three fibronectin domains as well as the cytokine receptor module, and it binds G-CSF with high affinity in a 2:2 stoichiometry. G-CSF is the physiological regulator of neutrophil production, stimulating the proliferation and differentiation of committed neutrophil progenitor cells without affecting other granulocytic lineages, and its levels are increased during infection. It also synergizes with IL-3 or SCF to stimulate the proliferation and differentiation of primitive multipotent hematopoietic progenitor cells. However, mice lacking G-CSF or its receptor, while neutropenic, possess some nature neutrophils and have only a mild reduction in committed granulocyte-macrophage progenitor cells, suggesting that G-CSF is not necessary for neutrophil lineage committment. It also enhances the survival of mature neutrophils and may prime their functional responses [3].

TPO and the TPO Receptor (TPOR)

Structurally, TPO can be divided into two domains, an amino terminal portion with EPO homology and a carboxy terminal domain that is widely species divergent and lacks homology with other known proteins. It is synthesized primarily by hepatocytes, endothelial cells, fibroblasts, and the proximal tubule cells of the kidney. The TPOR gene, *c-mpl*, was originally identified as the cellular counterpart of *v-mpl*, the oncogene carried by the murine myeloproliferative leukemia virus, and is restricted in its expression to megakaryocytes, platelets, and primitive hematopoietic cells. TPO affects all aspects of megakaryocyte and platelet development, including stimulation of megakaryocyte progenitor cell proliferation and differentiation, and stimulation of platelet release. It has also been shown to stimulate proliferation of hematopoietic stem cells. Mice lacking either TPO or TPOR possess an identical phenotype and exhibit an 80–90% decrease in platelets with no effect on numbers of other differentiated blood cells. However, despite their selective thrombocytopenia, both mutant mice also exhibit a 60% reduction in multipotent and committed myeloid progenitors and are deficient in stem cells capable of long-term repopulation [11].

Cytokines Signaling through Receptors with a Common β subunit

The three cytokines, interleukin-3 (IL-3), granulocyte/macrophage CSF (GM-SCF), and interleukin-5 (IL-5), which signal through receptors sharing a common β subunit, are members of a subfamily of cytokines that appear to share a common ancestry despite their lack of homology at the amino acid level. Their genes have a similar structure and map closely together on chromosome 5. The cytokines signal through receptors comprised of a cytokine-specific α chain, which alone exhibits low affinity for the cytokine, and a larger, shared β chain, βc, that can interact with any of the three α chain–cytokine complexes to generate a specific, high-affinity complex (Fig. 3C). Ligand-induced disulfide bonding between an α and β subunit and dimerization of the βc subunit are required for signaling, producing a complex consisting of two receptor α chains, two βc chains, and two ligand molecules. The shared βc leads to competition between IL-3, GM-CSF, and IL-5, although the biological significance of this is not well understood. Their α subunit ECD cytokine receptor module structures are similar to that of EPOR (Fig. 2), and the divergent ICDs are required for proper receptor function. The 120-KDa β subunit contains two ECD cytokine receptor motifs and a long cytoplasmic tail that is required for proliferative signaling. Box 1 and box 2 motifs in the ICD contain the docking sites necessary for the association of JAK2 and recruitment of other signaling molecules to the activated receptor resulting in rapid tyrosine phosphorylation of several cellular proteins, including the β subunit of the receptor itself (Fig. 3C) [12].

IL-3 AND THE IL-3 RECEPTOR (IL-3R)

IL-3 is a secreted monomeric glycoprotein that is synthesized almost exclusively by T cells in response to antigen stimulation. The IL-3R consists of a unique IL-3-specific, low-affinity α subunit and the βc subunit. However, a duplication of the βc gene has occurred in the mouse to produce βIL-3, which interacts with IL-3Rα to create an additional low-affinity receptor for IL-3. IL-3 is a pleiotropic hematopoietic cytokine supporting the proliferation and differentiation of both primitive multipotent progenitor cells and committed myeloid progenitors. Working alone, it stimulates primitive hematopoietic cells to form multilineage colonies, comprised of neutrophils, basophils, eosnophils, monocytes, and megakaryocytes. In combination with SCF, IL-1, and CSF-1, IL-3 stimulates the proliferation and differentiation of even more primitive precursors and, in concert with other late-acting hematopoietic cytokines, it stimulates multipotent cells to specific lineage commitment. For example, the combination of IL-3 and CSF-1 allows the proliferation of primitive cells that do not respond to CSF-1 alone to give rise to committed, CSF-1R-expressing macrophage progenitors. Once committed, progenitor cells and their progeny lose expression of IL-3R and their ability to respond to IL-3. Since IL-3 synthesis is highly restricted and regulated and IL-3-deficient mice display only a diminished delayed hypersensitivity with no obvious steady-state hematopoietic phenotype, it seems to be important only during hematopoietic demand [3].

GM-CSF AND THE GM-CSF RECEPTOR (GM-CSFR)

GM-CSF, a monomeric glycoprotein, is constitutively synthesized by macrophages, endothelial cells, and fibroblasts and inducibly expressed in a variety of cells, especially T cells. The GM-CSFR α subunit has three alternative transcripts used to produce either the main form, a soluble form, and an alternative membrane-spanning subunit with an elongated C terminus. All are functional, but their relative physiological significance is not yet well understood. GM-CSF has been shown to have a broad range of biological effects, acting on both progenitor cells and mature, terminally differentiated cells. The survival, proliferation, and differentiation of all stages of cells in the neutrophil, macrophage, and eosinophil lineages are supported by GM-CSF and it appears to be important in the activation and enhancement of function of mature cells in these lineages. However, at most stages of development, each lineage also requires the presence of the other lineage-specific cytokines, G-CSF, CSF-1, and IL-5. The antigen-presenting dendritic cells appear to be derived from myeloid lineage and possibly also from lymphoid lineage cells. GM-CSF, with IL-4 and TNFα, stimulates myeloid-derived predendritic cells to differentiate into mature dendritic cells. GM-CSF also synergizes with EPO and TPO on primitive hematopoietic cells to generate erythroid and megakaryocytic progeny, respectively, and *in vivo* GM-CSF administration increases the number of circulating neutrophils, monocytes, and eosinophils and the number of tissue-fixed macrophages. Nevertheless, since GM-CSF-deficient mice have normal granulocyte and macrophage production in both steady-state and stressed conditions, GM-CSF apparently does not play an important role in blood cell production [13].

IL-5 AND THE IL-5 RECEPTOR (IL-5R)

Biologically active IL-5 is a dimer that is secreted predominantly by antigen-stimulated T lymphocytes, but also by NK cells, mast cells, B cells, eosinophils, and endothelial cells. However, only eosinophils and basophils, precursors of both lineages, and some B cells express both subunits of the IL-5R, so IL-5 has a restricted biological activity. It is the primary late-acting cytokine for eosinophil proliferation and differentiation, for mature eosinophil survival and activation, and *in vivo*, in the development of eosinophilia. Complete abrogation of the development of eosinophilia secondary to parasitic infections is observed in IL-5-deficient mice.

In combination with GM-CSF, IL-3, and IL-4, IL-5 stimulates the survival, proliferation, and differentiation of the basophil-mast cell lineage [14].

Cytokines Signaling through Receptors with a Common γ Subunit

Five hematopoietic cytokines, IL-2, IL-4, IL-7, IL-9, and IL-15, signal through a high-affinity receptor comprised of a cytokine-specific α chain and a common γ chain (γc), in a manner similar to IL-3, GM-CSF, and IL-5 and their common βc, in that the cytokine binds the α subunit with low affinity and requires further binding of the γc for high-affinity $\alpha \cdot \gamma$c complex signaling (Fig. 3D). Additionally, IL-2R and IL-15R, which probably have a common ancestry and constitute a separate cytokine receptor subfamily, also share a common β subunit, IL-2Rβ (Fig. 2), which can directly bind both ligands and, thus replace the α chain but is most usually included in a high-affinity $\alpha \cdot \beta \cdot \gamma$c complex (Fig. 3D). The signaling pathways described for all of these receptors are similar and, because the ICDs of each receptor lack catalytic activity, involve associated JAKs. Upon ligand binding and assembly of the high-affinity receptor–ligand complex, JAK1, constitutively associated with the α and β subunits, and JAK3, constitutively associated with the γc subunit, become activated and transphosphorylated. The JAKs phosphorylate the receptor chains, leading to recruitment and phosphorylation of phosphotyrosine-associating signaling intermediates, including STATs 1, 3, 5A, 5B, and 6 that dimerize and move to the nucleus to direct transcription. The MAPK pathway is activated via Shc interactions with the α or β subunit and PI3 kinase activation is brought about by recruitment of the Src-family kinases and/or insulin receptor substrates 1 and 2 to the receptors. Spontaneous mutations in the genes for γc and JAK3 produce, respectively, an X-linked and an autosomal severe combined immunodeficiency (SCID) syndrome in man, and mice lacking either γc or JAK3 are also severely immunodeficient, indicating the importance of these shared signaling pathways in immune function. Because the primary role of IL-2, IL-7, and IL-15 is regulation of the development and/or function of lymphoid lineage cells, only a very brief summary of their actions is outlined, while IL-4 and IL-9 have more pleiotropic effects and will be covered in more detail [15].

IL-2, IL-7, IL-15, and Their Receptors

IL-2 stimulates both a prompt T-cell immune response to antigens and a subsequent, rapid dampening of this response. Because IL-2-deficient mice die from autoimmune disorders yet mount a normal response to infections, the primary role of IL-2 appears to be the promotion of apoptosis of excess activated T cells. IL-15 exhibits considerable functional overlap with IL-2 in their regulation of innate immunity, through the two shared receptor subunits, IL-2Rβ and γc. However, in contrast to IL-2/IL-2Rα, IL-15Rα alone binds IL-15 with high affinity and IL-15R signaling has distinct effects on activated T cells. Rather than promoting apoptosis of excess T cells as does IL-2, IL-15 stimulates the survival of memory T cells for the secondary immune response. IL-7 plays a critical role in the early development of both B and T cells in mice and of T cells only in man. Deletion of the gene for either IL-7 or IL-7R produces severe B- and T-cell lymphopenia.

IL-4 and the IL-4 Receptor (IL-4R)

IL-4 is a pleiotropic cytokine that is secreted by T_H2 and NK1.1+T cells, basophils, mast cells, and eosinophils. The high affinity IL-4R is widely expressed in hematopoietic and nonhematopoietic cells and its expression is upregulated by IL-4 itself. In hematopoietic cells, the IL-4R consists of the IL-4Rα and γc subunits and requires JAK3 for STAT6 activation. Nonhematopoietic signaling by IL-4 utilizes the IL-4Rα subunit in combination with the IL-13R rather than γc. Consequently, IL-4 can act on many cell types and can modulate cytokine production by a variety of tissues. Its primary role, however, is in the regulation of T helper cell differentiation to T_H2 cells and in B-cell Ig switching. Apart from its effects on lymphocytes, IL-4 inhibits CSF-1-induced macrophage colony formation and megakaryocyte colony formation, but enhances G-CSF-induced granulocyte colony formation and IL-3-induced basophil, mast cell, and eosinophil generation. Along with GM-CSF and TNFα, IL-4 induces the differentiation of myeloid lineage cells into dendritic cells. It also regulates the production of inflammatory mediators in a pattern consistent with its anti-inflammatory role.

IL-9 and the IL-9 Receptor (IL-9R)

IL-9 is a pleiotropic growth factor that appears to be important in the pathogenesis of asthma. It is predominantly secreted by T cells, especially T_H2 cells, but also by mast cells and eosinophils, particularly those derived from asthmatic subjects. IL-9R is expressed by T and B cells, eosinophils, and neutrophils and IL-9 has been shown to stimulate early T-cell development, B-cell Ig production, including IgE, mast cell survival, proliferation, and cytokine release as well as up-regulation of eosinophil IL-5R expression. Hence, IL-9 influences many aspects of the inflammatory process that underlies asthma.

Cytokines Signaling Through a Common gp 130 Subunit

At least three hematopoietic cytokines, IL-6, IL-11, and leukaemia inhibitory factor (LIF), and three nonhematopoeitic cytokines, oncostatin M (OSM), ciliary neurotrophic factor (CNTF), and cardiotrophin-1 (CT-1), comprise the IL-6-type cytokine group. All share at least one subunit, gp130, in their receptor complexes and, as a result, functional overlap is frequent. IL-6R and IL-11R each contain a unique, cytokine-specific α chain, whereas the LIFR-specific subunit is much larger (Fig. 2) and also is more promiscuous because it is incorporated in receptor complexes for OSM, CNTF, and CT-1. The signaling complexes for this group of cytokines also vary in their stoichiometry. Il-6, IL-6Rα, and gp 130 bind together in a 2 : 2 : 2 molecular ratio to produce a high-affinity hexameric complex. A similar situation exists for the

IL-11R while LIF complexes in a 1 : 1 : 1 stoichiometry with LIFR and gp130. Gp130 signal transduction is mediated via activation of the constitutively associated JAKs, JAK1, JAK2, and TYK-2, which activate STAT1, STAT3, and STAT5 in turn. The MAPK pathway is also activated through SHP-2 binding to a juxtamembrane phosphotyrosine on gp130. Gp130-deficient mice die *in utero* due to severe hematopoietic and cardiac problems [16].

IL-6 AND THE IL-6 RECEPTOR (IL-6R)

IL-6 is a multifunctional cytokine produced by both lymphoid and nonlymphoid cells and its receptor is also expressed on a wide range of cells. In hematopoiesis, IL-6 acts synergistically with a number of cytokines, including SCF, IL-3, FL, CSF-1, and TPO, to stimulate primitive multipotential hematopoietic cell proliferation, myelopoiesis, and megakaryocyte production as well as lymphopoiesis. IL-6-deficient mice have reduced primitive multipotent progenitors, megakaryocyte progenitors, and neutrophils, and they are severely defective in their responses to tissue damage or infection.

IL-11 AND THE IL-11 RECEPTOR (IL-11R)

IL-11 is similar to IL-6 in many respects apart from its shared receptor subunit. Both IL-11 and its receptor are widely expressed in hematopoietic and nonhematopoietic tissues. Its expression is induced by pro-inflammatory and anti-inflammatory cytokines, and hematopoietic cell targets include those of the myeloid, erythroid, and megakaryocytic lineages. Although the effects of IL-11 on hematopoiesis are very similar to those of IL-6 and largely synergistic, adult IL-11Rα nullizygous mice are hematopoietically normal, highlighting the significant redundancy in this family of cytokines.

LIF AND THE LIF RECEPTOR (LIFR)

LIF is also multifunctional and is produced by monocytes and stromal cells in response to activating stimuli such as IL-1β and LPS. The hematopoietic effects of LIF are similar to those of IL-6 and IL-11 and, as for IL-11, LIF-deficient mice have no obvious hematopoietic phenotype.

Conclusions

The maintenance of the normal complement of hematopoietic cells requires the complex interaction of a large number of cytokine signaling pathways, some of which are redundant while others are critically important. Moreover, this complexity is increased by the actions of a number of other cytokines that also regulate hematopoiesis but primarily act elsewhere and which are therefore not covered in this review. They include IL-1 and IL-18, which do not signal through cytokine receptor motif-containing receptors, IFNγ, TGFβ, TNFα, and TNFβ, which commonly inhibit rather than promote hematopoiesis, several chemokines, and the mammalian Notch paralogs and their ligands, which regulate primitive hematopoietic cells.

In an extension of their physiological roles, some hematopoietic cytokines are used extensively in the clinical setting, usually to correct deficiencies of specific hematopoietic lineages such as EPO for anemia, TPO for thrombocytopenia, and GM-CSF and G-CSF for granulocytopenias.

References

1. Till, J. E. and McCulloch, E. A. (1980). Hemopoietic stem cell differentiation. *Biochim. Biophys. Acta.* **605**, 431–459.
2. Harrison, D. E. (1992). Evaluating functional abilities of primitive hematopoietic stem cell populations. *Curr. Top. Microbiol. Immunol.* **177**, 13–30.
3. Stanley, E. R. (2001). The hematopoietic cytokines. In K. F. Austen, M. M. Frank, J. P. Atkinson, and H. Cantor, Eds., *Samter's Immunological Diseases*, 6th ed., pp. 175–193. Lippincott Williams and Wilkins, Philadelphia.
4. Ihle, J. N. (2001). Signal transduction in the regulation of hematopoiesis. In G. Stamatoyannopoulos, P. W. Majerus, R. M. perlmutter, and H. Varmus, Eds., *The Molecular Basis of Blood Diseases*, 3rd ed., pp. 103–125. W. B. Saunders, Philadelphia.
5. Kaushansky, K. (2001). Hematopoietic growth factors and receptors. In G. Stamatoyannopoulos, P. W. Majerus, R. M. Perlmutter, and H. Varmus, Eds., *The Molecular Basis of Blood Dieseases*, 3rd ed., pp. 25–54. W. B. Saunders, Philadelphia.
6. Galli, S. J., Zsebo, K. M., and Geissler, E. N. (1994). The kit ligand, stem cell factor. *Adv. Immunol.* **55**, 1–96.
7. Lyman, S. D. and Jacobsen, S. E. W. (1998). *c-kit* ligand and flt3 ligand: Stem/progenitor cell factors with overlapping yet distinct activities. *Blood* **91**, 1101–1134.
8. Stanley, E. R. (2000). CSF-1. In J. J. Oppenheim and M. Feldmann, Eds., *Cytokine Reference: A Compendium of Cytokines and Other Mediators of Host Defence*, pp. 911–934. Academic Press, London.
9. Bourette, R. P. and Rohrschneider, L. R. (2000). Early events in M-CSF receptor signaling. *Growth Factors* **17**, 155–166.
10. Constantinescu, S. N., Ghaffari, S., and Lodish, H. F. (1999). The erythropoietin receptor: Structure, activation and intracellular signal transduction. *Trends Endocrinol. Metab.* **10**, 18–23.
11. Kaushansky, K. (1995). Thrombopoietin: The primary regulator of platelet production. *Blood* **86**, 419–431.
12. Bagley, C. J., Woodcock, J. M., Stomski, F. C., and Lopez, A. F. (1997). The structural and functional basis of cytokine receptor activation: Lessons from the common β subunit of the granulocyte-macrophage colony-stimulating factor, interleukin-3 (IL-3), and IL-5 receptors. *Blood* **89**, 1471–1482.
13. Kastelein, R. A. and Shanafelt, A. B. (1993). GM-CSF receptor: Interactions and activation. *Oncogene* **8**, 231–236.
14. Adachi, T. and Alam, R. (1998). The mechanism of IL-5 signal transduction. *Am. J. Physiol.* **275**, C623–C633.
15. He, Y. W. and Malek, T. R. (1998). The structure and function of gamma c-dependent cytokines and receptors: Regulation of T lymphocyte development and homeostasis. *Crit. Rev. Immunol.* **18**, 503–524.
16. Bravo, J. and Health, J. K. (2000). Receptor recognition by gp 130 cytokines. *EMBO J.* **19**, 2399–2411.
17. Townsend, M. J., Fallon, P. G., Matthews, D. J., Smith, P., Jolin, H. E., and McKenzie, A. N. J. (2000). IL-9-deficient mice establish fundamental roles for IL-9 in pulmonary mastocytosis and goblet cell hyperplasia but not T cell development. *Immunity* **13**, 573–583.

CHAPTER 350

Regulation of Bartlett Endogenous Stem Cells in the Adult Mammalian Brain: Promoting Neuronal Repair

Rodney L. Rietze and Perry F. Bartlett
Institute for Brain Research, The University of Queensland, Brisbane, Queensland, Australia

Adult Neurogenesis Revealed

A central dogma of neuroscience for greater than a century was that the brain was a static nonrenewable organ. The centerpiece of this dogma was that no new neurons were produced in the central nervous system (CNS) following development. Following a histological examination of the developing CNS, investigators as famous as Ramón y Cajal [1,2] and His [3] concluded that owing to its complex architecture and elaborate connections, the idea of the brain changing at all after its formation could not be seriously entertained. While reports describing mitotic figures in the cerebrum of the adolescent rat [4] and lining the walls of the lateral ventricles [5,6] surfaced in the early 1900s, these early reports of mitotic activity were basically overlooked and rarely cited owing to the prevailing hegemony and the crude nature of the detection methods used.

It was not until the late 1950s with the introduction of [^{3}H]-thymidine autoradiography that a serious challenge was mounted to the dogma. Smart [7] was the first to apply this technique to study proliferation in the adult brain and while he reported that neurogenesis and gliogenesis occurred in the 3-day-old postnatal mouse, he was unable to detect evidence for such an event in the adult. It was not until several years later that Altman [8] discovered putative adult neurogenesis through a series of experiments designed to test the kinetics of glial cell proliferation following brain trauma. However, even though Altman [9–14] continually reported [^{3}H]-thymidine autoradiographic evidence for new neurons in the neocortex, dentate gyrus, and olfactory bulb of adult rats, a lack of definitive immunocytochemical markers allowed this also to be dismissed. Indeed, it took the better part of another two decades before two teams—Michael Kaplan and colleagues [15,16] working on adult rats and Fernando Nottebohm and colleagues [17–21] investigating the seasonal production of learned song in birds—succeeded in unambiguously demonstrating production of new neurons in the adult brain. Although these studies documented neurogenesis in the adult CNS, they suggested this phenomenon was localized to two distinct brain regions: the olfactory bulb and the hippocampus.

Another major step was combining BrdU labeling techniques with specific monoclonal antibodies that recognized cell-type-specific molecules [22], which ultimately enabled the unambiguous identification of newly generated cell types. This technological achievement had three immediate effects: It provided indisputable evidence of the incidence and location of neurogenesis and gliogenesis in the adult CNS, it facilitated the stereological estimation of the total number of specific cells being generated in any given region at any given time, and it engendered a more detailed search for the precursor cell population that continued to give rise to new cells in the adult brain. It was against this background that the race to identify and isolate and the cell responsible for producing these neurons in adult brain began.

Isolation and Culture of Neural Stem Cells

Traditionally, stem cells were thought to be located only in tissues where differentiated cells were most susceptible to loss and the need for replacement great, such as the skin [23], intestinal epithelia [24], and the blood [25]. Indeed, the best known example of an adult stem cell is the hematopoietic stem cell (HSC), which is found in the bone marrow and is ultimately responsible for the generation of all blood cell types throughout the life of the animal [25–27]. Because the adult CNS was thought not to exhibit a significant amount of neuronal death and to have no regenerative capacity, the existence of neural stem cells seemed both unlikely and unnecessary. However, in 1992 two independent groups successfully demonstrated the presence of putative stem cells in the adult mammalian CNS.

In our study, cells from various regions of the adult mouse forebrain were dissociated and cultured at high density in the presence of serum and basic fibroblast growth factor (FGF-2), allowed to proliferate, and then differentiated into neurons using a conditioned medium from an astrocyte-like cell line (Ast-1) [28]. The FGF-2 system was ultimately refined to show that single precursors could be grown *in vitro* to give rise to clonal progeny [29]. Reynolds and Weiss [30] employed epidermal growth factor (EGF) in a serum-free culture system. which resulted in the death of the majority of cell types harvested from the periventricular (PV) region within 3 days of culture, but allowed a small population (<0.01%) of the starting population to enter a period of active proliferation, even at very low cell densities. By using such a system, they were able to demonstrate that a single adult CNS cell could proliferate to form a ball of undifferentiated cells they termed a *neurosphere*, which in turn could be dissociated to form numerous secondary spheres, or induced to differentiate, generating the three major cell types of the CNS. This was a major finding because it demonstrated for the first time that the precursor cell exhibited the three cardinal attributes of a bona fide stem cell: proliferation, self-renewal, and the ability to give rise to a number of differentiated, functional progeny [24,31].

Just as the advent of [^{3}H]-thymidine autoradiography, BrdU immunocytochemistry, and hybridoma technologies facilitated a more detailed understanding of adult neurogenesis, the neurosphere culture system provided investigators with a valuable tool for assaying neural stem cell activity. Although the neurospheres provided the means to begin to study neural differentiation and search for stem cell activity in different parts of the nervous system, it told us nothing about the identity, number, or factor regulation of the stem cell itself. The neurosphere assay did, however, provide a means to begin to identify and purify this stem cell using this assay. Surprisingly, it was almost another 10 years before this was achieved.

Because no specific stem cell markers were available, we began the purification process in 1995 using a totally empirical approach. Single cells harvested from the PV region of adult mice were examined for their ability to bind a number of fluorescently tagged ligands or antibodies using a fluorescent activated cell sorter (FACS). Populations with different binding properties were sorted and then assayed for stem cell activity using the neurosphere assay. After examining more than 500 different ligands and antibodies, we were able to identify a small, distinct population of adult stem cells whose cell surface was characterized by the expression of minimal or undetectable levels of peanut agglutinin receptors (PNA)^{lo} and mCD24 (also known as heat-stable antigen, HSA)^{lo}. This PNA^{lo}HSA^{lo} subpopulation (Fig. 1A, lower boxed region), a discrete population representing $0.27 \pm 0.07\%$ of the unsorted population, was comprised almost exclusively of NSCs. When cultured under clonal conditions, in which one PNA^{lo}HSA^{lo} cell was plated per well, approximately 80% of the cells (1:1.28) gave rise to neurospheres [32]. Furthermore, the PNA^{lo}HSA^{lo} population contained 63.2% of the overall NSC activity, strongly suggesting that the vast majority of NSCs in the periventricular region of adult mice were of this phenotype.

Although FACS analysis suggested that all the putative NSCs were similar, a more precise characterization of the differentiation and self-renewal capacity of the PNA^{lo}HSA^{lo} cells was undertaken to confirm that bone fide neural stem cells had been purified [32]. To determine this, clonally derived neurospheres generated from individual PNA^{lo}HSA^{lo} cells were differentiated by growth factor withdrawal and the addition of serum, and then assessed by immunocytochemistry (ICC) for the presence of neurons and glia. In all cases, differentiated clonally derived spheres (a total of 201 in $n=3$ experiments) contained glial fibrillary acidic protein positive (GFAP+ve) astrocytes, β-tubulin type III positive (β-tubulin+ve) neurons, and O4 positive (O4+ve) oligodendrocytes. To assay for self-renewal capacity, clonally derived spheres were once again generated from freshly isolated PNA^{lo}HSA^{lo} NSCs, then after 7 days *in vitro*, each primary sphere was dissociated into a single cell suspension and plated into individual 96-well plates. In all cases (348 spheres across $n=3$ experiments) secondary spheres were formed from each dissociated primary sphere. These clonal experiments demonstrated that each cell could proliferate, self-renew, and give rise to a family of differentiated progeny, the hallmark characteristics of a neural stem cells [24, 33]. To determine whether neurospheres derived from sorted cells retained these properties when passaged, three clonally derived cultures from individual PNA^{lo}HSA^{lo} cells were generated and passaged every 5–7 days *in vitro* for 3 months. When individual spheres from each culture were transferred to differentiating conditions (116 spheres in total), GFAP+ve, O4+ve, and β-tubulin+ve cells were present in every sphere examined. Furthermore, in 100% of the cases, individual passaged spheres retained the ability to give rise to more than one subsequent sphere, suggesting that they have all of the characteristic *in vitro* properties previously described for NSCs [30,34,35].

Recently, several different neural cell types have been purported to be capable of behaving as neural stem cells under a variety of conditions *in vitro* [26,36–40]. That is,

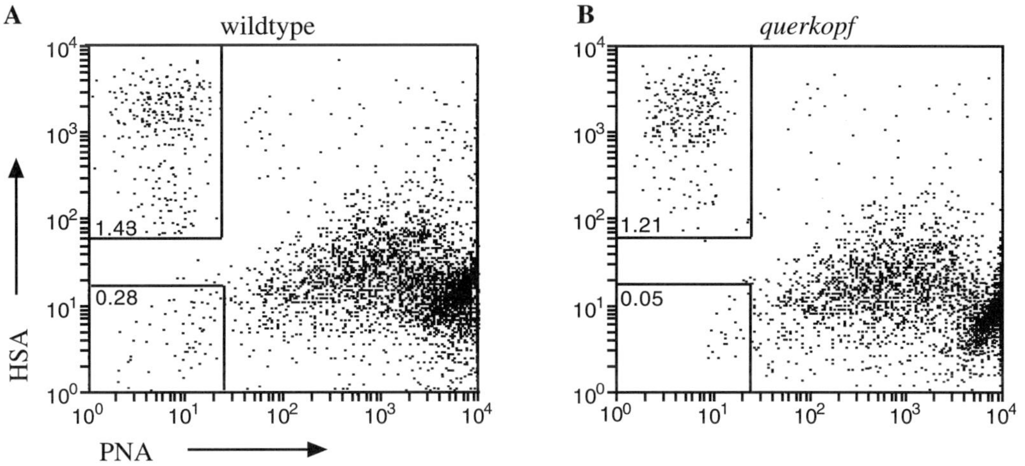

Figure 1 Querkopf mice are deficient in $PNA^{lo}HSA^{lo}$NSCs. Viable CNS cells were harvested from individual adult *querkopf* mutant mice and wildtype littermates and sorted for PNA and HSA binding. Flow cytometric analysis revealed similar sorting profiles between all mice sampled, except for the $PNA^{lo}HSA^{lo}$ NSC populations, which showed an approximately 6 fold diminution (mean 6.17 + 1.36 SEM) in the percentage of cells of this phenotype in the mutant *querkopf* compared to wildtype littermates. Of note, was the dramatic loss of the ultra-low PNAHSA population. Values in boxed regions represent the % of total viable cell number.

they have shown the ability to generate clusters of cells *in vitro*, which are reminiscent of multipotent neurospheres. Unfortunately, it is unclear whether the formation of such clusters represents bona fide neurospheres (i.e., contain multipotent NSCs), because high cell densities and protracted times in culture makes such an interpretation difficult. We sought to investigate whether subpopulations additional to the $PNA^{lo}HSA^{lo}$ in the PV contained latent stem cell activity (as measured by neurosphere formation) by culturing sorted cells in serum free media (NS-A) supplemented with growth factor(s) other than EGF and bFGF. Thus, cells were harvested from adult CBA mice and sorted into $PNA^{lo}HSA^{lo}$, $PNA^{lo}HSA^{mid\text{-}hi}$, $PNA^{mid\text{-}hi}HSA^{lo}$, and $PNA^{mid\text{-}hi}HSA^{mid\text{-}hi}$ fractions. Each fraction was cultured for a total of 21 days *in vitro* at equivalent densities, with new media added every 7 days *in vitro*. The factors and respective concentrations employed were PDGF alone (10 ng/ml), LIF alone (10,25, 50 ng/ml), BMP-2 alone (25,50,100 ng/ml), growth hormone alone (50,100,500 ng/ml), stem cell factor (200 ng/ml) alone, and in combination with IL-3 (2 ng/ml), IL-6 (20 ng/ml), and G-CSF (20 ng/ml). This latter combination of factors was employed to assay for hematopoietic stem cells. In all of the preceding cases, after three independent rounds of experimentation, in the absence of EGF or bFGF costimulation, no latent stem cell activity was detected in any of the various subpopulations. Furthermore, we also followed culture conditions described by Kondo and Raff [40] to specifically screen for oligodendrocyte precursor cell (OPC) activity, but failed to observe the formation of neurospheres in any of the sort fractions. Of note in this regard, the $PNA^{lo}HSA^{lo}$ cells do not share the phenotype of OPCs because they are negative for the A2B5 antigen, which is present on all OPCs [40–42].

The fact that we assayed for latent stem cell activity (sphere formation) in other subpopulations using a variety of growth factors and had not detected spheres up to 3 weeks after stimulation suggested that we had not missed a population of stem cells, which reinforced our original observation that there is a single predominant stem cell type in the PV region. This is a very important point, because the periventricular region not only contains the highest frequency of EGF- or FGF-2-responsive NSCs in the CNS [30,43–47], but, more importantly, the NSCs from this region of the adult telencephalon have been shown to have the capacity to generate neurons and glia under a variety of in both the normal and diseased animals [48–51]. It is these stem cells that continue to generate neuroblasts, which migrate along the rostral migratory stream and ultimately differentiate into new neurons in the olfactory bulb throughout adulthood [48–50,52], underlining their functional significance in neuronal production under normal physiological conditions. It is also these stem cells that have been shown to generate neurons in the hippocampus and cortex after injury [53,54].

We sought to verify that the NSC population we had identified corresponded to the functional stem cell population. We did this by examining the stem cell pool of *querkopf* mutant mice [55], which show significant lack of neurons in the region of the brain that continually acquires new neurons in the postnatal and adult animal: the olfactory bulb. We reasoned that if $PNA^{lo}HSA^{lo}$ NSCs generated new neurons in the olfactory bulb under normal conditions, the neuronal deficiency in *querkopf* may be reflected by a selective deficiency in the $PNA^{lo}HSA^{lo}$ stem cell population. Examination of the FACS dot-plot profiles revealed this to be the case: sorting profiles from mutant and wild-type mice were indistinguishable, except for the size of the $PNA^{lo}HSA^{lo}$ compartment, where mutant mice exhibited an approximate sixfold reduction in cells of this phenotype. Indeed, closer examination of the $PNA^{lo}HSA^{lo}$ region (lower boxed area; Fig. 1B) revealed a loss of an as yet unrecognized PNAHSA ultra-lo population. Whether this is indeed a distinct population of cells has yet to be determined, because the reduction in $PNA^{lo}HSA^{lo}$ cell numbers did not affect the frequency of stem cell activity contained within this compartment, when compared to littermate controls. Because no change in stem cell activity was detected in other sorted populations, these results cannot be explained by a shift in stem cell phenotype. Furthermore, the studies of Thomas *et al.* [55] showed that there was no change in the proliferation rate or cell death rate observed in these mice during development and postnatally, further supporting the conclusion that the main defect is the lack of stem cell activity. Preliminary experiments have since revealed that when multipotent neurospheres are generated and differentiated from *querkopf* mice, they exhibit a significant reduction in the percentage of neurons produced *in vitro* as compared to littermate controls, providing further support for an underlying stem cell defect. Taken together, the results discussed demonstrate a high degree of correlation between the size of the $PNA^{lo}HSA^{lo}$ population and neuronal production, suggesting that this population acts as a functional stem cell *in situ* generating new neurons under normal physiological conditions.

Recently, two publications appeared in *Science* [56,57] claiming that bone marrow cells had the ability to generate new neurons *in vivo*. The strategy those researchers had employed to purify stem cells from the bone marrow was based on the selective exclusion of the DNA-binding dye Hoechst 33342 [58,59]. Sorting based on Hoechst 33342, a dye that specifically but reversibly joins with DNA [60,61], seems to rely on the differential ability of cells to exclude the dye, and in murine bone marrow defines a small (0.07%) homogeneous population, termed SP (side population) cells. This SP population, which extrudes the dye very efficiently, expresses low or undetectable levels of CD34, lacks markers of mature blood cell types (i.e., Lin^-), and provides long-term hematopoietic reconstitution in lethally irradiated mice [59], suggesting it represents a bona fide hematopoietic stem cell population. Because it has been suggested that the HSC and the NSC represent the same cell in different parts of the body, and the Hoechst 33342 technique has been used to isolate nonhematopoietic (muscle and epidermis) stem cell types [58,62], we sought to determine if these two populations were similar by performing further immunocytochemical examination of the $PNA^{lo}HSA^{lo}$ NSCs and by seeing if Hoechst exclusion could reliably predict neural stem cell activity.

Accordingly, PV cells were harvested from adult mice and processed so as to isolate the $PNA^{lo}HSA^{lo}$ population and finally incubated in media containing 5 µg/ml Hoechst 33342 for 60 min at 37°C before sorting occurred. As shown in Fig. 2, two discrete regions of PNA/HSA labeled periventricular cells (Figs. 2A and 2C), were analyzed for the distribution of the Hoechst 33342 staining patterns (Figs. 2B and 2D). The Hoechst staining profile of cells within the $PNA^{lo}HSA^{lo}$ compartment (Fig. 2B) revealed that the vast majority (approximately 90%) of the $PNA^{lo}HSA^{lo}$ cells were brightly labeled, while the majority of cells in the $PNA^{lo}HSA^{mid\text{-}hi}$ compartment were Hoechst low (Fig. 2D). Culturing of the Hoechst low population shown in Fig. 2D under conditions known to generate neurospheres did not increase NSC frequency more than simply selecting for $PNA^{lo}HSA^{mid\text{-}hi}$ cells. To further identify which populations of cells excluded the dye and confirm that Hoechst 33342 labeling was ineffective to enrich for neural stem cell activity, PNA/HSA labeled periventricular cells were first analyzed on the basis of Hoechst staining. As illustrated in Fig. 2F, three distinct populations (Hoechst-bright, Hoechst-mid, and Hoechst-low) of PV cells can be distinguished on the basis of Hoechst 33342 staining. Examination of the distribution of PNA/HSA labeled cells within the Hoechst low population revealed that the majority (86.6%) of Hoechst 33342 low cells were contained within the $PNA^{lo}HSA^{mid\text{-}hi}$ compartment (Fig. 2E). Thus, in contrast to previously published work demonstrating that HSC activity is unambiguously contained within a specific Hoechst "low" compartment, Hoechst 33342 labeling could not be used to enrich for NSC activity.

Furthermore immunocytochemical analysis of $PNA^{lo}HSA^{lo}$ cells revealed no expression of cell-surface antigens associated with endothelial and hematopoietic stem cells. NSCs were negative for the following markers:

1. CD90.2, which reacts with the Thy 1.2 alloantigen on thymocytes and peripheral T cells [63];
2. CD31, also known as PECAM-1 (platelet endothelial cell adhesion molecule), which is expressed constitutively on adult endothelial cells and weakly on peripheral platelets and leukocytes [64–66];
3. CD117, which reacts with the mouse c-Kit receptor and is expressed on virtually all hematopoietic progenitor cells [67,68];
4. CD135, which reacts with Flk-2/Flt-3, a receptor protein tyrosine kinase closely related to c-Kit, and is expressed in hematopoietic stem cells and also primitive progenitor cells [69,70];
5. CD34, a cell-surface protein of unknown function that has in the past been used to positively select for long-term repopulating HSCs [26,27];
6. CD45, a cell-surface tyrosine phosphatase present on all nucleated hematopoietic cells including HSCs, but not found on any nonhematopoietic cell types [71]; and
7. Ly-6A/E, also known as Sca-1, which is expressed on multipotent, long term-engrafting, HSCs in mice [26,27].

Thus, the data strongly suggest that the $PNA^{lo}HSA^{lo}$ NSC is unrelated to the bone marrow-derived stem cell.

Regulation of Stem Cell Differentiation into Neuron

During the past decade an increasing number of studies have been directed toward understanding the growth factor regulation of nervous system development [72–75]. Of interest here is that many of the factors that play critical roles during early neural development are thought to modulate

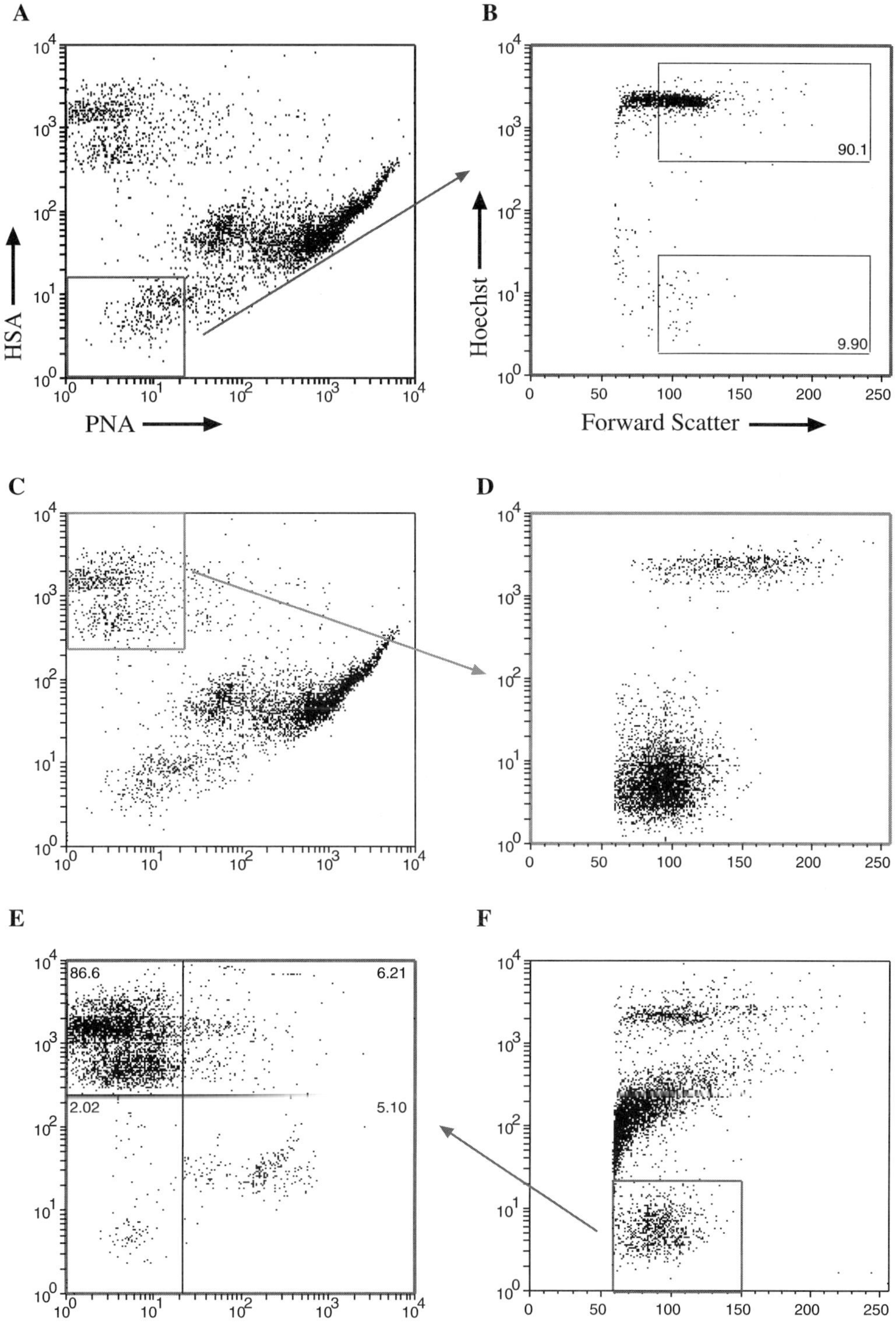

Figure 2 Hoechst staining profile of PNA/HSA labeled periventricular cells. Two populations [$PNA^{lo}HSA^{lo}$(A) and $PNA^{lo}HSA^{mid\text{-}hi}$(C)] were analyzed from the same suspension of PNA/HSA labeled periventricular cells to determine the distribution of Hoechst 33342 stained cells within each compartment. Approximately 90% of cells within the $PNA^{lo}HSA^{lo}$ compartment (box, A) were brightly labeled with the DNA-binding dye Hoechst 33342 (upper box, B). In comparison, the majority of cells within the $PNA^{lo}HSA^{mid\text{-}hi}$ compartment (box, C) were not Hoechst bright (D). Panel (E) illustrates the Hoechst staining pattern of a population of freshly isolated >7μm periventricular cells. When the Hoechst low compartment (box, F) was analyzed for the distribution and percentage of PNA/HSA labeled cells, the majority of cells (86.6%) were located in the $PNA^{lo}HSA^{mid\text{-}hi}$ compartment (E). Numerical values represent the % of cells located within that region (B) or quadrant (E) from individual representative experiments. (A color representation of this figure is available on the CD version of the *Handbook of Cell Signaling*.)

various aspects of adult neurogenesis, including extrinsic cues such as cytokines signaling through the leukemia inhibitory factor (LIF) receptor complex, such as LIF and ciliary neurotrophic factor (CNTF) [37,76] and growth factors such as members of the FGF family [74], the bone morphogenic protein (BMP) family [77], and platelet-derived growth factor (PDGF) [36,78] as well as intrinsic cues, including neurogenic basic helix–loop–helix (bHLH) genes such as *NeuroD* and *Neurogenin* [79–83].

While CNTF, interleukin-6 (IL-6), oncostatin-M, and LIF all signal through the LIF receptor and its associated receptor subunit gp130, they exhibit a wide range of effects on many different cell types. For example, in the developing CNS, both CNTF and LIF have been shown to play a vital role in the differentiation of neural precursors into astrocytes [36–38, 84–87] or oligodendrocytes [85,86,88–91]. Somewhat paradoxically, LIF and CNTF have also been shown to promote neurogenesis under some conditions [76,85,92–95]. To determine its role on neural stem cells, Shimazaki and colleagues [96] analyzed the number of multipotent neurospheres that could be derived from LIFR-deficient (heterozygous) and wild-type mice. They report that both the number of NSCs and the number of tyrosine hydroxylase (TH) positive neurons in the olfactory bulb (implied progeny of NSCs) were significantly reduced [96], suggesting the importance of the LIFR in NSC regulation. Of interest though, when the authors investigated the number of spheres that could be generated from E14 $LIFR^{-/-}$ mice versus wild-type controls, they found no significant difference. However, when assayed for the ability to continue to generate spheres upon passaging (i.e., self-renewal) they found that unlike wild-type controls, cultures derived from $LIFR^{-/-}$ mice could not be expanded beyond seven passages, suggesting that the LIFR played a more specific role in the long-term maintenance of stem cells in culture. Consistent with this hypothesis, the authors report a 37% reduction in the number of sphere-producing cells harvested from adult $LIFB^{+/-}$ versus wild-type mice. In addition, when CNTF, which also signals through the LIFR/gp130 complex, was injected into the lateral ventricle of adult mice for 6 days, these authors report a 24% increase in the number of EGF-responsive sphere forming cells, confirming the importance of the LIFR, but also suggesting that CNTF may play a role in regulation of the NSC. Given that CNTF and its specific receptor subunit are expressed in the developing CNS and, most interestingly, in the periventricular region of the adult forebrain [97,98], and that it has been shown in the past to enhance the survival of a variety of neural cell types [75], these results are not totally unexpected. Having confirmed that LIF and CNTF are indeed candidate molecules for the maintenance of NSCs, the challenge now is to more clearly define the role(s) and mechanism of action of each cytokine that signals through the LIFR in regulating the activity of the NSCs.

Although we know that the interaction between external influences such as cytokines and internal cues causes the differential expression of neurogenic genes and hence regulates neural stem cell fate [99–101], what is less understood is how a particular cell (a stem cell, for example) can process and integrate the variety of simultaneous external signals it receives (which in turn active different signal transduction pathways) and ultimately achieve an appropriate biological outcome (i.e., differentiation into a neuron). One possible explanation is that under some circumstances, neural stem cells have the capacity to overcome signals that normally inhibit neuronal differentiation. In the case of LIF and CNTF, which both signal through the LIF receptor and employ the JAK/STAT pathway [102], work in our lab suggests that the differential regulation of JAK/STAT signaling via suppressor of cytokine signaling (SOCS) molecules is important in determining the differentiation of NSCs into neurons versus astrocytes.

SOCS molecules are key regulators of LIF receptor signaling and are able to inhibit the JAK/STAT pathway downstream of cytokines such as IL-6 or LIF [103–105]. At least eight members of the SOCS family (SOCS1–7 and CIS) [106] are differentially expressed in a variety of tissues and can be up-regulated by a range of cytokines and growth factors. Recently we have shown that SOCS1, SOCS2, and SOCS3 are expressed throughout neural development, with SOCS2 being the most highly expressed.

A most surprising finding was that astrocytes expressed SOCS1 and SOCS3, whereas neurons expressed SOCS2 exclusively. This pattern of expression was not altered by treatment with a variety of cytokines, although cytokines regulated the level of individual SOCS gene expression. Our observation that SOCS2 was restricted to the neuronal lineage was reinforced by its *in situ* expression pattern, which showed SOCS2 expression in neurons of the developing cortex and hippocampus [107]. Both the *in situ* observations and the Northern analysis [107] showed that the level of SOCS2 mRNA is significantly up-regulated as neurons are generated in the developing forebrain and down-regulated postnatally. Surprisingly, SOCS2 was not expressed in the adult multipotent neural stem cells isolated by FACS, but was expressed in progenitors and neurons but not astrocytes, suggesting a role for SOCS2 in the regulation of neurogenesis.

Results from experiments in which SOCS2 was overexpressed or deleted in neural cells support this concept. Overexpression of SOCS2 in neurospheres resulted in significant induction of neuronal differentiation. Deletion of SOCS2 expression from CNS neural precursors resulted in decreased numbers of neurons being generated. *In vitro*, the neural stem cells from $SOCS2^{-/-}$ mice generated significantly fewer neurons but more astrocytes compared to stem cells from wild-type mice, a result that was also confirmed in neuroepithelial cells from E10 mice. Because no difference was seen in the total number of cells generated in culture and the total number of differentiated cells was not affected, the results support the idea that SOCS2 plays a role in neurogenesis by promoting neuronal and inhibiting glial cell fate and not in cell survival or proliferation. The *in vivo* findings supported the *in vitro* observations.

We found a 30% reduction in the number of NeuN positive neurons across all layers of the cortex in adult SOCS2$^{-/-}$ mice compared to wild-type mice, but found no reduction in the total cell density in the mutant cortex, indicating that non-neuronal cells, glia, had been generated in proportionally higher numbers, similar to our observations *in vitro*. This decrease of the neuron-to-glia ratio is similar to that observed *in vitro* for neural precursor cells differentiated in the presence of LIF or CNTF, in which the neuron-to-glia ratio is also decreased [37,87]. Thus, expression of SOCS2 appears to promote the ability of a precursor cell to differentiate into a neuron and inhibit its differentiation into an astrocyte. Conversely, a lack of SOCS2 expression appears to favor astrocyte differentiation.

Because the level of SOCS2 expression appears to regulate neural precursor cell differentiation, the key questions are what regulates SOCS2 expression and what does SOCS2 regulate? Growth factors that signal through the LIF receptor complex appeared to be the most likely candidates for both regulating and being regulated by SOCS2. We have shown that they up-regulate SOCS2 expression in neuroepithelial cells and SOCS2 has been shown, albeit weakly, to regulate downstream signaling by LIF in some cells [108]. Furthermore, LIF and related molecules have been shown to exert effects on neural precursors similar to that observed here with changes in SOCS2 expression levels: either an increase in neuronal differentiation [76] or promotion of astrocyte differentiation [37,38,87]. Addition of LIF to neurosphere cultures profoundly increased the number of astrocytes while inhibiting neuron differentiation; however, this effect was the same in both SOCS2$^{-/-}$ and wild-type mice. Thus, it appears that SOCS2 is not a major regulator of biological signaling through the LIF receptor complex in neural precursor cells. This finding is in accordance with SOCS2$^{-/-}$ mice whose phenotype of gigantism appears to be due to dysregulation of GH and/or IGF1 signaling [109]. GH now also appears to be a major regulator of neuronal differentiation. Although a role for GH in neural stem cell differentiation has not previously been described, GH is expressed in whole fetal rat brain during neural development, from E10, with a peak of expression before birth [110]. Moreover, GH production has been shown in cultured brain cells [111], and somatostatin is produced by embryonic cortical neurons [112]. Together with our demonstration here that one form of the GH receptor is expressed as early as E10 in the mouse brain, these studies further support a role for GH, modulated by SOCS2, in negatively regulating neuronal differentiation of neural precursor cells.

How Does SOCS2 Regulate Neuronal Differentiation?

The main role of SOCS2 appears to be regulation of STAT5 activation downstream of GH signaling. SOCS2$^{-/-}$ mice exhibit an enlarged growth phenotype consistent with hyperresponsiveness to GH signaling [109]. The overgrowth phenotype of the same line of SOCS2$^{-/-}$ mice used in this study requires STAT5b for expression of the phenotype [113]. In addition, both STAT5a and STAT5b activation is prolonged in cells from these mice in response to GH due to their inability to effectively down-regulate the activation [113]. Regulation of STAT5 activation by SOCS2 has also been demonstrated in overexpression studies [114,115]. Although the precise mechanism by which SOCS2 regulates STAT5 activation remains to be determined, it appears to involve SOCS2 competitively binding to the STAT5 binding sites on the GH receptor and thus inhibiting STAT5 activation [115,116].

Although SOCS2 regulates STAT5 activation, how this translated to regulation of neurogenesis was unclear. One possibility was that GH inhibits expression of neurogenic genes, such as *Neurogenin*, similar to our recent demonstration that members of the FGF family regulate *Notch* and *Delta* expression to inhibit neuronal differentiation [101]. Expression of SOCS2 would therefore be required to overcome GH inhibition of neurogenic gene expression. The level of expression of neurogenic genes is important in determining neuronal versus glial cell fate. Neural progenitor cells from Neurogenin-2 (Ngn2)/Mash1 double knock-outs showed decreased neuronal and increased astrocyte differentiation of cortical progenitor cells [83], whereas overexpression of Neurogenin-1 (Ngn1) promoted neurogenesis and inhibited gliogenesis by competing with the JAK/STAT pathway for the transcriptional coactivators CBP/p300 as well as directly inhibiting STAT activation [83].

Given that SOCS2 appears to act by regulating STAT5 signaling, it may as a consequence regulate the availability of the JAK/STAT pathway to compete with Ngn1 for CBP/p300. We found not only that the levels of Ngn1 were decreased in SOCS2-cells compared to wild-type cells *in vitro* and *in vivo*, but also that addition of GH acutely decreased Ngn1 levels in wild-type cells *in vitro* by 50–60%, to a level comparable to that in SOCS2$^{-/-}$ cells. This thus provides a potential mechanism for the decreased neurogenesis and increased gliogenesis observed in the SOCS2$^{-/-}$ cells. However, unlike SOCS2-mice, Ngn1 null mice do not have a robust cortical phenotype. This is presumably due to compensation by other bHLH genes, such as Ngn2 and Mash1, and double knock-outs are required before a phenotype is observed [82,117]. Therefore, it is probable that SOCS2 does not affect Ngn1 alone but also affects other bHLH genes, although this remains to be determined.

Thus, our data suggest a hypothesis whereby SOCS2, via its regulation of GH signaling, regulates levels of Ngn1 expression and thus the ability of a neural stem cell to differentiate into a neuron or an astrocyte. Although we have shown that SOCS2 regulates the biological effects of GH signaling in neural precursor cells, we have also shown that only LIF and related cytokines, not GH, regulate SOCS2 expression in these cells. We suggest therefore that LIF-like molecules inhibit a precursor's response to GH by up-regulation of SOCS2 expression, while at the same time promoting astrocyte differentiation by activation of STAT3 [38], a process that is possibly regulated by SOCS3 as part of a normal negative feedback loop for LIF signaling [108]. SOCS2 would, therefore, be acting as an integrator of multiple signal transduction pathways and its level of expression would be important in determining final biological outcome of signaling by multiple stimuli.

References

1. Ramon y Cajal, S. (1906). *The Structure and Connexions of Neurons.* Elsevier, New York.
2. Ramon y Cajal, S. (1913). *Degeneration and Regeneration of the Nervous System,* Oxford University Press.
3. His, W. (1889). Die neuroblasten und deren entstehung im embryonalen mark. *Abh. Math. Phys. Cl. Kgl. Sach. Ges. Wiss.* **15**, 313–372.
4. Allen, E. (1912). The cessation of mitosis in the central nervous system of the albino rat. *J. Comp. Neurol.* **22**, 547–569.
5. Hamilton, A. (1901). The division of differentiated cells in the central nervous system of the white rat. *J. Comp. Neural* **11**, 297–320.
6. Sugita, N. (1918). Comparative studies on the growth of the cerebral cortex. *J. Comp. Neural* **29**, 61–117.
7. Smart, I. (1961). The subependymal layer of the mouse brain and its cell production as shown by radioautography after tritiated thymidine injection. *J. Comp. Neurol.* **116**, 325–347.
8. Altman, J. (1962). Are neurons formed in the brains of adult mammals? *Science* **135**, 1127–1128.
9. Altman, J. (1963). Autoradiographic investigation of cell proliferation in the brains of rats and cats. *Anat. Rec.* **145**, 573–591.
10. Altman, J. and Das, G. D. (1965). Autoradiographic and histological evidence of postnatal hippocampal neurogenesis in rats. *J. Comp. Neurol.* **124**(3), 319–335.
11. Altman, J. (1966). Proliferation and migration of undifferentiated precursor cells in the rat during postnatal gliogenesis. *Exp. Neurol.* **16**(3), 263–278.
12. Altman, J. and Das, G. D. (1966). Autoradiographic and histological studies of postnatal neurogenesis. I. A longitudinal investigation of the kinetics, migration and transformation of cells incorporating tritiated thymidine in neonate rats, with special reference to postnatal neurogenesis in some brain regions. *J. Comp. Neurol.* **126**(3), 337–389.
13. Altman, J. (1969). Autoradiographic and histological studies of postnatal neurogenesis. IV. Cell proliferation and migration in the anterior forebrain, with special reference to persisting neurogenesis in the olfactory bulb. *J. Comp. Neurol.* **137**(4), 433–457.
14. Altman, J. (1969). Autoradiographic and histological studies of postnatal neurogenesis. 3. Dating the time of production and onset of differentiation of cerebellar microneurons in rats. *J. Comp. Neurol.* **136**(3), 269–293.
15. Kaplan, M. S. and Hinds, J. W. (1977). Neurogenesis in the adult rat: Electron microscopic analysis of light radiographs. *Science* **197**, 1092–1094.
16. Kaplan, M. S. and Bell, D. H. (1984). Mitotic neuroblasts in the 9 day old and 11 month old rodent hippocampus. *Journal of Neuroscience* **4**, 1429–1441.
17. Nottebohm, F. *et al.* (1976). Central control of song in the canary, *Serinus canarius. J. Comp. Neurol.* **165**(4), 457–486.
18. Burd, G. D. and Nottebohm, F. (1985). Ultrastructural characterization of synaptic terminals formed on newly generated neurons in a song control nucleus of the adult canary forebrain. *J. Comp. Neurol.* **240**(2), 143–152.
19. Goldman, S. A. and Nottebohm, F. (1983). Neuronal production, migration, and differentiation in a vocal control nucleus of the adult female canary brain. *Proc. Natl. Acad. Sci. USA* **80**(8), 2390–2394.
20. Nottebohm, F. (1985). Neuronal replacement in adulthood. *Ann. N.Y. Acad. Sci.* **457**, 143–161.
21. Paton, J. A. and Nottebohm, F. N. (1984). Neurons generated in the adult brain are recruited into functional circuits. *Science* **225**(4666), 1046–1048.
22. Kohler, G. and Milstein, C. (1975). Continuous cultures of fused cells secreting antibody of predefined specificity. *Nature* **256**(5517), 495–497.
23. Huelsken, J. *et al.* (2001). beta-Catenin controls hair follicle morphogenesis and stem cell differentiation in the skin. *Cell* **105**(4), 533–545.
24. Potten, C. S. and Loeffler, M. (1990). Stem cells: Attributes, cycles, spirals, pitfalls and uncertainties. Lessons for and from the crypt. *Development* **110**(4), 1001–1020.
25. Morrison, S. J. *et al.* (1995). The biology of hematopoietic stem cells. *Annu. Rev. Cell Dev. Biol.* **11**, 35–71.
26. Weissman, I. L. (2000). Stem cells: Units of development, units of regeneration, and units in evolution. *Cell* **100**(1), 157–168.
27. Weissman, I. L. (2000). Translating stem and progenitor cell biology to the clinic: Barriers and opportunities. *Science* **287**(5457), 1442–1446.
28. Richards, L. J. *et al.* (1992). De novo generation of neuronal cells from the adult mouse brain. *Proc. Natl. Acad. Sci. USA* **89**(18), 8591–8595.
29. Kilpatrick, T. J. *et al.* (1995). The regulation of neural precursor cells within the mammalian brain. *Mol. Cell. Neurosci.* **6**(1), 2–15.
30. Reynolds, B. A. and Weiss, S. (1992). Generation of neurons and astrocytes from isolated cells of the adult mammalian central nervous system. *Science* **255**(5052), 1707–1710.
31. Hall, P. A. and Watt, F. M. (1989). Stem cells: The generation and maintenance of cellular diversity. *Development* **106**(4), 619–633.
32. Rietze, R. L. *et al.* (2001). Purification of a pluripotent neural stem cell from the adult mouse brain. *Nature* **412**(6848), 736–739.
33. Loeffler, M. A. P. and C. S. (1997). Stem cells and cellular pedigrees-a conceptual introduction. In Potten, C. S., Ed., *Stem Cells*, pp. 1–28. Academic Press, San Diego.
34. Bjornson, C. R. *et al.* (1999). Turning brain into blood: A hematopoietic fate adopted by adult neural stem cells *in vivo* [see comments]. *Science* **283**(5401), 534–537.
35. Gritti, A. *et al.* (1999). Epidermal and fibroblast growth factors behave as mitogenic regulators for a single multipotent stem cell-like population from the subventricular region of the adult mouse forebrain. *J. Neurosci.* **19**(9), 3287–3297.
36. Johe, K. *et al.* (1996). Single factors direct the differentiation of stem cells from the fetal and adult central nervous system. *Genes. Dev.* **10**, 3129–3140.
37. Bonni, A. *et al.* (1997). Regulation of gliogenesis in the central nervous system by the JAK-STAT signaling pathway. *Science* **278**(5337), 477–483.
38. Rajan, P. and McKay, R. D. (1998). Multiple routes to astrocytic differentiation in the CNS. *J. Neurosci.* **18**(10), 3620–3629.
39. Vescovi, A. L. and Snyder, E. Y. (1999). Establishment and properties of neural stem cell clones: Plasticity in vitro and in vivo. *Brain Pathol.* **9**(3), 569–598.
40. Kondo, T. and Raff, M. (2000). Oligodendrocyte precursor cells reprogrammed to become multipotential CNS stem cells [see comments]. *Science* **289**(5485), 1754–1757.
41. Eisenbarth, G. S. *et al.* (1979). Monoclonal antibody to a plasma membrane antigen of neurons. *Proc. Natl. Acad. Sci. USA* **76**(10), 4913–4917.
42. Raff, M. C. *et al.* (1983). A glial progenitor cell that develops *in vitro* into an astrocyte or an oligodendrocyte depending on culture medium. *Nature* **303**(5916), 390–396.
43. Morshead, C. M. *et al.* (1994). Neural stem cells in the adult mammalian forebrain: a relatively quiescent subpopulation of subependymal cells. *Neuron* **13**(5), 1071–1082.
44. Craig, C. G. *et al.* (1996). In vivo growth factor expansion of endogenous subependymal neural precursor cell populations in the adult mouse brain. *J. Neurosci.* **16**(8), 2649–2658.
45. Gritti, A. *et al.* (1996). Multipotential stem cells from the adult mouse brain proliferate and self-renew in response to basic fibroblast growth factor. *J. Neurosci.* **16**(3), 1091–1100.
46. Weiss, S. *et al.* (1996b). Multipotent CNS stem cells are present in the adult mammalian spinal cord and ventricular neuroaxis. *J. Neurosci.* **16**(23), 7599–7609.

47. Tropepe, V. *et al.* (1997). Transforming growth factor-alpha null and senescent mice show decreased neural progenitor cell proliferation in the forebrain subependyma. *J. Neurosci.* **17**(20), 7850–7859.
48. Doetsch, F. and Alvarez-Buylla, A. (1996). Network of tangential pathways for neuronal migration in adult mammalian brain. *Proc. Natl. Acad. Sci. USA* **93**(25), 14895–14900.
49. Lois, C. and Alvarez-Buylla, A. (1993). Proliferating subventricular zone cells in the adult mammalian forebrain can differentiate into neurons and glia. *Proc. Natl. Acad. Sci. USA* **90**(5), 2074–2077.
50. Lois, C. *et al.* (1996). Chain migration of neuronal precursors. *Science* **271**(5251), 978–981.
51. Luskin, M. B. *et al.* (1993). Neurons, astrocytes, and oligodendrocytes of the rat cerebral cortex originate from separate progenitor cells: An ultrastructural analysis of clonally related cells. *J. Neurosci.* **13**(4), 1730–1750.
52. Doetsch, F. *et al.* (1999). Subventricular zone astrocytes are neural stem cells in the adult mammalian brain. *Cell* **97**(6), 703–716.
53. Magavi, S. S. *et al.* (2000). Induction of neurogenesis in the neocortex of adult mice. *Nature* **405**(6789), 951–955.
54. Nakatomi, H., Kuriu, T., Okabe, S., and Yamamoto, S. (2002). Regeneration of hippocampal pyramidal neurons after ischemic brain injury by recruitment of endogenous neural progenitors. *Cell* **110**, 429–441.
55. Thomas, T. *et al.* (2000). Querkopf, a MYST family histone acetyltransferase, is required for normal cerebral cortex development. *Development* **127**(12), 2537–2548.
56. Brazelton, T. R. *et al.* (2000). From marrow to brain: Expression of neuronal phenotypes in adult mice. *Science* **290**, 1775–1779.
57. Mezey, E. *et al.* (2000). Turning blood into brain: Cells bearing neuronal antigens generated *in vivo* from bone marrow. *Science* **290**, 1779–1782.
58. Gussoni, E. *et al.* (1999). Dystrophin expression in the mdx mouse restored by stem cell transplantation. *Nature* **401**(6751), 390–394.
59. Goodell, M. A. *et al.* (1997). Dye efflux studies suggest that hematopoietic stem cells expressing low or undetectable levels of CD34 antigen exist in multiple species. *Nat. Med.* **3**(12), 1337–1345.
60. Lalande, M. E. *et al.* (1981). Hoechst 33342 dye uptake as a probe of membrane permeability changes in mammalian cells. *Proc. Natl. Acad. Sci. USA* **78**(1), 363–367.
61. Smith, P. J. *et al.* (1988). A mammalian cell mutant with enhanced capacity to dissociate a bis-benzimidazole dye-DNA complex. *Carcinogenesis* **9**(3), 485–490.
62. Dunnwald, M. *et al.* (2001). Isolating a pure population of epidermal stem cells for use in tissue engineering. *Exp. Dermatol.* **10**(1), 45–54.
63. Ledbetter, J. A. *et al.* (1980). T cell subsets defined by expression of Lyt-1,2,3 and Thy-1 antigens. Two-parameter immunofluorescence and cytotoxicity analysis with monoclonal antibodies modifies current views. *J. Exp. Med.* **152**(2), 280–295.
64. Vecchi, A. *et al.* (1994). Monoclonal antibodies specific for endothelial cells of mouse blood vessels. Their application in the identification of adult and embryonic endothelium. *Eur. J. Cell Biol.* **63**(2), 247–254.
65. Vanzulli, S. *et al.* (1997). Detection of endothelial cells by MEC 13.3 monoclonal antibody in mice mammary tumors. *Biocell* **21**(1), 39–46.
66. DeLisser, H. M. *et al.* (1994). Molecular and functional aspects of PECAM-1/CD31. *Immunol. Today* **15**(10), 490–495.
67. Valent, P. (1994). The riddle of the mast cell: kit(CD117)-ligand as the missing link? *Immunol. Today* **15**(3), 111–114.
68. Ikuta, K. and Weissman, I. L. (1992). Evidence that hematopoietic stem cells express mouse c-kit but do not depend on steel factor for their generation. *Proc. Natl. Acad. Sci. USA* **89**(4), 1502–1506.
69. Orlic, D. *et al.* (1993). Purification and characterization of heterogeneous pluripotent hematopoietic stem cell populations expressing high levels of c-kit receptor. *Blood* **82**(3), 762–770.
70. Matthews, W. *et al.* (1991). A receptor tyrosine kinase specific to hematopoietic stem and progenitor cell-enriched populations. *Cell* **65**(7), 1143–1152.
71. Trowbridge, I. S. and Thomas, M. L. (1994). CD45: An emerging role as a protein tyrosine phosphatase required for lymphocyte activation and development. *Annu. Rev. Immunol.* **12**, 85–116.
72. Mehler, M. F. *et al.* (1997). Bone morphogenetic proteins in the nervous system. *Trends Neurosci.* **20**(7), 309–317.
73. Semkova, I. and Krieglstein, J. (1999). Neuroprotection mediated via neurotrophic factors and induction of neurotrophic factors. *Brain Res. Brain Res. Rev.* **30**(2), 176–188.
74. Bartlett, P. *et al.* (1998). Regulation of neural stem cell differentiation in the forebrain. *Immunol. Cell Biol.* **76**(5), 414–418.
75. Turnley, A. M. and Bartlett, P. F. (2000). Cytokines that signal through the leukemia inhibitory factor receptor-beta complex in the nervous system. *J. Neurochem.* **74**(3), 889–899.
76. Richards, L. J. *et al.* (1996). Leukaemia inhibitory factor or related factors promote the differentiation of neuronal and astrocytic precursors within the developing murine spinal cord. *Eur. J. Neurosci.* **8**(2), 291–299.
77. Li, W. *et al.* (1998). Neuronal differentiation of precursors in the neocortical ventricular zone is triggered by BMP. *J. Neurosci.* **18**(21), 8853–8862.
78. Williams, B. P. *et al.* (1997). A PDGF-regulated immediate early gene response initiates neuronal differentiation in ventricular zone progenitor cells. *Neuron* **18**(4), 553–562.
79. Lee, J. *et al.* (1995). Conversion of *Xenopus* ectoderm into neurons by NeuroD, a basic helix–loop–helix protein. *Science* **268**(5212), 836–844.
80. Ma, Q. *et al.* (1996). Identification of neurogenin, a vertebrate neuronal determination gene. *Cell* **87**, 43–52.
81. Cai, L. *et al.* (2000). Misexpression of basic helix–loop–helix genes in the murine cerebral cortex affects cell fate choices and neuronal survival. *Development* **127**(14), 3021–3030.
82. Nieto, M. *et al.* (2001). Neural bHLH genes control the neuronal versus glial fate decision in cortical progenitors. *Neuron* **29**(2), 401–413.
83. Sun, Y. *et al.* (2001). Neurogenin promotes neurogenesis and inhibits glial differentiation by independent mechanisms. *Cell* **104**(3), 365–376.
84. Hughes, S. M. *et al.* (1988). Ciliary neurotrophic factor induces type-2 astrocyte differentiation in culture. *Nature* **335**(6185), 70–73.
85. Murphy, M. *et al.* (1997). Cytokines which signal through the LIF receptor and their actions in the nervous system. *Prog. Neurobiol.* **52**(5), 355–378.
86. Gadient, R. A. *et al.* (1998). Effect of leukemia inhibitory factor (LIF) on the morphology and survival of cultured hippocampal neurons and glial cells. *Brain Res.* **798**(1–2), 140–146.
87. Koblar, S. A. *et al.* (1998). Neural precursor differentiation into astrocytes requires signaling through the leukemia inhibitory factor receptor. *Proc. Natl. Acad. Sci. USA* **95**(6), 3178–3181.
88. Mayer, M. *et al.* (1994). Ciliary neurotrophic factor and leukemia inhibitory factor promote the generation, maturation and survival of oligodendrocytes in vitro. *Development* **120**(1), 143–153.
89. Gard, A. L. *et al.* (1995). Astroglial control of oligodendrocyte survival mediated by PDGF and leukemia inhibitory factor-like protein. *Development* **121**(7), 2187–2197.
90. Barres, B. A. *et al.* (1996). Ciliary neurotrophic factor enhances the rate of oligodendrocyte generation. *Mol. Cell Neurosci.* **8**(2–3), 146–156.
91. Marmur, R. *et al.* (1998). Differentiation of oligodendroglial progenitors derived from cortical multipotent cells requires extrinsic signals including activation of gp130/LIFbeta receptors. *J. Neurosci.* **18**(23), 9800–9811.
92. Ip, N. Y. *et al.* (1991). Ciliary neurotrophic factor enhances neuronal survival in embryonic rat hippocampal cultures. *J. Neurosci.* **11**(10), 3124–3134.
93. Oppenheim, R. W. *et al.* (1991). Control of embryonic motoneuron survival *in vivo* by ciliary neurotrophic factor. *Science* **251**(5001), 1616–1618.
94. Martinou, J. C. *et al.* (1992). Cholinergic differentiation factor (CDF/LIF) promotes survival of isolated rat embryonic motoneurons *in vitro. Neuron* **8**(4), 737–744.
95. Galli, R. *et al.* (2000). Regulation of neuronal differentiation in human CNS stem cell progeny by leukemia inhibitory factor. *Dev. Neurosci.* **22**(1–2), 86–95.

96. Shimazaki, T. *et al.* (2001). The ciliary neurotrophic factor/leukemia inhibitory factor/gp130 receptor complex operates in the maintenance of mammalian forebrain neural stem cells. *J. Neurosci.* **21**(19), 7642–7653.
97. Seniuk-Tatton, N. A. *et al.* (1995). Neurons express ciliary neurotrophic factor mRNA in the early postnatal and adult rat brain. *J. Neurosci. Res.* **41**(5), 663–676.
98. Ip, N. Y. *et al.* (1993). The alpha component of the CNTF receptor is required for signaling and defines potential CNTF targets in the adult and during development. *Neuron* **10**(1), 89–102.
99. Solecki, D. J. *et al.* (2001). Activated Notch2 signaling inhibits differentiation of cerebellar granule neuron precursors by maintaining proliferation. *Neuron* **31**(4), 557–568.
100. Nakashima, K. *et al.* (2001). BMP2-mediated alteration in the developmental pathway of fetal mouse brain cells from neurogenesis to astrocytogenesis. *Proc. Natl. Acad. Sci. USA* **98**(10), 5868–5873.
101. Faux, C. H. *et al.* (2001). Interactions between fibroblast growth factors and Notch regulate neuronal differentiation. *J. Neurosci.* **21**(15), 5587–5596.
102. Stahl, N. *et al.* (1995). Choice of STATs and other substrates specified by modular tyrosine-based motifs in cytokine receptors. *Science* **267**(5202), 1349–1353.
103. Endo, T. A. *et al.* (1997). A new protein containing an SH2 domain that inhibits JAK kinases. *Nature* **387**(6636), 921–924.
104. Naka, T. *et al.* (1997). Structure and function of a new STAT-induced STAT inhibitor. *Nature* **387**(6636), 924–929.
105. Starr, R. *et al.* (1997). A family of cytokine-inducible inhibitors of signalling. *Nature* **387**(6636), 917–921.
106. Hilton, D. J. *et al.* (1998). Twenty proteins containing a C-terminal SOCS box form five structural classes. *Proc. Natl. Acad. Sci. USA* **95**(1), 114–119.
107. Krebs, D. L. and Hilton, D. J. (2001). SOCS proteins: Negative regulators of cytokine signaling. *Stem Cells* **19**(5), 378–387.
108. Greenhalgh, C. J. *et al.* (2002). Growth enhancement in suppressor of cytokine signaling 2 (SOCS-2)-deficient mice is dependent on signal transducer and activator of transcription 5b (STAT5b). *Mol. Endocrinol.* **16**(6), 1394–1406.
109. Reynolds, B. A. and Weiss, S. (1996). Clonal and population analyses demonstrate that an EGF-responsive mammalian embryonic CNS precursor is a stem cell. *Dev. Biol.* **175**(1), 1–13.
110. Nicholson, S. E. *et al.* (1999). Mutational analyses of the SOCS proteins suggest a dual domain requirement but distinct mechanisms for inhibition of LIF and IL-6 signal transduction. *EMBO J.* **18**(2), 375–385.
111. Krulich, L. *et al.* (1968). Stimulatory and inhibitory effects of purified hypothalamic extracts on growth hormone release from rat pituitary *in vitro*. *Endocrinology* **83**(4), 783–790.
112. Hojvat, S. *et al.* (1982). Growth hormone (GH), thyroid-stimulating hormone (TSH), and luteinizing hormone (LH)-like peptides in the rodent brain: non-parallel ontogenetic development with pituitary counterparts. *Brain Res.* **256**(4), 427–434.
113. Ram, P. A. and Waxman, D. J. (1999). SOCS/CIS protein inhibition of growth hormone-stimulated STAT5 signaling by multiple mechanisms. *J. Biol. Chem.* **274**(50), 35553–35561.
114. Delfs, J. *et al.* (1980). Somatostatin production by rat cerebral neurones in dissociated cell culture. *Nature* **283**(5748), 676–677.
115. Lobie, P. E. *et al.* (1993). Localization and ontogeny of growth hormone receptor gene expression in the central nervous system. *Brain Res. Dev. Brain Res.* **74**(2), 225–233.
116. Ram, P. A. and Waxman, D. J. (2000). Role of the cytokine-inducible SH2 protein CIS in desensitization of STAT5b signaling by continuous growth hormone. *J. Biol. Chem.* **275**(50), 39487–39496.
117. Hansen, J. A. *et al.* (1999). Mechanism of inhibition of growth hormone receptor signaling by suppressor of cytokine signaling proteins. *Mol. Endocrinol.* **13**(11), 1832–1843.

Index

A

B

D

E

F

G

H

I

J

M

N

O

P

Q

R

S

T

U

V

W

U

V

W

ISBN 0-12-124549-7
90038
9 780121 245498